Proceedings

of the Fifth International Conference on the Bearing Capacity of Roads and Airfields

AF739993

Norwegian University of Science and Technology
TRONDHEIM, NORWAY, 6 – 8 JULY 1998

Editors:

Rasmus S. Nordal,
Norwegian University of Science and Technology
Geir Refsdal, Norwegian Public Roads Administration

VOLUME I

© **Faculty of Civil and Enviromental Engineering**
Department of Road and Railway Engineering

ISBN 82-519-1346-2
Printed by TAPIR

Supporting Organizations
Norwegian University of Science and Technology
Norwegian Civil Aviation Administration
Norwegian Public Roads Administration

National Organizing Committee
Ivar Horvli (Chairman), Norwegian University of Science and Technology
Kjell Levik (Co-Chairman), Norwegian Public Roads Administration
Bjørn Finborud, Berdal Strømme as
Øyvind Hallquist, Norwegian Civil Aviation Administration
Geir Lange, Norwegian Civil Aviation Administration
Rasmus S. Nordal, Norwegian University of Science and Technology
Geir Refsdal, Norwegian Public Roads Administration
Olav Reine, Veidekke ASA
Astrid Bye (Conference Secretary), SEVU

International Advisory Committee
Kaare Flaate (Chairman), Norwegian Public Roads Administration
Ivar Horvli (Co-Chairman), Norwegian University of Science and Technology
David Bennet, Australian Road Research Board Ltd.
Jean-Francois Corté, Laboratorie Central des Ponts et Chaussées, France
Rudolf Floss, Technische Universität München, Germany
Les Hawker, Highways Agency, United Kingdom
Michel Hovan, Federal Aviation Administration, USA
Byron N. Lord, Federal Highway Administration, USA
El Hussein Mohamed, National Research Council of Canada
M.I. Pinard, Ministry of Works, Botswana
Xinglong Wang, Heilongjiang Institute of Highway and Transport, China
Vladislav M. Yumashev, State Highway Research Institute, Russia

Technical Advisory Committee
Geir Refsdal (Chairman), Norwegian Public Roads Administration
Per Magne Noss (Co-Chairman), The Norwegian Asphalt Contractors Association
Bjørn Finborud, Berdal Strømme as
Åsmund Knutson, Norwegian Public Roads Administration
Geir Lange, Norwegian Civil Aviation Administration
Helge Mork, Norwegian University of Science and Technology
Tore Slyngstad, Norwegian Public Roads Administration
Jostein Myre, Norwegian Public Roads Administration

Conference Secretariat
SEVU, Congress Department
NTNU, Gløshaugen
N-7034 Trondheim, Norway

INTRODUCTION

Road and air transport with their primary role in the transportation system are drawing considerable public and political attention all over the world. The steady growth of traffic volume and weight reduces the service life of existing pavements, causing escalating maintenance costs as well as increasing safety and environmental problems. To meet these challenges, there are great needs for further development of road and airport networks , and for cost-effective rehabilitation and strengthening of existing pavements.

The BCRA conferences have been organized every fourth year since 1982 when the first one was held in Trondheim, Norway. The second was held in Plymouth, England in 1986, the third in Trondheim in 1990 and the fourth in Minneapolis, Minnesota , USA in 1994. These conferences have focused on the monitoring of the bearing capacity , the performance of pavements and the evaluation and use of this information for maintenance, strengthening and load regulations.

The themes of the BCRA conference in 1998 cover important aspects of pavement performance, pavement condition surveys, evaluation of structural capacity, properties of pavement material, design and pavement rehabilitation and strengthening. The contributions are documented in these proceedings containing 180 papers from all over the world.

The conference will provide a platform for the presentation and discussion of these research results and new developments, for the exchange of experience from current practice and for the identification of further research and development needs.

I would like to express my gratitude to the authors and the chairmen, the members of the International Advisory Committee, the Technical Advisory Committee, the Conference Secretariat and the editors of the proceedings.

Ivar Horvli

Chairman of the Organizing Committee

VOLUM 1

Table of contens

Airfield Pavements

Pavement Condition Surveys

Backanalysis of Deflection Measurements

Structural Evaluation and Design

VOLUM 2

Structural Evaluation and Design

Full Scale Testing

Asphaltic Materials

VOLUM 3

Granular Materials and Subgrade Soils

Use of Reinforcement

Pavement Rehabilitation

Case Histories, Airfields

Case Histories, Roads

Continuing from page 114

Fifth International Conference on the
BEARING CAPASITY OF
ROADS AND AIRFIELDS

AN INTERNATIONAL VIEW OF PAVEMENT ENGINEERING

F. C. RUST

CSIR Transportek

Pretoria, South Africa

JOE P. MAHONEY

University of Washington

Seattle, Washington, USA

JAMES B. SORENSON

Federal Highway Administration

Washington, D.C., USA

ABSTRACT

The authors have attempted to show that local differences in pavement practices and climate are typically not all that different (with the exception of countries such as South Africa). Further, the transfer of technology is difficult given the current transfer venues. To address this issue, two major new approaches are proposed. The International Pavement Guide is not being developed and requires much discussion if such an effort is ever undertaken. An international technology workshop is currently being planned for 1998. Possibly, this joint South Africa/USA activity can serve as a model for this type of venue.

INTRODUCTION

Virtually all pavement design procedures (either new construction, reconstruction, or rehabilitation) are based on either empirical or mechanistic-empirical principles. Further, similar materials are used whether it be asphalt concrete, portland cement concrete, crushed stone, etc. Unfortunately, local differences in how such principles and materials are used make sharing and transfer of pavement-oriented knowledge far more difficult than it should be. Local differences in pavement design, construction, evaluation and rehabilitation are somewhat understandable due to variations in available materials, climate, traffic, and budget; however, improved technology transfer of known pavement practices is needed.

This paper will be used to first overview general road and pavement statistics followed by more specific pavement practices for a selection of countries. Such information will be used to propose improved mechanisms for sharing pavement technology among countries. The issue of the potential benefit of such technology sharing will also be addressed. As the authors will show, there are generally broad similarities in pavement practices with a few countries having somewhat different practices (such as South Africa). Such information (including both the similarities and differences) supports the need for improved technology sharing.

GENERAL STATISTICS

Road and pavement related data from a number of countries have been reported earlier[1], some of the information and its relation to technology transfer will be discussed here. One group of countries are used to show broad world trends followed by more specific data from various countries (mostly Europe and the USA). First, a selection of countries were chosen to represent Africa, the Americas, Asia, and Europe. In Table 1, these countries are shown along with information on size, kilometers of motorways and total road network, percent paved, and road density. Most countries shown for Europe, Asia, and the USA have a substantial percentage of the road network paved (greater than 50 percent). Of the world's 24 least developed countries with national route systems with more than 10,000 km of highways (United Nations[2]), the average percentage of roads paved was 13 percent. These countries had a Gross Domestic Product of only US$319 per capita in 1994. This compares to US$21,875 for developed market economies (a difference by a factor of about 70).

Table 2 shows various traffic statistics (where available) in terms of annual vehicle-kilometers per capita and ton-kilometers per capita. As expected the figures vary considerably from country to country. Table 3 provides a view of specific climate statistics. As one would expect, the countries near the equator have little variation between the coldest and warmest months. Some of the largest differences occur in the USA, South Korea, and Europe. The majority of the selected cities have substantial precipitation. Depth of freeze of course influences pavement design and performance strongly - possibly more so than any other single climate measure.

The information shown in this section of the paper reveals that:

- Most "developed" countries have about 50 percent or more of their roads paved.
- For countries such as Denmark, Estonia, Latvia, Lithuania, Japan, and Sweden, about 300 E80s are applied to each kilometer of road (on average) for each person in those countries. The USA has a value about double that figure.
- Yearly mean and highest mean monthly temperatures for selected cities in various countries shown exhibit only modest differences. The mean temperature for the coldest months reveal significant differences.
- Most cities studied have measurable precipitation days for at least 30 percent (or more) of a typical year.
- Numerous countries have very warm summers accompanied by significant freezing depths due to cold winters.

PAVEMENT TYPES AND PRACTICES

This section will be used to overview some of the available information about the pavement types used in various countries along with various pieces of design-oriented information.

Table 4 shows the relative percentages of pavement types shown for various European countries (the selection of countries was influenced by an earlier version of the paper presented in Estonia[1]). Flexible pavements surfaced with asphalt concrete tend to dominate (as is true with most countries throughout the world). This view is supported by statistics from the USA (FHWA[3]) which reveals that 94 percent of the public road system is flexible pavement and the remaining six percent is rigid.

Table 5 is a summary of work done by Nissoux et al.[4] for PIARC. It shows rigid pavement practices for 11 European countries and Japan. Basically, plain jointed concrete pavement tends to dominate. For heavily trafficked roads, the slab thicknesses range between 220 to 260 mm (a bit thinner for CRCP).

One country which, in general, designs and constructs their flexible pavements quite differently than most countries is South Africa. About 70 percent of its pavements are constructed with thin asphalt bound surfaces (40 mm or less) placed on top of high quality crushed stone bases. These bases are placed on either unstabilized or cement-stabilized subbases. The design and construction requirements are unique and result in high performance pavements which can accommodate high E80 loadings. Further, South Africa has adopted advanced pavement design and test methods (such as a mechanistic-empirical pavement design procedure and in-situ material testing devices such as the Dynamic Cone Penetrometer (DCP)). The details associated with its pavement practices are important and, as such, not necessarily straightforward to transfer to another country. Evidence of this fact is that South African engineers have published technical papers widely in proceedings associated with international pavement conferences; however, there is little evidence, in the view of the authors, that South African practices have been adopted elsewhere.

The question may be asked whether it would be worth while to embark on significant activities to transfer South African technology to the USA. In order to demonstrate the possible effect such technology transfer may have, a pilot project to compare the relative cost of pavement structures from the two countries was conducted. The purpose of this work was to calculate the relative cost of three typical flexible pavement structures designed using the South African mechanistic pavement design method, the Californian Department of Transportation (Caltrans) pavement design method, and the Washington State DoT pavement design method. The following conditions were selected :

- pavement to be a flexible structure;
- three levels of traffic (1 million, 10 million and 30 million equivalent standard axles or E80s), and
- two subgrade types (weak with a CBR of 5 and strong with a CBR of 15).

One of the problems encountered was that the high quality aggregate bases often used in South Africa, had never been constructed in the USA and therefore it was difficult to estimate construction costs for these bases in the USA. For the purpose of a relative analysis, the construction costs of the various layers as currently prevalent in South Africa was therefore used. A more accurate answer could be obtained using USA costs once high quality granular bases have been constructed.

Table 6 shows the structures obtained from the various design processes as well as their relative costs. The data is shown graphically in Figures 1 and 2. It is obvious that the design philosophy to use high quality granular bases supported by a cemented subbase and covered with a relatively thin wearing course yield more cost effective designs than that utilizing relatively thick asphalt layers on weaker granular layers. It is evident that, should one be able to construct these materials cost-effectively in the USA, a significant initial cost saving should be effected. The saving on initial cost could be somewhere between 50 and 80 per cent. This initial, somewhat superficial calculation shows that it may well be worth while to transfer the materials and design technologies from South Africa to the USA (or elsewhere) for use in areas with similar climatic conditions.

TECHNOLOGY SHARING VENUES

The preceding data suggests that many countries have broadly similar pavement practices with the exception of countries such as South Africa; however, this is not to say that pavement performance is necessarily similar among countries. This suggests that potential exists in understanding both the similarities and differences in country pavement practices. The need for technology transfer and training of staff in the USA was highlighted by Carter and Rochon.[5,6] Further, there exist numerous activities and processes through which exchanges of pavement practices are made (or enhanced). These include, but are not limited to, international conferences (such as the BCRA), study tours, agency agreements, technical reports, books and short courses. These are briefly discussed below.

International conferences

International pavement-oriented conferences have become quite common. Typically, proceedings are published which become widely available. Pros include the fact that they provide "packaged," recent technical information, that they provide a venue for meeting colleagues and that they provide a written record of proceedings. Cons include the fact that they are expensive to attend when all expenses are totaled, and that information presented may not be applicable or appropriate for the attendee.

Study tours

Pavement study tours have been commonly done for the last 100 or so years (a study tour to Europe by Washington State engineers in 1900 had a profound impact on the Northwestern part of the USA and the "good roads" movement). Until recently, structured study tours by US personnel were rather rare; however, this changed in 1990 with the European Asphalt Study Tour (with representatives from AASHTO, FHWA, SHRP, TAI, and TRB) and the 1992 U.S. Tour of European Concrete Highways (with representatives from AASHTO, ACPA, FHWA, PCA, SHRP, and TRB). Organized tours from various countries have been commonly observed in the USA. Pros include the fact that substantial pavement-orientated information is exchanged in a short period of time, a trip report (documentation) is often produced, and face-to-face meetings/introductions take place. Cons include the fact that benefits of such trips tend to accrue primarily to those who participate in the study tour, and that results are a bit slow in being placed in practice though notable exceptions have occurred in the USA (SMA wearing courses being one).

Agency agreements

Agency agreements enable a continuing collaboration between two or more agencies. Such agreements often include provisions for exchanging information such as research data and reports, personnel visits, round-robin laboratory testing, etc. Two such agreements the authors are aware of exist between the State of California and South Africa and between the State of Minnesota and Finland. One of the major pros is the fact that detailed exchanges of information take place over an extended period of time. A down side is that fact that information exchange can be slow and primarily resides with the parties to the agreement.

Technical reports

Technical reports are widely available today which document various studies and/or pavement practices. This has been a common venue for sharing technical information for over the last 100 years or so. Pros include careful documentation of results and the fact that most major studies receive such documentation. Cons include the fact that the distribution of technical reports is limited, there is a mixture of languages, terminology, and units between countries which slows and impedes technology sharing, and technical reports generally are used to document specific studies - the scope of which may be of limited use to the reader.

Books

Books on pavement practices are actually not all that common. Pros include the fact that books can provide a careful, complete explanation of the practices being described, and that they can be used in a variety of instructional venues. Cons include the fact that book development can lag behind pavement practice development, i.e., the information can be "dated" by the time the book is published, and that updates are not timely and typically require the purchase of a new, revised book.

Short courses

Short courses have been used for some years as a way of packaging current pavement practices and presenting such practices to a small group (generally 25 to 30 attendees per course). Pros include the fact that they provide recent, relevant information by competent instructors, and that question and answer format is a plus in learning new information. Cons include the fact that they are expensive on a cost per attendee basis (typically US$500 to over $1,500 per day if all costs considered), that details associated with implementation are often lacking, and that national courses may not address local practices and needs.

Clearly, of the areas which are commonly used for exchange of evolving pavement practices, all have positive and negative aspects (as noted via the listed "pros" and "cons"). Are there other venues which might improve the pavement technology transfer process and, specifically, the "effectiveness" of such transfers?

NEW DIRECTIONS

Thus far, this paper has been a mixture of very general statistics for a selection of countries followed by a "snapshot" of selected pavement practices, followed by a brief discussion of some commonly used technology sharing practices. All of the technology sharing practices have pros and cons. Possibly one of the single largest deficiencies is in the area of material specifications and construction practices (for which very limited international information exists). It seems, in the opinion of the authors, that these areas are rarely adequately addressed and are crucial to the proper performance of any pavement project. Recent work in technology management (Rust and Vos) has indicated that technology transfer is more effective if it is managed as part of a holistic, systems approach to technology management. This implies that planning for technology development should include the essential elements of technology transfer and training up-front rather than "post planning" after technology has already been developed. This approach often calls for "reverse thinking", forcing researchers to think of the implementation of solutions first before developing them, thus facilitating the eventual transfer of such technology to practice. It also calls for using the appropriate "delivery systems" in order to ensure ease of technology transfer. In the case of transferring technology from one country to another after it had been developed and implemented in the former, one has to take into consideration that the technology package may have to be modified to suit the conditions, practices, material availability and material cost in the "target country". The most appropriate delivery system will also depend on the specific technology package and the circumstances in the target country.

The following "picture" emerges from the preceding information:

- Local (country) differences are apparent in usage of pavements; however, broad similarities are evident with respect to use of pavement types, materials, traffic, and importantly, climate. This runs a bit counter to the general view which seems to be that local conditions are somewhat unique requiring unique practices. Further, even modest differences in pavement practices make technology sharing more difficult. It is apparent that some countries, such as South Africa, build rather unique pavement structures. The authors have not attempted to address the issue of pavement performance.
- Those countries which experience substantial freezing weather do have a special set of design and performance issues to deal with; however, these are common among all countries north or south of 40° latitude (or high elevations). Designs for these climate regions appear to have a common solution as well, i.e., thicker pavement structures and specification of layer materials.
- The common venues for sharing knowledge about pavement practices all have pros and cons.
- In the complex case of transferring technology from one country to another, care must be taken to adjust technology packages and to select delivery systems to suit the conditions in the target country.

Given the above "observations," what are potential, improved pavement technology "delivery systems"? Two concepts will be briefly described which may have merit. The first is an International Pavement Guide, and the second is the concept of international technology workshops.

INTERNATIONAL PAVEMENT GUIDE

To bring together the substantial and relevant information that most "pavement people" would like to have, a document which reflects the "best" pavement design, rehabilitation, maintenance, and construction practices is desirable. A much more modest attempt at this was completed two years ago by the Washington State Department of Transportation (WSDOT) which resulted in the three-volume WSDOT Pavement Guide (Volume 1 (Pavement Policy), Volume 2 (Pavement Notes), and Volume 3 (Case Studies and Software Users' Guides)). This WSDOT document is over 800 pages in length and combines the important policy issues along with general to very specific pavement design, evaluation, and construction issues. One of the difficulties with this document is its length and the difficulty associated with locating specific information. Further, a document of this size is difficult to update and the addition of photographs (particularly, color) and other descriptive information is expensive. To address this issue, a CD-ROM containing this information, along with easy "links" to definitions, articles, photographs, design equations, etc. was recently completed for WSDOT. The CD-ROM format has an additional benefit; each copy costs only about US$10 (as opposed to US$40 for each copy of the paper Guide).

The International Pavement Guide would be far more ambitious. Unfortunately, there are several potential impediments to its development and use. These include:

- Development costs
- Lack of an international organization to sponsor and support the effort
- Implementation difficulties

INTERNATIONAL TECHNOLOGY WORKSHOPS

The second concept is to use international technology workshops as a means to enhance and accelerate the sharing of relevant technology between countries. An example of this type of workshop is currently being planned for the USA during 1999. The planned workshop, sponsored largely by the FHWA, will feature South African technology which has potential applications in the USA. The workshop will include sessions on materials, structural design, pavement maintenance, accelerated pavement testing, and practical field demonstrations. Documentation of development history and how these procedures or equipment were used will be fully included in a set of notes. As appropriate, equipment will be demonstrated in "field" applications (such as the semi-automatic DCP). All of this information will be presented by those South Africans most familiar with the specific technology being presented. Naturally, information on how to obtain specific reports, software, or equipment will be fully described at the workshop. How additional information or service can be obtained for any of the covered technologies will also be provided. It is expected that this type of activity can put detailed information into the hands of practitioners quickly and in a cost effective manner.

CONCLUDING REMARKS

Although environmental and institutional conditions vary greatly between countries, the authors are of the opinion that much benefit can be obtained from the transfer of technology packages with proven track records from one country to another. However, the magnitude and cost of such activities should not be underestimated. In addition, such activities require advanced planning using a holisitc approach rather then fragmented singular efforts.

An earlier version of this paper was presented at the 23 rd International Road Conference - Baltic Road Council during August 1997 in Parnu, Estonia. This version contains updated and improved information and is presented in the spirit of continuing the dialogue on international pavement practices.

REFERENCES

1. MAHONEY, JP, SORENSON, JB, RUST, FC. (1997), "An international view of pavement engineering", 23 rd International Road Conference - Baltic Road Council, Parnu, Estonia, 1997.

2. United Nations (1996), "The Least Developed Countries - 1996 Report," United Nations, New York, February 1996.

3. Federal Highway Administration (1995), "Highway Statistics 1994," Report No. FHWA-PL-95-042, FHWA, Washington, D.C., October 1995.

4. NISSOUX, J, GOUX, M, SOMMER, H. (1994), "Synoptic Table on Standards and Practices for Concrete Roads - International Revue," Seventh International Symposium on Concrete Roads, Vienna, Austria, October 3-5, 1994.

5. CARTER, E.C. and ROCHON, S.M. (1996a), "Development and Training of Pavement Engineering Personnel, "Final Report, Project 20-7(66), National Cooperative Highway Research Board Research Program, Transportation, Washington, D.C., February 1996.

6. CARTER, E.C. and ROCHON, S.M (1996b), "Compendium of Training Related to Pavement Engineering, "Final Report, Project 20-7 (66), National Cooperative Highway Research Board Research Program, Transportation, Washington, D.C., February 1996.

7. IRF (1994), "World Road Statistics 1989 - 1993, "International Road Federation, Washington, D.C.

8. World Bank (1993a), "Estonia - The Transition to a Market Economy," The World Bank, Washington, D.C., March 1993.

9. Organization for Economic Cooperation and Development (1993), "Road Strengthening in Central and Eastern European Countries, "Report Prepared by an OECD Scientific Expert Group, OECD, Paris.

10. RUST, FC and VOS, RM. "A proposed holistic framework for managing the development of road infrastructure technology in South Africa." Paper presented at the Annual Transportation Research Board Conference, Washington DC, 1998.

Table 1 : Country Areas and Road Kilometers

Country/State	Area (sq. km)	Motorways (km)	Total Network** (km)	% Paved	Total Roads/ Area (km/sq. km)
Africa					
Kenya	582,646	-	62,573	13	0.1
South Africa	1,123,226	1,953	182,329	30	0.2
Americas					
Chile	756,945	-	79,593	14	0.1
USA	9,809,418	86,818	6,277,859	59	0.6
Washington State	115,509	*1,229	97,090	68	0.8
Asia					
Japan	337,801	5,410	1,130,892	72	3.3
South Korea	99,392	1,602	61,296	85	0.6
Europe					
Denmark	43,094	747	71,111	100	1.7
Estonia	45,100	62	14,771	55	0.3
Hungary	93,030	441	158,711	44	1.7
Iceland	104,000	-	11,279	24	0.1
Sweden	411,114	1,044	135,920	71	0.3

- 1993 statistics (mostly)
- Source: IRF [7] and Library of Congress (1992)

* Interstate highways only

** Public roads only

Table 2 : Calculated Country Traffic Statistics

Country/State	Roads Only		Total**
	Vehicle*-km / Capita	Tonne-km / Capita	Tonne-km / Capita
Africa			
Kenya	183		
South Africa	2,161	910	
Americas			
Chile			
USA	13,764	4,564	13,306
Washington State	14,531		
Asia			
Japan	5,421	2,251	4,446
South Korea	1,198	1,084	2,269
Europe			
Denmark	7,441	2,083	2,604
Estonia		2,608	11,648
Hungary	2,219	576	3,603
Iceland			
Sweden	6,464	2,871	5,053

* Includes cars, buses, trucks

** Includes road, water, rail modes combined

Table 3 : Calculated Climate Statistics

Location	Temperatures		Annual Days with Precipitation (%)
	Hottest Month Minus Coldest Month	Coefficient of Variation (%)	
Africa			
Nairobi, Kenya	2°C	6	32
Johannesburg, S. Africa	9°C	25	19
Americas			
Santiago, Chile	11°C	27	8
San Jose, Costa Rica	3°C	5	47
Washington, D.C., USA	25°C	69	34
Seattle, WA, USA	14°C	45	41
Spokane, WA, USA	23°C	89	31
Asia			
Tokyo, Japan	21°C	57	29
Seoul, South Korea	30°C	100	31
Europe			
Copenhagen, Denmark	18°C	78	47
Tallinn, Estonia	22°C		49
Budapest, Hungary	24°C	82	37
Reykjavik, Iceland	11°C	80	58
Riga, Latvia	23°C	180	53
Vilnius, Lithuania	24°C		
Stockholm, Sweden	21°C	114	45

Table 4 : Pavement Types for Various Central, Eastern European, and Baltic Countries (after OECD[9] and World Bank[8])

Country	Pavement Types (%)						
	Asphalt Concrete	Portland Cement Concrete	Asphalt Macadam	Sett Paving	Bitumen	Gravel	Other
Bulgaria	96.6	0.0	1.0	0.6			1.8
Czech Republic	95.9	2.6	0.0	1.5			0.0
Hungary	96.1	1.9	2.0	0.0			0.0
Lithuania	55.0	1.0	44.0	0.0			0.0
Poland	100.0	0.0	0.0	0.0			0.0
Romania	90.0	10.0	0.0	0.0			0.0
Russia	87.0	8.0	0.0	0.0			5.0
Ukraine	55.0	1.5	13.0	9.5			21.0
Estonia	17.4	0.0	?	?	27.9	46.7	8.0

Table 5 : Rigid Pavement and Design Practice - Twelve Countries*
(after Nissoux et al. [4])

Rigid Pavement Types	**Percentage**
• JPCP	75%
• JRCP	25%
• CRCP	50%
Slab Thickness (typical)	**mm**
• JPCP	220-260
• JRCP	unknown
• CRCP	200-230
Maximum Allowable Axle Loads	**tonnes**
• Single	
• Mean	11.25
• Range	10-13
• Tandem	
• Mean	19
• Range	18-20
Transverse Joint Spacing (typically with dowel bars)	**m**
• Mean	5.0
• Range	3.5-6
Dowel Bars	**mm**
• Diameter (range)	16-30
• Length (range)	250-700
• Spacing (range)	300-500

* Includes Austria, Belgium, France, Germany, Great Britain, Italy, Japan, Netherlands, Norway, Spain, Sweden, Switzerland

Table 8 : Cost of comparable South African and US pavement designs

South African Designs									
Million ESALS	**1 to 3**			**3 to 10**			**30**		
Subgrade type	**Layer type**	**Thick-ness**	**Cost**	**Layer type**	**Thick-ness**	**Cost**	**Layer type**	**Thick-ness**	**Cost**
Weak, CBR = 5	AC	40	$3.18	AC	40	$3.18	AC	50	$3.98
	G2	150	$0.52	G2	150	$0.52	G1	150	$3.17
	G5	150	$0.89	C3	250	$3.19	C3	300	$3.83
	G5	300	$1.78	G5	300	$1.78	G5	300	$1.78
		Total :	$6.37			$8.67			$12.75
Strong, CBR = 15	AC	40	$3.18	AC	40	$3.18	AC	50	$3.98
	G2	150	$0.52	G2	150	$0.52	G1	150	$3.17
	G5	150	$0.89	C3	250	$3.19	C3	300	$3.83
		Total :	$4.60			$6.90			$10.97
Caltrans Designs									
TI Class	**9**			**12**			**13.5**		
Million ESALS	**0.8 to 1.27**			**9.5 to 13.5**			**26.1 to 35.6**		
Subgrade type	**Layer type**	**Thick-ness**	**Cost**	**Layer type**	**Thick-ness**	**Cost**	**Layer type**	**Thick-ness**	**Cost**
Weak, CBR = 5	AC	152	$12.13	AC	183	$14.56	AC	213	$16.98
	G5	107	$0.63	G5	198	$1.17	G5	213	$1.26
	G6	259	$1.53	G6	351	$2.07	G6	396	$2.35
		Total :	$14.29			$17.80			$20.59
Strong, CBR = 15	AC	137	$10.92	AC	183	$14.56	AC	183	$14.56
	G5	137	$0.81	G5	198	$1.17	G5	107	$0.63
	G6	107	$0.63	G6	107	$0.63	C3	168	$2.14
		Total :	$12.36			$16.36			$17.33
WSDoT Designs									
Million ESALS	**1**			**10**			**30**		
Subgrade type	**Layer type**	**Thick-ness**	**Cost**	**Layer type**	**Thick-ness**	**Cost**	**Layer type**	**Thick-ness**	**Cost**
Weak, CBR = 5	AC	105	$8.36	AC	105	$8.36	AC	105	$8.36
	G3	280	$3.69	BC	135	$8.82	BC	230	$15.02
				G3	170	$2.24	G3	135	$1.78
		Total :	$12.04			$19.41			$25.15
Strong, CBR = 15	AC	105	$8.36	AC	105	$8.36	AC	105	$8.36
	G3	75	$0.99	BC	60	$3.92	BC	115	$7.51
				G3	105	$1.38	G3	135	$1.78
		Total :	$9.34			$13.66			$17.64

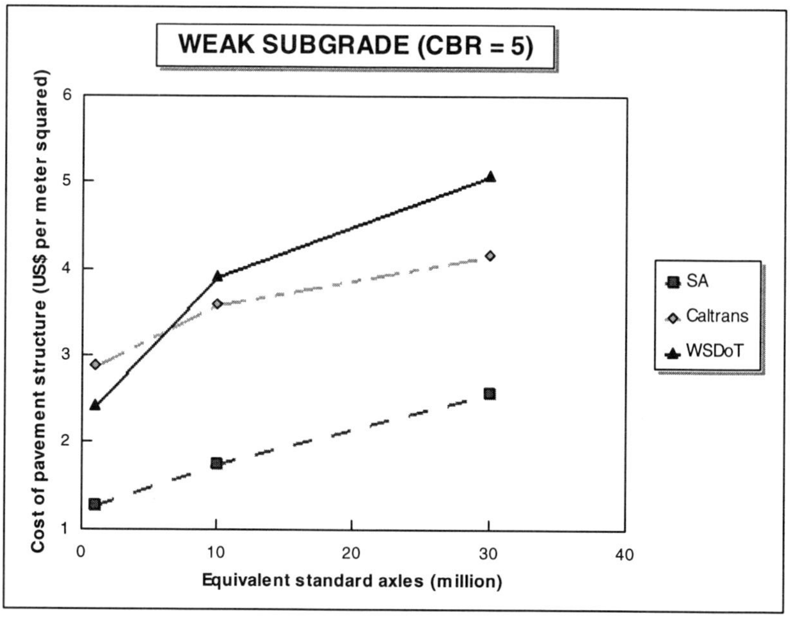

FIGURE 1 : COST OF COMPARATIVE DESIGNS ON A WEAK SUBGRADE (CBR = 5)

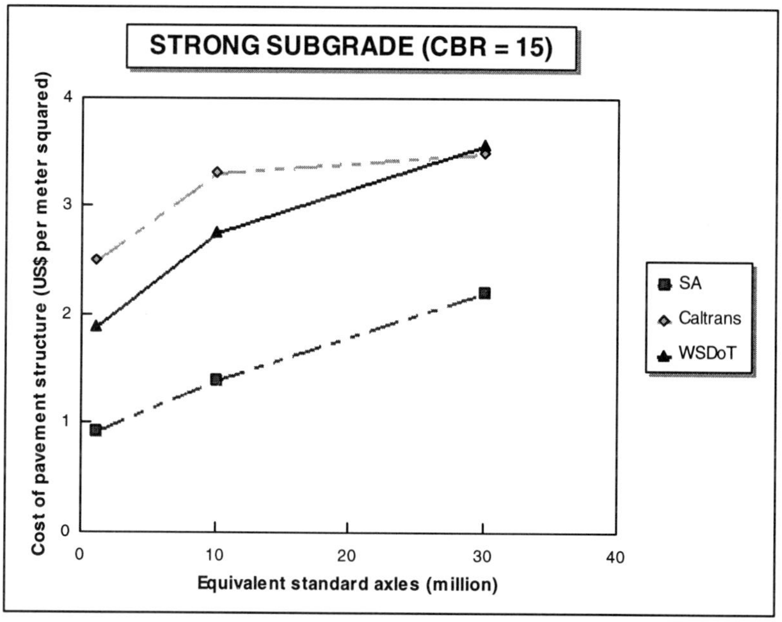

FIGURE 2 : COST OF COMPARATIVE DESIGNS ON A STRONG SUBGRADE (CBR = 15)

Fifth International Conference on the
BEARING CAPASITY OF
ROADS AND AIRFIELDS

FROM RATIONAL TO SUSTAINABLE PAVEMENT MANAGEMENT

F.J. KONINGS[1]
A. TIMMERMAN[2]
A.A. VAN DE GRIEND[3]

1,2: students, Polytechnic 's-Hertogenbosch, 5201 AS 's-Hertogenbosch, the Netherlands, fax 0031-73-6295205, e-mail @hsbos.nl
3: MSc civil engineering, Tauw bv, 3526 AB Utrecht, the Netherlands, fax 0031-30-2889484, e-mail u.tmcb@tauw.nl

ABSTRACT

During the last 25 years, a uniform system has been developed in the Netherlands for rational pavement management. In this system, a maintenance technique is selected based only on its cost and the expected life. No other aspects are taken into account.

Recently in the Netherlands, interest has been growing in sustainable construction. This concerns such aspects as maintenance-free (re)design, maximum reuse/recycling of materials, and minimisation of adverse effects on the environment and/or humans. Thus, sustainable construction not only is a matter of finances and civil engineering but also takes into account environmental and social aspects.

Although technical and financial aspects remain key issues, environmental and social aspects also play an important--if not decisive--role in the final selection of measures. Therefore, an investigation was conducted into possibilities for integrating these aspects into the current pavement management system. The new system takes aspects into account that result in sustainable construction by carrying out a so-called multi-criteria analysis.

Key words:

Pavement management
Sustainable construction
Multi-criteria analysis

INTRODUCTION

During the last 25 years, a uniform system has been developed in the Netherlands for rational pavement management, as detailed in issue No. 20 of the 'Information and technology center for transport and infrastructure' (CROW) – "Rationeel Wegbeheer (Rational Pavement Management)" (1). Several pavement management software-programs have become available which can be used to determine residual lives, maintenance measures and costs based on the above issue. In this, maintenance measures are usually compared using financial and civil engineering criteria.

Recently in the Netherlands, interest has been growing in sustainable construction, meaning "dealing sensibly with scarce means". This concerns such aspects as maintenance-free (re)design, maximum reuse/recycling of materials, and minimisation of adverse effects on the environment and humans. Thus, sustainable construction not only is a matter of finances and civil engineering but also takes into account environmental and social aspects.

As part of a final project carried out at Polytechnic 's-Hertogenbosch, an investigation was conducted into possibilities for integrating sustainable construction into the current pavement management systems and in particular into the current pavement management software. In other words, it was investigated how in the field of pavement management in addition to financial and civil engineering also environmental and social aspects could be taken into account. The aim of this article, although it discusses only asphalt pavement, is to initiate the forming of opinion in this regard.

RATIONAL PAVEMENT MANAGEMENT

Most pavement management software-packages used in the Netherlands are of similar design. First, for each type of damage a residual life is calculated based on figures derived from a visual inspection and life expectancies models developed for the type of damage concerned. The shortest residual life is usually considered indicative of the road section involved. Next, the maintenance measure to be taken is determined. In this, a selection is usually made from an existing set of maintenance/reconstruction measures, where for each measure the price per square metre is known. The most suitable measure is established based on the so-called principle of cost-effectiveness: for each measure, the quotient of its cost and the resultant residual life (i.e. its effect) is determined. In this, the measure (or combination of measures) with the lowest quotient is usually selected. Consequently, the current generation of pavement management software-packages is mainly of a financial/civil-engineering nature.

Although technical and financial aspects will remain key issues, other aspects also may play an important--if not decisive--role in the final selection of measures. Such aspects may be not only of an operational or political nature, but also of an environmental or social one, as mentioned above.

In most current pavement management software-packages, such aspects are not included, partly because it is hard or even impossible to predict them. However, they could be included at a later stage by manually adapting the calculated results. Furthermore, some of the above aspects are predictable and therefore could be integrated into the current pavement management software.

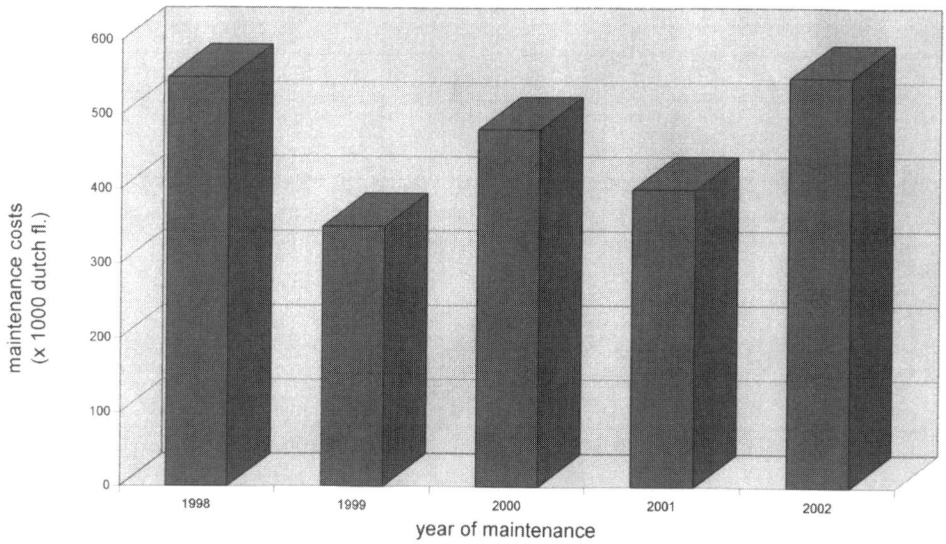

SUSTAINABLE CONSTRUCTION

Sustainable construction may be expressed in three lines of policy:

- Integral life-cycle management:
  reduction of the consumption of raw materials by adapting production processes, and the reduction, separation and reuse/recycling of construction and demolition waste;
- Energy saving:
  reduction of energy consumption in order to save the supply of finite fossil fuels and to decrease the emission of CO_2 and acidifying substances;
- Quality improvement:
  aiming at the optimum quality of the construction's surroundings by lengthening its residual life and increasing its reparability and possibilities for reuse, and by reducing (noise) nuisance.

(Source: Dutch Ministry of the Environment)

Although these lines of policy apply to all social sectors, road construction is one of the most important ones because of its massive use of (natural) raw materials. Therefore, this sector should especially use means as efficiently as possible and should gear supply and demand to each other to the utmost extent. For this purpose, building materials should be dealt with efficiently in the stages of design, execution as well as management.

To make the concept of sustainable construction more quantifiable and thus make it easier to relate it to pavement management, the three above lines of policy were subdivided into the following five aspects:

- Consumption of raw materials:
  = the consumed quantity of finite raw materials;
- Energy consumption:
  = the consumed quantity of energy;
- Nuisance:
  all consequences experienced by humans as being negative;
- Waste:
  = the quantity of released materials that cannot be reused/recycled;
- Emissions:
  = the emission of substances or sound into the surroundings.

INTEGRATION OF SUSTAINABLE CONSTRUCTION INTO THE PAVEMENT MANAGEMENT SYSTEM

In the current pavement management systems, measures are selected in a number of successive steps, namely:

1. Input of data derived from visual inspection;
2. Calculation of the residual life based on the inspection figures, the pavement life, and life expectancy models or warning and norm limits existing for each type of damage;
3. Calculation of the quotient of the cost and effect (i.e. lengthening of the pavement life) of each measure based on an existing set of maintenance/reconstruction measures;
4. Selection of the most cost-effective maintenance measure (in the case of longer planning periods, this may also be a combination of measures).

This schematic representation shows that a measure is selected after the cost and effect of each measure from a set of measures have been established. Thus, in that selection stage new selection criteria could be added, such as the consumption of raw materials, energy consumption, nuisance, waste and emissions. Qualification/quantification could occur as follows:

- Consumption of raw materials:
  this is normally expressed in kg (of material). Because the consumption of e.g. one kg of sand is not as harmful to the environment as that of one kg of bitumen, equivalence factors would have to be used to sum up the effects of various materials. For example,
  a "kilogram equivalent" (kgE) could be introduced as a reference unit;
- Energy consumption:
  this is expressed in kWh;

- Nuisance:
  it is harder to quantify this aspect. It may be considered to assess measures in this regard using + and -. However, some elements could be quantified, e.g. "hours of delay" (as a result of traffic jams or making a detour);
- Waste:
  this can be quantified similarly to the consumption of raw materials;
- Emissions:
  this aspect could also be quantified using + and -. In addition, the emission of substances could be quantified in e.g. g of CO_2, and that of sound in dBa's (which include a correction for the duration and for the moment when the sound-pressure level concerned occurs).

To determine the extent (amount) of each aspect, for each measure the entire process should be considered, i.e. from extraction to demolition/recycling.

Some of the above aspects are interdependent (e.g. energy consumption generally is related to the consumption of raw materials). However, in this investigation they were considered independent of each other.

For three imaginary maintenance measures, the cost per year of life extension (the criterion of cost-effectiveness) as well as five criteria connected with sustainable construction are given in Table 1. For each measure, a total of six criteria were quantified/qualified. In this, the original figures were converted into figures per m^2 of pavement, as given in the table.

Table 1 Resumé of selection criteria for the determination of measures

maintenance measure	cost per year of life extension (*f*)	consumption of raw materials (kgE)*	energy consumption (kWh)*	nuisance (h.o.d.)	waste	emissions
a	1	60	20	20	-	-
b	2	80	20	10	- -	-
c	4	100	30	5	0	0

(the figures are imaginary)
* kgE = kg equivalent
h.o.d. = hours of delay

0 = neutral
\- = unfavourable
\- - = highly unfavourable

To make a properly weighted selection, a multi-criteria analysis could be carried out. Such an analysis may include both quantifiable and non-quantifiable aspects. Furthermore, it would allow various selection criteria to be compared. For example, it could check whether a different measure would be preferable if highest priority were given to the aspect of nuisance. During this project, a type of multi-criteria analysis developed by TauwMabeg was carried out, which is based on the EVAMIX method.

To give an impression of what result could be attained, the above measures a, b and c were subjected to such an analysis. In this, the following order of priority was used, from highest to lowest: 1. nuisance, 2. emissions, 3. waste, 4. cost, 5. consumption of raw materials, and 6. energy consumption. Table 2 shows that in that case measure c would be preferable, despite its relatively high cost.

Table 2 Result of multi-criteria analysis

position	maintenance measure	percentage of preference
first	c	95
second	b	5
third	a	0

CONCLUSIONS

The above design for extension of the current pavement management system is in keeping with the higher priority nowadays given to sustainable construction. To put this idea into practice, the structure of the current pavement management software would have to be adapted. As a result, the selection of maintenance measures, which currently is usually based only on the principle of cost-effectiveness, would then also be based on a number of environmental and social aspects.

To work out this subject, a deeper insight would have to be gained into the effects of an adapted pavement management system on the chosen maintenance measures, the feasibility in practice, the degree to which the selection criteria are interdependent, and weighting factors would have to be established and missing data collected (e.g. on the consumption of materials, energy consumption, and nuisance).

REFERENCES

1. Stichting Centrum voor Regelgeving en Onderzoek in de Grond-, Water- en Wegenbouw en de Verkeerstechniek (CROW), Rationeel Wegbeheer, Volumes A, B and C, Ede, April 1989

Fifth International Conference on the
BEARING CAPASITY OF
ROADS AND AIRFIELDS

A NEW APPROACH TO MODELLING FLEXIBLE PAVEMENT RESPONSE

DARREN MERRILL[1], ANDREW BROWN[1], ANTHONY LUXMOORE[2] and STEPHEN PHILLIPS[3]

[1]Transport Research Laboratory, Crowthorne, UK
[2]University of Wales Swansea, Swansea, UK
[3]Highways Agency, UK

ABSTRACT

Recent research at TRL has indicated that the assumed forms of deterioration used in pavement design, fatigue cracking at the underside of the roadbase layer and rutting originating in the subgrade, are no longer the primary forms of deterioration in thick, flexible pavements. Cracking initiated at the surface and rutting caused by deformation in the surfacing are more commonly observed.

Existing pavement models were reviewed and were found to be unsuitable for predicting these forms of deterioration. This paper describes an approach to modelling pavement deterioration based on a Finite Element response model. Finite Element techniques are able to model the three dimensional contact stress distributions that have been predicted to have a large influence in the deterioration of the surface layers. A three dimensional Finite Element model has been constructed using ABAQUS. A comparison of this model under a uniform, vertical contact stress distribution and a non-uniform, three dimensional contact stress distribution is given.

INTRODUCTION

Current flexible pavement design methods are based on the assumption that the primary forms of pavement deterioration are fatigue cracking at the underside of the roadbase and structural rutting resulting from an accumulation of deformation in the subgrade. Pavement response models are used to calculate stresses and strains at critical locations in the pavement structure and then transfer functions are used to relate these responses to pavement deterioration. By altering the pavement thickness or material stiffness, the level of calculated stress or strain can be controlled to ensure that the primary forms of deterioration are prevented during the design life of the pavement.

Research has indicated that the assumed forms of deterioration are not the primary forms of deterioration in thick, well-constructed, flexible pavements. Therefore, a new approach is required to model the deterioration observed. This paper outlines the results of recent research undertaken at the Transport Research Laboratory. Site investigations were carried out on a number of trunk roads and motorways in the UK to establish the manner in which they were deteriorating. Available modelling methods were then reviewed to assess whether they could be used to predict the types of pavement distress observed and if not, to identify how improvements in pavement modelling can be made.

REASONS FOR NEW APPROACH

Flexible pavement deterioration in the UK

Flexible pavements are designed to minimise the risk of two modes of structural deterioration over the design life of the road. These are fatigue cracking initiating at the underside of the asphalt layer and structural rutting resulting from an accumulation of deformation in the subgrade.

Leech and Nunn (1997) investigated deterioration in well constructed, thick, fully flexible pavements. Extensive structural investigations were carried out at eight motorways and four trunk roads throughout England showing deterioration. These investigations failed to detect evidence of fatigue at the underside of the asphalt layer which suggested that this form of deterioration is less important than previously thought. However, deterioration was found to have taken place at the surface in the form of cracking and rutting. Nishizawa et al. (1997) came to similar conclusions stating that 'the failure modes of asphalt pavements are rutting and surface cracking' and also that 'rutting is usually due to plastic deformation in the asphalt layer, not due to failure of the subgrade'.

Data from TRL experimental roads reported by Nunn et al. (1997) shows that, at a thickness of greater than 200mm of asphalt, the rate of rutting was independent of thickness. This suggested that conventional structural rutting was not occurring and that rutting was originating in the surface layers. Examples of this type of rutting have also been observed by Dawley et al. (1990) and Paul (1996). The occurrence of non-structural rutting is likely to depend on asphalt mix design, compaction, climatic conditions and traffic loading.

Leech and Nunn (1997) found that structural deterioration within the lower layers of the pavement only occurred where there was poor construction practice or surface deterioration had been left untreated for a long period. A large number of cores were taken at each site directly on cracks and elsewhere but no evidence of cracking in the lower roadbase was found. Cracks appeared to have initiated at the surface and propagated downwards.

Surface cracking can develop as the wearing course loses its ability to withstand tensile stresses. This is probably due to the binder becoming progressively embrittled at the surface of the wearing course with exposure to solar radiaton and to the atmosphere. Consequently this process, which is generally known as ageing, was examined. Rheological testing of binders extracted from wearing course samples confirmed that most ageing occurs at the top of the wearing course layer and diminishes rapidly with depth. Ageing was also observed further down in the pavement structure but this is better described as curing as it can have a beneficial effect on pavement performance. This gradual hardening of the binder increases the elastic stiffness of the mixture and thus improves the load spreading ability of the layers with time.

While there was evidence of surface cracking being present to some degree in the majority of older pavements, the investigation did not identify the conditions conducive to surface cracking and the mechanism itself.

It can be concluded from the review of pavement deterioration that fatigue of the roadbase layers is not a significant problem for thick, fully flexible pavements (Leech and Nunn,1997). If fatigue occurred, there would either be cracks propagating from the bottom of the pavement or that stiffness of the roadbase would have reduced. Wu (1992) failed to detect any evidence of fatigue damage in a detailed study carried out on behalf of TRL, in which the residual fatigue lives of four heavily trafficked motorways in the UK were measured. Leech and Nunn (1997) cored many sites and found no evidence of fatigue damage. Additionally, cores taken through surface cracks generally showed that they did not propagate to the bottom of the roadbase. While investigating the structural performance of SHRP-NL test sections in The Netherlands, Schmorak and Van Dommelen (1994) also concluded that 'fatigue initiating at the underside of the asphalt will rarely or never be the predominant failure mechanism and that surface cracking is a major cause of structural distress'.

For thick, long life pavements in the UK, cracks can initiate in the pavement surfacing and propagate with time into the lower layers. Longitudinal cracks, occurring close to the wheelpaths, are related to traffic loading whereas transverse cracks are more likely to be related to temperature changes: these forms of deterioration are classified as load associated and non-load associated respectively. However, thermal effects may be largely responsible for propagating both types of cracking.

Existing pavement response models

A review of pavement response models was carried out. These models can be classified into two groups; models based upon the Burmister's solution, referred to in this paper as Multi-Layered elastic Theory (MLT), and models based upon the Finite Element (FE) method. Whilst FE models have been in existence for many years, MLT models are used almost exclusively (Nunn and Merrill,1997) in pavement design. This is because they are particularly suited to calculating the responses thought responsible for the assumed forms of deterioration in flexible pavements. Also, these types of models do not require large computing resources which are essential for many finite element models.

MLT models assume that the pavement layers are homogeneous and extend infinitely in all horizontal directions with the foundation assumed to be a half-space. Most commonly loading is applied statically at the surface by a vertical wheel load which is assumed to be circular and uniformly distributes the contact pressure over the loaded area. This type of model will generally calculate horizontal tensile strain to be a maximum at the base of the asphalt layer but will suffer from mathematical singularities when calculations are required close to the edge of the applied load. The assumed discontinuity between loaded and unloaded regions is responsible for this and it prevents the calculation of realistic solutions. As a result, MLT models are considered unsuitable for investigation of the pavement responses close to the pavement surface.

The development of FE based pavement response models has increased in recent years. The attraction of using FE methods is their flexibility because in general, they are not limited by geometry and, can accept complex loading and material behaviour. A major advantage of finite element models is that responses can be calculated close to the wheel load because the effects of the assumed load discontinuity caused by the wheel load can be controlled.

Despite the general flexibility of the FE models, most pavement FE models are restricted to axi-symmetric analysis to reduce the computation time and simple material behaviour is often assumed. Also, the prescription of load is generally restricted to the simple uniformly distributed case used in MLT models and is considered to be adequate for pavement design purposes, given the current assumptions concerning pavement deterioration.

DETERMINATION OF THE REQUIREMENTS FOR A NEW MODEL

The review of available pavement design models together with the investigations of pavement deterioration have exposed a contradiction. The existing models are not suitable for modeling pavement responses close to the surface although it is clear the primary forms of deterioration are located in this region. A new improved approach is needed which is able to reasonably predict responses close to the pavement surface.

Investigation of significant factors

During the review of available models BISAR-PC, an MLT model, was used to assess the effects of changes in selected material parameters. BISAR-PC is a restricted version of BISAR (de Jong et al.,1973) for use with personal computers. The elastic stiffness of the layers appeared to have a large effect on the calculated responses, particularly at the surface.

Jacobs (1995) used BISAR's facility of accepting horizontal loads and its multiple load feature to simulate the effects of more realistic tyre contact stresses. Particular care was taken to avoid singularities by not defining calculation points in the pavement close to the wheel load. The introduction of a more realistic wheel load was shown to have a dramatic effect on responses close to the surface. In particular, large tensile stresses were calculated under the edge of the loaded area. This effect diminishes rapidly with depth.

Requirements for a new model

A set of basic requirements have been defined for a new model and particular attention was paid to the deterioration observed in thick pavements in the UK. For a pavement response model capable of calculating the responses close to the surface which are likely to influence the appearance of load-associated surface cracking the following features must be included:

- a realistic tyre contact stress distribution and,
- values of stiffness for pavement materials which accurately reflect changes in material properties with depth due to ageing and temperature.

To include these features in a model, an approach based on Finite Element analysis appears to be suitable. This approach preserves the possibility of including other features such as complex material behaviour.

CONSTRUCTION OF A NEW MODEL

ABAQUS/Standard, a general purpose finite element software, was used for the finite element calculations in this study. This package enables plane strain, axi-symmetric and full three dimensional models to be constructed and run, if required, using complex material behaviour and dynamic analysis routines. Plane strain models were constructed in this work but found to be unsuitable for pavement analysis.

Creation of the finite element models

An axi-symmetric FE model was produced because using this type of geometry enabling close comparison MLT models. However, differences will always be present because MLT models provide 'closed form' exact solutions of a particular simplified system whereas FE methods are always approximate. If the output from both types of model are comparable, the FE approximation can be viewed as accurately representing the simple MLT system.

For a model to precisely replicate MLT solutions, the model geometry should extend infinitely in all horizontal directions and be infinite in one vertical direction. This is not possible using finite elements, which intrinsically have boundaries at a finite distance from the load and therefore, the boundaries will always have an effect. However, the positions of the boundaries can be chosen such that their effects are not large. Duncan et al. (1968) state that the vertical boundaries of a FE model must be at least 12 times the radius of the load contact area, with the horizontal lower boundary at a depth of at least 50 times the load radius. Under these conditions a finite element model will be cumbersome when considering the small area of interest close to the wheel load is a fraction of the overall mesh size.

To reduce the overall mesh size, infinite elements were used. These infinite elements have been successful in modelling Boussinesq's problem of calculating the deflection resulting from a point load applied to an infinite half-space (ABAQUS/Standard,1996). The finite elements at the outer edges of the mesh were substituted for infinite elements both on the base and the vertical boundaries. An example of the type of mesh used is given in Figure 1. The finite elements are purposely small close to the loaded region because of the expected high stress gradients.

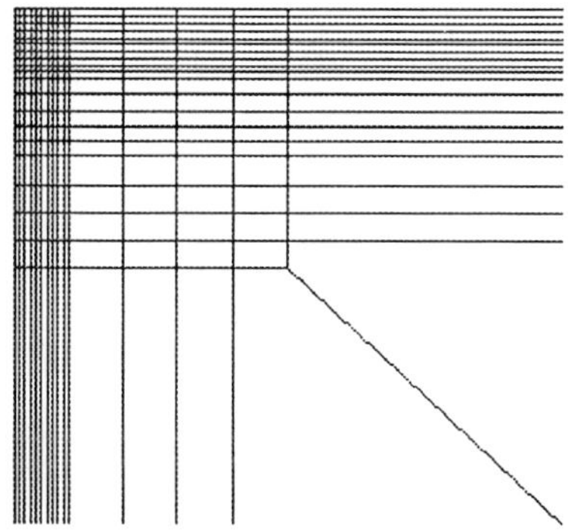

Figure 1. Example of the finite element mesh

A three dimensional FE model was produced, based upon the mesh used in the axi-symmetric model. Three dimensional models require extensive computer resources to cope with the large number of nodes and elements in the mesh. To reduce the computation effort, the model was constructed as a quartile by applying rules of symmetry along the two vertical planes. Thus, only 25% of the total number of elements in the whole model are required, substantially reducing memory requirements and computation times. Infinite elements were again used on the bottom and vertical boundaries at the extremities of the mesh.

Loading is specified through point loads rather than distributed across a face of a loaded element. In ABAQUS, distributed loads can only be specified normal to a particular element face whereas point loads can be defined in all three coordinate directions which is essential for the model to accept realistic contact stress distributions.

Comparison of FE models with MLT models

To compare the output of the axi-symmetric model and the MLT solution, strain was calculated at two positions for a particular pavement construction under identical loading conditions. The calculations were carried out at positions where MLT models are known to produce reliable results; directly under the wheel load at the bottom of the asphalt layer and also on the surface of subgrade. A similar comparison was done to check the output of the three dimensional model. A summary of the calculated strains is given in Table 1. The results showed that the calculated strains using the axi-symmetric model to be within 5% of the MLT solution while the 3D FE model was within 10% of the MLT solutions.

Response	MLT	AXI FE	3D FE
ε_{11} at asphalt base	8.61E-05	8.42E-05	7.75E-05
ε_{22} at asphalt base	-9.79E-05	-9.60E-05	-9.85E-05
ε_{11} at top of subgrade	8.59E-05	8.09E-05	7.8E-05
ε_{22} at top of subgrade	-1.99E-04	-2.05E-05	-2.13E-05

Table 1. Calculated strains using MLT and FE methods

The greater difference with the 3D FE model is consistent with the findings of Chen et al. (1995) that axi-symmetric methods, both MLT and FE, overestimate the strains produced using 3D FE analysis. The agreement could be improved using more elements in the mesh, or by utilising quadratic type elements in the mesh. Both these methods are known to improve the finite element approximations but the desire to improve the solution must be balanced with the need for a practical model with tolerable computation times. For these reasons, the output obtained from this model is considered reasonable by the authors.

REALISTIC CONTACT STRESS DISTRIBUTIONS

For pavement design purposes, the contact stress distribution between tyre and pavement is assumed to apply a uniformly distributed vertical pressure over a circular area. This is analogous to modelling the tyre as a structureless balloon in contact with a smooth surface. In reality the contact stress distribution is more complex due to the rigidity of the rubber and the internal structure of the tyre.

The measurement of contact stress between tyre and pavement surface is a difficult problem. A successful method has been developed by CSIR in South Africa using the Vehicle-Road Surface Pressure Transducer Array system or VRSPTA (de Beer,1994). Using an array of pins, with one row instrumented, measurements are taken as a wheel rolls across the pins.

Forces are measured in three directions so that the full three dimensional stress distribution can be observed. Examples of these measurements are given below.

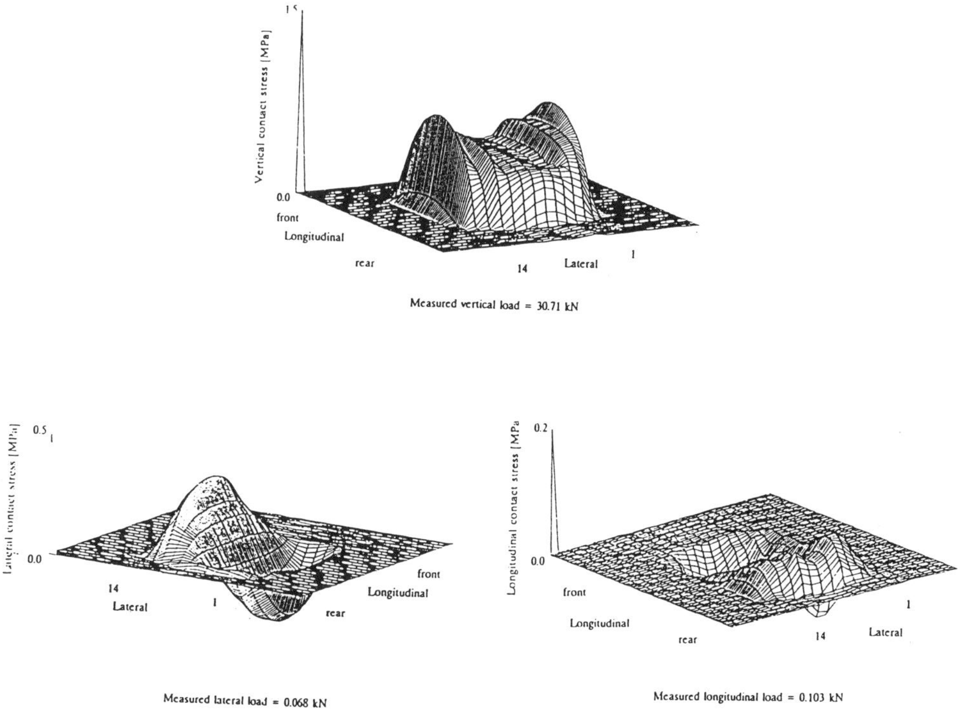

Figure 2. Vertical, lateral and longitudinal contact stress distributions of a 40kN bias-ply truck tyre inflated to 620kPa, de Beer (1994)

The contact stress distribution is highly non-uniform in the vertical direction and horizontal components also exist. The peaks of vertical contact stress close to the tyre edge are caused by the rigidity of the tyre walls. These peaks are not evident at low load levels. The contact pressure at the centre of the tyre is approximately proportional to the inflation pressure. As expected the peak values of vertical stress are directly below the centre of the tyre in the moving direction. The measured horizontal force normal to the direction of travel is, in general, inward toward the centre line of the tyre. The contact stresses in the longitudinal direction, in the moving direction, peak close to the front of the tread, becoming neutral in the centre with the opposite peak at the rear.

EFFECT OF THE THREE DIMENSIONAL CONTACT STRESS DISTRIBUTION ON PAVEMENT RESPONSE

Three dimensional vertical and transverse contact stress distributions were generated based on data measured by de Beer (1994). The longitudinal stresses were neglected because under a free rolling wheel they should not be large compared to vertical stresses (Mante et al.,1995) and therefore should not have a great effect on the output. Also, including longitudinal stresses would have invalidated one of the assumed axes of symmetry in the FE mesh since the stresses are not symmetrical about the central axis, doubling the number of elements required for a solution.

The output of FE analysis using a three dimensional, non-uniform, generated load and a conventional uniform load under similar conditions were compared. Figure 3 shows the transverse strain response close to the surface. Under the uniform distribution, no tension is seen under the edge of the tyre whereas the 3D distribution produces significant tension in this area. Deeper in the pavement structure the responses are similar suggesting that the 3D distribution has little effect. Displacement of the surface under both types of load are similar, confirming that equivalent wheel loads were applied. These results are important because tensile transverse strains in the surfacing are a likely initiator of load associated surface cracking.

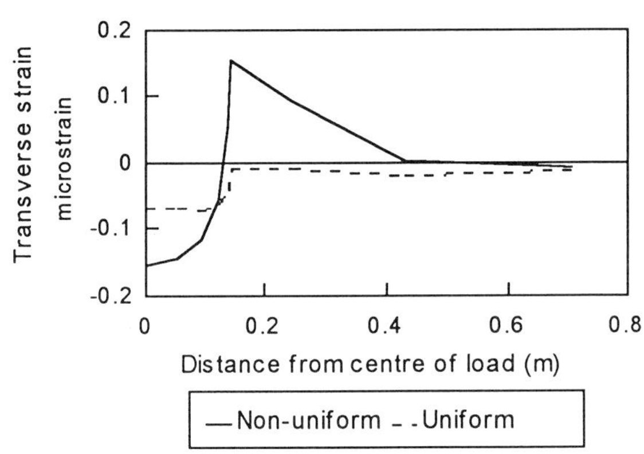

Figure 3. Transverse strain at the surface

CONCLUSIONS

- Fatigue cracking at the underside of the asphalt layer and structural rutting are not significant forms of deterioration in thick, well constructed flexible pavements.
- Cracks initiating at the surface and non-structural rutting are observed on thick, flexible pavements.
- Conventional pavement response models do not predict the necessary conditions for the initiation of surface cracking.

- The three dimensional nature of the tyre contact stress distribution has an important bearing on the predicted pavement response close to the surface and it must be taken into account in any model that attempts to predict pavement behaviour close to the surface.

- The effect of the three dimensional contact stress distribution diminishes rapidly with depth. For analyses at positions deep within the pavement structure, a simplified circular load can be used. The conventional MLT models may be justified for this purpose.

- Axi-symmetric and plane strain type finite element models are not suitable for modelling pavement responses close to the surface.

- A three dimensional finite element model has been produced which can represent the behaviour of the pavement close to the surface. This model uses simple linear elastic material behaviour and neglects the effects of temperature. Development of the model may include these features as necessary in the future.

ACKNOWLEDGMENTS

Copyright Transport Research Laboratory 1998.

This research was sponsored by the Highways Agency, UK. This paper is published by permission of the Chief Executive of the Highways Agency and the TRL. The assistance of the Civil Engineering Department of the University of Wales Swansea is gratefully acknowledged. The views expressed are not necessarily those of the Department of the Environment, Transport and the Regions.

REFERENCES

1. LEECH, D and NUNN, M.E. Deterioration mechanisms in flexible roads. Proceedings of the 2nd European Symposium on Performance and Durability of Bituminous Materials. Leeds, UK, 1997.

2. NISHIZAWA, T., SHIMENO, S. and SEKIGUCHI,M. Fatigue analysis in asphalt pavements with thick asphalt mixture layer. 8th International Conference on Asphalt Pavements, Seattle, USA, 1997.

3. NUNN, M.E, BROWN, A., WESTON, D. and NICHOLLS, J.C. Design of long-life flexible pavements. TRL Report 250. Transport Research Laboratory.

4. DAWLEY, C.B, HOGEWEIDE C.L. and ANDERSON K.O. Mitigation of instability rutting of asphalt concrete pavements in Lethbridge, Alberta, Canada. Proceedings of the Association of Asphalt Paving Technologists, Vol 59, 1990.

5. PAUL, H.R. Case history:I-10 Pavement distress. Proceedings of the Association of Asphalt Paving Technologists, Vol.65, 1996.

6. WU, F. Assessment of residual life of bituminous layers for the design of pavement strengthening. PhD thesis, The Polytechnic of Wales, 1992.

7. Van DOMMELEN, A and SCHMORAK, N. Verification of design method. Delft University of Technology, 1990.

8. NUNN, M.E. and MERRILL, D.B. Review of flexible and composite pavement design methods. COST Action 333. TRL Paper PA/3298/97, 1997

9. de JONG, D.L., PEUTZ, M.G.F. and KORSWAGEN, A.R. COMPUTER PROGRAM BISAR- Layered systems under normal and tangential surface loads. AMSR.0006.73 Shell research BV.

10. JACOBS, M.M.J. Cracking in asphalt mixes. PhD thesis, The Delft University of Technology, The Netherlands, 1995.

11. DUNCAN, J.M, MONISMITH, C.L and WILSON, E.L. Finite element analyses of pavements. HRR228, Soils, stresses and pavement analyses. Washington : Highway Research Board, 1968.

12. ABAQUS/Standard Users Manual Version 5.6, Hibbit, Karlsson and Sorenson Inc, 1996.

13. CHEN, D-H, ZAMAN, M, LAGUROS, J and SOLTANI, A. Assessment of computer programs for analysis of flexible pavement structure. Transportation Research Record 1482, TRB, National Research Council, Washington, USA, 1995.

14. de BEER, M. Measurement of tyre/pavement interface stresses under moving wheel loads. Pavement Engineering Technology, Division of Roads and Transport Technology, CSIR, Pretoria, 1994.

15. MANTE, B.R, MOLENAAR, A.A.A and GROENENDIJK, J. Stresses and strains in pavements. Unpublished report. Delft University of Technology, The Netherlands, 1995.

Fifth International Conference on the
BEARING CAPASITY OF
ROADS AND AIRFIELDS

ANALYSIS OF THE NORWEGIAN PAVEMENT DESIGN MANUAL WITH A LIFE-CYCLE COST PERSPECTIVE

EVEN K. SUND

Research Engineer, SINTEF Civil and Environmental Engineering,
Highway Engineering, N-7034 Trondheim, Norway
Fax: +47 73 59 70 20
E-mail: Even.K.Sund@civil.sintef.no

ABSTRACT

A method for calculating the life-cycle cost of road pavements considering the user costs in a way that makes them comparable over a defined analysis period is presented. The method has been used together with a mechanistic analysis to analyze the life-cycle costs of pavements for low traffic volume designed using the Norwegian Pavement Design Manual (PDM) [1]. The length of the design period was increased from 10 to 20 years when this design manual was introduced in 1992, leading to an increase in initial pavement thickness and costs. The question whether this was economically wise was raised. The analyses presented in this paper are based on the work presented in the thesis submitted by the author for the partial fulfillment of the requirements for the degree of Dr.ing. at the Norwegian University of Science and Technology (NTNU)[2]. Based on these analyses there are no savings to be made in highway agency costs by deviating from the new Norwegian PDM. User costs were calculated using the Norwegian Pavement Management System. Based on the total life-cycle costs the pavement design thickness should be increased for roads with low traffic volume.

INTRODUCTION

The Norwegian PDM [1] prescribes the use of a 20-year design period for pavements on main roads. This was an increase from the previous version of the PDM [3], which had 10 years as a standard design period. This increase resulted in new roads being designed with thicker layers than before, resulting in higher construction costs. The question that naturally arose was whether this increase in design thickness was economically sound. Or more generally stated: are pavements designed using the latest version of the Norwegian PDM the most economical ones, i.e. with the lowest life-cycle costs? It is impossible to give a general answer to this question. Many assumptions have to be made in order to assess the life-cycle costs of the different pavement designs. If one or more of these assumptions are changed, the answer to the question may also change. Therefore it is essential to include sensitivity analyses to any investigation of pavement life-cycle costs.

Wear from studded tires is critical for most Norwegian pavements with an AADT above 2000-3000 vehicles/day. For these pavements the analyses have to be carried out in a slightly different manner than for those pavements where this damage mechanism is not critical, i.e. the pavements with low-volume traffic. The analyses presented in this paper are for pavements with a low traffic volume (AADT=1000 vehicles/day).

The mechanistic analyses were performed using the VESYS 3A-M program. The use of a mechanistic approach using different primary response or damage models could have given different results.

Analyses of the economic aspects of the new design manual have been conducted earlier [4], but these were not based on a mechanistic analysis to establish the relative pavement lives and did not include the user costs.

METHOD USED FOR LIFE-CYCLE COSTING

In this section I present a theoretical model of life-cycle costing for highway pavements, including a method for correcting the marginal user costs for comparability over a defined analysis period. The life-cycle costs of a pavement are traditionally expressed as:

$$LCC_j = C_j + \left[\sum_{y=0}^{N} \left(M_{jy} + U_{jy} + E_{jy} \right) \cdot (1+i)^{-y} \right] - (R_{jC} + R_{jM}) \cdot (1+i)^{-N} \qquad (1)$$

LCC_j : Life-cycle costs of pavement alternative *j*.
C_j : Construction costs of pavement alternative *j* (above the foundation level).
M_{jy} : Maintenance costs for pavement alternative *j* in year *y*
U_{jy} : Accumulated marginal user costs[1] for pavement alternative *j* in year *y*
E_{jy} : Accumulated environmental costs for pavement alternative *j* in year *y*
R_{jC} : Residual value of the construction costs for pavement alternative *j* at the end of the analysis period (year N)
R_{jM} : Residual value of maintenance costs for pavement alternative *j* at the end of the analysis period (year N)
y : year in analysis period
i : Real discount rate (%) - common for all alternatives
N : Analysis period (years) - common for all alternatives

This model does not take into consideration the fact that the user and environmental costs (seldom included in practical applications) of different alternatives may not be directly comparable. The highway agency costs of alternatives with different number of complete service life cycles are made comparable by the use of a residual value. The user costs do not

---

[1] Marginal user costs are defined as the portion of the total user costs that can be directly attributed to pavement condition.

have a residual value, but nevertheless a correction should be done to make them comparable for different alternative pavement strategies. An illustration of the importance of this is illustrated in Figure 1, where the accumulated present value of user costs for a 20-year analysis period is shown as a function of expected pavement service life when no correction is made.

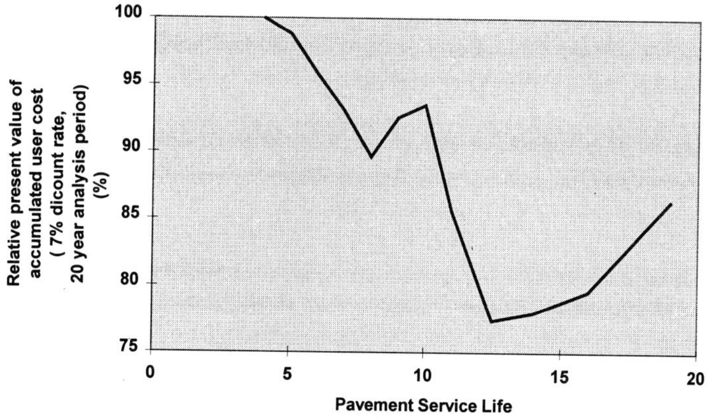

FIGURE 1 Relative present value of accumulated user costs (20-year analysis period, 7% discount rate) as a function of expected pavement service life (calculated using the Norwegian PMS)

As seen in Figure 1 the accumulated user costs have a peak value when the expected service life is 10 years. This is simply because 10 years goes two times into 20 years, which is the length of the analysis period. This alternative has two complete service life-cycles included in the analysis period, where as a pavement with a predicted service life of e.g. 13 years has only 1.5 complete cycles, and gains an extra "advantage" because of this. This is illustrated in Figure 2.

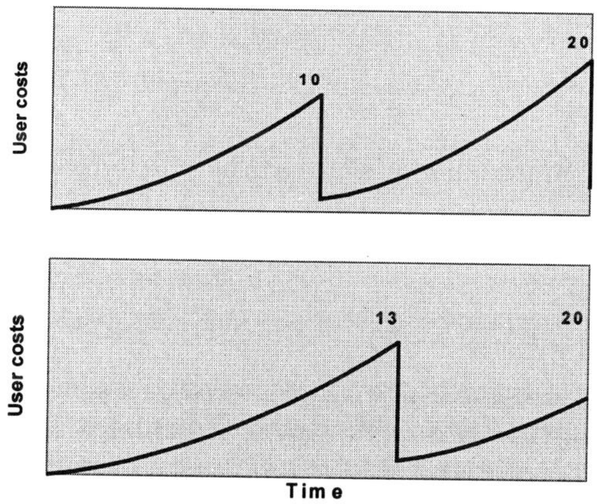

FIGURE 2 Illustration of the user costs as a function of time for pavement service lives of 10 and 13 years (2% annual traffic increase)

The accumulated user costs of the alternative with an expected life of 13 years are not directly comparable with those of the alternative with 10 years expected life. To make the accumulated user costs directly comparable they have to be corrected. The methodology for doing these corrections in the accumulated discounted user costs is briefly described below.

1. The annualized user costs are calculated for the maximum whole number of service life-cycles within the analysis period (called the annuity period). This value is called the *Representative Annuity,* and is illustrated in Figure 3.

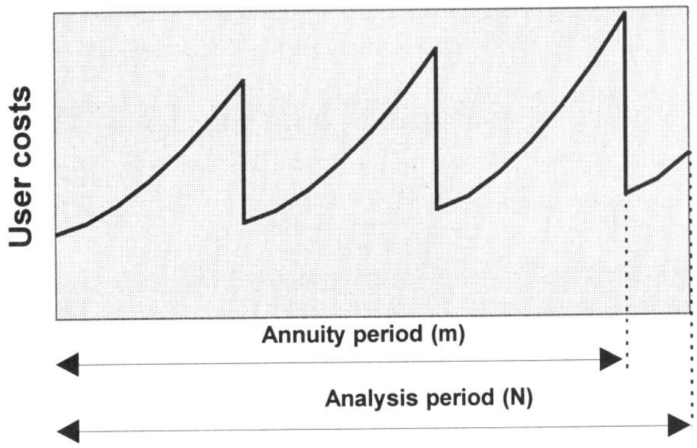

FIGURE 3 The annuity period

$$RA\ (Representative\ Annuity) = \left(\sum_{y=0}^{m} (U_y + E_y) \cdot (1+i)^{-y} \right) \cdot \left(\frac{i(1+i)^m}{(1+i)^m - 1} \right) \quad (2)$$

2. The representative annuity is transformed into an equivalent present value for the whole analysis period (N), by multiplying the representative annuity by the inverse annuity factor for the whole analysis period. The *RA*-value is thereby made representative of the user costs for the whole analysis period.

$$RPV\ (Representative\ Present\ Value) = RA \cdot \left(\frac{(1+i)^N - 1}{i(1+i)^N} \right) \quad (3)$$

3. To take into consideration the expected annual traffic growth in the period between the end of the annuity period (m) and the end of the analysis period (N), the present value is corrected as follows:

$$C_{Traf} = \frac{\textbf{averge traffic over the analysis period (N)}}{\textbf{average traffic over the annuity period (m)}} = \frac{(1+v)^N - 1}{(1+v)^m - 1} \cdot \frac{m}{N} \quad (4)$$

where v = expected annual traffic increase in decimals

$$CPV\ (Corrected\ present\ value) = c_{traf.} \cdot RPV \quad (5)$$

By using this method of correction the accumulated user costs as a function of predicted pavement service life is altered from the situation illustrated in Figure 1 to that shown in Figure 4.

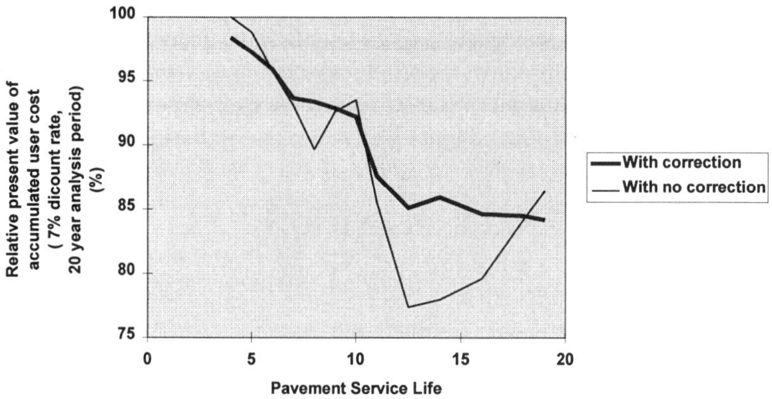

FIGURE 4 Relative present value of accumulated user costs (20-year analysis period, 7% discount rate) as a function of expected pavement service life, after correction because of different number of service life-cycles in the analysis period

The user cost function now has a more logical shape[2], and the user costs calculated for a fixed analysis period are comparable for alternatives with different expected service lives. There are still some minor "irregularities" on the curve. This is mainly due to that the PMS-program calculates user costs using whole years as the basic time unit, which causes some minor differences in calculating user costs for 1/2 years in years when maintenance takes place.

A more appropriate model than the one represented by equation (1) would be:

$$\begin{aligned} \mathrm{LCC_j} &= \mathrm{C_j} + \left[\sum_{y=0}^{N}\left(M_{jy}\right)\cdot(1+i)^{-y}\right] - (\mathrm{R_{jC}} + \mathrm{R_{jM}})\cdot(1+i)^{-N} \\ &+ \left[\sum_{y=0}^{m}\left(U_{jy} + E_{jy}\right)\cdot(1+i)^{-y}\right]\cdot \mathrm{A_m}\cdot \mathrm{A_N}\cdot \mathrm{c_{traf}} \end{aligned} \qquad (6)$$

where m = the period of time for the maximum number of whole service life cycles for alternative j in the analysis period N

A_m = annuity factor for transforming a present value to an equivalent annuity over the period m $\left(\dfrac{i(1+i)^m}{(1+i)^m-1}\right)$

A_N = summation factor for transforming an annuity over the period N to an equivalent present value $\left(\dfrac{(1+i)^N-1}{i(1+i)^N}\right)$

---

[2] *When a polynomialcurve is fitted to the calculated points the* $R^2=0.95$

This model makes the user- and environmental costs comparable between alternatives with different life-cycle duration when using a fixed length of the analysis period. As a series of maintenance actions during the analysis period may consist of varying types (and thus costs), a simple annuity calculation over one maintenance cycle would not yield the correct answer.

ANALYSIS AND RESULTS

Procedure

The procedure used in the analysis is outlined in Figure 5.

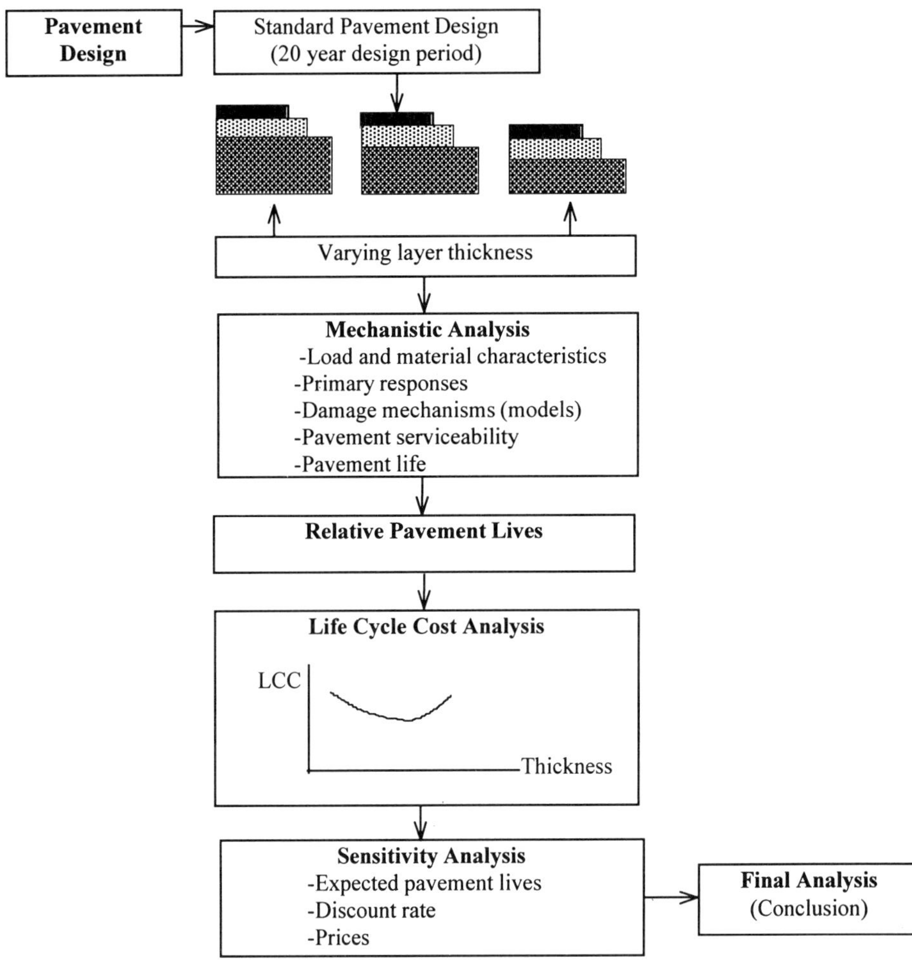

FIGURE 5 Outline of the procedure used for analyzing the Norwegian PDM with a life-cycle cost perspective

The mechanistic analysis was performed using the VESYS 3A-M program developed by the Federal Highway Administration (FHWA) of the United States. The program was only used to calculate the relative pavement lives, not absolute pavement lives, as a function of varying pavement thickness. The VESYS program calculates the primary responses under a circular load assuming linear elastic, isotropic and homogeneous response. Typical values of the E-modulus and Poisson's ratio were used in these analyses. The primary responses are used in models for the following damage mechanisms: permanent deformation, cracking (fatigue) and roughness. These are transferred to a performance model that uses the Present Serviceability Index (PSI)

Three pavement configurations with different materials in the surface and base layers were analyzed. All pavements were analyzed for good, medium and poor subgrade conditions, as they are defined in the Norwegian PDM. For each of the three pavement types the thickness for each layer as prescribed by the Norwegian PDM were found. The base and/or subbase layer thickness was varied around these values, yielding different relative pavement lives from the VESYS analysis. The PDM-designs were assumed to have normal pavement lives, as documented in the report from the BUAB-project [5][3]. The VESYS analysis gave quite large effects on pavement lives, even with moderate variations in base and/or subbase layer thickness. For the pavements thinner than the PDM-design a comparison between the VESYS analysis and curves given in the BUAB-report was made. The F_{Diff}-value expresses the "strengthening need" of a pavement relative to the PDM design. Figure 6 shows the result of the comparison for one of the pavement configurations for varying subgrade conditions (B2=good, B4=medium and B6=poor).

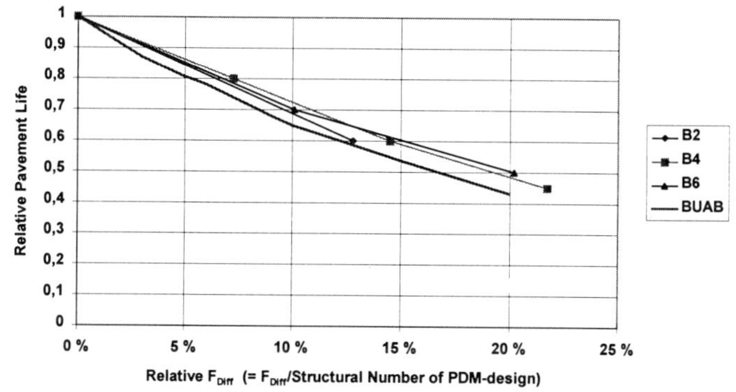

FIGURE 6 Comparison of relative pavement lives from BUAB and the VESYS analysis for three different subgrade conditions using a relative F_{Diff} (Pavement 1)

The reduction in pavement lives for this pavement is of the same magnitude for the VESYS analysis as was found in the BUAB project.

Results

Prices used in the analysis were the same as in the Norwegian Pavement Management System. Normal maintenance activities were assumed at the end of the pavement lives. The results of the life-cycle cost analysis are shown in Figure 7

---

[3] BUAB - "Better Utilization of the Bearing Capacity of Roads" was a research project carried out by the Norwegian Road Research Laoratory in the period 1990-94

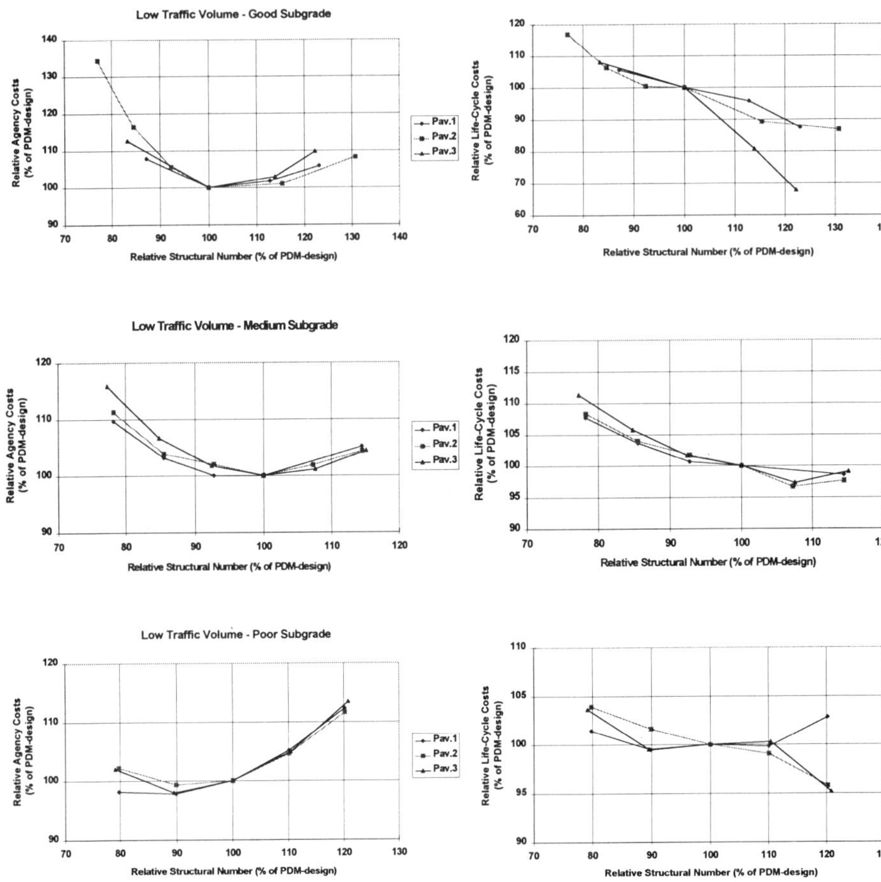

FIGURE 7 Results of the life-cycle cost analysis, using a 7% p.a. real discount rate and a 20-year analysis period

For good and medium subgrade conditions the analysis shows that the PDM-design is optimal with respect to highway agency costs. For poor subgrade the thinner pavements have approximately the same or slightly lower calculated highway agency costs then the PDM-design. For all subgrade types, except one case for poor subgrade conditions, the calculated total life-cycle costs increase when the pavements are designed thinner than the PDM prescribes. For pavements designed thicker than PDM-level, the tendency is opposite. The calculated highway agency costs increase, while there is a tendency towards lower total life-cycle costs. This is mainly due to the large predicted increase in pavement life in VESYS when the pavement thickness is increased. A lower estimated pavement life increase would mean higher life-cycle costs for these pavements.

Sensitivity analysis

An extensive sensitivity analysis was carried out. How the input parameters in VESYS affected the relative pavement lives was analyzed before deciding on which values to use in the further analyses. The results of the life-cycle cost analysis were also subjected to sensitivity analysis of the most important parameters, which are the expected pavement lives, the discount rate and the prices of materials.

The discount rate is an important parameter than can significantly influence the conclusions. Generally a high discount rate favors alternatives with a low initial investment and higher maintenance costs. A discount rate of 7% p.a. is relatively high. In these analyses the discount rate influenced the conclusions based on the highway agency costs, but did not alter the ranking based on the total life-cycle costs. Changing the life expectancies of all the pavements by +/-20% in some cases marginally shifted the conclusion based on both highway agency and total life-cycle costs. Varying the material prices by +/-25% in some cases marginally shifted the conclusion based on highway agency costs.

CONCLUSIONS AND RECOMMENDATIONS

The LCC-analyses presented in this paper indicate that in most cases no savings in highway agency costs can be expected by deviating from the designs prescribed by the Norwegian PDM for pavements with low traffic volume. Designing thinner pavements than the PDM prescribes can in a few cases give lower highway agency costs, but these are more than outweighed by the increase in marginal user costs. Designing stronger pavements than prescribed by the PDM increases the highway agency costs for all analyzed pavements. However, because of the quite substantial increases in predicted pavement lives, the total life-cycle costs were in most cases lower than for the PDM designed pavements. Based on this the conclusion for these pavements are that thicker pavement designs should be used for roads with low traffic volume.

Life-cycle costing should be made an integrated part of pavement design. Even though the basis for conducting such analyses can still be greatly improved, LCC-analyses should still be used to evaluate different pavement design alternatives. A computerized method should be integrated with the existing and future systems for pavement design. In Norway there are existing computer programs for flexible pavement design using design levels 1(simple empirical design) and 2 (semi-empirical design). Design level 3 (mechanistic design) has not yet been developed. When a mechanistic system is developed, a life-cycle costing module should be made a part of it. In this way different technical solutions can be generated with varying thickness, maintenance strategies etc., and the one with the lowest life-cycle costs can be identified. There is also a general need to improve the models used for calculating the user costs.

Further analyses using different methods of mechanistic analysis than those implemented in VESYS should be conducted in order to verify the conclusions. The influence that pavement thickness has on resulting pavement life is a key input when doing life-cycle cost analysis of alternative pavement designs.

ACKNOWLEDGEMENTS

The thesis on which this paper is based was carried out at the Norwegian University of Science and Technology, Department of Road and Railway Engineering. Professor Rasmus

S. Nordal was my supervisor throughout the work with this thesis. I would like to thank him for his support and guidance. I would also like to thank everybody else at the Department of Road and Railway Engineering and SINTEF Highway Engineering for keeping my spirits up and providing an excellent working environment.

I thank the Norwegian University of Science and Technology for granting me a scholarship. The Norwegian Public Roads Administration has also supported my work, both financially and with background material for the analyses.

REFERENCES

1. Norwegian Public Roads Administration, Directorate of Public Roads — *Normaler, Vegbygging* No. 018 in the Public Roads Administration series of manuals, Oslo 1992

2. SUND, E. K. — *Life-Cycle Cost Analysis of Road Pavements* Dr.ing. thesis, Department of Road and Railway Engineering, The Norwegian University of Science and Technology, Trondheim, 1996

3. Norwegian Public Roads Administration, Directorate of Public Roads — *Normaler, Vegbygging* No. 018 in the Public Roads Administration series of manuals, Oslo 1980

4. BERTELSEN, D. SKOGLUND, R. — *Levetidskostnader for vegkonstruksjoner* SINTEF Highway Engineering, Report No. STF61 A93013, Trondheim, 1993

5. Norwegian Public Roads Administration, Directorate of Public Roads, Road Research Laboratory — *Sluttrapport fra etatsatsingsområdet Bedre utnyttelse av vegens bæreevne* Publication No.75, Oslo 1994

Fifth International Conference on the
BEARING CAPASITY OF
ROADS AND AIRFIELDS

INVESTIGATION OF THE RELATIONSHIP BETWEEN PAVEMENT REHABILITATION STRATEGIES AND LIFE CYCLE COSTS

LI NINGYUAN, RALPH HAAS, AND MICHEL HUOT

Department of Civil Engineering
University of Waterloo
Waterloo, Ontario
Canada N2l 3E5

ABSTRACT

This paper presents a new approach to determining the optimal maintenance and rehabilitation (M&R) treatments for small and middle sized pavement networks. This approach integrates pavement performance prediction models with rehabilitation treatment effects and impacts on the rate of pavement future deterioration and maintenance needs. The rehabilitation treatments include full-wide and full length asphalt concrete thin and thick overlays with recycle of one or more existing pavement layers. Each selected maintenance or rehabilitation project is evaluated on a year-by-year basis as is identified as being economically feasible as a whole. Inputs required for this program includes the pavement network inventory data, available M&R treatment alternatives and their associated costs, plus the available budget for each programming year. Outputs from the program gives a list of recommended M&R projects for each pavement section throughout the program years. Finally, the paper provides an example of applying the cost-effectiveness based rehabilitation program to a small road network in order to illustrate the working system and procedures. The investigation described in this study was primarily concerned with integration of the performance prediction model, the cost-effective M&R treatment and the network optimization. The principle and methodology developed in this study can be applied to a regional level of pavement network management system.

Keywords: Pavement, rehabilitation, cost analysis, life cycle, performance

INTRODUCTION

Paved road infrastructure of a country or region plays a very important role in its economic development and the welfare of its people. Preservation of the existing road network has lead to a considerable expenditure on pavement maintenance and rehabilitation (M&R) requirements. Due to other competing infrastructure demands on government finance and the shrinking budgets of many agencies who maintain or own various pavements, it is becoming increasingly important to minimize these expenses. Determination of future needs and the optimal investments for the preservation of pavement network is concerned with cost-effective M&R treatment alternatives and priority programming, which may be carried out through pavement life-cycle analysis.

As the vehicle operating costs are strongly influenced by road condition, the deterioration and rehabilitation of paved roads have become the major concerns of many highway agencies, particularly in the developed countries where a large number of paved road networks have been established with a huge amount of previous investments.

The concept of pavement performance modeling and various measuring methods has gained widespread recognition around the world as a useful way of depicting general deterioration of a pavement from good condition to poor and the effects of rehabilitation improving the serviceability (1).

When a pavement section reaches the minimum acceptable service level specified, it becomes a need. In identifying deficient pavements and estimating future needs in a network, both the existing pavement performance and treatment effect on the pavement future deterioration are considered by using a comprehensive prediction model, i.e., integrating pavement deterioration with treatment actions. Figure 1 shows two different deterioration curves of a pavement with each one corresponding to an alterntive M&R program and some contstraints.

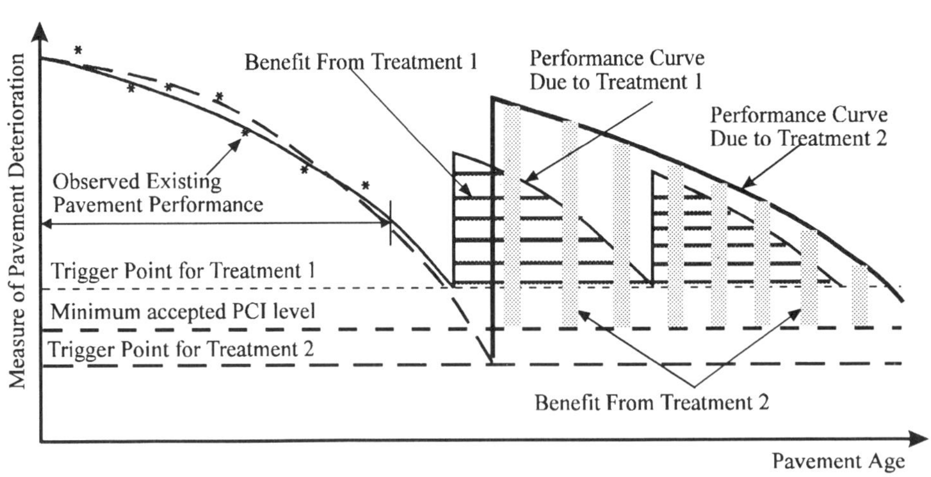

Figure 1 Performance defined by the treatment strategies and benefit with time

Figure 1 indicates that if a pavement performance analysis period goes beyond the initial service life of the alternative being considered, then the performance prediction should include a separate deterioration model to consider the effects of the selected "M&R" treatment(s) within the program period. In a previous research an integrated performance prediction model with regular treatments was developed to predict the pavement serviceability-age relationship over the entire analysis period, including the prediction model for the initial pavement structure and the modified prediction models for the pavement structure after implementing each of the preservation actions.

GENERAL DESCRIPTION OF THE OPTIMIZATION MODEL

The general formulation developed by Li (2) for the network optimisation, as shown in Equations [1] to [5], is a multi-year integer programming on a year-by-year basis. In the process of programming both the rehabilitation treatment effects and their potential impacts on the rate of pavement future deterioration are considered in the prediction model. This optimisation model is designed to maximise the total value of effectiveness/cost ratio for a pavement network in which there are totally S sections, M standardised M&R alternative treatment options, and T years of programming period. The objective function of the optimisation models is, given budget limitations and other constraints, to maximise the total effectiveness/cost ratio of the network. The M&R treatment projects that produce the greatest cost-effectiveness of the network in each programming year are considered as the optimal M&R treatment actions, and therefore are given to the first priority as compared to all other alternative M&R programs. The total cost of implementing all the optimal M&R treatment actions assigned to the network in each programming is calculated as the present worth. The effectiveness credited to each pavement section in considered as the product of the area between the predicted pavement performance curve and the minimum acceptable level, multiplied by length and traffic volume of the pavement, and sum of the effectiveness for all of the pavements in the network is the total effectiveness gained in that programming year. It should be noted that effectiveness calculated by using this method can be a positive value (or credit) or a negative value (or debit), depending on the predicted pavement performance curve (i.e., the predicted pavement condition state) is above or below the minimum acceptable serviceability level.

As concerning negative effectiveness, it means that pavement surface condition or serviceability is below an economic evaluation based standard level or minimum acceptable serviceability level, which is related to speed, roughness and safety, etc. In other words, if pavement condition state is below the minimum acceptable level, it means extra user costs, including travel delay, extra gas consuming, fast vehicle depreciation, uncomfortable or unsafe driving environment, etc.

Generally, the objective function of the network optimisation model is:

$$\text{Maximise: } \sum_{s=1}^{S}\left\{\sum_{m=1}^{M} X_{stm} \times \left[\frac{(PCS_{stm} - A_{st}) \times L_{st} \times ESALs \times D_{st}}{L_{st} \times W_{st} \times C_{stm} \times (1+R)^{t}}\right]\right\}, \qquad \forall t \qquad [1]$$

and the objective function is subject to the following conditions or constraints:

$$\sum_{m=1}^{M} X_{stm} = 1, \qquad \forall\, s,\ t \qquad [2]$$

$$X_{stm} = \begin{cases} 1 & \text{if maintenance alternative } m \text{ is selected for section } s \text{ in year } t \\ 0 & \text{otherwise} \end{cases} \qquad [3]$$

$$\sum_{s=1}^{S} \sum_{m=1}^{M} X_{stm} \times \left(L_{st} \times W_{st} \times C_{st}\right) \le B_t, \text{ for } t = 1,\ 2,\ 3,\ \ldots,\mathrm{T} \qquad [4]$$

$$\sum_{m=1}^{M} PCS_{st} + (X_{stm} \times \Delta PCS_m) \le PCS_{\max imum}. \qquad \forall\, s,\ t \qquad [5]$$

where

PCS_{st} = generalised Pavement Condition State (such as Pavement Condition Index, PCI, Present Serviceability Index, PSI, etc.) for section s (of S total sections in a road network) at year t (of T years of analysis period)

A_{st} = the minimum acceptable level of PCS required for pavement section s at year t, and (PCS_{st} - A_{st}) can be either positive or negative value,

L_{st} = length (km) of pavement section s in year t,

$ESALs$ = Equivalent Single Axle Load Applications on pavement section s in year t

D_{st} = number of service days for traffic flows by pavement section s in year t if treatment alternative strategy m is selected,

W_{st} = width (m) of pavement section s in year t,

C_{stm} = unit cost ($ / per square meter) of a standardised M&R treatment alternative strategy m is applied to pavement section s in year t,

R = discount rate for calculating present value of future cost,

B_t = budget limit for all the M&R actions in the network in programming year t

ΔPCS_m = treatment effect of a standardised M&R action, which is defined as an amount of PCS that can be recovered, from the existing Pavement Condition State, by the M&R action alternative m.

$PCS_{max.}$ = Maximum value of pavement condition state defined for a pavement. For examples: if PCS is defined by PSI in one case, which is measured on a scale of 0 to 10, with 10 being perfect, then the highest level of the PCS is 10, i.e., $PSI_{max.}$ is 10; if PCS is defined by PCI in another case, which is measured on a scale of 0 to 100, with 100 being perfect, then the highest level of the PCS is 100, i.e., $PCI_{max.}$ is 100. Considering the construction quality, reliability and many other factors, the maximum PCI of a pavement is usually defined as 95 in stead of 100 in real situations.

LIFE CYCLE OF DETERIORATION AND REHABILITATION

Figure 2 illustrates a typical life cycle of a pavement, which depicts the relationship between pavement deterioration-rehabilitation treatment in (a) and the costs in (b). The purpose of this example is to explain the comprehensive prediction of asphalt pavement performance with standardized rehabilitation treatments and their impacts on the rate of pavement future deterioration. The standardized rehabilitation provides the highway agency with the ability to analyze quantitatively the cost-effectiveness and therefore perform economic analysis of each rehabilitation strategies.

As shown in Figure 2, within the analysis period of 24 years, two standardized rehabilitation treatments, i.e., thin and thick surface overlays with recycle of exiting asphalt concrete pavement overlayer(s). The first treatment recovers the pavement PCI by 2.5 in the 9th year, while the treatment recovers PCI by 3.5 in the 15th year. After implementation of the two rehabilitation treatments, the pavement structural thickness is increased to 808 mm after 9 years and 868 mm after 15 years, respectively.

Consequently, three separate pavement performance prediction models have to be developed. Each one corresponds to one of three different pavement structures, initial pavement condition states and traffic volumes, as shown in the upper part of the figure. In this way, both the pavement treatment effects and their impacts on the rate of pavement future deterioration can be considered in the time-related performance prediction models.

The input data for the pavement structure described in this example is taken from Highway 25 which is located in southern Ontario, Canada. The initial pavement thickness above the subgrade, He, is 748 mm in terms of equivalent granular base layer thickness, and its initial pavement condition index (PCI) is 9.0.

Rehabilitation treatments such as thin or thick overlay with asphalt concrete or recycled bituminous surfacing have the immediate effects of reducing the roughness to approximately the level of new pavements and the surface distress to zero, and the future effect of reducing the rate of deterioration through strengthening the pavement.

The associated cost-streams, shown in (b) of the figure, when combined with the volume and growth of traffic and discounted to present worth, provides the basis for determining the economic benefits and returns of a given rehabilitation strategy. Typically, the total vehicle operating costs outweigh the road costs by large factors in the order of ten to twenty so that small savings yield high returns.

(a) Cycle of Deterioration and Rehabilitaton

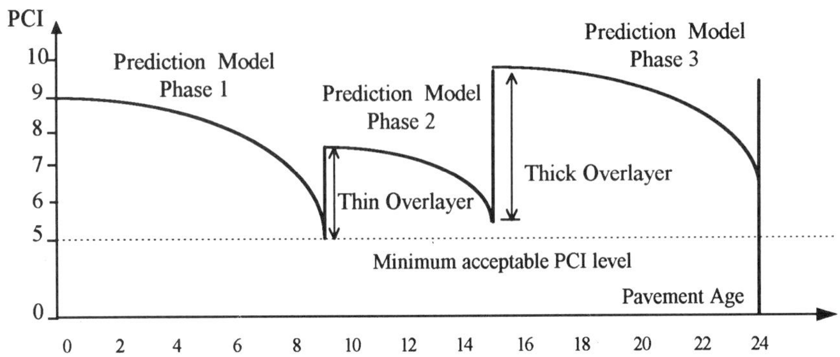

(b) Road Construction/Rehabilitation Costs and Vehicle Operating Costs

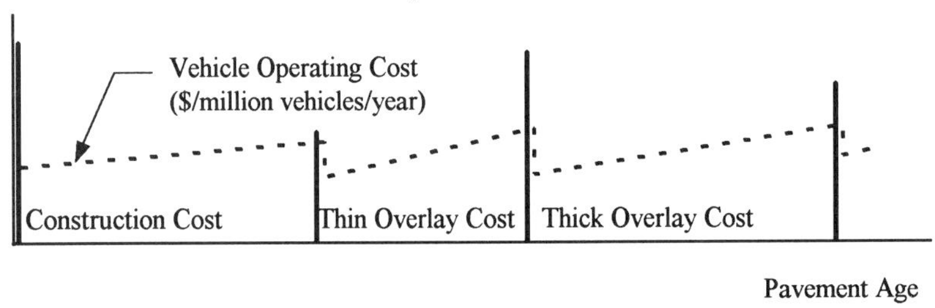

M&R Action Time	Year 9	Year 15
Type of M&R Treatment	Thin overlay (30 mm)	Milling, 50% Recycling, and resurfacing (30 mm increase in thickness)
M&R Treatment Effect	PCI is raised by 2.5 (or ΔPCI=2.5)	PCI is raised by 3.5 (or ΔPCI = 3.5)
Structural Changes		
Initial Structure H_e = 748 mm	After treatment 1 $H'_e = 748+30\times2 = 808$ (mm)	After treatment 2 $H''_e = 828+30\times2 = 868$ (mm)
Prediction Model		
Phase 1 ($PCI_0 = 9.0$)	Phase 2 ($PCI_0 = 7.5$)	Phase 3 ($PCI_0 = 9.0$)
Traffic AADT in year 1 = 6500	AADT in Year 9 = 9200	AADT in Year 15 =12,000

* Equivalent granular base layer thickness

Figure 2 Integrated Multi-Year Prediction of Pavement Performance

EXAMPLE OF AN OPTIMAL REHABILITATION PROGRAM

To describe explicitly how the integer programming works, consider the following example, which involves five asphalt pavement sections and three standardized M&R strategies. The pavement condition state is determined by the time-related Markov probabilistic prediction model, which has been described in detail in reference (3). Entries in Table 1 are section number, length and width, the number of 80 kN equivalent single axle loads (ESALs), condition state of the current year and predicted pavement condition states in the next 5 years. The prediction of pavement future condition state is conducted by using Ontario Pavement Analysis of Costs (OPAC) model, considering the following factors: 1) the total number of 80 kN equivalent single axle loads (ESALs), 2) pavement structural thickness and 3) subgrade soil strength.

Table 1 Predicted PCI Values of Pavement Sections in the Next Five Years

Section Number	Length/Width of Section (m)		Traffic Volume Annual 80 kN ESALs*	PCI**					
				Current Year	Programming Year 1	2	3	4	5
1	1500	15	2.9×10^5	79	74	68	63	55	50
2	1200	7.5	2.1×10^5	61	57	54	50	46	42
3	900	7.5	0.8×10^5	49	44	40	36	32	29
4	1400	12	1.4×10^5	89	84	79	74	70	64
5	1100	7.5	2.2×10^5	83	79	74	69	63	57

* 80 kN Equivalent Single Axle loads (ESALs)

** PCI values are calculated by Markov process-based prediction model.

The additional information needed includes the available treatment strategies, treatment effect and the cost of applying each alternative. Table 2 lists three standardized M&R treatment strategies for the small pavement network. Each of the strategies includes: 1) treatment effect in terms of raising the existing pavement condition state by a certain amount of PCI points, 2) treatment impact on the existing pavement in terms of an increase in pavement thickness as compared with the existing pavement and, 3) unit cost for each of the M&R actions.

Table 2 An Example showing Feasible M&R Treatment Strategies

No. of Treatment Strategy	Treatment Effect and Impact	Cost ($/M²)
No.1 Routine Maintenance	• Raise PCI by 0 • No structural change to pavement	0.3
No.2 Major Maintenance	• Raise PCI by 15 • Correct all pavement surface distresses (crack sealing, distortions and pothole patching, etc.) • Thin coat, or spraying • No structural change to pavement	9
No.3 Minor Rehabilitation	• Raise PCI by 30 • Milling old surface and resurfacing • 40 mm asphalt overlay resurfacing • Increase structural thickness by 40 mm of asphalt concrete surface layer	15

To simplify the example, assume that the minimum acceptable PCI level is 45 for all of the pavement sections and an annual budget scenario is $200,000, $150,000, $350,000, $150,000 and 100,000 for the programmed years 1 through 5. In addition, annual traffic growth rate for all pavement sections is 4 % and discount rate is 5%. The solution obtained using the integer programming model is summarized as follows:

Optimal Rehabilitation Treatments for the Network Preservation Table 3 gives a five-year program developed for the small pavement network, together with the predicted PCI and the treatment effect for each pavement section in each programming year. As indicated previously, the PCI at the first column in each programming year is predicted by the time-related Markov prediction model. The second column is outputs of the optimal rehabilitation treatment strategies produced by the integer programming, which is based on the greatest yearly total effectiveness/ratio. The ΔPCI at the third column is the treatment effect corresponding to the selected M&R action in the programming year.

Table 3 Program of Multi-Year M&R Treatment Strategies in the Next Five Years

Section Number	Predicted PCI, Selected Optimal M&R Actions for Programming Years 1 through 5																	
	Current year			Year 1			Year 2			Year 3			Year 4			Year 5		
	PCI	M&R	ΔPCI	PCI	M&R	ΔPCI	PCI	M&R	ΔPCI	PCI	M&R	ΔPCI	PCI	M&R	ΔPCI	PCI	M&R	ΔPCI
1	79	1	0	74	1	0	68	**3**	**30**	93	1	0	89	1	0	84	1	0
2	61	**2**	**15**	72	1	0	67	1	0	62	**3**	**30**	87	**2**	**15**	95	1	0
3	49	**3**	**30**	74	1	0	70	1	0	66	1	0	61	1	0	57	1	0
4	89	1	0	84	1	0	79	1	0	74	1	0	70	1	0	64	1	0
5	83	1	0	79	**3**	**30**	95	1	0	92	1	0	89	1	0	85	1	0
* Pavement Condition Index (PCI) predicted in each year.																		

Cost-Effectiveness Analysis of the M&R Program Information about costs and benefits in terms of effectiveness is summarized in Table 4 and Table 5, respectively. Entries in Table 4 include pavement section number, area of pavement surface of each section and allocation of the annual budget to the network. In the column of each programming year are the selected optimal M&R actions and the corresponding costs for all the pavement sections. In addition, the annual budget limitation and actual spending in each programming year are listed in the last two rows of the table. The total cost of M&R program for each pavement section during the five years analysis period is provided in the last column of the table.

The example has illustrated the rehabilitation program for the five pavement sections in the period of five analysis years with 3 alternative treatment. In such a way, the cost for each M&R treatment assigned to each pavement section is shown in Table 4 for the programming years 1 through 5. Similarly, the effectiveness of the selected optimal M&R projects for each pavement section can also be calculate on the basis of predicted PCI and traffic in each year.

Table 4 M&R Treatment Actions and Costs in Each Programming Years

Section Number	Section Area LxW(m²)	Allocation of the Annual Investment to the M&R Program										
		Year 1		Year 2		Year 3		Year 4		Year 5		Total Cost
		M&R	$ ×10³	M&R	$× 10³	M&R	$ ×10³	M&R	$ ×10³	M&R	$ ×10³	$ (×10³)
1	22500	1	6.7	1	6.7	3	338	1	6.7	1	6.7	364.8
2	9000	2	81	1	2.7	1	2.7	3	135	2	81	302.4
3	6750	3	101	1	2	1	2	1	2	1	2	109.4
4	16800	1	5	1	5	1	5	1	5	1	5	25.0
5	8252	1	2.5	3	124	1	2.5	1	2.5	1	2.5	134.0
Annual Budget (×10³)		200		150		350		150		100		950
Annual Spending (×10³)		196.2		140.4		350.2		151.2		97.2		935.2

Table 5 Pavement PCI and Effectiveness of Each Section in Each Programming Year

Section Number	Section Length (M)	ESALs (×10⁵)	PCI and Effectiveness Based on the M&R Program										
			Year 1		Year 2		Year 3		Year 4		Year 5		Total
			M&R No.	Effec. (×10⁸)	M&R No.	Effec. (×10⁸)	M&R No.	Effec. (×10⁸)	M&R No.	Effec. (×10⁸)	M&R No.	Effec. (×10⁸)	Effective-ness (×10⁸)
1	1500	2.9	1	126.15	1	100.05	3	208.80	1	191.40	1	169.65	796.05
2	1200	2.1	2	68.04	1	55.44	1	42.84	3	105.84	2	126.00	398.16
3	900	0.8	3	20.88	1	18.00	1	15.12	1	11.552	1	8.64	74.19
4	11400	1.4	1	76.44	1	66.64	1	56.84	1	49.00	1	37.24	286.16
5	1100	2.2	1	82.28	3	121.00	1	113.74	1	106.48	1	96.80	520.30
Total Effectiveness (×10⁸)			373.79		361.13		437.37		462.24		438.33		2074.86

CONCLUSIONS AND RECOMMENDATIONS

This paper presented a new integrated multi-year optimization model developed for the preservation of pavement network through cost-effectiveness analysis. The optimization model can be used to produce the optimal multi-year M&R program under the constraints of annual budget limitation and requirements of pavement serviceability. The decisions made on the M&R treatment strategies for each pavement section are based on the integration with the following three sub-systems:

1. The time-related Markov probabilistic prediction model.
2. A set of standardized M&R treatment strategies to be used in the network preservation, which consider treatment effects of the M&R actions and their impacts on future deterioration or pavement future M&R needs.
3. Cost-effectiveness based integer programming of multi-year M&R actions for the pavement network under the constraints of available funds, minimum acceptable pavement serviceability level and programming requirements.

The output from this program are a list of M&R program for individual pavement improvements within the programming analysis period. All feasible combinations of M&R treatment strategies for each pavement section along the programming years are compared in the process of integer programming, and the best one are selected for implementation on the basis of their economic consequences.

The developed multi-year integer programming uses the concept of maximizing the total effectiveness/cost ratio of the network as a objective function. Although a minimum acceptable serviceability level is defined for each pavement, a M&R project may be assigned to the pavement that has not reached, or even fairly higher than, its minimum acceptable serviceability level. This is different from many existing priority programs which will not consider a major M&R action for the pavement until it reaches the minimum acceptable serviceability level. In other words, under certain circumstance, it may be more economical to apply a major maintenance or a rehabilitation treatment to a pavement before its terminal serviceability level is reached.

In conclusion, the optimization methodology developed in this study provides a very powerful tool for selecting the optimal M&R treatment strategies for a pavement network. It has been shown to be an efficient means of solving the pavement management optimization problems. The results from the model are promising, but more work should be conducted to make the procedure directly implemented by provincial departments of transportation for their PMSs.

REFERENCES

(1) Feighan, K.J., Shahin, M.Y., and Sinha, K.C. A Dynamic Programming Approach to Optimization for Pavement Management Systems. *Proceedings, Second North American Conference on Managing Pavements*, Toronto, Ontario, Canada, Vol. 1, pp. 195-206, November 1987.

(2) Li, Ningyuan, Development of A Probabilistic Based, Integrated Pavement Management System, Ph.D. Thesis, Department of Civil, Engineering University of Waterloo, Canada, 1997.

(3) Li, Ningyuan, Ralph Haas and Weichau Xie, Development of a New Asphalt Pavement Performance Prediction Model", *Canadian Journal of Civil Engineering*, Vol. 24, No. 4, August 1997.

Fifth International Conference on the
BEARING CAPASITY OF
ROADS AND AIRFIELDS

NEW CALCULATION METHOD PROPOSAL FOR ROAD FLEXIBLE PAVEMENT

GIUSEPPE ABBRESCIA[1] E PASQUALE COLONNA[2]

[1]Professor†, Polytechnic, 70125 Bari, Italy
[2]Professor, Polytechnic, 70125 Bari, Italy, 039-80-5460329, colonna@teseo.it

ABSTRACT

Empirical and rational calculation methods of the flexible road surfaces generally assume an overall thickness of the structure, in other words by setting the thickness of the various layers, they enable them to be checked according to the tensions and the deformations present in the layer and to their interfacing.
This work proposes to directly carry out the determination of the thickness of the individual layers of pavement in accordance with the characteristics of the material and the traffic, beginning with the results of simple CBR tests.
The method which was originally taken up by prof. Abbrescia, who was unable to see it through to the end, was completed by the other author in terms of both the theoretical part and the information application which was essential to resolve the necessary repetitions.
Key words: pavements, CBR.

INTRODUCTION

Calculation of flexible road pavements is one of the most difficult sectors for modern road infrastructure engineering. Actually it is well-known that the problem is a difficult one to solve due to the influence of several factors, the main ones of which are the difference of the materials, their high sensitivity on the one hand to high temperatures and on the other to the freezing cold. Then there is the substantial difference of the quality of the materials locally available, also in terms of resistant support, the difficulty of determining a correct rheological model, the contemporary presence of several layers, some of which have a binder and others which do not, the uncertainty of determining the tension distribution laws, the deformation and the temperatures and finally the uncertainty and variability of the load cycles in terms of both the entity and spatial distribution.
This theme has been one which the Dipartimento di Vie e Trasporti of the Polytechnic of Bari has been particularly concerned with for quite some time by means of theoretical and experimental investigations. Professor Abbrescia began to propose an appropriate rheological model in 1964 (1) for bituminous conglomerates, whereas the idea of using the

CBR test on them, to determine the thickness of the pavements came in 1978 (2). In 1991 (3) he came up with an organic proposal even though it was not easy to implement and later set about perfecting his studies, when the efforts of research were abruptly interrupted by an unhappy epilogue of a sudden illness.
The other author of this paper, a former student of Professor Abbrescia was working on the same theme (4) even though it was from different viewpoints and he decided to continue the work which had been started, even with the difficulties connected with the impossibility of establishing a dialogue.
The theoretical schematization of the method was perfected as will be shown later and it was adapted to the possibility of using an appropriate calculation programme which was later actually carried out. The results obtained underwent control by means of an initial control with an Italian pavement catalogue and then by means of a (BISAR) calculation programme which uses the algorithms of a rational method.
The work cannot be considered completely finished because some aspects can be investigated further and further lines of development can be hypothesised, as will be mentioned later. It was, however, considered to be the right moment to publicise the point the research has reached, in order to testify the use and the continuity of research into flexible road pavements undertaken by the Dipartimento di Vie e Trasporti of the Polytechnic of Bari.

CALCULATION METHODS OF THE FLEXIBLE PAVEMENTS

It is well known that the calculation methods of flexible pavements can generally be separated into two wide categories, empirical ones and rational ones.
Empirical methods are the result of observations and appropriate experiments carried out over the years which have been catalogued in an orderly manner and have made up useful data banks whose reasonable interpretation has represented the basis for the construction of each method. They are used by consulting booklets or catalogues in which it is possible to find the solution which is closest to the needs of the planner, starting from the materials available, from the type and amount of traffic and from the surrounding conditions. The limit of these methods is represented by the fact that they are rather "rigid", they run into problems when innovative materials are proposed and anyway offer a rather limited range of solutions.
The rational methods, on the other hand, develop theoretical models which are aimed at determining the tensions and deformations induced by loads within a planned pavement, leaving the planner with the job of comparing these values with the admitted ones. These methods also require pre-dimensioning and therefore basically become control methods, reaching the best solution generally after numerous attempts. It is thus obvious that the reliability of these procedures is connected to experimental control and to calibration of the algorithms on the one hand, and to the real capacity of considering all the factors in play on the other (environmental and heat conditions for instance.).
In any case it would seem to be almost definitively ascertained that any rational method for the calculation of the flexible pavements must take the following aspects into account:

1) the capacity of the pavement to oppose itself to the progressive accumulation of plastic deformations (rutting);
2) the need to contrast the fatigue phenomena, originating from the traction forces to the layer interface;

3) the capacity to withstand heat and environmental excursions with particular reference, on the one hand to high temperatures and on the other to freezing temperatures.

THE NEW METHOD

The new method proposed has the ambitious objective of really presenting itself as a project procedure. It will thus be able to directly provide the thickness of the pavement by means of an appropriate algorithm once the traffic and material characteristics have been assigned. The method can doubtless be considered one of the most rational, because it comes from a theory which examines the equilibrium conditions of an elementary body inside the pavement. At the same time, however, it has some empirical characteristics because at its basis there is the way the pavement reacts to a completely conventional test like the CBR test.

BASIC CONCEPTS

Before going on to the description of the theoretical model adopted, we shall describe the basic concepts which have determined the approach of the methodology, in its later stages.

The CBR test as the model of a pavement under load stress. The surface of flexible road pavement which undergoes the action of traffic is subject to a distributed load which acts on an almost circular contact area with any tyre. This load tends to punch the pavement instant by instant which reacts as a consequence.
As is well known, the CBR test determines the resistance of a road material inserted into a cylindrical metal container which is subjected to a specific punching action, performed by a circular section piston in the middle of the upper surface of the test sample.
There is a clear analogy between the two phenomena. There are also other aspects which lead us to give credit to the CBR test as a representative of the real case of a flexible pavement under load stress, particularly in terms of its resistance capacity to permanent deformations.

A) In the CBR test, just like in the real situation, the pavement has a side border which is, however, difficult to quantify. Triaxial tests carried out in the laboratory of the Dipartimento di Vie e Trasporti of the Polytechnic of Bari (5) have shown that the increase in resistance that a test sample of bituminous conglomerate is subjected to with this bordering is mainly obtained as soon as the bordering activity is carried out, it then varies very little with increases in this action.

B) Studies carried out by AASHO Road Test had already indicated that the main cause of the rutting mechanism is shear strain rather than stiffening, but this was definitively confirmed in 1987 (6) and finally during the SHRP investigation.

C) If a mixture is subjected to shear strain it tends to expand. If expansion is impeded (as is the case with the pavement which has material surrounding it) border tensions are generated which oppose the shear strain. This mechanism (along with the rigidity of the binder) is the thing which gives most resistance to the permanent deformations in a bituminous conglomeration (7).

D) In the CBR material deformation test the shear strain of the material itself and the

$$H_1 = C_1 \cdot \ln\frac{D_1 - E_1}{(CBR)_{inf} - E_1} \cdot \left(1 - \frac{\ln\frac{n}{n_0}}{\ln\frac{(CBR)_1 - E_1}{(CBR)_{min} - E_1}} \right) \quad [10]$$

with

$$C_1 = \frac{r_{s1}}{2 \cdot k_1^2 \cdot tg\phi_1 \cdot \frac{R_1}{\eta_1} \cdot 250} \quad [11]$$

$$D_1 = \frac{q \cdot \frac{R_1}{\eta_1} \cdot n_0 \cdot \frac{\pi}{2} \cdot r_{s1}}{v} \quad [12]$$

$$E_1 = \frac{p_0 \cdot \frac{R_1}{\eta_1} \cdot n_0 \cdot \frac{\pi}{2} \cdot r_{s1}}{v} - c_1 \cdot \frac{1 - 2 \cdot tg\left(45° - \frac{\phi_1}{2}\right) \cdot tg\phi_1}{tg^2\left(45° - \frac{\phi_1}{2}\right) \cdot tg\phi_1} \quad [13]$$

n = the number of load passages Q which travel, with velocity v, on the circular area of radius r_S;
n_0 = fixed number of load repetitions regarding the average value of the CBR index considered.
Moreover, for the layer of thickness H1, the bottom print radius r_{S2} and top print radius r_{S1} are linked by:

$$r_{s2} = r_{s1} \cdot \sqrt{e^{\frac{H1}{C1}}} \quad [14]$$

AUTOMATIC CALCULATION PROGRAMME

Because the calculation procedures are rather laborious, an appropriate programme has been set up to speed up the process and make it more reliable.
In this programme, firstly the physical-mechanical characteristics of the pavement are assigned along with those which define the type of traffic.
Then the operator attributes arbitrary values to the thickness of the layers, while the programme takes care of firstly calculating the thickness of the foundation and later also defines the thickness of all the other layers. The procedure foresees the continuation of the calculation for consecutive repetitions with the aim of making the load conditions congruent on the interface surfaces.
The process stops when the individual thickness of the layers regarding a certain calculation cycle differ from those regarding the previous cycle by any quantity however little it may be.

CONTROL OF THE METHOD

The method underwent a double control procedure, comparing the results obtained firstly with those of another “empirical” methodology and then with those of another “rational”

method.
First of all the Road Pavement Catalogue drawn up by the CNR in Italy in 1994 was considered. After all the necessary connecting procedures were carried out to make the input data homogeneous in terms of the pavements and of the vehicle loads, it was possible to compare the two project proposals. The results were satisfactory, with maximum differences of about a centimetre for the surface layers and higher ones (3 - 5 cm) for the deepest layer, limited to secondary roads outside the town. The new method seems to be more prudent in the latter circumstance, giving values which were actually higher for the thicknesses.
The second control was carried out using the BISAR calculation programme used at the Dipartimento di Vie e Trasporti in the Polytechnic of Bari. The programme, which is based on the elastic multi-layer model resting on an infinite semi-space, allows tensions, deformations and movements to be calculated.
All the information regarding each project utilized for comparison with the Italian Catalogue is inserted into the programme. The compatibility of the results is then checked with the known formulas usually used to evaluate the fatigue damage caused by a number of load repetitions N (Van Dijk, etc.). Tensions and deformations were found along the vertical plane passing through the centre of the individual wheel print and in the interface planes between the various layers. All the results were satisfactory without any exceptions.

POSSIBLE FUTURE DEVELOPMENTS OF THE METHOD

The method has shown itself to be quite reliable, however, some aspects must be pointed out.

a) Traction stress on the base of the layers is not taken into consideration whatsoever. As is well-known, they are the prime cause of fatigue damage. This problem must be resolved, even though many researchers believe that the fatigue cracking comes from the energy dissipation of the energy accumulated before they appeared, which in turn depends on the slope of the permanent deformations curve in accordance with the time in logarithmic scale (8) and, therefore with the results of CBR test.
b) The characteristics of bituminous conglomerates are strongly influenced by temperature, so the choice of making the thicknesses of the layers depend on tests (CBR, cut) to be carried out at 20°C, without a thermostat may be disputed. However, carrying out the tests at different temperatures, or with a similar gradient to the real one, could be an interesting way of developing the research project.
c) The problem of freezing has not been specifically considered, even though appropriate tests could be carried out, conditioning the materials.
d) The procedure is particularly appropriate to evaluate the resistance characteristics and, therefore, to plan the thicknesses of pavements which use innovative materials. This is because the possible behaviour in a real situation can be determined from the CBR test that can simulate real behaviour quite effectively as has already be explained.
e) A further interesting research development may be represented by carrying out CBR tests with a container provided with pressure cells, in order to experimentally determine the real diffusion capacity of the loads with variations in the materials.
f) If [1], [2] and [3] are correctly processed they allow both the deformations and the elasticity module and the coefficient of viscosity of the layers to be calculated. The procedure can, therefore, favour new investigation possibilities of these parameters.
g) It should also be pointed out that the automatic calculation procedure allows numerous

planning tests to be carried out very rapidly, by varying the material and the number of layers, permitting the most appropriate choice from all points of view.

CONCLUSIONS

A new planning methodology for the thickness of flexible road pavements has been described. It is based on the great analogy existing between the mode of the CBR test and the real behaviour of the pavements themselves which are cyclically punched by the vehicle loads and thus forced to react and this compromises the capacity for developing sideways actions.
After having described the theoretical model and having set out the main resolving formula, we showed the controls carried out by comparing the results of a purposed made calculation programme with the results of two other planning models one of which is empirical and the other rational.
Finally we outlined the possible future research development lines.

REFERENCES

1. ABBRESCIA, G., Un nuovo metodo di calcolo degli spessori delle pavimentazioni flessibili a mezzo dell'indice di CBR, La rivista della strada, n.447, 1978

2. ABBRESCIA, G., Sviluppi di un nuovo metodo di calcolo degli spessori delle pavimentazioni flessibili a mezzo dell'indice di CBR, Le Strade, n.1248, 1988

3. ABBRESCIA, G., Il proporzionamento delle pavimentazioni flessibili, Edizioni Fratelli Laterza, Bari, 1991

4. COLONNA, P., Studio del comportamento strutturale dei manti bituminosi drenanti ai fini della determinazione di un criterio di progetto, Le Strade, n.1271, 1990

5. BONVINO, U., Contributo alla definizione di un criterio di resistenza dei conglomerati bituminosi drenanti, Le Strade, n.1275, 1991

6. EISENMANN and HILMER, Influence of wheel load and inflation pressure on the rutting effect at asphalt pavement. Experiments and theoretical investigations, Proceedings of sixth international conference on the structural design of asphalt pavements, 1987

7. SOUSA J.B., and SOLAIMAIAN, Abridged procedure to determine permanent deformation of asphalt concrete pavement, TRB annual meeting 1994

8. SCHAPERY., A theory of crack growth in viscoelastic media, ONR Contrect No, Technical report, n.2, 1973

Fifth International Conference on the
BEARING CAPASITY OF
ROADS AND AIRFIELDS

Analysis on Long Term Performance Investigations in Germany

Michaela Friedrich

Dipl.-Ing. Univ. (civil engineering)
Technische Universität München
D-80290 München, Germany
Fax: ++49 89 28923800
E-mail: michaela.friedrich@lrz.tu-muenchen.de

Abstract

The data collected on many heavily trafficked road segments for the research project "long term performance investigations" during the 70's and 80's all over Germany shows that during the last decades the traffic was increasing. The number of heavy vehicles was growing and also the weights of the axles were increasing. Considering that the load is the most important factor for the fatigue life of a structure, road design should not be depending on the number of vehicles but on the value of the axle loads. A probability calculus with the collected information confirms the results of the AASHO-Road-Test for German road and traffic conditions. The axle loads have to be valued by the rule of the third to fourth power in case of the bituminous layers and by the rule of the third to fifth power in case of the concrete pavements.
Keywords: long term investigations, traffic loads, increase rates, impact on road structure, damage progression

1 Introduction

In former years numerous projects were in progress to determine a relationship between the traffic load impact and the long term performance of roads. One of the best known tests was the two year AASHO-Road-Test, which took place at the end of the fifties in the Ottawa, Illinois, USA. The empirical result of this project is the rule of the fourth power. Because the traffic loads and pavement constructions in Germany are different than those of the AASHO-Road-Test, the Bundesanstalt für Straßenwesen BASt (Federal Highway Research

can be calculated. On Autobahns an average number of 500 to 7.000 vehicles per day is received, on Bundesstraßen this value ranges between 200 and 2.000 daily trucks and on Landesstraßen it varies between 20 and 700 vehicles per day. A comparison of the proposed and the actual traffic in Germany shows that the exact estimation of the development of heavy traffic for a period of 20 or 30 years is very difficult. The mean increase rates of the average number of daily heavy vehicles in the 70's and 80's show a growth of more than 4% at Autobahns, of about 3% at Bundesstraßen and of approximately 1,5% at Landesstraßen. These real increase rates exceed the values, which are given in the German design standards to take into account the rise in traffic.

3 Valuation of the traffic load data

The traffic is characterized by a wide range of different axle loads as well as different axle and wheel configurations. To simplify further investigations one has to modify these varying combinations. The mixed traffic has to be converted in a cumulative number of passages of reference axle loads. Two methods should be mentioned to describe the heavy traffic in an easier way. First, the real combination can be transformed to the counted number of an average valued equivalent axle load. Second, the actual mixed traffic can be destribed as a cumulative number of equivalent axle cycles of a reference axle load. This reference axle will be an isolated axle with 100 kN and single wheel. People in highway research doubt that the rule of the fourth power (result of the AASHO-Road-Test) is valed for Germany, because the road constructions and the type of traffic in this country is different to the test conditions. For this reason further studies will value the traffic with the rule of the first to the sixth power.

3.1 The average valued equivalent axle load

The traffic data recorded by the research project "Langzeitbeobachtungen" is the basis to calculate a non valued arithmetical mean μ and the variation coefficient v of all axle loads for each location. Formula 1 and 2 show the proceeding. As already mentioned, axle loads with less than 20 kN are neglected.

$$\mu = \frac{\sum_{i=2}^{k}(\overline{x}_i \cdot n_i)}{\sum_{i=2}^{k} n_i} = \frac{\overline{x}_2 \cdot n_2 + \overline{x}_3 \cdot n_3 + \ldots + \overline{x}_7 \cdot n_7}{n_2 + n_3 + \ldots + n_7} = \frac{3 \cdot n_2 + 5 \cdot n_3 + \ldots + 13 \cdot n_7}{n_2 + n_3 + \ldots + n_7} \qquad \text{Formula 1}$$

$$v = \frac{12 - \mu}{\lambda \cdot \mu} \qquad \text{Formula 2}$$

μ	average axle load
n_i	number of cycles in load class i
$\overline{x}_i$	arithmetical mean of load class i
v	variation coefficient
λ	factor of the Gaussian normal distribution

At Autobahns the average axle load is 5,6 tons, at Bundesstraßen the mean is 5,3 tons and at Landesstraßen it is 5,1 tons. High axle loads have a stronger impact on road constructions than lower weights. This effect has to be considered when calculating the valued equivalent axle loads. Following the valuation of the dynamic variation of wheel-loading, the mean and the deviation of the load distribution is necessary to calculate the valued equivalent axle loads. Depending on the rule of the first to the sixth power the following equations (Formula 3.1 to 3.6) are used to get the valued equivalent axle loads.

Rule of the first power: $\mu_{equi(1)} = \mu$ Formula 3.1

Rule of the second power: $\mu_{equi(2)} = \left[\mu^2 \cdot \left(1 + v^2\right)\right]^{1/2}$ Formula 3.2

Rule of the third power: $\mu_{equi(3)} = \left[\mu^3 \cdot \left(1 + 3 \cdot v^2\right)\right]^{1/3}$ Formula 3.3

Rule of the fourth power: $\mu_{equi(4)} = \left[\mu^4 \cdot \left(1 + 6 \cdot v^2 + 3 \cdot v^4\right)\right]^{1/4}$ Formula 3.4

Rule of the fifth power: $\mu_{equi(5)} = \left[\mu^5 \cdot \left(1 + 10 \cdot v^2 + 15 \cdot v^4\right)\right]^{1/5}$ Formula 3.5

Rule of the sixth power: $\mu_{equi(6)} = \left[\mu^6 \cdot \left(1 + 15 \cdot v^2 + 45 \cdot v^4 + 15 \cdot v^6\right)\right]^{1/6}$ Formula 3.6

v variation coefficient (Formula 2)

μ average axle load (Formula 1)

From these formulas and the traffic load data of the "Langzeitbeobachtungen" the following results are equated apart for each type of road. Tables 4, 5 and 6 show respectively the mean, the minimum and the maximum values of the equivalent axle load.

	1st Power	2nd Power	3rd Power	4th Power	5th Power	6th Power
Mean	**5,60 t**	**6,34 t**	**6,87 t**	**7,33 t**	**7,73 t**	**8,10 t**
Minimum	4,94 t	5,77 t	6,26 t	6,66 t	7,02 t	7,34 t
Maximum	6,36 t	6,91 t	7,35 t	7,73 t	8,07 t	8,47 t

Table 4: Average equivalent axle load at Autobahns

	1st Power	2nd Power	3rd Power	4th Power	5th Power	6th Power
Mean	**5,34 t**	**6,13 t**	**6,65 t**	**7,10 t**	**7,50 t**	**7,86 t**
Minimum	4,21 t	5,09 t	5,63 t	6,10 t	6,51 t	6,87 t
Maximum	6,82 t	7,37 t	7,81 t	8,20 t	8,56 t	8,88 t

Table 5: Average equivalent axle load at Bundesstraßen

	1st Power	2nd Power	3rd Power	4th Power	5th Power	6th Power
Mean	**5,24 t**	**6,06 t**	**6,61 t**	**7,09 t**	**7,50 t**	**7,88 t**
Minimum	4,43 t	5,28 t	5,80 t	6,21 t	6,57 t	6,90 t
Maximum	6,74 t	7,30 t	7,75 t	8,15 t	8,50 t	8,87 t

Table 6: Average equivalent axle load at Landesstraßen

The higher the power, the higher the value of the average equivalent axle load is. Valued with the rule of the fourth power one can get an average equivalent axle load of 7,3 tons at Bundesautobahnen, 7,1 tons at Bundesstraßen and at Landesstraßen. A relevant increase compared to the non-valued mean axle loads is to be seen. This rise is even higher when using the rule of the fifth or the sixth power.

3.2 Number of equivalent axle cycles of 10 tons

Apart from the valued equivalent axle load, the heavy traffic can also be transformed to a calculated number of equivalent passages of 10 tons. In this case, the effect that heavy loads impact the road construction more than the low ones has to be respected, too. This happens with the valuation by the first to the sixth power. The cycles of load class 1 are neglected. By the following formulas the equivalent number of passages of 10 tons can be equated:

$$n_{10t\,(N)} = \sum_{i=2}^{7} \left(\frac{\bar{x}_i}{10}\right)^N \cdot n_i = \left(\frac{3}{10}\right)^N \cdot n_2 + \ldots + \left(\frac{13}{10}\right)^N \cdot n_7 \qquad \text{Formula 4}$$

or:

$$n_{10t\,(N)} = \left(\sum_{i=2}^{7} n_i\right) \cdot \left(\frac{\mu_{equi(N)}}{10}\right)^N \qquad \text{Formula 5}$$

$n_{10t\,(N)}$	equivalent number of passages of 10 tons (100 kN)
$\bar{x}_i$	arithmetic mean of load class i
n_i	number of cycles in load class i
N	value of power (1 to 6)
$\mu_{equi\,(N)}$	valued equivalent axle load

Tables 7, 8 and 9 contain the minimum and maximum numbers of 10 ton charges which were calculated seperately for all Autobahns, Bundesstraßen and Landesstraßen that were part of the research project "long term performance investigations".

	1st Power	2nd Power	3rd Power	4th Power	5th Power	6th Power
Minimum	228.000	149.000	107.000	86.000	75.000	69.000
Maximum	3.701.000	2.672.000	2.152.000	1.875.000	1.744.000	1.715.000

Table 7: Number of 10 ton charges at Autobahns

	1st Power	2nd Power	3rd Power	4th Power	5th Power	6th Power
Minimum	45.000	30.000	22.000	19.000	18.000	18.000
Maximum	1.056.000	758.000	607.000	531.000	497.000	494.000

Table 8: Number of 10 ton charges at Bundesstraßen

	1st Power	2nd Power	3rd Power	4th Power	5th Power	6th Power
Minimum	7.000	5.000	3.000	3.000	2.000	2.000
Maximum	431.000	340.000	294.000	275.000	290.000	324.000

Table 9: Number of 10 ton charges at Landesstraßen

Taking into account an increasing value of the power the number of equivalent cycles of 10 tons goes down. According to this analysis there are about 90.000 to 190.000 equivalent numbers at Bundesautobahnen, 20.000 to 550.000 equivalent cycles at Bundesstraßen and 3.000 to 275.000 equivalent numbers at Landesstraßen by valueing the traffic with the rule of the fourth power.

4 Fatigue behaviour of road structures

The equivalence relations between different axles are equivalence values of unit damage. They are derived from calculations of stress in the pavement structure and take into account the fatigue behaviour of the materials. Equivalency factors depend on the material, the type of damage, and the pavement structure. To get to know which factor has to be used to value the increase of an axle load, the theoretical fatigue life has to be calculated.

4.1 Fatigue behaviour of asphalt pavements

The following investigations are based on the real traffic data and the actual road structures. There are three different types of bituminous pavement constructions (A, B and C) which were designed according to the German standards in the 60's and 70's: Type A: Bituminous pavement with asphalt-mixture bases; Type B: Bituminous pavement with cement aggregate bases; Type C: Bituminous pavement with unbound aggregate bases.

With the following two methods the fatigue life of asphalt structures can be calculated. The first method (method A) takes into consideration an axle load of 100 kN and the equivalent number of passages of 100 kN. For the second method (method B) the valued equivalent axle load and the counted number of load cycles are taken into account. The calculation of the theoretical fatigue life is based on bending stresses at the bottom of the bounded pavement layer. By equating the results of both methods, a statement concerning the power, which is relevant for the damage progression, can be made.

4.1.1 Method A

To calculate the theoretical fatigue life one needs the stress and the permissible strength as well as the deviation of both. The bending stress μ_S at the bottom of the asphalt layer is the result of a wheel load of 50 kN. According to a study of Dr. Wüst the variation coefficient v_S for stress is 20% [4]. The permissible strength μ_R (resistance) depends on the equivalent number of axle load cycles of 100 kN and the fatigue behaviour of the bituminous material. Depending on the power, which is used to value the traffic volume, there are six different equivalent numbers of load cycles. For this reason there are also 6 different values for the permissible strengths. The variation coefficient v_R (resistance) for strength is estimated with 10%. Considering that stress and strengh are distributed as the Gaussian normal distribution (Figure 1) the local probability of structural failure $P_f(k)$ and the time variant probability of structural failure Z(k) can be calculated with the following formulas (Formulas 6, 7 and 8):

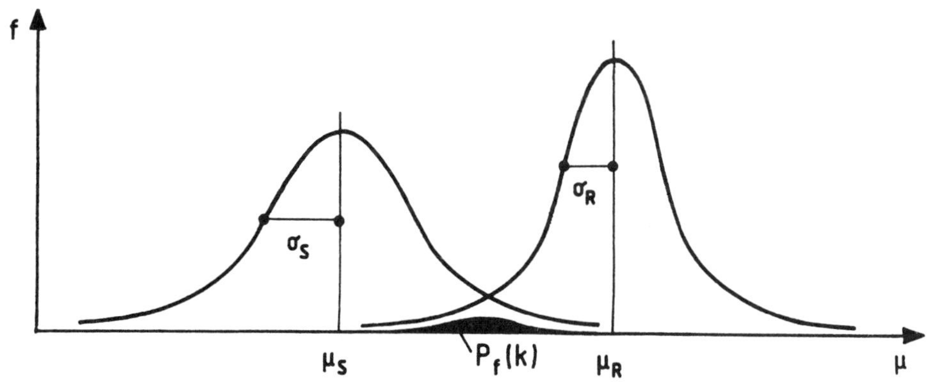

Figure 1: Distribution of stress and strength

$$\text{ß} = \frac{\mu_R - \mu_S}{\sqrt{\sigma_R^2 + \sigma_S^2}}$$ Formula 6

$$P_f(k) = \frac{1}{\sqrt{2 \cdot \pi}} \int_{-\infty}^{\text{ß}} e^{-\frac{t^2}{2}} dt$$ Formula 7

$$Z(k) = 1 - \prod_{k=1}^{K} (1 - P_f(k))^{n(k)}$$ Formula 8

μ_S	mean stress	
μ_R	mean strength	
σ_S	$= v_S \cdot \mu_S$	deviation of stress
σ_R	$= v_R \cdot \mu_R$	deviation of strength (resistance)
$P_f(k)$	local probability of failure	
$Z(k)$	time variant probability of failure	
k	variable for period	
K	total number of periods	
n(k)	number of load cycles in period k	

Figure 2 shows as an example the time variant probabilities of structural failure, which depend on the rule of the first to the sixth power [5].

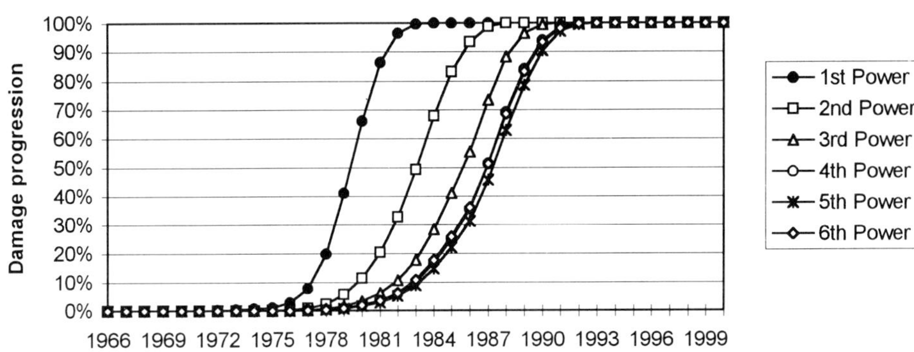

Figure 2: Damage progression according to method A (axle load: 100 kN)

There are six different values for the permissible strengths, so there are also six different courses of structural deterioration.

4.1.2 Method B

The conversion of mixed traffic to an average equivalent axle load and the counted number of passages is the basis to get the damage progression according to method B. The valued equivalent axle load causes stress μ_S at the bottom of the bituminous layer. Depending on the rule of the power, six different values of loads give rise to six different values of stresses. The variation coefficient v_S is 20% according to method A. The permissible strength μ_R depends on the counted number of axle loads and the fatigue behaviour of asphalt. In constrast to method A the result is just one value, which describes the resistance. The variation coefficient v_R is 10% (see method A). Formulas 6, 7 and 8 provide the calculation of the theoretical fatigue life. Figure 3 shows the courses of the time variant probabilities of failure depending on the rule of the first to the sixth power [5].

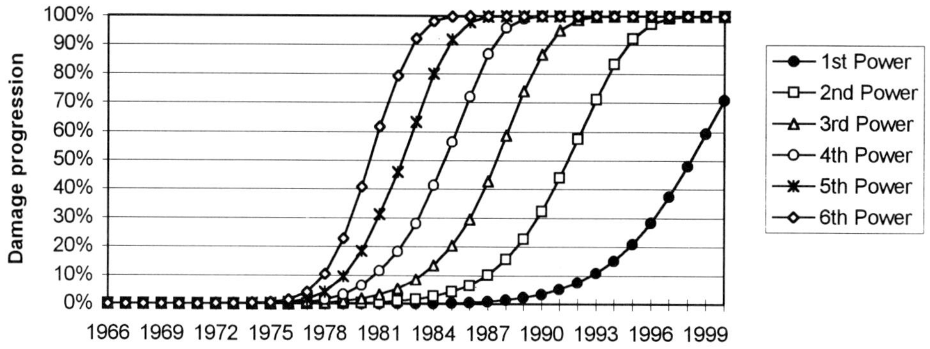

Figure 3:Damage progression according to method B (axle load: equivalent average axle load)

4.1.3 Determining of the decisive rule

There are six different courses of the damage progression in both methods A and B; but they are different. According to method A the fatigue life of the road structure is longer when the power has a higher value. This is the result of the increasing permissible strengths when using a higher power (Figure 2). In contrast to this the fatigue life according to method B is longer when using a lower value for the power. The reason is the lower stress at the bottom of the layer by calculating the valued equivalent axle load with a rule of a lower power (Figure 3). The power, which causes the same fatigue life at both methods, is the relevant one to value correctly the impact on the road structure by an increasing load. Figure 4 shows how the value of the power can be estimated. The point of intersection of both courses is the searched factor [5].

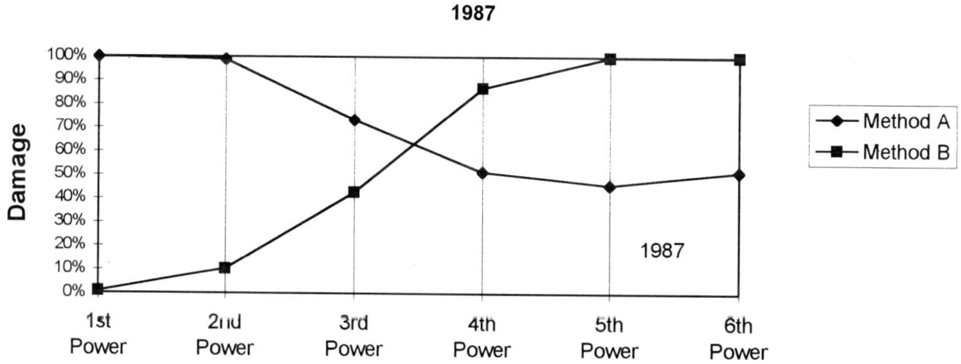

Figure 4: Determination of the decisive rule

As a result of studying selected road segments, axle loads must be valued by the rule of the third to the fourth power in case the of bituminous layers.

4.2 Fatigue behaviour of concrete pavements

Concrete pavements have a clear limit of permanent strength. Stresses which are less than this strength do not have any impact on the structure of the road. Therefore axle loads with less than about 8 tons are not damaging to the concrete slabs, if they are built according to the German design standards. In contrast to the bituminous pavements, the detrimental traffic can no longer be described with a Gaussian normal distribution. For this reason the two methods have to be modified. The result however is very similar to those of the asphalt pavements. An increase of the axle loads must be valued by the rule of the third to the fifth power to estimate the influence of increasing loads on the damage progression of concrete slabs.

References

1 Kalisch, H.: "Einbau und Betrieb von 146 Achslast- und Achsmengenzählgeräten für das Forschungsvorhaben 'Langzeitbeobachtungen'", Straße und Autobahn, Heft 11, 1976

2 Eisenmann, J., Friedrich, M.: "Analyse und Quantifizierung der Verkehrsbelastung an ausgewählten Straßenabschnitten", Forschungsbericht des Prüfamtes für Bau von Landverkehrswegen, TU München, im Auftrag der BASt, Juni 1995 (nicht veröffentlicht)

3 Zahnmesser, W.:"Auswirkungen verschiedener Lastzugkombinationen auf die Straßenbeanspruchung", Forschung Straßenbau und Straßenverkehrstechnik, Der Bundesminister für Verkehr, Heft 327, Bonn,1981

4 Wüst, W.: "Langzeitverhalten von Asphaltstraßen - Eine Untersuchung mit statistischen und wahrscheinlichkeitstheoretischen Methoden", Mitteilungen des Prüfamtes für Bau von Landverkehrswegen der Technischen Universität München, Heft 68, 1995

5 Friedrich, M.: "Analyse des Schwerverkehrs und Quantifizierung seiner Auswirkungen auf die Straßenbeanspruchung mit Hilfe der Potenzregel", Mitteilungen des Prüfamtes für Bau von Landverkehrswegen der Technischen Universität München, Heft 73, 1998

Fifth International Conference on the
BEARING CAPASITY OF
ROADS AND AIRFIELDS

MECHANISTIC EVALUATION OF PAVEMENT PERFORMANCE

MAGDY Y. MIKHAIL[1] AND MICHAEL S. MAMLOUK[2]

[1]Research Engineer, Nichols Consulting Engineers, 1885 Arlington, Reno, NV 89509.
[2]Professor, Dept. of Civil and Envir. Eng., Arizona State University, Tempe, AZ 85287-5306.

ABSTRACT

The factors affecting dynamic load variability are studied and statistically analyzed with regard to vehicle-pavement interaction.

Available pavement performance models are either based on empirical relations or statistical predictions. In either case, the produced models are not adaptable to new conditions such as different material types, vehicle types, or climatic conditions. No completely mechanistic performance models are currently available that do not depend on previous observations or questionable assumptions.

In this paper, a concept for mechanistic-based performance models for flexible pavements was developed which considers the interaction between vehicles and pavement. A dynamic vehicle model was used to estimate the dynamic wheel forces, and a three-dimensional finite element nonlinear dynamic pavement model was used to determine the dynamic pavement response. The study considers the visco-elastic nature of asphalt concrete and the nonlinearity and plasticity of granular and subgrade materials. The accumulation of plastic deformation in the wheel path and the increase in pavement roughness due to repeated traffic loads were estimated. The effect of pavement roughness on vehicle bouncing and the effect of vehicle bouncing on the progression of pavement roughness were investigated under different roughness levels and suspension types. The number of equivalent 80-kN single axle load repetitions to cause a decrease in serviceability from one level to another was estimated under different conditions without the need for empirical observations. If failure is defined by a certain serviceability level, the number of load repetitions to cause failure can be estimated.

Several performance models were developed using the prescribed concept for different truck suspension types. The reverse of this analysis can be used to design the pavement section to sustain a certain number of load repetitions before failure using mechanistic procedure.

INTRODUCTION

The study of pavement response and performance is complex since it involves many factors such as dynamic loading, nonlinear material properties and environmental Conditions. Most of the previous studies assume that pavements are loaded statically, while in reality pavements are subjected to dynamic loads. The forces applied by vehicles vary instantaneously above and below the static weight due to the interaction effect between vehicles and pavement. Several factors contribute to this load variation such as road roughness, vehicle configuration, suspension type, tire type, and vehicle speed. Increasing dynamic forces increases the damage rate to pavements and shortens their service lives.

OBJECTIVE AND APPROACH

The main objective of this study is to develop mechanistic-based pavement performance models that can predict the pavement condition at different ages and provide good estimates of the pavement service lives under different conditions. The factors affecting dynamic load variability are investigated with regard to pavement-vehicle interaction. Furthermore, the study considers the visco-elastic nature of asphalt concrete and the nonlinearity and plasticity of granular and subgrade materials. The accumulation of plastic deformation and the increase in pavement roughness due to the repeated traffic loads is also investigated. In this study a vehicle dynamic program is used to predict the dynamic loads generated by moving traffic. The structural response of pavement is analyzed using a three-dimensional, dynamic finite element program. The progression of pavement roughness due to the dynamic forces of vehicles and the reciprocal effect of the increased pavement roughness on increasing the dynamic forces of vehicles are used to develop mechanistic-based performance models under different conditions.

LITERATURE REVIEW

Vehicle Dynamic Models

Vehicle dynamic models have been developed by previous researchers to predict the dynamic forces produced by different axle and wheel configurations at different locations on the pavement surface. Vehicle dynamic models represent the truck with a number of masses, springs, and dashpots. The models are typically implemented in time-domain computer simulation programs. Among the vehicle dynamic computer models is the Florida COMprehensive Pavement Analysis System (COMPAS) (1 & 2). The COMPAS model is capable of generating the dynamic force profile of wheel loads for different combinations of vehicle types, suspension types, vehicle speeds, and levels of pavement roughness. The COMPAS program simulates vehicles using a pitch-plane model with a combination of masses, springs, and dashpots.

Pavement Dynamic Model

Several pavement dynamic models have been developed in the last few decades.

ABAQUS is one of the available pavement dynamic models, which is a versatile three-dimensional finite element program and can be used in many applications. ABAQUS has the capability to simulate moving loads and offers several methods for dynamic analysis (3).

Several researchers have studied the dynamic structural response of pavement using ABAQUS. Zaghloul and White (4, 5) found close agreement between the results from ABAQUS and field measurements. They also found that the speed of the vehicle has a significant effect on both the elastic and plastic pavement responses. In addition, the predicted pavement response to the falling weight defelectometer load obtained from ABAQUS was found to match the measured deflection (5).

Vehicle-Pavement Interaction

The subject of vehicle-pavement interaction considers the effect of pavement roughness on the vehicle dynamic forces and the effect of these forces on the pavement response and performance. The roughness of the pavement surface excites and bounces traveling vehicles. Increasing surface roughness or vehicle speed increases the bouncing of vehicles and consequently increases the dynamic forces applied on the pavement. The forces applied by vehicles vary instantaneously above and below the static weight due to the interaction effect between vehicles and pavement. Several factors contribute to this load variation such as road roughness, vehicle configuration, suspension type, tire type, and vehicle speed. On the other hand, increasing dynamic forces increases the damage rate to the pavement and shortens its service life (6).

VEHICLE DYNAMICS AND PAVEMENT RESPONSE

Estimation of Vehicle Dynamic Forces

In this study the Florida COMPAS was used to determine the dynamic tire forces under various conditions. The following factors and levels were studied:

1. Truck Type: Three types were used following the FHWA classification; 2-axle tractor, axle trailer (2-S2), 3-axle tractor, 2-axle trailer, (3-S2), and 5-axle multitrailer (2-S1-2)
2. Suspension Type: air bag, leaf spring, and walking beam
3. Speed: 20, 75, and 130 km/h
4. Roughness: Three levels with values of PSI 4.6, 3.5, and 2.4.

The combinations of the different truck types, suspension types, speeds, and pavement roughness levels produce different dynamic tire forces. These factors and levels resulted in 81-factor combinations. The gross vehicle weights for the three truck types were chosen so that the axle weights on rear axles were approximately equal to 80 kN for the purpose of comparison. Numerical outputs were generated for all the above combinations. The results show the variation of dynamic tire forces with distance and time under various conditions.

Figure 1 shows an example of the dynamic tire force for the second axle versus distance for the different types of suspension of 2-S1-2 truck at a speed of 130 km/h for an average present serviceability index of 2.4. The figure shows that the walking beam suspension causes more dynamic load variation than the other two suspension types.

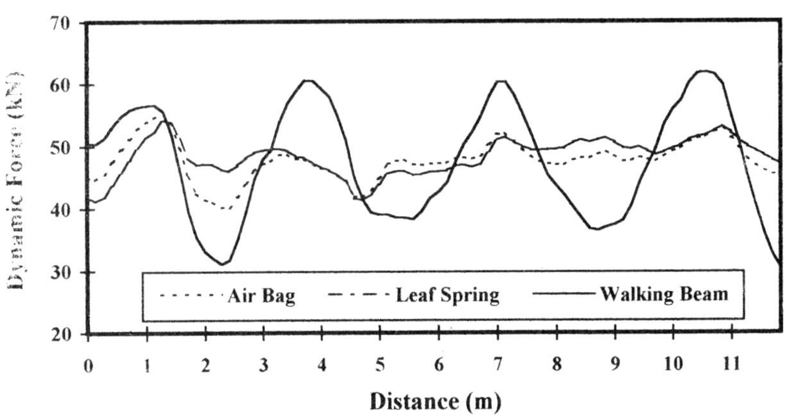

Figure 1. Dynamic Tire Forces Versus Distance for 2-S1-2 at a Speed of 130 km/h and PSI of 2.4

Dynamic Load Coefficient

The dynamic load coefficient (DLC) is used to characterize the variability of the wheel force and is defined as :

$$DLC = \frac{\text{Standard Deviation of Dynamic Wheel Force}}{\text{Mean Dynamic Wheel Force}} \quad (1)$$

The results showed that the air bag suspension produces the least DLC, whereas the walking beam suspension produced the highest DLC. Furthermore, increasing the truck speed and/or the pavement roughness increase the DLC for all types of suspension.

A full factorial experimental design was used for the above factors and levels. An analysis of variance (ANOVA) was performed on the results to test the significance of the different factors. All the main effects were found to be significant at a significance level of 0.05.

Finite Element Model

The pavement sections used in the study were assumed to consist of three layers; asphalt concrete surface, granular material base, and cohesive soil subgrade. One set of pavement layer thickness was used as shown in Table 1. A three-dimensional finite element model using the ABAQUS program was developed to represent the pavement structure as illustrated in Figure. 2. Due to symmetry, the figure represents one half of the pavement section on one side of the wheel path. A 12.2 m (40 ft) long wheel path was used with two 12.2 m unloaded zones before and after the wheel path to reduce the end effect.

In the finite element model brick elements with 8 nodes were used. In the region of the wheel path a fine mesh was used. The area of each element in the wheel path was taken so that it would equal to the tire-pavement contact area. Both the asphalt concrete surface and the base course were modeled as single-element layers, while the subgrade was modeled as a multi-element layer. Infinite elements were used at the bottom of the subgrade. The mesh dimensions were selected to obtain appropriate aspect ratios of the elements and to achieve the desired degree of accuracy. Movement was restrained at the plane of symmetry in the perpendicular direction.

TABLE 1. Layer Thickness for Pavement Section Used

Layer	Medium-Thickness (mm)
Asphalt Concrete	100
Granular Base	150
Subgrade	1500

Material Properties

Asphalt concrete was modeled as a visco-elastic material. The response of this type of material is time dependent. The time dependent properties were represented by the ratio of instantaneous shear modulus to the long term shear modulus (G-ratio) which can be expressed as follow (3):

$$\text{G-ratio} = 1 - \frac{\text{Long Term Shear Modulus}}{\text{Instantaneous Shear Modulus}} \qquad (2)$$

The elastic-plastic Drucker-Prager model was used in the analysis to model both granular base and subgrade materials. Table 2 shows the material characteristics used for this model (4, 5).

Loading Cycles

Moving wheel loads were simulated in the analysis using the ABAQUS program. For each passage of a wheel load both vehicle speed and the variable dynamic loads due to truck bouncing were simulated in the analysis. In each finite element the load was assumed to increase linearly to a peak value then decrease linearly to zero with a duration matching a speed of 75 km/h. This loading pattern was repeated for the different elements in the wheel path at different times depending on the vehicle speed. Different load magnitudes were applied on each element to simulate the variable dynamic loads obtained from the Florida COMPAS program.

Pavement Response

The ABAQUS program was run to evaluate the structural response of pavement under different conditions. The program was installed on an HP9000 model J200 workstation with 128 Mbytes RAM, 9 Gbytes hard disk, and 100 MHz PA-RISC 7200 CPU. Each computer run for a one wheel pass took an average of 3 hours. Multiple wheel passes increased the run time approximately linearly. For example, 20 wheel passes took about 60 hours of run time.

The permanent displacements at the pavement surface due to the compressibility of all layers were evaluated using the ABAQUS program for different roughness levels using the appropriate dynamic loads obtained from the Florida COMPAS program.

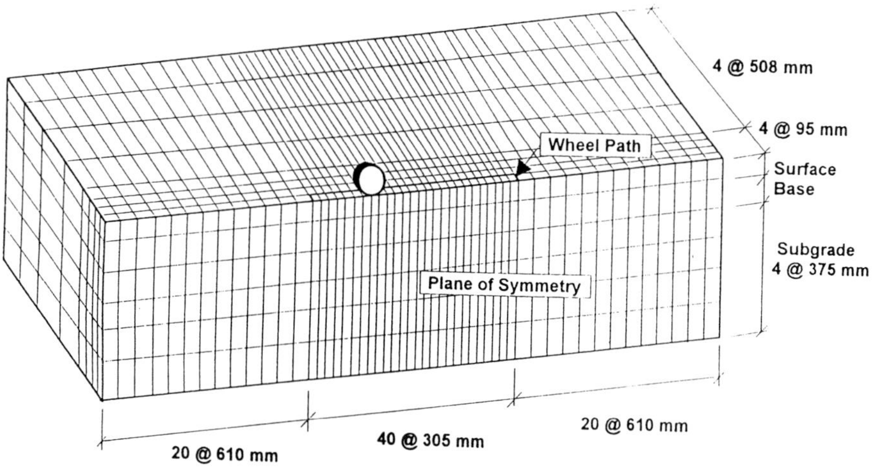

Figure 2. Finite Element Mesh Used in the Analysis

TABLE 2. Material Properties

Layer	Parameter	Medium Stiffness
Asphalt Concrete	Modulus Of Elasticity (MPa)	2.75
	Poisson's Ratio	0.35
	G-Ratio	0.8
	Damping Coefficient	0.05
	Density (Mg/m^3)	2.403
Granular Base	Modulus of Elasticity (kPa)	415
	Poisson's Ratio	0.4
	Yield Stress (kPa)	0.243
	Damping Coefficient	0.05
	Density (Mg/m^3)	2.163
	Angle of Internal Friction (degrees)	38
	Cohesion	0
Subgrade	Modulus of Elasticity (kPa)	52
	Poisson's Ratio	0.45
	Yield Stress (kPa)	0.138
	Damping Coefficient	0.05
	Density (Mg/m^3)	1.922
	Angle of Internal Friction (degrees)	7.3
	Cohesion (Pa)	76.66

DEVELOPMENT OF PERFORMANCE MODELS

Pavement performance models were developed to relate pavement condition to the cumulative number of ESAL repetitions. In these models it is assumed that pavement condition is measured by surface roughness. Thus, it is implicitly assumed that other forms of distress will develop at the same rate as that of roughness. It is also assumed that roughness is developed through the development of different levels of permanent deformation at different locations within the wheel path due to vehicle dynamic loading.

Starting with a smooth surface of a new pavement (PSI = 4.2) the dynamic load profile obtained from the COMPAS program was input to the ABAQUS program to estimate the permanent displacement on the pavement surface at 0.3 m (1 ft) intervals after different numbers of load repetitions. Since each load repetition nonlinearly contributes to the increase in the pavement roughness, a large number of load repetitions has to be simulated in the analysis. After many trial runs it was found that the first load repetition resulted in the largest permanent displacement on the pavement surface due to material nonlinearity. Further load applications resulted in reduced permanent displacements. To reduce the computational time, 19 load repetitions were used in each case. Also, the permanent displacement due to the first load application was not considered in the development of the performance models assuming that this displacement was achieved during construction.

Using the computed permanent deformation after 19 load applications at a certain PSI level the new slope variance (SV) was computed using the following equation (7):

$$SV = \frac{\sum Y^2 - (1/n)(\sum Y)^2}{n-1} \qquad (3)$$

where Y is the difference in elevation of two points 0.3 m apart and n is the number of readings along the road profile. The slope variance was used to estimate the new PSI value according to the following equation (7):

$$PSI = 5.03 - 1.9 \log (1+SV) - 0.01\sqrt{C+P} - 1.38\, RD^2 \qquad (4)$$

where

SV = slope variance,
C = linear feet of major cracking per 1000 ft^2 area,
P = bituminous patching in ft^2 per 1000 ft^2 area, and
RD = rut depth in inches.

The terms C, P and RD are neglected since 95 % of the variation of the PSI value is described by the slope variance (8).

Because of the time limitation of running the ABAQUS program, the number of load applications to go from one PSI level to the next level was extrapolated linearly from the results of the 19 load repetitions. This method is justified since most of the nonlinearity occurs in the first few load applications.

The new PSI level was used to compute the new dynamic load profile using the Florida COMPAS program. This process was repeated until the failure PSI of 2.5 was reached.

Performance models were developed starting with a high PSI value of 4.2 then goes down to 3.86, 3.55, 3.0, and 2.5. The number of load repetitions required to go from one level to the next level was estimated for each PSI interval. Figure 3 shows a flow diagram for the development of performance models.

Performance models were developed for three suspension types. Truck 3-S2 type was used with a single axle load of 80 kN and an average truck speed 75 km/h in all cases.

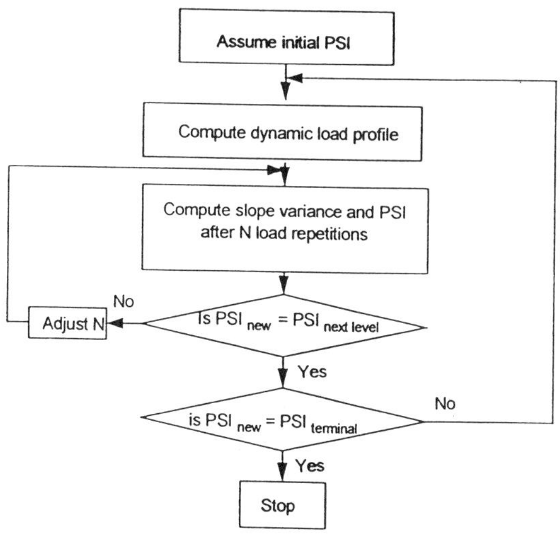

Figure 3. Flow diagram of the Model Developed

Figure 4 shows the PSI versus the number of load repetitions for the different suspension types. It is clear from the figure that the air the bag suspension gives the highest number of load repetitions until failure, whereas the walking beam suspension produces the lowest number of load repetitions. This indicates that the air bag suspension has the least rate of pavement failure as compared to other suspension types.

It should be noted the slope of the performance curve stays flat for a large number of repetitions after which it increases sharply in all cases. This indicates that when the pavement is smooth it can carry a large number of load repetitions before the PSI level drops appreciably. On the other hand, when the pavement is rough large truck bouncing occurs which results in a faster rate of pavement deterioration.

Fifth International Conference on the
BEARING CAPASITY OF
ROADS AND AIRFIELDS

MODERN CONCEPT FOR REFINING RUSSIAN METHOD OF FLEXIBLE ROAD PAVEMENT DESING

V.D.KAZARNOVSKY, Dr.Sc., Prof, Academ. of Transp. Academy of RF
V.M.YUMASHEV, Ph. D., Academ. of Transp. Academy of RF
M.L. POPOV, Eng.
State Road Research Institute (Soyuzdornii)
79, Chaussee of Enthusiasts, Balashikha - 6, Moscow region,
143900, Russia
Fax: (095) 521 18 92

ABSTRACT

The climatic conditions in Russia vary extensively and in this country it is to a great extent necessary to use local soils and road-building materials. These materials are variable in composition and condition, and are characterized by properties which vary over a wide range. For these reasons the pavement structural design in this country is different from empirical methods commonly used abroad. The structural design in Russia was traditionally based on physically substantiated theoretical schemes developed in road pavement mechanics applying general principles of structural mechanics.

In the current Russian manual on flexible pavement design (BCH 46-82) provision is made for the design according to two conditions: based on frost resistance and based on strength. The design according to strength is subdivided into three independent designs, taking into consideration various kinds of pavement deterioration.

Work is now going on to improve the interpretation of the «road pavement strength» concept and to verify the physical sense of the strength criteria. Consideration is also given to the different deterioration mechanisms by decreasing the load distributive capacity (plate effect) of bound layers when cracks appear. Accumulating permanent deformations under repeated loads, accompanied by increasing surface unevenness with time, is also considered. When considering the pavement deterioration mechanisms under multiple load applications, in this sense, modulus of elasticity, Poisson's ratio, fatigue index, and dynamic creep index are regarded as design parameters of the structural layer materials.

An assessment of the pavement thermal crack resistance is applied as an additional characteristic of the structure.

Key words: shape-changing permanent deformation, pavement structure, repeated short-term loading, fatigue process.

INTRODUCTION

An initial version of the Russian method for flexible road pavement design was drawn up as a standard method during the 1950's (Prof N.N.Ivanov), though its development was actively under way during the 1930's and 1940's.

The method was developed under certain circumstances characterized by the following.

- high diversity of natural conditions over the former USSR territory ranging from subarctic to subtropical;
- undeveloped road network and a very scarce experience in the road construction and operation in a number of regions;
- centralization of road science and its low development in regions;

- isolation of the road industry and the road science in particular from abroad.

The above circumstances had their negative effects, but, along with this, also created prerequisites for rather original and deep scientific research and developments. One such example is the development of the Russian method for road pavement design. Unlike empirical design methods in common use in the West (for example, those based on bearing CBR), this method is based on strict theoretical design schemes applying theoretically substantiated parameters of strength and deformation properties of soils and material which are generally accepted in the field of construction.

The initial version of the design method started from the assumption that soils and road-building materials behaved as elastic-viscous-plastic materials. Visco-plastic properties influence the stress distribution in a structure and predetermine in all cases the accumulation of permanent deformations in the structure under the action of multiple loads. Deformation moduli of soils and materials, assessed with consideration of a specified admissible value of total deformation of the structure, were used as a basic characteristic of their mechanical properties. The design was brought to the observance of condition that the accumulated deformation of the pavement surface to the end of the life should not exceed the admissible one. Practically, this condition meant that the total (so-called «equivalent») deformation modulus of the structure should not be lower than the required one, specified by standards for roads of a given category.

AN UPDATED DESIGN METHOD

In the 1960's, under the guidance of Prof. A.M.Krivissky, complex investigations were carried out as a result of which the design method was updated. The following fundamentals, proved in the process of research, formed the basis for updating /1/:

1. Stress distribution in present-day flexible road pavement structures is in compliance with the theory of elasticity for layered systems, and, therefore, it is precisely the theory that shall be applied and there is no need for any correction coefficients when determining stresses in the structure.

2. In some stressed state of the structure, permanent deformations under loads practically speaking do not accumulate. It is proposed to use the conditions of local ultimate shear equilibrium at the most stressed point of the structure as conditions under the observance of which no accumulation of permanent deformations will occur. If there is no accumulation of permanent deformations, the structure behaves as an elastic system (the performance in the elastic stage), and the question on forecasting the deformation accumulation no longer arises, i.e. the design is conducted to exclude permanent deformations.

3. An additional condition, governing the structural strength, is non-exceeding an admissible value of flexural tensile stresses in the bound layers.

According to the above items, a nomenclature of design soil and material characteristics of the structural layers substantially changes. There appear in the design: moduli of elasticity and Poisson's ratios as deformation characteristics; angle of internal friction and adhesion as strength characteristics, and additionally flexural tensile strength for bound layers.

The fundamentals listed above are used in the design methodics that is prescribed by a current normative document /2/.

Along with the strength design, the structure is checked for frost resistance, and the test method in the current instruction is not different from that in the first version of the normative document. However, this question is not considered within the context of this report.

More than 15year experience in the use of the last version of the design method has shown that at present, the instruction itself needs certain updating, some of the fundamentals included.

First of all, it is found that the condition of local ultimate equilibrium by no means ensures avoiding the accumulation of permanent deformations in the structure. This is explained by the fact that non-exceeding the shear resistance does not guarantee against the shear deformation accumulation. Under certain conditions soils and road-building materials can accumulate permanent deformations even at tangential stresses that are much lower than the shear resistance.

In this connection it is possible to speak about the road pavement structure performance in the elastic stage only in a sense of the compliance of the stress distribution with the theory of elasticity for layered systems. Permanent deformations, to a certain extent, can appear almost always, and their magnitude, negligible for a single load application, can be considerable for load applications measured in thousands and hundreds of thousands. Depending on soil and material properties as well as on level of the stressed state, governed by the pavement structure, the intensity of deformation accumulation may be higher or lower, may damp out or may not, etc.

Thus, there appears a problem of improving both the forecast of conditions and the evaluation of permanent deformation accumulation intensity, that is, to a certain sense, the necessity of some return to the old version of the design method.

A LATE REVISION OF THE METHOD

To solve this problem, in the 1980's, Soyuzdornii carried out a great number of investigations. A number of works in this field was performed in MADI (TU) under the supervision of Prof. A.P.Vasiliev.

Along with this, large works were conducted (particularly, by Prof B.S.Radovsky and his co-workers /3/ in Kiev) on studying the second process, proceeding in the road pavement structure with bound layers under multiple load applications - the process of fatigue.

The results of all these studies indicate that the process, that brings the road pavement to final failure under multiple load applications, is a combination of independent simple (elementary) processes, conditions for creating and developing of which are governed by their own laws.

This process as a whole seems to be as follows:

1. When the structure is subjected to a single short-term action of the external load, there appears a stressed state described by the theory of elasticity. Under specified conditions for the external load, stress distribution and stress magnitude at different points in the structure depend on geometric parameters (layer thicknesses), material elastic properties (modulus of elasticity and Poisson's ratios) as well as on mutual positions of these layers.

2. Under the repeated short-term action two processes are in progress: the fatigue process in bound layers, the intensity of which is dependent upon the ratio of acting tensile stresses to material fatigue strength, and the process of accumulating permanent deformations. As it is found, the permanent deformations in the structure, the degree of compaction of which conforms the standards, are mainly (about 85%) shape-changing deformations developing under the action of tangential stresses.

3. The fatigue process develops in any case and finally manifests itself in (rupture) microcracking due to the fatigue failure of the bound layer under the action of flexural tensile stresses. Thereby, the bound distributive capacity (plate effect) of the bound layer decreases. In the limiting case, when the bound layers completely fail in this way, a picture of the stress state in the structure approaches the one, occurring in homogeneous semispace or discrete medium. Such failure is indicative of the deterioration of the basic quality of the bound layers - the increased stress distribution over the underlying courses. In this case, the stressed state of the structure is transformed in such way that the stresses in the layers, underlying the bound ones, and in the subsoil increase.

When describing the fatigue process, the following relationship may be applied:

$$N_p = \left(\frac{R}{r}\right)^m \tag{1}$$

where: N_p = number of load applications resulting in failure of the bound layer material;
r = acting tensile stress;
R = laboratory tensile strength of the material under single load application (with the same loading rate as that for stress r);
m = index of material fatigue properties depending on material type and design state.

Tentative values of index «m», established as a result of special investigations and recommended by Prof B.S.Radovsky et al /3/, are given in Table 1.

Table 1

ą	Bound layer material	m
1	Tar concrete and cold asphalt concrete	2.0-3.0
2	Hot asphalt concrete with bitumen of medium viscosity	3.5-5.0
3	Ditto, with viscous bitumen	5.0-6.3
4	Cement-or lime-stabilized soil	12.0-20.0
5	Cement concrete	20.0 - 37.0

4. The process of accumulating permanent shape-changing deformations may develop regardless of the fatigue process. Its intensity is dependent on the stress state of the layers and the subsoil as well as properties of the materials in these layers and of the soil. The development of the fatigue process, with an increase in stresses in the underlying courses and soil, stimulates a growth in intensity of the permanent deformation accumulation for the portion of the structure and the soil which underlie the bound layer. The process of permanent deformation accumulation eventually results in the pavement evenness decrease (rutting, etc.) and, in the long run, making the structure inoperative (the state of refusal) according to the condition of surface evenness.

The process of accumulating permanent shape-changing deformations is described on the basis of the relationship:

$$\lambda_{per.z} = \frac{(\sigma_1-\sigma_2)-(\sigma_1-\sigma_2)^*}{\eta_{N\sigma}} \tag{2}$$

where:

$\lambda_{per.z}$ = permanent vertical deformation of an elementary layer at depth z along the axis of loading area;

$(\sigma_1 - \sigma_2)$ = stress deviator;

$(\sigma_1 - \sigma_2)^*$ = deviator at which the accumulation of permanent deformations begins (so called «threshold of accumulation»);

η_{N_σ} = parameter taking account of the material properties to accumulate permanent shape-changing deformations under the action of multiple loads, which depends on a level of the stress state (σ) and number of load applications (N). The latter is determined from soil or material tests under conditions of the triaxial stress state at the application of multiple load with specified parameters. This parameter is proposed to be referred to as rheodynamic viscosity.

Thus, the threshold of accumulation and the rheodynamic viscosity are the characteristics of soils and materials that reflect their capability to accumulate permanent deformations under the action of multiple loads (so called «rheodynamic characteristics»).

DISCUSSION

The complexity of the process of making the road pavement structure inoperative under multiple load applications points to the necessity to apply in this case another approach to the notion of the strength or «road pavement bearing capacity» than that accepted in this field by analogy with the design of bases and foundations for static load.

The strength or bearing capacity of the road pavement should be understood as its property to with-stand deterioration, more or less intensively, under multiple load applications due to fatigue failure of bound layers and / or permanent deformation accumulation.

To adjust the value of road pavement strength in the above sense, it is not sufficient to know a value of the elastic deflection of the structure, but it is necessary to have an apparatus and required indices of properties of soils and materials in structural layers in order to have a possibility to predict a period of the structure performance to refusal, i.e. the structure resource.

In this case, the notion «strength» or «bearing capacity» of the structure will reflect conditions of its performance under multiple loading or in time, i.e. will reflect its durability. The structure that can withstand more load applications can be called in this case a more durable one.
Taking into account the above mechanism of the pavement performance will allow to verify the design method and to reveal during computation more efficient structures, with due regard for peculiarities of the road pavement performance under multiple load applications.

FURTHER DEVELOPMENTS

Vast part of the territory in Russia is characterized by low winter temperatures and high temperature gradients. This is one of the reasons why the problem of road pavement crack resistance is urgent. The appearance of temperature cracks in the structure creates sites of origin for its failure just because in the vicinity of the crack the structure stressed state changes and the process of permanent deformation accumulation speeds up under the action of loads, not to mention the consequences of water ingress and freezing in the cracks, the contamination of the latter and the appearance of additional temperature deformations, etc. That is why, when evaluating the structure strength in a broad sense, it is required to consider its resistance to temperature cracking and, if this is inevitable, to consider the influence of this process on accumulating permanent deformations under transport loads. At present studies are under way on evaluating the structure crack resistance and methods of its improving. This aspect of the road pavement design is not covered in the existing Russian method of design and requires further development.

The realization of the above directions in the development of the Russian method for flexible road pavement design will allow to substantially refine the method. However, at the same time, this realization will involve more thorough and complicated tests of soils and road-building materials as well as the need of inclusion of additional parameters, not determined before. Along with this, this approach will make it possible to reduce values of different safety factors, which are now introduced due to the absence of knowledge of actual mechanism on the process and which often reduce to zero the rigorism of applied design schemes themselves.

REFERENCES

1. Иванов Н.Н., Кривисский А.М.и др. Конструирование и расчет нежестких дорожных одежд. -М.: Транспорт, 1977.-102 с.

   IVANOV, N.N. and KRIVISSKY, A.M. et al. Flexible Road Pavement Structural Design. M.: Transport, p. 102, 1977.

2. Инструкция по проектированию дорожных одежд нежесткого типа. ВСН 46-83. Минтрансстрой. -М.: Транспорт, 1985.-157 с.

   Flexible Road Pavement Design Instruction. VSN 46 - 83. Mintransstroy. M.: Transport, p. 157, 1985.

3. Радовский Б.С., Супрун А.С., КозаковИ.И. Проектирование дорожных одежд для большегрузных автомобилей. -К.: Будивэльник, 1989.-168 с.

   RADOVSKY, B.S., SUPRUN, A.S. and KOZAKOV, I.I. Road Pavement Design for Heavy Vehicle Traffic. K.: Budivelnik, p. 168, 1989.

Fifth International Conference on the
BEARING CAPASITY OF
ROADS AND AIRFIELDS

LOADING CHARACTERISTICS OF MODERN TRUCK TIRES AND THEIR EFFECTS ON SURFACE CRACKING OF ASPHALT PAVEMENTS

Reynaldo Roque, Leslie A. Myers, and Byron E. Ruth

Associate Professor, Graduate Research Assistant, and Professor
University of Florida, Department of Civil Engineering,
Gainesville, Florida, USA

ABSTRACT

Surface-initiated wheel path cracking of asphalt pavements has become increasingly more prevalent in recent years. Various explanations have been proposed for this phenomenon, including dissipated energy concepts and the potential effects of lateral contact stresses under truck tires. However, until recently, limited, if any, measured tire-pavement contact stresses have been obtained for different types of truck tires and for use in pavement stress analysis. Tire-pavement contact stresses (vertical, longitudinal, and transverse) were measured for both bias ply and radial truck tires in order to clearly demonstrate the differences in their surface loading characteristics and in their resulting effects on near-surface pavement stresses. A transducer-based system originally developed for tire companies to improve tire wear characteristics was used to obtain the measurements, which indicated that transverse stress distributions under bias-ply tires were significantly different than those under radial tires. Pavement stress analyses indicated that significant transverse tensile stresses, which would explain the development of surface-initiated longitudinal wheel path cracks, develop under the treads of radial tires. Surface tensile stresses did not develop when the pavement was subjected to the bias-ply contact stresses. Tire surveys have indicated that use of radial tires has increased dramatically, while bias-ply tires are almost no longer used. This fact, combined with the results of this study indicate that transverse contact stresses induced by radial tires appears to explain the development of surface-initiated cracks. Tire-pavement contact stresses need to be better defined and must be considered in modern pavement and mixture design procedures.

Key Words: Surface Cracking, Asphalt Pavement, Tire Stresses

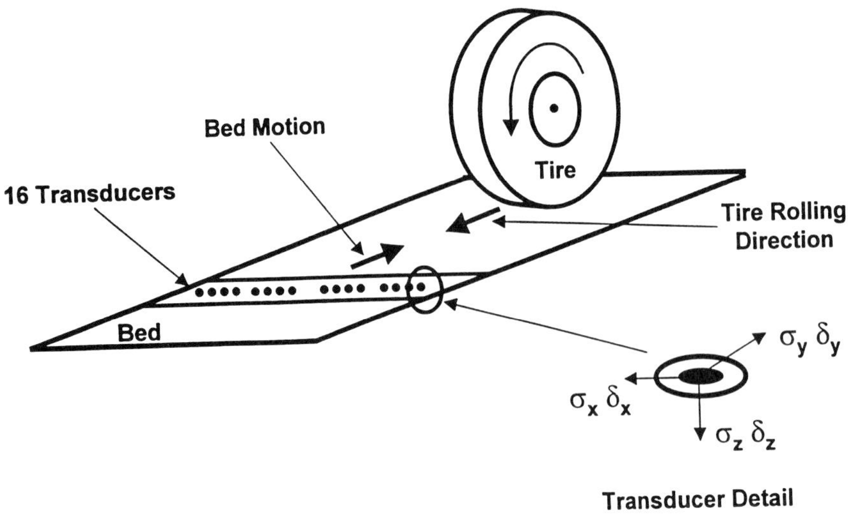

Figure 2. Schematic of System Used to Measure Tire Contact Stresses

The process is repeated with the tire at different transverse positions to obtain a very detailed pattern of the stresses under the tire. In fact, force measurements are made every 2.54 mm longitudinally and every 3.56 mm in the transverse direction. This results in about 4000 stress measurements (vertical, transverse, and longitudinal) under a typical truck tire with five treads and gross contact area of 225 x 200 mm.

Results A description of the tires used in this study, including the load and inflation pressures used to obtain the data, is presented in table 1. As described in the previous section, about 4000 vertical, 4000 transverse, and 4000 longitudinal stress measurements were obtained under each tire. A thorough evaluation of these data indicated that there were significant differences between the stresses under radial tires and stresses under bias ply tires, particularly the transverse stresses.

Table 1. Details of Tires Used For Measured Contact Stress Evaluation.

Tire Type	Contact Area	Number of Treads	Load (kN)	Inflation Pressure (kPa)
Radial, R24.5	22.5 x 200mm	5	25.0	793
Bias Ply, 11-22.5	22.5 x 200mm	5	24.5	793

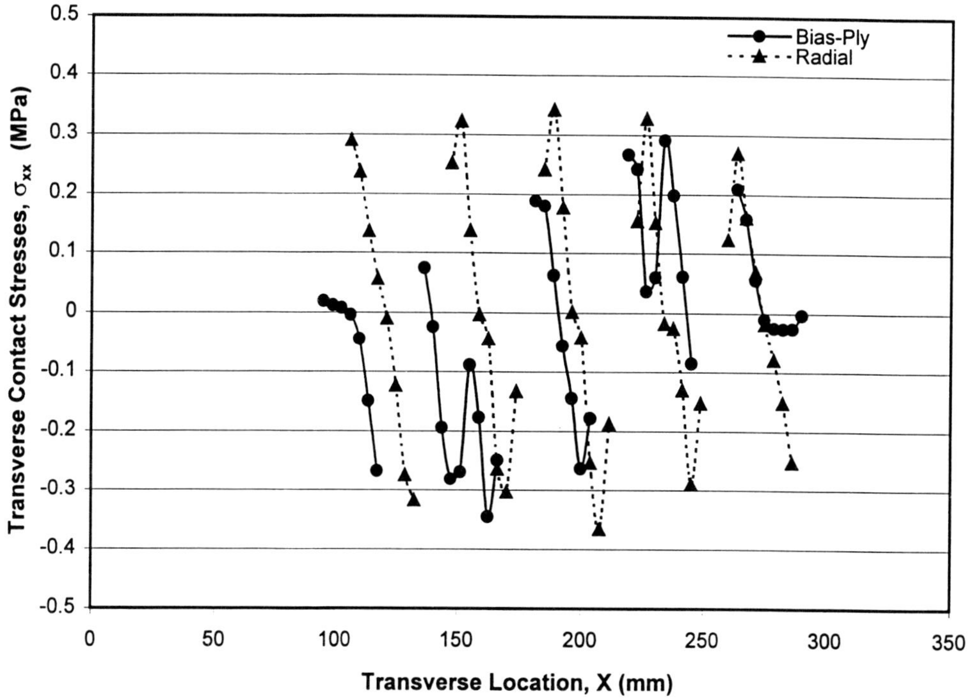

Figure 3. Measured Transverse Contact Stresses Under Truck Tires

Figure 3 shows a comparison of the transverse stresses between the two tires obtained at one longitudinal location near the center of the contact area. It should be noted that the patterns were similar at other longitudinal locations. A detailed review of the stresses presented in figure 3 indicate the following:

- The stress distribution is very similar underneath all the treads of the radial tire, where the transverse stresses reverse from positive on one side of the tread to negative on the other. The effect is that stresses are pushing away from the center of the tread, which will result in a tensile stress on the surface of the pavement immediately under each tread of the radial tire.

- Unlike the radial tire, the transverse stress distribution for the bias-ply tire changes significantly from one side of the tire to the other. The stresses under the left-most tread are predominantly negative (pulling to the right), while the stresses under the right-most tread are predominantly positive (pulling to the left). The overall effect is that the bias-ply tire will tend to pull the pavement in toward center of the tire. Although the stresses do vary within individual treads, the effect is much smaller than for the radial tire. Furthermore, since except for the middle tread, the stresses are predominantly of one orientation, they will not result in a tensile stress on the surface of the pavement.

The effects observed in figure 3 are explained by differences in tire construction. Bias-ply tires have more rigid walls, but less rigid tread structures than radial tires. This implies that the radial tire tread structure will deform much less when loaded, which results in a more uniform stress distribution accross the width of the tire. This has been attributed to the longer life of radial tires than bias-ply tires, but as indicated below, it may have the effect of reducing the cracking performance of pavements.

It should be noted that the tire contact stress measurements obtained in this study corresponded well with measurements obtained by de Beer et al.(11) using a system developed in South Africa.

PAVEMENT STRESS ANALYSIS

Analytical Model The measurements presented above indicated that a fairly detailed representation of tire-pavement contact stresses may be critical for the proper determination of pavement surface stresses. Thus, vertical, transverse, and longitudinal contact stresses determined from measured data under the five tire treads were applied to over 145 uniformly circular loads were used in the BISAR elastic layer computer program (12). Figures 4 and 5 show the circular loads that were used to replicate the stress distributions under the bias-ply and radial tires, respectively. The figures illustrate the effects described earlier in conjunction with figure 3, which indicate that the pavement experiences an overall "pulling inward" effect induced by the bias-ply tire, whereas the radial tire induced a "pushing outward" effect under each individual tread. The stress analysis would indicate whether either distribution may result in tensile stresses at the pavement's surface.

The standard pavement structure used for the analysis was a 200 mm asphalt concrete layer (E_1 = 5500 MPa), underlain by a 300 mm base (E_2 = 300 MPa) and subgrade (E_3 = 100 MPa). Various other pavement structure combinations were evaluated to determine their effects on surface stress distributions, but pavement structure was found to have a negligible effect on surface stresses. Therefore, only the results of the standard pavement are presented below. The total applied loads corresponded to the loads presented in table 1.

Results Figures 6 and 7 show the transverse stresses on the surface of the pavement predicted at one longitudinal position for the bias-ply and the radial truck tires. These figures clearly show that bias-ply tires did not result in tensile stresses at the pavement's surface, while the radial tire induced very significant tensile stress under the left-most tread. There are two reasons why stresses were critical under this particular tread: 1) this was the widest tread for this particular tire, which resulted in higher transverse stresses than the thinner, right-most tread (see figure 3); and 2) the effect of surface compression induced by pavement bending during loading is less on the outer edge of the tire than at the center of the tire. The higher compressive stresses caused by bending is the reason that tension did not develop under the center tread, even though the transverse stresses in the center tread were as high as for the left-most tread (see figure 3). The results in figure 6 appear to indicate that the compressive stresses caused by pavement bending overwhelmed any tensile stresses that were developed by the bias-ply tire pulling the pavement in towards the center of the tire. Analyses were also conducted to determine the effects of pavement structure (e.g.,

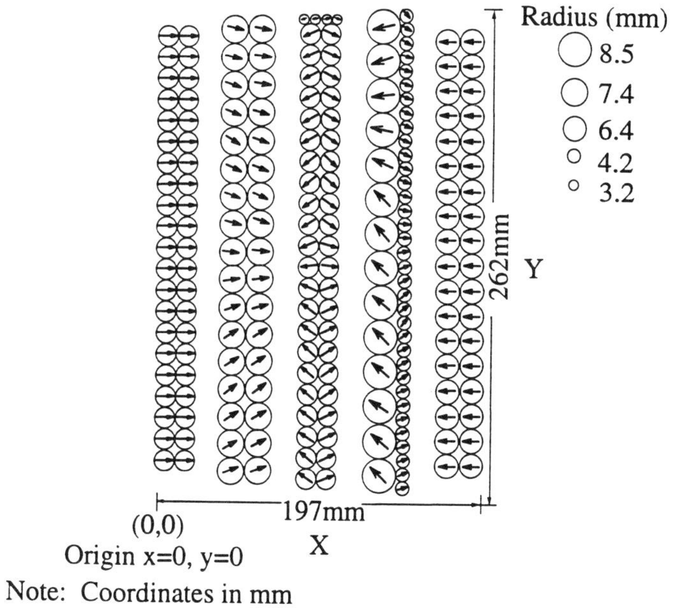

Figure 4. Loading Configuration Used in BISAR to Represent Measured Stresses Under Bias-Ply Truck Tire

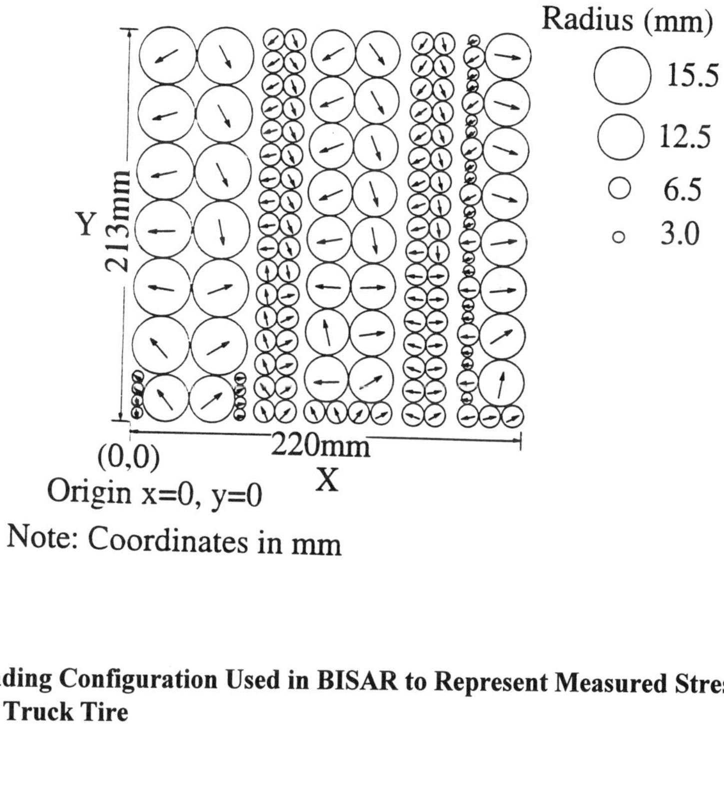

Figure 5. Loading Configuration Used in BISAR to Represent Measured Stresses Under Radial Truck Tire

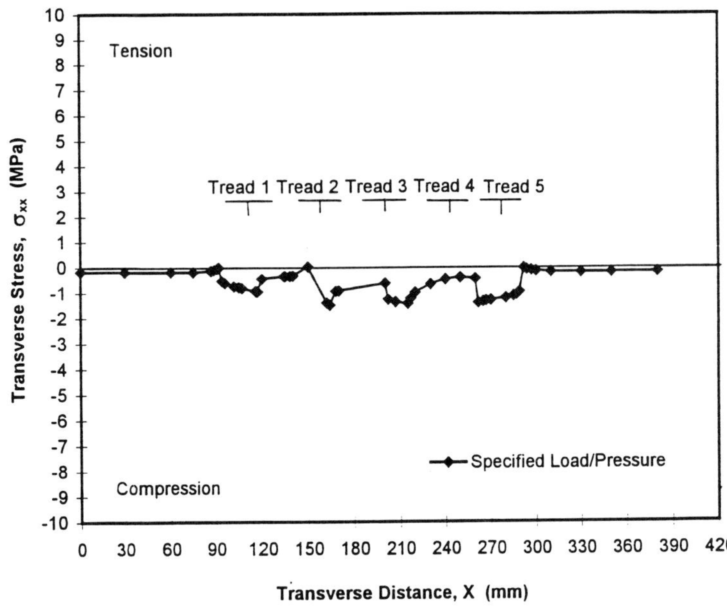

Figure 6. Predicted Transverse Stress Distribution Along Pavement Surface: Bias-Ply

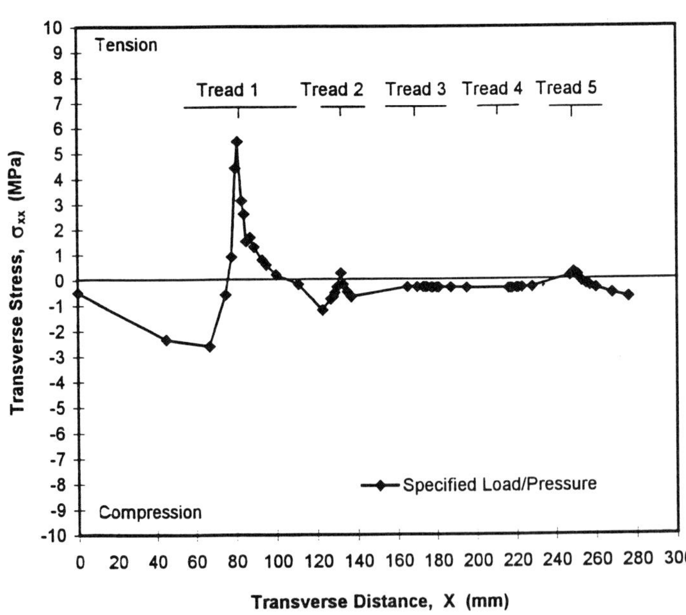

Figure 7. Predicted Transverse Stress Distribution Along Pavement Surface: Radial

layer thickness and stiffness) on surface tension. Results demonstrated that the same high tensile stresses develop at the surface of the pavement, regardless of the surface modulus, base modulus, or layer thickness combinations used (i.e. these surface stresses were relatively independent of pavement structural characteristics).

CONCLUSIONS

The results of the measurements and analyses presented in this paper indicate that tensile stresses generated by radial truck tires appear to explain the development of surface-initiated longitudinal wheel path cracks. Furthermore, the fact that bias-ply tires do not appear to induce surface tensile stresses provides additional support for this explanation, since this type of cracking has become more prevalent in recent years and thus coincides with the increased use of radial tires. These results of this investigation also indicate that measured tire-pavement interface stresses, particularly lateral stresses, must be more clearly defined and considered for proper evaluation of pavement cracking performance and mixture design.

REFERENCES

1. CROW - Center for Research and Contract Standardization in Civil and Traffic Engineering. Report of the Working Group. Surface Cracking in Asphalt Layers. The Netherlands, March, 1990.
2. HUANG, Y. Pavement Analysis and Design. Prentice-Hall, Inc., 1993.
3. LYTTON, R., ROQUE, R., UZAN, J., HILTUNEN, D., FERNANDO, E., AND STOFFELS, S. Performance Models and Validation of Test Results. Final Report to Strategic Highway Research Program; Asphalt Project A-005, July, 1993.
4. MARSHEK, K., CHEN, H., CONNELL, R., AND HUDSON, W. Experimental Determination of Pressure Distribution of Truck Tire-Pavement Contact. Transportation Research Record No. 1070, Transportation Research Board, pp. 9-13, 1986.
5. MARSHEK, K., CHEN, H., CONNELL, R., AND SARAF, C. Effect of Truck Tire Inflation Pressure and Axle Load on Flexible and Rigid Pavement Performance. Transportation Research Record No. 1070, Transportation Research Board, pp. 14-21, 1986.
6. CHEN, H., MARSHEK, K., AND SARAF, C. Effects of Truck Tire Contact Pressure Distribution on the Design of Flexible Pavements: A Three-Dimensional Finite Element Approach. Transportation Research Record No. 1095, Transportation Research Board, pp. 72-86, 1986.
7. HUGO, F. AND KENNEDY, T.W. Surface Cracking of Asphalt Mixtures in Southern Africa. Proceedings of Association of Asphalt Paving Technologists, Vol. 54, pp. 454-496, 1985.
8. ROQUE, R., HARDEE, H., AND RUTH, B.E. Thermal Rippling of Asphalt Concrete Pavements. Proceedings of Association of Asphalt Paving Technologists, Vol. 57, pp. 464-483, 1988.
9. ROQUE, R. AND RUTH, B.E. Mechanisms and Modeling of Surface Cracking of Asphalt Pavements. Proceedings of Association of Asphalt Paving Technologists, Vol. 59, pp. 396-431, 1990.

10. JACOBS, M.M. Crack Growth in Asphaltic Mixes. Ph.D. Dissertation, Delft Institute of Technology, The Netherlands, 1995.
11. DE BEER, M., FISHER, C., AND JOOSTE, F. Determination of Pneumatic Tyre/Pavement Interface Contact Stresses Under Moving Loads and Some Effects on Pavements With Thin Asphalt Surfacing Layers. Eighth International Conference on Asphalt Pavements, Seattle, Washington, pp. 179-226 , 1997.
12. MYERS, L.A. Mechanism of Wheel Path Cracking That Initiates At the Surface of Asphalt Pavements. Master's Thesis, University of Florida, 1997.

Fifth International Conference on the
BEARING CAPASITY OF
ROADS AND AIRFIELDS

THE LIFTING OF AXLE LOAD RESTRICTIONS DURING SPRING THAW - A NORWEGIAN EXPERIMENT

GEIR REFSDAL

Public Roads Administration, Buskerud
P.O.Box 2265 Stromso, 3003 Drammen, Norway
E-mail: geir.refsdal @ statens.vegvesen.no

ABSTRACT

On the Norwegian road network of 53.000 km of main and secondary roads, more than 50 % of the main road network and about 80 % of the secondary road network have spring thaw axle load restrictions imposed for approx. 8 weeks every year. This was the situation in 1994.

In 1994 a four year research project, BUAB [1)] («better utilisation of the bearing capacity of the road network»), was completed /1/. One important aspect of this project was to evaluate the economic effects of a possible lifting of all axle load restrictions during the spring thaw period.

It was estimated that the annual extra costs for the road owner would be approx. 145 mill NOK The profit for the road users would at the same time amount to approx. 330 mill NOK, thus resulting in a socio-economic profit of 185 mill NOK, provided the surfacing service life was maintained at its present level, i.e. in average a 10 years service life for main roads and 15 for secondary roads. A potential loss of 210 mill NOK was estimated if the extra costs to maintain the roads surfacing service lives were *not* allocated.

In 1995 *all* axle load restrictions were lifted and the required annual funds to maintain the surfacing service life were allocated.

Experience
It was considered that the calculated extra costs for the road owner should at least be in the correct magnitude. After four years it can be concluded that

- the extra allocations to counteract the sudden and very local collapse in the road structure and the reduced surfacing service life seems to have been adequate.
- the first period with sudden and very local collapse in the road structure seems to be over.

The future may still come up with very severe spring thaw effects, but again - the *extra* costs for this related to the lifting of the axle load restrictions - should be taken care of within the present budget.

1) The BUAB project, "Better utilisation of the bearing capacity of the road network", was carried out from 1990 - 1994 with the following project team: Par Games, Åsmund Knutson, Terje Moen, Geirmund Nordal, Geir Refsdal, Paul Senstad (project coordinator), Tor-Sverre Thomassen and Torkild Thurmann-Moe.

NORWEGIAN AXLE LOAD POLICY

Since the 1940's new road pavements have been designed for 10 tons axle loads in Norway. Prior to 1979, roads were allowed for 10 tons axle load only if the road could tolerate this axle load also in the spring thaw period. Very few roads were therefore allowed for 10 tons axle load. This axle load policy was however left in 1979. The roads which earlier had been allowed for an all year axle load of 8 tons were then lifted to 10 tons, but with load restrictions to 8 tons during the spring thaw period. This represented a much better utilisation of the bearing capacity of the road network. In 1994 almost 90 % of the main road network was open for 10 tons axle load. However, 50% of these roads had axle load restrictions during the spring thaw period.

During the winter period the allowable axle loads for 8 tons roads have generally been increased to 10 tons during the winter period, provided the bridges have been up to this standard.

As one of the very few countries in Europe, Norway also has tyre pressure restrictions. To day the allowable tyre pressure is 9,0 bar. Up to 1978 it was 7,0 bar.

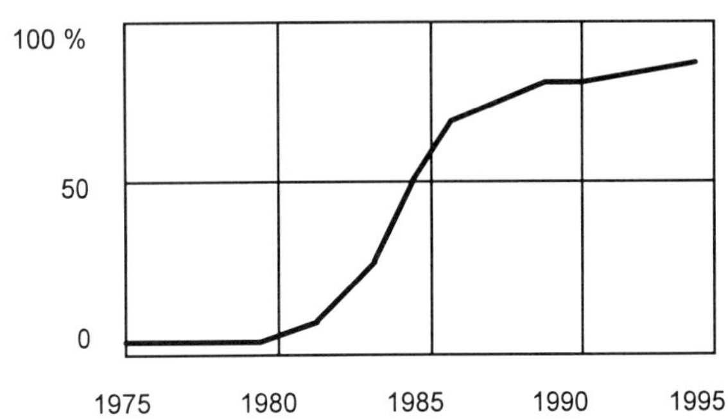

Fig.1. Development in 10 tons allowable axle loads on the main road network 1975 - 1994

AXLE LOAD POLICY DURING SPRING THAW

1994: 50 % axle load restrictions

Norwegian axle load policy has to a large extent been linked to the bearing capacity during spring thaw. During the years 1981 to 1994 more than 50 % of the main road network of 26000 km had axle load restrictions during spring thaw. The restrictions were generally imposed when the thawing was at a depth of 20 to 25 cm and lifted at a thaw depth of 100 to 125 cm. In general the allowable single axle load in this period was reduced with two tons on the main road network, from 10 to 8 tons, and for a period of approx. 8 weeks.

On the secondary road network of 27000 km, the general bearing capacity situation is worse, and the major part of the roads have their allowable single axle load reduced from 8 to 6 tons.

Allowable axle load summer / spring thaw	10/10	10/8	10/7	10/6	8/8	8/7	8/6	8/4	6/6
Main road network	44,4	40,8	0,7	3,4	2,7	1,1	6,8	0	0
Secondary road network	5,1	23,0	0,9	6,9	15,1	5,1	42,0	1,6	0,4

Fig.2 Allowable axle loads (%) for summer and spring thaw on the Norwegian road network, January 1994.

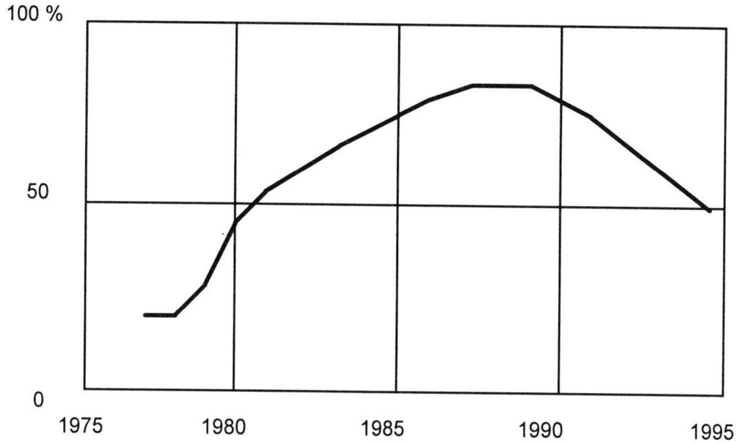

Fig.3. Extent of axle load restrictions (%) during spring thaw on the Norwegian main road network, 1977 - 1994

THE BEARING CAPACITY PROJECT (BUAB) 1990 - 1994

In 1990 the Norwegian Directorate of Public Roads decided to look into several aspects of bearing capacity on the Norwegian road network through the BUAB project. An important part of this was to investigate the bearing capacity cost implications for both the road owner and the road user related to a general increase in the allowable axle load to 10 tons for the complete road network, as well as lifting all axle load restrictions during the spring thaw period.

ROAD OWNER COSTS

Reduced surfacing service life

The typical service life for a Norwegian asphalt surfacing on the main road network is 10 years, and on the secondary road network typically 15 years. The spring thaw axle load restrictions, normally down two tons from 10 to 8 tons, contribute to these service lives. If the axle load restrictions should be lifted, the 10 years surfacing service life on the main road network would be reduced from 10 to 8 years. On the secondary road network, the lifting of axle load restrictions on a 8 tons road from 6 to 8 tons in the spring thaw would reduce the surfacing service life in average from 15 years to 11 years.

Locally the lifting of spring thaw axle load restrictions, may not affect only the surfacing service life. In some cases, but generally for very short sections only, a need for a full reconstruction may come up, due to a sudden collapse in the pavement structure during spring thaw. This should also be taken into account when the road owner costs are calculated.

Extra annual costs per km

If all spring thaw axle load restrictions on the Norwegian main and secondary road network should be lifted, the surfacing service life would be reduced. This would have its price tag. For the individual road section, the price would depend on the allowable axle load in the spring thaw today, and after a lifting, on the type of road (main road / secondary road), on traffic (AADT) and on surfacing type and road width.

The calculated extra annual maintenance cost per km (NOK per km and year) is shown below.

MAIN ROADS
10/8 to 10/10 tons: k x 5.300 NOK/km/year
10/6 to 10/10 tons: k x 20.300 NOK/km/year
8/6 to 8/8 tons: k x 7.200 NOK/km/year

SECONDARY ROADS
10/8 to 10/10 tons: k x 2.700 NOK/km/year
10/6 to 10/10 tons: k x 10.400 NOK/km/year
8/6 to 8/8 tons: k x 3.700 NOK/km/year
8/4 to 8/8 tons: k x 12.900 NOK/km/year
(1$ = approx. 7.50 NOK)

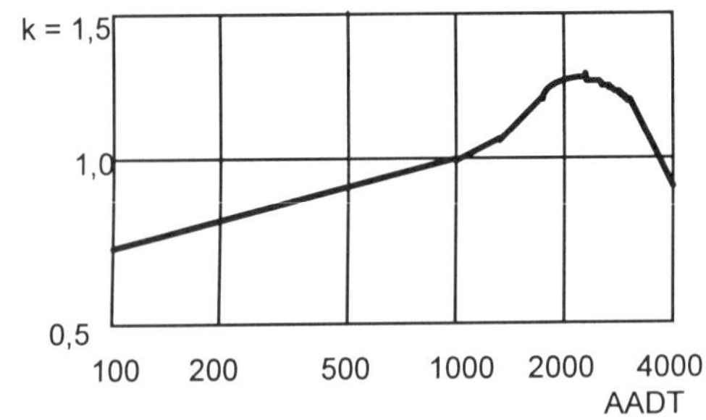

Fig.4. Increase in average annual maintenance cost related to a more frequent surfacing, depending on road type, AADT and allowable axle load

Extra costs for the complete road network - 145 mill NOK/year

The extra annual road owner costs will come to its full value when a full cycle of road surfacing has been completed. The first year we will hardly notice it, except for the occasional breakdown of a weak road section, which will have to be reconstructed rather than resurfaced. Especially this could be the case for the first 2-4 years.

The calculation was linked to the general pavement design table for Norwegian roads /2/. The design includes both allowable axle load, AADT and ESA's. The effect of lifting axle loads for two months was derived from the design table and transferred to reduced surfacing service life. In this way also the cost implications could be calculated. The situation prior to a potential lifting was ignored completely in the calculation as it reflects the present maintenance situation only .

The extra annual costs for the complete road network is calculated to 145 mill NOK, with 80 mill NOK for the main road network and 65 mill NOK for the secondary road network.

No spring thaw restrictions below 8 tons

During spring thaw, axle load restrictions below 8 tons are often introduced. One step towards a complete lifting of all spring thaw axle load restrictions could be to have a minimum of 8 tons during this period. The price tag for this has been calculated to 85 mill NOK, with 25 and 60 mill NOK for the main road network an the secondary road network respectively.

Change in road class	**Annual average cost increase, NOK (km)**	
	Main road network	Secondary road network
from 10/8 tons to 10/10 tons	51,0 mill (10620 km)	14,0 mill (6210 km)
from 10/7 tons to 10/10 tons	1,7 mill (188 km)	1,1 mill (246 km)
from 10/6 tons to 10/10 tons	16,3 mill (883 km)	15,2 mill (1855 km)
from 8/7 tons to 8/8 tons	0,6 mill (288 km)	1,3 mill (1368 km)
from 8/4 tons to 8/8 tons	10,2 mill(1778 km)	30,4 mill (11347 km)
from 8/4 tons to 8/8 tons	0 mill (0 km)	4,3 mill (418 km)
total	79,8 mill	66,3 mill

Fig.5. Increase in average annual maintenance costs related to a more frequent resurfacing as a result of lifting of all spring thaw restrictions in allowable axle loads, for all road classes and the complete road network

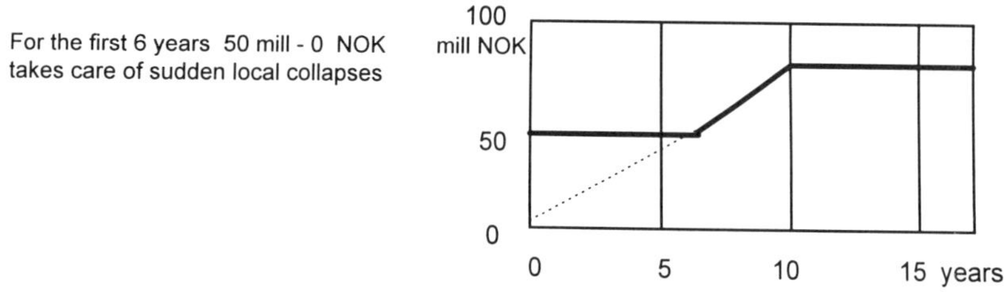

Fig.6. Increase in future road owner costs from year 1 following the lifting of spring thaw axle load restrictions on all main roads

ROAD USER COSTS

As a consequence of the spring thaw axle load restrictions, the heavy vehicles are not able to take advantage of their loading capacity, resulting in a reduced load for each trip as well as a need for an increased number of trips. The load restrictions also influence the possible utilisation of production equipment, the quality of products which have to be stored, and to some extent also the localisation of industries and the practising of forced leave of absence..

The road user costs are also influenced by the road surfacing condition. The socio-economical profit related to a lifting of axle load restrictions may therefore be lost if the surfacing condition is not maintained at its present level.

Annual road user profit: 330 mill NOK

The annual road user profits can be split into 1) direct profits. 2) indirect profits and 3) time savings. The following profits have been calculated:

Direct profits:	140 mill. NOK.
Indirect profits:	70 mill. NOK
Time savings:	120 mill. NOK
Total socio-economic profit:	**330 mill. NOK**

Extra cost for the road owner	**Profit for the road user**	**Socio-economic profit**
145 mill NOK per year - main road network: 80 mill - secondary road netw: 65 mill	330 mill NOK per year - main road network: 190 mill - secondary road netw: 140 mill	185 mill NOK per year - benefit / cost = 2,3

Fig.7. The socio-economic profit, provided the present road surfacing condition is maintained at its present level following a lifting of the axle load restrictions during the spring thaw

Warning: a potential annual loss of 210 mill NOK

Without an increased maintenance budget, a reduced surfacing quality will be the result, with a potential annual loss for the road users of 210 mill. NOK. This loss can be related mainly to increased petrol costs and increased accident rate.

Increased comfort costs for all road users, calculated to 4000 mill NOK annually, have not been considered as the present road choice hardly will be influenced by a reduced surfacing condition.

Extra cost for the road owner	**Profit for the road user**	**Socio-economic loss**
0 mill NOK per year, i.e. the road owner will resurface at the present frequency, and the road surfacing condition will fall	*210 mill NOK per year* - main road network: 120 mill - secondary road network: 90 mill	*210 mill NOK per year*

Fig.8. The socio-economic loss provided the present road surfacing condition is not maintained following a lifting of the axle load restrictions during the spring thaw

1. JANUARY 1995: ALL AXLE LOAD RESTRICTIONS LIFTED

The BUAB Project Report /1/ was published in November 1994. A political decision was taken almost immediately to lift all spring thaw axle load restrictions from 1. January 1995, and the maintenance budget was increased accordingly to take care of the following reduced surfacing service life.

Four spring thaw periods experienced since 1995
Since 1995 four spring thaw periods have been experienced. Each of the 19 District Road Administrations have received in average 8 mill NOK per year (from 1 to 21 mill) to offset the effects of the lifted axle load restrictions. It is clear that the effect of the spring thaw periods may vary considerably, both from one district to another as well as from one year to another.

It was considered that the calculated extra costs for the road owner should at least be in the correct magnitude. After four years it can be concluded that

- the extra allocations to counteract the sudden and very local collapse in the road structure and the reduced surfacing service life seems to have been adequate.
- the first period with sudden and very local collapse in the road structure seems to be over.
- the future may still come up with very severe spring thaw effects, but again - the *extra* costs for this related to the lifting of the axle load restrictions - seems to be taken care of within the present budget.

References
/1/ Better utilisation of the roads bearing capacity. Final report. Publication No. 75 from NRRL, Nov. 1994. (*In Norwegian*).
/2/ Pavement design and materials manual. Publication no. 018, Norwegian Public Roads Administration. Dec. 1991 (*In Norwegian*), ISBN 82-7207-305-6
/3/ Road owners extra costs related to a lifting of the present spring thaw restrictions. Calculation method and results. NRRL, Internal Report No. 1539, Feb. 1993 (SINTEF, NTNU). (*In Norwegian*).
/4/ An administrative lifting of axle load restrictions - increased road deterioration. The District Road Administrations opinions of extra costs. NRRL, Internal Report No. 1590, Feb. 1993. (*In Norwegian*).
/5/ Axle road restrictions on roads. An examination of international practise. NRRL Internal Report No. 1588, Feb. 1993. (*In Norwegian*).

The Pavement Team – Integrated project delivery

for the Fifth International Conference on the Bearing Capacity of Roads and Airfields Trondheim, Norway

George Cook, *General Manager, Pavement Team, BAA plc*
John Moss, *Managing Director, Southern Region, AMEC Civil Engineering*

Abstract

The Pavement Team is a 5-year partnering agreement between BAA plc and AMEC Civil Engineering for the project management, design and construction of major airfield works. The Framework commenced in November 1995 and has since completed or has in progress 32 projects valued at close to £70m. BAA plc, the UK based airport owner operator, decided to pursue strategic partnering as the means of breaking the mould and moving away from traditional contract procurement into long term agreements targeted to achieve substantial improvement over traditional methods.

The paper describes the Pavement Team as a case study for the successful development of a long term partnering agreement for major airfield works, giving the client and contractor views of traditional methods, their aspirations from partnering and the three stage selection process. With the award made, both parties were pitched into a new way of working for which no one had the blue print and new methods and processes had to be developed. This early period was characterised by some quick wins but also some frustration at road blocks which proved difficult to move. The paper describes how these challenges were overcome and also the tangible performance benefits that are now bring delivered. Having just passed the mid-way point the paper concludes by summarising the issues facing the team and how it is intended to take the business forward.

Key Words

Process improvement, metrics, attitude, waste reduction, shared benefits.

What was wrong with the old way?

Project performance was mixed – some contracts went well, others less so. Good performance tended to be associated with good working relationships, though the tendering process (with almost invariably the lowest bid selected) meant that these relationships were seldom carried forward. Even during the bid process the contractor would identify potential areas for future claims and discount the bids accordingly. Thus the scene was set for conflict to enable the contractor to get back to a profit making position. Project mobilisation was slow with the need for the contractor to request passes, airside driving licences, etc., and familiarise themselves with the airport environment. On the client side the "oh – they're just another contractor" attitude was characterised by low trust and expectation. The contract invariably required approval of subcontractors, suppliers, materials, test information and concrete/asphalt mixes which all contrived to impede a rapid start on site.

Clients felt they 'engaged' the construction industry almost unwillingly as theonly means to achieve their business goals. Certainly the client understood very little about how the construction industry worked and it was very difficult to understand costs from the way that tenders were assembled. Contractors were adept at "Bill structuring" to maximise value rather than truly reflecting the cost of an operation.

Tenders were won and the contractual game commenced, which demanded considerable resources from both sides. Formal written communication and rather less verbal dialogue was the norm and generally both parties collected data to use in contractual disputes. A great deal of effort went into building claims or providing a defence against them. The work itself could become secondary to contract engineering and the individual personalities involved. Subcontractors were employed on a pay when paid basis thus creating another interface where conflict flourished. While there was much formal contract administration, costs increased due to delayed responses and issues left to fester rather than being resolved as quickly as possible. Both parties felt they were smarter than the other and there was seldom much open, honest, dialogue.

What did we believe partnering could offer?

For the client organisation, the move towards partnering was driven by recognition that traditional procurement methods offered very little scope to improve value for money. The Latham Report (1) published in 1994 highlighted the adversarial nature of traditional tendering and advocated a move to new forms of contract with the emphasis on dispute resolution to maintain a clear focus on the project goals. Previously there was a belief that traditional tendering would always offer the best value, almost in a Darwinian concept of the survival of the fittest, being able to win more and sustain their business. If that was the theory, the practice was different with tender bids having little reference to out-turn costs.

From the contractors point of view, partnering offered the opportunity to develop a relationship that would improve cashflow (it often took in excess of 2 years to complete final accounts) and to provide a small but secure profit instead of the unpredictable gains and losses of traditional contracts. Figure 1 below shows the situation on five contracts AMEC were running at the time the BAA partnering arrangement was first suggested.

	1994		**1995**			
	Turnover	Profit	Turnover	Profit	Tender	Outturn
Contract A	22m	(292)	24m	(1200)	17m	30m
Contract B	9.7m	(1126)	14.6m	(817)	12m	16m
Contract C	4.4m	(497)	6.9m	(400)	5.6m	6.9m
Contract D	5.6m	(2572)	17.2m	(4875)	20m	34m?
Contract E	17.6m	(-)	41.4m	(218)	36m	60m?

It would also allow investment in Research and Development and equipment that had been impossible on one off contracts with little chance of repeat business. The opportunity to work in a strategic partnering relationship also gave the contractor the ability to establish some product differentiation in the way construction was executed. A key thought was that if the potential for contractual conflict was reduced so that both parties could align in the knowledge that both would benefit, the focus of effort could move to improving the basic processes and achieving sustainable benefits.

There was a perception that the construction industry had failed to move forward as other industries. Womak and Jones (2) identified how the car industry had changed as led by Japanese manufacturers, but could the same approach be taken in construction? One thing was clear that the industry could not change from within, but would respond to sufficiently enlightened clients.

The Selection Process

The three stage selection process was initiated by a call for completion in the European Journal in November 1994. The scope of works was described as " Framework agreement for aircraft pavements and ancillary works' at Heathrow, Gatwick, Stansted and Southampton Airports. Each part of the process was marked against pre-defined criteria with companies achieving above a cut-off invited to continue to the next stage.

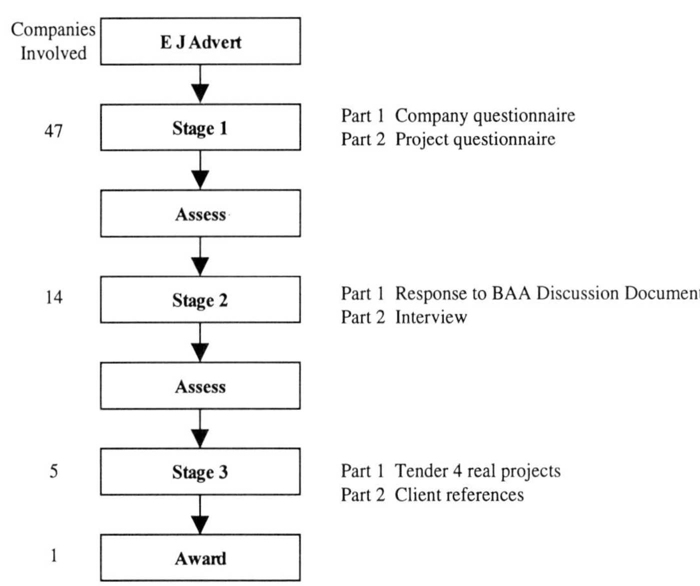

Selection Process

The scoring from all stages was aggregated as follows:-

Combined Stages 1 and 2		40%
Client references		10%
Stage 3 Tender	- Hard costs	40%
	- Soft issues	10%

The total selection process took 12 months and AMEC recognised the need throughout this period to develop the attitude and soft skills that would be required to successfully execute the framework agreement.

What were the early results?

The selection process was long and had required much effort from all parties. Perhaps as a consequence little thought had gone into how the framework would operate and be managed. To compound matters, the target commencement of the framework had been known well in advance and a backlog of work had built up. In addition, a large additional project had 'popped up' which was required to be completed by the following Spring.

In effect, we had a partnering arrangement but no real idea of how the process should operate. While some staff were enthusiastic, others saw the framework as a threat and were unwilling to move from traditional working procedures. In retrospect, perhaps this was inevitable and the fact that there was plenty of work to progress, enabled some working relationships to build but essentially these early projects were business-as-usual under a different name. AMEC had well-developed contract procedures that had served them for many years. However, in this new environment with a wish not to refer to the contract, much of the working discipline was lost, this gave the sceptical further possible evidence that partnering might not work as envisaged. This gap between aspiration and practice was recognised and a series of partnering workshops, led by Susanne Palmer of the Palmer Partnering Group, was set up in March 1996.

Three partnering sessions were held, one for key stakeholders and two vertical groups. These sessions produced the Team Charter, encompassing the mission statement and values. The workshops also gave the framework its logo and its name, the Pavement

Team. Looking back these workshops marked the 'birth' of the Pavement Team and it is from this point that a new team culture began to grow, linked to but different from either parent company culture.

There were clearly early successes and it was recognised that the 'pop up' job could never have been completed in the same timescale, or at the same cost using traditional methods. There was also some scepticism and belief that the framework was being used to paper over the cracks in the system without addressing the real issues.

Moving forward

Development of shared goals and a single culture whilst cementing the core also brought some of the unresolved issues into sharper focus. There were still Resident Engineer's teams on site and while the Pavement Team was seen differently, all their systems and procedures were separate and related to the previous way of working. Some staff were able to make the transition into a single team while others could not. Often the dissenters were most vociferous in pointing out weaknesses. Typically from the client this would be as criticism of any quality, safety or management failures, the contractor would complain about late or poor design information and late discussion making. In nearly all cases, performance was still better than normal but those resistant to change were expressing their frustration through their traditional blame culture.

The creation of a single team also highlighted other areas, for instance, where both had financial reporting functions drawing information from different sources. Bringing these teams together and using the same base data simplified our processes and procedures. The emphasis on improvement led to the identification of 5 key processes – project management, design, production, financial and support administration. Process mapping allowed these processes to be described at high level as a means of explaining the activities of various parties.

Continuing on page 1847

Fifth International Conference on the
BEARING CAPASITY OF
ROADS AND AIRFIELDS

STRUCTURAL MODELING CONCEPTS FOR THE DESIGN OF AIRPORT PAVEMENTS FOR HEAVY AIRCRAFT

NAVNEET GARG[1], EROL TUTUMLUER[2] & MARSHALL R. THOMPSON[3]

[1]Post-Doctoral Research Associate, e-mail: n-garg@uiuc.edu
[2]Assistant Professor, e-mail:tutumlue@uiuc.edu
[3]Professor Emeritus, e-mail: mrthomps@uiuc.edu
Department of Civil Engineering, University of Illinois @ Urbana-Champaign
205 North Mathews, Urbana, IL 61801-2352, U.S.A
Phone: (217)333-3930 FAX: (217)333-1924

ABSTRACT

The advent of heavier aircraft like the Boeing 777 with a tridem gear configuration necessitates the need for reconsidering structural modeling concepts for airport pavement design. Current mechanistic design methods such as LEDFAA (FAA-AC:150/5320-16, 1995), Corps of Engineers (TM 5-825-2-1/AFM 88-6, 1989), and the Asphalt Institute (AI MS-1, 1982), use Elastic Layer Programs (ELPs) to predict airfield pavement responses to gear loads. In these procedures, the stress dependent behavior of cohesive subgrade soils is neglected and granular layer moduli are assigned as a function of subgrade CBR. In the ongoing research at the FAA Center of Excellence (COE) for Airport Pavement Research at University of Illinois, two Finite Element Programs (FEPs), ILLI-PAVE (isotropic material properties) and GT-PAVE (anisotropic material properties), are used to predict pavement responses under a single wheel of the Boeing 777 tridem gear. Both ILLI-PAVE and GT-PAVE consider the non-linear stress dependent modulus behavior of granular bases and subgrade soils. GT-PAVE predicts higher critical pavement responses than ILLI-PAVE because of the reduced horizontal stiffness of the granular base. When granular base residual stresses, which occur due to compaction during construction and repeated traffic loading, are included in the GT-PAVE analyses, surface and subgrade deflections show good agreement with ILLI-PAVE. Sensitivity analyses were conducted to study the effect of varying asphalt concrete (AC) thickness, AC modulus, granular base thickness, and subgrade resilient modulus on critical pavement responses (surface deflections, tensile strain at the bottom of the AC layer - related to AC fatigue, and vertical strain/stress at the top of subgrade - related to subgrade rutting). The results from the sensitivity analyses are presented and discussed.

KEY WORDS: Boeing-777, anisotropic, AC strain, subgrade stress, residual stress

INTRODUCTION

In mechanistic-empirical (M-E) flexible pavement design, load-related pavement responses (stresses, strains, deflections) are used to predict the development of pavement distress through the use of transfer functions. In flexible pavements, the tensile strain at the bottom of asphalt concrete (AC) layer is related to the fatigue life of the pavement. Reduced strain corresponds to increased fatigue life. The deviator stress at the top of the subgrade is related to subgrade rutting potential. A subgrade stress ratio (SSR = deviator stress/unconfined compressive strength) criterion is used (at the University of Illinois) to consider cohesive soil subgrade rutting. The landing gears for large aircraft include multiple wheels. The main landing gear of the Boeing-777 is unique - six wheels are arranged in a tridem configuration. Pavement response is the resultant of individual wheel load responses.

Elastic layer programs (ELP's) like BISAR, JULEA, and LEDFAA can accommodate multiple wheel loads. However, these programs do not consider the non-linear stress-dependent behavior of paving materials/subgrade soils (unbound granular materials exhibit stress-hardening, and fine-grained cohesive soils exhibit stress-softening behavior) nor do they include any paving material/subgrade failure criteria. In the ongoing research at the FAA Center of Excellence (COE) for Airport Pavement Research at University of Illinois, two Finite Element Programs (FEPs), ILLI-PAVE (isotropic material properties) [1] and GT-PAVE (anisotropic material properties) [2], are used to predict pavement responses under a single wheel of Boeing 777 tridem gear. Both ILLI-PAVE and GT-PAVE consider the non-linear stress dependent modulus behavior of granular bases and subgrade soils. However, ILLI-PAVE and GT-PAVE pavement responses correspond to a single wheel load.

A three dimensional finite-element program like ABAQUS can accommodate multiple wheel loads, non-linear stress-dependent pavement material models, and failure criteria. Relative to ILLI-PAVE and GT-PAVE, this approach is more complex and "computer intensive" (machine capability/computational time). Three dimensional finite-element pavement programs are in a developmental stage and are not currently used in M-E flexible pavement design procedures.

An "engineering approach" for estimating flexible pavement responses to multiple wheel gear loading is being developed at the University of Illinois. Single wheel responses are calculated using ILLI-PAVE. Two options were considered to approximate multiple wheel gear loading responses from ILLI-PAVE calculated single wheel responses - (1) Utilize ILLI-PAVE final iteration granular base and subgrade moduli (for a single wheel of the gear) as modulus inputs to an Elastic Layer Program that can accommodate multiple wheel loading. Thompson (one of the authors) has utilized this procedure for several years and presented it in a US Air Force Pavements Seminar at Tyndall AFB, FL, in 1986. The base and subgrade moduli are for various "layers" of the pavement section. (2) Postulate that the principle of superposition can be used to predict pavement responses for a multiple wheel gear configuration. Thus, ILLI-PAVE "single-wheel" loading responses (deflections, stresses, strains) can be used to "sum up" the interactions of the various wheel loads. Previous studies [3, 4, 5] have shown that the principle of superposition is reasonably valid for certain known conditions for flexible pavements except at depths near the surface. The known conditions are that (1) the stress and deflection basins under a single-wheel load can be either correctly measured or correctly computed, and (2) the load per wheel for the single wheel and the multiple wheels are the same.

ILLI-PAVE and GT-PAVE predict realistic flexible pavement structural responses [1, 2, 6, 7]. A sensitivity analysis was conducted to study the effect of varying asphalt concrete (AC) thickness, AC modulus, granular base thickness, and subgrade resilient modulus on critical pavement responses (surface deflections, tensile strain at the bottom of the AC layer - related to AC fatigue, vertical strain/stress at the top of subgrade - related to subgrade rutting). The results from the sensitivity analysis are presented and discussed.

RESULTS FROM ILLI-PAVE ANALYSES

In ILLI-PAVE [6] the pavement is considered an axisymmetric solid of revolution. Nonlinear, stress-dependent resilient modulus material models and failure criteria for granular materials and fine-grained soils are incorporated into ILLI-PAVE. The principal stresses in the granular and subgrade layers are modified at the end of each iteration so that they do not exceed the strength of the materials as defined by the Mohr-Coulomb theory of failure.

The asphalt concrete surface is modeled as a constant modulus material. The granular base layer is characterized using the minor principal stress (σ_3) model and the K-Theta (bulk stress) model. The arithmetic resilient modulus-deviator stress (stress-softening) model is used for characterizing the cohesive soils. Three new models, Uzan's model [8], log resilient modulus-deviator stress model, and log resilient modulus-log deviator stress model, were incorporated into ILLI-PAVE as part of the FAA Center of Excellence Program. Uzan's model describes the descending behavior of resilient modulus in sand subgrades [7] with axial strain because it considers the effect of shear stress on resilient modulus.

A sensitivity analysis was conducted to study the effect of varying asphalt concrete (AC) thickness, AC modulus, granular base thickness, and subgrade resilient modulus on critical pavement responses (surface deflections, tensile strain at the bottom of the AC layer - related to AC fatigue, vertical strain/stress at the top of subgrade-related to subgrade rutting). Boeing-777-200A aircraft load characteristics (wheel load = 42,500 lbs, tire pressure = 182 psi, radius of loaded area = 8.62 in.) were used.

In the first analysis, AC surface thickness was varied to establish its effect on critical pavement responses (tensile strain @ bottom of AC layer & subgrade deviator stress). Granular base was modeled using K-Theta and Uzan models. The 43.5-in. thick granular base layer properties were: for K-θ model, M_R (psi) = $2253*\theta^{0.7}$; and for Uzan's model, M_R (psi) = $2318*\theta^{0.64}*\sigma_d^{0.065}$, friction angle ϕ = 40°. The cohesive subgrade properties were: Unit Weight = 120-pcf, Poisson's Ratio = 0.45, Cohesion = 15-psi, E_{Ri} = 6000-psi, Modulus at Failure = 3000-psi. AC surface thicknesses of 8, 9.5, 11, and 12 in. were considered. Figures 1 and 2 show the variation in AC strain and subgrade deviator stress with increase in AC thickness. The K-Theta and Uzan models produced similar results.

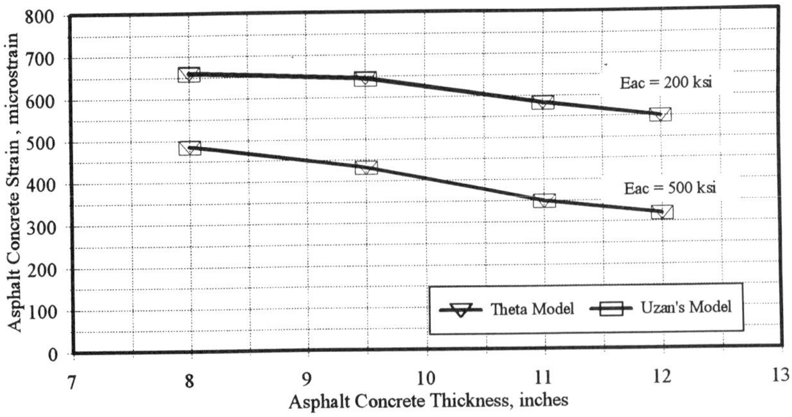

Figure 1. Variation in Strain at the Bottom of AC Layer with AC Thickness

Variations in surface deflection and subgrade deflection with increase in AC thickness are shown in Figure 3. The subgrade deflections are about 48-percent of the surface deflection for E_{ac} = 500-ksi, and 42-percent for E_{ac} = 200-ksi.

In the second analysis, AC thickness was constant at 8-in., and the granular base thickness was varied from 20-in. to 43.5-in.. Granular base was modeled using the K-Theta model. The aircraft loading, granular base and subgrade used were similar to those used in the first analysis. The objective was to study the effect of varying granular base thickness on critical pavement responses (tensile strain @ bottom of AC layer & subgrade deviator stress).

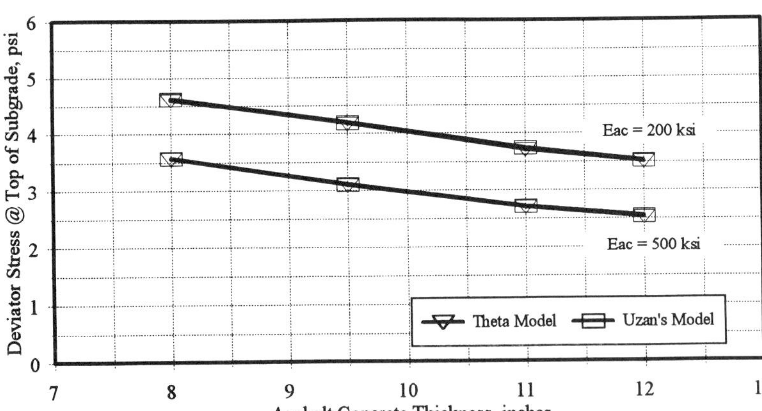

Figure 2. Variation in Subgrade Deviator Stress with AC Thickness

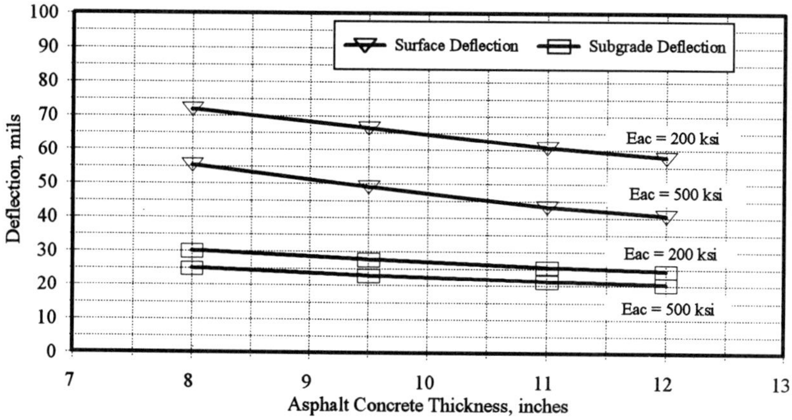

Figure 3. Variation in Deflections with Change in AC Thickness

Figures 4 and 5 show the variation in AC strain and subgrade deviator stress with increase in granular base thickness (T_{GR}). An increase in T_{GR} does not significantly affect the tensile strains at the bottom of asphalt concrete layer. Subgrade responses (stresses, strains, deflections) decrease with an increase in T_{GR}. Surface deflections are not significantly affected by increase in granular base thickness (not shown here). The contribution of granular base layer to the surface deflection increases with the increase in granular base thickness. For T_{GR} = 20-in., the contribution of granular base layer to the surface deflection was about 14-percent. Increasing T_{GR} to 43.5-in. increased the granular base layer contribution to about 18-percent. The rate of decrease in the subgrade responses diminishes for granular base thicknesses greater than T_{GR} = 34-in.

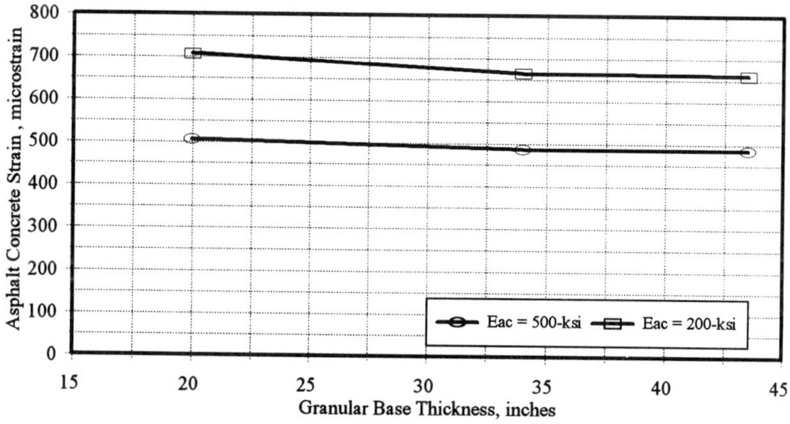

Figure 4. Variation in Strain at the Bottom of AC Layer
with Granular Base Thickness

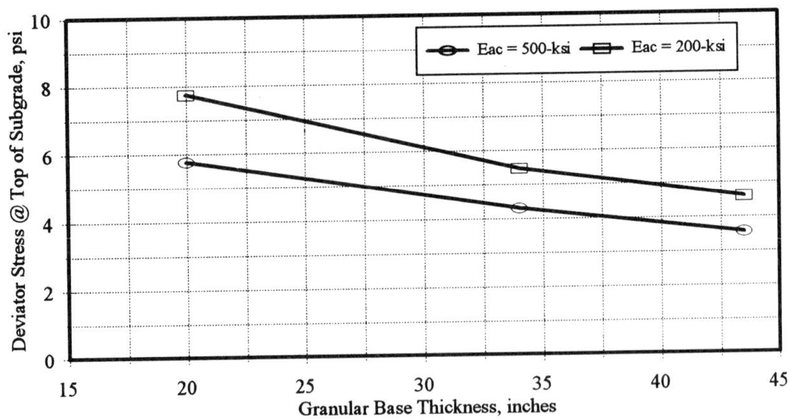

Figure 5. Variation in Subgrade Deviator Stress with Granular Base Thickness

RESULTS FROM GT-PAVE ANALYSES

The recently developed nonlinear finite element program GT-PAVE [2] was used to employ the cross-anisotropic model in granular bases of conventional flexible pavements. GT-PAVE runs on a PC, considers the cross-anisotropic behavior exhibited by unbound aggregate bases/subbases and also handles locked-in residual stresses due to compaction and repeated traffic loading. Isoparametric 8-node quadrilateral elements are used in the program to analyze a flexible pavement consisting of either linear or nonlinear elastic layers. The details on the nonlinear solution technique and the base and subgrade characterization models used in GT-PAVE are described elsewhere [2, 9].

The first analysis involved varying the AC surface thickness and studying its effects on critical pavement responses (tensile strain @ bottom of AC layer & subgrade deviator stress). A 43.5-in. thick granular base layer was modeled as a cross-anisotropic material. Three different cohesive subgrades (soft, medium, and stiff) were considered. The results are shown only for the medium cohesive subgrade which is similar to the one used for ILLI-PAVE (Unit Weight = 120-pcf, Poisson's Ratio = 0.45, Cohesion = 15-psi, E_{Ri} = 6000-psi, Modulus at Failure = 3000-psi). AC surface thicknesses of 8, 9.5, 11, and 13 in. were considered. Figures 6 and 7 show the decreasing trends in AC strain and subgrade deviator stress with an increase in AC thickness.

Figure 8 shows the variation in surface and subgrade deflections with an increase in AC thickness. Due to the low horizontal stiffnesses computed by GT-PAVE in the anisotropic granular base, the subgrade deflections are only about 40-percent of the surface deflections.

The AC thickness was kept constant at 8-in., and the granular base thickness was varied from 20-in. to 43.5-in. The aircraft loading, and the granular base and subgrade properties used were the same as used in first analysis. Figures 9 and 10 show the variation in AC strains and subgrade deviator stress respectively with increase in granular base thickness (T_{GR}). An increase in T_{GR} increase does not significantly affect the tensile strains at the

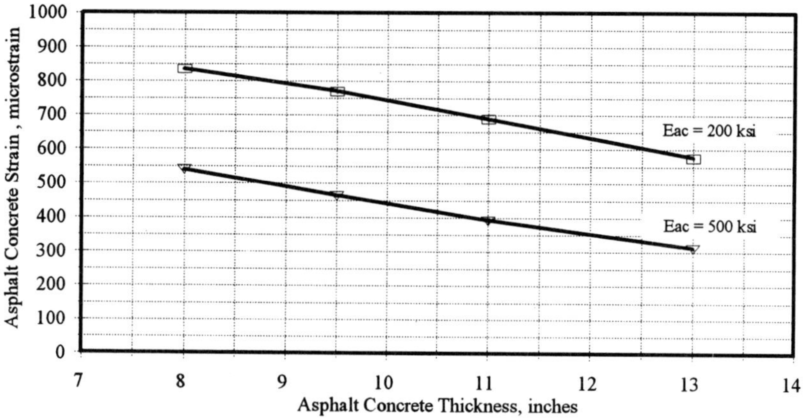

Figure 6. Variation in Strain at the Bottom of AC Layer with AC Thickness

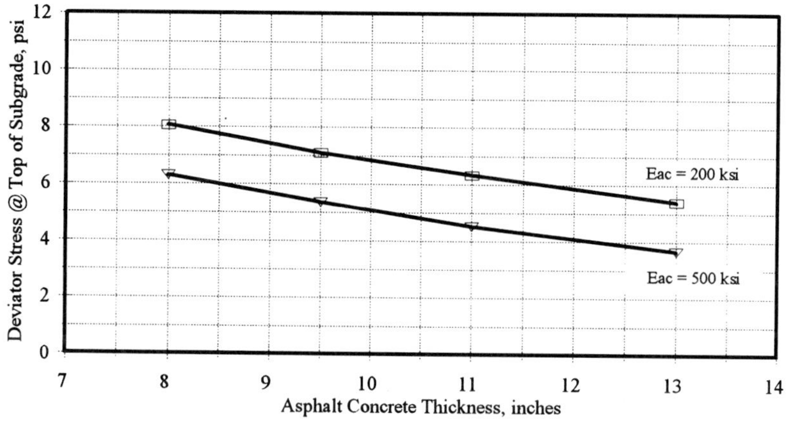

Figure 7. Variation in Subgrade Deviator Stress with AC Thickness

bottom of asphalt concrete layer. Subgrade responses (stresses, strains, deflections), however, reduce considerably with an increase in T_{GR}. Surface deflections are not significantly affected by an increase in granular base thickness for soft and medium subgrades (not shown here). However, for the stiff subgrade, the surface deflections increased by about 14-percent for Eac = 500-ksi, and 10-percent for Eac = 200-ksi. The contribution of granular base layer to the surface deflection increased as the granular base thickness increases.

Earlier studies have shown that residual stresses develop in granular base due to compaction during construction and subsequent traffic loading. These residual stresses were considered in the GT-PAVE analysis. Residual stresses of 2, 3, and 5-psi were used for analysis of the pavement cross-section with 8-in. AC surface, 43.5-in. thick granular base layer, and medium strength subgrade. Figure 11 shows the effect of residual stresses on surface and subgrade deflections. Including granular base residual stress in the analysis

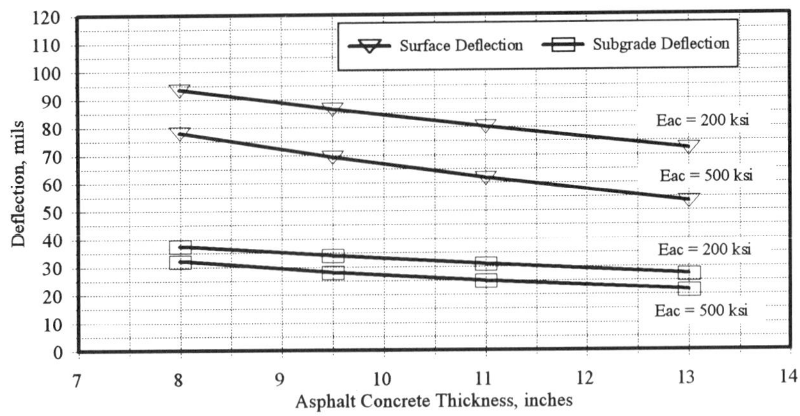

Figure 8. Variation in Deflections with Change in AC Thickness

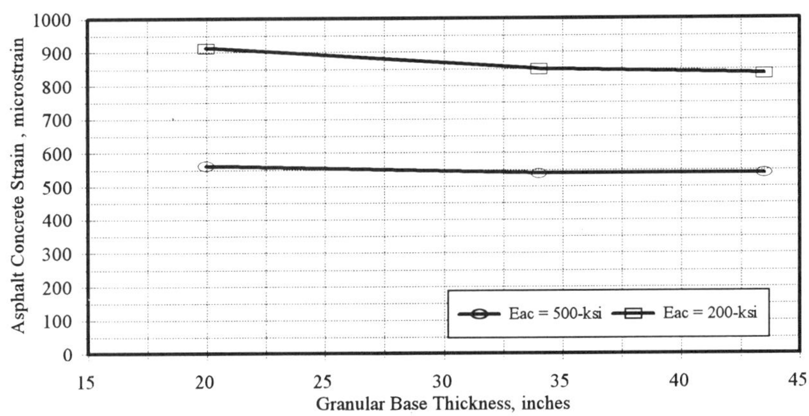

Figure 9. Variation in Strain at the Bottom of AC Layer with Granular Base Thickness

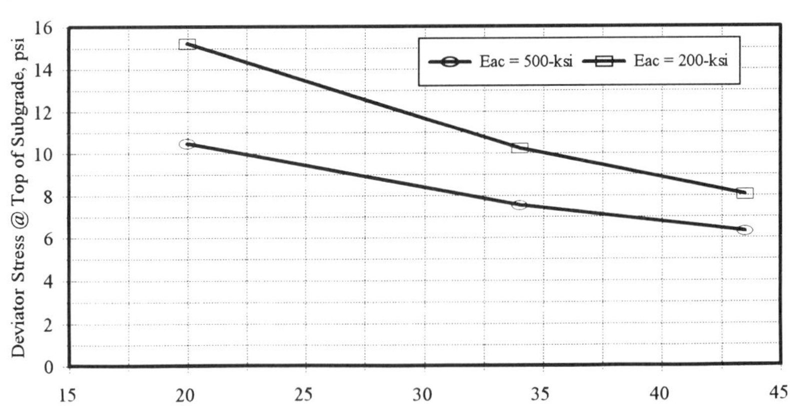

Figure 10. Variation in Subgrade Deviator Stress with Granular Base Thickness

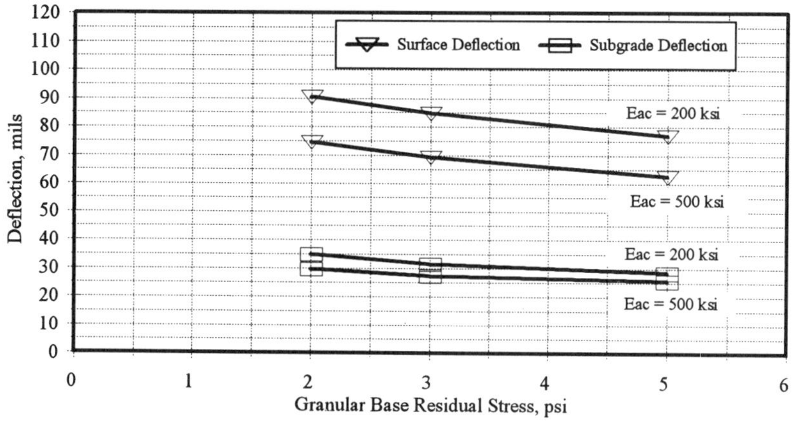

Figure 11. Variation in Deflections with Change in Granular Base Residual Stress

reduces the surface and subgrade deflections since the residual stress in horizontal direction reduces the vertical compressive strain and stress in the pavement. With proper confinement maintained from residual stresses, subgrade and surface deflections predicted by the two programs, ILLI-PAVE and GT-PAVE show good agreement. The anisotropic moduli help predict the stress state more accurately under the wheel load where dilative behavior is observed due to high major to minor principal stress ratios.

SUMMARY & CONCLUSIONS

An "Engineering Approach" is being developed at the University of Illinois Federal Aviation Administration Center of Excellence for Airport Pavement Research to estimate critical pavement responses and their locations for multiple wheel gear loads (i.e. Boeing-777 aircraft). In the ILLI-PAVE approach, single wheel responses are used along with the principle of superposition. Also, ILLI-PAVE final iteration granular base and subgrade moduli (for a single wheel of the gear) are used as modulus inputs to an ELP (BISAR) that can accommodate multiple wheel loading. The base and subgrade moduli are for various "layers" of the pavement section. The tensile strain at the bottom of the asphalt concrete layer and subgrade deviator stress are critical pavement responses for conventional flexible pavements and full-depth asphalt concrete pavements. Reasonable estimates of these critical pavement responses are achieved if the stress and deflection basins under a single-wheel load can be either accurately measured or computed. ILLI-PAVE and GT-PAVE have been found to predict realistic flexible pavement structural responses. This paper summarizes the results from sensitivity analyses that were conducted to study the effect of varying asphalt concrete (AC) thickness, AC modulus, granular base thickness, and subgrade resilient modulus on critical pavement responses. Increasing granular base thickness does not have any significant effect on the tensile strains at the bottom of AC layer and surface deflections. With increased granular base thickness, the contribution of granular base layer to the surface deflection increased. However, granular base thickness increase resulted in reductions in subgrade deviator stress, deflection, and strain. GT-PAVE predicted critical pavement responses are higher than ILLI-PAVE predicted responses, because of reduced horizontal stiffness of anisotropic granular materials computed by the nonlinear models in GT-PAVE. However, when residual stresses are included in the analysis, GT-PAVE responses approach ILLI-PAVE responses.

ACKNOWLEDGEMENTS / DISCLAIMER

This paper was prepared in a study conducted at the University of Illinois Federal Aviation Administration Center of Excellence (COE) for Airport Pavement Research. Funding for the COE is provided in part by the FAA under Research Grant No. 95-C-001. The contents of this paper reflect the views of the authors who are responsible for the facts and accuracy of the data presented within. The contents do not necessarily reflect the official views and policies of the FAA. This paper does not constitute a standard, specification, or regulation. The authors would also like to acknowledge graduate research assistants Umit Seyhan and Brian Heil for their help.

REFERENCES

1. Raad, L., and Figueroa, J. L., “Load Response of Transportation Support Systems”, Journal of Transportation Engineering Division, ASCE, Vol. 106, No. TE1, January 1980.
2. Tutumluer, E., “Predicting Behavior of Flexible Pavements with Granular Bases”, Ph.D. Dissertation, Georgia Institute of Technology, Atlanta, GA, 1995.
3. Chou, Y. T., Ledbetter, R. H., “The Behavior of Flexible Airfield Pavements Under Loads - Theory and Experiments”, M.P. S-73-66, U.S. Army Engineer Waterways Experiment Station, Vicksburg, Mississippi, July 1973.
4. “Investigations of Pressures and Deflections for Flexible Pavements”, Report 1, “Homogeneous Clayey-Silt Test Section”, Report 4, “Homogeneous Sand Test Section”, Technical Memorandum No. 3-323, March 1951 (1), December 1954 (4), U.S. Army Engineer Waterways Experiment Station, Vicksburg, Mississippi.
5. Ahlvin, R. G. et al., “Multiple-Wheel Heavy Gear Load Pavement Tests Technical Report S-71-17, Vol. I-IV, November 1971, U.S. Army Engineer Waterways Experiment Station, Vicksburg, Mississippi.
6. Thompson, M. R., “ILLI-PAVE Based Conventional Flexible Pavement Design Procedure”, Proceedings, Vol. 1, Seventh International Conference on Asphalt Pavements, Nottingham, U.K., 1992.
7. Garg, Navneet, “Mechanistic-Empirical Evaluation of the Mn/ROAD Low-Volume Road Test Sections”, Ph.D. Dissertation, University of Illinois @ Urbana-Champaign, IL, 1997.
8. Uzan, J., “Characterization of Granular Materials”, Transportation Research Record 1022, Transportation Research Board, 1985.
9. Tutumluer, E., and Barksdale, R.D., “Behavior of Pavements with Granular Bases-Prediction and Performance”, In Proceedings of the UNBAR4 Symposium, July 17-19, Nottingham, UK, Edited by Dawson A. R. and Jones R. H., 1995, pp. 173-183.

Fifth International Conference on the
BEARING CAPASITY OF
ROADS AND AIRFIELDS

CONSTRUCTION AND ENLARGEMENT OF GERMAN AIRPORTS WITH REFERENCE TO STRUCTURAL AND GEOTECHNICAL ASPECTS

RUDOLF FLOSS

Univ.- Prof. Dr.-Ing., Technical University of Munich,
D-81245 Munich, Germany

Abstract

The contribution gives an insight into design requirements and construction experiences arising from special structural and geotechnical problems in case of the new Munich Airport and the current enlargements of the airfields in Leipzig and Dresden.

For instance special requirements are related to watershed management, pavement design and geotechnical problems coming from underground conditions in watertight building pit enclosures and foundation. Additionally in case of the Munich Airport geosynthetic sealing mats were used in a technique employed for the first time which was designed to protect the groundwater from chemical deicers.

Quality assurance with regard to both the protection of the groundwater as well as to earthworks for the runways and for foundation work is based on specially developed testing procedures.

Key words

Runways, taxiways design regulations, groundwater protection, groundwater control, treatment of sewage, de-icing, restoration of runways, geosynthetic clay liner, dynamic compaction control.

1. Introduction

The Federal Republic of Germany has a decentralized airport system. One of the features of this system is the location of airports in centres of population and industry, ensuring the integration of these areas into the worldwide network of air routes and thus giving them additional competitive advantage.

This paper gives an insight into the design principles and some construction experiences arising from special structural and geotechnical problems in case of the new Munich airport and the current enlargements of the airports in Leipzig and Dresden.

2. Fundamentals for the design and construction of German airport runways

2.1 Regulations and guidelines

For the design and construction of runways and taxi areas the specifications and guidelines for roads are used to a large extent. [1 to 4].

The recommendations of the International Civil Aviation Organization (ICAO) for permissible inclinations and cross falls according to flight operational factors are applied for the surface drainage of air traffic areas.

Military airfields are subject to special guidelines issued by the Federal Ministry of Defence [5].

2.2 Design loads

The dimensioning is carried out for the number of load cycles to be expected from the aircrafts used, based on an pavement in-service life of 20 years. The dimensioning can be based either on the actual landing-gear geometry or on the corresponding equivalent single-wheel loads, using the ACN/PCN load systems[1)].

Reference values for the loading of individual air traffic areas are shown in Table 1. According to present experiences, they are sufficient for the design and are based on the design load which is decisive for the pavement.

[1)] ACN = Aircraft Classification Number
PCN = Pavement Classification Number

The operation areas are used by vehicles (e. g. tankers, tractors, special winter vehicles) with wheel loads up to 125 KN and wheel pressures up to approx. 1 N/mm².

Tab. 1: Loading frequency and additional loads

2.3 Pavement design

Normally, the air traffic areas are designed for the particular instance. The bearing capacity of the subsoil and subgrade has to be considered, including any seasonal changes. The technical requirements set up for roads including bearing and frost protection layers are decisive for the bearing capacity and for the degree of compaction of subsoil and subgrade.

The aprons in the taxiway and parking areas are designed for static loads, the runway heads and taxiways for single lane loads of the different landing gears. The mid-section of the runway is a 25 m wide strip along the runway centre line and gets more than 95 % of the load cycles. The thickness of the pavement can be reduced within this mid-section, because there is none rut traffic. This is also applicable to the small loading on the 10 to 18 m wide outer area strips.

Dynamic loads and additional loads resulting from weight displacements are taken into account by impact load factors (Table 1). Normal braking forces are calculated with a coefficient of friction $\mu = 0.3$, emergencies – for example slow down take-offs – with values up to $\mu = 0.6$. When aircrafts are being turned on aprons and parking areas, loads develop between the wheels and the surface of the pavement resulting from the lateral skid forces of the wheels, which is considered by coefficients of friction up to $\mu = 0.7$.

Theoretical design principles are increasingly used. The calculated stresses are compared to the working stresses of the building material depending on the number of load cycles. The most widely known design principle for concrete aprons is based on the Westergaard's theory. Corresponding computer programmes are available by means of which the loads of any landing gear configurations can be calculated. For the dimensioning of asphalt pavements empirical and theoretical design principles are usual. The most common methodes are: CBR-method (California Bearing Radio), the FAA method (Federal Aviation Administration), the US-Navy method and the McLeod method.

3. Munich Airport: Measures to protect the groundwater

3.1 Construction measures and buildings

The Munich Airport came into operation in 1992. It is located 29 km north-east of Munich at an altitude of 448 m above sea level. The airport has two runways each with a length of 4000 m and a width of 60 m. They are running at a displaced centre of 2300 m in the east-west direction and are offset against each other by 1500 m.

Geologically, the new airport is located in the Northern part of the gravel plain belonging to the foothills of the Alps. There are fluvio-glacial deposits of the devensia ice age as well as the holocene. The tertiary layers of the upper freshwater molasse (very sandy gravels with low permeable silts, clays and marls) are below the approx. 9 m thick quaternary gravels. The original tertiary surface was approx. 50 to 100 m higher than present and was subject to erosion especially in the ice age.

The tertiary surface is not very permeable, so that there is an upper quaternary and a tertiary groundwater reservoir. The tertiary groundwater is used as a drinking water supply for the surrounding communities and must be protected against pollution. Thus, the water-supplying interests and effects on the groundwater balance were very important for the designing and construction of the airport.

Altogether, more than 2 million m^2 paved air traffic and transport areas as well as more than 80 buildings, halls, tunnels, tanks and bridges were built-up [6]. A multitude of the buildings had to be permanently founded in the groundwater. These deeply founded buildings were designed and carried out as water impermeable troughs, safe to floating. Apart from water pumping and cut off walls, special construction methods with narrow tight walls and bulkheads as well as underwater concrete layers and horizontal sealing layers were allowed in the particular case.

The authorities required a proof regarding the effects of the structures on the groundwater regime. A groundwater dam-up of maximum 10 cm was allowed at the buildings and for sheetings and impermeable walls remaining in the ground.

3.2 Groundwater control and waterway rearrangement

The permanent and safe trafficability of the runways, taxiways and aprons required a certain distance between the groundwater level and the substructure of the paved areas. Furtheron the groundwater level must not rise above the frost depth which is laid down at 1 m.

By means of a drainage ditch system a permanent lowering of the water table by approx. 1.5 m in the inside area is achieved. The groundwater quantity removed from the drainage ditches is lead back to the groundwater by a seepage plant. The seepage plant consists of a pump station with a performance of 120 to 450 l/s, an approx. 6 km long distribution conduit and 44 control shafts as well as 132 seepage wells each with about 10 m depth.

Furthermore, another hydrological measure became necessary: A multitude of small waters crossing the airport site served to cultivate the areas and to maintain a favourable groundwater level for agriculture. These waters had to be rearranged.

Fig. 1: Munich airport: Groundwater management

3.3 Removal and treatment of sewage

The different kinds of sewage are collected and removed by a combined system of sewage, mixed water and rainwater pipes. The sewage is treated in the purification plant.

The sewage of the aircrafts is also conducted into this sewage system. The waste water incurring from washing the aircrafts is pretreated by physico-chemical methods before being conducted into the sewage system. Contaminated rain water is being filtered by means of a special separator.

The rain water draining off from the runways and taxiways is normally not contaminated and can seep into the ground at the sides or it is lead to drainage trenches via open ditches.

To enable trouble-free traffic in winter the areas must be free of snow and ice, so that their gripping capacity ensures safe take-offs, landings and taxiing of the aircrafts. Besides the mechanical clearing, chemical de-icing materials consisting of glycol and technical ureas are used. The procedure used for the de-icing of the air traffic areas must not detrimentally contaminate the streams and the groundwater. For this purpose, a composite system of operational and constructional measures was devised. The de-icing waste water from the runways and taxiways is collected in basins and for purification it is conducted to the sewerage-treatment plant together with the local sewage.

In the area of the 16 km long taxiways which are only de-iced with glycol, the contaminated melted snow and ice and rain water is purified by means of a bacteriological method (Fig. 2). At the sides of the taxiways the contaminated water flows through a 20 m wide gravel and sand body with transverse fills of sand. At a depth of about 1.50 m this area is sealed with a geosynthetic clay liner to protect the groundwater beneath. Within a period of 5 to 20 days the naturally existing bacteria dissect the main component of the de-icing glycol solution into harmless water and carbon dioxide. The water which seeps into the ground at the end of the sealing area is not detrimentally contaminated and can be conducted into the groundwater without danger. This system allows the direct transport of the purified melt-water into the groundwater without using the sewerage-treatment plant [7].

Fig. 2: Munich airport: Bacteriological treatment system for contaminated water from taxiways

4. Leipzig/Halle Airport: Construction and enlargement [8]

The current enlargement of the Leipzig/Halle airport by means of constructing and operating a new runway at a length of 3600 m with an adequate taxiway and apron system and terminals as well as further infrastructure considerably changes the existing airport. (Fig. 3). In 1990 the airport showed serious structural and functional faults because of the insufficient length and location of the runways, the taxiway system and the old apron.

For the new development the functional connection of the existing structure with the planned extension northern of the federal motorway A 14 and the new railway track of the Deutsche Bahn AG was an important matter of concern. The north extension is built to complement the existing airport, whereas the functional conception developed up to now for the connection of rail, road and air traffic shall also be continued.

The new runway North shall be 3600 m long and 60 m wide with two parallel taxiways of 30 m width each. Furtheron, it is planned to connect the existing traffic areas in the South to the new runway North by bridges for the aircrafts, which have to cross the federal motorway at a height of 4.70 m and the new railway track at a height of 7.40 m. In addition, the public roads running parallel to the motorway in the North and South have to be crossed.

Fig. 3: Flughafen Leipzig-Halle: MASTERPLAN AIRPORT 2000

The whole system of the air traffic areas has to be planned so that the permissible inclinations of the runways, taxiways and aprons are not exceeded. Altogether, for this reason and for sufficient drainage of the areas by a ditch system 2.830.000 m^3 earth has to be moved. To improve the bearing capacity of the subsoil, extensive soil stabilization has to be carried out.

The airport is located in the Leipzig lowlands and is part of a ground moraine. Due to several glacier movements a heterogeneous stratigraphical sequence had been created. The complicated bedding conditions are characterized by an irregular distribution of boulder clays, glacial marls and melt-water sands. The deeper lying tertiary soils are also characterized by a multiple stratigraphical sequence of sands, silts, clays and lignite beds.

The enlargement of the airport has an effect on the balance of nature and the landscape. New groundwater creation is influenced by sealed surfaces (approx. 160 ha altogether) and soil compaction. Contamination of the groundwater has to be avoided when soil layers are exposed. This can be achieved by an expert treatment of the soils, a sufficient cover, a careful handling of groundwater damaging materials and a controlled drainage system.

5. Dresden Airport: Restoration and renewal of the runway

The extension of the Dresden airport comprised as short-term measure the restoration of the existing runway. The renewal of the runway combined with a lengthening of 300 m is planned for the medium and long term. The existing runway, the taxiway system and the apron show considerable structural as well as functional problems concerning its levelness, length and position.

The following report describes the restoration of the existing runway which is 51 m wide, 2507 m long and has 14.5 m wide safety stripes at its both sides. The subsoil near to the surface consists of sand and gravel fill as ground levelling and in the northern part mainly of decomposed granodiorite rock.

The old runway, built in 1935, was unpaved until 1950 and then extended and constructed for public air traffic by an unreinforced concrete apron of 2 x 20 cm respectivly 25 cm layer thickness with an intermediate insulating card board. Because of spallings, cracks and a too low gripping capacity of the concrete surface, in 1973 this runway was restored by installing on top of it a bituminously bound superstructure (2.5 cm leveling course, 17 cm bearing layer, 2.5 to 3.5 cm wearing layer). But extensive spallings of the wearing layer, reflection cracks in the asphalt surface and water outlets required another restoration after a short time.

In 1988/89 the pavement was again built over by an unreinforced concrete layer without any expansion joints. Longitudinal joints were arranged at a distance of 4.25 m and cross cuts at a distance of 5.0 m both sealed with bituminous material or joint tape. Since this renewal, damages of the concrete layer have been developed increasingly, mainly starting with splittings at the joint edges. In addition, cracks developed since 1991 which could lead to extensive failures of the slab edges and softening of the concrete. Most of the sealing material swelled and squeezed out of the joints.

Detailed tests were carried out (measurements of PCN values and the slab deflection and eveness, taking up of cracks, investigations of drill cores) with the following results: There were voids under the slab edges (3.5 mm on average) as far as they had not yet cracked, whereas the centre of the slabs was firmly connected to the asphalt layer base. The cracks running diagonally to the slab edges initially occurred mainly in the areas where the aircrafts were landing, later on they extended over larger areas. The cracks accumulated in special sections of the runway and at special distances

almost perpendicular to these (orientated 12-30). These runways serve about 260,000 operations per year - the 2 parallel runways being the most used (about 95 percent of the operations). Copenhagen Airports is served by aircraft from all ICAO categories, most of the operations being in category B, C and D (Typical aircraft in each category are B 737, A320 and B767).

This paper deals with the evaluation of Runway 04L-22R because of the importance of this runway combined with the age and condition of the pavement, which has been measured regularly since 1989. Runway 04L-22R has a total length of 3,600 metres, and a width of 45 metres (excl. shoulders). The runway is a flexible pavement, and has today the following structure (including age of each layer):

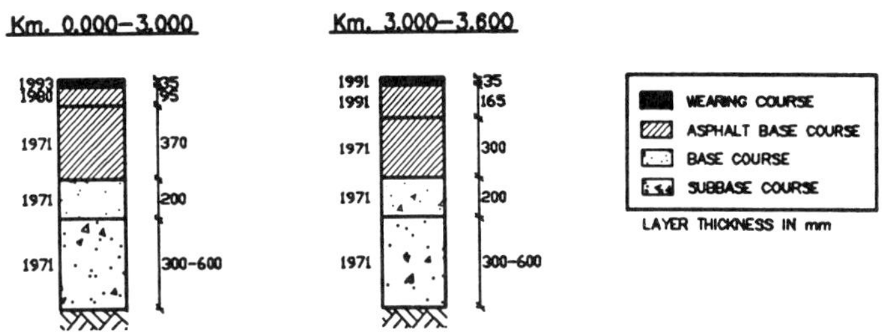

Figure 1 Pavement structure

In 1996 Runway 04L-22R carried about 120,000 operations, 55 percent of these operations being take-off. The distribution (ICAO categories) is shown in the table below:

ICAO Category	A	B	C	D	E
Operations	534	8,395	100,597	6,225	420

Table 1 Operations on 04L-22R in 1996

It is important for Copenhagen Airports to be able to predict the short-term condition as well as the long-term condition of the pavements at the airport. The short-term condition, i.e. the condition during the following 2 years, shall be predicted with such a degree of certainty, that no other major maintenance than the predicted maintenance is needed. This is due to economic as well as traffic planning. The long-term condition, i.e. the condition during the following 10 years, shall be predicted with a such degree of certainty that long term investment in rehabilitation work can be planned properly.

The evaluation is done by the Periodic Measurement Decision Making System (PMDM System), which consists of periodic measurements of the structural condition (bearing capacity) and the functional condition. The intervals between measurements are shortened as the classification and/or the condition of the pavement necessitates more intensive surveillance.

The bearing capacity is evaluated by measurements with a Heavy Weight Deflectometer (HWD), and the functional condition by measurements with a Road/Runway Surface Profiler (RSP) and a Safegate Friction Tester (SFT). Further, systematic visual inspections of the surface of the pavement are carried out yearly (BELREG).

RAMBØLL´s Department of Roads and Airports is assisting Copenhagen Airports in the evaluation of the pavements. The experience of RAMBØLL in this field consists of similar work in Copenhagen Airports and Aarhus Airport on runways, taxiways and aprons, and general work experience on other airports in Denmark and overseas.

MEASUREMENTS

As mentioned previously the following measurements are carried out: HWD, RSP and Safegate Friction Tester. Additionally visual inspections are carried out.

In the period 1990-1993 the Falling Weight Deflectometer (FWD) was used. Since 1993 the bearing capacity of the runway has been measured by the HWD, a specially developed FWD for use in airports. To minimise the influence of environmental factors on the calculations, the FWD/HWD measurements are carried out under the same weather conditions (temperature and rainfall) every year (except in 1990 where measurements were carried out during the summer). HWD drops are done in three lines along the full length of the runway. In two lines, symmetrically located about the centre line (± 4,5 meters, line A and B - corresponding to the area serving the most part of the operations), the interval between drops is 60 meters. In the third line, line C, placed 10 meters from the centre line, drops are done each 120 meters. In areas where the pavement is evaluated to deteriorate faster than the surrounding areas the distance between each drop is halved. In this paper, only results from lines A and B are discussed, but the same discussions apply to line C.

The RSP measurement has been carried out since 1993 to evaluate the roughness and the rut depth of the pavement. The measurements are carried out in four lanes along the full length of the runway. The lanes are placed symmetrically about the centre line ±3-5 meters and ±5-8 meters respectively. The traffic is mostly served in the inner lanes. The roughness is calculated in terms of International Roughness Index (IRI - given in mm/m), and the rut dept in mm is calculated, assuming a 3 meter straightedge.

The friction is tested, normally in lines 5 metres symmetrically placed about the centre line, whenever there is suspicion of low friction, and is thus not carried out in direct connection with the periodic measurements, but the results are part of the overall condition (structural and functional) of the runway. However the friction of the runway has no interest in connection with the bearing capacity, as it only gives information about surface conditions without any correlation to the structural condition.

To support the evaluation of the measured results, visual inspections are carried out. For this purpose a special form has been developed for use at Copenhagen Airports, enabling an evaluation of the structural, functional and mechanical condition of the pavement (the mechanical condition is the condition of the wearing course).

To verify the measured results, and for further evaluation of the condition of the asphalt, core samples are taken and analysed in RAMBØLL's laboratory. The cores give detailed information about the condition of each asphalt layer in the pavement, which then can be compared to the condition based on the HWD measurements.

RESULTS

In 1990 our first FWD measurements were carried out showing that the pavement needed up to 350 mm of new overlay in order to be able to carry the traffic for the next 20 years.

Based on the knowledge of the pavement in connection with the known limitations of the backcalculation models, this result was decided to be checked again with new measurements after which decisions would be taken based on the rate of change in the FWD/HWD results. This is how the Periodic Measurement Decision Making System was developed.

According to this system, the description of the results is naturally divided into three groups:

- Yearly measurement results
- Periodic measurement results
- Supplementary investigations (Cores)

Yearly Measurement Results

From the HWD testing the elasticity moduli of the layers are calculated using the backcalculation program ELMOD. The backcalculation is carried out assuming a two layer system; an asphalt layer of 500 mm thickness and an unbound semi-infinite layer. The unbound layer thus consists of the base course, the subbase course and the subgrade (figure 1). Based on the calculated elasticity moduli and the estimated traffic the residual service life of the pavement and the needed overlay are calculated. Further the PCN and the ACN (B747-400) values are calculated.

The composition of the runway results in a division of the runway into two main sections (figure 1). One of the main sections is again divided into 3 subsections due to the variation in condition, as can be seen from the HWD results in table 2. This reveals two subsections with significantly less calculated bearing capacity than the rest of the runway, as can be seen from figure 2, showing the variation in the surface modulus of the runway.

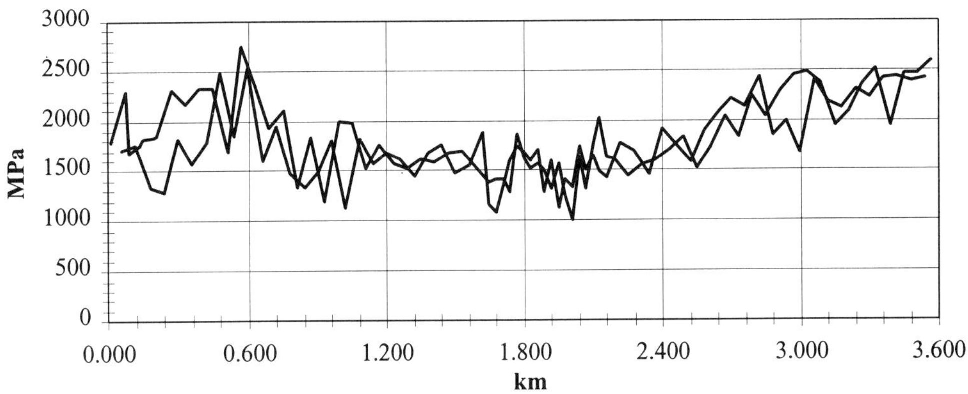

Figure 2 Surface modulus, 1997 HWD measurements at 9°C in two lines

Based on the distribution of traffic as given in table 1, and Copenhagen Airports' own analysis of the growth in operations, the design traffic over the next 20 years has been estimated. This design traffic includes the B 777-200, which is not yet operating in Copenhagen Airports, but is expected in the future.

The results from the two subsections km 1.650-1.740 and km 1.950-2.070 show lower elasticity moduli of both the asphalt layer and the unbound layer. The result is a considerably shorter ELMOD calculated residual service life (6 and 2 years respectively in 1994 and 3 years in 1997) compared with 13 years in 1994 and 1997 in the rest of the main section, and correspondingly higher need of overlay (116 mm and 250 mm respectively in 1994 and 99 mm and 153 mm respectively in 1997) compared to the rest of the main section (43 mm in 1994 and 48 mm in 1997). Further it should be noticed that the main section (km 3.000 - 3.600), which was partly reconstructed in 1991, has a service life of more than 20 years. The differences in the 1997 and the 1994 results are explained below in "periodic measurements".

		Main sections		Sub sections		
Section	Unit	0.000-3.000	3.000-3.600	1.650-1.740	1.950-2.070	0.000-1.650 1.740-1.950 2.070-3.000
$E_{0, Surface}$	MPa	1,673	2,324	1,338	1,389	1,726
$E_{Asphalt}$	MPa	2,677	3,932	2,004	2,126	2,783
$E_{0, Unbound}$	MPa	255	306	184	163	269
PCN[1]	-	118	120	112	107	119
ACN[2]	-	61	61	61	62	61
Residual life[3]	Year	12	20	3	3	13
Overlay	mm	61	6	99	153	48
Roughness	mm/m	1.7	2.2	1.6	1.6	1.7
Rut depth	mm	1.0	1.1	0.9	1.1	1.0
Friction	-	0.6	0.7	0.6	0.6	0.6

Table 2 Calculated average results of the HWD, RSP and SFT measurements in 1997

Based on these results, maintenance should have been applied before 1996 (1994 results) or shall be applied before 2000 (1997 results). This difference clearly demonstrates the necessity of a decision making system. Further, based on experience in heavy duty asphalt pavements it is the assessment of the authors that the condition of the pavement of all subsections is better than that implied by the calculated residual service life and needed overlay. This applies both to the 1994 and the 1997 results.

This assessment is further confirmed by the functional condition expressed by roughness and rut depth, which is no worse in the two subsections than in other sections (table 2).

Periodic Measurement Results

For a better evaluation of the actual condition and the development in the condition of the pavement, the trends of the calculated results in the structural and functional parameters over the years are analysed.
The structural parameters chosen are the calculated elasticity moduli (figure 3). To support the evaluation based on the development of the elasticity moduli, the development in the calculated residual service life and needed overlay are also observed (figure 4).

Figure 3 clearly demonstrates the different levels in the backcalculated elasticity moduli of each main and subsection. The two subsections with the lowest bearing capacity are seen to have the lowest calculated elasticity moduli. From figure 3 it can be seen:

- That there is a significant difference between the two main sections after 1993. The reconstruction of the upper 200 mm of main section km 3.000 - 3.600 in 1991 has improved the actual structural condition by about 30%. No measurements were carried out on this section in 1992.
- That there seems to be an improvement in the structural condition of the asphalt pavement from 1990 to 1992. This is due to the ELMOD temperature correction as

[1] In the PCN calculations a maximum value of 120 is chosen
[2] The ACN is given for B 747-400
[3] In the residual service life calculation a maximum service life of 20 years is chosen

stated below in "Supplementary Investigations (cores)". The 1990 measurement temperature was 18°C, while the rest of the measurements have been carried out in colder conditions (5-9°C). A correction of the 1990 measurement for a temperature condition similar to that for the other measurements will raise the calculated elasticity moduli of the asphalt pavement to moduli at the same order of size as the 1992 measurements (figure 6).

- That the recycling of the wearing course (35 mm) of main section km 0.000-3.000 between the measurements of 1992 and 1993 has no influence on the actual structural condition. It can not be stated whether the new wearing course has influenced the rate of deterioration as sufficient measurements were not taken before the recycling was carried out.
- That the asphalt pavements of main section km 0.000-3.000 has had an approximate rate of deterioration of about 240 MPa per year since 1992.
- That the two worse sections (km 1.650-1.740 and km. 1.950-2.070) have a consistently worse pavement condition than the rest of the main section (km 0.000-3.000). If the deterioration of the asphalt pavement continues with a constant rate of change, the two worse subsections will have a service life 3 years shorter than the rest of the section.
- That the rate of deterioration of main section km 3.000-3.600 is very low over the period 1993 to 1996. In 1997 there is a deterioration in the structural condition of the asphalt pavement. The coming years' measurements will show whether this deterioration is the beginning of a continuous process.
- That the elasticity modulus of the unbound layer has minor changes (up and down) over the years with no sign of fatigue weakening. The fluctuation of the modulus reflects natural changes connected to dry and wet periods at the time of measurements.
- That the two subsections of worse structural condition have lower elasticity moduli of the unbound materials than the rest of the main section, which has an elasticity modulus of the same size as the other main section.

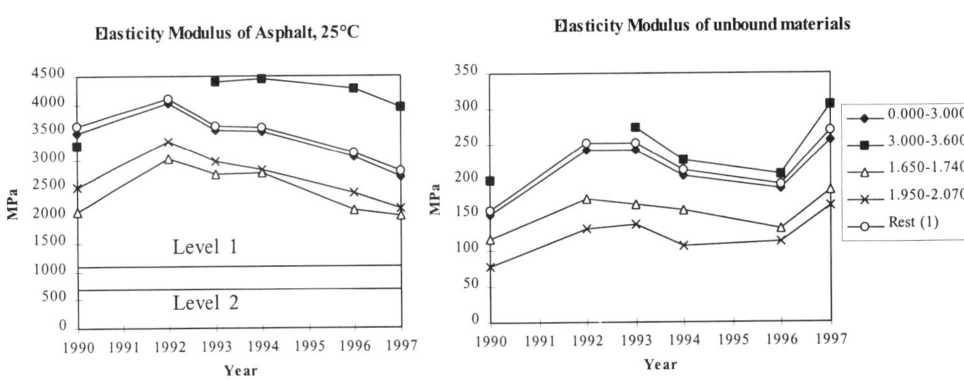

Figure 3 HWD backcalculated elasticity moduli. At the critical levels there is 2 years until execution of overlay (level 1) or reconstruction (level 2)

Based on the design traffic the results of the ELMOD calculations of the residual service life and the needed overlay can be seen in figure 4.

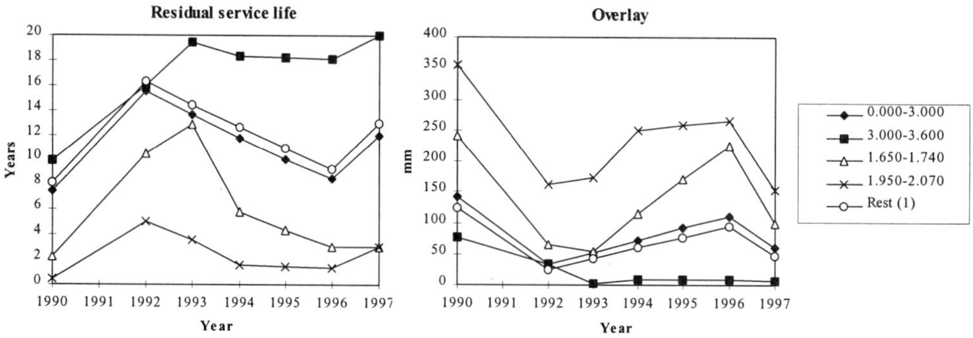

Figure 4 Development in residual service life and needed overlay for 20 years traffic, 1990-1997 calculated by ELMOD

Basically the residual service life and the needed overlay express the same results, as they are closely connected. From figure 4 it can be seen:

- That, the rise in residual service life and fall in needed overlay from 1990 to 1992 and from 1996 to 1997 is due to the similar rise in elasticity modulus of the unbound layers observed in all subsections, thus giving an incorrect picture of the rates of change.
- That the rates of change in the residual service life have abated in the recent years in the subsections with lower bearing capacity and in the rehabilitated main section km 3.000-3.600. The rest of the runway has a continuous fall in service life is assessed.
- That figure 4 in view of these consideration confirms the statements of figure 3.

The functional parameters used to support the structural condition are the roughness and the rut depth. These parameters have been measured since 1993. As these results still show a pavement with a very high functional condition in 1997 with no significant changes over the years, only the 1997 results are demonstrated (table 2).

It is our assessment, that the functional condition will decline as the structural condition declines but, as can be seen from the results, obviously not before the structural condition has reached a certain level, beyond the level of the structural condition of the pavement in 1997. If such a coherence could be established before the structural condition has reached the critical level, it would enable an inexpensive and quick method for detecting areas with lower bearing capacity, and thus further investigations could be focused in these areas of the pavements.

Supplementary investigations (Cores)

To investigate what causes the low elasticity moduli of the asphalt in the two weaker areas 9 cores have been taken and analysed in the laboratory in early spring 1998 (figure 5).

3 cores have been taken from each of the two worse subsections (km 1.650-1.740 and km 1.950-2.070), and 3 cores from the good subsections adjacent to the worse subsections. The cores have been taken at points where HWD measurements have been carried out. For stiffness modulus testing in the laboratory (Indirect Tensile Test (ITT) by means of the Nottingham Asphalt Tester) 3 cores from good subsections and 3 cores from worse subsections were selected. The stiffness modulus has been measured at both the temperature of the HWD measurements (9°C) and at the selected ELMOD standard temperature (25°C).

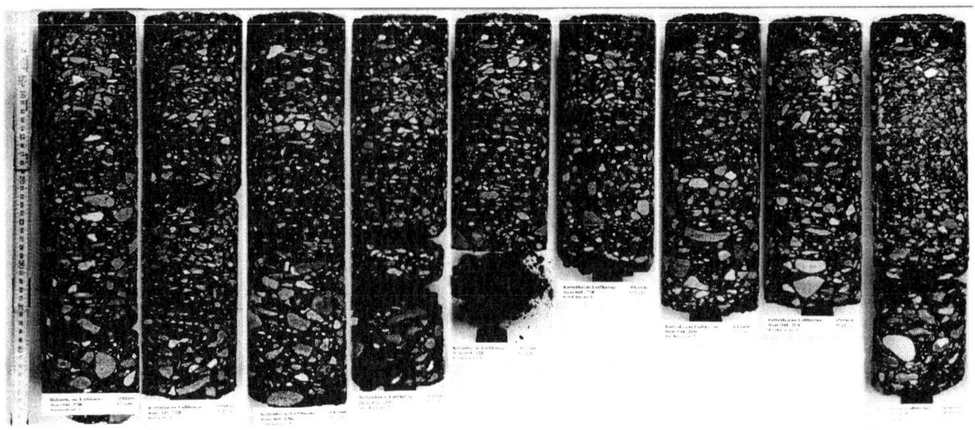

Figure 5 Core samples, 1998

The ITT testing has been carried out upon different samples of homogeneous layers of the cores. The sampling of the measured stiffness moduli into one stiffness modulus representing the core has been carried out with Odemark´s equation, [1], assuming the core was one whole intact core. The primary results of the analysis of the cores can be seen from table 3.

Core number			Unit	1	3	4	5	7	9
Subsection				Weaker	Weaker	Stronger	Stronger	weaker	Stronger
Station			km	1.650	1.710	1.800	1.860	2.010	2.220
Placement			m	4.5 Left	10 Right	4.5 Left	4,5 Right	4.5 Left	4.5 Left
Height			mm	482	462	418	277	347	428
HWD		9°C	MPa	4,960	2,518	6,437	6,191	4,774	7,027
modulus		25°C	MPa	2,218	1,080	2,761	2,656	2,048	3,015
Bulk Density	Top		kg/l	2.620	2.671	2.644	2.606	2.693	2.630
	Top ABS[4]		kg/l	2.237	2.177	2.218	2.273	2.269	2.290
	Middle ABS		kg/l	2.260	2.253	2.320	2.246	2.270	2.271
	Bottom		kg/l	2.231	2.226	2.248	2.302	2.228	2.270
Stiffness	Top	9°C	MPa	10,490	9,810	10,100[5]	10,520	9,430	10,430
Modulus		25°C	MPa	2,020	2,251	-	2,550	1,880	1,790
Indirect	Top ABS	9°C	MPa	7,250	5,920	5,380	8,880	7,550	8,630
Tensile		25°C	MPa	2,510	1,100	1,140	1,490	1,340	1,560
Strength	Middle ABS	9°C	MPa	3,940	2,380	4,810	5,560	3,710	5.680
(ITT)		25°C	MPa	-	340	750	1,070	-	1,000
	Bottom	9°C	MPa	4,170	2,810	2,400	6,600	1,690	4,270
		25°C	MPa	730	410	-	1,340	-	560
Stiffness, Core	Odemark	9°C	MPa	5,130	3,340	4,190	6,620	4,640	6,140
Cracks	Rupture			2	1	0	0	1	1
	Cohesion			0	0	3	1	0	0
Intact core[6]				+	+	÷	+	+	÷
Intact bottom				+	+	÷	÷	÷	÷

Table 3 Laboratory analysis of core samples, 1998.

[4] Asphalt Base Course (ABS)

[5] Calculated as the mean of the rest of the stiffness moduli of the wearing course

[6] Cracks are assumed to be due to the drilling

In the measurement of the stiffness modulus some cores have been divided in until 8 samples for the testing. Because of cracks and disintegration of the adhesion between asphalt layers and in some cases even disintegrated material it has only been possible to drill out three whole core of the asphalt layer.

The variation in the stiffness modulus measured at 9°C and 25°C when tested in the ITT, shows poor correlation with the one used in the ELMOD calculations (figure 6). The investigation shows that the correlation properly should depend on the level of the elasticity modulus.

For instance where the ELMOD makes a correction from 4.170 MPa at 9°C to 1.790 MPa at 25°C, the correction should be to 800 MPa at 25°C. This is important to know when overlay design and PCN values are calculated. This emphasises the importance of carrying out material analysis in order to have the right properties of the materials before judgement and redesign of pavements. This is in conjunction with results found by Said, [2].

Further it can be seen from the results of the analysis of the cores, that the worse sections are the sections where the bulk density of the lower asphalt layers are the lowest, indicating that the worse section could be due to insufficient compaction of the lower asphalt layers compared to the asphalt outside these section.

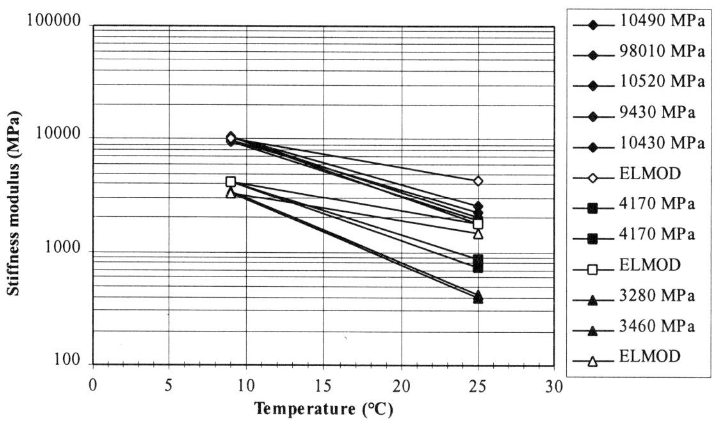

Figure 6 Temperature correction by ITT measurements and by ELMOD calculations

CRITICAL LEVELS AND RESIDUAL SERVICE LIFE

Based on the ITT results from the cores, it is possible to calculate critical levels of the backcalculated elasticity modulus using BISAR, [3], based on elasticity theory and a fatigue design life of two years and the Danish fatigue criterion, [4].

For the 2 types of maintenance stated below the following critical levels have been calculated:

- Level 1: Overlay, where the top 150 mm of the existing asphalt pavement is used as an asphalt base course. In this case a critical value of 1.100 MPa has been calculated.
- Level 2: Reconstruction, where the old asphalt layer is replaced by a new asphalt layer. In this case a critical value of 700 MPa has been calculated.

The residual service life of the asphalt pavement can be calculated by extrapolation of the curves in figure 3, based on the assumption that the rate of deterioration does not change and based on the assumption that the critical level given above does not change.

The residual service life given in the table below is calculated as the years until the interception of the curves in figure 3 and the critical levels plus the 2 years of design life.

		1.650-1.740 1.950-2.070	0.000-1.650 1.740-1.950 2.070-3.000
Level 1	Years	5	8
Level 2	Years	7	10

Table 4 Residual service life assessed using the PMDM System

The critical levels shall be confirmed by further testing of cores from asphalt layers in order to improve the ratter insubstantial basis.

CONCLUSIONS AND RECOMMENDATIONS

The decisions based on results of the periodic measurements are:

- The structural condition of the pavement is decreasing due to decrease of the structural condition of the asphalt layer.
- The structural condition of the asphalt layer is acceptable as long as the rate of deterioration is constant and above the critical level.
- The critical levels of the backcalculated elasticity modulus is 1.100 MPa when overlay is used as maintenance and 700 MPa when reconstruction is used.
- The residual service life of the runway is 5-7 years depending on the chosen type of maintenance of weaker subsection and 8-10 for stronger subsections.

The Periodic Measurement Decision Making System (PMDM System), as presented in this paper, includes the following:

- Carrying out periodic HWD measurements in the same points and as close to the same conditions as possible each time. The frequency of the measurements depends on the level of bearing capacity and on changes in the rate of deterioration.
- Determination of the critical levels from core analysis.
- Localising of weak areas and extend the measurements in these areas if necessary.
- Carrying out supplementary testing when needed, such as core analysis, testing of functional condition and others.

The measurements at Copenhagen Airports will be continued and will be extended to other parts of the paved areas (taxiways and aprons). Further the HWD measurements will be supplemented by other measurements and other surveillance methods such as georadar measurements, when they prove to be satisfactory measurements.

How future measurements will influence this method of evaluation will be announced in future papers.

Further, Copenhagen Airports is developing a pavement management system (named BELREG) for all pavements. This software will include the Periodic Measurement Decision Making System, which will influence the calculated residual service life of the individual subsections of the different pavements. The initial program will be completed in July 1998 and will likewise be announced in a future paper.

REFERENCES

[1] Ullidtz, P. Pavement Analysis. Elsevier Science Publishes B. V., 1987
[2] Said. S. F. Variability in Roadbase Layer Properties Conducting Indirect Tensile Test. ISAP Conference, Seattle, 1997
[3] BISAR-PC, Shell Pavement Design Method. User Manual. Shell Bitumen (1988)
[4] Bärenholt E. et al. Dimensioneringsdiagrammer for fleksible vejbefæstelser. Laboratorierapport 26, Statens Vejlaboratorium, 1976 (in Danish)

Fifth International Conference on the
BEARING CAPASITY OF
ROADS AND AIRFIELDS

DUTCH STRUCTURAL EVALUTION METHOD FOR CONCRETE AIRFIELD PAVEMENT

MARC STET[1], EGBERT BEUVING[2] AND BERT THEWESSEN[3]

[1]Senior Consultant, KOAC•WMD Dutch Road Research Laboratory, Schumanpark 43, NL-7336 AS Apeldoorn,The Netherlands, Fax: +31 55 543 3111, E-mail: KOACOA@PI.NET
[2] Project Manager, CROW (Information and Technology Centre for Transport and Infrastructure), PO Box 37, NL-6710 BA Ede, The Netherlands, Fax: + 31 318 621 112, E-mail: E_Beuving@compuserve.com
[3]Senior Consultant, Witteveen + Bos Raadgevende Ingenieurs BV, PO Box 233, NL-7400 AE Deventer, The Netherlands, Fax: +31 570 697 555, E-mail: H.Thewessen@witbo.nl

ABSTRACT

In order to harmonize FWD condition surveys on concrete pavements in the Netherlands, CROW's working group "Uniform Evaluation of Concrete Pavements (UEC)[1] has developed a structural evaluation procedure for jointed concrete airfield pavements. The committee has developed procedures for in-situ testing and models for determining of the mechanical pavement properties based on FWD measurements. With respect to the foundation of the slab-on-grade model, the Pasternak foundation was chosen as an attractive alternative for the Winkler foundation. The introduction of Pasternak's shear constant in Winkler's model allows for the removal of the discrepancies between Westergaard's theory and multilayer theory, while great advantages of Westergaard's model (edge and corner loadings) can be maintained. Codes for backcalculation and a stress model have been programmed. Consideration is given to the evaluation of varying pavement components such as slab thickness, support conditions, load transfer efficiency, material properties and fatigue behaviour. The remaining structural life can be derived from traffic data and material strength using mechanical pavement properties of a slab-on-grade. Insight is given in the relationship between the number of tests and the reliability of the structural life determination.

[1] The following persons have contributed to the evaluation method: E. Beuving of CROW, mrs. M.E. van den Bol-de Jong of the Road and Hydraulic Engineering Division (RHED), F. Van Cauwelaert of Febelcem, J.W. Frénay of VNC, L.Heleven of the Flemisch Community, L.J.M. Houben of the Techical University of Delft, A.C. Pronk of the RHED, M.J.A. Stet of the Dutch Road Research Laboatory, B. Thewessen of W+B and J.P. Verbeek of the Ministery of Defence

Keywords: Falling Weight Deflectometer, Structural Evaluation, Concrete Pavement, Pasternak Foundation.

INTRODUCTION

Rigid pavement evaluation normally consists of a serviceability survey and a structural condition survey to determine whether the pavement can adequately withstand the expected traffic loads. Structural pavement condition is an essential part in the assessment of the maintenance strategy. For the assessment of the structural condition, non-destructive pavement testing by means of falling weight deflectometer, coring and additional material testing have become popular in evaluating in-service pavements. The structural condition is deduced from the response in terms of vertical displacements due to the impact of a load, simulating actual traffic. FWD tests have a good load and deflection repeatability and can be performed at low costs and minimum traffic disruption. Whilest measurements are performed in a relatively standardized manner, many interpretation methodologies have been developed in the last decades. In order to harmonize and standardize the conventional measuring practice and evaluation procedures currently being applied in the Netherlands, CROW instigated a working group to develop an uniform evaluation method for jointed concrete pavements. The method has to be as uniform as practical and should be such that it can predict the remaining structural life with sufficient accuracy. The method developed by CROW's working group "Uniform Evaluation of Concrete Pavements""UEC"on is outlined and discussed in the next chapters.

STRUCTURAL EVALUATION PROCEDURE

In structural evaluation procedures, choices in evaluation techniques, analytical models and structural performance models have to be made.The working group UEC started with a literature survey, evaluating available slab-on-grade models, joint models, structural evaluation procedures, methods to determine cement concrete strength, fatigue models and probabilistic evaluation models [1]. The literature survey resulted in a methodology in which several models were chosen. Other models such as the slab-on-grade model were developed by the working group. The mechanistic based evaluation procedure comprises of:

- Guidelines for performing FWD measurements, PCI condition surveys and coring.
- An analytical pavement model for a translation of deflections into structural pavement parameters allowing to determine bending stresses in a concrete pavement due to individual and combined effects of wheel loading (and thermal curling). The model should enable to take into account edges, slab support conditions and load transfer to an adjacent slab;
- The governing properties of cement concrete are the Young's modulus and the modulus of rupture. The parameters can be derived from laboratory tests on specimen extracted from the pavement by coring. The Young's modulus and the slab thickness are input variables for the assessment of structural pavement parameters.
- A performance model taking into account fatigue accumulation and stress analysis for

lateral wander of air traffic based on the Palmgren-Miner hypothesis. Based on this performance model a residual life has to be estimated.

- Statistical technigues to adequately determine the residual pavement life as a function of the number of tests taking into account the interaction of variable pavement components such as slab thickness, load transfer, material strength as a structural system.

The evaluation method is outlined in figure 1. This paper discusses items 1, 4 and 5. More information can be found in [2];

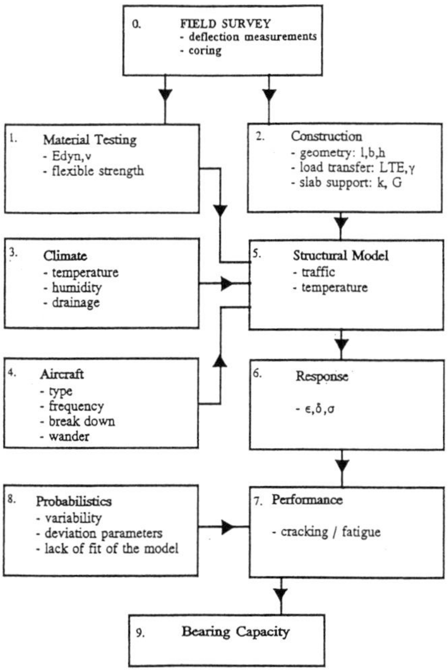

Figure 1. Outline of the UEC Method for assessing concrete pavement strength

STRUCTURAL PAVEMENT MODEL

The basis of the structural method is the structural pavement model. Measured deflections have to be translated into stresses, by backcalculation of the Young's modulus of the concrete slab and the foundation characteristics. Most pavement models rely either on the well known Westergaard solutions, on Burmister multi-layer theories or on finite element codes. The latter appears to be elaborate and time consuming in use. The former two models have their own intrinsik advantages and disadvantages. Since the limited slab dimensions and the support conditions of subgrade and joints are vital parts of the mechanical evaluation, it was decided to develop a model that was capable of combining all advantages. The model should enable backcalculation of mechanical parameters of slab and grade, and take into account edges, support conditions, load transfer. Besides this it should enable the calculation of stresses due to external loads. A rigid pavement can be satisfactorily modelled as a slab-on-grade system. The Pasternak foundation was chosen as an attractive alternative for the classical Winkler foundation. The introduction of a horizontal linkage, Pasternak's shear constant, in Winkler's model is a remedy for the discrepancies between Westergaard's theory and the multilayer theory, while the great advantages of Westergaard's model (edge and corner loadings) can be maintained. Although the capabilities of this model were recognized earlier, it was Van Cauwelaert who solved the mathematical problems for rectangular interior and edge loads in 1997 [3].

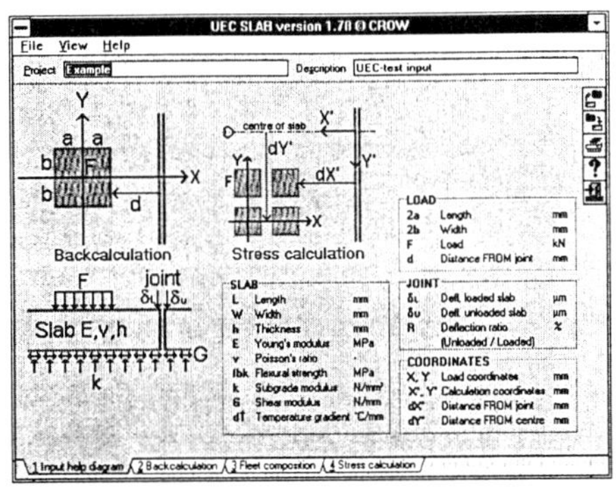

Figure 2. Van Cauwelaert's slab-Pasternak two-parameter foundation model

The model can be considered as an improvement of Westergaard's slab on a dense liquid foundation. The constant, k, that defines the subgrade and/or foundation in classical works of rigid pavements is the modulus of subgrade reaction. The reaction of the foundation on the slab is k times the deflection, and represents the reactive pressure to a load induced deflection. It can be determined through plate loading tests at a standard plate diameter of 30 inches. The Pasternak Foundation model can be considered as a linked spring system. The boundary conditions at the joint edge presume a difference in deflection between loaded and unloaded joint sides, allowing for some shear transfer through the joint (shear force left of the slab is equal to shear force of right slab side) and assumes no moment transfer through the joint (M=0). The parameter G of Pasternak can be seen as a horizontal spring, which is connected to Winkler's vertical k-spring. In case G=0, one arrives at a Winkler type foundation. Based on Van Cauwelaert's closed form integral approach codes for backcalculating and determining stresses have been programmed. The program UECSLAB can be used to backcalculate deflections in evaluating Falling Weight Deflectometer surveys. UECSLAB also computes the vertical deflection on top of a slab submitted to interior and/or edge loads, the vertical stress on the subgrade and the bending stresses at the bottom of the slab.

Backcalculation of Slab-on-Grade variables

The mechanical parameters are assessed in an iterative computer based technique in which a deflection bowl is calculated from Young's modulus of concrete (E), Poisson's ratio (υ), slab thickness (h), modulus of subgrade reaction (k) and shear modulus (G) and deflection ratio

(R). The deflection ratio is defined as the deflection of the unloaded slab edge to the adjacent loaded slab edge. An indication of the shear transfer (γ) at the edge is obtained from the delection ratio defined as the ratio of deflection of the unloaded edge to the loaded slab edge. The shear transfer at the slab edge is:

$$\gamma = \frac{2R}{1+R} \tag{1}$$

For R=0 (γ=0) there is no load transfer and for R=1 (γ=1) there is full load transfer. The backcalculation is based on laboratory derived values for the Young's modulus and Poisson ratio of the cement concrete. This dynamic modulus is asssessed from resonance tests according to NEN 5979 or NBN B15-230. Both methods relate the dynamic Young's modulus and the longitudnal (compression) velocities α and the transversal velocity β in an infinite medium:

$$E_d = \rho * \alpha^2 * \frac{(1+\mu)*(1-2\mu)}{1-\mu} \tag{2}$$

and

$$E_d = \rho * \beta^2 * 2\ (1+\mu) \tag{3}$$

In which:

E_d	=	dynamic Young's modulus [Pa].
ρ	=	material density [kg/m^3]
α	=	longitudinal velocity [m/s]
β	=	transversal velocity [m/s]
μ	=	Poisson ratio.

The resonance measurements are performed on cores extracted from the pavement. The Poisson ratio is determined from combined resonance and Betontester tests. From derived values for slab thickness, concrete modulus and Poisson ratio and deflection ratio, a deflection bowl can be calculated and compared to the characteristic profile obtained from site measurements. If the calculated deflections at each point and SCI matches the measured values to within a fit of 2%, no further refinement is required. The pavement characteristics obtained this way are considered to be representative for the pavement structure of the evaluated slab.

The main screen of the backcalculation module of UECSLAB is given in figure 3. Although the module is self-explainatory, several help-boxes are made in which several aspects of a specific topic are explained. The help-boxes give ,when activated, a so called 'suggestion value' for the parameter involved. This (calculated) suggestion value can be used as a guidance for a first estimate, but can be ommitted or overruled by the user. The backcalculation module computes maximal 10 displacements due to a rectangular load. The load can be arranged at user defined slab positions. The geophones can be placed perpendicular (direction X) or parallel (direction Y) to the edge or at given positions on the slab.

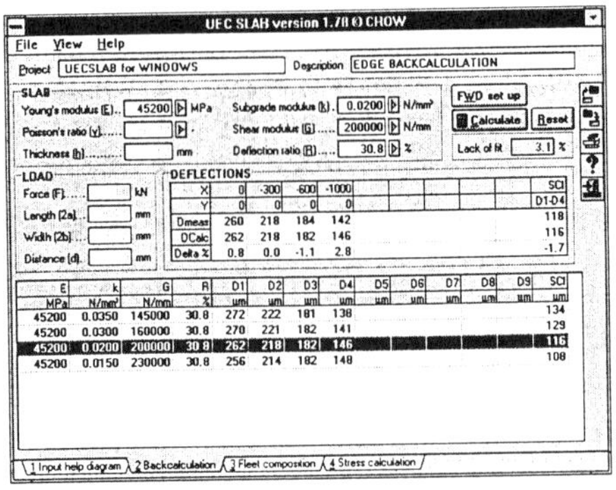

Figure 3. Screen for backcalculation structural slab-on-grade variables

Stress determination of external wheel loads

The fact that the edge stress is the controlling stress is generally accepted. Flexural bending stresses can be derived in the stress-calculation module of UECSLAB. This module computes the response of slabs with a finite edge due to external aircraft undercarriage loads. The closed form integral solutions of Van Cauwelaert enable the calculation of stress response at any given position on the slab. The response is expressed as horizontal bending stress at the bottom of the slab (σx and σy), vertical stress on top of the foundation (σz), and the vertical displacement (δ). The undercarriages of more than 50 aircrafts are predefined. The program also enables the user to add his own fleet to the database.

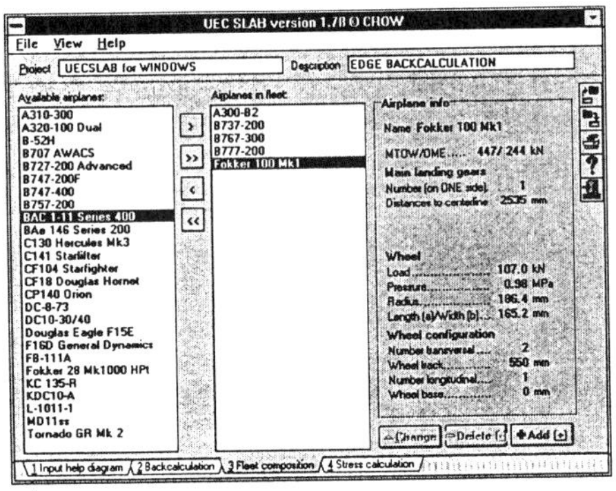

Figure 4. Selecting aircraft for fleet to be evaluated

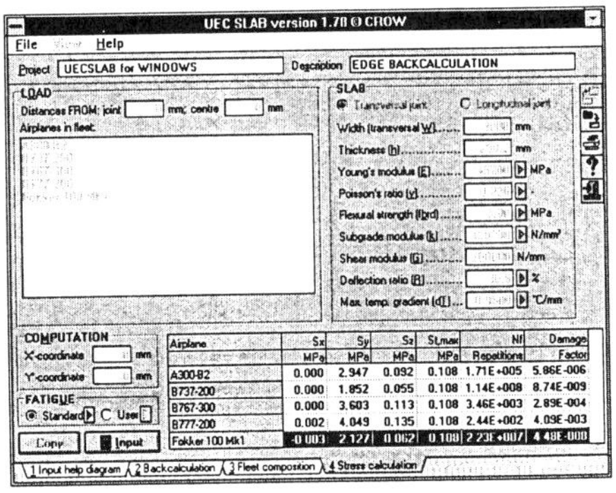

Figure 5. Example of stress determination

Thermal stresses

Temperature variations induce flexural stresses into the pavement. Due to daily variations a slab can either curl or warp. The critical stress condition occurs in daytime when the warping stress is due to positive thermal gradients. The temperature stresses are determined by taking into account a linear thermal gradient distribution by means of a VNC-modified Eisenmann's approach [4]. The maximum thermal stress occurs at the center of the slab edge. This stress decreases to the opposite edge end. Since Van Cauwelaert's slab model is essentially a finite slab model with a single edge, it was necessary to determine a thermal stress distribution at fictitious crossing slab edges along the pavement.

RESIDUAL LIFE ASSESSMENT

Performance model

Like other structural materials, pavement materials are subject to fatigue. In structural evaluation procedures of concrete pavements, choices in evaluation techniques, analytical models and structural performance models have to be made. In general a structural performance model comprises of a fatigue model and a damage hypothesis. Fatigue failure induced by traffic load repetitions and temperature variations is one of the basis structural distresses which affect the performance of pavements. In pavements, the fatigue behaviour of plain concrete is governed by flexural tensile stresses. In general, the fatigue behaviour is dependent on the types of (external) loads, their sizes and duration as well as on the number of repeated loads. The ratio between the maximum stress state (seviceability state) and the static tensile strength of plain concrete (ultimate state) appears not to be a single constant. The number of cycles to failure can be empircally described by a classical Wöhler relation. The choice of the fatigue model is an important one, which can influence the evaluation result tremendously. CROW working group UEC has performed a study into concrete fatigue models [5]. The working group suggests a specific fatigue relationship [for Dutch concrete mixes]. This relationship originates from the laboratory, but has been verified by field observations. The relationship is a conventional field-calibrated log N_f-σ relationship with given variations. The following relationship is proposed:

$$\log N_f = \frac{12.903\ (0.995\ -\ \frac{\sigma_{max}}{f_{bd}})}{1.000\ -\ 0.7525\ \frac{\sigma_{min}}{f_{bd}}} \quad with\ 0.50 \leq \frac{\sigma_{max}}{f_{bd}} \leq 0.833 \qquad (4)$$

Where f_{bd} is the modulus of rupture of the concrete which can be assessed in laboratory tetst on specimen extracted from the pavement, σ_{max} is the sum of the traffic and thermal stress and σ_{min} is the thermal stress. The relationship is valid for Dutch concrete and has a 50% reliability. The deviation of this relationship at an 95% reliability interval is log $N_f \pm 1{,}25$ or $\sigma_{max} \pm 0.08$.

Traffic analysis

For the calculation of the residual life, it is necessary to have detailed information on the traffic which has used the pavement in the past and to forecast the use of the pavement in the future. Lateral wander effects are addressed with a normalized beta-distribution. The complex frequency distribution and different gear loads must be analytically transferred into a fatigue damage distribution of the pavement. The traffic is carefully examined by means of a lateral wander analysis to determine the actual number of coverages at several positions on the slab.

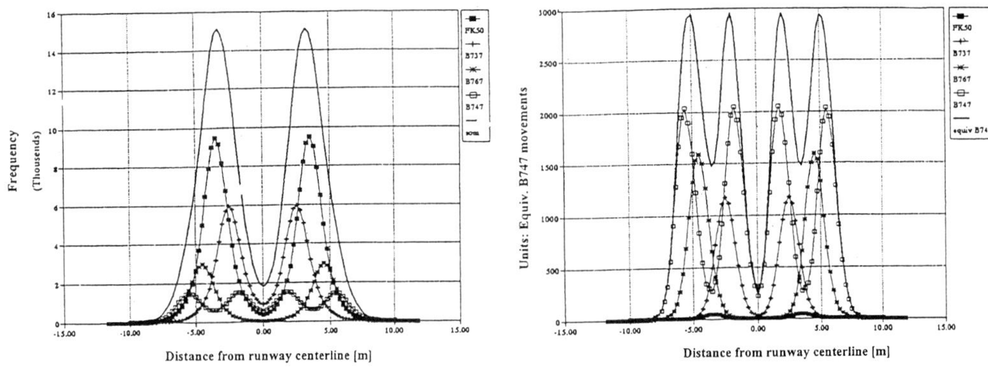

Figure 6 Lateral wander and damage analysis

The closed form integral solutions of Van Cauwelaert enable the calculation of stress response on any given position on a slab (edge). This enables the expression of the fleet into a number of historic and future coverages of a design aircraft on the basis of a fatigue damage concept. The pavement life is determined by cumulating the fatigue damage for each aircraft along different tracks and positions on the slab. Accumulation of the effects of the number of load repetitions is made on the basis of Miner's damage hypothesis. Based on the historical traffic data, present structural condition and traffic forecast, insight in future structural deterioration is gained.

Residual life

The residual pavement life is assessed by deducting the historical load repetitions from the allowable number of load repititions corrected for lateral wander. The accumulated pavement damage is assessed by adding the damage of the aircraft, taking into account the position on the slab edge. Wander effects are addressed with a normalized beta-distribution. The damage is the reciproke of the allowable number of aircraft passages. According to the Palmgren-Miner hypothese, the structural pavement life is consumed when the cumulative damage number is equal to 1 and the residual life is nil. If the allowable number exceeds the number of historical repetitions, the residual life can be calculated based on the difference of allowabel and past traffic, and the future traffic.

Statistical analysis

The residual life determination procedure is based on an analysis of individual slabs results.

The FWD measurements, coring, material testing and backcalculation are performed per slab. Hence, the reliability of the residual life of the pavement depends on the number of random tests, called observations that are performed. Figure 6 shows that the accuracy to determine the residual life depends on the number of measurements observations. For instance, a number representing 10% failure (90% sound condition) can be assessed with 20 observations with a 95% reliability interval and a bandwidth of $\pm 10^{2.0}$ repetitions. A total number of 30 observations reduces the bandwidth to $\pm 10^{1.5}$ and $\pm 10^{1.2}$ respectively. The UEC method gives recommendations for a minimum of 30 measurements and allows for an in-situ verification based on the standard deviation of FWD measurements.

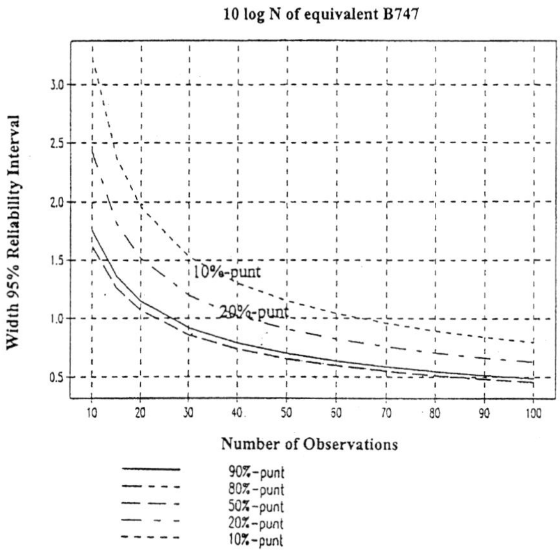

Figure 7. Width of 95%-reliability interval for percentile points of log Nf based on a lognormal distribution of the residual life

CLOSURE

The objective of this paper is to document on the progress of UEC. Their report will be published by the end of 1998. The "UEC method' presents guidelines for the evaluation of concrete pavements. This method has several advantages over other methods, e.g.:

- the method is based on a slab-Pasternak model which encompasses the disadvantages of the Westergaard model and Burmister model;
- the method uses a revised fatigue relationship for plain concrete;
- the method is uniform as well as practical and presents guidelines for field-measurements, laboratory testing, evaluation of field data and gives guidelines for the assessment of the remaining pavement life;
- the method enables to evaluate a pavement for different levels of probability since the origin of the evaluation method is a probabilistic one. The method uses a statistic procedure for the assessment of the required number of tests and the reliability of the evaluation procedure.

AKNOWLEDGEMENTS

This study was sponsored by the Information and Technology Centre for Transport and Infrastructure (CROW), the contributors of the Dutch Study Group for Airport Pavements and the Belgian Federation of Cement Industry. Their support is greatefully acknowledged.

DISCLAIMER

The views expressed in this paper are those of the Dutch working group UEC. The working group consists of representatives of the Dutch airport and Aviation Industry and pavement experts. The contents do not nessessarily reflect official views or ploicies of the Netherlands Directorate General of Civil Aviation.

REFERENCES

1. STET, M.J.A. (1996), "Evaluatiemodellen betonverhardingen; State-of the Art van literatuur, CROW, Ede, 105 p;

2. "Guidelines and Evaluation Method for Jointed Concrete Airport Pavements", CROW, Ede (in Dutch; to be published in 1998);

3. CAUWELAERT, F. Van (1997), "The limited rectangular slab on a Pasternak foundation", Doc. UEC-98-006, Brussels, October 20th, 34 p;

4. EISENMANN, J. (1970), "Bemessung von Zementbetondecken", Forchungsarbeiten aus den Strassenwesen, no. 82, Kirschbaum Verlag

5. FRENAY J.W. and M.J.A. STET (1997), "Fatigue Properties of Concrete Pavements", 3rd Airport Pavement Workshop, CROW, Ede, pp 163-178;

6. BOL, M.E. van den (1998), "Reliability and minimum number of observations", Doc. UEC-98-005, 5 p.

Fifth International Conference on the
BEARING CAPASITY OF
ROADS AND AIRFIELDS

EFFECTS OF DE-ICING AGENTS ON AIRFIELD ASPHALT PAVEMENT

JOHANN ROLLEN[1] and GEIR LANGE[2]

[1] Senior Engineer, Civil Aviaion Administration, PO Box, 53, 157 39 Stocholm Arlanda, SWEDEN, Fax: +46 797 67 45, E-mail: johan.rollen@lfvteknk.lfv.se

[2] Senior Engineer, Civil Aviation Administration, PO. box 8124 Dep. N-0032 OSLO, NORWAY, Fax +47 22 94 23 90, E-mail geir.lange@caa.no

ABSTRACT

The Swedish Civil Aviation Administration (Swedish CAA) is currently running an R&D project which aims to optimise the de-/anti-icing of runways and aircraft considering aviation safety, environmental impact and financial aspects. Items discussed are, i.a., runway slipperiness, degradation of materials, remote de-icing sites and requirements for aircraft and runway de-/ anti -icers.

In 1997 Civil Aviation Administration in Norway and Sweden experienced severe problems with some asphalt pavements on runways were the de-icing had been done by using new chemicals instead of sand or Urea. At Sundsvall Airport the surface had to be remixed after only five year and at Oslo Airport, Fornebu the porous friction course in October 1997 had to be removed due to such problems.

In November the same year CAA, in Norway and Sweden, started a joint venture program in order to investigate these problems. The main objective was to develop a new test method (or use an existing method) to measure the effects various de-icing agents have on asphalt pavements. In the AMS specifications AMS 1431 and 1435 (solid and fluid de-icing agents) there is no reference to asphalt pavements, only PCC surfaces.

The different de-icing agents are: Acetate, Formeate, Urea, Glykol+acetate and water, as a reference.

After the weatherometer a number of existing test methods will be carried out in order to measure the same effect as we get in the weatherometer.

INTRODUCTION

The Swedish Civil Aviation Administration (Swedish CAA) is currently running an R&D program which aims to optimise de-/anti icing of runways and aircraft considering aviation safety, environmental impact and financial aspects. This paper describes two of the projects under this R&D program.

1. Specifications for de-icing agents
2. Runway friction

1. SPECIFICATIONS FOR AIRCRAFT- AND RUNWAY DE/ ANTI-ICING AGENTS

The existing specifications for aircraft- and runway de/ anti -icers (AMS, ISO) do not fully consider all aspects that can be of interest. Areas not fully involved are environmental impact, electrical equipment, snow removal equipment, bitumen, joint sealers etc. In these areas it is necessary to add requirements. Among other things, a high rate of degradation of asphalt pavements has been noticed on some runways in Sweden and Norway. The bitumen in the pavement surface has dissolved, causing contamination of runway lights and on some runways the degradation has caused FOD-risk due to stone loss, which can entail aviation safety problems as well as financial consequences.

A workgroup run by the Swedish and Norwegian Civil Aviation Administrations has been established to improve existing specifications. In the group there are also participants from the Swedish National Road and Transport Research Institute (VTI), SINTEF Civil and Environmental Engineering and the Norwegian Building Research Institute.

The group has started to develop test methods for determine the effect of de/ anti-icers on bitumen. In this work, a weatherometer located in Trondheim has been central in verifying the test methods used.

LABORATORY STUDIES

This study use a large scale weatherometer in order to accelerate these de-icing effects. Five samples form four runways with different asphalt mix and binders have been collected and 4 de-icing agents is sprayed on these samples twice a day. This exposure seams to be similar to 3-4 de-icing seasons in Oslo and 7-8 de-icing seasons at Gothenburg Airport. A total of 120 de-icing treatment are going to be applied on the samples using same amount as we use on the runway.

The weatherometer at NBI normally runs test samples of bituminous roofing membranes, which consists of similar materials and are exposed to similar climatic conditions as an asphalt pavement on a runway.

The 20 samples of different asphalt mixes are in the weatherometer exposed to:

- IR heating
- UV
- Water
- Freezing
- De-icing agent
- Temperature variations Monday - Friday - 10°C to +30 °C with de-icing
- Temperature variations Sutreday - Sunday - 10 °C to +60 °C with de-icing

Water samples is collected every 18th day (5 times during the test) to measure if there is any bitumen in it.

After 12 weeks cores from the exposed asphalt are going to be compared with cores just exposed to water to measure the effect in a number of existing test methods.

Furthermore bitumen and aggregate are going to be tested separately in order to measure any reduced adhesion properties.

Figure 1 : Asphalt samples and electrical components in the weatherometer

The group will continue its work with improving the specifications by :

- development of a full scale test method for measuring the effect of de/ anti-icing materials on i.e. bitumen;
- setting up the criteria for how to make life cycle analysis of de/ anti-icing materials;
- setting up the criteria for how to assure that an aircraft de/ anti-icing fluid is recoverable.

2. RUNWAY FRICTION

Runway friction is affected by a number of parameters. To understand the mechanisms behind this, the Swedish Civil Aviation Administration has performed two series of friction measurements during the winter of 1996 on Stockholm-Arlanda and Gothenburg-Landvetter airports. The equipment used was a SAAB friction tester The measurements were performed on the runways during days with aircraft de-icing going on.

At every measurement, external data such as time, runway in use, air and surface temperature, precipitation, runway de/ anti-icing and amount and type of aircraft de-icing fluid used during the day were recorded. Since these are very complex circumstances, it is difficult to sort out the effect from the different parameters. Nevertheless, factors or phenomenon which seem to effect runway friction are:

- chemical deposits on runway - in particular Type 2 (Non-Newtonian) de/ anti-icing fluids from departing aircraft;
- rubber deposits in touch down zones;
- polished stone material on the surface;
- de/ anti-icing chemicals accumulate on the runway;
- quick temperature rises, causing accumulated chemicals to rise to the surface and make the runway slippery.

The group will continue its work with;

1. evaluation of different methods for de/ anti-icing or snow removal on runways;
2. evaluation of methods to improve surface characteristics regarding friction;
3. study of the effect of Type-4 fluid on runway friction;
4. follow the Joint NASA/ Transport Canada Friction Testing Program.

FIELD STUDIES

CAA Norway has constructed a friction test track located at Oslo Airport Gardermoen. This test-track consists of 8 different section with different surface texture. Along one side a drainage channel has been constructed in order to allow the use of all different chemicals and run tests to measure the effect from these chemicals on the asphalt pavements. Field Studies on this test track will be carried out in late May 1998.
CAA Sweden have measured different runway in order to compare asphalt with and without de-/ anti-icing agents.

CONCLUSIONS

These two projects will be finished in June 1998 and therefor the conclusions will be summarised at the Conference in July. Further examples of differences in wet friction properties from different runways in Sweden will be presented.

PAVEMENT SURVEYING WITH EURADAR : HIGH SPEED MEASUREMENTS OF LAYER THICKNESS AND DEFECTS, EXPERIENCES ON ROADS AND AIRFIELDS IN HOLLAND

ADRIAN J. VAN LEEST MSc. C.Eng.

Executive manager, Infratech Pavement Consultants,
P.O. Box 2477, 5202 CL 's-Hertogenbosch, The Netherlands,
Telephone (+)31 - 73 623 21 00, Telefax (+)31 - 73 623 12 32

ABSTRACT

Variation in layer thickness or lack of historical constructiondata and evaluation of the quality of interfaces between layers make it necessary for a pavementmanager to perform time consuming and thorough core-drillings. Such core sampling is costly and their number is therefor necessarily limited, this will increase the uncertainties regarding the determination of these basic parameters. Let us add to this that attempting to evaluate the bonding of interfaces by means of core samples which, by their very nature, disturb the sampled materials significantly, is often an overly optimistic endeavour.

Highway and airfield engineers, who have long wished to see through their pavements almost by transparency and continuously, without having to break them up, or at least only at the most suitable locations, are now (almost) seeing their wishes come true. Nowadays it is possible to obtain an almost continuous image of the layer thickness of a pavement in a very short time, at a speed of approx. 70 to 80 km/hour. At lower speeds it is even possible to find defects in a pavement, such as cavities and delamination.

AUSCULT' from France has developed a new, portable, light weight, easy to handle radarsystem, the EURADAR, which has been introduced by INFRATECH in the Netherlands.
Lately Infratech Pavement Consultants has performed a large number of EURADAR measurements on several types of pavements like roads, bridges and runways.
The results, price-quality ratio, comparison with traditional cores and future applications are very positive for daily use of the EURADAR.

Keywords: radar, high speed pavement surveying, layer thickness, defects

INTRODUCTION

Since some time Infratech has been dealing with the search for a suitable radar to be used in pavement surveys. Until recently in the Netherlands only radar techniques based on the so called ground penetrating radar were used. These were developed to search for different materials in deeper soil and were not high speed and not very accurate.

Besides doing visual condition inspections and bearing capacity tests, collecting layer thicknessdata is very important for the pavementmanager, and this for many reasons:

- inventory of the material layers of existing pavements as a basic parameter for a pavement management system;
- essential information for the processing of deflection measurements and insight in the variation in thickness and therefor necessary for an adequate maintenance plan;
- determination of the total quantity of materials to be removed in case of reconstruction of the pavement, which can be under estimated when there is a lack of thicknessdata.

Infratech was looking for a radarsystem with the following specifications:

- contactless, high speed technique capable to drive in normal traffic;
- especially designed for pavement constructions;
- proved accuracy;
- operational system without developing problems;
- user friendly presentation of results, suitable for further analysis;
- if possible, a by government authorities accepted way of testing.

Reason for this is that with this kind of radar, collection of thicknessdata will be very rapidly, without disturbing the trafficflow too much or leave (road)workers too long in unsafe conditions under traffic.

In 1996 Infratech has found the answers to these questions, the so called EURADAR. The EURADAR consists of an electromagnetic wave transmitter and receiver system whose design and development, undertaken by AUSCULT' from France in the mid 1990s, have resulted in an operational, high-performance system for the qualitative and quantitative evaluation of pavement structures. AUSCULT' has brought into this radarsystem many years of research and experiences with using radar techniques in surveying pavements.
The results of the radarmeasurements are presented by means of an interactive softwareprogram and sophisticated graphic presentation modes.

PERFORMANCE OF EURADAR

The EURADAR generates a monostatic or bistatic puls with a wide frequency band centred on 1000 MHz. It is composed of a microwave electronic system and a horn antenna. It has been especially designed for the surveying of pavement constructions like roads and airfields. Owing to its improved gain and dynamic range, it allows surveying of pavements to a sufficient depth.
The principle of radartechnology is shown in the following figure.

PRINCIPLE OF MEASUREMENT

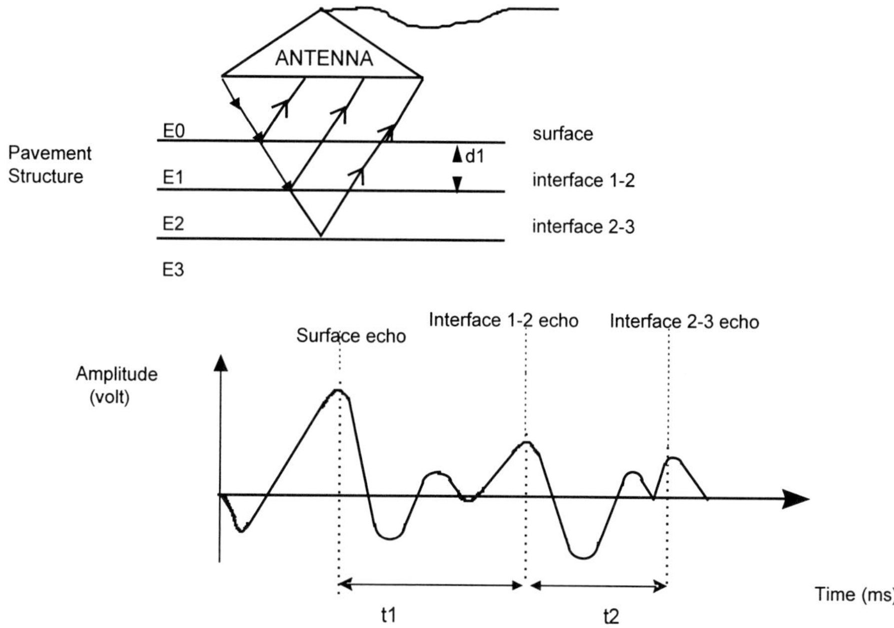

Signal received by the antenna

$$d1 = \frac{t_1 c}{2\sqrt{E_1}} \quad \text{with}$$

t_1, t_2 = propagation times
C = celerity ($3*10^8$ m/s)
E_1 = permittivity
d_1 = thickness of the layer

The antenna radiates an electromagnetic wave towards the structure to be analysed. With each change of medium, part of the energy is reflected and then picked up by the antenna; the other part continues its propagation by transmission through the host medium.
The antenna restores an electric signal containing mainly the echoes indicative of reflections at the interfaces of encountered media. In this way also the timedelay between various reflections is recorded. After conversion from microwave to audio frequencies, the signal becomes compatible with conventional acquisition, recording and processing equipment.

The EURADAR can be mounted on any kind of vehicle equipped with an antenna positioning system and an energy source. Also an incremental encoder of distances covered should be mounted and furtheron a DAT-recorder and oscilloscope will be installed in the vehicle.
More recently the EURADAR was completed with software designed for qualitative analysis (portable microcomputer, high-definition screen, colour printer) making it possible to obtain radarscans in colour. Also (semi-)automatic calculation of layer thicknesses is possible with this software.
With only the recording equipment on board, because of the as simple as possible configuration, processing and interpretation of measuring results will be done at the office and surveydata will be available in a few days if required.

Two of the projects in the Netherlands with EURADAR were a Dutch highway and a Dutch airport runway. Before reporting about the results of these projects, some characteristics of EURADAR will be explained.

Depth of penetration by the antenna
The depth of penetration with the EURADAR (1000 MHz.) depends on:
* the loss of wave-energy in the materials;
* the dielectric contrast between layers.

These two parameters are determined by the chemical compound and the water content of the materials. Also the compaction can play an important role.

Bituminous binders usually have little effect on the dielectric characteristics of the asphaltmix, compared with the basic ingredients like sand and crushed rock. On the contrary hydraulic binders (cement, blast furnace slag, fly-ash) can have a significant effect, especially when applied in high contents.
The water content has a negative as well as a positive effect because on the one hand the investigation depth is limited but on the other hand the contrast between layers will be bigger.

Specifically, radar evaluation is in principle possible on pavement structures of the following type:
a large investigation depth (about 1 m.) is possible in:
- bituminous layers, asphalt concrete as well as bitumen-treated base layers;
- unbound granular base layers with a low water content.

A medium investigation depth (about 0.8 m.) is possible in:
- cement-treated sand and granular base layers;
- granular base layers with binders like fly-ash, blast furnace slag.

A small investigation depth (about 0.5 m.) is possible in:
- cement concrete pavements, especially in early age;
- granular base layers of blast furnace slags;
- unbound granular base layers with a very high water content;
- clayey materials.

Penetration will be nil in metals or fine-meshed metallic fabrics.

Discernment of layer thickness
The minimum thickness that may be measured is about 30 to 50 mm, depending on the nature of the materials.

Surveying speed
Surveying speed depends essentially on the required measurement intervals, type of pavement, aim of the investigation and category of road.
For the measurement of thicknesses (quantitative measurements):
- 70 to 80 km/hour on highways, interstates and runways;
- between 30 and 60 km/hour on other roadcategories and taxiways.

For the investigation of defects (qualitative problems):
- 10 to 30 km/hour on pavements of roads and airfields;
- 5 to 10 km/hour on bridges, tunnels etc. because of the small length.

Measurement interval
The interval of measurement depends on the time needed for transmitting, receiving and processing of the radarsignals during measurement and therefor on the travelling speed. In practice the interval of measurement can be varied between a few centimetres and several meters.

Accuracy of measurement
The accuracy is plus or minus 5 mm for bituminous layers which are homogeneous and plus or minus 20 mm for base layers. For a qualitative good radarsignal homogeneous material should be present in a width of about 80 cm.

Applications
As long as there is a clear and adequate dielectric contrast between two layers of a pavement structure, it is possible to pick up almost continuously an echo allowing the location of interfaces. This phenomenon is used for the quantitative measurement of thicknesses.
On the other hand, dielectric irregularities which may be limited in space result in a change of the shape and/or the amplitude of the signal, which are processed qualitatively.

The major applications of the EURADAR-system are:
<u>In quantitative analysis of pavements:</u>
- thickness measurement of different layers of a pavement structure;
- measuring total thickness of the bituminous layer on bridges in order to calculate the total permissible weight on the construction and to control a milling machine;
- commissioning measurement of a newly build pavement structure, layer by layer;
- indication of thickness of various materiallayers in an existing road construction for evaluating reinforcement requirements.

<u>In qualitative analysis of pavements:</u>
- detecting and locating structural defects in layers and on the interface of materiallayers like cavities, moisture zones, debonded layers, segregation of mixes;
- colour displaying of radar images to investigate variations in layer thickness or determination of homogeneous sections, even in a poorly known pavement structure.

The results in terms of layer thickness are available in digital format which means statistical calculations can easily be made. Besides the data can be imported in any pavement management system.

Projects that have been performed by Infratech in the Netherlands in the last three years are: determination of layer thickness on several highways, cityroads, rural roads and agricultural roads, locating voids under concrete slabs at a junction of two highways, thickness (and variation) of an industrial pavement, search for debonding layers on three bridges, one of which was 5 km long. Commissioning the thickness of an asphalt overlay of a reconstructed runway of a national airport.
The projects on one of the highways and on the airport runway will be described and the results will be presented.

EXPERIENCES ON A DUTCH HIGHWAY

Ordered by the Dutch Highway Authorities (Rijkswaterstaat) Infratech performed measurements with EURADAR on a Dutch highway in order to determine the thickness of the asphaltic layers and if possible the base layer of cement stabilised sand. In this way of working it is possible to search for weak spots in the pavement construction because the measurements allow an almost continuous scan of the road.

In this case the objective was to prepare for a total reconstruction of the road, that's why layer thickness has been determined between the wheeltracks of the shoulder, slow lane and fast lane on both carriageways. According to the authorities one might expect an uniform construction in each lane, so sufficient information can be gathered in this way of measuring. The road was originally constructed with about 30 cm asphalt on 20 cm cement stabilised sand, this construction should be the same over the total length of the road.

Performing measurements
In May 1995 the radarmeasurements were performed under traffic at a speed of about 70 to 80 km/hour. Extra security cars will be necessary when this speed causes an obstacle for normal traffic. At this speed a lot of data in a short time can be collected without disturbing traffic flow. The total length to be measured was 30 km of lane.
Performing an equal amount of core drillings would cost a lot of time and probably a trafficjam.

In order to increase accuracy and verification of materials and thickness a small coredrilling with a 13 mm diameter was carried out after which the materiallayers were investigated by an endoscope.

Processing of radardata
Collected radarsignals have been processed semi-automatically into layerthicknesses. If there is a clear reflectionsignal, calculation can be done automatically. When the reflection is weak or of bad quality, manual processing will be necessary.

After a first interpretation it can be concluded that the asphaltlayer in this project is clearly visible but there is a slight problem in seeing the baselayer. The cause of this is a bad contrast in dielectric constant due to mixing of materials. Thicknesses of 12 to 15 and sometimes 20 to 25 cm are calculated.
In order to verify the thicknesses of asphalt as well as base layer and to improve accuracy, the authorities have given a limited number of core drillings of some locations.
Together with this information the asphalt thickness can be very well calculated. On first sight there is a good comparison between all lanes on one carriageway. Although the asphalt thickness is about 30 cm over a long distance, large variations can be present. For this road asphalt thickness varies between 15 and 80 cm.

If in a single location there is no registration of layer thickness because an interface between layers can not be detected, it is statistically allowed to interpolate between two calculated values due to the large number of data.

A graphical presentation of the layer thickness is given in the following figure.

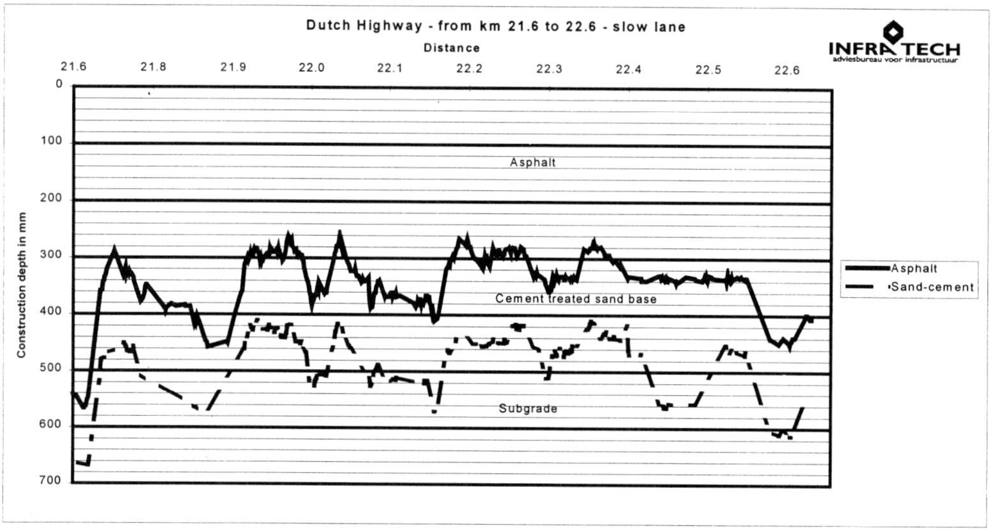

EXPERIENCES ON A DUTCH RUNWAY

In cooperation with the Dutch Ministry of Defense, the manager of this civil/military airport, and the contractor who applied the asphalt overlay Infratech performed measurements with EURADAR to determine the thickness of the new overlay, the old asphalt layer and the total thickness of the asphalt pavement. The aim of this investigation was to check whether the thickness of the overlay was according to the specifications.

Performing measurements

In june 1997 measurements have been performed in 5 longitudinal tracks: centerline, 7. m. and 13 m. left and right parallel to the centerline. The speed of measurement was 20 to 30 km/h which leads to an interval of 0.5 m. The total length of the runway is about 3000 m. including two runway-ends (runovers) of 400 m. each, the width of the runway is 45 m, of which the center part of 30 m. is the most important. Total time on the runway, including 4 calibrations next to cores was about 2 hours. During measurements there was no airtraffic on the airport.

Processing of radardata

Besides displaying the results in a graph, the measured thicknesses for each 0.5 m. have been statistically processed by calculating the mean value and standard deviation for each section of 50 m. The results of the centerline track are shown in the following table.

Layerthickness measurements by EURADAR on National Airport
The Netherlands
Runway 04 - 22, measured at the centre line

INFRATECH

Statistical processing of calculated layerthicknesses

		new asphaltlayer			old asphaltlayer			old+new
begin section	end section	mean value in cm.	number of counts	stand.dev. in cm.	mean value in cm.	number of counts	stand.dev. in cm.	mean value in cm.
0 + 0	0 + 49	26,6	100	1,25				26,6
0 + 50	0 + 99	28	98	0,32				28
0 + 100	0 + 149	28,6	100	0,29				28,6
0 + 150	0 + 199	28,1	99	0,79				28,1
0 + 200	0 + 249	27,1	98	0,63				27,1
0 + 250	0 + 299	25,9	100	0,4				25,9
0 + 300	0 + 349	24,7	96	0,72				24,7
0 + 350	0 + 399	24,3	93	0,78				24,3
0 + 400	0 + 449	22,5	76	0,74	29,1	72	1,14	51,6
0 + 450	0 + 499	21,9	69	0,31	29,4	69	0,7	51,3
0 + 500	0 + 549	22,1	99	0,31	27,9	92	2,22	50
0 + 550	0 + 599	22,2	97	0,28	27,3	95	1,04	49,5
0 + 600	0 + 649	22,1	96	0,37	28,7	90	1,23	50,8
0 + 650	0 + 699	22,4	100	0,27	27,7	98	0,89	50
0 + 700	0 + 749	22,5	100	0,3	27,9	78	0,86	50,4
0 + 750	0 + 799	22,3	90	0,28	28,1	86	2,06	50,4
0 + 800	0 + 849	22,4	100	0,19	28,1	97	0,8	50,6

Both runway-ends (run-overs) have cracked and seated concrete slabs underneath the asphalt overlay, which is the reason no old asphaltlayers have been found there. The mean overlay thickness of the runway is 22.45 cm. (based on 3880 counts in 2,200 m.). When we make a frequency distribution of all thicknesses of the overlay on the runway, we can see that in the centerline the overlay is thicker than 205 mm., but no value is thicker than 240 mm. The thickness class of 220-225 mm, containing the overall mean value, contains 35 % of the thickness values.
If the pavement manager had set his specifications on a mean thickness with a certain standard deviation or reliability interval he could have calculated easily whether the contractor has been working according to specifications.

In this specific project evenness of the runway after overlay was more important than layerthickness itself, so the contractor was asked to apply a total asphalt thickness of 19 cm excluding leveling layer to obtain a good bearing capacity.

COMPARISON BETWEEN THICKNESSES BY RADAR AND CORES

In the highway project the results of a radarmeasurement, in this case the layerthicknesses, were compared with the results of cores on the same lane. From this comparison it appears that 85 % of the radar thickness is exactly the same as the cores. Comparison of the homogeneous sections with mean value and standard deviation of the radarmeasurements with a single observation from the core leads to the conclusion that the continuous image of the radar is statistically more reliable than the single values from cores. It is generally known that the reliability of a certain variable is better when the number of counts is bigger.

COST OF RADAR VERSUS COREDRILLINGS

One of the aspects of performing radar measurements are the costs. Per thickness count the cost by radar (about 1000 measurements per kilometre) will be lower than by normal coredrillings (3 to 5 per kilometre). Eventually the total cost may be higher, but by radar a lot of extra information along the length of the pavement is gathered which will be very useful, e.g. in total reconstruction of pavements, cold recycling of existing pavementlayers, milling pavements at a greater depth. Especially in these cases exact and sufficient data about thickness, and the length to which a certain thickness extends, are very important, if not indispensable. Cost are of course depending on the details of investigation, the necessary security measures and the allowable time for investigations (during daytime or night-time, in one or more phases etceteras).
A global comparison for Dutch circumstances on highways has shown that one layer thickness by radar is about 400 times cheaper than one by core drilling. In other circumstances other comparisons should be made, but in many cases radar will be cheaper per thickness count.
Besides the cost per count, there is a lot more information about the variation of thickness which is continuously visible with radar and only at large intervals with cores.

CONCLUSIONS AFTER THE HIGHWAY AND RUNWAY PROJECT

The traditional way of working of the Dutch authorities in the highway project has been: drilling cores on the shoulder in order to obtain a layerthickness every 500 m. Already from these cores it appeared a large variation in thickness over the total road length could be expected. This was exactly the reason to perform radarmeasurements. At last cores on specific locations have been drilled, based on the radardata.
In future similar projects Infratech advises to follow the next order in this kind of projects: first of all performing radarmeasurements and determining quantitative homogeneous sections, so on each homogeneous section a core can be drilled for verification of material layers. After this layer thicknesses can be calculated. If necessary extra cores can be drilled on specific locations lateron.

It is very important to ensure exact locations of radar and cores in order to make an adequate calibration and accurate calculation of thicknesses. Radar and cores should be on the same spot, in the longitudinal as well as in the transversal cross section.

If any information about the history, constructiondata or quality of the pavement is already available, this should be studied before doing radar measurements because it might influence the way of working. When a heterogene construction is expected one has to search more intensively for variations. This means a much lower speed than 80 km/hour should be applied.
In the runway project it was possible to perform acceptance tests on thickness when the pavement was commissioned by the contractor. Because of the large number of data in each track, variations in thickness could easily be found. All thickness data is available in digital format which gives the opportunity to automate the processing, so reliable statistical calculations can be made. In this project thickness in a longitudinal track was

rather constant because the evenness of the old runway was rather good.
The recently developed thickness measurement equipment called EURADAR, the sophisticated software for analysis and presentation of results, and the quality of evaluated results have led to the availability of particularly effective pavement surveying equipment.

The last conclusion is that radar will never replace coredrillings but radar is an effective tool to determine homogeneous sections based on a lot of data. Radar enables the still required drillings to be limited en better located. The result will be a more reliable prediction of layerthickness and not just being a lucky guess.

PROPOSITIONS:

1. Nowadays it is possible to determine layer thicknesses of a pavement almost continuously by means of radar measurements at high speed. The accuracy of the calculated thicknesses can be increased by using cores on the same spots and can be considered as acceptable related to the large number of measuring points compared to cores in a certain interval.
It is always finding a compromise between drilling of some cores with a higher accuracy or performing radarmeasurements almost continuously with a somewhat lower accuracy.

2. It is also possible to use radar when commissioning overlaid or new pavements when the accuracy is within the boundaries of specifications, for example base and asphalt can be measured layer by layer. It depends on the specifications of the contract whether radar can prove if the thicknesses are correct or not. It is a matter of combining accuracy of thickness with the large number of counts.
An example how commissioning measurement is done in a runway project in The Netherlands is given in the next figure.

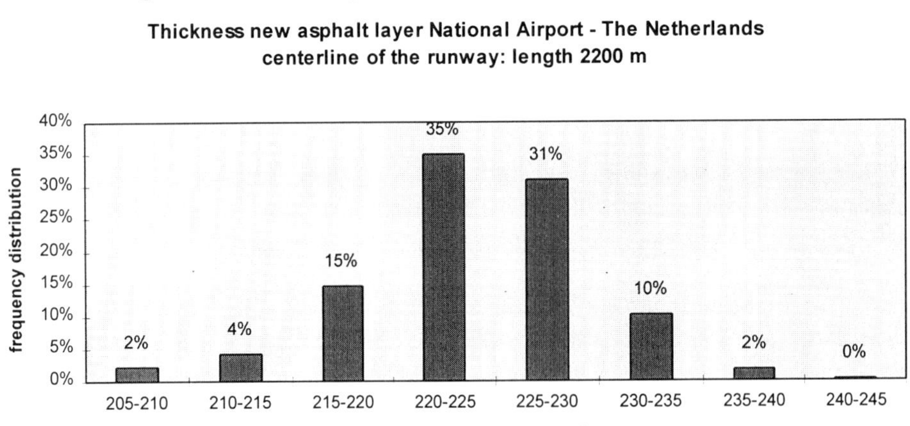

Figure: presentation results radar measurements on commissioning of an asphalt runway, in this project over a length of 2.2 km 3880 counts were collected.

3. Radar technique is an especially effective and suitable tool in collecting thicknessdata and locating homogeneous sections over a roadlength or detecting defects in a pavement. Radar will never replace coredrillings but can, as a pre-investigation, better point out the locations which are interesting to drill and as such limit the number of costly and time consuming coredrillings.
That's why Infratech recommends: radar first, specific drillings next.

4. Analysis of radar scans, if not automatically, should be done by experienced and thoroughly trained technicians. Qualifying the bonding between layers is a more difficult problem which will require many investigations, along with other methods, in order to validate interpretations which are always complex. Unfortunately radarsignals on old, sometimes historically build pavements can not always be clearly analysed with the state of the art radar analysis. When there is a mess in the construction, there is a mess on the radar!

Fifth International Conference on the
BEARING CAPASITY OF
ROADS AND AIRFIELDS

DYNAMIC INVESTIGATIONS IN ASSESSING THE STRUCTURAL CONDITION OF PAVEMENTS

J-M. SIMONIN[1], P. MAISONNEUVE[2]

[1] Research Engineer, LCPC, 44340 Bouguenais, France, ☎(33) 2 40 84 58 29; Fax: (33) 2 40 84 59 92; E-mail: Jean-Michel.Simonin@lcpc.fr

[2] Technician Expert, LCPC, 44340 Bouguenais, France, ☎(33) 2 40 84 58 45; Fax: (33) 2 40 84 59 92; E-mail: Patrick.Maisonneuve@lcpc.fr

ABSTRACT

The behavior of pavements, and consequently their bearing capacity, is obviously influenced by their structural distresses, in the form of cracks or loss of cohesion. Deflection measurement represents a basic approach to determining the structural condition of the road base layer, especially with flexible pavements. However, on semi-rigid pavements, as well as on thick bitumen pavements, deflection tends to provide delayed information about the ongoing processes of structural deterioration. Such is true in the case of a lack of bond between the wearing course and the road base. Furthermore, only severe cracks accompanied by a loss of cohesion in the road base actually induces significant changes in deflection.

In contrast, these types of deterioration induce rapid modifications in the dynamic response of the pavement. For more than thirty years, the French National Public Works Laboratory has been investigating the application of dynamic tests in assessing the structural condition of pavements, especially with a treated road base layer. An original testing device, called "Colibri", was developed at the beginning of the 1990's. A mechanical impulse is applied to the surface of the pavement, at the test point, and its response is measured by a set of accelerometers. Frequency analysis of the signals makes it possible to assess various aspects of the pavement's structural condition at each test point. Since 1992, this device has been extensively used in many maintenance studies for a variety of road projects. The results, from a number of experiments, are presented in this paper and revealed the following:

- The lack of bond between the upper layers of a pavement results in a significant change in the dynamic response of the pavement above 1 kHz,
- A loss of cohesion in the upper part of the base layer induces even more obvious changes, within the 300 to 500-Hz frequency band, and
- "Colibri" provides valuable information for monitoring the evolution of a wearing course or an overlay.

The paper's conclusion focuses on the ability of this device to track and diagnose the structural condition of more recently treated road bases.

INTRODUCTION

For over 30 years, France's National Public Works Laboratory ("LCPC") has been working on the use of mechanical waves in performing diagnostic assessments of roadway pavements. A dynamic investigation consists of applying an impulse to the surface of the pavement and then measuring the response at various distances. Special attention is given to the propagation of Rayleigh waves, without necessarily ignoring the other waves.

The initial systems provided an impulse to the pavement on an adjustable single-frequency band. In order to span an adequate frequency band, it was necessary to repeat the test a number of times, which extended the experimentation time to several hours.

Towards the beginning of the 1990's, various studies demonstrated that it was indeed possible to apply a shock-type stress [1]. Hence, during a single test, a wide spectrum of frequencies could be spanned, thereby reducing the time of experimentation to approximately one minute. In 1994, a prototype, called "COLIBRI", was developed based on this principle [2]. At roughly the same time, an effort was undertaken to establish a better codification of the dynamic investigation test. Following various validation phases, the "COLIBRI" prototype was successfully used in 1996 and 1997. It has served to:

- Identify bonding defects at the interface between the wearing course and the underlying layer;
- Characterize the cohesion or the fragmentation of the base layer; and
- Evaluate and monitor the evolution in the wearing course's mechanical characteristics.

PRINCIPLE OF DYNAMIC INVESTIGATION

A dynamic impulse, x(t), is induced in the body of a roadway pavement from different types of waves. Dynamic investigation consists of measuring the propagation of these waves at various points on the pavement's surface (pavement response y(t)).

In considering the pavement as a mechanical system, the Fourier transform of x(t) and y(t), can be defined as $X(\omega)$ and $Y(\omega)$, along with their complex conjugate functions $X^*(\omega)$ and $Y^*(\omega)$.

The spectral analysis serves to define the following set of spectral functions:

$$G_{xy} = X(\omega)\,Y^*(\omega)$$
$$G_{xx} = X(\omega)\,X^*(\omega)$$
$$G_{yy} = Y(\omega)\,Y^*(\omega)$$

This analysis also serves to define both the transfer function $T(\omega)$ and the coherence function $\gamma(\omega)$.

$$T(\omega) = \frac{Y(\omega)}{X(\omega)} = \frac{G_{xy}}{G_{xx}} \qquad \gamma(\omega) = \sqrt{\frac{|G_{xy}|^2}{G_{xx}\,G_{yy}}}$$

The coherence function lies between 0 and 1. It enables verifying that the response recorded does indeed result from the loading x(t). The transfer function T(ω) is representative of the mechanical system and thus of the pavement's structure.

PRINCIPLE BEHIND THE "COLIBRI" MEASUREMENT SYSTEM

The "COLIBRI" device consists of a trailer assembly that includes:

- a measurement beam with three accelerometers and an impact separator;
- two screws jack enabling the lowering and raising of the beam as well as the placement and removal of the sensors required carrying out the measurement;
- An electronic assembly that makes it possible to guide the various devices composing the trailer system as well as to process the signals;
- An electric group for supplying the trailer assembly with electricity; and
- The trailer's signaling system.

This trailer is pulled by a vehicle, which contains, in particular, an acquisition chain and a computer for both coordinating the various measurement phases and storing the measurement results.

The principle of the test performed by "COLIBRI" is depicted in Figure 1 below:

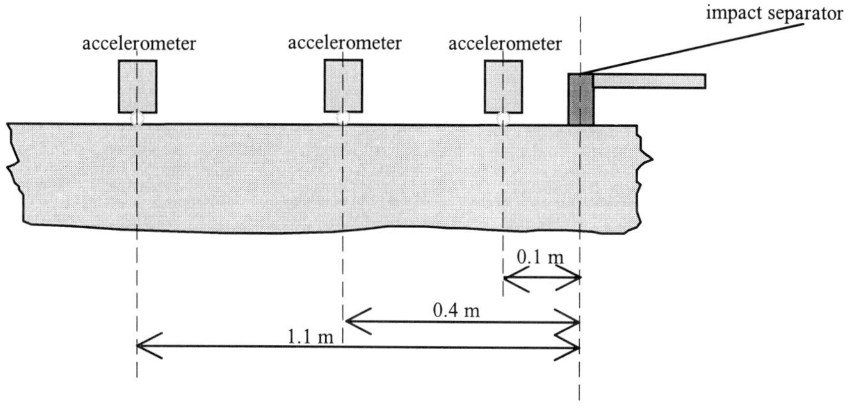

Figure 1: Principle of the impulse method

Three accelerometers are positioned vertically on the pavement and held in place by means of a spring-mass system. A wide-band dynamic impulse is then applied to the pavement by an impact separator fitted with a force sensor. The pavement's response curves are measured by the accelerometers. The four signals (impulse and responses) are digitized and recorded by a computer system. Based on the recorded signals at each measurement point, one of the three dynamic investigation methods may be applied. The test lasts about one minute, which includes the lowering and raising of the beam.

READING THE DYNAMIC INVESTIGATION MEASUREMENTS

Three methods are available for reading the dynamic investigation measurements. The first two focus on the value of the modulus of the transfer function at a frequency of either 300 or 1,500 Hz all along a given route. They allow obtaining information on the cohesion or cracking of the road base (AD1 method) as well as on the state of the interface between the wearing course and the underlying layer (AD2 method). The third method is aimed at identifying the relationship existing between the waves' propagation speed and their wavelength for the entire frequency spectrum at a given measurement point. This method enables obtaining information on the mechanical characteristics of the wearing course.

1) AD1 and AD2 methods

a) Testing procedure

These two methods are applicable to zones where the pavement exhibits a homogeneous structure. Measurements are made at a succession of regularly spaced measurement points at an interval of no more than 10 meters. At each measurement point, an impulse is applied to the pavement, and the response at the pavement surface is measured at a distance of 10 cm. At each measurement point, the value of the transfer function between the response and the impulse at frequencies of both 300 Hz (AD1 method) and 1,500 Hz (AD2 method) is determined. The measurement is then validated if the level of coherence between signals proves to be sufficient.

b) Utilization of measurement readings

The investigated **zone** is to be divided into **sections**, which contain a set number of consecutive measurements (normally 20).

Over the investigated **zone**, the following are to be computed:

- The 20% threshold, $\mathbf{Z}_{20}^{300}$ of the set of valid measurements taken at 300 Hz; and
- The 20% threshold, $\mathbf{Z}_{20}^{1500}$ of the set of valid measurements taken at 1,500 Hz.

On each predetermined **section**, the following are to be computed:

- The standard deviation, σ_{300}, of the set of valid measurements taken at 300 Hz; and
- The threshold at 80%, $\mathbf{S}_{80}^{1500}$ of the set of valid measurements taken at 1,500 Hz.

For the investigation carried out at 300 Hz, the indicator $I_{300} = \frac{1}{2} \times \frac{\sigma_{300}}{Z_{20}^{300}}$ is computed for each section of 20 measurements. For the investigation at 1,500 Hz, the indicator $I_{1500} = \frac{1}{2} \times \frac{S_{80}^{1500}}{Z_{20}^{1500}}$ is computed for each section of 20 measurements.

Note: The expression "threshold at X%", $\mathbf{E}_X$, refers to the **entire** set of measurements, whereby the value $\mathbf{E}_X$ is such that X% of the results from the measurements under consideration turn out to be less than this value $\mathbf{E}_X$.

c) Characterization of the cohesion or the fragmentation of the road base layer

For a pavement with a fragmented base layer, the waves propagation in the pavement varies enormously with the position of the measurement point. Therefore, both high and low measurement values are in fact obtained. The route diagram winds up being rather chaotic. In order to confirm this observation, it would be necessary to compute the threshold at 20%, $\mathbf{Z}_{20}^{300}$, of the entire set of measurements over the investigation **zone**, along with the standard deviation s_{300} of each predetermined **section's** specific set of measurements. The indicator $I_{300} = \frac{1}{2} \times \frac{\sigma_{300}}{Z_{20}^{300}}$ can now be computed. The histogram of this indicator's variation then is superimposed on the route diagram (Figure 2). The values obtained are representative of the pavement structure condition. The interpretation of this indicator's value can be read on the following table:

Value of the indicator I_{300}	Interpretation
Less than 0.8	Road base in a good state of repair
Between 0.8 and 1.2	Uncertain road base quality, requires monitoring
Greater than 1.2	Road base either deteriorated or cracked

Figure 2 displays the example of a pavement whose road base layer is in good condition. The measurements recorded reflect a homogeneous level. The indicator at 300 Hz, which remains less than 0.8, would indicate a satisfactory cohesion for the road base.

ABSTRACT

The structural performance and condition of road pavements has traditionally been measured by external observation or mechanical testing. This paper describes a feasibility study, carried out for the UK Highways Agency, to investigate whether technology has advanced to the point where reliable sensors are available at moderate cost to enable the performance and condition of pavements to be monitored (remotely) without the need for attendance or traffic management.

It was recognised that price was a key element if the concept was to be widely adopted. The evaluation of a range of existing instruments, new applications and new concepts is described, in particular one based on a single layer deflectometer, and the other a simple method of measuring stress in the pavement foundation. Neither requires cables in the pavement, and the deflectometer can be remotely monitored from a moving vehicle.

The existing concept of a strain gauged core, for installing into an existing pavement, was further developed into an instrumented core, incorporating signal conditioning, micro-controller, wireless transmitter and battery components. The study clearly demonstrated the benefits of wireless, remotely monitored retrofitted instruments to the concept of 'Smart' pavements. The integrated 'Smart' sensor, which removes the relatively high expense of manually reading/downloading data or performing labour intensive surveys and investigations, is considered to be the preferred method for pavement monitoring in the future.

Key words: Instrumentation, Smart sensors, Wireless telemetry, Monitoring, Trial pavements

INTRODUCTION

Highway pavement designs are traditionally based on a combination of mathematical modelling, materials characterisation and monitoring of pavement performance. Pavement trials come in many forms [1] and may include:-

- full scale test tracks loaded by trucks (eg. WESTRACK, MNROAD [2]);
- trial sections loaded by accelerated test equipment (eg. CAPTIF, HVS [3]);
- smaller scale laboratory test facilities.

Generally, within a trial pavement, it is possible to control many of the pavement, loading and environmental variables, and this makes data interpretation easier. However construction of pavement trials, and the associated monitoring of performance under these controlled conditions, are relatively expensive.

Another source of performance data is from routine monitoring of existing pavements on a highway network (eg. Deflectograph, Falling Weight Deflectometer). However, due to the large lengths of pavement to be covered, it is unusual for any particular section to be tested at frequencies greater than 3-5 years. High speed survey equipment (eg. HRM, ARAN, Laser RST) may cover a network every 1 - 2 years, but they only collect parameters measurable on the surface, giving little indication of the changes in load spreading ability occurring within the pavement and subgrade, especially throughout the year as temperature and moisture conditions vary.

In order to overcome the drawbacks of expense (for specially constructed trials) and infrequent data (for routine testing on existing networks) a third scenario was considered; the use of low cost, yet reliable, instrumentation fitted into existing pavements.

The following sections describe a feasibility study that was carried out by Scott Wilson Pavement Engineering, in conjunction with the Department of Civil Engineering and Department of Electrical and Electronic Engineering at the University of Nottingham. The research, funded by the UK Highways Agency (HA), was to determine whether sensor technology had developed to the stage where instruments were sufficiently accurate, reliable, robust and cost effective for general use on a highway network.

The study incorporated:-

- a literature search and market review;
- laboratory testing and calibration of selected instruments;
- development of new sensor prototypes;
- trials in a half scale Pavement Test Facility.

The research reviewed the instruments currently available, and their ease of installation and use, predominantly for flexible pavements. A number of new concepts were developed to fill gaps in the existing instrumentation. The importance of wireless telemetry for downloading of data was demonstrated and the research team considered how the new ideas could be implemented in practice. With the possibility of increased reliability in diagnosis of pavement behaviour, it is anticipated that considerable savings could result through better design and more effective maintenance.

REVIEW OF EXISTING TECHNOLOGY

In an ideal world a pavement incorporating 'Smart' sensors is able to inform on its structural integrity. In order to achieve this objective, there are a number of significant issues that need to be considered, including:-

- what are the measurement criteria?
- should static or dynamic readings be taken?
- are single or multiple instruments required per site?

- what is the optimum instrument spacing?
- what is the precision of measurements?
- should "continuous" logging or only "threshold" values be stored?

From discussions held between the collaborating organisations, experts around the world, and the HA, it was agreed that the measurement criteria were ideally:-

- horizontal strain (at various locations within bound bituminous materials);
- stress/pressure (at top of subgrade);
- permanent deformation (within bituminous materials);
- temperature (within bituminous materials);
- moisture/ground water level (in unbound materials and subgrade);
- loading magnitude/position.

It was apparent from the literature review that instruments currently designed for pavement applications were intended to be installed during construction of the pavement, and were generally not readily adaptable for insertion into an existing pavement [4]. Furthermore, few of these could be regarded as low cost.

The whole concept of retrofitting 'Smart' sensors into existing pavements relies on cost effectiveness. Existing pavements are much more variable than specially built trials, and so a number of sensor installations would be required to obtain a representative view of performance. Furthermore, it was accepted that, over a period of years, some of the installed sensors would be damaged, removed (during maintenance) or would otherwise cease to function. Loss of sensors would not be significant if they were comparatively cheap to replace.

Nevertheless the cost of detailed instrumentation at many hundreds of installations around a network would be prohibitive even with low cost sensors, so it was suggested that a two level approach was adopted.

Level I: A limited number of sites, probably adjacent to existing Weigh-in-Motion (WIM) installations, where most of the above criteria would be measured to a reasonable level of precision. Sensor duplication would be considered along with dynamic measurement of strain/stress/deflection and "continuous" logging. Other criteria like pavement temperature and ground water levels would be logged at appropriate intervals.

Level II: A much larger number of sites, perhaps with monitoring aimed at identifying either deteriorating trends in the readings from an instrument, or the exceeding of a threshold value that could be used to simply classify the pavement condition, for example as poor, satisfactory or good.

Three different types of strip gauge for strain measurement were identified for the laboratory tests and trials; manufactured by Dynatest, Techni Measure and Measurements Group. Only the first was specifically designed for use in bituminous materials; the other two being developed for use in or on concrete. The Techni Measure gauge cost less than $20 compared to around $600 for some of the other makes. In addition, the use of conventional foil strain gauges, glued to a cylinder of bituminous material that could be installed into a core hole [5] was identified as a suitable procedure for further investigation.

Two pressure cells were tested, designed for measurement of subgrade vertical stress (by measuring strain changes on a diaphragm on which the pressure acts); the Kulite and the Nottingham (developed in previous research at the University [6]). Thermocouples and thermistors are available for measuring temperature (generally very reliable and relatively cheap), and both were investigated in this study.

Instrumentation for measuring permanent deformation of pavement layers did exist, in the form of the Multi-Depth Deflectometer (MDD, [7]), but, costing in excess of $4000, it would not be appropriate for general use in 'Smart' pavements. Similarly, Time Domain Reflectometry (TDR, [2]) is commonly used to measure moisture in trial pavements. Although the TDR probes are not expensive, the reading equipment is, and requires calibration for a particular soil type. Measurements are relatively labour intensive, which has cost implications.

The aspect of measuring vehicle loading was not specifically addressed in the study, since WIM technology is being developed by several different companies. However, it is envisaged that 'Smart' sensors will often be installed alongside WIM sites, so that pavement response can be related to vehicle loading.

It is interesting to note that, although measuring different parameters, existing instruments have much in common, since they all rely on cabling, an electrical circuit, and the measurement of current, voltage or resistance (which are of course interrelated).

NEW SENSOR CONCEPTS

In order to fulfil the objectives of the study, the research team developed a number of new devices (and also some new applications of existing concepts) for pavement instrumentation, which can be installed into an existing pavement (ie. retro-fitted), namely:-

- Retrofit Stress Gauge;
- Retrofit Asphalt Deflectometer;
- Piezoelectric Subgrade Sensor.

Retrofit Stress Gauge, RSG

The RSG, Figure 1, is designed to measure subgrade stress, once installed in a horizontal hole drilled from the side of the pavement into the subgrade. It consists of a flat brass recessed disc and a ring, trapping a circular rubber membrane above the recess. A long brass tube is fitted through the side of the disc to allow fluid to pass into the cavity. For the prototype, de-aired water was used and a bleed screw was fitted to ensure that the cavity was completely filled with water. An adaptor on the end of the tube provided a facility for the application of a back pressure, and also measurement of pressure with a proprietary transducer. This pressure is generated by stresses developed in the subgrade when loaded.

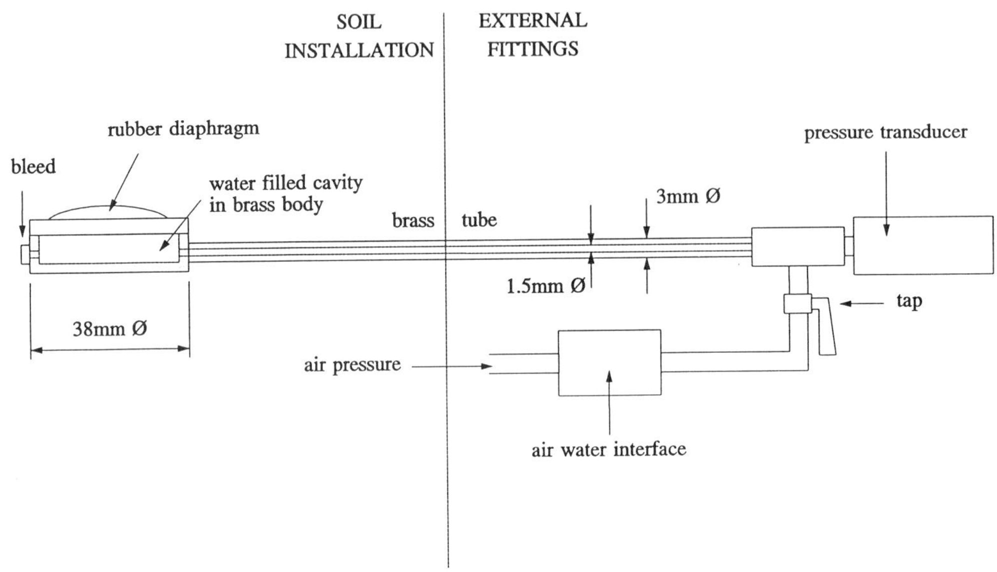

Figure 1 : Retrofit Stress Gauge Concept

The RSG is suitable for installation on embankments or via a prepared pit at the roadside, but requires special care to ensure that the diaphragm is in contact with the surrounding soil. This equipment requires a source of power for the pressure transducer and the prototype must be read manually, so it is only suitable for the better instrumented Level I sites.

Retrofit Asphalt Deflectometer, RAD

The RAD is shown in Figure 2 and consists of a wire coil, a sliding collar and a microswitch. The circuit is closed by the microswitch after a fixed surface displacement has occurred relative to the sliding ring, which is a fixed datum relative to the bottom of the layer being monitored. The closed circuit is tuned so that it can be detected by a metal detector on a passing vehicle. The design of the contact arrangement permits the displacement to continue beyond the pre-set limit.

The RAD can be easily installed into a small diameter hole cored from the surface. A typical installation consists of RAD's of varying length (eg. 50mm, 100mm, 150mm), set to varying levels of permanent displacement (eg. 5mm, 10mm, 15mm). An array of RAD's can be exploited for multi-depth use to give more detailed information about change in layer thickness (and hence the least rut resistant material) than is possible from surface profile measurement. Costing around $15 each, and requiring no external power source, the RAD offers a cheap alternative to the MDD, which is particularly suitable for Level II sites.

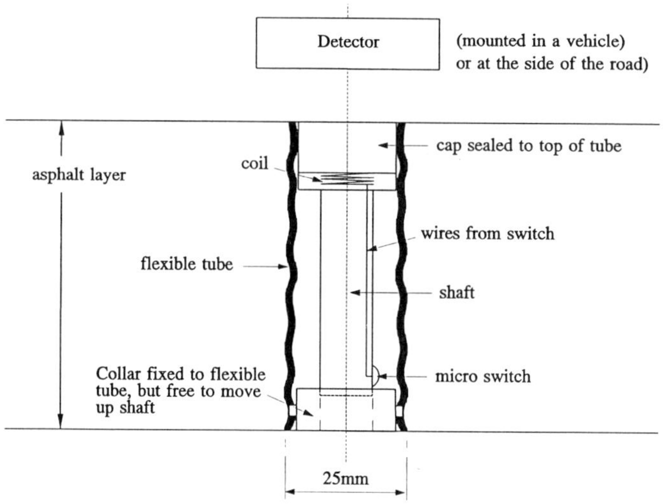

Figure 2 : Retrofit Asphalt Deflectometer

Piezoelectric Subgrade Sensor, PSS

The PSS uses a piezoelectric cable to measure stress. This is similar in approach to the measurement of vehicle loading using piezo sensors inserted into the pavement surface. When the piezoelectric material undergoes a dynamic change in shape, it generates a charge. By having the outer braided sheath and inner core connected directly to the piezo material, this charge can be monitored and converted to a voltage in a charge amplifier, which, with selection of a time constant appropriate to the duration of the loading, can be calibrated to give a stress measurement.

In order to ensure encapsulation of the cable by the subgrade soil, the PSS is ideally installed during construction/reconstruction, although retrofitting is possible. The equipment does not require a source of power, but does require monitoring equipment. The cable can be substituted by regular (and cheap) co-axial cable with appropriate amplification, so could be suitable for Level II sites.

To simplify data handling and analysis, the PSS does not have to discriminate between vehicle types, but can simply be calibrated to give a reading above a critical subgrade stress for the type of material in which it is installed. This threshold value would depend on the pavement structure and number of load applications for which the pavement was designed.

PILOT SCALE TESTING

Instruments selected from the review, together with those developed by the research team, were calibrated in the laboratory and then installed in a half scale Pavement Test Facility (PTF) at the University of Nottingham. The PTF permits moving wheel loads to be applied to pavement sections 5m long by 2.4m wide, in a test pit 1.4m deep. The pavement consisted of a 100mm thick Hot Rolled Asphalt (HRA) bituminous surfacing, a 200mm thick granular Type 1 sub-base layer and a 1.1 metre thick Keuper Marl subgrade layer, the latter with a CBR value of 5%.

The selected instruments, together with some of their characteristics, are summarised in Table 1.

Instrument	Concept	Measurement Parameter	When Installed
Nottingham Pressure Cell x 3 no.	Existing	Subgrade vertical stress	Prefixed
Kulite Pressure Cell x 1 no.	Existing	Subgrade vertical stress	Prefixed
Retrofit Stress Gauge	New concept	Subgrade vertical stress	Retrofit
Piezoelectric Subgrade Sensor	New application	Subgrade all round stress	Retrofit
Dynatest Strip Gauge	Existing	Asphalt tensile strain	Prefixed
Techni Measure Strip Gauge	New application	Asphalt tensile strain	Prefixed
Measurements Group Strip Gauge	New application	Asphalt tensile strain	Prefixed
Strain Gauged Cores x 3 no.	Existing	Asphalt strain	Retrofit
Retrofit Asphalt Deflectometer	New concept	Asphalt deformation	Retrofit
Coaxial cable	New application	Asphalt strain/cracking	Prefixed
Thermocouples x 3 no.	Existing	Temperature	Prefixed/Retrofit

Table 1 : Instrumentation for Pilot Scale Testing

Note : Most instruments also suitable for installation beneath or within rigid and composite pavements.

The Keuper Marl subgrade is essentially a permanent installation in the PTF made from compacted wet unfired clay bricks, but if the test requires a change in bearing capacity an upper layer can be replaced with wetter or drier material. The pressure cells were installed just below the surface of the subgrade. The sub-base grading complied with the requirements of the Department of Transport Specification for Highway Works [8]. The embedment strain gauges and coaxial cable were installed on the surface of this layer.

The HRA wearing course material was a 30% stone/14mm nominal size material complying with Column 3/2 of BS594 [9]. The RAD and the thermocouples were retrofitted to the pavement by drilling through this layer. Finally, the RSG and the piezoelectric cable were retrofitted into the subgrade layer via an access pit at the side of the pavement. A section through the instrumented pavement is shown in Figure 3.

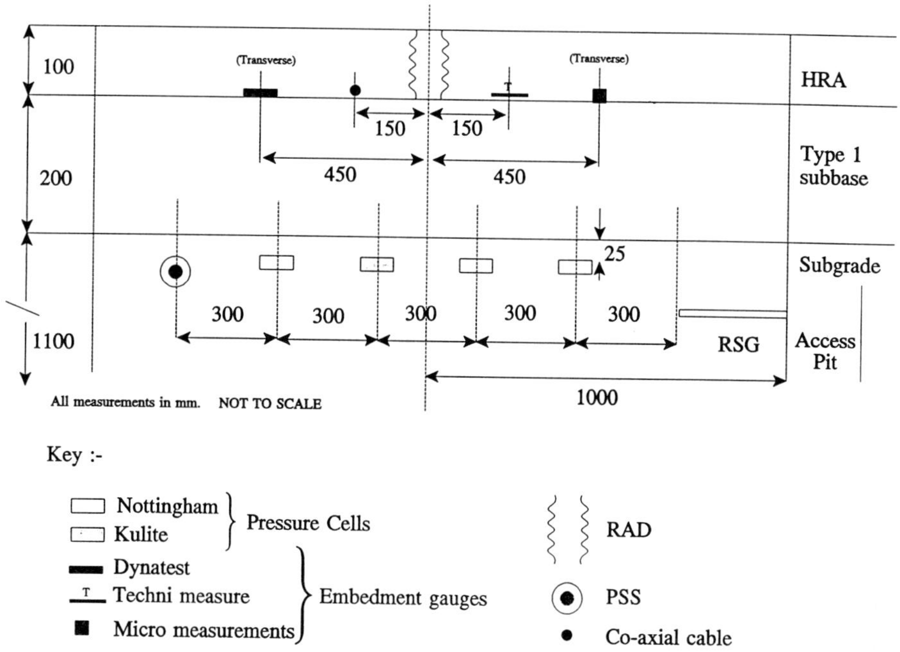

Figure 3 : Section Across Instrumented PTF

Trafficking was performed for 50,000 passes under a 9kN loaded wheel, and data was recorded and analysed using a purpose designed data logger with a dedicated laptop computer. After completion of trafficking, the Strain Gauged cores were installed (these had not been ready prior to trafficking) and a Falling Weight Deflectometer (FWD) was used to obtain surface deflections under a measured load. Pavement layer stiffnesses were then calculated using a back-analysis procedure, the stresses and strains computed from the model obtained, and compared with the values measured by the in situ sensors. The instruments were then rated according to cost, ease of installation, accuracy and durability.

The combination of calibration data, instrument duplication and pavement analysis from FWD data gave some confidence that readings close to the true field stress and strain were being measured. The best performers were the RAD and RSG, with high ratings also given to conventional thermocouples (temperature), the PSS (stress), the Techni Measure gauge designed for concrete (strain) and the Nottingham Pressure Cell (stress).

Subsequent to the pilot scale testing, the research team reviewed the available methods for measurement of moisture in unbound materials. Three different instruments were assessed, all based on the TDR principle. The cheapest to purchase, and easiest to use, was found to be the Theta Probe. In common with all TDR's it requires calibration, a power source and data logging facilities, but was considered a suitable candidate for Level I sites, especially in combination with the RSG and/or the PSS. (All of which could be inserted from a pit excavated at the side of the road in the case of an existing pavement).

Technological developments take place rapidly, and a continuous review was performed by the team to take advantage of relevant advances. One area that was positively identified as potentially benefitting the introduction of 'Smart' sensors relates to data collection. Conventional pavement instrumentation technology requires cabling and a physical interface with a data logger and computer. Experiences in this project, and related work carried out on 'live' sites, demonstrated how the installation of cabling is incompatible with the construction process. The RAD and RSG were developed with this in mind; neither requires cables in the pavement and the RAD can even be remotely monitored from a vehicle. The further benefits of wireless, remotely monitored, retrofitted instruments to the concept of 'Smart' sensors is clear when the costs of manually reading instruments on site are taken into account. The research team therefore turned their attention to this aspect.

INSTRUMENTED CORES

In order to overcome the problem of installing cables in the surface of a pavement, the research team developed the idea of using a strain gauged cylindrical specimen (installed in a core hole), to become an "instrumented core", incorporating wireless telemetry for downloading of data.

Some of the considerations addressed were as follows:-

- Size - miniaturisation is not necessary; a 90mm diameter cylinder 200mm long could be installed in a pavement element (instrumented core) giving space for batteries;

- Processor - a lower cost programmable processor may be more suitable than a hard coded integrated circuit, since it would give more flexibility, for different instrument combinations;

- Power - low power consumption, long life is required (for both sensors and telemetry);

- Sampling Rate - continuous sampling (1000 measurements/second) is technically possible, with post processing off site; however this uses power, reducing life;

- Storage - alternatively the sensor could be designed to switch on and take readings at infrequent intervals only, with data stored for later downloading;

- Cost - long life may not be absolutely necessary if sensor cost (including installation) can be made sufficiently cheap. A certain amount of redundancy should be expected in any case;

- Wireless Telemetry - transmission range <10m may be adequate if a roadside receiver/data logger is installed. Effectiveness of transmission through pavement materials required assessment.

Description

After consideration of these aspects, a prototype "instrumented core" (Figure 4) was developed comprising four elements:-

- Bituminous material;
- Sensors;
- A-D Converter/Microcontroller/Wireless Transmitter;
- Batteries.

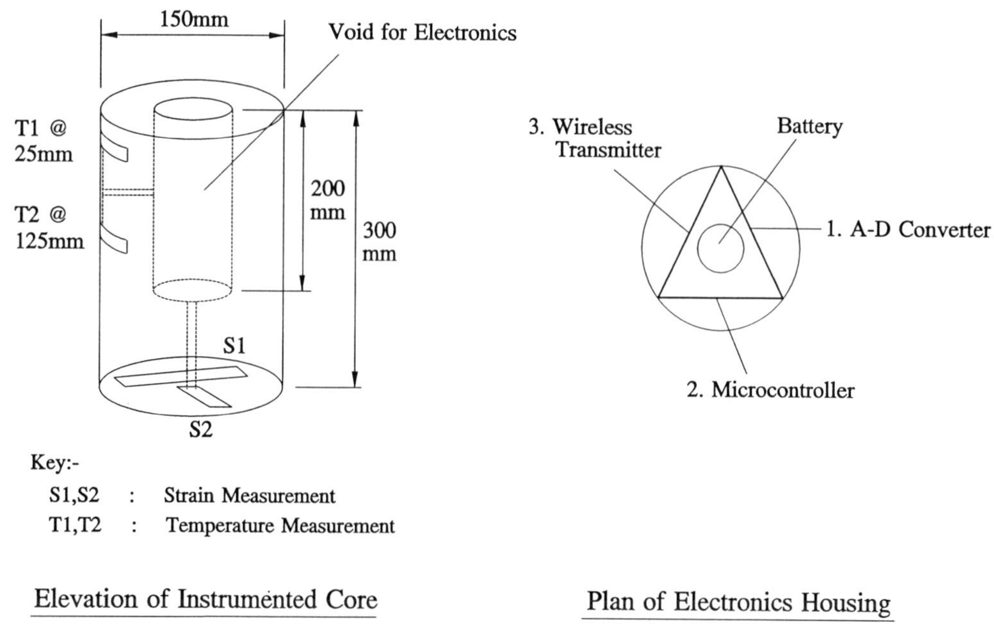

Figure 4 : Instrumented Core

The bituminous material is used to house the other components, and is the pavement element on which measurements will be taken once installed. It is a cylindrical specimen manufactured in the laboratory, typically to a 150mm external diameter. Varying depths of specimen can be made to suit the likely pavement construction at a particular installation site. (Ideally, cores would be taken from site, instrumented in the laboratory, then reinstated. However, the coring process cuts away approx. 5mm of material, and previous experience has shown that such a reinstated core, glued into place, does not act in a homogeneous fashion, so does not produce the correct data).

The sensors shown in Figure 4, comprise two linear thermistors; one at 25mm depth from the surface (to measure diurnal changes) and the other at a depth of 125mm (where changes will mostly occur over several months). These are wired to produce voltages in the range 1-5V, over the range of -30 to +50°C.

profilometer which strictly complies with the accuracy and reproducibility specified in the Method, i.e. that has been approved for it. The reference test is specified in a conventional but extremely precise manner: it is performed with a high definition laser transverse profilometer which has been calibrated on a test bench immediately prior to each use, it is implemented in line with an extremely rigorous and demanding procedure, and finally analysis involves use of a particular so-called "reference" algorithm. The same factors are also present in the measurement of surface deformability (4) for example.

Table 1: LPC Test Methods, as at 31 January 1996

Test	Stage
Surface distress survey (including a catalogue of surface defects)	Approved
Measurement of surface deformability (deflection, curvature)	Approved
Structural profile (1 - core sample method)	Approved
Structural profile (2 - radar testing method)	Approved
Measurement of transverse evenness	
Measurement of longitudinal evenness	Draft
Measurement of geometrical characteristics	Draft
Measurement of texture	Draft
Measurement of friction	Draft
Measurement of in situ performance (Ovalization))	Approved

An advantage of this method is that it guarantees the quality of data collection while at the same time opening it up to different devices and even to different service providers. In addition, because they do not impose particular devices the LCP Methods are not at the mercy of the industrial policy of a particular equipment manufacturer.

The approval of devices The approval of a type of device in relation to a Test Method provides a means of ensuring that its design and construction meet, in normal operation, the performance requirements of the Method. The data collection devices in use in France have generally been approved at the same time as the Methods which implement them. We shall now briefly illustrate the approval process by referring to two examples - the approval of deflectographs and transverse profilometers.

The various deflectographs which are used in France (types "02 - short frame", "03 - long frame" and "04 - elongated beam") were approved at the same time as the surface deformability method, in Autumn 1994 (9, 10, 11). This operation involved all the units which are in use on French road networks - and not only one representative of each type- so it was possible not only to approve these devices but also to identify, on some units, the malfunctions which might interfere with their approval. These observations and the analysis of their causes have been used to specify the checking procedure which is to be applied periodically to all devices. This procedure makes it possible to be certain that each unit is capable of correctly carrying out the Test Method.

In practical terms, various test sections were selected which were roughly representative of the different pavements on which surface deformability measurements need to be made. These sections were approximately 200 m in length. Each deflectograph made ten runs over each section, so it was possible to assess its repeatability and, by comparing the results of all the devices, evaluate the reproducibility of results. A sensor was attached to a measurement point on each section to allow reference tests to be performed there. The accuracy of each device was thus evaluated. Figures 1 and 2 show some of the results obtained during this process - i.e. the accuracy and repeatability of each "03 long frame" deflectograph. By comparing these results with the specifications of the Method (which are shown on the Figures) it can be seen that each of the devices has been approved for untreated pavements or pavements with thick bituminous bases (sections 4, 5, 6, 7 and 8) but not for pavements with semi-rigid bases. Table 2 shows the general scope of application for each type of device.

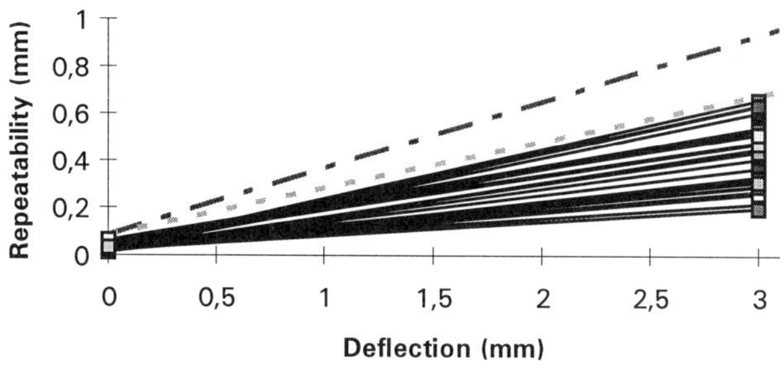

Figure 1 : Repeatability of "03 - long frame " deflectographs - from (9)

Several units failed this approval process, although their type passed. When these units were tested it was observed either that the measurement beam was distorted or that there were electrical connection problems between the beam and the data collection system in the deflectograph control cab. Once these defects were corrected the devices were approved after retesting. The problems which were discovered are now among the points on these devices which are periodically checked .

In the same way the approval process for transverse profilometers for the LPC Method for transverse profile measurement was carried out at the same time that the Method was approved, in April 1995. Here too, all the transverse profilometers in use of French roads underwent the approval procedure in order to identify the aspects which require periodic checking in order to make sure that each unit is operating correctly. In this case, as all the units were recent none of them failed the approval procedure.

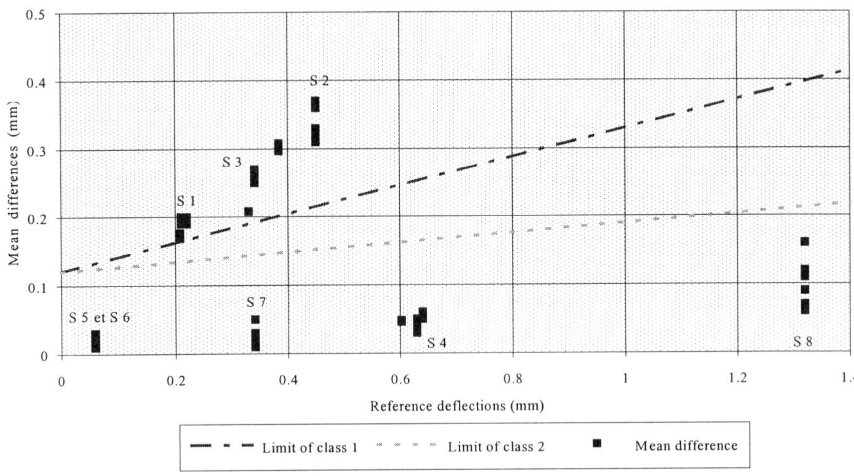

Figure 2 : Precision of "03 - long frame " deflectographs - from (9)

Type of deflectograph	Approved for
02 - Short frame	Untreated pavements
03 - Long frame	Untreated pavements and thick bituminous pavements
04 - Elongated beam	All pavements

Table 2 : Result of the approval process for Deflectographs

The authorization of staff The opinions of operators play an important role in some Methods, including visual surface distress surveys. When possible it is preferable to replace these types of surveys with more objective measurements (1). However, when no operational techniques are available for this purpose the training of test operators should be extremely thorough and backed up by authorization (12, 13, 14).

Visual distress surveys are currently an important part of the activities of the Laboratoires des Ponts et Chaussées, involving more than 25 teams. Particular attention has therefore been given to ensuring that, in the absence of total objectivity in the surveys, there is a high level of uniformity between the teams. An LPC Distress Survey Method was drafted which is now linked to a catalogue which codifies the different types of distress and their degrees of severity. Above all it was decided to train 40 employees to a high level in this area, and these workers now have an authorization to perform this type of survey.

Between 1992 and 1995 all the employess carrying out visual damage surveys were brought together for two day training sessions three times a year. During these sessions each employee performed a distress survey on several test sections. The results were compared, and the divergences and differences of opinion were discussed by all participants. A large number of ambiguities were identified in this way.

Since 1995 employees have been subject to an annual authorization procedure. They must each attend a session at the beginning of every year, during which their level of skill is assessed. They then survey a reference section. The results of this survey are compared to an average survey. The employee receives an authorization certificate on condition that he has attended at least three training or authorization sessions during the last three years and surveyed at least 100 km of road in the last year (in an accompanying role in the case of new candidates).

The competence of workers who perform a test is always an important factor in the quality of the results, even if the human factor has less influence on the result than is the case with visual surface distress surveys, for example. It is therefore preferable for the workers who are competent to conduct a test to be identified by name, if only because they are familiar with the Test Method and have a certain amount of experience in its use. This is the current aim of the LPCs. Authorization of a new worker involves both training him in the Test Method and his acquisition of practical experience through accompaniment.

QUALITY FROM NEW DEVICES

Sometimes the procedures outlined above are not sufficient to ensure that pavement condition tests produce results of an adequate quality. For example, in the case of measurement of surface deformability, only "04 - elongated beam" deflectometers are approved for all types of pavement. However, their systematic use is impossible, firstly because their size makes them unusable on most secondary roads, and secondly because they are difficult to use even on major roads. Under these circumstances, the Method specifies that combinations of different types of deflectographs should be used according to the specific conditions. In practice this is extremely difficult, and therefore expensive, to implement and can therefore lead to the use of non-approved devices.

In such situations, the way of achieving high quality is to develop new devices. The relevant Test Method is an integral part of the specification for the necessary development work. The following section illustrates the contribution made by the development of a new Deflectograph, Flash to the quality of pavement condition tests.

The Flash deflectograph was developed in order to obtain a deflection measurement device which is approved for all types of pavement but which is much less cumbersome than the "04 - elongated beam" deflectograph. This device was intended to replace all the types of deflectograph currently used by the LPCs. The objective was to be able to make high quality surface deformability measurements under "reasonable" conditions. Furthermore, this device was expected to increase the speed of measurement by a considerable margin in order to reduce costs.

The Flash deflectograph consists of a two axle truck with a wheel base of approximately 5 m whose rear axle can bear a load of 13 tonnes (15). As with a conventional deflectograph, the deflection is measured at each pair of rear axles by two articulated sensor arms on a reference beam by means of two circular sensors. The principal innovative features which permit the device to achieve its aims are as follows:

- the new geometry of the beam which allows accurate measurement on all types of pavements with a highly manoeuvrable short frame type truck,
- the technology used to construct the beam. Carbon fibre is used, which results in a considerable reduction in mass and considerably speeds up the return motion between each measurement. As a result the overall duration of the measurement sequence can be reduced by a factor of three.

When developed, the deflectograh was subjected to a specific approval procedure for the LPC Surface Deformability Measurement Method and was accepted for all types of pavement. Furthermore, the measurement speed of the device is three times faster than current deflectographs. The ultimate objective of the LPCs is now to replace their existing 25 long frame and short frame deflectographs by about ten Flashes.

CONTRIBUTIONS TO INTERNATIONAL ACTIVITIES

It is significant that the main international co-operation activities in the area of pavement condition surveys relate to the harmonization of the data collection procedures and comparisons between measurement devices. The activities of the LPCs, which we have described above (see § 2.), are therefore completely in line with the concerns of the international community. The participation of the LPCs in various working groups is a natural means of extending these activities and gaining recognition for them.

PIARC has already run, or is in the process of organizing, a number of experiments in order to compare techniques for pavement condition surveys. The Laboratoires des Ponts et Chaussées have, in particular, taken part in experiments which compare different devices for measuring the skidding resistance of road surfaces and the texture parameters which determine it. They intend to play an even greater part in comparative tests of transverse and longitudinal profilometers which are currently being organized.

In recent years the European Community has sponsored an increasing number of work programmes whose essential aims are to harmonize pavement condition survey procedures. The LPCs are participating, for example, in the programmes given below:

- COST 324, which has an intermediate objective of producing a European method for monitoring reference road sections and which has now been extended as part of the European PARIS project.
- COST 336, whose objective is to draw up a European guide for correct usage of the "Falling Weight Deflectometer" (FWD).

CONCLUSIONS

The quality of the data from pavement condition surveys is assessed on the basis of its repeatability, reproducibility and accuracy and has a major influence on the value of the analysis to which field data is subjected. The French network of Laboratoires des Ponts et

Chaussées has paid a great deal of attention to achieving this quality. The work has two main components.

About ten LPC Test Methods have been completed or worked on in the last three years. Five of these have now been approved and are therefore operational. The formalization of these Methods has made it possible to approve the devices which implement them. Approval procedures of this type have been organized for deflectographs and transverse profilometers. Concurrently, the staff employed to conduct the tests have been trained in the LPC methods and are now subjected to periodic authorization procedures, in particular in the case of those tests where the human factor still dominates (for example, visual pavement surface distress surveys). Several years from now, all the test methods, approval and testing procedures for devices, staff training and authorization will be incorporated into a Standard Quality Plan for pavement condition tests.

This formalization and organization work is still insufficient to ensure that surveys are of satisfactory quality in all cases. New development, for example for the Flash deflectograph, has resulted in improvements in the quality of measurements and the rate and cost-effectiveness of testing.

REFERENCES

1. LIVNEH M. et al. On the repeatability and reproductibility of manual pavement distress surveys methodes. Proceedings of the Third International Conference on Managing Pavements, San Antonio, 1994.

2. LEPERT PH. and JOUBERT P. Recent Developments in the Pavement Management System in France. Proceedings of the 4th International Conference on Managing Pavement, Durban, 1998.

3. LCPC. Relevé des dégradations de surface des chaussées - Méthode d'essai LPC n° 38. 1993.

4. LCPC. Exécution et interprétation de la mesure de déformabilité de surface - Méthode LPC ° 39. 1997.

5. LCPC. Ovalisation, Exécution et exploitation des mesures - Méthode LPC n° 41. 1995.

6. LCPC. Application d'un radar pulsé monostatique à l'auscultation des chaussées - Méthode LPC n° 42. 1996.

7. LCPC. Exécution et exploitation des carottages - Méthode LPC n° 43. 1997.

8. LCPC. Mesure et interprétation du profil en travers - Projet de Méthode LPC. 1998 .

9. SIMONIN J.M. and RIOUALL A. Qualification des Déflectographes. Bull. liaison des LPC, n° 208, 1997.

10. LEPERT PH., LEROLLE Y. and SIMONIN J.M. Evaluation conjointe CEBTP/LPC des performances du Curviamètres MT15. Bull. liaison des LPC, n° 208, 1997.

11. LEPERT PH. and KOBISCH R. Le FWD : conditions de mise en oeuvre et performances - exemples d'utilisation en France. Bull. liaison des LPC, n° 208, 1997.

12. LEPERT PH. and GOUX M.T. Evaluation du Réseau français de routes nationales basée sur le relevé de dégradations de surface, Proceedings of the 4ème Congrès International de la route, Rabat, 1994.

13. LEPERT PH. et al. An evaluation of the French national highway network based on surface damage surveys. Proceedings of the 3rd International Conference on Managing Pavements, San Antonio, USA, 1994.

14. LEPERT PH. Surface Damage to Evaluate 34 000 km of Highways. Symposium on Highway Maintenance, Nottingham, UK, 1993.

15. SIMONIN J.M. et al. Le Déflectographe Flash. Bull. liaison des LPC, n° 208, 1997.

Fifth International Conference on the
BEARING CAPASITY OF
ROADS AND AIRFIELDS

COMPARATIVE INVESTIGATION OF SEVERAL MEASURING DEVICES AND TEST METHODS TO DETERMINE BEARING CAPACITY

WILFRIED SCHELLENBERGER

Dipl.-Ing., Laboratory for quality control of the State agency for road construction of the state of Thüringen, Königstraße 22,
D- 98553 Schleusingen, Tel. und Fax: 036841/ 40052,
Federal Republic of Germany

ABSTRACT

The possibilities to measure the bearing capacity especially of asphaltconstructions have increased considerably in the last couple of years. But what has not been done yet are sufficient comparative investigations with relative simple and cheap test-methods and -devices in order to find correlations, which are exact enough to allow the use of these devices when measurements are necessary on short notice.

To determine this, measurements were taken at 11 selected test sections in asphalt roads with the following instruments:

- Benkelman-beam (simple and)
- Radius of curvature
- Falling- Weight- Deflectometer (FWD)

The thickness of the asphalt (asphalt concrete) varied between 4 and 40 cm.

Evaluating the results the following relations were found:

- vertical displacement: Benkelman-beam- FWD
- Radius of curvature - FWD- Benkelman-beam (modified)
- Dependence of vertical displacement and radius of curvature from the thickness of the asphalt concrete
- Influence of cracks on the results
- Intensity of the bending- trough depending on the thickness of the asphalt concrete as well as the influence of these results on the accuracy af the measurement with the Benkelman-beam.

Further more a new Benkelman-beam was developed with a variable elecronic 3-point-measurement at a distance from the centre from 0 to 750 mm. With this device only the not vertical displacement in the centre can be measurement.

Using the variable adjustment of the chord between the centre and the following inductive-displacement-device, a fast adjusting to the stiffness of the road construction is possible. Using a third inductive-displacement-device and having the absolute vertical displacement in the centre, weak spots in the asphalt construction and below it can be found and localized.

keywords: curvature measuring instrument
= Krümmungsmesser

radius of curvature
= Krümmungsradius

Fifth International Conference on the
BEARING CAPASITY OF
ROADS AND AIRFIELDS

COMPARATIVE INVESTIGATION OF SEVERAL MEASURING DEVICES AND TEST METHODS TO DETERMINE BEARING CAPACITY

WILFRIED SCHELLENBERGER

Dipl.-Ing., Laboratory for quality control of the State agency for road construction of the state of Thüringen, Königstraße 22, D- 98553 Schleusingen, Tel. und Fax: 036 841/400 52, Federal Republic of Germany

1.0 INTRODUCTION (MEASURING DEVICES FOR BEARING CAPACITY)

In 1961 the laboratory for quality control in Schleusingen built the first Benkelman-beam in Germany (former German Democratic Republic GDR) and with it measured the bearing capacity of a couple of thousand kilometres of roads during the next couple of years. The further development created a very light and quickly to operate device with very thin-walled pipes made of precision steel (picture 1, left).

During the next years the reconstructed roads had to be strenghthend again because of the increase of heavy tuty traffic. It showed that the Benkelman- beam did not reach the necessary differentiation in every case because of the increased stiffness of the construction. This led to the development of a device to measure the curvature, the so- called curvature measuring instrument (cmi). Having variable chordal lengths (between 0,25 and 0,55 m) this device allowed a detailed judgement using the calculated radius of curvature. This characteristic value reacts much more sensitive to variations concerning the geo-hydrological circumstances and the thicknesses of the bound layers (picture 1, right and picture 2).

To achieve a better relation between the vertical displacement in the centre of the load and the connected radius of curvature a second dial gauge was added to the Benkelman- beam in a minimal distance of 500 mm from the centre of the load (picture 3).
It was also tried to find a relation between the now establishing falling weight deflectometer and the Benkelman- measurement as well as to the cmi-measurement. Furthermore it should be checked if there are additional possibilities to foresay something concerning the quality of the construction and the condition of the bearing capacity above the formation level. Some additional tests were also done with the French Lacroix device.

picture 1: Benkelman- beam (left) and curvature measuring instrument - cmi- (right)

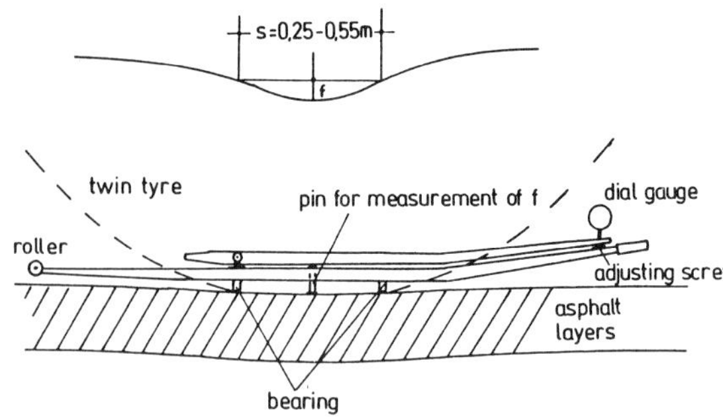

picture 2: principle of the cmi

2.0 SELECTION OF TEST SECTIONS

11 test sections were chosen after a number of cores were taken before. Number 6 later had to be deleted, because it schowed inhomogenities.

Table 1: Summary of the test sectiones

Section Number	Thickness of bound layers (average) $\overline{h_{asphalt}}$ in cm	Geological circumstances at foundation
1 Hibu - Wiedersbach	4,1	cohesive sand / -stone (Trias)
2 Brattendorf - Wiedersbach	40,0	clay (Keuper)
3 Gerhardtsgereuth - Schleusingen	19,3	cohesive sand / -stone
4 Themar - Henfstädt	15,4	sediment of river
5 L1 134 - Geisenhöhn	7,0	cohesive sand
7 Römhild - Mendhausen	31,9	clay (Keuper)
8 Haina - Exdorf	18,0	clay - silt
9 Vachdorf - Leutersdorf	10,4	lime marl
10 Hinternah - Waldau	11,0	cohesive sand / - stone
11 Heckengereuth - Ratscher	12,9	cohesive sand

The 40 cm of test section no. 2 can be explained as follows: The lower 10 centimetres of the asphalt base layer cracked under the heavy construction traffic so that additionally 30 centimetres had to be built up extra. A 25 cm thick bound construction could not be found for this measuring program.
All kind of soils in the road foundations were more or less of high water sensitivity.

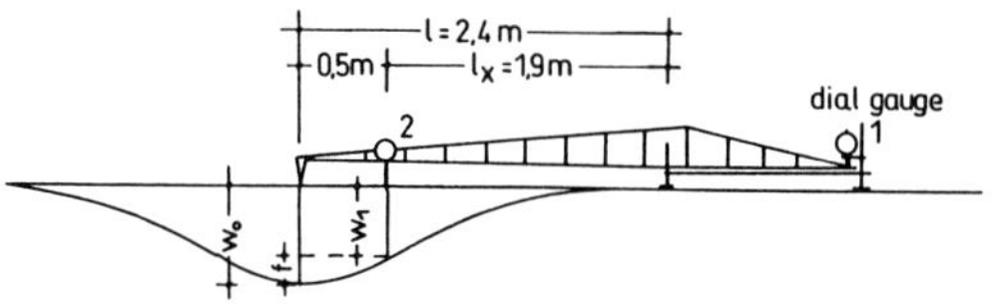

$$w_1 = w_o \frac{l_x}{l} - A \qquad A = \text{vertical shifting at dial gauge 2}$$

$$f = w_o - w_w = w_o - \left(w_o \frac{l_x}{l} - A\right) = w_o \left(1 - \frac{l_x}{l}\right) + A =$$

$$w_o \left(1 - \frac{1,9}{2,4}\right) + A = 0,2083\ w_o + A$$

picture 3: modified Benkelman- beam

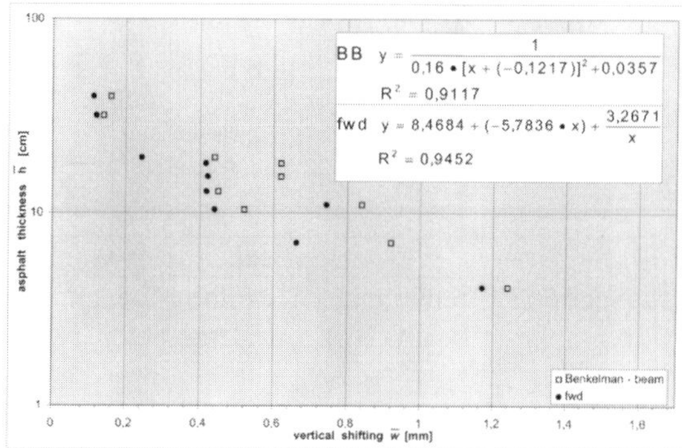

picture 4: vertical deformation with Benkelman- beam and fwd in dependency of thickness asphalt

3.0 CONDUCTING AND EVALUATING THE MEASUREMENTS

The distance of the measuring points were exactly 5 m. Before the commencement of the measurements the points were marked. A major problem in conducting the measurements was to exactly use the same point with the following measuring device. It was strongly observed that the distance from the middle point of the mark was less than 5 cm. But a part of the variation of the measurements may be already a result from those little differences. The measurements took two days during which the sky was overcast so that the temperature on the road surface almost did not change at all and were around 18 °C. The measurements per section took about 1 hours for all measuring devices.

3.1 Influence if the thickness of the bound layers

3.1.1 Comparison Benkelman-beam to falling weight deflectometer

The vertical displacements of both the Benkelman- and the falling weight deflector-measurements show a straight line in a half logarithmic scale (picture 4). The quotient between both lines is $w_{Benkelman\text{-}beam} / w_{fwd} \sim 1,2$.

While the result of Benkelman- measurement is the elastic shifting , the result of the fwd-measurement is the summary of elastic and semielastic shifting. That`s why the absolute relation is higher.
This relation does also not change when the construction shows cracks. But a small parallel movement is seen when the formation of the road shows extreme differences like sections no. 3 and 8. Section 2 had the strongly cracked lower 10 centimetre in the asphalt base corse. They act basicly like a noncohesive layer.

3.1.2 Comparison with the Lacroix divce for orientation

Exactly one year later on almost all sections the measurements were repeated with the Lacroix device. The results are shown in table 2 for orientation. However it must be mentioned, that the measuring points where not identical. The distance between the Lacroix-measurement point where exactly 6 metres.

Table 2: Comparison with the Lacroix- measurements for orientation

Section Number	Vertical shifting w (mm)			Quotient	
	Benkelman-beam (Bb)	falling- weight-deflectometer (fwd)	Lacroix (lac)	$\frac{Bb}{lac}$	$\frac{fwd}{lac}$
1	1,24	1,170	0,736	1,68	1,59
2	0,16	0,112	0,082	1,95	1,37
3	0,44	0,244	0,389	1,13	0,63
4	0,62	0,421	0,384	1,61	1,10
5	0,92	0,662	0,590	1,56	1,12
7	0,14	0,120	0,083	1,69	1,45
8	0,62	0,416	0,478	1,30	0,87
9	0,52	0,440	-	-	-
10	0,84	0,743	0,597	1,40	1,24
11	0,45	0,418	0,271	1,65	1,54

The values of the Lacroix- device are between those of the Benkelman- and the falling-weight- deflector- measurements. This is realistic because the shorter test arm of only 1,65 metres. The thickest asphalt concrete layers give the highest factors, because the measuring base is in the bending line.

But interesting are the quotients of measuring sections 3 and 8, where the foundation was very inhomogenious. Here the values are relatively low compared to the other measurement methods.

3.2 Settlement tray at Benkelman- measurements

The Benkelman-measurement is often said to have larger measurement errors because both the front and the back measurement-bases are too much positioned into the settlement tray. Sometimes the usefulness of this measurement method was doubted in general.

During these investigations measurements were also taken at all points with a second dial gauge and further three Benkelman- beams in the following distances from the centre of the load (picture 5 and 6).

- 0,50 m — 2nd dial gauge at the Benkelman-beam (1/100 mm accuracy of dial gauge)
- 1,00 m — 2nd Benkelman- beam (1/100 mm)
- 2,40 m — 3rd Benkelman- beam at the front of measurement- base (1/1000 mm)
- 3,60 m — 4th Benkelman- beam at the rear of measurement-base (1/1000 mm)

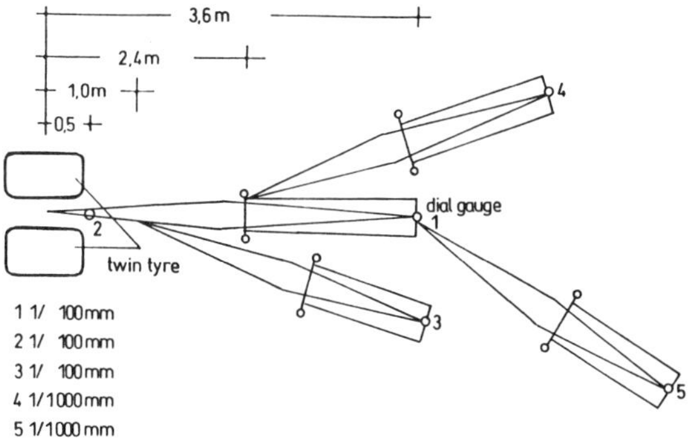

picture 5: layout of the measurement with 4 Benkelman- beams

picture 6: measurement on the road with 4 Benkelman- beams

Table 3: Arithmetic means of the Benkelman- measurements

Section Number	$\overline{h_{asphalt}}$ cm	vertical shifting $\overline{w}$ (mm) in a distance from the centre of the load (m) 0	0,5	1,0	2,4	3,6
1	4,13	1,24	0,88	0,05	0,016	0,004
2	39,95	0,16	0,07	0,11	0,02	0,001
3	19,30	0,44	0,28	0,11	0,01	0,02
4	15,35	0,62	0,37	0,15	0,06	0,01
5	7,0	0,92	0,66	0,11	0,01	0,009
7	31,88	0,14	0,06	0,12	0,02	0,01
8	18,03	0,62	0,43	0,09	0,01	0,008
9	10,40	0,52	0,34	0,10	0,003	0,009
10	11,00	0,84	0,64	0,06	0,01	0,02
11	12,88	0,45	0,30	0,07	0,002	0,01
		0,14 - 1,24	0,06 - 0,88	0,05 - 0,15	0,002 - 0,06	0,002 -0,02

Following conclusions can be drawn:

- near by the front measuring base (spacing 2,4 m) the average vertical shiftings are 0,002 - 0,02 mm. If test section no. 4 with it peaty foundation is deleted, only $\overline{\overline{w}}$ = 0,011 mm will remain as an absolut average.

- at the rear measuring base (spacing 3,6 m) the deflection lessens to $\overline{\overline{w}} \approx$ 0,01 mm.

- the influence of the asphalt concrete thickness is unexpectedly small and decreases in accordance with the increasing age of the construction.

Accordingly sufficient precise deflection measurements can be carried out with the Benkelman- beam.
Other influences like the other twin tyre of the lorry in a distance of about 1,60 m, the strong slope of the gradient or the differences between the arbitrary chosen measuring points are of much higher implication.

3.3 Curvature measurements

The radius of curvature is an important parameter to calculate stresses and strenghthening of flexible pavements.
The radii of curvature have been calculated by the test results with fwd`s, cmi and Benkelman- beam with the second dial gauge in a distance of 500 mm from the centre of load:

curvature measuring instrument:

$$R = \frac{s^2}{8f} \cdot 3{,}5\ (m)$$

s = chordal length (m)
f = segmental height (m)
3,5 = magnifying factor at the instrument

During the fwd- measurements (radius of the contact area r = 150 mm) the second geophon has been arranged as near as possible to the edge of the base place.
The distance r from the centre of load was 185 (fwd of the GSA Kaiserslautern) respectively 200 mm (fwd of the Federal Road Research Laboratory, BASt).

$$R = \frac{w_1}{w_o} \cdot \frac{r^2}{2(w_o - w_1)}\ (m) \qquad \text{(french model)}$$

w_1 = vertical shifting within the spacing r from the centre of load (m)
w_o = vertical shifting in the centre of load (m)

In the course of 25 years using the cmi it turned out that the radii of curvature are distributed in a completely logarithmic manner. Therfore the geometric mean $\overset{o}{R}$ of the single test sections and not the arithmetic mean $\overline{R}$ has been calculated

$$\overset{o}{R} = \sqrt[n]{R_1 \cdot R_2 \cdot \ldots R_n} \quad \text{or} \quad \lg \overset{o}{R} = \frac{\sum_{i=1}^{i=n} \lg R_i}{n}$$

Out of the fwd- measurements you have to use the arithmetic mean.
As mentioned before the differences are resulting not only by the minor differences regar-ding the accuracy of the measuring point but also mainly by the changing load conditions of twin tyres and circle place or camber in between twin tyres. Here crack formations are

effecting extremely the results. As expected the radii of curvature are increasing dependent on the asphalt thickness.
The mean values are shown as a diagrammatic figure on pictures 7 - 10.

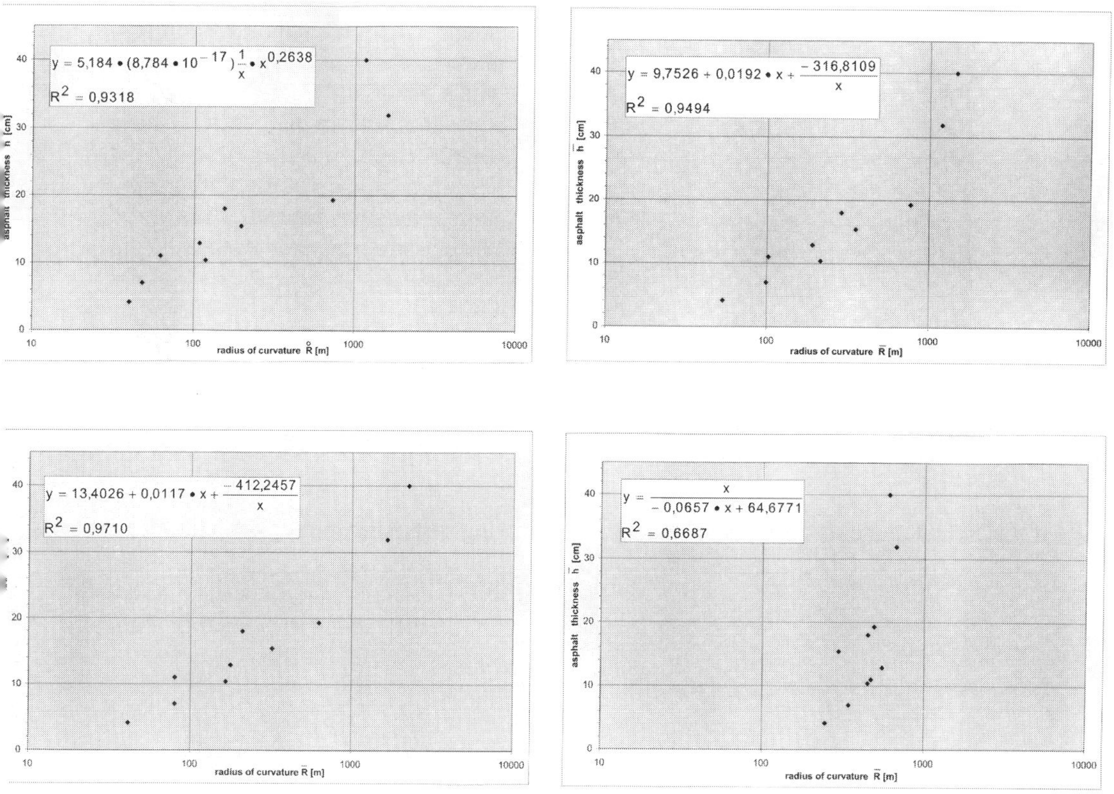

picture 7 - 10: radius of curvature in dependency of thickness asphalt

picture 7: cmi

picture 8: fwd (BASt)

picture 9: fwd (GSA Kaiserslautern)

picture 10: modified Benkelman- beam

Between the radii of curvature of both fwd- devices differences were noted. The second geophon of the GSA was situated a little to near to the load plate.
The optimum should be about 5 cm distance.
However a correlation was found between the radii of the cmi- and fwd- measurement. I admit, that there is a high dependency on the layer thickness of asphalt (picture 11).
Furthermore you have to consider that the cmi- measurements take into account only the elastic and the fwd- measuremants additional the semi- elastic shiftings.
Constructions with crack- formations didn`t match the statistical connex.
However as the radii of curvature regarding Benkelman- measurement with a second dial gauge are very unprecise to useless in this technology of measuring. The second dial gauge needs to have an indication accuracy of 1/1000 mm and should be easier manually readable.

Referring to this you can expect essentially more accurate test results by the measuring system according to Dr. Köhler.

On the basis of the further evaluation of the radii of curvature along with an extended chordal length by the integration of more distant geophones with s = 0,70 m; 1,0 m; 2,0 m the correlation shown on picture 11 have been received.

Yet a principle derives: to minimize the chordal length is correct, if the vertical shiftings are registered exactly.

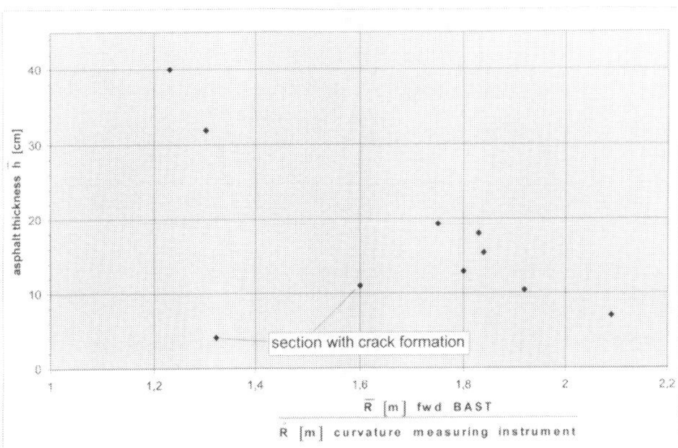

picture 11: correlation of the radii between cmi- and fwd- measurement (BASt)

If the french model is used to calculate R the following surprising findings can be stated:

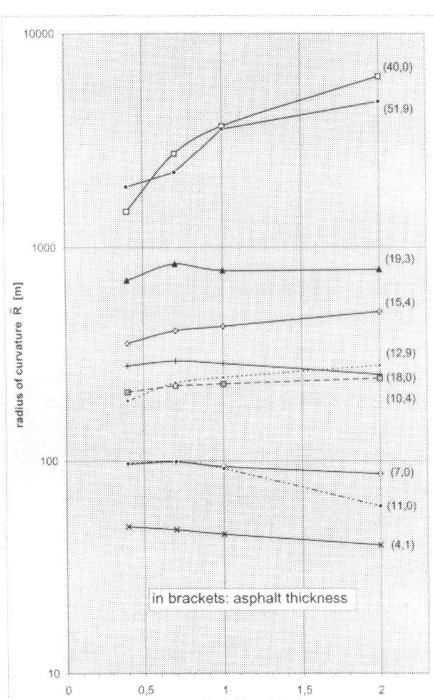

picture 12: radius of curvature with variable chordal length (fwd BASt)

- at R ≤ 100 m the radii of curvature with s = 1,0 instead 0,4 m were approximately 10 % lower ($\overline{h_{asph}}$ ~≤ 15 cm)
- at R ~ 100 to ~ 750 m there is an increase up to 20 % ($\overline{h_{asph}}$ ~ 15 - 25 cm)
- at more than R ~ 750 larger chordal length than ~ 0,4 m are resulting in useless radii of curvature (increase to ~ 350 % with $\overline{h_{asph}}$ ≥ 25 cm)

Furthermore an influencing of the results exists due to the bearing capacity properties of the unbound layers and the road foundation. The problem „ time history" is of fore- grounded character regarding larger chordal length of the fwd- measurement.

4.0 BENKELMAN- BEAM ON ELECTRONIC MEASURING BASE

Evaluating all this tests results along with the radii of curvature a modified Benkelman-beam was constructed and installed (see picture 12; in the meantime the guide- heads of ITW 2 and 3 are supplied with rolls).
This device, which couldn`t be used yet for these comparative measurements, is able to combine the properties of the Benkelman- beam and the cmi, i.e. it is possible to measure the vertical shifting in the centre of load as well as to calculate the radius of curvature by the second inductive linear variable differential transducer (LVDT).
The special advantage is, that the second LVDT can be shifted without any control in the spacing of 100 to 700 mm to the centre of load. Therewith an adjustment to the stiffness of the construction is possible without any problem.

Finally the third LVDT, it spacing can be changed from ~ 130 to 750 mm, enables to record the course of the settlement depression. If the total construction thickness is well known a fundamental statement is furthermore possible, whether the weak- points regarding bearing capacity are caused by a possibly not frost resistant road construction or by the foundation see picture 13).

picture 13: Benkelman- beam with 3 inductive transducer (LTVI)

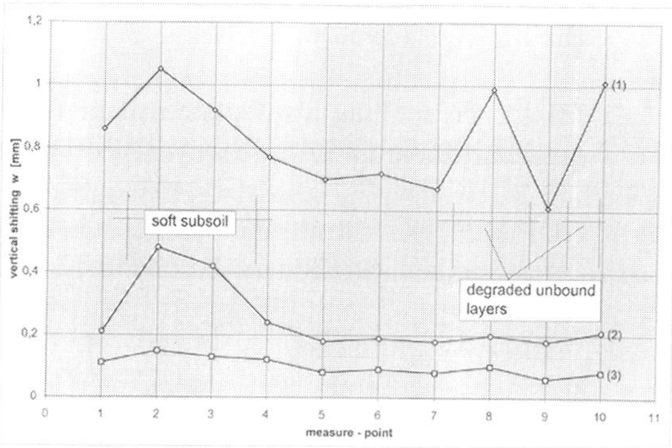

picture 14: vertical deformation in distance from the centre of the load

The upper curve (1) shows the vertical shifting in the centre of load, the centre curve (2) in the actual case in the distance of 500 mm, as the construction thickness was ~ 50 cm too. Curve (3) with a distance of 1000 mm should indicate which bearing capacity properties are existing at the zone ~ 0,5 m below the formation level.
If the curves (1) and (2) are of the same tendency it is assumed that the road construction consists of frost-resistant construction material. The failing is caused in the formation level area (sub-soil). If the same tendency is continuing downwards, even the lower-lying zone shows deficiencies regarding bearing capacity. If there are vertical shiftings only at curve (1), you can be sure that there are deficiencies in the unbound construction layers.These knowledges are especially important regarding pavement design.

5.0 SUMMARY

The results of the deseribed comparative measurements can be summarized as follows:

- there is a functional relation of the vertical shifting between the Benkelman- and the fwd-measurements.
- the influence of the settlement tray on the result of the Benkelman- measurement is only small.
- there is a relation of the radii of curvature between cmi- and fwd- measurements, which is influenced by the thickness of the asphalt-concrete.
- the increase of the chordal lengths up to 1,0 m gives approximation results still under 25 cm of asphalt- concrete, but the resolution of the dial gauge must be 1/ 1000 mm.
- the measuring of the vertical shifting outside of the centre of load gives important informations about the state of the whole construction.

6.0 REFERENCES

1. MÜLLER, F.: Krümmungsmessungen an der Oberfläche von Straßenbefestigungen. Die Straße 9 (1969)6, S. 285
2. MÜLLER, F.:Tragfähigkeitsmessungen als Grundlage der Verstärkung von Asphalt-straßen. Straße und Autobahn 42 (1991), Heft 4
3. KÖHLER, M.: Ermittlung der Steifigkeitskonstellation vorhandener flexibler Fahrbahn-befestigungen. Straße und Autobahn 47 (1996) Heft 3
4. WEIST, W.; PLEHM, T.; SPAHR, H.: Einsatz des Deflectographen „Lacroix" in Brandenburg. Straße und Autobahn 5 (1995) S. 277-282

Fifth International Conference on the
BEARING CAPASITY OF
ROADS AND AIRFIELDS

ROLLING DEFLECTOMETER METER DATA STRATEGY DOS AND DON'TS

Carl A. Lenngren

Researcher, Vägverket Konsult,
Box 1200, S-462 28 Vänersborg, Sweden
Fax +46521279124, E-mail carl.lenngren@swipnet.se

ABSTRACT

Pavement management input data in Sweden are primarily based on surface characteristics. These data are good for assessing user costs but they are of less value for estimating maintenance and reconstruction costs. Structural assessment is very important and the falling weight deflectometer can be used for this purpose. It has proven to be an excellent tool for overlay design. For this purpose its simplicity and straightforwardness are well documented. However, the relatively low capacity per unit makes this device less attractive for pavement management network level use. In addition the falling weight deflectometer stationary operation mode requires lane closures and high user costs on high volume roads. The Swedish National Road Administration decided that a high-speed rolling deflection tester should be tried using lasers sensors mounted on a heavy truck. The Swedish Road and Transportation Institute had a working system ready by the end of 1997. Forty sensors are mounted on a heavy truck to determine two transverse profiles. One profile constitutes an unloaded case. The other profile just behind the rear wheels of the vehicle constitutes the loaded case. The high sampling rate is adequate for filtering the macro texture of the pavement. The present paper discusses some different strategies for optimizing the rolling deflection tester usage.

Keywords: bearing capacity, deflection, high-speed non-destructive testing, laser

INTRODUCTION

The Swedish National Road Administration (SNRA) pavement management system input data are primarily based on surface characteristics. These data are assessed annually at an

	PCC	Base	Subgrade
Mean Modulus [MPa]	42400	180	160
Coefficient of Variation	.230	.235	.148

Table 2. Portland Cement Concrete FWD moduli

There were a few slabs showing substantially lower stiffness. At 60 m 18000 MPa was backcalculated. Apart from those slabs the variability was small, and most FWD center deflections were in the range from .15 to .20 mm at the 50 kN load level.

The RDT deflection at four different speeds is plotted in Figure 4. The results look very surprising as there is heaving except for the slowest speed at 10 km/h. There are a few possible causes for this to occur.

- The surface wave amplitude being greater than the recorded deflection
- A reflected wave counteracting the recorded deflection
- The loaded transverse profile is too close to the wheel and away from the maximum deflection
- Time dependent behavior of the soil, possibly from pore pressures building up

The surface wave travels much faster than the vehicle. Even if they in fact affect the sensor readings, it would not be consistent with speed. A reflection wave could however reduce the deflection in such an orderly fashion. Depending on layer thickness it is hard to determine if this is the case without pavement instrumentation. Accelerometers and LVDT:s in the pavement would help to resolve the issue.

The lag of deflection behind the wheel is usually due to the viscoelastic properties of asphalt concrete. The PCC response should be much more instantaneous and is unlikely to explain the effect.

The pore pressures in the soil may be a plausible explanation to the speed effect. It is a well-known fact that less permeable base layers often lead to faulting and other distresses on PCC pavements. The highest heaving effect coincides with the part of the runway displaying the most surface distress and lower slab modulus. Further the ground water table in the area is known to be rather high.

Power spectrum density diagrams were also plotted for the two profiles. A firsthand evaluation showed that loaded profiles generally are noisier than the unloaded ones. A spike at the seven meter wavelength corresponds to the slab length. The subsequent analysis will study the feasibility of using this type of evaluation for testing joint load transfer efficiency.

The results found are going to change the subsequent evaluation of the RDT data. Just correlating results with the FWD is not enough to account for dynamics. Instrumented pavement sections would solve some of the issues as would surface wave analysis. Even if the results are disappointing from a layer elastic analysis point of view, the potential of locating areas with high pore pressures is certainly intriguing.

Figure 4. PCC RDT deflection

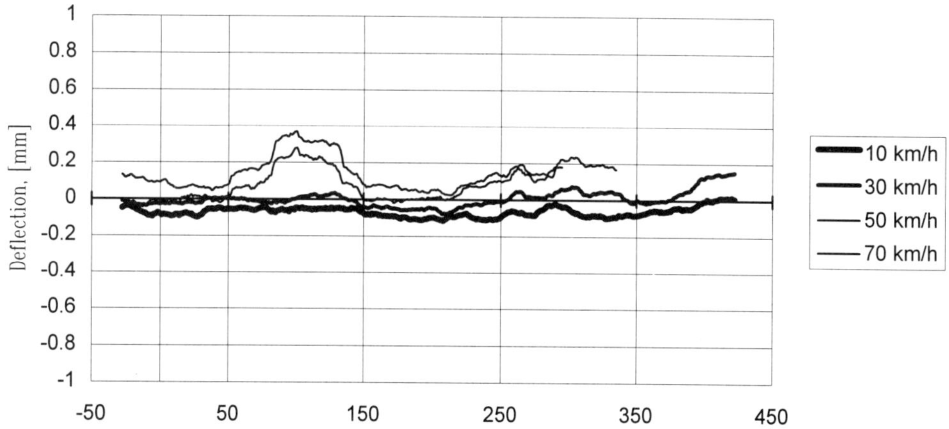

STRATEGIES

Given that the data assessment system is working reliably the strategies for using RDT data for pavement management systems (PMS) would be as follows.

1. Comparative studies by repetitive tests.
2. Simple Approach
3. Linear elastic solutions
4. Green's function approach
5. Full DEM/FEM model

Comparative studies by repetitive tests

The idea is to gather data in such a way that the average deflection is stored every 400 - 1000 meters. This way the coordination between measurements error will be limited. This method should be at hand during 1998. It potential for PMS network level use is fair.

Simple Approach

Regression equations between deflection data based on Do and the so-called area parameter and strain criteria. Equations are available for FWD deflectometers. There is a need to calibrate towards RDT-use. Implementation is expected in 1999. It potential for PMS network level use is good; program level use is fair.

Linear elastic solutions

As with the simple approach this method needs to be calibrated. It is not known at the present time how useful this approach is when deflections are averaged over long distances. For overlay purposes one knows that averaging deflection basins for more than 20 meters often results in a conservative design. Provided data can be averaged for 20 meters only, this approach may be implemented in the year 2000. Its potential for PMS program level use is good. The FWD will still be used for project level use, however the RDT continuously collected data have a potential of complementing the FWD tests.

Green's function approach

Green's function approach is a model solving the dynamics of the moving load. It is based on superpositioning the wheel. It will be helpful in determining whether there are standing waves present or not, due to a stiff layer somewhere in the structure. The methodology has also been used when considering frozen layers at shallow depths. The Green's function approach is going to be applied for structural response research for the RDT. In the future it may be applied for production work possibly at the project level.

Full DEM/FEM backcalculation model

In accordance with the Green's function approach it is also possible to build an optimization backcalculation algorithm that uses complete response-time histories to recover pavement layer thicknesses and stiffnesses. This model will require a lot of computer resources and it is only intended for research. However, 5-10 years ahead it may well be used for production work also at the project level and for overlay design as well.

DOS AND DON'TS FOR SENSORS

Resolution and signal to noise ratio

As the resolution of the laser sensors is not quite on par with stationary types, strong structures yielding small deflections from any given load should be a cause for concern. The lack of resolution is somewhat compensated for by the high sampling rate, 16 or 32 kHz. At a speed of 16 m/s (58 km/h), 100 or 200 samples would be taken per decimeter length on the road. As comforting as this sounds, two sensors are compared, so the anticipated error (99 % confidence) is in the range of 2 - 4 thousands of a millimeter. This is provided the deflection does not change over the length being sampled, the readings are all adequate and that no out of range condition prevails. In addition the outermost lasers are being angled outward, so the best possible resolution is about 5 - 10 thousands of a millimeter if a long base line involving the outer sensors is used. For a quantitative deflection analysis the resolution seems to be adequate, however for backcalculation of moduli analyses the resolution is near the limit for what is needed, and may not always work. Structurally sound pavements will only yield small deflections; the maximum could be less than .010 mm. Considering these facts, it is probably best not to use angled sensors.

Surface wave interference

Surface waves travel much faster than the vehicle and should therefore not interfere with the signal. However, certain wavelengths could coincide with the sampling rate. The reference profile may be affected by a dip in the surface wave, and the profile under load may be affected by the top of a surface wave. The amplitude of the wave is likely to be small and the probability of a coinciding sampling rate is also small. The Green's function approach should be able to model this phenomenon so that the effect could be ruled out.

Reflected wave interference
In FWD analysis the sensors far away from the load often yield unrealistically small deflections. There is some concern that these sensors may be affected by a wave being reflected back to the surface counteracting the deflection. However, this is usually occurring for those sensors being about 1.5 m or more away from the load. By using a shorter base line than 2 m the problem should be all but eliminated, except for AC on PCC pavements.

Deflection lag
The lag of the deepest point behind the tires is varying, depending on the speed of the vehicle, as experimentally confirmed by DeBeer.[2] At high speeds there is a risk of missing any developed deflection completely. This is of great concern for placing the sensor bar behind the rear axle. The sensor bar does not have to be placed precisely at the maximum deflection point as long as you know where the maximum deflections occur. Hence, the RDT prototype was equipped with four sensors placed longitudinally behind the wheel. This helped in establishing where the maximum was located. Over the hundred sites investigated about 25 percent were outside the range of these four sensors. Five or six sensors will most likely be adequate. A scanning device placed longitudinally will also serve this purpose.

Out of range conditions
Rough roads would undoubtedly lead to more out of range situations; i.e. more road has to be traveled before adequate sampling is achieved. By lowering the speed of the vehicle to 10 or 15 m/s the influence of the roughness can be controlled. After all, trucks go slow on very rough roads anyway. By lowering sensor resolution, a wider standoff range can be achieved. However, as mentioned above it is not a good idea to skip on the resolution.

Inadequate brightness and scattering
A problem with new asphalt pavements is the black surface. It reduces the reflection of the laser spot quite remarkably and hence reduces the number of samples per traveled length quite substantially. At one of the test sites at a brand new road, the pavement in addition to the dark surface also had a very open macrotexture. This type of texture will further slow down the drying process of a wet surface. The pavement has to be dry, as otherwise present water will scatter the light. At the site rainy weather did not permit the open macrotexture to dry properly which led to loss of data while sampling. Unfortunately, the results from this site could not be used. A feedback system should tell the operator whether the data could be used or not. In rainy and wet climates one has to expect long delays and plan for it accordingly.

DOS AND DON'TS FOR OPERATION

The coordination of the vehicle involves the operation of the tests and keeping a protocol of which sections and how they have been measured.

Longitude
For subsequent measurements it is important that the vehicle measures in the same track and that there is no offset. The experience from the L-RST has shown that good

repeatability can be achieved in the longitudinal direction. However, manual systems do require that the operators be trained. Automated triggering should be used for all research and other critical work.

Latitude

The bearing capacity of a roads can vary not only lengthwise. Several tests made across a road transversally with an FWD show considerable variation. Side support is beginning to influence the result when the load is closer than 60 cm from the edge. Wheel path compaction is another very strong reason why the stiffness varies transversally on old roads. Thickness variation is another reason. Rut repair may contribute to a thicker layer right in the wheel path. To handle the situation RDT drivers must be trained to follow the wheel paths at all times. The lateral position should be videotaped at all times for reference reasons.

Crabbing-Angle Errors may occur due to cross slope of the pavement. It can be shown that as long as the crabbing angle stays less than 3.5° for a 9-m long baseline the effect is small[3]. In most cases for highways the angle would be considerably less. However, when driving in curves the deviation between profiles may be more than five centimeters and the introduced error too large. The radius of curvature can be sampled at all times though so it should be possible to determine if the data are valid or not due to turning motions. As mentioned above there may be a bias in the signal due to the macrotexture. That problem is more serious as rut repair may cause two very different textures to be close and thus affect the sensors differently. By assessing the texture for each pair of sensors on the reference and loaded profile as well, one may exclude incompatible data where the texture differs.

Global Positioning System

The Global Positioning System (GPS) could help considerably in determining the exact location of the RDT at all times. As the vehicle is moving it is not possible to achieve an adequate accuracy at the time of testing. However, it is possible to store the signal with the data stream being collected. Thus, positions would be documented. The method as such may require setting the system clock within .05 seconds from accurate Universal Time though.

Speed control

Controlling the speed of the vehicle would be utmost important. All asphalt pavements are more or less susceptible to the loading time. Many soils, especially cohesive soils react very differently under different loading conditions too. In effect this means that the vehicle should be traveling at a constant speed when measuring, perhaps 20 m/s (72 km/h) would be a representative and yet safe speed to test at. Minor variations from the norm could be permitted. As mentioned above 10 or 15 m/s would be appropriate speeds on rough roads.

CONCLUSIONS

The Road Deflection Tester is capable of recording deflections caused by a moving load at high speeds. However, the dynamic behavior of materials being used in pavements and the natural soils make interpretation of the results very difficult without additional information about the sites tested. It is obvious that a traditional elastic model cannot be used at higher measuring speeds. Asphalt concrete modulus is highly influenced by temperature. It is also influenced by the loading rate. A better understanding of this property may be achieved by testing at different speeds. High pore pressures in the soil under PCC slabs may have lessened deflection with speed. An elastic or even viscoelastic model would not account for such behavior. However, potential of faulting could perhaps be detected if handled properly. The problem may be handled by always testing at a number of different speeds in the range of 10 - 90 km/h. Actually, this may be a good way of identifying sections with high pore pressures. The elastic model may perhaps be used at the lowest speed.

The system needs a higher sensor resolution and/or higher loads on heavy PCC pavements for obtaining a better signal to noise ratio. Joint transfer load efficiency may be detected by evaluating power spectrum densities. The prototype axle load of 9 tons was near the resolution limit at least for relatively strong structures. The new RDT higher load would increase the signal to noise ratio. However, all things considered the RDT is probably better suited for flexible pavements.

REFERENCES

1. Arnberg, Å. Holen, G. Magnusson. "The High-speed road deflection tester", Heavy vehicles and roads, Proceedings of the third international symposium on heavy vehicle weights and dimensions organised by the University of Cambridge, Edited by D. Cebon and C.G.B. Mitchell, pp.176-181, Thomas Telford, London, 1992.

2. De Beer, "Measurement of tyre/pavement interface stresses under moving wheel loads", CSIR Research Report DPVT 224, ISBN 0-7988-5443-X, 1994.

3. Rish, J.W. et alia ”Electro-optical approach to pavement deflection management” Proceedings ”Nondestructive Evaluation of Aging Aircraft, Airports, Aerospace Hardware, and Materials”, Oakland, CA 1995 pp. 363-372. SPIE Volume 2455

Fifth International Conference on the
BEARING CAPACITY OF
ROADS AND AIRFIELDS

FLASH : A NEW DEFLECTOGRAPH

J-M SIMONIN[1] H. VIALLETEL[2]

[1]Research engineer, LCPC, 44340 Bouguenais, France, ☎ (33) 240845829 ;
Fax : (33) 240845992 ; Email : Jean-Michel.Simonin@lcpc.fr
[2] Civil engineer, "Project manager", CECP 49100 Angers.

ABSTRACT

Deflection measurement is a basic element to evaluate the bearing capacity of roads. In 1993, the French laboratories decided to develop a new deflectograph: Flash. This new device had to meet four requirements:

- To ensure a good accuracy of measurement even on the rigid pavement;
- To increase the speed of the truck during the measurement;
- To increase the driveability compare to the last version of the deflectograph;
- To simplify operation of the device, particularly for secondary functions (putting up and down the measurement beam, its protection against crushing).

The new deflectograph, Flash, uses the same measurement principle as its predecessors, but it is based on more modern design, construction, servocontrol, and metrological technologies.

The beam is made of composite materials. To protect it from the risk of crushing, a system of guides has been developed. The beam is moved by a linear rail system. An absolute coder is used to measure deflection so that no calibration procedure is necessary. Flash provides a high and adjustable measurement rate: two measurement points (centerline and edge) every 4 meters at 3.5 km/h, every 10 meters at 8 km/h.

In 1995, a test program was carried out to verify the metrological performance of the Flash deflectograph. In 1996 and 1997, 1000 km of measurements allowed us to test the Flash prototype. In 1998, after the last modifications, the French laboratories started to be equipped with this new deflectograph which is more accurate, faster, easier to drive and not more expensive than its predecessor.

DEFLECTION MEASUREMENT BY DEFLECTOGRAPHS

For nearly 30 years, the French national network of Laboratoires des Ponts et Chaussées has been using devices known as deflectographs [1] to measure the vertical deformation of a pavement under a truck rear axle loaded to 13 tons. The long experience acquired has made it possible to develop a vast domain of use for these devices. The main use is the survey of the highway network, and more particularly the evaluation with time, primarily under the effect of traffic. This use allows locating weak areas for which overlays can be planned. The devices are then used to check the effectiveness of the overlays. Less often, they are used for a « point zero » on a new pavement, which will then become the reference for subsequent investigations. Moreover, construction contractors may request that the device be used on subgrades, so that weak spots can be detected during construction and receive special attention after finishing the pavement.

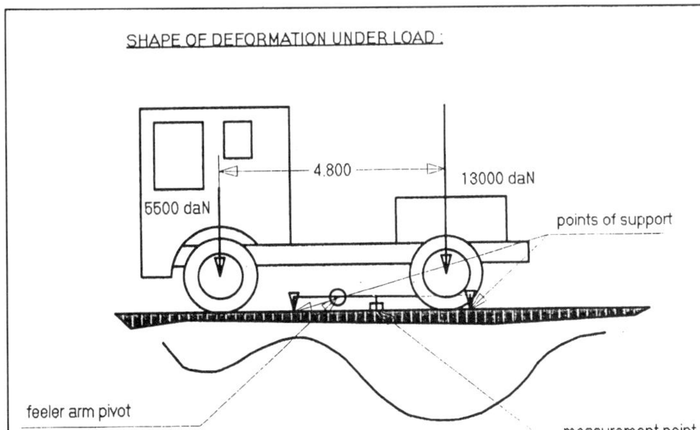

Figure 1 : Principle of deflectograph measurement

Theorical and experimental [2] work has shown that three parameters are needed for all these checks. One is the maximum deflection under the rear axle, another is the measured or calculated curvature (i.e.radius of curvature) of the deflection basin centered on its summits. The third one is the temperature of the pavement when the other two parameters are measured. After processing, these parameters can be used to interpret the condition of the pavement. Over time, deflectographs have incorporated all these parameters, and in addition they have been automated and computerized.

DESCRIPTION OF EXISTING EQUIPMENT

The existing deflectographs are type 02, 03 or 04. When making measurements, the vehicle travels at a constant speed of about 3.5km/h (~1m/s) and a measurement is made every 4 meters. The device consists of a carrier, a beam and two feeler arms, a system to pull the beam and orient and at least, an electronic data-processing system which records the measurements. The device measures a deflection every 2 cm up to a position a few centimeters behind the rear axle. The deflection bowl is so measured on a 1.50-meter length.

It was decided to develop a new generation of deflectograph for many reasons. The three points of support of the beam are used as reference plane for the measurement. The measurements made assume that this reference plane is not affected by the deflection basin. However, it is not always the case, depending on the structure of the pavement investigated [3]. With flexible pavements, this is not very important; the maximum deflection is rather large and the deflection basin is small, and so the reference may be regarded as valid. However, on rigid pavements, where the maximum deflection is small and the zone of influence large, the reference may cease to be altogether satisfactory. Creating three types of deflectographs, 02, 03 and 04 [3] solved the problem. The 03 and 04 generations should have superseded the 02 generation, but for reasons of maneuverability, the 02 still remains quite useful. Obviously, it entails some duplication of resources for the operating authorities, necessity of having two rather expensive devices to make measurements of the same type.

As we stated earlier, the measurement is made at a constant speed, about 3.5 km/h on average. This speed is a constraint on measurement campaigns covering many kilometers of pavement because it entails a rather long investigation time. In addition, the slow speed means that many safety precautions must be taken, which adds to the measurement cost.

The speed is imposed by the very principle of the device : a beam is made to slide along the pavement, inducing vibrations in it. The vibrations may be comparable in magnitude to the measured deflection and so alter the results. So the beam must be moved as smoothly as possible. In addition, the front stop, even after a deceleration, causes shock that also produces vibrations, which must be limited. Finally, the beam must be rapidly immobilized on the pavement to ensure a regular interval between measurements. All these constraints limit the speed between 3 to 4 km/h, depending on the pavement.

Accordingly the objectives set for the new generation were to have only one device, to guarantee a sufficient measurement accuracy on any type of pavement, to increase the system's speed to 8 km/h, while taking some liberties with the measurement interval, increased from 4 to 10 meters.

The next step then was to look for solutions capable of attaining the assigned objectives. Many approaches were considered, but we shall describe only the solutions actually adopted.

INNOVATIVES FEATURES OF THE FLASH DEFLECTOGRAPH

The first point studied was the optimization of the positions of the points of support to guarantee the precision of the measurement on any type of pavement. Our study led us to the geometry shown on figure 2.

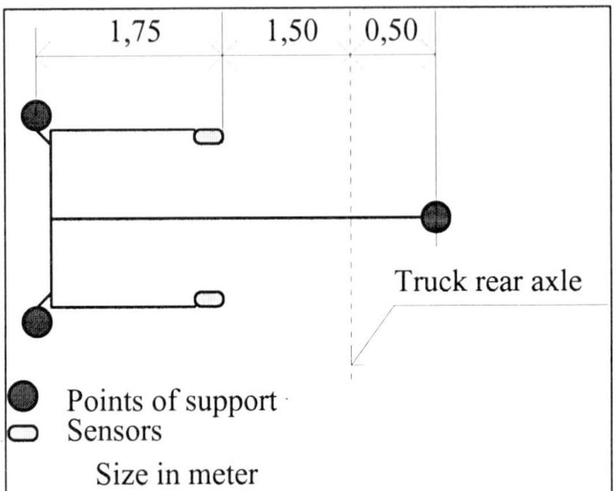

Figure 2 : Geometry of Flash beam

The second part of our study concerned the measurement speed of the carrier vehicle. Our objective was to reach 8 km/h (2.23 m/s) with a measurement interval of 10 meters, while retaining the possibility of making a measurement every 5 meters at a speed of 4 km/h . To attain this performance level, we first had to think of protecting the beam from the risk of crushing which was a problem encountered in the previous generation. A system of guides and stops was developed to solve this problem. As for the beam itself, we worked to reduce its weight by half without sacrificing any rigidity. For this purpose, we used composite materials, which combine these two features of light weight and rigidity. The skids, which support the beam, were chosen for a good compromise between the coefficient of friction for stopping, the damping coefficient, and abrasion resistance. Here, plastics meet our needs best. Similarly, the traction system has been replaced by a linear rail system.

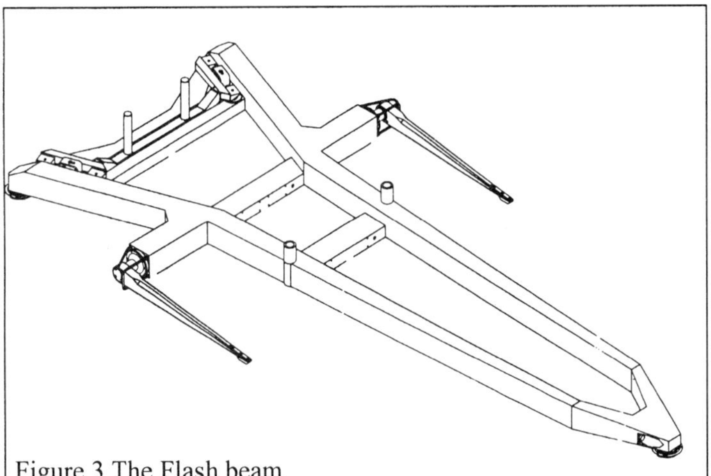

Figure 3 The Flash beam

Finally, to simplify the deflection measurement displacement sensor part, we turned to an absolute rotary coder, which will spare operators the need to go through a calibration procedure. Moreover, the electric cables are now moved by the beam traction system, which makes it possible to eliminate practically all the forces applied to the beam.

Flash also includes an inclinometer, which measures directly the radius of curvature [4]. This device exists in prototype form; it has been through a series of tests and satisfied the experimenters. Its incorporation in the Flash deflectograph will ensure that it is manufactured and distributed.

Finally, Flash retains the possibility of making measurements in a bend, and all the measurement control and acquisition has been revised.

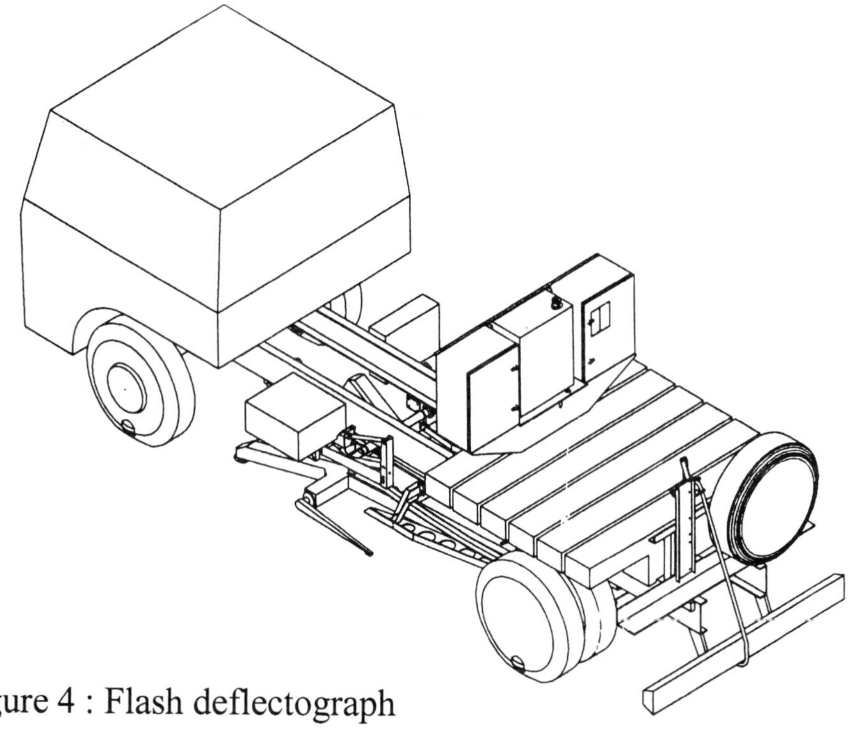

Figure 4 : Flash deflectograph

EXPERIMENTS

In 1994, mechanical and electronic tests allowed us to optimize the control system and general mechanism.

In early 1995, the Blois laboratory received the prototype and tested it. This gave the operators time to familiarize themselves with the new device. After these measurement campaigns, operators made remarks and suggestions to improve the prototype.

During the summer of 1995, the last modifications have been carried out. In November, the Flash metrological performances have been evaluated. The measurement speed was 7 km/h with a measurement interval of 10 meters. The evaluation method was the same method used for the other equipment for deflection measurements [5] [6] [7], in accordance with the French LPC method for surface deformability tests [8]. Flash made 1092 deflection and curvature measurements; 96 % of the deflection measurements and 92 % of the curvature measurements were validated. The percentage of valid measurements is higher than those obtained with the other deflectographs, especially for curvature [5]. The measurement accuracy of deflection is evaluated by comparison between Flash measurements and reference measurements (a displacement sensor anchored at a depth of 6 meters). Figure 5 shows the deviation between Flash and reference measurements versus reference measurements. These deviations are lower than the class 2 limits defined in the French LPC method of deformability measurement [8].

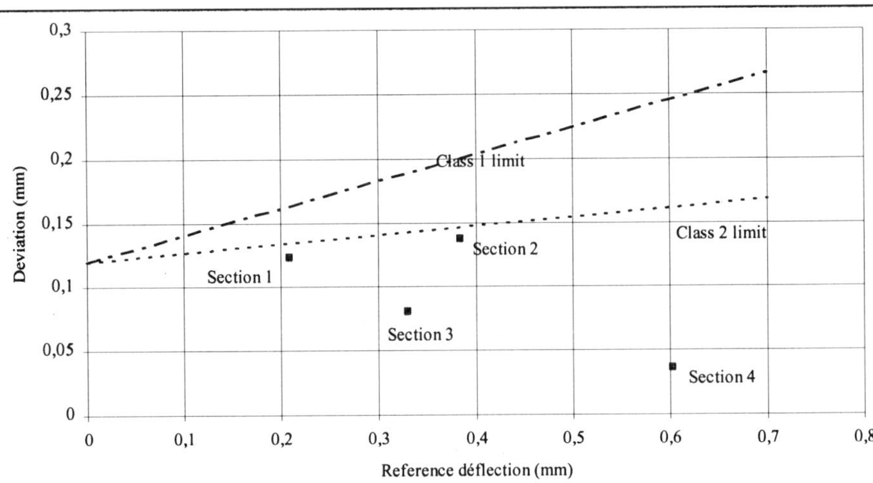

Figure 5 : Deviation between Flash and reference measurements versus reference measurement

Flash repeatability is illustrated by figure 6. It presents the Flash measurements executed on one section where Flash repeated 10 times all the measurements. This shows us the good repeatability of the measurements. Flash has class 2 approval for every type of pavement.

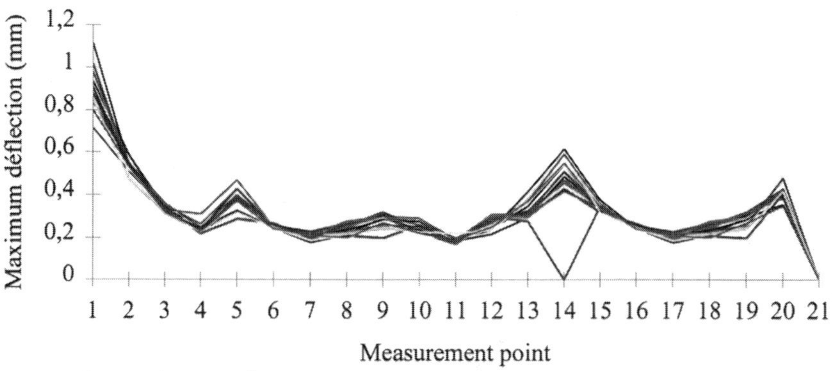

Figure 6 : Ten times Flash measurements on one section

Since 1996, the Blois laboratory has used Flash for operational studies. Therefore, 1000-km of measurements with a 10 meters interval have been made. These measurement campaigns, from 1 to 50 km long, have been made on different types of pavements. The results show that it would be possible to use only one deflectograph by laboratory. This systematic use allows reducing the measurement cost, and to propose competitive prices. Its driveability on small working sites and winding roads has been appreciated. Its capability to reduce speed is an advantage for the survey of base courses where measurements are more difficult.

Today, the last adjustments are carried out ; they include recent electronics and computer evolutions. In 1999, Flash will be distributed.

CONCLUSION

Flash deflectograph is a new device for deformability measurements which associates measurement speed and metrological performance. Flash offers a good opportunity in the competitive French market. With its large capabilities, it will become the standard equipment for deformability measurement. Several French laboratories ask for Flash deflectograph.

REFERENCES

[1] J. Lacroix ; « Le déflectographe : pour une auscultation rapide des chaussées » ; Bulletin de Liaison des Laboratoires des Ponts et Chaussées ; N° 3 ; Sept 1963 ; pp. 1-16 (in french)

[2] R Kobisch and R Tricki ; « Determination of tree-layer pavement moduli from the analysis of deflection basins leasured with the 03 / 04 Lacroix deflectograph » ; BCRA Trondheim ; Jul.1990

[3] De Boissoudy and al. ; « Le deflectographe 04 » ; Bulletin de Liaison des Laboratoires des Ponts et Chaussées ; N° 129 ; Jan.1984 ; pp. 81-98 (in french)

[4] M Rengarth and J-P. Roche ; « Mesure du rayon de courbure des déflexions sur chaussées - Utilisation d'un inclinomètre de précision » ; ; Bulletin de Liaison des Laboratoires des Ponts et Chaussées ; N° 97 ; Sept. 78 ; pp.195-198 (in french)

[5] J-M Simonin and A.Riouall ; « Evaluation desf deflectographes » ; Bulletin de Liaison des Laboratoires des Ponts et Chaussées ; N° 208 ; Avril.1997 ; pp. 39-47 (in french)

[6] Ph. Lepert, G. Aussedat and J-M Simonin ; « Evaluation du curviamètre MT 15 » ; Bulletin de Liaison des Laboratoires des Ponts et Chaussées ; N° 209 ; Mai.1997 ; pp. 3-11 (in french)

[7]Ph. Lepert, J-M Simonin and R. Kobisch ; « Le FWD : performances, utilisation en France et en europe » ; Bulletin de Liaison des Laboratoires des Ponts et Chaussées ; N° 209 ; Mai.1997 ; pp. 19-28 (in french)

[8] J-M Simonin and al. ; « Méthode LPC d'exécution et d'exploitation de la déformabilité de surface » ; LCPC ; Collection techniques and méthodes ; N°39 ; Avril 1997

FINITE ELEMENTS INVESTIGATION OF PAVEMENT - FWD INTERACTION

A. SCARPAS[1], R. AL-KHOURY[1], C. VAN GURP[2]

[1]Delft University of Technology, Stevinweg 1, 2628 CN Delft, The Netherlands

[2]KOAC-WMD, Schumanpark 43, 7336 AS, Apeldoorn, The Netherlands

ABSTRACT

A computational study is presented of the interaction between falling weight deflectometer (FWD) devices and pavement structures. Wave generation and propagation due to an FWD pulse in pavement structures have been simulated by means of dynamic nonlinear finite element analyses. Constitutive modeling of the rate dependent asphaltic material response has been formulated on the basis of a comprehensive laboratory investigation.

The contribution focuses on the interaction between instrument characteristics (e.g. load pulse duration, rubber pad characteristics etc.) and pavement material characteristics. The numerical results compare favourably with available field data and enable the interpretation of the discrepancies between field measurements obtained from different instruments. Understanding of the causes of these discrepancies provides the key towards the development of harmonization and FWD device calibration protocols at network level.

Keywords: dynamic finite element analyses, FWD wave generation, FWD wave propagation, FWD calibration

INTRODUCTION

The Falling Weight Deflectometer (FWD) test has been developed in an attempt to obtain the material characteristics of road and airfield pavement layers. Because of its nondestructive nature, fast testing procedure and its resemblance to moving traffic action, the FWD test has become one of the most popular evaluation tests for the development of maintenance strategies for road and airfield pavements.

Many types of FWDs are available in the market. However, instruments from different manufacturers often produce different results when used at same locations of a pavement. As a consequence, deflection data generated by one FWD cannot be compared directly to those produced by another FWD. This is particularly pronounced when testing a pavement with soft subgrade and/or at high temperatures. Up until now there is no clear harmonization methodology capable of correlating measurements obtained from different FWD devices. Consequently, no objective calibration procedure is yet available for instrument maintenance and certification purposes.

Without such a procedure, inaccurate and/or biased data can lead to incorrect conclusions about the structural condition of the pavement and serious errors in estimates of the properties of the pavement layers. Estimates indicate that errors of 5% in FWD measurement results can lead to errors of up to 25% in pavement life prediction.

In addition, the lack of fundamental understanding of the reasons causing the above measurement discrepancies has hindered the development and adoption of objective calibration procedures for instrument maintenance and state certification purposes.

In view of the above, an extensive project of investigation has been initiated in Netherlands between the Delft University of Technology, KOAC□□WMD bv. and the Dutch Center for Information & Technology for Transport & Infrastructure (CROW) for the purpose of developing a methodology for harmonization of FWD measurements obtained from instruments of different types.

In the preliminary phase of the investigation, the results of which are reported in this contribution, three dimensional dynamic non-linear finite element analyses were performed by means of the Finite Element System CAPA-3D. The simulations focussed on the interaction between instrument characteristics and pavement material properties.

FINITE ELEMENTS MODELING

The finite elements system CAPA-3D [1] was utilized for the simulation of the dynamic non-linear response of pavement structures subjected to FWD load pulses. In the following several aspects of the finite elements modeling will be discussed.

Pavement geometry

The overall dimensions of the simulated pavement were 6 m wide and 15.4 m deep. The finite elements mesh consisted of 1040, 20-noded brick elements. A detail of the loaded region is shown in Fig. 1. Because of symmetry, only a quarter of the pavement was simulated.

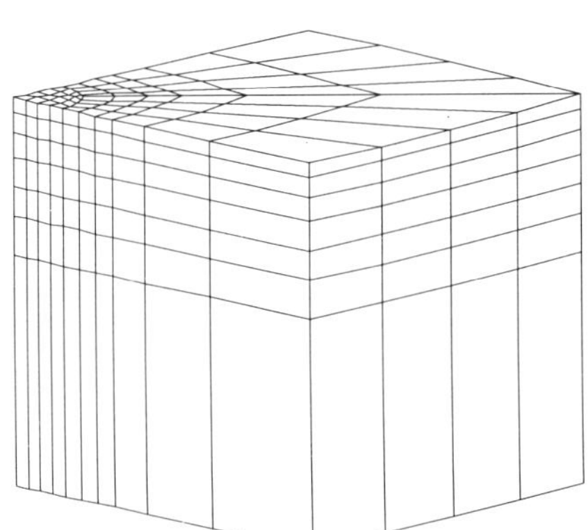

Fig. 1 Detail of finite elements mesh

The pavement model was assumed to be consisting of an asphalt concrete layer 150 mm thick with E=1000 MPa, a granular sub-base layer 250 mm thick with E=200 MPa and a sub-grade layer. Two different sub-grade types were examined. One representing a stiff sand layer with E=100 MPa and another representing a soft sand layer with E=25 MPa.

Pavement materials characteristics

A pre-requisite for utilization of the finite element method (FEM) is the availability of material constitutive models capable of describing the triaxial material response in the linear and the non-linear ranges. For the sub-grade and sub-base materials it was felt that adequate experimental information was available from the literature.

For asphalt concrete, a project of experimental and analytical investigation was undertaken by the authors. The goal was the development of a constitutive model for characterization of the material response under triaxial conditions. The model was implemented in CAPA-3D and was utilized for simulation of the dynamic non-linear response of asphalt concrete.

The theory of dynamic plasticity has been chosen as the most suitable constitutive framework for modeling of the strain rate dependent response of asphalt concrete. In similarity to inviscid plasticity, the plastic rate is defined as:

$$\dot{\varepsilon}_p = \dot{\lambda} \cdot \frac{\partial f}{\partial \sigma} \qquad \text{with:} \qquad \dot{\lambda} \geq 0 \ , \ f \leq 0 \ , \ \dot{\lambda} \cdot f = 0 \qquad (1)$$

Evolution of plastic flow is determined by the consistency condition $\dot{f}(\sigma, \dot{\delta}, \kappa) = 0$ in which, $\dot{\delta}$ is the deformation rate and κ is some measure of hardening/softening.

The flow surface proposed by Desai [2] has been chosen for simulation of the hardening response of the material. In stress invariant space it is defined as:

$$f_D = \frac{J_2}{p_a^2} - \left[-\alpha \cdot \left(\frac{I_1 + R}{p_a} \right)^n + \gamma \cdot \left(\frac{I_1 + R}{p_a} \right)^2 \right] \cdot (1 - \beta \cdot \cos 3\theta)^{-1/2} = 0 \qquad (2)$$

in which I_1, J_2 and J_3 are the stress invariants, θ is the Lode angle, p_a is the atmospheric pressure and R is the triaxial tensile strength.

Determination of the model parameters was carried out on the basis of the results of laboratory uniaxial compression and uniaxial tension tests. The tests were performed under displacement control on cylindrical specimens of an asphalt concrete mix of type 0/16 with 6% bitumen 80/100. The specimens were 100 mm high and 100 mm in diameter.

The imposed displacement was increased monotonically until complete annihilation of the strength of the specimen. Because of the significant influence of strain rate and temperature on the response of asphalt concrete, tests were performed at various levels of strain rate and temperature.

As expected, increasing strain rates and/or lower temperatures resulted to higher material strengths. Typical results are shown in Fig. 2. Details can be found in [3].

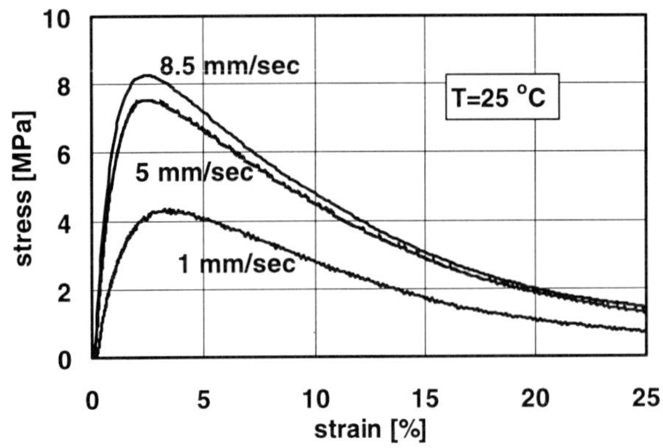

Fig. 2(a) Uniaxial compression test [3]

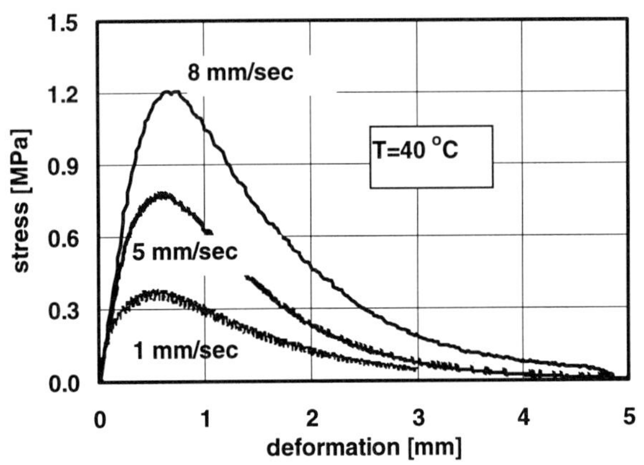

Fig. 2(b) Uniaxial tension test [3]

The Desai surface in inviscid formulation was also utilized for simulation of the non-linear response of both, the granular sub-base and the sand sub-grade. The non-linear material parameters are shown in Table 1.

Material	γ	β	n	R
Asphalt concrete	0.087	0.	2.	0.16
Granular sub-base	0.060	0.	2.5	0.
Sand sub-grade	0.028	0.	2.5	0.

Table 1 Non-linear material parameters

Loading foot plate

In order to investigate any possible interaction effects between the load foot plate and the pavement, parametric analyses were performed in which the geometric and material properties of the foot plate were included explicitly, Fig. 3. Various degrees of lateral restraint between the foot plate and the pavement surface were simulated by the introduction of an interface layer. The individual characteristics of each of the simulated foot plate materials are shown in Table 2.

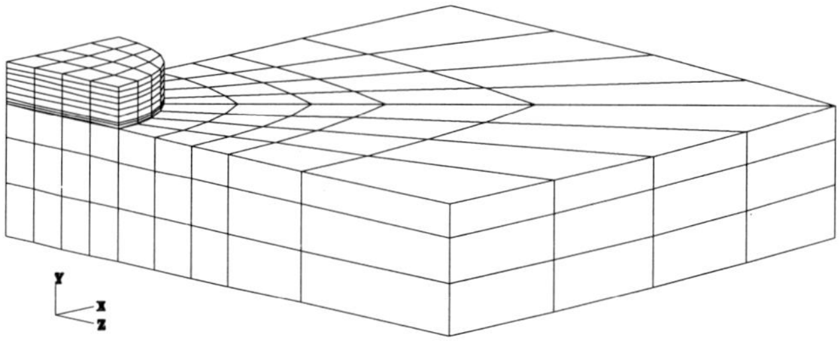

Fig. 3 FEM detail of loading foot plate

Material	thickness [mm]	E [MPa]	ν
Steel disk	19.2	200000	0.2
PVC	22.4	1000	0.3
Rubber pad	6.	5, 50, 250, 1000	0.49
Interface	0.1	1, 10	-

Table 2 Loading foot plate characteristics

FWD pulse shape

Two different pulse shapes were applied. One representing a half sine and one consisting of two successive peaks as shown in Fig. 4. The pulse amplitude was 0.707 MPa and was applied over an area of 150 mm radius. Pulse durations of 25 ms and 50 ms were examined.

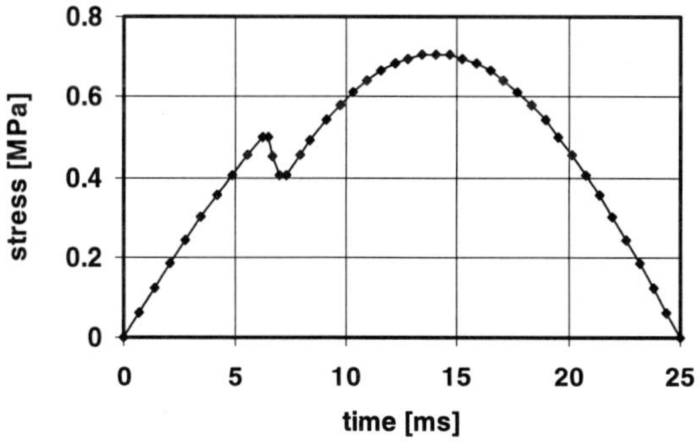

Fig. 4 Two peaks pulse shape

DYNAMIC PAVEMENT RESPONSE SIMULATION

Dynamic non-linear finite element analyses can provide invaluable information about the dynamic phenomena that occur within the body of the pavement as a result of a load impulse. By means of in-time integration of the equations of the dynamic motion of the system, the generation, propagation and internal reflection of the waves caused by the FWD pulse can be traced in time. Also, the influence of rate effects on the response characteristics of the materials of the pavement layers can be accurately simulated. This is of paramount importance for viscoplastic materials like asphalt concrete in which sensitivity to strain rate can alter substantially their engineering properties.

Analyses have been carried out for several combinations of the above mentioned pavement material properties and pulse characteristics. Some of the main conclusion of these analyses will be reviewed in the following.

Pulse duration vs. sub-grade stiffness

In Fig. 5, time snapshots of surface deflections are shown for two pavements, one with sub-grade stiffness E=25 MPa and another with E=100 MPa, subjected to a half sine 50 ms FWD pulse. Propagation of the wave front along both, the vertical and the horizontal directions can be observed. At any given instant of time, the differences between the actual deflection profile and the classical "static deflection bowl" utilized in back-calculation procedures are apparent.

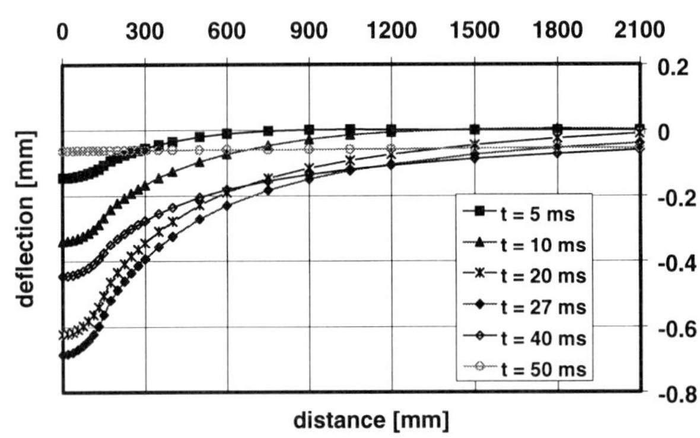

Fig. 5(a) Time snapshots of surface deflections for stiff subgrade (E=100 MPa)

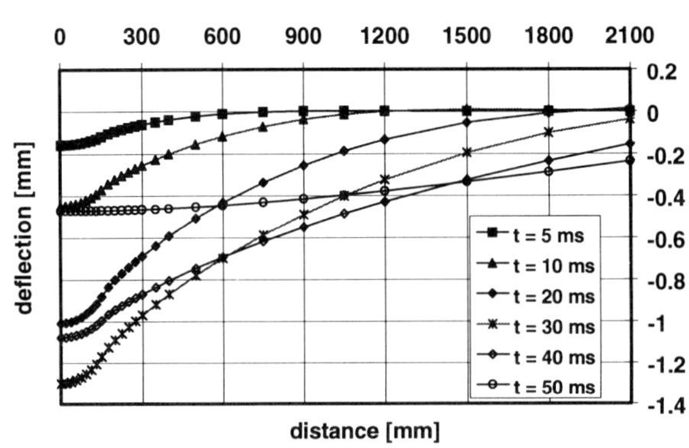

Fig. 5(b) Time snapshots of surface deflections for soft subgrade (E=25 MPa)

In Fig. 6, the surface deflections are compared when the pavements are subjected to a 25 ms and to 50 ms half sine FWD pulses. In agreement with field experience, the following observations can be made:

In the case of the stiff sub-grade pavement:

(a) the vertical surface deflection profile is not influenced by the duration of the load pulse, Fig. 6(a)

(b) although the vertical displacement profiles for both pulse durations are almost the same, the amount of energy transported in the body of the pavement as manifested by the surface Rayleigh waves is different, Fig. 7(a).

In the case of the soft sub-grade pavement:

(a) the vertical surface deflection profile is influenced by the duration of the load pulse. The deflection during the 50 ms pulse duration is about 20% higher than that during the 25 ms pulse, Fig. 6(b)

(b) there are only slight differences in the surface Rayleigh waves between the 25 and 50 ms pulse durations, Fig. 7(b).

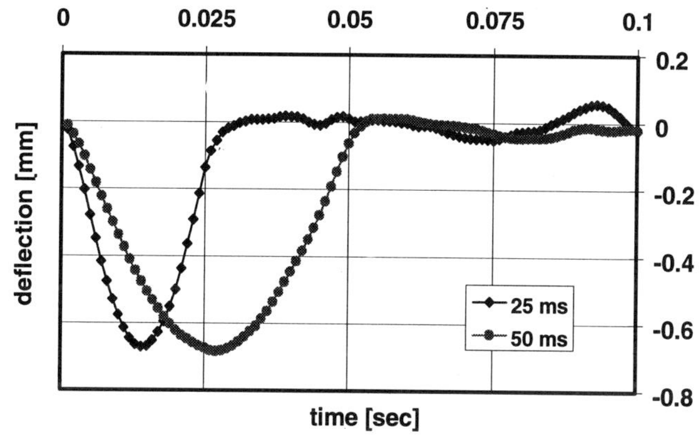

Fig. 6(a) In-time response of stiff (E=100 MPa) sub-grade

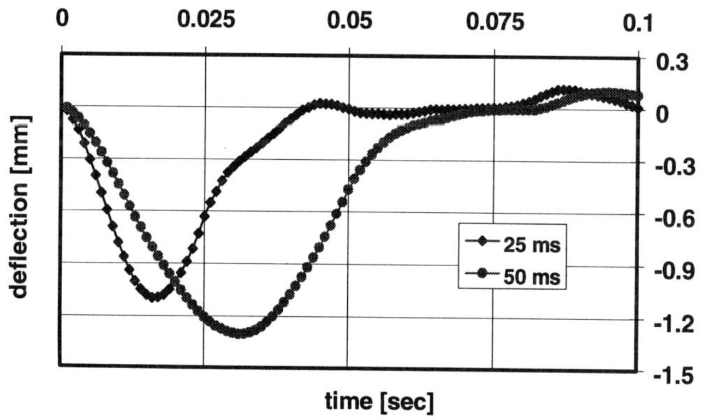

Fig. 6(b) In-time response of soft (E=25 MPa) sub-grade

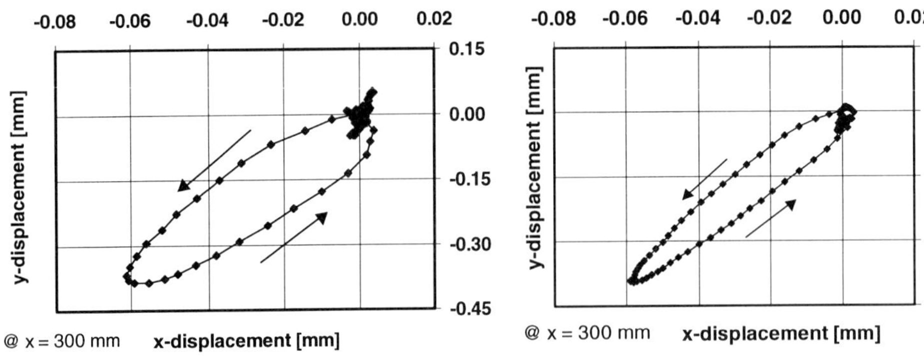

Fig. 7(a) Rayleigh waves in stiff sub-grade (E=100 MPa) pavement

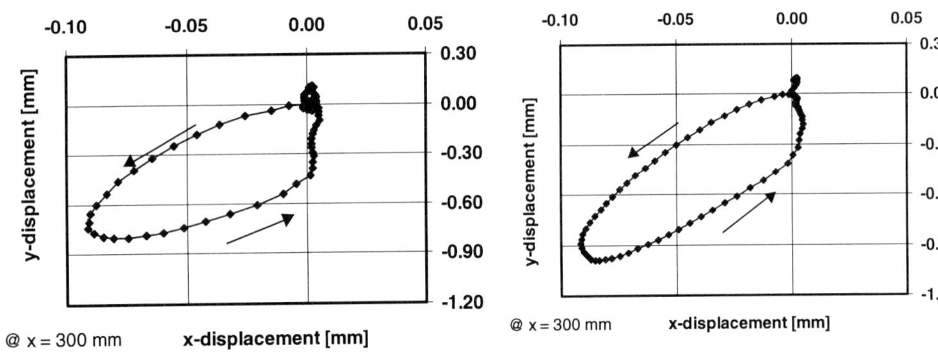

Fig. 7(b) Rayleigh waves in soft sub-grade (E=25 MPa) pavement

Influence of pulse characteristics

In order to clarify the issue whether the load pulse characteristics have any influence on pavement response two different pulse shapes were examined. A 25 ms half sine and one containing two peaks, Fig. 4. The case of a pavement with sub-grade E=100 MPa was considered. As shown in Fig. 8, no influences was observed on the magnitude of the maximum pavement surface deflection.

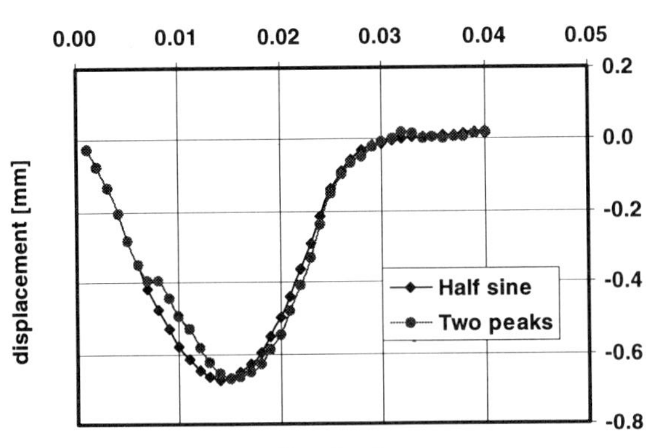

Fig. 8 Pavement surface deflections under different pulse shapes

Pulse shape vs. pad stiffness

The influence of the material properties of the rubber pad on the characteristics of the generated FWD pulse was investigated by including the foot plate in the finite elements mesh, Fig. 3. The analyses focussed exclusively on the material properties of the rubber pad, Table 2. Two different pulse shapes were examined. A 25 ms half sine pulse and the pulse of Fig. 4.

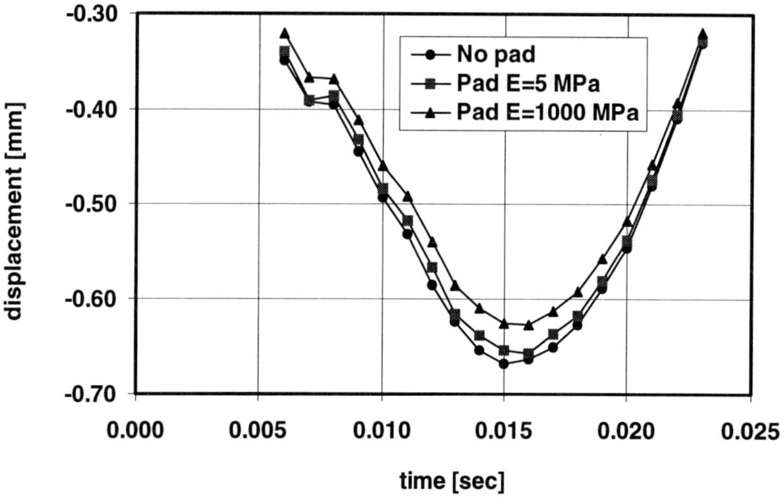

Fig. 9 Influence of rubber pad properties on pavement surface response

In the case of the half sine pulse, the influence of the rubber pad stiffness on pavement response was found to be negligible.

In the case of the two peaks pulse, for increasingly higher pad stiffnesses, some influence was observed on the overall surface deflections, Fig. 9, however, for all examined stiffnesses, the discontinuity between the load pulse peaks can also be traced in the generated response. Additional pulse shape characteristics are currently under investigation in order to clarify the issue further.

CONCLUSIONS

Nonlinear dynamic finite element analysis was capable of simulating the salient features of pavement response when subjected to an FWD load pulse. Understanding of the wave generation and propagation phenomena that occur is of vital significance before any attempt can be undertaken for the development of an FWD instrument calibration procedure. Such an attempt constitutes the second phase of this investigation and will be reported in the future.

This first phase of the investigation focused on identifying the interrelations between the various parameters affecting FWD-pavement interaction. This study has shown that the pulse duration (frequency) has no influence on the surface deflections of pavements with stiff sub-grades and has rather significant effect on pavements with poor sub-grades.

For the range and characteristics of the pulses that were examined no significant effect was found on the overall pavement behaviour.

Explicit inclusion of the foot plate geometry and material characteristics in the finite elements model was found to have no influence for the case of the half sine pulse. For the two peaks pulse, the stiffness of the rubber pad had some minor influence on the magnitude of the overall surface deflections. Additional future work is necessary for clarification of this issue.

ACKNOWLEDGEMENTS

CAPA-3D has been developed with partial financial assistance from the Dutch Foundation for Technical Sciences (STW). The authors are grateful to the members of the Study Committee P8, "Calibration of Falling Weight Deflection Equipment" of the Dutch Center for Information & Technology for Transport & Infrastructure (CROW) and in particular to its Chairman, ir. E. Beuving, for their support.

REFERENCES

1. Scarpas A. CAPA-3D Finite Elements System - User's Manual, Parts I, II and III, Faculty of Civil Engineering, TU-Delft, 1992.
2. Desai C.S. Modeling and Testing: Implementation of Numerical Models and their Application in Practice. CISM Courses and Lectures 311, Springer Verlag, New York, 1990.
3. Scarpas, A., Al-Khoury, R.I.N., van Gurp, C.A.P.M. and Erkens, S.M.J.G. Finite Elements Simulation of Damage Development in Asphalt Concrete Pavements. 8-th International Conference on Asphalt Pavements, University of Washington, Seattle, 1997, pp.673-692.

Fifth International Conference on the
BEARING CAPASITY OF
ROADS AND AIRFIELDS

GUIDELINES FOR CALIBRATION OF FALLING WEIGHT DEFLECTOMETERS

RENE CLEMEN, PRODUCTION MANAGER
Fuglesangsallé 16, 6600 Vejen
Denmark
Tel.: +45 7536 1111
Fax.: +45 7536 0986
E-mail: r.clemen@phoenix.dk

ABSTRACT

The interest and need for systematic pavement maintenance has been growing tremendously over the last decade. This has increased the demand for specialised equipment for planning and maintenance in practice and for pavement condition survey equipment.

The FWD is today the most used non-destructive testing (NDT) device for testing of load bearing capacities of road and airfield pavements.

The FWD equipment has undergone an impressive development and is a true state-of-the-art device using the latest electronics and computer technology.

In Europe and the USA research work has been done in order to harmonise test methods and calibration procedures for the FWD. In Europe this work is done by the Centre for Research and Contract Standardisation in Civil and Traffic Engineering (C.R.O.W.), the Netherlands. A special group - the Study Committee P7 "Falling Weight Deflection Testing" - has developed preliminary guidelines for Falling Weight Deflectometer calibration.

The calibration protocols are partially developed by the Study Committee P7 and partially with amendments from the protocols developed by Strategic Highway Research Program (SHRP), National Research Council at Washington, DC.

Different levels of the calibration programme have been developed, ranging from simple to difficult, and from low-priced and convenient to more expensive and comprehensive.

This paper will discuss new equipment which has been developed to comply with the new calibration tests performed according to the new protocols at verification centres.

INTRODUCTION

In Europe there are 3 manufacturers of Falling Weight Deflectometers FWD, namely PHØNIX Pavement Consultants, KUAB and Dynatest. All have their own FWD calibration programmes. Normally deflection transducers, load cell and temperature sensors are calibrated.

PPC has as the only manufacturer chosen to base its calibration programme on CROW (Preliminary Guidelines for Falling Weight Deflectometer Calibration). First of all this programme was chosen because it has been developed for FWDs operating with load pulse widths within the range of 20-35 msec. and our opinion is that this procedure gives the best calibration result - and it is an European standard. CROW operates with the following calibration procedures (protocols A-G):

A. Relative Calibration Verification of FWD deflection sensors.
B. FWD short-term repeatability verification
C. FWD long-term repeatability verification
D. Reference LVDT calibration procedure (Linear Variable Differential Transducer)
E. FWD deflection sensor calibration verification
F. FWD group field calibration procedure
G. FWD field calibration

When PPC calibrates FWDs this is done on the basis of Protocols A, B, D and E, which normally gives a good an secure calibration result. (Protocols C, F, G are for more comprehensive analysis programmes in which PPC takes part every time the Study Committee P8 and CROW arrange comparison tests among various FWD brands).

In stead of "just" using Protocol A, B, D and E, PPC has taken a step further in the endeavour to obtain the optimum calibration and thus the most precise and correct results.

PPC Calibration Equipment
PPC has developed a calibration equipment which works with the newest technology and principles within this sphere. Furthermore the equipment has a state-of-the-art design which makes it possible to bring along the equipment to the customer to be used at site.

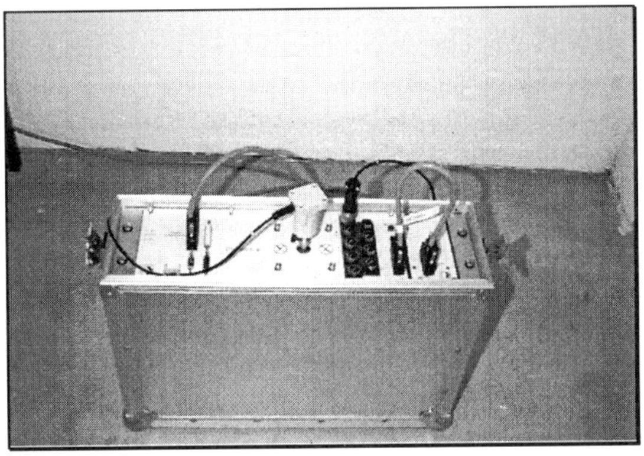

BASIC THEORY

When a geophone (deflection sensor) is calibrated the problem is that the curve is not linear in the low frequency range from 7-0 Hz.

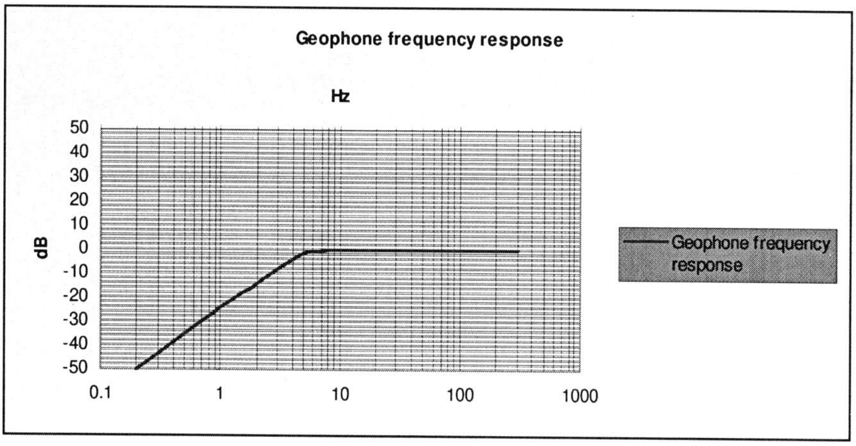

This problem may be compensated for by adding a fast filter at a natural frequency of 4.5 Hz (- 3 dB) which is similar but in opposite direction of the geophones own frequency characteristics (see below figure).

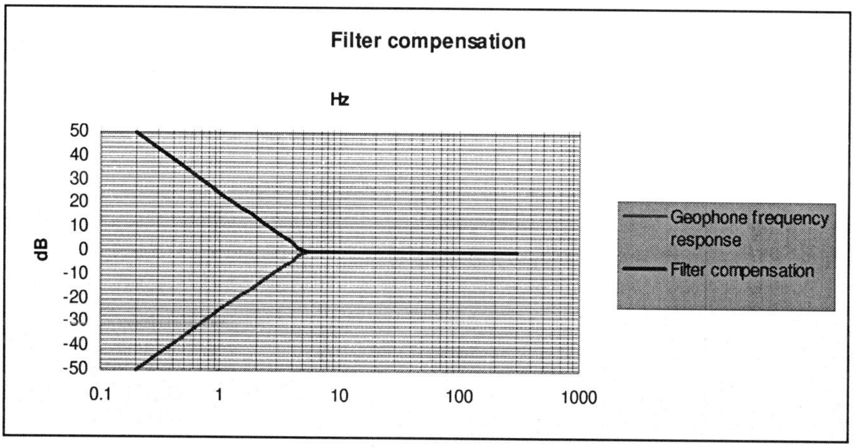

In this way a resultant characteristic is obtained which is linear from approx. 0 Hz to 300 Hz. To compensate for geophone unlinearity within the low frequency area in this way is not a new method but a method which has been used for many years.

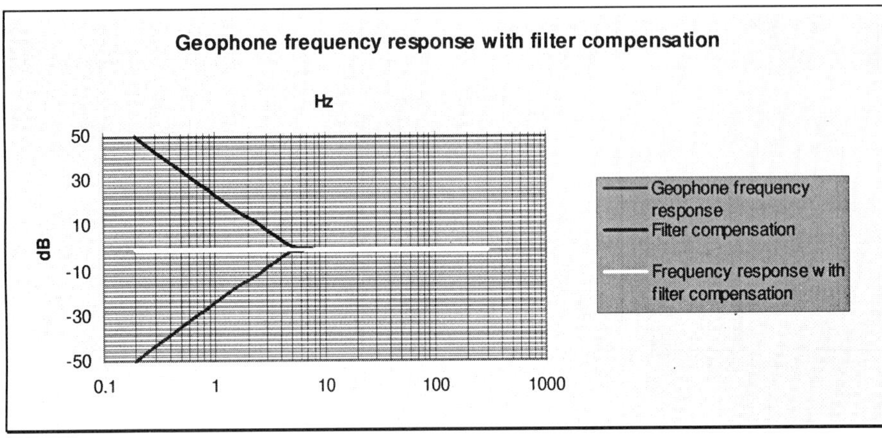

PPC differs from other FWD manufactures in that PPC provides for the fact that geophones are not uniform in their characteristics. The natural frequency has according to the geophone manufacturer a tolerance of +/- 0.5 Hz. This means that it may vary from 4 to 5 Hz. It is very important to consider this tolerance as it may give an inaccuracy on the peak value of up to 5%, which will again result in a large inaccuracy when back calculating.

If the geophone has a natural frequency of 4 Hz and a filter compensation of 4.5 Hz is used, this means that the frequency characteristic and thus the geophone output will be considerably wrong. That is up to 5% too high.

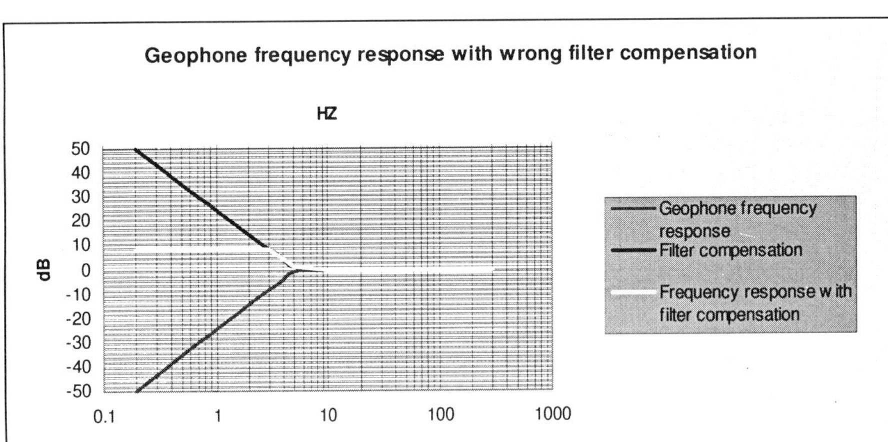

If a geophone has a natural frequency of 5 Hz and the filter compensation is 4.5 Hz, this will mean that the frequency characteristic and thus the output of the geophone will be noticeably wrong. That is up to 5% too low compared to the reference.

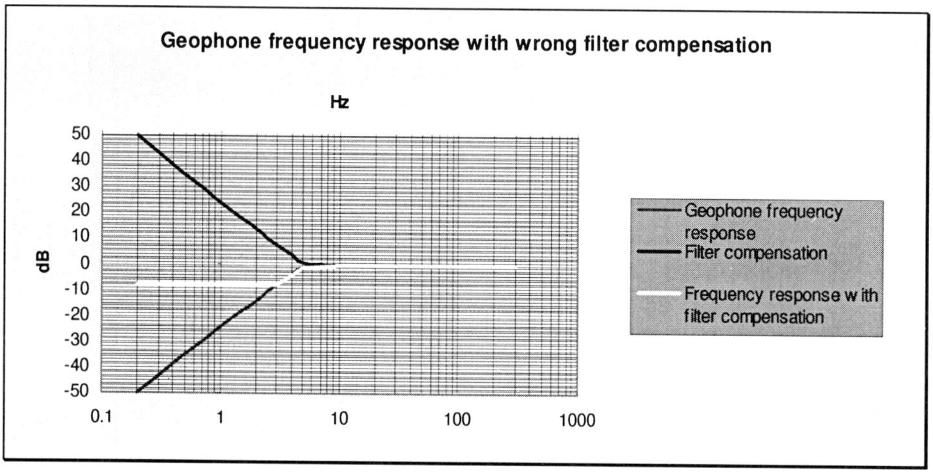

EQUIPMENT FOR ABSOLUTE CALIBRATION

When PPC makes an absolute geophone calibration, a special developed calibration equipment is used that is the newest technique within this area.

The system consists of a flight case which is easy and handy to transport worldwide and a note book Personal Computer with Windows 95 software.

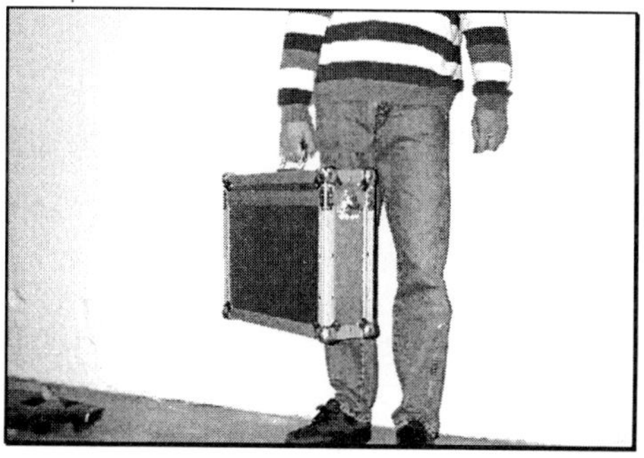

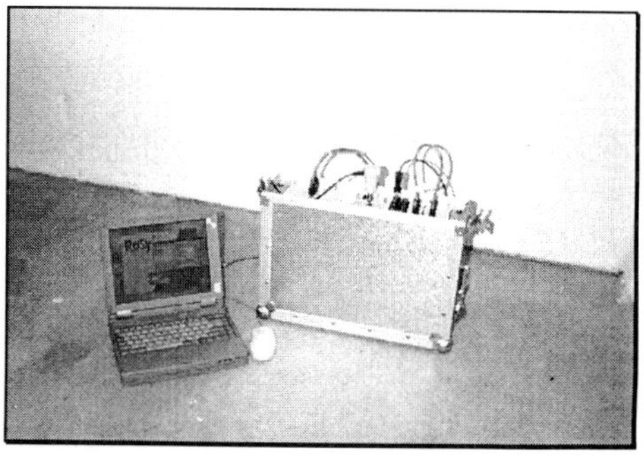

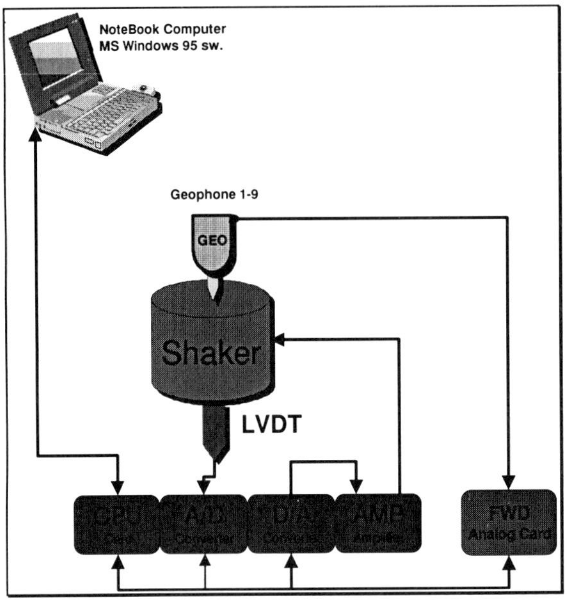

The flight case is constructed around a mini-shaker which is capable of generating pulses within the entire geophone frequency and displacement ranges. A highly accurate LVDT is mounted on the mini shaker which is used for reference and for control of the shaker's fluctuation.

The control of the computer and the calibration process is taken care of by a CPU built into the flight case.

PPC CALIBRATION PROCEDURE

When PPC calibrates geophones (displacement transducers), three different calibration parameters are used:

- Off set
- Cut off frequency
- Slope

The off set and the natural noise which influences the analog card and the channel and for which the system compensates automatically. This means that in practice the value is 0.
Cut off frequency is exactly the frequency where the system shall compensate for the non-liniarity in the low frequency range.
Slope is the amplification required to make geophone and calibration reference show the same peak value.
These three software values have the great advantage that they do not change over time and the current development of the calibration may be followed.

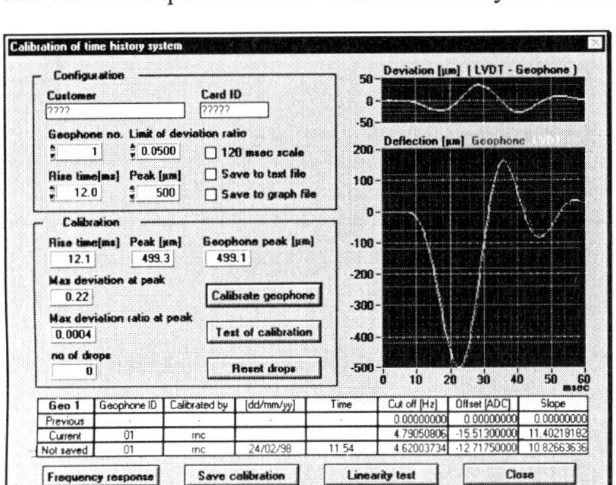

Calibration is done at a given rise time (frequency) and peak value which may be changed by the operator as required.

When the system has automatically found offset, cut off frequency and slope, a test drop is made with the found calibration data. The operation may now decide whether the found values shall be accepted.

Normally the values will be within the below range:

	Min.	Max.	
OFFSET	- 100	+ 100	ADC
CUT OFF FREQ.	4.0	5.2	Hz
SLOPE	10	12	gg

The found calibration values may now be tested in various ways to find out if they are acceptable. A number of test drops may be made to check that "max. deviation ratio at peak" does not exceed a fixed value. Normally the deviation from the reference is below 0.1% This test does however not indicate whether the cut off frequency is correct but only whether the slope is correct. In order to check the cut off frequency a frequency response test must be made.

Frequency response

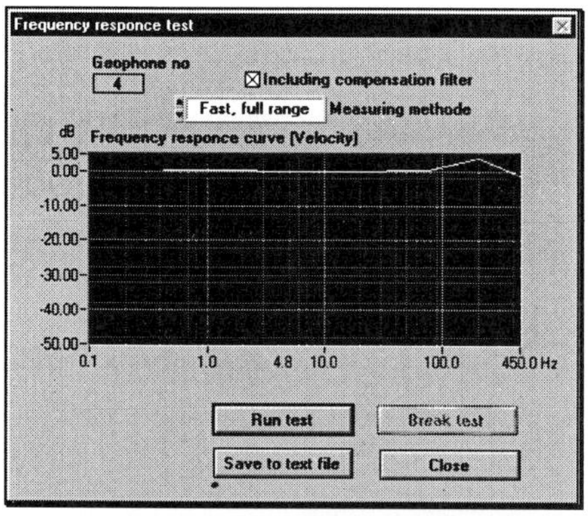

After a correct calibration a frequency response should look like this response.

Under frequency response it is possible to see which frequency characteristic the geophone has within the range of 0.4-450 Hz.

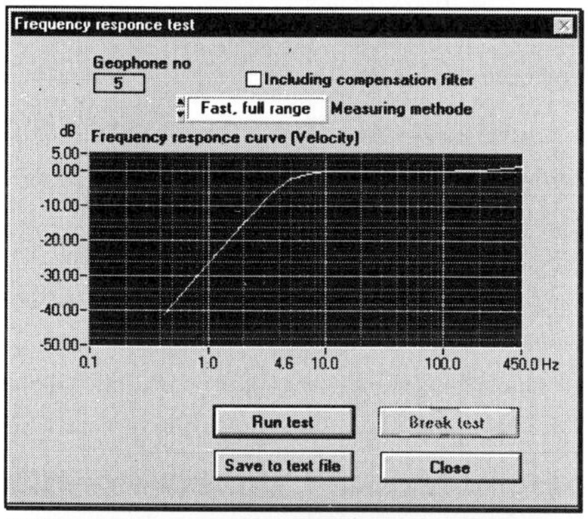

Here is the the same geophone shown with the same calibration parameter (offset and slope) but without compensation filter.

As it clearly appears, the geophone frequency response is no longer linear.

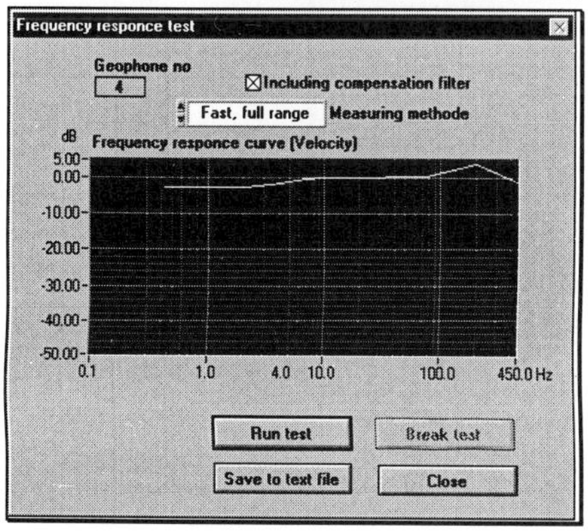

Here is the same geophone again with the same calibration parameter. However the cut off frequency has been changed with -0.5 Hz.

As it appears, the geophone becomes unlinear from approx. 7 Hz.

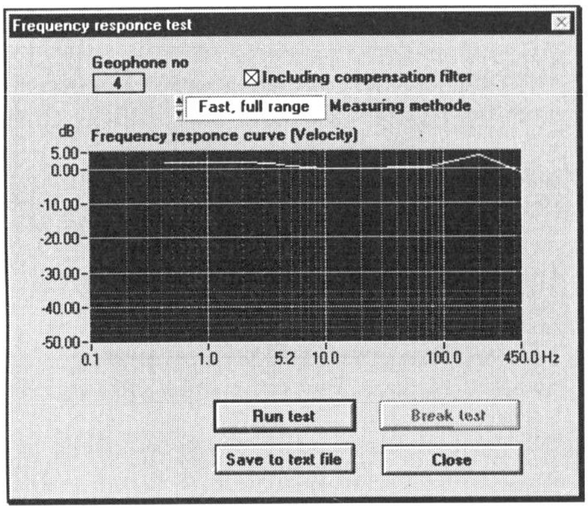

Here is the same geophone again with same calibration parameter. However cut off frequency has been changed with +0.5 Hz.

As it appears, the geophone is unlinear from approx. 7 Hz but this time in the opposite side.

Peak value
In the below table it appears how large an influence a correct cut off frequency has on the peak value.

Peak value test at diff. cut off frequencies

Cut off frequency Hz.	Drop. No.	REFERENCE µ m.	GEOPHONE µ m.	DEVIATION µ m.	DEVIATION %
Correct cut off frequency at 4.75 Hz.	1	497	497.1	-0.1	-0.02
	2	501	500.9	0.1	0.02
	3	500.6	500.2	0.4	0.08
	4	499.4	498.8	0.6	0.12
	5	500.4	500.4	0	0.00
	6	499.1	498.7	0.4	0.08
	7	500	500.6	-0.6	-0.12
	8	500.2	500.3	-0.1	-0.02
	9	500.7	499.8	0.9	0.18
	10	500.7	499.9	0.8	0.16
					0.05
Incorrect cut off frequency at 4.0 Hz.	1	499.4	478.3	21.1	4.23
	2	499.6	478.7	20.9	4.18
	3	499.4	478.7	20.7	4.14
	4	498.9	478.3	20.6	4.13
	5	500.7	479	21.7	4.33
	6	500.4	479.3	21.1	4.22
	7	500.6	478.7	21.9	4.37
	8	499.4	478	21.4	4.29
	9	499.8	478.1	21.7	4.34
	10	499.9	478.4	21.5	4.30
					4.25
Incorrect cut off frequency at 5.2 Hz.	1	500.5	512.6	-12.1	-2.42
	2	499.2	511.4	-12.2	-2.44
	3	499.4	511.5	-12.1	-2.42
	4	500.6	512.4	-11.8	-2.36
	5	499.2	512.3	-13.1	-2.62
	6	499.6	512.4	-12.8	-2.56
	7	501.1	513.6	-12.5	-2.49
	8	500.3	512.6	-12.3	-2.46
	9	500.3	512.5	-12.2	-2.44
	10	498.8	511.6	-12.8	-2.57
					-2.48

Linearity test

After the frequency test the system automatically performs a linearity test in accordance with the conditions set forth in C.R.O.W. protocol E. The linearity test gives a good picture of the quality of the calibration as more combinations of rise time and amplitude for the pulse are tested.

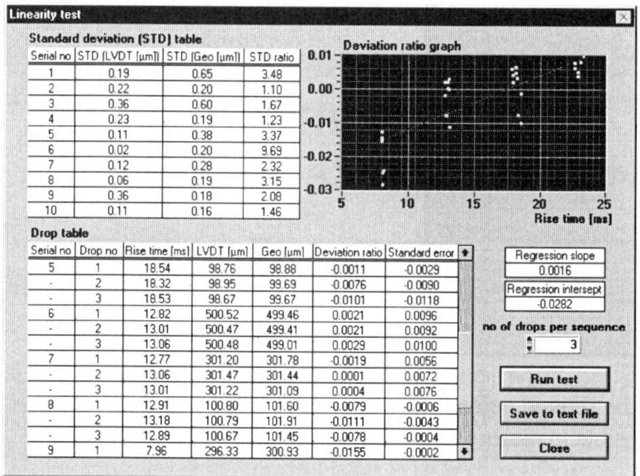

Standard deviation (STD) table

Serial no	STD (LVDT [µm])	STD (Geo [µm])	STD ratio
1	0.19	0.65	3.48
2	0.22	0.20	1.10
3	0.36	0.60	1.67
4	0.23	0.19	1.23
5	0.11	0.38	3.37
6	0.02	0.20	9.69
7	0.12	0.28	2.32
8	0.06	0.19	3.15
9	0.36	0.18	2.08
10	0.11	0.16	1.46

Drop table

Serial no	Drop no	Rise time [ms]	LVDT [µm]	Geo [µm]	Deviation ratio	Standard error
5	1	18.54	98.76	98.88	-0.0011	-0.0029
-	2	18.32	98.95	99.69	-0.0076	-0.0090
-	3	18.53	98.67	99.67	-0.0101	-0.0118
6	1	12.82	500.52	499.46	0.0021	0.0096
-	2	13.01	500.47	499.41	0.0021	0.0092
-	3	13.06	500.48	499.01	0.0029	0.0100
7	1	12.77	301.20	301.78	-0.0019	0.0056
-	2	13.06	301.47	301.44	0.0001	0.0072
-	3	13.01	301.22	301.09	0.0004	0.0076
8	1	12.91	100.80	101.60	-0.0079	-0.0006
-	2	13.18	100.79	101.91	-0.0111	-0.0043
-	3	12.89	100.67	101.45	-0.0078	-0.0004
9	1	7.96	296.33	300.93	-0.0155	-0.0002

Data analysis software

PPC has also developed an advanced user friendly software which can be used for data analysis and manipulation.

For many years it has not been quite clear what time history data could be used for. So far only peak/max. values have been used for back calculation. However a tool is now available, which makes it easy and quick to make FFT analyses (Fast Fourier Transformation - frequency analyses) and in this way see what the frequency content of a drop and the peak value.

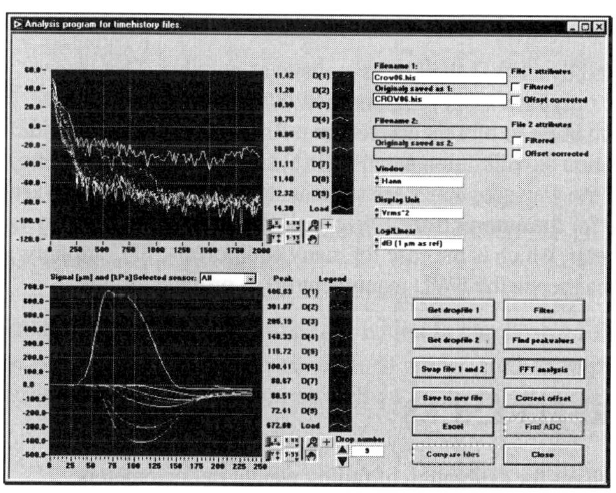

Another facility of the tool is low and high pass filter. These filters may among other things be used for noise filtering of time history signals or for comparison with data from other FWD manufacturers using filter manipulation in data collection software (see below).

INTRODUCTION

Currently, among the methods of overlay design of pavements that are used in the world, prevail and have a great popularity the procedures based on measurements of deflection, given it relatively simple of its application. However, among some groups of Mexican technicals devoted to these disciplines, anxieties exist, due to unfortunate experiences in the use of these methods, whose improper application not only it would not solve the problem, but would aggravate it given the scarce available resources in the country; for the previously exposed, in this work is made a series of reflections in relation to the use of measures of deflection, in its meaning as in the application of its results, finally, it is offered a series of recommendations for the overlay design of asphalt pavements.

CHARACTERISTIC OF THE DEFLECTION MEASUREMENTS

Relationship between maximum deflection and structure

It does not exist a clear relationship between the design of a pavement and the deflection measurement on the surface, since it can exist an infinity of different pavements that have the same deflection but a different structure and behavior.

The previous fact has been described in different articles, (1,2 and 3), as is observed in Figure 1, for the test road 1, 2 and 3, which some of the highest deflections correspond to thick pavements while pavements quickly failed, gave apparently acceptable values; only in the road 1 is observed the logic trend, to smaller thickness pavement, greater value in the deflection measured with Benkelman beam.

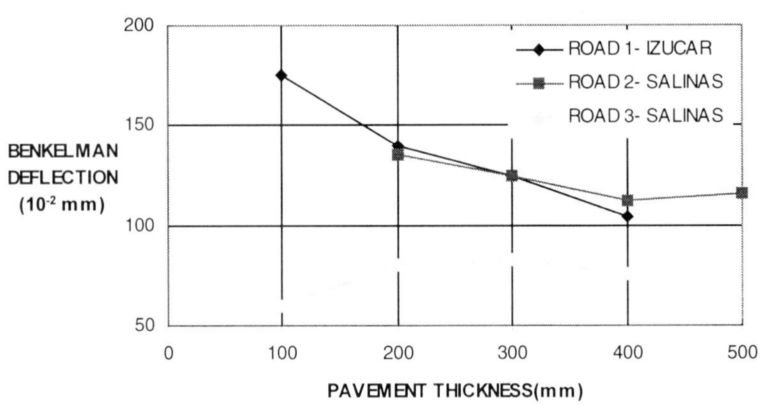

Figure 1. Results of deflection tests on test roads of Izucar and Salinas (3).

These results are important due to the fact that each test road has the same quality of subgrades and pavement materials in all its sections, being the only one variable, the pavement thicknesses.

Evolution of the deflection

Many studies show that the evolution of the deflection depends fundamentally on the type and structure of the pavement, as well as of the traffic and climatic conditions, some of the typical trends of the evolution of these measurements are described below.

A typical graph of the performance of the deflection for a well designed pavement, with asphalt concrete surface, is shown in Figure 2 (3), where it is appreciated three phases:

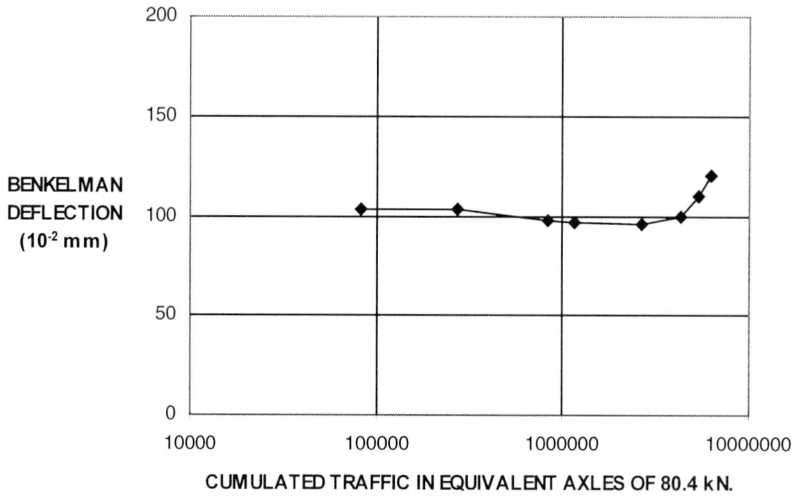

Figure 2. Relationship between Benkelman deflection and cumulated traffic.

a) The initial, immediate to the construction, when the pavement structure is consolidated, the surface course is toughened and the deflection shows a rapid decrement.
b) The functional, when the pavement performs service to the traffic and the deflection stays constant or is increased slightly.
c) The failure stage, when the destructive effect is accelerated by the traffic and climatic agents while the deflections grow significantly.

The graph of Figure 2 corresponds to a test road in Mexico, with pumiceous subgrades and a pavement composed by 240 mm of base and subbase layers, with surface course of 60 mm of thickness, whose service life was for more than 10 years, a cumulated traffic of 5.2 million of equivalent axles of 80.4 kN and whose failure was by fatigue cracking.

Some asphalt pavements, with thin surface courses, sustain adequately the effects of the traffic and environment, showing a reasonable increase in the magnitude of the deflection up to a point where, as is observed in Figure 3 (1), due to effect of the over compaction of the materials, begins to show slowly decrements in the magnitude of the deflection up to the functional failure.

The pavements shown in Figure 3, belong to three sections of the tests ring number 3, tested in the circular track of the Engineering Institute of the UNAM, whose pavements had granular base layer from 400 to 600 mm and thin surface courses. The subgrades formed by poor quality silty or sandy clays, ML - CL with saturated CBR of 7%.

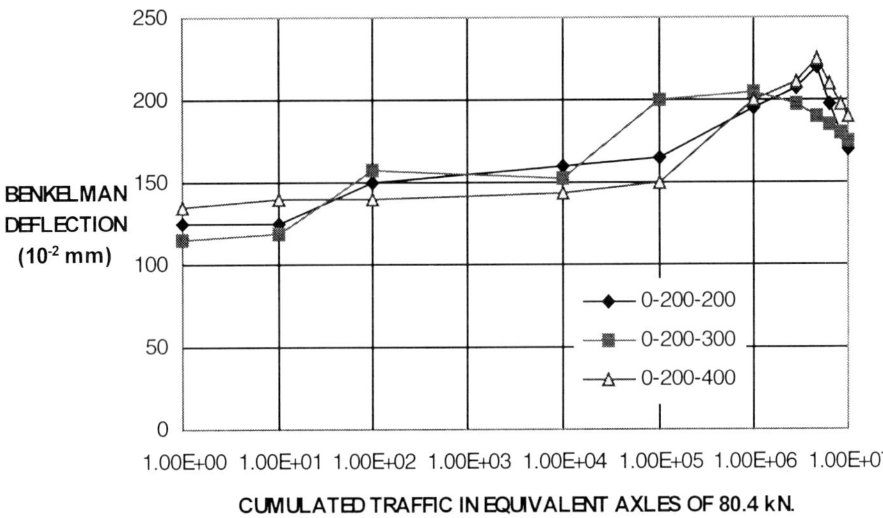

Figure 3. Benkelman deflection versus cumulated traffic.

In the same Figure 3, it is also appreciated that the variation of the deflection with the cumulated traffic is random, since the order of magnitude of the deflection varies according to the volume of cumulated traffic when the test is accomplished, for example the section with design 0- 600 has the greatest initial measurement, to 100 applications the least and for 1,000,000 , is the intermediate.

Experiences in other countries (4), have shown that pavements with a rigid superior layer (stabilized with portland cement, asphalt, etc.), exhibit a constant level of deflection, up to a point in where show sudden increases in the magnitude of the deflections due to the fact that such layer begins to suffer structural cracking.

In some countries, the climatic effect causes meaningful variations in the deflection, as is the case of tropical countries where the rain season causes weakening of the roads for effect of the moisture increment in the materials. Also, in the case of the Nordic countries, where the freezing and thaw effect weakens notably the resistance of the roads. In the case of Mexico, this effect has been studied with detail in the central zone of the country, the densest region in population and highways, by means of 52 test sections with pavements from 200 to 700 mm of thickness, subgrades formed by plastic fine materials that vary from regular to poor quality, in highways with more than 10 service years.

In the development of the study during 4 years (1972-1975), two annual measurements with equipment Dynaflect were accomplished, the first between the months of April and May, before the beginning of the summer and the second in the months of September and October, final part of the rains season when we expected that the materials that form the road structure had the greatest moisture of the year. In these conditions the analysis of the deflections (Figure 4) showed that the seasonal variations are of scanty importance and though the data dispersion is considerable, a sensitive change in the resistance is not appreciated, therefore, the climatic effect for most of the highways of Mexico is uniform; though different isolated cases will have to be analyzed, as the roads located in the mountainous zones of the north of the country or some tropical areas with very high rainfall.

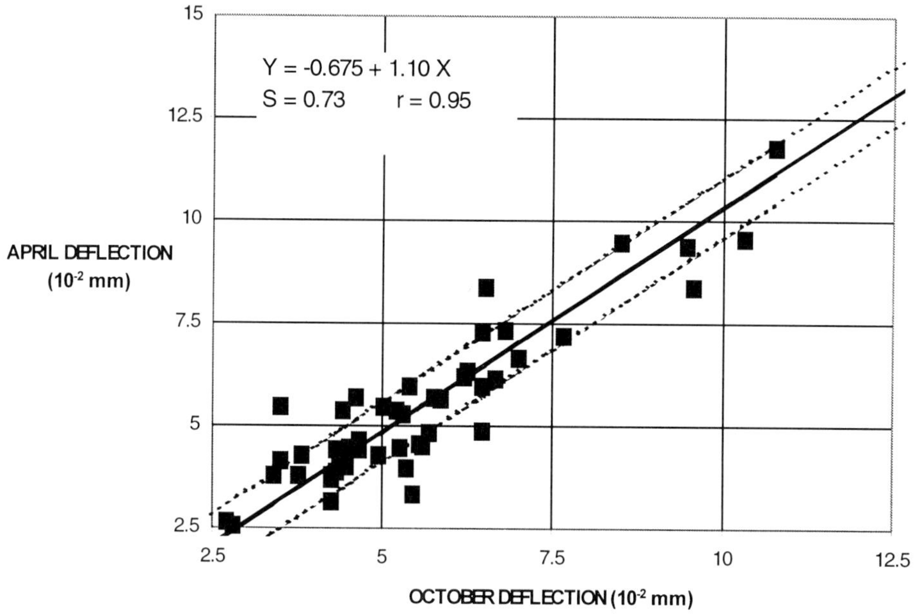

Figure 4. Correlation between two series of tests with Dynaflect Device.

Those same seasonal trends have been found with the use of the falling weight deflectometer in several highways located in the central part of the country (5).

In relation to the variation of the deflection with the temperature, it can be said that it has been very studied and only it is necessary to correct the value of the deflections when the asphalt surface course has a greater thickness to 50 mm or when the temperature of the surface course during the measurement defers in more than a Celsius degree of the reference, and in the case of using equipment that take readings at different distances from the apply load, only the deflection maximum and those found within a radio of 500 mm, should be corrected.

As conclusion, is observed that the evolution of the deflection depends on, as was established at the beginning of this discussion, the type and structure of the pavement, the traffic and environmental conditions, and the studied period of service life.

Relationships between deflection and service life

Currently, one of the principal applications of the measures of deflection is the use of them to overlay design of pavements in service, on the base that the remaining life is in function of the level of deflection; with that principle, several organizations have developed their methods, based on experiments more or less vast according to the resources or their own needs; but as already it has been seen that the deflection by itself is not a complete measure of the bearing capacity, it can not be used to predict the service life of a pavement in any condition.

As a rule, from the experiences of organizations as the Road and Transportation Research Laboratory of England, the California Department of Transportation, The Asphalt Institute and others, is accepted that for certain specific conditions in each case (structures, traffic, environment), exists a relationship between the initial deflection of a pavement, and its remaining life, but these correlations vary, as was said, according to the specific conditions of each case.

In Figure 5 are presented the different correlations between initial deflection and service life, that use some of the specialized organizations including one developed in Mexico (3); the examination of Figure, shows the so notable dispersion in the forecasts of the service life, for example for a deflection of 0.7 mm, the remaining life of the pavement can vary from 65,000 to 3,000,000 of equivalent axles of 80.4 kN, according to the criterion that is used.

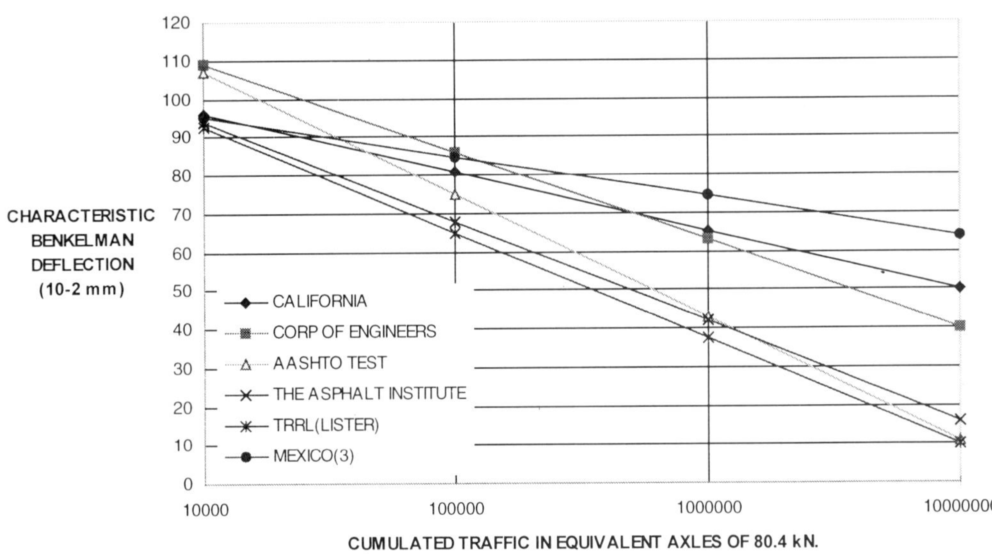

Figure 5. Comparison of different correlations between Benkelman deflection and service life.

Prado(7), in the test circular track of the Engineering Institute of the UNAM, found that for pavements with thin surfaces courses with the same subgrades exists a relationship between the initial deflection of the pavement and its service life, when the quality of the base layer is similar.

The correlation among measures of deflection and life of the pavement resulted different for the types of base layers (crushed basalt, volcanic slags), giving an inferior duration the pavements with base of slags, even when they had the same initial deflection. On the other hand, it was not found correlation defined between the deflection measurements after 100,000 repetitions with the remaining life of the pavement.

The same described limitations have been found in different studies, propitiating the development of techniques for the calculation the layer moduli, in order to characterize the pavement structures (4, 8, 9).

As conclusion of this clause, it can be said that for certain specific conditions (type of pavement, quality of materials, environment), relationship between the initial deflection and the service life of pavement has been found, but this relationship changes notably, with any variation in material as well as in structural or environmental conditions; but furthermore not finding relationship between the measurements of deflection in any moment and the remaining life, even for the case of conditions very controlled as in the test circular track, hinders more the establishment of forecasts for purpose of overlay design.

METHODS OF OVERLAY DESIGN OF ASPHALT PAVEMENTS, BASED ON DEFLECTION MEASUREMENTS

The development of the methods of pavement reinforcement, it has been adjusting to the era and the needs of the moment, the traditional approach until before 1960s, were based on layer analysis and experience; from those years, the not destructive procedures, based on measurements of superficial deflection, began to be used and developed with a wide acceptance, so that currently they are still used, in spite of that from 1970s, analytical methods based on mechanical models more or less supported by experiment have appeared.

The principal reasons of the popularity of the methods based on deflections, are their rapidity and relative economy.

Among these methods, those which have a greater experimental support, are found the corresponding to Road and Transportation Research Laboratory of England, California Department of Transportation, The Asphalt Institute, etc., among others. These methods have several common characteristics among them: to handle a permissible level of deflection, to suppose that upon reinforcing a pavement is reduced the level of deflection at a permissible level, etc.

In spite of the coincidences in several of these methods, none established a criterion of universal application and furthermore, none of these systems are interchangeable for the strong differences in the correlations between deflection and performance (Figure 5), that can be attributed to variations in type and structure of pavements, materials, environmental conditions and measurement technics.

In the last years, the efforts have been increased to determine the structural properties of the pavements on base of the tests of deflection, with the use of falling weight deflectometers, measuring the deformation at least in five points, obtaining the maximum value from the deflection, as well as the curvature from the distorted zone. With the previously exposed and knowing the layer thicknesses and depth of the rocky layer, if exists, the layer moduli are calculated by computer programs based on elastic layers theories or finite elements. The process ends when the best adjustment are obtained between the deflection measures and the calculated.

Currently many difficulties are found that do not permit to use these methods extensively, among them, that in many cases, there is not unique solution, in other words, that different sets of layer moduli can result in a same deflection profile, situation that is complicated when the pavement has very thin surfaces courses or cement treated base layers. (10).

To confirm the previously exposed, in an accomplished study (10) to measure the precision of these methods, data of 26 measurements to 13 specialized agencies were given; the results showed a great variation in the modules calculated by each one of the agencies, though five of them achieved reasonable results, what implies that such agencies have personal with most experience in the layer moduli calculation than the others.

Here , is pointed out the importance of the complementary information concerning layer thicknesses, characteristic of resistance of the pavement materials, etc; for that it can be used the radar equipment, borings and laboratory tests, all to calculate adequately the layer moduli, in whose analysis is reiterated, it is necessary the good judgement and the experience of the pavements engineer.

Currently, it is developing prudently the application of the results of the investigations, in order to produce methods for the design of pavements reinforcement, efficient and reliable.

CONCLUSIONS

The deflection for itself does not permit to characterize adequately a pavement structure, therefore it is necessary to bind it with additional information as curvature radio of the distorted zone, characteristic of the materials, existing structure, environmental conditions, type of traffic and other.

The measures of deflection are very useful, since by the use of them can be obtained general resistance conditions, to locate homogeneous highway sections (zoning) or singular points (of high or low resistance); accomplished in periodical form, permit to detect trends in the highways structural behavior, seasonal resistance variations, etc.

Other practical and valuable utilization of these measurements, using its rapidity and simplicity, it is the control of the degree of compaction of the materials during the construction, which is practiced in Mexico since 1973 (6, 11).

The methods of overlay design based on measurements of maximum deflection, are limited to the particular conditions for those which were established the correlations.

The methods of overlay design based on the layer moduli calculation are in process of development, since this design is currently more an art than a technique.

RECOMMENDATIONS

In Mexico, to improve the use of the measurements of deflection, is complemented the information with the accomplishment of a condition survey and borings in points elected after of analyzing the results of the deflections. In each boring are measured layer thicknesses, are extracted material for tests of quality in the laboratory and in some few borings, are accomplished plate bearing tests at three levels to characterize more adequately the layer moduli and therefore, to accomplish a overlay design more reliable.

To give idea of the work volumes required for the overlay design, in a section of 36.0 km of length, were accomplished 360 deflection measurements and 12 borings which in 4 were made plate tests.

With the described method, 12 overlays have been designed approximately, that are in service, all they currently in very good conditions, but since all have less than two years in operation, it is not possible to extract conclusions.

REFERENCES

1. Corro S. y Prado G. "Diseño Estructural de Carreteras con Pavimento Flexible", Informe No. 325, Instituto de Ingeniería, U.N.A.M., México, D.F.,1974.
2. Padilla, E. y González, E. " Estudio del Comportamiento de Pavimentos Flexibles en Tramos Representativos de la Red Nacional, Etapa 1974", Informe interno entregado a la SOP, Instituto de Ingeniería, U.N.A.M., México, D.F.,1974.
3. Padilla, E. "Reflexiones Sobre el Uso de las Medidas de Deflexión Superficial en Pavimentos Flexibles", 10ª Reunión Nacional de Ingeniería de Vías Terrestres, Guanajuato, Guanajuato, México, 1992.
4. Jeuffroy, G. "Introductory Paper Session no. VI", International Simposium on Bearing Capacity of Roads and Airfields, Proceedings, Vol.2, Trondheim Norway, 1982.
5. Orozco, R.V. y Torres, R. "Evaluación de la Capacidad Estructural de Pavimentos a Partir de Mediciones de Desplazamientos Verticales con el Deformómetro de Impacto", 11ª Reunión Nacional de Ingeniería de Vías Terrestres, Morelia, Michoacán, México, 1994.
6. Padilla, E. "Pavement Design in Pumice Subgrades, A Method that Combines Layered Elastic Theory and Plate Bearing Tests", Second Transportation Research WorkShop, Phoenix, Arizona, U.S.A., 1982.
7. Prado, G. "Datos Complementarios de la Ponencia de México, Referentes a Medidas de Deflexión con Viga Benkelman", XV Congreso Mundial de Carreteras, Memoria AI PCC, México , D.F., 1975.
8. Ullidtz, P. and Stubstad, R.N. "Why Bother about Layer Module", Third International Conference on Bearing Capacity of Roads and Airfields, Proceedings Vol.2, Trondheim Norway, 1990.
9. Molenaar, A.A. "State of the Art of Pavement Evaluation", Fourth International Conference on Bearing Capacity of Roads and Airfields, Proceedings Vol.2, Minneapolis, Minnesota, U.S.A., 1984.

10. Lytton, R.L., Germann, F.P., Chou Y. and Stoffels S.M. "Determining Asphaltic Concrete Structural Properties by Non Destructive Testing". National Cooperative Highway Research Program, Report 327, Transportation Research Board, Washington, D.C., U.S.A., 1990.
11. Padilla, E. "Use of Nondestructive Tests for Construction Control of Roads", Fourth International Conference on Bearing Capacity of Roads an Airfields, Proceedings Vol.2, Minneapolis, Minnesota, U.S.A., 1994.

Fifth International Conference on the
BEARING CAPASITY OF
ROADS AND AIRFIELDS

LOAD TRANSFER MEASUREMENTS ON IN-SITU CONCRETE PAVEMENTS

DR. KIERAN FEIGHAN

Managing Director, PMS Pavement Management Services Ltd, 226 Harolds Cross Road, Dublin 6W, Ireland. Phone: +353-1-4979760. Fax: +353-1-4979300. kfeighan@iol.ie.

ABSTRACT

A major rehabilitation project is currently underway to widen one of the most heavily trafficked national routes in Ireland. The original roadway was constructed over 30 years ago, and consists of 2 lanes of reinforced concrete slabs in each direction, with a flexible hard shoulder. The rehabilitation project involves widening of the roadway to three lanes in both directions with construction of a new lane (flexible construction), with repair and strengthening to the original lanes using a bituminous overlay typically 150 mm in thickness.

Visual inspection of the existing lanes showed significant faulting at a large number of joints. The original construction was typically of concrete slabs on grade, without a significant sub-base layer. There were concerns about the stability of the joints, and at the possibility of premature reflection cracking in the bituminous overlay layer.

Extensive Falling Weight Deflectometer (FWD) testing was carried out to determine joints with excessively high deflections and/or joints showing excessively poor load transfer. Criteria were developed to indicate joints which would benefit from sub-sealing of the joints. Joints thus identified were sealed by the contractor using pressure grouting techniques, and further FWD testing was carried out seven days later to determine the success of the maintenance. Joints showing unacceptable results were re-grouted, and tested again a further seven days later. Criteria for identification of joints which are likely to benefit from grouting are put forward based on the results of the various testing phases.

INTRODUCTION

The N7 National Primary route links the Republic of Ireland's capital city, Dublin with the 2nd and 3rd largest cities, Cork and Limerick. The section of the route closest to Dublin was constructed as a dual carriageway section over 30 years ago. Each carriageway consisted of 2 lanes of reinforced concrete slabs on a variable quality sub-base, with a flexible hard shoulder. The slabs were typically 20 to 25 metres in length, 4.8 metres in width, and 225 mm in thickness. The sub-base provided varied in thickness and quality. As a result of large increases in traffic, both inter-urban and commuter, it became necessary to widen the roadway to three lanes in each direction. It was decided to construct a new, fully flexible third lane using the existing hard shoulder and adjacent land, and to upgrade the concrete lanes to carry the additional traffic. A 150 mm bituminous overlay was to be applied to strengthen the concrete lanes, and to provide a uniform appearance and ride quality across all three lanes.

Visual examination of the existing concrete lanes showed that the concrete was generally in very good condition. However, there were clear problems at many of the joints, with faulting at many of the joints, pumping and rocking of the slabs in the vicinity of many joints clearly visible, particularly in rainy conditions, and transverse cracks within 2-3 metres of many of the joints. There were concerns about the stability of the joints, and at the possibility of premature reflection cracking in the bituminous overlay layer.

FWD TESTING AND GROUTING CRITERIA

It was decided to use Falling Weight Deflectometer (FWD) tests to determine the joint condition, and to select joints for sub-sealing using pressure grouting techniques. A Dynatest model 8081 FWD was used in the evaluation. A nominal 70 kN load was specified to ensure sufficiently high deflections. A geophone setup shown in Figure 1 was used, with sensors placed 200 mm and 300 mm from the centre of the load plate on both sides of the load plate. Assuming the FWD was moving in the direction of traffic, the testing sequence at each joint was as follows. The load plate was placed on the Approach side of the joint, with the D2 and D3 sensors straddling the joint. A 3 drop test sequence was carried out, with typical loads of 55, 77 and 70 kN. The FWD was then moved so that the plate was on the Leave side of the joint, with the D8 and D9 sensors straddling the joint. Another 3 drop sequence was then carried out. The difference between the D2 and D3 readings (or D8 and D9 readings for the Leave side) was used to determine the load transfer, and the suitability of the joint for grouting, as described below.

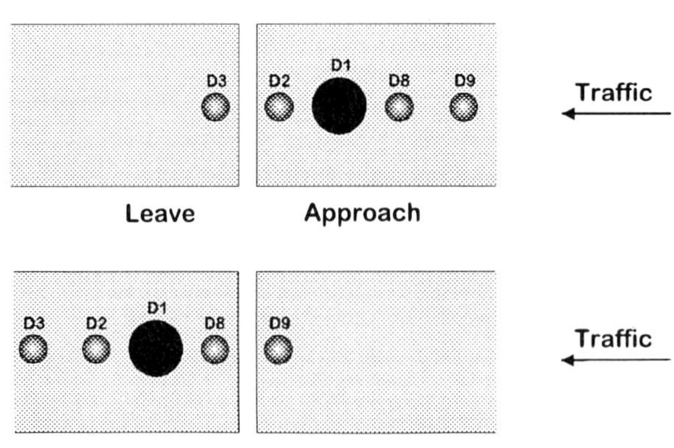

Figure 1

Tests were carried out on a preliminary basis on a number of joints to determine the range of deflections to be encountered. Limited trials were carried out using a number of different grouting techniques. Deflections were measured before and after grouting to determine the likely effect of grouting on the deflection levels measured. On this basis, criteria were developed and incorporated into the tender documents for the contract as follows:

The criteria for selecting slab joints for pressure grouting under a 70 kN nominal load were:

(i) Joints where D1 was greater than 250 microns.
OR
(ii) Joints where (D2 - D3) was greater than 150 microns.

The grouting mixture was standard cementitious grout with a water/cement ratio of 0.45 to 0.6, inserted under a nett pressure of 3 bars. A drilling pattern was specified, typically with holes at 500 mm or 1 metre transverse centres, extending back to 2 metres from the joint on the approach and leave sides. Grouting was to continue at each hole until one of the criteria below was met:

(i) Refusal under nett pressure of 3 bars.
(ii) Uplift of the slab exceeding 2 mm.
(iii) An uncontained grout leak.

Seven days after grouting was completed, further FWD testing was carried out using the same testing arrangement on all grouted joints. The Contractor was not to allow vehicles to traffic the transverse joints during this seven day period. On the basis that where grouting was used, the joint condition should have improved, the following criteria were specified:

(i) Joints where D1 was greater than 200 microns.
OR
(ii) Joints where (D2 - D3) was greater than 75 microns.

Joints which failed either of these criteria were to be regrouted.

Testing was carried out on the Westbound side of the dual carriageway. Both Slow and Fast lanes were tested on the Approach and Leave sides as shown in Figure 1. Slabs were typically 20 to 25 metres in length, with dowelled joints present in intact joints. A number of joint repairs, some with mini-slab replacements at the joints, some with full-depth bituminous patching, had been carried out over the years at deteriorated joints, and load transfer devices were typically not in place at these locations. Testing was carried out in the inner wheelpath (approx 1 to 1.5 metres from the inside edge of the slab) on both traffic lanes. The total length of the site was approx. 5 kilometres, and due to traffic management and construction requirements, the site was divided into three sections for testing and grouting. Testing and grouting was carried out at various times between March and July, 1997.

In carrying out the grouting on the first one kilometre section, it was noted that many of the joints took very little grout. It was not clear if this was as a result of poor technique, or an absence of voids. In any case, the grout was allowed to cure, and was retested 7 days later. The After criteria described before (D1 > 200 microns, (D2-D3) > 75 microns) were applied to the grouted joints, and a number of the joints failed to pass these criteria. These joints were re-grouted, left for a further 7 days and re-tested. About half the joints showed improvement, many showed no improvement, and a few showed disimprovement. The joints that failed the post-grout criteria were again re-grouted, left for a further 7 days and re-tested.

Clearly, a tradeoff has to be made between attempting to ensure that all joints are suitably stabilised, and the delay that ensues from the re-grouting and curing period. As a result, considerable analysis was carried out on the results from the first kilometre to determine parameters that would indicate when further grouting was unlikely to yield further improvement. These "ballpark" parameters were then available on the remaining sections, as grouting took place, to determine which joints should be re-grouted for the second time. Further analysis has subsequently been carried out, as described below, and may be useful to others engaged in similar testing activities.

ANALYSIS OF RESULTS

All concrete joints were tested initially, and the results from over 200 joints in each lane were collated and analysed. Joints were initially selected for grouting based on the criteria shown above. If a failure occurred on one side of the joint, both sides of the joint were grouted, and re-tested. As a result, there are results available from many joint sides which did not fail the initial criteria. In addition, it was decided to grout a number of joints which had <u>not</u> failed either criteria to determine if grouting would have any effect. Typically, these joints were adjacent to failed joints, e.g. a number of locations where the Slow lane joint failed, but the Fast lane joint passed had both joints grouted and re-tested. In the end, results were available for Before and After grouting, from 323 locations made up of 161 Approach and 162 Leave locations.

Some relevant overall statistics are useful at this stage. All deflections are in microns, corrected to a standard 70 kN load. The average D1 deflection before grouting on the Approach side was 184 microns; on the Leave side the average was significantly higher at 240 microns. Similar calculations on the D2-D3 criterion showed an overall average on the Approach side of 101 microns, and 156 microns on the Leave side. The overall average values are particularly relevant when compared with the grouting criteria, with the <u>average</u> results on the Leave side being very close to the stated grouting criteria.

Of the 323 locations which were actually grouted, 22 failed on the Approach side (based on the Before grouting results) due to D1 deflections in excess of 250 microns, while 70 locations failed on the Leave side using this criterion. Applying the specified (D2-D3) criterion of 150 microns, 39 failed on the Approach side, and 77 on the Leave side. Overall, 47 locations on the Approach side failed under one or both criteria, while 85 locations failed on the Leave side. Clearly, on average, conditions were significantly worse on the Leave side, and these results confirm other studies which indicate that if testing is to be carried out on one side of the joint only, the Leave side should be tested.

Analysis of the results after grouting showed a very significant overall improvement, particularly on the Leave side. The average D1 deflection on the Approach side was 161 microns, and the average on the Leave side was 162 microns. The average D2-D3 value was 67 microns on the Approach side, and 71 microns on the Leave side. Clearly, the grouting was successful in overall terms in reducing the deflection, increasing the load transfer at the joints and producing much more uniform conditions on both sides of the joints.

PREDICTIVE PARAMETERS

As outlined above, it was decided to post-analyse the data to determine if parameters could be established which would identify which joints would improve, and which joints would have little additional improvement from grouting. For each joint side, the D1 and D2-D3 parameters were already available. Two further parameters were calculated for each joint side. When testing was carried out, deflections were measured at three different load levels, typically 55, 77 and 70 kN. The three points for each location can be plotted and values for the slope and intercept of the least-squares best-fit line joining the three points can be easily calculated. Previous research has shown some relationship between the magnitude of the intercept, and the likelihood of the presence of significant voids [1]. It was decided to examine if the slope and/or intercept would be useful in identifying "improvable" joints.

The following procedure was adopted. The Before and After deflection results at all of the joint locations were compared, and each location assigned to one of three categories;

Category 1: Improvement (D1 or (D2-D3) decreased by at least 50 microns)
Category 2: No Change
Category 3: Disimprovement (D1 or (D2-D3) increased by at least 50 microns)

A database table was set up, containing the following fields:

Pavement Lane (Fast/Slow)
Joint Number
Side (Approach/Leave)
D1 (microns)
D2-D3 (microns)
Slope (microns/kN)
Intercept (microns)
Category (1, 2 or 3)

tended to separate improvable and non-improvable locations. This was used in combination with various intercept parameters, as shown in the table. In general terms, the results were quite similar to those obtained from using absolute deflection parameters, with a clear ability to distinguish improvable from non-improvable locations. Again identification of the optimum balance is subjective; I would suggest 20/3 (intercept > 20 microns or slope > 3 microns/kN) or 10/3 as possible criteria. Finally, an examination of the data using intercept alone was carried out. This also provided a good basis for distinguishing between improvable and non-improvable locations, with an intercept of 10 microns or above being the most promising parameter.

A similar analysis was carried out on the After Grouting and After Re-Grouting data. Obviously, the number of data points available was much lower, as most of the grouted joints showed satisfactory performance. A total of 50 joint locations were available, and of these 20 were identified as improving significantly after re-grouting, 25 locations did not show any significant improvement, and 5 locations showed significant disimprovement. As previously described, it is even more critical that as few "disimprovers" as possible are identified for re-grouting using the parameter criteria. The reasons why some joint locations disimproved varied on this site. It is believed that some additional voids may have been created under adjacent joints or corner locations during re-grouting, either by disturbance of the sub-base or lifting of the slab. In addition, it is possible that the sub-joint support provided by the first grouting process was disturbed by the re-grouting process.

Examination of Table 2 shows that absolute deflection criteria of D1 > 250 and D2-D3 > 100 microns appears to provide the best combination, with all "improvable" joints identified, and only 2 "disimprovers" identified. Application of the stated post-grouting criteria of D1 > 200 and D2-D3 > 75 microns led to identification of all "improvable" joints, but also to identification of a number of additional locations which did not improve, and to selection of all 5 joints which subsequently disimproved under re-grouting.

Using the Intercept/Slope combinations, the best combination appears to be 10/3, with all but one of the "improvable" joints selected, and only 2 "disimprovers" identified. Using the intercept parameter alone, an intercept of > 10 microns identified 16 of the 20 "improvable" joints, and only selected 1 of the "disimprovers". Overall, if consistent and unchanging selection parameters are required for selection of joints before grouting, and for re-grouting, the criteria best suited on this site appear to be D1 > 250 microns or D2-D3 > 100 microns, or alternatively, Slope > 3.0 microns/kN and Intercept > 10 microns. Further testing is currently being carried out in very different temperature conditions, and will be used to determine if the slope/intercept parameters are applicable over a range of temperatures.

SUMMARY

Joint undersealing using pressure grouting was carried out on a large number of joints prior to bituminous overlay on a heavily trafficked national route in Ireland. Falling Weight Deflectometer tests were carried out at various stages to monitor the joint condition. The grouting was clearly successful in reducing the absolute deflection, increasing the load transfer, and producing more uniform support on both sides of the joints. Absolute deflection criteria, based on D1 and (D2-D3) parameters, were specified for selection of joints for grouting, and re-grouting. The criteria were successful in identifying "improvable" joints, while minimising the number of "unimprovable" joints selected for grouting.

Alternative criteria, derived from testing at different load levels, were developed. These criteria, based on the slope and intercept of the best-fit line to the deflections measured at various load levels, proved equally successful at distinguishing between improvable and unimprovable joints. It is felt that the alternative criteria may be less temperature-susceptible than the absolute deflection criteria, and should be considered either separately, or in conjunction with the absolute deflection criteria, on future performance specifications for pressure grouting.

REFERENCES

1. AASHTO Guide for Design of Pavement Structures. American Association of State Highway and Transportation Officials. Washington D.C. 1993.

2. Design Manual for Roads and Bridges, UK Department of Transport. Volume 7, Section 3, Part 2, HD 29/94. London. 1994.

Fifth International Conference on the
BEARING CAPASITY OF
ROADS AND AIRFIELDS

NONDESTRUCTIVE EVALUATION OF D-CRACKING IN ASPHALT-OVERLAID PCC PAVEMENTS

By ROBERT K. KUMAPLEY[1], LESLIE TITUS-GLOVER[1], EMMANUEL B. OWUSU-ANTWI[2], AND ERIC ODURO-KONADU[3]

[1]Project Engineer, ERES Consultants, Inc., 505 West University Avenue, Champaign, Illinois 61820-3915, tel.: (217)-356-4500, fax: (217)-356-3088
[2]Assistant Professor, Arizona State University, Dept. of Civil Eng. and Environmental Eng., P.O. Box 5306, Tempe, Arizona 85287-5306
[3]Director (Materials), Ghana Highway Authority, P.O. Box 1641, Accra, Ghana

ABSTRACT

Currently, the only way to assess D-cracking in the portland cement concrete (PCC) layer of composite asphalt concrete-overlaid PCC pavement (AC/PCC) is by the examination of full-depth cores taken through the pavement. Coring operations to identify and track D-cracking can be expensive and time-consuming, particularly in situations where D-cracking is widespread but not visible at the surface. This paper presents a simple nondestructive procedure for evaluation of D-cracking problems in AC/PCC. The procedure involves the use of the ratio of deflections at the slab edge to the deflections at the slab center, DEF_{ratio}. The deflections are obtained from nondestructive testing using a falling weight deflectometer (FWD). The procedure is a good supplement to visual condition data, because it highlights areas that, although they appear satisfactory visually, actually have durability problems and structural deficiencies.

Keywords D-cracking, DEF_{ratio}, Nondestructive testing, Visual condition assessment, Falling weight deflectometer, Asphalt-overlaid PCC pavement.

INTRODUCTION

Portland cement concrete (PCC) pavements are rehabilitated by resurfacing with asphalt concrete (AC) more commonly than by any other method (1). With the growing number of AC-overlaid PCC (AC/PCC) pavements in the United States (U.S.) and worldwide, there is a need for simple, efficient, repeatable, and economical tools for early detection and monitoring of structural problems in these pavements.

The most common visual condition assessment procedures for AC-surfaced pavements, the Long-Term Pavement Performance (LTPP) *"Distress Identification Manual for the Long-Term Pavement Performance Project (SHRP-P-338)"* (2) and the pavement condition index (PCI) procedure described in ASTM D5340 (3), do not account for D-cracking in the PCC layer of AC/PCC pavements. The sole dependence on visual condition data for management of maintenance and rehabilitation (M&R) resources is likely to result in erroneous decisions and schedules. Currently, the most acceptable way to assess D-cracking in the PCC layer of AC/PCC pavements is by the examination of full-depth cores taken through the pavement. This operation can be expensive, particularly in situations where D-cracking is widespread.

This paper presents a nondestructive procedure that was used to evaluate D-cracking in the PCC layer of AC/PCC airfield pavements in the U.S. The procedure involves the use of the ratio of the normalized deflection—from heavy falling weight deflectometer (FWD) testing—at the slab joint, edge, or crack (where D-cracking always starts) to the slab center, DEF_{ratio}. The results and procedures presented in this paper may be useful to pavement engineers faced with widespread D-cracking problems. DEF_{ratio} values, visual inspection results, and the condition of PCC cores are compared to demonstrate the effectiveness of the DEF_{ratio} procedure as a tool for evaluating the soundness of PCC in AC/PCC pavements.

PREVIOUS RESEARCH

DEF_{ratio} is defined as follows (4; 5):

$$DEF_{ratio} = \frac{DEF_{edge}}{DEF_{center}} \qquad (1)$$

where

DEF_{ratio} = Ratio of deflections at slab edge and center of slab.

DEF_{edge} = Deflection under the FWD load plate at the edge of slab, normalized to the maximum operational load (typically 223 kN [50,000 lb]).

DEF_{center} = Deflection under the FWD load plate at the center of slab, normalized to the maximum operational load (typically 223 kN [50,000 lb]).

Pavement deflection data were collected at the joints (at midslab in the main gear path) and center of the PCC slabs using a heavyweight FWD (4; 5). Target FWD loads were selected to cover the range of single wheel loads of the expected operational aircraft.

The DEF_{ratio} procedure was originally developed to investigate the progression of D-cracking in a rehabilitated D-cracked PCC pavement at Forbes Field, Kansas (4). Even though the findings from that study were based on very limited data, the potential use of the technique as a nondestructive tool for detecting D-cracking was promising. With the availability of more data from six major airfields, Kumapley et al. (5) further investigated the potential of the DEF_{ratio} procedure for assessing D-cracking in jointed plain concrete pavements (JPCP) with the following conclusive results:

- DEF_{ratio} values for sound PCC will always be less than 3, even for the worst load transfer efficiency (LTE) case of 0 percent.
- DEF_{ratio} values substantially greater than 3 indicate potential D-cracking problems.

- The DEF_{ratio} procedure is an effective tool for early detection, assessment, and monitoring of D-cracking in airfield pavements irrespective of PCC thickness, site conditions, and other characteristics.

The above findings and conclusions were based on a regression model with DEF_{ratio} and LTE as the dependent and independent variables, respectively. The model obtained with a coefficient of determination, R^2, of 63 percent is as follows (5):

$$DEF_{ratio} = 2.9565 - 0.0186 \times (LTE) \quad (2)$$

LTE was computed as the ratio of the unloaded slab's deflection to the loaded slab's deflection. LTE is a key factor that influences the deflections obtained from the slab edge. Figure 1 shows a plot of the data used in developing the model. The model developed for sound PCC covers LTE values from 17 to 99 percent .

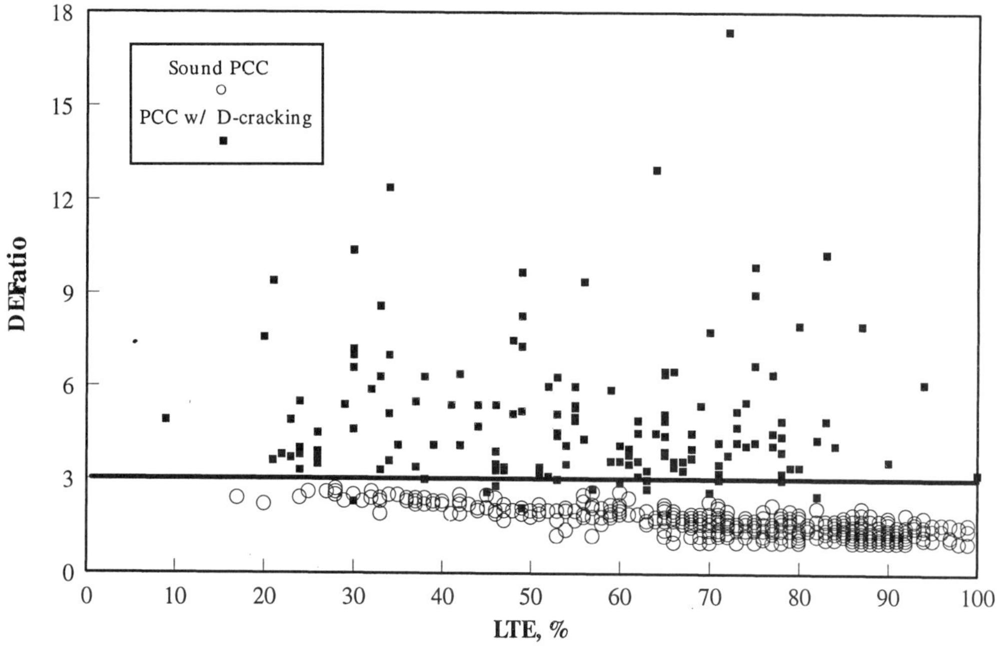

Figure 1. A plot of DEF_{ratio} versus LTE (5).

As can be seen in figure 1, there is a significant difference in DEF_{ratio} values for sound PCC and PCC with D-cracking. Kumapley et al. also noted that all the DEF_{ratio} values for sound PCC (with no D-cracking) are consistently less than 3, irrespective of other pavement properties and features such as PCC thickness, subgrade type, and load transfer (5). Kumapley et al. also noted a strong linear relationship between LTE and DEF_{ratio}.

Based on the above findings and conclusions, this study was conducted to investigate the potential and effectiveness of the DEF_{ratio} procedure, originally developed for investigating D-cracking problems in bare PCC pavements, for investigating D-cracking in the PCC layer of AC/PCC pavements.

DATA COLLECTION

The airside pavement network at Dayton includes 3 runways, 25 taxiways, 8 aprons, and emergency and access roads. Figure 2 shows the location of the AC/PCC pavements.

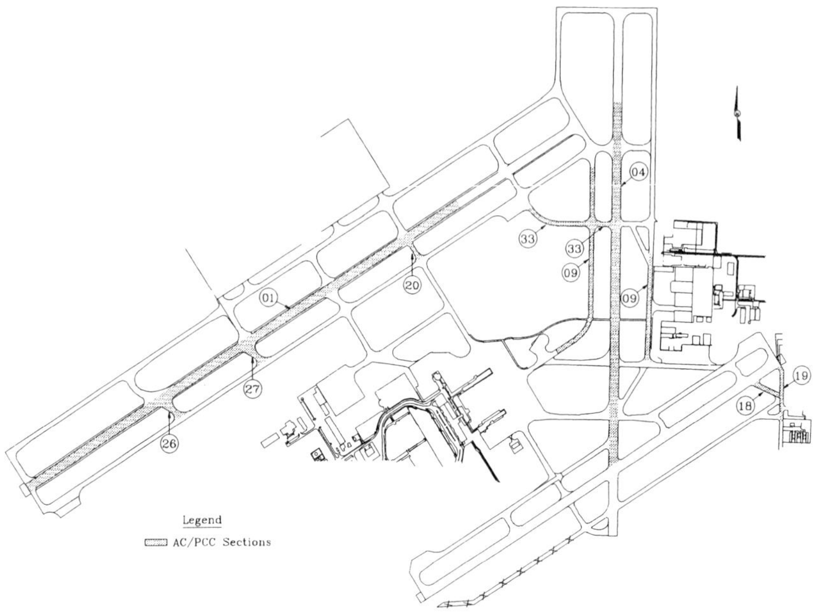

Figure 2. Pavements evaluated at Dayton International Airport.

The PCI procedure described in ASTM D5340 was used to obtain distress data. Pavement deflection data were collected at the joints (at midslab in the main gear path) and center of the PCC slabs using a Dynatest heavyweight FWD. The presence of reflective longitudinal and transverse cracks in the AC layer clearly identified the joint locations in the underlying PCC. Three target load magnitudes of 133, 178, and 223 kN (30, 40, and 50 kips) were applied to the pavements at each test location. The target loads were selected to cover the range of single wheel loads of expected operational aircraft at Dayton and were used to detect loss of support (voids) underneath the slabs. Loss of support may influence DEF_{ratio} results. The pavement surface temperature during FWD testing was between 21 and 32 °C (70 and 90 °F).

DATA ANALYSIS

DEF_{ratio} Analysis Using equation 1, the DEF_{ratio} values were computed for all the AC/PCC pavement sections. A summary of the DEF_{ratio}, PCI, and ratings results for the AC/PCC pavements at Dayton are presented in table 1. The ratings for the average PCI were adopted from the PCI field survey manuals for asphalt-surfaced pavements.

Table 1. Summary of AC/PCC pavement condition data.

Facility (Section no in figure 2)	Average void depth, x 10^{-4} mm	PCI information		DEF_{ratio}		
		Average, %	Rating	Average	COV, %	Rating
RW 6L-4R (01)	305	76	Very good	1.3	10	Excellent
RW 18-36 (04)	450	80	Very good	1.2	12	Excellent
TW A (09)	434	70	Very good	2.6	30	Poor
TW C (12)	406	58	Good	1.9	36	Fair
TW J (18)	330	97	Excellent	1.2	4	Excellent
TW K (19)	381	100	Excellent	1.1	16	Excellent
TW M (20)	279	85	Very good	1.2	7	Excellent
TW S (26)	318	80	Very good	1.4	13	Very good
TW T (27)	300	75	Very good	1.3	16	Excellent
TW Z (33)	250	70	Very good	2.6	31	Poor

Note: a) COV = coefficient of variation.
b) RW and TW = runway and taxiway, respectively.

Because the PCI procedure for AC-surfaced pavements does not specifically account for D-cracking in the underlying PCC, the PCI database for Dayton did not contain any D-cracking data for the AC/PCC, even though there were signs of D-cracking.

As can be seen in table 1, the average DEF_{ratio} values are always greater than 1. This is due to higher deflections at the slab edge compared to the deflections at the center of the slab. DEF_{ratio} values close to 1 are expected for stiffer pavements in excellent condition, and values of 3 or greater indicate potential structural problems, such as D-cracking, at the edges of the slabs (4; 5).

Void Detection Analysis Because the presence of voids may influence the DEF_{ratio} results, a loss of support analysis was conducted. The procedure used in this study, developed by Crovetti and Darter (6), gives an indication of the existence of voids at test locations using load versus deflection plots. As required by the procedure, the loads and corresponding deflections at each test location were used to estimate the void depth. Void depth greater than .0762 mm (3 mils) is likely to influence DEF_{ratio} results significantly. A summary of the results of this analysis is presented in table 1. The average void depth for the facilities evaluated did not exceed 0.045 mm (1.8 mils). This finding is consistent with the results of the visual survey that did not give any indication of loss of support problems such as pumping and faulting. Therefore, no significant influence is expected from voids at the test locations.

Load Transfer Analysis Currently, there is no widely accepted procedure for evaluating load transfer at reflected/joints in AC/PCC pavements; thus, no load transfer analysis was conducted for the pavements in this study. As shown in equation 3, LTE is computed for bare PCC as the ratio of the unloaded slab's deflection to the loaded slab's deflection.

$$LTE = \frac{\text{Unloaded slab deflection x 100}}{\text{Loaded slab deflection}} \qquad (3)$$

Equation 3 gives misleading results for AC/PCC pavements due to the significant compression in the AC layer underneath the FWD load plate (i.e., the loaded slab deflection).

There is ongoing research into the use of the DEF_{ratio} procedure for evaluating load transfer in AC/PCC pavements. The research is based on the concept that it should be possible to assess the load transfer for two similar AC/PCC sections or adjoining slabs by comparing their DEF_{ratio} values. For example, Taxiway J (section 18) has an average DEF_{ratio} value of 1.2 and coefficient of variation (COV) of 4 percent, indicating minimal variation in the DEF_{ratio} values along the taxiway. COV values less than 20 percent indicate uniform conditions. The Taxiway J data therefore suggest that the PCC layer in the AC/PCC pavement along the taxiway is sound, with good load transfer, and that these conditions are uniform along the taxiway. Taxiway C, on the other hand, has average DEF_{ratio} and COV of values of 1.9 and 36 percent, respectively. The high COV for Taxiway C suggests that some sections along the taxiway may be experiencing structural problems, triggering the need for a more detailed investigation.

DEF_{ratio} RATINGS

To simplify interpretation of the results, each pavement at Dayton Airport was rated as "Excellent," "Very Good," "Good," "Fair," Poor," "Very Poor," or "Failed" based on average DEF_{ratio} values. The ranges of DEF_{ratio} values for which these ratings apply are shown in table 2.

Table 2. Ratings for average DEF_{ratio} values.

Rating	DEF_{ratio} range
Excellent	1.0 - 1.3
Very good	1.3 - 1..5
Good	1.5 - 1.7
Fair	1.7 - 2.0
Poor	2.0 - 3.0
Very poor	3.0 - 4.0
Failed	> 4.0

The ratings were developed in this study to enable the comparison of the structural condition and soundness of the PCC layer in each facility in the pavement network at Dayton Airport. The ratings were developed based on the cores taken at Dayton for verification of the DEF_{ratio} results, results from previous studies that used DEF_{ratio} values for investigating D-cracking (4;5), and numerous evaluations of airfield PCC pavements (7; 8; 9). Laboratory strength tests were also performed on selected cores from midslab to confirm the soundness of the concrete in the pavement sections without D-cracking. Average tensile strength for cores without D-cracking ranged from 2758 to 3792 kPa (400 to 550 psi). The sections from which the cores were taken, together with the condition of the cores with respect to durability, can be found in table 3.

Table 3. Visual inspection of PCC layer core samples.

Facility (section)	Average DEF_{ratio}	PCC core		
		Location	Condition	Is D-cracking present?
RW 6L-4R (01)	1.3	Joint and Midslab	Excellent	No
RW 18-36 (04)	1.2	Joint and Midslab	Excellent	No
TW A (09)	2.6	Joint	Poor	Yes
		Midslab	Excellent	No
TW Z (33)	2.6	Joint	Poor	Yes

The results in table 3 show that the condition of the PCC layer in the AC/PCC runways is excellent. However, the cores taken from the joints in Taxiways A and Z were in poor condition. This is consistent with the results from previous studies, which showed that PCC with durability problems exhibit DEF_{ratio} values of 3 or greater, and DEF_{ratio} values are typically less than 1.5 for sound PCC (4; 5).

SUMMARY OF FINDINGS AND CONCLUSIONS

The DEF_{ratio} procedure presented in this paper has been shown to be a simple, efficient, repeatable, nondestructive, and economical tool for early detection and monitoring of structural problems in the PCC layer of AC/PCC pavements. The results of this study suggest that the DEF_{ratio} procedure originally developed for bare PCC is capable of highlighting structural problems in the PCC layer in AC/PCC pavements. Large DEF_{ratio} values obtained for AC/PCC pavements may be due to reasons other than D-cracking in the PCC layer (such as the presence of voids underneath slabs and poor load transfer at joints and cracks).

This study revealed that AC/PCC pavements with DEF_{ratio} values greater than 2.5 may be experiencing structural problems and may be in need of detailed investigation to identify the cause of the problem. Although the ratings presented in this paper are specific to Dayton International Airport, similar ratings can be developed for other airfields following the procedures presented in this paper. The ongoing research on this subject and the availability of more data from other airfields will enable a more general rating system to be developed.

A major advantage of the DEF_{ratio} procedure is that numerical DEF_{ratio} values enable direct comparison of the extent of structural problems in different pavements in a network, without additional pavement closures to collect the data needed. The procedure is a good supplement to visual condition data because it highlights areas that, although they appear satisfactory visually, actually have durability problems and structural deficiencies.

ACKNOWLEDGMENT

The results presented in this paper are from work conducted for the Department of Aviation, City of Dayton, at Dayton International Airport. The authors wish to express their appreciation to the Department of Aviation of the City of Dayton and ERES Consultants, Inc. The opinions and conclusions expressed in this paper are those of the authors and in no way reflect those of any of the aforementioned agencies.

REFERENCES

1. HALL, K.T., DARTER, M.I., and STEEL, D.A. *Guidelines for Evaluation of Asphalt-Overlaid Concrete Pavements*, Transportation Engineering Series N0. 79. Transportation Cooperative Highway Research Program. University of Illinois, Champaign-Urbana, Illinois, 1995.

2. *Distress Identification Manual for the Long-Term Pavement Performance Project (SHRP-P-338)*, Strategic Highway Research Program. National Research Council, Washington, D.C., 1993.

3. AMERICAN SOCIETY FOR TESTING AND MATERIALS. *Annual Book of ASTM Standards, Road and Paving Materials; Vehicle-Pavement Systems,* Vol. 04.03, 546-593, 1997.

4. KUMAPLEY, R.K. AND OWUSU-ANTWI, E.B. "Performance of a Rehabilitated D-Cracked Jointed Plain Concrete Airfield Pavement," *Transportation Research Record No. 1568: Pavement Rehabilitation and Design*, National Research Council, Washington, D.C., pp. 148-154, 1997.

5. KUMAPLEY, R.K., OWUSU-ANTWI, E.B., AND TITUS-GLOVER, L. "Nondestructive Technique for Evaluating D-cracking in Airfield Pavements," *Proceedings of the 6th International Purdue Conference on Concrete Pavement, Volume 2*, pp. 139-154, Purdue University, West Lafayette, Indiana, 1997.

6. CROVETTI J.A. AND DARTER, M.I., "Void Detection Procedures" Appendix C to Final Report, National Cooperative Highway Research Program, Transportation Research Board, National Research Council, Washington D.C., 1985.

7. ERES CONSULTANTS, INC. "Airport Pavement Evaluation, Runway 13-31, Forbes Field, Topeka, Kansas," Final Report for Bartlett & West Engineers, Inc., Topeka, Kansas 1993.

8. ERES CONSULTANTS, INC. "Evaluation of New Portland Cement Concrete Pavement in Taxiways E and G at Chicago's O'Hare International Airport, ERES Project No. 94-097-S1," Final Report for Howard Needles Tammen and Bergendoff (HNTB), Chicago, Illinois, 1995.

9. ERES CONSULTANTS, INC. "Airport Pavement Evaluation and Management Study, Volumes I, II, and III, ERES Project No. 94-064-S1", Final Report for The City of Dayton, Department of Aviation, Dayton International Airport, Vandalia, Ohio, 1996.

Fifth International Conference on the
BEARING CAPASITY OF
ROADS AND AIRFIELDS

STABILITY PROCESSING EVALUATION METHOD FOR THE SUBGRADE USING THE HFWD

NAGATO ABE[1],ETSUO SEKINE[2],MASAKI KAMIURA[3]
TERUHIKO MARUYAMA[4] AND YASUO MIURA[5]

[1]Senior Research Engineer, Toa Doro Kogyo Co., LTD. Research Laboratory, 232-0033, Yokohama, Japan, abe@msb.biglobe.ne.jp
[2]Senior Research Engineer, Railway Technical Research Institute, 185-0034, Kokubunji, Japan, kei@rtri.or.jp
[3]Professor, Hokkai Gakuen University, Department of Civil Engineering, 064-0926, Sapporo, Japan, kamiura@hokkai-s-u.ac.jp
[4]Professor, Nagaoka University of Technology, Department of Civil and Environment Engineering, 940-2188, Nagaoka, Japan, terum@nagaokaut.ac.jp
[5]Sub chief, Japan Freight Railway Co., LTD., Department of Transport and Technology, 112-0004, Tokyo, Japan

ABSTRACT

Sampling on the site, in-site CBR test and the strength examination such as plate bearing tests are difficult in a weak subgrade. Then, the strength evaluation method of the ground in a weak subgrade at the silty sand and cohesive soil was examined by using Handy Falling Weight Deflectometer (HFWD).

The plate bearing test is 7 times the radius of the loading plate in the vertical direction and 3 times the radius in the horizontal direction. In the dynamic loading test, including FWD and HFWD tests, where impact load is applied, the range of loading effect is 2 to 3 times deeper and wider than in the static loading test.

The above-mentioned investigations and studies confirm that FWD and HFWD can be used to take measurements of elastic behavior of the subgrade and the base course, thereby contributing to the evaluation and design of a stabilization process.

FWD and HFWD equipment is not only useful for structural evaluation of pavement but also applicable to the construction management, design and evaluation of a stabilization process for the subgrade.

Key words: HFWD, K value, elastic modulus, stabilization, subgrade

INTRODUCTION

Sand soil and granular crushed stone are the main materials for the subgrades and the base courses of roads and railways. The strength of these base courses and subgrades is managed through in-situ density tests, in-situ CBR tests and plate bearing tests.

Yet problems with these types of tests lie in the fact that only a limited number of points on roads can be measured in a day and that a considerable amount of reaction force equipment is required for measurement. In view of such constraints, CBR tests and plate bearing tests are normally conducted at the intervals of between 50 m and 100 m [1]. For the proper management and control of the degree of compaction, however, it is necessary to take measurements at a larger number of points by employing a relatively accurate yet simple and quick testing method, so that distribution patterns of bearing capacity in longitudinal and traverse directions can be more precisely identified.

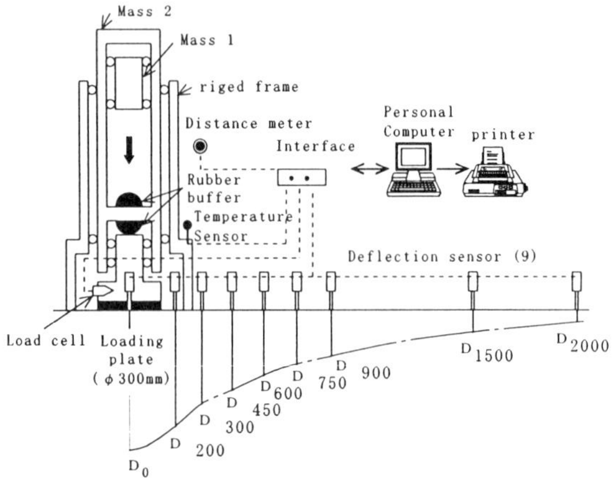

Fig.1 Outline of Falling Weight Deflectometer (FWD)

When pavements and tracks are constructed on soft and weak subgrades, their structures should be made thicker. This leads to a higher cost, in addition to an increased degree of settlement due to repeated loads. In order to circumvent these problems, the construction of a pavement or a track is usually preceded by the preparation of a load-bearing layer by means of a subgrade stabilization process or through the replacement of materials in the upper layer of the subgrade with granular crushed stone or sand.

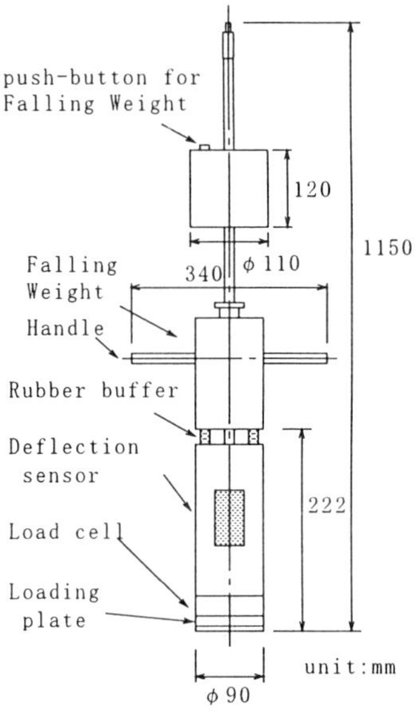

Fig.2 Outline of Handy FWD

A simple method is being sought for determining the bearing capacity of a soft subgrade. Also, methods for calculating the thickness of the layers which require a stabilization process or material replacement have yet to be established. Equally, it is necessary to find a method for properly evaluating increases in the strength of such layers after stabilization process.

As an alternative to the in-situ CBR test and the plate bearing test that facilitate the investigation of soil bearing capacity, the present authors have explored the applicability of the Falling Weight Deflectometer (FWD: Fig.1) and its small and compact version, i.e., the Handy Falling Weight Deflectometer (HFWD: Fig.2), to the compact management of subgrade soil and base courses. As the names indicate, these types of equipment are used to measure deflection resulting from the impact caused by a weight that has been dropped.

A series of experiments and researches revealed close relationships between the compact density ratio, the air space ratio, and the K30 value, all of which represent the degree of compaction, and between the K value and the elastic modulus that are determined through FWD and HFWD testing.

Furthermore, the modulus of deformation obtained from cyclic triaxial tests was found to coincide with the elastic modulus determined through FWD testing when a strain level measured by FWD is taken into account. In view of these positive findings, FWD equipment began to be used for in-situ tests for the purpose of managing the degree of compaction [2].

This paper describes the investigations which have been carried out to verify that FWD and HFWD methods are applicable to the evaluation of soft subgrade, the thickness design for implementing the subgrade stabilization process, and the measurement of increases in strength after stabilization.

TEST METHOD

Test Equipment. Table 1 lists the features of the FWD equipment manufactured by PHØNIX in Danish, and the HFWD equipment made by Freesia Macross in Japan. An FWD is built in a VW vehicle, and nine geophones are used for the measurement of deflection. An HFWD, on the other hand, weighs merely 15 kg and therefore is portable, so that measurement can be made at points unaccessible to vehicles. In the HFWD, there are three options for the falling weight: 8 kg, 10 kg, and 15 kg, and also for the diameter of the load bearing plate: 9 cm, 20 cm and 30 cm.

Table 1 Compared with FWD and HFWD device

	FWD	HFWD
Dia. of Loading Plate (mm)	300	90, 200
Hardness of Rubber Buffer	42	61
Falling Weight (kg)	130	8, 10, 15
Falling Height (mm)	50,100,200,400	50,150,300
Number of Deflection Sensor	9	2
Maximum Deflection (mm)	2	2

Results of Bearing Capacity Tests. Three types of soil structures with different degrees of compaction were prepared in a test pit by using mechanically stabilized crushed stone. These structures were then subjected to FWD and HFWD testing to measure the impact load applied and the deflection caused by the load, thereby determining the K value which represents the bearing capacity coefficient of the soil structures. Because the load bearing plate of the HFWD is as small as 9 cm in diameter, linear correction was made for HFWD tests in accordance with a diameter ratio against the plate whose diameter was 30 cm, the size usually employed for the plate bearing test and the FWD test [2].

Fig. 3 compares kv and other values obtained from FWD and HFWD tests with K30 values which were made available through plate bearing tests. As the figure illustrates, K values from the HFWD test were approximately twice as large as K30 value, while those from the FWD test were two and a half times larger. The kv values, which were calculated using tangent gradients observed in the plate bearing tests, stood at 3.5: 1 against K30 value. Although the K values from the HFWD indicate the results after correction with a diameter ratio of 9 cm to 30 cm, the relationship between these values from the HFWD tests and those from the FWD tests is close to the y=x relationship. The fact that the K values determined through HFWD are smaller than those determined through FWD appears to reflect differences in ranges affected by impact load and also in strain levels at the time of measurement. Because the number of deflection sensors for HFWD was limited to two, the elastic moduli of mechanical stabilized crushed stone in the HFWD tests were calculated through Boussinesq's equation[3]. In the FWD tests, six pieces of deflection data out of nine were used for determining elastic moduli through multi-layer elasticity calculations.

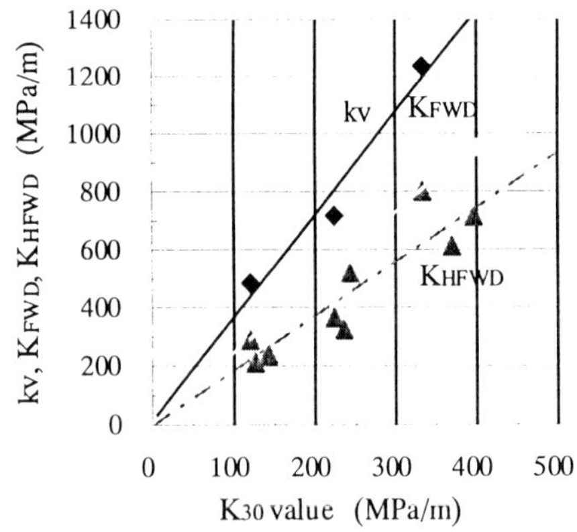

Fig. 3 Compares kv, KFWD, KHFWD with K30 value

Table 2 Compares elastic moduli of 3 compaction level

degree of compaction	Elastic moduli (MPa) FWD	HFWD
96%	181	171
91%	119	116
86%	79	77

Table 2 compares elastic moduli obtained through the measurements of the three test structures consisting of mechanically stabilized crushed stone. Since the values from the HFWD and FWD were close enough, it can be said that the multi-layer elasticity analysis is valid as long as the effects of impact load on the structures are confined to elastic deformation. Disparities in K values between HFWD and FWD as shown in Fig. 3, in contrast to the values of elastic moduli in Table 2, can be attributed to differing levels of loading energy of the equipment, which in turn give rise to variances in ranges affected by such loading.

Evaluation and Design of Subgrade Stabilization Process

For roads in general, The Asphalt Pavement Manual specifies that the value 3 or higher in subgrade design CBR should be ensured by means of the CBR test method. The said manual also mandates the construction of subgrade when subgrade design CBR is less than 3. For railways, on the other hand, the K30 value on the roadbed surface should be 68.7 MPa/m. When bearing capacity is inadequate, the top layer, of 30 cm to 50 cm in thickness, needs to be subjected to a stabilization process with lime of cement, or the materials of the layer must be replaced with mechanically stabilized crushed stone or crusher run. In view of the above-mentioned criteria and standards, a design approach should be established to make up for the low bearing capacity of soft subgrade and to secure a certain level of the bearing capacity of subgrade, giving consideration to total construction costs.

In the following subsections, comparisons are made among the results from the CBR test, the plate bearing test, and the FWD test, while investigating design methods for improving thickness. By using the subgrade consisting of volcanic cohesive soil and silty sand, the CBR tests for measuring unconfined compressive strength were implemented at the Utsunomiya Freight Terminal and at Sumidagawa Station of the Japan Freight Railway Co., Ltd. before and after the stabilization of the subgrade.

Table 3 Grading characteristics of subgrades

	Utsunomiya Freight Ter	Sumidagawa Station
Soil type	Volcanic cohesive soil	Silty sand
Gravel (%)	0	28.5
Sand (%)	29.5	32.9
Fine grain (%)	70.5	38.6
Coefficient of uniformity	7.44	103
Coefficient of curvature	0.65	0.23

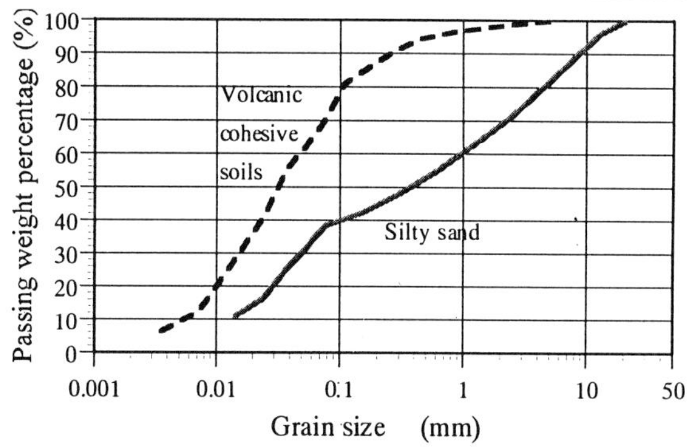

Fig. 4 The grain size accumulation curves of the 2 materials

Characteristics of in-situ subgrade.

Table 3 and Fig. 4 show the grading characteristics and grain size distribution of the subgrade of the two stations, respectively. Volcanic cohesive soil, with a uniformity coefficient smaller than 10 and a curvature coefficient smaller than 1, belonged to a category defined as "poor grade size distribution". Silty sand, in contrast, was characterized by the almost equal percentages of gravel fraction, sand, and fine-grained fraction, and its grain sizes were fairly well distributed.

Table 4 The physical characteristics and unconfined compressive strength of the subgrade soil

		Utsunomiya Freight Ter		Sumidagawa Station	
		No.1	No.2	No.1	No.2
Before stabilization	Soil type	Volcanic cohesive soil		Silty sand	
	Water content (%)	123.7	118.3	35.2	31.1
	ρt (g/cm^3)	1.311	1.323	1.791	1.620
	ρd (g/cm^3)	0.586	0.606	1.325	1.236
	CBR Value (%)	2.6	2.9	3.7	3.1
	σ compress (kPa)	12.3	13.5	34.5	24.9
After stabilization	Solidifying agent	Lime-based solidifying		Cement	
	additional rate (%)	23.2		11.6	
	add volume (kg/m^3)	253.0		202.7	
	σ compress (kPa, 7day)	328		332.2	
	Failure strain (%)	4.25		2.84	
	Water content (%)	73.6		29.5	
	ρt (g/cm^3)	1.456		1.886	
	ρd (g/cm^3)	0.839		1.456	

The physical characteristics and unconfined compressive strength of the subgrade soil were improved through the stabilization process indicated in Table 4. Since volcanic cohesive soil with poor grain-size distribution contained a high percentage of water, a lime-based solidifying agent was used for the soil in order to absorb water and reduce the moisture content. For stabilizing silty sand, cement was added in view of its favorable grain-size distribution and its water content, which is lower than 50%.

In the mix design of the stabilization process, CBR values are used most widely for roads, while unconfined compressive strength is pursued in constructing railways. The relationships between CBR values and unconfined compressive strength are represented by the following equation in "Design and Construction Guide for Geotechnical Structures" [3].

$$CBR = qu / a \qquad (1)$$

Where, CBR: CBR in laboratory test (%)

qu: unconfined compressive strength (kPa)
a: conversion coefficient (15 to 20)

The authors of this paper also relied on unconfined compressive strength tests and CBR tests in mix design for the stabilization process. The comparative analysis of results from the two types of testing brought a conversion coefficient of 17, demonstrating that stabilization process CBR values of not less than 20% could be maintained regardless of conversion coefficients, as long as unconfined compression strength (7 days curing) was 400 kPa or more.

Relationships Between K Values and CBR Values. While the conversion curve developed by Middlebrooks et al is often used in defining the relationship between the value from in-situ CBR testing and K_{30} values from the plate bearing test, the following equation by Ueshita et. al[5] is well known in Japan.

$$K_{30} = 5.66 \times CBR + 22.4 \qquad (2)$$

Where, K_{30}: Modulus of subgrade reaction (MPa/m)
CBR: in-situ CBR (%)

In general, K_{30} values are calculated under the condition of 0.125 cm settlement in the plate bearing test. In this particular case, the relationship between load and settlement is almost linear, and the range of settlement does not exceed 1% in terms of strain. Accordingly, the effect of plastic deformation is not likely to be reflected in the results of measurement using the FWD or HFWD.

Fig. 5 illustrates the relationship between K_{HFWD} values after diameter correction and K_{30} values in the plate bearing test, for both of which volcanic cohesive soil was used. Based on this relationship, the conversion coefficient from K_{HFWD} to K_{30} values is determined to be 1.8.

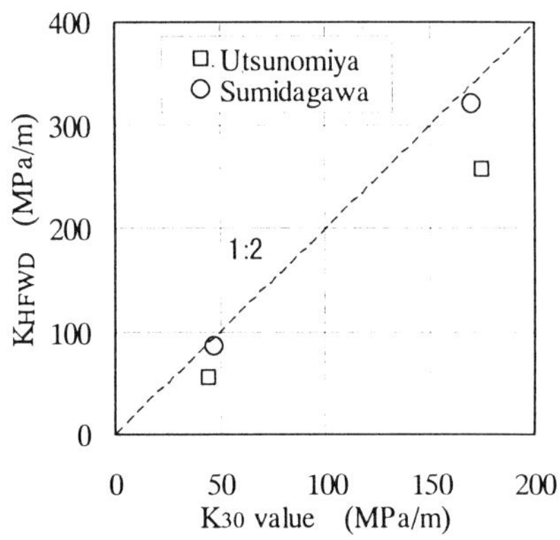

Fig. 5 The relationship between K_{HFWD} values and K_{30} values

The relationship between CBR values and K_{30} values as identified by Ueshita and the Portland Cement Association (PCA) of the United States[6] is seen in Fig. 6. According to our observations in a series of tests for this paper, their relationship turned out to be closer to Ueshita's when the strength of the subgrade was lower, but it became similar to PCA's as the strength increased to a certain level. This could be explained by the fact that the relationship put forth by Ueshita pertains to the measurement of the subgrade only, whereas the one identified by PCA

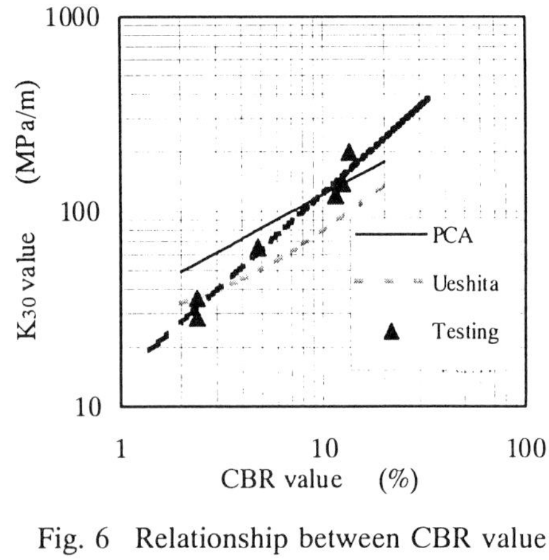

Fig. 6 Relationship between CBR value and K_{30} value

includes the measurement of the base course as well.

The regression curve in the figure is represented by equation (3). The correlation coefficient R^2 of the regression curve is 0.966.

$K_{30} = 14.16 \times CBR^{0.938}$ (3)

Where,

K_{30}: Modulus of subgrade reaction (MPa/m)

CBR: in-situ CBR value (%)

In reference 2), the existence/non-existence of crushed stone and dry density exerted strong influence on the conversion coefficient of mechanically stabilized crushed stone, which was found to be as large as 2.5 in this particular case.

Elasticity Analysis by HFWD. The elasticity coefficient of two layers can be determined by measuring deflection using the HFWD and then comparing it with the deflection calculated according to the multi-layer elasticity theory.

In our research, the subgrade was assumed to be of a two-layer structure, with a view to investigating the possibility and feasibility of using the HFWD so that CBR values of respective layers and the unconfined compressive strength of the stabilization process can be determined in a simple and convenient manner.

Esg was calculated through equation, which was based on Timoshenko's theory of elasticity. The calculation result was then compared with the result from the tests involving the two layers in order to determine the depth to which the bearing capacity could be calculated for a given radius of the loading plate.

The average CBR of the subgrade after improvement ranges from 7% to 10%. Table 5 compares the results from various types of strength tests, including the unconfined compression test and the plate bearing test, with those measured by the HFWD.

For the measurement by the HFWD in Table 5, a loading plate with a 9 cm diameter was used, so the range for evaluation should have been down to a depth of 60 cm. The evaluation through the HFWD method is considered to have addressed only the stabilization process in the upper part of the subgrade, without incorporating the effect of the lower part which was soft and weak. There were similar findings as to the stress status in the test pit where gravel was used. In consideration of these results, it is valid and feasible to determine the range of stress propagation in the direction of depth according to the ratio of the depth to the diameter of a loading plate.

Table 5 Compares the results from various types of strength tests

	Utsunomiya Freight Ter.		Sumidagawa Station	
	before	after 60cm	before	after 50cm
CBR Value (%)	2.6	10.3	2.8	7.3
σ compress (kPa)	38.2	333.3	35.3	328.4
K_{30} value (MPa/m)	46.9	182.6	49.5	177.2
K_{HFWD} (MPa/m)	56.1	266.2	89.0	331.5
Estimated K_{30} (MPa/m)	31.2	147.9	38.0	207.4
Estimated CBR (%)	1.5	22.1	2.7	21.5

Conversion Coefficient from K_{HFWD} to K_{30} Values. The next step in our research involved investigations into the conversion coefficient for translating K values into K_{30} values on the basis of the relationships of the K values obtained through the FWD and HFWD tests and the K_{30} values identified in the plate bearing tests. These different types of tests were conducted on the soil structures containing volcanic cohesive soil and silty sand, as described earlier, and also on embankments consisting of crusher run and mechanically stabilized crushed stone.

According to the K values from the FWD and HFWD and the K_{30} values from the plate bearing tests, both of which are seen in Fig. 6 and Table 5 for the subgrade before and after the stabilization process, the ratio of K_{30} values to K values is close to 1:1 when K_{30} values are smaller, but it changes to 1:2 or 1:3 when K_{30} values increase.

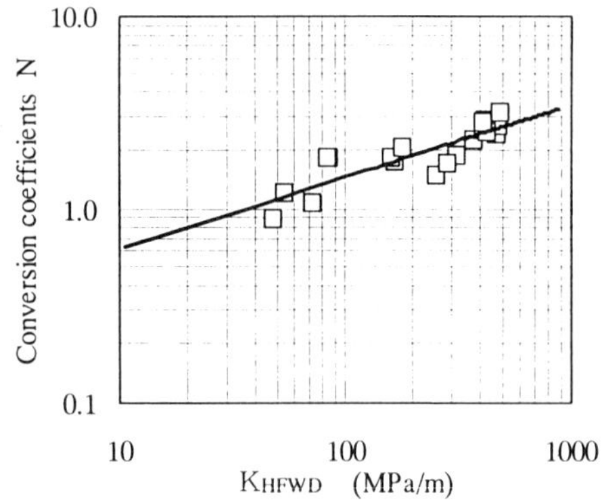

Fig. 7 relationship between the conversion coefficients and KHFWD

A conversion coefficient related to the strength of soil is required for the purpose of precisely calculating K_{30} values based on the measurement results from the FWD and HFWD. Fig. 7 illustrates the relationship between the conversion coefficients, which represents the ratio of K_{HFWD} to K_{30} values, and K_{HFWD}, when these values were determined through the FWD and HFWD testing for subgrade soil and crushed stone.

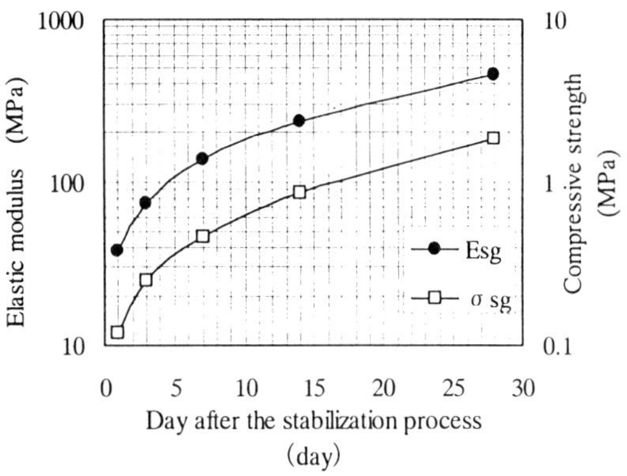

Fig. 8 The relationship between elastic modulus and unconfined compressive strength

As the linearity in the figure shows, the conversion coefficients and K_{HFWD} are highly correlated. Such relations make it possible to determine values corresponding to K_{30} by using the conversion coefficient N, on the basis of measurement results from the FWD and HFWD. The following are the equation (4),(5),(6) for deriving K_{30} values from the HFWD.

$$K_{30} = N \times K_{HFWD} \quad (4)$$

$$K_{HFWD} = p / d \times 0.3 \quad (5)$$

$$N = 0.260 \times K_{HFWD}^{0.373} \quad (6)$$

Where , K_{30}: Modulus of subgrade reaction (MPa/m)

K_{HFWD}: Modulus of subgrade reaction by FWD (MPa/m)

p: contact pressure (MPa)

d: deflection (deformation) (m)

N: conversion coefficient based on K_{HFWD}

ρ d: dry density (g/cm^3)

Based on the above-mentioned outcome of the investigation, Fig. 8 illustrates the relationship between subgrade elastic modulus and unconfined compressive strength over elapsed time (days) after the stabilization process. As the figure shows, decreases in deflection after the subgrade stabilization process, increases in the elastic modulus, and increases in unconfined compressive strength are almost in parallel with each other. In the case of silty sand, unconfined compressive strength after 7 days exceeded 500 kPa, indicating sufficient strength. The elastic modulus of the subgrade after 28 days also became 10 times larger than the modulus on the first day.

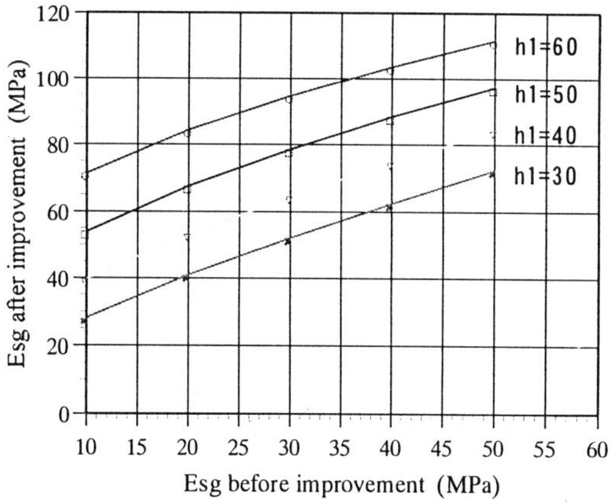

Fig. 9 Design chart of the subgrade stabilization process layer

Thus the relationship between the elastic modulus of the subgrade and unconfined compressive strength can be represented by the following equation.

Esg (MPa) =250 × qu (MPa)　　(7)

Where, Esg: elastic modulus of the subgrade　(MPa)
qu: unconfined compressive strength　(MPa)

A comparative study was carried out of the elastic modulus of the subgrade before improvement and the target elastic modulus after improvement, when the CBR value of the stabilization process layer was 20% (E=200MPa) and h_1 as the layer thickness ranged from 30 cm to 60 cm. Given the results presented in Fig. 9, it is possible to determine the elastic modulus of the subgrade through the FWD measurement on the un-improved subgrade and then calculate the required thickness of the layer to be improved, so that the target CBR obtained from the Esg=100 CBR relationship can be satisfied.

The above-mentioned investigations and studies confirm that the FWD and HFWD can be used to take measurements of elastic behavior of the subgrade and the base course, thereby contributing to the evaluation and design of a stabilization process.

Conclusion

This paper can be summarized as follows.

1) Elastic moduli of the subgrade as determined through the FWD and HFWD are very close to each other. Both single-layer elasticity analysis and multi-layer elasticity analysis are valid as long as deflection takes the form of elastic deformation.

2) The range of loading effect in the static loading test, such as the plate bearing test, is 7 times the radius of the loading plate in the vertical direction and 3 times the radius in the horizontal direction. In the dynamic loading test, including FWD and HFWD tests, where impact load is applied, the range of loading effect is 2 to 3 times deeper and wider than in the static loading test.

3) In FWD tests with impact load, the range affected by loading is larger than in the static loading test, and the bearing capacity coefficient K becomes smaller as the bearing capacity decreases. For this reason, the range of loading effect and the existence or non-existence of a soft and weak layer must be taken into consideration in taking measurements or in analyzing the subgrade.

4) The relationship between in-situ CBR and K_{30} values is closer to the relationship identified by Ueshita when the bearing capacity of the subgrade is smaller, and it becomes similar to PCA's as the bearing capacity increases.

5) As a method for converting the results of bearing capacity measurement from the FWD and HFWD into K_{30} values, a proposal was made for conversion coefficients using K values obtained from the FWD and HFWD.

6) The elastic modulus of the subgrade as derived from FWD and HFWD measurement, is approximately 250 times larger than unconfined compressive strength.

7) FWD and HFWD equipment is not only useful for the evaluation of pavement structure but also applicable to the construction management and the design and evaluation of a stabilization process for the subgrade and the base course.

REFERENCES

1. Japan Road Association, Manual of Asphalt Pavement, Japan, 1986.

2. Nagato Abe, Masaki Kamiura, Etsuo Sekine and Teruhiko Maruyama, Evaluation of Bearing Capacity of Base Course and Subgrade by Falling Weight Deflectometer, Third International Conference on Road & Airfield Pavement Technology, Beijing, P. R. China, April, 1998.

3. Railway Technical Research Institute, Design and Construction of Soil Mechanistics, Japan, 1994.

4. Etsuo Sekine, Katsumi Muramoto, Makoto Sunaga, Nagato Abe and Futoshi Nakanishi, Structural evaluation of Railway roadbed using the FWD, JGEA, The 32th annual meeting of soil and foundation engineering, pp.2213-2214, 1996.7.

5. Kaname Ueshita, Mechanics of the pavement, Pavement, Kensetsu-tosho, Vol.8, No.6, pp.33-38, 1973.

6. Portland Cement Association, Thickness Design for Concrete Pavements, U.S.A., 1966.

Fifth International Conference on the
BEARING CAPASITY OF
ROADS AND AIRFIELDS

ALFRED, a Norwegian Road Monitoring System.

TOR ARNE HETLAND,

Senior Consultant, Datainstrument AS, P.O.Box 64,
N-5035 Bergen, Norway

ABSTRACT

Measuring cross and longitudinal profiles forms a basis from where essential road and pavement characteristics can be calculated. Repeated measurements over several years disclose development trends and form a basis for effective pavement maintenance.

In Norway cross and longitudinal profiles are measured for all main highways and county roads every year. 13 measuring equipments are owned and operated by the local Public Roads Administration in the counties. Around 100.000 lane kilometres are measured every year during the season from April/May through September.

The measuring equipment consists of an extendable 2 m wide beam with 17 ultrasonic sensors measuring the cross profile, and a laser-based sensor unit measuring the longitudinal profile. A common instrument computer performs data acquisition, calculations, and storing. Operator interface and final data export are utilized in a conventional lap top PC, operating the instrument computer through a network connection.

Recorded data are stored for each metre. From these 'raw'-measurements rut depth and rut area, cross fall, horizontal radius, and IRI among others are calculated. In a postprocessing operation, average values for 20 m sections are calculated and stored in the Central Road Data Bank.

A quality assurance system is built around the operation and maintenance of the measuring equipment. Thus operation by non-engineers with moderate training is possible.

Datainstrument AS operates a service- and support- system for each equipment to ensure maximum operational effectiveness.

Key words: Measuring equipment, cross profile, longitudinal profile.

counties the vehicles are also used for other tasks like recording of road furnishing and friction measurements.

The measuring season for pavement conditions is short in Norway. In the southern part they start in April/May, or in a few cases even before. This depends on the weather conditions. In the three northern counties they usually start in late May or June. Due to frost heave, many roads can not be measured until this effect has vanished, unless they want to measure it explicitly as some counties are doing on certain roads. During September, most measuring programmes are completed. In years with good weather conditions, the programmes may be completed earlier, other years the time may become too short, especially in the western and northern parts of the country.

Since the 13 measuring equipments are locally situated, they are able to adapt the measuring programme to the local conditions and combining the various tasks in an optimal way. The equipments are increasingly used for repeated follow up on new and reworked road segments. Thus, it is important to be able to insert a short break in the ordinary programme to measure certain road segments immediately after finish / rework has been conducted and at certain intervals afterwards.

Since the width of the equipment is less then the limits where an accompanying warning vehicle is required, the equipment is operated in normal traffic, however , usually at a reduced speed of about 50 km/h.

The cost of the measuring equipment is around 1 Million NOK, vehicle excluded. The cost is so moderate that most counties can afford to own their own equipment and have the benefits!

MEASURING EQUIPMENT DESCRIPTION.

The measuring equipment consist of

- a beam of ultrasonic transducer measuring the cross profile
- a sensor unit carrying gyros, accelerometers, inclinometers and a laser distance sensor
- a high precision tacho
- an instrument computer
- an operators computer

The beam is mounted at the front of the vehicle as shown in Figure 4. It actually consists of two halves (9 and 8 sensors), which are mounted in parallel to give a cross profile length of 2 m and 12.5 cm between each profile point. An 18th sensor is placed on top of the beam measuring a fixed distance to obtain a reference for distance calculation on the other 17 sensors. It is possible to adjust the two beams with respect to each other to obtain a profile length of up to 3.75 m, but the option is seldom used since an accompanying warning vehicle then would be required.

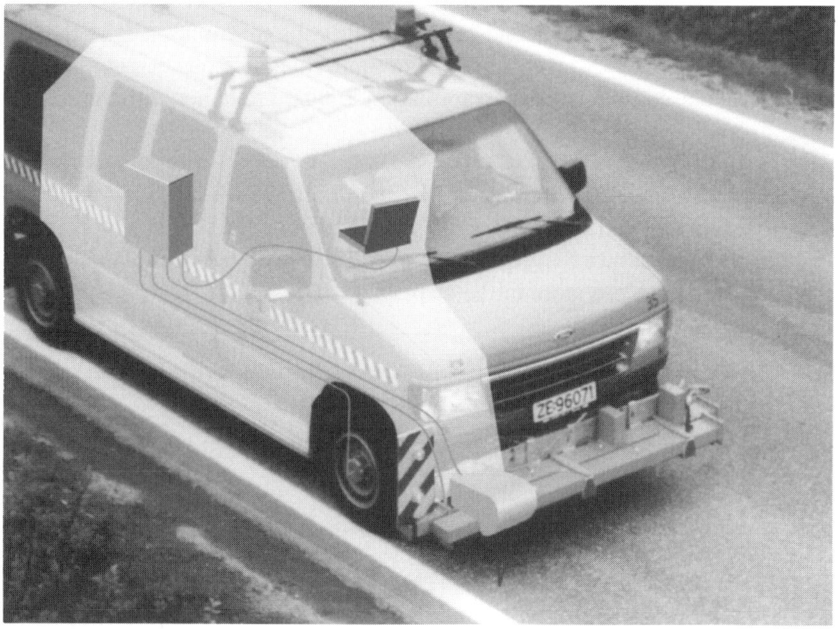

Figure4. Measuring equipment and vehicle

The sensor unit is mounted on top of the beam, on the right side and positioned so that the laser beam will measure in the middle of the wheel track. The unit is heated to minimize temperature drift in the sensors.

The tacho is attached to the front wheel on the right side of the car as shown.

The instrument computer is actually a standard industrial PC, equipped with the necessary interface electronics for the sensors, and a network connection for operation by the operator computer which is a standard lap top PC. The software in the instrument computer reads the sensors, and calculates

- cross profile,
- cross fall,
- horizontal radius,
- longitudinal profile and
- travelled distance
- travelling speed

The data are stored on file usually every 1 spatial m. 0.5 and 0.25 m interval could also be selected. The longitudinal profile will always have 0.25 m between each height.

Also calculated are IRI, rut depth, rut depth position and sensor statistics to provide running information to the operator. This information is also used to present average values at the end of each measured road segment to allow the skilled operator to make a brief judgement of the quality of the recorded data.

The data are positioned along the actual road by connecting information from the road network file to the travelled distance from one or more reference positions along the travelled route. The used road network information is stored together with the recorded data.

During operation, the operator observes the screen in Figure 5 which shows some typical data. To some extent the operator will use the information on the screen to confirm the driving pattern and also assure that significant irregularities do not occur which will result in excessive values in one ore more of the displayed data.

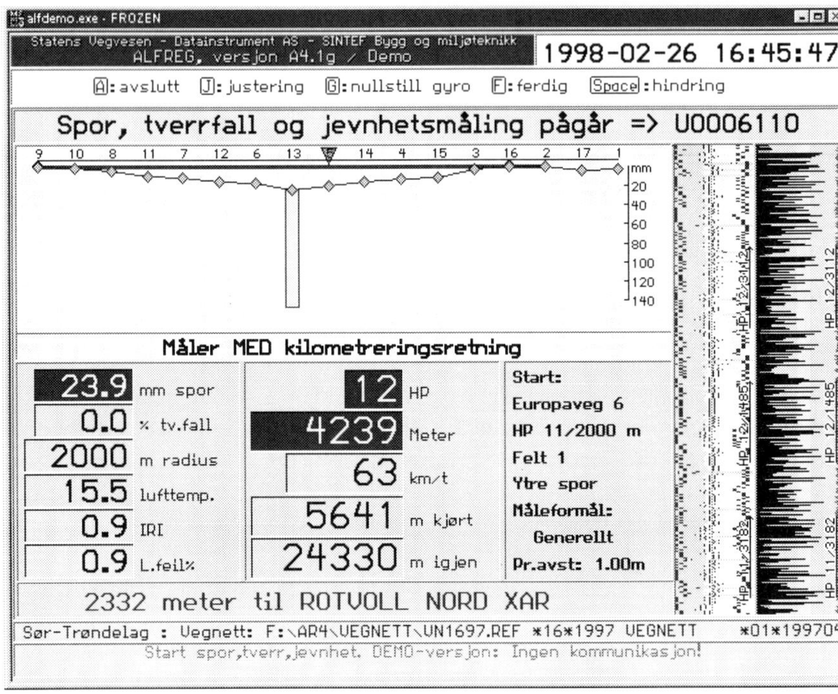

Figure 5. Operators screen during recording of pavement data.

DATA FLOW AND POST PROCESSING.

Figure 5 shows the essential data flow in the system. The data recorded and stored in the vehicle are labelled 'raw data'. These data are permanently stored and archived at the county Public Roads Administration. Post processing programs (from SINTEF and the Directorate of Public Roads) perform calculation of 20 m values which are submitted to the Central Road Data Bank. This operation is usually executed by the vehicle operator when convenient. From the Road Data Bank the data are then available for pavement management and budgeting to apply work and money in the right priority.

From the raw data it is also possible to extract and present detailed information on a per m basis which is of special values for the local administration.

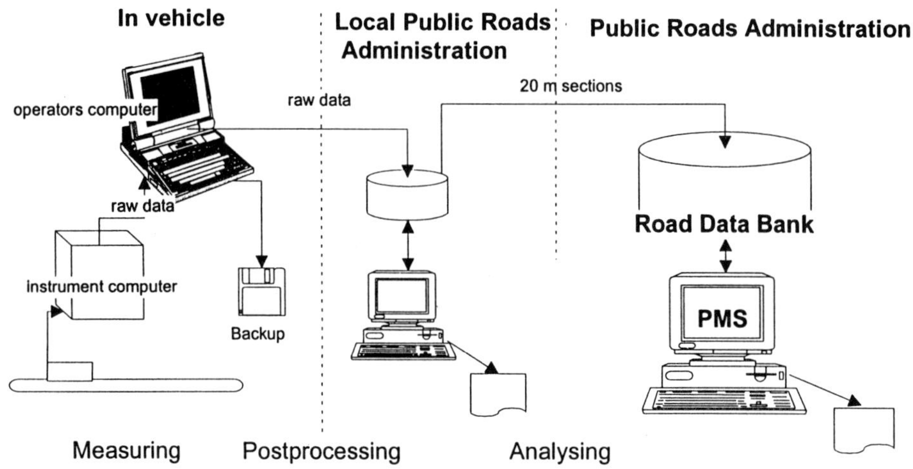

Figure 6. Main data flow of the Alfred Road Monitoring System.

QUALITY ASSURANCE SYSTEM.

The quality assurance system is based upon a set of procedures covering

- yearly control and calibration of equipment,
- establishing local control road sections,
- checking of equipment against control road sections
- education of operators

and further for daily use by the operators ,detailed procedures covering

- daily control and calibration,
- recording data
- processing of Road Data Bank files
- transmission to the Road Data Bank

The sensor unit is tested / calibrated in the laboratory each winter and each vehicle and measurement equipment is carefully checked at the start of a new measuring season. At least one control road section is established in each county. This section is travelled several times for control of repeatability and on at least 6 positions the cross profile is checked against manually measured data. Where possible the control sections are located so that it could be reached by neighbouring counties too and thus obtain a 'cross-vehicle' control. Several times each season the equipment shall be checked against the control section by the operator.

At the beginning of each season the Directorate of Public Roads gather several vehicles in 'regional tests' where one or more road sections are measured several times by all and the results carefully analysed to confirm equal results and disclose any errors in equipment, driving pattern etc.

During the measuring season Datainstrument AS provides telephone as well as field assistance whenever needed. There is also a frequent communication with the operators to secure that important information is gathered and distributed. A full stock of spare parts is always available to secure short downtime if something breaks down or is damaged.

After the season all operators and most of the pavement management people who are the primary users of the recorded data are gathered at the yearly 'pavement condition measurement' conference to discuss the experience from the last season and propose necessary improvements and fixes to equipment and procedures.

FUTURE DEVELOPMENTS.

The selection of pavement characteristics measured and calculated is and has always been closely matched to the users (Norwegian Public Roads Administration) needs and requirements. Thus other characteristics that would be more or less easily implemented will wait until the data becomes relevant for the users.

One such characteristic is macro texture which will be made available in 1998. The calculation will conform to the CEN standard. Other parameters to describe longitudinal unevenness has been investigated, however, it has been decided to continue with the IRI.

Use of GPS as position reference may become relevant in the future and the equipment has been used to supply short time position during GPS fall outs with good results.

The system will continuously provide current vehicle position along the road network. This information could be made available as a service on the internal network and the equipment will then provide the position reference for other kind of measuring equipment that could be connected to the vehicle and operated in parallel or in separate run, i.e. video recording of the road, recording of items along the road etc.

On the postprocessing part a major revision is planned to replace old programs and obtain a higher integration of the post processing, reporting, book keeping and data transfer parts.

CONCLUSION.

When conclusions shall be drawn from the development and running of the Alfred system, a main point is the importance of dealing with the whole process including equipment, operators, users, service, as one system which serves the one purpose of providing the user, i.e. the Norwegian Public Road Authorities with the data they require. Thus, establishing an overall quality system and an open dialogue between a stable group of operators, users and developers are equal important as the measuring equipment itself.

EVALUATION OF LONGITUDINAL ROUGHNESS AND RIDE QUALITY OF AIRPORT PAVEMENTS

Marc Stet[1] and Tony Gerardi[2]

[1] KOAC•WMD Pavement Research & Consultancy,
Apeldoorn, The Netherlands, E-mail: koacoa@pi.net

[2] APR Consultants, Inc. Medway Ohio, USA

ABSTRACT

This paper describes a method for routinely evaluating airfield pavement surface roughness from an aircraft ride quality perspective. The method is designed to be as compatible as practical with existing pavement management system data collection and reporting processes. Ride quality has always been recognized as a major factor in a pavement's serviceability. Now it can be measured and tracked on a consistent basis.

Key words: Pavement roughness, Ride Quality, Airport pavements, Pavement management.

INTRODUCTION

Smooth airport pavements last longer and cause less aircraft fatigue damage. Most airport owners & operators think about runway roughness only when complaints from pilots and passengers start occurring. However, there is a recent interest in adding runway ride quality as a part of Pavement Maintenance/Management Systems (PMMS). By maintaining pavements to a "not to exceed" level of roughness, pavement life will be extended. PMMS are becoming more attractive as airport authorities are discovering that the up front investment of time and money has significant long term payoffs. As with any relatively new science, PMMS are continually being refined. This paper describes a method for routinely evaluating airfield pavement surface roughness from an aircraft ride quality perspective. The method is designed to be as compatible as practical with existing pavement management system data collection and reporting processes. Aircraft *Ride quality* is the best measure of pavement roughness. Ride quality is generally measured as

condition [4]. It's structural load carrying capability is another measure of its goodness. ICAO's ACN-PCN reporting method is commonly used as a criterion for safeguarding the pavement's integrity and obtaining assurance of optimum service life [5]. An airport pavement could, in theory, have a pavement condition index (PCI) of 90 to 100, and a high load carrying capability, but be unusable because of poor ride quality. Although this is an extreme example, it does emphasize the importance of including "rideability" as a necessary part of any pavement management system.
Pavement engineers have been aware of this for many years, but until recently, there was no practical way of quantifying or tracking pavement rideability for airport pavements. Some airport managers rely on pilot and passenger feedback, but this type of a report of a suspected rough area lacks definition. It doesn't specify where or how rough it is.

Simulation of Aircraft Response

A roughness evaluation requires a profile elevation survey along the centerline of the runway over the reported rough areas, plus survey lines along the same area where the main gear track would normally be - about 3 to 3.5 meters on each side of the runway centerline. High speed profilers are available and simulation technology is mature and practical. Pavements can be monitored for ride quality just as easily able than any other parameter considered in modern airport pavement management systems. With high-speed profilers such as APR's automated rod and level device or KOAC•WMD's ARAN vehicle, a 3,600m long runway strip can be measured within two hours, getting profile measurement every 10 cm [6].
For analysis use can be made or the PC program *TAKEOFF*, developed and refined by Gerardi [7]. The refinements include the modeling of the simulated aircraft for the taxi-in, taxi-out, takeoff, landing and high speed turn out operations and includes provisions for the automated scan of these ground operations to search out the most severe aircraft response for each of the simulated operations. The aircraft simulation program library contains aircraft parameters for designated aircraft classes. The aircraft simulation program computes pilot station acceleration (PSA), aircraft center of gravity acceleration (CGA), dynamic tire load, and dynamic pavement load as a function of location on the measured runway profile input. The PSA and CGA for a FAA class D11 aircraft (wide-body, 2 wing mounted and 1 tail mounted engine) during takeoff from Amsterdam Airport Schiphol runway 24-06 are plotted in figure 1. as a function of runway stationing. The dynamic loading of the nose gear and main landing gear to the pavement is plotted in figure 2.

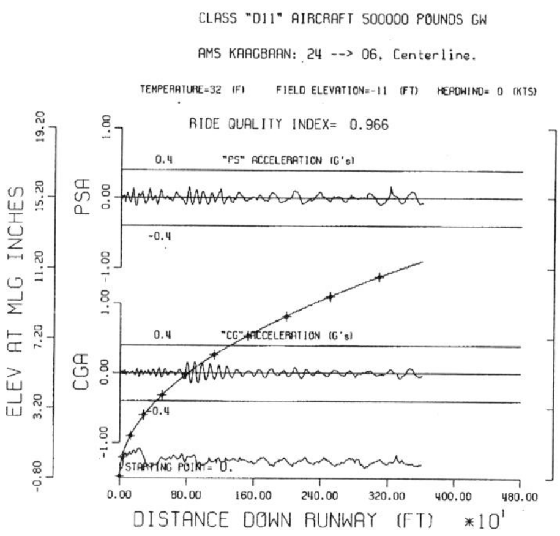

Figure 1: Time history of Takeoff simulation

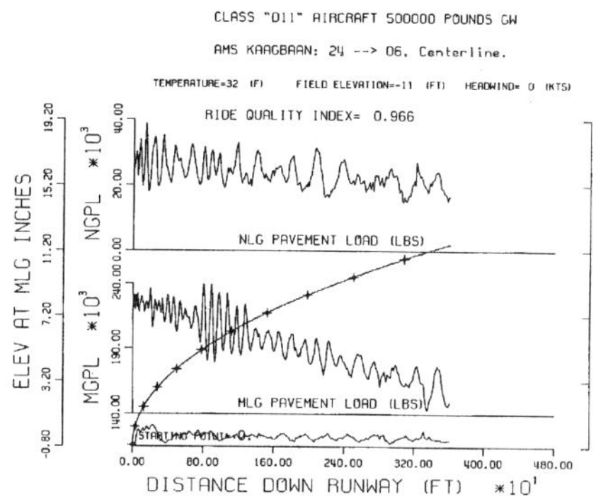

Figure 2: Dynamic tire and pavement loading during Takeoff.

The plotted bottom trace in figure 1(& 2) is the measured runway profile input to the simulation. The plotted trace rising from left to right, is the computed aircraft velocity which starts at zero at the start of takeoff on the left and increases to the aircraft rotation velocity on the right. The top trace in figure 1 is a plot of the pilot station acceleration (PSA) and the center trace is a plot of the aircraft center of gravity accelerations (CGA). The dynamic loading of nose gear and landing gear to the pavement is plotted in figure 2. The traces show the dynamic tire load and dynamic pavement load as a resonse on the measured runway profile input. Based on the vertical accelerations, the program provides a single overall measure of the ride quality for the entire pavement lane. This ride quality index (RQI) will be different for different aircraft and therefore differs on the same pavement. The Ride Quality Index is derived by calculating a normalized root mean square (RMS) of the peak pilot and C.G. vertical accelerations and combining them. The roughness can be ranked as smooth, average or excessive.

Adding rideability to PMMS

Simulation of aircraft response to a runway profile provides a practical method for accurately locating rough sections of pavement and quantifies the level of roughness. The vertical acceleration at the pilot's station and aircraft c.g. is predicted as well as the pavement dynamic loading from the main and nose landing gear. This technology has been successfully used on multiple occasions to accurately locate reported bumps on the runway.

This same technology is also a practical method for adding *"rideability"* to existing pavement management systems. However, the strategy in isolating a bump in a reported rough area on a pavement is different than that required to quantify the rideability of a pavement in a pavement management system. The approach suggested is to measure the pavement profile, break it into sections and calculate the "Ride Quality" of each section

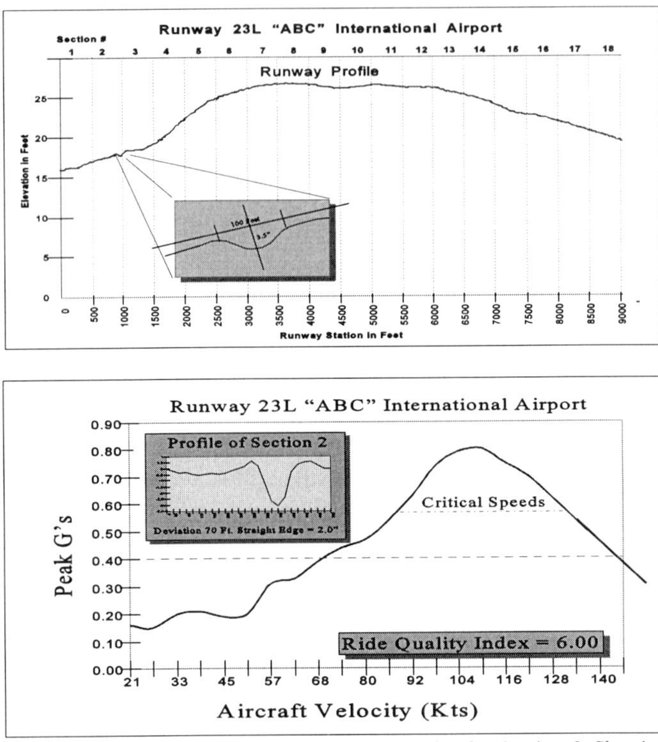

Figure 3: Runway Profile and Velocity Plot for Section 2 Showing Resonant Speeds (Normally in Color)

using aircraft simulations for all possible velocities. The paper presents the results of the application of this technique to an in-service runway, referred to as runway 23L "ABC International Airport". The speed at which an aircraft traverses a section of pavement is a key factor of its response. In theory, any section of pavement could be traversed at any speed, e.g. takeoff, landing, taxi or aborted takeoff.

The analysis approach is to consider the pavement in sections; 152.4 meters (500 feet) for example. Each section is evaluated in the following manner: Simulate an aircraft taxiing at a constant speed of 12 knots (20 feet per second) over that section of pavement. Somewhere during that 12 knot simulation, a peak acceleration occurred for the pilot's station and for the aircraft's C.G. These values are stored for later use and plotting.

For the next simulation, the speed is increased to 14.9 knots (25 fps) and the same section of pavement is evaluated at that speed. Once again, the peak accelerations are stored for later use. This process is continued in 2.97 knot (5 foot per second) increments until the aircraft's rotation velocity is reached. This computer program is called "*VSWEEP*". This method does not consider the effect residual motion imparted from a previous section of pavement or aircraft horizontal acceleration in the pavement section being analyzed. It examines each pavement section autonomously and at constant velocities. Aerodynamic lift is included as well as all other non-linearities simulated in the *TAKEOFF* computer program. Once this *"velocity sweep"* is completed and peak vertical accelerations have been stored for the pilot's station and center of gravity, a *"velocity plot"* can be generated. An illustration of a velocity plot is shown in figure 3. The value of this type of plot is that it immediately points out the aircraft velocities that cause resonance between the natural frequencies of the aircraft and the bumps and dips in that section of the pavement. The peak C.G. response curve will reflect the speeds which excite the aircraft's *"heave"* mode of vibration for the bumps and dips on that section of pavement. The pilot's station response will reflect those that excite the aircraft's pitch mode of vibration. The *"sharpness"* of the peaks will reflect the energy content for that section of pavement. The ride comfort criteria line of .4g is drawn on the plot as a reference point. However, it should be noted that a .4g exceedance here is not as significant as it would be in a *TAKEOFF* or *LANDING* simulation because of the resonance *"search"* being conducted in this analysis. An example of the results of a *VSWEEP* analysis for Runway 23L of ABC International Airport, the pavement management ride quality analysis program analysis are shown in Figure 4. Figure 4 shows the calculated Ride Quality Index (RQI) for each 500 foot section of pavement. This type of graphical information could be very

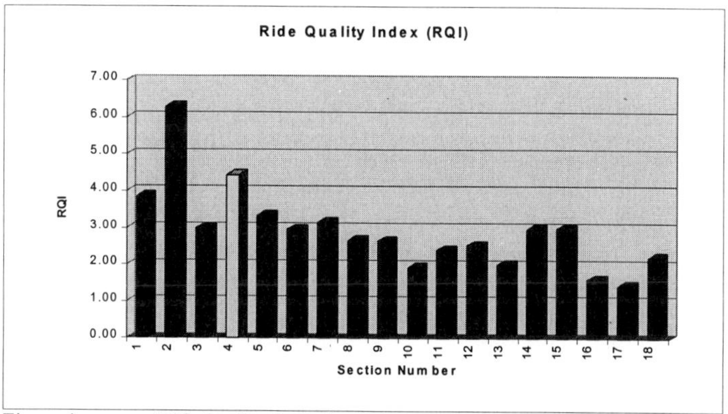

Figure 4: Ride Quality Assessing Using VSWEEP, a Pavement Management Tool (Normally in Color).

useful for managers in making informed decisions regarding pavement maintenance and management. Note that section 2, where a 3.5 inch dip is located, produced an RQI over 6. Experience has shown that sections having an RQI greater than 4.0 can generally be considered rough. A value of 6.0 indicates severe roughness. Again, the *VSWEEP* program looks at all aircraft velocities to calculate the RQI. If this dip had been located further down the runway, it could have stopped commercial jet operations until repairs were made. This process could be used as a criteria for determining when rehabilitation is needed, and as acceptance criteria for new and rehabilitated pavements.

Periodic evaluation of the pavement in this manner will allow the operator to track pavement roughness *"growth"*. (Roughness growth is a term used to illustrate increases in pavement roughness levels). The airport owner will be better equipped to forecast when and where rehabilitation will be required. The change in pavement smoothness and ride quality will be a reflection of the traffic that operates on it and the subgrade motion caused by frost heave, settlement, and the like.

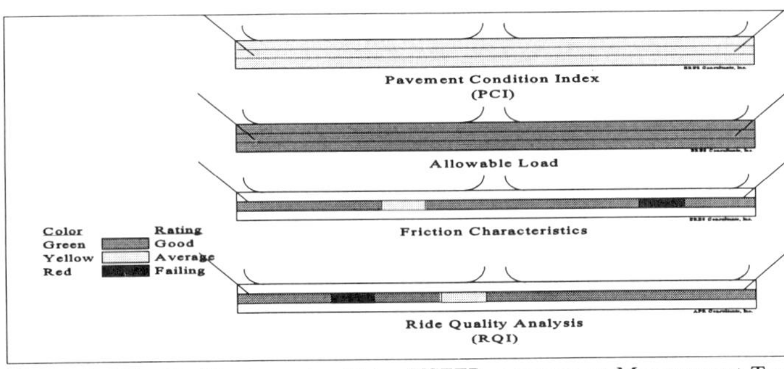

Figure 5: Ride Quality Assessing Using VSEEP, a pavement Management Tool (Normally in Color)

Figure 5 illustrates how *Aircraft Ride Quality* as well as *Friction Characteristics* could be incorporated into existing pavement management/maintenance systems. One chart can be used to depict the overall condition of the runway, PCI, Structural, Friction and Ride Quality. It is designed to help managers make informed decisions regarding maintenance based on solid engineering analyses. It allows direct comparisons year after year on a consistent and repeatable basis, and can be used to track roughness growth as well as aid in predicting when major rehabilitation will be required.

CONCLUSIONS AND RECOMMENDATIONS

1. A method has been developed that will access airport pavements for aircraft ride quality. The method is designed to be compatible with existing pavement maintenance-management systems and provide a practical, user friendly and consistent way of evaluating a pavement for its ride quality.

2. Pavement profile measuring equipment and aircraft simulation technology are available to make "ride quality" assessment a practical pavement management parameter.

Maintenance decisions can be based on hard data and solid engineering. Reported rough areas can be isolated and optimized repairs can be made. For example, instead of overlaying several thousand feet of runway, it may be possible to repair only eighty to one hundred feet. The most practical way to accomplish this is to measure the runway profile and simulate an aircraft operating on it to determine its aircraft "ride quality". Profiling devices are available and simulation technology is mature and practical. A runway pavement can be evaluated within two hours with easy-to-ship equipment. Pavements can be monitored for ride quality just as easily as they can for structural integrity. If fact, it may be more repeatable than any other parameter considered in modern airport pavement management systems.

3. Aircraft simulation technology is a mature science that will assess airport pavements for ride quality. *VSWEEP* is a computer program designed to be used with existing pavement maintenance/management systems and provide a practical, user friendly and consistent way of evaluating a pavement for its ride quality. Simulation of aircraft response to a runway profile provides a practical method for accurately locating rough sections of pavement and quantifies the level of roughness. The vertical acceleration at the pilot's station and aircraft c.g. is predicted as well as the pavement dynamic loading from the main and nose landing gear. This technology has been successfully used on multiple occasions to accurately locate reported bumps on the runway. The speed at which an aircraft traverses a section of pavement is a key factor of its response. In theory, any section of pavement could be traversed at any speed, e.g. takeoff, landing, taxi or aborted takeoff. Consequently, a ride quality analysis used as a pavement management tool, should consider all aircraft velocities.

4. There is a need for the establishment of runway roughness criteria based on aircraft response. Rough runways and taxiways cause aircraft fatigue damage, they can be an aircraft safety issue, can reduce pavement life, can reduce aircraft stopping ability and cause pilot and passenger complaints. Aircraft simulation technology is a mature science that will assess airport pavements for ride quality. When used as a pavement management tool, the simulations will *"flag"* areas of the pavement that should be evaluated for roughness in more detail and perhaps repaired.

5. Smooth pavements last longer, cause less fatigue to aircraft and smooth pavements are safer. Yet, there are currently no official criteria that tells an airport operator when existing pavements have become too rough. There are, however, some unofficial criteria that are used by the industry; ICAO, the Boeing Co, and a military specification all call for "not to exceed" bump and dip amplitudes verses wavelength. While these do provide useful guidance, they do not account for multiple events and the complex interactions between the runway and the aircraft. Another unofficial criteria used in the industry is a "Human Comfort" vibration criteria. This type of criteria is more applicable to the needs of the industry since it reflects the aircraft response to the runway profile. Recently, ICAO

initiated a study on aircraft ride quality. A committee of peers comprised of industry, government and contractors is looking into implementing guidelines or criteria of this nature. Many members of the ACNSG (Airfield Classification Number Study Group) are using the software and are recommending it for pavement evaluation.

CLOSURE

The objective of this paper is to document on the importance of pavement roughness and airport pavement ride quality as an airport safety issue. Rideability should be integrated in PMMS. KOAC•WMD is a representative and can provide roughness evaluation software, high speed profile measurements and roughness evaluations. KOAC•WMD also performs structural pavement evaluations, PCN pavement strength classifications, automated visual condition surveys and helps you with your PMMS. We aim for co-partnering rather than for working on a project basis only. We can be reached by Email via koacoa@pi.net.

REFERENCES

[1] Tony Gerardi and Wayne Johnson (1994), "Dynamic Loading of Airfield Pavements Caused by Aircraft Operations on Rough Runways", Presented at the IATC in Washington DC June 24.

[2] Edward Gervais (1991), "Runway roughness measurement, qualification and application; The Boeing Approach", Proc: ASTM Conference on Airport-Pavement Interaction, pp. 121-131;

[3] Marc Stet, Marien Voorwinde and Kars Drent (1995), "Schiphol's Airport Maintenance Management Approach",Proc. 2nd ICPT, Singapore;

[4] Mo Shahin (1994), "Pavement Management for Airports, Roads, and Parking Lots", Chapman and Hall;

[5] Marc Stet (1991), "Airport Classification Systems for Structural Pavement Strength", 2nd CROW Workshop on Airport Pavements, Amsterdam Airport Schiphol;

[6] Marc Stet and Tony Gerardi (1997), "Ride Quality Assesement of Airport Pavements", 3rd CROW Workshop on Airport Pavements, Amsterdam Airport Schiphol;

[7] Tony Gerardi (1997), "Takeoff 4.0 and Landing 3.0", Aircraft Simulation for IBM Compatible PC's Runway Interaction.

Fifth International Conference on the
BEARING CAPASITY OF
ROADS AND AIRFIELDS

A FRESH APPROACH TO RUT PREDICTION IN ASPHALT

NICHOLAS THOM[1], ANDREW COLLOP[2] AND RICHARD ELLIOTT[3]

[1,2]Lecturer, University of Nottingham, NG7 2RD, UK, fax 44 115 9513898

[3]Senior Engineer, Scott Wilson Pavement Engineeering, NG7 2QP, UK,

fax 44 115 9431302

ABSTRACT

Experiences over recent summers in the UK have concentrated minds on the problem of asphalt rutting and its prediction. In this paper, the mechanisms at work in rut development are explored in some detail and a predictive approach is presented which takes account of the contribution from both the aggregate skeleton and the bituminous binder in a fairly fundamental way. The model presented deals with shear flow in an asphalt. Essentially the asphalt is modelled as a granular material, whose permanent deformation rate is severely restrained, sometimes dominated, by the viscous contribution of the bitumen. To prove the capabilities of the model, it has been applied to a range of results from the Repeated Load Axial Test (RLAT) and it is seen to be capable of simulating each of them with a reasonable degree of accuracy. Each of the parameters in the model is seen to have a particular meaning in terms of the physical properties of the asphalt. The model is then applied to the prediction of rutting in an accelerated wheel tracking facility. The required adjustments and corrections are pointed out and the fact that it is not possible to deduce the angle of internal friction from a conventional RLAT is highlighted as a particular problem. None the less rut predictions are of the right order of magnitude and the possibility of using a confined version of the RLAT is seen as a way forward, enabling much more confident prediction of ruts based on that test.

INTRODUCTION

The principle which has guided the direction of this paper is that the permanent deformation of a bituminous material can only be understood as a combination of the properties of an unbound aggregate together with the viscous nature of bitumen. An unbound aggregate has its own peculiar nature in the way it deforms under cyclic load. Researchers have shown (1,2) that the closer the applied stresses are to the limiting shear stress ratio for the material, the more rapidly permanent shear deformation will occur. Beyond the limiting stress ratio, or if tension is applied to the material, deformation resistance disappears. However, if the material is bound with bitumen, it would be logical to assume that the bitumen has the effect of greatly slowing down the deformation which would otherwise have occurred. On a particulate scale, one aggregate particle can only slip relative to another if it can overcome the viscous resistance of the bitumen which is surrounding the whole contact region.

This paper will commence with a discussion of the detailed mechanisms of rut development before giving a description of the new modelling approach, with reasons for its selection. It will then outline the way in which the aggregate and bituminous components of the model have been combined, before introducing its use in the interpretation of laboratory data and looking forward to the assessment of rutting under idealised and then real conditions.

TYPES OF RUTTING IN ASPHALT

Densification It is certainly the case that inadequate compaction of a material will result in additional compaction by traffic, leading to an increase in the density of the material and consequent surface deformation. On the UK trunk road network, the 'percentage refusal density' (PRD) test (3) is used to assess the degree of compaction in a material, 100% representing full compaction relative to a laboratory technique. It would be logical therefore to expect that a material with only 95% PRD might densify under traffic, eventually perhaps achieving around 100% PRD. A 100mm layer of asphalt would therefore lose around 5mm in thickness purely from this effect, leading to a 5mm rut.

However, modern compactive techniques rarely result in 95% PRD values other than at isolated locations, 98 to 100% being much more common. It is therefore considered here that densification, though a real effect, would rarely account for more than a few millimetres of rut at most.

Shear Flow In contrast to densification, which allows only a certain limited degree of surface deformation, shear flow is a mechanism which has the potential to generate large and almost indefinitely increasing ruts. As is well known, it manifests itself clearly by the formation of shoulders either side of a rut and it represents the most damaging form of deformation in asphalt. It is also not easy to model nor, if evidence from many relatively recently constructed highways in the UK is considered, to predict. Experience reveals that two materials, both meeting the same specification and both placed and compacted satisfactorily, can develop quite different levels of deformation and it is the understanding of this effect which is the principal subject of this paper.

MECHANISMS OF SHEAR IN ASPHALT

Shear necessarily involves the movement of certain particles of aggregate relative to others. Three elements which could affect this movement are as follows:

a) Aggregate shape;
b) Friction at aggregate contact zones;
c) Viscosity of the bitumen.

Aggregate Shape In order for large shear deformation to occur, it will be necessary for some particles to move quite considerable distances relative to others and this movement will be made more difficult the more angular and rough the aggregate particles are.

Friction at Aggregate Contact Zones In the case of an unbound aggregate direct particle to particle contact clearly means that the frictional properties of the aggregate are important. In the case of an asphalt, it is usually considered that a film of bitumen is present between aggregate particles although this is an area worthy of further investigation. It may be that sufficient particles still touch directly to significantly influence behaviour but it is also certain that the contact zone will be surrounded by bitumen and that this bitumen may influence the effective friction at the contact.

Viscosity of the Bitumen This is considered as a separate effect from any influence the bitumen has on friction. The bulk of the bitumen will not be in the immediate contact zones but, as slip occurs, it will still be expected to flow and change its shape as the void spaces change. This will inevitably slow down the deformation process.

MODELLING OF INDIVIDUAL COMPONENTS

Granular Material Perhaps two key facts emerge from the work of those who have investigated the nature of permanent deformation accumulation in granular material, namely:

a) strain is often approximately logarithmically dependent on the number of load cycles (4),
b) strain is also closely tied to proximity of the applied stress to failure (1).

The second fact surely needs no explanation and will surprise no-one. The reason for the first is that, on each loading, there will be a few particle contact points where the forces are sufficient to induce irreversible slip. Following these slippages, the particles have to re-orientate themselves slightly, meaning that on the next load application there will again be a few cases among these re-orientated particles where forces induce slip. Thus the process is one of repeated slip and re-arrangement, with the mass of aggregate usually progressing towards an ever more stable state.

The research described in this paper uses a relatively fundamental model for shear deformation in granular material, described in detail in (2). The resulting equation, relating shear strain to the number of cycles of load, is as follows:

$$\gamma = A \log_e [(\tau_f - \tau)/\tau_f - B\tau \log_e N] \quad (1)$$

where:
γ = shear strain
τ = applied shear stress
τ_f = shear stress to cause failure (depends on other stresses)
N = number of load applications
A and B are constants

The derivation of equation 1 comes primarily from the two observations listed at the start of this section. As may be seen, it contains the logarithmic dependence on N as well as the proximity of the applied stress to failure, in the term $(\tau_f - \tau)/\tau_f$. It also has the merit of fitting laboratory data reasonably well.

The model parameters may be seen in terms of the factors affecting deformation. The constant A is directly related to the strain required to induce failure, which does not vary greatly from one aggregate to another. It is a function of grading and particle shape. B depends on the frictional resistance at particle contacts. The failure shear stress τ_f depends on stresses in other directions and is affected principally by particle shape.

Bitumen The properties of bitumen are of course temperature and loading rate dependent. However, at a particular temperature, it is a reasonable approximation to assume linear viscous flow characteristics, which would enable the following equation to be used:

$$d\gamma / dt = \tau / \eta \quad (2)$$

where:
γ = shear strain
t = time
τ = shear stress
η = shear viscosity

In the case of an aggregate / bitumen mixture, a shear stress on the whole material becomes an infinite range of different stresses at different locations within the bitumen. Similarly, the shear strain in the material as a whole represents an infinite variety of shear strains within the bitumen. The result is that, if the above equation is applied to the viscous flow aspects of a bituminous mixture, then the "effective" shear viscosity which applies will be very different from, and much greater than, the shear viscosity of the bitumen itself.

DEVELOPMENT OF COMPOSITE MODEL

The principle of the composite model is that an applied shear stress pulse of magnitude τ can be divided into an aggregate affecting component τ_a and a bituminous component τ_b, such that $\tau_a + \tau_b = \tau$. To ensure compatibility, the shear strain in any one load application, due to each component, is the same. The key quantity to be expressed is therefore $d\gamma/dN$.

Granular material Differentiating equation 1 with respect to N, and replacing τ by τ_a, the component of shear stress acting on the aggregate skeleton:

$$d\gamma/dN = -(A\,B\,\tau_a / N) / [(\tau_f - \tau_a)/\tau_f - B\,\tau_a \log_e N] \quad (3)$$

Unfortunately, equation 3 is only valid for the case where τ_a remains constant throughout, whereas in this model the relative proportions of τ_a and τ_b vary with N. It is therefore necessary to re-express the equation without reference to N.

Rewriting equation 1: $N = \exp(-e^{\gamma/A}/B\tau_a + (\tau_f - \tau_a)/B\tau_f\tau_a)$

and: $[(\tau_f - \tau_a)/\tau_f - B\,\tau_a \log_e N] = e^{\gamma/A}$

Substituting both expressions into equation 3:

$$d\gamma/dN = -A\,B\,\tau_a \exp(e^{-\gamma/A} + e^{\gamma/A}/B\tau_a - (\tau_f - \tau_a)/B\tau_f\tau_a) \quad (4)$$

Equation 4 gives the amount of shear strain expected in a given load application when the total shear strain has reached γ and the shear stress acting on the aggregate skeleton at that time is τ_a.

Bitumen To solve for the viscous component it is necessary to assume a stress pulse shape. For this it is suggested that the most appropriate course of action is to replace the true pulse shape by an equivalent square wave form with the same maximum stress and the same area, as illustrated in Figure 1.

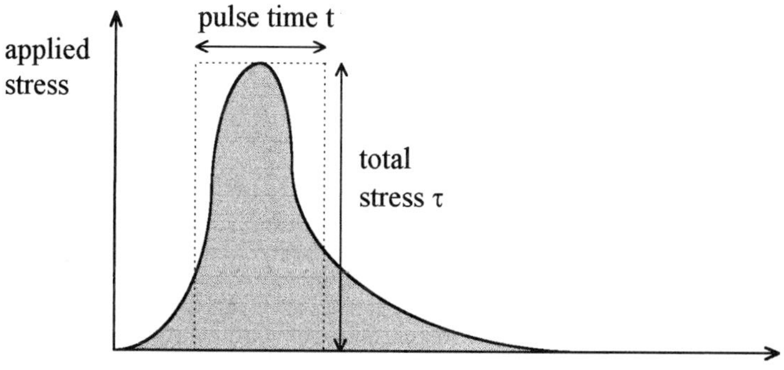

Figure 1: Idealisation of a Real Stress Pulse

Equation 2, written in terms of the "effective" mixture viscosity, therefore becomes:

$$d\gamma / dN = t\,\tau_b / \eta_{eff} = t\,(\tau - \tau_a) / \eta_{eff} \quad (5)$$

Equations 4 and 5 can now be equated, giving:

$$t\,(\tau - \tau_a)\,/\,\eta_{eff} + A\,B\,\tau_a \exp(e^{-\gamma/A} + e^{\gamma/A}/B\tau_a - (\tau_f - \tau_a)/\,B\tau_f\,\tau_a) = 0 \qquad (6)$$

Equation 6 represents the relationship between the aggregate component of shear stress, τ_a, and the shear strain in the mixture, γ, and is the key equation presented in this paper. It can be seen that, provided material properties A, B, τ_f and η_{eff} are known and the loading is specified, equation 6 can be solved numerically for τ_a in terms of γ. From equation 5, $d\gamma/dN$ can then be determined as a function of γ, and numerically integrated to give γ as a function of N. Alternatively, given an experimental curve of γ versus N, equations 5 and 6 can be used to determine the material parameters A, B, τ_f and η_{eff} that give a theoretical curve that best fits the experimental results.

USE OF THE MODEL

Repeated Load Axial Test (RLAT) The RLAT (5) is now commonly used in the UK for discriminating between the permanent deformation susceptibilities of bituminous materials. A repeated vertical load of 100kPa, of approximately square wave form, is applied, with a stress pulse time of 1 second followed by a 1 second rest period, and with zero applied horizontal stress. From a consideration of the Mohr circle of stress, the maximum shear stress, on 45° planes, is 50kPa.

As a check on the validity of the proposed model, it was decided to try to fit it to examples of RLAT data. Three examples are shown in Figure 2. The lines shown are the best predictions found and the points represent actual data from tests on (rolled asphalt) wearing course and binder course materials, carried out at 40°C. Figure 3 shows four further examples, this time on (dense graded) laboratory prepared specimens, the same material in each case, but with different levels of compaction.

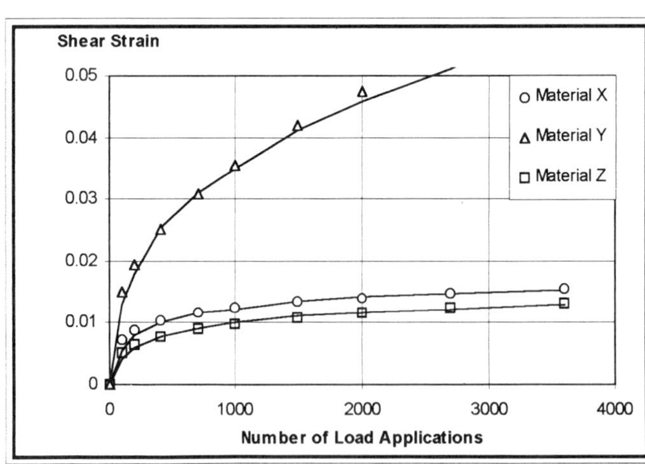

Figure 2: Measured and Predicted RLAT Results from Site Samples

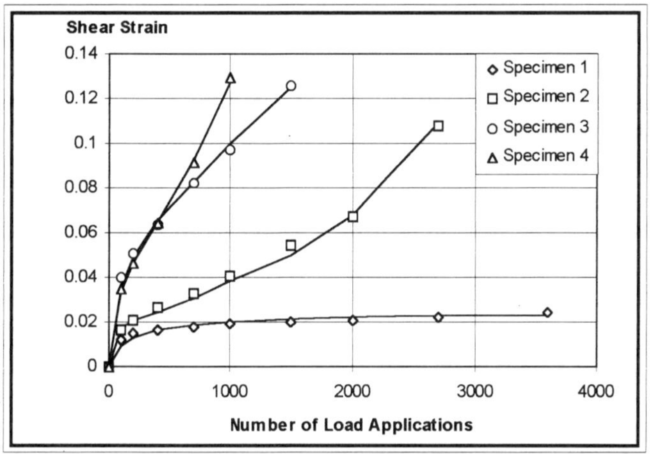

Figure 3: Measured and Predicted RLAT Results from Laboratory Samples

It is clear from Figures 2 and 3 that the model is generally capable of describing the shapes of the experimental curves. The model parameters for the above curves are given in Table 1.

Table 1: Parameters Deduced from RLAT Results

	A	$B (m^2/N)$	η_{eff}(Pa.secs)	τ_f(kPa)
Site materials:				
X	$-5.0\ 10^{-3}$	$1.4\ 10^{-7}$	$4.5\ 10^{8}$	55.5
Y	$-1.05\ 10^{-2}$	$2.4\ 10^{-8}$	$2.0\ 10^{8}$	49.5
Z	$-5.0\ 10^{-3}$	$1.8\ 10^{-7}$	$9.0\ 10^{8}$	57.5
Laboratory specimens:				
1	$-8.0\ 10^{-3}$	$1.0\ 10^{-7}$	$2.5\ 10^{8}$	65.0
2	$-5.5\ 10^{-3}$	$1.0\ 10^{-7}$	$4.6\ 10^{7}$	51.0
3	$-2.0\ 10^{-2}$	$1.0\ 10^{-7}$	$5.5\ 10^{7}$	49.0
4	$-1.2\ 10^{-2}$	$1.0\ 10^{-7}$	$3.7\ 10^{7}$	48.0

It is noted that the order of magnitude of the parameters A, B and τ_f is the same as has been found previously for unbound materials (2). The values of η_{eff} are much higher than typical bitumen viscosities at the test temperature (40°C), which would be expected to be around 10^3 to 10^4 Pa.s, but are of the order expected from mixture viscosity tests.

Materials X, Y and Z are similar in composition, although X is a wearing course and Y and Z are binder courses. Y was extracted from an area where significant rutting had taken place. All parameters are rather different for Y than for the others. A lower τ_f and higher A value indicate a weaker aggregate structure, whereas a higher B and lower η_{eff} suggest easier inter-particle movement. In this particular case, it may be that lack of air voids has led to pressure in the bitumen, decreasing particle contact forces and therefore reducing strength, as well as easing inter-particle movement.

If the loading times are adequately chosen (and if a correlation exists with the truck rate of loading and the rate of loading of the RSST-CH test) than the relationship previously determined and presented in Quadrant "Shif Factors" would be able to predict the number of ESALS actually travelled over the sections.

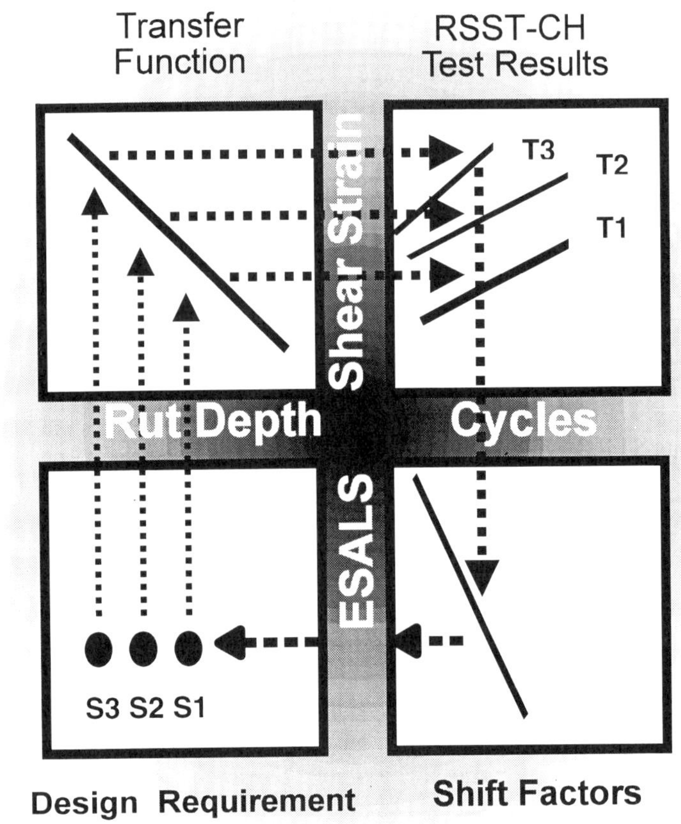

Figure 2 – Schematic representation of the permanent deformation validation concept

TEST PROGRAM

Cores from each cross section were obtained from between the wheel path. Specimens were obtained from the top 5 cm of each core. RSST-CH testing was undertaken with different loading times, namely 0.1, 0.2, 0.3 and 0.4 s haversine loading and in all cases with a rest period of 0.6 seconds. For each test condition 3 replicates were tested at 50 ºC. For each cross section tests were conducted at all these loading times. The magnitude of the loading pulse was 70 KPa. Data presented this far is only based on test results from cores from the second cross section.

TEST RESULTS

It can be observed in Figure 3 the typical effect of loading rate observed in all cross sections. With an increase in loading times, an increase in the rate of accumulation of permanent deformation is observed.

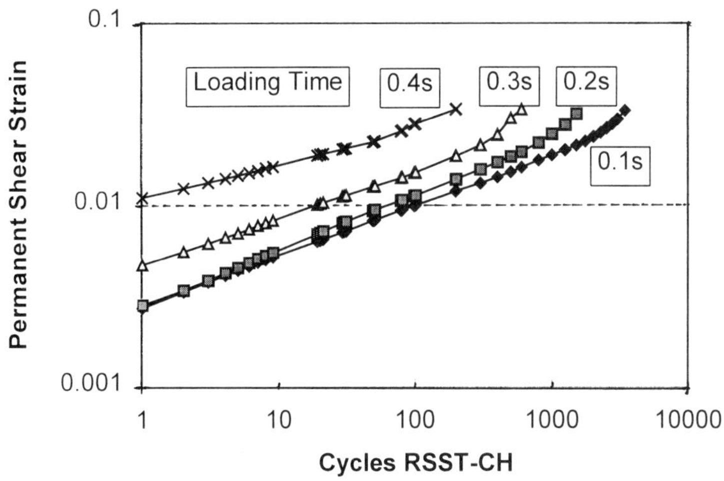

Figure 3 – Effect of loading time on the accumulation of permanent deformation in the RSST-CH test at 50 °C Temperature.

Tables 3 and 4 present a summary of the results obtained for each cross section. It must be noted, that the number of cycles RSST-CH cycles presented in these tables are the number of cycles required to reach the permanent shear strain obtained from the field, at the corresponding cross section. The conversion to ESALs was obtained based on the following relationship:

$$\log(\textbf{Cycles}) = -4.36 + 1.24 \log(\textbf{ESAL})$$

Figures 4 and 5 show the predicted ESALs to measured rut depth as function of load time.

Two approaches were followed to validate the effect of loading time; A and B.
In Approach A, for a given cross section the loading time that corresponds to the expected number of ESALs was obtained using a logarithmic interpolation between consecutive load times, that are above and below of the target value. In this case, it would be assumed that the relationship ESAL-RSST-CH cycles was valid, and it would vary the loading time to obtain the loading time that would yield the best relationship. The loading time was searched to reach the measured rut depth.
In section test number 103, for cross section number 2, 3 and 4, the load time was obtained by the load time that is closer of the expected number of ESALs, respectively, 0.1, 0.1 and 0.3 seconds.

Tables 5 and 6 present, respectively for ST64 and ST103, the results for approach A and B and the corresponding differences to loading time function of speed, to approach A, and 8E+06 ESALs, to approach B.
It can be observed that the differences for approach A, in both section tests, are very low, indicating that the RSST-CH also predict well the rut depth for sloped roads. In terms of approach B, although some discrepant values have been found, the overall approach appears to yield reasonable results.

Table 5 – Summary results of the analysis made to identify the effect of loading time for ST64

		Approach A		Approach B	
Cross Section Number	Loading Time F(vel)	Interpolated Time	Dif. to Loading Time F(vel) (%)	ESALs Assuming correct Loading Time f(vel)	Dif. to 8E6 ESALs (%)
1					
2	0.200	0.175	-12.44	17 274 351	115.93
3	0.267	0.246	-7.74	22 271 702	178.40
4	0.320	0.229	-28.59		

Table 6 – Summary results of the analysis made to identify the effect of loading time for ST103

		Approach A		Approach B	
Cross Section Number	Loading Time F(vel)	Interpolated Time	Dif. (%)	ESALs Assuming correct Loading Time f(vel)	Dif. to 8E6 (%)
1				3 574 780	-55.32
2	0.114	0.1	-12.50	4 237 169	-47.04
3	0.160	0.1	-37.50	5 426 836	-32.16
4	0.178	0.3	-68.75	16 319 966	104.00

It can be concluded that the loading time is an important factor in the correlations made. Small variations in loading time cause significant effects in the predictions made. On the other hand, it can be assumed that errors in measuring truck speed (or the effect caused by trucks going at lower speeds than expected) may be the cause of some dispersion in comparisons between observed and predicted data.

A representation of predicted ESALs to measured rut depth assuming correct loading time function of speed, approach B, is represented in Figure 6, where it can be observed that the results are mostly with half and order of magnitude of the actual number of ESALs. For site 103 the results have less dispersion than for site 64. It is to be expected high data scatter in site 64, because the variation of rut depth function of measured truck speed is not as expected (see Table 1). Material from cross section 2 may not be identical to that of other cross sections.

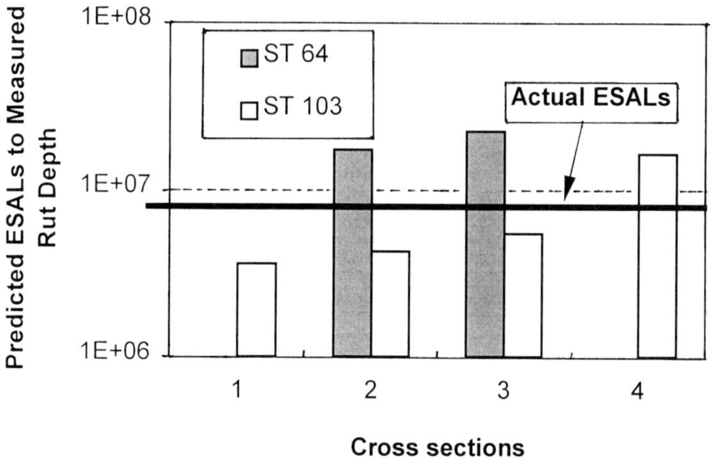

Figure 6 – Predicted ESALs to measured rut depth assuming correct loading time function of speed (Approach B).

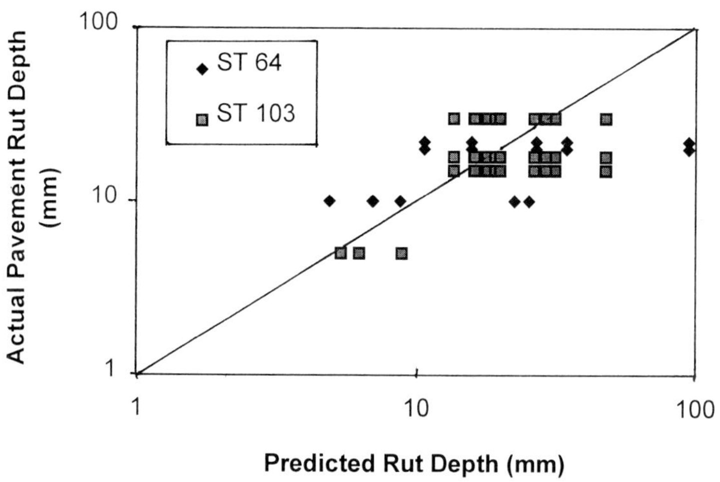

Figure 7 – Comparisons between predicted versus actual rut depth.

Figure 7 presents the comparison of predicted (using approach B) versus observed rut depth, based on the actual results for each of the tests. It can be observed that test results for site 103 generally are very good. This graph is presented to show the effect of test variability on actual rut depth predictions, when compared with observations. Normally, average all test results would be used for predictions. In this case the average of all the values would be just below the equality line, indicating that the concepts proposed are slightly conservative.

High level of heterogeneity and variability of the most of associated characteristics requires the probabilistic approach to the conclusions based on the data analysis results with wide application of statistical methods. Some of the methods show efficient applicability aimed at the analysis of the data under consideration and their interpretation as well as at practical solutions of pavement related problems.

SOME NOTES REGARDING DATA OBTAINED

Taking into account peculiarities of the behavior and properties of the measurements, it should be pointed to the fact that particularities of characteristics, measurement errors and other factors have significant influence on the adequate description of the object [3].

Analysis of actual deviations of road profiles as well as various corresponding characteristics has been made on the basis of application of statistical methods so that to describe in some details peculiarities of the data under circumstances. This approach in such conditions helps overcome the difficulty and increases the reliability of estimates and conclusions in analysis of pavement surface roughness data. Also, analysis of the road roughness data for different road sections has been prepared [4], and the results of the analysis should be taken into consideration in estimating a type of the roughness for road sections.

The actual data consist of some characteristics for 32 Israel road section profiles with different levels of roughness. The profilometer device created in the Materials and Research Division of the Israel Public Works Department has been used. The profilometer records deviations with smooth component eliminated. In other words the measured deviations are filtered with respect to main profile of the road. Roughness measurements of the road surface can be recorded with continuous or discontinuous presentation of the measured deviations. Characteristics for measured deviations of each road section profile are presented in Table 1 that summarizes available data and particularly includes measurement interval (*INT*, *m*), number of measurements *N*, mean value and standard deviation (Std) in mm for recorded deviations, contribution (*EU*) of the random component variance in general variance of the measured deviations, and the first autocorrelation $r(1)$ estimated from autocorrelation functions for sequences of roughness measurements for various measurement intervals. In addition, the table also includes estimated type of roughness for road sections determined on the basis of statistical analysis of the data obtained. The results of the analysis are presented in current investigation.

Table 1. Summary of the Available Data for Road Sections With Roughness Type Estimations

Road, Sec.	INT, *m*	N	Std, *mm*	Mean, *mm*	EU	r(1)	Road Type
3P	0.18	40	3.165	0.434	0.854	-0.366	Rough
3P	0.36	20	2.806	1.001	0.804	-0.206	Rough
3M	0.18	40	3.137	0.022	0.885	-0.097	Rough
3M	0.36	20	3.389	-0.072	0.966	-0.18	Rough
4P	0.18	56	3.476	0.776	0.837	-0.343	Rough
4P	0.36	28	3.654	0.698	0.989	-0.062	Rough
4P	0.54	19	3.776	1.792	0.992	-0.086	Rough
4M	0.18	56	3.631	0.235	0.856	-0.268	Rough
4M	0.36	28	3.839	0.441	0.933	-0.096	Rough
4M	0.54	19	3.688	0.902	0.997	0.006	Rough
383B	0.06	75	3.364	0.573	0.474	0.574	Rough
383B	0.12	38	3.487	0.416	0.682	-0.02	Rough
383B	0.18	25	3.628	0.726	0.585	-0.523	Rough
383B	0.24	19	2.815	0.648	0.643	-0.597	Rough
383AB	0.06	132	2.674	0.455	0.508	0.569	Transitional
383AB	0.12	66	2.61	0.535	0.742	-0.016	Transitional
383AB	0.18	44	2.886	0.515	0.679	-0.479	Transitional
383AB	0.24	33	2.208	0.843	0.917	-0.267	Transitional
383AB	0.30	27	2.318	0.611	0.998	0.046	Transitional
383AB	0.36	22	2.625	0.696	0.855	0.219	Transitional
353P	0.06	120	1.148	0.567	0.524	0.672	Smooth
353P	0.12	60	1.14	0.568	0.938	0.245	Smooth
353P	0.18	40	1.125	0.534	0.949	0.125	Smooth
353P	0.24	30	1.238	0.552	0.953	0.217	Smooth
353P	0.30	24	1.045	0.625	0.979	0.012	Smooth
353P	0.36	20	1.098	0.672	0.963	-0.193	Smooth
383A	0.03	114	1.442	0.41	0.504	0.699	Smooth
383A	0.06	57	1.313	0.301	0.618	0.504	Smooth
383A	0.09	38	1.51	0.359	0.821	0.347	Smooth
383A	0.12	29	1.384	0.281	0.882	-0.001	Smooth
383A	0.15	23	1.461	0.282	0.869	-0.016	Smooth
383A	0.18	19	1.484	0.237	0.946	-0.228	Smooth

Table 2. The Results of Canonical Discriminant Analysis (Part 1)

Canonical Discriminant Analysis

32 Observations	31 DF Total
5 Variables	29 DF Within Classes
3 Classes	2 DF Between Classes

Class Level Information

TYPE	Frequency	Weight	Proportion
Rough	14	14.0000	0.437500
Smooth	12	12.0000	0.375000
Transitional	6	6.0000	0.187500

Squared Distance to TYPE

From TYPE	Rough	Smooth	Transitional
Rough	0	73.71596	12.06946
Smooth	73.71596	0	26.42675
Transitional	12.06946	26.42675	0

F Statistics, NDF=5, DDF=25 for
Squared Distance to TYPE

From TYPE	Rough	Smooth	Transitional
Rough	0	82.12388	8.73995
Smooth	82.12388	0	18.22535
Transitional	8.73995	18.22535	0

Prob > Mahalanobis Distance for
Squared Distance to TYPE

From TYPE	Rough	Smooth	Transitional
Rough	1.0000	0.0001	0.0001
Smooth	0.0001	1.0000	0.0001
Transitional	0.0001	0.0001	1.0000

Table 3. The Results of Canonical Discriminant Analysis (Part 2)

Univariate Test Statistics

F Statistics, Num DF= 2 Den DF= 29

Variable	Total STD	Pooled STD	Between STD	R-Squared	RSQ/ (1-RSQ)
INT	1.3313	1.2605	0.6446	0.161359	0.1924
Std	1.0095	0.2653	1.1769	0.935385	14.4762
Mean	0.3260	0.3267	0.0969	0.060761	0.0647
EU	0.1671	0.1719	0.0203	0.010161	0.0103
r(1)	0.3402	0.3079	0.1983	0.233812	0.3052

Univariate Test Statistics

Variable	F	Pr > F
INT	2.7899	0.0780
Std	209.9048	0.0001
Mean	0.9380	0.4030
EU	0.1488	0.8624
r(1)	4.4249	0.0210

Average R-Squared: Unweighted = 0.2802953
Weighted by Variance = 0.4185578

Class Means on Canonical Variables

TYPE	CAN1	CAN2
Rough	3.867701068	-0.071891604
Smooth	-4.718085046	-0.056511199
Transitional	0.411534267	0.280769476

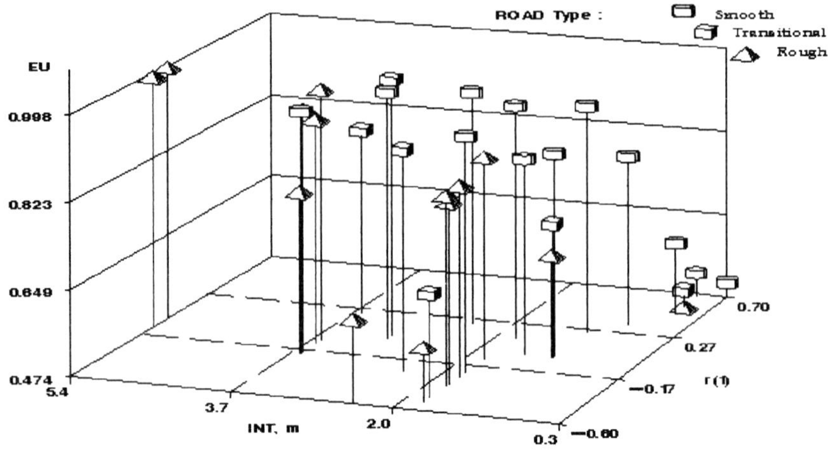

Figure 3. Three-Dimensional Scatter Plot Using *EU* and *r(1)* Values for Various Measurement Intervals (*INT*), in m, with Smooth, Transitional and Rough Road Roughness Types Delineated

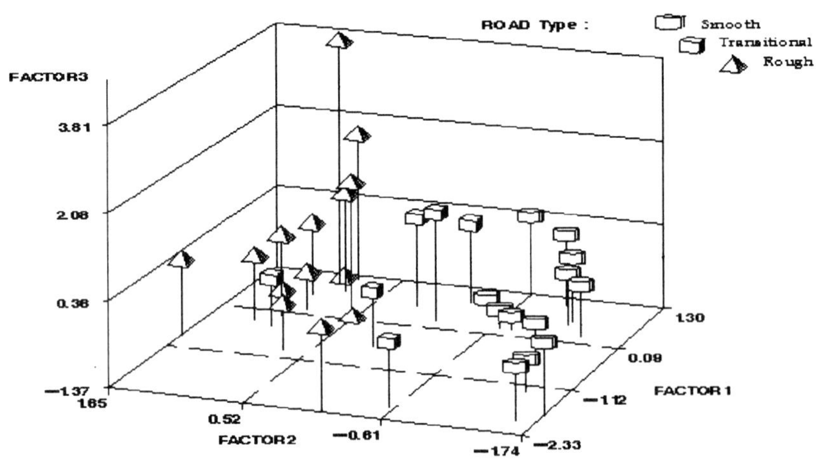

Figure 4. Three-Dimensional Scatter Plot for Three Factors from the Results of Factor Analysis with Smooth, Transitional and Rough Road Roughness Types Delineated

CONCLUSIONS

Elaboration of the methods aimed at estimating the type and level of the road roughness has significant meaning in pavement surface investigations with great influence of road roughness characteristics on pavement performance, riding comfort for the road users, etc.

The problem of dividing the road profiles into statistically homogeneous units has noticeable connection with the subject under consideration. The problem is of significance in order to describe and investigate road roughness characteristics and peculiarities in their behaviour. All this has meaning influence on the accuracy of statistical model construction as well as correct description of the object under study.

Analysis of the road roughness measurements for various actual road sections by employing some statistical procedures in order to estimate the type of the road roughness and to compare results has been fulfilled. Results of the analysis show efficiency and practical applicability of the applied methods in solution some evaluating and distinguishing problems in road roughness data investigations.

REFERENCES

1. PATERSON,W.D.O. *Road Deterioration and Maintenance Effects: Models for Planning and Management.* Johns Hopkins Univ. Press: Baltimore and London, 1987.

2. SHAHIN, M.Y. *Pavement Management for Airports, Roads and Parking Lots.* Chapman & Hall: NY & London, 1994.

3. FERRONSKY, V.I. AND DIVINSKY, M.L. Probabilistic Analysis of Cone Penetration Testing Data: Some Problems and Solutions. *Penetration Testing 1988 (ISOPT-1), Orlando, USA/* Rotterdam, pp. 751-755, 1988.

4. DIVINSKY, M. AND NESICHI, S. Road Roughness Analysis by Using the SAS System. *Proceedings of the SUGIS'95*, M.I.A. Ltd., Tel Aviv, pp. 113-119, 1995.

5. SAS/STAT. *User's Guide, Version 6, Fourth Edition, Vol. 1.* Cary, NC: SAS Institute Inc., 1989.

Fifth International Conference on the
BEARING CAPASITY OF
ROADS AND AIRFIELDS

THE NEW FRENCH CATALOGUE OF PAVEMENT SURFACE DISTRESSES MORE THAN A DICTIONARY!

PH. LEPERT[1], M. SIFFERT[2], L . BERTRAND[3]

1. Research Director, head of the « Data collection management » section, Laboratoire Central des Ponts et Chaussées, 44340 Bouguenais, France
2. Engineer, head of the « Co-ordination Centre », Laboratoire Régional des Ponts et Chaussées, 78 190 Trappes, France
3. Engineer, Deputy Director, Laboratoire Régional des Ponts et Chaussées, 69674 Bron Cedex

ABSTRACT

Deflection measurement is a basic approach of the structural condition of the roadbase, especially on flexible pavements. But the surface distresses, as they can be recorded by visual or automated surveys, provide a pertinent and useful information to complete the deflection measurements, **if this information is reliable**. It has become obvious, in the five last years, that distresses surveys can be reliable only if :

1 - the distresses, and their levels of severity, were rigorously defined,
2 - stringent surveys procedures were applied for data collection, and
3 - the surveys was automated, as far as possible.

The French guide of pavement surface distresses is an attempt to meet the first requirement. Therefore it is firstly designed as a dictionary, which precisely defines all the distresses which are or may be observed, today, on the French roads networks (motorways, highways, secondary roads, etc.). The different levels of severity of these distresses are illustrated with photographs. In this instance, the guide is a basic tool to support rational visual as well as automatic surveys procedures.

Furthermore, this guide also provides information which makes it possible to understand each distress : its causes, its evolution, the parameters which govern it, etc. This information was derived from a synthesis of 30 years of experience of the engineers of the French Public Works Laboratories (Laboratoires des Ponts et Chaussées). It also take into account results and conclusions of a number of experiments on test sections as well as on the French Accelerated Loading Facility. Thank to this «state of knowledge» content, the guide is a very useful document for pedagogy.

INTRODUCTION

The bearing capacity of a pavement is a useful indicator of its mechanical condition, and therefore its ability to carry traffic, in particular heavy traffic, at a given moment. It cannot however be interpreted on its own. It is obvious that the structure of a pavement (whether its base is treated with a bituminous or hydraulic binder or not) determines to a considerable extent what constitutes acceptable levels of deflection for it. In the same way distresses which can be observed on the surface of the road provide valuable information on the probable state of damage of the pavement, and therefore assist interpretation of measured deflection. For example, it is not surprising to find large deflections (for the structure in question) on a pavement which is severely deformed and/or cracked. Indeed, in the maintenance study process a surface distress survey is often carried out before deflection measurements and these are used to decide between several hypotheses of structural behaviour. The measurement of bearing capacity and visual damage surveys should not therefore be presented as two alternative competing approaches in maintenance studies. On the contrary, these two types of investigation complement one another very closely.

The considerable and unceasing efforts which have been made over many years to improve the standard of bearing capacity measurements (1,2,3) needed to be matched by attempts to make damage surveys more rigorous and precise. This appears particularly necessary if one considers that the subjective judgement of personnel plays a role in visual distress surveys. Ultimately, the only possible way of achieving truly objective data collection is to replace visual condition surveys by physical distress measurements. While waiting for this technical advance, and also to prepare the ground for it, activities have been focused on two aspects:

- distress survey procedures have been rigorously formalised. These are presented in other papers (4, 5) ;
- to provide a basis for these procedures a pavement distress catalogue has been produced which precisely defines distresses and their levels of severity on the basis of an analysis of their cause and development.

OBJECTIVES OF THE CATALOGUE

Harmonising distress surveys When several teams cover a road in the same way, their visual perception of the distresses they observe is not generally very different. However, there is considerable diversity in the way surveys are conducted as regards the detailed classification of surface distresses and, in particular, the assessment of their "severity".

As we have already seen, the first stage in the harmonisation of surveys is to specify procedures which can be applied uniformly by all the teams engaged in a maintenance study. In each case the procedure will vary depending on the nature of the study. It will be fairly general, therefore rapid, for network studies. It will be much more detailed and meticulous for project studies, for example. In France these procedures were formalised some time ago and have been regularly updated. They are published, in particular, in the form of "Test Methods" (6). Defining such procedures is not however on its own sufficient to harmonise surveys. It is also essential for all field personnel to code identical or sufficiently similar observations in the same way. The technicians who interpret these surveys must also know and use the same coding as the field personnel. There is therefore a very strong need for a single coding system which is suitable for on-site observations and realistic with reference to the conditions under which surveys are conducted. The catalogue of surface distresses aims primarily to satisfy this need and it is therefore a supplement to the Test Method (6).

Defining a suitable common language for maintenance studies In addition to codifying visual observations, the catalogue helps to define a common language for engineers who need to exchange data and knowledge about pavement condition or performance. In order for such exchanges to take place effectively the technical language must identify the various types of distress which are observed on a road and their different levels of severity by using precise "standardised" terms. In addition to this, it is extremely important that :

1) the proposed definitions and classification correspond precisely to on-site observations ;
2) the described levels of severity express the different stages in the development of the defect.

If these conditions are not met, the terminology will be inappropriate for communication and will therefore not be used.

Developing codification which is both accurate and applicable in the field (i.e. which is realistic as regards survey conditions) was the most important issue in production of the catalogue.

DESIGN OF THE CATALOGUE

A functional definition of distresses and their severity A study of pavement performance was at the origin of the catalogue of distresses. Pavement deterioration processes were listed, classed on the basis of type of structure and/or type of surfacing, and analysed. The causes of these processes were identified, as were the constructional or environmental factors which influence their development. The sequence of distresses that each process produces on the pavement surface was described in detail, as was the change in the appearance of these distresses. This first phase of work assembled the knowledge required to list pavement distresses and to construct a functional classification on the basis of type of pavement structure and type of wearing course, which involved linking these distresses to clearly identified processes of deterioration.

A systematic attempt to find suitable photographs Concurrently, photographs were collected on sites all over the country. Technicians from the Laboratoires Régionaux des Ponts et Chaussées tried to find examples of each distress at different levels of severity on the national and local road networks. In some cases this was quite difficult, in particular for distresses whose consequences are such that they must be repaired without too much delay and which are therefore rarely encountered. Each distress was photographed under strictly specified conditions, in particular (see the example on figure 1):

1) a camera with a 50 mm lens was used,
2) a 1.5 metre rule, with 10 centimetre graduations, was placed on the carriageway in the field of view,
3) three photographs were taken, one from above, one from the verge, and a third parallel to the axis of the road.

It was at times quite difficult to take these photographs, in particular parallel to the axis of the road for which specific protection was required on account of heavy traffic. More than 700 photographs were nevertheless taken, of which about 200 were used in the catalogue.

Figure n° 1 : Longitudinal crack, non specific to wheelpath, severity level 2
Example of photograph - from (8)

Production of the catalogue The catalogue combines the knowledge which was assembled during the behavioural study and the photographs obtained in the field. Teams who conduct damage surveys in the field were also brought in to ensure that the document, in addition to being rigorously accurate as regards current theoretical knowledge, was of a form which was easy to use in the field.

Finally, in addition to definitions of distresses and their levels of severity, illustrated by photographs, the catalogue provides full information on the structural behaviour of pavements and wearing courses. It contains a considerable amount of information on distress processes. Thus, the catalogue is not only an illustrated dictionary but can be used by technicians everywhere (in laboratories, consultancies and maintenance services) to provide an initial guide to pavement behaviour.

THE STRUCTURE OF THE CATALOGUE

A technical guide and an illustrated dictionary The first part of the catalogue contains a summarised but full definition of the pavement distresses which are dealt with in the catalogue. Some definitions are illustrated with simple diagrams, such as those on figures 2 to 4.

The second part is a description of the pavement and surfacing structures used in France. This section used terminology drawn from technical documents, of which the most important is the "Guide Technique de Conception et de Dimensionnement des Chaussées"

(7). The various distress modes for each structure and surfacing are described, as they have been observed over the last thirty years, with their causes and the factors which affect their development.

The third part of the catalogue consists of an information sheet for each distress. Each sheet recapitulates a general definition of the distress and describes its different levels of severity. Each level is illustrated by a photograph which backs up the accompanying text definition (figure 1). Each sheet also describes the possible development of the distress, when necessary for different types of structure, as well as its causes and the factors which affect its development. These sheets are arranged in five major categories :

* structural distresses of granular-based, bitumen-based, or semi-rigid pavements,
* structural distresses of concrete slab pavements,
* structural distresses of continuously reinforced concrete pavements,
* distresses of pavements with surface dressings,
* lastly, distresses of AC wearing courses.

A document and a CD-Rom The catalogue is currently published in document form. A CD-Rom version is being prepared which will include educational software in addition to the information in the catalogue and software which allows this to be readily accessed. The educational software is designed to help in the training of various categories of staff, from technicians conducting distress surveys to engineers who are in charge of maintenance studies.

THE TYPES OF DISTRESSES CONTAINED IN THE CATALOGUE

The following information is taken from the catalogue and illustrates its contents. Pavement distresses are divided into three main separate categories: permanent deformations (mostly of longitudinal profile), cracking (transverse, longitudinal or alligator cracking, crazing, etc.), surface distresses (bleeding, ravelling, delamination). Distinctions have been made between different distresses within these categories, when they indicate or are caused by different structural behaviours. Finally, we have endeavoured to characterise the stage of development of each distress by means of a severity indicator.

Permanent deformations As regards deformations, the catalogue distinguishes between rutting, edge settlement and settlement in other parts of the road. Rutting is divided into two types, structural rutting and surface rutting, both of which involve permanent longitudinal deformation which is produced by the passage of wheels. However, the first type of rutting is caused by a structural weakness and involves deformations of both base layers and wearing course. On the other hand, surface rutting is caused by creep in the wearing course when the characteristics of the bitumen are not suited to the climate and traffic. It is important to distinguish between these two types of rutting because their causes, and hence their remedies, are fundamentally different. In the same way, edge settlement, not to mention other types, generally occurs because the strength of the edge of the pavement is inadequate for the traffic it carries. This type of distress frequently occurs on pavements with unbound bases, particularly inside the bends, when the effectiveness of the drainage ditch is inadequate. On the inside lane of the bend, vehicles tend to stay close to the edge of the pavement which has been weakened by the presence of water. The granular materials creep and settlement occurs at the edge of the pavement. The treatment for this too is specific and can involve constructing a shoulder using bound material.

The severity of permanent deformations depends on their depth. This is simply due to the fact that when deformations develop, they become deeper. In France, it is often considered that a deformation is undesirable (severity 1) if its amplitude exceeds 15 millimetres, and that it is severe (severity 2) if its depth exceeds 30 millimetres. These thresholds are however only guides and may be modified depending on the nature of the road involved.

Cracks A crack is a line of breakage which appears on the pavement surface. There are many types of cracks and they have widely differing causes and appearances.

Longitudinal cracks are amongst the best known. These are of two types: those which occur only in the wheel tracks (figure 2), and those which also occur elsewhere (figure 3). The first type is caused by traffic-generated pavement fatigue. The second type can appear and propagate at any point on the pavement surface (outside and sometimes inside the wheel tracks) and is usually caused by construction problems (a poor joint in the base appearing on the surface) or climatic phenomena (differential swelling caused by frost).

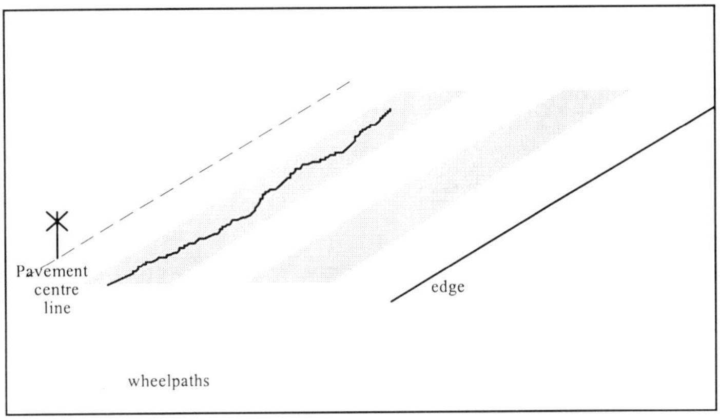

Figure n° 2 : Longitudinal crack in wheelpath (diagram illustrating the definition) - from (8)

The same group also contains another type of cracking, known as "transverse". These cracks usually only appear on pavements with bases treated with hydraulic binders (semi-rigid). They are roughly perpendicular to the axis of the pavement and can cover all or part of its width. They are caused by the well-known phenomenon of concrete shrinkage. During setting, the bound material contracts and the layer breaks at fairly regular intervals of between 3 and 10 metres depending on the nature of the aggregate (siliceous or calcareous) and the binder (cement or blast furnace slag). These cracks are not distresses in themselves as they are the result of normal behaviour of the pavement. However they require regular maintenance (sealing) and should therefore be recorded in a distress survey.

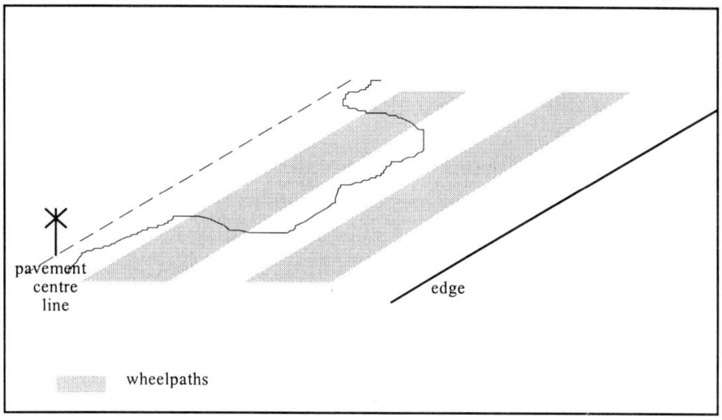

Figure n° 3 : Longitudinal crack, non specific to wheelpath (diagram illustrating the definition - from (8))

Longitudinal or transverse cracks almost always evolve in the same way: branching appears, then crazing occurs along the cracks and/or varying amounts of aggregate is lost from the edges of the cracks. The severity of cracks is classified on this basis: simple (severity 1), branched (severity 2), cracks with spalling and/or crazing (severity 3). It should be noted that when longitudinal cracks develop on semi-rigid pavements they can ultimately cut across transverse cracks. This results in a characteristic "block cracking" (see figure 4) which is typical of advanced damage in this type of pavement.

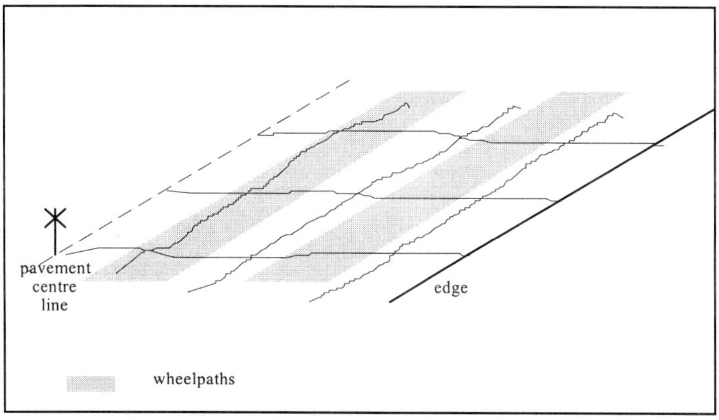

Figure n° 4 : Block cracking (diagram illustrating the definition) - from (8)

Crazing is also included in the category of cracks. As in the case of longitudinal cracks, a distinction is made between crazing which is specific to wheel tracks (often called « alligator cracking ») which results from traffic-generated pavement fatigue and crazing which is not specific to wheel tracks (which can occur both within and outside them) which is mainly caused by climatic phenomena. The severity classes are determined on the basis of both the size of the pattern in the surfacing and the amount of material which is lost as crazing tends to evolve on the one hand by subdivision of the existing pattern and on the other by loss of material from the surfacing layer.

Surfacing distresses The surfacing of a pavement often tends to wear, particularly in the wheel tracks. Either excess binder covers a large proportion of the aggregate particles (bleeding), or the aggregate particles are polished. Both of these result in a reduction in skidding resistance. This is the first group of surfacing distresses.

On the other hand, a poorly constructed wearing course may be damaged by raveling. In the case of surface dressings, the most common type of distress is "combing". This consists of the more or less total loss of aggregate in bands which run parallel to the axis of the road. This distress can be caused by an incorrectly adjusted bitumen jet or an excessively low laying temperature. It is dangerous as it can lead to extremely large differences in skidding resistance under the different wheels of a vehicle. When the interface between a bituminous concrete wearing course and the base course has been poorly constructed, particularly when there is no bonding between these two courses, fatigue rapidly occurs in the surfacing which is removed in large plates. This distress is known as delamination. The severity of these distresses depends on the amount of material lost.

CONCLUSIONS

The primary aim of the catalogue is to be a tool for harmonisation in the area of pavement surface distresses. It defines a precise and relevant vocabulary for all users which covers the behaviour of pavements in line with more than 30 years of on-site observations. It supplements the existing "Méthode de Relevé des Dégradations des Chaussées" improving the uniformity of field data and ensuring that this data is more comprehensible and therefore better exploited by pavement maintenance technicians. This need for harmonisation, which was first sensed at national level, has now become an international concern, in particular in Europe. It is significant that the work performed in order to develop this catalogue has also contributed to the European COST 324 and PARIS Projects.

In addition to this role as a dictionary, the catalogue contains basic information about the behaviour of pavements which means that it constitutes a genuine introductory guide to this topic. It allows not only the teams engaged in field surveys but also engineers in charge of maintenance studies and road network managers to use the same terms for the same concepts.

Finally, because it precisely defines a terminology it provides an invaluable tool to enable highway engineers to explain what they expect from an automatic pavement distress survey system which we know is awaited with so much impatience.

REFERENCES

1. Exécution et interprétation de la déformabilité de surface - Méthode d'essai LPC n° 39, Coll. Technique et Méthodes des LPC, LCPC, Paris, France, Juillet 1996

2. RIOUALL A. and SIMONIN J.M. Evaluation des déflectographes, Bulletin de Liaison des LPC n° 208, 39-47, 1997

3. Harmonisation of the use of the Falling Weight Deflectometer on pavement - part 1 : Harmonisation of FWD measurements and data processing for flexible road pavement evaluation, FEHRL report n° 1996/1, 1996

4. PH. LEPERT and M.-T. GOUX Evaluation du réseau français de routes nationales basées sur le relevé de dégradations de surface. 4ème Congrès International de la Route, Rabat, 2 Juin 1994.

5. PH. LEPERT et al. An evaluation of the french national highway network based on surface damages surveys, 3rd International Conference on Managing Pavement, May 21-26, San Antonio, USA, 1994

6. Relevé des dégradations de surface des chaussées - Méthode d'essai LPC n° 38-2. Coll. Techniques et Méthodes des LPC, LCPC, Paris, France, Juillet 1997

7. LCPC - SETRA Conception et dimensionnement des chaussées - Guide technique. Paris, Décembre 1994.

8. Catalogue des dégradations de surface des chaussées - Annexe à la Méthode d'essai LPC n° 38-2. Coll. Techniques et Méthodes des LPC, LCPC, Paris, France, Février 1998

Fifth International Conference on the
BEARING CAPASITY OF
ROADS AND AIRFIELDS

CHARACTERIZATION OF MANUAL DISTRESS DATA VARIABILITY: APPARENT BIAS AND PRECISION

GONZALO R. RADA[1], Ph.D., P.E., CHUNG-LUNG WU[2], Ph.D., P.E., GARY E. ELKINS[2], P.E., AND RAJESH K. BHANDARI[3]

[1]Chief Engineer, [2]Senior Engineer, and [3]Project Engineer

PCS/LAW, A Division of Law Engineering and Environmental Services, Inc.

12104 Indian Creek Court, Suite A, Beltsville, Maryland 20705

ABSTRACT

This paper presents the findings of an effort undertaken by the Long-Term Pavement Performance (LTPP) program to assess variability of distress data collected using manual procedures in the field. The primary source of data used to characterize the variability of manual procedures is the results from nine LTPP distress rater accreditation workshops conducted between 1992 and 1996. This data source provided the distress ratings of 11 to 16 individual raters per workshop who performed the ratings on the same day on the same test section; each workshop included an asphaltic concrete (AC) and a portland cement concrete (PCC) pavement test section. In addition, a reference rating of the distress present on the same test sections was performed using a consensus rating method by a group of expert raters immediately prior to the workshop; data from these reference surveys were used as a surrogate of "ground-truth". Based on the analyses of these data, both the apparent bias and precision for the common distress type-severity level combinations were quantified. General trends were also studied as well as the impact of distress data variability on single composite numbers or indices such as the PCI value.

Key words: pavement distress, distress variability, distress surveys, bias and precision

INTRODUCTION

Pavement distress is the primary performance indicator used in highway and airfield pavement maintenance and rehabilitation decision making. Through the use of composite indices such as the Pavement Condition Index (PCI), distress serves as a tracking mechanism and ultimately a trigger for maintenance and rehabilitation actions. The process of collecting pavement distress data, analyzing it, and basing decisions for maintenance and rehabilitation outlays, i.e. pavement management, is an economic essential in operating a pavement facility.

Reliable forecasting must be based on knowledge of performance deterioration (increasing amounts of distress) measured periodically over time. These data, coupled with mathematical modeling techniques, provide a picture of the performance of pavement types which may be used to predict future conditions. This would incorporate the effects of different routine maintenance activities performed during the pavement service life and therefore would provide rational estimates of expected performance of future treatments critical to evaluating economic benefits of different alternative treatment strategies.

Accurate, reliable distress data are critical to sound modeling and forecasting. Bad data will produce bad models but the most accurate models will provide poor results if the distress data are not consistent, accurate and repeatable. The primary means of distress data collection has been the individual human being visually detecting and recording distress according to some written guidelines; distress definitions and methods of evaluating the recorded data. Because distress types can be challenging to describe in clear, unambiguous terms the level of variability in distress data collection is quite high. As a consequence, the survey methodologies are of necessity geared to softening the impact of large variations in the data while still providing an adequate response to conditions; the proper threshold of condition where some construction activity is warranted.

To promote uniformity and consistency of distress data collection, one of the early Long-Term Pavement Performance (LTPP) efforts was to develop standard definitions, measurement procedures and data collection forms. These guidelines are contained in "Distress Identification Manual for the Long-Term Pavement Performance Project"[(1)] , which is typically referred to as the Distress Identification Manual (DIM).

The LTPP manual distress survey procedure contained in the DIM is based upon a single rater's interpretation of distresses, preparation of maps indicating the location and nature of all distresses within the monitoring portion of the test section, and summarization of the extent and severity of all distresses found. In typical practice, copies of the previous manual distress survey maps are used. The rater adds any additional distress identified or notates differences in distress severity or interpretation from the last survey.

The data quality control and assurance function for manual distress surveys consists of distress rater accreditation workshops, office review of all manual data by a senior staff member, and data logic and range checks performed after entry of the data into the LTPP Information Management System (IMS). Accreditation workshops were developed to help promote quality and consistency in manual distress data collection. All surveys entered into the LTPP IMS must have been conducted by a rater accredited within the last two years.

Although extensive efforts were taken in the development of the DIM to define distress types and severity levels in objective terms, the application of these definitions still requires subjective interpretation by the rater. It is this subjectivity through which the variability between raters is created, although some variations can be created by mistakes in the arithmetic summary process. Recognizing that rater variability exists, this study was undertaken to quantify the level of variability so that researchers and/or analysts using manually collected LTPP distress data could incorporate, or at least recognize, this important feature in their analyses.

DATA SOURCE

Results from the nine LTPP rater accreditation workshops held to date were used in this study. This data source provides the distress ratings of 11 to 16 individual raters per workshop who performed the ratings on the same day on the same test section.

There are two types of workshops, accreditation and re-accreditation. The accreditation workshop provides an extensive introduction to distress definitions and rating procedures contained in the DIM. They are targeted at individuals who are being accredited for the first time, while re-accreditation workshops are conducted for accredited raters who are familiar with LTPP distress survey methods but require the two-year re-accreditation.

The accreditation and re-accreditation process consists of two major parts: a written examination and a two-part field survey examination. The written examination is intended to test the rater's knowledge of LTPP distress definitions and procedures. The field survey examinations are intended to measure each rater's capabilities in observing, interpreting and recording distress data. The field examinations are conducted on two 150-m long pavement sections, one AC surfaced and one jointed PCC pavement (JCP). Each rater is given three hours to complete an independent distress rating of each section following all LTPP procedures including preparation of a detailed scaled map and reduction of mapped quantities onto the summary distress forms.

Prior to the workshops, the sections were surveyed in detail using a consensus procedure by the workshop instructors, and in some cases, other knowledgeable personnel. The results of these consensus surveys were used as the reference values against which the individual raters' results were compared for accreditation. These reference values are considered to be a surrogate of "ground truth" for the distress actually present. To assess the variability in the LTPP manual rating methods, only the summary distress data from the exam accreditation ratings were used. The results of the practice sessions were not used.

GLOBAL TRENDS

To gain a general understanding of the magnitude of variability associated with manually collected LTPP distress data, plots of distress quantity at each severity level and total across all severity levels for a distress type were developed for each of the common distress types identified at the nine accreditation workshops. An example plot showing the trends for total transverse cracking in both AC and PCC pavement sections is presented in Figure 1.

In this figure, the letters “R” and “I” along the X-axis denote the values pertaining to the reference and the group of individuals, respectively. As stated earlier, the reference values are quantities of distress, for each distress type and severity level, determined by the consensus survey conducted by the workshop instructors immediately prior to the workshops. Average, minimum and maximum values of the distress quantities recorded by the individual raters for each workshop are also shown in Figure 1. A complete set of global trend plots are presented in a report being prepared by LAW/PCS for the U.S. Federal Highway Administration (FHWA) to be published later this year.

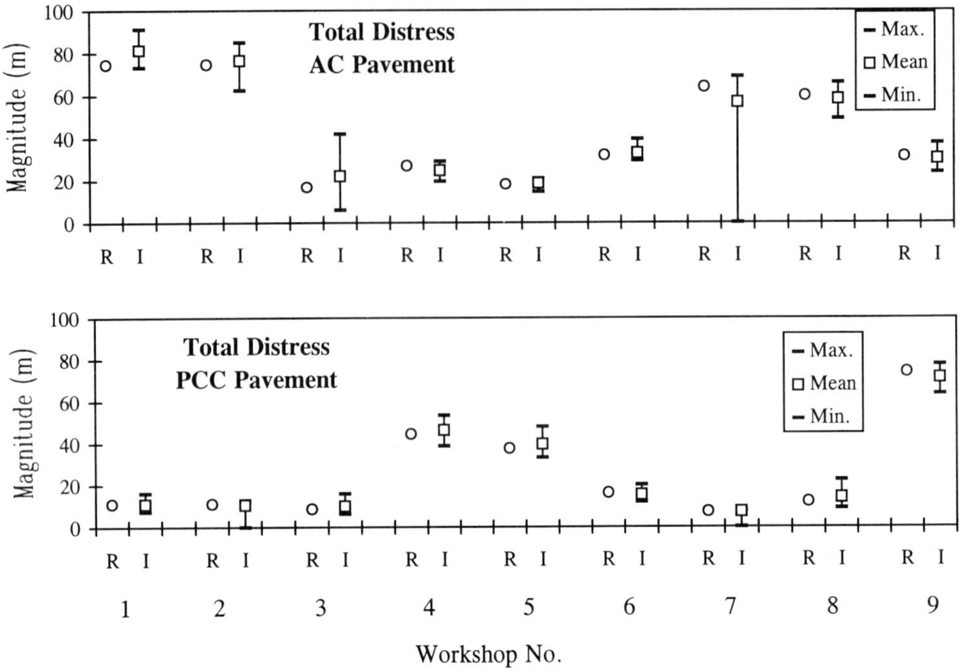

Figure 1 Global Trends for Transverse Cracking (meters)

The following observations were made based on the information presented in those plots:

- Although the magnitude of variability for any given distress type and severity level varies from workshop to workshop, in general the variability is large and the scatter of data tends to increase in magnitude as the quantity of distress increases.
- For total distress summed across all severity levels for each distress type, the group means are generally closer to the reference value and the between rater scatter is smaller than for the individual distress severity levels. This is indicative of the greater variability in classification of severity level for a distress type than that associated with distress type identification.
- For closely related distress types, such as fatigue cracking and longitudinal cracking in the wheel path, compensatory differences between the group ratings and reference values were observed, i.e. group ratings indicated higher quantity of fatigue cracking and lower quantity of longitudinal cracking as compared to the reference values.
- There does not appear to be a significant positive or negative bias in the data; i.e., no tendency to consistently rate all distress type and severity level combinations higher or lower.
- Due to the relatively high variability between distress rater values, the reference value is almost always within the range of data scatter for all distress type and severity level combinations.

These observations are consistent with those reported in an earlier study.[(2)]

PAVEMENT CONDITION INDEX

Another approach in evaluating distress rating variability is to use a composite distress statistic typical of those used in pavement management systems. For this comparison, distress data from the individual raters at each workshop were used to compute the well established and widely recognized Pavement Condition Index (PCI) value according to the procedure developed by the U.S. Army Corps of Engineers.[3] Due to differences in the definition of distress types and severity levels between the LTPP DIM and PCI methods, this exercise was limited to AC pavements and even then, distress types had to be combined (e.g., longitudinal cracking in the wheel path and not in the wheel path) and a few minor assumptions made (e.g., severity levels assumed to be the same for a few distress types).

To compute the PCI, deduct values are computed for each distress/severity level. In computing these deduct values, weighing curves are used to transform the extent for each distress severity level. These weighing functions act to suppress the variability in individual distress ratings whose extent are less than the trigger levels that indicate the need for application of corrective treatments. Thus, it is expected that the PCI will result in a lower between rater variability and bias from the reference values.

The statistics of the PCI values for the nine workshops (i.e., AC pavements) are summarized in Table 1. The major observations and/or conclusions derived from these data are as follows:

Table 1. Summary Statistics of Pavement Condition Index (PCI) for AC Pavements

Workshop			GROUP STATISTICS					
No.	Location	Date	Refer.	Min.	Max.	Mean	Stdev.	COV
1	Reno	6/9/92	43	44	52	48	2.56	5.33
2	Reno	7/14/92	43	45	74	56	7.12	12.69
3	Minnesota	8/10/93	44	41	61	46	5.07	11.03
4	Buffalo	5/3/94	50	58	71	64	3.89	6.05
5	Houston	10/4/94	82	80	86	83	1.68	2.03
6	Reno	4/4/95	59	60	68	65	2.22	3.41
7	Minnesota	10/3/95	64	47	76	63	7.92	12.50
8	Reno	7/30/96	30	32	61	42	7.88	18.69
9	Champagne	9/17/96	50	43	61	51	6.94	13.71

- As expected, there is generally an improvement in the agreement between the individual raters, the group mean and the reference PCI value. The difference between the group mean and reference value ranges from 1 to 14 PCI. It is interesting to note that the reference PCI is always less than the group mean, except for Workshop No. 7 where the reference PCI is 1 greater than the group mean.

- Variability between individual raters also improved as compared to those for individual distress/severity levels. The coefficient of variation ranges from a low of 2.0% to a high of 18.7%.

BIAS AND PRECISION

One of the main objectives of the study was to quantify the bias and precision associated with LTPP manual distress data to form the basis for future pavement design and performance evaluation. Towards this end, data obtained from the distress rater accreditation workshops were manipulated and analyzed quantitatively.

Coefficient of Variation and RMSE Evaluation

The coefficient of Variation (CV), defined as standard deviation divided by mean value, is a statistical term normally used for representing variability associated with experimental data. However, the CV can be misleading when dealing with small magnitudes, as if often the case with distress data. A small amount of distress can inflate the CV tremendously; however, that variation in data (high CV) is not as much of a concern when dealing with small distress amounts. To overcome this deficiency, an alternative approach was used for determining the CV associated with manual distress data.

Specifically, the coefficient of variation was determined by plotting standard deviation versus mean for each distress type-severity level combination and fitting the best line through those data (y-intercept was forced through 0). The slope of this best fit line (in percentage terms) is a measure of the CV, assuming a linear increasing relationship between standard deviation and mean. An example of the derived CV for total transverse cracking in AC and PCC pavements is shown in Figure 2. The regression line marked as STDEV1 represents the derived CV for the data in question.

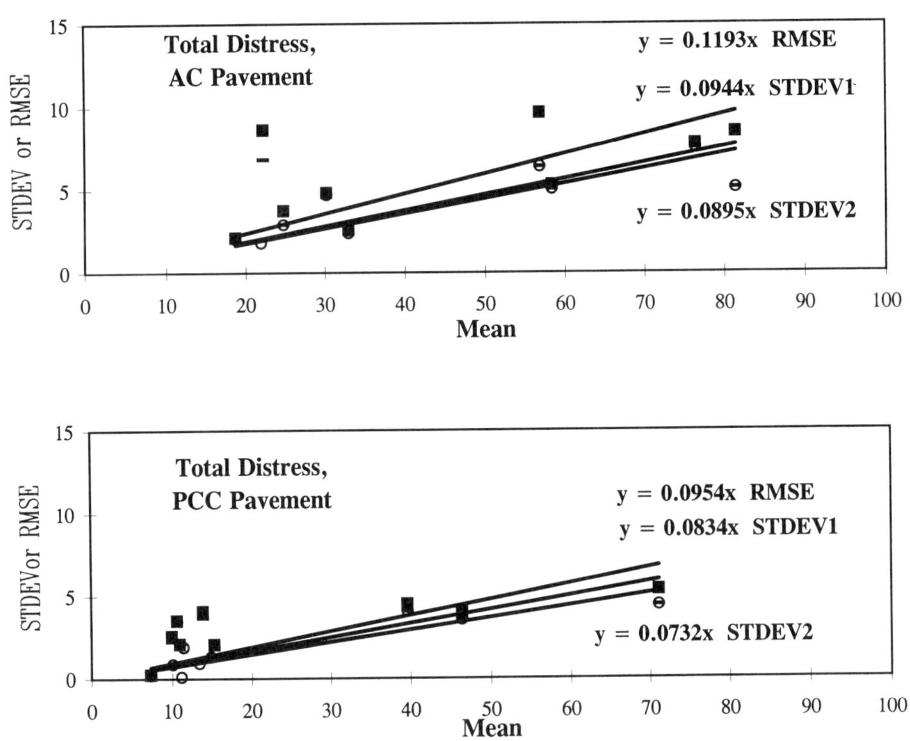

Figure 2. Standard Deviation and RMSE vs. Group Mean for Transverse Cracking (m)

Also shown in this figure are regression lines of RMSE and STDEV2 vs. Mean. RMSE (root mean square of error) is defined as the square root of the summation of the squared difference between the reference and the individual rater values divided by the number of raters in the workshop. This term combines the variability and bias associated with the individual rater data relative to the reference values.

The STDEV2 regression lines were developed using the standard deviation and mean values from the nine workshops after rejecting observations that were statistically considered as outliers. Individual data points were considered outliers, and thus eliminated from this analysis, if the difference between the individual distress values in each workshop and their group means were greater than three times the standard deviation derived from the STDEV1 regression lines discussed previously. This exercise was only performed on total distress. Once a data point was identified as an outlier, it was also dropped from further analysis at the individual severity level (low, moderate and high).

The CV for total distress data, without rejection of outliers, ranged from 9 to 38% for AC pavements. For PCC pavements, the CV ranged from 8 to 22% for cracking related distresses. The CV were relative high for joint spalling of PCC pavements, ranging from 25 to 71%. Except for fatigue cracking of AC pavements and transverse joint spalling of PCC pavements, no appreciable differences were observed between the STDEV1 and RMSE regression lines, which indicates that the difference between the group means and their corresponding reference values is small. The elimination of outliers did not seem to affect that observation, except for joint spalling of PCC pavements, which appears to be a more difficult distress to quantify as compared with others.

General Assessment of Bias and Precision

To gain a general understanding of the distress data, indicators of overall bias and precision were calculated based on data from the nine workshops; refer to Table 2 for sample results. Pooled reference and group mean values shown in this table represent average values derived from the nine workshops for each distress type-severity level combination. The difference between the pooled reference and group mean values was used to characterize the apparent bias for each distress type-severity level combination. The CV, pooled group standard deviation, and RMSE shown on this table were obtained using the procedures described in the previous section.

The data presented in Table 2 and standard deviation/RMSE plots such as the one shown in Figure 2 led to the following observations and/or conclusions relative to bias and precision:

- The apparent bias for most distress type-severity level combinations is small as illustrated in Figure 2. For most distress types, the RMSE and STDEV1 regression lines were relatively close to each other. In addition, there does not appear to be a uniform bias in the data; i.e., no tendency to consistently rate all distress type and severity level combinations higher or lower.

- Precision of distress data relative to the group mean appears to be acceptable, especially for total distress quantities where the CV ranges from 9 to 38% for AC pavements and from 8 to 22% for PCC pavement cracking related distresses.

Quantification of precision for joint spalling in PCC pavements appears to be more difficult; however, the CV was significantly reduced after rejection of data outliers.

- Precision for total distress quantities is generally much better than that for the individual severity levels, which is consistent with an earlier observation regarding difficulties in distinguishing severity levels.

Table 2. Indicators of Bias and Precision for AC and PCC Pavement Distresses

				Pooled values with outliers					Pooled values w/o outliers			
				Group					Group			
Distress Type	Unit	Distress Severity	Pooled Ref.	Mean	Stdev.	RMSE	COV (%)	Bias	Mean	Stdev.	COV (%)	Bias
AC Section												
Fatigue Cracking	Sq. Meters	Low	8.22	8.85	4.24	6.06	47.9	0.63	8.25	3.66	44.3	0.03
		Moderate	5.29	7.04	4.09	4.40	58.1	1.75	6.58	3.67	55.8	1.29
		High	0.72	0.60	1.02	1.25	169.9	-0.12	0.46	0.70	151.1	-0.26
		Total	14.23	16.49	6.21	9.46	37.7	2.26	15.29	5.00	32.7	1.06
Long. Cracking WP	Meters	Low	8.82	8.56	2.27	2.95	26.5	-0.26	8.70	2.24	25.8	-0.12
		Moderate	4.99	7.99	4.25	6.66	53.2	3.01	7.94	4.25	53.5	2.95
		High	4.63	1.71	2.42	5.07	141.9	-2.93	1.52	2.00	131.9	-3.12
		Total	18.44	18.26	6.02	7.43	33.0	-0.18	18.16	5.66	31.2	-0.29
PCC Section												
Corner Breaks	No.	Low	0.44	0.77	0.38	0.48	49.8	0.32	0.76	0.38	50.2	0.32
		Moderate	2.78	2.07	0.75	1.01	36.4	-0.70	2.10	0.75	35.8	-0.67
		High	0.78	0.85	0.54	0.65	63.2	0.07	0.86	0.39	46.1	0.08
		Total	3.89	3.68	0.51	0.56	13.9	-0.20	3.72	0.49	13.1	-0.17
Long. Cracking	Meters	Low	3.59	3.55	1.10	1.21	31.1	-0.03	3.45	0.80	23.2	-0.14
		Moderate	3.18	2.98	1.08	1.13	36.4	-0.20	3.05	1.04	34.0	-0.12
		High	0.78	0.50	0.50	0.69	99.2	-0.28	0.47	0.47	99.5	-0.30
		Total	7.54	7.04	1.56	1.83	22.1	-0.51	6.98	1.21	17.3	-0.57

- As shown in the regression line plots, the standard deviation appears to increase as distress magnitude increases; however, data bias does not seem to be affected by distress magnitude.

- The significant reduction in CV resulting from elimination of outliers appears to suggest that the precision of PCC joint spalling may be improved through additional rater training.

Quantification of Bias and Precision

The analyses conducted in the previous section indicated that the difference between group mean and corresponding reference values is generally small. A logical implication from this observations is that the group mean may be considered as an unbiased estimate of the reference value. In this section, a more vigorous analysis of the data was undertaken to better quantify the bias and precision associated with manual distress data.

It has been shown that, for all distress data, the standard deviation increases as the mean or magnitude of distress increases. A straight-line regression technique was used in the previous section to define the relationship between these two parameters. The resulting lines gave a general indication of data variability; however, to better quantify bias and precision, analyses were conducted to establish regression functions that better define that relationship. Three different functions were evaluated: straight line, logarithm transformation and square root transformation of the mean values.

It was determined from this effort that regression equations with the square root transformation generally produced a better fit between standard deviation and mean for the three functions considered, and thus was used in the remaining analyses. Figure 3 shows the square root function relationship between standard deviation and mean for total fatigue cracking in AC pavements and corner breaks in PCC pavements. The general form of the equation is: Standard Deviation = C * Sqrt(Mean), where C is the regression constant.

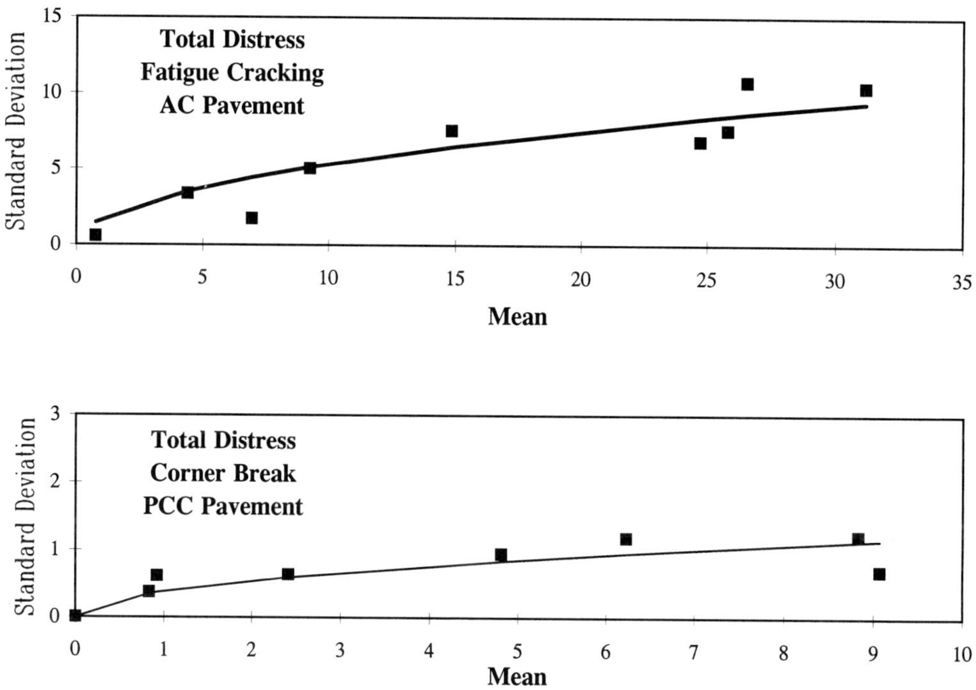

Figure 3. Standard Deviation vs. Mean, Square Root Function

Next, an outlier analysis similar to that in the previous section was conducted. Individual data were considered outliers and eliminated from further analyses if the difference between the individual value and group mean was greater than three times the standard deviation derived from the respective regression line. However, the standard deviation defined by the square root function was used instead of the straight line function. A new set of regression equations was then developed using the revised data.

Finally, a bias analysis was conducted for all distress type and severity level combinations in each workshop. A bias was considered to exist if the absolute difference between the workshop mean and reference value was outside the range of three times the derived standard deviation divided by the square root of the number of raters in the workshop.

Partial results (total distress only) from this bias analysis are shown in Table 3 for both AC and PCC pavements.

Table 3 also shows the number of raters before and after the outlier analysis, reference values, workshop means, computed standard deviation from individual data points, derived standard deviation from regression lines, regression equation constants and their coefficient of determination (R^2); very few data points were identified as outliers. Three symbols are used under the bias column: "N" indicates no bias between group mean and reference value, "Y" indicates that bias does exist, and "NA" indicates that there were no observations.

Table 3. Results of Bias Analysis Based on Square Root Transformation for AC and PCC Pavements

Distress Type	Unit	Section ID	Distress. Sev.	No. of Rates Original	No. of Rates Outliers	Ref.	Mean	Stdev	Derived Stdev	Bias	Slope C	R^2
Long. Cracking WP	Meters	AC # 1	Total	12	12	0.00	0.00	0.00	0.00	NA	1.8485	0.720
		AC # 2	Total	13	13	0.00	0.18	0.34	0.79	N		
		AC # 3	Total	15	15	16.00	15.83	8.57	7.35	N		
		AC # 4	Total	11	11	46.40	36.56	13.50	11.18	N		
		AC # 5	Total	12	12	36.00	41.06	5.16	11.84	N		
		AC # 6	Total	13	13	36.00	27.81	10.64	9.75	Y		
		AC # 7	Total	12	12	7.70	8.24	5.25	5.31	N		
		AC # 8	Total	16	15	21.90	31.17	14.01	10.32	Y		
		AC # 9	Total	6	6	2.00	1.20	1.10	2.02	N		
Transverse Cracking	Meters	AC # 1	Total	12	12	74.60	81.36	5.08	6.06	Y	0.6716	0.655
		AC # 2	Total	13	13	74.60	76.37	7.53	5.87	N		
		AC # 3	Total	15	13	16.90	21.91	1.80	3.14	Y		
		AC # 4	Total	11	11	27.10	24.74	2.91	3.34	N		
		AC # 5	Total	12	12	18.20	18.72	2.04	2.91	N		
		AC # 6	Total	13	13	32.00	32.93	2.39	3.85	N		
		AC # 7	Total	12	12	64.00	56.79	6.39	5.06	Y		
		AC # 8	Total	16	16	59.90	58.33	5.04	5.13	N		
		AC # 9	Total	6	6	31.30	30.18	4.70	3.69	N		
Corner Breaks	No.	PCC # 1	Total	12	12	9.00	8.83	1.21	1.14	N	0.3834	0.792
		PCC # 2	Total	14	14	9.00	9.07	0.70	1.15	N		
		PCC # 3	Total	14	13	2.00	0.92	0.62	0.37	Y		
		PCC # 4	Total	11	11	0.00	0.00	0.00	0.00	NA		
		PCC # 5	Total	12	12	3.00	2.42	0.64	0.60	Y		
		PCC # 6	Total	13	13	7.00	6.23	1.19	0.96	N		
		PCC # 7	Total	11	11	0.00	0.00	0.00	0.00	NA		
		PCC # 8	Total	16	16	5.00	4.81	0.95	0.84	N		
		PCC # 9	Total	6	6	0.00	0.83	0.37	0.35	Y		
Transverse Cracking	Meters	PCC # 1	Total	12	12	11.10	10.95	2.09	1.82	N	0.5500	0.772
		PCC # 2	Total	14	13	11.10	11.42	1.90	1.86	N		
		PCC # 3	Total	14	13	8.70	9.48	1.57	1.69	N		
		PCC # 4	Total	11	11	44.60	46.41	3.53	3.75	N		
		PCC # 5	Total	12	12	37.70	39.62	4.04	3.46	N		
		PCC # 6	Total	13	13	16.20	15.30	1.85	2.15	N		
		PCC # 7	Total	11	11	7.40	7.28	0.22	1.48	N		
		PCC # 8	Total	16	15	12.10	13.32	3.04	2.01	N		
		PCC # 9	Total	6	6	74.10	71.10	4.40	4.64	N		

Regression equation: Standard Deviation = $C * (\text{Mean})^{0.5}$

In general, no bias was observed for AC pavement distresses, except for fatigue cracking where five of the nine workshops indicated some bias. For PCC transverse joint spalling, differences seem to exist between the group mean and reference values; however, these differences often resulted from the small magnitude of mean and reference values, or the small variation (standard deviation) in distress data. Thus, from an engineering viewpoint, one can conclude no bias was observed for PCC pavement distresses.

From the above analyses, it can be concluded that the group mean may be viewed as an unbiased estimate of the reference value. Therefore, assuming a normal distribution and using a 95% confidence level, one can state that the true value is bound by the measured value ± 2*standard deviations, where the standard deviation can be calculated using the regression equations discussed earlier in this section. For example, for AC pavement transverse cracking (meters), the derived standard deviation for a measurement of 30 meters is 3.68 (=0.6716*sqrt(30)). Or, with a 95% confidence level, the true value is within the range of 22.64 and 37.36 meters (=30±2*3.68).

SUMMARY AND CONCLUSIONS

One of the primary purposes of the LTPP surface distress monitoring efforts is to provide detailed, distress-specific condition data for use in the development of distress specific performance prediction models. Because manual distress surveys are conducted by individual raters whose biases can lead to variability between raters, it was hypothesized that distress data variability existed and that it could potentially be quite large. Thus, the purpose of the study presented in this paper was to quantify manual distress data variability, with special emphasis on the bias and precision of the data.

Results from nine LTPP rater accreditation workshops conducted during the period of 1992 to 1996 were used as the only source of data in this study. From the analyses, it was concluded that, in general, difference between group means and reference values are small and can be ignored. The group means for different distress type and severity level combinations can be regarded as unbiased estimates of the corresponding reference values (surrogate of ground truth). Other important conclusions are summarized below:

1. Individual rater variability for any given distress type-severity level combination is typically large. Also, there is a decrease in variability with an increase in the magnitude of the mean as captured in the coefficient of variation.

2 Both apparent bias and precision for the common distress type-severity level combinations were quantified. The apparent bias is small and not uniform; i.e., no tendency to consistently rate all distress type and severity level combinations higher or lower. However, the associated precision or variability is very sensitive to the magnitude and range of distress quantities present on a section. The coefficient of variation ranges from less than 10% to well in excess of 100%, although it generally decreases with increased distress quantities. Large CV values were observed only where low quantities of a given pavement distress are present. Thus, the large variability indicated by these values may not necessarily indicate poor precision.

3 There is a significant improvement in precision for total distress quantities when compared with individual severity levels. The CV in some instances is still high if

viewed individually but, as noted before, that is primarily attributable to low distress magnitude and therefore may not necessarily indicate poor precision.

4 When all distress type-severity level combinations are viewed in terms of a single composite number such as the PCI value, there is excellent agreement between the individual raters, the group mean and the reference value. The individual rater variability is also small when viewed in terms of this composite value.

REFERENCES

1. "Distress Identification Manual for the Long-Term Pavement Performance Project", Strategic Highway Research Program, National Research Council, Publication No. SHRP-P-338, Washington, D. C., May 1993.

2. Rada, G. R., Bhandari, R. K., Elkins, G. E., and Bellinger, W. Y., "Assessment of Long-Term Pavement Performance Program Manual Distress Data Variability: Bias and Precision," Transportation Research Record 1592, TRB, Washington D. C., pp. 151-168, 1997.

3. Shahin, M. and Kohn S., "Pavement Maintenance Management for Roads and Parking Lots," Final Report CERL-TR-M-294, U. S. Army Construction Engineering Research Laboratory, Champaign, Illinois, 1981.

Fifth International Conference on the
BEARING CAPASITY OF
ROADS AND AIRFIELDS

FIELD MEASUREMENTS OF CRACK ACTIVITY AND LABORATORY SIMULATION OF CRACK REFLECTION PHENOMENON IN PAVEMENTS OVERLAYS

JORGE C. PAIS[1], PAULO A. A. PEREIRA[2]
JORGE B. SOUSA[3]

[1]Research Assistant, University of Minho, Guimarães, Portugal, Fax: +351-53-510-217, Email: jpais@eng.uminho.pt

[2]Associate Professor, University of Minho, Guimarães, Portugal, Fax: +351-53-510-217, Email: ppereira@ci.uminho.pt

[3]General Director, CONSULPAV, Portugal, Fax: +351-1-510-2296, Email: jmbsousa.consulpav@taguspark.pt

ABSTRACT

ABSTRACT: This paper shows a field measurement of the crack activity of longitudinal and transversal cracks with a Crack Activity Meter. The Crack Activity Meter was used to measure load associated differential vertical and horizontal movement between the two sides of a crack in pavements before overlay. The crack movements after overlay were calculated with a finite element model. The crack activity before overlay was used to estimate, with a finite element model, the crack activity after overlay function of overlay thickness and material stiffness.

The laboratory simulation of reflection cracking was undertaken with the Reflective Cracking Device. The Reflective Cracking Device used with the shear machine introduced by SHRP program, was employed to simulate in laboratory the crack activity patterns measured in the field and the crack activity movements calculated with the finite element model.

Key words: Crack activity, Reflective cracking, Fatigue life,

INTRODUCTION

Overlays are the most commonly used method for pavement maintenance. However, they often do not perform as desired due to the existing cracks, which propagate through the new pavement layers. This type of cracking is known as reflective cracking and it is a result of vertical and horizontal movements at the crack tip. Those movements, caused by thermal stresses, traffic loads or by a combination of these two mechanisms, induce stress concentrations in the overlay. Thus existing cracks of the old pavement propagate through the new pavement layer. It is usual to see the overlays cracked after few years of service.
This paper presents an approach to evaluate the reflective cracking in pavement overlays based on laboratory tests with the Reflective Cracking Device (RCD) designed by Sousa et al (1996). Old pavement crack differential movements were measured in road with the Crack Activity Meter (CAM) designed by Rust (1987) and produced by CSIR in South Africa.

CRACK ACTIVITY

Load associated reflective cracking is governed simultaneously by an horizontal opening and/or closing and a vertical shearing at the crack zone. The simulation of this process must consider the simultaneity of these two modes of opening in the analysis of this problem.
The CAM is composed of two LVDT (Linear Variable Differential Transformers), one placed vertically and the other placed horizontally allowing the measurement of both differential movements. Figure 1 shows a schematic representation of the CAM placed over the crack.

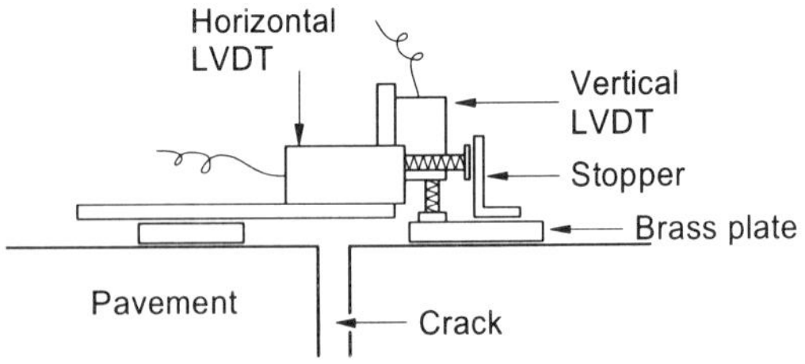

Figure 1. Crack Activity Meter.

Crack activity measurements

The evaluation of the crack activity before overlay was made in 16 cracked cross sections, with 500m long, in the Portuguese road network and only flexible pavements were studied. 13 transversal cracks and 11 longitudinal cracks were analyzed.
The typical crack activity for a longitudinal crack is shown in Figure 2, and represents the crack activity of the cracked pavement used for this study. Figure 3 shows the typical crack activity for transversal cracks. Horizontal positive values represent opening of the crack while negative values represent crack closing.

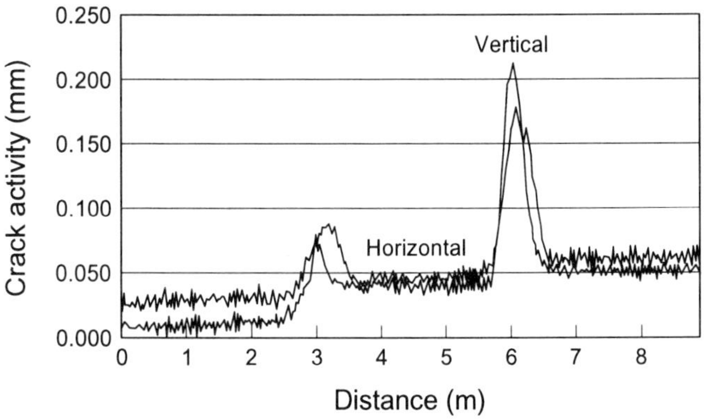

Figure 2. Crack movements of the longitudinal crack before pavement overlay.

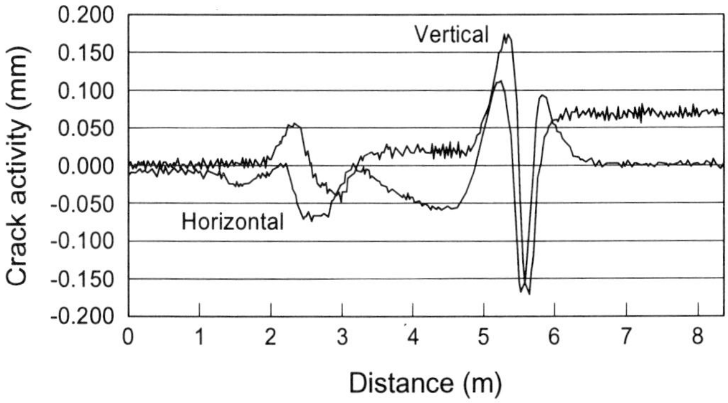

Figure 3. Crack movements of the transversal crack before pavement overlay.

Statistical calculations were made to correlate the measured crack activity with pavement and crack characteristics. To evaluate pavement characteristics, 11 FWD measurements were made in each cross section. The pavement layers stiffness was the average of the 11 measurements. For each pavement the considered stiffness was the bituminous layer stiffness and the thickness was the equivalent thickness for the bituminous layer stiffness according Equation 1. A thickness for the subgrade, defining a rigid layer, was considered.

$$\frac{E_1 h_1^3}{12(1-\nu_1^{\ 2})} = \frac{E_2 h_2^3}{12(1-\nu_2^2)} \qquad (1)$$

The models to predict the crack activity before overlay as function of pavement characteristics are expressed by the Piecewise model with breakpoint. This model is a two tends linear regression with break point. Equation 2 shows the model to predict the horizontal crack activity before overlay for the transversal cracks.

$$AH = a_1 + b_1 * af + c_1 * h_t + d_1 * E_t \ , \ AH < AH_0 \qquad (2)$$
$$AH = a_2 + b_2 * af + c_2 * h_t + d_2 * E_t \ , \ AH \geq AH_0 \quad R^2 = 0.951$$

where AH is horizontal crack activity before overlay (e-6), af is the crack width (mm), h_t is pavement thickness (cm), E_t is pavement stiffness (MPa), and a_i, b_i, c_i, d_i are statistical coefficients. The vertical crack activity before overlay can be estimated using the Equation 3.

Table 1. Coefficients for Equation 2.

a_1	b_1	c_1	d_1	a_2	b_2	c_2	d_2	AH_0
-64.66	12.23	1.002	0.007651	-58.11	40.54	1.394	0.008957	123.9

$$AV = a_1 + b_1 * af + c_1 * h_t + d_1 * E_t \ , \ AV < AV_0 \qquad (3)$$
$$AV = a_2 + b_2 * af + c_2 * h_t + d_2 * E_t \ , \ AV \geq AV_0 \quad R^2 = 0.951$$

where AV is vertical crack activity before overlay (e-6).

Table 2. Coefficients for Equation 3.

a_1	b_1	c_1	d_1	a_2	b_2	c_2	d_2	AV_0
137.1	13.16	-1.528	0.006864	-1312	217.6	8.304	0.05128	174.4

Table 3. Coefficients for horizontal crack activity before overlay for longitudinal cracks.

a_1	b_1	c_1	d_1	a_2	b_2	c_2	d_2	AH_0
92.33	-1.703	-0.1844	-0.001347	55.93	-88.43	9.044	-0.09681	103.9

Table 4. Coefficients for vertical crack activity before overlay for longitudinal cracks.

a_1	b_1	c_1	d_1	a_2	b_2	c_2	d_2	AV_0
173.7	3.567	-0.6103	-0.01501	185.4	47.59	0.5914	-0.02649	115.6

Figure 4 shows a typical comparison between observed and predicted crack activities for the horizontal activity of transversal cracks.

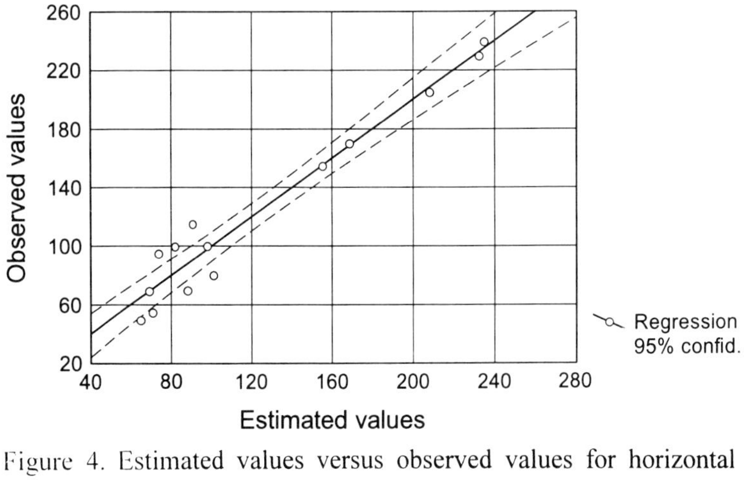

Figure 4. Estimated values versus observed values for horizontal activity of transversal cracks.

Estimation of crack activity after overlay

The crack activity measured in the pavement before the overlay is different from the crack activity after overlay since it depends on the overlay thickness and material stiffness. The crack activity after overlay can be estimated using a finite element analysis and knowing the old pavement properties such as the one shown in Figure 5.

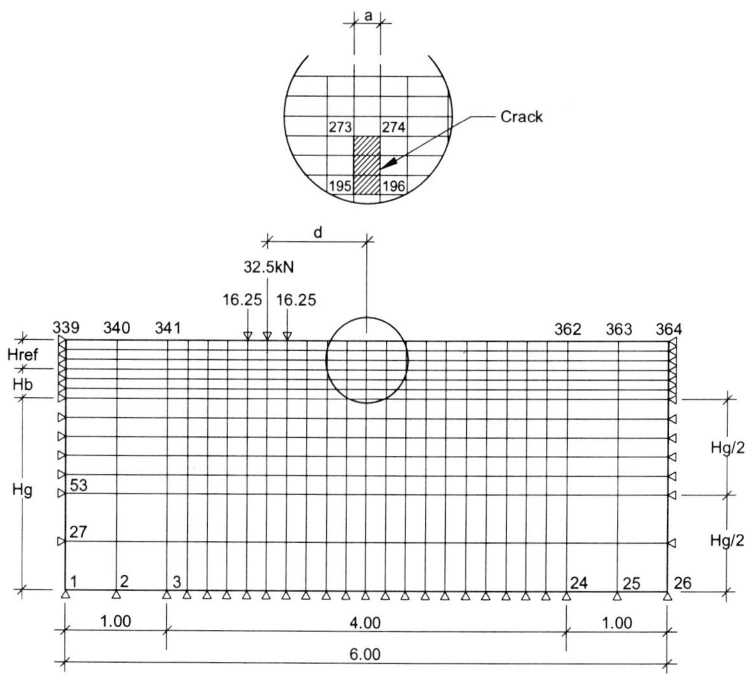

Figure 5. Finite element model for estimation of crack activity after overlay.

The influence of the overlay thickness in the crack activity after overlay, for 5000 MPa overlay stiffness, is presented in Figure 6.

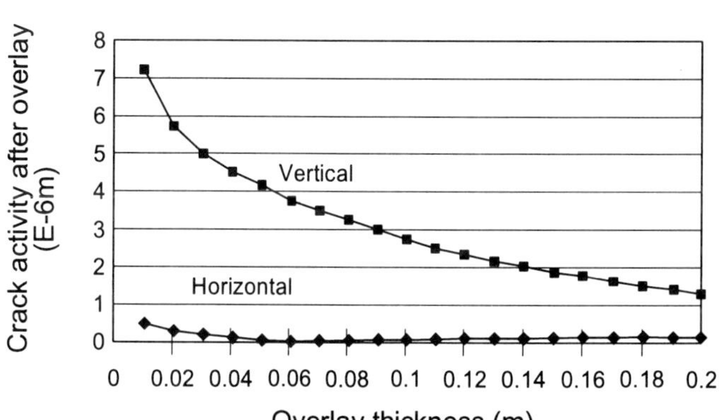

Figure 6. Crack activity after pavement overlay as function of overlay thickness.

EXPERIMENT DESIGN

The laboratory simulation of reflective cracking was made on bituminous mixtures produced in laboratory. The bituminous mixture used in this study was a dense graded bituminous mixture with a maximum aggregate size of 25 mm manufactured according the Portuguese standards and used in binder layers. The bitumen used to produce the bituminous mixture was a conventional 50/70-penetration grade.
For these tests the crack width of the reflective cracking was set to 6 mm, corresponding to three times the crack width measured in pavement. The test frequency was 10 Hz and the test temperature was 20 °C.
To simulate the reflective cracking through overlays, the crack activity after overlay, calculated using the finite element model, was applied to the specimen by the shear machine through the RCD. Because this crack activity is very low, test would take several days. Therefore a test program was undertaken at higher crack activity levels involving the use of a fatigue law to extrapolate the results to the desired crack activity (Pais et al, 1998).

TEST RESULTS

To evaluate reflection cracking through overlays Sousa et al (1996) suggested that, an equivalent stiffness in the crack zone could be used to define the fatigue life for this problem. This equivalent stiffness is defined as the average stiffness in specimen volume above the crack and is defined as follows:

$$S = \frac{F/L.H}{\delta/w} \qquad (5)$$

where S = equivalent stiffness in crack zone; F = measured force; δ = applied displacement; L = crack length (i.e. specimen length); H = specimen height and w = crack width.

The AASHTO TP8-94 standard test defines the fatigue life for flexural beam specimens when the specimen stiffness is reduced to 50% of the initial stiffness. For this study, the fatigue life was defined as the number of load cycles to reach 50% of the initial equivalent stiffness in crack zone.
A reflective cracking fatigue model can be established using a linear regression between logarithm of fatigue life, *log* N, and the logarithm of displacement, log Δ, defined as follows:

$$N = a\left(\frac{1}{\Delta}\right)^{b} \qquad (6)$$

where N is the fatigue life; Δ is the crack activity after overlay and a, b are experimentally determined coefficients.

Using the equivalent stiffness in the crack zone to analyze the reflective cracking, two types of stiffness evolution were found during the test. The most usual stiffness evolution found in these tests was the logarithmic law (Figure 11) but in some tests a exponential law (Figure 12) was found.

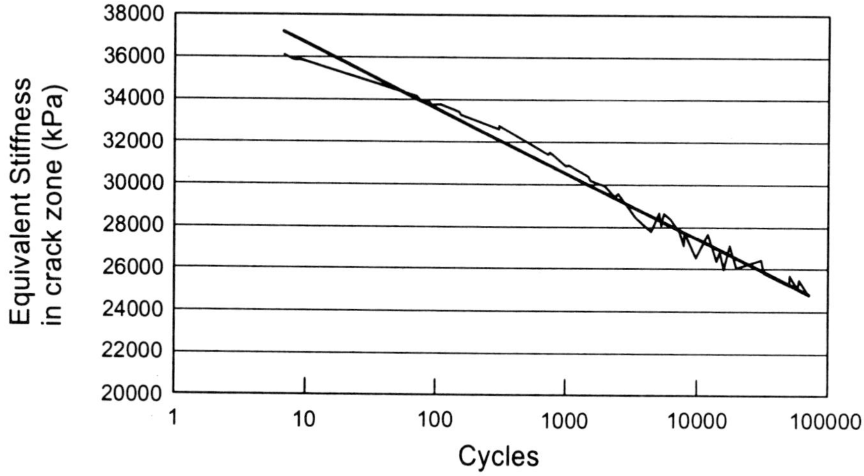

Figure 11. Logarithmic evolution of equivalent stiffness in crack zone.

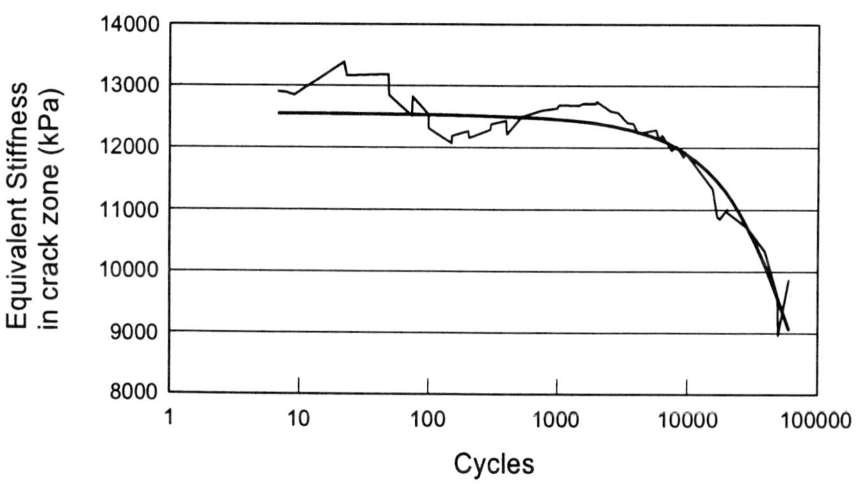

Figure 12. Exponential evolution of equivalent stiffness in crack zone.

The reflective cracking fatigue life as function of crack activity and specimen thickness for longitudinal cracks is shown in Figure 13 where one can conclude that the increase of crack activity decreases the fatigue life and the increase of specimen thickness increases the fatigue life.

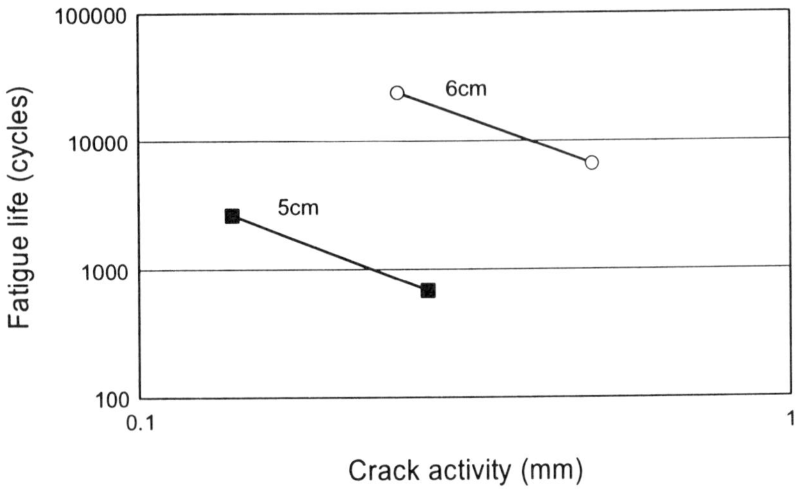

Figure 13. Fatigue life laws for longitudinal cracks.

Those two thicknesses and three crack activities can be used to extrapolate the reflective cracking fatigue life for other thicknesses and crack activities. However, further conditions need to be evaluated to demonstrate the validity of these assumptions.

CONCLUSIONS

To evaluate the reflective cracking potential, the crack activity must be measured. This crack activity before overlay can be used to estimate the crack activity after overlay. Both crack activities can be well correlated with old pavement and overlay characteristics.
After overlay, the horizontal crack activity in the bottom of new layers is almost null and the vertical activity is highly reduced with the overlay thickness and follows a logarithmic law. These crack activities are influenced by the overlay properties, namely the overlay thickness and material stiffness.
The laboratory study with the reflective cracking device showed that the stiffness behaviour during test follows either an exponential law or a logarithmic law.
The reflective cracking fatigue life is influenced mainly by the crack activity after overlay and overlay thickness. The increase of crack activity decreases the fatigue life and the increase of specimen thickness, representing the overlay thickness, increase the fatigue life.
This approach appears to open promising avenues and further research should be made in this area.

ACKNOWLEDGEMENTS

The authors acknowledge and are thankful to the insight and support of Dr. M. Conceição Azevedo of APORBET, Portugal.

REFERENCES

1. Pais, J.C., P. Pereira, M. Azevedo & J.B. Sousa, 1998. The reflective cracking in thin surfacings, Proceeding of the First International Conference on Highway Surfacing.
2. Rust, F. C., 1987. Load associated crack movement and aspects of the rehabilitation of reflection cracking in cemented pavements, Faculty of Engineering, University of Pretoria, South Africa.
3. Sousa, J.B., S. Shatnawi & J. Cox, 1996. An approach for investigating reflective fatigue cracking in asphalt-aggregate overlays, Proceeding of the Third International RILEM Conference.

Fifth International Conference on the
BEARING CAPASITY OF
ROADS AND AIRFIELDS

A PRACTICAL SYSTEM FOR CRACK IMAGE COLLECTION AND AUTOMATIC IDENTIFICATION

BRIAN FERNE[1], LES HAWKER[2] AND JOHN PYNN[3]

[1]Project Manager, Transport Research Laboratory, RG45 6AU Crowthorne, UK, +44 1344 770686, ferneb@h@trl.co.uk

[2]Client, Highways Agency, SE1 0TE London, UK, +44 171 921 4041, les.hawker@highways.gov.uk

[3]Senior Researcher, Transport Research Laboratory, RG45 6AU Crowthorne, UK, +44 1344 770686, pynnj@h@trl.co.uk

ABSTRACT

Recent evidence has demonstrated that the deterioration of thick, well constructed flexible pavements is not structural and that deterioration generally starts at the surface in the form of cracking and rutting. Such behaviour of long life roads emphasises the need for accurate repeatable and economical methods of monitoring and correcting these defects before they affect the structural integrity of the pavement.

This paper describes the development of a system for the automatic detection of surface cracking. Automatic analysis depends on obtaining uniformly bright images of high quality. These are collected at survey speeds of up to 80 km/h using novel linescan video cameras operating with artificial illumination. Surveys can thus proceed either by day or night. Purpose built hardware onboard the survey vehicle assembles the line images into video frames prior to automatic processing to identify cracks. Firmware is used to carry out part of this processing in real time, the remainder of the processing being carried out off-line on a PC.

In such a system the correct application of defect information depends on accurate knowledge of its location. In the UK such referencing is currently provided by accurately measuring the distance travelled with respect to automatically recognised studs on the road surface. An improved system using GPS is also being investigated.

Such a system provides enormous benefits by reducing considerably the need for hazardous and relatively expensive slow speed visual condition surveys as well as eliminating the inconsistency and subjectivity of manual interpretation. This development will be a major contribution to the effective monitoring and maintenance of flexible pavements.

KEYWORDS: Visual condition surveys, Crack detection, Image processing, DSP

INTRODUCTION

Recent research has concluded that the deterioration of thick, well constructed flexible pavements initiates at the surface in the form of rutting and surface cracking. Such behaviour of these long life pavements emphasises the need for accurate repeatable and economical methods of monitoring and correcting these defects before they affect the integrity of the structure. As these pavements normally carry large amounts of traffic, monitoring should take place at normal traffic speed to avoid disruption. In the UK depth of rutting is measured every two years using the High-speed Road Monitor (Jordan and Still, 1982), which operates at speeds up to 80 km/h. This paper describes a new system to monitor surface cracks at similar survey speeds. The development has been financed by the UK Highways Agency and carried out by TRL, who deployed a multi-disciplinary team of scientists, engineers and software writers, as reported by Ling (1998). The development has taken 5 years, involving several prototypes. One of these prototypes has been described by Hawker (1997).

THE CRACK MONITORING SYSTEM

System requirements

The requirements of the crack monitoring system are to automatically identify and locate surface cracks, of widths 2 mm and greater, over 80% of the width of a traffic lane during surveys carried out at speeds up to 80 km/h. The surveys are to be carried out using a specially equipped vehicle, with a survey team consisting of a driver and operator. A further requirement is to provide a facility for inputting the crack information into UK pavement management systems. The target performance is to identify at least 80 % of the cracks recorded that would have been identified by manual inspections with fewer than 5% false cracks being found.

System outline

Images of the road surface are collected by three linescan cameras that are mounted in a survey vehicle and look straight down at the road surface. The cameras operate under artificial illumination to avoid the shadowing that can occur with natural lighting. Linescan cameras are used because it is easier to obtain uniform lighting over the narrow strip of road surface viewed by the linescan camera, rather than the much larger rectangular area that would be needed for a conventional video frame camera. In this application the illumination is angled at 30 degrees to the surface to increase the shadowing produced by cracks.

The camera looks vertically downwards at the road surface and images a narrow strip of the surface measuring 1.024 m in the transverse direction and 2 mm in the direction of travel. The camera provides a black and white line image in digital form that consists of a string of 512 pixels representing the variations brightness of the surface along the length of the narrow strip. Each pixel therefore covers a 2 mm square area of road surface. A two dimensional image or frame is created by triggering the camera to output a line image every 2 mm of forward travel and storing the line image as a row of pixels in the digital memory of a purpose built Scan Data Converter, or SDC. After 512 rows have been stored a frame consisting of 512 by 512 pixels, which represents an image of approximately a 1.024 metre square area of the road surface, is complete. Note that this sampling rate provides the necessary resolution to detect 2 mm wide cracks.

Distance units attached to the rear wheels of the survey vehicle are used to trigger the cameras and provide the location of the frame from the start of the road section being surveyed.

Consecutive frames are output from the SDC in analogue form as videoframes to a SVHS videotape recorder to provide a permanent record of the survey. Location information is recorded in the hidden lines within the video frame by a purpose built Video Data Correlator, or VDC. The frames, each with their associated location information, are also output in digital form to a real time processing unit that undertakes the initial, or primary, stage of image processing.

Primary processing entails removing the unwanted variations in image brightness produced by imperfections in the lighting system, and then carrying out statistical analysis on the image to determine appropriate threshold levels for crack detection. The unit then thresholds the image using these levels, and stores the thresholded image on removable hard disk for subsequent off-line processing to identify and classify the cracks. It should be noted that the thresholding process reduces the size of the image data and the rate at which it is produced by a factor of four. This enables the reduced image to be written to conventional hard disk in real time when surveying at 80 km/h.

Three TDI cameras, SDC's, VDC's, videotape recorders, and primary processing units are provided in order to monitor a 2.9 m width of lane. This width allows a small amount of overlap at the sides of the images collected by adjacent cameras.

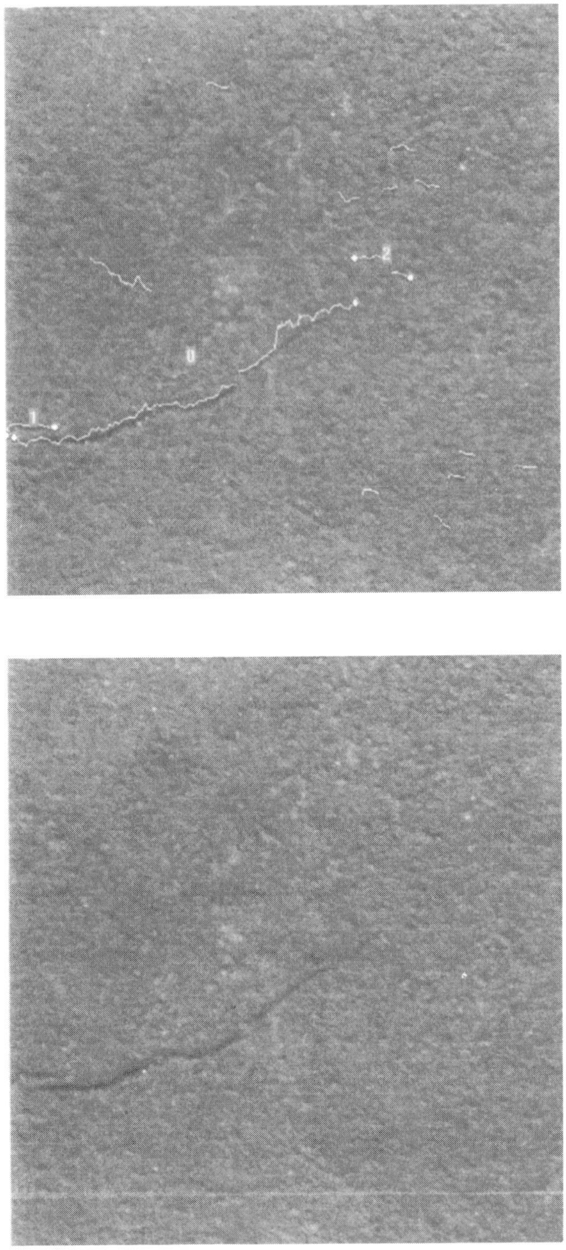

Figure 1 Illustration of crack detection, and example of an image of the road surface

The off-line, or secondary, processing software identifies cracks in the images, as shown in Figure 1, locates the cracks and classifies them according to their length and direction. The results are presented as three crack lists obtained from the images collected by the nearside, middle and offside cameras. Further software is also provided to input the information in the crack lists into the UK CHART (1986) maintenance management system.

Figure 2 Survey Vehicle

Equipment details

The survey vehicle is a Mercedes 811D Panel Van, the exterior of which is shown in Figure 2. The three linescan cameras are mounted in the rear compartment on a transverse rigid beam and view the road surface through slots in the floor of the compartment. The lighting units are mounted inside the brushed skirt hung under the rear of the vehicle. The vehicle incorporates a 'pod' on either side at the rear to extend the camera's field of view and the illumination outside the vehicle's width, in order to image a maximum 2.9 m width of traffic lane.

Time Delay and Integration (TDI) linescan cameras are used, as shown in Figure 3. The conventional linescan camera has a single line of CCD sensors, whereas the TDI camera contains many parallel lines of sensors which are synchronised so that, in turn, each line of sensors images the same narrow strip of road surface as the survey vehicle moves forward. The effect is to amplify the image, resulting in a reduced lighting requirement. In this application TDI cameras with 96 lines, with each line containing 512 CCD sensors, are used. Each time the camera is triggered, namely every 2 mm of forward travel, the camera outputs a single line image, which is the sum of the 96 line images of the same narrow strip of road surface.

Figure 3 Rear of Vehicle showing TDI Linescan cameras

The TDI camera is connected via a camera interface to the SDC. The interface supplies electrical power, controls the camera's auto iris lens, and delivers the camera trigger pulses supplied by the distance units.

Lighting of the 2.9 m wide strip of road surface is provided by twenty-six halogen fog lamps which are butted together to form a lighting array 13 lamps long and 2 lamps high, as shown in Figure 4. The lamps are mounted close to the road surface and angled down to provide the illumination at 30 degrees to the road surface. Power for the lighting and image capture hardware is provided by a generator mounted under the vehicle.

Figure 4 Lighting Array

A front facing shuttered video camera is mounted on the driver's dashboard to provide images of the road ahead. These images are stored on SVHS video cassette. This camera provides a visual locational record of the survey, and can be used to obtain information on the general state of the road surface.

The SDC's and VDC's and front camera are connected to a computer network to enable the survey operator to control the image collection from a PC mounted on the bench in the operator's cabin, see Figure 5. A list containing the identifiers of the sections of road to be surveyed during a run is entered into the PC and down loaded into the VDC's prior to

1) the backcalculated mould must fit the measured one perfectly
2) there is only one combination of parameters which fits the mould
(only one definite solution)

To get a first impression of the sensitivity of back calculation a simple system is proposed. A halfspace and an asphalt overlay may represent a two layer system. If there will occur misjudging in asphalt layer thickness from +20 to - 20 % the backcalculated e- modulus of this layer vary in the way presented in figure 1. It's obvious that thin layers are more sensitive to errors.

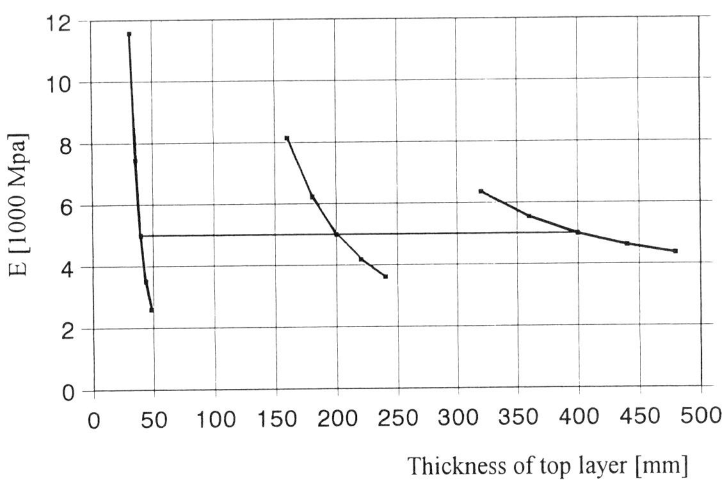

Figure 1: Backcalculated e- modulus of top layer of a two layer system influenced by estimation errors of layer thickness

CONVERGENCE AND DEFINITENESS OF BACK CALCULATION

There seems to be a considerable sensitivity to input parameters.
Several attempts have been done to refine back calculation procedures. Some examples are given in literature [1], [2]. Despite of excellent convergence sometimes there are some doubts necessary about the results obtained.
As typical example of this behaviour let's have a look at figure 2, representing the back calculation results of a three layer system. After a certain number of iterations the derivation from measured to calculated mould minimizes nearly to zero and seems to be excellent.

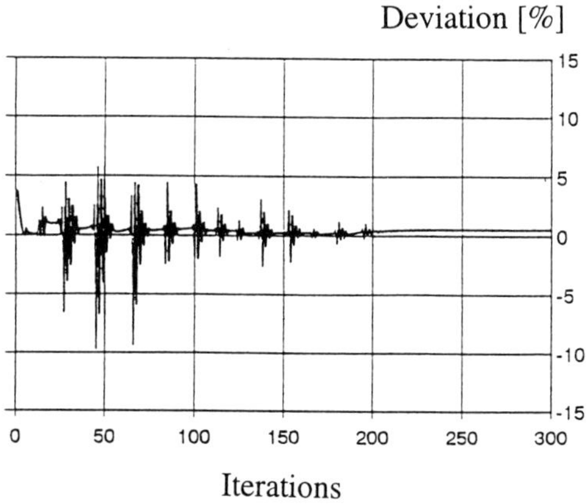

Figure 2: Convergence behaviour of a critical three layer system
(Deviation from measured to calculated mould)

But it's also necessary to look at the same time at figure 3, where the backcalculated e-modulus of each layer are presented.

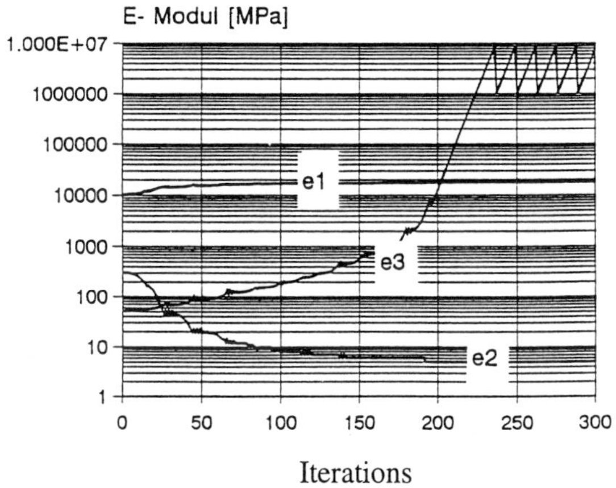

Figure 3 Backcalculated modulus of a critical three layer system

At the latest from iteration number 200 the results can be described as nonsense. Despite of excellent convergence there are several combinations of layer modulus possible, which result in the same deflection mould. Definiteness of three layer systems is not existing. There are systems which behave uncritically, but you can't prevent the critical ones.

In the next step a systematical checking of results of backcalculation procedure is presented based on the assumption that materials behave linear elastic.
Experience has shown that systems containing asphalt top layer thicknesses of 200mm and an unbound second layer of 500mm behave stable and don't tend to critical behaviour.

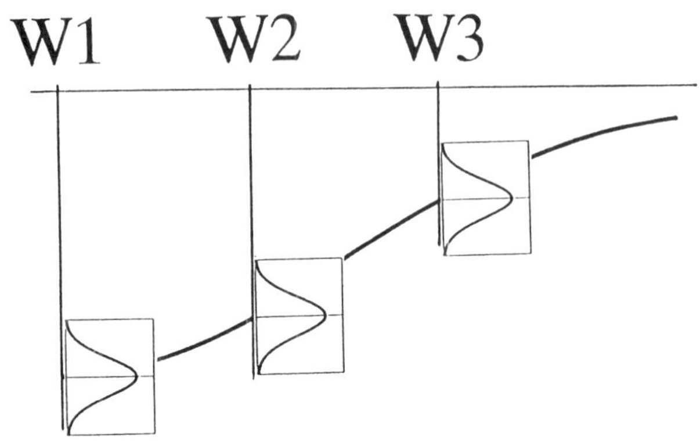

Figure 4: Scheme of deflection mould with normally distributed readings

It is assumed that three deflections are measured, which are influenced by normally distributed errors, which occur in the range from +5 to -5%. The connected results of back calculation for top layer are shown in figure 5.

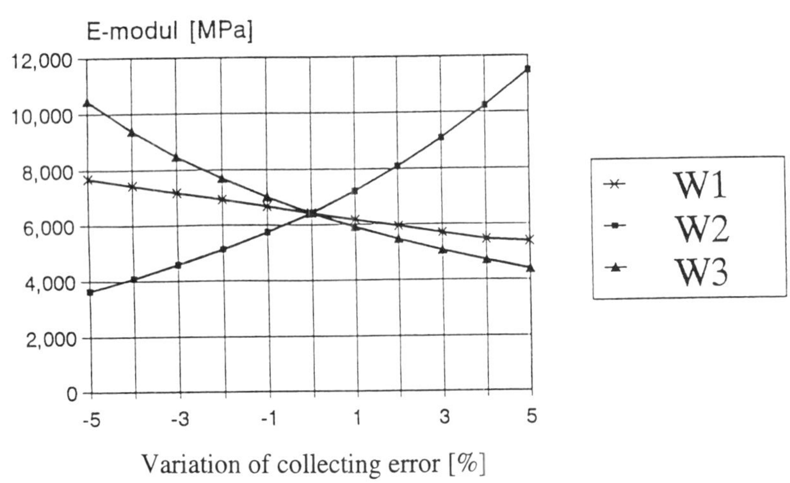

Figure 5: Sensitivity to variation in reading errors

The deflection w1 is the most insensitive to errors in collecting, in contrary to deflection w2 which results in considerable differences in obtained e- modulus from 3500 to 11500 N/mm² (error free value is 6500 N/mm²).
The consequences of these remarkable results are demonstrated in figure 6, where error combinations were made by chance. Each deflection is connected with collecting errors from +3 to -3%. Backcalcuted modulus are taken in multi layer calculation to determine maximum tensile strain at the bottom of top layer. This strain is considered to be the central value for service life considerations.

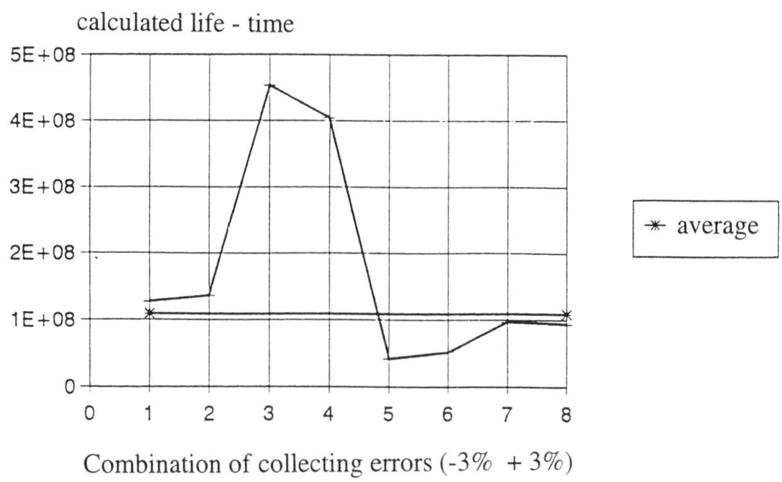

Figure 6: Calculated service life influenced by recording errors

There is no doubt about the necessity to require high accuracy to get reasonable results. But is to keep in mind that for example the accuracy of geophones in FWD is in the range from +3 to -3%.

The characteristics of the uncritical system (which is a quite normal system) mentioned above point out the extreme sensitivity to collecting errors (or processing errors as well).

moduli appear quite reasonable. In most cases the base is in good condition and the problem lies in the lack of discrimination between these layers in backcalculation computer programs.

The shape and magnitude of the deflection basin is highly dependent upon pavement stiffness and subgrade support. Even today some agencies rely upon maximum deflections to identify sub-standard structural support. The use of surface curvature index (SCI) or radius of the deflection basin has a greater relationship to pavement structural response and stress levels than maximum deflection. A very stiff pavement over weak, compressible soils may have exceptionally high maximum deflections but very low stresses due to the large radius of the NDT deflection basin. Consequently, the shape of deflection basin relates much more realistically to the pavement's structural behavior than magnitude of deflections.

Considerable emphasis has been placed on temperature effects and ways to correct to a standard temperature. This type of approach may be useful for a general appraisal of a highway network and to locate pavement sections of dubious structural strength. However, it may be totally inadequate when trying to identify specific causes for pavement distress or structural deficiencies. The most definitive approach is to selectively perform an NDT evaluation at different pavement temperatures and during different seasons of the year. This approach can improve our ability to determine the significance of temperature and climatic changes in relation to pavement stresses and longevity. Temperature effects may be fairly minimal in pavements constructed with non-temperature susceptible granular materials regardless of asphalt concrete pavement thickness. However, full-depth asphalt pavements are temperature susceptible in their response characteristics as measured by NDT methods.

Another facet of NDT and pavement temperature involves cracked pavements. Normally it is desirable to test between cracks to minimize the potential for abrupt changes in the deflection response. This does not necessarily eliminate the problem when determining layer moduli. Testing conducted at high pavement temperatures reduce the moduli of the asphalt concrete, minimizes the influence of cracks and allows for more realistic determination of moduli for the base course and underlying layers. This approach was used successfully in the evaluation of a cracked airfield taxiway (1).

Seasonal testing of pavements can be beneficial in assessing if a critical condition exists which can produce high load-induced stresses and early distress in the pavement. High water table and subsurface seepage zones that occur during the spring or rainy season can sometimes drastically alter a pavement's response to NDT. As an example, a pavement section with cypress ponds on both sides of the road was found to have maximum deflections in the wet season that were three times greater than during dry periods.

Conventional FWD configurations space the geophones (sensors) at 30 cm intervals in a straight line from the load. This is convenient from the standpoint of automatically raising and lowering the sensors for testing but it does not provide for good definition of the stiffness of the asphalt concrete layer. The Dynaflect was initially studied and the sensors' spacing altered longitudinally and sensors were placed next to the loading wheels. Layer moduli prediction equations developed on this basis appeared to give fairly reliable results. Subsequently an analytical investigation was performed using a dual loading with spacings of 61 and 102 cm between the 18.1 cm diameter loading plates. Layer moduli prediction equations were developed using a rigid-plate approximation system in the elastic layer program BISAR (2).

Since an FWD with dual loading was not available for pavement testing it was decided to develop prediction equations based upon the superposition principle. Therefore, conventional FWD equipment with offset sensors could be used to simulate the dual loading configurations (3). As illustrated in Figure 1, dual loading produces double bending in the pavement which allows for improved measurements relating to stiffness of the asphalt concrete layer. A major deficiency in the superposition approach is that no replication of deflections under the plate is obtained as compared to a dual load.

The ensuing segments of this paper present the FWD sensor configuration, layer moduli prediction equations, and analyses of data recently collected on interstate pavements for an investigation of longitudinal cracking.

FWD SENSOR CONFIGURATION AND LAYER MODULI PREDICTION EQUATIONS

The configuration of deflection sensors relative to the FWD's 0.3 m diameter load plate is illustrated in Figure 2. These sensor spacings were selected in an attempt to maximize sensitivity for layer moduli determinations using the superposition approach. Replicate deflection readings are obtained in the transverse direction (e.g., sensors 6-7, 8-9, and 10-11). In the testing of very stiff pavements, sensors 10-11 will provide more reliable predictions of the composite modulus (E_{12}) of the asphalt concrete and granular base than sensors 8-9. The moduli prediction equations for a four-layer system are as follows:

A. Adaptation of Standard FWD System with 0.3 m Diameter Plates at 0.61 m Load Spacing (8,000 lbs total)

1. Asphalt concrete layer modulus E_1:

$$E_1 = 78.2254(t_1)^{0.5554}(D_{y/0} - D_{y/10})^{(-0.7966-19.1332/t1)} *(D_{y/0} - D_{x/8})^{17.4791/t1} \qquad \text{Eqn. 1}$$

$$R^2 = 0.968$$

2. Composite layer modulus E_{12} (composite of E1 and E2):

$$E_{12} = 1171.774(t_1)^{-0.4086}(t_2)^{-0.4927} * [(D_{y/0} + D_{y/24})/2 - D_{x/12}]^{-0.475-1.3034/t1-1.7208/t2} \qquad \text{Eqn. 2}$$

$$R^2 = 0.946$$

where $E_{12} = [(E_1^{1/3} t_1 + E_2^{1/3} t_2)/(t_1 + t_2)]^3$

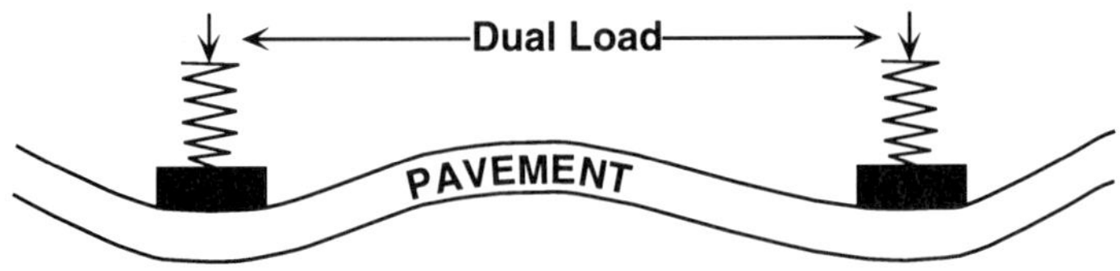

Figure 1 Double Bending to Improve Sensitivity to Asphalt Concrete Pavement Stiffness

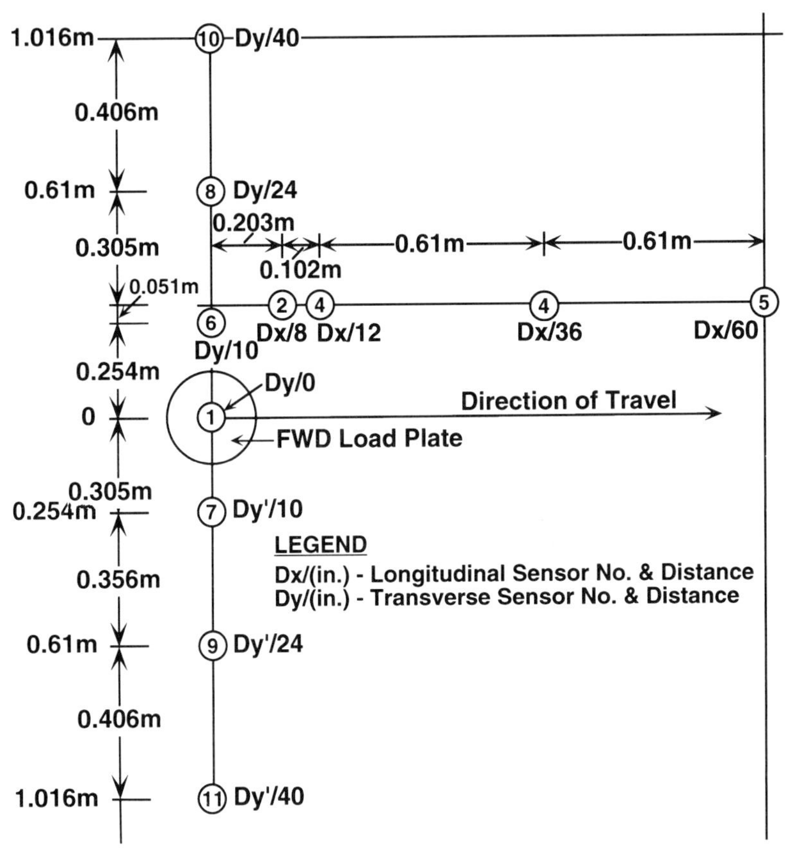

Figure 2 Schematic of FWD Sensor Configuration for Superpositioned Dual Load

3. Subgrade Modulus E_4:

$$E_4 = 36.334\,(D_{x/60})^{-1.015} \qquad R^2 = 0.996 \qquad \text{Eqn. 3}$$

4. Subbase Modulus E_3:

$$E_3 = 105.81136(t_2)^{-1.0785}\,(D_{x/36} - D_{x/60})^{-6.02523 + 2.4888/(D_{x/60})}$$

$$*\,[(D_{y/0} + D_{x/12}) - 1.15D_{x/36}]^{2.1609 - 1.6202/(D_{x/60}) - 5.302/t_2}$$

$$*\,(D_{x/60})^{3.6706 + -0.0498t_1 - 0.686t_2 - 3.0998/(D_{x/60})} \qquad \text{Eqn. 4}$$

$R^2 = 0.904$

[All thicknesses (t) in inches
All deflections (D) in mils (0.001 in.)
Conversion E (MPa) = 6.90 (ksi)]

B. Adaptation of Standard FWD System with 11.82 in. Diameter Plates at 40 in. Load Spacing (8,000 lbs total)

1. Asphalt concrete layer moduli E_1:

$$E_1 = 140.3731(t_1)^{0.3149}(D_{y/0} - D_{y/10})^{(-0.901 - 10.2687/t_1)}$$

$$*\,(D_{y/0} - 1.05D_{y/24})^{8.8724/t_1} \qquad \text{Eqn. 5}$$

$R^2 = 0.962$

2. Layer moduli E_{12} (composite of E1 and E2):

$$E_{12} = 2102.548(t_1)^{-0.4575}t_2^{-0.578}(D_{y/0} - 1.02D_{y/10})^{-2.9971/t_1}$$

$$*\,[(D_{y/0} + D_{y/40})/2 - 0.97D_{y/24}]^{-0.878 + 2.5454/t_1} \qquad \text{Eqn. 6}$$

$R^2 = 0.967$

where $E_{12} = [(E_1^{1/3}\,t_1 + E_2^{1/3}\,t_2) / (t_1 + t_2)]^3$

The results indicated that the regression models developed for the adaptation of the standard FWD system were similar to those previously developed for the FWD dual load systems. These equations can be evaluated by performing field tests to collect deflection data, which can be used to indirectly assess the relative accuracy of the FWD dual-load, rigid-plate system.

There are limitations in the use of these equations. Table 1 gives the range of parameters used in developing the predictive equations and the suggested limitations for layer thicknesses.

Table 1 Moduli and Thicknesses Used in Development of Prediction Equations

Layer	Modulus (ksi)	Thickness (in.)	Suggested Limiting Layer Thickness Range
Asphalt Concrete	150, 300, 600, 1200	3, 6, 8	2 to 10 in. (5-25 cm.)
Base Course	42.5, 85, 120	8, 12, 16	6 to 18 in. (15-45 cm.)
Subbase	15, 45, 75	12	8 to 16 in. (20-40 cm.)
Subgrade	5, 10, 20, 40	infinite	—

PRELIMINARY EVALUATION OF PREDICTIVE CAPABILITY

Table 2 presents layer moduli computed from FWD deflections measured on five different highway sections. The equations for the two different configurations for superpositioned dual loads generally gave different results of similar magnitude. The greatest differences occurred between US 301A and US 301B. The value for E_3 (16.28) is definitely questionable considering the exceptionally high value for E_4. However, the predicted deflections compare favorably with the measured values except for $D_{y/0}$ and $D_{y/10}$ for US 17B. Table 3 compares the measured deflections with those predicted using the moduli in Table 2. In general, the predicted deflections are extremely close to the measured values.

Table 2 Layer Moduli Predictions Using the Adaptation of FWD Prediction Equations

Test Road	Predicted Moduli (ksi)							
	E_1^a	E_1^b	E_{12}^a	E_{12}^b	E_2^a	E_2^b	E_3	E_4
SR 24	67.49	61.31	46.06	69.03	41.53	48.22	40.34	59.35
US 301A	178.61	198.85	99.84	89.99	75.44	60.08	16.28	82.22
US 301B	379.32	421.78	160.28	144.38	104.44	82.02	44.66	66.80
US 17A	127.85	123.04	86.89	95.87	64.52	79.73	39.24	32.52
US 17B	151.97	161.29	111.69	111.69	83.43	70.10	41.22	35.61

[a] Equation 1 for E_1, 8.2 for E_{12}

[b] Equation 5 for E_1, 8.6 for E_{12}

Table 3 Comparison of Measured and Predicted FWD Deflections

Test Road	Type	Deflection (mils)							
		$D_{y/0}$	$D_{y/10}$	$D_{y/24}$	$D_{y/40}$	$D_{x/8}$	$D_{x/12}$	$D_{x/36}$	$D_{x/60}$
SR 24	Measured	11.52	5.20	1.59	0.86	3.07	2.17	0.89	0.62
	Predicted	12.07	4.90	1.56	0.87	2.99	2.42	0.92	0.58
US 301A	Measured	9.15	5.40	1.64	0.80	3.71	3.11	0.87	0.45
	Predicted	8.56	5.01	1.76	0.68	3.49	2.89	0.75	0.37
US 301B	Measured	6.12	3.96	1.55	0.82	2.90	2.29	0.94	0.55
	Predicted	5.45	3.45	1.51	0.83	2.52	2.15	0.89	0.52
US 17A	Measured	10.61	7.87	3.97	1.77	5.51	4.37	2.12	1.14
	Predicted	10.86	7.19	3.37	1.78	5.52	4.80	1.91	1.07
US 17B	Measured	8.58*	5.27*	2.24	1.29	3.97	3.38	1.41	1.06
	Predicted	6.41*	3.66*	2.11	1.57	2.83	2.55	1.63	1.10

* Excessive difference between measured and predicted.

Recently FWD tests were performed on sections of interstate highway. Layer thicknesses were approximately the same for each of these pavements, i.e., 6, 10.5, 12 inches for the asphalt concrete, granular base, and subbase, respectively. The subgrade was assumed to be 5.3 m thick overlying bedrock. A total of 7 or more locations were tested on each section of pavement using both the standard and modified FWD sensor positions.

Analyses were performed five different ways as stipulated for each of the following cases:

Case 1: 3-layer analyses using BISDEF's default seed moduli (350, 30, 15 ksi)

Case 2: 4-layer analyses using BISDEF's default seed moduli (350, 30, 30, 15 ksi)

Case 3: 3-layer analyses using predicted moduli as seed moduli

Case 4: 4-layer analyses using predicted moduli as seed moduli

Case 5: 4-layer analyses using predicted moduli as seed moduli keeping the subbase seed moduli constant and allowing BISDEF to determine the moduli of the other layers.

Moduli for Cases 1 and 2 are given in Table 4. Asphalt concrete and base course moduli are drastically different between the two cases. The Case 1 base course moduli were unrealistically high for I-95. This problem was eliminated using the 4-layer, Case 2, analysis but the subbase moduli consistently exceeded the base moduli by a factor of 2 to 20 except for I-10 Duval W.B. Similar results were obtained in Cases 3 and 4 (see Table 5) using predicted moduli as seed moduli in BISDEF. Comparison of Cases 1 and 3 and Cases 2 and 4 indicate most values are in the same order of magnitude. Only I-10 Duval W.B. gave consistently rational values regardless of seed moduli or number of layers in the analysis.

Table 4 Comparison of 3 and 4 Layer BISDEF's Analyses Using Default Values

Location	Layer Moduli-ksi							
	AC		Base		Subbase		Subgrade	
	Case 1*	Case 2*	Case 1	Case 2	Case 1	Case 2	Case 1	Case 2
I-95 St. Johns	263	678	275	69	—	143	28	23
I-95 Duval	180	488	501	58	—	1072	35	24
I-10 Duval E.B.	72	155	80	31	—	282	36	33
I-10 Duval W.B.	302	398	64	41	—	54	29	28
I-10 East Madison	265	484	109	38	—	310	38	33
I-10 West Madison	257	540	90	26		547	37	31

*Case 1: 3-layer analyses using BISDEF's default seed moduli
Case 2: 4-layer analyses using BISDEF's default seed moduli

Table 5 Comparison of 3 and 4 Layer BISDEF Analyses Using Predicted Values as Seed Moduli

Location	Layer Moduli-ksi							
	AC		Base		Subbase		Subgrade	
	Case 3*	Case 4*	Case 3	Case 4	Case 3	Case 4	Case 3	Case 4
I-95 St. Johns	397	736	223	56	—	225	28	23
I-95 Duval	163	487	668	48	—	850	34	24
I-10 Duval E.B.	102	154	67	31	—	228	37	33
I-10 Duval W.B.	297	373	65	45	—	46	29	28
I-10 East Madison	262	479	110	37	—	270	38	33

Case 3: 3-layer analyses using predicted moduli for seed values

Case 4: 4-layer analyses using predicted moduli for seed values

Table 6 gives the predicted and BISDEF Case 5 modulus values. Again the moduli for I-10 Duval W.B. are essentially the same as for Cases 2 and 4. The base moduli for the other sections are reasonable with the exception of I-95 Duval. It is obvious that 4-layer BISDEF analyses are questionable, particularly in these analyses where the subbase moduli was considerably greater than the limestone base moduli. The use of a constant subbase value tends to minimize this problem.

Table 6 Comparison of Predicted Moduli with 4-Layer BISDEF Analyses with Predicted Subgrade Modulus as Constant (Fixed) Value

Location	Layer Moduli-ksi							
	AC		Base		Subbase		Subgrade	
	P*	5**	P	5	P	5	P	5
I-95 St. Johns	240	499	150	123	60	60	35	25
I-95 Duval	220	190	150	454	60	60	40	32
I-10 Duval E.B.	70	105	90	63	40	40	45	37
I-10 Duval W.B.	110	352	90	50	40	40	40	28
I-10 East Madison	150	308	110	78	60	60	48	36

* Predicted values

** Case 5 with subbase constant using predicted modulus

In summary, all but two of the pavement sections had significant pavement cracking. No information was available on uniformity or stratigraphy of the foundation soils. Without this information it is impossible to properly evaluate layer moduli by elastic layer analyses. However, estimates of layer moduli can best be obtained using the prediction equations or by using these predictions as seed moduli and by holding the subbase modulus value constant.

CONCLUSIONS

Pavement layer moduli obtained from the developed prediction equations appears to give fairly reliable and realistic values. In general the predicted deflections closely match those measured using the modified sensor configuration. Backcalculated layer moduli using BISDEF 4-layer analyses were not adequate for seven of the eight interstate pavements evaluated in this study because of extremely high and unrealistic subbase moduli. This problem was minimized when the subbase modulus was fixed using the predicted value and allowing BISDEF to solve for the moduli of the other layers.

In general the major deficiency in the analysis of FWD data is the lack of information on stratigraphy of foundation soils. The prediction method does not guarantee accurate results but it does provide a quick means of estimating moduli for evaluation of pavements.

REFERENCES

1. Badu-Tweneboah, K., Manzione, C.W., Ruth, B.E., and Miley, W.G., "Prediction of Flexible Pavement Layer Moduli from Dynaflect and FWD Deflections," STP1026, American Society for Testing and Materials, 1988, pp. 245-277.

2. Roque, R., Romero, P., and Ruth, B.E., "Evaluation of a Dual-Loading Nondestructive Testing System to Better Discriminate Near-Surface Layer Moduli," Transportation Research Record 1355, TRB, 1992, pp. 52-58.

3. Ruth, B.E., Tia, M., and Shen, X., Flexible Pavement Evaluation/Design-Phase II, Final Report, Project No. 450433812, Department of Civil Engineering, University of Florida, 1993, pp. 1-147.

Fifth International Conference on the
BEARING CAPASITY OF
ROADS AND AIRFIELDS

STRUCTURAL MODELING OF FLEXIBLE PAVEMENTS IN THE LVR TEST SECTIONS AT Mn/ROAD PROJECT

NAVNEET GARG[1] AND MARSHALL R. THOMPSON[2]

[1]Post-Doctoral Research Associate, e-mail: n-garg@uiuc.edu
[2]Professor Emeritus, e-mail: mrthomps@uiuc.edu
Department of Civil Engineering, University of Illinois @ Urbana-Champaign
205 North Mathews, Urbana, IL 61801-2352, U.S.A
Phone: (217)333-3930 FAX: (217)333-1924

ABSTRACT

In a mechanistic-empirical (M-E) pavement design procedure, pavement responses (stresses, strains, deflections) are used to predict pavement distresses through the use of transfer functions. FWD (Dynatest Model 8000) equipment was used at Mn/ROAD to measure pavement surface deflection response. The field measured pavement responses are compared to the ILLI-PAVE (finite element program) predicted pavement responses. The material properties used for the ILLI-PAVE runs correspond to the field conditions at the time of FWD testing. The asphalt concrete layer was modeled as constant modulus material in ILLI-PAVE. Backcalculation algorithms were used to estimate asphalt concrete modulus. A relationship was established between the backcalculated asphalt concrete modulus and the pavement temperature (BELL's temperature). The granular base material was modeled as a stress hardening material. Granular base material and subgrade sand properties were determined from the repeated load triaxial tests. The K-θ model was used to estimate modulus values. In the sand subgrade, when the confining stress is higher than the deviator stress, it exhibits descending behavior of resilient modulus with axial strain. Since the K-θ model does not describe the descending behavior of resilient modulus with axial strain, Uzan's model was used for the sand. The arithmetic model was used to characterize the stress softening behavior of cohesive soils. Subgrade "breakpoint" modulus (E_{Ri}) values were estimated using backcalculation algorithms. Estimated pavement responses are fairly accurate for pavement cross-sections where the asphalt concrete surface behaves like a structural layer (asphalt concrete thickness higher than 5-inches). For thinner asphalt concrete surfaces, a less favorable deflection match is obtained, but the predicted deflection basin geometry is comparable to the measured one.

KEY WORDS: flexible pavement response, FWD testing, ILLI-PAVE, deflection basin

INTRODUCTION

In a M-E design procedure, pavement responses (stresses, strains, deflections) are used to predict pavement distresses through the use of transfer functions. At Mn/ROAD (Minnesota Road Research Project), approximately 4500 sensors were embedded in different pavement layers to measure the pavement responses to different loading and climatic conditions. Mn/ROAD is the largest and most technologically advanced pavement research facility in the world. It is located parallel to Interstate 94 (I-94) in Otsego, Minnesota. The 500-feet long pavement sections are designed so that different combinations of materials, layer thicknesses, and design details can be evaluated. Six different dense-graded granular materials were used as base and subbase materials, and two different subgrades (cohesive soil and sand) were used. In aggregate classes CL-3sp and CL-4sp, crushed particles were not allowed. Ten to fifteen percent crushed particles were required for CL-5sp, and 100-percent crushed particles were required for CL-6sp. The pavement cross-section details for conventional flexible pavements and full-depth asphalt concrete pavements in the low-volume road loop are given in Table 1.

TABLE 1. Low-Volume Road Test Sections at Mn/ROAD Project

Cell No.	AC Thickness, inches	Marshall AC Mix Design	Granular Base (GB)	GB Thickness, inches	Subgrade Type
24	3	35 Blows	CL-6sp	4	Sand
25	5	50 Blows	--	--	Sand
26	6	50 Blows	--	--	Cohesive
27	3	35 Blows	CL-6sp	11	Cohesive
28	3	50 Blows	CL-5sp	13	Cohesive
29	5	50 Blows	CL-4sp	10	Cohesive
30	5	75 Blows	CL-3sp	12	Cohesive
31	3	75 Blows	CL-5sp, CL-3sp*	16	Cohesive

** 4-inch CL-5sp Base & 12-inch CL-3sp Subbase*

FWD tests were conducted to measure the surface deflection response of the pavement. In this study, the field measured pavement responses are compared to ILLI-PAVE (finite element program) predicted pavement responses.

FWD TESTING AT Mn/ROAD

The Dynatest Model 8000 FWD is used at Mn/ROAD. The testing pattern for each section included ten test points spaced at 50-feet intervals with four longitudinal offsets from the centerline (-9.5-feet - outer wheel track westbound lane, -6.25-feet - between wheel tracks westbound lane, +6.25-feet - between wheel tracks eastbound lane, +9.5-feet - outer wheel track eastbound lane) tested. Pavement surface deflections under the center of the 12-inch diameter loading plate (D0), 12 inches from the load center (D1), 24 inches from the load center (D2), and 36 inches from the load center (D3) were recorded. FWD test dates were selected to cover a range of pavement climatic conditions.

ILLI-PAVE MODELING OF LVR TEST SECTIONS AT Mn/ROAD

ILLI-PAVE [1] is a finite element computer program for plane strain analysis of elastic solids with stress dependent properties. The pavement is modeled as a three-dimensional pavement section by using a two-dimensional half space of a finite solid of revolution. By symmetry, the solution of the three-dimensional solid may be specified in terms of a plane radial section of rectangular configuration. Loading is specified in terms of the surface contact pressure and radius of loaded area. Only one load can be accommodated. ILLI-PAVE includes stress-strain characterization models that realistically represent the nonlinear stress dependent resilient behavior of granular materials and fine-grained soils. The failure of granular material and cohesive soils is incorporated into the analysis by Mohr-Coulomb failure criteria. The Mohr-Coulomb failure criteria allows for the stress redistribution in the elements so that the granular materials do not achieve unrealistic stress values (i.e. tensile stress at the bottom of the unbound granular layer). The stresses in the granular layer are modified so that the stresses lie within the Mohr-Coulomb failure envelope.

ILLI-PAVE runs were conducted for test sections to compare the predicted pavement responses with the FWD measured pavement responses. The ILLI-PAVE material properties correspond to the field conditions at the time of FWD testing. The asphalt concrete layer is modeled as a constant modulus material in ILLI-PAVE. Backcalculation algorithms were used to estimate asphalt concrete modulus [2]. A relationship was established between the backcalculated asphalt concrete modulus and the pavement temperature (BELL's temperature, described elsewhere [3]). The asphalt concrete modulus was estimated from the following relationship:

$$\text{Log}\, E_{AC} = 3.3804 - 0.04771 * T \qquad \text{[1]}$$

No. of observations = 755, $R^2 = 0.89$, SEE = 0.146

where E_{AC} is asphalt concrete modulus (ksi), and T is temperature (degree Centigrade).

Asphalt concrete moduli were estimated from the FWD test temperature. The granular base material is modeled as a stress hardening material. The K-θ model was used to estimate modulus values. The 'K' and 'n' values, used in ILLI-PAVE analysis, for the K-θ model (for different granular materials used) are listed in Table 2 [4].

TABLE 2. Results from Resilient Modulus Testing on Mn/ROAD Granular Materials

Material	Moisture Content, %	Dry Density, pcf	THETA Model K	n	R^2
CL-3 sp	8.0	128	2012	0.65	0.99
CL-3 sp	6.8	127	2707	0.60	0.99
CL-4 sp	9.4	132	6490	0.20	0.89
CL-4 sp	7.9	130	2410	0.60	0.94
CL-5 sp	6.8	137	3827	0.45	0.88
CL-5 sp	7.7	139	5358	0.32	0.93
CL-6 sp	6.3	134	2583	0.64	0.99
CL-6 sp	5.4	133	2807	0.64	0.99
CL-6 sp	7.3	131	3206	0.62	0.99
CL-6 sp	6.3	139	4597	0.57	0.99

The subgrade soil "breakpoint" modulus (E_{Ri}) was estimated from the FWD data using the following backcalculation algorithms:

Conventional Flexible Pavements [5]:

$$\text{Log } E_{Ri} = 1.51 - 0.19 * D3 + 0.27 * \text{Log}(D3) \quad \text{[2]}$$

$R^2 = 0.99$ SEE = 0.05

Full Depth Asphalt Concrete Pavements [5]:

$$\text{Log } E_{Ri} = 24.7 - 5.41 * D3 + 0.31 * D3^2 \quad \text{[3]}$$

$R^2 = 0.98$ SEE = 0.64

where E_{Ri} is subgrade "Breakpoint" Modulus (ksi), and D3 is FWD deflection at 36-inch offset (mils).

Table 3 summarizes the properties used in the ILLI-PAVE runs. Asphalt concrete moduli were obtained from Equation 1. The coefficient of lateral pressure at rest and Poisson's ratio values for asphalt concrete, granular material, and subgrade soil were obtained from Reference 1.

TABLE 3. Summary of Material Properties for ILLI-PAVE Runs

Property	Asphalt Concrete			Subgrade			
	40 °F	70 °F	100 °F	Stiff	Medium	Soft	Very Soft
Unit Weight, pcf	145	145	145	125	120	115	110
Lateral Pressure							
Coefficient at Rest	0.37	0.67	0.85	0.82	0.82	0.82	0.82
Poisson's Ratio	0.27	0.4	0.46	0.45	0.45	0.45	0.45
Unconfined Compressive Strength, psi				32.8	22.85	12.9	6.21
Deviator Stress, psi							
Upper Limit				32.8	22.85	12.9	6.21
Lower Limit				2.0	2.0	2.0	2.0
K_1, psi/psi				-1.11	-1.11	-1.11	-1.11
K_2, psi/psi				-0.178	-0.178	-0.178	-0.178
Deviator Stress @ Breakpoint, psi				6.2	6.2	6.2	6.2
E_{Ri}, ksi				12.34	7.68	3.02	1.0
E Failure, ksi				7.605	4.716	1.827	1.00
E Constant Modulus, ksi	1475	240	38				
Friction Angle, degrees				0	0	0	0
Cohesion, psi				16.4	11.43	6.45	3.11

The two subgrade types (cohesive soil and sand) in the LVR loop were modeled differently. Two test sections (Cell-24 & Cell-25) have sand subgrades. Cell-24 is a conventional flexible pavement with 3-inch asphalt concrete and 4-inch granular base. Cell-25 is a 5-inch full-depth asphalt concrete section. Sand is a granular material and exhibits stress-hardening behavior (increased modulus with increased stresses). At the 3-foot depth in the sand subgrade, the confining stresses exceed the deviator stresses (Figure 1).

When the confining stress exceeds the deviator stress, granular material exhibits descending behavior (Figure 2) of resilient modulus with axial strain [6]. The K-θ model does not describe the descending behavior of resilient modulus with axial strain because it neglects the effect of shear stress on resilient modulus. Uzan's model considers the effect of shear stress on resilient modulus. Figure 2 shows the comparison in the behavior of subgrade sand predicted by K-θ model and Uzan's model. Uzan's model was recently (April 1997) incorporated into the ILLI-PAVE program. For Cells 24 and 25, the resilient behavior of the top 6-feet of subgrade sand was characterized using Uzan's model. At a depth of 6-feet (72-

inches) in the sand subgrade, the magnitude of shear strain is 80-microstrain and it reduces as the depth increases (Figure 3). Iwasaki et. al. [7] showed that as the magnitude of shear strain reduces (lower than 100-microstrain) G/Gmax (G is shear modulus and Gmax is shear modulus at very low or zero shear strain) approaches 1. Since the magnitude of shear strain

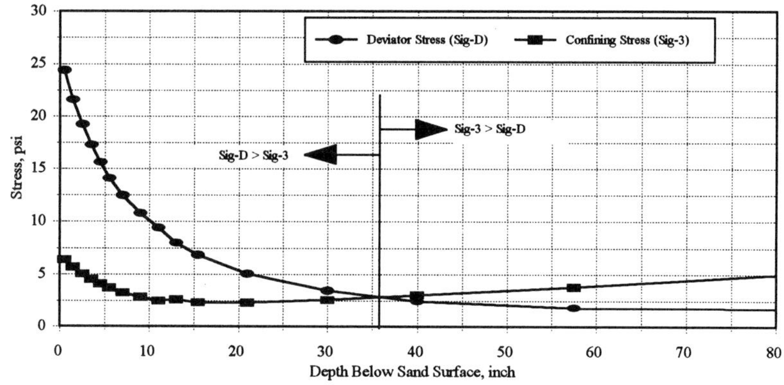

Figure 1. Stresses in the Sand Subgrade with Depth
[Cell-25: 5-inch full-depth asphalt concrete pavement]

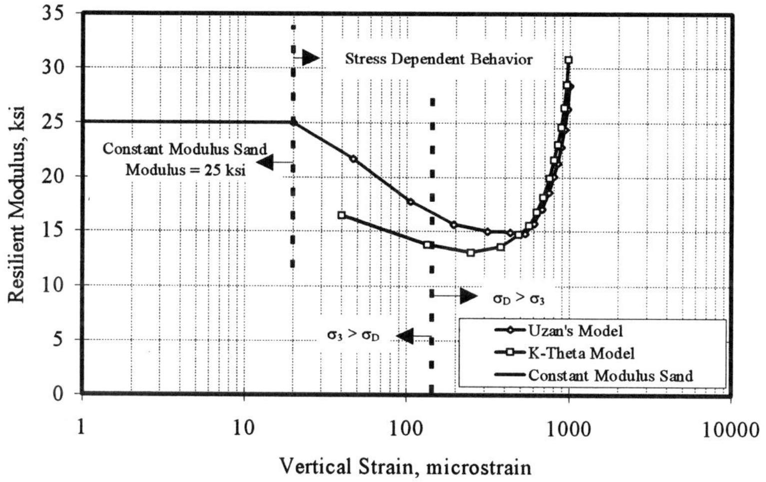

Figure 2. Behavior of Subgrade Sand Predicted by K-Theta Model and Uzan's Model
[Cell-25: 5-inch full-depth asphalt concrete pavement]

is very low (lower than 80-microstrain) as the depth in sand increases, the resilient modulus of sand predicted by Uzan's model at the six-foot depth (resilient modulus @ 6-feet depth = 25-ksi) was assigned to the sand at depths greater than 6-feet. Thus, the subgrade sand below the 6-feet depth was modeled as a constant modulus material with a resilient modulus of 25-ksi.

The remaining test sections have cohesive subgrades. The arithmetic model (Figure 4) was used to characterize the stress softening behavior of cohesive soils. In the arithmetic model, the resilient modulus at the breakpoint in the bilinear curve, E_{Ri}, is a good indicator of a soil's resilient behavior. The slope values, K_1 and K_2, display less variability and influence pavement structural response to a smaller degree than E_{Ri}.

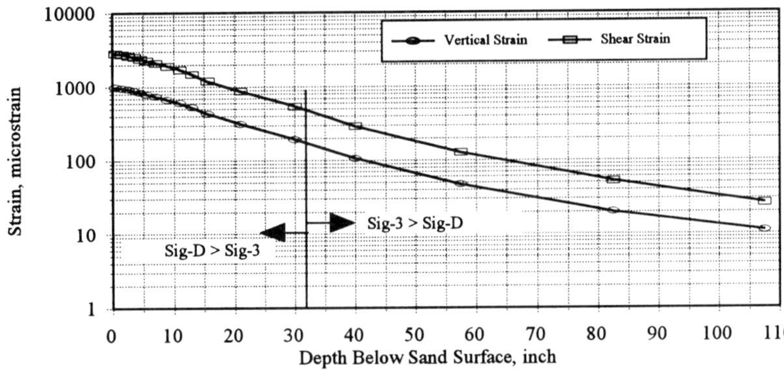

Figure 3. Vertical and Shear Strains in the Sand Subgrade with Depth
[Cell-25: 5-inch full-depth asphalt concrete pavement]

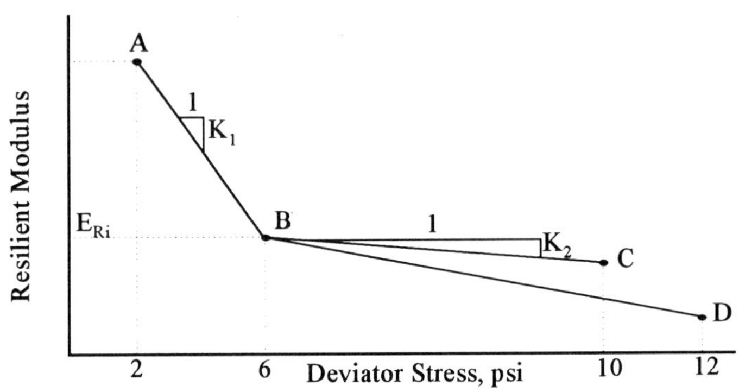

Figure 4. Arithmetic Stress Dependent Model for Fine-Grained Soils

COMPARISON BETWEEN ESTIMATED (from ILLI-PAVE) & MEASURED (from FWD Testing) PAVEMENT RESPONSES

Each FWD deflection value in the following figures is the mean of ten FWD tests performed on a particular test date.

Full-Depth Asphalt Concrete Pavements

Cell-25 is a 5-inch full-depth asphalt concrete pavement over a sand subgrade. The subgrade sand is characterized using Uzan's model. Comparisons between FWD deflections and predicted deflections (from ILLI-PAVE) are shown in Figure 5. The peak center deflections (D0) were overpredicted by about 1-7 mils (2-33 percent). The difference between measured and predicted D1 ranged from 1-2 mils (1-16 percent). ILLI-PAVE predicted deflection D2 were 1-2 mils (4-28 percent) lower than the measured values. The difference between the measured and predicted D3 ranged from 1-2 mils.

Cell-26 is a 6-inch full-depth asphalt concrete pavement over a cohesive subgrade. The arithmetic model is used to characterize the stress-softening behavior of cohesive subgrade.

Comparisons between FWD and ILLI-PAVE deflections are shown in Figure 6.

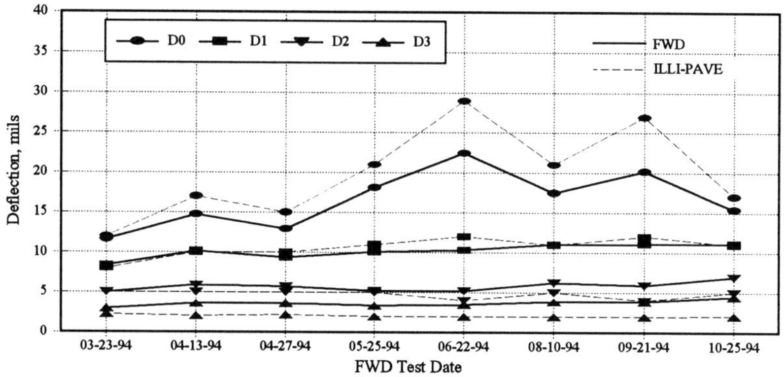

Figure 5. Comparison Between FWD Deflections and ILLI-PAVE Predicted Deflections for Cell-25

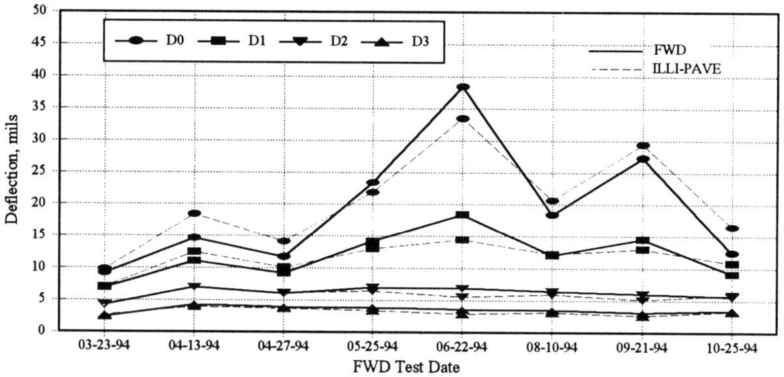

Figure 6. Comparison Between FWD Deflections and ILLI-PAVE Predicted Deflections for Cell-26

There is variability in the field measured deflection values because of the inherent variability in the materials and the variation in the layer thickness. The ILLI-PAVE predicted deflections are within two standard deviation of the field measured values. The predicted values are fairly accurate. Figure 7 shows measured and predicted AREAs. "AREA" is a geometric property of the deflection basin combining all measured deflections in the basin. Combining the different sensor deflections into one number minimizes the effect of a possible sensor malfunctions. AREA is defined as:

$$AREA = 6*[1 + 2*D1/D0 + 2*D2/D0 + D3/D0] \qquad [4]$$

AREA is a function of sensor location and has units of length (inches in this case). Stiffer pavements show larger AREAs.

Conventional Flexible Pavements

Cell-30 has a 5-inch asphalt concrete surface, 12-inch granular base, and a cohesive subgrade. ILLI-PAVE predicted deflections are within two standard deviations of the field measured values. Comparisons between FWD deflections and ILLI-PAVE deflections are shown in Figure 8. The difference between the predicted and measured D0 values is 1-18 percent. In the case of deflection D1, the difference is about 2-7 percent. Deflections D2 differed by 2-10 percent. The differences for deflection D3 were less than 1-mil.

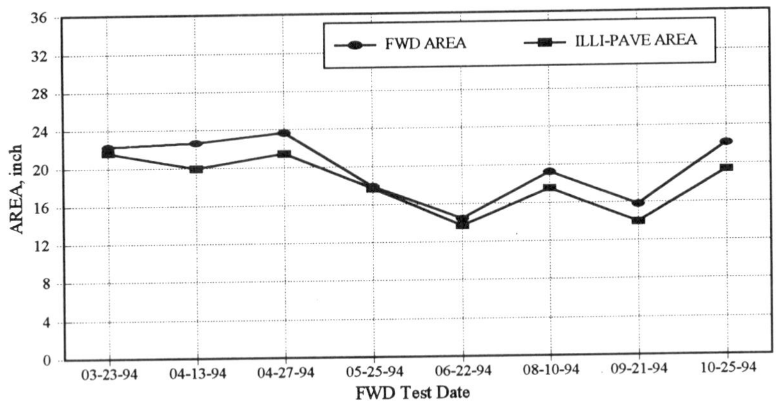

Figure 7. Comparison Between FWD AREA and ILLI-PAVE Predicted AREA for Cell-26

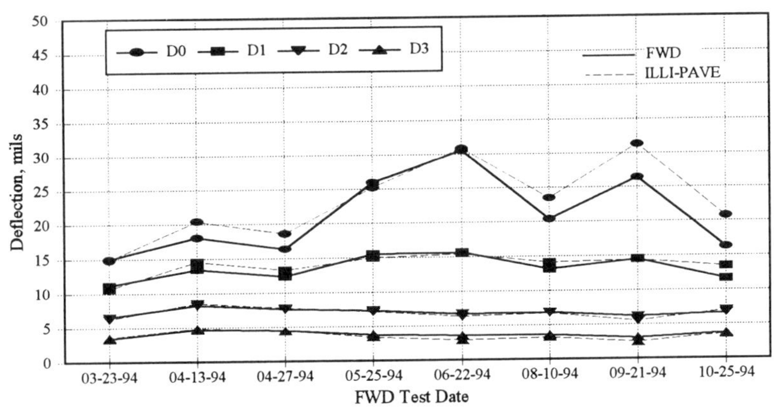

Figure 8. Comparison Between FWD Deflections and ILLI-PAVE Predicted Deflections for Cell-30

Cell-31 has a 3-inch asphalt concrete surface, 4-inch granular base (CL-5sp), 12-inch granular subbase (CL-3sp), and a cohesive subgrade. The comparison between FWD deflections and ILLI-PAVE deflections is shown in Figure 9. Deflection D0 values are 12-29 percent higher than the measured values. A similar pattern is observed for deflection D1. The predicted values were 7-17 higher than the measured values. In the case of deflection D2, the difference between the predicted and measured values ranged from 1-7 percent. The predicted values were within two standard deviations of the field measured values. The difference in predicted and measured deflection D3 values were less than 1-mil. Figure 10 shows the comparison between measured and predicted AREAs.

DISCUSSIONS

An important step in a mechanistic-empirical pavement design procedure is the accurate estimation of critical pavement responses. Strain at the bottom of asphalt concrete layer, vertical stress/strain at the top of the subgrade, and subgrade stress ratio (SSR = repeated

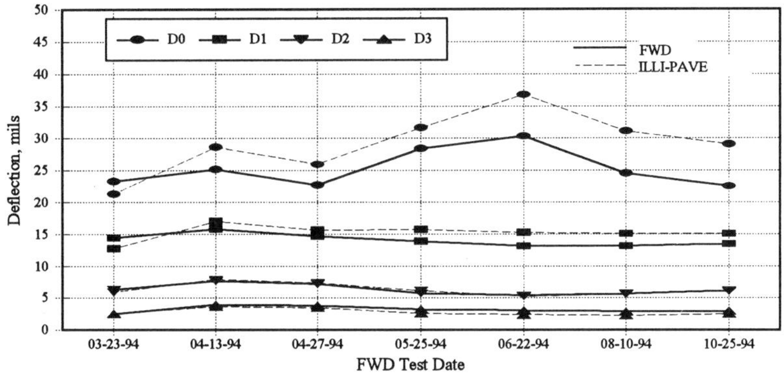

Figure 9. Comparison Between FWD Deflections and ILLI-PAVE Predicted Deflections for Cell-31

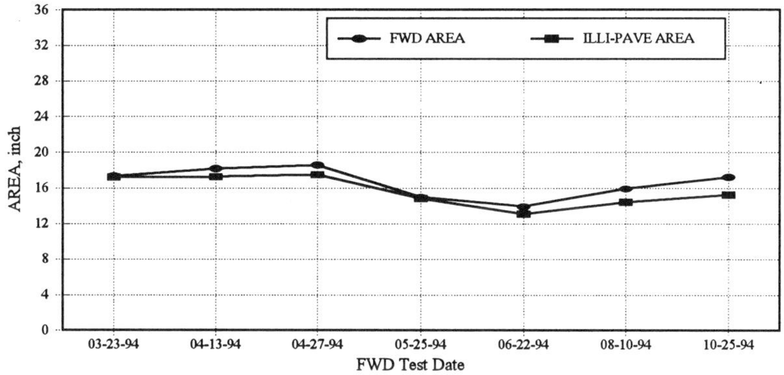

Figure 10. Comparison Between FWD AREA and ILLI-PAVE Predicted AREA for Cell-31

deviator stress/unconfined compressive strength of soil) are typically considered. The strain at the bottom of asphalt concrete layer is related to the fatigue life of the pavement; higher strains are associated with reduced fatigue life. The stress/strain at the top of subgrade and SSR are related to the permanent deformation accumulation in the subgrade. Higher stress/strain and SSR are associated with higher permanent deformation accumulation. Pavement thicknesses should be adequate to limit the strain at the bottom of asphalt concrete surface, and the stress at the top of the subgrade. A stress criteria (based on shear stress and shear strength) should be used to consider granular base rutting.

SUMMARY & CONCLUSIONS

ILLI-PAVE was used to predict Mn/ROAD LVR flexible pavement deflections and the results were compared with the field FWD responses. The analyses showed that more accurate pavement response predictions are achieved for thicker asphalt concrete surfaces (5-inches or higher). The behavior of thin asphalt concrete surfaces (3-inch in the case of Mn/ROAD test sections) is between a surface treatment and a structural layer. Thin asphalt concrete surfaces exhibit a 'membrane' type behavior. This makes the modeling of stress states occurring in the pavement cross-section difficult. Predicting exact deflections in a basin is only one aspect of pavement response. If the geometry (e.g. AREA, curvature, etc.) of the deflection basin can be predicted with a reasonable amount of accuracy, critical pavement responses like strain at the bottom of the asphalt concrete surface can be predicted. This study showed that the finite element program ILLI-PAVE can be used to reliably estimate pavement responses. Pavement response estimates are fairly accurate for pavement cross-sections where the asphalt concrete surface behaves like a structural layer (asphalt concrete thickness higher than 5-inch). For thinner asphalt concrete surfaces, the exact match may not be obtained on the deflections, but the predicted geometry of deflection basin was comparable to the measured one.

ACKNOWLEDGEMENTS / DISCLAIMER

This study was sponsored by the Illinois Department of Transportation (IDOT)-Division of Highways as part of IHR-535 (COOPERATIVE EVALUATION OF Mn/ROAD TEST RESULTS TO ILLINOIS CONDITIONS). IHR-535 is a cooperative effort between the Minnesota Department of Transportation (Mn/DOT), the Illinois Department of Transportation (IDOT), and the Department of Civil Engineering - University of Illinois @ Urbana-Champaign to provide broad utilization of the data obtained from the Minnesota Road Research Project (Mn/ROAD). The cooperation is facilitated through a MEMORANDUM OF UNDERSTANDING between Mn/DOT and IDOT. The contents of this paper reflect the views of the authors who are responsible for the facts and accuracy of the data presented. The contents do not necessarily reflect the official views or policies of the IDOT or the Mn/DOT.

REFERENCES

1. Elliott, R. P., and Thompson, M. R., "Mechanistic Design Concepts for Conventional Flexible Pavements", Civil Engineering Studies, Transportation Engineering Series No. 42, University of Illinois @ Urbana-Champaign, 1985.
2. Garg, N., "Mechanistic-Empirical Analysis of Mn/ROAD Low-Volume Road Test Sections", Ph.D. Thesis, Department of Civil Engineering, University of Illinois @ Urbana-Champaign, June 1997.
3. Stubstad, R. N., Baltzer, S., Lukanen, E. O., and Ertman-Larsen, H. J., "Prediction of AC Mat Temperatures for Routine Load/Deflection Measurements", Proceedings, 4th International Conference on Bearing Capacity of Roads and Airfields, Minneapolis, MN, July, 1994.
4. Garg, N, and Thompson, M. R., "Triaxial Characterization of Mn/ROAD Granular Materials", Transportation Research Record 1577, Transportation Research Board, 1997.
5. Thompson, M. R., "ILLI-PAVE Based NDT Analysis Procedures", ASTM International Symposium on Non-Destructive Testing of Pavement Structures, ASTM Annual Meeting, Baltimore, MD, June 1988, Published in ASTM STP 1026, 1989.
6. Uzan, J., "Characterization of Granular Materials", Transportation Research Record 1022, Transportation Research Board, 1985.
7. Iwasaki, T., Tatsuoka, F., and Takagi, Y., "Shear Moduli of Sand Under Cyclic Torsional Shear Loading", Soils and Foundations, Vol. 18, No. 1, March 1978.

Fifth International Conference on the
BEARING CAPASITY OF
ROADS AND AIRFIELDS

APPLICATION OF FINITE ELEMENT DYNAMIC ANALYSIS FOR IN SITU MATERIAL CHARACTERIZATION OF PAVEMENT SYSTEMS

WAHEED UDDIN[1]

[1]Assistant Professor, Department of Civil Engineering
The University of Mississippi, University, MS 38677, USA
cvuddin@olemiss.edu

ABSTRACT

This paper presents some results of the three-dimensional finite-element computer simulations conducted to study the effects of pavement cracking, voids, falling weight deflectometer (FWD) dynamic loading, truck loading, and thermal gradient on the structural responses of a jointed plain concrete pavement and to analyze deflection data collected on asphalt pavement-subgrade systems in northern Mississippi. The use of three-dimensional finite element dynamic analysis to backcalculate pavement moduli from the measured surface deflection response of the pavement-subgrade model subjected to a standard FWD load is demonstrated. In situ layer moduli for uncracked asphalt and concrete pavements, for the concrete pavement having cracks in both the concrete and base layers and for a cracked pavement with voids beneath the concrete layer, are backcalculated by matching finite-element simulation results to field measurements of FWD deflections. Reasonable good agreement is shown between the backcalculated and laboratory moduli. Effect of viscoelastic material characterization of asphalt layers on pavement deflection response and backcalculated moduli is investigated. These simulation studies show that advanced computational procedures are necessary to accurately analyze and design pavements. These analyses are not possible with static elastic layered analysis and two-dimensional finite element procedures traditionally used for pavements.

Keywords: Pavement, finite element, simulation, falling weight deflectometer, dynamic analysis

BACKGROUND

Objectives During the last decade massive amount of data have been collected from the long-term pavement performance (LTPP) and asphalt research studies in North America and the LTPP pavement evaluation data collection is continuing. However, valid pavement performance models and improved pavement design procedures have not been developed. Without accurate pavement modeling and three-dimensional dynamic analysis, correct pavement responses may not be calculated. Subsequently, this will result in inadequate modeling of pavement performance, design, and nondestructive evaluation. This paper shows that it is imperative to use advanced three-dimensional finite element pavement analysis procedures for correctly simulating uncracked and cracked or deteriorated pavements subjected to dynamic loads produced by moving traffic and nondestructive evaluation equipment. Moduli for uncracked pavement and for pavement having cracks are backcalculated by matching finite-element simulation results to field measurements. Further, three-dimensional simulation models and backcalculated moduli are used to study the deflection response of asphalt pavements.

Pavement nondestructive evaluation using current static analysis procedures Pavement nondestructive evaluation (NDE) is performed by analyzing surface deflection measurements under a known dynamic load. Pavement deflection response can be analyzed using an appropriate structural response analysis model to calculate the in situ Young's modulus for each layer in the pavement-subgrade system (1,2). The use of multilayered linear elastic theory for structural analysis of a pavement-subgrade system subjected to a falling weight deflectometer (FWD) or moving wheel load is based on the assumption that the pavement-subgrade system behaves as a linearly elastic system. Other key assumptions include: static loading and no consideration of loading mode and duration, infinite horizontal extent of pavement layers, and homogenous and isotopic material properties. Each layer is characterized by its Young's modulus and its Poisson's ratio. Assuming a semi-infinite subgrade and infinite lateral boundaries, unique values of surface deflections at specified distances from the load can be theoretically predicted.

The modulus backcalculation procedure involves an iterative application of the multilayered elastic theory. Surface deflections are predicted using assumed values of the Young's modulus and the Poisson's ratio of the pavement layers. Calculated surface deflections are matched with measured deflections until the percentage of error is reduced to an acceptably low value and the final pavement moduli are considered as the values of effective in situ Young's moduli of the pavement layers (2). With the above assumptions, the linearly elastic response of a pavement-subgrade model may be reasonable, in the absence of pavement discontinuities and nonlinearities. The possibility of nonuniqueness of the backcalculated moduli is a serious limitation of the iterative backcalculation procedures because more than one combination of moduli can produce deflections which may match the measured deflections. During the development of the PEDD1 backcalculation program this problem was recognized and the uniqueness of backcalculated moduli was ensured by using predictive equations for "seed" moduli. The "seed" moduli are uniquely related to measured peak FWD force, deflections and radial distances of FWD sensors from the load center, layer thicknesses and types of layer materials (1,2).

On the other hand, the finite-element method allows for the dynamic analysis of pavements and the consideration of finite or infinite dimensions of the physical pavement structure. Two-dimensional finite-element programs have been developed exclusively for flexible pavement analysis, e.g., ILLIPAVE (3). The KENSLABS and ILLISLAB finite element programs have been developed for rigid pavements (4, 5). These programs are capable of performing only static analyses and can not model three-dimensional boundaries of a pavement-subgrade system and can not simulate dynamic loading conditions.

THREE-DIMENSIONAL FINITE ELEMENT SIMULATIONS

The finite-element method enables the evaluation of the state of stress and strain in a continuum by transforming the continuum into an assemblage of finite elements. The elements are interconnected at their common nodes. For example, the finite element codes ABAQUS (6) and LS-DYNA3D (7) are available for comprehensive structural pavement response analysis procedures such as static and dynamic analysis (impulsive, steady-state vibratory forces, and moving wheel loads), a variety of material models (linear elastic as well as nonlinear elastic and viscoelastic material constitutive models), and problems involving thermal gradient conditions and body-to-body contact. Effects due to cracks, voids and material degradation can be analyzed. Mesh size and configuration is an important part of finite-element modeling; precise mesh refinement being necessary in regions of high stress intensity. A pre- and post- processor such as PATRAN (8) is used to visualize the initial model and deformed model after the analysis. Some examples of three-dimensional finite element (3-D FE) dynamic simulations are presented in the following sections.

EVALUATION OF JOINTED CONCRETE PAVEMENT

Field study US Highway 78 near Holly Springs in northern Mississippi, a jointed plain concrete pavement, was built in 1989. The project study area in Marshall county extends 11 miles from State Highway 7 near Holly Springs to the county line near Desoto County. The 254 mm (10 inches) thick and 4.2 m (14 ft) wide Portland cement concrete (PCC) slab has 6.1 m (20 ft) transverse joint spacing. The base consists of 152.4 mm (6 inches) cement treated base (CTB), constructed from in-place sand clay topping material. The transverse joints are constructed using 457 mm (18 inches) long dowel bars at 305 mm (12 inches) spacing.

Figure 1. PCI data summary for the concrete pavement tests sections

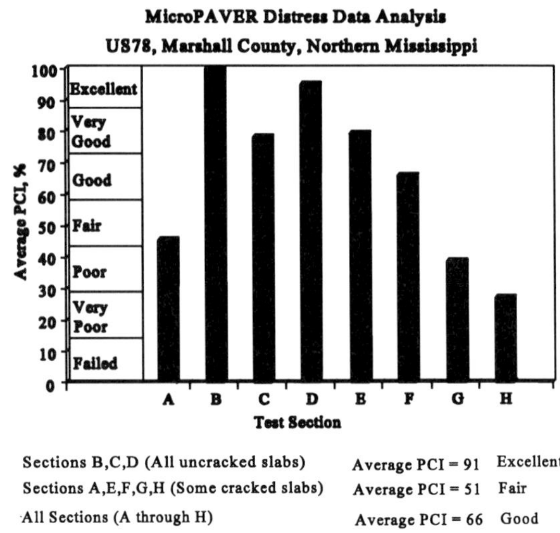

Table 1. A comparison of laboratory and backcalculated moduli for uncracked slab E3

Backcalculation Method	Backcalculated Modulus, MPa (ksi)		
	Concrete *	CTB +	Subgrade
LEEP-WESDEF Static Analysis	47,010 (6,823)	9,977 (1,448)	196 (28.4)
MODULUS-4 Static Analysis	57,669 (8,370)	4,975 (722)	164 (23.8)
PEDD1 Static Analysis	44,785 (6,500)	2,081 (302)	134 (19.4)
ABAQUS Dynamic Analysis	24,074 (3,494)	4,685 (680)	168 (24.4)
Laboratory Modulus * +	24,570 (3,566)	5,854 (850)	-

Visual distress survey, FWD deflection data and other nondestructive test data using ground penetrating radar and thermographic survey were collected from nine test sections on US Highway 78 in May 1994. Figure 1 shows the pavement condition index (PCI) summary of the test sections based on the MicroPaver procedure.

At first, the FWD deflection data on an uncracked pavement section were analyzed using the static layered elastic analysis incorporated in the PEDD1, MODULUS and WESDEF backcalculation programs. The measured deflections were normalized to 40 kN (9,000 lbs) FWD peak load before comparison. The backcalculated moduli are shown in Table 1.

Limiting ranges of moduli were not set in these static analyses. The modulus values backcalculated from the PEDD1 static analysis were more reasonable than the other two backcalculation programs, therefore, these values were used as initial estimates of layer material properties for backcalculation using dynamic analysis. Table 1 also shows the laboratory modulus values based on laboratory compressive strength tests on concrete and CTB cores extracted from uncracked slab E3.

3-D FE Pavement models The US Highway 78 pavement was modeled as a three-layered linear elastic system that consists of the three layers and subgrade (9). The PATRAN code was used to develop these models for ABAQUS analysis. The models were developed for FWD loading respectively in the mid-slab position and the approach and leave slab positions of used for the transverse joint testing. A refined mesh was developed for the vicinity of the loaded area and the transverse joint. Transverse joints with dowel bars were modeled using gap and beam elements. Further, a model was developed for a cracked section with voids beneath the concrete layer. Mesh size and configuration is an important part of finite-element modeling, precise mesh refinement being necessary in regions of high stress intensity. A refined mesh was developed for the vicinity of the loaded area and the transverse joint. Falling weight deflectometer loading was used in the simulation study. The center of the load was located at a distance of 1. 5m (5 feet) from the edge of the outside shoulder and 3m (10 feet) from the transverse joint. The model has a subgrade depth of 12m (40 feet). The dimensions and boundary conditions for the three-dimensional finite-element pavement models were established earlier and reported in detail by Uddin et al, (9). The dowel bars in the transverse joint were modeled using beam elements. Gap elements were used to model the contact between the two faces of the crack in the concrete slab at the transverse joint (9).

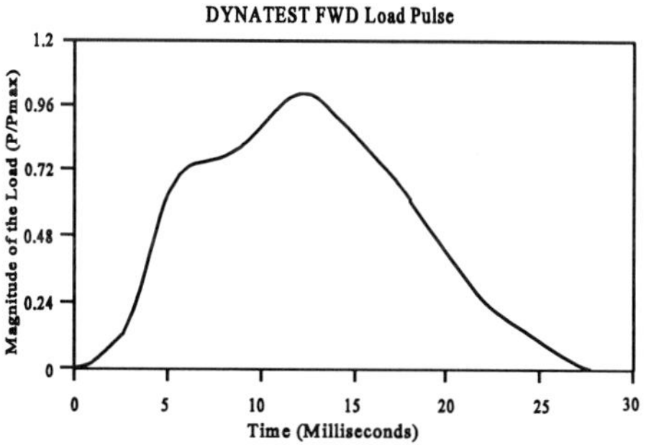

Figure 2. Dynatest FWD pulse

Moduli backcalculated from dynamic analysis Some results of three-dimensional finite element analysis were reported earlier (9). The FWD load time history based on a typical FWD pulse of 33 msec duration, measured on the US Highway 78 test pavement, was used to simulate the FWD impact load. Figure 2 shows the measured FWD pulse. The peak FWD load was taken as 9000 lbs (40 kN) distributed on an area of 109.4 sq. inch (706 sq.

cm). The computed dynamic deflections were compared with the FWD measured deflections and modulus values were adjusted until a close agreement was achieved between computed and measured dynamic deflections. The computed and measured deflections for the uncracked test slab E3 are shown in Figure 3. The backcalculated modulus values for the uncracked concrete, the uncracked CTB base, and the subgrade are listed in Figure 3. These backcalculated modulus values of concrete and CTB layers from the dynamic analysis are in reasonable agreement with the laboratory moduli shown in Table 1.

Figure 3. Moduli of uncracked pavement and measured versus computed deflections

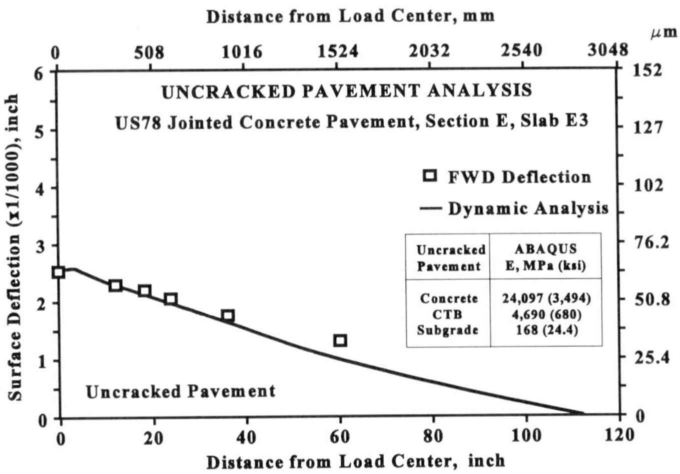

Modulus backcalculation for cracked pavements Three-dimensional finite-element models of cracked and deteriorated pavements were made to conduct dynamic analyses. Cracks and voids in the pavement were modeled using gap elements. A longitudinal crack in slab F3 initiated from the transverse joint at a distance of 1.5 m (5 feet) from the pavement edge. A core extracted from this test section indicated cracking in the concrete layer only (9). This distress was simulated in the finite-element model and the calculated dynamic deflections were substantially higher than the corresponding deflection calculated for the uncracked pavement. After a few iterations, a lower concrete modulus value yielded dynamic deflections which agree reasonably well with the measured FWD deflections. The backcalculated modulus values for the cracked concrete was 12,759 Mpa (1,850 ksi). Using the FWD data, distress data, and core data collected in this study, slab Gl, in test section G, was found to have cracking in concrete and cement treated base layers. The reduced moduli of concrete and CTB (2,069 Mpa or 300 ksi) were successfully used to match the predicted and measured FWD deflections for slab G1.

Dynamic response of pavement with cracks and voids Voids beneath some slabs were observed at some pavement sections of US Highway 78 by the thermographic equipment, as mentioned before. High severity longitudinal crack was recorded on Slab H2 in section H. This slab also showed a possible void underneath the surface in thermograph image. This was verified by a core taken from this slab after slab grouting and undersealing work. The core extracted from this test section indicates cracking in both concrete and cement treated base layers, as well as a thin layer of grout material. The longitudinal crack through the concrete and CTB layers was simulated in the finite element model of this cracked slab H2. The depth of the simulated void is assumed as 5.1 mm (0.2 inch). The modulus values of both cracked layers (concrete and CTB) and the unchanged subgrade modulus were used in the preliminary iterations. After further reduction of the subgrade modulus the computed dynamic deflections agreed reasonably well with the measured FWD deflections, as shown in Figure 4. The modulus values from the final iteration are also listed in Figure 4. These results are indicative of the importance of using dynamic finite-element analysis to backcalculate appropriate values of in situ effective pavement modulus for cracked layers.

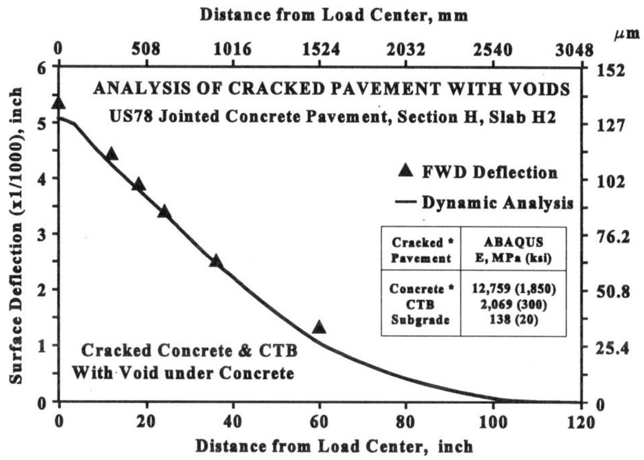

Figure 4. Measured versus computed FWD deflections for slab H2 with cracked concrete, cracked CTB and voids

MODULUS BACKCALCULATION FOR ASPHALT PAVEMENTS

The PEDD1 backcalculated moduli for a good asphalt pavement section of the Lancaster Drive site in Oregan compared reasonably well with the laboratory moduli and the measured FWD deflections matched well with the deflections computed from the ABAQUS dynamic analysis using these moduli. These results and some other results of cracked asphalt pavements are described in detail by Uddin and Pan (10).

As a part of on-going asphalt pavement performance study for the Mississippi Department of Transportation, FWD deflection data were collected during 1995-97 on selected highway sections. The deflection data were measured at a peak load of around 4,083 kg (9,000 lb) in the outer wheelpath, about 1 m (3 ft) from the pavement edge. The deflection data were collected in late fall on MS Highway 6 East on a SHRP-LTPP section in northern Mississippi. The analysis of this data was performed at in situ temperature condition without making any correction for temperature. The pavement was in good condition and no cracks and rutting were observed in the test area. Table 2 presents the pavement structure and in situ effective elastic moduli backcalculated using the PEDD1 program. These modulus values were used to compute surface deflections of the 3-D FE model of the pavement subjected to the FWD dynamic load pulse. The LS-DYNA3D code on Windows95 computer station was used for these analyses. Figure 5 shows the 3-D FE model of this pavement.

Table 2. Layers and backcalculated moduli for the asphalt pavement of MS 6 East

Layer	Thickness mm (inches)	PEDD1 Backcalculated Modulus MPa (ksi)
AC Surface (Asphalt)	139.7 (5.5)	3,259 (473)
Asphalt Base	127.0 (5.0)	4,134 (600)
Cement Treated Subbase	152.4 (6.0)	395 (57)
Subgrade	Semi-infinite	300 (43.6)

Figure 5. 3D-FE Model of MS 6 East

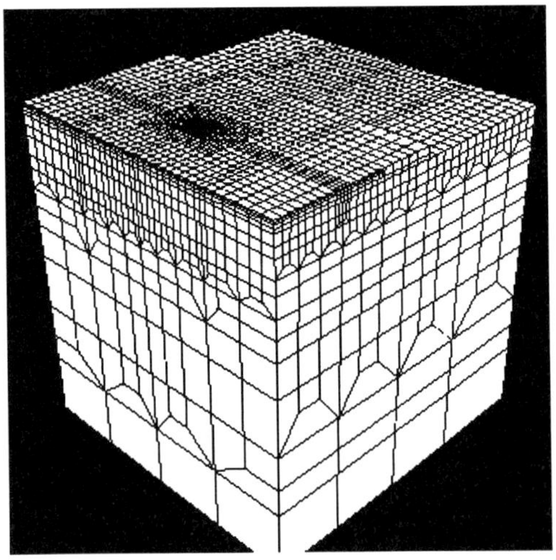

For this good condition asphalt pavement the dynamic deflections computed using the PEDD1 backcalculated moduli match reasonably well with the measured FWD deflections. Figure 6 shows the peak measured FWD deflection and computed deflection-time history under the load center assuming linear elastic material properties. This analysis indicates that the elastic moduli backcalculated by the PEDD1 program are the best estimate of the in situ moduli based on 3D-FE dynamic analysis for this good condition pavement, similar to the findings of the 3-D FE dynamic analysis conducted for the asphalt pavement section of Lancaster Drive. The FWD data of MS Highway 6 East was further analyzed using linear viscoelastic material parameters for Material type 6 from the LS-DYNA3D library of material models for asphalt layers. The peak deflection-time history from the viscoelastic analysis is plotted in Figure 6 to compare with the peak measured FWD deflection and the deflection-time history from linear elastic analysis. The maximum deflection is 15 percent larger than the measured FWD deflection and the corresponding calculated deflection assuming a linear elastic model. After increasing the bulk modulus by 16 percent, a better match of computed and measured FWD deflections was achieved. This corresponds to the instantaneous Young's modulus of 3,780 MPa (549 ksi). The results are preliminary, however, these results demonstrate that viscoelastic material modeling and analysis of asphalt pavement using three-dimensional finite element pavement models subjected to dynamic load pulses are within the practical limits of easily available computing stations.

Figure 6. Matching of measured FWD deflections and deflections computed from 3D-FE analysis

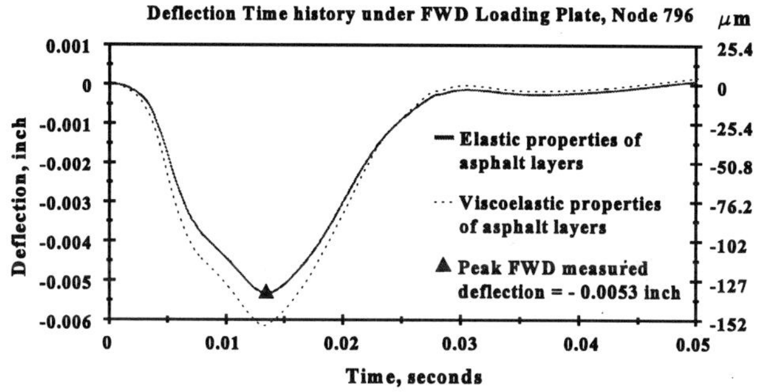

SUMMARY AND CONCLUSIONS

Structural response analyses of selected concrete and asphalt pavements have been conducted using static analysis and three-dimensional finite element dynamic simulations. For jointed concrete pavements, lower effective in situ moduli were backcalculated compared to the moduli backcalculated from the static analysis. However, for good condition asphalt pavements the effective in situ moduli backcalculated from the PEDD1 static analysis provided good match of measured FWD deflections with the deflections computed by the three-dimensional finite element dynamic analysis. This study demonstrates the extensive usefulness of three-dimensional finite element dynamic analysis to simulate the effects of pavement geometry, cracking, viscoelastic material moduli and FWD load-time history which is not possible with traditional layered elastic analysis, as well as other two-dimensional finite element programs which allow only static analysis. Work is currently underway to analyze FWD deflection data from polymer-asphalt pavement sections and evaluate the FWD-pavement contact simulation models.

The research reported herein was supported by the Mississippi Department of Transportation. The LS-DYNA3D finite element simulations were conducted with the assistance of Dhafer Marzougui of the National Crash Analysis Center of the George Washington University. The author acknowledges the efforts of other contributors including A.B. Crawley and R.M. Hackett. The contents of this paper reflect the views of the author who is responsible for the facts, findings and data reported herein.

REFERENCES

1. UDDIN, W., MEYER, A.H., and HUDSON, W.R. Rigid Bottom Considerations For Nondestructive Evaluation of Pavements. *In Transportation Research Record 1070*, TRB, National Research Council, Washington, D.C., 1986, pp. 21-29.
2. UDDIN, W. A Nondestructive Evaluation and Rehabilitation Design Methodology for Asphalt and Concrete Pavements. Proceedings, Second International Conference on Road and Airfield Pavement Technology, Singapore, 1995, Vol 2, pp. 610-618.
3. ELLIOTT, R.P. and THOMPSON, M.R. *Mechanistic Design Concepts for Conventional Flexible Pavements*. Transportation Engineering Series No. 42, University of Illinois, Urbana, 1985.
4. HUANG, Y.H. Pavement Analysis and Design. Prentice-Hall, Inc. 1993.
5. TABATABIE, A.M. and BARENBERG, E.J. Finite Element Analysis of Jointed or Cracked Concrete Pavements. In *Transportation Research Record 671,* TRB, National Research Council, Washington, D.C., 1978.
6. ABAQUS, Theory Manual, Version 5.3. Hibbitt, Karlsson, & Sorenson, Inc., 1993.
7. LS-DYNA3D, A Nonlinear, Explicit, Three-Dimensional Finite Element Code for Solid and Structural Mechanics. User Manual, Version 932, Livermore Software Technology Corporation, May 1995.
8. PATRAN User's Manual. PDA Engineering, PATRAN Division, 1993.
9. UDDIN, W. HACKETT, R.M., JOSEPH, A., PAN, Z., and CRAWLEY, A.B. Three-Dimensional Finite-Element Analysis of Jointed Concrete Pavement Having Discontinuities. In *Transportation Research Record 1482,* Presented at the 1995 Annual Meeting of the Transportation Research Board, National Research Council, Washington, D.C., 1995.
10. UDDIN, W. and PAN, Z. Finite-Element Analysis of Flexible Pavements with Discontinuities," *Proceedings, ASCE Transportation Congress*, American Society of Civil Engineers, September 1995, Volume 1, pp. 410-423.

Fifth International Conference on the
BEARING CAPASITY OF
ROADS AND AIRFIELDS

PAVEMENT ASSESSMENT INCLUDING BONDING CONDITION: CASE STUDIES

BACHAR AL HAKIM[1], ROBERT ARMITAGE[2] AND NICHOLAS THOM[3]

[1]Pavement Engineer, [2]Director, Scott Wilson Pavement Engineering, Nottingham, NG7 2QP, UNITED KINGDOM
[3]Lecturer, Nottingham University, Nottingham NG7 2RD, UNITED KINGDOM

ABSTRACT

The estimation of pavement properties with sufficient accuracy is an important part of pavement assessment. Therefore a new back-analysis method is suggested to assess the bonding condition between bituminous layers in addition to the layer stiffnesses, from Falling Weight Deflectometer (FWD) test results.

Case studies are detailed describing the evaluation of three recently constructed flexible pavements using the new back-analysis process. The results showed good quality bituminous materials, poor bonding between bituminous layers, weak sub-base and good subgrade for most pavements.

Good correlation was found between the weak sub-base and the poor bonding condition for the investigated pavements. The latter is believed to be the main reason for the de-bonding failure since inadequate support was provided to the lower bituminous layer during construction.

Keywords: Pavement assessment, FWD, Back-analysis, Bonding condition.

INTRODUCTION

A cost effective pavement maintenance and rehabilitation process requires an accurate knowledge of existing insitu properties. Therefore non-destructive testing, such as the Falling Weight Deflectometer (FWD), has become widely used for pavement structural evaluation.

The FWD applies an impact load to the pavement surface, in order to simulate the action of moving traffic, and the pavement response in terms of deflections is measured at seven radial locations. Optimising the best set of pavement properties to fit the measured deflections are then carried out in a process known as back-analysis. The back-analysed properties are utilised to calculate critical pavement strains under a standard 80kN axle load. Finally the pavement life is predicted, employing appropriate failure criteria.

The pavement structure is usually modelled as a layered linear elastic system, where each layer is described by its stiffness and Poisson's ratio. The pavement construction type and the layer thicknesses are determined by a coring survey, whereas an appropriate Poisson's ratio is assigned for each material by experience. Conventionally, only the layer stiffnesses are back-analysed, assuming full bonding exists between different layers.

However weak bonding between bituminous layers may develop during construction due to poor quality control, lack of tack coat, contamination of the lower layer or laying in cold weather (1). Furthermore a poorly compacted sub-base may cause bituminous layer debonding, since sufficient support is not be provided during construction. Therefore, a new back-analysis method has been developed (2), to predict both the pavement layer stiffnesses and the bonding condition between bituminous layers, from FWD test results.

BONDING CONDITION BETWEEN PAVEMENT LAYERS

A numerical solution of a multilayer system requires knowledge of the boundary conditions between layers in order to predict the structural response to surface loading. These conditions can be either full bond, with the same shear stresses on both sides of the interface, or complete de-bonding, where no shear stress will transfer between the layers. A general fundamental model for the interface, assuming that the shear stress is a function of the difference in horizontal displacements of the layers above and below the interface, can be used, i.e.

$$\tau = K_s (u_2 - u_1)$$

where,

τ, shear stress,
K_s, horizontal shear reaction modulus at the interface,
u_2, u_1, horizontal displacements on both sides of the interface.

This equation represents Goodman's constitutive law to describe the interface behaviour (3).

Numerical analysis to identify the range of bonding condition (K_s) has been carried out (2). The results showed that K_s varies from 10 MN/m^3 (complete de-bonding) to 10^5 MN/m^3 (full bonding), and beyond these two limits there is no significant change in pavement response such as strains and displacements.

Furthermore, the state of adhesion between pavement layers can be divided into three parts as follows (2):

i)	$K_s \leq 100$ MN/m^3	(de-bonding),
ii)	$100 \leq K_s < 10000$ MN/m^3	(intermediate case),
iii)	$K_s \geq 10000$ MN/m^3	(approximately full bonding).

THE NEW BACK-ANALYSIS PROCEDURE

A brief description of the new back-analysis process is presented in this paper; however the details of its development and validation can be found elsewhere (2). Simple static analysis for the pavement as a multilayer linear elastic system is adopted. The back-analysis process is performed to calculate the shear reaction modulus (K_s) between bituminous layers as a pavement parameter, in addition to the layer stiffnesses (E) from the measured deflections.

The shape of the deflection bowl is known to be influenced by different layer properties, for example the upper pavement parameters affect the deflections close to the load centre, whereas the outer deflections are mainly influenced by the subgrade (4).

Firstly a deflection database is built for a combination of layer stiffnesses and bonding conditions for each pavement construction type. A search is then performed for the best equations for the pavement parameters as dependant variables, with the deflections as independent variables. Several measures of the shape of the deflection bowls are used as variables in the equations. Finally, the measured deflections at each pavement location are input into the equations to predict the pavement parameters.

THE INVESTIGATED PAVEMENTS

Pavements from three newly constructed road sections were investigated. The FWD survey was performed at 10m spacing, a nominal contact pressure of 1MN/m^2 was applied and the deflections were measured at 0, 300, 600, 900, 1200, 1500 and 2100mm from the FWD load centre. The pavement temperatures at 100mm depth below the surface were recorded during the testing. Average pavement temperatures of 6, 10 and 12°C were found for roads A, B and C respectively.

Table 1 summarises the pavement layer thicknesses as extracted from the construction drawings. Road A includes three different sections within Lane 1, whereas roads B and C consist of one section where both Lanes 1 and 2 were tested. Bituminous layer de-bonding was suspected since the conventional back-analysis procedure produced low bituminous stiffnesses (when combining the two layers) for these new materials.

Table 1: Summary of Pavement Layer Thicknesses

Road	Section	Layer Thickness (mm)		
		Upper Bituminous	Lower Bituminous	Sub-base and Capping
A	1, 2 and 3	110	190	490
B	Lanes 1 and 2	110	130	625
C	Lanes 1 and 2	110	155	535

All the test points were included in the back-analysis procedure to determine the layer stiffnesses and the bonding condition between the bituminous layers. A four-layer pavement structure was used to model the two separate bituminous layers, the unbound layer and the subgrade. The results of the back-analysis are presented in Figure 1 for Road A, Figures 2 and 3 for Lanes 1 and 2 of Road B respectively, and in Figures 4 and 5 for Lanes 1 and 2 of Road C respectively. A summary of the 15 and 50 percentile layer stiffnesses and bonding conditions at the site temperature is presented in Tables 2 and 3.

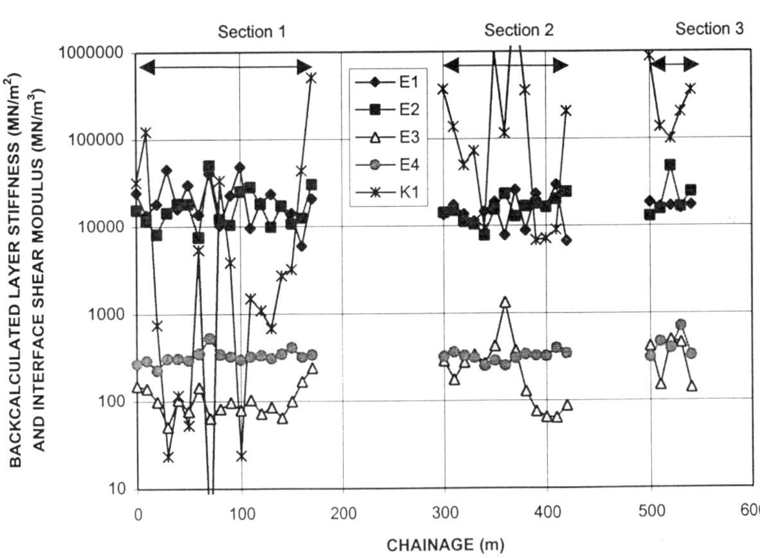

Figure 1: Profiles of the back-analyzed Pavement Parameters for Road A

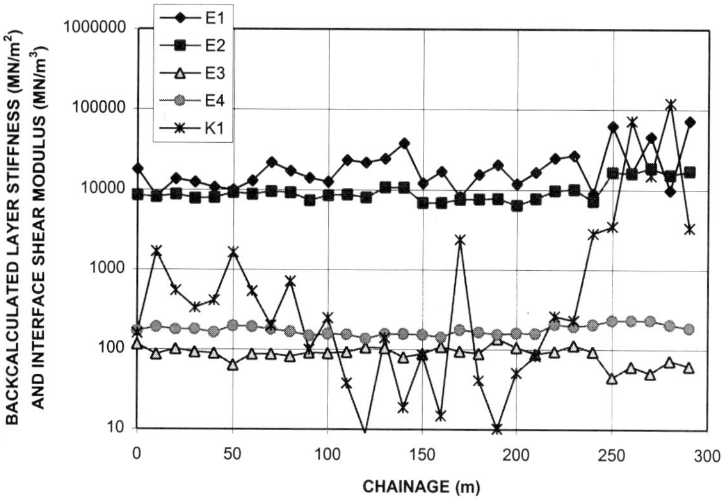

Figure 2: Profiles of the Back-analyzed Pavement Parameters for Lane 1 of Road B

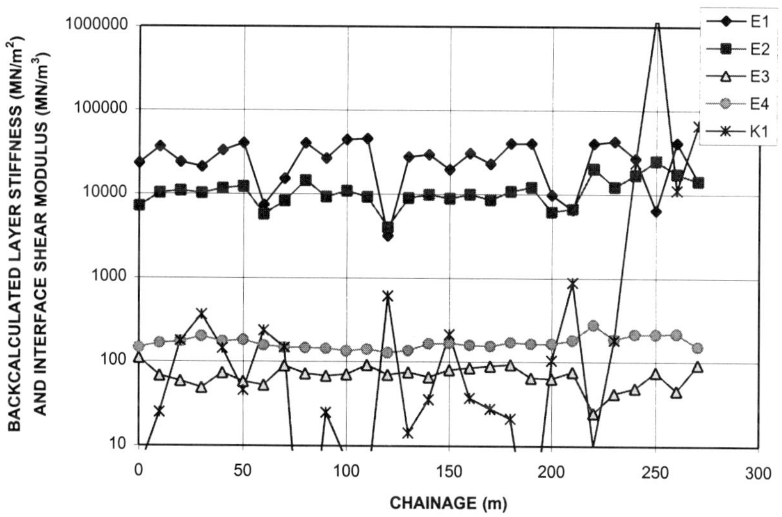

Figure 3: Profiles of the Back-analyzed Pavement Parameters for Lane 2 of Road B

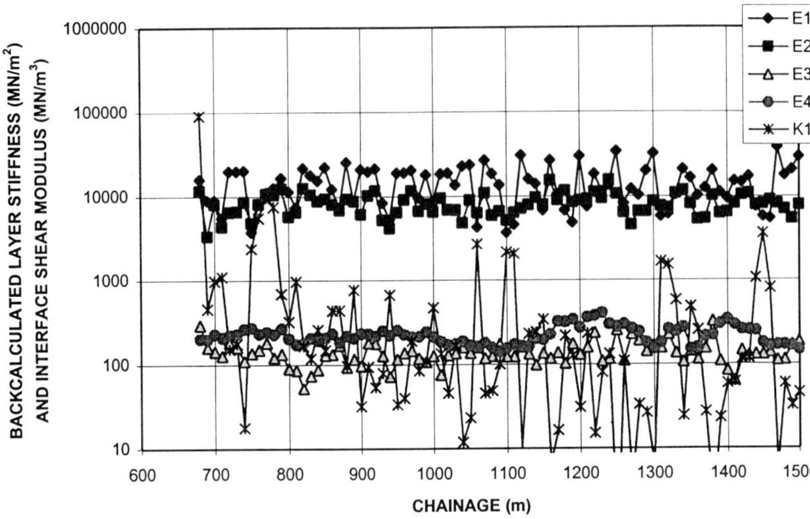

Figure 4: Profiles of the Back-analyzed Pavement Parameters for Lane 1 of Road C

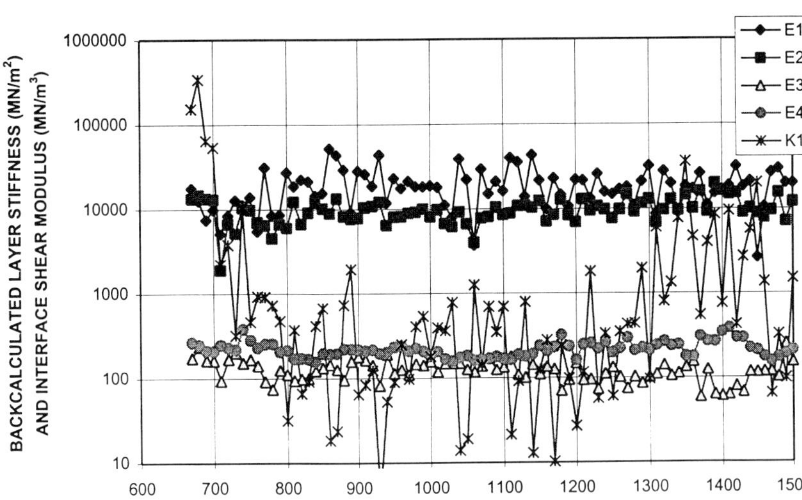

Figure 5: Profiles of the Back-analyzed Pavement Parameters for Lane 2 of Road C

Table 2: Summary of the 15 Percentile Back-analysed Pavement Properties

Road	Section	K_1 (MN/m^3)	E_1 (MN/m^2)	E_2 (MN/m^2)	E_3 (MN/m^2)	E_4 (MN/m^2)
A	1	40	12000	10200	70	290
A	2	7000	8600	11000	75	280
A	3	80000	16500	14600	145	325
B	Lane 1	40	10300	7500	70	155
B	Lane 2	10	10000	7200	50	144
C	Lane 1	30	7000	6000	95	170
C	Lane 2	170	10000	6800	90	175

Table 3: Summary of the 50 Percentile Back-analysed Pavement Properties

Road	Section	K_1 (MN/m^3)	E_1 (MN/m^2)	E_2 (MN/m^2)	E_3 (MN/m^2)	E_4 (MN/m^2)
A	1	2000	18500	15000	100	320
A	2	100000	14700	15500	270	325
A	3	100000	17000	16600	290	400
B	Lane 1	250	16400	8700	90	175
B	Lane 2	75	20000	10300	70	165
C	Lane 1	130	14000	8000	130	220
C	Lane 2	200	18000	10000	120	215

Bonding condition between bituminous layers. The shear reaction modulus (K_1) results presented in Tables 2 and 3 indicate poor bonding between the bituminous layers in section 1 and good bonding in Sections 2 and 3 of Road A. However full de-bonding was found in both lanes of Roads B and C, with the 15 percentile interface shear modulus below 100 MN/m^3. The poor bonding results were confirmed by a coring survey in Section 1 of Road A, where the two bituminous layers separated during coring.

Bituminous materials. The stiffness results shown in Tables 2 and 3 (E_1 and E_2) indicate adequate quality bituminous materials for all roads taking account of test temperature. This was expected for the newly constructed pavement sections.

Unbound layer. Excepting Section 3 of Road A, the sub-base layer stiffnesses (E_3), based on the 15 percentile values, are lower than 100 MN/m^2. A value between 100 and 200 MN/m^2 is typical for a good granular sub-base. It is interesting to notice the good correlation between the sub-base stiffness and the bonding condition, where the weaker the sub-base stiffness the poorer the bituminous layer bonding.

Subgrade. A very good subgrade condition was found in all roads with stiffness values (E_4) above 100 MN/m^2.

Pavement life assessment. The bituminous stiffnesses were adjusted to the appropriate design temperature and the 15 percentile back-analyzed pavement parameters were used to calculate;

i) the maximum tensile strain at the bottom of both bituminous layers (since in the case of de-bonding these two layers will act independently),

ii) the maximum compressive strain at the top of the subgrade.

The resultant strains were then input into pavement performance models to predict the life based on the failure criteria, i.e. fatigue cracking of the bituminous materials and permanent deformation of the subgrade. Table 4 shows the critical strains and the pavement residual lives for the different sections. It is not possible to ascertain whether the standard performance models are applicable to a de-bonded pavement, however this approach enables a relative comparison between sections.

Fatigue cracking is the predicted failure mechanism for the above pavements, where the maximum tensile strain is located at the bottom of the lower bituminous layer. The pavement residual lives presented in Table 4 are calculated based on the actual bonding condition predicted from the back-analysis. The adhesion properties may improve under the effect of future traffic loading which would result in higher lives.

Table 4: Critical Pavement Strains and the Residual lives

Road	Section	ε_{t1}* Strain x 10^{-6}	ε_{t2}** Strain x 10^{-6}	ε_z+ Strain x 10^{-6}	Life msa++
A	1	50.3	58.8	41.6	27.2
A	2	23.5	46.5	31.6	66.2
A	3	5.2	26.0	25.8	>100
B	Lane 1	83.1	84.0	88.7	12.9
B	Lane 2	90.6	92.9	88.1	8.5
C	Lane 1	84.0	87.5	95.6	10.9
C	Lane 2	69.8	76.1	86.5	19.4

Note : * Tensile strain at the bottom of the upper bituminous layer.
** Tensile strain at the bottom of the lower bituminous layer.
\+ Compressive strain at the top of subgrade.
++ Millions of standard 80 kN axles.

Where bonding is poor, the strain at the bottom of the upper layer is slightly less than at the bottom of the lower layer. Assuming the crack propagation time could be reduced for the thinner upper layer, the actual failure mode could involve cracking in this layer only. In order to determine whether this is the case, controlled trials with 'fixed' levels of bonding would need to be performed.

COMPARISON WITH THE CONVENTIONAL ASSESSMENT METHOD

The two bituminous layers were then combined in the back-analysis process for the above pavement assuming full bonding between the various layers, as commonly found in the conventional methods. Low bituminous material stiffnesses were found to compensate for the layer de-bonding, as shown in Table 5. The stiffness values for the combined bituminous layers are apparent results, which include the effect of bonding conditions. The remaining back-analysed stiffnesses (for sub-base and subgrade) are slightly different from those presented in Table 2 due to the analysis process.

Table 5: Summary of the 15 Percentile Back-analysed Pavement Stiffnesses at site Temperature Using the Conventional Method

Road	Section	E_1 (Combined bituminous layers) (MN/m^2)	E_2 Sub-base (MN/m^2)	E_3 Subgrade (MN/m^2)
A	1	7500	50	240
A	2	10000	65	250
A	3	14500	230	320
B	Lane 1	4500	50	190
B	Lane 2	4300	50	200
C	Lane 1	4000	60	230
C	Lane 2	5000	80	250

CONCLUSIONS AND RECOMMENDATIONS

Assessment of the bonding condition between bituminous layers in FWD back-analysis is suggested, in order to predict pavement life based on more accurate data. This may help in understanding the failure mechanism on pavements where de-bonding is suspected.

Three newly constructed pavements were investigated using a new back-analysis method. The results show good quality bituminous layers in all sections, poor bonding in two of the tested pavements, weak unbound sub-base materials and good subgrade.

A weak sub-base layer may be the main reason for the de-bonding failure, since good correlation was found between these properties for the investigated pavements.

A weak sub-base layer may be the main reason for the de-bonding failure, since good correlation was found between these properties for the investigated pavements.

Since the variation of bond between bituminous layers with time and traffic loading is not well established, pavement re-testing in the future, or monitoring of controlled trials, should be performed in order to determine the failure mechanisms that occur in such pavements.

REFERENCES

1. Lepert, P. H. Poilane, J.P. and Bats-Villard, M. Evaluation of Various Field Measurements Techniques for the Assessment of Pavement Interface Condition. Proceedings of the 7th Int. Conference on Asphalt Pavements, Vol. 3, Nottingham, 1992, pp224-237.

2. Al Hakim, B. An Improved Backcalculation Method to Predict Flexible Pavement Layer Moduli and Bonding Condition Between Wearing Course and Base Course. PhD Thesis, Liverpool John Moores University, 1997.

3. Goodman, R. E. Taylor, R. L. and Brekke, T. L. A Model for the Mechanics of Jointed Rocks. J. of Soil Mechanics and Foundation Division, ASCE, Vol. 94, No. SM3, 1968, pp 637-659.

4. Brown, S. F. Tam, W.S. and Brunton, J.M. Structural Evaluation and Overlay Design: Analysis and Implementation. Proc. of the 6th Int. Conf. on the Struct. Design of Asphalt Pavements, Vol. 1, Ann Arbor, 1987, pp 1013-1028.

Fifth International Conference on the
BEARING CAPASITY OF
ROADS AND AIRFIELDS

JOINTED CONCRETE PAVEMENT BACKANALYSIS USING A SIMPLE REGRESSION APPROACH

by

Teruhiko MARUYAMA
Professor, Department of Civil and Environment Engineering
Nagaoka University of Technology,
Fax: (81)-258-47-9613 E-mail: terum@nagaokaut.ac.jp
Nagaoka, Niigata, 940-2188, Japan

Boming TANG
Associate Professor, Department of Civil and Environment Engineering
Nagaoka University of Technology,
Fax: (81)-258-47-9626 E-mail: tang@vos.nagaokaut.ac.jp
Nagaoka, Niigata, 940-2188, Japan

Kazunori MANABE
Research Engineer, Toa Road Industrial Corporation
Fax: (81)-45-251-4213
Minami-ku, Yokohama, 232, Japan

Hiroji KOSEKI
Research Engineer, Research Lab. of Obayashi Road Corporation
Fax: (81)-48-866-6564 E-mail: koseki@obayashi-roda.co.jp
Urawa-Shi, Saitama, 336, Japan

Prepared for presentation at the Fifth International Conference on the
Bearing Capacity of Roads and Airfields
Norwegian University of science and Technology
Trondheim, Norway July 6-8, 1998

ABSTRACT

By applying regression analysis to FEM calculation results, a simple in-situ backcalculation approach was developed, which can be conveniently used to estimate the foundation elastic modulus E_0 (or reaction modulus k) and slab elastic modulus E_c under slab interior and joint edge FWD loading conditions. The evaluation results from several in-service pavements were presented to investigate the validity and accuracy of the proposed procedures. Furthermore, the reasonableness of the moduli derived from two different foundation models was discussed.

Key Words: concrete pavement, back-analysis, foundation model, radius of relative stiffness

1. INTRODUCTION

FWD testing and backcalculation (BC) is now widely recognized as a useful method for pavement structural evaluation and rehabilitation design. In the last 20 years, numerous backcalculation methods have been developed. For various optimization programs, the basic BC procedure is to adjust the set of moduli until an acceptable match between the theoretical and measured deflection is obtained[1)]. An exception to the optimization procedures is the ILLI-BACK program based on the closed-form approach, which can be applied to rapid analysis of plate-on-liquid model at slab interior loading case[2,3)]. Due to the performance of concrete pavement is largely controlled by the behavior of its joints, the joint conditions should be included in the backanalysis procedures.

In this study, the theoretical deflection basin characterizations of jointed concrete pavement are comparatively analyzed by using the finite element method for plate-on-liquid and plate-on-elastic solid foundation. It is interesting to note that a unique relationship also exists between the Spreadability Index (SI) of deflection basin and the radius of relative stiffness of the pavement system for jointed edge/corner loading conditions. These relationships are similar to that of interior condition as described in ILLI-BACK, only that the AREA or SI is defined by the unloaded side deflection basin.

By applying regression analysis to FEM calculation results, a simple in-situ evaluation approach has been developed in this paper, which can be conveniently used to estimate the foundation elastic modulus E_0 (or reaction modulus k) and slab elastic modulus E_c under slab interior and joint edge FWD loading conditions. Finally, the evaluation results from several in-service pavements are presented to demonstrate the validity of the proposed back-analysis procedures, and discuss the reasonableness of the moduli derived from two different foundation models. The theoretical analysis and BC results demonstrated that the Ec derived from liquid model should be always greater than that is from solid assumption, and the unreasonable high Ec-values are mainly due to the deviation between theory model and actual pavement structure, especially, the deflection basin shapes.

2. DEVELOPMENT OF BACK-ANALYSIS PROCEDURE

2.1 Modeling of jointed concrete pavement

In general, two types of structural models are widely used for evaluating the load deformation characteristics of concrete pavements. These are plate-on-liquid model and plate-on-elastic solid assumption. The finite element analysis computer program

KENSLABS[4)] was used to perform the computations in this study. The KENSLABS program has the capability to model the foundation as liquid and/or solid assumption, and the load transfer across a joint is modeled by means of shear spring.

In Japan, the typical jointed mesh-reinforced concrete slab is 8.0~10.0m in length, and 4.0~5.0m in width. For these large size slabs, the joint load transfer efficiency has only a very little effect on the deflections at slab interior FWD load case. So that, one-slab was considered for the interior load, and two-slab system for joint load case, as shown in Fig.1.

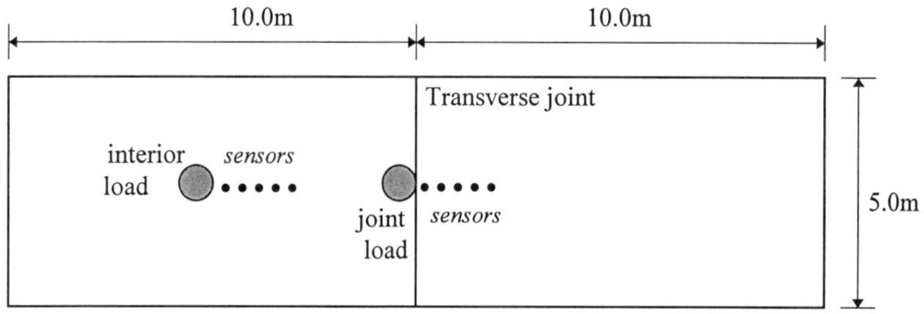

Fig.1 Structure model for deflection analysis of slab interior and joint load case

By considering the popularly used FWD device in Japan, a magnitude load of 100kN with a 30cm diameter loading plate, and six sensors spaced 30cm apart was selected as standard plate-sensor arrangement in this study. For edge deflections measured across joints, the load transfer efficiency is defined as:

$$LT = \frac{D_{30}}{D_0} \times 100 \tag{1}$$

in which LT is the load transfer efficiency, D_0 and D_{30} refer to deflections at the loaded and unloaded slabs, respectively.

Using the finite element method as described above, numerous theoretical deflection basins were generated for combinations of pavements structural parameters of Table 1, which covered wide ranges of factor levels.

Table 1 Combinations of pavements structural parameters deflection analysis

Pavement structural parameters		Range of values
Slab thickness h (cm)		20, 23, 25, 28, 30, 33, 36
Concrete elastic modulus E_c (10^5 kg/cm^2)		2, 3, 4, 5, 6, 7
Shear spring constant of joint C_w (kg/cm^2)		10^3, 10^4, 10^5
Foundation modulus	reaction modulus k (kg/cm^3)	3, 7, 10, 15, 20, 25, 30
	elastic solid modulus E_0 (kg/cm^2)	500, 1000, 2000, 5000, 7000

2.2 Characterizations of theoretical deflection basins

For concrete pavement structural evaluation and stress analysis, the radius of relative stiffness of the slab-foundation system is an important factor. As defined in Eq.2 & 3, the radius of relative stiffness has linear dimension, which is significantly relative to the deflection basin shape of pavement structure under a load.

For liquid foundation $\ell_k = \left[\dfrac{E_c h^3}{12(1-\mu_c^2)k} \right]^{\frac{1}{4}}$ (2)

For solid foundation $\ell_e = \left[\dfrac{E_c h^3 (1-\mu_o^2)}{6(1-\mu_c^2)E_0} \right]^{\frac{1}{3}}$ (3)

where μ_0 is Poisson ratio of solid foundation; μ_c is Poisson ratio of concrete slab; other factors have been defined in Table 1.

As stated, the famous BC program of ILLI-BACK demonstrated that unique relationships exist between deflection basin AREA and the radius of relative stiffness, however, the concept of AREA is restricted to analysis of interior load case. In this study, the Spreadability Index (SI) of deflection basin is expressed in a dimensionless form as followings:

For slab center load case $SI_1 = (D_0 + D_{30} + D_{60} + D_{90} + D_{120} + D_{150}) / D_0$ (4a)

$SI_2 = (D_0 + D_{30} + D_{60} + D_{90}) / D_0$ (4b)

For joint edge load case $SIJ_1 = (D_{30} + D_{60} + D_{90} + D_{120} + D_{150}) / D_{30}$ (5a)

$SIJ_2 = (D_{30} + D_{60} + D_{90}) / D_{30}$ (5b)

where D_0, D_{30}, D_{60}, D_{90}, D_{120}, D_{150} refer to deflections at 0cm, 30cm, 60cm, 90cm, 120cm and 150cm from the load center. The Spreadability Index SI_1 and SI_2 are corresponded to 6 (long) & 4 (short) sensors arrangement.

Since the deflection basin is discontinued curve at joint load condition, the Spreadability Index is defined by the unloaded side deflection basin only, not including the deflection under the center of load plate.

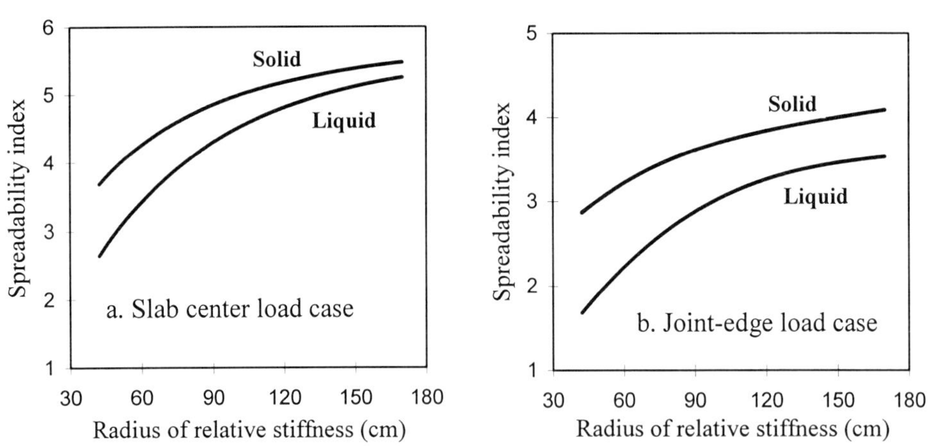

Fig. 2 Relationship between radius of relative stiffness ℓ and spreadability index (SI) of deflection basin

As shown in Fig.2, there is a one-to-one relationship between SI and ℓ_k or ℓ_e, which is not only for the interior loading, but also exists for joint-edge load case. It is also noted that for the same SI-values of the deflection basin, the corresponded ℓ_k is always greater than ℓ_e. This difference is caused by the basic assumptions of two foundation models. However, because the radius of relative stiffness is proportion to the concrete elastic modulus E_c-values, the high ℓ_k-values mean that the E_c backcalculated from liquid model may be not the same as that derived from elastic solid foundation.

2.3 Regression analysis

In light of the above investigations, the independent variables are studied in various ways in an attempt to determine the best form of BC model. It was found that the maximum deflection (D0) at the load center, SI of deflection basin (represented the radius of relative stiffness of the slab-foundation system), slab thickness, and load transfer efficiency are the significant variables. By applying regression analysis to FEM calculation results, a simple in-situ backcalculation approach has been developed as given in Table 2.

Table 2 Regression analysis results for 100kN FWD load*

load	Model	Regression equations	n	R^2	SEE
slab center	liquid	$k=\exp(3.93+24.30D_0^{-0.056}-15.34SI_1^{0.18})$	168	0.99	0.2
		$E_c=10^5\left\{\dfrac{\exp(5.20+0.113SI_1^{2.40})}{h^{3.0}}k\right\}$	168	0.98	12300
	solid	$E_0=10^3\exp(3.22+23.56D_0^{-0.058}-16.01SI_1^{0.138})$	168	0.99	96
		$E_c=10^5\left\{\dfrac{\exp(7.86+0.004SI_1^{4.16})}{h^{3.0}}\left(\dfrac{E_0}{10^3}\right)\right\}$	168	0.98	13000
joint edge	liquid	$k=\exp(-23.95+17.86D_0^{-0.088}+18.92SIJ_1^{-0.215})LT^{-0.58}$	504	0.99	0.2
		$E_c=10^5\left\{\dfrac{\exp(-26.75+39.15D_0^{-0.020}+0.177SIJ_1^{2.39})}{h^{3.0}}k^{0.28}\right\}$	504	0.97	15000
	solid	$E_0=10^3\exp(-26.47+24.01D_0^{-0.575}+12.92SIJ_1^{-0.239})LT^{-0.84}$	504	0.99	77
		$E_c=10^5\left\{\dfrac{\exp(-30.76+52.87D_0^{-0.046}+0.023SIJ_1^{3.41})}{h^{3.0}}\left(\dfrac{E_0}{10^3}\right)^{-0.89}\right\}$	504	0.98	16000

*For other load level, only the deflection D_0 need to be modified according to the linear relationship between deflection and load magnitude.

It can be seen that both the correlation coefficient and standard error of the estimate are quite good. Theoretically, these equations can be conveniently used to rapidly estimate the foundation elastic modulus E_0 (or reaction modulus k) and slab elastic modulus E_c under slab interior and joint edge FWD loading conditions. The regression equations shown in Table 2 are for SI_1 and SIJ_1 of six sensors arrangement, the similar evaluation equations for four sensors arrangement of SI_2 and SIJ_2 were also developed.

2.4 Compared with ILLI-BACK

For 8 typical test sections of SHRP as given in Table 3, the backcalculation results obtained from ILLI-BACK were presented in reference[2], with which the validity of the proposed BC equations can be demonstrated. Table 4 shows that the results obtained with ILLI-BACK and this analysis compare very well.

Table 3 Structures of test section of SHRP

Test section	Slab thickness h (cm)	The material of base course	FWD Load (kN)	Deflection (μm)				
				D_0	D_{30}	D_{60}	D_{90}	D_{150}
I	27.3	Asphalt treaded base	74	88	78	64	52	38
J	32.6	Crushed stone base	75	67	60	56	52	43
K	28.5	Cement-aggregate mix base	82	77	71	66	60	49
L	26.4	Crushed stone base	75	182	171	154	133	85
M	22.7	Gravel base	77	201	170	137	110	69
N	24.0	Cement-aggregate mix base	70	136	123	104	90	65
O	24.0	asphalt interlayer + JRCP	75	85	79	72	69	56
P	24.0	asphalt interlayer + CRCP	80	94	79	67	55	39

Table 4 Comparison of backcalculated results

Section	ILLI-BACK				This Study			
	Liquid model		Solid model		Liquid model		Solid model	
	k(kg/cm³)	E_c(kg/cm²)	E_0(kg/cm²)	E_c(kg/cm²)	k(kg/cm³)	E_c(kg/cm²)	E_0(kg/cm²)	E_c(kg/cm²)
I	14.1	4.3×10^5	4360	2.9×10^5	15.0	4.1×10^5	4800	2.8×10^5
J	7.5	8.8×10^5	3510	7.1×10^5	8.3	9.6×10^5	4100	6.3×10^5
K	7.4	12.3×10^5	3370	9.8×10^5	8.6	14.5×10^5	3900	9.7×10^5
L	4.1	4.1×10^5	1620	3.1×10^5	4.1	4.0×10^5	1800	2.6×10^5
M	8.3	3.2×10^5	2320	2.1×10^5	8.6	2.4×10^5	2600	1.8×10^5
N	6.8	4.9×10^5	2320	3.6×10^5	7.6	4.8×10^5	2500	3.1×10^5
O	5.6	18.0×10^5	2670	14.5×10^5	8.2	20.1×10^5	3000	14.4×10^5
P	16.3	5.6×10^5	4850	3.8×10^5	16.8	5.4×10^5	5200	3.8×10^5

3. FIELD EVALUATION RESULTS

In this section, the BC results from several in-service pavements will be presented to evaluate the proposed back-analysis procedures, and to discuss the reasonableness of the moduli derived from two different foundation models.

3.1 For slab center load case

Field evaluations were recently conducted on six test sections as given in Table 5. These pavements represented typical designs and climatic conditions in Japan. For an example, the backcalculated results of section-A are presented in Fig.3, and the results from all six sections are summarized in Table 6.

It can be noted that pavements with cement-treated base layers generally have higher k-values than pavements with untreated bases. Meanwhile, there is a linear relationship between the estimated foundation reaction modulus k-values and subgrade elastic modulus E_0 (see Fig.4). The E_c backcalculated on the basis of the liquid model is approximately

1.2~1.3 times as which is derived from the solid model, and the backcalculated ℓ_k is about 25% higher than the corresponded ℓ_e-values.

Table 5 Structures of six in-service pavements

Test section	Slab thickness h (cm)	The thickness & material of base course	Construction years	FWD test years
A	25	15cm cement stabilization base	1986	1994
B	30	15cm cement stabilization base	1982	1994
C	30	15cm cement stabilization base	1974	1994
D	32	22cm crushed aggregate base	1994	1995
E	25	4cm asphalt interlayer + 25cm granular base	1976	1996
F	25	4cm asphalt interlayer + 25cm granular base	1976	1996

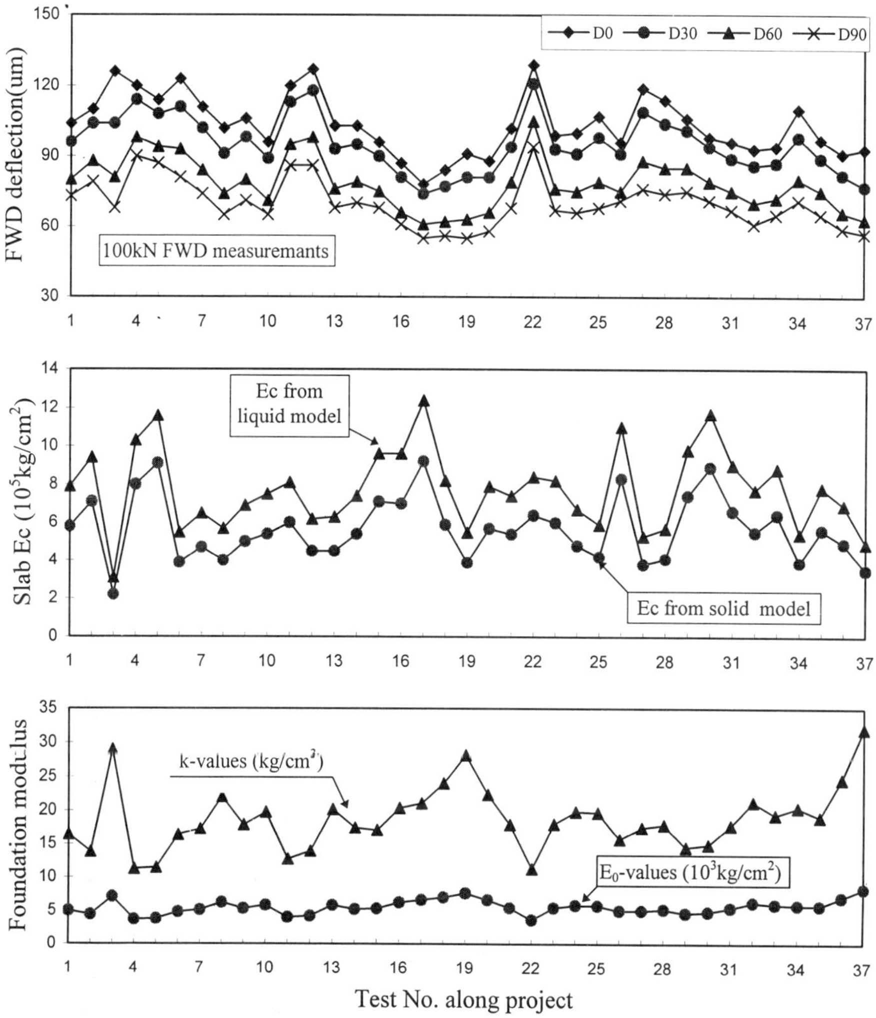

Fig.3 FWD measurements and BC results of test section-A

Table 6 The summarized BC results from all six sections

Test section	100kN FWD measurements			Slab-on-liquid model			Slab-on-solid model		
	Number of test	Deflection D_0 (μm)	Spreadability Index SI_2	k (kg/cm³)	E_c(10^5 kg/cm²)	ℓ_k (cm)	E_0 (kg/cm²)	E_c(10^5 kg/cm²)	ℓ_e (cm)
A	37	96.0	3.35	18.8	7.7	87	5500	5.7	63
B	25	84.4	3.37	19.0	6.1	89	6160	4.0	65
C	40	127.5	3.47	11.6	5.3	102	3600	4.2	78
D	24	210.7	3.49	5.0	3.7	102	1800	2.5	78
E	27	343.0	3.48	4.1	3.3	103	1300	2.6	78
F	34	342.7	3.55	3.7	4.1	112	1220	3.3	88

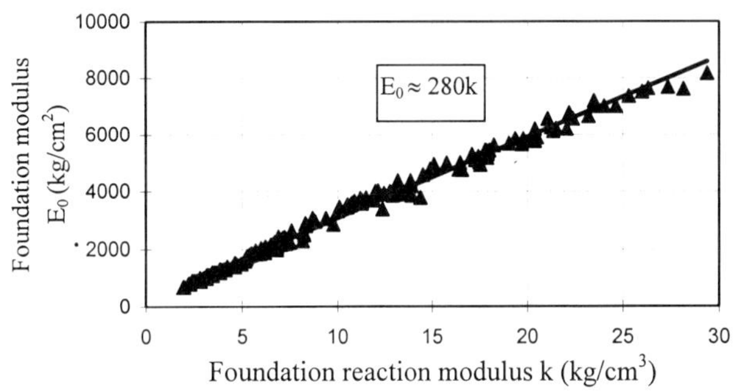

Fig.4 Relationship between backcalculated foundation reaction modulus k-values and elastic modulus E_0

3.2 For joint-edge load case

Current concrete pavement back-analysis procedures are usually for slab center FWD load case. However, for jointed concrete pavement structural design and evaluation, the critical load positions are joint-edge or slab corner. To investigate edge load case, the transverse joint edge deflection testing of section E & F was also conducted. The FWD measurements are shown in Fig.5, and the evaluated results from joint-edge load case are given in Table 7.

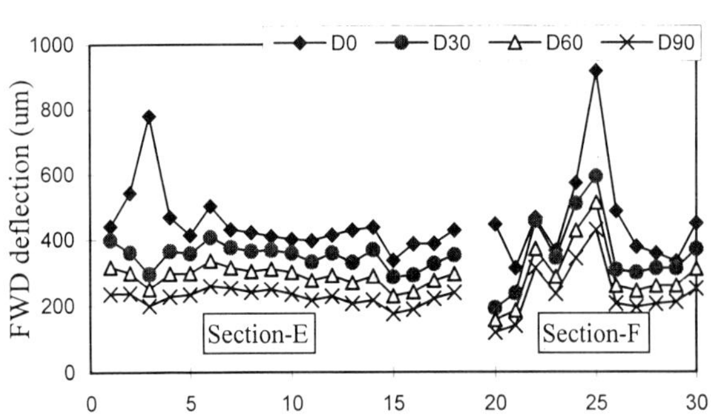

Fig.5 Joint-edge deflection under 100kN FWD load case

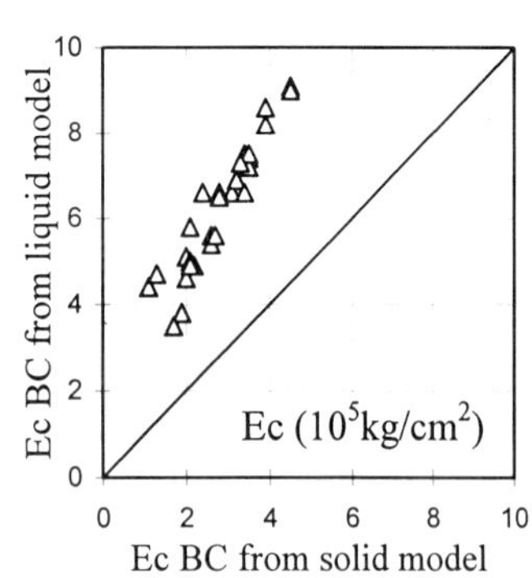

Fig.6 Backcalculated Ec-values

Table 7 Back-analysis results from joint-edge FWD load case

Test section	100kN FWD measurements				Slab-on-liquid model			Slab-on-solid model		
	Number of test	D_0 (μm)	Index SIJ_2	LT (%)	k (kg/cm^3)	E_c(10^5 kg/cm^2)	ℓ_k (cm)	E_0 (kg/cm^2)	E_c(10^5 kg/cm^2)	ℓ_e (cm)
E	17	446	2.47	80.7	3.4	6.0	124	1307	2.7	78
F	10	463	2.50	72.5	3.4	6.7	129	1350	3.0	81

Compared with the slab center results of Table 6, it can be observed that for solid model the BC results from joint-edge load case are nearly same as that from slab center condition. However, for liquid model, the concrete slab elastic modulus E_c-value and radius of relative stiffness ℓ_k derived from joint-edge load case is obvious higher than that is estimated from slab center load case. It also may be noted in Fig.6 that the slab modulus backcalculated from liquid model is approximately two times as which is determined using the solid model, and the former is considerably higher than what would normally be expected.

3.3 Discussion of the results

Back-calculation of pavement moduli is an inverse problem solution. The possibility of BC is depended to a very large extent on the information that is contained in the deflection basins. In fact, all theory models are not closely enough related to the realities of a pavement, albeit of differing degrees[5,6].

When a load is placed on a concrete pavement, the magnitude and shape of the deflection basin is a function of the thickness and elastic modulus of the slab, and foundation modulus. In liquid model, the deflection at a node depends solely on the subgrade reaction modulus k-values, at the node but not elsewhere. In contrast to liquid model, elastic solid model takes into account the effect of shear interaction between adjacent foundation elements, the deflection at a node depends not only on the elastic modulus E_0, but also on the deflections at other nodes. Because of the difference in the basic assumption of two types of supporting model, the shapes of the theoretical deflection basins of the concrete slab are different between the liquid foundation and the elastic foundation.

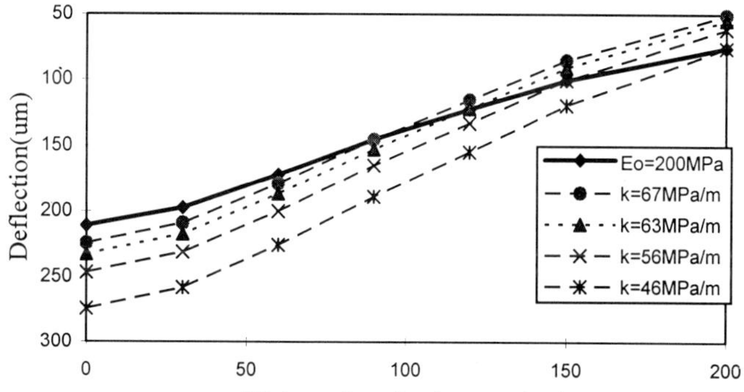

Fig.7 Comparison of deflection basin shape between liquid and solid foundation model

In Fig.7 the E_0 was assumed to be 200MPa for slab on elastic assumption, the corresponding values of subgrade reaction modulus k, used for pavements on liquid foundation were determined by matching deflections at 90cm, 120cm, 150cm and 200cm from load center, respectively. It can be recognized that the deflection basin is much steeper in the liquid subgrade than in the elastic subgrade. When the deflection at the farther point is in agreement, the deflections closer to the load for the case of liquid subgrade are consistently greater than those for a solid subgrade, and as the agreement point farther from load center, much greater difference exists in load center deflections between two models. This demonstrated that the Ec derived from liquid model should be always greater than that is from solid assumption.

4. SUMMARY AND CONCLUSION

The back-analysis procedures described in this paper provide a rapid method to determine the in-situ concrete slab modulus-E_c and foundation modulus E_0 & k-values. These procedures can be conveniently used for either slab interior or joint-edge FWD loading conditions.

The field backcalculated results and theoretical comparison of deflection basin shape between liquid and solid foundation model demonstrated that the E_c derived from liquid model is much greater than that from solid assumption, and the unreasonable high E_c-values are mainly due to the deviation between theory model and actual pavement structure, especially, the deflection basin shapes.

5. REFERENCES

1) P.Ullidtz, and N.F.Coetzee, Analytical procedures in Nondestructive Testing pavement evaluation, Transportation Research Record 1482, TRB, 1995, pp61-66.
2) A.M.Ioannides, Concrete pavement backcalculation using ILLI-BACK3.0, Nondestructive Testing of Pavement and Backcalculation of Moduli (Second Volume), ASTM STP 1198, 1994, pp103-124.
3) M.I.Dater, K.D.Smith and K.T.Hall, Concrete pavement backcalculation results from field studies, Transportation Research Record 1377, TRB, 1992, pp7-16.
4) Y.H.Huang, Pavement Analysis and Design, Prentice-Hall, Inc., Englewood Cliffs, 1993.
5) J.Uzan and R.Briggs, Backcalculation of design parameters for rigid pavements, Transportation Research Record 1377, TRB, 1992, pp107-114.
6) B.Tang and T. Maruyama, Evaluation of back-calculated results of rigid pavement, Journal of Pavement Engineering, JSCE, Vol-1, Dec.1996, pp95-101.

Fifth International Conference on the
BEARING CAPASITY OF
ROADS AND AIRFIELDS

THE BEARING CAPACITY ANALYSIS OF RIGID PAVEMENTS

IVAN GSCHWENDT[1], RUDOLF STAŇO AND EDUARD ŠTEFANEC[2]

[1]Professor, Slovak Technical University Bratislava
[2]Research workers, Slovak Technical University Bratislava
813 68 Bratislava, Slovakia

ABSTRACT

The paper deals with the FWD measurements generally used to determine pavement bearing capacity. The KUAB - FWD was used for test of rigid pavements on airport runways: structure (A) with one cement concrete slab and structure (B) with two cement concrete slabs. In order to ascertain the interaction of slabs and load transfer across transverse and longitudinal edges, the plate bearing tests were performed on edges.
Keywords: Airfield pavements, Bearing capacity, Deflection measurements.

INTRODUCTION

The pavement management systems requires repeated measurements and evaluation of the serviceability and bearing capacity of pavement structures. Deflections measured at plate bearing tests are fundamental for the bearing capacity evaluation. When conducting the plate bearing tests and deflection measurements, the deflections near the joint on the loaded and unloaded slabs can be measured simultaneously. The deflection difference can be used to evaluate the joint performance and the load transfer.

DEFLECTION MEASUREMENTS

The technical specification for testing and evaluation of a bearing capacity of the cement concrete pavements on airport runways suggest that the falling weight deflectometer (FWD) is the preferred device based on its ability to measure the load - induced deflection and deflection profile (bowl). For described tests the KUAB 2m-150 model has bean used. This device produced the max load approximately 150-153 kN and contact pressure of 0,95 MPa. (The diameter of the circular loading plate is 450 mm). One deflection sensor (absolute deflection transducers) is located in the center of the loading plate, the others 300, 450, 600, 900, 1200 and 1500 mm from the center. The sensor configuration is shown in Figure 1.

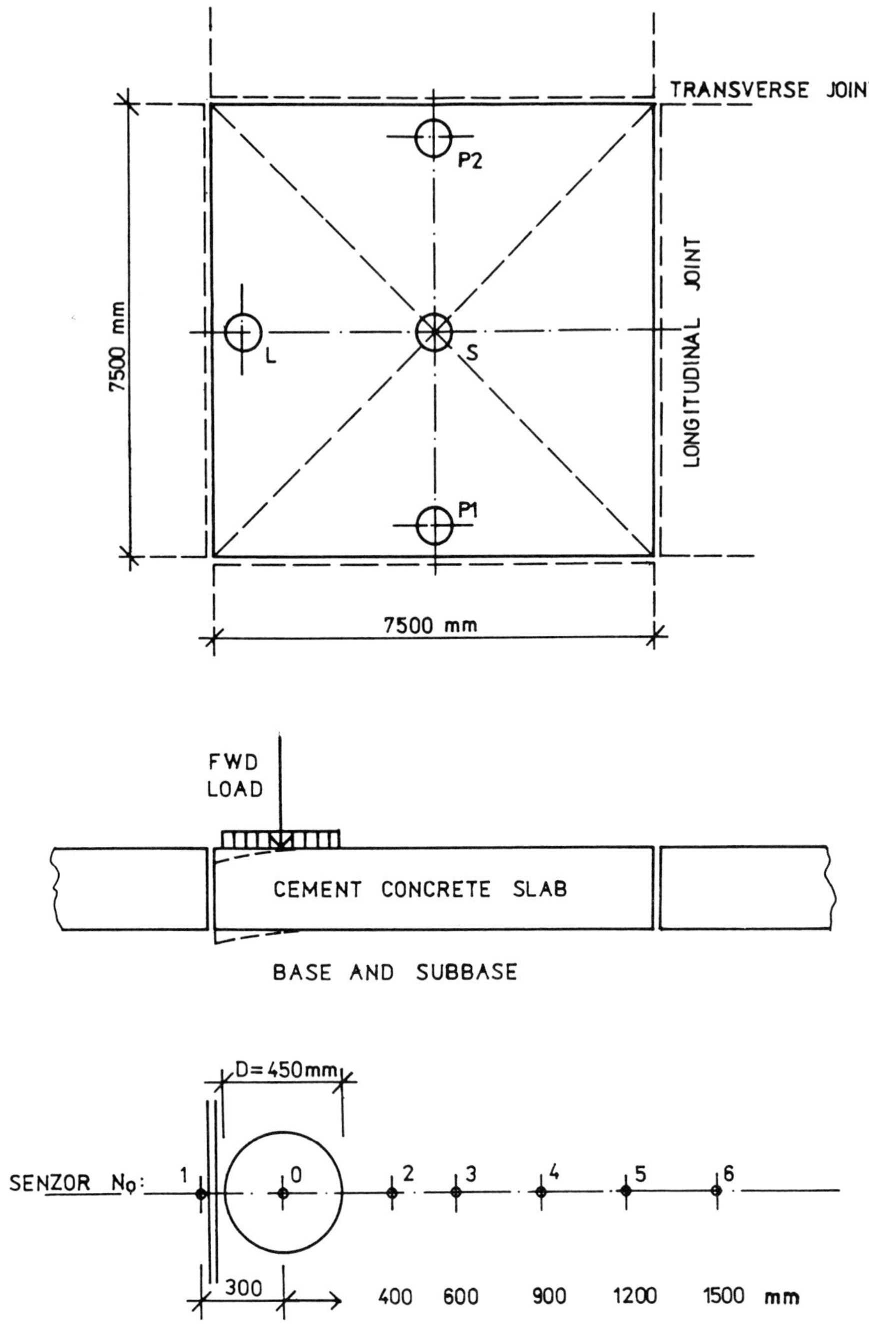

FIG.1 Loading test arrangement and deflection measurement

Sensor No 1 was placed on the edge of the unloaded slab.

Max. Surface deflection and deflection profile under dynamic load were measured on two type of cement concrete pavements:

- pavement structure type A and type B. Structure type A has slab 300 mm thick, 40 mm layer from open graded asphalt mix and cement stabilization base course of 200 mm. The total thickness is 840 mm. Structure type B has upper slab 220 mm thick, lower slab 240 mm thick (interlayer 40 mm open graded ashalt mix), cement stabilization base course of 150 mm. The total thickness is 900 mm. The dimensions of most of the slabs are 7500 x 7500 mm.

The loading tests with deflection measurements were performed on slabs in the geometric centre and on edges, along the transversal and longitudinal joints. To analyse and evaluate of pavement structure with one slab (A) and structure with two slabs (B), the tests were performed in two series: first on the upper slab and the second on lower slab or stabilized base layer. The vertical temperature differential, e. g. the difference between the temperatures of the top and bottom of a concrete slab, has a strong influence on the slab curl and conditions for deflection test. Therefore the time of testing (period of year and daytime) becomes critical. For the limitation of temperature influence on deflection measurements, it is recomended to test under condition when the temperature gradient is in range ± 0,013°C/mm.

Part of the results of the deflection measurements on pavement structures with one concrete slab (structure A) and on structures with two slabs (structure B) in series I. and series II. are in table 1 and table 2. [1]

The variability of the deflection in the centre of the concrete slabs on structure A and B as well as more characteristics from the deflection measurements are in table 3. The stiffness of structure A or B is characterized by the base (subgrade) effective moduli of reaction:

$$k_{ef} = P / V \qquad [MN \, . \, m^{-3}]$$

where V is the volume of the deflection basin under dynamic load P. The value of r_0 is the radial distance from the loaded area, where the deflection d_r is equal to zero.

Table 1

Max. deflection d_0 mm . 10^{-3}, SERIES I

- in centre of slab ... S
- on edges / transversal joints 1,2 ... P1, P2

Slab No	Test point	Load P kN	Defl. d_0 mm . 10^{-3}	Section
F 83	S	154,5	325	
	P 1	154,5	398	
	P 2	154,4	165	A
E 83	S	152,4	208	
F 312	S	156,2	318	
	P 1	152,0	408	
	P 2	151,5	291	B
E 312	S	153,2	245	

Table 2
Max. deflection d_0 mm . 10^{-3}, SERIES II, on base of slabs No 83, 312
- under centre of slab ... S
- under transversal joints 1,2 ... P1, P2

Slab No	Test point	Load P kN	Defl. d_0 mm . 10^{-3}	Section
F83	S	149,1	899	
	P 1	147,2	1213	**A**
	P 2	145,1	1681	
F 312	S	153,8	454	
	P 1	150,3	488	**B**
	P 2	151,0	363	

Table 3

SLAB No	SERIES	LOAD kN	DEFL. d_0 10^{-3} . mm	DISTANCE r_0 mm	MODULI k_{ef} MN . m^{-3}
		PAVEMENT STRUCTURE A			
F 83	I	153,3	269	3574	41,68
	II	148,6	895	2264	36,29
F 84	I	152,9	331	2982	48,32
	II	146,6	1052	2246	31,30
F 86	I	152,9	282	3080	50,15
	II	148,2	1039	2255	30,72
F 87	I	152,5	299	3160	46,06
	II	147,2	1555	1928	33,98
		PAVEMENT STRUCTURE B			
F 311	I	156,2	387	3281	38,98
	II	153,2	406	3103	35,75
F 312	I	156,3	380	3375	40,77
	II	153,8	454	3207	29,92
F 314	I	153,7	426	2665	51,52
	II	153,7	392	3095	38,93
F 315	I	153,3	394	2665	52,75
	II	153,0	394	2965	39,73

The plate bearing tests with deflection measurements to evaluate the joint performance and load transfer at joints were conducted on 60 concrete slabs, pavement structures K 1, K 2 and K 3 on sections 1, 2 and 3 of airfield runway. The temperature conditions during the plate bearing tests are characterized by the "zero temperature gradient" and mean absolute temperature 9,0 - 9,5°C(very close to average annual air temperature).
Two basic characteristics of the load transfer between the joints [1] were calculated on the basis of a large number of deflection measurements:

a) load transfer

$$LT = (d_u / d_l) . 100$$

where d_u is deflection of the unloaded slab,
d_l - deflection of the loaded slab

b) load transfer efficiency

$$LTE = 2d_u / d_u + d_l$$

The results of the LT and LTE calculations and their mean values for three runway structures are in table 4. The characteristics were calculated separately for transverse and longitudinal joints.

Table 4

Pavement structure	Transverse joints		Longitudinal joints	
	LT	LTE	LT	LTE
K 1	58,63	0,727	25,62	0,396
K 2	80,89	0,887	51,03	0,651
K 3	83,79	0,907	43,15	0,588

EVALUATION OF MEASUREMENTS

The deflections measured by the KUAB 2m-150 device are recorded and printed in the standard form in compliance with the software "KUAB". The record contains a number of data, from which we chosed the max. deflection for P = 150 kN or characteristics of deflection profile.
A very simple comparison the maximum deflection of the slabs geometric centre gives the differential evaluation the structures A and B: the ratio is 1,35. On the base course of the structure B is the mean deflection 400 . 10^{-3} mm, on the base of the structure A is a deflection from 1000 up to 1500.10^{-3} mm. The influence of temperature on the deflection of concrete slabs 7500 x 7500 mm under dynamic load P = 150 kN can be followed by the mean values in table 5.

Table 5

Structure	Temperature gradient	Deflection d . 10^{-3} mm	
		in geom. centre	on edge
A	0	188	214
	+ ΔT	475	214
B	0	254	235
	+ ΔT	437	235

The ratio between the midslab deflections of structure A at zero temperature gradient and under gradient ΔT is 2,2, but of structure B is 1,86 only.
Remarkably different are the deflection basin shapes for stiff and weak pavements [2]. In Fig. 2 we have deflection profiles of the F 83 slab (section A), then F 312 (section B). In Fig. 3 there are deflection profiles of the base layer on section A: under slab F83, of the lower concrete slab on section B: F 312 and for comparison two profiles F 312 and F 83 in

To improve the results from the Dynaflect, NRRL initiated a research program to study alternative geophone configurations for the device, and also if possible find more optimal sensor configurations for the FWD. An other part involved the development of an analysis program enabling the computation of "Bearing Capacity" from backcalculated layer moduli.

BACKCALCULATION PROCEDURE

One of the goals of the BUAB project was to develop a tool for interpreting bearing capacity based on deflection and condition measurements. As part of this, the DimTo program was developed. This is a program for structural design and strengthening of road pavements according to level two of the Norwegian Design Manual (NDM). Included in DimTo and the backcalculation program EMODUL is a backcalculation procedure developed by SINTEF. This procedure uses a corrected CHEVRON version for forward deflection computations.

Layer handling

As the DimTo program is linked to the Norwegian Road Data Bank, there should be a possibility to include a great number of layers. Due to limitations in CHEVRON and some possible modifications during the process, the number of layers to be identified by the user is limited to 12, the subgrade included. At a maximum, the procedure can backcalculate moduli for four layers for FWD and for three layers for Dynaflect. Of these, maximum two can be defined within the same category. The remaining layers must be given a constant modulus.

The materials are defined within different groups, according to the division into the categories *Pavement*, *Base layer*, *Subbase* and *Subgrade* in the NDM. For each material there is defined a load distributing coefficient a. The elastic modulus [MPa] of a material can be calculated according to the bending theory for plates as $E = 111.1 \bullet a^3$. For bituminous materials the modulus is adjusted to the current surface temperature T as measured by the deflection device.

In the procedure there is an automatical joining of materials that belong to the same category. Equally, base and subbase layers are joined if the pavement is to be regarded as a three-layer system. Further, there is a joining of layers if the layers are thinner than certain minimum values, and for certain combinations of bituminous pavement and base materials. For such joined layers a combined modulus is computed according to the index system developed after the AASHTO field test. When the adjusted modulus for the combined layer is distributed to the included individual layers, this happens according to the same principle.

Starting modulus equations

The backcalculation includes an iterative process where starting moduli are based on the combined modulus and thickness for the layers in a three-layer linear elastic system with pavement, base and subgrade. Starting moduli are calculated separately for Dynaflect (DFL) and FWD, and as an example the starting modulus for the subgrade is expressed as:

$$E_{Sg}^{St} = C_1 + C_2 \bullet (\delta_r^m)^{C_3}$$

where δ_r^m: measured deflection at distance r [cm], $r \in$ [90,120,150]

C_1, C_2, C_3: constants; for r=120: C_1=÷6.0, C_2=13.0, C_3=÷0.95 for FWD (δ_r^m: [mm])

C_1=÷5.0, C_2=1029, C_3=÷0.96 for DFL (δ_r^m: [µm])

These equations do all have a regression coefficient $R \geq 0.98$. For Dynaflect geophones 60 cm or closer to the mid-point, the equations include pavement and base layer thicknesses, too.

For the pavement and base layers are developed formulas which include two or three deflection differences over the inner parts of the deflection basins. For base layer thickness greater than 30 cm, all these equations do have an $R > 0.90$, and for no one R is less than 0.84. These equations represent a further development of the later mentioned regression analyses.

Missing or incorrect deflection measurements

In version 2 of the procedure is included a routine to exclude possible erroneous deflections (1). As a main rule, measured deflections should decrease with increasing distance from the load, but for Dynaflect 3 % or 1 μm less variation from this is accepted. There are rules for disregarding sensors with either extremely great or extremely small deflections, too.

To compute the starting moduli if deflection values at the predefined distances for these equations are excluded, there is included a routine to predict estimates for measured deflections at these distances based on measured values at the current ones. For this are used routines for cubic spline from the mathematical library NUMLIBC, developed at the former Department of Mathematics at the Norwegian Institute of Technology. Normally, a very close estimate for the deflection at possible missing or erroneous sensors can be found (1).

The backcalculation process

The backcalculation process runs for a maximum of 15 iterations for each deflection basin to be analyzed, and operates on the current non-erroneous deflections. The iteration process stops if no moduli are changed or if the calculated deflection deviations are within pre-set limits. Towards the end of the process these limits are expanded in two steps if necessary. Both for Dynaflect and FWD are included both absolute and relative tolerances for deflection deviation at individual sensors, and the most gentle of these is adopted in each case. Besides, there are criteria for the mean relative deviation for sensors.

The routine for adjusting the combined layers' moduli is except for the subgrade modulus based on deflection *differences.* The moduli are adjusted according to a log-log deflection (difference) to modulus connection, like the process used for the –DEF series of programs (2).

DimTo can do a simplified analysis of deflection measurements. The starting modulus equations are then used directly to estimate moduli for joined layers, and the modulus of the joined layer is distributed to the included layers according to the index system.

When the backcalculation process is finished, the modulus of bituminous materials is adjusted from the current surface temperature T to the standard temperature of 25 °C. Thereafter, load distributing coefficients a are computed as $a_i = 0.208 \bullet (E_i^{25})^{1/3}$ where E_i^{25} is replaced by E_i for non-bituminous materials.

Calculation of bearing capacity

When bearing capacity according to structural data is calculated in the DimTo program, the load distributing coefficients from the backcalculated moduli are used. The method is based on the index system, where current index value above a certain depth is compared to the related index value demand for the current material. Based on the maximum difference between these two values at a certain measurement point, the bearing capacity at the point can be calculated. For a stretch of the road, the bearing capacity is calculated by a statistical analysis of the individual point values as the 10 % percentile value.

SENSOR CONFIGURATION

A brief literature survey was carried out to find used sensor configurations and possible research work on sensor configuration (3). This survey also should give a background for the sensor configuration optimalization which was to be carried out.

Geophone configurations for Dynaflect

The standard geophone set-up for Dynaflect is two to five geophones in a row along the symmetry line midway between the steel wheels with an internal geophone spacing of 30 cm.

A comprehensive research work concerning optimal geophone configuration for Dynaflect has been conducted in Florida (4). The geophone configuration shown at figure 1 was found to be optimal, and is used by Florida Department of Transportation (FDOT).

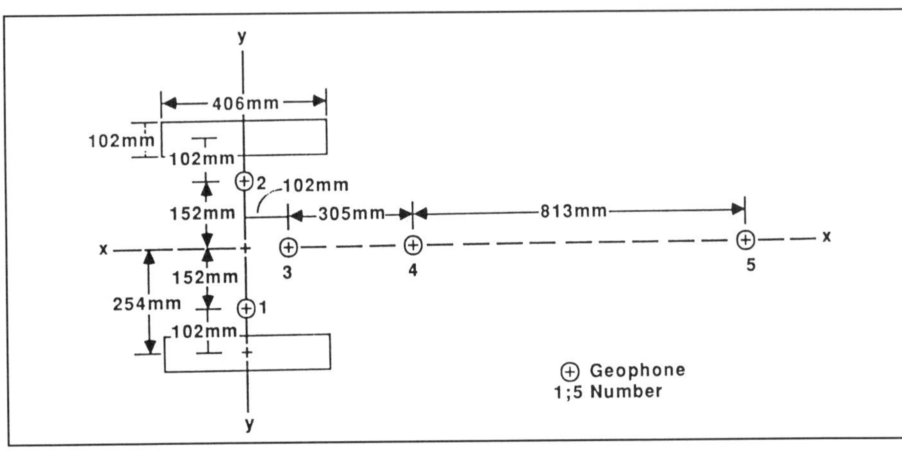

FIGURE 1 Modified Dynaflect geophone configuration used by FDOT (5)

Sensor configurations for FWD

The number and type of deflection sensors used for different FWD's vary with producer, model and agency, but six or seven sensors is normal. Table 1 summarizes the found sensor configurations for measurements on roads with bituminous surface layer(s).

TABLE 1 Overview over several FWD sensor configurations in use (summary from (3))

Agency	Radial distance from load centre [cm]														
	0	20	30	45	50	60	65	80	90	120	150	160	180	210	250
RST, Sweden	X	X	X	X		X			X	(x)					
Denmark	X	X	X	X			X		X	X					
Nottingham	X		X			X			X	X	X			X	
the Netherlands	X	X	X			X			X	X			X		
Japan	X	X		X		X			X		X				X
Florida	X	X	X		X			X		X		X			
Texas	X		X			X			X	X	X		X		
SHRP	X	X	X	X		X			X		X				

A research work in Texas on optimal sensor configuration for secondary roads concluded that sensors should be placed as close as possible to the load axis (6). Physical limitations connected to the load plate, gave the following sensor configuration recommendations:

0-20-27.5-35-45-75-120 cm

Nevertheless, it was suggested to study the possibility of modifying the load plate to allow placing of sensors 2 and 3 at distances 7.5 and 15 cm from the load centre.

A Nottingham study concerning four-layer structures recommended to place the two innermost sensors at distances 0 and 20 cm from the load centre (7). To backcalculate the subbase, it was recommended to place the third sensor at a distance where the deflection ratio δ_i/δ_0 is 0.8. Two sensors should be placed 60 and 120 cm from the third sensor. The remaining two sensors should be placed to fill "holes" in the deflection basin.

REGRESSION ANALYSES

Statistical analyses were carried out to study possible optimal sensor configurations for FWD and Dynaflect (3), based on results from CHEVRON computations for several hundred three-layer structures. A selection of 360 of these structures was used for the analyses.

Analyses for Dynaflect

For the subgrade, the modulus equations were found as function of deflections at far offset. For the top (asphalt concrete) and intermediate (called base) layers, the equations were found for deflection differences close to the load.

For the subgrade modulus, the statistical significance increased slightly with increasing distance. For the top layer, the innermost geophone was placed as close to one of the wheels as possible. The statistical significance decreased as the distance to geophone no. 2 increased, but for geophone no. 2 closer than 5 cm along the symmetry line, R was greater than 0.9. For the base layer, three geophone placings with their deflection differences had to be considered. It was found to be favourable to place a geophone along the line between the wheels, while the second geophone should be placed 0-10 cm along the symmetry line for thin base layer and 20-30 cm along the symmetry line for thick base layers. The third geophone should be placed as close to the second as possible for thin base layers and 40-50 cm along the symmetry line for thick base layers.

Analyses for FWD

For FWD, the analyses were carried out for one sensor in the centre of the loading plate and with radial offsets for the others varying from 20 to 150 cm.

For the surfaçe course, the most significant connections were found for sensors 0, 20 and 30 cm or 0, 25 and 30 cm from the load centre, both with an R of 0.99. The three configurations with highest R for the base course modulus was found to be the following:

- 30 - 45 - 75 - 90 cm: $R = 0.964$
- 30 - 60 - 75 - 90 cm: $R = 0.957$
- 30 - 45 - 60 - 90 cm: $R = 0.944$

For the subgrade, the statistical significance increased with increasing distance to the sensor, and both for $r = 120$ cm and $r = 150$ cm was found $R > 0.99$.

FIELD MEASUREMENTS

Practical measurements

The field measurements were carried out at totally 27 test sites with a wide variety of subgrades, including both old and fairly new roads. Subbases were granular, but both granular, bituminous and cementitious base courses were present. 60 % of the roads had asphalt concrete and 40 % soft bituminous pavement layers. Total pavement thickness for the new roads varied from 30 cm to 120 cm, and bituminous thickness varied from 2-3 cm to 16 cm (8).

Measurement points were situated in outer wheel path and between wheel paths. Surface layer temperature was measured at two or three levels and samples were taken to the laboratory for determination of layer thickness and for indirect tensile test. At 13 sites the structure was dug up to determine also unbound layer thickness and to do several standard laboratory tests.

Sensor configurations

The field measurements included three sensor configurations for each device. The standard and FDOT set-ups were chosen as two of the Dynaflect configurations. Based on the regression analyses, physical and practical considerations, configuration 3 from table 2 was chosen as the third. The geophones were manually moved around to fulfil the geophone set-ups.

TABLE 2 Chosen Dynaflect geophone configurations with distances in cm from the midpoint between the steel wheels (negative distances are towards the wheel(s))

Geophone configuration	Geophone no.				
	1	2	3	4	5
1 (Standard)	0	30	60	90	120
2 (FDOT)	-15 (Right)	-15 (Left)	10	40	120
3	-15 (Right)	0	15	14	120

The SHRP set-up was chosen as standard FWD sensor configuration. The 0-20-30 cm set-up was maintained for the three innermost sensors. The remaining positions were chosen based on the statistical analyses. The FWD configurations shown at table 3 were finally used. No. 2 has non-increasing spacing outwards.

TABLE 3 Chosen FWD sensor configurations with distances in cm from the load centre

Sensor configuration	Sensor no.						
	1	2	3	4	5	6	7
1 (SHRP)	0	20	30	45	60	90	150
2	0	20	30	45	70	90	150
3	0	20	30	45	70	110	150

BACKCALCULATION OF LAYER MODULI

For each device, deflections from two representative points, one in outer and one between wheel paths, at 12 of the sites with digging, were chosen for backcalculation. The same point was used for all devices and configurations, and the point with deflections closest to the mean values for the current line and sensor set-up was chosen. For different points for different devices, the FWD's were paid most attendance. Six of the 12 sites had clayey/silty subgrade, the remaining non-cohesive subgrade.

The EMODUL program was used for the analyses. The individual results are not reported here. Details can be found elsewhere (9), but as some revisions were done to the program (1), the analyses were re-done with the new version. For FWD, this is only done for KUAB. The analyses are carried out both with varying and constant pavement layer modulus, the constant value is the tensile stress value found in the laboratory at the current temperature.

Pavement layer modulus is backcalculated for Dynaflect geophone set-ups 2 and 3, but not for standard Dynaflect configuration. For that case, the standard modulus for the current pavement material is adjusted for temperature deviation from 25 °C and used as a constant value.

Accuracy

RMS (root mean square (10)) is a measure of deflection accuracy, even though this parameter in this context has some drawbacks. A summary of RMS and other accuracy parameters for Dynaflect is given in table 4, and for KUAB FWD in table 5.

TABLE 4 Summary of accuracy and program problems for Dynaflect

Configuration	Cases without problems for two or three config.			Cases with too great deviations			Cases with program failure			Cases without problems for any config. (13 cases)	
	Mean RMS [%]	Cases	Cases with the least RMS	Total	Cold mixed/thin base	Rock	Total	Cold mixed/thin base	Rock	Mean RMS [%]	Cases with the least RMS
1	8.1	20	6	1	1		2	-	2	8.3	1
2	6.6	18	4	4	1		2	2	-	5.7	4
3	5.6	19	11	4	1		-	-	-	4.5	8

TABLE 5 Summary of accuracy and program problems for KUAB FWD

Configuration	Cases without too great deviations for two or three config. Rock at shallow depth ...				Cases with too great deviations			Cases without too great deviations for any config. (13 cases)	
	included			excluded					
	Mean RMS [%]	Cases	Cases with the least RMS	Mean RMS [%]	Total	Cold mixed/thin base	Rock	Mean RMS [%]	Cases with the least RMS
1	3.1	16	7	2.9	4	2	1	2.3	5
2	3.5	17	3	3.2	2	-	1	2.4	3
3	2.5	14	6	2.5	6	2	2	2.2	5

Comparison of different configurations

The mean relative modulus values for different configurations for accepted cases are summarized in table 6 and 7 for Dynaflect and KUAB, respectively. The results are compared for both varying and constant pavement layer modulus to exclude any influence from the very high pavement layer modulus for Dynaflect set-up 1 due to low temperature (all except one are below + 10 °C), possibly giving a too low base layer modulus.

TABLE 6 Mean relative modulus values for different Dynaflect configurations; for pavement modulus given both without (in brackets) and with outlying values

	Configuration							
	2 compared to 1				3 compared to 1 (for Pavement: 3 compared to 2)			
Pvmt modulus	Varying		Constant		Varying		Constant	
Wheel path	Outer	Between	Outer	Between	Outer	Between	Outer	Between
Pavement					0.96 (1.12)	1.37 (1.51)		
Base	1.17	0.99	1.14	0.98	1.09	0.97	1.09	1.00
Subgrade	0.90	1.00	0.94	1.00	0.945	1.01	0.93	1.01

TABLE 7 Mean relative modulus values for different KUAB FWD configurations

	Configuration							
	2 compared to 1				3 compared to 1			
Pvmt modulus	Varying		Constant		Varying		Constant	
Wheel path	Outer	Between	Outer	Between	Outer	Between	Outer	Between
Pavement	1.04 (1.09)	1.01 (1.06)			1.03	1.02		
Base	0.93	0.94	0.99	1.01	0.94	0.98	1.00	0.99
Subbase	1.06	1.07	1.01	1.03	1.01	0.99	0.97	1.02
Subgrade	0.99	1.01	0.99	1.00	1.00	1.01	1.00	0.99

The subgrade modulus is approximately the same for different set-ups. With very few exceptions, all deviations are within ±15 % for Dynaflect and within ±5 % for FWD. Three of the pavements are for FWD backcalculated with four layers. For these, the subbase modulus is higher for set-up 2 than for 1. For the base layer, Dynaflect set-ups 2 and 3 give higher moduli than set-up 1 for both varying and constant pavement layer modulus, while FWD configurations 2 and 3 both give less base layer moduli than configuration 1. The pavement modulus is greater for Dynaflect set-up 3 than for 2 for all cases except two in outer wheel path, and for all cases except one between wheel paths. For the outlyers, set-up 3 gives almost just half the value of set-up 2. For KUAB, the pavement modulus deviations vary widely, but as the “positive” deviations are greater and more numerous than the “negative”, the main impression is greater pavement modulus for set-up 3 and 2 than for 1.

Modulus values for varying and constant pavement layer modulus are compared in table 8 and 9 for Dynaflect and KUAB, respectively. The rather small mean deviations for pavement modulus for set-up 1 for Dynaflect indicate that the temperature correction works fairly well. The greater deviations for set-ups 2 and 3, especially in outer wheel path, show that there are certain difficulties in backcalculating pavement layer modulus from the Dynaflect. For all sites except one, KUAB pavement layer moduli are less than the value found from the laboratory test. This indicates either that the pavements are not as intact as the samples tested in the laboratory tell, or points towards a systematic weakness in the backcalculation procedure. As table 9 shows, this “underestimating” of pavement modulus is followed by a greater base modulus. For the other layers there are just minor differences between the configurations.

TABLE 8 Mean relative modulus values for varying against constant pavement modulus for different Dynaflect configurations

	Configuration 1		Configuration 2		Configuration 3	
Wheel path	Outer	Between	Outer	Between	Outer	Between
Pavement	1.13	1.13	0.63	0.97	0.67	0.91
Base	1.02	1.03	1.02	1.05	1.00	1.05
Subgrade	1.01	1.00	1.02	1.01	1.02	1.00

TABLE 9 Mean relative modulus values for varying against constant pavement modulus for different KUAB configurations

	Configuration 1		Configuration 2		Configuration 3	
Wheel path	Outer	Between	Outer	Between	Outer	Between
Pavement	0.56	0.64	0.68	0.64	0.62	0.71
Base	1.20	1.14	1.15	1.17	1.15	1.12
Subbase	0.90	1.03	0.94	1.05	0.94	0.97
Subgrade	1.00	0.99	0.99	1.00	1.01	1.00

Comparison of Dynaflect and KUAB

The comparison between Dynaflect and KUAB are based on set-ups 1, 2 and 3 for Dynaflect relative to configuration 1 for KUAB. The mean results are shown in table 10, where the values for the base layer are based on sites where a three-layer structure is used for KUAB. As can be seen, Dynaflect gives greater moduli than the KUAB FWD for all configurations and layers, and partly the deviations are fairly great.

TABLE 10 Mean relative modulus values for Dynaflect against KUAB for varying pavement modulus

	Configuration 1		Configuration 2		Configuration 3	
Wheel path	Outer	Between	Outer	Between	Outer	Between
Pavement			1.19	1.39	1.19	1.57
Base	1.62	1.38	1.95	1.37	1.46	1.32
Subgrade	1.84	1.81	1.53	1.45	1.68	1.57

CONCLUSIONS

There is developed a backcalculation procedure for use with deflection data for Dynaflect and FWD stored on a standardized Norwegian Road Data Bank format. For FWD the deflections can be measured with almost a free sensor set-up. For Dynaflect, both the standard and the FDOT geophone set-up with some variants of these can be analyzed.

The procedure uses linear elastic theory for the iterative process. When good estimates exist on the layers in the pavement, the procedure in most cases succeeds in the backcalculation. Nevertheless, for thin cold-mixed bituminous layers in combination with a thin base layer, certain problems may occur. However, some of the problems disappeared when pavements with non-cohesive subgrades were backcalculated without a stiff layer in the subgrade. It is found to be some problems connected to measurements carried out at sites with rock at shallow depth, too. For Dynaflect, this is especially connected to the standard geophone set-up, where deflections close to or like zero may occur at several geophones.

Analyses of measurements carried out with three different sensor configurations for Dynaflect and FWD, show that the used configurations are equal as far as subgrade modulus is concerned. On the contrary, there are great deviations between subgrade modulus for Dynaflect and FWD, with 65 % greater value for Dynaflect as an average. Analyses for Dynaflect give greater moduli than FWD for the other layers, too, but there is a tendency towards decreasing relative value as layers higher in the structure are analyzed.

By a comparison of backcalculated pavement layer modulus with values found in the laboratory, analyses for the FWD give as a mean 35 % lower value. Analyses for Dynaflect give a similar but less pronounced tendency. This "underestimation" of pavement modulus is for FWD compensated for by a higher base layer modulus.

No major differences could be found between the different sensor configurations for FWD, neither by looking at the moduli nor by looking at the deflection deviations. The backcalculation procedure had somewhat greater problems by the iteration process for configuration 3 than for 1 and 2. However, when the deflections passed the deviation tests, the deviations as a mean were marginally the least for configuration 3.

For Dynaflect, geophone set-ups 2 (FDOT) and 3 give marked less deflection deviations than the standard set-up. Set-up 3 seems to have a bit less average deviations than 2, and the least deviations for most cases. Set-up 2, on the other hand, has the advantage of having a geophone by each wheel of the device, giving the possibility to operate on a mean value of these two. Besides, this set-up gives the closest correlation between backcalculated pavement layer modulus and values from the laboratory. Dynaflect geophone set-up 2 (FDOT) therefore was recommended as the new standard set-up.

From 1996 a variant of the FDOT geophone set-up has been used for Dynaflect within the Norwegian Public Roads Administration. This variant includes the four innermost geophones, but has no demand for the outermost sensor. The SHRP configuration is adopted as the standard set-up for FWD measurements.

REFERENCES

1. MORK, H.: EMODUL ved ulike sensorplasseringar, SINTEF-report STF61F94009 (In Norwegian), Trondheim, 1994

2. VAN CAUWELAERT, F. J.; ALEXANDER, D. R.; WHITE, T. D.; BARKER, W. R.: Multilayer Elastic Program for Backcalculating Layer Moduli in Pavement Evaluation, First International Symposium on Nondestructive Testing of Pavements and Backcalculation of Moduli, ASTM STP 1026, Baltimore, 1989

3. MORK, H.: Optimal sensorplassering for fallodd og Dynaflect, SINTEF-report STF61F93005 (In Norwegian), Trondheim, 1993

4. RUTH, B. E.; TIA, M.; BLOOMQUIST, D.; BADU-TWENEBOAH, K.: Structural Characterization and Stress Analysis of Flexible Pavement Systems, Department of Civil Engineering, University of Florida, Gainesville, 1987

5. BADU-TWENEBOAH, K.; MANZIONE, C. W.; RUTH, B. E.; MILEY, W. G.: Prediction of Flexible Pavement Layer Moduli from Dynaflect and FWD Deflections, First International Symposium on Non-destructive Testing of Pavements and Backcalculation of Moduli, ASTM STP 1026, Baltimore, 1989

6. NAZARIAN, S.; BRIGGS, R. C.: Determination of Structural Integrity of Secondary Roads Using Falling Weight Deflectometer Device, Third International Conference on Bearing Capacity of Roads and Airfields, Trondheim, 1990

7. BROWN, S. F.; TAM, W. S.; BRUNTON, J. M.: Structural Evaluation and Overlay Design: Analysis and Implementation, Sixth International Conference Structural Design of Asphalt Pavements, Ann Arbor, 1987

8. MORK, H.: Nedbøyingsmålingar med ulike sensorplasseringar for fallodd og Dynaflect i Sør-Trøndelag hausten 1991. Felt- og laboratorieundersøkingar, SINTEF-report STF61F93004 (In Norwegian), Trondheim, 1993

9. MORK, H.: Nedbøyingsmålingar med ulike sensorplasseringar for fallodd og Dynaflect i Sør-Trøndelag hausten 1991. Etterrekning av E-modular, SINTEF-report STF61F93006 (In Norwegian), Trondheim, 1993

10. IRWIN, L. H.; YANG, W. S.; STUBSTAD, R. N.: Deflection Reading Accuracy and Layer Thickness Accuracy in Backcalculation of Pavement Layer Moduli, First International Symposium on Non-destructive Testing of Pavements and Backcalculation of Moduli, ASTM STP 1026, Baltimore, 1989

Fifth International Conference on the
BEARING CAPASITY OF
ROADS AND AIRFIELDS

EFFECT OF FWD SENSOR LOCATIONS ON BACKCALCULATED LAYER MODULI

[1]SHUICHI KAMEYAMA, [2]KUNIHITO MATSUI, [3]ATSUSHI KASAHARA, [4]KENJI HIMENO AND [5]TERUHIKO MARUYAMA

[1]Associate Professor, Hokkaido Institute of Technology, 7-15, Maeda, Teine-ku, Sapporò 006-8585, JAPAN, Fax +81-11-685-0780, E-mail kameyama@hit.ac.jp
[2]Professor, Tokyo Denki University, Ishizaka, Hatoyama, Hiki, Saitama 350-03 JAPAN, Fax +81-492-96-6501, E-mail matsui@g.dendai.ac.jp
[3]Professor, Hokkaido Institute of Technology, 7-15, Maeda, Teine-ku, Sapporo 006-8585, JAPAN, Fax +81-11-685-0780, E-mail kameyama@hit.ac.jp
[4]Professor, Chuo University, 1-13-27, Kasuga, Bunkyo-ku, Tokyo 112-8551, Japan, Fax: +81-3-3817-1795, E-mail himeno@civil.chuo-u.ac.jp
[5]Professor, Nagaoka University of Technology, 1603-1, Kamitomioka, Nagaoka 940-21, Japan, Fax +81-258-41-9613, E-mail terum@vosccnagaoka.ac.jp

ABSTRUCT

In this study, we investigated the effect that measurement errors have on the backcalculated moduli, and we developed a method using a genetic algorithm that can determine the location of sensors that will minimize this effect, i.e., the optimum sensor locations. Also, in order to verify the validity of the optimum sensor locations, we performed backcalculation simulation with the sensors set in the optimum locations and in the conventional locations.

Key words : Backcalculation, Backcalculated Moduli, Measurement error, Genetic Algorithm, Optimum Sensor Locations

INTRODUCTION

In order to carry out a maintenance and rehabilitation strategy based on Pavement Management System (PMS), appropriate evaluation of the functional and structural capacities of the pavement is needed. The functional capacity of the pavement is evaluated according to surface characteristics such as roughness, rutting and cracking. The structural capacity of pavements is determined by non-destructive deflection tests such as FWD.

Backcalculation is widely used as a method to evaluate the structural capacity of pavements from FWD-measured deflection. This method involves estimating backcalculated moduli of the pavement indirectly from measurements of the surface deflection. The least residual method is widely used as a backcalculating technique. In this method, the summation of square values of the difference between measured and calculated deflections is used as the evaluation function, and the backcalculated moduli which minimizes this is estimated.

When layer moduli are backcalculated from a set of surface deflections measured by FWD, it has been known that the backcalculated moduli are affected by measurement errors as well as model link error[1]. Our previous studies have shown that the effect of measurement error on the backcalculated moduli greatly depends on the number of FWD deflection sensors and the position of the sensors[2]. However, despite the fact that sensor locations is a very important factor in improving the reliability of the backcalculated moduli, there have been very few studies on the appropriate positions for deflection sensors. Deflection sensors are usually set at intervals of 300mm; however, there has been no investigation of whether this is the optimum interval for deflection measurement.

In this study, we investigated the effect that the measurement error has on the backcalculated moduli, and we developed a method using a genetic algorithm[3] that can determine the sensor locations that will minimize the effect of measurement error, i.e., the optimum sensor locations. Also, in order to verify the validity of the optimum sensor locations, we performed backcalculation simulation with the sensors set in the optimum locations and in the conventional locations.

CONDITIONS OF ANALYSIS

Three-layered pavement structures were used in this study. The layer thickness, layer moduli, and Poisson's ratios are shown in Figure 1. It was assumed that the deflection sensors could be set at 50-mm intervals from the center (0 mm) of the loading plate to 2,000 mm in the radial direction except for the points within the radius of the loading plate (50mm, 100mm and 150mm from the center of the loading plate) at which deflection can not be measured. Thus, deflection sensors could be set at a total of 38 positions.

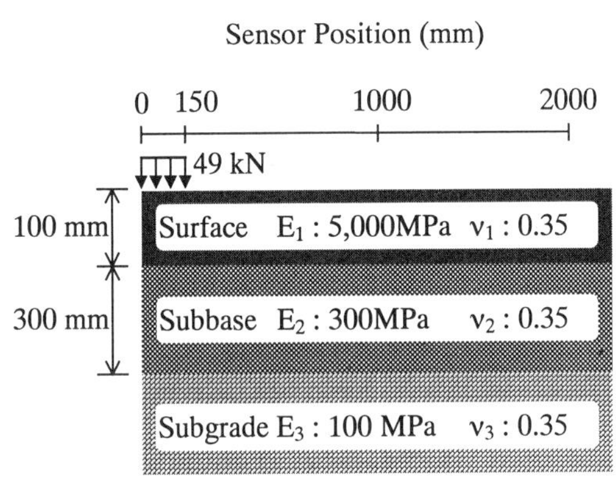

Figure 1. Three layer pavement

APPLICATION OF GENETIC ALGORITHM

We used a genetic algorithm (GA) as a method for optimizing the sensor positions. The concept of GA was derived from the evolutionary theory of biological species. The flowchart of the GA is shown in Figure 2, and each stage of the algorithm is explained below.

Method for expressing genes It is assumed that one chromosome has 38 genes as shown in Figure 3. The number of genes corresponds to the number of points at which deflection sensors can be set. Each gene is expressed by 0 or 1; 1 means a deflection sensor has been set.

For example, if the gene is 1000110..., it means deflection sensors are set at 0, 350, 400 mm, and so on. In this analysis, the number of measurement points was varied from 5 to 10, the range in which the number of sensors currently used in most FWD measurements falls due to operational and economic considerations.

Initialization of genes As shown in Figure 4, n chromosomes containing genes are generated randomly to form the initial group. Although the appropriate number of chromosomes varies according to the optimization problem, and there is no standard, generally 30 to 150 chromosomes are considered to be a sufficient number[4]. In this study, 50 chromosomes were used in consideration of the calculation time.

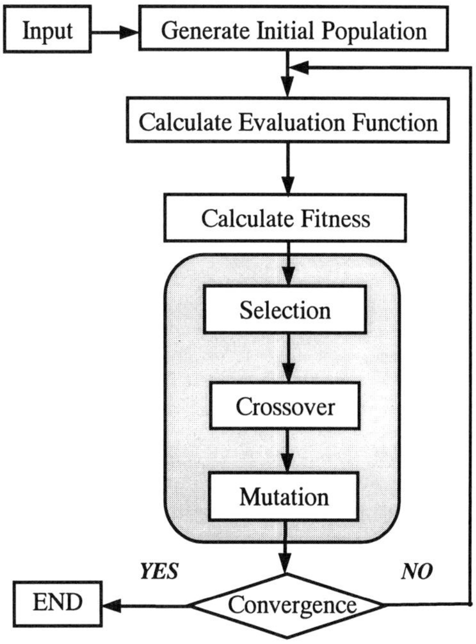

Figure 2. Flow of genetic algorithm

Calculation of the evaluation function

Assume that surface deflection of an M-layered pavement structure was measured at N points using an FWD. The Gauss-Newton method is one of the iterative methods that repeatedly adjust the layer moduli and call a mechanistic analysis program until a suitable match between the measured and calculated deflection basins is obtained[5].

If we assume that the deflection reading error at each sensor is random and that there is no correlation between measurement errors, then the sensitivity of the backcalculated moduli with respect to deflection reading error can be calculated by Equation [1] (1).

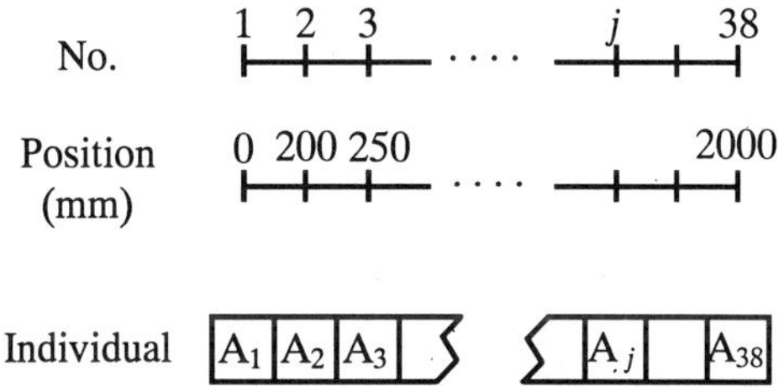

Where, $A_j \begin{cases} 0 & \cdots \text{ Sensor is set.} \\ 1 & \cdots \text{ Sensor is not set.} \end{cases}$

Figure 3. Genotype

$$\sum_{j=1}^{M}\left(\sum_{i=1}^{N}\frac{\partial z_i}{\partial E_j}\frac{\partial z_i}{\partial E_k}\right)\lambda_{jl} = \frac{\partial z_l}{\partial E_k} \qquad (1)$$

$$(l = 1,\ldots, N),\ (k = 1,\ldots, M)$$

Here, λ_{jl} is $\partial E_j / \partial u_l$, which is the sensitivity of the j th layer modulus with respect to deflection reading error at measurement point l.

If we let us be the true value of deflection at each point and Δu_i the error, the layer modulus E_j of layer j will be affected by errors at all measurement points, and therefore Equation (1) can be written as Equation (2):

$$E_j(u_0+\Delta u_0) = E_j(u_0) + \sum_{i=1}^{N} \frac{\partial E_j}{\partial u_i} \Delta u_i \qquad (2)$$

$$(j = 1,..., M)$$

Where,

$$u_0 = (u_{01}, u_{02}, ..., u_{0N})^T$$
$$\Delta u_0 = (\Delta u_{01}, \Delta u_{02}, ..., \Delta u_{0N})^T$$

If σ_i^2 (i = 1,..., N) is the variance of deflection at point i, the variance of deflection at point i for R number of measurements is σ_i^2/R, and, from Equation (2), the variance and covariance matrix of backcalculated moduli can be written as:

$$\left[\sigma_{EE}^{u}\right] = \left[\frac{\partial E}{\partial u}\right]^T \begin{bmatrix} \ddots & & \\ & \sigma_i^2/R & \\ & & \ddots \end{bmatrix} \left[\frac{\partial E}{\partial u}\right] \qquad (3)$$

Where,

$$\left[\frac{\partial E}{\partial u}\right] = \begin{bmatrix} \frac{\partial E_1}{\partial u_1} & \frac{\partial E_2}{\partial u_1} & \cdots & \frac{\partial E_M}{\partial u_1} \\ \frac{\partial E_1}{\partial u_2} & \frac{\partial E_2}{\partial u_2} & \cdots & \frac{\partial E_M}{\partial u_2} \\ \vdots & \vdots & & \vdots \\ \frac{\partial E_1}{\partial u_N} & \frac{\partial E_2}{\partial u_N} & \cdots & \frac{\partial E_M}{\partial u_N} \end{bmatrix} \qquad \begin{bmatrix} \ddots & & \\ & \sigma_i^2/R & \\ & & \ddots \end{bmatrix} = \begin{bmatrix} \sigma_1^2/R & & & 0 \\ & \sigma_2^2/R & & \\ & & \ddots & \\ 0 & & & \sigma_N^2/R \end{bmatrix}$$

The trace of this variance and covariance matrix is the evaluation function J. The measurement accuracy of the FWD deflection sensor is within 2μm, and it can be assumed that this value does not depend on the deflection[6]. Thus, in this study, we assumed that the standard deviation (σ_i) of deflection measured by all sensors was 2μm, and we calculated the evaluation function J for each chromosome by Equation (4).

$$J = Trace\left(\left[\sigma_{EE}^{u}\right]\right) = \sum_{j=1}^{M} \sum_{i=1}^{N} \left(\frac{\partial E_j}{\partial u_l}\right)^2 \left(\frac{\sigma_i^2}{R}\right) \qquad (4)$$

To obtain the minimum value by GA, the following expression converts the evaluation function J into fitness F:

$$F = U - J \tag{5}$$

where U is a constant such that $U > J$ for all chromosomes. As U may not be available a priori, generally the maximum J of past generations is used for U.

Sampling mechanism First, we list the chromosomes in the group in the order of fitness. Next, the chromosomes to be used for reproduction are selected. This selection is based on the fitness of each chromosome: chromosomes with a high fitness are likely to be used for reproduction, while those with a low fitness are likely to be exterminated. The exterminated chromosomes in the group are supplemented by offspring from the reproduction between the surviving chromosomes which have a high fitness. For crossover, two chromosomes are randomly selected from a group, and these are used to produce one offspring. These two parent chromosomes were allowed to participate in the production of offspring many times. The method of crossover in this study was uni-crossover[7].

Mutation has the effect of bringing variance to the chromosomes produced by the genes as described above. Even if the group remains stationary at a local optimum, mutation will enable the group to proceed from there.

Calculations are repeated until the maximum generation number G_{max} is reached, and the gene of the best chromosome obtained within G_{max} is taken as the solution. In this study, as a decrease in the evaluation functions was not seen after 2,000 generations even with changes in the number of sensors, G_{max} was taken to be 2,000.

OPTIMUM SENSOR LOCATIONS

The locations for 5~10 sensors at which the evaluation functions are minimized were obtained by a GA. Figure 4 shows the relationship between the number of sensors and optimum sensor locations. The optimum sensor locations are distributed in three ranges: center of the loading plate (0mm), 200~400mm, and 1,000~1,200mm. Even if the number of sensors is increased, there are no optimum sensor locations outside the ranges of 200~400mm, and 1,000~1,200mm.

We calculated the evaluation functions when 10 sensors were set at the optimum locations, case 1 (0, 200, 400, 600, 800, 1,000, 1,200, 1,400, 1,600 and 1,800mm) and case 2 (0, 200, 300, 450, 600, 750, 900, 1,200, 1,500 and 2,000mm). In case 1, the sensors were set at 200-mm intervals

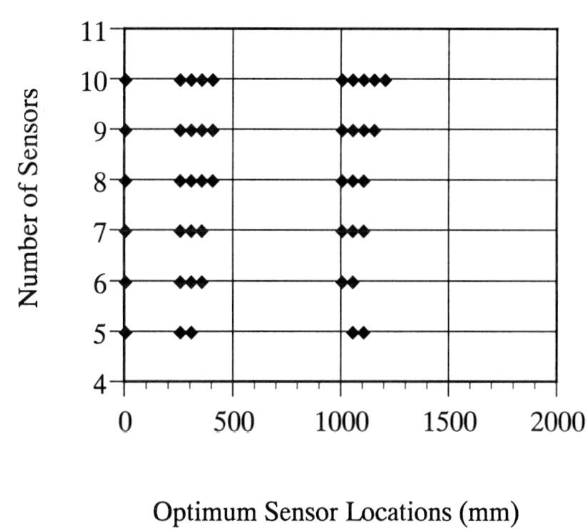

Figure 4. Relationship between the number of sensors and optimum sensor locations

As was the case in E_1, the optimum sensor locations shift with changes in E_2 and E_3, and the effect of the layer moduli is greatest the further away the sensors are from the loading plate.

CONCLUSIONS

The following conclusions were drawn from this study.

1) The effect of errors in the measurement of surface deflection on the back-calculated moduli were evaluated, and the locations of deflection sensors that minimize this effect were determined. The optimum sensor locations for a three-layered pavement structure are distributed in three ranges: 0mm, 200 ~400mm and 1,000~1,200mm. These ranges are the same for any number of deflection sensors in the range of 5~10.
2) Monte Carlo simulation was used to verify the validity of the optimum sensor locations. The standard deviations of backcalculated moduli of the surface layer and subbase estimated from deflections measured at the optimum sensor locations are much smaller than the standard deviations of backcalculated moduli obtained from deflections measured at the conventionally used sensor locations. Thus, the accuracy of backcalculated moduli of the surface layer and subbase can be improved by appropriate selection of the location of deflection sensors. This would mean a reduction in the required number of FWD measurements.
3) The effects of changes in the layer moduli of the surface, subbase and subgrade on optimum sensor locations were investigated. It was found that the locations of all of the sensors except for those in the center of the loading plate moved further away from the loading plate with increases in the surface layer modulus and moved

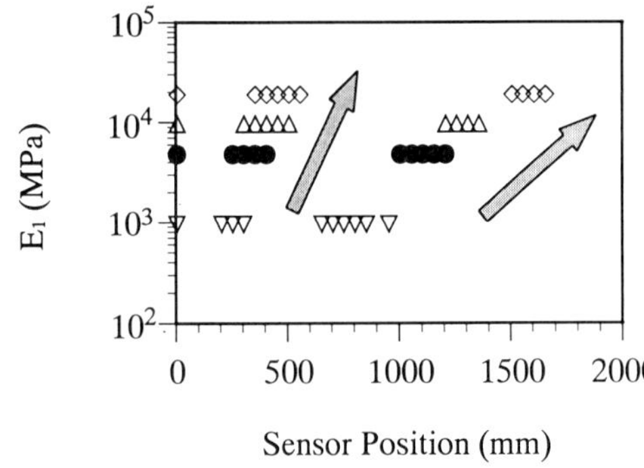

Figure 9. Relationship between optimum sensor locations and E_1

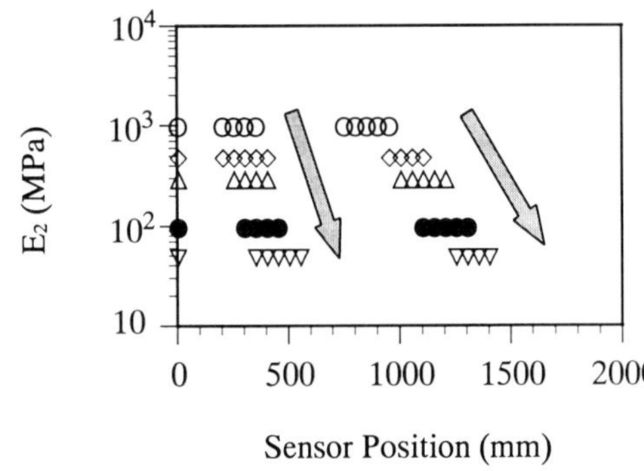

Figure 10. Relationship between optimum sensor locations and E_2

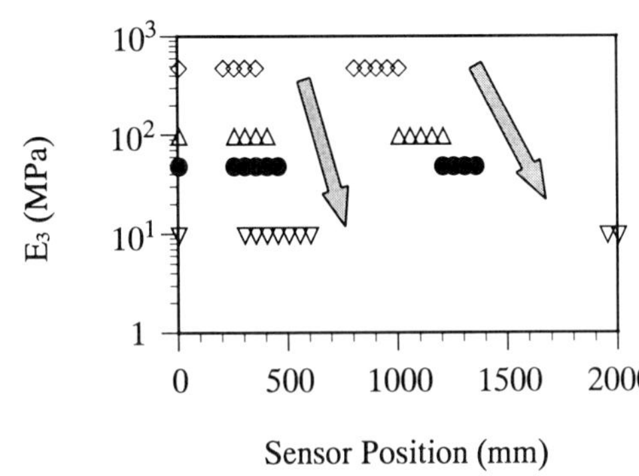

Figure 11. Relationship between optimum sensor locations and E_3

closer to the loading plate with increases in subbase and subgrade layer moduli.

Since it has been shown that the locations of deflection sensors greatly affect the backcalculated moduli, the development of an FWD apparatus that can adjust the sensor locations would be desirable.

REFERENCES

1. Matsui, K., Inoue, T., and Sampei, T. Reflection of Model and Measurement Errors on Stiffness Estimates. Proceeding of the 7th International Conference on Asphalt Pavements, Volume Two, pp.263-276, 1992.
2. Matsui, K., Kasahara, A., and Okada, K. Effect of Errors in Deflection Readings and Pavement Model on Backcalculated Layer Moduli. Proceeding of JSCE, Concrete Engineering and Pavements, No.526/V-29, pp.55-62, 1995 (in Japanese).
3. Gordberg, D. E. Genetic Algorithms in Search, Optimization, and Machine Learning. Addison-Wesley Publishing Company, Inc., 1989.
4. Grefenstette, J. J. Optimization of Control Parameters for Genetic Algorithms. IEEE Trans. On Systems, Man, and Cybernetics, Vol.SMC-16, No.1, 1986.
5. Himeno, K. Maruyama, T. and Kasahara, A. Development of Pavement Evaluation System Using The Falling Weight Deflectometer. Proceeding of 2nd International Symposium on Pavement Evaluation and Overlay Design, pp.4.6.1-4.6.20, 1989.
6. Irwin, L. H., Yang, W. S. and Stubstad, R. N. Deflection Reading Accuracy and Layer Thickness Accuracy in Backcalculation of Pavement Layer Moduli. Nondestructive Testing of Pavements and Backcalculation of Moduli, pp. 229-244, ASTM 1989.
7. Michalewicz, Z., Genetic Algorithms + Data Structures = Evolution Programs, 2nd Extended Edition. Springer-Verlag Berlin Heidelberg New York, 1994.
8. Himeno, K., Maruyama, T. and Hayashi, M. The use of FWD Deflection Data in Mechanistic Analysis of Flexible Pavements. Proceedings of 3rd International Conference on Bearing Capacity of Roads and Airfields, Vol.1, pp.401-410, 1990.

Fifth International Conference on the
BEARING CAPASITY OF
ROADS AND AIRFIELDS

BACK-CALCULATION OF FLEXIBLE PAVEMENTS WITH CONSIDERATION OF TEMPERATURE GRADIENTS

JOÃO ROCHA DE ALMEIDA

Associate Professor, Dept. Civil Eng., New University of Lisbon, 2825 Monte da Caparica, Portugal (Fax: + 351-1-2948398 ; E-mail: jr@mail.fct.unl.pt)
Researcher, ICIST, Av. Rovisco Pais, 1096 Lisboa Codex, Portugal

ABSTRACT

In general, the temperature inside the bituminous layers of a pavement structure varies with depth. Therefore, these layers are subjected to temperature gradients which sometimes are of significant magnitude. As it is well established that the elastic properties of the bituminous mixes are strongly dependent on temperature, attention should be given to thermal gradients when back-analysing flexible pavements. In fact, some guides for field testing operations prescribe the measurement of temperatures at different depths in the pavement. Nevertheless, temperature gradients are very seldom taken into account in current back-calculation procedures.

A simple approach for considering thermal gradients in pavement evaluation is presented herein. In this approach, the pavement structure is modelled as a multi-layered elastic system. The bituminous material is divided into several sub-layers, each with a different temperature corresponding to the values measured "in situ". The back-calculation process is then carried out assuming that the elastic moduli of the bituminous sub-layers are correlated according to appropriate relationships between elastic modulus and temperature obtained experimentally.

In order to assess and validate the methodology proposed, a case study is analysed, consisting of the interpretation of deflection data collected using the Falling Weight Deflectometer (FWD) on a survey in which temperatures were taken at various depths. The results of the back-calculation are discussed, giving particular emphasis to the implications of the possible options of dealing with the temperature data.

Keywords: back-calculation, layer modulus, temperature, FWD.

INTRODUCTION

The deflections on flexible pavements are very dependent on the temperature of the bituminous layers. The asphalt binder is responsible for this temperature sensitivity, for its stiffness decreases sharply when temperature rises. Consequently, monitoring of pavement temperatures along with the adoption of techniques for quantifying the influence of temperature on the elastic properties of bituminous materials forms a fundamental part of a pavement back-calculation procedure.

"In situ" pavement testing may be carried out throughout the year, spanning a wide range of temperatures. Hence, it is necessary that the results obtained from a field survey (under particular temperature conditions) can be compared with results obtained under other conditions or converted to standard conditions appropriate for design. In thick bituminous layers, the pavement temperature is not generally uniform in depth. Thus, it may be useful to dispose of a back-calculation technique capable of taking into account the effect of thermal gradients in the bituminous layers of a pavement structure.

ACQUISITION OF TEMPERATURE DATA

In a field survey, it is essential to know not only the air temperature but also the pavement surface temperature at the time of testing. The surface temperature can be readily determined if a measuring device is installed on the falling weight. In this manner, the surface temperature can be obtained simultaneously with the recording of deflections. For this purpose, use can be made of infra red sensors.

Although the temperature inside the pavement can be estimated with reasonable accuracy from the knowledge of the air or surface temperatures, it is strongly recommended to measure it directly using mercury thermometers or electric thermocouples. To enable execution of temperature measurements at depth, holes with diameter of usually 5 to 7 mm are drilled into the pavement at various locations.

After a hole is drilled, a small quantity of fluid is poured into the hole so that the sensitive tip of the measuring instrument is covered. The fluid (glycerol or mineral oil) serves as a contact medium between the bituminous mix and the thermometer. Since heat is released during drilling, the pavement temperature must not be taken immediately after drilling but only after a minimum of 15 minutes.

Very often, all holes are drilled to the same depth beneath the pavement surface (e.g., 40 mm, as prescribed by TRL [1]). Nevertheless, particularly for pavements with thick bituminous layers, it is advisable to follow the procedure proposed by the US Strategic Highway Research Program (SHRP) [2] and also adopted in the Netherlands [3]. According to this procedure, the temperature is taken at a depth of 25 mm, at the mid-depth of the total bituminous thickness and at 25 mm above the bottom of the bituminous layers, as shown in Figure 1. A different hole must be drilled for every depth at which a temperature measurement is required. The distance between holes must be at least 0.5 m. Since the exact layer thicknesses are probably not known when the temperature is measured, it may be required to make use of data from specifications or other records to estimate the thickness of the bituminous layers. The temperature sensors must have a read-out resolution of 0.1°C and the times of measurement must be recorded to the minute, being also important to measure accurately the actual depth of each drill hole.

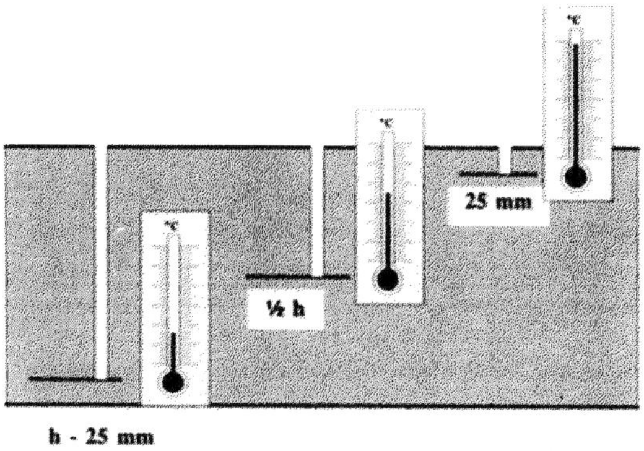

Figure 1 – Measurement of pavement temperatures at several depths (after CROW [3])

RELATIONSHIPS BETWEEN MODULUS AND TEMPERATURE

In order to model the effect of temperature gradients on the structural behaviour of a flexible pavement, use can be made of relationships for characterising the variation of the elastic modulus of the bituminous mixes with temperature.

A relationship of this kind was developed by Ullidtz and Peattie [4], derived from investigation of the stiffnesses of typical bituminous mixes using the Shell procedure [5], together with analysis of deflection data from the AASHO road test. That relationship is:

$$\frac{E_T}{E_{15}} = 1 - 1.384 \log_{10} \frac{T}{15} \qquad (1)$$

Where: E_T, E_{15} = modulus at a temperature T (>1°C) and at a reference temperature of 15°C.

Equation 1 may be expressed in a different form, relating the moduli at two different temperatures, T_1 and T_2 (in °C), as follows:

$$\frac{E_{T_1}}{E_{T_2}} = \frac{2.6277 - 1.384 \log_{10} T_1}{2.6277 - 1.384 \log_{10} T_2} \qquad (2)$$

Where: E_{T_1}, E_{T_2} = modulus at temperatures T_1 and T_2, respectively.

Another expression for modelling the variation of stiffness with temperature, presented by Rada et al. [6] and adopted in the AASHTO Design Guide [7], is:

$$\frac{E_{T_1}}{E_{T_2}} = 10^{3.245\text{x}10^{-4} (T_1^{1.798} - T_2^{1.798})} \qquad (3)$$

Where all symbols have the same meaning as in Equation 2 but temperatures are in °F.

Antunes [8], based on the analysis of back-calculated moduli obtained from FWD data collected at different temperatures, proposed the following linear relationships between elastic modulus and temperature for two types of bituminous mixes:

Asphalt concrete: $$\frac{E_{T_1}}{E_{T_2}} = \frac{1.635 - 0.0317\, T_1}{1.635 - 0.0317\, T_2} \qquad (4)$$

Bituminous macadam: $$\frac{E_{T_1}}{E_{T_2}} = \frac{1.795 - 0.0398\, T_1}{1.795 - 0.0398\, T_2} \qquad (5)$$

Where all symbols have the same meaning as in Equation 2.

The variation of modulus with temperature according to each of the relationships presented is shown in Figure 2. In this figure, the reference temperature is taken equal to 20°C, a value adopted for design in many procedures.

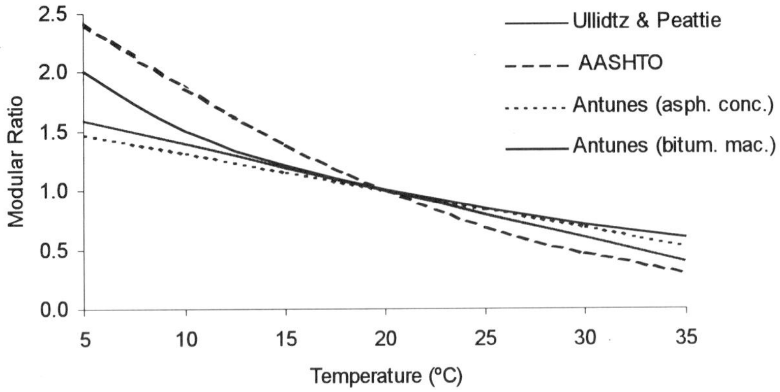

Figure 2 – Relationships between modulus of bituminous mixes and temperature

Although the AASHTO guide does not refer how its relationship was formulated, it possibly resulted from a predictive procedure in which the stiffness of a typical bituminous mix is estimated in function of its composition, temperature and loading time. This scheme is based on the hypothesis that the back-calculated "in situ" moduli vary with temperature at the same ratio as shown in the design curves. It was found that the relationships established using this approach normally exhibit a greater variation of the elastic modulus with temperature [8]. This could explain the greater temperature sensitivity displayed by the AASHTO relationship in Figure 2.

In any case, the degree of susceptibility to temperature should be assessed on the basis of the characteristics of the bituminous mixes. Thus, whenever possible, laboratory testing should be performed on samples taken from the road in order to select the relationship most adequate for a particular type of bituminous mix.

BACK-CALCULATION PROGRAM

The back-calculation program used in this study is LEAD (Layered Elastic Analysis of Deflections), developed by Almeida [9] for the estimation of the elastic properties of the pavement layers given a measured deflection basin. This program performs the interpretation of nondestructive data collected by the FWD.

LEAD follows an iterative scheme whereby the elastic moduli of the paving materials are successively adjusted so as to achieve the best possible match between the computed set of deflections and the surface deflections measured using the FWD. For the calculation of pavement deflections, LEAD makes use of the program BISTRO [10], in which the pavement structure is modeled as a semi-infinite, axi-symmetric and multi-layered system.

Knowing the surface deflections measured at n different radial locations, LEAD determines the values of m elastic parameters X which lead to the best fit between the measured and the computed deflections. The solution is found through minimization of an objective function F(X), given by:

$$F(X) = \sum_{i=1}^{n} W_i (u_i - u^*_i(X))^2 \tag{6}$$

Where: n = number of sensors where deflections are measured.
u_i = measured deflection at sensor i.
u^*_i = computed deflection at sensor i.
W_i = weighting coefficients ("weights").

Depending on the user's choice, the function F(X) can represent:
- the sum of squares of the absolute deflection differences (if $W_i=1$).
- the sum of squares of the relative deflection differences (if $W_i=1/u_i^2$).
- any other weighted sum of squares of the deflection differences, with user-defined weights.

The Gauss-Newton algorithm [9] is used to find the set of elastic parameters that reproduce more closely the measured data. According to this method, in each iteration an adjustment $\{\Delta X\}$ is added to the previous estimate of elastic parameters, $\{X\}$, in order to minimize the function F(X). The adjustment $\{\Delta X\}$ is obtained by solving the following system of simultaneous equations:

$$[S]\{\Delta X\}=\{R\} \tag{7}$$

Where: $S_{ij} = \sum_{k=1}^{n} W_k \frac{\partial u^*_k}{\partial X_i} \frac{\partial u^*_k}{\partial X_j}$ and $R_j = \sum_{k=1}^{n} W_k (u_k - u^*_k) \frac{\partial u^*_k}{\partial X_j}$

Indices i and j vary both from 1 to m

This method requires the evaluation in each iteration of the derivatives $\partial u^*_i/\partial X_j$, i.e., the sensitivities of the computed deflections u^*_i with respect to the unknown parameters X_j. Each sensitivity factor is obtained using a forward finite difference scheme, i.e., by increasing the value of X_j by 1% while keeping all the remaining parameters unchanged and determining the corresponding change in the value of u^*_i.

The iterations continue until the reduction in the value of the error function between successive iterations becomes very small (lower than a predefined limit).

LEAD can model non-linear behaviour of unbound granular materials and soils. However, since this research is primarily concerned with the characterization of bitumen bound layers, all materials are assumed as linear elastic in the subsequent analyses presented herein.

CONSIDERATION OF TEMPERATURE GRADIENTS

A simplified methodology commonly adopted involves the determination of an equivalent temperature for the whole bituminous layer that can, in an approximate way, reproduce the effect of temperature gradients. Some procedures take for equivalent temperature the value measured at a given depth (say, one-third or half of the asphalt thickness) [7]. However, if temperatures are available at various depths, it is more appropriate to define the equivalent temperature, T_{eq}, as [8]:

$$T_{eq} = \frac{\int_0^h T(z)\,dz}{h} \qquad (8)$$

Where: h = overall thickness of the bituminous layers.
z = vertical co-ordinate.
T(z) = diagram of temperature with depth.

The equivalent temperature is not used during the iterative process of back-calculation, being only necessary for performing the adjustment of the back-calculated elastic moduli from the "in situ" temperature to a standard temperature.

A more elaborated approach should be able to model a non-uniform temperature distribution. In fact, the temperature varies with depth into the pavement and so does the elastic modulus of the bituminous mixes. Therefore, it would appear logical to perform the back-calculation considering a series of independent bituminous sub-layers, each with a different modulus. However, this approach is not feasible in practical terms, due to the following reasons:

- The number of elastic moduli to be determined could become very large in comparison to the number of measured deflections. This would turn the back-calculation in a problem numerically ill conditioned or even undetermined if the number of unknown elastic parameters would exceed the number of deflections.
- Estimation of the elastic properties of very thin layers or adjacent layers with similar characteristics poses great difficulties. In these situations, it is not normally possible to determine accurately the contribution of each individual layer to the pavement structural behaviour, either because it is small (case of thin layers) or because it is difficult to discriminate (case of similar layers). The sets of moduli obtained in these cases are often unrealistic.

To overcome these limitations, an alternative way for taking into account thermal gradients in pavement back-calculation was developed. The technique proposed relies on the assumption that the dependence of the elastic stiffness of bituminous materials on temperature can be described by a relationship such as the ones given by Equations 2 to 5, requiring also the knowledge of the temperature at different depths within the pavement.

When modelling the pavement structure, the bituminous material is divided into several sub-layers. The thicknesses of these sub-layers are such that the mid-depth of every sub-layer coincides with a depth where the temperature has been measured. To each sub-layer is then assigned a constant temperature equal to the value recorded "in situ" at the corresponding mid-depth.

The back-calculation is performed using the algorithm described previously. However, only one of the bituminous sub-layer moduli is taken as an unknown parameter to be estimated in the back-calculation. In fact, at any stage of the computation, given a certain value of this parameter, the remaining sub-layer moduli are automatically determined using the relationship adopted between modulus and temperature.

This approach has been incorporated in a modified version of program LEAD, in which Equations 2 to 5 were implemented for evaluating the variation in modulus induced by temperature.

CASE STUDY

To illustrate the technique proposed for dealing with temperature gradients, an example is presented concerning the interpretation of FWD data collected in Portugal by the Laboratório Nacional de Engenharia Civil (LNEC), from a full-scale test section at the Alcácer do Sal by-pass to the EN5 and EN120 roads [11].

The pavement analysed was built in 1992 and is composed of:
- Wearing course of asphalt concrete with a thickness of 5 cm.
- Binder course (for pavement levelling) of dense bituminous mix with a thickness of 5 cm.
- Road base of bituminous macadam with a thickness of 10 cm.
- Granular base of crushed limestone with a thickness of 20 cm.
- Sub-base of well-graded sand with a thickness of 20 cm, laid directly on the subgrade.

FWD testing was carried out both at the offside wheel path and at the lane centre, with a spacing of 20 m between consecutive test points. In order to assess the effect of temperature on the structural behaviour of the bituminous layers, several surveys were conducted at different times of year. In each survey, the pavement temperature was measured periodically at the surface and at depths of 25, 50, 75, 100, 125 and 150 mm.

The tests were carried out at three load levels: 20, 40 and 65 kN (the two latter values correspond to standard wheel loads commonly adopted in European countries). The loads were transmitted to the pavement by a circular plate with a diameter of 30 cm. The deflection sensors were positioned at distances of respectively 0, 30, 45, 60, 90, 150 e 250 cm away from the centre of the loading plate.

Only data from a survey carried out in September of 1994 is analysed here, for the greatest temperature gradients were registered on that survey. The deflections exhibited a reasonable degree of uniformity throughout the section. In view of that and of the reduced length of the section (200 m), no division into sub-sections was necessary. From coring, it was found that the actual thicknesses of the bituminous layers were slightly lower than the nominal values. The average values obtained were: 44 mm for the wearing course, 45 mm for the binder course and 87 mm for the road base.

To avoid problems that usually occur in back-calculation when thin layers or similar layers are considered separately (as discussed above), the various bituminous courses were treated as an homogeneous layer. For the same reason, the granular base and sub-base were grouped into a single layer. Hence, the pavement structure to be back-calculated was:
- Bituminous layer 176 mm thick.
- Granular layer 400 mm thick.
- Semi-infinite subgrade.

As a single bituminous material was considered in the back-calculation, it had to be decided which relationship between modulus and temperature presented by Antunes should be selected, the one for asphalt concrete or the one for bituminous macadam. From laboratory testing on cores taken from the pavement, it was concluded that the dense bituminous mix and the bituminous macadam had similar properties in terms of volumetric composition and grading [11]. Since the combined thickness of these two courses was much larger than the wearing course thickness, it was decided to adopt the relationship between modulus and temperature developed for bituminous macadam.

For the objective of taking into account temperature gradients, the bituminous material was partitioned into seven sub-layers, each one corresponding to a temperature reading taken at a different depth, as shown in Figure 3.

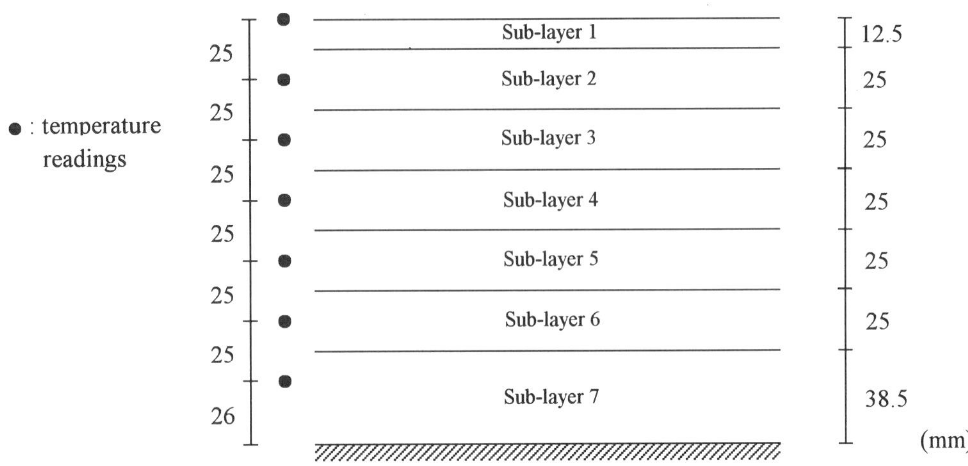

Figure 3 – Temperature readings and corresponding subdivision of bituminous material

A Poisson's ratio of 0.35 was assumed for all pavement layers. For each load level and loading path, the deflection basin selected for analysis was the one that more closely resembled the fictitious basin resulting from the average of deflections measured in all test points.

The outer deflection sensor had been positioned at 250 cm away from the load, a radial distance greater than usually adopted in FWD surveys. Hence, the corresponding deflections measured at this position were very small, particularly for the lower load level, often reaching values of the same order of magnitude as the errors of the measuring device. Furthermore, it is known that temperature variations only induce significant changes in the top pavement layers, having hardly any effect on the furthest deflections. For these reasons, the deflections at a radius of 250 cm were removed from the back-calculation process.

The back-calculation was carried out using program LEAD. Two approaches were considered: approach 1, where the bituminous material is divided into several layers, and approach 2, where only one bituminous layer is required due to the equivalent temperature concept. Both approaches rely on relationships between temperature and bituminous layer modulus, although in approach 2 these relationships are only required for correcting the back-calculated value to a standard temperature. The solution is obtained by minimising the sum of squares of the absolute deflection differences. This option leads to a better match of the deflections closer to the load, enabling to get more accurate estimates of the bituminous layer moduli, as these moduli only have a significant influence on the inner deflections [9].

Figure 4 shows an example of profile of the pavement temperature in depth, together with the respective moduli obtained using approach 1. As would be expected (cf. Figure 2), the AASHTO relationship is responsible for the greatest change in stiffness with temperature, yielding the lowest modulus for higher temperatures (at the top) and the highest modulus for lower temperatures (at the bottom). This variation is less pronounced when using Antunes equation and becomes even smaller when using Ullidtz equation.

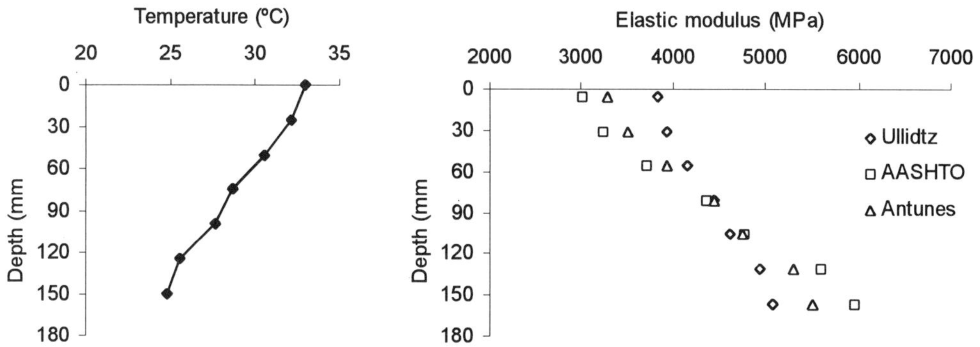

Figure 4 – Temperature and corresponding back-calculated moduli (65 kN load, wheel path)

Table 1 summarises the back-calculation results. For every load level and loading path, is presented the solution found using each approach and relationship considered for modelling temperature gradients. In this table, the elastic modulus of the bituminous material is corrected to a temperature of 25°C, a value adequate for Portuguese design conditions. The back-calculated moduli of the granular layer and of the subgrade are also displayed, together with the average differences between measured and computed deflections.

Table 1 – Back-calculated moduli and deflection differences (Alcácer do Sal by-pass)

		20 kN load				40 kN load				65 kN load			
		wheel path		lane centre		wheel path		lane centre		wheel path		lane centre	
		appr.1	appr.2	appr.1	appr.2	appr.1	appr.2	appr.1	appr.2	appr.1	appr.2	appr.1	appr.2
Ullidtz relation	E_{bit}(MPa)	4367	4385	4945	4963	4686	4707	4721	4747	5046	5072	5278	5316
	E_{gr}(MPa)	209	208	200	198	205	203	239	236	277	274	264	260
	E_{sg}(MPa)	234	234	296	296	222	221	236	236	222	222	234	234
	def.dif.(%)	3.74	3.77	3.81	3.82	2.76	2.78	4.71	4.73	2.50	2.53	5.91	5.92
AASHTO relation	E_{bit}(MPa)	5195	5209	5893	5896	5576	5592	5611	5639	5869	5896	6189	6183
	E_{gr}(MPa)	213	208	202	198	208	203	243	236	282	274	266	260
	E_{sg}(MPa)	234	234	296	296	222	221	235	236	222	222	234	234
	def.dif.(%)	3.73	3.77	3.82	3.82	2.75	2.78	4.71	4.73	2.49	2.53	5.87	5.92
Antunes relation	E_{bit}(MPa)	4800	4742	5442	5366	5152	5089	5183	5133	5460	5420	5705	5681
	E_{gr}(MPa)	213	208	202	198	208	203	242	236	280	274	268	260
	E_{sg}(MPa)	234	234	296	296	222	221	235	236	222	222	234	234
	def.dif.(%)	3.75	3.77	3.84	3.82	2.77	2.78	4.74	4.73	2.51	2.53	5.93	5.92

Compared to typical values, the subgradc moduli arc quitc high, for these represent overall values that are likely to be affected by the increase in stiffness normally observed in the deeper soil. Hence, the stiffness of the upper subgrade is probably overestimated. For an improved modelling of the subgrade, a rigid bottom should be prescribed at a certain depth or a stress-dependent constitutive law should be adopted. However, for simplicity, this was not tried, since the study mainly focused on the properties of the bituminous layers.

It can be noticed than an increase in load corresponds generally to an increase in the stiffness of the granular layers and to a decrease in the subgrade modulus. These variations are in agreement with both the stress hardening and the stress softening behaviour normally attributed to granular materials and clayey soils, respectively [9]. With respect to the load path, the layer moduli obtained at the wheel path and at the lane centre are fairly similar.

For each deflection basin analysed, the moduli of the unbound layers and the average deflection match remain practically unchanged regardless of the formulation followed. However, some differences occur concerning the modulus of the bituminous layer. In fact, when using Ullidtz relationship, this modulus is consistently lower than when applying Antunes equation, which in turn yields values always lower than the ones obtained using the AASHTO equation. It is worth noting that the moduli derived from approaches 1 and 2 are almost identical for all basins, which suggests that the equivalent temperature approach may be satisfactory for modelling the effect of temperature gradients in the bituminous layers.

CONCLUSIONS

The existence of temperature gradients can have a relevant effect on the resilient response of flexible pavements. To evaluate this effect, temperature readings should be taken at various depths into the pavement and use should be made of relationships between temperature and elastic modulus. This information enables to take temperature gradients into account, by means of two types of approach: one, where the bituminous material is modelled as a series of layers with modulus ratios given by their temperatures, and other, in which an equivalent temperature is used for characterising the overall behaviour of the bituminous material.

From analysis of FWD data, it was found that both approaches lead to similar results but the back-calculated modulus of the bituminous layer can be significantly affected by the type of equation adopted for describing the temperature sensitivity of bituminous mixes. This may have consequences in terms of the fatigue cracking design criterion and underlines the need for selecting relationships that can accurately represent the influence of temperature on the resilient response of bituminous materials.

REFERENCES

1. Kennedy, C.L., "Pavement deflection: operating procedures for use in the UK", Laboratory Report 835, Transport and Road Research Laboratory, Crowthorne, 1978.
2. SHRP, "SHRP-LTPP manual for FWD testing: version 1.0", 1989.
3. CROW, "Deflection profile - not a pitfall anymore" (in Dutch), C.R.O.W., Ede, 1995.
4. Ullidtz, P.; Peattie K.R., "Programmable calculators in the assessment of overlays and maintenance strategies", Proc. 5th Int. Conf. on the Structural Design of Asphalt Pavements, Vol.1, Delft, pp.671-681, 1982.
5. Shell International Petroleum Co. Ltd., "Shell pavement design manual", London, 1978.
6. Rada, G.R.; Witczak, M.W.; Rabinow, S.D., "Comparison of AASHTO structural evaluation techniques using nondestructive deflection testing", Transportation Research Record 1207, Transportation Research Board, Washington DC, pp.134-144, 1988.
7. AASHTO, "Guide for design of pavement structures", Washington DC, 1986.
8. Antunes, M.L., "Pavement bearing capacity evaluation using dynamic non destructive tests" (in Portuguese), Ph.D. Thesis, IST, Lisbon, 1993.
9. Almeida, J.R. de; Brown, S.F.; Thom, N.H., "A pavement evaluation procedure incorporating material nonlinearity", ASTM STP 1198, pp.218-232, 1994.
10. Peutz, M.G.F.; Van Kempen, H.P.M.; Jones, A., "Layered systems under normal surface loads: computer program BISTRO", Koninklijke/Shell-Laboratorium, Amsterdam, 1968.
11. LNEC, "By-pass to EN5 and EN120 in Alcácer do Sal: observation of full-scale test pavement behaviour" (in Portuguese), LNEC, Lisbon, 1995.

Fifth International Conference on the
BEARING CAPASITY OF
ROADS AND AIRFIELDS

DETERMINATION OF VISCOELASTIC PROPERTIES OF ASPHALT CONCRETE LAYER USING FALLING WEIGHT DEFLECTOMETER(FWD) TESTS

JONG RYEOL KIM[1] and LESLIE TITUS-GLOVER[2]

[1] Assistant Professor, University of Hawaii, Honolulu, Hawaii, USA
[2] Research Engineer, ERES Consultants, Inc., Champaign, Illinois, USA

ABSTRACT

A simple dynamic analysis procedure for the falling weight deflectometer (FWD) test on airfield asphalt concrete pavements is presented. Several nondestructive tests were performed along asphalt pavements using Dynatest™ FWD equipment. The time histories of the surface deflections due to the dynamic load were recorded by several receivers. The asphalt concrete layer was assumed to be viscoelastic (Maxwell model), and the others were elastic. This assumption was verified by the hysteretic loop method. The viscoelastic concepts were employed to explain the time-delayed response. The dynamic properties (complex modulus and complex compliance) of the asphalt concrete layer and the material constants for the Maxwell model were backcalculated based upon the test results.

INTRODUCTION

Knowledge of pavement material properties and subgrade is essential for evaluating pavement structural capacity. The material properties in existing pavements can be obtained through numerous laboratory tests and in-situ nondestructive tests. These material properties, together with variables such as the geometry and magnitude of future wheel loads and pavement thickness, are used with mechanical and empirical methods to assess remaining life and future pavement response with respect to stress, strain, and deflections in the pavement layers.

The falling weight deflectometer (FWD) test is the most widespread nondestructive test (NDT) to evaluate airfield pavement capacity because of its simplicity and ability to impose high amplitude dynamic loads. In this test, the dynamic load is applied to the pavement and the surface deflections are recorded by several sensors on the pavement surface. The process, which uses layer thickness information and theoretical models to determine the material response parameters for each layer within the pavement structure from deflection basin measurements, is known as backcalculation.

Most of the tools currently used for backcalculation of pavement moduli are computer programs based upon the multi-layer elastic theory. Examples of these programs include BISDEF, CHEVDEF, ELSDEF, EVERCALC, and WESDEF. In these programs, the viscous property is ignored and only the peak load and the peak deflection are taken into account to calculate the elastic modulus for each structural layer such that the difference between the measured and predicted peak deflection basins is minimal. The results of these programs may be questionable when they analyze the FWD test on asphalt concrete pavement because asphalt concrete is known to be viscoelastic. Recently, some attempts have been made to account for the viscoelastic properties (Stubbs et al, 1994).

This paper presents a dynamic analysis of the falling weight deflectometer (FWD) tests using the viscoelastic theory. Some considerations are presented that may assist in analyzing FWD tests. The theoretical considerations are supplemented with a series of FWD tests performed on asphalt concrete pavement.

LOAD-DEFLECTION HISTORY OF FWD TESTS

In general, the loading history of the FWD test is the same as one step of a haversine load (Fig.1), and its magnitude $F(t)$ can be written as:

$$F(t) = F_0 \, (1 - \cos \omega t) \tag{1}$$

where t is the time, F_0 is the load amplitude, ω is the angular frequency for the load ($\omega = 2\pi/T$), and T is the load period. Equation 1 shows that the haversine load is the summation of a constant load, F_0, and a sinusoidal load, $F_0 \cos \omega t$.

When a material is subjected to an oscillatory (haversine) load, the deflection response is also oscillatory, and its magnitude depends on the material properties and locations. For example, if the haversine load or the FWD test subjects on the linear elastic halfspace, the vertical deflection $u(t)$ can be calculated by the solution of Ahlvin and Ulery (1962) by changing the static load to the haversine load (the load is uniform on the plate):

$$u_0(t) = \frac{F(t)(1 + \nu)}{\pi a E} [\frac{z}{a} A + (1 - \nu) H] \tag{2}$$

where a is the radius of the loading plate, ν is the Poisson's ratio, E is the elastic modulus, z is the vertical distance from the center of the loading plate, and A and H are functions a, z, and the horizontal distance from the center of the loading plate r, respectively. For the vertical surface deflection at the center of the loading plate $u_0(t)$, $A = 0$ and $H = 2$, Eq.2 becomes:

$$u_0(t) = \frac{2 F(t)(1 - \nu^2)}{\pi a E} = \frac{2(1 - \nu^2)}{\pi a E} F_0 (1 - \cos \omega t) \tag{3}$$

Eqs.1 and 3 show that, theoretically, when the load history is haversine the deflection history is also haversine with the same frequency on the linear elastic halfspace; $u_0(t)$ increases with

$F(t)$ and reaches the peak when $F(t)$ is the maximum. Then, $u_0(t)$ decreases with $F(t)$ and vanishes when $F(t)=0$ (Fig.1a). The trend of the deflection history is similar to Fig.1a for the linear elastic layered system due to the haversine load.

For a viscoelastic material, however, the deflection response lags behind the load history due to the viscous property. The difference between the peak load and the peak deflection is known as a phase angle, $\delta(\omega)$. The deflection history of a linear viscoelastic material due to a haversine load can be expressed as:

$$u(t) = u_c(t) - u_a \cos(\omega t - \delta) \tag{4}$$

where $u_c(t)$ is the creep deflection and u_a is the deflection amplitude. The creep deflection is due to the constant load, F_0, and is related to the creep function, and the sinusoidal deflection, $u_a \cos \omega t$, is due to the sinusoidal load, $F_0 \cos \omega t$.

The vertical surface deflection history at the center of the loading plate for the viscoelastic halfspace can be obtained from Eq.3 using the elastic-viscoelastic correspondence principle. The details of this principle can be found elsewhere (Lee, 1955; Flugge, 1967; Findley et al., 1989).

Fig.1b shows the load and the vertical surface deflection at the center of the loading plate for the viscoelastic halfspace (viscoelastic solid). The deflection recovers fully if the material is solid when the recovery time is infinite. Otherwise, if the material is liquid, the deflection response does not recover fully. Fig.1c is another illustration of the relationship between $F(t)$ and $u_0(t)$ for the viscoelastic halfspace (viscoelastic liquid). The trend of the deflection history will be similar to Fig.1b or 1c if the pavement is a linear viscoelastic layered system or a combination of linear elastic-viscoelastic layered systems.

DYNAMIC PROPERTIES OF LINEAR VISCOELASTIC MATERIAL

When a sinusoidal stress, $\sigma(t)=\sigma_0 \cos \omega t$, is applied to a linear viscoelastic material, the strain is sinusoidal at the same frequency, but it lags by a phase angle, δ, $\epsilon(t)=\epsilon_0 \cos(\omega t-\delta)$, where σ_0 is the stress amplitude and ϵ_0 the strain amplitude. The relationship between stress and strain of the material is therefore a function of σ_0, ϵ_0, and δ.

Fig.2 presents the relationship bctwccn thc sinusoidal stress and its strain response. When the material is linearly viscoelastic, this relationship leads to an elliptical hysteretic loop. The details of the elliptical hysteretic loop for the linear viscoelastic material can be found in the literature (Kim, 1995).

A convenient way of expressing this relationship is in terms of an exponential relation and complex notation. Since $e^{i\omega t} = \cos \omega t + I \sin \omega t$, where I represents the imaginary axis of the complex variable, the sinusoidal stress, $\sigma(t)$, and its strain response, $\epsilon(t)$, can be expressed by using the real part of $e^{i\omega t}$, $\cos \omega t$, as follows:

$$\sigma(t) = \sigma_0 e^{i\omega t} \tag{5}$$

$$\epsilon(t) = \epsilon_0 e^{i(\omega t - \delta)} = \epsilon_0 e^{i\omega t}(\cos\delta - i\sin\delta) \tag{6}$$

Then, the relationship between stress and strain can be expressed with the complex compliance, $J^*(\omega)$:

$$J^*(\omega) = \frac{\epsilon_0}{\sigma_0}(\cos\delta - i\sin\delta) = J_1(\omega) - iJ_2(\omega) \tag{7}$$

where $J_1(\omega)$ is the storage compliance and $J_2(\omega)$ is the loss compliance. The magnitude of the complex compliance, $|J^*(\omega)|$, is defined as:

$$|J^*(\omega)| = \left|\frac{\epsilon_0}{\sigma_0}\right| = [J_1^2(\omega) + J_2^2(\omega)]^{1/2} \tag{8}$$

Similarly, the complex modulus, $E^*(\omega)$, and the magnitude of the complex modulus, $|E^*(\omega)|$, can be defined as:

$$E^*(\omega) = \frac{\sigma_0}{\epsilon_0}(\cos\delta + i\sin\delta) = E_1(\omega) + iE_2(\omega) \tag{9}$$

$$|E^*(\omega)| = \left|\frac{\sigma_0}{\epsilon_0}\right| = [E_1^2(\omega) + E_2^2(\omega)]^{1/2} \tag{10}$$

where $E_1(\omega)$ is the storage modulus and $E_2(\omega)$ is the loss modulus. Comparison of Eqs.7 and 9 shows that $J^*(\omega)$ and $E^*(\omega)$ are reciprocal:

$$J^*(\omega)\, E^*(\omega) = 1 \tag{11}$$

Also, from Eqs.8 and 10 it can be shown that:

$$|J^*(\omega)|\,|E^*(\omega)| = 1\,, \qquad \tan\delta = \frac{J_2(\omega)}{J_1(\omega)} = \frac{E_2(\omega)}{E_1(\omega)} \tag{12}$$

EXPERIMENTS

Several in-situ tests were performed at Robins Air Force Base in Georgia in September 1995 using the Dynatest™ Falling Weight Deflectometer (FWD). The tested section has a macadam layer (15.25 cm) between the asphalt concrete layer (22.85 cm) and the subgrade. The subgrade was classified as SM (silty sands) and was presumed to be present to a substantial depth.

The deflection and load histories were used to determine the structural characteristics of each layer in the pavement. The FWD is a loading device that creates a dynamic load by dropping a mass onto a spring connected to a loading plate. By varying the mass and the height from which the mass is dropped, the magnitude of the load can be changed. The pavement deflection history is measured by seven deflection transducers, one located at the center of the loading plate ($u_0(t)$) and six additional sensors spaced at intervals of 30.5 cm (u_1, u_2, ... , u_6). A frequency of 100/3 Hz was used. Fig.3 is a typical result of a FWD test.

DATA ANALYSIS

Backcalculation Using Elasticity

Elastic layer theory was used to backcalculate the elastic moduli of the asphalt concrete, base, and subgrade layers. The maximum deflection of each location and peak load were used. A two-layer analysis was done initially to obtain the elastic modulus of the asphalt concrete separate from the composite modulus of the underlying layers. Three-layer analyses were then done to obtain the elastic modulus of the macadam base separate from that of the subgrade. The backcalculation results and the assumed Poisson's ratio are shown in Table 1.

Table 1. Summary of Backcalculation Using Linear Elastic Layer Theory

Layer	AC	Base	Subgrade
Elastic Modulus	1038 MPa	345 MPa	241 MPa
Poisson's Ratio	0.35	0.4	0.45

Backcalculation Using Coupling Elasticity and Viscoelasticity

Since the result of the FWD test shows that the surface deflection at the center of the load plate lags behind the load history (Fig.4), the pavement could be assumed as viscoelastic. To verify this assumption, Fig.5 was constructed. The sinusoidal load was plotted as a function of the sinusoidal deflection. The pavement structure was assumed to be a Maxwell model. The sinusoidal load and the sinusoidal deflection were obtained by Eqs.1 and 4, respectively. Because the hysteretic loop is elliptical, the pavement structure or asphalt concrete layer can be assumed to be linear viscoelastic.

A rigorous viscoelastic layer theory could be used to determine the material properties in each layer (Ashton and Moavenzadeh, 1967; Huang, 1967). For simplicity, however, this was not attempted in this study. Instead, a simple coupling elastic-viscoelastic analysis was used to backcalculate the material properties of the asphalt concrete, base, and subgrade layers. The elastic modulus of each layer was determined using elastic layer theory with FWD test results (using the peak load and the peak deflections). If a pavement layer was viscoelastic, the elastic modulus of this layer based on the elastic theory was assumed to be the magnitude of its complex modulus at a given frequency (Eq.10).

In this study, the asphalt concrete layer was assumed to be viscoelastic, and the others were elastic. The following procedure was used to determine the material properties of the pavement:

1. The elastic modulus of each layer, E_i, and the equivalent elastic modulus of the pavement system, E_{sys}, are determined using the elastic layer theory (the peak load and the peak deflection are used).
2. The elastic modulus is assumed to be the magnitude of the complex modulus at a given frequency, $E_i = |E^*(\omega)_I|$, if the layer is viscoelastic. Also the equivalent elastic modulus of the pavement system, E_{sys}, is assumed to be equal to the magnitude of the complex modulus of the pavement system, $|E^*(\omega)_{sys}|$, at a given frequency.
3. The phase angle of the pavement system, $\delta(\omega)_{sys}$, is determined from the load history and the deflection history at the center of the loading plate (difference between the peak load and the peak deflection).
4. Once $|E^*(\omega)_{sys}|$ and $\delta(\omega)_{sys}$ are known, the other viscoelastic properties for the pavement system are calculated as discussed before.
5. The storage modulus of the asphalt concrete layer, $E_1(\omega)_{AC}$, can be backcalculated using elastic layer theory (it is assumed that the storage modulus of each layer, $E_1(\omega)_I$, obeys the elastic layer theory).
6. Once $|E^*(\omega)_{AC}|$ and $E_1(\omega)_{AC}$ are known, the other viscoelastic properties of the asphalt concrete layer are calculated.

In the Maxwell model, a spring and a dash-pot are connected in series. Then, the storage modulus, $E_1(\omega)$, and the loss modulus, $E_2(\omega)$, of the Maxwell model can be written as:

$$E_1(\omega) = \frac{\eta^2\omega^2/R}{1 + \eta^2\omega^2/R^2}, \qquad E_2(\omega) = \frac{\eta\omega}{1 + \eta^2\omega^2/R^2} \tag{13}$$

where R is the elastic (spring) constant and η is the viscous (dash-pot) constant.

For the tested section, $|E^*(\omega)|_{sys}$ is 517 MPa and $\delta(\omega)_{sys}$ is 40° when $\omega = 100/3$ Hz. The storage modulus, $E_1(\omega)$, the loss modulus, $E_2(\omega)$, the complex compliance, $|J^*(\omega)|$, the storage compliance, $J_1(\omega)$, and the loss compliance, $J_2(\omega)$, at a given frequency and temperature can be calculated with Eqs.7 through 12. The elastic constant, R_{AC}, and the viscous constant, η_{AC}, for the asphalt concrete layer can be calculated with Eq.13. Some properties of the pavement system and the asphalt concrete layer are summarized in Table 2.

Table 2. Properties of Pavement System and Asphalt Concrete Layer

Properties	Value	Properties	Value
$\delta(\omega)_{sys}$	40°	$\delta(\omega)_{AC}$	33.6°
$\|E^*(\omega)\|_{sys}$	517 MPa	$\|E^*(\omega)\|_{AC}$	1038 MPa
$E_1(\omega)_{sys}$	396 MPa	$E_1(\omega)_{AC}$	861 MPa
$E_2(\omega)_{sys}$	332 MPa	$E_2(\omega)_{AC}$	571 MPa
R_{AC}	1245 MPa	η_{AC}	56.4 MPa-sec

CONCLUSIONS

Based upon this study, the following conclusions can be made:

1. The results of the FWD tests show that asphalt concrete pavement exhibits viscous properties.
2. The asphalt concrete layer can be assumed to be linear viscoelastic. This assumption can be verified by the elliptical hysteretic loop method.
3. The equivalent elastic modulus of the pavement system, E_{sys}, can be assumed to be equal to the magnitude of the complex modulus of the pavement system, $|E^{*}(\omega)|_{sys}$, at a given frequency if E_{sys} is calculated using the peak load and the peak deflections.
4. The phase angle, $\delta(\omega)_{sys}$, can be determined directly from the load history and the deflection history at the center of the loading plate.
5. The dynamic properties of the asphalt concrete layer can be backcalculated from the results of FWD tests (load history and deflection history) using viscoelastic theory.

REFERENCES

Ahlvin, R.G., and H.H. Ulery (1962), "Tabulated Values for Determining the Complete Patter of Stresses, Strains and Deflections Beneath a Uniform Circular Load on a Homogeneous Half Space," *Bulletin 342*, Highway Research Board, pp. 1-13.

Ashton, J.E., and F. Moavenzadeh (1967), "Analysis of Stresses and Displacements in a Three-Layered Viscoelastic System," *Proc. of 2nd Inter. Conf. on the Struct. Design of Asph. Pave.*, University of Michigan, pp. 209-219.

Findley, W.N., Lai, J.S., and Onaran, K (1989), "Creep and Relaxation of Nonlinear Viscoelastic Materials," Dover Publishing Company.

Flugge, W. (1967), "Viscoelasticity," Blaisdell Publishing Company.

Huang, Y.H. (1967), "Stresses and Displacements in Viscoelastic Layered Systems under Circular Loaded Areas," *Proc. of 2nd Inter. Conf. on the Struct. Design of Asph. Pave.*, University of Michigan, pp.225-224.

Kim, J.R. (1995), "Viscoelastic Analysis of Triaxial Tests on Asphalt Concrete," Ph.D. Thesis, University of Minnesota.

Lee, E.H. (1955), "Stress Analysis in Viscoelastic Bodies," *Quarterly of Applied Mathematics*, vol. 13, pp. 183-190.

Stubbs, N., V.S. Torpunuri, R.L. Lytton, and A.H. Magnuson (1994), "A Methodology to Identify Material Properties in Pavements Modeled as Layered Viscoelastic Halfspaces (Theory)," *Nondestructive Testing of Pavements and Backcalculation of Moduli*, vol.2, ASTM STP 1198, American Society for Testing and Materials, Philadelphia, pp. 53-67.

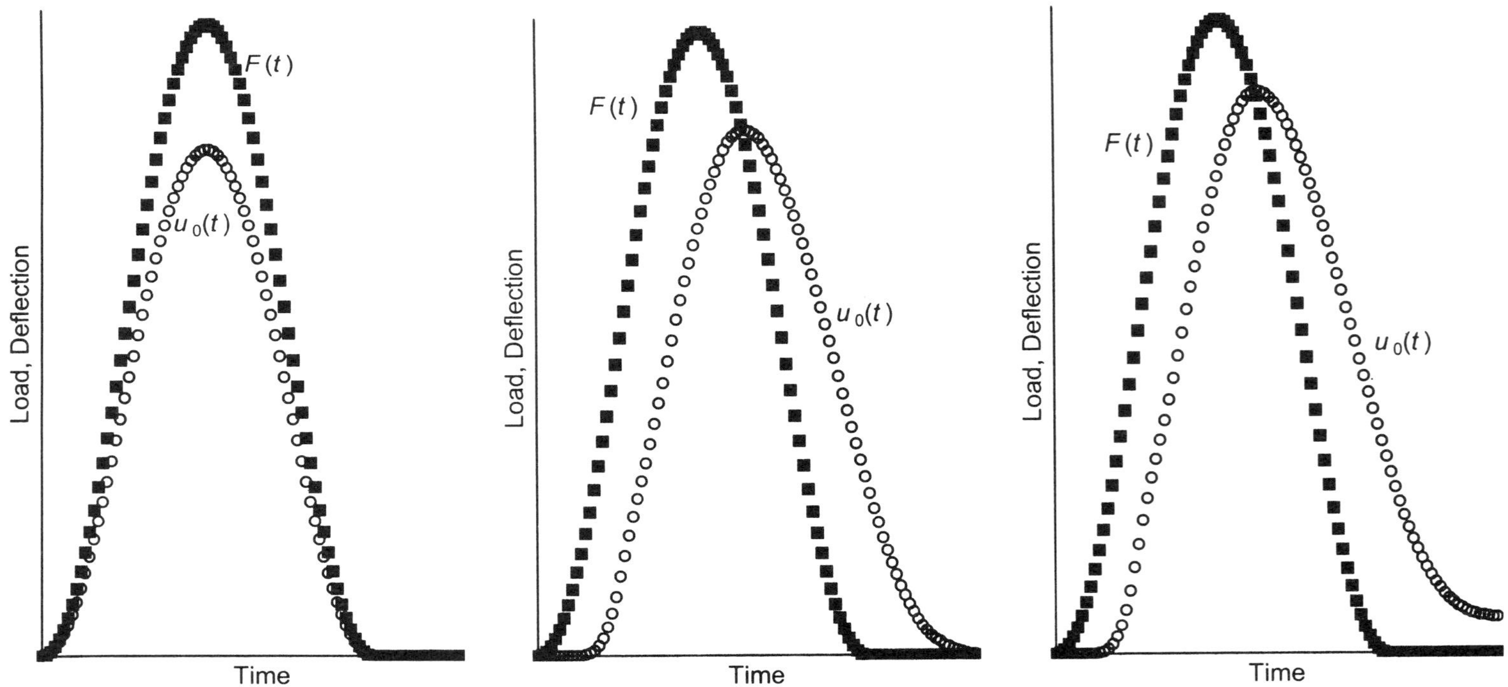

(a) Elastic (b) Viscoelastic Solid (c) Viscoelastic Liquid

Figure 1. Load and Deflection Histories in FWD Tests.

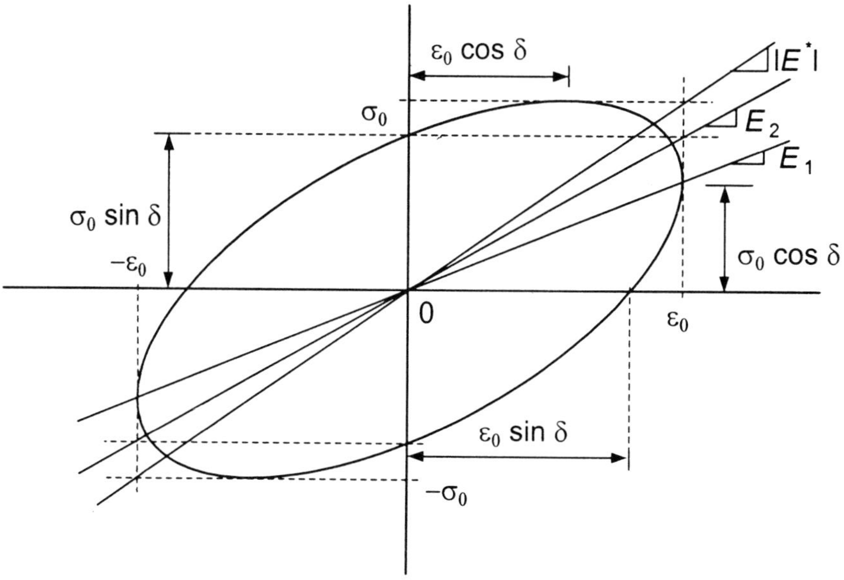

Figure 2. An Elliptical Hysteretic Loop for the Linear Viscoelastic Material

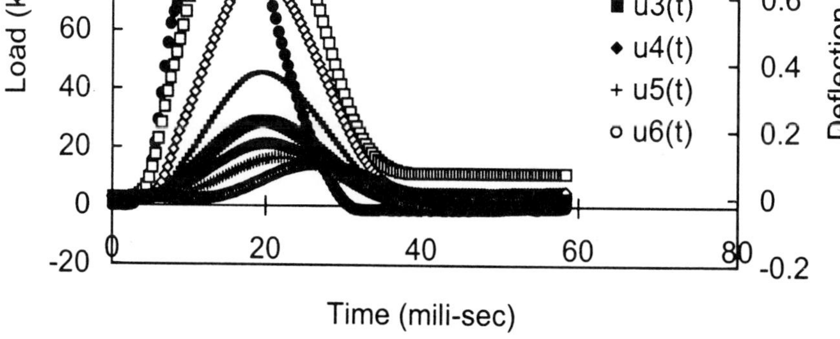

Figure 3. A typical Results of FWD Tests.

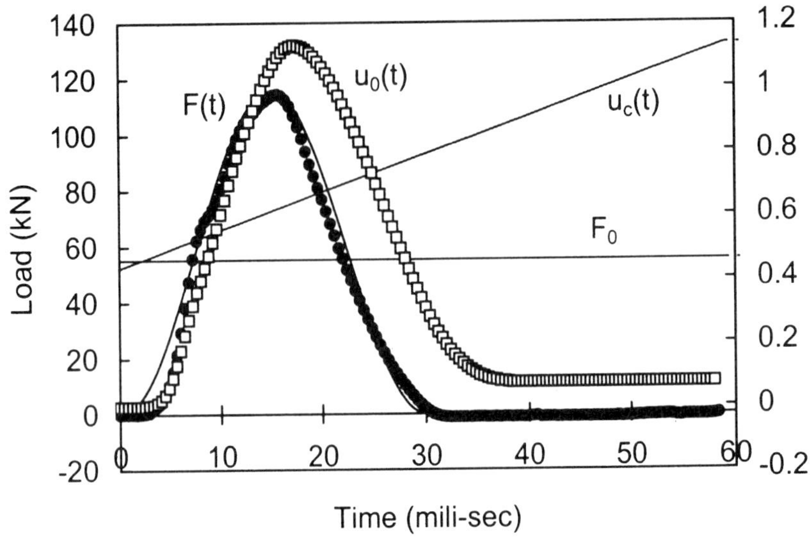

Figure 4. Load and Deflection Histories in for the FWD Test Site.

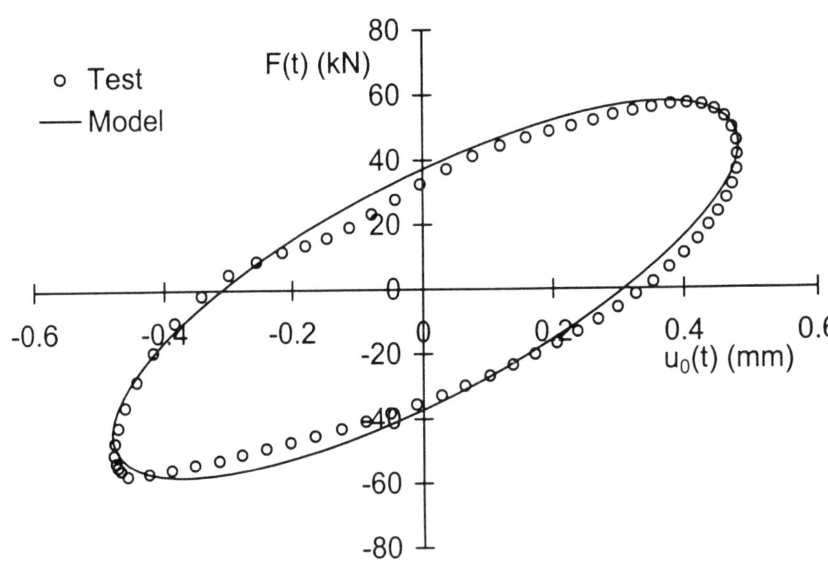

Figure 5. An Elliptical Hysteretic Loop for the FWD Test Site.

Fifth International Conference on the
BEARING CAPASITY OF
ROADS AND AIRFIELDS

ESTIMATION OF SUBGRADE THICKNESS FROM DEFLECTION BASIN ANALYSES

JACOB UZAN

Professor of Civil Engineering, TECHNION, Haifa, ISRAEL 32000,
Tel. (972) 429-3325, Fax (972) 422-5176, e-mail: cvrjaco@tx.technion.ac.il

ABSTRACT

The current procedure included in MODULUS computer program for estimating the subgrade thickness is sensitive to measurement errors and may be inaccurate. A new procedure based on back calculating the subgrade thickness from the deflection bowls using available computed data-bases of deflection bowls is proposed. Problems associated with global and local minimum are discussed.
Five test sections in Texas are analyzed using both the current (MOD4-2) and the new (MOD4-3) procedures. It is found that the results with MOD4-2 and MOD4-3 are in some cases different.

Key words: Flexible pavements, back calculation of moduli, subgrade thickness, surface deflection bowls.

INTRODUCTION

Several problem areas in the field of back calculation of pavement layer moduli need to be addressed and solved urgently in order to get reliable and realistic results (Uzan, 1993). The paper addresses one problem area related to knowledge of the depth to a rigid layer or the thickness of the subgrade.

The parametric study conducted by Rohde, 1990 on 12 sites in Texas concentrated on the effect of the depth to a rigid layer on the back calculated moduli. It is shown clearly that as the thickness of the subgrade increases, the back calculated modulus of the subgrade increases and that of the base decreases. In some cases, the use of an infinitely thick subgrade resulted in an inverse pavement structure (i.e. the back calculated base modulus is lower than the subgrade modulus). This result has, in turn, a predominant effect on the structural evaluation of the pavement, the rehabilitation strategy and overlay thickness.

The paper presents the methodology currently used in back calculation programs such as MODULUS (Uzan et al, 1988, Uzan et al, 1989). In view of the limitations of the existing procedure, a new procedure for determining the subgrade thickness is presented and

and discussed. Results of back calculation with the current (MOD4-2) and new (MOD4-3) procedures are presented and discussed.

BACKGROUND OF THE EXISTING METHODOLOGY

Current procedures for estimating the depth to a rigid layer are based on the Boussinesq equation for the surface deflection under a concentrated point load at the surface (Rohde, 1990). The intercept of the deflection (d) versus inverse of radial offset (1/r) function with the 1/r axis is a measure of the depth to a rigid layer. A smaller intercept corresponds to a deeper rigid layer (or thicker upper subgrade). In the case of a uniformly distributed load over a circle and a multilayered system, the loading system, the thickness and the stiffness of the upper layers also affect the intercept. The approach taken by Rohde is to develop a regression equation relating the depth to a rigid layer with the intercept and basin shape factors SCI (surface curvature index), BDI (base damage index) and BCI (base curvature index). These factors are defined as the difference between the surface deflections in mils at 0.m and 0.3m for SCI, at 0.3m and 0.6m for BDI and at 0.6m and 0.9m for BCI. These factors represent in global terms the thickness and stiffness of the upper layers. Four separate equations for different AC layer thickness ranges were developed from 1080 runs of the BISAR computer code. These equations were included in the program MODULUS to estimate the depth to a rigid layer. It is worth mentioning that the regression equations were developed for a rigid layer that is 100 times stiffer than the subgrade.

Three major drawbacks to the above procedure are noted:
(1) The procedure requires that four sensors be placed at 0m, 0.3m, 0.6m and 0.9m.
(2) The maximum intercept is sensitive to measurement errors, material segregation and cracking which greatly affects the slope of the segments d-1/r. Figure 1 presents results of 6 deflection basins in the d (or normalized deflection)-1/r space. These deflection basins were obtained from section 11, station 5 at different periods (Rohde, 1990). It is seen that 4 basins are smooth while the other 2 basins have a "jump" caused by a measurement error at the sensor located at 1.2m (1/r = 1/1.2 = 0.833 1/m). This measurement error led to a larger intercept (as compared to the intercept corresponding to a smooth basin) and thus to a smaller depth to rigid layer.
(3) The calculated depths to rigid layer change substantially when two of the four regression equations are used (i.e. using the AC thickness corresponding to the transition from one equation to the other). This result is an indication of the reliability of the regression equations that may be not accurate enough to predict depth to rigid layer. Figure 2 presents results of predicted depth to rigid layer using the regression equations and generated (artificial) deflection bowls. These bowls were obtained using the computer program WESLEA (Van Cauwelaert et al, 1989) and pavement layer thicknesses and moduli within the range of values for which the regression equations were developed (Rohde, 1990). It is seen that the regression equations are not good in predicting the depth to bedrock (for artificial bowls).
In view of the above limitations, a new procedure was developed to back calculate the thickness of the upper subgrade.

PROPOSED METHOD FOR SUBGRADE THICKNESS EVALUATION

Background

The subgrade thickness is usually unknown and not clearly defined. In some cases, which are easy to deal with, bedrock made of very stiff material may exist. In other cases, the material below the subgrade may be only slightly stiffer than the upper subgrade. In the cases, which are difficult to deal with, the same subgrade material (such as sandy subgrade material) exists to very large depth below the surface, however its stiffness increases with depth. In fact, any subgrade thickness evaluation should refer to the cases where the material changes from one kind or physical state to another kind or physical state. The case of sandy subgrade material is a misapplication of the approach: The increase in stiffness with depth can be correctly handled by using nonlinear models. However since most back calculation schemes use linear theories, this case will be approximated as if a fictitious stiff layer exits at some depth below the subgrade (Rohde, 1990).

If the depth to a rigid layer is known, the subgrade thickness is input to the back calculation procedure (like the other pavement layers) and the layer modulus evaluated. However this situation is not common and the subgrade thickness is usually unknown. Theoretically, including two additional unknowns to the back calculation procedure, namely the modulus and the thickness of the upper subgrade, can solve this situation. The increase in the number of unknowns may lead to non-unique solution or/and excessive sensitivity of the solution and will require much longer time to run the program. Therefore, the proposed procedure uses an a-priori knowledge of the relative stiffness of the upper and lower subgrade layers (or subgrade and bedrock layers) to reduce the number of unknowns to minimum and to improve the sensitivity of the system.

Other questions pertinent to the choice of the solution methodology. What degree of accuracy is needed to evaluate the subgrade thickness? What change in subgrade stiffness is relevant for the structural evaluation of the pavement and for rehabilitation purposes? It is the writer's opinion that the most important back calculated parameter is the subgrade modulus. It should therefore be evaluated within less than 20-50 percent, depending on the level of reliability to be achieved. As the back calculated subgrade modulus corresponding to a given measured deflection bowl depends strongly on the subgrade thickness, it is of utmost importance to evaluate the subgrade thickness. This will insure or at least help obtaining "realistic" back calculated subgrade modulus.

Description of the proposed procedure

On the basis of the above discussions, a new procedure (named MOD4-3) for evaluating the subgrade thickness is proposed. It is based on the following:
(1) The modular ratio between the upper subgrade and the lower "stiff" infinite subgrade layer may be assigned the following values: 5, 10, 100 and 1000. The user must make a choice based on his knowledge of the subsurface condition, i.e. whether rock exists or not. In the cases where no outcrops are seen in the vicinity of the pavement or no other information concerning the existence of a stiff layer is available, a ratio value of 5 is

recommended. This ratio value may also represent stiffening of the material due to overburden. Four databases are made available to the user for the four different modular ratios.
(2) The subgrade thickness is evaluated using the back calculation scheme included in MODULUS. The above mentioned databases were generated for four layer systems (two pavement layers, a finite subgrade and an infinite subgrade layers) and include a wide range of parameters. The Poisson's ratios were assumed 0.35 for all layers. The databases include surface deflections at nine radial distances for 21840 cases.
The systems are restricted to only three pavement systems (two pavement layers and the finite subgrade). The user must combine judiciously any adjacent layers of similar stiffness to come up with a total of three layers including the finite subgrade. For example, a treated base may be combined with the surface course and a granular subbase with the granular base or subgrade. These layer thicknesses are required only for evaluating the subgrade thickness. They can be changed later for a more accurate evaluation of pavement layer moduli.
(3) After the user entered the modular ratio of his choice and the pavement layer thicknesses, the program uses the spline interpolation technique to generate a working database for the specific upper-lower subgrade modular ratio and pavement layer thicknesses. The database is then used to determine, for each deflection bowl, four unknowns: the thickness and modulus of the subgrade and the moduli of the two pavement layers.
(4) The computation of the subgrade thickness and layer moduli is an error minimization problem that is handled using the search pattern in MODULUS. The objective function (mean squared error) is neither convex nor concave. Therefore, it is anticipated that the solution may reach a local instead of a global minimum. In order to reduce the magnitude of this problem and to take care of real conditions, two steps were added to the current search process. First the minimization process is initiated at two different initial values of the subgrade thickness. These initial values are chosen to correspond to a thin subgrade (subgrade thickness of 4.64 and less times the radius of the loading area) and thick subgrade (subgrade thickness of more than 4.64 times the radius of the loading area) respectively. Two solutions are then obtained. The second added step is to eliminate unrealistic solutions. As in real conditions, the measured deflection bowl may be distorted due to cracking, the back calculation process may tend to introduce a weak thin subgrade layer. This situation is taken care by disregarding solutions with a back calculated subgrade modulus less than 35 MPa or a subgrade thickness of less than 300 mm. (These values can be adjusted according to the site conditions to eliminate unrealistic solutions). When two realistic solutions are obtained, the solution with the small squared error is selected for further analyses. It is worth mentioning that this procedure introduces a substantial improvement in solving the minimization problem.
(5) When several deflection bowls are analyzed, an average value is computed to represent the pavement section. Like in the MODULUS program, outliers are disregarded based on the 0.10 percentile (i.e. when their value exceed the average by more than 1.3 times the standard deviation). This procedure is found useful to detect and disregard unwanted results due to several reasons such as longitudinal or transversal cracking in the vicinity of the test.

The proposed procedure is a compromise between accuracy and running time. The main drawback is the running time which is much longer than in the existing procedure

(Rohde, 1990). The limitation to 3 layer systems for flexible pavements and the use of interpolation will have a mild effect on the subgrade modulus results.

Evaluation of the proposed procedure

The proposed new procedure for back calculating subgrade thickness was evaluated from several aspects:
(1) The interpolation scheme and the variable spacing in the database. In order to evaluate the accuracy of the interpolation scheme, one generated database was compared with the results obtained directly from the forward program WESLEA (i.e. without interpolation). The results presented in Figure 3 were obtained for a pavement section with 254 mm asphalt concrete (AC), 279 mm base, a modular ratio of upper and lower subgrade of 100 and wide ranges of subgrade thickness and modular ratios of upper (AC and base) and subgrade layers. Figure 3 show the error between surface deflections obtained with and without interpolation versus the normalized deflection (where d=deflection, E=upper subgrade modulus, p=pressure and a=radius of loaded area). It is seen that the error in absolute terms is smaller than 0.006 and in most cases less than 0.002 normalized deflection. This value corresponds to about 5 microns error in the surface deflection. This level of error is good. It is worth mentioning that the error comes from two interpolation steps, one for the AC layer thickness and the second for the base layer thickness. Because of this error, surface deflections smaller than about five microns will not be used in the back calculation.
(2) Accuracy of prediction using artificial deflection bowls. For illustration purposes, 140 runs were generated using four different values of depth to rigid layer. Figure 4 presents a comparison between the depth to rigid layer used in generating the bowls and the back calculated one using the proposed procedure. It is seen that the prediction is very good except for one case at 0.7 m. For artificial bowls, one should get a perfect match, without outliers. The round off error is the reason for the outlier case in which the match is not perfect. The generated deflection bowls were stored in mils with 2 digits after the decimal point (i.e. with an accuracy of one hundredth of a mil or 4 microns). When the deflection data was stored with four digits after the decimal point, perfect match was obtained for all 140 cases.

On the basis of the above analyses, It appears that the proposed procedure is capable of predicting the depth to a rigid layer for artificial deflection basins with good numerical accuracy. However, the procedure requires some broad knowledge of the modular ratio of the upper to lower subgrade layer for choosing the appropriate modular ratio among the four values of 5, 10, 100 and 1000. The restriction that very small deflections should not be used could create a real problem, especially for rigid pavements. In the following verifications and applications, deflections smaller than 8 mils were eliminated and not used in the prediction of the depth to rigid layer.

VERIFICATION

In this section, results of analyses using the existing procedure in MODULUS (Rohde, 1990) and the proposed one are presented. The verification is conducted by comparing the predicted rigid layer depths from back calculation on 5 pavement sections (7, 8, 9, 11 and 12) to that obtained through drilling and cone penetration. It should be emphasized

FHWA/TX-88-1123-1, Texas Transportation Institute, Texas A&M University, College Station, Texas, Sept. 1988.

Uzan, J., Lytton, R.L. and German, F.P., "General Procedure for Backcalculating Layer Moduli", Nondestructive Testing of Pavements and Backcalculation of Moduli, ASTM STP 1026, Albert J. Bush and Gilbert Y. Baladi, Eds., American Society for Testing and Materials, Philadelphia, 1989.

Uzan, J., Zollinger, D.G. and Lytton, R.L., "The Texas Flexible Pavement System (TFPS)", Research Report FHWA/TX-91/455-1, Texas Transportation Institute, Texas A&M University, College Station, Texas, Nov. 1991.

Uzan, J., "Advanced Backcalculation Techniques", Nondestructive Testing of Pavements and Backcalculation of Moduli, ASTM STP 1198, Harold L. Von Quintus, Albert J. Bush and Gilbert Y. Baladi, Eds., American Society for Testing and Materials, Philadelphia, pp. 217-228, 1993.

Van Cauwelaert, F.J., Alexander, D.R., White, T.D. and Barker, W.R., "Multilayer Elastic Program for Backcalculating Layer Moduli in Pavement Evaluation", Nondestructive Testing of Pavements and Backcalculation of Moduli, ASTM STP 1026, Albert J. Bush and Gilbert Y. Baladi, Eds., American Society for Testing and Materials, Philadelphia, pp. 171-188, 1989.

Table 1: Results of back calculation for 5 pavement sections.

	Modular ratio	Site 7	Site 8	Site 9	Site 11	Site 12
Depth to rigid layer, m.	100 Mod4-2	3.17	5.09	4.62	4.00	7.62
	100 Mod4-3	6.50	7.62	6.43	7.62	7.62
	10 Mod4-3	5.00	7.62	5.36	7.62	7.62
	5 Mod4-3	3.24	7.62	4.76	7.62	7.62
Subgrade modulus, MPa.	100 Mod4-2	80	126	77	108	130
	100 Mod4-3	132	145	82	139	130
	10 Mod4-3	123	150	81	143	128
	5 Mod4-3	105	156	81	148	131
Granular base modulus, MPa.	100 Mod4-2	402	721	293	448	159
	100 Mod4-3	142	601	292	367	159
	10 Mod4-3	161	568	292	357	205
	5 Mod4-3	226	536	269	348	191
Asphalt concrete modulus, MPa.	100 Mod4-2	6559	4841	7483	#	#
	100 Mod4-3	7352	5214	4262	#	#
	10 Mod4-3	7303	5331	4276	#	#
	5 Mod4-3	7090	5448	4614	#	#
Error per sensor, percent	100 Mod4-2	1.7	2.6	4.9	4.6	5.2
	100 Mod4-3	1.1	2.2	2.6	2.7	5.2
	10 Mod4-3	1.1	2.0	2.8	2.4	5.8
	5 Mod4-3	1.2	1.8	2.8	2.2	5.6

AC thickness is 25 mm., the modulus was not back calculated.

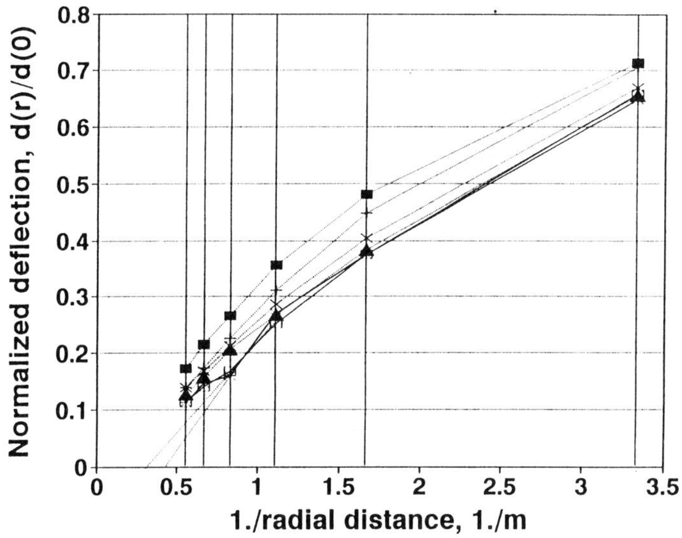

Figure1: Deflection basins in Normalized deflection - Inverse radial distance space.

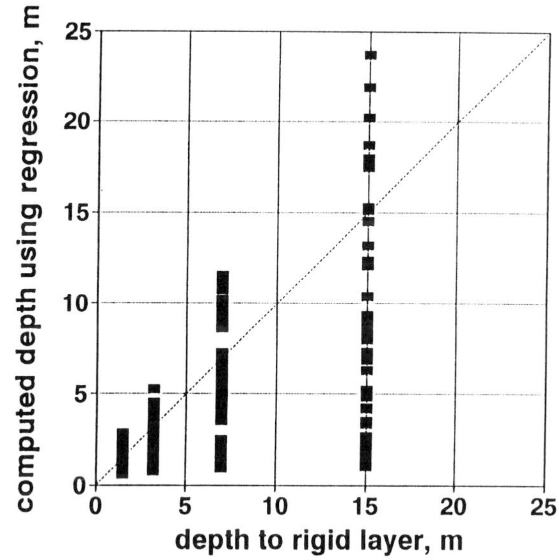

Figure 2: Predicted using regression equations (in MOD4-2) versus computed depth to bedrock.

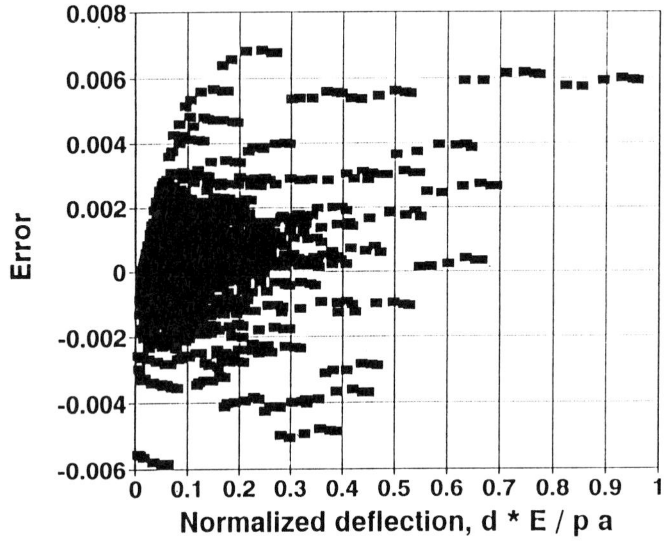

Figure 3: The absolute error in the interpolation of normalized surface deflections in MOD4-3.

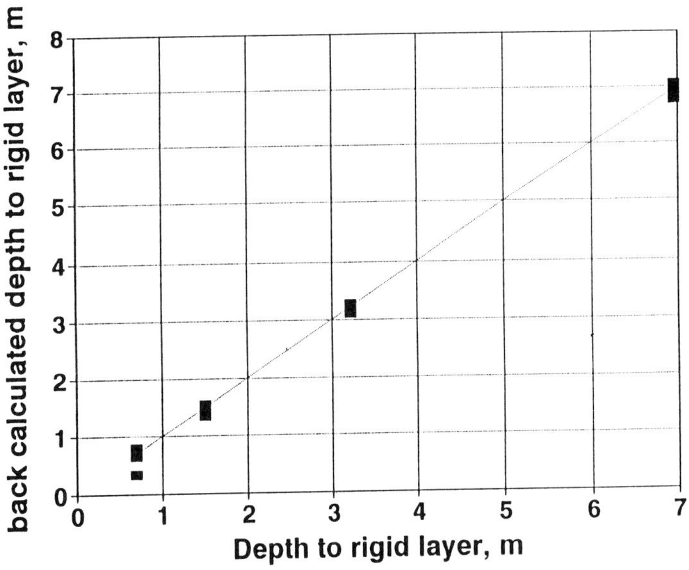

Figure 4: Predicted using MOD4-3 versus computed depth to bedrock.

Fifth International Conference on the
BEARING CAPASITY OF
ROADS AND AIRFIELDS

FULL STRUCTURE DIAGNOSIS OF LOW-VOLUME ROADS

CHRIST VAN GURP[1], ALBERT BODE[2],
MARTIN WENNINK[3] AND ANDRÉ MOLENAAR[4]

[1]Head Research and Consultancy, KOAC•WMD, Schumanpark 43,
NL-7336 AS Apeldoorn, Fax: +31 55 5433111, Email: koacoa@pi.net
[2]Government Service of Land and Water Management, P.O. Box 20021,
NL-3502 LA Utrecht, Fax: +31 30 2858999, Email: a.g.bode@lbl.agro.nl
[3]Pavement Consultant, KOAC•WMD, Schumanpark 43,
NL-7336 AS Apeldoorn, Fax: +31 55 5433111, Email: koacoa@pi.net
[4]Professor, Delft University of Technology, P.O. Box 5048,
NL-2600 GA Delft, Fax: +31 15 2783443, Email: a.molenaar@ct.tudelft.nl

ABSTRACT

This paper presents results of a study on the development of design and structural evaluation procedure for low-volume roads. The research programme has been conducted within the framework of the activities of the Working Group Design Low-volume Roads (WOVP). The developed methodology has been calibrated with the use of test and analysis data of 46 test sections. The load-carrying capacity of the pavement structure is expressed by the so-called Modified Structural Number, SNC. This parameter can be predicted from data collected with Falling Weight deflection measurements. The methodology uses the strain at the bottom and surface of the asphalt layer, the strain at the bottom of the bound road base, and the vertical compressive strains at the top of the unbound road base and subgrade as design criteria. An especially developed software tool enables easy use of the multitude of equations to verify design and to determine the required strengthening.

Keywords: Falling Weight Deflectometer, Design, Structural Evaluation, Low-volume Road.

INTRODUCTION

Most worldwide used design and structural evaluation procedures for roads are mainly developed for highway pavement structures. These methodologies are not fully applicable to low-volume roads since:

- load-carrying capacity of these roads may be very poor,
- traffic intensities are low, but traffic loads may be comparatively high,
- structures have been built historically leading to ample variation in layer thicknesses and material usage,
- often thin asphalt layers are applied in combination with high stiffness road bases,
- high groundwater tables are frequently measured,
- roads may be very narrow resulting in edge distress.

OBJECTIVES AND STRUCTURE OF STUDY

A two-stage approach was used in the development of the design procedures. In 1993 a feasibility study was conducted on 12 selected pavement sections of 300 m each (see shaded row in Table 1). The following tests were performed in situ: geometry, distress identification survey, dynamic cone penetrometer test, FWD test, coring. The asphalt cores were investigated in the laboratory on aspects as: degree of compaction, composition and gradation, stiffness vs. temperature.

Estimates of pavement lives were made on the basis of the CROW standard procedure for processing of FWD data [1]. This CROW approach has the option of field calibration of computed pavement lives to observed extent of distress. Good correlation was found between the pavement lives and the Modified Structural Number, SNC. This number is a measure for load-carrying capacity (see later in this paper.)

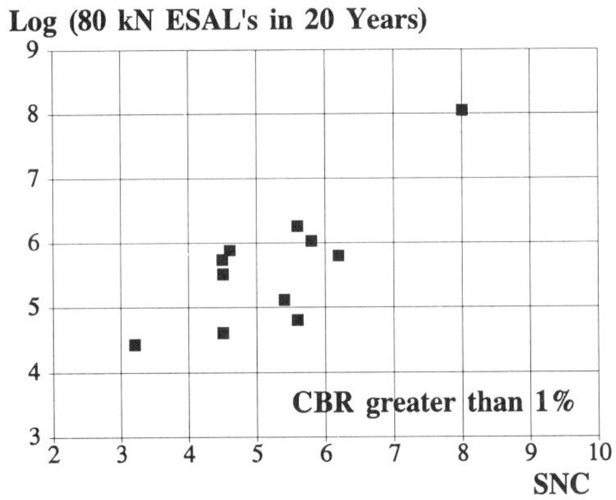

Figure 1 Prediction of structural pavement life by SNC

The second stage of the study started in 1994. Table 1 shows the factorials of the pavement sections. Based on the experiences of the first stage a different approach of selection of the pavements structures was applied. In the first stage a weighted rating of a set of distress types was used whereas in the second stage only cracking was used as selection criterion.

Table 1 Factorials (Number of pavement sections per cell)

Pavement Width (m)	Structural Support of Subgrade				
	Peat	CBR < 5%		5% ≤ CBR ≤ 10%	
	Cracking 5 - 15%	Cracking 5 - 10%	Cracking > 10%	Cracking 5 - 10%	Cracking > 10%
< 3.00	2	2	2	2	2
3.00 - 3.99	2	0	0	0	2
3.00 - 3.99	3	3	3	3	0
4.00 - 4.99	2	2	2	2	2
5.00 - 5.99	2	2	2	2	2

Since the current generation of backcalculation codes have great difficulty with accurate estimation of layer moduli in situations with thin asphalt layers, development of an overall measure of structural capability was considered as desirable. A set of additional conditions was formulated to guarantee widespread use of the new method.

- In case of structural evaluation and strengthening, critical strain and stress values should be inferred from the deflection basin and if needed some extra, easy-to-collect data.
- Conversion of truck traffic intensities to repetitions of equivalent standard axles (ESAL's) should be avoided since intensities are easier and more accurate to assess than the number of ESAL's.
- Since most low-volume roads are narrow, pavement edge distress is frequently observed. The use of thin asphalt layers leads to higher stresses in the underlying layers. Consequently critical stresses and strains in these layers should be addressed as well. On some low-volume roads surface initiated cracking could be observed. This phenomenon was assumed to be due to the rigid road base.

  The following criteria were included in the new design and evaluation approach:
  - horizontal tensile strain at the bottom of the asphalt layer
  - horizontal tensile strain at the pavement surface near the tyre rim
  - horizontal tensile strain at the bottom of the bound road base
  - compressive vertical strain at the top of the unbound road base
  - compressive vertical strain at the top of the sub-base and subgrade

Various relationships have been developed amongst these criteria, the deflection basin and the parameter SNC, and layer thickness. These relations have been extended to predictive models of pavement life per criterion.

MEASURE OF LOAD-CARRYING CAPACITY

Since many low-volume roads are actually thin asphalt roads, much of the load-carrying capacity is provided by the road base and the underlying substructure. Besides that current backcalculation codes are not sensitive to thin asphalt layers, in other words, inaccurate layer moduli are found for thin asphalt pavements. This all pleads for usage of a measure of load-carrying capacity in which all pavement layers play their role. The Modified Structural Number, SNC [2] has been selected for this purpose. This parameter consists of two components, i.e. the pavement structure and the subgrade.

$$SNC = SN + SN_{SG} \tag{1}$$

$$SN = 0.04 \sum_{j=1}^{N} a_j h_j \tag{2}$$

$$a_j = a_s \sqrt[3]{\frac{E_j (1 - v_s^2)}{E_s (1 - v_j^2)}} \tag{3}$$

$$SN_{SG} = 3.51 \log CBR - 0.85 (\log CBR)^2 - 1.43 \tag{4}$$

where
- SNC = Modified Structural Number
- SN = Structural Number of layered system
- SN_{SG} = Structural Number of subgrade
- h_j = Thickness of layer j (mm)
- a_j = Structural coefficient of layer j
- a_s = Structural coefficient of standard material = 0.14
- E_j = Stiffness modulus of layer j (MPa)
- E_s = Stiffness modulus of standard material = 200 MPa
- v_j = Poisson's ratio of layer j
- v_s = Poisson's ratio of standard material = 0.35
- CBR = California Bearing Ratio of subgrade (%)

Contribution of layers to the determination of SN is limited to a depth of 700 mm. All layers deeper need to be accounted for in the subgrade. In case the total thickness of pavement layers is less than 700 mm, the remaining thickness must be supplied from the subgrade. In this study SN is computed for a reference asphalt temperature of 20 °C.

CALIBRATION OF METHODOLOGY

In common fatigue life calculations, crack initiation at the bottom of the asphalt layer defines the end of the structural pavement life. Since this distress phenomenon is difficult to validate, another definition of pavement life is used in the study presented. In case of asphalt strain being the predominant distress criterion, pavement life is defined as the time needed for development of cracking in the representative wheel path to an extent of 25 percent. Neither distinction is made between alligator cracking and longitudinal cracking, nor are various classes of severity of cracking used.
To model the pattern of crack growth visible at the pavement surface, basic information of the research projects FORCE [3] and LINTRACK [4] was retrieved. In these projects crack growth was meticulously mapped. A crack growth performance model was developed from these data.

In case deformation played a more distinct role, calibration of the methodology was based on projects with extensive records of development of secondary rutting as a function of time or load repetitions. The WOVP sections did not provide this information, but again the LINTRACK project [4] did. In this latter project the Shell subgrade fatigue line proved to be very accurate in predicting the number of load repetitions needed for formation of a rut depth of 18 mm. Beware that in this case rutting or similar forms of permanent deformation need to be deep seated. In other words, the deformation must be primarily found in the layers beneath the asphalt layer and not in the asphalt layer itself.

Based on the two above mentioned approaches, calibration factors were computed for all WOVP test sections and design parameters. Results of pavement life computations need to be multiplied by these factors to achieve the final prediction of actual pavement life. Application of this approach and the calibration factors revealed that pavement life computation can be based reliably on FWD data and truck traffic intensities. Conversion of traffic intensities to numbers of equivalent standard axles appears to be redundant. Material characteristics seem to be of secondary importance.

DESIGN

Because of the limited number of pages available for description of the design and evaluation methodology, only the basic steps will be presented. For a detailed description one is referred to [5].

The first step in the design process is to kick off with a set of layer thicknesses and layer stiffnesses. This set is converted to the Modified Structural Number, SNC. In a next step SNC is used to predict the magnitude of the critical design parameters. Equation (5) supplies an example, i.e. for the bottom asphalt strain. The equations for the other design parameters are similarly structured, obviously with other regression coefficients.

$$\log \epsilon_{r1,o} = 4.2408 - 3.2399 \log SNC + 0.2110 \log h_1 \quad (5)$$

where $\varepsilon_{r1,o}$ = Maximum strain at bottom of asphalt layer (μm/m)
h_1 = Thickness of asphalt layer (mm)

The asphalt strain is defined as the strain computed under a typical FWD load pulse of 50 kN. Relationships between FWD deflection basin and normal and shear forces in the truck tire pavement contact area have been developed to predict the critical strain at the surface of the pavement [5].

With the estimates of truck traffic intensities, material to be used, provisional pavement structure and the road width as input, the magnitude of all critical design parameters is determined. Subsequently all these values are checked against the allowable values. If one or more values exceed the allowable value, layer thickness and/or layer stiffness of the provisional pavement structure needs to be adjusted. This stepwise approach must be continued until all specifications are complied with. A spread sheet program has been developed to facilitate this calculation process. The attractive point of this software tool is that no complicated multi-layer programmes and unreliable determination of ESAL's are needed to determine layer thicknesses of a low-volume road.

STRUCTURAL EVALUATION

In designing a pavement structure we try to fit a combination of layer materials that meet all design criteria. Under assumption of certain resistance against fatigue or deformation, we can determine the required layer thicknesses. Once the materials have been selected and layer thicknesses have been determined, we can compute the Modified Structural Number, SNC. However, in structural evaluation of in-service roads, this information is usually lacking. Layer thickness can still be determined by coring or Ground Penetrating Radar, but material characteristics need to be determined in another way. This involves that we cannot compute SNC according to the procedure shown in Eq. (1). Since SNC and structural support of the pavement system are closely related, SNC can easily be predicted from the FWD deflection basin directly. Layer thickness is often included as an extra variable to increase accuracy (see next equation)

$$\begin{aligned}\log SNC = {} & 1.82472 + 0.03344 \log h_1 + 0.11832 \log BCI + \\ & - 0.16207 \log BDI + 0.12659 \log d_0 - 0.57878 \log d_{900} + \\ & + 0.19996 \log d_{1800} - 0.19829 \log SCI_{300}\end{aligned} \quad (6)$$

where SNC = Modified Structural Number
d_i = Deflection at offset i mm (μm)
SCI_{300} = Surface Curvature Index : $d_0 - d_{300}$ (μm)
BDI = Base Damage Index : $d_{300} - d_{600}$ (μm)
BCI = Base Curvature Index : $d_{600} - d_{900}$ (μm)

h_1 = Thickness of asphalt layer (mm)

Accordingly critical strains can directly be inferred from the FWD deflection basin. The following equation presents the relationship between the critical strain at the pavement surface and the FWD deflection basin (FWD load = 50 kN).

$$\epsilon_{r1,b} = 194.895 - 20.7769 \sqrt{SCI_{300}} \quad \mathbf{(7)}$$

where $\varepsilon_{r1,b}$ = Maximum horizontal strain at pavement surface (μm/m)
SCI_{300} = Surface Curvature Index : d_0 - d_{300} (μm)

Data on truck traffic history provides the necessary for determination the residual pavement life. Eq. (7) and the equations for the other re-design criteria are used for this purpose.

$$\log\epsilon_{r1,o} = -1.06755 + 0.56178 \log h_1 + 0.03233 \log d_{1800} + + 0.47462 \log SCI_{300} + 1.15612 \log BDI - 0.68266 \log BCI \quad \mathbf{(8)}$$

$$\log\epsilon_{r2,o} = 0.0931 + 0.4011 \log d_0 + 0.3243 \log d_{1800} + 0.4504 \log d_{300} + - 0.9958 \log d_{900} + 0.8367 \log BDI \quad \mathbf{(9)}$$

$$\log \epsilon_{v4} = 2.48589 + 0.34582 \log SCI_{300} + 0.16638 \log d_{1800} + - 0.68746 \log (h_1 + h_2 + h_3) + 0.47432 \log BDI \quad \mathbf{(10)}$$

where $\varepsilon_{r1,b}$ = Max. horizontal strain at pavement surface (μm/m)
$\varepsilon_{r2,o}$ = Max. horizontal strain at bottom of bound base (μm/m)
ε_{v4} = Max. vertical compressive strain at top of subgrade (μm/m)

Prediction of the residual pavement life with the use of all design criteria has the major benefit that the source of any insufficiency is disclosed. If the surface asphalt strain governs the residual life, than most likely the load-carrying capacity of the whole structure still complies with the requirements; only the wearing course and maybe another asphalt layer need to be renewed. If the subgrade strain is the most critical strain, than the whole structure must be assessed to determine the most feasible rehabilitation strategy. The WOVP approach satisfies these requirements.

Although the approach has been set up for use on low-volume roads, some first attempts of usage of the approach on the arterial motorway network appeared to be very successful. These types of roads are usually designed with the strain at the bottom of the asphalt layer as critical design parameter. Evaluation however, often reveals that cracking observed at the pavement surface of motorways typically is surface initiated. The equations of the WOVP methodology accurately distinguish between roads where cracking was bottom initiated and surface initiated. Coring data validated the predictions.

APPLICATION

Structural evaluation All equations and steps needed for the structural evaluation are presented in detail in [5]. Usage of all steps and equations can only be conducted with sufficient accuracy and speed, if a dedicated software code is written for this purpose. Figure 2 shows the basic lay-out of a spread sheet program written as a first step towards computerization. The spread sheet requires deflection data, temperature, extent of cracking, layer thickness and pavement width as input. Additionally data on daily truck traffic intensities, age of pavement and type of asphalt used need to be provided.

Deflection Basin

Deflector	Deflection
mm	μm
0	842
300	685
600	541
900	445
1200	394
1500	350
1800	304
Temperature (°C)	10.7
Cracking	64%

Temperature Normalisation

Parameter	Factor	Value
	TNF	μm
d0	0.96	874
SCI300	0.90	175
SCI600	0.92	328
BDI	N.A.	144
BCI	N.A.	96

Pavement Structure

Layer Number	Thickness (mm)	Width (m)
1	171	2.80
2	150	
3	0	
Bottom	14000	

Correction Factors

Pavement Width	1.00
Lateral Wander	0.60
Edge Effect	1.00

SNC: 4.64

Design Parameters	Required	Actual	Years	Label	dSNC
Asphalt Strain bottom	150	298	0.8	Beware	1.10
Asphalt Strain surface	N.A.	-80	N.A.	O.K.	0.00
Strain Bound Base	80	334		Beware	
Strain Sub-base	845	945	19.2	Beware	0.23
Strain Sub-base (weak inter-layer)	N.A.	841	N.A.	O.K.	N.A.
Strain Subgrade	845	945	19.2	Beware	0.23

Traffic

Average Daily Truck Traffic Intensity	11
Age (+ Re-design Period)	30
CUVA	4015

Strengthening

Stiffness Modulus Overlay (MPa)	5000
Thickness Overlay (mm)	67

Figure 2 Example of WOVP approach

The traffic data are summarized into the parameter CUVA. This parameter is defined as the number of truck passages per year.

Deflections are normalized to a reference temperature of 20 °C [6]. The extent of cracking has some effect on this normalization process. Additionally critical strains are determined based on deflections and layer thicknesses (see column "Actual"). Type of material used, traffic data, pavement width and age determine accordingly the (residual) pavement life.

The computed pavement lives are checked against the desired outcome. This value is entered under the heading "Age (+ Re-design Period)". If the pavement life fails to meet this value the label "Beware" is printed, otherwise "O.K." is presented. The spread sheet makes clear which and how many criteria are really critical.

Strengthening In case of inadequate residual pavement life the pavement structure must be strengthened. The WOVP approach uses for this purpose exactly the same set of equations as in the structural evaluation. The example of Figure 2 demonstrates how much the pavement structure needs to be strengthened in terms of SNC. In this case this strengthening equals 1.03. This figure would be equivalent to an overlay thickness of 63 mm when the overlay stiffness equals 5000 MPa. However, the lack of load-carrying capacity is mainly due to inadequate behaviour and performance of the sub-base (see Figure 2). Application of an overlay is consequently not the correct way of strengthening. It would be much better if we decided for a rehabilitation strategy in which the sub-base is structurally improved.

CONCLUSIONS

The WOVP design, evaluation and strengthening methodology for low-volume roads enables quick estimation of required design or strengthening. No additional software codes as BISAR and backcalculation programmes are required.
The WOVP methodology uses the asphalt strain at the bottom of the asphalt, the asphalt strain near the tyre rim, the horizontal strain at the bottom of the bound base, the vertical compressive strains at the top of unbound base, sub-base and subgrade as design and evaluation criteria.

The load-carrying capabilities of a pavement structure can adequately be expressed in the so-called Modified Structural Number. Extra use of layer thicknesses increases predictive accuracy of estimation of critical strains.

Reliable relationships have been developed amongst FWD deflections, parameter SNC and the design criteria.

REFERENCES

1 Deflection Basin no Pitfall any longer (in Dutch). Publication 92, CROW. Ede, the Netherlands. August 1995.

2 PATERSON, W.D.O. Road Deterioration and Maintenance Effects; Models for Planning and Management. The Highway Design and Maintenance Standards Series, World Bank / John Hopkins, University Press. Washington, DC / Baltimore, MD, USA, 1987.

3 OECD Full-scale Pavement Test. Proceedings, OECD, Paris, 1991.

4 GROENENDIJK, J., MIRADI, A., MOLENAAR, A.A.A., VOGELZANG, C.H., DOHMEN, L.J.M., and MAAGDENBERG, A.M., Pavement Performance Modelling using LINTRACK. Proceedings 8th Int. Conf. Asphalt Pavements, Vol. 2, University of Washington, Seattle, WA, pp. 1505-1526, 1997.

5 VAN GURP, C.A.P.M. and WENNINK, P.M., "Design of Pavements of Low-volume Roads - Design, Evaluation and Strengthening" (in Dutch). Report E96007, KOAC•WMD Dutch Road Research Laboratories. Apeldoorn, 1997.

6 VAN GURP, C.A.P.M., Characterization of Seasonal Influences on Asphalt Pavements with use of Falling Weight Deflectometers. Ph.D.-thesis. Delft University of Technology. Delft, June 1995.

Fifth International Conference on the
BEARING CAPASITY OF
ROADS AND AIRFIELDS

STRUCTURAL PERFORMANCE OF CONCRETE PAVEMENTS IN CHINA

Xiao FENG[1], Boming TANG[2] and Tingyuan HAN[3]

[1]Shanghai Pudong Municipal Engineering Construction Department
(633 Xiexu Road, Shanghai, 200023, CHINA, E-mail: fengxiao@yahoo.com)
[2]Dr. of Eng., Associate Prof., Dept. of Civil & Environmental Eng., Nagaoka University of Technology
(Nagaoka, Niigata, 940-21, JAPAN, E-mail:tang@nagaokaut.ac.jp)
[3]Highway Administration Dept., Tangshan
(Tangshan, Hebei 063000, CHINA)

Abstract

The general development and technology background of portland cement concrete (PCC) pavements in China are introduced. Authors have participated in the researches on the structural and service performance evaluation of concrete pavement in Hebei, Zhejiang and Guangdong provinces in China. Based on the data collected in these typical regions by a means of field survey, systematic analysis and evaluation have been made. As concerned with the subgrade condition, the subgrade stability coefficient is discussed and used for the first time. Three-parameter prediction models and recommended evaluation standard of structural performance, including ***P***ercentage of slab cracking distress (P), ***CR***acking level (CR) and ***P***avement ***C***ondition ***I***ndex (PCI) are developed in this paper. Then, the prediction equations of those parameters are suggested, too. At the end, an example for application is presented.

Key words: PCC, pavement condition, structural performance, prediction model

1. Introduction

Extensive concrete pavement construction was launched in the early 1970's in China. At the very beginning the mileage of portland cement concrete pavements was only about 200km, which took a small share in China's paved highways. There has been, however, a rapid growth of concrete pavement from then on, particularly in heavy-duty highways. The design and construction specifications of PCC pavement were issued in the middle 1980's. At the end of 1990, the mileage of PCC pavement was 11,373km that formed 4.4% of the total high-type pavements. By the end of 1995, the mileage of concrete pavements constructed and put to use had been 30,000km. It was estimated that the mileage of new concrete pavement would be 2,500 kilometers each year.

At present, concrete pavements in China have not been used for a long time. Early distress is common due to the lack of experience in construction, management and maintenance. Some concrete pavements have stepped into the stage of rehabilitation and reconstruction. Considering the mileage of PCC pavement built each year, any deficiency in their design and construction can result in continuous maintenance problems of significant proportions. Therefore, it is very important for design, construction and management to study the laws of pavement structural performance.

During recent years, the efforts were focused on the inspection and evaluation of concrete pavements worldwide. Many nations and districts have established evaluation systems according to their demands and conditions. Among them the Concrete Pavement Evaluation System (COPES) may be the most influential. It shows a bright future in systematically analyzing concrete pavement distress causes and improving the level in design, construction and management.

A lot of researches have been carried out in China. Some provinces or cities have developed regional pavement management systems (PMS). The data amounts are not very large and affecting factors are not thoroughly considered because it's not very long after they were built. Meanwhile, these systems may be only suitable for those particular areas. Therefore, there has a certain distance from actual application.

Authors of this paper participated in several researches on structural performance of concrete pavement. Thus, it becomes possible to overcome those defects mentioned above based on the rich data gathered from long-term and extensive surveys. Some universal laws are developed in which an important road structural parameter is introduced.

2. Technical Background of Concrete Pavements in China

China has a vast area, a variety of climate types, and most of the geological types. This situation demands special consideration on environmental factors and employs different measures according to various circumstances. First of all, typical areas should be selected. Hebei, Zhejiang and Guangdong provinces, which are located separately in the northern, middle and southern China, are areas where concrete pavements were constructed relatively in advance. Some old concrete pavements have been in service for more than two decades. Natural conditions and technical situation of concrete pavement in these areas are representative. Hence, the nation-wide state of affairs about PCC pavement can be learned by studying those regions' situation.

- Hebei province lies in the north of Chinese mainland. The monthly average highest temperature in the area is 24.4 °C while the lowest is -5.5 °C. It is dry with an annual rainfall of 600mm. Concrete pavements in this region were mostly reconstructed over old roads in the early 1980's. For this reason, subgrades were widened for the most part. Typical slab dimension is 5.0m in length and 3.0m or 3.75m in width. Roadway with four lanes is 12~15m in width. Designed thickness of slab is in the range of 20~22cm with a bending strength of about 5.0MPa. Longitudinal joints are often plain joints with tie bars. Contraction joints are usually formed by a means of joint pressing-in method. Small-sized machines and manual work are the main methods used in construction.

- Zhejiang province is located in the eastern coastal area of China. The monthly average highest temperature in the province is 29 °C occurred in July while the lowest is 5 °C in January. It's rainy in summer with an annual rainfall of about 1500mm. Consequently the air in this area is very moist and the subgrade is weak. Concrete pavements in this area were mostly widened and reconstructed over old roads in the late 1970's and the early 1980's. Typical slab dimension is 5.0m in length and 3.5m or 4.5m in width with a roadway about 6~9m. Designed slab thickness is 18~20cm with a bending strength of only 4.5MPa because there were not many heavy load vehicles at that time. Longitudinal joints are often plain joints without tie bars. Contraction joints were usually formed by joint pressing-in method. Small-scale machines and manual work are mainly applied in construction. As the numbers of heavy load vehicles increased staggeringly, pavement distress appeared quickly and slab cracking distress was particularly serious.

- At the end of 1990, Guangdong province ranked first of all provinces in China with 3515km concrete pavements. With the development of economy, the province planed to construct more concrete pavements as the high-quality pavements. Hot and moist climate is a distinguishing feature of this area. Common asphalt pavement is unable to prevent water from permeating the ground. In this case, mechanic property and stability of concrete pavement is in an advantageous position. Designed slab thickness, which is up to 25~27cm in some peculiar districts, is often 20~25cm with a bending strength of only 4.5MPa. Longitudinal joints are often plain joints or tongue-and-groove joints with tie bars. Contraction joints were usually formed by mechanic saw cut.

3. Main Content of Long-term Inspection about Structural Performance

We selected more than forty stretches in the areas mentioned above to conduct long-term extensive investigation by a means of field survey in order to analyze and conclude the major types and causes of concrete pavement distress and the deteriorating regularity of structural performance and to appraise the key measures adopted in design, construction and management. The principal steps and content in the researches are illustrated in Figure 1 and are described in detail in the following section.

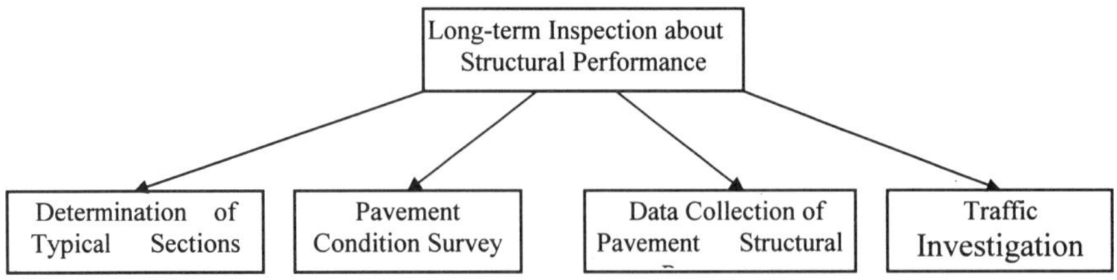

Figure 1. Main Content of Long-term Inspection About Structural Performance

3.1 Determination of Typical Stretches

Several factors should be taken into consideration when determining representative stretches. They include not only the representation of stretch distribution, time of use, environment, traffic volume, construction quality, pavement structure and pavement situation, but also the possibility of investigation. More than 40 typical stretches were chosen for continuous investigation, each of which varies from 1km to 3km in length and has two lanes on each side.

3.2 Data Collection of Pavement Structural Parameters

The thickness and strength of concrete slab are measured by sampling cores from the roadways. Ten cores randomly distributed along the main lanes of the stretches are sampled. The bending strength can be determined by indirect tension test and the thickness can be accurately measured with those samples.

The equivalent resilient modulus on the top of the subgrade can be backcalculated by measuring the resilient deflections in the middle of longitudinal side of a concrete slab.

3.3 Traffic Investigation

Traffic load is the chief factor affecting pavement performance. In order to learn the accumulated standard axle loads (100KN) of each stretch, investigations as follows have been made:

- Collect traffic volume data over the years. Numbers of middle and heavy vehicles which have strong influences on pavement performance should be in special consideration.
- Investigate the composition of loading so that the standard axle loads can be determined.
- Determine the horizontal force distribution coefficient.

3.4 Pavement Condition Survey. Data Bank Establishment

Field survey procedure is conducted according to the distress categories. Distress identification standardization is provided in Ref. 1 in detail.

Distress data on the inside and outside lanes are recorded respectively at every stretch. That means that there are more than 160 substretches representing all concrete pavements in those three areas. While in statistics, the total slab numbers and areas of every substretch are used as the basic units. It should be noted that only the most significant distress can be included in statistics in a slab. The slab position and direction in survey are illustrated in Figure 2:

left		right	
out	in	in	out

Figure 2. Slab Position and Direction in Survey

4. Analysis and Establishment of Three-Parameter Prediction Model

4.1 Data Bank of Pavement Condition Based On Field Survey

Data bank of pavement condition is established after continuous investigation in different stretches. It's impossible to list the structural situations of all stretches due to the space limitation. So it is only the data of Hebei province that is introduced in this paper. But the data of the others are also included when considering and presenting basic laws. Investigations in the region were conducted three times in different years. Original Data of the two middle heavy load lanes, i.e., two substretches, are added up respectively so the simplified average data showed in Figure 3 is easy to understand. Developing trends of three parameters (P, CR, PCI) in ten stretches included in the data bank are indicated as follows:

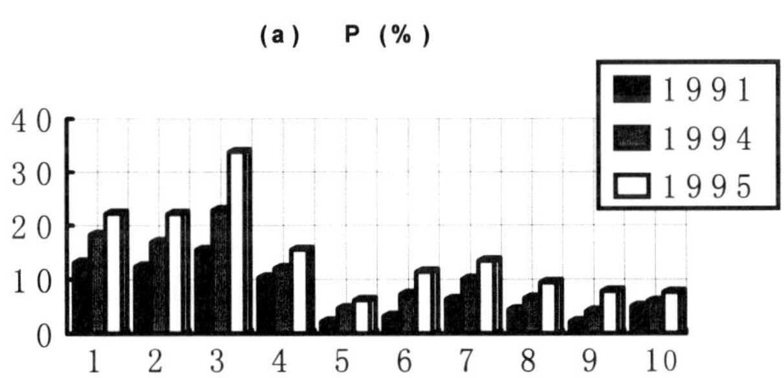

4.4.2 Analysis and Prediction of Cracking Level (CR)

The prediction model of cracking level (CR, $m/1000m^2$) is developed as follows:

$$CR=10.68(N*Y)^{0.85}(S^{3.37}+0.125\times SUB) \quad (4)$$

$(n=118, \quad R^2=0.7, \quad SEE=8.99)$

Analysis shows that N, S and SUB have similar influences on CR as on P. The service life of stretches are 22 years (SUB=0) and 15 years (SUB=1) if that CR reaches $100m/1000m^2$ is regarded as the condition of the end of service life. It can be seen that the unsteady subgrade will cause one third loss in service life. Therefore, special measures should be taken to ensure the stability of subgrade if old road is to be reconstructed.

4.4.3 Analysis and Prediction of Pavement Condition Index(PCI)

Regression analysis indicates that there is no significant relationship between PCI and S. Therefore, a model is developed as follows:

$$PCI=95(1-0.0245(N*Y)^{0.555}(1+0.679\times SUB)) \quad (5)$$

$(n=117, \quad R^2=0.69, \quad SEE=4.19)$

Note: The original PCI value is defined as 95 because of early cracking and surface defects before use owing to construction quality. Figure 4 is plotted based on equation 5:

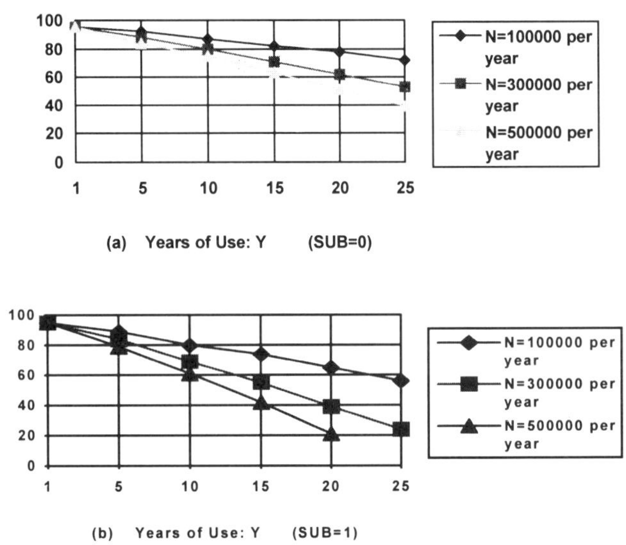

Figure 4. Illustration of PCI Loss Over Traffic Load and Subgrade Stability (a and b)

Figure 4 shows that PCI decreases steadily with years of use. That means that there has no evident defect in design and construction stages of pavements. The service life of stretches which subgrades are steady (SUB=0) will be 25 years with an average loss of 1~2% in PCI every year if that PCI decreases down to 50 is regard as the condition of the end of service life. Meanwhile the life of other stretches that have unstable subgrades (SUB=1) will only be about 15 years with an average loss of 3~4% in PCI every year. It can also be seen that the unsteady subgrade will even make up two fifths loss in service life.

4.5 Simple Analysis of Main Deterioration causes on PCC Pavement Structural Performance

There are a great deal of failure modes and mechanisms of concrete pavement. Some are due to traffic loading, and the others are due to non-traffic factors, such as construction quality. They can be briefly divided into three categories according to actual data and practice:

- Design Defect (Insufficient pavement structural strength). It's obvious that the old concrete pavements that have thin slab thickness and low strength are confronted with severe structural failures. Construction defect may also cause such failures.
- Construction Defect. It's a key factor affecting the service life of concrete pavement, such as material quality, mixture quality, control and management level and late joint pressing-in, etc.
- Environmental Effect. Uneven subgrade strength resulting in loss of support may be a major reason. Several types of pavement distress such as pumping, faulting and blow-up may be attributed to the uneven subgrade. Of course, big rainfall and tardy maintenance may fasten the developing process.

5. Application of Prediction Models

5.1 Recommended Three-Parameter Evaluation Standard of Distress Condition

Based on detailed analysis and calculation on distress data as well as the experiences in maintenance and management, the evaluation standard of distress condition is recommended in Table 1.

Table 1. Recommended Three-Parameter Evaluation Standard of Distress Condition

	Excellent	Good	Fair	Poor	Very Poor
P(%)	<5	5~15	15~30	30~50	>50
CR(m/1000m^2)	<10	10~50	50~100	100~200	>200
PCI	>90	90~75	75~60	60~40	<40
Recommended Measures	Daily Maintenance	Minor Remedy, Medium Remedy	Medium Remedy, Major Remedy	Major Remedy, Reconstruction	Reconstruction

5.2 Prediction of Three Parameters of Every Stretch

By the models developed in this paper, the predictions about three parameters of the investigated stretches in the next decade are showed in Table 2:

Table 2. Prediction About the Three Parameters of Stretches

Stretch	Current Situation (1995)			After 2 Years (1997)			After 5 Years (2000)			After 10 Years (2005)		
	P	CR	PCI	P	CR	PCI	P	CR	PCI	P	CR	PCI
01	34.0	96.6	65.3	40.0	130.0	57.3	50.0	200.0	45.3	68.0	300.0	25.3
02	24.3	74.5	68.9	30.0	110.0	55.0	40.0	170.0	35.9	55.0	270.0	23.0
03	45.7	130.2	55.9	53.0	200.0	48.0	66.0	300.0	35.9	85.0	430.0	15.9
04	15.7	39.8	68.7	20.8	60.0	64.7	27.8	90.0	54.8	45.8	140.0	44.8
05	7.3	17.5	80.4	10.3	27.5	78.0	15.0	43.0	75.0	22.3	70.0	70.0
06	12.0	19.3	75.8	20.0	30.0	72.0	32.0	45.0	65.0	52.0	70.0	55.8
07	18.0	37.9	49.5	26.0	58.0	40.0	38.0	90.0	30.0	58.0	140.0	15.0
08	10.8	21.5	77.3	15.0	28.0	70.0	20.7	37.0	60.0	30.7	55.0	40.0
09	8.8	21.9	80.9	15.0	32.0	79.0	24.0	47.0	75.9	38.8	72.0	70.0
10	8.2	22.3	74.6	11.3	30.0	71.5	16.0	42.0	67.0	27.0	62.0	59.6

The department responsible for the pavements can be aware of the situation of all stretches according to the predictions and Table 1. Thus, suitable maintenance and rehabilitation measures can be taken. For instance, the current pavement condition of stretch 03 is poor according to Table 1. It will become very poor within 1~2 years if proper rehabilitation activities are not taken immediately. Rehabilitation methods other than overlay is much more expensive and traffic interrupting than overlay methods because structural failures are extensive and severe in this stretch. Therefore, rehabilitation methods with overlays are strongly recommended such as asphalt concrete overlay. Meanwhile, pre-overlay repair is necessary as the existing pavement is severely deteriorated.

6. Conclusions

Latest research results about structural performance evaluation of concrete pavement in China are introduced in this paper. Base on systematic data from long-term inspection, the three-parameter prediction models are developed. Combined with the recommended three-parameter evaluation standard based on practice, current structural condition can be learned and suitable measures can be determined by pavement managers.

Uneven stability of subgrade can't be neglected because it has a significant influence on pavement structural of concrete pavement. It is an important factor for design and reconstruction, especially over weak subgrade.

Of course, new train of thought and results presented in this paper can be taken in establishing pavement management system. They can also be followed for reference by other pavement users of different areas in the world.

REFERENCES

(1) Specification of Cement Concrete Pavement Design for Highway, JTJ 012-94, P.R.China, Department of Transportation, 1994.
(2) Tang B.M. and Z.K. Yao. Long-Term Structural Performance Evaluation of Cement Concrete Pavement. China Journal of Highway and Transportation, Vol 9, No.2, Jun. 1996, pp.20-27.

Fifth International Conference on the
BEARING CAPASITY OF
ROADS AND AIRFIELDS

THE DEVELOPMENT OF MULTI-DIMENSIONAL TRANSFER FUNCTIONS FOR THE PERMANENT DEFORMATION OF PAVEMENT STRUCTURES

HECHTER THEYSE

Project Leader, Division of Roads and Transport Technology, CSIR,
P O Box 395, Pretoria, 0001, South Africa, +27 12 841-2690,
htheyse@transprt.csir.co.za

ABSTRACT

This paper describes recent developments in the permanent deformation modelling of pavement structures in South Africa. The current South African Mechanistic Design Method (SAMDM) is a critical layer approach and the aim is to convert it to a system approach for at least the permanent deformation modelling component. A conceptual model of permanent deformation has been formulated. The characteristics of the model has been investigated and mathematical functions that should be suitable to describe the model has been identified. The model was calibrated for different pavement layers of varying material quality with permanent deformation data from the Accelerated Pavement Testing (APT) programme that has been ongoing in South Africa with Heavy Vehicle Simulators (HVSs) for more than 20 years. Although the models have been calibrated with data applicable to South African conditions, the basic concepts remain universally applicable.

Key words: Mechanistic-empirical design, permanent deformation, system approach.

INTRODUCTION

The South African Mechanistic Design Method[1,2,3] (SAMDM) has been used for pavement design in South Africa in combination with other design methods for a number of years. This method is based on a critical layer approach to pavement design whereby the layer life is predicted for each pavement layer and the most critical layer controls the pavement life. The layer life is predicted based on a number of failure mechanisms for different material types each associated with a particular critical stress parameter. The layer lives are, however, predicted from a condition of no distress to a condition of terminal distress defined in terms of the specific mode of distress for a particular material type. There is therefore only a start and an end condition with no simulation of distress during the pavement life and no combined contribution from all the pavement layers to a distress manifestation such as surface rut. The ultimate goal of the current research conducted at the Division of Roads and Transport Technology is to convert the current SAMDM into a comprehensive mechanistic-empirical design method with a permanent deformation component for all layers and a fatigue component for bound layers. The permanent deformation of all layers will therefore contribute to the surface rut of the pavement and a simulation of rut with increasing load repetitions will be possible.

To be able to achieve this, a general permanent deformation model was formulated for the empirical component of the design procedure. The design method will therefore remain a mechanistic-empirical method where the working stresses and strains will be calculated by an independent structural analysis technique. The working stresses and strains will then be entered into the permanent deformation models and a calculation of the total permanent deformation will be possible at any number of load repetitions. This paper will not discuss the structural analysis technique and the accompanying material models that should be used but will focus on the characteristics and calibration of the permanent deformation models.

Data from a number of sources such as the repeated load triaxial test or Accelerated Pavement Testing (APT) may be used to develop these permanent deformation models. It was decided to use the APT data from the Heavy Vehicle Simulator (HVS) test programme that has been running for more than 20 years in South Africa to develop the permanent deformation models for the SAMDM.

The proposed permanent deformation models have got unique features that should also allow load history to be introduced in permanent deformation modelling.

HVS INSTRUMENTATION AND MEASUREMENTS

The definitions of permanent deformation concepts in this document are closely related to HVS instrumentation and measurements, as the HVS is the main source of data used in the development of the models described in this document. HVS test sections are normally instrumented with Multi-depth Deflectometers (MDDs) consisting of a stack of Linear Variable Displacement Transducers (LVDTs). The LVDTs are housed in a module that can be fixed at a predetermined depth in the pavement structure, usually at an interface between two layers. A reference core runs through the LVDT modules and is anchored at a depth of 2,5 to 3 m.

Two types of measurements are obtained from the MDD system. Firstly, a set of depth deflection influence lines can be measured as the HVS wheel load passes over the MDD stack. These influence lines are sampled at 256 points. A typical example of a reduced data set is illustrated in Figure 1(a). By extracting the peak deflections from these influence lines, and plotting each peak against the depth at which the MDD module for that reading was installed, a depth deflection profile as shown in Figure 1(b) may be obtained.

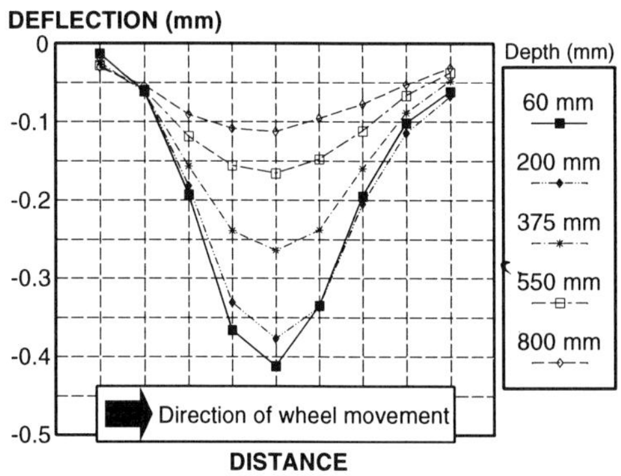

(a) Reduced sample of depth deflection data

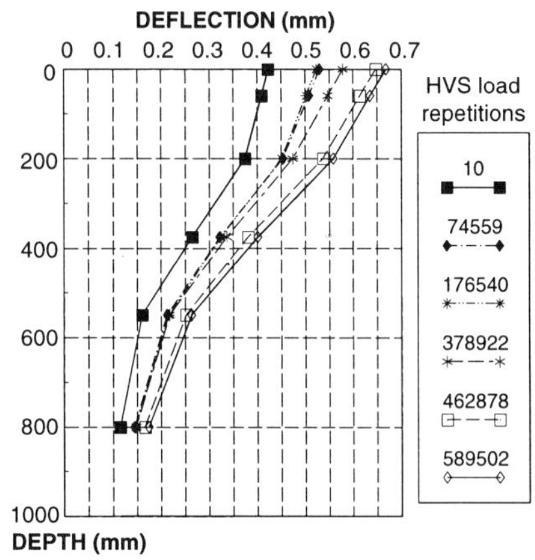

(b) Depth deflection profile of maximum deflection

FIGURE 1: TYPICAL DEPTH DEFLECTION RESULTS FROM AN HVS-TEST

Secondly, the permanent vertical displacement of each MDD module may be recorded at regular intervals during a particular test. The data obtained in this manner may be plotted as a trace of the permanent downward MDD displacement against the number of load repetitions as shown in Figure 2.

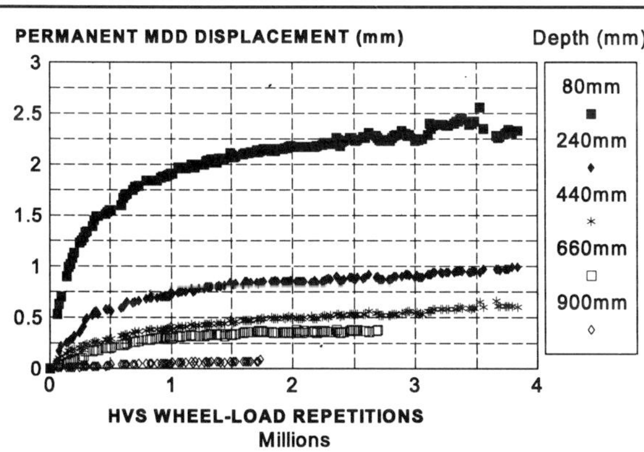

FIGURE 2: PERMANENT DOWNWARD DISPLACEMENT OF MDD MODULES WITH INCREASING LOAD REPETITIONS

$$\epsilon_p = AN^b \qquad \text{.........................} \textbf{(1)}$$

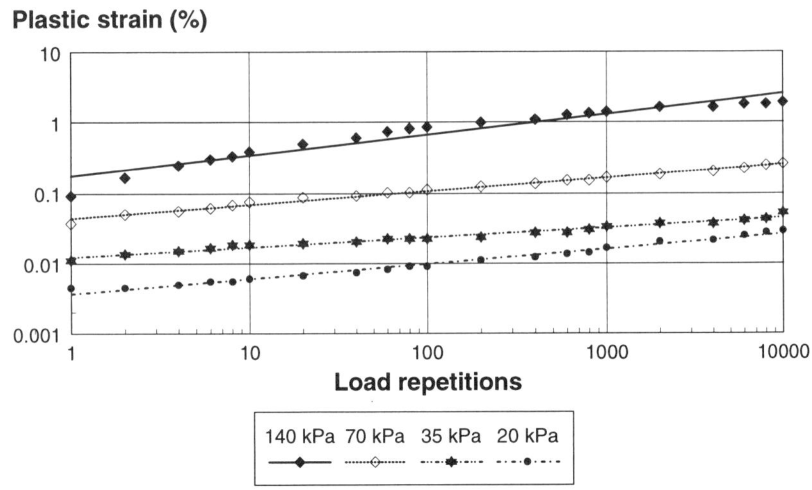

FIGURE 4: PERMANENT STRAIN AGAINST THE NUMBER OF LOAD REPETITIONS AT DIFFERENT STRESS LEVELS FROM THE WORK OF MONISMITH ET AL[5]

The permanent strain measured at 140 kPa deviates slightly from the straight-line relationship and this deviation is quite large on a normal scale. This function is, however, still used extensively[6,7,8,9] to describe the relationship between permanent strain and the number of load repetitions. Leahy and Witczak[8] listed a few other functions that may also be used for this purpose.

Wolff[10] used the data from APT testing in South Africa to establish the relationship given in Equation 2 between the plastic strain of base and sub-base layers and the number of load repetitions at a constant stress condition. This function also fits the permanent displacement of MDD-modules (Figure 2) during HVS-testing well (Theyse[11]). A typical result of this function fitted to the permanent MDD-displacement data from an HVS-test measured at the top of a pavement foundation, is shown in Figure 5. The physical meaning of the regression coefficients "m" and "a" are also shown in the figure.

$$\epsilon_p = (mN + a)(1 - e^{-bN}) \qquad \text{...................} \textbf{(2)}$$

where ϵ_p = accumulated permanent strain
N = number of load repetitions
m,a,b = regression coefficients
e = base of the natural logarithm

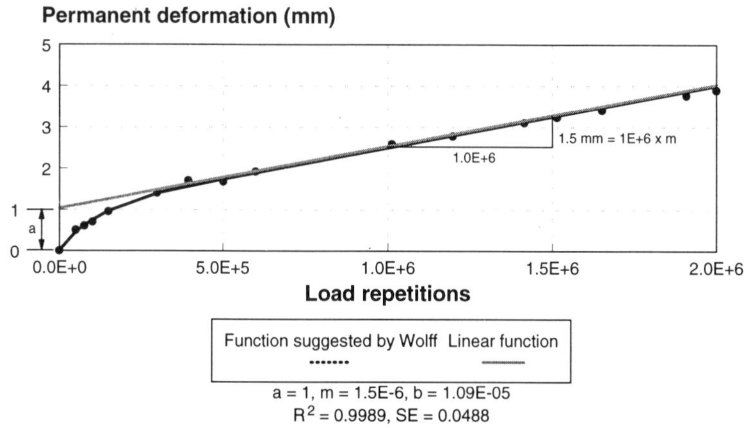

FIGURE 5: PERMANENT DEFORMATION AS A FUNCTION OF LOAD REPETITIONS AT A CONSTANT STRESS LEVEL UNDER ACCELERATED PAVEMENT TESTING

Figure 6 shows two examples of permanent strain data reported by Monismith et al[4] and Maree[12] for subgrade and high quality crushed stone base layer material respectively. Compared to permanent MDD displacement data from APT testing, the initial rate of increase of the laboratory data is very high, reducing to almost zero after the initial settling-in phase while the MDD displacement data from APT tests keep on increasing at a steady rate. The general pattern does, however, seem the same with an initial rapid increase in permanent deformation reducing to a steady state increase afterwards. It must also be kept in mind that the MDD displacement data reflects the superposition of the steady state increase in permanent deformation of all the material layers below the position of the MDD.

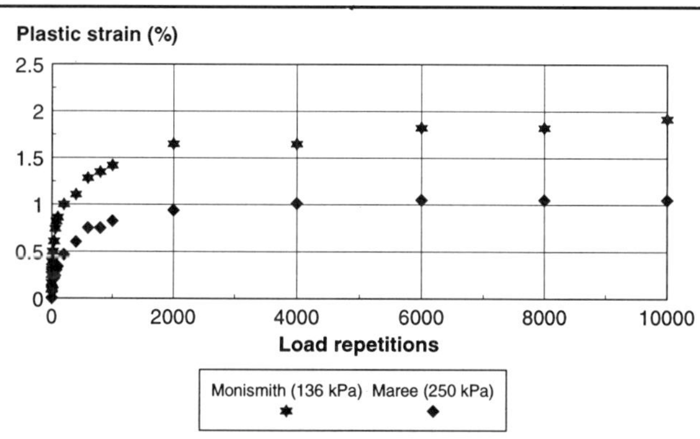

FIGURE 6: PERMANENT DEFORMATION DATA FROM REPEATED LOAD TRIAXIAL TESTS

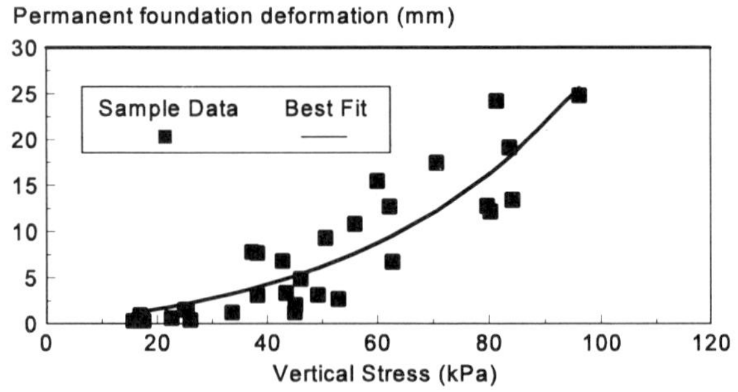

FIGURE 9: PERMANENT PAVEMENT FOUNDATION DEFORMATION AS A FUNCTION OF THE VERTICAL STRESS AT THE TOP OF THE PAVEMENT FOUNDATION

By repeating the regression shown in Figure 9 for a range of load repetition values, the author[11] could establish a relationship between the regression coefficients "A" and "B" of Equation 4 and the number of load repetitions. The coefficient "A" was independent of material type and a linear function of the number of load repetitions on a log-log plot while "B" was independent of the number of load repetitions but had different values for different pavement foundation material qualities. This led to the development of a single equation to describe the permanent pavement foundation deformation (PD) as a function of the vertical stress (σ_v) and the number of load repetitions (N) for three different material quality groups given in Equation 5[11].

$$PD = 18\times10^{-6} N^{0,813}\left[e^{B\sigma_v} - 1\right] \quad \ldots\ldots\ldots\ldots\ldots\ldots \textbf{(5)}$$

with B = 0,01 for CBR > 80 %
= 0,117 for 25 > CBR > 80 %
= 0,993 for 7 > CBR > 25 %

CONCLUSIONS AND RECOMMENDATIONS

A conceptual 3-dimensional model has been formulated for permanent deformation and the advantages of such a model have been highlighted. The model was calibrated with permanent pavement foundation deformation data from HVS-tests done in South Africa over the past 20 years. The model was calibrated for the effect of the two primary independent variables namely stress level and the number of load repetitions and one secondary independent variable namely the material quality. The model could not be calibrated for any other secondary independent variable such as moisture content and initial density as these parameters were not always recorded in detail during previous HVS-tests.

By considering laboratory test results published by other researchers it was possible to confirm the similarity in the basic trends between laboratory and APT permanent deformation data. Hopefully these similarities, if better understood, will enable the development of the permanent deformation models in the laboratory followed by calibration of these models with limited

APT-tests. The laboratory environment will allow better control and recording of all the independent variables.

REFERENCES

1. Van Vuuren D J, Otte E and Paterson W D O. *The Structural Design of Flexible Pavements in South Africa*. Proc. 2nd Conference on Asphalt Pavements in South Africa, Durban, 1974.

2. Walker R N, Paterson W D O, Freeme C R and Marais C P. *The South African Mechanistic Pavement Design Procedure*. Proceedings of the fourth International Conference on the Structural Design of Asphalt Pavements, University of Michigan, Ann Arbor, Michigan, U S A, 1977.

3. Theyse H L, de Beer M and Rust F C. *Overview of the South African Mechanistic Pavement Design Method*. Transportation Research Record 1539, Transportation Research Board, Washington, D.C., 1996.

4. Barksdale R D. *Laboratory Evaluation of Rutting in Base Course Materials*. Proceedings of the third International Conference on the Structural Design of Asphalt Pavements, University of Michigan, Ann Arbor, Michigan, U S A, 1972.

5. Monismith C L, Ogawa N and Freeme C R. *Permanent Deformation Characteristics of Subgrade Soils in Repeated Loading*. Paper presented at the 54th Annual Meeting of the Transportation Research Board, Washington, D.C., 1975.

6. Molenaar A A A. Personal communication.

7. Kenis W J. *Predictive Design Procedures, a Design Method for Flexible Pavements using the VESYS Structural Subsystem*. Proceedings of the fourth International Conference on the Structural Design of Asphalt Pavements, University of Michigan, Ann Arbor, Michigan, U S A, 1977.

8. Leahy R B and Witczak M W, *The Influence of Test Conditions and Asphalt Concrete Mix Parameters on Permanent Deformation Coefficients Alpha and Mu*. Journal of the Association of Asphalt Paving Technologists, Volume 60, 1991.

9. Owusa-Antwi E B, Titus-Glover L and Khazanovich L. Paper presented at the 77th Annual Meeting of the Transportation Research Board, Washington, D.C., 1998. (TRB preprint 981119)

10. Wolff H. *Elasto-Plastic Behaviour of Granular Pavement Layers in South Africa*. Ph D thesis, University of Pretoria, Pretoria, South Africa, 1992.

11. Theyse H L. Mechanistic Empirical Modelling of the Permanent Deformation of Unbound Pavement Layers. Proceedings of the eighth International Conference on the Structural Design of Asphalt Pavements, University of Washington, Seattle, Washington, U S A, 1997.

12. Maree J H. Ontwerpparameters vir Klipslag in Plaveisels. (Design Parameters for Crushed Stone in Pavements). M Eng thesis, University of Pretoria, Pretoria, South Africa, 1978.

13. Kondner R L. *Hyperbolic stress-strain response: Cohesive soils*. Proceedings of the American Society of Civil Engineers, Vol. 89 No. SM1, Paper 3429, 1963.

14. Kondner R L and Zelasko J S. *A Hyperbolic Stress-Strain Formulation for Sands*. Proceedings of the Second international conference of Soil Mechanics and Foundation Engineering, Vol. I, Brazil, 1963.

15. Duncan J M and Chang C Y. *Non-linear analysis of stress and strain in soils*. Journal of the Soil Mechanics and Foundation Division, American Society of Civil Engineers, 1970.

Fifth International Conference on the
BEARING CAPASITY OF
ROADS AND AIRFIELDS

ON THE EFFECT OF DYNAMIC ROUGHNESS INDEX TO ALLIGATOR CRACKING

VESA MÄNNISTÖ[1] AND SEPPO JÄRVINEN[1]

[1]Inframan Ltd, Päiväntaite 8, 02210 Espoo, Finland, vesa@inframan.fi

ABSTRACT

The effect of road unevenness on road users is commonly expressed using International Roughness Index (IRI). The Technical Research Centre of Finland (VTT) has developed another measure, Dynamic Roughness Index (DRI), which describes how the pavement feels the unevenness of the road due to dynamic axle loads of heavy vehicles.

Reliable road deterioration models are key elements of any pavement management system (PMS). The PMS-systems used in Finland are partly lacking good models for prediction of distresses, especially for prediction of alligator cracks, which existence is due to high traffic loads and low bearing capacity. One solution to improve these models would be to utilise the DRI.

This paper concentrates on modelling of existence of alligator cracks on soft emulsion pavements in Finland, using logistic regression function to predict the probability of existence of alligator cracks as a function of independent variables, such as traffic, bearing capacity or DRI. The logistic function is of form

$$P(y_i = 1 \mid x_{ij}) = \frac{e^{\alpha+\beta xj}}{1+e^{\alpha+\beta xj}}$$

where y_i is a Bernoulli distributed variable indicating existence of alligator cracks, x_{ij}:s are values of independent variables, and α and β are vectors of unknown parameters. A logistic classification method is used to classify road sections to two classes: with or without alligator cracks.

The results of this study show, that dynamic roughness index, base curvature index, and pavement age are all statistically significant variables to explain the existence of alligator cracks. The models were able to recognise "no alligator cracks" situation in 80 percent of cases, and "alligator cracks" in 70 percent of cases.

Keywords: Dynamic Roughness Index, Pavement Management Systems, logistic regression.

INTRODUCTION

Pavement management systems (PMS) have become widely used in preparation of long-term and short-term strategies for pavement maintenance and rehabilitation (1). Reliable road condition data and deterioration models are key elements of any pavement management system.

The PMS-systems used in Finland are partly lacking good models for prediction of distresses, especially for prediction of alligator cracks, which existence is due to high traffic loads and low bearing capacity. One solution to improve these models would be to utilise the dynamic roughness index (DRI).

The effect of road unevenness on road users has been commonly expressed using International Roughness Index (IRI). The Technical Research Centre of Finland (VTT) has developed another measure, Dynamic Truck Index (or Dynamic Roughness Index), DRI, which describes how the pavement feels the unevenness of the road due to dynamic axle loads of heavy vehicles (2).

The main indicators for road deterioration in northern areas are climate, thaw, traffic and time, all these related to construction and maintenance standards. Visual distresses like alligator cracks progress usually non-linearly in several phases: cracks in the old pavement layer reflect during the first one to three years. After this phase progression of cracks smoothens down until road and pavement structures start to deteriorate. The third phase consists of accelerated deterioration, until the road comes to its final stage, when it collapses.

The standard approach would be to model the area of alligator cracks as such, but it has proved to be extremely difficult, due to low accuracy of visual inspection of cracks, and also due to very skewed distribution of this variable. A more robust method is to model the probability, whether a road section has alligator cracks or not. This method has been tested in identification of low bearing capacity in Finland (3).

An interesting question is, how this type of statistical model would be able to discriminate already cracked road sections from good ones. This information could be used in the PMS to predict the need of treatments. Another application would be planning of distress inventories; these models could rule out non-cracked sections from measurement programs.

A corollary problem would be: how many groups are needed to classify alligator cracks? The current solution is for two classes, but a clustering procedure could be used to find an optimal number of classes for alligator cracks.

METHODS

The modelling of existence of alligator cracks can be done using a logistic regression function (4) to predict the probability of existence of alligator cracks as a function of independent variables, like traffic, bearing capacity, pavement age or DRI. The logistic function is of form

$$P(y_i = 1 | x_{ij}) = \frac{e^{\alpha+\beta xj}}{1+e^{\alpha+\beta xj}}$$

where y_i is a Bernoulli distributed response variable indicating existence of alligator cracks, x_{ij}:s are values of independent variables (either continuous or discrete), and α and β are vectors of unknown parameters. Furthermore, a logistic classification method is used to classify road sections to two classes: with or without alligator cracks (5). The limiting probability of 0.5 is normally used in the analysis.

DATA

The data used in this analysis was collected from the road condition data bank of the Finnish National Road Administration. The data set consists of condition measurements executed in 1996 on soft emulsion pavements only, and where the last periodic treatment had been an overlay. All the data is handled as 100 m sections, which is the standard procedure in Finland.

The data set was classified to two classes according to traffic volume to test the effect of road construction standard. The final set had 13 489 observations, of which 8893 were of low traffic (ADT < 800) and the rest 4596 of traffic > 800 v.p.d.

The following variables were used:

Dynamic Roughness Index (DRI) is a 4th power function of dynamic and static axle load forces on road, i.e.

$$DRI = \frac{100}{N} \times \sum_{i=1}^{N} \left(\frac{F_{D_i}}{F_S}\right)^4,$$

where F_D = dynamic axle load
F_S = static axle load
N = observations, where $F_D > F_S$

Values of DRI vary usually between 100 (absolutely even road) and 300 (high value in practice). DRI is measured with a Road Surface Monitoring –vehicle (2).

Utilisation of deflection bowl information has traditionally been constrained to D0-deflection, although it does not correlate well with other road condition variables. In this study, **Base Curvature Index (BCI)** and **Surface Curvature Index (SCI)** are utilised instead of D0. The definitions of BCI and SCI are as follows:

$$BCI = D_{900} - D_{1200}$$
$$SCI = D_0 - D_{200}$$

where D_x = deflection at offset x.

Pavement age is probably the best indicator of both traffic loading and effect of time and climate on pavement.

Alligator cracks, measured with visual inspection are one of the most common distress type on soft emulsion pavements in Finland. For modelling purposes, this variable is classified into two classes:

0 = area of alligator cracks is zero
1 = area of alligator cracks is greater than zero

RESULTS

The logit-models were estimated separately for three situations:

- All roads
- Traffic less than 800 ADT
- Traffic more than 800 ADT

Standard linear regression models and discriminant models were also tested, but they appeared to be of less significance.

All roads

The model is as follows:

$$P(Y = 1/X) = -5.85 + 0.0275 * DRI + 0.0132 * BCI + 0.179 * AGE,$$

where		
	P(Y=1|X)	logistic probability function for existence of alligator cracks
	DRI =	dynamic roughness index
	BCI =	base curvature index
	AGE =	pavement age
	N =	number of observations (=2697)

The coefficient of determination (R^2) is 0.80, and all the independent variables are significant at risk level $p < 0.01$.

Table 1 shows the classification power of the model. This model is able to recognise "no cracks" situation with 79 percent accuracy, and "cracks" situation with 68 percent.

Table 1. Classification power for all roads

	Model	
Data	**No cracks**	**Cracks**
No cracks	79	21
Cracks	32	68

Higher traffic roads (ADT > 800)

The model is as follows:

$$P(Y=1/X) = -5.35 + 0.023 * DRI + 0.26 * AGE\ ,$$

where		
	P(Y=1|X)	logistic probability function for existence of alligator cracks
	DRI =	dynamic roughness index
	AGE =	pavement age
	N =	number of observations (=919)

The coefficient of determination (R^2) is 0.82, and both the independent variables are significant at risk level $p < 0.01$. In this case, the base curvature index appeared not to be a significant factor, which is likely due to better construction standard of higher trafficked roads.

Table 2 shows the classification power of the model. This model is able to recognise "no cracks" situation with 80 percent accuracy, and "cracks" situation with 70 percent.

Table 2. Classification power for higher traffic roads

	Model	
Data	**No cracks**	**Cracks**
No cracks	80	20
Cracks	30	70

Low traffic roads (ADT < 800)

The model is as follows:

$$P(Y=1/X) = -6.38 + 0.0321 * DRI + 0.0138 * BCI + 0.165 * AGE\ ,$$

where		
	P(Y=1|X)	logistic probability function for existence of alligator cracks
	DRI =	dynamic roughness index
	BCI =	base curvature index
	AGE =	pavement age
	N =	number of observations (=1778)

The coefficient of determination (R^2) is 0.80, and again all the independent variables are significant at risk level $p < 0.01$.

Table 3 shows the classification power of the model. This model is able to recognise "no cracks" situation with 80 percent accuracy, and "cracks" situation with 70 percent.

Table 3. Classification power for higher traffic roads

	Model	
Data	**No cracks**	**Cracks**
No cracks	79	21
Cracks	31	69

CONCLUSIONS AND RECOMMENDATIONS

The results of this study show that dynamic roughness index is a very significant factor in prediction of existence of alligator cracks. Only base curvature index of the bearing capacity variables was found to be significant in this data set. The effect of traffic load is reflected through pavement age, because these soft emulsion pavements are rather low trafficked in Finland.

The estimated logit models were able to classify road sections into two classes extremely well; in "no alligator cracks" situation, the classification was successful in about 80 percent of cases, and in "alligator cracks" situation in about 70 percent of cases, respectively.

These models could be used in prediction of alligator cracks in the PMS-systems, but also in programming of visual condition surveys, where the models can indicate, whether it pays off to execute a more precise visual inspection.

The importance of dynamic roughness was compared to normal roughness (IRI). These two variables are so highly mutually correlated, that due to statistical reasons they can not be put into same models.

These models could be improved in the future by

- adding more independent variables to the models
- estimation of models for asphalt concrete pavements
- testing autoregressive models, which would diminish the effect of dependent residuals
- applying other new methods, like LAD (Least Absolute Deviation), or TOBIT-models (for mixed-distributed response variables)

REFERENCES

1. TAPIO, R. and MÄNNISTÖ, V. Highway Performance Monitoring Systems, Final Report. World Road Association, Paris, 1997.

2. HUHTALA, M. *et al.* Roughness Measurement Devices and Dynamic Truck Index. Proceedings of the Fourth International Conference on Bearing Capacity of Roads and Airfields, Minneapolis, 1994.

3. MÄNNISTÖ, V. and TAPIO, R. Estimation of Road Bearing Capacity with Other Automatically Measured Road Condition Variables. Proceedings of the Third International Conference on Bearing Capacity of Roads and Airfields, Trondheim, 1990.

4. McCULLAGH, P. and NELDER, J.A. Generalised Linear Models, Second Edition. Chapman and Hall, London, 1989.

5. SEBER, G.A.F. Multivariate Observations, Wiley & Sons, New York, 1984.

Fifth International Conference on the
BEARING CAPASITY OF
ROADS AND AIRFIELDS

THE REMAINING LIFE FACTOR FOR OVERLAY DESIGN

MOSHE LIVNEH

Harry Oppenheimer Professor of Engineering, Transportation Research Institute, Technion City, Haifa 32000, Israel, Tel: +972 4 8292362
Fax: +972 4 83610919, E-mail: moshliv@tx.technion.ac.il

ABSTRACT

The overlay design process generates a basic debate over whether a determination of the remaining life parameter is required at all; and, if so, how should this parameter be taken into consideration to account for the total pavement and its subgrade fatigue. That this debate exists at all cannot be understood, as it seems obvious that the true concept of the overlay design should be associated with the "past memory" (fatigue) effect of the pavement. The future life span of a pavement from a given point in time no doubt depends on the volume of traffic that has already passed; thus, it cannot be stated that the pavement's current (deteriorated) material data at any point in time were essentially "created" for that given point in time rather than having been of a deteriorated nature arising from previous traffic-induced stress. Unfortunately, this "correct" concept of the residual life phenomenon is not used in rehabilitation design methods. This may perhaps, be ascribed to the difficulty involved in determining the residual life parameter for a given pavement; thus, it is "convenient" to ignore it. It is demonstrated here, however, that the use the "correct" concept of residual life in the design or evaluation process is both important and decisive. The paper suggests a convenient method for including this residual life parameter within the design process for flexible pavement rehabilitation.

PROBLEM DEFINITION

In order to demonstrate the issue of the residual life factor in the overlay design, the following question should raised: Does the future life span of a pavement at a given point in time depend on the volume of traffic that has already passed (i.e., is the future behavior of a pavement affected by the pavement's "past memory")? Or is the future life span of the pavement at the given point in time independent of the volume of traffic already passed (i.e., is the future behavior of a pavement unaffected by "past memory")? In other words, is it correct to state that at any point in time, the pavement's current (deteriorated) mate1rial data are essentially of an initial nature--i.e., "created" for that given point in time--rather than being of a deteriorated nature arising from stress induced by previous traffic.

The following example illustrates the meaning of the above questions. According to the TRRL design chart (Ref. 1), a deflection of 35×10^{-2} mm indicates the end of the life of pavement *A* and the start of the life of pavement *C* (Fig. 1). This means that a measured value of the central deflection alone cannot physically predict the remaining life of a given pavement structure. In the case of pavement *A*, the remaining life value was 0%; and in the case of pavement *C*, the remaining life value was 100%. In order to predict correctly the remaining value, the past traffic should be introduced to the chart. In the Asphalt Institute (Ref. 2) approach, however, no past traffic is needed. The prediction of the pavement bearing capacity in terms of allowable repetitions of future traffic is carried out with the criterion of the central deflection alone.

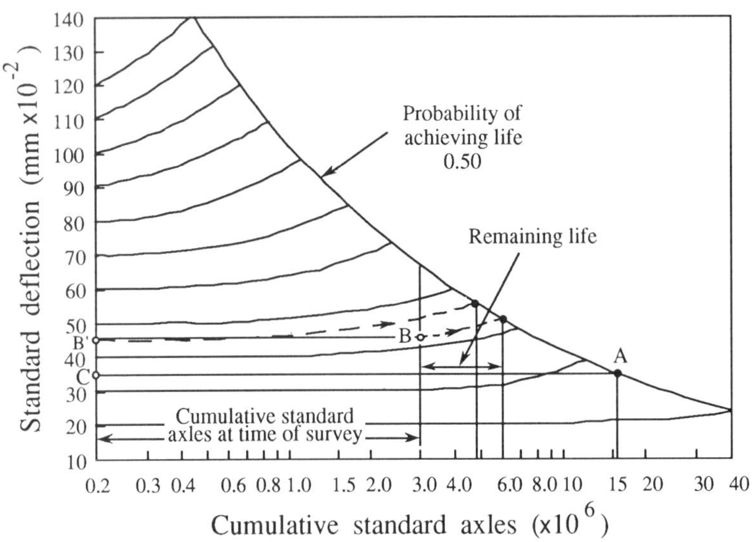

Fig. 1: Relationship Between Standard Deflection and Life of Pavements with Granular Bases. Source: Kennedy and Lister (Ref. 1)

The latter approach assumes that after its rehabilitation (even if of a very light nature), the pavement may be considered "brand new"; and its stiffness may thus be determined again by measuring the deflection bowl at the surface of the wheel's critical track without considering its past traffic. In other words, this approach makes no allowances for the remaining life factor as though future rehabilitation action will wipe out all past fatigue effects on the pavement structure and its subgrade. To clarify this point, it is worthy to quote Shell's statement, cited by Koole (Ref. 3): *"If, after a certain proportion of service life has been consumed, an overlay is applied, the permanent deformation due both asphalt and subgrade deformation will automatically be eliminated and the original design is restored."* Or, *"Any overlay will automatically eliminate all ill effects."*

It should be mentioned here that the above approach is limited, though only for the total structure and its subgrade fatigue. As for the rehabilitation of the upper asphaltic layers alone, almost all methods do take into consideration their fatigue effects. This is done by addressing their remaining life parameter, as described recently by Livneh (Ref. 4).

To sum up, the two approaches discussed above indicate that for the total structure and its subgrade, a basic debate exits over whether a determination of the remaining life parameter is required at all in the overlay design process; and, if so, how this parameter should be taken it into consideration in the overlay design process. That this debate exists at all, however, cannot be understood, as it seems obvious that the true concept of the overlay design should be associated with the "past memory" effect. This paper will show how this concept should be taken into account in the overlay design process. Naturally, the presentation is limited only to the total structure and its subgrade criteria.

THE FUNDAMENTAL APPROACH OF OVERLAY DESIGN

Prior to an elaboration of the remaining life issue, a brief description of the fundamental approach of overlay design is in order. According to AASHTO (Ref. 5), the effective structural number of the existing pavement (in mm), SN_{eff}, is

$$SN_{eff} = CF \times SN_O \qquad (1)$$

where SN_O is the original structural number (in mm) of the pavement (immediately after its opening for traffic) and CF is the condition factor (i.e., the reduction factor of SN_O from pavement deterioration). The above AASHTO guide makes it possible to determine the value of SN_{eff} directly through an analysis of the deflection bowl by using the following equation:

$$SN_{eff} = 0.0234 \times H_O \times (E_{PT})^{1/3} \qquad (2)$$

where H_O is the total thickness of all pavement layers above the compacted subgrade (in mm) and E_{PT} is the pavement modulus of elasticity at time t along the wheel's critical track (in Mpa).
When the two above equations are linked, the following relationship may be obtained:

$$CF = (E_{PT}/E_{PO})^{1/3} \qquad (3)$$

where E_{PO} is the effective modulus of elasticity of the original pavement (immediately following the opening of the pavement for traffic).

Now, if an asphaltic overlay is being placed for the purpose of structural improvement, the required thickness of this overlay, D_{OL}, is a function of the structural capacity required to meet future traffic demand and the structural capacity of the existing pavement. The required thickness to increase structural capacity to carry future traffic is determined by the following expression:

$$SN_{OL} = a_{OL} \times D_{OL} = SN_D - SN_{eff} \qquad (4a)$$

where SN_{OL} is the required overlay structural number (in mm), a_{OL} is the structural coefficient for the asphaltic overlay, SN_D is the structural number required to carry future (design) traffic (in mm). In the same manner, the rehabilitation thickness from any other design method, H_{OL}, is as follows:

$$H_{OL} = H_D - H_{ES} \qquad (4b)$$

where H_D is the total thickness required to carry future (design) traffic, and H_{ES} is the effective thickness of the existing (deteriorated) structure in terms of standard materials as defined by the given design method and as explained below.

Now, if Eq. 3 is used to calculate H_{ES}, no allowance is made to take into account the remaining life factor. To do so, the steps described in the following section are required.

MODIFIED DETERMINATION OF CONDITION FACTOR

The main drawback of the method described in the previous section is its avoidance of the "pavement memory" effect discussed earlier. This is illustrated in Fig. 2, which is based on the phenomenon associated with the low central deflection values in Fig. 1. The horizontal line represents the negligible change in the effective pavement modulus (zone I in Fig. 2), which occurs with the decrease in remaining life. Similarly the inclined line represents the severe change in the effective pavement modulus (zone II in Fig. 2). It is obvious that the required overlay thickness for an existing pavement, characterized by state *A* in Fig. 2, is by no means identical with that required for the pavement characterized by state *B* in Fig. 2, as suggested by the above equations.

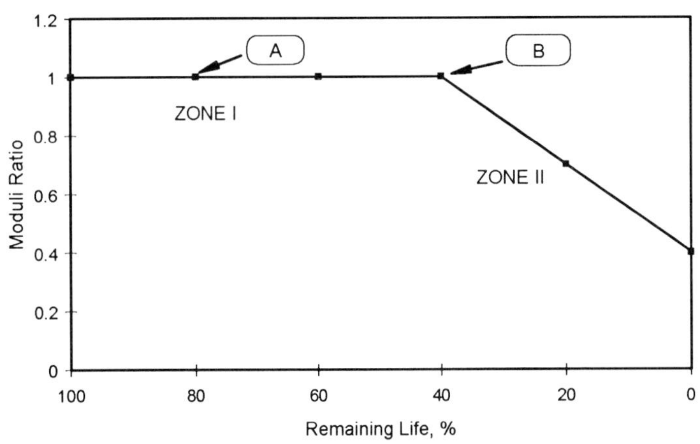

Fig. 2: Schematic Presentation of Moduli Ratio (E_{PT}/E_{PO}) Variation with Remaining Life Factor (*RL*)

It should be mentioned that because of these and other reasons , AASHTO (Ref. 5) states the following: *"Three alternative evaluation methods are recommended to determine effective structural capacity. (1) Structural capacity based on visual survey and materials testing....(2) Structural capacity based on nondestructive deflection testing....(3) Structural capacity based on fatigue damage from traffic....Because of the uncertainties associated with the determination of effective structural capacity, the three methods cannot be expected to provide equivalent estimates. The designer should use all three methods where possible and select the 'best' estimate based on his judgment. There is no substitute for solid experience and judgment in his selection."*

It can be shown that *CF* is related to *RL* through the pavement design equations. The expression obtained for the AASHTO design method of a new structure (Ref. 5) is

$$CF = (RL/100)^{0.165} \qquad (5)$$

It should be mentioned that it is possible to derive similar expressions to that of AASHTO (Eq. 5) for any other design method. A comparison of Eq. 5 with expressions derived from the extended CBR method is shown in Fig. 3. This extended CBR method was developed by Uzan (Ref. 6) for the Israeli Public Works Department. Fig. 3 shows that the use of Eq. 5 for local conditions leads to a favorable coefficient of safety. It should be noted that the pavement capacity at its "death" (for which *RL=0%*) still has some salvage value, which corresponds to an equivalent value of subbase material. Therefore, Eq. 5 and all other relationships given in Fig. 3 are limited to a range where, *RL* is higher than 5% for asphaltic materials and higher than 40% for granular materials. These conditions are introduced in order to limit the minimum values of *CF* to 0.5 in asphaltic materials and to 0.8 in granular materials. Thus the relevant expressions for the complementary range are these:

$$CF_A = 0.5+2.20\times(RL/100) \quad \text{for } RL<5\% \text{ in asphaltic materials} \qquad (6a)$$
$$CF_G = 0.8+0.15\times(RL/100) \quad \text{for } RL<40\% \text{ in granular materials} \qquad (6b)$$

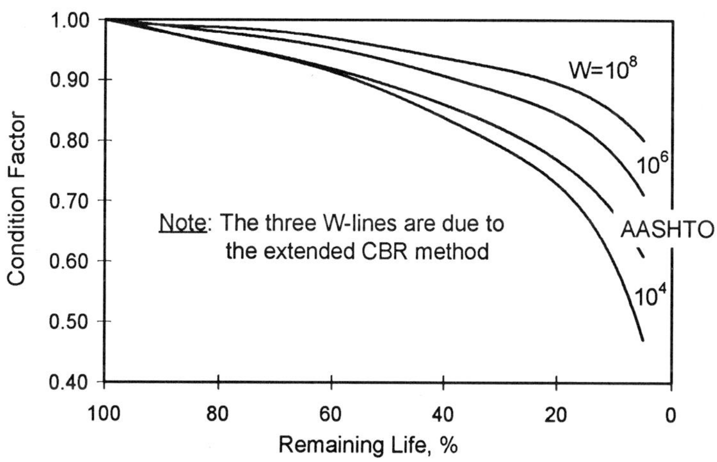

Fig. 3: Condition Factor (*CF*) versus Remaining Life Factor (*RL*) from AASHTO and Extended CBR Methods

Now, it can be shown that the use of CF $(=CF_A=CF_G)$ from Eq. 3 instead of CF $(=CF_A=CF_G)$ from Eq. 5 or CF_A from Eq. 6a and CF_G from Eq. 6b may lead to considerable errors in the calculated values of asphaltic overlay thickness. Fig. 4 indicates that for a pavement with a total thickness, H, of 80.0 cm (with 15.0 cm of the original asphaltic layers and 20.0 cm of the original granular base layer), the difference in the overlay thickness (i.e., the overlay thickness that is due to *CF* from Eq. 5 and Eq. 6 minus the overlay thickness due to *CF* from Eq. 3) can reach very high values, of up to 15.0 cm in extreme cases. Therefore, a very realistic determination of *RL* is essential for obtaining the correct values of overlay thickness. The method of a realistic determination of *RL* is dealt with in the following section.

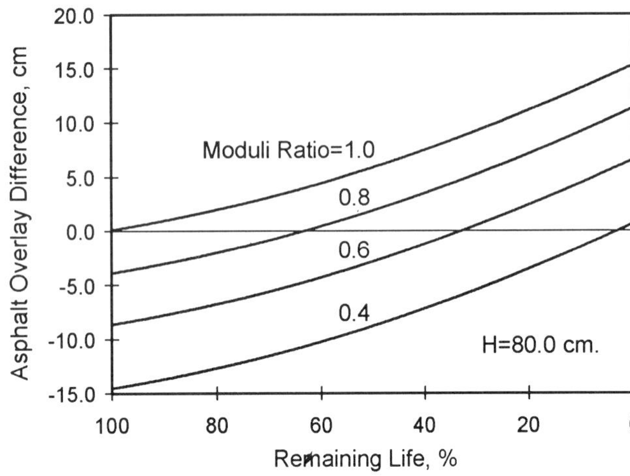

Fig. 4: Asphaltic Overlay Difference versus Remaining Life (*RL*) for Different Moduli Ratios (E_{PT}/E_{PO})

DETERMINATION OF REMAINIG LIFE

A realistic determination of *RL* calls for the involvement of the past memory of the pavement. This is done by using the TRRL method for defining the following past memory correction factor, ϕ:

$$\phi = W_{CF}/W_{UF} \qquad (7)$$

where W_{UF} is the uncorrected future standard axle capacity derived from Fig. 1 (i.e., the amount of standard axle capacity corresponding to the measured standard deflection and zero commutative standard axles at time of survey, W_P), and W_{CF} is the correct future standard axle capacity, (i.e., the amount of standard axle capacity corresponding to the measured standard deflection and to the real value of W_P). For example, pavement *B* in Fig. 1 yields a future standard axle capacity of 3.0×10^6 repetitions. This value of repetitions is the correct one, as it takes into account the past standard axles of the same (in this particular case) 3.0×10^6 value. Fig. 1 also enables a determination of the uncorrected future standard axles capacity for the same standard deflection measured in pavement *B*. For this uncorrected capacity, the past standard axle capacity is equal to zero (point *B'* in Fig. 1). Therefore, the uncorrected future standard axle capacity is equal to 5.0×10^6 repetitions. For these values, ϕ equals $3.0\times10^6/4.8\times10^6=0.6$.

The above correction factor is required in order to estimate the future standard axle capacity owing to any given design method, by considering the deteriorated pavement as a new pavement that possesses a reduced thickness with no past memory effect. This reduced thickness, H_{MS}, is calculated from the measured pavement modulus, E_{PT}, with the aid of the following equations:

$$H_{MS} = H_{OS} \times (E_{PT}/E_{PO})^{1/3} \tag{8a}$$
$$H_{OS} = H_O \times (E_{PO}/E_{PS})^{1/3} \tag{8b}$$

where E_{PS} denotes the standard pavement modulus as defined by the design method, and H_{OS} denotes the transformed original pavement thickness of the actual designed structure (H_O, which is compatible with E_{PO}) into an equivalent thickness with a standard structure compatible with E_{PS}. It should be noted that E_{PO} is backcalculated from deflection bowls measured on non-trafficked areas or, alternatively, is evaluated from E_{PT} by means of Fig. 1 and given relationships between FWD central deflections and British standard deflections. In this way, the value of the uncorrected future standard axle capacity derived from any CBR design method (for which, as mentioned before, W_P is considered to be zero) is attained, together with the subgrade CBR value evaluated from the backcalculated subgrade modulus,. Thus, the required value of RL may be derived from the following expression:

$$RL = \phi \times W_{UA}/(W_P + \phi \times W_{UA}) \tag{9}$$

where W_{UA} denotes the above uncorrected future standard axle capacity. In cases in which the correct value of ϕ is not taken into consideration, the overlay calculations are run, by default, with $\phi = 1.0$. This default value of ϕ leads to erroneous overlay thicknesses as shown in Fig. 5, in which the repetition factor denotes the following ratio: W_{UA}/W_P. According to the figure, the difference in the overlay design thickness in an 80.0 cm total thickness structure (H_{OS}), reaches a high value of about 7.0 cm. This demonstrates that a real value should be assigned to the ϕ factor.

In this context, the use of a pre-defined value of *0.5* is further suggested for the ϕ factor when tests do not provide the real in-situ value. This intermediate value leads to a less erroneous overlay thickness than the *1.0* default value, as demonstrated in Fig. 6. The 7.0 cm overlay thickness difference in Fig. 6 reduces here to only 4.0 cm.

Finally it is worthwhile noting again that H_{MS} does not represent the effective standard thickness H_{ES} required for the final overlay design. This final design is done by subtracting the H_{ES} value from the total thickness calculated for the new design traffic (i.e., the H_D value) as indicated by Eq. 4b. The difference between these two definitions can be evaluated through the following equations:

$$H_{ES} = CF_{(Eq.\,5\ \&\ Eq.\,6)} \times H_{OS} = CF_{(Eq.\,5\ \&\ Eq.\,6)} \times (E_{PO}/E_{PS})^{1/3} \times H_O \tag{10a}$$
$$H_{MS} = CF_{(Eq.\,3)} \times H_{OS} = CF_{(Eq.\,3)} \times (E_{PO}/E_{PS})^{1/3} \times H_O \tag{10b}$$

It is obvious that the more that H_{EM} differs from H_{MS}, the more important it is to take into account the remaining life factor.

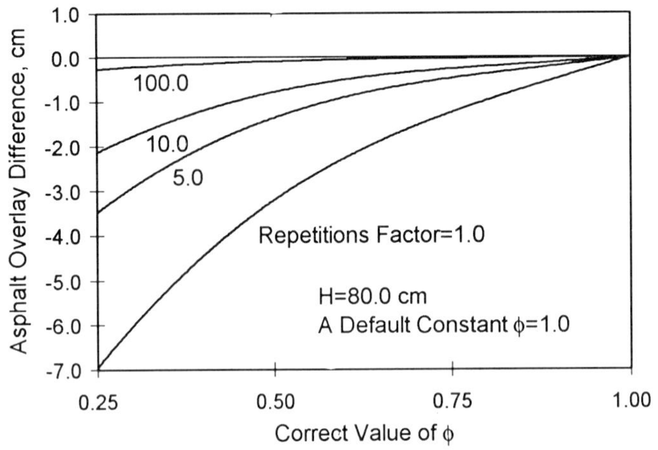

Fig. 5: Asphalt Overlay Difference owing to a Default Value of 1.0 for the Correction Factor (ϕ) versus Correct Value of Correction Factor

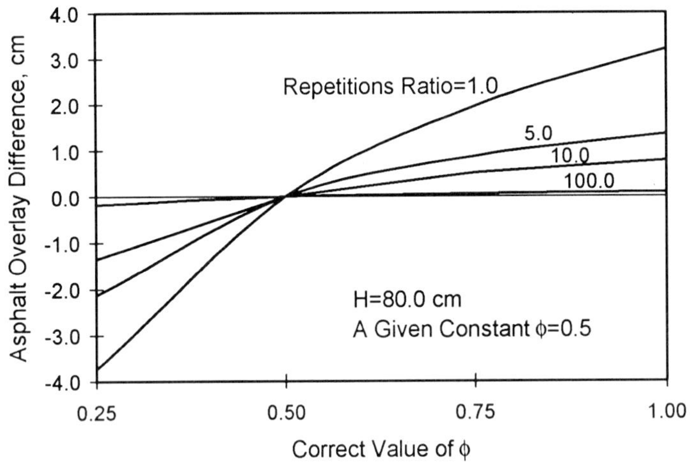

Fig. 6: Asphalt Overlay Difference owing to a Pre-Defined Constant Value of 0.5 for the Correction Factor (ϕ) versus Correct Value of Correction Factor

RESIDUAL LIFE OWING TO SURFAFE DAMAGES

As stated previously, the overlay design procedure also includes an evaluation of residual life through a visual survey of pavement surface damages. This is done by using the following relationships:

$$RL_A = -[\ln(1-0.01\times CCI)]/3 \leq 1.0 \qquad (11a)$$

$$RL_G = -[\ln(1-0.01\times OPI)]/3 \leq 1.0 \qquad (11b)$$

where *CCI* denotes the cracking condition index, and *OPI* is the overall pavement index. These two indices, which were defined by Livneh (Ref. 8), obviously are obtained from the visual survey. It should be added that Eq. 11a and Eq. 11b are based on the flexible pavement model (α=3) suggested by Van Gurp et al. (Ref. 9).

Recently it was shown to be possible to correlate a new RL_A equation from in-situ observations carried out by Cohen et al. (Ref.7). The observed data are shown in Fig. 7; the derived equation is

$$RL_A = 100\times[1.0-(FCA/21)^{0.28}]\leq 1.0 \qquad (12)$$

where *FCA* is the observed fatigue cracked area (in percentages). It should be noted that the area of the critical wheel track is about 25% of the total area. Thus, according to Eq. 12, the remaining life factor reduces to zero when the ratio of the fatigue cracked area to that of the critical wheel track increases to 75%. This value appears to be compatible with the zero structural strength of the asphalt layers.

The *SF* factor shown in Fig. 7 denotes the observed shift factor required to translate the asphalt fatigue model of the extended CBR design method into the observed in-situ model. Finally, it is suggested that the design RL_A value will be the minimum one derived from Eq. 9, Eq. 11a, and Eq. 12. In the same manner, the design RL_G value will be the minimum one derived from Eq. 9 and Eq. 11b. Knowing these two values enables, then, a calculation of the "correct" condition factor with the aid of Eq. 5, Eq. 6a, and Eq. 6b. Again, it should be noted that the calculation by means of Eq. 3 may lead to an erroneous value.

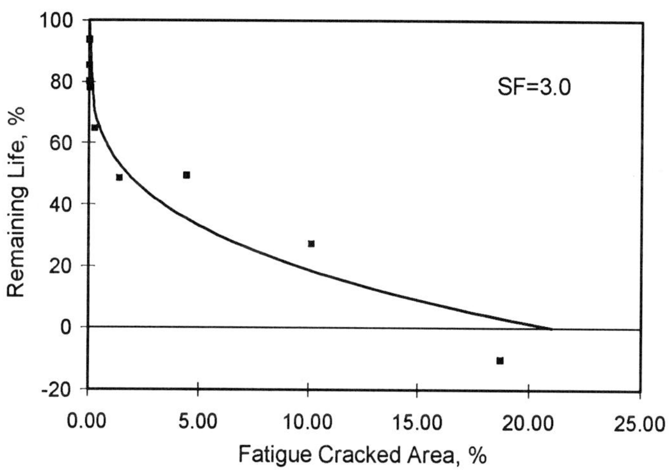

Fig. 7: Remaining Life versus Percentage of Fatigue Cracked Area as Derived from Local Observations.

CONCLUSIONS

Unfortunately, the concept of the residual life phenomenon is not used as a routine matter in rehabilitation design and structural evaluation methods. This may perhaps be ascribed to the difficulty involved in determining the residual life parameter for a given pavement; thus it is "convenient" to ignore it. This paper, however, demonstrates that using the "correct" concept of residual life in the design or evaluation process is both important and decisive. To conclude, this paper suggests a convenient method for including this residual life parameter within the design process for the rehabilitation of flexible pavement.

ACKNOWLEDGMENTS

This paper is partially based on a current study, Design Guides for Flexible Pavement Rehabilitation, which is being carried out for the Israeli Public Works Department, to whom thanks are due. The paper was prepared with the assistance of Mr. Arieh Aines, graphics editor of the Transportation Research Institute at the Technion, to whom thanks are also due.

REFERENCES

1. KENNEDY, C.K., and LISTER, N.W., "Prediction of Pavement Performance and the Design of Overlay." TRRL Laboratory Report No.833, Transportation and Road Research Laboratory, 1978.
2. ASPHALT INSTITUTE, "Asphalt Overlays for Highway and Street Rehabilitation." Manual Series No. 17 (MS-17), 1983.
3. KOOLE, R.C., "Overlay Design Based on Falling Weight Deflectometer Measurements." Transportation Research Record No. 700, TRB, 1979.
4. LIVNEH, M., "Determination of Residual Life in Flexible Pavements." Canadian Journal of Civil Engineering, Vol. 23, No. 5, Canadian Society of Civil Engineering, 1996.
5. AASHTO, "Guide for Design of Pavement Structures." American Association of State Highway and Transportation Officials, 1993.
6. UZAN, J., "A Pavement Design and Rehabilitation System." Transportation Research Record No. 1539, TRB, 1996.
7 COHEN, E., UZAN, J., RUBINSTEIN, I., and BRAUN-COHEN, S., "Development and Implementation of a Rehabilitation System." Preprint Paper No. 970280, TRB, 1997.
8. LIVNEH, M., "Local Criteria for Evaluating Pavement Surface Conditions." Research Report No. 97-255, Transportation Research Institute, Technion, Haifa Israel, 1997.
9. VAN GURP, C., MOLENAAR, A.A.A., VALK, H., and VAN VELZEN, F.J.M., "Testing the Delft University Pavement Management System." Transportation Research Record No. 977, TRB, 1984.

Fifth International Conference on the
BEARING CAPASITY OF
ROADS AND AIRFIELDS

ILSL2 — A FINITE ELEMENT PROGRAM FOR ANALYSIS OF CONCRETE PAVEMENTS AND OVERLAYS

By Lev Khazanovich[1] and Tom Yu[1]

[1] ERES Consultants, Inc., 505 West University Avenue, Champaign, Illinois 61820-3915, tel.: (217)-356-4500, fax: (217)-356-3088

ABSTRACT

Numerous finite element programs are available today for analysis of concrete pavements and overlays. The most widely used finite element computer programs - such as ILLI-SLAB, J-SLAB, FEACONS IV, and KOLA - utilize Westergaard's model, which treats the pavement system as a one- or two-layered slab resting on a Winkler or elastic solid foundation. This approach has several significant limitations, including those discussed in this paper. This paper also demonstrates how these limitations are overcome in the finite element program ILSL2, which is an extension of ILLI-SLAB.

Features that distinguish ILSL2 from all other programs include a wide selection of subgrade models, the ability to analyze the effects of independent action of the pavement layers, the ability to analyze the effect of linear and nonlinear temperature distribution throughout the pavement thickness, and the ability to analyze partial-depth cracks in the concrete layer. These features are discussed in the paper. The paper also presents the results of field verification of ILSL2 using data obtained from instrumented slabs. Examples of application of ILSL2 for analysis of rigid pavements and overlay are also presented. It is demonstrated that ILSL2 is an attractive alternative to general purpose 3D finite element packages.

Keywords PCC Pavements, Finite Element Analysis

INTRODUCTION

The performance of rigid pavements is dependent on the stresses and deflections imposed on the material through repeated traffic and environmental loadings. For example, jointed plain concrete pavement transverse cracking is governed by the maximum tensile stresses at the bottom of the concrete slab, and crack deterioration in jointed reinforced concrete pavements, and continuously reinforced concrete pavements can be related to the shear stresses in the reinforcement and at the crack surface. Therefore, reliable predictions of pavement responses are essential for a mechanistic-empirical design procedure. The structural model used for those predictions should satisfy the following requirements:

- The model should adequately describe the pavement structure (constructed layers and subgrade).
- The model should account for cracks and joints.
- The model should be able to analyze multi-wheel loading.
- The model should be able to analyze temperature curling and moisture warping.

Finite element methods permit development of structural models that satisfy all these requirements. A variety of finite element programs are available to a pavement engineer today. The programs developed specially for analysis of concrete pavement systems include ILLI-SLAB (1), WESLIQID (2), J-SLAB (3), FEACONS-IV (4), KENSLAB (5), and KOLA (6). Most of these programs are capable of analyzing multi-wheel loading of one- or two-layered medium thick plates resting on a Winkler foundation or an elastic solid (ILLI-SLAB, WESLIQID, KENSLAB). They can accommodate the effect of temperature curling and separation of the concrete slab from the subgrade. At the present time, however, most of finite element programs, including J-SLAB, FEACONS, and KOLA, can analyze only a single layer slab when a temperature gradient is introduced. WESLIQID and KENSLAB can analyze a two-layered slab, but only if the temperature gradient is linear and both layers have the same deflection profile.

Actual pavement behavior may be significantly different than the behavior of a slab resting on the dense liquid or elastic solid foundation. A real subgrade may exhibit behavior somewhere between the two extremes described by the Winkler and elastic subgrade models. If a concrete pavement system incorporates a stiff stabilized base or Portland cement concrete (PCC) overlaid by another PCC slab, separation may occur between constructed layers. Moreover, temperature distribution throughout the thickness of a concrete pavement may be highly nonlinear. At the present time, however, the majority of the pavement programs cannot account for these effects.

This paper summarizes the latest developments in the enhancement of ILLI-SLAB, referred to as ILSL2. The following features of ILSL2 distinguish it from all other programs:

- A wide selection of subgrade models.
- An ability to analyze the effect of independent action of two pavement layers.
- An ability to analyze the effect of linear and nonlinear temperature distribution throughout the pavement thickness.
- An ability to analyze partial-depth cracks.

The paper also presents examples of application of this program, as well as the results of field verification of the program using data obtained from instrumented slabs.

ADVANCED SUBGRADE MODELS

Winkler or dense liquid (DL) and elastic solid (ES) subgrade models (figures 1 and 2, respectively) are used most often in pavement modeling. ILLI-SLAB, WESLIQID, KENSLAB are capable of analyzing both of these foundation models. However, neither the DL nor the ES idealization is entirely adequate when applied to real soils, and the predictions from both exhibit discrepancies with observed, in situ behavior. The DL model assumes that shear resistance of the subgrade is negligible compared to shear capacity of the subgrade. It models the foundation as a set of independent springs, represented as the foundation parameter, k. On the other hand, the ES model attributes to the foundation a higher degree of shear interaction than is usually available in the field. Use of ES models results in infinite stress predictions under the edges and corners of a plate resting on it. Both models fail to describe subgrade behavior beyond the slab edge; soil deflections vanish more rapidly than predicted by the ES but not nearly as fast as assumed in the DL formulation. Khazanovich and Ioannides (7) identified a number of subgrade models that promise to provide more realistic characterization of a PCC slab support. The finite element formulations for the following models were developed and incorporated into ILSL2 in addition to DL and ES models:

- Two Parametric (TP) model (figure 3).
- Zhemochkin-Sinitsyn Shtaerman (ZSS) model (figure 4).
- Kerr-Vlasov (KV) model (figure 5).

Also known as the Pasternak or Vlasov subgrade, tthe TP model offers an attractive alternative to the ES continuum in providing a degree of shear interaction between adjacent soil elements. One of the possible visualizations of this model is a combination of a shear layer resting on top of the spring layer (figure 3). The deflection profile predicted by TP vanishes much faster than the corresponding ES basin and may be a better approximation of the deflections observed in a real foundation of finite depth. Comparisons between DL and TP presented by Pronk (8) suggest that the latter can be considered a logical improvement of DL.

A finite element formulation of TP for ILLI-SLAB was first developed by Ioannides, et al. (9). That solution, however, failed to account properly for behavior of the subgrade beyond the slab edge and, therefore, was limited to the case of interior loading. Khazanovich and Ioannides (10) developed an improved, mathematically rigorous finite element formulation, which was incorporated into ILSL2. Semi-infinite elements in one or two horizontal dimensions that properly describe the model behavior near slab edges and corners were introduced. Appropriate terms are added to the stiffness matrix to reflect the energy of soil deformation beyond the slab.

Despite its advantages, TP cannot be used for the estimation of subgrade stresses because, like ES, it predicts infinite stresses under a free slab edge. The subgrade idealization proposed by Zhemochkin and Sinitsyn and by Shtaerman overcomes this weakness. ZSS represent an in-series combination of the DL and ES models (see figure 4). This model allows a limited degree of interaction between adjacent foundation elements, yet prevents the generation of infinite edge stresses. Khazanovich and Ioannides (10) developed a finite element formulation for ZSS using available formulations for DL and ES models. This formulation was incorporated into ILSL2.

KV model was proposed by Vlasov (11) for account for a soft layer placed on a stiffer subgrade. Kerr (12) proposed another physical interpretation of this model as a spring layer placed above a TP foundation (see figure 5). He and his colleagues obtained several analytical solutions for several practically important boundary value problems and of KV. In contrast to ZSS, KV leads to a banded stiffness matrix, without predicting infinite soil pressures under the free edge of the slab, as ES and TP would. Although KV requires twice as many nodes as ZSS, the fact that the stiffness matrix is banded allows considerable savings in computational effort. The disadvantage of this model is its requirement for three material parameters, k_U, G, and k_L, where k_U and k_L are the upper and lower spring stiffnesses, respectively.

Figure 1. Slab on Winkler foundation.

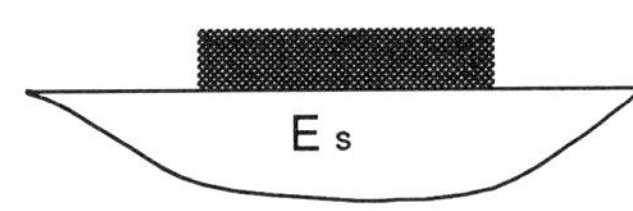

Figure 2. Slab on ES foundation.

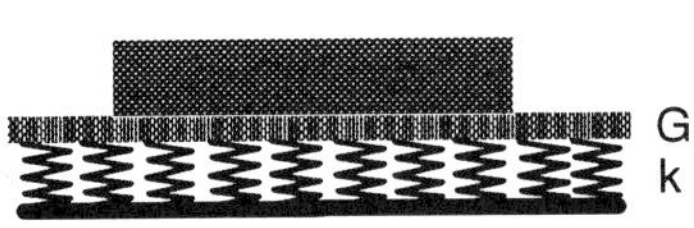

Figure 3. Slab on TP foundation.

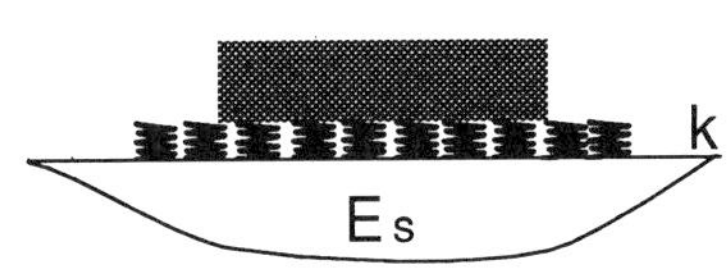

Figure 4. Slab on ZSS foundation

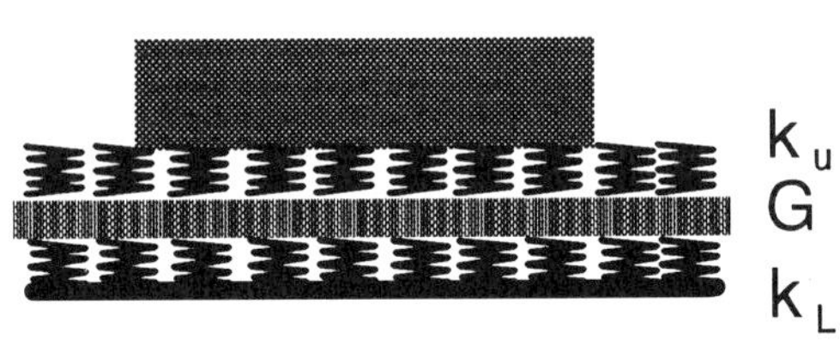

Figure 5. Slab on KV foundation.

Comparison of the Subgrade Models

In this study, the models were compared based on their ability to match the results obtained from the other models. A 200-mm thick, 3.6-m wide, and 4.8-m long slab was analyzed. The applied pressure of 0.7 MPa was distributed over a square with a size of 133 mm. These conditions correspond to a typical highway pavement slab loaded by a Falling Weight Deflectometer (FWD) load. Three cases were considered: interior loading, edge loading, and corner loading. The parameters of the models were selected to match closely maximum deflections and slab stresses from an ES foundation, whose modulus of elasticity, E_s, was fixed at 50 MPa. The priority was given to matching of the deflections from the interior loading, since an FWD center slab test is a common method for obtaining information for in situ pavement material properties. The result of the analysis are summarized in table 1.

Table 1. Effect of Subgrade Model

Model	Interior Loading		Edge Loading		Corner Loading	
	Deflection, mm	Stress, kPa	Deflection, mm	Stress, kPa	Deflection, mm	Stress, kPa
ES	0.36	1560	0.64	2734	1.06	1433
DL	0.36	1727	1.08	3260	2.62	1878
TP	0.36	1566	0.56	2435	0.73	1420
KV	0.36	1572	0.66	2760	1.21	1405

Notes: DL model-k=19.0 kPa/mm; TP model-G=16.6 kN/mm, k=8.15 kPa/mm; KV model: k_U=272 kPa/mm, G=34.9 kN/mm, k_L=7.33 kPa/mm.

Since DL foundation is a single parameter model, varying the coefficient of subgrade reaction, k, allowed the matching of the interior deflections only. The interior slab stresses, as well as edge and corner slab deflections and stresses, are higher for the DL foundation than for the ES foundation.

Varying the parameters of the TP model permitted matching of both maximum plate stresses and deflections from the interior loading, but failed to match edge and corner loading responses. On the other hand, the set of parameters of the KV model, which provided the match of the maximum plate interior stress and deflection and the maximum edge deflection to the corresponding responses for the ES model, also provided a good match for all other maximum responses. This supports the recommendations of Kerr (12) and Kefali (13) that the KV model can be substituted for the ES model without changes in slab responses. Consequently, the KV model can be used in place of the ZSS model, which provides intermediate subgrade modeling between DL and ES models. Because the KV model is more computationally efficient than ES and ZSS models, we can conclude that this model will be widely used in the near future.

TOTSKY MODEL

ILSL2 represents a significant improvement over other finite element programs for the analysis of curling of two unbonded pavement layers. Curling of unbonded PCC overlays or PCC slabs constructed on a stabilized base is a difficult phenomenon to analyze because curling can cause the upper pavement slab to lift off the underlying pavement or stabilized base. Many finite element programs for PCC pavements allow analysis of two-layered systems; however, in almost all cases, this is accomplished by converting the two-layer system to a structurally equivalent single-layer system. This conversion is feasible only if the two layers are assumed either fully bonded or fully unbonded and they assume the same deflection profile. Because the two pavement layers are not actually modeled as two separate layers, most finite element programs, including ILLI-SLAB, cannot analyze the independent actions of the two layers.

Until recently, the separation between the slab and the base could only be modeled using 3-D finite element programs. ILSL2 incorporates a new approach to analyzing the layer separation problem (14, 15), developed by a Russian researcher Totsky (16). This approach models the multi-layered pavement system resting on subgrade as a series of springs and plates. The plate elements model the bending, whereas the springs accommodate the direct compression occurring in such a system.

Implementation of Totsky's approach in a finite element code required the introduction of a special 8-noded (24-degree-of-freedom) element. The first four nodes are placed at the neutral axis of the upper plate, while the other four nodes are placed at the neutral axis of the lower plate. In view of the potential for separation during temperature analysis, it is assumed for simplicity that the stiffness matrices for both the spring interlayer and subgrade are dependent only on nodal displacements, and not on nodal rotations. The stiffness of the interlayer spring may be specified by the user, or the program will calculate it from the plate parameters using recommendations developed by Khazanovich and Ioannides (14).

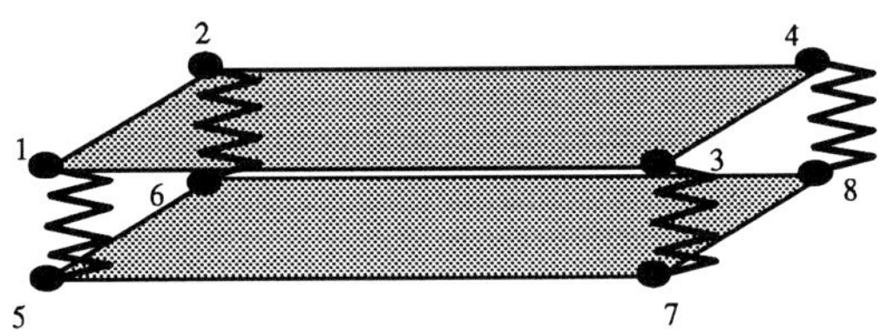

Figure 6. 8-nodded finite element for Totsky Model.

The curling problem in the Totsky model is solved iteratively. The analysis begins with all of the interface springs in compression (compression due to the self-weight of the slab), then the pavement layers are allowed to curl. If any of the springs are in tension at the end of the first iteration, those springs are removed and the system reanalyzed. The iteration continues until an equilibrium condition has been reached. The interface springs that have been removed during the solution process represent the layer separation. The use of the Totsky model in ILSL2 allows very accurate modeling of the curling problem.

Field Verification of ILSL2

In a recent study conducted by ERES Consultants, Inc. for the Colorado Department of Transportation, the structural response of jointed concrete pavement slabs was evaluated using data obtained from instrumented slabs (17). The instrumented 295-mm thick PCC slabs were a part of newly constructed jointed plain concrete overlay on an existing 80-mm thick stiff asphalt concrete (AC) pavement on I-70, near the Kansas-Colorado border. The instrumentation consisted of dial gauges for measuring curling deflections at the slab corner and longitudinal edge and surface-mounted strain gauges for measuring load strains at the longitudinal edge at midslab. Temperature profile through the pavement slab was also measured on 30-minute intervals during field testing. A thorough analysis was performed on the collected data to determine whether the measured structural responses of PCC pavements are consistent with the analytically obtained values.

Several anomalies were observed in the comparison of the measured pavement responses with the responses calculated using ILLI-SLAB with either bonded or unbonded interface conditions (no separation) between the PCC slab and the base. The measured deflections were about 40 percent higher than the calculated values, but the measured strains were only about 50 percent of the calculated values. If a 180-mm AC base is assumed fully bonded to the PCC surface, the calculated strain values are in line with the measured values; however, in that case, the calculated deflections would be lower, which would lead to an even greater discrepancy between the calculated and measured deflections.

The only way to explain the seemingly conflicting measured responses appears to be that under certain loading conditions, the PCC slab can separate from the stabilized base but under other loading conditions, the pavement system may behave as if the base is bonded to the slab. The high curling deflections measured at the slab corners and at the longitudinal edge are possible only if the slab is free to lift off the base and if the slab is initially curled up (i.e., the slab is curled up at zero temperature gradient due to differential shrinkage, moisture gradient, or built-in curling). Therefore, the ability of ILSL2 to analyze this type of problems is very important.

Application of ILSL2 for Analysis of Mismatch Joints in Unbonded PCC Overlays

Transverse joints in unbonded concrete overlays are often deliberately mismatched with those in the underlying pavement to improve faulting performance. A minimum offset distance of 0.9 m between the joints in the overlay and the underlying joints or cracks is usually recommended (18). By placing the joint in the overlay before the joint in the underlying pavement, a "sleeper slab" effect is provided that is expected promote good load transfer across the joints. No documented data or analytical examinations are available to support this recommendation.

A factorial of ILSL2 runs were performed for the above JPCP to determine the effect of joint mismatching on pavement responses caused by the combined effect of corner loading and nighttime temperature gradient (19). The overlay thickness was varied from 150 to 350 mm, the underlying slab thickness was varied from 200 to 300 mm, and the temperature gradient was equal to -0.437 or -0.656 °C/cm. The resulting maximum bending stresses at the transverse edge in the overlay are given in table 2. It can be observed that joint mismatching significantly reduced bending stresses in the overlay in a few cases. In the majority of the cases considered, the resulting stresses are close for matched and mismatched joints. The maximum corner deflections in the pavement with mismatched joints are significantly lower than in the corresponding pavement with matched joints, as shown in table 2. Therefore, it appears that joint mismatching leads to better protection of the overlay and underlying pavement and subgrade from traffic and climatic loading by reducing corner deflections and some critical stresses.

Table 2. Maximum bending stresses and corner deflections in unbonded PCC overlays.

H_{OL}, mm	H_{SL}, mm	Temperature Gradient, °C/cm	Maximum Bending Stress, kPa		Maximum Deflection, mm	
			Matched Joints	Mismatched Joints	Matched Joints	Mismatched Join
150	200	-0.437	1867	1396	2.03	0.89
150	200	-0.656	2344	1958	1.93	0.86
150	250	-0.437	1656	1352	1.93	0.86
150	250	-0.656	2218	1952	1.85	0.84
150	300	-0.437	1550	1340	1.88	0.86
150	300	-0.656	2155	1977	1.83	0.84
250	200	-0.437	1551	1482	1.63	0.94
250	200	-0.656	1907	2031	1.55	1.02
250	250	-0.437	1659	1651	1.65	0.91
250	250	-0.656	2183	2321	1.57	0.97
250	300	-0.437	1807	1807	1.70	0.91
250	300	-0.656	2407	2542	1.63	0.94
350	200	-0.437	1048	1110	1.57	1.12
350	200	-0.656	1279	1362	1.57	1.22
350	250	-0.437	1254	1336	1.60	1.09
350	250	-0.656	1568	1566	1.57	1.14
350	300	-0.437	1456	1502	1.63	1.07
350	300	-0.656	1833	1896	1.55	1.09

Note. PCC overlay and underlying slab: Modulus of elasticity=28 GPa; coefficient of thermal expansion =3.3×10^{-6} 1/°C; Poisson's ratio=0. Subgrade k value = 27 KPa/mm. AC interlayer thickness = 50 mm; corresponding interlayer spring stiffness = 27 Mpa/mm.

EFFECT OF NONLINEAR TEMPERATURE GRADIENT

The nonlinearity in the temperature distribution through a concrete slab was theoretically predicted and was subsequently proven by experimental data presented by numerous investigators. Janssen (20) concluded from the results of field testing, laboratory testing, and computer modeling that significant drying in concrete pavement slabs usually occurs only at rather shallow depth. This causes shrinkage that is nonlinear throughout the slab thickness and affects concrete pavement in a manner similar to a nonlinear temperature distribution. Choubane and Tia (4) analyzed numerous field test data and concluded that a quadratic function can adequately describe in situ temperature distributions in many cases. They stated that replacing an actual distribution by a linear approximation leads to increased maximum stresses for daytime conditions and to reduced maximum stresses in the slab for nighttime conditions.

Korenev and Chernigovskaya (21) proposed splitting any arbitrary temperature distribution throughout the slab thickness into three components: one causing only slab expansion or contraction, another causing curling (i.e., slab bending), and a third tending to cause distortion of the cross-section, thereby giving rise to self-equilibrating stresses resisting this distortion. This four-step approach was implemented into ILSL2 for analysis of two-layered slabs:

1. Split the nonlinear temperature distribution into its three components (the part that cause constant throughout the slab thickness strain, the part that causes linear throughout the slab thickness strain, and the part that causes nonlinear strain part).
2. Using available finite element formulations, determine the deflection profile and bending stress distribution due to the applied traffic loads and the linear strain-causing component of the original nonlinear temperature distribution.
3. Using a closed-form analytical solution, evaluate the normal components of the self-equilibrating thermal stress distribution due to the nonlinear strain-causing component of the temperature distribution.
4. Superimpose the bending stresses from Step 2 and the self-equilibrating thermal stresses from Step 3 to obtain the resultant stress distribution.

A detailed description of the each step and examples of application of this procedure can be found elsewhere (15).

ANALYSIS OF PARTIAL-DEPTH CRACKS

The ability to analyze partial-depth cracks is important for analysis of both new and overlaid pavements. ILSL2 allows for the analysis this problem (22). The partial-depth crack is treated like a conventional aggregate interlocked joint with infinite shear load transfer, but with the ability to change the moment load transfer at the crack. Figure 7 shows how the moment load transfer at a crack changes with crack depth to thickness ratio. This relationship was obtained using an available fracture mechanics solution (23).

A distinguishing feature of the partial-depth crack is the presence of infinite stresses near the crack tip induced by virtually any applied load. To characterize the singularity in the stress field at the crack tip, ILSL2 calculates the stress intensity factor and J-integral.

Consider an unrestrained highway slab with dual wheel loading (22). The pavement is 3.7 m by 4.6 m with a k-value of 27.1 MPa, concrete elastic modulus of 27,560 MPa, and a Poisson's ratio of 0.15. The loaded area of a single tire is 413 cm^2 with a tire pressure of 620 kPa. Figure 8 shows that the crack depth where the maximum stress intensity factor occurs is dependent on the slab thickness. The 203-mm slab had the highest stress intensity factor at a crack depth of 0.75*h. To propagate an existing crack, assuming linear elastic fracture mechanics, the applied K_I must be greater than or equal to the critical stress intensity factor (K_{IC}) of the concrete. Mindess and Young (24) quotes typical ranges for concrete K_{IC} as 440 to 1429 $kPa/m^{1/2}$. For this range of K_{IC}, it can be concluded that crack propagation may occur in all the slabs if LEFM assumptions are used. Crack propagation will most likely occur in the 203-mm slab especially for crack depths between 0.5*h and 0.90*h. Once the crack depth approaches the thickness of the slab (0.95*h), the mode of failure becomes shear, not bending. This is demonstrated in the stress intensity curves for the bending mode, which decrease after reaching a maximum value and then go to zero for crack depths approaching the slab thickness.

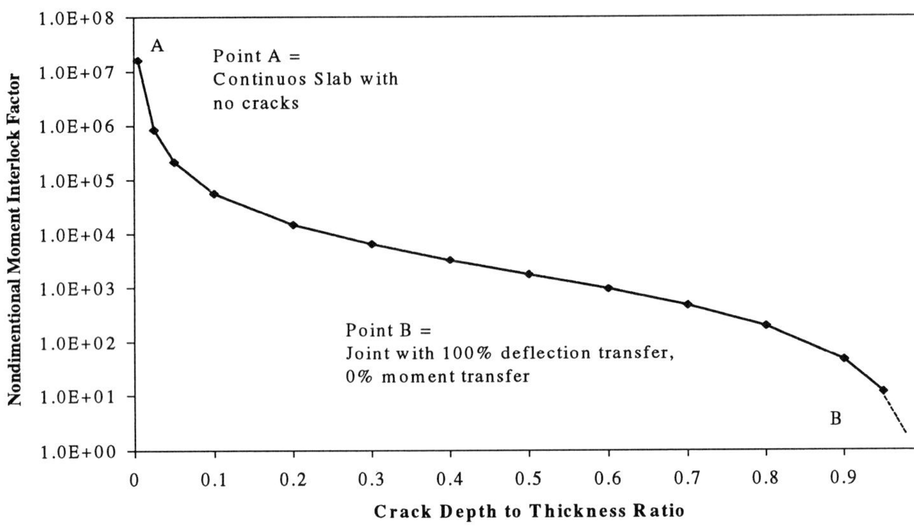

Figure 7. Moment interlock factor versus crack depth to thickness ratio.

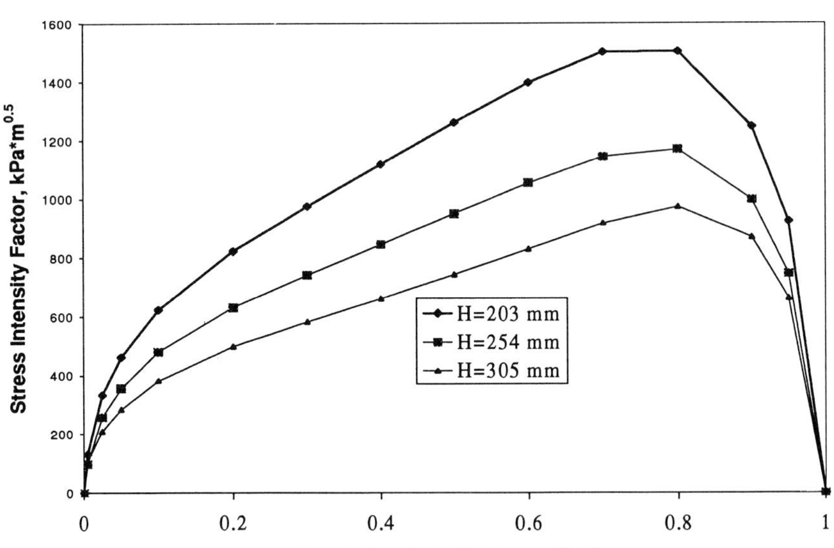

Figure 8. Effect of slab thickness and crack depth on stress intensity factor.

CONCLUSIONS

This paper presented a finite element code, ILSL2, that represents an extension of the ILLI-SLAB program. Results from numerous ILSL2 runs demonstrated the potentials of the new version. Due to its ability to account for different subgrade models, separation between constructed layer, nonlinear temperature gradient throughout the pavement depth, and effect of partial-depth cracks, ILSL2 is a very powerful analytical tool and attractive alternative to the general purpose 3D finite element packages.

Nevertheless, ILSL2 has several limitations. Being an extension of an old-generation engineering program, ILLI-SLAB, ILSL2 is cumbersome to use; however, more seriously, there are problems associated with the core of the ILLI-SLAB code itself which were inherented by ILSL2. Throughout its development history, the task of maintaining a clean and efficient code was not a high priority for ILLI-SLAB. Consequently, the current code contains numerous inefficiencies and redundancies.

At the present time, ERES Consultants, Inc., in cooperation with Michigan and Minnesota Departments of Transportation, Michigan Technical University, University of Michigan, Michigan State University, and University of Minnesota, is working on a new finite element code. The new program will retain all positive features of ILSL2 but will be more computational efficient and provide a user-friendly interface. The maximum allowable number of nodes will be significantly increased, and computation time will be decreased. The increased analysis capacity is a very important improvement to enable more accurate analysis of many common problems, especially when significant slab curling is involved. The interface will be a Microsoft Windows application that provides several on-screen forms that users fill out to enter the analysis input. Various options will be provided for defining the finite element mesh. Graphical displays will be used to enable users to verify the input data and to view analysis results. The graphical display will also be used to show contour plots of stress, strain, or deflection. The new program is expected to be released in March 1999.

REFERENCES

1. Tabatabie, A.M., and E.J. Barenberg. "Structural Analysis of Concrete Pavement Systems." ASCE, *Transportation Engineering Journal.* Vol. 106, No. 5, 1980.
2. Chou, Y.T. *Structural Analysis Computer Programs for Rigid Multicomponent Pavement Structures with Discontinuities- WESLIQID and WESLAYER.* Technical Report GL-81-6, U.S. Army Engineer Waterways Experiment Station, May, 1981.
3. Tayabji, S.D., and B.E. Colley. "Improved Pavement Joints." *TRR 930*, 1983.
4. Choubane, B., and M. Tia. "Nonlinear Temperature Gradient Effect on Maximum Warping Stresses in Rigid Pavements." *Transportation Research Record No. 1370*, 1992.
5. Huang, Y.H. *Pavement Analysis and Design.* Prentice Hall, 1993.
6. Kok, A.W.M. "A PC Program for the Analysis of Rectangular Pavements Structures." *Proceedings,* Second International Workshop on the Design and

Rehabilitation of Concrete Pavements, Sigüenza, Spain, 1990
7. Khazanovich, L., and A.M. Ioannides. "Glasnost and Concrete Pavement Engineering: Insights Concerning Slab-on-grade from the Former Soviet Union." *Proceedings*, Fifth International Conference on Concrete Pavement Design and Rehabilitation 1993.
8. Pronk, A.C. "The Pasternak Foundation: An Attractive Alternative for the Winkler Foundation." *Proceedings,* Fifth International Conference on Concrete Pavement Design and Rehabilitation,1993.
9. Ioannides, A.M.et al. "Finite Element Analysis of Slabs-on-Grade Using a Variety of Support Models." *Proceedings,* Third International conference on Concrete Pavement Design and Rehabilitation, Purdue University, 1985.
10. Khazanovich, L., and A.M. Ioannides. "Finite Element Analysis of Slabs-on-Grade Using Improved Subgrade Soil Models." *Proceedings*, ASCE Specialty Conference 'Airport Pavement Innovations—Theory to Practice,'1993.
11. Vlasov, V.Z., and N.N. Leont'ev. "Beams, Plates and Shells on Elastic Foundations." NASA-NSF, NASA TT F-357, TT 65-50135, Israel Program for Scientific Translations, 1960 (translation date: 1966).
12. Kerr, A.D. "Elastic and Viscoelastic Foundation Models." *Journal of Applied Mechanics, Transactions*, ASME, Vol. 31, No. 3, pp. 491-498, 1964.
13. Kneifati, M.C. "Analysis of Plates on a Kerr Foundation Model." *Journal of Engineering Mechanics, ASCE*, Vol. 111, No. 11, 1985.
14. Khazanovich L. and A.M. Ioannides. "Structural Analysis of Unbonded Concrete Overlays under Wheel and Environmental Loading."*Transportation Research Record No.* 1449, 1994.
15. Khazanovich, L. *Structural Analysis of Multi-Layered Concrete Pavement Systems.* Ph.D. Thesis, University of Illinois, Urbana, IL, 1994.
16. Totsky, O.N. *Behavior of Multi-Layered Plates on Winkler Foundation.* Stroitel'naya Mekhanika I Raschet Sooruzhenii, No. 6, Moscow, pp.54-58, 1981 (in Russian).
17. Yu, H. et al. *Field and Analytical Evaluation of the Effects of Tied PCC Shoulders and Widened Slabs on Performance of JPCP*. Final Report. Denver, CO: Colorado Department of Transportation, 1995.
18. American Concrete Pavement Association, "Guidelines for Unbonded Concrete Overlays," *Technical Bulletin TB-005.0 D* (1990).
19. ERES Consultants, Inc. *Evaluation of Unbonded Portland Cement Concrete Overlays*. Draft Final Report. Transportation Research Board, 1997.
20. Janssen, D.J. "Moisture in Portland Cement Concrete."*Transportation Research Record No 1121*, 1987.
21. Korenev, B.G., and E.I. Chernigovskaya. *Analysis of Plates on Elastic Foundation.* Gosstroiizdat, Moscow, 1962 (in Russian).
22. Roesler J., and L. Khazanovich. "Finite Element Analysis of PCC Pavements with Cracks."*Transportation Research Record No.* 1568, 1997.
23. Rice, J.R. "The Line-Spring Model for Surface Flaws." *The Surface Crack: Physical Problems and Computational Solutions*, ed. J. L. Swedlow, ASME 1972.
24. Mindess, S. and Young, J. F., *Concrete*, Prentice-Hall, Inc., 1981.

Fifth International Conference on the
BEARING CAPASITY OF
ROADS AND AIRFIELDS

FLEXIPAVE: A FINITE ELEMENT BASED PROGRAM FOR THE ANALYSIS OF FLEXIBLE PAVEMENTS

MUHAMMAD N.S. HADI[1] AND AHMED A. MALIK[2]

[1]Lecturer, Department of Civil, Mining & Environmental Engineering,
University of Wollongong, Wollongong 2522, Australia,
Fax: +61 2 42 21 3238, email: m.hadi@uow.edu.au
[2]PhD Candidate, Department of Civil, Mining & Environmental Engineering,
University of Wollongong, Wollongong 2522, Australia

ABSTRACT

The aim of this paper is to describe FlexiPave, a program that has been developed for the structural analysis of flexible road pavement structures by the Finite Element method. The pre-processor of FlexiPave generates an appropriate Finite Element model while reducing the user inputs to only the number of layers along with their respective thickness and material properties. The frontal method of finite element matrices is used for the analysis. The post-processor of FlexiPave shows the deformed shape of the pavement cross-section with marked critical locations along with responses at these points. An additional feature of FlexiPave is its ability to act as a pre- and post-processor for the general-purpose finite element programs, STRAND6 and MSC/NASTRAN. The main reason for this is to allow users to take advantage of these well-known and tested programs, which are widely used. FlexiPave was developed using Visual Basic 5.0 ActiveX technology. Two ActiveX controls were used, one for the finite-element analysis and the other for communicating with STRAND6 and MSC/NASTRAN. A number of examples were run to evaluate the developed system and they proved that FlexiPave is a versatile program that allows the engineering user to analyse flexible pavements easily without the need to refer to user manuals. FlexiPave provides pavement engineers the opportunity to explore several pavement designs and to adopt the most economical one.

KEYWORDS: FlexiPave, Flexible Pavements, Finite Elements, Pavement Analysis, ActiveX.

THE DESIGN OF FLEXBILE PAVEMENTS

The design of flexible pavements was traditionally based on using procedures based on empirical equations and graphs that were developed over the years reflecting the experience of road engineers. There are several of these procedures and it looks like that different organisations like AASHTO, US Corps of Engineers and others, tend to adopt their own method of design. This mainly leads to confusion among pavement engineers especially when different methods lead to different designs.

In the last ten years or so, pavement engineers decided to adopt the "Mechanistic Design Method". In this method strains or stresses at critical sections are to be calculated due to vehicular loading. These responses (strains or stresses) are used in fatigue criteria suggested by the design standards, to estimate the allowable number of vehicular load repetitions. In turn this will be compared with the expected number of repetitions and eventually a decision is made on whether the design is accepted or a redesign is necessary.

CALCULATING RESPONSES OF PAVEMENTS

The steps involved in calculating the responses of flexible pavements to the vehicular loading is rather complex.

Boussinesq (1) developed equations, based on the theory of elasticity, for calculating both the vertical and horizontal stresses due to a point load on the surface of a semi-infinite, homogeneous, isotropic, elastic half-space. Zienkiewicz (2) used the finite element method to determine induced stresses by applying it to the Boussinesq problem and using tetrahedral elements. Desai and Christian (3) also applied the method to solve this problem, as did Ioannides and Donnelly (4) who have also used three-dimensional elements.

Bowles (5) produced computer programs, based on Boussinesq equations, which subdivided the load into a number of point loads and then sums the induced stresses at selected points due to each of the point loads. Booker and Small (6) presented a finite layer method for the analysis of pavements subjected to horizontal loading. AUSTROADS (7) recommends using the finite difference based program CIRCLY which is developed by Wardle (8). Hadi and Symons (9) reported that the stresses and strains induced in a pavement due to a standard vehicular load calculated using CIRCLY, STRAND6 and MSC/NASTRAN are of comparable magnitudes.

However, in spite of the availability of different general purpose finite element programs along with their user friendly pre-processors the user still has to go through the difficult task of creating the finite element mesh of the pavement structure. Moreover, the user needs to reconstruct the finite element mesh each time layer thickness or wheel dimensions are changed. This fact limits the use of finite element programs for pavement structures. FlexiPave is a user-friendly Windows based program developed in Visual Basic 5.0, which generates and analyses a three dimensional finite element mesh representing a pavement structure. The user needs only to input the number of layers together with their material properties and FlexiPave will generate the finite element mesh. Wheel load and dimensions are set at default in line with the recommendations of AUSTROADS. The user has the option of changing these default values to suit the pavement to be designed.

In addition to its own solver, FlexiPave acts as a pre- and post-processor for STRAND6 and MSC/NASTRAN.

STRAND6 is a general-purpose finite element system developed by G+D Computing (10) in Australia. It includes interactive graphical modelling and post-processing capabilities, and a large library of elements. These can be applied to a wide range of problems including static and dynamic, linear and non-linear structural analysis and potential flow/field problems such as heat transfer. Although STRAND6 has its own pre- and post-processors but still the need of FlexiPave was felt mainly due to the complexity involved in defining the mesh compared to the relatively simple geometry of the model.

Figure 1: Typical Finite Element Model Viewed in FlexiPave

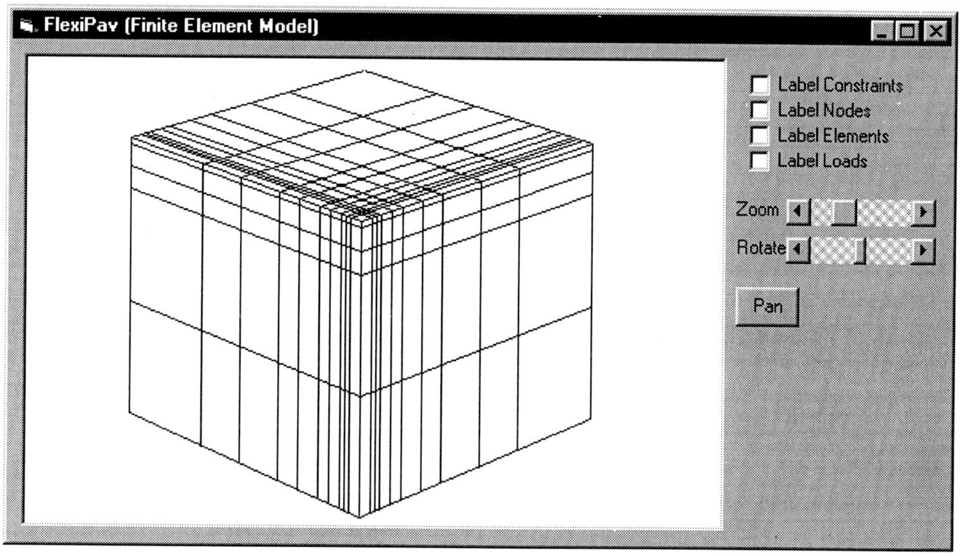

MSC/NASTRAN FlexiPave can also generate input data files to be used in MSC/NASTRAN. This program was chosen since it is well known all over the world and has been tested and used extensively. The generated data files are standard MSC/NASTRAN files that can be processed by MSC/NASTRAN directly or be read into the pre-processor for MSC/NASTRAN for Windows, FEMAP. In the former case, the user needs to issue the command for running MSC/NASTRAN, for example "*nastran file_name*", and MSC/NASTRAN in turn will analyse the model and produce the results files, ie. files with the extension *.f04* and *.f06*. In the latter case, Windows case, the user needs to run FEMAP, then to read the data file by issuing the "translate" command, then the "analysis" command needs to be issued to run the analysis. After completing the analysis, the user needs to issue the command "translate" again to generate the results files, ie. *.f04* and *.f06* files. FlexiPave can read the results file with the extension *.f06* which contains the displacements, strains and stresses at nodes and elements.

FINITE ELEMENT MODEL

After considering the various possibilities regarding type of mesh and type of elements, it was concluded that a simple cube shaped model, as shown in Figure 1, with Hexahedral (brick) finite elements proves to be the best choice for flexible road pavements. Due to the symmetry, only a quarter of the loaded portion of the pavement structure was considered along with the necessary adjustments to the translational constraints of all the nodes lying in the planes of symmetry.

The size of the brick elements is gradually increased in such a manner that smaller elements are used in regions of stress concentration i.e., under and near the load. This ensured continuity in geometry and more accuracy in the results obtained in the region of interest. However an option is also provided for the user to change the default sizes of the elements.

The load used in the finite element model was set as a default pressure with an intensity of 640 kPa applied on the surface of two brick elements having an area of 62.5 x 125 mm^2 each. However the user can change these parameters by providing different wheel specifications. Each element possesses the parent property of the layer of material or the sub layer of a material in which it is present. The property of any layer of material can be set as either isotropic or orthotropic along with their respective required parameters.

FLEXIPAVE

Following is a brief description of the various features of FlexiPave:

Menu Bar

FlexiPave menu bar comprises the ***File***, ***Export***, ***Output***, ***Options*** and ***Help*** menus (Figure 3). The ***File*** menu contains the typical, *New, Open, Save, Save As* and *Analyse* sub menus. The *Analyse* sub-menu, after checking any missing inputs or errors, performs the actual Finite Element Analysis.

The ***Export*** menu contains *Save As STRAND, Save As MSC/NASTRAN, Save As AutoCAD, Save Model Picture and Save Pavement Picture* sub-menus. The *Save As STRAND* sub-menu allows to export the generated FE model in STRAND6 as an ASCII file. The model can then be loaded in STRAND6 by using the ASCII IN option in STRAND6. The *Save As MSC/NASTRN* has similar action but for MSC/NASTRN. The *Save AS AutoCAD* sub-menu allows to save the three dimensional model as an AutoCAD script file, which can then be loaded in AutoCAD using the SCRIPT command.

The ***Output*** menu contains the *Read STRAND Output, Read MSC/NASTRN Output, Read FlexiPave Output, Save Output, View Output, Compare Outputs* and *Report* sub-menus. These allow to save and read different outputs generated after performing different analyses on a pavement by varying certain parameters. The *View Output* sub-menu allows to view the numerical values of responses at any desired locations. These results can then be graphically compared using the *Compare Outputs* sub-menu. The *Report* sub-menu generates a text report summarising the loading characteristics, pavement material properties and layer dimensions along with stresses, strains and displacements at user selected critical locations. Information regarding the finite element model can also be included in the report. The image of deformed pavement cross-section, the three-dimensional Finite Element model, and the graphical comparisons of results can be saved as BITMAP picture or captured for pasting in documents.

Figure 2: FlexiPave Preferences Screen View

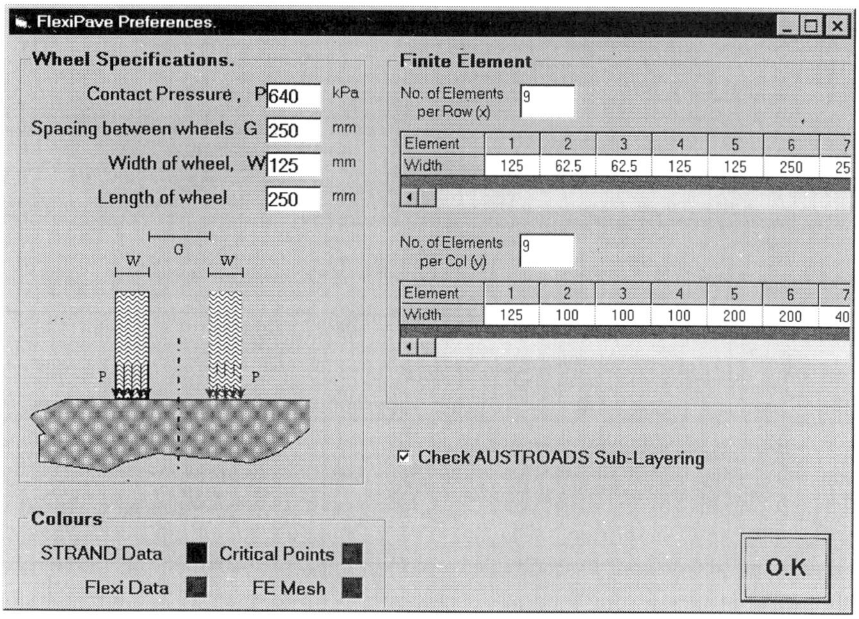

The ***Options*** menu contains the *Loading, FE Model* and *Preferences* sub-menus. The *Loading* sub-menu allows changing the default loading characteristics, which includes the load magnitude, rectangular contact area and the wheel spacing. The *FE Model* sub-menu allows changing the default element sizing in the Finite Element mesh depending upon the desired accuracy (Figure 2). The *Preferences* sub-menu can be used to change different environment settings of the program itself.

The ***Help*** menu contains the *Contents* and *About* sub-menus. The *Contents* sub-menu displays a detailed description of all the menus and features of the program. A Windows 95 style Tool-Tip help system is also provided, which is activated whenever the mouse pointer pauses over a control. The *About* sub-menu simply tells about the people behind the software and its version number.

User Input

If all the default settings are acceptable then the user is left to supply only the number of material layers and their respective values of modulus of elasticity and Poisson's ratio as shown in figure 3. If a material layer is opted to have an orthotropic behaviour then values of modulus of elasticity and Poisson's ratio, both in horizontal and vertical directions are required. In case the default settings are not acceptable then the user has the option to change the wheel specifications, sizing of the finite brick elements and the number of sub-layers of each material along with the respective properties of each sub-layer.

Figure 3: Typical Input parameters for a Flexible/Recycled Pavement

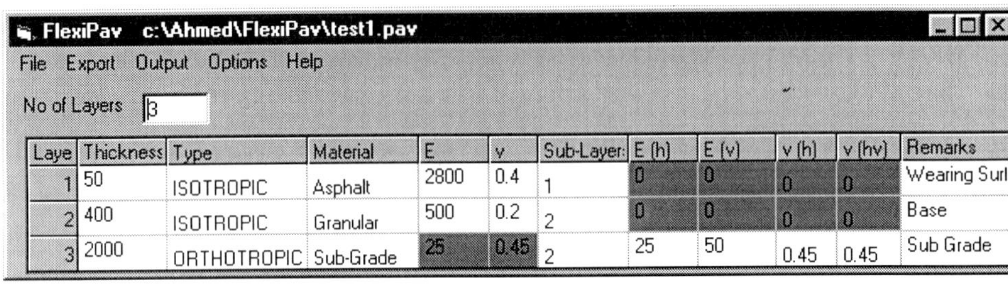

Laye	Thickness	Type	Material	E	v	Sub-Layer	E (h)	E (v)	v (h)	v (hv)	Remarks
1	50	ISOTROPIC	Asphalt	2800	0.4	1	0	0	0	0	Wearing Surl
2	400	ISOTROPIC	Granular	500	0.2	2	0	0	0	0	Base
3	2000	ORTHOTROPIC	Sub-Grade	25	0.45	2	25	50	0.45	0.45	Sub Grade

Pre-Processing

After checking possible errors in the small user supplied information, sub-layering is done for granular material placed directly over the subgrade with each sub-layer having varying values of modulus of elasticity according to the specifications of AUSTROADS (1992). After this, depending on the desired accuracy of the results, each layer or sub-layer is further divided into layers having the same properties. Every layer is discretised into small prismatic elements having the same horizontal layout. The information regarding every brick element comprises the tag numbers of its nodes and a property number to which the element belongs. For orthotropic property, the 9 terms of the stress-strain matrix [C] are automatically calculated from the supplied values of E (Modulus of Elasticity) and ν (Poisson's ratio) in the horizontal and vertical directions. More details about this can be found in the user manual of STRAND6 (10). The Gauss Elimination method is used for calculating the inverse of the stress-strain matrix. The generated finite element model can also be saved as an ASCII file whose format is recognised by STRAND6 or MSC/NASTRAN.

Processing

The linear finite element analysis is carried out using the program for the 3-D Hexahedral finite element analysis, originally written by Tirupathi and Ashok (11). In order to reduce the analysis time and storage space, the Frontal method is used to reduce the size of the matrices involved. In this method, the order of element numbering plays a more important role than the order of node numbering. The Frontal method relies on the fact that a degree of freedom (dof) can be eliminated as soon as all stiffness values in rows and columns for that dof are complete. Irons (12) observed that all of the degrees of freedom for a node can be eliminated when it appears for the last time as elements are assembled in an ascending order. The solver of FlexiPave was developed using ActiveX technology.

Figure 4: Flexible Pavement deformed cross-section

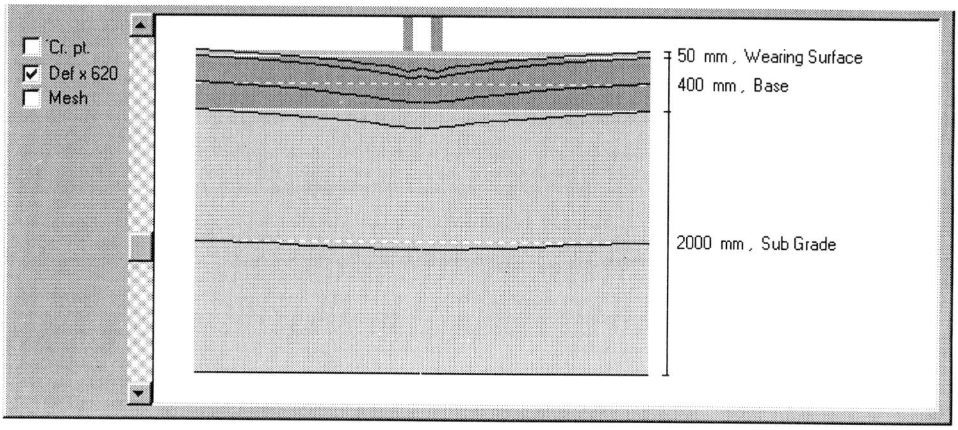

Post-Processing

The post-processor of FlexiPave presents the output generated after the finite element analysis, in the form of a deformed pavement cross-section along with responses at specified critical locations only. The default critical points are considered under and between the wheels lying on the interface of the different material layers as specified in AUSTROADS. This helps in focussing the attention to what is important from the design point of view. In addition to critical locations, numerical values of stresses, strains, and deflections at any other location, and stress or strain contours can also be viewed.

FlexiPave also has the ability to post-process the ASCII output file created by STRAND6 and MSC/NASTRAN. Moreover graphical comparisons between outputs from different models or between different analyses of the same model can also be made.

Figure 5: Profile of Normal Stress in Horizontal-axis between the Wheels

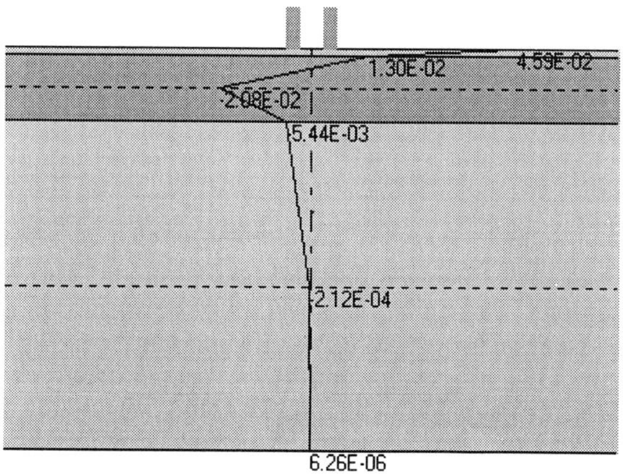

Graphic Preview

The graphic preview screen as shown in Figure 4 shows the pavement cross-section along with properties regarding every layer while differentiating them with different colours. The doted lines indicate the sub-layering of a material layer. The full 3D finite element model can also be viewed from within FlexiPave by selecting the "FE Model" option. After performing the analysis or reading the ASCII output file generated by STRAND6 or MSC/NASTRAN, FlexiPave can also show the deformed shape of the cross-section with marked critical locations along with the responses at these points (Figure 5). The user can look at the different stress, strain and displacement profiles or contours by choosing the options menu (Figure 6).

Figure 6: Screen View of Options for Profile and Contour Selection

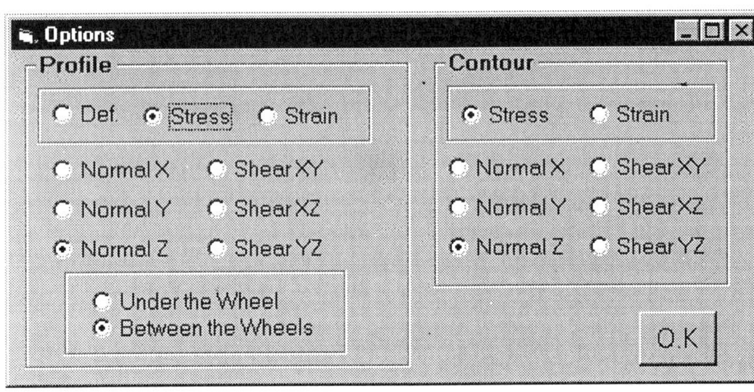

EVALUATION

A number of examples were run to test the integrity of the system. Here only one example is shown, in which, a three layered pavement structure, as shown in figure 7, was analysed using FlexiPave, STRAND6 and MSC/NASTRAN. The first layer had a depth of 50 mm representing a bituminous wearing surface with isotropic properties. The second layer had a depth of 400 mm representing a granular base material, with isotropic properties. The third layer had a depth of 2000 mm representing a sub-grade with orthotropic properties. The sub-grade has been assumed to have a finite depth to limit the modelling requirements. Input to FlexiPave is shown in Figure 3.

The loading was kept as the default of 640 kPa contact pressure (20 kN force by each tyre). The finite element model of the pavement in figure 7 was also exported to STRAND6 and MSC/NASTRAN and thus analysed. The results obtained from both STRAND6 and MSC/NASTRAN showed agreement with the results of FlexiPave.

The pavement structure was also analysed using CIRCLY. The choice of this program was made due to its wide acceptance amongst pavement engineers and being the recommended program by AUSTROADS for analysing pavement structures. Results showed a wide agreement. Specifically values of horizontal strain at the bottom of the asphalt layer and the vertical compressive strain at top of the subgrade were compared and those in turn revealed comparable magnitudes of strains.

Figure 7: An Example of a 3 Layered Pavement Structure

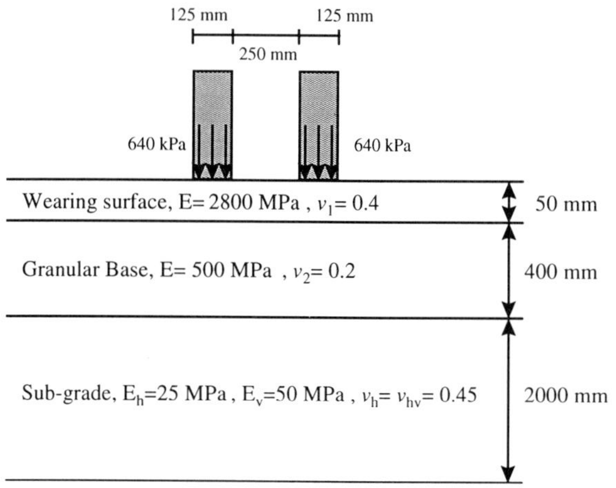

CONCLUSION

The design of flexible and recycled pavements according to AUSTROADS (7) is a lengthy process that requires a lot of calculations and referral to the different tables and figures included in AUSTROADS. In addition, the design process is an iterative process, where the pavement engineer should assume a trial thickness and then by using the recommendations will decide whether the design is adequate or not. Due to the complexity of the design process and the cost involved in conducting a single iteration of the design, most design engineers will stop after two or three iterations.

FlexiPave allows pavement engineers to investigate several designs quickly thus giving them the opportunity of achieving a cost effective design quickly. In addition, achieving an optimum design will cut the cost of building road pavements as well as reducing the amount of materials required in building them.

A number of examples were run to evaluate the developed system and it can be concluded that FlexiPave is a versatile program that allows the user to efficiently analyse flexible pavement structures using the Finite Element method, without the need to refer to user manuals of general purpose Finite Element programs and to know about the concepts of Finite Element method.

FlexiPave reduces the time and effort required in creating or changing a finite element model considerably, thus providing an efficient mean of design. Apart from keeping the Finite Element analysis in background, FlexiPave filters the huge generated output to only what is necessary form a design point of view. Moreover, the option to use the external finite element solvers of STRAND6 and MSC/NASTRAN gives an additional advantage to verify and compare the results more easily by using these well-known and tested programs.

The results obtained from STRAND6 and MSC/NASTRAN were compared with the results of FlexiPave analysis and they showed almost no difference in displacements and stress/strain distribution pattern in the structure. However, slight differences in the magnitudes of stresses and strains were observed. These slight differences were mainly due to truncation errors. Moreover, these results were compared with those obtained from CIRCLY which proved to be in agreement.

The ease in changing different parameters of a pavement and the built-in ability to graphically comparing different output sets makes FlexiPave an ideal tool for the pavement engineer.

Finally, FlexiPave bridges the gap by providing Finite Element software for the analysis of flexible pavement structures according to the current Australian code of practice.

REFERENCES

1. BOUSSINESQ, M.J. Application Des Potentiels, 'a l'Etude de l'Equilibre et du Mouvement Des Solides Elastiques. Gauthier-Villars, Paris, (1885).
2. ZIENKIEWICZ, O.C. The Finite Element Method, Third Edition, McGraw-Hill Book Company, UK, (1977).
3. DESAI, C.S. and CHRISTIAN, J.T. Numerical Methods in Geotechnical Engineering, McGraw-Hill Book Company, New York, USA, (1977).
4. IOANNIDES, A.M. and DONNELLY, J.P. Three-Dimensional Analysis of Slab on Stress-Dependent Foundation", Pavement Evaluation and Rehabilitation, Transportation Research Board, National Research Council, 1196, USA, 1988.
5. BOWLES, J.E. Foundation Analysis and Design. Fourth Edition. McGraw-Hill Book Company, USA, (1988).
6. BOOKER, J.R. and SMALL, J.C. Finite Layer Analysis of Layered Pavements Subjected to Horizontal Loading, Numerical Methods in Geomechanics. Innsbruck, pp. 2109-2113, 1988.
7. AUSTROADS. Pavement Design - A Guide to the Structural Design of Pavement, Sydney, (1992).
8. WARDLE, L.J. Program CIRCLY User's Manual, Geomechanics Computer Program - Number 2, Division of Applied Mechanics - CSIRO, Australia, 1977.
9. HADI, M.N.S and SYMONS, M.G. Computing Stresses in Road Pavements Using CIRCLY, MSC/NASTRAN and STRAND6, Australian Civil/Structural Engineering Transactions. Vol. CE38, No. 2,3 & 4. pp. 89-92, 1996.
10. G+D COMPUTING. STRAND6 Reference Manual and User Guide, Ultimo NSW, Australia, (1993).
11. TIRUPATHI, R.C. and ASHOK, D.B. Introduction to Finite Elements in Engineering, Prentice Hall, New Jersey, (1997).
12. IRONS, B.M. A frontal solution program for finite element analysis. Int. J. for Numerical Methods in Engineering, Vol. 2, 5-32, 1970.

Fifth International Conference on the
BEARING CAPASITY OF
ROADS AND AIRFIELDS

THE NOAH SOFTWARE

BERNARD ECKMANN
AB NYNAS Petroleum, Group Competence Center, S-14982 Nynäshamn, Sweden

ABSTRACT

The very purpose of the NOAH software is to enable the user to derive the maximum benefit from "multi-layer elastic stress/strain" computation programs in the frame of rational pavement design methods. This is achieved by following major features :

- All the various design parameters (structural, environmental, loading) can be treated as "*Design Variables*" which are assigned several successive values.
- The software includes a "*Formula Generator*" by which the user can enter, directly from the keyboard, all kinds of mathematical relationships which may be used at various stages :
  * to describe material properties (e.g. stiffness) as a function of environmental parameters (e.g. temperature).
  * to define "*Performance Indicators*" relating to calculated stresses and strains. Performance Indicators are in fact characteristics of the material and depend on the type of distress which is investigated (rutting, fatigue,...).
- The NOAH software also allows to address probabilistic pavement design via the so-called "*ROSENBLUETH approximation*". In this technique, the continuous statistical distributions of the input parameters are replaced by discrete equivalent "Point-estimate" distributions.

Beside these key features, the programmation under Windows® and many other facilities such as an extended Database function make NOAH a highly flexible and powerful tool.

Keywords. Pavement design, probabilistic design, ROSENBLUETH, computation software, NOAH.

INTRODUCTION

Although pavement design engineers are getting more and more familiar with the principles of multi-layer stress and strain calculations, they do rarely take advantage of their full potential. A key element of pavement design should be to evaluate the sensitivity of the tentative pavement structure to the variations of all the numerous parameters involved. Such an analysis does however request a large number of individual stress/strain calculations and additional data treatment. This explains why it is seldomly done ! Although certain softwares already permit to automatically perform such simulations, they are often tailormade to specific needs and thus restricted in use. This has urged NYNAS to develop a software tool of such a level of flexibility that it can easily be used or adapted by pavement engineers working according to different design methods.

CONCEPTS USED BY NOAH

The NOAH software (developed under WINDOWS) is based on automated computation loops which allow to investigate the effect of all relevant design parameters on the structural performance of a road. To this end it uses several key concepts which are briefly explained hereafter :

Structural Parameters. They describe the pavement as such and include :

- Number of layers.
- Layer thickness.
- Mechanical properties of the constitutive material (**E** : Stiffness, **V** : POISSON's ratio,...).
- Interface conditions between two layers (partial or full bond).

Environmental Parameters. Those are independent factors which may change the value of the structural parameters. A good example is given by the temperature which has obviously an impact on the stiffness of bituminous paving materials. But there may be other parameters than temperature, such as for instance load application time or frequency (bituminous materials), moisture (soils), age,....
A major asset of the NOAH software is the fact that the user may himself define and introduce the Environmental Parameters he believes to be relevant !

Loading Parameters. They describe the load(s) which are applied on top of the pavement structure. The NOAH software assumes circular imprints and only vertical forces (no tangential shear forces). But there are no restrictions as to the number of imprints !

Design Variables. Each of the parameters having an impact on the final performance (Structural, Environmental and Loading Parameters) can be treated as a *Variable*, successively taking several *States*. It is of course possible to simultaneously declare several variables within the same NOAH computation case. Each combination of the states of these variables will then lead to a separate single computation run (full matrix).

Distribution. This is a particular type of *Variable* which is to be used in the case of *Probabilistic Pavement Design*. The aim is then to evaluate the effect of the natural probabilistic distribution of a design parameter around its nominal value. *Distribution* variables are supposed to be bounded and defined by :

- the lower (B-) and upper (B+) boundaries of the distribution,
- the mean (X_0) and standard deviation (S) values.

Performance Indicator. This is a calculated magnitude which quantifies the response (hence the "Performance") of a given road structure to a single set of environmental and load data. The simplest *Performance Indicator* is, for instance, the tensile strain calculated at the bottom of a certain pavement layer. But *Performance Indicators* may be much more sophisticated. For instance, an experimentally established mathematical formula giving the number of cycles for failure for the material of this layer when repeatedly subjected to this tensile strain (Fatigue Law).

The ROSENBLUETH approximation. This is the particular technique which the NOAH software uses to determine, with a minimum number of computations, the statistical distribution of a *Performance Indicator* as a result of the statistical distribution of one or several *Design Parameters*.

In the following, we will try to illustrate how NOAH works on the basis of a practical computation case.

EXAMPLE COMPUTATION CASE

Input data. Figure 1 shows an example pavement structure of which the "Fatigue Life" is supposed to be solely dependent on the tensile strain ε generated at the bottom of the bound materials (layer 1).

Furthermore, the stiffness of these materials is dependent on temperature according to the relationship written in Figure 1. The tensile strain is thus not only a function of the applied loads (Q) but also a function of the layer temperature !

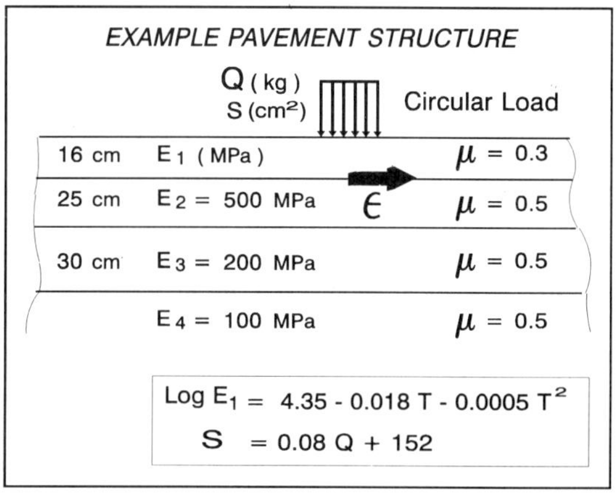

Figure 1

To calculate the total number of admissible traffic loads (N), one needs to take into account that the pavement is subjected to both a frequency histogram (spectrum) of axle loads and a frequency histogram of temperatures. For the sake of simplicity, we consider in this example that the traffic spectrum is identical at every considered temperature step.

The total number of admissible axle loads is usually calculated according to MINER's rule which supposes that the damages generated by different loadings are simply cumulative, irrespective of the order in time at which these damages occur. The procedure is outlined in Figure 2 which also indicates the Fatigue Law used in this example.

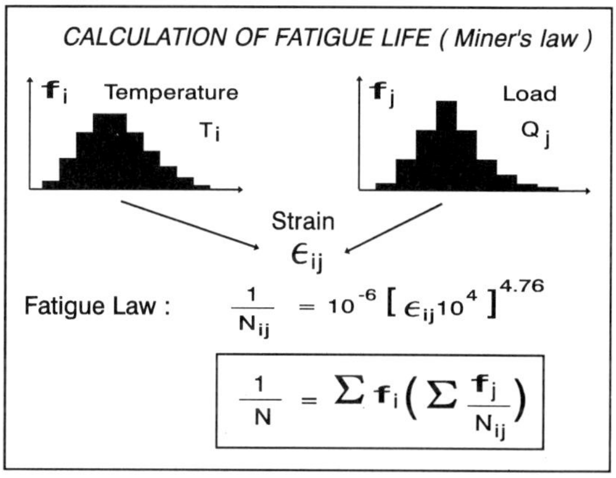

Figure 2

Design Problem. The example pavement being subjected to the above described traffic and temperature histograms, one wants to determine the **statistical distribution** of the total number of admissible traffic loads (**N**) as a function of the statistical distribution of the thickness of Layer 1 (asphalt layers). The nominal thickness of Layer 1 is equal to 16 cm. One assumes that the statistical distribution which represents the expected natural "scatter" below and above this mean value is a bounded distribution (so-called "Beta-Distribution") described by :

- A mean value of 16 cm and a standard-deviation of 0.8 cm
- A lower and an upper bound respectively equal to 14 cm and 17.5 cm

Before illustrating how such a problem would be handled within NOAH, we will however also use this example to briefly explain the theoretical background for the ROSENBLUETH approximation.

The creation of a Reference Set is greatly facilitated by the possibility to access the *NOAH Database* which, among others, includes following items :

Typicals. These are known pavement structures, environmental and loading conditions ("Catalogue" Structures, recorded climatic conditions, "standard" axle types and loads) which the user can enter himself so as to progressively set up a "Library of Typicals". In this particular example, a typical belgian structure has been retrieved from the Database since its characteristics are close to those of the desired reference structure. It is then enough for the user to adjust only those data which differ from those of the chosen Typical (e.g., in this case, to change 22 cm into 16 cm - see Figure 6).

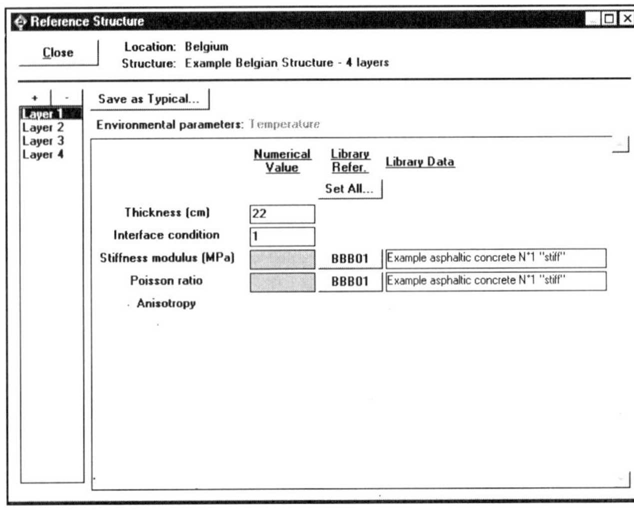

Figure 6

Materials. Structural parameters relative to mechanical properties (E, ν , ...) are defined by either fixed values or by mathematical formulas which relate these structural parameters to one or several environmental parameters. It is one of the strong assets of the NOAH software that such formulas can be freely defined by the user himself and saved in a Database. In this example, the "BBB01" reference refers to a mathematical formula expressing stiffness in relation to temperature (TE) (see Figure 7). By accessing the Database, the user may himself modify the mathematical formula via a built-in "Formula Generator" which provides all necessary operators and mathematical functions. As a matter of fact, new "materials" (i.e. new mathematical relationships) can as easily be created from "scratch" or via a "Copy-Paste-Modify" sequence.

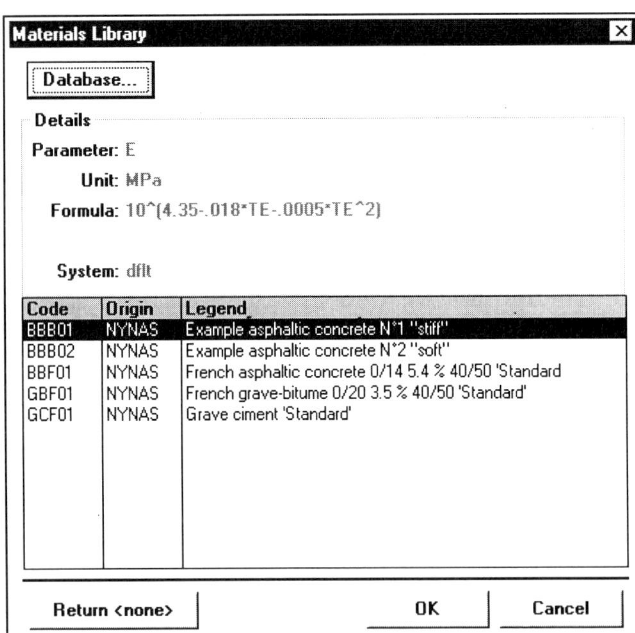

Figure 7

Load Configurations. To speed-up the definition of loading conditions, especially in the case of multiple imprints (e.g. airplane carriages !) loading geometries (coordinates of imprints) can be saveguarded as "*Load Configurations*".

Definition of Design Variables. Next, the user defines those design parameters (structural, environmental or loading factors) which he wants to treat as Variables. In the present case, three *Design Variables* are to be defined :

- Temperature of Layer 1 and Loading (they will allow to calculate the Fatigue Life according to the procedure given in Figure 2),
- the *Distribution Variable* describing the statistical distribution of the thickness of layer 1.

Figure 8 illustrates the case of the *Loading Variable.* As indicated on Figure 1, a particular relationship between the load and the tyre pressure has to be observed. This is possible by declaring a "*Macro-Variable* " with two "components". For each "*State*" of the "*Loading Macro-Variable*", two values (one for the total load and one for the pressure) have to be entered !

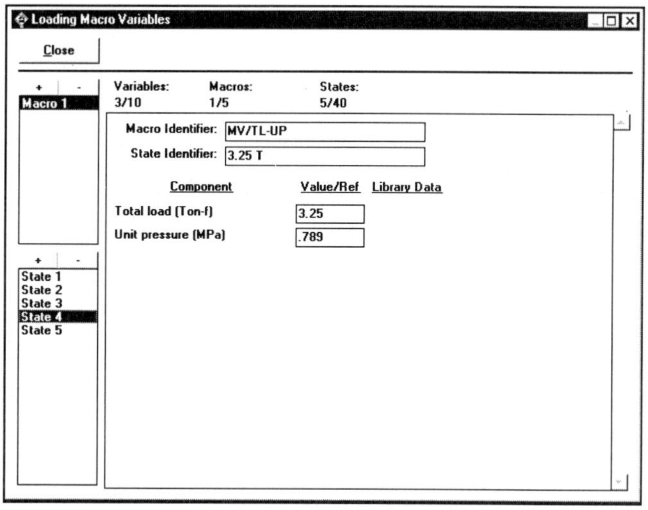

Figure 8

Figure 9 illustrates the case of the *Distribution Variable* representing the thickness of layer 1 (D/H1). The user is requested to enter the corresponding Mean, Standard Deviation, Minimum and Maximum values. From these data, the software calculates the characteristics of an equivalent 3 points ROSENBLUETH distribution (i.e. 3 data points : x_-, x_0, x_+ and the associated probabilities P_-, P_0, P_+) according to the relationships given in Figure 3.

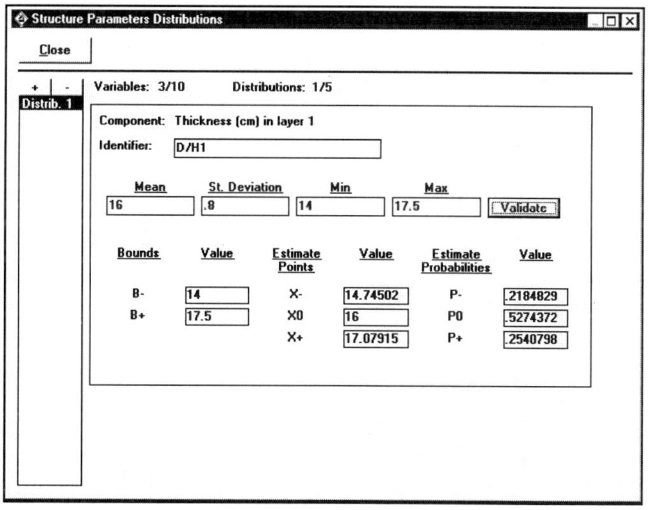

Figure 9

Computation runs. NOAH automatically computes a fixed set of stresses and strains corresponding to all the combinations of the various *States* of all the defined *Design Variables*. The computation program (STRESS) presently incorporated into NOAH systematically calculates :

- component stresses and strains (orthogonal reference system)
- vertical displacement and maximum shear stress in the horizontal plane
- maximum and minimum normal stresses and strains in the horizontal plane.

Result Analysis. The basic objective of *Result Analysis* is to generate a two-dimensional plot in which the calculated values of a given "*Performance Indicator*" are shown as a function of the different *States* of one of the *Design Variables*. To define such a plot, two steps are necessary :

Definition of a Performance Indicator. These are mathematical formulae which relate computed stresses and strains to a specific aspect of pavement performance (e.g. resistance to fatigue). Such formulae are directly entered from the keyboard (via the *"Formula Generator")* or retrieved from the Database. The formulas are entered *layer per layer* and may include, beside the elementary results, all the structural and environmental data pertaining to the layer. On Figure 10 (layer 1), one recognizes the formula for elemental fatigue damage already given in Figure 2.

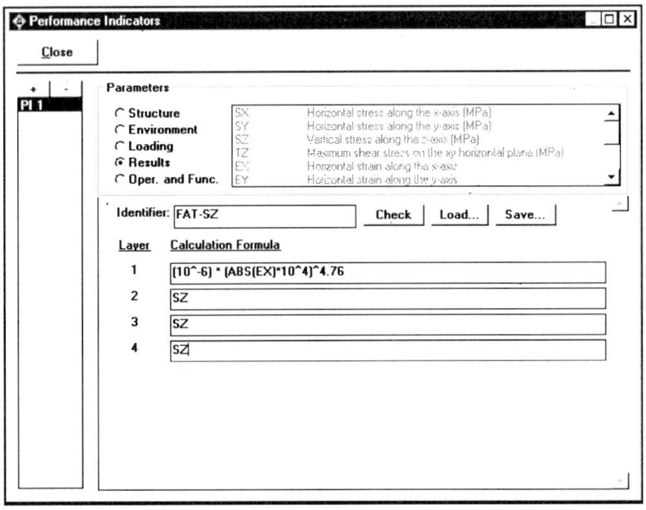

Figure 10

Contraction of Variables. In the final graphical output, the X-axis corresponds to one of the Design Variables (in this case, it is the D/H1 *Distribution*) whereas the calculated *Performance Indicator* is presented on the Y-axis. If the computation case treats several design parameters as variables, then the calculated *Performance Indicator* values belong to an "hyper-space", each variable accounting for one dimension of this "hyper-space". To end-up with a two-dimensional plot, it becomes then necessary to "*contract*" all the variables other

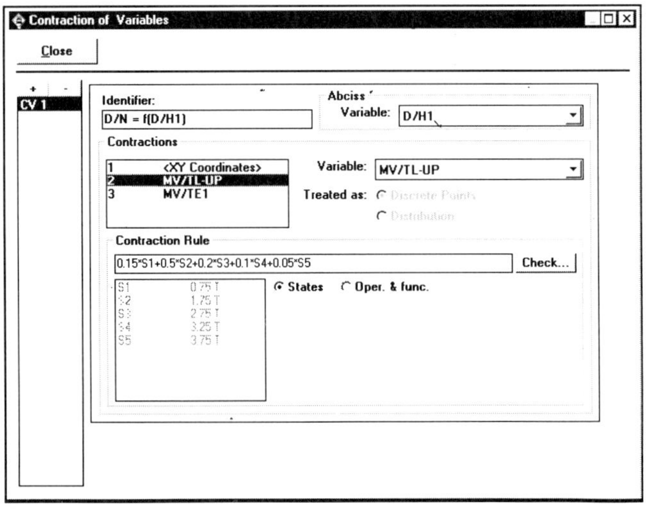

Figure 11

than the one selected on abciss. This has to be done, for each of these variables, by either selecting a particular state or by combining the results obtained for several states via a mathematical formula. In the case of the treated example, there are two such contraction steps which correspond in fact to the weighted average reflecting Miner's Law (see Figure 2). On Figure 11, the "*Contraction Rule*" written for the *Loading Variable* (MV/TL-UP) reflects the traffic spectrum frequencies. A similar rule would be written at the next contraction level for the Temperature histogram.

Master Graphs. The combination of a *Performance Indicator* and a *Contraction* defines a so-called "*Master Graph*". This term reflects the fact that there is a remaining degree of freedom provided by the "Z- components" (depths) of the *Performance Indicator*. Each of these "Z- components" can be referred to as a a "*Master Serie*" defining one particular curve. Figure 12 shows the *Master Graph* display corresponding to the treated example. At the first calculation depth (bottom of layer 1), one recognizes the number of permissible axle loads, together with the corresponding probabilities, in relation to the bounds (B_- and B_+) and the point estimates (x_-, x_0 and x_+) for the thickness distribution. The other *Master Series* are related to the vertical stress (see also Figure 10 for the definition of the *Performance Indicator* per layer).

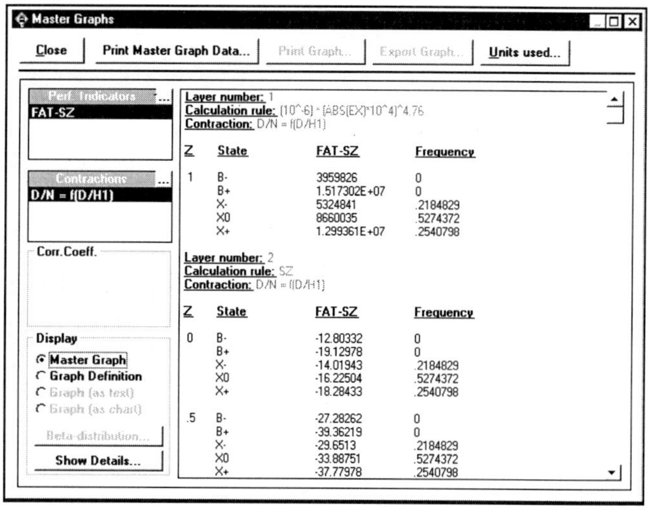

Figure 12

Graphs. The final graphical output is defined by the selective extraction of one or several *Master Series* from the *Master Graph.*

In the case of the treated example, the final step would consist in converting the point estimate of the resulting Fatigue Life distribution into an equivalent continuous distribution ("Step 4" of the ROSENBLUETH process). This is triggered by the "*Beta-Distribution*" command (see Figure 12). The software then determines coefficients α and β. The resulting distribution is displayed in a normalised form

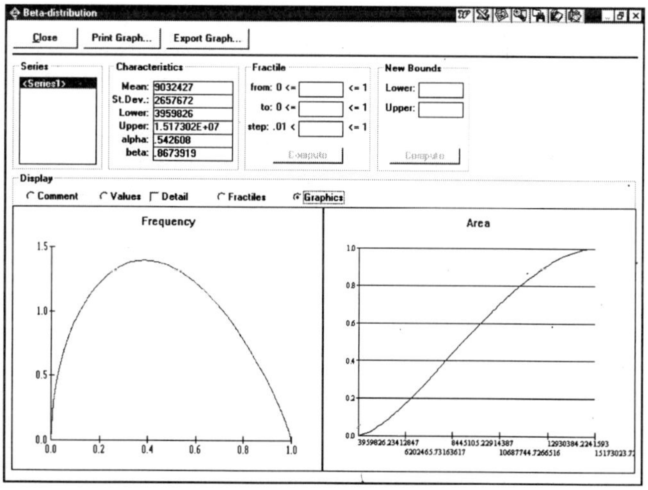

Figure 13

together with a cumulated frequency curve. On the same screen, the user can also retrieve Fractiles, i.e. the "level of risk" related to any given value of the *Performance Indicator* (see Figure 13).

Graphs can be displayed on the screen but the software does not provide extended graphical possibilities. Instead, Graphs can be "exported" , which enables the user to work with its own graphical software. All results, also at intermediary stage, can of course be printed in an extensive or selective way.

CONCLUSION

It is essential to underline that the NOAH software does not propose a new pavement design method. Its prime ambition is rather to be a very versatile and powerful computation tool, usable in the frame of any rational pavement design method based on calculated stresses and strains. To achieve this goal, NOAH has been provided with many specific and genuine facilities such as :

- the possibility to perform *Probabilistic Calculations* based on the ROSENBLUETH point estimate approximation,
- The "*Formula Generator*" which can be used at various stages (Description of material properties, definition of "*Performance Indicators*" and "*Contraction Rules*").
- The *Database* function which, beside the storage of Material Properties and "*Typicals*", offers many additional facilities. The user may, for instance, define its own systems of units as well as its own set of *Environmental Parameters*.
- Finally, the software has been programmed in such a way that future evolutions such as for instance the switch to a different type of calculation program, the addition of new design parameters,.... will be possible without major changes to the existing program structure.

Its extended possibilities, especially with regard to sensitivity studies and probabilistic calculations, let us hope that this new tool will foster new advances in the field of rational pavement design and support the development of flexible pavement structures.

Technical data. The present version of NOAH has been developed under WINDOWS 3.1 and VISUAL BASIC 3.0. It is compatible with WINDOWS 95. It requires a 486/33 or greater IBM compatible computer, a minimum of 5 MB free disk space, 8 MB RAM and a VGA colour monitor.

ACKNOWLEDGEMENTS

The author is highly endebted to Dr.Ir.F. Van Cauwelaert who, as a Consultant to NYNAS NV, has provided the mathematical input and back-up to the developments presented in this paper. Special thanks go also to Mr. C. Danheux for his dedication to the NOAH programming work.

REFERENCES

1. VAN CAUWELAERT, F. Use of the ROSENBLUETH theory for the determination of the Frequency Distribution of a function of Random Variables. Proceedings of the C.R.O.W. Workshop "Probabilistiek in het verhardingsontwerp", Ede, The Netherlands, 1994.

2. ECKMANN, B. New Tools for Rational Pavement Design. Proceedings of the 8th Int. Conf. on Asphalt Pavement, Seattle, Vol. I, pp. 25-42, 1997.

Fifth International Conference on the
BEARING CAPASITY OF
ROADS AND AIRFIELDS

NON LINEAR FINITE ELEMENT ANALYSIS OF RECYCLED ROAD PAVEMENTS

MUHAMMAD N S HADI[1] and HAMID R ZARGHAMPOUR[2]

[1]Lecturer, Department of Civil, Mining and Environmental Engineering, University of Wollongong, Wollongong, 2522, Australia, Fax: +61 2 42 213238, email: m.hadi@uow.edu.au
[2]Technical Development Officer, Department of Technical Development, Swedish National Road Authority, Sweden

ABSTRACT

An important step in the mechanistic design method for flexible and recycled pavements is calculating responses of pavements to vehicular loads. These responses are traditionally calculated assuming linear elastic media. Although such an assumption makes the problem easier to deal with, it leads usually to inaccurate results. Several researchers have attempted to model pavements as non linear material, but so far there is no unique solution that pavement engineers agree on. This research attempts to analyse recycled pavements assuming non linear elasto-plastic material properties which incorporate Drucker-Prager failure model. A pavement structure composing of two layers, 375 mm recycled layer resting on subgrade, was considered. For each of the four subgrade materials considered, the top recycled layer was assumed to have six different properties, henceforth each of the resulting model was analysed using three techniques, CIRCLY, linear finite elements and non linear finite elements. The runs of CIRCLY were to bench mark the other solutions. Graphs are drawn to highlight the viability of recycling and to assess the correctness of the different analysis techniques. Results from the three methods were to some extent diverse having different values of critical strains. However, they all proved that recycling is viable and considerable gain in load repetitions is achieved.

Key Words: Non linear analysis, Finite Elements, Flexible Pavements, Recycled Pavements.

INTRODUCTION

Pavement engineers are under pressure to build and maintain the road network with minimum cost. They have to come up with methods of stretching the available funds to the maximum limit and at the same time satisfy the expectations of the community for safety and comfort in driving. This is coupled with the cuts imposed on the funds available for infrastructure maintenance and construction.

The majority of roads in Australia are near the end of their useable life spans. Rebuilding these roads using traditional methods is an expensive exercise that requires significant resources (both materials and funds). This necessitates the need of employing other methods of building roads that minimise the cost and reduce the waste in materials. These methods should also be quick to accomplish in order to reduce the impact on flowing traffic. One of these methods is *pavement recycling* which is environmentally friendly, cost effective and quick method. Pavement recycling basically consists of reclaiming a certain depth of an existing pavement. This material is then mixed with predetermined percentages of cement, slag and/or lime together with water, and the produced material is placed back on the road and an adequate compaction subsequently applied.

For the last eight years or so, pavement recycling has been successfully used in Australia, as it has proved to cut the cost of maintaining the road network as well as using a minimum amount of new materials. The depth of recycling ranges from shallow lifting to depths of 200 to 300 mm down to deep lifting to depths of 600 to 700 mm.

An important step in the design of recycled and flexible pavements is calculating the response of the pavement to a standard vehicular load. Such response has been characterised to be a non linear problem in continuous media. A large number of investigations during recent decades, indicate that there are considerable disagreements between the conventional solutions of the theory of elasticity and the actual measurements of pavement responses. The reason for these disagreements appears to be caused by the simplifying but necessary assumptions of linearity in soil materials. Therefore, the effect of non linearity is of added interest in the procedure of pavement analysis and design. The inclusion of this important factor in the procedure of analysis demands an understanding of the material's characteristics and their behaviour under moving loads. This paper attempts to tackle this problem.

PAVEMENT RECYCLING

A number of trials of pavement recycling were conducted in the early 1990's and proved to be a viable method. Details of these tests are documented in several conference proceedings for example, a conference organised by the Roads and Traffic Authority, RTA of the New South Wales state held in 1993 (1). Recycling has proved to offer many improving factors to the reconstruction process of pavements. These include fast construction, improving the strength of the pavement, providing economical solutions, and reducing the maintenance cost and time. Before recycling is carried out, laboratory testing is to be conducted to determine the mechanical properties of the pavement material. This will determine the percentage and type of binder to be added during the recycling process. In addition, the amount of water will be determined at this stage to achieve the best performance. The recycling process involves the following steps:

- Assessment of the geometrical design of the road. Accordingly, a redesign of the geometry may be required. This will be followed by reconstructing the road. This may involve the grade and horizontal and vertical curves.

- By using recycling equipment (stabiliser), the pavement material is reclaimed up to a certain depth.
- Water and the binder as determined from the laboratory test are added and mixed to the reclaimed pavement material. During the mixing process the material is pulverised to its natural grading.
- The new mixed material is replaced on the pavement and the necessary compaction is applied to achieve the required strength of the pavement.
- The final step in the process is curing to prevent rapid evaporation of moisture from the surface and to allow the binder to harden fully. A light coat of bituminous material is commonly used.

A number of binders have been used successfully. These include: Portland cement, lime (quicklime or hydrated lime), cement with waste materials, such as fly ash and granulated blast furnace slag, and bituminous products. The most suitable and the percentage of the binder depends of the properties of the existing pavement.

DEVELOPMENT OF PAVEMENT DESIGN

Introduction Generally, flexible and recycled pavement design methods are categorised into five groups:

Empirical Methods based on the soil classification system provided by Hogentogler and Terzaghi (2). *Limited Shear Failure* based on Terzaghi's (3) formula for bearing capacity and developed by Barber (4) and McLeod (5). *Limited Deflection* first introduced by Kansas State Highway commission (6). This method refers to determining the thickness of pavement without allowing the deflection in pavement beyond a specified limit. *Methods Based on Road Tests* such as the ASSHTO Road Test (7). Finally, *Mechanistic Empirical Methods*, due to the importance of this method, it is described in the next section.

The Mechanistic Empirical Method This method is based on the evaluation of pavement responses to applied loading under different conditions. For these evaluations, data from field investigation and laboratory testing are needed. The obtained pavement responses from these testings are then used to predict possible distresses in the pavement. This method of design requires a failure criterion. The finally designed pavement is then checked against the defined failure criterion. If the condition for the assumed failure criterion is not satisfied, some changes in the designed pavement structure is then necessary. In general, it is difficult to assume a criterion that covers all the possibilities of failure. However, a few of the suggested failure criteria during the years are presented here.

Kerkhoven and Dormon (8) first suggested the use of vertical compressive strain on the surface of subgrade as a failure criterion to reduce the permanent deformation. The choice of this parameter as a failure criterion can be explained as, the plastic strains in the paving materials is proportional to their elastic strains. The limitation of elastic strains in the subgrade controls the same factor in the other components above the subgrade also. Consequently, the permanent deformation on the surface of the pavement will also be controlled. Saal and Pell (9) suggested the use of horizontal tensile strain at the bottom of the asphalt layer as a failure criterion for minimising the fatigue cracking. Both the above mentioned failure criteria have been adopted in the mechanistic design of pavement by the Shell Petroleum International as presented by Claussen et al. (10) and Asphalt Institute as reported by Shook et al. (11). Moreover, AUSTROADS (12) has adopted the asphalt fatigue criterion, published by Shell, to develop charts for the determination of pavement thickness. The advantages in adopting the mechanistic empirical methods in the design of pavement structure are: improvement and reliability of design and feasibility of usage of the limited field and laboratory data.

The fatigue criteria for subgrade, cemented materials and asphalt are based on experience over the years. The fatigue criterion for subgrade recommended by AUSTROADS was developed by Youdale (13) by back-analysing 25 pavements selected from the NAASRA (14) granular thickness chart using CIRCLY (15) to calculate the compressive vertical strains at top of subgrade. This chart is the basis of the design chart recommended by AUSTROADS which is produced in Figure 8.4 of AUSTROADS. The origins of the NAASRA and hence AUSTROADS design charts can be traced back to the charts developed by the California State Highway Department back in the 1940's. During the years the original design chart was modified to suit the Australian road conditions and considering the research on road technology carried out by the different Australian road authorities. A comprehensive review of the origins of AUSTROADS design procedure in presented by Jameson (16). It is worth noting here that Potter et al. (17) has reported that about ten percent of the pavement engineers who responded to a recent survey indicated their dissatisfaction with this failure criterion. Other design standards have included similar failure criteria which relate the allowable number of standard load repetitions, N and the vertical compressive strain at the surface of the subgrade, ε. These are listed in equation (1) and values of the constants f_4 and f_5 are presented Table 1. Figure 1 presents the relation between the subgrade strain and the load repetitions using these relations.

$$N = f_4(\varepsilon)^{-f_5} \qquad (1)$$

Table 1: Factors for Subgrade Fatigue Criteria

	f_4	f_5
AUSTROADS (12)	$1.66*10^{-15}$	7.14
Asphalt Institute, AI (18)	$1.365*10^{-9}$	4.477
Shell (19)	$6.15*10^{-7}$	4.0
TRRL (20)	$6.18*10^{-8}$	3.95

It is clear from Figure 1 that although the factors seem to be different, the resulting relation between the strain and the load repetitions is almost the same, except the AUSTROADS relation which seems to be slightly high, thus giving more allowable load repetitions. In this work the recommendations of AUSTROADS were adopted due to its relevance to Australia.

Figure 1: Subgrade Failure Criteria

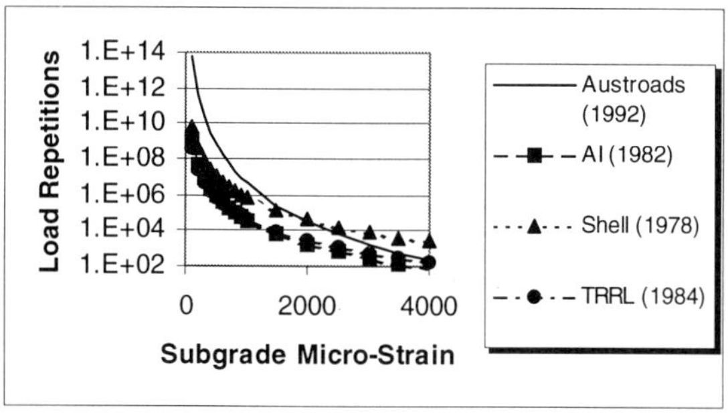

MODELLING, ANALYSIS AND RESULTS

Introduction In this work, a road pavement recycled to a depth of 375 mm has been analysed, using the finite element program MSC/NASTRAN running on a SUN Ultra E3000 supercomputer. The pavement was assumed to consist of two layers. The top recycled layer was assumed to be composed of a linear elastic material. Material properties of the lower layer were assumed to be linear in the linear analysis and non linear in the non linear analysis.

In the linear analysis, the pavement responses were studied when the granular material in the lower layer was assumed to have linear elastic behaviour in line with the theory of elasticity and the layered theory. In the non linear analysis, the non linear behaviour of the granular materials used in building the pavement structure was studied. There have been numerous attempts to correctly model and include the constitutive behaviour of soil in the analysis during the years. In most finite element formulations employed to solve these types of problems, the modulus of elasticity E is expressed as a function of stress and/or strain level, while Poisson's ratio is kept constant. For the studied model in this work, an elasto-plastic relationship was set up by using a recommended model for description of the non linear behaviour of granular material. This was utilised by using Drucker-Prager model (21). In both analyses a three dimensional finite element model was used which is a 3 metre cube. Due to symmetry, this cube represents quarter of the analysed pavement structure. The model is composed of 810 three dimensional hexahedral elements with different sizes, smaller under and near the load and becoming larger away from the load. Strains at the critical locations as specified in AUSTROADS were computed from both analyses and later on were compiled and compared. These strains were compared to those obtained by analysing the same pavement structures using CIRCLY.

The thickness of the first (recycled) and second layers are 0.375 m and 2.625 m, respectively. The total thickness of the structure is 3 m. A total of forty eight runs were conducted, twenty four for the linear and a further twenty four for the non linear. Four basic materials are considered for the second layer. The first representing inorganic clay, the second heavy clay, the third clayey sand and the fourth sand. These materials are represented by defining their basic properties, viz. cohesion, c, angle of internal friction, ϕ, modulus, E and Poisson's ratio, ν. Values for these properties were adopted from (22) and are presented in Table 2. For each basic type of soil, six material properties are considered for the top recycled layer. These properties, for the first recycled layer, are described by modulus of elasticity E and Poisson's ratio ν. The values of E were 500, 1000, 2000, 3000, 4000 and 5000 MPa. The value of Poisson's ratio ν is kept constant at 0.2. For the linear analysis, the material models for the lower layer was limited to the modulus and Poisson's ratio. Figure 2 illustrates the physical model of the pavement.

Figure 2: Cross Section of the Two Layered Recycled Pavement System

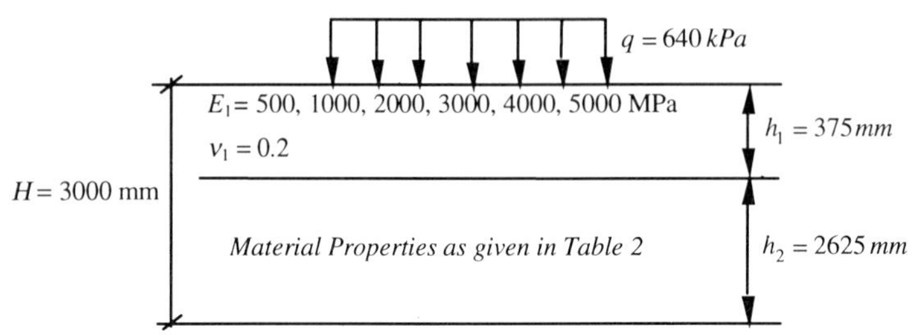

Table 2: Properties of Subgrade

Type of Soil	Inorganic Clay	Heavy Clay	Clayey Sand	Poorly Graded Sand
Cohesion, kPa	11	76.5	11	2
Angle of Internal Friction, φ	19	7.3	31	37
Modulus, MPa	150	31	50	250
Poisson's Ratio, ν	0.4	0.4	0.4	0.4

Figure 3 Shows the finite element model used in both the linear and no linear analyses. The model is composed of 810 elements that are called hexahedral solid elements, CHEXA. Each of these elements has eight grid points and each grid point has six possible components of displacement, *ie* three translations (in the x, y, and z directions) and three rotations (about x, y, and z-axes). The total number of grid points in the model is 1100. All grid points in the model were assumed to be fixed to rotate in all directions. This was done in order to comply with the finite element theory definition of CHEXA elements. Moreover, all grid points at the lower face of the model were fixed to displace in the vertical direction and all grid points in the faces of symmetry (east face and west face) were fixed to translate in the direction normal to their planes. This was done to simulate the bigger model of the pavement, *ie* 6 by 6 by 3 metre pavement structure.

The size of the CHEXA elements in the model varies in such a manner that smaller elements are used in the region of stress concentration. The further from this region the larger elements are used. This also ensures continuity in the geometry and more accuracy in the results obtained in the region of interest for the problem. The variation of the side length of these prismatic elements are 62.5, 125, 250, 500 and 1000 mm respectively. The load used in the finite element model consists of a pressure with an intensity of 640 kPa. This load is applied on the surface of the second and third solid elements off the finely meshed corner of the model (Figure 3). The horizontal dimensions of the area are 125 x 125 mm^2, which gives a total force 10 kN. This represents a quarter of the full loading recommended by AUSTROADS.

Figure 3: General View of the Finite Element Model

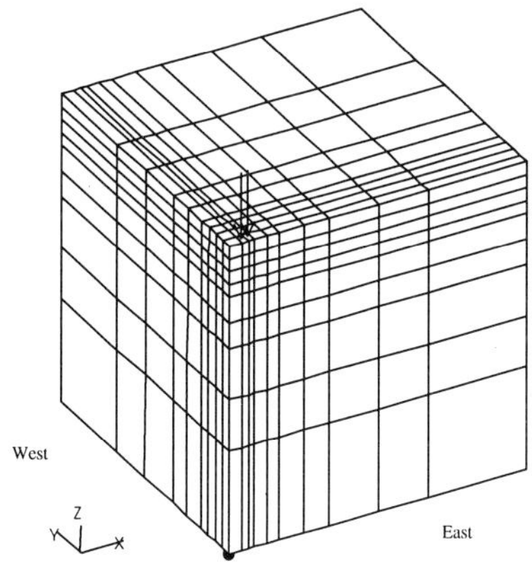

In order to bench mark the results of the finite element linear and non linear analyses, the finite difference based program, CIRCLY was used to analyse the pavement structures. The model for CIRCLY has infinite horizontal dimensions and the thickness of the subgrade is also infinite, however the thickness of the top layer is kept the same, 375 mm. Material properties are assumed similar to those in the linear finite element analysis. The CIRCLY analyses were conducted on a PC.

Results MSC/NASTRAN running on a SUN Ultra E3000 supercomputer was used to conduct the linear and non linear analyses and CIRCLY running on a PC was used for comparison purposes. The vertical compressive strain at top of the subgrade directly under the wheel load were extracted from the finite element solutions and CIRCLY runs (Figure 4). These values of strain were used as a basis to calculate the allowable load repetitions (ESL) as recommended in the subgrade fatigue criterion of AUTSROADS (Figure 5).

The analysis of the 500 MPa top layer was assumed to be the deteriorated pavement and the other pavements were assumed to be the recycled one. Accordingly, the gain in the load repetitions was calculated for each of the moduli, viz. 1000, 2000, 3000, 4000 and 5000 MPa based on the results of the 500 MPa analysis. These values are graphically presented in Figure 6.

It is to be noted that the strains from CIRCLY are considerably higher than the other analyses and hence the lower load repetitions. The strains from the linear analysis were the least, yielding the highest load repetitions. The gain in load repetitions using the linear analysis was the highest compared to CIRCLY which gave the lowest gain.

Figure 4: Variation of Subgrade Strain with Recycled Layer Modulus

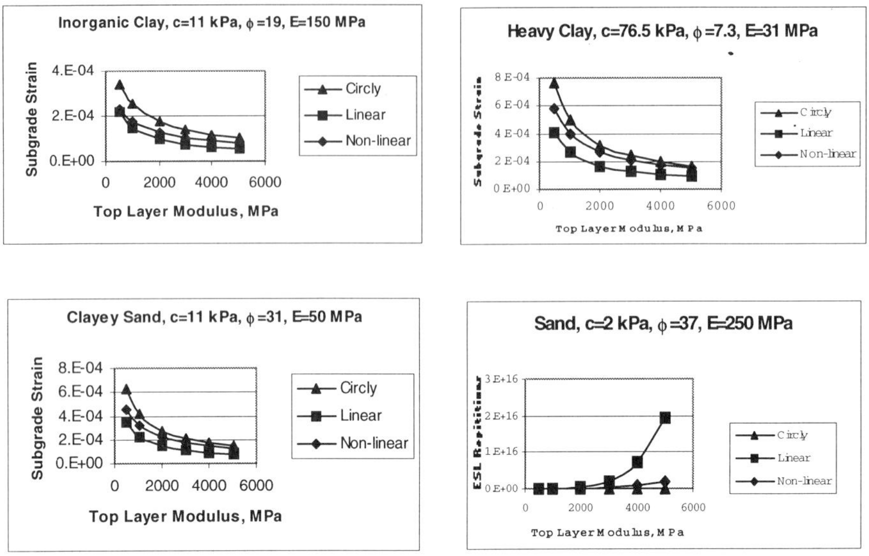

Figure 5: Variation of Load Repetitions with Recycled Layer Modulus

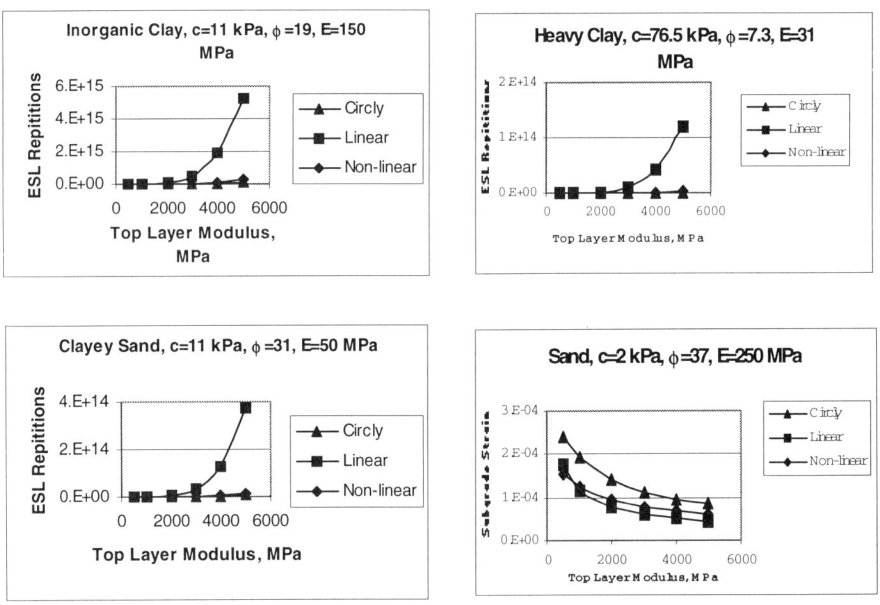

Figure 6: Variation of Gain in Load Repetitions with Recycled Layer Modulus

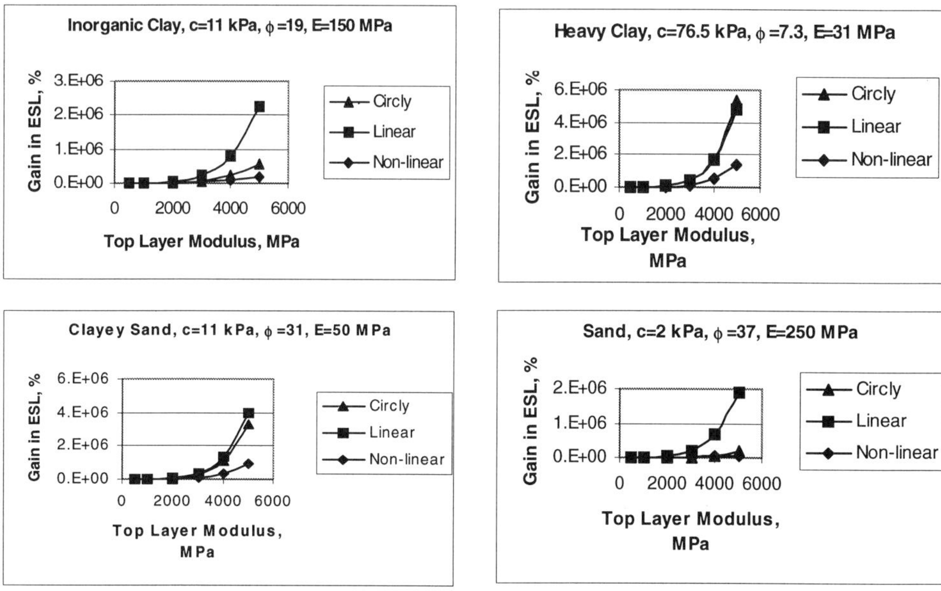

CONCLUSION

The current design method of recycled and flexible pavements recommended by AUSTROADS (12) is based on the mechanistic design method. This design standard recommends using the finite difference based program, CIRCLY. Although CIRCLY is versatile and easy to use, it is believed that the time is ripe to move to more sophisticated analysis methods such as the finite element. Since these programs offer a wide range of material types and solution techniques, thus representing the actual pavement more realistically and considering all the different possibilities in the pavement structure and loading pattern.

In this work a non linear analysis of a pavement structure was conducted together with a linear analysis. The results of these analyses were compared with the results obtained from CIRCLY. The pavement was assumed to be recycled and these results were compared with a deteriorated pavement. The three analyses revealed that the results of CIRCLY were more conservative compared to the non linear analysis. Also, the gain in the load repetitions was the highest in the linear analysis. All analyses revealed that recycling is viable and by increasing of the modulus of the top layer a considerable increase in the number of load repetitions is yielded, this is in the ten fold range. This proves that recycling is a viable method for road stabilisation.

REFERENCES

1. RTA. Proceedings of the Workshop on Deep-Lift Recycling of Granular Pavements. Roads and Traffic Authority of New South Wales, Sydney, Australia, (1993).
2. HOGENTOGLER, C.A., TERZAGHI, C. Interrelationship of Load, Road and Subgrade, Pub Roads, May. pp. 37-64, 1929.
3. TERZAGHI, C. Theoretical Soil Mechanics, Wiley, New York, USA, (1943).
4. BARBER, E.S. Application of Triaxial Compression Test Results to the Calculation of Flexible Pavement Thickness, Proceedings, Highway Research Board, pp. 26-39, 1946.
5. MCLEOD, N.W. Some Basic Problems in Flexible Pavement Design. Proceedings, Highway Research Board, pp. 91-118, 1953.
6. KANSAS STATE HIGHWAY COMMISSION. Design of Flexible Pavement Using the Triaxial Compression Test, Bulletin 8, Highway Research Board, 1947.
7. AASHTO. AASHTO Interim Guide for Design of Pavement Structures. American Association of State Highway and Transportation Officials, (1972).
8. KERKHOVEN, R.E., DORMON G.M. Some Considerations on the California Bearing Ratio Method for Design of Flexible Pavement. Shell Bitumen Monograph No.1, 1953.
9. SAAL, R.N.J., PELL, P.S. Kolloid-Zeitschrift MI, Heft 1, pp. 61-71, 1960.
10. CLAUSSEN, A.I.M, EDWARDS, J.M., SOMMER, P., UGE, P. Asphalt Pavement Design-The Shell Method. Proceedings, 4th International Conference on the Structural Design of Asphalt Pavements. 1:39-74, 1977.
11. SHOOK, J.F, FINN, F.N., WITCZAK, M.W., MONISMITH, C.L. Thickness Design of Asphalt Pavements - The Asphalt Institute Method. Proceedings, 5th International Conference on the Structural Design of Asphalt Pavements. 1:17-44, 1982.
12. AUSTROADS. Pavement Design, A Guide to the Structural Design of Pavement. AUSTROADS, Sydney, Australia, (1992).
13. YOUDALE, G.P. Review of Limiting Subgrade Strain Criterion. Submission to NAASRA Working Group on the Revision of NAASRA Interim Guide to Pavement Thickness Design (IGTPTD), April, 1984.
14. NAASRA. Interim Guide to Pavement Thickness Design. NAASRA Sydney, (1979).
15. WARDLE, L.J. Program Circly User's Manual. Geomechanics Computer Program - Number 2. CSIRO - Division of Applied Mechanics, Australia, 1977.
16. JAMESON, G.W. Origins of AUSTROADS Design Procedure for Granular Pavements. ARRB, Transport Research, ARR 292, Australia, 1996.
17. POTTER, D.W, JAMESON, G.W, MAKAROV, A, MOFFATT, M.A, CROPLEY, S.M. A Basis for Incorporating Reliability in the AUSTROADS Pavement Design Procedures. ARRB TR WD TI96/014, Australia, 1996.
18. AI. Research and Development of the Asphalt Institute's Thickness Design Manual. (MS-1), 9th ed, Research Report 82-2, Asphalt Institute, USA, 1982.
19. SHELL. Shell Pavement Design Manual. Shell International Petroleum Company. London, UK, (1978).
20. TRRL. The Structural Design of Bituminous Roads. Laboratory Report 1132, Transport and Road Research Laboratory, Department of the Environment, Department of Transport, UK, 1984.
21. CHEN, W.J. AND MIZUNO, E. Non Linear Analysis in Soil Mechanics – Theory and Implementation. Elsevier, (1990).
22. HAUSMANN, M.R. Engineering Principles of Ground Modification. McGraw-Hill, (1990).

Fifth International Conference on the
BEARING CAPASITY OF
ROADS AND AIRFIELDS

ANALYSIS OF THE PAVEMENT BEARING CAPACITY ON THE DEFLECTION BOWL BASIS

JÁN ČELKO[1] AND JOZEF KOMAČKA[2]

[1]Assoc. Prof., University of Žilina, 010 26 Žilina, Slovak Republic,
Fax 00421 89 43351, E-mail celko@fstav.utc.sk
[2]Assist. Prof. , University of Žilina, 010 26 Žilina, Slovak Republic

ABSTRACT

Design of the pavement overlay was realised only on the base of maximal value of pavement deflection in the past. The deflection was measured by diagnostic equipment but alone value of deflection did not characterise a pavement structure sufficiently. This source of mistakes can be removed by using of whole deflection bowl. The article described the new method used in Slovakia, which is based on foreign experiences. The created computer program analyses whole deflection bowl measured by equipment KUAB or PHÖNIX. The values of deformation characteristics of pavement layers are calculated by iteration computing. The present values of characteristics are found by comparison of computed and measured values. The values are used for computing of residual life of pavement, a critical layer, and for overlay, too. In conclusion the examples of calculation for different homogeneous method are presented.
Key words: pavement, bearing capacity, deflection, deformation.

INTRODUCTION

The pavement bearing capacity is the important factor for a road administrator from global state of the pavement point of view. The demands are necessary for maintenance, reparation and reconstruction of the road and highway network in relation to financial and material demands. However the bearing capacity is not important for road users because only consequences of the insufficient bearing capacity are interesting for them.

A problem of the bearing capacity evaluation relates to the Pavement Management System (PMS) implementation in Slovak Republic. The bearing capacity is one from four parameters, which are used in Slovak PMS. The pavement bearing capacity was evaluated by using of the nomographic charts in the past. Nomograms were created on the base of empirical experiences and the bearing capacity was evaluated according to maximal deflection value. This mode of the evaluation is insufficient and inaccurate because only a maximal deflection provides few information about pavement layers and subgrade condition. In addition, almost same values of the maximal deflection is possible to obtain for different pavement structures and even on the same pavement structure for different values of the layers modulus. The purchasing modern testing devices KUAB and PHONIX apply for calculation of pavement bearing capacity firm software of these facilities. The software was created by design methods and boundary conditions which are valid in foreign countries, but for the evaluation only a maximal deflection or difference between maximal deflection and deflection at distance 45 cm from axle of the load was used. Therefore the computer program named CANUV which uses all deflection bowl measured by KUAB or PHÖNIX was developed in author's department. The program respects the Slovak method for design of the flexible pavement [1].

DESCRIPTION OF THE PROGRAM

The program is created for PC computers. The program modules are connect to one complex unit, which contents these parts:

1. data conversion and input file preparation
2. layers modulus backcalculation
3. pavement residual life calculation
4. overlay design

Data conversion and input file preparation

The text output file from deflectometer KUAB 2m - 150 or PHÖNIX ML 5000 contains a lot of data but not all are necessary for the calculation of the layers modulus, pavement residual life and overlay design. Therefore only data about measured points are selected and the text output file is conversed to a database file, which for each measured point contains:

- identification data: a) road number
  b) initial and terminal node of a section
  c) chaining within the node section
  d) date of the measuring
- deflection bowl values (seven deflection values for KUAB or six for PHÖNIX)
- pavement temperature
- load force value
- radius of the FWD loading plate

Input file for the calculation is prepared after this conversion for each measured point. The file preparation is automatic but previous data must be completed about:

- number, type and thickness of the asphalt layers
- type of the subbase layers: a) unbounded
  b) bounded by cement

c) bounded by bitumen

- total thickness of the subbase layers
- initial elastic modulus values for the surface layers, subbase and subgrade
- traffic load and its annual growth
- required residual life of the pavement

Layers modulus backcalculation

The pavement structure from input file is transformed to three-layered system (asphalt layers, subbase, subgrade) in the first step. The equivalent modulus of the asphalt layers at the temperature 11° C (middle from three values, which characterise climatic conditions in Slovak Republic) is then calculated by equation

$$E_{EKV} = \frac{\sum_{i=1}^{n} E_i \times h_i}{\sum_{i=1}^{n} h_i} \qquad (1)$$

where E_i = elastic modulus value of the asphalt layer „i" used for new layer at the temperature 11°C [MPa]
h_i = thickness of individual asphalt layer [cm]

The total thickness of the asphalt layers for the calculation is sum of the individual asphalt layers thickness.

After this transformation subgrade is divided into three parts. Two layers with 50-cm thickness and third infinite vertically. It is assumed that second layer has elastic modulus higher than first about 30 MPa and third about 60 MPa higher then first. The then program calculates theoretical surface deflection values in this layered system at positions corresponding to deflection sensors' location. The modulus values for the asphalt layers, subbase and subgrade defined in the input file preparation are used for the first calculation. If the differences between measured and calculated deflections after the first calculation are not within allowance the initial modulus values are changed. The searching of new modulus values continues in pursuance of the defined algorithm [2] while the criterion of the allowance is attained. The finding is limited by 100 iterations otherwise measured point is not evaluated.

Theoretical deflections are calculated by program OPMEKO [3]. This program uses the same linear elastic multi-layered assumption in their formulation as programs BISTRO or CHEVRON, i.e.:

- the applied load is to be acting uniformly over a circular area and this area is to be fully flexible in contact with pavement surface
- all layers above the bottom layer have know thickness but they are infinite horizontally
- the bottom layer is infinite vertically and horizontally

- each layer is elastic, isotropic, linear and homogenous in behaviour
- the material properties of each layer are characterised by Young's modulus and Poisson's ratio which are uniform throughout the whole layer
- variation of the interface bonding degree between layers is possible

The example of the deflections calculated on one pavement type by CHEVRON, BISTRO and OPMEKO is on fig.1. Deflections calculated by OPMEKO and BISTRO produce excellent agreement and it is valid for other pavement structures too.

The computed elastic moduli of the asphalt layers are calculated for the asphalt layers temperature at the time of the deflection measuring. Then they are recalculated to the temperature 11°C by equation

$$E_i = E_{EKV} + \left(E_{CAL} - \left(7717.04 - 225.29 \times T + 1.94 \times T^2 - \left(5500 - E_{EKV}\right)\right)\right) \quad (2)$$

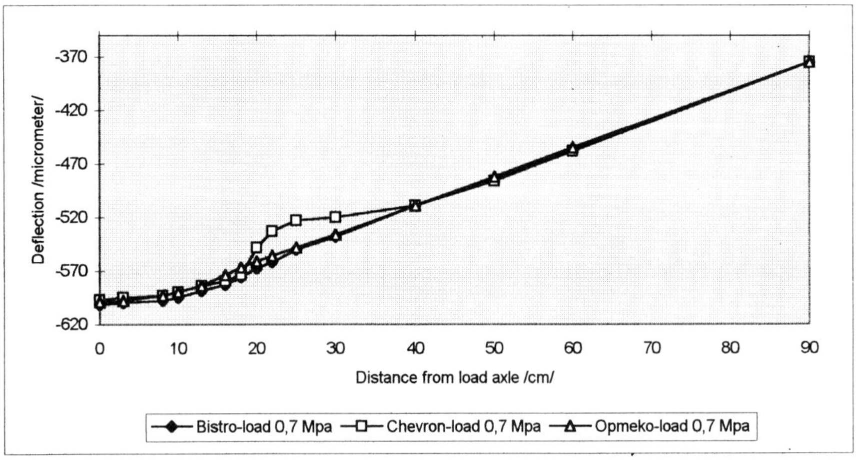

Fig.1 Comparison of the deflections calculated by different programs

where E_{11} = equivalent elastic modulus for temperature 11° C [MPa]
E_{EKV} = equivalent modulus calculated by equation (1) [MPa]
E_{CAL} = equivalent calculated modulus for the asphalt layers temperature at the time of the deflection measuring [MPa]
T = asphalt layers temperature at the time of the deflection measuring [°C]

Pavement residual life calculation

The elastic modulus of asphalt layers recalculated to 11° C temperature and elastic modulus of the subbase and three layers of the subgrade from backcalculation are used in this part.

The residual life of the pavement is calculated by two criterions. The first is expressed by formula

$$\frac{\sigma_{r,i}}{S_N \times R_{i,i}} \leq 1 \qquad (3)$$

where $\sigma_{r,i}$ = radial stress on the bottom of the pavement bounded layer „i“ [MPa]

$R_{i,i}$ = maximal tensile stress produces failure after one loading cycle of the pavement layer „i“ [MPa]

S_N = fatigue coefficient of the pavement layer material

Because S_N is depended on the number repetitions of standard axle load (S_N=f(log N)), it is possible from this dependence to calculate the residual life of bounded layers by formula:

$$\log N_i = \frac{A \times R_i - \sigma_{r,i}}{B \times R_i} \qquad (4)$$

where N_i = residual life of layer „i“ in standard axles load [S.A.L.]

$\sigma_{r,i}$ = computed radial stress on the bottom of pavement bounded layer „i“ [MPa]

A,B = parameters of layers:

- for bituminous layers A=0.95 B=0.12
- for cement bounded layers A=1.0 B=0.07

Second criterion is

$$\frac{\sigma_z}{\sigma_{z,m}} \leq 1 \qquad (5)$$

where σ_z = vertical stress on the top of subgrade [MPa]

$\sigma_{z,m}$ = vertical stress limit of the subgrade [MPa]

$\sigma_{z,m}$ depends on the number repetitions of standard axle load and elastic modulus of subgrade ($\sigma_{z,m} = f(\log N, (E_{SUBGR}))$. This relation is used for the calculation of subgrade residual life that is computed by formula

$$\log N_{SUBGR} = \frac{0.00346 \times E_P - \sigma_z}{0.7 \times \sigma_z} \tag{6}$$

where σ_z = computed vertical stress on the top of subgrade [MPa]

E_{SUBGR} = elastic modulus of subgrade determined by backcalculation [MPa]

N_{SUBGR} = residual life of subgrade in standard axles load [S.A.L.]

The fewest computed values from N_i and N_{SUBGR} determine a "critical layer" of the pavement. For this layer a total residual life of the pavement is solved by follow formula:

$$N_{RL} = N - \sum_{i=1}^{n} N_N \times \delta_i \tag{7}$$

For first $N_{RL} < 0$ is RL = i - 1 years,

where	N_{RL}	= residual life expectancy after i - year	[S.A.L.]
	RL	= residual life of pavement	[years]
	N	= total residual life of critical pavement layer	[S.A.L.]
	N_N	= traffic load per year in design section	[S.A.L.]
	δ_i	= annual increases of the traffic load	[%]

If a computed residual life of the pavement is less then the residual life required in the input data file preparation, overlay thickness is calculated.

Overlay thickness design

In the process of overlay design are used elastic modulus of the pavement layers obtained by backcalculation and thickness of the layers from the input data file preparation. The other layer (overlay) is joined to layered system that is used for the backcalculation and pavement residual life calculation. The initial thickness of the overlay is calculated from the maximum deflection value in the measured point. In this new layered system are computed values of radial stress on the bottom of each bounded layer and vertical stress on the top of subgrade.

The criterions for design of the overlay thickness are same as for the calculation of residual life (formula (3) and (5)). Stress - resistance ratio from left part of formula (3) is calculated for each bounded pavement layers and stress ratio of subgrade from left parts of formula (5), too. If the maximum value from these ratios is not from 0.85 to 0.95 then the initial value of overlay thickness is decreased (for ratio less then 0.85) or augmented (for ratio greater than 1.0) by one centimetre. This procedure is repeated until the maximum value from these ratios is from 0.85 to 0.95.

If all measured points are evaluated, homogenous sections are created. The overlay thickness determines the division to these homogenous sections. Statistical method or maximal 3-cm difference between the overlay thickness in one section is used for the creating of the sections. It is necessary to say that there are differences in homogenous sections created by these two methods.

PRACTICAL OUTPUTS

The outputs of the program are possible in text file (.TXT) or database file (.DBF). A database file is required for the using in Pavement Management System.

The program has been used by Slovak Road Databank for the roads bearing capacity evaluation from 1996. The example of the text output file is gets below.

ROAD NUMBER I/73 (E 71)

Initial node	Terminal node	Chaining [m]	Elastic Modulus E [MPa] Asph.	Subbase	Subgrade	Critical layer	RL [year]	Overlay [cm]
2833B00300	2833A01900	0	3863,5	135,0	74,0	1	0,0	5,0
2833B00300	2833A01900	100	4476,1	270,0	93,0	1	2,0	5,0
2833B00300	2833A01900	200	4476,8	200,0	188,0	1	0,0	7,0
2833B00300	2833A01900	300	5254,5	208,0	549,0	1	1,0	6,0
2833B00300	2833A01900	400	5425,8	167,0	451,0	1	0,0	9,0
2833B00300	2833A01900	500	7384,0	194,0	100,0	1	1,0	5,0
2833B00300	2833A01900	600	4131,5	153,0	232,0	1	0,0	9,0
2833B00300	2833A01900	700	5977,8	131,0	151,0	1	0,0	10,0
2833B00300	2833A01900	800	4479,8	112,0	196,0	1	0,0	8,0
2833A01900	2833A02000	29	5390,7	115,0	232,0	1	0,0	12,0
2833A01900	2833A02000	129	5926,5	237,0	230,0	1	0,0	10,0
2833A01900	2833A02000	229	4813,7	71,0	219,0	1	0,0	12,0
2833A01900	2833A02000	329	4230,8	96,0	233,0	1	0,0	11,0
2833A01900	2833A02000	429	5639,6	98,0	203,0	1	0,0	13,0
2833A01900	2833A02000	529	3221,8	81,0	150,0	1	0,0	15,0

Fig.2 Example of the text output file

Section	Initial node	Terminal node	Chaining [m] Starting	Ending	Length of section [m]	Thickness of overlay [cm]
1	2833B00300	2833A01900	0	100	100	5,0
2	2833B00300	2833A01900	100	300	200	7,0
3	2833B00300	2833A01900	300	400	100	9,0
4	2833B00300	2833A01900	400	500	100	5,0
5	2833B00300	2833A01900	500	800	300	9,0
6	2833B00300	2833A02000	800	29	29	12,0
7	2833A01900	2833A02000	29	129	100	10,0
8	2833A01900	2833A02000	129	429	300	12,0
9	2833A01900	2833A02000	429	529	100	15,0

Fig. 3 Homogeneous sections created by statistical method

Section	Initial node	Terminal node	Chaining [m] Starting	Ending	Length of section [m]	Thickness of overlay [cm]
1	2833B00300	2833A01900	0	100	100	5,0
2	2833B00300	2833A01900	100	300	200	7,0
3	2833B00300	2833A01900	300	400	100	9,0
4	2833B00300	2833A01900	400	500	100	5,0
5	2833B00300	2833A01900	500	800	300	9,0
6	2833B00300	2833A02000	800	429	429	12,0
7	2833A01900	2833A02000	429	529	100	15,0

Fig.4 Homogenous sections for 3-cm difference

CONCLUSIONS

On the basis of the practical experiences with the using of this program by Central Road Databank it is possible to state that:

⇒the theoretical assumptions for backcalculation of the pavement layers modulus were verify because if we know the layer thickness of the pavement structure approximately 95% of the measured points were evaluated

⇒the values of elastic pavement layers modulus obtained by backcalculation are in the real range of the possible values

On the other way it is necessary to say that some obtained results indicated some problems connected with this evaluation method. They are above all information about thickness and types of pavement layers and their changes, instantaneous resistance in rupture bounded pavement layers, temperature differences between top and bottom of the asphalt layers. Answers to these problems will contribute to the improvement of pavement bearing capacity evaluation by this method.

REFERENCES

1. The design and appraisal of flexible pavement. The technical condition. Research institute of civil engineering. Bratislava 1976. In Slovak
2. Komačka, J., Čelko, J.: CANUV - Computer program for analysis of the bearing capacity on the deflection bowl basis. Studies of University of Transport and Communications in Žilina, Volume 19, pp. 63-70. Žilina 1996
3. Novotný, B.: OPMEKO - computer program for design of flexible pavement. Bratislava 1993.

Fifth International Conference on the
BEARING CAPASITY OF
ROADS AND AIRFIELDS

THE NEW AUSTRIAN STANDARD FOR PAVEMENT DESIGN

R. BLAB[1], J. LITZKA[2] AND C. MOLZER[1]

[1] Senior Research Assistant; [2] Professor, Head; Technical University of Vienna, Institute for Road Construction and Maintenance, A-1040 Vienna, Gusshausstrasse 28, Fax: ++43-1-504 42 09, E-mail: sekr@istu.tuwien.ac.at

ABSTRACT

The Austrian specification for pavement design was first published in 1986. In the light of changes in heavy traffic composition and heavy vehicle technology and after more than 10 years of experience with standardised pavement types for flexible and concrete roads, the specification has been revised comprehensively with a special focus on the following objectives:

- establishment of an improved method for the assessment of design traffic based on extensive experimental and theoretical research
- extension of the range of standardised pavement types to include pavements made in part of recycled material
- introduction of a new load class for very heavy traffic situations.

The standardised pavements are set out in matrix-type tables showing six traffic load classes and six different construction types. These tables are the result of analytical design calculations taking into account traffic, structural and environmental data. The paper presents the design tables and describes the methods used in traffic assessment and design calculation.

Keywords: Pavement design, analytical design calculation, design traffic, standardised pavements

1. INTRODUCTION

After more than 10 years of experience with the Austrian specification for standardised pavement design, RVS 3.63 (1, 2), a need was felt for a comprehensive review. The most obvious changes made were the introduction of a new load class as a reaction to a substantial increase in heavy traffic and admission of the use of recycled materials in standardised pavements. On the basis of previous research, new vehicle equivalency factors were moreover specified for use in the determination of design traffic. In addition - and not immediately recognisable for the user - the analytical design method for the specification of standard structures was also modified, which resulted in minor changes in the thicknesses of bound layers as compared with previous specifications.

2. DESIGN CATALOGUE

2.1 Structure

The two design tables forming the core of the revised specification RVS 3.63 (2), are shown in Figures 1 (bituminous pavements) and 2 (concrete pavements). The two tables are organised in an analogous manner. Determination of the design traffic volume (see Section 3) is followed by classification in one of 6 load classes (S, I - V). For each load class, one out of 6 different construction types can then be chosen:

- Construction types 1 - 3: bituminous base and wearing course on unbound base and subbase (flexible pavement)
- Construction type 4: bituminous base and wearing course on cement-treated base and unbound subbase (semi-rigid pavement)
- Construction type 5: concrete pavement on unbound subbase (rigid pavement)
- Construction type 6: concrete pavement on cement-treated subbase (rigid pavement)

The standards to be met by the materials of each pavement course (see keys to Figures 1 and 2) are specified in the relevant Austrian specification (RVS) (2). Beside these standard materials, the following proportions of reclaimed asphalt concrete RAC (crushed or milled granulated asphalt) may be used:

– subbase	≤	50 % RAC	
– base course	≤	5 % RAC	
	≤	50 % RAC	for load classes III - V only
		100 % RAC	for construction type 3, suitable only for load classes III - V

An important precondition for use of the standardised pavements is adequate loadbearing capacity at formation level. In view of the requirements defined in the applicable guideline RVS 8.24 (2), the deformation modulus E_{V1} (plate-bearing test) must be a minimum of 35 N/mm^2. If the modulus is not high enough, suitable action (stabilisation, soil exchange) must be taken to ensure the required minimum bearing capacity. If on the other hand the formation is reliably proven to have substantially higher bearing capacity, use of a standardised pavement would in fact result in overdesign. In such a case one may consider departing from the standardised design and opting instead for customised (special) design taking into account the higher bearing capacity of the formation.

Fig.1: Design table for flexible and semi-rigid pavements, RVS 3.63 (2)

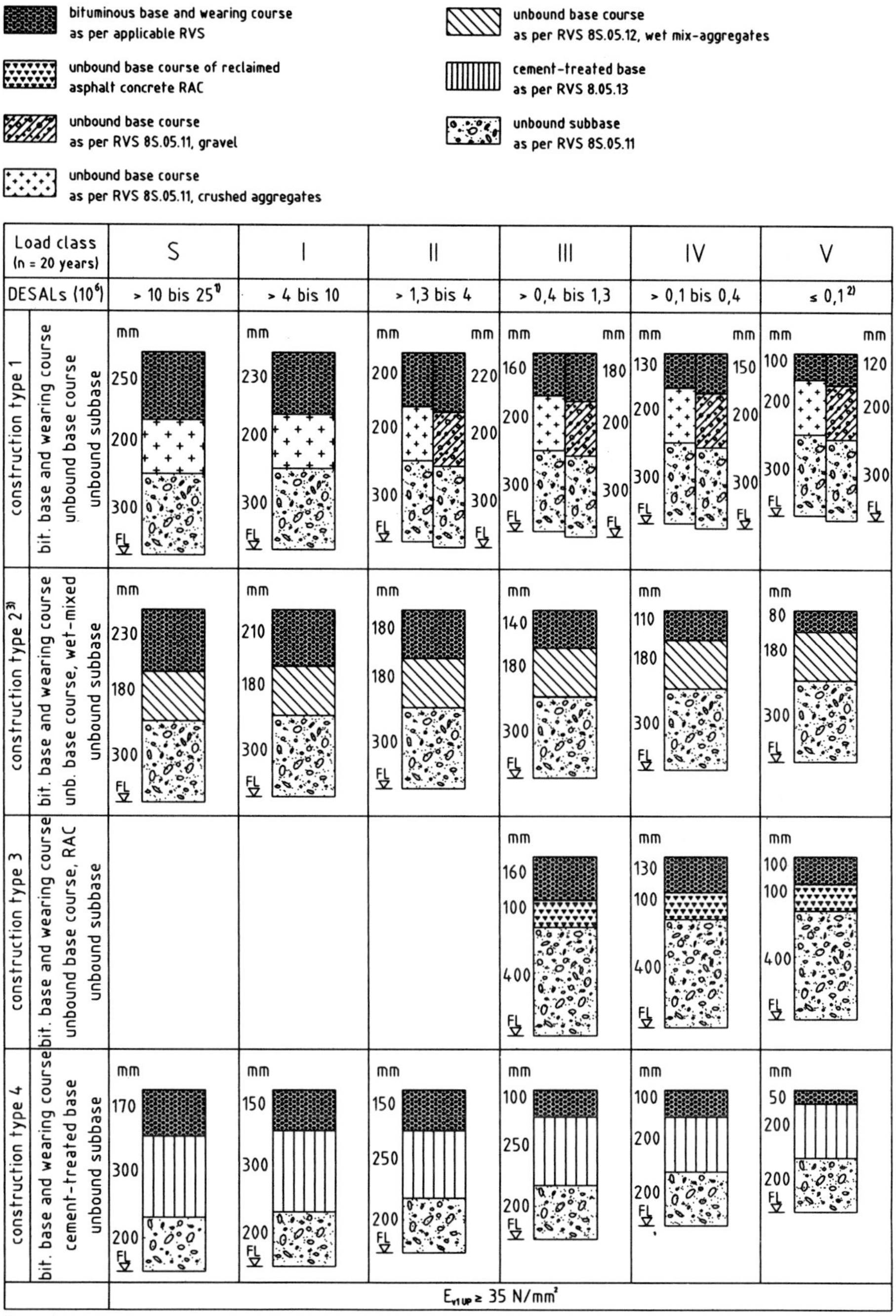

Comments:

1) In case of higher traffic volumes a customised structural design is required.

2) Further differentiation is possible for secondary local and rural roads.

3) Thickness specifications for asphalt package valid only, if $E_{v1} \geq 120$ N/mm² is reached on wet mixed unbound base course. Otherwise the asphalt thickness of construction type 1 shall be used.

Fig.2: Design table for concrete pavements, RVS 3.63 (2)

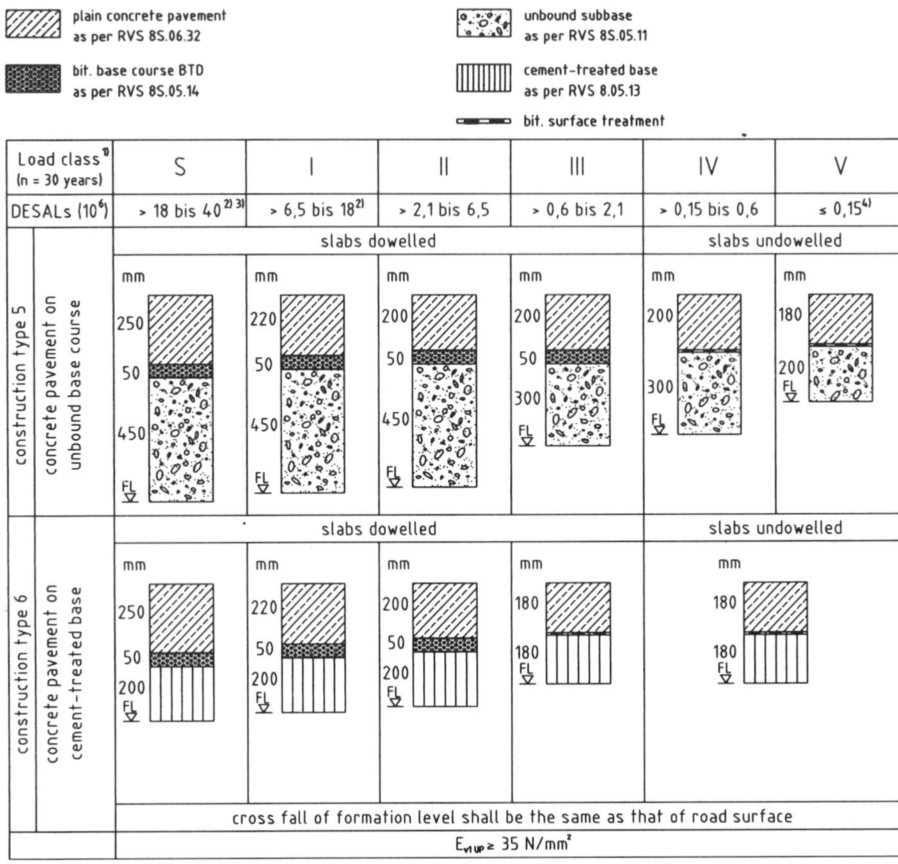

Comments (for slab geometry also see RVS 8S.06.32):

1) In the presence of permanent traffic close to the slab edge, frequent crossing of the free slab edge and numerous manholes in the slab concrete thickness shall meet the specifications of the next higher load class.

2) If traffic load is higher than 10 Million DESALs, maximum slab length shall not exceed 25 times the slab thickness.

3) If slab length is limited to 6 m, a design load of up to 70 Million DESALs is allowed. For higher design loads a customised structural design shall be used.

4) Further differentiation is possible for secondary local and rural roads.

3. DETERMINATION OF DESIGN TRAFFIC VOLUME

3.1 The principles

One of the most significant changes to previous procedures is the derivation of new equivalency factors for translation of the heavy vehicles universe to Design ESALs (equivalent single axle loads) based on the 100 kN standard axle load. Neither the 4th power law nor AASHO axle load equivalency factors were employed. The calculations were made using data material available from a previous study (3). Extrapolations performed in this study yielded detailed information on the shares of different types of vehicles in the heavy traffic, the distribution of axle loads and tyre types with estimated tyre pressures for the entire federal road network in Austria. These data were used to compute new vehicle equivalency factors for different universes of heavy vehicles and different heavy vehicle types by relating these to equivalent fatigue damage caused in the bound pavement courses (4, 5, 6) (also see Section 4).

Another change involved revision of the factor used to take into account lateral wheel path distribution. This was based on the results of a comprehensive study of heavy vehicle track behaviour as determined by lane width, speed and rutting (7, 8). These data were used to compute the shift factors incorporated in Table 1.

The substantial increase in heavy traffic also prompted introduction of the new load class S. When the old specification entered into force in 1986, only 0.3% of the entire Austrian road network was exposed to a traffic load above the limit specified for load class I. In 1995, this upper limit was already being exceeded on more than 20% of Austrian motorways. The upper limit of the newly introduced load class S now covers 99% of motorways and all federal highways.

3.2 Practical implementation

Traffic data on heavy traffic (annual average daily traffic of commercial vehicles - $AADT_{cv}$) are available for Austria's entire federal road network and large parts of the provincial road networks. In most cases, detailed data are also available for different types of heavy vehicles (trucks, trucks with trailers or semitrailers, busses). Routine application of RVS 3.63 is therefore based on these data.

If, in some exceptional cases, precise data are available on the existing range of axle loads, the specification recommends use of customised analytical pavement design.

Generally the relevant traffic load - expressed in Design ESALs (= DESALs) - is determined by means of the following formula:

$$\mathbf{DESAL = ESAL_{day} \,.\, R.V.S.\, 365.\, n.z}$$

where

$ESAL_{day}$ average daily standard load applications (passages of the 100 kN standard axle load) across the entire cross-section at the time the road is opened to traffic

These are calculated as follows:

- if the annual average daily traffic $AADT_i$ of vehicle type i across the entire cross-section is known at the time the road is opened to traffic:

$$ESAL_{day} = \sum_i (AADT_i . A_i)$$

where A_i = the mean equivalency factor of vehicle category i as shown in Table 1;

- if data from traffic counts do not differentiate between vehicles of different vehicle categories:

$$ESAL_{day} = AADT_{cv} \cdot A_{cv}$$

where

$AADT_{CV}$ annual average daily traffic of commercial vehicles (trucks, etc. and busses within a 24-hour period, all days) across the entire cross-section at the time the road is opened to traffic; determined from counts or on the basis of estimates of the AADT level and the share of heavy traffic

A_{cv} mean equivalency factor of the $AADT_{cv}$ universe for the road category concerned as shown in Table 2.

R Direction factor reflecting the distribution of heavy traffic across the two traffic directions (0.5 if heavy traffic is distributed evenly across the two directions).

V Factor taking into account the distribution of heavy traffic across several carriageway lanes 1.0 with 1 or 2 lanes
0.9 with 3 or more lanes

S Shift factor taking into account the distribution of vehicle tracks across a lane as shown in Table 3

n Design period in years (generally 20 years for bituminous pavements and 30 years for concrete pavements).

z Growth factor reflecting an annual growth rate p (%).
The growth factor is generally derived from the results of traffic counts or traffic forecasts. If no counts or forecasts are available, a mean annual growth rate of p = 3% is assumed for motorways and p = 1% for other roads.

Table 1: Average equivalency factors for different categories of heavy vehicles

Vehicle category i	A_i
Trucks	0.70
Trucks with trailer, semitrailer	1.20
Busses	0.60

Table 2: Average equivalency factors for $AADT_{CV}$-universe

Road category	A_{CV}
Motorways	1.00
Other roads	0.90

Table 3: Shift factor S

b_f	3.00 m	3.25 m	3.50 m	3.75 m	≥ 4.00 m
S	0.90	0.85	0.80	0.75	0.70

4. DESIGN METHOD EMPLOYED

As already explained, the specifications in the guideline for standardised pavements were derived by analytical calculations based on the elastic multilayer theory (bituminous pavements) or the slab theory (concrete pavements). For the different pavement courses, the following assumptions were used (see als (1)).

Formation level: For computation purposes, the static deformation modulus (E_{V1} = 35 N/mm^2) was transformed into a dynamic modulus of elasticity (E_{dyn} = 4.E_{V1} = 140 N/mm^2). Seasonal variations in subgrade strength were taken into account by dividing the year into four different periods with a decrease in the E_{dyn}-value by 50 % during the spring thaw period.

Unbound bases and subbases: Depending on the thickness of each layer and the material used (gravel, crushed material, etc.), its dynamic elasticity modulus was stated as a multiple of the modulus of the unbound supporting layer (factor 1.5 to 3.0).

Bituminous layers: The bituminous base course and the surface course were regarded as one layer. The dynamic elasticity modulus of this bituminous layer at a given temperature was derived from a diagram in the Shell Design Manual (9) as no representative values were available from laboratory tests. In order to take into account different temperature situations, the year was split into six significant periods. For each such period a further distinction was made between day and night. Thus a total of twelve situations were taken into account, differing in terms of the temperature of the bituminous layers and the bearing capacity at formation level. For calculation of the relevant stresses and strains an improved strength hypothesis for asphalt was applied (10, 11). Permissible load repetitions to fatigue damage were calculated using the fatigue law based on the VESYS program (12).

Cement-treated base course: Following measurements in the field and in the laboratory and the performance of backcalculations, the modulus of elasticity of these layers was stated as a constant 5,000 N/mm^2 under operating conditions, e.g. in the presence of micro cracks. The permissible load repetitions to fatigue damage were calculated by means of a fatigue law derived from backcalculations and comparisons with existing pavement structures (4).

Concrete layers: Calculations were based on Westergaards theory as modified by Cauwelaert (13), assuming an elasticity modulus of 30,000 N/mm^2. Permissible stresses or load repetitions were determined by using Smith´s diagram (14), with warping stress as a permanent stress. The periods during which stress from traffic loads is augmented by maximum warping stress was estimated as 5 % of the year. For the remaining 95 % of the time only traffic-related stresses were taken into account (14). For concrete pavements the year was thus divided into two different periods.

Stresses and strains in bituminous and semi-rigid pavements were calculated by means of the BISAR program (15), those occurring in concrete pavements by means of the SLAB-program (13). For representative traffic situations (based on data available for the whole Austrian federal road network) and the climatic conditions described calculations were made for different thicknesses of bound layers. The calculation procedure is shown in the flow chart in Fig. 3. From these results design curves were derived and used in developing a scale of pavement thicknesses for standardised pavement types suitable for the different load classes.

Figure 3: Methodology of analytical design calculation (4)

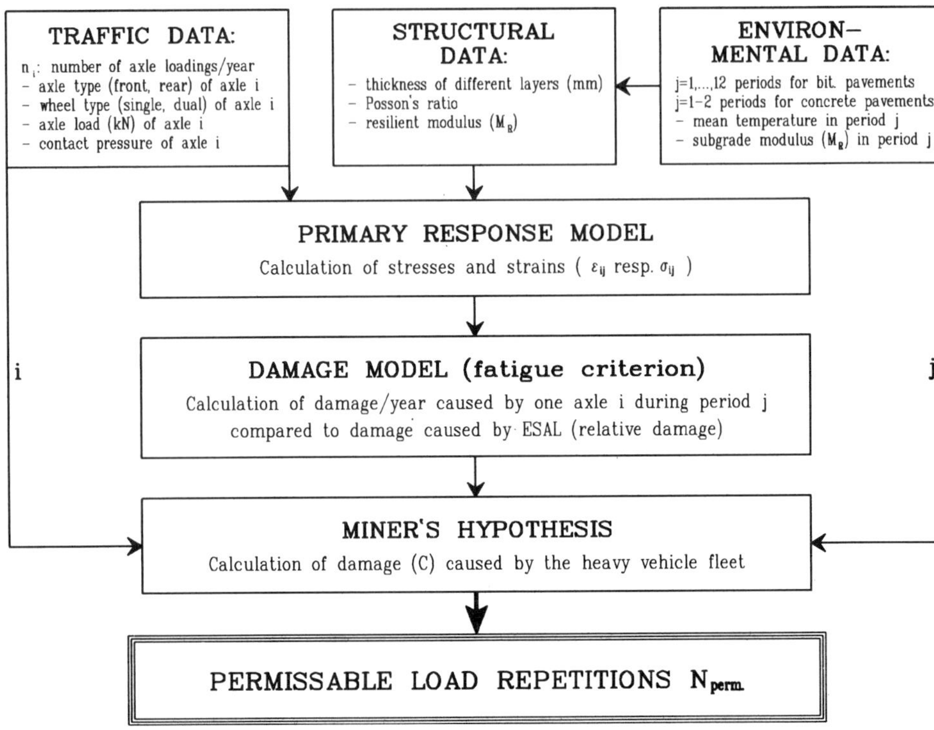

5. APPLICATION OF THE DESIGN STANDARD

In routine cases, the user only has to determine the design traffic volume as described in Section 3 and, thereby, the relevant load class. Then he can select the appropriate variant from among 6 technically equivalent standardised pavements (construction types) and, if desired, compare the economics of two or three eligible variants. If in a specific case complete information is available on the full range of axle loads, special design calculations may be carried out as described in Section 4.

For a correct application of the specified standardised pavements, however, the following additional conditions must also be observed:

- The thicknesses specified for bound pavement courses apply only in the case of (completely) new construction, not in the case of stage by stage construction. In the latter case, material fatigue occurring during the first loading phase must be taken into account in designing the second construction stage.

- The fundamental conditions regarding the strength of the subgrade and the unbound courses must be strictly complied with.
- Breakdown of the asphalt package in construction types 1 to 4 into a bituminous base course and a surface course must be carried out in conformity with the thickness specifications contained in the relevant asphalt specifications (2).
- Special attention must be paid to ensuring full bonding of pavement courses in bituminous structures as calculations are based on the assumption of one compact asphalt package.
- On sections exposed to slow-moving heavy traffic (before intersections, on uphill sections) the material of the asphalt courses must conform to special requirements (deformation resistance).

It is sometimes argued that standardised pavement design is uneconomical. This may occasionally be true but it must be noted that the thickness increments selected (30 - 40 mm) are quite adequate in practice. Design reliability depends in any case greatly on the accuracy of the traffic count or the traffic projection. If actual conditions differ from the general assumptions underlying the standard, special design calculations must be made.

ACKNOWLEDGEMENTS

The authors wish to thank the Austrian Federal Ministry for Economic Affairs for sponsoring this project and the members of the working group "Pavement Design" within the Austrian Association of Transportation Engineers, FVS, for their fruitful discussions and inputs.

REFERENCES

1. LITZKA J. and HERBST G.: A new specification for the structural design of pavements in Austria. Proceedings International Conference on Bearing Capacity of Roads and Airfields, Plymouth, 1986

2. RVS, Richtlinien und Vorschriften für den Straßenbau (Guidelines and Specifications for Road Construction). Austrian Research Association for Transportation Engineering, Vienna, Austria.

3. MOLZER C., FUSSEIS W., LITZKA J. and STEIERWALD G.: Auswirkung von Achslasterhöhungen auf das Bundesstraßennetz (Effects of increased axle loads on the federal road network). BMwA, Schriftenreihe Straßenforschung, H. 450, Wien, 1995.

4. LITZKA J., MOLZER C. and BLAB R.: Modifikation der österreichischen Methode zur Dimensionierung des Straßenoberbaues (Modification of the Austrian method for pavement design). BMwA, Schriftenreihe Straßenforschung, H. 465, Wien, 1996.

5. MOLZER C. and BLAB R.: Untersuchungen zur Einführung einer neuen Lastklasse für höchstbelastete Straßen in der österreichischen Bemessungsrichtlinie (Investigations for the application of a new load class for highly traffic roads in the Austrian Design Guide). Straße und Verkehr, Heft 7, 1996.

6. BLAB R., MOLZER C. and LITZKA J.: Modification of the Austrian guideline for standardised Asphalt Pavements, Proceedings of the 8th ISAP-Conference, Seattle, 1997.

7. BLAB R. and LITZKA J.: Measurement of the Lateral Distribution of Heavy Vehicles and its Effects on the Design of Road Pavements. 4th Intern. Symposium on Heavy Vehicle Weights and Dimensions, Ann Arbor, Michigan, 1996.

8. BLAB R.: Die Fahrspurverteilung als Einflussgröße bei der Bemessung des Straßenoberbaues (Lateral wheel path distribution as influence parameter for pavement design). Dissertation an der Technischen Universität Wien, Fakultät für Bauingenieurwesen, Ph.D.Thesis, Wien, 1995.

9. SHELL Pavement Design Manual: Asphalt Pavements and Overlays for Road Traffic. Shell International Petroleum Comp. Ltd., London, 1978.

10. HAGEMANN R.: Ein Verfahren zur Beurteilung flexibler Fahrbahnbefestigungen unter Berücksichtigung von Festigkeitshypothesen für Asphalte (An assessment procedure for flexible pavements applying strength hypotheses). Dissertation an der Universität Hannover, Ph.D.Thesis, Hannover, 1980.

11. MOLZER C.: Modifikation der österreichischen Bemessungsmethode für Oberbauten mit bituminösen Befestigungen (Modification of the Austrian design method for flexible pavements), Dissertation an der Technischen Universität Wien, Fakultät für Bauingenieurwesen, Ph.D.Thesis, Wien, 1995.

12. KENIS W.J., SHERWOOD J.A. and MC MAHON T.F.: Verification and application of the VESYS structural subsystem. 5th International Conference on the Structural Design of Asphalt Pavements, Delft, 1982.

13. CAUWELAERT F. VAN: A rigorous analytical solution of a concrete slab submitted to interior and edge loads with no, partial and full shear transfer at the edge. Proceedings of the 5th International Conference on Concrete Pavement Design and Rehabilitation, Vol. 1, Purdue Universtiy, Indiana USA; 1993.

14. EISENMANN J.: Betonfahrbahnen (Concrete pavements) Handbuch für Beton-, Stahlbeton und Spannbetonbau; Entwurf, Berechnung, Ausführung. Verlag Wilhelm Ernst u. Sohn, Berlin, München, Düsseldorf, 1979.

15. DE JONG D.L., PEUTZ M.G.F. and KORSWAGEN A.R.: External report computer program BISAR, Layered systems under normal and tangential surface loads. Koninklijke/Shell - Laboratorium Amsterdam, Shell Research B.V., 1979.

Fifth International Conference on the
BEARING CAPASITY OF
ROADS AND AIRFIELDS

VARIABILITY OF MECHANISTIC-EMPIRICAL FLEXIBLE PAVEMENT DESIGN PARAMETERS

DAVID TIMM[1], BJORN BIRGISSON[2] AND DAVID NEWCOMB[3]

[1]Research Fellow, [2]Research Associate, [3]Associate Professor
University of Minnesota, Department of Civil Engineering
500 Pillsbury Dr SE, Minneapolis, MN 55455, U.S.A.
Tel: (612) 626-1749 Fax: (612) 626-7750

ABSTRACT

As mechanistic-empirical (M-E) flexible pavement design procedures gain greater acceptance, there is a need to investigate the design input parameters and assess their influence on the predicted life of the pavement structure. Specifically, the incorporation of reliability into the design procedure requires the determination of the variability of the associated design input parameters. Reliability-based design is advantageous since it allows for a rational accounting of the statistical variability in the design parameters, may aid in the optimization of resources and provides a common basis when comparing different designs. As a first step toward reliability-based M-E design, the variability associated with the primary design inputs were investigated and quantified. The inputs included: layer modulus, layer thickness, Poisson's ratio, tire configuration, axle load, and tire pressure. The sources of information used to statistically characterize each of the variables included data from the existing literature and data from the Minnesota Road Research Project. Monte Carlo simulation, a method of artificially generating and combining each of the distributions, was used to demonstrate the effect of variability on design reliability. It was found that reducing the input variability (i.e., better quality control) can lead directly to improved design reliability for the same mean design input values.
Keywords: flexible pavement design, reliability, Monte Carlo, input variability

5. HADLEY, W.O., IRICK, P. and ANDERSON, V. Materials and Construction Variability Based on SHRP-LTPP Data. Proceedings, 4th International Conference on the Bearing Capacity of Roads and Airfields, Vol. 2., 1994, pp. 863 - 882.
6. ALLEN, D.L. and GRAVES, R.C. Variability in Measurement of In-Situ Material Properties. Proceedings, 4th International Conference on the Bearing Capacity of Roads and Airfields, Vol. 2., 1994, pp. 989 - 1005.
7. NOURELDIN, A. S. Influence of Stress Levels and Seasonal Variations on In Situ Pavement Layer Properties. Transportation Research Record No. 1448, Washington, D.C., 1994, pp. 16-24.
8. NEWCOMB, D.E., CHADBOURN, B.A., VAN DEUSEN, D.A. and BURNHAM T.R. Initial Characterization of Subgrade Soils and Granular Base Materials at the Minnesota Road Research Project. Final Report, Report No. MN/RC – 96/19, Minnesota Department of Transportation, December, 1995.
9. RADA, G. and WITCZAK M.W. Comprehensive Evaluation of Laboratory Resilient Moduli Results for Granular Material. Transportation Research Record No. 810, Washington, D.C., 1981, pp. 23-33.
10. NOURELDIN, A.S., SHARAF, E., ARAFAH, A. and AL-SUGAIR, F. Estimation of Standard Deviation of Predicted Performance of Flexible Pavements Using AASHTO Model. Transportation Research Record No. 1449, Washington, D.C., 1994, pp. 46-56.
11. LUKANEN, E.O. Pilot Program for Evaluation of Structural Adequacy of Flexible Pavements for Counties and Municipalities. Investigation No. 650, Minnesota Department of Transportation, 1980.
12. YODER, E. and WITCZAK, M.W. <u>Principles of Pavement Design</u>, 2nd Edition, John Wiley & Sons, Inc., New York, 1975.
13. SHAH, N. An Investigation of Overloads on Highway Bridges. MS Thesis, Illinois Institute of Technology, Chicago, IL, 1989.
14. LIU, W. D., CORNELL, C.A. and R.A. IMBSEN. Analysis of Bridge Truck Loads, in <u>Probabilistic Methods in Civil Engineering</u>, Edited by P.D. Spanos, ASCE, 1988, pp. 221-224.
15. CHIA-PEI, J.C. and CHING, C. Truck Load Distribution and Its Impact on Vehicle Weight Regulations in Taiwan. Transportation Research Record No. 1501, Washington, D.C., 1995, pp. 87-94.
16. KIM, O. and BELL, C.A. Measurement and Analysis of Truck Tire Pressure in Oregon. Transportation Research Record No. 1207, Washington, D.C., 1988, pp. 100-110.
17. MIDDLETON, D.R., ROBERTS, F.L. and Chira-Chavala, T. Measurement and Analysis of Truck Tire Pressures on Texas Highways. Transportation Research Record No. 1070, Washington, D.C., 1988, pp.1-8.
18. GERRARD, C.M. and HARRISON, W.J. A Theoretical Comparison of the Effects of Dual-Tandem and Dual-Wheel Assemblies on Pavements. Proceedings, Fifth Conference, Australian Road Research Board, 1970.
19. KULKARNI, R.B. Rational Approach in Applying Reliability Theory to Pavement Structural Design. Transportation Research Record No. 1449, Washington, D.C., 1994, pp. 13-17.
20. DEVORE, J.L. <u>Probability and Statistics for Engineering and the Sciences, Fourth Edition</u>, International Thomson Publishing Inc., 1995.

Fifth International Conference on the
BEARING CAPASITY OF
ROADS AND AIRFIELDS

ESTIMATED FACTORS OF SAFETY OF THE AASHO ROAD TEST FLEXIBLE PAVEMENT SECTIONS BASED ON LIMITING EQUILIBRIUM METHODS

TOMMY C. HOPKINS
Head of Geotechnology

MIKHAIL E. SLEPAK
Senior Research Engineer

Kentucky Transportation Center, College of Engineering
University of Kentucky, Lexington, Kentucky, USA 40506-0281

ABSTRACT

During 1959 and 1960 near Ottawa, Illinois, the American Association of State Highway Officials (AASHO) conducted an extensive roadway test involving many flexible pavement sections of various thicknesses. Trucks of different axle configurations and loads were driven over the roadways for several months and pavement conditions were noted periodically. Although this road test has been analyzed in various ways, a different type of analysis is presented. A generalized, mathematical stability model developed by Hopkins in 1991 was used to calculate the factors of safety against failure of 215 flexible pavement sections of the road test and to examine the viability of using limiting equilibrium approaches for analyzing the bearing capacity of flexible pavements. A Prandtl-type failure surface is assumed in the analyses. Shear strength parameters of the paving materials were estimated based on testing of similar materials and values reported in published sources. Factors of safety of all sections ranged from 0.63 to 1.81. Factors of safety were very small for sections that deteriorated rapidly, or had short pavement lives. Average factors of safety of 53 percent of the sections (present serviceability index, P_t, equal 2) ranged from 0.95 to 1.14; values of weighted, 18-kip (8,172 kg), equivalent single-axle loads (ESAL) were less than 0.5 million. Average factors of safety of 13.5 percent of the sections, which survived the road test ($Pt > 2$), ranged from 1.51 to 1.80 while (ESAL) applications exceeded 8 million. Average factors of safety of about 33.5 percent of the sections ranged from 1.3 to 1.4; ESAL values ranged from 0.5 to 8 million.

INTRODUCTION

The problem of tire-flexible pavement-subgrade interaction has traditionally been treated as a problem of a multilayered linear elastic media (1). However, depending on the tire pressure, thickness and temperature of the asphalt pavement, and subgrade strength properties, tire-

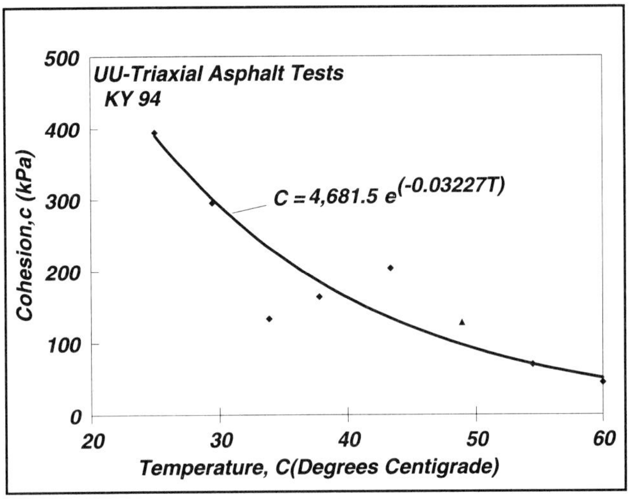

Figure 4. ***Variation of cohesion with temperature***

value of ϕ increases from a value of about 26 to 44 degrees. The value of c decreases from about 386 kPa (56 psi) to 51 kPa (7.4 psi).

BEARING CAPACITY ANALYSES

Example of Typical Data Set-Up and Types of Analysis. Analyses of the flexible pavement sections of loops 3, 4, 5, and 6 (lanes 1 and 2) of the AASHO Road Test (6) were made using two different of pavement temperature assumptions. In the first analysis, the surface temperature (T_s) and the average air temperature (T_a) were assumed to be 60° and 27.3° C, respectively. Surface temperatures of pavements in this part of the country can reach values of 60° C. To account for temperature variation (and variation of shear strength) with depth in the asphalt pavement, relationships given by Southgate, et al (11) were used. For assumed surface and air temperatures of 60° and 27.3° C, the temperature, T, for any depth, D (in cm) may be determined from their relationships. For this particular case, the relationship is (in cm)

$$T = 58.1 - 1.825D + 0.0334D^2. \qquad (4)$$

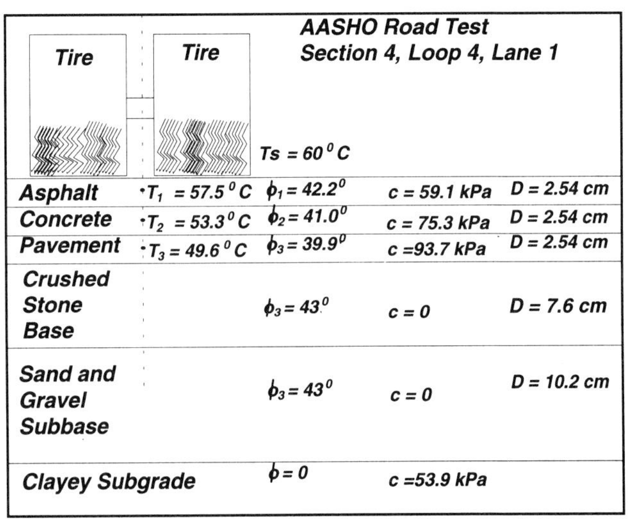

Figure 5. ***Typical data set-up***

Each pavement section was divided into 2.54-cm layers (Figure 5, pavement section 599, loop 4, lane 1). The temperature at each mid-point of the layer was calculated from Equation 4. Using equations given in Figures 3 and 4 and the temperature of each asphalt concrete layer, shear strength parameters, ϕ and c, of each finite layer were determined. A second set of analyses was performed using a surface temperature of 25° C. Temperatures were assumed to be 25° C throughout the full depths of the asphalt concrete pavements.

Results of Bearing Capacity Analyses. In Figure 6, the factors of safety of the pavement sections (T_s equal 60° C) are shown as a function of weighted, 18-kip, equivalent single-axle loads (ESAL). Factors of safety of some 29 pavement sections of a total of 215 sections are not shown in this figure because the serviceability index never reached a value of 2.0 during the road test. The factors of safety of those 29 sections ranged from 1.19 to 2.00 and averaged 1.51. ESAL values were greater than eight million. Factors of safety of all sections ranged from 0.63 to 1.8. As shown in Figure 6, the slope of the factor of safety-ESAL curve increases rapidly up to a factor of safety of about 1.3 and an ESAL value of one million. When the factor of safety is greater than 1.3, or the ESAL value is greater than about one million, the slope of the curve tends to decrease, become flatten, and approach a safety factor of about 1.5. At that stage, the ESAL value approaches eight million. When the factor of safety exceeds a value of 1.5, the ESAL values are greater than eight million.

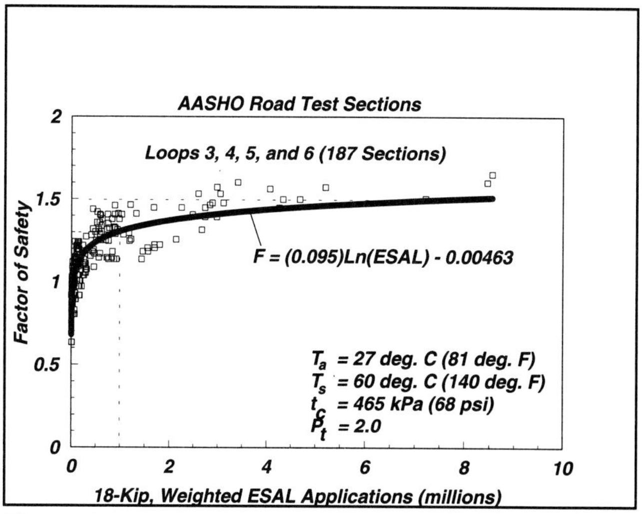

Figure 6. ***Factor of safety as a function of 18-kip, weighted ESAL applications***

The curve in Figure 6 was based on a surface temperature of 60° C and an average air temperature of 27.3 ° C. As a means of examining the effect of temperature on the factor of safety (Figure 7), the analyses were repeated using a surface temperature of 25° C and assuming that the temperature of the asphalt pavement of each section was equal to a constant value of 25°C throughout. Decreasing the surface temperature

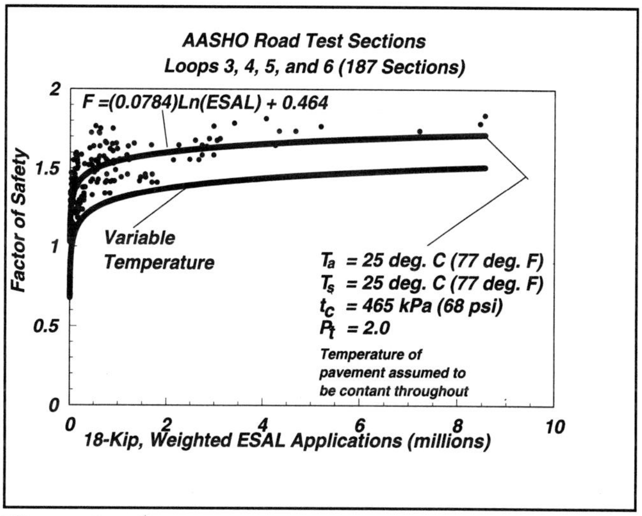

Figure 7. ***Comparison of ESAL-factor of safety curves based on different asphaltic pavement temperatures***

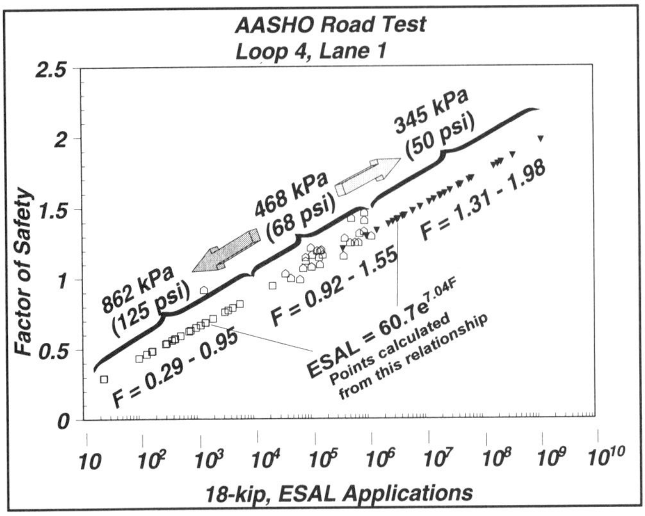

Figure 8. ***Influence of the magnitude of tire stress***

of the asphalt pavement from a value of 60° to 25° C causes a slight increase in the factor of safety. For ESAL values ranging from 1 to eight million, the differences in factors of safety of the two curves range from 0.23 to 0.20, as shown in Figure 7.

Data shown in Figure 6 may be illustrated in another form. Values of ESAL were sorted in ascending order, grouped (arbitrarily), and the average factor of safety of each group was computed. As the average factor of safety increases, the value of ESAL for each group increases, as shown in Table 1. The average factor of safety of 53 percent of all sections ranged from only 0.95 to 1.14 and ESAL values were less than only 0.5 million (P_t equal 2.0). The serviceability index of some 13.5 percent of the sections did reach a value of 2.0 during the duration of the road test and the ESAL values exceeded 8 million. The average factor of safety of this group generally exceeded 1.5.

Table 1. Average factor of safety of ESAL groups

ESAL Group (P_t = 2.0)	Average Factor of Safety, F	% Total Pavement Sections
<50,000	0.95	14.9
50,000-200,000	1.08	26.0
200,000-500,000	1.14	12.1
500,000-1 million	1.30	17.2
1 million-8 million	1.40	16.3
>8 million	1.51	13.5

Influence of the magnitude of contact tire stresses. Bearing capacity analyses were also performed to examine the effect and significance of varying the unit tire contact stress on the factor of safety and value of ESAL. Analyses were performed on sections of Lane 1 of Loop 4. As shown in Figure 8, when a unit tire stress of 468 kPa is used, the factors of safety of the sections range from 0.92 to 1.55. Analyzing the same sections of Lane 1 of Loop 4, but using a unit tire stress of 862 kPa, factors of safety ranging from 0.28 to 0.95 are obtained. These values are much lower than those obtained when a unit tire contact of 468 kPa is used. To compare ESAL values for the two different unit tire contact stresses, a regression analysis was performed using all loop sections and where the factor of safety is treated as the dependent variable and the parameter, ESAL, was treated as the independent variable. The equation relating ESAL and factor of safety is shown in Figure 8. For the unit contact stress of 468 kPa, the factor of safety ranged from 0.92 to 1.55 while the values of ESAL ranged from 39,361 to 3.3 million (Figure 7). When a unit tire contact stress of 862 kPa is used, factors of safety ranging from 0.29 to 0.95 are obtained; ESAL values only range from 75 to 48,613. Hence, considerable reduction in values of ESAL , or pavement life occurs. However, when the unit tire contact stress is decreased from 468 to 350 kPa, factors of safety of the sections increase and range from 1.31 to 1.98. Pavement life increases dramatically, or the ESAL values increase to a range of 612,337 to 68.3 million. These data are summarized in Table 2.

Table 2. Influence of tire contact stress on pavement life

Unit Tire Contact Stress (kPa)	Factor of Safety, F	$ESAL = 60.7e^{7.0372}$
350	1.31 — 1.98	0.61 - 68.3 million
468	0.92 — 1.55	0.04 - 3.3 million
862	0.29 — 0.95	0.00008- 0.05 million

SUMMARY AND CONCLUSIONS

A relationship between weighted, 18-kip (8,172 kg) equivalent single-axle load applications and factor of safety was established by analyzing 215 flexible pavement sections of the AASHO Road Test (6). Results illustrate that reasonable values can be obtained from the limiting equilibrium approach. Factors of safety of all sections ranged from 0.63 to 1.81. Sections that had large pavement lives, or large values of ESAL, had large factors of safety. Where small factors of safety were calculated, pavement lives of those sections were small. The average factor of safety of 53 percent of all sections ranged from only 0.95 to 1.14 and ESAL values were less than only 0.5 million (P_t equal 2.0). Increasing the tire contact stress from 468 kPa to 862 kPa causes significant decreases in the values of ESAL that a flexible pavement may sustain. That is, the life of the pavement decreases significantly. Decreasing the tire contact stress from 468 kPa to 350 kPa causes an increase in the values of ESAL and therefore increases, significantly, the life of pavements. Results of the analysis indicate that design thicknesses obtained from the original AASHO Road Test design procedure yield correct results only when the design contact tire pressure is about 468 kPa--the value used in the test. If significantly larger tire contact stresses are anticipated, then larger thicknesses would be required than those obtained from the test equations. The limiting equilibrium approach could be used to check this situation.

ACKNOWLEDGMENTS

The research reported herein was funded by the Federal Highway Administration and the Kentucky Transportation Cabinet through the University of Kentucky Research Foundation.

The authors are deeply grateful to those sponsors.

REFERENCES

1. BURMISTER, D. M.; (1943), "The Theory of Stresses and Displacements in Layered Systems and Applications to the Design of Airport Runways," Highway Research Board Proceedings, Vol. 23.

2. ENGINEERING MANUAL (July 1951), Corps of Engineers, U.S. Army, Part XII, Chapter 2.

3. McLEOD, N. W.; (1953), "Some Basic Problems in Flexible - Pavement Design," Highway Research Board, Proceedings, Thirty-Second Annual Meeting.

4. PRANDTL; (1921), "Hautaufsatze Uber die Eindringungsfestigkrit (Harte) plasticher Baustoffe und die Festigkeit von Schneiden," Zaitschrift Fur Angewandte, Mathematik und Mechanik, Vol. 1,No. 1.

5. HOPKINS, T. C.; (June 1991); "Bearing Capacity Analyses of Pavements," Research Report KTC-91-8, University of Kentucky Transportation Center, College of Engineering.

6. The AASHO Road Test (May 1962), Proceedings, Special Report 73, Publication No. 1012, National Academy of Sciences - National Research Council, Washington, D.C.

7. HOPKINS, T. C.; (January 1986); "A Generalized Slope Stability Computer Program: User's Guide for HOPK-I," Research Report UKTRP-86-2, University of Kentucky Transportation Center, College of Engineering.

8. JANBU, N.; (1954), "Application of Composite Slip Surface for Stability Analysis, European Conference on Stability of Earth Slopes," European Conference on Stability of Earth Slopes, Stockholm, Sweden.

9. BISHOP, A.W.; (1955), "The Use of the Slip Circle in the Stability Analysis of Slopes," Geotechnique, Vol. 5.

10. MORGENSTERN, N. R. and PRICE, V. E.; (1965), "The Analysis of the Stability of General Slip Surfaces," Geotechnique, Vol. 15.

11. SOUTHGATE, H. F.; SHARPE, G. W.; DEEN, R. C.; and HAVENS, J. H.; (August 1982), "Structural Capacity of In-Place Asphaltic Pavements from Dynamic Deflections," Fifth International Conference on the Structural Design of Asphalt Pavements, Proceedings, Vol. 1, The Delft University of Technology, The Netherlands.

Fifth International Conference on the
BEARING CAPASITY OF
ROADS AND AIRFIELDS

PAVEMENT DESIGN FOR ULTRA-HEAVY MINE TRUCKS

ROGER J THOMPSON[1] AND ALEX T VISSER[2]

[1]Assoc Professor of Mining Engineering
University of Pretoria, Pretoria 0002
Republic of South Africa
Tel +12 420 3195, Fax +12 362 5173
E-mail r.thompson@eng.up.ac.za

[2]Professor of Civil Engineering
University of Pretoria, Pretoria 0002
Republic of South Africa
Tel +12 420 3168, Fax +12 362 5218
E-mail avisser@postino.up.ac.za

ABSTRACT

The structural design of an unsurfaced mine haul road relates to the ability of the road to carry the imposed loads without undue deformation which can lead to excessive maintenance or premature rehabilitation. Current surface mine haul road structural design techniques are empirical and based primarily on the previous experience of personnel assigned to pavement design. This has the potential for unwarranted expenditure for too thick a structure, or conversely, premature deformation leading to the need for excessive expenditure on maintenance. The aim of this paper is to present a mechanistic design method for ultra-heavy loads and to provide an insight into the material behaviour under such loads.

Pavement deflections generated by ultra-heavy trucks were studied on a number of opencast mine road sites in South Africa. Multi-depth deflectometers were installed in the various pavement layers encountered and pavement response data were analysed to back-calculate layer elastic moduli using multi-layer elastic theory. Typical modulus values were deduced for the various types of material commonly used in mine road construction, together with an indication of their tendency to exhibit stress-sensitivity. The current state of mine haul road structural design was assessed through the analysis and quantification of existing pavement structural performance and by comparative analysis, the associated limiting design criteria and optimal mechanistic design were deduced. Finally, the benefits of the proposed mechanistic design method were verified by comparison with existing haul road designs.

KEYWORDS: Haul road, structural, design, hauling, truck

INTRODUCTION

The South African coal mining industry produces in excess of 106 million tons of coal per annum from surface mines which require, inter alia, the transport of raw coal from the pit to the loading or transfer point. Ore, and to a lesser extent waste, are both transported by large haul trucks running on unpaved haul roads that have, at best, been empirically designed with little or no recognition of the consequences of inadequate design on cost per ton hauled, operational efficiency or safety.

In addition to the mechanical engineering challenges involved in the design and construction of large dump trucks, specific civil engineering problems arise, primarily as a result of the gross vehicle mass (GVM) and the resultant loading applied to the pavement. Taking the Caterpillar 789 as a typical example of a 170 ton capacity rear dump truck (single front and dual rear wheel axle), the GVM of 317,4 ton results in a maximum dual wheel single axle load of 2 086 kN and a contact radius of 0,513 m (using 37.00R57 tyres inflated to 640 kPa).

The operating performance of a pavement or road structure system can be subdivided into three distinct design categories; structural, functional and maintenance scheduling and management. When applied to the design of mine haul roads, each of these components is addressed individually and then combined to produce specific and individual haul road designs based on a comprehensive general design strategy, applicable to most surface mining operations. The structural design component of the total strategy is addressed in this paper.

Aim and Scope of Paper

The aim of this paper is to present a mechanistic design method for ultra-heavy loads and to provide an insight into the material behaviour under such loads. The current state of haul road structural design is briefly introduced and through an analysis of the independent variables predominantly controlling the structural performance of a haul road, the experimental design is described.

Twenty seven independent variable combinations were assessed at nine mine test sites to enable a valid comparison to be made between existing and proposed haul road designs. Existing designs at each site were assessed as a basis for comparison with results from the mechanistic analysis. The derivation of limiting design criteria and the optimal equivalent mechanistic design is described. Finally, a comparative analysis was conducted to verify both the utility and associated construction cost benefits of the proposed mechanistic haul road design technique.

CURRENT STATE OF SURFACE MINE HAUL ROAD STRUCTURAL DESIGN

The findings of a United States Bureau of Mines (USBM) study[1] were that haul road structural design techniques were entirely empirical, mine operators often electing to forego the use of sub-base materials and accept infringements on mobility in the interest of economics. The USBM work was one of the first annotated applications of the California

Bearing Ratio (CBR) cover thickness design technique for mine haul roads.

CBR Cover Curve Structural Design Technique

The CBR method[2] has been widely applied to the design of surface mine haul roads in which untreated materials are used. The CBR method was based on empirical results relating to the design of asphalt-surfaced airfield pavements for wheel-gear loads up to 4 400kN. When aggregate-surfaced mine haul roads are considered in conjunction with stabilised bases, albeit at similar load levels, the same approach is of questionable validity.

South African Mechanistic Design Method

The South African Mechanistic Design Method (SA-MDM) is based on a theoretical model of pavement layer response parameters (stresses, strains and deflections in each layer) by assuming a linear-elastic multi-layer system. Empirically derived limiting design criteria are then used with which to assess the pavement under the specific loading conditions, thereby determining the level of service and in turn, the time at which some maintenance or rehabilitation would be required. The method has been applied mostly to the design of asphalt pavements in the public domain (freeways, collector and access roads) subject to equivalant 80kN axle loads[3].

Little published data exists concerning design criteria for mine haul roads. The approach adopted was therefore to identify those damage parameters applied in the design of roads subject to standard axle loadings and by means of categorising haul road test section structural performance, deduce acceptable limiting design criteria for haul road structural design. The development of empirical limiting fatigue or distress criteria applicable to mine haul road structural design can then be used to evaluate the structural performance of a pavement and, if necessary, to evaluate the efficacy of corrective measures or new structural designs for much heavier haul trucks.

ANALYSIS AND QUANTIFICATION OF EXISTING PAVEMENTS

Experimental design

In order to determine the suitability of the existing haul road structural design techniques, existing mine pavements were assessed in terms of a number of independent variables (factors) which were recognised as predominantly controlling the structural performance of a haul road. These factors were:

(i) Applied load/stress

(ii) Subgrade strength
(iii) Structural thickness and layer strengths

To fully characterise the structural performance of existing haul road designs, each factor combination was analysed at various levels using a designed factorial experiment. Nine test sites were identified at three mines which enabled 27 factor combinations to be assessed. The utility of each design was assessed over a range of factor levels thus enabling a comparative analysis to be completed for all combinations of haul truck load, construction material, structural thickness and layer strengths.

As a precursor to the analysis of the structural performance of haul roads and the derivation of limiting design criteria, a classification of haul road structural performance was conducted to indicate in broad terms the adequacy of the various designs encountered at each mine test site. This was achieved by categorising the structural performance of each site, together with a short summary of the structural defects observed or reported by mine personnel. In addition, the maximum deflection recorded by the MDDs in the structure is depicted for each site as an aid to classification. Table 1 summarises this data. The in-situ CBR was measured by means of a Dynamic Cone Penetrometer (DCP)[4].

CBR Analysis

Results relating the required cover thickness and CBR value for a particular vehicle and the actual (as constructed) cover were compiled for each test site and are summarised in Table 1. Results ranged from under-design of the road (inadequate cover) to probable over-design (excessive cover). In the case of indicated under-design, a correlation was seen between poor structural performance and inadequate pavement layer CBR values. Recourse to mechanistic analysis is therefore indicated to quantify the extent of any over- or under-design recognised from the CBR approach since the latter technique cannot assess each layer independently in terms of its load bearing capability.

Mechanistic Analysis

An array of six linear voltage differential transducers (LVDTs), anchored at depths of 5 to 9 m, were used in this study to determine individual pavement layer deflections resulting from an applied load[5]. Typical Multi-Depth Deflectometer (MDD) generated pavement maximum deflection profiles are presented in Figure 1 in which it is seen that no load induced deflections were recorded below 3,2m depth, assuming no deflection of the anchor occurred. This fact enables an "equivalent stiff layer" to be included in the multi-layer elastic model of the pavement below this depth.

The solution technique adopted with the multi-layer elastic analysis program ELSYM5A[6]

TABLE 1 Summary of Structural Analysis Results for Performance Classification, CBR and Mechanistic Evaluation.

MINE	SITE	LAYERS	PERFORMANCE CLASSIFICATION			CBR EVALUATION	MECHANISTIC EVALUATION			
			Structural defect description	Performance index Poor = 0 Excellent = 10	Max deflection (mm) Dual wheel load 2x430kN	(Based on cover for largest truck operated)	Layer	Max vert strain (microstrain) For dual wheel loads of 2x430kN	Safety factor layer 1	Safety factor layer2
1	1	4	Deformation and rutting on a large scale, poor structural performance	1	8,7	Layer 3 approaching minimum cover requirements	2 3 4	1950 5450 4300	5,0	2,9
	2	4	Deformation and rutting, poor structural performance	2.5	6,4	Road exceeds minimum cover requirements	2 3 4	8400 Rock layer 2750	3,1	5,2
	3	4	Excellent structural performance	8	1,6	Road exceeds minimum cover requirements	2 3 4	Stabilised 1350 900	1,7	No data
2	1	4	Localised deformation and rutting, good structural performance	6	2,5	Road exceeds minimum cover requirements	2 3 4	950 2600 1200	9,0	4,3
	2	4	Localised excessive rutting, reasonable structural performance	4	2,9	Layers 3 and 4 less than minimum cover requirements	2 3 4	1350 3000 1500	11,9	6,3
	3	3	Localised deformation and rutting, good structural performance	6	2,3	Layer 3 less than minimum cover	2 3	1400 1900	9,1	6,1
3	1	4	Excellent structural performance	9	No data	Road exceeds minimum cover requirements	No data	No data	No data	No data
	2	3	Deformation and rutting, good structural performance	6	3,4	Road exceeds minimum cover	2 3	1800 1050	6,3	8,7
	3	3	Reasonable structural performance, rutting	4	4,6	Layer 3 approaching minimum cover	2 3	4650 950	6,3	8,5

program involved the manual iteration from some assumed seed moduli for each layer until calculated deflections matched those measured by the LVDTs. The goodness of fit was determined for a particular LVDT (usually located centrally within the layer) to within 2% of the actual recorded deflection. Figure 1 illustrates the conformity achieved between actual and modelled deflections for a particular mine road test site.

Since no mechanistic limiting design criteria exist relating to ultra-heavy axle loads on mine haul roads, two design criteria were identified from conventional road design.

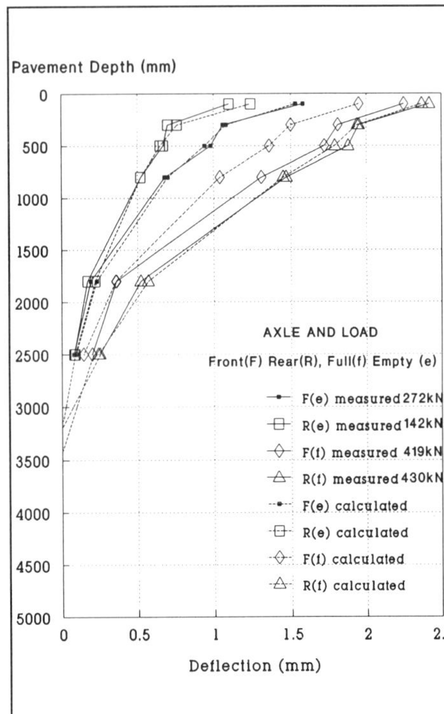

Figure 1 Typical haul truck front and rear (laden and unladen) actual and back-calculated deflection profiles

Elastic Vertical (Compressive) Strain

Since much of the structural deterioration of a pavement is attributable to the stresses or strains developed in individual pavement layers these parameters are important when considering limiting design criteria: Vertical strains in the top of sub-base and sub-grade layers are associated with rutting and shear deformation whilst horizontal strains in the upper stabilised layers with cracking(7). Limitations are placed on the permissible compressive vertical elastic strains at the top of subgrade and selected layers to prevent rutting and subsequent deformation of the road surface. The design criteria adopted for the layers below the wearing course was that of vertical compressive strain.

Results of Analysis - Vertical Strain Design Criterion

Figure 2 illustrates the vertical strains recorded for a 2x430kN dual wheel load in each pavement layer for the three typical test sites. By comparing these data with the performance classification it was found that the vertical strain criterion correlates well with performance of the road; those sites exhibiting poor performance and an associated excessive deformation/maximum deflection were seen to be associated with large vertical compressive strain values in one or more layers. From analysis of the data it was found that an upper limit of 2 000 microstrain should be placed on layer vertical strain values for the typical load repetitons found on the mines. Strains exceeding this value have been shown to be

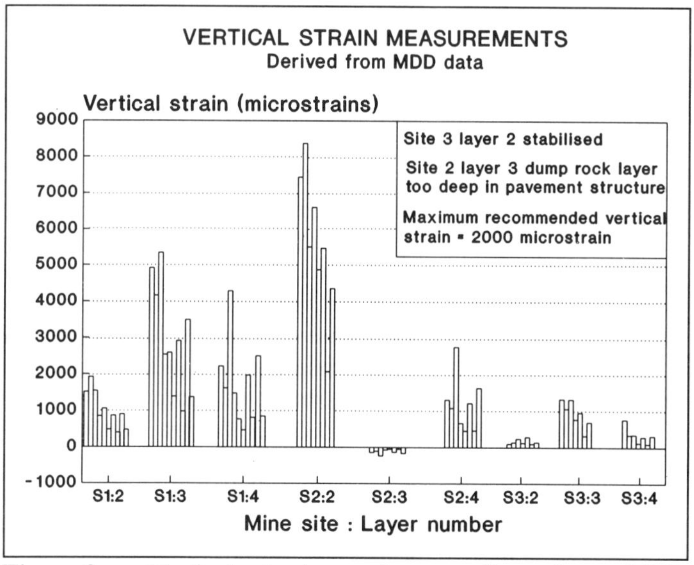

Figure 2 Vertical elastic strain recordings from MDD deflection data for various test sites under wheel loads of 430kN

associated with unacceptable structural performance.

If the CBR evaluation data is compared with the vertical strain data from the mechanistic analysis and the performance classification, it is seen that the outcome of the CBR technique is under-design of the pavement (the maximum allowable vertical strain criterion is exceeded) despite the actual cover exceeding predicted minimum requirements. Several of the pavement designs included rock or stabilised layers at various depths in the structure. Although vertical strains are much reduced, a deep rock layer does not appear to improve performance of the road as seen in Table 1 and Figure 2. When a 240mm thick stabilised layer was located higher in the structure it was seen that the road performed well and was not susceptible to the effects of high axle loads in the upper layers, primarily due to the load carrying capacity of the stabilised layer. This result had important implications in terms of the mechanistic structural design of mine haul roads.

Safety Factor

Granular materials exhibit distress through cumulative permanent deformation or inadequate stability. Both forms of distress are related to the ultimate shear strength of the material and to prohibit shear failure or excessive gradual shear deformation in the layer, the traffic generated shear stresses must be limited. Safety factors vary widely with depth within a layer, typically between compressive and tensile conditions. In this analysis, safety factors were calculated under single or at the centre of dual wheel assemblies at the mid-depth of the layer, following Maree[8].

Results of Analysis - Safety Factor Design Criterion

With regard to the FOS design criteria for the upper layers, since the applied stresses were

MECHANISTIC DESIGN CASE STUDY

The optimal mechanistic structural design of a surface mine haul road embodies the selection of target effective elastic modulus values for the construction materials available and the placement of those materials such as to optimise their performance both as individual layers and over the entire structure. Performance has been analysed in terms of minimum wearing course thickness and compaction and the limiting design criteria of vertical strain in the base, subbase and subgrade layers. In addition, of the various design options analysed at each mine test site, the inclusion of a rock layer immediately below the wearing course proffered the structure increased stiffness to the applied loads without recourse to excessive structural thickness. Rock is typically a cheap and readily available waste material on the mines.

For comparative purposes, two design options were considered for a road under construction; a conventional design based on the CBR cover curve design methodology, and the mechanistically designed optimal equivalent. It is assumed that in-situ and road construction material properties remain the same irrespective of the structural design technique adopted. For both options a minimum wearing course thickness of 200 mm, compacted to a density of 98% Mod. AASHTO is adopted. An Euclid R170 truck is used to assess the response of the structure to applied loads generated by a fully laden rear dual wheel axle and the assumption is made of no load induced elastic deflections below 3 000 mm. The various design options are summarised in Table 3 and Figure 4.

Table 3 CBR and Optimal Mechanistic Structural Design Case Study Data

Layer	Layer Thickness (mm)	Mod. AASHTO compaction (%)	CBR (%) achieved	Target effective elastic modulus (MPa)[1]	Material description
CBR Design					
1	200	98	90	150	Selected ferricrete
2	150	95	50	100	Selected ferricrete
3	150	93	35	100	Selected ferricrete
4	300		25	3 000	Selected sandstone ex-pit, <300mm block size
5	In-situ		17	85	In situ compacted
Optimal Mechanistic Design					
1	200	98	90	150	Selected ferricrete
2	500		>200	3 000	Selected sandstone ex-pit, <300mm block size.
3	In-situ		17	85	In situ compacted

Notes
1. Maximum wheel load 429 kN, dual wheel single rear axle, maximum tyre pressure 640 kPa.

CBR Cover Curve Design

Structural design for the CBR design data are given in Table 3. To determine the likely structural performance of the road in the light of the critical design factors previously identified, a mechanistic analysis was performed by assigning target effective elastic modulus values for each layer.

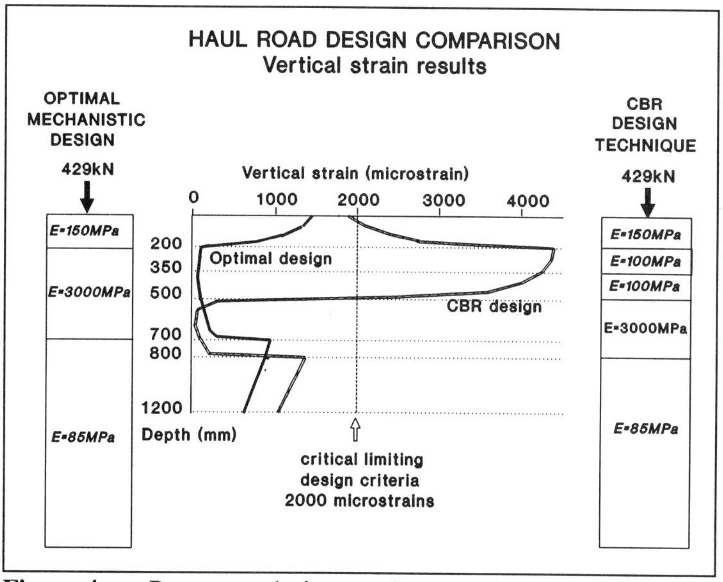

Figure 4 Pavement designs and results in terms of vertical compressive elastic strain criteria for CBR and mechanistic optimal design comparison

Figure 4 relates to the results of the mechanistic analysis of the CBR derived cover curve design. It is evident that excessive vertical compressive strains were generated in the top of layers 2 and 3. Strains in excess of 2 000 microstrain have been associated with an unacceptable amount of rutting and pavement deformation as evidenced by the performance ranking referred to in Table 1. Surface deflections generated by the applied load of 3,65 mm do not appear excessive but when accompanied with the severe load induced strains, will eventually initiate structural failure. The comments made regarding the inapplicability and under-design apparent with the CBR design technique are borne out by these results, specifically the large vertical strains developed in the pavement as the design layer strengths approach the cover curve line.

Optimal Haul Road Design

The design adopted is depicted in Figure 4 and described in Table 3. The results of the mechanistic analysis of the optimal design showed that no excessive vertical compressive strains were generated in the structure, primarily due to the support generated by the shallow rock layer. Maximum vertical strain values did not exceed the limiting design value of 2 000 microstrain and maximum surface deflections were approximately 2 mm. The proposed optimal design thus provides a better structural response to the applied loads than the thicker CBR based design and, in addition, does not contravene any of the proposed design criteria.

Cost Implications of Optimal Design

A cost comparison was compiled based on contractor tender unit costs for the construction of a road designed according to the CBR cover curve technique and one constructed according to the mechanistically derived optimal design. The variable costs taken into account were those of the volume and area of materials required and the associated costs of placing and compaction. An equivalent cost of the optimal mechanistic design was calculated from the actual construction costs of the conventional design. It was found that a 29% variable cost saving is realised, by virtue of the reduced material volumetric and compaction requirements associated with the mechanistic optimal design. In terms of total construction cost (including preliminary and general costs) a 17% total cost saving realised. Subsequent to this analysis several roads were constructed according to the mechanistic design and during an extremely wet 1996 summer, superior performance was reported, compared with the existing roads.

CONCLUSIONS AND RECOMMENDATIONS

- Existing empirical haul road design methods are deficient when ultra-heavy axle loads are considered in conjunction with multi-layer structures. The potential for under-design associated with multi-layer structures limit the utility of the CBR cover curve technique when applied to mine haul road structural design.
- Of the two design criteria proposed, only the vertical strain criterion correlated with structural performance of the road and an upper limit of 2 000 microstrain was proposed for layer strain values for the number of load repetitons found on the mines. For the wearing course, a 200 mm layer of compacted (95-98% Mod. AASHTO) good quality gravel (field CBR>80) was recommended.
- Target effective elastic modulus values were proposed for typical construction materials. A modulus of 150-200 MPa was proposed for G4-G6 gravels when used in wearing course and 75-100 MPa when used in a base or sub-base layer.
- Modulus values for in-situ sub-grade material were site dependant and it was recommended that DCP derived CBR values are used together with recommended modulus values for this material.
- The optimal structural design of surface mine haul roads incorporates a dumprock layer immediately below the wearing course layer, the thickness of which was dependant on the underlying in-situ material effective elastic modulus, applied load and contact stress.
- The proposed optimal design provided an improved structural response to the applied loads and, in addition, did not exceed any of the proposed design criteria, as opposed to the conventional CBR design.
- As a result of the reduced structural thickness, a potential construction cost saving of 17% over the conventional design could be realised. Superior performance was reported for roads built according to the mechanistic design.

ACKNOWLEDGEMENTS

The permission of AMCOAL to publish this paper, together with their financial support for the research, is gratefully acknowledged, as is the advice and co-operation of mine personnel.

REFERENCES

1. KAUFMAN, W W & AULT, J C. 1977. The design of surface mine haul roads manual. USDOI Information circular 8758, USBM, pp 2-7, 19-30.
2. AMERICAN SOCIETY OF CIVIL ENGINEERS. 1950. Development of the CBR flexible pavement design methods for airfields. Symposium, Transactions, vol 115, ASCE.
3. SOUTH AFRICAN ROADS BOARD. 1994. The South African mechanistic pavement rehabilitation design method. Research report RR 91/242, Department of Transport, South Africa.
4. KLEYN, E G, MAREE, J H & SAVAGE, P F. 1982. Application of a portable pavement dynamic cone penetrometer to determine the insitu bearing properties of road pavement layers and subgrades in South Africa. ESOPT II, Amsterdam, Holland.
5. DE BEER, M, HORAK, E & VISSER, A T. 1988. The multi-depth deflectometer (MDD) system for determining the effective elastic moduli of pavement layers. 1st Int Symp on Nondestructive Testing of Pavements and Back calculation of Moduli ASTM STP 1026, Baltimore, USA.
6. ELSYM5A. 1985. Interactive version 5A users guide. FHWA.
7. PATERSON, W D O & MAREE, J H. 1978. An interim mechanistic procedure for the design of asphalt pavements. National Institute for Transportation and Road Research, technical report RP/5/78, CSIR, Pretoria, South Africa.
8. MAREE, J H, VAN ZYL, N J W & FREEME, C R. 1982. The effective moduli and the stress dependence of pavement materials as measured in some HVS tests. Session H of Transportation Infrastructure, Annual Transportation Convention (ATC), Pretoria, South Africa.
9. COMMITTEE OF STATE ROAD AUTHORITIES. 1985. Technical recommendations for highways (TRH) 14: Guidelines for road construction materials. NITRR TRH14, Pretoria, South Africa.
10. MAREE, J.H. & FREEME, C.R. 1981. The mechanistic design method as used to evaluate the pavement structures in the catologue of the draft TRH4 (1980). National Institute for Transportation and Road Research, NITRR technical report RP/2/81, CSIR, Pretoria, South Africa.
11. THOMPSON, R J & VISSER, A T. 1994. The structural design of surface mine haul roads. Haul roads research project, Stage 1 contract report for AMCOAL, University of Pretoria, Pretoria, South Africa.

Author Index

T

U

V

v

W

X

Y

Z

Ø

Å

INDEX

NOTE A reference is generally given only once, even if the subject may appear several times in the same paper.

A

B

C

Q

R

S

T

U

V

W

Z

Proceedings

of the Fifth International Conference on the Bearing Capacity of Roads and Airfields

Norwegian University of Science and Technology
TRONDHEIM, NORWAY, 6 – 8 JULY 1998

Editors:

Rasmus S. Nordal,
Norwegian University of Science and Technology
Geir Refsdal, Norwegian Public Roads Administration

VOLUME II

© Faculty of Civil and Enviromental Engineering
Department of Road and Railway Engineering

ISBN 82-519-1346-2
Printed by TAPIR

Supporting Organizations
Norwegian University of Science and Technology
Norwegian Civil Aviation Administration
Norwegian Public Roads Administration

National Organizing Committee
Ivar Horvli (Chairman), Norwegian University of Science and Technology
Kjell Levik (Co-Chairman), Norwegian Public Roads Administration
Bjørn Finborud, Berdal Strømme as
Øyvind Hallquist, Norwegian Civil Aviation Administration
Geir Lange, Norwegian Civil Aviation Administration
Rasmus S. Nordal, Norwegian University of Science and Technology
Geir Refsdal, Norwegian Public Roads Administration
Olav Reine, Veidekke ASA
Astrid Bye (Conference Secretary), SEVU

International Advisory Committee
Kaare Flaate (Chairman), Norwegian Public Roads Administration
Ivar Horvli (Co-Chairman), Norwegian University of Science and Technology
David Bennet, Australian Road Research Board Ltd.
Jean-Francois Corté, Laboratorie Central des Ponts et Chaussées, France
Rudolf Floss, Technische Universität München, Germany
Les Hawker, Highways Agency, United Kingdom
Michel Hovan, Federal Aviation Administration, USA
Byron N. Lord, Federal Highway Administration, USA
El Hussein Mohamed, National Research Council of Canada
M.I. Pinard, Ministry of Works, Botswana
Xinglong Wang, Heilongjiang Institute of Highway and Transport, China
Vladislav M. Yumashev, State Highway Research Institute, Russia

Technical Advisory Committee
Geir Refsdal (Chairman), Norwegian Public Roads Administration
Per Magne Noss (Co-Chairman), The Norwegian Asphalt Contractors Association
Bjørn Finborud, Berdal Strømme as
Åsmund Knutson, Norwegian Public Roads Administration
Geir Lange, Norwegian Civil Aviation Administration
Helge Mork, Norwegian University of Science and Technology
Tore Slyngstad, Norwegian Public Roads Administration
Jostein Myre, Norwegian Public Roads Administration

Conference Secretariat
SEVU, Congress Department
NTNU, Gløshaugen
N-7034 Trondheim, Norway

INTRODUCTION

Road and air transport with their primary role in the transportation system are drawing considerable public and political attention all over the world. The steady growth of traffic volume and weight reduces the service life of existing pavements, causing escalating maintenance costs as well as increasing safety and environmental problems. To meet these challenges, there are great needs for further development of road and airport networks , and for cost-effective rehabilitation and strengthening of existing pavements.

The BCRA conferences have been organized every fourth year since 1982 when the first one was held in Trondheim, Norway. The second was held in Plymouth, England in 1986, the third in Trondheim in 1990 and the fourth in Minneapolis, Minnesota , USA in 1994. These conferences have focused on the monitoring of the bearing capacity , the performance of pavements and the evaluation and use of this information for maintenance, strengthening and load regulations.

The themes of the BCRA conference in 1998 cover important aspects of pavement performance, pavement condition surveys, evaluation of structural capacity, properties of pavement material, design and pavement rehabilitation and strengthening. The contributions are documented in these proceedings containing 180 papers from all over the world.

The conference will provide a platform for the presentation and discussion of these research results and new developments, for the exchange of experience from current practice and for the identification of further research and development needs.

I would like to express my gratitude to the authors and the chairmen, the members of the International Advisory Committee, the Technical Advisory Committee, the Conference Secretariat and the editors of the proceedings.

Ivar Horvli

Chairman of the Organizing Committee

VOLUM 1

Table of contens

Bearing Capacity, Policies, Concepts and Costs

Effect of Axle Loads and Tyre Pressures

Airfield Pavements

Pavement Condition Surveys

Backanalysis of Deflection Measurements

Structural Evaluation and Design

VOLUM 2

Structural Evaluation and Design

Full Scale Testing

Environmental Effects

Asphaltic Materials

VOLUM 3

Granular Materials and Subgrade Soils

Use of Reinforcement

Pavement Rehabilitation

Case Histories, Airfields

Case Histories, Roads

Continuing from page 114

Fifth International Conference on the
BEARING CAPASITY OF
ROADS AND AIRFIELDS

DESIGN AND PERFORMANCE MODELS FOR CONCRETE BLOCK PAVEMENTS

M. HUURMAN [1] L.J.M. HOUBEN [2] AND A.A.A. MOLENAAR[3]

[1]Consultant Netherlands Airport Consultants, the Hague
former Research Engineer Delft University
[2]Senior Research Engineer Delft University
[3] Professor of Highway Engineering
Delft University of Technology
Faculty of Civil Engineering and Geo Sciences
P.O. Box 5048, 2600 GA DELFT, The Netherlands

ABSTRACT

It is a well known fact that concrete blocks are an extremely useful pavement material for urban areas on soft soils. This is because this pavement can be easily relayed without the need of using new material and because of the fact that utilities that are burried in the road construction can be reached easily. However also on heavily loaded area like container terminals and airport aprons, concrete block pavements have been used very successfully. One of the problems with concrete block pavements is however their sensitive to the development of uneveness.

Transverse (rutting) and longitudinal unevenness are the most well known defects on this pavement type and it is therefore essential that block pavements are designed and constructed in such a way that the occurence of these defects is overcome as well as possible. This paper describes the development of design performance models for concrete block pavements by taking fully into account dynamic loads caused by traffic driving over an uneven pavement surface, the non-linear stress dependent behaviour of unbound materials used in the base and the subbase, and the non continuous characteristics of the block layer.

INTRODUCTION

Traditionally a large part of the Dutch road network is paved with small elements like concrete blocks and clay bricks. Figure 1 e.g. gives an overview of the pavement types used in the city of Rotterdam. Block pavements are not only attractive to urban areas but also for heavily loaded areas like container terminals and airport aprons, this paving material has been successfully used. The dominant damage type on block pavements is deformation either in the transverse direction (rutting) or in the longitudinal direction. These deformations can be contributed to the deformations in the bedding sand layer, the unbound base and subbase and the subgrade. The development of permanent deformations is accelerated by the fact that by nature, the concrete block pavement (c.b.p.) has not a very smooth surface. This means that significant dynamic forces can be generated by traffic which accelerate the deformation process. Especially when block pavements are laid manually by means of paviours, the initial smoothness is certainly not as high as that of e.g. asphalt pavements.
In designing concrete block pavements the dynamic traffic load has therefore to be taken into account. Furthermore such design models should take into account the fact that the non-homogeneous block layer is not capable of taking any bending moments. The load is transferred from one block to the other by vertical shear forces in the joint and horizontal compressive joint forces. Also the fact that the unbound materials used for base and subbase are non-linear should carefully be modelled especially since locations underneath the blocks are loaded almost to shear failure which means that a redistribution of stresses takes place.

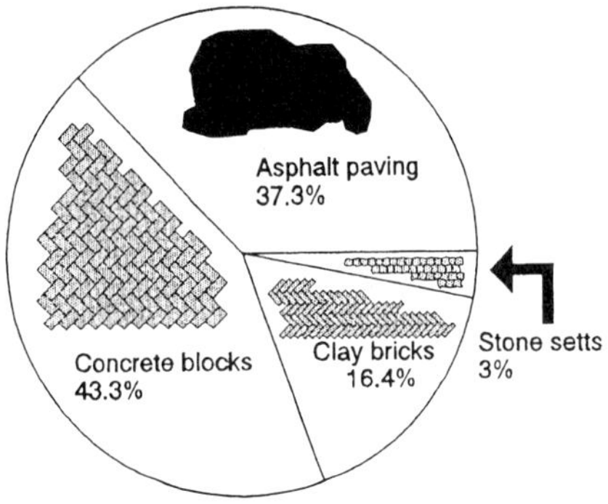

Figure 1: Paved road surfaces in Rotterdam.

Since the currently available design methods don't take into account these specific features, a research program was started at the Delft University to develop improved models. This paper describes some of the main results and findings of that research program.

DEVELOPMENT OF AN ANALYTICAL TOOL

Figure 2 is a schematic representation of a c.b.p.. As mentioned before, the block layer is a non-continuous, non-homogeneous layer where load transfer takes place in the joints. Only shear and compressive forces can be generated by the jointing sand. This is schematically represented in figure 3.

Since no finite element codes were available that allowed an accurate analysis of this behaviour, that is so specific for a concrete block layer, to be made, much effort had to be invested in the development of such models. Finally the model shown in figure 4 was developed.

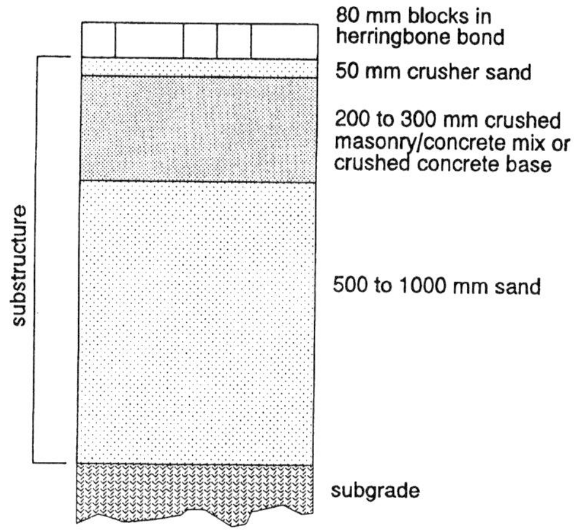

Figure 2: Principle of a concrete block pavement.

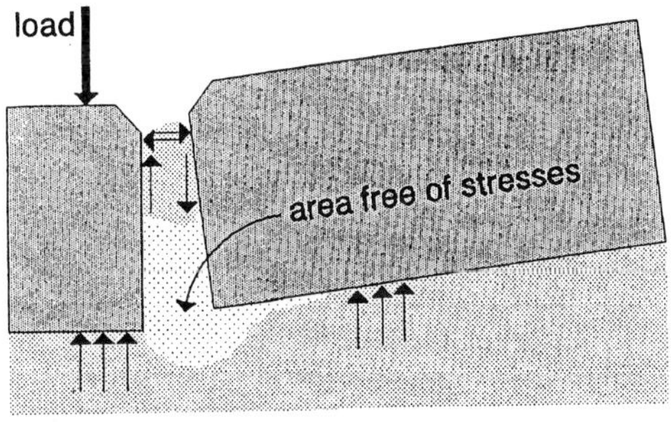

Figure 3: Transfer of loads in and near joints.

This representation allows a stiff block (dark gray part) to translate and rotate. As soon as tensile forces are generated in the horizontal springs, representing the joints, these springs are de-activated. Also precautions have been given to the fact that no vertical tensile stresses can be generated in the bedding sand layer.

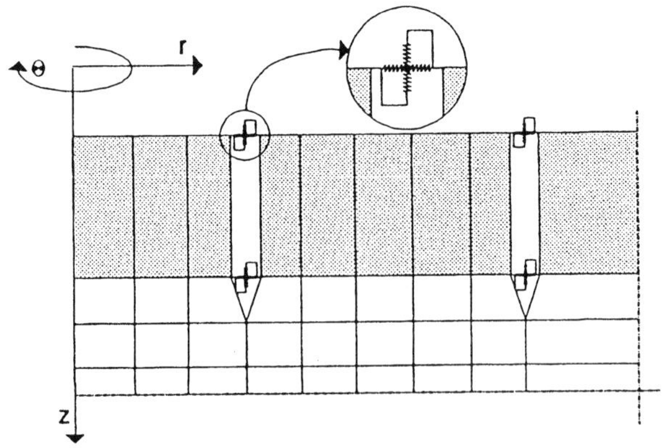

Figure 4: Modelization of a joint.

In order to be able to make finite element analyses, information on the values to be used for the joint spring stiffnesses need to be known. In order to obtain these values, F.W.D. deflection measurements were done on a number of pavements and the spring stiffnesses were back calculated from the measured deflection bowls using the developed finite element code.

Figure 5a shows the development of the maximum deflection d_0 in time of a specific pavement. As one will observe, d_0 gradually decreases in time and some seasonal influence is visible. Figure 5b shows the development of the curvature index $IDK_{0.3}$ in time. This value is the difference between the maximum deflection and the deflection measured at 0.3 m from the loading centre. A slight decrease is observed as well as some seasonal influences.

Both figures show the progressive stiffening trend that is observed on many c.b.p.'s.

It is obvious that a range of values for the spring stiffnesses was obtained; the average value for the horizontal spring stiffness appeared to be 5500 N/mm per mm joint length while the mean shear stiffness took a value of 550 N/mm mm. Both values are valid for a block thickness of 80 mm.

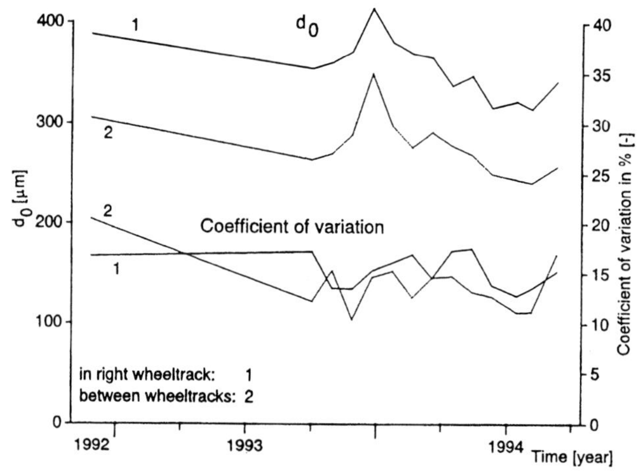

Figure 5a: Change of this maximum FWD deflector in time.

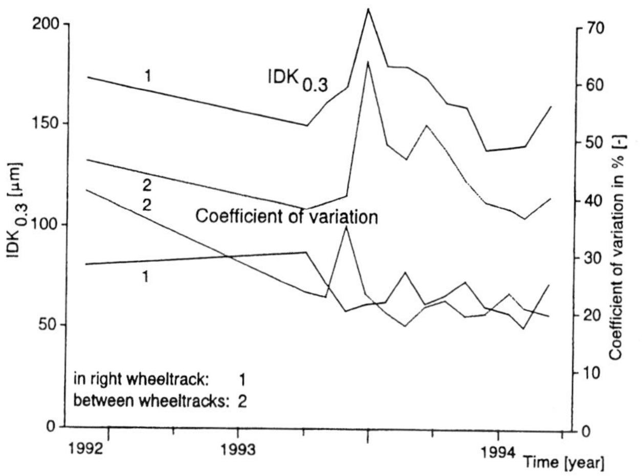

Figure 5b: Change of the FWD surface curvature index in time.

MATERIAL MODELLING

As has been mentioned before, localized high stresses can be generated underneath the blocks. At these locations the stresses can become so high that shear failure is likely to occur. On the other hand the well known M_r - θ model predicts high modulus values at high stress levels. For the sake of completeness it is recalled that this model is written as:

$$M_r = k_1 \theta^{k2} \qquad \text{..........(1)}$$

where: M_r = resilient modulus
θ = sum of principle stresses
k_1, k_2 = constants

The M_r - θ model has another consequence which is that high stresses not only generate high moduli but also that these locations of high stiffness "attrack" the stresses so that the moduli at these locations are becoming even higher. It is obvious that this in severe contradiction with practice which means that material models are needed that allow softening under high stresses allowing for a redistribution of the stresses. In order to derive such a model repeated load triaxial tests were performed on a number of subbase sands and unbound granular base course materials. Figure 6 shows the results of resilient modulus testing on a particular sand. As one can observe the M_r - θ model describes only a general trend but is absolutely incapable in describing the behaviour at the various combining levels. Based on experimental results like those shown in figure 6, it was concluded that the resilient modulus is dependent on the confining stress and θ. If θ increases at a constant combining stress level, the modulus will decrease if the vertical stress comes close to failure conditions. In order to take this behaviour into account, the following model was derived:

$$M_r = k_5 \, (\sigma_3 / \sigma_0)^{k6} \, (1 - k_7 \, (\sigma_1 / \sigma_{1,f})^{k8}) \quad \ldots\ldots\ldots(2)$$

where: M_r = resilient modulus
σ_3 = combining pressure
σ_1 = vertical pressure
$\sigma 1,f$ = vertical pressure at which failure occurs at the given δ_3
σ_0 = reference pressure = 1kPa
k_5, k_6, k_7, k_8 = constants

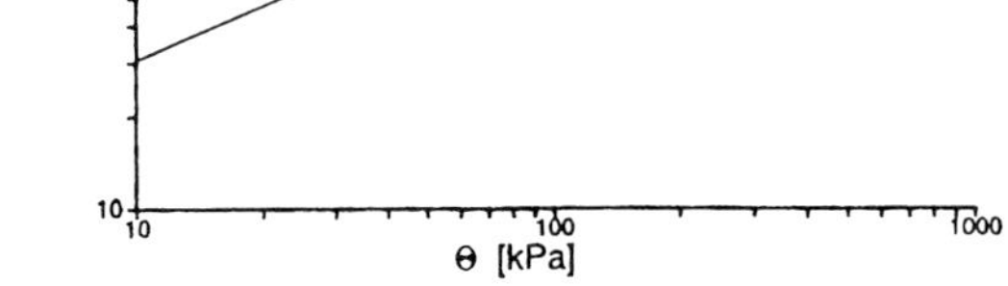

Figure 6: Resilient modulus behaviour of a sand characterised by means of the M_r-θ model.

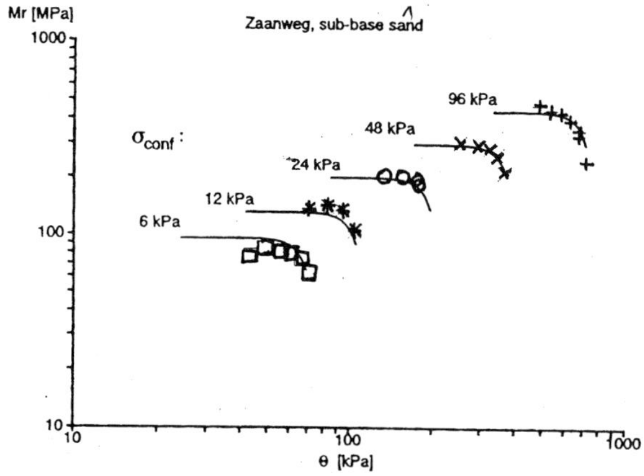

Figure 7: Resilient modulus behaviour of a sand characterised by means of the improved model.

As shown in figure 7 this model nicely describes the observed behaviour.

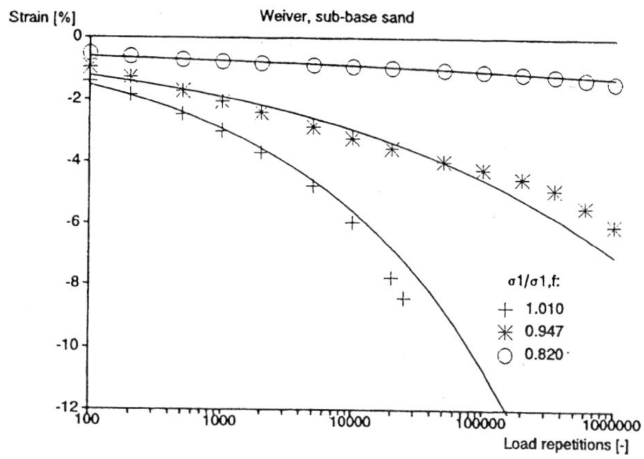

Figure 8: Permanent deformation behaviour of a sand.

Furthermore the observed permanent deformation behaviour (figure 8) was modelled using:

$$\varepsilon_p = A\,(N/1000)^B + C\,(e^{D(N/1000)} - 1) \qquad \text{..........(3)}$$

where: ε_p = permanent strain

N = number of load repetitions

A, B, C, D = material parameters

LONGITUDINAL UNEVENNESS

The longitudinal profiles of a number of roads were measured using the ARAN of the Ministry of Transport. The measurement results were used to derive Power Spectral Density relations, examples of which are given in figure 9. This type of information was used together with a half vehicle model in order to be able to analyse the behaviour of various types of vehicles on road surfaces with various degrees of roughness.

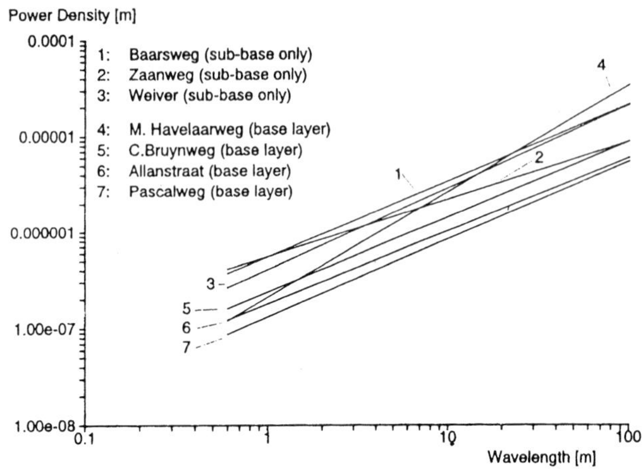

Figure 9: PDS relations for some concrete block pavements.

DESIGN MODEL

If the initial profile of a road is known, then the dynamic loadings due to a particular vehicle can be calculated. If the loads are known, as well as the resilient and permanent deformation characteristics of the materials used, then the stresses in the various layers of a c.b.p. can be calculated as well as the resulting permanent deformations, and the resulting longitudinal profile. Then the entire procedure is repeated again and as a result the development of the rutting and longitudinal unevenness as a result of the number of load repetitions is obtained. It is obvious that such an integrated models requires extensive computation time and therefore simplifications were made which overcome the need to do finite element analyses and vehicle response studies for each vehicle passage. A large amount of finite element calculations was e.g. made for different types of structures and varying material characteristics in order to determine the development of permanent deformation due to a static load. These results were then used to develop a regression equation which gave similar results. Several studies were made using the model and the results of one are shown in figures 10 and 11.

Figure 10 shows the development of the development of the rutdepth due to a truck with a 100 kN rear axle load and a 71 kN front axle load. The "standard" case shows the result with standard values for the stiffnesses of the tyres, springs and shock absorbers, while "2xk"indicates a two times higher tyre pressure and "C/3" indicates the condition of worn out shock absorbers.

From this figure it is clear that not well maintained cars result in accelerated pavement deterioration. This is also indicated in figure 11 where the standard deviations of the front and rear axle load are given for the three conditions considered.

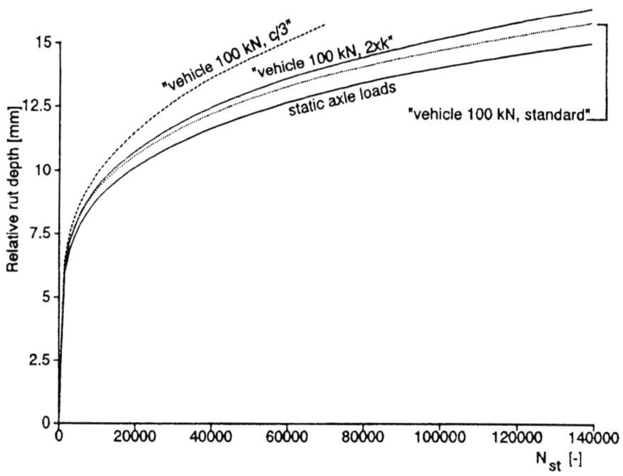

Figure 10: Rut depth development of a c.b.p. in relation to vehicle characteristics.

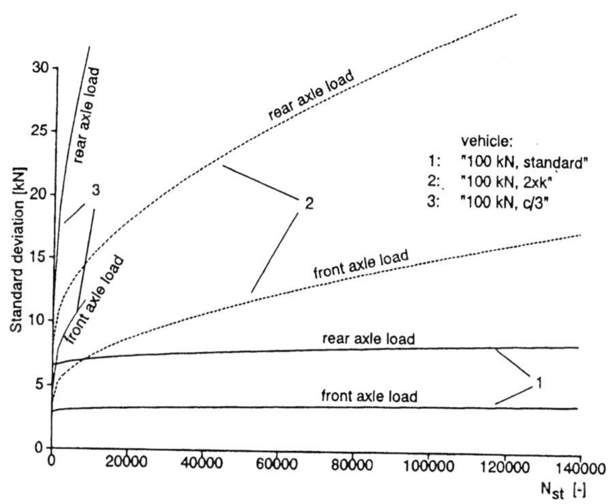

Figure 11: Standard deviation of the front and rear axle as a function of the vehicle characteristics and roughness development.

CONCLUSIONS

Based on the research that was performed, the following conclusions have been drawn.

- In the analysis of c.b.p.'s it is essential that the block layer as well as the characteristics of the unbound base, subbase and subgrade are correctly modelled. Application of linear elastic layer/theory is not justified neither the use of the M_r - θ models.
- Dynamic loads should be taken into account when designing c.b.p.'s. This is because even the initial roughness levels are such that is relatively high dynamic loads will develop from the very beginning of the pavement life.
- In order to reduce pavement maintenance, the truck fleet should be well maintained.
- Too narrow lanes and traffic regulating measures like artificial unevenesses accelerate pavement deterioration.

REFERENCES

1. HUURMAN, M. Permanent deformation in concrete block pavements. Dr.Sc. dissertation; Delft University of Technology; Delft – 1997

2. HUURMAN, M.: HOUBEN, L.J.M.; Kok, A.W.M. Development of a three dimensional finite element model for concrete block pavements. Proceedings 4th Int. Conf. On Concrete Block Paving; Auckland – 1992

3. VAN NIEKERK A.A. AND HUURMAN, M. Establishing complex behaviour of unbound road building materials from simple material testing. Report 7-95-200-16; Road and Railroad Research Laboratory; Delft University of Technology – 1995

Fifth International Conference on the
BEARING CAPASITY OF
ROADS AND AIRFIELDS

POTENTIAL APPLICATIONS OF GEOMETRICAL MODELING TO STRESS - STRAIN BEHAVIOR OF ASPHALTIC CONCRETE LAYERS IN FLEXIBLE PAVEMENT

NII O. ATTOH-OKINE, Ph.D., P.E.

Dept. of Civil and Environmental Engineering, Florida International University, FL 33199, Miami, U.S.A., Fax: (305) 348-4057, E-mail: okine@eng.fiu.edu

ABSTRACT

The paper presents the use of Geometrical Modeling in the stress-strain behavior of asphaltic layers in flexible pavement structure. Asphaltic concrete can be assumed as a granular material (collection of aggregates and fillers) bonded with a binder. In asphaltic concrete mixtures in which the binder/matrix constitutes a small percentage of the total volume, traffic (repètitive) are usually transmitted through particle (aggregate-aggregate) contacts or aggregate-binder-aggregate contacts.

Therefore the use of contact mechanisms and computational techniques like the Discrete Elements Method can be rational to some extent if there are no aggregate-binder-aggregate contacts. Unfortunately, in asphaltic pavements the binder matrix introduces additional forces and load transmissions. The paper will provide a framework for realistic application of Geometrical Modeling in stress-strain in asphaltic concrete mixtures taking into consideration the aggregate-binder-aggregate interaction and binder matrix load. Both 2D and 3D assemblies of asphaltic mixtures will be investigated.

INTRODUCTION

Asphaltic concrete (AC) layers in flexible pavements can be perceived as a collection of particles (aggregates) bonded with a binder. Therefore, AC materials belong to a class of two-phase media consisting of particles (aggregates) and the asphalt binder matrix. The force/load transmission due to traffic loads can directly act through aggregate-aggregate contacts and aggregate-asphalt binder-aggregate contacts. Chang and Meegoda (1997) named the above contacts as Type I and Type II contacts, respectively.

Traditionally, two approaches have been used as a solution for stress-strain behavior in two-phase media - the continuum mechanical approach and microstructural approach. The most widespread is the microstructural approach. The microstructural approach is to find macro-level state variables that are based on micro-variables such as aggregate contact forces, aggregate displacement and local geometrical characteristics [Bagi, 1996]. Generally, these microscopic models are based on Discrete Element Modeling (DEM) [Cundal and Strack, 1979]. The DEM is a numerical scheme used to simulate the response of granular-media by modeling the dynamic behavior of the assemblies of the model. The method makes simplifying constitutive assumptions for each aggregate and then applies Newtonian mechanics to determine translational and rotational motion of each aggregates in the assemble.

Cambou et al. (1995) introduced the homogenization techniques in the microstructural approach. In this technique, the macroscopic properties of two-phase media or three-phase media are assumed to be related to a basic structure, and the properties of the constituents and their interactions. Based on this assumption, the overall behavior of such media can be derived from the local properties.

EMERGING APPLICATIONS

Zhu et al. (1997) develops a technique, which uses binder-contact law and basic cells to formulate a stress-strain equation. Zhu et al. (1997) assumed that the particle aggregates are hexagonal arrayed and the binder particles behave elastically. Zhu et al. (1997) analyzed AC layer in three levels: the micro-level, the meso-level and the macro. Zhu et al. (1997) demonstrated that the classic stiffness coordinate transformation method is generally not applicable at the meso-level.

Chang and Meegoda (1997) developed a micromechanical simulation of hot mix asphalt (HMA) using DEM. Figure 1 shows the micromechanical system of HMA used. The aggregates in contact are represented by a mass with a moment of inertia connected by visco-elastic elements in the normal and tangential directions at each contact.

Bagi (1996) applied geometrical modeling to analysis of stress-strain in granular assemble. The author discussed how Voronoi - Tessalation, Dirichlet Tessalation and Delaynay network can be applied in stress-strain application in granular assemble.

Voronoi, System is suitable for the analysis of granular assemble, The Dirichlet analysis appears to be suitable for any particles with a smooth convex shape and assemblies having a well-defined and special regular geometry. If Dirichlet cells of two aggregates have a common side, when one connects the two aggregate's centers by straight line, the connecting lines form the Delaunay network of the assembly.

APPLICATION

Modification was introduced to Bagis geometrical model to "binder" – "aggregate-binder-aggregate" – to effectively discuss this, Bagi (1996) introduced the concept of a material cell and space cell system. The material cell is the basis of stress definition and the space cell forms the basis for strain definition. Other terms introduced are branch vectors and complementary areas.

If A1 and A2 are aggregates, with aggregates with real or virtual contact c, the branch vector is

$$l_i^a = v_i^{1c} - v_i^{2c}$$

As shown on figure 2.

Assign vector b_i^k to each face of the aggregate in the following way:

- The magnitude of b_i^k is equal to the area of the face.

- The direction of b_i^k is normal to the face, poiňtivity outwards.

Another vector a_i^k can be defined as: $$a_i^k = -\frac{1}{D} b_i^k$$

Where D – nodes. Those vectors are shown on Figure 3. The complementary area vector is defined as: $$d_i^{12} = \frac{1}{(D+1)} \sum_{t=1}^{T} (a_i^{1(t)} - a_i^{2(t)})$$

t - is the number of cells

The Volume of the space cells: $$V = \sum_c l_i^c d_i^c$$

The stress can be expressed as: $$\sigma_{ij} \frac{1}{V} \sum_{(c)} l_i^c F_j^c = \frac{1}{V} \sum_L V^L \sigma_{ij}^{-L}$$

c – index of the faces of material cell system

F_j – contact force between neighboring aggregates

l_i^c – branch vector

The displacement gradient tensor

L – index of material cell

$$\bar{e}_{ij} = \frac{1}{V} \sum_{K} V^K \bar{e}_{ij}^K = \frac{1}{V} \sum_{(c)} \Delta u_i^c d_j^K$$

K – index of space cells

c – index of the edges of space cell system

Δu_i – relative translation of neighboring aggregates

d_j – complementary area vector

SPECIAL CASES

The geometrical interpretation of granular assembly presented by Bagi (1996) was modified

- In case of type I, the branch vector should remain the same as presented by Bagi (1996)
- In the case of Type II, contacts the branch vector should be modified

$$l_i^c = \frac{v_i^{ic} - v_i^{2c}}{\alpha_i} \qquad \alpha_i < 1$$

α_i is a function of the binder properties, environmental conditions, temperature and traffic loads. This will consequently affect the volume computation hence, the stress and the strain.

SUMMARY

This paper attempts to modify the branch vector use in geometrical modeling of granular assemblies. The modification of this vector can make a method applicable to aggregate-binder-aggregate contacts. Experimental methods coupled with numerical modeling is required to establish the various values of α_i for the modification. It will be appropriate if initially α_i is model for different mechanical models of HMA to establish the ranges in which models will be both physically and geometrically interpretable. The concept can be good but more work is required.

REFERENCES

1. CHANG, K.G. and MEEGODA J. Micromechanical Simulation of Hot Mix Asphalt. ASCE Journal of Engineering Mechanics, Vol 123, No 5, pp 495-503, 1997.

2. BAGI, K. Geometrical Modeling of granular Assemblies. Acta Technica Acad Sci. Hung. Vol 107 No (1-2) pp 1-16, 1995-96

3. CUNDAL, P.A. and STRACK O.D.L. A Discrete Numerical Model for Granular Asssemblies. Geotechnique Vol 29, pp 47-65, 1979

4. CAMBOU, B. et al. Homogenization for Granular Materials. Euv. J. Mech., A/Solids Vol 14 No 2, pp 255-278, 1995

5. ZHU, H., RISH, J.W. and DASS, W.C. Constitutive Relations for Two-phase Particulate Materials with Elastic Binder. Computer and Geotechnics, Vol 20, No 3 / 4 pp 303-322, 1997

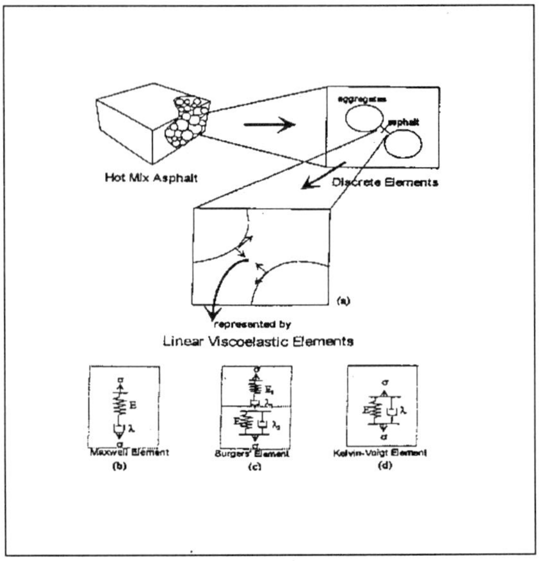

FIG. 1. Micromechanical System of HMA (Chang and Meegoda, 1997)

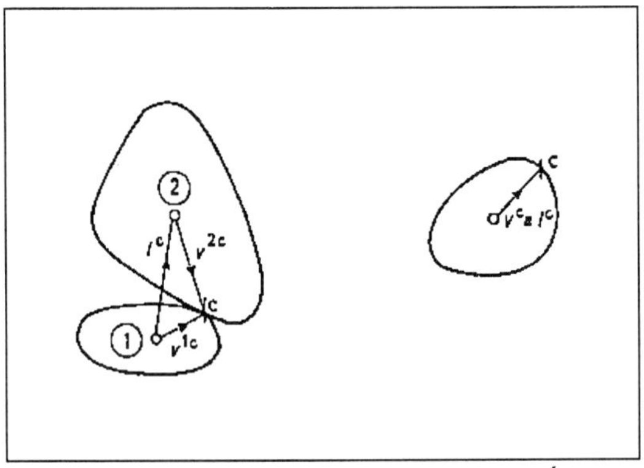

FIG. 2. Branch Vector (Bagi, 1996)

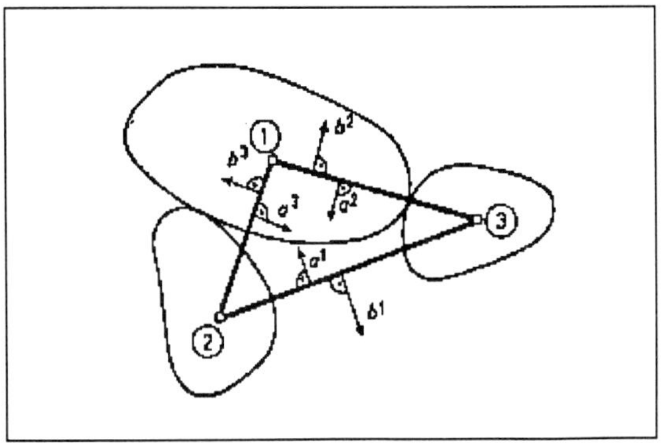

FIG. 3. Complementary Area Vector (Bagi, 1996)

Fifth International Conference on the
BEARING CAPASITY OF
ROADS AND AIRFIELDS

WINKLER - PASTERNAK - KERR FOUNDATIONS

- NOTES ON BOUNDARY CONDITIONS -

AD C. PRONK & MARION E. van den BOL

Road and Hydraulic Engineering Division
Directorate-General of Public Works and Water Management
P.O. Box 5044, 2600 GA Delft, The Netherlands
E-mail: A.C.Pronk@DWW.RWS.MINVENW.NL

ABSTRACT

In cement concrete pavement design models the application of a Winkler (elastic spring) foundation is more commonly used than the application of a Boussinesq (elastic solid) foundation. The last one is more used in the design of flexible pavements (Burmister multi-layer theory: BISAR). By adopting the thin plate theory for the response of the cement concrete pavement on top of the Winkler foundation a simple design model is obtained. Compared to the Burmister model the big advantage of this model is the opportunity to calculate stresses at the edge and corner of the concrete plate. This can be done by using the formulas as given by Westergaard or by closed integral solutions as developed by Van Cauwelaert. However, the simulation of the foundation behaviour by a one parameter model like the Winkler foundation turns out to be quite unsatisfactory. A better simulation is obtained by using the Pasternak foundation (a two parameter model by applying a shear layer on top of the Winkler foundation).
The latest development is an extension of the Pasternak model by inserting a second Winkler model on top of the shear layer. This model is called a Kerr foundation. For the determination of the stresses at the edge (and corner) of a plate, boundary conditions are needed in order to solve the problem. This paper deals with the problems which are encountered in the formulation of these boundary conditions, the advantages and disadvantages of the foundation models and the viability of (prescribed) boundary conditions. This is illustrated by calculations for thin plates (with a joint) on a Pasternak foundation using closed integral solutions for a rectangular load which are developed by F. van Cauwelaert.

Key words: Pasternak Foundation, Thin Plate Theory, Edge Loading,
Boundary Conditions

SYMBOLS

a	=	Half width of the rectangular load area	[m]
b	=	Half length of the rectangular load area	[m]
d	=	x coordinate of the joint	[m]
g	=	Dimensionless parameter	
h	=	Thickness of (concrete) plate	[m]
p	=	Normal homogeneous stress distribution within load area	[Pa]
k	=	Modulus of subgrade (foundation) reaction	[N/m^3]
s,t	=	Dummy integration variables	
x,y	=	Cartesian coordinates	[m]
$A\{s\},A_s$	=	Integration constants depending on s	
$B\{s\},B_s$	=	Integration constants depending on s	
$C\{s\},C_s$	=	Integration constants depending on s	
$D\{s\}$	=	Integration constant depending on s	
E	=	Young's modulus of (concrete) plate	[Pa]
G	=	"Shear" modulus of shear layer in foundation	[N/m]
L	=	Elastic length: $L^4 = N/k$	[m]
$M_{\perp}\{x,y\}$	=	Bending moment M_{xx} at coordinates x,y (normal to $x=x_o$)	[N]
$M_{\parallel}\{x,y\}$	=	Bending moment M_{yy} at coordinates x,y (parallel to $x=x_o$)	[N]
$M_{\varnothing}\{x,y\}$	=	Twisting moment M_{xy} at coordinates x,y (parallel to $x=x_o$)	[N]
N	=	"Bending Stiffness" of concrete plate	[Nm]
$T_x\{x,y\}$	=	Shear stress in the concrete plate at x,y	[Pa]
$S_x\{x,y\}$	=	Shear stress in the foundation shear layer at x,y	[Pa]
$W_{inf.}\{x,y\}$	=	Deflection at coordinates $0<x<d$,y for infinite plate	[m]
$W_{left}\{x,y\}$	=	Solution of homogeneous differential equation ($x \to -\infty$)	[m]
$W_{right}\{x,y\}$	=	Solution of homogeneous differential equation ($x \to +\infty$)	[m]
$W\{x,y\}$	=	Total deflection	[m]
$W_{i,j}$	=	Differentiation of deflection with respect to x (i) and y (j)	[m^{1-i-j}]
$\alpha;\gamma$	=	Parameters	
$\sigma_{xx;yy}$	=	Bending stresses (x and y direction) in the plate	[Pa]
μ	=	Poisson's ratio for (concrete) plate	
Δ	=	Laplace operator $\{\partial^2/\partial x^2 + \partial^2/\partial y^2\}$	[$1/m^2$]

INTRODUCTION

In the design of concrete pavements the Winkler foundation is next to the Boussinesq model the most commonly used model for the response of a foundation. The combination of the Winkler foundation (one parameter model) with a thin plate replacement for the concrete layers is known as the Westergaard model in case of circular and elliptical loads [1,2]. A more advanced model is the thin plate resting on a Pasternak foundation (two parameter model) [3,4]. Solutions for point loads are given by Kerr [5,6,7,8] and for rectangular loads by Van Cauwelaert [9]. This last model is also referred to as the Van Cauwelaert model [10]. The most recent model for the response of a foundation is the Kerr foundation [7]. The main advantage of the spring type foundations in combination with a thin plate on top of it is the applicability of calculating stresses at the edge and corner of the concrete plate with rather simple analytical formulas [11,12,13]. With exception of finite element methods (FEM) this is not yet possible in the linear (visco-) elastic multi layer models [14]. Also there are no analytical formulas known for the edge and corner loading of a thin plate resting on an elastic solid (Hogg and Jeuffroy models; [15,16]).

If boundaries (joints) are present in the plate, so called boundary equations are needed in order to meet the requirements at the boundary. There are two types of requirements. The first type deals with the requirements with respect to the equilibrium of forces and moments and the second one deals with assumptions with respect to the continuity in deflections. The requirements with respect to the equilibrium of forces and moments are such that the shear force (including the derivative of a possible twisting moment) and the moment normal to the boundary are continuous at the boundary. The shear force and moment are zero if no foundation and plate (completely free boundary) are present at the other side of the boundary. If the boundary still allows a transfer of shear force and/or moment two other boundary equations are necessary for the solution. In this paper no moment transfer between the plates is assumed while shear force transfer will take place through the Pasternak foundation (and plate in case of physical contact between plates). Therefore only one extra boundary equation is needed. In the case of a Winkler foundation (no shear force transfer possible in the foundation) a difference in deflections at the boundary is widely accepted but in the case of a Pasternak foundation equality in deflections is pre-assumed [17]. However, in the view of the present authors this is a too stringent requirement. Any fourth (and third in case of moment transfer) boundary equation is viable if it doesn't conflict with the requirements with respect to the equilibrium of forces and moments [18]. In this paper the needed fourth boundary equation is postulated as a ratio γ of the deflections to the left and right side of the boundary at $x = d > 0$. For $\gamma=1$ the boundary requirement as postulated by A. Kerr [16] is obtained. The load area on the plate is represented by a rectangle (sides 2a an 2b) with its centre at $x=0$ and $y=0$. For this system solutions of the homogenous differential equation are available in the form of (single) closed integrals [9].

FOUNDATION MODELS

Winkler foundation model

The most simple representation of a foundation response is the Winkler foundation which consists out of an infinite set of uncoupled springs (figure 1) with a spring constant k (modulus of subgrade reaction). The differential equation for a loaded thin plate on top of such a foundation is given by:

$$N.\Delta.\Delta.W + k.W = p \qquad (1)$$

There is no (horizontal) load transfer possible through the foundation.

Pasternak foundation model

In case of the Pasternak foundation model the foundation is capable of transferring loads in horizontal direction through an (artificial) shear layer on top of the Winkler foundation (figure 1). It should be marked that the shear layer is not a real elastic layer with a finite thickness but a mathematical representation for the behaviour of the subgrade reaction. The implication of this note will be highlighted later on. The differential equation is:

$$N.\Delta.\Delta.W - G.\Delta.W + k.W = p \tag{2}$$

Kerr foundation model

Although the Pasternak foundation model is already a better simulation for the response of a foundation there are still some disadvantages. A model which overcomes most of these disadvantages is the Kerr foundation model [6]. The response is modelled by inserting a second Winkler foundation on top of the Pasternak foundation. The main disadvantage of this model is the fact that instead of a fourth order differential equation the governing equation for a plate on this foundation is of the sixth order. Only analytical solutions for **point** loads at the edge are known up till now. In this paper this foundation model will not be elaborated. However, notice that by the appearance of the second Winkler foundation on top of the shear layer it is possible to obtain a difference in deflections without disturbing the integrity of the shear layer. Also the shear force in the plate and the shear force in the shear layer become two independent values.

Figure 1.

Winkler and Pasternak foundations

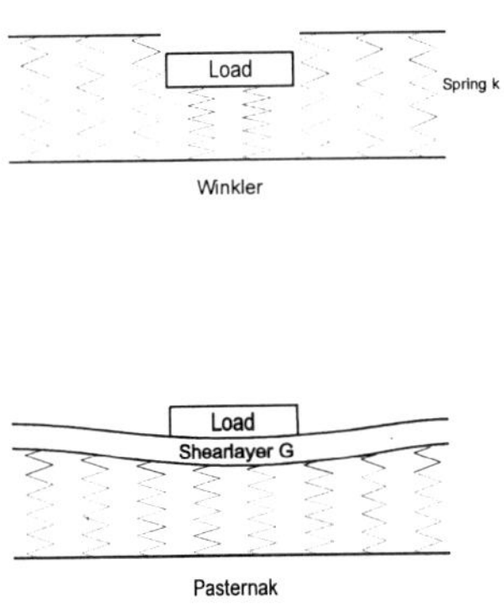

Figure 2.

Boundary condition: continuity in deflections

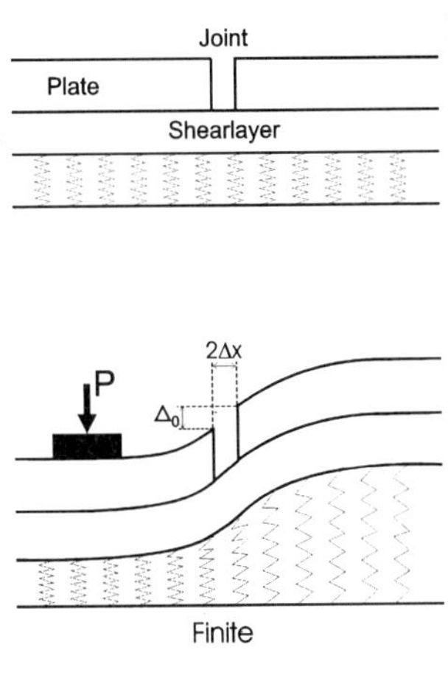

Figure 3.

Vertical interface in the shear layer

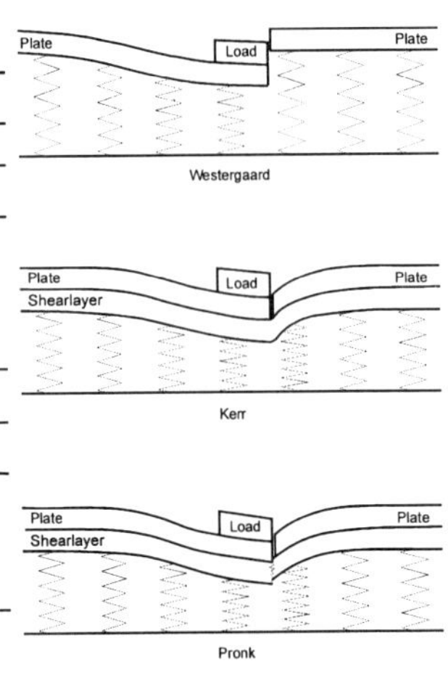

Observed foundation response in practice

Even with a finite (free) joint between plates, it is observed that if one plate is loaded the adjacent plate will deflect also. This observation cannot be modelled in case of a Winkler foundation model. It is possible to model the observed difference with a Pasternak foundation model by adopting a finite distance between the two plates (figure 2). Because there is no dead weight on or confinement of the foundation between the two adjacent plates it might well be that the foundation parameters k and G will be different for the foundation between the plates. Taking into account the complexity of the problem: two fourth order differential equations, one second order differential equation and six necessarily boundary equations this presentation seems not suitable for practical purposes. Therefore the Pasternak foundation model is adopted with two adherent plates (no finite joint between the two plates) which may show a difference in deflections (figure 3). To solve the boundary problem taking into account a possible difference in deflections a further investigation was needed of the formulation of the boundary conditions.

FORMULATION OF BOUNDARY CONDITIONS

General

The problem of a loaded thin plate resting on a Pasternak foundation (Winkler foundation is obtained for G=0) is described by a fourth order differential equation in x and y (equation 2). The complete solution exists out of a particular solution (satisfying the load condition) and four solutions of the homogenous differential equation (p=0). These last solutions are needed to meet the "boundary" requirements at the crossing of the loaded area and the boundary requirements at infinity (infinite plate) or at a joint (finite plate) [13]. The boundary requirements at infinity are that the deflection and all derivatives go to zero. If the load consists out of an area with a constant load (e.g. the Westergaard model), the solution for the deflection of an infinite plate is often represented in a single closed integral form. However, the integral can be written as the combination of the particular solution: W=p/k (inside the loaded area) and W=0.(outside the loaded area) and four solutions of the homogenous differential equation. The combination must satisfy the requirements at the crossing of the loaded area [13]. While the solution for the loaded infinite plate satisfies these requirements it can be used as the particular solution in situations with a finite loaded plate. At a joint between two finite (or semi-infinite) plates normally four boundary equations are needed to solve the two differential equations. If the loaded plate has a **completely free joint** only two boundary equations are necessary for the loaded plate. The word completely free implies that next to zero moment transfer also no load transfer takes place at the joint to the adherent plate. In case of a Pasternak foundation this means that the shear layer in the foundation also stops at the joint.
A **free joint between two plates** doesn't imply that load transfer through the foundation will not take place. Only in the case of a Winkler foundation these formulations are the same because no load transfer is possible in the foundation (uncoupled vertical springs).
The boundary conditions can be divided in equations following from the equilibrium situation at the joint and equations with respect to the continuity of the deflection at the joint. The equilibrium situation is defined by the moments normal to and shear forces parallel to the joint. The boundary conditions for the moments and shear forces (including a possible twisting moment) are that both values are continuous at a joint. Only in the case of a **completely free joint** these boundary conditions alone will lead to the necessarily four boundary equations ($M_{\perp}=0$; $T_x+S_x-M_{\varnothing,y}=0$ at both sides of the joint). In all other cases one or two equations more are needed. This is illustrated by the "boundary"

requirements at the crossing from the loaded area to the unloaded area. Because no joint (physical disturbance in the plate and foundation) is present at this crossing, two equations more are needed: continuity in deflection and its first derivative normal to the crossing. It seems a little bit obvious to formulate or deduce the boundary requirements in this way, but it should be emphasized that in principle only the boundary requirements with respect to the equilibrium situation have a fundamental character. In the following paragraph this observation will be worked out for an infinite plate resting on a Pasternak foundation with a joint (disturbance) at x=d. The plate left to the joint carries a rectangular constant load (sides 2a an 2b) with its centre in x=0 and y=0 (Van Cauwelaert model).

Two adherent semi-infinite plates on a Pasternak foundation with a joint at x=d

In this paper only the case of a free joint between the two adherent plates will be dealt with. This implicates no moment and shear force transfer through the plates. However, due to the presence of a shear layer directly under the plate, shear force (including the twisting moment) transfer will occur in this (artificial) layer. Therefore three boundary equations are obtained from the equilibrium requirements:

$$M_{\perp x\uparrow d} = M_{\perp x\downarrow d} \equiv 0 \;\; ; \;\; [T_x + S_x - \frac{\partial}{\partial y} M_{\varnothing}]_{x\uparrow d} = [T_x + S_x - \frac{\partial}{\partial y} M_{\varnothing}]_{x\downarrow d} \tag{3}$$

Thus one equation is still needed in order to solve the boundary problem. If the shear layer in the foundation is considered to be a real elastic shear layer which retains its integrity it is obvious that the deflection must be continuous (equation 4). This fourth boundary equation is postulated by A. Kerr [7,17]. This solution rules out a difference in deflections and will lead only to a change in the first derivative of the deflection at the joint. However, in the derivation of the differential equation for the Pasternak foundation the shear layer is in fact only a artificial elastic layer without a thickness and represents only in a mathematical way the shear transfer capability of the foundation. Therefore there are no physical objections to adopt another boundary equation as long as this fourth boundary equation doesn't conflict with the boundary equations which follow from the equilibrium point of view [18]. This will lead to a general formulation for the fourth boundary equation as given in equation 5.

$$W_{x\uparrow d} = W_{x\downarrow d} \tag{4}$$

$$F\{\, W, W_x, T_x + S_x - M_{\varnothing, y}, M_{\perp} \}_{x\uparrow d} = F\{\, W, W_x, T_x + S_x - M_{\varnothing, y}, M_{\perp} \}_{x\downarrow d} \tag{5}$$

In this paper two cases will be dealt with. In the first one (I) a difference between deflections is assumed and which can be represented as a ratio (equation 6):

$$\gamma . W_{x\uparrow d} = W_{x\downarrow d} \;\; ; \;\; 0 < \gamma < 1 \tag{6}$$

In the second one (II) a relationship between the differences in deflections and the total shear force at the joint is assumed (equation 7). This type of relationship is also used in linear elastic multi-layer programs as a representation of the interface condition between two layers. For α=0 the deflections must be equal (Kerr's boundary condition) which represents in the linear elastic multi-layer system the condition of full friction between the layers. The other extreme case is obtained for α=1 which implicates that the shear force is zero at the joint (completely free boundary). In the linear elastic multi-layer system this value represents the condition of complete slip between the layers.

$$\frac{1}{L^3}.(1-\alpha).\left[W_{x\uparrow d}-W_{x\downarrow d}\right]=\alpha.\left[T_x+S_x-M_{\varnothing,y}\right]_{x=d} \quad (7)$$

CALCULATIONS

In Annex I the formulas are given which are used for the calculation with the software package Mathematica. Further details about the final equations are given in reference [20]. Based on backcalculation of FWD measurements on an airfield [10], the following input was used in the calculations: E= 46,300 MPa, h= 0.265 m, p= 1.41544 MPa, μ= 0.215 [-], G= 44 MN/m, k= 20 MN/m^3, a= 0.1329 m, b= 0.1329 m, d= 0.133 m, α,γ: variables [-] ranging from 0 to 1. The centre of the loaded rectangle (sides 2a and 2b) was located at x=0 and y=0. Deflections (W) and stresses (σ_{xx} and σ_{yy}) are calculated along the x axis (y=0). For the first fourth boundary restriction (case I) the results have been verified with the program UEC slab which was developed by Van Cauwelaert [9,10]. These last results are also in agreement with calculations with a FEM model [9,10]. The normal stress σ_{xx} at the joint vanishes while the stress σ_{yy} increases with decreasing γ and α values (figures 5 and 8).

For case I the deflections and the bending stresses σ_{yy} at the joint and in the centre are given in the figures 4 and 5 as a function of the ratio γ. The calculated values for case II are presented in the figures 7 and 8. In figures 6 and 9 the deflection and bending stress σ_{yy} are given as a function of the distance x from the centre for γ=0.7 (case I) and α=0.1 (case II). Although the deflection profiles are nearly equal in figures 6 and 9, the bending stresses σ_{yy} differ quite a lot. Therefore the boundary equations 6 and 7 will lead to different results.

Figure 4. The deflections at the joint and in the centre as a function of the ratio γ

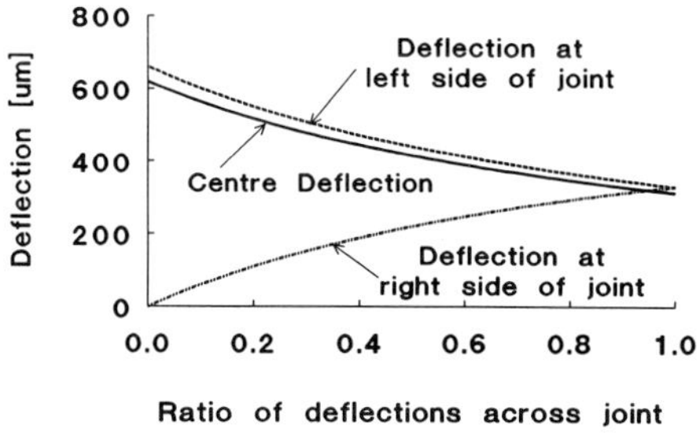

Figure 5. The stresses σ_{yy} at the joint and in the centre as a function of the ratio γ

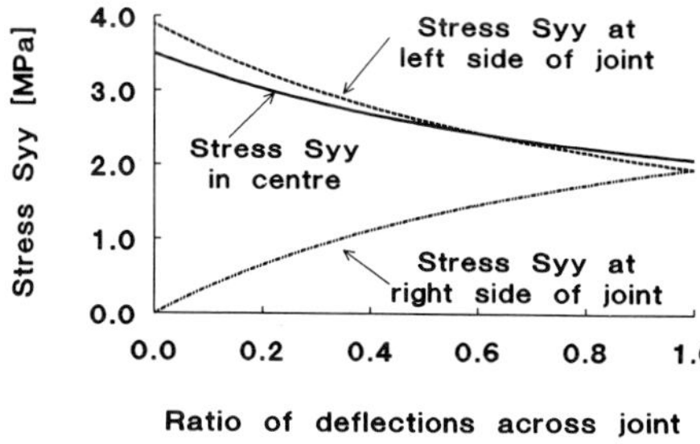

Figure 6. The deflection and the stress σ_{yy} as a function of the distance x for γ=0.7 (case I)

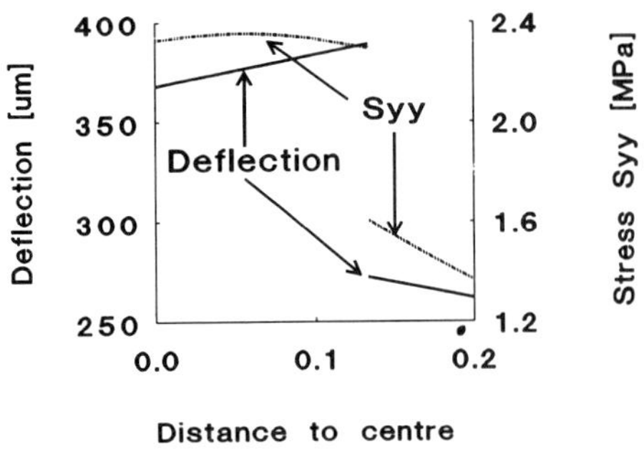

Figure 7. The deflections in the centre and at the joint as a function of the parameter α

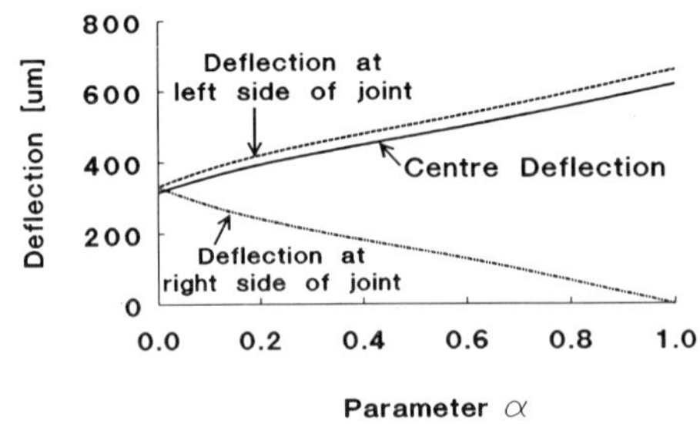

Figure 8. The stresses σ_{yy} at the joint and at the centre as a function of the parameter α

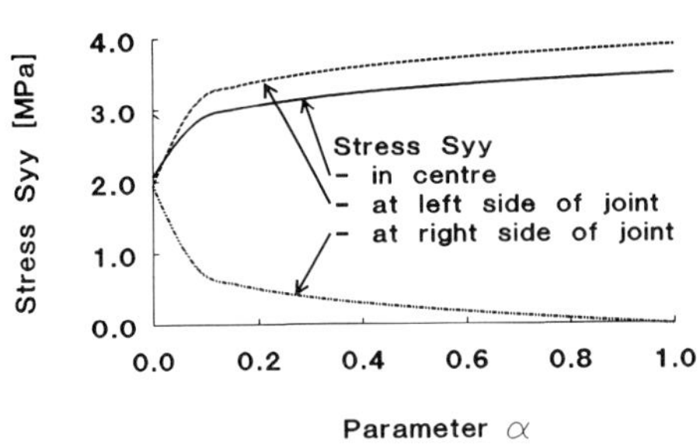

Figure 9. The deflection and the stress σ_{yy} as a function of the distance x for α=0.1 (II)

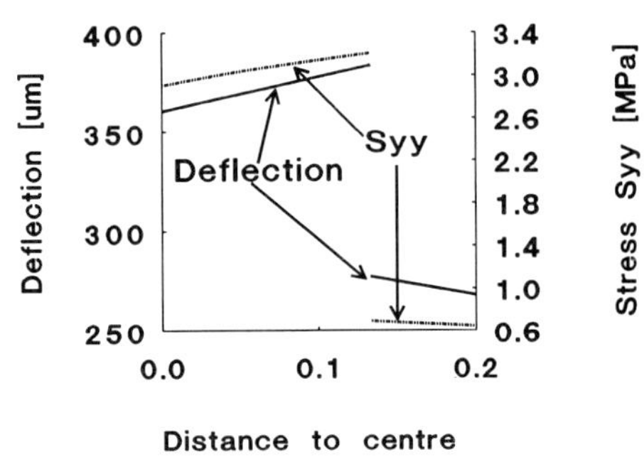

DISCUSSION

The calculations with the modified fourth boundary requirements give realistic values. From a physical point of view the restriction for case II seems more logical. However, measurements on airfield pavements show that the ratios of the deflections at both sides of the joint remain constant if measured along the joint. This observation is in favour of the restriction as mentioned in case I. If the shear modulus G goes to zero, the Pasternak foundation should degenerate into the Winkler foundation. The formulation of the boundary requirement with respect to the shear force transfer leads to a system in which the two adherent plates still make contact (see equation 3). The parameters γ (case I) and α (case II) of the fourth boundary requirement resemble the amount of contact. For γ=0 or α=1 the free plate condition in the Westergaard model is obtained. It should be noticed that in case I the ratio of the bending stresses σ_{yy} (S_{yy} in the figures) at both sides of the joint equals the ratio of the deflections. Furthermore the sum of the deflections (or stresses σ_{yy}) at both sides of the joint is a constant and doesn't depend on the ratio γ. Also in case I the load transfer coefficient (LTC) is directly related to the ratio γ: $LTC=2\gamma/(1+\gamma)$ [20]. For the given material and construction values these sums are respectively 661 μm and 3.50 MPa. However, the maximum bending stress σ_{yy} will occur at a distance x which depends on the ratio γ. For $\gamma=0$ the maximum value occurs at the joint and for $\gamma=1$ in the centre of the load.

CONCLUSIONS

- The Pasternak foundation represents a better simulation of the foundation response in practice.
- By adopting an interface condition in the shear layer of the foundation at the joint it is possible to use a Pasternak foundation model in connection with a difference in deflections at both sides of a joint.
- The adoption of a vertical interface condition allows for a variety of viable boundary restrictions as long as they are not conflicting with the boundary conditions based on the equilibrium of forces and moments. Based on measurements the boundary condition of a constant ratio in deflections at both sides of a joint looks favourable.

REFERENCES

1. WESTERGAARD, H.M., "New formulas for stresses in concrete pavements of airfields", Transactions, 1948, ASCE, Vol. 13, pp. 425-439

2. WESTERGAARD, H.M., "Stresses in concrete pavements computed by theoretical analysis", Public Roads, Washington 7, 1926, no. 2.

3. PASTERNAK, P.L., "On a new method of analysis of an elastic foundation by means of two foundations constants", (in Russian), Gasudarstvennoe Izdatelstvo Literaturi po Stroitelstvui Arkhitekture, Moscow, USSR, 1954.

4. PRONK, A.C., "The Pasternak foundation - An attractive alternative for the Winkler foundation", Proc. of the 5th Int. Conf. on Concrete Pavement Design and Rehabilitation, Vol. 1, Purdue Univ., West Lafayette, Indiana, 1993.

5. KERR, A.D., "Elastic and visco-elastic foundation models", Journ. of Appl. Mechn., ASME, Vol. 31, 1964, pp. 491-498.

6. KERR, A.D., "A study of a new foundation model", Acta Mechanica, Vol. 1, 1965, pp. 135-147.

7. KERR, A.D., "On the formal development of elastic foundation models", Ingenieur-Archiv 54 (1984), pp. 455-464.

8. KERR, A.D., "The evolution of foundations models and analyses for concrete pavements", Proc. of the 3rd Int. Workshop on Theoretical Design of Concrete

Pavements, Krumbach, Austria, 1994, pp. 61-70, CROW publication, Ede.

9. CAUWELAERT, F. van, "The limited rectangular slab on a limited Pasternak-Kerr foundation", Internal report, UEC, 1997

10. UEC Report, Final report of the CROW Workgroup UEC on the "Unification of Evaluation methods for Cement concrete pavements", CROW, Ede. UEC-97-070, 1998, (in Dutch).

11. CAUWELAERT, F. van, "Integral solution for limited slabs submitted to uniform distributed load", Proc. of the 2nd Int. Workshop on Theoretical Design of Concrete Pavements, Siguenza, Spain, 1990, pp. 219-227, CROW publication, Ede.

12. CAUWELAERT, F. van, "A rigorous analytical solution of a concrete slab submitted to interior and edge loads with no, partial and full shear transfer at the edge", Proc. of the 5th Int. Conf. on Concrete Pavement Design and Rehabilitation, Vol. 1, Purdue Univ., West Lafayette, Indiana, 1993, pp. 255-276.

13. PRONK, A.C., "The wedge-segment (pie) load case in the Westergaard model", Proc. of the 3rd Int. Workshop on Theoretical Design of Concrete Pavements, Krumbach, Austria, 1994, pp. 45-53, CROW publication, Ede, The Netherlands.

14. BURMISTER, D.M., "The theory of stresses and displacements in layered systems and applications to design of airport runways", Highway Res. Proc. 23, (1943), pp. 126-148.

15. HOGG, A.H.A., "Equilibrium of a thin plate, symmetrically loaded, resting on an elastic foundation of infinite depth", London, Edinbergh and Dublin, Philosophical Magazine and Jour. of Science, Vol. 25, (168), 1938.

16. JEUFFROY, G., "Considérations théoriques sur le calcul des chaussées en béton", Ann. Ponts et Chaussées, 129, (1959), no.3.

17. KERR, A.D., "On the derivation of well posed boundary value problems in structural mechanics", Int. Journ. Solids and Structures, Vol. 12, 1976.

18. PRONK A.C., "Some notes on the formulation of boundary requirements for a thin plate on a Pasternak foundation", Paper to be presented at the 4th Int. Workshop on Theoretical Design of Concrete Pavements, Portugal, 1998

19. HOLL, D., "Thin plates on elastic foundation", Proc. 5th Int. Congr. for Appl. Mechn., Cambridge, Maas, 1938, John Wiley & Sons, Inc., New York, 1939, pp. 71-74.

20. PRONK, A.C., "Equations for the problem of a thin finite plate on a Pasternak foundation", W-DWW-98-010, Road & Hydraulic Eng. Div., Delft, 1998

ANNEX I CLOSED INTEGRAL SOLUTIONS

Infinite Plate

The differential equation for a thin plate resting on a Pasternak foundation is given by:

$$N.\Delta.\Delta W - G.\Delta W + k.W = p \qquad \textbf{(8)}$$

The Winkler foundation is obtained for G=0. Abbreviations: $G/N=2g/L^2$ and $N/k=L^4$.
If the load is represented by a rectangular area and a constant stress p within the area the pavement model is called a Van Cauwelaert model [10]. F. van Cauwelaert developed for this case the solution of the loaded infinite plate in the form of a closed integral [19] starting from equation 9:

$$W = \int_0^\infty \int_0^\infty \frac{4pL^4}{\pi^2 N} \cdot \left[\frac{\cos\left(\frac{xt}{L}\right) \sin\left(\frac{at}{L}\right) \cos\left(\frac{ys}{L}\right) \sin\left(\frac{bs}{L}\right)}{ts(t^4 + 2t^2 s^2 + 2gt^2 + 2gs^2 + s^4 + 1)} \right] ds\, dt \qquad \textbf{(9)}$$

In reference [10] the dominator is split into two parts. Here we follow a slightly different reduction by splitting the dominator term $1/(t^4+2t^2s^2+2gt^2+2gs^2+s^4+1)$ into three parts:

$$A_s + \frac{B_s t^2}{t^2 + z_1^2} + \frac{C_s t^2}{t^2 + z_2^2}$$

$$\text{with } z_1^2 = s^2 + g + \sqrt{g^2 - 1} \text{ and } z_2^2 = s^2 + g - \sqrt{g^2 - 1}$$

$$A_s = \frac{1}{z_1^2 z_2^2} \; ; \; B_s = + \frac{z_2^2}{z_1^2 - z_2^2} A \; ; \; C_s = - \frac{z_1^2}{z_1^2 - z_2^2} A$$

Due to symmetry we only have to deal with the solution for the first quadrant (x,y ≥ 0).
The partial integration with respect to t of the terms within the brackets leads to:

$$\text{for } a > x > 0 : W = \int_0^\infty \frac{2p \cos\left(\frac{ys}{L}\right) \sin\left(\frac{bs}{L}\right)}{\pi ks} \times \qquad \textbf{(11)}$$

$$\left[A_s + B_s e^{-z_1 \frac{a}{L}} \cosh\left(z_1 \frac{x}{L}\right) + C_s e^{-z_2 \frac{a}{L}} \cosh\left(z_2 \frac{x}{L}\right) \right] ds$$

$$\text{for } 0 < a < x : W = \int_0^\infty \frac{2p \cos\left(\frac{ys}{L}\right) \sin\left(\frac{bs}{L}\right)}{\pi ks} \times$$

$$\left[0 - B_s e^{-z_1 \frac{x}{L}} \sinh\left(z_1 \frac{a}{L}\right) - C_s e^{-z_2 \frac{x}{L}} \sinh\left(z_2 \frac{a}{L}\right)\right] ds \tag{12}$$

Integration of the term which contains A_s with respect to s and repeating the same type of splitting for the dominator leads to:

$$b > y > 0 : \frac{p}{k}\left[+1 + \frac{u_2^2}{u_1^2 - u_2^2} e^{-\frac{b}{L}u_1} \cosh\left(\frac{y}{L}u_1\right) - \frac{u_1^2}{u_1^2 - u_2^2} e^{-\frac{b}{L}u_2} \cosh\left(\frac{y}{L}u_2\right)\right]$$

$$y > b > 0 : \frac{p}{k}\left[+0 - \frac{u_2^2}{u_1^2 - u_2^2} e^{-\frac{y}{L}u_1} \sinh\left(\frac{b}{L}u_1\right) + \frac{u_1^2}{u_1^2 - u_2^2} e^{-\frac{y}{L}u_2} \sinh\left(\frac{b}{L}u_2\right)\right] \tag{13}$$

$$\text{with } u_1^2 = g + \sqrt{g^2 - 1} \text{ and } u_2^2 = g - \sqrt{g^2 - 1}$$

Note that this deduction leads to the formulation of the deflection in terms of the particular solution (p/k for a>x>0 and b>y>0 and 0 for x>a>0 and y>b>0) and solutions of the homogenous differential equation. The solution above is valid for the first quadrant (x>0 and y>0).

Semi-infinite plate with a joint at x=d

Taking into account the symmetry, for solving the boundary problem at x = d > a > .0 the needed solutions of the homogenous differential are given by equations 14 and 15.

$$W = \int_0^\infty \frac{2p \cos\left(\frac{ys}{L}\right) \sin\left(\frac{bs}{L}\right)}{\pi ks} \times$$

$$\left[A\{s\} e^{+z_1 \frac{x}{L}} \sinh\left(z_1 \frac{a}{L}\right) + B\{s\} e^{+z_2 \frac{x}{L}} \sinh\left(z_2 \frac{a}{L}\right)\right] ds \tag{14}$$

$$\text{for } x < d \;\; ; \;\; A\{s\} \text{ and } B\{s\} \text{ are functions in } s$$

$$W = \int_0^{\infty} \frac{2p\, cos\left(\frac{ys}{L}\right) sin\left(\frac{bs}{L}\right)}{\pi\, ks} \times$$

$$\left[C\{s\}\, e^{-z_1 \frac{x}{L}} sinh\left(z_1 \frac{a}{L}\right) + D\{s\}\, e^{-z_2 \frac{x}{L}} sinh\left(z_2 \frac{a}{L}\right) \right] ds \qquad \textbf{(15)}$$

for $x > d$; $C\{s\}$ *and* $D\{s\}$ *are functions in* s

Boundary requirements

For a free plate boundary condition the boundary requirements with respect to the equilibrium are given by equations 16 and 17.

$$-N.\left[\frac{\partial^2}{\partial x^2} + \mu \frac{\partial^2}{\partial y^2}\right] W \equiv 0 \text{ for } x \uparrow d \text{ and } x \downarrow d \qquad \textbf{(16)}$$

$$-N.\left[\frac{\partial^3}{\partial x^3} + (2 - \mu)\frac{\partial^3}{\partial x \partial y^2} - \frac{2g}{L^2}\frac{\partial}{\partial x}\right] W = \text{continous at } x = d \qquad \textbf{(17)}$$

If the artificial shear layer in the foundation remains its integrity at the boundary, the fourth boundary equation is the continuity of the deflection itself [17]:

$$W = \text{continous at } x = d \qquad \textbf{(18)}$$

However an interface (disruption) in this artificial layer is also viable, which will lead to a difference in deflections at both sides of the boundary. For thin finite plates on a Winkler foundation a commonly adopted equation is:

$$W_{x \downarrow d} = \gamma . W_{x \uparrow d} \qquad \textbf{(19)}$$

Another boundary equation could be equation 20. This restriction supposes a relationship between the difference in deflections and the shear force present at the joint. This type of relationship is also used in linear elastic multi layer programs for describing the amount of adhesion or friction between the separate layers.

$$(1 - \alpha).\frac{W_{x \uparrow d} - W_{x \downarrow d}}{L^3} = -\alpha.\left[\frac{\partial^3}{\partial x^3} + (2 - \mu)\frac{\partial^3}{\partial x \partial y^2} - \frac{2g}{L^2}\frac{\partial}{\partial . x}\right] W_{x=d} \qquad \textbf{(20)}$$

For α=1 the complete free boundary condition is obtained and for α=0 the condition as postulated by Kerr (equality in deflections) is obtained.

Fifth International Conference on the
BEARING CAPASITY OF
ROADS AND AIRFIELDS

SUBGRADE FAILURE CRITERIA

VINCENT JANOO,[1] LYNNE IRWIN,[2] ROBERT EATON[1]
AND CHERYL RICHTER[3]

[1]Research Civil Engineer, U.S. Army Cold Regions Research and Engineering Laboratory, Hanover, NH USA 03755
[2]Director, Cornell University Local Roads Program, Ithaca, NY 14853
[3]Highway Research Engineer, Federal Highway Administration, Turner-Fairbank Highway Research Center, McLean, VA 22101-2296

ABSTRACT

An international group of researchers is developing a generalized subgrade failure criterion for use in mechanistic designs/evaluation of pavements. This paper presents the preliminary results from accelerated pavement testing using the heavy vehicle simulator (HVS) in the Frost Effects Research Facility (FERF) at the U.S. Army Cold Regions Research and Engineering Laboratory (CRREL). Tests are being conducted on four soils, each tested at several moisture contents. The test sections are instrumented with stress, strain, moisture and temperature sensors. In addition, surface deformation is obtained periodically. This paper presents preliminary results from the first two (out of twelve) test sections.

INTRODUCTION

Pavement researchers have been refining mechanistically based design methods since the middle 1960s. While the mechanics of layered systems are well developed, there is much work to be done in the areas of material characterization and failure criteria. With respect to asphalt concrete pavements, the current failure criterion used is the horizontal tensile strain at the bottom of the asphalt concrete layer and the vertical strain at the top of the subgrade. The current subgrade failure criterion—limiting subgrade strain—was developed based on limited data from the AASHO road test.

As a result of the work of Dormon and Metcalf,[1] the current design criteria for pavements stipulate that subgrade failure is predicted by the level of vertical compressive strain at the subgrade interface. However, the limiting subgrade strain criterion was based on a single soil (A-6) at the AASHO road test during spring thaw (saturated) conditions. This failure criterion may not be applicable for the full range of materials types (gravel, sand, silt or clay) in the subgrade, nor for a partially saturated moisture condition of the material. Peattie[2] recognized the limitations of the data from the AASHO road test and cautioned against the generalized use of the limiting stress or strain criteria to other soil types or environment. At the end of the AASHO road test, there was an urgent call for trial sections for developing relationships between damage to the subgrade and axle loads as a function of soil type.[2]

As work continues in an effort to provide a mechanistic basis for the AASHTO pavement design method, it is important to refine the subgrade performance criteria. By investigating the performance of full-scale pavements involving several types of subgrade soils, tested at several moisture conditions, this study will provide new insight into the subgrade rutting phenomenon. An international group consisting of the U.S. Federal Highway Administration, Turner-Fairbank Highway Research Center (TFHRC), the U.S. Army Corps of Engineers Cold Regions Research and Engineering Laboratory (CRREL), the Cornell University Local Roads Program, the Danish Road Institute (DRI), the Minnesota Department of Transportation (MinnRoad) and the Technical Research Centre of Finland Road, Traffic and Geotechnical Laboratory (VTT) has been formed to conduct the research program. This paper presents a brief description of the test program and some of the preliminary results from the first two test sections at CRREL.

TEST SOILS

The soils selected for the project reflected the typical subgrade materials found in various parts of the U.S. In terms of the AASHTO soil classification system, the subgrade soils selected for testing in the FERF were A-2-4, A-4, A-6 and A-7-6 (Fig. 1). In terms of the USCS soil classification system, the soils range from SM, to CH. In addition to the sieve and hydrometer analysis, Atterberg limits, specific gravity, standard Proctor density and the California bearing ratio tests were performed on the soils. The results of the basic tests are shown in Table 1. Detailed information on the soil types and test results can be found in Janoo et al.[3]

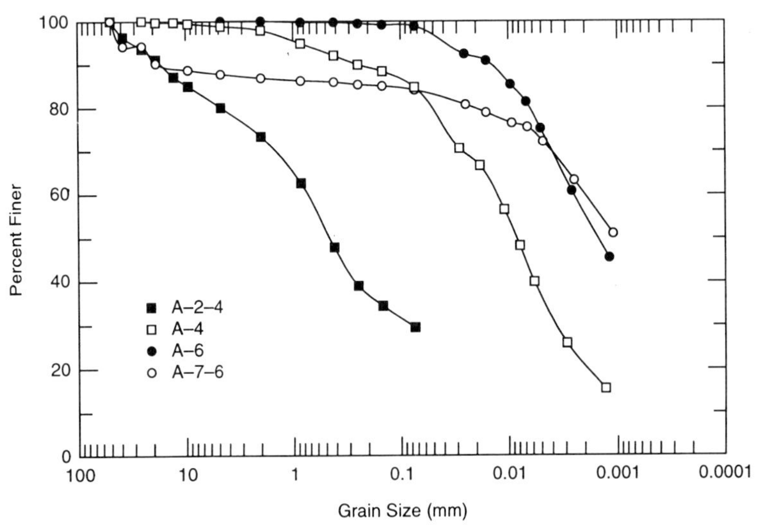

Figure 1. Grain size distribution of test subgrade soils.

Table 1. Properties of test subgrade soils.

Classification		Percent passing No. 200	Atterberg limits		Specific gravity	Proctor compaction	
AASHTO	*ASTM*		*Liquid limit*	*Plasticity index*		*Optimum moisture*	*Max. density (kg/m³)*
A-2-4	SM	29.9	30	3	2.72	10.0	1949.5
A-4	ML	84.7	28	8	2.73	16.5	1824.5
A-6	CL	87.1	33	15	2.79	16.1	1770.0
A-7-6	CH	84.2	57	37	2.80	22.6	1534.6

DESCRIPTION OF TEST SECTION

Two subgrade test sections are in place at one time in the FERF. Each test section is 20 m long by 7.5 m wide by 3 m deep. The nominal cross section is 75 mm of asphalt concrete, 225 mm of crushed base over 3.5 m of subgrade soil. Each test section has concrete walls on two sides, and a concrete floor below. Six individual test cells are created in each test section. These test cells are 7.5 m long by 1 m wide. The center-to-center distance between the test cells is 1.3 m. This allows us to load the pavement to either three strain levels with one replicate at each level or six strain levels. The first two subgrades tested were the A-2-4 and the A-4 soils at optimum moisture content.

The subgrade was compacted in 150-mm lifts. On each layer the Clegg hammer and dynamic cone penetrometer (DCP) was used to measure the uniformity of density and strength of the subgrade. Rod and level surveys were also done to determine layer thickness and locations of instrumentation. Falling weight deflectometer (FWD) measurements were taken at each subgrade level, on top of the base and on the surface.

DESCRIPTION OF ACCELERATED LOADING DEVICE

The various strain levels applied to the top of the subgrade in each test cell are done with an accelerated loading device. The accelerated loading device in the FERF is the South-African-made heavy vehicle simulator (HVS). The HVS (Mark IV) delivered to CRREL was a modification of the previous MK III. The modifications included increased speed, computer and/or manual control, and an electric motor to drive the test carriage.

The HVS is approximately 23 m long, 3.7 m wide and 4 m high. It weighs about 50 tons. For this test program, Michelin dual truck tires at 690 kPa pressure are used and applied in the unidirectional mode. The constant-velocity test cell length is 7 m. The test speed is 13 km/hr. Wheel wander over the width of the test cell (1 m) in increments of 50 mm is used.

INSTRUMENTATION

Instrumentation for measuring temperature, water content, dynamic stress and strain, and permanent strain is installed in each test section. Subsurface stresses, strains and permanent displacements are measured in the vertical and in the two perpendicular horizontal directions.

For temperature measurements, two thermocouple strings are installed in each test section from the surface to a depth of 3 m. Temperatures are collected on an hourly basis. In-situ moisture content is measured with the Vitel Hydra Probe on an hourly basis. Calibration curves were developed for each of the soil types since the calibration curves provided by the manufacturer were found to provide erroneous results for the A-2-4 soil.

The soil stresses in the x, y and z directions are measured using Dynatest soil stress cells. (X is in the direction of wheel travel, y is transverse to the direction of wheel travel, and z is in the vertical direction below the surface of the pavement structure.) Two ranges of stress cells are installed, mostly at 150 mm below the top of the subgrade. In some test cells an additional set is installed 450 mm below the top of the subgrade. Stress cells with a range of 10 to 200 kPa are installed to measure the horizontal stresses, and cells with a range of 100 to 800 kPa are installed to measure the vertical stresses. In addition, Geokon stress cells are used to measure the stresses in the base and at the bottom of the subgrade. Typical dynamic measurements in the vertical direction are shown in Figure 2.

Soil strain measurements are made with the εMU inductive coils strain system. The εMU system is used for measuring the dynamic deformation between pairs of 100-mm-diameter inductive coils. The nominal spacing between the two coils is 150 mm. The εMU gages are capable of measuring dynamic strain as well as permanent displacement. Pairs of coils are installed in three orientations to measure strains in the vertical and in two perpendicular horizontal directions. The coils are also installed starting in the base in columnar stacks to a depth of 1220 mm from the surface. A typical vertical strain measurement is shown in Figure 3.

Two special techniques had to be used in order to be able to use the εMU gages to measure dynamic strain. Electronic noise on the output signal was bothersome. To overcome it an elliptical low-pass filter with a cutoff frequency of 40 Hz was used. The second problem occurred because the coil gages produce a signal due to the moving metal of the HVS wheel carriage. Thus when the loaded wheel moves over the pavement, the εMU signal is the

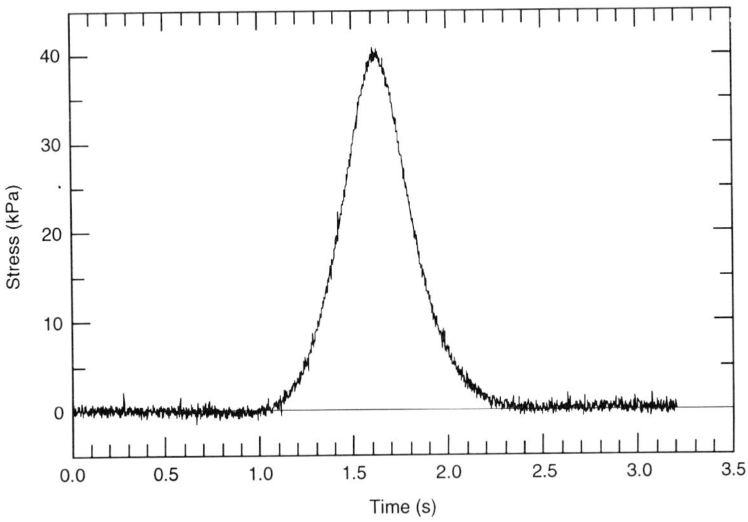

Figure 2. Typical vertical dynamic stress measurement from a Dynatest stress cell.

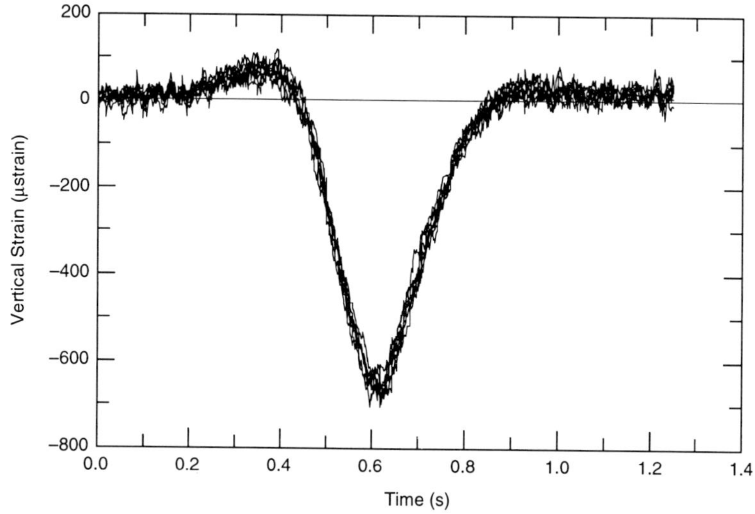

Figure 3. Typical vertical dynamic strain measurement from the εMU system.

combined effect of the dynamic strain and the moving metal. To correct for this a baseline measurement was made by moving the wheel over the pavement at zero load. This signal was subtracted digitally from the combined signal, thereby leaving only the dynamic strain response.

Surface rutting is measured with a laser profiler. The profiler uses a horizontal metal beam approximately 3 m in length with a stepper motor/cable system that moves a small carriage on the beam. Mounted on the carriage is a small infrared laser range finder. The computer program controls the speed of the carriage and collects 256 data points over the length of the run (approximately one point every 9 mm) as the carriage transverses the test section. Twenty-four cross section measurements are taken 305 mm apart and stored in a file for transfer to the main data system.

RESULTS

The analysis planned is on a hierarchical basis, the simplest being an elastic analysis. The analysis becomes more complex as one climbs up the hierarchical ladder, leading to plastic and elastic plastic characterization of the subgrade. The preliminary results from the A-2-4 and A-4 subgrade presented in this paper are based on linear elasticity. Failure is defined as 12.5 mm of vertical surface rutting. The results are for the test soils constructed at the optimum moisture content.

Strain results

Figure 4 shows the elastic vertical strains measured in the various test cells as a function of load passes. In addition, the limiting subgrade failure criteria as developed by Shell, the Asphalt Institute and the Corps of Engineers are shown for comparison.

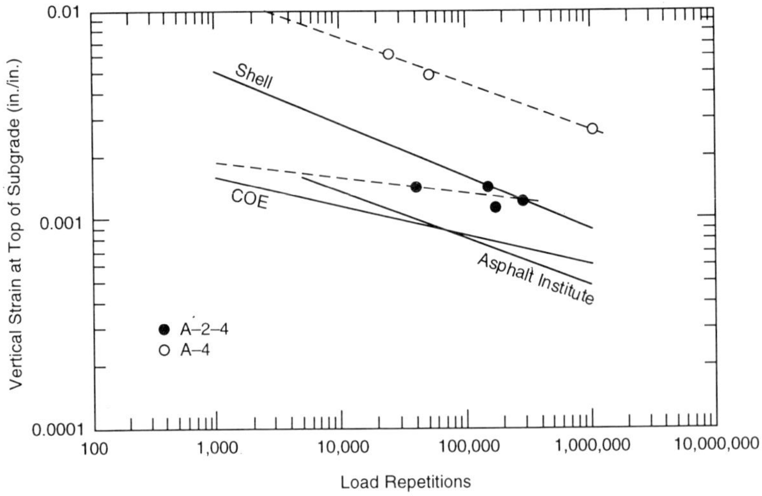

Figure 4. Vertical elastic strain as a function of passes.

Figure 5 shows the dynamic elastic strains measured in the horizontal x direction. The x direction is in the direction of wheel movement. The data were fitted with a power relationship similar to that used by Shell and the Asphalt Institute. The constants are shown in Table 2.

$$N_d = f_4(\varepsilon_z)^{-f_5}$$

where N_d = number of load repetitions (to 12.5 mm rut depth)
ε_z = vertical subgrade strain (μstrain)
f_4, f_5 = constants.

The R^2 values for the power equations are shown in Table 2. For the A-4 soil, the f_5 coeffi-

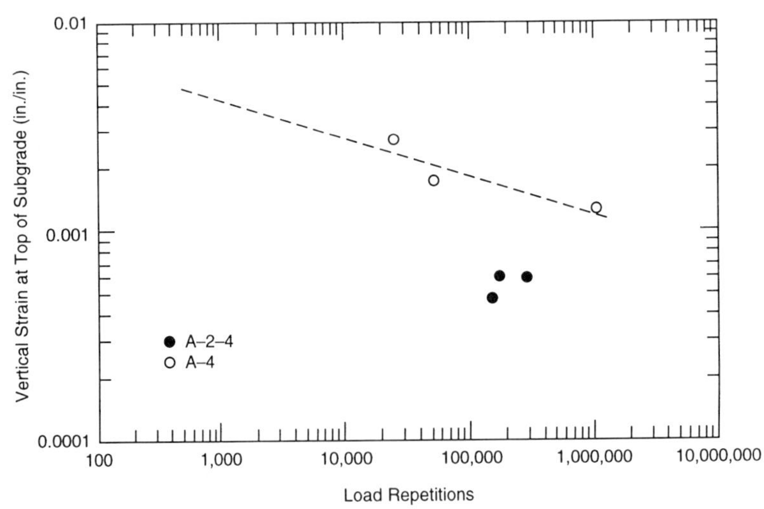

Figure 5. Horizontal elastic strain as a function of passes.

Table 2. Subgrade strain criteria used by various agencies.

Agency	f_4	f_5	*Rut depth (mm)*
Asphalt Institute	1.375E–9	4.45	12.5
Belgian Road Research Center	3.05E–9	4.35	10.2
Shell (revised 1985)	6.15E–7	4.00	—
50% reliability			
FERF Test Sections			
Z direction			
A-2-4 (Optimum moisture)	2.94E–4	11.86 ($R^2 = 0.40$)	12.5
A-4 (Optimum moisture)	1.95E–5	4.63 ($R^2 = 0.99$)	12.5
X Direction			
A-4 (Optimum moisture)	6.66E–5	5.55 ($R^2 = 0.67$)	12.5

cient (i.e., the slope of the line) for the vertical strain failure criterion is similar (4.6, versus 4.0 to 4.4) to those developed previously from the AASHO road test data (note that the AASHO road test subgrade was an A-6 soil). However, the f_5 coefficient for the A-2-4 soil is substantially different. And the f_4 coefficients (i.e., the intercept) are all quite different.

The AASHO road test results, and the results from the CRREL FERF tests on A-4 and A-2-4 soils are shown together in Figure 4. Taken together, the lines represent the results for three soils, an A-6 in a nearly saturated condition, and an A-4 and A-2-4 in a partially saturated condition. For a given level of strain, say 1500 microstrain, the predictions of allowable load repetitions would span from about 1000 to greater than 10,000,000. This leads to the preliminary conclusion that soil type and perhaps also degree of saturation may both have an effect on the subgrade vertical strain failure criterion.

These preliminary findings are based on very limited data. The conclusions may possibly change after additional testing over a broader range of soil type and moisture conditions is completed.

Stress results

The dynamic vertical and horizontal stress failure criteria are presented in Figures 6 and 7. The stresses were measured at a depth of 150 mm beneath the surface of the subgrade. The vertical-to-horizontal stress ratio associated with failure is in the range of 3 to 7.

Using a power law model similar to that used for strain, the following relationship was developed for the vertical stress in the A-2-4 soil:

$$\sigma_z = 100{,}000\ N^{-0.61} \qquad (R^2 = 0.96)$$

And for the horizontal stress in the A-2-4 soil:

$$\sigma_x = 100{,}000\ N^{-0.74} \qquad (R^2 = 0.85)$$

There was not sufficient data to develop any relationships for the A-4 soil, but the data are plotted in Figures 6 and 7.

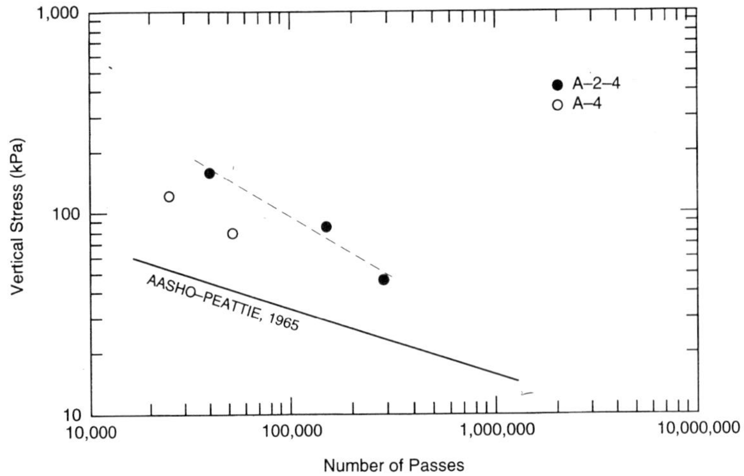

Figure 6. Vertical dynamic stress as a function of load passes.

The vertical stress relationship is compared to the AASHO road test result in Figure 6. Again it is observed that the AASHO A-6 soil, the A-2-4 and the A-4 soil appear to fall on different lines. The slopes for the A-2-4 and the AASHO lines are similar, but not identical. It is too soon to attempt to make any conclusions based on such limited data.

There are insufficient data for the A-4 soil to develop any reasonable relationships.

CONCLUSIONS

Based on limited preliminary data from tests on full-scale pavements utilizing the HVS in the CRREL Frost Effects Research Facility, the following conclusions appear warranted.

1. Pavement permanent deformation (rutting) does appear to be related to the level of verti-

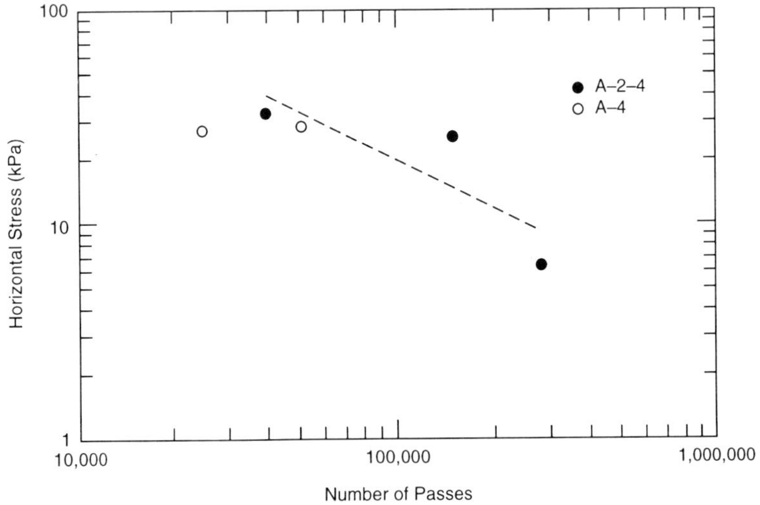

Figure 7. Horizontal dynamic stress as a function of load passes

cal compressive *strain* on the subgrade, as was originally postulated by Dorman and Metcalf [1]. However, early indications suggest that soil type may be an important consideration in the failure criterion. It is also possible that soil moisture content may be a factor.

2. Rutting may also be related to the level of vertical compressive *stress* on the subgrade, as was originally suggested by Peattie[2]. Again, early indications suggest that soil type may be a consideration in the failure criterion.

3. Both vertical and horizontal dynamic strains are being measured with the eMU system. Special techniques had to be employed (noise filtration and signal summing) in order to get good quality data. A value of Poisson's ratio computed from the data indicates that the data are reasonable.

4. Both vertical and horizontal dynamic stresses are being measured using Dynatest soil stress cells. The principal stress ratio at failure computed from the data seems to be reasonable.

The results reported herein are based on the testing of only two pavement sections, one with an A-2-4 subgrade and one with an A-4 subgrade. Overall plans for the study call for testing a total of four subgrade soil types (the two cited, plus an A-6 and an A-7-6). Each soil is planned to be placed at three different levels of moisture content. Thus twelve soil type - moisture content combinations are envisioned. Periodically, as additional opportunities to report project results occur, updates on the results will be made.

REFERENCES

1. DORMON, G.M. and METCALF, C.T. Design curves for flexible pavements based on layered system theory, Highway Research Record, No. 71, 1965.

2. PEATTIE, K.R. Design curves for flexible pavements based on layered system theory, Discussion, Highway Research Record, No. 71, 1965.

3. JANOO, V., IRWIN, L., EATON, R., KNUTH, K. and MARTINSON, C. Subgrade failure criteria, Volume 1, CRREL Report, in preparation, 1998.

Fifth International Conference on the
BEARING CAPASITY OF
ROADS AND AIRFIELDS

DETERMINATION OF IN-SERVICE ASPHALT PAVEMENT SUBGRADE SUPPORT FOR CONCRETE PAVEMENT DESIGN

By Trent J. McPeak[1], Lev Khazanovich[1], Michael I. Darter[1], and Curt A. Beckemeyer[1]

[1]ERES Consultants, Inc., 505 West University Avenue, Champaign, Illinois 61820-3915, tel.: (217)-356-4500, fax: (217)-356-3088

ABSTRACT

To evaluate the structural capacity of any pavement, certain information is required, including the knowledge of the strength properties of the underlying subgrade. Currently, generally accepted methods exist for the backcalculation of asphalt (AC) pavement and portland cement concrete (PCC) pavement layer stiffness characteristics using data collected with a falling weight deflectometer (FWD). Backcalculation programs based on elastic layer theory are used for AC pavements. Plate theory solutions for a slab on a dense liquid (Winkler) foundation are used for PCC pavements. There does not currently exist a widely accepted procedure for the backcalculation of subgrade support for flexible pavements resting on a dense liquid foundation, a required input for concrete pavement design.

The DIPLOMAT pavement evaluation program provides the capability to model a pavement structure that contains a mix of layers that behave as elastic or plate materials. The pavement subgrade can be modeled as either an elastic solid or a dense liquid. This program is especially useful for this study by modeling AC pavements treating the subgrade as a dense liquid foundation. A neural networks based backcalculation program, DIPLOBACK, has recently been.coded based on the DIPLOMAT algorithms.

The result is a reliable method for determining the subgrade support conditions that will exist under a concrete-reconstructed asphalt pavement, as well as the structural properties of the asphalt layer, through FWD testing the in-service asphalt pavement. Also presented is the simple closed-form solution for estimation of these structural parameters.

Keywords Backcalculation, Subgrade, Falling Weight Deflectometer, Pavement

INTRODUCTION

For several years now, nondestructive deflection testing has been used to determine the structural characteristics of an in-place pavement. Impulse loading devices, such as the Dynatest, KUAB, Jils, and Phonix FWDs, apply a load to the pavement surface and measure the resultant deflections. The data collected by the FWD are useful in determining the structural characteristics of a pavement cross section. This type of information is essential in evaluating the remaining life and required overlay thickness of a pavement selected for rehabilitation design.

ERES Consultants, Inc. (ERES) was contracted to provide pavement analysis and design services for a 10-mile, flexible pavement section of the Kansas Turnpike System. As part of this study, each feasible rehabilitation alternative for the section was considered and compared based on a life cycle cost analysis. The problem encountered was simply, "How to determine the subgrade support required for the PCC designs, k-value, directly from the FWD measurements?" The k-value should not be backcalculated using plate theory, as for rigid pavements, because of the deviation from plate theory behavior that a flexible system experiences, namely vertical compression of the upper layers.

This paper details the solution to this problem and presents a simple, theoretically sound method for the determination of the subgrade k-value as backcalculated from deflection data measured on a flexible pavement. This is an important tool because any pavement rehabilitation design study must examine every possible reconstruction option in order to find the most economical alternative.

ELASTIC LAYER THEORY

The elastic layer model for pavement structures (Burmister[1]) has been used successfully for a number of years to predict pavement deflections, stresses, and strains in AC pavements. The theory is based on the following assumptions:

1. A pavement is considered to be a multi-layered linear elastic system.
2. Each layer is of uniform thickness and infinite dimensions in horizontal directions, resting on a semi-infinite elastic and isotropic domain.
3. Each layer is assumed to consist of a homogeneous, isotropic, elastic material whose constitutive equation is governed by Hooke's law.
4. In the absence external loading, the system is free of stresses and deformations.
5. There are no body forces acting on the system, i.e. all layers are weightless.

The upper layers of the system are defined by thickness, elastic modulus, and Poisson's ratio of each layer. The subgrade is defined by elastic modulus and Poisson's ratio. The subgrade is assumed to be semi-infinite in depth, as described above, although several models allow a rigid layer to be placed beneath a subgrade of a user-defined finite depth.

Attractive features of this approach include the ability to represent a layered pavement structure with variable interface conditions between consecutive layers. Elastic layer theory allows the determination of displacement, stress, and strain fields, including through-the-thickness vertical stresses and displacements, thereby accounting for pavement layer compressibility.

PLATE THEORY

Plate theory has been used for many years in the analysis of PCC slab-on-grade pavement systems. Conventional analysis and mechanistic-based design procedures for PCC pavement systems make use of closed-form solutions developed over the last 80 years on the basis of quite restrictive assumptions. The assumptions used in the original development of the slab on dense liquid foundation model (Westergaard[2]) are as follows for the constructed (plate) layer:

1. Lines normal to the middle surface before deformation remain straight and normal to the middle surface of the deformed plate.
2. There is no deformation in the mid-plane of the plate, i.e. this plane remains neutral during bending.
3. The normal stress in the transverse direction can be disregarded.

The Winkler, or dense liquid, foundation (Winkler[3]) is the simplest foundation model. This model idealizes the subgrade as a system of closely-spaced, independent springs. No shear interaction exists between adjacent springs. Only one parameter is required to define the structural characteristics of this model, the modulus of subgrade reaction, k. This is the proportionality constant between the applied pressure and a load plate deflection at any point. The deflections are assumed to be elastic and recoverable upon removal of the load.

The slab is defined by a thickness, elastic modulus, and Poisson's ratio; the subgrade is characterized by a modulus of subgrade reaction, k. A very important assumption associated with plate theory is no compression of the upper layer. All of the deflection is attributed to compression of the subgrade and bending of the plate.

ELASTIC LAYER ON SPRINGS

The formulation of the isotropic, elastic layer on a springs foundation was originally developed by Van Cauwelaert[4]. The theory has been coded into a computer program called DIPLOMAT (Khazanovich[5], Khazanovich and Ioannides[6]). Also presented in the literature is the mathematical formulation of other combinations of pavement system cross-sections. The program is capable of analyzing a pavement system of up to five layers, which may be isotropic, plate, or spring layers. To accommodate multi-wheel loading, superposition is used, and Cartesian coordinates (x, y) are adopted, instead of polar coordinates (r, ϕ).

DIPLOMAT output results include horizontal and vertical stresses, strains, and deflections, as well as shear stresses and strains, at user-specified locations in the pavement system. Knowledge of these critical pavement responses facilitates the prediction of pavement performance under repeated loading. Modeling of pavement deflection at the surface of the system allows the formulation of backcalculation techniques. The remainder of this study will focus on backcalculation of isotropic-over-springs pavement systems as it applies to AC pavements.

BACKCALCULATION THEORY

Pavement deflections measured with an FWD in interior regions of an AC pavement (away from cracks and the pavement edge) may be used to backcalculate the structural parameters of the pavement layer and foundation. These backcalculation results may be examined to draw some conclusions about the degree of structural deterioration in the pavement layers, the support conditions of the in-place subgrade materials, and the expected remaining life of the overall pavement structure. In addition, the backcalculated moduli may be used for design of future structural overlays for the pavement.

The general principle of backcalculation is based on the assumptions that FWD-measured deflections are dependent on the following parameters:

- Thickness and stiffness of pavement layers.
- Stiffness of the foundation.
- Magnitude of the applied load.
- Radius of applied load.
- Rate of loading.
- Position of the deflection sensors.

If the applied load, layer thickness, and deflection positions are known, the measured pavement deflections may be used to solve for the stiffness of the pavement layers and foundation. The measured deflections are fit to theoretically predicted deflections by various methods. In this study, the isotropic layer on springs foundation solutions presented above were used to predict deflections, using the DIPLOMAT computer program.

DIPLOBACK

The pavement backcalculation computer program, DIPLOBACK, was originally developed for the backcalculation of composite (AC-overlaid concrete) pavements (Khazanovich and Roesler[7], McPeak and Khazanovich[8]). Both constructed layers were modeled as isotropic elastic layers, whereas the subgrade was modeled as a Winkler foundation. The advantage of this approach is, unlike plate theory, compressibility throughout the thickness of the constructed layers (especially the AC layer) is not ignored, and the subgrade is characterized in a conventional manner for PCC pavement design (k-value). The same statement holds true for AC pavements on a springs foundation. Compression is allowed in the AC layer and the typical subgrade model for PCC pavements is used.

The computer program DIPLOMAT was used for forward calculations to determine predicted pavement deflections based on a range of typical input values for the pavement structure. Another distinguishing feature of DIPLOBACK is that it uses neural networks to speed up the backcalculation process. This makes DIPLOBACK an attractive alternative for AC pavement backcalculation with the subgrade modeled as springs. The neural networks concept is still relatively young and has not attained wide-spread acceptance, therefore a closed-form solution using simple equations, that may be entered into a spreadsheet, was developed as described in the next section.

AREA-BASED BACKCALCULATION

The AREA parameter was first proposed by Hoffman and Thompson[9] for the backcalculation of flexible pavements with an isotropic, elastic subgrade characterization. Backcalculation solutions for PCC pavement structures based on the AREA method have been developed based on Westergaard's[2] plate theory solutions with a dense liquid subgrade characterization (ERES[10], Ioannides[11], and Hall[12]).

In this study, the standard deflection sensor configuration used in the LTPP program was used, placing sensors at 0, 203, 305, 457, 610, 914, and 1524 mm (0, 8, 12, 18, 24, 36, and 60 in) from the center of the load plate of 150 mm (5.9 in) radius. It was found that a unique relationship, with relatively small variation, exists between AREA and l_k when the first 2 sensors, at 0 and 203 mm, were excluded from the analysis, due to compression in the AC layer. Hall, et al[13] found that excluding the sensors at 0 and 203 mm improved backcalculation for composite pavements. AREA for remaining sensors is calculated as follows:

$$AREA5 = 3 + 6\left(\frac{d_{18}}{d_{12}}\right) + 9\left(\frac{d_{24}}{d_{12}}\right) + 18\left(\frac{d_{36}}{d_{12}}\right) + 12\left(\frac{d_{60}}{d_{12}}\right)$$

where di are the measured deflections at i-inches from the center of the applied load. AREA has units of length because each of the deflections is normalized with respect to one of the deflection sensors. This normalization is done to remove the effect of load level and to restrict the range of possible values. The AREA term may be easily defined for any number and arrangement of deflection sensors, however, this study will concentrate on the LTPP sensor configuration described above.

The AREA term is valuable in PCC backcalculation, using plate theory, due to the fact that there is a unique relationship between AREA and the radius of relative stiffness, defined as:

$$l_k = \sqrt[4]{\frac{E_c h_c^3}{12\left(1-\mu_c^2\right)k}}$$

where l_k is the radius of relative stiffness, E_C is the PCC elastic modulus, h_C is the PCC slab thickness, μ_C is the PCC Poisson's ratio, and k is the modulus of subgrade reaction. A study was conducted to determine a sensor configuration for which there exists a unique relationship between the AREA and radius of relative stiffness using AC-specific parameters, as opposed to PCC-specific parameters, along with the Winkler subgrade characterization.

The DIPLOMAT program was used to calculate pavement surface deflections for a range of input parameters. The input ranges used were:

- E_{AC} ranged from 100,000 to 1,000,000 psi.
- h_{AC} ranged from 6 to 16 in.
- k ranged from 20 to 350 psi/in.
- μ_{AC} was 0.35.
- Load was 10,000 lb.
- Load radius was 5.9 in.

Deflections at positions corresponding to the SHRP sensor spacing were output and calculations were performed to determine the radius of relative stiffness, l_k, and AREA term for each theoretical deflection basin. A statistical regression was done to determine the AREA versus l_k relationships. Figure 1 shows this relationship for the AREA definition, AREA5, as described above. The relationship between AREA5 versus l_k was found to be:

$$l_k = -8 \times \ln\left[1.21 \times 10^{-7} \times \ln(48 - AREA5)^{11.33}\right]$$

Nondimensional Deflections

According to plate theory (Westergaard[14]), the deflection at the plate surface on a dense liquid foundation, at any radial distance from the applied load, is a function of a nondimensional deflection coefficient, d*, the load magnitude, and the pavement structure parameters for the slab and foundation:

$$d_r^* = \frac{d_r k l_k^2}{P}$$

The nondimensional deflection is therefore a function of load radius a, the radial distance r, and l_k. For a constant load radius (5.9 inches), d* for a given radial distance is therefore a function solely of l_k. The theoretical deflections computed earlier by DIPLOMAT, elastic layer on springs model with asphalt-typical input values, were used to develop nondimensional deflections versus l_k relationships. Figure 2 shows plots for these relationships.

Regression equations were developed for the nondimensional deflection versus l_k relationships at the appropriate radial distance from the applied load. The equations were developed for each sensor spacing considered in this study using the following form. The coefficients A, B, C, D, and E are given in table 1.

$$d^* = Al_k^4 + Bl_k^3 + Cl_k^2 + Dl_k + E$$

Table 1. Regression coefficients for the d* versus l_k equations.

Distance, in	A	B	C	D	E	R^2
12	-2.78E-8	5.18E-6	-3.56E-4	1.09E-2	-8.93E-3	96.6
18	-2.71E-8	5.22E-6	-3.77E-4	1.25E-2	-5.21E-2	99.6
24	-1.88E-8	3.90E-6	-3.09E-4	1.16E-2	-6.91E-2	99.9
36	6.30E-9	-4.59E-7	-4.55E-5	5.57E-3	-5.12E-2	99.9
60	3.17E-8	-5.47E-6	3.14E-4	-5.07E-3	2.31E-2	99.9

Solution for k and E_{AC}

With l_k determined from the AREA5 relationship, and d* computed for various sensors from l_k, the subgrade k-value may be determined for each sensor at 12, 18, 24, 36, and 60 inches from the applied load by the following equation:

$$k = \frac{Pd^*_{*r}}{d_r l_k^2}$$

where P is the applied load, and d_r is the deflection measured at a distance, r, from the applied load. The k-value for the FWD test may then be determined by averaging the individual values obtained from the various deflection measurement locations. The elastic modulus of the asphalt surface layer may then be obtained easily from the following relationship:

$$E_{AC} = \frac{12(1-\mu_{AC}^2)kl_k^4}{h_{AC}^2}$$

RESULTS OF FIELD TESTING

FWD testing was conducted by ERES Consultants on an asphalt-surfaced section of the Kansas Turnpike in July, 1997. The FWD testing was conducted between mileposts 89 and 94 in the southbound traffic lane using loads ranging from about 8,000 to 16,000 lb and the sensor spacing described earlier. The asphalt layer thickness was approximately 13 in and was relatively uniform along the length of the project, based on the results of pavement coring.

The resulting data were backcalculated using DIPLOBACK neural networks, the AREA5 closed-form method, and WESDEF elastic layer program. Figure 3 shows a comparison of DIPLOBACK and the AREA5 method. It can be seen that these two methods agree quite well. Figure 4 shows a comparison of the AREA5 results versus WESDEF backcalculation of the subgrade elastic modulus. The trends are nearly the same for the two methods.

CONCLUSIONS

Presented in this paper is a theoretically sound, closed-form procedure for the backcalculation of the subgrade modulus of reaction, k, for flexible pavement systems. A neural networks-based procedure, DIPLOBACK, was developed based on deflections predicted by the DIPLOMAT pavement evaluation computer program. Also, a closed-form solution based on the popular AREA method was developed. The AREA5 method allows the user to perform this analysis without the need for the DIPLOMAT program or a neural network system.

The results of DIPLOBACK and the AREA5 method were compared and close agreement was found. The AREA5 method was also compared with the results of the WESDEF elastic layer backcalculation program. The trends observed for the two methods were nearly identical.

In any pavement evaluation and reconstruction design, all feasible alternatives must be considered. Previously, for existing AC-surfaced pavements, the subgrade support for reconstruction design with PCC could not be evaluated directly with FWD data. The new backcalculation methods allow for the first time direct calculation of the modulus of subgrade reaction, a necessary input in concrete pavement design, for flexible pavement systems.

REFERENCES

1. Burmister, D.M. (1943). "The Theory of Stresses and Displacements in Layered Systems and Application to the Design of Airport Runways," Proceedings, Highway Research Board, Vol. 23, National Research Council, Washington, D.C., pp. 126-144.
2. Westergaard, H.M. (1926). "Stresses in Concrete Pavements Computed by Theoretical Analysis," Public Roads, Vol. 7, No. 2, April, pp. 25-35.
3. Winkler, E. (1867). "Die Lehre von der Elastizitat und Festigkeit," (Theory of Elasticity and Strength), H. Dominicus, Prague (in German).
4. Van Cauwelaert, F. (1990). "'Westergaard's Equations' for Thick Elastic Plates," Second International Workshop on the Theoretical Design of Concrete Pavements, Siguenza, Spain, October, pp. 165-175.
5. Khazanovich, L. (1994). "Structural Analysis of Multi-Layered Concrete Pavement Systems," Doctoral thesis presented to the University of Illinois, Urbana.
6. Khazanovich, L., and A.M. Ioannides (1995). "DIPLOMAT: An Analysis Program for both Bituminous and Concrete Pavements," Transportation Research Record No. 1482, TRB, National Research Council, Washington, D.C., pp. 52-60.

7. Khazanovich, L., and J. Roesler (1997). "DIPLOBACK: A Neural Networks-Based Backcalculation Program for Composite Pavements," Transportation Research Record No. 1570, TRB, National Research Council, Washington, D.C., pp. 143-150.
8. McPeak, T.J., and L. Khazanovich (1997). "Backcalculation of Asphalt-Overlaid Concrete Pavements," Aircraft/Pavement Technology In the Midst of Change, ASCE, Seattle, WA, August, pp. 107-124.
9. Hoffman, M.S., and M.R. Thompson (1981). "Mechanistic Interpretation of Nondestructive Testing Deflections," Civil Engineering Studies, Transportation Engineering Series No. 32, Illinois Cooperative Highway and Transportation Research Program Series 190.
10. ERES Consultants, Inc. (1982). "Nondestructive Structural Evaluation of Airfield Pavements," prepared for U.S. Army Corps of Engineers Waterways Experiment Station, Vicksburg, MS.
11. Ioannides, A.M. (1990). "Dimensional Analysis in NDT Rigid Pavement Evaluation," Journal of Transportation Engineering, ASCE, Vol. 116, No. 1, New York, NY.
12. Hall, K.T. (1991). "Performance, Evaluation, and Rehabilitation of Asphalt-Overlaid Concrete Pavements," Doctoral thesis presented to the University of Illinois, Urbana.
13. Hall, K.T., M.I. Darter, T.E. Hoerner, and L. Khazanovich (1997). "LTPP DATA ANALYSIS Phase I: Validation of Guidelines for k-Value Selection and Concrete Pavement Performance Prediction, FHWA-RD-96-198, Federal Highway Administration, Washington, D.C.
14. Westergaard, H.M. (1939). "Stresses in Concrete Runways of Airports," Proceedings, Highway Research Board, Vol. 19, National Research Council, Washington, D.C.

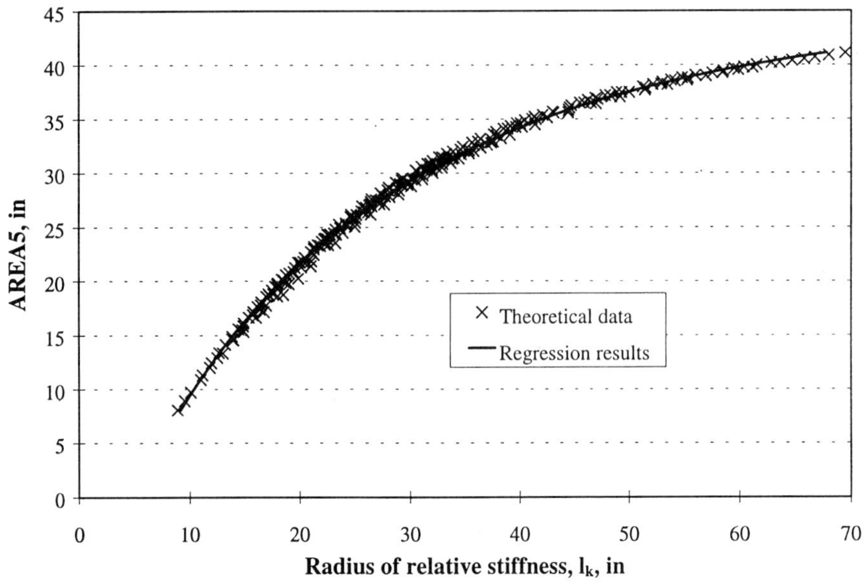

Figure 1. Regression results for AREA5 versus l_k.

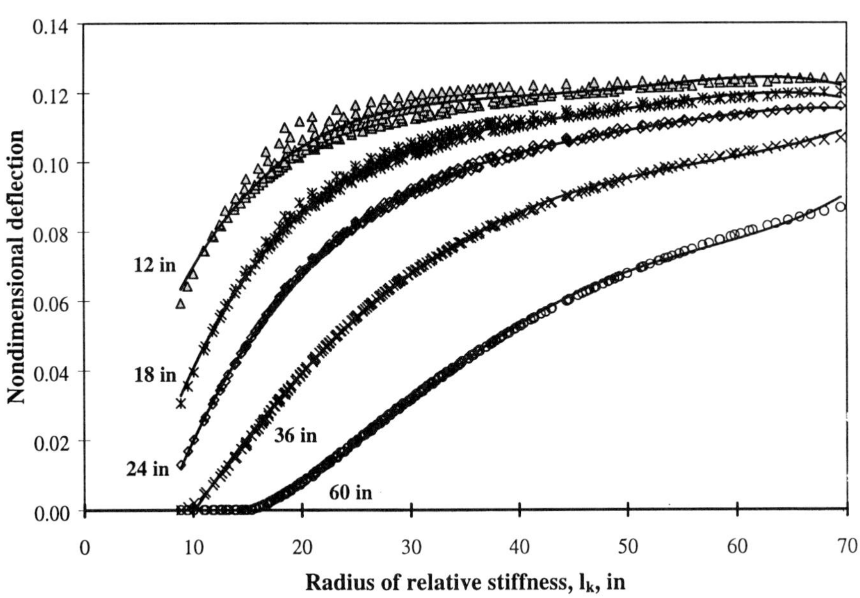

Figure 2. Nondimensional deflection coefficients versus l_k.

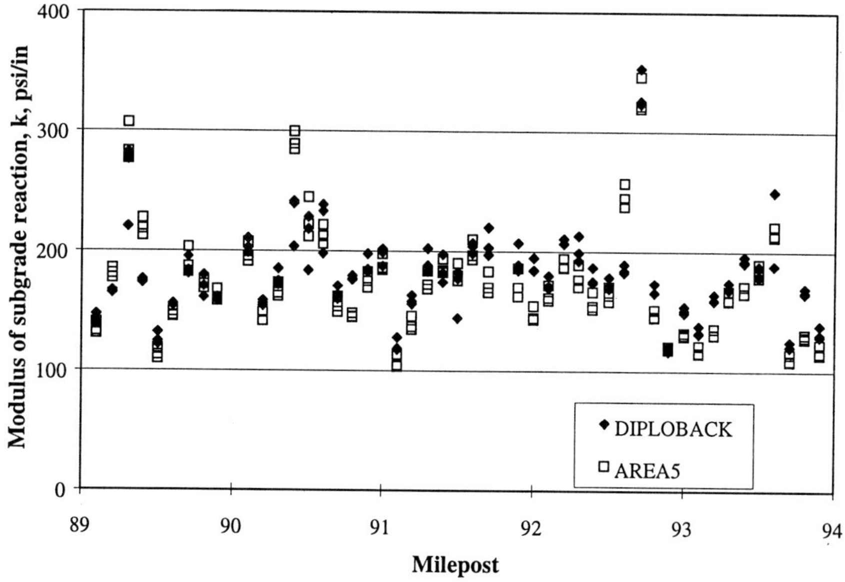

Figure 3. Comparison of k from DIPLOBACK and AREA5.

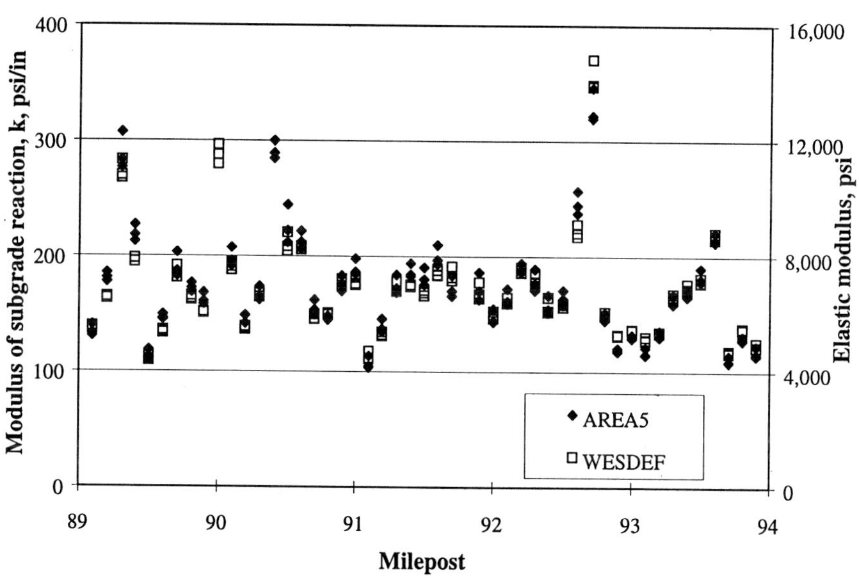

Figure 4. Comparison of k from AREA5 and E from WESDEF.

Fifth International Conference on the
BEARING CAPASITY OF
ROADS AND AIRFIELDS

CONSOLIDATION INDUCED STRAINS IN ASPHALT CONCRETE PAVEMENTS

K. ONITSUKA [1]   J. N. SHEN [2] ,   S. R. SHEN [3]

[1] Professor, Saga University, Saga 840, Japan

[2] Graduate Student, Ditto, 97ts51@edu.cc.saga-u.ac.jp

[3] Graduate Student, Ditto

ABSTRACT

Behaviors of pavement structures constructed on soft ground will be influenced by consolidation of the soft ground. Building pavements with low embankment is one of the practical ways that can eliminate the influence of the consolidation on the pavement behaviors. In the study, construction of pavements with low embankment on the soft foundations was simulated by CRISP, a finite element method (FEM) program. The pavement behaviors in consideration of the consolidation of the soft ground were discussed. Parametric studies to reduce the influences of consolidation of the soft ground on pavement behaviors were also dealt with finally.
Keywords: asphalt concrete, pavement, consolidation, finite element method

INTRODUCTION

Pavement structures now are faced to build on soft ground due to the limitation of high quality lands. There are commonly two different ways to build pavements on the soft ground: with low embankment (the height of embankment including the pavement structure is less than 3m) and high embankment. In Saga, Japan, where there is a soft ground of Ariake clay with a depth of about 10m to 20m, it has been preferably adopted that pavements are constructed with low embankment on soft ground. Fig.1 is a typical pavement structure with low embankment on soft ground. The reason for the use of low embankment is that it has fewer problems such as slip failure and settlement than high embankment. The traffic load induced stresses will be, however, transferred to the soft foundation. The soft ground will consequently generate consolidation under the pavement self-weight and traffic loads because the soft ground have high compressibility with high water content and low strength. The pavement structure must be paved strong enough to sustain traffic loads and the additional strains induced by consolidation of the soft ground. Therefore, it is important and necessary to analyze the influence of consolidation of the soft ground on the pavement structure.

In convention, analyses of pavement structures are mostly based on a multi-layer system, either linear or non-linear elastic. The influence of the consolidation of soft ground on pavement structures can not be taken into account by this theory. Fortunately, finite element method (FEM) has been used to analyze pavement structures since 1960 (Duncan et al 1968). Now, a lot of programs based on FE are available to perform pavement analysis.

In the study, the construction of pavements with low embankment was simulated by CRISP FE program. The influences of consolidation of the soft ground on pavement behaviors are discussed. The pavement behaviors concerned are settlement of the pavements, vertical compressive strains on the top of subgrade and horizontal tensile strains at the bottom of asphalt concrete layer as well. These are related with analytical design methods (Huang, 1993). Pavement structures were modeled as an axis-symmetric problem. Elements with coupled displacement and pressure are adopted for the soft ground. Cam-clay model was adopted for the soft ground.

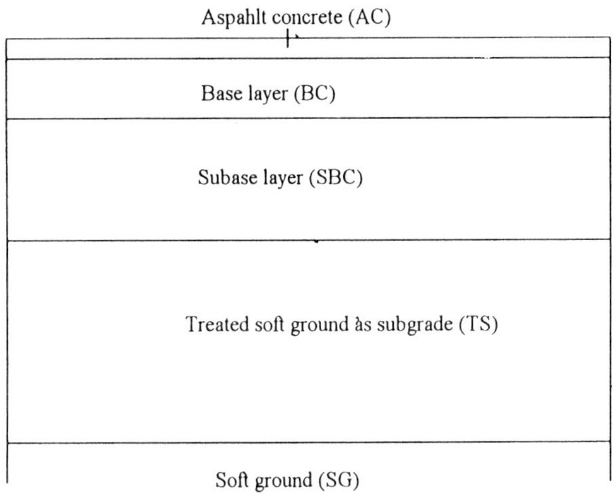

Fig.1 A typical pavement structure

ANALYSIS MODEL

Asphalt concrete pavements are chosen with priority on soft ground areas. The pavement discussed here is a typical asphalt concrete pavement, which is used in Saga. It consists of a 5 cm asphalt concrete as surface layer, a 15 cm aggregate base layer, a 30 cm cement treated subbase and treated subgrade. A 10 m deep soft ground supports the pavement structure.

A pavement structure acted by one wheel traffic load has been modeled in highway engineering as a multi-layer half space acted by a uniformly distributed circular load. This is a typical axisymmetric problem. In the study, an axisymmteric (2D) model was discussed. The geometry of the 2D model is shown in Fig. 2.

Here it should be noted that the width of the model influences the calculating results of the pavements, especially on the vertical compressive strains along the top of the subgrade. But the calculating results of strains in pavements will be convergent if the width is over 6 m wide. When consolidation of the soft ground is concerned, the width should be wider. It depends on the depth of the soft ground discussed in the consolidation. The model discussed here is lager than traditionally used model that only ten times the radius of loaded area was assumed, and also different from the suggestion made by Duncan et al (1968). Duncan et al pointed analysis should extend to 20 times the radius horizontally. The model is 25 m wide 11 m deep. It is to simulate a 10m deep soft ground which supports the pavements with thickness of 1m.

Meshing of pavements is another important factor influencing the calculation results. Studies (Shen, 1997) shows the convergent results, not only the tensile strains but also the compressive strains from the FEM can properly agree with the results from multi-layer elastic theory. In the study, asphalt concrete has one element in depth, two elements in base, three elements in subbase and subgarde layers. Element sizes in horizontal direction vary from 0.05 cm near loading the center to 5 m at the right boundary. Elements in all layers change in same size horizontally. Totally, there are 176 elements.

A static analysis was performed in the study. Loads considered here in the analysis are not only traffic load, but also pavement self-weight. Pavement structures are treated as external loads that acted on the soft foundation. The four pavement layers are divided into three load blocks. Each of the blocks is put on the soft foundation with in 10 days. Thus, the pavements are expected to complete in 30 days. The loading speed affects the settlement of pavements on soft ground. It should not be too fast.

After construction of the pavements is completed, then traffic load is exerted on the pavement surface. The traffic load is standard traffic load, having a single axle with dual wheel each side. The total weight is 100KN. Thus the load in the 2D model is 0.795MPa on 10 cm.

In-situ stresses were included in the study. The water table was set up at the top of subgrade. The boundary conditions were applied in such a way that no vertical displacement was prescribed to the bottom of the model, no displacement to the right side horizontally and no excess pore pressure along the right side of the model In case of consolidation, see Fig 2 also.

MATERIAL MODELS

Pavement materials are traditionally described as linear or nonlinear elastic. This is not always the case, especially in case that consolidation of the soft ground are considered. The

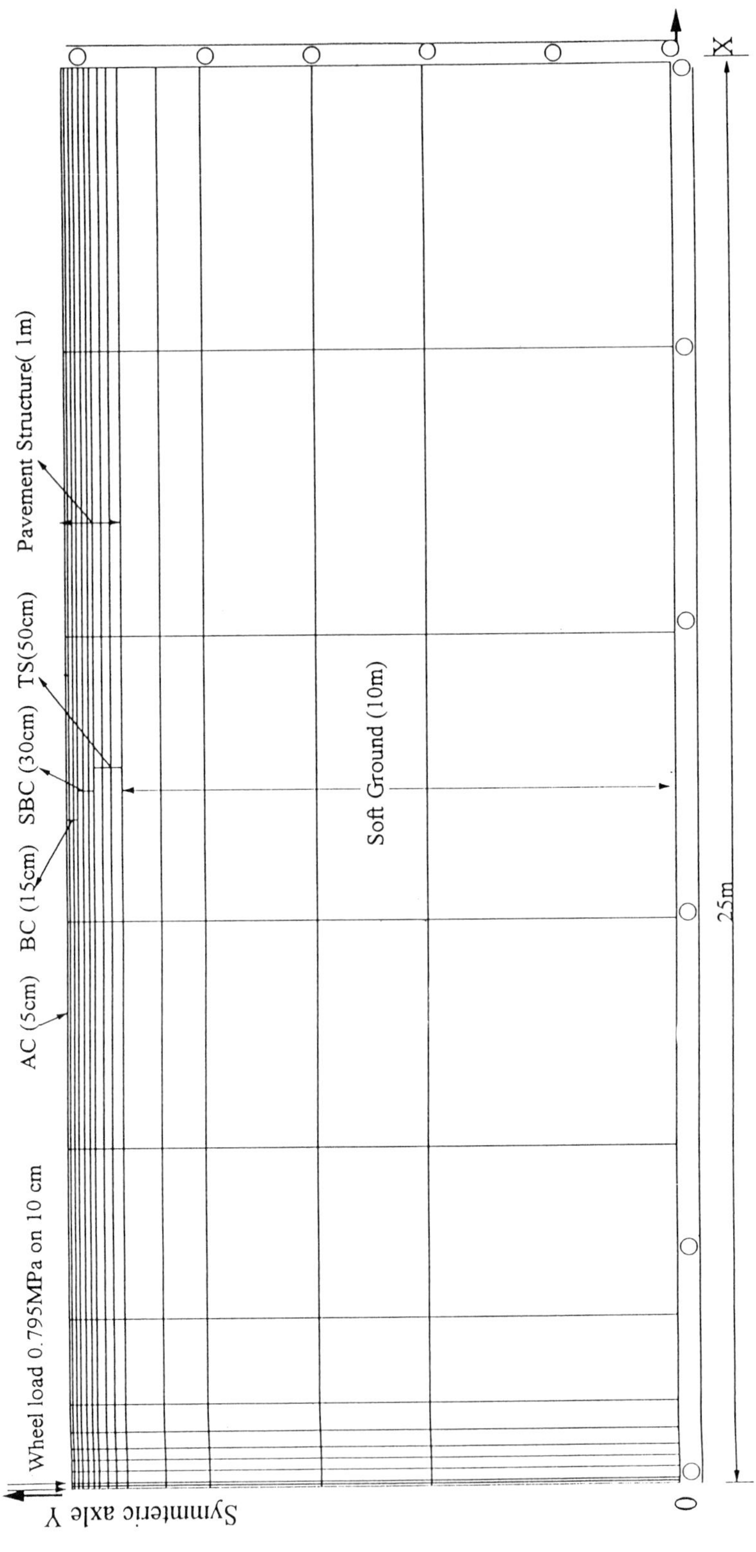

Fig.2 Model and meshing of pavement structure

settlement of the pavements due to the consolidation can be as much as 10 to 40 cm in Saga, Japan (Fujikawa, 1996). The behavior of the soft ground will be by far no elastic. Therefore, critical state machanics could be proper to discuss the properties of the soft ground under the pavements. This can be illustrated by a comparison of the measured (Fujikawa, 1996) and the predicted settlements of a five-layer pavement with time presented in Fig.3. This calculated settlements are based on two different models. The results from elastic model were based on elastic theory for all pavement layers with values of E1-E4 and E5 for the soft ground. The results from Cam-clay model were based on elastic model for all pavement layers with values of E1-E4 and Cam-clay for the soft ground with values of κ, M, λ. It is apparent that the measured values agree better with the calculating settlements from Cam-clay than those from elastic model, Specially, after some days of consolidation, soil of the soft ground reaches critical state, consequently the settlements obtained from the Cam-clay are larger than from elastic model.

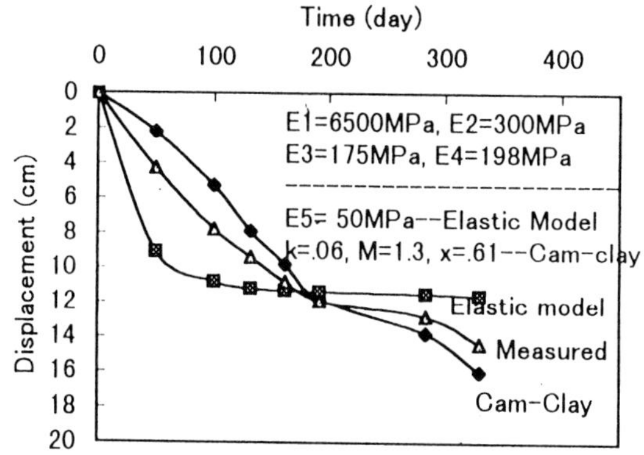

Fig. 3 A comparison of settlements by different models

Cam-clay model was utilized therefore in the study. Following is a brief introduction of the model. It is a critical state model of soil.mechanics. A state of soil under triaxial test can be described by three parameters, p', q, V. When the triaxial test proceeds, the sample yields at a point when the value of q is given by the following equation:

$$q = M p' (\Gamma + \lambda - \kappa - V - \lambda \ln (p'))/ (\lambda - \kappa) \quad (1)$$

where M is a constant for a particular soil, Γ, another constant for a particular soil, λ and κ, slops of the virgin compression, swelling lines in (ln (p'), V) plot, q, the mean normal effective stress, p , the deviator stress, V, the specific volume.

Equation (1) describes a surface in (p', V, q). When the state of a soil specimen can be represented by a point below the surface, then the soil is elastic. Soils states on the surface indicate yield, and it is impossible for soil samples to exit in states equivalent to points above the surface. A traditional formation of the yielding surface can be expressed as:

$$q = M p' \ln (p_c' / p') \quad (2)$$

where p_c' is the isotropic pre-consolidation stress for a soil sample lying on a particular κ-line.

The characteristic values used for the CRISP analysis are modules of elastic (E) and Poisson's ratio (μ) for elastic model, and κ, λ, Γ, M for Cam-clay model. The values adopted in the study are listed in Tab1.

Table1 Parameters of pavement materials

Layers	E(MPa)	μ	κ	λ	Γ	M	γ (kN/m³)	Kx (m/d)	Ky(m/d)
AC(E1)	6500	0.35					23	1.E-5	0.5E-5
BC (E2)	300	0.35					19	1.E10-2	0.5E-2
SBC(E3)	160	0.35					19	1.E10-2	0.5E-2
TS (E4)	150	0.35					19	1.E10-2	0.5E-2
SG		0.25	0.06	0.61	2.5	1.3	15	1.E10-3	0.5E-3

NUMERICAL RESULTS

Fig.4.1-4.2 presented are the numerical results. Responses of the pavement construction calculated are settlement at the loading center, the horizontal tensile strains at the bottom of the asphalt concrete layer and the vertical compressive strain on the top of subgrade as well. The beginning of the consolidation assumed in Fig.4 is the time just after the pavement structures are open to the traffic loads.

Settlement at the loading center could increase about 10 cm when the degree of consolidation reaches about 90 % in a little more than one year after pavements are paved and open to traffic loads. The vertical compressive strains on the top of subgrade also change with the occurrence of the consolidation of soft ground. The stable strain is about 25% larger than the initial one when the consolidation starts. In other words, the vertical strain will be underestimated without considering the influence of consolidation by using the traditional analysis method such as multi-layer elastic theory. Obviously, the effect of increasing vertical strain will prompt the occurrence of rutting on the pavement surface, because the rutting are controlled by the vertical strains on top of subgrade.

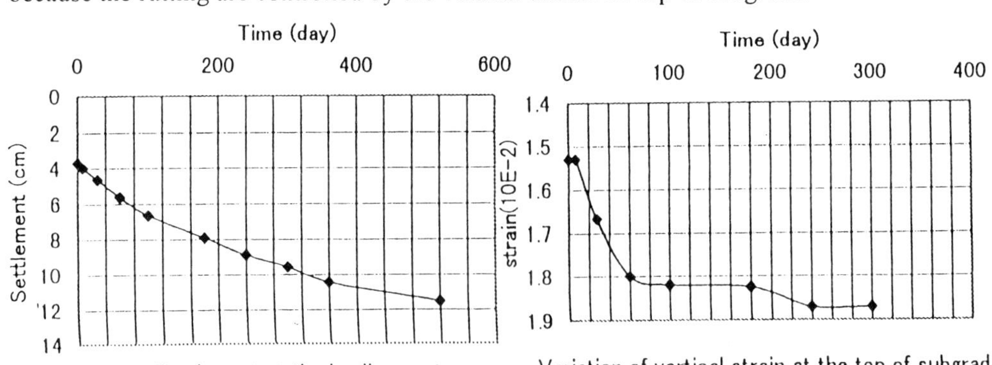

Fig. 4.1 Effects of consolidation of soft ground on asphalt concrete pavement

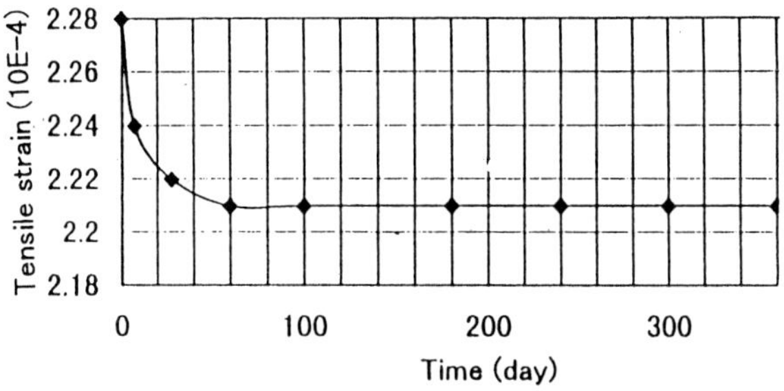

Variation of tensile strain at the bottom of AC

Fig. 4.2 Effects of consolidation of soft ground on asphalt concrete pavement

The horizontal tensile strains at the bottom of the asphalt concrete are also influenced with the time by the consolidation of soft ground. However, the strains are decreasing a little with time. The stable strain is about 5% lower than the initial one. Although the decreasing ratio is small, the effect on the pavement performance tends to be safe.

Strains induced by consolidation of the soft ground are in essence affected by the differential settlement In order to investigate the variation of the differential settlement of soft ground shapes of settlement at different time will be discussed. Figs5 presented below are plots of the distribution of settlement of soft ground at two different times. It is clear that the shape of settlement at the beginning is steeper near the loading area. With time, it becomes more and more flat. However, the differential settlement between the loading center and the point at 6m away from the loading center increase 3 times during the period of 2 years after consolidation. As we know, the horizontal tensile strain at the bottom of pavement layers is more directly related with the curvature of the distribution of the settlement, therefore it will decrease with the consolidation. On the other hands, the vertical strain on the top of subgrade seems more likely related with the differential settlement. This is why the vertical compressive strain increases with the consolidation.

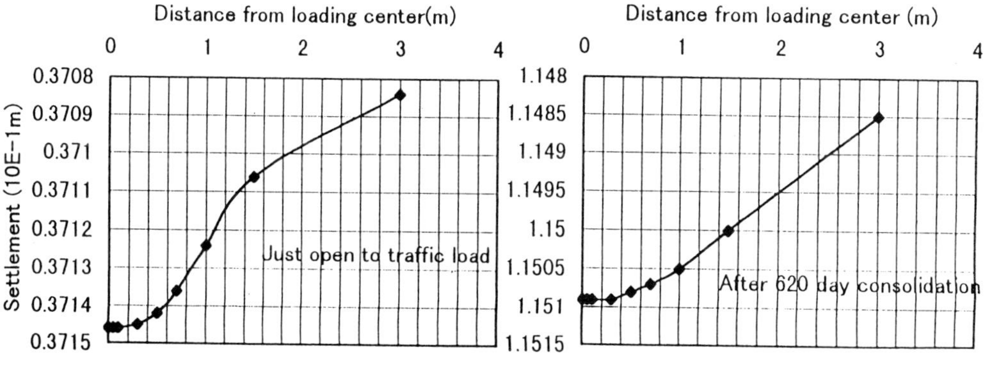

Fig.5 Distribution of settlement of soft foundation with consolidation

PARAMETRIC STUDIES

Usually, It is necessary to carry out some treatments of soft ground before pavements are built. There are currently two main approaches to treat the ground. One is to enhance the stiffness and strength of the soft ground as subgrade layer. The other is to reduce the self-weight of the treated layers. In order to model these two methods, a parametric study was made in the study.

The modules of the treated subgrade adopted are from 20MPa to 150MPa. The influences of modules of elasticity of the treated subgrade are show in Fig6. The influence of the module of elasticity on reducing the settlements of the pavements is insignificant. However, it is rather efficient to eliminate the increasing of the vertical compressive strains on the top of subgrade. The horizontal tensile strains at the bottom of asphalt concrete will be more stable with the improvement of the subgrade.

As for the unit weight of the treated subgrade, no significant influences on the pavement response was obtained with its values from 14-20 kN/m

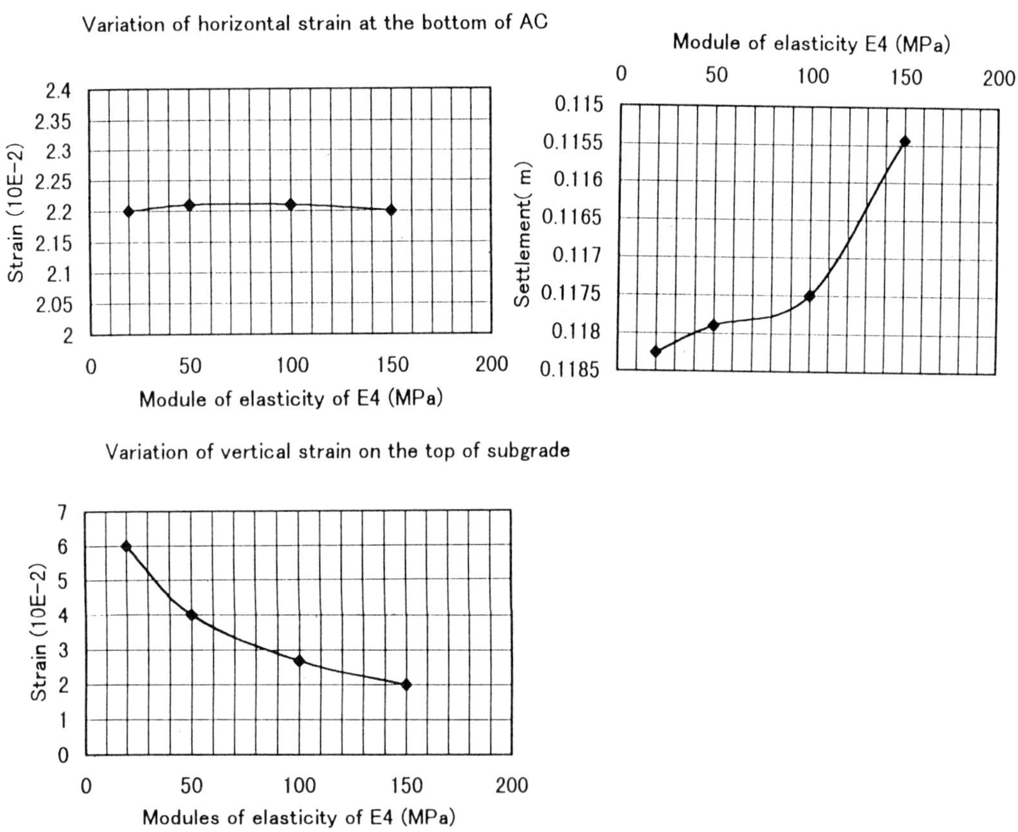

Fig.6 Influences of modules of elasticity E4 on pavement behaviors

CONCLUSION

In the study, asphalt concrete pavements with low embankment on soft foundation was simulated by CRISP FE program. The influences of consolidation of the soft ground on the pavement were discussed. Pavements were modeled as an axisymmetric model. Elastic model was adopted for the top four pavement layers, Cam-clay model for the soft ground. The following results has been observed:

1) Cam-clay model is better than elastic model to predict settlement on asphalt concrete pavements constructed on Ariake soft ground.

2) The influence of the consolidation on the vertical compressive strains on the top of the subgrade is significant. A 25% increasing than the initial one when it is stable has been observed in the study. So it is serious without considering the consolidation with respect the compressive strains. In contrast, influence of the consolidation on the horizontal tensile strains is relatively small and about 5% decreasing than the initial one was observed.

3) A study on the distribution of settlement of the soft ground shows that the curvature of the curve becomes more and more flat, but the differential settlement increase with time. Thus, the strain distributions in pavements are accordingly affected by the changing of the distribution of the settlement.

4) Improvements of the subgrade, either enhancing the stiffness or reducing the unit weigh, have no significant effect on the reduction of pavement settlements but can affect the strain distribution in the pavements. The pavement settlement is mainly controlled by the properties of the soft foundation. But, improvements with increasing the stiffness of subgrade can change the strain distributions in the pavements, especially for the vertical strains on the top of subgrade.

ACKNOWLEDGEMENTS

The authors are grateful for the help given by Dr. Jinchun Chai, an Associate Professor, Department of Civil Engineering, Saga University.

REFREERENCE

1. Duncan, J.M., Monismith, C.L. & Wilson, E.L.(1968), Finite element analysis of pavements, Highway Res. Record No. 228, Highway Research Board, Washington. DC.
2. Huang, Y.H. (1993), Pavement analysis and design. Printice-Hall, Englewood Cliffs, NJ, USA
3. Britto, A.M., Gunn, M.J.(1978), Critical state soil mechanics via finite elements, Elis

Horwood Limited
4. Wood, D.M.(1990), Soil behavior and critical state soil mechanics Cambridge University press
5. Fujikawa, K. (1996), A study on design of pavements with low embankment on soft foundations considering traffic induced settlement, Dissertation, Saga university, (in Japanese)
6. Brow, S.F. (1996), Soil mechanics in pavement engineering. Geotechnique 46,No. 3, 383-426

Fifth International Conference on the
BEARING CAPASITY OF
ROADS AND AIRFIELDS

Sensitivity Analysis of Predicted Stress-dependency and Plastic Behavior for Load Zoning

Seong-Wan Park[1] and Emmanuel G. Fernando[2]

[1] Graduate Research Assistant, Texas Transportation Institute, Texas A&M Univ., College Station, Texas 77843, U.S.A.
[2] Associate Research Engineer, Texas Transportation Institute, Texas A&M Univ., College Station, Texas 77843, U.S.A.

ABSTRACT

Most of the load-zoned pavements in Texas are presently posted using gross vehicle weight (GVW) limits to prevent or minimize the development of permanent deformation due to excessive truck loads. In reality , however, pavement damage is more related to axle loads and axle configuration rather than GVW. Thus, load restrictions need to be related to axle loads, axle type, pavement structure, and material characteristics. For this reason, a nonlinear elasto-plastic finite element program using the Universal Soil Model was developed to evaluate the structural adequacy of load-zoned pavements. This program can assess pavement response based on a bearing capacity approach using the Mohr-Coulomb failure criterion. The stress-dependent behavior of pavement materials is also considered. A sensitivity analysis is conducted to identify the significant factors in the pavement structural model. The predicted trends of resilient and plastic behaviors from this analysis are presented.

Key words: Resilient modulus, load-zoned pavements, stress-dependency, plastic strains.

INTRODUCTION

Texas has approximately 28,175 km of load-zoned pavements. About 98 percent of these are posted with a gross vehicle weight restriction of 260 kN. While load limits based on gross vehicle weight are simple to implement, the procedure is fundamentally flawed since the gross load from a vehicle is transmitted through the axle tires. Thus, the tire loads and the geometric arrangement of the tires comprising the axle are the factors that more directly influence the response of the pavement to the vehicle rather than its gross weight. Therefore a need exists to determine the load limits on the basis of axle load and axle configuration. To achieve this objective, it is important to identify the materials related-factors that significantly affect predicted pavement response. This paper presents the findings from the sensitivity analysis of predicted stress-dependency and plastic behavior of load-zoned pavements.

NUMERICAL ANALYSIS

An incremental nonlinear 2-D elasto-plastic finite element program was developed to establish the sensitivity of predicted pavement response to changes in pavement material parameters [1]. To model the pavement response under the standard 80 kN single axle load, a 40 kN single wheel load was assumed in the finite element analysis. A tire pressure of 690 kPa acting on a circular tire footprint of 136-mm radius was used. For the boundary conditions, the sides of the finite element mesh were restrained in the horizontal direction, while the bottom of the mesh was restrained in both the vertical and horizontal directions to represent the rigid layer.

In the sensitivity analysis, each of the K_1 to K_3 and strength parameters was varied one at a time from its assumed base value, while the other parameters were held at their corresponding base values. The change in the predicted pavement response due to a change in a given parameter was evaluated using the finite element program. Table 1 shows the base values assumed in the sensitivity analysis for the stress-dependent and Mohr-Coulomb strength parameters, respectively. Each parameter was varied ±30 percent from its base value except for the surface layer, K_3, and the subgrade, K_2, which were fixed at zero. Specifically, the induced stresses at the edge and center of the 40 kN wheel load were evaluated at three different depths corresponding to the bottom of the surface and base layers, and the top of the subgrade. These stresses were used to determine the predicted plastic strain under loading at each evaluation position.

Table 1. Base Levels of Factors Evaluated in Sensitivity Analysis.

Factor	Surface	Base	Subgrade
K_1	50,000	700	400
K_2	0.1	0.6	0.0
K_3	0.0	- 0.3	- 0.3
C (kPa)	1721	69	90
ϕ (degrees)	0	50	30

SENSITIVITY OF STRESS-DEPENDENT MODULUS

The material parameters considered in the sensitivity analysis are the K_1 to K_3 coefficients of the Universal Soil Model [2,3,4] that govern the stress-dependency of the resilient modulus of each layer. This model is given by the relation:

$$E_r = K_1 \, Atm \left(\frac{\theta}{Atm} \right)^{K_2} \left(\frac{\tau_{oct}}{Atm} \right)^{K_3} \tag{1}$$

where

E_r	=	resilient modulus.
Atm	=	atmospheric pressure (100 kPa).
θ	=	first stress invariant = $\sigma_1 + \sigma_2 + \sigma_3$.
σ_i	=	principal stresses.
τ_{oct}	=	octahedral shear stress.

The coefficients, K_1, K_2, and K_3, are determined from resilient modulus tests. Since the calculated stresses are normalized with respect to the atmospheric pressure, these coefficients are dimensionless. From results of the study conducted by Jooste and Fernando [5], the coefficient, K_1, was found to have the most influence on the predicted resilient modulus. In general, the higher the K_1, the higher the predicted resilient modulus. This is illustrated in Figure 1, which shows predicted resilient moduli for a granular base material at three different values of K_1. The data shown were calculated assuming a pavement with a 100-mm thick asphalt concrete surface layer and a 200-mm thick granular base layer.

For a given curve, it is observed that the resilient modulus increases with increasing wheel load, illustrating the hardening effect of increasing confinement on the predicted resilient modulus. This hardening effect is associated with the K_2 term,

$$K_2 \; term = \left(\frac{\theta}{Atm} \right)^{K_2} \tag{2}$$

As the wheel load increases, the confining pressures also increase resulting in higher predicted values for the resilient modulus. It is noted that the octahedral shear stress also increases with increasing wheel load, which will tend to decrease the resilient modulus. However, for the pavement and range of wheel loads considered in Figure 1, the increase in confinement with higher wheel loads more than compensates for the softening effect of the octahedral shear stress. Thus, the resilient modulus is predicted to increase with higher wheel loads in the figure shown. However, the opposite trend may be obtained for other pavements (such as thin pavements), where the softening effect of the octahedral shear stress may be more pronounced.

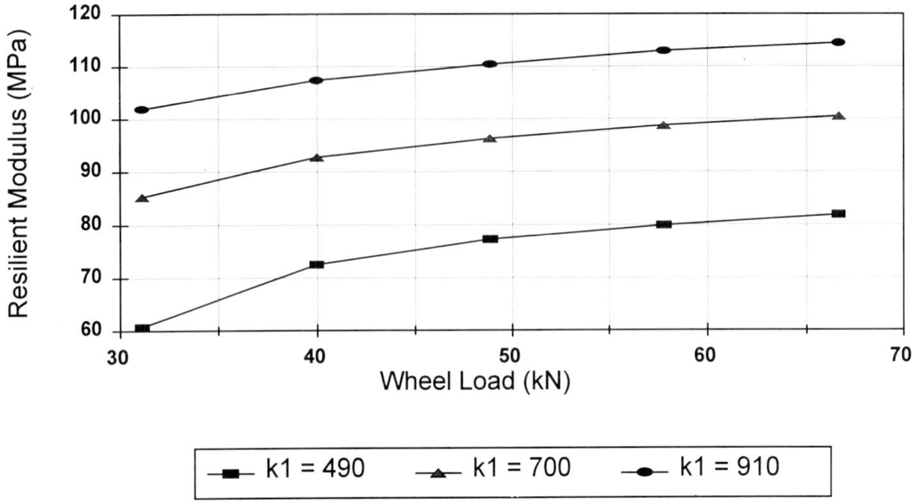

Figure 1. Variation in Resilient Modulus With Parameter, K_1 [5].

The hardening effect of higher confinement and the softening effect of higher octahedral shear stress can be discerned from Figure 2. The K_3 term in the figure is equal to,

$$K_3\ term = \left(\frac{\tau_{oct}}{Atm}\right)^{K_3} \tag{3}$$

As the wheel load increases, the K_2 term increases because of higher confinement. However, the octahedral shear stress also increases so that the K_3 term diminishes with higher wheel loads. Consequently, while the effect of the higher K_1 is generally to increase the predicted resilient modulus, the effects of K_2 and K_3 depend on the interactions between these coefficients, the applied loads, and the pavement geometry. The tendency of a material to stiffen with increasing confinement, θ, is related to K_2. However, this tendency is counteracted by the softening effect under increasing shear, as controlled by the coefficient, K_3. The greater the tendency of a material to stiffen under increasing confinement, the higher the effect of K_2. Similarly, the greater the tendency of a material to soften under shear, the higher the effect of K_3. The effects of these coefficients on the resilient modulus are also affected by the applied loads and pavement geometry due to the effects of these latter factors on the induced stresses. The coefficients, K_1, K_2, and K_3, are also used in evaluating the stress-dependency of the Poisson's ratio based on the relationship developed by Uzan [6].

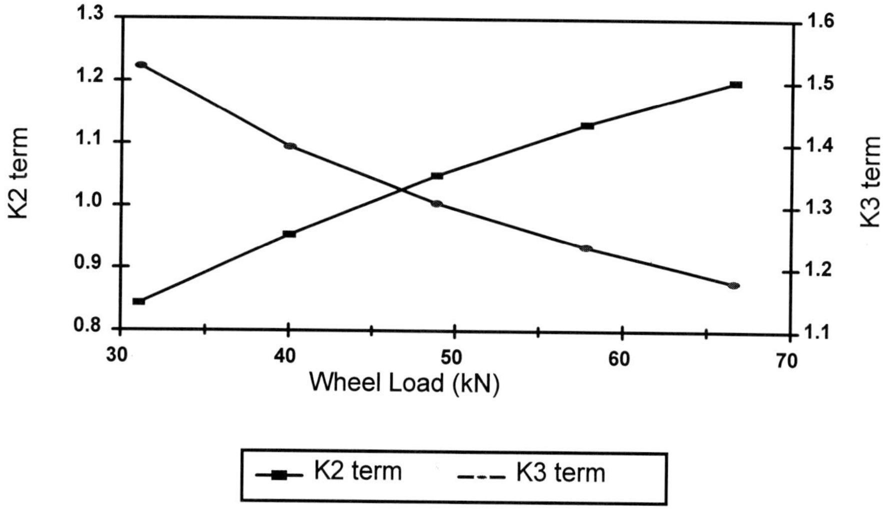

Figure 2. Illustration of the Hardening Effect Due to θ and the Softening Effect Due to τ_{oct} [5].

SENSITIVITY OF PREDICTED PLASTIC STRAIN

The results are based on the predicted plastic strains from modeling of the post yield behavior of a thin pavement with a 25-mm thick asphalt concrete surface layer and a 150-mm thick granular base layer. Figures 3 to 7 show the results of the analysis. Plotted in these figures is the percent change in the predicted plastic strain due to a corresponding change in a given parameter. The higher the percent change in the predicted plastic strain, the higher the bar corresponding to that change and the more important a given parameter is. Since no yielding was predicted at the subgrade, the figures show the results based on the predicted plastic strains at the bottom of the asphalt surface and base layers. In all cases, the predicted plastic strains at the center of the wheel load were the most critical.

The predicted plastic strain is most sensitive to the parameter, K_1. Figures 3, 5, and 7 show that the predicted plastic strain changes significantly whenever K_1 for a given layer is varied. Figure 3 shows that increasing the surface K_1 by 30 percent led to a higher predicted plastic strain at the bottom of the surface layer. This is due to the higher induced stresses resulting from a stiffer surface brought about by an increase in K_1. It is noted that a stiffer material would show a higher cohesion, which would counteract the effect of higher predicted stresses. However, since the parameters were varied only one at a time, this increase in cohesion with increase in K_1 was not considered in the sensitivity analysis. Thus, it is conceivable that the opposite effect may be observed in practice, depending on the relationships between stiffness, cohesion, and induced stresses for a given material.

The effect of varying the surface K_1 on the predicted plastic strain at the bottom of the base is relatively small compared to the corresponding effect on the predicted plastic strain at the bottom of the surface layer (Figure 3). It is observed that the higher the surface K_1, the lower the predicted plastic strain at the bottom of the base. This is due to the lower stresses predicted in the base because of the stiffer surface.

The cohesion of the surface layer has a significant effect on the predicted plastic strains as shown in Figure 4. The lower the cohesion, the higher the predicted plastic strain at the bottom of the surface. It is also observed that the predicted plastic strain at the bottom of the base increases when the surface cohesion is decreased by 30 percent. This reflects the redistribution of the stresses after initial yielding within the surface layer. Because of this, the base starts to carry more of the load, resulting in higher stresses at the bottom of the base and greater plastic strains.

The biggest change in the predicted plastic strain at the bottom of the asphalt layer occurred when the base K_1 was varied. Figure 5 shows that the predicted plastic strain in the asphalt layer increased by about a 110 percent when the base K_1 was reduced by 30 percent. This is due to the higher tensile stresses predicted in the surface layer as a result of a softer base providing diminished support to the overlying material. Figure 5 also shows that the predicted plastic strain at the bottom of the base increased when the base K_1 was raised by 30 percent. Again, the observed effect is due to the higher induced tensile stresses at the bottom of the base due to the higher stiffness. Note, however, that the base cohesion was not varied at the same time that the base K_1 was increased.

The effect of base K_2 was most pronounced for the surface material. Figure 5 shows that the predicted plastic strain at the bottom of the surface increases by more than 50 percent when the base K_2 is reduced by 30 percent. The lower this parameter, the lower the base stiffness, leading to higher tensile stresses at the bottom of the surface and to higher predicted plastic strains at this same location. The effect of base K_2 is relatively less when the predicted plastic strain at the bottom of the base is considered. Note that when the base K_2 is reduced by 30 percent, the predicted plastic strain at the bottom of the base is also reduced owing to the decrease in the predicted tensile stresses as a result of a softer material. Again, no corresponding change was made in the cohesion of the base when K_2 was reduced by 30 percent. Figure 5 also shows that the predicted plastic strains at the bottom of the asphalt and base layers are not sensitive to changes in the base K_3.

The effects of changes in the base cohesion and angle of internal friction are shown in Figure 6. It is observed that both parameters have a significant influence in the predicted plastic strains. The lower the values of the parameters, the higher the predicted plastic strains at the bottom of the base and surface layers.

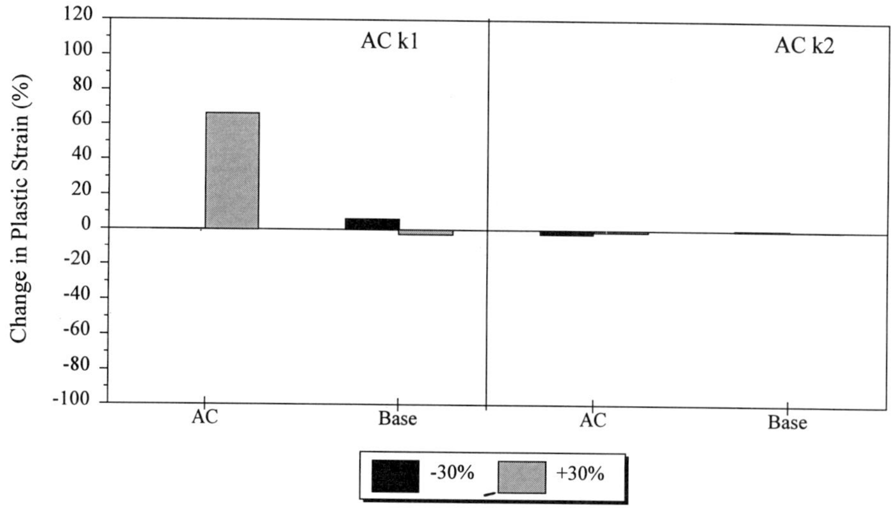

Figure 3. Sensitivity of the Plastic Strain to Changes in k1 and k2 Parameters of the Asphalt Concrete (Center of Load, Thin Pavement).

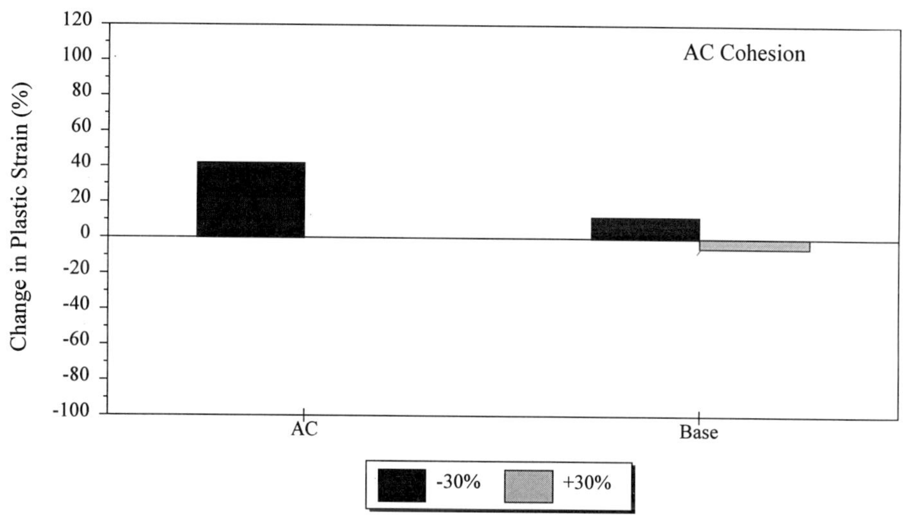

Figure 4. Sensitivity of the Plastic Strain to Changes in Cohesion of the Asphalt Concrete (Center of Load, Thin Pavement).

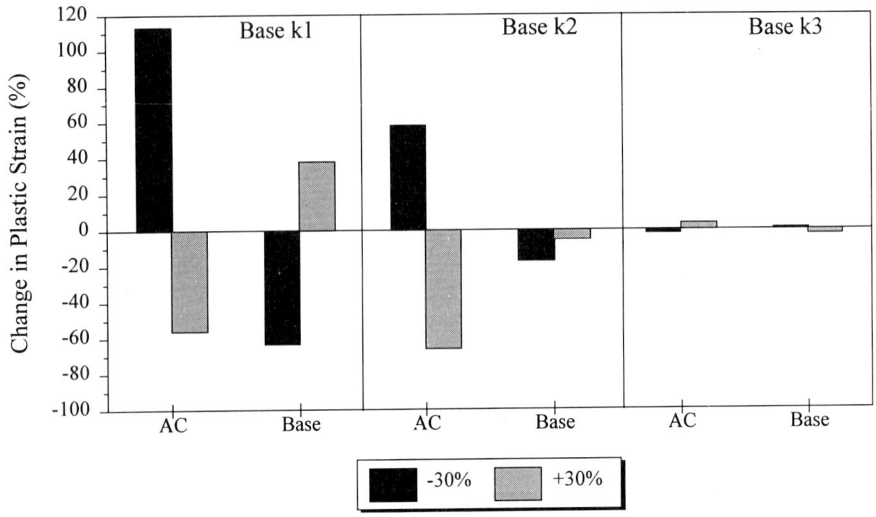

Figure 5. Sensitivity of the Plastic Strain to Changes in k1, k2, and k3 Parameters of the Base (Center of laod, Thin Pavement).

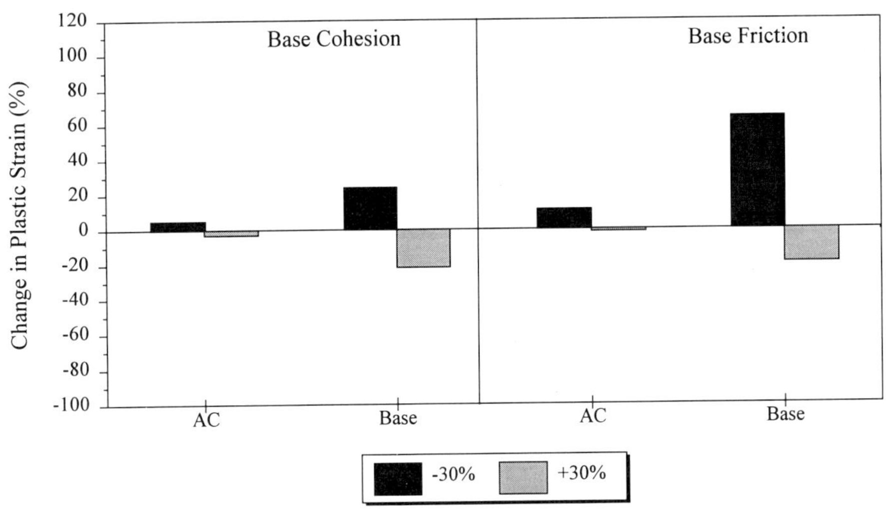

Figure 6. Sensitivity of the Plastic Strain to Changes in Cohesion and angle of Friction of the Base (Center of Load, Thin Pavement).

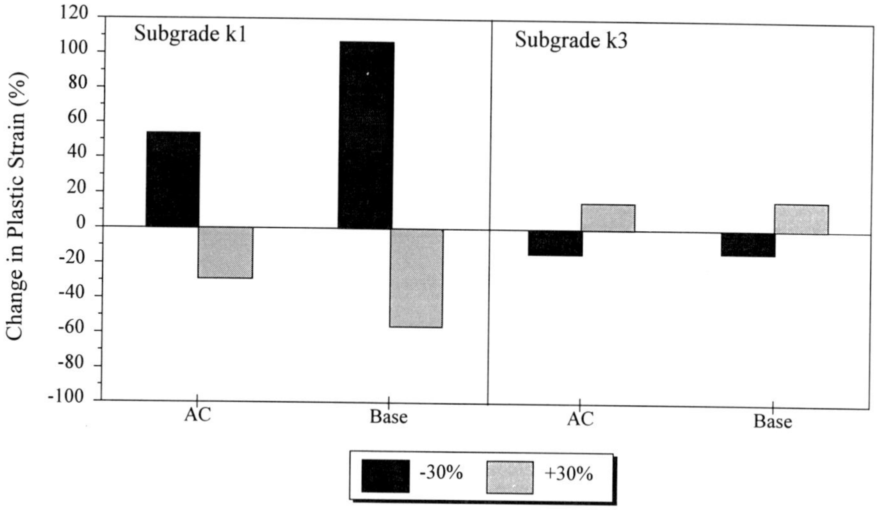

Figure 7. Sensitivity of the Plastic Strain to Changes in k1 and k3 Parameters of the Subgrade (Center of Load, Thin Pavement).

Figure 7 shows that the subgrade K_1 significantly influences the predicted plastic strains in the asphalt and base layers. The lower the subgrade K_1, the higher the predicted plastic strains owing to the higher tensile stresses in the overlying materials because of a softer subgrade. The effect of subgrade K_3 is also significant but to a lesser extent compared to the effect of subgrade K_1. Note that the subgrade K_3 is negative. Since the predicted values for θ and τ_{oct} at the top of the subgrade are less than the atmospheric pressure, the stiffness of the subgrade decreases when the parameter, K_3, is increased by 30 percent. Thus, the predicted plastic strains at the bottom of the surface and base layers go up with increase in this parameter.

CONCLUSIONS

Based on the sensitivity analysis of predicted stress-dependency and plastic behavior using the Universal Soil Model and Mohr Coulomb yield criteria, the following observations are noted.

- The effect of the higher K_1 is generally to increase the predicted resilient modulus and the effects of K_2 and K_3 depend on the interactions between these coefficients, the applied loads, and the pavement geometry.
- Accurate estimates of K_1 to K_3 parameters are important, and of the parameters, the K_1 parameter is the most influential.
- For the surface and base, cohesion and friction angle have a significant influence.
- Largest changes in predicted plastic strain occurred when base K_1 was varied.

- Variations in subgrade K_1 parameter resulted in significant changes to the predicted plastic strain, followed to a lesser extent by the subgrade K_3 parameter.
- Base and subgrade are important to pavement load carrying capacity. Good subgrade support will mobilize confining pressures in base layer and reduce bending effect in upper layers.

REFERENCES

1. Park, S. W. and E. G. Fernando. *Development of a Methodology for Posting Load Limits on Load-Zoned Pavements*. Research Report 1701-1. Texas Transportation Institute, Texas A&M Univ., College Station, Texas, 1997.

2. Witczak, M.W. and J. Uzan. *The Universal Pavement Airport Design System Report I of IV: Granular Material Characterization*, University of Maryland, College Park, Md., September 1988.

3. Uzan, J. Granular Material Characterization, In *Transportation Research Record* 1022, TRB, National Research Council, Washington, D.C., 1985, pp 52-59.

4. Lytton, R.L., J. Uzan, E.G. Fernando, R. Roque, D. Hiltunten, and S.M. Stoffels. *Development and Validation of Performance Prediction Models and Specifications for Asphalt Binders and Paving Mixes*. Report SHRP-A-357. Strategic Highway Research Program. National Research Council, Washington, D.C., 1993.

5. Jooste, F. J. and E. G. Fernando, *Development of A Procedure for the Structural Evaluation of Superheavy Loads Routes*, Research Report 1335-3F, Texas Transportation Institute, Texas A&M Univ., College Station, TX, 1995.

6. Uzan, J. Resilient Characterization of Pavement Materials. *International Journal for Numerical and Analytical Methods in Geomechanics*, Vol. 16, 1992, pp 435-459.

ACKNOWLEDGMENTS

This work was conducted in a study funded by the Texas Department of Transportation, and the Federal Highway Administration, U.S. Department of Transportation.

DISCLAIMER

The contents of this paper reflect the views of the authors who are responsible for the facts and accuracy of the data presented. The contents do not necessarily reflect the official views or policies of the Texas Department of Transportation or the Federal Highway Administration.

Fifth International Conference on the
BEARING CAPASITY OF
ROADS AND AIRFIELDS

SELECTION OF INTERLAYER SYSTEMS TO PREVENT REFLECTIVE CRACKING

A.H. DE BONDT[1], A.A.A. MOLENAAR[2] AND A. SCARPAS[2]

[1]Ooms Avenhorn b.v.
P.O. Box 1, 1633 ZG Avenhorn, the Netherlands
[2]Delft University of Technology
Faculty of Civil Engineering and Geo Sciences
P.O. Box 5048, 2600 GA DELFT, The Netherlands

ABSTRACT

Several types of interlayer systems are available nowadays which are effective in retarding or even preventing reflective cracking. Since reflective cracking can have various causes, it is essential to select the right type of interlayer system. Should the system e.g. behave like a SAMI where relatively large displacements are allowed or should the interlayer system behave more like a reinforcing system. This paper will present a number of guidelines for the selection of the most appropriate interlayer system given the conditions that prevail on a particular road. The guidelines are based on the results of an extensive research project that has been carried out at the Delft University.

INTRODUCTION

In the last decades several interlayer systems, such as geogrids, geotextiles and steel reinforcements, have been introduced in the market which all claim to give a significant improvement to the resistance of overlays to reflective cracking. Basically these systems can be discriminated into two major groups which are:

a. systems which allow the overlay to behave independently from the structure below and in this way retard the penetration of cracks in the overlay,

b. systems which reinforce the crack resistance of the overlay.

Furthermore there is a group of products which seem to have a more or less hybride behaviour, since their properties are such that they should be classified somewhere in between the two major groups mentioned above.

It is quite obvious that the choice on the type of interlayer system depends on the mechanisms responsible for the crack development and propagation. Is the cracking e.g.

driven by temperature movements or traffic loads. Are the loading conditions such that they produce just tensile forces or tensile and shear forces or mainly shear forces. Furthermore it is important to know whether the loads applied in a fast rate or a slow rate or one is dealing with a combination of both. Recognition of the forces driving the crack initiation and progression phase is therefore essential in order to be able to design or select the most appropriate interlayer systems. The paper therefore will start with a description of the tools that are available to monitor crack behaviour. Secondly the principles of the major groups mentioned above will be discussed and finally it will be discussed how the most appropriate interlayer system can be selected.

CRACK EVALUATION

The chance that existing cracks in a pavement structure will propagate through an overlay which is placed on top of it is very much dependent on the amount and type of movement at a crack. Figure 1 e.g. gives an indication of the amount of horizontal movement in relation to the change in temperature that was measured at a specific transverse crack in a highway structure consisting of a 250 mm thick bituminous top layer on top of a 400 mm thick cement treated base [1].

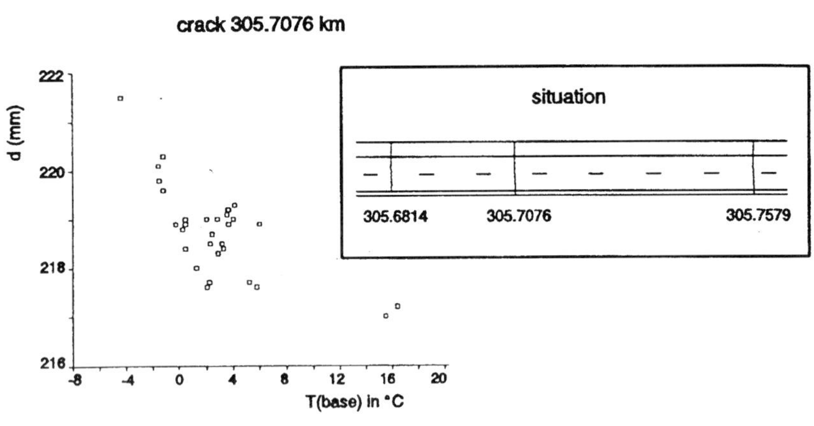

Figure 1: Example of crack movement caused by temperature cycles

Figure 2 shows the deflections that were measured by means of a falling weight deflectometer on a crack in the same structure. The figure shows that the loading plate of the F.W.D. was placed at various distances from the loading centre. Figure 2a shows the deflections as measured on a fairly warm summer day when the transverse crack was closed, while figure 2b shows the deflections measured on the same crack during a relatively cold day.

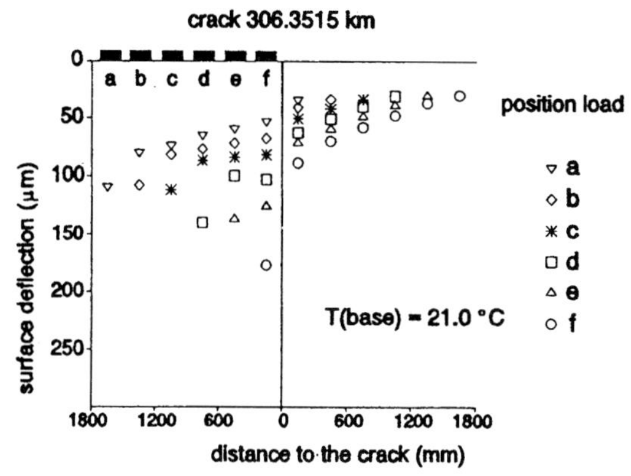

Figure 2a: Surface deflections in summer

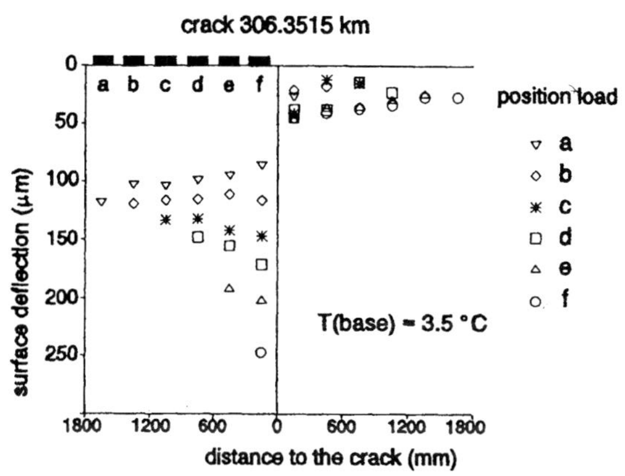

Figure 2b: Surface deflections in winter

From figures 1 and 2 it is clear that especially the winter period will be very harmful to an overlay which is placed on top of the pavement since under those conditions, the overlay will be subjected to horizontal tensile stresses due to the low temperatures and shear stresses because of vehicles passing the cracked zone. The temperature stresses however are applied at a much slower rate than the traffic related ones.

This particular example clearly shows that only taking into account the thermal movement would probably result in a wrong selection of the interlayer system. In that case one would most probably select a SAMI type of interlayer which allows the overlay to move freely over the original structure. It will be obvious however that a SAMI type solution will not be the most appropriate one for solving the problems induced due to the shear movement which are significant in this case.

In this particular example the interlayer layer system has to cater movements in the horizontal as well as vertical direction. In cases where the pavement is relative thin and the subgrade is of good quality, the shear movement in wintertime might be much less than the

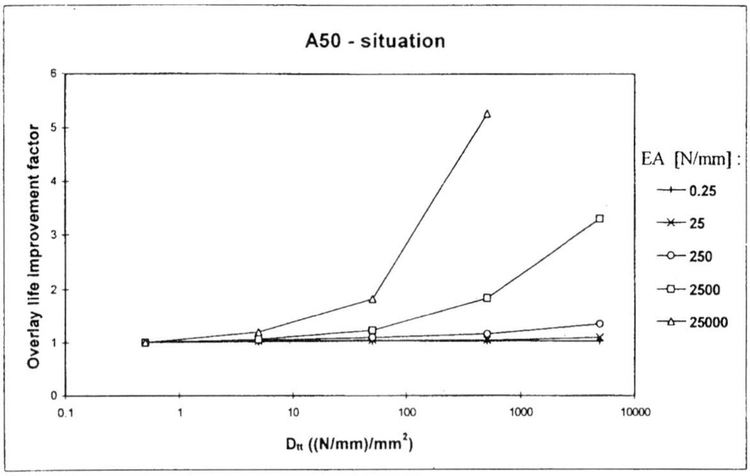

Figure 4: Influence of reinforcement bond on overlay life improvement factor

	Thickness [mm]	E [MPa]	v	D_{tt} [N/mm^3]	EA [N/mm]
Overlay	100	4500	0.35	-	-
Interface layer	1	-	-	various	-
Reinforcement	-	-	-	-	450 or 27000
Interface layer	1	-	-	various	-
Existing asphalt	248	4500	0.35	-	-
Bound base	400	10000	0.2	-	-
Subgrade	-	40	0.35	-	-

Table 1: Properties of the analyzed structures

In the analyses the bound base and existing asphalt layer were assumed to be cracked and crack propagation through the overlay was simulated.

When no reinforcement was applied, the total thickness of the interface was 2 mm. The factor D_{tt} represents the stiffness of the bond between the layers. The factor EA is the stiffness of reinforcing system per unit width. When no reinforcement is applied, low D_{tt} values represent SAMI conditions. High values of D_{tt} represent perfect bond between the layers. The results of two of the analyzed cases will be presented here. The first case is the situation where crack opening develops because of the thermal contraction of the cement treated base. The second case is the situation where bending and shearing develops because of a 50 kN load placed directly next to the crack.
Figures 5a, 5b and 5c show the development of the stress intensity factor K_I in relation to the crack length for the thermal loading case in case no reinforcement is applied (figure 5a), and when a stiff (figure 5b) or soft reinforcement (figure 5c) is used.

From figure 5 it is clear that a low D_{tt}, representing a SAMI, gives much lower K values in the unreinforced case than when the overlay is firmly 'glued' to the existing pavement (high D_{tt}). This means that a SAMI is effective in reducing crack reflection due to thermal loading. Comparing figures 5b and 5c with figure 5a shows that application of a reinforcement has no use when the stiffness of the bond is low. In practical terms this means that a reinforcement will have little effect when used together with a SAMI. Figures 5b and 5c also show that the type of reinforcement is of less importance in reducing the K values due to thermal movement. The stiffness of the bond is of much greater importance.

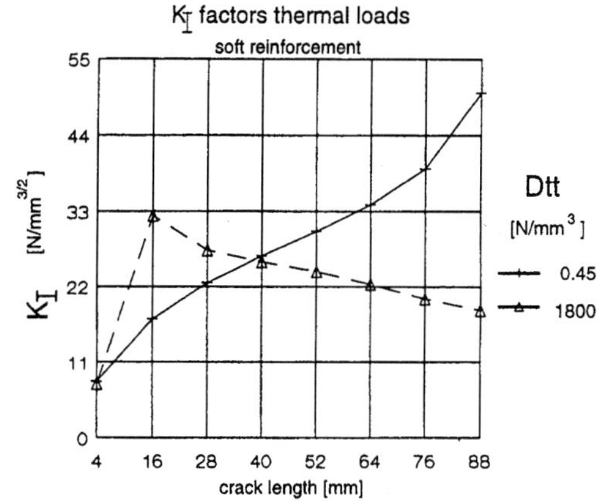

Figure 5a: Effect of bond stiffness on K_I due to thermal movements (no reinforcement)

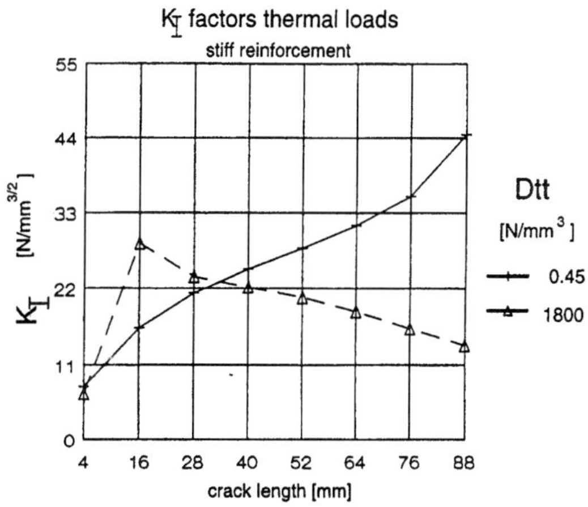

Figure 5b: Effect of bond stiffness and stiff reinforcement on K_I due to thermal movements

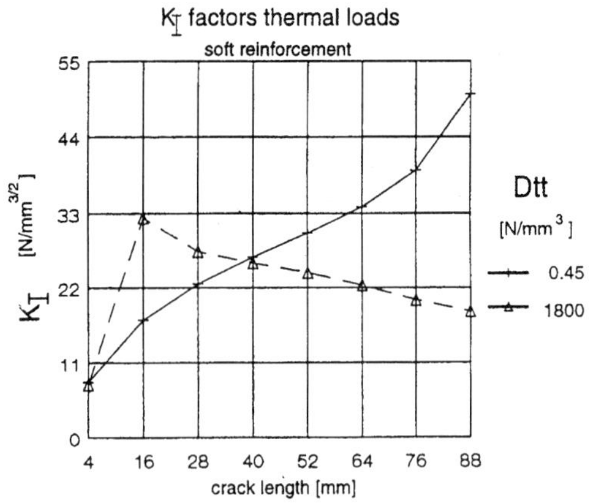

Figure 5c: Effect of bond stiffness and soft reinforcement on K_I due to thermal movements

It is recalled that K_I is used to calculate the crack propagation rate following:

$$dc/dN = A K^n$$

where	dc/dN	=	increase in crack length per load cycle
	K	=	stress intensity factor
	A, n	=	material parameters

Figures 6a, 6b and 6c show the effect of the bond stiffness and presence of a reinforcement on the K_I due to bending caused by the traffic load. Figures 7a, 7b and 7c show the effect on the K_2 due to the shearing caused by the traffic load.
Figure 6a shows that in the unreinforced case a high bond stiffness results in higher K_I values when the crack length is between 4 and 40 mm, so a SAMI helps in reducing the K_I due to traffic loads. Figure 7a however shows, as expected, that the bond stiffness has no effect on K_2 which is caused by the shearing action of the traffic load.
When figures 6b and 6c are compared with figure 6a then it becomes clear that the reinforcement has a positive effect on reducing K_I. Again one can notice that the stiffness of the bond is of greater importance than the stiffness of the reinforcement. One should realize however that e.g. a polypropylene with a lower EA is usually attached to the existing pavement by means of e.g. a tack coat which has a relatively low D_{tt}, while a steel net with a high EA is quite often attached to the existing pavement by means of nailing; this results in a relatively high D_{tt}. All in all this indicates that one has to consider the effectiveness of the entire system, that is the reinforcement and the way how it is attached to the pavement.
Although a reinforcing has a positive effect in reducing K_I, figures 7b and 7c show that the most significant effect is in reducing K_2. In this particular case the stiffness of the reinforcement and the bond stiffness are of limited importance; the presence of the reinforcement itself is important.

With respect to the analysis results presented here, it should be noted that they should not be generalized. The type of structure as well as the thickness of the cracked bound layers and the thickness of the overlay have an important effect on the magnitude of the K values and the effect of bond and reinforcement stiffnesses. All this means that generalized solutions are hard to give and that a proper analysis using a finite element program like CAPA [6] is needed in order to be able to make a proper selection of the most effective interlayer system.

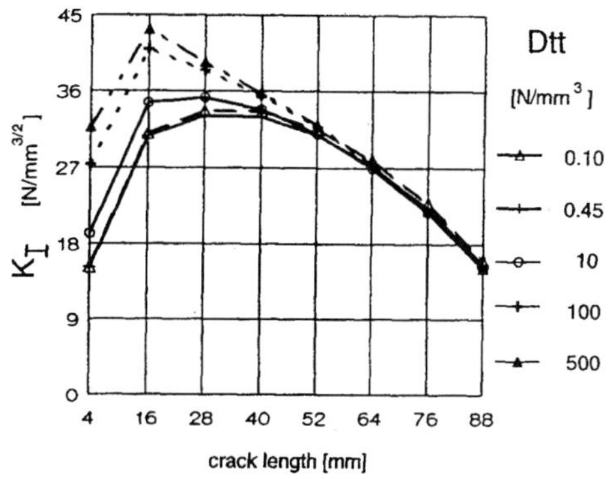

Figure 6a: Influence of bond stiffness on K_I in unreinforced overlay due to traffic movements

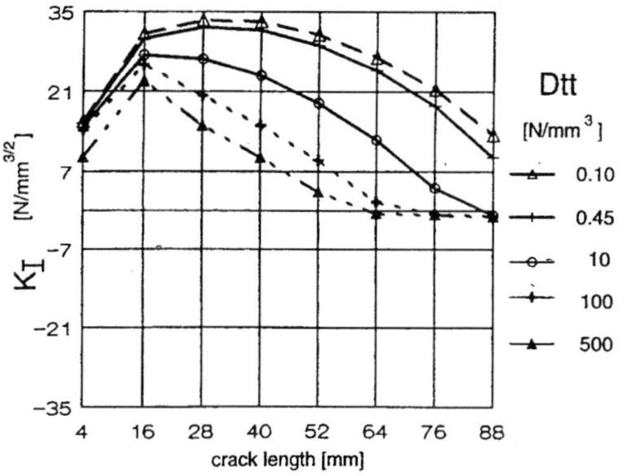

Figure 6b: Influence of bond stiffness on K_I in overlay with stiff reinforcement due to traffic movements

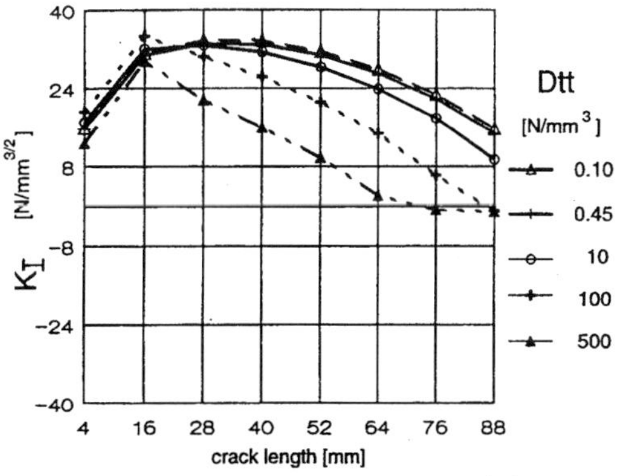

Figure 6c: Influence of bond stiffness on K_I in overlay with soft reinforcement due to traffic movements

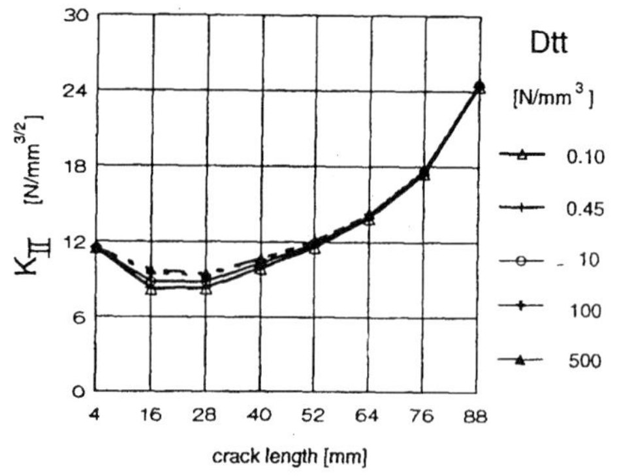

Figure 7a: Influence of bond stiffness on K_2 in unreinforced overlay due to traffic movements

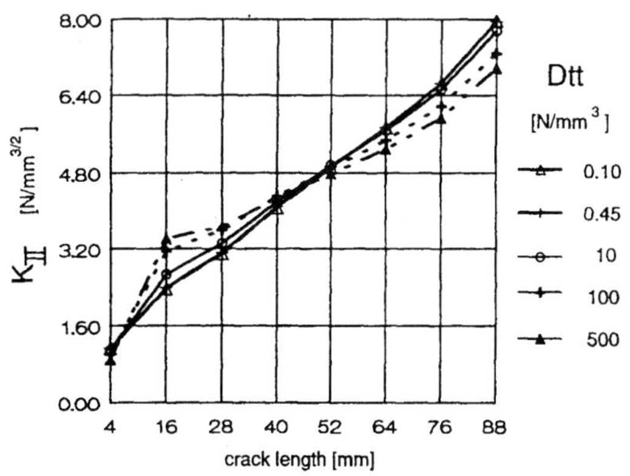

Figure 7b: Influence of bond stiffness on K_2 in overlay with stiff reinforcement due to traffic movements

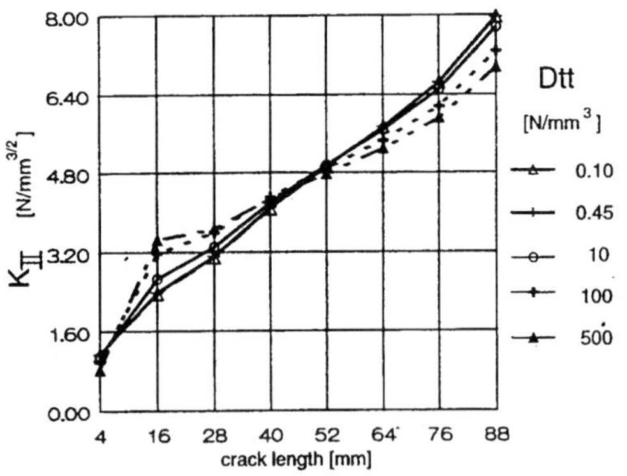

Figure 7c: Influence of bond stiffness on K_2 in overlay with soft reinforcement due to traffic movements

SELECTION OF THE INTERLAYER SYSTEM

From the discussion given in the previous sections it is clear that no generalized solutions can be given with respect to the selection of interlayer systems to retard the propagation of cracks through an overlay. Nevertheless some general guidelines can be given.
If measurements indicate that a significant amount of shear movement can occur then a SAMI type of interlayer system, or a non woven soaked with emulsion or bitumen, will not be very effective; this is clearly indicated by figures 7a, 7b and 7c. A significant amount of shear movement can be expected in cases where the pavement consists of relatively thick cracked bound layers. Especially when wide cracks occur, high shear deformations can occur in these cases. For these conditions reinforcements with a high EA and a high bond stiffness are preferred when both temperature movements and traffic loads occur. When only traffic loads occur, the requirement with respect to the EA of the interlayer product can be less strict. In any case a high bond stiffness is required.
In cases where measurements show that shearing movements are not such an issue, SAMI type interlayers can be very effective. These conditions occur e.g. when the pavement consists of a relatively thin cracked asphalt layer on top of a good unbound base and subbase.
In all cases, appropriate measurements and finite element analyses are strongly recommended in order to arrive to the most appropriate interlayer system.

CONCLUSIONS

The following conclusions can be drawn.

a. It is essential to perform a proper evaluation of the cracked pavement in order to determine the causes and nature of the cracks and to assess their reflective potential.
b. No general solution can be given with respect to which type of interlayer system is the most appropriate given the prevailing conditions.
c. Nevertheless it can be stated that a SAMI type of interlayer system seems to be the most appropriate if shearing movements near cracks are small.
d. In all other cases a reinforcing type of interlayer seems to be the most appropriate. However when a reinforcement is applied a high bond stiffness is necessary in order to let the reinforcement be effective.

LITERATURE

1. DE BONDT, A.H.; STEENVOORDEN, M.P.
Reinforcement test sections A50 (A6) Friesland. Report 7-95-209-21; Road and Railroad Research Laboratory; Delft University of Technology; Delft – 1995

2. VANELSTRAETE, A.; FRANCKEN, L. (EDITORS)
Prevention of reflective cracking in pavements. RILEM Report 18; E. & F.N. Spon; London – 1997

3. VANELSTRAETE, A.; FRANCKEN, L.
Laboratory testing and numerical modelling of overlay systems on cement concrete slabs. Proc. 3rd RILEM Conf. On Reflective Cracking in Pavements; Maastricht - 1996

4. SCARPAS, A.; DE BONDT, A.H.; GAARKEUKEN, G.
Reflective cracking control via reinforcing systems: FE modelling of reinforced overlays. Proc. 3rd RILEM Conf. On Reflective Cracking in Pavements; Maastricht – 1996.

5. BRUIJN, N.; SCARPAS, A.; DE BONDT, A.H.
(Un)-reinforced interfaces in asphalt concrete pavements. Report 7-94-203-19; Road and Railroad Research Laboratory; Delft University of Technology; Delft – 1994.

6. SCARPAS, A.; BLAAUWENDRAAD, J.; DE BONDT, A.H.; MOLENAAR A.A.A.
CAPA: a modern tool for the analysis and design of pavements. Proc. 2nd RILEM Conf. On Reflective Cracking in Pavements; Liege – 1993.

Fifth International Conference on the
BEARING CAPASITY OF
ROADS AND AIRFIELDS

PERFORMANCE AND AXLE LOAD EQUIVALENCY OF UNBOUND GRANULAR PAVEMENTS

B.T. VUONG[1], K.G. SHARP[2] AND J.R. JOHNSON-CLARKE[3]

[1]Senior Research Scientist, ARRB Transport Research Ltd
[2]Principal Research Scientist, ARRB Transport Research Ltd
[3]Senior Research Scientist, ARRB Transport Research Ltd
E-mail: [1]binhv@arrb.org.au, [2]kierans@arrb.org.au and [3]jimjc@arrb.org.au

ABSTRACT

This paper presents the results of trials of an unbound crushed rock conducted as part of a series of axle load equivalency trials conducted using the Accelerated Loading Facility (ALF). Dual wheel loads of 40 kN, 50 kN and 60 kN were used for trafficking the pavements in an attempt to determine the relative damaging effects of axle load to be examined. The crushed rock section tested in the axle load equivalency trial was uniform enough to allow axle load equivalency factors of the crushed rock pavement to be determined without the need to correct the data for the effects of asphalt temperature, and pavement moisture conditions. The load damage exponent (LDE) value determined for this pavement was about 7.5, which is much higher than the LDE of 4 obtained from the AASHO Road Test, but similar to the "subgrade damage" exponent of 7.14 recommended in the Austroads Pavement Design Guide.

INTRODUCTION

Current Australian reviews of vehicle mass limit regulations are hampered by the lack of relevant performance data on the impact of axle load, tyre type and tyre pressure, especially with respect to chip seal pavements, which form the bulk of the Australian sealed road network. There is a need to investigate the implications of increased axle loads and related parameters on the pavement design procedures currently recommended in the Austroads Pavement Design Guide (1).

Accordingly, it was proposed (2, 3, 4) that field trials be conducted, using the Accelerated Loading Facility (ALF), to investigate the performance and axle load equivalency of high quality crushed rocks. The first trial was conducted on a chip seal crushed rock pavement at a site at Beerburrum, Queensland. However, because of problems related to seal disintegration and variation in moisture condition during the trial, there was uncertainly about the reliability of the findings obtained from this trial (5). The second trial was conducted on a typical "municipal" or "arterial" granular pavement with thin asphalt surfacing at a site in Dandenong, Victoria. This crushed rock pavement was uniform enough to allow axle load equivalency factors of the crushed rock pavement to be determined without the need to correct the data for temperature, pavement moisture and subgrade condition.

This paper describes the results of the analysis of the field performance data from the ALF experiments conducted on the crushed rock pavement at Dandenong.

BACKGROUND

The original load equivalency concept developed from the AASHO Road Test was used to convert the number of passes of a particular axle load and configuration which causes pavement damage to a terminal condition into the number of equivalent standard axle loads (P_s) which would cause the same terminal damage. This concept was expressed in the form of a "power law":

$$LEF|_{p_t} = (N_s/N_i)|_{p_t} = (P_i/P_s)^{LDE}|_{p_t} \quad (1)$$

where LEF = Load Equivalency Factor,
p_t = pavement terminal condition (e.g. a terminal rut depth of 20 mm),
N_s = number of Standard Axles of load P_s to the terminal condition p_t,
N_i = number of axles of load P_i to the terminal condition p_t, and
LDE = load damage exponent.

Because a LDE value of 4 was obtained from the AASHO Road Test analysis, this law has been called the "Fourth Power Law" and for many years it has been used to simplify the determination of the number of repetitions, N, of a standard load, P_s, from mixed traffic for the purpose of determining pavement thickness at a given design pavement condition. However, many critics (6, 7) have pointed out that this equation may not be appropriate in all situations unless the environment, traffic, materials, pavement type and pavement construction methods were the same as, or similar to, those in the AASHO Road Test.

Accordingly, it was proposed (2, 3) that a field trial using three different ALF axle loads (namely 40 kN, 60 kN and 80 kN dual-wheel loading) be conducted to determine the load equivalency of a typical chip seal crushed rock pavement at a site at Beerburrum, Queensland. The pavement consisted of a nominal 150 mm crushed rock base and 150 mm of crushed rock sub-base over a silty sand subgrade having a soaked CBR of about 10. The crushed rock layers were compacted to 104-106% standard maximum dry density (or 102-104% modified maximum dry density). Initially the pavement was sealed with a single seal coat composed of 10 mm aggregate and a bitumen spray rate of $1.4 L/m^2$. However, during the first two experiments using the 40 kN and 60 kN axle loads, the thin seal disintegrated under ALF loading during periods of rain. Therefore, it was decided to use a double seal in the final experiment using the 80 kN axle load. Unexpected record rainfall during the 60 kN experiment also created problems in maintaining similar moisture conditions in the pavement layers during the three experiments.

Consequently, the adjustment method proposed by Vuong (5) was used in the analysis of the data to correct the effects of pavement moisture condition on surface deformation and to assess the damaging effects of different axle loads. In this method, a regression rut depth model, having a predetermined mathematical form as shown in eqn (2), is used to fit all the deformation data obtained from the experiments.

$$RD = A.\ N^{B} \qquad (2)$$

where RD is the rut depth or permanent deformation,
N is the number of applied load cycles, and
A and B are constants obtained from the fitted curve.

This relationship was derived from laboratory repeated load triaxial testing on unbound materials and subgrade under the same repeated loads (8, 9, 10). Therefore, it is assumed when using the above relationship that the pavement has been subjected to a selected reference load.

Pavement damage due to different axle loads (at the same pavement condition) can be adjusted using eqn (3).

$$N_{equivalent} = N_{applied}(L_{applied}/L_{reference})^{LDE} \qquad (3)$$

where $N_{applied}$ = number of cycles of load to a given deformation under the moisture conditions operating during testing,
$N_{equivalent}$ = equivalent number of cycles of load to a given deformation at the reference moisture condition,
$L_{applied}$ = axle load used in the experiment,
$L_{reference}$ = reference axle load (standard axle load of 40 kN adopted in this study), and
LDE = Load Damage Exponent.

Pavement damage due to different pavement conditions (at the same axle load) can, on the other hand, be adjusted using eqn (4).

$$N_{equivalent} = N_{applied}(D_{applied}/D_{reference})^{EDE} \qquad (4)$$

where $N_{applied}$ = number of cycles of load to a given deformation under the moisture conditions operating during testing,
$N_{equivalent}$ = equivalent number of cycles of load to a given deformation at the reference moisture condition,
$D_{applied}$ = measured maximum deflection corresponding to a given deformation,
$D_{reference}$ = reference maximum deflection at the reference moisture condition, and
EDE = Environmental Damage Exponent.

Using this method, EDE and LED values can be selected based on the criterion of closeness of the agreement between the corrected data. *Fig. 1* shows that, for the Beerburrum ALF experiments on the chip seal crushed rock pavement, the criterion was best met for EDE values in the range of 8-10 and LED values in the range of 6-8, and that it would take about 2 $x10^6$ to 6 $x10^6$ cycles of the 40 kN ALF axle load to cause a maximum pavement deformation of 20 mm. However, there was uncertainty about the reliability of the findings because of: (i) the large corrections of the measured deformation (in the 60 kN experiment) required to take account of variation in moisture condition, and (ii) the deformation measured in the 40 kN experiment was small (10 mm maximum) and affected by seal disintegration.

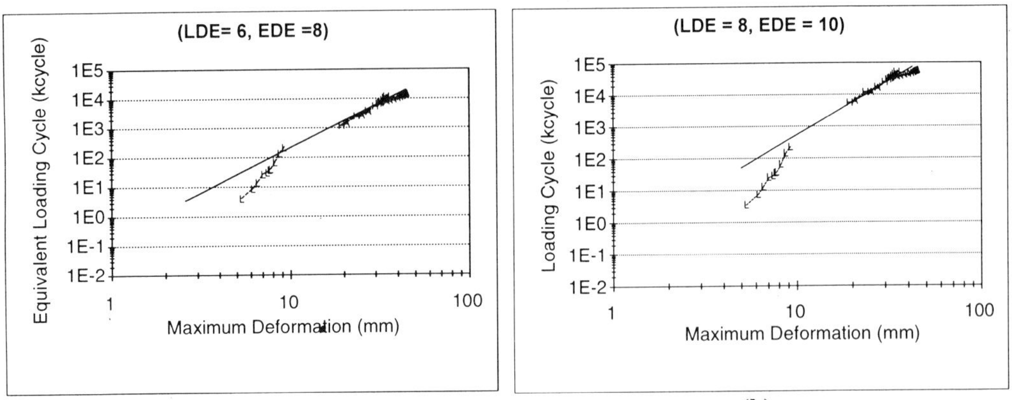

(a) (b)

Fig. 1 - Comparison of deformation obtained from ALF experiments on the crushed rock pavement at Beerburrum (5)

Accordingly, it was proposed (4) that typical "municipal" or "arterial" granular pavements, which had a nominal pavement thickness of 200 mm over a subgrade having an unsoaked CBR of about 10 and a soaked CBR of about 3-4 be tested to allow for expected pavement design lives of the unbound pavements in the range of 10^3-10^5 ESAs. The trial was subsequently conducted at a site at Dandenong, Victoria. Details of the trial and performance data obtained from the trial follow.

DETAILS OF THE DANDENONG LOAD EQUIVALENCY ALF TRIAL

Details of the trial are briefly described below. A more detailed report on the construction of the test sites is contained in Sharp (11). Performance data obtained from the trial are given in Vertessy (12) and the analysis of the data is reported in full in Vuong and Sharp (13).

Structural Details of Test Pavement

The test pavement consisted of:

- an asphalt surfacing (nominally 30 mm thick),
- a base crushed rock (nominally 200 mm thick),
- an imported clay (nominally 400 mm thick),
- a free-draining, permeable layer (nominally 100 mm thick),
- the natural clay subgrade stabilised to a nominal depth of 300 mm, and
- the natural subgrade.

The asphalt surfacing was a dense-graded mix consisting of 7 mm aggregate and Class 320 binder. The surface of the base layers was primed prior to the application of the asphalt.

The crushed rock was a 20 mm non-plastic (PI = 0) material. It was an angular, rough, hard, blue-grey material containing some dolomite siltstone and some blue-grey sand silt. The results of classification tests (Plasticity Index, Liquid Limit, Linear Shrinkage, laboratory

CBR) and Modified compaction testing (Maximum Dry Density (MDD) and Optimum Moisture Content (OMC)) conducted on the crushed rock are given in *Table 1*. Grading of the crushed rock complied with the VicRoads (1990) specifications for Class 1 base.

Table 1
Laboratory Properties of Base and Subgrade Materials

Material	Plasticity Index (PI)	Liquid Limit (LL)	Linear Shrinkage (LS)	Maximum Dry Density (t/m^3)	Optimum Moisture Content (%)	Unsoaked CBR (%)	Soaked CBR (%)
Crush Rock Base	0	21-23	1-3	2.2*	6.8*	166	118
Imported Subgrade	4-5	16	1-3	2.08**	8.6**	10	3

* Modified compaction.
** Standard compaction.

The field density and moisture content of the crushed rock base in the test section (S1-1) was measured with the nuclear density gauge in back-scatter mode after placement and after each experiments. The results indicated that, after placement, the crushed rock was quite uniform and had a field relative density and relative moisture content of 100% and 50% respectively. However, the final moisture contents measured after the ALF experiments were much higher, having an average moisture content of about 70% of OMC_{mod}. An investigation of the causes of this increase in moisture content indicated that the pavement had wet up after heavy rainfall before the site was sealed and also from the adjacent shoulder.

The imported subgrade was a Silurian clay, typically found in the Melbourne area, of nominal CBR 8-10 (dry) and a soaked CBR of about 3. Results of classification tests and Standard compaction tests conducted on the subgrade are also given in *Table 1*. Field subgrade CBRs measured with depth soon after the completion of ALF trafficking were the range 2-4, indicating that the subgrade in this section was fully soaked.

ALF EXPERIMENTS AND RESULTS

Accelerated Loading Facility

The Accelerated Loading Facility (ALF) is a mobile, relocatable road testing machine which applies full-scale rolling wheel loads to a test pavement 12 m long at a speed of 20 km/h (see *Fig. 2*) The load, which can be varied in 10 kN increments between 40 and 80 kN, tracks linearly and is guided by the main frame. The dual-wheel or single-wheel half single axle loading can either be channelised or applied over a normal transverse distribution about 1.4 m or 1 m wide. The load is applied in one direction only, with the wheel lifted off the pavement at the end of the cycle and supported by the main frame on its return. Approximately 380 load cycles can be applied each hour.

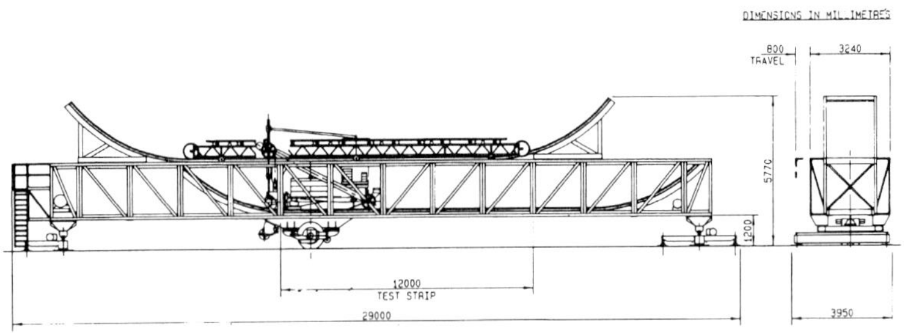

Fig. 2 - Schematic diagram of ALF

ALF Experiments

Three Experiments were conducted on the crushed rock pavement using ALF axle loads of 40 kN, 50 kN and 60 kN respectively. The number of loading cycles applied in each experiment is given in *Table 2*. For reasons described shortly, the 40 kN test was divided into two experiments during data analysis.

Table 2
Condition of the Crushed Rock Section S1-1 During ALF Experiments

Exp.	ALF Load (kN)	Offset (m)	No of Cycles (kcycles)	Average Initial Maximum Deflection # (mm)	Average Pavement Moisture Content (% of OMC)	Average Subgrade CBR	Average Asphalt Temp. (°C)
1a	40	10-14	23.5	2000	70	3	20
1b	40	15-20	23.5	1700	65	3	20
2	50	22-26	6.4	2000	70	3	25
3	60	29-32	1.7	2000	73	3	27

\# Measured with FWD contact stress of 500 kPa.

In the 40 kN Experiment, extensive surface cracking appeared after only 20,000 cycles, followed by an increase in the rate of deformation and shoving at the edges of the test section. The appearance of this cracking was a cause for concern because, at the prevailing low ambient temperatures, the predominant failure mode was fatigue of the thin asphalt surfacing rather than the expected deformation of the crushed rock base or subgrade. A pavement heating system developed for an earlier trial (14) was therefore used after 44,000 cycles, and in the subsequent experiments, the temperature in the asphalt layer maintained at 25°C in an attempt to prevent any fatigue cracking in the asphalt. The results of pavement condition testing (deflection, temperature and moisture contents) and pavement performance monitoring (cracking and rutting) are briefly described below.

Pavement Deflection, Temperature and Moisture Condition

Table 2 also summarises the average FWD maximum deflection (measured just prior to testing commencing), subgrade moisture content and subgrade CBR (measured at the conclusion of each ALF experiment), and asphalt temperature (measured during each ALF experiment) for the three crushed rock sections. Referring to *Table 2*, there were three sub-sections that had similar FWD maximum deflections under the FWD load of 500 kPa (about 2000 µm), similar subgrade moisture contents (about 70% of OMC) and similar subgrade CBR values (about 3). These were:

- offsets of 10-14 m (40 kN Experiment);
- offsets of 22-26 m (50 kN Experiment); and
- offsets of 29-32 m (60 kN Experiment).

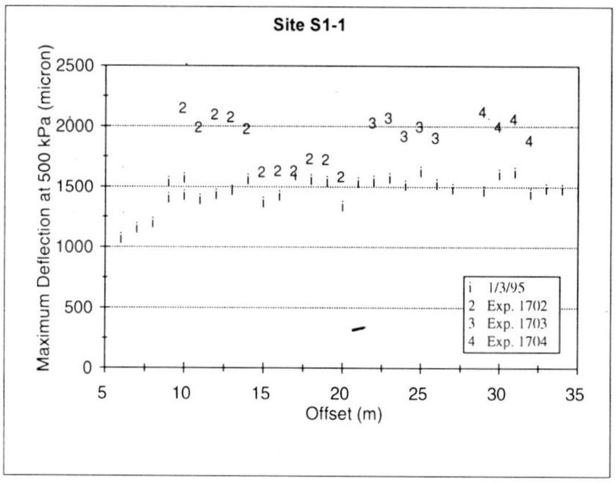

Fig. 3 - Variation in FWD maximum deflection along the pavement length

In addition, as shown in *Fig. 3*, there were only small variations in FWD maximum deflection along the length of each sub-section, i.e. in the range of 1900-2100 µm for all of the experiments shown in *Table 2*. It was considered reasonable, therefore, that the performance of these uniform sub-sections could be used to determine a load equivalency factor for the crushed rock pavement, without any corrections for variations in pavement moisture content being required.

With respect to asphalt temperature, the results in *Table 2* indicate that the temperature of the asphalt in the 40 kN Experiment was lower than that in the other experiments. However, as discussed below, the effects of temperature on the mode of failure and pavement deformation were not significant, particularly during the early loading stage before cracking occurred.

Cracking

Table 3 summarises the development of cracking. In all cases, cracking was first observed on the surface well before the pre-determined termination point of the test, viz. an average maximum deformation of 20 mm. This indicated that, despite the use of the pavement heating system to maintain an asphalt temperature at about 25°C, the initial mode of failure in the sections was fatigue of the thin asphalt surfacing rather than the expected permanent deformation of the base layer and/or subgrade.

As discussed below, because heaving was observed at the two edges of the ALF loading area throughout the loading history (before and after cracking occurred), and because longitudinal cracks were first observed at the two edges of the ALF loading area (where heaving occurred) well before the development of transverse cracks, it was thought that the cracking was due shearing and heaving rather than fatigue.

Table 3
Crack Development in Crushed Rock Section S1-1

Experiment	ALF Dual-Wheel Load (kN)	Loading Cycle (kcycle)	Crack Length (m/m^2)	Maximum Deformation (mm)
1a	40	20	>1	>10
2	50	6	>1	>15
3	60	1.7	>1	>15

Maximum Deformation and Rutting

Figs 4a, 4b and *4c* compare the maximum deformation and loading cycles for the 40 kN, 50 kN and 60 kN experiments conducted on the crushed rock pavement. The results indicate that there were variations in deformation along the length of the pavement section used in the 40 kN experiment, whereas the deformation which occurred along the length of each sub-section tested in the 50 kN and 60 kN experiments was very uniform. It could be argued that the variation in deformation along the length of the sub-section tested in the 40 kN experiment (see *Fig. 4a*) was possibly due to the effects of stiff sections at both ends of the sub-section concerned (i.e. the sections between chainages 0-10 m and 15-20 m). On that basis, it was decided to use the maximum deformations that occurred in the middle of the sub-section (between offsets 12-13 m) as characteristic values for this sub-section.

Heaving was observed at the two edges of the ALF loading area throughout the loading history (before and after cracking occurred), indicating that shear failure was occurring (see *Fig. 5a)*. The deformation and heave near the pavement edge were much higher than that on the other side of the wheelpath due to the higher moisture contents near the pavement shoulder. *Fig. 5b* shows the relative deformations of the base, sub-base and subgrade at the end of the ALF experiments. These results indicate that at least 70% of the total pavement surface deformation occurred in the crushed rock layer rather than in the subgrade. In other words, subgrade deformation may not always be the most critical factor influencing the life of granular pavements. This observation is similar to that observed in previous ALF experiments as reported in Vuong *et al.* (5).

It should also be noted that, because of the heaving, and because the narrow ALF loading distribution (0.9 m wide) was used in the experiments, the rut depth, which is calculated from the measured transverse deformation data assuming a 1.2 m straight edge, was substantially greater than the measured deformation. Referring to *Fig. 6*, a maximum deformation of 15 mm corresponded to a rut depth in the range of 18 mm to 22 mm. The differences between the measured deformation and calculated rut depth increased with increasing axle load.

According to Vuong *et al.* (5), and also intuitively, these differences would have been smaller had the wider (1.4 m) ALF transverse loading distribution been adopted. Because the rut depth was related to the amount of heave under the narrow ALF loading distribution, which is not presentative of real traffic, the maximum deformation was therefore preferred over the rut depth as the performance parameter of interest in this study, particularly in the derivation of axle load equivalencies.

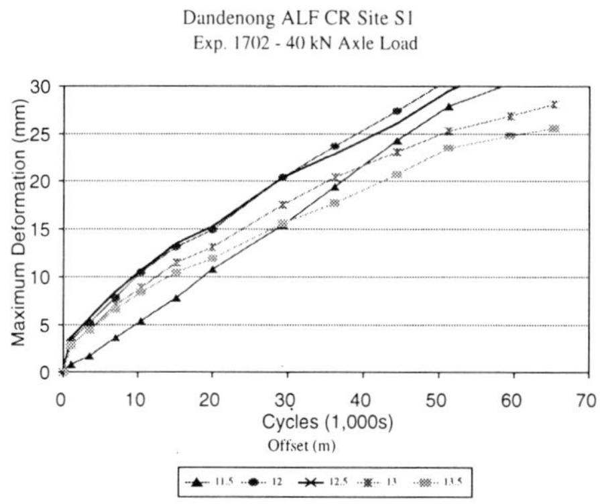

(a)

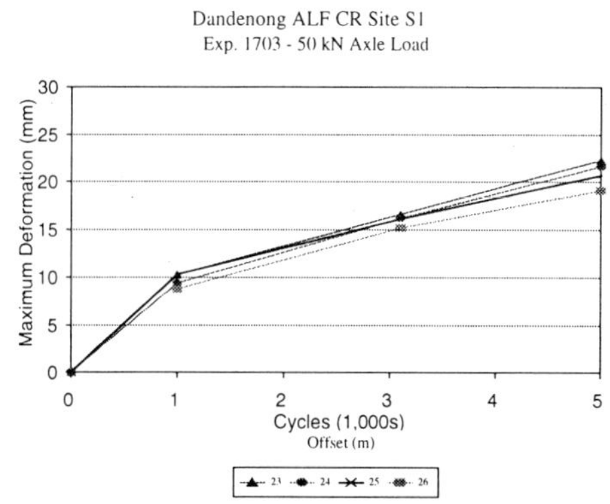

(b)

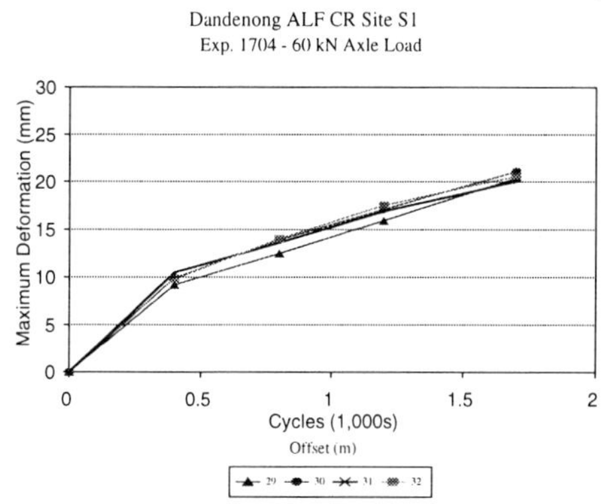

(c)

Fig. 4 - Maximum deformation of the crushed rock pavement at Dandenong

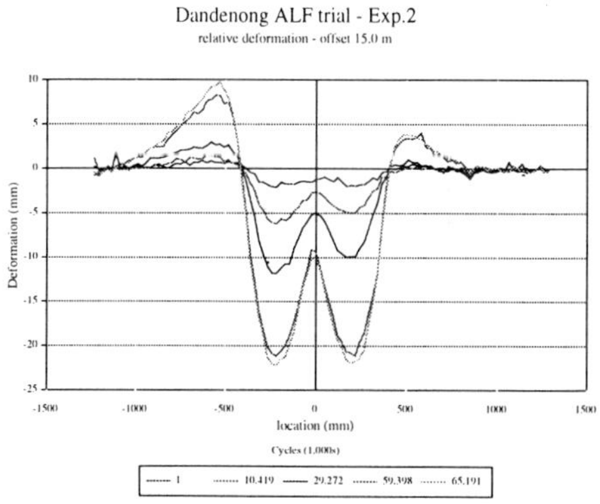

(a) Mean maximum surface with cycles

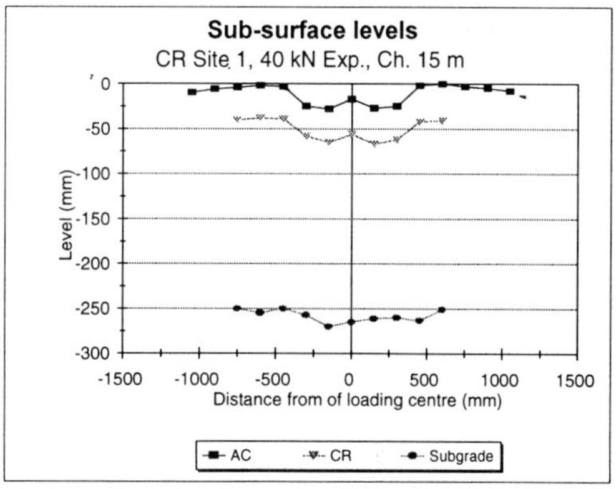

(b) Sub-surface deformations

Fig. 5 - Surface and sub-surface deformations in crushed rock pavement

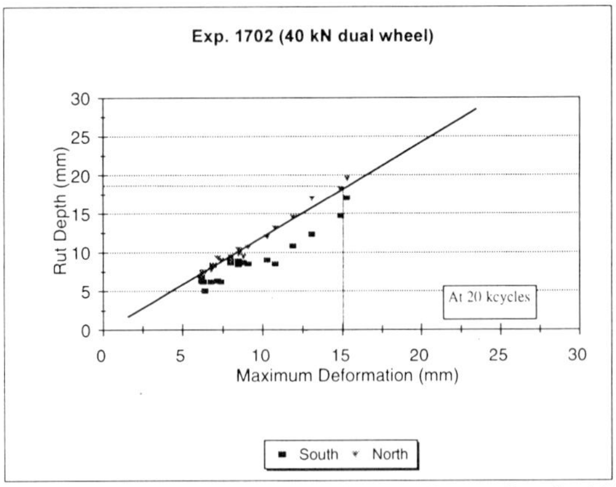

Fig. 6 - Comparison of maximum deformation and rut depth (40 kN Experiment)

DEFORMATION LIFE AND AXLE LOAD EQUIVALENCY

Based on the results already discussed above, it was decided to select a terminal deformation criterion of 15 mm (or terminal rut depth of 20 mm) in order to reduce the many uncontrolled variables in the trials, including:

- the unexpected early fatigue failure of the thin asphalt surfacing; and
- the changes in pavement moisture condition after cracking of the asphalt surfacing occurred.

Another concern was the effect of asphalt temperature on asphalt modulus and hence pavement deformation in the 40 kN Experiment. However, it is contended that the effects of asphalt modulus (and hence temperature) on pavement deformation were not significant for the following reasons:

- the asphalt layer was very thin (30 mm thick);
- there was sufficient pavement deformation in the 40 kN experiment before the appearance of cracking, which was first observed after 20 kcycles of loading, to justify the use of the permanent deformation data in the analysis; and
- the average temperature in the principal loading stage of the 40 kN Experiment (about 20°C) was not much lower than that in the 50 kN and 60 kN Experiments, where the asphalt temperatures were about 25°C and 27°C respectively. Moreover, previous ALF trials have indicated that below 25°C the deformation of asphalt is minor.

Therefore, in the comparison of pavement lives and determination of axle load equivalency using performance data obtained from the above experiments, no correction for moisture condition and asphalt temperature was considered to be necessary.

Based on the terminal deformation criterion of 15 mm, the deformation lives of the selected pavement sub-sections tested with the 40 kN, 50 kN and 60 kN axle loads are given *Table 4,* together with the calculated LEF and LDE values for the 50 kN and 60 kN axle load as compared with the reference load of 40 kN using eqn (1). It is apparent that in each loading case the LDE value was greater than 7, and that the LDE value of 4 obtained from the AASHO Road Test analysis was not appropriate in this case.

It was also possible to determine the values of LDE for different terminal deformation criteria less than 15 mm by using eqn (3) to adjust the surface deformation results obtained from the 50 kN and 60 kN experiments to the case of 40 kN using a Load Damage Exponent (LDE). *Fig. 7* shows the generalised curve of maximum deformation against equivalent ALF cycles ($N_{equivalent}$) produced by the reference axle load (standard axle load of 40 kN adopted in this study). The curve fitted to the data has the same form as eqn (2), i.e. linear relationship between logarithm of maximum deformation and logarithm of loading cycles. By inputting different values of Loading Damage Exponent, LDE, in eqn (3) to convert the 50 kN and 60 kN axle loads to the equivalent 40 kN axle load, a value of LDE was derived which produced the closest generalised relationship between surface deformation and loading cycles for all experiments. As shown in *Fig. 7*, when a LDE value of 7.5 was used, the results of surface deformation converted from the 50 kN and 60 kN results were very close to the generalised deformation curve for the 40 kN load. The LDE value derived using this method was similar to those derived using eqn (1) and with a terminal maximum deformation of 15 mm (see *Table 4*). The closeness of the agreement between the uncorrected data shown in *Fig. 7* also suggests that eqn (3) can be used to adjust the surface deformation of existing pavements for the effects of axle load, if the values of LDE of the pavements concerned can be determined.

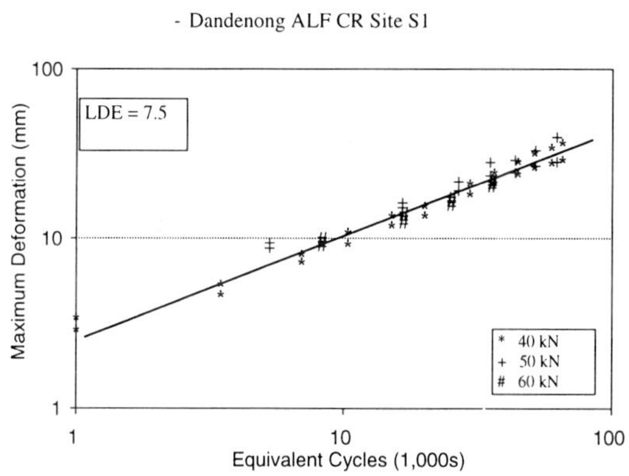

Fig. 7 - Generalised deformation curve for crushed rock pavement at Dandenong

Table 4
Pavement Lives of Crushed Rock Section S1-1 Under Different Axle Loads

Load (kN)	Exp. No.	Number of cycles to 15 mm maximum deformation (10^3 cycles)	LEF	LDE
40	1a	20	1	1
50	2	2.7	7.4	8.9
60	3	1	20	7.4

CONCLUSIONS AND RECOMMENDATIONS

In summary, the crushed rock pavement tested in the Axle Load Equivalency ALF trial was typical of a pavement used in arterial road applications (i.e. total pavement thickness of 200 mm on a subgrade CBR 3). It had a relative short life, of about 20,000 ESAs for the termination maximum deformation of 15 mm (or 30,000 ESAs for the termination maximum deformation of 20 mm). The crushed rock section was uniform enough to allow axle load equivalency factors to be determined without the need to correct the data for temperature, pavement moisture and subgrade condition. The load damage exponent (LDE) value determined for this pavement was about 7.5, which is similar to the LDE values of 6-8 estimated from the previous Beerburrum Axle Load Equivalency ALF trial, much higher than the LDE of 4 obtained from the AASHO Road Test, but similar to the "subgrade damage" exponent of 7.14 recommended in the Austroads Pavement Design Guide.

It is recommended that further ALF trials using dual-wheel and wide single-wheel loads on instrumented pavements of different material types and configurations be conducted to determine the load equivalency of the pavement materials. Performance data obtained from these ALF trials can also be used to establish the relationships between results obtained from laboratory characterisation testings and field performance.

ACKNOWLEDGMENT

The contribution of the ARRB Transport Research ALF staff (David Fossey, Nick Vertessy, Danny Cheesman, Peter Armstrong and Phillip Smith) is gratefully acknowledged.

REFERENCES

1. AUSTROADS (1992). Pavement Design - A Guide to the Structural Design of Road Pavements. (Austroads: Sydney.).

2. QUEENSLAND TRANSPORT (1991). Final proposal and first progress report: ALF Beerburrum II trial. Report RP21001A, August.

3. VUONG, B. T. (1992). ALF Beerburrum II trial - proposed analysis approach. Australian Road Research Board, Working Document WD RI92/007.

4. SHARP, K.G. (1995). Proposal for an Accelerated Loading Facility Trial on In Situ Stabilisation Treatments and Axle Load Equivalency. APRG Document 95/06(RE).

5. VUONG, B.T., JAMESON, G.W., SHARP, K.G., BARAN, E. and JOHNSON-CLARKE, J.R. (1995). Performance and Axle Load Equivalency of a Crushed Rock Pavement Under Accelerated Loading. ARRB Working Document WD RI95/006, February.

6. OECD ROAD TRANSPORT RESEARCH (1991). OECD full-scale pavement test. (OECD. Paris.)

7. OECD ROAD TRANSPORT RESEARCH (1992). Dynamic loading of pavements.

(OECD. Paris.).

8. KENIS, W.J., SHERWOOD, J.A. and McMAHON, T.F. (1981). Verification and application of the VESYS structural sub-system. Proc. 5th Int. Conf. on Structural Design of Asphalt Pavements.

9. MONISMITH, C.L. (1976). Rutting prediction in asphalt concrete pavements. Trans. Res. Rec. 616, pp. 2-8.

10. VUONG, B.T. (1992). Influence of density and moisture content on dynamic stress-strain behaviour of a low plasticity crushed rock. Road & Transport Research 1(2), June.

11. SHARP, K.G., VERTESSY, N.J.(1997), MOFFATT, M.A., ARMSTRONG, P. and YEO, R. (1997). Axle Load Equivalency and In Situ Stabilisation ALF Trials: Construction Report. APRG Document No. 97/06 (MA), July 1997.

12. VERTESSY, N.J. (1997). Crushed Rock Axle Load Equivalency and In-situ Stabilisation ALF Trial: Data Report on Testing of Unbound Crushed Rock and Marginal Materials. APRG Document No. APRG 97/02 (MA). February 1997.

13. VUONG, B.T. and SHARP, K.G (1996). Crushed Rock Axle Load Equivalency and In-situ Stabilisation ALF Trial: Performance and Axle Load Equivalency of Unbound Granular Pavements. APRG Document No. APRG 97/02 (MA). February 1997.

14. OLIVER, J.W.H., JAMESON, G.W., SHARP, K.G., VERTESSY, N.J., JOHNSON-CLARKE, J.C. and ALDERSON, A.J. (1997). Evaluation of Non-resistant Properties of Asphalt Mixes Under Field and Laboratory Conditions. Transport Research record 1590, pp. 53-61.

Fifth International Conference on the
BEARING CAPASITY OF
ROADS AND AIRFIELDS

THERMO-VISCO-ELASTIC MODELLING OF PAVEMENTS BEHAVIOUR AND COMPARISON WITH EXPERIMENTAL DATA FROM LCPC TEST TRACK

J.V. HECK*, J.M. PIAU**, J.C. GRAMSAMMER**,
J.P. KERZREHO**, H.ODEON**

*Ph'D student from civil engeenering laboratory of Nantes - St Nazaire,
Ecole Centrale de Nantes - Université de Nantes, France
**Department of Road Materials and Structures, LCPC, Nantes, France

ABSTRACT

The French national laboratory for Civil Engineering (LCPC) is equipped with a circular test track for the study of pavement behaviour under accelerated loading conditions. LCPC also works on laboratory characterisation and theoretical modelling of the behaviour of pavement materials. The aim of this paper is to link the thermo-visco-elastic behaviour of asphalt mixes (AM) measured in the laboratory, with field observations made on the test track. We describe the fitting of experimental data through Huet & Sayegh's constitutive model, as well as the finite element program developed for the calculation of visco-elastic pavements under traffic load.

Key words : Asphalt material, test track, thermo-visco-elasticity, modelling, finite elements.

INTRODUCTION

The accurate determination of stress and reversible strain fields in roads under traffic load is a question of major concern for the understanding of most pavement distress problems such as rutting, fatigue or cracking. The basic model usually used for that purpose is the analytical Burmister model for elastic multilayers [1]. However, as shown by the LCPC circular test track [2], due to the viscous properties and the thermal susceptibility of asphalt concretes (AC), the reversible strain signals appear to be very changing in nature with time, within a given pavement, depending mainly on the temperature and traffic velocity conditions. This makes the choice of an equivalent elastic modulus in the Burmister model somewhat cumbersome. Moreover, the asymmetry of the experimental strain signals under the passing of a load may sometimes become very pronounced, whereas elastic models predict symmetric strain histories. An analytical extension of Burmister's model for visco-elastic pavements has already been proposed by Hopman [3]. For the same problem we have developed a numerical finite element approach, which can deal at the same time with non-linear elasticity of unbound granular materials if needed [4]. In the near future, the same numerical tool should also include the modelisation of permanent strains in road materials (rutting of AC or/and unbound layers).

This paper only deals with visco-elastic aspects. It begins with the laboratory characterisation of AM behaviour, from measurement of complex modulus (from alternate flexural tests) or secant modulus (from monotonic tensile tests), determined at different temperatures. In both cases, these data are used to fit parameters of Huet & Sayegh's model [5,6].

Then, we derive the set of equations for the calculation of a visco-elastic pavement under a moving wheel load. The problem is solved in the case of uniform speed, using the reference frame attached to the wheel. The time dependency of the constitutive law thus disappears, replaced by a spatially non-local stress/strain relationship. The program developed in the 3D finite element code CESAR-LCPC [7] to achieve the solution, is presented.

Finally the paper illustrates some first applications of this chain from lab-to-field, to the modelling of reversible strain signals obtained on the test track during some experiments, including high and (relatively) low temperature conditions. The model naturally accounts for the qualitative and quantitative changes of signals with the temperature, as well as for the asymmetry of signals before and behind the wheel.

MODELLING OF ASPHALT CONCRETE BEHAVIOUR IN LAB

In the small strains domain $\left(\varepsilon \leq 5.10^{-4}\right)$ and for small numbers of load cycles, AM behaviour can be assumed as linear thermo-visco-elastic, with a good approximation. According to the french pavement design method, AM behaviour is usually identified in France through the measurement, at different temperatures, of the complex modulus $E^*(\omega,\theta)$, derived from alternate flexural tests, or the measurement of the secant modulus $S(t,\theta)=\sigma(t,\theta)/\varepsilon(t)$, derived from monotonic tensile tests.

Regarding the interpretation of complex modulus measurements, Huet & Sayegh have introduced in the 60ies a constitutive law which, since then, has always been confirmed [8]. Let us recall it briefly.

Huet & Sayegh's model

The representation of experimental data of complex modulus in classical Cole & Cole and Black planes led Huet & Sayegh to propose the following dependance of E* with ω and θ :

$$E^*(\omega\tau(\theta)) = E_0 + \frac{E_\infty - E_0}{1 + \delta\,(i\omega\tau(\theta))^{-k} + (i\omega\tau(\theta))^{-h}} \qquad (1)$$

with : E_0, E_∞ = limits of complex modulus for $\omega = 0$ or $\omega = +\infty$; h, k = exponents such that $1 > h > k > 0$, related respectively to the ratio E_{imag}/E_{real} when ω tends to 0 (respectively to infinity) ; δ = dimensionless constant. $\tau(\theta)$ is a function of temperature, which accounts for the classical equivalence principle between frequency and temperature. Huet & Sayegh suggest to approximate it by an Arrhenius or Eyring type law, $\tau(\theta) = A\exp(-B/T)$ with $T = 273° + \theta$. In fact, as regards the limited range of temperature observed in pavements, the following convenient exponential-parabolic law can also be used :

$$\tau(\theta) = \exp(A_0 + A_1\theta + A_2\theta^2) \tag{2}$$

Fitting from alternate flexural test

The classical french alternate flexural test is performed on trapezoïdal specimens under imposed sinusoïdal strain for crossed conditions : [1Hz, 3Hz, 10Hz, 30Hz, 40Hz] ⊗ [-10°C, 0°C, 10°C, 20°C, 30°C, 40°C]. The norm and the phase angle of the complex modulus is measured for each case. First, taking these data in the Cole & Cole (E_{imag} *vs* E_{real}) and Black (*phase angle vs* $|E^*|$) planes easily yields E_0, E_∞, h, k and δ coefficients, owing to their explicit geometrical significance. Then $\tau(\theta)$ can be fitted from complex modulus isotherms. A program based on this principle has been developed for an automatic matching. Such a global fit, showing a good correlation, is presented on figs. 1 and 2 for the reference AM generally used on the test track (0/14mm classical AM containing 5.7% of 50/70 bitumen). For this material, E_0 = 70 MPa, E_∞ = 30000 MPa, h = 0.53, k = 0.18, δ = 2.9, $\tau(\theta) = \exp(3.022 - 0.389\,\theta + 0.00198\,\theta^2)$.

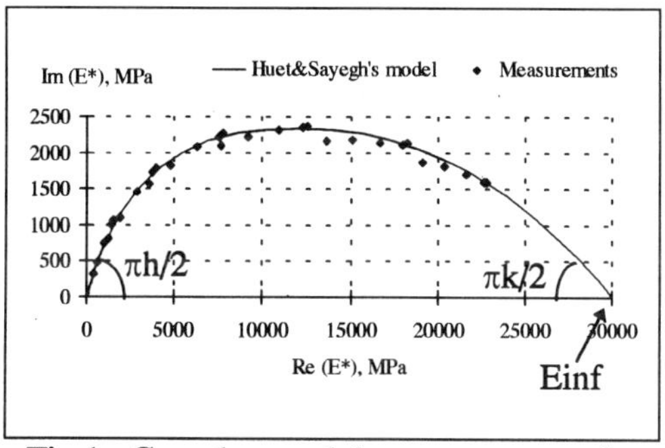

Fig.1 : Complex modulus master curve for the reference AM in Cole & Cole axes

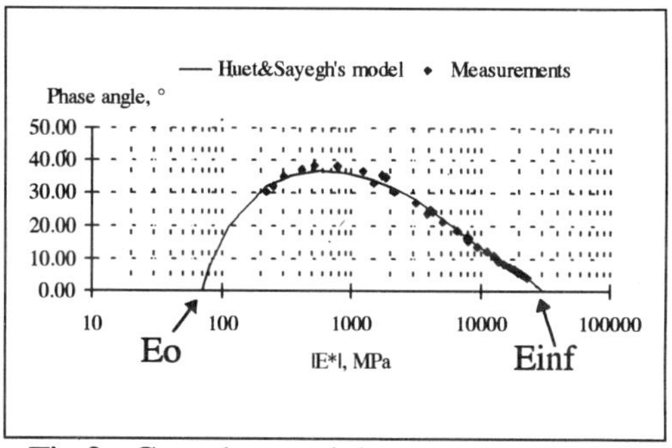

Fig.2 : Complex modulus master curve for the reference AM in Black space

Modelling of AM transient behaviour

Before presenting identification of Huet & Sayegh parameters from monotonic tensile tests, we need to show how this model can be used to predict the transient behaviour of AM, that is how to compute σ(t) (resp. ε(t)) when ε(t) (resp. σ(t)) is given. This is also an essential step for the following section relative to pavement modelling under traffic load.

In fact assuming linearity and isothermal conditions, stress and strain can be simply related through Boltzman equations,

$$\sigma(t) = R(t) * \varepsilon_{,t}(t) = \int_{-\infty}^{t} R(t-u)\,\varepsilon_{,t}(u)\,du \tag{3}$$

$$\varepsilon(t) = F(t) * \sigma_{,t}(t) = \int_{-\infty}^{t} F(t-u)\,\sigma_{,t}(u)\,du \tag{4}$$

with, R(t) = relaxation function , F(t) = creep function , $f_{,t}$ = derivative of f with time and $*$ = convolution operator.

But the creep and relaxation functions are mathematically related with the complex modulus. In particular, Huet & Sayegh have shown that their equation for the complex modulus is related to the following analogical model of figure 3 with two branches : one, with a spring and two parabolic dashpots corresponding to instantaneous and delayed elasticity (branch I), the other one, with the spring E_0 ($<< E_\infty$) representing the static or long term behaviour (branch II).

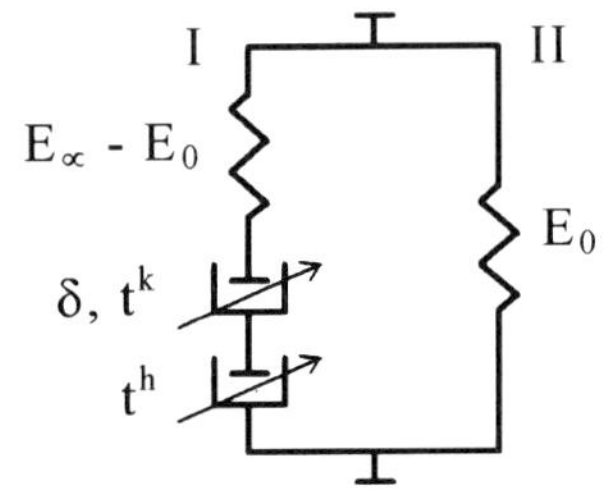

Fig.3 : Analogical representation of Huet & Sayegh's model

A simple analytical expression of the creep law of branch *I* can be derived :

$$F^I\left(\frac{t}{\tau(\theta)}\right) = \frac{1}{E_\infty - E_0}\left(1 + \delta\,\frac{(t/\tau(\theta))^k}{\Gamma(k+1)} + \frac{(t/\tau(\theta))^h}{\Gamma(h+1)}\right) \quad \textit{(with } \Gamma = \textit{Euler function)} \tag{5}$$

No such simple formula exists for the whole model, but using the decomposition of fig.3, the constitutive law of Huet & Sayegh's model in time domain can be stated under the system :

$$\begin{cases} \varepsilon(t) = F^I(t/\tau(\theta)) * \sigma^I_{,t} \\ \sigma(t) = \sigma^I(t) + \sigma^{II}(t) = \sigma^I(t) + E_0\,\varepsilon(t) \end{cases} \tag{6}$$

This one can be solved numerically, either to pass from σ(t) to ε(t) or from ε(t) to σ(t), once Huet & Sayegh's parameters are known.

In order to validate this theoretical work, several unidimensional simulations of laboratory tests have been carried out on AM, in the time domain (direct tensile tests, shear tests...), starting from alternate flexural test results. The good agreement generally obtained between our simulations and experimental results (difference < 5%) seems to validate the linear behaviour of AC (at least for small reversible strains). This fundamental assumption underlies the approach.

Note that in the time domain the equivalence principle between frequency and temperature converts into an equivalence principle between time and temperature, as can be seen on the expression of F^I which depends on the ratio t / τ(θ). Note also that the previous model can be easily converted into tensorial frame, assuming a constant and real Poisson's ratio, around 0.35 [9].

Fitting from direct tensile test

The direct tensile test is carried out on cylindrical specimens under constant strain rate conditions. The maximal strain is generally chosen as 50 μdef . A classical experimental campaign includes the different test duration [0.5s, 1.5s, 5s, 15s, 50s, 150s], in conjunction with temperature conditions [0°C, 10°C, 20°C]. The results are expressed in terms of secant modulus.

Assuming that the tensile tests can be accurately described by Huet & Sayegh's model in time domain, we have developed a procedure to derive parameters $E_0, E_\infty, h, k, \delta$ and the function $\tau(\theta)$ from secant modulus campaigns. The secant modulus master curve is first built for a given temperature θ_0. Then we numerically look for the Huet & Sayegh parameters, including $\tau(\theta_0)$, which minimise the difference between the experimental master curve and the theoretical one calculated from the equations (6). Finally keeping $E_0, E_\infty, h, k, \delta$ constant, we seek the function $\tau(\theta)$ of type (2), leading to the best fit of experimental secant modulus isotherms. A numerical program has been developed on this principle to perform the whole identification.

The application of the two identification modes on AM, which are both tested in frequency and time domains, generally leads to slightly different sets of Huet & Sayegh parameters. Nevertheless, when applied to field calculations such as those presented hereafter, the models behave "tangentially" in the contemplated range of temperatures and times and finally give a close response (difference < 4%).

MODELLING OF VISCO-ELASTIC PAVEMENT UNDER TRAFFIC LOAD

Problem formulation

Let us consider a standard pavement structure, composed of elastic layers (subgrade, bound or unbound granular materials,...) and AC layers, responding to the constitutive law (6), with possibly different coefficients in each asphalt layer. Our aim is to calculate the response of the road under the passing of a vehicle at constant speed, V , along the Ox axis (possibly inclined in the case of a slope). Small strains are assumed.

In the reference frame $\Re' = (O', X, y, z)$ attached to the wheel load $(X = x+Vt)$, the mechanical fields (displacements, strains, stresses) appear to be static, depending only on X, y, and z. Then, the introduction of time steps can be avoided. Moreover, working in $\Re'$ leaves classical equilibrium and kinematic equations unchanged, as well as time independent constitutive laws such as elasticity. As regards visco-elastic constitutive laws, just a slight change occurs, since time convolution turns into spatial convolution with respect to X . Thus system (6), now expressed in tensorial form, becomes :

$$\begin{cases} \varepsilon\,(X,y,z,\theta) = \int_{-\infty}^{X} F^I\left(\dfrac{X-X'}{V.\tau(\theta)}\right)\left[(1+\upsilon)\,\sigma^I - \upsilon\; tr\left(\sigma^I\right)\mathrm{I}\right]_{,X}(X',y,z,\theta)\; dX' \\ \sigma\,(X,y,z,\theta) = \sigma^I(X,y,z,\theta) + 2\mu_0\; \varepsilon\,(X,y,z,\theta) + \lambda_0\; tr(\varepsilon\,(X,y,z,\theta))\,\mathrm{I} \end{cases} \quad (7)$$

This system can be seen as defining a linear but non local constitutive law. The strain at a point (X, y, z) appears to depend not only on the stress at the same point, but also on stresses at all points (X', y, z) of half-axis $-\infty < X' < X$, which represent the "past" of point X (fig. 4).

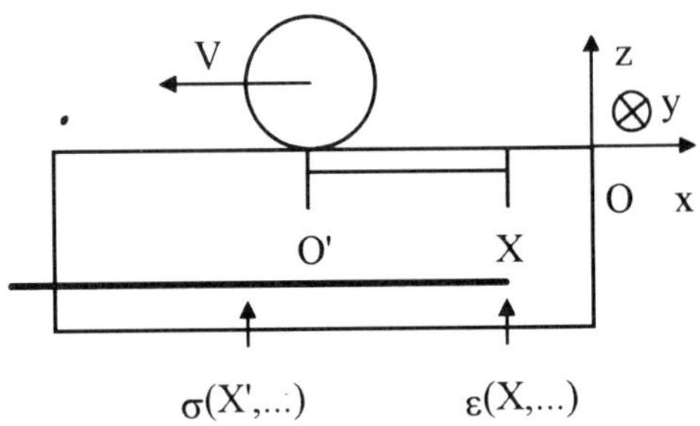

Fig.4 : Principle scheme for the calculation of visco-elastic pavements

Finite element solving

Solving of the previous problem has given rise to a dedicated program, called CVCR, in the finite element code CESAR of LCPC. Its development derives from classical discretization rules applied to the *Virtual Work Principle* including constitutive law (7). Finally, the algebraïc linear system is obtained :

$$K^{ve}U = F \quad (8)$$

where the visco-elastic stiffness matrix K^{ve} is found to have a much larger band-width than a usual elastic matrix, because of the non-local nature of system (7). In order to avoid too much storage and too long calculation times, especially in 3D, (8) is solved through the following iterative algorithm, chosen for the sake of simplicity :

$$K^{ea}\Delta U_n = F - K^{ve}U_{n-1} \qquad \text{with} \qquad \Delta U_n = U_n - U_{n-1} \quad (9)$$

where K^{ea} is now a classical elastic stiffness matrix, associated with some arbitrary elastic modulus E^a of asphalt layers. The term $-K^{ve}U_{n-1}$ is determined at the finite element level, avoiding the explicit calculation of K^{ve}.

The code was validated against a straightforward calculation of stresses and strains within a homogeneous single-layer pavement, using the *correspondance principle* and *elastic* finite element calculation. The code was also successfully compared with the program VEROAD developed by Hopman [3]. More details about CVCR can be found in [4]. Note that recent developments in CVCR permit to deal with a vertical temperature gradient in asphalt layers and elastic non-linearity of unbound layers [10].

COMPARISON BETWEEN TEST TRACK SIGNALS AND SIMULATION

A first comparison of our lab-to-field approach with results gained from the LCPC test track is shown in [4]. It deals with the flexible pavement structure of fig. 5. AC properties of the surface layer are derived from a complex modulus campaign in the lab.

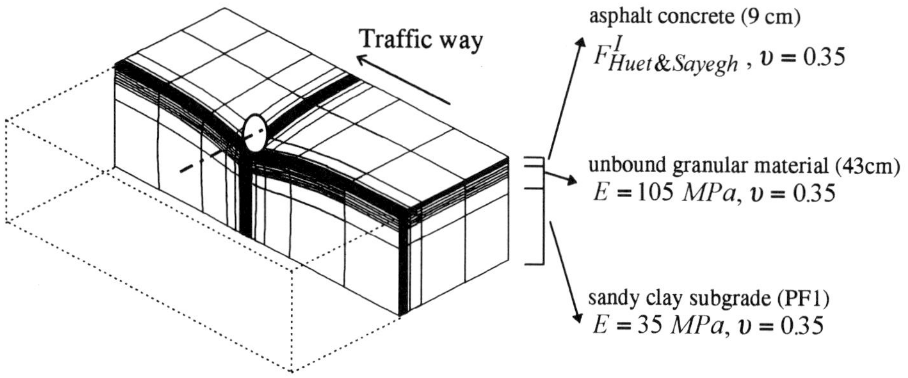

Fig.5 : Flexible pavement simulation with CVCR : stucture and deformed mesh

Figure 6 compares the measured and computed surface deflection curves obtained at 11°C, under one wheel of a dual-wheels axle (V = 72 Km/h).

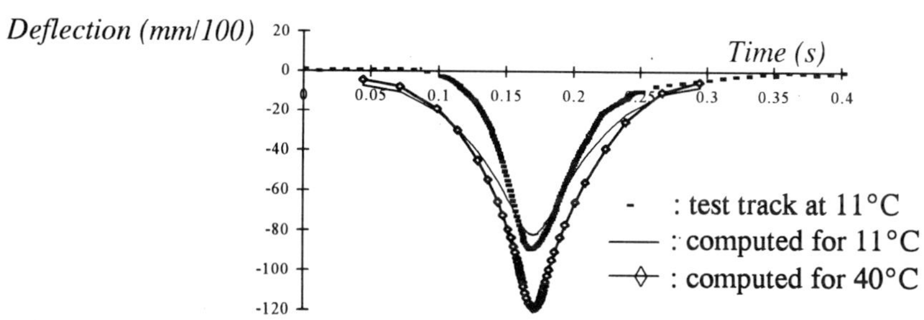

Fig. 6 : Vertical displacement under wheel path

The extent of the experimental deflection bowl is less than the computed one, but the accordance of the peak values is fairly good. It is believed that a part of the difference between the bowl shapes comes from the elastic non-linear behaviour of unbound materials, not taken into consideration here. Besides the computed deflection curve at 40°C comes to illustrate the theoretical effect of AM thermo-susceptibility on pavement behaviour. The increase of the deflection found with the temperature rise (30% here) corroborates the observations done on the test track. [4] also contains some discussion about the measured and computed strains for the same test track experiment.

Figures 7 to 18, relative to research on rutting, are another example which illustrates the potentiality of our lab-to-field approach. Figures 7, 9, 13, 15 show experimental horizontal strain signals measured during a test track rutting experiment, at the bottom of a 8 cm thick wearing course. By convention, ε (resp. σ) is taken positive for contraction (resp. compression). Strains of figs. 7, 9 were obtained for a pavement temperature around 20°C, those of figs. 13, 15 around 38°C. Remark the important qualitative and quantitative changes of the signals between the two situations. Fig. 13 for instance, shows a large extension peak, which does not exist on fig. 7.

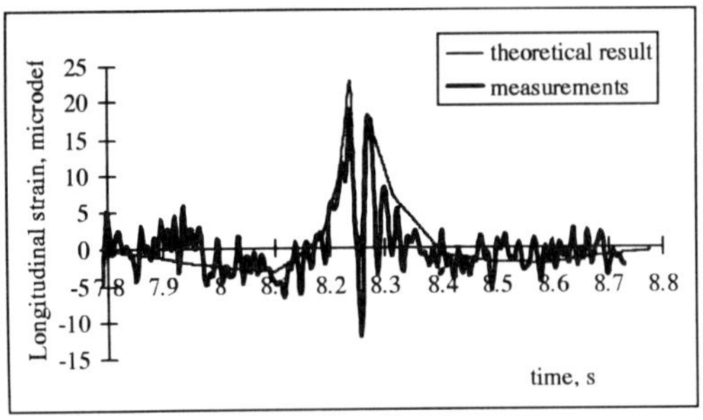

Fig.7 : Longitudinal strain, low θ

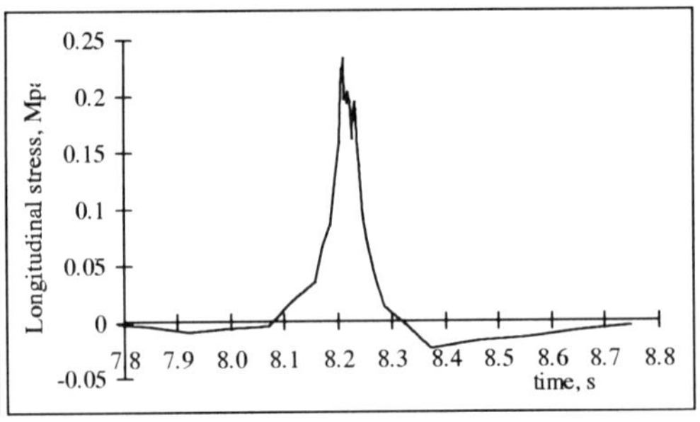

Fig.8 : Longitudinal stress, low θ

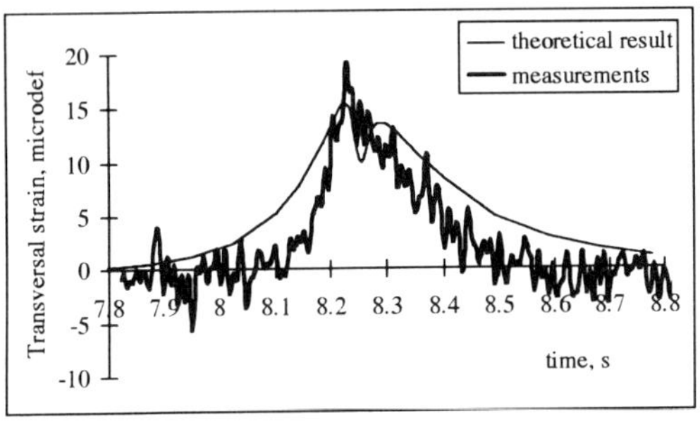

Fig.9 : Tranversal strain, low θ

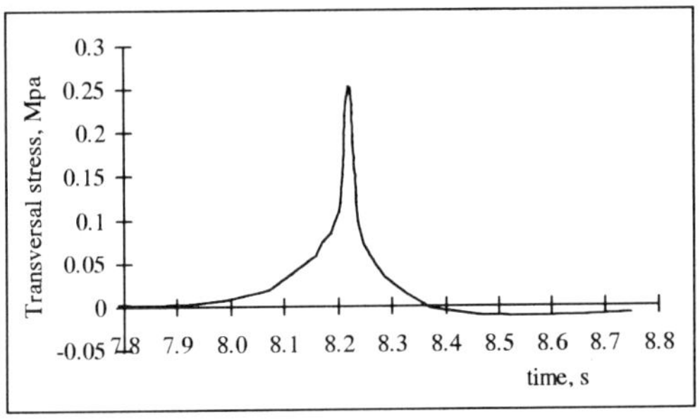

Fig.10 : Tranversal stress, low θ

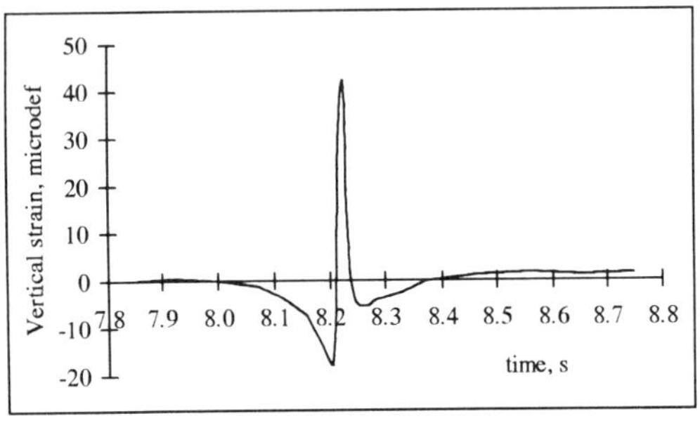

Fig.11 : Vertical strain, low θ

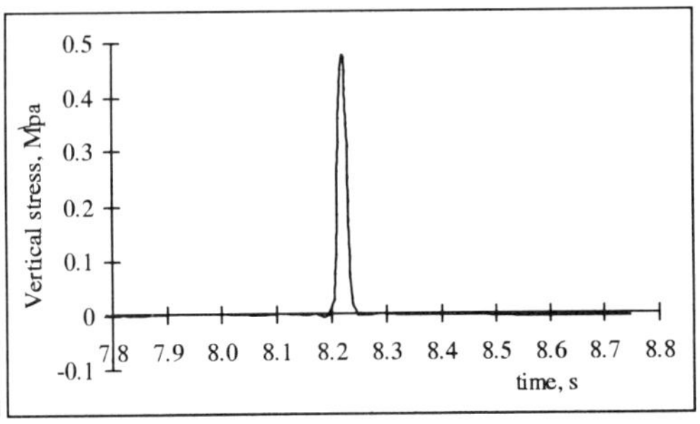

Fig.12 : Vertical stress, low θ

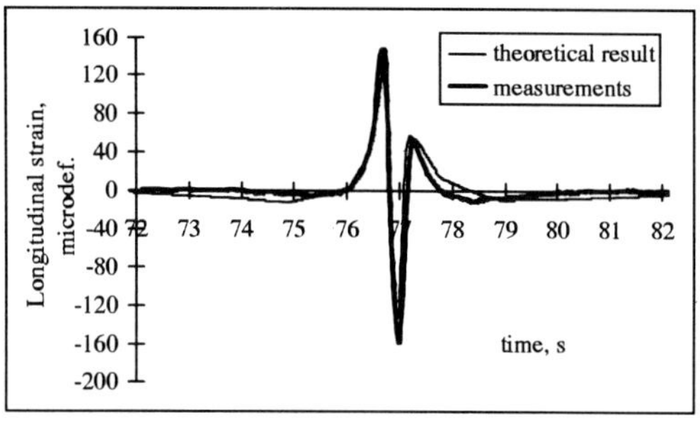

Fig.13 : Longitudinal strain, high θ

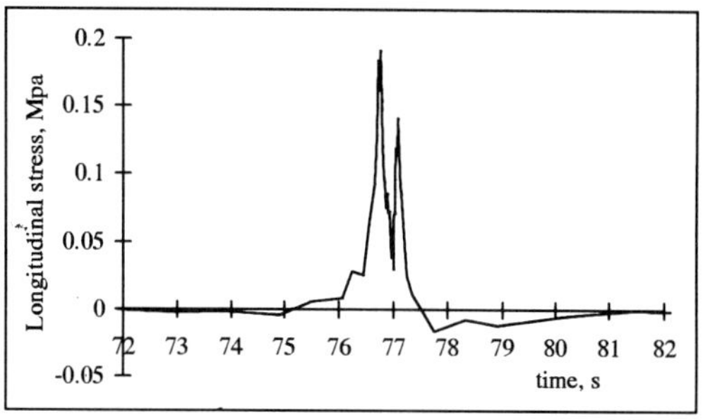

Fig.14 : Longitudinal stress, high θ

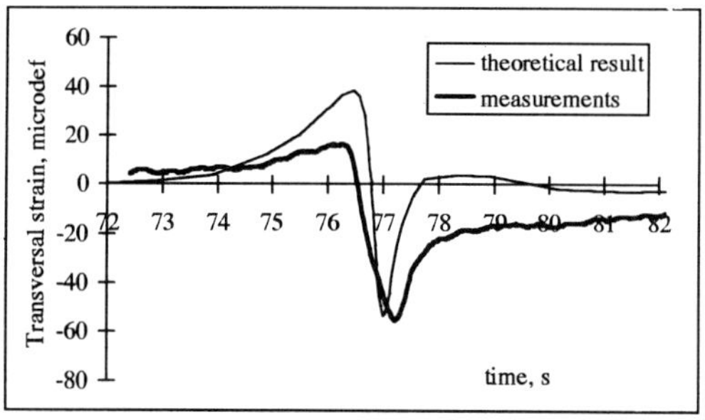

Fig.15 : Tranversal strain, high θ

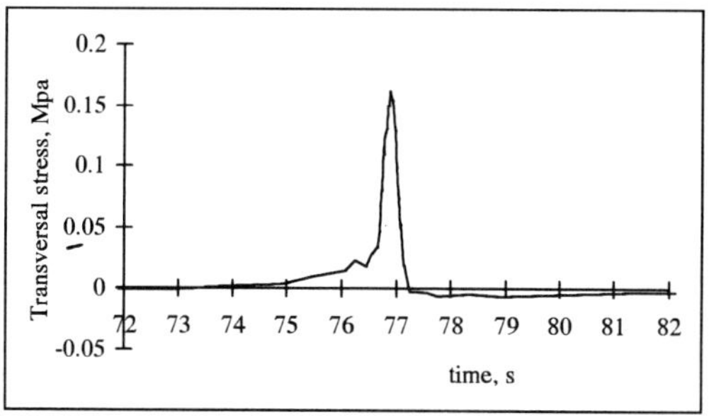

Fig.16 : Tranversal stress, high θ

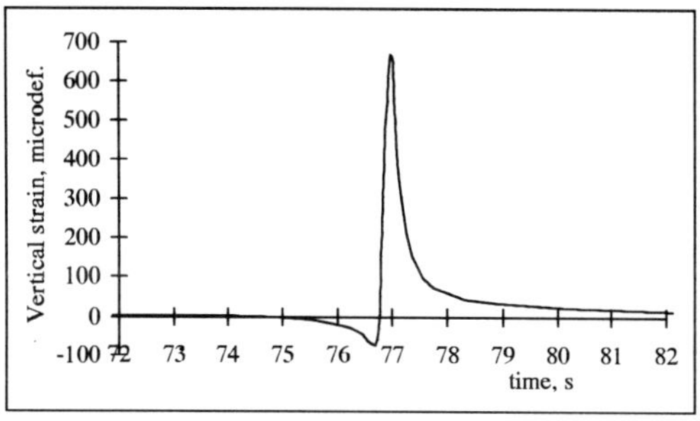

Fig.17 : Vertical strain, high θ

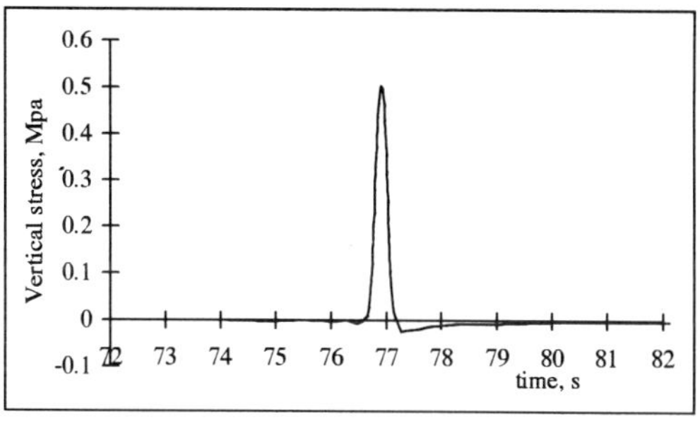

Fig.18 : Vertical stress, high θ

The computed signals of figures 7, 9 and 13, 15, derived from the complex modulus data and modelling with CVCR, show that the simulation is able to reproduce the wide variability of field signals. The other non measured, but computed data such as vertical strains and stresses can then be taken with confidence to help in the physical analysis of the whole situation. For instance, one can see that, under the wheel, the largest strain is in fact the non measured vertical one (contraction). Moreover, despite the extension peak of the horizontal strain of fig. 13, the corresponding horizontal stress is compressive. The situation can be compared to that of a triaxial test showing Poisson's effect, under classical stress conditions, $\sigma_{vertical} = \sigma_{axial} > \sigma_{horizontal} = \sigma_{confining} > 0$. More detailed analyses of such data are obviously of great help in the understanding of physical phenomena, such as rutting.

CONCLUSION

A lab-to-field approach for the visco-elastic calculation of pavements under traffic load has been presented. The input data for the characterization of AM behaviour can be either the complex modulus or secant modulus values deduced from current lab campaigns. In both cases, these data are interpreted through Huet & Sayegh's thermo-visco-elastic constitutive law. Finally the program CVCR developed within the finite element code CESAR-LCPC permits to achieve structural calculations. The comparisons performed until now between the test track results and CVCR simulations are rather encouraging. They allow to get confidence in the whole set of computed data such as the reversible strain and stress fields. Thus, CVCR is believed to provide a useful tool for the understanding of more complex pavement behaviour, such as rutting or fatigue. Introduction of rutting in CVCR is currently investigated.

REFERENCES

[1] BURMISTER, D. M., "The theory of stresses and displacements in layered systems and applications of the design of airport runways", Proceedings of the Highway Research Board, vol.23, November 1943.

[2] AUTRET, P., DE BOISSOUDY, A., GRAMSAMMER, J.C., "Le manège de fatigue du LCPC - Premiers résultats", Bulletin de liaison des laboratoires des Ponts et Chaussées, n°155, p 33-45, 1988.

[3] HOPMAN, P.C., "VEROAD: A linear viscoelastic multi-layer program for the calculation of stresses, strains and displacements in asphaltic road constructions", Delft University of Technology, ISSN-0169-9288-7-93-500-6, December 1993 (part I), ISSN-0169-9288-7-94-500-7, November 1994 (part II).

[4] PIAU, J.M., HECK, J.V., GRAMSAMMER, J.C., ODEON, H., " Modélisation viscoélastique des chaussées et comparaison avec comportement observé sur manège routier ", Colloque scientifique EC'97: comparaison entre résultats expérimentaux et résultats de calcul, Strasbourg, 22 et 23 mai 1997.

[5] HUET, C., "Etude par une méthode d'impédance du comportement viscoélastique des matériaux hydrocarbonés", Thèse de Docteur-Ingénieur, Faculté des sciences de Paris, France, 1963.

[6] SAYEGH, G., "Contribution à l'étude des propriétés visco-élastiques des bitumes purs et des bétons bitumineux", Thèse de Docteur-Ingénieur, Sorbonne, France, 1965.

[7] HUMBERT, P., "CESAR-LCPC: un code général de calcul par éléments finis", Bulletin de liaison des Laboratoires des Ponts et Chaussées, n°160, p112-115, 1988.

[8] DE LA ROCHE, C., "Module de rigidité et comportement en fatigue des enrobés bitumineux. Expérimentations et nouvelles perspectives d'analyse", Thèse de l'Ecole Centrale de Paris, 1996.

[9] DOUBBANEH, E., "Comportement mécanique des enrobés bitumineux des petites aux grandes déformations", Thèse de Docteur-Ingénieur de l'INSA de Lyon, 1995.

[10] HORNYCH, KAZAÏ, A., P., PIAU, J.M., "Study of the resilient behaviour of unbound granular materials", BCRA'98, Trondheim, Norway, 1998.

Fifth International Conference on the
BEARING CAPASITY OF
ROADS AND AIRFIELDS

Full-scale pavement test with hydraulic bound materials using 50% granulate asphalt

M. Horat[1], O. Özkul[1], M. Caprez[2]

[1]Dipl. Ing. ETH, IVT Swiss Federal Institute of Technology, ETH
[2]Dr. Dipl. Ing. ETH, IVT Swiss Federal Institute of Technology, ETH
Hönggerberg, CH-8093 Zürich, Switzerland, Fax +41 1 633 10 39

Abstract

In late 1994 construction work started for a test on a circular test facility for full scale trials. The test track was divided into four different sections in order to meet the goals, which can be summarized as follows:

- to study the performance of a hydraulic bound sub base layer consisting of approximately 50% granulate asphalt and 50% gravel.
- to study the effects of a stress absorbing interlayer (SAMI) located between a hydraulic bound layer and a bituminous bound layer.
- to compare the performance of bituminous hot mix layers with a different grading and different binders.
- to compare the performance of stone mastic asphalt wearing courses with a different grading and different binders.

As it eventually turned out, the trial had- so far - to be divided into two parts:
1. About 875'000 loadings corresponding to an axle load of 100 kN
2. Replacement of the surface layers and about 1'275'000 loadings corresponding to an axle load of 115 kN.

The analysis includes periodical observations, laboratory testes, field measurements and back-calculation. The measurements include rut depth, longitudinal profile, static and dynamic deflections, temperatures at different depths and strain measurements between different layers as well as comparisons with test sections on real trafficked roads. The test is planned to end in 1998, while first results will are presented in this paper In its entirety the test should help to answer questions concerning design, service life, maintenance and recycling.

Key words: hydraulic bound layers, stress absorbing membrane interlayer, recycling, full-scale tests

Introduction

In many countries people involved with road design, construction and maintenance are facing similar challenges. A few examples chosen at random are, for example: increasing traffic loads, increasing maintenance tasks, increasing lack of high quality materials, increasing need to recycle road construction materials. In 1994 a full-scale test was started that touches on some of these topics. The initial obejctives of the test may be described as follows:

1. The main task was to study the performance of a hydraulic sub base layer consisting of approximately 50 % granulate asphalt and 50 % gravel. There were several reasons for this approach. Generally, hydraulic bound layers have a higher layer coefficient than, for example, gravel layers. Therefore, a hydraulic bound layer might not only be used for new roads, but also for structural reinforcement of existing roads. Additionally, hydraulic bound layers at the current prices often cost less than comparable bituminous bound layers.
2. The use of asphalt granulate in a hydraulic bound layer has different reasons, as well. One is that there are increasing amounts of rubble from old roads and there is a lack of disposal sites, which leeds to a increasing need to recycle. But a lot of old roads had asphalt layers bound with tar bitumen and for health and environmental reasons are, therefore, not allowed to be recycled unbound or to be heated for reuse in a hot mix layer. They can be recycled in a cold mix layer, which has its own advantages and disadvantages [1], or, as in this test, they can be used as asphalt granulate in a hydraulic bound layer. Since hydraulic bound sub base layers often tend to cause problems with shrinkage cracks, it is an additional assumption that the relative elasticity of the bituminous or tar bitumen bound parts of the asphalt granulate might reduce this problem.
3. As mentioned above, in the past there were often problems with shrinkage cracks of hydaulic bound layers causing reflective cracking on other layers. Therefore, the effects of a stress absorbing membrane interlayer (SAMI) placed between the hydraulic bound sub base and a bituminous bound base layers was tested as a possible solution.
4. The available length of the test track should be divided in different sections in order to compare the performance of bituminous hot mix layers (HMT) with different gradings and different binders as well as the performance of stone mastic asphalt (SMA) wearing courses with different gradings and different binders.
5. Whereas in previous tests the loading system initially was set corresponding to an axle load of 100 kN (maximum according to the Swiss standard for single axles), this time the system should be corresponding to an axle load of 115 kN (maximum according to the Swiss standard for drive axles). As will be seen below, not all af these initial ideas could be achived in the imagined manner and the goals of the experiment had partly to be changed during the course of the test.

Test Facility

The test was carried out on the circular test facility that has already been used for different previous tests. The fatigue test track consists of an annular concrete pit. The width on top is 4.5 m, on the bottom 2.5 m, the depth is 2 m and the medium diameter is 32 m. With these dimensions full-scale pavements can be applied. The loading system consists of three arms, each fitted with two truck wheels and driven by a 55 kW direct current motor. The load applied by the wheels can be varied from 50 kN to 80 kN using water containers. For trucks this would mean axle loads between 10 und 16 tonnes. Tyre pressure is set in accordance with axle load. The transversal distribution of the loads can be controlled up to a width of 1.4 m by hydraulic rams. For the test it was programmed corresponding to the transversal distribution of truck wheel paths as measured on heavy trafficked roads. The speed of rotation of the arms can be held constant or varied according to a sinusoidal function by making use of a frequency generator. The maximum speed is about 80 km/h. It is also possible to control the ground water. Consequently, to leave aside influences due to level changes, the ground water level was set at the same depth during the whole test.

Structures

Four sections out of five were chosen in order to conduct the test presented in this paper. The remaining section was used for other purposes, i.e. to study the performance of sett-pavings. Intitially the structures were planned to be constructed as follows with the bituminous bound layers intentionally beeing relatively thin.

Table 1: Initial structures as designed

<table>
<tr><th>Section 61</th><th>Section 62</th><th>Section 63</th><th>Section 64</th></tr>
<tr><td>35 mm Stone Mastic Asphalt 11,Binder 1</td><td>35 mm Stone Mastic Asphalt 11,Binder 2</td><td>25 mm Stone Mastic Asphalt 8,Binder 1</td><td>25 mm Stone Mastic Asphalt 8,Binder 2</td></tr>
<tr><td>70 mm Bit. Hot Mix Base 0/22, Binder 3</td><td>70 mm Bit. Hot Mix Base 0/22, Binder 4</td><td>50 mm Bit. Hot Mix Base 0/16, Binder 3</td><td>50 mm Bit. Hot Mix Base 0/16, Binder 4</td></tr>
<tr><td colspan="4">15 mm Stress Absorbing Membrane Interlayer</td></tr>
<tr><td colspan="4">250 mm Hydraulic Bound Sub Base</td></tr>
<tr><td colspan="4">Silty Sand Subsoil</td></tr>
</table>

After construction the quality of some of the bituminous layers was not as good as expected. Still, it was decided to start the test in order to see what happens to these layers and then to replace the wearing courses of all sections at a later stage. After milling about 40 mm - 70 mm of the existiting surface to an even surface, the replacement was 35 mm stone mastic asphalt with the same binder for all sections. This meant that the test had to be conducted in two parts.

Table 2: Theoretical initial structures for part two

<table>
<tr><th>Section 61</th><th>Section 62</th><th>Section 63</th><th>Section 64</th></tr>
<tr><td colspan="4">35 mm Stone Mastic Asphalt 11, Binder 5</td></tr>
<tr><td>≈65 mm Bit. Hot Mix Base 0/22, Binder 3</td><td>≈65 mm Bit. Hot Mix Base 0/22, Binder 4</td><td>≈35 mm Bit. Hot Mix Base 0/16, Binder 3</td><td>≈35 mm Bit. Hot Mix Base 0/16, Binder 4</td></tr>
<tr><td colspan="4">15 mm Stress Absorbing Membrane Interlayer</td></tr>
<tr><td colspan="4">250 mm Hydraulic Bound Sub base</td></tr>
<tr><td colspan="4">Silty Sand Subsoil</td></tr>
</table>

Construction

Construction work started at the end of November 1994 with the preparation of the formation. Then some plate tests were done. The hydraulic bound layer was constructed and compacted on the fifth of December for all four sections. On that day it was raining in the morning and the formation was very wet. While the layer at first made a very inhomogenuous visual impression it looked much better after a few days. Thus, it was decided to construct the bituminous layers for all four sections on December the 13th. At first the modified bituminous binder of the SAMI was sprayed on the surface of the hydraulic bound layer. Then the already wraped and warmed up aggregate was added and compacted while the surplus aggregate was vacuumed up. After that the other bituminous layers of the different sections were constructed. As it turned out, the circumstances of this day finally mounted up to unsatisfying results concerning material properties and also the longitudinal eveness of some of the sections. With hindsight this was probably due to a combination of different causes:

- There were some unnecessary delays because some of the materials were not ready in the right quantity at at the right time.
- Although it was sunny in the afternoon and relatively warm in the afternoon, the temperatures were not always according to the Swiss standards.
- The construction equipment may not have been ideal

The construction work for the second part of the test (replacement of the surface layers) was done in April 1996. This time the weather conditions were better, there were no delays and heavier construction equipment was used. As a result, the new surface layer had satisfying characteristics.

Figure 1: ground plan of the test track at the beginning

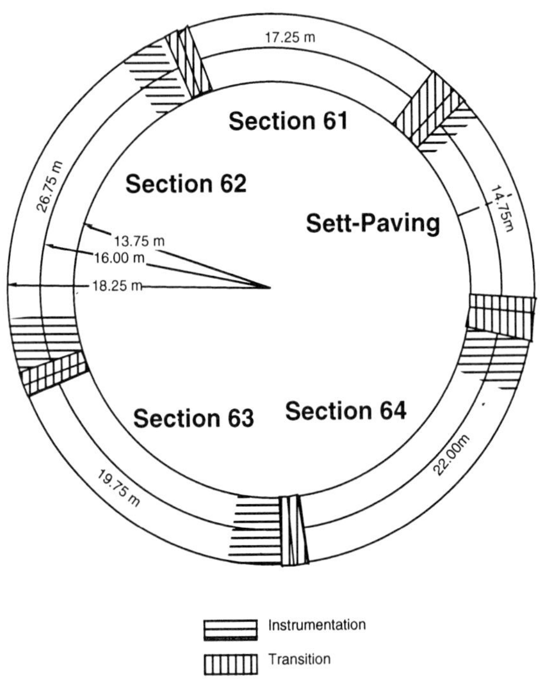

Instrumentation

At various levels sensors for measuring temperatures were inserted during construction. This allowed to have constant temperature data ranging from -2 cm to - 2 m on each section along with air temperature. At the same time, a total of 44 strain gauges for measuring transversal and longitudinal strains between the different layers were installed in the instrumentation zone of each section.

Measurements

Regular standard measurements on marked spots of each section during the test included:
- Longitudinal and transversal profiles
- „Statical“ and „dynamical“ deflection measurements

Among other things there were also :
- Laboratory tests on cores (taken at various stages of the test) and material samples
- Plate tests
- Nuclear isotopic soil penetrometer
- Strain measurements along with opto-electronical deflection measurements at various speeds and wheel positions
- Collection of climate datas

Conduct of the test

Load application started in March 1995. It was possible to apply about 875’000 loadings during the first part (loads corresponding to an axle load of 100 kN and about 1’275’000 loadings during the second part (loads corresponding to an axle load of 115 kN). Speed was usually varying between 20 and 35 km/h. Interruptions took only place in case of measurements, repair work or during wintertime, when there was snow on the test track.

Figure 2: Loadings per month

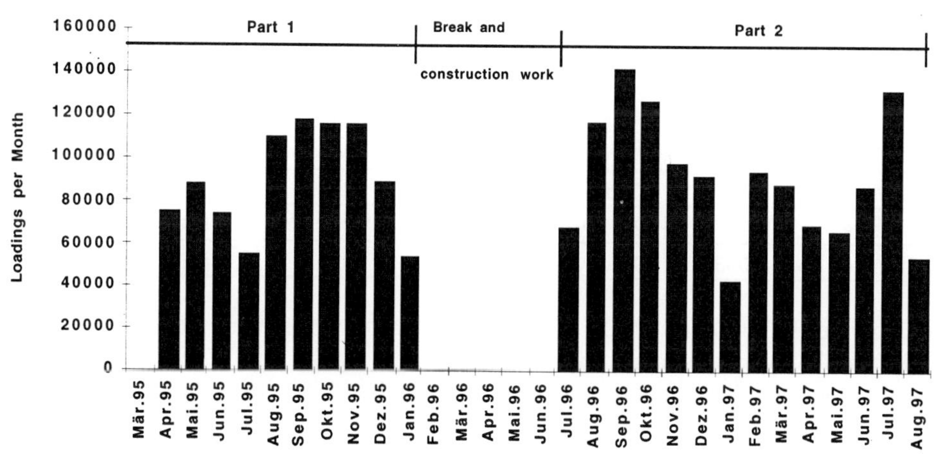

First Results

Material properties and specifications At different stages of the project cores and material samplings were investigated. Some of the characteristics of the materials are presented here. The hydraulic bound layer was bound with a mortar binder (70kg / m^3) and consisted of 38 % asphalt granulate, 57 % gravel, 4 % concrete granulate and 1 % diverse aggregate.

Figure 3: Grading analysis of the sub base according to the Swiss Standards [2]

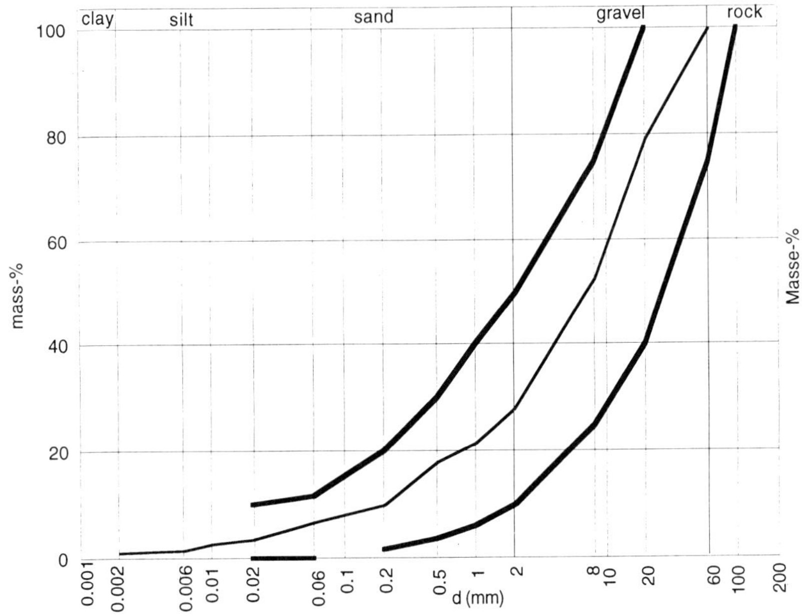

After construction the degree of compaction as measured with a nuclear isotopic soil penetrometer had an average of 97.9 % and the medium water content was 6.4 %. The mean compression strength of nine samples was as follows.

Table 3: Compression strength of the hydraulic bound material

9 samples	weight (kg)	age	unit mass (t/m^3)	ultimate load (kN)	compr. strenght (N/mm^2)
mean value	2.055	7 days	2.20	32	3.95
mean value	2.057	28 days	2.21	52	6.45
mean value	2.053	92 days	2.20	66	8.24

Laboratory tests from cores taken after part one and samples taken at the end of 1997 showed that the hydraulic bound layer seemed to be in a good state. There were no visible cracks on these samples. Cross beams taken at the end of 1997 made the same impression. The next two tables (4 and 5) show a selection of mean values from seven cores taken in each section after construction of part one and a sample of the Swiss standards [2]. A comparison of the two tables illustrates why the construction of part one was not satysfactory. Especially the void contents were too high. Six cores from each section - taken after part one - indicate how much the void contents changed after 875'000 loadings (Table 6). The SMA 11 surface layer of part two had good values after construction. Mean values of six cores from each section taken after part two had also comparable satysfactory results.

Table 4: A few of the characterisics of the bituminous bound materials (mean values)

Section Nr.	Material	Binder content (M-%)	Density maximum (g/cm³)	Void content (Vol-%)	Degree of compaction (%)	Softening point (°C)	Penetration (1/10 mm)
61	SMA 11	6.12	2.456	4.5	99.2	58.9	37
	HMT 22	4.28	2.537	5.5	102.3	61.3	33
62	SMA 11	6.00	2.467	3.5	100.9	68.3	38
	HMT 22	4.44	2.542	4.4	102.8	67.0	23
63	SMA 8	6.36	2.437	14.4	89.1	58.3	39
	HMT 16	4.99	2.525	8.2	97.3	59.9	37
64	SMA 8	6.19	2.451	7.0	98.6	67.9	38
	HMT 16	5.13	2.511	9.0	96.5	64.5	26

Table 5: Exerpt from the Swiss standards for mean values of these materials [2]

Belagstyp	Binder content (M-%)	Void content (Vol-%)	Degree of compact. (%)
SMA 8 S	–	–	–
SMA 11 S	6.5.....7.5	3.0.....4.0	98
HMT 16 S	4.3.....4.9	3.0.....6.0	98
HMT 22 S	4.0.....4.6	3.0.....6.0	98

Table 6: A few of the properties of the bituminous bound materials after part one

Section Nr.	Material	Binder content (M-%)	Density maximum (g/cm³)	Void content (Vol-%)	Softening point (°C)	Penetration (1/10 mm)
61	SMA 11 S	6.12	2.456	5.7	62.7	32
	HMT 22 S	4.28	2.537	5.3	63.9	32
62	SMA 11 S	6.00	2.467	3.5	69.4	37
	HMT 22 S	4.44	2.542	5.2	67.4	21
63	SMA 8 S	6.36	2.437	4.7	61.1	35
	HMT 16 S	4.99	2.525	5.0	58.9	36
64	SMA 8 S	6.19	2.451	6.4	68.8	36
	HMT 16 S	5.13	2.511	7.8	66.1	24

Deflection The following table shows results of the deflection measurements as measured with the Benkelman beam. These are mean values of several measurements during each part.

Table 7: Deflections, Benkelman beam

Part	Pav. temp. (°C)	Air temp. (°C)	Sect. 61		Sect. 62		Sect. 63		Sect. 64	
			Defl. (1/100mm)	C_V (%)	Defl. (1/100mm)	C_V (%)	Defl. (1/100mm)	C_V (%)	Defl. (1/100mm)	C_V (%)
I (6 Measurements)	17.4	12.8	22	26	27	34	29	30	32	42
II (8 Measur.)	18.8	12.8	37	30	43	39	46	27	47	40

The values are not corrected with temperature. Therefore, the mean values of the temperatures during those measurements are also given (it is by coincidence that the mean air temperaure turned out to be the same for both parts). The deflections of part two were generally higher during all seasons, probably mainly because of the thinner structures. During both parts the change of deflections over time appeard to be rather parallel for all the sections.

Surface distress There were no visible cracks during both parts of the test. There was some creeping in the transition zone of section 61 during part one. But this was exactly at a place where cores had been taken.

Evenness Figures 4 and 5 show the longitudinal profiles. The difference in eveness between the two parts is quite obvious (Section 64 was shortened for part 2 due to construction reasons).

Figure 4: Longitudinal profiles of the first and the last measurement, phase one

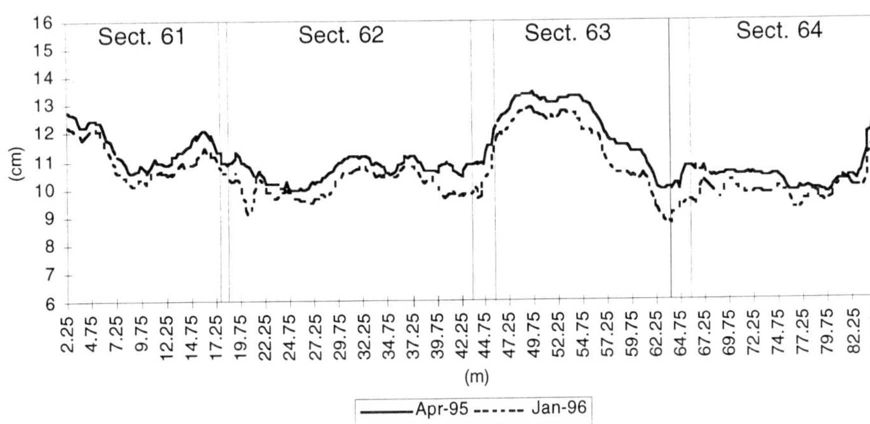

Figur 5: Longitudinal profiles of the first and the last measurement, phase two

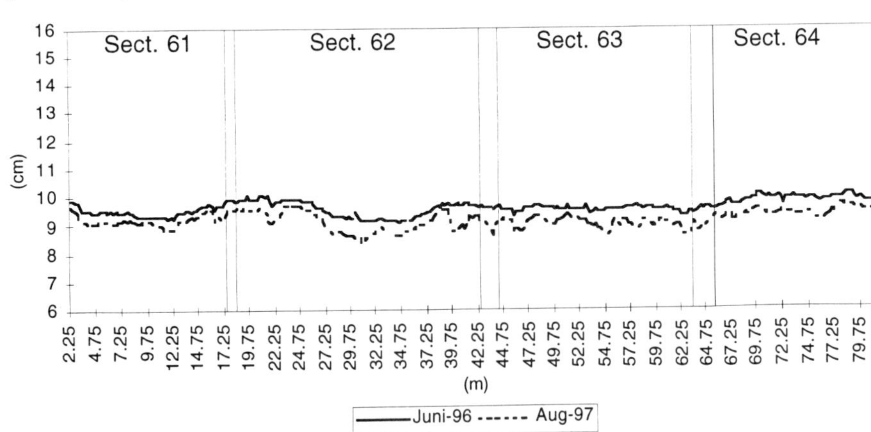

The rut depths of phase one and two are presented in figure 6 using the traffic eqivalence factors according to the AASHTO Guide [3]. The rut depths show a typical increase during high temperature periods, but at the beginning of part one they also refer to additional compaction due to the high void contents of some of the layers. At least for part two the rutting of the sections with HMT 16 (63 and 64) and the sections with HMT 22 (61 and 62) each have similar tendencies.

Figure 6: Rut depths (2 m bar)

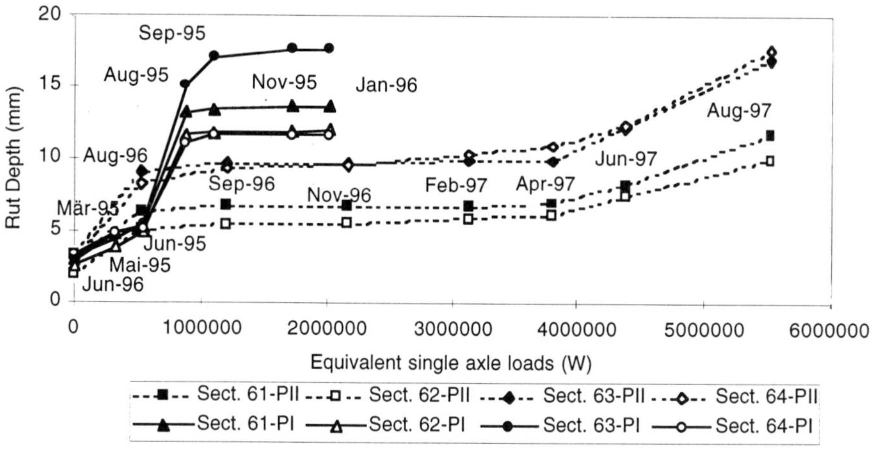

Conclusions

Part one with about 2'021'000 equivalent single axle loads showed impressivly the effects of partly inhomogenous hot mix base layers and stone mastic asphalt layers with partly too much void content, and ,therefore, also stressed the importance of careful construction work under the right conditions.
Part two with about 5'520'000 equivalent single axle loads justified an increasing use of stone mastic asphalt layers in Switzerland.
After three years and about 7'542 '000 equivalent single axle loads (factors according to AASHTO) the combination of the hydraulic bound layer and the stress absorbing membrane interlayer performed very satisfactory. Cores and cross beams of the whole pavement structures showed no problems with slip or cracks. Backcalculation of the layer coefficient (according to AASHTO) of the hydraulic bound layer with approcimately 50% asphale granulate resulted in values that are in accordance to the layer coefficients of conventional hydraulic bound layers.

Outlook

After a break in August 1997, it is planned to carry on with the test for a few month in 1998, partly also in order to investigate wether hydraulic bound sub bases are a good option for sub base layers that could be designed for more then the theoretical 20 years of

current designing methods in Switzerland. Comparisons with test sections on real trafficked roads are also planned to go on in 1998.
Additionally, laboratory tests with hydraulic bound layers made with 100% asphale granulate already showed some promising results [4].

Acknoledgements

The following institutions contributed to this test: Institut für Geotechnik, ETH Zürich; Tiefbauamt des Kantons Zürich, Bundesamt für Strassenbau, Ingenieurbureau Heierli AG.

References

1. Beligni, M., Horat, M. and Caprez, Dr. M. Heiss- und Kaltmischfundationsschichten aus recycliertem Ausbauasphalt, , Schlussbericht zum Forschungsauftrag 14/91, Forschungsberichte des Eidg. Departementes des Innern, Nr. 351, Bern, August, 1995

2. Hrsg. Vereinigung Schweizerischer Strassenbaufachleute (VSS), Schweizer Normen (SN), Zürich

3. AASHTO Interim Guide for the Design of Pavement Structures 1972, AASHTO, Washington D.C., 1972

4. Shojaati, M. Einsatz von zementstabilisiertem Asphaltgranulat in Fundationsschichten, Schlussbericht erscheint 1998

Fifth International Conference on the
BEARING CAPASITY OF
ROADS AND AIRFIELDS

RELATING ACCELERATED PAVEMENT TESTING TO IN SERVICE PAVEMENT PERFORMANCE

THOMAS D. WHITE[1] AND JEFF HUA[2]

[1]Professor, Purdue University, School of Civil Engineering, West Lafayette, IN, 47907, USA, Fax 765-496-1364, E-mail tdwhite@ce.ecn.purdue.edu
[2]Research Assistant, Purdue University, School of Civil Engineering, West Lafayette, IN, 47907, USA, Fax 765-496-1364, E-mail jiangfen@ce.ecn.purdue.edu

ABSTRACT

Purdue University has developed an accelerated pavement test (APT) facility. This facility is being utilized to evaluate Superpave asphalt mixtures and criteria. In conjunction with this testing a model is being developed for predicting the performance of in service pavements from APT tests. Factors have been identified that have an effect on rutting magnitude. These factors include pavement geometry, boundary conditions, materials and loads; test conditions; construction variables and material variables. Determining the effect of these factors will provide a means of predicting in service pavement performance from APT tests. APT test data are presented for tests where truckloads of mixture for a Superpave project were diverted for placement in the APT facility. These data were utilized to calibrate a creep rate model. Subsequently, the model was utilized to predict the effect of loading speed, contact pressure and wander.
Key Words: APT, rutting, model

INTRODUCTION

Rational pavement design generally uses magnitude of tensile strain at the bottom of the lowest asphalt layer and vertical strain at the top of the subgrade as criteria for pavement thickness determination. Conceptually, the asphalt tensile strain is related to fatigue cracking and the subgrade vertical strain is associated with permanent rutting. However, modern asphalt pavements have asphalt surfaces that are relatively thick or are full depth asphalt pavements. Older pavements, asphalt or concrete are likely to have multiple overlays. These overlays are used to increase structural capacity or restore serviceability. Rutting from heavy, high volume truck traffic is one component of pavement performance. The rutting leads to roughness and loss of serviceability. Perhaps more importantly, rutting leads to a concern for safety. With rutting a driver's ability to steer a vehicle is affected.

Also, if ruts are deep enough water accumulates, creating hydroplaning potential. In winter, the accumulated water may freeze in the wheel paths. Such rutting is primarily in the asphalt layers near the surface.

Asphalt mixture rutting is time dependent and rutting of a well-designed asphalt mixture used on an in service pavement may take years to develop. This is also the case for evaluating the effect of using new materials or of making modifications to mix design procedures, specifications, and construction practices. Accelerated pavement testing can compress the time to obtain asphalt pavement performance. Several factors can be controlled to magnify damage per load application. Recognizing the potential of accelerated pavement testing, a number of agencies in the United States have or are developing accelerated pavement testing facilities.

In large part, application of accelerated pavement testing results has been limited to a relative comparison of materials or pavement sections or empirical analysis. These approaches limit application of APT results. Usefulness of accelerated pavement testing would be enhanced through rational analysis of the time dependent material properties, loading and physical geometry of the pavement and boundary conditions. Such an analysis would make possible extrapolation of the accelerated pavement test results to predict performance of in service pavements.

THEORY FOR RUTTING

Asphalt concrete is a time, temperature, and stress dependent material. Asphalt mixture subjected to repeated loading exhibits elastic, plastic, visco-elastic, and visco-plastic responses. The elastic properties do not contribute to permanent deformation and are therefore modeled by modulus of elasticity and Poisson's ratio. Plastic properties contribute to permanent deformation, which is cumulative under repeated loading. A creep model (1) can be used to characterize the permanent deformation properties of the asphalt mixtures.

$$\dot{\varepsilon} = A\sigma^n t^m$$

Where:

$\dot{\varepsilon}$ = Creep rate
σ = Mises equivalent stress
t = total time
A,m,n = parameters related to material properties

The goal in application of the above creep model is to predict performance of in service pavements from accelerated pavement tests. However, there are several factors that influence rutting as shown in Figure 1. They should be accounted for in the predictive model. Of these factors vehicle speed and contact pressure are represented directly in the creep model as time and normal pressure, respectively Once the creep model is calibrated, layer thickness is accounted for in the model geometry. The creep model and simulation of load sequence can also account for wander. Temperature, asphalt mixture characteristics and construction quality have to be determined by tests and are represented in the values of the constants in the creep model.

Continuing research at Purdue University is concerned with applying the creep model in a finite element method analysis to predict in service pavement rutting. Tests have been conducted in an accelerated testing facility to evaluate some of the factors under controlled conditions.

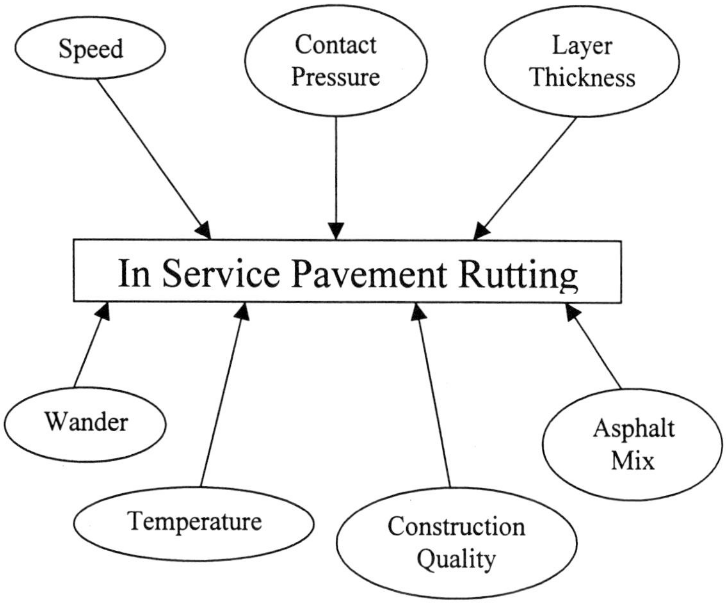

Figure 1 Rutting Factors

INDOT/Purdue APT Facility

An accelerated pavement testing facility was developed by Purdue University (2) for the Indiana Department of Transportation (INDOT). Prototype scale pavement sections can be installed in a test pit in the facility. The Accelerated Pavement Tester (APT) loading system has the capability of applying moving wheel loads to pavement test sections installed in the facility. The APT is designed to apply up to a 20,000-lb load on a dual wheel or super single assembly. The load carriage travels at 5 mph. There are many miles of concrete pavement with asphalt overlays in Indiana. As a result, tests in the APT Facility were initiated with asphalt overlays constructed on a concrete slab. As a result, a heating system was developed for the asphalt overlay that uses hot water pumped through hoses embedded in the concrete. Air heating is also utilized to minimize air pavement temperature differential. High test temperature and slow speed results in significant damage to a pavement structure for each wheel pass. The APT device is shown in Figure 2.

Typically, two paving lanes (each lane is ten feet wide) are constructed in the APT facility. Test lanes are five feet wide. The concrete slabs on which the asphalt is placed are also five feet wide and twenty feet long. Each slab can be heated separately. As a result, various factors related to mixture performance can be studied. Examples are quality of construction (high and low density), tire pressure, temperature and asphalt content.

Figure 2 INDOT/Purdue University APT

In initial studies of asphalt mixture rutting performance (2) the asphalt layer was modeled resting on a rigid surface. A finite element mesh was created to represent the overlay. Four factors control the finite element mesh geometry:

1. Pavement geometry, which controls the general size of the mesh.
2. Load configuration, such as dimensions and spacing of contact area.
3. Degree of detail, i.e., locations where response parameters will be predicted.
4. Desired accuracy.

The general mesh is shown in Figure 3.

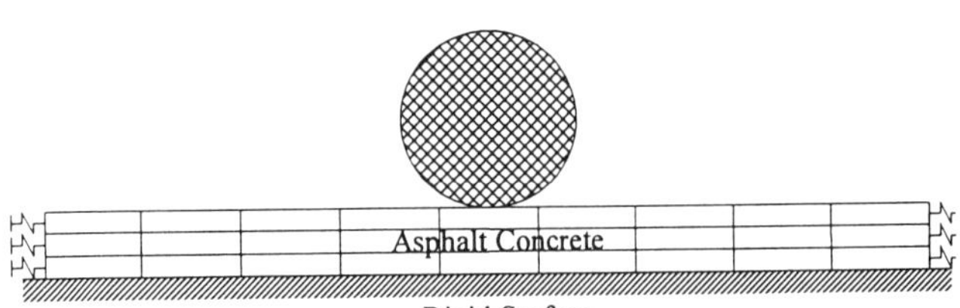

Figure 3 Side View of Finite Element Model (2)

APT TESTS

The Indiana Department of Transportation

(INDOT) paved approximately forty Superpave projects in the 1997. Arrangements were made to ship mixture from one of the projects located approximately thirty miles from the APT facility. In preparation for placement of the mixture the in-slab heating system was turned on the night before. A light tack coat was applied to the concrete surface. Two truckloads of mixture for the project were diverted and sent to the APT facility in West Lafayette. Conventional laydown and compaction equipment was utilized to place two ten-foot wide paving lanes. The mixture was constructed three inches thick. An attempt was made to achieve different compaction levels in the two lanes. This attempt was not as successful as desired.

The mixture placed was a 12.5-mm Superpave mixture. Design gradation, asphalt content and volumetric properties are given in Table 1. Laboratory compaction (N_{design}) was selected for 3 to 10 million ESALs.

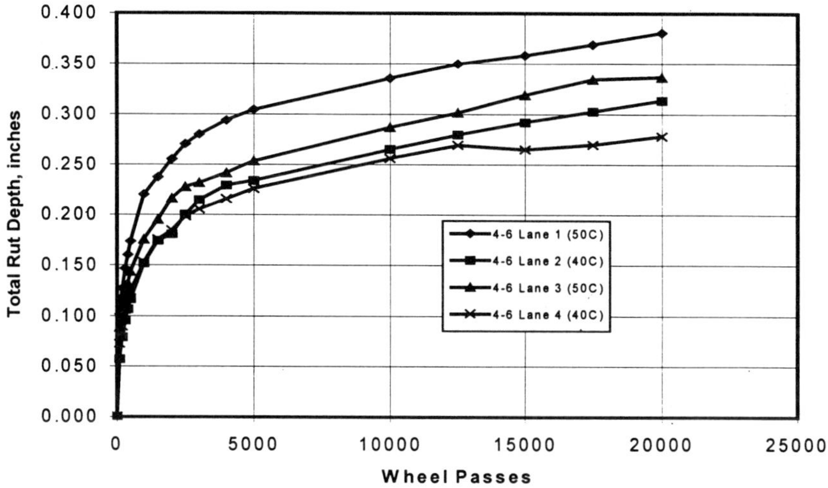

Figure 4 APT Rutting

After construction temperature sensors were embedded in the asphalt. These sensors provide feedback to the heating control system. Nine transverse sections are marked at two-foot intervals on the pavement surface and are used as references for measuring transverse profiles at various increments of traffic. A transverse profile device was designed and fabricated for this purpose. The transverse profile device consists of a beam spanning the test section. A small wheel attached to the beam is lowered to the pavement surface and rolled across the surface. Transducers are used to determine the wheels horizontal and vertical movement. This data is recorded electronically. An initial profile is measured. The difference in the initial profile and subsequent profiles at various levels of wheel passes is used to compute rutting. Rutting in these tests is taken as the difference in the lowest point on a profile and an imaginary straight edge resting on the test lane surface and spanning the wheel paths. Because there are end effects the center three transverse profiles are averaged and used as a single result.

As noted above, the original approach to the tests involved an attempt to construct the two paving lanes at different densities. This approach was not completely successful. Each paving lane includes two five-foot test lanes. The test plan included tests of both levels of density at forty and fifty degrees centigrade. All tests were conducted with 9000 lbs. on dual wheels and a tire pressure of 90 psi.

A total of twenty thousand load cycles were applied to each test lane. All loads were applied in a single wheel path and in one direction. Single wheel path loading was selected to be compatible with previous tests. Single wheel path loading also includes a component of damage acceleration. The APT has the capability of applying normally distributed random traffic. Traffic can also be applied in both directions. Each test lane requires approximately seven days for preparation, traffic application and sampling.

A decision was made to conduct the tests at two temperatures. One test lane of each paving lane was trafficked at 40°C and the other at 50°C. Data for each test lane is plotted in Figure 4. This data is the straight-line rutting described above.

ANALYSIS

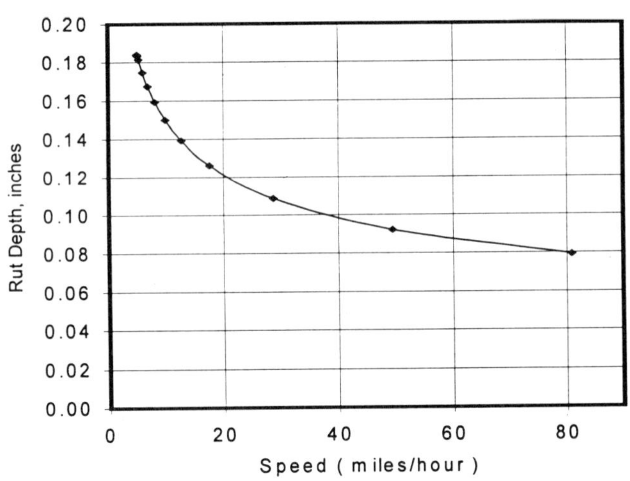

Figure 5 Effect of Speed on Rutting

An analysis was conducted using the actual data to calibrate the creep rate model. This is a trial and error procedure but with experience only two or three trials are required to achieve reasonable agreement between the measured and predicted rutting. The agreement is very good up to 5000 wheel passes. After 5000 wheel passes the predicted and measured rutting do not agree as well. Consequently, the following analysis is made for up to 5000 wheel passes.

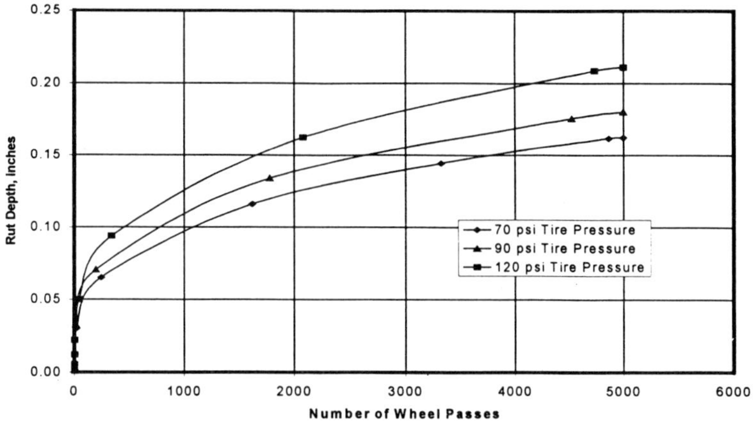

Figure 6 Effect of Tire Pressure on Rutting

Time of loading is one of the variables in the creep rate model. Time of loading for one wheel pass depends on the length of the wheel print and vehicle speed. Consequently, time of loading varies with vehicle speed. Using this simple model for time of loading the rutting test results at five miles per hour can be extrapolated to higher traffic speeds. These results are shown in Figure 5.

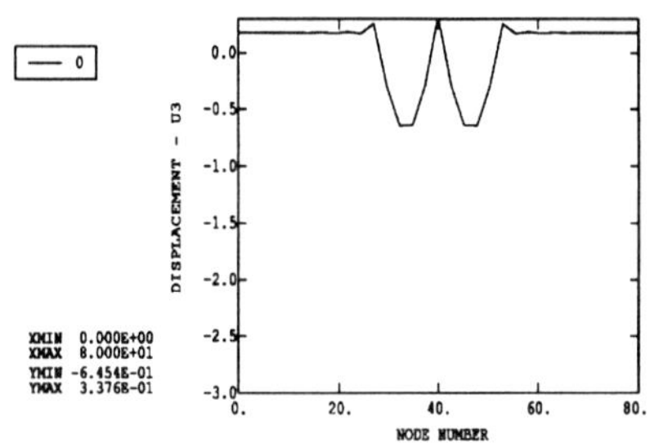

Figure 7 Rutting with Zero Wander

Tire pressure is also a variable in the creep rate model. Actual tire pressure was 90 psi in the APT tests. Gross contact pressure in the analysis was computed based on wheel load and measured gross tire print area (2). The gross tire contact pressure was found to approximately equal the tire pressure. As with speed, rutting at 90 psi contact pressure can be extrapolated

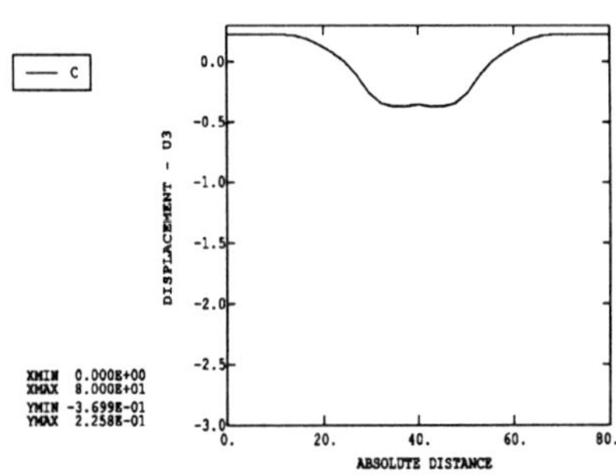

Figure 8 Rutting with Fifteen Inches of Wander

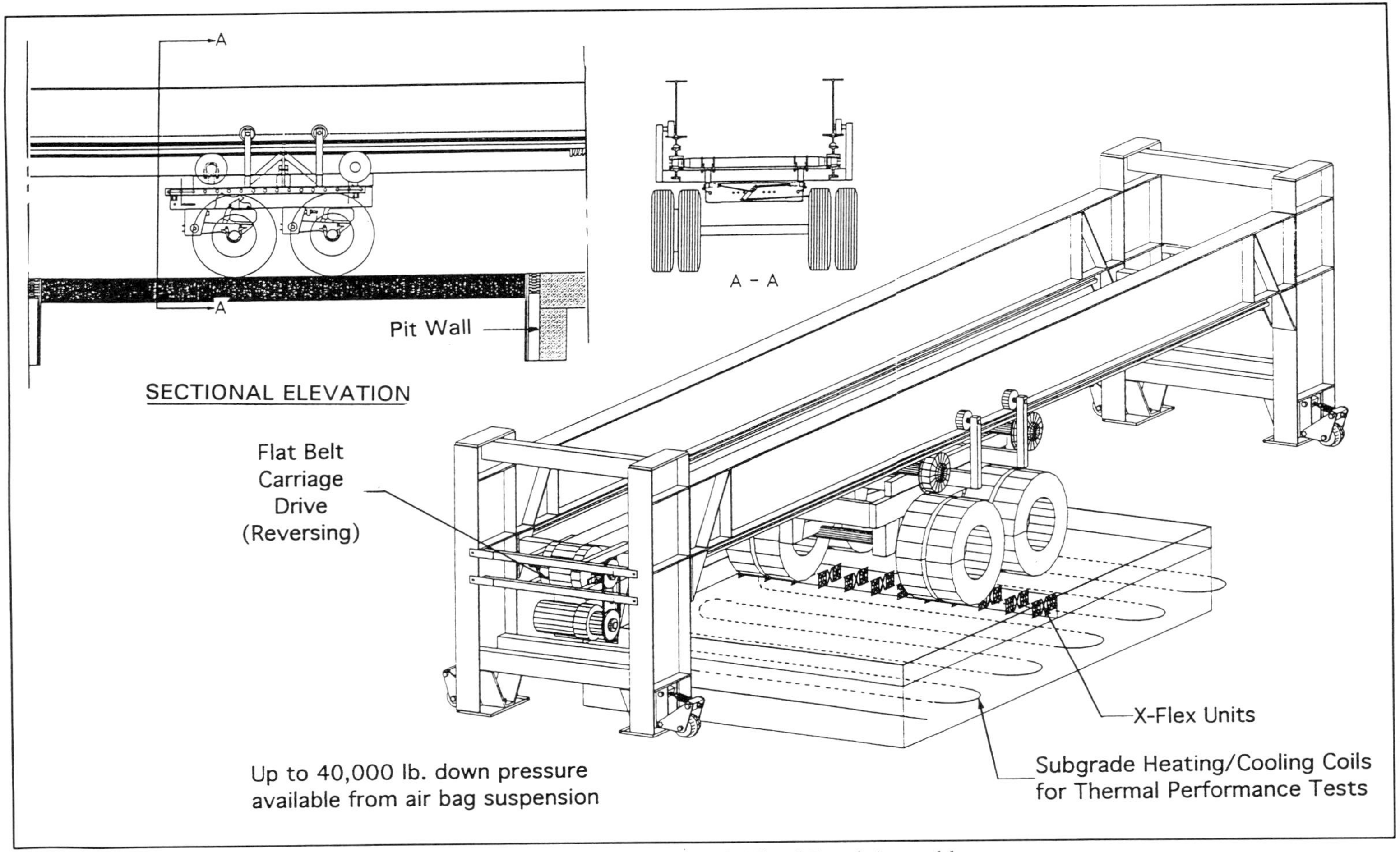

FIGURE 1 Test Frame and Wheel Load Assembly
(Design Provided by Cardwell International, Ltd.)

The test frame and loading devices were designed and fabricated by Cardwell International, Ltd., of Newton, Kansas. The arrangement allows the system to load one or both axles as desired. One or more pairs of tires may be replaced by a super-single if a test requires so. Normally the system would be loading in both direction as the wheel assembly moves back and forth. However, one-way traffic simulation can be achieved through a hydraulic system that can lift the wheel axles either manually or automatically. The automatic mode will cause the eight wheels to be lifted off the ground when the carriage reaches the end of the track until it goes back to its initial position and start a new load cycle. The manual mode is used when the whole test frame needs to be moved off the specimen or across the laboratory space. The frame is moved by pulling it using an overhead crane. Accurate positioning is achieved manually with a pry-bar.

The TAM is moved back and forth along the track using a flat conveyor belt driven by a 20 HP variable speed electric motor which reverses direction every time the carriage reached one end or the other of the reaction frame. The fastest safe operating speed achieved is 300 cycles per hour, or 600 load applications per hour for the two-way passage operation. At this rate, the average speed of the wheels axles is 5.6 km/h (3.5 mph) over the total travel distance of 9.1 m (30 ft) whereas the speed is about 11.3 km/h (7 mph) constant over the middle portion of the track where the specimens are usually placed in the pit.

Pulse Load System Fatigue tests of jointed Portland Cement Concrete Pavements (PCCP) are performed by applying cyclic loads (thumping) across the joint using an MTS system with hydraulic actuators. Generally it takes several million load cycles to fail PCCP joints, as opposed to few hundred thousand to fatigue AC pavement. This allows comparing the behavior of different shear transfer devices at a joint. Using this setup a speed of 9000 applications per hour have been achieved which is 15 times faster that the rolling axles. With this sytem, more cycles can be applied in a reasonable duration of time.

The same reaction frame carrying the wheel load assembly is used to support the pulse load system. The wheel carriage (bogie) is moved to one end of the frame (East end) to make room for the pulse load system at the midspan of the frame. The pulse load system consists of a two actuators (cylinders) activated by a hydraulic power supply (MTS pump) and positioned across the PCCP joint such that when one actuator is down, the other one is up. This results in a rocking action on the two sections of the pavement simulating the passage of a truck over the joint even though it does not exactly replicate the rolling wheel effects.

Heating and Cooling System The temperature control equipment consists of a boiler for heating and a chiller system for cooling or freezing. They both use the same 55% ethylene glycol-based fluid that circulates in the thermal coils to heat/cool the pavement slab and/or subgrade. A set of control valves on the supply and return lines directs the solution to either the boiler or the chiller depending upon the test requirements. Temperature ranges between -23°C and +66°C (-10°F and +150°F). The depth of the pipes in the subgrade determines how much of the soil underneath the pavement is thermally controlled. To add more flexibility to the testing facility, the elevation of the heat exchanger (piping network) was made variable by having six lines of holes at 8 inches intervals (vertically) so that when soil is removed from the pit, the pipes can be re-installed at a different depth. In the present position, 17 U-tubes are installed at the third row of holes from the top, i.e. at 914 mm (36") below the lab floor level, and evenly spaced horizontally at 203 mm (8 in.) intervals.

The capability of heating and cooling from the top through thermal panels was added to duplicate the usual natural conditions and allow the flexibility to control the temperature gradient through the pavement surface and the subgrade. This is accomplished by circulating the glycol fluid through thermal panels (stainless steel plate coils) connected with flexible pipes so that pavement can be cooled/heated from the top down. This allows to specify the temperature of both the pavement surface and the subgrade, and hence the ability to vary the temperature gradient across the pavement depth. A strip on the surface of the specimen can be exposed to allow for the wheel load to travel on the heated/cooled pavement while it is being heated or cooled. Each of the lower and upper system can be operated independently. Through manipulating the appropriate set of valves, it is possible to have the top and bottom of the pavement both cooled, both heated, top heated/bottom cooled, or vice versa. Thermocouples are installed on the tubes but also below and above both the base and the pavement. Surface heating is also achieved by using radiant heat lamps (infrared heater). This best simulates exposures of AC pavement to the sun and is specially used in rutting tests.

Water Table and Soil Moisture Control A sprinkler/drainage system is installed in both the middle pit and the environmental pit. The inside wall of the concrete pit are coated with a water proof sealant, and the inside walls of the environmental pit are covered with a water-tight rubber membrane to prevent water from getting into its thermal insulation. The water sprinkler consists of three soaker hoses buried right under the surface of the aggregate base, transverse to the direction of wheel rolling. The location of the hoses is at 0.6 m (2 ft), 3.1 m (10 ft) i.e. at the middle, and 5.5 m (18 ft) spacing from one end or the other of the pit. The middle soaker hose was mostly used during the concrete pavement test (as it lays under the PCCP joint) to achieve the most serious soil erosion and pumping action at the joint.

Three water collectors are placed horizontally at the bottom of each of the pits consisting of 102 mm (4 in.) perforated black plastic pipes (corrugated) leading into a metal riser tube of 254 mm (10 in.) diameter installed vertically at one corner of each pit. A layer of 279 mm (11 in.) of UD-1 or pea gravel was laid at the bottom of both pits to facilitate water drainage. Filter fabric was then placed on top of the gravel. With the riser tubes, the water level can be observed from the surface and excess water can be removed by dropping a sump pump to the bottom when necessary. TDR (Time Domain Reflectometry) are used to measure the soil moisture at the beginning of testing and at regular interval as water is added to the subgrade.

TESTING ACTIVITY

ATL-Exp #1 The purpose of the experiment was to validate the performance of cold in-place recycled (CIPR) asphalt pavement when topped with an asphalt overlay. This experiment was conducted in the large pit (see Figure 2). No heating or cooling was used for this test.

FIGURE 2 Wheel Load Testing of AC Overlay
(Photo taken by KSU Photographic Services)

The pavement structure consisted of a 3.7 m (12 ft)-wide by 6.1 m (20 ft)-long test section and 76 mm (3 in.)-thick asphalt overlay (KDOT mixture BM-2C) over a 102 mm (4 in.)-thick CIPR base stabilized with fly ash. The *asphalt overlay* has a dense aggregate gradation and is typically used as a binder course on heavily traveled asphalt pavements in Kansas. The *CIPR layer* was constructed with reclaimed asphalt pavement (RAP) and 10% class C fly ash (by weight) was spread uniformly on the RAP. The *granular subbase* is crushed limestone placed in loose lift of 152 mm (6 in.) thickness and compacted to 95% of the laboratory maximum dry density (MDD). The optimum moisture content of this aggregate base (KDOT AB-3) material was approximately 10%.

Subgrade soil of 1220 mm (48 in.)-thick was placed and compacted on top of the 279 mm (11 in.) of pea gravel that were put at the bottom of the pit for drainage. It is a typical silty soil (AASHTO A-4) from a borrow pit placed in approximately 152 mm (6 in.) lifts and compacted to 90% of MDD. Its liquid limit, plasticity index, sieve analysis, MDD, and optimum moisture content were calculated in the Materials lab at KSU. The density obtained was monitored with a nuclear gage. The top 457 mm (18 in.) of the subgrade were compacted to 95% of MDD. The in-place moisture content of the subgrade was measured with TRD waveguides as 15.5% during construction and decreased to about 10.5% three months later.

Testing was conducted between January 10, 1997 to March 29, 1997. Load was applied using the wheel carriage (tandem axle with dual wheels loaded to 151 kN (34,000 lbs) and tire pressure of 620 kPa (90 psi). A monitoring plan was developed to measure the extent of cracking, rutting, deflection and roughness before loading and every 10,000 load repetitions.

Deflection data were collected on the section at five different locations on each wheel path with a Dynatest 8000 FWD at 4 m (13 ft) intervals. The transverse profiles at three different locations on the test section and the longitudinal profile on each wheel path were measured with a Face Dipstick.

Severe rutting (up to one inch) was observed after 30,000 repetitions of the K-ATL wheel loading. The ruts were repaired with hot mix asphalt (SM-2C and SM-1T). Loading continued up to 145,000 repetitions. Then a new thin patching material was placed at the beginning and at the end on both load paths where the wheels enter and leave the specimen. At 168,000 load repetitions, the maximum rut depth had again reached 1 in. and the patch had cracked. The section was then removed although no cracking was visible in the pavement. A detailed description of this experiment is given in Ref.[2]. Analysis of the test results including mechanistic response analysis of the deflection data, back-calculated layer moduli, stress-strain states, and transverse profile data are summarized in Ref. [3]. A design sensitivity of the CIPR pavement layer is also included.

ATL-Exp #2 The purpose of this experiment was to compare three different types of shear transfer devices in jointed PCCP. The pavement of Exp #1 was removed and three 6.1 m (20 ft)-long strips (lanes), 229 mm (9 in.)-thick were constructed side-by-side with joints at mid-length on top of the existing aggregate base. The test setup is shown in Figure 3.

FIGURE 3 Testing PCCP joints with the Pulse Load System

The north lane has 6 epoxy-coated steel dowels 30 mm (1-3/16 in.) diameter, the middle lane has 6 FiberCon dowels, and the south lane has 11 *X-FLEX* units fabricated by Cardwell Concrete Systems, a subsidiary of.Cardwell International, Ltd. All three types of shear transverse devices are 457 mm (18 in.)-long. FiberCon is a proprietary product of Concrete Systems, Inc., and *X-FLEX* is a shear transverse device patented by KSU professors [4].

Testing started on May 11,1997 and was completed on August 22, 1998. The pulse loading assembly (thumper) was used to pound the different types of PCCP joints, and a total of 4.7 million load repetitions were applied during this time period. Test results are proprietary to Cardwell International, Ltd., and may be published elsewhere.

ATL-Exp #3 The purpose of this experiment is to compare Superpave SM-2C mix design with Marshall BM-2C mix design when used as asphalt concrete overlays over existing PCCP. During the application of the axle loads, the pavement was constantly heated at 50°C (122°F) from the surface with the infrared radiant heaters. The asphalt overlays were about 102 mm (4 in.) thick and were placed as three adjacent lanes of about 1.9 m-wide each (76 in.) on top of the previously tested PCCP sections of ATL-Exp #2. Each of the three 6.1 m-long (20 ft) strips were divided into two sections placed parallel to the direction of the rolling wheels. The two sections consist of mixes prepared with optimum and optimum +0.5% binder content. The PCCP under the south lane had a failed joint with X-FLEX as shear transfer device, the one under the central lane has an unbroken joint with FiberCon, and that under the north lane had an unbroken joint with standard epoxy-coated steel dowels.

The strips were tested in pairs such that two adjacent lanes were each loaded with one half tandem axle. The design tandem axle load was 151 KN (34,000 lbs) for this test. The testing frame was moved laterally every so many cycles to simulate wander of the traffic on the pavement. The lateral movement is applied at seven discrete intervals with a number of cycles in each interval that represent the random (normal) distribution of the traffic. The carriage and consequently the center of each pair of wheels were therefore moved according to this distribution within a 3 ft-wide range (±1.5 ft from central position). This has proven to be essential for asphalt pavement tests to avoid early excessive rutting that would be obtained if the wheels were traveling on the same fixed path.

The central and north lanes were tested first for up to 80,000 cycles. At that stage rutting in the north lane had exceeded 1 inch. Therefore the tandem axle was moved over the central and south lanes that were tested for an additional 20,000 cycles. The north lane therefore had 80,000 cycles, while the central lane had a total of 100,000 cycles and the south lane had 20,000 cycles. Test data are currently being thoroughly studied and analyzed to summarize the results and draw final conclusions of the experimental investigation.

ATL-Exp #4 Preparation for this experiment is currently in progress. Its objective is to compare composite versus epoxy coated steel dowels in PCCP under adverse thermal conditions. Two 200 mm-thick (8 in.) slabs separated by a construction joint have been placed side-by-side in the environmental pit.

Both slabs will be tested simultaneously. The total load on both strips will be 178 kN (40,000 lbs). The edges of the PCCP sections have been tied down at the two ends perpendicular to the direction of traffic. The thermal system will be used to cause warping of the slab and joint movement. The pulse load device (thumper) used earlier and shown in Figure 3 will be used to apply the cyclic loads on the test specimens.

CONCLUSION

An new accelerated pavement testing laboratory has recently been constructed at KSU. A tandem axle assembly moving at an average speed of 5.6 km/h (3.5 mph), or a pulse load system with hydraulic actuators, can apply up to 178 kN (40,000 lbs) on flexible or rigid pavement specimens placed on soil in a 6.1 m×6.1 m×1.8 m test pit. Heating and cooling between -23°C and +66°C of the pavement surface or the soil/subbase under the pavement can be achieved in a second 6.1 m×3.7 m×1.8 m environmental pit. Within about 13 months of operation three experiments were successfully conducted on both AC and PCCP using a combination of test features. Preparation of a fourth experiment is presently underway. Two other experiments are scheduled for 1998.

ACKNOWLEDGMENTS

The building construction was made possible thanks to gifts from KSU alumni. The test frame and loading devices were contributed by Cardwell International, Ltd., Newton, Kansas, and a matching grant from the Kansas Technology Enterprise Corporation (KTEC). The recommendation of the Advisory Committee are much appreciated. Upcoming testing activity is supported by the Midwest States Pooled Funds Accelerated Testing Program.

REFERENCES

1. METCALF, J.B. Application of Full-Scale Accelerated Pavement Testing. NCHRP Synthesis 235, Transportation Research Board, National Research Council, Washington, D.C., (1996).

2. MELHEM, H.G. Development of an Accelerated Testing Laboratory for Highway research in Kansas. Final Report, Kansas Department of Transportation, Topeka, Kansas, in print (1997).

3. MELHEM, H.G., HOSSAIN, M, XU, J.C., and GISI, A.J. Testing of Cold In-Place Recycled Pavement at the Kansas Accelerated Testing Laboratory. To be submitted to the International Conference on Accelerated Pavement Testing, to be held in Reno, Nevada, October 18-20, 1999.

4. HU, K.K., KIRMSER, P.G., SWARTZ, S.E., HOSSAIN, M., and GAN, L. Experimental Evaluation of the X-FLEX System for Shear Transfer Device to Couple the Portland Cement Concrete Pavements. Final Report prepared for Cardwell International, Ltd., Department of Civil Engineering, Kansas State University, Manhattan, KS, (1994).

Fifth International Conference on the
BEARING CAPASITY OF
ROADS AND AIRFIELDS

PAVEMENT PERFORMANCE MODELLING USING APT

P.D. BHAIRO[1], J. GROENENDIJK[2], A.A.A. MOLENAAR[2], A. VAN DOMMELEN[3], A. MIRADI[2], C.H. VOGELZANG[2]

[1]KOAC/WMD Road Consultancy
[2]Delft University of Technology
[3]Road and Hydraulics Engineering Division, Ministry of Transport
Corresponding address: Delft University of Technology
Faculty of Civil Engineering and Geo Sciences
P.O. Box 5048, 2600 GA DELFT, The Netherlands

ABSTRACT

Accelerated pavement testing (APT) is considered to be an essential tool in the development of design and performance prediction models. Also in the Netherlands and APT device called LINTRACK is used for these purposes. This paper describes some of the results and findings that have been obtained from these tests. Emphasis will be placed on the performance of two sections with respect to their cracking and permanent deformation behaviour. Also attention will be paid to observations that have been made with respect to some damage that is most probably related to construction but which has a significant influence on performance.

INTRODUCTION

It is well recognised that theory based design and evaluation procedures should be calibrated to real life performance in order to be able to apply them with confidence. During the last decades the understanding has grown that this can be accomplished to a very large extent by means of accelerated pavement testing (APTT) since this allows performance to be evaluated as it develops in time under conditions which are much better defined than the performance of sections which are subjected to real traffic. For that reason an APT device, called LINTRACK, was also built in the Netherlands and is used for performance modelling purposes as mentioned above. This paper describes the most important results as obtained on two full depth asphalt pavements which were loaded by means of 75kN wheel loads using wide based tyres. Especially this tyre type was used since about 80% of the towed truck axles in the Netherlands are equipped with it.

LINTRACK

The APT device used by the Road and Hydraulics Engineering Division (RHED) of the Ministry of Transport and the Road and Railroad Research Laboratory (RRRL) of the Delft University of Technology, is a linear tracking device called LINTRACK (figure 1).
Details of the device are not given here; the interested reader is referred to [1]. The two pavement structures that were tested were full depth asphalt structures on a sand subgrade.
Structure I had a total thickness of 150 mm consisting of two layers, an 80 mm bottom and a 70 mm top layer, both made of gravel asphaltic concrete (GAC). GAC was a mixture that was used until recently as a base course mixture.
Structure V_a consisted of a single layer of 75 mm of GAC. Structure I was subjected to 4.2 Mcycles of a 75 kN single wheel with a wide base tyre; structure V_a carried 625 kcycles of the same 75 kN load. In both cases the tyre pressure was 850 kPa.
During the tests regular strain measurements were taken as well as falling weight deflectometer (F.W.D.) measurements. Also the visual condition was systematically observed by means of crack mapping; furthermore the rut depth development was regularly measured. Figure 2 gives an overview of the measurements performed as well as of the locations where they were taken. The F.W.D. measurements were taken at locations where the strain gauges were installed.

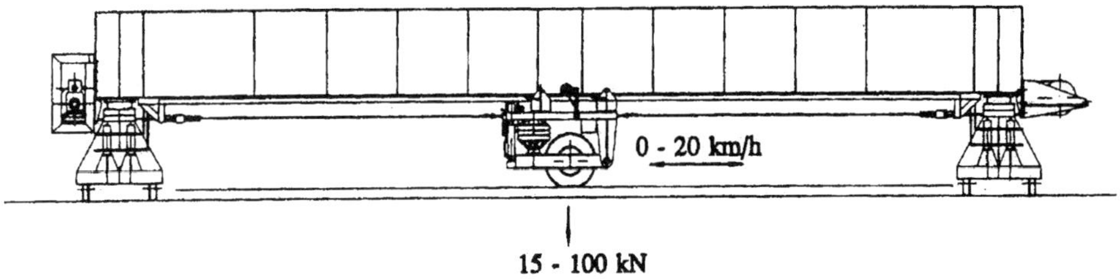

Figure 1: LINTRACK

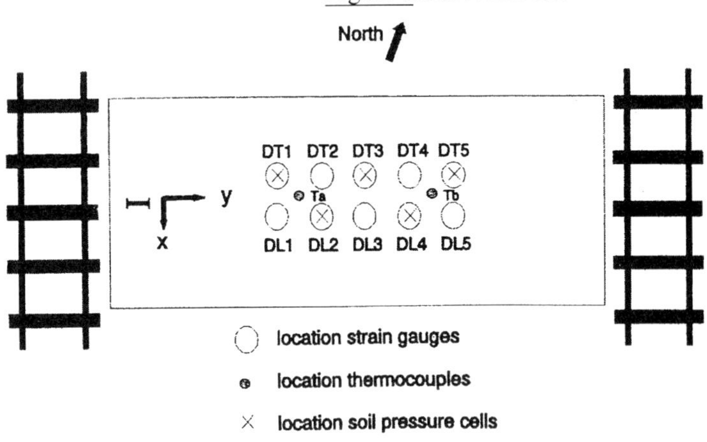

Figure 2: Location of LINTRACK measurement devices

TEST RESULTS

Strain Measurements

In this section examples will be given of the most relevant test results. The development of the measured tensile strain will be discussed as well as the decrease of the asphalt stiffness in time as observed by means of deflection measurements.

Furthermore some results regarding the visual condition and the development of permanent deformation will be presented.

A typical example of the development of the tensile strain in the transverse direction with the number of load repetitions as observed on section I is given in figure 3. Special attention is called for the strain relaxation that is clearly visible after about 400 kcycles and 2 Mcycles.

The relaxation took place during a period in which LINTRACK was not functioning, because of maintenance. It is believed that some of this recovery can be assigned to healing but that some of it should be assigned to effects like thixotropy etc.. However the observed phenomenon is not fully understood neither from a qualitative nor from a quantitative point of view.

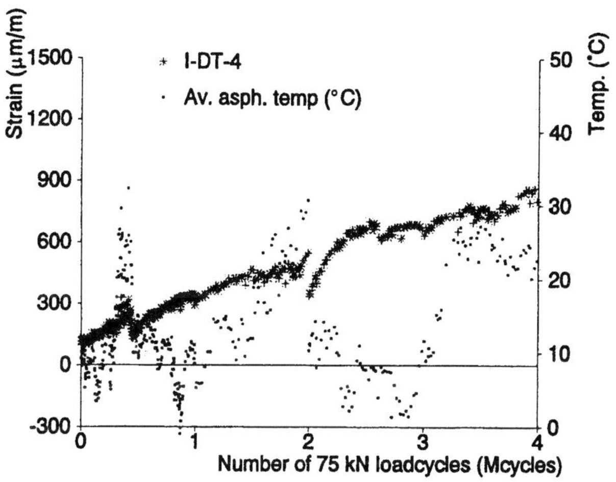

Figure 3: Strain peak development DT4

Deflection measurements, back calculated moduli

Figure 4 shows, for section I, the decrease of the asphalt modulus in relation to the applied number of load repetitions. The moduli values were back calculated from the measured deflection profiles. Although a significant amount of scatter is present, a clear trend can be observed which is a decrease in time of the back calculated modulus. Also a slight increase in subgrade modulus is observed.

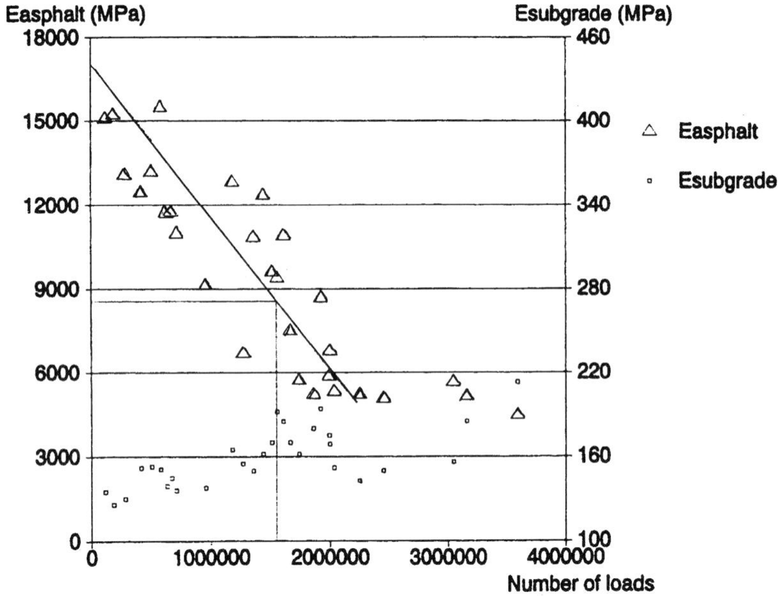

Figure 4: Development of pavement stiffness with loading I-DT4 (uncorrected for temperature)

Visual condition

Figure 5 shows the increase in the amount of cracking, as observed on section I, in relation to the applied nr. of load repetitions together with results obtained on section V_a. As mentioned before the visual condition was analyzed by means of mapping the visible cracks.

For qualification purposes, the sections were covered by a grid and a grid cell was considered to be failed as soon as it contained a crack. It will be obvious that the fraction of the area cracked is dependent on the gridsize. The larger the gridsize the higher the chance that crack will hit a grid cell. The results shown are valid for a grid size of 100 mm in length and 2400 mm wide. This means that the entire width of the trafficked area was covered by each cell and that the length of the trafficked area was divided in strips of 100 mm long. This way of characterizing the amount of cracking in fact means that the percentage of lane length is affected by cracking is monitored.

With respect to the type of cracking it should be mentioned that the majority of the cracking was transverse cracking and that the majority of the cracks appeared to be surface cracking in the top few centimetres of the top layer.

Figure 6 shows the development of the observed cracking in relation to the ratio of applied nr. of load repetitions n to the totally applied nr. of load repetitions N. It is remarkable to note that, although both sections were quite different from each other and although the cracking development was also quite different, their performance is quite similar when represented on a non dimensional basis. It is believed that this information is extremely helpful in the development of performance models based on visual condition which are often used in pavement management systems.

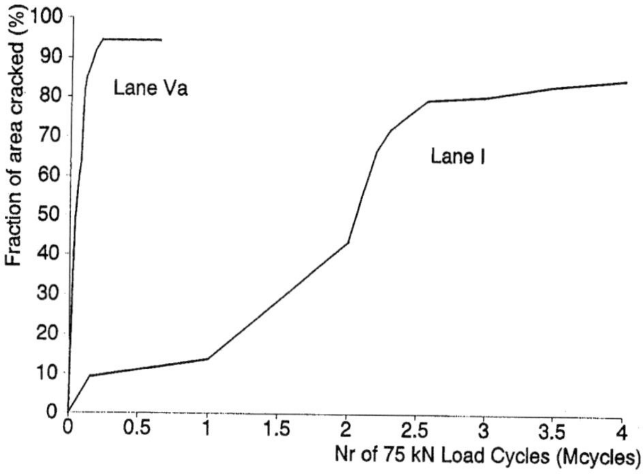

Figure 5: Crack development of LINTRACK lanes I van V_a

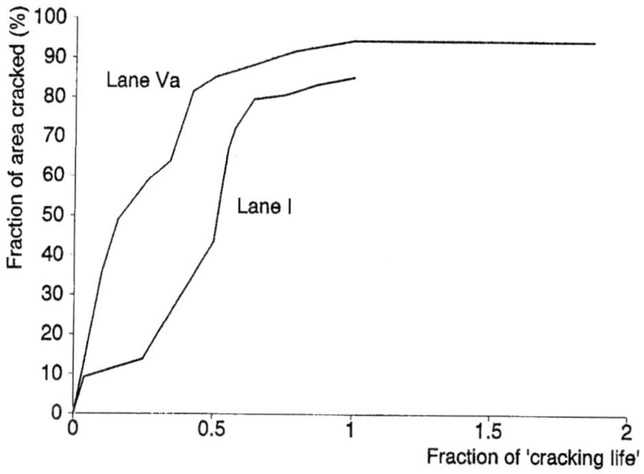

Figure 6: Normalized crack development of LINTRACK lanes I and V_a

Permanent deformation

The permanent deformation was measured by means of profile measurements across the entire width of the test sections. Figure 7 shows how the deformation developed on sections I and V_a with respect to the number of load repetitions. The "dip" in the rut development that is observed on section I around 1 – 1.6 Mcycles coincides with the Winter of 1992 – 1993. In the Netherlands, the warning level with respect to the rutdepth is set at 18 mm. The "rutting" life of section I is about 2.5 Mcycles while for section V_a it is not more than 38 cycles. When the rutdepth development is represented in a semi-normalized way by using on the horizontal axis the ratio of applied number of load repetitions to the number of load repetitions at which a rutdepth of 18 mm is reached figure 8 is obtained. Again a fairly similar behaviour is observed when the results are plotted on a semi non dimensional scale.

Coring revealed that hardly any permanent deformation had occurred in the asphalt layer so all deformation could be attributed to the sand subgrade.

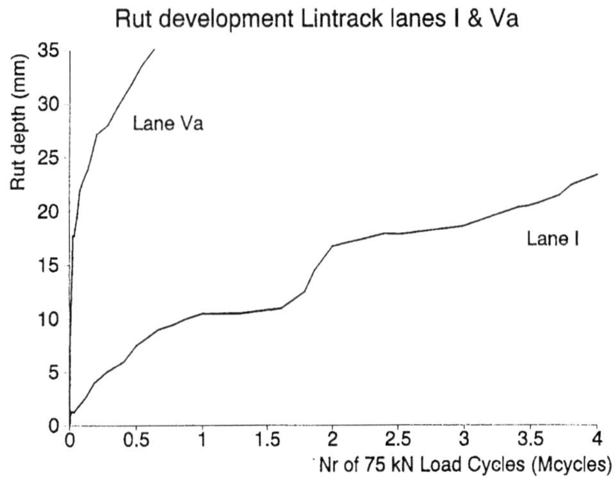

Figure 7: Rutdepth as function of the number of load cycles

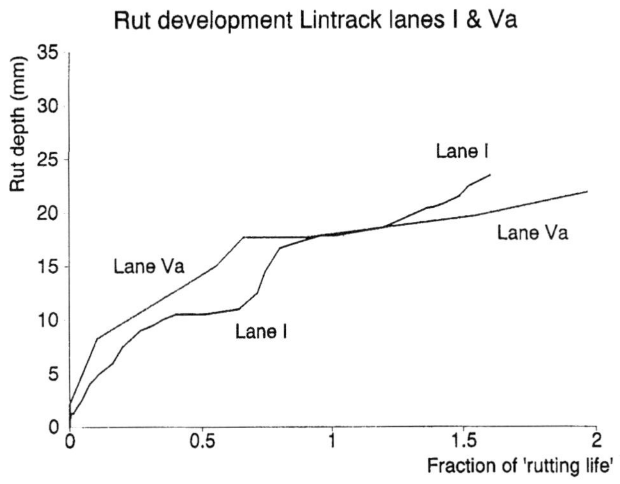

Figure 8: Rutdepth as function of the normalized number of load repetitions

ANALYSES

Asphalt fatigue life Prediction

As has been mentioned before, calibration of design methods is one of the objectives of APT programs. Therefore the pavement life of sections I and V_a was predicted using the well known multi-layer linear elastic approach. The calculated strains which are a result of the super single wheel load are introduced in appropriate fatigue relations to obtain pavement life. The problem of such an approach however is, that the physical condition that is related to the end of life definition as used in design procedures is quite often not defined. The fatigue relations e.g. that are used in the Netherlands for asphalt fatigue analyses, are all based on constant displacement fatigue relations where the end of life is defined as the number of load repetitions at which the effective stiffness of the asphalt layer has decreased to 50% of its original value. What this means in terms of visible damage is absolutely unclear.

Furthermore one has to realize that lab. fatigue relations cannot immediately be used for field conditions. This implied that the lab. results had to be transformed to LINTRACK conditions. For this a special procedure has been developed, which is described in [2]. Finally effects like lateral wonder and healing have to be taken into account but it should be recognized that only limited information is available to quantify the latter effect; especially the effect of long restperiods (several weeks), which occurred because of maintenance of LINTRACK, was very difficult to quantify.

First of all the pavement life was estimated by determining the number of load repetitions at which the modulus of the asphalt layer, as back calculated from the deflection measurements, had reduced to 50% of the initial value.

Secondly the number of load repetitions has been determined at which the strain level had increased to the level which coincides with a decrease in modulus of 50%. For the 150 mm thick section I this meant that the number of load repetitions to a strain increase of 1.7 times the initial strain had to be determined. For section V_a this meant that the number of load repetitions to a strain increase of 1.5 times the initial strain had to be established. The results of these analyses are given in table 1.

Test section	N_{fat} strain gauge [Kcycles]		N_{fat} FWD [Kcycles]
	$\varepsilon_n = 1.5\ \varepsilon_i$	$\varepsilon_n = 1.7\ \varepsilon_i$	$E_n = 0.5E_i$
I		1400	1850
V_a	20		25

Table 1: Pavement life as determined from strain increase and modulus decrease.

From figure 6 it can be derived that that amount of cracking at the pavement lives mentioned in table 1 is around 20% in both cases (for section I $N_{cracking}$ = 4.2 Mcycles, for section V_a $N_{cracking}$ = 625 k cycles).

In both cases the calculated pavement life appeared to be a factor of 2 – 4 shorter than the observed pavement life even by taking into account the positive effects of lateral wander, healing and difference in thickness between the lab. fatigue beam and the actual pavement thickness. Information on how this thickness effect was quantified can be found in [2].

Permanent deformation

Since it was observed that the permanent deformation that had developed on both sections should not be attributed to deformation of the asphalt but the deformation of the subgrade, the number of load repetitions to a rutdepth of 18 mm was related to the initial subgrade strain, since in this way a subgrade strain criterion could be developed.
The result of this analysis is shown in figure 9. In this figure also the Shell subgrade strain criterion is shown and one can concluded that this criterion is very good applicable for the particular type of sand subgrade underneath the test sections. Since this type of sand is used quite often under the heavily loaded highways in the western part of the country, the Shell subgrade strain criterion can be used with confidence for design purposes.

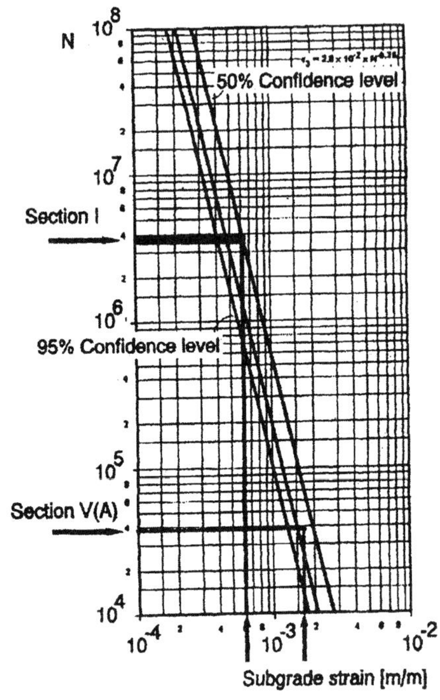

Figure 9: Shell subgrade criterion with 85 and 95 percent confidence limits, together with LINTRACK results

SPECIAL OBSERVATIONS

When pavements are designed one normally assumes that the layers are built of homogeneous and isotropic materials. It is a well known fact however that this assumption is far from reality and observations made on the test sections pointed out that real life is indeed significantly different from the assumptions. The first observation was that on a few locations of test section I deterioration started at the bottom of the top layer (which is around the neutral axis) because of the fact that some segregation occurred during the laying and compaction process. This damage had a significant effect on the damage development since it progressed from these locations. Normally the effects of these locally poor conditions on overall performance are not taken into account in the design and analysis which complicates the comparison of observed with predicted performance. The second observation that was made is that the density of the asphalt layer showed quite some variations over the width of the test sections. The variation that was observed on section V_a is shown in figure 10 [3].

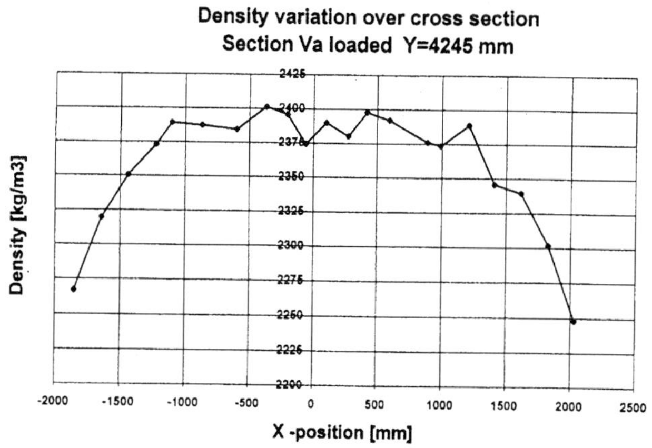

Figure 10: Variation in density in transverse direction

A similar pattern was observed on section I [1]. It is obvious that such variations will also occur in practice and since the areas with the lower density have a higher sensitivity for cracking, permanent deformation and ravelling, it is most likely that deterioration will initiate there when subjected to loading. These effects however could not be studied in the LINTRACK tests since the wheel loads were applied on the central area and because the test sections were not subjected to the direct influence of climate since the LINTRACK is covered by a hull. Nevertheless it is believed that the initial construction flaws as described here have a significant influence on pavement performance, and that special attention has to be given to such details in order to achieve that pavements have the lives as designed.

CONCLUSIONS

The following conclusions have been drawn based on the performance of two test sections.

a. It seems realistic to assume that a full depth asphalt pavement is failed if the asphaltic modulus has decreased to 50% of its original value.

b. It seems that the failure condition coincides with a 20% cracked area. It should be noted however that this percentage is dependent on how the condition survey is performed.

c. Pavement life predictions made by linear elastic theory appear to be conservative when compared to in situ performance.

d. The Shell subgrade strain criterion appears to be very applicable for subgrade sands as used in the Netherlands.

LITERATURE

1. GROENENDIJK, J.; VOGELZANG, C.H.; MIRADI, A.; MOLENAAR, A.A.A. AND DOHMEN, L.J.M.
   Linear tracking performance tests on full depth asphalt pavement.
   Transportation Research Record 1570; TRB, Washington D.C. – 1997

2. SABA, H.; GROENENDIJK, J. AND MOLENAAR A.A.A.
   Estimation of crack growth parameters and fatigue characteristics of asphalt mixes using simple tests. Report 7-95-209-35 M; Delft University of Technology, Delft – 1995

3. BHAIRO, P.D.
   Comparison of the predicted observed pavement life of LINTRACK test lane V_a
   MSc-thesis; Delft University of Technology, Delft - 1997

Fifth International Conference on the
BEARING CAPASITY OF
ROADS AND AIRFIELDS

FULL-SCALE ACCELERATED PAVEMENT TESTING IN FINLAND AND SWEDEN

JARI PIHLAJAMÄKI[1] AND MATTI HUHTALA[2]

[1]Senior Research Scientist, Jari.Pihlajamaki@vtt.fi
[2]Chief Research Scientist, Matti.Huhtala@vtt.fi
Technical Research Centre of Finland, PL 19031, FIN-02044 VTT, Finland,
fax. +358 9 463 251,

ABSTRACT

A full-scale pavement testing facility (HVS) was bought in 1997 by Finnish and Swedish partners together on 50/50 bases. Pavement life can be estimated with HVS tests which simulates pavement deterioration within two or three months while real traffic over a period of several years. HVS is mobile so it can be easily moved from test site to an other. HVS has a heating/cooling system and thus the pavement temperature can be held constant which is necessary to compare the performance of different road structures. It has been started six years combined Finnish Swedish project on accelerated pavement testing with HVS NORDIC. The purpose of this project is to maximise the use of HVS for both countries.

We were in the beginning of the learning process of accelerated pavement testing and there were also a lot of teething problems at HVS. The harmonisation of test procedure including instrumentation and measurement plans is under development. The first tests have lasted longer than expected. The measured data will be analysed more detailed later when the common data base system is ready to be used.

Key words: pavement performance, accelerated pavement testing, response measurements

INTRODUCTION

The nordic countries have had traditionally close co-operation in road research. The Nordic countries had for instance commonly a person as a loaned staff at SHRP during the whole research program. Co-operation happens usually in the Nordic Road Association (PTL), especially in its permanent committees. PTL had also an ad hoc group on accelerated pavement testing in late 1980's; no real co-operation within a common facility was economically possible at that time and the result was a published report.

The Finnish National Road Administration (Finnra) and Technical Research Centre of Finland (VTT) started discussions with the Swedish National Road Administration (SNRA) and Swedish Road and Transport Institute (VTI). They had shown interest for an accelerated testing facility earlier. The result from the negotiations was positive; a Full Scale Accelerated Testing Facility will be bought together on 50/50 bases.

The researchers of VTT have participated in international research projects where accelerated pavement testing has been used and came to the conclusion that the testing facility should be linear and mobile and it should have complete temperature control. The bids for offers were sent to all potential manufacturers.

A full-scale pavement testing facility was bought in 1997 by Finnish and Swedish partners, the Finnish National Road Administration (FinnRA), the Swedish National Road Administration (SNRA), Technical Research Centre of Finland (VTT) and Swedish National Road and Transport Institute (VTI). In Finland it is used for the Road Structures Research Programme 1994 - 2001 (TPPT) funded by Finnra. In Sweden there is no similar national research programme but it will be used for pavement research in separate projects.

The facility was manufactured in the Republic of South Africa and is called Heavy Vehicle Simulator (HVS) Mark IV. In Finland and in Sweden it is called as HVS NORDIC. In principle it is one year in Finland, next year in Sweden and so on. The facility is illustrated in Figure 1.

Figure 1. The facility HVS NORDIC.

THE AIM OF ACCELERATED PAVEMENT TESTING

Life cycle cost analysis is one of the most important parts of TPPT. For life cycle cost analysis the life time of pavement must be determined. This can be done with HVS tests which simulates pavement deterioration within two or three months while real traffic over a period of several years. The aim of accelerated pavement testing with HVS is:

- to modell the deterioration process
- to check laboratory test results
- to compare different structures
- to check the reliability of the dimensioning system
- to investigate the influence of the repair techniques on the life time

THE PROPERTIES OF HVS

The facility is totally computer controlled. The loading is linear, one can say it simulates traffic loading more realistically compared to carousel types.

HVS is mobile so it can be easily moved from test site to an other test site or test road. HVS NORDIC is the only mobile APT facility in Europe. It can be run over short distance by itself at walking speed, in practice only at the same test site. Over longer distances it can be moved as a semitrailer to other test areas or to VTI in Sweden. The speed is 50 km/h and special permits are needed. Because it has steering wheels, it can in spite of its long length turn even on relative sharp corners.

Temperature varies in Finland very much in different seasons let's say from -30 to +30 °C. It is important to test different road structures in the same circumstances. That's why HVS has a heating/cooling system and thus temperature can be held constant. The air inside the box is heated or cooled and controlled in order to keep the pavement temperature constant. The temperature of pavement can be between -5 and +35 °C, when ambient temperature is +25 °C. The standard pavement temperature is selected to be +10 °C. HVS NORDIC is the only mobile APT facility in the world with full temperature control. Figure 2 illustrates HVS NORDIC with heating/cooling box.

Figure 2. HVS NORDIC with heating/cooling box.

HVS's length is 23 m, width 3.5 m, height 4.2 m and weight 46 t. HVS can be run by diesel or electric motor. Diesel motor provides power also for the heating/cooling system and thus it is independent of external power.

The loading can be uni- or bidirectional. HVS-NORDIC has dynamic loading property. The loading can be varied dynamically +/- 20 % of static load in sinusoidal mode. As far as we know there are no possibilities for dynamic loading in any other APT facility.

Loading wheel can be dual or single. The lateral movement of loading wheel can be 0.75 m with 0.05 m steps (loading beam can have 16 different lateral positions). The wheel load can be selected from 20 kN to 110 kN (corresponding axle loads 40...220 kN). The speed can be selected from 1 to 12 km/h, for short period up to 15 km/h. The test wheel can be dual or single wheel.

It is possible to get 25.000 load repetitions per day (24 h) when loading is bidirectiopnal.

THE SIX YEARS COMMON RESEARCH PROJECT

The organisation of the project. It has been started six years combined Finnish Swedish project on accelerated pavement testing with HVS NORDIC. The purpose of this project is to maximise the use of HVS for both countries. The organisation of the project consists of three groups; streering, program and operative ones.

The function of steering group is to make principle decisions of research focusing and financing. It consists of the directors of partners and has meeting once a year.

The programme group is responsible for research programme. It watches the quality of scientific research and that research will be get through according to plans. It makes different proposals for steering group to make decisions.

The group is responsible for that tests will get through in standardised way so it is possible to compare results and experiences with those of other parties.

The group watches development of research area and stimulates co-operation between this project and other parties. It is responsible for marketing of HVS and distribution of research results. The group consists of the members of all four parties.

The operative group is responsible for detailed design and plans of test structures, instrumentation and measurements. The group makes decision of measuring devices, instruments, gages and data acquisition system.

The group is responsible for that the personnel of HVS has enough training to work with HVS. The group makes proposals to improve the facility itself and the test method.

The group is responsible for that reports will be written and distributed. The members of the group are from VTT and VTI.

A subgroup for operative group, Data base group, was set up. It is responsible for that data collection, transfer and storing will happen in reasonable way.

Test structures. The test program for six years has been completed. The program consists of more than 30 structures but is preliminary one and will be modified based on experiences and needs. The first test structures in Finland have thin bituminous layer and unbound layers.

The first test runs in Sweden will be on conventional Swedish structures with well-documented and well-known performance from studies in the field e.g. survey and test sections on the road network. All test sections will be built on fine sand subgrade of the same type as was used in the first test runs in Finland

There are plans at three levels; a plan for one test, a plan for one year and a plan for six year, the whole project.

Test site. In Finland the test site is located a few hundred metres from the office of VTT personnel. The site includes two test pits. One is made of concrete, the walls are insulated and there is complete water table regulation. Its length is 36 meters including a 16 meter long slope. Its depth is 2.5 meters and width 4 meters at the top and 3 meters at the bottom. Two or three test pavements can be constructed in a test pit.

The second test pit is parallel to the first one. It is excavated mainly in rock and it has no water table regulation. It is of the same size as the test pit no 1.

The test pits are outside but there is also a tent which covers nearly the whole test pit and which is big enough to house HVS NORDIC.

In Sweden HVS will be located at VTI in Linkoping. At VTI there is an indoor full-scale pavement test facility where pavements can be constructed by ordinary road construction machines. The size of the test pits are 3 m in depth, 5 m in width and 15 m in length. Two test pits will be used for the HVS-experiments, which means that one test-section can be constructed while the test is running on the other.

Instrumentation. Instrumentation for response measurements is mainly based on the experience which has been attained at the Virttaa test site during last 15 years. The position of the sensors is presented in Figure 3.

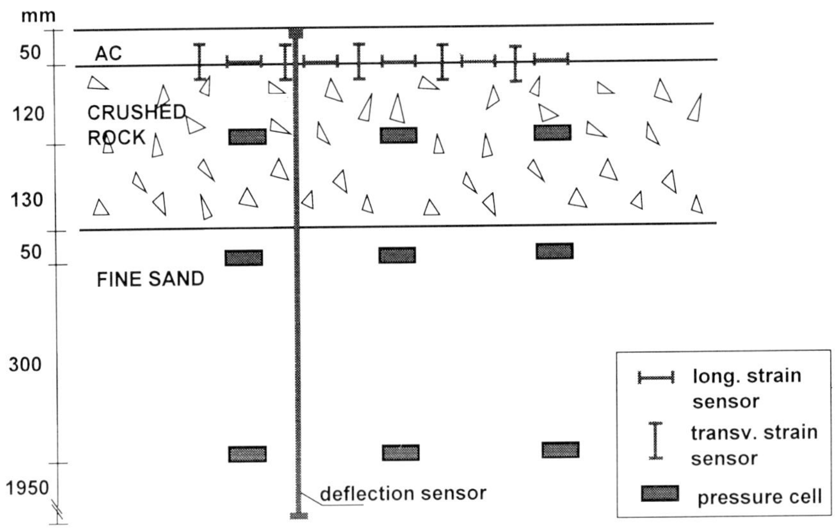

Figure 3. Instrumentation plan.

The basic instrumentation is based on strain gauges which are installed at the bottom and on the surface of the bituminous layers. VTT uses retrofit strain gauges. The most common set of gauges is four longitudinal and four transverse gauges at the bottom and three longitudinal and three transverse gauges on the surface.

Stress in unbound layers are measured with pressure cells which have been bought from the University of Nottingham. They are installed at three levels in the structure and three at the same level. The height levels are in the middle of the base course, 100 mm and 350 mm in the subgrade.

Deflection under the moving wheel load is measured by a deflection rod which is anchored at the bottom of subgrade near the concrete or rock bottom.

In Sweden also subgrade deformation will be measured by a plate connected to a steel rod which goes through the bottom of the pit.Vertical strains in unbound materials and in subgrade are not measured at the present time. Several systems are under consideration for near future.

The temperature of bituminous layers is measured always at three depths. In thaw tests also other specific measurements like water content, pore water pressure, suction, heat-flux, frost depth, radiometric water content and density measurements have been made. In situ plate bearing tests in subgrade have been done, too.

Deflection before and after the test is measured by falling weight deflectometer. That cannot be done under HVS but it must be driven away which means also that there is no temperature control during the measurements.

During the test deflections are measured by a modified Benkelman beam. The present system is not good and better system is under consideration. It will be based on laser.

Measurements. A test is started first with prerun in order to relax possible residual stresses and cause some post-compaction. After the prerun so called zero measurements are made. They include considerable amount of response measurements: strain, stress and deflection measurements at different wheel loads, tyre inflation pressures and temperatures. Both wide base tyre and dual tyres are used. The same measurements are repeated later during the test but not so complete.

The zero measurement program lives still but at least five wheel loads from 30 kN to 90 kN and five tyre pressures from 500 to 900 kPa have been used. The highest loads cannot be used if the pavement is weak. Speeds have been 1, 4, 7, 10 and 12 km/h.

All measurements are made at pavement temperatures +5, 10, 15 and 20 °C. The change of the temperature by five degrees takes a few hours. Basically the same program is made later during the test but in much simplified fashion.

The standard loading parameters are dual wheel, 60 kN, 12 km/h and asphalt temperature 10 °C. The lateral distribution of loadings is fixed, too.

During the test response measurements are made about twice a week but only with standard loading parameters.

Transverse profiles are measured at five locations with a laser profilograph constructed by VTI for HVS NORDIC. The measurement of longitudinal profile is under consideration. Cracks are drawn on paper with an aid of 1 m by 1 m grid which is divided to 100 mm by 100 mm squares.

RESULTS OF FIRST TESTS

The first test structures have been tested since the beginning of July 1997. It was decided to start with traditional pavements, thin asphalt layer (50 mm) and unbound base layer on sand subgrade. It was loaded by 60 kN twin tyre bidirectionally with tyre pressure 800 kPa. It was estimated that it will last less than half million load repetitions. However, it lasted 1,6 million repetitions without cracks, only 25 mm rutting was found. To accelerate the test the water table level was raised from the original level (2.3 m from surface) to 0.3 m from surface. Only 0.1 million load repetitions were needed to increase rutting from 25 mm to 50 mm and also some transverse cracks appeared to road surface. As the pavement was opened one could see very clearly that the rut was due to rutting in the subgrade and the base course was badly deformed. The development of rutting can be seen in Figure 4.

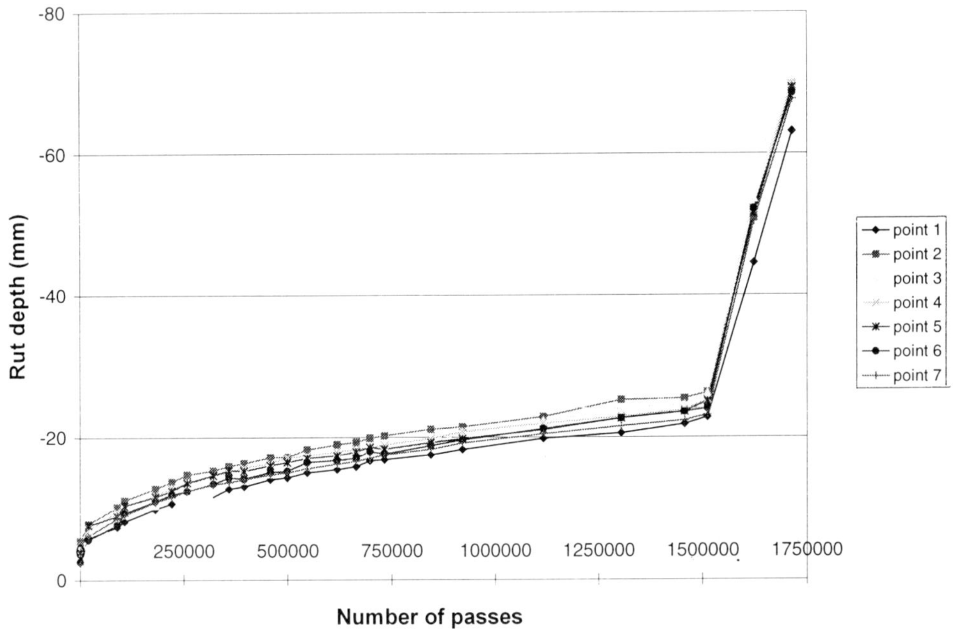

Figure 4. Rut depths during the test 1.

In the beginning of the test initial measurements were made with different wheel loads. When the tyre pressure was the same, the increasing of wheel load had no effect to strains at the bottom of asphalt layers. The reason is the thin asphalt layer compared to the imprint area of the tyre.

Next test was made with single tyre. Wheel load was 70 kN, tyre pressure 850 kPa and loading bidirectional. First transverse cracks were found after 0.8 million load repetitions. The test was continued to 1.4 million load repetitions but no more cracks appeared. Besides the cracks were very thin, hair cracks and did not get larger after further 0.6 million load repetitions.

Because of very few cracks compared to number of loadings the next test was decided to load with same load parameters but unidirectionally which is situation in normal traffic. By now the number of loadings is 300.000 but no cracks can be seen.

The next three tests are so called thaw tests. A thin pavement (500 mm) with 50 mm asphalt concrete lies on frost susceptible subgrade (silt). The depth of frost was 1.3 m from road surface. The loading with HVS will start in the end of April when the level of frost is 0.8 m from road surface. It has been estimated to make these tests within four ... five weeks.

The next two tests pavements are designed for heavy trafficed roads. The thickness of bituminous layers will be 150 mm on crushed rock bed.

CONCLUSIONS

We were in the beginning of the learning process of accelerated pavement testing and there were also a lot of teething problems at HVS. Even we were prepared for some unexpected problems there were very many technical difficulties with HVS NORDIC. It seems that they all can be solved but it has made effective running time shorter and taken too much time of the personnel and caused problems in test schedule.

HVS will be taken to Sweden in September 1998 and the huge amount of data can really handled after that. The handling of data will be easier after the common database is really ready to be used.

The accelerating effect can be attained by greater number of loading in a time period, increasing wheel load and making the pavement weaker.

In our case the bituminous courses were thin, only 50 mm. The strain measurements revealed that the strains in thin bituminous pavement do not increase with load and thus the performance of bituminous layers is much better than expected in the present thinking concerning bituminous pavement design. This phenomenon sets limits to the testing of thin bituminous mixtures and the test must be carefully planned. The value of response measurements increases.

The effect of thin bituminous layers means that if the pavement will be made weaker in order to accelerate failure and make the test shorter bituminous mixture itself cannot be part of the test but it is only a protecting layer. The tests concerning unbound layers may be, however, at the same time realistic.

Planning of the tests must be done very carefully. Even the personnel of VTT has experience from several APT facilities that has not really understood. It means for instance that the goal of each test must be carefully studied. The previous experience of VTT is from circular tracks and thus both smaller speed and total temperature control are new.

Each test produces considerable amount of data. It will give new ideas. The APT are not simply performance tests or that only the life-time of different pavements are compared.

Very important part are the other measurements especially response measurements. They give the final comparison.

Fifth International Conference on the
BEARING CAPASITY OF
ROADS AND AIRFIELDS

ACCELERATED LOAD TESTING OF THE RTM2 TEST PAVEMENT IN THE DANISH ROAD TESTING MACHINE

R.A.MACDONALD [1] AND W.ZHANG [2]

[1]Research Engineer, Danish Road Institute (DRI)
Elisagaardsvej 5, P.O. Box 235, 4000 Roskilde, Denmark
Fax: +45 46 300105, E-mail: rmd@vd.dk
[2]Post Doctoral Research Assistant, Technical University of Denmark
2800 Lyngby, Denmark
Fax: +45 45 936412, E-mail: wei@ivtb.dtu.dk

ABSTRACT

Since 1994, two flexible test pavements, instrumented to measure stresses and strains in three primary axes within the upper 420 mm of the subgrade, have been constructed and load tested in the Danish Road Testing Machine (RTM). One objective of this research, which is part of the international Pavement Subgrade Performance Study, is to develop an improved mechanistic subgrade failure criterion. The first test pavement was load tested between December and April, 1996. This paper describes RTM2, the second test pavement, which was load tested between June and December, 1997. Simulated traffic loads were applied to both test pavements by a dual tyre load cart within a climate chamber maintained at a temperature of approximately 25°C. Instruments installed in both test pavements included: Strain measuring sensors (Soil Strain Deformation Transducers and Asphalt Strain Gauges), Soil Pressure Cells, Thermocouple Probes and Pore Pressure Sensors. Routine monitoring of instrument responses, measurement of surface profiles with a Profilometer, and structural testing with FWD/LWD (Falling/Light Weight) Deflectometers were undertaken at regular intervals during the construction and load testing programmes. This paper describes the RTM2 test pavement currently being load tested in the Danish Road Testing Machine: test pavement materials; types of instruments installed in the test pavement; instrument responses; surface profile monitoring and structural testing undertaken during the construction and load testing programmes. The paper describes some experiences and conclusions from both test programmes, presents the results of preliminary analyses and draws initial conclusions.

Keywords: Flexible pavement, instrumentation, subgrade, full-scale, Denmark

INTRODUCTION

An international project sponsored by the Federal Highway Administration (FHWA), USA, entitled the Pavement Subgrade Performance Study, which was formulated to investigate the failure mechanisms of a range of subgrade soils, commenced in 1994. The primary objectives of the study are to research subgrade deformation under the influence of loading and climatic factors, and to develop new and improved subgrade failure criteria and performance models.

The Pavement Subgrade Performance Study will investigate the performance of full-scale pavements involving several different types of subgrade soils, tested at several moisture conditions. Full-scale pavements are being constructed and tested in the Frost Effects Research Facility (FERF), at the Cold Regions Research and Engineering Laboratory in Hanover, New Hampshire, USA. The fuller objectives of the study are to:

- Develop an improved mechanistic subgrade failure criterion (elastic or plastic) for new and reconstructed pavements.
- Develop a foundation for research on seasonal effects of environment on pavements.
- Investigate the seasonal variability of pavement materials.
- Foster international co-operation in pavement research.

As part of this international study, and to initiate the empirical study, the Danish Road Institute (DRI) undertook the task of constructing and load testing a preliminary test pavement in the Danish Road Testing Machine (named RTM1) during 1995/96. The objective was to evaluate different instruments, installation and testing procedures of interest to the full-scale test sections, and to provide a preliminary pavement deterioration model. Accelerated loading of the first test pavement was stopped prematurely during 1996, well before failure, due to the adverse soil reinforcing effect of a central column of instruments, which markedly affected the pavement deformation mode [1]. Consequently, with experience from the first test pavement, during 1996/97 the Danish research team constructed a second test pavement with similar pavement materials and structure (named RTM2). This paper describes the second test pavement.

In the RTM1 study, accelerated loading at two dual wheel load levels (20 kN and 40 kN) for 50,000 load repetitions each were undertaken. Structural and functional performance models, including relationships for IRI (International Roughness Index) and RD (Rut Depth), for the subgrade material were developed, and are described in detail in the literature [3, 4]. These relationships have not been verified for the current RTM2 study, as yet, though this will be investigated in the data analysis stage during 1998.

The Danish Road Testing Machine and Test Pavement

The Danish Road Testing Machine is a linear accelerated pavement testing facility, with a width of 2.5 m and a length of 27 m. The central 9 m is the instrumented test section, which is 2 m deep. Details of the RTM are shown in Figure 1.

Response instruments are installed within the central six metres of the full length of the test section, in two parallel lines on either side of the longitudinal centreline of the test pavement (i.e., under each of the dual tyre wheelpaths). The RTM1 study found that responses from

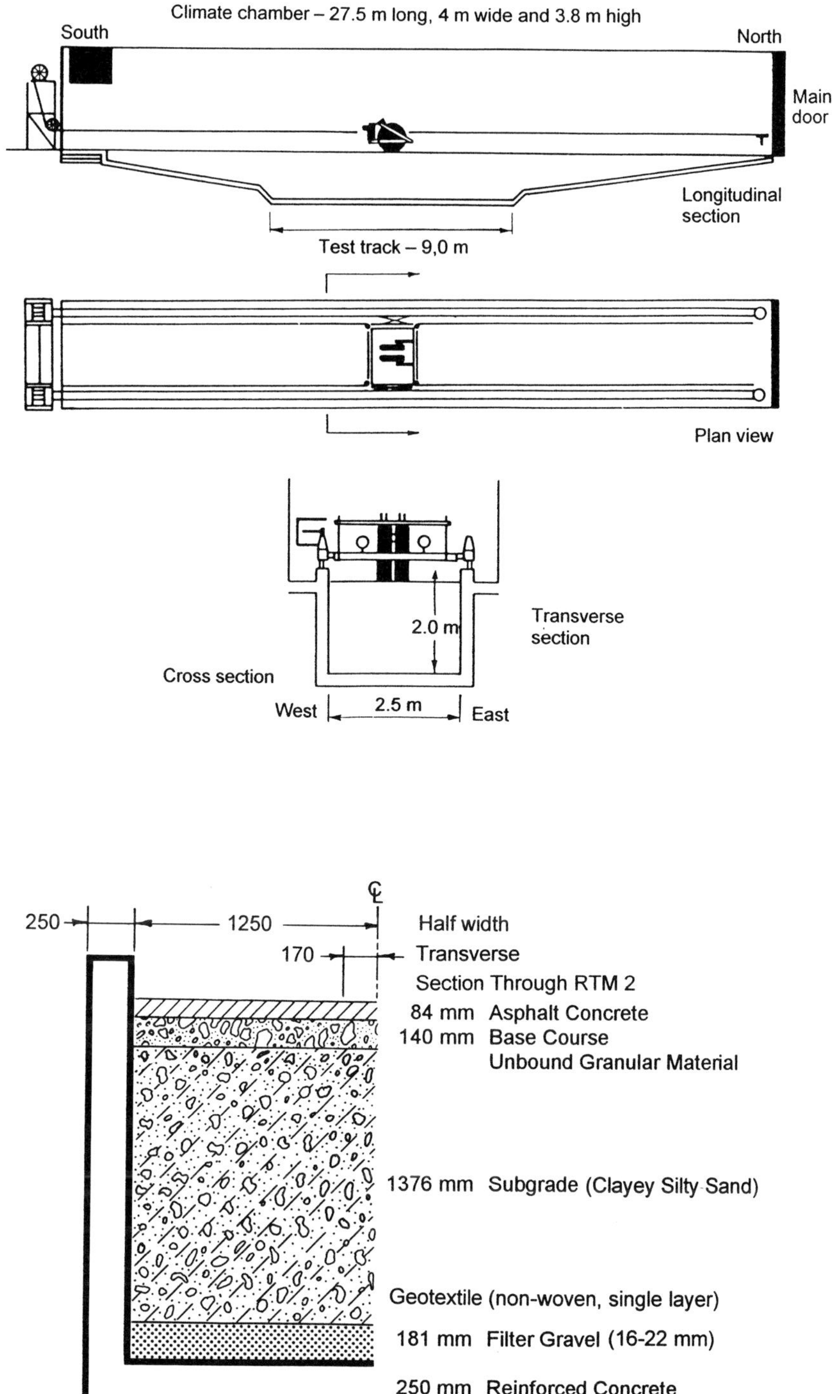

Figure 1. The Danish Road Testing Machine, RTM2.

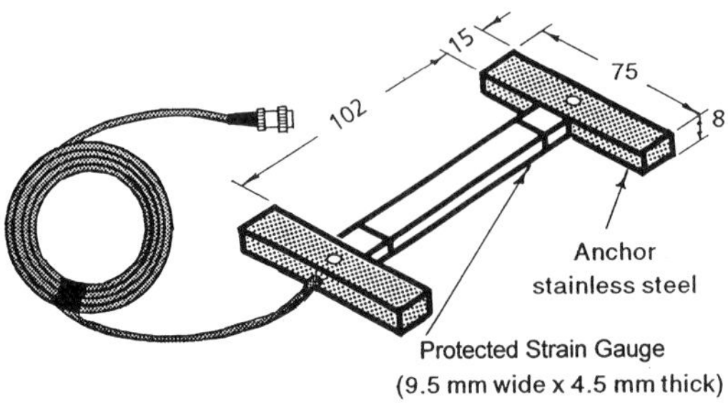

(a) Asphalt Strain Gauge

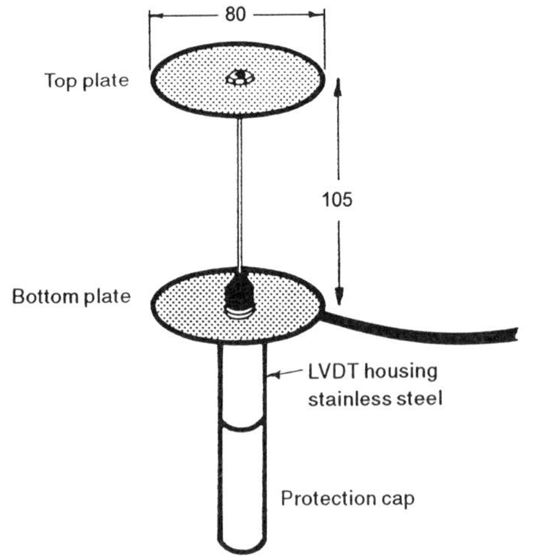

(b) Soil Deformation Transducer (STD) (Vertical Strain Measurement).

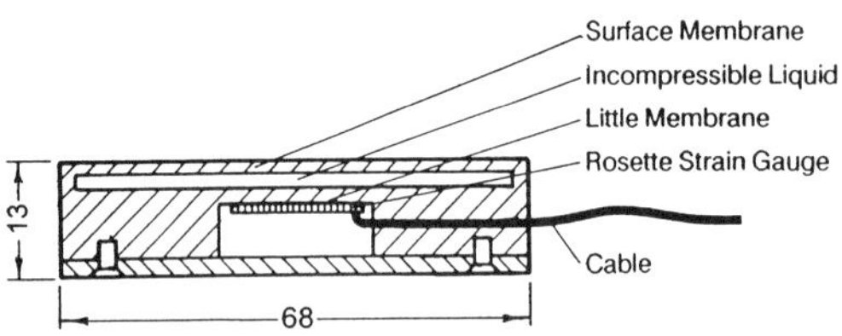

(c) Cross section of a Soil Pressure Cell (SPC).

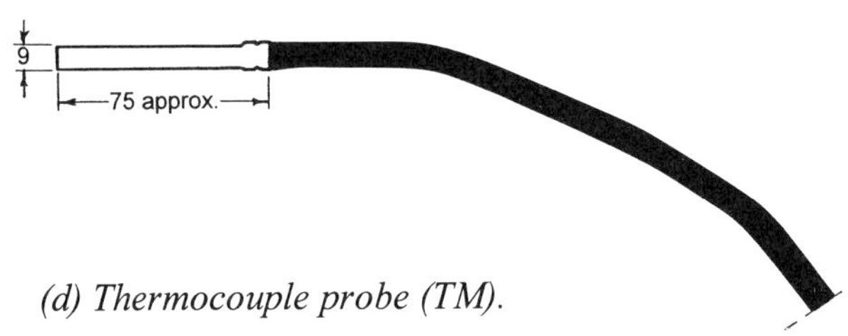

(d) Thermocouple probe (TM).

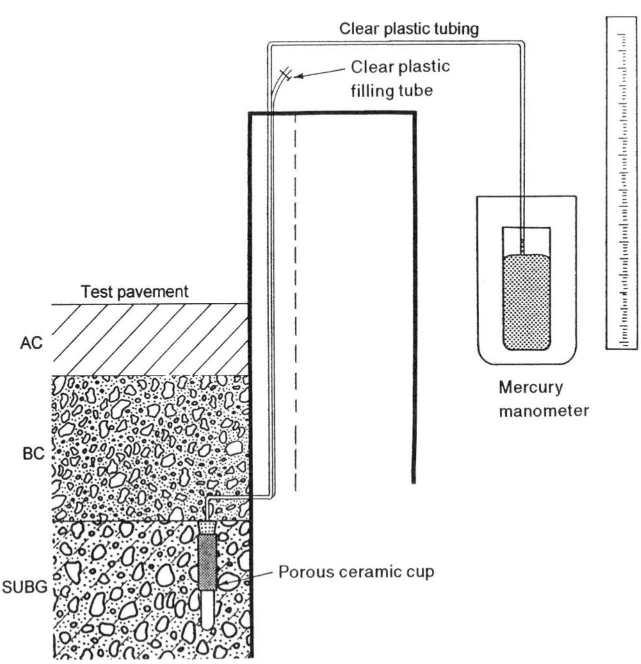

Figure 2.
Instruments installed
in the RTM2 Test Pavement.

so only rolling compaction was applied. Mean compaction densities ranged between 92 % and 100 % (of the maximum dry density determined in the standard Proctor laboratory compaction test), and average moisture contents ranged between 9 % and 11 %.

The base course material is relatively coarse grained and non-plastic, so drying to a suitable moisture content was easily achieved. Compaction was carried out by numerous passes of the same type of dual drum roller. No vibratory compaction was applied. An average compaction density of 96% (variation between 94 and 97%) and an average moisture content of 4% were recorded.

Before commencing the main AC surfacing work, Asphalt Strain Gauges were laid on the base course on bedding layers of sand asphalt (a mixture of sand and cold bitumen). An emulsion prime coat was sprayed over the whole surface of the base course and allowed to cure. The AC surfacing was constructed in one layer in a continuous laying operation performed by a small paving machine. Ten cores were taken from the pavement, close to, but beyond the instrumented test section, and to complete the surfacing work, three layers of a proprietary compound (for protecting the bituminous surfacing against hydraulic oil spillage) were applied.

For every constructed layer, the level of the surface at a systematic number of positions was recorded with the **Profilometer**. These and other point measurements form the basis for the determination of layer thicknesses and instrument positions. The **Profilometer [1, 2, 5]** is a digital profiler consisting of a wheeled T-frame running on the same rails as the loading cart. For vertical profile measurements of the pavement surface to micron accuracy, a small metal wheel, coupled to a high precision LVDT, is run on the pavement surface; it is connected to a digital transducer that records the position of the wheel in the three axes. The Profilometer measures plan distances in the X (transverse) and Y (longitudinal) axes to an accuracy of ±0.6 mm. It is also equipped with a manually operated laser distance measuring unit (dmu) for measuring pavement layer surface levels to an accuracy of ±3 mm.

ACCELERATED LOADING AND TESTING

Accelerated loading was applied at three dual wheel loading levels at a surface temperature of 25°C approximately, as follows:

- 40 kN at 600 kPa tyre pressure for 50,000 (50 k) load repetitions
- 50 kN at 700 kPa tyre pressure for 50 k load repetitions
- 60 kN at 800 kPa tyre pressure for 50 k load repetitions

The main programme of simulated traffic loading was applied by dual wheel loading at regular load repetition increments for each load level: 0 k (start), 1 k, 3 k, 10 k, 20 k, 30 k, 40 k and 50 k load repetitions. After each load increment, stresses and strains were measured under a single wheel, dual wheel and FWD loading. Under single and dual wheel loading, stresses and strains were recorded twice at each load repetition stage: immediately before completing an increment of load repetitions, and immediately after commencing the next increment of load repetitions. Pavement layer temperatures were monitored and recorded at the same time as the stress and strain measurements. Instrument responses were also measured under FWD loading, which was moved incrementally along the two instrument lines at intervals of 75 mm.

Structural deterioration was monitored by FWD testing. These standardised FWD tests were carried out longitudinally along the two instrument lines at 300 mm intervals at certain intervals of load repetitions. Surface profile measurements using the Profilometer were conducted after each load repetition increment for each load level; on each measurement, five longitudinal and 31 transverse profiles were recorded. From the longitudinal profiles, Slope Variance (SV), Present Serviceability Index (PSI) and International Roughness Index (IRI) were computed; permanent surface deformation (rutting) was calculated from the transverse profiles. To monitor soil moisture development, soil suction was also observed on a regular basis.

SOME PRELIMINARY RESULTS

To study the influence of load type on pavement responses, stresses and strains in the pavement layers were measured under both single and dual wheel loading at load cart speeds of approximately 20, 16, 10 and 8 km/h, as tabulated in Table 1. The data summarised in the table indicate that as speed of the applied loading decreases, stresses and strains induced in the pavement layers increase. Asphalt is a viscoelastic material, so a lower speed of applied loading causes a reduction in asphalt resilient modulus, thereby increasing stresses and strains in the underlying pavement layers.

Table 1. Measured stresses and strains (average values) under 50 kN dual wheel loading at varying speeds of the load cart. SGL 1 to 3: refer to the upper SubGrade Layers 1 to 3.

		No. Of Gauges	20 km/h (5.56m/s)	16 km/h (4.44 m/s)	10 km/h (2.78 m/s)	8 km/h (2.22 m/s)
Vertical stresses in the subgrade (SPCs) (kPa)	SGL 1	2	120	125	132	140
	SGL 2	2	103	106	110	112
	SGL 3	2	57	58	60	62
Vertical strains in the subgrade (SDTs) $\mu\varepsilon$	SGL 1	4	2165	2286	2418	2512
	SGL 2	4	1245	1296	1393	1429
	SGL 3	4	1587	1634	1712	1762
Horizontal strains at the bottom of the AC (ASG) $\mu\varepsilon$						
	X	4	-309	-320	-329	-337
	Y	4	-493	-520	-576	-598

SGL 1: 0 – 128 mm of the subgrade
SGL 2: 128 – 260 mm of the subgrade
SGL 3: 260 – 400 mm of the subgrade
X and Y : the X and Y axes, respectively.

AC = Asphalt Concrete
ASG = Asphalt Strain Gauge
SPC = Soil Pressure Cell
SDT = Soil Deformation Transducer

Stresses and strains induced in pavement layers are determined by the pavement structure, the properties of pavement materials and the loading. Boussinesq's equations express the distribution of stresses and strains within a homogeneous, isotropic, linear elastic infinite halfspace under a point load. If the test pavement is assumed to be a single-layer system, one of Boussinesq's equations can be adapted to estimate the equivalent resilient modulus under the rolling wheel load when travelling at different speeds. The form of the Boussinesq equation (Eq. 1) used in this paper is given here.

$$\varepsilon_z = \frac{(1+\mu)*p}{2*\pi*E}\left(\frac{3*z^3}{(r^2+z^2)^{2.5}} - \frac{2*\mu*z}{(r^2+z^2)^{1.5}}\right)$$

Where:

μ	is Poisson's ratio for the material: in this paper, 0.35 is assumed,
p	is the amount of the applied load: in this case, 25 kN single wheel load,
E	is the elastic modulus of the material,
r	is the radial distance to the load, and
z	is the vertical distance to the load (depth.).

Analyses undertaken using a spreadsheet, with εz, p, r and π as assumed or known values (developed in Microsoft Excel 97), gave results tabulated in Tables 2 and 3. The inbuilt solver function (a standard MS Excel add-in tool) is invoked to estimate E and z to fit the rising (on left side) arm of the calculated curve to the measured strain values using a least squares error approximation procedure. Thus, by iteratively adjusting depth z and modulus E, calculated strain distributions can be adjusted to match measured strain distributions precisely, as depicted in Figure 3.

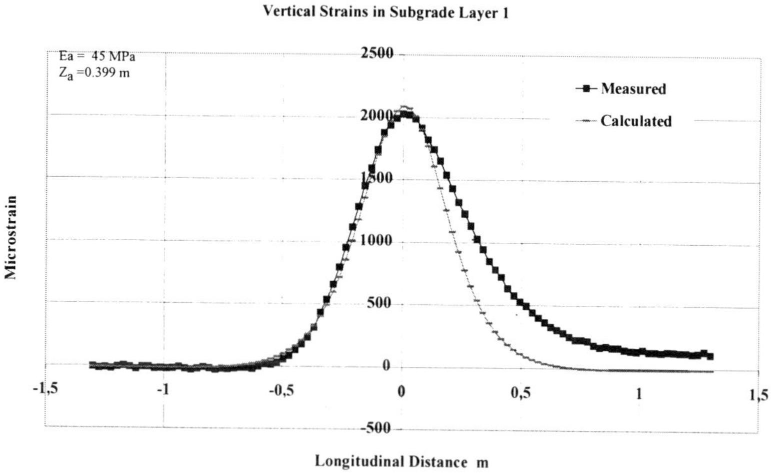

Figure 3. Typical measured and calculated vertical strain distributions.

Figure 3 shows that a good fit has been achieved between the measured and estimated strains as the wheel load approaches the SDT (the rising arm of the curve). However, as the wheel load departs the SDT (to the right of the peak), the measured strains are larger than the estimated values because these are composed of elastic, plastic and viscoelastic strains, which can only be approximately modelled by this form of Boussinesq equation.

Using this Boussinesq equation, an SDT (i.e., a strain measuring instrument) located in the layered road pavement can be considered as equivalently located in an infinite halfspace with a resilient modulus of Ea and at a depth of z_a. Table 2 tabulates estimated resilient moduli and equivalent depths determined by applying this Boussinesq equation approach at different speeds of applied loading.

This analysis shows that not only equivalent depth, but also resilient modulus changes with the speed of the applied wheel load. Variations in the estimates of equivalent depth are due to the influence of applied load speeds on the asphalt and the base course layers, whilst variations in the resilient moduli emphasise the significant degree of viscosity of the subgrade materials (which in the RTM2 subgrade is a clayey silty sand).

Table 2. Equivalent depths and resilient moduli for 25 kN single wheel loading at different load cart speeds computed using Boussinesq's equation. SGL 1 to 3 are SubGrade Layers 1 to 3, as described in Table 1.

		Resilient Modulus, MPa				Equivalent depth, mm			
	No. of Gauges	20 km/h	16 km/h	10 km/h	8 km/h	20 km/h	16 km/h	10 km/h	8 km/h
SGL1	4	44	45	32	33	457	443	480	467
SGL2	4	56	58	42	45	580	564	609	575
SGL3	4	30	30	21	23	734	722	792	755

Table 3 compares base and subgrade strains measured under FWD loading and single wheel loading at 20 km/h (4.44 m/s). Subgrade strains measured under dynamic wheel loading vary from 2.1 to upwards of 2.7 times the magnitude of the strains measured under FWD loading. This is due to the dynamic response of the pavement materials, since, whereas the speed of the single wheel loading is 20 km/h, FWD loading is approximately equivalent to dynamic pneumatic tyred loading applied at a speed of 60 - 80 km/h (i.e., for a load pulse period of 40 – 25 ms).

Table 3. Strains measured under single wheel loading and FWD loading.
Single wheel load: 30 kN at 800 kPa tyre pressure; FWD load: 28.3 kN at 400 kPa.
SGL 1 to 3 are SubGrade Layers 1 to 3, as described in Table 1.

Strains measured	Pavement Layer	No. of Gauges	Single wheel Loading $\mu\varepsilon$	FWD Loading $\mu\varepsilon$	Single wheel / FWD Ave. %
Vertical	BASE	2	1836	875	112
Vertical	SGL 1	3	1669	618	170
Vertical	SGL 2	3	823	360	129
Vertical	SGL 3	4	1047	385	172
Longitudinal	SGL 1	1	-433	-163	166
Transverse	SGL 1	2	-608	-183	231

CONCLUSIONS

This paper briefly describes the construction, instrumentation and load testing of the second test pavement constructed in the Danish Road Testing Machine (RTM2) in connection with an international Pavement Subgrade Performance Study. The paper also presents some preliminary results and interesting methods of analysis being investigated and developed in the current ongoing study.

The Danish Road Institute has extensive experience in fabricating and installing response-measuring instruments. In the second test pavement constructed in the Danish Road Testing Machine (RTM2), vertical stresses and strains are being measured only in the top 400 mm of the subgrade and in the base course layer (to monitor deformation in the base course). Horizontal stresses and strains, however, are only being measured only in the upper 130 mm of the subgrade (as their values are relatively much smaller than the corresponding vertical responses). Pore Pressure Sensors are being used to record negative pore pressure in the subgrade, in order to obtain indications of the fluctuations in moisture contents; and thermocouple probes in both the bound and unbound pavement layers are being used to monitor upper pavement temperatures.

To study the influence of load type on pavement responses, stresses and strains in the pavement layers have been measured under stationary FWD loading and dynamic wheel loading moving at different speeds. The data indicate that as the wheel load speed decreases, stresses and strains in the pavement layers increase, as may be expected, due in part to the viscoelastic nature of the subgrade material.

By adapting and applying one of Boussinesq's equations, a comparison of estimated and measured vertical strains in the subgrade under single wheel loads at different speeds, is also presented. A good degree of fit is observed between the measured and estimated strains on wheel load approach to the response instrument, but not on departure, indicating that a significant degree of elastic, plastic and viscoelastic straining is occurring within the subgrade material.

Accelerated Load Testing Programme

The first phase of accelerated loading of the current RTM2 test pavement has applied 150,000 load repetitions at three load levels (40 kN, 50 kN and 60 kN dual wheel loading). To date, the pavement has developed an average depth of permanent deformation (rutting) of 10 mm. It is intended to induce subgrade failure during further accelerated load testing, and to carefully observe the deterioration of the test pavement's functional condition and the responses of its constituent materials. Accelerated loading will therefore continue during 1998, preceded by adverse climatic treatment to initiate subgrade weakening. On completion of the climatic treatment, accelerated loading will be continued with between 40 kN to 60 kN halfaxle loading applied by the dual wheel load cart.

It is expected that structural failure of the RTM2 test pavement will occur during 1998, primarily due to permanent deformation (rutting) of the subgrade. Analysis of the structural and functional performance of the test pavement up to failure will be undertaken and reported during 1998.

ACKNOWLEDGEMENTS

The research partnership of the Danish Road Institute and the Technical University of Denmark, wish to acknowledge and thank the Federal Highway Administration (FHWA), USA, and a number of research colleagues for their help, good advice and willing assistance, and in particular to thank:

- Prof. Lynne Irwin, Cornell University Local Roads Program, Cornell University, 416 Riley-Robb Hall, Ithaca, NY 14853-5701, U.S.A.

- Dr. Vincent C. Janoo, US Army Cold Regions Research and Engineering Laboratory, 72 Lyme Road, Hanover, NH 03755-1290, U.S.A.

REFERENCES

1. DANISH ROAD INSTITUTE. TRB Annual Meeting 97. DRI Paper Presentation at Session 13, Pavement Instrumentation, Part 1. Report 84, 1997.

2. DANISH ROAD INSTITUTE. Subgrade Performance Study. Part 1: Materials, Construction and Instrumentation. DRI Report 85, 1997.

3. ERTMAN LARSEN, H.J. and ULLIDTZ, P. Pavement Subgrade performance Study in the Danish Road Testing Machine, Eighth International Conference on Asphalt Pavements, Seattle, Washington, USA, 1997.

4. ERTMAN LARSEN, H.J. and ULLIDTZ, P. Development of Improved Mechanistic Deterioration Models for Flexible Pavements, 4th International Conference on Managing Pavements, Durban, South Africa, Theme 2: Implementation of PMS Outputs, Tuesday, 19th May, 1998.

5. KRARUP, J. Bearing Capacity and Water, Part I: Materials, Construction and Instrumentation, Note 328, Danish Road Institute (DRI), 1992.

Fifth International Conference on the
BEARING CAPASITY OF
ROADS AND AIRFIELDS

COMPARISON OF SOME STRUCTURAL ANALYSES METHODS USED FOR THE TEST PAVEMENT IN THE DANISH ROAD TESTING MACHINE

S. BALTZER[1], W. ZHANG[2], R. MACDONALD[1] AND P. ULLIDTZ[3]

[1]Research Engineers, Road Directorate, Danish Road Institute, 4000 Roskilde, Fax: + 45 46 30 01 05, E-mail: sub@vd.dk, rmd@vd.dk
[2]Post Doctoral Research Assistant, Technical University of Denmark, 2800 Lyngby, Fax: + 45 93 64 12, E-mail: wei@ivtb.dtu.dk
[3]Associate Professor, Technical University of Denmark, 2800 Lyngby, Fax: + 45 93 64 12, E-mail: pullidtz@ivtb.dtu.dk

ABSTRACT

A flexible test pavement, instrumented to measure stresses and strains in three primary axes within the upper 400 mm of the subgrade, has been constructed and load tested in the Danish Road Testing Machine (RTM). One objective of this research, which is part of the international Pavement Subgrade Performance Study, is to develop an improved mechanistic subgrade failure criterion.

Simulated traffic loads were applied between June and December 1997 by a dual tyre loading cart within a climate chamber maintained at a constant temperature. Three load levels (40 kN, 50 kN and 60 kN) were applied. In all three test series, 50,000 load repetitions were applied at each load level, and the test chamber temperature was maintained at approximately 25°C.

Instruments installed in the test pavement included: Strain measuring sensors (LVDT Soil Strain Deformation Transducers and Asphalt Strain Gauges), Soil Pressure Cells, Thermistors and Pore Pressure Sensors. Routine monitoring of instrument responses and surface profiles with a Profilometer and FWD/LWD structural testing were undertaken at regular intervals during the construction and load testing programmes.

This paper compares various structural analysis methods used for the RTM test pavement with data from FWD testing undertaken after the construction and loading programmes. Multilayer linear elastic forward and backcalculation methods, a finite element program and MS Excel spreadsheet based methods are compared.
Keywords: Pavement Design, Instrumentation, Stress, Strain, E-moduli.

INTRODUCTION

A number of different models (or programs) can be used for backcalculation of layer moduli of a pavement structure. The problem with backcalculation is that the solution may not be unique. It depends on how the pavement structure is modelled and which program is used. One way of verifying the sets of layer moduli is to use them in forward calculation of stresses and strains and then compare these values to stresses and strains measured in a test pavement.

During 1996, a test pavement was constructed in the Danish Road Testing Machine (RTM). This pavement ("RTM2") is instrumented with LVDT Soil Strain Deformation Transducers (SDTs) to measure strains in unbound materials. Asphalt Strain Gauges (ASGs) to measure strains at the bottom of the asphalt and Soil Pressure Cells (SPCs) to measure stresses in unbound materials. These instruments have been developed over the past 25 years and have been used with success in previous studies [1-2]. Simulated traffic loads were applied in series of 50,000 load repetitions. Stresses and strains in the pavement have been measured both under rolling wheel loading at planned load repetitions and FWD loading.

This paper focuses on analysing deflections, strains and stresses measured under FWD loading. Several structures are used to model the test pavement in this approach, and programs based on Elastic Layer Theory (ELT), Finite Element Method (FEM) and Method of Equivalent Thickness (MET) are utilised, both in backcalculation to obtain layer moduli, and in forward calculation to obtain stresses and strains. Non-linearity of the subgrade is taken into consideration for some models. The calculated stresses and strains are compared to the measured values. For the more promising models, strain distributions are also compared.

TEST PAVEMENT

The Danish Road Testing Machine, RTM, is a linear pavement testing facility enclosed in a climate chamber. The actual test section has a width of 2.5 m and a length of 9 m. A test pavement, with a structure as shown in Figure 1, has been constructed and instrumented.

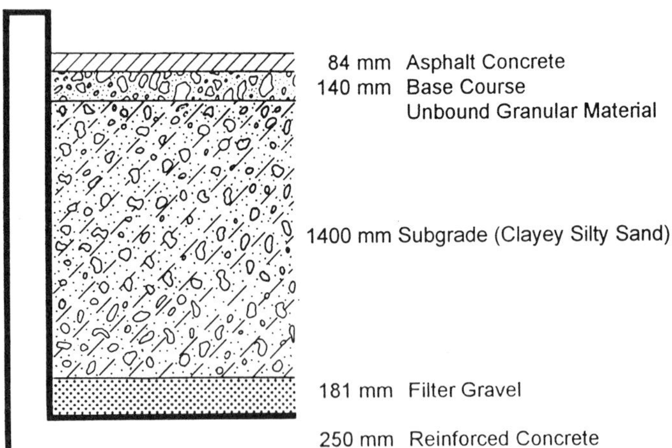

Figure 1: Test Pavement Structure.

A densely graded asphalt concrete is used for the asphalt concrete layer. The maximum aggregate size is 16 mm. The AC mix is composed of crushed granite aggregate, sand and fillers with a 60 pen. bitumen binder. Mix tests indicated a bitumen content of 5.3 percent and a Marshall density of 2420 kg/m^3.

The base course is a graded crushed natural aggregate with an aggregate particle size ranging from 0-32 mm. It has an optimum moisture content of 6.9 % and a maximum dry density of 2190 kg/m^3. The level of compaction varies considerably throughout the pavement between values of 94-97 % of maximum dry density. The relatively low compaction level achieved is due to a low moisture content of around 4 %, and to the fact that the 150 mm granular layer is placed directly on a relatively soft subgrade material.

A Danish "Moraine Clay" classified as a clayey silty sand (AASHTO classification A-4(0)) is being used as the subgrade. It has an optimum moisture content of 9 % and a maximum dry density of 2045 kg/m^3. It is constructed in nine layers of approx. 150 mm thickness each. The two uppermost subgrade layers are fairly well compacted, at approx. 97 % compaction (moisture content approx. 10.5 %), while the third subgrade layer is only compacted to a degree of approx. 92 % (moisture content approx. 9 %).

The reinforced concrete bottom was tested with FWD before construction of the filter gravel layer. Based on the previous studies carried out in the RTM, the concrete bottom has a modulus of 27,500 MPa and the drainage layer has a modulus of 250 MPa.

Instruments: Eight ASGs are installed at the bottom of the asphalt concrete layer, four in the longitudinal direction and four in the transversal direction. Nineteen SPCs and 18 SDTs are installed in the unbound materials to measure strains and stresses in the vertical and horizontal directions. Instruments are laid out along six metres of the full length of the test section, in two lines coincident with the centrelines of the wheelpath of the double wheel used for the accelerated loading. The current structural design criteria applied in pavement design are horizontal strains at the bottom of the asphalt and the vertical strain at the top of the subgrade. Therefore, this paper concentrates on analysis of vertical stresses and strains recorded with SPCs and SDTs and horizontal strains recorded with ASGs.

BACKCALCULATION OF LAYER MODULI

The deflections used in the backcalculation were measured before the accelerated loading started and after 50,000 load repetitions of 40 kN. The contact stresses were 400 kPa and 600 kPa, respectively. FWD tests were carried out in both instrument lines at intervals of 300 mm. Three programs based on ELT, FEM and MET were used to backcalculate layer moduli. In each program several structures were set up to model the test pavement. Figure 2 shows the structures used in backcalculation. In all five structures, the asphalt- and base course layers are modelled the same way. The subgrade layer varies from being semi-infinite to being finite either including or excluding the drainage layer. The backcalculated layer moduli would be more comparable if e.g. a 4-layer model was set up and used in all programs. But the programs capable of calculating model structures of many layers, also obtain the best deflection fits using more layers. In respect of the different programs, it has therefore been chosen to model several different structures.

3-layer structure

AC	84 mm	$\nu_{AC} = 0.35$	$E_{AC} = ?$
BC	140 mm	$\nu_{BC} = 0.35$	$E_{BC} = ?$
SG	∞	$\nu_{SG} = 0.40$	$E_{SG} = ?$

4-layer structure

AC	84 mm	$\nu_{AC} = 0.35$	$E_{AC} = ?$
BC	140 mm	$\nu_{BC} = 0.35$	$E_{BC} = ?$
SG	1580 mm	$\nu_{SG} = 0.40$	$E_{SG} = ?$
Con	∞	$\nu_C = 0.15$	E_C= 27,500 MPa

5-layer structure

AC	84 mm	$\nu_{AC} = 0.35$	$E_{AC} = ?$
BC	140 mm	$\nu_{BC} = 0.35$	$E_{BC} = ?$
SG	1580 mm	$\nu_{SG} = 0.40$	$E_{SG} = ?$
Con	250 mm	$\nu_C = 0.15$	E_C= 27,500 MPa
NS	∞	$\nu_{NS} = 0.40$	$E_{NS} = ?$

6-layer structure

AC	84 mm	$\nu_{AC} = 0.35$	$E_{AC} = ?$
BC	140 mm	$\nu_{BC} = 0.35$	$E_{BC} = ?$
SG	1400 mm	$\nu_{SG} = 0.40$	$E_{SG} = ?$
DL	180 mm	$\nu_{DL} = 0.35$	E_{DL}= 250MPa
Con	250 mm	$\nu_C = 0.15$	E_C = 27,500 MPa
NS	∞	$\nu_{NS} = 0.40$	$E_{NS} = ?$

7-layer structure

AC	84 mm	$\nu_{AC} = 0.35$	$E_{AC} = ?$
BC	140 mm	$\nu_{BC} = 0.35$	$E_{BC} = ?$
SG1	300 mm	$\nu_{SG1} = 0.40$	$E_{SG1} = ?$
SG2	1100 mm	$\nu_{SG2} = 0.40$	$E_{SG2} = ?$
DL	180 mm	$\nu_{DL} = 0.35$	E_{DL}= 250 MPa
Con	250 mm	$\nu_C = 0.15$	E_C = 27,500 MPa
NS	∞	$\nu_{NS} = 0.40$	$E_{NS} = ?$

Legend:

AC	Asphalt Concrete
BC	Base Course
SG	Subgrade
DL	Drainage Layer
Con	Reinforced Concrete
NS	Native Soil
ν	Poisson's Ratio
E	E-modulus

Figure 2: Pavement model structures used in backcalculation.

Elastic Layer Theory (ELT) is the most widely used method for pavement design. Elastic theory is based on the assumption that the load is static and all materials are homogeneous, isotropic and linear elastic. Layer moduli of the RTM test pavement have been calculated using the backcalculation program MODCOMP3 [3]. MODCOMP3 can also take non-linearity of the pavement layers into consideration, but for that purpose more than one load level FWD testing is required. The deflections used in the backcalculation were measured under one load level FWD testing, so only linear elastic analysis has been carried out.

MODCOMP3 can analyse pavement structures composed of up to 15 layers. The RTM test pavement has been modelled both as a 6-layer structure and a 7-layer structure, in which the subgrade is divided into an upper and a lower layer as shown in Figure 2. The representative layer moduli are given in Table 1.

Table 1: Representative layer moduli backcalculated using MODCOMP3 (ELT).

	Before accelerated loading		After 50,000 load repetitions	
	6-layer	7-layer	6-layer	7-layer
E_{AC} (MPa)	6715	6245	7015	5710
E_{BC} (MPa)	90	125	40	115
E_{SG1} (MPa)	--	130	--	95
E_{SG2} (MPa)	160	170	140	145
E_{NS} (MPa)	280	265	250	240

Finite Element Method (FEM). The basic concept of the finite element method is that a continuum structure can be treated as an assembly of finite elements, which are connected with finite "joints". The finite element still assumes the strains to be compatible, except for special elements that may allow for discontinuities. Analysing deformations, forces, stiffness of elements and conditions of joints, a matrix of equilibrium equation for the whole structure can thereby be established. In solving the equilibrium equation matrix, stresses and strains in the structure are obtained. In the finite element method, a layered pavement is divided into rows and columns, e.g., elements. Non-linearity of materials can be taken into consideration.

The model used for non-linearity of the subgrade is based on the major principal stress, and has the following format:

$$E = C \cdot \left(\frac{\sigma_1}{\sigma'}\right)^n$$

Where C and n are material constants, σ_1 is the major principal stress, and σ' is a reference stress (atmospherical pressure).

Three structures comprising 3-, 4- and 5-layers, are used in the backcalculation. Table 2 shows the representative results from the calculation.

Table 2: Representative layer moduli backcalculated with FEM for three structures.

	Before accelerated loading			After 50,000 load repetitions		
	3-layer	4-layer	5-layer	3-layer	4-layer	5-layer
E_{AC} (MPa)	5000	4600	6400	5950	5390	6960
E_{BC} (MPa)	260	280	120	139	145	95
C_{SG} (MPa)	64	100	35	46	78	21
n_{SG}	-0.32	0	-0.75	-0.375	0	-0.8
E_{NS} (MPa)	--	--	150	--	--	150

INTRODUCTION

Labour intensive construction of roads on large-scale projects has been revived in South Africa because of the demand for job creation. A lot of research effort has been directed towards the social benefits and management of these projects. Little attention has, however, been paid to the technical aspects relating to the quality of the construction and the bearing capacity of pavements constructed in this manner. A trial project was therefore initiated by the Gauteng Department of Transport and Public Works (GAUTRANS) in South Africa to investigate the feasibility of labour intensive construction in terms of logistics, supervision, management and the quality of the end-product. The project consisted of the labour intensive construction of an emulsion treated natural gravel base layer on 3,5 km of road, combined with the labour intensive construction of several experimental sections of various base layer types for the purpose of Accelerated Pavement Testing (APT) and Long-Term Pavement Performance (LTPP) studies. The accelerated testing is being done with the Heavy Vehicle Simulator (HVS) that has been in the service of GAUTRANS for about 20 years.

The experimental sections consist of 3 lanes with a uniform cross section, one lane for each traffic direction and an additional lane for accelerated testing. The traffic lane adjacent to the HVS-lane will serve as the LTTP-sections. This paper will concentrate on the technical aspects related to the construction and accelerated testing of the experimental sections.

Figure 1 gives detail regarding the layout of the experimental sections. Only the base layers (excluding the crushed stone base layer) were constructed labour intensively. The rest of the structure up to sub-base level was constructed using conventional machine construction. The materials used in the base layers were:

- Emulsion treated natural gravel (ETB). A sandstone natural gravel was treated with 1% emulsion (0,6% residual binder) and 1% ordinary Portland cement (OPC) by mass.

- Crushed stone. A graded crushed stone from a commercial supplier.

- Composite macadam (CM). A 53 mm aggregate was placed in a single or double layer. This large stone layer was filled ⅔ of its depth with a fine filler. A layer of 26 mm stone was placed on top of the 53 mm layer. Interlock between the small and large stone layer was achieved in the top third of the large stone layer. The 26 mm stone and the top third of the 53 mm stone layer was penetrated with a slurry.

- Waterbound Macadam (WM). A 53 mm aggregate was placed at the full depth of the base layer and filled with a fine filler.

- Ash (ASH). A clinker ash waste product.

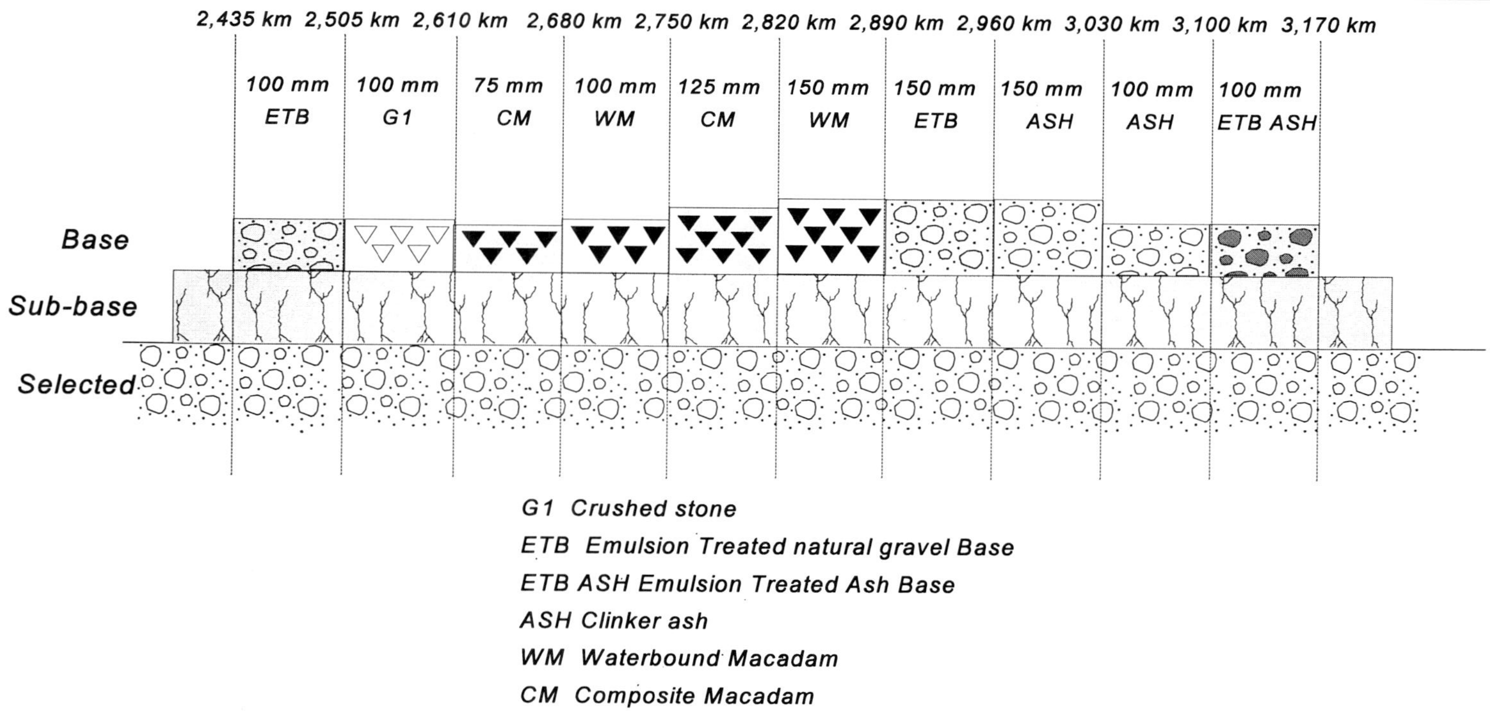

FIGURE 1: LAYOUT OF THE EXPERIMENTAL SECTIONS

Emulsion treated ash (ETB ASH). A clinker ash was treated with 2,5% emulsion (1,5% residual binder) and 1% ordinary Portland cement (OPC) by mass.

Compaction of the natural gravel ETB and Macadam type base layers was done with small pedestrian rollers and plate vibrators while the compaction of all ash base layers was done with conventional heavy construction rollers. A 25 mm asphalt wearing course was placed to improve riding quality.

MATERIAL TESTING DURING THE CONSTRUCTION OF THE EXPERIMENTAL SECTIONS

Accelerated pavement testing requires detailed knowledge regarding the composition of the pavements to be tested before commencing with the APT-tests. This knowledge will help identifying potentially weak layers in the pavement and will assist in explaining some of the behaviour of the test pavements. For this reason a detailed laboratory and field testing programme was completed during the construction of the experimental sections.

The subgrade of the road consists of an existing gravel road. In addition to the required fill material, selected material with a thickness of 300 mm was imported from local sources. The material for the selected layers from km 2,435 to 2,780 is the same good quality sandstone material as that used in the subbase and emulsion treated natural gravel base layers. From km 2,780 to 3,170 the selected layer consists of a poor quality ferricrete material. The subbase consists of a cemented sandstone material.

Tables 1 and 2 give a summary of the average compaction and material strength values obtained. The support for the base layers may be divided into two sections, the first section stretching from km 2,400 to 2,780 and the second one from km 2,780 to 3,180. On the first section the selected layer consists of the relatively good quality sandstone material with a strong cement treated sub-base layer. On the second section the selected layer consists of the poor quality ferricrete material with a slightly weaker cement treated sub-base.

DEFLECTION TESTING BEFORE COMMENCING THE ACCELERATED PAVEMENT TESTING

Deflection testing consisted of Road Surface Deflectometer (RSD) measurements which is an electronic Benkelman beam deflection instrument, Deflectograph measurements which is a moving Benkelman beam deflection instrument and a Falling Weight Deflectometer (FWD). The road surface was marked with a white paint dot at 1m intervals. It was therefore possible to do measurements with the RSD and FWD on exactly the same points. The positions at which the Deflectograph readings were taken may not exactly coincide with the positions at which the other two readings were taken but should be in close proximity.

Table 1: Average compaction results				
Layer	Material	Layer thickness	Average compaction (% mod AASHTO)	Specification (% mod AASHTO)
Selected	Sandstone	300 mm	99,6	93
	Ferricrete	300 mm	95,1	93
Subbase	Cement treated sandstone	150 mm	96,5	95
Base	Crushed stone	100 mm	98,6	100 - 102
	Ash	100 mm	106	100 - 102
		150 mm	102	100 - 102
	ETB Ash	100 mm	105	97
	ETB	100 mm	97	97
		150 mm	97	97
	Waterbound Macadam	100 mm	80 % of solid density	88 - 90 % WM1 86 - 88 % WM2
		150 mm	77 % of solid density	

Table 2: Average material strength results				
Layer	Material	Laboratory strength*	Field DCP strength*	Specification*
Selected	Sandstone	111 % @ 100 %	253 % @ 99,6 %	15 % @ 93 %
	Ferricrete	2 % @ 96 %	62 % @ 95 %	15 % @ 93 %
Subbase	Cement treated sandstone km 2,400 - 2,780	2486 kPa @ 100 % 1940 kPa @ 97 % 1576 kPa @ 95 %	1683 kPa @ 96,5 %	1500 - 3000 kPa @ 100 % 1000 - 2000 kPa @ 97 %
	Cement treated sandstone km 2,780 - 3,180	1702 kPa @ 100 % 1267 kPa @ 97 % 977 kPa @ 95 %	861 kPa @ 96,5 %	1500 - 3000 kPa @ 100 % 1000 - 2000 kPa @ 97 %
Base	Crushed stone	111 % @ 98 %	238 % @ 98,6 %	80 % @ 98 %
	Ash	99 % @ 98 %	121 % @ 104 %	80 % @ 98 %
	ETB Ash	1529 kPa @ 100 % 1328 kPa @ 97 %	1089 kPa @ 105 %	1500 - 3000 kPa @ 100 % 1000 - 2000 kPa @ 97 %
	ETB	550 kPa @ 100 % 509 kPa @ 97 %	919 kPa @ 97 %	750 - 1500 kPa @ 100 % 500 - 1000 kPa @ 97 %

* CBR (%)or UCS (kPa) at a percentage mod. AASHTO compaction

Figure 2 shows the deflection readings from the different instruments. The correlation between the FWD and the RSD readings is very good from km 2,435 to 2,960. There seem to be a slightly bigger difference between these two instruments on the ash sections from km 2,960 to 3,180. There is a culvert at km 2,770 to 2,775 which was picked up by all the instruments registering a low deflection at that point.

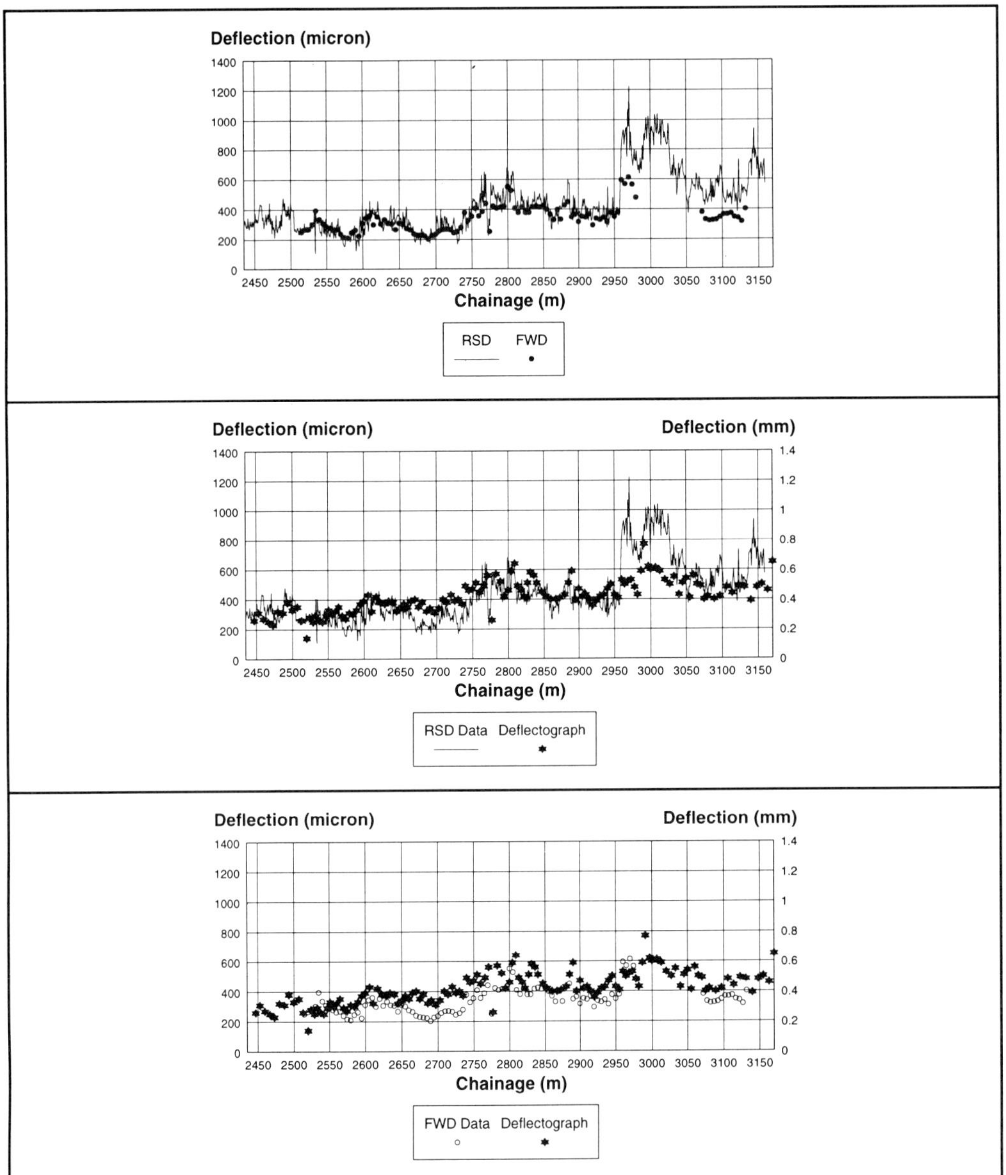

FIGURE 2: DEFLECTION DATA FOR THE EXPERIMENTAL SECTIONS MEASURED WITH VARIOUS DEFLECTION DEVICES

ACCELERATED PAVEMENT TESTING

The progress of the HVS-testing at the time of writing this paper is shown in Table 3. Limited results from the accelerated tests will be presented for the following types of measurements taken during the test:

- Straight-edge surface rut. A manual rut reading is taken from a straight-edge on a fixed grid along the HVS-test section. The maximum rut is normally reported at cross-section positions along the HVS-section or the average rut for all the points on the grid is reported. These readings are repeated at regular time intervals during the HVS-test.

- Laser profilometer surface rut. The road profile is measured with a moving laser on the same cross-sections at which the manual rut readings are taken. The data are recorded at regular intervals during the HVS-test and reported in the same way as the manual rut readings.

- RSD road surface deflections. Road surface deflections are normally measured with an electronic Benkelman beam at all points along the centre line and 200 mm to either side of the centre line. These deflections are measured over a range of wheel loads not exceeding the test wheel load.

- Multi-Depth Deflectometer[1] (MDD) deflection and permanent deformation results. The HVS-sections are instrumented with a stack of Linear Variable Displacement Transducers (LVDTs) normally fixed at the layer interfaces and anchored at a depth of 2,5 to 3m. Resilient deflections under a moving wheel load and the permanent downward displacement of the MDD-modules with increasing wheel load applications may be recorded during the test.

Table 3: Progress of HVS-testing at the time of writing this paper

Base type and thickness	Tests programme		
	Dry test done at equilibrium moisture content		Wet test
	200 000 repetitions @ 40 kN dual wheel load	200 000 repetitions @ 70 kN dual wheel load	Test @ 70 kN to failure
150 mm ASH	completed	completed	completed
100 mm ASH	completed	completed	completed
100 mm ETB ASH	completed	completed	completed
100 mm ETB	completed	completed	completed
100 mm crushed stone	completed	completed	ongoing

The test program consisted of 200 000 load repetitions of a 40 kN dual wheel load after which the load was increased to 70 kN for another 200 000 repetitions. After the first 400 000 load repetitions 87 mm diameter holes were drilled through the asphalt surfacing and base layer next to one half of the test sections and filled with water. The water level was kept at a constant height in these holes and water therefore seeped into the base of the test section.

Surface rut Figure 3 shows the surface rut data obtained from the laser profilometer measurements for the HVS-tests listed in Table 3.

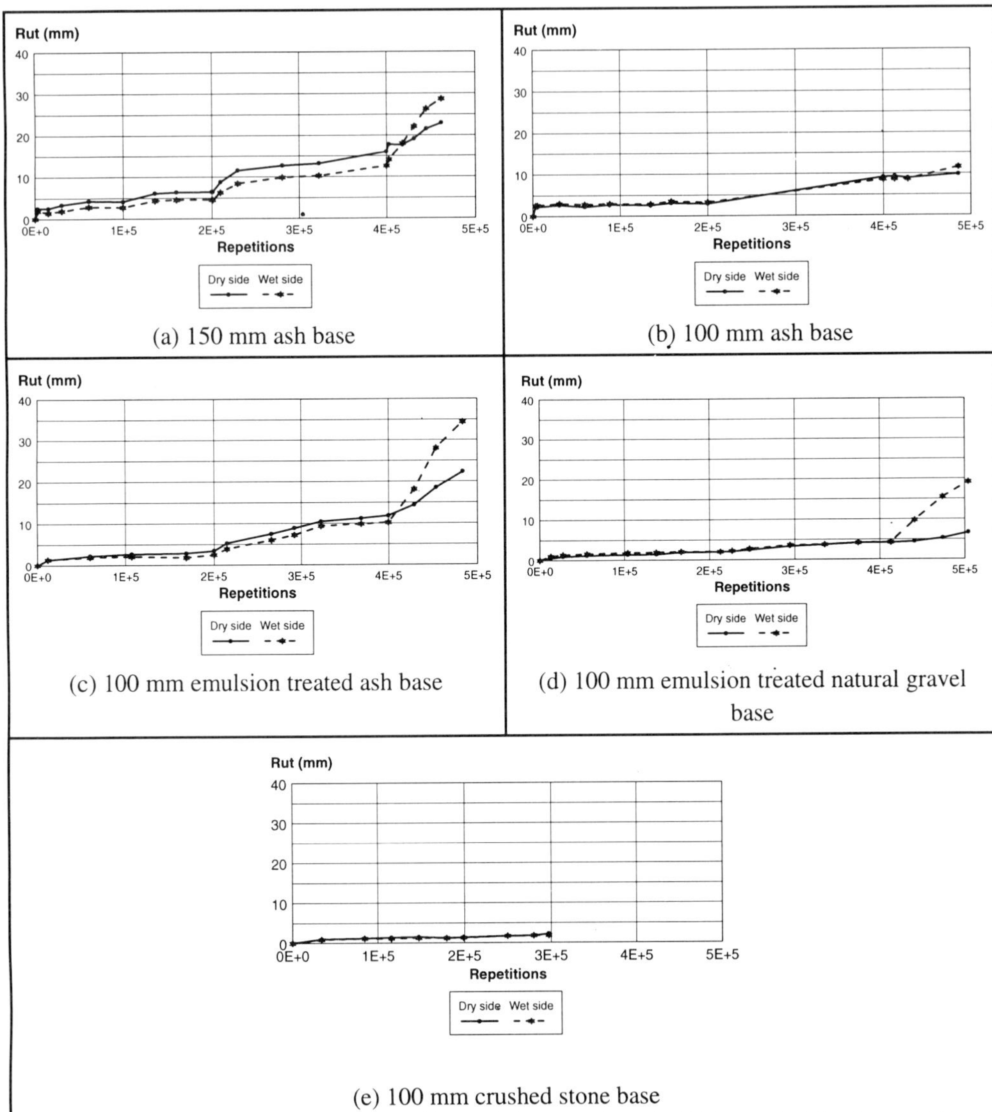

(a) 150 mm ash base

(b) 100 mm ash base

(c) 100 mm emulsion treated ash base

(d) 100 mm emulsion treated natural gravel base

(e) 100 mm crushed stone base

FIGURE 3: SURFACE RUT FOR THE HVS-TESTS LISTED IN TABLE 3

The average rut is shown for the two halves of the HVS-section. The surface rut generally increased after 200 000 load repetitions when the test load was increased from 40 to 70 kN and increased again after 400 000 repetitions when water was introduced into the base layer on one half of the test section. The rutting on the dry side of the test sections is generally lower than on the wet side, although some increase in rut rate did occur even on the dry side during the wet test. This was caused by water being pumped under the asphalt surfacing to the dry side of the test section.

Surface deflection Figure 4 contains the RSD deflection results for the HVS-test sections listed in Table 3. The average deflection for the dry and wet halves of the test sections are shown.

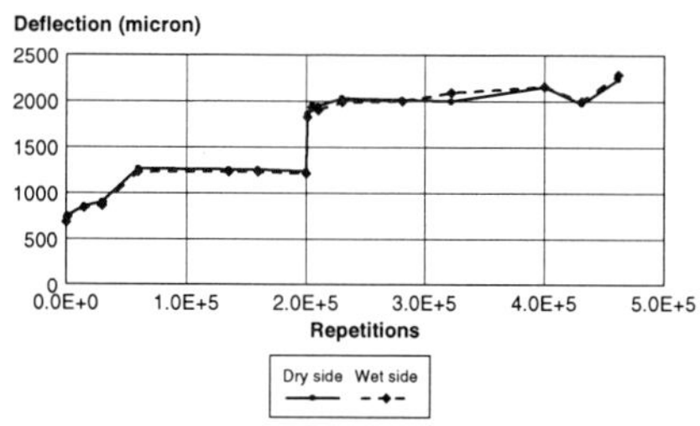

(a) 150 mm ash base

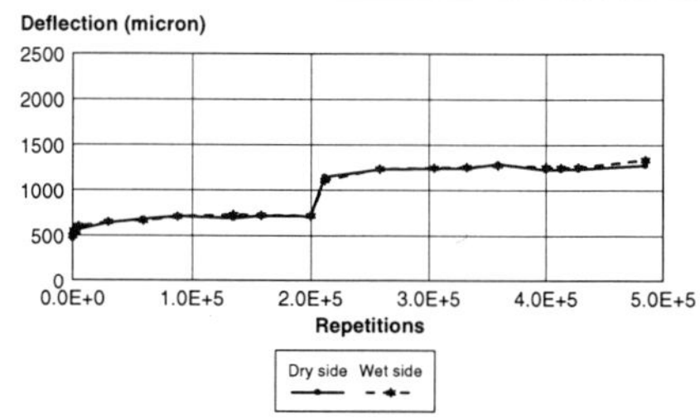

(b) 100 mm ash base

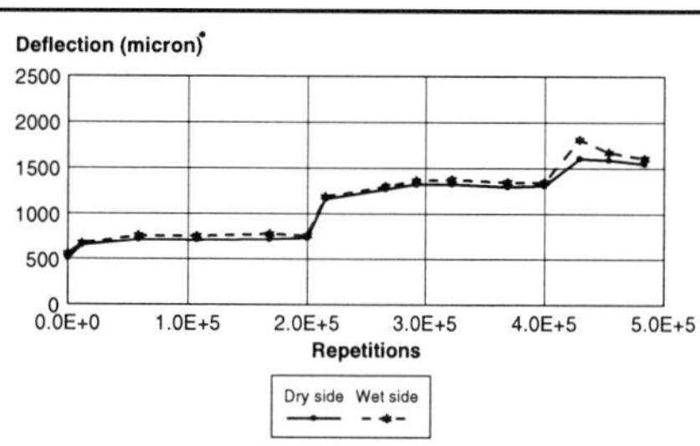

(c) 100 mm emulsion treated ash base

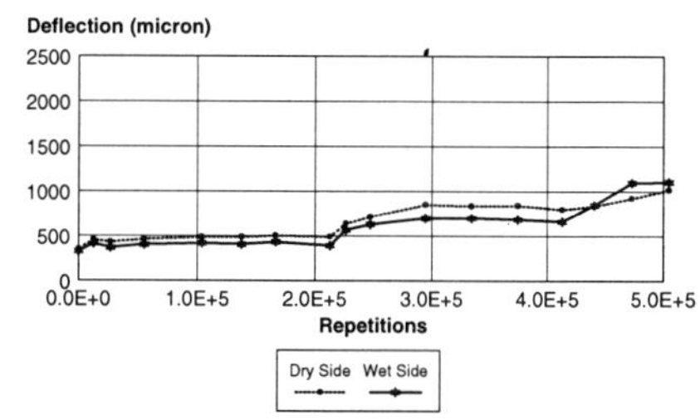

(d) 100 mm emulsion treated natural gravel base

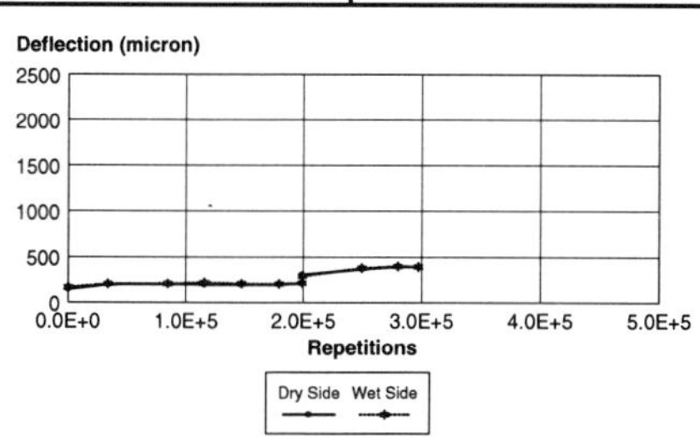

(e) 100 mm crushed stone base

FIGURE 4: SURFACE DEFLECTION FOR THE HVS-TESTS LISTED IN TABLE 3

CLASSIFICATION OF DESIGNS

Figure 5 illustrates the permanent MDD-displacement results for the duration of the dry test on the section with the 150 mm ash base layer at three positions where MDDs were installed. Similar results were obtained for the section with the 100 mm ash base and the 100 mm emulsion treated ash base layer. The function listed in Equation 1[2] was fitted to the surface MDD-displacement data for these three test sections for the duration of the 40 kN dual wheel load test (200 000 repetitions).

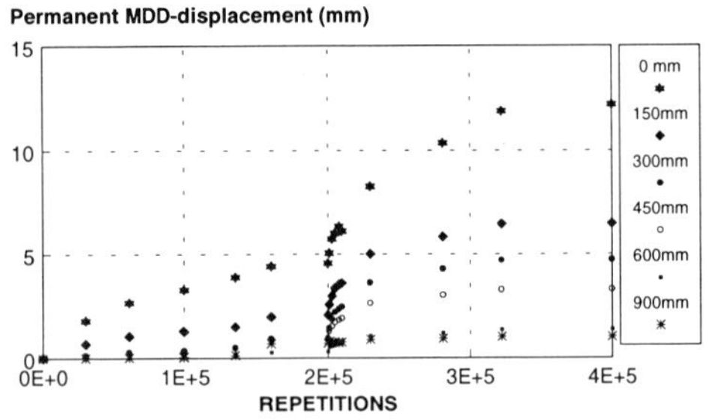

(a) 150 mm ash base, MDD 4

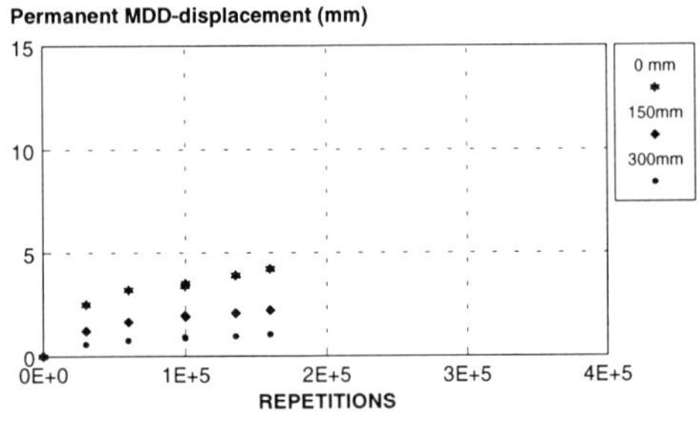

(b) 150 mm ash base, MDD 8

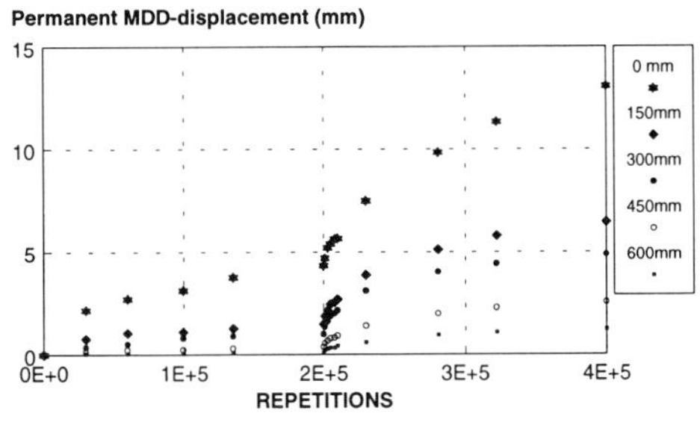

(c) 150 mm ash base, MDD 12

FIGURE 5: PERMANENT MDD-DISPLACEMENT RESULTS FOR THE SECTION WITH THE 150 mm ASH BASE

$$MDD\text{-}displacement = (mN + a)\left(1 - e^{-bN}\right) \quad \text{.......... Eq 1}$$

where N = Number of load repetitions
a,m,b = regression coefficients
e = base of the natural logarithm

The regression coefficient "m" represents the rate of increase in the permanent MDD-displacement during the linear phase, after the initial bedding-in phase. For the surface MDD-modules "m" is therefore equal to the rate of rutting after the bedding-in phase. Table 4 gives a summary of the linear rut rates for three of the test sections.

Using Equation 1, the number of load repetitions to 20 mm surface rut could be predicted for the 150 mm ash base section, the 100 mm ash base section and the 100 mm emulsion treated ash base section. These results are summarised in Table 5.

Table 4: Summary of the linear rut rate (mm per million load repetitions) results for three test sections

Test section	MDD Nr 4	MDD Nr 8	MDD Nr 12
150 mm ash base	11,9	10,5	11,6
100 mm ash base		4,2	3,2
100 mm emulsion treated ash		8,8	6,2

Table 5: Summary of the number of load repetitions to 20 mm surface rut for three test sections

Test section	MDD Nr 4	MDD Nr 8	MDD Nr 12
150 mm ash base	$1,49 \times 10^6$	$3,06 \times 10^6$	$1,55 \times 10^6$
100 mm ash base		$4,48 \times 10^6$	$5,88 \times 10^6$
100 mm emulsion treated ash		$2,17 \times 10^6$	$3,09 \times 10^6$

Based on the information in Tables 4 and 5 combined with the profilometer rut data from the previous section, the following statements can be made regarding the sections with the 150 mm ash base, the 100 mm ash base and the 100 mm emulsion treated ash base:

- The 100 mm ash base section performed best with and estimated life of about 4,5 to 5,5 million standard (40 kN) axle loads for a terminal condition of 20 mm rut.

- The 100 mm emulsion treated ash base section was second best with an estimated life of about 2 to 3 million standard (40 kN) axle loads for a terminal condition 20 mm rut.

- The 150 mm ash base section performed worst with and estimated life of about 1,5 million standard (40 kN) axle loads for a terminal condition 20 mm rut.

4. RESULTS FROM LOADING TESTS

The test sections were constructed in October 1991 and the first test series was carried out in December 1991. This has been followed by several more test series, most of which were in the period till November 1994 (1992: May, June, September and December, 1993: April and August, 1994: April and November). The general procedure of the test series consisted of ordinary FWD tests for determination of the deflection basin, FWD tests over the instruments to determine the pavement response (stresses and strains) under FWD loading, and measurement of stresses and strains under a dual wheel truck loading.

4.1 Methods for back-analysis of FWD tests

A number of difficulties were encountered in analysing the deflection basins from the FWD tests. Two of the pavements are nominally 4-layer systems, with the asphalt layer being thicker than the granular base layer. Section III is nominally a 3-layer system but could only be analysed as a 4-layer system, with the granular base being divided into two layers, each 300 mm thick. For all of the pavements however the subgrade is layered. In the top 1.5-3 m of the moraine subgrade the original lime content has been partly washed out, resulting in a softer material than further down where the lime content is preserved.

Several different back-analysis methods for determination of Young's modulus were attempted:

- **a linear elastic analysis** of a 4- or 5-layer system (divided subgrade) using the Waterways Experiment Station's program WES5 for the forward analysis,
- **non-linear elastic analysis** of a 4-layer system (with the subgrade treated as non-linear) using Odemark-Boussinesq's method (also known as the Method of Equivalent Thickness', MET), a modified version of WES5 (incorporating a non-linear subgrade), and a Finite Element Program for the forward analysis, and
- **a simplified dynamic analysis** based on Odemark-Boussinesq, but not including inertial effects.

All these methods are approximate in the sense that the assumptions on which the forward analysis routines are based are idealisation of reality. None of the methods consider the particulate nature of the pavement materials, nor the plastic, viscous and viscoelastic deformations related to this particulate nature.

The "information" contained in the deflection basins is not sufficient to determine the five unknowns (moduli and/or non-linearity). Several combinations of moduli and non-linear parameters may produce theoretical deflection basins that match the measured basins. None the less some trends appear from the analysis. One is that the modulus of the subgrade remained fairly constant.

The regression coefficient "m" represents the rate of increase in the permanent MDD-displacement during the linear phase, after the initial bedding-in phase. For the surface MDD-modules "m" is therefore equal to the rate of rutting after the bedding-in phase. Table 4 gives a summary of the linear rut rates for three of the test sections.

Using Equation 1, the number of load repetitions to 20 mm surface rut could be predicted for the 150 mm ash base section, the 100 mm ash base section and the 100 mm emulsion treated ash base section. These results are summarised in Table 5.

Table 4: Summary of the linear rut rate (mm per million load repetitions) results for three test sections

Test section	MDD Nr 4	MDD Nr 8	MDD Nr 12
150 mm ash base	11,9	10,5	11,6
100 mm ash base		4,2	3,2
100 mm emulsion treated ash		8,8	6,2

Table 5: Summary of the number of load repetitions to 20 mm surface rut for three test sections

Test section	MDD Nr 4	MDD Nr 8	MDD Nr 12
150 mm ash base	$1{,}49 \times 10^6$	$3{,}06 \times 10^6$	$1{,}55 \times 10^6$
100 mm ash base		$4{,}48 \times 10^6$	$5{,}88 \times 10^6$
100 mm emulsion treated ash		$2{,}17 \times 10^6$	$3{,}09 \times 10^6$

Based on the information in Tables 4 and 5 combined with the profilometer rut data from the previous section, the following statements can be made regarding the sections with the 150 mm ash base, the 100 mm ash base and the 100 mm emulsion treated ash base:

- The 100 mm ash base section performed best with and estimated life of about 4,5 to 5,5 million standard (40 kN) axle loads for a terminal condition of 20 mm rut.
- The 100 mm emulsion treated ash base section was second best with an estimated life of about 2 to 3 million standard (40 kN) axle loads for a terminal condition 20 mm rut.
- The 150 mm ash base section performed worst with and estimated life of about 1,5 million standard (40 kN) axle loads for a terminal condition 20 mm rut.

The performance of these sections under HVS-testing is believed to be closely related to the density of the base layers. The 100 mm ash base was compacted to 106% mod. AASHTO compaction, the 100 mm emulsion treated ash base to 105% mod. AASHTO compaction and the 150 mm ash base 102% mod. AASHTO compaction.

These predictions are, however, only valid for the moisture conditions experienced during the dry test. The moisture conditions during the dry test are believed to be representative of the equilibrium pavement moisture conditions for the region where the tests were conducted.

CONCLUSIONS

Using the Heavy Vehicle Simulator (HVS) it was possible to assess the performance of several pavement structures constructed labour intensively. The duration of each test on a particular pavement design lasted about six weeks. During this period is was possible to collect enough data to make predictions about:

- The expected behaviour and estimated life of these pavements under dry conditions,
- the expected behaviour of these pavements under wet conditions and
- the load sensitivity of these designs although this point is not addressed in this paper.

The compaction level of the base layer on three of the pavements investigated had a major impact on the rutting behaviour of these pavements and ultimately the estimated life of these pavements.

REFERENCES

1. De Beer M, Horak E and Visser A T, *The Multi-Depth Deflectometer (MDD) System for Determining the Effective Elastic Moduli of Pavement Layers.* Nondestructive Testing of Pavements and Back-calculation of Moduli, ASTM STP 1026, A.J. Bush III and G. Y. Baladi, Eds., ASTM, Philadelphia, 1989.

2. Wolff H. *Elasto-Plastic Behaviour of Granular Pavement Layers in South Africa.* Ph D thesis, University of Pretoria, Pretoria, South Africa, 1992.

Fifth International Conference on the
BEARING CAPASITY OF
ROADS AND AIRFIELDS

FULL-SCALE TESTING OF PAVEMENT RESPONSE

by

PER ULLIDTZ
Associate professor Dr. Techn.
Technical University of Denmark, DK-2800 Lyngby, Denmark
Phone: +45 45 25 15 18, Fax: +45 45 93 64 12
E-mail: pullidtz@ivtb.dtu.dk

PETER EKDAHL
MSc in Civil Engineering
Lund University, Box 118 S-221 00 Lund, Sweden
Phone: +46 46 22 29 893, Fax: +46 46 12 32 72
E-mail: peter.ekdahl@tft.lth.se

ABSTRACT

Three pavement sections in southern Sweden were instrumented in late 1991. Each section have instruments for measuring the strain at the bottom of the asphalt layer, vertical stress and strains in the subgrade and temperatures at different depths. The purpose was to evaluate different theoretical methods for determining stresses, strains and deflections in pavement structures.

Falling Weight Deflectometer (FWD) tests were done on the test sections, and stresses and strains were measured both under FWD loading and under a rolling wheel load. Different back-analysis procedures were used to derive layer moduli from FWD testing. The forward calculated stresses and strains were then compared to the measured values. The tests were done under different climatic conditions, the latest in November 1996, where almost all of the gauges were still functioning.

Of the different methods used for back-analysing FWD data and for calculating the stresses and the strains, the simple method based on Odemark's transformations and Boussinesq's equations appeared to be as good as, or better than, the more sophisticated methods using layered elastic theory or the Finite Element Method.

KEYWORDS

Asphalt pavements , pavement response, instrumented road sections, FWD tests

4. RESULTS FROM LOADING TESTS

The test sections were constructed in October 1991 and the first test series was carried out in December 1991. This has been followed by several more test series, most of which were in the period till November 1994 (1992: May, June, September and December, 1993: April and August, 1994: April and November). The general procedure of the test series consisted of ordinary FWD tests for determination of the deflection basin, FWD tests over the instruments to determine the pavement response (stresses and strains) under FWD loading, and measurement of stresses and strains under a dual wheel truck loading.

4.1 Methods for back-analysis of FWD tests

A number of difficulties were encountered in analysing the deflection basins from the FWD tests. Two of the pavements are nominally 4-layer systems, with the asphalt layer being thicker than the granular base layer. Section III is nominally a 3-layer system but could only be analysed as a 4-layer system, with the granular base being divided into two layers, each 300 mm thick. For all of the pavements however the subgrade is layered. In the top 1.5-3 m of the moraine subgrade the original lime content has been partly washed out, resulting in a softer material than further down where the lime content is preserved.

Several different back-analysis methods for determination of Young's modulus were attempted:

- **a linear elastic analysis** of a 4- or 5-layer system (divided subgrade) using the Waterways Experiment Station's program WES5 for the forward analysis,
- **non-linear elastic analysis** of a 4-layer system (with the subgrade treated as non-linear) using Odemark-Boussinesq's method (also known as the Method of Equivalent Thickness', MET), a modified version of WES5 (incorporating a non-linear subgrade), and a Finite Element Program for the forward analysis, and
- **a simplified dynamic analysis** based on Odemark-Boussinesq, but not including inertial effects.

All these methods are approximate in the sense that the assumptions on which the forward analysis routines are based are idealisation of reality. None of the methods consider the particulate nature of the pavement materials, nor the plastic, viscous and viscoelastic deformations related to this particulate nature.

The "information" contained in the deflection basins is not sufficient to determine the five unknowns (moduli and/or non-linearity). Several combinations of moduli and non-linear parameters may produce theoretical deflection basins that match the measured basins. None the less some trends appear from the analysis. One is that the modulus of the subgrade remained fairly constant.

4.2 Subgrade moduli

The subgrade modulus has been determined in five different ways. **1/** During the installation the moduli varied between 64 and 114 MPa, calculated from FWD-measurements made on the subgrade. **2-3/** Based on measured stresses and strains in the subgrade under both rolling wheel load and FWD-loading the moduli were calculated (σ/ε). In general the FWD-loading gave slightly higher moduli than those under a rolling wheel load. **4-5/** Both linear and non-linear analysis were done based on deflection basins from the FWD-tests. The linear analysis gave moduli almost twice as big as those obtained from the non-linear analysis. Dynamic linear analysis gave quite similar results as static linear analysis.

Table 2 Subgrade moduli (MPa) from FWD test and measured stresses and strains

Section	FWD Installation	Wheel σ/ε	FWD σ/ε	FWD linear	FWD non-linear
I	114	105-125	120-140	150-180	80-110
II	64	50-80	50-80	130-170	50-80
III	77	100-120	150-190	110-140	70-90

4.3 Layer moduli

In all back-calculations the modulus of the asphalt layer was found to be temperature dependent. If written in the format:

$$E_t = E_{ref} \times \exp(a \times \left[t - t_{ref}\right]) \qquad (1)$$

where

E_t is the modulus at temperature t
a is a constant and
t_{ref} is a reference temperature (in this case 10 °C)

the following values are obtained

Table 3 Values from temperature calculations according to equation 1

Test section	a	E_{ref}, MPa	R^2
I	-.08079	5678	0.915
II	-.08823	5884	0.950
III	-.07875	4545	0.938

No systematic changes in the **asphalt modulus** over time have been observed during the past five years. Neither would any structural deterioration have been expected with the rather low amount of heavy traffic. More surprisingly a similar variation- with temperature- of the modulus of the base layer (macadam), as for the asphalt, was observed in some of the back-calculations. This was most pronounced for test section I using the linear elastic back-analysis, i.e. an increasing asphalt modulus led to an increasing modulus of the macadam layer. Whether this variation in modulus is a real variation or due to the problems of determining too many unknowns in the back-calculation, is not certain.

The **modulus of the macadam base layer** appeared to be increasing over time. For test section I from an initial value of about 500 MPa to a final value of 650 MPa, for test section II from 250 MPa to 700 MPa and for test section III from 200 MPa to 350 MPa (for the upper 300 mm of the macadam layer).

The modulus of the subbase layer, and of the lower 300 mm of the macadam layer for test section III, was very dependent upon the method chosen for back-calculation. Approximate values using a 5-layer linear elastic analysis and a 4-layer system with a non-linear subgrade are given below:

Table 4 Modulus of the subbase layer with 5 layer linear and 4 layer non-linear systems

Test section	5-layer linear	4-layer non-linear
I	80-120 MPa	110-140 MPa
II	60-100 MPa	100-140 MPa
III	70-120 MPa	100-140 MPa

The moduli are rather lower than what would normally have been expected for these types of materials, especially for the linear elastic analysis. It appears that the modulus of the subbase is underestimated and that of the subgrade overestimated with the linear elastic back-analysis. The subbase moduli remained relatively constant over the test period.

4.4 Stresses and strains

Vertical stresses at the top of the subgrade, longitudinal strains at the bottom of the asphalt layer and vertical strains at the top of the subgrade were measured under FWD loading and under a rolling wheel load. By testing at short length intervals in the line of the gauges an influence line for the FWD load could be obtained. A close fit to the influence line could be obtained using Boussinesq's equations for a point load on a semi-infinite space, by selecting the appropriate depth for stress, and depth and modulus for strain. Some examples from test section II, November 1994, are shown below:

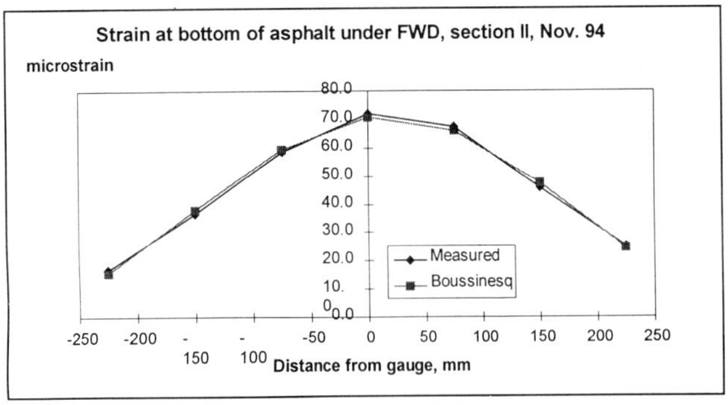

Fig. 2 Influence line for one asphalt strain gauge.

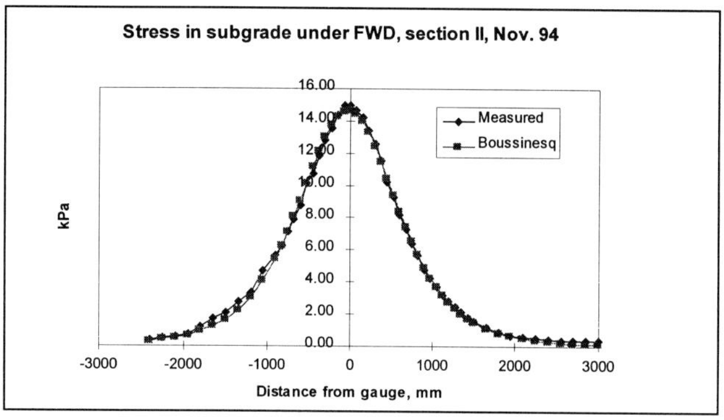

Fig. 3 Influence line for one soil pressure cell.

If the stresses are axially symmetric then the total force on the plane of the pressure cell can be calculated and compared to the known input force from the FWD. The pressure on the plane of the pressure gauge will vary from point to point, and the pressure at the position of the gauge is not necessarily the average value on the plane. With this procedure the measured stress may be adjusted to the average value. The adjustment factors range from 0.69 to 1.38, with a mean value of 0.92 for the nine pressure cells.

For test section II, November 1994, an excellent agreement with Boussinesq's equations for a point load on a semi-infinite space was found using the following fictive depths (Odemark's method) and moduli (Poisson's ratio equal to 0.35):

Asphalt strain:

Gauge	**1**	**2**	**3**	**4**
Depth	496 mm	474 mm	448 mm	471 mm
Modulus	327 MPa	402 MPa	410 MPa	356 MPa

The modulus should be compared to the macadam base modulus given in Table 5 below. The best agreement is obtained with the MET. The equivalent thickness [Odemarks's method, $h_e=h_1(E_1/E_2)^{1/3}$] of the asphalt is 420 mm.

Stress in subgrade:

Gauge	**1**	**2**	**3**
Depth	1249 mm	1233 mm	1201 mm

The equivalent depth to the stress gauges is about 1200 mm.

Strain in subgrade:

Gauge	**1**	**2**
Depth	1146 mm	1143 mm
Modulus	76 MPa	101 MPa

Backcalculation using the Finite Element method (FE), the Method of Equivalent Thickness (MET, using ELMOD4) and Waterways Experiment Station's layered linear elastic program (WES) gave the following moduli:.

Table 5 Calculated layer moduli using different methods

Layer	FE	MET	WES
Asphalt	7312 MPa	6996 MPa	8028 MPa
Macadam	826 MPa	386 MPa	824 MPa
Subbase	95 MPa	205 MPa	58 MPa
Subgrade	69 MPa	54 MPa	235 MPa

With FE and MET the subgrade was treated as a non-linear elastic material. The modulus given corresponds to the stress level at the top of the subgrade during the FWD test. The "surface" or mean modulus would be about 100 MPa.

In the following figures the measured stresses and strains have been compared to theoretical values, using the same routines as for the backcalculation. All values have been adjusted to a contact stress of 500 kPa.

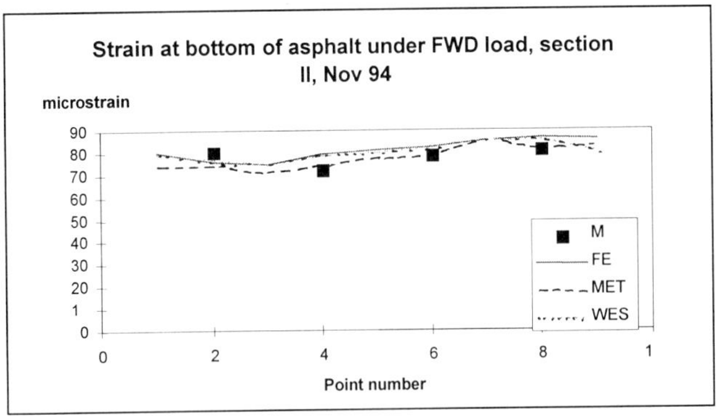

Fig. 4 Measured and calculated asphalt strains.

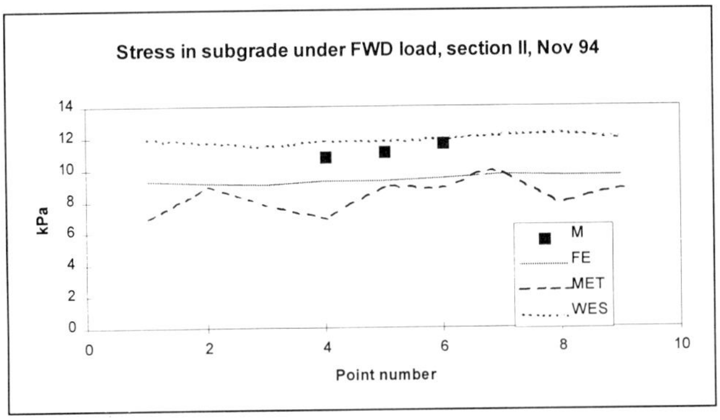

Fig. 5 Measured and calculated subgrade stress.

The strain at the bottom of the asphalt layer is predicted reasonably well by all three methods. For the stress in the subgrade WES provides the best prediction, whereas it underestimates the strain in the subgrade by a factor of 2 to 3.

These results show that the linear elastic theory of elasticity cannot be used for the evaluation of this pavement structure. The fact that WES gives the best prediction of the subgrade stress is of little consolation, because the stress alone is not sufficient to predict future damage. The modulus of the subgrade is also needed, and this is overestimated by a factor of 2 to 3 by WES.

The two non-linear methods give a better prediction of the subgrade strain (and modulus), but they both underestimate the subgrade stress. A possible explanation for this could be the "stress concentration" in granular materials noted earlier by Frölich (1934) (4), Ververka (1973) (5) and Ullidtz et al. (1996) (6).

The average ratios of calculated and measured stresses and strains, for all three test sections and the first seven test series, are given in the table below (WES was done for a 5-layer system):

Table 6 Ratios of calculated/measured stresses and strains

	FE	MET	WES
Asphalt strain	1.10	1.03	1.18
Subgrade stress	0.73	1.04	0.82
Subgrade strain	0.99	1.04	0.54

On an average MET gives the best agreement, but for the asphalt strain, the coefficient of variation is larger with MET than for the other two methods (40% versus approximately 20% for all other values).

5. CONCLUSIONS

Of the different methods used for back-analysing FWD data and for calculating the stresses and the strains, the simple method based on Odemark's transformations and Boussinesq's equations appeared to be as good as, or better than, the more sophisticated methods using layered elastic theory or the Finite Element method. Apparently the differences between the theoretical assumptions of continuum mechanics and the real conditions of particulate pavement materials are so important that no improvement is obtained through the use of more sophisticated methods. Until methods based on more appropriate assumptions, like the Distinct Element Methods, become available for pavement analysis, the simple methods are to be preferred.

ACKNOWLEDGEMENTS

The test sections are financed by NCC, Malmoe municipality, BFR and SBUF. The authors are indebted to Finn Ole and Jan Sjoelin (Technical University of Denmark) who installed most of the instruments and carried out the measurements.

REFERENCES:

1 Zhang, J, Ebersohn, W. & Selig, E.T. "Evaluation of Earth Pressure Cells For Pavement Subgrade Performance Study", Geotechnical Report No. CRR95-427F, Cold Regions Research and Engineering Laboratory, U.S. Army Corps of Engineering, Hanover, NH, 1995

2 Askegaard, V. "Pressure Cell Calibration Tests", Department of Structural Engineering and Materials, The Technical University of Denmark, Series S, No. 9508, 1996

3 Wahlman, T. Stahle, F. "Öppna överbyggnader och dränerande bärlager. Provväg Borrebackevägen Malmö 1991-1994". SBUF project 0115, BFR project 901035-0

4 Frölich, O.K. "Drukverteilung im Baugrunde", Springer-Verlag, Wien 1934.

5 Veverka, V. "Modules, contraintes et déformations des massifs et couches granulaires", Rapport de Recherche No. 162/VV/1973, Centre de Recherches Routières, Bruxelles, 1973

6 Ullidtz, P, Askegaard, V. & Sjølin, F.O. "Normal Stresses in a Granular Material Under Falling Weight Deflectometer Loading", Transportation Research Record 1540, Washington, D.C. 1996

Fifth International Conference on the
BEARING CAPASITY OF
ROADS AND AIRFIELDS

STUDY IN REFLECTIVE CRACKING ON COMPOSITE PAVEMENT

SATOSHI TANIGUCHI[1] AND TAKUYA IKEDA[2]

[1]Researcher, Pavement Division, Public Works Research Institute,
[2]Head, Pavement Division, Public Works Research Institute,
Ministry of Construction, Government of Japan
305-0804, Tsukuba, Japan

ABSTRACT

This paper presents results of study on preventing reflective cracking by crack prevention sheets and crack inducing joints. In 1990, test pavement was constructed at the Pavement Test Field of the Public Works Research Institute. The test pavement consisted of a 5cm asphalt concrete layer on 15cm cement concrete slabs, which had 11 joints with 6cm depth at 5m intervals. Crack prevention sheets between the asphalt layer and the concrete slabs were installed on 4 joints. Crack inducing joints were cut in the asphalt layer with 1.5cm deep, just above 3 joints of the concrete slabs. No special measures was taken for other 4 joints. The test pavement received 390,000 times of loading for six years by Loading Vehicles with 98kN rear axle load. In order to study effects of sheets and inducing joints, cracking on the pavement surface was surveyed at every 20,000 times of loading. In 1996, detailed survey was conducted including core boring. Reflective cracking on the joints with no measures grew immediately after its occurrence. Cracking on the joints with crack prevention sheets grew slowly. However, the crack prevention sheets did not comletely prevent occurrence of reflective cracking. Since the shape of cracking is jagged and branched, the cracking lead to further deflects such as spalling and faulting. On the contrary crack inducing joints properly lead reflective cracking from joints on cement concrete slabs. Crack inducing joints did not cause spalling and other fatal deterioration. Therefore, it is effective to construct crack inducing joints where reflective cracking is expected.

Key Words: composite pavement, reflective cracking, crack prevention sheets, and crack inducing joints

INTRODUCTION

Composite pavement, consisting of an asphlat layer on cement concrete slabs, is a durable pavement structure with good riding comfortability in comparison with ordinary asphalt pavement and cement concrete pavement. However, except when the asphalt layer is constructed on continuously reinforced cement concrete slabs or semi-flexible pavement, the cement concrete slabs have joints at constant intervals in order to absorp expansion and contraction by change of temperature. Therefore, reflective cracking in the asphalt layers is a crucial problem for composite pavement structures.

In order to study the effect of crack prevention sheets and crack inducing joints, test composite pavement was constructed at the Pavement Test Field in the Public Works Research Institute. The test section received repeated loading by real-sized large vehicles. This paper presents the result of the survey for six years.

STUDY PROCEDURE

Pavement structure and joint installation Figure 1 shows the pavement structure of the test pavement. In 1990, the test pavement was constructed at a part of a circular test track (628 meter long in total), consisting of a 5cm asphalt layer on 15cm cement concrete slabs. The cement concrete slabs were roller compacted cement concrete pavement, which had 11

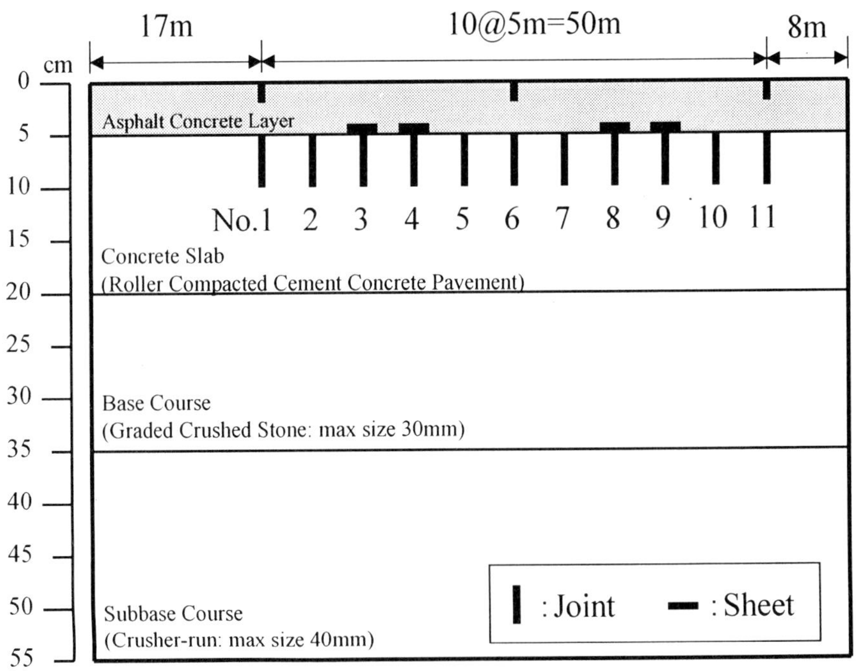

Figure 1 Pavement Structure

contraction joints with 4mm width and 6cm depth at 5m intervals. Crack prevention sheets were installed between the asphalt layer and the concrete slabs to cover 4 joints on the slabs. Crack inducing joints were cut in the asphalt layer with 4mm width and 1.5cm depth, just above 3 joints of the slabs. No measures were taken for other 4 joints.

Loading test by 'Loading Vehicles' The test pavement received 390,000 times of loading by 'Loading Vehicles' (Photograph 1) with 44.1kN front axle load and 98.0kN rear axle load. Loading by the heavy vehicles began in 1990 and ended in 1996. Table 1 shows the loading times of each year. At every 20,000 times of loading, location of surface cracking on the test pavement was sketched. Through this survey, occurrence and progress of each cracking were identified.

Photograph 1 Loading Test by the 'Loading Vehicle'

Table 1 Loading times of each year

Year	Loading Times
1990	70,000
1991	70,000
1992	80,000
1994	80,000
1995	80,000
1996	10,000
TOTAL	390,000

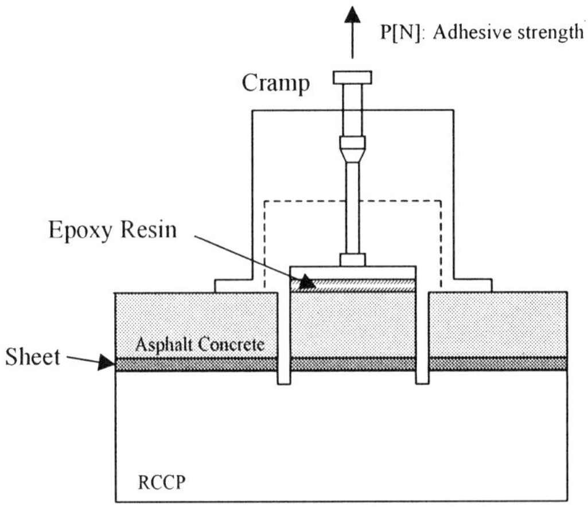

Figure 2 Adhesive Strength Tester

Core boring After loading by Loading Vehicles finished, detailed crack survey and core boring were conducted. Cores were cut out in 15cm diameter and 20cm depth to observe the cracking occurrence in the asphalt layer and the concrete slabs.

Adhesive strength test Adhesive strength between the asphalt layer, the crack prevention sheets and the cement concrete slabs was measured using adhesive strength tester shown on Figure 2.

RESULT OF THE SURVEY

Crack occurrences Figure 3 shows crack occurrence on the test pavement at 40,000 times of loading in 1991. This figure indicates that a crack expanded continuously on No.10 joint with no special measures. On the contrary, a crack appeared discontinuously on No.9 joint with a crack prevention sheet and no crack appeared at No.11 crack inducing joint.

Figure 4 shows cracking after 390,000 time of loading and location of core boring. This figure indicates that cracking finally expanded continuously on entire joints with crack prevention sheets. Besides this, longitudinal cracks appeared on the inner tread. However no cracking appeared at all crack inducing joints.

Figure 5 shows progress of cracking on representing 6 joints. Cracking at joint No.5 with no special measures appeared and expanded entirely between 40,000 and 60,000 times of loading in 1991. Cracking at joint No.4 with a crack prevention sheet appeared between 40,000 and 60,000 times of loading in 1991. However, the cracking needed another 160,000 times of loading before the cracking reached entire width

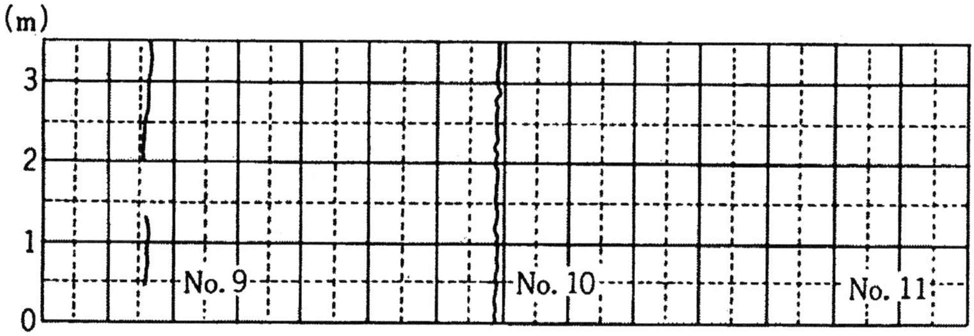

Figure 3 Cracking at 40,000 times of loading in 1991

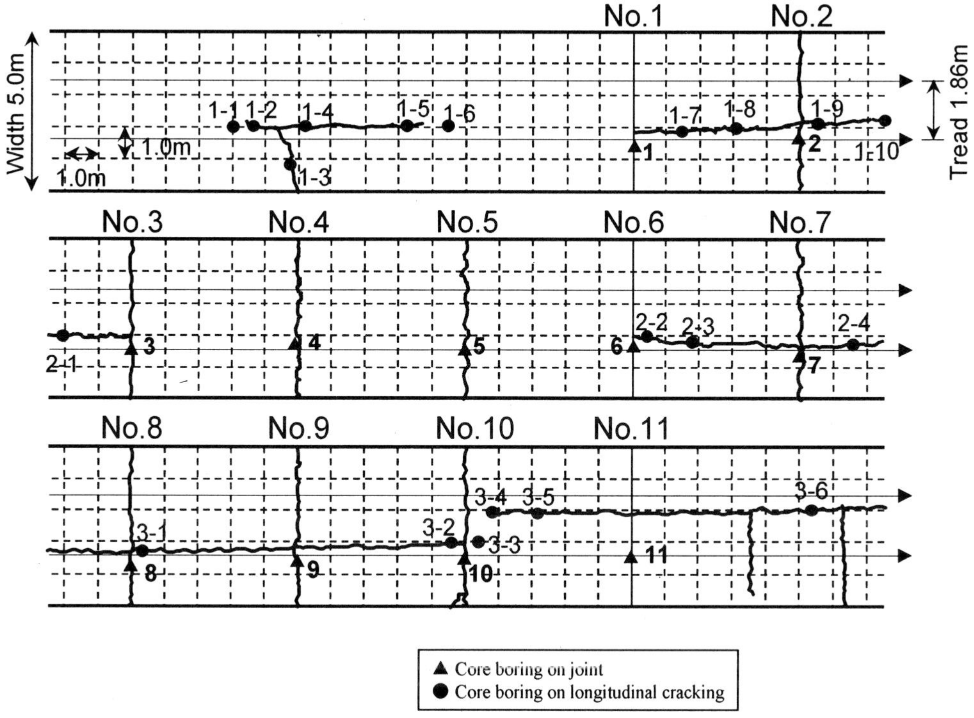

Figure 4 Cracking after entire loading test and location of core boring

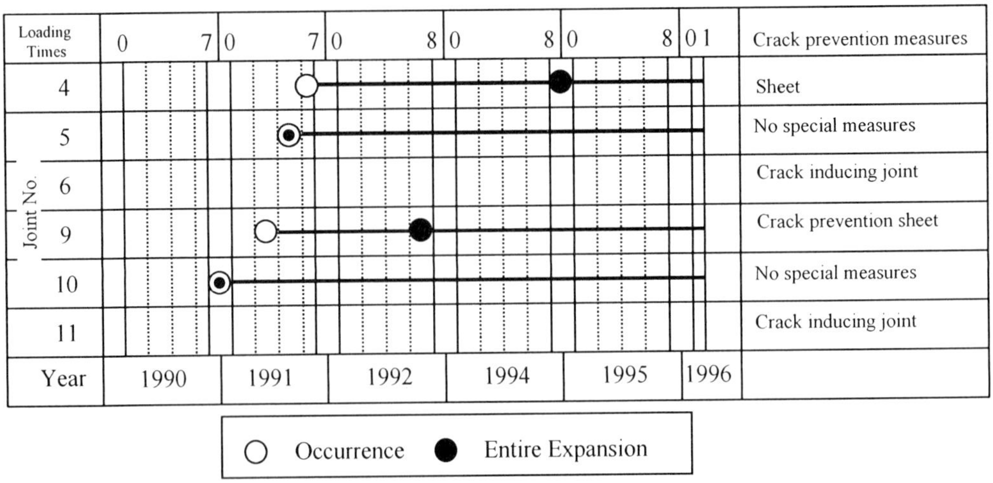

Figure 5 Crack occurrence chart

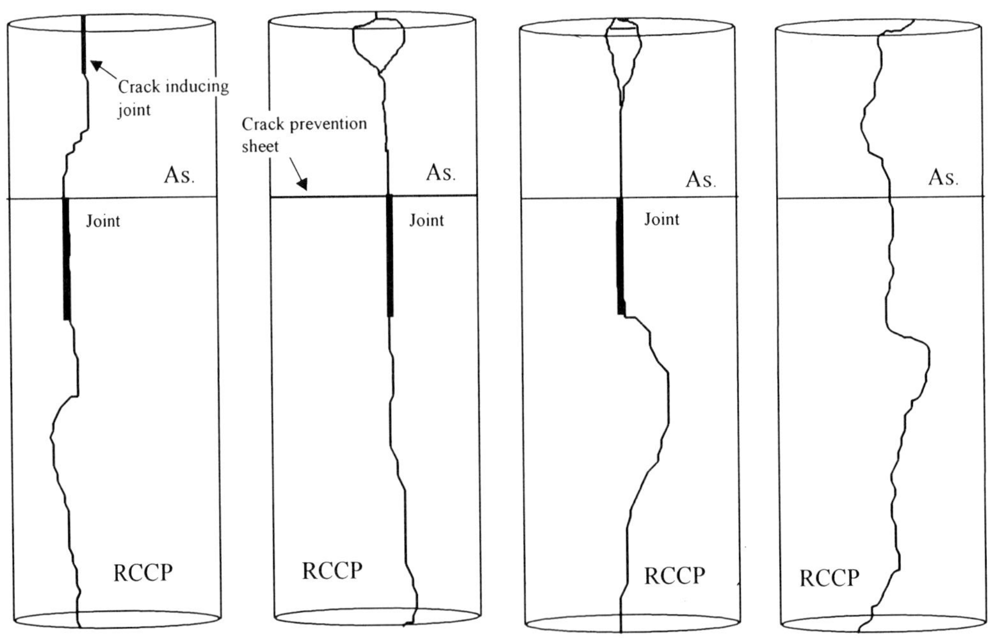

(1) Core No.6 (2) Core No.9 (3) Core No.10 (4) Core No.2-3

Figure 6 Cracking on cores

Photograph 2 Survey on longitudinal cracking

Core Borings Figure 6 shows examples of cores obtained from the test pavement. Core No.6 indicates that cracking on the joint properly connected to the crack inducing joint. However, cracking on cores No.9 and No.10 reached to the surface and branched out. As shown on the core No.2-3, the cause of longitudinal cracking at the inner tread of the surface course was cracking in the cement concrete slabs.

Survey on longitudinal cracks In order to identify the cause of the longitudinal cracking, the asphalt layer was cut partially on longitudinal cracks (Photograph 2). It was confirmed that cracking on the surface course occurred above the longitudinal cracking, which occurred after construction, in cement concrete slabs.

Adhesive strength The test pieces were detached between the sheets and the cement concrete slabs. Adhesive strength between the sheets and the asphalt layer was higher than that between the sheets and concrete slabs.

ANALYSIS

Effect of crack prevention sheet Since cracking on crack prevention sheets expanded slowly, effect of crack prevention sheets was confirmed. However, the sheets did not prevent occurrence of refelective cracking. Cracking on joints with no special measures appeared earlier than cracking on joints with crack prevention sheets. In addition, cracking expanded immediately to the entire width on the joint with no special measures, while cracking expanded slowly on the joint with crack prevention sheets. Furthermore, cracking in cores on joints with sheets appeared on the whole thickness and had branches. Since the shape of cracking was jagged and branched, the cracking caused spalling as well as cracking on joints with no special measures.

Effect of crack inducing joint Since crack inducing joints lead progress of cracking properly, no spalling and faulting was observed on the surface course. Reflective cracking is considered to grow from bottom to surface on cracking or joints of cement concrete slabs. When it grows, cracking traces weak points of the layer. When crack inducing joints are installed, the position is thinner and weaker than other positions. Therefore, cracking connects to the crack inducing joints. The crack inducing joint can be considered as an artificially installed cracking. In case of natural cracking, its shape is jagged and branched, and may cause spalling and faulting. Since the shape of artificail cracking is rigid and sharp, it does not lead spalling and faulting.

When crack inducing joints were installed, workers tried to cut the joints just above the joints in cement concrete slabs. Howerver, construction errors were not inevitable. When cores were observed, crack inducing joints were installed within 5 to 10mm range on joints in cement concrete slabs. This amount of errors does not affect proper crack induction.

CONCLUSION

The results of the test paving and the survey were summarized as follows.

1) Crack prevention sheets delay occurrence and progress of reflective cracking. However, they do not prevent reflective cracking.
2) Crack inducing joints properly lead reflective cracking.
3) Crack inducing joints do not cause spalling and faulting.

In this study, the depth of crack inducing joints was constant at 1.5cm. In order to introduce this technique to actual sites, standard depth and width of the joints should be decided. The range of construction errors is also to be identified. Furthermore, proper treatment such as joint sealing after installation of crack inducing joints should be studied.

ACCELERATED LOAD TESTING OF EXPERIMENTAL TEST SECTIONS TO FAILURE

ROHAN W. PERERA[1], CHUCK A. GEMAYEL[2], and STARR D. KOHN[3]
[1] Senior Engineer, [2] Senior Associate, and [3] Vice President
Soil and Materials Engineers, Inc.
43980 Plymouth Oaks Boulevard
Plymouth, Michigan 48170, U.S.A.

ABSTRACT

The Michigan Department of Transportation (MDOT) constructed a series of flexible pavement test sections for the Specific Pavement Studies experiment (SPS 1) in the Long Term Pavement Performance program. Falling Weight Deflectometer (FWD) tests conducted before the pavement was opened to traffic indicated excessive deflections on a portion of the roadway that encompassed two of these test sections. MDOT decided to reconstruct the pavement within the limits where excessive deflections were noted. Prior to reconstruction, an accelerated load test was performed to fail these test sections. The load test was conducted by driving loaded trucks over the test sections. Rut depths were measured at regular intervals during the period the pavement was subjected to traffic. FWD measurements were used to backcalculate pavement layer moduli. These moduli were used to calculate the strain on top of the subgrade. Existing rutting models were then used to theoretically predict the load applications that each test section could undergo before failure. The predicted load applications were then compared with the actual load applications.
Keywords: FWD Tests, Rutting, Pavement Performance, Backcalculation

INTRODUCTION

The Michigan Department of Transportation (MDOT) constructed a series of flexible pavement test sections for the Specific Pavement Studies experiment (SPS 1) in the Long Term Pavement Performance program. These test sections were constructed at St. John's, Michigan. Falling Weight Deflectometer (FWD) testing conducted prior to opening the pavement to traffic indicated excessive deflections on a portion of the roadway that encompassed two of these test sections. As these test sections were located on a highway that is expected to receive heavy traffic, MDOT decided to reconstruct the pavement within these limits. Prior to reconstruction, an accelerated load test was performed to fail these test sections. The load test was conducted by driving loaded trucks over the test sections. The load test on one section was terminated after the application of 2,300 Equivalent Single Axle Loads (ESALs), at which stage the average rutting in the outer wheel path of the section was 17.4 mm. The load test on the other section was terminated after the application of 6,180 ESALs, at which stage the average rutting on the outer wheel path of the test section was 1.7 mm. Falling Weight Deflectometer tests were conducted on the outer wheel path of these test sections prior to the application of loading. Rut depths were measured at regular intervals during the period the pavement was subjected to traffic. Existing rutting models were then used to theoretically predict the load applications that each test section could undergo before failure. The predicted load applications were then compared with the actual load applications. The approach taken in this analysis was to simulate a real-life situation that can be experienced by an engineer. The engineer has the FWD data in possession, and from this data the engineers has to determine how many load applications a pavement section can undergo before failure. The engineer can then use the results from the analysis to take whatever action is necessary to remedy the situation. This paper examines how close the ESALs to failure predicted by the different models agreed with the results observed in the field.

TEST SECTIONS

The two sections analyzed in the study consisted of an asphalt concrete surface placed over a dense graded aggregate base. Table 1 presents the layer thickness in the two test sections. These test sections were constructed on a clay subgrade.

Table 1 Pavement Layer Thickness of Test Sections

Section Number	Layer Thickness (mm)	
	Asphalt Concrete	Aggregate Base
13	100	200
14	175	300

FIELD TESTING

FWD Tests: Falling Weight Defletometer (FWD) tests were conducted on the outer wheel path of the test sections prior to the application of traffic loading. The tests were conducted at 15 m intervals. The load plate had a radius of 150 mm. Deflections were measured at distances of 0, 200, 300, 450, 600, 900 and 1500 mm from the center of the load plate. Figures 1 and 2 show the deflection profiles along test sections 13 and 14, respectively for a normalized load of 27 kN. Each figure shows the deflection measured at the seven deflection sensors for each test station. Temperature holes were used to measure the pavement

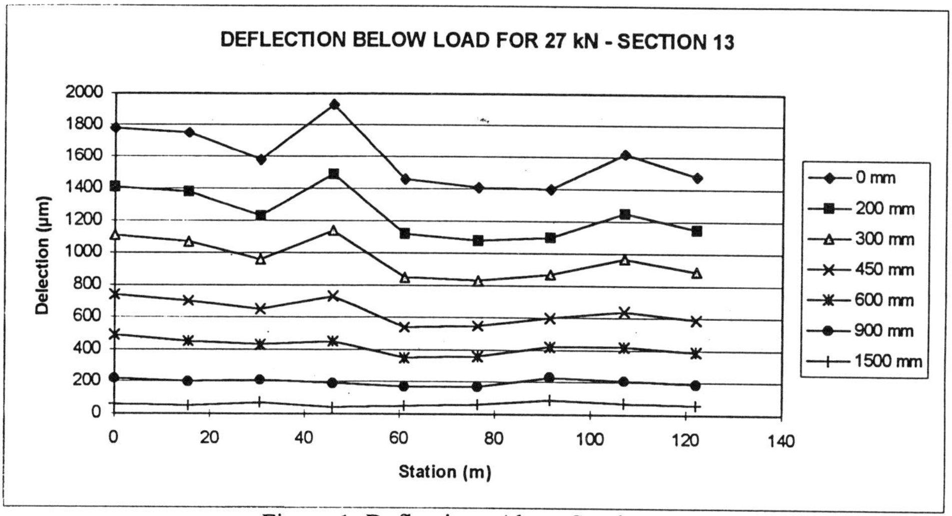

Figure 1 Deflections Along Section 13

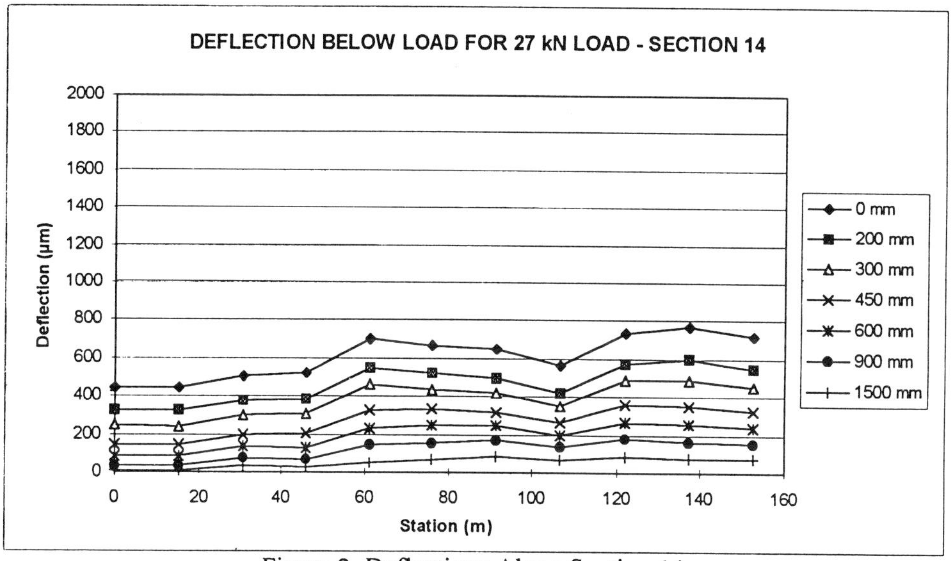

Figure 2 Deflections Along Section 14

temperature at the time of the FWD tests. In section 13, three holes were drilled to depths of 25 mm, 50 mm, and 75 mm. In test section 14, three holes were drilled to depths of 25 mm, 87 mm, and 150 mm. The average temperature of the pavement determined by averaging the temperature obtained from the three holes were 34 °C for both sections.

Traffic Load Applications: Traffic loadings were applied using two types of trucks: single axle and tandem axle. The single axle trucks consisted of two single axles, and comprised 85% of the total traffic volume. These loaded trucks generally weighed 134 kN. The weight carried by the steering axle was 53 kN and the weight carried by the rear axle was 80 kN. The tandem axle trucks consisted of two axles, a steering axle and a tandem axle, and comprised the remaining 15% of the traffic. The loaded tandem axle trucks weighed 196 kN, with the weight carried by the steering axle and the tandem axle being 53 kN and 142 kN, respectively. Traffic loading on test section 13 was terminated after the application of approximately 2,300 ESALS, at which stage the average rut depth of the test section in the outer wheel path was 17.4 mm. During latter stage of load application on section 14, multiple axle trucks were used to apply traffic loading. The loaded weight of these trucks were generally 668 kN and each truck consisted of 11 axles. Traffic loading on section 14 continued until the cumulative ESALs that were applied reached 6,180, at which stage the accelerated load test was terminated. The average rut depth in the outer wheel path at test section 14 at the time the test was terminated was 1.7 mm. The axle weights were used to compute the ESAL applications at the test sites.

Rut Depth Measurements: Rut depth measurements were conducted at the test sections at regular time intervals at 15 m spacing. The rut depths were measured using a strangling with a calibrated wedge. The stringline was stretched across the lane, and then a calibrated wedge was pushed between the wire and the pavement. The wire is connected to an electrical system, and emits a beep when the wedge comes in contact with the wire. The rut depths measured in the field are presented in Table 2 and 3 for sections 13 and 14, respectively.

Dynamic Cone Penetration Tests: Dynamic cone penetration tests were conducted at selected locations to estimate the strength of the subsurface material. The Dynamic Cone Penetrometer (DCP) was developed by the US Army Corps of Engineers. The DCP consists of a 16 mm diameter steel rod with a steel cone attached to one end which is driven by means of a sliding dual mass hammer. The depth of cone penetration is measured at selected penetration or hammer drop intervals. DCP tests were conducted at two locations within section 13, and at one location just outside section 14.

DATA ANALYSIS

Objective: The objective of the data analysis was to compare the load applications to failure of the test sections determined from available methods in the literature and then to compare them with the results obtained from the load tests. The data analysis approach that was adopted attempted to simulate a real-life situation that would be experienced by an engineer. The engineer has FWD data of a pavement section, and wants to get an estimate of the expected traffic loading that the pavement can undergo before failure. The engineer looks at available information that is available in the literature that can be used to do this analysis; then makes use of these methods with the FWD data to get an answer.

Table 2 Rut Depths at Section 13

Cumulative ESALs	Rut Depth (mm) Station (m)								Average (mm)
	0	15	30	46	61	76	91	107	
353	4.4	3.2	1.9	10.2	2.5	2.5	1.3	3.8	3.7
806	8.9	6.4	6.4	16.5	5.7	5.7	4.4	7.0	7.6
1596	15.9	11.4	10.2	25.4	7.6	7.0	7.6	7.6	11.6
1916	21.0	13.3	12.7	31.8	9.5	9.5	10.2	12.1	15.0
2300	24.1	14.6	15.2	34.9	9.5	11.4	12.1	17.1	17.4

Table 3 Rut Depths at Section 14

Cumulative ESALs	Rut Depth (mm) Station (m)											Average (mm)
	0	15	30	46	61	76	91	107	122	137	152	
353	0.0	0.0	0.6	0.0	0.0	0.0	0.0	0.0	0.0	0.0	0.0	0.1
2752	0.6	0.0	1.9	1.9	1.9	1.9	1.9	1.3	2.5	3.2	2.5	1.8
3653	0.6	0.0	1.9	1.3	1.9	1.9	1.9	0.6	2.5	3.2	1.9	1.6
5157	0.6	0.0	1.9	1.9	0.6	1.3	2.5	1.3	2.5	3.8	2.5	1.7
6180	0.6	0.0	1.3	1.9	2.5	1.9	1.9	1.3	1.9	3.2	1.9	1.7

Rutting Models: Rutting occurs due to permanent deformation in any pavement layers or the subgrade. This deformation can occur because of consolidation or the lateral movement of the material due to traffic loads. Rutting in the asphalt layer can occur due to plastic movement of the asphalt concrete mix during hot weather, or can also be caused by consolidation of the layer (lacking compaction during construction). A literature review indicated that the models that are available to predict rutting can be divided into two categories; empirical and mechanistic-empirical. For this analysis, only the mechanistic-empirical models were reviewed. Mechanistic analysis together with observed pavement performance have been used to develop these models. The available models can be divided into two categories. The first category of models calculate the total accumulated deformation on the pavement surface based on the permanent deformation properties of each pavement layer and the subgrade. The other category of models assume that the rutting occurs only in the subgrade.

The first category of models compute the deformation that occur in each pavement layer (asphalt concrete surface and base layer) and the subgrade using layer properties. All these deformations are added together to give the overall deformation of the pavement. Examples of such models are the VESYS (1), the Ohio State Model (2), and the model presented by Owusu-Antwi et al (3).

In the second category of models, it is assumed that rutting will not occur in the asphalt concrete and the base layer and that all the rutting will occur in the subgrade layer. In these models, it is assumed that proper mix design and material selection criteria will be used so that rutting will not occur in the asphalt concrete layer and the base layer. The Asphalt Institute (4), Shell Model (5), AUSTROADS model (6) and model developed by Pidwerbesky (7) are examples of models that fall into this category. In all these models, the strain on top of the subgrade is related to the allowable number of load repetitions.

The rutting models that were selected for evaluation in this study were the models that predicted the ESALs based on the limiting subgrade strain criterion. These models assume that deformations will not occur in the asphalt concrete and the base layer, and that all deformation will occur in the subgrade. The assumption that rutting will occur only in the subgrade can be justified if the following two criteria are satisfied: (1) the mix design properties of the asphalt concrete and quality control tests conducted during construction indicate that that rutting is not expected to occur in the asphalt concrete layer, (2) prior experience with the aggregate base and the quality control tests during construction indicate that rutting is not expected to occur in the aggregate base, (3) test pits performed at the completion of the load test confirmed that the rutting occurred in the subgrade.

The models that were selected for evaluation were the Asphalt Institute Model (4), the Shell Model (5), AUSTROADS model (6) and model developed by Pidwerbesky (7). In all these models, the allowable ESAL applications is based on the strain at the top of the subgrade. The selected models are presented in Table 4.

Table 4 - Rutting Models Selected for Evaluation

Source	Model
Asphalt Institute Model - Shook et. al (4)	$\varepsilon_c = 0.0105\ N^{-0.223}$
Shell Model - Claussen et al (5)	$\varepsilon_c = 0.028\ N^{-0.25}$
Pidwerbesky et al (7)	$\varepsilon_c = 0.012\ N^{-0.145}$
AUSTROADS (6)	$\varepsilon_c = 0.0085\ N^{-0.14}$

In the above table, N is the allowable ESALs, while ε_c is the vertical compressive strain on top of the subgrade

The Asphalt Institute Model (4) gives the allowable number of ESALs a pavement structure can support based on a failure rut depth of 12.5 mm. The Shell Model (5) gives the allowable ESALs a pavement structure can support before it reaches a Present Serviceability Level of 2.5. This model was developed from data collected at the AASHO road test. The model developed by Pidwerbesky et al (7) was based on data obtained from five pavements in New Zealand and present the subgrade strain criteria that will prevent excessive deformations from occurring in pavements. The AUSTROADS (6) strain criteria is used in the design of pavements.

The layer moduli at selected FWD test locations within the test sections were determined by backcalculation. These moduli were used to compute the strain on top of the subgrade and the rutting criteria that were described earlier were used to predict the allowable load applications. The predicted load applications were then compared with the load applications that were applied.

Backcalculation of Layer Moduli: An examination of the DCP test profiles performed in section 13 indicated that the subgrade was extremely weak to a depth of approximately 0.45 m below the aggregate base. The DCP test in section 14 did not indicate a weak subgrade layer below the aggregate base.

The backcalculation of the layer moduli for test section 13 was performed using the MODCOMP program. A five layer system was used to backcalculate the moduli. The layers that were considered were: (1) asphalt concrete layer of 100 mm thickness, (2) aggregate base layer of 200 mm thickness, (3) a subgrade layer of 450 mm thickness, (4) a subgrade layer whose thickness was determined based on the depth to the stiff layer estimated by MODCOMP, and (5) a stiff layer. The moduli of all five layers were assumed to be unknown, and backcalculated using MODCOMP. Backcalculation was carried out earlier by fixing a moduli value for the fifth layer. It was observed that allowing the program to backcalculate the value of the stiff layer resulted in lower error values, and yielded moduli values that were reasonable. The moduli that were backcalculated at selected locations are presented in Table 5.

Table 5 Backcalculated Modiuli for Section 13

Section	Station (m)	Backcalculated Moduli (MPa)				
		Asphalt Concrete	Aggregate Base	Subgrade Layer 1	Subgrade Layer 2	Stiff Layer
13	15	1509	28	7	27	110
13	30	1213	65	6	28	93
13	61	1144	68	7	29	113
13	76	1068	84	6	31	87
13	91	1433	76	7	21	326
13	107	1261	51	7	24	119
13	122	1288	61	8	26	79

Backcalculation of section 14 was also carried out using MODCOMP. A similar approach to that used in section 13 was used in the backcalculation of layer moduli. Although the program was able to backcalculate moduli values with a low error value, the obtained moduli did not agree with the observations from the DCP test. The program predicted very low moduli values for the aggregate base layer, that did not agree with the observations from the DCP tests. The backcalculation of layer moduli at this section was then performed using ELMOD. A four layer system was used to backcalculate the moduli: (1) asphalt concrete layer of 175 mm thickness, (2) aggregate base layer of 300 mm thickness, (3) subgrade layer of 450 mm thickness, (4) and a subgrade layer underlain by a stiff layer, with the thickness of the subgrade layer being determined by the program. This program gave moduli that were reasonable and that generally agreed with the observations from the DCP test. The moduli that were backcalculated at selected locations are given in Table 6.

Table 6 Backcalculated Modiuli for Section 14

Section	Station (m)	Backcalculated Moduli (MPa)			
		Layer			
		Asphalt Concrete	Aggregate Base	Subgrade Layer 1	Subgrade Layer 2
14	0	944	90	55	83
14	76	827	76	41	28
14	152	668	76	41	28

COMPARISON OF PREDICTED AND ACTUAL LOADING TO FAILURE

In this section, the ESALs to failure predicted by the four models are compared with the observations from the field tests. In order to use these models, the strain at the top of the subgrade has to be computed. The traffic loading at the test sections were applied during the daylight hours in August. Based on the air temperatures during the period in which traffic loading were applied to the site and the pavement temperatures measured in the holes (34 °C), the asphalt modulus was assumed to be 1378 MPa. This asphalt concrete modulus was used together with the moduli values presented in Table 5 and 6 to compute the strain on top of the subgrade. A load of 40 kN, and a tire pressure of 689 kPa was used to compute these strains using the ELSYM 5 program. The computed strain values on top of the subgrade as well as the ESALs predicted by each of the rutting criteria for pavement failure are presented in Table 5. The ESALs that were applied in the field to obtain 10 mm of rutting and 12.5 mm of rutting are also shown in this table. The rut depths presented in Table 7 for 10 mm and 12.5 mm rutting was obtained by linearly interpolating the values presented in Table 2. The backcalculated moduli presented in Table 5 in section 13 show that generally low values were obtained for the aggregate base. These low moduli may cause some rutting to occur in the aggregate base. However, the test pits did not show any signs of rutting ocurred in the base course.

The results presented in Table 7 show that out of the evaluated criteria, the Shell criteria gave values that best matched the results obtained from the field data. The values predicted by Asphalt Institute criteria were very conservative, and were about 30 times lower than the values predicted by the Shell criteria. The values predicted by the AUSTROADS criteria were higher by a factor ranging from 2 to 3.5 over the Asphalt Institute Criteria. The values predicted by Pidwerbesky model were generally close but lower than the values predicted by the Shell Criteria.

Table 7 Comparison of Allowable ESALs

Section	Station (m)	Vertical Compressive Strain on Subgrade	Allowable ESALs to Failure				Field Data ESALS for	
			Asphalt Institute Criteria	Shell Criteria	Pidwerbesky	AUSTROADS	10 mm Rutting	12.5 mm Rutting
13	15	0.00463	39	1338	712	77	1375	1781
13	30	0.00414	65	2092	1540	170	1554	1964
13	61	0.00389	86	2684	2366	266	>2298	>2298
13	76	0.00374	102	3142	3103	352	2016	>2298
13	91	0.00372	105	3210	3220	366	2016	> 2298
13	107	0.00416	64	2052	1490	165	1891	>2298
13	122	0.00379	97	2979	2832	320	1770	1947
14	0	0.00058	436,696	Note 1	Note 1	Note 1	>6179	>6179
14	76	0.000679	215,410	Note 1	Note 1	Note 1	>6179	>6179
14	152	0.000679	215,410	Note 1	Note 1	Note 1	>6179	>6179

Note 1: Greater than 1,000,000 ESALs

Traffic loading on section 14 was terminated after 6,180 ESAL applications. The average rut depth of that section obtained by averaging the rut depths measured at 15 m spacing was 1.7 mm when traffic loading was terminated. As shown in Table 7, the Asphalt Institute criteria, which was the most conservative criteria predicted over 215,000 ESAL before failure for the test section. The ESALs predicted for station 0 was double the value predicted for stations 76 and 152. In the field, when the traffic loadings were terminated, the rut depths at station 0 was 0.6 mm, while the rut depths at stations 76 and 152 were both 1.9 mm.

CONCLUSIONS

Two test sections (sections 13 and 14) in Michigan were subjected to traffic to determine the loading required to fail the test sections. Both test sections were constructed over a clay subgrade. The traffic loading on the first test section was terminated after the application of approximately 2,300 ESALs at which stage the average rutting in the outer wheel path of the section was 17.4 mm. Traffic loading on the other test section was terminated after approximately 6,180 ESALs at which stage the average rutting of the outer wheel path of the test section was 1.7 mm.

An analysis was performed using moduli obtained by backcalculating FWD deflections and four rutting criteria to theoretically compute the ESALs to failure. The rutting models that were evaluated were the Asphalt Institute criteria (4), Shell criteria (5), AUSTROADS model (6) and model developed by Pidwerbesky (7). All these models are based on a limiting strain criteria on the subgrade. A comparison of the predicted values using these criteria with field values at section 13 indicated that the Shell criteria gave values that agreed closely with values observed from the field test. The values predicted by the Asphalt Institute criteria were very conservative, and were about 30 times lower than the values predicted by the Shell criteria. The values predicted by the AUSTROADS criteria were higher by a factor ranging from 2 to 3.5 over the Asphalt Institute Criteria. The values predicted by the Pidwerbesky model were generally close but lower than the values predicted by the Shell Criteria. For section 14 that had an average rut depth of 1.7 mm in the outer wheel path, the Asphalt Institute criteria, which was the most conservative criteria predicted ESALs to failure ranging from 215,000 to 436,000. The ESAL applications on section 14 when the traffic loading was terminated was 6180, with an average rut depth of 1.7 mm, and no conclusions could be drawn about the long-term performance of the section.

REFERENCES

1. FHWA, Predictive Design Procedures, VESYS User Manual, Report No. FHWA-RD-77-154, Federal Highway Administration, 1978.
2. Majidzadeh, K. Implementation of a Pavement Design System, Research Project Number EES 579, Ohio State University, Columbus, 1981.
3. Owusu-Antwi, E.B., Titus-Glover, L., and Khazanovich, L. A Rutting Model for Implementation in Mechanistic-Based Design Procedures. Presented at 77th Annual Meeting, Transportation Research Board, Washington, D.C., January 1998.
4. Shook, J.F., F.N. Finn, M.W. Witczak, and C.L. Monosmith. Thickness Design of Asphalt Pavements - The Asphalt Institute Method, Proceedings, 5th International Conference on the Structural Design of Asphalt Pavements, Delft, Netherlands, Vol. 1, pp 17-44, 1982.
5. Claussen, A.I.M., J.M. Edwards. P. Sommer, and P. Uge. Asphalt Pavement Design - The Shell Method, Proceedings, 4th International Conference on the Structural Design of Asphalt Pavements, Ann Arbor, Michigan, Vol. 1, pp. 39-74, 1977.
6. AUSTROADS, Pavement Design: A Guide to the Structural Design of Asphalt Pavements, AUSTROADS, Sydney, 1992.
7. Pidwerbesky, B.D., Steven, B.D., and Arnold, G. Subgrade Strain Criterion for Limiting Rutting in Asphalt Pavements, Eighth International Conference on Asphalt Pavements, Seattle, Washington, Vol. 1., 1529- 1544. 1997.

Fifth International Conference on the
BEARING CAPASITY OF
ROADS AND AIRFIELDS

FULL SCALE TEST ON AIRFIELD PAVEMENT

GEIR LANGE

Senior engineer, Civil Aviation Administration, N-0032 Oslo, Norway

ABSTRACT

This paper describes the construction of a full scale test for airfield pavements. The test track was located at the new main airport outside Oslo, which were established in order to investigate and evaluate 10 different pavement constructions by simulating heavy aircraft take off weights (TOW). The test track consists of 8 flexible, 1 PCC and 1 gravel section with different total pavement thickness. Each material used in the base course of these 10 sections were measured during construction. DCP measurements was conducted before construction and samples were sent to a laboratory to measure the CBR value. On each compacted layer, plate bearing tests and Falling Weight Deflectometer (FWD) were used to measure E- modulus and k-values.
After 12 months of heavy traffic, materials from part of the section were investigated in order to identify any deterioration. The traffic obtained could be compared to a DC 10/ MD 11 with 25 metric tons wheel loads and tire pressure 1,9 MPa.
Further more, 3 different transitionramps between the PCC slab and the flexible pavement were built, the effect of a geotextile between sub-base and base course was investigated. This paper summarises the findings from the test track.

INTRODUCTION

Background

Existing design models for airfield pavements are based on experiences from different contries with different materials and established practises. Calculation of the thickness of each layer is determined with different safety factors in order to make sure that the construction is sufficient for extreme conditions that can occur during its design life.

The new main airport in Norway - Oslo Airport, Gardermoen - is designed for 250.000 movements pr. year in the year 2020, of which 11% are expected to be wide-body aircrafts. To choose a design method was not easy. All available design methods (except one) gave different results compared to the method that has been used for Norway's 55 Airports. The only design method that gives an equal pavement construction is the Transport Canada method (1) and also the construction of the new runway at Hong Kong International Airport - scheduled to be opened during the summer of 1998 (design method unknown).

In Norway PCC pavement has only been used on one runway (Bodø Airport, constructed in 1955) and on the apron at 7 airports. Out of approximately 10 million m^2 airfield pavement in Norway, only some 500.000 m^2 is PCC.

A new apron at Oslo Airport Fornebu was in 1988 constructed with a PCC pavement. In 1991 severe D-cracking occured in the pavement at all 5 gates. To gain experiences from

these problems, a field study was carried out in 1992 in order to get some answers of why the PCC cracked like it did. Two slabs were removed and Benkelmann beam and FWD were used to measure the bearing strength on each different layer.
The results were devastating:

Layer # *From top*	*Material*	*Thickness* cm	*E-modules* MPa	*Load-distribution*
1	PCC	30	-	-
2	Crushed rock 0-63mm	10	120	1,04
3	Crushed rock	85	225	1,28
4	Sand	100	270	1,36
5	Sub-base -silt	-	160	1,14

Figure 1 Results from Fornebu 1992 (2)

The bearing strength increases for each layer downwords in the construction !

Existing methods
Stopway and apron area were decided to be constructed with PCC at Gardermoen. In order to calculate the thickness of the PCC pavements 7 different institutes or design methods were asked / used to calculate and recommend the pavement thickness based upon the same given parameters (design aircraft, concrete quality, aggregate quality, bearing capacity of the base course etc.).

The results were as follows:

Pavement	*Recommended thickness*	**Norwegian practise**
PCC	310 - 470 mm	300 or 350 mm
Asphalt - total thickness	125 - 540 mm	Max. 150 mm

Figure 2 Recomanded pavement thickness

Aircraft loads
An international airfield is normally trafficked by size aircrafts, from Cessna with a maximum TOW of 1.043 kg, to Boeing 747 with a MTOW of 396.000 kg. MTOW appears on the pavement when the aircraft is taxing from the gate and out to the runway end and in the first ¼ of the runway. Touch down is not as critical as many seams to think.
The design aircraft for a specific airport is the heaviest aircraft that can be expected at the airport in ordinary scheduled traffic during its design life.

The difference between airports and road construction (in Norway) is shown in figure 3:

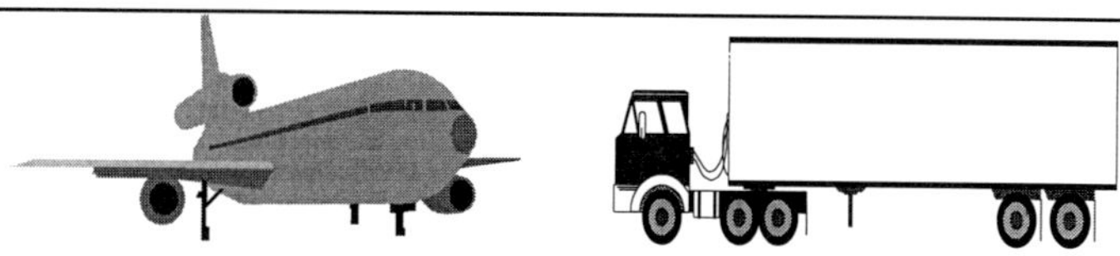

TOW:	396 tons	50 tons
Wheel-load:	25 tons	5 tons
Tire pressure:	1,9 MPa	0,9 MPa
Departures:	100.000/year	50.000 /days

Figure 3 Loads on airport verses roads

Figure 4 shows typical data for some of the usual aircraft's (and the heavy vehicle used at Gardermoen).

Aircraft	*Wheel-configuration*	*TOW*	*Wheel-load*	*Load pr. leg*	*Tire-pressure* MPa	kg/cm^2
B737	Dual	58	13	26	1,08	11,0
B747	Complex	373	21	84	1,30	13,2
B777	Complex	310	23	136	1,38	14.0
DC 10/MD 11	Complex	285	27	110	1,41	14,4
DC 9/ MD 80	Dual	64	15	30	1,17	12,0
IL 62	Dual Tandem	162	?	60	1,65	16,8
Dash 8	Dual	20	4,5	9	0,74	7,5
Truck	Super single	50	4	12	<0,9	9,2
Vehicle at GM	Dual	150	25	50	1,9	19,3

Figure 4 Aircraft data.

Why the full scale test track?
In order to give the best possible background to develop the «correct» pavement construction for the new airport at Gardermoen and to finally examine the Norwegian practice with «real» aircraft wheel loads, it was decided in 1994 to construct a full scale test section for different airfield pavement constructions. As shown in figure 4 the wheel-load and tire-pressure of the heavy vehicle used on the Gardermoen site is similar to a DC10/MD11 which are the usual design aircraft on international airports.

CAA joined in 1993 the joint venture project «Quality of crushed rock and gravel». This program will be presented in 9 different papers on BCRA `98. The Gardermoen test track was one of 4 full scale test tracks in this project.

The main objectives for the Gardermoen full scale test track were :

- Construction of 10 different pavement structures on top of the same sub-grade material
- Monitoring of permanent deformation
- Investigate different compacting methods of sub-grade
- Documentation of the crushed rock bearing strength
- After 2 years of traffic, investigate the quality of the used material. (The period was decreased to 12 months)

FIELD STUDIES

The Test Track
The test track was constructed at Gardermoen Airport during the summer of 1995 and included 10 different pavement structures, with a total length of 215 m and width of 8 m. Maximum thickness on section 7 - 10 was 86 cm. The track was located just outside the apron area to be a part of the transport route between the rock crushing plant and the construction site for the new runway. The weight of the «heavy vehicle» that was used for mass-removal varied from 110 - 150 metric tons on a total of 6 wheels (wheel load 25 tons and tire pressure 1,9 MPa)

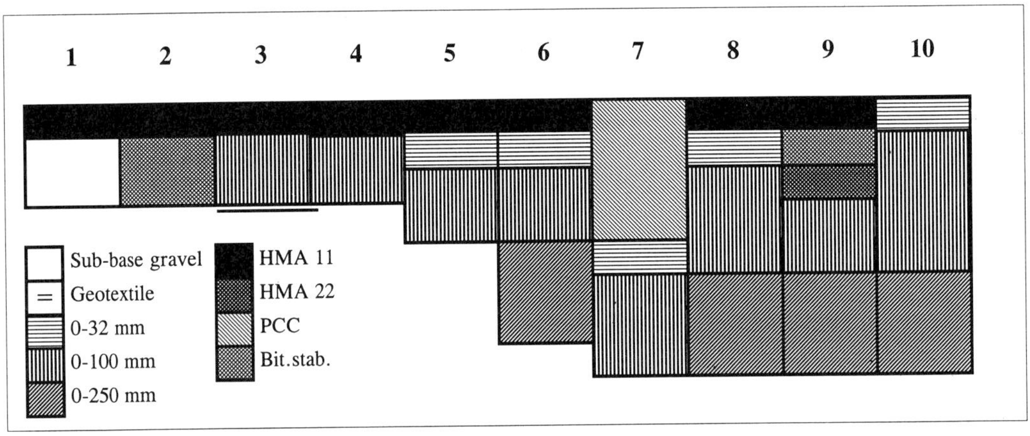

Figure 5 10 different pavement constructions

Each section
The length of each section varies from 7,5 to 27,5 m.

- The asphalt pavement used was the same for all sections; 11 cm thick asphalt concrete with 6,0% of binder (performance grade PG 64-34) and 0,3% cellulose fibres .

- Section 1 - 4 were constructed in order to investigate the use of different materials in the base-course for small airports and different compacting methods for the sub grade. One half was compacted with roller, the other half by wheel from a lorry.

- Section 3 and 4 were identical exept the use of a geotextile between the sub-grade and the base-course. The effect of using geotextile were tested.

- Section 5 and 6 were constructed in order to investigate the effect of using a 40 cm base-coarse with 0-250 mm crushed rock.

- Section 7 had a 370 mm thick PCC pavement.

- Section 8 and 9 was the two different constructions recommended for the Gardermoen airport.

- Section 10 contained crushed rock 0 - 32 mm as top layer (same material as base course in section 5 to 8)

The program contained 4 phases:

In phase 1 - Before construction - all materials to be used were tested. Los Angeles- and CBR-values of the material, and Marshall for the asphalt pavement were measured.

In phase 2 - During construction - DCP-measurement was carried out on sub-grade before compacting, after 5 and 8 passes with the roller, and finally on wheel compacted. After the compacting of each different layer the Benckelmann beam was used to measure the k-value and E-modules. A FWD was also used to measure the theoretical bearing strength on top of the base-course.

In phase 3 - Test period - the number and weight of all passes were determined, and the rut-depth measured every second week during the first 3 months, and once every 2nd months until the end of the Summer of 1996.

Phase 4 - Investigation - after 12 months of traffic the different materials used in the construction were investigated to determine if there had been any changes in property.

RESULTS

Sub-grade

Homogenizing of the sub-grade

The sub-grade material was removed from section 5-10 in order to get sufficient depth of each construction. This material was used as a sub-base course which gave the necessary homogenity of the sub-grade. The coeffisient of uniformity (Cu number) increased from 3,0 (measured before construction) to 4,5 on section 1-4 after action had been taken. The requirements for the sub-grade was >4,0.

Different methods of compacting

A test program was carried out in order to measure the sufficient number of passes and methods of compacting sub-grade material at Gardermoen. DCP, Bomag Terrameter (on the vibrate-roller) and a Troxler were used to measure the effect of the compacting.

On section 1-4, a 16-tons vibrate-roller was used on one half (respectively 8 and 10 passes) and wheel-compaction on the other half (lorry with total weight of 55 tons).
During compaction it was observed that the surface of the wheel-compacted materials was far more firm than the other. Wheel-compaction is a time consuming method compared to useing a roller. Figure 6 shows the different k -values and E-modules.

Section	Method	k-value	E_2/E_1	E-modules -E_2
1	Wheel	42	6,0	59
1	Roller	57	4,0	69
4	Wheel	49	4,0	61
4	Roller	51	3,6	61

Figure 6 - quality of sub-grade

In order to determine the sufficient number of passes by the roller, measurements from section 7 and 8 were used.

Section	Passes	k-value	E_2/E_1	E-modules E_2
7	2	17	10,6	45
7	5	20	10,7	54
7	8	28	5,0	57
7	10	27	6,6	80
8	2	15	13,5	49
8	5	19	8,8	49
8	8	32	5,8	66
8	10	27	5,7	59

Figure 7 - Compaction control by number of passes

Measurements of k-value showed differences from section 1-4 which was far more firm than section 7-8. The recommended number of passes was determined to be 8. This number was also the results when Troxler and DCP measurements were used. The compaction increases most between 5 and 8 passes, while above 8 passes no significant changes were noticed.

The best compaction of the sub-grade at Gardermoen occur when we used 4 passes with vibrating-roller followed by 4 passes with wheel compaction.

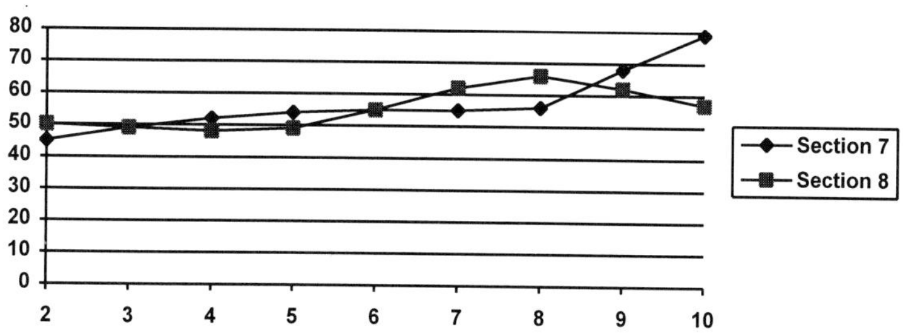

Figure 7 - E-modules as a function of passes

CBR-variations on the sub-grade

The design value for Gardermoen, and the maximum CBR value on subgrade according to ICAO - Annex 14 (3) used for calculating ACN/PCN, is 15%. From 4 samples taken from the test section, Cu-number and CBR value were measured in laboratory.

Section	**At depth in meters**	**Cu-number**	**CBR-value**
2	0,5	2,9	34%
4	0,5	2,7	24%
6	0,9	2,6	26%
8	1,1	3,2	34%

Figure 8 - 'Cu-number and CBR-values

The average CBR-value at Gardermoen was determined to 30%.

After compaction each section were measured with DCP and CBR-values calculated:

Section	Compaction methods	CBR	k-value
1-4	Wheel	44	54
1-4	Roller	35	46
7-8	Roller	32	25

Figure 9 - CBR and k-value

Base-course

Effect of Geotextile between subgrade and base-course

Section 3 and 4 had the same construction, except the geotextile used in section 3. As shown in figure 10 and 12 the geotextile gave a slightly higher k-value but a lower E2 modulus.

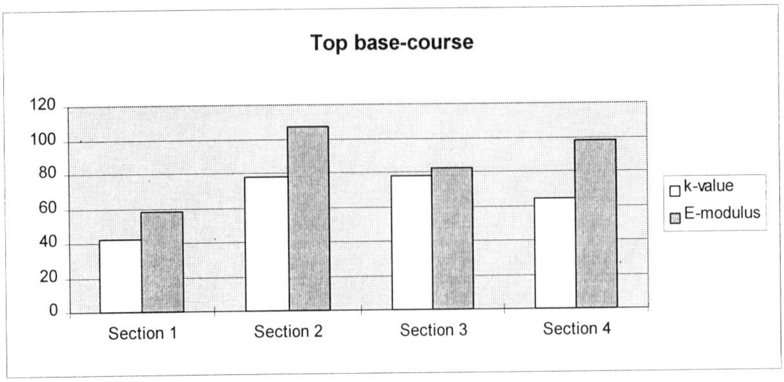

Figure 10 - k-value and E_2 modules section 1, 2, 3 and 4 with same compaction methods

3 different quality 0-32 mm crushed rock

Section 8 was constructed with 3 different 0-32 mm crushed rock materials in order to investigate the differences on crushed rock and crushed gravel. Fk-Garderfjell (Crushed rock) were produced at the Gardermoen site. Gk-Hovinmoen was an imported crushed gravel. In phase 4, these materials were investigated to determine any changes of LA-value and grading curves. Figure 12 shows the grading curves from 1995 and 1997 for two of them.

Section	Material	Passes	E_2 on top base-course	k-value	LA-value Phase 1	LA-value Phase 4
8	Fk - Garderfjell	1	135	70		
8	Fk - Garderfjell	2	135	92	24	27
8	Fk - Garderfjell	3	123	92		
8	Gk - Hovinmoen	1	113	56		
8	Gk - Hovinmoen	2	130	78	21	21
8	Gk - Hovinmoen	3	141	90		

Figure 11 - 2 different 0-32 mm material

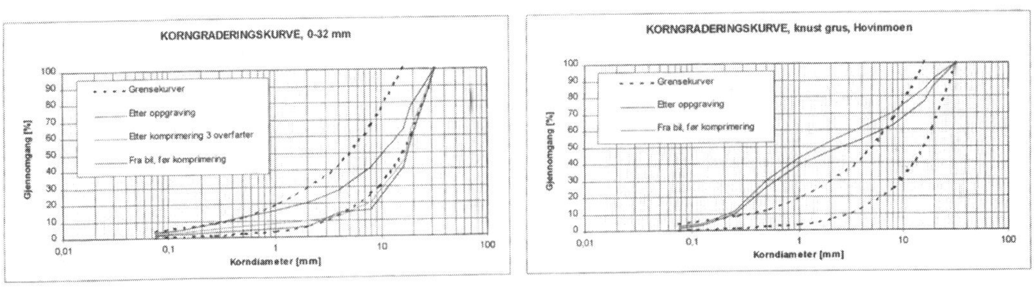

Figure 12 - grading curves phase 1 and phase 4

Amount of fines

In the tender paper for the new airport the different materials used in the base-course were 0-250, 0-100 of 0-32 mm crushed rock produced on site. The requirement for the amount of fines (< 0,075 mm) were set to be respectively 7, 6 or 5%. In theory this gives the best base course, but during construction, a lot of separation was registered in the different layers. It looks like the contractor had some difficulties in getting a homogeneous layer. Nearly all the fines were located in the centre of the test track which resulted in initial deformation in the asphalt pavement before opening the test track to traffic.

CBR on each layer

Based on measurement by Benkelmann Beam CBR-value were calculated.

Section	Top base - course	CBR-value
1	Sand/gravel	36
3	0-100 mm with geotextile	121
4	0-100 mm without geotextile	113
5 and 6	0-32 mm	179

Figure 13 - CBR-value on top base-coarse

Section	Layer	CBR-value
8	0-32	179
8	0-100	73
8	0-250	59
8	Subgrade	32

Figure 14 - CBR value for different layer in section 8

PCC pavement

Section 7 contain a PCC slab, 370 mm thick and 7,5 meter in square. On each end different transition ramps were constructed.

Surface

The surface smoothness was measured just after paving. The rutting was then measured every 2nd week during the first 3 months followed by measurements every 2nd month until the end of the test period. Figure 15 shows initial deformation, and after 4 and 6 weeks, finally after 7 months.

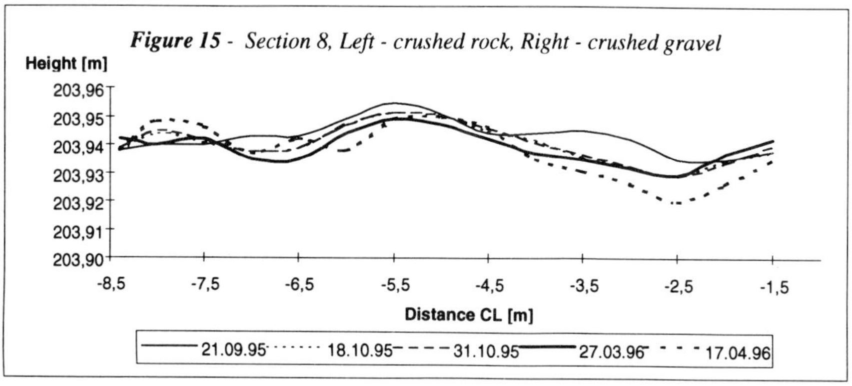

Figure 15 - Deformations on section 8

Figure 16 - Picture of the transition ramp between section 6 and 7 with initial deformations

Traffic intensity
The two types of vehicles that were used had a maximum weight of 110 and 150 tons, all contained six wheels with maximum tire pressure of 1,9 MPa. As mentioned earlier the maximum tire-pressure on airfields is 1,9 MPa (fighters) but the overall design aircraft MD 11 has a tire-pressure of 1,41 MPa and wheel-load of 25 and 27,5 tons. In other words this test section get the same loads as an airport.
Each vehicle was weighed and registered in order to get the actual number of passes and total load. The test track was a one way route and only full-loaded vehicle passed the section.

The total number of passes is shown in figure 17:

Months	Number of passes	Netto weight tons	Brutto weight tons	Average passes/day	Average tons/passes
3.10.95 - 30.10.95	2 627	133 000	247 000	139	94
27.11.95 -15.5.96	18 225	973 000	1 798 000	154	99
Sum	20 852	1 106 000	2 045 000		

Figure 17 - total number of passes

There was no heavy traffic on the section after May 1996.

In order to illustrate the effect of the number of passes each day the new airport is designed based on 250.000 movements each year in the year 2020. 11% of the traffic is wide body aircrafts (ICAO code D and E) which is equal to 37 movements /day. This means that the average number of passes on the test section was 3 times higher than the expected daily traffic at Gardermoen in the year 2020.

CONCLUSIONS AND RECOMMENDATIONS

- The results show that wheel compacting gives a more firm sub-grade than a roller-compacted does on this type of material.
- Geotextile has no effect -as assumed- on this kind of sub grade materials.
- The base-course layer 0-250 mm crushed rock gave no significant increased bearing strenght at Gardermoen.
- The requirement of fines content had to be decreased in order to avoid initial deformations.
- Quality Control are extremely important when using this kind of materials.
- Compare the E-modules from the Gardermoen and Fornebu site; - We are talking of two different E-modules - Why?

ACKNOWLEDGEMENTS

I would like to address a special thanks and many thoughts to my late colleague Mr. Arnold Hagen who was the force in establishing this project.

The investigation during construction in 1995 was handled by two graduate students; Fredrik Røtter and Øyvind Salomonsen

The investigation of the materials in 1997 was handled by two other graduate students; Olav Aune and Arve Vassbotn

REFERENCES

1. «Manual of Pavement Structural Design» Transport Canada - ASG-19 July 1992
2. CAA Norway - Investigation of PCC apron at Oslo Airport Fornebu
3. Test track - Airfield pavements Røtter/Salomonsen 1995
4. Test track - Investigations Aune/Vassbotn 1997

Fifth International Conference on the
BEARING CAPASITY OF
ROADS AND AIRFIELDS

ANALYSIS OF SEASONAL VARIATIONS IN MOISTURE STATE USING LTPP DATA

Y. JANE JIANG, Senior Project Engineer
SHIRAZ D. TAYABJI, Ph.D., P.E., Regional Vice President
HESHAM ALI, Ph. D., Senior Project Engineer

ERES Consultants, Inc.
9030 Red Branch Road, Suite 210, Columbia, Maryland 21045, USA
Phone: 410-997-6181; Fax: 410-997-6413

and

CHERYL ALLEN RICHTER, P.E.

Federal Highway Administration, Pavement Performance Division
6300 Georgetown Pike, McLean, Virginia 22101, USA
Phone: 703-285-2183; Fax: 703-285-2767

ABSTRACT

As part of the United States Long Term Pavement Performance (LTPP) program, a comprehensive study is underway to develop data needed to obtain a fundamental understanding of the characteristics and the effects of the seasonal variations in the moisture and temperature conditions (specifically the freeze state) within the pavement layers. As part of this study, referred to as the Seasonal Monitoring Program (SMP), moisture and temperature conditions are being monitored at sixty-two instrumented test sites across the United States and Canada. The instrumentation consists of a 3-prong time domain reflectometry (TDR) probes for measuring moisture content and electrical resistivity and temperature probes for estimating the freeze state.

As part of the current LTPP data analysis effort, procedures are being established to transform the instrumentation data into engineering parameters. The soil dielectric constants determined from the TDR data are being used to compute the volumetric and gravimetric moisture content. The soil electrical resistivity data, in conjunction with the temperature data, are being used to estimate the freeze state.

The SMP data collection effort also includes deflection testing, profile testing, and distress data collection on a seasonal basis (12 to 14 times per year). Recently, a study was initiated to determine the in-situ moisture contents using the TDR data and to determine the in-situ freeze state using the electrical resistivity data. This paper presents the preliminary results of the study related to the TDR data study. Key issues related to data interpretation are discussed. Also, preliminary results of analysis of the effect of the in-situ moisture variability on the structural behavior are presented.

INTRODUCTION

Moisture and temperature conditions of pavement materials can significantly affect pavement performance. Increase of the moisture in subgrade and unbound base material can weaken these material and therefore affect the pavement's response to traffic loading and ultimately, pavement service life. Also, variation of moisture in subgrade can change the volume of the swelling soil which may result in detrimental deformation of the pavement structure. In order to better understand the environmental factors and their effects on pavement performance, the LTPP SMP was initiated several years ago. A total 64 LTPP sections were designated for inclusion in the SMP. These sections are monitored on a much more frequent basis (12 to 14 times a year) to acquire comprehensive data on seasonal variation in pavement conditions and in pavement response. As part of this effort, TDR sensors are used to allow determination of moisture content in unbound base and subgrade materials. Other data that are also collected at the SMP sites include freeze state, deflection, profile, distress, and weather data.

TDR DATA COLLECTION

TDR data are being collected using the 3-prong TDR probes developed by the Federal Highway Administration (FHWA) (1). A total of 10 TDR probes are used at each section, with the first 8 probes spaced at 152 mm (6 in) and the last 2 probes spaced at 302 mm (12 in).

TDR probes have been used to determine the in-situ moisture content in soils since the late 1970s, with the initial use in the agricultural field. In the TDR method, an electromagnetic wave is transmitted along a coaxial metallic cable which acts as a waveguide and through the granular medium. The velocity of the pulse through the granular medium is influenced by the dielectric constant of the material (which is also affected by the moisture changes in the material). Once the pulse reaches the end of the probe, a portion of the signal is reflected back through the shielding of the coaxial cable of the Tektronix unit. A drop in reflected voltage is seen on the display of the Tektronix unit when the signal reaches the beginning of the TDR probe due to the increased resistance of the smaller path in the printed circuit board and a large rise in the reflected voltage is noticed when the signal reaches the end of the probe. This waveform is recorded in an ASCII file using a personal computer. The horizontal distance between the initial and final inflection points of the TDR trace response is the travel time of the signal, which represents the apparent length (La) of the probes. The bulk dielectric constant, Ka, can then be calculated from the apparent length.

The dielectric constant of water is approximately 80, that of dry soil is between 3 and 8 (depending on soil type and density), and that of wet soil is between 3 and 60 (depending on the moisture content). The volumetric moisture content can be estimated from the measured dielectric constant using regression type equations such as Topp's equation (2) or semi-empirical soil mixing formulations such as Roth's model (3). The gravimetric moisture content is then determined using standard formulations that require the knowledge of dry density of the soil.

DETERMINATION OF THE SOIL BULK DIELECTRIC CONSTANT

The determination of the dielectric constant requires the computation of the apparent length of the TDR probe from the TDR probe's return signal. Several methods exist for computation of the apparent length. The Method of Tangents and the Method of Peaks are widely used by researchers.

The Method of Tangents was found to be the most accurate method by comparing the TDR results and the laboratory-determined moisture content in a recent study conducted by Klemunes (5). This conclusion was verified in this study using Klemunes' calibration database. The Method of Tangents is illustrated in figure 1. The initial inflection point is determined by locating the intersection of the horizontal and negatively sloped tangents at the trace's local maximum value, and the final inflection point is located at the intersection of the horizontal and positively sloped tangents at the trace's local minimum value. The distance between the two inflection points is the apparent length of the TDR trace.

However, the Method of Tangents was also found not to be applicable in some situations, such as very dry soil TDR traces. For these situations, the Method of Peaks has been found to be more reliable and is proposed for use. The Method of Peaks is

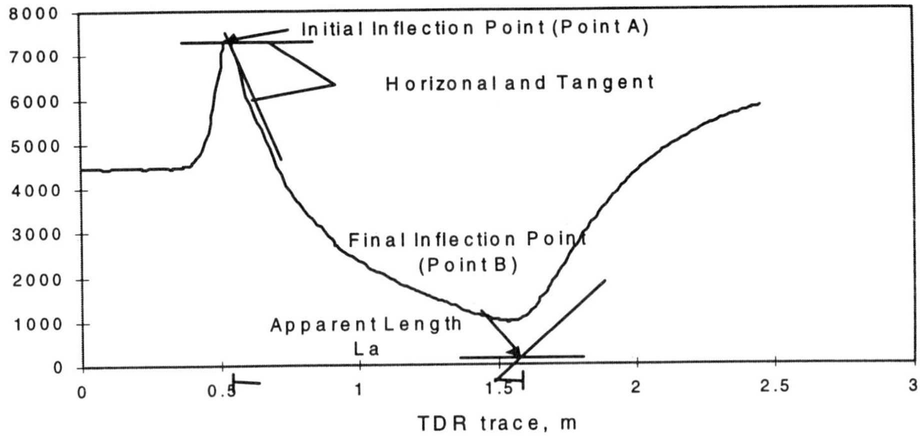

Figure 1. Illustration of the Method of Tangents (Section 51_113).

illustrated in figure 2. The initial inflection point is determined by locating the intersection of the tangents drawn on both side of the first inflection point. The final inflection point is located at the intersection of the tangents drawn on both sides of the final inflection point.

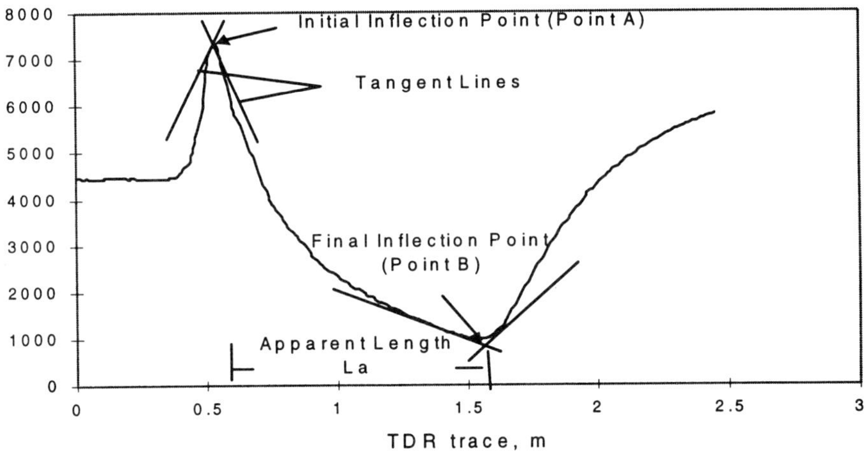

Figure 2. Illustration of the Method of Peaks (Section 51_113).

Factors influencing the accuracy and precision of the TDR interpretation method include extremes in soil mixture conductivity, salinity, and possibly temperature effects. Saline-like conditions in the soil can cause a short circuit in the TDR probes, which makes the final inflection point of the TDR trace difficult to locate. In the proposed trace interpretation procedure, an error code is assigned to flag such conditions.

Once the apparent length is determined, the dielectric constant of the soil mix can be calculated using the following standard equation (1):

$$Ka = \left(\frac{La}{(L * Vp)} \right)^2$$

where:

	Ka	=	Dielectric constant of the soil or granular base
	La	=	Apparent length of the TDR trace, m
	L	=	Actual length of the TDR probe, m
	Vp	=	Propagation velocity of the TDR wave, normally 0.99

DETERMINATION OF THE VOLUMETRIC MOISTURE CONTENT

In-situ moisture content is a very important condition that affects pavement behavior and long-term performance. More specifically, moisture condition has a significant influence on the moduli of unbound pavement materials. Resilient modulus of the soil typically decreases with the increase in the moisture content. Many laboratory studies have been conducted to address this issue. The modulus of a fully saturated material may be as much as 50 percent lower than that of the same soil in a partially saturated condition, although some researchers have arrived at different conclusions.

Two approaches have been used to relate the bulk dielectric constant to soil volumetric moisture content (Vw%): the empirical and the semi-empirical approach. The most accurate models are those that have been calibrated in laboratory test. In the empirical approach, the apparent dielectric constant, Ka, is related to the volumetric moisture content directly using empirical regression functions. No physical justification of the model form is considered in this approach. The functional form most frequently used is the third-order polynomial. Topp was one of the first researchers who obtained the third-order polynomial empirical model relating Ka to Vw% (2). Topp's equation is the most widely accepted empirical model used to determine volumetric moisture content directly from the apparent dielectric constant. It is a generic model applicable to all soils; therefore, it is not very accurate in all cases. The advantage of the empirical model is its ease of use, since no other soil properties are required to calculate Vw%. When calibrated to a specific soil type, this approach can also give accurate results.

The semi-empirical approach involves derivation of a fundamental equation from the soil dielectric mixing models. The soil mixing models relate the composite dielectric constant of a multiphase mixture to the dielectric constant and volume fractions of its constituents (soil, water, and air). The following three mixing models are most widely used in the semi-empirical approach:

- 3-phase model (5) which includes soil, water, and air.
- 4-phase model (3) which includes soil, free water, bound water, and air.
- Maxwell-De Loor model (7) which includes soil, free water, bound water, and air.

Compared to the empirical models, the mixing models are more physically sound and have the potential to describe the relationship between the apparent dielectric constant and the volumetric moisture content because they account for the influence of the individual constituents. However, when developing the sample specific semi-empirical models using the actual TDR traces and the laboratory soil tests, these models sometimes yield physically meaningless estimates of some parameters (e.g., negative soil dielectric constant).

Klemunes (5) recently performed a comprehensive study for developing calibrated mixing models for specific soil samples. Twenty-eight soil samples were obtained from 28 LTPP SMP sites. For each soil sample, TDR traces and dry density and gravimetric moisture content were obtained at several moisture levels and up to six compaction levels in a laboratory. A total of 415 data points were obtained, with measurements including specific gravity, dry density, gravimetric moisture content, volumetric percentage of the soil mixture, and the corresponding TDR traces. After eliminating the outliers, 375 data points remained in the database. This provides a good database for calibrating the models relating the soil bulk dielectric constant with the soil volumetric moisture content and was used as the project database.

It was observed from the project database that fine- and coarse-grained soils have very different behavior in relating the apparent dielectric constant to the volumetric moisture content. The dielectric constant of fine grained soil is about 4, whereas the coarse-grained soil may have a dielectric constant of about 8. Water in the soil can be further separated into free water and bound water. Free water has a dielectric constant of about 80, but bound water's dielectric constant is in the range of 3 to 5. This may affect the accuracy of the computed water content when the volumetric moisture content is low and a large portion of the water is bound.

Hierarchical empirical models were developed under this study. In developing the empirical models, third-order polynomial models were tried first. However, in some cases, the developed models produced a functional form of Vw%=f(Ka) that did not monotonically increase with Ka. This is not physically rational. For these cases, a power model with a phase shift was used. These empirical models are comparable to the 3-phase mixing models. Two of the 28 site-specific models were dropped because of unsatisfactory data. The empirical regression equations developed under this project using the project database are given below:

Third-Order Polynomial Models:

$$Vw\% = a_0 + a_1\ Ka + a_2\ Ka^2 + a_3\ Ka^3$$

Power Models:

$$Vw\% = c_0\left(Ka - c_1\right)^{c2}$$

where a_0, a_1, a_2, a_3 = Regression coefficients as given in table 1
c_0, c_1, c_2 = Regression coefficients as given in table 2
Ka = Bulk dielectric constant

The R^2 and the standard error of the models are also given in tables 1 and 2. The 95 percent ($\alpha = 0.05$) confidence limit of the mean model prediction can be obtained by multiplying the standard error by Z_α of 1.96. However, since another independent data set of the measured moisture content, TDR traces, and other laboratory soil test data is not available, the precision and bias of these models cannot be established at this time.

Table 1. Coefficients for the 3rd-Order Polynomial Models.

Model	Level	Soil Class	Model Vw range, %	a0	a1	a2	a3	n	R2, %	Standard Error, Vw%
161010	1	A-2-4	2.8-29	3.5	-1.533	0.4032	-0.0135	10	96	2.0
251002	1	A-3	5-22	-17.4	7.308	-0.6367	0.0236	12	99.5	0.4
501002	1	A-7-5	8-38.3	-0.2	2.248	-0.005	-0.0009	17	72	4.7
A-2-4	2	A-2-4	2.8-34	-6.8	4.015	-0.1877	0.0037	127	90	2.7
A-3	2	A-3	4.4-35	-9.2	4.198	-0.1965	0.0042	83	95	1.6
A-4	2	A-4	6.7-50	-0.5	3.027	-0.082	0.0011	60	65	6.5
A-7-5	2	A-7-5	8-42.3	-4.9	3.453	-0.087	0.0007	35	81	4.0
A-7-6	2	A-7-6	11.3-30.6	10.8	0.930	-0.0168	0.0004	26	60	3.5
Coarse	3	Coarse	2.8-35	-6.1	3.526	-0.1428	0.0026	237	91	2.4
All soil	4	All soil	2.8-50	-0.2	2.285	-0.0433	0.0003	375	75	4.6

Note: Under the Model column, the six digit numbers refer to the test section identification number as used in the LTPP program. Only a few examples of site specific models are given.

Table 2. Coefficient for the Power Empirical Models.

Model	Level	Soil Class	Model Vw range, %	c0	c1	c2	n	R2, %	error, Vw%
271018	1	A-1-b	9.6-23	7.4	4.594	0.4905	17	84	1.9
404165	1	A-2-4	4.5-28	6.7	3.142	0.4809	10	89	3.0
460804	1	A-6	22-47.5	16.5	6.601	0.2838	13	74	4.4
481068	1	A-7-6	19-48	0.0	-70.574	2.4700	9	49	2.7
A-1-b	2	A-1-b	6.5-23	5.2	3.649	0.6148	27	87	2.0
A-6	2	A-6	21-47.3	6.1	-5.610	0.5262	17	80	4.2
Fine	3	Fine	6.7-50	7.7	1.807	0.4940	138	65	6.1

Note: Under the Model column, the six digit number refers to the test section identification number as used in the LTPP program. Only a few examples of site specific models are given.

Site-specific models are recommended for calculation of the volumetric moisture content for the LTPP SMP sites whenever possible. If site-specific models are not available, AASHTO soil class-specific models should be used. If the AASHTO soil class of the section is not known, then the soil type (coarse or fine) models or the generic model (applicable to all soils, but is less reliable) can be used. These calibrated empirical models give comparable accuracy compared to the calibrated 3-phase mixing models. The site-specific, laboratory-calibrated empirical or soil mixing models give the most precise relationship between the bulk dielectric constant and volumetric moisture content.

Gravimetric moisture content can be calculated from volumetric moisture content using the dry density information.

STUDY RESULTS

Soil dry density information is very limited in the LTPP database. Work is underway to identify or estimate the dry density for each SMP site at various depth where TDR probes are installed.

For this study, only the empirical model that is applicable to all the soils was used in this study. A Window based software program, Moister, was developed as part of this study to view and process TDR traces using the Method of Tangents and the Method of Peaks. The research team used the Moister program and processed a total of 19,663 TDR traces from the SPS SMP sites. The TDR waveforms were obtained from the LTPP Information Management System (IMS) database in

December 1997 All traces were processed and reviewed semi-manually. The apparent length, dielectric constant, and volumetric moisture content were calculated for each trace. The newly developed site-specific universal empirical model and Topp's universal equation were used to calculate the volumetric moisture content from the soil dielectric constant. Work is underway to compute gravimetric moisture contents.

Time history plots can be generated to examine the seasonality of the in-situ moisture content in the soil. Figure 3 shows an example time history plot for LTPP section 83_1801 in Manitoba. It should be noted that the TDR method can only be used to estimate the liquid moisture content. As shown, the soil moisture content does vary significantly throughout the year. During winter, the volumetric moisture content dropped to a very low level which may indicate a frozen soil situation since this section is in a deep freeze zone. Furthermore, the soil volumetric moisture con-

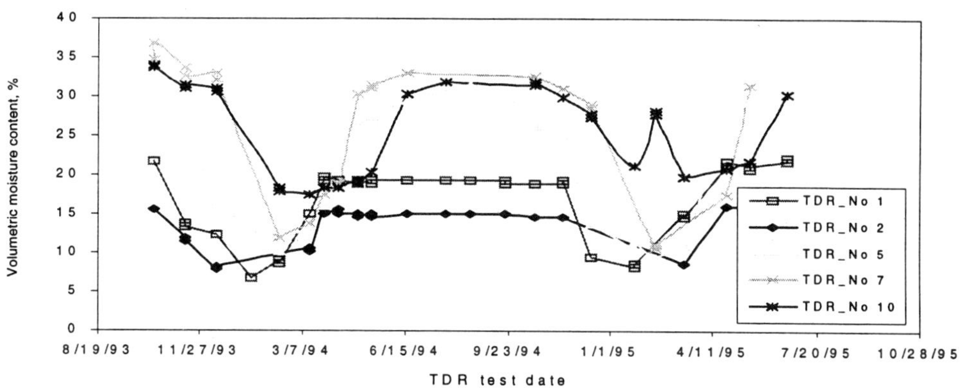

Figure 3. Example moisture content time history plot for section 83_1801 in Manitoba.

tent vary considerably by depth also. This may be because of many factors, such as precipitation, soil permeability, groundwater level, etc. Figure 4 gives another example of time history plot for LTPP section 01-101 in Alabama. It is clear that the seasonal variation of the computed moisture content of this section is much less than that of the Manitoba section. Moreover, there is no sharp drop in moisture content during the winter time for this section since it is in a nonfreeze area.

Sensitivity of the TDR Data Interpretation Procedure

The sensitivity of the TDR data interpretation procedure was assessed by reviewing the month to month variations in the subgrade. The data for the top four TDR sensors were averaged for each time of the seasonal testing (about once a

ACKNOWLEDGMENTS

This work was conducted as part of LTPP Data Analysis Technical Support Contract, FHWA Contract Number DTFH61-96-C-00003.

REFERENCES

1. Rada, G.R., G.E. Elkins, B. Henderson, R.J. Van Sambeek, and A. López, Jr. LTPP *Seasonal Monitoring Program: Instrumentation Installation and Data Collection Guidelines*. Report FHWA-RD-94-110. FHWA, U.S. Department of Transportation, April 1994.

2. Topp, G.C., J.L. Davis, and A.P. Annan, *Electromagnetic Determination of Soil Water Content: Measurements in Coaxial Transmission Lines*, Water Resources Research, Vol. 16, No. 3, June 1980.

3. Roth, K., R. Schulin, H. Flφhler, and W. Attinger, *Calibration of Time Domain Reflectometry for Water Content Measurement Using a Composite Dielectric Approach*, Water Resources Research, Vol. 26, No. 10, October, 1990.

4. Jiang, Y. Jane, S.D. Tayabji, *Preliminary Proposed Procedure for Determination of In-Situ Moisture Contents*, Technical Memorandum, LTPP Data Analysis Technical Support Contract, FHWA, U.S. Department of Transportation, February 1997.

5. Klemunes, J. A. Jr., *Determining Soil Volumetric Moisture Content Using Time Domain Reflectometry*, M.Sc. thesis, University of Maryland, 1995.

6. Bohl, H. and K. Roth, *Evaluation of Dielectric Mixing Models to Describe the θ(_)-Relation*, Symposium on Time Domain Reflectometry in Environmental, Infrastructure, and Mining Applications, September 7-9, 1994.

Fifth International Conference on the
BEARING CAPASITY OF
ROADS AND AIRFIELDS

QUANTIFICATION OF SEASONAL VARIATION EFFECTS OF SUBGRADE SOIL MOISTURE AND PAVEMENT TEMPERATURE ON PAVEMENT PERFORMANCE USING LTPP DATA

FOUAD M. BAYOMY[1], CHERYL A. RICHTER[2] AND ARAMIS A. LOPEZ[2]

[1]Associate Professor of Civil Engineering, University of Idaho, Moscow, Idaho, 83844-1022 USA
FAX (208) 885-6608, E-mail: bayomy@uidaho.edu
[2]Research Engineer, Pavement Performance Division, HNR-30, Federal Highway Administration, McLean, VA 22101-2296, USA

ABSTRACT

The Seasonal Monitoring Program (SMP) of the Long Term Pavement Performance (LTPP) of the Federal Highway Administration (FHWA) has been developed to address, comprehensively, seasonal variation effects on pavement performance. This paper focuses on analyzing data collected at a LTPP seasonal site in the State of Idaho, USA over the period (1989-1997). The comprehensive database for this site includes information on pavement bearing capacity as determined by nondestructive testing using FWD, surface distress, cross and longitudinal profiles, moisture content and temperature. The database includes also data on frost penetration and weather data (precipitation and air temperature) as well as traffic data. Analysis was limited to temperature and moisture effects on performance. Results indicate that no significant moisture variation was observed along the year in the pavement system. Temperature has more profound effect and it controlled the variation of the pavement structural support over the time. The impact of environmental changes appears in the incremental weakening of the pavement system which lead to faster deterioration with time once damage has started.

INTRODUCTION

There are many parameters that affect performance of in service pavements. It may be convenient to categorize these parameters into four groups: a group of parameters related to the pavement structure itself; another group related to the environment; a third group related to the level of use of the pavement which includes traffic loads; and a fourth group

related to the history of pavement improvements including maintenance and rehabilitation activities. The Seasonal Monitoring Program (SMP) of the Long Term Pavement Performance (LTPP) of the Federal Highway Administration (FHWA) has been developed to address, comprehensively, seasonal variation effects on pavement performance (1). There are 64 pavement sites in the SMP experiments. Each site is instrumented with Time Domain Reflectometry (TDR) as a moisture-measuring device, thermistors to measure temperature and resistivity props to measure resistivity as a frost indicator. A local weather station is also installed to measure air temperature and rainfall intensity. The Falling Weight Deflectometer (FWD) is used to test the pavement structure capacity on nearly monthly basis (14 times a year). Distress and roughness measurements are collected quarterly. The SMP sites are part of the General Pavement Studies (GPS of LTPP. For details on the LTPP program, the reader is referred to the original Strategic Highway Research Program (SHRP) research plans (2) and the FHWA-LTPP program reference guide (3). Other data, which are normally collected for the GPS sites, are also collected for the SMP sites, which include inventory, rehabilitation, maintenance, material testing and traffic. LTPP data is stored in a relational database referred to as the LTPP Information Management System (LTPP-IMS) (4).
The scope of this paper is confined to address "environmental" impact on "performance". Two environmental factors are considered: temperature as a major influential factor on asphalt layers; and moisture as a major influential factor on the unbound layers. Performance is addressed from functional as well as structural viewpoints. The Functional aspect is measured by roughness and distresses while the structural aspect is addressed by non-destructive testing (i.e. FWD testing).

SELECTED SITE

The site selected is the SHRP-LTPP site designated as 16-1010. This site is designated as a GPS-1 experiment in Cell 1-130 (2) which was assigned to LTPP in 1989. It is located on the south bound of U.S. Interstate I-15 approximately 1.5 miles south of the city of Robins near Idaho Falls. A small lake is located approximately 0.8 Km away from the site. The test section is located on a divided highway consisting of two traffic lanes in each direction. The outside of the test section is stripped 3.96 m wide. The outside shoulder is approximately 2.7 m wide. The pavement section is composed of 277-mm asphalt concrete over an average of 137-mm granular aggregate base. The subgrade is gravelly-silty sand. The material properties determined from the GPS laboratory and documented in the site installation report (5) are presented in Table 1. The climatic characteristics of the site location based on 11-year data are: Freezing Index is 688 °C-days, Precipitation is 305 mm, number of days above 32 °C is 15, Days below 0 °C is 185, number of freeze-thaw cycles is 136, and number of wet-days is 114. The AADT traffic in 1989 was 3160 (two-way) of which 25% truck traffic. The GPS lane carried approximately 46% of the total AADT. The estimated annual 80 kN Equivalent Single Axle Loads (ESALs) were 147,000. The site was instrumented on September 30, 1993 and initial SMP data was collected next day on October 1, 1993. Other GPS data (Distress, roughness, ...etc. have been collected since October 1990. The 16-1010 site is now de-assigned from the GPS and SMP experiments due to major rehabilitation, which was placed in August 1997.

Table 1. Material Properties of the Pavement Section at Site 16-1010 (After Ref. 5)

Description	Surface Layer	Base layer	Subgrade
Material	AC	Aggregate	Gravely-Silty Sand
Thickness (mm)	277	140	∞
In-Situ Dry Density (kg/m^3)		2047	1877
In-Situ Moisture Content (%)		3.1	6.8
Lab Max Dry Density (kg/m^3)		2243 @ 5% MC	1826 @ 11% MC
Liquid Limit		NP	NP
Plastic Limit		NP	NP
% Passing 75-μm (#200) sieve		8.0	10.7

ENVIRONMENTAL FACTORS

The data collected in the SMP include subsurface temperature, subsurface moisture, and frost conditions, level of ground water table, and air temperature and precipitation. Temperature and moisture data are discussed and used in the performance analysis.

Moisture Variation: Time Domain Reflectometry was used to determine the moisture conditions at ten depths below surface. The first TDR#1 was placed approximately in the middle of base course layer (0.356 m deep from the pavement surface) while the other nine TDR#2 through 10 were placed in the subgrade layer each 0.152 m except for the bottom two where they were spaced 0.305 m. The volumetric moisture content is calculated form the TDR traces (6). A computer program "Moister" was used to automate this process. Analysis of the TDR data revealed that the moisture variations in the top one-meter of the pavement system are somewhat larger than at deeper depths, as shown in Figure 1.

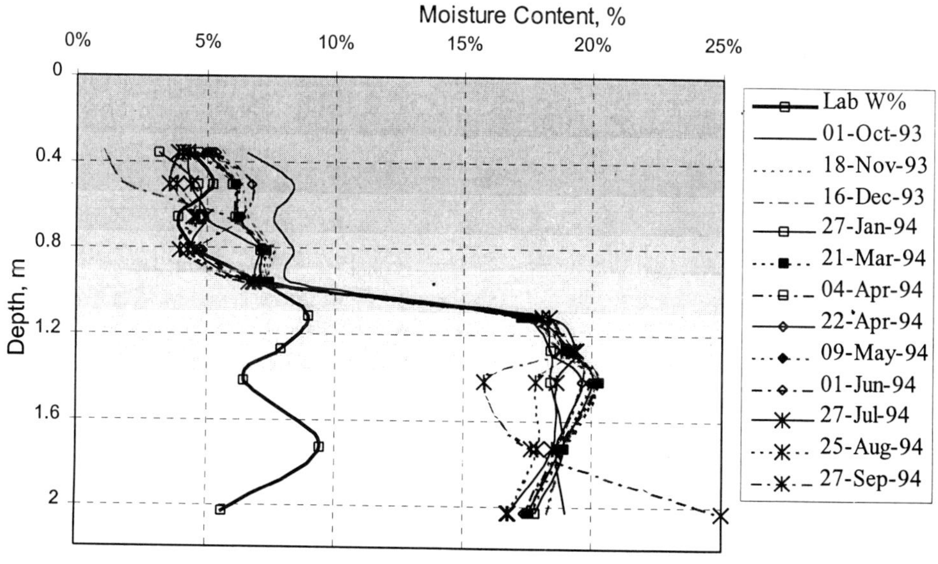

Figure 1. Variation of Moisture Content During One Year Cycle

Below one meter, the soil is over saturated and moisture measured by TDR stayed constant. It is noticed that moisture content, W% measured next day of installation indicates a jump after one meter below the surface. This jump was not evident in the laboratory measured moisture content from the samples extracted at different depths during installation. Data in Figure 1 shows that the jump in moisture values existed all the time. This could be due to the closeness of a small lake.

Temperature Variation: Subsurface temperatures are measured on hourly basis at 18 depths using MRC thermistor sensors. Depths considered are at 0.013 m below surface for the first sensor and goes as deep as 2.138 m for sensor No. 18. Details of installation are found in Ref. (5). In addition, pavement near-surface temperatures were measured at three depths (T1 at 25 mm, T2 at 140 mm, and T3 at 250 mm) using manual temperature probes during the FWD testing.

Temperature distribution for a one-year cycle is shown in Figure 2. The IMS database has temperature data for every hour of the year. The daily statistical average was selected for a daily representation. It is to be noted that the first day of the month was selected arbitrary. The range of temperature change is not expected to deviate far from that shown in Figure 2. Notice that the temperature distribution of Nov. 1, 94 is very close to that of Nov. 1, 93. It is also noticed that the temperature did not reach below freezing at depths deeper than 0.70 m from the surface. This may indicate non-frost conditions in the deep subgrade layers.

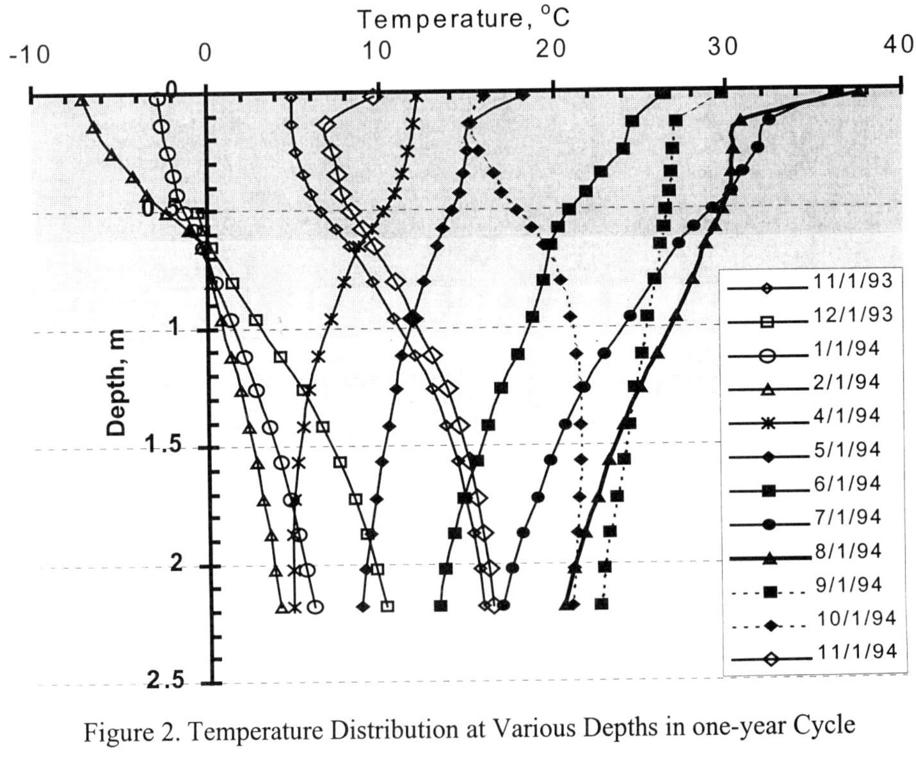

Figure 2. Temperature Distribution at Various Depths in one-year Cycle

PERFORMANCE INDICATORS

Roughness and surface distresses (particularly fatigue cracking and rutting) were considered in the performance analysis. The structural capacity of the pavement system was assessed by non-destructive testing using FWD.

Roughness is measured by Profilometer and Mays meter. International Roughness Index (IRI) in the left and right wheel paths, and the slope variance were calculated and stored in the IMS database.

Surface distresses are collected based on the LTPP distress identification manual (DIM) (8). While many distress elements were considered, data collected for this site have shown that fatigue cracking and rutting were the most dominant distresses. Fatigue cracking was measured by the area of cracked surface. Either a dipstick or photographic survey-based method was used in rut-depth determination.

Structural performance is determined by means of FWD testing. Deflection data measured by FWD were collected every site visit along with pavement temperatures. Modulus backcalculation program (9) was used to estimate the moduli values for the considered site. While results of such analysis were marginally acceptable, they were not acceptable for frozen seasons where deflections were relatively small. Therefore, the structural performance as related to seasonal variations in this site would be assessed based on the change in deflection data rather than the change in moduli values.

ANALYSIS

Fatigue Cracking: Figure 3 shows the progression of fatigue cracking with pavement life. Data shows that for this section cracking did not start until summer of 1991. This means the pavement has served about 22 years (8/69 to 8/91) before it started to crack. It took the asphalt surface of the pavement about six years to fail below an acceptable level of serviceability where a major rehabilitation was needed. The progression of cracking continued in a non-linear form until the pavement was rehabilitated in August 1997.

Rutting: Figure 4 shows the variation of the cross-profile at the instrumented location. The increase in rut depth that is observed does not show significant increase in rut-depth. It could be argued that the 12-mm maximum rut depth shown was mainly due to primary rutting and occurred in the early time the pavement service life. It is also noticed that the rut depth in the inner wheel path was larger than that of the outer wheel path. This could be attributed to non-uniformity of the pavement section or the subsurface conditions across the lane width. However, data collected using photographic method showed less rutting than that measured by dipstick method. Either method showed no progress in rutting during the monitoring period.

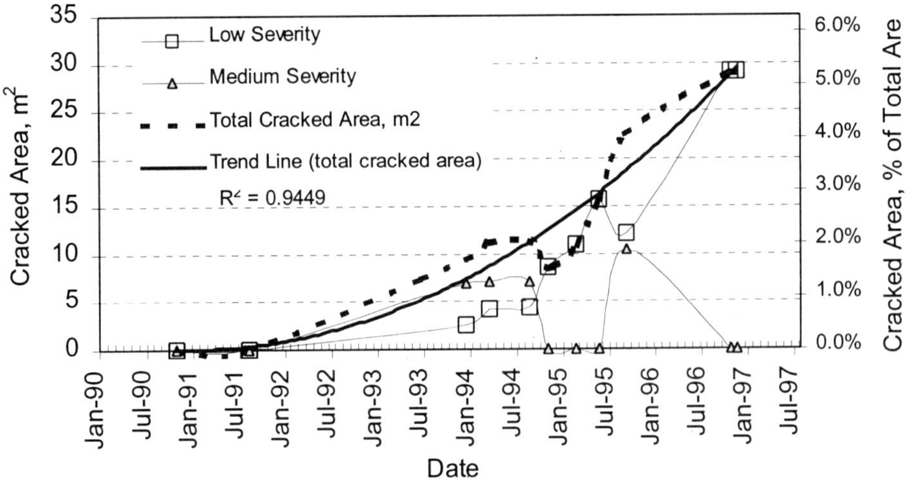

Figure 3. Propagation of Fatigue Cracking

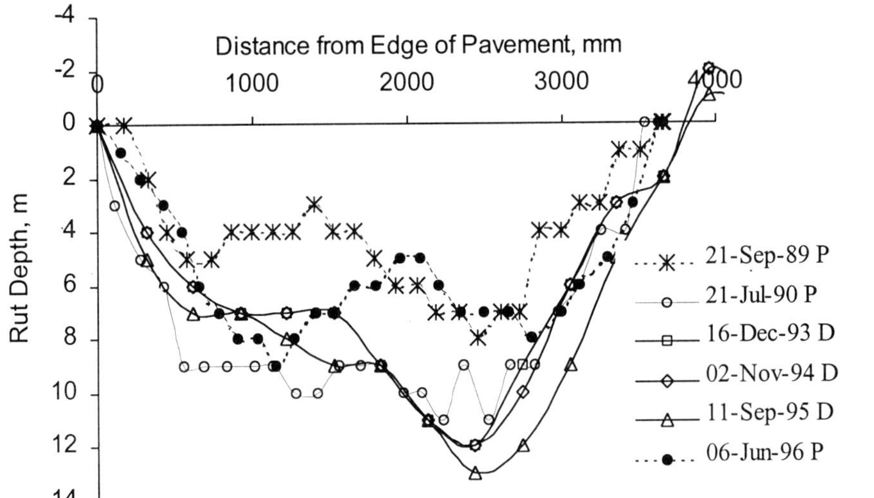

Figure 4. Cross Profile at the Leave End of Section 16-1010

Roughness: Figure 5 shows the change in IRI, Mays meter measurements and the slope variance. It is evident that the pavement roughness increased during the study period (1989 to 97). The IRI values measured by the Profilometer were about 1.5 times of those measured by the Mays meter. The successive increase in roughness indicates the effect of both environmental factors as well as traffic.

Structural Support: Deflection data are plotted versus temperature and moisture in Figure 6. Deflections D1–D7 represent those measured at FWD sensors 1-7. It is evident from

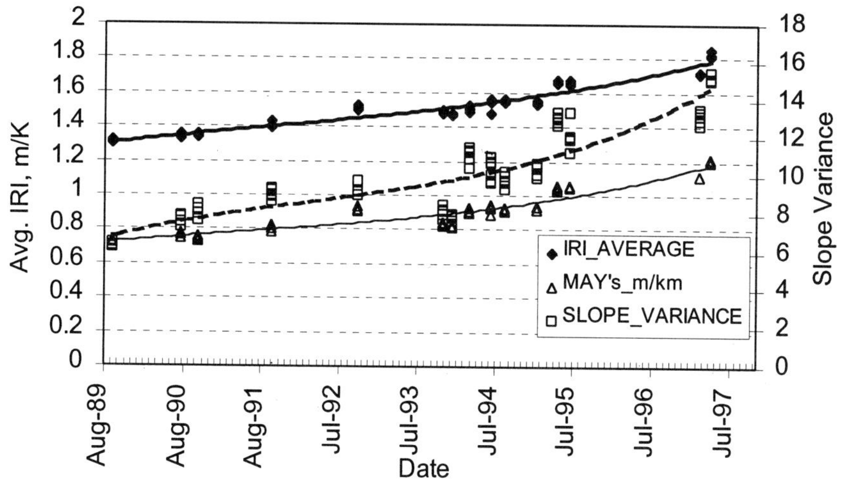

Figure 5. Roughness Measurements

Figure 6 that the variation in deflection patterns along the time (season) followed very closely the variation of the temperature patterns. This validates the great dependency of the structural support on temperature. However, for moisture, there was no great variation that influenced the deflections. It is also noticed that for this location, the low deflections occurred during November-February period. The transition period is quite small and the data collected was not close enough to capture variation in deflections during (summer-winter) or (winter-spring) transitions. However, from the temperature data (Figure 2), it seems that frost depth is no more than 0.60 m. More close analysis was performed by calculating the layer moduli using the MODULUS backcalculation software (9). Results for a one-year cycle are presented in Figure 7. E1 (AC surface layer modulus) was presented as a function of time and temperature while E2 and E3 for the granular base and subgrade soil were presented as a function of time and moisture. TDR#1 presents moisture in the base while TDR#2-9 were selected for different depths of the subgrade. Results show that the variation in E1 can go through three time-regions: high during winter (approximately Oct-Jan), medium during spring (approximately Feb-May), and low during summer (approximately June to Sept). The borders of these ranges are not very definite due to the relatively large gap in data collection. For subbase and subgrade moduli, the time-regions were different. It is observed that two regions may be obtained: high or frozen modulus during winter (Nov-Jan) and normal during the rest of the year. From year to year, it seems that the cycle is repeated without much deterioration in the bearing capacity of the pavement system. This is observed by the deflection distributions in Figure 6. Note that in Round 2 deflections are almost the same. This was also associated by similar distribution of the temperature and moisture. It is therefore concluded that, for this site, progression of failure, functional as measured by roughness and structural as indicated by fatigue cracking is not directly related to environmental changes. However, from roughness and fatigue cracking data, it is believed that the environmental changes cause incremental weakening in the system that cause the damage to progress at a faster rate with time.

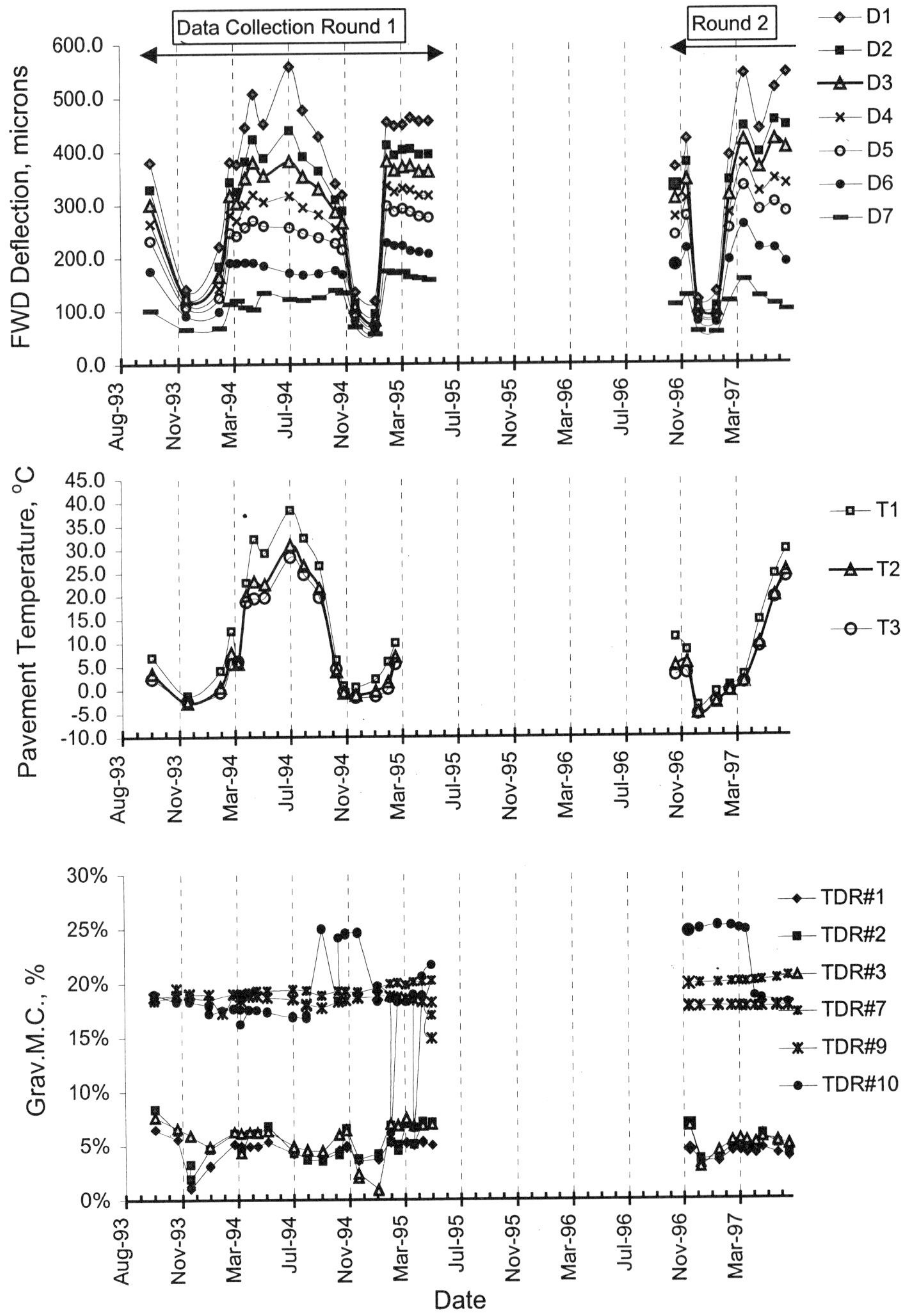

Figure 6. FWD Deflections as related to Temperature and Moisture

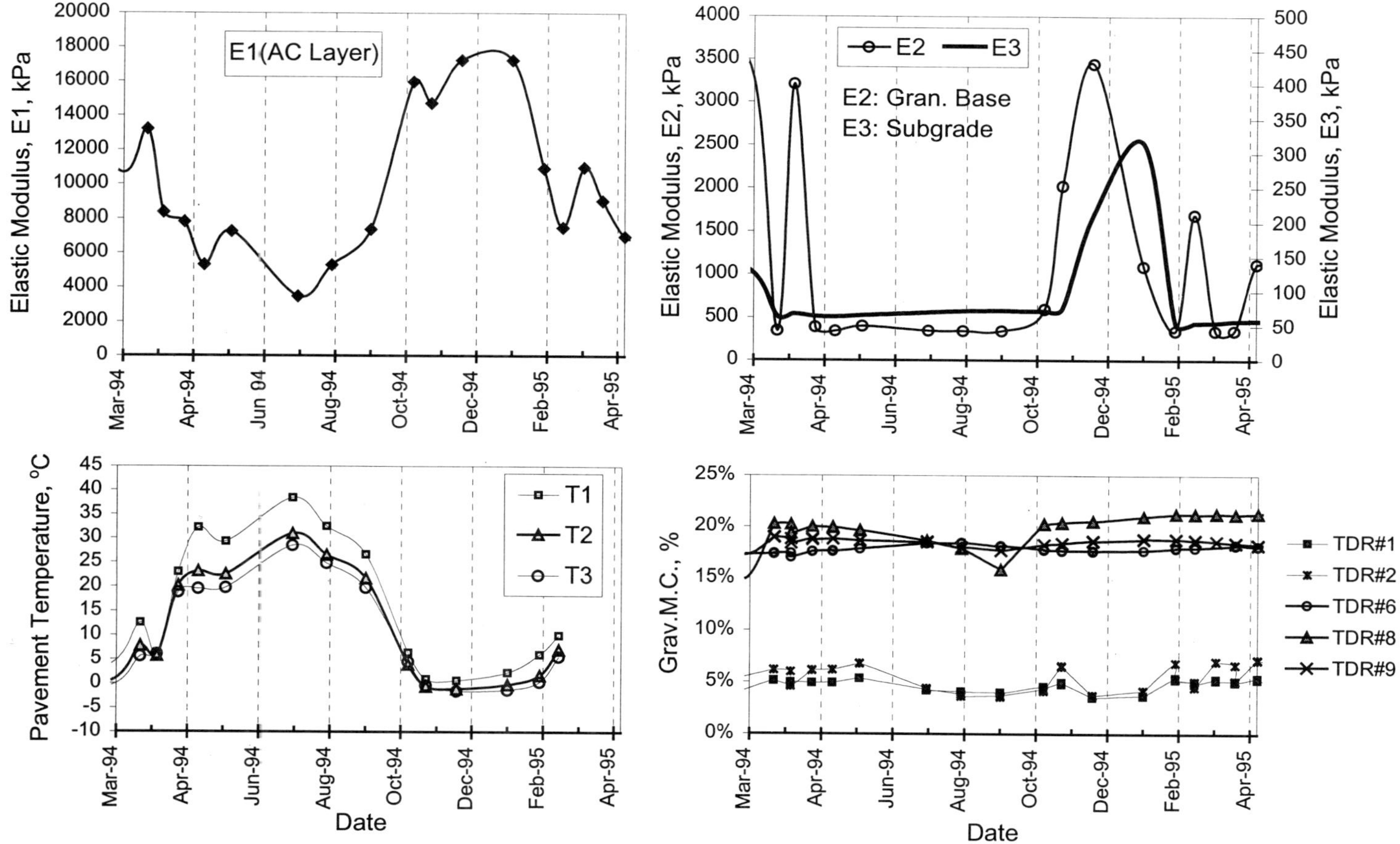

Figure 7. Seasonal Variation of Elastic Moduli (Backcalculated from FWD) with Temperature and Moisture

CONCLUSIONS

Data analyzed at the selected site have indicated:

- There is no significant variation in the moisture content in the subgrade and the base along the year. Accordingly moisture did not influence structural support variations.
- Structural capacity of the pavement system, measured by FWD, varied in a fashion similar to that of the pavement near-surface temperatures (within asphalt bound layers).
- Fatigue cracking was not evident in the initial period of monitoring. However, once it started, it took about 6 years for the pavement to be rehabilitated. Roughness as well increased rapidly during that period, apparently due to increase of fatigue cracking.
- It is believed that the environmental changes cause incremental weakening in the system that cause the damage to progress at a faster rate with time.

ACKNOWLEDGEMENT

This study was performed during an IPA contract between the U.S. Department of Transportation, Federal Highway Administration, Pavement Performance Division and the University of Idaho. Support provided by these agencies is greatly appreciated.

REFERENCES

1) Rada, G., R., Elkins, G.E., Henderson, B., Van Sambeek, B., and Lopez, A. Jr., "LTPP Seasonal Monitoring Program: Instrumentation Installation and Data Collection Guidelines," FHWA-RD-94-110, April 1994.
2) "Strategic Highway Research Program: Research Plans," Final Report, Transportation Research Board, Washington, D.C., May 1986
3) "Program Reference Guide, Version 1.0," LTPP, Federal Highway Administration, Turner-Fairbank Highway Research Center, McLean, VA, March 1996.
4) Ostrom, B., Walker, D., Harris, S., and Rowshan, S., "Long-Term Pavement Performance Information Management System: Data Users Reference Manual," FHWA-RD-97-001, May 1997.
5) "LTPP Seasonal Monitoring Program: Site Installation and Initial Data Collection, Section 161010, Idaho Falls, Idaho", FHWA 16-1010, Federal Highway Administration, Turner-Fairbank Highway Research Center, McLean, VA, February 1994.
6) Klemunes, J., "Determining Soil Volumetric Moisture Content Using Time Domain Reflectometry," FHWA report No. 97-139, Federal Highway Administration, Turner-Fairbank Highway Research Center, McLean, VA, February 1998.
7) ERES Consultants, Inc., "Analysis of Time Domain Reflectometry Measurements: Work Order No. 7, Task 2 – Interim report I," Federal Highway Administration, Turner-Fairbank Highway Research Center, McLean, VA, October 1997.
8) Distress Identification Manual for the Long-Term Pavement Performance Project, SHRP-P-338, Strategic Highway research Program, National Research Council, Washington, DC, 1993.
9) MODULUS Backcalculation Program, Version 4.20, Texas Transportation Institute, College Station, TX, 1993.

Fifth International Conference on the
BEARING CAPASITY OF
ROADS AND AIRFIELDS

CALCULATION OF AC LAYER TEMPERATURES FROM FWD FIELD DATA

R.N. STUBSTAD[1], E.O. LUKANEN[2], C.A. RICHTER[3], AND S. BALTZER[4]

[1]President, Dynatest Engineering, P O Box 865, Ojai, CA 93024, USA; Fax: +1 805 646-7228; E-mail: stubstad@dynatest.com

[2]Principal, Braun Intertec, 6875 Washington Ave. So., Minneapolis,MN 55439, USA; Fax: +1 612 942-3059; E-mail: elukanen@brauncorp.com

[3]Research Engineer, FHWA, 6300 Georgetown Pike, HNR-30, McLean,VA 22101, USA; Fax: +1 703 285-2767; E-mail: cheryl.richter@fhwa.dot.gov

[4]Research Engineer, Danish Road Institute, 4000 Roskilde, Denmark; Fax: +45 46 30 01 05; E-mail: sub@vd.dk

ABSTRACT

The temperature of asphaltic concrete (AC) is a very important data item that is required for analysis of pavement deflection data. Determining the temperature which represents the AC layer at the time of FWD testing is currently accomplished in many ways. Some methods utilize the surface temperature, the air temperature, and/or the previous 5-day mean air temperature as inputs to figures or equations which determine the representative temperature of the AC. Other methods utilize a single manual measurement within the AC layer itself, by drilling a hole in the pavement and manually measuring the temperature at an appropriate depth *in situ.*

The original "BELLS" equation was described in a paper presented at the 1994 BCRA conference and proceedings. This equation was used to determine the temperature of the AC at a given depth, where the 5-day mean air temperature, the infrared surface temperature, the thickness of the AC layer, and the time of day were used as inputs to the equation. A new BELLS equation has now been developed, called "BELLS3", which may be used for normal, routine FWD "production" testing.

The Standard Error of the Estimate for the large BELLS3 data set is ±1.9°C, based on over 10,000 FWD data points, from the Long Term Pavement Performance Program's Seasonal Monitoring Program (SMP), taken throughout the USA and southern Canada. BELLS3 is based on the previous average 1-day air temperature, not the 5-day average normally used.

Key Words: BELLS; Asphalt Pavement Temperature; Pavement Temperature Prediction, Falling Weight Deflectometer; FWD

INTRODUCTION

The correct determination of the representative temperature of asphaltic concrete (AC) pavement layers has always been somewhat of an issue for non-destructive pavement testing, because the deflections and/or moduli derived from these tests are highly temperature-dependent. At the same time, determining the asphalt layer (mat) temperature at each and every Non-Destructive Testing (NDT) test point is generally impractical unless some automated method can be employed. Oftentimes, this problem is addressed in the field by periodically drilling holes in the AC pavement, measuring the temperature manually, and using this local temperature measurement for the following series of NDT test points, say for the next one or two hours until a new manual measurement is made, etc.

Based on earlier work by Southgate *(Ref. 1)*, a new method of determining the temperature within the AC, at each NDT test point, without delaying the testing procedure, was presented by Baltzer, Ertman-Larsen, Lukanen and Stubstad, called "BELLS", in the Proceedings of the Fourth International Conference on the Bearing Capacity of Roads and Airfields *(Ref. 2)*. This method was based on the following input data: the 5-day mean air temperature, the infrared (IR) temperature of the AC surface, the depth, and the time of day. Subsequent data analysis and quality assurance checks have now revealed, however, that the IR temperature gauge used in the first BELLS study was not working properly over the full range of temperatures studied, thus the equation derived only works well on data gathered using that specific IR device. As a result, temperatures derived using the BELLS(1) equation, with a properly functioning and calibrated IR device, will be essentially correct if the mat temperature is in the 15-25°C range, but can be off by as much as ±5°C if the mat temperature is some 20°C above or below these nominal temperatures (ie. around freezing the BELLS1 equation will *over* predict the mat temperature by 5°C while around 40°C BELLS1 will *under* predict by 5°C).

As a result, and also due to the somewhat limited database used in the BELLS1 study, a follow-up study was undertaken as part of the Federal Highway Administration's (FHWA) Long Term Pavement Performance (LTPP) program data analysis effort. This study used data from the LTPP Seasonal Monitoring Program (SMP) where Falling Weight Deflectometer (FWD) tests were conducted at approximately monthly intervals throughout the year *(Ref. 3)*. This study initially used data collected in 1994 and 1995 from some 3,300 FWD test points spread across 25 SMP sites, and was later expanded to include 16 additional sites tested in 1995 and 1996 for a total of over 10,000 FWD test points.

THE USE OF INFRARED TEMPERATURE DEVICES

Using the SMP database from the first 25 sites, an initial investigation was conducted as to the validity of the IR surface temperature data for the eight LTPP FWDs used in this study. This was accomplished by comparing the regression analyses of the upper-most manual AC temperature readings in the database with the IR readings obtained through FWD measurements. The manual temperature readings were taken through predrilled holes in the pavement, concurrent with each FWD test round. The IR readings were obtained through the use of commercially available IR sensors mounted on the FWDs used in the study.

This initial correlation study clearly revealed that there were significant differences between the IR sensors mounted on the FWDs. Further analysis found that the differences were associated with the brand of IR sensor and the calibration of these sensors by the equipment operators. In particular, it was discovered that the temperature compensation circuitry in the particular IR device used for the original BELLS1 data set was faulty and the readings were incorrect unless the temperature of the sensor itself was around 20°C.

In the end, only four of eight IR devices used for the first 25 sites were found to be reliable. These four were relatively new and were of a different manufacturer than the four older IR devices. The newer IR manufacturer is Raytek, which (currently) appears to provide the best product on the market for the purposes at hand. Furthermore, it was found that the factory calibrations provided with the four Raytek devices were consistently better than the "corrected" settings arrived at through periodic operator calibration. Therefore, all four data sets were modified to reflect the IR temperature which would have been recorded using Raytek's factory IR calibration constants, not the operator-modified constants.

It will be shown below that the IR reading is the most sensitive parameter in the BELLS equations. Therefore, to sum up the conclusions about automated IR devices, it is recommended that periodic calibration of IR sensors should be carried out by FWD operators; however, if the constants derived are reasonably close to the factory settings, the factory settings should be maintained, not altered. If they are significantly different from the factory settings, the suspect IR device should be returned to the manufacturer for an electronic and mechanical check, and a new calibration.

THE INITIAL SMP DATABASE

By the end of 1995, there were already several thousand FWD test points in the North American SMP database, spread out over 25 pavement sections. These sections were all tested with the FWD (with concurrent IR and pavement temperature measurements), at various times of the year. These sections included a wide variety of pavement thicknesses, geographic location, climatic variation, and design details.

In order to develop a reliable and sufficiently accurate equation which can be used to predict AC temperatures using easy-to-obtain data, without interrupting the FWD testing procedure and also using little or no destructive measuring techniques, the following database was extracted from existing information:

1) The nominal thickness of the AC layer at each SMP site.
2) Temperature vs. depth measurements within the AC mat at each FWD test site and at each time of test.
3) The air temperature at the time of FWD measurement at each site, along with the high and low temperatures for each of the previous five days and nights.
4) The IR surface temperature using a properly calibrated and functioning IR sensor mounted on the FWD, at each test point and testing time.
5) The time, date, and location of each input temperature in the database.

Before an attempt was made to recalculate BELLS, a comparison between two controlled in-depth temperature measurements at each SMP site was conducted. These two methods of pavement temperature measurement were: *a)* manual readings taken through pre-drilled holes in the pavement, at three depths within the pavement, and *b)* thermistor temperature readings from pre-installed pavement sensors, also at three depths in each pavement section. This comparison is shown in *Figure 1*

As can be seen in *Fig. 1*, even with fairly well controlled in-depth temperature measurements, some scatter in the data is evident. This was true even though about 5% of the available data was "culled" due to very large discrepancies, indicating some obvious (but unspecified) errors in the data. The scatter observed is greatest at higher temperatures. Although the intercept and slope of the regression were close to zero and unity, respectively, the Standard Error of the Estimate (SEE) was approx. ±1.3°C. Naturally, the correlation coefficient (R^2) was very good, at approx. 0.99.

Since the manual temperature data resulted in a better correlation with all the BELLS models subsequently attempted, these temperature measurements were selected as the most correct independent variable data set available, and they were thus used throughout in this study.

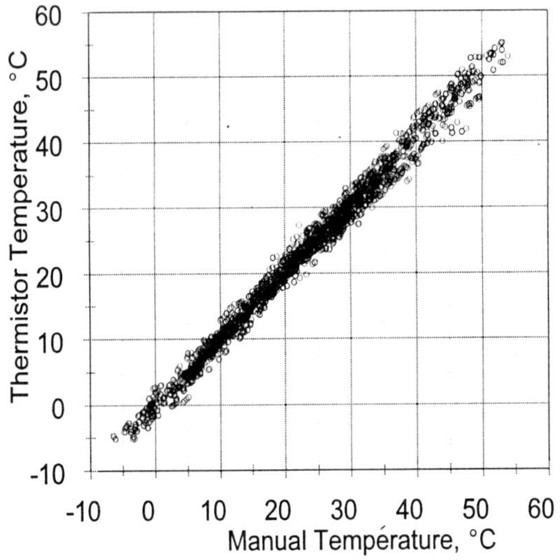

Figure 1 — *Comparison of manual and thermistor temperature data*

BELLS2

Based on the complete pre-1996 LTPP Seasonal Monitoring Program (SMP) data set, and using all properly functioning and calibrated IR devices mounted on the various FWDs located in the four coast-to-coast LTPP regions, a new "BELLS" equation was derived, called "BELLS2". Due to an improvement in the form of the BELLS2 equation and the fact that much more data were available, the SEE for BELLS2 was only ±1.6°C, as compared with ±1.8°C for BELLS1, based on data within a temperature range of 0°- 40°C in both cases.

The obtained value of ±1.6°C is thus considered quite satisfactory indeed, especially considering the fact that the SEE for the two types of in-depth temperature measurements was ±1.3°C, as indicated above and as shown in *Fig. 1*. The plot of the BELLS2 results vs. the as-measured in-depth pavement temperatures is shown in *Fig. 2*.

As mentioned above, there was a calibration or function problem with the IR sensor used to generate the BELLS1 equation. A direct comparison of calculations made with the BELLS1 equation vs. the BELLS2 equation, using the BELLS2 data set, was conducted. The results of this comparison are shown in *Fig. 3.*

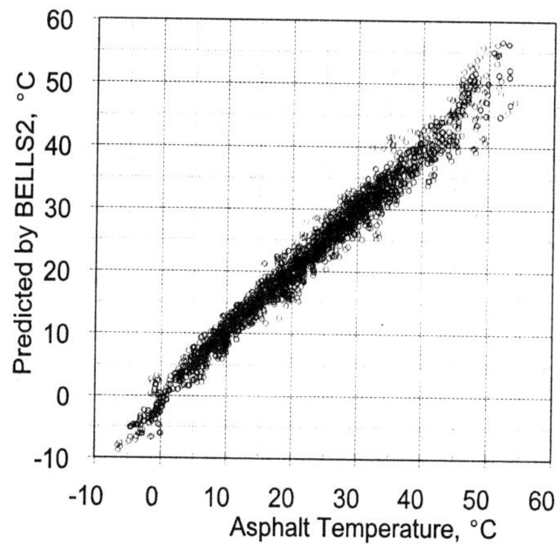

Figure 2 — *Comparison of BELLS2-derived temperatures vs. in-depth temperature*

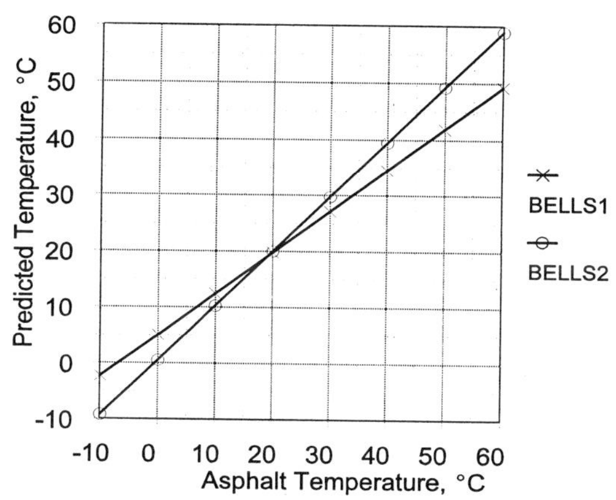

Figure 3 — *BELLS1 vs. BELLS2 calculated in-depth pavement temperatures*

THE EFFECT OF SHADING ON IR TEMPERATURES

Another issue was revealed during this study. It was noticed that the FWD testing procedure involves a large number of drops of the FWD weight at each SMP test point (21 points per pavement section). Each FWD is equipped with a cover which tends to shade the test point. In addition, the tow vehicle is parked over the next test point (~7.6 m ahead) during a test, such that—during sunny or partly cloudy days—each test point is shaded by the testing procedure for a total of approx. six minutes.

In essence what this means is that both BELLS1 (even without the IR problem outlined above) and BELLS2 are only valid for a research-type testing procedure such as LTPP's. For most FWD "production" test procedures, the FWD remains parked for less than one minute, thus on clear or partly cloudy days, the IR temperatures automatically recorded during routine testing will tend to be higher than those measured with the SMP procedure.

In order to adjust the BELLS2 results to correspond to routine FWD production testing, it was necessary to adjust the SMP database considering the effect of shading on the measured IR temperatures.

In order to accomplish this, limited tests were conducted in Florida and California on AC pavements, where the IR temperatures were monitored as a function of shading time. This was carried out, in both locations, on both sunny, partly cloudy, and cloudy days. In so doing, it was noticed that—even on completely cloudy days with no precipitation—some shading still takes place. In summary, the following corrections were applied to the recorded SMP IR temperature readings:

Sky Cover Reported	**Added to IR Temperature**
Sunny	+4°C
Partly Cloudy	+3°C
Cloudy	+1.5°C

To develop the BELLS3 equation, below, these temperatures were added to all SMP IR temperatures, when the sky condition was noted as being in any of the three above categories, which was every test point in the database except where "rain/precipitation" was noted.

BELLS3

Meanwhile, the SMP database had one more year under its belt, resulting in a grand total of over 10,000 data points. The number of test sections had also increased, from 25 to 41, representing a larger geographical and climatic variation as well as more rounds of SMP testing on the original pre-1996 test sections. A map of these locations is shown in *Fig. 4.*

First, the BELLS2 regression was checked by calculating the temperatures in all new SMP sections using the BELLS2 regression coefficients, and comparing these with the measured in-depth temperatures, both at mid-depth and at the third-point depth. It was found that there was no practical difference in the calculated vs. the measured temperatures, thus the BELLS2 model was verified with a large volume of new test data. In this case, the average difference between the predicted and measured temperature is +0.16°C with a standard deviation of the differences of ±1.85°C.

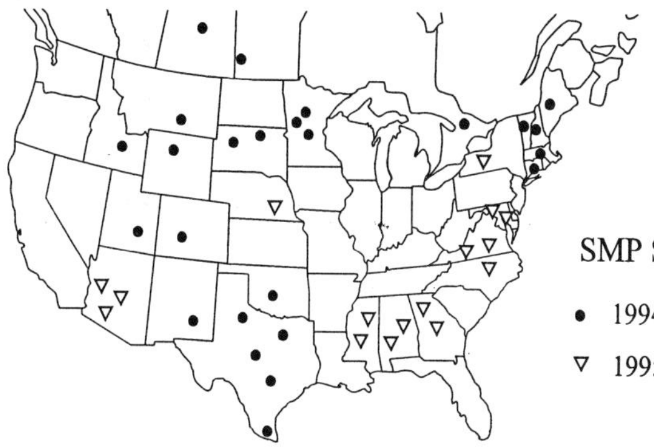

Figure 4 — *Location of final SMP study sections*

As mentioned earlier, both the BELLS2 and BELLS3 models were somewhat different from BELLS1 in terms of the form of the regression equation used, in several respects, as follows:

1. It was found that the previous one- and two-day average air temperatures were better correlated to the existing pavement temperatures than the previously-recommended 5-day mean air temperature. For simplicity, the previous day's average air temperature was used throughout, ie. the average of the previous day's high and low temperatures measured at the weather station on site.
2. Instead of the six-variable multiple regression form used in BELLS1, a better five-variable form was found, and used throughout.
3. Instead of using two shifted 24-hour sine waves to represent the temperature changes in-depth within the AC mat, a so-called "18-hour" sine wave, with a flat temperature representing an early morning "temperature reversal" time period of 6 hours was employed. This was based on a better correlation with actual measured temperatures, in-depth. The two shifted 18-hour sine waves are shown in *Fig. 5.*

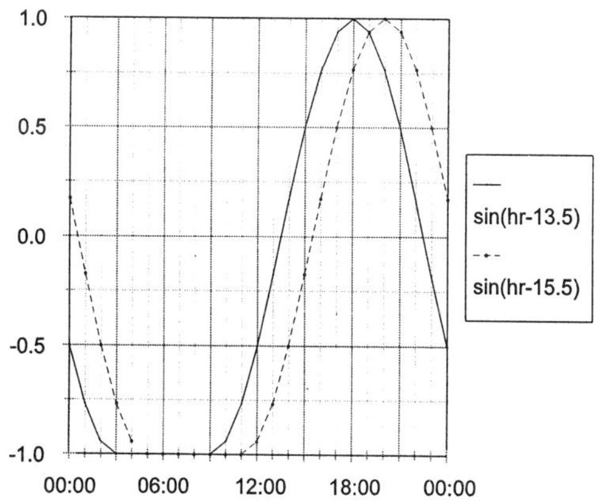

Figure 5 — *18-hour cycle in-depth temperature sine function*

The resulting BELLS3 regression equation is as follows:

$$T_d = 0.95 + 0.892IR + (\log(d) - 1.25)(-0.45IR + 0.62(1\text{-}day) + 1.83\sin(hr_{18} - 15.5)) + 0.042\sin(hr_{18} - 13.5)$$

where:

T_d	=	pavement temperature at desired depth d, ^{o}C
IR	=	infrared surface temperature, ^{o}C
log	=	base 10 logarithm
d	=	depth (eg. mid-depth) at which mat temperature is to be predicted, mm
1-day	=	average of the previous day's high & low air temperature
sin	=	sin function over an 18-hour period, and 2π radians equal to one 18-hour cycle
hr_{18}	=	time of day in the 24-hour system, but calculated using an 18-hour AC temperature rise and fall time, as explained in the notes below

Notes: When using the $\sin(hr_{18}-15.5)$ decimal function, only use times between 11:00 hrs and 05:00 hrs the following morning. If the actual clock time is not within this time range, then calculate the sin as if the time is 11:00 hrs (ie. where the sin = –1). If the time is between midnight and 05:00 hrs, add 24 to the actual decimal time. Then calculate as indicated in the following example: Actual clock time is 13:15 hrs; in decimal form, 13.25 – 15.50 = –2.25; –2.25/18 = –0.125; –0.125*2π = –0.785 radians; sin(–0.785) = –0.707. [Note that an 18-hour sin function is assumed, with a flat "–1" segment between 05:00 and 11:00 hrs in the morning.]

When using the $\sin(hr_{18}-13.5)$ decimal function, only use times between 09:00 hrs and 03:00 hrs the following morning. If the actual clock time is not within this time range, then calculate the sin as if the time is 09:00 hrs (ie. where the sin = –1). If the time is between midnight and 03:00 hrs, add 24 to the actual decimal time. Then calculate as indicated in the following example: Actual clock time is 15:08 hrs; in decimal form, 15.13 – 13.50 = 1.63; 1.63/18 = 0.091; 0.091*2π = 0.569 radians; sin(0.569) = 0.539. [Note that an 18-hour sin function is assumed, with a flat "–1" segment between 03:00 and 09:00 hrs in the morning.]

All 10,304 data points resulted in the following BELLS3 statistics: SEE = ±1.9°C and R^2 = 0.975. For ease of use, a *BASIC* program listing of the BELLS3 algorithm is presented in *Fig. 6.*

A plot of the data set of the measured temperatures (at both the mid- and third-depths in the AC mat) vs. the BELLS3 calculated values is plotted in *Fig. 7.*

```
'Program to illustrate the implementation of the BELLS3 equation.
'***************************************************************

CLS

INPUT "Input surface temperature, °C "; ir
INPUT "Input hour of test, 24 hour system "; hr
INPUT "Input minutes past the hour "; min
INPUT "Input the depth for predicting the asphalt temperature, mm "; d
INPUT "Input average air temperature for the day before the test date, °C "; air

decimal.hrs = hr + min / 60

IF decimal.hrs > 11 OR decimal.hrs < 5 THEN
     IF decimal.hrs < 5 THEN decimal.hrs = decimal.hrs + 24
     sine15.5 = SIN(2 * pi * (decimal.hrs - 15.5) / 18)
  ELSE
     sine15.5 = -1
END IF

IF decimal.hrs > 9 OR decimal.hrs < 3 THEN
     IF decimal.hrs < 3 THEN decimal.hrs = decimal.hrs + 24
     sine13.5 = SIN(2 * pi * (decimal.hrs - 13.5) / 18)
  ELSE
     sine13.5 = -1
END IF

td = 0.95 + .892 * ir
logdepth = LOG(d) / LOG(10) - 1.25
firstbracket = -.45 * ir + .62 * air + 1.83 * sine15.5
last.term = .042 * ir * sine13.5
td = td + logdepth * firstbracket + last.term

PRINT "The predicted temperature is "; td
```

Figure 6 — *BASIC program listing for the BELLS3 algorithm*

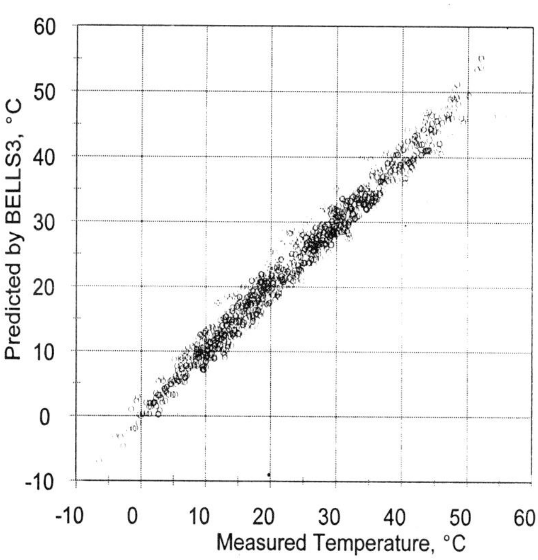

Figure 7 — *Comparison of BELLS3 temperatures vs. in-depth temperatures*

CONCLUSIONS AND RECOMMENDATIONS

Use of the BELLS3 equation for predicting in-depth asphalt pavement temperatures is recommended because existing methods are decidedly worse than the above, other than perhaps very careful and time-consuming & destructive manual measurements *in situ* in-depth, on a point-by-point basis. Even if this is done, problems often exist because of the heat generated by the drill bit, plus the measured temperatures are significantly affected by the amount of liquid poured into the hole and/or by how long the equipment operator leaves the manual temperature probe inserted into the pavement, etc.

Some "fine tweaking" of the BELLS3 equation is recommended at some point in time, because the shading adjustments made to the SMP data set are based on a small amount of data. Further research could clear up this uncertainty.

It should, however, be remembered that the BELLS3 equation was developed using daytime FWD and IR data only, thus extrapolation into the nighttime hours may not be accurate. The range of AC thicknesses employed should also be adhered to, ie. BELLS3 should only be used for AC pavement thicknesses between 45 and 305 mm. Also, IR readings taken on pavements which are at or below freezing can negatively affect the accuracy of the IR temperature readings, thus below-freezing use of IR sensors on AC pavements, and the subsequent use of BELLS, is not recommended.

Most importantly, the accuracy and calibration of the IR sensor is vital, because the BELLS3 equation is mostly affected by this one reading alone, with a t-stat = 292. [The next-most important t-stat, = 92, is the previous 1-day mean air temperature.] Similarly, BELLS2 is most influenced by the IR temperature reading as well.

ACKNOWLEDGEMENTS

This work was sponsored by the U.S. Federal Highway Administration as a part of the LTPP data analysis effort. Participation of the FHWA is gratefully acknowledged.

Details of the above findings and conclusions, with much more additional information and research results, are published by the FHWA in *Ref. 3.*

REFERENCES

1. SOUTHGATE, H.F. "An Evaluation of Temperature Distribution of Asphalt Pavements and its Relationship to Pavement Deflection". MSCE Thesis, University of Kentucky, Lexington, Kentucky, 1968.

2. BALTZER, S., ERTMAN-LARSEN, H.J., LUKANEN, E.O. and STUBSTAD, R.N. "Prediction of AC Mat Temperatures for Routine Load/Deflection Measurements". Proceedings of the 4th International Conference on the Bearing Capacity of Roads and Airfields, Minneapolis, Minnesota, pp. 401-410, Vol. 1, 1994.

3. LUKANEN, E.O., STUBSTAD, R.N. and BRIGGS, R.C. "Temperature Predictions and Adjustment Factors for Asphalt Pavements". FHWA Report No. FHWA-RD-98-085, 1998.

SEASONAL VARIATION OF PAVEMENTS RESPONSE TO FALLING WEIGHT DEFLECTOMETER

MARIA DE LURDES ANTUNES[1], ANTÓNIO PINELO[2] AND ANTÓNIO GOMES CORREIA[3]

[1]Research Officer, Laboratório Nacional de Engenharia Civil, P-1799, Lisboa, Portugal
[2]Principal Research Officer, Laboratório Nacional de Engenharia Civil, P-1799, Lisboa, Portugal
[3]Professor, Technical University of Lisbon/IST/DECivil, P-1096, Lisboa, Portugal

ABSTRACT

This paper presents some results of a research study performed by LNEC concerning the seasonal variation of pavements response to Falling Weight Deflectometer. An airport pavement was instrumented with thermocouples inside the bituminous layers and tensiometers in the subgrade. The study comprised the observation of the pavement's behaviour during a period of more than one year, through Falling Weight Deflectometer testing, and measurement of temperatures in the bituminous layers and pore water pressure in the subgrade. These *in situ* tests were complemented with site investigation and laboratory tests aiming at the mechanical characterisation of the materials.

This work allowed to get indications about the variability of the soil's in situ modulus with the variations of the pore water pressure. Regarding the bituminous layers, relationships between the moduli and temperature were derived. For both cases, there was a reasonable agreement between the in situ behaviour and the laboratory test results.

INTRODUCTION

It is acknowledged that climatic conditions can have a significant influence on the structural behaviour of flexible pavements. Besides frost, the most frequently considered climatic effect is the influence of temperature in the behaviour of bituminous layers. In Portugal, this is usually taken into account in pavement design, by using relationships between temperature and bitumen stiffness, such as the ones established by Shell (1). The elastic modulus of the bituminous materials is calculated from the volumetric mix composition and the bitumen stiffness corresponding to the design temperature. The effects of hydrologic conditions on the performance of unbound granular layers and soils are indirectly taken into account by using appropriate construction practices.

In studies concerning bearing capacity evaluation of in-service pavements using non-destructive tests, such as the Falling Weight Deflectometer (FWD), climatic conditions during testing must also be taken into account. The elastic moduli of the pavement's layers and subgrade derived from the test results must be "corrected" in order to take into account the difference between the climatic conditions during testing, and the ones that are considered as "representative" for design purposes. However, the variation of the modulus of in-service bituminous layers with temperature can be different from the relationships considered in pavement design, due to numerous effects, such as ageing, for example. On the other hand, the sensitivity of the subgrade modulus to hydrologic conditions can be important, and this is very often not taken into account in pavement's bearing capacity evaluation.

This paper presents some results of a research study performed by LNEC concerning the seasonal variation of pavements response to FWD (2). The study comprised the observation the behaviour of an airport pavement during a period of more than one year, through FWD testing, and measurement of temperatures in the bituminous layers and pore water pressure in the subgrade. These *in situ* tests were complemented with site investigation and laboratory tests aiming at the mechanical characterisation of the materials, for comparison with elastic moduli derived from back-analysis of FWD test results.

SITE DESCRIPTION

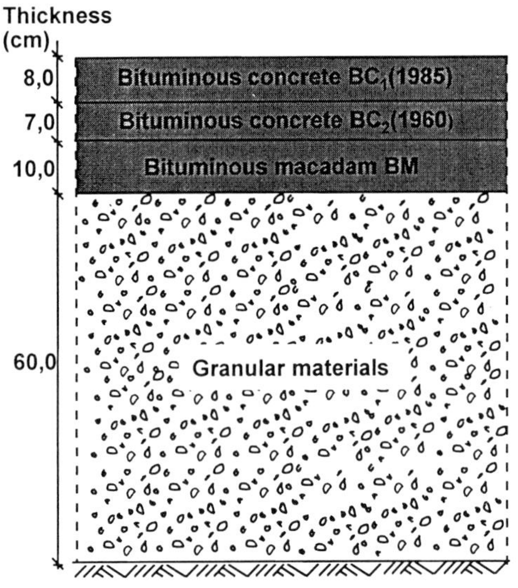

Figure 1 – Pavement's structure

The pavement under study is a stretch of a taxiway at Lisbon airport, built in 1960, and later reinforced with a bituminous overlay. The pavement's structure, represented in figure 1 consists of unbound limestone granular materials, with a total thickness of 0.60 m, a layer of bituminous macadam (BM), 0.10 m thick, and two layers of bituminous concrete. The lower layer (BC_2), which is 0,07 m thick, was the original wearing course, and the upper layer (BC_1), 0.08 m thick, was placed in 1985 as an overlay. Table 1 presents the composition of the bituminous mixtures and the characteristics of the recovered binder. These characteristics were determined from testing cores extracted from the pavement.

From the results of table 1, it can be concluded that the bottom layer, "bituminous macadam", has a rather low bitumen content, compared to the two upper layers. Cores taken from the pavement also showed that the material from that layer looked very heterogeneous.

The subgrade is formed by an upper layer of clayey sand, classified as SC, according to ASTM D 2487-85. The *in-situ* compaction characteristics and water content, determined at the beginning of the study, is presented in table 2, together with the results from compaction tests (AASHTO T 180-86). From geophysical site investigations (3) it was concluded that there is a rock layer (limestone) at a depth of approximately 2 m.

Table 1
Composition of the bituminous mixtures and characteristics of the recovered binder

Layer	ρ_{mix} (g/cm^3)	P_b (%)	V_v (%)	V_b (%)	P_{200} (%)	Pen_{25} (0,1 mm)	T_{rb} (°C)
BC_1	2,39	*4,5*	4,1	10,3	9,2	29	60
BC_2	2,42	*3,9*	4,3	9,2	8,5	44	53
MB	2,42	*2,0*	7,5	4,7	4,3	39	56

Symbols:

ρ_{mix} - Density;
V_v - Void content by volume;
V_b - Binder content by volume;
P_{200} - Percent aggregate passing N° 200 ASTM sieve.
P_b - Binder content by weight;
Pen_{25} - Penetration of recovered binder;
T_{rb} - Rec. binder softening point;

Table 2
Compaction characteristics of the subgrade soil

Compaction test (1)		*In situ* characteristics (2)		
$\rho_{d\,max}$ (kg/m^3)	ω_{opm} (%)	$\rho_{d\,o}$ (kg/m^3)	ω_o (%)	S_r (%)
1890	13,9	1800	16,8	90,1

Symbols:

$\rho_{d\,max}$ - Maximum dry density
ω_{opm} - Optimum moisture content
(1) - Modified Proctor
$\rho_{d\,o}$ - *In situ* dry density
ω_o - *In situ* moisture content
S_r - Saturation degree
(2) - Determined in February1992

LABORATORY TESTS

This section presents laboratory tests performed on materials extracted from the pavement and subgrade, aiming at the determination of E-moduli of the materials, to be compared with results derived from back-calculation of FWD test results. These tests were performed on samples from the bituminous bound layers – bituminous concrete and bituminous macadam – and from the subgrade soil. The samples from bituminous materials were obtained through coring in the pavement, whereas the soil samples were obtained from site investigations near the pavement's shoulder. Due to airport restrictions, it was not possible to perform test pits inside the pavement, therefore, no samples were collected from the granular layers.

Mechanical characterisation of bituminous mixtures – Cores extracted from the pavement were used for preparation of three sets of 6 cylindrical specimens representing the three existing bituminous layers. Each of these specimens was used for indirect tensile tests at three temperatures - 0°C, 20°C and 30°C - for determination of the E modulus of the material. These tests were performed with a servo-hydraulic equipment existing at LNEC, which is able to induce repeated loads with numerous shapes, in controlled stress or controlled strain.
The applied load was a haversine load pulse, with a duration of 0.05s, in order to be as similar as possible to the load pulse applied by the KUAB FWD. This load configuration was used with the purpose of comparing the moduli determined in laboratory with the

moduli derived from FWD test results. Each load cycle was followed by a rest period of 0.45s. The tests were performed in controlled stress. The load amplitude applied was chosen to correspond to approximately 30% of the tensile strength of the material, also determined in indirect tensile tests. The E moduli of the materials were determined from the recovered vertical displacement after 200 load cycles at each temperature. Table 3 presents the average modulus obtained for each type of mixture.

Table 3

Average moduli (E) obtained in indirect tensile tests

Temperature	0°C		20°C		30°C	
Layer	No of tests	E (MPa)	No of tests	E (MPa)	No of tests	E (MPa)
BC1	5	13990	6	8970	6	6820
BC2	4	12180	6	8370	6	5840
BM	5	8520	6	5230	5	4040

Mechanical characterisation of the soil – The mechanical characterisation of the subgrade soil was performed through resonant column tests on laboratory compacted specimens (4). In these tests, a cylindrical soil specimen is submitted to shear dynamic loads covering a range of frequencies, until resonance is achieved. From the natural frequency of the sample thus determined, the dynamic shear modulus (G) and the damping (D) can be obtained. The corresponding Young modulus (E) can be calculated from the expression:

$$E = 2G(1+\nu)$$

The test specimens were prepared with the same compaction and water content as the ones measured *in situ* (see table 2). The tests were performed for several confining stresses, covering a range of applied shear deformations which was considered representative of the effect of vehicle loads. Figure 2 presents the results, corresponding to the average of results obtained from two specimens.

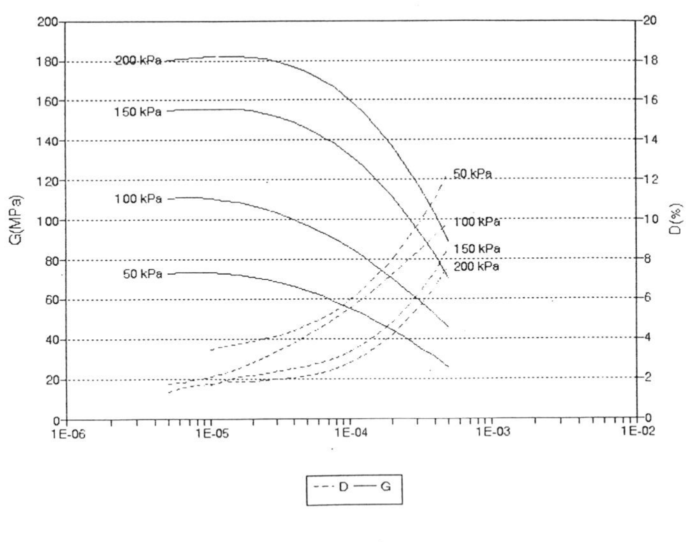

Figure 2 – Resonant column test results obtained with soil specimens from the subgrade

IN-SITU TESTS

Instrumentation – The pavement was instrumented with the purpose of measuring the temperature inside the bituminous layers and the pore water pressure in the soil.

Resistance thermometers were installed inside the pavement at 0.02, 0.05, 0.09, 0.13, 0.18 and 0.23 m from the surface, connected to a data logger which automatically recorded the temperature for each channel every 30 minutes. Thus, the temperature profiles through the full depth of the bituminous layers (see examples in figure 6) were obtained. Air temperature was also recorded using the same equipment.

Three tensiometers were installed in the subgrade, with the porous cell approximately 0.60 to 0.70 m below the top of the subgrade. These apparatus allow for assessing the pore water pressure in the soil, through the measurement of the pressure of the water contained in the porous cell, once this is in equilibrium with the soil water (5) (6). The pressure was measured with a mercury column, every month, together with FWD tests. In addition, data concerning monthly precipitation was collected from meteorological records.

FWD tests – During a period of 18 month, FWD tests were performed monthly, at ten fixed locations, covering an area 50 m long and 12.5 m wide, around the instrumented section. The equipment used was a KUAB 150, with a 450 mm diameter loading plate, and 7 deflection transducers located at 0, 0.30, 0.45, 0.60, 0.90, 1.5 and 2.5 m from its centre. At each testing point, three heights of fall were used, corresponding to nominal peak loads in the order of 50, 100 and 150 kN.

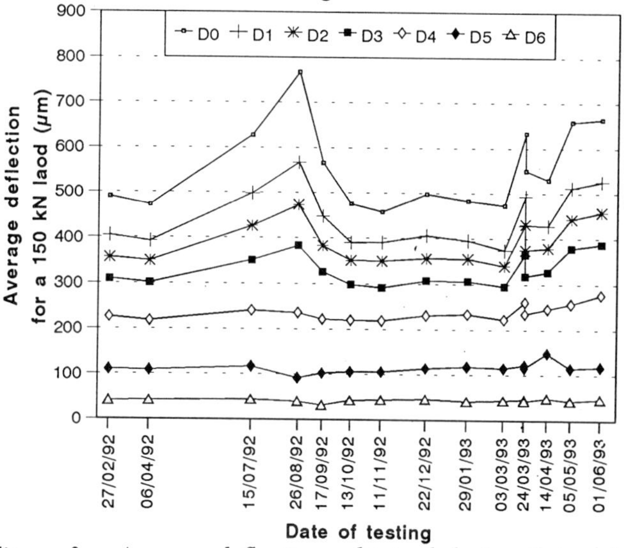

Figure 3 – Average deflections obtained during the study period induced by 150 kN load

Figure 3 shows the variation of the average deflections obtained at the third height of fall during the study period, normalised for 150 kN.

INTERPRETATION OF TEST RESULTS

Temperature of bituminous layers – One of the issues addressed in this study is the variation of the back-calculated elastic moduli of the bituminous layers as a function of the layer's temperature, measured as described in the previous section. The back-calculation of elastic moduli was performed considering two layers of bituminous materials: the first one comprising the two bituminous concrete layers (referred previously as BC_1 and BC_2) and the second comprising the bituminous macadam (BM), which is significantly different from the previous ones, in terms of composition.

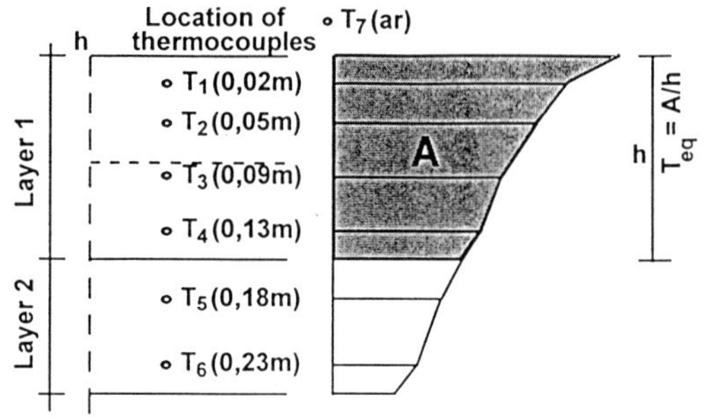

Figure 4 – Calculation of "equivalent temperature"

For each set of FWD tests, performed at least once a month, a representative temperature was calculated for the two layers considered in the back-calculation. This was the weighted average of measured temperatures with depth (figure 4), designated as "equivalent temperature" (T_{eq}).

The choice of this equivalent temperature was supported by a number of calculations of theoretical deflections induced by FWD loads, using a Finite Element program (2), for two types of pavement structures: structure 1, where the bituminous materials were subdivided in a number of layers corresponding to the number of temperature measurements, and structure 2, where only one layer was considered, figure 5.

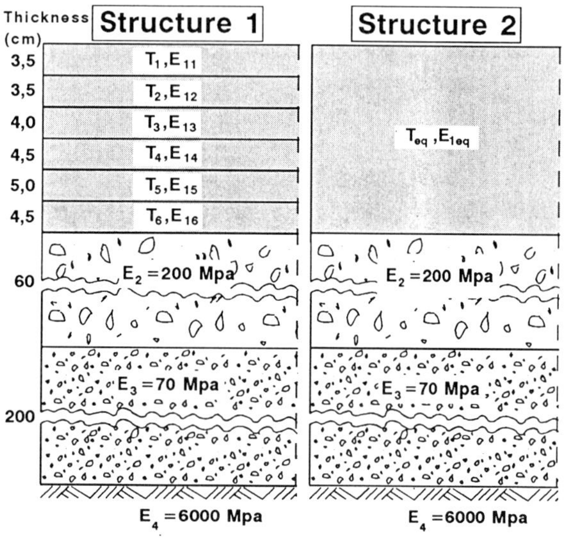

Figure 5 – Structures analysed for supporting the calculation of equivalent pavement temperatures

In structure 1, an elastic modulus was assigned to each individual bituminous layer (E_{1i}), as a function of temperature (t_i). This elastic modulus was calculated for the average mix composition of layers BC1 and BC2 (table 1), using the Shell nomographs (1). In structure 2, the same procedure was used to calculate an equivalent modulus (E_{1eq}) corresponding to the equivalent temperature (T_{eq}).

Figure 6 presents two examples of results obtained for two extreme cases of measured temperature profiles: one corresponding to high temperatures, and another corresponding to low temperatures.

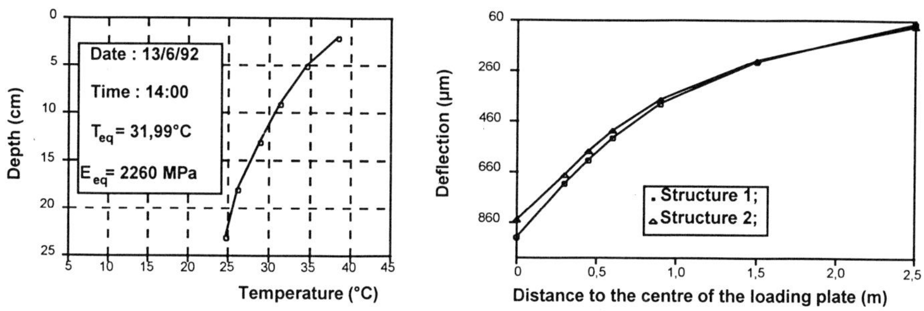

Figure 6 a) – Comparison of deflections obtained for structures 1 and 2, for a situation of high temperature

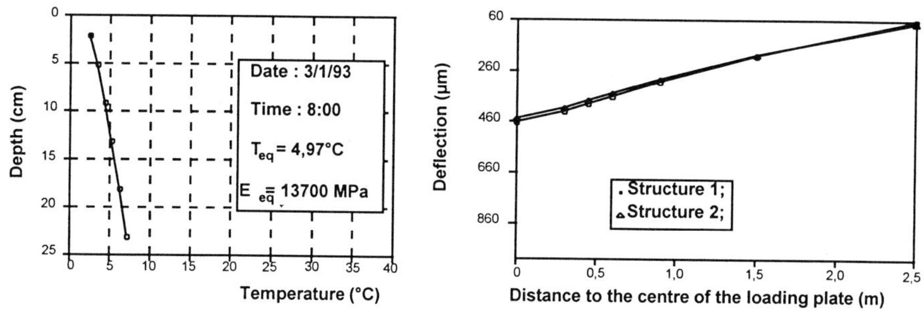

Figure 6 b) – Comparison of deflections obtained for structures 1 and 2, for a situation of low temperature

Back-calculation of layer moduli – The back-calculation of layer moduli from FWD test results was performed using a finite element computer program specially developed for linear elastic analysis of pavement structures under dynamic loads (2).

The pavement was considered as a three layer structure: the first layer comprising the two upper layers of bituminous concrete, with a total thickness of 15 cm, the second consisting of the 10 cm thick bituminous macadam, and the third being the 60 cm thick unbound granular layers. According to the results from geophysical investigations, the subgrade was sub-divided in 2 layers: an upper layer of soil, 2m thick, and a rock layer underneath, with a fixed E-modulus of 2000 MPa. Typical values were fixed for the materials Poisson coefficients, as well as for the densities and percent of damping.

Table 4 presents an example of back-calculated layer moduli obtained from tests at three different load levels. The moduli obtained for the soil are compared with the results obtained in the resonant column tests, for the same shear strain levels, and for a confining stress of 50 kPa, which is considered to be in the order of the in situ confining stress, taking into account the structure's self weight.

The results presented in table 4 show a reasonable agreement between subgrade E-moduli derived from FWD tests, and moduli obtained from laboratory tests. These results also illustrate the stress dependency of the behaviour of the soil and granular material.

Table 4

Layer moduli obtained from tests at three different load levels

Date of test	Peak load (kN)	Bituminous concrete		Bituminous macadam		Granular material	Subgrade (upper layer)		
		T_{eq} (°C)	E_{FWD} (MPa)	T_{eq} (°C)	E_{FWD} (MPa)	E_{FWD} (MPa)	E_{FWD} (MPa)	γ_{rz} (10^{-6})	E_{RC} (MPa)
03.93	48.4	27	4000	20	2900	320	150	130	145
	96.8		4400		2960	230	130	280	110
	133.3		4500		3230	190	120	420	90

Symbols:

T_{eq} - Equivalent temperature of bituminous layers

E_{FWD} – Moduli back-calculated from FWD test results

γ_{rz} - Shear strain in the subgrade corresponding to FWD load

E_{RC} - Moduli calculated from resonant column test results

The results obtained during the study period for the E-moduli of the bituminous materials are presented in figures 7 and 8, respectively for bituminous concrete and bituminous macadam, as a function of the layer's equivalent temperature at the time of testing. These results are compared with the results obtained in laboratory indirect tensile tests.

From the results presented in figures 7 and 8, it seems reasonable to consider a linear relationship between the E-modulus of bituminous materials and the temperature, for the range of temperatures that occurred during this study. This can also be concluded from the indirect tensile test results, also in figures 7 and 8, although there is a shift between the moduli obtained in laboratory and the moduli obtained from FWD tests. Equations 1 and 2 express the relationship obtained between the FWD back-calculated modulus at temperature t (in °C), E_t, and the corresponding modulus at 20°C, E_{20}, respectively for the bituminous concrete and for the bituminous macadam.

$$E_t^{BC} = (1.635 - 0.0317)E_{20}^{BC} \quad (1)$$

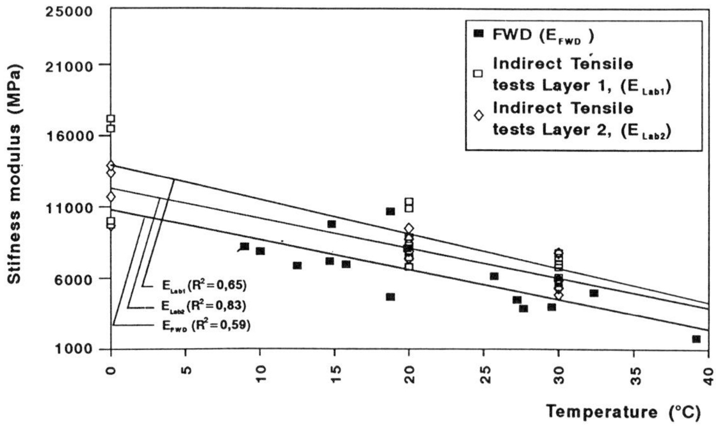

Figure 7 – Variation of the modulus of the bituminous concrete with temperature

The results obtained are not in agreement with the temperature dependency usually considered for the stiffness of bituminous materials (1), but it should be stressed that the relationships derived from this study refer to "in service" materials, for which a certain hardening of the binder has already taken place (see table 1). Furthermore, it must be pointed out that the bituminous macadam has a very low binder content, which presumably makes it less sensitive to temperature than other mixtures.

$$E_t^{MB} = (1.795 - 0.0398t)E_{20}^{MB} \qquad (2)$$

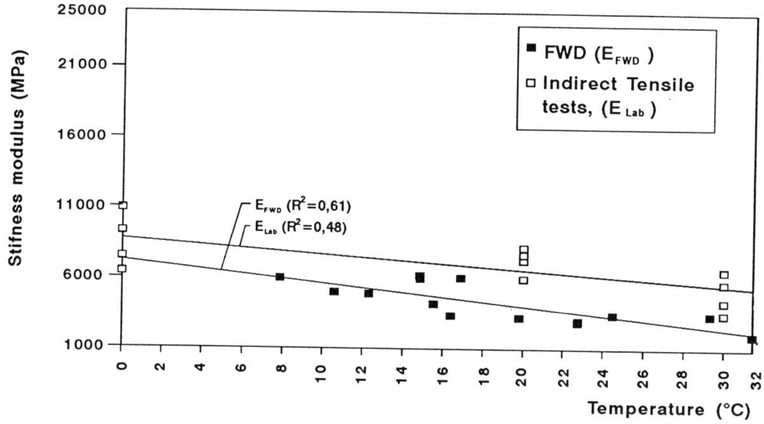

Figure 8 – Variation of the modulus of the bituminous macadam with temperature

Figure 9 presents the seasonal variation of the subgrade modulus, together with the variation of pore water pressure in the soil and monthly precipitation.

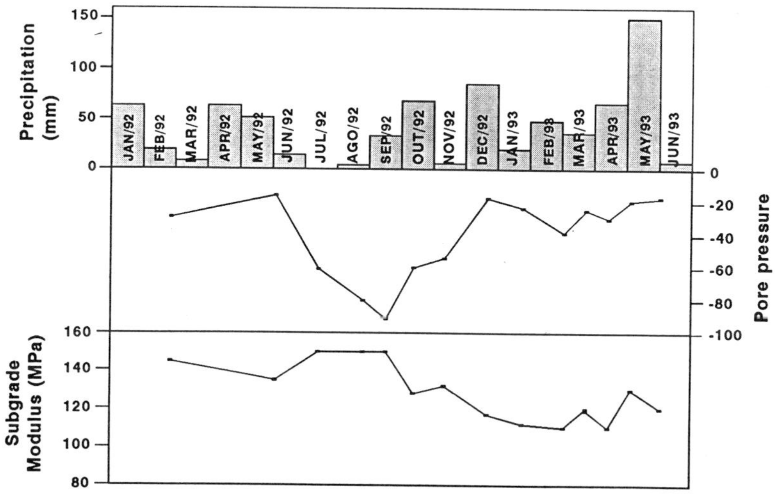

Figure 9 – Seasonal variation of subgrade modulus, soil water pressure and precipitation

The results presented in figure 9 illustrate the influence of hydrological conditions on the subgrade E modulus. It can be seen that there is a direct relationship between the E modulus and the pore water pressure. The seasonal variation of pore water pressure is directly related to precipitation, as expected, although with a slight delay in time. The most unfavourable situation takes place in the Spring, when there is a reduction of approximately 30% of the E modulus, with respect to the maximum value, which takes place in the Summer.

CONCLUSIONS

The results presented above stressed the importance of climatic effects on the structural behaviour of a pavement in terms of the layers E-moduli, back-calculated from FWD test results.

For the bituminous layers, it was found that the average of the temperature profile through the layer's depth, designated as "equivalent temperature", is suitable for representation of the effect of different temperature profiles in terms of deflections. Relationships between the back-calculated E-moduli and the layers average temperatures were derived for the two main types of bituminous mixtures in the pavement. Similar relationships were also obtained from indirect tensile laboratory tests performed on specimens cored from the pavement.

The results obtained from FWD tests at several peak loads have shown the stress dependency of the behaviour of the unbound granular material and the soil. There was a good agreement between the back-calculated E moduli of the soil and the results from laboratory resonant column tests. It was verified that the seasonal variation of back-calculated E moduli is very closely related to the variation of pore water pressure in the soil.

REFERENCES

1. SHELL Shell pavement design manual – asphalt pavements and overlays for road traffic. London, 1978.

2. ANTUNES, M.L. Pavement bearing capacity evaluation using dynamic non destructive testing (in Portuguese). PhD Dissertation, Technical University of Lisbon, 1993.

3. OLIVEIRA, M.P.; RODRIGUES, L.F. Seismic refraction survey, near the S3 taxiway at Lisbon airport (in Portuguese). LNEC Report No 118/92, 1992.

4. GOMES CORREIA, A.; SANTOS, J.A.; CARVALHO, A.T. Dynamic properties of a clayey sand from the subgrade of a pavement at Lisbon airport (in Portuguese). Technical University of Lisbon, 1993.

5. PAUTE, J.L. Utilisation des tensiomètres dans les chaussées – pose des capteurs et réalisation des mesures. Bulletin de Liaison des Ponts et Chaussées 168, 1990.

6. GOMES CORREIA, A. Contribuition a l'ètude mécanique des sols soumis a des chargement cycliques. These presentée pour l'obtention du diplome de Docteur Ingenieur, École Nationalle des Ponts et Chaussées, Paris, 1985.

Fifth International Conference on the
BEARING CAPASITY OF
ROADS AND AIRFIELDS

MOISTURE MODEL AND CRITICAL HEIGHT OF HIGHWAY SUBGRADES

[1]XINGLONG WANG, HUIMIN DAI, XUEFANG QIN, DONGXIU SHEN AND [2]JIAYI SHEN

[1]Heilongjiang Institute of Highways and Transport, Harbin 150080, China
[2]Heilongjiang Communications Department, Harbin 150000, China

ABSTRACT

Based on the observed data from three years' experiments of the test roads and laboratory tests, the authors developed the moisture model for and critical height of silty subgrades. These achievevments act as a scientific basis for further complementary revision to the related contents in the design specifications of highway subgrades, and can be applied to solve highway frost damages in both economic and rational way to ensure the bearing capacity of roads in cold regions.

Key words: moisture model, critical height, subgrade, seasonal frost.

INTRODUCTION

To solve the distortion and faulting of highway pavements due to frost heave of soils so as to realize or guarantee the anticipated bearing capacity, Frost Heave Susceptibility of Highway Subgrade Soils in Seasonal Frost Regions, a major research project assigned by Heilongjiang Communications Department was undertaken by Heilongjiang Institute of Highways and Transport in cooperation with Lanzhou Institute of Glaciology and Geocryology, Chinese Academy of Sciences. In view of the characteristics of frost heave susceptibility of highway subgrade soils, test roads were constructed in Qing-an Frozen Ground Experimental Station, Heilongjiang Institute of Highways and Transport. Simulated tests were carried out in the Frozen Ground Laboratory, Lanzhou Institute of Glaciology and Geocryology, Chinese Academy of Sciences. Part of the research findings related to determine the moisture model and critical height of subgrades are presented in this paper.

FIELD STUDIES

Basic Conditions of Experimental Field

Qing-an Frozen Ground Experimental Station, 46° 53′ N, with an area of about 6,000 square meters is located in the center of Heilongjiang Province and at the western foot of Xiaoxing-anling Mountain. The soils here are silty and the silty soil strata is 4 - 6 m thick from the lower level upward along the beach of a river called Gemuke River. Annual mean water content of soils in corresponding specified sections varies between 28.8 and 37.0 percent. Annual mean groundwater level is correspondingly 20 to 200 cm. Annual mean air temperature is 1.7℃, annual mean freezing index 2,362 degree-days, and the lowest air temperature -42℃. Concerning the three essential factors for frost heave, i.e. soil, water and temperature, their combination here is nearly the optimum. Therefore the frost heave susceptibility is rather strong here (at the lowest section, annual mean frost heaving reaches 397mm, the highest 437 mm; annual frost heaving ratio[1] is 34.4%, the highest is 38.2%). Characterized by typical seasonal frost soil, the station is a perfect place for experimental studies on the prevention of engineering frost damages.

An Outline of the Test Roads

To experimentally investigate the moisture model and critical heihgt of subgrades in cold regions, silty soils (silty clay of intermediate liquid limit exactly), which are more commonly adopted and the most frost susceptible for road construction in Heilongjiang Province, were chosen to construct 7m wide and 200m long test roads (50 m long for each section of road) at Qing-an Frozen Ground Experimental Station. Under different levels of ground water and surface water, the test roads were constructed to be of both variable and identical depths of fill. The compactness is 93 per cent from 0 to 80 cm of the subgrades and 90 per cent below 80 cm. Basic physical properties of the soil are summarized in Table 1.

Table 1 Basic Physical Properties of the Subgrade Soils

Grain-size analysis				L.L (%)	P.L. (%)	PI	Optimum moisture content (%)	Optimum compactness (t/m3)	Desig. of soils
2 ~ 0.05	0.05 ~ 0.01	0.01 ~ 0.002	< 0.002						
9.8	55.0	22.73	12.47	41.6	22.0	19.6	18.4	1.83	(1)
9.15	52.65	29.05	9.15	38.0	21.0	17.0	16.1	1.82	(2)

Note: Both (1) and (2) are silty clay of medium L.L..

Items and Procedures of the Tests and Observations

The tests and observations cover mainly the items of air temperature, subgrade temperature, depth of frost penetration, total frost heave, subdivision frost heave, uneven frost heave, surface water, ground water, and water content of subgrade soils.

Air temperature is read from thermometers and thermocouples directly for the depths of 0, 40, 80, 160 and 240 cm of the subgrades. Depth of frost penetration is determined with Danilin gauge specially for frozen ground, the total frost heave with frost heave marker and frost heave nail (3 as a group, at the interval of 40 cm). Frost heave marker and ringlike frost heave gauge are employed for determining the subdivision frost heave to the depths of 20, 40, 60, 80, 100, 120, 140, 160, 180, and 200 cm. Uneven frost heave is observed with an accurate surveyer's level to measure frost heave nails which are at the interval of 50 cm and arranged in the form of plum blossoms. Surface water is observed with an surveyer's level, ground water with an surveying rod or an surveying rope to detect the water level in a steel pipe buried in the subgrades (as inlets of ground water many holes in the shape of plum blossoms were drilled on the part of pipe below earth surface, and covered with wire gauze as filter screen). And the moisture content is determined using a thermostatic oven.

As for the above items, air temperature and subgrade temperature are observed twice a day (at 7 a.m. and 2 p.m.), depth of frost penetration and ground water once a day, total frost heave, subdivision frost heave and uneven frost heave once every five days, and moisture content of subgrade soils three times every experimental year, namely before freezing, at time maximum depth of frost penetration appears, and in spring-thaw period.

LABORATORY INVESTIGATIONS

Aiming at supplementing and perfecting the experiments on the test roads, systematic experiments on silty soils were also conducted under different moisture content, density, freezing rate and water supply; in the Frozen Ground Laboratory of Lanzhou Institute of Glaciology and Geocryology, Chinese Academy of Sciences.

The experiments continue to use the method that boundary temperature of soil samples is controlled by refrigeration cyclic bath of high precision (±0.02℃). Steady loading is achieved using an untomatically adjusted lever loading device for frost heave tests. Test data are monitored and stored (at 100 points per second with temperature drifting kept to ±0.025 ℃) by HP 3045S automatic data acquisition system of high accuracy (±1 ηV). And a computer is applied for statistical analysis of the test data.

TEST RESULTS AND THEIR ANALYSIS

Moisture State and Critical Height of Subgrade

Moisture state of a subgrade is usually determined by comparing Wx the arithmetic mean of relative moisture content of the upper 0 - 80 cm subgrade soil with W the proposed values (see Table 2) for critical

Table 2. Proposed values for critical relative moisture content

Natural zoning	Soil group												
	Sandy soil				Clayey soil				Silty soil				
	Critical relative moisture content												
	W0	W1	W2	W3	W0	W1	W2	W3	W0	W1	W2	W3	
II1,2,3 II1a II2a	0.45	0.70	0.75	0.80	0.45 0.50	0.50 0.55	0.60 0.65	0.70 0.75	0.50	0.55 0.60	0.60 0.65	0.70 0.75	Clayey soil: the denominator is only for II1,2,3; silty soil: it is only for II2a

Note: In table 2 W0 is the lower-bound relative moisture content for subgrade soils to be in dry state; W1 is the limit relative moisture content for subgrade soils to be in dry state or mid-moist state; W2 is that for them to be in mid-moist state or moist state; and W3 is that for them to be in moist state or over-moist state.

relative moisture content, according to Bing[2], 1990. If Wx < W1 for instance, the subgrade is then supposed to be in dry state.

The moisture state of subgrades is the principal factor deciding the bearing capacity of roads and of vital importance for determining the design strength of subgrades, the type of pavement structure, and the depth of frost protection course. And It is dominated by the ambient environment including the rainfall and evaporation in Summer and Autumn, the freezing index and temperature gradient in Winter, especially the ground water level and long-standing surface water before freezing. Therefore, to insure the bearing capacity of roads it is very necessary to make investigations on the dependence of moisture state upon in-situ ground water and long-standing surface water. It is known to all that with regard to highways in seasonal frost regions the moisture content adopted for the determination of subgrade moisture state should be that observed in spring-thaw period. The best approach to get this moisture content is to apply in-situ test. But for considerable highways to be designed one can by no way apply in-situ test. What one can do is only to estimate the moisture content of subgrades on the basis of routine investigations (to those items including hydrogeological type, natural water content of soil, ground water level, drainage situation, geology, geomorphology, vegetation and others) of construction sites[8].

In order to realize correct prediction to the moisture state of subgrades in seasonal frost regions, it is crucial to establish the moisture model and then the critical height of subgrades. In studies we seized on one of the principal factors, i.e. the influence of ground water on moisture state of silty subgrade. By analysis with mathematical statistics to the test data the moisture model for silty subgrade under the action of ground water was developed, as expressed by Eq. 1.

$$W_{ms} = 2.371\ Hu^{-0.240} \qquad (1)$$

$$R = 0.91$$

$$F = 298.58 > F_{0.05} = 4.02$$

in which Wms refers to the mean relative moisture content of the upper 0 - 80 cm subgrade soil in spring-thaw period; Hu is the distance from road bed to ground water level before freezing in cm; and R is correlation coefficient.

By Eq. 1 critical height of silty subgrade in different moisture states, under the action of ground water is calculated and proposed (see Table 3).

Table 3 Critical height Wms (cm) of subgrades in different moisture states, under the action of ground water

Moisture state	Calculated	Recommended
0.55	439.69	440
0.60	306.04	310
0.70	161.05	170

Note: 0.55 is the limit relative moisture content for dry state or mid-moist state of subgrade soil; 0.60 is that for mid-moist state and moist state; and 0.70 for moist state and over-moist state.

And based on the test data on another principal factor, i.e. the influence of long-standing surface water on the moisture state of silty subgrade, the moisture model for silty subgrade under the action of long-standing surface water at varied distances was developed with regression analysis, as expressed by Eq. 2.

$$Wms = 1.761\ Hs^{-0.198}\exp(-0.018L) \qquad (2)$$

$$R = 0.95$$

$$F = 82.19 > F_{0.05} = 3.59$$

where Hs is perpendicular distance of road bed to long-standing surface water level before freezing in cm; 'L' implies the minimum horizontal distance from road bed to long-standing surface water in m; other symbols are as above.

Applying Eq. 2 the critical height of silty subgrade under the circumstance of with long-standing surface water is calculated and proposed (see Table 4), corresponding to different moisture states.

Table 4 Critical height Hs of subgrades in different moisture state, under the action of long-standing surface water

L(m) / Hs(cm) / Wms	2.5		5.0		7.5		10.0		12.5		15.0	
	a	b	a	b	a	b	a	b	a	b	a	b
0.55	443	440	352	380	280	300	222	240	177	190	140	150
0.60	275	300	218	240	174	190	138	160	110	120	87	100
0.70	119	130	94	100	75	80	60	60	47	50		

Notes: "a" represents the calculated values and "b" the recommended.

Clearly from Table 4, it is implied that to increase the horizontal distance from road bed to long-standing surface water would yield the same effect as to increase the critical height of subgrades. This plays a positive role for the improvement of subgrade moisture state, the mitigation of subgrade frost heave, and the prevention and control of highway frost damages. This is of practical significance for the economical and resonable determination of the height (which refers to the distance from road bed to underground or surface water level) of subgrades, especially for the guarantee of the bearing capacity of roads.

CONCLUSIONS

The moisture model developed for silty subgrade in seasonal frost regions were established through the processing of data from 3 years' observations of the test roads and laboratory tests under the conditions of different ground water and lon-standing surface water levels. Therefore the results are reliable. The critical height from the models for silty subgrade in different moisture states are in line with engineering reality and can be put into practice.

The influence of long-standing surface water upon the moisture state decreases with the increase of distance to the road bed. Therefore by increasing the horizontal distance from the subgrade to the long-standing surface water appropriately the moisture state of subgrades can be improved, which is of practical signigicance for the performance of subgrade strength, for the prevention of highways from frost damages, and further for guarantee of the bearing capacity of roads in cold regions.

ACKNOWLEDGEMENTS

The authors are indebted to Mr. L.M. Wang of the Qing-an Frozen Ground Experiment Station, Heilongjiang, P.R. China for his assistance with field data collection and logistics. The support and guidance by Prof. H.X. Liu of Heilongjiang Institute of Cold Regions Buildings, P.R. China, is very much appreciated.

REFERENCES

1. DAI, H.M., WANG, X.L. Frost heave suceptibility of highway bridge foundation soil in seasonal frost region. Cold Reg. Sci. Technol., 20: 141 - 146, 1992.

2. BING, W.S. Proposed values for critical relative moisture content. Highway Subgrade Engineering, W.S. Bing (Ed.), Harbin University of Industries Press, Harbin, P.R. china, pp. 27-29, 1990.

3. No. 2 Highway surveying & Designing Institute, Ministry of Communications, P.R. China. JTJ 013-86 Standard Specifications of Highway Subgrade Designs. 46pp., 1986.

insulation layer and limit frost heave to 6.0 cm for the design winter, which corresponds to a freezing index of 1200 °C.days. This allowable heave was calculated based on the width of the pavement subjected to differential heave.

Using "FROST-1D" with a 50-mm thick extruded styrofoam layer placed at a depth of 0.40 m below the pavement surface with the parameters derived from the basic data collected prior to rehabilitation gives the results presented in Figure 4 and a maximum surface heave of 5.3 cm for the design winter. Since it is a low volume road, the thermal design section was not verified from a structural point of view.

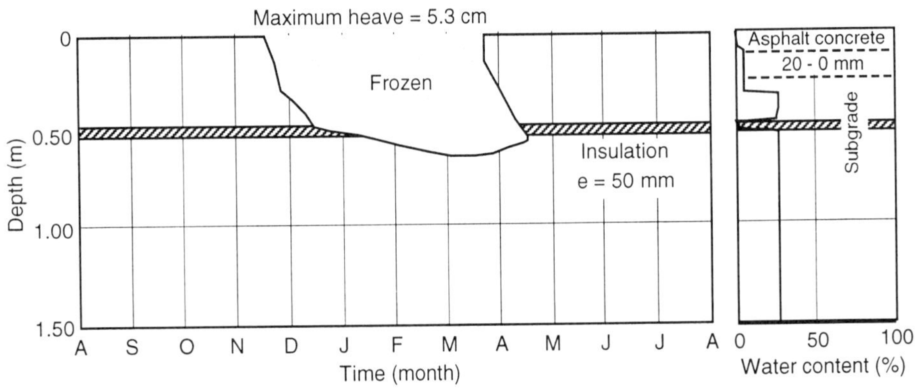

Fig. 4 - Thermal analysis of rehabilitation (FI = 1200 °C•d)

Observations and field performance

Figure 5 illustrates the design solution proposed for the rehabilitation of Montbray street. In order to provide a transition towards the utility trench, which displays little surface heave owing to less frost-susceptible fill material and towards the south side sloping terrain, which also has little surface movement owing to the presence of a snow bank, the thickness of the insulation layer was reduced to 25 mm over a width of 2.4 m.

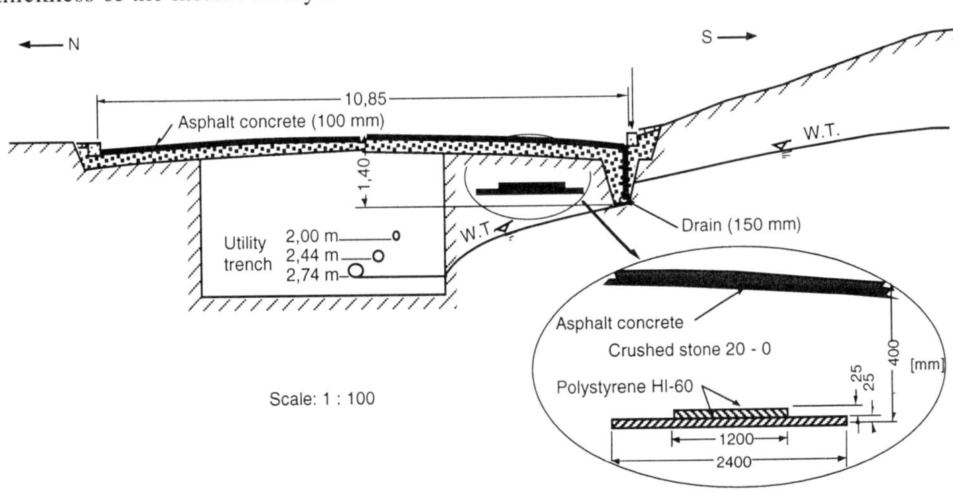

Fig. 5 - Rehabilitation of Montbray street

The 1996-97 winter was relatively severe with a freezing index of 1140 °C.days. Frost heave data and frost depth with time are shown in Figure 6 for the road prior to rehabilitation (1995-96) and for the rehabilitated design in 1996-97. Surface heave measurements shown in Figure 7 confirm that deformations were relatively uniform and that maximum differential heave did not exceed 6 cm, i.e. the design allowable heave.

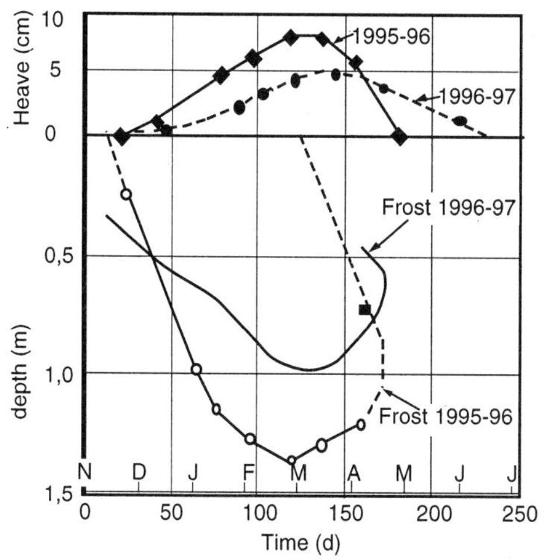

Fig. 6 - Observations prior and after rehabilitation

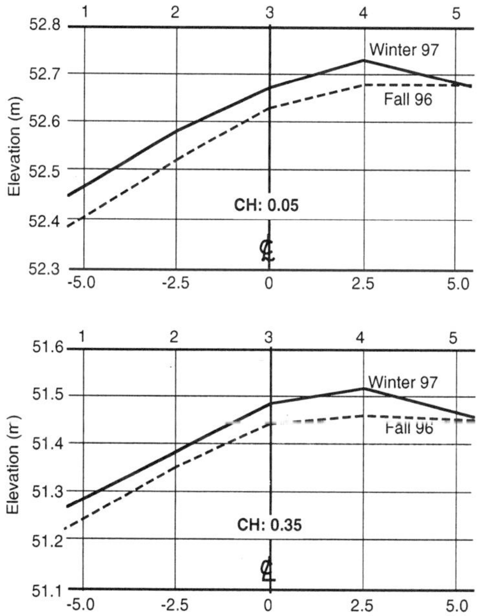

Fig. 7 - Surface heave after rehabilitation at two locations

So far, i.e. after two years of observations, no longitudinal cracks appeared in mid lane between the utility trench and the sloping terrain.

CONCLUSIONS

A design methodology for urban roads in cold regions was developed and evaluated on 12 test sites in the Quebec region. Essentially, the design methodology consists of four steps and puts a significant emphasis on collecting basic data related to climate, environment and soil characteristics. Another important feature is the back-calculation of the segregation potential from field observations such as surface heave and frost depth. Finally, the design of the rehabilitation ensures that surface heave does not exceed an allowable value that needs to be defined for each project. The design methodology is cost-effective since it focusses on actual soil and traffic conditions rather than designing by empiricism or catalogues.

ACKNOWLEDGEMENTS

The authors gratefully acknowledge the continued support of the engineers of the municipalities involved in this project, particularly to P. Bernier. Thanks are also extented to F. Thivierge, H. Lansac, R. Juneau and F. Gilbert for their active involvement in each sub-project. Financial support for this research project was obtained from "Travaux d'infrastructures Canada-Québec", the municipalities and from NSERC research grants of the researchers.

REFERENCES

1. DYSLI, M. The radiation index. VTT Symposium 94. Frost in geotechnical engineering. Vol. 1.233-238, 1989.

2. KERSTEN, M.S. Thermal properties of soils. Univ. of Minnesota, Engineering Experiment Station, Bull. 28, 227 p, 1949.

3. KONRAD, J.-M. Sixteenth Candian Geotechnical Colloquium: Frost heave in soils: concept and engineering, Can. Geotech. J.31, 2, pp.233-245, 1994.

4. JESSBERGER, H.L. and JAGOW, R. Determination of frost susceptibility of soils. VTT Symposium 95 "Frost in geotechnical engineering" Saariselka, Vol 2, 449-469, 1989.

5. SHEN, M. and KONRAD, J.-M. Dam construction in northern environment: A numerical study. 8th International Specialty Conf. on Cold Regions Engineering. ASCE, 724-735, 1996.

Fifth International Conference on the
BEARING CAPASITY OF
ROADS AND AIRFIELDS

PREDICTION OF PAVEMENT RESPONSE IN COLD REGIONS

ERIK SIMONSEN [1], VINCENT C. JANOO [2], AND ULF ISACSSON [3]

[1]PhD Candidate, Royal Institute of Technology, 100 44 Stockholm, Sweden. [2]Res. Civ. Eng. Cold Regions Research and Engineering Laboratory, 03755-1290 Hanover, USA. [3]Prof., Royal Institute of Technology, 100 44 Stockholm, Sweden.

ABSTRACT

Although the effects of climate on pavement structures are recognized as a major contributor to the deterioration of cold region pavements, only a few predictive pavement response models concerned with seasonal freezing and thawing have been developed. This paper presents a procedure for predicting response on pavements subject to seasonal freezing and thawing. The adopted procedure is based on a climatic effect analysis and a structural response analysis. The climatic effect analysis, calculating soil moisture and temperature profiles, is performed using a coupled mass and heat transfer model, FROSTB, developed by the Cold Regions Research and Engineering Laboratory (CRREL). Based on results obtained, a structural response analysis is performed using ABAQUS, a commercially available all-purpose finite element (FEM) computer code. The results from the climatic effect and structural response analysis are compared with field data from an extensively instrumented test road located in Vermont, USA. In the structural response analysis surface deflections and relative change in pavement stiffness, indicated by the Subgrade Strength Index, are predicted accurately. Although a time lag between maximum measured and predicted surface deflection is observed during thawing, the procedure is found to be promising and further research is warranted. (Keywords: freezing, thawing, bearing capacity, finite element)

INTRODUCTION

The road infrastructure represents a major national investment and constitutes a primary condition for domestic and international commerce. All-season service and long term pavement performance are essential for continued economic growth. Consequently, the prediction and assessment of the structural condition of pavements is of primary importance. In cold regions, pavement analysis encounters special difficulties. In such areas, the deterioration of pavement structures is caused not only by commercial vehicles, but to a large extent by seasonal freezing and thawing. The main body of pavement analysis software has been developed by adopting soil mechanics techniques and, in particular, multilayer techniques [1]. Due to the nature of moving loads, pavement geometry and material response, these techniques are not ideally suited for pavement analysis [2]. Also, it is generally impossible to account for material non-linearity resulting from seasonal variations in temperature and soil moisture. However, using the finite element method (FEM), the above mentioned shortcomings can often be eliminated.

This paper presents an initial attempt to implement ABAQUS [3], a commercially available FEM program in an evaluation procedure for pavements subject to seasonal freezing and thawing. This computer code has previously been used with success in structural analysis of pavements [4, 5, 6]. A two step procedure is adopted for the analysis. A coupled heat and mass transfer model, FROSTB [7], is initially used to simulate the effects of climate on the pavement structure. Based on the results obtained, a structural response analysis calculating daily variations in critical parameters, such as stresses, strains and surface deflections, is performed using ABAQUS. The results from the structural analysis are compared to field data collected from an extensively instrumented test road located on Berlin Town Highway 27 (BTH27), Montpelier, Vermont, USA. The results presented in this paper emphasize on the structural response analysis.

METHOD AND MATERIALS

The BTH27 test site, is instrumented with a variety of sensors to monitor seasonal variations in climate and pavement performance. BTH27 consists of a 6 cm asphalt concrete layer, a 46 cm silty sand with gravel base, and a sandy silt subgrade. Further details on BTH27 are contained in [8, 9].

The FROSTB model has primarily been developed for frost heave simulation, but also calculates thaw settlement, temperature and moisture distribution, and pore water tensions under various environmental conditions. Further details on the model can be found in [7]. In this study, FROSTB was used to calculate parameters critical to thaw weakening. Soil moisture and temperature are predicted on a daily basis and compared with field data collected from BTH27. Further details on the climatic effect analysis are contained in [9, 10]. Figure 1 shows the FEM mesh used during the structural response analysis. The total lateral and vertical dimensions of the mesh are 10.8 and 13.6 m, respectively. Further details on FEM model verification can be found in [11]. In order to account for the daily variation in material properties within a single computer run, field variables were introduced in the analyses. The field variables are not directly involved in the analysis but are used as a flag to trigger material properties. The field variables are applied to all the nodes in the model and represent user-defined parameters varying with time. Within each analysis step, the program checks the state of the field variables at the nodes and interpolates the values to the material points. The program then selects, for each element, material properties corresponding to the current values of the field variables. Consequently, the material properties may vary continuously over the extent of the model. In this way, the material property will not only vary between different material layers but also within each element layer. For each new analysis step, new field variables are applied to the nodes and new material properties are selected for each element. The procedure is then repeated as many times as desired.

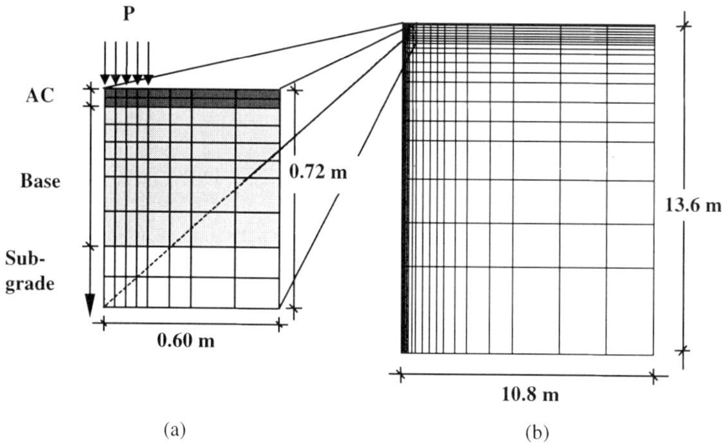

Figure 1 BTH27 finite element mesh. (a) Detail of critical region close to load center, and (b) entire finite element mesh.

In this study, the daily temperature and moisture distribution from the one-dimensional FROSTB climatic effects analysis were applied as field variables to all the nodes in the two-dimensional ABAQUS structural analysis. Temperature was used to alter the resilient modulus of the asphalt layer as well as for defining the interface between frozen and unfrozen conditions. The moisture distribution was used to establish the degree of saturation, which in turn alters the resilient modulus of the unbound materials. Within each one-day analysis step, a linear elastic procedure was used. However, during the analysis the resilient modulus varies continuously at all depths corresponding to the state of stress and the field variables (temperature and soil moisture) applied.

A uniform pressure at 690 kPa was applied at the surface of the four elements, corresponding to the area of a FWD loading plate, at the upper left corner. During the analysis, the model was loaded 198 times, from 23 October 1992 to 8 May 1993, corresponding to the number of days in the FROSTB climatic effects analysis. The simulation starts 16 days before and 47 days after the occurrence of negative temperature in the asphalt surface. The simulation ends approximately 35 days after the pavement is completely thawed and 20 days after maximum water table elevation. A combination of two regression equations was used to describe the seasonal variation in asphalt modulus. For temperatures in the interval - 26 to 12 °C, an equation presented by Schmidt [12] was used. At high temperatures, the Schmidt equation tended to give values of the resilient modulus which were considered unreasonably low. To compensate for this, a combination of the Schmidt equation and an equation presented by Ullidtz [13] was used at higher temperatures. The Ullidtz equation is based on back-calculated moduli from FWD measurements. However, back-calculated asphalt moduli often tend to give relatively high values compared to laboratory determined resilient. For this reason, the modulus at higher temperatures (12 °C and above) was calculated as the mean value of the Schmidt and Ullidtz equations. The resulting asphalt modulus as a function of temperature is presented in Figure 2.

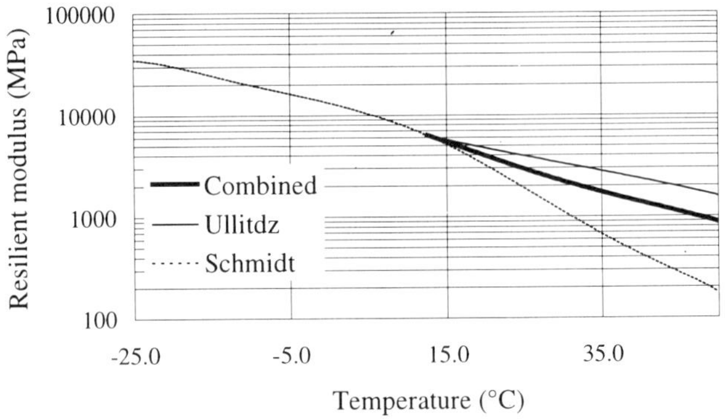

Figure 2 Asphalt concrete modulus as a function of temperature.

Over several years, a large material database has been collected at CRREL for different types of unbound materials [14, 15]. The equations, representing a frozen and unfrozen state, are given in Table 1.

During 1992-93, measurements with a FWD were performed on BTH27 at several occasions. The measurements were performed with a Dynatest 9000 FWD at four load levels. All drops were then normalized to a 690 kPa load level. This load level corresponds to that used in the structural analysis. The surface deflection was measured with seven sensors (S_1 - S_7). To quantify the bearing capacity of BTH27, and as tools for comparing the pavement response with response calculated in the FEM analysis, several structural measures were used. In this paper only the Total Basin Area Ratio (A_{TotR}) and the Subgrade Strength Index (SSI) will be mentioned. The Total Basin Area Ratio is defined as the ratio of current area relative to a basin area measured at a reference date prior to freezing. The Subgrade Strength Index is based on a similar concept as Basin Area Ratio, but calculated as the ratio of current fourth deflection basin area relative to a reference area measured prior to freezing. Further details on this matter can be found in [11].

Based on the assumption that the number of load applications to cause failure is proportional to resilient elastic strain, the cumulative damage is calculated using Miner's rule and the Shell strain criteria for fatigue cracking and rutting. The number of load applications at each strain level was calculated using traffic information from a weight-in-motion (WIM) installation at BTH27.

Table 1 Resilient modulus, M_r (kPa), for unbound BTH27 materials (after [14,15])

	Frozen	Not frozen/Thawed
Base	2000000	$1.9 \cdot 10^6 \cdot f(S)^{-0.876} \cdot f(\sigma)^{0.164}$
Subgrade	2500000	$0.2 \cdot 10^6 \cdot f(S)^{-0.885} \cdot f(\sigma)^{0.417}$
$f(S) = S/S_0$	$S_0 = 1.0$ (%) $\sigma_0 = 1.0$ (kPa)	S = degree of saturation σ_d = deviator stress σ_3 = minor principal stress
$f(\sigma) = \dfrac{9\sigma_3^2 + 6\sigma_3\,\sigma_d}{\sqrt{2}\,\sigma_d\,\sigma_0}$		

RESULTS AND DISCUSSION

Results from the strain calculations are shown in Figure 3. For reference, measured temperatures at the surface and 107 cm below the surface, as well as frost and thaw depth, predicted by FROSTB, are shown in the top two graphs, respectively. Daily precipitation in mm, presented as bars, is also included in the uppermost graph. The horizontal lines in the frost-depth graph represent the layer interfaces. The two lowermost graphs illustrate the vertical strain at the top of the subgrade and horizontal strain at the bottom of the asphalt layer, respectively. Tensile strain is defined positive and compressive strain negative. Cumulative damage for fatigue cracking and rutting is also included in the two lowermost graphs, respectively.

Throughout the freezing season, the subgrade compressive strain decreases constantly, closely following the shape of the frost penetration curve, until the freezing front reaches the subgrade. At this point the subgrade strain suddenly drops and remains at a near constant level for the remainder of the winter. The minimum winter subgrade strain is about 5 % of the level prior to freezing (18 November). On approximately 5 April, the subgrade strain starts increasing rapidly, reaching a maximum on 8 April. The maximum subgrade strain occurs approximately nine days after the thaw front reaches the subgrade and five days after the pavement is completely thawed. The maximum subgrade strain is substantially greater than the winter strain and about 70 % higher than the strain level prior to freezing. The subgrade strain starts falling off approximately on 9 April, which is about one week after the pavement is completely thawed. This period, commonly referred to as the recovery period, is the period during which surplus water is draining from the pavement structure. After a couple of days, the strain descending rate starts leveling out. However, the subgrade strain does not fall back entirely to the corresponding fall level. The shape of the asphalt horizontal strain curve shows much the same patterns as the subgrade strain curve. However, some important differences may be seen. The asphalt strain appears to be more directly linked to the asphalt surface temperature than the subgrade strain, which is more related to moisture conditions and thaw depth. The asphalt strain rapidly decreases to an almost constant level when the asphalt temperature falls below 0 °C. This level is about 15 % of the pre-freezing level (18 November). On 22 March, the pavement starts thawing from the surface. The data shown in Figure 3 do not support a relationship between increasing asphalt strain and the advance of the thaw plane. In contrast, the asphalt strain closely follows the temperature fluctuations at asphalt temperatures above 0 °C. This has the effect that the maximum asphalt strain coincides with the time of maximum asphalt temperature, rather than any time during the thawing period. The time of maximum surface temperature and horizontal asphalt strain (1 May) is well after the pavement is completely thawed. The calculated maximum asphalt strain is about 90 % higher than the pre-freezing reference level (18 November).

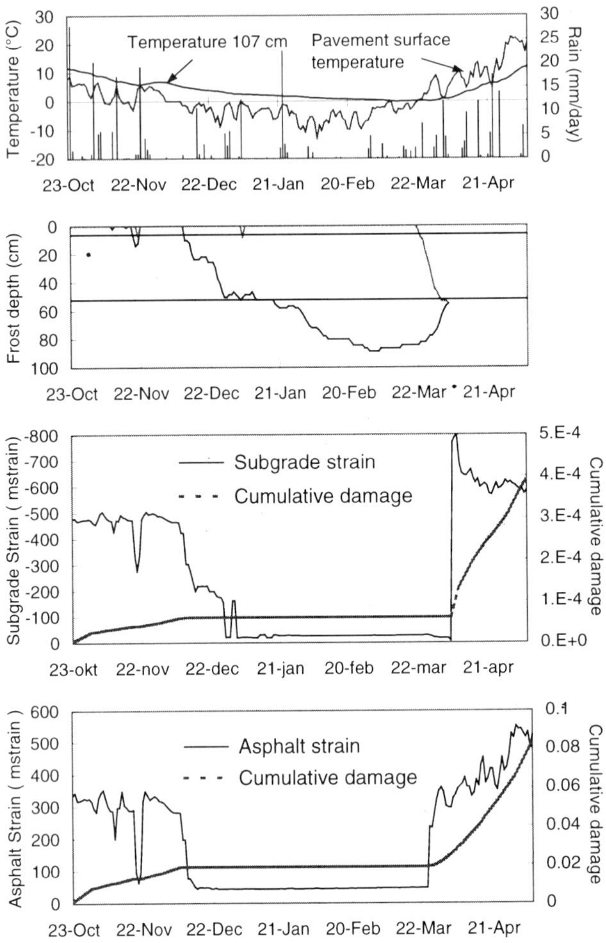

Figure 3 Measured temperature and rainfall, together with calculated frost/thaw penetration, cumulative damage and strain on BTH27, 1992-93.

The damage calculations, shown in the two lowermost graphs, clearly indicate the thaw period as critical during the simulation period. Approximately 75 % of the total cumulative damage related to fatigue cracking and 85 % damage related to rutting occurs after the start of thaw. However, due to the very low traffic volume (approximately 3500 ESAL per year) practically no damage, calculated using the Shell criteria, occurs during the period. Little or no damage during the simulation period was also confirmed by visual inspection of the BTH27.

In Figure 4, the three lowermost graphs illustrate measured and calculated surface deflections on BTH27 during 1992-93. For reference, measured and calculated moisture contents at 76 cm depth are included in the uppermost graph, and calculated frost/thaw depth as well as measured water table elevations in the second graph. For simplicity, only the first (S_1) and fourth (S_4) sensor deflection, measured at stations 1CL (center of lane) and 1WP (outer wheel path), are shown. The sensors are located at the load center and 92 cm from the load center. The simulated deflections are calculated at nodes located at the same distances as the FWD sensors. For simplicity, these nodes are referred to as "sensors" in the same manner as the FWD sensors.

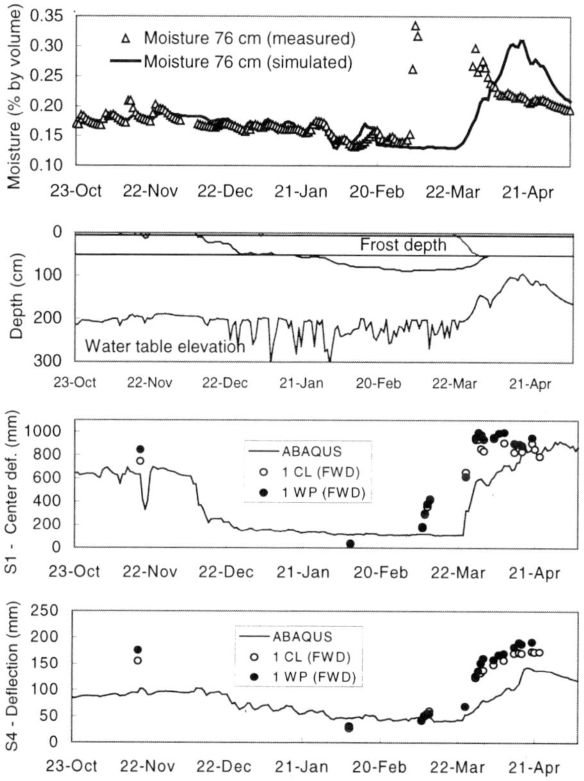

Figure 4 Moisture content, water table and frost depth together with measured and calculated deflections on BTH27 1992-93.

Inspection of the figures shows that the predicted deflections are generally somewhat lower than the measured deflections. On 20 November, a relatively large drop in calculated deflection may be observed at the center sensor (S_1). This is caused by a sudden drop in surface temperature and subsequent freezing of the asphalt and the uppermost parts of the base layer. The duration of the low temperatures is short and the center deflection soon returns to the normal pre-freezing level. The fourth (S_4) sensor is not observably affected by the sudden freezing of the asphalt and the uppermost parts of the base layer. However, this is reasonable since the deflections at this distance from the load center is normally considered to reflect the displacement at corresponding depth below the surface, where the soil remains unaffected by the surface temperature drop. On the pre-freezing reference day, the calculated deflections are under predicted in the structural analysis. The under-prediction is most severe at the fourth sensor, indicating that the error in estimated stiffness increases with depth.

In Figure 4, only a marginal reduction in center deflection occurs after the freezing front continues below the unbound base layer, revealing a large contribution from the base and asphalt layer to the overall pavement stiffness. The calculated winter deflection is about 20 % of the deflection on the pre-freezing reference day. The maximum calculated center deflection is approximately 95% of the corresponding measured value. The rate of decrease is significantly less for the fourth sensor deflection (S_4) than for the center deflection (S_1). The minimum calculated deflection relative to the pre-freezing reference day for the fourth (S_4) deflection is about 50 %. In the figure, a time lag is observed between the measured and predicted deflections. This time lag is most visible for the center deflection. The day of maximum measured center deflection seems to coincide with the time of maximum measured moisture content (at 76 cm depth) as well as the approximate time when the freezing front is penetrating the top of the subgrade. This observation indicates the subgrade as the critical region. However, since the progress of the freezing front is in fact calculated, some uncertainty exists for the exact time it reaches the subgrade. In contrast, the time of maximum calculated deflection coincides with the maximum calculated moisture at 76 cm. Since the unbound material properties are based on a relationship between moisture content and resilient modulus, it might be expected that the variation in calculated deflection would closely follow the predicted moisture content variation. This should at least be the case if the contribution of the asphalt to the overall stiffness is marginal. Based on this, the observed time lag could solely be explained by the fact that the in climatic effects analysis a time lag occurred between measured and predicted maximum moisture content and temperature during thawing [9]. In the climatic effect analysis, the predicted maximum moisture coincided with the time of maximum water table elevation, giving a time lag of about two weeks.

Structural measures such as Basin Area, Fourth Deflection Basin Area, Total Basin Area Ratio and Subgrade Strength Index have become popular tools for indicating pavement strength during spring thaw [16]. In Figure 5 the Total Basin Area Ratio (A_{TotR}) and Subgrade Strength Index (SSI) are shown. The pre-freezing reference day is 18 November. Except for that the predicted stiffness at greater depths is exceeding the actual stiffness, the relative change in pavement response is predicted well during the thaw period.

Structural analysis of cold regions pavements encounters special difficulties. The seasonal change in material properties makes it extremely difficult to predict pavement response. The procedure adopted in this study shows promising performance in predicting pavement response during freezing and thawing. However, more research is needed in the validation of this procedure. In particular, research is needed on the structural behavior of pavement materials during thawing, as well as on the ability to predict seasonal variations in critical climatic variables such as temperature, soil moisture and pore water pressure.

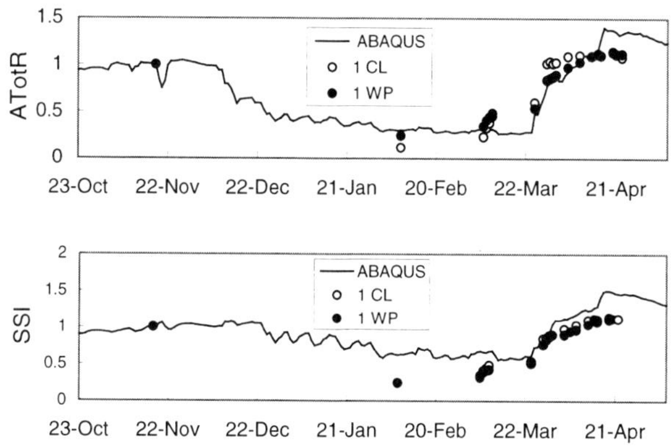

Figure 5 Deflection basin structural measures (Basin Area, Fourth Deflection Basin Area, Total Basin Area Ratio and Subgrade Strength Index) on BTH27, 1992-93.

CONCLUSIONS

During thaw weakening, the calculated subgrade strain increased by approximately 70 % compared to a pre-freezing level. The calculated horizontal asphalt strain did not increase to a maximum during the thaw period but corresponded to the time of maximum surface temperature. The strain exceeded the pre-freezing level by approximately 90 %. The surface center deflection was predicted well, as was the maximum calculated center deflection (approximately 95 % of the measured). However, a time lag was observed between measured and calculated center deflection. This time lag was mainly result of a time lag predicted in the climatic effect analysis. The result indicates that a relatively large contribution of asphalt stiffness to the overall pavement stiffness may have further extended the center deflection time lag. The relative change in pavement stiffness, indicated by Total Basin Area Ratio and Subgrade Strength Index is predicted well. The pavement analysis procedure adopted in this study shows promising results. However, further research is needed in characterizing material properties during thaw, as well as predicting the climatic effects on pavement structures critical to the thaw weakening process.

REFERENCES

1. SCARPAS, A. (1995) *Finite element modeling of pavement structures*, In: Bituminous Pavements, Materials, Design and Evaluation, Oulu 25-28 April.

conditions is also provided.

Keywords: bearing capacity, freeze-thaw effect, weakening, cusion course, thermal transmission.

INTRODUCTION

Increasing the bearing capacity of pavement is an important research project to ensure high speed driving of vehicles and to guarantee the service life of pavement; it is much more important in frost region. Except some roads in dry road sections, weakening of bearing capacity is more serious in the frost region in China than in non-frost region. In addition to climatic factor, the weakening of pavement bearing capacity is also related to repetitive loading, fatigue-resistant property of pavement materials, subgrade strength, hydrogeological condition, and pavement structure. Freeze-thaw process of subgrade soils is the more harmfull period of frost action. The bearing capacity of pavement is dependant on the moisture content and density of subgrde soils after thawing. Based on test data and engineering practice, the weakening causes of bearing capacity and the strength weakening rules of each structural layer were discussed. Some technical measures to increase the bearing capacity of pavements were developed in the paper. To prevent pavement and subgrade from frost heaving and thaw settlement so as to increase the bearing capacity of roads, a calculation method for the depth of frost protective cusion course developed by the authors were introduced in detail in the paper.

CAUSES FOR WEAKENING OF BEARING CAPCITY IN COLD REGIONS

According to geometric and hydrogeological conditions of subgrades, the weakening (or attenuation or decreasing the same) of bearing capacity can be divided into several situations as followed.

Weakening of Bearing Capacity of Pavements on High Embankments

In general, when the depth of fill exceeds 2m, the quantity of migrated moisture due to frost effects and capillary effects will not be large. However the rain-water may sometimes largely come into the subgrades via pavement joints and other places. When the water comes into pavement layers especially into subgrades, and when the subgrades are subjected to frost effects, the bearing capacity will decrease.

Weakening of Bearing Capacity of Pavements on Low Embankments

For the embankments with depth of fill less than 2m, the weakening of bearing capacity is directly affected by the accumulated surface water at both sides of the roads and/or by the ground water. If there is also rain-water coming into the roads simultaneously, the weakening will be more serious than the above situation.

Weakening of Bearing Capacity of Pavements on Zero-fill or Cut Subgrades

In addition to the influence of the above water sources, weakening in this situation is often resulted from water permeation from side ditch to road bed, because the level of ditch bottom is high above that of road bed. It is very difficult to drain off the water in the road bed, therefore the subgrades will be over moist. The weakening in this situation is much more serious.

ANALYSIS TO RULES FOR WEAKENING OF BEARING CAPACITY IN COLD REGIONS

Strength Variation of Cement Concrete Pavement in Test Section

To analyse the strength variations of pavement structural layers, moduli of test road A constructed in 1993 were measued layer by layer in the spring-thaw period in 1994. The test road is in unfavorable hydrogeological condition; depth of fill is about 1m. Calculations based on test data show that moduli of structural layer containing lime soils and subgrade decrease 15 - 20% on average, and moduli of sandy gravel cusion course and cement stabilized sandy gravel basecourse decrease 10 - 15%. It can be seen that the extent to which the bearing capacity of pavement can decrease depends on the hydrogeological condition of subgrade and the materials of structural layer.

Analasis to the Bearing Capacity of Existing Roads

To analyse further the variation of bearing capacity, investigations and measurements were carried out in the spring-thaw period in 1996 to a cement concrete pavement, which were constructed in 1989 and so has been open to traffic for 7 years. We drew the following conclusions. For the non-frost-heaving subgrade with depth of fill more than 2m, even if there is some accumulated water (say about 10 - 20cm deep) at both sides of the subgrade, the bearing capacity could only decrease by 1 - 2% or so, and for frost heaving subgrade the bearing capacity could drop 2 -5% on average. For the subgrade with depth of fill less than 2m and in unfavorable hydrogeological condition, bearing capacity for pavement on non-frost-heaving subgrade decreases 3 - 5% approximately on average; for pavement on frost heaving subgrade decreases 5 - 8%. For zero-fill or cut subgrade, bearing capacity in non-frost-heaving and hydrogeologically unfavorable road section averagely drops 5 - 10%; bearing capacity of pavement on frost heaving subgrade drops 10 - 12%, and decreases upto 15% for individual road sections. For some other road sections where sugrades are in dry state, there are basically no changes in the bearing capacity. Decrease could be within 1 - 2% if there are changes.

Weakening Coefficents for Structural Layers of Pavements

The weakening of pavement bearing capacity results mainly from strength decrease of each structural layer of pavement. Especially, the strength decrease of subgrade has more obvious influence to pavement bearing capacity. The extent to which the strength of each structural layer could decrease relates not only with types of materials, hydrogeological condition of subgrade, but also with the position of each layer. But it should be noted that with the year-to-year repetitive action of vehicle loads and repetitive freeze-thaw in structural layers, in certain suitable conditions, the moisture content and density of subgrade and the strengths of structural layers tend to reach stable values, and at this time the bearing capacity of pavement would maintain at a certain level. Based on the above analysis to the weakening of bearing capacity and relevant observed data, the weakening coefficients for material strengths of structural layers in hydrogeologically unfavorable road section are proposed (see Table 1), for reference in design.

Table 1 Proposed Weakening Coefficients for Structural Layers in %

Structural Layers and Materials		Mid-moist Bm=1-0.75	Moist Bm=0.75-0.5	Note
Base-course	Stablized granular type	3 - 5	5 - 10	①
	Stablized soil type	5 - 10	10 - 15	
Cusion Course	Natural sandy gravel	5 - 10	10 - 15	②
	Sand	5 - 10	10 - 15	
Subgrade	Non-frost heaving soil	10 - 15	15 - 20	③
	Frost heaving soil	15 - 20	20 - 30	

Note: ① Apply upper limit for the cement concrete pavements affected by water permeation through cracks. ② Apply upper limit when more soils are contained. ③ Apply upper limit in case of large frost penetration. ④ Bm = (Wl - Wx) / (Wl - Wp), where Wx is the natural moisture content of subgrade soils, Wl is liquid limit of soils, and Wp is plastic limit of soils.

TECHNICAL MEASURES FOR PREVENTING THE BEARING CAPACITY OF PAVEMENT FROM DECREASING IN COLD REGIONS

It can be seen from the analysis to the weakening causes and rules for the bearing capacity of pavements that to prevent the unfavorable effects of water on subgrades and pavements and to control the extent of frost effects on roads are the key to increase the bearing capacity of roads. Based on research findings and summary of engineering experiences, the following technical measures are put forward to prevent from decreasing of pavement bearing capacity.

In Design For road section that may be influenced by ground water and accumulated ground suface water, depth of fill should be increased properly to keep subgrade in dry state. Apply weakening coefficients of strength and increase depth of structural layer properly in pavement design. Frost resistant cusion course of a certain depth is arranged to prevent from frost heaving of pavement and thaw settlement of subgrade. Highlight drainage design of structural layer.

In Construction Execute special treatment to zero-fill and cut subgrade in hydrogeologically unfavorable road section, including a. adopt permeation-proof measures or make deep excavation of side ditch to let the ditch bottom lower than the bottom of road bed, in order to prevent from water permeating from side ditch to road bed, b. move outward the ditch to leave a 1.5m or more distance between the ditch and foot of slope, and c. in hydrogeologically unfavorable road section, replace the upper part (60 - 80cm) of subgrade soils with sandy gravel or increase the compactness of back filled soils. Increase the compactness of fill subgrade. Try not to employ structural layer that contains soils. Apply geotextile products.

In Maintenance Repair pavement cracks in time to prevent water permeating into structural layers. Drain off the accumulated water in side ditch to decrease water effects upon subgrade.

Calculation of the Depth of Frost Resistant Cusion Course
To decrease frost heaving failure of pavement and the displacement of of thaw settlement of subgrade so as to increase bearing capacity of pavement, design of frost resistant cusion course must be done properly

and matrials of good thermal insulating properties be adopted as far as possible. Calculation of the depth of frost resistant cusion course is heat issue as well as mechanical issue. A calculation method for the depth of frost resistant cusion course is developed as followed.

He, the depth of frost resistant cusion course can be calculated based on the principle of heat condution. Assume that heat condution in the pavement is one-dimensional and stable and that pavement body is thermal insulating materials. Hence, Qp the heat emitted through pavement structure should be less than or equal to Qg the heat emitted through general ground surface, i.e., $Qp \leqslant Qg$. Make further substitutions (process omitted here), we obtain the calculation formula for He (see Equ. (1)).

$$He = (((-Tg)/(2(1-0.17Tg)\lambda g) - ((H/\lambda p)+(Hu/\lambda u)))\lambda e = \alpha \lambda e \qquad (1)$$

in which α is coefficient for cusion course, Tg pavement temperature in ℃ (average minus local temperature in the coldest month), λg mean thermal conductivity for the frozen and unfrozen soils below ground surface in W/m·℃, λp weighted mean of thermal condutivity for materials of all structural layers in W/m·℃, Hu the allowable depth of frost penetration in m (calculations were made in relevant papers by the authors), λu thermal conductivity of subgrade soils in W/m·℃, H pavement depth in m based on strength calculations, and λe thermal conductivity for the materials of frost resistant cusion course W/m·℃.

To simplify calculations a list of α the cusion course coefficent were made (see Table 2). Find λe in Table 3 after materials for frost resistant course is determined, then He the depth of the frost resistant cusion course can be worked out by Equ. (1).

Table 2 Proposed α the Coefficient of Frost Resistant Course

Maximum Depth of Frost Penetration	Type of Basecourse	α Subgrade Soils Non-frost Heaving	Frost Heaving
50 - 100cm	A	0.10 - 0.15	0.15 - 0.20
	B	0.09 - 0.12	0.12 - 0.15
	C	0.07 - 0.10	0.10 - 0.12
100 - 150cm	A	0.15 - 0.20	0.20 - 0.25
	B	0.12 - 0.15	0.15 - 0.20
	C	0.10 - 0.12	0.12 - 0.15
150 - 200cm	A	0.20 - 0.25	0.25 - 0.35
	B	0.15 - 0.20	0.20 - 0.30
	C	0.12 - 0.15	0.15 - 0.25
More than 200cm	A	0.25 - 0.30	0.30 - 0.45
	B	0.20 - 0.25	0.25 - 0.40
	C	0.15 - 0.20	0.20 - 0.30

Note: 1. A denotes sand and stone type, B lime and flyash type, and C slag type. 2. Apply bigger values to cement concrete pavement. 3. Apply bigger values to moist subgrade.

Table 3 List of λe the Thermal Conductivity for Common Materials Used in Frost Resistant Course, in W/m·℃

Designation of Materials	Thermal Conductivity	Note
Sand and gravel type	1.5 - 2.0	Bigger values for materials of large moisture content
Industrial wastes	0.8 - 1.2	
Stablized soils	1.2 - 1.5	

Fifth International Conference on the
BEARING CAPASITY OF
ROADS AND AIRFIELDS

MODELING ANALYSIS ON THE HEAT STABILITY OF EMBANKMENT FOR THE CONTINUOUS PERMAFROST DISTRICT OF CHANG SHITOU MOUNTAIN IN HUA SHIXIA VALLEY

LI DONGQING[1], FANG JIANHONG[2], AND LI YICHI[2]

[1]Doctor,State Key Lab. of Frozen Soil Eng., Lanzhou Inst. of Glaciology and Geocryology, Chinese Academy of Sciences,730000,Lanzhou,China,
dqli@public.lz.gs.cn
[2]Engineer,Highway Research and Survey Design Institute Of Qinghai Province, 810001,Xinin,China

ABSTRACT

The heat stability of embankment in the continuous permafrost district of Chang Shitou Mountain in Hua Shixia Valley is modeled. Critical heights of embankment for asphalt, concrete and gravel road surface along the National Road of Qinghai-Sichuang are figured out by computers, according to the geographical and geological conditions of the Hua Shixia Valley. The modeled results are in good agreement with the field measurements on the Qinghai-Xizang Highway. The conclusion is that the critical heights are 4.3, 1.4 and 1.0 m for asphalt, concrete and gravel road surface, respectively.
Key Words: Continuous Permafrost, Heat Stability, Embankment Heights

INTRODUCTION

The Qinghai-Sichuang Highway, i.e. National Road 214, crosses the eastern part of the Qinghai-Xizang Plateau, China. It is one of the most important highways connecting the three provinces of Qinghai, Sichuang and Xizang. It runs between north latitudes ranging from 33.5° to 36° and between east longitudes 96° and 99°, with elevations ranging from 3,000 to 5,000 *m*. Because of the high elevation and low mean annual air temperatures, there are many large areas underlain by permafrost. The highway crosses about 300 *km* of permafrost, the thickness of which has been decreasing in this region since the 1960s. Zhu Linnan (1995) and Li Dongqing (1996) thought the decrease could continue 30-40 years into the future. Thus, there is a prerequisite to study and to predict the decreasing trend of the permafrost along the Qinghai-Sichuang Highway under the conditions of warmer climate. This is particularly important for road fill and surfacing specifications in the Hua Shixia Valley area.

HEAT AND MASS TRANSFER MODEL

Since Harlan (1973) presented the first model for the coupled heat and mass transfer in freezing soils, his strategy has been referred to and further developed by many other researchers (Guyman and Luthin (1974), Outcalt (1977), Taylor and Luthin (1978), Jame and Norum (1976) and Nixon (1975)). These authors also applied two partial differential equations, in terms of temperature and water (ice) content or water pressure, to describe the transient processes of heat and water flows in freezing soils. In this section, the two equations are coupled through the characteristic relationship between the unfrozen water content and the temperature. The essential ideas of heat and mass transfer model will be discussed theoretically.

Assuming that vapour transfer and heat convection are omitted, the application of the laws of heat balance to saturated or unsaturated soil and the mass conservation of water in soil pores results in the following two equations:

$$\frac{\partial}{\partial x}(\lambda_x \frac{\partial T}{\partial x}) + \frac{\partial}{\partial y}(\lambda_y \frac{\partial T}{\partial y}) = C\rho \frac{\partial T}{\partial t} - L\rho_i \frac{\partial \theta_i}{\partial t} \quad (1)$$

$$\frac{\partial}{\partial x}(K_x \frac{\partial \psi}{\partial x}) + \frac{\partial}{\partial y}(K_y \frac{\partial \psi}{\partial y}) = \frac{\partial \theta_u}{\partial t} + \frac{\rho_i}{\rho_w} \frac{\partial \theta_i}{\partial t} \quad (2)$$

Here ρ_w, ρ and ρ_i are the densities of water, soil and ice, respectively; t, x and y denote the time and two space coordinates; θ_i and θ_u are the volumetric ice and unfrozen water content; $\lambda_x(\lambda_y)$ and $K_x(K_y)$ are the thermal and hydraulic conductivities in the x and y directions; T is the temperature; C is the specific heat; L is the potential heat ; ψ is the total potential of water in soil; $\psi = \varphi + Z$, where φ and Z denote the volumetric potential of water in the soil and the gravitational potential, respectively. In general, Z can be omitted and therefore $\psi = \varphi$.

According to the characteristic relationship $\theta_u = f(T)$ between the unfrozen water content θ_u and the temperature T, we obtain the following differential relationship:

$$\frac{\partial \theta_u}{\partial t} = \frac{\partial \theta_u}{\partial T} \frac{\partial T}{\partial t}$$

Substituting this differential relationship into equation (1) and reducing (1), we can derive:

$$(C\rho + L\rho_w \frac{\partial \theta_u}{\partial T})\frac{\partial T}{\partial t} =$$

$$\frac{\partial}{\partial x}(\lambda_x \frac{\partial T}{\partial x}) + \frac{\partial}{\partial y}(\lambda_y \frac{\partial T}{\partial y}) + L\rho_w\left[\frac{\partial}{\partial x}(\mathrm{K}_x \frac{\partial \psi}{\partial x}) + \frac{\partial}{\partial y}(\mathrm{K}_y \frac{\partial \psi}{\partial y})\right] \quad (3)$$

Equations (2) and (3) can be solved when the function ψ is given and the characteristic relationship between the unfrozen water content and the temperature is determined by testing. Therefore, in terms of the different types of potential functions, we can derive different types of calculation models.

Taylor (1976,1978) investigated the volumetric potential of unfrozen water in frozen soil, and found that the volumetric potential could be determined using the characteristic relationship between the unfrozen water content and the temperature, similar to the volumetric potential of unfrozen soil, when the action of ice in the frozen soil is omitted, i.e. the characteristic relationship of moisture for unfrozen soil is suitable for frozen soil.

By introducing the new differential relationship:

$$\frac{\partial \psi}{\partial T} = \frac{\partial \psi}{\partial \theta_u}\frac{\partial \theta_u}{\partial T}$$

and by using the term of the differential water capacity $c = \frac{\partial \theta_u}{\partial \psi}$ and the soil-water diffusivity $D = \frac{\mathrm{K}}{c}$, we reduce equation (3) to:

$$C(T)\frac{\partial T}{\partial t} = \frac{\partial}{\partial x}(\beta_x(T)\frac{\partial T}{\partial x}) + \frac{\partial}{\partial y}(\beta_y(T)\frac{\partial T}{\partial y}) \quad (4)$$

where

$$C(T) = C\rho + L\rho_w \frac{\partial \theta_u}{\partial T}$$

$$\beta_x(T) = \lambda_x + L\rho_w D_x \frac{\partial \theta_u}{\partial T} \text{ and}$$

$$\beta_y(T) = \lambda_y + L\rho_w D_y \frac{\partial \theta_u}{\partial T}$$

Since the characteristic relationship between the unfrozen water content and the temperature can be determined by testing, the nonlinear equations (2) and (4) can be solved using different methods. The implicit finite difference method was used to derive the results reported in this paper.

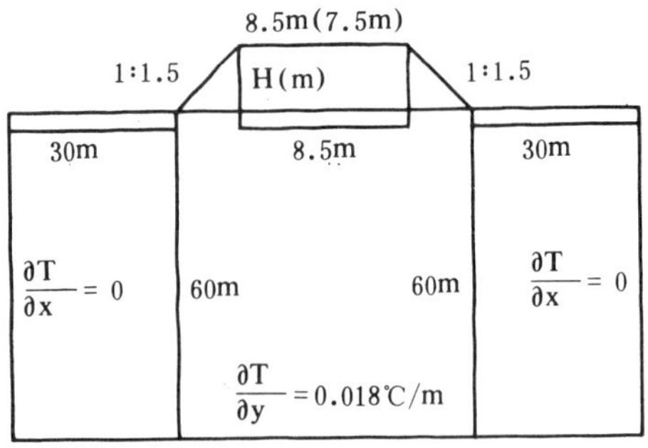

Fig. 1 The profile of embankment for calculations

BOUNDARY CONDITIONS

The mathematical model for the calculation is given above. According to the Chinese standards for designing and building of national highways, the widths of road surface and shoulder is 8.5 and 0.5 *m,* respectively. In the calculations, the height of embankment can be varied, the slope gradient is 1:1.5 and the level of the bottom boundary is located at 60 meters depth. In the Hua Shixia Valley, the ground temperature gradient at the lower boundary at 60-meter-depth was assumed to be $\frac{\partial T}{\partial x} = 0.018$, i.e. 0.018°C per meter. Based on the relevant drilling records on temperature and its gradient, the heat flux at the bottom of the permafrost is found to be a constant value. On the other hand, the lateral boundary (30 *m* from the slope corner) is situated so far from the road that the temperature gradient there is regarded as $\frac{\partial T}{\partial x} = 0$ in horizontal direction, see figure 1. The temperature of embankment slope is 0.3-0.5°C lower than the upper boundary temperature at the road surface, because the latter is often influenced by traffic. Using meteorological information of air stations from the Hua Shixia Valley and Ma Duo County, and by regression analysis of the sine function, we can reduce the upper boundary temperature at the road surface to the following sine function:

$$T = T_s + g(t) + 12.2\sin\left(\frac{2\pi}{8640}t + \frac{4\pi}{3}\right)$$

Here, $g(t) = At$, and T_s and A are the mean annual temperature at the ground surface (Wu Ziwang, 1988) and the annual rate of temperature increase, respectively. In terms of the principles on the adhere layer, T_S is +2.0, 0.5 and 0.0°C for asphalt, concrete and gravel road surface, respectively, and A is taken as 0.04°C/a. As for the continuous permafrost region in the Hua Shixia Valley (Drill Hole No.2), the hydrogeological conditions are as follows:

(1)0~1.20*m*, organic clayey soil, the water content was 40.0%, the dry density 0.9 g/cm^3; the heat capacities of unfrozen and frozen soils are 1.1704 and 1.34178 $J/cm^3 \cdot °C$, and the heat conductivities of that are 13.974 and 16.72 $J/cm \cdot h \cdot °C$, respectively.

(2)1.20~3.30*m*, clayey soil, the water content was 178.5%, the dry density 0.355 g/cm^3; The heat capacities of unfrozen and frozen soils are 1.0336 and 0.88937 $J/cm^3 \cdot °C$, and the heat conductivities of that are 26.36 and 77.404 $J/cm \cdot h \cdot °C$, respectively.

(3)3.30~7.60*m*, gravel, the water content was 24.5%, the dry density 1.52 g/cm^3; the heat capacities of unfrozen and frozen soils are 2.8242 and 2.23007 $J/cm^3 \cdot °C$, and the heat conductivities of that are 40.5848 and 56.9024 $J/cm \cdot h \cdot °C$, respectively.

(4)Below 7.60*m*, mantle rock, the water content was 4%, the heat capacity of soil is 2.284464 $J/cm^3 \cdot °C$, the conductivity of soil is 97.2 $J/cm \cdot h \cdot °C$.

(5)Fill materials of embankment contain clayey soil with coarse-grained gravel, the water content was 8.0-10.0%, the dry density 2.0 g/cm^3. The capacities and conductivities of unfrozen and frozen soils were taken as that in 3.3~7.60 *m* deep gravel soil.

The relationship between the unfrozen water content, temperature, total moisture and the heat parameters are taken from the book of the experimental research on moisture transfer in frozen soil (Xu Xuezu, 1991).

Because traffic is increasing with time, a rebuilding of the highway is required and a high-standard pavement will gradually replace the existing gravelly road surface. To keep the road surface even and stable, first of all, the fill stability must be maintained. Firstly, to prevent thaw settlement, the permafrost table under the roadbed must be maintained and not lowered, secondly the frost deformation of the active layer above the permafrost must be controlled to a given value to maintain an even road surface, that value is about 1 %. The calculations referred in the paper is done due to these two principles. For the different types of the road surface and embankment heights, the changes in the level of the natural permafrost table as the age of the road runs are analyzed by computer modeling, and the critical design heights of the embankment for a twenty year period after construction are determined.

CALCULATION RESULTS

The computer analyses are summarized in Fig. 1 to Fig. 4. Fig.2(a), Fig.3(a) and Fig.4(a) show for the asphalt, concrete and gravel road surface respectively, the relations between the heights of embankment and the changes of the level of natural permafrost table under the middle sections of road in the twentieth year after construction on the moisture sections in Hua Shixia Valley. On the other hand, Fig.2 (b), Fig.3 (b) and Fig.4 (b) show the relations between the heights of embankment and the artificial permafrost table (changed permafrost table under the road surface) under the middle sections of road, i.e. the critical design heights have a relationship with the age of the road after construction.

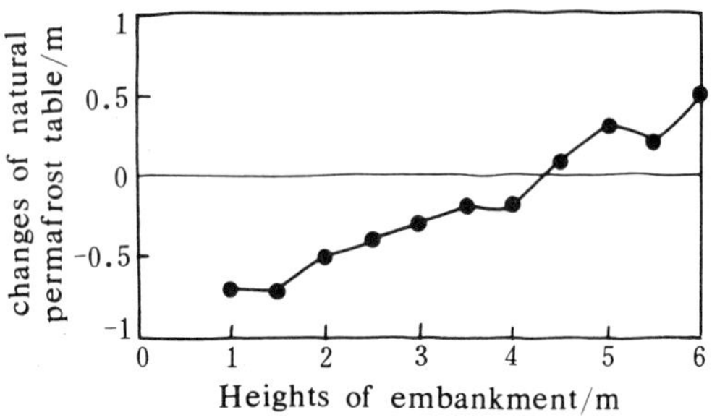

Fig. 2 (a) Relation between the height of embankment and the changes of natural permafrost table under roadbed (Asphalt)

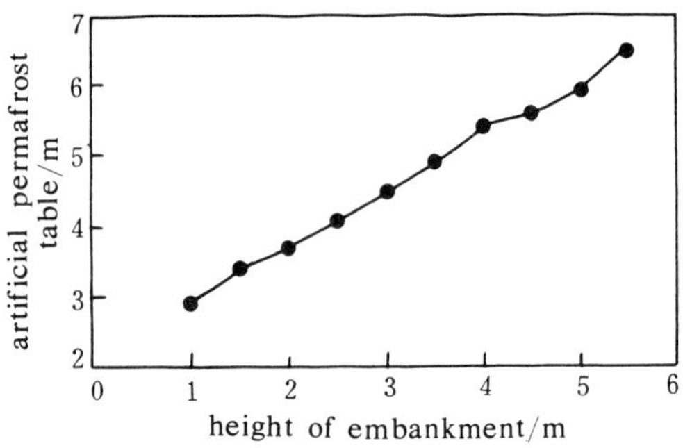

Fig. 2 (b) Relation between the height of embankment and the artificial permafrost table of embankment (Asphalt)

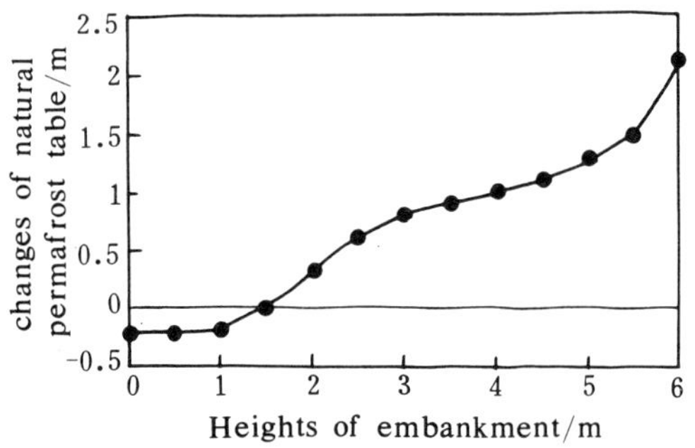

Fig. 3 (a) Relation between the height of embankment and the changes of natural permafrost table under road (concrete)

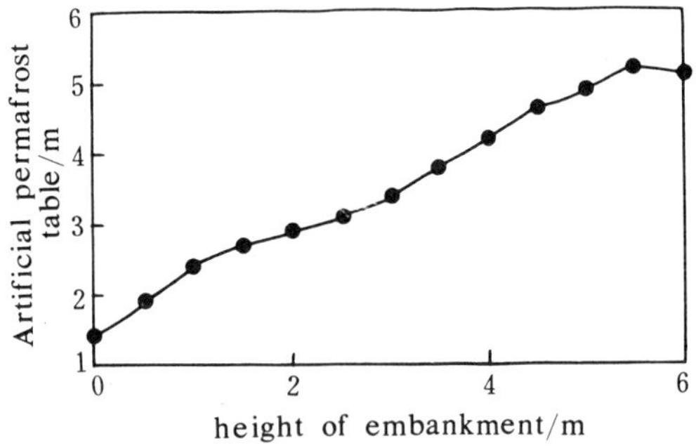

Fig. 3 (b) Relation between the height of embankment and the artificial permafrost table of embankment (Concrete)

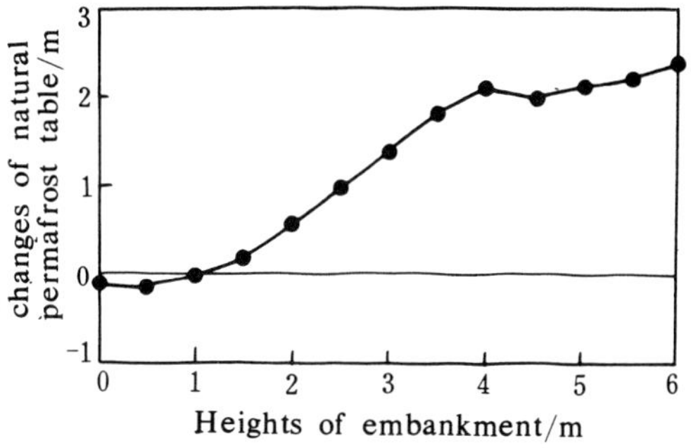

Fig. 4 (a) Relation between the height of embankment and the changes of natural permafrost table under road (Gravel)

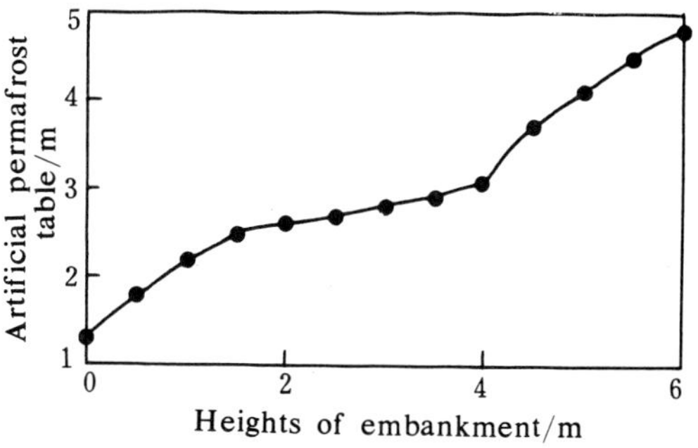

Fig. 4 (b) Relation between the height of embankment and the artificial permafrost table of embankment (Gravel)

From Fig. 2 to Fig. 4, to a certain extent for embankment height lower than 4.0 *m*, it is seen that there are almost linear relations between the heights of embankment and the changes of natural permafrost and artificial permafrost under the road in the twentieth year after construction. This is in correspondence with the observed results on the Qinghai-Xizang Highway for thirty years or more (Wu Ziwang, 1983).

From regression analysis, the heat-stable critical heights of embankment for asphalt, concrete and gravel roads are figured out as 4.30, 1.40 and 1.00 m in the twentieth year after construction. Besides it is found from the calculation result that when the height of embankment is more than 5.0 m for asphalt road, there is an upper thawed course (Huan Xiaoming, 1983) above the artificial permafrost table in the embankment later than ten years after construction, although the artificial permafrost table under the road is located above the natural permafrost table for the whole analysis period. The frost deformation of the embankment therefore would be increased, and can't reach the design deformation requirement. But for concrete and gravel road there is no upper thawed course in the embankment for embankment heights less than 5.0 m.

CONCLUSIONS

(1) It is important that the running time is introduced into the determination of the critical heights of embankment in cold regions, particularly more important in regions with decreasing permafrost thickness.

(2) The heat-stable critical heights of embankment in Chang Shitou Mountain are

4.30, 1.40 and 1.00 m for a twenty year analysis period after construction for asphalt, concrete and gravel road, respectively.

ACKNOWLEDGEMENTS

This research was supported by the Natural Fund of Sciences, China. Authors wish to thank professor Wu Ziwang for help and guide and all other researchers of the exploring team for their cooperation in the field.

REFERENCES

1. ZHU, L.N. et al. (1995). Permafrost Degeneration in the East of Tibetan Plateau, Journal of Glaciology and Geocryology, 17(2): 120-124, Science Press, Pecking, China.
2. LI, D.Q. et al. (1996). Numerical Calculation on Permafrost Degradation. In Proceedings of the Fifth Chinese Conference on Glaciology and Geocryology, Vol.2, Gansu Culture Press, China, pp. 1082-1086.
3. HARLAN, R.L. (1973). An analysis of coupled heat-fluid transport in partially frozen soil. Water Resources Research. 9, pp.1314-1323.
4. GUYMAN, G.L. and LUTHIN, J.N. (1974). A coupled heat and moisture transport model for arctic soils, Water Resources Research, 10, pp.1314-1323.
5. OUTCALT, S. (1977). Numerical modelling of the ice lensing processes, Int. Symp. On Frost Action in Soils, Lulea, Sweden, Feb 1977, Proc., Lulea University of Technology, Vol.2, pp.75-91.
6. TAYLOR, G.S. and LUTHIN, J.N. (1978). A model for coupled heat and moisture transfer during soil freezing. Canadian Geotech.J., 15: 548-555.
7. JAME, Y.W. and NORUM, D.I. (1976). Heat and mass transfer in freezing unsaturated soil in a closed system. In Proc. 2nd Conference on Soil Water Problem in Cold Region, Edmonton, Alba., 1-18.
8. NIXON, J.F. (1975). The role of convective heat transport in the thawing of frozen soils. Canadian Geotech. J., 12(4): 425-429.
9. WU, Z.W. et al. (1988). Road Engineering of Frozen Soil, Lan Zhou University Press, Lan Zhou, China.
10. XU, X.Z. et al. (1991). The Experimental research on Moisture Transfer in Frozen Soil, Science Press, Pecking, China.
11. HUAN, X.M. (1983). Determination of the Critical Height of Embankment for Highway in Permafrost Regions of Tibetan Plateau. In Proceedings of Second National Conference on Permafrost (Selection), Gansu People's Publishing House, China, pp. 391-397.

Fifth International Conference on the
BEARING CAPASITY OF
ROADS AND AIRFIELDS

EVALUATION TO THE TYPICAL STRUCTURES OF CEMENT CONCRETE PAVEMENTS IN HEILONGJIANG

[1]XINGLONG WANG, HUIMIN DAI, XUEFANG QIN, GUOCHANG XUE
[1]Heilongjiang Institute of Highways & Transport, Harbin 150080, China
[2]JIAYI SHEN, [3]JINGMAN JI AND [4]GUANGXIN LIU
[2]Heilongjiang Communications Department, Harbin 150000, China
[3]Heilongjiang Road Administration, Harbin 150000, China
[4]Heilongjiang Communications College, Harbin 150000, China

ABSTRACT

In recent years more and more cement concrete pavements have been constructed in Heilongjiang Province, China. Due to lack of construction experiences plus the existence of rigid cold climate, we are facing with increasing problems such as decreasing of bearing capacity, premature failures and others. To cope with, a project entitled Study on Structures and Stability of Basecourse and Cusion Course of Cement Concrete Pavement in Seasonal Frost Regions was carried out. Some of the research findings are shown in the paper.

Based on extensive survey of existing roads, in-situ experiments and laboratory tests, the minimum frost-resistant depths for cement concrete pavements of typical structures were proposed. Rules for strength decreasing for the common base courses, cusion courses and subgrades are analyzed in the paper. The typical structures were evaluated in terms of both structural stability and strength stability.

In addition to traditional items for test road observations, a new approach, which we call Pre-installed Bearing Plate method was adopted to monitor the variations in resilient modulus of each structural layer.

Keywords: resilient modulus, evaluation, typical structure, cement concrete pavement.

WATER STABILITY AND FROST STABILITY OF COMMON MATERIALS FOR THE BASECOURSE AND CUSION COURSE IN HEILONGJIANG

Early in the province, cement concrete pavements were mainly adopted for the construction of urban roads. Structural arrangement for basecourse and cusion course then was different from each other, depending on sourses of lacal materials and the state-of-art at that time. Since the late 1980's, with the rapid development of high-grade highways, a typical structure with cement stabilized sandy gravel or cushed stone as basecourse, cement or lime-flyash stabilized granular materials as subbase course, and natural sandy gravel or other coarse granular materials as cusion course has been formed for the cement concrete pavements in the province.

To be clear of both the water stability (W.S.) and frost stability (F.S) of the common materials for basecourse and subbase course in the province, laboratory investigations were carried out in the institute, the results of which are summarized in Table 1. In the test, mixture ratios for lime flyash and lime-flyash stablized sandy gravel (or crushed stone) are 1:4 and 10:20:70 respectively. Lime content in lime stabilized clayey soil is 10 per cent. Number of freeze-thaw cycles is 5, with temperature down to -20℃.

From Table 1 it can be seen that the water stability and frost stability for the common materials of basecourse and subbase course are satisfactory, according to laboratory tests. The results for lime-flyash or cement stabilized sandy gravel (or crushed stone) are obviously desirable. But results for lime soil are a little bit too inferior. This is one of the main reasons that lime soil has been less applied as basecourse and cusion

Table 1 Water Stability and Frost Stability for Common Materials

Desig. of Materials		A		B		C	D	E	F
Age in d		30	60	30	60	30	30	30	30
Compressive strength in 10^{-1}MPa	Unsaturated	17.4	20.2	45.3	87.4	23.9	44.2	49.6	48.0
	Saturated	9.9		39.8	77.8	13.6	41.5	44.7	43.7
	Freeze-thaw	8.9		37.2	76.5		40.2		
Water stability		0.57		0.88	0.89	0.57	0.94	0.90	0.91
Frost stability		0.51		0.82	0.88		0.91		

course materials in construction of high-grade highways. However, under favorable conditions of water and temperature, it is still an economical selection to use the lime soil as subbase course materials. As cusion course materials lime soil is not favorable because of its unsatisfied water stability and frost stability; as an alternative lime-flyash stabilized industrial slags can be applied. Even so, when it is permitted lime-flyash or cement stabilized granular materials is the first choice.

Although for the cement concrete pavement the cement stabilized sandy gravel or other half-rigid basecourses have been arranged, and cusion course of coarse granular materials may be added if necessary, such questions are still difficult to answer by laboratory tests as how the stability of basecourse and cusion course structures will be and if their strengths will be decreased under natural effects of dry-moist and freeze-thaw cycles, etc. in in-situ conditions.

OBSERVATIONS TO STRENGTH VARIATIONS IN BASECOURSE, CUSION COURSE AND SUBGRADE

To know about strength variations in the typical basecourse, cusion course of cement concrete pavement in the province, i.e., to reconfirm strengths and stabilities of typical basecourse and cusion course, test roads with the typical stuctures were constructed in Qing-an Frozen Ground Expermental Station in 1993. Moduli of resilience at tops of subgrade and each structural layers were monitored with pre-installed bearing plates, which were placed on top of each structural layer and connected with steel rods. The bearing plate is 30cm in diameter. From observations in nearly 3 years it is concluded that the method is convenient and successful for continuous observations at fixed points. Data of resilient modulus in MPa from in-situ measurements are listed in Table 2.

Table 2 Strength Variations in Structural Layers at Different Priods

Measured layer and point no.		Autumn 1993	Autumn 1994	Winter 1994	Winter 1995	Spring 1994	Spring 1995
Cement stabilized sandy gravel	1	425.2				340.1	
	1		303.1		1088.6		265.2
	5		578.3		403.7		240.3
Lime soil	2	217.6				93.2	
	2		149.9		365.4		62.8
	6		128.1		539.4		83.1
Sandy cusion course	3	54.5				41.4	
	3		40.2		153.8		27.0
	7		42.8		165.3		36.2
Subgrade	4	46.8				27.5	
	4		36.6		188.9		21.2
	8		40.0		227.4		17.2

Based on the observed resilient moduli at top of certain layer and the layer below, the resilient modulus of materials of this layer was then backcalculated with calculating method proposed in the Design, Construction and Maintenance of Cement Concrete Pavement1 and the design specifications in use for cement concrete pavement. And based on the observed and backcalculated moduli of resilience the decreasing percentages of moduli were calculated and listed together in Table 3.

Table 3 Decreasing Percentage of Observed and Backcalculated Moduli

Measured layer and point no.		Autumn-Autumn Obser.	Autumn-Autumn Backc.	Spring-Spring Obser.	Spring-Spring Backc.	Autumn-Spring Obser.	Autumn-Spring Backc.
Cement	1	28.7	24.1	22.0	-	20.0	-
stabilized						12.5 (37.6)	
sandy gravel	5					58.4	
Lime	2	31.1	29.1	32.6	34.8	57.2	-
soil						58.1 (71.1)	
	6					35.1	58.5
Sandy	3	26.2	26.7	34.8	43.1	24.0	15.5
cusion						32.8 (50.5)	34.4 (51.9)
course	7					15.4	13.9
	4	21.8	21.8	22.9	22.9	41.2	41.2
Subgrade						42.1 (54.7)	42.1 (54.7)
	8					57.0	57.0

Note: Data at numerator denotes decreasing ratio from the first year to the second year, at denominator the second year to the third year, and in brackets the first year to the third year.

It should be noted that because the calculation methods and models could not be in perfect accordance with the actural boundary conditions, the theoretically backcalculated results must be subjected to certain errors. Therefore, the data here based on backcalculations are only for reference in analysis. From the strength variations in structural layers at different periods, it can be seen that strength of material itself or strength at top of each layer realizes the highest value in the year of construction completion or in the first year after completion. After suffering from dry-moist cycles and frost-thaw cycles in different seasons, strengths decrease considerably comparing with the initial values.

From the authors' point of view the basecourse and cusion course should be of certain stability in strength, that is to say the variation and decreasing extent of·both strength at top of each layer and strength of material should not be too large. Especially the strength at top of basecourse must meet the specifications' requirements for equivalent modulus of resilience. The observations on the test roads of typical structures show that although the strength at top of each structural layer and the strength of material of each structural layer have all decreased by large extent, the equivalent modulus of resilience at top of basecourse are still two times and above of the lower limit stipulated in specifications in use. This implies the typical structure of cement concrete pavement in Heilongjiang has relatively more abundant storage of strength. Therefore, instead of strength stability the authors consider more attention should be payed to structural stability, which refers to the characteristics for materials of structural layer to keep their own structures. This is the main reason that frost resistant depth of pavement is also studied at the same time.

RECONFIRMATION OF FROST RESISTANT DEPTH SUITABLE FOR TYPICAL CEMENT CONCRETE PAVEMENTS IN HEILONGJIANG

Based on the observations on test roads for two winters, the following correlation exists between the measured overall frost heaving quantity of pavement and the depth of pavement. See Table 4 and Equ. 1.

Table 4 Pavement Depth (Hp) vs. Overall Frost Heaving Quantity (Hf)

Hp, cm	89.3	114.0	64.5	70.7	76.9	83.1	89.3	95.4	114.0	64.5	70.7	76.9	83.1
Hf, mm	29.7	21.3	60.0	56.3	47.7	42.7	37.7	44.7	36.3	61.7	52.3	40.7	29.3

$$Hp = 129.286 - 1.06Hf \qquad (1)$$

$$R = 0.80,\ F_0 = 19.00 > F_{0.05}(1, 11) = 4.84$$

Based on the upper limits for control of overall frost heaving quantity, which were put forward for cement concrete pavement by the authors in earlier researches, the minimum frost resistant depth required by typical structure is calculated by Equ. 1, as listed in Table 5.

Table 5 Minimum Frost Resistant Depth (Min. FRD) for Pavements with the Typical Structure in Heilongjiang

Grade of highway	Grade II and above	Below Grade II
Min. FRD in cm	67 - 88	40 - 67

Figures in Table 5 are basically in harmony with those stipulated in the specifications in use. Difference lies in that for highways below Grade II the allowable frost heaving value here gets a little bit larger, and the lower limit of minimum frost resistant depth gets smaller. The lower limits in the table apply to road sections with favorable water and temperature conditions, upper limits to road sections with unfavorable water and temperature conditions.

SUGGESTIONS ON THE SELECTION OF TYPICAL BASECOURSE AND SUBBASE COURSE STRUCTURE FOR CEMENT CONCRETE PAVEMENTS IN HEILONGJIANG

According to the above analysis and in connection with State-of-art Investigations on the Existing Cement Concrete Pavements in Heilongjiang, a list for selection of common materials for basecourse and subbase course of cement concrete pavement in Heilongjiang is given, as illustrated by Table 6.

Table 6 List of Common Materials for Selection for Basecourse and Subbase Course of Cement Concrete Pavements in Heilongjiang

Desig. of materials	Basecourse	Subbase course	Cusion course
1	A	A	B
2	C	B	C
3	C	A	A
4	C	A	A
5	C	B	B
6	C	B	C
7	C	C	A

Note: 1 refers to cement or lime-flyash stabilized sandy gravel (or crushed stone), 2 soil bounded crushed stone, 3 cement-lime stabilized soil, 4 lime stabilized flyash, 5 lime-flyash soil, 6 lime soil, 7 sand, natural sandy gravel or steel slag and others, A not recommended for selection, B conditionally selected when economically feasible or under favorable water and temperature conditions, and C suitable for selection.

CONCLUSIONS

The typical basecourse and cusion course structure for cement concrete pavement consisting of cement stabilized sandy gravel (or crushed stone) as basecourse, cement or lime-flyash stabilized granular materials as subbase course, and natural sandy gravel and other coarse granular materials as cusion course can meet the actual demands and therefore is practical in Heilongjiang.

The strengths of the typical basecourse and cusion course in use are of abundant storage. Due to the inferior stability of lime soil, in road section in moist or nearly moist state it is not suitable for use as materials of structural layers. Lime-flyash or cement-lime stabilized soils may be considered to use in subbase course or cusion course. See Table 6 in the text for detail selections.

For the typical cement concrete pavement structure in use in Heilongjiang, the proposed minimum frost resistant depth for highways of Grade II and above is 67 - 88cm, for highways below grade II 40 - 67cm. For roads with cut sections where ground water level is higher, measures should be taken to improve dry-moist state of the upper 80cm subgrade, to keep the part of subgrade in mid-moist or nearly mid-moist state. When in difficulty to change, stabilizing treatment must be conducted.

For the natural conditions in Heilongjiang, the water stability and the frost stability for the basecourse and cusion course of cement concrete pavement are of the same importance. On the basis of proper selection of basecourse and cusion course, drainage in basecourse, surface course and shoulder and in the interface of them must be highlighted. This is because the cement concrete pavements are charaterized by high rigidity and multiple joints. Furthermore, it is also needed to point out that in design of basecourse and cusion course for cement concrete pavement, their connection and also matching of strength and deformation with road shoulder should be stressed. Otherwise, even if a cement concrete pavement is of good strength and stability, its performance and service life will still be seriously affected by water permeation, frost heaving, and displacement.

REFERENCES

MINISTRY OF COMMUNICATIONS, P.R. CHINA. Design, Construction and Maintenance of Cement Concrete Pavement. 1992.

Fifth International Conference on the
BEARING CAPASITY OF
ROADS AND AIRFIELDS

ACCELERATED AGE HARDENING OF ASPHALT TO SIMULATE LONG TERM EFFECTS

BYRON E. RUTH[1] AND MANG TIA[1]

Professor, University of Florida, Dept. of Civil Engineering, PO Box 116580, Gainesville, Florida 32611-6580, USA

ABSTRACT

Testing equipment were designed, constructed, and evaluated for its effectiveness in producing accelerated age hardening of thin asphalt films. The concept followed in the development of the test equipment was to provide a method for the investigation of asphalt age hardening that simulates the thin asphalt films and temperatures found in asphalt paving mixtures.

A spray coating unit was constructed to coat the inside of 150 mm dia. by 597 mm long PYREX™ glass cylinders. The coated cylinders were used in the forced draft age hardening unit that was built to achieve accelerated aging of the asphalt films.

Keywords: Asphalt, Aging, Testing Equipment

INTRODUCTION

The hardening or aging of asphalt binders is recognized as occurring in two distinct phases; the first being a short term effect resulting from hot mix production operations and the second being a long term effect produced by environmental exposure over the service life of the pavement. Both of these aging effects result in hardening of the binder which in turn increases the viscosity and stiffness of the asphalt, and asphalt paving mixture, eventually reducing durability and resistance to cracking.

It has been observed that excessive hot-mix temperatures (e.g., 180C) has little effect on the long term hardening of certain asphalts whereas other asphalts exposed to high mixing temperatures exhibit a substantial increase in rate of hardening similar to polymerization (1). Although short term hardening may be primarily related to volatile losses, the long term loss of oily components may also occur as a result of syneresis and separation (2).

In general, long term changes in the rheological properties of asphalt binders are attributed to oxidative hardening. Pavements with air void contents in excess of 5 percent will likely experience accelerated aging due to its greater permeability which provides greater exposure to atmospheric oxygen (3). The properties of an asphalt, such as stiffness and viscosity, are influenced by the formation and dissolution of polar bonds. In a dispersed structure, the colloids are highly mobilized and oxidation can occur more easily. However, in a highly associated asphalt, the colloids are highly immobile particularly at low temperatures which inhibits oxidation (4). At high temperatures the microstructure begins to dissociate and the molecules oxidize more freely.

The influence of temperature on the rate of age hardening was very pronounced when comparing binder viscosity-age data from different climatic zones (5). Desert areas in the southwestern US exhibited hardening rates that were two to four times greater than those encountered in the Southeast and Midwest. A longer duration of high pavement temperatures obviously contributed substantially to the greater rate of aging.

At present, the recommended PAV test procedure for the "Superpave" asphalt binder specification requires a testing time of 20 hours at 90, 100, or 110C at a pressure of 2.0 MPa using 50 g. of RTFOT residue (6). Although this test method may provide a realistic approximation of long term aging (e.g., 5-8 years), the influence of pressure and above normal temperatures on the relative changes in properties between different asphalts is unknown. Consequently, a test method that exposed thin asphalt films to temperatures approximating those occurring in asphalt pavements would logically be better for simulation of in situ age hardening.

OBJECTIVES

The primary goal of the research was to develop equipment for the accelerated aging of asphalt binders which would be more indicative of field conditions. The main objectives of the study were:

1. To produce specimens having a film thickness less than 5 micron with sufficient uniformity to minimize variability of exposure.
2. To have sufficient exposed surface area to produce at least 50 g. of age hardened residue for use in subsequent rheological tests.
3. To construct a forced draft oven capable of providing adequate exposure of the asphalt film to produce an accelerated aging effect at typical pavement temperatures.
4. To evaluate the effect of temperature and exposure time on the viscosity and Asphalt Aging Index (AAI).
5. To compare the effectiveness of the process with other established procedures, particularly the PAV.

DEVELOPED AGING EQUIPMENT CHARACTERISTICS

The concept of thin film, forced draft, aging of asphalt resulted in the selection of 150 mm dia. (I.D. 140 mm), 597 mm long, PYREX™ glass cylinders which were to be spray coated on the inside and exposed to hot air moving through the cylinder. A spray coating unit and an age hardening unit were designed, constructed, and modified to achieve the desired operational characteristics.

Spray Coating Unit. Figure 1 shows an illustration of the spray coating unit and its key components. The unit performs two main functions: 1) the pressurized nozzle applies the heated asphalt as an atomized spray onto the inside surface of the glass cylinder; and 2) the cylinder is rotated and a servo unit moves the cylinder past the tip of the spray nozzle to achieve a thin, uniform, asphalt coating. The oil bath heated asphalt reservoir was pressurized to 70 kPa and the air pressure for the atomizing nozzle was adjusted to achieve a mist spray. An oil bath temperature of 171C was necessary to achieve a suitable spray to compensate for the air cooling effect of the atomization air at the nozzle.

Age Hardening Unit. Figure 2 illustrates the basic operational components and functional characteristics of the age hardening unit. A 1400 cfm blower produced an air velocity of approximately 18 m/s through the glass cylinders. Dual duct (ring) heaters installed within the 31.4 cm dia. duct provided sufficient heating capacity to maintain temperature slightly over 110C. An adjustable fresh air inlet and hot air exhaust provided make-up air; however, the most effective operation occurred when the system is closed to outside air. A DC motor/chain drive connected to the roller rods supporting the glass cylinders provided slow rotation during the operation of the unit.

MATERIALS AND TEST PROCEDURES

A preliminary evaluation of the new equipment was conducted to evaluate the need for modifications in equipment and procedures. This was followed by the testing of two different AC-30 materials, designated OR-2 and OR-3, that were aged in the new aging unit at 85C. The asphalt residue was subjected to viscosity tests at 5, 25, and 60C. Subsequently a temperature of 110C was used to determine the relative influence of temperature at the various exposure times.

Short and Long Term Aging. Short term aging of the asphalts was performed by Thin Film Oven Test (TFOT) according to ASTM D1754. The TFOT residue was applied to the inside surface of the glass cylinders using the spray coating unit. Typically two cylinders were coated with 45 to 60 g. of asphalt and processed in the age hardening unit of different exposure times (i.e., at 85C for 0, 48, and 72 hrs and at 110C for 0, 20, 48, and 72 hrs.). Upon completion of the accelerated (long term) aging, the asphalt residue was removed without the use of solvents. Liquid nitrogen was dispensed into a non-shattering container. Initially a large

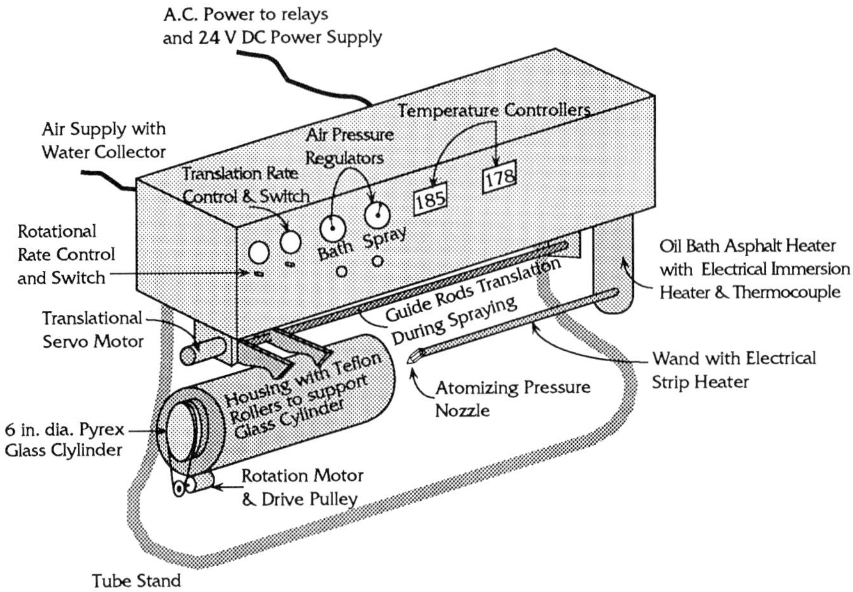

Figure 1 Spray Coating Unit

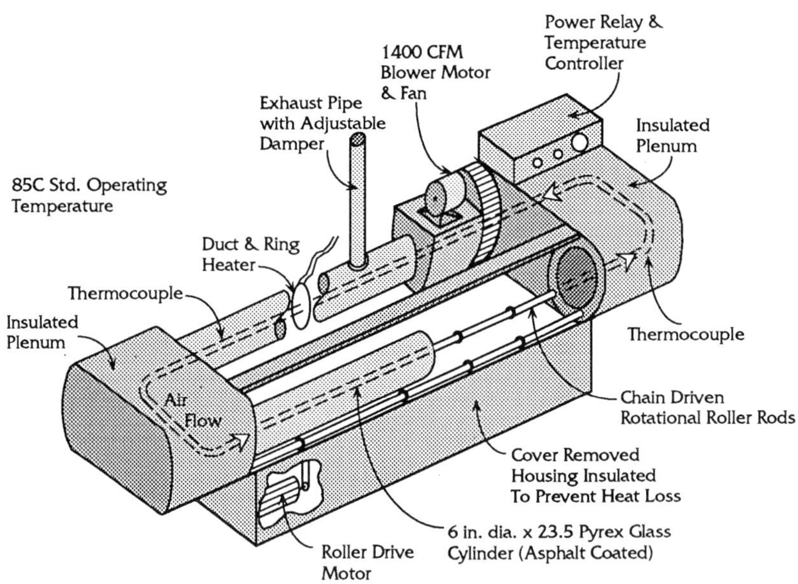

Figure 2 Age Hardening Unit

styrofoam cooler was used and the asphalt coated glass cylinder rotated to freeze the entire sample. Then a curved putty knife was used to scrape the asphalt from cylinder into a pan. A dewar flask was tried but only 1/2 of the sample could be exposed and removed at a time. It did prove more effective in reducing evaporation loss of the liquid nitrogen.

The removed asphalt residues were placed in sealed sample containers for subsequent viscosity tests. Upon heating of the samples prior to pouring of the asphalt, it was observed that air had been trapped in the samples. This required some stirring and a longer oven heating time to dissipate the air. The best procedure to minimize this problem is to add asphalt shavings gradually to a heated container immediately after freezing and removal from the glass cylinder. The asphalt residue from the 110C aging was then tested at 15C, 20C, 22.5C, 25C and 30C in the Schweyer Rheometer using D, E, and F capillaries and at 60C in the Brookfield Rheometer.

The Pressure Aging Vessel (PAV) was used to process the TFOT residue for 20 hours at 2069 kPa and at temperatures of 90, 100 and 110C. The residue from the PAV was tested at 60C using the Brookfield DVIII (Thermocell) rheometer and the Cannon Schweyer Constant Stress Rheometer at 25C. The constant power viscosity was (η_j) computed from the Schweyer Rheometer test results using the shear susceptibility (complex flow) and viscosity at 1.0 sec^{-1} (7).

The purpose of using a constant power viscosity (η_j) is to minimize extrapolation errors incurred by computing a viscosity at a shear rate substantially outside the shear rate of the test data. Although a constant power of $J = 100\ W/m^3$ was used by H. E. Schweyer, the test equipment, procedures, and constant power value selected is intended to closely approximate the stress and shear rate obtained at different test temperatures. If this is accomplished, the (η_j) versus log K plot will be linear from 60 C to low temperatures approaching the glass transition temperature. Also, the Aging Index (AI) was computed as follows to show the relative change in viscosity due to aging:

$$AI = \frac{\eta_{100}\ (\text{Aged})}{\eta_{100}\ (\text{Original})} \qquad \text{eqn. 1}$$

EVALUATION OF ASPHALT VISCOSITY DATA

Viscosity Trends. The investigation initially addressed the hardening produced at 85C. This temperature was selected as representative of Dubai pavements during extreme hot weather conditions. Viscosity tests performed on the aged residue was evaluated at 5, 25, and 60C after 48 and 72 hrs. of exposure in the accelerated age-hardening unit. Regression analyses were performed to evaluate the relative reliability of the viscosity (η_{100}) - temperature data. Subsequently, aging was conducted at 110C but the amount of viscosity data was increased to accommodate the greater hardening rate.

The results of these regression analyses are given in Table 1. The coefficients of determination (R^2) indicated an excellent fit to the 85C and 110C data. The values predicted by the regression equations were essentially the same as measured viscosities except for OR-3 after 72 hrs. where the 25C predicted viscosity was about 30 percent greater than the measured values.

Table 1
Regression Analyses of Viscosity data for TFOT Asphalt Residue
Processed in the Accelerated Age Hardening Unit

A.C. Code	Temp. °C	Process Time, hrs.	Regression Information[a]			
			n	R^2	Constant, a	Constant, b
OR-2	85	48	3	0.998	182.9376	-71.2398
		72	3	0.997	191.5532	-74.6511
	110	20	6	0.999	174.1847	-67.7085
		48	9	0.998	174.8280	-67.9108
		72	5	0.998	167.6144	-64.9897
OR-3	85	48	3	0.987	189.5249	-73.8963
		72	3	0.986	194.3812	-75.7867
	110	20	9	0.992	155.1410	-60.0479
		48	11	0.998	169.4364	-65.7223
		72	13	0.991	164.9669	-63.8698

[a] Regression equation conformed to a power law model using a constant power viscosity of 100 watts/m^3 and temperature in K.

$$\text{Log}\ \eta_{100} = a + b\text{Log}\ (K)$$

The viscosity test results for the PAV processed residue are given in Table 2. The TFOT was used to simulate short-term hardening instead of the current SHRP procedure that requires the RTFOT. This was done to minimize handling of the residue prior to insertion into the PAV.

Table 2
Viscosity Test Results for TFOT-PAV Residue

A.C. Code	η_{100}(PAV, 90C, 20 hr.)		η_{100}(PAV, 110C, 20 hr.)	
	V_{25} - Pa.s	V_{60} - Pa.s	V_{25} - Pa.s	V_{60} - Pa.s
OR-2[a]	1.13E6	2.55E3	2.04E6	6.20E3
	1.10E6	2.23E3	2.73E6	6.25E3
OR-3[b]	1.35E6	2.69E3	3.84E6	11.85E3
	1.77E6	2.75E3	5.12E6	13.16E3

[a] OR-2 original A.C.: V_{25} = 2.20E5 Pa.s and V_{60} = 304 Pa.s
[b] OR-3 original A.C.: V_{25} = 1.85E5 Pa.s and V_{60} = 309 Pa.s

The viscosity-temperature relationships for the PAV and accelerated age hardening unit for asphalts OR-2 and OR-3 are illustrated in Figures 3 and 4, respectively. The effect of aging time and process temperature appears logically to result in a parallel shift in the viscosity-temperature relationships. The shift from the original AC-30 relationship for OR-3 was substantially greater than for OR-2, thereby indicating greater age hardening potential. Also, OR-3 was more susceptible to the higher process temperature in the aging unit as suggested by the shift in viscosity trends between 85C and 110C for the 48 hr. exposure time. Similarly, the difference in PAV results between 90C and 110C indicated a greater change in hardening for OR-3.

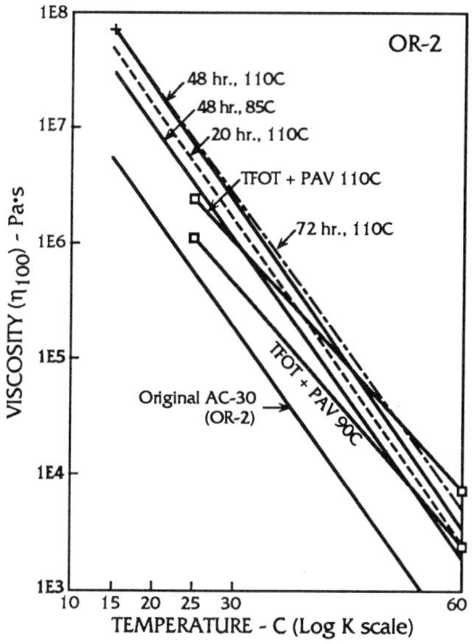

Figure 3 Viscosity-Temperature for OR-2.

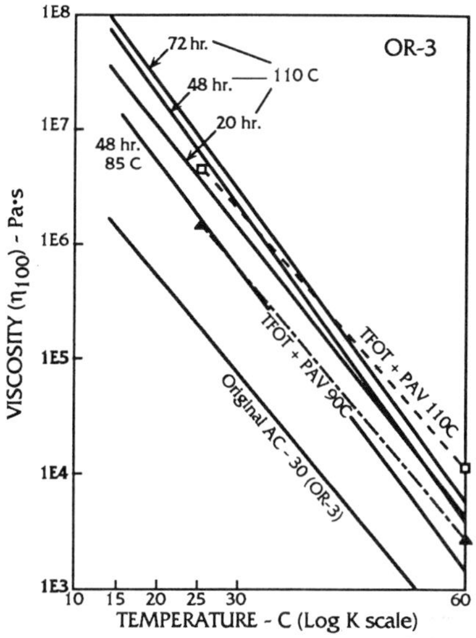

Figure 4 Viscosity-Temperature Trends for OR-3.

Asphalt Aging Index (AAI) values obtained from viscosity tests on the residue from the PAV and age hardening unit are given in Tables 3 and 4, respectively. Comments relating to the interpretation of these results are as follows:

1. On a comparative basis, approximately 15 to 50 hours in the accelerated asphalt aging unit produced AAI values, based upon V_{25} equivalent to those from the PAV. The PAV results based on V_{60} exceeded the viscosity and AAI values obtained from the aging unit.
2. The use of high pressure in the PAV was based on preventing morphological changes in polymer modified asphalts. However, these test results indicate that rheological properties at various test temperatures may be different than if aging was performed at atmospheric pressure.
3. The effect of short-term aging temperature on long-term aging has not been assessed. Some asphalts are very susceptible to higher asphalt hot-mix temperatures, resulting in greater initial hardening (loss volatiles?) and a much greater rate of long-term hardening (1).
4. Accelerated aging tests conducted at temperatures above those normally encountered in the pavement may produce changes in composition and rheological properties in certain asphalts that is not representative of their in-service behavior.

Comparison of AAI Values. California Tilt Oven (CTO) AAI values for asphalts OR-2 and OR-3 were obtained from an investigation conducted by C. T. Chiu (8). These values along with those from PAV and accelerated age hardening unit were used to prepare bar chart comparisons.

Table 3
PAV Asphalt Aging Index Values

A.C. Code	Asphalt Aging Index - AAI			
	PAV, 90C, 20 hr.		PAV, 110C, 20 hr.	
	60C	25C	60C	25C
OR-2	7.9	5.1	20.0	10.6
OR-3	8.8	8.4	40.0	24.2

Table 4
Asphalt Aging Index Values for Age Hardening Unit

A. C. Code	Time Hrs.	85C Process Temperature		110C Process Temperature	
		60C	25C	60C	25C
OR-2	0*	2.9	2.8		
	20	--	--	8.4	20.7
	48	4.7	6.0	11.5	28.8
	72	5.8	8.3	16.4	29.7
OR-3	0*	2.8	1.5		
	20	--	--	13.6	20.1
	48	6.0	5.0	16.1	36.2
	72	7.4	6.3	20.2	39.7

* After TFOT

Figures 5 and 6 compare the 85C age hardening unit values with CTO and PAV data. The 5C data were obtained from a study by T. S. Wattleworth (9). The OR-2 data shown in Figure 5 for 48 and 72 hours were similar to the 90C PAV and somewhat less than the 72 hr. CTO at 111C. The OR-3 data in Figure 6 are similar to OR-2 except the 5C values at 48 and 72 hrs. are considerably greater than the 25C and 60C data. The greatest differences between the two asphalts occurs at the 110C PAV and the 168 hr. CTO. This is further illustrated in Figures 7 and 8 where the 48 and 72 hr. 25C AAI values for the 110C age hardening unit are comparable to the 110C PAV and the 168 hr. CTO. However, the 60C values from the 110C aging unit were considerably lower than either the CTO or 110C PAV.

On the basis of the 25C AAI, the thin films in the accelerated age hardening unit can be age-hardened in less time than the CTO. However, the results for the 60C AAI indicate that the 168 hr. PAV and 110C CTO produce greater hardening than the aging unit.

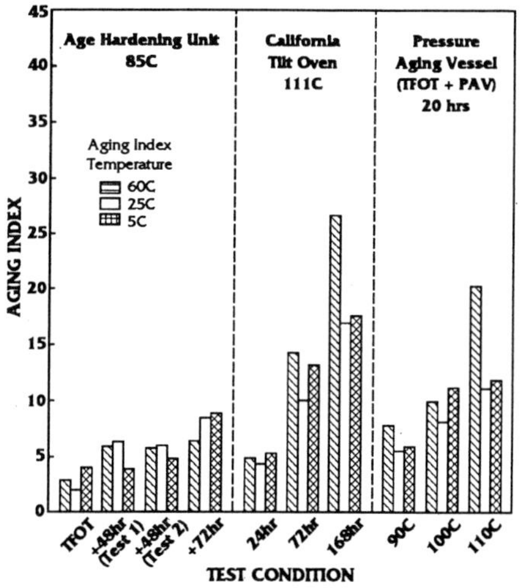

Figure 5 Asphalt Aging Index Values: OR-2, 85C

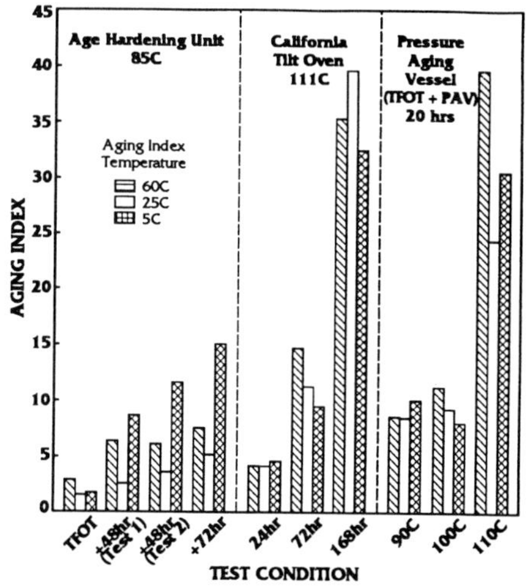

Figure 6 Asphalt Aging Index Values: OR-3, 85C

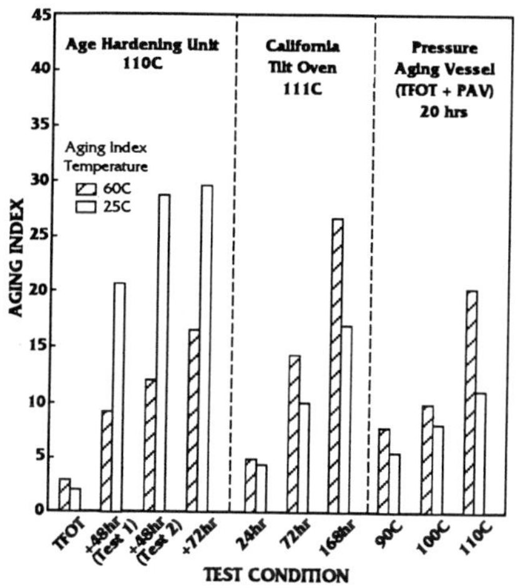

Figure 7 Asphalt Aging Index Values: OR-2, 110C

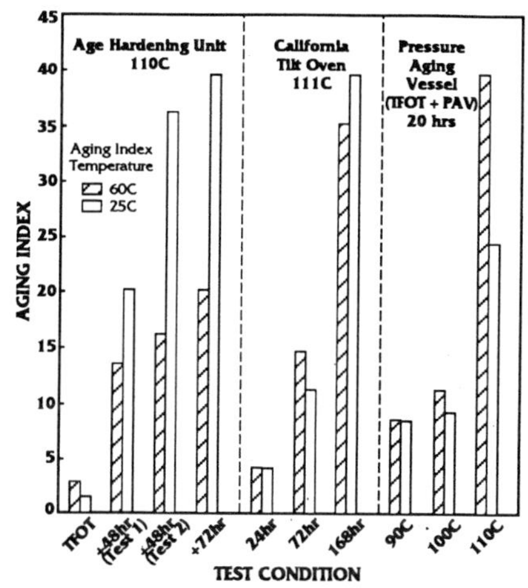

Figure 8 Asphalt Aging Index Values: OR-3, 110C

SUMMARY AND CONCLUSIONS

An accelerated age hardening unit has been designed, constructed and evaluated for its effectiveness in producing an asphalt residue comparable in viscosity to the pressure aging vessel (PAV). The results of this preliminary investigation indicate that the aging unit produces asphalt residues that are similar to the PAV and California Tilt Oven depending upon the time and temperature used in the processes.

Specific conclusions derived from the study are:

1. The rate of hardening in the aging unit is highly dependent upon process temperature, where extending the exposure time beyond 48 hours produces relatively small increases in the degree of hardening.

2. Asphalt OR-3 was selected because it exhibited greater hardening than other previously tested asphalts. The results from the PAV and aging unit showed only slightly greater hardening than OR-2 except for the 110C PAV viscosities and asphalt aging index values.
3. The comparison of PAV and aging unit indicated greater hardening after 20 hrs in the PAV based on viscosity tests at 60C. This trend reversed at 25C, resulting in greater hardening with the aging unit after 20 hours of heating to temperatures of 85C and 110C. These results suggest that the high pressure used in the PAV may alter the composition and consequently, change the rheological behavior of the asphalt residue.
4. The California Tilt Oven (111C for 168 hrs) generally produced greater hardening than the aging unit. However, in consideration of effectiveness, the aging unit produced a higher rate of hardening than the CTO.

In summary, the accelerated aging unit has the capability of exposing relatively thin films (130 microns) of asphalt to high temperatures representative of those encountered in asphalt pavements. The temperature combined with the high rate of air flow in the forced draft system appears to be very effective in hardening asphalt cements. This equipment is deemed suitable for research investigations into the age hardening of asphalt binders in pavements.

REFERENCES

1. Page, G.C., B.E. Ruth, and R. Roque, "Asphalt Binder Hardening-Causes and Effects," *Proceedings*, the Association of Asphalt Paving Technologists, Vol. 54, 1985, pp. 140-167.
2. Roberts, F.L., P.S. Kandhal, E.R. Brown, D. Lee, and T.W. Kennedy, "Hot Mix Asphalt Materials, Mixture Design, and Construction," 1st Ed., NAPA Education Found., Lanham, MD, 1991.
3. Tuffour, Y.A., and I. Ishai, "The Diffusion Model and Asphalt Age-Hardening," *Journal of the Association of Asphalt Paving Technologists*, Vol. 59, 1990, pp. 110-127.
4. Petersen, J.C., J.F. Branthauer, R.E. Robertson, P.M. Harnsberger, J.J. Duvall, and E.K. Ensley, "Effects of Physicochemical Factors on Asphalt Oxidation Kinetics," *Transportation Research Record 1291*, Transportation Research Board, 1992, pp. 1-10.
5. Guan, L., and B.E. Ruth, "Age Hardening Trends and Predictions," *Proceedings*, First Materials Engineering Congress on Serviceability and Durability of Construction Materials, ASCE, Vol. 1, 1990.
6. *Performance Graded Asphalt Binder Specification and Testing*, Superpave Series No. 1 (SP-1), Asphalt Institute, Research Park Drive, P.O. Box 14052 Lexington, KY 40512-4052.
7. Tia, M., and B.E. Ruth, "Basic Rheology and Rheological Concepts Established by H.E. Schweyer," STP941, American Society for Testing and Materials, 1987, pp. 118-145.
8. Chiu, C.T., "Investigation of Laboratory Aging Processes for Development of an Effective Procedure to Characterize Asphalt Durability," Doctoral Dissertation: Univ. of Florida, Gainesville, FL, May 1994.
9. Wattleworth, T.S., "Development and Evaluation of a Testing Device for Accelerated Age Hardening of Bituminous Binders," Masters Thesis: Univ. of Florida, Gainesville, FL, Aug 1994.

Fifth International Conference on the
BEARING CAPACITY OF
ROADS AND AIRFIELDS

AGING OF ASPHALT AIRFIELD PAVEMENTS IN NORWAY - A field and laboratory study

JORALF AURSTAD[1], EIVIND O. ANDERSEN[2] and GEIR LANGE[3]

[1,2]Research Engineer, SINTEF Civil and Environmental Engineering
N-7034 Trondheim, Norway, Fax: +47 73591478
E-mail: [1]joralf.aurstad@civil.sintef.no [2]eivind.o.andersen@civil.sintef.no
[3]Senior Engineer, Norwegian Civil Aviation Administration
Box 8124 Dep., 0032 Oslo, Norway, Fax: +47 22942390
E-mail: geir.lange@caa.no

ABSTRACT

Asphalt airfield pavements in Norway are often exposed to extreme and rapidly changing climate, causing severe distresses and shortened pavement life. It has been shown that grooving of the pavement surface in order to avoid hydroplaning, actually accelerates this deterioration.

The pavements on most of these airfields have traditionally been a standard dense-graded asphalt concrete (AC 11 mm) with a 180pen binder. In 1991, the Norwegian Civil Aviation Administration initiated a pavement research and development program, where the main objective was to develop more lasting pavements with sufficient wet friction properties.

As a part of this project, bituminous binders and asphalt mixes have been studied by accelerated aging in the laboratory. A large-scale program for accelerated aging of asphalt mixes in a weatherometer has been conducted. SHRP aging procedures for binders have also been tried out. The properties have been evaluated using both traditional and new Superpave binder tests. The different accelerated aging procedures have been compared.

Some Norwegian airfields have also been reconstructed/repaved with "new" asphalt mixes (different aggregate gradings, modified binders etc.). These new pavements are now closely monitored. Core samples for laboratory investigations are collected regularly to study the mix and binder properties over time.

This paper gives a state-of-the-art; the in-field behaviour and aging of different mixes are discussed. Experiences so far by using accelerated aging methods are also discussed.

Key words: Airfields, Asphalt pavements, Aging, Weatherometer, Superpave

1 INTRODUCTION

Norway has a rapidly changing climate. Considerable differences are recorded from south to north and between coastal and inland locations. Air temperatures can range from -40°C to +35°C for selected airfields. Combined with other unfavourable climatic factors (large amounts of precipitation, many freeze-thaw cycles), construction of lasting and resistant asphalt pavements in Norway is a big challenge.

The pavements on most Norwegian airfields have traditionally been a standard dense graded asphalt concrete (AC 11mm) with a 180pen binder. The pavement surfaces on all civil airfields have been grooved in order to avoid hydroplaning and ensure sufficient wet friction conditions.
In 1990 the Norwegian Civil Aviation Administration (NCAA) carried out a distress identification program on the main route airports. These investigations revealed severe pavement distresses on a number of airfields. Cracking and ravelling were developing at a faster rate than NCAA's plans and budgets for maintenance and rehabilitation.
On the basis of these results, the NCAA in 1991 initiated a pavement research and development program, where the main objective was to develop longer lasting asphalt pavements.

2 TYPES OF INVESTIGATION

The program was divided into four different sub-projects:

I. ***Literature review:*** The literature study took advantage of the investigations carried out in the SHRP project.
II. ***Laboratory investigations of existing/old pavements:*** A number of samples were collected from selected airfield pavements representing different ages, climates and types of distresses (chapter 3).
III. ***Accelerated aging tests:*** Aging of both binders and asphalt mixes have been studied in the laboratory. For asphalt mixes, an extensive investigation was carried out using a large weatherometer facility. Aging of binders has been studied using the SHRP specifications (RTFOT/ PAV). The different accelerated aging procedures have been compared (chapter 4).
IV. ***New asphalt pavements:*** A number of Norwegian airfields have been reconstructed/repaved with "new" asphalt mixes (different aggregate gradings, modified binders etc.). These new pavements are now closely monitored. Core samples for laboratory studies are collected regularly to study the mix and binder properties over time. The friction properties are also monitored (chapter 5).

3 STUDIES ON EXISTING/OLD PAVEMENTS

The initial investigations conducted by the NCAA indicated that many flexible airport pavements have premature deterioration, resulting in shortened pavement life.

To verify (and quantify) these observations, a number of samples were collected from selected airfield pavements. By correctly identifying distresses, gathering historical data (pavement age, mix characteristics, traffic loads etc.) and performing laboratory analyses, the following conclusions could be drawn:

Durable pavements with satisfactory performance (lifetime>20 years) were characterised by

- High binder contents (> 5.5 %)
- Low air void contents (< 4.5 %)
- High degree of uniformity/homogeneity

Pavements with unsatisfactory performance showing early distresses were characterised by

- ♦ Very stiff binders (low penetration values)
- ♦ Inhomogeneous surfaces
- ♦ Grooving

These findings are in agreement with international literature on the subject. In this sense, the results did not represent any major new findings. Nevertheless, some essential factors influencing long time asphalt pavement performance were confirmed. And it was obvious that grooving gives the pavement densely spaced points of weakness.

The final conclusions from these initial laboratory studies were:

- ♦ There is a strong need for development of more climatic resistant pavements.
- ♦ Pavement management and preventive maintenance techniques should be emphasised.
- ♦ Grooving should be avoided (given that alternative pavement surfaces can provide sufficient skid resistance).

4 ACCELERATED AGING IN LABORATORY

4.1 Scope

In order to study the influence of different mix properties in the laboratory, an accelerated aging test program was conducted. A large number of samples were mixed and compacted in the laboratory. Selected mix parameters were varied, the project was designed to study different aggregate gradings, binder stiffnesses, binder contents and ***general*** effects of polymer modifications and fibre additions.
Identical twin samples were made of each mix. One sample was exposed to climatic influences, the other was used as a reference for comparison (mix and binder properties before and after aging). Some field samples were also included.
A total of ***50 different mixes*** were included in the project (41 lab-mixed + 9 field samples), in two consecutive test programs. For further details, see /1/ and /2/.

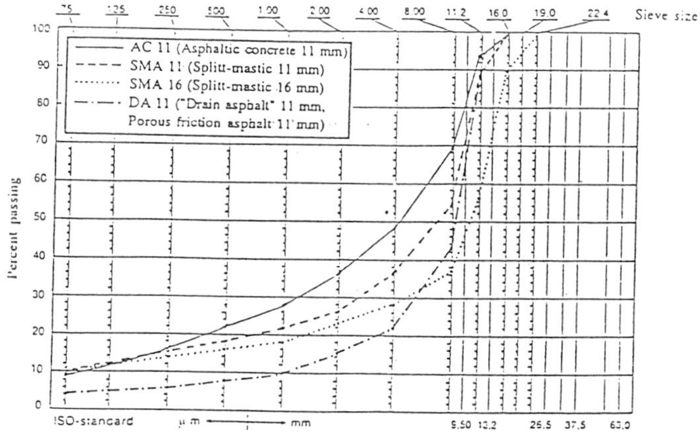

FIGURE 1 Different aggregate gradings for the accelerated aging program /1/

4.2 Apparatus

The accelerated aging test apparatus utilised in the project was developed by the Norwegian Building Research Institute. This is a large-scale climatic chamber (weatherometer) able to simulate a number of climatic influences. In this case, the test samples were exposed to a 6-month program with a precisely defined 24-hour cycle of varying climate; heating (+60°C), UV-radiation, precipitation and surface freezing down to -10°C (figure 2). This program has earlier been shown to equal 10-15 years of outdoor aging on bituminous roofing membranes in central Norway.

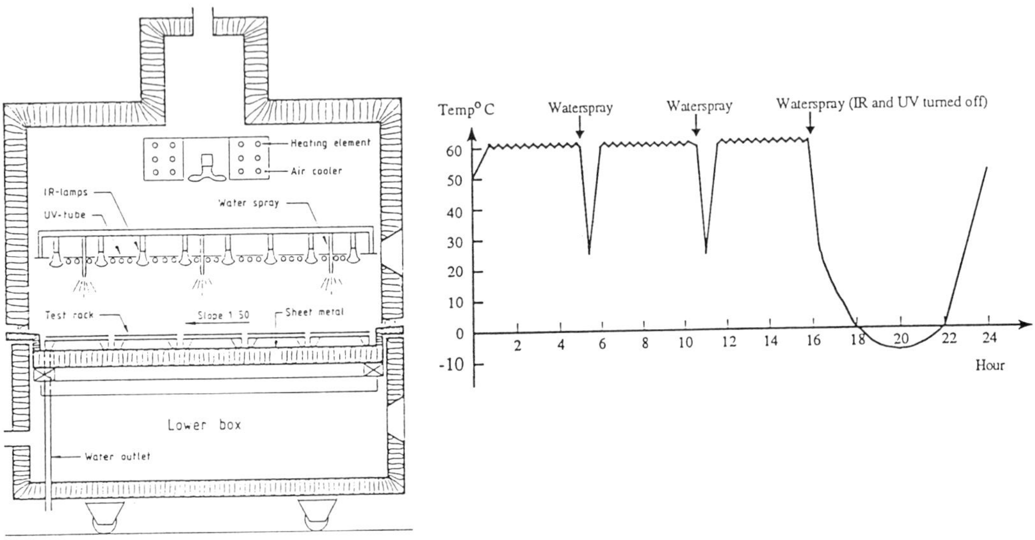

FIGURE 2 Section of apparatus for accelerated aging (weatherometer) /1/ and surface temperature of samples during one cycle (24 h) of the weatherometer test /3/

4.3 Results from the weatherometer test

As mentioned previously, a total of 50 samples were aged in the weatherometer. Due to the relatively large sample size (44x44x5 cm), the different mixes could be evaluated with regards to:

- Visual effects (deformations, changes in surface characteristics)
- Mix properties (testing/analyses of cores from aged and unaged samples)
- Binder properties (testing/analyses of original binders and recovered binders from aged and unaged samples)

By ***visually observing*** the samples during the test period, it could be seen that the different mix variables had significant effects;

- High (unmodified) binder contents (> 6%) and soft binders (pen >180) lead to extensive deformation and poor mix properties. Bleeding, binder hardening and incipient "micro-cracking" appeared.

- Polymer modification and/or fibre additives had a surprisingly large effect; both deformations and surface distresses were strongly reduced or eliminated. Especially the samples with PmB maintained a highly satisfactory and appealing surface during the test, even at high binder contents (> 6%).

Analyses of ***mix and binder properties*** also revealed interesting results (binder analyses were run on binders recovered from the upper 10 mm of the specimens). Some examples of the findings are:

- Mix stiffness (indirect tensile strength, Marshall stability) increased significantly during the aging test for mixes with conventional binders. In the polymer modified samples these properties remained almost unchanged.
- SMA and DA mixes revealed greater changes in both mix properties and binder stiffness than AC-mixes.
- Binder tests confirmed reduced aging (drop in pen. value) with increasing binder content.
- PmBs were less effected than unmodified binders on the basis of pen values.

Further details, photos, figures etc. are given in /1/ and /2/.

4.4 Comparison to SHRP laboratory aging methods

Laboratory aging using this kind of weatherometer can hardly become routine tests because of time, costs etc. However, the procedure has during this project showed to be an interesting approach when larger samples ("pavement areas") are to be studied.

For detailed studies, for instance of aging properties of specific binders, simpler and quicker conditioning tests must be used. The SHRP tests for simulating short term (RTFOT/TFOT) and long term (PAV) aging are well suited for this purpose.

During 1994-1998 a project aiming to study and evaluate the new Superpave binder technology has been carried out in Norway /4/, /5/. By combining some of the results from this study with analyses of recovered binders from the weatherometer tests, a simple comparison has been made between the accelerated aging in the weatherometer and the SHRP aging procedure, see table 1. The results are presented in figures 3-6 (unmodified binders only).

TABLE 1 Different aging procedures

	Short-term aging	Long-term aging
Weatherometer approach	Laboratory mixer (batch 30 kg), 5 min at spec. mixing temp.	Weatherometer, 24 weeks spec. program
SHRP procedure	RTFOT, 163°C 75 min	PAV, 100°C 20 h

Figure 3 shows that the SHRP short-term aging procedure (RTFOT) leads to an over-all higher penetration drop (more severe aging) than the mixing procedure used for the weatherometer samples.

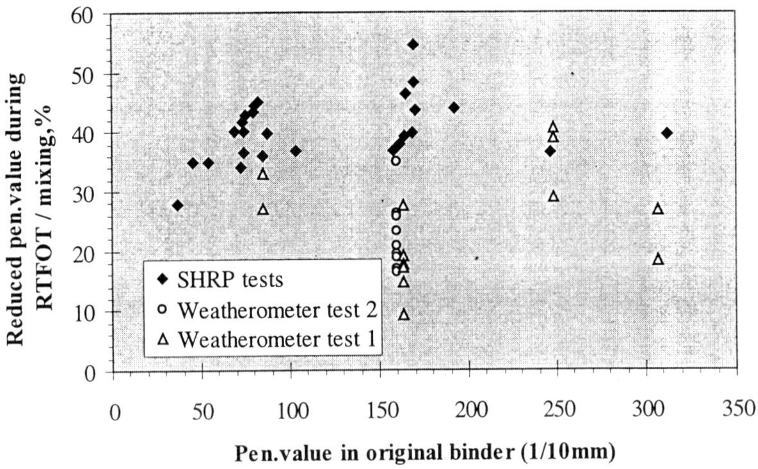

FIGURE 3 Short-term aging (expressed as reduced penetration values) during laboratory mixing of weatherometer samples compared to RTFOT (unmodified binders).

Data for the weatherometer-aged mixes in figure 4 are based on recovered binders from the upper 10 mm of the samples. Here we see that 24 weeks in the weatherometer gives a more severe aging than the PAV.

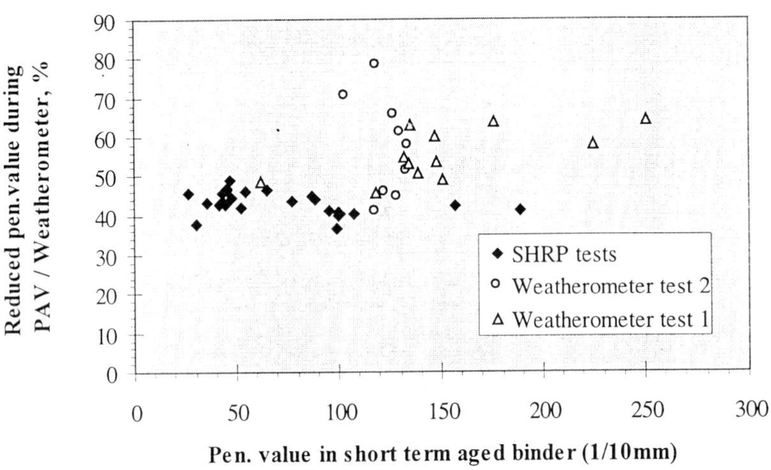

FIGURE 4 Long-term aging (expressed as reduced penetration values) during weatherometer tests compared to PAV (unmodified binders only).

Figure 5 shows the ***total aging*** (short-term + long-term) of the different procedures. It seems like the total aging effects of these completely different procedures are very much comparable. Based on penetration, the binders show (more or less) ***similar hardening***!

The same comparison is made in figure 6, here using the SHRP low temperature criteria. We see that the weatherometer samples fit very well into the results from the SHRP tests.

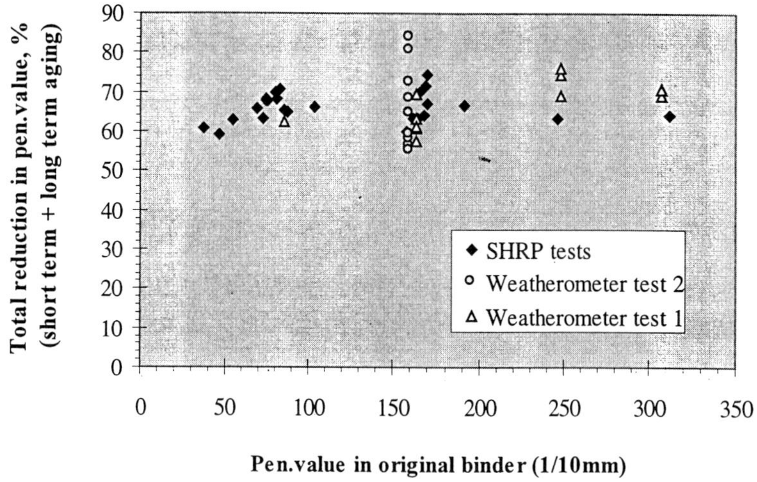

FIGURE 5 Total aging (expressed as reduced penetration values) during mixing + weatherometer tests compared to RTFOT + PAV (unmodified binders only).

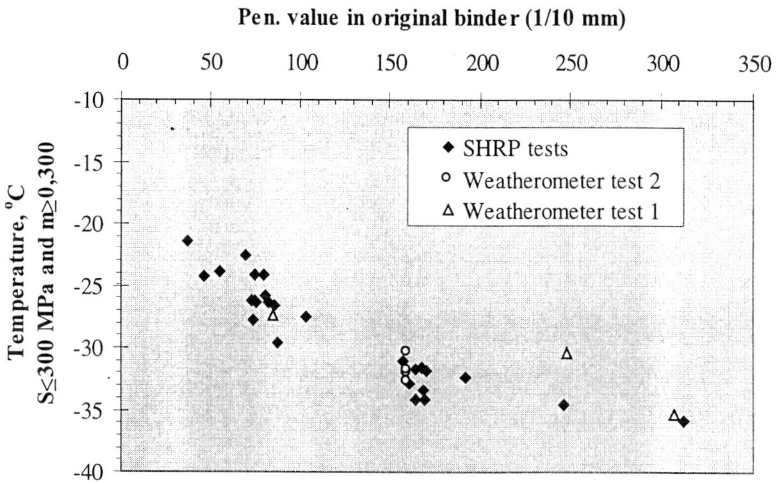

FIGURE 6 Low-temperature properties measured with Bending Beam Rheometer on RTFOT+PAV-aged binders and binders recovered from weatherometer-aged samples (unmodified binders only).

5 NEW ASPHALT RUNWAY PAVEMENTS – FIELD TRIALS

The weatherometer program showed that high binder contents (thick bitumen films on the aggregate), fibre additives and polymer modification had favourable effects on aging. Regarding mix alternatives, SMA pavements were of specially interest due to deformation resistance and good friction properties (macro-texture).

As a result of the accelerated aging program, a number of airports have been repaved since 1992, using other types of mixes than the standard AC11 180pen concept. Some of these pavements are now closely monitored to evaluate the functional properties over time.

5.1 Friction

The friction properties for some of the pavements have been compared. Figure 7 shows an example using one of the Pennsylvania State University models. Based on macro-texture values and measured skid resistance (SAAB Friction tester, 1-mm water-film) at specific driving speeds, speed-friction relationships have been calculated.

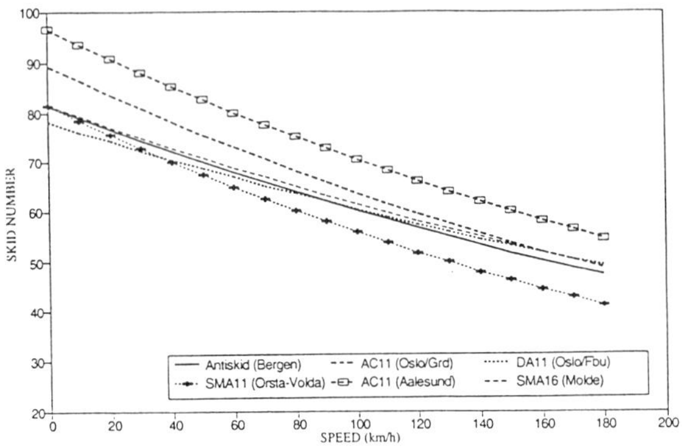

Skid Number $SN = SN_{60} \exp [-0.041*(V - 60)*MD^{-0.47}]$

FIGURE 7 Comparison of skid numbers as a function of speed (input values; macro-texture (MD) and measured skid resistance at 60 km/h), 1 mm water-film /6/

All these pavements are above ICAO's recommended "maintenance level" for wet friction. And perhaps the most promising result; SMA16-mm pavements show very high friction values, without grooving!

5.2 Mix and binder hardening (aging)

The weatherometer test also indicated that mixes with "open-graded" surfaces like SMA might have a faster rate of aging/binder hardening than dense AC pavements, and thereby might need more extensive use of preventive surface treatments (rejuvenators, fog seals etc.).

In order to study this in the field, a monitoring program with distress identification and core sampling is now being carried out on four Norwegian airports, representing different climates and different asphalt mixes, see figure 8.

Some preliminary results from binder analyses are shown in figure 9. As for the weatherometer samples, these data refer to binders recovered from the upper 10 mm of the asphalt pavement. (Mixes are shown in figure 8.)

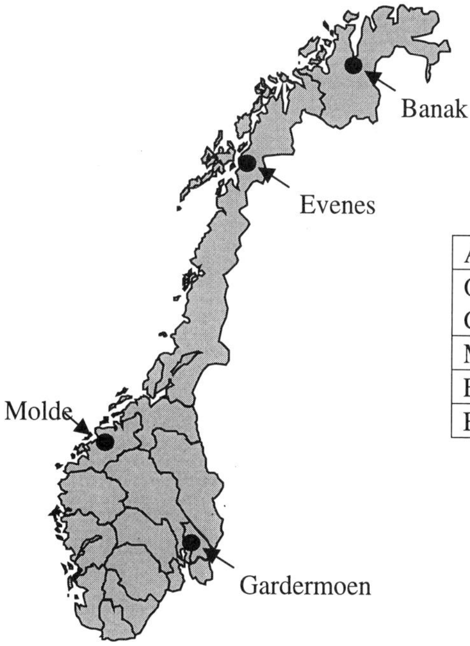

Airfield	Age	Pavement
Oslo/ Gardermoen	1989	AC11 mm, 180pen bit. (grooved surface)
Molde	1993	SMA16 mm, 180pen bit.
Evenes	1994	SMA16 mm, 180pen bit.
Banak	1993	SMA11 mm, PmB

FIGURE 8 Airfields that are being studied closely with regards to long-term asphalt pavement performance

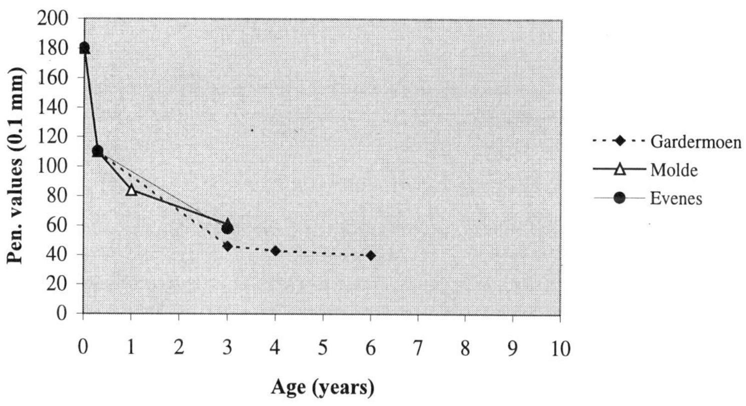

1st value: Original binder pen. 2nd value: Binder pen. after compaction (short-term aging)

FIGURE 9 Aging on different Norwegian airfields (expressed as 25°C pen. values), preliminary results

The figures show that even in Nordic climates a substantial age hardening can be observed. When comparing the results, we see that the grooved AC (Gardermoen) and the coarse SMA (Molde, Evenes), all with 180pen bitumen, more or less have the same rate of hardening. To see whether pavements with PmB (Banak) have less severe field aging will be very interesting (that was actually indicated in the weatherometer program).

This work will be supplemented with other analyses, measured mix and binder properties will be compared to the functional behaviour of the pavements (distress development). NCAA also intends to utilise Superpave technology in future work on this project.

6 CONCLUSIONS

The preliminary results are promising, hopefully this work will result in more resistant and longer lasting pavements. The following conclusions can be drawn:

- Laboratory studies of aging, using both SHRP methods and weatherometer/climatic chamber, can give valuable results. It has been shown that high binder content, low void content and the use of PmB can reduce aging and the effects of aging.
- Further, it has been shown that the chosen program for weatherometer testing gives the same over-all aging effects as the SHRP procedures (RTFOT + PAV).
- Recent field studies have revealed substantial age hardening on Norwegian airfields. Compared to these field studies, a preliminary conclusion is that the laboratory long-term aging procedures (PAV, weatherometer) simulate a shorter time period than the prospected 8-10/10-15 years.
- High skid resistance makes SMA a good alternative to grooved AC pavements. Compared to dense asphalt mixes, the project has revealed more severe surface aging on SMA pavements due to the coarse macrostructure. We believe that this can be controlled by preventive surface maintenance.

ACKNOWLEDGEMENTS

This project is funded by the Norwegian Civil Aviation Administration. The authors want to address special thanks to project manager Arnold Hagen, NCAA for his enthusiasm and contribution. Arnold Hagen died 28 July 1997.

REFERENCES

1. AURSTAD, J., ANDERSEN, E.O. and HAGEN, A. Evaluations of long-term asphalt pavement performance on Norwegian airfields. Proceedings of the 4th BCRA Conference; Minneapolis 17-21 July 1994
2. AURSTAD, J., ANDERSEN, E.O. Aging of asphalt pavements on airfields, final report. SINTEF Report STF22A97520, 1997 (*in Norwegian*)
3. GJELSVIK, T. Large scale tests facilities for durability studies in Trondheim. Norwegian Building Research Institute 1985
4. ANDERSEN, E.O., SKOGLUND, R. Experiences and results using Superpave binder tests. Proceedings of the 5th International RILEM Symposium Lyon, 14-16 May 1997
5. ANDERSEN, E.O. Adjustments of the Superpave binder technology to Norwegian conditions. SINTEF Report STF22A98452, 1998 (*in Norwegian*)
6. AURSTAD, J. Pavement friction on airfields. Report STF61A95014, SINTEF Civil and Environmental Engineering 1995

Fifth International Conference on the
BEARING CAPASITY OF
ROADS AND AIRFIELDS

AGEING AND DEGRADATION OF ASPHALT PAVEMENTS ON LOW-VOLUME ROADS

BJØRN OVE LERFALD

Doctoral student, Norwegian University of Science and Technology,
N-7034 Trondheim, Norway, fax +47 73 59 14 78,
E-mail; bjorn.o.lerfald@civil.sintef.no

ABSTRACT

Asphalt pavement research in Norway has until now mainly focused on problems related to wear caused by studded tyres on high-volume roads. However, about 95 % of the Norwegian road network consist of low-volume roads. These roads are primary affected by other types of distress mechanisms. This paper gives an overview of different types of defects occurring on asphalt pavements on low-volume roads in Norway. Results from a field investigation on pavement conditions in central Norway is also given.

INTRODUCTION

This work is a part of a new research programme at SINTEF and the Norwegian University of Science and Technology. The programme focuses on Life Cycle Management and the service life of buildings and infrastructure. Ageing and deterioration of asphalt pavements on low-volume roads is one of four doctoral studies connected to this research programme. This paper summarises some of the work carried out at the start of this study.

About 95 % of the Norwegian road network consist of low-volume roads. In this case, an annual daily traffic (AADT) of 2000-3000 is defined as low-volume roads. On these roads, other distress mechanisms than wear from studded tyres dominate. These factors are plastic deformations, ageing and fatigue. Ageing due to climatic exposures are among the most important factors influencing the degradation of asphalt surfaces on these roads.
The ageing properties of asphalt mixtures are related to factors such as binder stiffness, binder content, aggregate mineralogy, void content and production.

VARIABLES IN SPECIFIC GRAVITY OF AGGREGATES

The specific gravity and absorption test methods for coarse aggregate (ASTM C127) and for fine aggregate (ASTM C128) should be scrutinized because of its potential affect upon the computed values of VMA that impacts the design of asphalt mixtures. Only the apparent specific gravity (G_{sa}) test is not dependent on the subjective test procedure for a saturated surface-dry condition. The amount of water absorbed by coarse aggregate (CA) requires a visual assessment of the saturated surface-dry condition. The use of the cone to determine when the fine aggregate (FA) is saturated surface-dry often gives variable results. The variability in specific gravity values can be due to testing and to differences in aggregate gravity.

Tables 1 and 2 summarize specific gravity test results for granite CA and FA obtained from two different sources. The values given in these tables represent one to three samples tested each week over five or more weeks. The CA mean specific gravity values are essentially identical for each of the three sample groups. These results imply that the aggregate gravities are fairly constant and variability is primarily due to testing. In Table 1 the G_{sa} of CA and FA for granite No. 1 are 2.732 and 2.723, respectively. These differences are slightly greater for granite No. 2 as given in Table 2.

Table 1 Specific Gravity Values for Granite No. 1

	No.	Mean Specific Gravity			% Water Absorption		
	Samples	G_{sb}	G_{SSD}	G_{sa}	Mean	Min.	Max.
CA[(a)]	14	2.701	2.713	2.734	0.44	0.3	0.7
	31	2.693	2.706	2.727	0.45	0.3	0.6
	57	2.689	2.706	2.736	0.65	0.4	1.0
	$\bar{\bar{x}}$	2.694	2.708	2.732			
FA[(b)]	24	2.677	2.694	2.723	0.64	0.3	1.3

(a) typical standard deviation of coarse aggregate = 0.014, (b) std. deviation FA = 0.023

Since the mean G_{sa} and absorption values for the CA and FA are essentially identical for granite No. 1, it is questionable that water permeable aggregate pores exist. Water retention on the surface of aggregate particles rather than absorption is a more plausible explanation.

The standard deviations for the specific gravity values given in Tables 1 and 2 are given in Table 3. Values of $G_{sb} \pm 2$ S.D. were selected and used to calculate the extreme range in VMA. Granites No. 1 and No. 2 had VMA differences of 2.3 and 4.2, respectively. The variation in G_{sb} value for the FA were substantially greater than for the CA, particularly for granite No. 2. This further exemplifies the potential for substantial differences in VMA values as a result of G_{sb}/water absorption variations.

Table 2 Specific Gravity Values for Granite No. 2

	No. Samples	Mean Specific Gravity			% Water Absorption		
		G_{sb}	G_{SSD}	G_{sa}	Mean	Min.	Max.
CA[a]	21	2.736	2.755	2.789	0.69	0.4	1.0
	30	2.737	2.755	2.788	0.68	0.5	1.1
	53	2.745	2.763	2.796	0.67	0.4	0.9
	$\bar{\bar{x}}$	2.739	2.758	2.791			
FA[b]	50	2.666	2.701	2.763	1.33	0.6	2.9

(a) typical standard deviation of coarse aggregate = 0.026, (b) std. deviation FA = 0.040

Table 3 Standard Deviation Values for Aggregate Specific Gravity

	No. Samples	Standard Deviation			Diff. in VMA[a]
		G_{sb}	G_{sa}	% Absorption	
Granite No. 1-CA	14	0.017	0.017	0.098	
	31	0.012	0.010	0.091	
	57	0.015*	0.014	0.140	
FA	24	0.023*	0.029	0.255	2.3
Granite No. 2-CA	21	0.026*	0.029	0.166	
	30	0.040	0.041	0.140	
	30	0.011	0.015	0.125	
FA	50	0.044*	0.036	0.370	4.2

(a) Diff. based upon ± 2 S.D. using 5.5% AC, G_{mb} = 2.400, and G_{sb}* values.

Groups of three replicate samples that conformed to the ASTM precision requirements were selected, the means of each three replicates computed, and the high and low mean G_{sa} values selected along with corresponding mean values of G_{sb} and absorption. The mean G_{sb} values were verified using G_{sa} and absorption to calculate a G_{sb}. The blend of CA and FA were used to calculate the G_{sb} of the blend. The VMA values were then computed using the G_{sb} values by assuming a bulk mix density of 2.400 and an asphalt content of 5.5 percent. The results in Table 4 indicate that the VMA determined from three replicate tests performed each week could vary as much as 0.8 percent. Another comparison was made using the mean values in Tables 1 and 2 to compute the G_{sa} of the blend and the typical min.-max. absorption values (CA & FA) to calculate the G_{sb} and corresponding VMA values given in Table 5. The purpose of this comparison is to emphasize that absorption variations have a very significant affect on G_{sb} and VMA if we assume the G_{sa} is constant. Again, the reliability of specific gravity tests is questioned. The test results may be precise but are they accurate in depicting absorption into aggregate pores and for use in VMA calculations?

Table 4 Difference in G_{sb} and Its Effect on VMA

	CA		FA		Blend 60/40		Diff.
Granite No. 1	Min.	Max.	Min.	Max.	Min.	Max.	Max-Min
$G_{sa}^{(a)}$	2.711	2.742	2.714	2.727	—	—	—
$G_{sb}^{(a)}$	2.673	2.707	2.678	2.683	2.675	2.697	0.022
$VMA^{(b)}$					15.2	15.9	0.7
Granite No. 2							
$G_{sa}^{(a)}$	2.765	2.798	2.737	2.792	—	—	—
$G_{sb}^{(a)}$	2.729	2.725	2.652	2.714	2.696	2.722	0.026
$VMA^{(b)}$					15.9	16.7	0.8

(a) Min-max of groups of 3 replicates conforming to ASTM precision requirements.
(b) Calculated using 5.5% AC and 2.400 for bulk density (G_{mb}) of mix.

COMPARISON OF EFFECTIVE, APPARENT, AND BULK SPECIFIC GRAVITY

Mixtures were prepared using two different aggregate blends. Values for the CA and FA were used to calculate the specific gravities and absorption of the aggregate blend. Table 6 gives a comparison of the test values for Mixtures N619 and G319. In general, the G_{sa} values were approximately equal to the effective gravity (G_{se}) of the aggregate as calculated from the Rice density test (1). This indicated that the absorption of water and asphalt is either the same or zero. Low absorptive aggregates (e.g., Mix N619) may retain moisture or asphalt equally because of retention due to surface texture rather than aggregate pores. Aggregates with high water absorption usually absorb only about one third to one half the amount of asphalt. However, Mix G319 deviated from the expected since the G_{se} was almost the same as the G_{sa} even though the asphalt absorption based on G_{sb} was 2.27 percent.

Table 5 Extreme Variation in BSG and VMA Due to Absorption

Aggregates Type/Source	G_{sa} Blend	% Water Absorption	Calculated G_{sb}	Computed Mix VMA, %$^{(a)}$
Granite No. 1	2.728	0.3	2.706	16.2
	2.728	1.3	2.635	13.9
	Diff. in values	1.0	0.071	2.3
Granite No. 2	2.780	0.5	2.742	17.3
	2.780	1.0	2.705	16.2
	2.780	1.7	2.655	14.6
	Max. Diff. in values	1.2	0.087	2.7

(a) VMA calculated using G_{sb} and assumed mix density of 2.400 with 5.5% asphalt content.

Table 6 Comparison of Test Parameters

Parameters	Asphalt Content - %			
	Mix N619			Mix G319
	5.0	6.0	7.0	7.0
G_{mm}, measured (Rice density)	2.455	2.423	2.396	2.424
G_{se}, calc. using G_{mm} [(a)]	2.647[(a)]	2.650[(a)]	2.659[(a)]	2.698[(a)]
G_{sa}, meas. CA & FA, calc. for blend[(a)]	2.650[(a)]	2.650[(a)]	2.650[(a)]	2.708[(a)]
Absorption water, %[(b)]	0.61	0.61	0.61	
G_{sb}, Meas. CA & FA, calc. for blend	2.608	2.608	2.608	2.536
Absorption AC, % (CA & FA)[(b)]	0.54	0.59	0.71	2.27

(a) $G_{se} \cong G_{sa}$ (b) % Water Absorption ≅ % AC absorption.

EFFECT OF VARIABILITY ON QC TEST PARAMETERS - A CASE STUDY

Data collected from QC testing of 338 samples for seven different mixtures were checked and analyzed to determine the major factors affecting measured and computed volumetric parameters. Tables 7 and 8 give the standard deviations for mixtures produced in two states under slightly different QC test procedures. The hot-mix samples collected from trucks were split to prepare samples for compaction (G_{mb}), Rice density (G_{mm}) testing and gradation. The values in Table 7 represent heating of the split samples to bring them up to compaction temperature, usually requiring about 45 m, prior to performing a Rice test and compaction of two specimens in the Superpave™ gyratory. Also, compaction was carried out to meet the number of revolutions corresponding to N_{design}. Table 8 is based on a somewhat similar approach except that split samples were heated for four hours prior to compaction and performing Rice tests. This produced a very uniform coating of the aggregate so that the S.D. of G_{mm} was very low. In this case the bulk densities (G_{mb}) of the compacted specimens at N_{design} were based upon the measured value at N_{max} and the gyratory specimen height values were used to compute G_{mb} at N_{design}. Also G_{se} values from mix design were used in binder content calculations. The pay schedule was different for the two states. In the first case, Table 7, the bid cost was based on asphalt content from sticking of asphalt tanks whereas in the second state (Table 8) it was based on tons of hot-mix. Asphalt contents determined from tank sticking were computed using temperature/specific gravity corrections and total produced tonnage of hot-mix. The estimates of asphalt content at the end of each day's production were used to compute the estimated G_{se} according to the following equation:

$$G_{se} = \frac{100 - \%\,AC}{\left(\dfrac{100}{G_{mmA}} - \dfrac{\%\,AC}{G_b}\right)} \qquad \text{Eqn. 1}$$

Table 7 Standard Deviation for QC Test Parameters Using N_{design}

Parameters	Standard Deviation (S.D.)					
	19.0 mm Mixtures			12.5 mm Mixtures		
	No. 1	No. 2		No. 3	No. 4	
	Drum*	Batch*	Drum*	Drum*	Batch*	Drum*
P_b	0.243	0.235	0.199	0.286	0.218	0.158
G_{mm}	0.008	0.010	0.008	0.008	0.008	0.005
G_{se}	0.008	0.002	0.004	0.006	0.004	0.006
G_{mb}	0.014	0.018	0.015	0.015	0.019	0.010
V_a	0.724	0.847	0.635	0.768	0.772	0.440
VMA	0.494	0.637	0.545	0.396	0.729	0.375
VFB	4.877	5.156	3.736	5.168	4.041	2.432
* No. Samples	16	45	57	33	42	45

Table 8 Standard Deviation for QC Test Parameters Using N_{max}

Parameters	Standard Deviation (S.D.)		
	25.0 mm Mixture	19.0 mm Mixture	12.5 mm Mixture
	No. 5 Drum*	No. 6 Drum*	No. 7 Drum*
P_b	0.077	0.103	0.103
G_{mm}	0.003	0.004	0.004
G_{se}	0.000[(a)]	0.000[(a)]	0.000[(a)]
G_{mb}	0.014	0.011	0.010
V_a	0.637	0.479	0.429
VMA	0.505	0.437	0.404
VFB	3.842	2.788	2.415
* No. Samples	17	47	31

(a) G_{se} from Design used (constant)

where: G_{se} = Estimated effective specific gravity of aggregate
% AC = Computed from tank sticking
G_b = Specific gravity of binder from mix design
G_{mmA} = Daily average G_{mm} (Max. specific gravity of mixture)

The accumulated average daily G_{se} values during production were used to calculate the asphalt contents reported in QC reports. Equation 2 was used for this calculation.

$$P_b = \frac{100\,(G_b)\,(G_{se} - G_{mmA})}{G_{mmA}(G_{se} - G_b)}$$ Eqn. 2

where: P_b = Calculated binder content, %
G_{se} = Effective specific gravity from Eqn. 1

The SD of G_{mm} values in Table 8 for Mixtures 5, 6, and 7 were considerably less than Mixtures 1 through 4. This is undoubtedly due to the four hour oven heating of samples prior to Rice testing. The computation of asphalt content is similar to that in Eqn. 2 except the input parameters are different.

$$P_b = \frac{100\,G_b\,(G_{se} - G_{mm})}{G_{mm}(S_{se} - G_b)}$$ Eqn. 3

where: P_b = Calculated binder content, %
G_b = Specific gravity of the binder
G_{se} = Effective specific gravity of aggregate from mix design (a constant)
G_{mm} = Maximum specific gravity measured by the Rice test.

A cursory inspection of the binder content (P_b) results in Table 9 suggests that the plant QC binder contents for Mixtures 5, 6, and 7 were substantially more consistent than for Mixtures 1 through 4. However, this is misleading since G_{se} was not computed during production of Mixtures 5, 6, and 7 and variability in G_{mm} was minimized by oven heating of samples for four hours prior to testing.

Table 9 Comparison of QC Binder Contents to Values Computed with Constant G_{se} and G_{mm} ± 2 S.D.

		Range in Binder Content (P_b), %								
		Mix 1	Mix 2B	Mix 2D	Mix 3	Mix 4B	Mix 4D	Mix 5	Mix 6	Mix 7
Plant QC	max.	5.03	5.58	5.68	6.16[b]	6.04	5.87	4.88	5.76	6.11
	min.	4.03	4.60	4.72	4.67	5.07	5.22	4.65	5.22	5.83
	Δ	1.00	0.98	0.96	1.49	0.97	0.65	0.23	0.54	0.28
95% Conf.[a]	max.	4.84	5.71	5.68	5.45	5.89	5.74	4.93	5.55	6.23
	min.	4.04	4.71	4.87	4.65	5.09	5.24	4.60	5.22	5.77
	Δ	0.80	1.00	0.81	0.80	0.80	0.50	0.33	0.33	0.46

(a) P_b calculated using mean G_{se} and mean G_{mm} ± 2 SD values. (b) Extreme value

The binder contents computed on the basis of the mean G_{se} and G_{mm} ± 2 S.D. are also given in Table 9 for comparison to the maximum and minimum values obtained during plant production. Similar P_b values and differences (Δ) indicate that the primary cause of fluctuation in binder content was due to variation of G_{mm}, regardless of the method used in calculation of P_b. It is obvious that both methods have deficiencies and their computed P_b values may not be totally representative of true values.

G_{se} values should not remain constant when the binder content changes unless the amount of asphalt absorption remains constant. The time, temperature, and asphalt content can effect G_{mm} and G_{se} values. Consequently, variation of these parameters during production may have a noticeable influence on computation of P_b, asphalt absorption, effective asphalt content, VMA, and air void content.

The standard deviation for G_{mb} in Tables 7 and 8 is about twice those for the G_{mm}. Since the reliability and methods of obtaining the bulk density (G_{mb}) of mixtures is in question, the mean G_{mb} was used to calculate the acceptable range in G_{mb} using 95% confidence and the mean of other parameters involved in the calculation of V_a and VMA. Table 10 presents these values and the extreme range in values obtained during production of each mixture. The results indicate that G_{mb} is the major source of variation in V_a and VMA. Furthermore, the density of

Table 10 Comparison of V_a and VMA Plant Values with Computed Values to Identify the Effect of Variation in G_{mb}

		Range in Air Void and Mineral Aggregate Void Contents								
		Mix 1	Mix 2B	Mix 2D	Mix 3	Mix 4B	Mix 4D	Mix 5	Mix 6	Mix 7
V_a, Plant	max.	4.6	5.4	5.0	4.5	6.4	5.1	5.1	4.9	4.7
	min.	2.1	2.0	2.4	1.1	2.4	3.2	3.0	3.0	2.7
	Δ	2.5	3.4	2.6	3.4	4.0	1.9	2.1	1.9	2.0
V_a, @ 95%	max.	4.3	2.5	5.0	4.4	5.9	5.1	5.2	5.0	5.7
Conf.(a)	min.	2.2	5.3	2.6	2.1	2.9	3.5	3.0	3.2	4.1
	Δ	2.1	2.8	2.4	2.3	3.0	1.6	2.2	1.8	1.6
VMA, plant	max.	14.0	15.9	16.1	15.2	16.9	16.2	13.3	14.2	15.0
	min.	12.3	12.9	13.2	13.0	14.1	14.6	11.6	12.5	13.5
	Δ	1.7	3.0	2.9	2.2	2.8	1.6	1.7	1.7	1.5
V_a, @ 95%	max.	14.1	15.8	15.9	15.4	16.7	16.2	13.7	14.9	15.2
Conf.(b)	min.	12.2	13.3	13.6	13.3	14.1	14.9	11.7	13.3	13.7
	Δ	1.9	2.5	2.3	2.1	2.6	1.3	2.0	1.6	1.5

(a) $V_a = \left(\frac{\overline{G_{mm}} - \overline{G_{mb}} \pm 2\text{ S.D.}}{\overline{G_{mm}}} \right) 100$

(b) $VMA = 100 - \frac{(100 - \overline{P_b})(\overline{G_{mb}} \pm 2\text{ S.D.})}{G_{sb}}$

cores taken from pavements for QC or QA purposes are not directly comparable to densities obtained from QC tests at the plant. Smooth sides of cores versus the texture of gyratory compacted samples will produce differences in density regardless of whether sample weight and volume in the mold or water emersion is used to determine G_{mb}. In a relative sense, the density, V_a and VMA values are good for general information but do not constitute high reliability for QC during production. In addition, the variable nature of G_{sb} value contributes to potential errors in VMA computations.

One of the major disadvantages in using either of the two methods for QC of hot-mix asphalt production is the lack of a direct measure of asphalt content. Recent studies indicate that the ignition oven test (2), if properly calibrated, gives reliable P_b values having low standard deviations. This is a relatively fast test as compared to conventional extraction tests (3) with their ash correction. The combination of P_b with G_{mm} measurements provide direct computation of G_{se}. Consequently, all three parameters could be used for QC during production. In the event G_{se} varied excessively it would be prudent to oven heat additional samples in an oven for 4 to 8 hours to establish whether or not variation in G_{se} was due to differential absorption. If G_{se} is constant, the G_{mm} value can be computed for different asphalt contents as a guide for comparison to the measured G_{mm}.

The precision of each Rice test may be questioned unless two or more replicate samples are tested. A testing protocol requiring two values of G_{mm} within 0.011 would probably be beneficial.

SUMMARY: CONCLUSIONS AND RECOMMENDATIONS

The variability in G_{sb} of aggregates and the bulk density of compacted asphalt mixtures can result in substantial changes in the computed VMA. The absorption of water into dense aggregates may be similar or almost equal to the absorption of asphalt. Since the saturated surface-dry condition is required for G_{sb} determinations and the G_{sb} is used to calculate asphalt absorption, it is obvious that variations in G_{sb}, a subjective test, affect both VMA and asphalt absorption values. It has been shown that angular fine aggregate has much greater G_{sb} testing variability than the coarse aggregate.

Analysis of gyratory compacted mix densities during QC testing indicated that variability in V_a and VMA was primarily due to fluctuations in bulk densities (G_{mb}). Similarly, methods for computing the asphalt content proved to be susceptible to variations in G_{mm}. The primary problems in QC testing are the conflict between using tests and computations that minimize time for making an assessment versus more lengthy and involved testing which may defeat the purpose of QC. Consequently, the following alternatives are recommended to facilitate QC and to simplify the parameters needed for evaluation of the uniformity of asphalt mixture production:

1. Monitor asphalt content (P_b) using the ignition oven or any other method that provides speed and accuracy.
2. Perform two replicate Rice tests to determine G_{mm} and check for conformance to required precision (i.e., values with an acceptable range of 0.011).
3. Compute the G_{se} value using P_b and G_{mm}. If desired for QC only, use G_{se} instead of G_{sb} to evaluate variability in VMA.
4. Establish criteria for acceptable QC limits for G_{se} considering that both P_b and G_{mm} can vary. As a supplementary standard procedure, two additional replicates could be heated in an oven for 4 to 8 hours in an effort to obtain uniform G_{se} values.
5. Do not include G_{sb}, G_{mb}, or VMA in QC parameters but rather for general information only.

REFERENCES

1. Standard Test Method for Theoretical Maximum Specific Gravity and Density of Bituminous Paving Mixtures, D2041-95, American Society for Testing and Materials, Vol.04.03, 1997.

2. Standard Test Method for Asphalt Content of Hot-Mix Asphalt by the Ignition Method, PS 90-97, American Society for Testing and Materials.

3. Standard Test Method for Quantitative Extraction of Bitumen from Bituminous Paving Mixtures D2172-95, American Society for Testing and Materials, Vol.04.03, 1997.

Fifth International Conference on the
BEARING CAPASITY OF
ROADS AND AIRFIELDS

EFFECT OF NEW DIRECT TENSION TEST PROTOCOL ON THE SUPERPAVE LOW-TEMPERATURE SPECIFICATION FOR BITUMEN BINDERS

RAJ DONGRÉ[1] AND JOHN D'ANGELO[2]

[1]Consultant, FHWA/GPI, 6300 Georgetown pike

McLean, Va-22209, Fax: (703) 285-2950

E-mail: Raj.Dongre@fhwa.dot.gov

[2]Asphalt Team Leader – FHWA/OTA

400 7th Street, SW, HTA-20, Washington D.C. – 20590

Fax: (202) 366-7909

E-mail: John.D'Angelo@fhwa.dot.gov

ABSTRACT

This paper reports the results of a study conducted to investigate the cause of poor reproducibility in failure properties measured using the new Superpave direct tension test. Several causes were discovered. It was found that the silicone rubber molds used to manufacture specimen lowered the failure properties of bitumen. It was also found that alcohol, which is used as a cooling medium, also lowers the failure properties. Neither effect is seen on the stress-strain curve of the bitumen but only on the failure point. The impact of these and other effects related to specimen geometry on the Superpave specification criteria of 1% strain to failure is discussed.

INTRODUCTION

The Federal Highway Administration (FHWA) is leading in implementing the new Superpave system for the United States. The Superpave system, developed during the Strategic Highway Research Program (SHRP), consists of three parts: a bitumen binder specification; a volumetric hot-mix design; and hot-mix analysis tests and performance prediction software (*1*). The new Superpave direct tension test (DTT) device is an integral part of the Superpave binder specification (*2*). The DTT device is used to determine low temperature failure properties of bitumen binders, such as the stress and strain (σ_f and ε_f) at failure. The specimen geometry is shown in Figure 1.

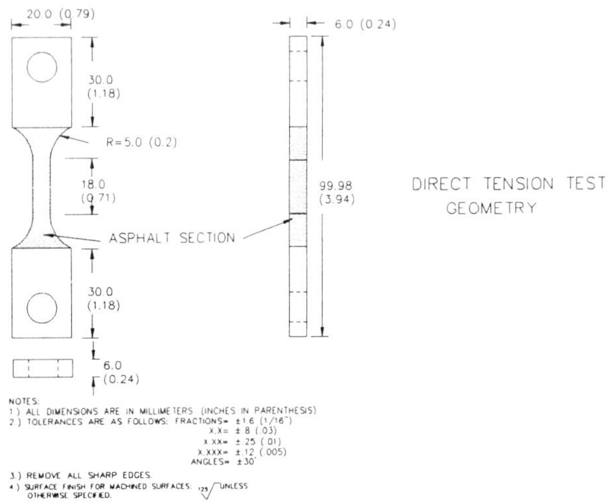

Figure 1. Superpave Bitumen Test Specimen Geometry

The strain at failure, ε_f, is used in Superpave to determine the lowest temperature at which the bitumen will perform with minimal thermal cracking (*3*). In the current Superpave low-temperature specification the failure strain limit of 1%, determined at an elongation rate of 1 mm/min, is used to control thermal cracking. However, in its current form, this limit is only applicable when the stiffness, (S(60)), of bitumen binders at 60s, determined using the Bending Beam Rheometer (BBR), is between 300 MPa and 600 MPa. When stiffnesses are in this range and the bitumen binder can resist a strain of 1% or more without fracture, then the maximum creep stiffness requirement of 300 MPa is waived. The 'm $\geq$ 0.3' criteria (m is the slope, at 60s, of the log stiffness versus log time curve) still must be met. This requirement was incorporated in the specification to accommodate the polymer modified bitumen binders which tend to be stiffer. The validation of this specification approach is beyond the scope of this paper. The Expert Task Group set up by the FHWA has assigned the validation and development of new specification to the low-temperature task group.

This paper focuses on the implications of recent findings about the testing protocol and data reduction used in the new DTT.

MAJOR FINDINGS

The majority of the published data on the DTT test results, has been developed using silicone rubber molds and air chilled specimen. Also, more recent data using the silicone rubber molds and the alcohol fluid bath. A recent round–robin between several laboratories, made it evident that the between–laboratory variability was unacceptably high. Several experiments were undertaken to investigate the cause of poor reproducibility. It was found that the age of the silicone

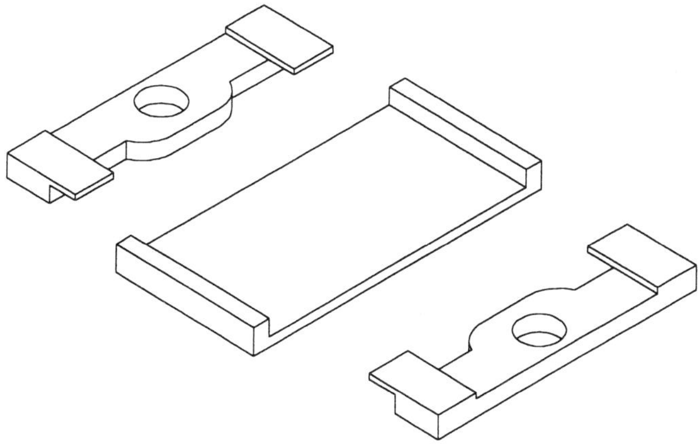

Figure 2. Specially Designed Aluminum Mold for Making Bitumen Specimen.

rubber molds and the alcohol cooling medium in the new DTT significantly affected the failure properties of the bitumen binders as shown in Tables 1 and 2. In Table 1, the seasoning time refers to the curing time in oven for the silicone rubber molds. The molds were cured by filling the specimen cavities with asphalt and heating at 160°C for the time specified in Table 1. The molds were cured or "seasoned" to simulate the effect of multiple use and physical aging experienced by the silicone rubber molds during service.

After these findings, a complete re-evaluation of the new DTT test method and related calculations was undertaken. This gave rise to three significant changes in the protocol for using the new Superpave DTT.

1) The silicone rubber molds used for producing bitumen test specimen have been replaced by specially designed Aluminum molds, Figure 2.
2) Alcohol was replaced by a mixture of 43% Potassium Acetate and 57% de-ionized water as the cooling medium.
3) The effective gage length, L_e, needed to compute accurate strains had to be changed to 33.8 mm from the current value of 26.66 mm based on the finite element analysis of the specimen geometry. These calculations were conducted by The Instron Corporation at their Boston, MA facility. A compliance value for the plastic end tabs was determined and included in calculations.

This paper discusses the effect of the new metal (Aluminum) molds, coolant fluid, specimen geometry, and compliance corrections on the current Superpave low-temperature specification parameter i.e. on ε_f=1%.

Table. 1. Effect of Molds on Failure Strain of Bitumen.

Mold Type	Seasoning Time, h	No. of Specimen	F. Strain %
Silicone Mold	0	24	1.05 (CV=25%)
Silicone Mold	8	24	3.31 (14.0%)
Silicone Mold	24	24	3.12 (19.0%)
Silicone Mold	48	24	2.73 (23.0%)
Metal	**0**	**> 24**	**3.05 (21.0%)**

Table 2. Effect of Cooling Medium on Failure Strain of Bitumen.

Cooling Medium	Mold Type	F. Strain, %
Ethanol-Denatured	Silicone Rubber	~0.6%
Methanol 100%	S. Rubber and Metal	~0.8%
Eth. Glycol+Water+Methanol (60:25:15)	S. Rubber and Metal	~0.7%
Methanol+Water (45:55)	S. Rubber and Metal	~2.0%
KOAc Deicing Fluid (Brine)	Metal	3.0%
AIR	**METAL (Aluminum)**	**3.05%**

MATERIALS AND METHODS

A total of 11 unmodified bitumen were used in this study. Six out of the 11 bitumen were selected from the Strategic Highway Research Program (SHRP) core set of bitumens. All bitumen were RTFOT+PAV aged according to Superpave protocols (*4*). Table 3 shows the bitumen along with the crude source. The experiment design, Table 4, consisted of determining failure stress and the corresponding strain at failure for the 11 bitumen binders. Four elongation rates (strain rates) and five temperatures were investigated for each bitumen.

Table 3. Bitumen and Crude Sources Used in This Study.

Bitumen/Origin	Crude Source
AAA-1/SHRP CORE SET	Llyodminster
AAB-1/SHRP CORE SET	Wyoming Sour
AAD-1/SHRP CORE SET	CA Coastal
AAF-1/SHRP CORE SET	West Texas Sour
AAK-1/SHRP CORE SET	BOSCAN
AAM-1/SHRP CORE SET	West Texas Intermediate
Imperial 85/100	Cold Lake
Husky Blend 100/150	Sasketchwan
Imperial 300/400	Cold Lake
Imperial 150/200	Cold Lake
Imperial 200/300	Cold Lake

Table 4. Experiment Design Used in This Study.

Temperature, °C	Strain Rate, %/min			
	0.375	1.0	2.0	3.75
T_{BBR} + 6	4[2]	4	4	4
T_{BBR} + 3	4	4	4	4
T_{BBR}[1]	4	4	4	4
T_{BBR} - 3	4	4	4	4
T_{BBR} - 6	4	4	4	4

[1]Note: T_{BBR} is the temperature at which BBR test is conducted (Low temperature grade + 10°C).
[2]Note: Number of Replicates in each cell is 4.

RESULTS

The dependence of failure strain on temperature and strain rate is shown for bitumen AAA-1 in Figure 3. Figure 4 shows the failure stress as a function of temperature and strain rate for the same bitumen. The failure strains were shifted using the rheological shift factors (calculated based on rheological complex modulus data). Figures 5 and 6 show the shifted failure strain for bitumen AAA-1 and AAM-1 respectively. Failure stress and strain for all 11 bitumens were successfully shifted within experimental error (25%) using this technique.

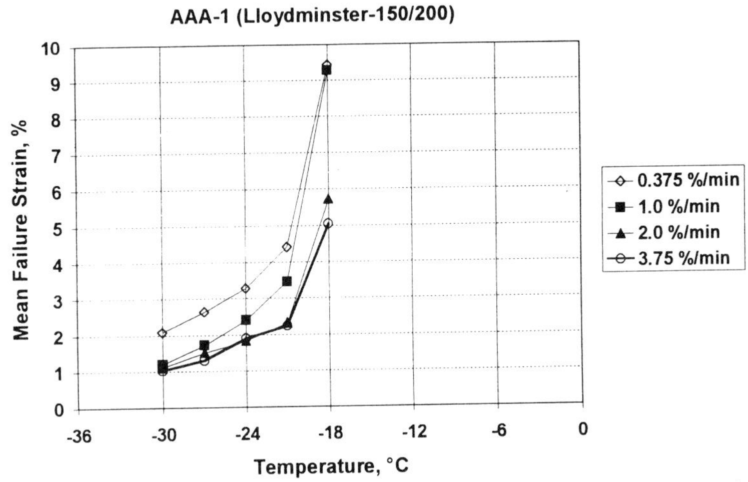

Figure 3. Temperature Dependence of Failure Strain at Various Strain Rates for Bitumen AAA-1 in PAV Aged Condition. (Le = effective gage length =26.66mm, No Compliance Correction Applied.)

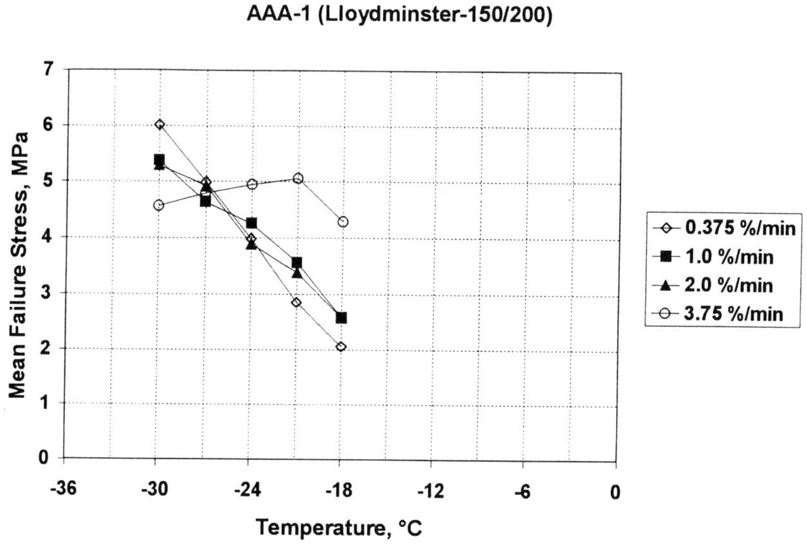

Figure 4. Temperature Dependence of Failure Stress at Various Strain Rates for Bitumen AAA-1 in PAV Aged Condition. (Le = effective gage length =26.66mm, No Compliance Correction Applied.)

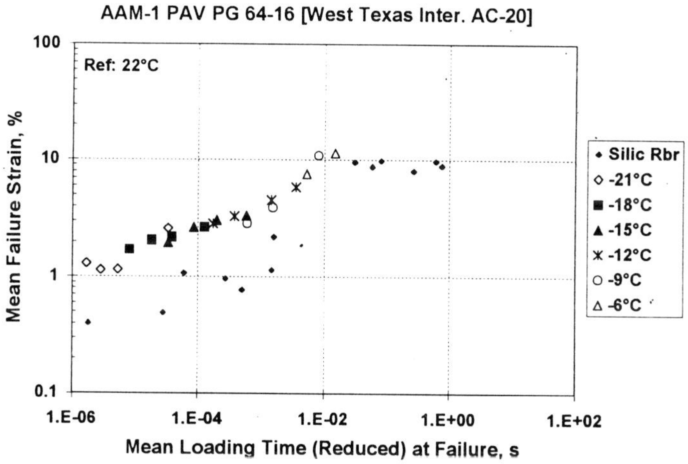

Figure 5. Failure Strain Master Curve for Bitumen AAA-1. (Le = effective gage length =26.66mm, No Compliance Correction Applied.)

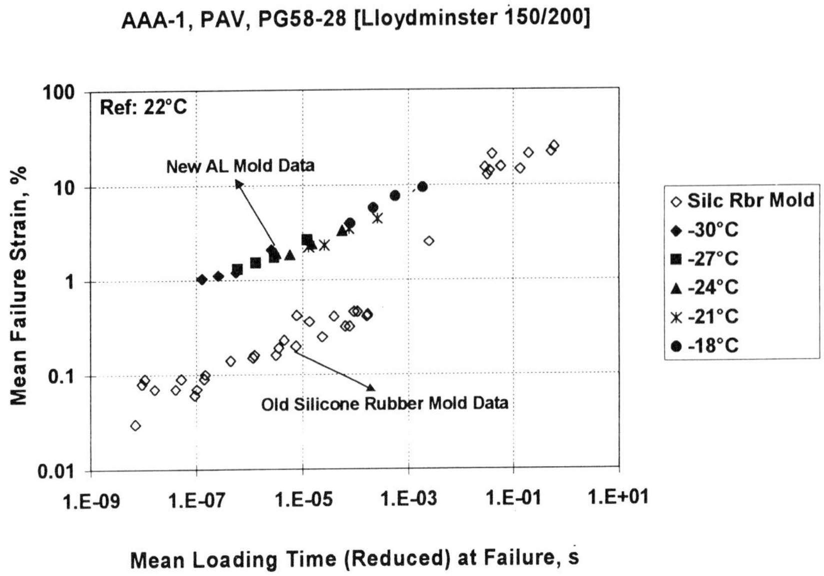

Figure 6. Failure Strain Master Curve for Bitumen AAM-1. (Le = effective gage length =26.66mm, No Compliance Correction Applied.)

DISCUSSION

Effect of Silicone Rubber Molds

The effect of silicone rubber molds on failure properties was found to be significant (Table 1). The silicone rubber mold, however, does not affect the stress-strain curve but only the failure (i.e., the breaking point), Figure 7. As shown in Figures 5 and 6 the effect of silicone rubber molds is greater on bitumen AAA-1 than on AAM-1. AAM-1 has a higher molecular weight. One possible explanation is that the silicone rubber imbibes (acts as a sieve for) the low molecular weight species. When hot bitumen (165°C) is poured into the heated silicone rubber mold (110°C) the low molecular weight species get absorbed and hence produce surface flaws on the specimen leading to premature failure. To validate this hypothesis the silicone rubber molds were cured (seasoned) with bitumen at 165°C for various periods of time. It was found that at a seasoning time of 8 h, the effect on failure strain was negligible (Table 1). At times greater than 48h the rubber starts deteriorating. Although the method of curing

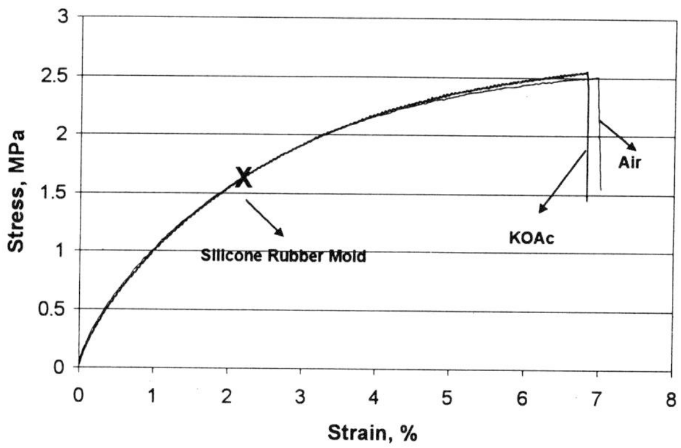

Figure 7. Stress-Strain Curve Showing the Effect of Molds and Cooling Medium.

the rubber eliminates the effect on failure properties, it is not a consistent method and hence of limited use in a specification test. We have observed (not shown here) that polymer modified bitumen are affected less by the silicone rubber molds with crumb rubber modified bitumen being the least affected. The validity of our hypothesis for the mechanism with which silicone rubber mold affects the failure properties of the bitumen is still being evaluated.

Effect of Alcohol

The original version of the new Superpave DTT was designed with alcohol as the cooling medium. However, it was found that alcohol affects failure properties of bitumen to the same extent as did the silicone rubber mold. The possible explanation for this effect is that alcohol is an environmentally active fluid for bitumen and therefore promotes environmental stress corrosion cracking (ESC). This phenomenon is known to occur in glass forming polymers, however, the mechanism is not completely understood (*5*). The effect of alcohol on bitumen under stress is dependent on its concentration. Table 2 shows the effect of various kinds of alcohol on the failure strain of bitumen. While investigating alternative cooling fluids for the new DTT, it was found that an aqueous mixture of Potassium Acetate (KOAc) and water (43% KOAc + 57% Water) did not affect the failure properties of bitumen, Figure 8. Similar data collected using a air cooled machine was used for comparison. It was assumed that air has no effect on the failure properties of bitumen.

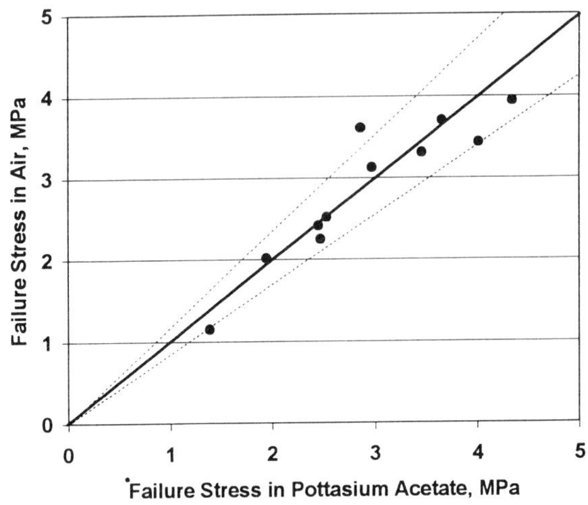

Figure 8. Comparison of Failure Strength for Various Bitumen Tested in Air Versus Potassium Acetate.

This mixture is commercially available in 50 KOAc / 50 Water concentration as a de-icing fluid, which must be diluted for use in the new Superpave DTT. The dilution is obtained by adding 15% by weight distilled water to the commercially available 50:50 mix of KOAc+Water. The dilution lowers the viscosity and prevents spontaneous crystallization, which was observed with the 50:50 mix. The lower viscosity produced by dilution also helps in maintaining vigorous flow at the lowest temperatures which further reduces any existing temperature gradients.

BBR – DTT Comparison

The current Superpave criteria based on the finding that the Bending Beam Rheometer (BBR) S and m value are sufficient to minimize low-temperature cracking where unmodified bitumen binders are used *(4)*. This is because for unmodified binders the failure strain was found to be unique function of stiffness at a loading time of 2 h (*4*). For modified bitumen binders, however, this correlation breaks down. Therefore, for modified bitumen a combination of criteria derived from the low-temperature rheology and failure are required. The Superpave assumption requires that for unmodified bitumens the critical cracking temperatures (Superpave low-temperature grade) determined using the BBR and the DTT criteria must agree within experimental error. Because of the problems with silicone molds and the alcohol bath, test data to verify this assumption had to be generated again. To develop new data the critical cracking temperatures were determined using the failure strains from the DTT. Initially, 1.75% strain to failure was chosen to determine the critical temperature.

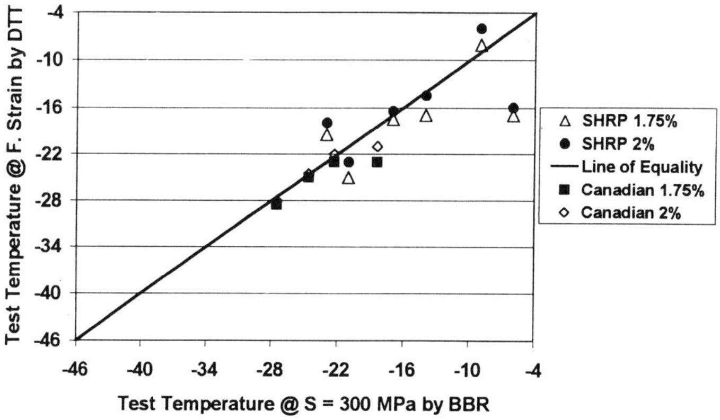

Figure 9. Comparison Between BBR and DTT Test Temperatures for all Asphalts.

A second value of 2% strain to failure was then added to evaluate the effect of a slightly higher value which is more suitable as an integer specification number. These values were selected for two reasons. First, the Superpave 1% strain to failure criteria can no longer be used, as the failure strains are higher because Aluminum molds are being used. The silicone rubber mold effect is no longer present. The second reason is that, at these strain levels the bitumen binders are in the middle of the brittle-ductile region. Selecting strains in the brittle-ductile region is desirable because they are just enough to produce reasonably repeatable results without appreciable change in the cross section (no necking). Using these values the temperature at which these strains are obtained were determined from the failure strain versus temperature plot (Figure 3). As shown in Figure 9 the agreement between the Superpave low-temperature grading test temperature (i.e. low-temperature grade + 10°C) determined using the BBR and the DTT criteria is reasonably good. Note that there is no significant difference between the 1.75% and the 2% criteria. Therefore a 2% strain to failure appears to be a reasonable specification criteria.

Effect of the New Effective Gage Length and Compliance Correction

So far all the failure strain data presented were computed using an effective gage length, L_e, of 26.66 mm. The effective gage length corrects test data for variations in the geometry. Details of the concept of effective gage length can be found elsewhere (*2 and 6*). An error in test data calculations presented so far is that the correction for compliance due to the bending of plastic end tabs and the loading pins. This has been neglected in the data presented. During the recent re-evaluation of the specimen geometry and the effective gage length calculations it was found that the correct value for the effective gage length should be 33.8 mm instead of 26.66 mm. A discussion of why 26.66 mm was chosen earlier can be found elsewhere (2 and 6).

The effect of the new 33.8 mm effective gage length is straightforward. For example the new recommended 2% strain to failure criteria discussed earlier will become 1.58% and similarly 1.75% becomes 1.38%. The effect of compliance of the plastic end tabs, however, is load dependent. Compliance is also temperature dependent because plastic is temperature dependent, however, the effect of elongation rate is negligible. The exact algorithm used in this study to correct for compliance of the plastic end tabs is proprietary and at this time cannot be made public. However, a rule of thumb (approximate) value for compliance correction, that does not include the small but important effect of temperature, may be given as follows. For every MPa of stress applied the correction to the strain value would be equal to the applied stress in MPa divided by 10. The rule of thumb correction is in % strain units. For example, at 1 MPa applied stress the correction subtracted from the computed strain value, at that stress, would be 0.1%. When all corrections are applied the 1.75% failure strain will change to 1.08% and 2% failure strain reduces to 1.18%. The failure stress in both cases is 4.0 MPa.

CONCLUSION

The results of this study show that the silicone rubber molds lower bitumen failure properties without affecting the stress-strain curve (i.e. stiffness or slope is not affected). It is therefore recommended that metal molds be used to manufacture the bitumen DTT specimens. A schematic for the molds, which was successfully used in this study, is also presented. It was also found that alcohol is an environmentally active fluid for bitumen and causes stress corrosion cracking, which results in lower failure strains. Therefore alcohol must not be used as a cooling medium when testing failure properties of bitumen. Like the silicone rubber molds alcohol only affects the failure properties and not the stress-strain curve. A 43% Potassium Acetate and 57% water mixture was found as a suitable medium for temperature control when testing to determine failure properties of bitumen. And finally, a finite element study of the specimen geometry revealed that the effective gage length for the Superpave DTT specimen must be changed to 33.8 mm from the current value of 26.66 mm. It was also found that compliance correction is required to account for the elongation produced in the plastic end tabs when load is applied. Both the new effective gage length and the compliance correction have been experimentally verified.

When all the effects are added together, however, the overall effect on the Superpave criteria of 1% failure strain is only minimal. This is due to fact that the amount of increase in strain obtained due to the use of new metal molds is almost negated by the reduction caused by the change in effective gage length and the compliance correction. If the strain rate is changed to 3.0 %/min from the current 3.75 %/min to maintain the elongation rate at 1.0 mm/min (i.e., (33.8 * 3.00/100) ~ 1.0 mm/min) the specification criteria will change to 1.30 % (from 1.75 %) and 1.6% (from 2.0%). This is because only the compliance correction of 0.4% (4 MPa) is applied to the calculated strain in each case.

ACKNOWLEDGEMENT

This work was funded by the Federal Highway Administration's Office of Technology Applications (FHWA-OTA). The authors would like to thank FHWA-OTA for their continued support. The authors also wish to thank Mr. D. J. Heidler, and Mr. Charles King of FHWA-OTA and Mr. Richard Heath of Dongre Laboratory Services Inc. for conducting the tests. Special thanks are offered to Mr. Steve McMahon of The Instron Corporation for help with finite element calculations and Mr. Gerry Reinke of MTE Services Inc. for numerous helpful discussions.

REFERENCES

1. *Superpave Performance-Graded Asphalt Binder Specification and Testing*, Publication SP-1. The Asphalt Institute, Lexington, Kentucky, 1994.

2. Dongre, Raj, D'Angelo, John, and McMahon, S. DEVELOPMENT OF THE SUPERPAVE DIRECT TENSION TEST DEVICE, Transportation Research Record, Transportation Research Board, TRB, Washington, D.C. 1996.

3. Anderson, D.A., and T.W. Kennedy. Development of SHRP Binder Specification. *Journal of The Association of Asphalt Paving Technologists*, Vol. 62, 1993, pp. 481-507.

4. Anderson, D.A., et al., Volume IV, Final Report for SHRP Project A002A, Binder Characterization and Evaluation - Test Methods, Strategic Highway Research Program, Washington, D. C., 1993.

5. Kilwon, C., Thomas, S. and Kumaran, M. G. Environmental stress cracking of rubber-modified styrenic polymers in Freon vapour, POLYMER, Vol. 38, No. 18, 1997.

6. Dongre, R. Development of Direct Tension Test to Characterize Failure Properties of Asphalt Cements. Ph.D. dissertation. Pennsylvania State University, University Park, PA, 1994.

Fifth International Conference on the
BEARING CAPASITY OF
ROADS AND AIRFIELDS

SUITABILITY OF THE SUPERPAVE BINDER TECHNOLOGY TO NORWEGIAN CONDITIONS

EIVIND O. ANDERSEN[1] AND LEIF J. BAKLØKK[2]

[1]Research Engineer, SINTEF Civil and Environmental Engineering, 7034 Trondheim, Norway, Fax: +47 73 591480, E-mail: eivind.o.andersen@civil.sintef.no
[2]Research Engineer, SINTEF Civil and Environmental Engineering, 7034 Trondheim, Norway, Fax: +47 73 591480, E-mail: leif.j.baklokk@civil.sintef.no

ABSTRACT

This paper discusses how the Superpave binder technology and specification applies to Norwegian conditions. Although the climate in Norway varies significantly from south to north and from coast to inland, climate is not well enough considered when selecting binder for a particular project. The Superpave binder specification was developed to select binders with respect to climate. This paper compares the current Norwegian practice for selection of binders for bituminous mixes to a similar selection using the Superpave procedures.

Some limited adjustments to the Superpave technology are proposed to make it applicable to Norwegian conditions. The most important adjustment is a new equation to predict critical high pavement temperatures. The Superpave-equation has been shown non-valid in Norway.

Binders selected using the Superpave procedures for intermediate to high traffic volume roads is considered a great improvement compared to our Norwegian practice. Superpave does not take into account the problems of seasonal variations in bearing capacity and other problems often associated with low volume roads. For low volume roads, Superpave may suggest binders which are much stiffer than those of the Norwegian recommendations. The Superpave procedures may therefor in many cases not be suitable for low-volume roads in Norway.

Key words: Bituminous binders, Superpave, Norwegian specifications, comparison

INTRODUCTION

In late 1994, a Norwegian project was started to evaluate the new Superpave binder technology. The project, which is now being finished, has been funded by Norwegian road and airfield authorities, asphalt paving contractors, oil companies operating on the Norwegian bitumen market and the Norwegian Industrial and Regional Development Fund.

During this project, SINTEF Civil and Environmental Engineering dep. of Highway Engineering and the Norwegian University of Science and Technology, have equipped their joint laboratory and built up expertise to perform binder tests according to the Superpave specification. A large number of binders (total of 40) have been tested using both traditional binder tests and the new Superpave tests. One has also looked at how this new technology applies to Norwegian conditions and how certain adjustments can be made to improve the overall applicability to Norwegian conditions.

This paper deals in particular with work that has been done to study how this new technology applies to, and how it can be adjusted to better apply to Norwegian conditions. The work and conclusions presented in this paper, are only meant to apply for unmodified bitumen. Polymer modified materials have not been considered and are beyond the scope of this investigation.

CURRENT PRACTICE IN NORWAY

The current binder specification in Norway is a typical traditional specification, based on tests such as penetration, viscosity, Fraass breaking point and ductility. Binders for bituminous mixes are chosen mainly on the basis of traffic volume and to a certain extent type of mix. Although the climate in Norway varies significantly from south to north and from coast to inland (illustrated by the temperature map in figure 1), climate is not well enough considered when selecting binder for a particular project.

The Superpave binder specification was developed to select binders with respect to climate. For these reasons, it is very interesting to see which binders are suggested in Norway using Superpave procedures, and how this compares to our current practice.

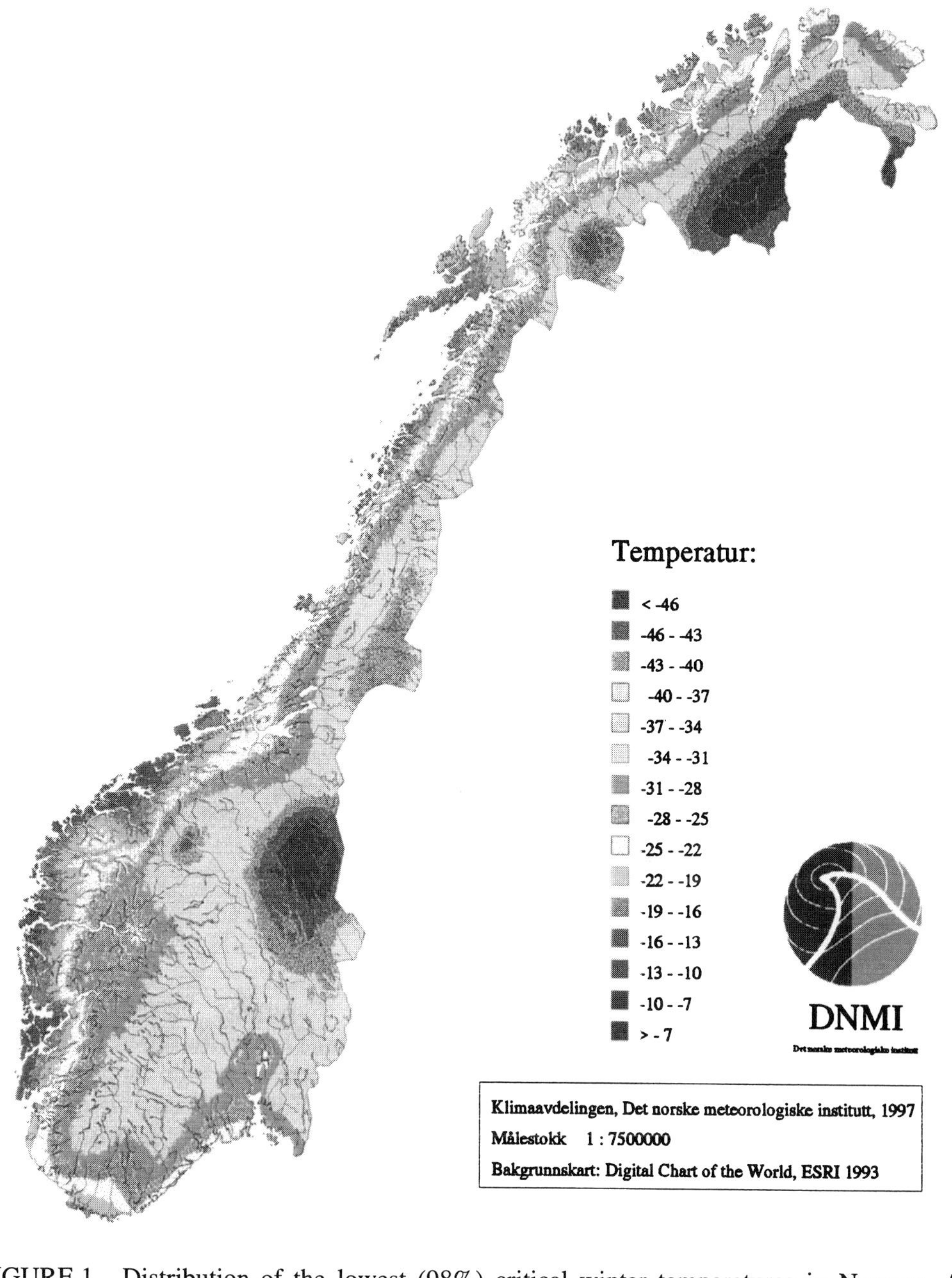

FIGURE 1 Distribution of the lowest (98%) critical winter temperatures in Norway. A similar map has also been made for the highest (98%) critical summer temperatures.

When a binder is selected with regards to rutting according to Superpave procedures for a specific project, two requirements for the binder must be met at the critical pavement temperature (equations 1 and 4):

- $G^*/\sin\delta \geq 1{,}0$ kPa in original binder (measured with a Dynamic Shear Rheometer)
- $G^*/\sin\delta \geq 2{,}2$ kPa in RTFOT-aged binder (measured with a Dynamic Shear Rheometer)

To make the comparison, a relationship between penetration and $G^*/\sin\delta$ has been established in this Norwegian project. To simplify, we neglect the $G^*/\sin\delta$-requirement after RTFOT-aging. Through our project, we have experienced that this will not affect our conclusions.

Figure 2 shows the relationship between penetration and the exact temperature where $G^*/\sin\delta = 1{,}0$ kPa for 26 unmodified binders of varying stiffness.

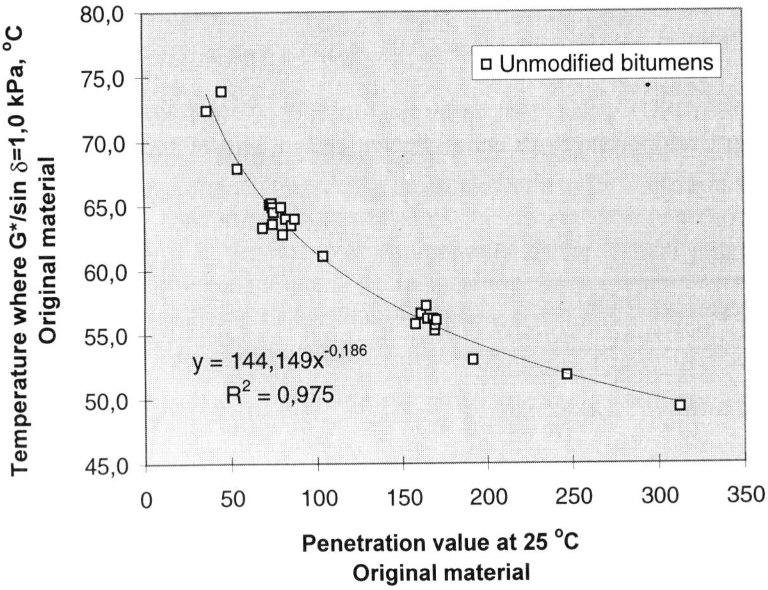

FIGURE 2 Relationship between penetration at 25°C and temperature where $G^*/\sin\delta = 1{,}0$ kPa in original binders.

Rearranging the equation in figure 2, gives the following relationship between critical pavement temperature ($G^*/\sin\delta = 1{,}0$ kPa) and penetration value:

$$Pen.value = 4{,}0412 \cdot 10^{11} \cdot Crit.pavem.temp.^{-5{,}3763} \quad \text{Equation 5}$$

Comparison for climate extremities in Norway To compare the two binder selection practices, two climates in Norway have been chosen that represent extremities with respect to pavement temperatures. These climates are typical for the Oslo and Tromsø areas.

Critical air and pavement temperatures (equation 4) and the corresponding penetration values (equation 5) are given in table 2. Critical pavement temperatures are adjusted for 50 km/h speed as described previously. Binder selection for traffic volumes exceeding 3,0E+07 ESALs is also included.

TABLE 2 Binders selected using Superpave procedures for two Norwegian climates (extremities) expressed by penetration values.

Location	Critical air temp., °C	Critical pavement temp., °C		Penetration value, 0.1 mm	
		ESAL≤1,0E+07	ESAL≥3,0E+07	ESAL≤1,0E+07	ESAL≥3,0E+07
Oslo	32	59,8	65,8	113	68
Tromsø	24	47,0	53,0	414	217

In our comparison, we select a 2-lane road and a design life of 20 years. From table 1, the AADT-values corresponding to 1,0E+07 and 3,0E+07 ESALs is 6.550 and 19.651 respectively.

In figure 3, the binder selection from Superpave (table 2) is compared to the recommendations in the Norwegian Public Roads Administration's norms for road building (4). As seen from figure 3, a change from our current Norwegian practice to Superpave would lead to a radical change in the use of binders in Norway.

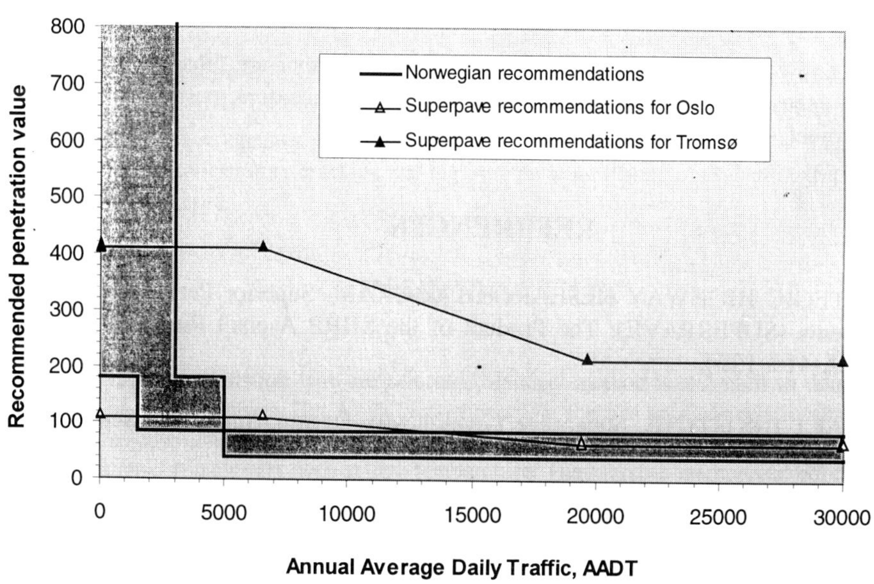

FIGURE 3 Comparison of Superpave and Norwegian recommendations for binder selection.

pavements and the asphalt mixtures should be considered in the mix design to ensure these failures will not take place in the asphalt pavement.

In this paper, through the analysis failure types of asphalt pavement and climate feature in China, the new design method of asphalt mixture is established.

MATERIALS

The properties of the asphalt used in this study are shown in Table 1.

The limestone aggregate is selected which properties are shown in Table 2.

The 5 kinds of gradations are selected for the study from the main mixture types of the asphalt pavement construction specification used in China and included in the following. The upper value and the middle values of the gradation limits of AC-16-I type, the middle values and the lower values of AC-16-II type, and middle values of the gradation limits of the open graded friction course (OGFC) are shown in Table 3[2].

SPECIMEN PREPARATION

The asphalt mixtures are tested by Marshall method to measure the structural parameter which results are shown in Table 4. According to the determined asphalt content and aggregate gradation, 30×30×6 cm and 30×30×5 cm slab specimens are prepared with the rolling wheel compactor. 30×30×5 cm specimens are used in rutting test and 30× 30×6cm specimens sawed into 30×35×250 mm and 40×40×250mm prism specimens which are used in the creep test at low temperature and fatigue test, respectively.

PAVEMENT PERFORMANCE OF ASPHALT MIXTURE

Skid Resistance Test of Asphalt Mixture The friction number test results of asphalt mixture are shown in Table 5. It is noted in Table 5 that friction number of asphalt mixtures increases with increasing of air void, when the air void varieties in a limited values.

Marshall Retained Stability Test of Asphalt Mixtures Marshall retained stability test results of asphalt mixture are shown in Table 6.It is shown in Table 6 the Marshall retained stability of asphalt mixture decreases with increasing of air void and reaches the minimum value at the air void of about to 10 percent.

Rutting Test of Asphalt Mixture Rutting test results of asphalt mixture in Table 7 shows that the dynamic stability of asphalt mixture approaches the maximum value with an optimum air void which is about 8 percent.

Creep Test of Asphalt Mixture at 0 °C Creep test results of asphalt mixture in Table 8 display that the creep rates of the mixtures increase with the decreasing of air voids. The higher the creep rate, the greater the deformation resistance, the better the performance of mixture at low temperature.

Fatigue Test of Asphalt Mixture at 15 °C The fatigue test results of asphalt mixture are shown in Figure 1. It is shown in Figure 1 that the fatigue life gradually get longer with the change of the air void from large to small, when air void becomes larger, there are many micro-cracks in asphalt mixture so that the fatigue life of asphalt mixture gets shorter.

NEW CONCEPT OF ASPHALT MIXTURE DESIGN

Features of Climate Regions for Asphalt Mixture Design As the regions in China vary greatly in climate from North to South, its performance of asphalt mixture should be considered differently. Therefore it is necessary to classify the technical indexes of asphalt mixture according to climate feature in every region. Climate regions for asphalt mixture design are categorized into three levels, based on the average highest air temperature in July, and each level is categorized into four sublevels, based on the yearly extreme low temperature. The climate in China is categorized into four regions according to the annual precipitation. Climate regions are shown in Table 9. It is very difficult that one asphalt mixture can meet the criterias for different climate types, so when asphalt mixture design is to be conducted. We must consider the most serious failure types of asphalt mixture associated with climate condition. Based on these points, the climate regions can be categorized as following:

a. When yearly extreme low temperature is lower than -21.5 °C, and the average highest air temperature in July is lower than 30 °C, low-temperature cracking of asphalt pavement is the main failure type in these regions, so the index of anti-cracking at low temperature is regarded as the control factor of asphalt mixture design.

b. When the average highest air temperature in July is higher than 30 °C, and yearly extreme low temperature is higher than -21.5 °C, the permanent deformation of asphalt pavement is the main failure type in these regions. The index of high temperature stability is regarded as the control factor for asphalt mixture design.

c. For other regions, both high temperature stability and anti-cracking at low temperature are considered as the control factor for asphalt-mixture design.

Method of Asphalt Mixture Design Based on the Pavement Performance According to climate feature in China, the method of asphalt mixture design can be divided into three types.

a. Design method of asphalt mixture based on rutting resistance.

The design flow chart of asphalt mixture on the base of high temperature stability is shown in Fig.2. When performance of asphalt mixture can not meet the rutting depth requirement, it can be adjusted with following method:

(1)Select higher stiffness asphalt or modified asphalt;

(2)Select the gradation type containing more coarse aggregate, but be careful filler content;

(3)Select angular aggregate of rough surface.

b. Design method of asphalt mixture of low temperature cracking resistance

The design flow chart of asphalt mixture of low temperature cracking resistance is shown in Fig. 3.

When performance of asphalt mixture can not meet the requirement of low temperature cracking, it can be perfected as following:

(1)Select softer asphalt or modified asphalt

(2)Select gradation with low air void or gradation near to the upper limit of the specification

c. The design method of asphalt mixture based on pavement performance

The design flow chart of asphalt mixture based on pavement performance is shown in Fig. 4.

When performance of asphalt mixture can not meet the requirement, it can be perfected as following:

(1)When water stability of asphalt mixture does not meet the requirement, anti-stripping agent can be added to asphalt.

(2)when high temperature stability of asphalt mixture does not meet the requirement, since there are some conflict between the high temperature stability and the low temperature cracking resistance, it can be perfected by adjusting the gradation of aggregate and the modification technology of asphalt. It is strongly recommended to use the gap gradation type such as stone mastic asphalt mixture (SMA).

Determination of the Optimum Gradation of Asphalt Mixture According to the climate feature in China and the relationship between pavement performance of asphalt mixture and air void, optimum gradations of asphalt mixture are put forward, which can meet the different requirement of pavement performance. The results are shown in Table 10. The results of fatigue resistance and skid resistance of asphalt mixture are shown in Table 11.

CONCLUSION

In this paper, the relationship between pavement performance of asphalt mixture and air void is studied systematically. For the different climate features in China, design method of asphalt mixture is proposed. Based on pavement performance of asphalt mixture, the optimum gradation of heavy duty asphalt pavement is recommended.

REFERENCES

1.Zhang Dengliang, Asphalt and Asphalt Mixture,Press of Communication (in China), 1993.
2.Neil. C. Kautz, Peter, E. Sebaaly, The Effect of Aggregate Gradation on Permanent Deformation of Asphalt Concrete, AAPT,Vol.63,1993.

Table 1 Properties of the Asphalt

Asphalt Source	Penetration 25 °C(1/10mm)	Ductility 15 °C ,(cm)	Soft Point (R&B), °C	Wax Content (%)
KLMY	92	>100	49.9	1.85

Table 2 Properties of Aggregate

Aggregate Type	SiO_2 Content (%)	Water Absorption (%)	Apparent Density (g/cm^3)		Bulk Density (g/cm^3)	
			>2.36mm	<2.36mm	>2.36mm	<2.36mm
Limestone	24.3	0.53	2.7824	2.7119	2.7488	2.6783

Table 3 Gradation of Asphalt Mixture

Mix Type		Percent Passing of Weight (%)										
		19	16	13.2	9.5	4.75	2.36	1.18	0.6	0.3	0.15	0.075
AC--16--I		100	95~100	75~ 90	58~78	42~63	32~50	22~37	16~28	11~21	7~ 15	4~ 8
Middle Value	A	100	100	90	78	63	50	37	28	21	15	8
AC--16--I		100	95~100	75~ 90	58~78	42~63	32~50	22~37	16~ 28	11~ 21	7~ 15	4~ 8
Middle Value	B	100	97	83	68	52	41	30	22	16	11	6
AC--16--II		100	90~100	65~ 85	50~70	30~50	18~35	12~26	7~ 19	4 ~ 14	3 ~ 9	2~ 5
Middle Value	C	100	95	75	60	40	26	19	13	14	6	4
AC--16--II		100	90~100	65~ 85	50~70	30~50	18~35	12~26	7 ~ 19	4 ~ 14	3 ~ 9	2 ~ 5
Middle Value	D	100	90	65	50	30	18	12	7	4	3	2
OGFC		100	/	90~100	/	11~35	8 ~ 25	/	5 ~ 17	4 ~ 14	3 ~ 10	2 ~ 7
Middle Value	E	100	/	95	/	23	17	/	11	9	7	5

Table 4 . The Structural Parameters of Asphalt Mixtures

Mix Type	A	B	C	D	E
Asphalt Content (%)	4.4	4.2	4.0	3.2	4.0
Density (g/cm^3)	2.4680	2.4560	2.3982	2.3210	2.2182
Asphalt Volume(%)	10.6	10.2	9.5	7.4	8.8
VMA(%)	13.6	14.0	16.3	18.4	22.7
Void Filled (%)	77.9	72.9	58.3	48.9	38.8
Marshall Stability(kN)	17.25	10.59	7.96	5.71	5.85
Flow Value(1/10mm)	28	31	37	27	24
Air Void(%)	3.0	3.8	5.8	9.0	13.9

Table 5. The Friction Number Test Results

Mix Type	A	B	C	D	E
Air Void (%)	3.0	3.8	5.8	9.0	13.9
Friction Number (BPN)	53	55	56	62	60

Table 6. Marshall Retained Stability Test Results

Mix Type	A	B	C	D	E
Air Void (%)	3.0	3.8	5.8	9.0	13.9
Retained Stability (%)	96.2	96.2	93.2	89.1	91.3

Table 7. Rutting Test Results at 60 °C

Mix Type	A	B	C	D	E
Air Void (%)	3.0	3.8	5.8	9.0	13.9
Dynamic Stability (cycle/mm)	600	867	3150	3316	818

Table 8. Creep Test Results at 0 °C

Mix Type	A	B	C	D	E
Air Void (%)	3.0	3.8	5.8	9.0	13.9
Creep Rate ($\times 10^{-6}$) (1/s)	2.44	2.34	2.04	/	/

Table 9. Climate Regions of Asphalt and Asphalt Mixture

Climate Type	Temperature(°C)		Precipitation(mm)
	the Average Highest Air Temperature in July	the Yearly Extreme Low Temperature	the Annual Precipitation
1--1--4	>30	< -37.0	<250
1--2--2	>30	-37.0 ~ -21.5	500 ~ 1000
1--2--3	>30	-37.0 ~ -21.5	250 ~500
1--2--4	>30	-37.0 ~ -21.5	<250
1--3--1	>30	-21.5 ~ -9.0	>1000
1--3--2	>30	-21.5 ~ -9.0	500 ~ 1000
1--3--3	>30	-21.5 ~ -9.0	250 ~ 500
1--3--4	>30	-21.5 ~ -9.0	<250
1--4--1	>30	> -9.0	>1000
1--4--2	>30	>-9.0	500 ~ 1000
2--1--2	20 ~ 30	< -37.0	500 ~ 1000
2--1--3	20 ~ 30	< -37.0	250 ~ 500
2--1--4	20 ~ 30	< -37.0	<250
2--2--1	20 ~ 30	-37.0 ~ -21.5	>1000
2--2--2	20 ~ 30	-37.0 ~ -21.5	500 ~ 1000
2--2--3	20 ~ 30	-37.0 ~ -21.5	250 ~ 500
2--2--4	20 ~ 30	-37.0 ~ -21.5	<250
2--3--1	20 ~ 30	-21.5 ~ -9.0	>1000
2--3--2	20 ~ 30	-21.5 ~ -9.0	500 ~ 1000
2--3--3	20 ~ 30	-21.5 ~ -9.0	250 ~ 500
2--3--4	20 ~ 30	-21.5 ~ -9.0	<250
2--4--1	20 ~ 30	> -9.0	>1000
2--4--2	20 ~ 30	> -9.0	500 ~ 1000
2--4--3	20 ~ 30	> -9.0	250 ~ 500
3--2--1	<20	-37.0 ~ -21.5	>1000
3--2--2	<20	-37.0 ~ -21.5	500 ~ 1000

Table 10. The Optimum Gradation of Heavy Duty Asphalt Pavement

Climate Regions	High Temperature Rutting(H)			Low Temperature Cracking Resistance(L)			Water Stability(w)			Meet HL	Meet HW	Meet LW	Meet HLW
	the Average Highest Air Tempera-ture in July	Dynamic Stability (cycle/mm) no less than	Grad-ation Types	The Yearly Extreme Low Temperature	Creep Rate (1/s×10^{-6}) no less than	Grada-tion Types	The Annual Precip-tation (mm)	Retained Stability (%) no less than	Grad-ation Types	Grad-ation Types	Grad-ation Types	Grad-ation Types	Grad-ation Types
1-1-4	>30	800	CDE	< -37.0	1.4	ABC	< 250	60	ABCDE	C	CDE	ABC	C
1-2-2	>30	800	CDE	-37.0~ -21.5	1.0	ABC	500~ 1000	70	ABCDE	C	CDE	ABC	C
1-2-3	>30	800	CDE	-37.0~ -21.5	1.0	ABC	250~ 500	65	ABCDE	C	CDE	ABC	C
1-2-4	>30	800	CDE	-37.0~ -21.5	1.0	ABC	< 250	60	ABCDE	C	CDE	ABC	C
1-3-1	>30	800	CDE	-21.5~ -9.0	0.7	ABC	> 1000	75	ABCDE	C	CDE	ABC	C
1-3-2	>30	800	CDE	-21.5~ -9.0	0.7	ABC	500~ 1000	70	ABCDE	C	CDE	ABC	C
1-3-3	>30	800	CDE	-21.5~ -9.0	0.7	ABC	250~ 500	65	ABCDE	C	CDE	ABC	C
1-3-4	>30	800	CDE	-21.5~ -9.0	0.7	ABC	< 250	60	ABCDE	C	CDE	ABC	C
1-4-1	>30	800	CDE	> -9.0	0.3	ABC	> 1000	75	ABCDE	C	CDE	ABC	C
1-4-2	20~30	600	ABCDE	> -9.0	0.3	ABC	500~ 1000	70	ABCDE	ABC	ABCDE	ABC	ABC
2-1-2	20~30	600	ABCDE	< -37.0	1.4	ABC	500~ 1000	70	ABCDE	ABC	ABCDE	ABC	ABC
2-1-3	20~30	600	ABCDE	< -37.0	1.4	ABC	250~ 500	65	ABCDE	ABC	ABCDE	ABC	ABC
2-1-4	20~30	600	ABCDE	< -37.0	1.4	ABC	< 250	60	ABCDE	ABC	ABCDE	ABC	ABC
2-2-1	20~30	600	ABCDE	-37.0~ -21.5	1.0	ABC	> 1000	75	ABCDE	ABC	ABCDE	ABC	ABC
2-2-2	20~30	600	ABCDE	-37.0~ -21.5	1.0	ABC	500~ 1000	70	ABCDE	ABC	ABCDE	ABC	ABC
2-2-3	20~30	600	ABCDE	-37.0~ -21.5	1.0	ABC	250~ 500	65	ABCDE	ABC	ABCDE	ABC	ABC
2-2-4	20~30	600	ABCDE	-37.0~ -21.5	1.0	ABC	< 250	60	ABCDE	ABC	ABCDE	ABC	ABC
2-3-1	20~30	600	ABCDE	-21.5~ -9.0	0.7	ABC	> 1000	75	ABCDE	ABC	ABCDE	ABC	ABC
2-3-2	20~30	600	ABCDE	-21.5~ -9.0	0.7	ABC	500~ 1000	70	ABCDE	ABC	ABCDE	ABC	ABC
2-3-3	20~30	600	ABCDE	-21.5~ -9.0	0.7	ABC	250~ 500	65	ABCDE	ABC	ABCDE	ABC	ABC
2-3-4	20~30	600	ABCDE	-21.5~ -9.0	0.7	ABC	< 250	60	ABCDE	ABC	ABCDE	ABC	ABC
2-4-1	20~30	600	ABCDE	> -9.0	0.3	ABC	> 1000	75	ABCDE	ABC	ABCDE	ABC	ABC
2-4-2	20~30	600	ABCDE	> -9.0	0.3	ABC	500~ 1000	70	ABCDE	ABC	ABCDE	ABC	ABC
2-4-3	20~30	600	ABCDE	> -9.0	0.3	ABC	250~ 500	65	ABCDE	ABC	ABCDE	ABC	ABC
3-2-1	<20	400	ABCDE	-37.0~ -21.5	1.0	ABC	> 1000	75	ABCDE	ABC	ABCDE	ABC	ABC
3-2-2	<20	400	ABCDE	-37.0~ -21.5	1.0	ABC	500~ 1000	70	ABCDE	ABC	ABCDE	ABC	ABC

Table 11. the Comparison ʻ Fatigue Resistance and Skid Resistance of Asphalt Mixture

Item	Fatigue Performance	Friction Number
Order	A>B>C>D>E	D>E>C>B>A

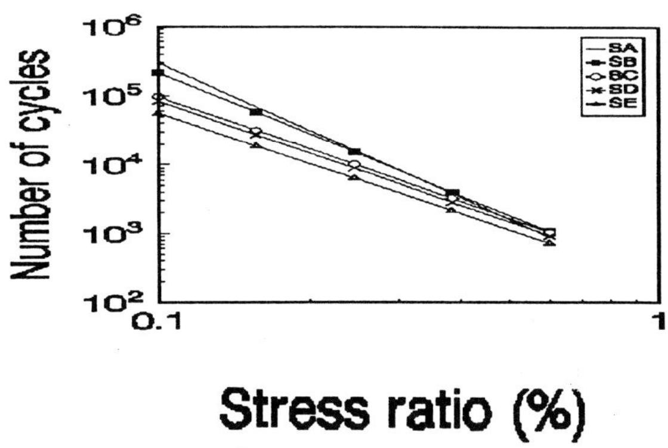

Figure 1. Fatigue Test Results

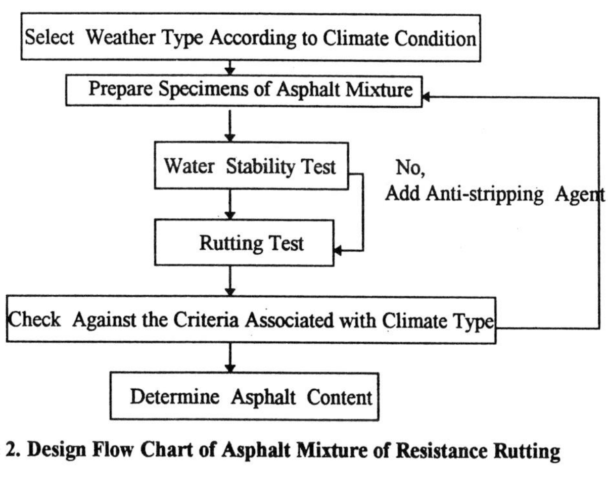

Figure 2. Design Flow Chart of Asphalt Mixture of Resistance Rutting

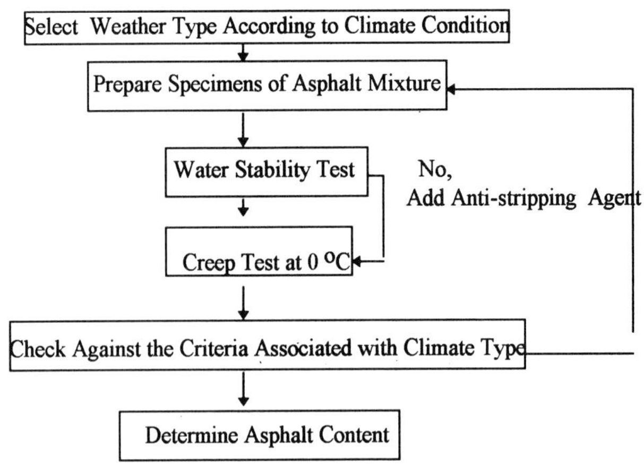

Figure 3. The Design Flow Chart of Asphalt Mixture of Low Temperature Cracking Resistance

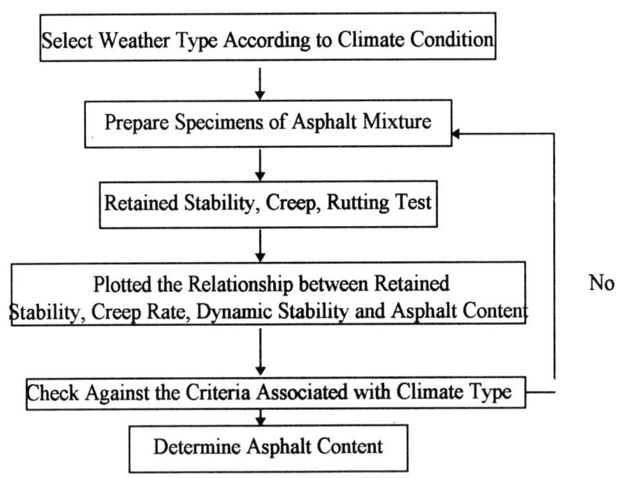

Figure 4. The Design Flow Chart of Asphalt Mixture Based on Pavement Performance

Fifth International Conference on the
BEARING CAPASITY OF
ROADS AND AIRFIELDS

FUNDAMENTAL PROPERTIES OF SMA AND CMHB MIXES

Louay N. Mahammad, Ph.D.[1], Zheng Z. Tan, P.E.[2], Baoshan Huang [3]
[1]Assistant Professor and Manager, [3]Research Associate
Louisiana Transportation Research Center, Louisiana State University, 4101 Gourrier Lane, Baton Rouge, LA 70808, U.S.A.

[2]Texas Department of Transportation, Houston, Texas, U.S.A.

ABSTRACT

A laboratory study was initiated to evaluate the performance of stone mastic asphalt, SMA, and coarse matrix high binder, CMHB, using locally available materials, as compared to a conventional Louisiana DOTD Type 8F wearing course.

Three mixture types were examined: SMA, CMHB, and conventional Type 8 WC mixture (serving as the control mix). For the SMA and CMHB, two types of modified asphalt cement were used: a polymer modified meeting PAC-40HG specifications and a chemically modified asphalt. In addition, for the SMA mixture, conventional AC-30 and a cellulose fiber was used. In order to evaluate the effect of aggregate type on SMA performance, three types (limestone, sandstone, and novaculite) were utilized for the SMA mixtures; the other mixtures incorporated limestone aggregates. Fundamental engineering tests such as indirect tensile strength and strain, resilient modulus, creep compliance, and fatigue life were used to evaluate the performance of these mixes.

In general, SMA mixes exhibited better performance than CMHB mixes. SMA mixes containing polymer modified asphalt and sandstone or novaculite aggregate provided lower axial creep slope than similar mixes with limestone aggregates. Conventional AC-30 and chemically modified binder exhibited lower asphalt draindown than polymer modified asphalt.

SMA mixes containing conventional AC-30 and cellulose fiber presented higher stiffness (resilient modulus), higher indirect tensile strength, and lower indirect tensile strain than similar mixes containing polymer modified or chemically modified asphalt.

Key Words: Asphalt Mixture Characteristics, SMA, CMHB

INTRODUCTION

Stone Mastic Asphalt (SMA)

Stone Mastic Asphalt (SMA) mixture was originated in Germany in the 1970s aimed at reducing permanent deformation in asphalt pavement [3]. Two German construction companies, Strabag and Teerbau led to the development of SMA. The philosophy of SMA mixture design is to use high quality coarse aggregate, through stone-on-stone contact, to form a structure that can carry heavy traffic loads. At the same time, the voids in the coarse aggregate are filled with a voidless mastic binder to ensure the durability and flexibility of the mix.

SMA is a gap-grade, dense (about 3 percent air voids in mix design), hot-mix asphalt (HMA) with a large proportion of coarse aggregate (passing 2 mm limited to about 20 percent, all aggregates are 100 percent

Fifth International Conference on the
BEARING CAPASITY OF
ROADS AND AIRFIELDS

FATIGUE AND PERMANENT DEFORMATION RESISTANCE OF ASPHALT MIXTURES CAN BE SPECIFIED

A.A.A. MOLENAAR[1] AND L.J.M. HOUBEN[2]

[1]Professor of Highway Engineering
[2]Senior Research Engineer
Delft University of Technology
Faculty of Civil Engineering and Geo Sciences
P.O. Box 5048, 2600 GA DELFT, The Netherlands

ABSTRACT

Beyond doubt, performance based and performance related specifications will be introduced gradually in the coming years. Essential however is that the test needed for quality control are fast and easy to perform. This implies that repeated testing is out of the question simply because of the fact that such testing takes too much time. Research is undertaken at the Delft University to come-up with such effective and simple procedures. The paper describes the developments so far with respect to the characterization of the fatigue and permanent deformation behaviour of asphalt mixtures.

INTRODUCTION

Although the transition from recipe based to performance based or related one makes sense from an engineering point of view, it might not be that easy to realize since performance testing requires repeated load testing which simply takes too much time. Therefore simplified procedures need to be developed which take away the burden of the time consuming repeated load tests especially those which are required for fatigue and permanent deformation characterization. This paper describes the developments that were made in this field by the Road and Railroad Research Laboratory (RRRL) of the Delft University. It will be shown that the fatigue characteristics can be estimated from simple indirect tensile tests combined with modulus testing. With respect to permanent deformation, it will be shown that estimation of the resistance should be possible by means of simple analyses.

ESTIMATION OF FATIGUE CHARACTERISTICS

Visco-elastic fracture mechanics to describe fatigue tests

Schapery [1, 2] has developed a theory which allows the crack propagation in visco-elastic media to be described. He has shown that the well known crack growth law which was developed by Paris and Erdogan [3] for elastic materials, can be used to describe the crack growth in visco-elastic materials as well. It is recalled that Paris' law can be written as:

$$dc/dN = A * K^{n} \quad \text{.......... (1)}$$

where: dc/dN= increase in crack length per load cycle
K = stress intensity factor at the tip of the crack
A, n = material parameters

Schapery showed that the constant A is dependent on parameters like tensile strength, fracture energy and modulus, while the constant n is dependent on the slope of the curve log (loading time) vs log (modulus). He also found that A was dependent on the shape of the load pulse and n was dependent on the type of test (constant load vs constant displacement type of fatigue testing).
The dependency of A and n on the above mentioned factors is shown in equations (2) and (3).

$$A = \frac{\pi}{6\sigma_m^2 I_1^2}\left(\frac{(1-v^2)\,D_2}{2\Gamma}\right)^{\frac{1}{m}} \int_0^{\Delta t} w(t)^{2\left(1+\frac{1}{m}\right)} dt \quad \text{.......... (2)}$$

$$n = 2/m \text{ for constant displacement tests .... (3a)}$$
$$n = 2 * (1 + 1/m) \text{ for constant load tests .... (3b)}$$

where: σ_m = tensile strength of the material given the prevailing temperature and loading conditions
I_1 = integration constant
v = Poisson's ratio
Γ = fracture energy at the prevailing loading time and temperature conditions
D_2 = 1/E at a loading time of 1 s and the prevailing temperature conditions
m = slope of the log t vs log E relation
w = shape of the stress intensity factor pulse

Work at the Delft University [4, 5, 6, 7, 8] has shown that it is indeed possible to use Schapery's theory to describe the fatigue behaviour of asphalt mixtures provided that some correction factors are applied.
In [4] the increase of displacement of direct tensile test specimens subjected to repeated loads was used to describe the increase in micro and macro damage. In [5] and [6] it was realized that the increase in overall displacement is indeed giving information on the damage development, but also that it is much better to determine the increase in damage by means of measurements at the opening of the notch that is usually made at the centre of the tensile specimens in order to promote cracking at that location. The reason for this is that the increase in displacement measured in this way can fully be contributed to the increase in

damage while the displacement measured over the entire specimen contains information on the increase in damage and the elongation of the undamaged part. Figure 1 gives a schematic overview of the type of tests that were performed. Figure 2 gives an example of the analyses that were made.

Figure 2a shows the increase in crack opening displacement (c.o.d.) as measured during the test. Figure 2b shows the effective crack length that was calculated by means of the finite element program CAPA [9], using the c.o.d. measurements as input. Figure 2c shows the stress intensity factor vs increase in c_{eq} as calculated by means of the finite element program. Figure 2d combines figures 2b and 2c.

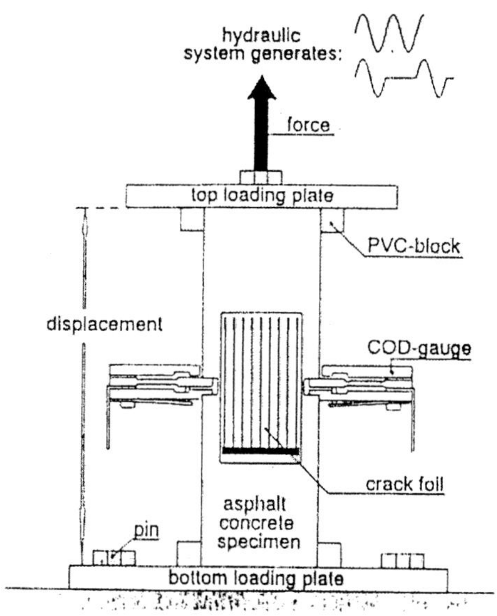

Figure 1: Direct tension test as used by Jacobs.

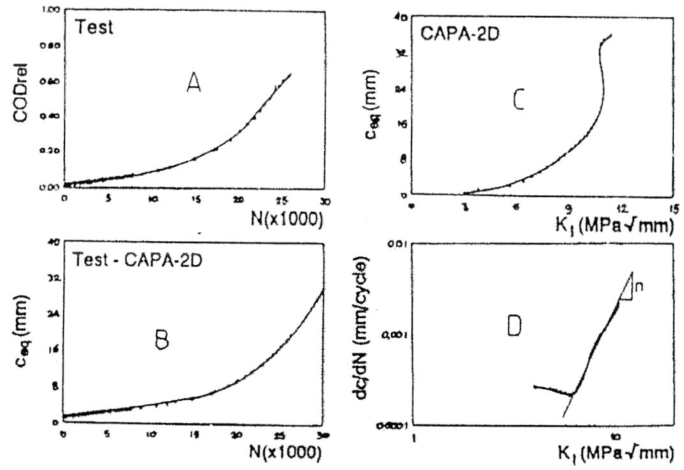

Figure 2: Procedure to calculate equivalent crack length, stress intensity factor and crack growth relation.

Mastercurves to estimate the modulus of the mixture and the slope of the crack growth relation

As indicated in equations (2) and (3), the constants A and n depend amongst other factors on the magnitude of the resilient modulus and the slope of the curve representing the relationship between the log of the loading time and the log of the modulus. This means that it is important to determine the master curve for the resilient modulus. Such a curve describes the relation between the loading time (or frequency) and the modulus for a wide range of loading times. In order to obtain such a master curve, the tests are normally performed at various temperatures at a limited number of loading times. Figure 3 shows how such a master curve is developed.

For a constant displacement type of test, the slope n of the fatigue relation

$$N = k1 * (1 / \varepsilon)^n \ldots\ldots (4)$$

where: N = number of load repetitions to failure
ε = applied strain
k1, n = material constants

is calculated using:

$$n = 2/(m * c) \ldots\ldots (5)$$

where: c = correction factor
m = slope of the modulus master curve at the loading time and temperature of interest

In reference [7] it has been shown that the correction factor c can be calculated using:

$$c = 0.8998 + 7.2814 \text{ E-5} * Smix + 7.1487 \text{ E-9} * (SMix)^2 \ldots\ldots (6)$$

where : Smix = resilient modulus of the mixture [MPa]

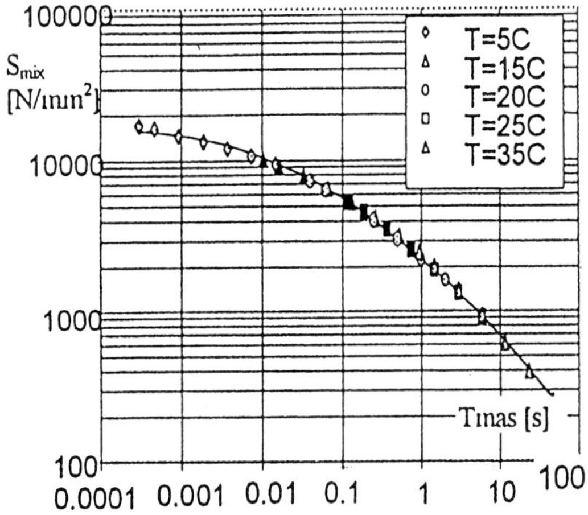

Figure 3: Mix stiffness mastercurve.

Mastercurves to determine the strength and the fracture energy of asphalt mixtures

Equation (2) indicates that in order to be able to estimate A, information on the tensile strength and fracture energy under the conditions of interest should be known. It has already been shown by Heukelom [10] that it is possible to present the tensile strength of asphalt mixtures in a non-dimensional way. The same procedure was followed in this research and the results are shown in figures 4 and 5.
Figure 4 shows the master curve for the tensile strength of gravel asphalt base courses.
Figure 5 shows the mastercurve that was derived in the same way for the fracture energy. These curves were obtained from direct tensile tests; the way in which the tensile strength, mix stiffness and fracture energy were determined is indicated in figure 6. This figure indicates that a continuous registration of force against displacement should be made. Furthermore it should be noted that the stiffness of the bitumen is backcalculated from the measured mix stiffness, the loading time to failure and the volumetric mix composition, using the equations that were developed by Francken [11].

Estimation of the crack growth parameter A

Jacobs showed in his study that values for A can best be estimated using equation (7). This equation was derived from available data using regression analysis. It is built up in the same way as equation (1) that was derived by Schapery.

log A = d - 2 * a * log (σ_m) - b * n * log (Γ) 2 - c * n * log (Smix) / 2.......... (7)

For base course mixes the following values for the various constants apply:
a = 0.39, b = - 0.04, c = 0.62, d = -1.29.

Estimation of the fatigue relation

When simulating a four point beam bending fatigue test by means of a finite element analysis, Sabha [7] showed that using the fracture mechanics approach the number of cycles to failure can be predicted with equation (8). In fact this equation is basically the same as equation (4).

$$N = \left[\frac{h}{AS^n_{mix}} \int_{c^o/h}^{c_f/h} \frac{1}{\left(\sqrt{cf(\frac{c}{h})}\right)^n} d(\frac{c}{h}) \right] \varepsilon^{-n} \qquad ......(8)$$

where: c_0 = initial crack length
c_f = track length at which specimen is considered to be failed
f(c/h) = equation describing the increase of the stress intensity factor when the crack progresses through the beam
h = beam height

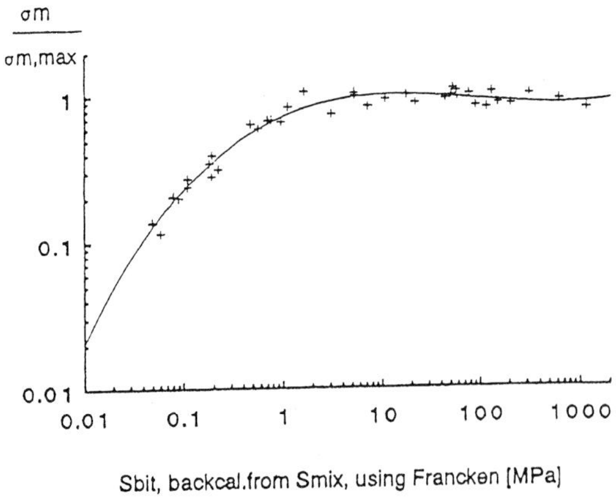

Figure 4: Mastercurve for the tensile strength of a base course mixture.

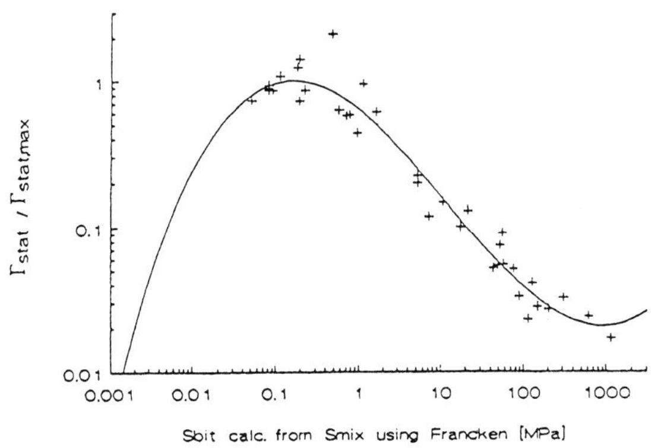

Figure 5: Mastercurve for the fracture energy of a base course mixture.

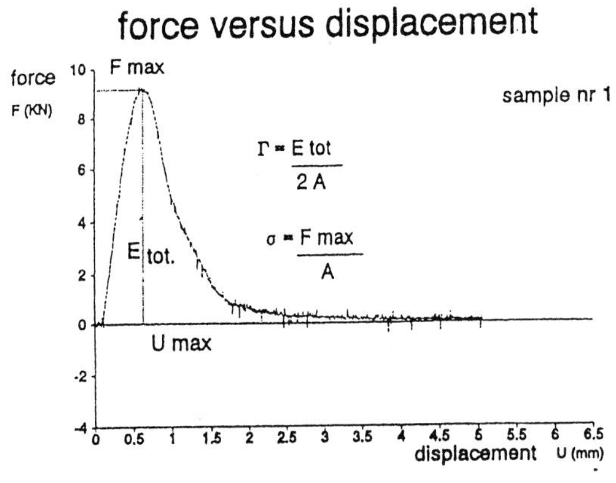

Figure 6: Example of how tensile strength and fracture energy are determined from the load vs displacement curve measured in a direct tensile test.

In a four bending displacement controlled test, the fatigue life is considered to be reached when the load that is needed to generate the displacement has decreased to 50% of its initial value. The finite element analysis showed that this coincides with a progression of the damage over 54% of the beam height.
Observing all these complicated equations might lead to the conclusion that the approach described here is far from practical. One should realize however that all the computations can easily be programmed into a pocket computer without the need of the technician having to understand how the calculations are performed. The beauty of this approach is in the fact that simple tests like the direct tensile test can be used to collect the data.

It might be argued that a direct tensile test is not a practical test to be used for quality control purposes. It is true that for control purposes, a test like the indirect tensile test (i.t.t.) is more appropriate. It has been shown [7] that reliable estimates for the modulus can be obtained using the resilient mode test in i.t.t. mode which is possible by means of relatively simple pneumatic driven equipment like the U.T.M. (marketed by ELE) or the N.A.T.. The tensile strength and fracture energy can be obtained by doing i.t.t.'s using a Marshall load frame. It is believed that in this way this type of testing comes within the possibilities of ordinary road materials testing laboratories which means that the procedure described here can be used very well for quality control purposes. A nice example of using this way of testing for quality control purposes is the work that was done at Amsterdam's airport Schiphol [12].

ESTIMATION OF THE RESISTANCE TO PERMANENT DEFORMATION

Permanent deformation in asphalt mixtures can be discriminated in plastic and viscous deformation. Viscous deformation is related to the "flow" in time of the bituminous mortar; it is a time dependent factor. Plastic deformation however is related to stresses; shear failure in e.g. an unbound granular materials is a nice example of plastic failure. In asphalt mixtures both phenomena can occur which makes the prediction of the resistance to permanent deformation rather complicated. Until now no nomographs are available to make such predictions. Therefore it is quite often believed that repeated loading will be needed to control whether the asphalt mixture has the required resistance to permanent deformation. This is considered to be a big disadvantage since repeated load testing doesn't fit into quality control systems because of the relatively long time that is needed to perform the tests. Work done at the R.R.R.L. of the Delft University suggests however that the resistance fo permanent deformation can be assessed from the volumetric mixture composition and some simple tests. In the sections here-after some details on this work will be given.

Relation between resilient and permanent deformation

In general the development of permanent deformation is described using equation (9).

$$\varepsilon_p = C * N^B \quad \ldots\ldots\ldots\ldots \quad (9)$$

where: ε_p = permanent strain
N = number of load repetitions
C, B = material constants

In the visco-elastic multilayer program VESYS [13], the constants C and B are written as:

$$C = \varepsilon_r * \mu /(1 - \alpha) \quad \text{.........} \; (10)$$

$$B = 1 - \alpha$$
$$\mu = C * B / \varepsilon_r$$

where: ε_r = elastic or resilient strain
a, μ = material constants

Equations (9) and (10) clearly indicate that mixes with a high resilient modulus will have a high resistance to permanent deformation. Since the resilient modulus (Smix) can be estimated from nomographs using as input the volumetric composition and the bitumen stiffness (Sbit), at least some of the resistance to permanent deformation can be assessed easily.

Influence of the aggregate skeleton and the type of filler

It is obvious that the aggregate skeleton and the type of filler have a significant influence on the resistance to permanent deformation of an asphalt mixture. Figure 7 [13] shows how the filler content and the void content of the filler influence the softening point of the bituminous mortar. It is obvious that a higher Tr&b results in a higher resistance to permanent deformation. A careful selection of the type and amount of filler is therefore essential to arrive to a deformation resistant mixture.
With respect to the influence of the aggregate skeleton it can be stated that the higher the resistance to shear failure is of the skeleton itself, the higher the resistance to shear failure of the asphalt mixture will be. It has been shown by van Niekerk e.a. [14] that the cohesion c and the angle of internal friction φ are dependent on the amount by which the actual gradation is different from the Fuller curve (quantified by means of the parameter Apsdo-fl) and the angularity or sharpness of the grains. Using these relations, the stress to failure can be calculated by means of the relationship given in figure 8.

Figure 8 shows how well the predicted stress at failure (predicted by means of the estimated values for c and φ agrees with the measured stress at failure. These equations at least allow potential granular skeletons for asphalt mixtures with high permanent deformation resistance to be selected.

Repeated load triaxial tests

As part of the "Heavy Duty Pavements Project", which is undertaken by the Ministry of Transport and the V.B.W. Asfalt, a large number of repeated load triaxial tests were performed in order to analyze the effects of mix composition, bitumen grade, temperature and loading conditions on the permanent deformation behaviour of number of wearing course mixtures (drain asphalt, s.m.a, dense asphaltic concrete), binder course mixtures and one base course mixture.

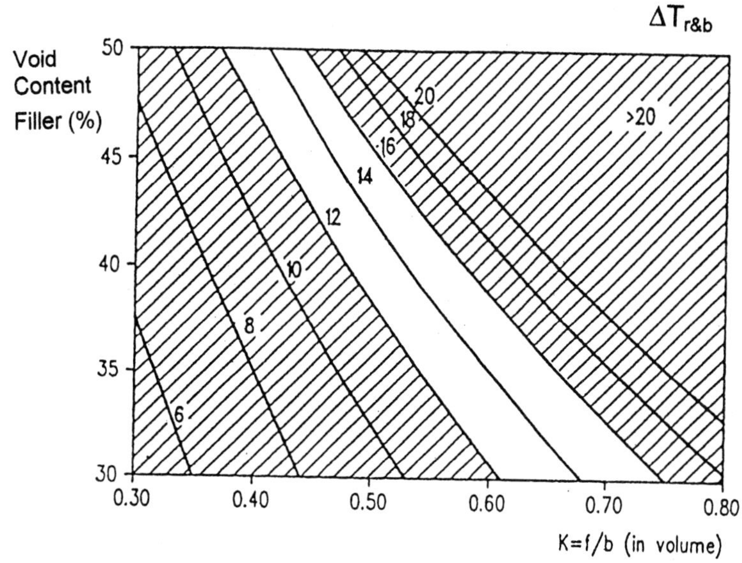

Figure 7: Influence of amount and type of filler on softening point.

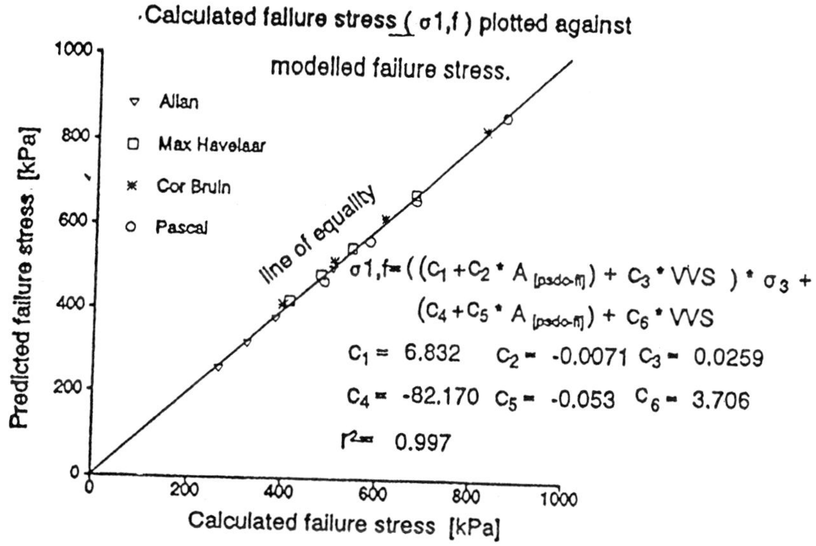

Figure 8: Comparison between predicted failure stress calculated with estimated C and φ and the failure stress calculated form the triaxial tests.

Some typical test results are shown in figure 9. It is quite clear from this figure that drain asphalt has no resistance to permanent deformation in case there is lack of confinement.

At higher confining stress levels, the drain asphalt mixture performed very good. This is important information for practice because these results show that it is not wise to place an overlay on top of a drain asphalt layer. This is because in those conditions the drain asphalt layer will get closer to the neutral axis which implies that the confinement stresses will decrease, resulting in a significant decrease of the deformation resistance.

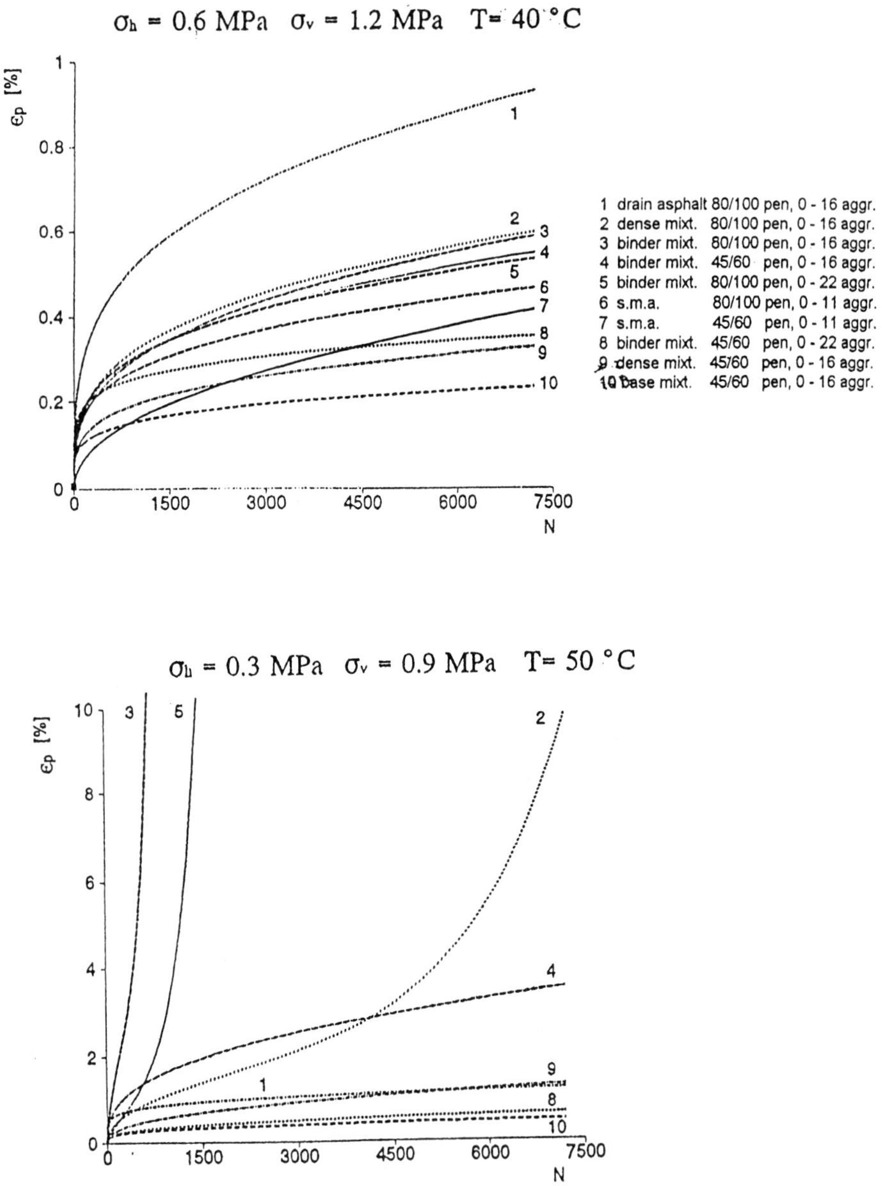

Figure 9: Examples of the repeated load triaxial test results.

Permanent deformation model

A model to predict the permanent deformation resistance potential of asphalt mixtures was developed by performing regression analysis on the repeated load triaxial test results [15]. The model which was obtained was:

$$C * B = k_o * SVSH^{k1} * (Smix * \%Rice)^{k2} .... (12)$$

where: C * B = the product of the intercept and slope value of equation (9)
SVSH = ratio vertical stress over horizontal stress
%Rice = density according to Rice
Smix = resilient modulus of the mix

The product C * B was selected since it gives an overall indication of the deformation resistance comparable with the product c * tg φ which is often used to characterize the shear resistance of unbound granular materials.
It is believed that equation (12) provides an easy way of characterizing the permanent deformation resistance of asphalt mixtures. The predictions can be made using the volumetric composition, the bitumen characteristics and the stress conditions as input. Because of its simplicity it can be used very well for performance related specifications and quality control systems.

CONCLUSIONS

From the work that has been presented in the previous sections, the following conclusions have been drawn.

a. In order to really use performance related specifications in practice, the tests needed to characterize the resilient modulus of asphalt mixtures as well as their resistance to cracking and permanent deformation must be effective, simple and above all fast.
b. It has been shown that the resistance to fatigue cracking as well as the resistance to permanent deformation can be assessed from the volumetric composition of the mixture, the characteristics of the bitumen and the temperature and loading conditions.
c. Relatively simple tests like the resilient modulus test, which can be performed by pneumatic driven equipment like the N.A.T. and the M.A.T.A., have shown to be important tests which should be incorporated in performance related quality control systems.
d. It is believed that the procedures presented here offer excellent possibilities for the introduction of performance related specifications.

REFERENCES

1. SCHAPERY, R.A. A theory of crack growth in visco-elastic media. Report MM 2764-73-1. Mechanics and Materials Research Centre, Texas A & M University – 1973.
2. SCHAPERY, R.A.
Non linear fracture analysis of visco-elastic composite materials based on a generalized J integral theory. Proceedings Japan - U.S.A. conference on composite materials. Tokyo – 1981.
3. PARIS P.C.; ERDOGAN, F. A critical analysis of crack propagation laws. Transactions of the ASME, Journal of Basic Engineering, series D, Vol. 85 No 3 – 1963.
4. MOLENAAR, A.A.A. Structural performance and design of flexible pavements and asphalt concrete overlays. Dissertation. Delft University of Technology; Delft – 1983.
5. SCARPAS, A.; JACOBS, M.; VERHOEVEN, K.M. Simulation of the crack growth process in asphalt pavements. C.R.O.W. publication 82, Ede – 1994.
6. JACOBS, M.M.J. Crack growth in asphalt mixes. Dissertation. Delft University of Technology, Delft – 1995.
7. SABHA, H.; GROENENDIJK, J.; MOLENAAR, A.A.A. Estimation of crack growth parameters and fatigue characteristics of asphalt mixes using simple tests. Report 7-95-209-35M. Road and Railroad Research Laboratory, Delft University of Technology; Delft – 1995.
8. ERKENS, S.M.J.G.; GROENENDIJK, J.; MORAAL J.; MOLENAAR, A.A.A.; JACOBS, M.M.J. Using Paris' law to determine fatigue characteristics - a discussion. Proceedings 8th International Conference on Asphalt Pavements, Seattle – 1997.
9. SCARPAS, A.; BLAAUWENDRAAD, J.; DE BONDT, A.H.; MOLENAAR, A.A.A. Capa a modern tool for the analysis and design of pavements. Proceedings 2nd International Conference on Reflective Cracking, Liege – 1993.
10. HEUKELOM, W. Observations on the rheology and fracture of bitumens and asphalt mixes. Proceedings A.A. P.T., Vol. 35 – 1966.
11. FRANCKEN, L. Complex modulus of bituminous mixtures. (in French) Bulletin de liaison des laboratoires des ponts et chaussees, special V. Laboratoire central des ponts et chaussees, Paris – 1977.
12. MOLENAAR, A.A.A.; VAN DER MEENT, A.; MUL, N.C.; NATARAJ, N.; DON, W. Design of polymer modified asphalt mixes for Amsterdam's Schiphol Airport. Proceedings 9th A.A.P.A. Int. Asphalt Conference, Brisbane – 1994.
13. FRANCKEN, L.; VANELSTRAETE, A. PRADO design of asphalt mixtures. (in Dutch). Belgian Road Research Centre, Brussels – 1997.
14. VAN NIEKERK, A.A.; HUURMAN, M. Establishing complex behaviour of unbound road building materials from simple material testing. Report 7-95-200-16. Road and Railroad Research Laboratory, Delft University of Technology – 1995.
15. KORTLEVER, M.; HOUBEN, L.J.M.; MOLENAAR, A.A.A. Permanent deformation of asphalt mixtures, part 1 - modelling by means of repeated load triaxial testing. (in Dutch). Report 7-96-114-5. Road and Railroad Research Laboratory, Delft University of Technology –1996.

Fifth International Conference on the
BEARING CAPASITY OF
ROADS AND AIRFIELDS

INFLUENCE OF LOADING PATTERN IN FOUR-POINT BENDING TESTS

JORGE C. PAIS[1], PAULO A. A. PEREIRA[2]

M. CONCEIÇÃO AZEVEDO[3]

[1]Research Assistant, University of Minho, Guimarães, Portugal, Fax: +351-53-510-217, Email: jpais@eng.uminho.pt

[2]Associate Professor, University of Minho, Guimarães, Portugal, Fax: +351-53-510-217, Email: ppereira@ci.uminho.pt

[3]Thecnical Director, APORBET, Portugal, Fax: +351-1-7931212

ABSTRACT

This paper presents the results of a study where the fatigue life of a dense bituminous mixture was evaluated for three different types of loading pattern. The first loading pattern was a sinusoidal wave, the second was a haversine wave and the third was a modified sinusoidal wave. The rest period was 50% of the load time. The mixes used in this study were produced and compacted in laboratory using a rolling wheel compactor.

The fatigue life of the mixes was compared based on the results of the four point bending test carried out at 20°C, under controlled strain mode of loading. Furthermore, comparisons were made between fatigue life predicted by the Shell, The Asphalt Institute, SHRP-A003A program procedures and the laboratory test results.

Key words: Bituminous Mixtures, Stiffness, Fatigue Life, Loading Pattern.

INTRODUCTION

The resistance to fatigue of a bituminous mixture is the ability to withstand repeated bending load without failure. This form of distress manifests itself by the appearance of cracking in the pavement surface. The fatigue of a bituminous mixture has been directly associated with the repeated application of tensile stresses or strains and it is generally accepted that it can be very well evaluated by a four point bending test (Azevedo, M.C., 1993). Fatigue tests are performed imposing strain or stress repetitively until failure occurs. The fatigue life is characterized by the relationship between the strain level and the number of repetitions to reach the mixture failure.

The mode of loading influences the mixture performance and controlled stress versus controlled strain is applied as a function of pavement structure (i.e., pavement thickness and stiffness). The repetitively applied loads usually follow a sinusoidal pattern although the traffic pavement response seems to be different.

The fatigue life of bituminous mixtures is influenced by several factors such as test temperature and frequency of applied loads. Significant differences are found when laboratory fatigue life is compared with fatigue life predicted by recognized fatigue models (Pereira et al, 1997). This can be explained by the fact that most of fatigue models were developed based on stress control tests, while most of actual laboratory tests are performed in strain control tests.

This paper shows a comparison of the laboratory fatigue performance for three type of loading pattern. The first loading pattern was the sinusoidal. This one is the most used in fatigue tests. The second pattern was the haversine once this wave is similar to the traffic pavement response. The last pattern was the modified sinusoidal simulating the delay between loads in the pavement.

The comparison among fatigue performance of these three load patterns allows for conclusions and recommendations about the definition of shift factors for fatigue models.

Stiffness

The bituminous mixtures exhibit linear-viscoelastic behaviour as such that their response is time of loading and test temperature dependent. This behavior is represented by the following relationship:

$$S_{mix} = \frac{\sigma}{\varepsilon}(t,T) \qquad (1)$$

where S_{mix} is the mixture stiffness, σ is the stress level, ε is the strain level, t is the time of loading and T is the test temperature.

The material stiffness measures the ability to spread the traffic loads over an area. This parameter represents an important rule in the design of pavements since the design strain level is a function of the material stiffness.

Fatigue life

Fatigue cracking is one of the major modes of distress considered in the flexible pavement design.

A fatigue model has generally been established using a linear regression between fatigue life logarithm, *log* N_f, and the initial strain logarithm, *log* ε_t, for controlled strain tests, as follows:

$$N = c\left(\frac{1}{\varepsilon_t}\right)^d \qquad (2)$$

where N is the number of repetitions to failure, ε_t is the tensile strain applied, c and d are experimentally determined coefficients. Sometimes this equation is expressed in terms of stress level instead of strain level.

Nowadays some organizations use the equation as follows:

$$N = K\left(\frac{1}{\varepsilon_t}\right)^a \left(\frac{1}{S_{mix}}\right)^b \qquad (3)$$

where S_{mix} is the stiffness, K is a factor that recognize the influence of bitumen content and degree of compaction. This equation shows an advantage in relation to the former one since the mixture stiffness has a great influence in the fatigue life. Stiffer mixes have a low fatigue response and softer mixes show a high fatigue response. The bitumen content or the voids filled with bitumen have a great influence in the fatigue response and thus some fatigue life models proposed by some institutions have applied this variable in such models.

The first well known model to predict the bituminous mixture fatigue life is the model proposed by Shell (1978) where the fatigue life is a function of strain level, bitumen content and mixture stiffness. This model is expressed by the following equation:

$$N_f = \left(\frac{\varepsilon_t}{(0.856V_b + 1.08)S_{mix}^{-0.36}}\right)^{-5} \qquad (4)$$

where N_f is the fatigue life, ε_t is the tensile strain, S_{mix} is the mixture stiffness and V_b is the bitumen content. This equation does not have a shift factor since it gives the design fatigue life. This equation was obtained from a work where bituminous slabs were tested in a wheel tracking machine under a rolling wheel and it was concluded that the crack patterns observed were very similar to those defined as alligator cracking in practice (Claessen et al, 1977).

The Asphalt Institute (1981) proposed a model to predict bituminous mixtures fatigue life, by the following equation:

$$N_f = S_f * 10^{(4.84(VFB-0.69))} * 0.004325 * (\varepsilon_t)^{-3.291} * (S_{mix})^{-0.845} \qquad (5)$$

where N_f is the fatigue life, S_f is a shift factor to convert laboratory results in field expected results, the recommended factor is 18.4 for a 10% cracked area, ε_t is the tensile strain applied, S_{mix} is the stiffness of a mix in PSI and VFB is the voids filled with bitumen.

Recently, the SHRP program (Tayebali 1994) proposed the following equation to predict the bituminous mixture fatigue file:

$$N_f = S_f * 2.738E5 * e^{0.077VFB} * \varepsilon_o^{-3.624} S_o''^{-2.720} \qquad (6)$$

where S_f is a shift factor to convert laboratory results in field expected results, the recommended factor is 10 for 10% cracked area and 14.0 for 45% cracked area, e is the base of natural logarithm, VFB is the percentage of voids filled by bitumen, ε_0 is the strain level and S''_o is the loss stiffness.

EXPERIMENT DESIGN

Materials

The bituminous mixture used in this study was a dense graded bituminous mixture with a maximum aggregate size of 25 mm, manufactured according to the Portuguese standards and used in binder layers. The aggregate was a crushed granit with the grading curve shown in Table 1. The specific unit weight of aggregate was 2.7 g/cm3, the water absorption was 0.97 % and the Los Angeles wear was 24%. The bitumen used to produce the bituminous mixture was a conventional 50/70-penetration grade. The bitumen penetration at 25°C was 62 (0.1mm) and the R&B temperature was 53.1° C.

Table 1. Aggregate grading curve

Sieve #	Percentage passing
1"	100
3/4"	94
1/2"	80.5
#4	52.5
#10	39
#40	21.5
#80	13.5
#200	7.5

Mixture design and specimens preparation

The bituminous mixtures were produced with the optimum bitumen content given by Marshall method (5% bitumen by total weight of aggregate).

To produce the specimens, bituminous mixture was produced and compacted in laboratory. The aggregates were heated at 179 °C and the bitumen was heated at 150 °C. After mixing, the mixture is placed in an oven at 135 °C during 4 hours to be subjected to the conditioning recommended by SHRP-A003A (Tayebali 1994).

The compaction was made with a lightweight vibratory steel roller in a steel-coated wood mold. From compacted slab, beams 38.1 cm long by 6.3 cm wide by 5.1 cm high were sawed (Figure 1).

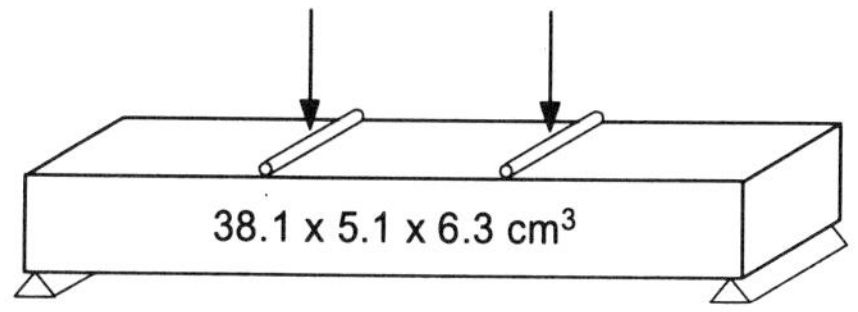

Figure 1. Four-point bending test on beam specimen.

Test apparatus

The test apparatus used in this study were a CS7800 Axial Testing System and a Flexural Beam Device (FBD), allowing 4 point bending tests. The FBD, identical to that one used during the SHRP program in the evaluation of the fatigue response, has free translation and free rotation at the reaction points and at load points. The machine is controlled by a microcomputer using a software that provides a feedback closed-loop control to the servohydraulic system, test temperature and data acquisition. This flexural beam device, in association with the software capabilities, has considerably improved the reliability of results compared with earlier devices.

TEST PROCEDURE

The test procedure utilized in this study usually includes a frequency sweep test for stiffness and phase angle characterization, as function of load frequency, and a fatigue test to evaluate the fatigue performance. These tests are performed for a given temperature in controlled strain mode of loading.
The three type of loading patterns were applied at 20°C, and the material dynamic response for the haversine and for the modified sinusoidal is shown in Figure 2 and 3, respectively. It has been observed that the material response to a sinusoidal wave without rest period is identical to the material response to the haversine wave.

Frequency sweep test

The frequency sweep test measures the stiffness and the phase angle of a mixture when subjected to different loading frequencies. Usually, these tests are executed on the same beams of the fatigue tests. All the frequency sweep tests of this study were executed only for sinusoidal wave without rest period at a strain level of 200 micro mm/mm, and at 10, 5, 2, 1, 0.5, 0.2 and 0.1 Hz. To minimize damage of the beam only 100 cycles were applied for the first four frequencies and for the other frequencies 10 cycles were applied.

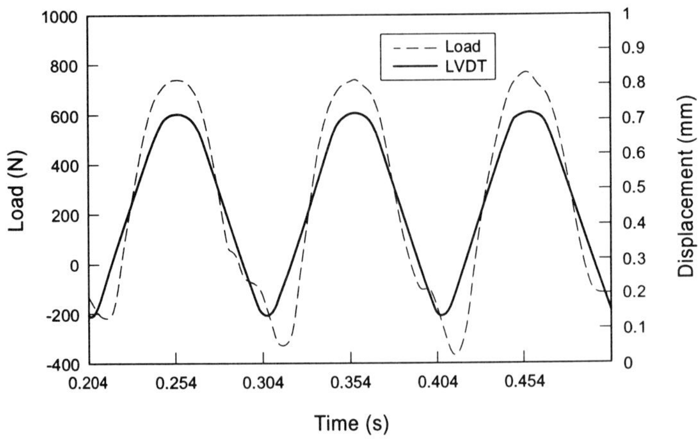

Figure 2. Applied haversine displacement and measured load.

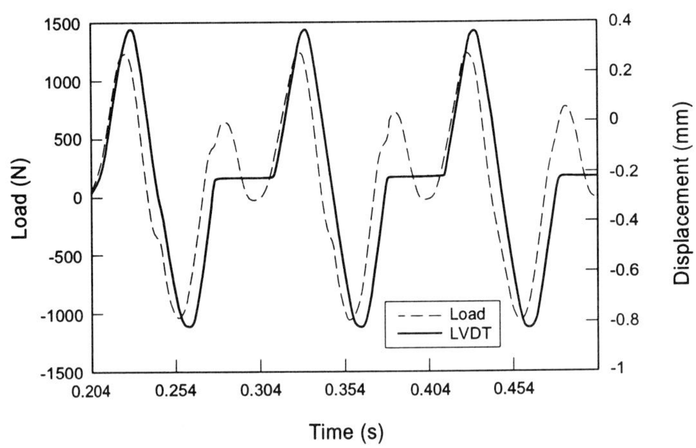

Figure 3. Applied modified sinusoidal displacement and measured load.

Fatigue test

The selected test strain levels were 200, 400 and 800 microstrain. The fatigue test was executed after the conclusion of the frequency sweep test and was done in control-strain mode. In this control mode the failure is defined at the point where the stiffness of the specimen decreases to 50% of the initial stiffness. The four-point bending tests were executed imposing a displacement to the center of the beam with a frequency of 10 Hz.

TEST RESULTS

Fatigue test analysis in controlled strain mode of loading requires the evaluation of the stiffness moduli during the test. In four point bending tests the material stiffness moduli is given by the following relationship:

$$E = \frac{Pa(3L^2 - 4a^2)}{4bh^3 d} \qquad (7)$$

where E is the stiffness moduli; P is the load; d is the applied displacement; L is the distance between support points; a is the distance between the load and the nearest support; b is the beam width and h is the beam depth.

In fatigue tests the stiffness exhibit an exponential behaviour during the test, but in the first cycles the decrease of the stiffness is due to the heating of the specimen (Di Benedetto et al. 1996) and so this part must be eliminated. The initial stiffness is given by extrapolating the curve-stabilized part until the first cycle. Once the frequency sweep tests were executed before fatigue tests, the heating of the specimen is produced only in the frequency sweep test and the cycles elimination must be made in this test.

Frequency sweep test results

Material characterization in terms of stiffness moduli and phase angle, obtained with frequency sweep tests, is shown respectively in Figure 4 and 5.

The stiffness or the phase angle behaviour, as function of loading frequency, can be expressed by a relationship as the follows:

$$E \text{ or } \delta = a.f^{\,b} \qquad (8)$$

where E is the stiffness moduli; δ is the phase angle; f is the frequency; a and b are experimentally determined coefficients.

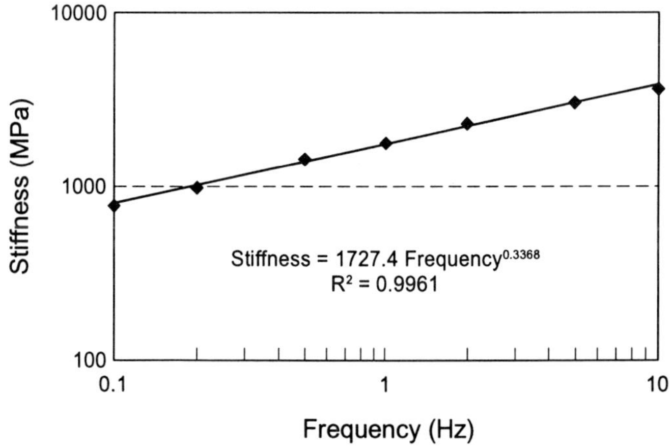

Figure 4. Stiffness as a function of load frequency.

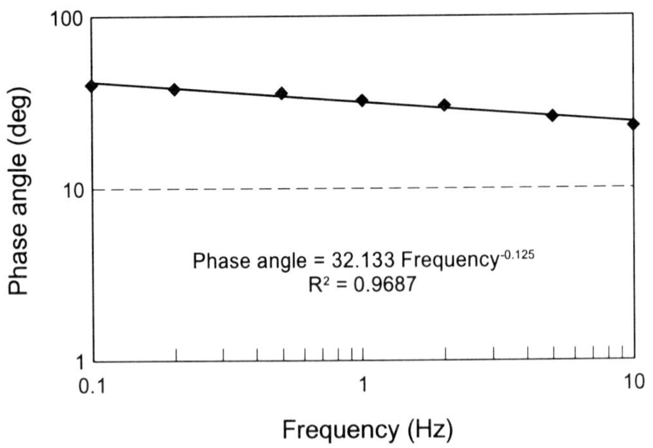

Figure 5. Phase angle as a function of load frequency.

Fatigue test results

Figure 6 presents the fatigue life as a function of tensile strain level for the three loading patterns used in this study and in Table 2 the coefficients for Equation 2 are shown, also with the R^2 coefficients for the different fatigue life laws.
The high R^2 coefficients indicate a good correlation between fatigue life and tensile strain level. Although, only three strain levels were applied.
For usual pavement tensile strain levels (i.e. about 100 microstrain), the fatigue life law for haversine loading pattern exhibit less fatigue life than the others. The modified sinusoidal loading pattern exhibits greater fatigue life.

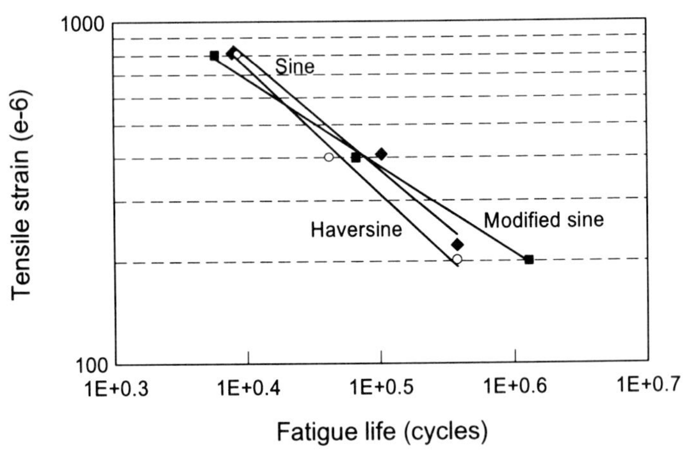

Figure 6. Fatigue curves for different loading patterns.

Table 2. Parameters of fatigue life laws.

Strain=a*N^b	Sine	Haversine	Modified Sine
slope (b)	-0.32708	-0.36459	-0.25605
intercept (a)	15865.07	20745.32	7163.243
R^2	0.976	0.991	0.997

When compared for the same tensile strain level, in this case 100 microstrain, the fatigue life for sine loading pattern is 2 times the fatigue life for haversine loading pattern and 30% the fatigue life for the same wave but with a 50% rest period. These results are shown in Table 3.

Table 3 also shows a comparison study between the design fatigue life given by the model proposed by The Asphalt Institute, the model proposed by Shell, the model proposed by SHRP and the values obtained for the loading patterns of this project.

Table 3. Comparison of fatigue life between predictive models and laboratory results.

Fatigue life for ε=100E-6	Sine	Haversine	Modified Sine
Laboratory	5 338 824	2 263 680	17 590 069
AI		2 403 355	
Shell		101 102 833	
SHRP		280 108 620	

From the comparison between laboratory test results and fatigue life models, one can conclude that there are some differences between these values. The fatigue life predicted by the Asphalt Institute model is the only one that can be compared with laboratory fatigue life.

CONCLUSIONS

This paper shown a study where three types of loading pattern were applied to fatigue bending tests and the laboratory results were compared with values predicted by design fatigue models.

The results indicate that sine and haversine loading pattern produce identical fatigue life for the same tensile strain level, while the sine loading pattern with 50% rest period increase 3 times the fatigue life.

Significant differences were found between the fatigue life prediction obtained by the different predictive methods.

ACKNOWLEDGEMENTS

The authors acknowledge and are thankful to the insight and support of Dr. Jorge B. Sousa of CONSULPAV, Portugal.

REFERENCES

1. The Asphalt Institute 1981. Thickness Design Manual (MS-1), 9th Edition, College Park, Maryland.
2. Azevedo, M.C. 1993. Características Mecânicas de Misturas Betuminosas para camadas de base de pavimentos (in portuguese), PhD Thesis, UTL, Lisboa.
3. Claessen, A.I.M., J.M. Edwards, P. Sommer, and P. Uge. (1977). Asphalt Pavement Design - The Shell Method, Proceedings, Fourth International Conference on the Structural Design of Asphalt Pavements, Vol. I, Ann Arbor, 39-74.
4. Di Benedetto, H, A. Ashayer Soltani & P. Chaverot 1996, Fatigue damage for bituminous mixtures: a pertinent approach, Journal of the Association of Asphalt Paving Technologists.
5. Pereira, P.A.A, J.C. Pais & J.B. Sousa. Comparison between laboratorial and field bituminous mixtures, Mechanical Tests for Bituminous Materials, Lyon, 1997.
6. Shell International Petroleum Company, Ltd 1978. Shell Pavement Design Manual, London.
7. Tayebali, A.A. 1994. Fatigue response of asphalt-aggregate mixtures. Report A404. SHRP A003A .

Fifth International Conference on the
BEARING CAPASITY OF
ROADS AND AIRFIELDS

THERMAL FATIGUE TESTING OF ASPHALT PAVEMENT OVERLAYS

A. SHALABY[1], A.O. ABD EL HALIM[2], AND S.M. EASA[3]

[1]Assistant Professor, Department of Civil and Geological Engineering, University of Manitoba, 15 Gillson St., Winnipeg, Manitoba, Canada R3T 5V6
Fax: +1(204) 474-7513, Email: a_shalaby@umanitoba.ca

[2]Professor, Department of Civil and Environmental Engineering Carleton University, 1125 Colonel By Dr., Ottawa, Ontario, Canada, K1S 5B6

[3]Professor and Head, Department of Civil Engineering, Lakehead University, 955 Oliver Rd, Thunder Bay, Ontario, Canada, P7B 5E1

ABSTRACT

The transverse cracking patterns of asphalt pavements suggest that there are two possible scenarios of failure. The first scenario, termed low-temperature cracking, considers that thermal strains incurred in the overlay have exceeded the failure strain. Due to a stress intensity at a material defect or cavity, a crack is initiated and propagated through the thickness of the overlay. The second scenario, termed thermal-fatigue, in which cyclic temperature variations cause a degradation of the material stiffness that also leads to transverse cracking. Based on the specific site temperature regime, one or a combination of both scenarios may occur. To extend the life of pavement overlays and reduce thermal cracking, research efforts have been heading towards two opposite extremes. The first is to reinforce the overlay, typically with a fabric or geogrid. The second is to precrack the overlay at short intervals and seal the cracks with a soft and flexible material that can tolerate high thermal strains. In this paper, an accelerated test method for thermal fatigue is proposed. Test specimens measuring 500 x 150 x 70 mm are saw-cut from field sections. The specimens are subjected to either a monotonic or cyclic tensile stress at a slow rate until failure. The test setup is enclosed in an environmental chamber, where the test temperature ranges from –20°C to +20°C. The cyclic loads are displacement-controlled and proportioned to produce inelastic strains in the tested specimen. Therefore, failure is incurred within the low-cycle fatigue Test results indicated that dissipated strain energy is a unique characteristic of the asphalt mixture. The results have shown that the test method is successful and suitable for examining thermal fatigue distress. The test method is also capable of handling pavement reinforcement materials.

Keywords: pavement cracking, thermal fatigue, asphalt testing, tensile stresses, cyclic loading

INTRODUCTION

Thermal cracking of asphalt pavements occurs in two basic modes. The first mode is associated with a severe temperature drop under restrained conditions, which causes thermal strains to exceed the failure tensile strain. The second mode is associated with cyclic temperature changes within the plastic zone causing irreversible damage and leading to the loosening of aggregate particles and eventually cracking after a relatively low number of cycles.

Thermal fatigue evolved to explain serious thermal cracks which were observed in pavements that do not normally experience low temperatures but are subjected to large thermal variations between maximum and minimum daily temperatures. There are two main approaches to thermal fatigue: the phenomenological approach which is based on the fatigue law, and the mechanistic approach which is based on crack propagation laws obtained from fracture mechanics. The research described in this paper is aimed at devising a laboratory test method to examine the behaviour of asphalt pavement under thermal cycling.

RATIONALE OF LOW CYCLE FATIGUE

Research on thermal fatigue was originally developed in the aerospace industry to study the behaviour of metallic alloys under high temperatures. The work has been extended to cover a variety of other applications including viscoelastic materials under low temperatures. Prediction of thermal fatigue lives is often based on the more established isothermal material behavior. It is generally recognized that the accuracy of thermal fatigue predictions is considerably poorer compared to the life predictions at isothermal conditions [Halford, 1987]. This problem is further augmented in low temperature conditions because of the relatively thicker and less conductive asphalt sections in comparison with thin metallic alloys.

The drawbacks of isothermal testing are that it cannot capture the differential thermal expansion between the various components and the change in strength and build-up (or decay) of strains during thermal cycling. A testing method for the isothermal cyclic loading is hereby proposed. The purpose of the test is to impose cyclic loads on asphalt samples within the elastic and inelastic ranges. The test focuses on the dissipated energy per cycle and the shape of individual loading cycles [Shalaby et al., 1996], [Shalaby et al., 1997a, 1997b]. The formation of thermal cracks requires a relatively low number of thermal cycles (less than 10000 cycles).

Data from year-round hourly temperature observations have shown that a pavement can be subjected to 120-200 cycles per year. These cycles have a maximum recorded temperature below +25°C, and have a temperature range larger than 12°C. Therefore, it is hypothesized that the failure will conform to the low-cycle fatigue mode. As such, large inelastic strains (in the order of elastic strains) lead to a form of premature failure.

It is interesting to note that, inconsistent with typical fatigue tests, it is more rational to perform the test at a constant speed. Under test conditions, the material response is largely affected by the rate of loading more than the frequency of loading. The latter varies with the amplitude of each cycle. The test method is termed Constant Speed Cyclic Test, CSCT. The test variables are the sample temperature, which is maintained constant during the test, the displacement amplitude, and the strain rate. The displacement amplitude and strain rate are arbitrarily selected such that a plastic deformations is incurred in the samples while it remains appreciably below the failure strain. The test setup consisted of a hydraulic actuator, a set of stationary and moving plates, and a data acquisition system connected to load cell and LVDT. The set up is shown in Figure 1.

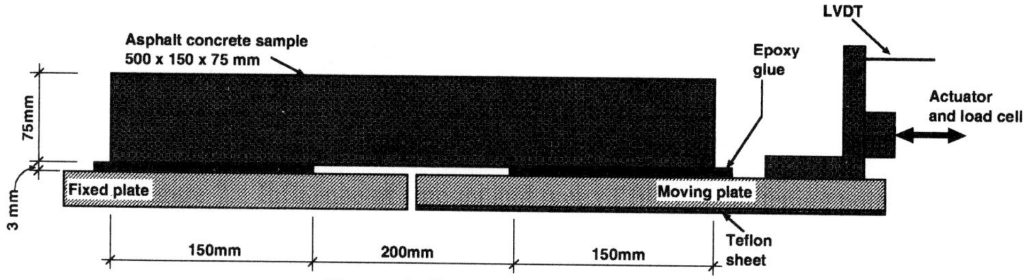

Figure 1: Sample setup for cyclic test

. MATERIALS AND TEST PREPARATIONS

Two asphalt mixes are used in this study. Both mixes are designed to meet HL3 aggregate gradation as specified by the Ontario Ministry of Transportation. The aggregate gradations are shown in Figure 2 with the control points and restricted zone recommended for Superpave mixes. It is noted that the gradation should avoid the restricted zone to improve rutting resistance by enforcing the S-shape gradation. This restriction has not been found relevant to thermal cracking. Aggregates and binders for both mixes are obtained from independent and unrelated sources. Large aggregates of both mixes were local HL3 stones. Two local paving contractors performed mixing and field compaction.

Table 1: Design material properties of Mix A and Mix B

Property	Mix A	Mix B
Asphalt cement content	5 %	5 %
Binder grade	85/100 (unmodified)	PG 58-34 (unmodified)
Air voids	4.3 %	4.1 %
Voids in mineral aggregates, VMA	16.1 %	15.2 %
Bulk relative density, BRD	2.436 gm/cm^3	2.393 gm/cm^3
Stability	13100 N	12262 N
Flow	8.9	9.2

A summary of mix properties is presented in Table 1. From the properties shown, it is reasonable to assume that both mixes have relatively stiff binders, and have similar aggregate type, and air voids ratio. The binder of Mix A is specified according to the older penetration-based method and that of Mix B is specified according to Superpave performance grades. The two mixes differ mainly in aggregate source, density, and gradations.

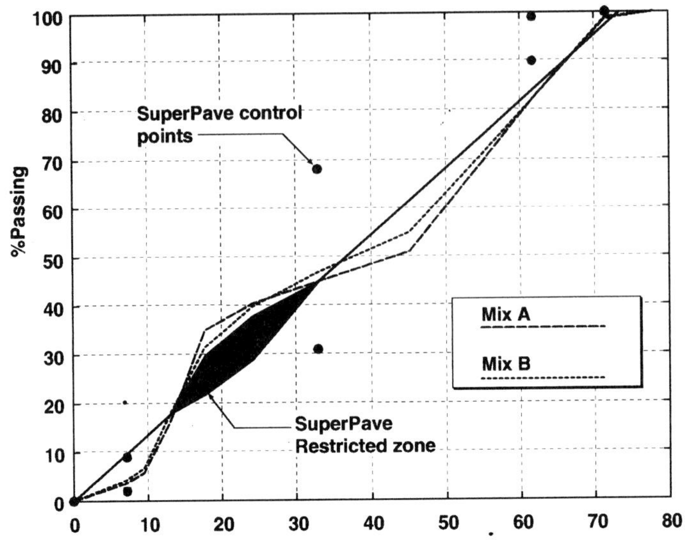

Figure 2: Aggregate gradation for Mixes A and B

The asphalt concrete mixes are laid into 1.2 x 2.4 m forms at the paving site and compacted with eight passes of a half-ton steel roller to achieve a pavement thickness of 75~mm. After seven days, each slab is saw-cut to manageable dimensions, properly marked, and transported to the laboratory. Slabs are then saw-cut to precise test sizes of 500 x 150 mm. Each test sample is glued to two 10mm steel plates at room temperature. The plates are aligned and positioned using a set of two removable side handles. A 24-hour epoxy-based adhesive is applied to bond the asphalt samples to the steel plates. The sample surface area in contact with the adhesive exceeded twice the area of the sample cross-section. Since the shear strength and tensile strength for asphalt concrete at low temperatures are approximately equal and the shear stresses never exceeded one half of the tensile stresses, the failure is dominated by tensile stresses. The test samples are kept in an environmental chamber at the desired test temperature for 24 hours prior to testing.

DAMAGE INDICES

A damage index is a reference to the degradation of a monitored material property. The selection of an appropriate property to monitor depends on the intended use of the information as well as the test method and set up. The following material properties and test characteristics are monitored to determine the propagation of damage in a given asphalt mix [Shalaby 1997]:

1. Maximum tensile and compressive stresses: the decrease in the maximum tensile stress during load cycling is a measure of the decay of the material. In cyclic tensile tests, the compressive stresses are indicative of plastic deformation and matrix rearrangement

during loading. The increase of compressive stresses is a measure of the resilience and hardening of the material.

2. Stress drop: in strain controlled tests, the difference between maximum and minimum stresses decreases with load cycling. Since the displacement is fixed, the drop in stresses reflects an equal drop in the secant stiffness modulus.

3. Strain energy: strain energy is an important parameter since it can capture both the elastic and plastic deformations. Results are compiled from several readings in each cycle, which is more meaningful and precise than selecting the highest or lowest reading.

4. Stiffness modulus: several stiffness moduli can be computed. The initial loading stiffness and the unloading stiffness of each cycle are the most significant. Due to the large applied strains, it is not recommended to use a single stiffness indicator since it will not reflect the inelastic behaviour.

5. Permanent deformation: the intercept of the unloading curve with the horizontal axis defines the permanent deformation on each cycle. This factor correlates to the amount of irrecoverable damage incurred to the test sample.

6. Number of cycles: the number of cycles to failure is not a good indicator since it can be affected by the various material and test conditions. In the absence of a clear failure criterion, the variability and repeatability of test results is questionable. In low cycle fatigue, it is not advisable to use this number as a performance measure since all tests will produce fairly consistent results within the same order of magnitude.

7. Hysteresis loop: the information gathered from the test includes the range of elastic deformation and the onset of plastic deformation, the tangential stiffness moduli, the comparison of loading and unloading paths, permanent deformation, and the shape of the loop

Comparison of indices is presented elsewhere, [Shalaby 1997]. The third index, strain energy, will be used in the following section to compare the performance of two asphalt mixes.

TEST RESULTS

The cyclic tests are terminated only after the sample shows significant (over 50%) deterioration. Each test is started with a damage-free sample. The analysis of results is performed using the mathematical software "Matlab". Typical test results depicting the hysteresis for both mixes at 0 °C and 1 % strain are shown in Figure 3. Special functions are developed to analyze the data, to obtain the maximum and minimum recorded stresses, and to compute the permanent deformation and the strain energy of each cycle. In addition, Matlab is also used for plotting the results.

The number of cycles required to induce particular damage levels according to the strain energy criterion is shown in Table 2. These results are based on the testing of 12 specimens. The tested samples are compared at two levels of damage, 25 % and 50 %. The comparison indicates that the second test group consisting of Mix B specimens has a lower extent of damage propagation due to cycling than the first test specimens of Mix A.

The results shown in Figure 3 and Table 2 suggest that Mix B will outperform Mix A under conditions of thermal fatigue similar to those applied during the accelerated testing. It should be noted that although each sample is referenced to its initial response in the first loading cycle, the comparison remains valid since the test specifically targets the relative damage propagation and not the absolute resistance to load or to cracking.

Table 2: Number of cycles required for damage propagation using strain energy data

Test Temperature	Number of loading cycles					
	0.25 % strain		0.50 % strain		1.00 % strain	
	Damage level*		Damage level*		Damage level*	
	25 %	50 %	25 %	50 %	25 %	50 %
Mix A						
0.0 °C	N/A	N/A	120	> 500	40	100
-20 °C	220	550	> 200	> 200	< 5	< 5
Mix B						
0.0 °C	N/A	N/A	> 500	> 500	60	110
-20 °C	> 2000	> 2000	> 200	> 200	< 5	< 5

*computed from drop of strain energy after first loading cycle

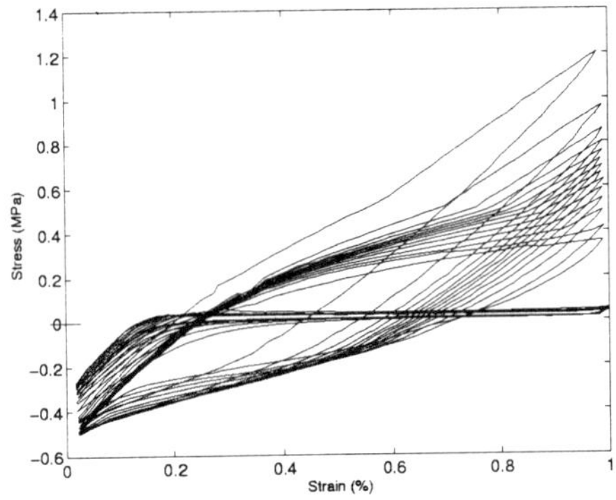

Mix A

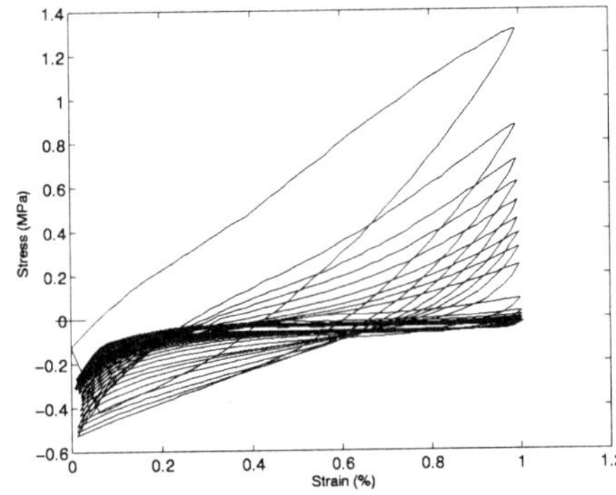

Mix B

Figure 3: Hysteresis loops depicting damage propagation at temperature of 0.0 ° C, strain rate of 0.1 mm/s, no rest period, and a strain amplitude of 1.0 %. A representative cycle is plotted for each 50-cycle interval.

CONCLUSIONS AND RECOMMENDATIONS

There are no specific test methods or design procedures to reduce thermal fatigue of asphalt mixes. While the CSCT method can provide indications of thermal fatigue potential as well as comparison of mix performance, it cannot be used to develop design guidelines or recommendations without significantly higher efforts being directed to this field.

In designing against thermal fatigue, it is necessary to test the material under repeated loads irrespective of the magnitude of stresses that are developed. Comparison across different mix types should be based on damage propagation rates, rather than maximum strength or limiting strains. The features of the CSCT testing method are:

1. It is more relevant to thermal loading of pavements including the two failure modes low-temperature cracking and thermal fatigue,

2. It is capable of modeling multi-layer pavement and reinforced pavement systems to study delamination and bond between layers,

3. The test outputs can be interpreted to evaluate damage, crack propagation and relevant mechanical properties,

4. It is designated to testing actual field extracted and relatively large samples,

5. The test can provide critical data on the aging of the mix as an input to mechanistic life prediction models, and

6. The TSRST is required to obtain the fracture temperature from a single cooling cycle and at a specified thermal gradient, say 10 ^{o} C/hour.

The CSCT approach can be used to construct a thermal fatigue database and develop prediction models based on thermal and material conditions in addition to crack type, length and location. Using the results of CSCT, it is possible to estimate the fatigue life under simulated field conditions, using one of several existing methods described by Ponniah and Kennepohl (1994). Combining results of high cycle and low cycle fatigue, relationships for the overall fatigue behaviour can be developed. The complex temperature cycling that occurs in field situations can be analyzed using small-interval components and probability density functions. The accumulated damage in the pavement structure can be estimated using Miner's hypothesis.

Mechanical properties of pavement materials are easily obtained at, or shortly after, the time of construction. Performance prediction models require information on material properties as a function of pavement life or number of loading cycles. The information is not readily available for a wide spectrum of mixes. In many cases, it is obtained indirectly from aged binders. In thermal cracking, the required properties are the tensile stiffness modulus and the tensile strength. The use of accelerated damage indices can provide a more precise estimate of stiffness and tensile strength at any number of cycles, and thus eliminate the need for aging models, from the viewpoint of thermal cracking only. Furthermore, this method can be applied to aged specimens to investigate the long-term effects of load cycling and aging.

In addition, this testing approach appears to be advantageous in testing of reinforced pavement sections for the following reasons:

1. Field extracted reinforced samples of relatively large size can be used; therefore, the size limitation of the TSRST test can be overcome.

2. Pavement reinforcement materials such as geogrids and are mobilized after considerable deformations and aggregate interlock, which can be simulated using the presented test set up.
3. Reinforcement materials often fail in bond with the pavement causing delamination or cracking. Therefore, repeated loading tests such as CSCT are suitable for testing the bond strength.

4. Load redistribution can cause stress concentrations around the highly stressed reinforcement strands. Repeated loading tests will highlight the damage propagation rates due to stress concentrations.

REFERENCES

HALFORD, G.R., (1987) "Low-cycle thermal fatigue" in Thermal Stresses II, (ed. Hetnarski, R.B.), Elsevier Science Publishers B.V., North Holland, The Netherlands, 329–428.

LEMAITRE, J., (1992). " A course on damage mechanics" Springer-Verlag, Berlin

PONNIAH, JOSEPH, and KENNEPOHL, GERHARD, (1994) "The economic benefits of sealing cracks in flexible pavements." Proceedings, Canadian Technical Asphalt Association, Vol. 39, pages 402–421.

PONNIAH, J., HAAS, R., and PHANG, W., (1987) "Thermally associated fatigue crack growth through asphalt overlays: an experimental investigation" Proceedings, Paving in Cold Areas/Workshop 3, Ottawa, Ontario, Canada, Vol. 1, 455–491.

SHALABY, A., (1997) "Analytical and experimental investigation of thermal cracking of asphalt pavements" doctoral thesis, Carleton University, Ottawa.

SHALABY, A., EASA, S.M., and ABD EL HALIM, A.O., (1997a) "Damage propagation in asphalt pavements at low temperatures" Annual Conference, Canadian Society for Civil Engineering, Sherbrooke, Quebec, Canada.

SHALABY, A., EASA, S.M., and ABD EL HALIM, A.O., (1997) "Accelerated test method for thermal fatigue of asphalt pavement overlays" Structural Faults and Repairs 97, Edinburgh, Scotland, Vol. 3, 151-158.

SHALABY, A., EASA, S.M., and ABD EL HALIM, A.O., (1996) "Effect of crack position on transverse cracking of asphalt pavements" Proceedings, Paving in cold areas (PICA-6) workshop, Niigata, Japan, Vol. 1, 297-313.

Fifth International Conference on the
BEARING CAPASITY OF
ROADS AND AIRFIELDS

STIFFNESS AND FATIGUE PROPERTIES OF AN ICELANDIC BITUMINOUS MIX

S. ERLINGSSON AND E. S. URBANCIC

Engineering Research Institute, University of Iceland, IS-107 Reykjavík, Iceland

ABSTRACT

A series of indirect tensile tests on a bituminous mix has been carried out. The purpose was to obtain stiffness values and fatigue properties for a commonly used mix in Iceland. Marshall specimen were used with aggregates consisting mostly of fresh igneous basalt with maximum aggregate size 16 mm. The grain size distribution was well graded with about 9 % fines content. The load applied in the tests was a symmetric haversine load pulse. For the stiffness testing, 30 – 50 load pulses were found necessary to precondition the specimens. Further is the resilient modulus found to be independent of stress level for low and moderate stress levels, but starts to decrease above a certain critical stress level. A simple relationship between the resilient modulus and temperature was established. In the fatigue tests, increase in accumulated deformation was very low in the beginning of the test but increased dramatically as a certain deformation level was exceeded. A fatigue life curve was determined for the mix.

Key words: Bituminous mix, mechanical properties, indirect tensile test, stiffness, fatigue.

INTRODUCTION

A series of indirect tensile tests on a bituminous mix has been carried out at the University of Iceland. The purpose was primarily to obtain stiffness values for a mix commonly used in Iceland and to determine a simple relationship between its resilient modulus and temperature. Additionally the fatigue properties of the mix were determined. In pavement design, stiffness and fatigue are very important properties. Stiffness quantifies the relationship between stress and strain in the mix and dictates the effectiveness of a layer to spread load. The fatigue strength of a bituminous material relates its ability to resist crack initiations and propagation under repetitive loading. The current test series represents the first steps in a newly intensified research field at the University of Iceland in making a total evaluation of mechanical properties of Icelandic bituminous mixes, with the aim to optimize the serviceability as well as the lifetime of commonly used surface courses.

A bituminous mixture is a complex non-linear viscoelastic material. Its response to external loading is therefore quite complex and highly temperature dependent (1). The stress condition in a bituminous road layer as a wheel load passes is aslo highly complex.

Sophisticated laboratory testing equipment is therefore needed to evaluate the mechanical properties of bituminous mixes in a realistic way (2). One commonly used method is the indirect tensile test, which has the great advantage of being a relatively simple test procedure. Furthermore, it is relatively easy to produce test specimen and both laboratory made specimens, such as Marshall specimens, and drilled cores from existing roads can be used (3).

TESTING EQUIPMENT

A Instron 8500 hydraulic testing system was used. It consists of a load frame, a digital control console, a hydraulic actuator and a computer. A data acquisition system was used to record load and displacement histories. A 10 kN electronic load cell was used and horizontal strain was measured with a high accuracy extensometer. The software package used, MAX, was written specifically for dynamic and fatigue testing by Instron. The temperature cabinet used, capable of maintaining a steady mean temperature within ± 1°C, was constructed at the University (4).

DESCRIPTION OF TEST SPECIMEN

Laboratory made specimens were fabricated in accordance with the Marshall method. They were stamped in a mold (102 mm in diameter, thickness about 60 mm) with 50 blows on each side. Mix samples for the specimens were taken from the production at the asphalt plant Höfði, Reykjavík. The aggregate consisted mostly of fresh igneous basalt with maximum aggregate size 16 mm. The grain size distribution was dense graded with about 9 % fines. The origin of the aggregates was from three quarries: aggregates 0 - 16 mm were crushed from the Seljadalur quarry, natural sand came from Núpar and filler from Leirvogur. A small amount of chalk was added to the mixture. The characteristics of the aggregates are given in Table 1.

Table 1. Aggregate characteristics.

Quarry	Ball mill value	Dorry abrasion	L.A. abrasion value	Petrographic analysis, 1. cl.	Crushing test value
Seljadalur	7.0	0.39	13.0	99%	27.5
Núpar	8.0	0.47	18.7	70-91%	29.7

The material characteristics of the bituminous mix are shown in Table 2.

Table 2. Characteristics of the bituminous mixture.

VMA	Specific gravity, G_s	Bitumen content	Bitumen penetration	Marshall stability	Marshall flow
16.0%	2.998	5.7%	89	10.5 kN	5.7 mm

INDIRECT TENSILE TESTS

In the indirect tensile test, a compressive load is applied across the vertical diameter of a cylindrical specimen. This results in a relatively uniform horizontal tensile stress along the vertical axial plane of the specimen and the resultant horizontal deformation can be measured during testing (5, 6, 7). The horizontal tensile stress resulting from the indirect tensile test can be expressed as (8):

$$\sigma_x = \frac{2P}{\pi t d}$$

where P is the applied repetitive load pulses, t is the specimen thickness and d is the specimen diameter. The resilient modulus, M_R of a bituminous mixture can be expressed in terms of the applied repetitive load pulses and the resilient horizontal deformation ΔH as (9):

$$M_R = \frac{P(\nu + 0.27)}{t\Delta H}$$

If Poisson's ratio is assumed to be equal to $\nu = 0.35$, then the maximum horizontal tensile strain ε_t in the middle of the specimen is:

$$\varepsilon_t = 2.1\frac{\Delta H}{d}$$

A well known (9) fatigue model for the relationship between the number of load repetitions to failue and the total horizontal tensile strain is given by the following power functions:

$$N_f = k_1\left(\frac{1}{\varepsilon_t}\right)^{k_2}$$

where N_f is the number of cycles to failure and k_1 and k_2 are regression constants. The tensile strain ε_t is either total or resilient strain, obtained during the loading period. However, the total and resilient strains are related, that is the total strain is the sum of the resilient strain and the permanent strain components, namely $\varepsilon_t^{tot} = \varepsilon_t^r + \varepsilon_t^{pl}$. Kim et al. (3) have concluded that fatigue life is better shown on plots using the resilient strain at the ordinate instead of the total strain.

TEST PROCEDURE

Before testing, specimen thickness was measured with 0.1 mm accuracy at three different locations. Two metal plates were glued onto the sides of the specimen at midheight opposite each other on the horizontal diameter. A deformation sensor was attached to the two metal plates for accurate deformation measurement. Compressive load was applied to the specimen through 13 mm wide metal strips attached to the loading surface. The inner surface of the strips was curved to match the surface curvature of the specimen (4).

Stiffness testing. A stress controlled test method was used where the applied load consisted of a series of symmetric haversine pulses, each followed by a rest period, during which a constant load corresponding to 10 % of the peak load was maintained. The haversine pulses had a risetime of 50 msec and the rest period lasted for 900 msec, resulting in a total cycle length of 1.0 sec. A haversine load pulse has been shown to give good agreement with measurements of pavement response due to traffic loading (10).

For preconditioning, 50 load cycles were applied in accordance with the ASTM D4123 standard, to ensure better contact between the specimen and the testing jig. In Figure 1 it may be seen how the resilient modulus varies as a function of the pulse number during the preconditioning part of the test. In most of the tests carried out, the modulus changed during the first 20 - 30 load pulses but became quite stable thereafter.

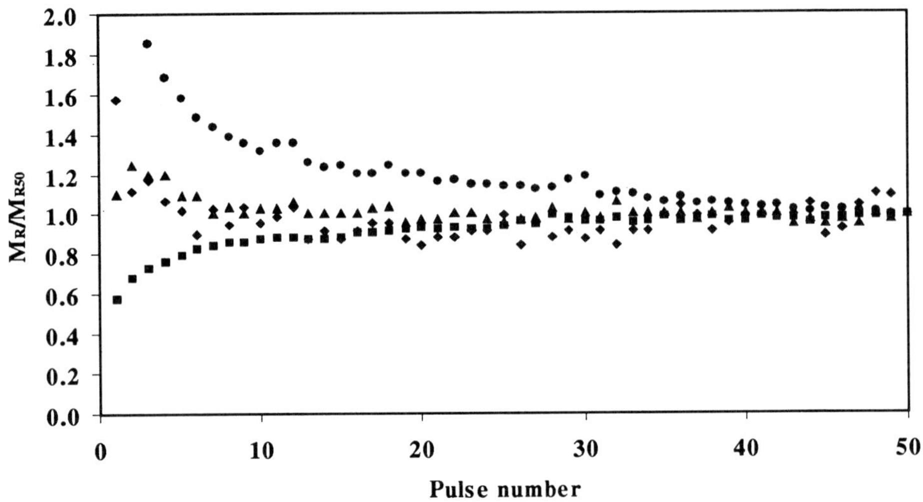

Fig. 1. Resilient modulus ratio as a function of pulse number during the preconditioning part of the test. Four different measuring sequences are shown. The measurements are normalized with the resilient modulus estimation at pulse number 50.

After preconditioning, 10 load cycles were applied for the actual stiffness estimation and the resilient modulus was determined as the average of these measurements. Figure 2 shows an example of the 10 applied load cycles used for the stiffness estimation and the resulting measured deformation.

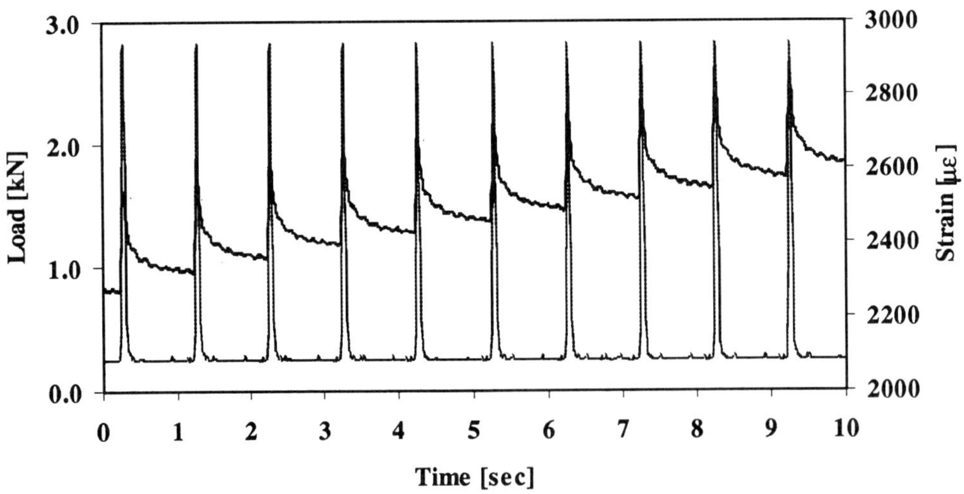

Fig. 2. Applied 10 load cycles used for stiffness estimation and the resultant strain.

Fatigue testing. The fatigue test was performed as a stress controlled test. The same load pulse was used as for the stiffness estimation, but the rest period was reduced to 400 msec and thus the total cycle length was reduced to 0.5 sec. For measuring the initial strain, 60 load cycles were first applied for preconditioning, thereafter 40 load cycles were applied where the initial total strain in each cycle was estimated (11). The initial strain was determined as the average of the resilient strain values obtained during these 40 load cycles. In all the measurements, the sustained load during the rest period was 10% of the applied load pulse. The test temperature was $T = 15°C$.

RESILIENT MODULUS TEST RESULTS

Influence of stress level on the resilient modulus. The effect of applied stress level on the stiffness modulus for the mixture was investigated. Four samples were tested at six different stress levels between 100 and 600 kPa at constant temperature, $T = 22°C$. At each stress level the specimen was tested twice; the second time after rotating the sample through 90 degrees. The results from these tests are shown in Figures 3 - 5.

In Figure 3 the resilient modulus estimation is plotted for each stress level. It can be seen that at low and moderate stress levels (σ_x = 100 - 400 kPa) the specimen stiffness is more or less independent of the applied stress level. However, at higher stress levels (σ_x > 400 kPa), the stiffness starts to decrease with increasing stress level.

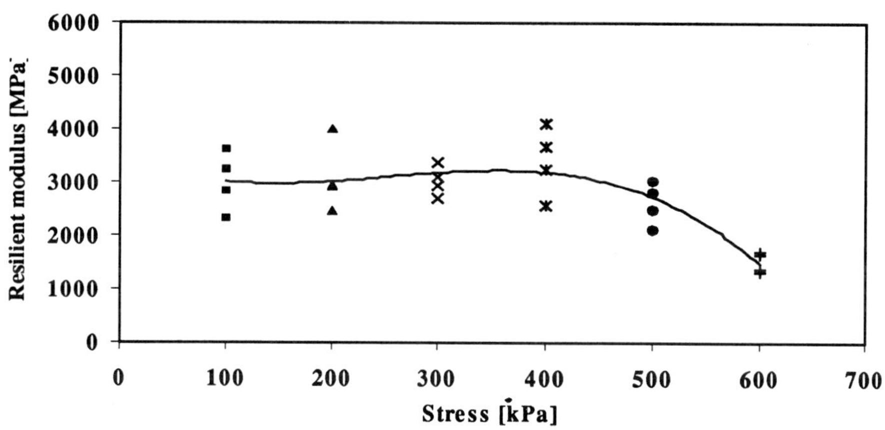

Fig. 3. Stress level influence on the resilient modulus at $T = 22°C$.

Measured horizontal tensile strain curves from one load cycle for each stress level are shown in Fig. 4. Clearly the strain increases with increasing stress level. The rate of strain increases as well with increasing stress level during the rise time of the load pulse, as it should, due to its viscoelastic nature, because the rate of the applied load is increasing. However, this does not cause an increase in the resilient modulus, see Fig. 3. Further it can be seen that during the load release as well as the rest time between the load pulses, the specimen shows a slow time dependent recovering deformation of a viscoelastic nature. This recoverable deformation is smaller than the deformation gained during the rise time of the load pulse and the difference is the permanent (plastic) deformation. Furthermore, as can be seen from the figure, the plastic deformation increases with increasing stress level.

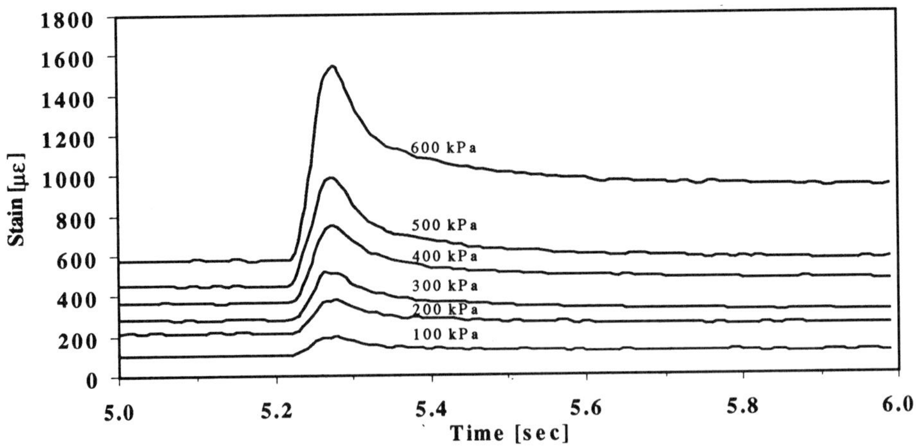

Fig. 4. Strain-time relationship for the six different stress levels. The actual values for the strain have been reduced by a constant to improve the visualization of the data.

To substantiate this further, a hysteresis loop has been plotted for all stress levels in Fig. 5. During the early loading portion of each hysteresis loops, the slopes are nearly parallel for low and moderate stress levels (σ_x< 500 kPa). Moreover these slopes have their counterparts in the unloading portion of each loop. This indicates that at low and moderate stress levels, the elastic effects are of primary importance during the early part of each path, gradually giving way to viscous and permanent deformations as the peak stress is approached. However, the curve at high stress levels (σ_x = 600 kPa) has a different slope during the early loading portion, indicating that viscous or even plastic behaviour are in ascendancy.

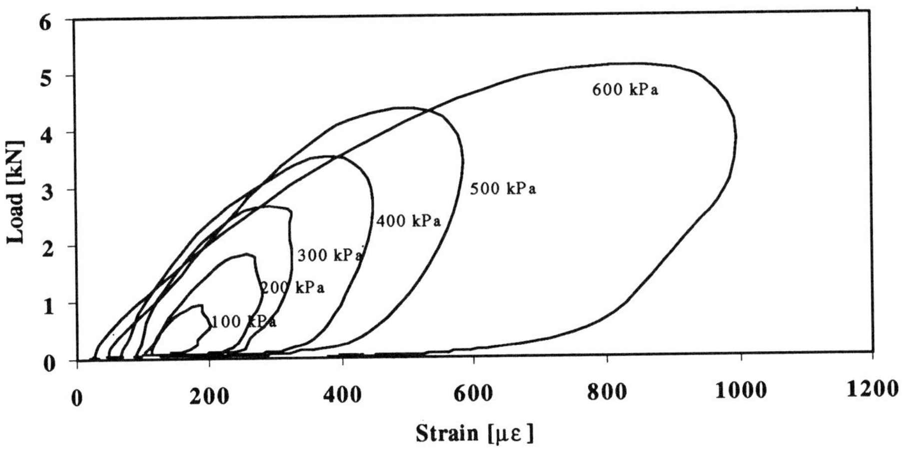

Fig. 5. Hysteresis loops showing the relationship between load and deformation during one load pulse for six different stress levels.

Resilient modulus versus temperature. To investigate the influence of temperature on the resilient modulus, deformation was measured at two peak loads. They were determined such as to give rise to a tension stress of 200 kPa and 300 kPa, respectively. Four samples were tested at both stress levels. The samples were tested at the following temperatures: 6, 10, 22, and 30°C, beginning with the lowest temperature and the smaller load. At each stress level and temperature the specimen was tested twice; the second time after rotation through 90 degrees. Typical response curves are shown in Figure 6.

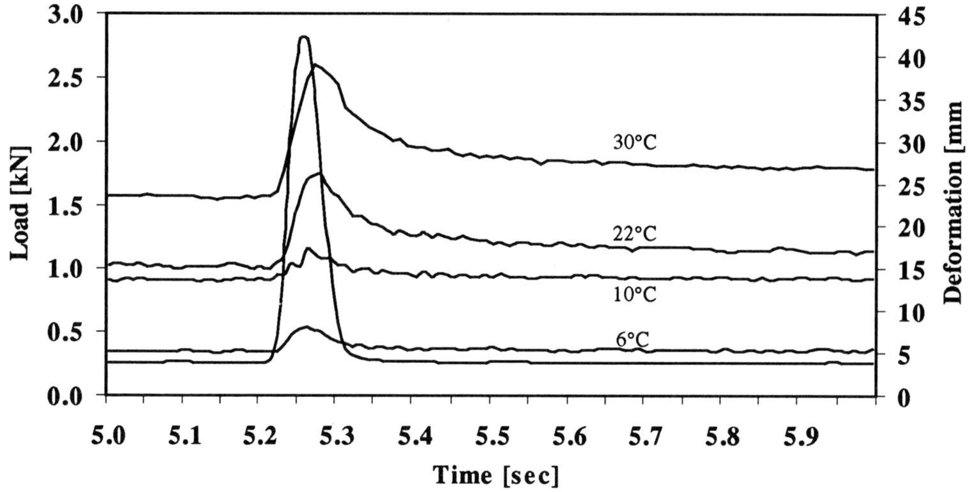

Fig. 6. A typical measured load pulse and horizontal deformation curves for four different temperatures.

Results from these resilient modulus estimations are shown in Fig. 7 along with 95% confidence limits for the mean value.

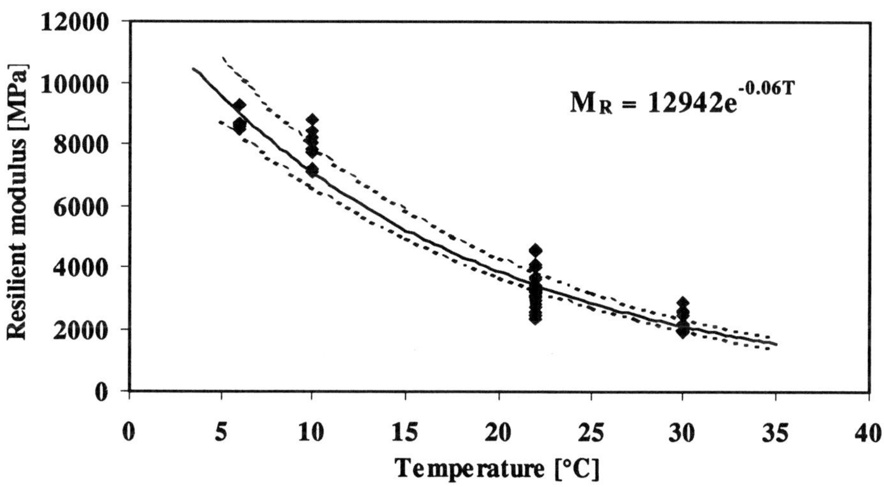

Fig. 7. Resilient modulus M_R as a function of temperature. The unbroken line is an exponential best fit curve and the dashed lines show the 95 % confidence limits for the mean value.

5. COOPER, K. E., BROWN, S. F., PRESTON, J. N. and AKEROYD, F. M. L.: Development of a Practical Method for Design of Hot-Mix Asphalt. Transportation Research Record no. 1317, pp. 42-51. National Academy Press, Washington D.C., 1991.

6. KHOSLA, N. P.: Effect of the Use of Modifiers on Performance of Asphaltic Pavements. Transportation Research Record no. 1317, pp. 10-22. National Academy Press, Washington D.C., 1991.

7. MAMLOUK, M. S.: Rational concept in relating lab testing to pavement analysis. 7th International Conference on Asphalt Pavements, pp. 124-140, Nottingham, 1992.

8. YODER, E. J. and WITCZAK, M. W.: Principles of pavement design, 2nd ed. John Wiley, New York, 1975.

9. SAID, S. F.: Tensile and fatigue properties of bituminous mixtures using the indirect tensile method. Ph.D. dissertation. Department of highway engineering, Royal Institute of Technology, Stockholm, 1989.

10. SAID, S. F. and CARLSSON, H.: Samband mellan fält- och laboratoriemoduler. (In Swedish) VTI särtryck nr. 235. VTI, Linköping, 1995.

11. ERLINGSSON, S. and URBANCIC, E. S.: Fatigue properties of an Icelandic bituminous mix estimated from indirect tensile tests. BUSL-report no. E-28, Reykjavík, 14 p., 1998.

Fifth International Conference on the
BEARING CAPASITY OF
ROADS AND AIRFIELDS

IMPROVEMENT OF PERMANENT DEFORMATION AND FATIGUE LIFE OF BITUMINOUS MIXTURES BY USING MODIFIED BITUMENS

PAULO A. A. PEREIRA[1], JORGE C. PAIS[2]
M. CONCEIÇÃO AZEVEDO[3]

[1]Associate Professor, University of Minho, Guimarães, Portugal,
Fax: +351-53-510-217, Email: ppereira@ci.uminho.pt
[2]Research Assistant, University of Minho, Guimarães, Portugal,
Fax: +351-53-510-217, Email: jpais@eng.uminho.pt
[3]Thecnical Director, APORBET, Portugal,
Fax: +351-1-793 1212, Email: aporbet@mail.telepac.pt

ABSTRACT

This paper describes the results of a study aiming the evaluation of pavement performance improvement by using a polymer modified bitumen as binder. Two dense graded bituminous mixtures were produced in laboratory. The first one was produced with conventional bitumen and the second one with polymer-modified bitumen.

Permanent deformation tests were carried out in a shear servo-hydraulic machine using the repetitive simple shear test at constant height. Fatigue life tests were carried out in an axial servo-hydraulic testing system using a flexural beam.

A comparison between these two mixtures was made for tensile strains calculated for a typical flexible pavement structure where these two mixtures were considered.

Key words: Bituminous Mixture, Permanent Deformation, Fatigue Life, Conventional Bitumen, Polymer Modified Bitumen.

INTRODUCTION

In order to increase the stiffness and the permanent deformation resistance of flexible pavements to high summer temperatures and to improve their resistance to crack initiation in winter conditions, the use of modified binders has begun in Portugal, namely in the North and Northest regions.

The purpose of this paper is to present information on the improvement of permanent deformation resistance and fatigue life of bituminous mixtures when using polymer modified bitumen (SBS).

In order to carriy out a comparison and to reveal the difference between a traditional mixture and a modified one, the mixtures to compare were subjected to the following tests:

- Repetitive simple shear test at constant height, according to SHRP program;
- Dynamic four point bending test.

ANALYSED MATERIALS

The bituminous mixtures of this study are a dense graded for a binder course with a maximum aggregate size of 25 mm according to the Portuguese standards. The aggregate used was a crushed granite with the grading curve shown in Table 1.

Table 1. Aggregate gradation

Sieve size (mm)	Percentage passing (%)
25.0	100
19.0	94
12.5	80.5
4.75	52.5
2.00	39
0.425	21.5
0.180	13.5
0.075	7.5

Two different bitumen types were used; a conventional 50/70-penetration grade bitumen and a polymer modified bitumen (SBS). The bitumen properties are presented in Table 2. The bitumen content was 5% by aggregate weight and the void content was 6%.

Table 2. Properties of bitumen

	Conventional	Modified (SBS)
Penetration, 25°C, 100 g, 5 seg., 0.1mm	62	61
Softening point, °C	53.1	58.0

SPECIMENS PREPARATION

The specimens were prepared in laboratory with the rolling wheel compaction method. So, the compaction was made with a lightweight vibratory steel roller in a steel coated wooden mould. The roller was repeatedly passed over the mixture placed in the mould. Rolling wheel compaction involves the use of weight and volume calculations to determine the mass of material to be compacted into a known volume to the desired void content.

After the mixture cooling, beams for the four point bending test and cylindrical specimens for shear tests were sawed.

Cylindrical test specimens with 15 cm diameter were cored from slabs compacted in laboratory for permanent deformation tests and prismatic specimens with 38.1 cm long by 6.25 cm wide by 5.0 cm high for fatigue tests.

It is interesting to note that the mixtures with polymer modified binder exhibited a better workability than the mixtures with unmodified binder.

TEST EQUIPMENT

The test equipment existing in th University of Minho, aiming the permanent deformation characterization, and used in this study, was a CS7400S shear servo-hydraulic machine, introduced during the SHRP program.

The machine testing capabilities and accommodation of add-on testing modules permits fatigue and stiffness testing as well as permanent deformation evaluation using axial and shear mode of loading (Pereira et al, 1997).

Specimens are glued to axial and shear plates such as represented in Figure 1.

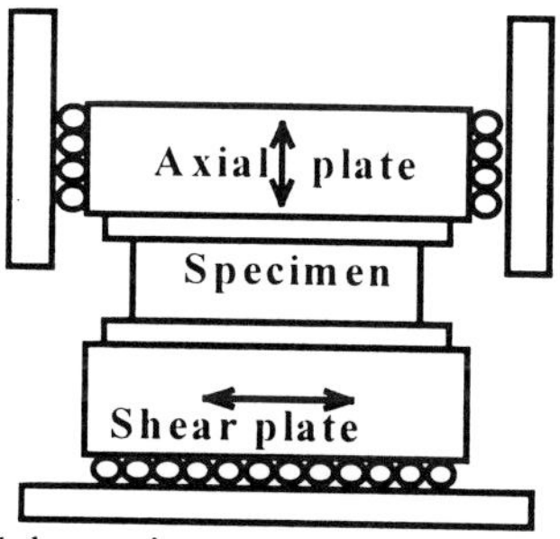

Figure 1. Schematic representation of shear apparatus.

Aiming the stiffness and fatigue life determination a servo-hydraulic equipment allowing four point bending test on beam specimens was used. The flexural beam device used to transfer the load to the specimen has a free translation and a free rotation at the reaction points and at load points (Pais et al, 1998).

Those equipments are controlled by a microcomputer using a software that provides a feedback closed-loop control to the servo-hydraulic system, test temperature and data acquisition. The flexural beam device, in association with the software capabilities, has considerably improved the reliability of results compared with earlier devices (Tayebali, 1994).

STIFFNESS AND FATIGUE PERFORMANCE

The four point bending tests were carried out under displacement control (strain control) at a 20°C temperature. For the stiffness evaluation, a cyclic load and frequency sweep, between 10 Hz and 0.1 Hz, were used.

For fatigue performance evaluation the same sinusoidal wave was used and the tests were performed for 3 strain levels and 10 Hz.

Figure 2 presents the stiffness of bituminous mixtures as a function of frequency, for the two types of mixtures analyzed.

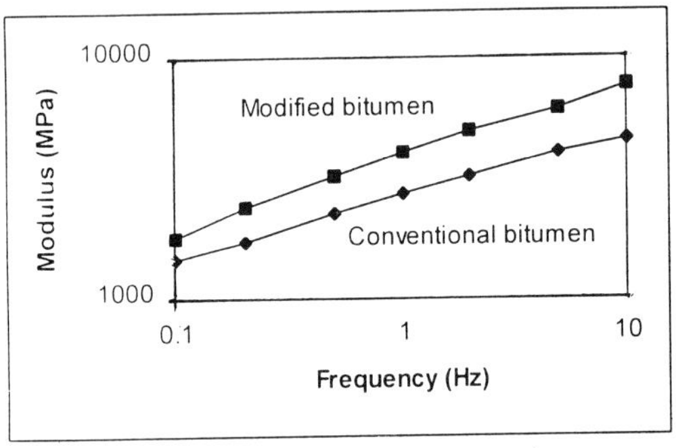

Figure 2 – Stiffness of bituminous mixtures as a function of frequency.

Figure 3 presents the fatigue life of the two bituminous mixtures.

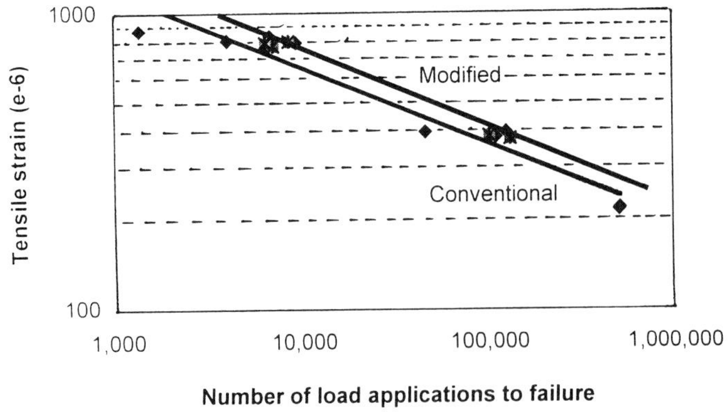

Figure 3 – Fatigue life laws for the two mixtures.

The comparative tests have explicitly shown the positive contribution of the SBS modified bitumen to improve the bituminous mixture behaviour, regarding fatigue life.

As a matter of fact, for an usual pavement tensile strain level of about 200 microstrain, the fatigue life of mixture with modified binder can be improved of about 49%.

On the other hand, a pavement design was performed, using a mechanistic approach, considering a typical pavement structure, within national conditions.

The pavement structure (whose response was calculated using a structural model where the materials were assumed as linear elastic) and the ALIZE computer program was set-up, assuming the materials properties presented in Table 3.

Table 3 – Layer modulus (E) and Poisson ratios (ν) adopted.

Surface course Bituminous mixture		Binder course Bituminous mixture		Base course Granular material		Sub-base course Granular material		Subgrade	
E1 (MPa)	ν1	E2 (MPa)	ν2	E3 (MPa)	ν3	E4 (MPa)	ν4	E5 (MPa)	ν5
3000	0.40	4000 a) 6000 b)	0.40	300	0.35	150	0.35	80	0.35
5 cm		12 cm		20 cm		20 cm			

a) Traditional mixture
b) Modified mixture

The stiffness of bituminous mixture for the binder course were the values obtained in this study (200 microstrain at 20 °C and 10 Hz), for the two types of mixtures analyzed.

The tensile strain calculated for a 80 kN load at the bottom of binder course was $185*10^{-6}$ and $147*10^{-6}$, for the traditional mixture (without modified binder) and modified mixture, respectively.

Using the laboratory fatigue life it can be observed that the number of load applications to failure is $1.4*10^{6}$ and $5.1*10^{6}$, for the traditional mixture and modified mixture, respectively.

Thus, the use of a polymer modified bitumen (SBS) can improve the bituminous fatigue life of about 2.6 times.

PERMANENT DEFORMATION AND FATIGUE CRACKING BEHAVIOUR

The repetitive simple shear tests at constant height were performed according to the SHRP program recommendations, as stated in AASHTO TP7 standard. These tests were undertaken at 60°C.

Figure 5 and 6 present the shear complex modulus and the phase angle as a function of frequency, for the two types of mixtures. Táble 4 shows the ESALs values obtained for the two types of mixtures.

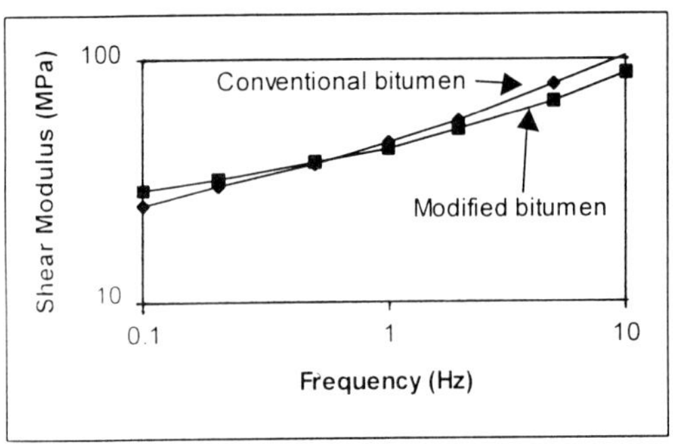

Figure 5 – Shear modulus as a function of frequency.

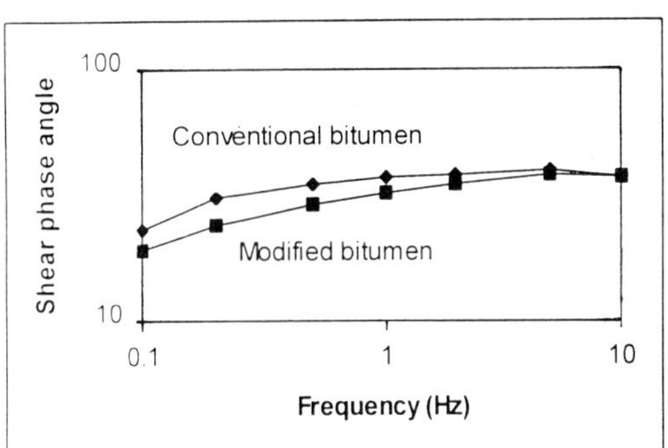

Figure 6 – Shear phase angle as a function of frequency.

Table 4 – ESALs obtained in shear tests

Specimen	Conventional ESALS to 12.5 mm	Modified ESALS to 12.5 mm
1	1 113 688	1 147 556
2	4 990 649	1 035 891
3	1 608 790	2 576 734
4	5 616 796	3 677 276
5		1 518 406
average	2 662 117	1 764 470

The permanent deformation behaviour of the two different types of mixtures have not revealed an expressive difference, such as the fatigue behaviour.

Nevertheless, the phase angle values of traditional bituminous mixtures are higher than for mixtures with modified binder, specially for the lower frequencies. As a matter of fact, for higher frequencies (2 to 10 Hz) the mean difference between the phase angle is about 5 to 10%, whereas the difference is about 20% for lower frequencies (0.1 to 0.5 Hz).

CONCLUSIONS

This study has compared fatigue response and permanent deformation behaviour of bituminous mixtures with conventional binder (50/70 pen) and polymer modified binder (SBS).
The use of elastomer modified binder was most expressively shown in the improvement of the fatigue behaviour. The service life of such materials can reach 2.6 times when compared with unmodified ones.

REFERENCES

1. Tayebali, *Stiffness of asphalt-aggregate mixes*, Report A388, SHRP A003A (1994).
2. Pereira, J.C. Pais and M.C. Azevedo, *Shear tests for bituminous mixture characterization*, (1997).
3. Pais, J.C., P. Pereira, J. Sousa, *The reflective cracking in thin surfacings*, First world Conference on Highway Surfacings, (1998).

Fifth International Conference on the
BEARING CAPASITY OF
ROADS AND AIRFIELDS

PERFORMANCE OF EMULSION-AGGREGATE MIXTURES DURING EARLY LIFE

H.E IBRAHIM[1] AND N.H THOM[2]

[1]Research Student, the University of Nottingham, NG7 2RD, England
[2]Lecturer, the University of Nottingham, England

ABSTRACT

Emulsion-aggregate mixtures are usually seen as one of the poorer bituminous materials, although much effort is being devoted to enhance the material to behave in a similar way to hot asphalt mixtures. Currently, there is no concensus for dealing with this type of slow curing mixture (up to 2 years). Some authorities characterize them mainly as asphalt mixtures. Others, on the other hand, are more cautious and treat the material as unbound. This paper describes a two phase study which has been carried out on the characterisation of mixtures during early life. The first phase was a study of the mechanical properties, the stiffness modulus of unsoaked and soaked specimens, measured in the Nottingham Asphalt Tester, both in indirect tensile and triaxial modes of testing, and the permanent deformation of both confined and unconfined specimens. The results showed high stress dependency, similar to that of unbound materials. Mixtures have generally shown high permanent deformation relative to comparable hot asphalt mixtures, but within acceptable levels. The second phase of the study was of performance in a pilot scale wheel tracking facility, for verification of the results from the first phase. Slabs of emulsion mixtures were laid-down on a base layer (sand or crushed stone) and/or a sheet of rubber. Measurements taken were surface rutting, vertical stress at the top of the base layer and tensile strain at the bottom of the emulsion mixture layer, by means of installed pressure cells and embedment strain gauges. The results showed the importance of the material non-linearity in stress and strain calculations and in back-calculation of moduli. In addition, fatigue cracking was found to be less significant than rutting.

Key Words: bitumen emulsion, emulsion mixtures, performance, slab testing.

INTRODUCTION

With modern technology for the production of bitumen-emulsion, industry is attempting to produce emulsion-aggregate mixtures with properties equivalent to those of hot bituminous mixtures. The material is generally characterised for stiffness in the indirect tensile mode of testing. However, many trials constructed to investigate material performance have shown a low curing rate and an equilibrium water content reported to be >1% (e.g. Leech[1], Robinson et al[2],). The first phase of this study (reported elsewhere[3]) was a laboratory work on specimens to investigate the mechanical properties of the material relative to that of hot mixtures. Measurements were taken in both indirect tensile and triaxial modes, in the Nottingham Asphalt Tester, NAT (Cooper and Brown[4]). The results highlighted 1) the stress dependency of the material in relation to curing time, emulsion type, residual bitumen content, and aggregate gradation, 2) poor fatigue resistance measured in the indirect tensile fatigue test, 3) higher permanent deformation compared to corresponding hot bituminous mixtures, dominated by the material response at early life. The behaviour and characteristics of the material are therefore different from those of hot mixtures. The objective of the work presented herein was to investigate the material failure characteristics. Therefore, performance has been studied in a pilot scale wheel tracking facility, in which a prototype representing the pavement layers under consideration is subjected to a simulated vehicle loading under controlled condition. Slabs of emulsion mixtures were laid down on a base layer (sand or crushed stone) and/or a sheet of rubber. Measurements taken were surface rutting, vertical stress at the top of the base layer and tensile strain at the bottom of the emulsion mixture layer, by means of installed pressure cells and embedment strain gauges. This test arrangement is considered suitable for assessing the characteristics of the material under a moving wheel and for design validation purposes. The effect of wheel load, mixture density, residual bitumen content and curing time are discussed. For validation of laboratory test results from the NAT in indirect tensile and triaxial modes of loading, comparisons are presented with back-calculated moduli using the slab measurements. In addition, possible failure mechanisms associated with this material are highlighted.

EXPERIMENTATION

Emulsion aggregate mixtures containing 4 % and 5 % residual bitumen content of cationic emulsion based on 62% of 100 pen bitumen and aggregate of 20 mm DBM grading (British Standard BS 4987: part 2) were used.

Testing used the NAT in the indirect tensile and, principally, the triaxial modes of loading. Figure 1 is a diagram of the triaxial test assembly. For triaxial testing, cylindrical specimens (101mm diameter × 160mm height) were fabricated in the gyratory compactor developed at the University of Nottingham. To allow the mounting of LVDTs on the specimens for vertical and horizontal deformation measurements, studs were inserted in the specimens during the compaction process through holes in the split moulds, and specially made screws were used. The fabricated cylindrical specimens were side wrapped with cling film after one day in the mould and left to cure at 20°C until the time of testing. The prepared cylindrical specimen was placed in a rubber membrane, which was attached to top and bottom steel platens by means of O-rings. The confining pressure was achieved by applying a vacuum inside the membrane through the top platen. During the test, the half height time of the repeated stress pulses was kept at 120 ms.

The equipment used for slab testing is known as the 'Slab Test Facility', STF (Figure 2), which applies a wheel load at a tyre pressure of 330 kPa. The facility may be considered to be approximately 1/3 scale. Two types of slab arrangement were planned. The first was of slabs laid on sand or crushed stone, with a rubber sheet underneath (see Figure 2). Measurements were taken of the horizontal strains in the surface layer and the vertical stresses on the base layer as well as surface rutting. The other test type was of slabs resting directly on the rubber sheet. In this arrangement, wheel load was controlled to a level leading to horizontal strains similar to that for slabs resting on unbound and rubber sheet layers. Measurements were of surface rutting and horizontal strains in the emulsion mixture layer. Table 1 presents a description of the slabs used in this study.

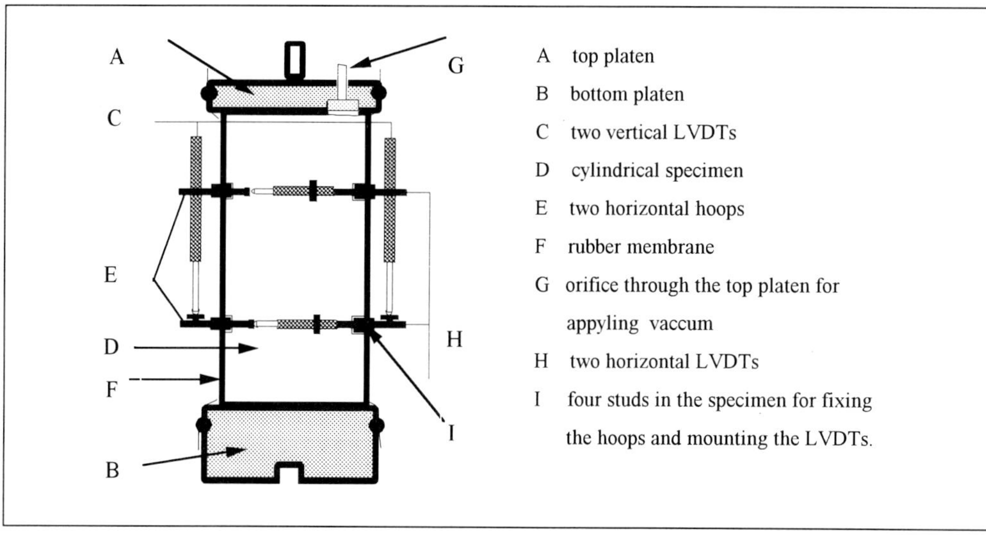

Figure 1 Schematic diagram of the NAT testing assembly in triaxial mode

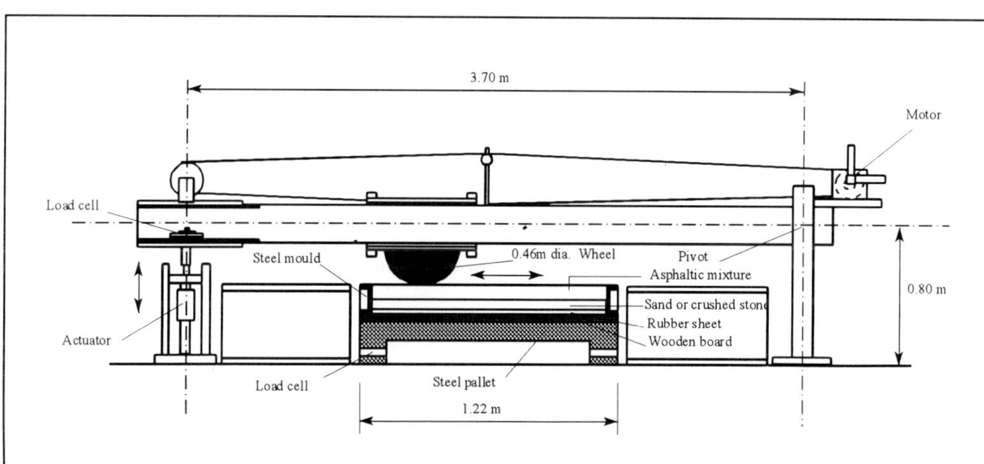

Figure 2 Slab test arrangement for testing in a pilot scale test

Table 1 Slab arrangements

Slab	Layers: Hot mixture	Layers: Emulsion mixture	Layers: Supporting layers	Dry density of emulsion mixture, g/cm^3	Curing at room temperature
S1	-	4 % RBC, 108 mm thick.	sand, 70 mm	2.21	10 days
S2	-	4 % RBC, 107 mm thick.	crushed limestone, 80 mm and rubber sheet 5.5 mm thick.	2.285	28 days
S3	-	5 % RBC, 105 mm thick.	crushed limestone, 80 mm and rubber sheet 5.5 mm	2.29	30 days
S4	(DBM-14 mm), 30 mm thick.	4 % RBC, 100 mm thick.	Sand 70 mm, rubber 5.5 mm	2.318	22 days
S5	-	5 % RBC, 101 mm thick.	Sand 70 mm, rubber 5.5 mm	2.225	34 days
S6	-	5 % RBC, 97 mm thick.	rubber 5.5 mm	2.34	37 days

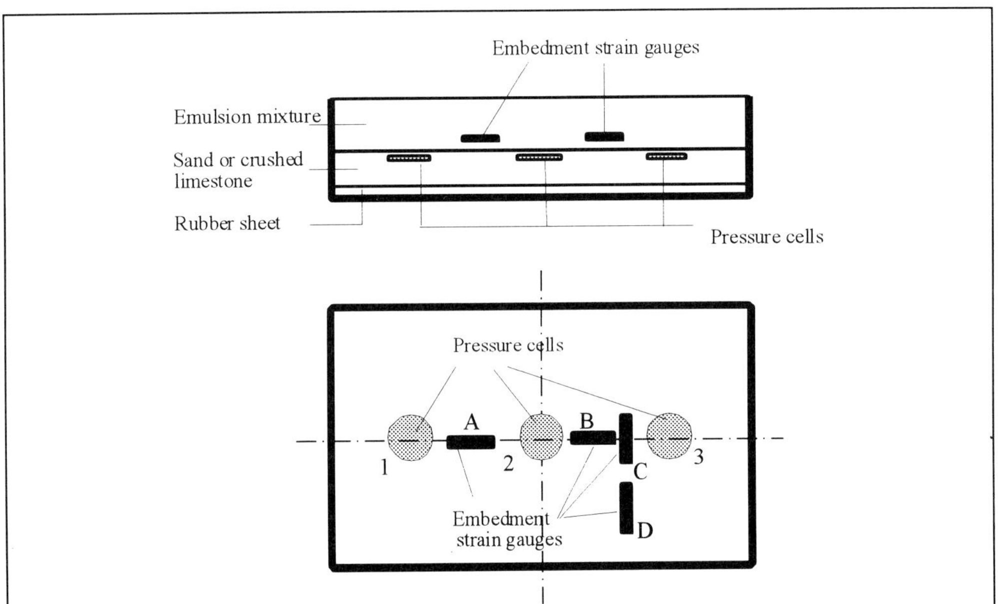

Figure 3 Location and orientation of pressure cells and embedment strain gauges

For measuring the vertical stresses on the top of the underlying layer and the horizontal strains at the bottom of the emulsion mixture layer, pressure cells and embedment strain gauges were used. Instrumentation was installed after laying and compacting each layer. Three cells were installed along the wheel path, 300 mm apart for a repeatability assessment. Embedment strain gauges were found to be a convenient tool for the measurement of the horizontal strains at the bottom of the emulsion mixture layer. Strain gauges with a gauge length of 60 mm were used. For better adhesion with the surrounding mixture, consideration was given to the surface roughness of the gauges. The location and orientation of the pressure cells and embedment strain gauges are shown in Figure 3.

TEST RESULTS AND DISCUSSION

Stress and Strain Measurements Typical strain measurements are presented in Figure 4. Generally, the transverse tensile strains were higher than the longitudinal strains for all tested slabs, in line with that reported by Huhtala et al[5].

The measured horizontal strains of various test slabs are presented in Table 2. The strains were almost constant over the test period, indicating no sign of local distress. However, longitudinal surface cracking occurred, accompanied by rutting. Dry sawing Slabs S1 to S3 for visual inspection showed cracks apparent over the top quarter height of the slabs, next to the wheel path. Generally, emulsion mixture layers seemed to crack due to shear, either because the tested slab was in direct contact with the wheel tyres and shear resulted from single tracking, or the applied wheel load was high.

Accordingly, two tests were carried out. The first (slab S4) was on a slab comprising similar layers (rubber sheet, sand, emulsion mixture) and covered by a hot asphaltic mixture, as described in Table 1. Longitudinal surface cracking, with rutting, and steady horizontal

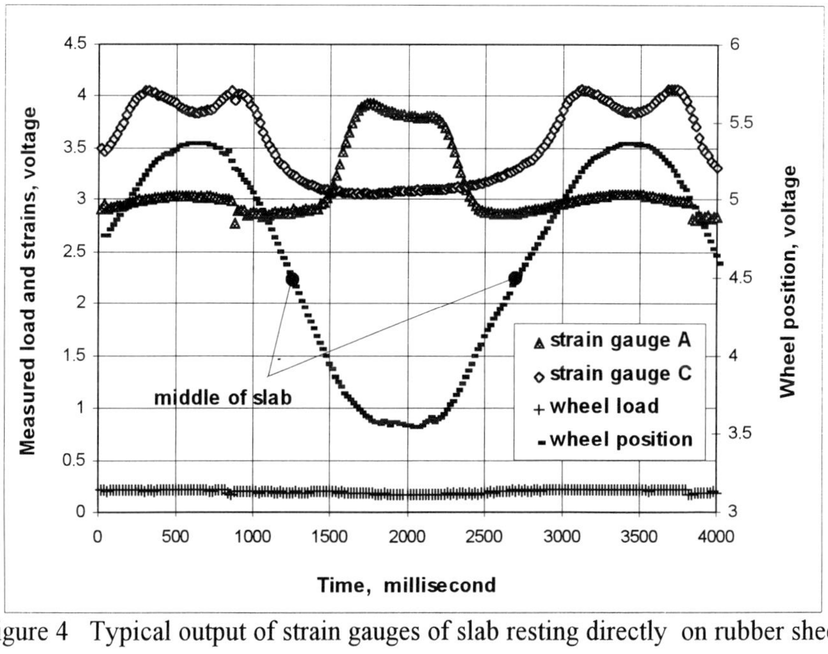

Figure 4 Typical output of strain gauges of slab resting directly on rubber sheet

Table 2 Test measurements.

Slab	Test temperature °C	Applied wheel load kN	Transverse horizontal strain microstrain	Vertical stress kPa
S1	21	1.5	186	128
S2	20	3.5	96	-
S3	23	3.2	128	85
S4	21	3.3	136	54
S5	27	1.0	97	-
		2.1	143	-
		3.0	215	-
		4.0	296	-
S6	20	2.15	227	-

strains over the test period resulted. However, on sawing the slab it was revealed that permanent deformation in the emulsion mixture layer was the key failure mechanism. A further test (slab S5) was carried out by applying different wheel load levels, 'multi-step loading', on a similar slab arrangement. A wheel load of 1 kN was applied for 35000 passes, then increased to 2.1 kN, as measured by the load cell, for 300 passes, 3 kN for 300 passes, and 4 kN for approximatelly 13000 passes, finally decreased to 3 kN until completion of the test, during which the horizontal strains, vertical stresses, and surface rutting were monitored. The measured strains were again constant for each applied load level. Rutting was moderate when a lower load level (up to 2 kN) was applied, but severe under the 4 kN wheel load.

It is thought that surface settlement due to material softening under the wheel load is a reason for the observed surface cracks. If damage occurred in the material due to fatigue, both the measured horizontal strains and the vertical stress would have changed as the test progressed. Re-applying a lower wheel load to the slab after testing at a higher load did not show any variation in the measurements.

However, a test on a slab of emulsion mixture resting directly on a rubber sheet (slab S6) was also carried out. It was aimed to apply a lower wheel load, resulting in a horizontal tensile strain level higher or at least similar to that of the slab resting on a base layer, when appyling 4 kN. A wheel load of 2.15 kN was used, resulting in a tensile strain of approximately 227 microstrain. The test in this arrangement is a wheel tracking fatigue test, as well as inducing rutting. Therefore, the tested slab was visually investigated for surface cracking, strain measurements were taken using transverse strain gauges and the number of wheel passes was recorded. The recorded number of wheel passes at commencement of cracking was approximately 146000. Although the resulting tensile strain at the bottom of the emulsion mixture layer due to the 2.15 kN wheel load was higher than in slab S5, the number of wheel load application causing cracking was also much higher. On the other hand, the resulting surface rutting was much less than that when high wheel load was applied.

It can be concluded that the cracking associated with applying a wheel load of 4 kN was not a fatigue phenomenon. Nor did it result from structural permanent deformation. It was a case of deformation within the layer surface due to material softening, which created high tensile stresses at the surface, next to the wheel path.

Comparison of Moduli The need for determination of a suitable method for characterizing the modulus of the material is paramount. For that purpose, the elastic modulus response of the material was studied in the wheel tracking pilot scale tests. Moduli of slab S3, described in Table 1, were back-calculated using both FENLAP (Finite Element Non-linear Analysis of Pavements) written in FORTRAN 77 language by Almeida[6], at the University of Nottingham and MPTRN (a multi-layer linear elastic program), and compared with those determined using the NAT in the indirect tensile and the triaxial modes of loading. Table 3 presents an example of the calculated values.

Assuming a linear response for the material, the calculated stiffness modulus which leads to a similar vertical stress at the top of the base to that measured was approximately 500 MPa, from which the calculated horizontal strain was 142 microstrain, based on the finite element program FENLAP. Assuming a non-linear response, the best estimates of parameters for the K-θ model (Hicks and Monismith[7]) were found to be k1= 9e4 and k2= 0.4 or k1= 6e4 and k2=0.5 (see Table 3). The measured and calculated values of stresses and strains using the finite element program are approximately similar. On the other hand, none of the calculated values using MPTRN is close to the measured values. Based on the linear program MPTRN, the calculated strains were higher than those measured; conversly, the calculated stesses were lower. Clearly, the finite element program is more suited for use in analysis of pavements incorporating an emulsion mixture layer, and more accurate than the linear elastic programs. Also, assuming the response of the material to be non-linear leads to more accurate values of stresses and strains.

Table 3 Calculated and measured stresses and strains in slab S3

mode of response		linear				non-linear k1/k2				
assumed stiffness		300	500	700	900	9e4 / 0.4	9e4 / 0.5	6e4 / 0.5	6e4 / 0.4	4e4 / 0.6
calculated stress	MPTRN	85.4	70.7	61.3	54.5					
	FENLAP	101	87.5	78.6	71.8	84	54	83	97	114
calculated strain	MPTRN	181	150	129	113					
	FENLAP	180	142	118	102	139	106	121	166	81
measured stress = 85 kPa, measured strain = 128 microstrain										
stiffness (ITSM)		approximately 1200 MPa								
stiffness (Triaxial)		k1 = 90000 and k2 = 0.4 after 30 days curing sided wrapped at 20°C								

As previously mentioned, an agreed test method for determination of emulsion mixture stiffness does not yet exist. Currently, it is characterised in the indirect tensile mode of test, following the procedure for hot asphaltic mixtures. Unfortunately, the behaviour of emulsion mixtures and hot asphaltic mixtures are completely different. However, calculated values of stiffness based on the pilot scale wheel tracking test measurements presented in Table 3 show proximity to the corresponding values from the triaxial mode.

Rutting For investigating rutting mechanisms in emulsion mixtures, transverse measurements of the surface levels of the above slabs were taken before running the test and at different numbers of wheel passes. The measured rutting profiles were then compared with permanent deformation measurements from the repeated load triaxial tests. A typical measurement of surface profile is presented in Figure 5.

Typically, rut profiles show an increase in slab surface level at the rut edges, indicating rutting occurring within the emulsion mixture layer. For slab S5, at a wheel load level of 1 kN, the maximum rut depth was 1.8 mm after applying 18230 passes. Generally, when applying a low level of wheel load, up to 2 kN, measured rutting depth indicated lower strains in comparison with those measured in the laboratory at similar stress levels on small size specimens. But, increasing the wheel load to 4 kN resulted in severe rutting. Measured rut depths were approximately 7 mm for slab S5 and 12 mm for slab S2. The measured ruts from slab S2 and the permanent deformation measured in the triaxial mode at a similar stress level can be compared from Figures 5 and 6. A steep strain rate was always associated with tests at applied load levels of 3 to 4 kN, and the measured rut depth is not in line with predictions from small size specimens. The thought was that severe rutting could be a result of severe settlement in the underlying layer, since formation of shoulders only occured at the rut edges, or rather only a small increase in the slab level was observed.

Of course, as a consequence of settlement in the supporting layer, the overlaid layer will bend and an increase in the horizontal strains occurs. As long as no change in the horizontal strain gauges was recorded, structural deformation was not occuring. Nevertheless, the slabs were sawn laterally after testing, for visual inspection of the crack formation, location, and orientation. In addition, thickness and levels of the layers were measured under and off the wheel path. Having investigated the sawn slabs, the following were observed:

- Cracks were visible at the edges of ruts, penetrating from the surface to a quarter depth of the slabs.
- No crack was observed at the bottom of the emulsion mixtures.
- Thickness reduction was only evident within the emulsion mixture layers.

It is therefore considered that the severe rutting in emulsion mixtures was due to softening in the material under the high applied wheel load which may be related to a lack of bond between aggregate particles. This result was confirmed by testing a slab resting directly on a thin rubber sheet (slab S6).

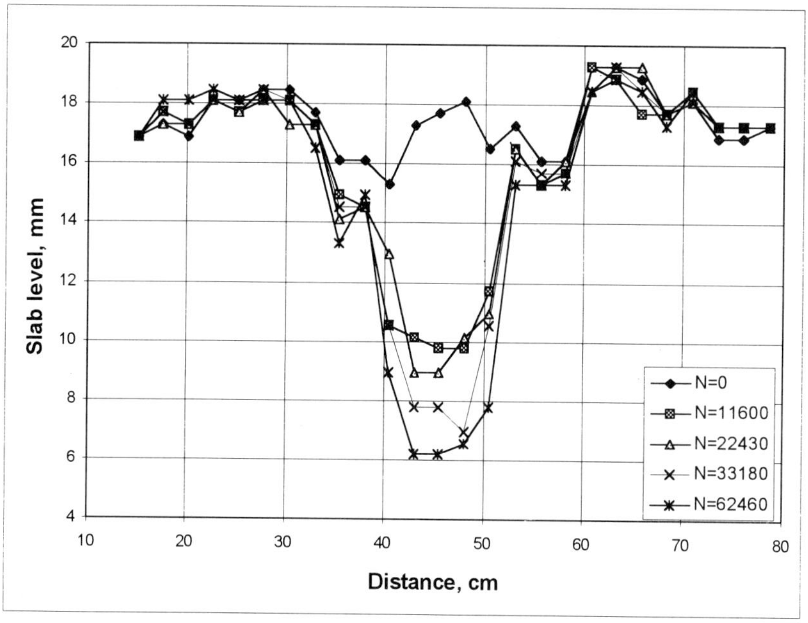

Figure 5 Rut profiles of wheel tracking in the STF on slab S2

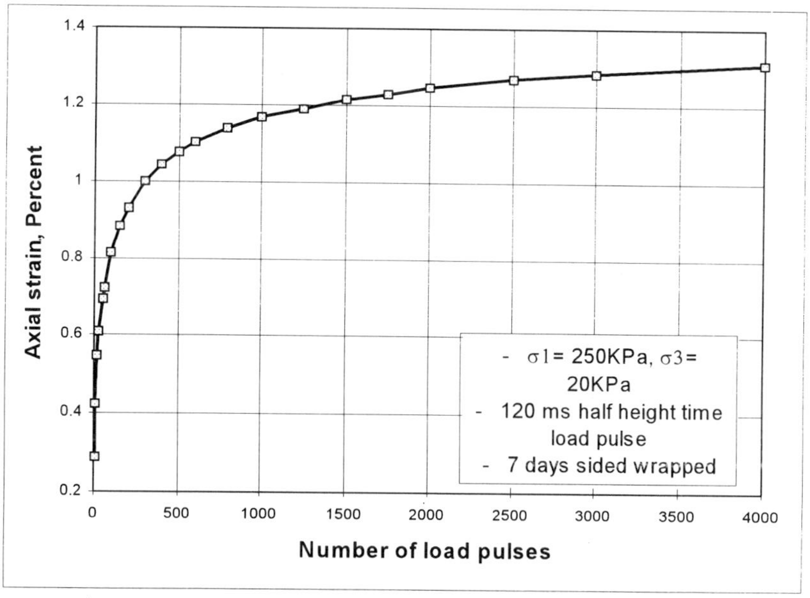

Figure 6 Permanent axial strain of specimens, containing 12.1 % void content, in the triaxial mode of the NAT.

CONCULSIONS

The level of wheel load is a significant factor affecting performance of this material. Excessive load might lead to deformation in the surface of the emulsion mixture layer which is considered to be due to a loss of cohesion, leading to less tensile stress resistance and occurrence of cracks at the layer surface. Two cases associated with emulsion mixtures can be distinguished. When applying a low wheel load, there is little rutting and reasonable fatigue resistance but the failure is prone to be in fatigue rather than in rutting . When applying high wheel load, excessive rutting is expected and the failure will be in rutting rather than in fatigue. It is apparent that emulsion mixtures behave in a similar way to granular materials rather than hot asphaltic mixtures, even after significant curing.

REFERENCES

1. LEECH, D. Cold-mix bituminous materials for use in the structural layers of roads. Transport Research Laboratory, Crowthorne, Project Report 75, 1994.

2. ROBINSON H., FORDYCE D., AND KHWEIR, R. Road maintenance using cold mix materials with hot mix properties. Eurasphalt & Eurobitume Congress, Strasbourg, 1996.

3. IBRAHIM, H.E. AND NICK, N.H. The effect of emulsion-aggregate mixture stiffness on both mixture and pavement design. Proceedings of the Second European Symposium on Performance and Durability of Bituminous Materials, Leeds, 1997.

4. COOPER, K.E. AND BROWN, S.F. Develoment of a simple apparatus for the measurement of the mechanical properties of asphalt mixes. Proceedings of the Fourth Eurobitume Symposium, Madrid, 1989.

5. HUHTALA, M., ALKIO, R., PIHLJAMAKI, J., PIENMAKI, M. AND HALONAN, P. Behavior of bituminous materials under moving wheel loads. Journal of the Association of Asphalt Paving Technologists, volume 59, pp. 422- 442, 1990.

6. ALMEIDA, J. Analytical techniques for the structural evaluation of pavements. Ph. D thesis, 1993.

7. HICKS, R.G. AND MONISMITH, C.L. Factors influencing the resilient response of granular materials. Highway Research Record, No. 345, pp. 15-31, 1971.

Fifth International Conference on the
BEARING CAPASITY OF
ROADS AND AIRFIELDS

PERFORMANCE RELATED TESTING WITH THE ASPHALT PAVEMENT ANALYZER

RONALD COLLINS[1] AND CYNTHIA LYNN[2]

[1]President, Pavement Technology Inc., Conyers, Georgia, USA
[2]Director of Materials, Pavement Technology Inc., Conyers, Georgia, USA

ABSTRACT

Permanent deformation, commonly called rutting, is a major pavement distress causing extensive damage to hot mix asphalt (HMA) every year. Heavy truck traffic, increased wheel loads, and the use of high pressure radial tires have increased the distress in recent years. After rutting has occurred on the roadway, it is expensive to correct. The laboratory prediction of rutting potential in HMA has been one of the greatest challenges facing pavement engineers.

The Asphalt Pavement Analyzer (APA) is a multi-functional loaded wheel tester that has been successfully used for evaluating permanent deformation, fatigue cracking, and moisture susceptibility of both hot and cold asphalt mixes. It features fully automatic, continuous measurement of rut profile under the wheel load. It can also test the asphalt specimens in dry conditions or submerged in water. The APA has the ability to develop the desired contact pressure, temperature, and moisture conditions during testing, and is capable of testing both rectangular and cylindrical shaped samples. Any number of compaction methods including gyratory, vibratory, kneading and Marshall may be used to prepare the samples. Roadway slabs and cores can also be tested.

The Georgia Department of Transportation began using a loaded wheel test in the mid eighties to predict rutting potential of mix. The APA is now required as proof testing for HMA placed on state highways. Several studies have proven the APA capable of recognizing mixes that are likely to fail under traffic.

INTRODUCTION

Approximately 93% of all the road surfaces in the United States are paved with hot mix asphalt (HMA). HMA combines bituminous binder and aggregate to give a pavement structure that is flexible over a wide range of climatic conditions. The ability of HMA to be produced from a wide variety of local aggregates and yet perform on a consistent basis makes it the pavement of choice throughout the United States and the rest of the world.

The vehicle miles traveled in America have increased approximately 75% in the past 20 years. The changing demographics in American society have also lead to many rural roads becoming high traffic roads as the population moves from urban to more rural areas. With 176 million automobiles on American roads today, there are as many cars as there are drivers. In the last decade, loads on our nation's highways have increased more than 60%. In addition to the increased loads, the increased distress due to the use of radial tires and higher tire pressures make it easy to see why some hot mix asphalt roads are rutting and cracking under these conditions.

PAVEMENT PERFORMANCE

When designing hot mix asphalt pavement, various weather conditions, traffic volumes and vehicle loadings are major considerations. These factors may change significantly over the life of the pavement. Often, highway engineers are forced to estimate these values with little predictive data available.

During the past 10 years, highway engineers have focussed their efforts on development of tests that predict performance. The main performance properties of interest are:

1. Permanent deformation
2. Fatigue cracking
3. Moisture induced damage

While some of the systems developed to predict these distresses have been adequate, most of them have fallen short under the wide variety of conditions incurred by the pavements.

There are almost an infinite number of variables in the different types of aggregates, combinations of aggregates, and the variety of binders used in making asphalt pavements. Asphalt roads have been produced out of limestone, granite, basalt, gravel, traprock, limerock, sandstone and other aggregates. The ability to recycle adds variables such as crushed concrete, roofing shingles, glass, plastic and tires to the combination of products that can be put into a HMA pavement. There are over 100 different crude oils around the world and each crude produces a slightly different asphalt binder. All these variables make modeling and accurate prediction of performance very difficult.

There is little similarity between the way mixes are produced in the lab, compacted in the lab, and tested in the lab, as compared to the way they are mixed, compacted, and placed in the field. After placement and compaction, the traffic situation and the distress that the pavements operate under are often considerably different than the test procedures utilized in the lab to predict performance.

Over the past 15 years the Europeans have developed wheel tracking units such as the Hamburg and French loaded wheel testers. In America, the Georgia loaded wheel tester and the Asphalt Pavement Analyzer (APA) have been developed to proof test the pavements and measure permanent deformation or rutting.

In 1985, the Georgia Department of Transportation initiated Research Project No. 8503, "Development of a Simplified Test Method to Predict Rutting Characteristics of Asphalt Mixes" intended to accomplish the initial development of the Georgia Loaded Wheel Tester (LWT). The prototype was a modified version of the Slurry Seal tester which operated under the following conditions:

1. Tire Pressure: 75 and 100 psi
2. Temperature: 95°F and 104 °F
3. Load: 50 lb, 75 lb., and 100 lb.

The loaded wheel tester demonstrated it could be successfully used in evaluating the rutting potential of HMA samples (Figure 1). It was further developed to accommodate three samples simultaneously and enable better control of testing temperature (Figure 2). This Modified Georgia Loaded Wheel Tester sparked much interest in the research community as a supplement to conventional mix design. Several projects were initiated to fully realize the capabilities of this concept. The latest version of the Georgia Loaded Wheel Tester is the Asphalt Pavement Analyzer (APA) (Figure 3).

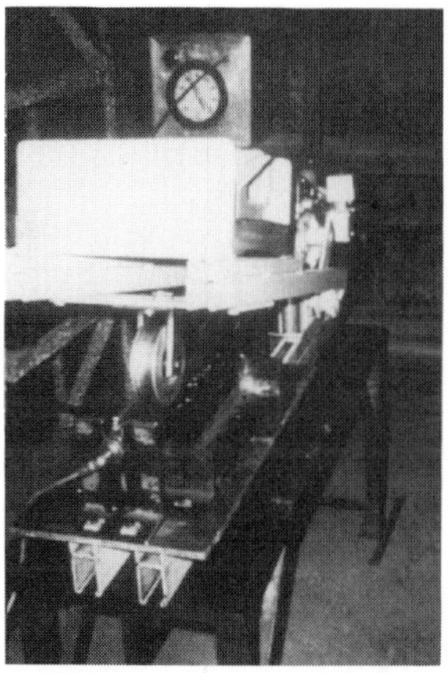

Figure 1: Original Georgia Loaded Wheel Tester

Figure 2: Georgia Loaded Wheel Tester

The APA is a multi-functional Loaded Wheel Tester that has been successfully used for evaluating permanent deformation, fatigue cracking, and moisture susceptibility of both hot and cold asphalt mixes.

The APA has the capability of testing both rectangular and cylindrical specimens. Laboratory mixed specimens, including those prepared by a gyratory compactor, or roadway cores can be tested (Figures 4 and 5). Figure 6 shows a field cut core cast in plaster to restore dimensional integrity. Permanent deformation tendencies can be determined simultaneously on three beam specimens, six cylindrical specimens, or a combination of both. Permanent deformation tests are conducted in a controlled temperature environment (30°C to 70°C)

Figure 3: Asphalt Pavement Analyzer

Figure 4: Beam Specimens

Figure 5: Cylindrical Specimens

Figure 6: Roadway cores cast in plaster

either dry or submerged. Each sample is loaded independently with up to 113 kg (250 lbs) which gives the option of testing each sample (set of two samples in the case of cylindrical specimens) at a different load. Tire pressures of up to 827 kpa (120 psi) can be simulated, with optional pressures up to 200psi available. An automated rut-depth measuring system, capable of plotting the deformation with respect to cycles or time, is also available.

The APA can determine fatigue resistance of a mixture with minimal change from the permanent deformation set-up. The rubber hose, used in the permanent deformation test, is replaced with a steel wheel. A beam is placed on supports at both ends with a rubber pad in the center to offset the dead weight of the beam. This prevents creep under its own weight. The test is conducted in a controlled temperature environment of 5°C to 30°C (41°F to 86 °F). Each sample can be loaded independently with up to 113 kg (250 lbs) and with contact pressure in excess of 827kPa (120 psi). A crack detection wire is placed on the bottom of each beam. When the wire is broken, meaning a crack has formed in the beam, the sample is unloaded automatically and the load cycle counter stops. When all three samples have failed, the machine shuts off automatically.

Moisture susceptibility of a mix can also be determined using the APA. This is accomplished by conducting the test for rutting susceptibility on both dry and pre-conditioned specimens. The pre-conditioned specimens are submerged in water during testing (Figure 7). This method compares favorably with AASHTO T-283 standard procedure for evaluating moisture susceptibility.

Figure 7: Specimens submerged in water

There have been numerous studies that used the LWT as a means to evaluate HMA parameters. Research has also evaluated the acceptability of the LWT. Some studies have verified these results with field performance. Fatigue testing and moisture susceptibility using the LWT have been researched and documented as well.

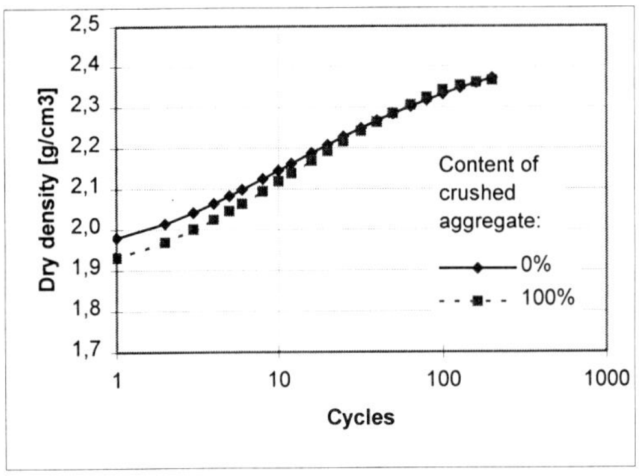

FIGURE 3a Dry density/log N curves for specimens with different content of crushed aggregate (4)

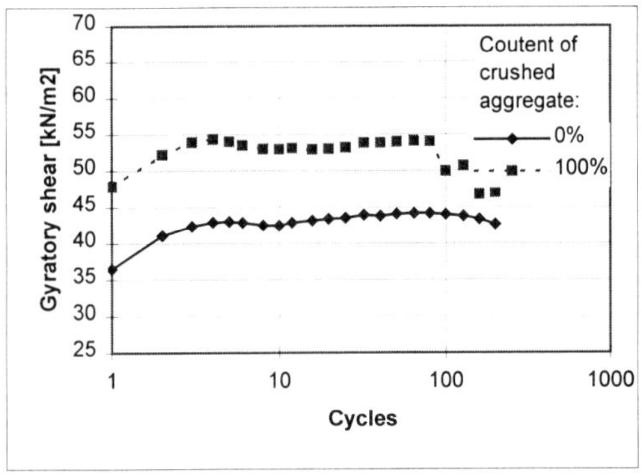

FIGURE 3b Gyratory shear/log N curves for specimens with different content of crushed aggregate (4)

Variations in gradation of aggregate show that densely graded aggregate gives specimens with higher density than open graded aggregate as expected. There is also a difference in gyratory shear resistance with dense graded aggregates giving shear resistance values 10 % or more above similar values for open graded aggregate and the AUT-gradation, figure 4a-c.

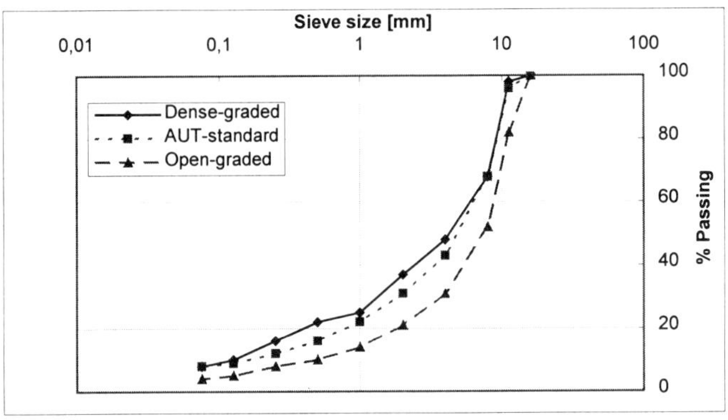

FIGURE 4a Gradation on the aggregate (4)

Fifth International Conference on the
BEARING CAPASITY OF
ROADS AND AIRFIELDS

GYRATORY TESTING OF COLD MIX ASPHALT

IVAR HORVLI[1] AND ANNE KARI TRØAN[2]

[1]Professor, Norwegian University of Science and Technology,
N-7034 Trondheim, Norway, Fax: +47 73 59 70 20,
E-mail: Ivar.Horvli@civil.sintef.no
[2]Research Engineer, SINTEF Civil and Environmental Engineering,
N-7034 Trondheim, Norway, Fax +47 73 59 14 78,
E-mail: AnneKari.Troan@civil.sintef.no

ABSTRACT

In Norway cold mix asphalt utilizing emulsified binders have been extensivly used especially for use on low volume roads. This paper presents results and main conclusions from recent research work to evaluate the use of cold mix asphalt for base course and wearing course by SINTEF Civil and Environmental Engineering, Division of Highway Engineering in Norway.

The gyratory compactor was used to simulate field compaction and to study the material characteristics. Different test procedures were used, changing the gyratory angle, axial stress and also the diameter of the mould. A number of asphalt mixes have been tested, with varying binder content, binder stiffness and amount of crushed material. For the recycled mixes, milled asphalt from different sources have been used. Density and gyratory shear modulus versus number of gyrations have been used to estimate the workability and field compaction characteristics. Reference values from the field have been derived from core specimens from the base course in one of the test roads in the Telemark Project.

Keywords: Gyratory compaction, Cold mix asphalt, Cold recycling of asphalt.

INTRODUCTION

In Norway cold mix asphalt using emulsifier binder is attracting a growing interest especially for use on low volume roads. Reasons for this interest is that cold mixes are simple to produce, are easily utilized in practice and may use 100 % recycled materials. The Telemark Project in Norway (AUT 1994-96) focused on the performance of cold mixes and on the development of suitable maintenance methods for cold mix asphalt for used for the improvement of the pavement standard for secondary roads. As a part of this research programme, SINTEF Civil and Environmental Engineering, division of Highway Engineering, was involved to determine

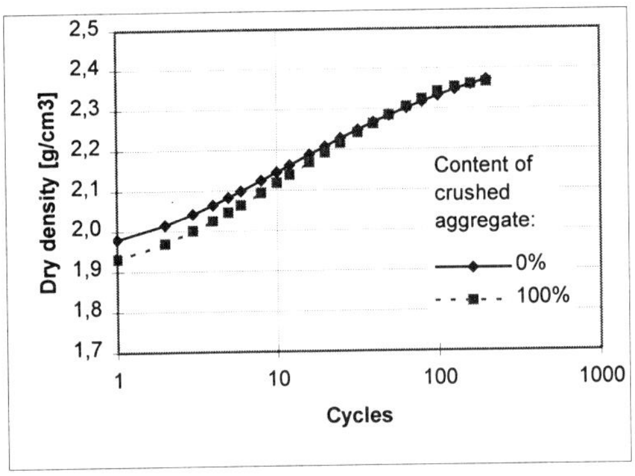

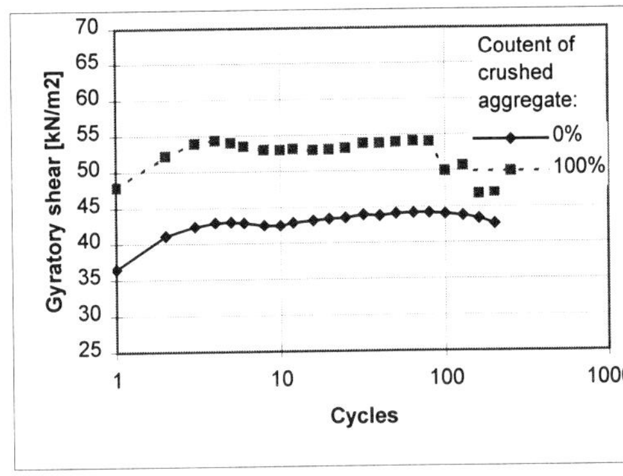

FIGURE 3a Dry density/log N curves for specimens with different content of crushed aggregate (4)

FIGURE 3b Gyratory shear/log N curves for specimens with different content of crushed aggregate (4)

Variations in gradation of aggregate show that densely graded aggregate gives specimens with higher density than open graded aggregate as expected. There is also a difference in gyratory shear resistance with dense graded aggregates giving shear resistance values 10 % or more above similar values for open graded aggregate and the AUT-gradation, figure 4a-c.

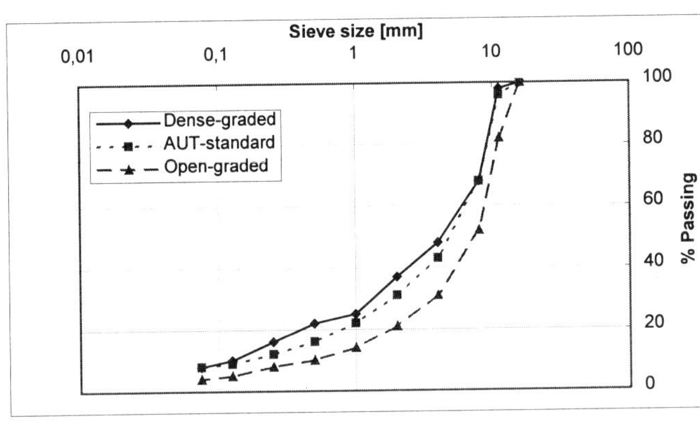

FIGURE 4a Gradation on the aggregate (4)

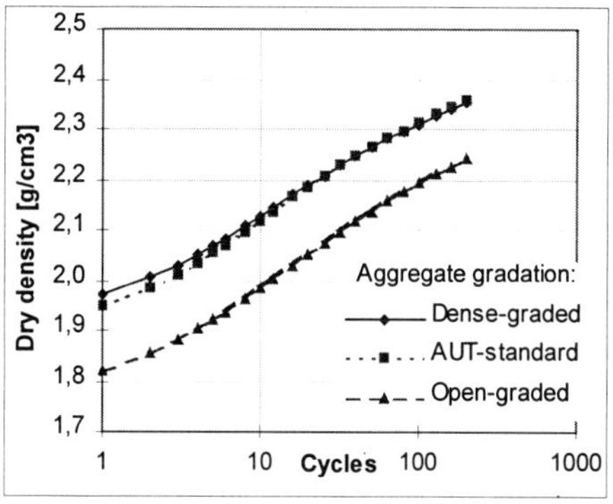

FIGURE 4b Dry density/log N curves for specimens with different aggregate gradation (4)

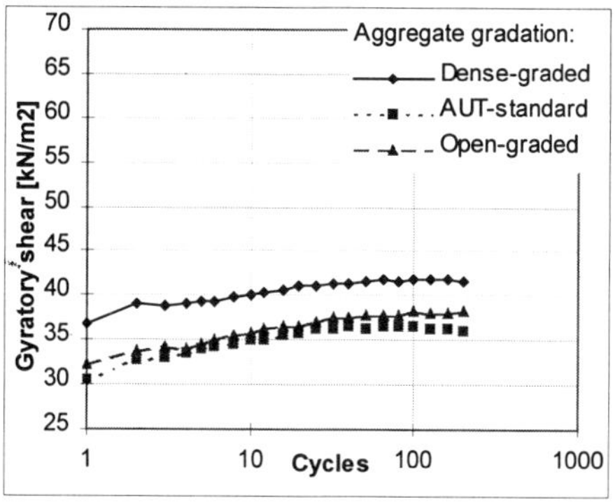

FIGURE 4c Gyratory shear/log N curves for specimens with different aggregate gradation (4)

Recycled asphalt

Cold mix recycled asphalt is more sensitive to binder content than binder type (MB1500/ MB2000). Mixture properties are in addition sensitive to aggregate gradations. Mixes with unsieved milled asphalt and the mixes with 70 % 8-16 mm and 30 % 0-8 mm (70/30) have very similar gyratory shear resistance/log N curves. Maximum gyratory shear resistance, however, occurs at different binder contents (2 % for unsieved milled asphalt and 1.5 % for 70/30). Mixes with 50 % course and 50 % fine milled fraction give significantly higher shear resistance for the same number of gyrations, N, see figure 5a-c.

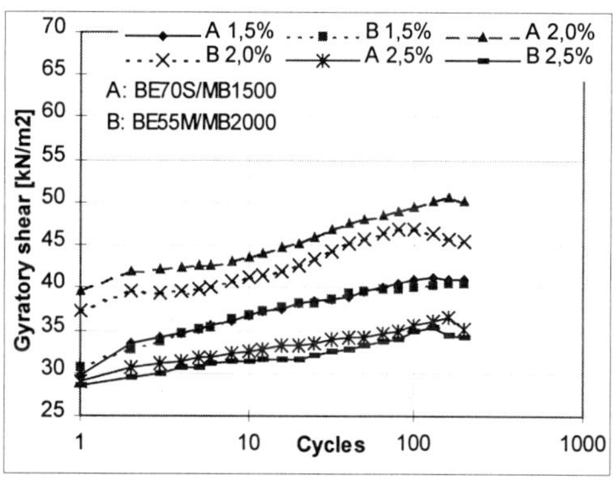

FIGURE 5a Gyratory shear resistance/ log N curves for cold mix asphalt with unsieved (one fraction) milled asphalt with different binder contents and stiffnesses (5)

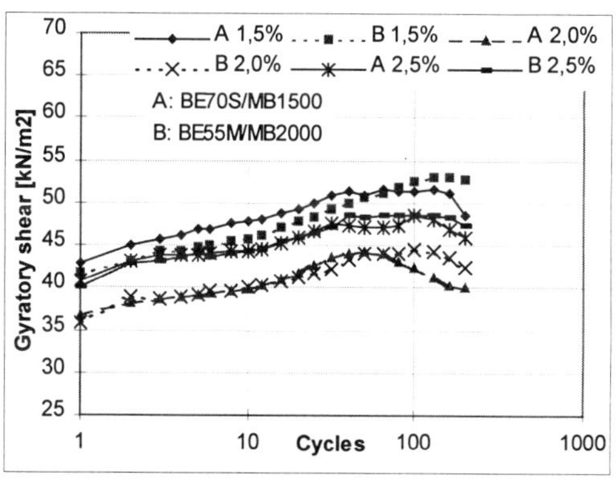

FIGURE 5b Gyratory shear resistance/log N curves for cold mix asphalt with 70 % 8-16 mm blended with 30 % 0-8 mm milled asphalt with different binder contents and stiffnesses (5)

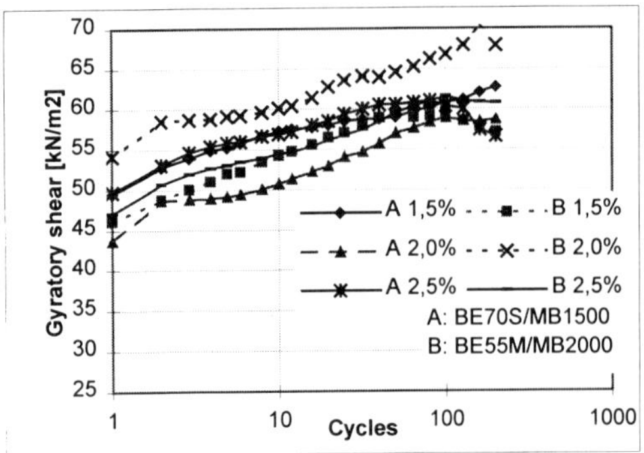

FIGURE 5c Gyratory shear resistance/log N curves for cold mix asphalt with 50 % 8-16 mm blended with 50 % 0-8 mm milled asphalt (5)

Within each type of aggregate, we find that binder content has a considerable effect on gyratory shear resistance, giving different optimum binder content with respect to shear resistance for each mixture (5, 6).

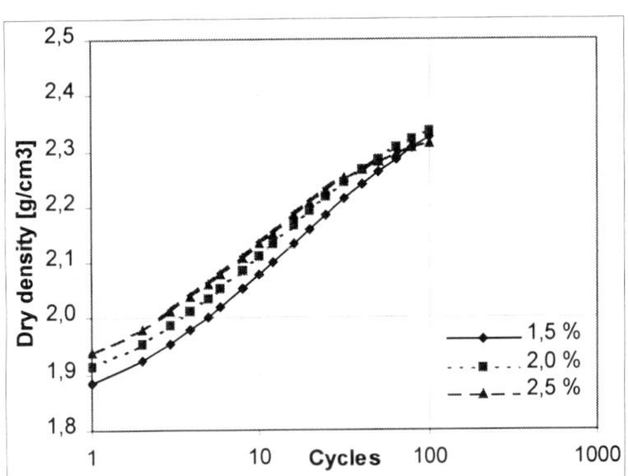

FIGURE 6a Dry density/log N curves for cold mix asphalt with unsieved milled asphalt and different binder content (6)

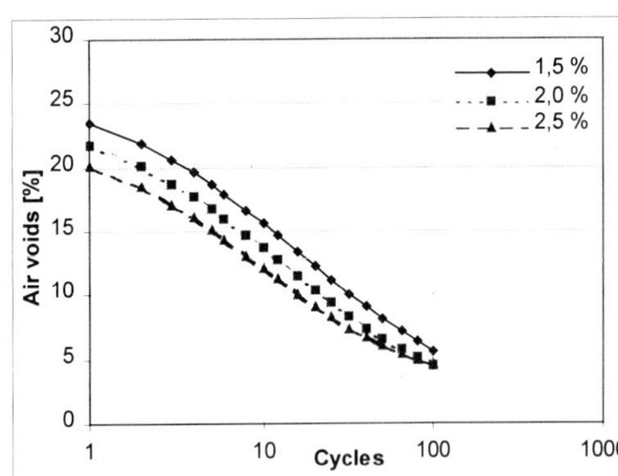

FIGURE 6b Air voids/log N curves for cold mix asphalt with unsieved milled asphalt and different binder content (6)

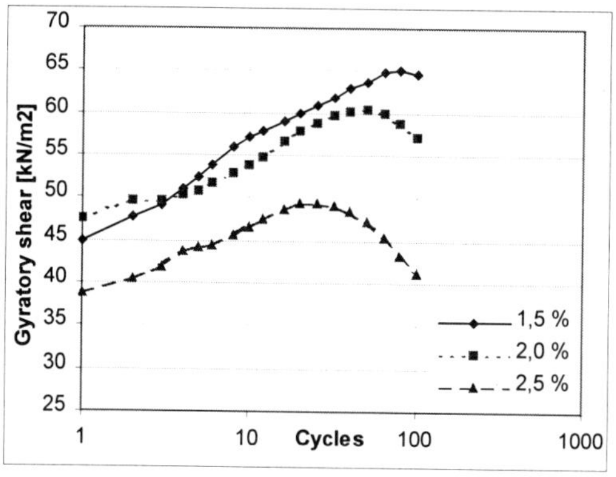

FIGURE 6c Gyratory shear resistance/log N curves for cold mix asphalt with unsieved milled asphalt and different binder content (6)

INTERPRETATION OF RESULTS FROM GYRATORY TESTING

The SUPERPAVE design method for hot mix asphalt (HMA), has specified interpretations of the relation of density versus number of cycles for the gyratory compactor. In principle the same interpretations can be used for cold mix asphalt, apart from the absolute values and design limits will be different.

Assume that N_{max} gives a compaction level that is equivalent to volume of traffic greater than the design traffic volume. To prevent high secondary compaction by traffic, a maximum specific gravity of 98 % of maximum theoretical specific gravity or a minimum volume of air voids of 2 % is recommended in the Superpave mix design specification for hot mix asphalt. Our laboratory investigations on cold mix and milled asphalt indicate critical (minimum) air void content in the range 2 – 4 % depending on the mixture properties (5, 6).

N_{ini} is used to evaluate the **workability** of the mixture:

- High density at N_{ini} indicates a mixture with good workability
- Low density at N_{ini} indicates a mixture with poor workability

The gyratory shear resistance, G, can also be used as an additional parameter in the interpretation of the workability.

- High G values imply a heavily workable mixture
- Low G values imply an easily workable mixture

If we consider the field situation during laying and compaction, a high degree of compaction at N_{ini} imply that relatively high degree of compaction is achieved after a few passes with a roller, and a low G value implies that the mixture can be easily handled with little resistance in the paver. Mixtures with a low degree of compaction at N_{ini} may need heavy rolling equipment. High G value confirms the need for heavy rolling equipment. High G also indicate large stresses on the screed in the paver due to high internal friction in the mixture.

The slope of the density-logN curve is used to evaluate the **compactibility** of the mixture.

- Large slope indicates an easily compactable mixture
- Small slope indicates a heavily compactable mixture

In addition to workability and compactibility, the **stability** of the mixture after compaction can also be evaluated using the G-logN curve in the region N_{design} - N_{max}:

- Shear modulus below a certain limit indicates an unstable mix. Our laboratory investigations on cold mixtures indicate a critical level at approximately 35-40 kPa
- Increasing or a constant curve indicates a stable mixture (increasing shear resistance with increasing load) if shear modulus is above the limit value.
- Decreasing curve indicates an unstable mixture (decreasing shear resistance with increasing load) if shear modulus is near or below the limit value.

ACKNOWLEDGEMENTS

The authors would like to thank the Norwegian Public Roads Administration (NPRA), the Public Roads Agency in Telemark and ICOPAL for co-operation in the Telemark Project (AUT, 1, 2). We also wish to thank NPRA and the Municipal Road Administration in Oslo and the Public Roads Agency in Sør-Trøndelag for co-operation in the research projects on recycled asphalt (5, 6). We also would like to acknowledge the assistance from Kwabena Asamoah, for complementing the test programme in AUT and making interpretations in his civil engineer thesis (4).

REFERENCES

1. HORVLI, I. and TRØAN, A.K. Gyratory test machine used to estimate workability and field compaction characteristics for cold mix asphalt (AUT - D2.1.5), SINTEF-STF22 F97508, Trondheim 1997 (in Norwegian)

2. TELLE, R., MYRE, J., HORVLI, I. et al. Mix design. Asphalt research project in Telemark (AUT). Norwegian Roads Administration, Report no. 87, September 1997 (in Norwegian)

3. HORVLI, I. Gyratory testing of cold mix asphalt. International Workshop on the use of Gyratory Shear Compactor, Nantes, December 1996

4. ASAMOAH, K. Evaluation of mix design properties and compactability of cold mix asphalt by use of gyratory compactor. Civil engineer thesis (master level) at Norwegian University of Science and Technology, Trondheim 1996

5. AURSTAD, J. and TRØAN, A.K. Cold Recycling of asphalt pavements. Examinations on milled asphalt from Huken gravel and asphalt plant, SINTEF - STF22 A97518, Trondheim 1997 (in Norwegian)

6. TRØAN, A.K. Recycling of milled asphalt in Sør-Trøndelag County, SINTEF - STF22 A97524, Trondheim, December 1997 (in Norwegian)

Fifth International Conference on the
BEARING CAPASITY OF
ROADS AND AIRFIELDS

CYCLIC LOAD TRIAXIAL TEST ON COLD MIX ASPHALT

IVAR HORVLI

Professor, Norwegian University of Science and Technology,
N-7034 Trondheim, Norway,
Fax: 47 73 59 70 20, E-mail: Ivar.Horvli@bygg.ntnu.no

ABSTRACT

Cyclic load triaxial tests have been carried out on cold mix asphalt compacted in the laboratory with a gyratory compacting device. Some tests were also carried out on core samples from field sites. A compacting procedure was selected that gave similar air void content for the laboratory specimens as found in situ. Different binder contents in the range from 3.5 to 5.5 % were selected.

The tests were run in a triaxial apparatus with continuous sinusoidal axial loading and a static confining pressure at different levels simulating total lateral stress including effects of traffic loads. The cyclic axial load was increased in steps until failure. Different failure criteria were used based on total permanent strain, Poisson's ratio and permanent deformation ratio. The load pulse frequency was varied within the range 3 to 20 Hz, and the resilient modulus was recorded at each frequency and stress level.

The resilient modulus of the laboratory compacted specimens and the core samples were found to be in the same range. Resilient moduli of cyclic load triaxial tests increased with increasing confining and axial stresses. The moduli from cyclic load triaxial tests and those from indirect tensile tests varied within the same range, but the mean values from the triaxial tests were slightly lower. A model for the resilient modulus as a function of the stress level is developed including the influence of varying frequencies.

BACKGROUND AND OBJECTIVES

On low traffic roads in Norway, cold mix asphalt utilizing emulsified binders have been more extensively used in the recent years. The growing interest in these asphalt mixes is due to their flexibility, the low energy consumption in production, the low production costs and the possibility using local gravel or milled asphalt as aggregate in an easy way.

The *load distribution factor* and *stability* are two parameters that are essential for the performance of road building materials. The load distribution factor can be expressed in terms of the *E-modulus*, and the stability properties can be expressed by *the permanent deformations* as a function of stress state (σ_1 σ_3).

This paper gives an overview of the results from dynamic triaxial tests that were carried out on laboratory compacted specimens as well as on core samples from field sites. In addition to the dynamic loading, some of the samples were subjected to static loading until failure after previous dynamic loading.

The main objectives of these experiments were to obtain more knowledge about the material characteristics of cold mix asphalt in order to utilise this type of material in a more optimal way. Then two key issues are:

- The relation between E-modulus and stress state
- The relation between permanent deformation and stress state

SAMPLES FOR TRIAXIAL TESTING

Two experimental methods were used to determine the E-modulus and the stability properties:

- Dynamic triaxial load test
- Indirect tension test (static)

Dynamic triaxial testing simulates traffic loads reasonably well and provides high quality information about the material parameters. The indirect tension test is used as a reference (standard test method used in Norway).

The test programme for dynamic triaxial and indirect tension test is described in ref. (1) which in addition gives an overview of fatigue tests and gyrator tests that are a part of the laboratory programme for the AUT project (Asphalt Development Project in Telemark, Norway (2)).

Core samples from the field
Core samples from field sites were used as a reference for determining the appropriate compaction level for laboratory-compacted specimens.

Two compaction methods were used in pilot tests for triaxial specimens:

- Kneeding compactor
- Kango compactor

In the main series, the kneeding compactor was chosen as the standard laboratory method. In addition, gyratory compactor was used to obtain more information about the material characteristics of the mixtures (6).

Core samples from two series from a road test site were used as a reference:

- 1 series from autumn 1995 (6 cores), the same year as produced
- 2 series from autumn 1996 (7 cores) after one winter season in service

Laboratory test programme
All the laboratory-compacted test specimens for dynamic triaxial tests were 10 cm in diameter and 20 cm in height. The specimens used for the indirect tension tests were cut in three by sawing to obtain a standard sample height of 6 cm. Core samples from the field for the indirect tension tests were also cut in three in the same way.

Two conditioning criteria were selected for the laboratory compacted specimens:

- Standard conditioning:
  7 days in an oven at a temperature of 40°C followed by 8 freeze / thaw cycles during a period of 48 hours

- Simplified conditioning:
  As standard, but without freeze / thaw cycles. The objectives with the simplified condition criteria were to see whether the freeze / thaw cycles had any effect on the properties of the mixture

During the tests the samples were maintained at a temperature of 25°C. More detailed results and data for the individual tests, are given in ref. (3) and (4).

DYNAMIC TRIAXIAL TEST

Test procedure
The test procedures for laboratory-compacted specimens and the core specimens from field samples are similar in order to be able to compare the results directly. Three core samples were tested with static loads until failure in order to compare static and dynamic failure criteria. Three testing procedures called A, B and H were used.

The samples were loaded by applying dynamic deviator load that was increased in steps until failure or large permanent deformations occurred. Continuous sinusoidal load pulses with a frequency of 10 Hz were used. This part of the test was characterised as the "step test". During the first steps, the load was applied with different frequencies in the range 5 to 20 Hz. This initial part of the test programme was defined as the "frequency test".

The static confining pressure was varied. In the first test series, the samples either have a lateral stress of 0 or 80 kPa. In the later series, different lateral stresses (80 – 40 - 0 kPa) were used for each specimen.

Stress conditions

Most of the tests were performed with a lateral pressure. In the field, the induced lateral stress will vary in relation to the dynamic axial stress $\sigma_3^{dyn} = K \cdot \sigma_1^{dyn}$. Simplifying the stress situation by using static lateral pressure, this should be the maximum values i.e.

1) $\sigma_3^{stat} = \max(\sigma_3^{dyn}) = K_{max}(\sigma_3^{dyn})$

Where σ_3^{dyn} = dynamic horizontal stress
σ_1^{dyn} = dynamic vertical stress
K = constant

In this case the maximum degree of mobilization for the material, 1/F, are the same as in the case with dynamic lateral pressure using the Mohr-Coulomb criteria, see Figure 1.

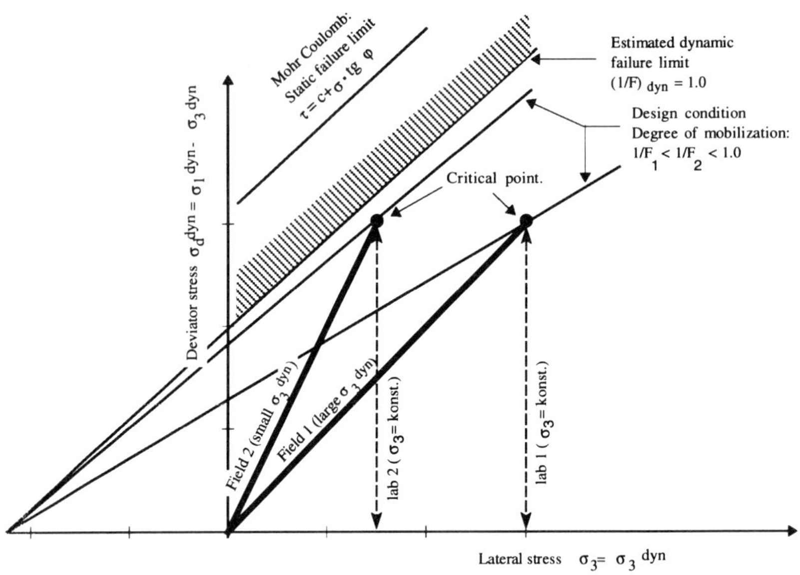

FIGURE 1 Stress state in field with dynamic traffic loads with different lateral stress respons (different ν) and equivalent simulation in a triaxial chamber with static lateral stress

Principles of interpretation

When material characteristics are stress dependent, it is essential to use relevant stress conditions in material testing procedures. E-modulus will be most relevant in a stress state in a design condition, i.e., when the stress situation is far from "failure" and permanent deformations are small. However, permanent deformations are interesting both in the design condition (corresponding to subsequent compaction by traffic) and for failure conditions (corresponding to an under-design case). Three alternative failure criteria or threshold limits were considered for the design condition:

- permanent deformation $\varepsilon_p \geq 5$ ‰
- the gradient of plastic flow development $d\varepsilon/dN > 7 \cdot 10^{-8}$
- Poisson's resilient coefficient $\nu \geq 0.4$ or 0.5

Laboratory results

Dynamic triaxial tests on cold mix asphalt from the Telemark project (AUT) show increasing resilient modulus / E-modulus with increasing frequency, being a well-known phenomena for asphalt mixes. When testing twice at the same frequency before and after a previous "step" with a higher stress level, the E-modulus was slightly reduced at the second time testing. This effect may be explained by a degradation of attraction in the material caused by stress over time.

The "frequency test" procedure was carried out on laboratory compacted specimens at different binder content and on core samples from the field at confining pressures of zero and 80 kPa. From these series the effect of frequency and confining pressure was found as expressed in Figure 2 and equation (1) in this paper.

The dynamic deviator stress was increased in steps until failure, and the dynamic E-modulus was plotted in diagrams as shown in Figure 3 and 4. We found that the E-modulus is not a unique function of stress, but is also affected of the stress history. In our interpretations, we therefor primarily used values from second time testing to determine correlation between E-modulus and stresses. In these interpretations we assumed a simple linear relation as shown in equation (1), and then derived the constants in this equation by changing one parameter at the time. We derived alternative expressions of the E-modulus as a function of the confining stress and the deviator stress as applied directly in the triaxial chamber, equation (1). We also reformulated this in a conceptual equation expressed by mean principle stresses, equation (2).

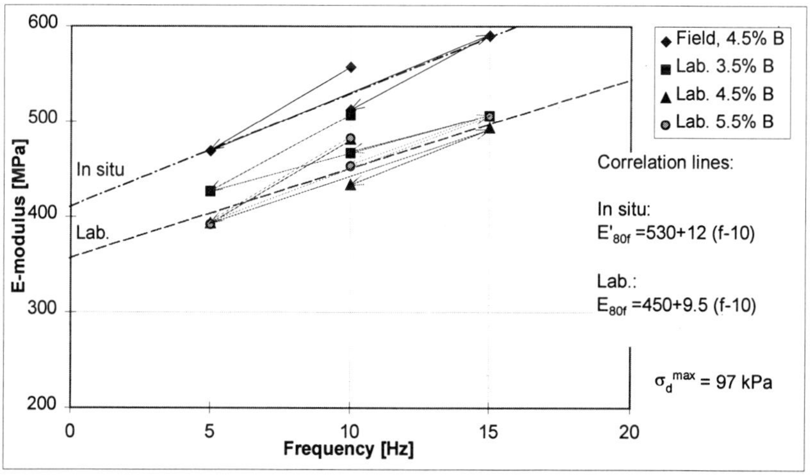

FIGURE 2 E-modulus versus frequency, mean values at 80 kPa confining pressure (1)

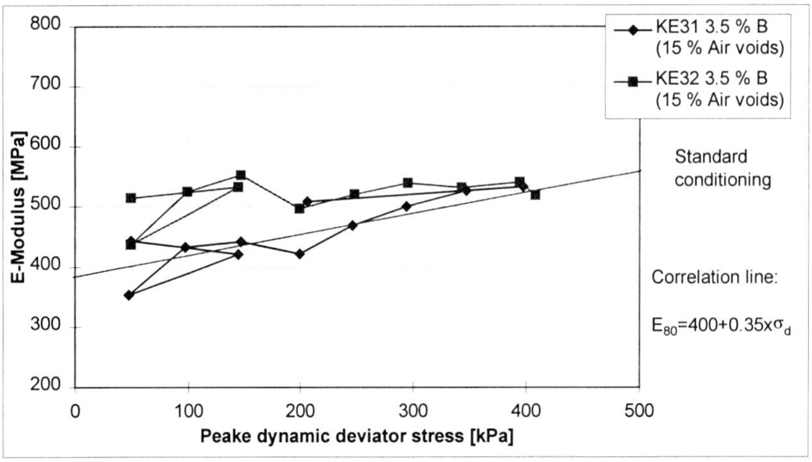

FIGURE 3 E-modulus versus deviator stress at a confining pressure of 80 kPa and binder content of 3.5 % (1)

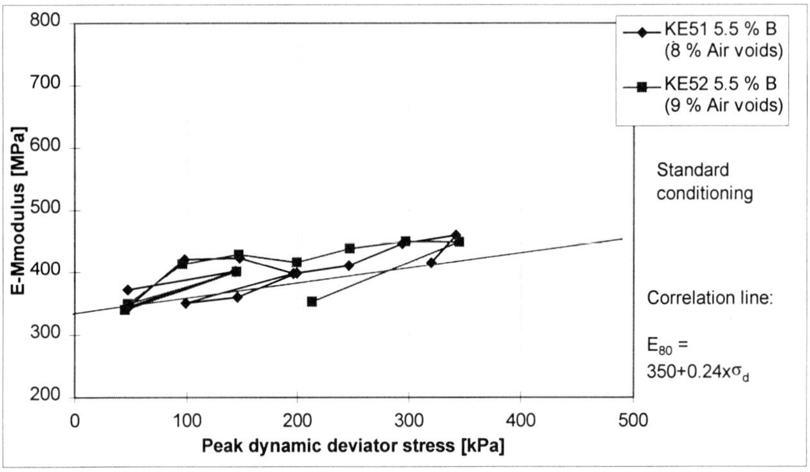

FIGURE 4 E-modulus versus deviator stress at a confining pressure of 80 kPa and binder content of 5.5 % (1)

Figure 5 shows that the E-moduli derived from indirect tensile stress test have a scatter in the same range as found from dynamic triaxial test. The mean value is however somewhat displaced.

Figure 6 shows the effect of the confining pressure on the maximum deviator stress at failure according to three alternative failure criteria. The results show a considerable effect of the lateral pressure on the maximum obtained failure stress level.

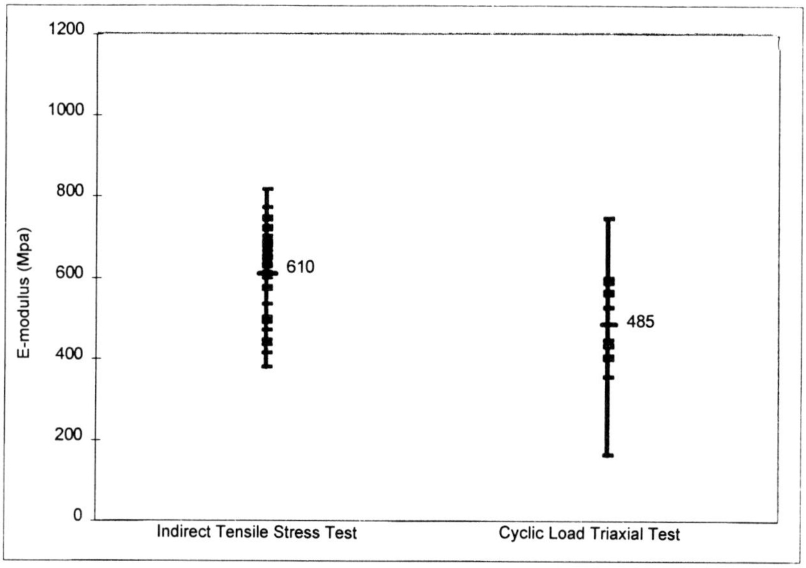

FIGURE 5 Span of variation in E-moduli from Indirect Tensile Stress Test and from Cyclic Load Triaxial Test on specimens with various bitumen contents and different stress states

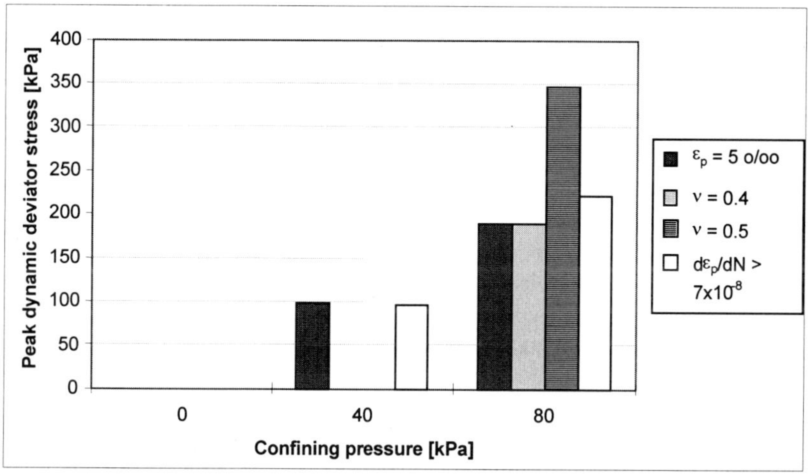

FIGURE 6 Permanent deformations at confining pressure of 40 kPa and 80 kPa for samples with 4.5 % binder content (1)

Load distribution factors

A well known approximate relation between E-modulus (Mpa) and the load distribution factor, a, is given by the equation:

$$a = 0.21\ \sqrt[3]{E}$$

Table 1 gives an overview of some E-moduli and the corresponding load distribution factors, a, derived from dynamic triaxial tests conducted at SINTEF (Foundation of Scientific and Industrial Research):

TABLE 1 E-modulus at 25°C and 10 Hz and load distribution factors, a, for several asphalt mixes used in Norway

Mix type cold mix			E-modulus (Mpa)	Load distr. factor	Ref.
Cold mix	Eg, B370	(emulsion)	400-600	1.5-1.8	(1)
Cold mix	Eg, MB1500	(emulsion)	200	1.2	(7)
Cold mix	Eg, MB6000	(emulsion)	400	1.5	(7)
Cold mix	Eg, B180	(emulsion)	1000	2.1	(7)
Cold mix	Asg, 4 % binder	(foam)	970	2.1	(8)
Cold mix	Asg, 4 % binder	(foam)	1420	2.4	(8)
Hot mix	Ab 16t, B85	(pen.bit.)	3200	3.1	(9)
Hot mix	Ag 16, B180	(pen.bit)	2300	2.8	(9)

SUMMARY AND CONCLUSIONS

The laboratory test programme have given us more insight in the behaviour of emulsion asphalt / cold-mix asphalt and its elastic and plastic properties. The main conclusions that might be drawn from these experiments are:

- The dynamic triaxial tests showed no unambiguous relationship between the E-modulus and the binder content

- No significant relationship was found between the E-modulus and the air void content· for dynamic triaxial tests as was found in the series of the indirect tensile stress. This is most likely due to the fact that the E-modulus from triaxial tests are highly dependent on the stress level (σ_{dyn}) which was varied within a considerable range in this test programme. However, the E - modulus from the dynamic triaxial test has a scatter within the same range as the results from the indirect tensile test, but with a somewhat lower mean value.

- The mean value of the E-moduli found by dynamic triaxial test conducted under a lateral stress of 80 kPa, is 120 kPa lower than the E-moduli from the indirect tensile test. There is, however, quite a large variation between the parallel tests. The lateral stress $\sigma_3 = 80$ kPa corresponds to a rather low stress level in a field situation for the base course. At $\sigma_3 \approx 200$ kPa (160-250 kPa), the results would coincide according to equation (1) with relevant values for the factor "b" found from this test programme.

- For mixes with low binder content (B = 3.5) the freeze / thaw conditioning (standard conditioning) gives a lower E-modulus in the triaxial test than without freeze / thaw cycles. For mixes with binder content B= 4.5 and 5.5 % there was no effect of the freeze / thaw cycles.

- There is very little difference between the mean values of E-moduli from the core samples from field in 1995 and 1996. E-moduli in 1996 seems to be a bit higher, but the scatter is rather high. On the other hand, the void content determined in situ seems to have been reduced somewhat with time.

- In general, we found that the dynamic E-modulus may be estimated as:

(1) $E = E^* + a \cdot \sigma_d + b \cdot \sigma_3 + c \cdot (f-10)$

where E*, a, b, c are constants
and f = frequency
σ_1 σ_3 = principle stress
$\sigma_d = \sigma_1 - \sigma_3$

The scatter was rather high, and typical values for the parameters a, b and c were found in the range:

a = 0.25 – 0.35
b = 0.7 – 1.4
c = 9 –12 in the frequency range 3 – 20 Hz

An other conceptual equation where the data from this test series fit pretty well is:

(2) $E = 360 + 0.9 \cdot \theta$
where $\theta = \sigma_m = 1/3 \cdot (\sigma_1 + 2 \cdot \sigma_3)$

RECOMMENDATIONS AND FURTHER INVESTIGATIONS

These investigations have given us a better basis for evaluating the load distribution factor and stability characteristics for emulsion gravel. There are however needs for further investigations to get a better knowledge of material properties of cold mix asphalt. In order to make further advance in understanding the material behaviour and to improve mix design for emulsion gravel, the following investigations would be of considerable interest:

- Conduct tests with more extensive consolidation procedures and with dynamic lateral stress in order obtain more uniform samples and to give a better simulation of lateral stress conditions in the field.
- Conduct dynamic triaxial testing on a number of core samples from the field obtained over a longer term time span in order to study in situ time-dependent effects such as healing, weathering and ageing

ACKNOWLEDGEMENTS

I would like to thank the Directorate for Pubic Roads, The Public Roads Agency in Telemark and ICOPAL for co-operation in the Telemark project (AUT). I would also like to thank Inge Hoff for performing sample compacting and testing, Stein Hoseth for preparing test specimens, Anne Kari Trøan for data presentations and Gudrun Foss Bakken for writing and layout.

REFERENCE LIST

1. HORVLI, I. and TRØAN, A.K. Dynamic triaxial test on emulsion gravel mixtures, summary report. SINTEF STF22 F97506, Trondheim, 1997 (in Norwegian)

2. TELLE, R., MYRE, J., HORVLI, I et. al. Mix design. Asphalt research project in Telemark (AUT). Norwegian Public Roads Administration, Report no 87, September 1997 (in Norwegian)

3. HORVLI, I. and TRØAN, A.K. Dynamic triaxial tests and indirect tensile tests on emulsion gravel, main investigation. SINTEF STF22 F97504, Trondheim, March 1997 (in Norwegian)

4. HORVLI, I. and TRØAN, A.K. Dynamic triaxial tests on emulsion gravel, supplementary investigation. SINTEF STF22 F97505, Trondheim, March 1997 (in Norwegian)

5. HORVLI, I. Material properties for cold mixed asphalt. NIF conference in Ålesund, Norway, 22-24. April 1997 (in Norwegian)

6. HORVLI, I. and TRØAN, A.K. Gyrator testing of cold mixed asphalt. BCRA-98, Trondheim 1998

7. SLYNGSTAD, T and HOSETH, S. Properties for gravel from Bolstadøyri, stabilised and unbound material. SINTEF, Project no 610234, preliminary paper Jan. 1987

8. BAKLØKK, L:J, DAHLHAUG, J.E, SLYNGSTAD, T. Dynamic material parameters for laboratory produced specimens of foamed asphalt, SINTEF STF61 83009, 1983

9. BAKLØKK, L.J. Function testing. Strength parameters for asphalt and unbound materials. NTNU division of Road and Railway Engineering, Paper no 953, March 1996

Fifth International Conference on the
BEARING CAPASITY OF
ROADS AND AIRFIELDS

PERFORMANCE OF HMA MIXES WITH HYDRATED LIME

Louay N. Mohammad, Ph.D.[1], Rana Gokman[2]
[1] Assistant Professor and Manager, [2] Graduate Research Assistant
Engineering Material Characterization and Research Facility
Louisiana Transportation Research Center, Louisiana State University
4101 Gourrier Lane, Baton Rouge, LA 70808, U. S. A.

ABSTRACT

The increasing volume of truck traffic combined with the hot and moist climate of Louisiana are identified as the major contributing factors to severe pavement distresses. Permanent deformation and moisture damage are known as the most common distresses of Louisiana pavements, resulting in the rutting and stripping of the asphaltic concrete. The use of hydrated lime is known to provide decreases in moisture susceptibility and increases in mixture stiffness.

A conventional asphaltic concrete mixtures and mixtures modified with hydrated lime were evaluated for their fundamental engineering properties as defined by permanent deformation characteristics and fatigue resistance. A typical Louisiana's low volume dense graded mixture was used. The test factorial included two aggregate types: limestone and gravel; and two asphalt cement types: a conventional AC-30 and a SB polymer modified. Hydrated lime was used as the mineral filler.

The results indicated that the addition of hydrated lime as mineral filler improved the permanent deformation characteristics and fatigue endurance of the asphaltic concrete mixtures. This improvement was particularly apparent at higher testing temperatures and/or with polymer mixes.

INTRODUCTION AND BACKGROUND

Construction personnel in Louisiana have pointed out that the performance of asphaltic concrete pavements has deteriorated over the last decade. Heavy loads coupled with the hot and moist climate of Louisiana are major contributing factors to the development of severe pavement distresses such as permanent deformation and loss of structure because of moisture damage. While reviewing the records of Hot Mix Asphalt concrete (HMAC) mixtures produced in the early 1960's [1] and today a major difference was identified as the lack of mineral fillers in today's mixes. These fillers increase film thickness, improve the cohesion of the binders and increase the stiffness of the mixtures. Thus, pavement performance and its rutting resistance can be improved with the use of mineral fillers.

The structure of hydrated lime consists of differently sized fractions. The larger size fraction performs as a filler and increases the stiffness of the bituminous mixture. The smaller size fraction increases binder film thickness enhancing viscosity of the binder, and improving the binder cohesion. This, in turn, leads the asphalt cement to coat the aggregate particles with a thicker film. Adhesion between the aggregate and binder increases [2], resulting in decreased mixture segregation.

It is reported [3-11] that hydrated lime could be used in many different applications in HMAC, and that these mixtures showed improved performance in the field. By maintaining a good adhesion between the aggregate and the asphalt cement, in the presence of water, hydrated lime is widely used as an antistripping agent. Its ability to reduce viscosity building polar components in the asphalt binder has enabled hydrated lime to show effect as an oxidation reducing agent. Also, its ability to increase mixture stiffness by filling air voids makes it an effective mineral filler.

The water sensitivity of bituminous mixtures treated with hydrated lime and antistrip agents was evaluated by researchers [6,7]. Results from these tests depended on the actual combinations of aggregate, asphalt, and additives. However, among the antistrip agents, hydrated lime (high calcium or dolomitic) was the most effective one overall. The wet-dry indirect tensile strength test conducted by Kennedy and Ping [6] revealed that lime was effective in increasing the tensile strength of the mix. The Texas freeze-thaw pedestal test performed by Kennedy et al. [7] further revealed that hydrated lime was effective in improving the water sensitivity of the mixture, thereby providing pavements that are highly strip resistant.

Shahrour [11] and Al-Suhaibani [4] evaluated the mineral filler properties of hydrated lime and other local fillers available in Saudi Arabia. The mechanical properties of the mixes were studied using tests such as the resilient modulus test, the indirect tensile strength test, Hveem stability, and Marshall criteria. The research results [11,4] revealed that the amount and characteristics of the mineral fillers can have an effect on the rutting susceptibility of flexible pavements, and that the use of hydrated lime can improve resistance of the mixes to rutting. The lime showed improved stiffening properties when incorporated into the mixture.

This paper presents the results of a comparative study conducted on conventional and modified HMA mixtures to determine if the use of hydrated lime as a mineral filler leads to any improvement in the fundamental engineering properties.

OBJECTIVE

The objective of this research is to evaluate the fundamental engineering properties of bituminous mixtures modified with hydrated lime. A Louisiana gravel mix and a limestone mix, both modified with hydrated lime, were used as comparison relative to conventional (unmodified) HMA mixtures. Fundamental tests considered include creep compliance and fatigue life. The Hamburg wheel tracking test device was also used in the performance evaluation.

SCOPE

A typical Louisiana's low volume dense graded mixture was used in this research. Two aggregate types were included in the test factorial: a limestone and a gravel, and two asphalt cement types: a conventional AC-30 and a polymer modified asphalt meeting Louisiana specifications for PAC-40HG, Table 1. The mineral filler included in this study was hydrated lime (HL). For each aggregate/asphalt cement combination, an unmodified and hydrated lime modified mixtures were considered. Thus, the number of mixtures utilized in this study was eight.

EXPERIMENTAL

Two aggregate types, that are commonly used in Louisiana, were included in this research: a limestone (from hereon referred to as stone), and a gravel. The stone was obtained from Vulcan Materials Company, Reed Quarry in Kentucky. The gravel, on the other hand, was obtained from ACME Gravel, Baywood, Louisiana.

Two types of asphalt cement were used in this study: a conventional viscosity graded AC-30 and a polymer modified asphalt cement meeting the LA DOTD specification for a PAC-40HG. The polymer modified asphalt cement is obtained by the modification of the asphalt cement with styrene-butadiene block co-polymer. Table 1 shows the specifications and properties of both asphalt cements.

The stone mixtures included 38 percent No. 78 and 37 percent No. 11 silicious lime stone, 18 percent coarse sand, and 7 percent fine sand. The proportions in the gravel mixture were 75 percent crushed gravel, 10 percent coarse sand, and 15 percent fine sand. For mixtures containing hydrated lime, the fine sand content is reduced to accommodate the hydrated lime. The hydrated lime is used at a rate of 1.5 percent of the total aggregate weight and was introduced as a slurry to the aggregate. The optimum binder content is determined from standard Marshall mix design. The asphalt contents for the gravel mixes and stone mixes are 5.2% and 4.7 %, respectively. Specimens were prepared using 75 blows per face of the Marshall hammer. Each specimen was 101.6 mm (4 in.) in diameter and about 63.5 mm (2.5 in.) high. The target air void level was 4 percent (± 0.5%). For the Hamburg wheel tracking device, slabs were prepared utilizing a linear kneading compactor.

At the target air void level, samples were statistically grouped to have similar mean air void levels. A triplicate was used for each test and mixture type.

MATERIALS CHARACTERIZATION

Indirect tensile creep and fatigue tests as well as a loaded wheel tracking test were conducted to characterize the eight mixtures and determine the effects of the hydrated lime mineral filler on the properties of HMACs. The fundamental tests were conducted using a fully automated MTS servo-hydraulic test system. The loaded wheel tracking test was conducted at Koch Materials laboratory, Terre Haute, Indiana. A detailed description of each test is given in References [12,13].

Wheel Tracking Test Utilizing the Hamburg Loaded Wheel Tracking Device (HWTD)

This test is considered a torture test. The Hamburg Wheel-Tracking Device produces damage by rolling a steel wheel across the surface of a slab that is submerged in water at 50°C (122°F) for 20,000 passes. The slabs have a length of 320 mm (12.6 inch), a width of 260 mm (10.2 inch), and a thickness of 80 mm (3.2 inch). They are secured in a reusable steel containers using plaster of paris, and are then placed into the wheel-tracking device.

The device tests two slabs at a time using two reciprocating solid-steel wheels. The wheels have a diameter of 203.5 mm (8 inch) and a width of 47 mm (1.85 inch). The load is 710 N (160 lb). The rolling speed is 1.1 km/h (0.68 mi/h). Each wheel rolls 230 mm (9.1 inch) before reversing direction. The device operates at 53 ± 2 passes/min. A maximum allowable rut depth of 4 mm at 20,000 passes is used in Hamburg, Germany. Colorado has the only specification for the HWTD test results in the United States. It considered the Hamburg specification to be too severe for their pavements and considers a rut depth of 10 mm after 20,000 passes to be more reasonable [14].

The following four indices are used to quantify the test results [15], Figure 1:

1. Post-Compaction Consolidation: A low post-compaction consolidation value (reported in mm) is desirable, because this would indicate that the compaction during laboratory fabrication was near optimum levels.
2. Creep Slope: The creep slope is the inverse of the rate of deformation in the linear region of the deformation curve. This linear region is the region after post-compaction effects have ended and before the beginning of stripping. The creep slope value is reported as passes/mm. The larger the value, the more energy is needed to deform the mix; therefore, the slower the rate of permanent deformation and the less sensitive the mix to rutting.
3. Stripping Slope: This slope is the inverse of the rate of deformation in the linear region of the deformation curve, after the beginning of stripping until the end of the test. The stripping slope is reported as passes/mm, The stripping slope is related to the severity of moisture damage, and for this reason a higher number of passes for each 1 mm of impression from stripping is the desirable value.
4. Stripping Inflection Point: The stripping inflection point (reported in number of passes) is the number of passes at the intersection of the creep slope and the stripping slope. The stripping inflection point is related to the resistance of the mixture to moisture damage and a high value for this index is desired. The higher the value for the stripping inflection point, the more energy is needed to cause stripping; therefore the less sensitive the mix is to stripping.

DISCUSSION OF RESULTS

A summary of the fundamental test results is presented in Tables 2 - 4. The results of these tests were statistically analyzed using the analysis of variance procedure provided in the Statistical Analysis System (SAS) program from SAS Institute, Inc. A multiple comparison procedure with a risk level of 5 percent was performed on the means. The tables for the grouping present the mean for the test results reported by mixture type. The group column contains the results of the statistical grouping reported with the letters A, B, C, and D. The letter A was assigned to the highest mean followed by the other letters in appropriate order. A double (or more) letter designation, such as A/B (or A/B/C), indicates that in the analysis the difference in the means was not clear-cut, and that the mean was close to either group. The mixtures evaluated are coded as follow:

SAC30N: Limestone Type 3 Binder Course with 4.7 % AC-30,
SAC30L: Limestone Type 3 Binder Course with 4.6 % AC-30 and 1.5 % HL
SPMACN: Limestone Type 3 Binder Course with 4.7 % PAC-40HG, and

SPMACL: Limestone Type 3 Binder Course with 4.6% PAC-40HG and 1.5% HL
GAC30N: Gravel Type 3 Binder Course with 5.8 % AC-30,
GAC30L: Gravel Type 3 Binder Course with 5.0 % AC-30 and 1.5 % HL
GPMACN: Gravel Type 3 Binder Course with 5.2 % PAC-40HG, and
GPMACL: Gravel Type 3 Binder Course with 5.2 % PAC-40HG and 1.5 % HL.

Permanent Deformation Tests

Effect on Indirect Tensile Creep Test

The effect of hydrated lime on the indirect tensile creep slope and time to failure of the limestone and gravel mixes is presented in Table 2. In this test flat slopes are desirable with long times to failure. Mixtures with a combination of low creep slope and long time to failure are considered to be less susceptible to rutting.

Table 2 shows that in general the addition of hydrated lime to the mixes (both limestone and gravel) decreased the creep slope values and increased the time to failure values. Also, the decrease in the creep slope and the increase in the time to failure results for SPMACL at 25°C (77°F) and 40°C (104°F) were significantly different from all other limestone mixes. The creep slopes for GAC30L at 25°C (77°F) and 40°C (104°F) were significantly lower, and the decrease in the creep slope and the increase in the time to failure results for GPMACL at 40°C (104°F) were significant.

These results indicate that, overall, hydrated lime worked well with both aggregate and asphalt cement types at both testing temperatures. However, the stone and gravel mixes modified with lime and containing PAC-40HG (SPMACL and GPMACL) showed the most improvement in creep properties, especially at the higher testing temperature and more for the gravel than for the limestone. This does seem to indicate that the polymer modified asphalt and the lime work well with the gravel aggregate to provide a HMAC that is less susceptible to rutting.

Effect on Hamburg Loaded Wheel Device Test Results

Table 3 presents the results for the tests conducted on the limestone and gravel slabs. For a rut and strip resistance mix higher creep slope and stripping slope values, as well as higher number of passes to 20mm rut depth and a higher stripping inflection point values, are desirable. The rut depth at 20,000 passes, however, is desired to be a low value.

The results show that the lime modified limestone mixes SAC30L and SPMACL showed better properties than the conventional mixes SAC30N and SPMACN, while all gravel mixes, in general, performed equally well. The results for the gravel mixture are somewhat limited regarding the stripping properties, because there was no onset of stripping during the test period. It is also important to note that the deformation curves were still in the linear region of the creep slope during the test period.

The addition of hydrated lime increased the creep slope for SAC30L, SPMACL, and GPMACL and the stripping slope for SAC30L and SPMACL. The lime addition also increased the stripping inflection point and the passes to 20 mm rut depth for SAC30L and SPMACL. The rut depth at 20,000 passes for SAC30L and SPMACL was lower than SAC30N and SPMACN, respectively. Among the stone mixes, SPAMCL had the most desirable rut properties (i.e. high creep slope value, low rut depth, and high number of passes). Also, compared with SAC30L, SPMACL did show a better performance. This may be due to the fact that the combination of PAC-40HG and lime worked well together, creating a mixture that is less susceptible to rutting as well as moisture damage. Also, the creep slope values and the rut depth at 20,000 passes are very close to each other for SPMACL and GPMAC mixes. A similar comparison was observed for SAC30L and GAC30Lmixes. However, the gravel mixes did not show any stripping tendencies and, in general, lower rut depths than the stone mixes.

When comparing the results of the two rut tests (indirect tensile creep test, and the HWTD test), it is noted that, overall, the addition of hydrated lime improved the performance indicators, and that the mixes with the polymer asphalt cement (PAC-40HG) in combination with the hydrated lime showed the most favorable results.

Effect on Indirect Tensile Fatigue Test

The results of the indirect tensile fatigue test are summarized in Table 4. While evaluating the fatigue resistance of HMACs, the fatigue slope and the number of cycles to failure are used as performance indicators. A low slope value and high number of cycles to failure are the desired properties.

Tables 4 shows the use of hydrated lime in the mixes improved the values for the fatigue slope of SAC30L, SPMACL, GAC30L, and GPMACL. The slope values for the stone mixes were very close to the gravel mixes. The decrease in the slopes were significant. The addition of hydrated lime also increased the number of cycles for SAC30L, SPMACL, GAC30L, and GPMACL. This increase was not significant for SAC30L and GAC30L, but was very significant for SPMACL and GPMACL. In general the number of cycles to failure were higher for the gravel mixes.

The results of the fatigue test indicate that the addition of hydrated lime as a filler increased the fatigue resistance of the mixes. Here again the mixtures containing a combination of hydrated lime and PAC-40HG showed the most improvement in fatigue properties. GPMACL was the mix that showed the highest endurance to the repeated cyclic loading of the fatigue test.

SUMMARY AND CONCLUSIONS

The performance of HMAC modified with hydrated lime mineral filler as defined by permanent deformation characteristics and fatigue endurance was evaluated. Four limestone and four gravel mixes were considered, one with AC-30 and the other a polymer modified binder, both modified with hydrated lime and were compared to mixes designed without lime. Generally, mixes with hydrated lime showed an improved performance relative to similar mixes with no lime. Also, mixtures containing the combination of hydrated lime and the polymer modified asphalt cement (PAC-40HG) showed the most improvement. This improvement was particular apparent at tests performed at higher temperatures. This can be attributed to the fact that the hydrated lime enhances the bond between the aggregate and binder and that the polymer modified binder adds elasticity to this bond. The following observations can be drawn from the data and analysis:

- The limestone and the gravel mixes with the lime filler (SAC30L, SPMACL, GAC30L, and GPMACL) showed an improvement in the indirect tensile creep slope at 25°C (77°F) and 40°C (104°F). The decrease in the slope values was more significant at the higher temperatures.
- The fatigue properties for the lime modified mixes of both aggregate types (SAC30L, SPMACL, GAC30L, and GPMACL) showed an improvement with a lower fatigue slope and higher number of cycles to failure. However, the most significant fatigue endurance was noted for the lime modified mixes with PAC-40HG (GAC30L and GPMACL).
- When comparing the results of the fundamental test with the HWTD results for the limestone mixtures it is apparent that the results are similar considering the improved resistance to permanent deformation of the mixtures with the hydrated lime additive.

ACKNOWLEDGMENTS

This work supported by the Louisiana Transportation Research Center in Cooperation with Louisiana Department of Transportation and Development and the Federal Highway Administration under state project number 93-3B. The authors wish to express their appreciation for this support. The authors would like to acknowledge Koch Materials for conducting the Hamburg loaded wheel tests.

REFERENCES

1. Shah, S. C., "Asphaltic Concrete Pavement Survey," Louisiana Transportation Research Center Report No. 9, Baton Rouge, Louisiana, 1963.
2. Dickinson, E. J., *Bituminous Roads in Australia,* Australia Research Board, April 1984, pp. 219, 253-259.
3. Metcalf, John B., "A Survey of the Use of Hydrated Lime as a Filler in Tar or Bitumen Road Surfacings," Chalk Lime & Allied Industries Research Association, Herts, October, 1959.
4. Al-Suhaibani, AbdulRahman, Al-Mudaiheem, Jamal, and Al-Fozan, Fahd, "Effect of Filler Type and Content on Properties of Asphalt Concrete Mixes," Effects of Aggregates and Mineral Fillers on Asphalt Mixture Performance, ASTM STP 1147, Richard C. Meininger, Editor, American Society for Testing and Materials, Philadelphia, 1992, pp. 107-130.
5. National Lime Association, *Lime, Handling, Application, and Storage,* Bulletin No. 213, Gutschick,

Kenneth A., and Lewis, Clifford J., Editors, Virginia, 1982.

6. Kennedy, Thomas W., and Ping, W. Virgil, "An Evaluation of Effectiveness of Antistripping Additives in Protecting Asphalt Mixtures from Moisture Damage," Annual Meeting of the Association of Asphalt Paving Technologists, March 4-6, 1991.

7. Kennedy, Thomas W., Roberts, Freddy L., and Lee, Kang W., "Evaluation of Moisture Susceptibility of Asphalt Mixtures Using the Texas Freeze-Thaw Pedestal Test," AAPT, Vol. 51, 1982, pp.327-341.

8. National Lime Association, "Hydrated Lime in Hot Mix Asphalt," Presentation Manual, 1993.

9. Kennedy, Thomas W., "Use of Hydrated Lime in Asphalt Paving Mixtures," National Lime Association, Bulletin No. 325, 1987.

10. Plancher, H., Green, E. L., and Petersen, J. C., "Paving Asphalts: Reduction of Oxidative Hardening of Asphalts by Treatment with Hydrated Lime - A Mechanistic Study," Federal Highway Administration, Report No. FHWA-RD-77-147, Washington, D.C., April 1977.

11. Shahrour, Mukhlis A., and Saloukeh, Bassam G., "Effect of Quality and Quantity of Locally Produced Filler (Passing Sieve No. 200) on Asphaltic Mixtures in Dubai," Effects of Aggregates and Mineral Fillers on Asphalt Mixture Performance, ASTM STP 1147, Richard C. Meininger, Editor, American Society for Testing and Materials, Philadelphia, 1992, pp.187-208.

12. Mohammad, L.N., and Paul, H.R., "A Parametric Evaluation of Fundamental Engineering Properties for Louisiana Hot Mix," Louisiana Transportation Research Center, Report No. 276, Baton Rouge, La., 1994.

13. Mohammad, L.N., and Paul, H.R., "Evaluation of the Indirect Tensile Test for Determining the Structural Properties of Asphalt Mix," National Academy of Science, Transportation Research Record No.1417, 1993, pp. 58-63.

14. Aschenbrener, Tim, "Evaluation of the Hamburg Wheel-Tracking Device to Predict Moisture Damage in Hot Mix Asphalt," Preprint, Transportation Research Board, 74th Annual Meeting, 1995, Paper No. 950475.

15. ..., "Hamburg Wheel Tracking Device," Technotes, A publication by Elf Asphalt Research, 1990.

Table 1. Asphalt Cement Specifications and Properties

SPECIFICATIONS		
Original Properties		
	AC - 30	LADOTD PAC-40 HG
Viscosity, 140°F (60°C), Poises	3000 ± 600	4000 +
Viscosity, 275°F (135 C), centiStokes	min. 350	max. 2000
Penetration, 77°F (2 °C), 100 g, 5 s, min.	50	50 - 75
Flash Point, Cleveland open cup, min. °F (°C)	450 (232)	450 (232)
Solubility in trichloroethylene, min., %	99.0	99.0
Separation of Polymer, 325°F (163°C), 48 hrs Difference in softening point from top and bottom, max., %	-	4.0
Force Ductility Ratio, (F_2/F_1, 39°F (4°C), 5 cm/min, @ 30 cm elongation, min.) F_2 = F @ 30 cm, F_1 = Peak Force	-	0.3
Test on residue from thin-film oven test		
Viscosity, 140°F (60°C), Poises Ductility, 77°F (25°C), 5 cm/min, min., cm Penetration Retention, min., % of original	15000 40 -	- - 50
Elastic Recovery, 25°C @ 10 cm elongation, % min.	-	60
PROPERTIES		
Original Properties		
Asphalt Cement Source	Marathon	Koch
Asphalt Cement Grade	AC - 30	PAC - 40 HG
Penetration @ 25 °C	57	59
Specific Gravity 25 / 25 °C	1.03	1.03
Absolute Viscosity @ 60 °C, Poises	3332	6629
Properties after R. T. F. O.		
Penetration @ 25 °C	30	38
Absolute Viscosity @ 60 °C, Poises	13448	18524

Table 2. Statistical Grouping of Indirect Tensile Creep Test Results

Limestone Mix								
Temp	25 °C		40 °C		25 °C		40 °C	
Property	CS		CS		TF		TF	
Mix Type	Mean	Group	Mean	Group	Mean	Group	Mean	Group
SAC30N	0.45	A	0.62	A	2264	B	37	B
SAC30L	0.39	A	0.57	A	3414	A	49	B
SPMACN	0.42	A	0.54	A/B	2125	B	37	B
SPMACL	0.29	B	0.43	B	3600	A	193	A
Gravel Mix								
Temp	25 °C		40 °C		25 °C		40 °C	
Property	CS		CS		TF		TF	
Mix Type	Mean	Group	Mean	Group	Mean	Group	Mean	Group
GAC30N	0.38	A	0.76	A	3600	A	59	B
GAC30L	0.35	A/B	0.68	B	3600	A	84	B
GPMACN	0.31	B	0.46	C	3600	A	116	B
GPMACL	0.31	B	0.36	D	3600	A	777	A

CS: Creep Slope for Calculated Indirect Tensile Creep Stiffness (kPa / sec)

TF: Time to Failure (sec)

Group: Statistical Grouping (columns with same letters indicate no significant difference)

Table 3. HWTD Test Indices for Test Mixtures

	Limestone Mixtures				Gravel Mixtures			
Test Indices/Air Voids	SAC30N	SAC30L	SPMACN	SPMACL	GAC30N	GAC30L	GPMACN	GPMACL
Air Void Levels (percent)	4.5	3.9	5.4	5.3	6.3	6.1	6.7	5.5
Post Compaction, mm	0.6	1.6	0.9	1.4	1.1	1.2	0.2	1.4
Creep Slope, passes/mm	2290	9800	4110	11350	18182	9756	10810	11765
Stripping Slope, passes/mm	385	1170	820	4350	[1]	[1]	[1]	[1]
Stripping Inflection Pt, passes	7385	9325	9815	14290	[1]	[1]	[1]	[1]
Rut Depth at 20,000 passes	[3]	16.0	11.8	4.1	2.6	3.8	2.2	3.9
Passes to 20 mm	13700	29600	23300	89000	[2]	[2]	[2]	[2]

[1] The mixes did not reach the state of stripping for the duration of the test.

[2] Sample reached 20 mm rut depth before 20,000 passes.

[3] The mixes did not reach a rut depth of 20 mm by the end of the test.

Table 4. Statistical Grouping of Indirect Tensile Fatigue Test Results

Limestone Mix				
Temp	25 °C			
Property	FS		N_f	
Mix Type	Mean	Group	Mean	Group
SAC30N	0.66	A	45118	B
SAC30L	0.58	A/B	77785	B
SPMACN	0.56	B	54985	B
SPMACL	0.47	C	476774	A
Gravel Mix				
Temp	25 °C			
Property	FS		N_f	
Mix Type	Mean	Group	Mean	Group
GAC30N	0.66	A	173652	B
GAC30L	0.59	A/B	327185	B
GPMACN	0.55	A/B	204851	B
GPMACL	0.47	B	1514986	A

FS: Fatigue Slope for Permanent Deformation (mm / sec)
N_f: Number of Cycles for Sample to Reach Failure
Group: Statistical Grouping (columns with same letters indicate no significant difference)

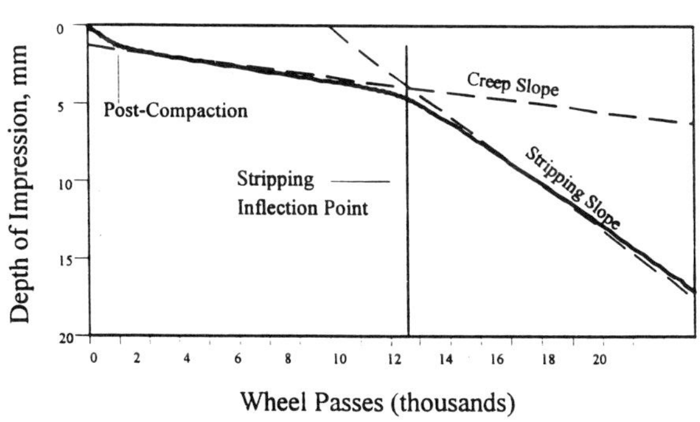

Figure 1. Data Analysis Indices for the Wheel Tracking Test

of hydrated lime to the aggregate before the production of HMA, this study was designed with the following objectives in mind;

1- To quantify moisture resistance of HMA specimens prepared in the laboratory of perfectly coated aggregate particles with hydrated lime, and specimen made with aggregate taken from an asphalt plant coated with hydrated lime using sprinkling method during cold feed.

2- To examine whether aggregate, treated with hydrated lime, produced in an asphalt plant can be duplicated in a laboratory setting, i.e. field simulation. The comparison is made through measuring the Indirect Tensile Strength (ITS) values of different specimens subjected to the same loading and environmental conditions to asphalt specimens made with the two aggregates.

3- To examine moisture resistance of specimens made with hydrated lime and those made with liquid antistripping agents.

INTRODUCTION

Moisture induced damage, i.e. loss of adhesion between asphalt and aggregate reduces the ability of some asphalt mixture to resist traffic loads and other environmental distresses. Pavements susceptible to moisture damage tends to strip, and ultimately yield inferior ride quality to motorists and higher cost. Several factors are known to cause stripping, such as pore pressure, surface polarity, presence of surface and subsurface moisture to name a few (1). Because of the complicated nature of these factors and the presence of interaction among them, it is impossible to eliminate the stripping problem without using an antistripping agent.

In a recent study (2) laboratory tests revealed that ITS and Tensile Strength Ratio (TSR) values for HMA specimens containing aggregate particles perfectly coated with hydrated lime were not significantly different from those ITS values produced by specimens containing liquid antistripping agents. In another recent study by the Virginia Transportation Research Council (3) it was concluded that hydrated lime showed no superiority over liquid antistripping additives. That conclusion was made after several overlay sections containing various antistripping agents were tested for TSR values. The use of liquid antistripping agents has been experienced by several public and private agencies with great success. The State of Minnesota, for example, used liquid antistripping agents in several projects some were built in 1990 such as US-169 near Belle Plaine, and T.H. 61 (N.B.) North of Winona. These two projects continue to perform very satisfactory as of the summer of 1997. Recently constructed projects include I-35 (S.B.) North of Owatonna in 1995, and I-94 (W.B.) Near Alexandria in 1996 also showed excellent field performance with no indication of any moisture damage after one year of service.

Although the development of tests to determine water sensitivity of HMA began in the 1930's, none of the tests procedures has emerged as acceptable over a wide range of conditions or materials, and none were performance related (4). The effort of five years of intensive research in asphalt binder and mixtures by the Strategic Highway Research Program (SHRP) resulted in some tests that can predict and evaluate moisture damage of HMA in the laboratory before construction (5,6,7). The new methods proposed by SHRP has not gained wide acceptance yet among asphalt researchers and technologists, and some of the existing methods like AASHTO T-283, and ASTM D-4867 continue to dominate the evaluation of moisture damage of HMA.

For an antistripping agent to be adopted by an agency it should have the following characteristics: **1-** Economical, **2-** Clean and easy to apply, **3-** Does not pose any health problems, **4-** Remain where it is needed the most, at the asphalt-aggregate interface, and **5-** Uniformly distributed on either aggregate surface or the asphalt binder mass.

Over the years asphalt technologists devised five different methods of incorporating hydrated lime into HMA. Some of these methods are crude and simple, while others are more effective but expensive. The five different methods are discussed in details in an NCAT publication (8).

The method used in this study is simply sprinkle the specified amount of hydrated lime on the coarse portion of the aggregate on the conveyor belt during cold feed and before the aggregate reaches the drum dryer. This method is by far the lease expensive and the most frequently used by the asphalt industry. In another treatment aggregate samples were obtained from an asphalt batch plant from the discharge end of the drum dryer during the production of lime treated HMA.

MATERIAL SELECTION

The materials shown in Table 1 were used in this study.

LABORATORY TESTING

Only two relevant tests were used to assess moisture damage to asphalt mixture specimens, namely ASTM D-4867, and AASHTO T-283. In a recently published report ,NCHRP # 373, it was concluded that ASTM is an effective laboratory test for evaluating moisture damage and additive effect (9). This conclusion was based on testing asphalt concrete mixture cores over 6- to 8-year period. Both Indirect Tensile Strength in psi or Kpa, and Tensile Strength Ratios were determined for each treatment.

The other test that was used in this study is AASHTO T-283 "*Resistance of Compacted Bituminous Mixtures to Moisture Induced Damage*"(10). In that test specimens are subjected to freeze/thaw (F/T) cycles then ITS is determined for each treatment. Although this test takes several days to complete, it is considered very reliable in assessing moisture damage for HMA under repeated F/T cycles.

Table 1: Materials used in the study

MATERIAL	NAME	SOURCE
1- Aggregate	Sioux Quartzite	Dell Rapids, S.D.
2- Binder	AC-20 = PG 64-22(SHRP)	Jebro Inc. Sioux City, IA.
3- Antistripping Agents	***I- Liquid*** a- REDICOTE 82-S b- Tall Oil Z-940	 AKZO Chemicals, Chicago, IL. Union Camp, Atlanta, GA.
	II- Hydrated Lime	SD Cement Plant, R. City, SD

In preparing specimens for laboratory tests, the following treatments were used:

1- Specimens containing perfectly coated aggregate with hydrated lime.
2- Specimens containing REDICOTE 82-S liquid antistripping agent.
3- Specimens containing Tall Oil Z-940 liquid antistripping agent.
4- Specimens made with aggregate treated with hydrated lime and obtained from asphalt plant (batch type).
5- Specimens made in the lab with aggregate mixed with hydrated lime in a manner similar to that used in the asphalt plant.

6- Specimens containing no antistripping agents of any kind. This treatment was considered a control treatment.

ANALYSIS OF TEST RESULTS

The determination of Indirect Tensile Strength (ITS) and Tensile Strength ratio (TSR) can be reviewed in either ASTM or AASHTO standards and no discussion on how the tests are conducted will be presented here. Test results of the six different treatments, however, are presented in Tables 2, and 3 for ASTM D-4867 and AASHTO T-283 respectively.

A- ASTM D-4867

This test evaluates the Indirect Tensile Strength (ITS) of dry (unconditioned) specimens and moisture conditioned specimens after 24 hours of soaking and no more than 80% saturation. The average ITS values for the moisture conditioned specimens are divided by the average ITS values of the dry (unconditioned) specimen to determine Tensile Strength Ratio (TSR). Most specifications sets the ratio at a minimum of 80 %, others might lower the ratio to a minimum of 70-75%. The data of Table 2 reveals that all treatments, with no exception passed the conservative 80% minimum. Therefore the data does not justify any further analysis.

It should be pointed out here that treatments with similar TSR values do not necessarily exhibit the same type of moisture resistance. For example treatments #1, and #6, both have the same TSR values but individual ITS values for treatment #6 are on the average higher than those of treatment #1 by 36%. Therefore caution must be exercised when evaluating TSR values for different treatments.

B- AASHTO T-283

Despite its lengthy time to conduct, and the complicated logistics associated with specimen handling through various stages of testing, this test method remain an effective means of evaluating long-term moisture damage to asphalt mixtures. In this test specimens are subjected to repeated freezing and thawing cycles (F/T). At the end of the specified number of F/T cycles, the Indirect Tensile Strength is determined. The results of this test can be used to predict long term stripping susceptibility of asphalt mixtures, and evaluating liquid antistripping additives which are added to the asphalt cement or pulverulent solids, such as hydrated lime, which are added to the mineral aggregate (AASHTO 1995).Table 3 shows results of the AASHTO T-283 test with zero, 3, and 6 F/T cycles.

Table 2: Tensile Strength Ratio for Different Treatments ASTM D-4867

TREATMENT	ITS (psi)		TSR (%)
	WET	DRY	
1- Perfectly Coated Agg. with Hyd. lime in the lab.	168.2, 159.7, 161.5	199.8,197.1, 192.4	83.1
2- REDICOTE 82-S	118.5, 120.2, 125.7	128.3, 139.6, 135.7	90.0
3- Tall Oil Z-940	126.7, 125.9, 115.8	147.7, 140.6, 134.9	87.o
4- Asphalt Plant Coating	139.1, 144.3, 129.7	158.9, 147.3, 151.5	90.1
5- Simulated Field Coating in the Laboratory.	162.3, 154.0, 158.9	173.3, 179.6, 177.2	89.6
6- No antistripping agents (Control Treatment)	117.2, 123.6, 120.0	142.7, 143.2, 148.1	83.1

Figure 1 represents a graphical illustration of data shown in Table 3. The most appropriate statistical tool to use to analyze the data of Table 3 is regression analysis due to the limited number of data point within each treatment. The results of Figure I suggested that the six treatments should be divided to three groups, the hydrated lime group (treatments 1, 4, and 5), the liquid antistripping group (treatments 2, and 3), and the control treatment (treatment 6). Therefore three regression applications were conducted for the previous three groups to examine whether the mean of each treatment, within a group, is similar or not to the mean of other treatments.

In each group a general model was developed for ITS values in terms of intercept values as well as coefficients for number of F/T cycles. Dummy variables were used to develop individual model for each treatment within the group. Individual models took a general shape like;

$$ITS = B_0 + B_1\, C + B_2\, C^2$$

I- The Hydrated Lime Group

The mean ITS value of treatments #4 and #5 were not significantly different at the 95% confidence level. Treatments #1 and #4 were also not significantly different at the 90% level. Therefore for all practical purposes the three treatments were not significantly different from each other.

Table 3: ITS values, in psi, for Different Treatments During AASHTO T-283 Test

TREATMENT	FREEZE/THAW CYCLES		
	NONE	3	6
1- Perfectly Coated Agg. with Hydrated lime in the lab.	181.6, 145.7 170.3, 163.1 **165.2**[a]	108.6, 101.3 107.2, 104.6 **105.5**	78.5, 75.1 73.8, 79.2 **76.7**
2- REDICOTE 82-S	123.2, 132.1 **127.7**	123.1, 111.3 **117.2**	109.2, 113.2 **111.2**
3- Tall Oil Z-940	123.1, 143.2 **133.2**	135.1, 103.2 **119.2**	98.8, 109.2 **104.0**
4- Asphalt Plant Coating	141.2, 145.1 **143.2**	103.7, 99.8 **101.8**	70.9, 69.1 **70.0**
5- Simulated Field Coating in the Laboratory.	150.6, 142.3 **146.5**	97.2, 103.7 **100.5**	67.6, 69.9 **68.8**
6- No antistripping agents (Control Treatment)	97.6, 104.5 102.1, 98.3 **100.6**	75.7, 61.0 69.1, 66.7 **68.1**	57.6, 54.8 57.2, 56.1 **56.4**

[a] Average of the group

II- The Liquid Antistripping Group

The mean ITS value for treatments# 2, and #3 were not significantly different at the 95% confidence level, implying that both liquid antistripping additives were very similar in yielding ITS values at different F/T cycles.

III- The Control Treatment

When treatment #1 was compared with the worst moisture resistant treatment #5 (examine Figure 1), the difference was statistically significant. This result confirms that the control treatment #6 is significantly different than all other treatments containing any type of antistripping additive.

CONCLUSIONS

1- Treatment #6 (control treatment without any antistripping agent) produced much lower ITS values than any of the other five treatments where one type of antistripping agent, liquid or Hydrated lime, was present in the HMA.

2- When treatments #2(containing liquid agent REDICOTE 82-S), and #3(containing liquid agent Tall Oil Campamid Z-940) were compared, there were no significant difference between the means. The model however could be improved if linear terms instead of quadratic terms were considered.

3- All treatments containing Hydrated lime behaved similarly in resisting moisture damage. Perfectly coated aggregate with Hydrated lime (treatment #1), aggregate mixed with hydrated lime in the asphalt plant (treatment #4), and aggregate treated with hydrated lime in the lab in a similar fashion of that practiced in the asphalt plant, i.e. simulated coating (treatment #5) all produced similar ITS values at different F/T cycles.

4- Treatments with hydrated lime showed higher initial ITS values at the intercept. Once F/T cycles were applied, ITS values dropped at a fast rate so that after six F/T cycles ITS values were significantly lower than those produced by treatments #2, and #3 which contains liquid antistripping agents.

5- It is believed that treatment #1 (perfectly coated aggregate in the laboratory) is the best way to coat aggregate with hydrated lime. Other methods such as Lime slurry can be as geed but not better. The results of this study showed that even with the best way of mixing hydrated lime with aggregate moisture resistance was lower, i.e. lower ITS values after few F/T cycles, than those produced by treatments containing liquid antistripping agents.

6- When liquid antistripping agents were incorporated into HMA, specimens produced higher ITS values than those specimens treated with hydrated lime after few F/T cycles.

7- The sprinkling method used in asphalt plants to mix hydrated lime with aggregate can be duplicated in a laboratory setting. This can assist in evaluating field performance of lime treated HMA before mass production for a particular job.

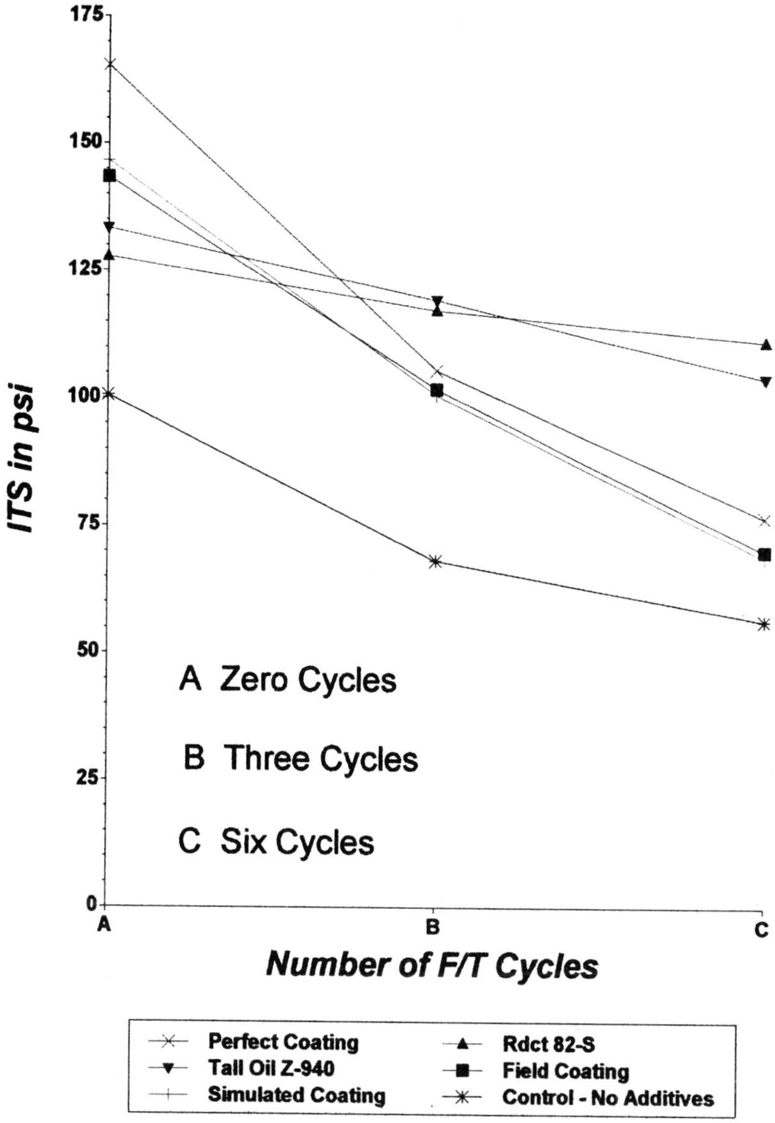

Figure 1 ITS Values at Different Freeze/Thaw Cycles

RECOMMENDATIONS

In light of the above conclusions it is recommended that the asphalt industry pays a closer attention to the use of liquid antistripping additives as an alternative to hydrated lime. Contractors could be given the choice to use either hydrated lime or a liquid antistripping agent.

REFERENCES

1- Selim, Ali A.,"Liquid Antistripping Additives vs. Hydrated Lime - Issue Revisited". Final report to the Quartzite Rock Association. August 1997.

2- Selim, Ali A., "Eliminating the Stripping Potential of the Sioux Quartzite by Means Other than Hydrated Lime". Final Report to the Quartzite Rock Association. October 1994.

3- Maupin, G. W.,"Follow-up Field Investigation of the Effectiveness of Antistripping Additives in Virginia". Virginia Transportation Research Council Project # 9398-010-940. Virginia DOT, January 1997.

4- Terrel, R. L.; Shute, J. W., "Summary Report on Water Sensitivity". SHRP-A/IR-89-0033, Strategic Highway Research Program, National Research Council, Washington .C. 1989

5- Curtis, C.; Epps, J.; Ensley, K. "Fundamental Properties of Asphalt-Aggregate Interactions Including Adhesion and Adsorption". SHRP-A-341, 1993.

6- Scholz, T; Bea, J.; Terrel, R.; Al-Joaib, "Water Sensitivity: Binder Validation". SHRP-A-402, 1994

7- Terrel, R.; Al-Sawalmi, S., "Water Sensitivity of Asphalt-Aggregate Mixes: Test Sections".. SHRP-A-403, 1994.

8- Hanson, D; Graves, R; Brown, E. "Laboratory Evaluation of The addition of Lime Treated Sand to HMA" NCAT, Report #94-0821, 1994.

9- Tunnicliff D.; Root, R., "Use of Antistripping Additives in Asphalt Concrete Mixtures -Field Evaluation". NCHRP Report # 373. Transportation Research Board, National Research Council. 1995.

10- AASHTO, "Standard Specifications for Transportation Materials and Method of Sampling and Testing". American Association of State Highway and Transportation Officials. Washington DC, 1995.

Fifth International Conference on the
BEARING CAPASITY OF
ROADS AND AIRFIELDS

DETERMINATION OF SHEAR STRENGTH OF ASPHALT LAYERS IN THE FIELD

By
ABD EL HALIM OMAR ABD EL HALIM[1]

[1]Professor, Civil and Environmental Engineering department
Carleton University
Ottawa, Canada

and

RAGAB ABDEL NABI[2]
ASSEM ABD EL ALEEM[3]

[2]Assistant Professor and [3]Professor, Civil Engineering Department
Zagazig University
Shoubra, Egypt

ABSTRACT

One of the main problems hindering the development of reliable models for the prediction of long term performance of asphalt pavements is the absence of reliable testing methods or procedures which can directly measure the mechanical properties of the different mixes in the field. At the present time, surface deflections under wheel load using the plate test method or the falling weight deflectometer are used to generate the elastic modulus of each of the pavement layers. The calculated values are then used as input for performance models.

Verification of such models indicated that many problems exist with the adoption of moduli values to predict the long term performance of asphalt pavements.

The need to provide more reliable information which reflect the actual performance of the asphalt layer as built in the field led to the development of a new testing facility termed "In-Situ Shear Strength Test ". The testing facility has been designed and built to determine the shear strength of asphalt mixes in the field. A comprehensive experimental program was carried out to verify the reliability of this new test facility. Also, the laboratory determined shear strength of asphalt cores recovered from the field were evaluated and compared with those measured in the field. This paper presents a description of the developed testing machine, outline of the testing program and test results showing the validity of using this new testing method in the field. In addition, the results of comparing the field and laboratory shear strength values showed that field strength is at least 50 % higher than laboratory determined values.
Key Words: Asphalt , Pavement, Testing, Shear, In-situ

INTRODUCTION

Evaluation of the performance of asphalt pavements in the field has been one of the most important issues in pavement industry. Information on the strength of the asphalt layer, surface distress and remaining life is very essential to manage the highways. At the present time, pavement performance is measured by carrying out non-destructive field tests or laboratory strength tests on asphalt specimens recovered from the field. However, strength of pavement could not be determined due to technical and economical difficulties.

The design of an in-situ test method for the evaluation of asphalt pavement in the field was selected to satisfy the following conditions: (1) The measured mechanical property should be significant to the overall performance of the asphalt layer, (2) The testing technique should be carried out with as little damage to the asphalt layer in the field as possible, (3) The test facility should provide comparative results about the shear strength at different points throughout the thickness of the asphalt layer, and (4) The test facility should be easy to handle, simple to operate and can be used under wide range of climatic conditions.

As a result, shear strength was adopted as the pavement performance indicator. This finding was further supported by the new guideline developed by the Strategic Highway Research

Program (SHRP). The authors adopted torsion as means of inducing shear stresses in the field due to both theoretical and practical reasons (1,2).

DEVELOPMENT OF THE IN-SITU SHEAR TEST FACILITY

Torsion was adopted as the source to induce pure shear to the asphalt test section. The development of the test facility, shown in Figure 1, considered the shape and dimensions of the test samples in the field,rate of loading, temperature during the test, recording of the test data and mobility of the test facility. The shape and dimensions of the samples in the field were selected to match laboratory prepared asphalt samples. As a result, the shape of the test specimens was selected to be a cylinder. The dimensions are the diameter of the test specimen and its depth. The diameter of the specimen is limited by the maximum torque that can be applied by the test facility. For a light weight test facility it was found that the maximum value of the diameter should not exceed 150 mm (6 inches). The depth of the sample will depend on the availability of a coring equipment. In the case of non-availability of a coring machine, the torsion can be applied directly on the surface of the asphalt (referred to as flush). On the other hand, if a coring machine is available, the shear test can be performed at any depth according to the preference of the engineer, (drilled). Also, cored samples from the test site (Figure 2) can be recovered and tested in the laboratory The speeds under which the test is performed is similar to the speeds of the laboratory test facility. Thus, comparisons can be made between shear strengths of field and laboratory tests. Due to the effect of temperature on the bonding agents used in the field during the testing procedure, field testing was performed only when temperatures are above 10 degrees. The test facility has been equipped with a recording data processor which can monitor and record up to 1000 data point. The recorded data include torque-deformation relationship, maximum torque and maximum deformation at failure. The test data can be retrieved later and fed to PC for further analysis. The design of the test facility ensured mobility through selection of light weight materials and the addition of four wheels for ease of movement.

TESTING PROGRAM

One of the advantages of the in-situ shear test facility is the ability to test the asphalt layer at different points throughout its thickness, or from the flush conditions to any drilled one. As a result, the pavement layer is partially drilled to a pre-determined depth. The testing program

Figure 1: Photo of the New In-situ Shear Test Facility

Figure 2: Photo of the In-situ Test Arrangement and the Recovered Cores

included 8 different standard Ontario asphalt mixes constructed on a number of access roads at the National Research Council of Canada (NRC). Comparing the measured in-situ shear strength of the partially drilled layer with the laboratory shear strength of full depth cores recovered from the same will provide is an important step to verify the in-situ test facility. Therefore, asphalt cores recovered from a number of the NRC test sections were tested in the laboratory. These cores were tested in full, one half, and one third their depth, respectively. The test results of the field cores were compared with results of shear strength tests performed on laboratory prepared specimens made from similar asphalt mix. The next step involved the comparison between the shear strength of field cores recovered from different test sections. The third and final stage dealt with the comparison between the laboratory measured shear strength and those measured in the field using the in-situ shear test facility.

FIELD TESTING

The in-situ shear test facility was verified using eight test sections with different asphalt mixtures constructed at the National Research Council in Ottawa. The tested sections were subjected to the same environmental conditions and were not opened to traffic. It was important to examine the ability of the test facility to recognize and/or distinguish between different asphalt mixes through their field measured shear strengths. Clearly, if the difference between the mixes is significant, then a strength test facility can be justified and used to rate and evaluate the performance of asphalt mixes in the field for the first time. The field tests were performed following two different techniques. The first technique was to carry out the field test without coring or drilling any of the asphalt surface (flush). The second technique was to drill about 20 mm deep before performing the shear test. It should be noted that the second technique is the one most simulated by laboratory shear tests. On the other hand the flush test is more representative of field conditions as will be discussed later.

The performance of the shear test in the field is simple and straightforward. The following steps constitute the procedure adopted:

1- Select the test section and determine the number and positions of the tests to be performed in the field

2- Bond the steel plates directly to the clean-dry surface using a proper epoxy. In case of drilling, the bonding of the plates may follow the drilling operation. Let the epoxy set (between 5 minutes and 24 hours) depending on the type of the epoxy, the traffic volume, time of the testing and the size of the project.

3- Connect the torsion shaft to the bonded plate and ensure that all connections are complete. Select the rate of loading and start the test.

4- At the end of each test ensure that the readings were recorded, take all necessary measurements (temperature, size of the plate, depth and mode of failure) and record any important observations. See Figures 1 and 2.

5- Repair the surface by adding cold mix and compact it.

Each shear test takes between 1 to 3 minutes depending on the speed of the load application and the size of the test section. Thus, a total of 12 test per each test road can be completed in less than one hour. It should be noted that very little damage is sustained by the pavement as a result of the test.

LABORATORY TESTING

Asphalt core samples were recovered from each of the eight field test sections. The core specimens were transported to the Civil and Environmental Engineering Laboratory at Carleton University. The laboratory testing program consisted of measurement of the densities and the determination of the shear strength of the core specimens. Shear tests were performed under room temperature and at speed similar to the one performed in the field. It is important to identify the main differences between the tests carried out in the field and those in the lab. Firstly, while the temperature in the field cannot be controlled the temperature during the laboratory testing was constant. Secondly, the testing in the field is performed on the surface without inducing any pre-damage to the asphalt. This advantage may be lost due to the coring, transporting and storing the recovered cores before testing in the laboratory. Thirdly, the laboratory test results can be directly used to calculate the shear strength of the asphalt. Test results obtained from the field, especially those performed without drilling, may require very complex mathematical equations. As a result of these differences, field tests were performed under temperatures closer to those observed during the laboratory testing. In addition, field tests were performed on the asphalt surface with and without drilling. The applied torque at failure is recorded and the shear strength can be calculated using Equation 1 :

Install Equation Editor and double-click here to view equation. (6.1)

where:

t : The calculated Shear Strength (MPa)

M : The applied Torque (N.m)

R : The radius of the tested Sample (m)

J : The polar moment of inertia (m^4)

Similarly, the shear strength of the field test samples was computed using Equation (1). Since both tests were performed on the same asphalt mix it was assumed that the difference between their results is due to the boundary conditions in the field in addition to the fact that some damage may have been inflicted on the laboratory tests samples during coring, transportation, storing, and preparing samples for testing. A total of 151 shear test were performed in this investigation. The available details of the used mixes provided from the National Research Council were published in earlier reports (1 to 4).

TEST RESULTS AND ANALYSIS

Length of Specimen

Tables 1 to 2 show the measured shear strength for specimens of different lengths for field cores and laboratory prepared specimens made of the same mix. As can be noted from Table 1, the shear strength of the laboratory prepared samples of mix a varies from 1.325 MPa for specimen length of 21 mm to 0.658 MPa for specimen length of 58 mm. Also, the mean value of each tested hight indicated that shear strength may increase as the depth decreases. The calculated means were 0.816, 0.824 and 0.966 MPa for full, half and third the hight of the sample, respectively. Table 2 shows similar results for the field asphalt cores of Mix a. The shear strength ranged from 1.225 MPa for specimen length of 19 mm to 0.940 MPa for specimen length of 54 mm. The mean strength values were 0.987, 1.084, and 1.100 MPa for full, half, and one third the depth of the specimens, respectively. Furthermore, the results shown in Tables 1 and 2 suggest that while the tests were performed on the same mix using the same test facility, a difference of up to 30% may exist between field and laboratory mixes due to different compaction, aging, and environmental conditions.

Table 3 shows shear strength results of laboratory tests performed on asphalt cores recovered from the field section of Mix b. It should be noted that Mix a has a larger maximum size aggregate than Mix b. The shear strength ranged from 1.517 MPa for specimen length of 16 mm to 1.155 MPa for specimen length of 54 mm. While the mean values of the shear strength of Mix b are higher than those of Mix a, the results of both Mixes suggest that the measured shear strength increases as the depth of the sample decreases. This result may explain why thinner asphalt layers may perform better to resist surface rutting than thicker ones. Similar results were obtained from tests performed on the other 6 mixes. As a result, it was concluded that the shear strength can be used to evaluate different asphalt mixes in the field. Also, it was apparent from the results that a thickness of 20 mm or more will be more representative of the average shear strength of the mix. Furthermore, it was established that a difference of at least 20 % is expected between the shear strength of field and laboratory mixes.

Laboratory and In-Situ Shear strength Test Results

The in-situ shear test facility was utilized to measure the shear strength of the 8 asphalt mixes used to construct the NRC access roads. Table 4 shows the mean and coefficient of variation (CoV for each test section. A minimum of 6 tests were carried out to obtain the mean values given in the table. Thus, for the laboratory and in-situ (drilled and flush) shear strength a total of at least 144 tests were carried out. In general, the test results were consistent as shown by the low values of the Coefficient Of Variation (C.O.V.). As shown in the table, the maximum calculated C.O.V. was 22% . Also, it was interesting to note that the lowest values of C.O.V were the ones calculated for the in-situ flush test method. This observation suggests that very little damage/unexplained variation have resulted from this test method. As can be seen in the tables the values of the C.O.V. were between 4% and 11%, while in the case of the drilled tests the maximum variation was 22%. This results suggest that coring, drilling and disturbing the field conditions can result in a higher C.O.V. and as a result less consistent measurements. The shear test results showed that the testing method can have a significant effect on the measured property. As shown in the table, the ratio between field and laboratory shear strength values was between 2.135 for Mix d to 6.659 for mix h for the flush test conditions.

In the case of a 20 mm drilling, the ratio was between 1.898 to 4.947. Furthermore, the test results shown in Table 4 support the finding that changing the boundary conditions in the field can result in a difference of up to 35 % as shown in the two in-situ test methods in the case of Mix h. These results were consistent for all the field asphalt mixtures.
In addition to the above results, the in-situ shear strength succeeded in identifying each mix with unique shear strength and coefficient of variation values. For example, Mix c is a modified and improved mixture of Mix a. The results in Table 4 support this observation. As seen in the table, there is between 10 % and 31 % improvement in the measured shear strength of both mixes. Finally, Mix h which has the highest in-situ shear strength values is a standard mix which is containing the highest portion of coarse aggregates in the 8 tested mixes.

CONCLUSIONS AND RECOMMENDATIONS

This paper presented the main results of laboratory and field testing program to justify the manufacture and subsequent use of a new in-situ shear test facility. The new facility was shown to be simple and reliable tool to determine an important engineering property of asphalt mixes as constructed in the field. Furthermore, the results presented in this paper support the following conclusions :

1- The developed In-Situ Shear Test Facility is a valid device for the determination and evaluation of asphalt mixtures in the field.

2- The shear strength can be utilized and adopted as a performance indicator. It is shown to be more reliable than the currently used asphalt density.

3- The in-situ test results showed more consistency than the test results obtained from laboratory tests. This would imply that field coring or disturbing asphalt layers in the field before testing may in fact induce damage to the test samples resulting in underestimating the actual strength of the pavement structure. This can lead to the conclusion that the pavement may require maintenance earlier than expected.

It is clear that the design and development of the in-situ shear test facility would provide pavement engineers and researchers with a much needed tool for asphalt management and research.

REFERENCES

1. Abdel Nabi, Ragab, "Effects of Highway Geometry on Pavement Distress", Ph.D. Thesis, Carleton University, Canada, 1995

2. Abd El Halim, A.O., Rickards, I., Haas, R.C., and Abd El Nabi, R.M., "The Development and Use of In-Situ Shear Test Device for Evaluation of Asphalt Performance", 8th ISAP International Conference on Asphalt Pavements, Vol. II, pp.1311 to 1328, Seatle, 1997.

3. Abd El Halim, A.O, R.Abd El Nabi, A. Abd El Aleem and Haas, R., " Effect of Highway Geometry and Construction Equipment on the Problem of Reflection Cracks", Proc. the Third International Conference ,Reflection Cracking in Pavements, Maastrichts, The Netherlands pp. 9 to 17, October 1996.

4. Abdel Nabi, R. M., A. O. Abd El Halim, A. M. Abd El Aleem and H. M. El-Hussein, "The Use of New In-Situ Test Machine to Evaluate the Shear Strength of Asphalt Layers" Annual Conference, Canadian Society of Civil Engineering, Volume II, Ottawa, 1995, pp. 89 to 98.

Table 1: Effect of Hight of Test Sample on Shear Strength of Laboratory Prepared Asphalt Specimens, Mix a

Size/location	H, mm	Specific Gravity	Shear Strength
full	58	2.184	0.658
full	58	2.286	0.897
full	60	2.323	0.893
H/2	32	2.226	0.785
H/2	35	2.236	0.847
H/2	30	2.233	0.847
H/2	31	2.253	0.818
top	25	2.201	0.796
middle	25	2.212	0.811
bottom	15	2.200	0.830
top	24	2.169	0.981
middle	15	2.205	0.919
bottom	23	2.190	0.833
top	21	2.271	1.325
middle	21	2.254	1.158
bottom	17	2.230	1.038

Table 2: Effect of Hight of Test Sample on Shear Strength of Cores Recovered from the Field, Mix a

Size/location	H, mm	Specific Gravity	Shear Strength
full	54	2.331	1.033
full	56	2.336	0.940
full	54	2.338	0.989
H/2	26	2.316	1.061
H/2	27	2.322	1.104
H/2	28	2.332	1.121
H/2	28	2.328	1.050
top	19	2.356	1.225
middle	18	2.356	1.165
bottom	18	2.361	0.995
top	18	2.331	1.181
middle	18	2.342	1.099
bottom	20	2.338	0.934

Table 3: Effect of Hight of Test Sample on Shear Strength of Cores Recovered from the Field, Mix b

Size/location	H, mm	Specific Gravity	Shear Strength
full	51	2.377	1.124
full	54	2.372	1.155
full	52	2.365	1.186
H/2	27	2.380	1.204
H/2	27	2.369	1.149
H/2	26	2.356	1.228
H/2	26	2.355	1.186
top	16	2.390	1.462
middle	24	2.377	1.308
bottom	16	2.278	1.517
top	18	2.400	1.484
middle	19	2.354	1.011
bottom	16	2.345	1.165

Table 4: Effect of Mix Type on In-situ and Laboratory Shear Strength

Mix Type	Laboratory, MPa (Coeff.of Variation)	In-situ Shear Strength	
		Drilled	Flush
a	0.616, (12%)	1.639, (6%)	2.374, (10%)
b	0.952, (20%)	2.308, (9%)	2.511, (8%)
c	0.758, (11%)	2.141, (13%)	2.608, (4%)
d	1.501, (14%)	2.849, (5%)	3.205, (4%)
e	0.865, (23%)	2.385, (5%)	2.691, (4%)
f	0.651, (21%)	1.655, (22%)	2.280, (11%)
g	0.507, (20%)	2.322, (17%)	2.541, (5%)
h	0.551, (12%)	2.726, (5%)	3.669, (5%)

Fifth International Conference on the
BEARING CAPASITY OF
ROADS AND AIRFIELDS

IMPROVING SHORT AND LONG TERM PERFORMANCE OF ASPHALT MIXES THROUGH THE USE OF AMIR COMPACTOR

ABD EL HALIM OMAR ABD EL HALIM [1] AND OSAMA OKAIL [2]

[1] Professor, Department of Civil and Environmental Engineering, Carleton University, Ottawa, Ontario, Canada
(613)520-3951, ahalim@ccs.carleton.ca
[2] Professor, Department of Civil Engineering
Ain Shams University, Cairo, Egypt
(011202)283-9496

ABSTRACT

Conventional rolling equipment, while capable of achieving a specified density, results in construction induced cracks, often visually apparent, and these are due to a mismatch between the geometry and relative rigidity of the roller and the asphalt mix. This led to the design and construction of a new compactor, the Asphalt Multi Integrated Roller (AMIR). In effect it replaces the cylindrical shape with a moving flat plate and results in a crack free asphalt layer and more uniform compaction along and across the mat while achieving the same if not better density with less passes than currently used rollers. A number of field trials and laboratory experiments have been carried out in Egypt, Canada and Australia. Extensive testing of cores, beams and slabs provided data on density, voids plus tensile strength, stripping and fatigue resistance. This paper briefly describes the development of the AMIR compactor. It then describes, in more detail, comparative test results performed on asphalt specimens from conventional and AMIR compacted sections. These results show significant improvements in strength, and fatigue life for the AMIR compacted specimens.

Key words: Asphalt, Compaction, Amir, Testing, Performance

INTRODUCTION

Construction induced cracks, known as "checking", have been observed to occur and pavement engineers attributed the causes of these cracks to properties of the asphalt mix, temperature during compaction, weak support or the use of inexperienced operators. The main causes of surface cracking of new asphalt layers during compaction have been well established through various field and laboratory investigations (1,2). In order to prevent this problem one must overcome the incompatibilities between the stiffness and geometry of the steel roller and those of the asphalt layer. There is very little one can do to increase the stiffness of the asphalt mix during compaction to bring the modulus ratio of the steel material and of the asphalt mix closer to unity. Therefore, the only solution is to decrease the stiffness of the steel drum. The second parameter is the relative geometry of the system. This will require increasing the diameter of the roller considerably to make the interface between the roller and the compacted asphalt surface as flat as possible.

The requirements for a crack free pavement system led to the development of an advanced compactor called AMIR (Asphalt Multi-Integrated Roller). Figure 1 shows the Australian prototype termed HIPAC (Hot Iron Pressure Asphalt Compaction). Both compactors have been designed and built to satisfy the main requirements for crack free asphalt surfaces. In addition to solving the problems of incompatibilities between the roller and the asphalt, the AMIR/HIPAC design provides better confinement and longer contact time. The larger contact area of the AMIR roller improves the confinement of the asphalt mat during compaction and reduces lateral movement of the mix.

ACTUAL CAUSES OF CONSTRUCTION CRACKS

The main causes of the phenomenon of construction cracks can be understood when the interaction between the roller and the asphalt mat during compaction is properly investigated. An analytical model was developed and supported with a laboratory simulation to show that the cracks are mainly a result of the geometry and material of the drum (3,4). Furthermore, it was concluded that the type of the asphalt mixture, strength of the structure under the asphalt layer, the temperature of the mix and the experience of the operator of the roller play very little, if any, role in the occurrence of the phenomenon of cracking. In order to verify these findings a number of field trials, where compaction of sand layer on top of a strong subbase layer, were carried out in Egypt in 1986, Canada in 1990, and Australia in 1996. The use of the sand layer was to show that the mix type, temperature of the mix and the degree of fluidity that may exist in an asphalt mix have no relation to the cracking of the surface during compaction. The Egyptian AMIR prototype was employed side by side with a vibratory steel roller. The Egyptian AMIR was built without an engine and therefore was pulled, through a system of pulleys, side by side by the vibratory roller which was used in the field test. Thus, the operation of both compactors was carried out by the driver of the steel vibratory roller. As a result of this arrangement the causes of the cracking related to the

Figure 1. Photo of AMIR/HIPAC

experience of the operator and/or speed of the roller were eliminated. The sand layer was 150 mm deep and 30 m long. The rollers made four passes before the test was completed. As soon as the vibratory roller passed over the sand layer, surface cracks similar to the ones described earlier, were observed on the compacted surface of the sand. The section compacted by the AMIR prototype was crack free. The Canadian self propelled AMIR compactor was used in Toronto and Ottawa and also in the Sydney trials in Australia. The Canadian tests were compared to a larger Dynapac vibratory compactor while the Australian trials employed a multi-pac Slovak roller which is considered one of the largest in the world. Figúres 2 and 3 show photos taken from the Australian sand test. The sand trials confirmed that : (a) Cracks will occur due to the use of steel drum roller, (b) The influence of temperature, roller operator and strength of subgrade on the initiation of surface cracks during compaction is questionable, since none was observed under the AMIR compacted section, (c) The AMIR concept is based on sound analytical basis, and (d) The AMIR compaction is superior to the present compaction methods in spite of the fact that its total weight is much less than the smallest vibratory steel roller utilized in the test.

EFFECT OF PNEUMATIC RUBBER WHEELED ROLLER

The AMIR compactors were used to compact standard Egyptian, Canadian and Australian hot asphalt mixes. The details of these field trials can be found in a number of published reports (1,5) and therefore will not be repeated here. The variables examined in the tests included the examination of the effect of pneumatic rollers on the induced cracks, criterion and objective of each test, thickness of the compacted layer, number of compacted layers, type of subgrade or subbase, season of compaction and the country where the tests were performed. However, in all the field tests the AMIR compactor and conventional rollers were used side by side on the same asphalt mix. Therefore, any difference in the end results of each field trial can only be explained by the difference in the compaction method.

The results of the asphalt field trials confirmed the observations reported from the sand tests. They showed that for the same asphalt mix construction induced cracks are caused by the steel roller. These results led to the consideration of the second question which addresses the ability of the pneumatic rubber roller to eliminate or cure these construction cracks. A number of field trials were planned and carried out where the pneumatic wheel roller was used after the passes of a static steel roller (Toronto) or a vibratory steel roller (Cairo, Ottawa and Sydney). The results of these field tests showed that the vibratory roller induced more surface cracks than the ones observed in the case of using a static steel roller only. The use of the pneumatic roller failed to eliminate any of the cracks resulting from the vibratory roller. In some cases the pneumatic roller made more than 14 passes on the same spot without any noticeable change in the condition of the cracks. Figure 4 shows the cracks after the pneumatic roller was used several times on the asphalt mat. The results of permeability tests, performed after the asphalt mixes were compacted using the pneumatic roller, showed that the construction induced cracks were present and that the water leaked through the surface with higher rates when compared to the rate on the sections compacted using the AMIR. Also, the AMIR compacted sections showed a remarkably crack free surface with tighter texture. In addition, the edge of the sections were clearly better compacted with no apparent cracks. Figure 5 shows a photo of the AMIR compacted section.

Figure 2. Vibratory Compacted Sand

Figure 3. AMIR Compacted Sand

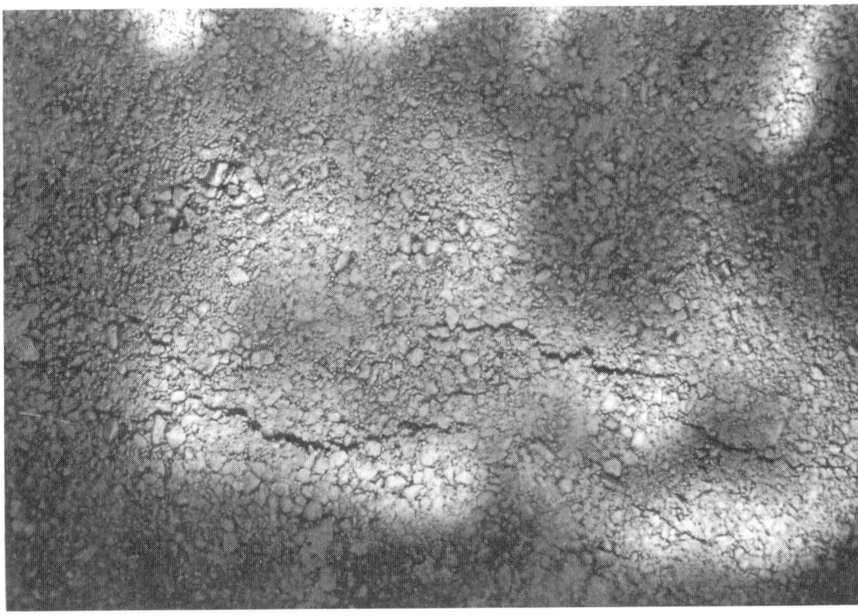

Figure 4. Conventionally Compacted Asphalt

Figure 5. AMIR Compacted Asphalt

QUALITY OF AMIR/HIPAC COMPACTION

The quality of the AMIR compaction method can be measured both in the short term and long term. Clearly the prevention of surface cracks during compaction is an added value to the quality of the compacted mix in the short term. This property was confirmed during the Australian field trial by measuring the in situ permeability test. It was reported (6) that the permeability test results taken on the AMIR compacted sections showed significant improvements when compared to the permeability of the same mixes when compacted using the Ottawa and Sydney). The conventional rollers. The AMIR compaction method has improved the permeability of the asphalt layer by a factor of 4. In some cases where the construction cracks, on the conventionally compacted mixes, were visible the tests could not performed due to the fact that the rate of water seepage was much faster than achieving the standard head. On the other hand, the AMIR compacted surfaces were so tight that on a number of occasions the standard head of 1000 mm was insufficient to cause the water to penetrate the surface. When the measurements were taken the average values showed that AMIR compacted sections had permeability of 0.034 and 0.033 ml/min, while the conventionally compacted section had 0.135 and 0.069 ml/min for AC 20 and AC20 High Bitumen mixes, respectively. Another short term improvement is the ability of AMIR compactor to achieve higher densities and a more uniform distributions. It should be noted that the weight of the AMIR compactor during the Australian tests was about 8.5 tons while the conventional compactors consisted of a steel vibratory compactor with a 16 tons (static), and a rubber wheeled roller with 13 tons weight. As shown in Tables 1 and 2, the mean values of the densities and the coefficients of variation (CoV) achieved in the Canadian as well as the Australian field tests support this observation. Clearly these results answer the question dealing with the ability of the AMIR compactor to achieve higher densities while having a much wider contact area and, therefore, lower pressure.

TABLE 1: MEAN VALUES OF SPECIFIC GRAVITY TEST RESULTS TORONTO SITE

	AMIR		CONVENTIONAL	
Thickness, mm	70	130	70	130
Mean Value	2.346	2.352	2.328	2.337
Std. Dev.	0.012	0.028	0.030	0.045
Coff.of Varatn.	0.50 %	0.84 %	1.29 %	1.93 %

TABLE 2: MEAN VALUES OF SPECIFIC GRAVITY TEST RESULTS SYDNEY SITE

	AMIR		Vib/Pnu	
Thickness, mm	100	80	100	80
Mix Type	AC20	AC20 HiBit	AC20	AC20 HiBit
Mean Value	2.269	2.312	2.262	2.289
Std. Dev.	0.046	0.050	0.064	0.093
CoV.	2.7 %	2.2 %	2.8 %	4.0%

The results given in Table 3 illustrate the improvement in the density achieved after each pass of the AMIR. It is important to note that the density reached a maximum of 2.329 only after 6 passes of the 8.5 ton AMIR while it took 13 passes of the 16 ton static steel roller, one additional vibratory which is about 23 ton and 6 more passes of the 13 ton pneumatic roller to reach a maximum of 2.305. Furthermore, increasing the number of passes by the rubber roller resulted in a drop of the maximum value to about 2.267; further supporting the assumption that pneumatic rollers do not close the construction induced cracks.

Two weeks after the completion of the Ottawa field trial, an interesting observation was reported. It was noted that cores recovered from the AMIR compacted section were fully bonded to the underlying old asphalt layers.On the other hand, cores recovered from the conventionally compacted section lacked any bond with the same old pavement, Figure 6. This observation was repeated on all other test sections where new mix was compacted using AMIR on top of an old pavement suggesting the possibility of achieving significant savings. Clearly, the developed bond will reduce the need for the use of tack coat.

TABLE 3: EFFECT OF NUMBER OF PASSES ON SPECIFIC GRAVITY SYDNEY SITE, MIX d

AMIR		Vibratory/Pneumatic			
No. of Passes	S. G.	Static	Vibr.	Pnuemat.	S.G.
1	2.144	3	1	---	2.292
3	2.227	9	1	---	2.235
4	2.314	9	1	3	2.301
6	2.329	9	1	6	2.305
N/A		11	1	12	2.267

Figure 6. Effect of Compaction on Bond

LONG TERM PERFORMANCE

The long term performance of asphalt mixes compacted using AMIR method was evaluated using a comprehensive laboratory investigation supported by field monitoring of the pavement conditions. The laboratory studies were performed on asphalt specimens recovered from test sections compacted in the field. The testing program included the study of the influence of the compaction method on the resistance of the mix to moisture damage, direct and indirect tensile strength, resistance to thermal cracking and indirect tensile and flexural fatigue tests. The following is a summary of some test results carried out in Canada and Australia.

Indirect tensile strength tests The first Ottawa field trials included two Ministry of Ontario standard asphalt mixes, Mix a and Mix b. The cores of Mix a were taken on two different dates. The first was one week after compaction while the other was one year after compaction. Thus, the test sections were left subject only to the severe conditions of the Canadian winter for one year. The cores of Mix b were recovered two weeks after compaction. The test results are given in Table 4. In all laboratory tests the cores were loaded with a specific reference to direction of rolling in the field. Thus, when a core is loaded parallel to direction of compaction it will indicate that construction cracks are subject to compressive stresses. In this case one should expect that the cracks will have the least effect on the measured property. Table 4 shows that AMIR compacted mixes will have higher tensile strength when compared with the strength of identical mixes compacted using steel drum rollers. Furthermore, the effect of rolling direction is more significant on steel compacted sections.Clearly, one should expect that the strength in the perpendicular direction is higher than the other direction. This is a defect in the drum compaction method which can have a serious effect on the occurrence of transverse cracking.

TABLE 4: SUMMARY OF INDIRECT TENSILE STRENGTH TEST RESULTS OTTAWA FIELD SITE, in kPa

	AMIR				Vibratory/Pneumatic			
Direction of rolling	Perpend.		Parallel		Perpend.		Parallel	
Type of Mix	a	b	a	b	a	b	a	b
Mean	1255	636	1158	616	715	533	827	597
Std. Dev	0.11	0.044	0.077	0.09	0.15	0.1	0.1	0.06
CoV in %	9	7	7	7	19	19	18	8

Fatigue testing Many highway agencies adopt fatigue of asphalt mixes as an indicator of the long term performance. Based on thisndirect tensile strength fatigue tests were performed on cores recovered from the Ottawa field sections. On the other hand, asphalt beams were cut from the Australian trials. While most of the Canadian fatigue data have been published (4,7), a summary is given in Table 5 of fatigue tests performed on 108 cores.

Table 5: MEAN FATIGUE LIVES OF ASPHALT CORES OTTAWA TEST SECTIONS

Mix type	AMIR			Conventional		
	Stress, MPa			Stress, MPa		
	0.206	0.275	0.413	0.206	0.275	0.413
Mix a, Summer	34171 (26%)	21126 (35%)	3856 (29%)	13473 (39%)	5158 (59%)	1001 (64%)
Mix a, Winter	33192 (13%)	12997 (20%)	2398 (20%)	17392 (27%)	6098 (38%)	1535 (30%)
Mix b, Summer	4080 (20%)	1916 (18%)	1013 (20%)	2252 (55%)	1312 (18%)	490 (32%)

The results shown in Table 5 indicate that the AMIR compacted sections will last at least 50 % longer than conventionally compacted mixes when subjected to severe stress and winter conditions as shown in the case of Mix a , winter and 0.413 Mpa applied stress. On the other hand, the ratio increased to 4 times when asphalt cores were tested within a week or two after compaction. Also, the AMIR compaction method resulted in longer fatigue lives for the two standard mixes shown in the table. Laboratory fatigue tests were performed on beams recovered from the Australian field trials. Similar to the cores used in the previous fatigue testing, beams parallel to and perpendicular to the direction of rolling were cut from the AC20 HB (Mix d) Australian field tests. It should be noted that the critical loading condition in the case of beams is the case when a beam is cut parallel to direction of rolling and loaded in flexure. This is in contrast to the case of loading a core in the direction of rolling. In the former case the cracks will be subjected to tensile stresses or strains as a result of the bending stresses while in the case of the core the crack will be closed as a result of the compressive stresses. The results of the Australian field trials are presented in Table 6. It should be mentioned that each value is the mean of at least 3 tests. Also, the data in the table are arranged to reflect the location of each set of beams in the field, for example; outside edge, middle and inside edge of the paved lane. Clearly, the results indicate that AMIR compacted beams taken from the edges (where the wheel path) have higher fatigue lives than the edges compacted by the conventional rollers.

TABLE 6: MEAN FATIGUE LIVES OF ASPHALT BEAMS SYDNEY FIELD TRIALS, MIX d

	AMIR		Vibr/Pnue	
Dir. of rolling	Perpnd.	Parallel	Perpnd.	Parallel
Mean (Std. Dev)	277426 (32%)	138465 (25 %)	101103 (42 %)	90582 (31 %)
Mean (Std. Dev)	57910 (30 %)	70940 (25 %)	502095 (30 %)	464710 (28 %)
Mean (Std. Dev)	351875 (42 %)	410460 (28 %)	290165 (16 %)	316290 (27 %)

CONCLUSIONS

The results of the field trials carried out in Egypt, Canada and Australia showed that the AMIR compaction method will provide comparable densities with better texture than those obtained by today's rollers. In addition, the AMIR compactor achieved better uniformity and comparable densities at lower number of passes. The results of the specific gravity tests showed that the use of the new compaction method can meet and exceed the current specifications and compaction standards. However, the main difference between the two compaction methods has been shown to be the presence of construction induced cracks. The test results of the indirect tensile strength showed consistently that the AMIR compacted asphalt concrete mixes will have higher strength. The results of the fatigue tests confirmed the assumption that construction induced cracks can affect the long term performance of asphalt layers. Furthermore, the results suggest that the influence of the direction of rolling is significant than the value of the air voids. Also, the selection of a proper compaction method can improve the fatigue life by a factor of 2 to 4 for the same asphalt mix.

ACKNOWLEDGEMENTS

The authors expresses their deepest appreciation to Mr. Ian Rickards of Pioneer Road Services, Australia, Dr. Svec of the National Research Centre of Canada, Dr. Haas of University of Waterloo and NSERC for providing part of the financial support.

REFERENCES

1. ABD EL HALIM, A.O., PHANG, W. and HAAS, R., An unwanted legacy of asphalt pavement compaction. Journal of Transportation Engineering, American Society of Civil Engineering, Vol. 119, No. 6, Nov./Dec., 1993, pp. 914-932.

2. ABD EL HALIM, A.O., PHANG, W. and HAAS, R.(1987). Realizing structural design objectives through minimizing of construction induced cracking. Sixth International Conference on Structural Design Of Asphalt Pavements, Ann Arbour, U.S.A. July 13-16, Vol.I., pp. 965-970.

3. ABD EL HALIM, A.O., Experimental and field investigation of the influence of relative rigidity on the problem of reflection cracking. Transportation Research Record #1060, pp. 88-98, U.S.A., 1986.

4. OKAIL, O., SHERIF, A. and ABD EL HALIM A.O., Effect of laboratory and field compaction on the shear and tensile strength of asphalt layers. Proceedings of Al-Azhar Engineering Fifth International Conference, Vol 4, pp. 527 to 549, Cairo, Egypt, Dec. 1997.

5. SVEC, O.J. and ABD EL HALIM, A.O. (1991). Field verification of a new asphalt compactor, AMIR. Canadian Journal of Civil Engineering, Vol. 18, No.3, pp. 465-471.

6. RICKARDS, AMIR-HIPAC trials in Australia. Proceedings of the Australian Research Board, November 1997.

7. ABD EL HALIM, A.O. and HAAS, R., Effect of Field Compaction Method on Fatigue Life of Asphalt Pavements. Transportation Research Record No. 1469, pp. 43 to 49, 1995.

Author Index

T

U

V

V

W

X

Y

Z

Ø

Å

INDEX

NOTE A reference is generally given only once, even if the subject may appear several times in the same paper.

A

B

C

D

E

F

G

H

I

J

K

L

M

N

O

P

S

T

U

V

W

Z

Proceedings

of the Fifth International Conference on the Bearing Capacity of Roads and Airfields

Norwegian University of Science and Technology
TRONDHEIM, NORWAY, 6 – 8 JULY 1998

Editors:

Rasmus S. Nordal,
Norwegian University of Science and Technology
Geir Refsdal, Norwegian Public Roads Administration

VOLUME III

© **Faculty of Civil and Enviromental Engineering**
Department of Road and Railway Engineering

ISBN 82-519-1346-2
Printed by TAPIR

Supporting Organizations
Norwegian University of Science and Technology
Norwegian Civil Aviation Administration
Norwegian Public Roads Administration

National Organizing Committee
Ivar Horvli (Chairman), Norwegian University of Science and Technology
Kjell Levik (Co-Chairman), Norwegian Public Roads Administration
Bjørn Finborud, Berdal Strømme as
Øyvind Hallquist, Norwegian Civil Aviation Administration
Geir Lange, Norwegian Civil Aviation Administration
Rasmus S. Nordal, Norwegian University of Science and Technology
Geir Refsdal, Norwegian Public Roads Administration
Olav Reine, Veidekke ASA
Astrid Bye (Conference Secretary), SEVU

International Advisory Committee
Kaare Flaate (Chairman), Norwegian Public Roads Administration
Ivar Horvli (Co-Chairman), Norwegian University of Science and Technology
David Bennet, Australian Road Research Board Ltd.
Jean-Francois Corté, Laboratorie Central des Ponts et Chaussées, France
Rudolf Floss, Technische Universität München, Germany
Les Hawker, Highways Agency, United Kingdom
Michel Hovan, Federal Aviation Administration, USA
Byron N. Lord, Federal Highway Administration, USA
El Hussein Mohamed, National Research Council of Canada
M.I. Pinard, Ministry of Works, Botswana
Xinglong Wang, Heilongjiang Institute of Highway and Transport, China
Vladislav M. Yumashev, State Highway Research Institute, Russia

Technical Advisory Committee
Geir Refsdal (Chairman), Norwegian Public Roads Administration
Per Magne Noss (Co-Chairman), The Norwegian Asphalt Contractors Association
Bjørn Finborud, Berdal Strømme as
Åsmund Knutson, Norwegian Public Roads Administration
Geir Lange, Norwegian Civil Aviation Administration
Helge Mork, Norwegian University of Science and Technology
Tore Slyngstad, Norwegian Public Roads Administration
Jostein Myre, Norwegian Public Roads Administration

Conference Secretariat
SEVU, Congress Department
NTNU, Gløshaugen
N-7034 Trondheim, Norway

INTRODUCTION

Road and air transport with their primary role in the transportation system are drawing considerable public and political attention all over the world. The steady growth of traffic volume and weight reduces the service life of existing pavements, causing escalating maintenance costs as well as increasing safety and environmental problems. To meet these challenges, there are great needs for further development of road and airport networks , and for cost-effective rehabilitation and strengthening of existing pavements.

The BCRA conferences have been organized every fourth year since 1982 when the first one was held in Trondheim, Norway. The second was held in Plymouth, England in 1986, the third in Trondheim in 1990 and the fourth in Minneapolis, Minnesota , USA in 1994. These conferences have focused on the monitoring of the bearing capacity , the performance of pavements and the evaluation and use of this information for maintenance, strengthening and load regulations.

The themes of the BCRA conference in 1998 cover important aspects of pavement performance, pavement condition surveys, evaluation of structural capacity, properties of pavement material, design and pavement rehabilitation and strengthening. The contributions are documented in these proceedings containing 180 papers from all over the world.

The conference will provide a platform for the presentation and discussion of these research results and new developments, for the exchange of experience from current practice and for the identification of further research and development needs.

I would like to express my gratitude to the authors and the chairmen, the members of the International Advisory Committee, the Technical Advisory Committee, the Conference Secretariat and the editors of the proceedings.

Ivar Horvli

Chairman of the Organizing Committee

VOLUM 1

Table of contens

Airfield Pavements

Pavement Condition Surveys

Backanalysis of Deflection Measurements

Structural Evaluation and Design

VOLUM 2

Structural Evaluation and Design

Full Scale Testing

Environmental Effects

Asphaltic Materials

VOLUM 3

Granular Materials and Subgrade Soils

Use of Reinforcement

Continuing from page 114

Fifth International Conference on the
BEARING CAPASITY OF
ROADS AND AIRFIELDS

MECHANICAL BEHAVIOUR OF UNBOUND GRANULAR MATERIALS FOR MODELLING OF FLEXIBLE PAVEMENTS

ANTÓNIO GOMES CORREIA[1] AND JOÃO ROCHA DE ALMEIDA[2]

1 Associate Professor, Dept. Civil Eng., IST, 1096 Lisboa Codex, Portugal (Fax: + 351-1-8497650 ; E-mail: acorreia@civil.ist.utl.pt)
2 Associate Professor, Dept. Civil Eng., New University of Lisbon, 2825 Monte da Caparica, Portugal (Fax: + 351-1-2948398 ; E-mail: jr@mail.fct.unl.pt) Researcher, ICIST, Av. Rovisco Pais, 1096 Lisboa Codex, Portugal

ABSTRACT

The structural analysis of pavements has been currently performed using the Burmister approach, in which the pavement is represented as an assemblage of linear elastic layers. However, the significance of the non-linear behaviour of unbound granular materials requires the use of other numerical tools, such as finite elements, which can accurately model material non-linearity.
In this paper, it is shown that the linear elastic approach can lead to solutions that violate the principles of mechanics. The differences between linear and non-linear analyses in terms of pavement design criteria are also discussed.

Keywords: aggregate, mathematical model, flexible pavement, finite elements.

INTRODUCTION

This paper is focused on the non-linear behaviour of granular materials and its significance on the response of standard pavement structures subjected to design loads. After presenting stress-dependent constitutive laws for resilient behaviour, established from cycling triaxial testing, attention is given to different approaches followed in structural analysis, namely the multi-layered elastic or Burmister's model and the finite element method. The latter allows taking into account the non-linear properties of the paving materials. Both techniques are used for analysing several standard pavement structures, enabling comparisons and discussion of the corresponding implications in terms of design criteria for flexible pavements.

RESILIENT BEHAVIOUR OF GRANULAR MATERIALS

Tests for determination of mechanical parameters Vehicle loading induces rotation of the principal stresses on the unbound pavement layers. To model this effect, complex laboratory testing is required. Nevertheless, cyclic load triaxial testing is widely accepted as the preferred technique for obtaining the mechanical parameters of soils and unbound paving materials, having been adopted by the Asphalt Institute [1], TRB [2], AASHTO [3] and CEN [4] as the standard test for determination of the resilient modulus for pavement design. Two main methods of cyclic load triaxial tests can be performed:

- CCP: testing under constant confining stress, σ_3 (only the deviator stress, $q = \sigma_1 - \sigma_3$, is cyclic).
- VCP: testing under variable confining stress (both the confining and the deviator stresses are cyclic).

Although VCP testing is obviously more realistic than CCP testing, the pre-standard European Committee for Standardisation [4] contemplates both procedures.

Constitutive laws From experiments carried out using CCP test method, the following relationship was established [5]:

$$E = K_1 \theta^{K_2} \tag{1}$$

Where E is the Young's modulus.
K_1, K_2 are material constants.
θ is the bulk stress ($\theta = \sigma_1 + 2\sigma_3$).

This relationship with a constant Poisson's ratio has proved to be inaccurate for predicting the real behaviour of granular layers under moving loads. To solve this drawback, other models have been proposed [6, 7]. One of these models, developed by Boyce [6], was subsequently modified based on experimental results derived from VCP triaxial tests conducted in France at the St. Brieuc LRPC [8], resulting in the equations below, expressed in terms of the bulk and shear moduli:

$$K = \frac{K_a \left(\frac{p}{p_a}\right)^{1-n}}{1 - \beta \left(\frac{q}{p}\right)^2} \tag{2}$$

$$G = G_a \left(\frac{p}{p_a}\right)^{1-n} \tag{3}$$

Where K and G are the resilient bulk and shear modulus, respectively.
$p = (\sigma_1 + 2\sigma_3)/3$ is the mean normal stress and $q = \sigma_1 - \sigma_3$ is the deviator stress.
K_a, G_a, n, β are material constants and p_a is a reference pressure.

From the theory of elasticity, K and G are easily converted into Young's modulus E and Poisson's ratio ν. The Boyce model and its variations have been incorporated in several finite element codes [9, 10, 11]. The program used herein adopts this model together with a plasticity criterion.

NUMERICAL METHODS FOR PAVEMENT ANALYSIS

The most usual numerical techniques for pavement analysis are based on the multi-layered elastic approach [12], whereby the pavement is modelled as a semi-infinite and axisymmetric system composed of a number of linear elastic layers. This technique requires less computational effort than the finite element method but it can not handle non-linear material behaviour. In contrary, finite elements are adequate for modelling the spatial variation of material properties. Furthermore, using finite elements, verification of yield criteria can be accomplished by means of an elastic-plastic formulation. Such features were implemented in the computer program FENLAP – Finite Element Non-Linear Analysis of Pavements [9], which was used in this study.

In FENLAP, the pavement is modelled as an axi-symmetric system, using 8-node isoparametric rectangular elements. An automatic mesh generator is built in the program. Different types of boundary conditions can be adopted, including infinite elements both in the horizontal and in the vertical direction. Various constitutive relationships for granular materials and soils are implemented in the program.

Input data required by FENLAP include the unit weight, the coefficient of earth pressure at rest, k_0, and the pore pressure for each layer. In fact, these values are necessary for evaluating the initial stresses, that affect the elastic response of stress-dependent materials.

Considering that the computed stresses may be incompatible with the material strength (e.g. high tensile stresses in unbound layers), FENLAP uses a simplified elastic-plastic algorithm in order to ensure that the stresses not only equilibrate the applied loads but also verify the yield criteria. This algorithm adopts the Drucker-Prager failure criterion [10], which in the p-q stress space is defined as:

$$q = M\,p + S \tag{4}$$

Where M and S are material constants.

According to this formulation, if the stresses at any point fall outside the yield surface, these stresses are projected towards this surface according to an associated rule. In order to preserve equilibrium, the removed stresses must be replaced by a set of statically equivalent nodal forces. Adding these nodal forces, the elastic analysis is repeated. The process is repeated until convergence is achieved.

Most unsaturated particulate materials exhibit apparent cohesion, represented by constant S in Equation 4. Hence, they can withstand tensile stresses up to the limiting value p*=M/S. This way, the modified Boyce model can be generalised to situations where the mean normal stress is tensile, providing that p is replaced by (p+p*). Introducing this modification in the failure criterion, the maximum stress ratio at failure becomes:

$$\left(\frac{q}{p + p^*}\right)_{max.} = M \tag{5}$$

LINEAR VERSUS NON-LINEAR ANALYSIS

Loading and pavement structure data Bearing in mind the objectives of this study, the analysis was focused on pavements where the granular layers (base and sub-base) play a significant role. The corresponding structures selected for analysis were chosen from a design catalog [13], and are defined in Table 1.

Table 1 – Pavement structures analysed and soil moduli

	Pavement 1	Pavement 2	Pavement 3	Pavement 4
$H_{bituminous}$ (m)	0.08	0.08	0.12	0.12
H_{base} (m)	0.16	0.16	0.16	0.16
$H_{sub\text{-}base}$ (m)	0.24	0.16	0.24	0.16
$E_{subgrade}$ (MPa)	30	60	30	60

For all structures, the Young's modulus, Poisson's ratio and unit weight of the bituminous layer were taken equal to 5000 MPa, 0.40 and 24 kN/m^3, respectively. For the subgrade, Poisson's ratio and unit weight were taken equal to 0.45 and 18 kN/m^3, whilst the Young's modulus corresponding to each structure is shown in Table 1. For the unbound granular materials, a unit weight of 22 kN/m^3 has been taken and the other mechanical parameters varied depending on the method followed in the analysis.
The wheel load adopted corresponds to a standard axle of 130 kN, being equivalent to an applied pressure of 676 kPa distributed over a circle of 0.175 m radius.

Linear elastic analysis To proceed with the linear elastic analysis, a constant Poisson's ratio of 0.30 was adopted for the granular layers. For each pavement structure, the Young's moduli of these materials were defined according to the following rules [14]:

$$E_{sub\text{-}base} = k\, E_{subgrade}, \quad \text{with } k = 0.2\, H_{sub\text{-}base}{}^{0.45} \quad (2 < k < 4) \tag{6}$$

$$E_{base} = \alpha\, E_{sub\text{-}base} \tag{7}$$

Where α is a coefficient that may vary, depending on the layer properties. In this study, values of α ranging from 1.5 to 2.5 were assumed.

The corresponding moduli obtained for each structure using these rules are shown in Table 2.

Table 2 - Moduli of the granular layers for linear elastic analysis (MPa)

	Pavements 1 and 3	Pavements 2 and 4
$E_{sub\text{-}base}$	71	120
E_{base} (α=1.5)	107	180
E_{base} (α=2.0)	142	240
E_{base} (α=2.5)	178	300

The linear elastic analysis was carried out using ELSYM5, a multi-layer program [15]. The corresponding values of the design parameters (ε_h - maximum tensile horizontal strain at the bottom of the bituminous layer; ε_v - maximum compressive vertical strain at the top of the subgrade) are presented in Table 3.

Table 3 – Linear elastic analysis: design parameters

	Pavement 1			Pavement 2			Pavement 3			Pavement 4		
	α=1.5	α=2.0	α=2.5	α=1.5	α=2.0	α=2.5	α=1.5	α=2.0	α=2.5	α=1.5	α=2.0	α=2.5
ε_h (με)	588	529	480	439	382	338	429	401	376	343	313	288
ε_v (με)	1530	1478	1428	1152	1108	1065	1065	1047	1026	811	794	774

Non-linear analysis (FENLAP) The generalized Boyce's relationship (Equations 2 and 3) and the Drucker-Prager failure criterion (Equation 4) were assumed for the granular layers. The corresponding parameters were extracted from Paute [8] and are the following:

K_a=109 MPa G_a=79.7 MPa n=0.326 β=0.216 p_a=0.1 MPa M=1.70 S=136 kPa

The coefficient k_0 in the granular layers was assumed equal to 1.0. For the remaining layers, the parameters were the same as used in the linear elastic. Pore pressure effects were not taken into account.

The finite element mesh had 484 elements and 1541 nodes. Infinite elements were placed at the lateral and lower boundaries of the domain, in order to achieve a better approximation of the semi-infinite domains considered when layered systems are used. The design parameters obtained from non-linear analysis are presented in Table 4.

Table 4 - Non-linear analysis: design parameters

	Pavement 1	Pavement 2	Pavement 3	Pavement 4
ε_h (με)	388	382	340	330
ε_v (με)	1370	1168	937	797

According to the results of the non-linear analysis, the Young's moduli of the granular materials vary with depth. This variation is more pronounced for pavement structures with thinner bituminous layers, as observed in Figure 1, referred to points at the load axis. It is also noticed that an increase in the subgrade stiffness conducts to a slight increase in the moduli of the granular layers (see Figure 2, also referred to points at the load axis). However, this increase is much smaller than what would be predicted from Equation 6.

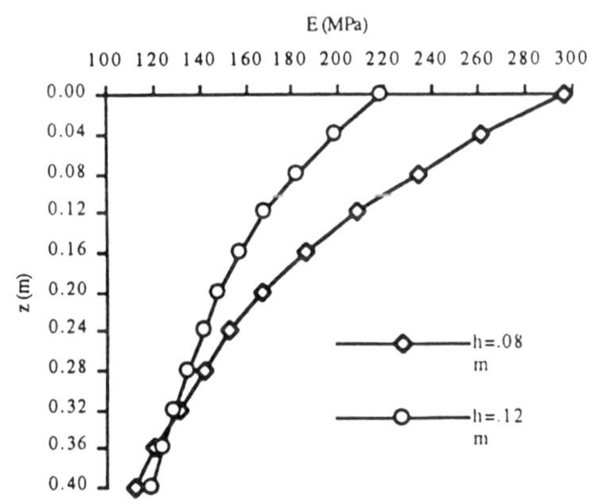

Figure 1 – Variation of E_{gr} with depth at the load axis (Pavements 1 and 3)

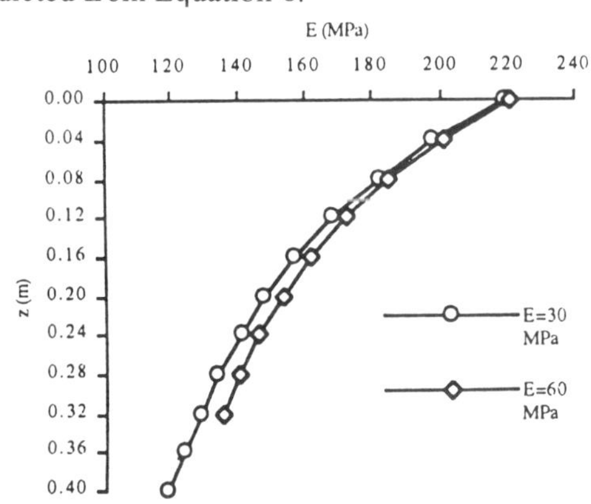

Figure 2 – Variation of E_{gr} with depth at the load axis (Pavements 3 and 4)

INTRODUCTION

In the past two decades research in the field of granular materials has at various research institutes and universities focused on more fundamental determination of the mechanical behaviour of these materials. These researches have learned that the mechanical behaviour of granular materials is more than anything stress dependent and that static and cyclic load triaxial testing provide a suitable though complex means of determining such stress dependent behaviour at stresses, densities, moisture contents and gradings representative for the materials in a pavement structure.
Research has also learned that finite element pavement design calculations based on stress dependent behaviour of the granular base and sub-base layers results in a much more realistic prediction of pavement performance than design calculations based on linear elasticity This is ofcourse especially the case for pavements in which a major structural contribution is required of the granular layers, i.e. thin pavements, small element pavements and unpaved roads.
A major impediment to the application of such a more fundamental approach in routine pavement design is however the complexity of the required static and cyclic load triaxial tests.
In this paper the results of a research are described in which it has successfully been attempted to quantitatively correlate the stress dependent failure-, resilient- and permanent deformation behaviour of a limited number of granular base and sub-base materials applied in the Netherlands to physical material properties such as the grading, the grainshape, the grain strength, etc.
The research involves four mixed granulates, mixes of crushed masonry and crushed concrete applied in the Netherlands as granular base material, one crushed sand and seven natural sub-base sands, on which a large number of triaxial tests were being conducted within the Ph.D. research conducted by Huurman [1], at the Delft University of Technology, into the development of traffic induced permanent deformation in concrete block pavements.
All triaxial tests on the sands and the mixed granulates have been conducted at specimen sizes (diameter*height) of 100*200 mm^2 and 400*800 mm^2 respectively.

ESTABLISHING OF PHYSICAL MATERIAL PROPERTIES

At first the physical material properties of which it was anticipated that they would be explanatory for the various types of material behaviour have been determined. These comprised of the particle size distribution, the specific density, the sharpness and the crushing resistance of the grains; where applicable all tests have been conducted in accordance with the relevant Dutch standards [2]. In the sections below a brief discussion is given of these tests; for the measured test results which cannot be included within this paper, reference is made to [1,3].

Particle size distribution

Extensive research has learned that the particle size distribution of unbound granular materials is probably the single most important property influencing the compactive behaviour and the shear strength. This research also supports the findings by others [4] that

Table 3 – Linear elastic analysis: design parameters

	Pavement 1			Pavement 2			Pavement 3			Pavement 4		
	α=1.5	α=2.0	α=2.5	α=1.5	α=2.0	α=2.5	α=1.5	α=2.0	α=2.5	α=1.5	α=2.0	α=2.5
ε_h ($\mu\varepsilon$)	588	529	480	439	382	338	429	401	376	343	313	288
ε_v ($\mu\varepsilon$)	1530	1478	1428	1152	1108	1065	1065	1047	1026	811	794	774

Non-linear analysis (FENLAP) The generalized Boyce's relationship (Equations 2 and 3) and the Drucker-Prager failure criterion (Equation 4) were assumed for the granular layers. The corresponding parameters were extracted from Paute [8] and are the following:

K_a=109 MPa G_a=79.7 MPa n=0.326 β=0.216 p_a=0.1 MPa M=1.70 S=136 kPa

The coefficient k_0 in the granular layers was assumed equal to 1.0. For the remaining layers, the parameters were the same as used in the linear elastic. Pore pressure effects were not taken into account.

The finite element mesh had 484 elements and 1541 nodes. Infinite elements were placed at the lateral and lower boundaries of the domain, in order to achieve a better approximation of the semi-infinite domains considered when layered systems are used. The design parameters obtained from non-linear analysis are presented in Table 4.

Table 4 - Non-linear analysis: design parameters

	Pavement 1	Pavement 2	Pavement 3	Pavement 4
ε_h ($\mu\varepsilon$)	388	382	340	330
ε_v ($\mu\varepsilon$)	1370	1168	937	797

According to the results of the non-linear analysis, the Young's moduli of the granular materials vary with depth. This variation is more pronounced for pavement structures with thinner bituminous layers, as observed in Figure 1, referred to points at the load axis. It is also noticed that an increase in the subgrade stiffness conducts to a slight increase in the moduli of the granular layers (see Figure 2, also referred to points at the load axis). However, this increase is much smaller than what would be predicted from Equation 6.

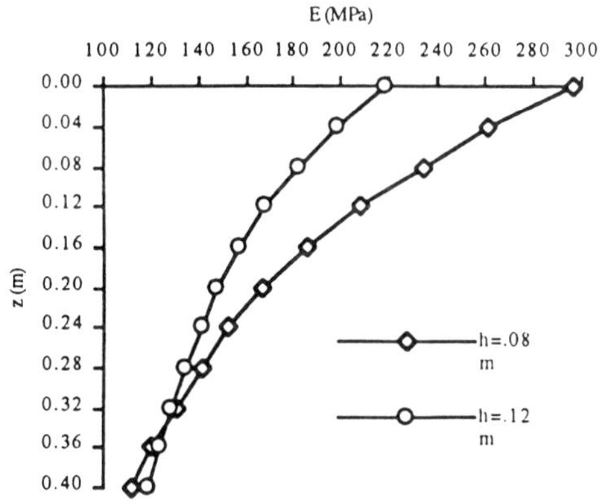

Figure 1 – Variation of E_{gr} with depth at the load axis (Pavements 1 and 3)

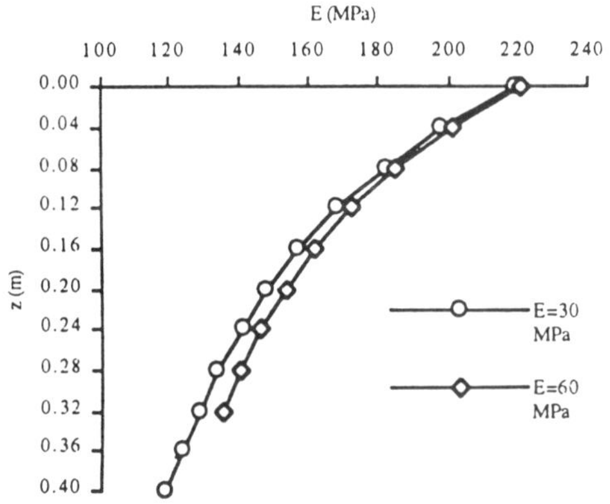

Figure 2 – Variation of E_{gr} with depth at the load axis (Pavements 3 and 4)

principles of rollability or outflow characteristics of the grains. Advantages of such an approach are its objectiveness and that it can take account of the different aspects to which the sharpness is related, i.e. the grainshape (rounded or angular), the grain surface (smooth or rough) and the grain diameter. In this research the outflow test introduced by Volders and Verhoeven, V.V. Sharpness (VVS), has been applied [1,3].

Specific density

The specific densities of the sands and the crushed granulates have been determined as they are required for the determination of the VVS. Due to differences in specific densities of on the one hand concrete and on the other hand masonry the bulk specific density also provides some insight in the composition of the mixed granulates, for the latter the apparent relative density is however a more appropriate measure.

Crushing resistance

Granular materials degrade as a result of among others mechanical forces introduced by compaction and repeated traffic loading. The resistance against mechanical degradation can be measured by different tests, introducing different degrading mechanisms as impact, abrasion and attrition. Examples of such tests are the Los Angeles abrasion test, the aggregate impact test, the micro Deval test and the Dutch crushing test. In this research the crushing resistance of the mixed granulates has been determined by means of the Dutch crushing test [2], in which the different coarse fractions are subjected to a gradually increasing load. The values found for the different mixed granulates ranged between 0.71 and 0.74. Typical crushing values for crushed masonry and crushed concrete range from 0.65 to 0.70 and 0.75-0.82, respectively.

ESTABLISHING OF MATERIAL BEHAVIOUR

Compactive behaviour and bearing capacity

By means of the proctor test [2] the moisture-density relations have been determined for the sands and the mixed granulates. From these relations the optimum moisture content (OMC) and the corresponding maximum proctor density (MPD) have been derived. At each moisture content also the California Bearing Ration (CBR) [2] has been determined.

Failure behaviour

Mohr-Coulomb's failure criterion describes the shear strength of (granular) materials by the cohesion (c) and the angle of internal friction (ϕ) as a function of the normal stress (σ_n). For the sands and the mixed granulates c and ϕ have been determined by means of static triaxial tests in which the axial stress at failure ($\sigma_{1,f}$) is determined for different levels of confining stress (σ_3). From the found combinations of $\sigma_{1,f}$ and σ_3 the so called Mohr circles can be plotted, c and ϕ can then be found as the intercept value and the slope of the line tangent to these circles. In terms of principal stresses Mohr-Coulomb's failure criterion can be rewritten to [1]:

$$\sigma_{1,f} = \frac{(1+\sin\phi)\sigma_3 + 2c\cos\phi}{(1-\sin\phi)} = A.\sigma_3 + B$$

$$with: A = \frac{(1+\sin\phi)}{(1-\sin\phi)}, B = \frac{2c\cos\phi}{(1-\sin\phi)} \qquad (2)$$

Resilient deformation behaviour

The stress dependent resilient deformation behaviour has been determined by means of cyclic load triaxial tests. To establish the stress dependency a sample is subjected to different cyclic deviatoric stress levels (σ_c) at 5 levels of confining stress (σ_3). This results in a few dozen stress combinations at which the axial and radial deformations of the specimen are measured. The resilient deformation behaviour can be described by means of the resilient modulus (Mr) and the Poisson's ratio (μ) according to:

$$Mr = \frac{\sigma_c}{\epsilon_{ax}} \; ; \; \mu = -\frac{\epsilon_{rad}}{\epsilon_{ax}} \qquad (3)$$

with:

Mr:	resilient modulus	{MPa]
σ_c:	cyclic deviatoric stress	[kPa]
ϵ_{ac}:	axial resilient strain	[-]
ϵ_{rad}:	radial resilient strain	[-]

A large number of different models exists relating the stress dependency of Mr to different stress components.

For the sands (s.) the stiffness, Mr, has been found to increase with increasing confining stress (σ_3) and to decrease at a given σ_3 level for high deviatoric stress levels. Figure 7 of the paper contributed by Huurman et al. to these proceedings shows the measured resilient behaviour for one of the sands together with the fitted Mr-$\sigma_3/\sigma_{3,0}$-$\sigma_1/\sigma_{1,f}$ model (eq. 4).

For the mixed granulates (m.g.) Mr has been found to mainly increase with an increase of the sum of principal stresses Θ, this can probably be attributed to the fact that these materials have not been tested at stress levels so close to failure as has been the case for the sands. The observed behaviour has been modelled as follows:

$$s.: \; Mr = k_1 \left(\frac{\sigma_3}{\sigma_{3,0}}\right)^{k_2} * \left(1 - k_3\left(\frac{\sigma_1}{\sigma_{1,f}}\right)^{k_4}\right) \; ; \; m.g.: \; Mr = k_1\left(\frac{\theta}{\theta_0}\right)^{k_2} \qquad (4)$$

with:

Θ:	sum of principal stresses Θ (Θ_0: reference stress = 1 kPa)	[kPa]
σ_3:	confining stress ($\sigma_{3,0}$: reference stress = 1 kPa)	[kPa]
σ_1:	first principal stress	[kPa]
$\sigma_{1,f}$:	maximum first principal stress on the basis Mohr-Coulomb (c, ϕ)	[kPa]
$k_1..k_4$:	material parameters found from regression analysis	[-]

Also the Poisson's ratio (μ) was found to be clearly stress dependent. For the sands (s.) and the mixed granulates (m.g.) μ was found to increase exponentially with the failure stress ratio, $\sigma_1/\sigma_{1,f}$, and the ratio σ_1/σ_3 respectively. This has hence been modelled as follows:

$$s.: \ \mu = v_0 + v_1 \left(\frac{\sigma_1}{\sigma_{1,g}}\right)^{v_2} \ ; \ m.g.: \ \mu = v_1 * \left(\frac{\sigma_1}{\sigma_3}\right)^{v_2} \tag{5}$$

with:

$v_0..v_2$: material parameters found from regression analysis [-]

Permanent deformation behaviour

The stress dependent permanent deformation behaviour has been determined by performing for each material three triaxial tests at three different stress levels. In these tests a sample is, at a certain stress level, loaded cyclicly either upto 10^6 load repetitions (N) or until it has accumulated more than 10% axial and/or radial strain. The occurring axial and radial deformations are measured at increasing N values (10^2, 10^3, ... ,10^6) from which the axial and radial strains ($\epsilon_{p,ax}$ en $\epsilon_{p,rad}$) can be determined.

The development of permanent strain with N was found to be strongly dependent on the failure stress ratio, $\sigma_1/\sigma_{1,f}$. It was observed that:

- the sands (s.) could be loaded to extreme ratios ($\sigma_1/\sigma_{1,f}$= 0.9-0.95) before considerable permanent strain develops,
- the mixed granulates (m.g.) already developed considerable permanent strain at lower ratios ($\sigma_1/\sigma_{1,f}$= 0.3-0.5).

It should however be borne in mind that due to higher c and ϕ values for the mixed granulates a certain $\sigma_1/\sigma_{1,f}$ -ratio corresponds to much higher absolute stress levels. Implying that a mixed granulate can indeed be more severely loaded than a sand before considerable permanent deformation develops. Figure 8 of the paper contributed by Huurman et al. to these proceedings shows the measured axial strain for one of the sands together with the fitted ϵ_p-N-$\sigma_1/\sigma_{1,f}$ model (eq. 6 and 7).

The observed permanent deformation behaviour has been modelled by means of the models below:

$$s.: \ \epsilon_p = A\left(\frac{N}{1000}\right)^B \ ; \ m.g.: \ \epsilon_p = A\left(\frac{N}{1000}\right)^B + C(e^{D\frac{N}{1000}} - 1) \tag{6}$$

with:

ϵ_p: permanent (axial or radial) strain [-]

N: number of load repetitions [-]

In which the stress dependency is taken account of in the coefficients A, B, C and D according to:

$$A = a_1 \left(\frac{\sigma 1}{\sigma 1,g} \right)^{a_2}, B = b_1 \left(\frac{\sigma_1}{\sigma_{1,g}} \right)^{b_2}$$

$$C = c_1 \left(\frac{\sigma 1}{\sigma 1,g} \right)^{c_2}, D = d_1 \left(\frac{\sigma_1}{\sigma_{1,g}} \right)^{d_2} \qquad (7)$$

with:

a_1.. d_2: material parameters found from regression analysis [-]

The model for the sands describes a linear relation between $\log(\epsilon_p)$ and log(N). In the model for the mixed granulates the second part allows for an exponential increase of $\log(\epsilon_p)$ with log(N) as has been observed for two of these materials at values of N > 50.000. Such an accelerated increase at high N-values was observed for the mixed granulates with the lowest crushing resistance.

By means of non-linear regression analysis the above discussed behaviour models have been fitted to the measured data, so obtaining the material parameters and the regression coefficients. The models were found to describe the measured behaviour well to very well. The goodness of fit between a model and the measured data can be evaluated from the regression coefficient, r^2, a value of $r^2=1$ indicates a perfect fit. Table 1 gives the values for r^2, averaged over the sands and the mixed granulates, as found for the chosen models. For the results of the individual materials reference is made to [1,3].

Material behaviour model			**r^2-averaged**	
			sands	**mixed-granulates**
$\sigma_{1,f}=A.\sigma_3+B$		(2)	> 0.990	0.967
$Mr\text{-}\sigma_3/\sigma_{3,0}\text{-}\sigma_1/\sigma_{1,f}$		(4)	0.970	-
$Mr\text{-}\Theta/\Theta_0$		(4)	-	0.725
$\mu\text{-}\sigma_1/\sigma_{1,f}$		(5)	0.733	-
$\mu\text{-}\sigma_1/\sigma_3$		(5)	-	0.853
$\epsilon_p\text{-}N\text{-}\sigma_1/\sigma_{1,f}$	axial		0.855	0.950
(6 and 7)	radial		0.865	0.928

Table 1: r^2-values for the different material behaviour models

CORRELATION OF MATERIAL BEHAVIOUR TO PHYSICAL MATERIAL PROPERTIES

In this phase an attempt has been made to quantitatively correlate, by means of non linear regression analysis, the various forms of material behaviour to the above discussed physical material properties, for the investigated sands and mixed granulates. This thus involved the MPDs and the OMCs and the model parameters of the different models applied for describing the stress dependent failure, resilient and permanent deformation behaviour, as given in table 1.
This has resulted in part in simple correlations, such as:

MPD and OMC:

sands:

$$MPD = (1702 * C_{cr}^{0.16}) - 5.6 * p_{fine} \quad (r^2=0.935) \quad (8)$$

$$OMC = (13.4 * C_{cr}^{-0.912}) + 0.91 * p_{fine} \quad (r^2=0.913) \quad (9)$$

mixed granulates:

$$MPD = 1319.1 + 6.0*C_e \quad (r^2=0.981) \quad (10)$$

$$OMC = 21.98 + 0.124*C_e \quad (r^2=0.959) \quad (11)$$

Failure behaviour for the mixed granulates: $\sigma_{1,f}=A.\sigma_3+B$

$$A = (6.83-0.0071*A_{psdc-fl}) + 0.0260*VVS \quad (r^2=0.998) \quad (12)$$

$$B = (-82.2-0.053*A_{psdc-fl}) + 3.71*VVS \quad (r^2=0.998) \quad (13)$$

Resilient modulus for the mixed granulates: $Mr=k_1.(\Theta/\Theta_0)^{k2}$

$$k_1 = 2.73*k_2^{-2.83} \quad (r^2=0.870) \quad (14)$$

$$k_2 = -2.02 + 1.93*C_u^{0.066} \quad (r^2=0.997) \quad (15)$$

Figures 1 and 2 (overleaf) give the relations between the coefficients k_1 and k_2 of the Mr-Θ model and between k_2 and the extension coefficient, d_{60}/d_{10}, respectively, together with the fitted models (eq. 14 and 15) for the mixed granulates.

For some correlations it was found not possible to straightforwardly correlate model parameters to relevant physical material properties. In these cases an indirect approach was adopted. An example is the model (5) for describing the Poisson's ratio (μ) for the mixed granulates, it proved not possible to correlate the model parameters v_1 and v_2 to physical material properties. Instead it was found very well possible to correlate μ at two different stress levels (σ_1/σ_3) to material properties, i.e. the grading parameter, d_{85}/d_{15}, and the grain sharpness VVS. The model parameters v_1 and v_2 can then easily be calculated from the values for μ at these two stress levels, which then constitutes the indirect approach.
Because of the size limitation of this paper, only a few of the established correlations between material behaviour and material properties are presented here. It is also considered sensible not to present these correlations numerically, for which reference is thus made to [3], but to discuss the nature or physical meaning of a number of these correlations. Below the nature of the established correlations for the compactive behaviour and the failure and resilient deformation behaviour for the sands and the mixed granulates will be discussed.

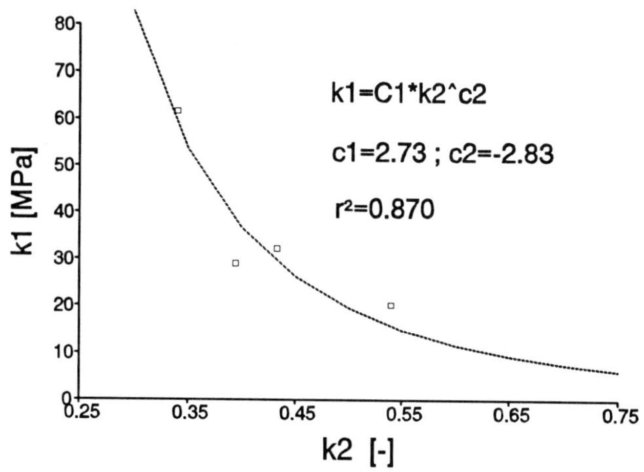

Figure 1: **Relations between the coefficients k_1 and k_2 of the Mr-Θ model together with the fitted model for the mixed granulates**

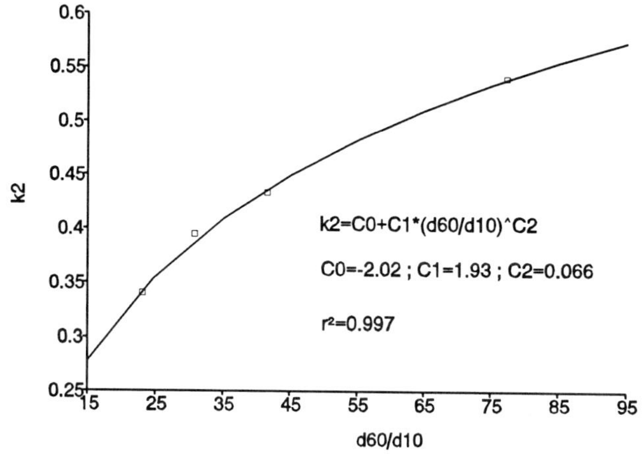

Figure 2: **Relations between k_2 Mr-Θ model and the extension coefficient, d_{60}/d_{10} together with the fitted model for the mixed granulates**

For the relatively more complex indirect correlations as have been established for the Poisson's ratio and the permanent deformation behaviour reference is made to [3]. For the mixed granulates it proved not possible to correlate the relatively complex models for describing the permanent deformation behaviour to material properties. This can be completely attributed to the limited size of the data set. As has been discussed it was however found that the crushing resistance clearly affected the observed accelerated accumulation of permanent strain at high numbers of load repetitions (>50.000).

For both the sands and the base materials the MPD was found to be inversely related to the OMC, such a relation could also be found for some 12 sands and mixed granulates investigated by Sweere et al. [5,6]. Such a relation seems sensible; a high MPD can be achieved by a material which can achieve a dense packing which corresponds to a small volume of voids. In a small volume of voids only limited moisture can be accommodated which thus corresponds to a low OMC.
For both these sands and mixed granulates the MPD could almost exclusively be explained from the grading of the materials. For the sands (8) by means of the coefficient of curvature ($= C_{cr}=d_{30}^2/(d_{60}*d_{10})$) and the percentage through the 0.063 mm sieve ($= p_{fine}$). A good grading, i.e. a high value for C_{cr}, results in grains in all fractions, so that voids left by grains of a certain size are filled in by grains of a next smaller fraction. Large relative amounts of fines, i.e. in excess of what is required for an optimum gradation, lead to a decrease of the MPD as one can imagine that a large grain, which has a higher weight than an equal volume of smaller grains, is displaced.
For the mixed granulates (9) with a much larger grading width (0/40 mm) than the sands (0/2 mm) the coefficient of extension ($= C_e=d_{85}/d_{15}$) proved adequate to quantify the grading. A high value of C_e, implying a well graded material, results in a high MPD.
The relative small differences in the specific densities and the apparent relative densities of both the sands and the mixed granulates seemed to have little influence on the found MPD values.

With respect to the failure behaviour the model parameters A and B for the sands were again found to correlate very well to the grading, described by the C_{cr} and p_{fine}.
A well graded material, i.e. high value for C_{cr}, will after compaction posses a high number of contact areas between its grains resulting in an increased friction and thus large resistance against shear failure ($\sigma_{1,f}$). In a well graded material the resistance against shear failure is to a limited extent negatively influenced by increasing amounts of fines. One can imagine that excessive fines may disturb a formed grain skeleton.
For the mixed granulates the parameters A and B could also be very well correlated to the grading (12 and 13), described by means of $A_{psdc-fc}$ (the area enclosed by the PSD curve and the Füller curve). The mechanism is the same as described for the sands.
The fact that the failure behaviour correlates to the same material properties as the MPD is a result of the fact that the failure behaviour is in fact determined by the density of the specimens and that all different materials have been tested for their mechanical behaviour at their respective MPDs. The same applies for the resilient deformation behaviour discussed in the next paragraph.
For the mixed granulates the sharpness of the grains (VVS) was also found to have a significant influence on the resistance against shear failure ($\sigma_{1,f}$), sharper grains result in increased interlocking between grains.

The stress dependency of the stiffness, Mr, has been described by the Mr-σ_3-$\sigma_{1,f}$ and the Mr-Θ models for the sands and the mixed granulates respectively. In the Mr-model for the sands the first term describes the increase of stiffness with the confining stress (σ_3).
The grading and the grain sharpness, described by $A_{psdc-fc}$ and VVS respectively, were found to be explanatory for the increase of Mr with σ_3. In a well graded material Mr can more strongly increase with σ_3 than in a poorly graded material. As a result of the larger number of contact areas at equal σ_3 the well graded material can thus take up a larger

deviatoric stress at equal deformation than a poorly graded material resulting in higher stiffness ($Mr = \sigma_c/\epsilon_{ax}$).
The same mechanism applies with respect to the grain sharpness (VVS). Between sharper grains more interlocking will occur than between less sharp grains, as a result again higher deviatoric stresses can be accommodated at equal deformation and confining stress.
The second term of the Mr-σ_3-$\sigma_{1,f}$ model describes both the degree and the gradualness of the stiffness decrease as observed when loading the sands upto failure ($(\sigma_1/\sigma_{1,f} \rightarrow 1)$). It has been found that a well graded and sharper sand experiences a more gradual relative stiffness decrease than a poorly graded more round sand.

For the more simple Mr-Θ model for the mixed granulates a strong inverse relation was found between the model parameters k_1 en k_2, (14), again such a relation also was found for sands and mixed granulates investigated by Sweere et al. [5,6]. The model parameter k_2, describing the stiffness increase with increasing stress levels (Θ), was found to relate very well (r^2=0.997) to the grading, described by the coefficient of uniformity (= $C_u = d_{60}/d_{10}$), (15). Similar as for the sands, the stiffness can develop more strongly in a well graded material as a result of the larger number of contact areas.

The reserach thus showed that the compactive behaviour and the different forms of mechanical behaviour could (very) well be directly or indirectly correlated to the physical material properties of the investigated materials. Such correlations can be applied to predict the material behaviour, required as input for non-linear elastic design calculations, from the physical material properties, which can be determined from simple (standard) tests, instead of having to establish the required behaviour from complex triaxial testing.

To conclude the study a series of design calculations for concrete block pavements have been performed by means of the axial symmetric finite element programme NOLIP, developed by Huurman [1]. Two sets of calculations have been performed, i.e. calculations in which the material behaviour has been implemented as:

- measured by means of the actual (triaxial) tests,
- predicted from the developed correlation models.

The calculations involved pavements with a concrete block toplayer and only a sand sub-base over the subgrade. From these design calculations the performance of the concrete block (c.b.) pavements in terms of the defect type rutting could be evaluated. Design calculations for c.b. pavements have been choosen as in this type of pavement, the unbound structural layers (the base and sub-base) are heavily loaded as a result of the discontinuous nature of the c.b. toplayer.
On the basis of these calculations one can evaluate whether prediction of material behaviour suffices for providing the input for such design calculations. Although the individual relations predict the different forms of material behaviour (very) well, from the rutting calculations one can evaluate the total effect as these calculations integrate all the forms of material behaviour.
Limitations in space do unfortunately not allow to describe these calculations in detail, for which reference is thus made to [3]. In brief these calculations demonstrated that:

- The identical pavements differing only with respect to the type of sand applied as sub-base, showed totally different performance. Two pavements showed excessive rutting after only a few thousand 50 kN wheel load repetitions, two other pavements reached the defect criterion of 15 mm rutting between 10.000 and 40.000 load repetitions and three of the pavements showed negliable or only little rutting even well in excess of 100.000 load repetitions. It needs to be pointed out that all the sands meet the current Dutch specifications [2] that apply for sands for sub-bases.

- With respect to the comparison between measured and predicted material behaviour five out of seven pavements showed a good to very good match. For the two pavements for which a poor match was found this could totally be attributed to too low prediction of only two parameters of the permanent deformation behaviour model (eq. 6 and 7).

CONCLUSIONS AND RECOMMENDATIONS

Conclusions

For the investigated sands and mixed granulates, models have been applied and/or developed which can describe the stress dependent failure, resilient and permanent deformation behaviour well to very well.

From the performed (triaxial) tests it has been found that both the sands and the base materials show very different compactive and mechanical behaviour.

Such mechanical behaviour can really only be correctly described by means of stress dependent material behaviour modelling and non-linear elastic calculations.
Such calculations, which have been performed as part of the research, on identical concrete block pavements, differing only with respect to the sub-base sand applied, showed completely different pavement performance with respect to the criterion of rutting. Pavements with certain sands incurred excessive rutting within a few thousand load repetitions while pavements with other sands showed negliable or only little rutting even after 100.000 load repetitions.

For all investigated sands and mixed granulates it was found very well possible to correlate the compactive, the failure and the resilient deformation behaviour to physical material properties such as the grading, the grainshape, the grain strength, etc.
With respect to the permanent deformation behaviour this was found well possible for the majority of the sands, the available dataset for the mixed granulates proved to be too small.

The approach described here of establishing such correlations not only provides qualitative insight in the influence of e.g. the grading, the grainshape, the grain strength, etc. on the compactive and mechanical behaviour, but it also provides a means to actually establish the required behaviour not from complex tests but from prediction from material properties which can be determined from simple tests which are often standard practice.

Such quantitative insight may also contribute to moving away from the conventional recipe based specifications towards more performance based specifications.

The current Dutch specifications as these apply to sands for sub-bases do not distinguish sands with respect to relevant mechanical behaviour.

Recommendations

The validity of the found correlations needs to be consolidated and extended through further research. Hereby special attention needs to be paid to:

- correlation of the more complex permanent deformation behaviour models for both sands and mixed granulates,
- incorporating the effect of density and moisture content, which are known to strongly influence the mechanical behaviour.

For unbound granular materials road engineering has to move away from the current material related, empirically based specifications towards more functional performance based specifications. Such functional specifications also make it possible to evaluate the suitability of alternative materials on the basis of e.g. the required mechanical behaviour.

REFERENCES

1. HUURMAN, M. Permanent Deformation in Concrete Block Pavements. Ph.D. dissertation, Delft University of Technology, Delft, 1997.

2. Centre R.O.W. Standard specifications for civil engineering works 1995. (in Dutch), Centre R.O.W, Ede, The Netherlands, 1995.

3. NIEKERK, A.A. VAN, AND M. HUURMAN. Establishing complex behaviour of unbound road building materials from simple material testing. Report 7-95-200-16, Delft University of Technology, Laboratory for Road and Railroad Research, Delft, 1995.

4. KOLISOJA, P. Resilient Deformation Characteristics of Granular Materials. Ph.D. dissertation, Tampere University of Technology, Tampere, Finland, 1997.

5. SWEERE, G.T.H. Unbound granular bases for roads. Ph.D. dissertation, Delft University of Technology, Delft, 1990.

6. LINSCHOTEN, E.P., HENSEN, E.B.J. AND SWEERE, G.T.H. Strength and deformation properties of six Dutch sands. (in Dutch),Report 7-83-205-1, Delft University of Technology, Laboratory for Road and Railroad Research, Delft, 1983.

Fifth International Conference on the
BEARING CAPASITY OF
ROADS AND AIRFIELDS

LABORATORY DETERMINATION OF BASE MATERIAL RESILIENT MODULI

MOHAMED BOUDALI[1] AND CLAUDE ROBERT[2]

[1]Professional Engineer, Ministère des Transports du Québec, G1P 3W8, Ste-Foy, Canada

[2]Professional Engineer, Responsable for soils and foundations section, Ministère des Transports du Québec, G1P 3W8, Ste-Foy, Canada
Fax.: (418) 646-6692, E-mail : crobert@mtq.gouv.qc.ca

Abstract

The resilient (elastic) modulus is the parameter that characterizes the elastic property of soils and granular materials. It is used in particular for sizing pavements and is the only parameter in the AASHTO Guide for Design of Pavement Structures that characterizes the mechanical behavior of materials. Direct measurement of the resilient modulus of granular materials is done in a laboratory using a repeated load triaxial apparatus. A testing program was carried out at the Ministère des Transports du Québec to determine the modulus for five MG-20 base materials of a different nature (granite, limestone, diabase, granitic gravel and Appalachian gravel) with six different particle size distributions falling within the MTQ specification grading range. The tests were done on samples with varying water content as well as on frozen samples.

The resilient modulus values are presented using the K-θ model in which the constant k_1 varies from 8,100 to 11,500 kPa, and where the constant k_2 is equal to 0.6. The limestone material presented the highest modulus while granite proved to be the lowest. These results demonstrate that the resilient modulus is a measure of the reaction of a material's skeleton or structure. One parameter that provides an indication of this structure is the void index; limestone had the lowest void index (closed structure) and granite had the highest (open structure). Finally, the resilient modulus values proved to be independent of sample particle size distribution provided that a variance of 25% around the average value was tolerated.

Keywords: resilient modulus, base materials, gradation, petrography.

1.0 Introduction

Pavement materials are normally designed for loading in an elastic range. Loads applied to the pavement produce stresses and deformations that can affect the various layers of pavement. Pavement layers must be sufficiently rigid to resist these stresses without deforming significantly. Base and subbase materials having high resilient moduli allow for greater dissipation of the stresses induced in the support soil. For granular materials with lower resilient moduli, base layer thickness must be increased.

The elastic or resilient modulus is the parameter that measures or characterizes the elastic property of materials. In the AASHTO (American Association of State Highway and Transportation Officials) [1] pavement design guide, the resilient modulus is the only parameter characterizing the mechanical behavior of the material that is needed for sizing a pavement.

Until today, a number of organizations continue to estimate the resilient modulus from empirical relations giving the resilient modulus in terms of the C.B.R. index, resistance index R (R_{value}) or other geotechnical parameters. However, these relations are not always valid and are specific to each type of material. A better way of determining the resilient modulus of a material is a test enabling direct measurement while still duplicating the actual loading conditions on the materials.

For this reason, a testing program was begun at the Ministère des Transports du Québec to determine the resilient moduli of various types of MG-20 base materials using a repeated load triaxial cell, and to study the modulus variation for different gradations within the specification grading range.

2.0 Resilient modulus definition

Resilient modulus M_r is the elastic modulus of a material. It is defined as the ratio of deviator stress applied to the soil over the resilient (elastic) vertical deformation:

$$M_r = \frac{q_r}{\varepsilon_r} \quad (2.1)$$

where q_r is the deviator stress applied to the sample and ε_r is the resilient axial deformation recovered after release of the deviator stress.

According to AASHTO, laboratory determination of this modulus should be done on representative samples (in terms of particle size distribution, dry density and water content), and under certain stress conditions simulating the actual site conditions encountered as closely as possible (layer thickness, repeated vehicle passage). The test principle consists of applying a repeated load to a material sample and measuring its deformations.

In this type of test, lateral stress (confinement) is generally held constant while repeated deviator stress is applied. Khedr [2] indicates that tests involving constant confinement stress do not simulate conditions on the ground where lateral stress varies simultaneously with the vertical stress applied by the vehicle load. In addition, according to the author, these tests can result in resilient modulus values that are higher than those obtained in tests where the lateral stress is cyclic. This is theoretically true (according to the generalized Hook's law) on condition that in both cases the resilient modulus is determined using the same relation (2.1). Nevertheless, Thompson [3] asserts that, for all practical purposes, the resilient modulus is approximately the same in both cases.

3.0 Modeling the resilient modulus for granular materials

Dunlap [4] indicates that the resilient modulus of coarse soil increases with confinement stress but that it is affected very little by the deviator stress since the latter does not cause excessive plastic deformation. He goes on to suggest that the resilient modulus be related to confinement stress through the following equation:

$$M_r = k_1 \bullet \sigma_3^{k_2} \tag{3.1}$$

where σ_3 is the confinement stress, and k_1 and k_2 are regression constants.

Seed et al. [5] suggest that the resilient modulus instead be a function of the sum of the principal stresses, i.e.:

$$M_r = k_1 \left(\frac{\theta}{p_a} \right)^{k_2} \tag{3.2}$$

where θ is the sum of the principal stresses ($\sigma 1 + \sigma 2 + \sigma 3$), k_1 and k_2 are regression constants, and p_a is a reference pressure (1 kPa). This relation implies that for a constant confinement stress, the resilient modulus should increase with the deviator stress. This increase in the resilient modulus with confinement and deviator stresses is mainly due to the dilatancy characteristics of these materials, according to various researchers. Under strong confinement stresses, the soil tends to resist shearing, resulting in smaller axial deformations, and consequently in a stronger elastic modulus. Typical values for the k_1 and k_2 parameters in relation (3.2) as applied to different unbound materials are given in Table 1.

In addition, the resilient modulus evaluated by May and Witczak [6] and Zaman et al. [7] for base materials varies from 51 to 195 MPa when the sum of the principal stresses varies from 100 to 690 kPa. These values are somewhat lower than the resilient moduli of granular materials as indicated by the Asphalt Institute [8] (M_r between 103 and 344 MPa).

May and Witczak [6] and Uzan [10] propose another model taking the deviator stress into account:

$$M_r = k_3 \bullet \left(\frac{\theta}{p_a} \right)^{k_4} \bullet \left(\frac{\sigma_d}{p_a} \right)^{k_5} \tag{3.3}$$

where θ is the sum of the principal stresses ($\sigma 1 + \sigma 2 + \sigma 3$), σ_d is the deviator stress ($\sigma_1 - \sigma_3$), k_3, k_4 and k_5 are regression constants, and p_a is a reference stress equal to 1 kPa. According to Santha [11], this equation conveniently describes the resilient modulus for soils.

The K-θ model (Equation 3.2), which relates the resilient modulus to the sum of the stresses applied to the sample, is the one most frequently used, while the Uzan model (equation 3.3) is more representative of the behavior of granular materials considering that it takes the effects of their shear strength into account. Several authors recommend using this model.

Table 1: k_1 and k_2 parameter values for various unbound base materials (from Zhou et al.) [9]

Authors	Materials studied	k_1	k_2
Hicks (1970)	Crushed gravel and crushed stone	1,600 - 5,000	0.57 - 0.73
Hicks et al. (1970)	Unbound base materials (San Diego Test Road)	2,100 - 5,400	0.61
Allen (1973)	Crushed gravel and crushed stone	1,800 - 8,000	0.32 - 0.70
Kalcheff & Hicks (1973)	Crushed stone	4,000 - 9,000	0.46 - 0.64
Boyce, Brown & Pell (1976)	Well-graded crushed aggregate	8,000	0.67
Monismith et al. (1972)	In-service base and subbase materials	2,900 - 7,750	0.46 - 0.65
Albright (1986)	Crushed aggregate (saturated) Crushed aggregate (opt. Water content)	1,300 - 2,000 2,000 - 2,600	0.69 - 0.78 0.70 - 0.73

4.0 Testing program

The testing program consists of determining the resilient modulus of various types of MG-20 base materials from a number of supply sources. The materials analyzed included a crushed granitic stone material, a crushed limestone and a crushed diabase with 100% fragmented particles each, a granitic gneiss gravel and an Appalachian gravel (comprised of sandstone and mudstone) containing about 30% unfragmented particles each.

For each of the materials, six different gradations were constituted, all within the MTQ specification grading range (Figure 1). Curves 1 and 6 follow the top and bottom of the range, respectively. Curve 5 follows the Talbot maximum density curve with a coefficient n = 0.5 (% passing d sieve = $((d/D_{max})^{0.5})$. The other curves (2, 3 and 4) were chosen to vary the C_c (coefficient of curvature), C_u (coefficient of uniformity) and % passing 0.080mm sieve.

Each sample of 150 mm diameter and 300 mm deep was compacted using a vibrating hammer to a maximum dry density at a water content approaching the modified Proctor optimum. The sample was compacted in seven layers and each layer was subjected to the vibrating hammer for one (1) minute.

Two regular tests were performed for each gradation: the first was done at the optimum water content; the second on the same sample after it was humidified to ensure saturation. Saturation of the samples was done by applying a low hydraulic gradient to the base of the sample (5 to 7 kPa). Drainage was done toward the top of the sample. Since the test is non-destructive, each one is repeated three (3) times on the same sample, thereby providing for a better result analysis. All tests were done at an ambient temperature of around 20°C.

FIGURE 1

Particle size distribution of analysed materials

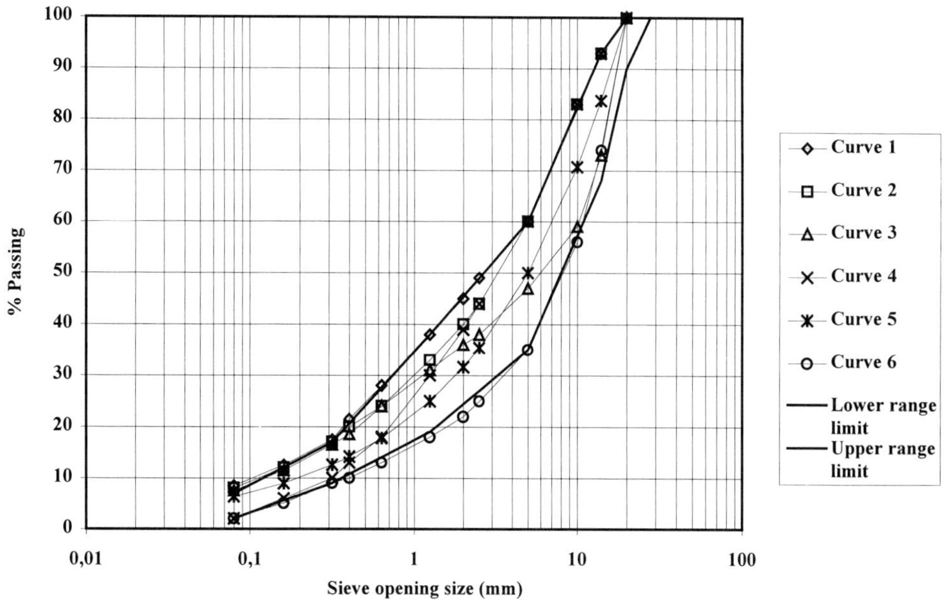

The tests were done in accordance with the SHRP-P46[12] procedure. Open drainage was maintained during the tests. The material is placed in an adapted triaxial cell and subjected to pre-loading of at least 200 cycles under confinement and deviator stresses of 103.4 kPa (15 psi). This conditioning is used to seat the sample and to eliminate plastic deformation as much as possible.

The test then consists of applying a repeated load to the sample defined by:

- a constant confinement pressure representing the minor principal stress σ_3. Five predefined confinement stresses are applied in succession (20.7, 34.5, 68.9, 103.4, and 137.9 kPa);
- an axial compression stress, σ_1, which produces the deviator stress. This stress varies in a cyclical manner from a minimum and maximum value. For each of the five confinement stresses used, three different deviator stresses are successively applied to the sample (20.7 to 62.1 kPa for the first confinement stress; 34.5 to 103.4 kPa for the 2nd; 68.9 to 206.8 kPa for the 3rd; 68.9 to 206.9 kPa for the 4th; and 103.4 to 275.8 kPa for the 5th).

The axial stress is applied by a hydraulic servo-system with a function generator capable of applying a Haversine-type pulse in which the signal crest corresponds to the load application time (0.1 seconds), followed by a relaxation period of 0.9 seconds. These two steps comprise one cycle. The axial load is applied for 100 cycles. It is assumed that, beyond the 95th cycle, the soil behavior is practically elastic. Consequently, only the results of the last five cycles are taken into consideration in calculating the resilient modulus. An average value for these last five cycles is used. The lateral deformation measurements are not made.

The apparatus used at the Ministère des Transports du Québec was manufactured by the American firm Structural Behavior Engineering Laboratories, Inc., located in Phoenix, Arizona. Considering the technical characteristics of the load cells and displacement sensors, the modulus measurement error is estimated at about 10% for small stresses and around 1% for large ones.

5.0 Presentation and interpretation of results

The results are presented and interpreted based on the K-θ model (Equation 3.2), a linear model between the logarithms of the resilient modulus and sum of the principal stresses. The Uzan model (Equation 3.3) is also applied to the data obtained, but for information purposes only.

5.1 Sample compaction

For all the materials studied, compaction using a vibrating hammer caused only minor particle breakage. This compacting method provides maximum dry densities equal to or slightly greater than those obtained with the modified Proctor test method (100-105%). Based on these results, we can conclude that compacting using a vibrating hammer is appropriate for sample preparation.

5.2 Resilient modulus

The average values of regression constants k_1 and k_2 of the K-θ model obtained for each of the six gradation curves are given in Table 2. These values include the results of tests performed on samples with optimum water content and on humidified samples. An analysis of the overall results indicates that saturated materials and those with an optimum water content behave similarly; in other words, the water content has little effect on the value of k_2 or on the value of constant k_1. The resilient modulus of the material with optimum water content differs only by about 10% from that in the saturated state. This result is in agreement with the findings of Cole et al. [13], Finn et al. [14], and Yeh and Su [15]. We can therefore say that for all practical purposes, when samples are drained and relatively permeable ($k \geq 10^{-5}$ cm/s), the water content has no influence on the resilient modulus, especially when it is greater than the optimum water content.

On the other hand, it was noted that any decrease in water content below the optimum causes an appreciable increase in the resilient modulus in which average increases of from 55 to 75% were recorded. The sample that produced a 55% increase had only undergone a 1.5% decrease in its initial water content, while the sample that showed an increase of 75% in its modulus had its initial water content reduced by 3.8%. This increase is caused by variation of constant k_1, while k_2 is relatively unaffected by a decrease in water content.

Table 2: **Results of repeated loads tests performed on granular base materials.**

Gradation curves	Dry density (kg/m³)	Void index	K-θ model			Uzan model			
			K_1 (kPa)	K_2	R^2	K_3 (kPa)	K_4	K_5	R^2
Crushed Granite									
1	2351	0.181	9777	0.540	0.931	6890	0.847	-0.296	0.991
2	2284	0.216	7308	0.596	0.935	4568	0.954	-0.330	0.993
3	2341	0.186	9147	0.614	0.973	7709	0.831	-0.227	0.996
4	2270	0.223	8566	0.588	0.960	6491	0.858	-0.264	0.995
5	2365	0.174	5305	0.663	0.969	3693	0.929	-0.245	0.996
6	2295	0.210	8732	0.611	0.963	6109	0.887	-0.255	0.996
Average		**0.198**	**8139**	**0.602**		**5910**	**0.884**	**-0.270**	
Crushed Limestone									
1	2303	0.135	12077	0.578	0.926	9023	0.861	-0.278	0.968
2	2333	0.121	11648	0.582	0.913	7280	0.982	-0.383	0.986
3	2362	0.107	19632	0.479	0.854	15223	0.844	-0.384	0.954
4	2229	0.173	10520	0.610	0.960	8071	0.884	-0.269	0.992
5	2326	0.124	9721	0.627	0.950	7393	0.873	-0.239	0.980
6	2262	0.156	16433	0.570	0.944	12937	0.843	-0.274	0.988
Average		**0.136**	**13339**	**0.575**		**9988**	**0.881**	**-0.305**	
Crushed Diabase									
1	2419	0.178	10243	0.585	0.972	8293	0.791	-0.200	0.993
2	2465	0.156	13231	0.539	0.960	9926	0.760	-0.206	0.990
3	2550	0.126	16471	0.516	0.906	10199	0.727	-0.176	0.951
4	2416	0.179	6085	0.665	0.959	4200	0.963	-0.281	0.993
5	2550	0.124	11414	0.580	0.917	7165	0.944	-0.344	0.991
6	2461	0.175	28184	0.391	0.737	23387	0.812	-0.472	0.927
Average		**0.156**	**14271**	**0.546**		**10528**	**0.833**	**-0.280**	
Granitic Gravel									
1	2355	0.122	9778	0.592	0.965	7773	0.823	-0.228	0.992
2	2338	0.130	8585	0.613	0.976	7032	0.794	-0.177	0.993
3	2357	0.122	12272	0.559	0.962	10295	0.771	-0.219	0.990
4	2260	0.170	11828	0.571	0.929	8559	0.907	-0.322	0.988
5	2328	0.134	15205	0.524	0.893	11121	0.906	-0.380	0.984
6	2309	0.142	13613	0.522	0.971	11466	0.692	-0.168	0.991
Average		**0.137**	**11880**	**0.564**		**9374**	**0.816**	**-0.249**	
Appalachian Gravel									
1	2202	0.167	13354	0.525	0.928	10473	0.807	-0.285	0.979
2	2193	0.172	11504	0.554	0.955	8259	0.790	-0.221	0.990
3	2308	0.115	13996	0.521	0.918	9596	0.835	-0.300	0.979
4	2211	0.162	9178	0.567	0.911	5950	0.930	-0.350	0.987
5	2273	0.132	25582	0.468	0.919	18642	0.702	-0.223	0.978
6	2226	0.158	24082	0.462	0.927	18383	0.681	-0.213	0.976
Average		**0.151**	**16283**	**0.516**		**11884**	**0.791**	**-0.265**	

No particular trend is evident from these results; low k_2 values are compensated for by higher k_1 values. For the crushed granite material, the data show that the k_2 constant varies from 0.54 to 0.66, giving an average value of 0.6. The k_1 constant values range from 5,300 to 9,800 kPa, giving an average value of 8,100 kPa. The maximum deviation of the results with respect to these average values is 25% (Figure 2). In such a case, it can be concluded that gradation (if constrained within the specification grading range) has no influence on the resilient modulus as observed by Zaman et al. [7].

For limestone, diabase, granitic gravel and Appalachian gravel, the average values of k_2 fall between 0.516 and 0.575, while the average values of k_1 are greater than those obtained for granite, varying from 11,880 to 16,280. Figure 3 shows the resilient modulus variation range for each of the materials analyzed using the relations expressed by the various values of k_1 and k_2 constants in the K-θ model. If, as in the case of the granite, we adopt a k_2 value equal to 0.6, the k_1 values for limestone, diabase, granitic gravel and Appalachian gravel become 11,500, 10,350, 9,600 and 9,900 respectively (Figure 4).

You will recall that the value of this constant for granitic material is 8,100 kPa. Therefore, the material that presents the strongest resilient modulus is the softest material (micro-Deval = 16%), i.e., the limestone, while the material with the weakest resilient modulus is the one thought to be the hardest - granite (micro-Deval = 14.8%). To understand these results, we must remember that the resilient modulus is the measurement of the reaction of a material's skeleton or structure (particles arrangement) and not the reaction of an isolated particle. This structure not only depends on the compacting force, but also on the particles shape (size, angularity, roundness, surface roughness). One element that measures or provides an indication of this structure is the void index. Table 2 shows that the average void index for crushed granitic material is the greatest (0.198) and that of limestone the smallest (0.136). Because of this, limestone presented the lowest coefficient of permeability. It is therefore imperative to balance a material's drainage with its support capacity.

5.3 Comparison between the K-θ and Uzan models

The relations proposed by the K-θ model (Equation 3.2) and the Uzan model (Equation 3.3) were applied to the experimental data obtained for the five materials analyzed. The coefficients of correlation indicate that the Uzan model equation is the one that best describes resilient modulus variation. Numerically, using the K-θ model gives an error (linearity due to the model and measurement errors) of ±25%. The Uzan model gives an error of ±16%. However, this model is more complex than the K-θ model since it is a function of two variables (θ and σ_d) and three regression constants, k_3, k_4 and k_5. It therefore cannot be presented on a two-dimensional graph. In addition, considering that the value of k_2 in the K-θ model can be considered constant and independent of any other parameter, the effect of certain parameters such as the suction and material structure (expressed by the void index) is factored in by constant k_1.

Pavement Rehabilitation

Case Histories, Airfields

Case Histories, Roads

FIGURE 2
Resilient Modulus of granitic material

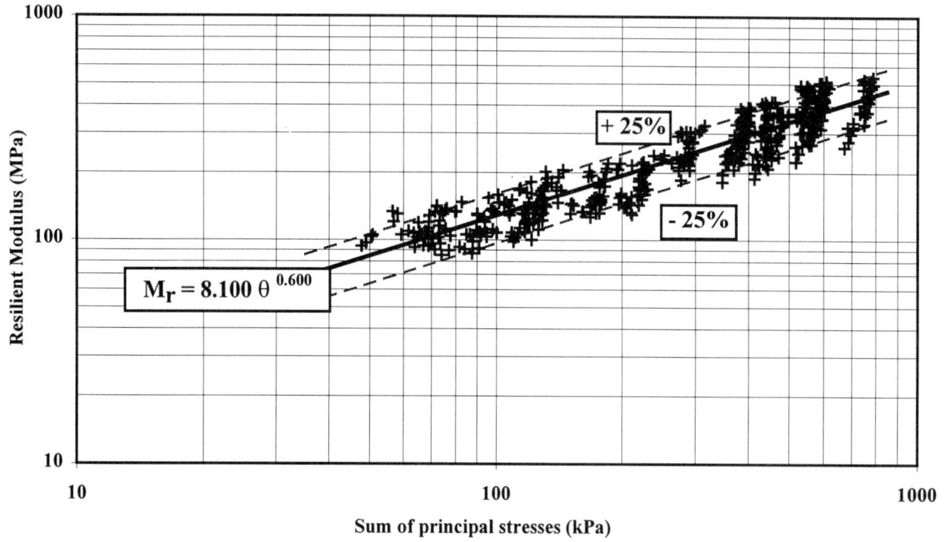

FIGURE 3
Resilient modulus various materials

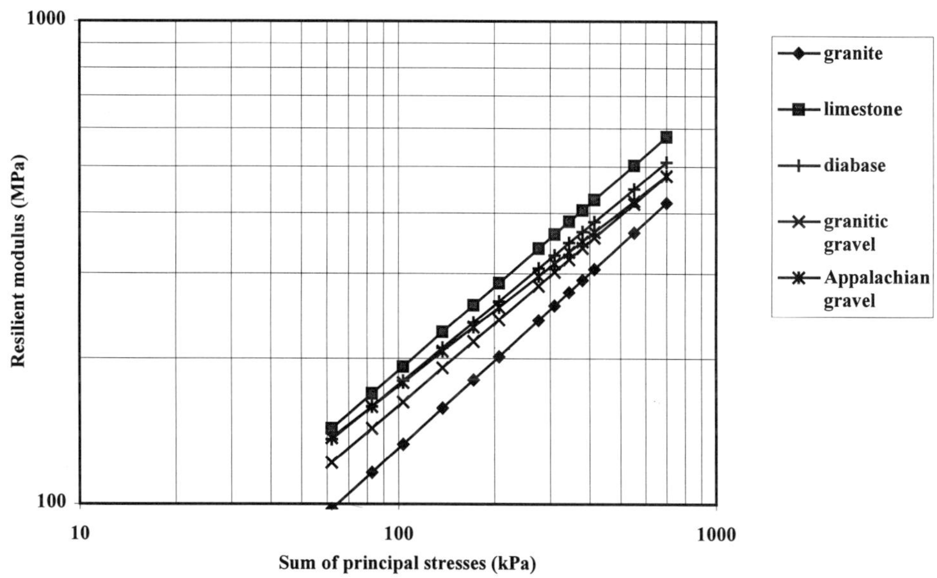

FIGURE 4

Corrected resilient modulus

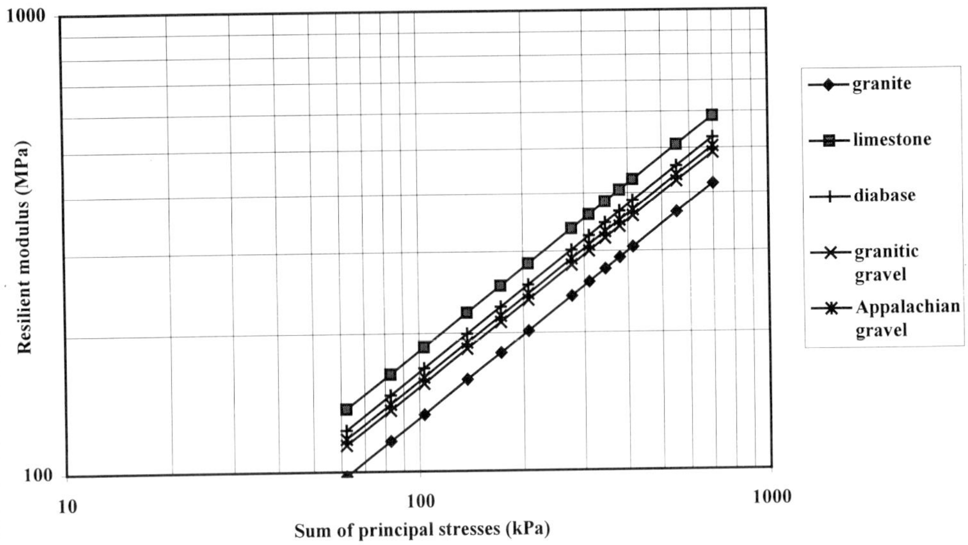

6.0 Conclusion

The resilient modulus was determined for five materials of a different nature, each in gradations falling within the MTQ's MG-20 specification grading range. The main observations are:

- if a maximum deviation of ±25% is tolerated, the gradation of the material has little impact on the resilient modulus. In such a case, the resilient modulus can be estimated using the K-θ model;
- the value of k_2 is virtually constant for unbound aggregate material (independent of the suction, geological origin, compacting, etc.), and is equal to 0.600; since this constant represents the stress sensitivity of the aggregate material, it could be related to a certain cohesion in the material ;
- constant k_1 varies, however. The results of the present study indicate that this constant varies from 8,100 to 11,500 kPa, mainly attributable to the material's void index (structure), rather than particle hardness (intrinsic characteristics);
- there is little difference between the resilient moduli of materials with optimum water content and those in a saturated condition because the tests were performed in the drained state; however, the resilient modulus increases when water content drops below the optimum water content.

ACKNOWLEDGEMENTS

We would sincerely like to thank those people who entrusted us with carrying out this study on the behavior of base materials subjected to cyclical loading. We would especially like to thank Mrs. Nelson Dubé, tech. and Guy Tremblay, ing. Our sincere thanks also to Mrs. Anne-Marie Leclerc, ing., head of the Service des matériaux des chaussées.

BIBLIOGRAPHY

[1] AASHTO (1993). Guide for Design of Pavement Structures, Washington, D.C.

[2] Khedr, S. (1985). Deformation Characteristics of Granular Base Course in Flexible Pavement. Transportation Research Record 1043, Transportation Research Board, National Research Council, Washington, D.C. pp 131-138.

[3] Thompson, M. (1989). Factors Affecting the Resilient Moduli of Soil and Granular Materials. Proc., Workshop on Resilient Modulus testing, State-of-Practices, Oregon State University, Corvallis.

[4] Dunlap, W. A. (1963). A Report on a Mathematical Model Describing the Deformation Characteristics of Granular Materials. Technical Report 1, Project 2-8-62-27, TTI, Texas A & M University.

[5] Seed, H.B., Mitry, F.G., Monismith, C.L. and Chan, C.K. (1967). Prediction of Pavement Deflection from Laboratory Repeated Load Tests. NCHRP Report 35.

[6] May, R.W. and Witczak, M.W. (1981). Effective Granular Modulus to Model Pavement Response. Transportation Research Record 810, Transportation Research Board, National Research Council, Washington, D.C., pp. 1-9.

[7] Zaman, M.M., Chen, D.-H., Laguros, J. (1994). Resilient Moduli of Granular Materials. ASCE, Journal of Transportation Engineering, vol. 120, No. 6, pp 967-987.

[8] Asphalt Institute (1981). Thickness Design-Asphalt Pavements for High Ways and Streets. Manual series No 1 (MS-1), Lexington, Ky.

[9] Zhou, H., Moore, L., Huddleston, J. and Gower, J. (1992). Free Drainage Base Materials Properties. HPR Project No. 5283, FWA-OR-RD-92-11.

[10] Uzan, J. (1985). Characterization of Granular Materials. Transportation Research Record 1022, Transportation Research Record 1022, Transportation Research Board, National Research Council, Washington, D.C., pp. 52-59

[11] Santha, B.L. (1994). Resilient Modulus of Subgrade Soils: Comparison of Two Constitutive Equations. Transportation Research Record 1462, Transportation Research Board, National Research Council, Washington, D.C., pp. 79-82.

[12] SHRP (1989). Resilient Modulus of Unbound Granular Base, Subbase Materials and Subgrade Soils. SHRP Protocol P-46, UG07, SS07.

[13] Cole, D., Bentley, D., Durell, G. and Johnson, T. (1986). Resilient Modulus of Freeze-Thaw Affected Granular Soils for Pavement Design and Evaluation. Part 1. Laboratory Tests on Soils from Winchendon, Massachussetts, Test Section.

[14] Finn, F.N., Saraf, C .L ., Kulkarni, R., Nair, K., Smith, W. and Abdulaah, A. (1977). Development of Pavement Structure Subsystems. Final Report NCHRP 1-10.

[15] Yeh, S .T. and Su, C.-K. (1989). Resilient Properties of Colorado Soils. Report No. CDOH-DH-SM-89-9.

Fifth International Conference on the
BEARING CAPASITY OF
ROADS AND AIRFIELDS

PERMANENT DEFORMATION IN GRANULAR MATERIALS - *State of the Art*

FREDRICK LEKARP[1] AND ULF ISACSSON[2]

[1]Ph.D. Student, Royal Institute of Technology, Highway Engineering, 100 44 Stockholm, Sweden, +46-8-108124, lekarp@infra.kth.se
[2]Professor, Royal Institute. of Technology, Highway Engineering, 100 44 Stockholm, Sweden

ABSTRACT

The unbound granular base and subbase layers play an essential role in the overall structural performance of a flexible pavement, especially when unsurfaced or thinly surfaced. Despite decades of research, the current knowledge concerning the behaviour of granular materials is limited. This paper discusses the plastic response of unbound granular materials under repeated loading. Based on a comprehensive literature survey, a *state of the art* is presented with regard to factors affecting permanent strain behaviour and different modelling techniques. The review of previous research shows that permanent deformation in granular materials is influenced by several factors: stress level, principal stress rotation, number of load applications, moisture content, stress history, density, grading and aggregate type. Computational modelling of the plastic behaviour is generally based on the gradual accumulation of permanent strain with number of load cycles and the impact of applied stresses. Over the years, different modelling techniques have been suggested. During the last two decades, more attention has been paid to the so-called shakedown theory and its potential for developing analytical models for road pavement design. Recent studies suggest that a shakedown approach may also be used to describe the permanent strain behaviour of granular materials under repetitive loading. (***Keywords: permanent strain; granular materials, plastic behaviour; shakedown***)

INTRODUCTION

The complex elasto-plastic behaviour of granular materials has been one of the main research topics in pavement engineering for many years. Many researchers have studied the performance of granular materials under repeated loading, using laboratory and *in situ* techniques. Most of the effort has been spent on understanding and modelling resilient behaviour, and considerable progress has been achieved. In comparison, less research has been done concerning plastic behaviour and permanent deformation development. The aim of this paper is to summarise the current understanding of the plastic behaviour of unbound granular materials under repeated loading, based on the literature available to date. The paper presents a *state of the art* in regard to factors affecting permanent strain behaviour and modelling techniques. A more detailed review of the literature is given elsewhere (1).

FACTORS INFLUENCING PERMANENT STRAIN BEHAVIOUR

In order to fully understand the plastic behaviour of granular materials, it is essential to discover the influencing parameters and their role in the material response. Research into this area has revealed the existence of several important parameters as described below.

Stress level: The literature available shows that stress level is one of the most important factors affecting the development of permanent deformation in granular materials. Early repeated load triaxial tests on granular materials, reported by Morgan (2) and Barksdale (3), showed clearly that accumulation of axial permanent strain is directly related to deviator stress and inversely related to confining pressure. Since then, several researchers have reported that permanent deformation in granular materials is principally governed by some form of stress ratio consisting of both deviatoric and confining stresses. Lashine *et al.* (4) and Brown and Hyde (5) reported that permanent axial strain settles down to a constant value directly related to the ratio of deviator stress to confining pressure. According to Brown and Hyde, similar test results are obtained with both constant and variable confining pressure, if for the latter case the mean value of the applied pressure is used in the analysis. Other researchers (6-10) have attempted to explain the permanent strain behaviour under repeated loading using the ultimate shear strength of the material. In this approach, the static failure line is considered as a boundary for permanent strain under repeated loading. This has been questioned by Lekarp and Dawson (11) who argue that failure in granular materials under repeated loading is a gradual process and not a sudden collapse as in static failure tests. Therefore, ultimate shear strength and stress levels that cause sudden failure are of no great interest for the analysis of the material behaviour when the increase of permanent strain is incremental.

Principal stress rotation: The effect of principal stress reorientation on permanent strain behaviour is not yet fully understood. This is probably due to the fact that repeated load triaxial testing, the most common means of reproducing traffic conditions in a laboratory, fails to provide for the continuous change in the direction of principal stresses. Nevertheless, tests in simple shear box have shown significant increase in permanent strain due to principal stress rotation (12,13). Similar observations have been reported by Chan (14) who conducted hollow cylinder tests both with and without the application of shear stress. Chan noted that when the applied shear stress was high in relation to the normal stress, there was a significant difference between the permanent strain obtained in these two cases. On the other hand, when the magnitude of the reversed shear stress was small compared to the normal stress, the difference in results was small. In the same study, Chan carried out a number of large-scale rutting tests, which confirmed the findings from the hollow cylinder tests.

Number of load applications: The growth of permanent deformation in granular materials under repeated loading is a gradual process during which each load application contributes to the accumulation of strain by a small increment. The number of load cycles is, therefore, one of the most important factors to consider in the analysis of the long-term behaviour of such materials. The significance of the number of load repetitions has been mentioned many times in the literature. Some researchers (2,3,15) have reported continuously increasing permanent strain under repeated loading. Others (10) have suggested that plastic behaviour of granular materials gradually stabilises, making it possible to define a limit value for the accumulation of permanent strain. Lekarp and Dawson (11) argue that stabilisation is achieved only when the applied stresses are low. High stresses, on the other hand, would result in continuous increase of permanent strain and gradual deterioration.

Moisture Content: The literature available reveals that researchers who have studied the effect of water content in granular pavement layers believe that the combination of a high degree of saturation and low permeability, due to poor drainage, gives rise to pore pressure, low effective stress and, consequently, low stiffness and low deformation resistance(16-20). Thom and Brown (17) reported that a relatively small increase in water content can trigger a dramatic increase in permanent strain rate. The moisture sensitivity of granular materials has been reported to increase with increasing fines content (18). The increase in the rutting potential of granular materials due to wetting has also been observed during *in situ* trials using Heavy Vehicle Simulator (20).

Stress history: The permanent deformation behaviour of soils and granular materials at any instant is directly related to the stress history, i.e. the order of the application of loads. Brown and Hyde (5) showed that permanent strain resulting from a successive increase in the stress level is considerably smaller than the strain that occurs when the highest stress is applied immediately (Fig.1). Even though the effect of stress history on permanent deformation behaviour has been recognised, very limited research appears to have been done to study this effect. In laboratory permanent deformation tests, the effect of stress history is eliminated by using a new specimen for each stress path applied.

Density: The effect of density, as described by degree of compaction, has been regarded by previous studies as being significantly important for the long-term behaviour of granular materials (3,16,21-24). Resistance to permanent deformation in these materials under repetitive loading appears to be highly improved as a result of increased density (Fig. 2). Holubec (16) suggested that the reduction of plastic strain due to increased density is particularly large for angular aggregates, providing there is no accompanying increase in the transient pore pressure during repetitive loading. For rounded aggregates, however, this decrease in strain with increasing density is not considered to be significant, as they are initially of higher relative density than angular aggregates for the same compactive effort.

Grading and aggregate type: Dunlap (25) discussed that if a change in grading produces an increase in relative density for the same compactive effort, the permanent strain will decrease. The effect of grading was also investigated by Thom and Brown (24) and said to be unclear, as it varied with the compaction level. When uncompacted, the specimens with uniform grading resulted in the least permanent strain. On the other hand, the resistance to plastic strain was similar for all gradings when the specimens were heavily compacted. This

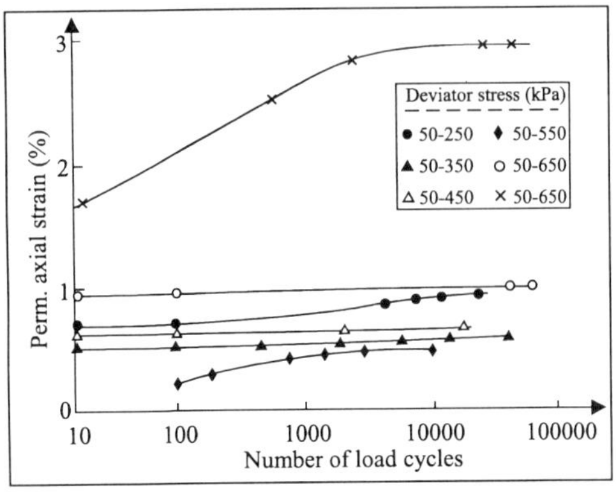

Fig. 1 Effect of stress history (after (5))

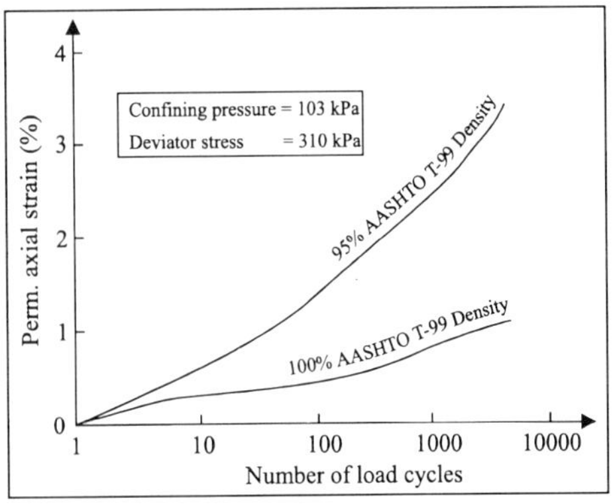

Fig. 2 Effect of density (after (21))

argument was questioned by Dawson *et al.* (19), who found the effect of grading on permanent deformation to be more significant than the degree of compaction, with the highest plastic strain resistance for the densest mix. The effect of fines content was investigated by Barksdale (3,21) and Thom and Brown (24), who concluded that permanent deformation resistance in granular materials is reduced as the amount of fines increases. Allen (22) related the difference in plastic strains between different aggregate types, of the same density, to the surface characteristics of the particles. It was concluded that angular materials, such as crushed stone, undergo smaller plastic deformations compared to materials like gravel with rounded particles. This was said to be the result of higher angle of shearing resistance in angular materials due to good particle interlock.

COMPUTATIONAL MODELLING OF PLASTIC RESPONSE

One of the main objectives of research into the long-term behaviour of granular materials is to establish a constitutive relationship which allows an accurate prediction of permanent strain. In such a relationship, it is essential to take into account the gradual accumulation of plastic strain as a function of the number of load applications and the important role played by stress condition. Different modelling techniques found in the literature can be divided into the following categories. The mathematical expressions are given in Table 1.

Static and cyclic response: Some attempts have been made to correlate repetitive and static loading test results (3,26). This approach has received mixed support since the behaviour of granular materials is generally regarded to be very complex and the repeated and static loadings do not necessarily induce the same structural response. Barksdale (3) made use of a complex hyperbolic expression, given by Duncan and Chang (27) for static triaxial tests, and related permanent axial strain in repeated load triaxial tests with applied stresses according to Eqn. [1]. Lentz and Baladi (26) suggested that, provided the specimens used in static and repeated load tests are identical in every respect, the permanent strain under repeated loading can be predicted by Eqn. 2, using the data from the static tests. Even though satisfactory results were reported by Lentz and Baladi, this method was based on the results with a single material and its validity was later rejected by Sweere (15).

Resilient and plastic response: Veverka (28) studied both resilient and plastic behaviour of granular materials and found a correlation between the two. He proposed a simple relationship between permanent and resilient strains given by Eqn. [3]. Sweere (15) studied the behaviour of unbound basecourse materials and sands, and could not confirm the equation suggested by Veverka.

Permanent deformation moduli: Most of the work on modelling permanent deformation behaviour deals with axial and horizontal stresses and strains. Jouve *et al.* (29) introduced a new approach by decomposing both stresses and strains into volumetric and shear components. In a similar manner to the elasticity theories, plastic deformation moduli were defined by Eqn. [4]. The experimental results were then used to express the permanent deformation moduli as functions of number of cycles as given by Eqn. [5].

Effect of number of load applications: Barksdale (3) suggested that, for a given stress condition, the accumulation of permanent axial strain can be defined by the log-normal expression given by Eqn. [6]. The long-term behaviour of granular materials was also investigated by Sweere (15) who found the log-normal approach to be inadequate and suggested a log-log expression for large numbers of load repetitions (Eqn. [7]). The applicability of this log-log model was later questioned by Wolff and Visser (30) who performed full-scale Heavy Vehicle Simulator (HVS) testing with several million load applications. Wolff and Visser suggested that the accumulation of permanent strain for very large numbers of load applications is best predicted by Eqn. [8]. It should be noted that Eqns. [6]-[8] all imply that accumulation of permanent strain with number of load cycles proceeds indefenitely. Yet several researchers (2,4,6,10,11,14,31) have reported that, at least for certain stresses, the growth of permanent strain eventually levels off resulting in an equilibrium state. Khedr (31) related the rate of permanent strain accumulation to the number of load repetitions using Eqn. [9]. While Khedr reported correlation coefficients quite close to unity for all tested samples, no other verification of this model was found in the literature. Paute *et al.* (10) expressed the accumulated permanent axial strain after an initial period of 100 cycles by Eqn. [10]. According to this equation, permanent strain goes towards a limit value (parameter A in the equation) as number of cycles, N, increases towards infinity. The parameter A is, therefore, considered as the limit value for total permanent axial strain. Lekarp and Dawson (11) showed that the equation suggested by Paute *et al.* is only valid at low levels of applied stresses.

Effect of stress condition: Lashine *et al.* (4) reported that permanent axial strain gradually settles down to a constant level, which is directly related to the ratio of the applied stresses (Eq. [11]). Pappin (6) expressed permanent shear strain rate as a function of the length of the stress path in p-q space and the applied shear stress ratio. He also calculated a shape factor for the variation of permanent strain with number of cycles, and expressed the total shear strain by Eq. [12]. Pappin stated that, unless the material was stressed close to the static failure limit, large permanent strains did not occur. However, the mathematical expression suggested by Pappin is not asymptotic to failure and predicts finite permanent strain even at or beyond the static failure stress. Shaw(32) attempted to apply the equation suggested by Pappin, but found it inadequate as the existence of stress paths with identical stress ratio but completely different strain rates was proved to be possible. Some researchers have suggested that the amount of permanent strain is determined by how close the applied

Table 1 *Computational models for permanent strain behaviour*

Expression	*Eqn. No.*	*Reference*
$\varepsilon_{1,p} = \dfrac{q/a\cdot\sigma_3^b}{1-\left[\dfrac{(R_f\cdot q)/2(C\cdot\cos\phi+\sigma_3\cdot\sin\phi)}{(1-\sin\phi)}\right]}$	[1]	(3)
$\varepsilon_{1,p} = \varepsilon_{0.95S}\cdot\ln(1-q/S)^{-0.15} + \left\{\dfrac{a\cdot(q/S)}{[1-b\cdot(q/S)]}\right\}\cdot\ln(N)$	[2]	(26)
$\varepsilon_{1,p} = a\cdot\varepsilon_r\cdot N^b$	[3]	(28)
$K_p(N) = \dfrac{p}{\varepsilon_{v,p}(N)}$, $G_p(N) = \dfrac{q}{3\varepsilon_{s,p}(N)}$	[4]	(29)
$G_p = \dfrac{A_2\sqrt{N}}{\sqrt{N}+D_2}$, $\dfrac{G_p}{K_p} = \dfrac{A_3\sqrt{N}}{\sqrt{N}+D_3}$	[5]	(29)
$\varepsilon_{1,p} = a+b\cdot\log(N)$	[6]	(3)
$\varepsilon_{1,p} = a\cdot N^b$	[7]	(15)
$\varepsilon_{1,p} = (c\cdot N+a)\cdot(1-e^{-bN})$	[8]	(30)
$\dfrac{\varepsilon_{1,p}}{N} = A_1\cdot N^{-b}$	[9]	(31)
$\varepsilon^*_{1,p} = A\cdot\left(1-\left(\dfrac{N}{100}\right)^{-B}\right)$	[10]	(10)
$\varepsilon_{1,p} = 0.9\dfrac{q}{\sigma_3}$	[11]	(4)
$\varepsilon_{s,p} = (fnN)\cdot L\cdot\left(\dfrac{q^\circ}{p^\circ}\right)^{2.8}_{max}$	[12]	(6)
$A = \dfrac{\dfrac{q}{(p+p^*)}}{b\cdot\left(m-\dfrac{q}{(p+p^*)}\right)}$	[13]	(10)
$\dfrac{\varepsilon_{1,p}(N_{ref})}{(L/p_o)} = a\cdot\left(\dfrac{q}{p}\right)^b_{max}$	[14]	(11)

Parameters	
$\varepsilon_{1,p}$	= accumulated permanent strain after N load repetitions
$\varepsilon^*_{1,p}$	= additional permanent axial strain after first 100 cycles
$\varepsilon_{1,p}(N_{ref})$	= accumulated permanent axial strain at a given number of cycles N_{ref}, $N_{ref} > 100$
$\varepsilon_{v,p}$	= permanent volumetric strain for $N > 100$
$\varepsilon_{s,p}$	= permanent shear strain for $N > 100$
ε_r	= resilient strain
K_p	= bulk modulus with respect to permanent deformation
G_p	= shear modulus with respect to permanent deformation
q	= deviator stress
p	= mean normal stress
q°	= modified deviator stress= $\sqrt{2/3}\cdot q$
p°	= modified mean normal stress= $\sqrt{3}\cdot p$
p^*	= stress parameter defined by intersection of the static failure line and the p-axis in p-q space
p_o	= reference stress
L	= stress path length
σ_3	= confining pressure
N	= number of load applications
S	= static strength
$\varepsilon_{0.95S}$	= static strain at 95 percent of static strength
C	= apparent cohesion
ϕ	= angle of internal friction
fnN	= shape factor
R_f	= a constant relating compressive strength to an asymptotic stress difference
A_1	= a material and stress-strain parameter given (function of stress ratio and resilient modulus)
A_2, A_3, D_2, D_3	= parameters which are functions of stress ratio q/p
m	= slope of the static failure line
a, b, c, A, B	= regression parameters (A is also the limit value for maximum permanent axial strain)

stresses are to the static failure stress. Gerrard *et al.* (33) plotted the applied stresses in a repeated load triaxial test on the Mohr circle and found the corresponding static shear strength. Permanent strain after certain number of load cycles were then drawn as contours for the corresponding ratio of the applied stress to the shear strength. Raymond and Williams (8) made use of the stress ratio $q_{max}/q_{failure}$, in which q_{max} is the maximum deviator stress and $q_{failure}$ the deviator stress at failure, to characterise the results of permanent deformation tests. Thom (9), on the other hand, suggested that permanent shear strain is better related to the stress ratio $(q_{failure}-q_{max})/q_{max}$. Paute *et al.* (10) defined a limit value for permanent axial strain and suggested that it varies with the maximum shear stress ratio, defined by using the static failure line, as shown in Eqn. [13]. This approach was later investigated and highly questioned by Lekarp *et al.* (34).

Shakedown approach: Some researchers (35-38) have developed computational procedures for pavement analysis based on the so-called shakedown theory. According to this theory, a pavement is liable to show progressive accumulation of plastic strains under repeated loading if the magnitude of the applied loads exceeds a limiting value, called the shakedown load. The pavement is then said to exhibit an incremental collapse or gradual failure. On the other hand, if the applied loads are lower than the shakedown limit, the growth of plastic strains will eventually level off and the pavement is said to have attained a state of shakedown by means of adaptation to the loads. The response of the pavement will then be elastic under additional load applications. The complex numerical models found in the literature deal with the response of the whole pavement as a single, unified structure. In a recent study, Lekarp and Dawson (11) suggested that the shakedown approach may also be employed in explaining the permanent deformation behaviour of granular materials. Using repeated load triaxial tests on different granular materials, a relationship was defined between the ratio of permanent strain after a given number of cycles, the stress path length and the maximum shear-normal stress ratio according to Eqn. [14]. At low stress ratios, the growth of permanent strain levelled off resulting in a final equilibrium state. High stress ratios, on the other hand, resulted in more progressive accumulation of permanent strain and gradual deterioration of the material. These results showed clear indications that the state of

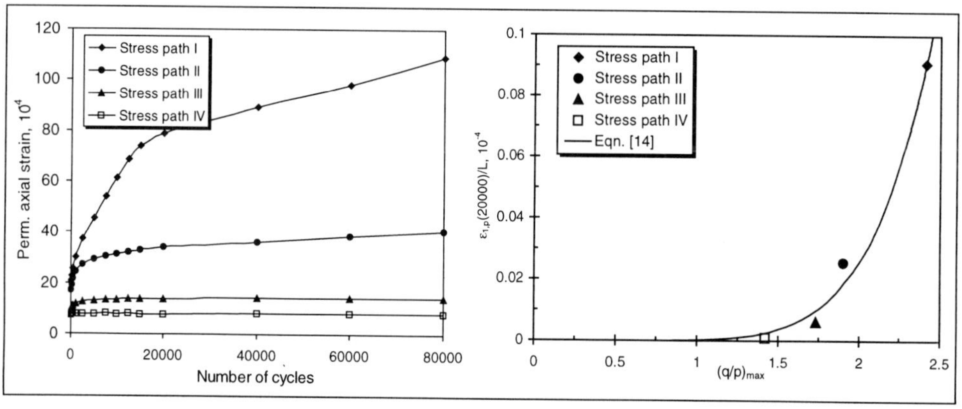

Fig. 3 Effect of no. cycles and stresses on permanent deformation development (after (1))

gradual collapse occurs beyond a certain threshold stress ratio, which could be a form of so-called shakedown limit. Lekarp and Dawson, however, pointed out that more research is needed to determine this shakedown limit.

CONCLUSIONS AND RECOMMENDATIONS

The literature survey revealed some of the complexities involving permanent deformation behaviour of unbound granular materials. The plastic behaviour of these materials is affected by several factors: stress level, principal stress rotation, number of load applications, moisture content, stress history, density, grading and aggregate type. While the impact of some of these factors, such as stress level and number of load cycles, has been investigated by many researchers, much less work has been done regarding the influence of reorientation of principal stresses, moisture content and stress history. There is clearly need for further studies in this area. Traditionally, the computational modelling of the permanent strain behaviour of granular materials has been based on the gradual accumulation of permanent strain with number of cycles and the impact of applied stresses. Several models have been introduced through the years. The review of the literature shows that these models are either based on continuously increasing plastic strain or assume that the growth of plastic strain eventually levels off resulting in an equilibrium state. A recent study (1,11) suggests the application of the so-called shakedown theory in modelling the long-term behaviour of granular materials. According to this approach, the response of the material stabilises if the applied stresses are below a certain threshold called shakedown limit. Stresses beyond this limit value would result in gradual deterioration of the material due to continuous increase in permanent strain. However, the possibility of using this approach needs yet to be confirmed.

REFERENCES

1. LEKARP, F. Permanent deformation behaviour of unbound granular materials. Licentiate Thesis, Royal Institute of Technology, Stockholm, 1997.
2. MORGAN, J.R. The response of granular materials to repeated loading. Proceedings, 3rd Australian Road Research Board Conference, Sydney, pp. 1178-1192, 1966.
3. BARKSDALE, R.D. Laboratory evaluation of rutting in basecourse materials. Proceedings of 3rd International Conference on the Structural Design of Asphalt Pavements, London, pp. 161-174, 1972.
4. LASHINE, A.K., BROWN, S.F., and PELL, P.S. Dynamic properties of soils. Report No. 2 Submitted to Koninklijke/Shell Laboratorium, Department of Civil Engineering, University of Nottingham, 1971.
5. BROWN, S.F. and HYDE, A.F.L. Significance of cyclic confining stress in repeated load triaxial testing of granular material. Transportation Research Record, No. 537, pp. 49-58, 1975.
6. PAPPIN, J.W. Characteristics of a granular material for pavement analysis. Ph.D. Thesis, Department of Civil Engineering, University of Nottingham, 1979.
7. BARRET, J.R. and SMITH, D.M. Stress history effects in base course materials. Proceedings, 8th Australian Road Research Board Conference, Vol. 8, pp. 30-39, 1976.

8. RAYMOND, G.P. and WILLIAMS, D.R. Repeated load triaxial tests on a dolomite ballast. Proceedings, The American Society of Civil Engineers, Vol. 104, No. GT7, pp. 1013-1029, 1978.

9. THOM, N.H. Design of road foundations. Ph.D. Thesis, Department of Civil Engineering, University of Nottingham, 1988.

10. PAUTE, J.L., HORNYCH, P., and BENABEN, J.P. Repeated load triaxial testing of granular materials in the French Network of Laboratories des Ponts et Chaussées. In Flexible Pavements, Edited by A.G. Correia, Proceedings, European Symposium on Flexible Pavements, Euroflex, pp. 53-64, 1993.

11. LEKARP, F., and DAWSON, A. Modelling permanent deformation behaviour of unbound granular materials. Construction and Building Materials, 1997 (*in press*).

12. YOUD, T.L. Compaction of sands by repeated shear straining. Proceedings of ASCE, Journal of Soil Mechanics and Foundation Division, Vol. 98, SM7, pp. 709-725, 1972.

13. ANSELL, P. and BROWN, S.F. Cyclic simple shear apparatus for dry granular material. Geotechnical Testing Journal, Vol. 1, No. 2, pp. 82-91, 1978.

14. CHAN, F.W.K. Permanent deformation resistance of granular layers in pavements. Ph.D. Thesis, Department of Civil Engineering, University of Nottingham, 1990.

15. SWEERE, G.T.H. Unbound granular bases for roads. Ph.D. Thesis, University of Delft, 1990.

16. HOLUBEC, I. Cyclic creep of granular materials. Department of Highways, Ontario, Report No. RR147, 1969.

17. THOM, N.H. and BROWN, S.F. Effect of moisture on the structural performance of a crushed-limestone road base. Transportation Research Record, No. 1121, pp. 50-56, 1987.

18. HAYNES, J.H., and YODER, E.J. Effects of repeated loading on gravel and crushed stone base course materials used in the AASHO Road Test. Highway Research Board Record, No. 39, Highway Research Board, 1963.

19. DAWSON, A.R., THOM, N.H., and PAUTE, J.L. Mechanical characteristics of unbound granular materials as a function of condition. In Flexible Pavements, Edited by A.G. Correia, Proceedings, European Symposium on Flexible Pavements, Euroflex 1993, Laboratório Nacional de Engenharia Civil, Lisbon, pp. 35-44, 1993.

20. MAREE, J.H., FREEME, C.R., VAN ZYL, N.J., and SAVAGE, P.F. The permanent deformation of pavements with untreated crushed stone bases as measured in Heavy Vehicle Simulator tests. Proceedings, 11th Australian Road Research Board Conference, Part 2, pp. 16-28, 1982.

21. BARKSDALE, R.D. The aggregate handbook. National Stone Association, Washington D.C., 1991.

22. ALLEN, J. The effect of non-constant lateral pressures of the resilient response of granular materials. Ph.D. Thesis, University of Illinois, Urbana, 1973.

23. MAREK, C.R. Compaction of graded aggregate bases and subbases. Proceedings, The American Society of Civil Engineers, Vol. 103, pp. 103-113, 1977.

24. THOM, N.H. and BROWN, S.F. The effect of grading and density on the mechanical properties of a crushed dolomitic limestone. Proceedings, 14th Australian Road Research Board Conference, Materials and Testing, pp. 94-100, 1988.

25. DUNLAP, W.A. Deformation characteristics of granular materials subjected to rapid repetitive loading. Ph.D. Thesis, Texas A & M University, 1966.

26. LENTZ, R.W. and BALADI, G.Y. Constitutive equation for permanent strain of sand subjected to cyclic loading. Transportation Research Record, No. 810, pp. 50-54, 1981.

27. DUNCAN, J.M. and CHANG, C.Y. Nonlinear analysis of stress and strain in soils. Proceedings, The American Society of Civil Engineers, Vol. 96, No. SM5, pp. 1629-1653, 1970.

28. VEVERKA, V. Raming van de spoordiepte bij wegen met een bitumineuze verharding. De Wegentechniek, Vol. XXIV, No. 3, pp. 25-45, 1979.

29. JOUVE, P., MARTINEZ, J., PAUTE, J.L., and RAGNEAU, E. Rational model for the flexible pavement deformations. Proceedings, 6th International Conference on Structural Design of Asphalt Pavements, Vol. 1, Ann Arbor, pp. 50-64, 1987.

30. WOLFF, H. and VISSER, A.T. Incorporating elasto-plasticity in granular layer pavement design. Proceedings, Instn. Civil Engineers Transp., No. 105, pp. 259-272, 1994.

31. KHEDR, S. Deformation characteristics of granular base course in flexible pavement. Transportation Research Record, No. 1043, pp. 131-138, 1985.

32. SHAW, P.S, Stress-strain relationships for granular materials under repeated loading. Ph.D. Thesis, Department of Civil Engineering, University of Nottingham, 1980.

33. GERRARD, C.M., MORGAN, J.R., and RICHARDS, B.G. An Approach to the design of flexible pavements for australian conditions. Australian Road Research Report 5 (8), 1975.

34. LEKARP, F., RICHARDSON, I. R. and DAWSON, A. Influences on permanent deformation behavior of unbound granular materials. Transportation Research Record, No. 1547, pp. 68-75, 1996.

35. SHARP, R.W. and BOOKER, J.R. Shakedown of pavements under moving surface loads. Journal of Transportation Engineering, Vol. 110, No. 1, pp. 1-14, 1984.

36. SHARP, R.W. Shakedown analysis and the design of pavements. Ph.D. Thesis, University of Sydney, 1983.

37. RAAD, L., WEICHERT, D., and HAIDAR, A. Shakedown and fatigue of pavements with granular bases. Transportation Research Record, No. 1227, pp. 159-172, 1989.

38. COLLINS, I.F., WANG, A.P., and SAUNDERS, L.R. Shakedown theory and the design of unbound pavements. Australian Road Research Board, Road and Transport Research Vol. 2, No. 4, pp. 28-37, 1993.

Mechanical properties of frictional materials

- An experimental study of load spreading and deformations

Ebrahim Parhamifar

MSc in Civil Engineering
Lund University, Box 118 S-221 00 Lund, Sweden
Phone: +46 46 222 91 34 E:mail_ Ebrahim.Parhamifar@tft.lth.se

ABSTRACT

There have been many attempts to establish analytical design methods.
Current pavement design models are often empirical and based on experiences from the field. Models are more or less sophisticated depending on which assumptions are used. Many design models are based on classical theory of elasticity, which imply an elastic material response.

Road construction materials are neither purely elastic nor homogenous. There are many reasons to question these methods as well as the empirical ones.

Frictional materials are viscoelastic in nature. This is reflected in both an elastic and a viscous response to external loading. In addition the response will be both size and time dependent. This complex behavior is not unique for frictional materials but it seems to be more pronounced than for other structural materials. There are several ways to consider a viscoelastic material behavior in a design but this requires appropriate experimental information where internal deformations are registered during external applied loads.

Deformations of granular materials used as roadbase materials under static and dynamic loading are studied in a box 800*800*800 mm. The thickness of the material layer is in accordance with the *Swedish National Road Administration* standards and the construction is instrumented at two levels to measure vertical responses. This means that- unlike triaxial measurements- deformation is measured inside the material.

The aim is to establish a relationship between applied stresses and accumulated deformations in granular materials. Six different materials are included in the tests. Materials are compacted properly and rested in two days before exposed loading.

KEYWORDS:
Frictional media, load spreading, viscoelastic, static and dynamic loading

1. THEORETICAL MODELING

Granular materials are special in the sense that they exhibit no or very little natural cohesion between its constituents as opposed to most other materials. Instead the friction between particles will be of significant importance. This difference in the internal structure compared to most other materials will have a pronounced influence on the mechanical behavior. This affects directly the strength characteristics which influences both the load spread from an external load as well as the deformation properties. Most analysis methods used for normal structural materials are less satisfactory and in some case completely inappropriate. This is in particular analysis methods based on the theory of elasticity. These methods normally require a homogeneous material with an internal tensile strength to be valid. Still, such methods often are used in absence of anything better because they have the advantage of being easily available. The problem with analyzing the load spreading in frictional media is directly connected to its unique characteristics. For this reason the first available formulations were not aimed at solving this particular problem. This is the case with the solution given by *Boussinesq (1885)* for the load spreading of a point load in a homogeneous half-sphere. Even if it is analytically tractable, as later solutions for almost any type of loading condition, it is based on assumptions, which are not in agreement with the properties of frictional media. The special appeal for these solutions stems from the variety of practical loading situations which can be accommodated, in easy to use charts, and, above all, the complete independence of vertical normal stresses and shearing stresses on any characterization or properties of the medium.

Comparisons and limitations of solutions obtained from the theory of elasticity in their application to particulate media are well documented in the literature *(Föpple and Föpple, 1897; Faber, 1933; Kögler and Scheidig, 1938; Turnball, Maxwell and Ahlvin, 1961; etc.).* Experiments have shown that deformations predicted by the theory of elasticity are incorrect in magnitude and distribution.

2. EXPERIMENTAL METHODS

One of the most important characteristics of road construction materials is its stress-strain behavior. When dealing with construction material, two concepts are primarily concerned. The first is the resistance of the material to stress and material failure. The second is the deferability of the material under stresses to which the material is subjected including the magnitude and the distribution and the rate of application. The second aspect is what explains by a stress-strain relation.

For a material, which is homogenous, isotropic and linearly elastic, it is possible to describe its stress-strain property by Young's modulus of elasticity (E) and Poisson's ratio. These could be obtained from a relatively simple test, where a bar of material is stretched, measuring the tension and resulting longitudinal and transverse deformation. With the elastic constants it is possible in an ideal media to determine the relationship between stresses and deformation for other type of loading than simple tension *(G.F.Sowers, 1988)*

Soils are not ideal materials, even though in some specific cases it is convenient to use a constant elasticity modulus and constant Poisson's ratio. In soils these are not constants, but variable which at best approximate the soil behavior for a given stress condition. They can change, sometimes radically, if the stress conditions change (changes in loading) or if the stresses are applied in a different way. Therefore, when elastic constants are applied to describe the soil behavior, they represent only the specific conditions for which measurements were taken or calculations made.

One of the most appropriate ways to find out the stress-strain characteristics of frictional media is to measure it directly either in laboratory or field. In real situations like considering a road structure, stress can be applied in many ways. All these different situations can not be represented in a single laboratory test. Therefore in order to collect reasonable data one has to conduct several different tests to represent the conditions in which the structure is being used. One the frequently used laboratory test is *Repeated Load Triaxial Apparatus* method.

By using this apparatus many different loading situations are easily simulated. There are a number of drawbacks involving triaxial testing *(Fredrik Lekarp 1997).*

Another way for measuring road constructions material responses is *in-situ* instrumentation *(Peter Ekdahl, 1997).* This method has been studied by several researchers worldwide and has given many good results.

Like any other methods *in-situ* testing of material properties has some advantages as well as disadvantages. One the most important drawbacks regarding in situ testing are high costs of instrumentation of the pavement structure.

In following pages a method which has been used to study the permanent deformations of granular materials under static and repeated dynamic loading is described and some of the results obtained from measurements are presented. It should be mentioned that this method is still in developing phase and some modifications are already made after this study.

One the most important and primary objective of this work is to study the behavior of granular materials under static loading. This kind of loading if often applied to pavements at bus stations and near traffic lights. Since the pavements are designed to carry dynamic loading, large plastic deformations due to static loading occur at these places. Study of stress-strain properties under static loading gives a better understanding of material behavior for pavement design especially for objects, which are constantly subjected to such loading situations.

Another objective of this study is to compare the results of repeated dynamic loading with results obtained by *Triaxial* testing apparatus. In order to do this comparison, one material is chosen and measurements under same conditions are conducted both by triaxial apparatus and testing box developed at Lund University. The results of the study show a very convincing tendency in measurements by *Testing Box* in comparison to an established method such as *Triaxial.*

3. TESTING BOX

The deformations in frictional materials are measured in a box 800*800*800 mm. The box is made of steel and the sides are together. One side is made of Plexiglas to make visual observations possible. To give the testing box a rigid support two beams of HE120A are used at the bottom according to figure bellow. To study and simulate the vertical deformations of frictions material under traffic loading a layer of crushed material is compacted and 10 transducers are placed at two different levels. A hydraulic pump applies desired loads to the construction and vertical displacements are registered.

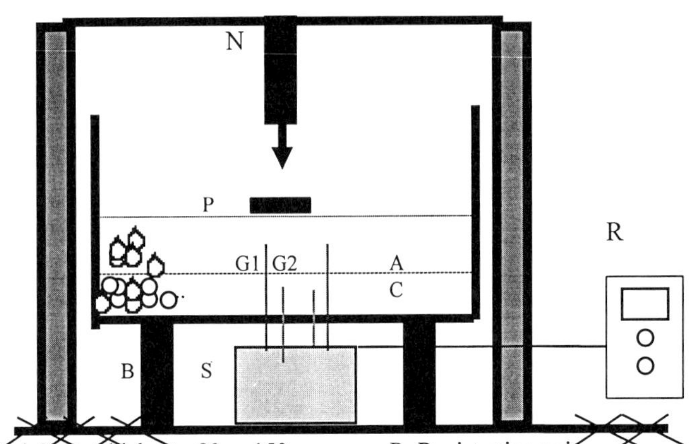

A: Granular material. Layer thickness 80 or 150 mm
B: Steel buck
C: 200 mm of 0-40 mm material
G1: Steel bars to transducers at middle of layer A
G2: Steel bars to transducers at bottom of layer A
R: Registration unit
S: Support to transducers
P: Loading plate
N: Loading unit

FIGURE 1 TESTING BOX AND LOADING UNIT

3.1 TRANSDUCERS

To measure movements of particles ten transducers are placed at two levels and five different locations according to figure 2.

Transducers have a circular shape with a diameter of 10 mm which easily could fit between real particles and do not disturb the material layer formation. The circles are connected to a measuring device outside the box by a stainless steel rod with two different lengths.

The measuring gauge has a length of 25 mm. To register the results of the measurements a program called *Datacapture* is used. The program is capable of registering 14 different signals. Two channels are used for registration of magnitude-applied loading and vertical movements of the loading unit. The accuracy of the loading unit is ± 0,02 mm. The transducers are very easy to replace and do not affect the position of steel bars.

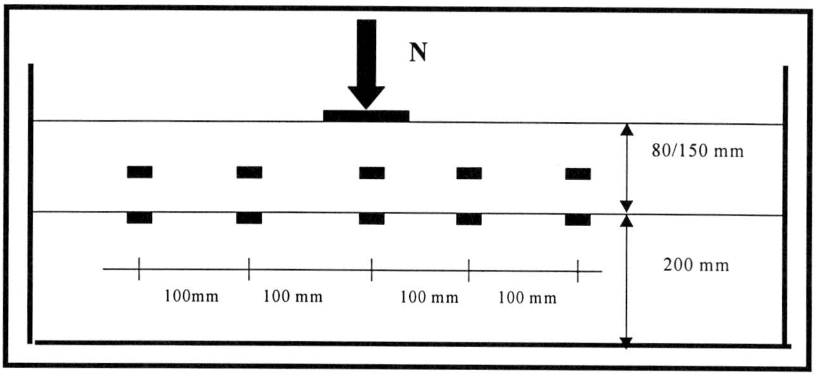

FIGURE 2 PLACEMENT OF TRANSDUCERS

3.2 LOADING UNIT

The loading unit is composed of a hydraulic pump controlled by a computer, which automatically gives all the signals to loading unit. Applied loads, which could go up to 100 kN are distributed to the material by a circular plate. The loading plate has a diameter of 100 mm with a thickness of 5 mm. A computer registers deformations and applied stresses at every second interval.

3.3 MATERIAL LAYER

To simulate a real site condition 200-mm of granular material are packed at the bottom of the box. One major function of this layer is to eliminate the effect of stiff plate at the bottom. Five transducers are placed at the top of this layer. Furthermore a layer of same crushed material, 80 or 150 mm thick recommended by *Swedish National Road Administration* is placed and packed at the top of previous layer. Five transducers are placed at the middle of the new layer. The distance between transducers is 100 mm.

3.4 COMPACTION OF THE MATERIAL

Compacting the material is a very important process prior to the measurements. The degree of compaction has a very major roll in how deformations are occurred in material. In order to get as high compaction as possible the following method is used:

- Determination of water content. This is done by taking 1000 gr. of the material and drawn it in *Ethano*l. This combination is then burned and when *Ethanol* is vaporized the stones are dried. One reason using this method is the time factor. The method is very quick and could be repeated several times to get a better result. The water content is then measured as the relationship between the stones before and after burning:

$WC\ \% = (m_1 - m_2) / m_1$

Where: m_1 is the weight before burning
m_2 is the weight after burning

- Mixing material and water. To approach a high degree of compaction recommended by *Swedish National Road Administration*, and to avoid separation of fine material from larger particles a mixer is used to mix the water and granular materials. The granular material is weighted and mixed with adequate water in a mixer at about 5 minutes. The material is weighted again before transferring to *Testing Box.*
- Compaction of material. The mixed material is compacted by a small compactor. The amount of material compacted every time is almost 40 mm thick. According to *Mark AMA 83* table C/5 the largest layer thickness should be 150 mm for this type of packing instrument (hand stamping 15 kg).

The layer construction with all transducers is covered with plastic and remains in 48 hours. This gives the construction possibility to stabilize and even water content is reached.

4. TRIALS FOR STATIC LOADING

To study permanent deformations caused by static loading six different type of material are chosen. Five materials named D1 to D5 are crushed materials and the last one is a natural material, which contains some crushed particles. These materials are regularly used as roadbase in Sweden. The aggregates are chosen from different places and have therefor-different properties.

The dry densities, water contents and other properties of aggregates used in trials are listed in table 1.

Table 1 Material properties

Material/Properties	**D1**	**D2**	**D3**	**D4**	**D5**	**D6**
Dry density (g / cm^3)	2,082	2,226	2,200	2,297	2,304	2,245
Moisture content (% weight)	7,7	5,3	7,0	6,2	5,7	6%
Particle density>5,6 mm	2,73	2,60	2,73	2,78	2,73	2,57
Flakiness ratio	1,41	1,48	1,42	1,46	1,45	1,37
Aggregate impact value	52	62	57	46	57	54
Nordic Ball Mill Value	23,5	24,8	20,6	15,3	15,7	25,1
% uncrushed material	-	-	-	-	-	39 /20

The recommended values for *Flakiness, Aggregate Impact Value* and *Ball Mill Value* according to *Swedish National Road Administration* are:

Nordic Ball mill Value<30
Aggregate impact value = No values are recommended
Flakiness ratio = No values are recommended

The applied loads vary from 3,75 kN to 15 kN. Loading speed is 0,25 kN per second. These four loading cycles are applied after each other. The loading program is listed in Table 2

Table 2 Loading program

Aggregate	**Thickness mm**	**Load / Cycles KN**			
D1	80 / 150	3,75 / 25	7,5 / 25	11,25 / 25	15,00 / 25
D2	80 / 150	3,75 / 25	7,5 / 25	11,25 / 25	15,00 / 25
D3	80 / 150	3,75 / 25	7,5 / 25	11,25 / 25	15,00 / 25
D4	80 / 150	3,75 / 25	7,5 / 25	11,25 / 25	15,00 / 25
D5	80 / 150	3,75 / 25	7,5 / 25	11,25 / 25	15,00 / 25
D6	80 / 150	3,75 / 25	7,5 / 25	11,25 / 25	15,00 / 25

5. RESULTS OF MEASUREMENTS

Static loading tests are conducted according to specifications in table 2. Some of the interesting results for material type D1, crushed material 0-40 mm and D6, uncrushed material 0-40 mm are shown in figures 1 and 3. In figure 2 a comparison between results from Triaxial and Testing box is illustrated. As it is seen in the figure the result of dynamic loading in *Testing Box* is very similar to *Triaxial* apparatus. The two transducers 5 and 8 are those placed directly under the center line of loading plate. Deformations illustrated in figure 1 and 3 are the difference in deformations for transducer 5 and 8 and shows therefore permanent deformation for a small specimen in that area. The same result is obtained for all other types of loadings and materials which indicates that Testing Box is capable of reproduce same result.

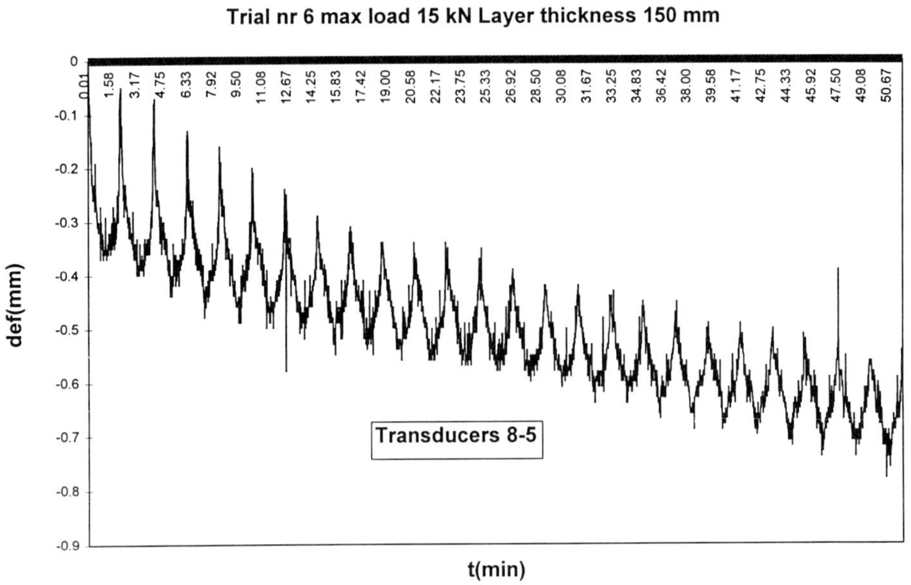

Diagram 1 **Permanent deformation registered for crushed material type D1**

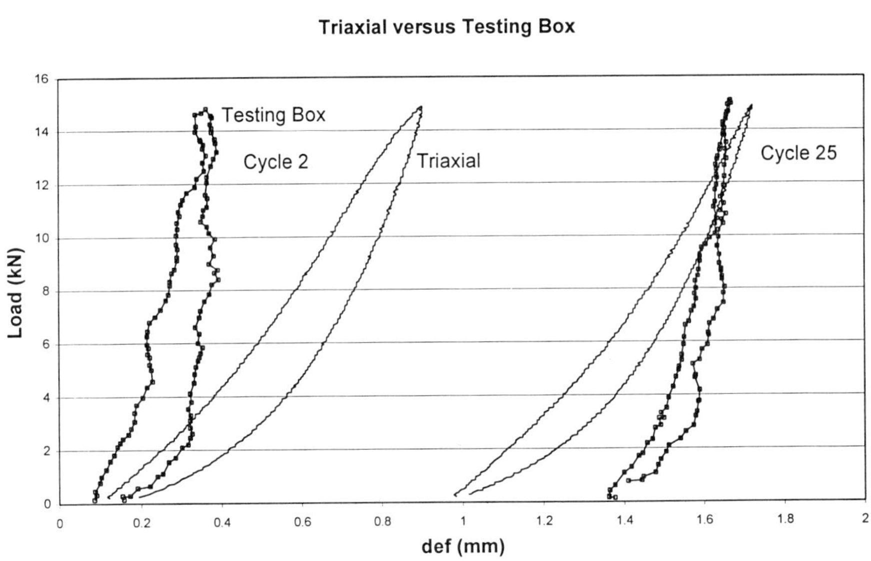

Diagram 2 **Triaxial and Testing Box**

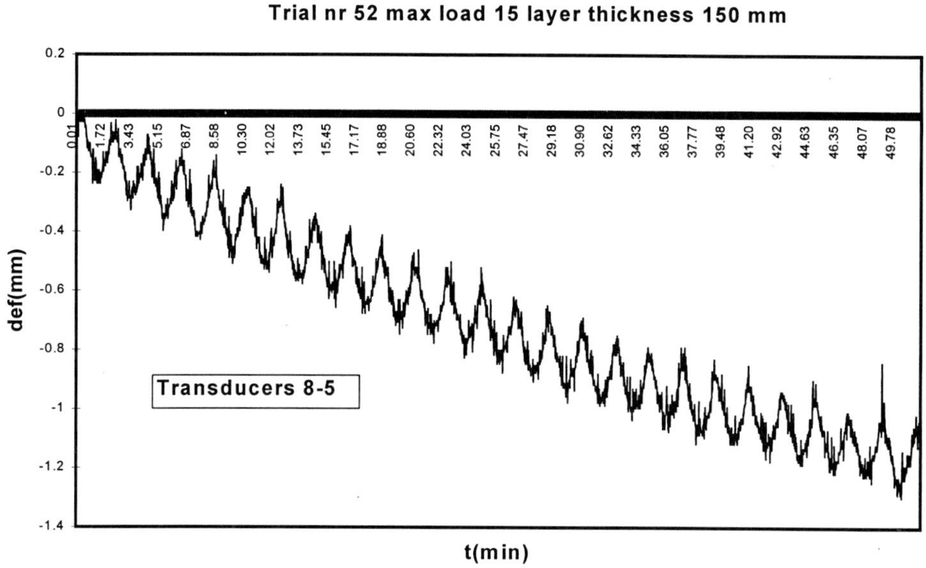

Diagram3 **Permanent deformation registered for uncrushed material type D6**

6. CONCLUSION

The results of dynamic loading measurements by *Testing Box* are very much similar to those conducted by *Triaxial.* The differences are very marginal as shown in diagram 2 and could be related to the volume subjected to stresses in *Testing Box* or number of particles in that area stresses are applied *(Sven Agardh & Thomas Lehrecke 1998).* In *Triaxial* testing, the volume of the material subjected to stresses is larger than the volume subjected to stresses in *Testing Box*. The diagram plotted after *Triaxial* testing shows the movements of the whole body of specimen subjected to applied stress. On the hand the deformations measured in *Testing Box* shows the behavior of a few number of particles due to external and internal changes and therefore the curves in diagram 2 are irregular . It should be mentioned however that the gauges used in *Testing Box* has not the same degree of accuracy as the *Triaxial* testing apparatus.

Results of static loading in *Testing Box* shown in diagram 1 and are for material type D1 and D2, which are a crushed respectively uncrushed material. The figures shows that magnitude of permanent deformations are increasing by time. This means that the material has *time softening* properties. But granular materials are known to have *time hardening* properties. This could be explained by the fact that in the beginning the materials are compacted due to applied stresses and behave like *time hardening* and later on and due to increase of stresses the particles crush or rotate and get a new position and this leads to *time softening* behavior.

The figures show the differences between deformations for crushed and uncrushed materials. The uncrushed material shows larger deformation than the crushed one.

The study shows also that granular materials behavior could not be only explained using theory of elasticity but more sophisticated stress-strain relationships has to be developed. This kind of modeling is probably possible by using results from different experiments.

Another interesting result is that the *Testing Box* is capable to produce measurements with the almost the same accuracy time after time.

ACKNOWLEDGEMENT

Construction Company Skanska in Sweden finances this study. The authors are indebted to Jim Bengtsson (Skanska Laboratory in Lomma) and Monica Berntman (University of Lund) and Sven Agardh (University of Lund) for all the help during the time.

REFERENCES:

1. Sowers G.F. '' Soil Mechanics in Highway Engineering'', Trans Tech Publications, 1988

2. Lekarp F. '' Permanent Deformations Behaviour of Unbound Granular Materials.Department of Infrastructure and Planing, Division of Highway Engineering, Royal Institute of Technology, 1997

3. Ekdal P. '' Full-scale Testing of Pavement Response'' University of Lund, Division of Road Construction, 1988

4. Agardh S. & Lehrecke T. '' Determination of Material Properties of Frictional Media'' University of Lund, Division of Road Construction, 1998

Fifth International Conference on the
BEARING CAPASITY OF
ROADS AND AIRFIELDS

RESILIENT DEFORMATION BEHAVIOUR OF AGGREGATES DETERMINED BY MEANS OF CYCLIC LOADING TRIAXIAL TESTS

PAULI KOLISOJA

Senior assistant, Tampere University of Technology,
P.O.Box 600, FIN-33101 Tampere, Finland
Fax: + 358 3 365 2884, E-mail: kolisoja@cc.tut.fi

ABSTRACT

The cyclic loading triaxial test is widely accepted to be a fundamental tool in determination of the basic mechanical properties of coarse grained granular materials typically used in unbound layers of road pavements. The test is especially well suited to the investigation of the resilient deformation behaviour of these materials due to its reasonable cost, straightforward applicability and accurate control of stresses and strains that it provides.

In the present study the results of more than 150 large scale - test specimen diameters 200 mm and 300 mm - cyclic loading triaxial tests have been analysed regarding the resilient deformation behaviour. The test series, which has mainly been commissioned by the Finnish National Road Administration and a consortium of Norwegian aggregate producers, Arbeidsfellesskapet KPG, have been performed in the Laboratory of Geotechnical Engineering at the Tampere University of Technology since 1992. The test materials have included a number of different types of Finnish and Norwegian aggregates whose grain size distribution has been varied from capping layer sand to materials corresponding to railway ballast.

After presenting the test materials and equipment that has been used in the research, this paper discusses the most important factors influencing the resilient deformation behaviour of granular materials and their particle level mechanisms of affection. As a practical result of the study the paper presents a method by means of which the values of resilient modulus can be approximated e.g. for analytical pavement design calculations. The method provides a reliable estimate of the modulus values on any practically feasible stress and density level if one cyclic loading triaxial test covering large enough range of stresses is performed and the maximum porosity value corresponding to the loosest state of the material is experimentally determined.

Key words: aggregate, cyclic, deformation, resilient, triaxial

INTRODUCTION

The availability of the coarse grained granular materials that have traditionally been used in construction of the unbound layers of road pavements has in many areas become scarce during the last few decades. Consequently the use of these materials should be optimised as much as possible and as large a share as possible should be replaced by various kinds of alternative construction materials i.e. industrial by-products, recycled material from old road pavements and other residual material resulting from road construction. The introduction of these alternative construction materials has, however, necessitated the development of new approaches to the design of road pavements, because the old empirical design methods are in most cases not applicable to the analysis and design of structures containing the alternative materials. Very often these new design approaches are called analytical or mechanistic design methods.

As it is well known, the key idea in mechanistic pavement design methods is to evaluate the stresses and strains under realistic traffic loads at the critical points of the pavement structure based on the analysis of stress-strain conditions of the entire structure. After which, the service life of the pavement can be estimated based on the obtained values of critical stresses and strains. As an input data the analysis needs of course reliable information on the deformation behaviour of the construction materials – also including the traditional ones. In this study the results of a considerable amount of large scale cyclic loading triaxial tests that have been performed with different types of Finnish and Norwegian aggregates have been analysed paying special attention to the particle level mechanisms of affection of the various factors that influence the resilient deformation behaviour of coarse grained granular materials.

AVAILABLE TEST RESULTS

Test equipment

A photograph of the test equipment that has been used in performing the test series referred to above is shown in Figure 1. The main components of the cyclic loading triaxial test facility developed in the Laboratory of Geotechnical Engineering at the Tampere University of Technology (TUT) at the beginning of this decade are a servo controlled hydraulic actuator, a loading frame, a constant confining pressure test cell enabling the maximum test specimen diameter of 300 mm and a micro computer based control and data acquisition system. Due to the fact that the main technical features of the testing facility have already been described in more detail in an earlier paper of this conference series /1/, the description is not repeated here.

Loading procedures

In performing the cyclic loading triaxial tests three different loading procedures have been adopted as follows:

1. A vast majority of the tests were performed according to the loading procedure suggested in the well-known American SHRP (Strategic Highway Research Program) protocol P46 /2/.

2. A number of tests were also performed using a test procedure that closely resembles a loading procedure that was recommended as a result of a reference test series performed at a few European laboratories and that is also likely to be included in the CEN standards /3/.
3. A couple of tests with some of the Norwegian test materials were additionally performed using a test procedure adopted at the Norwegian Geotechnical Institute NGI in cyclic loading triaxial testing of very large – test specimen diameter of 600 mm – aggregate specimens /4/.

Figure 1. Cyclic loading triaxial test facility at Tampere University of Technology.

Test materials

Most of the test specimens were prepared using aggregates whose grain size distribution had been prepared in a predetermined composition. Thereby the grading of the test materials was varied from capping layer sand to very open-graded materials corresponding to railway ballast. In addition to the grain size distribution the density, moisture content and aggregate type were also varied during the test series.

A summary of the grain size distributions of the Finnish test materials is presented in Figure 2a and that of the Norwegian test materials in Figure 2b. In both figures the letters NG, CG and CR stand for the aggregate types that have been tested i.e. natural gravel, crushed gravel and crushed rock respectively. In addition to the test materials possessing one of the grain size distributions of Figure 2 and being composed of newly produced aggregates some of the specimens were also prepared by using materials that had been taken out of existing road pavements.

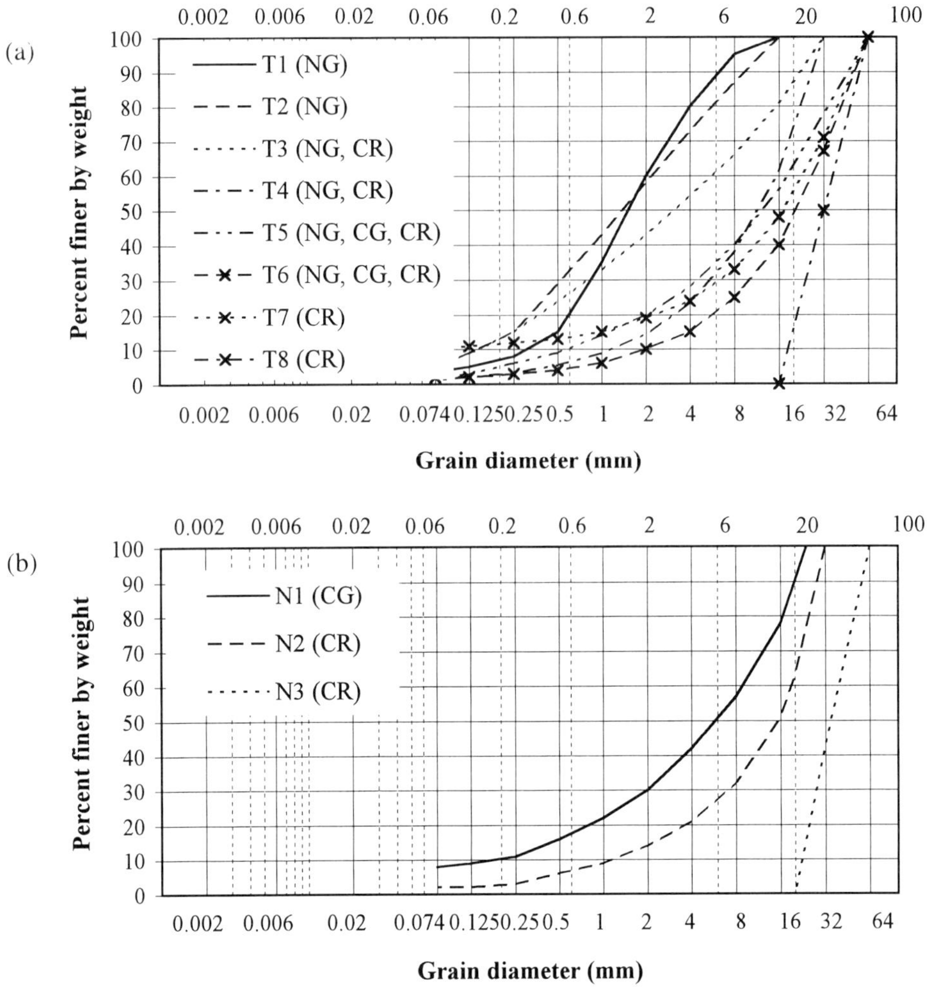

Figure 2. Objective gradings of (a) the Finnish and (b) Norwegian aggregates prepared to a predetermined grain size distribution.

FACTORS AFFECTING THE RESILIENT DEFORMATION BEHAVIOUR OF AGGREGATES

Mechanical model of a particle contact

As stated in the beginning, in the present study particle level considerations have been utilised in identifying the mechanisms of affection of the various factors influencing the resilient deformation behaviour of granular materials. In fact, several different versions of the mechanical model of the particle to particle contact are available in the literature presenting the results of the research work that has been made in micromechanical modelling of granular materials. Here the model suggested by Ting et al. /5/ is given as an example (Figure 3).

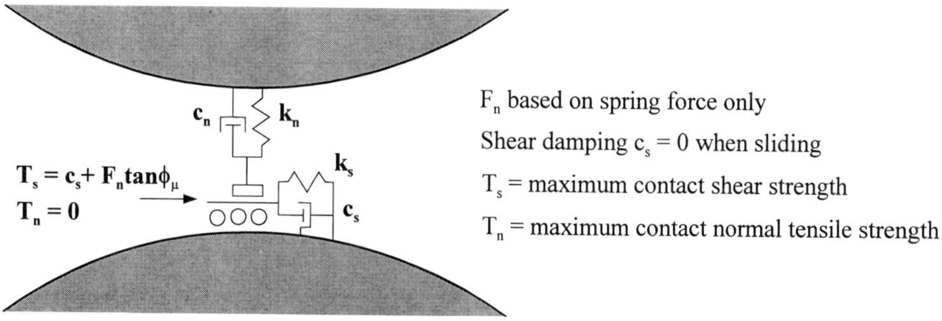

Figure 3. An example of a mechanical model of a particle contact /5/.

According to Figure 3 factors that obviously affect the resilient deformation behaviour of a dry granular material are at least the normal and shear stiffnesses of the particle contact i.e. the stiffness of the contacting particles and sliding resistance in the contact plane. Moreover, the possible existence of cementing at the particle contact and too low strength of the aggregate particles may affect the deformations that take place at the particle contact. In the case of a moist material the physical properties of water – such as surface tension and freezing – and the electro-chemical interactions between water and aggregate particles provide some additional influences on the behaviour.

Effect of stress state

Experimentally it has frequently been observed that an increase in the hydrostatic stress component increases the stiffness of a granular material. A straightforward particle level explanation for this can be given by means of Equation 1 which presents the displacement between particles 1 and 2 consisting of linear elastic material and being compressed centrally against each other:

$$\delta = \sqrt[3]{\frac{9F^2}{16}\frac{\left(\frac{1-\nu_1^2}{E_1}+\frac{1-\nu_2^2}{E_2}\right)^2(r_1+r_2)}{r_1 r_2}} \qquad \text{(Eq. 1)}$$

where

δ	is	displacement between particles
F	is	force compressing the particles against each other
$E_{1,2}$	are	Young's moduli of particles 1 and 2
$\nu_{1,2}$	are	Poisson's ratios of particles 1 and 2
$r_{1,2}$	are	radii of particles 1 and 2

According to Equation 1 the relation between the contact force and the displacement has the form $\delta = k \cdot F^{2/3}$ or $F = k^* \cdot \delta^{3/2}$, where k and k^* are coefficients that depend on particle size and material properties of the particles. This means that the relative increase in the displacement between the particles is the lower the higher the contact force. On the level of a larger particle system this means of course increasing stiffness alongside with increasing stress level. Moreover, the increase in the normal force at the particle contact increases the maximum shear force that resists the sliding of the particles (Figure 3) and is thus decreasing the deformations of the entire particle system.

The deviator stress component increase, in turn, has normally been observed to decrease the stiffness of a granular material. A natural particle level explanation for which is easily derived from the mechanical model of a particle contact presented in Figure 3. At a certain level of hydrostatic stresses each particle contact has a maximum value of shear force which they can transmit and if that value is reached the particle contact starts to slide. Now, as the deviator stress increases, the number of particle contacts that are sliding is continuously increasing and thus, larger and larger rearrangement of the particle system is needed before a stable arrangement against the imposed deviator stress is obtained.

Effect of density

In the available literature the importance of the effect of density on the resilient deformation behaviour of a granular material has sometimes been considered to be insignificant /6/ sometimes dependent upon the material /7/ but quite often fairly important (e.g. /8/). Among the cyclic loading triaxial test results available in this study the effect of density has been of the type shown in Figure 4. Figure 4 presents the values of resilient modulus as a function of the volume content of solid particles (i.e. 1-porosity) and the sum of principal stresses θ for a well graded crushed rock material with grain size distribution T5 of Figure 2.

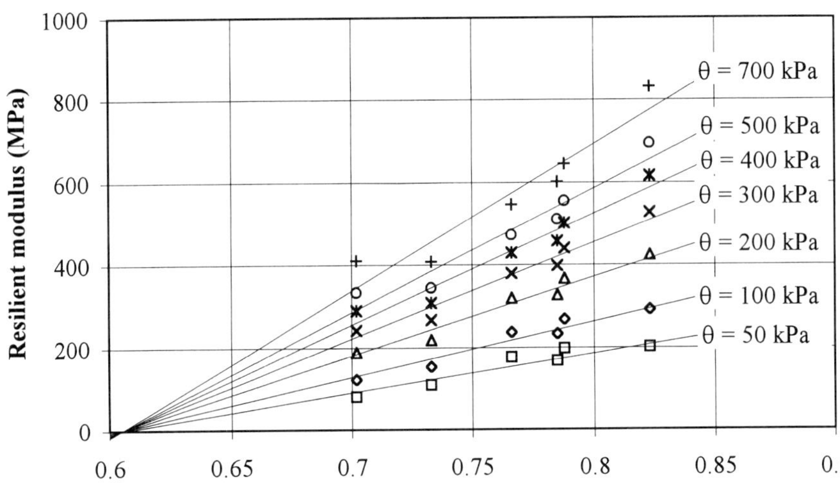

Figure 4. Variation of the resilient modulus of a well graded crushed rock material as a function of density and stress level.

Regarding particle level, the explanation for the marked effect of density on the stiffness of a granular material is also fairly straightforward: the denser the granular material is packed, the higher the co-ordination number (i.e. the average number of contacts per particle) and thus, the lower the average contact force and the smaller the deformation at the particle contact. The high number of particle contacts also efficiently prevents significant rearrangement of the particle system as the particles tend to slide in relation to each other. In fact, significant rearrangement of a densely packed particle system requires increase in the total volume of the material, i.e. dilation. Dilation, in turn, requires work to be done against the external load, which is from outside of the material observed as a higher stiffness value.

Effect of moisture content

The effect of moisture content on the resilient deformation behaviour of granular materials is fairly complex because it depends not only on the moisture content itself but at least on the grain size distribution and probably also on the electro-chemical properties of the material. According to the test results available in the present study the effect of moisture content was observed to be insignificant in open graded test materials (e.g. T4 and T6 of Figure 2) and modest in the well graded materials in which the highest values of resilient modulus were normally determined in a partly saturated condition /9/. The particle level explanation for this is obviously the negative pore water pressure induced by the surface tension of the water concentrated near to the particle contacts of a partly saturated material. However, in a very open graded material the loads in the particle skeleton are in any case mainly transmitted via the larger particles and thus the moisture content of the material in between the larger particles has hardly any effect on the macroscopic behaviour of the material.

The effect of moisture content on the mechanical behaviour of a granular material can be much more significant if a material with high fines content is near to full saturation and is exposed to rapid loadings. Due to the inadequate permeability of the material, excess pore water pressures may develop giving rise to the reduction of effective stresses between the particles and thus to much lowered stiffness values. Stiffness values on the other extreme may be observed if the material is first wetted and then dried to a fairly low moisture content. In the present study these kinds of tests were not performed, but in a latter test series the phenomena was observed to depend very much on the aggregate type as presented in another paper of this conference /10/.

Effect of grain size distribution

The effect of grain size distribution on the stiffness of a granular material was also observed to be somewhat complex and, as mentioned above, interrelated to e.g. moisture content of the material. As a general trend, however, it was observed that the value of resilient modulus was increasing as the maximum grain size of the material increased. The particle level explanation for this is most obviously based on the fact that the loads in the particle skeleton of a granular material are – as can be easily shown e.g. by means of numerical simulation models – mainly carried by chain like structures rather than being evenly distributed into the material. In this case, using Equation 1 it can be shown that the total deformation of the particle system is lower the larger the average particle size is i.e. via the fewer particle contacts the loads are transmitted /9/.

Effect of aggregate type

Physically the effect of aggregate type is most obviously reflected mainly from the mineralogical composition and internal structure of the aggregate because it affects both the strength and the stiffness of the aggregate particles, grain shape and surface roughness of the particles, electro-chemical properties of particles etc. Separation of the effects of these factors from each other was, however, at least with the present level of knowledge observed to be fairly impossible. On the other hand, the combined effect of all the factors depending on the aggregate type was unambiguous as will be shown in the next section.

APPROXIMATION METHOD OF THE RESILIENT MODULUS

The basic equations of the approximation method of the resilient modulus derived based on the available test results have a form analogous to the well known kθ /11/ and Uzan /12/ models as follows:

$$M_r = a_1(n_{\max} - n)\theta_0\left(\frac{\theta}{\theta_0}\right)^{0.5} \qquad \text{(Eq. 2)}$$

$$M_r = b_1(n_{\max} - n)\theta_0\left(\frac{\theta}{\theta_0}\right)^{0.7}\left(\frac{q}{\theta_0}\right)^{-0.2} \qquad \text{(Eq. 3)}$$

where

M_r	is	resilient modulus
n	is	porosity of material
$n_{\max}$	is	experimentally determined maximum value of porosity
θ	is	sum of principal stresses, $\theta = \sigma_1 + \sigma_2 + \sigma_3$
θ_0	is	reference stress, 100 kPa
q	is	deviator stress, $q = \sigma_1 - \sigma_3$
a_1, b_1	are	material parameters

As far as the effect of stress state is concerned it was observed that the moduli values of the available test materials could be described almost without any loss of accuracy even if the exponent values of Equations 2 and 3 reflecting the intensity of stress non-linearity of the modulus values were replaced by properly selected constant values /9/. To avoid the risk of slight over-estimation of the modulus values due to too low a shear strain level the use of a loading procedure similar to the CEN procedure and Equation 3 is obviously somewhat more recommendable than the use of the SHRP loading procedure and Equation 2.

The effect of density is taken into account in Equations 2 and 3 in linear relation to the difference between the experimentally determined maximum porosity value and the actual porosity value of the material. The approach was observed useful as was shown in Figure 4 in which the mathematical formulation of the set of straight lines is similar to Equation 2 as the sum of the principal stresses is given the values 50 kPa, 100 kPa etc.

The effect of factors other than stress level and density must be included in Equations 2 and 3 into the parameters a_1 and b_1 respectively. In the analysis of the available test results it was observed that e.g. the grain size distribution of the material has a remarkable, and in the case of one single aggregate type to some extent a predictable effect on the values of these parameters (Figure 5) but the variability between different aggregate types is far too large to be neglected. This is demonstrated in Figure 6 which compares the values of parameter a_1 of 10 different open graded Finnish and Norwegian aggregates.

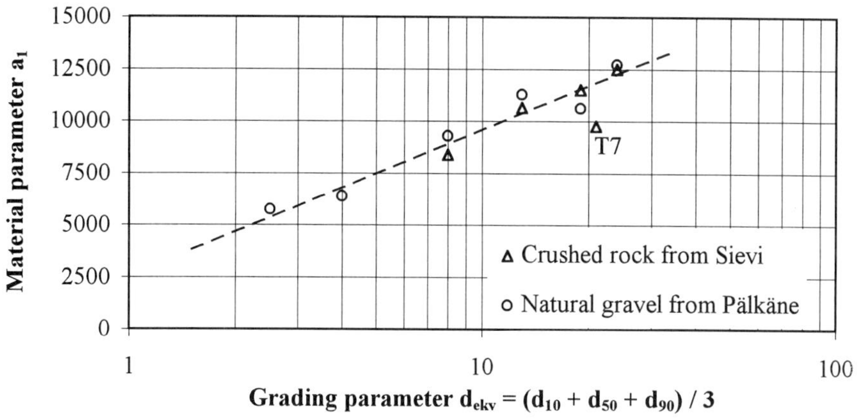

Figure 5. The effect of grading on the value of material parameter a_1 in Equation

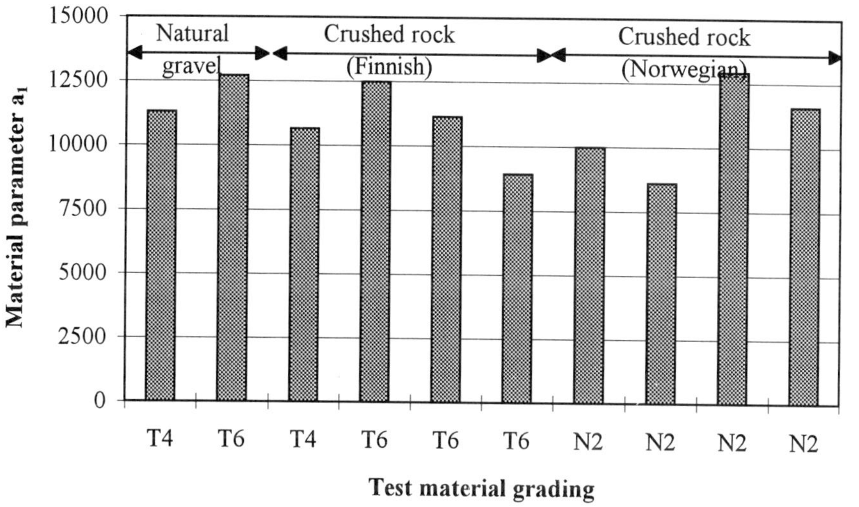

Figure 6. Values of parameter a_1 of 10 different open graded aggregates.

Consequently the most recommendable approach to apply the approximation method of resilient modulus presented by Equations 2 and 3 is to perform one cyclic loading triaxial test covering a large enough range of deviator stresses at a density state and moisture content relevant to the actual application. Following which the parameters a_1 and b_1 can be solved and an approximation of the resilient modulus at any combination of stress state and density can be obtained directly using Equations 2 and 3.

CONCLUSIONS

Based on the analysis of more than 150 large scale cyclic loading triaxial tests an approximation method was presented by means of which the number of laboratory tests needed to obtain reliable estimates of the resilient modulus of an aggregate at different stress levels and density states can be minimised. In addition to this, physically justified particle level explanations were found for the influencing mechanisms of the most important factors affecting the resilient deformation behaviour of granular materials.

ACKNOWLEDGEMENTS

The Finnish National Road Administration and the consortium of Norwegian aggregate producers, Arbeidsfellesskapet KPG, are gratefully acknowledged for financing the major part of the research projects that have output the experimental results forming the background material of this study. Furthermore the staff of the Laboratory of Geotechnical Engineering at TUT are warmly thanked for their assistance during the projects.

REFERENCES

1. KOLISOJA, P. (1994). Large scale dynamic triaxial tests with coarse grained aggregates. Proceedings of the 4th International Conference on the Bearing Capacity of Roads and Airfields. Minneapolis, USA. Vol. 2, pp. 883-897.
2. AASHTO T 294-92 I (1992). Interim method of test for resilient modulus of unbound granular base/subbase materials and subgrade soils - SHRP protocol P46. American Association of State Highway and Transportation Officials.
3. CEN prENV 00227413 (1997). Unbound and hydraulic bound mixtures for roads - Test methods - Cyclic load triaxial test. Draft. Brussels: CEN/TC227/WG4/TG2. 18 p.
4. VIK, G. (1995). Vakuumtreaksforsøk - Treaksialforsøk med syklisk belastning på naturgrus (0-32 mm). NGI report 531001-1. Oslo: Norwegian Geotechnical Institute. 24 p. + 29 Fig. *(In Norwegian)*
5. TING, J. M., CORKUM, B. T., KAUFFMAN, C. R. & CRECO, C. (1989). Discrete numerical model for soil mechanics. ASCE, Journal of Geotechnical Engineering, Vol. 115, No. 3, pp. 379-398.
6. THOM, N. H. (1988). Design of road foundations. PhD thesis, University of Nottingham. 307 p.
7. CHEUNG, L. W. (1994). Laboratory assessment of pavement foundation materials. PhD thesis, University of Nottingham. 223 p.
8. RADA, G. & WITCZAK, M. W. (1981). Comprehensive evaluation of laboratory resilient moduli results for granular materials. TRR 810, pp. 23-33.
9. KOLISOJA, P. (1997). Resilient deformation characteristics of granular metrials. Thesis for the degree of Doctor of Technology, Tampere Univ. of Technology. 214 p.
10. SAARENKETO, T., SCULLION, T. & KOLISOJA, P. (1998). Moisture susceptibility and electrical properties of base course aggregates. Proceedings of the 5th International Conference on the Bearing Capacity of Roads and Airfields, Trondheim.
11. BROWN, S. F. & PELL, P. S. (1967). An experimental investigation of the stresses, strains and deflections in a layered pavement structure subjected to dynamic loads. Proceedings of the 2nd International Conference on Structural Desing of Asphalt Pavements. Ann Arbor, USA. Pp. 487-504.
12. UZAN, J. (1985). Characterization of granular material. TRR 1022, pp. 52-59.

Fifth International Conference on the
BEARING CAPASITY OF
ROADS AND AIRFIELDS

STUDY OF THE RESILIENT BEHAVIOUR OF UNBOUND GRANULAR MATERIALS

PIERRE HORNYCH[1], ABDELLAH KAZAI[2] AND JEAN-MICHEL PIAU[3]

[1]Engineer, [2]PhD Student, [3]Engineer
Laboratoire Central des Ponts et Chaussées, Centre de Nantes,
BP 19, F 44340 Bouguenais, France.

ABSTRACT

This paper presents a study carried out to characterise the resilient behaviour of a 0/10 mm granular material. The study includes monotonic triaxial shear tests, and repeated load triaxial tests, used to determine its resilient behaviour, for different moisture contents. The resilient behaviour of the material is first described using the non linear elastic model of Boyce (1); then, the model is improved to take into account the anisotropy observed in most tests. Finally, the implementation of the model in the finite element code CESAR - LCPC is described, and a first comparison of the model predictions with results of a full scale experiment is presented.
Key words : Granular materials, cyclic triaxial tests, models, elasticity, anisotropy.

1 INTRODUCTION

Repeated load triaxial tests have been used for several years in the Laboratoires des Ponts et Chaussées, in France, to study the resilient behaviour of unbound granular materials, for pavement modelling applications. A standard test procedure, with loadings following different cyclic stress paths, has also been proposed (2). In this study, carried out on a 0/10 mm granular material, the standard test procedure was modified, by increasing the number of cyclic stress paths, in order to characterise better the stress-path dependent behaviour of the material. With the results obtained, it was possible to improve the modelling of the resilient behaviour.

2 TEST EQUIPMENT AND MATERIAL

2.1 Repeated load triaxial apparatus. This study was carried out using a triaxial apparatus developed at LCPC for testing subgrade soils. It is designed for specimens with a diameter of 76 mm and a height of 152 mm, and equipped with a servo-pneumatic loading system, capable of cycling both the axial stress and the confining pressure. The load signals used are sinusoidal, and the maximum loading frequency is 2 Hz. The triaxial cell is equipped with transducers measuring the axial and radial strains on the central part of the specimen.

2.2 Material. Because of the small specimen size imposed by the triaxial apparatus, the unbound granular material used for this study is a 0/10 mm crushed granite. Its grading curve, and its main properties, are summarized on figure 1. The grading curve corresponds to that of a standard 0/20 mm granular material reduced to the scale ½.

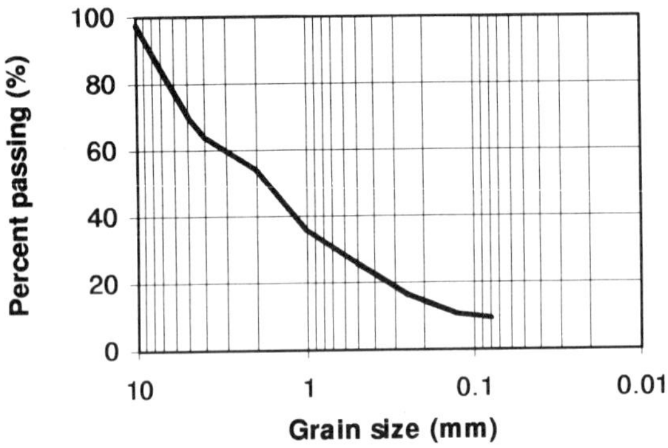

Unbound material properties	
Optimum dry density kg/m^3	2200
Optimum water content	6.3 %
Fines content	9 %
Los Angeles abrasion	20

Figure 1 : Grading curve and properties of the tested material

In the triaxial tests, the material was tested at a dry unit weight $\rho_d = 2134$ kg/m^3 $= 0.97\ \rho_{d\,OPM}$, corresponding to usual in situ conditions, and at several different water contents. The triaxial specimens were compacted in 4 layers, using a manual hammer.

2.3 monotonic triaxial tests - failure lines. To determine the failure envelope of the granular material, 3 series of triaxial shear tests were carried out, at 3 water contents : w = 4.3 %, w = 6.3 % and w = 8.5 %. The tests were performed in drained conditions, at a loading speed of 0.5 mm / minute. The corresponding failure lines are shown on figure 2.

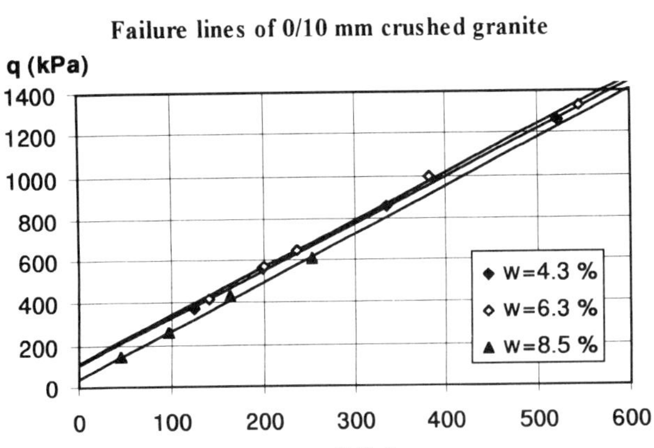

Failure lines	
w=4.3 %	q =2.22p+109 ϕ=54.1° c=65 kPa
w=6.3 %	q =2.26p+115 ϕ=55.1° c=50.6 kPa
w=8.3 %	q =2.29p+40 ϕ=54.3° c=17.9 kPa

Figure 2 : Failure lines obtained for different water contents

The failure lines are represented in p,q space, with $p = \frac{\sigma_1 + 2\sigma_3}{3}$ and $q = \sigma_1 - \sigma_3$. Each line is defined by its equation, of the form $q = Mp + S$ (M, S parameters), and also by the corresponding values of the Mohr-Coulomb parameters: friction angle ϕ and cohesion c. It can be noted that the friction angle is high (around 54°) and remains fairly constant, while the cohesion decreases when the water content increases. This is probably due to the fact that the material is unsaturated, and that the cohesion results from the negative pore pressure existing in the specimen. The values are in agreement with those obtained on 0/20 mm granular materials used for road bases.

3 REPEATED LOAD TRIAXIAL TESTS

3.1 Test programme. The objective of the tests was to study the resilient behaviour of the material, for various cyclic stress paths. The test procedure adopted is summarized in table 2 and on figure 3. The stress paths are linear in p,q space, and can be defined by initial static stresses p_0 and q_0, and cyclic stresses p_r and q_r. The test procedure includes 4 consecutive steps :

- a conditionning phase (20000 cycles) along a stress path with $q_r/p_r=3$ and $q_r=200$ kPa.
- a series of short loadings (100 cycles each), with stress ratios $q_r/p_r= 0$, 2.5 and 3;
- a second conditionning (20000 cycles) following a stress path with $q_r/p_r = 2$ and $q_r = 600$ kPa;
- finally, a series of short loadings (100 cycles) with ratios $q_r/p_r = 0$, 0.5, 1, 1.5 and 2.

loading	p_r kPa	q_r kPa	loading	p_r kPa	q_r kPa
1(1)	**66**	**200**	17	60	30
2	50	0	18	120	60
3	100	0	19	180	90
4	150	0	20	240	120
5	200	0	21	75	75
6	60	150	22	150	150
7	120	300	23	225	225
8	17	50	24	300	300
9	33	100	25	100	150
10	50	150	26	200	300
11	66	200	27	300	450
12(1)	**300**	**600**	28	400	600
13	50	0	29	90	180
14	100	0	30	180	360
15	150	0	31	250	500
16	200	0	32	300	600

(1) cyclic conditionning, lasting 20000 cycles. Other loadings last 100 cycles

Table 2 : Consecutive cyclic loading sequences applied in the repeated load triaxial tests

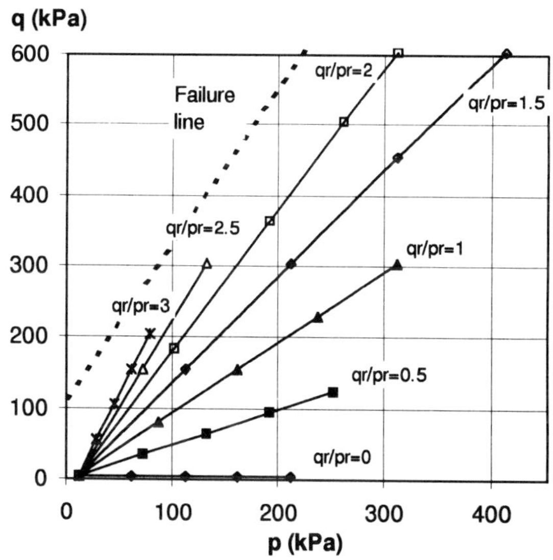

Figure 3 : Cyclic stress paths

The objective of the cyclic conditionnings is to stabilise the permanent strains of the material, and attain a practically elastic behaviour. Generally, only one conditionning is performed, with a stress level corresponding to the maximum stresses applied in the test. In these tests, because a large range of stress paths was applied ($0 \leq q_r/p_r \leq 3$), two sucessive conditionnings, with different stress paths ($q_r/p_r = 3$ and $q_r/p_r = 2$) were necessary to obtain a stable resilient behaviour for all the stress paths.

All the loadings were performed with initial static stresses $p_0 = 10$ kPa and $q_0 = 5$ kPa, at a frequency of 1 Hz, and in drained conditions. The test programme included 2 series of tests :

- first a series of 4 identical (repeatability) tests, at a water content w = 4.3 %;
- then, a second series of tests, at water contents of 3.3, 4.3, 5.3 and 6.3 %.

3.2 Results of conditioning. Figure 4 presents an example of the response of the material during one of the conditionings at $q_r/p_r = 3$ (axial stress-axial strain (σ_1 - ε_1) cycles). It can be seen that as the number of cycles increases, the permanent strains tend to stabilise and the response becomes more and more elastic (the hysteresis decreases). The elastic stiffness of the material also increases with the number of cycles.

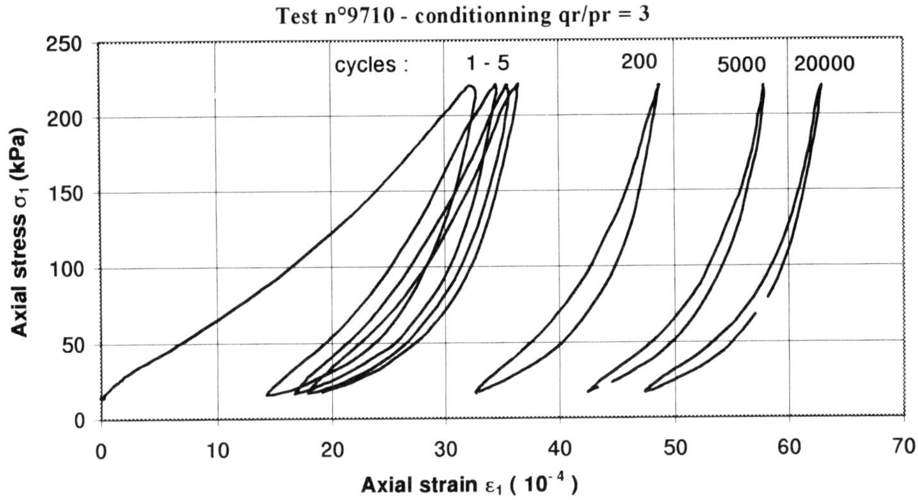

Figure 4 : stress - strain cycles obtained during a conditionning at $q_r/p_r = 3$

Figures 5 shows, for all the tests, the variations of the resilient modulus M_r (secant elastic modulus) during the conditioning at $q_r/p_r = 3$. It can be seen that :

- the resilient modulus increases rapidly during the first cycles, and generally becomes constant after about 1000 cycles of loading (more in some tests);
- the values of M_r are not very sensitive to water content; they are similar in all tests, (130 to 150 MPa) except in the test with the highest water content (9801 - w=6.3 %), where M_r is significantly lower (100 MPa).

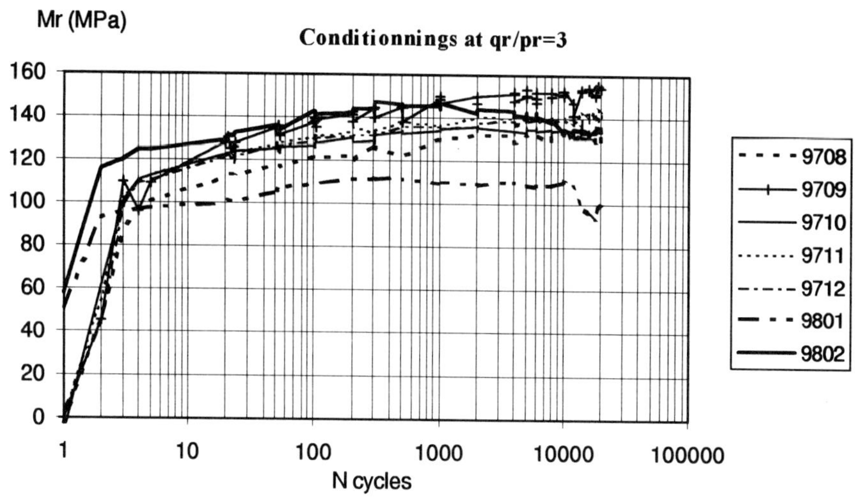

Figure 5 : Cyclic conditionnings at $q_r/p_r = 3$
Variations of the resilient modulus with the number of cycles.

3.3 Resilient behaviour. In the results presented here, the resilient strains are the strains during unloading; they are positive in compression. For each individual loading, the value of the resilient strain is the average of the last five load cycles.

Figure 6 shows a typical example of the resilient behaviour of the granular material. This figure represents variations of the resilient volumetric strains (ε_{vr}) and resilient shear strains (ε_{qr}) obtained for each cyclic loading with the mean stress p and the stress ratio q/p.

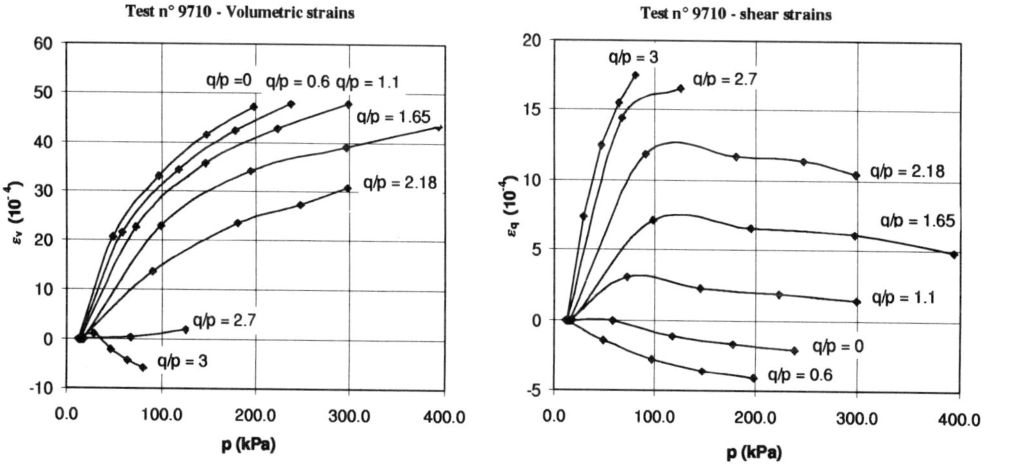

Figure 6 : Example of resilient behaviour;
variations of ε_{vr} and ε_{qr} with mean stress p and stress ratio q/p.

The response is strongly non linear and depends on the stress path :

– the resilient volumetric strain ε_{vr} increases when p increases, and decreases when q/p increases;
– ε_{vr} is positive for $q/p \leq 2.5$ (contractant behaviour) and negative for $q/p = 3$ (dilatant behaviour);
– the shear strain ε_{qr} increases when p increases, and also when q/p increases;
– ε_{qr} is negative for $q/p = 0$ and 0,5 , which indicates that the response of the material is anisotropic (for an isotropic loading ($q/p = 0$), we have $\varepsilon_{3r} > \varepsilon_{1r}$).

The anisotropic behaviour of the material (observed in all tests) is probably due to the compaction of the specimens in horizontal layers , and to the conditionning.

4 MODELLING OF THE RESILIENT BEHAVIOUR

4.1 Boyce model. Several studies carried out at LCPC (2) have shown that the resilient behaviour of unbound granular materials can be described using a non linear elastic model proposed by Boyce (1). This model derives from an elastic potential W, expressed by :

$$W = \frac{p^{n+1}}{p_a{}^{n-1}}\left[\frac{1}{(n+1)K_a} + \frac{1}{6G_a}\left(\frac{q}{p}\right)^2\right] \qquad (1)$$

where K_a, G_a and n are the parameters of the model, and p_a a constant equal to 100 kPa

This leads to the following expressions for the volumetric and shear strains :

$$\varepsilon_v = \frac{1}{Ka}\frac{p^n}{p_a{}^{n-1}}\left[1 + \frac{(n-1)K_a}{6G_a}\left(\frac{q}{p}\right)^2\right] \quad (2) \qquad \text{and} \qquad \varepsilon_q = \frac{1}{3Ga}\frac{p^n}{p_a{}^{n-1}}\frac{q}{p} \quad (3)$$

The Boyce model presents the advantage of being simple (3 parameters), and simulates correctly the effect of the stress path q/p on the resilient behaviour (dilatancy for high values of q/p).

Thus, the results of all the tests were interpreted using the Boyce model. The values obtained for the model parameters (K_a, G_a, n) are presented in table 3, and an example of the fit obtained is shown on figure 6. It can be seen that the fit is relatively good for ε_v, but less satisfactory for ε_q; in particular, the model, which is isotropic, cannot represent the negative values of ε_q obtained for $q/p < 1$. In addition, the values of the model parameters are often unrealistic (n close to zero, K_a sometimes negative - see table 3).

4.2 Generalisation of the Boyce model for a cross-anisotropic material. One of the possible reasons for the poor results obtained with the Boyce model was the anisotropic response of the granular material (characterised by negative values of ε_q for isotropic loadings). For this reason, an attempt was made to generalise the Boyce model to the case of a cross-anisotropic material (anisotropy between the vertical direction, and the horizontal ones).

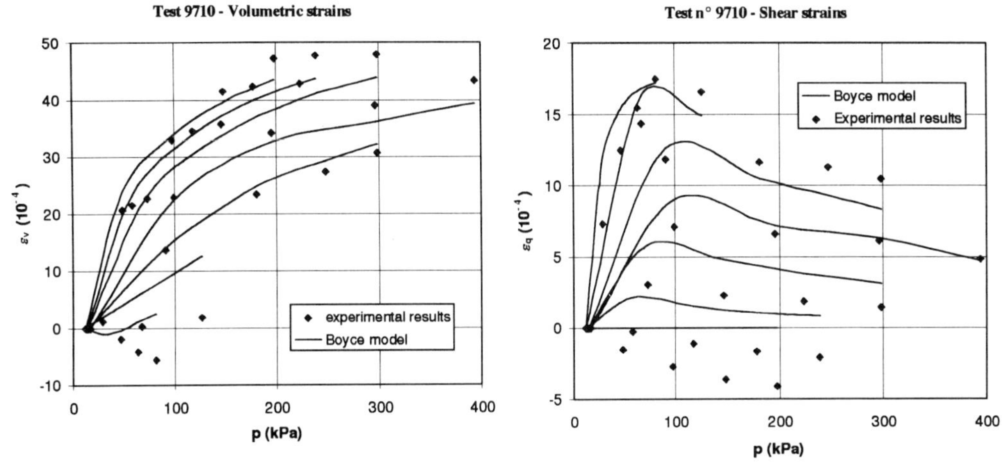

Figure 6 : Test n°9710 - adjustment of experimental results with the Boyce model

		Isotropic Boyce model				**Cross-anisotropic model**				
Test	**w %**	**K_a MPa**	**G_a MPa**	**n**	**$\rho^{(1)}$**	**K_a MPa**	**G_a MPa**	**n**	**γ**	**$\rho^{(1)}$**
9708	4.1	8.40	40.07	0.113	0.851	12.69	36.42	0.198	0.84	0.885
9709	4.2	-1.58	39.48	-0.021	0.724	10.74	30.21	0.207	0.64	0.913
9710	4.3	0.008	40.67	0.0	0.737	9.33	31.63	0.204	0.69	0.916
9711	4.1	-3.55	44.03	-0.056	0.641	8.29	29.79	0.202	0.61	0.912
9802	3.3	-11.95	49.47	-0.166	0.595	8.72	32.50	0.196	0.61	0.861
9712	5.6	-9.49	32.62	-0.174	0.568	5.88	20.88	0.192	0.50	0.935
9801	6.3	4.59	33.23	0.062	0.771	13.36	26.62	0.232	0.74	0.841

(1) ρ : correlation index

Table 3 : Values of model parameters obtained for all the tests

To introduce anisotropy in the model, the expression of the elastic potential was modified, by multiplying the principal stress σ_1 by a coefficient of anisotropy γ. The elastic potential then becomes :

$$W = \frac{\left[\frac{\gamma\sigma_1 + 2\sigma_3}{3}\right]^{n+1}}{p_a^{\,n-1}} \left[\frac{1}{(n+1)K_a} + \frac{1}{6G_a}\left(\frac{\gamma\sigma_1 - \sigma_3}{\frac{\gamma\sigma_1 + 2\sigma_3}{3}}\right)^2\right] \quad (4)$$

This leads to the following stress-strain relationships :

$$\varepsilon_v = \frac{p^{*n}}{p_a^{\,n-1}}\left[\frac{\gamma+2}{3K_a} + \frac{n-1}{18G_a}(\gamma+2)\left(\frac{q^*}{p^*}\right)^2 + \frac{\gamma-1}{3G_a}\left(\frac{q^*}{p^*}\right)\right] \quad (5)$$

$$\varepsilon_q = \frac{2}{3}\frac{p^{*n}}{p_a^{\,n-1}}\left[\frac{\gamma-1}{3K_a} + \frac{n-1}{18Ga}(\gamma-1)\left(\frac{q^*}{p^*}\right)^2 + \frac{2\gamma+1}{6Ga}\left(\frac{q^*}{p^*}\right)\right] \qquad (6)$$

with : $p^* = \frac{\gamma\sigma_1 + 2\sigma_3}{3}$ (7) and $q^* = \gamma\sigma_1 - \sigma_3$ (8)

All the results were reinterpreted using the anisotropic model . The values of the model parameters are given in table 4, and an example of the fit obtained is presented on figure 7. It can be seen that the anisotropic model describes both ε_v and ε_q quite well; the same is true for all the tests, as indicated by the good values of the correlation indexes in table 3. The values of the model parameters are also more consistent; for example, in the 4 tests at w = 4,3 %, the values of the model parameters K_a, G_a, n and γ are very similar.

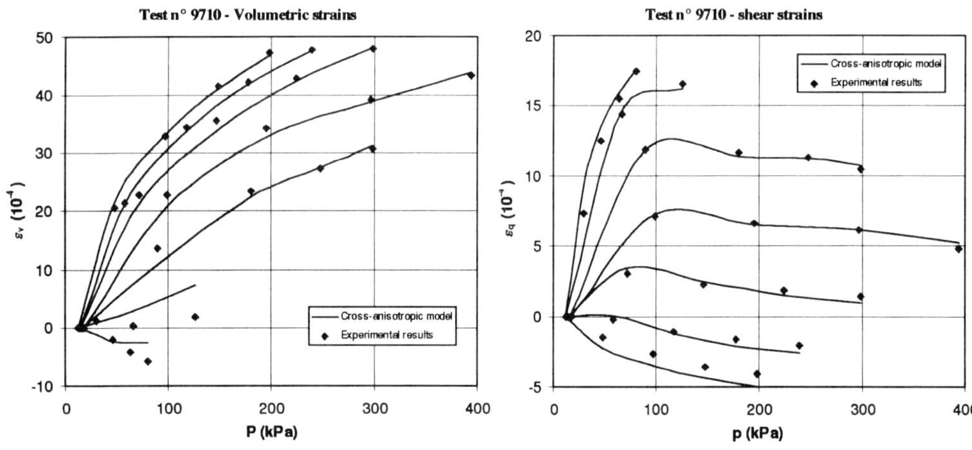

Figure 7 : Example of fit obtained with the cross anisotropic model

5 IMPLEMENTATION OF THE NON LINEAR MODEL IN THE FINITE ELEMENT CODE CESAR-LCPC

On of the objectives of the work was to apply the model developped above to pavement structure calculations. For this purpose, the model was implemented into the finite element code CESAR-LCPC. The case with given initial stresses σ° was considered, in order to account for stresses due to the weight of materials, pore pressures, initial compaction, or also rutting strains.

The implementation approach is the following. Let us note $\varepsilon = B^{-1}(\sigma)$, the tensorial relationship between ε and σ, which derives from the above anisotropic constitutive law. This law was first analytically inverted. In the case with initial stresses, it yields $\sigma = B(\varepsilon + B^{-1}(\sigma^0))$. Then using the Virtual Work Principle, the problem is stated under the form : find the field displacement u, such that for any (kinematic) virtual displacement field, δu :

$$\delta P_i + \delta P_e = 0 \qquad (9)$$

with : δP_i = virtual work of internal forces = $-\int_{\Omega} B\left(\varepsilon + B^{-1}(\sigma^0)\right)\delta\varepsilon$ (10)

δP_e = virtual work of external forces, Ω = geometry of the analysed structure, $\delta\varepsilon$ = virtual strain field, derived from δu.

Using the fixed point method, the previous problem can be solved by the following iterative process :

$$\int_{\Omega} E_a \Delta\varepsilon_n \delta\varepsilon = \delta P_e - \int_{\Omega} B\left(\varepsilon_n + B^{-1}(\sigma^0)\right)\delta\varepsilon \quad \text{with} \quad u_{n+1} = u_n + \Delta u_n \qquad (11)$$

where E_a is a constant arbitrary elastic (possibly isotropic) tensor, chosen to ensure the convergence of the algorithm.

Finally, after a classical finite element discretisation procedure, we obtain the iterative Newton-Raphson scheme, which was implemented in CESAR-LCPC :

$$\underline{\underline{K_a}}\,\underline{\Delta U_n} = F_e - F_i(\underline{U_n}) \qquad \underline{U_{n+1}} = \underline{U_n} + \underline{\Delta U_n} \qquad (12)$$

with $\underline{\underline{K_a}}$ = stiffness matrix, function of E_a, $\underline{U_n}$= nodal displacement vector, F_e= vector of external forces, $F_i(\underline{U_n})$= vector of internal forces, function of $\underline{Un}$.

Iterations are stopped when $\max\left(\left|\frac{\Delta U_n}{U_n}\right|\right)$ is lower than a specified value.

This algorithm is implemented into the « CVCR » module of CESAR-LCPC, which can deal at the same time with the thermo-visco-elastic behaviour of asphalt layers under moving loads (3),(4).

6 COMPARISON WITH RESULTS OF A FULL SCALE EXPERIMENT

6.1 Description of the experiment. In order to make a first validation of the model, an attempt was made to compare its results with data of a full scale experiment carried out previously at the Centre d'Experimentations Routières in Rouen.

The experiment consisted in applying cyclic plate loadings on an instrumented layer of unbound granular material. The tested structure is described on figure 8; it had a size of 4 × 6 meters. It consisted of a subgrade soil (a silty sand), covered by two layers of unbound granular material, each 20 cm thick. The granular material used was a 0/20 mm crushed granite, of the same origin as the one tested in this study. The cyclic loadings were applied using a 380 mm diameter steel plate. Two 42 mm thick polyamide plates were interposed between the steel plate and the granular material, to avoid direct contact with the very stiff steel plate. The structure was instrumented with 4 vertical displacement transducers (see figure 8) and 4 vertical pressure transducers. Three displacement transducers were placed on the top of the granular layer, on the center of the plate (DV1) and at 35 cm (DV2) and 80 cm (DV3) from it. The last displacement transducer (DV4) was placed on the top of the subgrade, under the center of the plate.

The loadings consisted of 200 000 load cycles, applied at a frequency of 0.5Hz, with a vertical load varying between 2 kN and 62 kN (vertical stress 18 kPa to 547 kPa). At the end of the experiment, several load cycles with different values of the load (16, 32, 62 and 98 kN) were also applied; only the results of these last cycles were used in the comparisons made here.

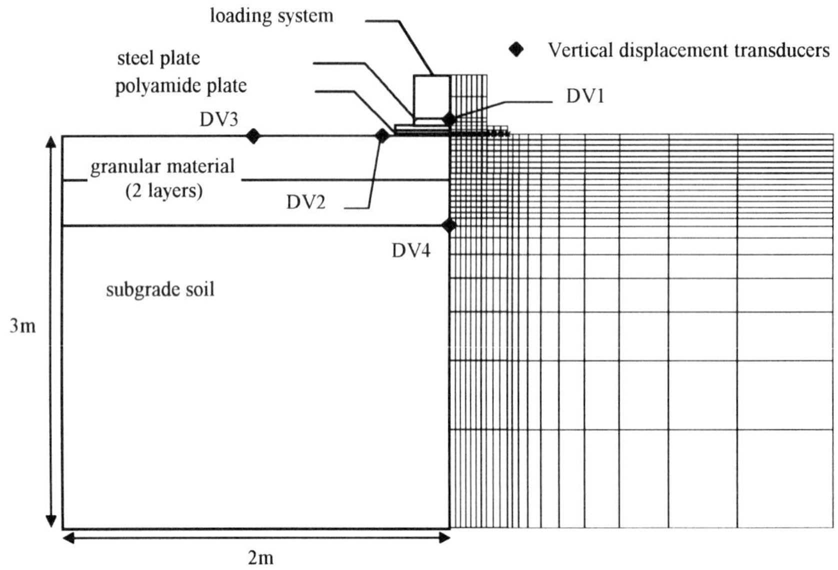

Figure 8 : Tested full scale structure and finite elements mesh.

The calculations with CESAR-LCPC were performed in 2D, in axisymetric conditions. The finite element mesh is shown on figure 8. The model used for the granular material was the Boyce model (isotropic or anisotropic), and linear elasticity was assumed for the soil.

Repeated load triaxial tests were performed on the granular material at the time of the experiment, giving the following parameters of the Boyce model for the water content and the density corresponding to the experiment : K_a=35.7 MPa , G_a=62 MPa and n=0.284 (anisotropic model parameters were not determined at that time). Plate loading tests were also carried out on the soil and indicated an elastic modulus E = 90 MPa.

6.2 Comparisons with the Boyce model . A first series of calculations was made using the standard Boyce model, with or without initial stresses . The values adopted for K_a, G_a and n were those obtained experimentally. Table 4 presents the results obtained for the load of 62 kN (comparison of the experimental and calculated displacements). It can be seen that :

-the Boyce model without initial stresses gives a relatively good value for the vertical displacement DV1, but overestimates the other displacements.
-The introduction of initial stresses (vertical stress equal to the weight of the structure and K_0 = 1) tends to reduce all the displacements, giving better results for DV2, DV3 and DV4, but underestimating DV1.

6.3 Comparisons with the anisotropic model. An attempt was also made to test the effect of anisotropy in the calculations. However, because no experimental values of anisotropic parameters were available for the material, only a parametric study could be conducted, trying to find the set of parameters giving the best fit with the experiment. The values retained were n=0.2, K_a = 15 MPa, G_a=30MPa, γ=0.85 and for the soil E = 95 MPa. As can be seen in table 4, the introduction of anisotropy improves significantly the predictions.

	Displacements			
Transducer	**Experimental** 10^{-5}m	**Boyce model** 10^{-5}m	**Boyce model + Initial stresses** 10^{-5}m	**Cross anisotropic model + Initial stresses** 10^{-5}m
DV1	-154	-128	-106	-143
DV2	-37	-56	-47	-43
DV3	0	-18	-12	-10
DV4	-38	-53	-39	-53

Table 4 : Comparison of experimental and calculated displacements

7 CONCLUSIONS

The main objective of this work was to improve the modelling of the behaviour of unbound granular materials. First, an experimental study was conducted, using repeated load triaxial tests, on a 0/10 mm granular material. A new test procedure, with a large variety of cyclic stress paths, was experimented. An anisoitropic response of the material was observed, and so the model adopted previously at LCPC to describe the resilient behaviour of granular materials (Boyce model) was modified to take into account this anisotropy. Good predictions were obtained with this anisotropic model.

Then, the new model was implemented in the finite element code CESAR-LCPC, and an attempt was made to compare its results with data from a simple experiment of loadings on a granular layer. This first comparison gave reasonably good predictions. A more detailed validation of the model, by comparisons with other laboratory and in situ results is planned.

REFERENCES

1. BOYCE, J.R. A non linear model for the elastic behaviour of granular materials under repeated loading. Int.Symposium on Soils under Cyclic and Transient Loading, Swansea, U.K., pp.285-294, Balkema (1980).

2. PAUTE J.L., HORNYCH P., BENABEN J.P. Repeated load triaxial testing of granular materials in the french network of Laboratoires des Ponts et Chaussées, Proceedings of the European Symposium Euroflex 1993, Lisbon, Portugal, pp. 53-64, Balkema (1996).

3. PIAU, J.M. HECK, J.V., GRAMSAMMER J.C., ODEON H., Modélisation viscoélastique des chaussées et comparaison avec le comportement observé sur manège routier, XVe Renc. Universitaires de Génie Civil, Strasbourg, France, pp.285-292 (1997).

4. HECK, J.V., PIAU J.M., Thermo-visco-elastic modelling of pavement behaviour and comparison with experimental data from the LCPC test track. Paper submitted for the BCRA conference, Trondheim (1998).

Fifth International Conference on the
BEARING CAPASITY OF
ROADS AND AIRFIELDS

ANISOTROPY OF UNBOUND PAVEMENT MATERIALS

KEMAS AHMAD ZAMHARI

Research Engineer, Institute of Road Engineering
Ministry of Public Works, Bandung 40294, Indonesia, Fax: + 62 22 780 2726
E-mail: Pusjal@melsa.net.id

ABSTRACT

This paper is concerned with the cross-anisotropic material related to analytically-based methods of pavement design and analysis. Some experimental evidence of material anisotropy is discussed. The elastic response of anisotropic material under repeated loading is reviewed. The effect of material anisotropic on elastic response of pavement and its consequence on the design life of pavement is presented. Mathematical techniques currently used to describe the behavior of cross-anisotropic materials are reviewed. The approaches are described and their effects on the laboratory procedure for determination of related elastic parameters are discussed. The application of the techniques by using the cubical repeated loading true-triaxial apparatus is demonstrated.

Keywords: Unbound material, anisotropy, true-triaxial.

INTRODUCTION

Current advancement in computers and numerical techniques have enabled increasingly realistic models of material behavior to be incorporated into the solutions of boundary value problems. Solutions, based on a simple linear isotropic elastic theory e.g. CHEV by the Chevron Research Company and BISAR by Shell have been supplemented by solutions embracing *nonlinear* isotropic elasticity e.g. Mich-PAVE, linear isotropic *viscoelasticity* e.g. VESYS and Kenlayer, and linear isotropic *elastoplasticity* e.g. Mechano-Lattice. Despite the inclusion of various factors such as nonlinearity and inelasticity, these solutions share a common feature in that all materials were assumed to be isotropic. In reality, however, pavement materials are not only inelastic and nonlinear but are also anisotropic.

The latter means that their characteristics differ according to direction. The simplest form of material anisotropy is cross-anisotropy where the material has an axis of symmetry of rotation. The elastic properties of such a material are equivalent in all directions perpendicular to the axis of symmetry but differ, in general, from those in a direction parallel to the axis. Gerrard and Mulholland (1), followed by Gerrard and Harrison (2), were among the first to consider the problem of cross-anisotropic half spaces subjected to circular loads of various type.

Their work fostered the development of a computer program, CIRCLY, by the Commonwealth Scientific and Industrial Research Organisation of Australia (CSIRO) (3). The program enables the analysis of the response of multilayered cross-anisotropic media under multiple complex circular loads to be calculated.

Later, NAASRA (now AUSTROADS) adopted the CIRCLY analysis in its structural design guide for flexible road pavements in Australia (4). This policy made Australia the first country in the world to adopt material anisotropy in its national road pavement design procedure. The policy was based on a fundamental assumption that all unbound materials are anisotropic. Indeed, as will be shown later, there is evidence that indicates that most pavement materials behave anisotropically rather than isotropically.

The AUSTROADS design guide progressed through many revisions. The final revision was established in 1992 but was first formally published in 1987. In that period many engineers became accustomed to assuming that unbound materials are anisotropic. Despite this, there have been few attempts to measure the true properties of anisotropic pavement materials. The work described in this paper intended to address this problem.

ANISOTROPY OF GEOMATERIALS

Compacted soils. Most of the studies on soil anisotropy have been restricted to naturally-deposited soils or soils which were prepared under a laboratory procedure used to simulate natural deposition soils. Studies on the anisotropy of compacted soils, such as in case of pavement structures, are few. This is perhaps because an analytical solution for a multilayer anisotropic pavement structure was not available at least until the CIRCLY analysis was developed. It can be argued, however, that the methods applied in the construction of a flexible pavement would promote anisotropy.

In flexible pavement construction, the loose materials are usually placed in a lift of 100 to 200 mm. The thickness of the lift is determined by the type of materials and the capacity of the compacting equipment. The material is then compacted by one or more items of compaction plant, usually road rollers, which travel back and forth on the surface of the lift.

Depending on the type of the compaction plant used, the equipment may apply a kneading action, vibrations, or combination of a kneading action and vibrations to the materials. Through this process, the materials under the compaction plant are densified in the vertical direction without any horizontal support except by the neighbouring materials themselves.

Studies on the effects of the methods of compaction have revealed that for clay soils, the compaction methods have a significant effect on the arrangement of clay particles. A kneading compaction at a moisture content related to the wet side of a compaction curve may yield clay particles arranged parallel to the surface. This particle arrangement is similar to a naturally-deposited clayey materials. In comparison, for granular materials, the compaction procedure may yield an effect similar to that produced by air-or-water-pluviated deposition of natural sand which also yield anisotropy.

The preceding discussion suggest that compacted pavement materials may, be expected to exhibit anisotropy. Some reports seem to confirm this (1, 5, 6). For the naturally-deposited soils, it is widely believed that the occurring anisotropy is of the cross-anisotropic type. This particular type of anisotropy has an axis of symmetry which coincides with the direction of deposition and isotropy in the horizontal planes. Five elastic parameters are required to describe materials with this type of anisotropy.

For pavement materials, the occurring anisotropy is probably closer to the orthotropic type because of the construction method and traffic loading. Material of this type has elastic symmetry with respect to three mutually perpendicular axes. In the case of the road pavement, the three directions may be the vertical, the horizontally parallel to the axis of road roller and/or traffic direction, and the horizontally perpendicular to the latter direction. Nine elastic parameters are required to describe orthotropic materials. However, as will be shown latter, the most comprehensive form of anisotropy that can be conveniently study at the present time is the elastic cross-anisotropic model.

Effects on elastic response of pavements. Since 1987 the AUSTROADS structural design guide road pavements recommends that unbound materials should be assume to be cross-anisotropic with a degree of anisotropy, *n*, equal to two. In other words, the material is considered to be two times as stiff in the vertical direction as in the horizontal direction.

In order to illustrate the effect of cross-anisotropy on pavement performance, an analysis was made on a prototype pavement as follows.

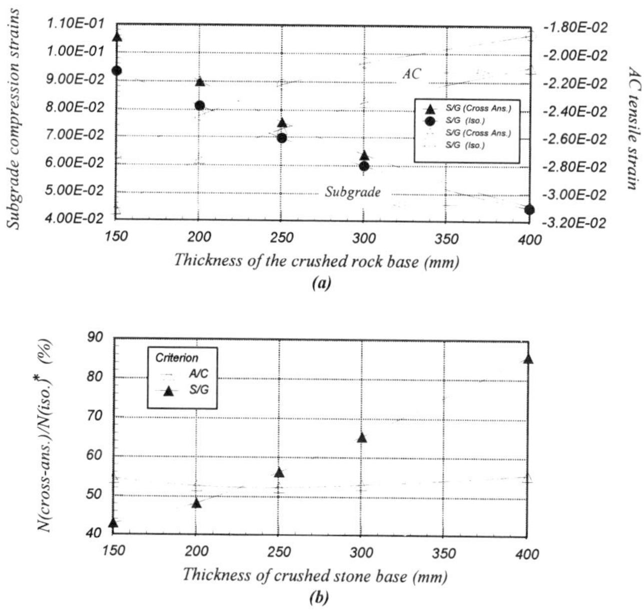

Figure 1 The effects of material anisotropy on (a) elastic strain and, (b) allowable cummulative ESA

The pavement was assumed to consist of a 100 mm layer of asphaltic concrete, laid on a crushed rock base whose thickness varied from 150 mm to 400 mm, and a fine-grained subgrade of 10% CBR. The response of this prototype pavement under the standard axle load was analysed using CIRLCY for both the isotropic and cross-anisotropic models. The AUSTROADS recommendations, which include the division of the granular base into several layers, the CBR-Elastic modulus relationship, and the degree of anisotropy, n, equal to two for cross-anisotropic materials were applied. By adopting AUSTROADS criteria on the subgrade vertical strain and asphaltic concrete tensile strain, both models were compared in terms of the allowable cumulative standard axle load. Figure-1 presents the results. This illustration shows that the assumption of cross-anisotropy might considerably reduce the number of cumulative standard axle loads that a pavement could withstand compared to that if the materials were isotropic. On the basis of the asphaltic concrete limited tensile strain criterion, for a given value of E_v, the pavement with cross-anisotropic materials, for which the degree of anisotropy $n = E_v/E_h = 2$, may only last to about 50% of the service life of the pavement with isotropic materials. On the basis of the subgrade vertical strain criterion, the service life ratio varied from as low as 40%, for a 150 mm crushed rock base, to 85% as the thickness of the base increased to 400 mm. It is apparent that these differences are large. The impact on the economics of the design is, of course, obvious.

THE DEFORMATION LAW

Basic equation. The constitutive equation of an elastic material in a three-dimensional space, known as the Generalised Hooke's Law, may be written in a general form as:

$$\{\sigma\}=[S]\{\varepsilon\} \qquad \text{Eq. 1}$$

Where $\{\sigma\}$ is the stress vector, $\{\varepsilon\}$ is the strain vector, and [S] is the elastic material-stiffness matrix. For the general anisotropic case, the elasticity stiffness matrix [S] consists of 36 independent elastic constants related to six stresses and six strain components. However, most of the materials possess some elastic symmetry, and permit the reduction of the number of independent elastic parameters to 21. Depending on the symmetric conditions, the number of elastic constant may be reduced further. A discussion on this aspect may be found elsewhere (7).

If a plane of isotropy is considered and the elastic symmetry is in the direction normal to the plane of isotropy then the material is considered to be cross-anisotropic. For this model the elastic constants are reduced to five. With reference to Cartesian axes xyz, in terms of the incremental stress-strain, the compliant matrix of Eq. 1 in engineering terms may be transformed as Eq. 2. This equation shows seven parameters used to describe material cross-anisotropy. These are, E_v, E_h , υ_{vh} , υ_{vh} , υ_{hh}, G_{vh}, G_{hv}. However, inter-relationships exist between the seven parameters so that there are five independent elastic constants. These are:

E_v : Modulus elasticity in vertical direction
E_h : Modulus elasticity in horizontal direction
υ_{vh} : Poisson's ratio, effects of vertical strain on horizontal strain
υ_{hh} : Poisson's ratio, effects of horizontal strain on complimentary horizontal strain
G_{vh} : Shear modulus of cross-anisotropic material

$$\begin{bmatrix} \delta\varepsilon_{zz} \\ \delta\varepsilon_{yy} \\ \delta\varepsilon_{xx} \\ \delta\gamma_{zx} \\ \delta\gamma_{zx} \\ \delta\gamma_{yz} \end{bmatrix} = \begin{bmatrix} \frac{1}{E_v} & -\frac{\nu_{vh}}{E_v} & -\frac{\nu_{vh}}{E_v} & & & \\ -\frac{\nu_{vh}}{E_v} & \frac{1}{E_h} & -\frac{\nu_{hh}}{E_h} & & & \\ -\frac{\nu_{vh}}{E_v} & -\frac{\nu_{hh}}{E_h} & \frac{1}{E_h} & & & \\ & & & \frac{1}{2G_{hv}} & & \\ & & & & \frac{1}{2G_{hv}} & \\ & & & & & \frac{1}{2G_{hh}} \end{bmatrix} \begin{bmatrix} \delta\sigma_{zz} \\ \delta\sigma_{yy} \\ \delta\sigma_{xx} \\ \delta\tau_{yx} \\ \delta\tau_{xz} \\ \delta\tau_{zx} \end{bmatrix} \qquad \text{Eq. 2}$$

Anisotropy and laboratory procedure. Eq. 2 shows that, in order to be able to determine the five elastic cross-anisotropic constants, a laboratory apparatus should have the capability to vary all three of the normal stresses as well as at least two shear stresses. In a conventional triaxial test the radial and tangential stresses are equal, and hence only two of the principal stresses can be varied independently.

$$\begin{bmatrix} \delta\varepsilon_{zz} \\ \delta\varepsilon_{yy} \end{bmatrix} = \begin{bmatrix} \frac{1}{E_v} & -2\frac{\upsilon_{vh}}{E_v} \\ -\frac{\upsilon_{vh}}{E_v} & \frac{1}{E_h}(1-\upsilon_{hh}) \end{bmatrix} \begin{bmatrix} \delta\sigma_{zz} \\ \delta\sigma_{yy} \end{bmatrix} \qquad \text{Eq. 3}$$

Eq. 3 is the transformation of Eq. 2 for a conventional triaxial test. Four out of the five elastic constants emerge but the constants E_h and υ_{hh} appear in combination, even though only two equations exist.

The most advanced triaxial test equipment produced to date is the true triaxial test apparatus. This type of apparatus may be broadly categorised into hollow-cylindrical and cubical true-triaxial.

In the hollow-cylindrical true-triaxial test the three normal stresses can be varied independently by varying the axial stress and the internal and external radial pressures. It is also possible to add an independently varying torsional shear stress to the ends of the cylinder. However, the radial stress, tangential stress, and shear stresses in particular, are heterogeneous. More and more knowledge regarding the distribution of the shear stresses in this type of apparatus has emerged in the last ten years. However, a characterisation of the material in accordance with the above equation requires the stress state to be clearly defined and, therefore, may invalided the application of the hollow cylindrical apparatus in the procedure.

In the cubical true-triaxial apparatus, three normal stresses can be varied independently but no shear stresses can be applied. If the axes of symmetry of the specimen coincide with the axes of the normal principal stresses, and no shear stresses are applied then for a cubical true-triaxial test, Eq. 2 becomes Eq. 4 which shows four constants appear in three equations. Both Eq. 3 and 4 demonstrate that none of the currently existing apparatus is capable to fulfilling the requirements for fully-characterised cross-anisotropic materials such as those given by Eq. 2. It is believed that, for this reason, most of the work in the past involved conventional triaxial tests performed on specimens cut at different orientations to the vertical.

$$\begin{bmatrix}\delta\varepsilon_1\\ \delta\varepsilon_2\\ \delta\varepsilon_3\end{bmatrix}=\begin{bmatrix}\frac{1}{E_v} & -\frac{\upsilon_{vh}}{E_v} & -\frac{\upsilon_{vh}}{E_v}\\ -\frac{\upsilon_{vh}}{E_v} & \frac{1}{E_h} & -\frac{\upsilon_{hh}}{E_h}\\ -\frac{\upsilon_{vh}}{E_v} & \frac{\upsilon_{hh}}{E_h} & \frac{1}{E_h}\end{bmatrix}\begin{bmatrix}\delta\sigma_1\\ \delta\sigma_2\\ \delta\sigma_3\end{bmatrix} \qquad \text{Eq. 4}$$

It appears that unless assumptions are made regarding further relationships between the five elastic constants it will be not possible, for the time being, to fully characterise cross-anisotropic materials. A comprehensive approach, involving all the five elastic parameters, was proposed by Graham and Houlsby (8).

The technique considers cross-anisotropy to be special case of isotropy, by multiplying the stiffness coefficients to change the stiffness of the material in the horizontal direction by an anisotropy factor $1/\alpha$. However, the principle of conservation energy requires the coefficient matrix to be symmetric. This condition is achieved by multiplying the second and third columns by $1/\alpha$ to give:

$$\begin{bmatrix}\delta\varepsilon_{zz}\\ \delta\varepsilon_{yy}\\ \delta\varepsilon_{xx}\\ \delta\gamma_{zy}\\ \delta\gamma_{zx}\\ \delta\gamma_{yz}\end{bmatrix}=\frac{1}{E^*}\begin{bmatrix}1 & -\frac{\nu^*}{\alpha} & -\frac{\nu^*}{\alpha} & & & \\ -\frac{\nu^*}{\alpha} & \frac{1}{\alpha^2} & -\frac{\nu^*}{\alpha^2} & & & \\ -\frac{\nu}{\alpha} & -\frac{\nu^*}{\alpha^2} & \frac{1}{\alpha^2} & & & \\ & & & \frac{(1+\nu^*)}{\alpha} & & \\ & & & & \frac{(1+\nu^*)}{\alpha} & \\ & & & & & \frac{(1+\nu^*)}{\alpha^2}\end{bmatrix}\begin{bmatrix}\delta\sigma_{zz}\\ \delta\sigma_{yy}\\ \delta\sigma_{xx}\\ \delta\tau_{zy}\\ \delta\tau_{zx}\\ \delta\tau_{yz}\end{bmatrix} \qquad \text{Eq. 5}$$

A comparison between Eqs. 2 and 5 establishes the following relationships:

$E_v = E^*$
$E_h = \alpha^2 E^*$
$\upsilon_{vh} = \upsilon^*/\alpha$
$\upsilon_{hh} = \upsilon^*$
$2G_{vh} = \alpha E^*/(1+ \upsilon^*)$

Eq. 5 shows a reduction in the number of elastic parameters from five to three. It is readily apparent that by solving the top-left 3 x 3 matrix of Eq 11, one may determine the three elastic constants E^*, υ^* and α.. In the cubical true-triaxial test where prinsipal stresses applied Eq. 5 becomes Eq. 6. The latter is a three-unknown-in-three-equation problem. This equation suggests that the use of a cubical true-triaxial presents a more straight forward interpretation of the technique proposed by Graham and Houlsby and thus is an ideal apparatus for determining the three unknown elastic parameters. The three elastic parameters, E^*, υ^*, and α, can be found by a substitution of the applied principal stresses and the related measured elastic principal strains into this expression.

$$\begin{bmatrix}\delta\varepsilon_1\\ \delta\varepsilon_2\\ \delta\varepsilon_3\end{bmatrix}=\frac{1}{E^*}\begin{bmatrix}1 & -\frac{\upsilon^*}{\alpha} & -\frac{\upsilon^*}{\alpha}\\ -\frac{\upsilon^*}{\alpha} & \frac{1}{\alpha^2} & -\frac{\upsilon^*}{E_h}\\ -\frac{\upsilon^*}{\alpha} & \frac{\upsilon^*}{\alpha^2} & \frac{1}{\alpha^2}\end{bmatrix}\begin{bmatrix}\delta\sigma_1\\ \delta\sigma_2\\ \delta\sigma_3\end{bmatrix} \qquad \text{Eq.6}$$

LABORATORY PROCEDURE AND TYPICAL RESULT

The true-triaxial test. Having considered of the advantages and disadvantages of variuos approaches, it was decided to select the solution of Graham and Houlsby as the preferred analytical solution for this experimental work. The elastic parameters presented hereafter were determined by employing this particular solution. For convenience, the five elastic parameters were represented by three elastic constants *Ev,* υ_{vh} and a vertical-to-horizontal stiffness ratio *n* equal to $1/\alpha^2$ in the Graham and Houlsby approach. The remaining two parameters may easily be transformed from these three constants.

A new cubical true-triaxial equipment was developed for this purpose. The equipment is a computer-controlled, pneumatic, repeated-loading system, all rigid boundary. The system has a facility for recording and displaying stresses, strains and temperature data for an extended period of time or a high number of loading applications. Instead of employing six loading devices in mutually-opposed pairs, the system employed only three platens loaded by three mutually-perpendicular hydraulic jacks. Three reaction walls were placed in the opposite side of each loading platen. Figure 2 presents the basic principle of this equipment and the photograph of the whole system.

Figure 2 (a) Loading platen arrangement; (b) True-triaxial test apparatus

Test sample. The material is a low plasticity, A-6 clay with Plastic Limit of 20% and Liquid Limit of 39%. The soil specimens prepared using the method known as the floating mould compaction. This method was developed by Shackel (9) for the preparation of cylindrical specimens. A variant of this method for making cubical samples was first used by Makiuchi (7). The uniformity of specimens prepared using this method was far better then that achieved using an ordinary static compaction procedure.

Stress state. With reference to Figure 3, the incremental principal stress, $\delta\sigma_{i,}$ is defined as the difference between the peak stress during each individual cyclic load, σ_i, and the constant static stress (σ_{cc}). The corresponding incremental principal strain, $\delta\varepsilon_{ir}$ (resilient), is defined as the difference between the peak strain, ε_{it}, and the remaining strain after the pulsing stress is relieved to constant static confining stress *(i.e.* $\delta\varepsilon_{ip}$, *permanent).*

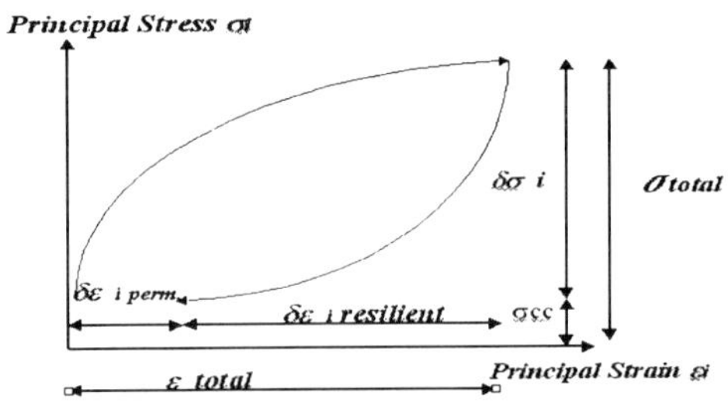

Figure 3 Definision of stress and strain

The static compression pressure for the whole test series was 20 kPa. This value is considered to be within the range of the overburden pressure normally experienced by subgrade soils in the field (10).

The stress invariant, in terms of octahedral shear stress, $\delta\tau_{oct}$, and octahedral stress ratio $k = \delta\tau_{oct} / \delta\sigma_{oct}$ were used to represent the stress state. The levels of stress magnitude were selected by examining various sources such as the literature, the strength of the soil, and the capacity of the experimental equipment. The general condition, for which all three principal stresses were compressive, occurred at a stress ratio, $k \leq {}^{1}/_{2}\sqrt{2}$. On the basis of the above information, the three levels of octahedral shear stress, $\delta\tau_{oct}$, were chosen to be 60 kPa, 90 kPa and 120 kPa. The three levels of octahedral stress ratio, k, were 0.1768, 0.3536 and 0.5304. These values are equal to 0.25, 0.50 and 0.75 times ${}^{1}/_{2}\sqrt{2}$ respectively.

In order to cover the effects of the intermediate principal stress, an intermediate principal stress parameter (i.e. the Nadai-Lode stress parameter) ζ was included, where,

$$\zeta = \frac{(2\delta\sigma_2 - \delta\sigma_1 - \delta\sigma_3)}{(\delta\sigma_1 - \delta\sigma_3)} \qquad \text{Eq. 7}$$

The Nadai-Lode stress parameter has limiting values of -1.0 for triaxial compression (i.e. $\delta\sigma_1 > \delta\sigma_2 = \delta\sigma_3$) and +1.0 for triaxial extension (i.e. $\delta\sigma_1 = \delta\sigma_2 > \delta\sigma_3$). In order to cover the whole range of the intermediate principal stresses from triaxial extension to triaxial compression, the three levels of the Nadai-Lode stress parameter, ζ, were initially selected to be +1.0, 0.0 and -1.0. However, further observations revealed that the selected method of analysis restricted the minimum level of the Nadai-Lode stress parameter to yield real solutions down to only -0.8. As a consequence, the levels of the Nadai Lode stress parameter were +1.0, 0.0 and -0.80.

Stress repetitions, pulsed-stress waveform, and frequency. Each test specimen was subjected to repetitive stress up to 10,000 applications or, alternatively, until any one of the cumulative principal normal strains exceeded 4%. As shown in the literature, subgrade soils tended to show a relatively stable elastic behaviour after about 3000 stress applications. Thus the 10,000 stress applications selected in this experiment were considered to be sufficient to represent the general conditions of subgrade soils in the field.

It was assumed that the intensity of the traffic-induced pulsed stresses in-situ varies with time according to a haversine function. For convenience, in this experimental work, the pulse loading time was chosen as 0.15 seconds. This value was considered to be representative of the stress pulses to be expected in-situ.

In this experimental work, a repetition frequency of one load pulse per second was selected. This value was within the range of the work reported by others and provided a feasible time for an experiment with a large number of cycles. Moreover, an observation on a preliminary test showed that this frequency provided sufficient time for most of the elastic strain to recover.

Typical result. Figures 4 presents typical results of the effects of stress states and the number of stress applications in terms of the elastic parameters E_v, υ_{vh} , and the degree of anisotropy, *n*. The figure shows the variation of the degree of anisotropy with the number of stress applications for ζ = -0.80. It suggests that the number of stress applications had some influence on the elastic parameters. Under k = 0.1768 and ζ= -0.80, for example, the vertical modulus, E_v, at 1000 stress applications was almost double its initial value. However, further increases of the number of stress applications practically had no effects on E_v.

Variation of the Poisson's ratio with the number of stress applications under the triaxial compression (ζ= -0.80) for both k = 0.1768 and 0.3536 (Figure 4b) were small. At k = 0.5304 the υ_{vh} was increased, however, after about 5000 repetitions the increment was also small.

The figure also suggests that the two-way interactions of the number of stress applications and the stress level may have some effects on the degree of anisotropy. The variation was likely to have been related to an alteration of the material structure such as a decrease or increase in the number of interparticle contacts in the material or a change in the orientation of clay fabrics caused by the repeated stressing. However, an analysis of variance of all test series on the effects of number of stress repetitions, shows that the number of stress applications either as an individual variable or in combination with other stress factors, had no statistically significant effects on the elastic parameters. It appears that the analysis of variance failed to indicate such effects. This discrepancy was probably because the effects, overall, were small. Nevertheless, the analysis indicates that individual variable of stress level and Nadai Lode stress parameter do have some effects on the elastic constants.

Although the effects of the number of stress applications are statistically insignificant, the number of stress applications are considered to be an important factor in road pavement analysis as a high number of stress applications is involved in a real pavement. Therefore, on the basis of the preceding discussion, it is sensible to restrict the observation of the elastic parameters to the conditions related to the highest number of stress applications (ie. 10,000 stress applications). This approach was used in further experiment of the effects of stress state on the cross-anisotropic elastic parameters (11).

CONCLUSIONS

Apart from a scarcity of evidence, the literature strongly suggest that pavement materials are likely to be anisotropic. The illustration presented in the paper indicates that material anisotropy may have a significant role in determination of the road pavement behavior. For a number of reasons, however, material anisotropy has been practically neglected in pavement engineering in the past.

The discussion suggested that by using the Graham and Houlsby approach, it was possible to characterise the cross-anisotropic subgrade material. The technique required the use of a cubical-true-triaxial test apparatus such as one that currently developed for this experimental work. However, the delineated cross-anisotropy is limited to a form in which five supposedly independent elastic parameters can be replaced by only three elastic parameters, E^*, υ^* and α. The last parameter, α, represents interrelationships between the five elastic parameters. Nonetheless, the assumptions implied in the Graham and Houlsby approach appear to represent a reasonable hypothesis regarding the material cross-anisotropy.

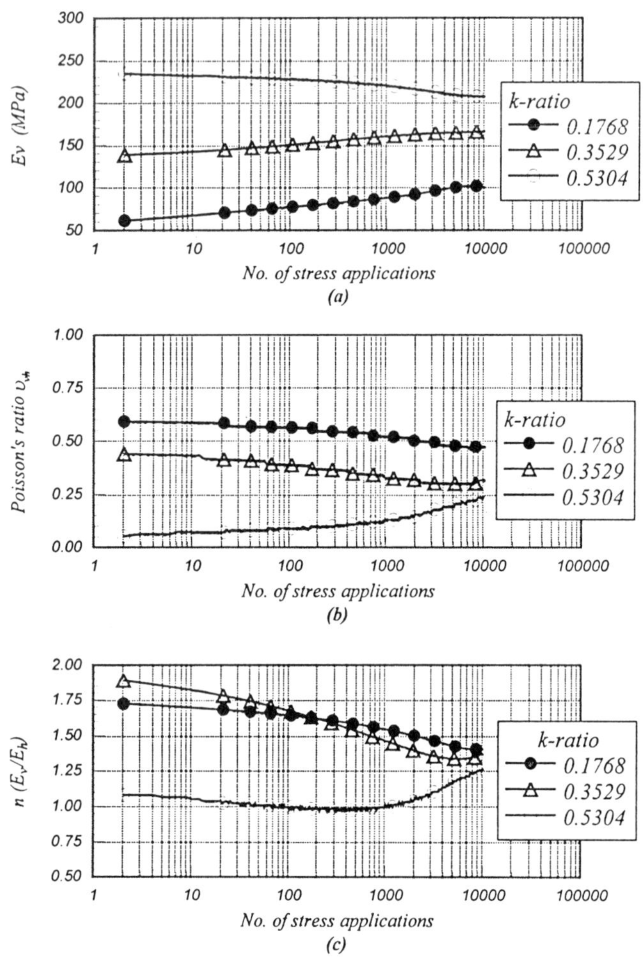

Figure 4 The effects of number of stress applications on elastic parameters ($\delta\sigma_{oct}$ = 60 kPa; ζ = -0.80))

The procedure introduced in the experiment has been proven to facilitate the study of material cross-anisotropy in unbound subgrade material subjected to repeated loading similar to those commonly encountered in pavements. The procedure enabled the determination of the elastic parameters required to describe cross-anisotropic materials.

ACKNOWLEDGEMENTS

It is gratefully acknowledged that the construction of the true-triaxial used in the experiment was financed by the Road and Transportation Authority of New South Wales, Australia. Most of the experimental work was conducted at the School of Civil Engineering, the University of New South Wales. The author expressed his sincere thanks to Associate Professor B Shackel for his guidance, critical appraisal and encouragement throughout the work.

REFERENCES

1. GERRARD, C.M., MULHOLLAND, P. 1966. "Stress Strain and Displacement Distributions in Cross-Anisocarpic and Two-Layer Isotropic Elastic System", Proc. Australian Road Research Board Vol. 3, pp. 1123-1158.
2. GERRARD, C.M., HARRISON, W.J., 1970. "Circular Loads Applied to a Cross-Anisotropic Half Space", CSIRO, Technical Paper No. 8 of Applied Geomechanics.
3. WARDELL, L.J., 1976. "Program CIRCLY - A Computer Program for the Analysis of Multiple Complex Circular Loads on Layered Anisotropic Media User's Manual", Geomechanics Computer Program No.2, CSIRO Division of Applied Geomechanics.
4. NAASRA, 1986. "Pavement Design. A Guide to the Structural Design Guide of Road Pavements", National Association of Australian Road State Authorities.
5. DEHLEN, G.L., 1966. "The Effects of Non-Linear Material on the Behaviour of Pavements Subjected to Traffic Loads", PhD Thesis, University of California.
6. MAKIUCHI, K. 1979. "A Study of Three Dimensional Constitutive Relationships for Subgrade Materials", PhD Thesis, The University of New South Wales.
7. GRAHAM, J., HOULSBY, G.T., 1983. "Anisotropic Elasticity of a Natural Clay", Geotechnique 33 No. 2, pp. 165-180.
8. LEKHNITSKII, S.G., 1963. "Theory of Elasticity of an Anisotropic Elastic Body", Holden-Day, Inc.
9. SHACKEL, B., 1976. "Factors to be Considered in Drafting of Compaction Specifications for Soils", Proc. 8th. Conf. ARRB, Vol. 8.
10. TRANSPORTATION RESEARCH BOARD, 1975. "Test Procedure for Characterizing Stress-Strain Properties of Pavement Materials", Special Report 162, 1975.
11. ZAMHARI, K.A., 1995. "Characterisation of Cross-Anisotropic Pavement Materials", PhD Thesis, The University of New South Wales.

Fifth International Conference on the
BEARING CAPASITY OF
ROADS AND AIRFIELDS

EFFECTS OF ANISOTROPY AND STRESS ON RESILIENT MODULUS OF UNBOUND GRANULAR MATERIALS

BINH T. VUONG

Senior Research Scientist, ARRB Transport Research Ltd,
Australia, Fax: 61 3 9887 8104, binhv@arrb.org.au

ABSTRACT

Ideally, the 3-dimensional (3-D) RTL test (viz. dynamic vertical stress and dynamic confining stress) should be used to closely simulate dynamic stresses in actual pavements because it would enable accurate determination of the resilient modulus and Poisson's ratio of granular unbound materials. However, the simple 1-D RLT test (viz. dynamic vertical stress and static confining stress) is currently adopted as the standard RLT test method for reasons of simplicity and availability of testing equipment. It is shown in this paper that the effects of stress and anisotropy on the resilient modulus of granular materials measured with 1-D and 3-D RLT testing are very significant. An improved understanding of these effects would assist in the interpretation of the measured resilient modulus from RLT testing and the development of relationships between field performance and laboratory results. **Key words:** Unbound granular materials, repeated load triaxial testing, resilient modulus, anisotropy, stress dependency.

INTRODUCTION

The use of the repeated load triaxial (RLT) test for determining resilient modulus and permanent strain of unbound materials has been increasing in Australia. AUSTROADS (1) have recently developed standard specifications for test equipment and testing procedures for the RLT test, and industry has also developed low-cost equipment and software which meet the required standards. Ideally, the 3-D RLT test (viz. dynamic vertical stress and dynamic confining stress) should be used to closely simulate dynamic stresses in actual pavements. However, simple 1-D dynamic loading (viz. resilient vertical stress and static confining stress) is used in the current standard RLT test method for reasons of simplicity and availability of testing equipment. It is anticipated that because granular materials often have anisotropic and stress-dependent characteristics, differences in stress state and stress path between field and laboratory conditions may result in significant differences between laboratory resilient modulus and field pavement modulus.

This paper discusses the effects of stress and anisotropy on the resilient modulus of granular materials measured with 1-D and 3-D RLT testing. Knowledge of these effects would assist in the interpretation of the measured resilient modulus from RLT testing and the development of relationships between field performance and laboratory results.

STRESS STATE AND STRESS PATH IN PAVEMENT AND IN RLT TESTING

Initial static insitu (at rest) stresses in basecourse layers of pavements are very small compared to the dynamic stresses generated by moving wheel loads. Initial insitu (at rest) stresses in basecourse layers consist of self-weight (gravitational) stresses and residual stresses resulting from the interlocking of particles after compaction. The self-weight stresses are usually less than 10 kPa, but the interlocking residual stresses can be higher, say in the range of 1-30 kPa, depending on material type. Moving wheel loads on the pavement result in dynamic compressive stresses in the pavement layers which can be as high as 600 kPa near the surface (depending on the tyre/road contact stress) and reduce with increasing depth.

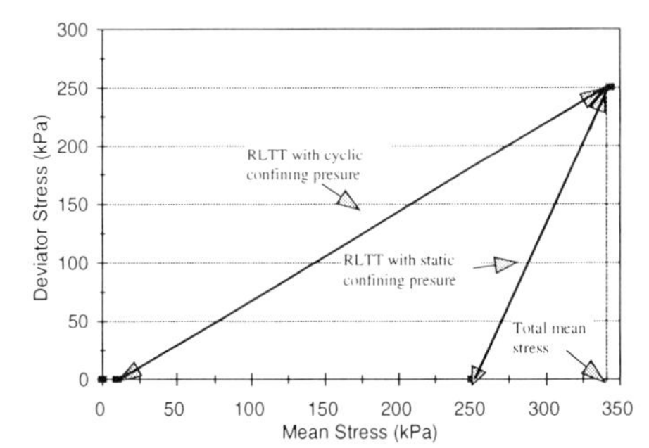

Fig. 1 - Differences in stress paths between 3-D and 1-D RLTT

Fig. 1 shows the *stress paths* used in the 3-D and 1-D RLT tests to achieve the same final *stress state* (represented by the total mean stress and deviator stress). Basically, the generalised Hook's law as given in Equations (1) and (2) is used to calculate resilient modulus (E_r) and Poisson's Ratio (ν_r) from the dynamic stresses and strains measured from the 3-D RLT testing.

$$E_r|_{\sigma_{ci}} = (\sigma_{1r}-\sigma_{3r}).(\sigma_{1r}+2.\sigma_{3r})/(\varepsilon_{1r}.(\sigma_{1r}+\sigma_{3r})-2.\varepsilon_{3r}.\sigma_{3r}) \quad (1)$$

$$\nu_r|_{\sigma_{ci}} = (\varepsilon_{1r}.\sigma_{3r}-\varepsilon_{3r}.\sigma_{1r})/(\varepsilon_{1r}.(\sigma_{1r}+\sigma_{3r})-2.\varepsilon_{3r}.\sigma_{3r}) \quad (2)$$

where σ_{ci} is the initial static confining stress representing the initial stress state (i.e. insitu stresses at rest),
σ_{1r} and σ_{3r} are the dynamic vertical and radial stresses; and
ε_{3r} and ε_{1r} are the resilient radial and vertical strains, respectively.

Because the Hook's law assumes that the material is isotropic (i.e. horizontal modulus = vertical modulus), the resilient modulus and Poisson's Ratio produced by the 3-D RLT testing (viz. using Equations 1 and 2) are considered as "equivalent 3-D isotropic".

The standard 1-D RLT testing is considered as 1-D vertical dynamic loading (i.e. $\sigma_{3r} = 0$, $\sigma_{1r} > 0$), but at static confining stresses much higher than the initial insitu stress. It produces 1-D vertical resilient modulus and Poisson's ratio at a higher static confining stress as given in Equations (3) and (4).

$$E_{vr}|_{\sigma_{ct}} = \sigma_{1r}/\varepsilon_{1r} = q_r/\varepsilon_{1r} \quad (3)$$

$$v_{vr}|_{\sigma_{ct}} = -\varepsilon_{3r}/\varepsilon_{1r} \quad (4)$$

where σ_{ct} is the static confining stress representing the combined insitu and dynamic horizontal stresses,
$q_r = \sigma_{1r} - \sigma_{3r} = \sigma_{1r}$ is the deviator stress; and
ε_{3r} and ε_{1r} are the resilient radial and vertical strains, respectively.

In the sections below, it will be shown, using experimental data, that granular materials are often anisotropic (i.e. vertical modulus can be much higher than horizontal modulus) and the effects of static and dynamic confining stresses on resilient modulus also differ.

EFFECTS OF ANISOTROPY ON RESILIENT MODULUS OF GRANULAR MATERIALS

For the purposes of demonstrating the effects of anisotropy on measured resilient modulus, 1-D RLT tests were conducted on specimens of unbound materials (20 mm nominal size), which had similar dry density, moisture content, but different degrees of anisotropy. The specimens were compacted using three different methods to produce different degrees of particle and/or particle-contact orientation, and hence different degrees of anisotropy (2, 3), viz.:

- static compaction, which assumes to produce a low degree of preferred particle orientation and/or preferred particle-contact direction;
- dynamic compaction, which assumes to produce a moderate degree of preferred particle orientation and/or preferred particle-contact direction; and
- vibratory compaction, which assumes to produce a high degree of preferred particle orientation and/or preferred particle-contact direction.

A study of the fabric (particle orientation and particle-contact direction) of the unbound pavement materials compacted with different methods was also conducted to confirm the above assumptions. This will be discussed in a separate report.

In order to demonstrate the effects of compaction method (or anisotropy) on material type, three different fine-grained granular materials (20 mm nominal size) were also included in the study: (i) a non-plastic crushed rock, (ii) a crushed gravel, and (iii) a flaky crushed rock. For each material, 12 specimens were prepared to specified dry density and moisture content using the three compaction methods described above.

The standard RLT testing equipment and test procedures specified in Standards Australia (1) were used in the testing program. In order to avoid early failure of the specimen and unstable results, particularly for the marginal materials, the stress paths with small step sizes as shown in *Fig. 2* were used. In the precondition stage, 1000 cycles were applied to the specimen at vertical and confining stresses of 350 kPa and 50 kPa, respectively. In other loading stages, 50 cycles were applied before the dynamic stresses and strains were measured.

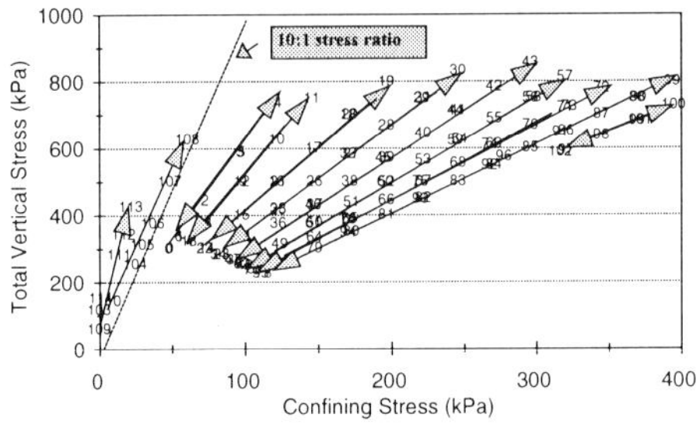

Fig. 2 - Stress paths used in the standard 1-D RLT testing

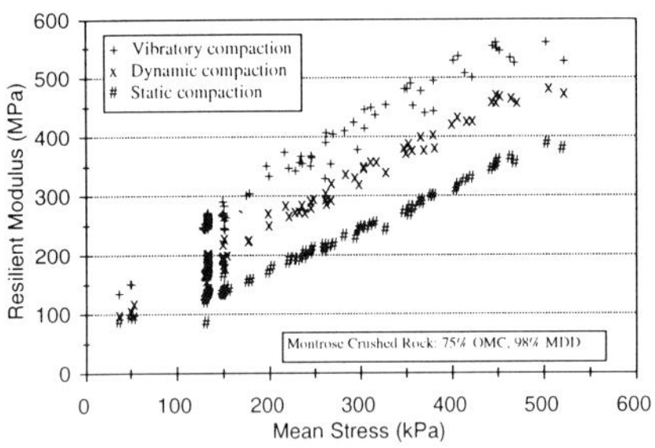

3-a - Boral Montrose non-plastic crushed rock

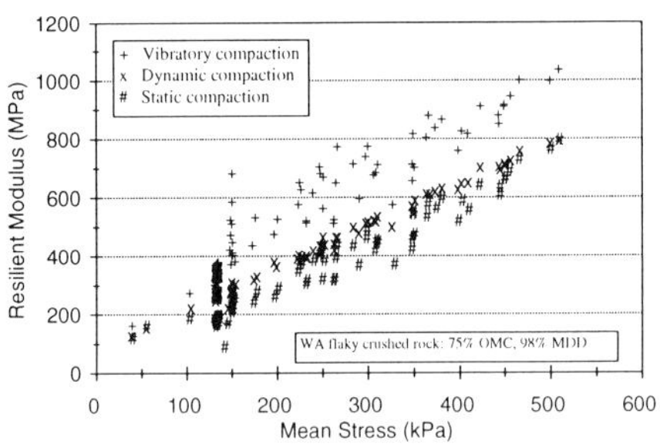

3b - Gosnells flaky crushed rock

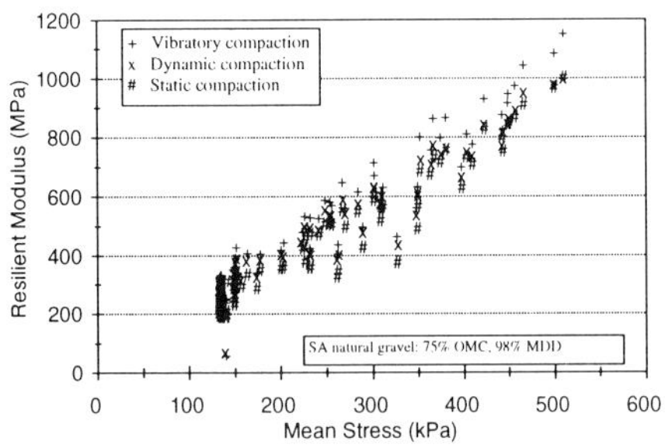

3c - Burra Morgan natural gravel

Fig. 3 - Results of 1-D resilient modulus obtained with various compaction methods

Typical results of resilient modulus obtained from standard 1-D RLT testing on the three materials (i.e. non-plastic crushed rock, the flaky crushed rock and the natural gravel) are shown in *Figs 3a, 3b and 3c,* respectively. The results are plotted in terms of resilient modulus versus mean stress, i.e. ($\sigma_m = 1/3(\sigma_1 + \sigma_2 + \sigma_3)$.

Referring to *Fig. 3*:

- for the non-plastic crushed rock (see *Fig. 3a*), the resilient modulus produced by the vibratory and dynamic compaction methods could be up to 75% and 40%, respectively, higher than that produced by the static compaction method;
- for the flaky crushed rock (see *Fig. 3b*), the resilient modulus produced by the vibratory and dynamic compaction methods could be up to 100% and 10%, respectively, higher than that produced by the static compaction method; and
- for the natural gravel (see *Fig. 3c*), the resilient modulus produced by the vibratory and dynamic compaction methods could be up to 15% and 5%, respectively, higher than that produced by the static compaction method.

These above results indicate that static compaction produced the lowest resilient modulus, whilst vibratory compaction produced the highest resilient modulus, and that the effects of anisotropy due to the sample preparation method on resilient modulus varied depending on material type. The effects were very significant for the flaky crushed rock and the non-plastic crushed rock, but not as significant for the high PI natural gravel. This may be due to the fact that the flaky materials and the non-plastic crushed rock had more flaky particles and lower PI than the natural gravel as was discussed in Vuong (3).

EFFECTS OF STATIC AND DYNAMIC CONFINING STRESSES ON RESILIENT MODULUS

For the purposes of demonstrating the effects of static and dynamic confining stresses (or stress state and stress path) on measured resilient modulus, both 3-D and 1-D RLT testing was conducted on specimens of the non-plastic crushed rock (20 mm nominal size) which had similar dry density, moisture content and degrees of particle orientation. The dynamic compaction method, which produced a moderate degree of anisotropy (as discussed earlier), was used to prepare the specimens.

Four specimens were tested in the study. Two specimens were tested with a series of stress stages using1-D RLT testing, and then with a series of stress stages using 3-D RLT testing. The other two specimens were tested with a reverse sequence of testing, i.e. with a series of stress stages using 3-D RLT testing, and then with a series of stress stages using 1-D RLT testing. In order to minimise the effects of loading history on the measured resilient modulus, only 16 stress stages were used in each test series, with the principal stress ratios being in the range of 2.5-10 and normal stresses in the range of 50-200 kPa.

Figs 4a and 4b compare the resilient moduli of the crushed rock obtained from both 3-D and 1-D RLT testing. The results in these Figures indicate that, for the crushed rock tested:

- at a given final stress state (represented by a mean stress and deviator stress), the standard 1-D RLT test (i.e. with static confining stress) produced a higher resilient modulus than the 3-D RLT test (i.e. with dynamic confining stress), indicating that the effects of static confining stress (or initial stress state) were different to those of dynamic confining stress; and

- at a given deviator stress, 1-D RLT testing with a higher static confining stress produced a much higher resilient modulus, whereas for 3-D RLT testing the resilient modulus was essentially independent of stress ratio or the level of dynamic confining stress.

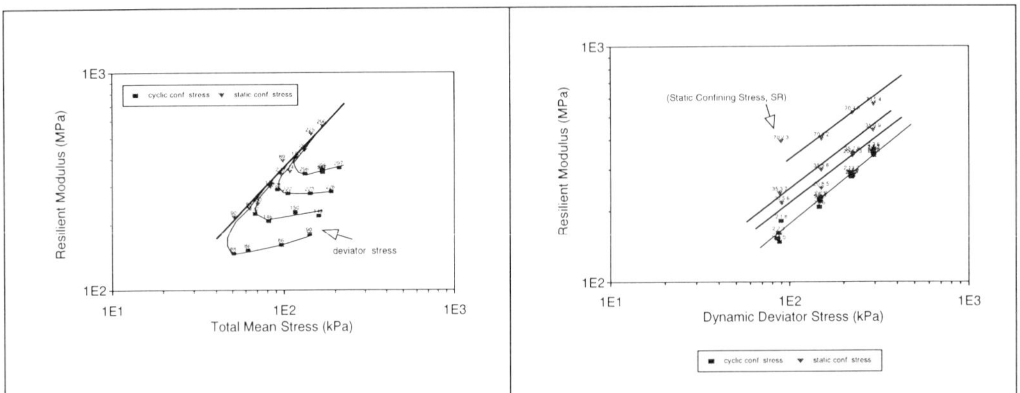

(a) Resilient modulus vs mean stress | **(b) Resilient modulus vs deviator stress**

Fig. 4 - Comparison of resilient modulus of non-plastic crushed rock determined with 1-D and 3-D RLT testing

INTERPRETATION OF 1-D AND 3-D RLT RESILIENT MODULI

Because the standard 1-D RLT test produces different stress conditions to those in actual pavements, and laboratory specimens (prepared by different compaction methods) can have different degrees of anisotropy compared to the field condition, care needs to be taken in interpreting the laboratory resilient modulus produced by RLT testing. *Table 1* summarises some of the common models for resilient modulus developed in the past using different stress parameters in the fitting of the test results obtained from different stress states and stress paths. They can be complicated or simple to suit research or practical applications.

By plotting the results obtained from both the 3-D and 1-D RLT testing against the stress parameters used in the models shown in *Figs 5(a, b, c, d, e* and *f)*, the ability of some simple models to take account of the effects of stress state and stress path on resilient modulus can be investigated. For the crushed rock studied, comments on the goodness of fit of the RTL data to the models are also given in *Table 1*.

The results shown in *Fig. 5* and *Table 1* indicate that it is difficult to develop a simple model that can adequately describe the test results obtained from both 1-D and 3-D RLT testing, possibly because of the many factors influencing the resilient modulus of unbound materials (e.g. anisotropy, stress dependency). It is expected that more comprehensive models can be developed for resilient modulus measured using 3-D RLT testing. However, because of the requirements of more complicated 3-D RLT testing equipment and test procedures to obtain the data, they may be not be suitable for practical use.

For the case studied, simple two-parameter models (Models 1, 2 and 3) closely fitted the results of the 1-D RLT testing, but did not fit the results of the 3-D RLT testing. Therefore, whilst the simple 1-D RLT testing has been considered to be more suitable for practical use, it may be more risky to apply the 1-D test results to real pavements because of the limitation of the test. The use of these models must be subjected to the stress conditions which produce the data that closely fit the models concerned.

The 3-parameter Model 5 can also be used as a descriptive model for resilient modulus derived from 1-D RLT testing. Compared to other 2-parameter models, it allows clearer quantification of the effects of both dynamic stress and static confining stress on the measured modulus.

Table 1
Models of resilient modulus of granular materials

Model	Equation	Reference	Comments
1	$E_r = K_2.\theta^{K1}$ $\theta = \sigma_1+2.\sigma_3$ = total stress	4, 5, 6	Does not take adequate account of the effects of static confining stress and stress ratio for both 1-D and 3-D resilient moduli; must be subjected to specified stress ratio and static/dynamic confining stresses (see Fig. 5a).
2	$E_r = K_1.(\sigma_{m,w}/\sigma_{ref})^{K2}.$ $\sigma_{m,w} = \sigma_{m,c} + (\sigma_{m,r}/2)$ $\sigma_{m,c}$ = initial static stress $\sigma_{m,r}$ = dynamic mean stress $= (\sigma_{1r} + 2\,\sigma_{3r})/3$	7	Could closely fit the results of 1-D modulus; does not fit the results of 3-D modulus; must be subjected to specified stress ratios (see Fig. 5b).
3a	$E_{r.} = [(A + B\, q_r)/q_r].\,\sigma_m$	8	Could closely fit the results of 1-D modulus; does not fit the results of 3-D modulus; must be subjected to specified stress ratio (see Fig. 5c).
3b	$E_r = K_1.(p'/q_r)^{K2}$ $E_r = K_1.(J_2/\tau_{oct})^{K2}$ $J_2 = (\sigma_1\sigma_3 + \sigma_1\sigma_3 + \sigma_1\sigma_3)$ = Second invariant $\tau_{oct} = 2/3(\sigma_1 - \sigma_3)$ = Octahedral shear stress p' = mean effective stress	9, 10, 11	As above
4	$E_r = K_1.\sigma_{oct}^{K2}.\,\tau_{oct}^{K3}$ $E_r = K_1.\theta^{K2}.q_r^{K3}$ $q_r = \sigma_1 - \sigma_3$ = deviator stress $\sigma_{oct} = (\sigma_1+2.\sigma_3)/3$ = Octahedral normal stress $\tau_{oct} = 2/3(\sigma_1 - \sigma_3)$ = Octahedral shear stress	12, 13, 14	Does not fit with all the results of 1-D or 3-D modulus; strong interdependence between normal stress and deviator stress (see Fig. 5d).

5	$E_r = K_1 . \sigma_c^{K2} . q_r^{K3}$ $E_r = K_1 . \sigma_c^{K2} . \theta^{K3}$	15, 16	Could fit to the results of 1-D resilient modulus; differentiates between the results of 1-D resilient modulus and 3-D resilient modulus; may not warrant a generalised model for both 1-D and 3-D resilient modulus (see Fig. 5e).

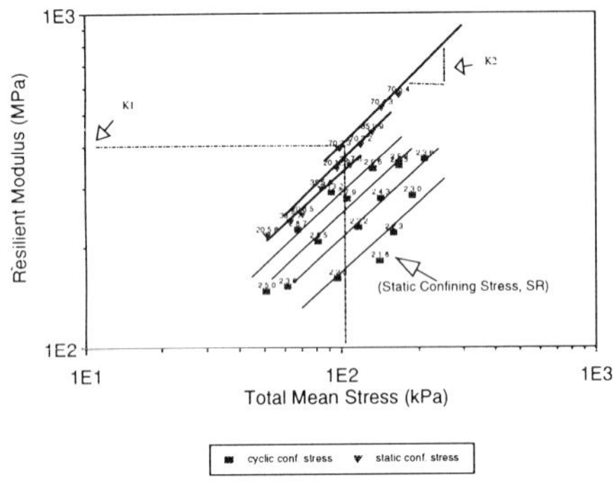

(a) Resilient modulus as a function of total mean stress

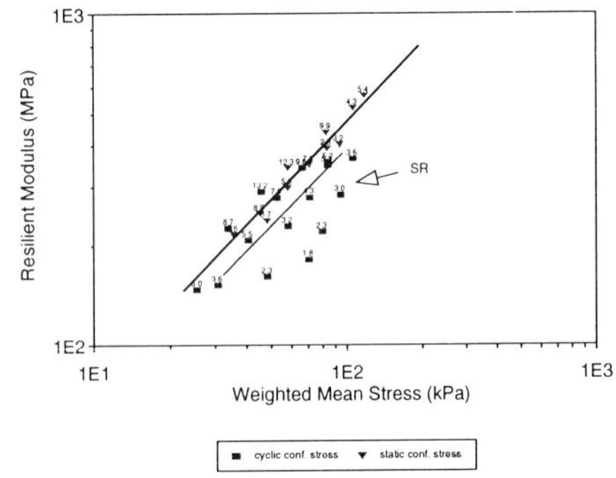

(b) Resilient modulus as a function of weighted mean stress

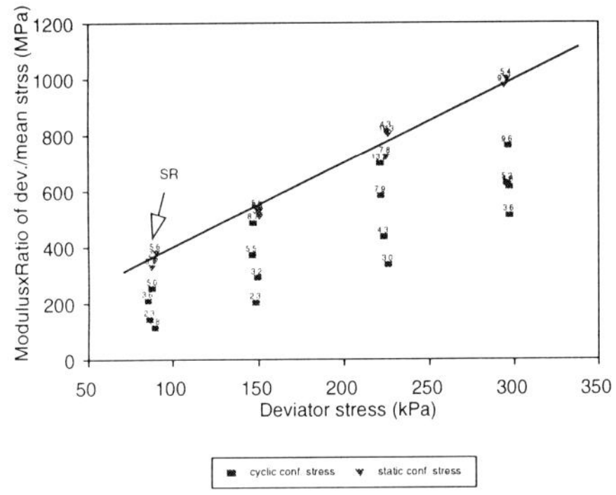

(c) Resilient modulus as a function of total mean stress/deviator stress

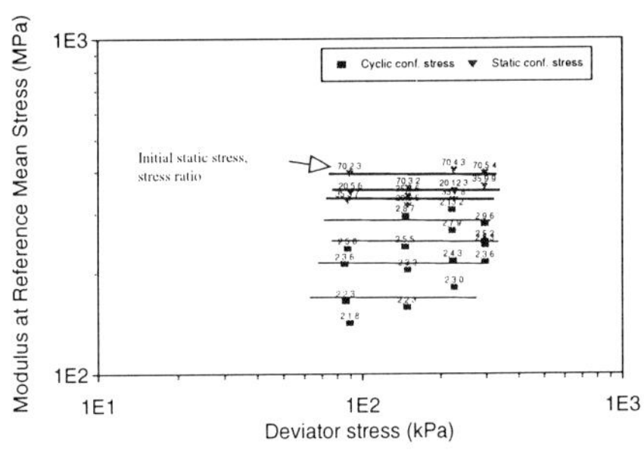

(d) Resilient modulus as a function of total mean stress and deviator stress

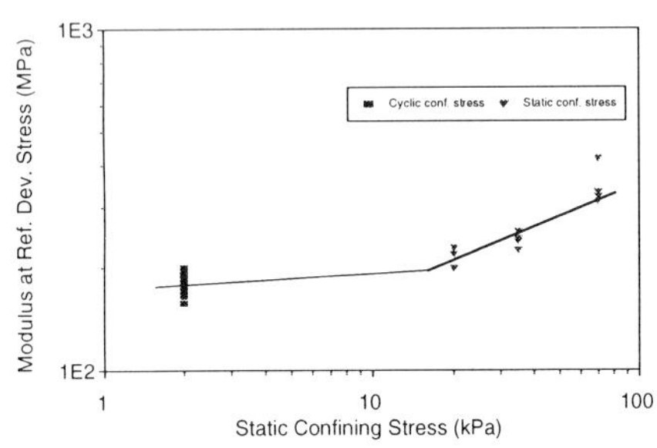

(e) Resilient modulus as a function of static confining stress and deviator stress

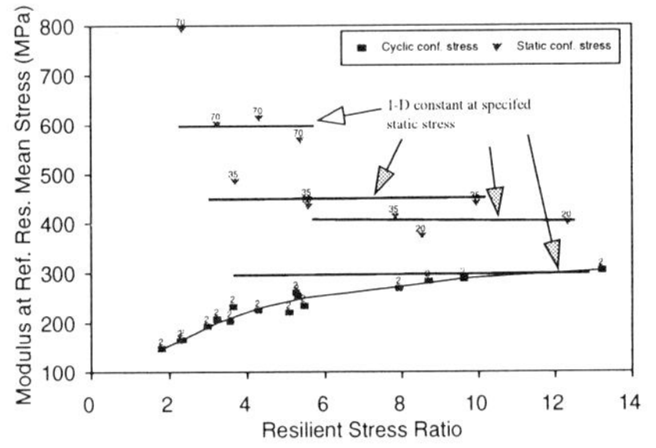

(f) Resilient modulus as a function of resilient mean stress and resilient stress ratio

Fig. 5 - Resilient modulus as a function of different stress parameters

INTERPRETATION OF 1-D RESILIENT MODULI OF VARIOUS MATERIAL TYPES

The suitability of the 2-parameter and 3-parameters models developed for resilient modulus measured with the standard 1-D RLT testing for various unbound material types is also examined below, viz. using these models to fit the results of 1-D resilient modulus for various unbound materials as shown in *Figs. 6, 7 and 8.*

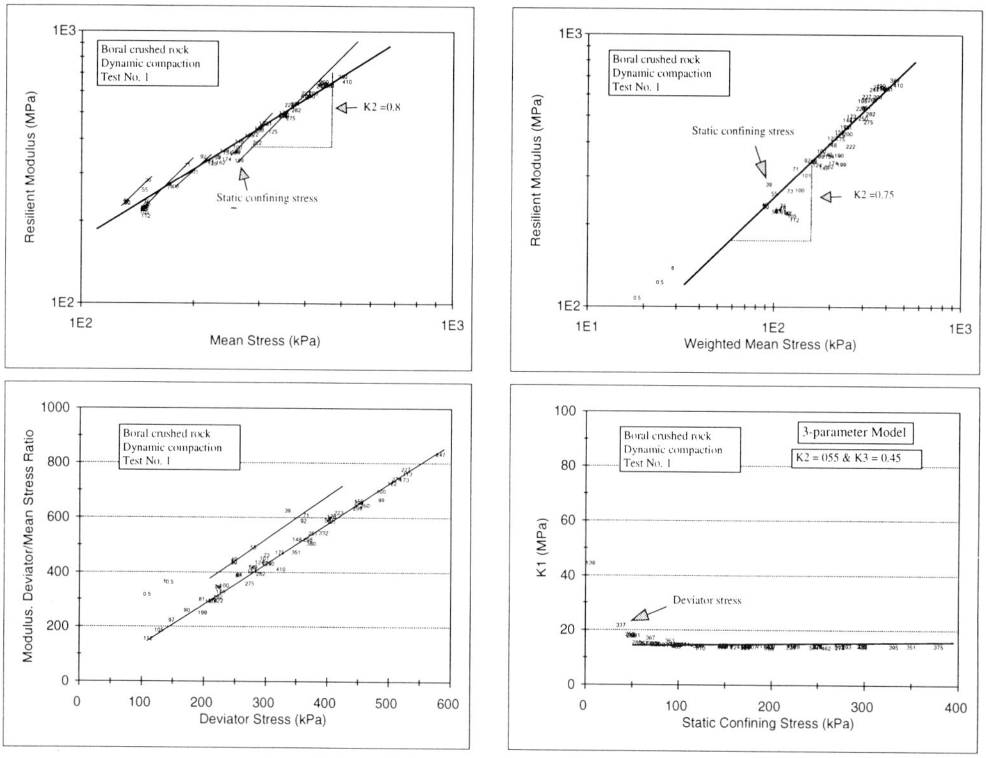

Fig. 6 - Interpretation of the non-plastic crushed rock modulus using different models

Referring to these Figures,:

- Model 1 appears to fit well with the test results for the non-plastic crushed rock (see *Fig. 6a)*, possibly because the slopes of the curves fitted to each static confining stress were very close to the slope of the curve fitted to all data points. However, it did not fit well with the results for the crushed gravel and flaky crushed rock (see *Fig. 7a* and 8a), possibly because the slopes of the curves fitted to each static confining stress were much lower than the slope of the curve fitted to all data points. Therefore, this model may not be suitable to describe 1-D resilient modulus for all unbound materials.
- Model 2 appears to fit well with the test results for the flaky crushed rock and the crushed gravel (see *Fig. 7b and 8b)*. However, it did not fit well with the results for the crushed rock (see *Fig. 6b*). Therefore, this model may not be suitable as a generalised model to describe the 1-D resilient modulus of various unbound materials.
- Model 3 appears to fit well with the test results for the flaky crushed rock (see *Fig. 7c)*.

However, it did not fit well with the results for the crushed rock and the crushed gravel (see *Figs 6c* and *8c*). Again, the results indicate that this model may not be suitable as a generalised model to describe 1-D resilient modulus of various unbound materials.

- Model 5 fitted well to the results of 1-D modulus obtained from RLT testing on the three different unbound materials (see *Figs 6d, 7d* and *8d*), particularly for confining stresses above 50 kPa. Therefore, the model appears to be suitable as a generalised model to describe the 1-D resilient modulus of various unbound materials.

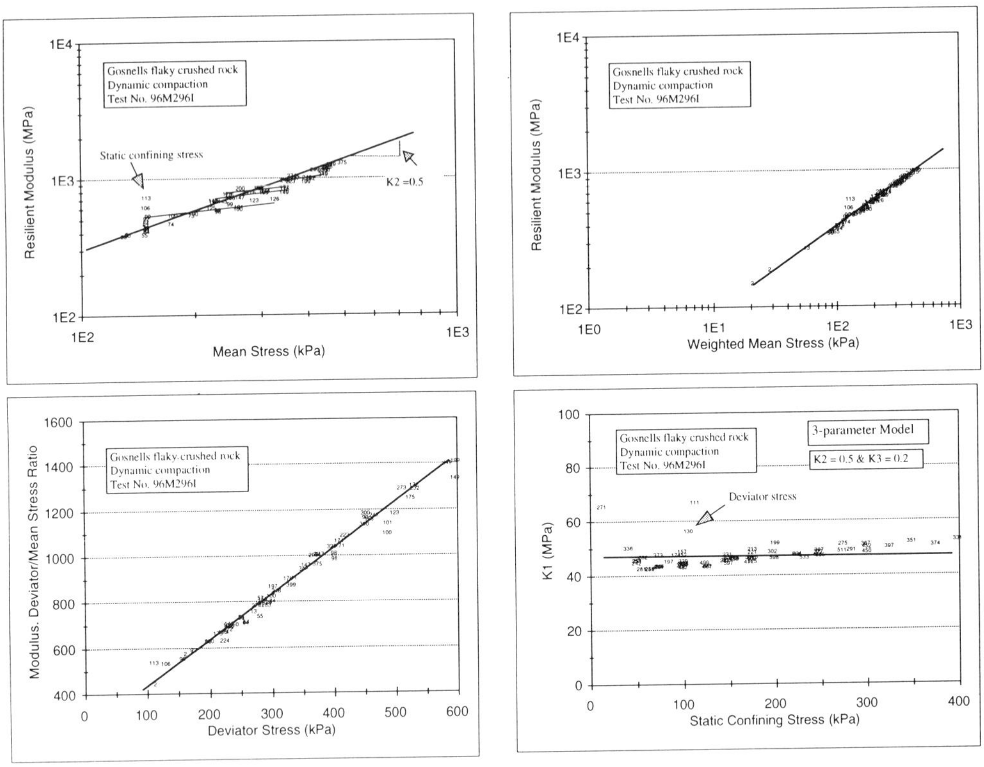

Fig. 7 - Interpretation of the flaky crushed rock modulus using different models

It was noted that, in using Model 5, different materials produced different parameters K_1, K_2, and K_3. These parameters may be effectively used to describe the stress-dependent characteristics of each material. For example:

- the crushed rock can gain significant stiffness under high static confining stresses (viz. a high K_2 value of 0.5) and high dynamic deviator stresses (viz. a high K_3 value of 0.45);
- the crushed gravel and flaky crushed rock can gain significant stiffness under high static confining stresses (viz. high K_2 values of 0.5 and 0.7, respectively), but may not gain significant stiffness under high dynamic deviator stresses (viz. a low K_3 value of about 0.2 for both materials).

Therefore, there is potential in using these parameters as material indices, particularly for the purpose of material ranking. This will be discussed further in a later report.

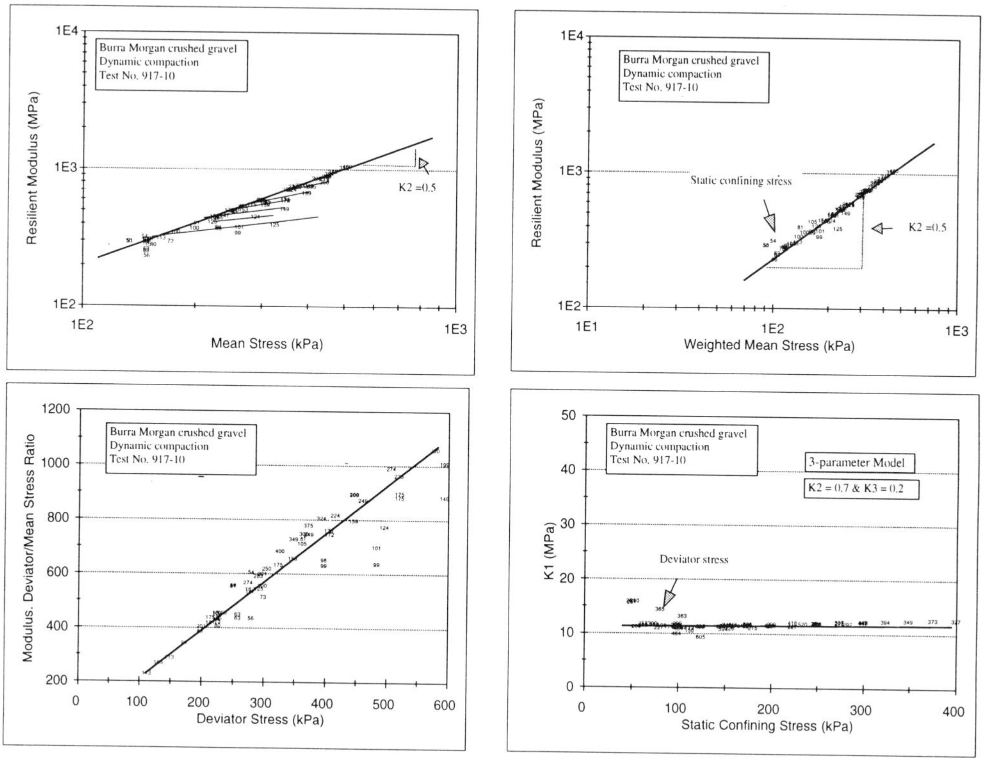

Fig. 8 - Interpretation of the crushed gravel resilient modulus using different models

CONCLUSIONS AND RECOMMENDATIONS

Initial static insitu (at rest) stresses, particularly in the top 100 mm of basecourse layers, are very small compared to the dynamic stresses generated by moving wheel loads. Therefore, the moduli obtained from current standard 1-D RLT testing (i.e. dynamic vertical stress and static confining stress) may be different to the actual pavement moduli (obtained with the 3-D dynamic loading).

It has been shown in this study that:

- Different compaction methods produced different values of 1-D resilient modulus. Because the different compaction methods produced different degrees of anisotropy (i.e preferred particle and particle-contact orientation), this may suggest that the effects of compaction method (or anisotropy) can be very significant for granular materials (i.e. the horizontal modulus can be much lower than the vertical modulus).
- Static/dynamic confining stresses can have a very significant effect on the measured resilient modulus of anisotropic granular materials. In other words, at the same levels of dynamic vertical stress, 1-D RLT testing (with static confining stresses) may produce higher resilient moduli than those produced by 3-D RLT testing (with dynamic confining stresses).

- It is difficult to develop a simple model that can adequately describe the test results obtained from both 1-D and 3-D RLT testing, possibly because of the many factors influencing the resilient modulus of unbound materials. It is expected that more comprehensive models can be developed for resilient modulus measured with 3-D RLT testing. However, because of the requirements of more complicated testing equipment and test procedures to obtain the data, they may be not be suitable for practical use.
- The simple 1-D RLT testing has been considered to be more suitable for practical use. However, because of the limitation of the test, it may be more risky to apply the 1-D test results to real pavement conditions. When using simple models to fit 1-D test results, the stress conditions which produce the data that closely fit the models concerned must be specified. There are also risks of producing significant fitting errors to experimental results when using simple models, because they may not fit well to RLTT results for different material types.
- The 3-parameter Model 5 can be used as a descriptive model for resilient modulus derived from 1-D RLT testing. Compared to other 2-parameter models, it allows clearer quantification of the effects of both dynamic stress and static confining stress on the measured modulus. Whilst it can closely fit the results obtained at static confining stresses greater than 50 kPa for different material types, work to date has suggested that the goodness of fit of the RTL data to the model may be poor at low confining stresses (<50 kPa) and improvements therefore need to be made to the model.

It is recommended that further studies be undertaken to establish relationships between test results obtained from the standard RLT testing and field performance. This will enable guidelines to be developed for the use of RLT testing results in material specifications and pavement design.

ACKNOWLEDGMENT

The contribution of the members of AUSTROADS Pavement Research Group of RLTT Users (particularly Mr Rod Brimle of Transport South Australia, Mr Frank Butkus of Main Roads Western Australia and Mr Peter Armstrong of ARRB TR) is gratefully acknowledged.

REFERENCES

1. STANDARDS AUSTRALIA (1995). Methods of testing Soils for engineering purposes. Method 6.8.1: Soil strength and consolidation tests - Determination of the resilient modulus and permanent deformation of granular unbound materials. AS 1298.6.8.1.

2. VUONG. B.T. (1979). Sand fabric and its influence upon the mechanical behaviour of sand. Monash University. Civil Engineering. Research Report No/1979. ISSN 0155-6282.

3. VUONG. B.T. (1997). Progress Report - A summary report on inter-laboratory precision study of resilient modulus and permanent strain testing. AUSTROADS Pavement Research Group Document No. APRG 97/01 (MA), February 1997.

4. HICKS, R.G. (1970). Factors influencing the resilient properties of granular materials.

Ph.D. Thesis, University of California, Berkeley.

5. ALLEN, J.J. AND THOMPSON, M.R. (1974). Resilient response of granular materials subjected to time dependent lateral stresses. Transportation Research Record TRR No. 510, pp. 1-13.

6. BROWN, S.F. (1974). Repeated load testing of a granular material. Journal of the Geotechnical Engineering Division. Proc. ASCE. 110 (GT7), pp. 131-138.

7. VUONG. B.T. (1991). Resilient modulus testing and application at ARRB. APRG Workshop on Elastic Characterisation of Unbound Granular Materials and Subgrades, Adelaide, Feb 1991. APRG Report No.3.

8. NATTATMADJA, A. (1992). Resilient modulus of granular materials under repeated loading. Proceedings of the 7th International Conference on Asphalt Pavements. Vol 1, pp. 172-185.

9. COLE, W.F. (1982). The performance of Deer Park basalt in triaxial testing. Proc. ARRB Conference, Vol. 1(3), pp. 52-59.

10. JOHNSON, T.C, BERG, R.L. AND DIMILLIO, A. (1986). Frost Action predictive techniques: an overview of Research results. Transportation Research Record TRR No. 1089, pp. 147-161.

11. TAM, W.A. AND BROWN, S.F. (1988). Use of the falling weight deflectometer for insitu evaluation of granular materials in pavements. Proc. ARRB Conference, Vol 14(5), pp.155-163.

12. SHACKEL, B. (1973). Repeated loading of soils - a review. Australian Road Research, Vol. 5(3), pp. 22-49.

13. UZAN, J. (1985). Characterisation of granular material. . Transportation Research Record TRR No. 537, pp. 49-58.

14. VUONG. B.T. (1985). Permanent deformation and resilient behaviour of a Victorian crushed rock using the repeated load triaxial test. Australian Road Research Board AIR 403-6.

15. PEZO R.F., KIM K.H, STOKOE I.I. AND HUDSON, W.R. (1991). Development of a reliable resilient modulus test for subgrade and non-granular subbase materials for use in routine pavement design. University of Texas at Austin, Research Report 1177-4f, Austin, Texas.

16. FELIBERTI, M., NAZARIAN, S., AND SRINIVASAN, T (1992). Critical evaluation of parameters affecting resilient modulus tests on subgrades. University of Texas at El Paso, Research Report 1177-2, El Paso, Texas.

Fifth International Conference on the
BEARING CAPASITY OF
ROADS AND AIRFIELDS

NEW HYPERELASTIC MATERIAL MODEL FOR GRANULAR MATERIALS IN PAVEMENT STRUCTURES.

INGE HOFF [1], STEINAR NORDAL[2] AND RASMUS S. NORDAL[3]

[1] Research Fellow, Norwegian University of Science and Technology
[2] Professor, Norwegian University of Science and Technology
[3] Professor Emeritus, Norwegian University of Science and Technology
N – 7034 Trondheim, Norway

ABSTRACT

To improve the analyses of stresses and strains induced by traffic loads in flexible pavement structures, a new material model based on hyperelasticity has been developed.

The proposed model includes the effect of resilient dilatancy in unbound granular materials. The three material parameters controlling the model is determined from repeated load triaxial test.

The hyperelastic model gives more realistic stress distribution in granular layers of a pavement structure than the use of the linear elastic model.

The hyperelastic model is being implemented in the general finite element programme ABAQUS and in a special tailored finite element programme.

Based on the predicted stresses the actual degree of shear strength mobilisation may be determined. This is used together with repeated load triaxial testing at corresponding stress levels in a draft design procedure to predict permanent deformations under traffic loading.

Keywords: Hyperelastic model, non-linear, permanent deformation, FEM, stress analysis.

INTRODUCTION

Traditionally the simple computer programmes for analysis of stresses and strains for pavement design have been based on the theory of linear elasticity. Well known and frequently used are the Chevron and the Bisar programmes. These analyses, which are based on the linear elastic model for granular materials, gives unrealistic values for the horizontal stresses in the lower layer of the pavement. For most typical structures considerable tensile stresses will be predicted in the lower layer of unbound granular materials. These predicted stresses are obviously wrong; as unbound materials have no tensile strength such stresses do in reality just indicate a failure condition.

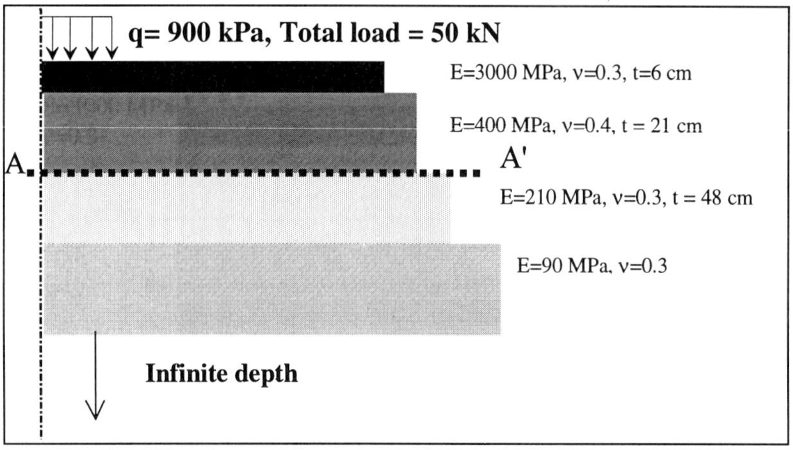

Figure 1 Typical pavement structure used for the computation of stresses

Stresses computed at level A – A' of the shown four-layer pavement structure are presented in figure 2. In the central area below the simulated wheel load there are tensile horizontal stresses.

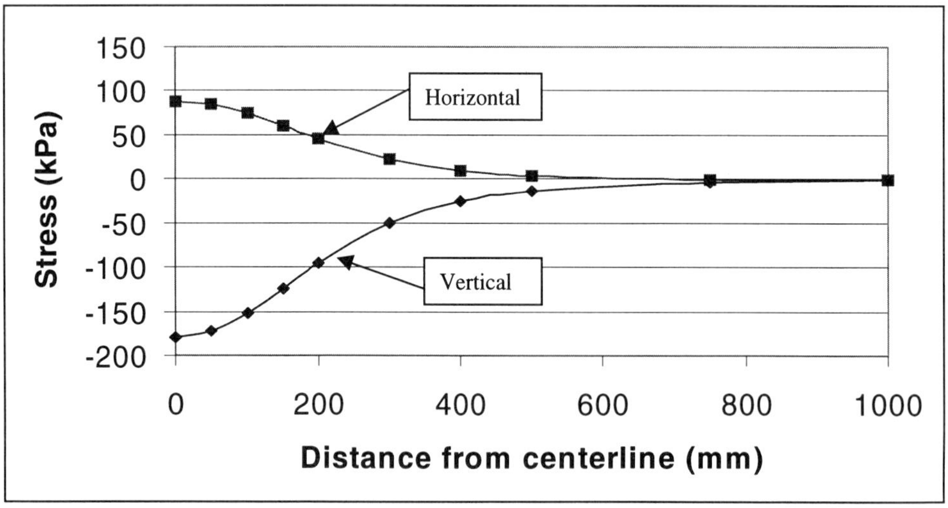

Figure 2 Horizontal and vertical stresses at the bottom of the second layer (A - A')

Different attempts to overcome the problem of unrealistic high tensile stresses have been made, involving:

1. Permanent horizontal compressive stresses from compaction. (Uzan /1/)
2. Using Poisson's ratio larger than 0.5 (Uzan et al/2/)
3. Anisotropic stiffness.
4. Stress adjustment using the Mohr-Columb failure criterion.

The four proposals lead to lower horizontal stresses, but require assumptions and material parameters that are not consistent with fundamental material behaviour as observed in triaxial tests.

NEW HYPERELASTIC MODEL

Theory of hyperelasticity

It is proposed to improve stress predictions by utilizing the theory of hyperelasticity in the stress-strain relationship for granular materials. The model will be formulated as a three parameter model, where the new parameter compared to conventional isotropic linear elasticity will couple shear and volumetric strain to model dilatancy during shear.

The theory of hyperelasticity is based on formulation of a strain energy function. The relation between stresses and strains is found from taking the derivatives of the energy function.

One dimensional bar

The basic principles of hyperelasticity may be illustrated by stretching of a one-dimensional simplified bar.

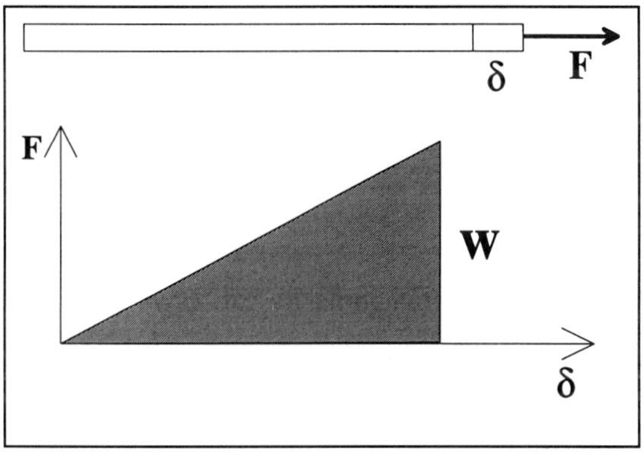

Figure 3 Work needed for the stretching of a one Dimensional linear elastic bar

The energy density is:

Where "E" is Young's modulus and "ε" is axial strain.
The stress in the bar is found by taking the derivative of the energy function with respect to the strain. The stress is then found to be:

$$\sigma = \frac{\partial U}{\partial \varepsilon} = E \cdot \varepsilon$$

We recognise this expression as Hook 's law for the uniaxial case.

Three dimensional continuum

For a three-dimensional continuum the strain energy expression is more complicated. One way to express it is to separate the volumetric and the shear the energy:

$$U = \frac{1}{2}K\varepsilon_v^{\ 2} + \frac{1}{2}G\gamma^2$$

Where K is bulk modulus, G is shear modulus; ε_v is volumetric strain and γ is shear strain.

By differentiation we obtain the conventional relationship from the theory of elasticity:

$$\begin{bmatrix} \sigma_m \\ \tau \end{bmatrix} = \begin{bmatrix} K & 0 \\ 0 & G \end{bmatrix} \cdot \begin{bmatrix} \varepsilon_v \\ \gamma \end{bmatrix}$$

Dilatation

In triaxial testing it is usually observed that a well-compacted granular material expands when subjected to shear strains. The explanation for this behaviour is due to the fact that the material is not a continuum, but consists of individual particles that can move relatively to each other.

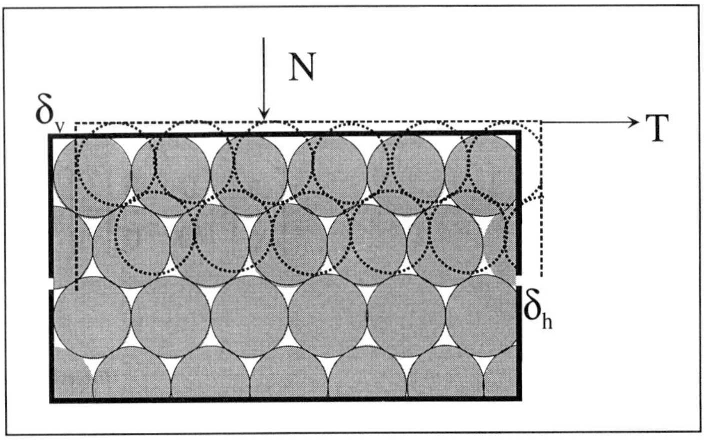

In a densely compacted granular material the particles are packed closely together leaving small pores between the grains. Shearing strain forces the particles to climb on each other, and the volume increases. Figure 4 illustrates this effect for equally sized spheres.

Figure 4 Dilatation effect by shear deformation of densely packed spheres.

Hyperelastic model including dilatation

To include the dilatancy we have added a new term to the strain energy function.

$$U = \tfrac{1}{2}K\left(I_1^\varepsilon\right)^2 + D \cdot I_1^\varepsilon J_2^\varepsilon + 2GJ_2^\varepsilon$$

I_1^ε is the first strain invariant (volumetric strain) and J_2^ε is the second deviator strain invariant. I_1^ε is the sum of the strains while J_2^ε is a function of squared strains.

The term "$D \cdot I_1^\varepsilon J_2^\varepsilon$" relates the volumetric strains to shear strains. From the energy function the stresses can be found by partial differentiation.

$$\sigma_1 = \frac{\partial U}{\partial \varepsilon_1}, \ \sigma_2 = \frac{\partial U}{\partial \varepsilon_2} \text{ and } \sigma_3 = \frac{\partial U}{\partial \varepsilon_3}$$

This gives the stress – strain relationship expressed by principal stresses and strains.

$$\begin{bmatrix}\sigma_1\\ \sigma_2\\ \sigma_3\end{bmatrix} = \begin{bmatrix} K+\frac{4}{3}\cdot G + D\cdot\varepsilon_1 & K+\frac{-2}{3}\cdot G - \frac{1}{2}\cdot D\cdot\varepsilon_3 & K-\frac{2}{3}\cdot G - \frac{1}{2}\cdot D\cdot\varepsilon_2 \\ K+\frac{-2}{3}\cdot G - \frac{1}{2}\cdot D\cdot\varepsilon_3 & K+\frac{4}{3}\cdot G + D\cdot\varepsilon_2 & K-\frac{2}{3}\cdot G_o - \frac{1}{2}\cdot D\cdot\varepsilon_1 \\ K-\frac{2}{3}\cdot G - \frac{1}{2}\cdot D\cdot\varepsilon_2 & K-\frac{2}{3}\cdot G - \frac{1}{2}\cdot D\cdot\varepsilon_1 & K+\frac{4}{3}\cdot G + D\cdot\varepsilon_3 \end{bmatrix} \cdot \begin{bmatrix}\varepsilon_1\\ \varepsilon_2\\ \varepsilon_3\end{bmatrix}$$

For the axisymmetric case where $\varepsilon_2 = \varepsilon_3$ this relationship could be simplified to:

$$\begin{bmatrix}\sigma_d\\ \sigma_m\end{bmatrix} = \begin{bmatrix} 3\cdot G & D\cdot\varepsilon_q \\ D\cdot\varepsilon_q\cdot\frac{1}{3} & K \end{bmatrix}$$

Where $\varepsilon_q = \frac{2}{3}(\varepsilon_1 - \varepsilon_3)$, $\varepsilon_q = \varepsilon_1 + \varepsilon_2 + \varepsilon_3$, $\sigma_d = (\sigma_1 - \sigma_3)$ and $\sigma_m = \frac{1}{3}(\sigma_1 + \sigma_2 + \sigma_3)$

It should be noticed that the stiffness matrix is not symmetric for the axisymmetric case, but it is symmetric in the general three-dimensional formulation. If the mean stress is tensile the material is in a failure state and the proposed model will not give reasonable results. In such a case the D-parameter is set to zero.

Comparison with triaxial tests

Dynamic triaxial tests performed on well-compacted, dense unbound granular materials show significant dilatation. Figure 5 shows in principle the results of back-calculation of a triaxial test, and the results from a test are presented in Figure 6.

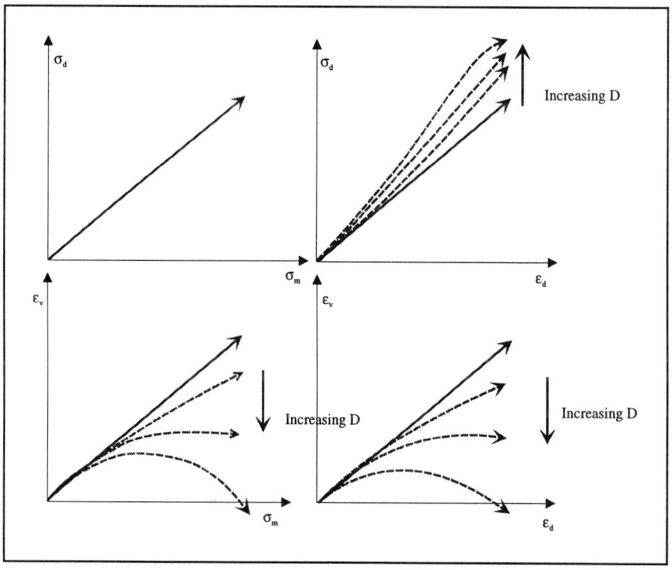

Figure 5 Dilatant behavior of granular materials illustrated by the hyperelastic model for simulated triaxial tests.

As can be seen from the last figure the sample starts out to behave linearly elastic at very low strains, but as the shear strains increase the dilatation effect increases. Depending on the confining pressure the sample will increase more or less in volume during testing. Such a volume increase is assumed to be mainly related to the shear induced dilatancy.

This effect can not be modelled by conventional theory of elasticity. The behaviour of a dilatant material is well described by the new hyperelastic model. The dilatancy parameter, D, may be determined from the results of repeated load triaxial tests as illustrated by Figure 6.

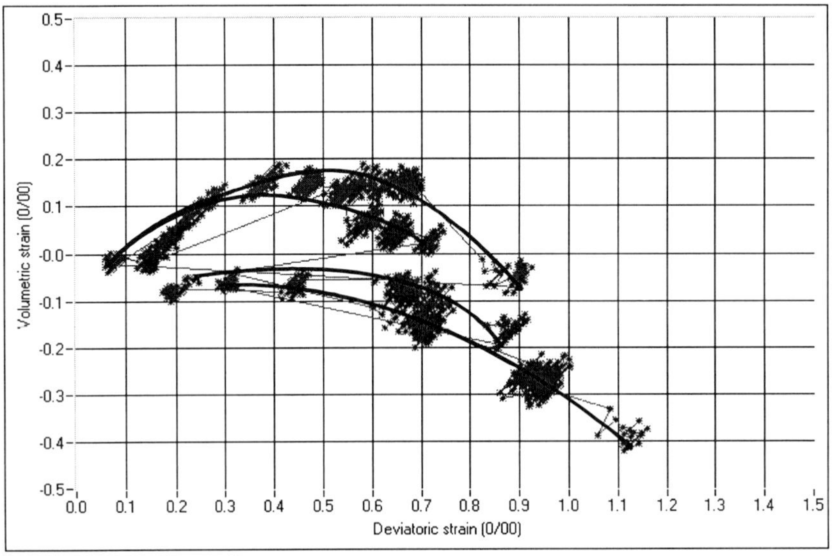

Figure 6 Result from repeated load triaxial testing of a dense granular material

Such laboratory curves will determine the dilatancy parameter D, (compare figure 4)

Example of application of the proposed model

The same four-layer pavement structure that was analysed with Chevron (Figure 2) has been recalculated with the new model, implemented in ABAQUS. All other conditions were kept the same.

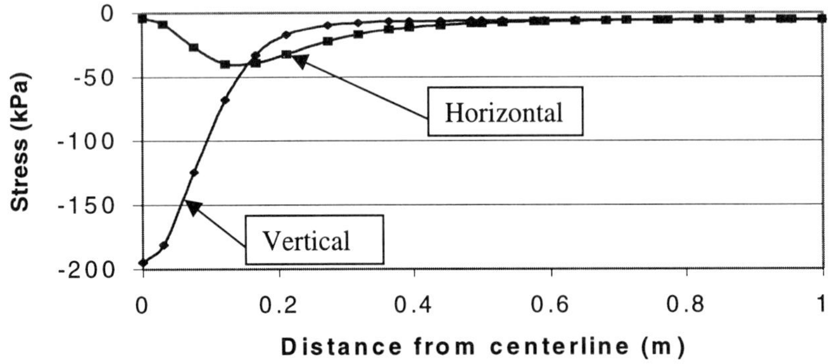

Figure 7 Horizontal and vertical stresses at the bottom of the second layer of the pavement in Fig.1 calculated with ABAQUS and the new material model.

The horizontal stresses in the pavement structure calculated with the new model are assumed to give a much more realistic stress situation. In the case studied, there occur no horizontal tensile stresses in the central load area in the unbound layer.

IMPLEMENTATION INTO FEM PROGRAM

The model is developed and implemented into the finite element programme ABAQUS. The model seems to be a powerful tool for the analysis of stresses and strains in granular layers of typical pavement structures.

ABAQUS is a very powerful tool for research purposes, but this programme is very large, quite difficult to use and rather expensive for non-academic applications.

Implementation into a specially tailored finite element program is also being carried out. This program will be made available from SINTEF during spring 1999.

TYPICAL VALUES OF PARAMETERS FOR THE HYPERELASTIC MODEL

Some typical parameter values are shown in the table below. A more detailed description of the equipment used is given by Hoff and Nordal /3/, and a presentation of the tested materials and the testing program is given by Bakløkk et al /4/.

Table 1 Example Values of Parameters for the Hyperelastic Model

Materials	K (MPa)	G (MPa)	D (Pa)
Well-graded crushed rock	600	75	$2.0 * 10^{11}$
Single sized crushed rock	200	100	$1.0 * 10^{11}$
Well graded natural gravel	300	110	$1.5 * 10^{11}$
Single sized natural gravel	400	200	$2.5 * 10^{11}$
Well graded crushed limestone	800	75	$4.5 * 10^{11}$
Single sized crushed limestone	400	75	$2.0 * 10^{11}$

The values presented in this table are based on a few tests on a limited number of samples and can not be used as "standard" values for design. Special repeated load triaxial tests must be carried out on the actual material to be used. In time increased testing and experience including field calibration might provide a basis for establishing a table of reference values of parameters for conventional materials for pavements.

PREDICTION OF PERMANENT DEFORMATIONS

The formation of wheel path ruts in flexible pavements is one of the most important distress mechanisms. The ruts are created by heavy traffic loads inducing high stresses that may lead to permanent deformations in the pavement layers and in the sub-grade.

The resistance against permanent deformations in granular layers is closely related to the materials characteristics, the state of stress and the degree of shear strength mobilisation. The basic deformation characteristics of a material may be examined most fundamentally by repeated load triaxial testing. A sample with representative density and grain structure is subjected to pulsating shear stresses, stepwise increasing at different levels of confining stress. Referring to Figure 8 the mobilised friction is defined as:

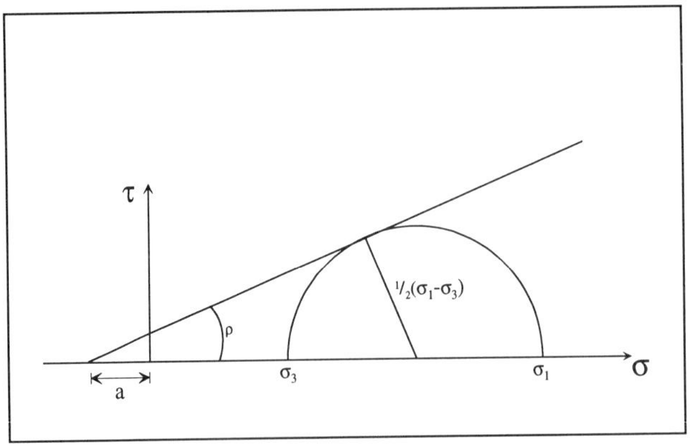

$$\sin\rho = \frac{\sigma_1 - \sigma_3}{\sigma_1 + \sigma_3 + 2a}$$

Where "a" is the apparent attraction for the material.

The degree of mobilisation of shear strength could also be expressed by a relative value; the degree of mobilised friction, S_m defined as:

$$S_m = \frac{\sin(\rho)}{\sin(\phi)}$$

where ϕ is the failure angle.

Figure 8 Mobilised friction

The permanent deviatoric strain observed by triaxial testing of a typical base-course material is presented in Figure 9 as a function of the degree of shear strength mobilisation. In this example practical limits for mainly elastic and plastic strain response are indicated by a grey and a dark line respectively. In the intermediate region some permanent strain develops initially, but the strain rate decreases with the number of load pulses.

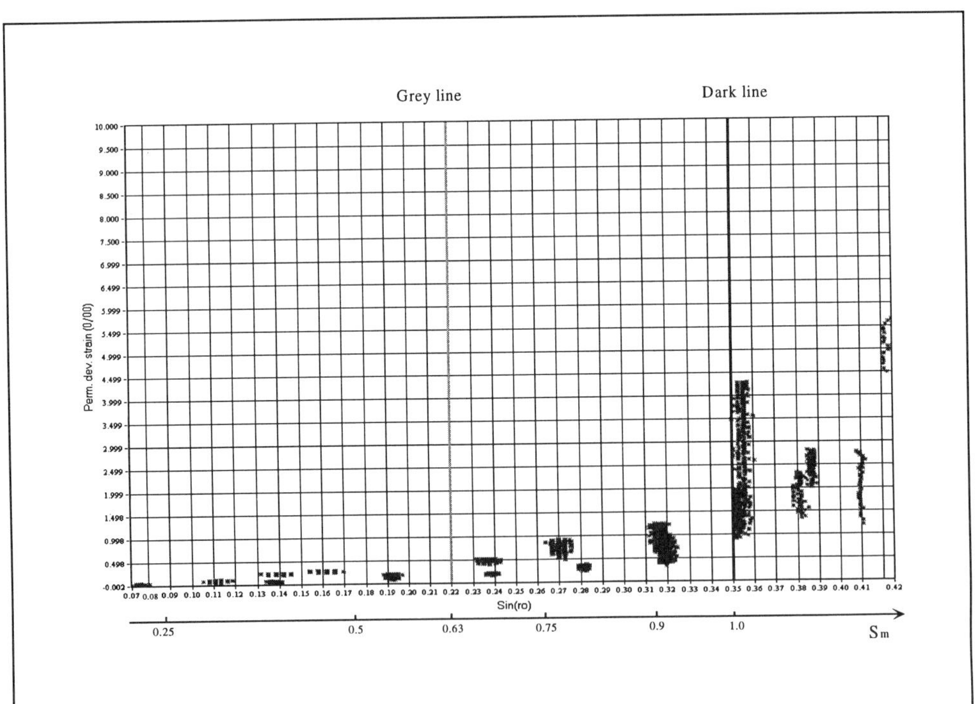

Figure 9 Permanent shear strain as a function of mobilised friction.

For this example the failure line is $\sin\rho_{failure} = \sin\phi = 0.35$ and the elastic limit line is $\sin\rho_{elastic} = 0.22$ and $S_m = \sin\rho_{elastic} / \sin\rho_{failure} = 0.63$.
Several complicated factors may strongly influence the development and amount of permanent deformations in a pavement. However, the results of the repeated load triaxial testing may first of all be used to assess the risk of developing excessive permanent deformations. Further it is believed that a rational prediction of the development of permanent deformation may also be obtained by the following design procedure:

1. Determine stress distribution and a corresponding map of degrees of mobilised shear strength using the computer program based on the new hyperelastic model. The material parameters should be derived from triaxial tests, but might also be obtained from a database with relevant specifications for recommended values.

2. Make an estimate of permanent deformations based on the degree of mobilised strength, and material parameters expressing the findings of figure 9. Some data do exist while much more testing and calibration work is needed to establish a table of recommended design parameters for typical materials.

3. Adjust the estimate for effects of the initial shake down under traffic, the sensitivity to rotation of induced traffic stress etc. based on field of full-scale test calibration data.

CONCLUSIONS AND RECOMMENDATIONS

Rational design of pavement structures must be based on a realistic analysis of stresses and strains induced by traffic loads. Frequently used computer programs based on linear elastic material models have serious shortcomings.

The authors propose the use of a new material model based on hyperelasticity for the unbound granular layers. This hyperelastic model takes account of the effects of dilatancy in well-compacted granular materials.

This hyperelastic model is believed to be an important contribution that will provide a basis for making more realistic stress analysis of pavement structures with unbound granular layers. The three necessary material parameters should preferably be determined by repeated load triaxial testing.

A draft design procedure for permanent deformations is proposed. It is based on the application of the concept of the degree of shear strength mobilisation, the use of the hyperelastic model and a procedure for repeated load triaxial testing.

More work is needed to complete the experimental basis for the hyperelastic model and to evaluate the implementation in various design procedures

ACKNOWLEDGEMENTS

The material model described in this paper has been developed during the Dr.Ing. study at NTNU sponsored by the Norwegian Public Roads Administration. The KPG-project, described by Hansen and Hansen /5/, has also financed some of the research work in this project.

REFERENCES

1 **Uzan, J.** Characterisation of Granular Material. Transportation Research Record. No 1022 pp 52 – 59.

2 **Uzan, J. Witczak, M. W., Scullion, T. And Lytton, R.L.** Development and Validation of Realistic Pavement Response Models. Proceeding of the 7th International Conference on Asphalt Pavements Vol. One (Design) pp 334 – 350

3 **Hoff, I. and Nordal, Rasmus. S.** Improvement of Equipment and Procedures for Triaxial Testing of Coarse Graded Aggregates. Proceeding, the 5th International Conference on the Bearing Capacity of Roads and Airfields.

4 **Bakløkk, L., Hoff, I. and Nordal, Rasmus. S.** Methods for Ranking Unbound Granular Materials for Pavement Construction. Proceeding, the 5th International Conference on the Bearing Capacity of Roads and Airfields.

5 **Hansen, E. K. and Hansen, G**. Quality of crushed rock aggregates for pavement structures. Proceeding, the 5th International Conference on the Bearing Capacity of Roads and Airfields.

6 **Nordal, Steinar** Lecture Notes for Ph.-D. Course in Soil Modeling.
Department of Geotechnical Engineering
NTNU 1994.

Fifth International Conference on the
BEARING CAPASITY OF
ROADS AND AIRFIELDS

QUALITY OF CRUSHED ROCK AGGREGATES FOR PAVEMENT STRUCTURES.

ERLING K. HANSEN, Project Manager,KPG
GUNNAR HANSEN, Chairman of Board, KPG

Arbeidsfellesskapet KPG
P.o.box 296, 1322 Høvik, Norway, Fax 63904603

ABSTRACT

Traditional Norwegian pavement structure design consists of subbase- and basecourse layers of unbound, mechanically stabilized gravel. Some areas are short of gravel of satisfactory quality and it is desirable to save available resources for other, more important purposes. Crushed rock materials have therefore more and more been used to substitute natural gravel in road pavement structures. According to the Norwegian Pavement Design Manual, however, crushed rock structures requires just the same as natural gravel. The crushed rock suppliers among many others, claim that crushed rock materials have many superior properties compared to natural gravel, such as better stability and bearing capability, especially in wet conditions. In order to document this, the producers in 1993 initiated a Joint Research Group - KPG, to carry out a study on the topic, including necessary laboratory and field tests. So far the project has resulted in a 7.5% theoretical reduction of necessary thickness by crushed rock basecourse layer in The Pavement Design Manual, and a 9.1% theoretical reduction in the subbase layer.
The remaining part of the study is concentrating on models for permanent deformation and pavement design.
The paper gives an overall picture of the project organization, the objectives and execution of the different tasks. The preliminary results so far will be presented in 8 separate papers in the Conference. Three Ph.D. candidates have been involved in different parts of the project. The project is planned to be completed by 1999.

Key words: Aggregate research, crushed rock, cyclic triaxial testing, permanent deformation.

INTRODUCTION

In the Norwegian Pavement Design Manual, the bearing capacity of crushed rock aggregates is ranked equal to that of gravel materials. Some crushed rock producers and users claim experience shows that crushed rock pavement structures produce less permanent deformation and exhibit better performance than gravel structures, especially in wet conditions. They claim crushed rock aggregates have at least 10% better bearing capacity than gravel materials.
To look closer into this assertion, 19 producers, 11 equipment dealers and 8 governmental agencies/technical organizations in 1993 established a Joint Research Group - KPG (Arbeidsfellesskapet KPG), and designed a research project on the topic. A research team and a research organization were established to perform the task.

OBJECTIVES

The overall goals of the research project for the crushed rock industry itself was:

1)To improve the road - and airfield pavement service life by better quality of the pavement structures by extensive use of crushed rock materials.
2)Increased usage of crushed rock materials in pavement structures in Norway and increased export for the same purpose to other countries.

The fulfillment of the goals should be reached through the following studies:

1) To develop or adopt laboratory test methods applicable to verify relevant properties, such as bearing capacity, stability and material coefficients of actual materials and layers in the pavement structure.
2) To run laboratory tests on typical material alternatives and field tests on corresponding structures.
3) Compare and verify the results from the tests by standard pavement structural design methods.
4) If possible, improve the capability of the design methods to incorporate and handle permanent deformations.
5) To produce information materials including a handbook on unbound materials in pavement structures.

PROJECT ORGANISATION

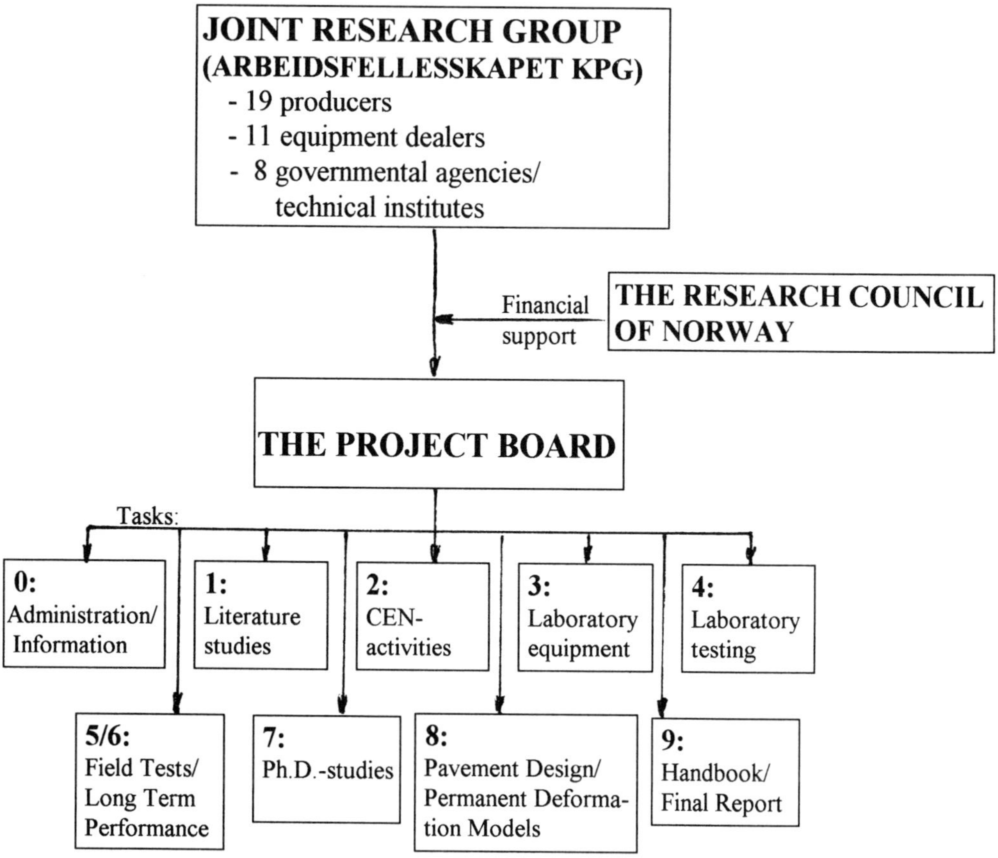

TASKS

Task 1, Literature studies: A thoroughly, international literature search on the field of interest was conducted by SINTEF - Civil and Environmental Engineering, Trondheim - and reported in the project's serial numbered reports. (1)

Task 2, CEN-activities: The main activity in this task has been to collect all available information relevant to the objectives of the project brought forward in the CEN-system (European Committee for Standardization), and to take this into account in the appropriate places in the project.

Task 3, Laboratory Equipment: The Los Angeles Test has never been a standard requirement in material testing in Norway. In an early stage of the project it become

clear that the Los Angeles Test Method was going to be adopted as standard within the CEN area. The project initiated an inter laboratory precision test on the method involving several material types and equipment makes (2).
Standardized methods for testing stability, bearing capacity and deformation resistance of coarse, unbound materials are also lacking. The project decided to choose dynamic triaxial tests as the basic test method for these properties. Equipment was designed and developed for material maximum sizes of 120 mm at the Norwegian Geotechnical Institute, NGI, Oslo (3).
Equipment and procedures for triaxial testing on smaller sized coarse Aggregates was developed by SINTEF (4).

Task 4, Laboratory testing: The geological composition, grading, mechanical properties etc. of all the material types in the project's program were tested by the established procedures, (1), (3), (4), (5).
As a tool for characterization of the mechanical strength of crushed rock materials, use of the ICT- Intensive Compaction Tester - was looked into in a separate study, (6), (7).
A small scale field test was undertaken to verify the laboratory test results in the field (8).
Dynamic triaxial tests was performed at NGI on samples of size 625 mm diameter and 1250 mm height, (3), (9), and on samples of size 300 mm diameter and 600 mm height, at Tampere University of Technology (4), (5).
Ranking pavement materials based on laboratory testing was object for another separate study, (10).

Task 5/6, Field Tests/ Long Term Performance: To construct field trial sections is important in order to:

1) Calibrate the methods developed for calculating pavement performance and service life.
2) Calibrate theories of material behavior and pavement designs.
3) Test certain material types and pavement designs.

Three road stretches and one airstrip section were constructed in connection to the project, all divided in several sub sections, (11), (12), (13), (14), (15).
These trial sections together with two earlier constructed trial sections are going to be monitored in several years, (16).

Task 7, Ph.D.- Studies: Three Ph.D.-candidates have been attached to the project. Special studies on mechanical properties of crushed rock materials together with stress, strains and deformation behavior, mechanisms and modeling have been elements in their programs supported by KPG. To date two of the candidates have completed their programs and earned their degrees, the third one are going to complete his study during this year, (17), (18) (19).

Task 8, Pavement Design/Permanent Deformation Models: Modern Pavement Structural Design Models are based on linear elastic layer theory. Input data comes from dynamic triaxial tests. In many cases the permanent deformation of the pavement surface rather than the fatigue represent the main cause of pavement failures. Permanent deformations rarely goes into the pavement design models. The task is aiming for an improved pavement design model incorporating permanent deformations. The NOAH Pavement Design Model has been chosen as the basic model for eventually improvements, (20), (21), (22) (23).

Task 9, Handbook/Final Report: The project is to be completed by a Handbook titled «Use of Unbound Crushed Rock Materials in Pavement Structures» and a Finale Report. The Handbook is to be published by this year. The Public Roads Administration is going to incorporate the Handbook in their own series of Technical Guidance. The Final Report is due at the completion of the project in 1999.

REFERENCES

(*= Paper on this Conferense)

1. BAKLØKK, L.J.: Steinmaterialer i vegkonstruksjoner. Litteraturstudium. Delprosjektrapport KPG 03, Arbeidsfellesskapet KPG, Oslo, 1995, (In Norwegian only).
2. NATVIK, J.: Orienterende ringtest Los Angeles. Delprosjektrapport KPG 01, Arbeidsfellesskapet KPG, Oslo 1995, (In Norwegian only).
3. VIK, G.: Sykliske Vakuumtreaksialforsøk på pukk og grus. Delprosjektrapport KPG 04, Arbeidsfellesskapet KPG, Oslo 1996, (In Norwegian only).
4. KOLISOJA, P.: Large Scale Dynamic Triaxial Tests I. Delprosjektrapport KPG 05, Arbeidsfellesskapet KPG, Oslo 1996.
5. KOLISOJA, P.: Large Scale Dynamic Triaxial Tests II. Delprosjektrapport KPG 07, Arbeidsfellesskapet KPG, Oslo 1996.
6. NATVIK, J.: Bruk av ITC (Intensive Compaction Tester) ved bestemmelse av steinmaterialers styrke. Delprosjektrapport KPG 10, Arbeidsfellesskapet KPG, Oslo 1997, (In Norwegian only).
7. *NATVIK, J.: Mechanical Properties of Crushed Rock Aggregates. Proceedings of the Fifth International Conference on the Bearing Capacity of Roads and Airfields, Trondheim, Norway, 6 - 8 July 1998.
8. NATVIK, J.: Feltforsøk i Vassfjell pukkverk. Delprosjektrapport KPG 13, Arbeidsfellesskapet KPG, Oslo 1997, (In Norwegian only).
9. *VIK,G., Holme, J., Høeg, K.: Cyclic Triaxial Tests on Gravel and Crushed Aggregates with 120 mm Maximum Grain Size. Proceedings of the Fifth International Conference on the Bearing Capacity of Roads and Airfields, Trondheim, Norway, 6 - 8 July 1998.
10. *BAKLØKK, L.J., HOFF, I., NORDAL, R.S.: Methods for Ranking Materials for Pavement Structures based on Triaxial-, CBR- and Gyratory Testing.

Proceedings of the Fifth International Conference on the Bearing Capacity of Roads and Airfields, Trondheim, Norway, 6 - 8 July 1998.
11. HAUSTVEIT, T.: Forsøksfelt Rv. 33, Lillo - Opsal, Oppland. ANLEGGSRAPPORT. Delprosjektrapport KPG 02, Arbeidsfellesskapet KPG, Oslo 1998, (In Norwegian only).
12. TRENGEREID. T.: Forsøksfelt Rv. 01, Matre - Nipetjønn, Hordaland. ANLEGGSRAPPORT. Delprosjektrapport KPG 08, Arbeidsfellesskapet KPG, Oslo 1998, (In Norwegian only).
13. NÅLSUND, R.: Forsøksfelt E 39, Bjørkøybekk på Hemnekjølen, Sør-Trøndelag. ANLEGGSRAPPORT. Delprosjektrapport KPG 09, Arbeidsfellesskapet KPG, Oslo 1998, (In Norwegian only).
14. LANGE, G.: Forsøksfelt på Gardermoen. Delprosjekt KPG 11, Arbeidsfellesskapet KPG, Oslo 1997, (In Norwegian only).
15. *LANGE, G.: Full Scale Test on Airfield Pavement Structure. Proceedings of the Fifth International Conference on the Bearing Capacity of Roads and Airfields, Trondheim, Norway, 6 - 8 July 1998.
16. *HAUSTVEIT, T., NÅLSUND, R., TRENGEREID, T., EVENSEN, R.: Crushed Rock Pavement Structure Trial Sections. Proceedings of the Fifth International Conference on the Bearing Capacity of Roads and Airfields, Trondheim, Norway, 6 - 8 July 1998.
17. KOLISOJA, P.: Resilient Deformation Characteristics of Granular Materials. Thesis for the Doctor of Technonlogy. Delprosjektrapport KPG 14, Arbeidsfellesskapet KPG, Oslo 1998.
18. NATVIK, J.: Vurdering av alternative testmetoder i undersøkelse av tilslagsmaterialer. Dr.ing.- avhandling. Delprosjektrapport KPG 15, Arbeidsfellesskapet KPG, Oslo 1998, (In Norwegian only).
19. HOFF, I.: Material Properties of Unbound Aggregates for Pavement Structures. Ph.D.-Thesis. Delprosjektrapport KPG 16. Arbeidsfellesskapet KPG, Oslo 1998.
20. HOLME, J.K.: Sykliske vakuumtriaksialforsøk på pukk og grus. Delprosjektrapport KPG 06, Arbeidsfellesskapet KPG, Oslo 1996, (In Norwegian only).
21. *KOLISOJA, P.: Deformation Behavior of Aggregates Determined by Means of Large Scale Cyclic Loading Triaxial Tests. Proceedings of the Fifth International Conference on the Bearing Capacity of Roads and Airfields, Trondheim, Norway, 6 - 8 July 1998.
22. *HOFF, I., NORDAL, S., NORDAL, R.S.: Predicting Safety Against Permanent Deformation Based on Triaxial Testing Using a New Material Model for Calculating Stresses in Pavement Structures. Proceedings of the Fifth International Conference on the Bearing Capacity of Roads and Airfields, Trondheim, Norway, 6 - 8 July 1998.
23. *ECKMANN, B.: Noah Software. Proceedings of the Fifth International Conference on the Bearing Capacity of Roads and Airfields, Trondheim, Norway, 6 - 8 July 1998.

Fifth International Conference on the
BEARING CAPASITY OF
ROADS AND AIRFIELDS

MECHANICAL PROPERTIES OF CRUSHED ROCK AGGREGATES

JOHN NATVIK

Norwegian University of Science and Technology, 7034 Trondheim, Norway, john.natvik@geo.ntnu.no

ABSTRACT

The paper describes and presents the results of laboratory and field tests of the mechanical strength of aggregates. In this work Intensive compaction testing (ICT) has been evaluated as an alternative test for determination of aggregates resistance to fragmentation. The laboratory results are evaluated by a field study. In this field study the fragmentation of aggregates in a road construction is defined.

The results from the laboratory testing show that Intensive compaction testing is well suited for determining aggregates resistance to fragmentation, and that it is sensitive to small differences in quality. The fragmentation in ICT depend on a number of parameters and these parameters are examined in the laboratory. The field test was carried out with 3 different materials and properties of both open (0-120 mm) and closed grading (0-32 mm). The results show that aggregate is fragmented dued to traffic load. In the section with open grading the aggregates shows different resistance to fragmentation. The results indicates a strong connection between observed fragmentation and the strength determinated in the laboratory, especially to the results from the Nordic abrasion test and the Norwegian abrasion value. The connection with the results from ICT is also strong. The Los Angeles test were not able to separate the materials as desired by results. Results from the section of closed grading do not indicate such a strong connection as the open grading. The resistance to fragmentation are almost the same for the aggregates. The strength of the aggregates, expressed by the results from the impact tests and ICT, are also almost the same. This lack of difference in strength makes it impossible to evaluate the connection between the results from ICT and the fragmentation in the road construction for this fraction.

Keywords: aggregate, mechanical strength, ICT.

3.3.2 Results, fraction 0-32 mm

The results of this fraction also indicates fragmentation. The amount passing the sieve size 16 mm is not increasing the first week, and it is a big difference in results for the to wheel lines (se figure 4). Calculated total increase (for 4 weeks) for the wheel line P1 were 7.8 to 8.9 weight %, and for P2 2.7 to 6.2 weight %. The amount of material passing the sieve size 75 μm was not increasing much the first week, except for the sample gravel 1. From 1 to 4 weeks there was an increase in amount < 75 μm. The total increase were from 1.3 to 2.1 weight %. The differences in fragmentation for the 3 materials are small.

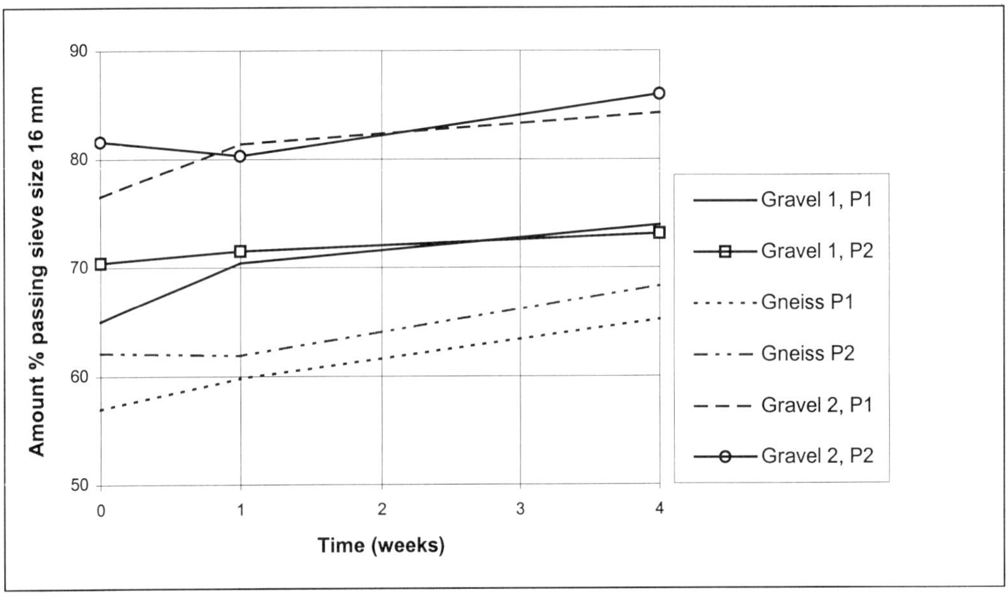

Figure 4: Amount (weight %) passing sieve size with opening 16 mm. Results from the fraction 0-32 mm.

3.4 Traffic load

The traffic load is evaluated by the maximum vertical stresses. These analysis is conducted for the main lorry used in the study for a defined road construction. Chevron analysis are used to calculate the vertical stresses, and the results indicates that the stresses are not as high as the stresses calculated for a worst case, but after all big. The vertical stresses were calculated to be 0.798 MPa on the front axle, and the maximum in the Norwegian standard for this lorry is 0.857 MPa. The total number of lorries passing the test road was 2881 /3/.

3.5 Discussion

This field study was done with only 3 different materials for each fraction. This makes it difficult to draw conclusions, but the results can be used as indications.

The results from the field study indicated fragmentation of aggregates dued to traffic load in a road construction. The traffic load expressed by vertical stresses are calculated to be high in the construction.

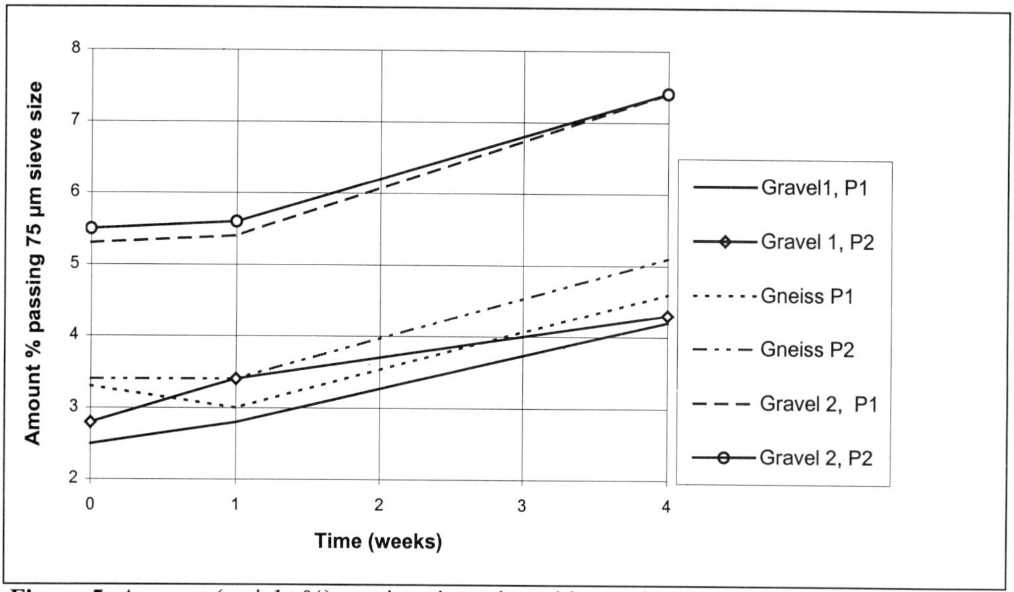

Figure 5: Amount (weight %) passing sieve size with opening 75 µm. Results from the fraction 0-32 mm.

In fraction 20-120 mm the increase in amount of material passing the sieve size 16 mm was 3.2 to 6.4 weight % after 4 weeks of traffic load. The increase in material passing the sieve size 75 µm was from 0.5 to 1.4 weight %. Regression analysis with results from this field study show a strong connection between results from laboratory tests and observed fragmentation in the road. Statistically the connection is strongest to the results from Nordic ballmill test and the Norwegian abrasion test. The connection is also strong to results from ICT, but the Los Angeles test were not able to separate the materials as desired by results.

For the fraction 0-32 mm the observed fragmentation showed a variation in results, and only small differences between the 3 tested materials. The increase in material passing the sieve size 16 mm was from 2.7 to 8.9 weight %, and the increase in material passing the sieve size 75 µm was 1.3 to 2.1 weight % after 4 weeks %. Regression analysis with results from this field study shows no strong connection between results from laboratory tests and observed fragmentation in road. The results showed only small differences in strength for the 3 materials as observed in the field study, but the connection is statistically weak. The strength of the aggregates, expressed by the results from the impact test and ICT, are also almost the same. This lack of difference in strength makes it impossible to evaluate the connection between the results from ICT and the fragmentation in the road construction. Separation in test materials may be the main cause for the variation in results.

4. CONCLUSIONS

Intensive compaction testing (ICT) is examined as an alternative test for determining aggregates resistance to fragmentation. The results show that Intensive compaction testing is well suited for this purpose, and that it is sensitive to small quality differences. Results from the fraction 8-14 mm show that ICT is able to separate samples better than standard aggregate tests. The results also show a strong connection to results from aggregate tests dominated by impact force. There are no such strong connection between results from the fraction 0-16 mm and results from standard aggregate tests. The fragmentation in ICT is depending by a number of parameters. The main parameters are pressure, angel of gyration, number of work cycles, water content, particle shape and particle size.

The connection between laboratory results from mechanical strength tests and results from the field, are studied by examining of the fragmentation in a road construction. Materials of both open (0-120 mm) and closed grading (0-32 mm) were examined. The results show that aggregate is fragmented dued to traffic load. In the section with open grading the aggregates showed different resistance to fragmentation. The results indicates a strong connection between observed fragmentation and the strength determinated in the laboratory, especially to the results from the Nordic abrasion test and the Norwegian abrasion value. The connection with the ICT-test is also strong. The Los Angeles test were not able to separate the materials as desired by results. Results from the section of closed grading do not indicate such a strong connection as the open grading. The resistance to fragmentation are almost the same for the aggregates.

5. REFERANSER

1. NATVIK, J. ICT (Intensive compaction testing) as a alternative test for determination of aggregates resistance to fragmentation, KPG project report number. 10, Department of Geology and Mineral Resources Engineering, NTNU, Trondheim, 1997 (In Norwegian).

2. NATVIK, J. The influence of blasting on aggregate quality, project report, Department of Geology and Mineral Resources Engineering, NTNU, Trondheim, 1997 (In Norwegian).

3. NATVIK, J. Field study. An evaluation of laboratory measured mechanical strength, KPG project report number. 13, Department of Geology and Mineral Resources Engineering, NTNU, Trondheim, 1997 (In Norwegian).

Fifth International Conference on the
BEARING CAPASITY OF
ROADS AND AIRFIELDS

Cyclic Triaxial Tests on Gravel and Crushed Aggregates with 120 mm Maximum Grain Size

By Gunnar[1] Vik, Jan K. Holme[1] and Kaare Høeg[2]

ABSTRACT

This article describes the equipment and laboratory procedures used for performing large scale cyclic triaxial tests on gravel and crushed rock aggregates with 120 mm maximum grain size. The six specimens tested had a diameter of 62,5 cm and a height of 125 cm. The confining stress was kept constant at 80 kPa during the testing by evacuating the inside of the specimen. Each specimen (except one) was tested in three sequences; phase 1 - moisture equal to construction practice; phase 2 - specimen soaked by water; and phase 3 - free water drained from specimen. The testing comprised packages of 1000 cycles at each deviatoric stress level which then was raised by an increment of 20 or 40 kPa to the next stress level. Some of the results are presented, and a comparison is made among the behaviour of the six different materials tested with respect to cyclic modulus (resilient modulus) and accumulated permanent deformations during the cyclic loading.

1. Introduction

In 1993 The Norwegian Aggregate Producer's Association together with the main user group (KPG) with financial support from the Research Council of Norway, started a research program on the use of crushed rock aggregates for road construction. One of the main project aims was to study how the different materials behaved under the cyclic loading immediately below a bituminous road pavement. One was especially interested in the behaviour of coarse aggregates with maximum grain size of 120 mm, for which very little laboratory data exists. NGI has previous experience from static triaxial tests on similar rock aggregates in connection with rockfill dam construction, and now the equipment was modified to enable cyclic testing of large cylindrical specimens, nominally 125 cm in height and 62,5 cm in diameter. The tests were performed in two series (1994 and 1995) with 3 tests each year.

1) Norwegian Geotechnical Institute (NGI), Oslo, Norway
2) University of Oslo and NGI

2. Six Materials Tested

The testing program included 6 different materials, two natural gneiss gravel materials (A and D) and four crushed rock aggregates (B, C, E and F) shown in Fig.1. Natural gravel material D was produced from A by removing some sand to produce a grain size distribution curve in accordance with the guidelines given by the Norwegian Public Road Administration. Figure 1 shows the grain size distributions for all materials. Crushed rock material B had a very open grading, 25-50 mm, similar to railroad aggregates, and material E had grading 20-120 mm. Materials C and F had both grading 0-120 mm, but were taken from different sites, as material C was made from biotic gneiss, and materials B, E and F were of granitic gneiss origin.

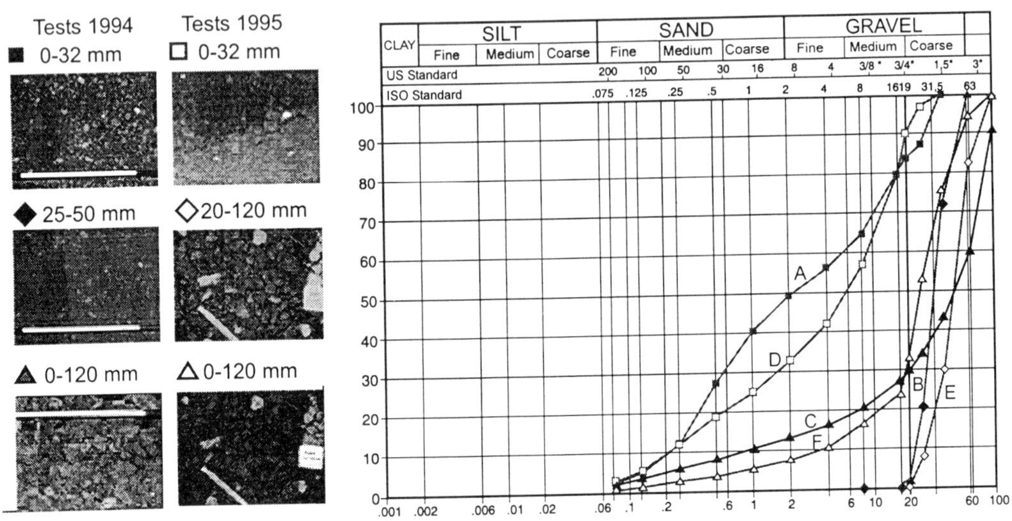

Fig. 1. Grain size distribution curves for the six different materials tested.

3. Testing Program

As the size of the specimens were so big (125 cm high and 62,5 cm diameter), the time spent on handling the material during specimen preparation, equipment assembly and specimen retrieval after testing, had major influence on testing costs.

3.1 Test Phases

To get as much information as possible out of each test specimen, each was tested in three phases with one phase following the other:

Phase 1: Material compacted with a water content equal to what would be the case during road construction, i.e. for gravel and well graded rock materials this refers to

Proctor Modified Optimum water content minus 1 per cent. For the open rock materials where the fines were removed, the rock particles were sprayed with water during compaction (materials B and E).

Phase 2: Water was introduced into the specimen to simulate the near saturation situation that occurs during spring thaw, when the drainage paths are blocked by ice and the bituminous road pavement collects sun radiation.

Phase 3: Drainage of all free water in the specimen to simulate the behaviour of a «mature road» base.

3.2 Applied Cyclic Loads

The loading period of 1 second consisted of a cosine loading pulse from 0 to maximum stress and back to 0 in 0.1 second followed by a resting period of 0.9 seconds as shown in Fig. 2. The testing in each phase comprised cyclic load packages of 1000 cycles at each deviatoric (σ_1 - σ_3) stress level which was then increased by 20 or 40 kPa to the next stress level. Between each package of 1000 cycles, 10 cycles of 20 kPa were run.

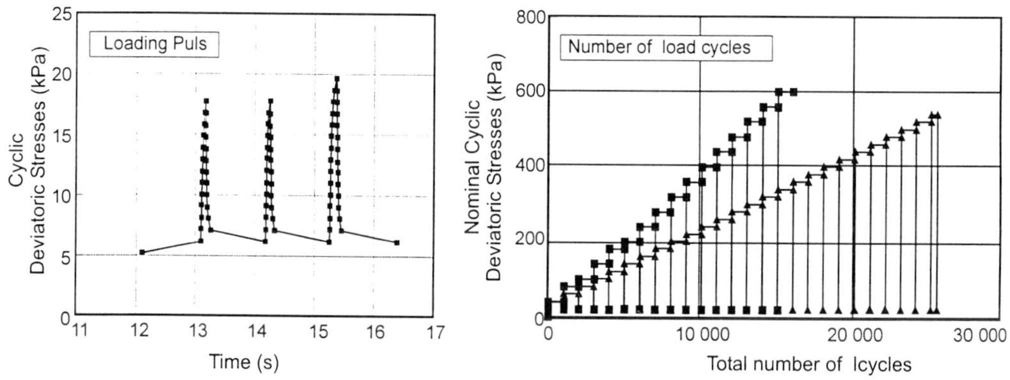

Fig. 2 Applied cyclic loads

For materials A, B and C (tested in 1994), the testing started with 10 cycles of 20 kPa deviatoric stress followed by 1000 cycles of 40 kPa deviatoric stress. The increment in deviatoric stress between each package of 1000 cycles was 20 kPa. Each phase was taken to 2.5% axial strain or 600 kPa maximum deviatoric stress.

For the tests performed in 1995, specimens D, E and F, a static triaxial test (monotonic loading) was performed before and after the cyclic testing program. The static test was unloaded after 1% axial strain was achieved, and then the cyclic testing was done in the same manner as for materials A-C, except that the increment in deviatoric stress between each package of 1000 cycles was 40 instead of 20 kPa to cut testing time. Each phase was taken to 1.5% axial strain or 600 kPa maximum deviatoric stress.

4. Specimen Preparation

The most demanding task in preparing large quantities of coarse grained aggregates is avoiding separation (segregation). Great care was therefore taken in order to reduce that problem as much as possible. The specimen preparation started with assuming the specimen density which would be achieved after compaction, and calculating the amount of material needed. The basis for this assumption was the natural gravel and well graded rock fill Modified Proctor Compaction tests on material < 32 mm. For each specimen the allocated material was mixed on the floor and split into eight batches, one for each layer to be compacted in the steel mould used to prepare the specimen. The water contents of the gravel materials were checked and water added to each layer the day before specimen compaction in an attempt to ensure homogeneity.

For specimen compaction one used a 4-column steel support frame and a split steel cylinder with overlapping external flanges for joining the two halves. First the bottom filter of the specimen was placed upon a centralising ring in the support frame. Then a custom made rubber membrane of 2,5 mm thick neoprene was fixed to the bottom filter. The filters were made by filling machined sections of the steel end platens with a mixture of epoxy and carborundum particles. Then the two halves of the steel scaffolding were erected and bolted to the support frame. The rubber membrane inside the cylinder was folded over the top edge.

Fig. 3 A test specimen in the split steel cylinder

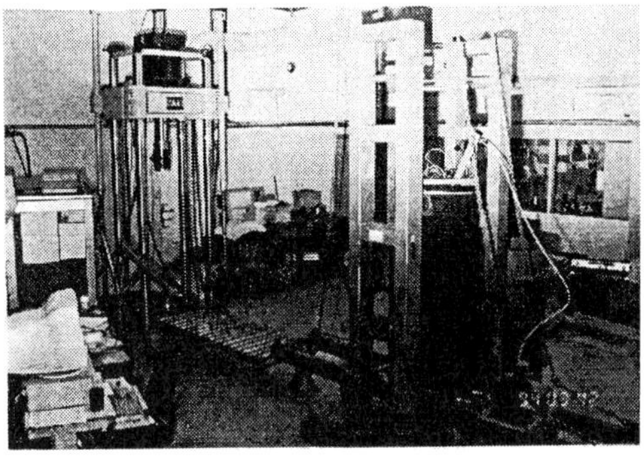

Fig. 4 An overview of the compaction frame with the finished specimen ready to be transferred by crane to the roller track and into the MTS testing machine.

Each of the eight specimen layers of material was levelled and compacted for a specified time. The compaction unit used was a 1 m long steel piston with an end surface diameter of 12 cm and a scaffold vibrator motor attached to the top. The open rock aggregate materials were washed with 5 litres of water on top of each layer with water draining freely from the bottom filter. The vibration duration varied from the bottom to the top layer in accordance with the experience that compaction of any layer also will influence the density of the layers below. Thus, the specimens A-C received 7 to 15 minutes at each layer from bottom to top, while the specimens D-F received 15-29 minutes of compaction work per layer.

Normally, the final layer had to be levelled with some extra material, or reduced according to the achieved dry density of the specimen. After final compaction the top cap was placed, and the rubber membrane was fixed to the top cap by hose clamps.

Consolidation started by closing all drainage and connecting the vacuum tube to the specimen top. Dial gages were used to check the consolidation process, and after achieving a consolidation stress of 50 kPa, the split scaffolds were taken off. After repairing any damages and leakages in the rubber membrane, the vacuum was increased to 80 kPa. A portal crane was attached to the top cap, and with the help of vacuum the whole specimen was transferred onto steel rails resting on rollers and rolled into position inside the MTS load frame. The MTS-piston was attached to the top filter plate, and the specimen was lifted up using vacuum. Then the roller table assembly was removed so the specimen could be lowered down to its final position on the MTS compression plate.

The initial measurements of consolidated specimen height and circumference were taken, instrumentation mounted and checked and testing started. Normally instrumentation and static testing (materials D-F) would take the first day. Test phase 1 would occupy day 2. The saturation process for phase 2 started immediately after finishing phase 1. Water was introducing into the specimen through the lower filter plate keeping a vacuum at the top filter. The vacuum was connected through a series of tanks. As water broke through in the upper filter, the water supply to the bottom filter was stopped. The volume of water introduced to the specimen was recorded during the process. The specimen would then be ready for phase 2 on day 3. After the end of testing phase 2, the storage tanks in the vacuum system were used to drain all free water in the specimen before starting the test in phase 3 with drained condition on day 4. Volume of drained water was recorded for calculation of saturation and porosity. Finally on day 5, the static test after cyclic loading (materials D-F) was performed, and the specimen dismantled.

5. Testing Equipment and Instrumentation

The tests were performed with a servo-controlled universal MTS testing machine. The MTS Teststar system was used for generating the loading pulses, counting number of pulses and data logging of 15 variables. The horizontal confining stress (σ_3) was generated by applying vacuum inside the specimens, using a compressed air-driven ejector pump which tolerated the moisture from the specimen. The compressor was supplied with a large storage tank and a diesel generator in case of power failure.

With the coarse grained materials tested, no reliable information on local cyclic lateral specimen strains could be expected, as the coarse particles would tend to rotate as well as displace against the rubber membrane. Therefore, priority was given to the measurement of vertical deformation, both by global and local dynamic measurements.

The local dynamic gauges consisted of two small LVDTs measuring the change of height over the central half of the specimen, one on each side of the specimen. The total measurement range for these transducers was +/- 1 mm, and as they approached the end of range due to permanent deformations during the cyclic loading packages, they were manually adjusted into range.

The global measurements were also over the central half of the specimen, but had a much larger range. The instruments were of the string potentiometer type with a range of 50 mm displacement. The instrument housing was attached to the base of the testing machine, and the strings were attached to extension wires up to the anchor points glued to the specimen membrane. Each measuring section consisted of two instruments, one attached at the upper ¼ point of the specimen and one attached at the lower ¼ point. The differential signal between these instruments gives the change in height over the middle half of the specimen. The overall change in specimen height was also monitored with the LVDT in the piston of the testing machine.

Load was measured with the MTS load cell between the top cap and the piston. The operating range of the machine was changed as the stresses increased to get good control in all parts of the tests.

All data were logged through MTS Teststar software, which was also used to set up testing parameters. However, logging and storing of data took significant computer processing time and affected the shape of the loading signal, especially the length of resting period between pulses. After a few trial cycles the parameters defining the wave shape were modified to keep the desired loading pulse.

To limit the amount of data, the logging was triggered by increments in deviatoric stress. The number of points logged per cycle was ≈ 25, and the increment in stress triggering data storage was 1/12 of the maximum deviatoric stress for each cycle. Still, one day of testing created easily 80 MB of data, as each of the cycles was logged.

6. Test Results

6.1 Cyclic E-modulus

In Fig. 5 the mean cyclic E-modulus (resilient modulus) for cycles 100-105 at each stress level is plotted versus the applied deviatoric stress.

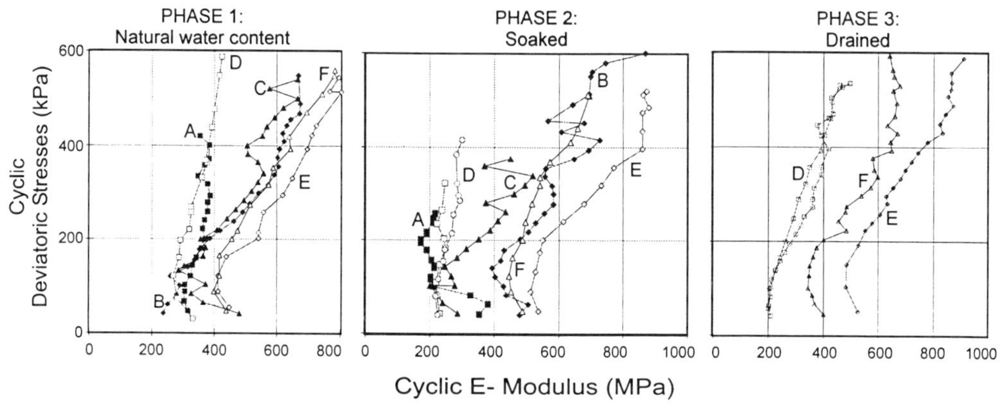

Fig. 5 *Mean cyclics E-modulus for cycles 100-105 at each stress level as a function of applied deviatoric stress for all three test phases and for all six materials (for legend, see Fig. 1).*

6.2 Permanent deformations

The development of permanent strain as function of deviator stress and number of load cycles is shown in Fig. 6 below.

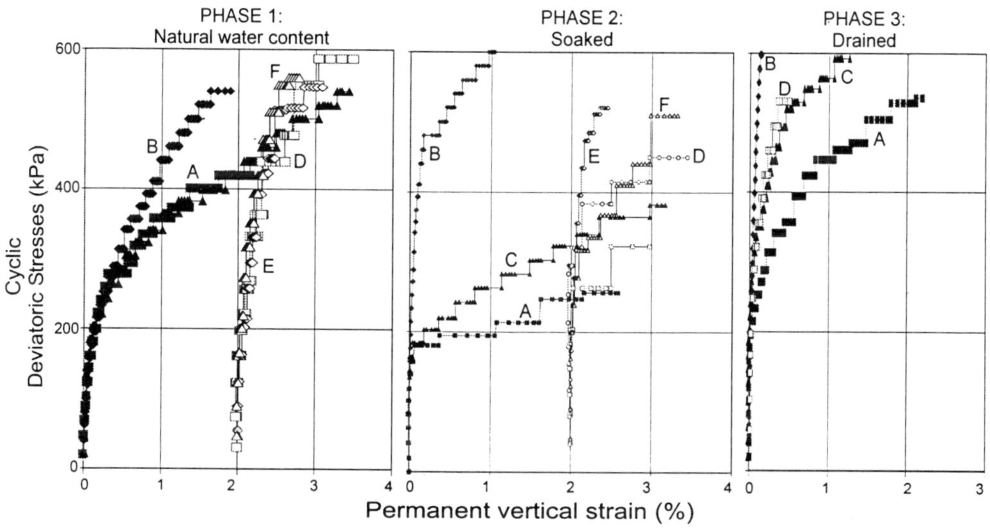

Fig. 6 Development of permanent strain as a function of deviator stress and number of cycles. At each stress level the permanent strain has been recorded after 200, 400, 600, 800 and 1000 cycles.

6.3 Discussion of results

The tests show that all materials develop higher cyclic E-modulus as the applied deviatoric stress increases. It should be noted that as the axial deviatoric stress increases, the mean stress also increases in these tests. From Fig. 5 it is apparent that the natural gravel materials A and D have lower moduli than the crushed rock aggregates at all stress levels, and that the differences in E-moduli between the two groups of materials increases as the applied stresses are increased. It can also be noted from Fig. 5 that the weakest of the rock aggregates is material C (0-120 mm), the biotic gneiss rock with a rather low static compressive strength, which produces a large fraction of fines during the crushing process The stiffest material is material B, fine grained granitic gneiss with grain size 25-50 mm.

The materials with substantial amounts of fines suffer the largest accumulated permanent deformations due to cyclic loading. In phase 2 (near saturated material) it may be implied from Fig. 6 that the development of internal pore pressure leads to a reduction of effective stress inside specimen A, and to some extent also in specimens C and D.

The material with the smallest permanent deformation due to cyclic loading was the open crushed rock aggregates represented by materials B and E (Fig 6).

In general, the crushed materials with open grading exhibit better properties in cyclic loading than the natural gravel and the crushed materials with fines.

Acknowledgement

The authors thank the Norwegian Aggregate Producer's Association, the main user group, KPG, and the Research Council of Norweay for their financial support and permission to publish the results of the work.

Referances

Andersen, K. H. og Høeg, K. (1994) Deformations of soils and displacements of structures subjected to combined static and cyclic loads, Proc. X ECSMFE, Vol 4 pp. 1147-1158, Florence, Italy.

Boyce, J.R.(1980) A non-linear model for elastic behaviour of granular materials under repeated loading. Int. Symp. on soils under Cyclic and Transient Loading, pp. 285-294, Swansea, UK.

Dawson, A.R., Correia, A.G., Jouve, P., Paute, J-L. and Galjaard, P.J (1994) Modelling resilient and permanent deflections in granular and soil pavement layers. Proc. 4. Int.Conf. on the Bearing Capacity of Roads and Airfields, Vol. 2 , pp 847-861 Minnesota, USA.

Kolisoja, P (1995) Large scale dynamic triaxial tests for Arbeidsfellesskapet KPG. Results of the SHRP tests. Tampere University of Technology, Laboratory of Geotechnical Engineering.

P. Kolisoja (1995) Large Scale Dynamic Triaxial Tests for Arbiedsfelleskapet KPG. Results of the SHRP tests. Laboratory report from Tampere University of Technology, Finland Laboratory of Geotechnical Engineering.

AASTHO T 294-92 I. Interim Method of Test for Resilient Modulus of Unbound Granular Base/Subbase Materials and Subbase Soils - SHRP Protocol P46. The American Association of State Highway and Transportation Officals. Washington D.C., 1992.

NGI project report no 940046-1: Triaxial tests on quarried and crushed rock (0 - 120 mm) subjected to cyclic loading. In Norwegian. NGI, 29 November 1994.

NGI project report no 531001-1:Triaxial tests with cyclic loading of natural gravel (0-32 mm). In Norwegian, NGI 4 April 1995.

NGI project report no 531001-2: Triaxial tests with cyclic loading of crushed rock aggregate (25 - 50 mm). In Norwegian, NGI 17 March 1995.
NGI project report no 531001-3: Triaxial tests with cyclic loading of three different gravel and aggregates. In Norwegian, NGI 12 February 1995.

NGI project report no 531001-4: Summary report. Cyclic vacuum triaxial testing of six different gravel materials and crushed aggregates. In Norwegian, NGI 1 February 1996.

Fifth International Conference on the
BEARING CAPASITY OF
ROADS AND AIRFIELDS

A STUDY TO IMPROVE EQUIPMENT AND PROCEDURES FOR TRIAXIAL TESTING OF COARSE GRAINED GRANULAR MATERIALS.

INGE HOFF [1] AND RASMUS S. NORDAL[2]

[1] Research Fellow, Norwegian University of Science and Technology (NTNU)
[2] Professor Emeritus, Norwegian University of Science and Technology
N – 7034 Trondheim, Norway

ABSTRACT

A new procedure and some equipment are developed to remove a sample of coarse grained materials from the compaction mould and to encapsulate it in a rubber membrane for triaxial testing. The samples seem to maintain the structure as compacted quite well.

The repeated load triaxial equipment used at NTNU/SINTEF has been modernised and improved. The equipment for registration and control has been changed and the measurement results are now more reliable.

A new loading procedure is developed to provide parameters for a new hyperelastic material model for the behaviour of granular materials. This model is believed to give a more realistic estimation of stresses and than obtained by the linear elastic model and it provides an improved basis for the prediction of permanent deformation in granular layers of pavements

Key words: Triaxial testing, Equipment, Unbound granular materials, Coarse grained.

INTRODUCTION

Repeated load triaxial testing to study the behaviour of granular materials in base- and sub-base courses has been carried out at several universities and research laboratories for a long time. This has lead to the development of a large variety of equipment and methods for

sample preparation, instrumentation and loading procedures. Steps to standardise triaxial testing methods are now being taken by AASHTO and CEN /1/. This may lead to the use of more standardised triaxial testing methods and procedures.
There will always be limitations in the capabilities of triaxial testing to reproduce the real behaviour of granular materials, subjected to the stress regimes induced by traffic loads. However, it is important to be able to produce samples that are highly representative for field conditions and to detect their behaviour under the most representative stress conditions generally obtainable by triaxial testing. Hence, there is still a need for further development of the triaxial testing for the study of the behaviour of layered pavement structures under traffic loads.

In this paper a brief report is given on some work in this area covering sample preparation, instrumentation and loading procedures with some conclusions and recommendations

The equipment and procedures presented in this paper will be described in more detail in a Dr.Ing. -thesis /2/, and some results from testing will be presented in another paper at the BCRA-conference /3/

SAMPLE PREPARATION

Figure 1 Gyrator compactor ICT 150 RD

Compaction

A well-suited compaction method is important to obtain samples with a density and a structure that are representative of actual field conditions. Various compaction methods based on static loading, pounding, kneading, vibrating and various combinations of these techniques have been utilized.
In this study we have used a gyratory compactor, ICT – 150 RD from Invelop oy, Finland. Figure 1 shows a picture of the gyratory compactor with a sample ready for compaction. The main reason for using the gyratory compactor for preparing samples for triaxial testing is the shear movements that the material experiences in the gyratory compaction. This movements is meant to be somewhat similar to the kneading actions under a moving roller in the field. Further the problem with separation during compaction is smaller than by use of vibrations. Some information about the material is also obtained through the compaction process by measuring the shear resistance and observing the development of density during the compaction process. The gyrator operates fast and it is easy and simple to use. It is very flexible and the compaction parameters can easily be adjusted.

This compaction method does have some drawbacks as well: the shear deformation varies from the periphery to centre of the sample; the density may vary with height, being lower in the central part of the sample; and the recommended 2:1 ratio of height to diameter for triaxial samples could not be obtained by this standard equipment. We have tried to solve the latter problem by preparing low friction ends as described below.

Transfer of sample from the compaction mould to the triaxial chamber

Most of our samples tested were made of coarse-grained materials and would fall apart outside the compaction mould if they were not confined in a rubber membrane under internal vacuum or external pressure. It is not possible to preserve a well functioning rubber membrane inside the mould during compaction. It will be penetrated by coarse angular grains after a few gyrations. To enable the transfer of the sample from the steel mould into a rubber membrane with top and bottom platens and place it in the triaxial apparatus, a special technique was developed.

The mould is fixed in the frame that is shown in Figure 2 and vacuum is applied. The sample is then slowly raised by a screw and the rubber membrane slides onto the sample as it raises. The bottom plate is supplied with an O-ring that is pressing airtight against the cylinder wall. This assures that the internal vacuum is maintained all through this process.

This method for removing of the sample from the compaction mould seems to entail very limited disturbance. Since the sample is kept under pressure all the time, the dry-density and the structure that the material has obtained during compaction is preserved until the sample is placed in the triaxial apparatus ready to be tested.

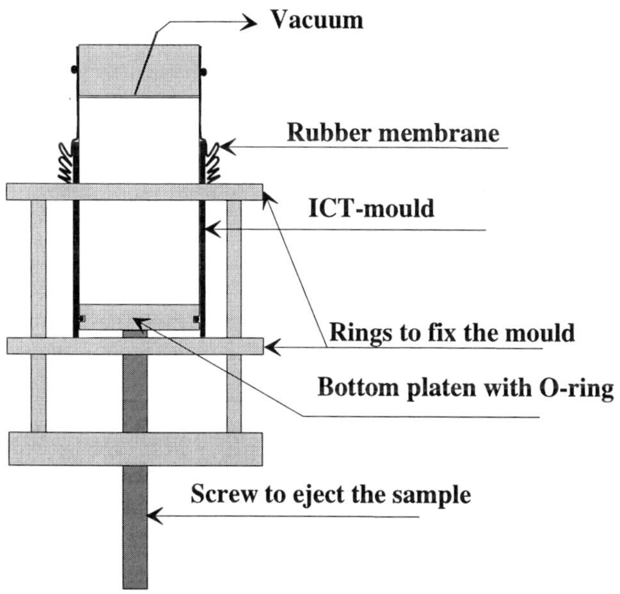

Figure 2 Sketch of equipment for removing of the sample from the compaction mould and encapsulating it in a rubber membrane under vacuum.

This method is practical, quite fast and very reliable. It is a bit laboriously to screw the sample manually. If a large number of tests are to be performed, motorization should be considered.

We have used the same technique for the new 300 and 400 mm triaxial testing equipment that is being built. However, these large samples have been ejected by a hydraulic actuator instead of by the screw shown here.

Low-friction ends

Samples from the gyratory compactor with a diameter of 150 mm can not be made higher than about 200 mm with the standard equipment. Usually a 2:1 relation between height and diameter is recommended for triaxial samples. This means that steps have to be taken to reduce end friction and thus obtain a practically uniform stress distribution in the entire sample during testing.

Figure 3 shows a sketch of the smoothening system used. For the coarse grained samples it was necessary to use several layers built up as an inverted filter with the coarser materials towards the sample.

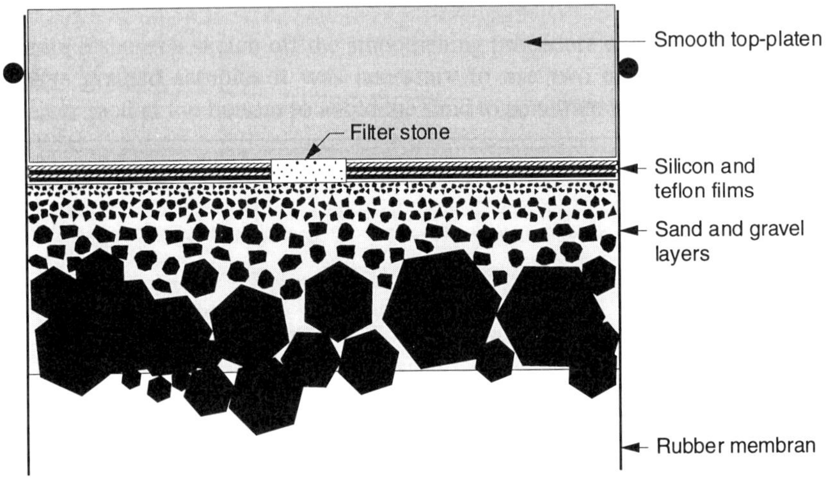

Figure 3 Smoothening of sample ends to reduce end friction

The sample form was observed during and after testing and the smoothening seemed to have functioned well. No signs of squeezing of the smoothening material or barrel shaping of the sample were observed for normal deformation levels. This method is a bit laborious, but it seems to function well.

INSTRUMENTATION

It is generally found important to place measuring devices inside the chamber and in direct contact with the sample. Measurements by a load cell placed outside the test chamber will be influenced by the friction in the bearings.

The instrumentation for the test performed in this work is shown in Figure 4. The radial strains are usually considered to be the most difficult ones to measure; however, the set-up presented seems to work well and give reasonable results.

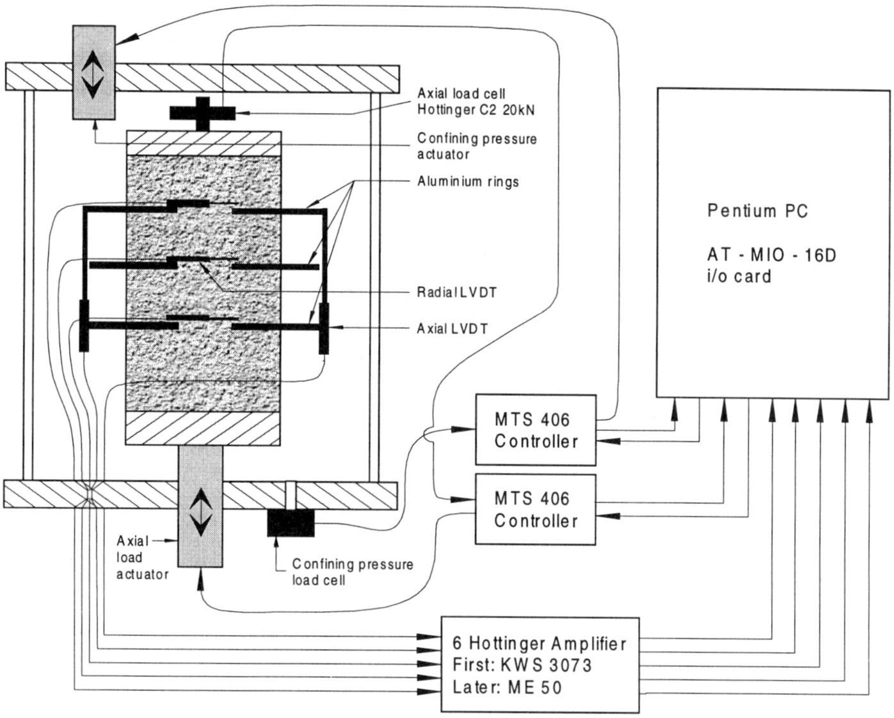

Figure 4 *Instrumentation for triaxial testing*

Load

The load is applied by a hydraulic actuator placed underneath the sample. It has a capacity of 20 kN force, which is sufficient for most unbound materials with sample diameter of 150 mm.

The load is measured by the load cell at the top of the sample and recorded by the computer. The loading system is of the closed-loop type: the control program will adjust the applied force until the measured signal equals the pre-set value and the load is varied with time in accordance with the selected load function.

The equipment also has the possibility to apply a varying confining pressure. A smaller hydraulic actuator is placed at the top of the cell and the piston works directly against the chamber fluid, which is water. The obtainable variation of the confining pressure depends on the volume capacity of the moving piston. It is also important to remove all air pockets to reduce the compressibility of the system. Obtainable variable confining pressure for this system is about 50 – 60 kPa with a plexiglas chamber and about 300 kPa when using the steel chamber.

Axial deformation

The axial deformation is measured by two LVDT's placed between an upper and a lower ring that are fixed to the rubber membrane 100 mm apart across the middle of the sample. This arrangement reduces the influence of non-uniform stresses that may occur close to the end platens. The use of LVDT's makes the calibration process rather simple.

Radial deformation

Realistic measurements of the radial deformation are the most difficult part of triaxial instrumentation. A comparison between triaxial test equipment at four different European laboratories /4/ showed a relative large scatter and especially were the differences for the radial strains large. Nevertheless is it important to measure the radial deformation to enable the calculation of volumetric strains. This again is an important prerequisite for the calculation and modelling of the resilient dilatation effects.

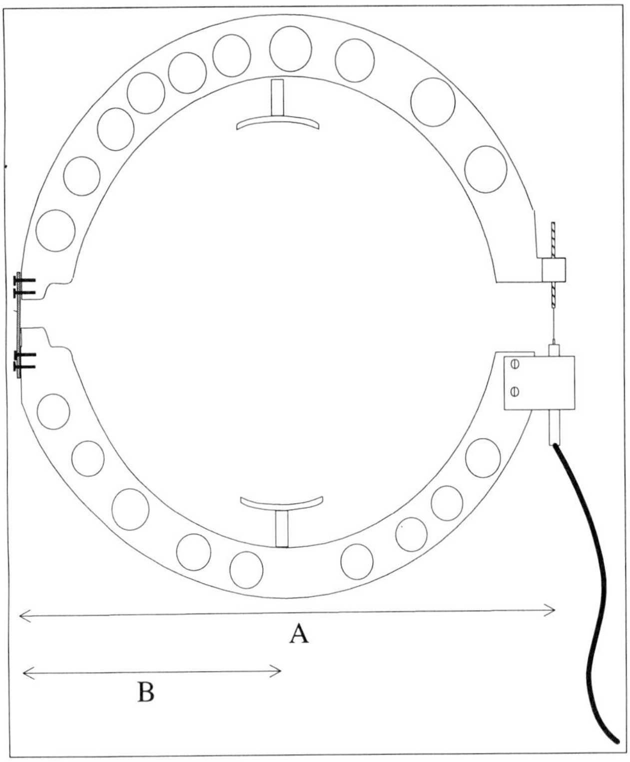

Figure 5 Caliper for measuring radial strains.

Figure 5 shows a sketch of the ring formed caliper used to measure the radial deformations. To make these calipers light they were made of magnesium. Holes were also drilled to reduce the weight.

Since the distance "A" is about the double of the distance "B" the deformation measured by the LVDT will also be about twice the deformation of the sample, improving the measuring accuracy.

The caliper is fixed to the sample by the pressure of a rubber band placed across the opening at the LVDT.

We have chosen not to use studs inside the sample because of the coarse grains, and the disturbance caused by large studs. For constant confining pressure tests the calipers contact points may be pressed against the rubber membrane, but this system would give large errors for varying confining pressure (VCP) tests due to the large movements of the rubber membrane in and out of the cavities formed by surface pores. For VCP-tests a new method for attaching the calipers to the sample was developed.

A small hole 1.2 mm is drilled through the rubber membrane and into one of the larger stone particles at the surface of the sample. Figure 6 shows a sketch of the set-up. A high speed (10 000 rpm or more) drilling device with a diamond drill is used. With this equipment a hole may be drilled while maintaining the vacuum inside the sample. A small bronze pin is then glued inside the hole, the membrane is glued to the fixture and a second membrane could also be used as shown.

then glued inside the hole, the membrane is glued to the fixture and a second membrane could also be used as shown.

This method has only been used for a small number off tests, but it seems to work satisfactorily.

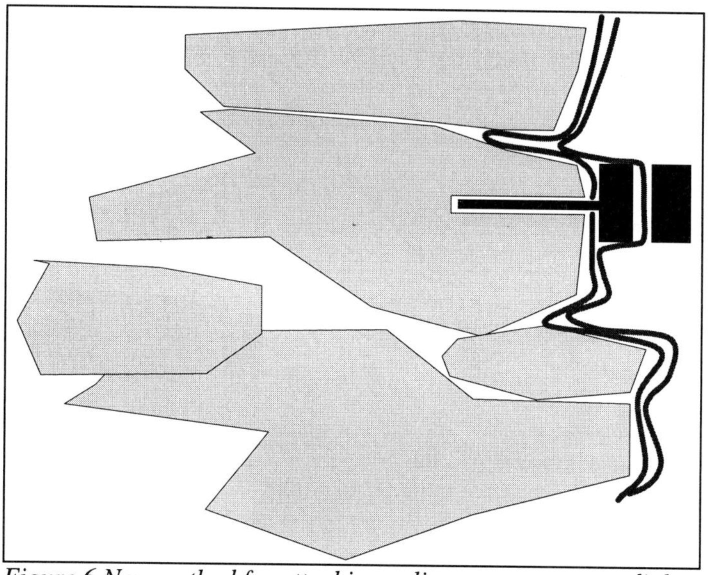

Figure 6 New method for attaching calipers to measure radial strains of coarse-grained triaxial samples.

The method requires relative large grains for the hole-size we used. It is also quite laboriously and is only relevant when it is impractical to use the more conventional studs.

LOADING PROCEDURE

Triaxial test can be performed for many purposes, and the loading procedures must be adapted for the specific type of load response to be studied. Traffic loading is rather complex, however if a standardised procedure could be utilised, it would be easier to compare result from different laboratories and make the results more useful for evaluation of experience and for application in practical design. Until recently the only standardisation in use was the SHRP standard /5/. Now a proposed new CEN-standard /1/ is already being taken into use by some laboratories.

Loading procedure for permanent deformation estimates

A special loading procedure was developed to determine parameters for the material model proposed in /6/. The static confining pressure is kept constant at four different levels and the deviator stress is increased in steps until the permanent deformation starts to develop rapidly. 3 000 cyclic load pulses are applied at each step. The cyclic load pulses are formed by a continuos haversine loading, with frequency of 10 Hz. An overview of the loading sequences is shown in figure 7. The confining pressure is set to 100 kPa to begin with, and the deviator stress is increased stepwise until "failure" is reached.

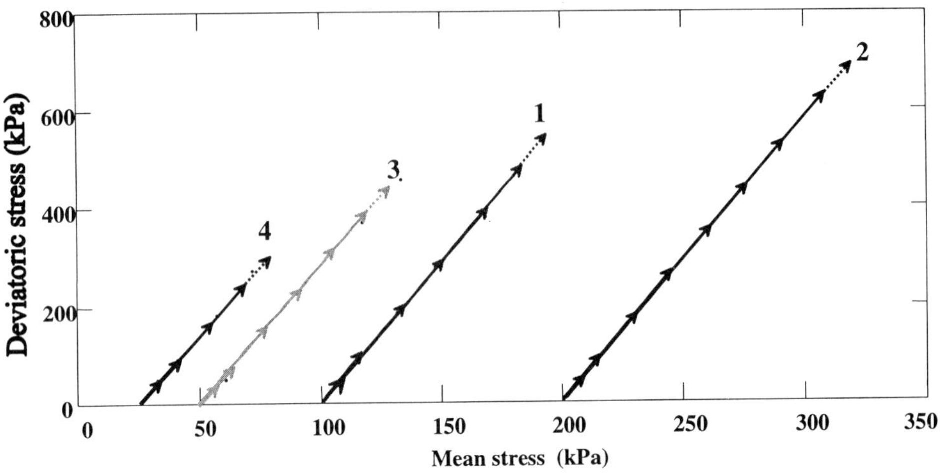

Figure 7 Loading procedure for the determination of parameters for the hyperelastic model

The total number of load steps applied depends on the material. A weak material will reach a deformation "failure" state sooner than a strong material that may endure many load steps and high stress levels before "failure" is reached. Figure 8 shows in principle how the permanent deformations develop during the test.

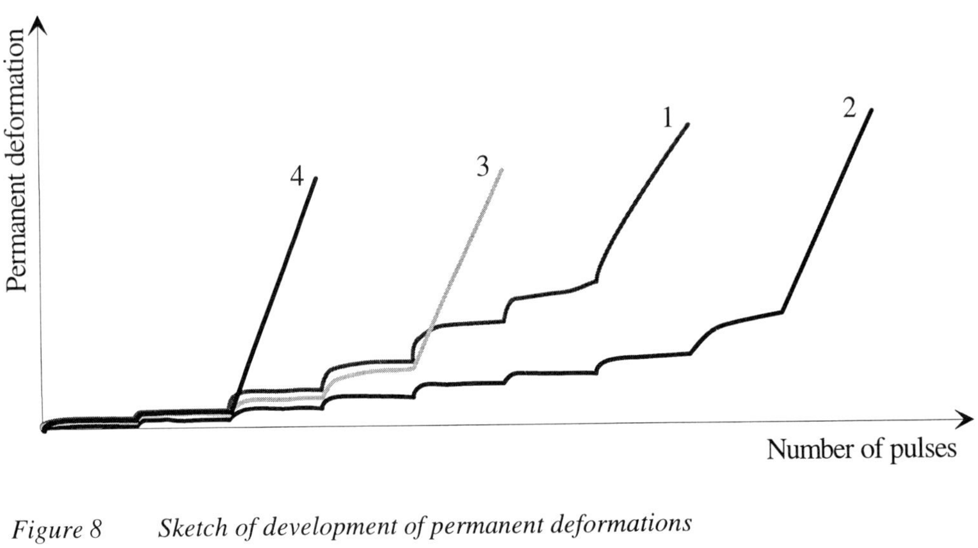

Figure 8 Sketch of development of permanent deformations

Relatively larger permanent deformations will develop during the first load series than in the following series for loads below the "failure-line"

It is not always easy to determine if a “failure” condition is reached. But we have used this definition: *Failure is reached when the permanent deformation continues and does not stabilise at the end of a load step.*

CONCLUSIONS AND RECOMMENDATIONS

The proposed procedure and techniques for sample preparation work well and give a large flexibility to produce samples with different densities. The smoothening procedure seems to reduce the friction enough to eliminate the problem with end friction for short samples.

The instrumentation set-up seems to give reliable measurements of both radial and axial strains for the load levels encountered. The proposed technique for attaching strain-measuring equipment to coarse-grained samples seems to work well and could eliminate some problems with studs that may disturb the sample.

There is a need for field calibration of triaxial test results obtained for gyratory compacted samples.

ACKNOWLEDGEMENTS

The development of equipment and procedures described in this paper has been carried out as part of the Dr.Ing.- study sponsored by the National Highway Administration. The KPG-project, described by Hansen and Hansen /7/, has also financed some of the research in this project.

REFERENCES

1 **CEN** Unbound and hydraulic mixtures for roads –
Test methods – Cyclic load triaxial test
TC 227 / WG 4 / TG 2

2 **Hoff, I.** Material Properties of Unbound Aggregates for Pavement Structures. Dr. Ing. Thesis. Department of road and Railway Engineering, Norwegian University of Science and Technology. Will be published in 1998.

3 **Bakløkk, L. J., Hoff, I. and Nordal, R. S.** Methods for Ranking Unbound Granular Materials for Pavement Construction.
Proceeding 5th International Conference on the Bearing Capacity of Roads and Airfields. Trondheim, Norway. July 1998

4 **Galjaard, P.D., Paute, J.L. and Dawson, A. R.** Comparison and Performance of Repeated Load Triaxial Test Equipment for Unbound Granular Materials
European Symposium of Flexible Pavements, Lisbon, September 1993.

5 **SHRP** Resilient Modulus of Unbound Granular Base/Sub-base Materials and Subgrade Soils. 1992 SHRP Protocol P-46

6 **Hoff, I., Nordal, S. and Nordal, R. S.** Permanent Deformation in Unbound Layers Predicted from Triaxial Testing using a new Methods for Calculating Stresses in Pavement Structures. Proceeding 5th International Conference on the Bearing Capacity of Roads and Airfields. Trondheim, Norway. July 1998

7 **Hansen, E. K. and Hansen, G**. Quality of crushed rock aggregates for pavement structures. Proceeding, the 5th International Conference on the Bearing Capacity of Roads and Airfields.

Fifth International Conference on the
BEARING CAPASITY OF
ROADS AND AIRFIELDS

A STUDY OF METHODS FOR RANKING UNBOUND GRANULAR MATERIALS IN PAVEMENT CONSTRUCTION

LEIF BAKLØKK [1], INGE HOFF [2] AND RASMUS S. NORDAL [3]

[1]Senior Research Engineer, SINTEF Civil and Environmental Engineering, Department of Highway Engineering
[2]Research Fellow, Norwegian University of Science and Technology (NTNU)
[3]Research Adviser, SINTEF Civil and Environmental Engineering, Department of Highway Engineering
N-7034 Trondheim, Norway

ABSTRACT

This paper describes methods for ranking of unbound granular materials according to their usefulness in load distributing layers of flexible pavement. The ranking system in Norway was originally developed on the basis of experience and a system for empirical pavement design, supplemented by some theoretical analysis, laboratory testing and results from the AASHTO Road Test. The draft ranking system presented here is based on laboratory investigations, and the material parameters are determined from triaxial tests and CBR^g -tests performed on gyratory compacted specimens.

The paper presents procedures for calculating load distribution coefficients (LDC), permanent deformation resistance coefficients (PDRC) and combined material indexes (CMI) for these materials.

The LDC (a) represents the relative load spreading capability of the material, and is based on the resilient properties of the materials determined by repeated load triaxial tests.

The PDRC (d) expresses the materials resistance against developing permanent deformation, and is based on results from triaxial tests, gyratory compaction test and CBR^g tests.

The CMI (m) gives a ranging of the materials, based on both the load bearing properties (resilient characteristics) and resistance against development of permanent deformation under repeated loading.

In recent years, rehabilitation of asphalt pavement is very often conducted in heavy traffic road. In this case, pavement is required to open for traffic as soon as possible immediately after the completion of construction to prevent traffic congestion leading to economic loss. For instance, the sequence of actual rehabilitation work is that: (1) starting subgrade construction being completed at midnight; (2) construction of subbase-course and base-course; one hour later (3) subsurface construction being completed at 5a.m., then at 6a.m. the pavement is made open for traffic; 7 days later, (4) the surface layer is placed on the subsurface layer; and one hour later (5) the pavement is made open for traffic. This leads to shorter or insufficient curing time for slag material. However, the effects of curing time upon the deformation behaviour of slag base-course material and more importantly upon the future performance of the pavement, when experienced traffic load in an early stage of hardening, are not fully understood. In this relation, the effect of curing time upon the deformation behaviour of stabilized subgrade and also the performance of asphalt pavement with stabilized subgrade have been studied and reported (Yoshida *et al.*, 1997 and 1998).

In this paper, the effects of stress condition and curing time upon the deformation characteristics of slag base-course material are first presented based on the results of repeated loading triaxial compression test. Then, the performance of asphalt pavement with slag base-course is discussed in terms of three indices, the present serviceability index proposed by American Association of State Highway Officials (1972), its modified version by Japan Road Association (1978) and the maintenance control index used by Japan Ministry of Construction (JSCE, 1992).

DEFORMATION CHARACTERISTICS OF SLAG BASE-COURSE MATERIAL

Test material

A compound slag was tested by means of repeated loading triaxial compression in this research and the deformation characteristics are investigated. The compound slag is a mixture of steel-making slag, blast-furnace slag and granulated slag in the ratio of 5: 3: 2 (weight). The physical properties of the compound slag are given in Table 1.

Repeated loading triaxial compression test

The test apparatus is an conventional dual-cell type triaxial compression one suitable for unsaturated soil, and the repeated axial load is applied by Bellofram cylinder through dual-direction electromagnetic valve (see Yoshida *et al.*, 1997). All the test samples were made by compacting the pre-screened material at its optimum water content in a mould and formed as 100mm in diameter and 200mm in height. Since the compound slag possesses favourable hydraulic-hardening nature, each specimen was cured in shallow ground, about 300mm below ground surface, prior to testing for certain period of time; 0, 0.5, 1, 3, 6 and 12 months.

In repeated loading triaxial compression test, each specimen was first isotropically consolidated under a pre-specified confining pressure, then a prescribed deviatoric stress was applied to the specimen repeatedly until the permanent strain increment becomes negligible. From this, permanent deformation characteristics are derived in terms of plastic axial strain. Subsequently, the deviatoric stress is reduced by a pre-specified magnitude and the reduced deviatoric stress is repeatedly applied to the specimen maintaining the mean principal stress constant, to obtain resilient deformation characteristics. The loading condition is summarized in Table 2. All the tests were performed under drained condition.

Resilient deformation characteristics

The resilient deformation modulus, M_r, is calculated as a ratio of repeated deviatoric stress to resulting resilient axial strain. Figures 1 and 2 shows the relationships between resilient

Table 1 Physical properties of compound slag

w_{opt} (%)	$\rho_{d,\,max}$ (g/cm^3)	ρ_s
9.3	2.293	3.151

Table 2 Loading condition

Number of load repetitions	Loading time (sec.)	Deviatoric stress (kPa)	Mean principal stress (kPa)
20000 - 30000	0.3	29 - 353	78, 157, 235

deformation modulus and deviatoric stress for curing time of 0 and 6 months, respectively. It is seen from these figures that the resilient deformation modulus decreases with deviatoric stress and that it becomes constant at a deviatoric stress greater than about 200kPa. It is also seen that the resilient deformation modulus is greater for 6 month curing than for 0 month curing due to development of cementation inherent in compound slag, and that the mean principal stress also affects the resilient deformation modulus; that is, the greater the mean principal stress is, the greater the resilient deformation modulus. This influence of mean principal stress upon resilient deformation modulus holds regardless of curing time as shown in Figure 3. Solid lines in Figures 1, 2 and 3 are regression lines given by

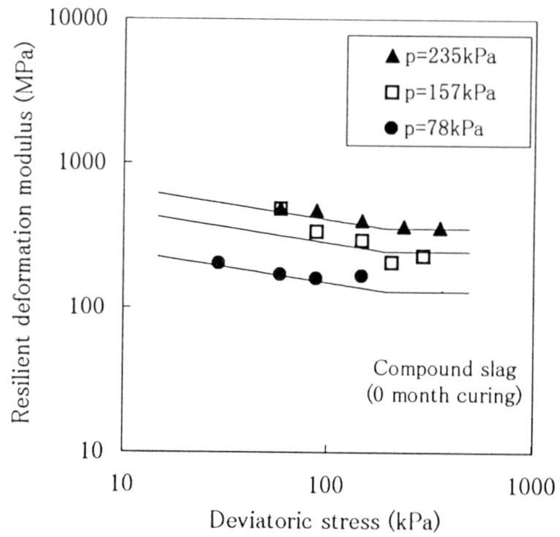

Figure 1 Relationships between resilient deformation modulus and deviatoric stress (curing time = 0 month)

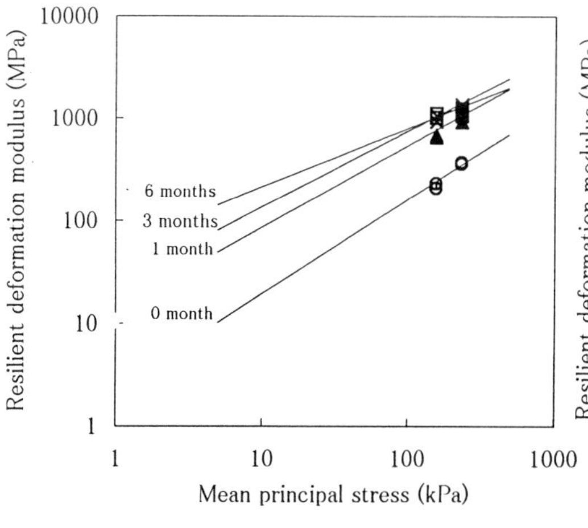

Figure 3 Relationships between resilient deformation modulus and mean principal stress

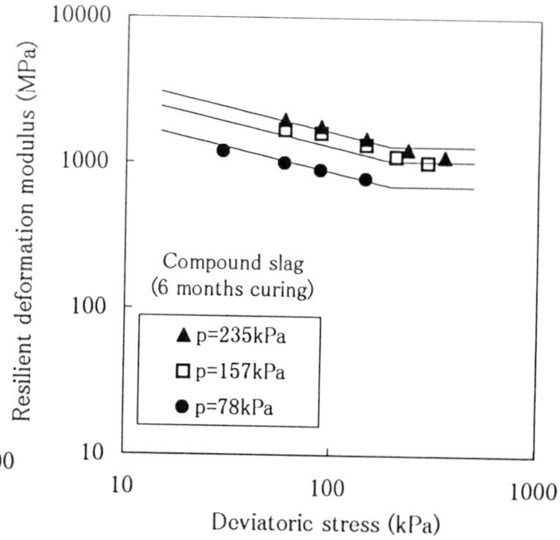

Figure 2 Relationships between resilient deformation modulus and deviatoric stress (curing time = 6 months)

Table 3 Material constants for resilient deformation modulus

Curing time (months)	K	M	N
0	2.381	0.916	0.214
0.5	9.250	0.869	0.242
1	13.89	0.794	0.242
3	24.93	0.740	0.261
6	57.76	0.571	0.327
12	76.25	0.525	0.363

Table 4 Material constants for resilient Poisson's ratio

Curing time (months)	A_0	A_1	A_2
0	0.3420	0.1530	-0.0520
0.5	0.1590	0.1220	-0.0251
1	0.1340	0.1300	-0.0269
3	0.0181	0.1160	-0.0293
6	0.0283	0.0768	-0.0201
12	0.0305	0.0	0.0

$$M_r = K\, p^M \, (200/q)^N \qquad [1]$$

where M_r is resilient deformation modulus in MPa, and p and q are mean principal and deviatoric stresses in kPa, respectively. K, M and N are material constants of which values are given in Table 3. Note that N is zero for $q \geq 200$kPa .

The resilient Poisson's ratio, ν_r , is calculated as a ratio of resilient lateral strain to resilient axial strain. Although not shown here, the relationships between resilient Poisson's ratio and stress ratio η (= q/p) are curve-fitted by

$$\nu_r = A_0 + A_1\eta + A_2\eta^2 \qquad [2]$$

where A_0, A_1 and A_2 are material constants of which values are given in Table 4. The resilient Poisson's ratio increases with stress ratio and decreases with curing time. Its dependency of stress ratio decreases with curing time, and at 12 months curing, the resilient Poisson's ratio becomes constant.

Permanent deformation characteristics

Figure 4 shows the relationships between the permanent axial strain and the number of load applications at a mean principal stress of 157 kPa for curing time of 0 and 6 months. It is seen that the permanent axial strain increases with the number of load applications rapidly at the beginning, then gradually at the number of load applications greater than about 2000. Note that, although in the figure there are no data at the early stage, it is reasonable guess that permanent strain has rapidly developed initially. Regression curves in the figure are given by

$$\varepsilon_p = \varepsilon_{p0} + (N - N_0) / \{ a + b(N - N_0) \} \qquad [3]$$

where ε_p is permanent axial strain, ε_{po} is permanent axial strain at N_o (= 2000), and a and b are material constants determined for each set of p, q and curing time. From this equation, the ultimate permanent axial strain is computed by setting N as infinite, and it is represented as

Table 5 Material constants for ultimate permanent axial strain

Curing time (months)	k (x 10^{-3})	u	v
0	0.8920	0.490	1.119
0.5	0.4690	0.288	1.050
1	0.2690	0.126	0.896
3	0.1270	0.311	1.012
6	0.1140	0.548	1.225
12	0.0979	0.288	1.106

$$\varepsilon_{p,ult} = k\,(q/p_0)^u / (p/p_0)^v \qquad [4]$$

where $\varepsilon_{p,ult}$ is ultimate permanent axial strain, p and q are mean principal and deviatoric stresses in kPa, respectively, p_0 is a reference pressure (= 98.1kPa), and k, u and v are material constants whose values are given in Table 5.

PERFORMANCE OF ASPHALT PAVEMENT WITH SLAG BASE-COURSE

Performance analysis

The performance of asphalt pavement with slag base-course is evaluated in terms of three indices, the present serviceability index (AASHO PSI), its modified version (JRA PSI) and the maintenance control index (MCI). These indices are given by

$$\text{AASHO PSI} = 5.03 - 1.91\log(1 + SV) - 0.01\,(C + P)^{1/2} - 0.21\,RD^2 \qquad [5]$$

$$\text{JRA PSI} = 4.53 - 0.518\log\sigma' - 0.371\,C'^{1/2} - 0.174\,D'^2 \qquad [6]$$

$$MCI = 10 - 1.48\,C'^{0.3} - 0.29\,D'^{0.7} - 0.47\,\sigma^{0.2} \qquad [7\text{-}1]$$

$$MCI_0 = 10 - 1.51\,C'^{0.3} - 0.30\,D'^{0.7} \qquad [7\text{-}2]$$

$$MCI_1 = 10 - 2.23\,C'^{0.3} \qquad [7\text{-}3]$$

$$MCI_2 = 10 - 0.54\,D'^{0.7} \qquad [7\text{-}4]$$

where SV is average of variance of longitudinal roughness, C is cracking ratio in $m^2/1000m^2$, P is patching ratio in $m^2/1000m^2$, RD is rut depth in cm, σ' is standard deviation of the longitudinal roughness in mm, C' is cracking ratio in %, D' is average rut depth in mm, σ is longitudinal roughness in mm. Note that the minimum value from MCI to MCI_2 is taken as the representative MCI. The overlay necessity level for asphalt pavement is usually considered 2.5 for AASHO PSI, 2.0 for JRA PSI or 4.0 for MCI.

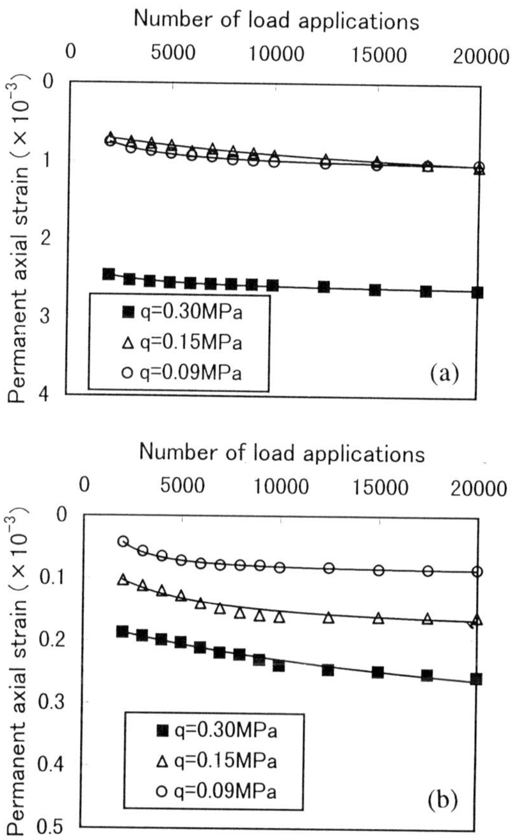

Figure 4 Relationships between the permanent axial strain and the number of load applications for curing time of 0 month (a) and for 6 months (b) (mean principal stress = 157 kPa)

$$\bar{x} = \sum_{1}^{n} x_i / n$$

$$\sigma = \sqrt{\frac{\sum_{1}^{n} (\bar{x_i} - x)^2}{n-1}}$$

Delete those data which are not in the range of $[\bar{x}-2\sigma, \bar{x}+2\sigma]$. Recount average and standard deviation till all data are in the range. They are the average and standard deviation of that group. Representative value can be calculated with following formula:

$$x = \bar{x} - t_\alpha \frac{\sigma}{\sqrt{n}}$$

Terms are defined as follows:

$\bar{x}$= average of the data

σ= standard deviation

n= number of samples

x_i= a single data

t_α= parameter in t—Distributing table that varies with reliability. When reliability is 97.5%, n=6,$t_\alpha/\sqrt{n}$=1.050

III Experiment Results and Analysis

1. Relationship Between Compression or Split Parameters and Curing Time

Statistical analysis of compressive strength and modulus, splitting tensile strength and modulus that obtained by cooperative units in different curing time under a curing temperature of 20±2 ℃ show the average and standard deviation of these parameters of the whole country. Recursive analysis of averages in different curing time shows the relationship between parameters and curing time. The statistical relations between compressive strength or modulus and curing time are shown in table 4 and table 5 while relations between splitting strength or modulus and curing time in table 6 and table 7.

Table 4 Compressive Strength Increasing Discipline of Semi-rigid Base Materials

Mixture	R=a+bd				$R=ad^b$			
	a	b	r	s	a	b	r	s
Lime Soil	1.7143	0.0048	0.5467	0.7034	0.8058	0.2392	0.8372	0.1182
Lime Flyash Stabilized Soil	1.3128	0.0303	0.9586	0.8569	0.3331	0.5777	0.9941	0.0475
Lime Flyash Bound Macadam	2.2945	0.0308	0.8927	1.4792	0.5246	0.5332	0.9764	0.0891
Cement Flyash Bound Macadam	1.6185	0.0093	0.8611	0.5228	0.7385	0.2901	0.9512	0.0712
Cement Bound Macadam	3.9221	0.0137	0.8781	0.7135	2.4369	0.1838	0.9875	0.0221
Cement Lime Flyash Bound Sand-gravel	1.2288	0.0343	0.9931	0.3819	0.4171	0.5369	0.9881	0.063
Cement Bount Sand-gravel	2.8[illegible]39	0.0108	0.7671	0.8644	1.4779	0.2327	0.9331	0.0679
Lime Cement Bound Sand-gravel	2.7521	0.0159	0.9884	0.2337	1.7784	0.2047	0.9333	0.0596

Notes: R= compressive strength (MPa); a,b= recursive coefficient; d= curing time (days); r= correlation coefficient; s= standard deviation .

All statistical relations in table 4 to 7 show high correlativity except that of lime soil. It also can be seen that exponential relations have higher correlativity than that of linear ones. So the strength and modulus of semi-rigid base course materials increase in an exponential double logarithm form along with curing time. The increasing rate decreases with curing time. Exponential relations have better correlation coefficient and lower standard deviation than that of linear relations.

Table 5 Compressive Modulus Increasing Discipline of Semi-rigid Base Materials

Mixture	Ep=a+bd				$Ep=ad^b$			
	a	b	r	s	a	b	r	s
Lime Soil	778.34	3.191	0.671	334.85	328.14	0.2858	0.895	0.107
Lime Flyash Stabilized Soil	486.06	9.983	0.940	343.12	111.24	0.5852	0.992	0.053
Lime Flyash Bound Macadam	581.99	13.67	0.993	146.15	191.82	0.5202	0.998	0.020
Cement Flyash Bound Macadam	933.84	9.965	0.939	344.60	331.69	0.4052	0.977	0.066
Cement Bound Macadam	1710.93	11.54	0.876	629.52	659.53	0.3306	0.970	0.062
Cement Lime Flyash Bound Sand-gravel	455.55	13.38	0.990	178.39	127.12	0.5913	0.999	0.016
Cement Bount Sand-gravel	1299.76	16.38	0.943	549.33	475.82	0.4147	0.959	0.091
Lime Cement Bound Sand-gravel	1547.04	0.22	0.991	100.08	1057.7	0.1842	0.934	0.052

Notes: Ep= compressive modulus (MPa)

Table 6 Splitting Strength Increasing Discipline of Semi-rigid Base Materials

Mixture	$\sigma_{sp}=a+bd$				$\sigma_{sp}=ad^b$			
	a	b	r	s	a	b	r	s
Lime Soil	0.1233	0.0007	0.817	0.0518	0.0463	0.3434	0.937	0.096
Lime Flyash Stabilized Soil	0.0416	0.0037	0.997	0.0231	0.0093	0.8332	0.997	0.040
Lime Flyash Bound Macadam	0.1018	0.0043	0.982	0.0773	0.0169	0.7683	0.995	0.054
Cement Flyash Bound Macadam	0.0906	0.0017	0.965	0.0448	0.0219	0.5642	0.989	0.061
Cement Bound Macadam	0.3078	0.0029	0.948	0.0925	0.1268	0.3577	0.997	0.017
Cement Lime Flyash Bound Sand-gravel	0.0916	0.0032	0.988	0.0467	0.0230	0.6405	0.999	0.019
Cement Bount Sand-gravel	0.3098	0.0029	0.940	0.0995	0.1219	0.3677	0.991	0.035
Lime Cement Bound Sand-gravel	0.1263	0.0020	0.994	0.0192	0.0580	0.3879	0.962	0.082

Notes: σ_{sp}= splitting strength (MPa)

Table 7 Splitting Modulus Increasing Discipline of Semi-rigid Base Materials

Mixture	Esp=a+bd				$Esp=ad^b$			
	a	b	r	s	a	b	r	s
Lime Soil	207.51	1.4783	0.812	100.72	93.665	0.3053	0.861	0.135
Lime Flyash Stabilized Soil	53.71	7.8843	0.994	49.28	23.341	0.6860	0.990	0.607
Lime Flyash Bound Macadam	156.29	8.0957	0.972	184.55	21.725	0.3413	0.995	0.062
Cement Flyash Bound Macadam	180.94	1.8400	0.917	75.95	66.050	0.4166	0.982	0.059
Cement Bound Macadam	361.64	5.1256	0.950	160.10	109.77	0.4740	0.995	0.035
Cement Lime Flyash Bound Sand-gravel	117.93	4.4906	0.986	69.584	25.343	0.6930	0.999	0.020
Cement Bount Sand-gravel	502.15	5.4201	0.993	60.178	262.16	0.3125	0.956	0.071
Lime Cement Bound Sand-gravel	196.96	3.2890	0.994	32.463	83.594	0.4134	0.981	0.060

Notes: Esp= splitting modulus (MPa)

Figure 1 show the compressive strength and modulus—curing time curves of cement bound macadam and lime flyash bound macadam. Their splitting strength and modulus—curing time curves are shown in figure 2. It is noted that the strength and modulus of cement bound macadam increase rapidly in its early 28 days curing, but slow down after that. Lime flyash bound macadam has a low strength and modulus in its early curing period. They increase slowly but continually especially splitting parameters. After 180 curing days, theirs splitting strength and modulus are all higher than that of cement bound macadam. It is concluded that lime flyash bound macadam has a good late strength.

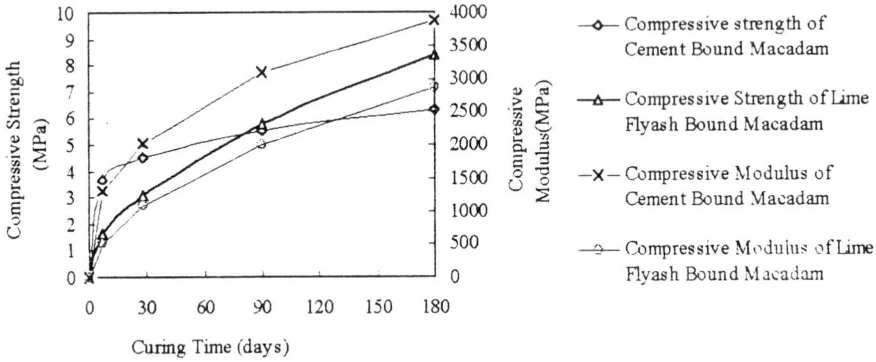

Fig. 1. Compressive Parameters vs. Curing Time

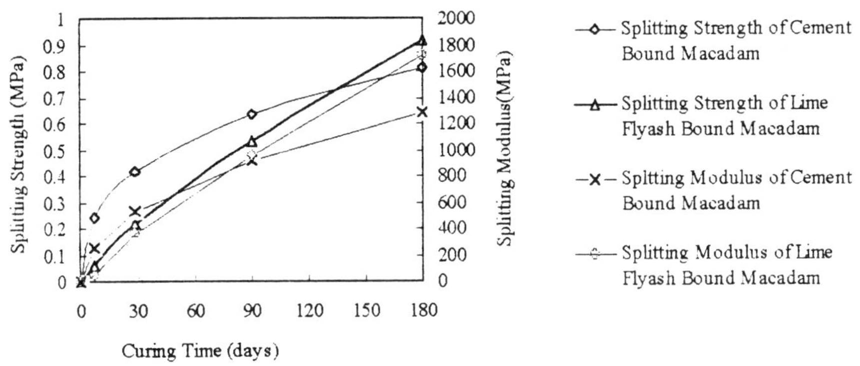

Fig. 2. Splitting Parameters vs. Curing Time

Using statistical relations in table 4 to table 7, we calculate strength and modulus of cement bound macadam and lime flyash bound macadam in different curing time and show them in table 8 and table 9 . Table 8 shows that strength and modulus of cement bound macadam at the 90th curing day attend to 80 percent of that at 180th curing day. Parameters at the 90th curing day can be used as design parameters in pavement design. As to lime flyash bound macadam, its strength and modulus at 90th curing day are only 60 percent of that at the 180th curing day. They still increase after 180 curing days(shown in figure 1 and 2). Lime flyash bound macadam take its strength and modulus at the 180th curing day as its design parameters for they can show its later character.

Table 8 Parameters of Cement Bound Macadam

Parameters(MPa) \ CuringTime (days)	28	90	180	Parameter Ratio between 90's and 180's(%)
Compressive Strength	4.49	5.57	6.33	87.99
Compressive Modulus	2092.9	3078.8	3871.8	79.52
Splitting Strength	0.4176	0.6341	0.8125	78.04
Splitting Modulus	532.7	926.4	1286.8	71.99
Average Value				79.39

Table 9 Parameters of Lime Flyash Bound Macadam

Parameters(MPa) \ CuringTime (days)	28	90	180	Parameter Ratio between 90's and 180's(%)
Compressive Strength	3.10	5.78	8.36	69.14
Compressive Modulus	1085.7	1993.0	2859.3	69.73
Splitting Strength	0.2186	0.5362	0.9133	58.71
Splitting Modulus	359.10	959.52	1719.7	55.79
Average Value				63.34

2. Relationship Between Modulus and Strength

The relationships between average or representative modulus and strength of various semi-rigid base course materials are obtained after tests through thousands samples and statistical analysis. The relationship between average compressive modulus and compressive strength is that:

$$\overline{E_p} = 958.9831\overline{R}^{0.5761} \quad (r=0.6088, \quad s=0.1285)$$

The relationship between representative compressive modulus and strength is that:

$$E_p = 839.5724R^{0.4976} \quad (r=0.5378, \quad s=0.1978)$$

The relationship between average splitting tensile modulus and strength is that:

$$\overline{E_{sp}} = 1743.4173\overline{\sigma_{sp}}^{1.172} \quad (r=0.9035, \quad s=0.0881)$$

The relationship between representative splitting tensile modulus and strength is that:

$$E_{sp} = 1484.4141\sigma_{sp}^{1.014} \quad (r=0.8990, \quad s=0.1329)$$

The above mentioned regression equations show that there is a higher correlativity between splitting tensile modulus and strength than that of compressive parameters. Materials with high splitting tensile strength also have high splitting tensile modulus. It dose not work on compressive parameters. All above formulae have reliability more than 90 percent.

3. Conversion Correlation among Compression, Split and Tension Parameters

1)Relationship Between Compression and Splittng Tension Parameters

Ten types semi-rigid base course materials which are cured under standard curing conditions are tested at their design curing time(90 days for cement-bound materials while

180 days for lime flyash-bound materials). All parameters are obtained using procedure controlled machine. The relationships between representative compressive and splitting tensile parameters are shown as follows:

$R=7.9027\sigma_{sp}^{0.8576}$ (n=10, r=0.9068, s=1.069)

$E_p=50.8979\,E_{sp}^{0.5280}$ (n=10, r=0.6525, s=0.1778)

2) Relationship Between Split and Tension Parameters

The research project, " Research on Simplify Test Procedure for Tensile Parameters of Pavement Materials ", which was finished by Communications Ministry did a large quantity of research on relationship between split and tension parameters. The results are shown in table 10

Table 10 Ratio between Split and Tension Parameters(Representative Value)

Mixture	S/σ_{sp}	Es/Esp
Lime Soil	1.57	5.08
Lime Flyash Bound Macadam	1.68	4.65
Lime Bound Macadam	1.67	11.76
Cement Bound Macadam	2.0	18.39

Notes: The S in table 10 represents tensile strength while Es represents tensile modulus.

4. Conversion Correlation between Curing Time and Curing Temperature

Semi-rigid materials form their strength through a serial of chemical and physical actions that require specified moisture and temperature. Proper moisture content provides essential crystallized water for $Ca(OH)_2$ crystal and volcanic ash action. Temperatures affect the speed that mixture form its strength. Strength grows slowly under a low temperature. When temperature is high, chemical actions act at a greatly accelerated pace and strength grows rapidly. Therefore research on semi-rigid base course materials should take temperature and curing time into account when moisture content is defined.

A large number of test data about strength forming of various semi-rigid base course materials are analyzed. It shows that strength grows in direct proportion with power of curing time (d^n) under standard curing conditions. When curing time is the same, strength grows in direct proportion with power of temperature (T^m). That is:

$R\propto(d^n,T^m)$

If strength under $d_1^nT_1^m$ equals to that under $d_2^nT_2^m$, $d_1^nT_1^m=d_2^nT_2^m$. Our study proved that the equation is tenable. So

$d_1/d_2=(T_2/T_1)^a$.

Using this equation, strength on design curing time under standard curing conditions can be obtained in much shorter time with higher temperature. It has great significance on pavement design and construction inspection. The two factors have lower correlativity when curing temperature is higher than 60 ℃. It is suitable to use 60 ℃ as standard temperature of quick curing.

In order to know the conversion correlation between curing time and temperature, Road and

Traffic Research Institute of Tongji University tested strength and modulus of two typical semi-rigid base course materials, lime flyash bound macadam and cement bound macadam. Regression analysis of representative strength and modulus of lime flyash bound macadam in different curing time and temperatures show their relation. Relationships between curing time and temperature for compressive parameters are shown as follows:

$R=0.021T^{1.259}d^{0.472}$

$E_p=21.023T^{1.001}d^{0.3667}$

Input $T_1=20$ ℃ and $T_2=60$ ℃, then

$d_{20}/d_{60}=18.74$ (for compressive strength)

$d_{20}/d_{60}=20$ (for compressive modulus)

Synthesize the two formulae and define $d_{20}/d_{60}=20$. It means that compressive strength and modulus of one curing day under 60 ℃ equal to that of twenty curing days under standard curing conditions.

Relationship between curing time and temperature for splitting parameters for lime flyash bound macadam are shown as follows:

$d_{20}/d_{60}=31.2$ (for splitting tensile strength)

$d_{20}/d_{60}=31.4$ (for splitting tensile modulus)

Define $d_{20}/d_{60}=30$ as the synthesis of the two formule.

In order to examine whether the relations work, a comparison between predicted parameters using quick curing and actual test parameters under standard curing conditions was done. Table 11 shows the comparison of splitting parameters of lime flyash bound macadam.

Table 11 Comparison between Predicted and Actual Values

20 ℃	Curing Time(d)	7	30	90	180	Curing Time(d)	7	30	90	180
	Actual Strength(MPa)	0.134	0.371	0.617	0.936	Actual Modulus(MPa)	111.0	525.1	1287	2608
60 ℃	Curing Time(d)	0.23	1	3	6	Curing Time(d)	0.23	1	3	6
	Predicted Strength(MPa)	0.138	0.339	0.667	1.022	Predicted Modulus(MPa)	113.4	473.8	1394	2754
Relative Error(%)		3	-8.6	8.1	9.2		2.2	-9.8	8.3	5.6

Relative errors between predicted value and actual test value in table 11 are less than 20%. The above relations can be used to predict design parameters in a short time.

Relationship between curing time and temperature for cement bound macadam parameters are shown as follows:

$d_{20}/d_{60}=7$ (for compressive parameters)

$d_{20}/d_{60}=9$ (for splitting tensile parameters)

Define $d_{20}/d_{60}=8$ for cement bound macadam parameters.

Other cooperative units obtain similar results with Tongji University. So test data under quick curing conditions can be used to predict design parameters if they are tested conforming Tést Procedure strictly.

IV Conclusion

The essay introduces the research results about mechanical properties of semi-rigid base course materials. They include following items: 1 Compressive (or splitting tensile) strength and modulus of semi-rigid base course materials increase with curing time. Cement bound materials can use their parameters at the 60th curing day as their design parameters while lime flyash bound materials use that of the 180th curing day. 2 Advance relationship between strength and modulus. 3 Give the conversion correlation among compression, split and tension parameters. 4 Put forward the quick curing method used to predict design parameters of semi-rigid base course materials in short term. All these results can be used as reference for design and construction.

References

1 Tongji University and Highway Planning Design Institute of Communications Ministry, Research on Material Design Parameters, 1995.
2 Communications Ministry P.R.China, Research on Simplify Test Procedure of Tensile Parameter of Pavement Materials,1988.

Fifth International Conference on the
BEARING CAPASITY OF
ROADS AND AIRFIELDS

EXPERIMENTAL STUDY ON THE IMPROVEMENT OF BEARING CAPACITY OF REINFORCED FLYASH SUBGRADE AND SOIL FLYASH MIX

J.N. JHA* A.K. CHOUDHARY** R. KUMAR*

* Lecturer(Civil) College of Engg. & Tech. Bathinda-151001 INDIA
** Lecturer(Civil), R.I.T. Jamshedpur-831014 INDIA

ABSTRACT

The paper presents some experimental investigation on the successful use of flyash in improving the load carrying capacity of expansive soil. The paper also presents the effectiveness of single layer of horizontal reinforcement in improving the bearing capacity characteristics of flyash subgrade, when placed below the footing. The results obtained are found to be quite encouraging.

Key Words:

Expansive soil, Flyash, Footing, Reinforcement, Bearing Capacity Ratio

INTRODUCTION

In India one fifth to one sixth part of surfacial deposits consists of expansive soil. Structures on these soil experience large scale damage due to heaving and shrinkage. At the same time construction engineers are apprehensive of using these soil as construction material. Extensive work has been done throughout the world and various solutions suggested based on different concept. Flyash an industrial waste, causing lot of environmental hazards and disposal problems if used as stabiliser may prove to be effective in modifying the undesirable properties of expansive soil, thus reducing the cost of construction. Earlier researchers(1,2,3) have reported the gainful utilization of flyash for pavement base and subbase course.

While constructing highways, low lying lands are quite often encountered along its root. It is a general practice to fill up these low lying areas by good quality soil and then improve / stabilize it up to the constructional requirement. In case of non-availability of good quality soil locally, it has to be transported from some other places, thus incurring a good amount of the project. To overcome this aspect, if low lying areas are filled up with flyash(Wherever available) and reinforcing horizontally its load carrying capacity can be improved considerably. The beneficial effect of geotextile in improving the load carrying capacity of subgrade is reported earlier also(4,5). Improvement in load bearing characteristics of flyash using geofabric was also reported (6). Keeping these aspects in mind two series of laboratory investigations have been conducted to asses the influence of

(i) Flyash as stabiliser for improving the undesirable properties of black cotton soil, thus improving the load carrying capacity and reducing the cost of construction.

(ii) Flyash as subgrade material after reinforcing it with geotextile.

EXPERIMENTAL INVESTIGATION

Tests were conducted using locally available soil procured from sector 20 , Rourkela and flyash from H.S.L. Rourkela, Tisco Jamshedpur.

Expansive Soil:

Soil which was investigated has the following basic properties.

Specific gravity:	2.50
Grain size distribution:	Sand-48%, Silt-39.5%, Clay 12.5%
Effective Size D_{10}:	0.001mm
Consistency limit:	Liquid limit-38%, Plastic limit-14.6%, Shrinkage limit-13.8%, Plasticity index-23.4%.

The soil classified as CI (as per IS 1498-1970) having low to medium degree of expansion as the differential free swell value was about 50%.

Flyash Properties:

Flyash procured from H.S.L., Rourkela and Tisco Jamshedpur has the following basic properties.

	Flyash	
	H.S.L. Rourkela	Tisco, Jamshedpur
Specific gravity:	2.50	2.09
Grain Size Distribution:		
Sand	28%	20%
Silt	33%	70%
Clay	39.1%	2%
Effective Size D_{10}	0.0007mm	0.013mm
Uniformity Coefficient, Cu	2.70	------
Coefficient of Curvature Cc	-----	3.85
Max. Dry Density	-----	0.934 gm/cc

Optimum Moisture Content ----- 48%

Reinforcing Material

In India, the availability and cost of reinforcing materials are very important factors. The manufacturers are also not aware of civil engg. requirements. With the result that the material produced donot often meet the specification. Reinforcement used in this investigation were geosynthetic and geotextile having thickness 0.01 and 0.02 cm respectively. These reinforcing materials were preferred because of its commercial availability and it happened to be cheap also. The only limitation in using these reinforcing materials were the non availability of its properties.

Laboratory Tests and Results

(i) Influence of Flyash on Expansive Soil:

The test materials were prepared by thoroughly mixing flyash with air dried expansive soil in proportion of 10, 30, 50, 70 & 90 percent by weight. Laboratory tests were carried out on different soil-flyash mix to find out the following properties. Tests were carried as per standard procedure.

(i) Atterberg's limit, (ii) Optimum moisture content, (iii) Maximum dry density, (iv) Shear strength parameters, (v) Bearing Capacity

Bearing capacity of different soil - flyash mix in both the cases (i.e. at O.M.C. and at Saturation) were determined by conducting model footing tests using cast iron model footings of size 3.81cmx3.81cmx1.0cm in a test tank of size 51cmx51cmx38cm. For load tests at saturation a sand layer, 2cm thick was provided at the bottom of the test specimen as well as sand drains of 8mm diameter were provided around the periphery of the specimen at a distance of 10 cm c/c, so that water percolates to the test specimen through the sand bed and sand drain. Atterberg's limit, O.M.C., maximum dry density, shear strength parameters and ultimate bearing capacity for different soil-flyash mixes have been tabulated in Table-1. Figure 1 and 2 shows the load - settlement curves for different soil - flyash mix both at O.M.C. and saturation respectively. The variation of ultimate bearing capacity with increasing flyash percentage has been shown in fig. 3. The experimental bearing capacity value have also been compared with theoretical value (7,8). Results shows that maximum increase in bearing capacity (At O.M.C.) is obtained for 30% flyash content, Where as for saturation it is only at 10% flyash content.

TABLE-1

Sr. No.	% Flyash	Moisture Content		Max. Dry Density (gm/cc)	Cohesion 'c' (kg/cm2)		Bearing Capacity(kg/cm2)		Angle of Internal Friction ϕ (degrees)	
		OMC	SAT.		OMC	SAT.	OMC	SAT.	OMC	SAT.
1.	0	14.65	15.20	1.935	0.47	0.25	6.20	2.06	22.0	6.0
2.	10	14.90	16.50	1.910	0.43	0.21	10.3	2.44	25.5	14.3
3.	30	16.00	20.00	1.810	0.35	0.14	12.6	1.88	29.4	20.5
4.	50	16.50	23.90	1.740	0.23	0.10	10.7	1.34	31.0	24.5
5.	70	17.50	24.75	1.675	0.15	0.08	6.8	1.28	32.0	26.5
6.	90	18.40	25.90	1.605	0.09	0.06	6.2	1.16	34.5	30.0

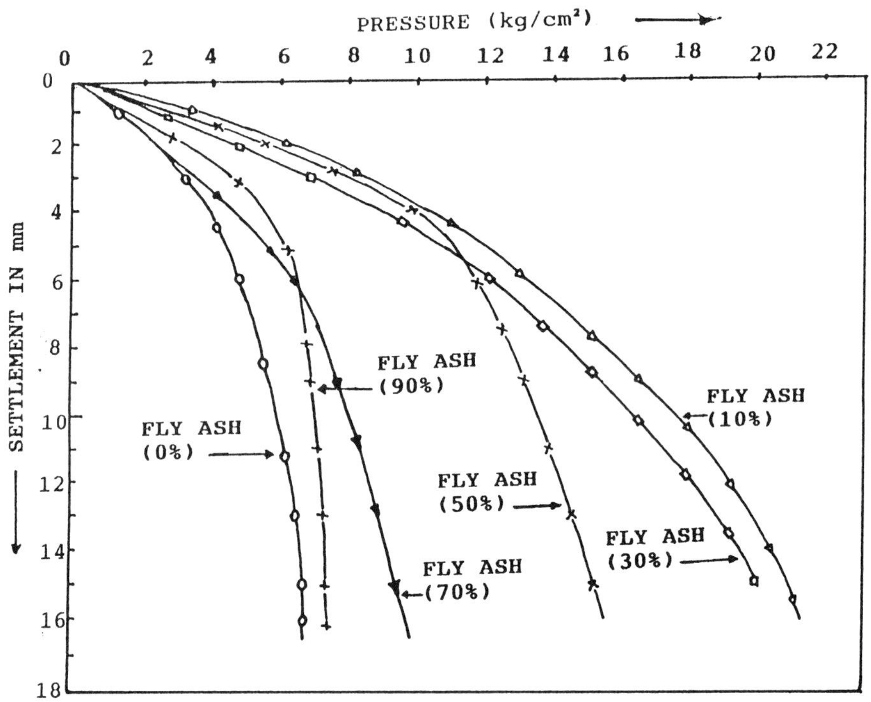

Fig. 1 - Pressure Vs. Settlement at OMC

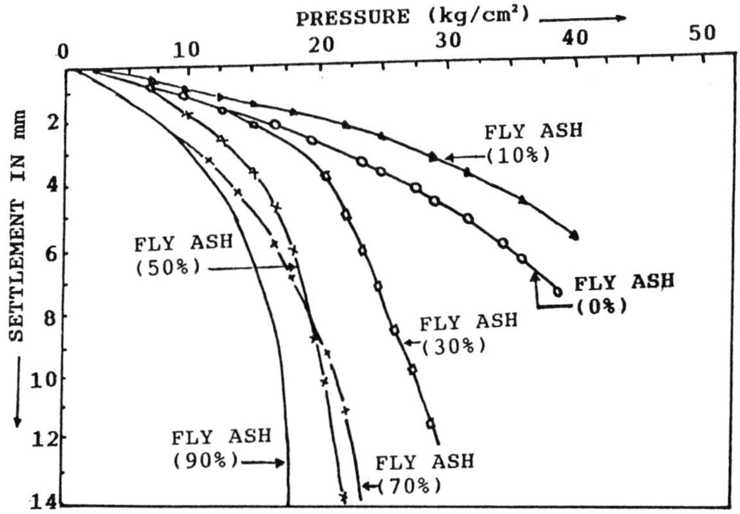

Fig. 2 - Pressure Vs. Settlement at Saturation

(ii) Influence of Geotextile on Flyash

Flyash can be used as an alternative to conventional borrow material for filling of low lying areas. Due to low unit weight of flyash, it becomes quite suitable for structural fills over soils with low pre consolidation pressure. But it has been found to have low bearing capacity. So to make it up to the constructional requirement, it was thought to reinforce the flyash with geotextiles/ geosynthetics, which have been found to be effective in other cases.

With this background in view, a series of two dimensional model footing tests were conducted in the laboratory using flyash as subgrade for both the cases i.e. unreinforced as well as reinforced case. The tests were conducted in an iron box of size 920mmx100mmx460mm using a model footing of size 100mmx100mm made out of well seasoned teak wood. The flyash bed was prepared by rainfall technique in two equal layers of six inch thickness and each layer being compacted by 10 number of uniformly distributed blows of standard proctor rammer over a wooden plank of size 920mmx100mmx40mm resting on the flyash subgrade. The relative density thus achieved was 34.0%, which was maintained constant for all the tests. The reinforcing materials used were geosynthetic and geotextile of thickness 0.1mm and 0.2mm respectively. In order to have an idea of effectiveness of friction on bearing capacity of the subgrade, tests were also conducted with geosynthetic coated with ennore as well as local sand for two different cases (Coated only on one side and on both sides).

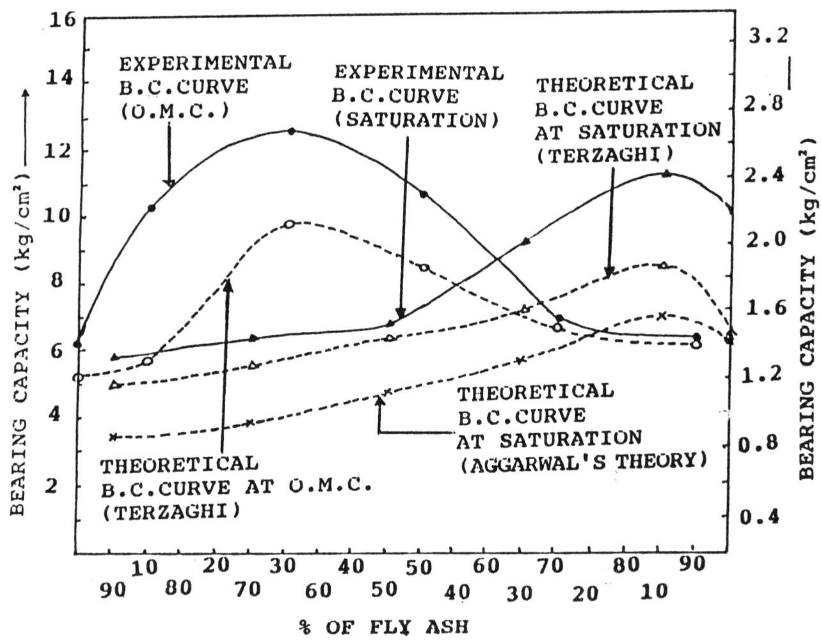

Fig. 3 - Bearing Capacity Vs. % of Flyash Mix

Parameters for study

(i) The ratio of depth of reinforcement layer below the footing(U) to width of footing(B), (i.e. U/B = 0.20, 0.60, and 1.0).

(ii) The ratio of length of reinforcement (L) to width of footing (B), (i.e. L/B = 4.0,6.0,8.0).

Table-2

Sr. No.	Type of Reinforcement	L/B	U/B	U.B.C.(KN/m^2)	BCR
1.	Unreinforced	-	-	78.50	-
2.	Geosynthetic	4	0.20	110.85	1.413
3.	-do-	4	0.60	127.53	1.625
4.	-do-	4	1.00	132.43	1.687
5.	-do-	6	0.20	118.70	1.512
6.	-do-	6	0.60	163.83	2.087
7.	-do-	6	1.00	122.63	1.562
8.	-do-	8	0.20	121.64	1.550
9.	-do-	8	0.60	114.78	1.462
10.	-do-	8	1.00	108.89	1.387
11.	Geotextile	4	0.20	144.21	1.837
12.	-do-	4	0.60	128.51	1.637
13.	-do-	4	1.00	137.34	1.749
14.	-do-	6	0.20	132.44	1.687
15.	-do-	6	0.60	146.17	1.862
16.	-do-	6	1.00	123.61	1.575
17.	-do-	8	0.20	103.99	1.325
18.	-do-	8	0.60	110.85	1.412
19.	-do-	8	1.00	130.47	1.662
20.	Geosynthetic coated with plain sand (one side)	6	0.60	215.82	2.749
21.	Geosynthetic coated with plain sand(both side)	6	0.60	223.67	2.849
22.	Geosynthetic coated with Ennore sand one side	6	0.60	196.20	2.499
23.	Geosynthetic coated with Ennore sand both side	6	0.60	204.05	2.599

The footing was placed centrally in the test tank and then loaded in small increments with the help of screw jack and settlement was recorded by the help of a set of two sensitive dial gauges placed diagonally opposite to each other(Fig. 4). The load settlement curve for each condition was obtained and the ultimate bearing capacity (UBC) was calculated using double tangent intersection method. The ultimate bearing capacity for unreinforced case (q_0) was also compared with that for

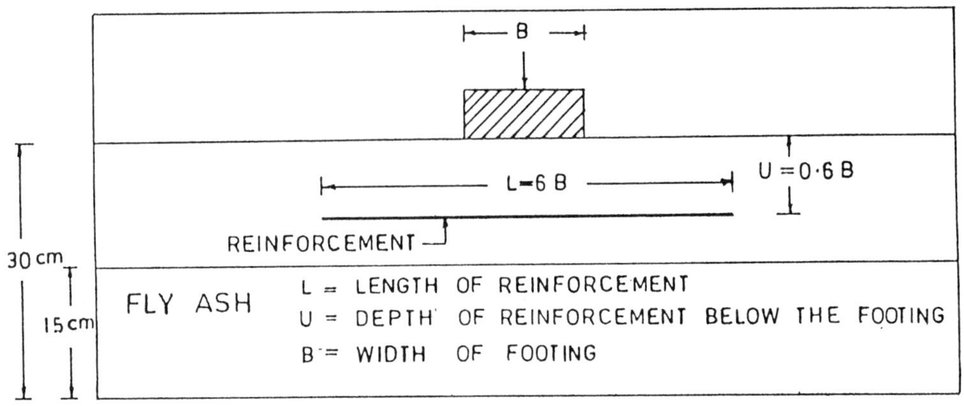

Fig. 4 - Placement of Reinforcement

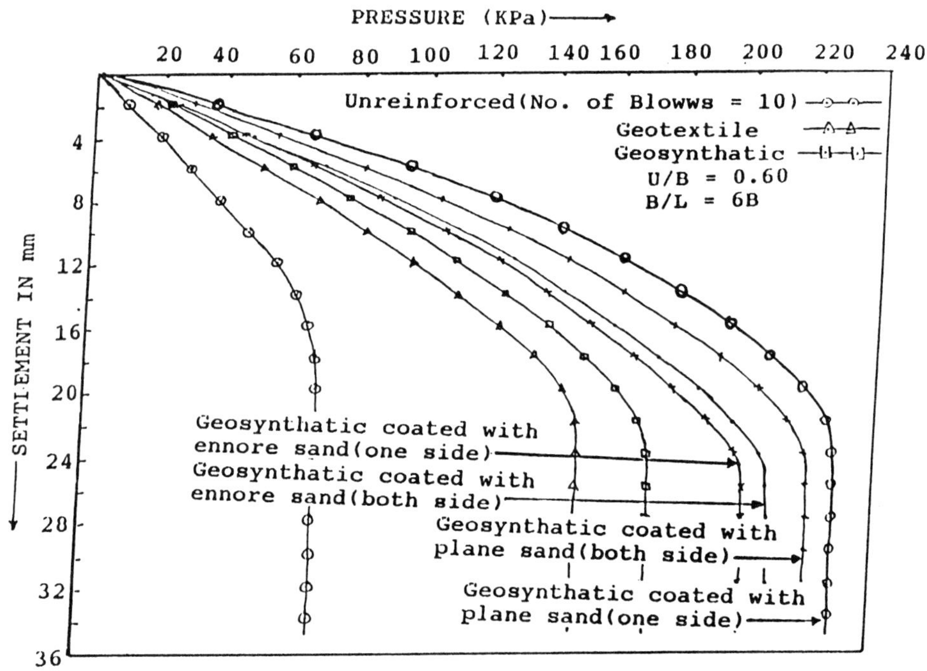

Fig. 5 - Pressure Vs. Settlement

the reinforced case (q) and the bearing capacity ratio (BCR) q/q_0 has been computed for all the tests. The bearing capacity value for different conditions have been given in Table -2. From Table 2 it can be seen that provision of single reinforcement having length (L) equal to 6B and placed at depth (U) equal to 0.60B nearly doubles the load carrying capacity of sub grades. Fig.5 shows the load settlement curves and optimum results obtained under different conditions.

It should be pointed out that there has not been a general consensus regarding the effect of depth ratio on bearing capacity of footing. The optimum depth ratios reported earlier are 0.25 and 0.175 for square foundation resting on reinforced sand(mild steel grids) and geogrid reinforced saturated clay respectively(9,10). The depth ratio becomes 0.3 when rectangular footing is supported over geogrid reinforced sand(11). In the present investigation the optimum depth ratio is 0.6. The disagreement with the results reported is most likely due to the difference in the material properties and geometrical dimensions of the reinforcement. Since the size, shape and spacing of geogrid bearing members together with the size of soil particles had a significant effect on geogrid soil reaction (12,13). Bearing capacity ratio generally increases with the increasing reinforcement size. The increase becomes more pronounced when reinforcement size varies from 3 to 8 times footing width(11,14,15,16). The present results also recommends the reinforcement length equal to six times the footing width, which is similar to other results reported.

CONCLUSIONS

Following are the main conclusions based on the findings of this study

(i) The cohesion `C' decreases and angle of internal friction (ϕ) increases for both the cases i.e. at O.M.C. and at saturation.

(ii) Soil flyash mix has the maximum bearing capacity at 30% flyash content in case of O.M.C. and at saturation it is only at 10% flyash content.

(iii) Flyash is quite suitable for structural fills over soils with low pre consolidation pressure and its load carrying capacity can be sufficiently improved by providing geotextiles/ geosynthetics reinforcement at suitable depth.

(iv) Roughness of reinforcing materials has marked influence on load carrying capacity of flyash subgrades.

(v) The trends observed for these conditions should be verified for larger footings. Additionally only surface footings were tested . The effect of footing embedment should also be included.

REFERENCES

1. **Dhavan,P.K.,Goswami,N.K. and Bhasin, N.K.**, Utilisation of flyash in Road construction. National workshop on utilization of flyash, Roorkee,pp. A.73- A.76,1988.
2. **Seehra, S.S and Arora, U.P.** Use of flyash in Road Construction. National workshop on utilization of flyash, Roorkee, pp. A.67 - A.72, 1988.
3. **Jain, K.K., Sahu, L. and Singh, S.K.** A study on the geotechnical properties of redmud and redmud flyash mixture. Indian Geotechnical Conference, Bombay, pp. 171 - 175,1990.
4. **Dembicki, E. and Alenowicz, J.M.** Influence of Geotextile on bearing capacity of two layer subsoil. First Indian Geotextile Conference on Reinforced Soil and Geotextile, Bombay, pp. A.61 - A.66, 1988.
5. **Perfetti, J. and Sangster, T.** The application for non woven needle punched Geotextile in low cost roads. First Indian Geotechnical Conference on Reinforced Soil and Geotextile, Bombay, pp. B.9 - B.14, 1988.
6. **Murtaza, G., Saha, S.S and Ahmad, M.** Improvement of load bearing characteristics of flyash using Geofabrics. First Indian Geotextile Conference on Reinforced Soil Geotextile, Bombay, pp. B.9 - B.14, 1988.
7. **Taylor, D.W.**, Fundamentals of Soil Mechanics, John wiley & Sons, New York.
8. **Agarwal, K.B.**, Bearing capacity in Expansive soils, Proceedings of Sixth International conference on Expansive Soils, New Delhi, pp 433-438, 1987.
9. **Singh, H.R.** Bearing Capacity of reinforced Soil beds. Ph.D. thesis, Indian Institute of Science, Bangalore, India, 1988.
10. **Mandal, J.N. and Saha, H.S.** Bearing Capacity tests on Geogrid reinforced Clay, Geotextile and Geomembranes, Vol. 11, Pp. 327, 1992.
11. **Yetimoglu, T., Jonathan, T.H. Wu, and Saglamer,A.** Bearing Capacity of rectangular footings on geogrid reinforced sand, Journal of Geotechnical Engineering Division, ASCE, Vol. 120, No. 12, pp. 2083 -2099, Dec. 1994.
12. **Jewell,R.A., Milligan, G.W.E., Sarsby, R.W. and Dubois, D.**, Interaction between soil and geogrids. Symposium polymer Grid Reinforcement in Civil Engineering, London, Pp. 18 - 30, 1984.
13. **Milligan, G.W.E. and Palmeria, E.** Prediction of bond between soil and reinforcement. Proceedings, International Conference in Geotechnical Engineering, Calgary, Alberta, pp. 147 -153, 1987.
14. **Omar, M.T., Das, B.M., Yen, S.C., Puri, U.K. and Cook, E.E.**, Ultimate bearing capacity of rectangular foundations on geogrid reinforced sand. Geotechnical Testing Journal, ASTM, Vol.16, No.2, pp. 246 - 252, 1993.
15. **Omar,M.T., Das, B.M.,Puri, V.K. and Yen,S.C.** Ultimate bearing capacity of shallow foundations on sand with geogrid reinforcement. Canadian Geotechnical Journal of geotechnical Engineering, Vol. 30,No.3, Pp. 545-549, 1993.
16. **Fragaszy, R.J. and Lawton, E.**, Bearing Capacity of reinforced sand subgrades. Journal of Geotechnical Engineering, ASCE, Vol. 110, No.10, Pp.1500- 1507, 1984.

Fifth International Conference on the
BEARING CAPASITY OF
ROADS AND AIRFIELDS

MOISTURE SUSCEPTIBILITY AND ELECTRICAL PROPERTIES OF BASE COURSE AGGREGATES

Timo Saarenketo[1], Tom Scullion[2] and Pauli Kolisoja[3]

[1]Roadscanners Oy Ltd, Valtakatu 21, FIN-96200 Rovaniemi, FINLAND,
Fax: +358 16 4200 511, e-mail: timo.saarenketo@roadscanners.fi
[2]Texas Transportation Institute, Texas A&M University,
College Station, TX 77843-3135, USA,
Fax: +1 409 845 0278, e-mail: t-scullion@tamu.edu
[3]Tampere University of Technology, Department of Civil Engineering,
P.O.Box 600, FIN-33101 Tampere, FINLAND
fax. +358 3 365 2884, e-mail: kolisoja @cc.tut.fi

ABSTRACT

This paper will report the results of a series of research projects performed in Finland and Texas concerning the relationship between electrical properties and strength and deformation properties of unbound base aggregates.

The research work has included measurements of dielectric value and electrical conductivity, suction tests and DCP and cyclic loading triaxial tests of known good and bad performing base materials from Finland and Texas. Tests have been performed in the laboratories of Texas Transportation Institute, Tampere University of Technology and the Road District of Lapland of Finnra.

The results have shown that suction properties can have a great effect on the moisture susceptibility of certain base materials. The resilient modulus of some base materials that were allowed to take on water were found to be less than 25% of the modli value measured when the material was dry. Results showed also that the known poorly performing base materials can have, due to high suction values, very high resilient modulus when they are dry or semidry. Important factors effecting the moisture susceptibility were the amount and quality of fines, i.e. cation exchange capacity, water adsorption properties and mineralogy of fines. These properties can be monitored in laboratory with a Tube Suction test developed during the research projects.

Key words: Ground Penetrating Radar, Dielectric Value, Electrical Conductivity, Base Aggregates, Bearing Capacity

INTRODUCTION

Ground Penetrating Radar (GPR) technique have been applied in Finland in road surveys for more than ten years mainly for measuring the thicknesses of structural layers and for locating damages in roads and finding possible causes. In North America the main applications in roads have mainly been measuring the thickness of the pavement and base layers for FWD back calculation purposes. However recently efforts have been made to apply GPR techniques to detect subsurface defects such as stripping in surfacing layers and voids beneath slabs or locating sinkholes beneath the highway structure /1/.

The two basic material properties affecting the propagation of GPR waves in road structures and subgrade soils are the dielectric value and electrical conductivity. The dielectric value of each road layer has to be estimated in order to calculate the thickness of the layer. As moisture content has a great effect both on the dielectric value and electrical conductivity there was interest in identifying the relationship between them and the actual volumetric or gravimetric moisture content of base materials.

Another reason for studying dielectric properties of base materials was the observation made in Finland that there seemed to be a correlation between low spring bearing capacity and a certain type of GPR reflection pattern from base courses. These relationships were noted even during the dry summer months when the road section was it's strongest. These reflection patterns could not be explained by any other reason than differences in the dielectric properties of the base material. There seemed to be also a very good correlation between the dielectric properties and strength and deformation properties of all types of soils and aggregates, i.e., the higher the dielectric value, the lower the spring bearing capacity.

This paper reports the results of several research project performed in Finland and Texas in order to study the relationship of dielectric value and electrical conductivity to the suction properties, moisture sensitivity, frost susceptibility, shear strength and resilient deformation response of different types of base aggregates from Finland and Texas.

FACTORS AFFECTING AGGREGATE STRENGTH AND DEFORMATION PROPERTIES

Besides the stress state, density, strength of the rock, particle shape and gradation the mechanical properties of all pavement materials and soils are to a greater or lesser extent affected by water. However the magnitude of these effects depends on the material properties, moisture content and even on the saturation history of the layer.

Water in soils and aggregates can be classified into 1. Adsorption water, which can be also called as hygroscopic water, 2. Viscous water, or capillary water and 3. Free water /2/. In adsorption water layer the water molecules closest to the mineral surface are systematically arranged toward the negatively charged mineral surface and the dielectric value of this water is thus low, close to the value of ice, 3.5 - 3.8. The bonding forces decrease with the distance from the mineral surface, and the outer water layers are very sensitive to the changes in pressure and temperature. The dielectric value of these water layers will be close to the dielectric value of free water, 81 /2/.

Both positive and negative pore-water pressures in a soil have a major effect on shear strength and volume change /3/. When discussing the negative pore-water pressure a term "Suction" is used for the thermodynamic quantity, Gibbs free energy which generates tension in the pore water between soil particles /4/. The total suction is composed of two components: 1. matric suction and 2. osmotic suction /5/.

In low moisture content base materials suction can increase the resilient modulus of aggregates /6, 7, 8/, but when the moisture content increases the effect of suction decreases /5/ and at high moisture contents positive pore water pressure can significantly decrease the base material resistance to recoverable and irrecoverable deformations /9/.

Hysteresis in the effect of wetting and drying, has an influence on the amount of suction in soils /5/. The matric suction is higher at a particular volumetric water content when the material is drying than at the same water content when it is wetting.

The freeze thaw process can also have a great influence on the recoverable and irrecoverable deformation of base materials. The grain size, pore size and total internal surface area of aggregates has a major influence not only on the freeze thaw durability, but also on the wetting-drying resistance and overall durability of aggregate /10/. The freezing process is also controlled by the amount of dissolved salts, relict salt and the byproducts of hydrolytic reactions, which effect the electrical conductivity of the base materials and which according to Kujala /11/ lowers the amount of free energy and thus lowers the freezing temperature. The freeze thaw process together with dynamic traffic loads can cause rapid disintegration of base course material with high specific surface area /12/ and there is also clear evidence of a correlation between thermal cracking on pavements and freeze thaw cycling properties of highly water adsorbing aggregates in Texas /13/.

According to Mitchell /14/ dielectricity can affect both attractive and repulsive forces in the interparticle system in soils. Dielectric values affect both the surface potential and diffuse layer thickness. The energy of repulsion is sensitive to changes in electrolyte concentration, cation valence, dielectric value, and pH; whereas the attractive energy is sensitive only to changes in the dielectric value and temperature.

ELECTRICAL PROPERTIES AND MOISTURE CONTENT OF BASE MATERIALS

Base material research started in 1993 in Road District of Lapland of Finnra in Finland with a study of the relationship between dielectric properties and moisture content of some Finnish aggregates and was continued in 1994-1995 at the Texas Transportation Institute in College Station, Texas, USA, where researchers used a special "bucket test" to obtain information on the dielectric value and electrical conductivity of eight Texas and three Finnish base aggregates at different moisture contents and densities /16/. Six of the tested materials were crushed from bedrock and five from gravel. To study the relation between dielectric and strength properties a DCP (Dynamic Cone Penetration) test was also performed.The results of the DCP-tests were transformed to CBR (California Bearing Ratio) values with the equation developed by US Army Corps of Engineers /17/. The CBR values range from 2 (very soft clay) to 100 (good quality base material). In order to study the effect of wetting and drying hysteresis on the dielectric and strength properties a special drying test was performed.

The results of Texas base material tests followed the results of the previous tests in Finland, ie. bad performing water sensitive aggregates had higher dielectrics and extremely water adsorbing aggregates, such as Iron ore gravel, had a lower dielectric value at the same moisture content than the good performers (Figure 1). The dielectric values increase as moisture content increases, but each material has its own characteristic curve. The explanation for this can be traced to the arrangements of water molecules around aggregate particles and to the amount of free water in the sample.

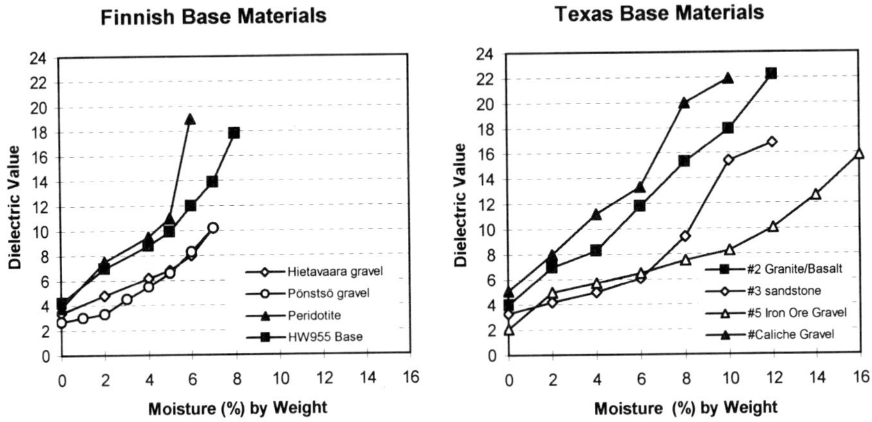

Figure 1. Correlation between gravimetric moisture content and dielectric value of Finnish (left) and Texas (right) base aggregates. Filled triangles and squares present problem aggregates

Figure 2 present the relationship of the mean CBR-value calculated from DCP-test results and the dielectric value and electrical conductivity. The measured CBR-values were relatively high with dielectric values less than 9 and after that stay almost at the same level until they reach the dielectric value of 16(see also figure 3). Changes in CBR-values are caused by the changes in the degree of compaction. Some of the test materials retained their strength properties even when the dielectric values were higher than 16, but in these cases, the material could not adsorb the added extra water and it was filtrating down to the bottom of the bucket. Thus, these conditions do not represent realistically in situ conditions inside a highway base course

When looking at the CBR-value vs. electrical conductivity ratio (Figure 2) CBR-value stayed relatively stable until electrical conductivity values became greater than 100 μS/cm, and then the strength properties of aggregates dropped. These results show that ion hydration might have a very important role in the strength properties of materials.

The hysteresis effect of wetting and drying to the strength and deformation properties of aggregates can be seen clearly in Figure 3, where the CBR-value vs. dielectric value of carbonate aggregates during wetting are plotted with the results measured during the drying cycle. The distributions of these results are the same as those published from the matric suction hysteresis /5, 4/ and explain how high resilient modulus values of the aggregates can be measured during the dry summer months.

The resilient modulus of four test aggregates were measured with dynamic triaxial test at two moisture contents in cooperation with the researchers of the Super Heavy Load Project in TTI /8/. The results of these tests followed the results of the bucket tests.

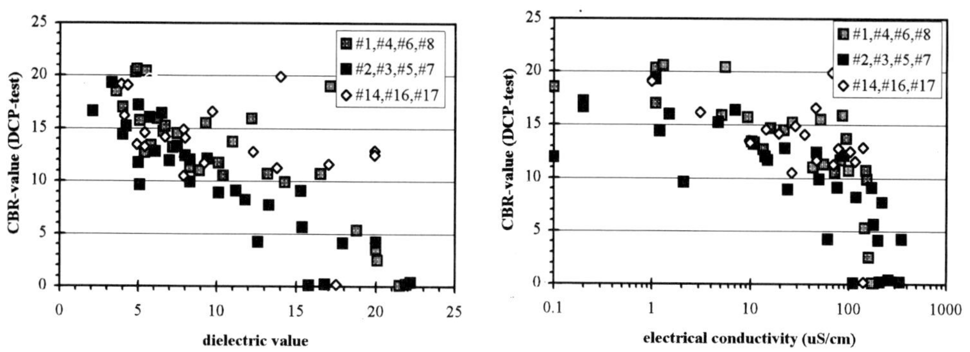

Figure 2. Correlation between CBR-value and dielectric value (left) and electrical conductivity (right) of Texas and Finnish base aggregates. Grey squares present Texas carbonate aggregates, black squares other Texas aggregates and open diamonds Finnish aggregates.

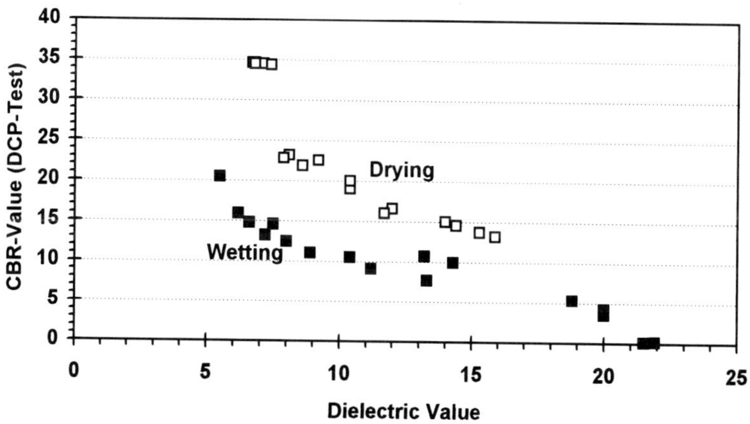

Figure 3. The effect of saturation and drying in the strength properties and dielectric value of Texas carbonate aggregates

TUBE SUCTION AND FREEZE-THAW TESTS

To estimate what is the equilibrium level of moisture in any base material, if water was available, researchers developed a special "tube suction test" /16, 17/. In this test base aggregates are compacted at optimum moisture content into a 200 mm high and 152 mm diameter plastic tube, which had 1 mm diameter holes about 10 mm above the bottom. Compacted samples were dried for 10 days at a temperature of about 40°C. After drying the dielectric value and electrical conductivity of the dry material were measured and the dried aggregate samples were then placed in a plastic tank with 20 mm deionised water. The water suction of the aggregate samples was measured by monitoring the changes in surface dielectric value over a 10 day period.

Tube Suction test performed both with Texas and Finnish aggregates have clearly shown that dielectric values of known bad performing base materials rise very rapidly and reach an asymptotic values above16, whereas the dielectric values of good performers does not markedly rise or if they rise they will stay under the value of 9 (Figure 4). Figure 5. shows the effect of fines content and fines quality on the Tube Suction value. These tests has been made with the Finnish aggregates at the Central Laboratory of the Road District of Lapland. Increases in the fines content effect the matric suction value and changes in fines quality i.e. cation exchange capacity (CEC), chloride content etc. influence the osmotic part of the total suction.

To evaluate what range of dielectric values should be expected if GPR surveys were performed in winter when the base course is frozen and what the effect of freezing and thawing has on these materials, a test was arranged to simulate freeze/thaw cycle in road bases. This freezing test was performed with the temperature of -5°C with the same samples that were used in the suction tests.

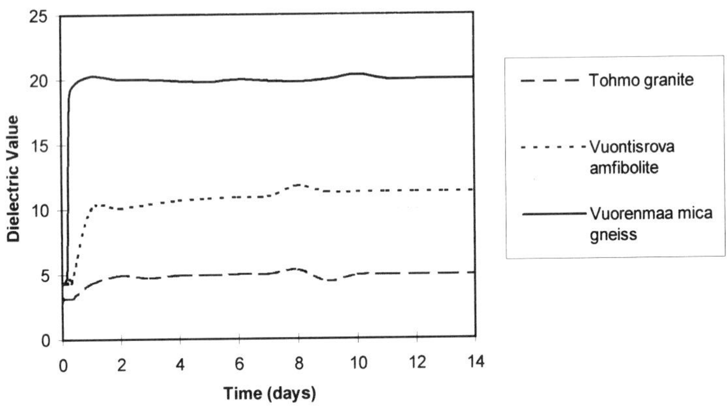

Figure 4. Typical Tube Suction test curves of good ,(granite), medium, (amfibolite) and bad (mica gneiss) performing Finnish base materials.

The results of freezing tests showed that dielectric values of frozen base samples started to rise from a value of 4, if the dielectric value of unfrozen base is higher than 9. This is an indicator of the presence of unfrozen water in frozen aggregates, which causes suction and is the greatest factor influencing the growth of ice lenses in pavement materials /11/. The dielectric limit value (9-10) for the frost susceptible aggregates are, according to these results, about the same as for the frost susceptible soils /19/.

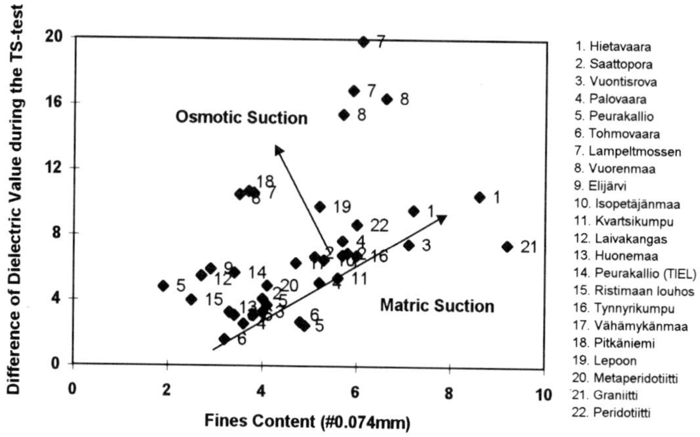

Figure 5. The effect of fines content (-200 mesh) to the difference of the dielectric value of wet base materials measured after the Tube Suction test and dry base material measured before the Tube Suction test. Measurements have been performed from 22 Finnish base materials at three fines content. Materials #3, amfibolite, #6, granite, and #8 mica gneiss are the same presented in figure 4. Materials higher from the matric suction line had higher CEC values and/or chloride contents.

RESILIENT MODULUS AND SUCTION TESTS

The resilient deformation response of the dry test materials/specimens and after they were allowed to suck water was determined using the large scale cyclic loading triaxial test facility available in the Laboratory of Geotechnical Engineering at the Tampere University of Technology. A photograph of the testing facility is shown elsewhere in these conference proceedings in connection with an article presenting the results of another series of large scale cyclic loading triaxial tests performed using the same equipment /20/. In addition to that article a more detailed description of the equipment can be found in some earlier publications /21,9/. The test specimen diameter that was used in the tests was 200 mm while the specimen height was about 400 mm. The selected test samples represented known good and bad performing base aggregates from the Road Districts of Lapland and Vaasa. Bad performers from Lapland were mainly mafic aggregates with high Mg, Ca content, and the problem aggregates from Vaasa were mainly gneisses with high mica content.

The resilient deformation response of the test specimen were tested according to the American SHRP protocol P46 /22/ which consist of 15 different combinations of constant confining pressure and cyclic axial load. As a result, 15 different values of resilient modulus M_r were obtained. Effect of suction value on the moisture sensitivity can be seen in figure 6. where resilient modulus values of granite and amphibolite base material are compared in the dry state to those obtained once the material has been allowed to suck in moisture and reach an equilibrium state. The difference between wet and dry resilient modulus values for the good performing aggregate is low typically less than 10 %. However for the known poor performer the difference was large, up to a 75 % reduction in resilient modulus was found.

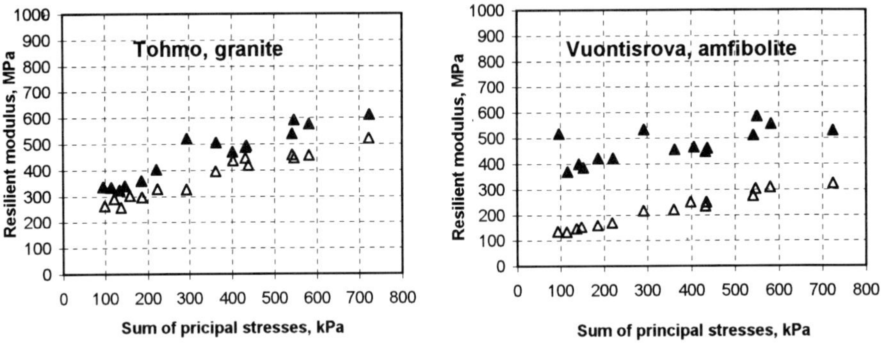

Figure 6. Resilient modulus test results of dry (filled triangle) and wet (open triangle) granite and amphibolite base materials. Materials are the same as presented in figures 4 and 5.

Some of the test materials showed extremely variable resilient modulus values when being dry which shows evidence of high suction value and cementation reactions during the drying phase. Some mica rich base materials from the area of the Road District of Vaasa on the other hand showed better resilient modulus values after the water suction than expected by the dielectric value measurements. The reason for that might be the measured low pH values of these aggregates, which prevents fines going into the water suspension under load repetitions and thus reducing high pore water pressure values. However pH does not improve material properties during the freeze-thaw cycles and in the field these aggregates have shown extremely bad performance during the spring thawing period.

CONCLUSIONS

Results of the surveys conducted have shown that dielectric value, which is a measure of how well the water molecules are arranged around and between the aggregate mineral surfaces and how much free water exists in materials, is a much better indicator of the strength and deformation properties of road aggregates than the moisture content. Furthermore, high dielectric values (calculated from GPR data for example) are good indicators of potential problems in the layer.

Tests have also clearly shown that suction has a great effect on the strength and deformation properties and moisture sensitivity of the base materials. Increasing fines content increases matric suction and increased cation content as a result of hydrolytic reactions and high chlorides content increase the osmotic part. Materials with high suction values have shown extremely high resilient modulus values when dry, but will loose their strength if the water is available. pH values seem to control these properties during the summer months but do not prevent the formation of segregation ice when the material freezes in moist conditions.

In general, dielectric values provide information about the position of the material in the aggregate moisture range, and electrical conductivity gives information about whether the material is on a drying or wetting cycle. High electrical conductivity values indicate loss of strength of a base course if the material has not low pH values. High dielectric values in base courses indicate always frost susceptible material. GPR and some other NDT methods can measure both of these values non-destructively and continuously. Water sensitive base aggregates can be identified even when they are dry from the dielectric dispersion properties. This can be done by using at least two GPR frequencies when measuring base course thickness and quality and/or comparing dielectric values or by studying changes in the reflected signal frequency response.

The results of these surveys can direct the strength and stiffness testing equipment, such as the Falling Weight Deflectometer, to the most important parts of the road and reduce the error when back calculating the layer moduli values.

ACKNOWLEDGMENT

This work has been conducted with the support of the Road District of Lapland and Vaasa of the Finnish National Road Administration, Texas Department of Transportation (TxDOT) and the Federal Highway Administration (FHWA).

REFERENCES

/1/ Saarenketo, T. and Scullion, T. 1994. Ground Penetrating Radar Applications on Roads and Highways. Research Report 1923-2F. Texas Transportation Institute, College Station, Texas. p. 36.

/2/ Saarenketo, T. 1996. Electrical Properties of Water in Soils. Proceeding of 6th International Conference on Ground Penetrating Radar (GPR'96). Sendai, Japan, p. 129-134.

/3/ Fredlund, D.G., Gan, J.K-M. and Gallen, P. 1995. Suction Measurements on Compacted Till Specimens and Indirect Filter Paper Calibration Technique. Transportation Research Board Paper 950891.

/4/ Lytton, R.L. 1994. Predicton of movement in expansive clays. Geotechnical Special Publications No. 40. "Vertical and Horizontal Deformations of Foundations and Embankments." Ed. A. Young, American Society of Civil Engineers.

/5/ Fredlund, D.G. and Rahardjo, H. 1993. Soil Mechanics for Unsaturated Soils, John Wiley & Sons, New York. p.517.

/6/ Thom, N.H., 1988. Design of Road Foundations. Th.D Thesis, Department of Civil Engineering, University of Nottingham.

/7/ Sweere, G.T.H. 1990. Unbound granular bases for roads. Ph.D. thesis, Delft University of Technology. Delft. 384 p.

/8/ Titus-Glover, L., and Fernando, E.G. 1995. "Evaluation of Pavement Base and Subgrade Material Properties and Test Procedures," Report 1335-2, Texas Transportation Institute, College Station, Texas.

/9/ Kolisoja, P. (1997). Resilient Deformation Characteristics of Granular Materials. Doctoral Thesis. Tampere University of Technology Publications 223. 188 p.

/10/ Hudec, P.P. and Achampong, F. 1994. Improving Aggregate Quality by Chemical Treatment. Report on the 2nd International Aggregates Symposium, Erlangen, Germany, October 22-27, 1990. ed. G.W. Luttig. p. 135-146.

/11/ Kujala, K. 1991. Factors affecting frost susceptibility and heaving pressure in soils. Acta Univ. Oul. C 58. p. 99.

/12/ Saarenketo, T. and Nieminen, P. 1989. Frost damage in road materials in Kittila, Northern Finland. Proceedings of Frost in Geotechnical Engineering. Vol 2. International Symposium, Saariselka, Finland, 13-15 March, 1989. Espoo 1989. ed. H. Rathmayer. p. 709-720.

/13/ Carpenter, S.H. and Lytton, R.L. 1977. Thermal Pavement Cracking in West Texas. Texas Transportation Institute Research Report 18-4F, Study 2-8-73-18. Texas A&M University, College Station, Texas. p. 232.

/14/ Mitchell, J.K. 1992. Fundamentals of Soil Behaviour. Second Edition. John Wiley & Sons, New York. p. 437.

/15/ Marjeta, J. 1993. Dielektrisyysmittarin käyttö murskeiden kosteuksen mittaukseen ja murskeiden laadun arviointiin. Insinöörityö, Rovaniemi Institute of Technology, Department of Construction technology. English Abstract: The Use of dielectric permittivity meter in moisture and qualitative determination of crushed gravel. p. 55.

/16/ Saarenketo, T. and Scullion, T. 1995. Using Electrical Properties to Classify the Strength Properties of Base Course Aggregates. Research Report 1341-2. Texas Transportation Institute, College Station, Texas.

/17/ Webster, S.L., Graw, R.H. and Williams, T.R.1992.Description and Application of Dual Mass Dynamic Cone Penetrometer" Report GL-92-3. USAE Waterways Expt Station, Vicksburg, MS 39180-6199

/18/ Scullion, T. and Saarenketo, T. 1997. Using Suction and Dielectric Measurements as Performance Indicators for Aggregate Base Materials. Transportation Research Record 1577, p. 37-44.

/19/ Saarenketo, T., 1995. The Use of Electric and Electrical Conductivity Measurements and Ground Penetrating Radar for Frost Susceptibility Evaluations of Subgrade Soils. Proceedings of the Symposium on the Application of Geophysics to Engineering and Environmental problems. Compiled by Ronald S. Bell. Orlando, Florida. pp.73-85.

/20/ Kolisoja, P., 1998. Resilient deformation behaviour of aggregates determined by means of cyclic loading triaxial tests. Proceedings of the 5th International Conference on the Bearing capacity of Roads and Airfields, Trondheim.

/21/ Kolisoja, P., 1994. Large scale dynamic triaxial tests with coarse grained aggregates. Proceedings of the 4th International Conference on the Bearing Capacity of Roads and Airfields. Minneapolis, USA. Vol. 2 pp. 883-897.

/22/ AASHTO T 294-92 I 1992. Interim method of test for resilient modulus of unbound granular base/subbase materials and subgrade soils - SHRP protocol P46. American Association of State Highway and Transportation Officials.

/23/ Tolla, P.,1986. Etelä-Pohjanmaan syväkivien soveltuvuudesta tien kantavan kerroksen materiaaliksi. Tampere Univeristy of Technology, masters thesis in civil engineering. 59 p.

Fifth International Conference on the
BEARING CAPASITY OF
ROADS AND AIRFIELDS

THE FAILURE MODEL OF ASPHALT PAVEMENT PAVING ON THE STABLIZED SOIL BASE COURSE

Wang Duanyi[1], Chen Yu[2], Wang Mao[3], Kou Xiaobo[4],
Lei Xiting[5] and Tian Shutian[6]

[1,2]Senior Engineer of Heilongjiang Transportation Research Institute, Harbin,China 150080
[3,4]Senior Engineer of Heilongjiang Communications Department, Harbin, China 150001
[5,6]Senior Engineer of Qiqihar Communications Bureau, Qiqihar, China 161001

ABSTRACT

Based on the strengthening principle of stabilized soil, the influence factors on the strength of stabilized soil were analyzed and the moisture situation of base course and subgrade in Heilongjiang province was investigated and evaluated. In the final part of this paper, the failure model of the asphalt pavement on the stabilized soil base course in Heilongjiang province was obtained.
KEY WORD Stabilized soil, Asphalt pavement, Failure model

Introduction

In the Heilongjiang province asphalt pavements have been built since the early 1960s. At that time and later, lime soil was our main base course for asphalt pavement. From 1970s to 1980s, cement and lime-cement treated soil were developed as base course for asphalt pavement. It should be said that the effect of these kinds of base course is positive to the highway transportation of the region, which was without sand and stone resources at that time and conditions. However, almost 40 years experience and failure investigation on asphalt pavements in the Heilongjiang province has indicated that asphalt pavements on stabilized soil base course are always exposed to premature failure and can not reach the assumed design life. For this reason, both direct and indirect losses are great.

Some results from the investigation indicate that the main reason to our province's asphalt pavement premature failure is lacking strength stability in the stabilized soil base course. The strength of the stabilized soil base course decrease after some years of use. In this paper, the failure model will be analyzed based on the data of laboratory and in situ investigation, as well as suggesting how to use stabilized soil material in our province's asphalt pavement structure.

Influence Factors on Strength of Stabilized Soil Material

The results of many studies have indicated that at a given time the strength of stabilized soil is a multiple various function as shown in equation (1),

$$S = F(a, b, c, d, e, f, g) \qquad (1)$$

where a is a mineral component parameter of soil; b is a parameter concerned in hydration component of soil; c is water content; d is the dose of mineral binder used; e is the curing time and age; f is curing condition; g is the initial strength of the stabilized soil material.

Analyzing construction data and condition, it can be said that the most important influence factor on the strength development of stabilized soil materials after construction is curing condition. Thus, a test was designed to examine if the stabilized soil is suitable for the base course in our area and how to use it, In the test, the following lime soil was adopted as an example.

Lime: Cao + Mgo content is 47.89%.

Soil: L.L. = 34%, Plastic Limit = 23.5%, PI = 10.5, belongs to silty clay loam.

The result of sieve test of soil used is shown in Table 1.

The test designed and results are shown as Table 2.

From Table 2 it can be seen that the saturation is the most obvious influence factor. Thus, we can reach the important conclusion, that if the stabilized soil after compacting is too wet, this will be unfavorable to the development of the strength of the material.

Evaluation of Wet Situation of Base Course

In general, the stabilized soil material should be at optimum water content after compacting. After paving, the asphalt pavement course blocks up the path of evaporation of water and pumping action during the day and night makes the stabilized soil material wetter and the water tends to saturation. Especially in the spring season, if there are ice crystal in the base course or subgrade, the material will be wetter than saturation. For example, in the Heilongjiang province the dry density of typical lime soil $r_d = 1{,}680\ \text{kg/m}^3$, the specific gravity of soil grain $g_s = 2.65$, the void of lime soil can be calculated with equation (2).

$$e = g_s r_w / r_d - 1 = 2.65 \times 1000 / 1680 - 1 = 0.58 \qquad (2)$$

where e is the void content of the material and r_w is the density of water.
If the optimum water content of the material is w = 18%, the saturation degree is

$$s = wgs / e = 18\% \times 2.65 / 0.58 = 82\% \qquad (3)$$

If water content of material approaches to saturation and the dry density is kept constant, the water content should be

$$w = se / g_s = 0.58 \times 100\% / 2.65 = 21.9\% \qquad (4)$$

If the volume of the material increases 1% (heave or other), the difference is

$$\Delta w = \Delta v\ r_w / r_d = 1\% \times 1000 / 1680 = 0.6\% \qquad (5)$$

where Δv is the difference of volume. At this time, the water content is w = 21.9% + 0.6% = 22.5. For most the area of the Heilongjiang province, the average freezing depth is 1.5 m to 2.0 m in the winter season. 1% of heave rate means that the surface of the road will heave 15 to 20 mm. This is possible if there is enough water supply and the soil is susceptible for freezing such as silty clay.

Table 3 shows water content investigation results of base course and subgrade in the spring season. Comparing Table3 to above calculated results, it is seen which saturation situation most base course and subgrades of Heilongjiang province's asphalt pavements are in.

The above calculation is based on the assumption that the dry density of the material is constant. If material for a long time is in an over-saturated situation, with the action of surface tension, the volume of voids will increase. This will result in decrease of the dry density. This difference can be calculated with equation (2) when g_s and r_w are kept constant and the difference of void rate is 1%.

$$\Delta r_d = -g_s\, r_w\, (1+e) - 2\Delta e = -\ 10.26 \qquad (6)$$

The decrease of dry density will influence on the development of the strength.

Influence of Freezing on Stabilized Soil Strength

The study indicate that after winter, even for road section, built in a dry situation, the strength of stabilized soil will also decrease. The resilience modules testing results with bearing plate conducted on Zhaodong test road of the Heilongjiang province are shown in Table 4.

From Table 4, it can be seen that:

(1) In the upper 20 cm of the pavement (the thickness of asphalt pavement is 4 cm), the decreasing strength is more obvious than in the sub-layer after winter.

(2) The lower strength before winter, the higher decrease strength after winter.

Conclusion

For road sections in wet conditions, with the action of freezing and water, the stabilized soil material can not provide enough strength and stability as base course as shown by the testing results conducted in the laboratory. Thus, the suggestion is that the stabilized soil material can not be used as base course for high class highways, and that it only can be used as subbase course. The action of freezing should also be considered when the stabilized soil material is used in a dry road section.

The failure model of asphalt pavement paving on stabilized soil base course in the Heilongjiang province can be proposed based on above analysis. It is

Asphalt pavement being up to standard
| over saturation of water, freezing and thawing
| and the action of traffic load
The strength of stabilized soil goes down
| circulation action
Asphalt pavement is failed

List of Reference

1. «Study on The Heave of Highway Subgrade in Seasonal Freezing Area»
   Heilongjiang Transportation Research Institute, 1993
2. «Investigating Report of Zhaodong Test Road»
   Heilongjiang Transportation Research Institute, 1985

Table 1 The result of sieve test of soil used

mm	<0.100	<0.050	<0.010	<0.005	<0.002
%	98.80	92.85	39.41	30.10	21.90

Table 2 The test designed and results

No.	Dose (%)	Curing	Freezing	Saturation	Age	Strength (MPa)
1	10	seal*	Y	Y	7	0.3
2	10	seal	Y	N	28	0.6
3	10	normal	N	Y	7	0.4
4	10	normal	N	N	28	0.5
5	12	seal	N	Y	28	0.34
6	12	seal	N	N	7	0.42
7	12	normal	Y	Y	28	0.34
8	12	normal	Y	N	7	0.41
I	1.52	1.66	1.66	1.38	1.53	
II	1.51	1.65	1.65	1.93	1.78	
R	0.01	0.01	0.01	0.55	0.25	

Note *: The specimen was sealed with emulsified asphalt.

Table 3 The investigation results of water content

Name of Road	Water Content (%)	
	Base course	Subgrade
Airport	22	23
Jingha	28	20
Yayu	25	30
Hatong	23	30
Heida	21	20
Hayi	16	19

Table 4 The resilience modules testing results

material	one month		
	dose (%)	thickness (cm)	average modules (MPa)
lime soil	11.5	15-16	3480
	8	12	4196
	11.5	39	2047
	8	34	2041
cement soil	10	12-14	4867
	10	33-38	1586
	spring season		
Lime soil	11.5	15-16	1567
	8	12	955
	11.5	39	2040
	8	34	610
Cement soil	10	12-14	1263
	10	33-38	603

Fifth International Conference on the
BEARING CAPASITY OF
ROADS AND AIRFIELDS

THE APPLICATION OF UNSATURATED SOIL MECHANICS THEORY TO THE DESIGN OF PAVEMENTS

S.Y. OLOO[1] AND D.G. FREDLUND[2]

[1]Dr., Department of Civil Engineering, University of Durban-Westville, Private bag X54001, Durban 4000, South Africa, Fax: 027 031 204 4755, E-mail: syoloo@pixie.udw.co.za
[2]Professor, Department of Civil Engineering, University of Saskatchewan, 57 Campus Drive, Saskatoon, Saskatchewan, Canada S7N 5A9, Fax: 306 966 5427 E-mail: Del_Fredlund@engr.usask.ca

ABSTRACT

Soil mechanics principles have been applied to the evaluation, measurement and prediction of resilient modulus and shear strength of subgrades and unbound pavement materials. Pavement layers are composed of compacted materials and often exist above the water table and are therefore unsaturated The lack of appropriate methods of defining the state of stresses in the unsaturated pavement materials has, for a long time, imposed limitations on the application of soil mechanics theories to the rational design of pavements. Recent advances in the theory of unsaturated soils allow for the incorporation of the net normal stress and matric suction as variables in the nonlinear response of unsaturated pavement materials to traffic loading. The influence of matric suction on the resilient modulus and shear strength of pavement materials as well as the incorporation of matric suction in saturated/unsaturated moisture flow models that can be used to predict equilibrium matric suction profiles under pavements are discussed in this paper.

Key words: Pavement design, matric suction, resilient modulus, shear strength

INTRODUCTION

Flexible pavements are made up of bound and unbound granular layers which distribute traffic stresses to the subgrade. The subgrade acts as the foundation for the pavement structure. The rational design of a pavement structure requires an understanding of the fundamental behaviour of both the pavements materials and the subgrade. Soil mechanics principles have been applied to the evaluation, measurement and prediction of resilient properties and shear strength of pavement materials.

Pavement layers are usually compacted and often exist above the water table and are therefore unsaturated. Lack of a suitable theoretical framework to deal with the state of stress in unsaturated pavement layers has, for a long time, imposed limitations on the application of soil mechanics theories to the rational design of pavements.

Matric suction and the net normal stress are now widely accepted as the stress state variable controlling the behaviour of unsaturated soils. Unsaturated soils theories allow for the incorporation of the net normal stress and matric suction as variables in the nonlinear response of unsaturated pavement materials to traffic loading.

This paper discusses the influence of matric suction on the resilient modulus and shear strength of pavement materials as well as the incorporation of matric suction in saturated/unsaturated moisture flow models that can be used to predict equilibrium matric suction profiles under pavements.

INFLUENCE OF MATRIC SUCTION ON THE RESILIENT MODULUS

Resilient deformations are the component of deformations that are recoverable after every loading cycle in a repeated loading test. They remain fairly constant even close to failure (1, 2). The resilient response of granular materials is different from that of fine grained soils. Granular soils exhibit a general increase in resilient modulus with increasing stress level (3). The general response of granular materials can be expressed as follows:

$$M_R = (\sigma_1 + \sigma_2 + \sigma_3)^n \qquad [1]$$

where k and n are material constants determined experimentally.

Fine grained soils tend to exhibit a decrease in resilient modulus at increasing levels of deviator stress. The relationship between resilient modulus and deviator stress can be expressed as follows (2):

$$M_R = k_2 - k_3(k_1 - \sigma_d) \qquad \text{for } k_1 > \sigma_d \qquad [2]$$

$$M_R = k_2 + k_3(\sigma_d - k_1) \qquad \text{for } k_1 < \sigma_d \qquad [3]$$

where k_1, k_2, k_3 and k_4 are material constants and σ_d is the deviator stress.

Prediction of strain in the pavement structure for pavement design purposes requires a knowledge of the resilient behaviour of the materials at conditions representative of those that may be expected to occur in-situ. These expected conditions will be controlled not only by the initial placement conditions but also by the environment in which the materials exist.

Moisture conditions in the pavement layers are controlled by climate and soil conditions. Russam (4) observed that water contents in subgrades reach equilibrium values under given environmental conditions. In many cases, the equilibrium water contents in subgrades do not reach saturation. Negative water pressures or matric suction should therefore be incorporated in the characterization of pavement materials for pavement design.

Fredlund and Morgenstern (5) demonstrated that matric suction, $(u_a - u_w)$, and the net normal stress, $(\sigma_1 - u_a)$, were the stress state variables controlling the behaviour of unsaturated soils. Repeated load triaxial test results on a glacial till reported by Sauer and Monismith (6) indicated a significant increase of resilient modulus with increasing matric suction.

Fredlund et al., (7) , on the basis of stress analysis, showed that the resilient modulus is a function of three stress variables:

$$M_R = f[(\sigma_1 - u_a), \sigma_d, (u_a - u_w)] \quad [4]$$

Repeated load triaxial tests conducted on compacted samples of a glacial till and reported by Fredlund et al., (8) confirmed the existence of a good correlation between resilient modulus and the stress variables. The relationship shown in Fig. 1, though derived from a series of cross plots, indicates a general increase of the resilient modulus with matric suction.

Jin et al., (9) carried out repeated load triaxial test on reconstituted soil specimens under controlled temperature, water content, dry density and stress conditions. The results indicated a decrease in resilient modulus with increasing water content. A general increase of resilient modulus with suction was inferred from the a soil-water characteristic curve developed by Sauer (10) for a glacial till.

Phillips and Cameron (11) determined the relationship between the resilient modulus and the soil suction for an expansive clay subgrade. At constant dry density the relationship between resilient modulus and suction was found to be linear. They modelled the resilient modulus of the samples tested using May and Witczak's equation (12):

$$M_R = k_o \left(\frac{\sigma_m}{p_a} \right)^{k_1} \left(\frac{\tau_{oct}}{\tau_{ref}} \right)^{k_2} \quad [5]$$

where σ_m = mean stress, p_a = atmospheric pressure, τ_{oct} = octahedral shear stress, τ_{ref} = reference octahedral shear stress and k_o, k_1, k_2 are regression constants not similar to the constant in Eq. 2 and 3.

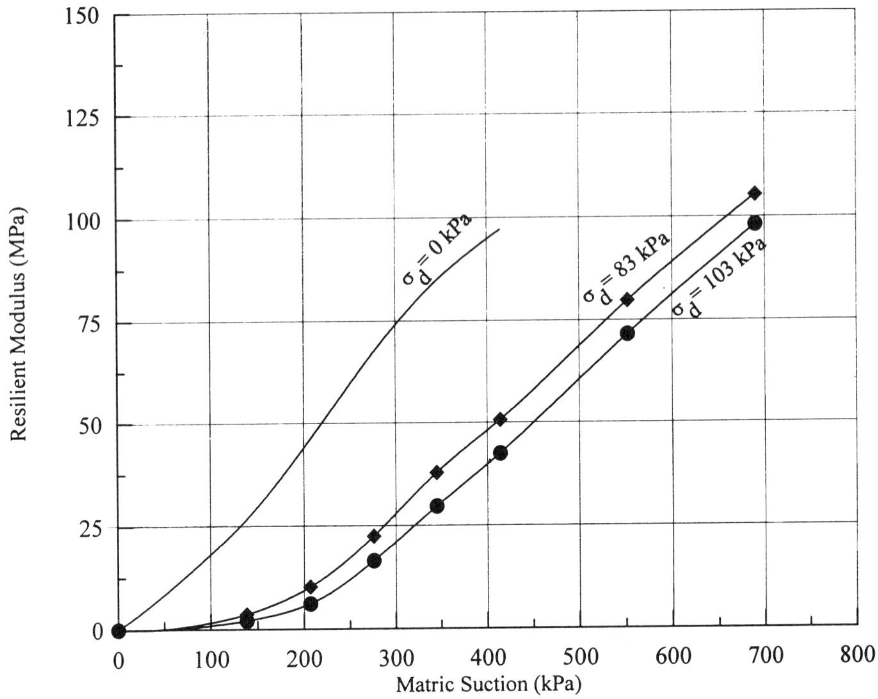

Fig. 1: The relationship between resilient modulus, matric suction and deviator stress (8).

By using statistical analysis to determine the regression constants k_o, k_1, and k_2, Phillips and Cameron (11) derived the relationship shown in Fig. 2 between the resilient modulus and the soil suction.

It is apparent from Fig. 2 and to a lesser extent from Fig. 1 that the form of the variation of the resilient modulus with suction is similar to that between shear strength and matric suction. The resilient modulus increases from a minimum value corresponding to a saturated soil. At high levels of matric suction, the rate of increase of resilient modulus with matric suction becomes negligible.

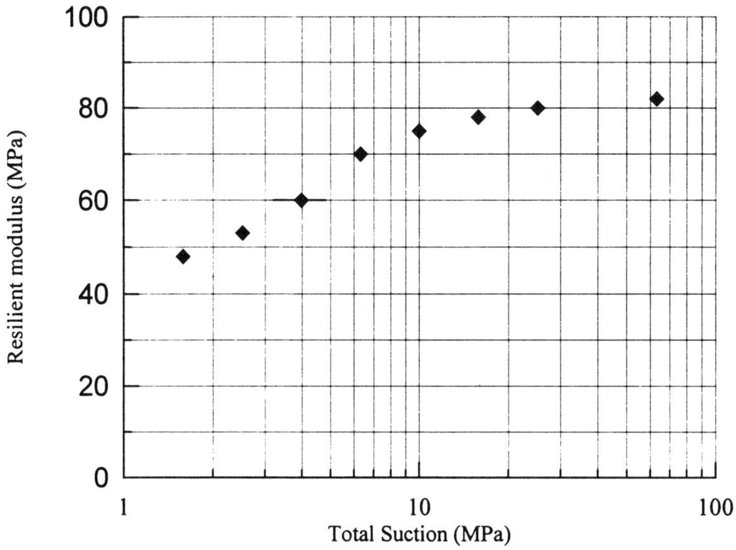

Fig. 2: The relationship between resilient modulus and soil suction at a constant deviator stress (11).

It follows, therefore, that the relationship between the resilient modulus, matric suction and deviator stress can be represented as follows for coarse grained soils:

$$M_R = k(\sigma_1 + \sigma_2 + \sigma_3)^n + k_s(u_a - u_w) \qquad [6]$$

And for fine grained soils:

$$M_R = k_2 - k_3(k_1 - \sigma_d) + k_s(u_a - u_w) \qquad \text{for } k_1 > \sigma_d \qquad [7]$$

$$M_R = k_2 + k_4(\sigma_d - k_1) + k_s(u_a - u_w) \qquad \text{for } k_1 < \sigma_d \qquad [8]$$

where k_s is a parameter that defines the rate of change of resilient modulus with matric suction, $(u_a - u_w)$. The parameter, k_s, is dependent on the soil type and similar in definition to $(\tan\phi^b)$ for the case of shear strength of unsaturated soils.

The above relationships define surfaces in three dimensional plots of resilient modulus, deviator stress and matric suction. An experimental program involving the measurement of the resilient modulus in a modified repeated loading triaxial apparatus that allows for the control of matric suction in addition to the applied deviator stress and confining pressure is under development to define the surfaces.

EFFECT OF MATRIC SUCTION ON SHEAR STRENGTH

Rutting in pavements is caused by permanent deformations and plastic flow of the pavement materials as a result of traffic loading. Permanent deformations of granular and subgrade materials depend on the magnitude, number, sequence of repeated principle stresses as well as on the shear strength of the materials.

The shear strength of a soil is defined in terms of cohesion, c and angle of friction, ϕ. To be consistent with traffic loading conditions, c and ϕ must be determined under repeated loading conditions. Data presented by Schimming et al., (13) and Seed and Chan (14) indicate that the shear strength of saturated cohesive soils is higher under transient loading than under static loading conditions. On the other hand, the shear strength of dry granular soils does not seem to be affected by the rate of loading (15, 16). The authors are not aware of reports in the literature of attempts to incorporate the effect of loading rate on the shear strength of pavement materials in design.

Raad and Figueroa (17) studied the influence of including the shear strength parameters c and ϕ on the response of pavements using the finite element method. A nonlinear finite element formulation with a Mohr-Coulomb failure criterion was utilised in the determination of stresses and strains in a 2-layer pavement structure consisting of a granular base and cohesive subgrade. The results indicated that the resilient deformations of the pavement were insensitive to variations in the shear strength parameters. However, the shear strength parameters played a significant role in the development of plastic zones under the loaded area. The development of plastic zones is directly related to permanent deformations in the pavement. The lower the shear strength, the higher the likelihood of large permanent deformations developing in the form of rutting. A number of improved structural design methods presently incorporate c and ϕ as well as a failure criterion in the prediction of rutting.

Because pavement layers are usually unsaturated, appropriate shear strength parameters should incorporate the influence of matric suction. The shear strength of the unsaturated soils are defined in terms of the two stress state variables, net normal stress, $(\sigma - u_a)$, and matric suction, $(u_a - u_w)$. Fredlund et al., (18) proposed the following equation for the shear strength of an unsaturated soil:

$$\tau_f = c' + (\sigma - u_a)\tan\phi' + (u_a - u_w)\tan\phi^b \qquad [9]$$

where: c' = effective cohesion of the soil, ϕ' = effective angle of friction, and ϕ^b = friction angle associated with the matric suction in the soil.

The relationship between shear strength and matric suction is nonlinear and takes the form shown in Fig. 3 at constant normal stress. The slope of the curve defines ϕ^b which decreases with increasing matric suction from a maximum of ϕ' for a saturated soil beyond the air entry value.

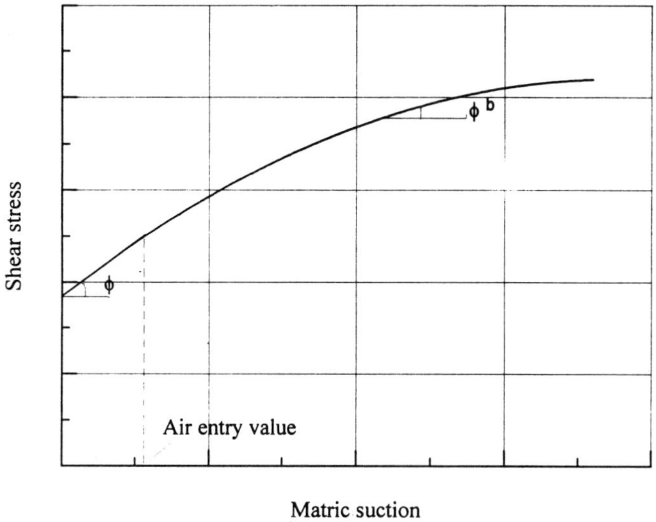

Fig. 3: Relationship between shear strength and matric suction at constant net normal stress

Matric suction can be visualized as contributing to the cohesion of the soil and Eq. 9 re-arranged as follows:

$$\tau_f = \left[c' + \left(u_a - u_w\right)\tan\phi^b\right] + \left(\sigma - u_a\right)\tan\phi' \qquad [10]$$

The influence of matric suction on the shear strength of pavement materials can be incorporated in design by using the following value for cohesion:

$$c = c' + \left(u_a - u_w\right) \tan\phi^b \qquad [11]$$

where $(u_a - u_w)$ is the matric suction in the pavement layer under consideration and ϕ^b corresponds to that level of suction. The friction angle to be used in analysis remains unchanged.

EQUILIBRIUM MATRIC SUCTION

The importance of incorporating matric suction into the resilient modulus and shear strength of unsaturated pavement materials has been discussed. A suitable value of matric suction to be used should correspond to the anticipated equilibrium water content under the pavement. Moisture conditions in the pavement layers are controlled by climatic and soil conditions and tend towards an equilibrium value under given environmental conditions. Aitchison and

Richards (19) and Loi et al., (20) have reported data on the development of matric suction beneath pavements. Studies reported by Crilly et al., (21) and Oloo and Fredlund (22) have also confirmed the existence of long term negative water pressures in subgrades.

The distribution of moisture in a layered soil depends on the relative hydraulic conductivities and the water retention characteristics of the soil in each layer. Saturated-unsaturated flow modelling that incorporates these parameters as well as flux boundary conditions (23, 24) are likely to give more insight into the problem of moisture distribution in layered soils. However, the determination of the surface fluxes to be used in the modelling requires the characterization of climatic and ground surface conditions and the solution of a system of coupled heat and mass transfer equations. Lytton et al., (23) presented a comprehensive model for the prediction of matric suction variations under pavements based on a finite difference solution of the coupled heat and mass transfer equations. Their model provides a means of characterizing climate as well. Solutions of the coupled heat and mass transfer equations have also been presented by Wilson (25) based on the finite difference method.

CONCLUSIONS

A relationship between resilient modulus, net normal stress and matric suction has been proposed based theoretical considerations and experimental observations of the behaviour of clay subgrades under repeated loading. Further experiments are planned to confirm the existence and form of the surface defining resilient modulus in a 3-dimensional space. The incorporation of matric suction in formulations of shear strength is fairly well understood and can be incorporated in pavement design as presented in this paper. Although basic research work has been done on saturated/unsaturated flow modelling and on the prediction of flux boundary conditions, further work is required to enable the theory to be used for the prediction of matric suction profiles under pavements.

REFERENCES

1. HAYNES, H.J., and YODER, E.J. Effects of repeated loading on gravel and crushed stone base coarse materials used in the AASHO road test. Highway Research Record, No. 39, Highways Research Board, Washington D.C., pp 82-96, 1963.

2. SEED, H.B., CHAN, C.K., and LEE, C. E. Resilience characterization of subgrade soils and their relation to fatigue failures in asphalt pavements. Proceedings 1st International Conference on Structural Design of Asphalt Pavements, Ann Arbor, Michigan, pp. 77-113, 1962.

3. KNUTSON, R. M., et al. Materials evaluation study - ballast and foundation materials research program. Technical Report No. FRA-OR&D-77-02. Federal Railway Administration, Washington, D.C, 1977.

4. RUSSAM, K. The prediction of subgrade moisture conditions for design purposes. Moisture Equilibria and Moisture Changes in Soils Beneath Covered Areas, Butterworth, Sydney, Australia, 1965.

5. FREDLUND, D. G., and MORGENSTERN, N. R. Stress state variable for unsaturated soils. ASCE Journal of the Geotechnical Engineering Division, **103**(GT5): 447-466, 1977.

6. SAUER, E.K., and MONISMITH, C. L. Influence of soil suction on behaviour of a glacial till subjected to repeated loading. Highway Research record, 215, TRB, Washington, D. C., pp. 18-23, 1968.

7. FREDLUND, D. G., BERGAN, A. T., and SAUER, E. K. The deformation characteristics of subgrade soils for highways and runways in Northern environments. Canadian Geotechnical Journal, 12, (2), pp. 213-223. 1975.

8. FREDLUND, D. G., BERGAN, A. T., and WONG, P. K. Relation between resilient modulus and stress conditions for cohesive subgrades. Transportation Research Record No. 642, TRB, pp. 73-81. 1977.

9. JIN M. S. and LEE, K. W. Seasonal variation of resilient modulus subgrade soils. Journal of Transportation Engineering, Vol. 120, No. 4, pp. 603-616. 1994.

10. SAUER, E. K. 1967. Application of geotechnical principles to road design problems. Dr. of Eng. Thesis, University of California, Berkeley, California, USA

11. PHILLIP, A.W. and CAMERON, D. A. The influence of soil suction on the resilient modulus of expansive soil subgrades. Proc. First International Conference on Unsaturated Soils, Vol. 2, Paris, France, pp. 171-176, 1995 .

12. MAY, R. W. and WITCZAK, M. W. Effective granular modulus to model pavement responses. Transportation Research Record, 810, TRB, pp. 1-19, 1981.

13. SCHIMMING, B.B., HAAS, H.J., and SAXE, H. C. Study of dynamic and static failure envelopes. Journal of Soil Mechanics and Foundations Division, ASCE, Vol. 92, SM2, pp. 105-124, 1966.

14. SEED, H.B., and CHAN, C.K. Clay strength under earthquake loading conditions. Journal of Soil Mechanics and Foundations Division, ASCE, Vol. 92, SM2, pp. 53-78, 1966.

15. HEALY, R.A. The response of soils to dynamic loading. Report II: Triaxial tests on saturated fine silty sand. U.S Army Corps of Engineers, Waterways Experimental Station, Vicksburg, Mississippi, 1962

16. WHITMAN, R.B., and HEALY, R.A. The response of soils to dynamic loads. Report 9: Shearing resistance of sands during rapid loading. U.S Army Corps of Engineers, Waterways Experimental Station, Vicksburg, Mississippi, 1962.

17. RAAD, L., and FIGUEROA, J. L. Load response of transportation support systems. Transportation Engineering Journal, ASCE, Vol. 106, No. TE1, pp. 111-129, 1980.

18. FREDLUND, D. G., MORGENSTERN, N. R., and WIDGER, R. A. The shear strength of unsaturated soils. Canadian Geotechnical Journal, **15**(3): 316-321, 1978.

19. AITCHISON, D.G., and RICHARDS, B.G. A broad-scale study of moisture conditions in pavement subgrades throughout Australia. Moisture Equilibria and Moisture Changes Beneath Covered Areas. Butterworth, pp. 184-225, 1965.

20. LOI, J., FREDLUND, D. G., GAN, J., K-M and WIDGER, R.A. Monitoring soil suction on an indoor test track. 7th Annual Meeting of TRB, Washington, D. C., 1992.

21. CRILLY, M. S., SHREINER, H. D. and GOURLEY, C. S. A simple field suction measurement probe. Proc. 10th Regional Conf. For Africa on soil Mech. Found. Eng. And 3rd Conf. On Tropical and Residual Soils, Maseru, pp. 291-298. 1991

22. OLOO, S.Y. and FREDLUND, D.G. Matric suction monitoring in an expansive soil subgrade in Kenya. Proceedings of the 1st International Conference on Unsaturated Soils, Paris, France, 6-8 September, pp. 631-636,1995.

23. LYTTON, R. L., PUFAHL, D. E., MICHALAK, C. H., LIANG, H. S., and DEMPSEY, B. J. An integrated model of the climatic effects on pavements. Final Report No. RF 7089. Texas Transportation Institute, Texas A&M University, Texas, 1990.

24. BARBOUR, S. L., FREDLUND, D. G., GAN, J. K. M., and WILSON, G. W. Prediction of moisture movements in highway subgrade soils. Proc., 45th Canadian Geotechnical Conference, Toronto, pp. 41A-1 - 41A-13, 1992.

25. WILSON, G. W. Soil evaporative fluxes for geotechnical engineering problems. Ph.D. Thesis, University of Saskatchewan, Saskatoon, Canada, 1990.

Fifth International Conference on the
BEARING CAPASITY OF
ROADS AND AIRFIELDS

THE NEW DYNAPLAQUE FOR THE CONTROL OF THE DEFORMABILITY OF PAVEMENT SUBGRADES

JEAN-FRANÇOIS CORTE[1], JEAN-MICHEL PIAU[1], PAUL CHASSAING[2], MICHEL FROUMENTIN[3]

[1]Laboratoire Central des Ponts et Chaussées, B.P. 19, 44340 Bouguenais, France
[2]Centre d'Etude et de Construction de Prototypes, 49000, Angers, France
[3]Centre d'Expérimentation Routière de Rouen, 76 ,Grand-Quevilly, France

ABSTRACT

Over the last twenty years, the deformability of pavement subgrades for new roads is being assessed in France by the use of the Dynaplaque apparatus. With the first generation of Dynaplaque, Young's modulus of the subgrade is estimated from the bounce height of a guided mass falling on a 60 cm diameter steel plate. A new type of Dynaplaque was built a few years ago by the Laboratoires des Ponts et Chaussées. With this new device, the force applied to the plate in contact with the soil is directly measured during the impact together with the vertical displacement of the plate. Changes in the instrumentation were motivated by the need to control the quality of cement stabilised subgrades and capping layers which are much stiffer than conventional unbound layers. The paper presents the new Dynaplaque 2 and a new algorithm selected for the calculation of the subgrade Young's modulus from the force and displacement signals. Examples of typical measured signals are given together with the comparison with the computed response of the Dynaplaque.

Key-words : subgrade, dynamic testing, plate loading test, algorithm

INTRODUCTION

For over 20 years, since 1976, the Dynaplaque has been used by the Laboratoires des Ponts et Chaussées for the assessment of the deformability of subgrades for pavement construction [1]. This apparatus has superseded traditional static plate loading tests since it permits to perform more than 20 tests per hour with only one technician in convenient and economical conditions. In the first generation of Dynaplaque loading is produced by the free fall of a cylindrical mass of 125 kg which bounces on a set of 3 springs attached to a stiff metallic plate 600 mm in diameter. With a correct choice for the height of fall, for the

where R is the radius of the base plate and α a dimensionless coefficient close to 1/2, according to Boussinesq's formula, for a rigid circular plate, lying on a homogeneous elastic half-space : $E = \frac{1-\upsilon^2}{2R} k$.

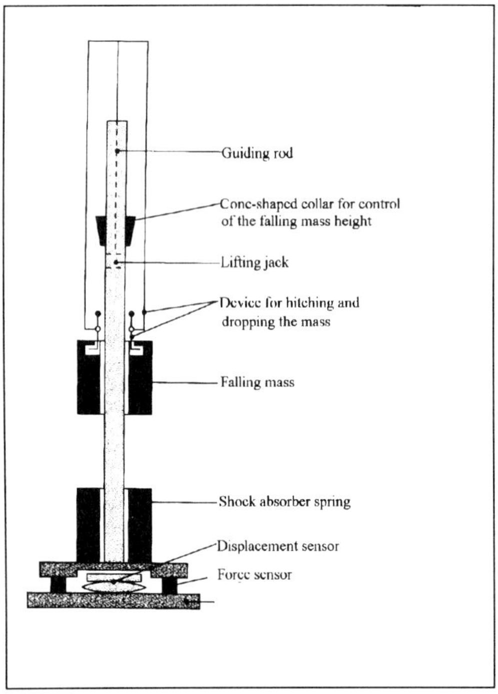

Fig.2 : Sketch of Dynaplaque 2

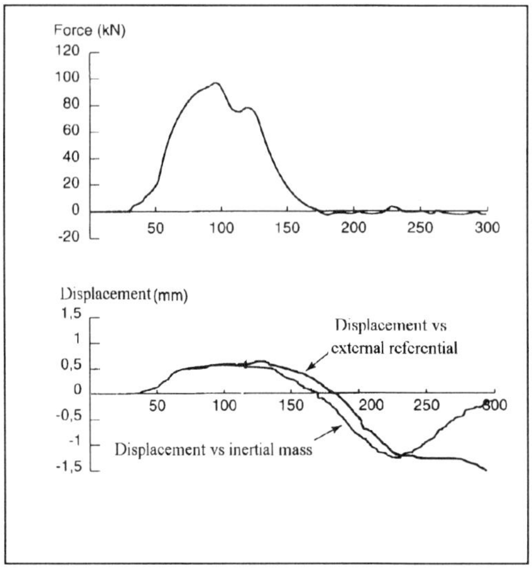

Fig.3 : Typical examples of the force and displacement signals obtained with Dynaplaque 2

As it can be seen from figure 5 which presents results of the subgrade modulus determined both with the old Dynaplaque from the coefficient of restitution and with the Dynaplaque 2 (here, modulus $E_{30/90}$) :

- it is possible with the Dynaplaque 2 to make an estimate of the modulus of very stiff supports (over 300 MPa) ;
- the correlation with static plate loading test appears fairly good over a large range of moduli although the way the dynamic modulus was computed could appear rather basic.

However as it will be shown in the further sections, in certain cases with very stiff subgrades, when decoupling could happen during the impact between the base plate and the soil, or on the contrary with very soft subgrades where a large phase lag would appear between the force and displacement signals, the previous mode of interpretation was not found satisfactory. This led to the search of a more mechanistic determination of the subgrade stiffness k .

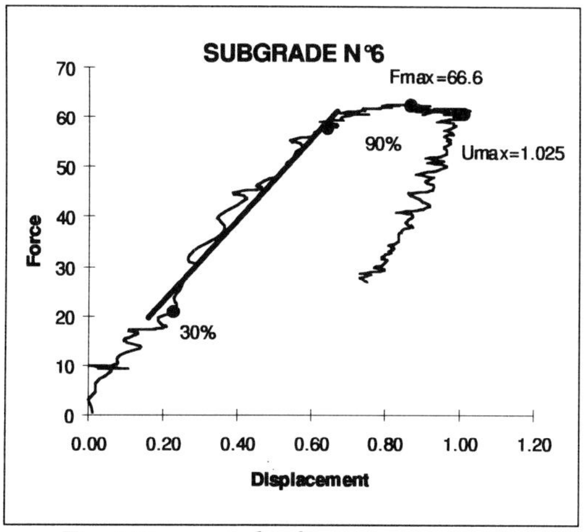

Fig.4 Force vs displacement curve for $E_{30/90}$ and $E_{fmax/umax}$ calculation

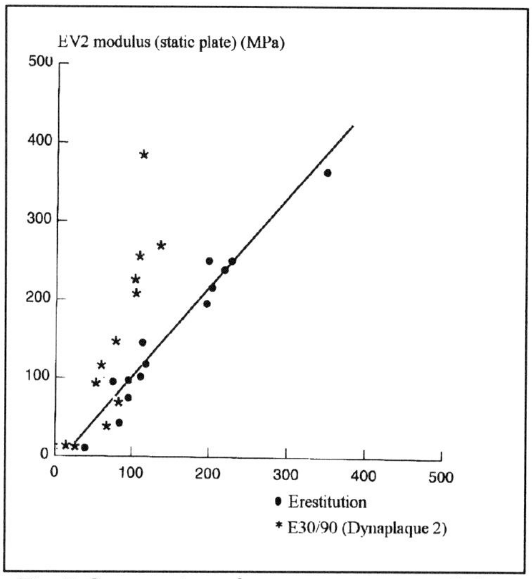

Fig.5 Comparison between static modulus, $E_{restitution}$ and $E_{30/90}$

MECHANISTIC INTERPRETATION OF DYNAPLAQUE 2 RESPONSE

Search for a reference value k_{ref}

Assuming the subgrade responses as a spring with stiffness k in parallel with a dash-pot of viscosity c, without added mass effect, the dynamic equation of the Dynaplaque 2 base-plate, taken alone and considered as a rigid solid of known mass m (38.5 kg) submitted to both the reaction of the subgrade and the impact force f(t) from the Dynaplaque reads :

$$m\ddot{u}(t) + c\,\dot{u}(t) + k\,u(t) = f(t) \tag{2}$$

with initial conditions : $u(0) = \dot{u}(0) = 0$.

u(t) and *f(t)* are both measured by the Dynaplaque 2. However due to imperfection of model (2) and experimental errors, these signals cannot be used directly to determine k and c.

Preferably k and c can be sought as the solution of the following least-square problem (P) :

" Find k_{ref} and c_{ref}, which minimize the function J(k,c) defined by :

$$J(k,c) = \int_0^T \left[u_{th}(t;k,c) - u(t)\right]^2 dt \tag{3}$$

where, T is the time duration of impact (typically T~ 20 ms) *and $u_{th}(t\,;\,k\,,\,c)$ the theoritical solution of equation :*

$$m\ddot{u}_{th} + c\,\dot{u}_{th} + k\,u_{th} = f(t) \tag{4}$$

with initial conditions : $u_{th}(t=0)=\dot{u}_{th}(t=0)=0$."

A computer program, based on an iterative gradient method, has been developed to solve this problem. It mainly relies on two subroutines : one for the direct calculation of u_{th} by Duhamel's integral for known values of t, k and c ; the other one for the numerical integration of J . At each iteration on k and c, these subroutines are used to evaluate J as well as its partial numerical derivatives. The process is stopped when no more improvement on J can be obtained. Numerous trials have shown that beginning the iterative process with different values of k and c , generally led to the same final set of values k_{ref}, c_{ref} or at least to close values, reinforcing the intrinsic meaning of k_{ref}.

The quality of these final data can be assessed through the corresponding value of J , or better by the visual comparison between the measured signal $u(t)$ and the computed response $u_{ref} = u_{th}(t;k_{ref},c_{ref})$. Such a comparison is presented on fig. 6, which shows quite a good match between u and u_{th}.

Simplified approach of k_{ref} value

For implementation of the algorithm in the microprocessor of the Dynaplaque 2, it was decided to look for a method which would approach the above satisfactory k_{ref} solution, but avoiding any iterative process. This goal was achieved using "weak" formulations of equation (2), such that measured data can be made "weak" solutions of (2).

More precisely, let's note $\{f(t_i);\ u(t_i)\}$ the force and displacement values measured at times $t_0 = 0,\ t_1 = \Delta t,\ ..t_i = i\ \Delta t,\ ...T = t_N = N\ \Delta t$, during a test impact (typically, N~300, $\Delta t = 0.07$ms). Be $\tilde{f}(t)$ and $\tilde{u}(t)$the piecewise linear functions of time, passing through these points.

Our aim will be to determine k and c such that $\tilde{f}(t)$ and $\tilde{u}(t)$can be related by equation (2), in a given way, which we are now to precise.

From classical integration rules, it can be shown that the search for solution of equation (2), is close to looking for function $u(t)$, checking $u(0)=\dot{u}(0)=0$and :

$$\int_0^T u(t)[m\,\ddot{g}(t)-c\,\dot{g}(t)+k\,g(t)]dt=\int_0^T f(t)g(t)dt \qquad (5)$$

for any *test function* $g(t)$, with first and second derivatives over [0,T] verifying, $g(T)=\dot{g}(T)=0$.

We call this problem (P^W). Note that in (P^W), $u(t)$ does not need to be as regular with time, as it were necessary to define equation (2). For this reason (P^W) is said to be a weak formulation of equation (2).

Now, simply by restricting the set of test functions $g(t)$, even "weaker" problems than (P^W) can be derived. In particular we can define problem (P^{Gn}) by imposing functions $g(t)$ to vary in a vectorial space G^n, with finite dimension n. Taking $n = 2$ will generally allow the previously defined function, $\tilde{u}(t)$, to be solution of such a problem with $f(t)=\tilde{f}(t)$, on condition that correct values of k and c are selected. Let $\{g_1(t),g_2(t)\}$ be a base of G^2; $\tilde{u}(t)$ will be a solution of (P^{G2}), if k_{appr} and c_{appr} are solution of the 2 x 2 linear system :

$$k_{appr} \int_0^T \tilde{u}(t)g_i(t)dt - c_{appr} \int_0^T \tilde{u}(t)\dot{g}_i(t)dt = \int_0^T \left[\tilde{f}(t)g_i(t) - m\tilde{u}(t)\ddot{g}_i(t)\right]dt \qquad (6_i)$$

for $i=1,2$. A short computer program was developed to calculate k_{appr} from (6), with G^2 being the polynomial set of smallest order with conditions $g(T)=\dot{g}(T)=0$. $\{g_1(t), g_2(t)\}$ were taken as, $\{(T-t)^2/2 , t(T-t)^2/2\}$. Integrals of equations (6_i) are calculated with Simpson's rule.

Since until now k_{appr} (resp. E_{appr}) has always shown to be close to k_{ref} (resp. E_{ref}) (see next section), this mode of interpretation of Dynaplaque 2 results looks as the most convenient one. Moreover the Dynaplaque's user can have a "visual" appreciation of k_{appr} and c_{appr} matching, versus field data, owing to the drawing on the same graph of measured displacement $u(t)$ and the computed signal $u_{appr} = u_{th}(t\ ; k_{appr}, c_{appr})$ (cf, fig.6)

COMPARISON BETWEEN THE DIFFERENT INTERPRETATION MODES

The 4 different previous modes of interpretation of Dynaplaque 2 were tested on a series of tests, including 9 different subgrades. Results are gathered in the following table.

Subgrade	**E_{ref} (MPa)**	$E_{30/90}$ (MPa)	$E_{fmax/umax}$ (MPa)	E_{appr} (MPa)
1	**24**	27	27	22
2	**38**	44	44	39
3	**65**	67	72	60
4	**81**	77	88	86
5	**97**	138	97	107
6	**123**	102	131	126
7	**313**	198	280	317
8	**10**	59	27	14
9	**10**	75	29	15

Table 1 : Comparison between different interpretation modes of Dynaplaque 2 results

The 4 modes of interpretation appear almost identical on the 6 first cases.

On case 7, corresponding to the stiffest subgrade, moduli E_{ref}, $E_{fmax/umax}$ and E_{appr} also give comparable values. The estimate given by $E_{30/90}$ leads however to a significantly lower value. We have in that case, $\left[E_{ref} - E_{30/90}\right] / E_{ref} \approx 35\%$.

Cases 8 and 9 correspond on the contrary to the softest subgrades. Now the $E_{30/90}$ modulus significantly overestimates the reference modulus E_{ref}. This is also the case with a less degree for $E_{fmax/umax}$. Besides modulus E_{appr} remains a good approximation of E_{ref}.

As an example, figure 6 presents the comparison between measured displacement u_{meas} and computed displacements u_{ref} and u_{appr}, obtained on subgrade n°7.

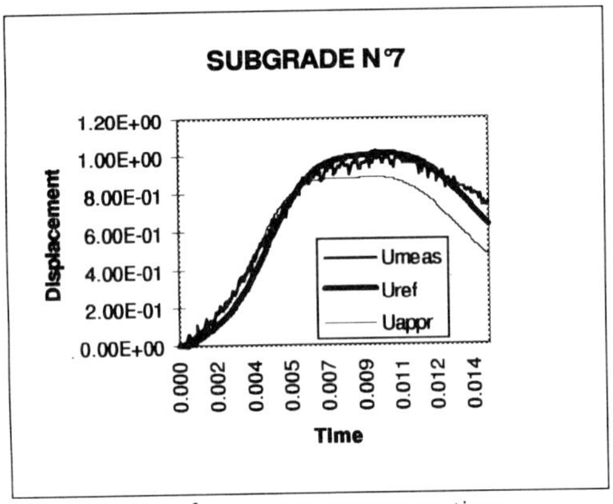

Displacement curves vs time

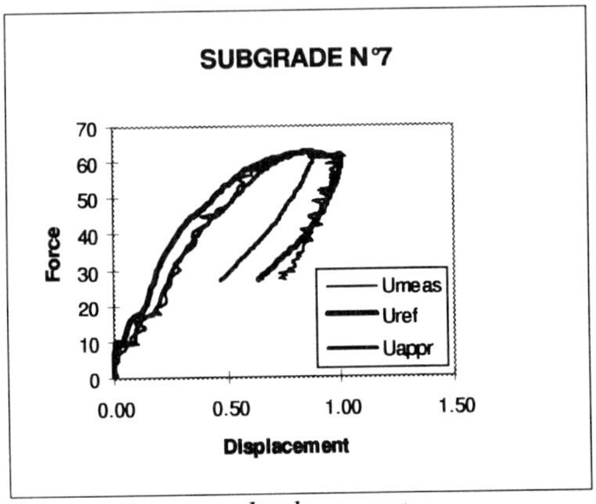

Force vs displacement curves

Fig6:Comparison between measured displacement and computed displacements with k_{ref} and k_{appr}

CONCLUSION

The modifications introduced in the design of the Dynaplaque 2 have significantly extended the field of testing of the former Dynaplaque. Now it can be used for testing of stiff subgrades made of cement treated soils or aggregates with a modulus up to 300 MPa. The measurements of the force and displacement of the base plate are of good quality. The corresponding temporal signals can be used for the determination of the subgrade modulus. Interpretation of the Dynaplaque response with the proposed algorithm permits to estimate a value of the subgrade Young's modulus close to that derived from the corresponding traditional static plate loading test, even in the case of soft subgrades.

REFERENCES

1. BENOIST J., SCHAEFFNER M. La Dynaplaque. Bulletin de liaison des Laboratoires des Ponts et Chaussées, n° 122, pp. 61-72. 1982.
2. CHASSAING P., CORTE J. F., FROUMENTIN M., SCHAEFFNER M. La Dynaplaque 2. Etude d'un nouveau matériel. Bulletin de liaison des Laboratoires des Ponts et Chaussées, n° 195, pp. 3-13. 1995.

Fifth International Conference on the
BEARING CAPASITY OF
ROADS AND AIRFIELDS

CHARACTERIZING SUBGRADE SOIL CONDITIONS THROUGH INSTRUMENTED PAVEMENTS AND LABORATORY EXPERIMENTS

S.T. DAI[1], C. SCHRADER[2], D. VAN DEUSEN[1],
G. JOHNSON[2] and D. RETTNER[3]

Office of Materials and Road Research
Minnesota Department of Transportation
1400 Gervais Avenue, Maplewood, MN 55109,USA

ABSTRACT

Field experiments were designed to study in-situ subgrade stiffness. First, the falling weight deflectometer tests (FWD) were conducted on three conventional flexible pavements at the Minnesota Road Research (Mn/ROAD) project. Then thin-walled tube subgrade samples were taken immediately after the FWD tests for the laboratory determination of resilient modulus. In addition, the depth to ground water table in each section was monitored by an open-stand pipe installed in each pavement. The purpose of the study is to 1) characterize subgrade modulus from FWD deflection basis using backcalculation process 2) obtain subgrade soil resilient modulus from laboratory experiments and compare backcalculated results with laboratory results. The analysis indicates that the backcalculated results using a stiff layer seem to be more reasonable than those without using a stiff layer for conventional flexible pavements. Also, the calculated depth to the stiff layer at water table matches well with measured depth to the stiff layer. Furthermore, it seems that material properties, such as plastic index and content of fine material, have great effect on resilient modulus. Both laboratory and backcalculated results show that the resilient modulus of subgrade soil with high plastic index is lower than that of subgrade soil with low plastic index.

INTRODUCTION

The resilient modulus is a widely used parameter to characterize the subgrade soil stiffness. Extensive studies have been conducted to investigate the resilient modulus (*1*). Research has primary focused on how to appropriately characterize in-situ subgrade stiffness from field experiments and laboratory test. Normally, FWD tests are conducted on pavement to evaluated pavement conditions. The pavement layer [1]moduli can be backcalculated from FWD deflection basin using a backcalculation process. Parker (2) backcalculated pavement layer moduli using FWD deflection basin and the software ELMOD. The results indicated that the backcalculated moduli of some fine-grained subgrade soils were essentially not dependent of the applied FWD load. Recently, Daleiden et al. (*3*) used SHRP-LTPP data to compare subgrade resilient moduli obtained from laboratory experiments with those obtained from the backcalculations using MODULUS 4.0. It was found that the ratio between the moduli determined from laboratory tests and the backcalculated moduli ranged from 0.01 to 10.34 with the mean value of 0.57. Based on FWD deflection basins, Thompson (4) proposed ILLI-PAVE based procedures to estimate in-situ subgrade resilient modulus. He pointed out that considerable in-situ moduli variability is typical and the backcalculated moduli using ILLI-PAVE based procedures are "reasonable engineering estimates." In addition, researchers have realized that the presence of a stiff sublayer has an effect on backcalculated pavement layer moduli (5-7). Traditionally, a layer, such as rock, was treated as a stiff layer. However, Mahoney et al. (8) pointed out that a saturated soil layer or ground water need also be treated as a stiff

1. Research Project Engineer
2. Research Scientist
3. Research Operations Engineer

layer during the backcalculation process. Newcomb et al. (9) considered saturated soil conditions in backcalculation of pavement layer moduli and illustrated that accounting for a saturated layer is important in the backcalculation process as well.

Resilient modulus of subgrade soil can also be evaluated from laboratory experiments under repeated load testing. Laboratory experiments showed that the resilient modulus of fine-grained cohesive soil was primarily affected by the deviator stress. Thompson and Robnett (10) showed that the resilient modulus could be described by a bilinear function of the applied deviator stress. When deviator stress is very low (less than a few kPa), the resilient modulus is a constant (4). Several procedures for laboratory experiments have been proposed by different agencies and researchers (11, 12, 13). It is apparent that the laboratory testing procedure on resilient modulus of soil is still under development. There is no unique procedure that can be used with precision by all material laboratories.

EXPERIMENT SETUP

Field Tests

In May of 1997, thin-walled subgrade samples were taken from three conventional flexible pavement sections at the Mn/ROAD project. The three sections are referred as section A, B and C (Fig. 1). The three sections have different pavement structures. Section A, referred as Cell 30 at Mn/ROAD, has a 127 mm asphalt concrete (AC) pavement and a 305 mm dense graded gravel base. Section B (Cell 22) contains a 200 mm AC pavement and a 460 mm crushed stone base. Section C (Cell 17) composes of a 200 mm AC pavement with a 710 mm dense graded gravel base. The subgrade soils of section B and C are classified as A-6 in the AASHTO classification, while subgrade soil of section A is classified as A-7. Some subgrade soil properties are listed in Table1. Also, an open standpipe is installed in each section to measure the depth to ground water table.

Table 1. Some subgrade soil properties.

Section	% passing #200 sieve	Liquid limit	Plastic Index	AASHTO Classification
A	65.7	45.4	27.7	A-7
B	58.6	37.7	18.4	A-6
C	57.8	35.5	17.4	A-6

During the sampling process, careful procedures were taken to preserve in-situ subgrade samples. First, the asphalt concrete surface was removed by drilling an approximately 25.4mm (10 inch) diameter core. Then the base material was removed using split-spoon method. Finally, a thin-walled metal tube was pressed into the in-situ subgrade soil to obtain the soil samples. The ends of the thin-walled tubes that contain the soil samples were immediately sealed using wax to prevent the soil from being disturbed and losing moisture. The thin-walled tube samples were later used for laboratory determination of resilient modulus.

Prior to the sampling, FWD tests were conducted on the top of the pavement. For each thin-walled tube sample, one set of FWD tests was conducted at the exact station where the sample was taken. In order to assess the uniformity of the subgrade soil, additional sets of FWD tests were performed at several offsets up to about 2.2 m away from the station where the thin-walled tube sample was taken. Fig. 2 gives a typical relationship between sample location and FWD test stations. For each set of FWD test, three different load levels were applied with three drops on each load level. The load ranged from about 22 kN to 70 kN. The obtained FWD deflection basins were used for backcalculation of pavement layer moduli.

Laboratory Experiments

The thin-walled subgrade samples were utilized for laboratory determination of resilient modulus. A closed-loop, servo-hydraulic and computer controlled material testing machine is used to conduct cyclic triaxial resilient modulus tests. The triaxial apparatus is instrumented with an internal load cell mounted in the middle of the base plate of the apparatus. The testing specimen directly seats on the top of the load cell through an aluminum platen. The advantage of using an internal load cell is that the friction between the loading rod and the top plate of the apparatus does not affect the load measurement. The internal load cell provides a feedback channel to control the applied cyclic load. Also, three internal LVDTs were mounted around the specimen at equal space and at the middle portion of the specimen to directly measure axial displacement of this portion of the specimen.

The resilient modulus tests were performed in general accordance with the Strategic Highway Research Program (SHRP) Protocol P-46 procedure. However, for some of the tests an additional series of measurements was conducted at a very low deviator stress (7 kPa). During the preparation of specimen, the ends of the specimen was carefully trimmed in an effort to make the two ends as parallel as possible. In general, the diameter of the specimen was about 71 mm (2.8 inches) and the height was about 152 mm (6 inches).

RESULTS OF FIELD TESTS

Resilient Modulus from Backcalculation

Backcalculations of pavement layer moduli were carried out based on the deflection basins of the FWD tests. The software EVERCALC5.0 was used for the backcalculations. The software is based on the linear elastic layered theory. The procedure in the software estimates the layer moduli using applied load and deflections measured by FWD sensors. The layer moduli were obtained by the best fit of the measured deflections. The best fit is achieved when the error between the calculated deflections and the measured deflections is minimized. The software has a stiff layer option, which means that pavement layer moduli can be backcalculated by assuming existence of a stiff layer. The software calculates the depth to the stiff layer using the scheme reported by Rhode and Scullion (14). In the study, two types of backcalculations were performed. One type is without stiff layer and another one is with stiff layer. The stiff layer modulus was selected as 345 MPa (14).

Fig. 3 shows the backcalculated resilient moduli of base and subgrade soil of pavement section A from the two types of backcalculations. It seems that the resilient moduli of both base and subgrade do not vary significantly with the stations, meaning that the base and subgrade materials can be considered uniform. The similar features have been also observed from the other pavement sections. The backcalculated results without accounting for a stiff layer shows that the resilient modulus of the base material is much lower than that of the subgrade soil. However, the results using a stiff layer demonstrate that the modulus of the base material is higher that that of the subgrade. In addition, the calculated depth to the stiff layer is compared with the measurements from the open standpipe. The calculated depth is about 1.58 m, while the measured depth is 2 m. The two depths are comparable. Furthermore, it seems that the backcalculated subgrade resilient modulus using a stiff layer matches reasonably with the laboratory result, which is shown later in the paper. Thus, it is apparent that the backcalculated results using a stiff layer seem more reasonable than those obtained without accounting for a stiff layer. Figure 3 also illustrates that the obtained resilient moduli of the both base and subgrade soil do not change significantly with the external load levels, which means that the subgrade soil may be viewed as a linear elastic material under the given load range.

Figures 4 and 5 give the backcalculated results of section B and C, respectively. Again, the subgrade resilient moduli calculated without using a stiff layer are higher than the base moduli. But the subgrade moduli are reduced drastically if a stiff layer is used in backcalculation process. For the section C, the calculated depth to the stiff layer is also comparable with the water table depth measured from the open standpipe. The calculated depth to the stiff layer from the pavement surface is 2.87 m and the measurement is 2.66 m. However, for the section B, it was found that there is a gravel layer under the subgrade at the depth of about 1.8 m. So, the gravel layer was also treated as a stiff layer during the backcalculation process. Again, the calculated depth to the stiff layer is 2.36 m, which reasonably matches with the depth to the gravel layer (1.8 m). In addition, the subgrade moduli of the both sections do not vary much with the external FWD loads, although a slight decrease of the subgrade moduli with the loads is observed. Furthermore, the backcalculated subgrade resilient moduli of section B and C using a stiff layer have similar values, which is expected since the two sections have the same type of subgrade soil (A-6).

Resilient Modulus from Laboratory Experiments

The thin-walled subgrade samples were tested in the laboratory to determine resilient modulus using SHRP P-46 with modification of adding a low deviator stress in the test. The results are given in Fig. 6 for the three sections. The resilient modulus decreases as the deviator stress increases. Also, It can be seen that the subgrade modulus of section A is much less than the moduli of section B and C. Notice that the subgrade of section A belongs to A-7 and the plastic index is much higher than the plastic indexes of the subgrade soils of section B and C. It may reflect that content of fine material in subgrade soil and plastic index of subgrade soil have significant effect on the resilient modulus and affect the subgrade response. This feature is consistent with the backcalculated results which also show that the subgrade modulus of section A is much less that those of section B and C (Fig. 3, 4 and 5). Furthermore, the backcalculated subgrade resilient modulus of section A matches with the laboratory results at a low deviator stress level. Also, it seems that the backcalculated subgrade moduli of section B an C are close to the laboratory results although the backcalculated moduli are slightly lower than the laboratory results.

SUMMARY AND CONCLUSION

A series of experiments, including FWD tests and thin-walled sampling, were conducted at the Mn/ROAD project in an effort to evaluate the in-situ subgrade resilient modulus. The backcalculation results show that the backcalculated subgrade modulus when accounting for a stiff layer seems to give more reasonable results than that without using a stiff layer. The calculated depth to the stiff layer is verified with the field measurement to coincide with the depth to the water table. Laboratory experiments are also conducted on the thin-walled samples to characterize subgrade soil stiffness through determining resilient modulus. The laboratory results illustrate that the modulus decreases as deviator stress increases. Also, it seems that plastic index and content of fine material in subgrade soil are important factors affecting the resilient modulus. The soil with higher plastic index and content of fine material shows low resilient modulus. This feature is also obviously dictated from the backcalculated resilient modulus values with accounting for a stiff layer in the backcalculation process. This means that the characterization of the subgrade soil stiffness under a conventional flexible pavement using the FWD backcalculation procedures with appropriate consideration of accounting for a stiff layer is consistent with the laboratory characterization of subgrade soil. The analysis suggests that backcalculation of pavement layer moduli for a conventional pavement using linear elastic layer theory should account for a stiff layer in the area where saturated subgrade soil or a stiff material layer underlying the subgrade is present.

ACKNOWLEDGMENTS

The article is based on the work supported by the Minnesota Department of Transportation and Local Road Research Board. The authors are extremely grateful to John Zollars and David Linell for assisting to conduct laboratory experiments. Special thanks are extended to Greg Larson for performing the field experiments.

REFERENCES

1. Thompson, M.R. Factors Affecting the Resilient Moduli of Soils and Granular Materials. Workshop on Resilient Modulus Testing, Oregon State University, 1989.

2. Parker, F. Jr. Estimation of Paving Materials Design Moduli from Falling Weight Deflectometer Measurements. Transportation Research Record 1293, 1991.

3. Daleiden, J.F., Killingsworth, B.K., Simpson, A.L. and Zamora, R.A. Analysis of Procedures for Establishing In Situ Subgrade Moduli. Transportation Research Record 1462, 1994.

4. Thompson, M.R. ILLI-PAVE Based NDT Analysis Procedures. Nondestructive Testing of Pavements and Backcalculation of Moduli, ASTM STP, 1026, pp.487-501, 1989.

5. Rohde, G.T., Smith, R.E. and Yang, W. Inclusion of Depth to a Rigid Layer in Determining Pavement Layer Properties. Proceedings of 3rd International Conference on the Bearing Capacity of Roads and Airfields, Trondheim, Norway, pp.475-486, 1990.

6. Uddin, W., Meyer, A.H. and Hudson, W.R. Rigid Bottom Considerations for Nondestructive Evaluation of Pavements. Transportation Research Record 1070, pp.21-29, 1986.

7. Rohde, G.T., Smith,R.E. and Scullion, T. Pavement Deflection Analysis on Sections Where the Subgrades Vary in Stiffness with Depth. Proceedings of 7th International Conference on Asphalt Pavements, Nottingham, England, pp.280-295, 1992.

8. Mahoney, J.P., Winters, N.C., Jackson, N.C. and Pierce, L.M. Some Observations About Backcalculation and Use of a Stiff Layer Condition. Transportation Research Record 1384, pp.8-14, 1993.

9. Newcomb, D., Van Deusen, D., Jiang, Y. and Mahoney J. Considerations of Saturated Soil Conditions in Backcalculation of Pavement Layer Moduli. Transportation Research Record 1473, pp.63-71. 1995.

10. Thompson M.R.and Robnett, Q.L. Resilient Properties of Subgrade Soils. *Transportation Engineering Journal*, ASCE, Vol.105, 1979.

11. AASHTO T292-91. Resilient Modulus of Subgrade Soils and Untreated Base/Subbase Materials. AASHTO, 1993.

12. NCHRP 1-28. Laboratory Determination of Resilient Modulus for Flexible Pavement Design. NCHRP. 1996.

13. SHRP Protocol P-46. Resilient Modulus of Granular Base\Subbase Materials and Subgrade Soils. June, 1993.

14. Washington State Department of Transportation. WSDOT Pavement Guide. Washington State Department of Transportation, February 1995.

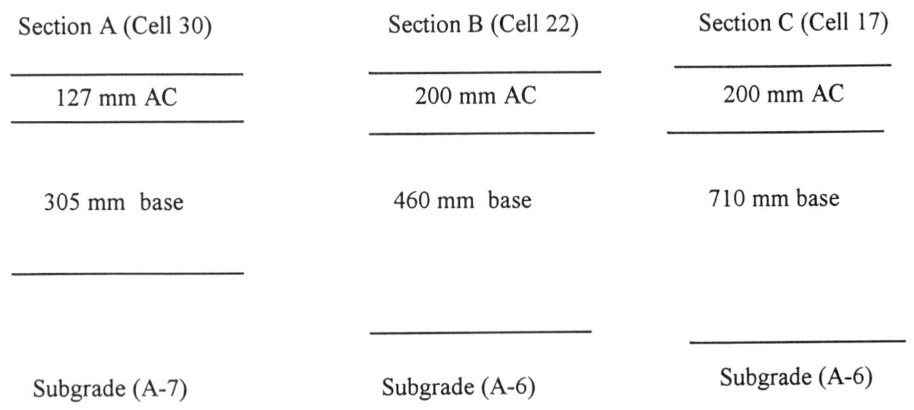

Fig. 1 Cross Sections of Tested Three Pavements.

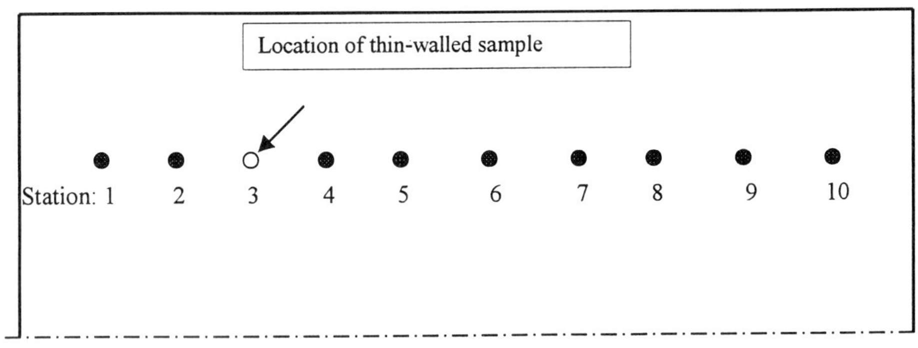

Fig. 2 Stations of FWD Tests and Location of Thin-walled Sample in a Typical Pavement Section.

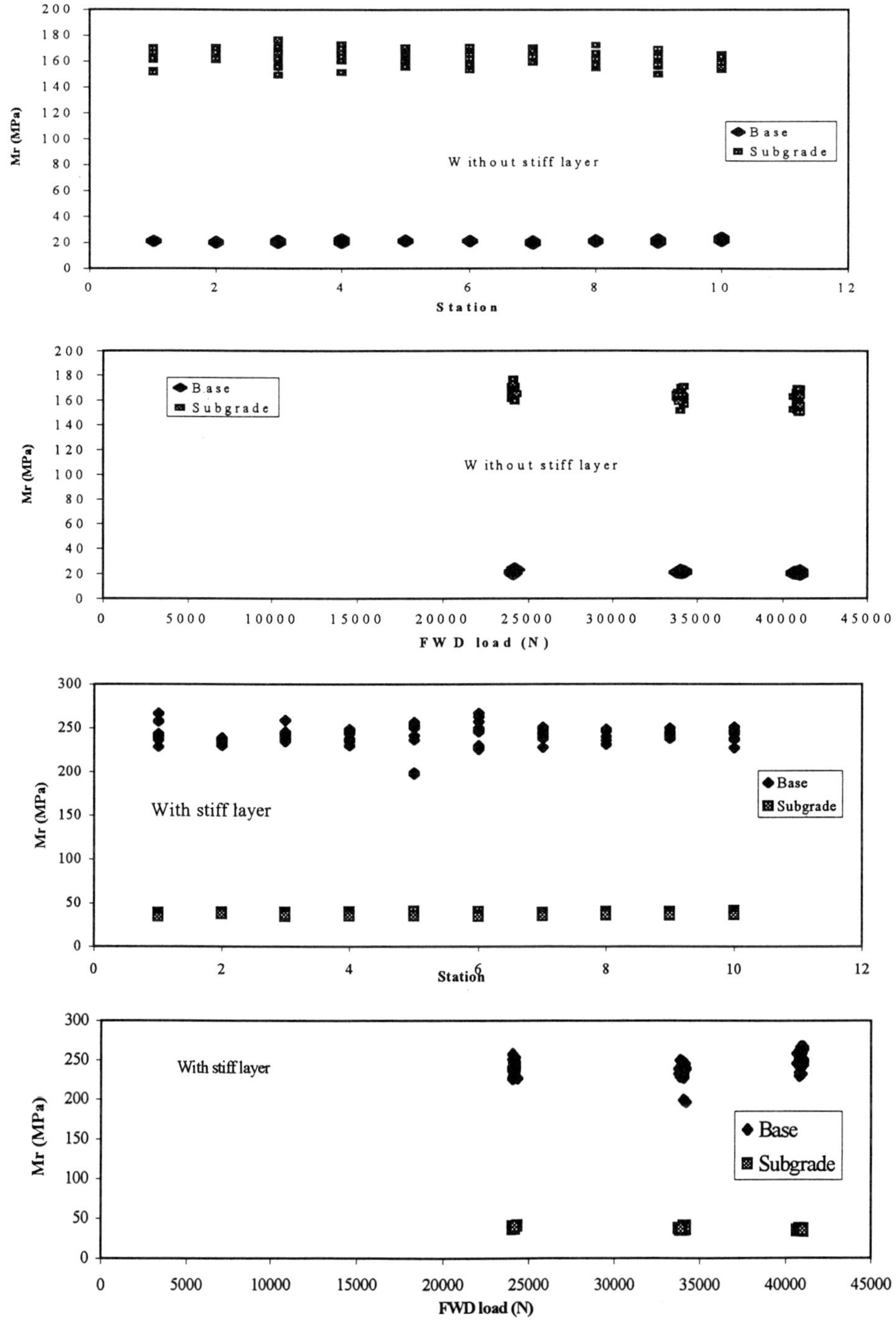

Fig. 3 Backcalculated Resilient Moduli of Section A with and without Stiff Layer.

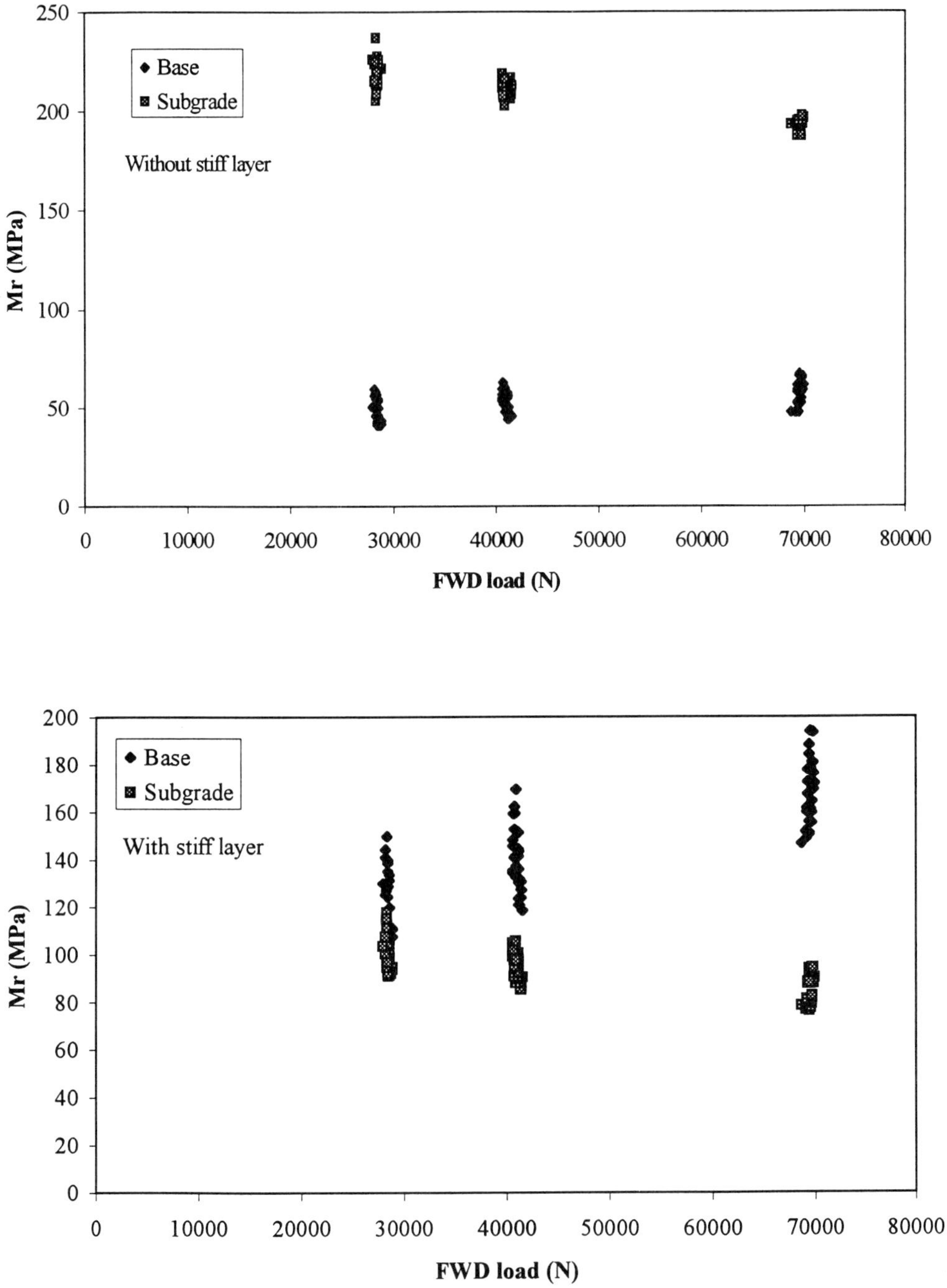

Fig. 4 Backcalculated Resilient Moduli of Base and Subgrade of Section B with and without a stiff layer.

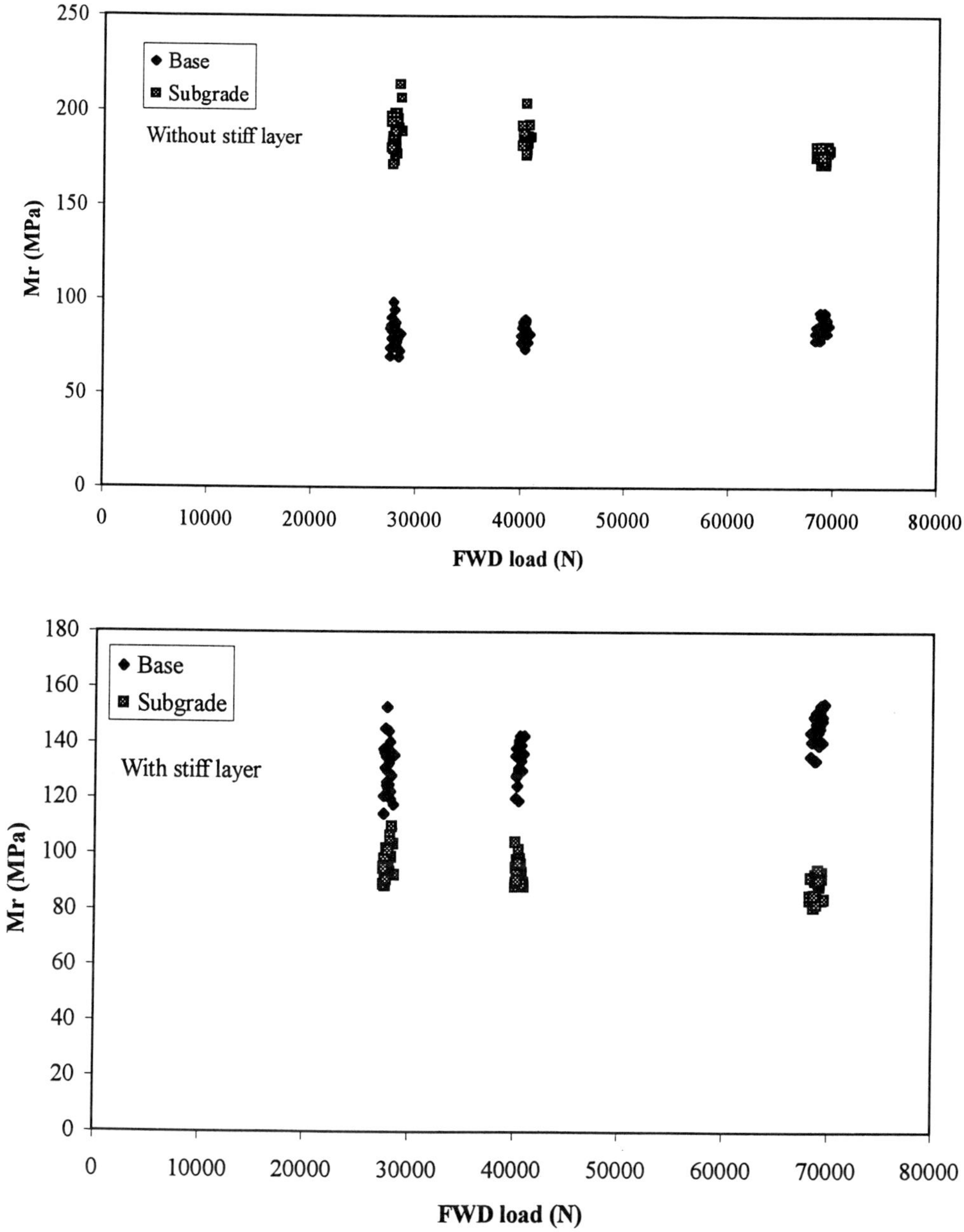

Fig. 5 Backcalculated Resilient Moduli of Base and Subgrade of Section C with and without a stiff layer.

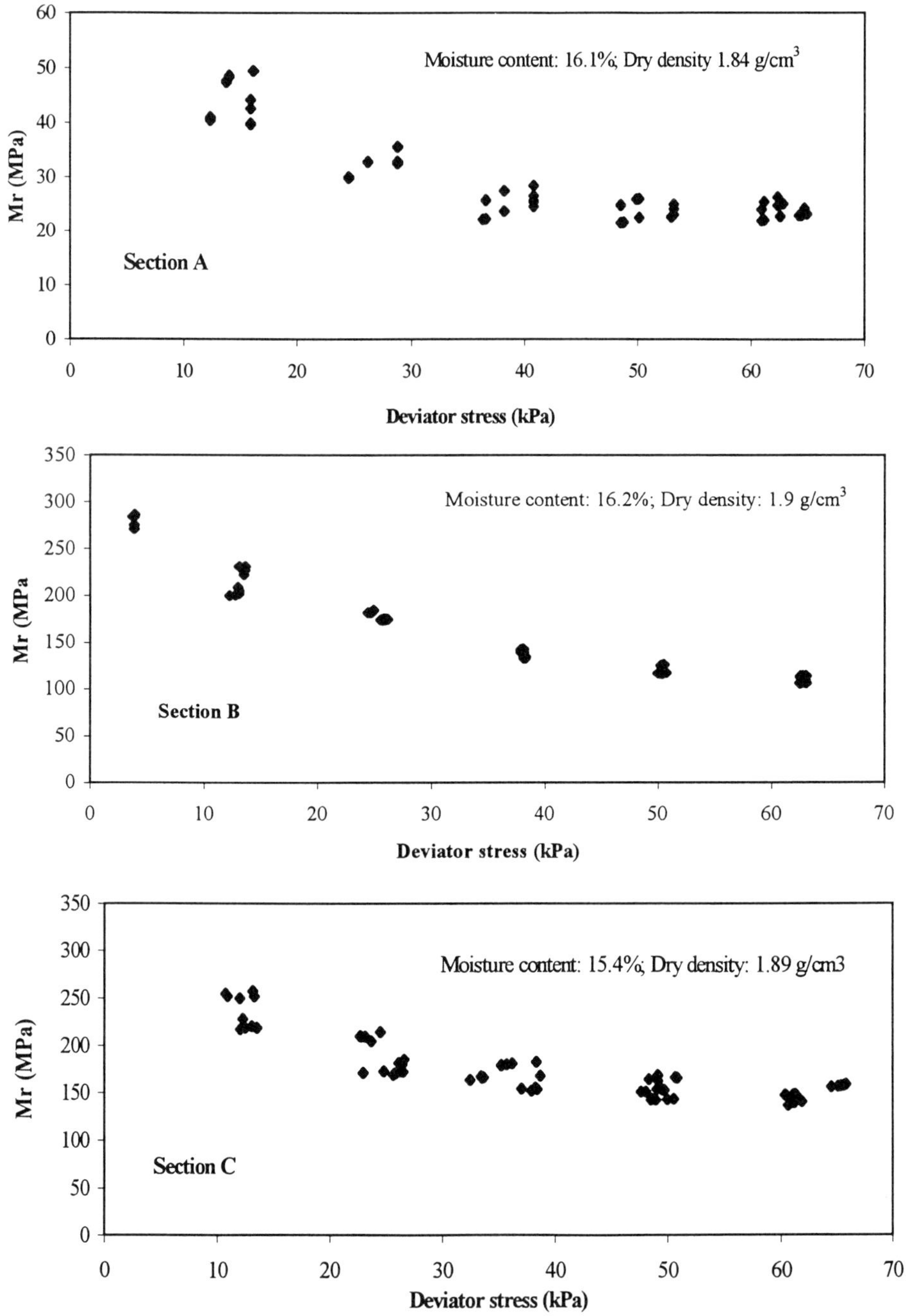

Fig. 6 Resilient Modulus of Subgrade Soil from Laboratory Experiments.

Fifth International Conference on the
BEARING CAPACITY OF
ROADS AND AIRFIELDS

THE USE OF SIMPLE SOIL TESTS IN ASSESSING SUBGRADE

LAM WAH CHEUNG

Senior Pavement Engineer, Scott Wilson Pavement Engineering Ltd., 9 Faraday Building, Highfields Science Park, University Boulevard, Nottingham NG7 2QP, England, United Kingdom.

Telephone: +44 115 9229098 Fax: +44 115 9431302
e-mail: lwc@swpe.co.uk

ANDREW R DAWSON

Senior Lecturer, Civil Engineering Department, University of Nottingham, University Park, Nottingham NG7 2RD, England, United Kingdom.

Telephone: +44 115 9513902 Fax: +44 115 9513898
e-mail: andrew.dawson@nottingham.ac.uk

ABSTRACT

A description of subgrade materials in terms of stiffness (resilient response) and resistance to permanent deformation is necessary because the resilient response governs the load spreading ability and enables a structural analysis to be performed, while the resistance to permanent deformation relates directly to the way in which pavement foundations fail by rutting. This paper discusses the abilities of simple soil test techniques in assessing the resilient response and resistance to permanent deformation of subgrade. Simple tests, including soil classification, composition, CBR, shear vane and suction tests, were carried out on three types of clay soils.

The permanent deformation characteristics of the soils under wheel loading were examined using a purpuse-built Soil Rut Testing Facility (SRTF), in which trays of soils were trafficked by a loaded wheel. The stiffnesses and permanent deformation data of the soils were obtained by Repeated Load Triaxial Tests (RLTT)

A comparison of the results from the simple tests and those obtained from the SRTF and the RLTT suggests that both the stiffness and the permanent deformation resistance can be related to the strength of the soil. Stiffnesses of soils increase non-linearly with the shear strength of the materials. For resistance to rutting, log-log relationships between soil strength and the rut depth achieved in the SRTF were observed. The results also showed that the relationships vary with soil types and stress conditions. It is, therefore, proposed that simple strength testing could be used for construction control purpose, but that specific site calibration would be required. An illustration of a typical application is given.

Key words: - subgrade, resilient behaviour, permanent deformation, strength, repeated load.

1. INTRODUCTION

Even before the completion of a road, the subgrade has already performed its most demanding role, with any overlying capping or sub-base, by providing a structural platform to bear direct loading from heavy construction plant. Then it supports loads from road traffic (transmitted through the upper pavement layers) throughout its design life.

A defective subgrade will result in a serious economic cost to society, both in repair and traffic congestion costs. On the other hand, pressure from limited resources and available budgets discourages any form of over-design of the pavement and under-estimation of the subgrade performance. For this reason, reliable subgrade characterisation is required so as to provide data for long term pavement design and to ensure that the standard of the design requirement is met during construction.

Failure of a subgrade of the pavement foundation may result from:-

1. Excessive resilience, which may cause distress in the overlaying layers;

2. Excessive permanent deformation under repeated traffic load, which induces local depression of the surface of a pavement;
3. Excessive deformation under single loading (leads to large localised shear distortion).

These failure modes can be represented by basic material parameters in terms of elastic stiffness, resistance to permanent deformation and shear strength respectively. For an adequately performing subgrade, the first two are the paramount requirements.

Ideally, the load response of the subgrade would be assessed by means of full-scale testing. However, time and financial constraints necessitate that pavement designs rely on simple assessment tests. This paper reports on the abilities of four simple soil assessment tests (plasticity, undrained shear strength, soil suction and the California Bearing Ratio). More sophisticated laboratory techniques (the Soil Rut Testing Facility (SRTF) and the Repeated Load Triaxial Test (RLTT) apparatus, were used to provide checks on the simple tests. Since much of the U.K. is covered by clayey soils, three types of cohesive soils, representing a range of plasticities, were selected for investigation.

2. THE SIMPLE TESTS

2.1 Plasticity Index

One of the methods to characterise fine grained soils is to find the plasticity of the materials using Liquid Limit (LL), Plastic Limit (PL) and Plasticity Index (PI) (B.S.1377, 1990). The liquid limit has been found to be directly proportional to the compressibility of a soil, whilst the plasticity index may serve as an indirect measurement of subgrade soil strength (Black and Lister, 1979).

2.2 Soil Suction

For cohesive soils Loach (1987) found that the resilient modulus was a function of soil fabric and suction. The term "soil suction" here refers to the negative pore water pressure, with respect to the atmospheric pressure, in a soil mass not subjected to any external stresses. The soil suction was measured by the application of a negative water pressure such that no net water inflow or outflow results (Dumbleton and West, 1968). This permits a rapid assessment.

2.3 Shear Vane Test

The shear vane test is at present one of the simplest tests to determine the undrained soil strength. The shear vane comprises a vane of four blades, which form a crossed-head, screwed into the end of a small diameter rod. The operating handle, with a proving gauge mounted on it, is attached to the other end of the rod. Readings of the proving gauge are calibrated to give the undrained shear strength (C_u) of the soil.

2.4 CBR

The California Bearing Ratio (CBR) test is a small scale shear test. During the test, a plunger is forced into the soil samples at a constant rate. The forces required to move the plunger 2.5mm and 5mm into the soil sample are expressed in terms of CBR values. Based on a low stress wave propagation technique, Heukelom and Klomp (1962) proposed a relationship between the resilient Young's modulus and CBR value. Another attempt to develop a single relationship between the resilient Young's modulus and CBR value was made by Powell et al (1984).

3. THE REFEFENCE TESTS

3.1 Soil Rut Testing Facility (SRTF)

The SRTF is a tool developed at the University of Nottingham to distinguish the resistance to permanent deformation of soils under traffic loading. It comprises a chain loop of six identical moulds of size 400mm by 280mm by 110mm, driven by means of a steel conveyor coupled to a motor. Single direction wheel trafficking at a test speed of 90m/hr and a contact pressure of 300kPa is allowed. Rut depth produced by the wheel was measured with a dial gauge, capable of reading to an accuracy of 0.01mm, at three locations near the middle of each tray.

3.2 Repeated Load Triaxial Test (RLTT)

The elastic and plastic behaviour of soils is highly stress dependent. The RLTT provides a simplified means of characterising this. The apparatus (Cheung, 1994) developed at the University of Nottingham is able to generate independent cyclic or other repeated axial and confining stresses by a pneumatic system. With the assistance of a signal generator, the frequency of load pulse can be controlled. The measurement of vertical deformation is made by a pair of strain loops mounted on the sides of the specimen whilst radial strains are measured by a pair of proximity transducers.

4. MATERIAL PREPARATION AND TESTING

4.1 Simple Tests and SRTF Test

The permanent deformation characteristics of soils under wheel loading were examined using the SRTF. Soils were compacted into the test tray of the SRTF in four layers. The compacted samples were then sealed to prevent loss of moisture and the samples were left for at least seven days in a high humidity room to allow variations in pore pressure to equalise before testing. Two layers of microwave plastic film were laid on the specimen to prevent change of moisture content during the test.

The tests were performed by running the soil specimens under the wheel of the SRTF for 1000 passes. The rut produced by the wheel was measured periodically. At the end of the wheel tracking test, the condition of the soil was examined for moisture content, for strength using a shear vane and by a soil suction test. Finally, materials were cut from the test specimen for CBR testing according to the procedure set in B.S. 1377 (1994).

4.2 Simple Tests and Repeated Load Triaxial Test

The size of each soil sample for the RLTT is 100 mm in diameter and 200 mm in height. Sample compaction effort, similar to the 'Proctor' requirement of B.S. 1377 (1994), was used for soil preparation. The specimens were tested at least 24 hours after the setting-up procedures had been completed. For all repeated loading tests, the load pulses were of a sinusoidual waveform and the loading frequency employed was 2Hz.

A multi-purpose testing procedure was developed to determine the resilient behaviour and the permanent deformation characteristics of the soil specimens. To study the permanent deformation characteristic of the soils, the samples were subjected to 1000 axial repeated loads at each of several, increasing, stress levels until the accumulated axial plastic strain was at, or just above 1%. The repeated deviator stress needed to cause this is here termed threshold stress, q_t. Each test was performed at zero confining pressure and in a drained condition. At its conclusion, a resilient test was carried out on the same specimen, covering 30 stress paths. For each of them, 50 cycles were applied and readings taken during the last 5 cycles. At the end of the repeated loading tests, soil strength was measured using a 19 mm shear vane. The moisture condition and the suction of the soils were then determined.

5. RESULTS OF TESTING

5.1 Soil Classification Tests

The test results for the three soil types are presented in Table 1. Among the soils, the London clay, L, (from Buckinghamshire, Southern England) is the soil with the highest clay content (74%), followed by the Bothkennar clay, B, (29%) (from Scotland) and Keuper Marl, K, (22%) (from Nottinghamshire, central England) together representative of a wide range of the cohesive soils in the U.K.

Table 1: Results of Soil Classification Tests

	London clay	Bothkennar clay	Keuper Marl
LL (%)	76.0	54.3	33.7
PL (%)	25.2	25.1	17.6
PI (%)	50.8	29.2	16.1
Plasticity	Very High	High	Low
Clay Content (%)	74	29	22

5.2 Soil Suction Tests

The wetting soil suction curves obtained from the three soils are presented in Figure 1. The suction is expressed in terms of pF, which is the logarithm of the pressure in centimetres of water. The results show that the suction of soils increased with decreasing moisture content. At the same moisture content, materials of higher plasticity and finer grading possess a higher suction value. It was also observed that the sensitivity of the suction to change in moisture content decreased with increasing plasticity of materials. Hence, the London clay was the least sensitive among the soils.

Figure 1 : Suction Curves of Soils

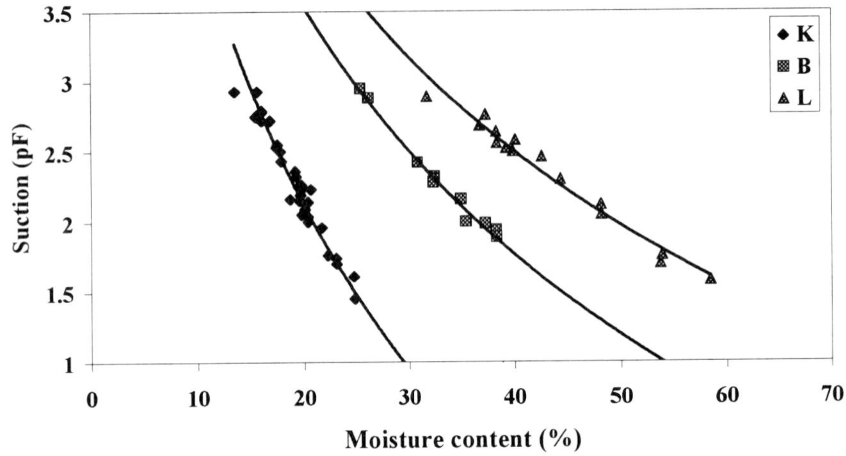

5.3 Strength Test

The strength of the samples was determined by shearing with a shear vane. The soil data are presented in graphic form (Figure 2). Although only limited data are available, straight lines radiating from the origin may be inserted onto the graph to represent the results. Hence, for a compacted soil the relationship between the soil suction and the shear strength can be expressed as :-

$$C_u = t*S \qquad \text{(Equation 1)}$$

Where:- t is a dimensionless constant.
S is the soil suction.
C_u is the undrained shear strength.

Figure 2 also indicates that the shear strength of soils of high plasticity was less sensitive to the change of soil suction value. Attempts were, therefore, made to normalise the soil suction values by the plasticity index of the materials. Equation 1 can be re-written as:-

$$Cu = T\frac{S}{PI} \qquad \text{(Equation 2)}$$

Where:- T is a dimensionless constant.

5.4 Wheel Tracking Tests

A total of twelve samples were tested in the SRTF. For soft samples, significant rutting was formed directly under the wheel and heaving at both sides of the wheel path could be observed. Table 2 summarises the test results showing the strength of each material and the

Figure 2 : Shear Strength and Suction Relationship

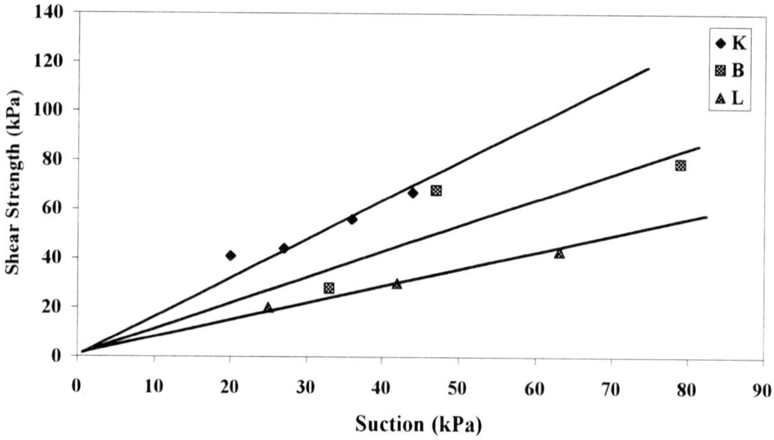

rut depths formed after 1000 passes (RD1000) of a rolling wheel. It was evident that the permanent deformation resistance of soils was a function of material strength and that CBR was a rather insensitive measure of rutting. The higher the strength of the soil, the lower the permanent deformation generated under repeated loading.

Figure 3 presents the rut against the number of passes. Except for the very soft samples of Keuper Marl, K5 and K6, a semi-log function between the accumulated rut depth and the number of passes may, generally, be observed

Table 2 : Results of SRTF and Associated Simple Tests

Sample	CBR (%)	Vane Shear Strength (kPa)	RD1000 (mm)
K1	13	>120	1.02
K2	5	>120	2.33
K3	2	91	5.19
K4	1	49	8.54
K5	0.5	26	16.51
K6	0.5	22	22.91
B1	1	41	9.34
B2	0.5	24	19.48
B3	0.5	22	22.92
L1	2	37	9.76
L2	2	37	10.17
L3	2	29	12.70

Acceleration of rutting was found in tests carried out on K5 and K6 of very low shear strength (being lower than 30 kPa). This might be partly explained by the observation made by Brown et al (1987) in their repeated load triaxial testing. They reported that when the deviator stress applied exceeded a certain limit, excessive permanent deformation development could occur. However, it was interesting to note that though the strength of samples B2 and B3 of Bothkennar clay and L3 of the London clay were all of low strength, below 30 kPa, acceleration of rutting was not found in these two soil types of higher plasticity. Hence, materials of a higher plasticity seem to perform better in resisting permanent deformation due to repeated loading.

Figures 4a and 4b compare the rut depths formed after 1000 passes with strength related parameters (undrained shear strength and CBR value). Linear log-log relationships are

Table 1. Summarised test results

Concrete Ring test

Road Base 0 - 32 mm	Mean % diff. from Ring + compacted fill			Repeatabilty Standard deviations		
	WD	DD	MC	WD	DD	MC
Method	Mean	Mean	Mean	St.dev.	St.dev.	St.dev.
ND - 1	-1.2%	-0.7%	-10.4%	1.1%	1.3%	4.3%
SR	-2.7%	-1.3%	-23.0%	1.4%	1.3%	4.0%
Ring + compacted road base (kg/m3 and %):						
Mean:	2224	2093	6.3%			
St. devn.:	3.4%	2.6%				
Concrete Aggregate 0 - 8 mm						
ND - 1	-1.9%	-0.6%	-21.8%	0.5%	0.6%	3.7%
SR	-2.3%	-1.5%	-13.3%	1.4%	1.4%	3.0%
Ring + compacted concrete aggregate (kg/m3 and %):						
Mean :	2036	1910	6.5%			
St. devn.:	2.6%	1.4%				

Field Density test

Road Base 0 - 32 mm	Mean % difference from SR		
	WD	DD	MC
Method	Mean	Mean	Mean
ND - 1	-2.1%	-0.9%	-55.8%
ND - 2	-1.6%	-0.6%	-45.1%
SR	0.0%	0.0%	-32.2%
MC test		**Mean :**	5.1%
Concrete Aggregate 0 - 8 mm			
ND - 1	-1.5%	-0.5%	-39.9%
ND - 2	-1.1%	-0.3%	-35.4%
SR	0.0%	0.0%	-20.5%
MC test		**Mean :**	5.1%

Key

ND - 1 Nuclear Densometer (ND) No. 1
ND - 2 ND No. 2

SR Sand Replacement test method
WD Wet Density
DD Dry Density
MC Moisture Content test (oven drying method)

% difference is determined from: (Measured - Actual) / Actual

Table 2. Comparison of results from this study with a previous DRI study [2].

Test Name	Test method	No. of groups	No. of test /group	Road Base 0 - 32 mm StDev.%	Road Base 0 - 32 mm dK%	Concrete Aggre- -gate 0 - 8 mm StDev.%	Concrete Aggre- -gate 0 - 8 mm dK%
Concrete Ring	ND-1	3	10	1.3%	-1.7%	0.6%	-2.4%
	SR	3	10	1.3%		1.5%	
Field Density Test	ND-1	1	20	1.7%	0.2%	1.3%	1.3%
	ND-2	1	20	2.4%	1.0%	1.1%	0.8%
	SR	2	20	2.1%		0.8%	
DRI Rpt 168	ND	25	min. 5	1.7%	2.7%	---	---
	SR	25	min. 5	1.7%		---	
DRI Rpt 168	ND	15	min. 5	Subbase :		1.6%	2.0%
	SR	15	min. 5	-- do --		1.6%	
DRI Rpt 168	ND	6	min. 5	Sandy Silt-Clay :		2.3%	1.7%
	SR	6	min. 5	-- do --		2.1%	

Table 3. Regression analysis of Wet Density measurements

Function : $Y = pr1 * X^2 + pr2 * X + pr3$ **kg/m3** **pr = regression parameter**

CONCRETE RING TEST

Y :	Base SR all	CAgg. SR all
X :	versus ND all	versus ND all
pr1	-0.00338	-0.00118
pr2	14.9463	6.0403
pr3	-14362.9	-5354.7
R^2 :	0.913	0.854
n	30	30

ALL DATA

Y :	Base SR all	CAgg. SR all
X :	versus ND all	versus ND all
pr1	-0.00420	0.00191
pr2	18.4350	-6.9793
pr3	-18057.6	8300.9
R^2 :	0.863	0.830
n	50	50

The regressed functions for Field Density (FD) test data alone are not tabulated.
The FD test correlations are very poor, since the R^2 values are less than 0.7.
The R^2 term expresses the proportion of the variation explained by the function.

Where:

- **ND** Nuclear Densometer (ND) type used in DRI Report No. 168 (1986)
- **ND-1** ND No. 1 used in both the Concrete Ring and Field Density tests
- **ND-2** ND No. 2 used in the Field Density test only
- **SR** Sand Replacement method (including oven dried moisture content method)

dK% = difference in the SR - ND characteristic compaction values for the material.

- **ND all** all ND test values
- **SR all** all SR test values
- **R^2 :** Regression coefficient
- **Base** Road Base 0 - 32 mm
- **CAgg.** Concrete Aggregate 0 - 8 mm
- **n** no. of test data values

of the radiation received by the detectors. Nuclear densometers should be suitably calibrated before use; routine annual maintenance checks by the manufacturer or supplier ensure that the radioactive sources, registers and electronics are properly maintained and adjusted for drift with time and use. A power-up calibration check is always carried out on a plastic block before commencing testing, and in some cases instruments may be calibrated against blocks of known density and moisture content. Suitable calibration block details are suggested in a number of references [1, 5, 6], and in certain situations it may also be appropriate to calibrate for individual site conditions [4]. The Nuclear densometers used in this study were only subject to routine annual servicing and standard power-up calibration checks before commencing each series of measurements.

TEST EQUIPMENT

Concrete pipe ring test. The main item was the concrete ring, as shown in Figures 1 and 2, which was coated internally with an impermeable membrane to inhibit absorption of moisture from the granular materials. The concrete pipe ring was weighed to an accuracy of ± 20 kg, and the internal dimensions measured at five diagonal positions to ± 1 mm; when empty, it weighed approximately 1827 kg and had an internal volume of approximately 1.2 m^3. When full with granular material, the weight of the concrete ring and compacted materials ranged between 4230 – 4600 kg, depending on the material type and degree of compaction.

Figures 1 and 2 show the plan views and sections through the concrete pipe ring. In Figure 1, the arrangement of the Nuclear densometer testing positions and the Sand Replacement test positions are indicated. For each filling of the concrete pipe ring with compacted granular material, ten tests of each type were carried out on the completed surface, commencing with the Nuclear densometer, and followed by Sand Replacement testing. Excavation holes for the Sand Replacement method were located within the area under each Nuclear densometer position. A level, well compacted foundation of granular materials was provided under the concrete pipe ring.

Field test area. No specialised equipment was required for this test, other than a level, well compacted foundation of granular materials. The plan area of the test area was approximately 8 m x 5 m.

CONSTRUCTION AND TESTING

Concrete pipe ring test. The granular materials were laid and compacted within the concrete pipe ring in three layers of compacted thickness ranging between 100 – 140 mm. The top layer was generally only 100 mm in compacted thickness. Compaction was applied by two types of frog rammer (for the two lower layers) and a vibratory plate compactor (for the top layer). During construction of each layer, approximately 8 kg of material was sampled from random positions in the layer and carefully stored in sealed plastic bags for moisture content determination. When the final layer of granular material was laid and compacted, the whole unit was weighed in a sling lifted by a shovel loader. An electronic weighing transducer with a range of up to 5000 kg and an accuracy of ± 20 kg, measured the total load. Nuclear densometer, followed by Sand Replacement testing was carried out at ten coincident positions radially distributed on the finished surface. Sand Replacement testing and moisture content determinations were, in all cases, carried out by experienced soils and materials technicians from the DRI's soil laboratory.

Field test area. The granular materials were laid and compacted in three layers of roughly 100 mm compacted thickness. Compaction was achieved with 5 to 10 passages of a 1300 kg dual drum vibrating roller for each layer of granular material. During construction of each layer approximately 16 kg of material was sampled from random positions in the layer and carefully stored in sealed plastic bags for moisture content determination. When the final layer of granular material was laid and compacted, Nuclear densometer testing, followed by Sand Replacement testing was carried out at twenty (20) coincident positions in a rectangular grid at 1 m centres on the finished surface.

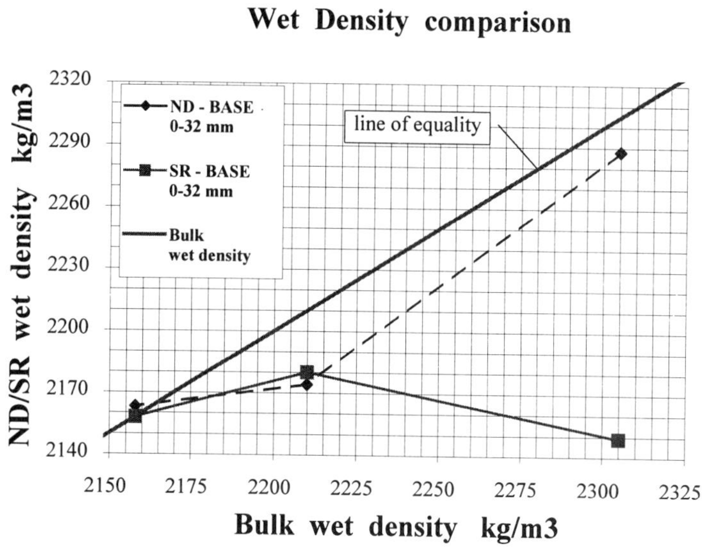

Figure 4 (a). Road Base (0 – 32 mm) material.

DISCUSSION

The summarised test results are tabulated in Table 1, which presents the results separately for the Concrete Ring test and for the Field Density test. The values tabulated for both test results are the mean results of the tests carried out, and are generally stated in terms of mean percentage difference from the reference value; for the Concrete Ring test, the reference is the actual bulk weights of the unit when filled with compacted granular materials, and in the case of the Field Density test, the reference is the Sand Replacement test values (for Wet and Dry Densities) and the oven dried Moisture Contents for the MC comparisons.

Table 2 presents a comparison of the results from this study with a previous DRI study. Subbase and Sandy Silt-Clay materials are included for comparison. The mean measurement variations recorded for Road Base material (DRI report 168) exceed the corresponding Concrete Ring values, but are closer to the Field Density test values. Also, for this study, the tabulated values indicate that test measurement variations are greater with increasing particle size and range, and that less measurement variation is observed with the Ring test.

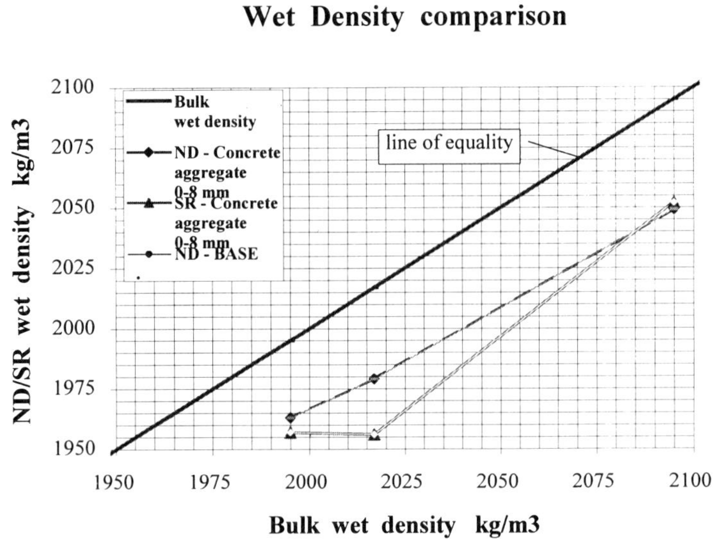

Figure 4 (b). Concrete aggregate (0 – 8 mm) material.
Figure 4. Comparison of bulk Wet Density and Wet Density determined by the ND and SR methods.

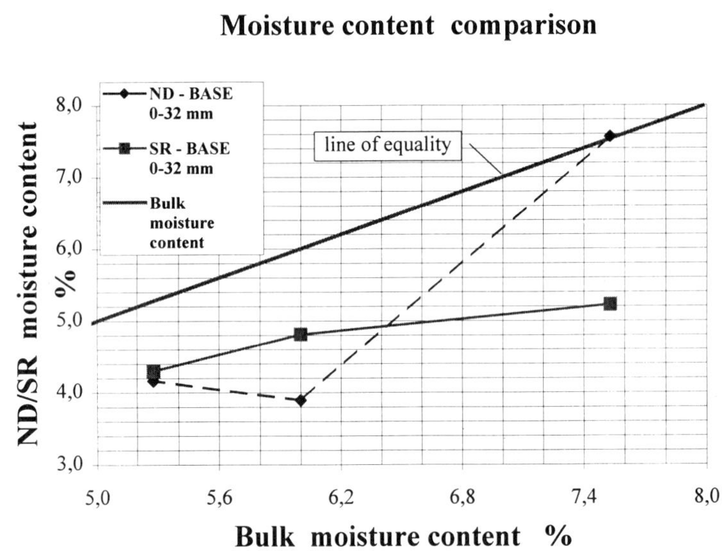

Figure 5 (a). Road Base (0 – 32 mm) material

Figures 4 and 5 show the mean results for the Concrete Ring test. Figure 4 displays the Wet Density (WD) comparison for the two granular materials, whilst figure 5 displays the Moisture Content comparison for the two materials. In all cases, with the exception of the lowest WD-Base data points (Fig. 4a), both the Nuclear densometer and Sand Replacement methods underestimate the actual Wet and Moisture Densities in the Concrete Ring test. Also, the Concrete Aggregate data values are far more consistent than the corresponding Road Base data values (i.e., measurement variations increase for the coarser material).

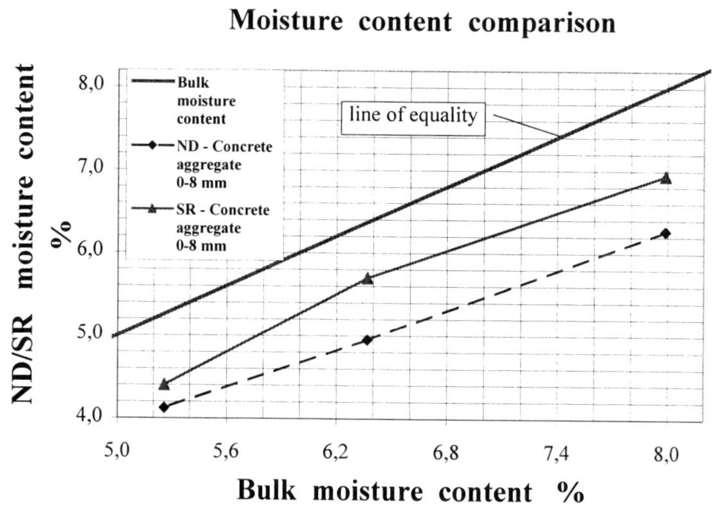

Figure 5 (b). Concrete aggregate (0 – 8 mm) material.
Figure 5. Comparison of bulk Moisture Content and Moisture Content determined by the ND and SR methods.

In Table 3, the results of regression analyses on the Wet Density test data values is given . A second order polynomial was chosen as the SR versus ND data values appeared to be distributed non-linearly. The R^2 coefficients, which express the proportion of the variation explained by the chosen function, are quite respectable. When both sets of data for the Concrete Reing test and the Field Density were combined, the R^2 regression term was 0.863 for the Road Base and 0.830 for the Concrete Aggregate materials, indicating that the functions explained 86% and 83% of the total variation in the data values for both materials.

CONCLUSIONS

This paper describes the tests carried out using two types of unbound granular materials to compare densities achieved after compaction in two separate experiments. Nuclear densometer and Sand Replacement methods were compared to each other and to the actual measured bulk densities of the compacted granular materials filling a controlled volume.

The experimental results indicated that on the basis of the concrete ring test, neither density measurement test method determined both moisture contents and dry densities of the unbound granular materials significantly better than the other. Both field test methods appear to underestimate actual densities and moisture contents to some degree. In most cases, the Nuclear densometer showed better repeatability and lower experimental deviation from the correct values for wet and dry material densities. For moisture contents determinations, the Sand Replacement method was more accurate, achieving 1% estimation errors, while the Nuclear densometer achieved overall estimation errors of 2%.

On balance, the results of this study indicate that the Nuclear densometer method is at least as accurate and reliable as the Sand replacement method for the two granular materials tested. The inclusion of other unbound road pavement materials in a wider study along the same lines, with more attention to calibration issues, would provide further useful information for highway engineers.

ACKNOWLEDGEMENTS

The Danish Road Institute wish to acknowledge and thank a number of persons and companies, and in particular Nymølle Stenindustrier a/s, for their help, good advice, willing assistance, and supply of granular materials during and throughout the course of the study. Mr Finn Hansen, from the Danish Road Institute (Consulting and Measurement Department) was responsible for the acquisition of construction materials and equipment and, assisted by the second author, carried out the sampling, filling, compaction and Nuclear densometer testing of the test materials.

REFERENCES

1. DANISH ROAD INSTITUTE. Komprimeringskontrol efter isotopmetode. Leveringsbetingelse og prøvningsmetoder 16 (in Danish). DRI, 1984.

2. DANISH ROAD INSTITUTE. Praktisk komprimeringskontrol med Isotopudstyr (in Danish), DRI Notat 168, 1986.

3. TRANSPORT RESEARCH LABORATORY. Soil stabilization with cement and lime (State–of–the–art Review), by P.T. Sherwood, London, HMSO 1995.

4. TRANSPORT RESEARCH LABORATORY. Compaction of Soils and Granular Materials: A review of research performed at the Transport Research Laboratory, by A. W. Parsons, London, HMSO 1992.

5. TRANSPORT RESEARCH LABORATORY. The measurement of the in-situ density of cement-bound materials, by P.T. Sherwood and A. J. Clark, TRRL Laboratory Report 1109, TRL, Crowthorne 1984.

6. American Society of Civil Engineers (ASCE). Nuclear Gage Tests on Soils Containing Various Sized Aggregates. M.A. Gabr, C.A. Gray, and H.M.Taylor. Geotechnical Special Publication No. 56, pages 35-47, ASCE, New York, 1995.

Fifth International Conference on the
BEARING CAPASITY OF
ROADS AND AIRFIELDS

INNOVATIVE COMPACTION TECHNIQUES FOR IMPROVING THE BEARING CAPACITY OF ROADS AND AIRFIELDS

M I PINARD

Director of Roads, Roads Department, Private Bag 0026,
Gaborone, Botswana

ABSTRACT

The transportation demands of the 21st century, particularly in terms of the continuing increase in the volume and weight of traffic using road and airfield pavements, have become increasingly difficult to meet with conventional 20th century construction technology. This has led to the comparatively recent development of a radically new type of compaction plant, the Impact Compactor, whose design is based on the concept of high energy impact compaction. Impact compaction technology offers a variety of new, innovative, cost-saving techniques which are increasingly being used to meet the construction challenges of the 21st century in a creative and cost-effective manner. This paper outlines the principles of high energy impact compaction and the characteristics of Impact Compactors. It then illustrates the use of the technology through a number of case histories in Southern Africa.

Key words: Compaction, Pavements, Roads, Airfields, Impact.

INTRODUCTION

Soil compaction is a critical aspect of the construction of road and airfield pavements since it affects their long-term performance and, ultimately, their maintenance and operational costs. In order to achieve the desired serviceability of such pavements at minimum cost, it is vital that the compaction process is carried out as effectively and efficiently as possible.

Traditional approaches to soil compaction have centred on the use of compaction plant which has relied on either static weight (e.g. smooth wheel rollers), kneading action (e.g. sheepsfoot rollers) or vibration (e.g. vibratory rollers) to achieve an increase in the strength of the material in order to create a load bearing capacity appropriate to the stresses to be imposed. However, the demand for more effective and efficient compaction has led to the development of a radically new type of compaction plant, the non-circular shaped, high energy, Impact Compactor. This revolutionary development has provided the construction industry with an improved compaction capability in terms of increased productivity, greater depth of compaction and reduced water requirements.

As with any new technology, new approaches are required to enable the unique features of the impact compaction plant to be fully exploited. Accordingly, this paper firstly reviews the historical development and principles of impact compaction and then outlines the characteristics and uses of Impact Compactors. The paper then highlights the shortcomings of traditional approaches to compaction and proposes more appropriate and rational approaches to impact compaction specification and control. Finally, the paper illustrates the application of impact compaction technology through the development of a low-cost road/airfield pavement system.

DEVELOPMENT AND PRINCIPLES OF IMPACT COMPACTION

Historical Development: Impact compaction is rooted historically in the various methods of dynamic compaction, such as the hand-ramming and tamping tools referred to by Roman historians. Other methods include the Ménard technique, which involves dropping a large weight from a height of up to 30 metres onto a soil mass and, more recently, the Impact Compactor which represents one of the latest developments in compaction technology.

The distinguishing feature of all Impact Compactors is their non-circular compacting masses which have a series of points alternating with flat compacting faces. The compactors vary in relation to their shape (3-, 4- or 5-sided) mass and configuration.

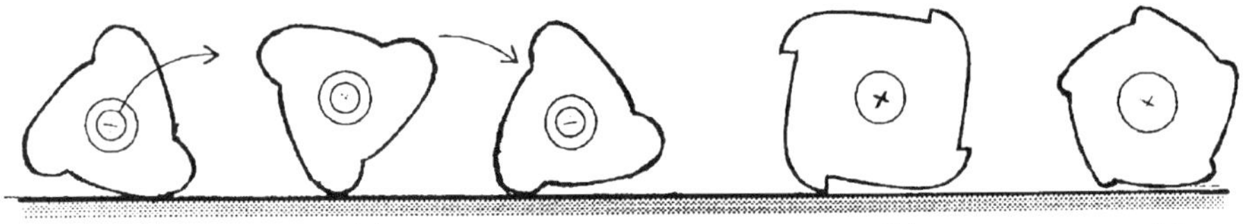

Figure 1 - Diagrammatic representation (not to scale) of (a) 3-sided Impact Compactor in operation, (b) 4-sided and (c) 5-sided Impact Compactor profiles

Fundamental Principles: The fundamental principles of impact compaction are embodied in the standardised laboratory compaction tests, such as the Proctor and Modified AASHTO tests, that are used to determine the density characteristics of soils. In these tests, energy is delivered to the soil at high amplitude and low frequency which, in practice, relates more to the process of impact compaction than to either static or vibratory compaction in which energy is delivered at low amplitude and high frequency.

Just as occurs when Modified AASHTO rather than Proctor laboratory compaction standards are followed, an increase in density and depth of soil compaction can be achieved if the following conditions prevail:

- The compaction force is applied over a large area;
- The force is dynamic and applied with high amplitude and low frequency, and
- The "parcel of energy" which produces the compacting force is large.

These principles of impact compaction are embodied in the design of Impact Compactors which was pioneered in South Africa (1). A modern, 3-sided Impact Compactor is illustrated in Photograph 1 below.

Photograph 1 - Profile of Compacting Masses of 3-sided Impact Compactor

Effectiveness of the Compaction Process: In general, the effectiveness of the compaction process depends primarily on three important, inter-related factors, namely:

- Soil type;
- Soil moisture content during compaction, and
- Nature and amount of the compactive effort.

For given soil conditions, the effectiveness of the compaction process is critically related to the nature, amount and manner of applying the compactive effort. As illustrated by the nail/hammer analogy below, due to the high load intensity and large surface area of the compacting masses, coupled with a high amplitude/low frequency mode of operation, Impact Compactors are able to achieve a depth of influence well in excess of that possible by conventional compaction plant.

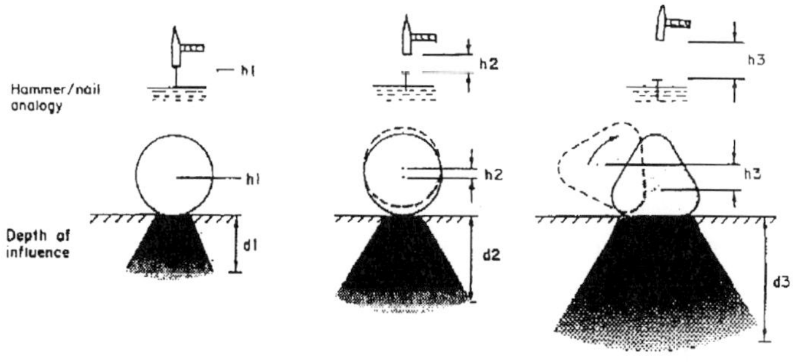

Figure 2 - Effectiveness of Compaction Process in Relation to the Nature and Manner of Applying the Compactive Effort

Soil Response to Compaction: During any compaction process, the stress waves generated by the compaction plant gradually overcome inter-particle soil friction. Initially, this causes plastic deformation (non-recoverable movement of soil particles) and an associated increase in density and, hence, strength, of the soil mass.

Further compactive effort results in further densification of the soil mass until it progresses from its elasto-plastic state (partially recoverable movement) to its elastic state (fully recoverable movement) when the soil is able to support the imposed load without permanent deformation. In this state, additional rolling has little, if any, effect on increasing soil density, i.e. the soil has been compacted to elastic restitution or "refusal". Further densification of the soil from its elastic state under a given compactive effort can then only be achieved by increasing the compactive "parcel of energy". In practice, this requires that the mass and/or amplitude and/or contact area of the roller be further increased, principles that are embodied in the design of Impact Compactors.

CHARACTERISTICS OF IMPACT COMPACTORS

Operational Features: During the compaction process, the compactor masses are towed along the ground by a wheeled tractor at a relatively high speed of approximately 12 km/hr. This operation causes the masses to rotate by contact with the ground surface and to alternatively rise up on their points and then to fall downward onto their compacting faces. This action produces a series of high amplitude, high impact blows delivered to the surface of the soil at a relatively low frequency of 90 to 130 blows per minute with the energy per blow varying between about 10 and 25 kilojoules, depending on the mass of the Impact Compactor.

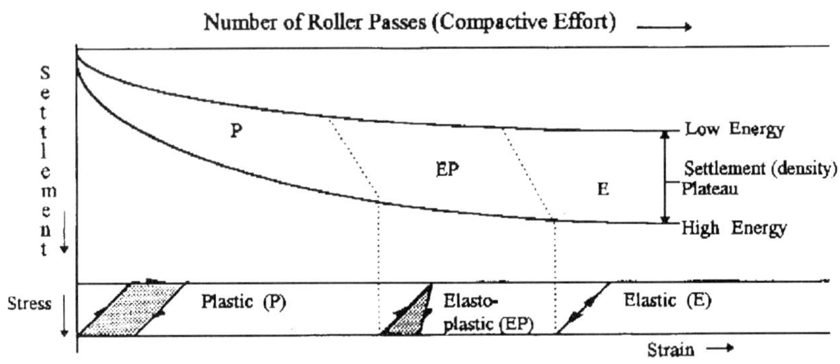

Figure 3 - Soil Response to Compaction in relation to Compactive Effort

As the compaction process proceeds towards elastic restitution, the full proof-stressing ability of the Impact Compactor is gradually realised. At this point, when the downward velocity of the compacting mass is arrested by the resistance of the soil, the dynamic compaction force would be of the order of 300-400 tonnes, depending on the type of compactor being used. The magnitude of such a force is well in excess of that imposed by vehicle or aircraft wheel loads which makes the Impact Compactor particularly well suited for proof compaction of subgrade soils.

Capabilities of Impact Compactors: By virtue of their unique design, Impact Compactors offer the following capabilities which enable many compaction operations to be carried out more efficiently and cost-effectively as compared with conventional compaction plant.

- Reduced water requirements (2);
- Deep/thick lift compaction (3);
- Effective compaction of rock fill
- In situ treatment of "collapsible" soils (4);
- Accelerated compaction of wet material
- "Super-proof" compaction

Selection and Use of Impact Compactors: The choice of Impact Compactor for a particular application is generally dictated by the soil type and associated surface shear strength characteristics, as well as the depth of compaction required. In this regard, careful consideration must be given to the choice of energy rating, shape and configuration of the Impact Compactor for which various models (3, 4 and 5-sided) are available.

The 3-sided Impact Compactor is normally best suited where deep, in-situ compaction is required. In contrast, 4- and 5-sided Impact Compactors are best suited for thick lift compaction of imported fill. The correct selection of machine will ensure that sufficient soil improvement is obtained to the required depth at the point of compaction refusal.

COMPACTION SPECIFICATION AND CONTROL

Traditional Approaches: Traditional approaches to compaction specification and control have generally evolved to suit the capabilities of conventional compaction plant for which relatively thin lift, layered construction and a related stepped density profile are normally specified. However, the enhanced compaction capabilities of Impact Compactors, particularly with regard to their greater depth effectiveness, has made traditional methods of specification and control generally inappropriate for use with this new technology. As a result, more appropriate methods of compaction specification and control have been developed to overcome the basic weaknesses inherent in traditional methods.

Conventional specifications generally indicate the procedure by which the compaction outcome is to be accomplished (method specification), the required quality of the compacted materials (end-result specification) or some combination of both the method and end-result specifications. The essential difference between these approaches depends on whether or not density requirements are specified.

Method specifications are generally established on the basis of a field trial which is used to establish moisture control, lift thickness, type, size and number of passes of the compaction plant. In contrast, end result specifications are generally expressed in terms of dry density achieved usually as a percentage of a laboratory maximum density (usually Proctor or Modified AASHTO).

Limitations of Traditional Approaches: Method specifications have the advantage of not requiring control testing of the finished product. However, the inherent variability of soils makes it difficult to establish general procedures that will be satisfactory throughout the project.

End result specifications also suffer from a number of inherent problems, not the least of which is the use of a criterion, usually dry density, which does not reflect the engineering properties of soils in roadway or runway service conditions. Other problems associated with the use of density as a method of compaction control include:

- The large ratio between volume of material tested to that compacted (typically 1: 100,000);
- Lack of correlation between laboratory and field compaction;
- Poor reproducibility of test results;
- Long duration of testing especially with deep/thick lift compaction;

In view of the above, new, more appropriate, methods of compaction specification and control are required to exploit the capabilities of Impact Compactors as well as to overcome the inherent difficulties of employing density as the primary measure for controlling compaction.

New Approaches: The new approaches to compaction specification and control are essentially a mixture of the traditional method and end-result specifications. In this new approach, the Impact Compactor is used as a "proof compactor" and direct measurement is undertaken of the engineering properties of the soil that are required for design purposes, i.e. elastic modulus (stiffness) rather than the conventionally specified proxy of dry density. Ideally, resistance to permanent deformation should also be assessed in situ. In this new approach, the inherent problems experienced with the conventional density approach are avoided and the advantages of using the Impact Compactor as a proof compactor, as described previously, can be realised to economic advantage. The advantages of this new approach are:

- Direct measurement of the engineering properties of the soil which provide an influential, quantitative input into any rational road or airfield pavement design procedure;
- Minimal use or possibly avoidance of time consuming laboratory testing as a primary measure of control of the compaction process;
- Immediate availability of results with minimum delay to compaction works;
- Comparatively simple control of the compaction even with heterogeneous materials.

The above methods of compaction specification and control place an additional onus on the site supervision team to ensure that sufficient compaction has been carried out. This requires a higher commitment of site personnel than the more traditional end-result approach which places the onus on the contractor to achieve a given result. However, this can be economically balanced by a reduction of testing resources on site and in the laboratory, both of which can be time consuming and costly.

APPLICATION OF IMPACT COMPACTION TECHNOLOGY

Underlying Philosophy: The underlying philosophy behind the application of impact compaction technology is to integrate the impact compaction and pavement design processes in a rational manner which allows optimum use to be made of available construction materials. In this approach the Impact Compactor is used as a highly efficient proof compactor based on the logic that:

- If, for a given moisture regime, the compactive effort and associated dynamic stresses induced by the Impact Compactor exceed those to be expected in service, and
- If the surface of the subgrade is compacted by as many passes as are required to transform the material from its given state to its elastic state,

then the compacted layer from the surface downwards will adapt to sustain the dynamic stresses imposed by traffic without deforming. As indicated in Figure 4, the vertical profile of this adaptation will not follow either the stepped density profile usually specified or the linear profile often assumed. Instead, the vertical profile will follow a strength-balanced density gradient which will be a direct response to the imposed stress and will be of a form similar to the way in which stress actually decreases with depth.

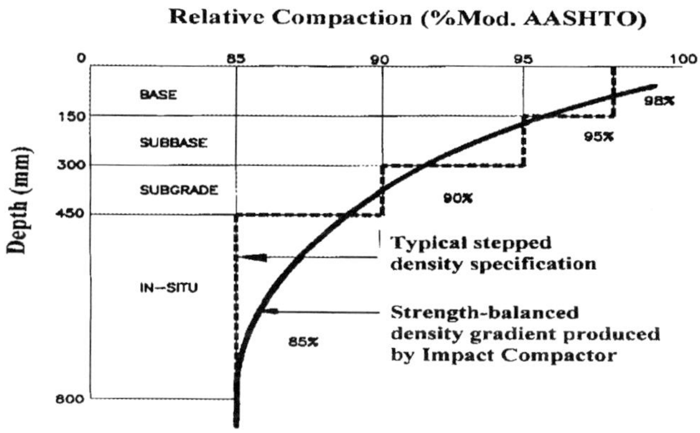

4 - Strength-balanced Density Gradient Response of Soil Layer to Impact Compaction

Key Elements of the Integrated Design System: The key elements of the integrated compaction and pavement design processes entail carrying out a representative impact compaction trial as a basis for determining the engineering properties of the soil as an input into an appropriate pavement design process. The compaction activities associated with the overall design process may be summarised as follows:

(a) Impact Compaction Trial: Carry out a field compaction trial on a representative test section(s) of the project site in order to establish the engineering properties of the soil for design purposes. This entails compacting the soil to refusal with an appropriate type of Impact Compactor chosen in relation to the soil type and type of compaction being undertaken (i.e. in-situ subgrade or embankment fill) in order to transform the soil into its elastic state. Establishment of the elastic state of the soil may be determined from measurement of settlement of the ground surface with each increment of say 10 compactor passes to a point where relatively little further settlement occurs. The output of a typical compaction trial is given in Figure 5.

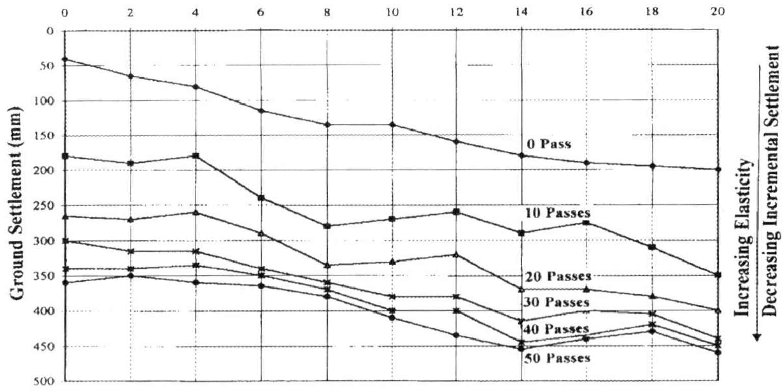

Figure 5 - Progressive Settlement of Ground Surface with Number of Compactor Passes

(b) Measurement of Engineering Properties of Soil: During the compaction trial, measure those fundamental engineering properties of the soil that critically influence its performance in service and are required for pavement design purposes. These properties include basic materials classification and in situ moisture content and, more importantly, soil stiffness (elastic modulus) and, desirably, resistance to permanent deformation for assessing long-term susceptibility to rutting. Other soil measurements would typically include the use of the Dynamic Cone Penetrometer (DCP) for determining the depth of influence of soil improvement. As indicated in Figure 5, the relationship between plate settlement, associated soil stiffness and number of compactor passes, provides a rational basis for choosing an economic compaction cut-off point, in terms of the number of compactor passes, at which the soil would have been transformed to its near elastic state.

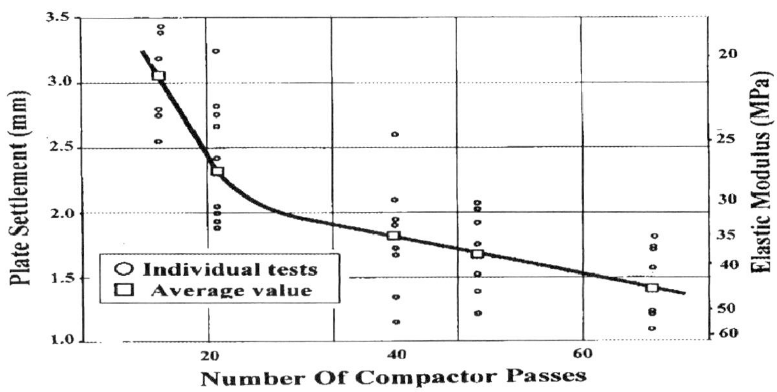

Figure 6 - Typical Output from Compaction Trial

Depending on the circumstances, various methods may be used for measuring the Elastic Modulus and Modulus of Deformation of the soil, including the Static or Dynamic Plate Load Test or the Falling Weight Deflectometer. If a change in moisture regime is expected to occur during service then, for design purposes, appropriate adjustments should be made to the value of the engineering properties of the soil measured during the compaction trial.

LOW COST ROAD SYSTEM

Materials typically make up about 70% of the cost of road construction of low-medium volume roads (traffic volumes up to about 2000 vehicles per day) which constitute a significant proportion of road networks in many countries. Substantial savings in new construction, upgrading or rehabilitation costs of such roads can be achieved if optimum use can be made of the relatively inexpensive in situ materials that often occur along their alignments.

Impact compaction technology has been used to advantage in the construction of a number of road projects in Southern Africa. Typically, an impact compactor of appropriate energy, shape and configuration, compatible with the soil type, is used as a "proof" compactor to compact and ultimately transform the soil to its elastic state. Under these conditions, the soil will have adapted to sustain the dynamic stresses imposed by the compactor and would exhibit a deep, well-balanced, profile which is comparatively less sensitive to traffic overloading than the shallower pavements of similar structural capacity (5).

As illustrated in Figure 7, a conventional pavement design would typically prescribe one or two imported layers over the subgrade. However, through the use of impact compaction, a deep pavement structure of at least equivalent structural capacity can be achieved in which in-depth compaction of the in situ material can be substituted for imported pavement layers, thus reducing costs quite significantly (of the order of 15 per cent). (6)

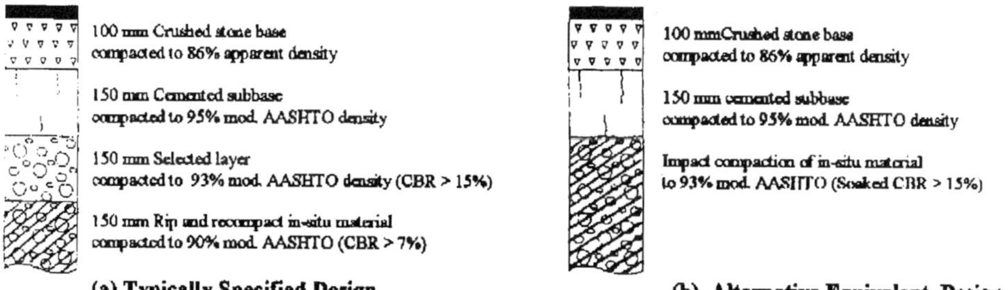

Figure 7 - Illustration of Low-cost Road System

SUMMARY AND CONCLUSIONS

Impact Compaction technology has evolved from the fundamental principles of standardised density tests and has led to the development of Impact Compactors of varying energy, shape and configuration. Compared to conventional compaction plant, Impact Compactors appear to be particularly well suited for achieving economies in the construction of roads and airfields in a variety of ways including: reduced water requirements, deep in-situ compaction, thick lift compaction, accelerated draining of wet material and proof rolling. These capabilities have provided the means of developing a low-cost road system which permits construction of stiffer, uniform subgrade conditions for minimum thickness pavements offering the potential for achieving significant cost savings.

Traditional specifications, which have evolved to suit the capabilities of lower energy compaction plant, are often not appropriate for application with Impact Compactors, particularly with regard to permissible layer thickness and the method of specifying quality control tests. This has engendered a need for new, more rational, methods of compaction specification and control, based on direct measurement of the engineering properties of the soil. Such approaches allow the potential of impact compaction technology to be exploited in a more efficient and cost-effective manner than is possible with more traditional approaches.

REFERENCES

1. Berrangé, A.R. South Africa takes the lead in the evolution of a new technology. Municipal Engineer. Johannesburg, Vol. 6, No. 2., 1975

2. Pinard, M.I and Ookeditse, S. Evaluation of high energy impact compaction techniques for minimising construction water requirements in semi-arid regions. Proc.14th ARRB Conf.Vol.7, pp 121-39, Canberra,1988.

3. Clifford, J.M. Impact Rolling and constr. techniques. Proc. 8th ARRB Conf. Vol. 8, pp21-5, 1976.

4. Pinard, M.I., Ookeditse, S. and Fraser, C. Evaluation of impact compaction trials on potentially collapsing sands in Botswana. Proc. Annual Transp. Convention, Vol IIIB, Pretoria, 1988.

5. Burgess, I.G and Joubert, J. A low-cost road system utilising in-depth compaction of in situ material. Proc. IRF Reg. Conf. Vol II, Paper No. 28, Johannesburg, 1995.

6. De beer, M, Kleyn, E.G and Savage, P.F. Towards the classification system for the strength-balance of thin surfaced flexible pavements. Proc. Annual Transp. Convention, Vol IIID, Pretoria, 1988.

Fifth International Conference on the
BEARING CAPASITY OF
ROADS AND AIRFIELDS

PRESSURE CELL APPLICABILITY

VAGN ASKEGAARD[1], ANDERS BRINK[1] and PER ULLIDTZ[2]

[1]Professor and Research Engineer, Department of Structural Engineering and Material, Technical University of Denmark, Building 118, DK-2800 Lyngby
[2]Dr. Techn., Department of Planning, Technical University of Denmark, Building 115, DK-2800 Lyngby

ABSTRACT

Structural analysis of pavements is based on analytical models, most often using continuum mechanics with linear elastic materials and static equilibrium. More advanced models may consider non-linear elastic, visco-elastic, elasto-plastic and/or anisotropic materials, dynamic loading or the particulate nature of materials. Acceptance of any of these models requires that experimental results make the theory probable. This again requires that the accuracy of the experimental results is known, a fact that is not always acknowledged.

In the case of measuring stresses and strains in particulate media this means estimating the accuracy of results obtained from tests with stress and strain cells mounted in or in contact with materials with loading history dependent properties and with not too well known loading histories.

The design of suitable cells is based on calculations involving approximation of cell geometry and of physical material properties of cell and particulate material. Design expressions for stress and strain cells placed in a material loaded in an arbitrary three-dimensional state of stress (also stress free strain state) are known. The expressions are based upon the cells being axially symmetric ellipsoids and the materials being linear elastic, homogeneous and isotropic.

These assumptions are not fulfilled in practice, but previous tests with thin cells (1:20 thickness/diameter) in sand, wheat and calcite powder have shown that pressure cell behaviour can be predicted with very good accuracy using the above assumptions.

In this paper the influence of thickness/diameter ratio on measuring accuracy has been studied. Local changes in homogeneity under load will happen in the particulate material near the cell. This is expected to be more pronounced with increasing thickness/diameter ratio, leading to less accurate measurements. Ratios of 1:20, 1:5 and 1:3 have been involved in calibration tests in sand, under very varying load histories.

It is concluded, as expected, that design expressions comply very well with actual pressure cell behaviour for the thin cell (1:20). Normal pressure under very varied load histories was determined with a coefficient of variation of about 0.04 with this cell. The design expressions also described the behaviour of the thicker pressure cell (1:5) very well, although the coefficient of variation is increased to about 0.08. The behaviour of the thickest cell (1:3) showed large load history dependency and could not be predicted, and it was not possible to determine the normal stress with an acceptable degree of accuracy. The thickest cell is, therefore, not suitable for verifying theoretical models.

Key words: Pressure cells, instrumentation, calibration, particulate media.

INTRODUCTION

Structural analysis of pavements is based on analytical models, most often using continuum mechanics with linear elastic materials and static equilibrium. More advanced models may consider non-linear elastic, visco-elastic, elasto-plastic and/or anisotropic materials, dynamic loading or the particulate nature of materials. Acceptance of any of these models requires that experimental results make the theory probable. This again requires that the accuracy of the experimental results is known, a fact that is not always acknowledged.

In the case of measuring stresses and strains in particulate media this means estimating the accuracy of results obtained from tests with stress and strain cells mounted in or in contact with materials with loading history dependent properties and with not too well know loading histories.

The designing of suitable cells is based on calculations involving approximations of cell geometry and physical material properties of cell and particulate material. Work in this field up to about 1972 is described in (1). Design expressions for stress and strain cells placed in a material loaded in an arbitrary three-dimensional state of stress (also stress free strain state) are given in (2). The expressions are based upon the cells being axially symmetric ellipsoids and the materials being linear elastic homogeneous and isotropic. Results from tests with thin cells in sand are mentioned in the same paper and correspond with the theoretically predicted.

Further experimental verification of the design expressions for a this normal stress cell and to a lesser degree for a thin shear stress cell is given in (3).The cells were placed in sand and wheat under very varied load histories involving permanent strain. The normal stress cell showed the actual normal stress to be nearly independent of nonlinearities, permanent strain, and type of particulate medium and stress state in the medium. The same was to a lesser degree the case for the shear stress cell.

In the present paper the influence of cell thickness-diameter ratio on accuracy and correspondence with theoretical predictions is studied. The cells are mounted in sand and very varied load histories are used to get an overall accuracy. The test programme was executed by an operator familiar with measurements in general but not with pressure cell measurements in particular. The operator was given a short introductory instruction and carried out a pre-test before the test series reported in this paper.

The test also included measurement of sensitivity under hydrostatic load and of compressibility.

CELL TYPES

Fig. 1. Cells used for calibration

Cell	Pressure range (KPa)	Geometry	Diameter of sensitive diaphragm (mm)	Thickness/ext. diameter ratio
I	0-200	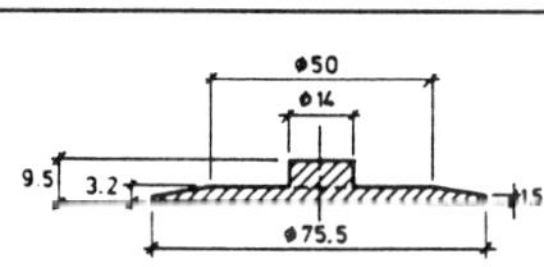	73	1:20
II	10-200	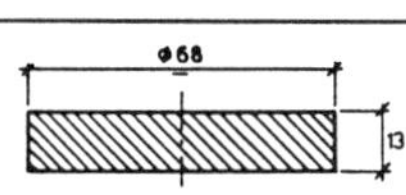	66	1:5
III	0-350	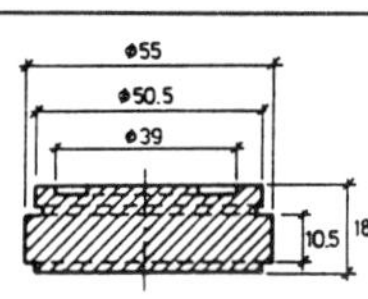	39	1:3

Table 1. Geometry of cells

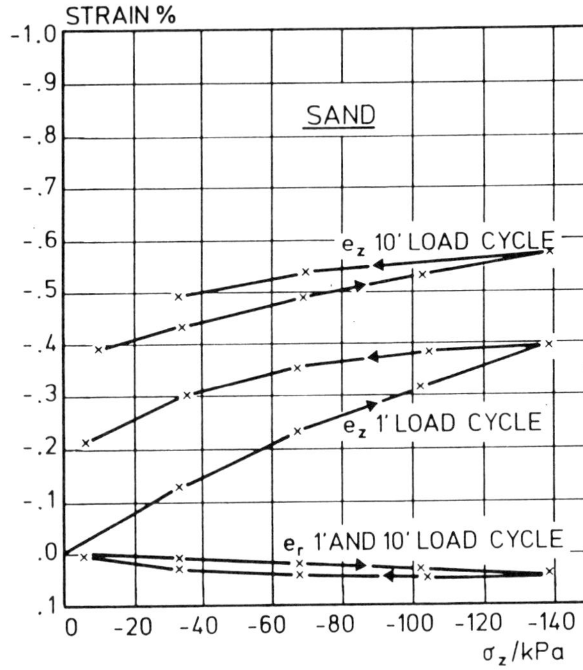

Fig. 6. Loading history.

The different maximum pressures were chosen so that the maximum signal from the stress cell was about the same independent of θ. The load cycles 2 - 8 were all taken from 20 to 140 kPa. The operator was given a short instruction in mounting the pressure cells in sand and made a pre-test himself. He then made two sets of tests with each of the three pressure cells giving a total of 18 individual tests. Each set included results from the 1st and 10th load cycle and for the three tilting angles. The best estimate of C and D for each cell is obtained by pooling the results for the two sets of tests on a cell (6 tests). The curves are shown in Fig. 7 in a co-ordinate system with the measured fluid pressure q (observation) as abscissa and $C\sigma_{33}^A + D(\sigma_{11}^A + \sigma_{22}^A)$ (estimate) as ordinate. The corresponding C and D values are also given in Fig. 7 for cell I and II but not for cell III, because these values and expression [2] do not describe the behaviour of this test cell which is load history dependent. This dependence is mainly attributed to the cells high thickness/diameter ratio which leads to a large load dependent change of homogeneity in the sand close to the cell and thereby a change of the stress distribution on the cell surface.

It must be remembered that pressure cells of the types described are used to determine the normal stress σ_{33}^A in the undisturbed material around the cell and acting normal to the pressure cell surface (see Fig. 2).

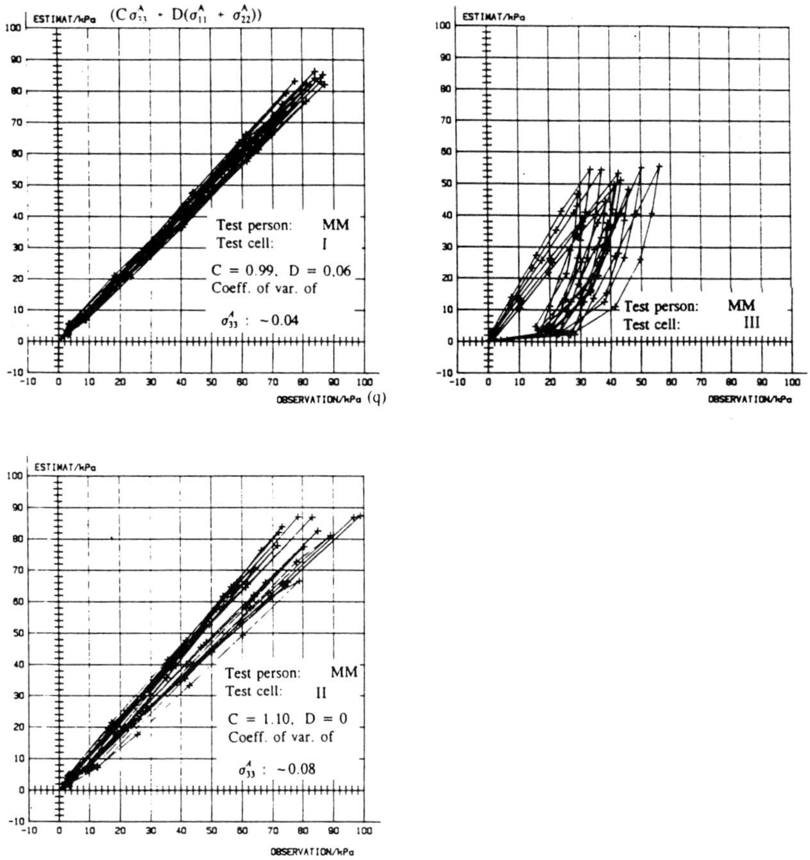

Fig. 7. Results of calibration tests.

From expression [2] it can be seen that if $\left(\sigma_{11}^A + \sigma_{22}^A\right)/\sigma_{33}^A$ can be well estimated or D is negligible, then σ_{33}^A can be found from the measured fluid pressure in the cell. These conditions are fulfilled in the tests with cells I and II in Fig. 7 and an estimate of the coefficient of variation of σ_{33}^A determined by using these cells under very varied test conditions can be found from this figure. The values found were 0.04 for the cell 1:20 increasing to 0.08 for the cell 1:5. The values are shown in Fig. 7. They include the effect of producing a homogeneous state of stress in the test chamber. This contribution is, however, the smallest part.

The measured and predicted (theoretical) C and D values and the coefficient of variation of σ_{33}^A are shown in Table 2. The correspondence between measured and predicted values is good.

Similar tests with identical cells of type I are reported in (7). An experienced operator obtained C = 1.01, D = 0.01 and a coefficient of variation of 0.05 for σ_{33}^A, while an inexperienced operator got C = 1.10, D = 0.03 and a coefficient of variation of 0.08. Although the values for C and D found by the different operators does not differ much a certain personal factor is, however, observed mainly caused by a slightly different mounting procedure. An operator should therefore make calibration tests to obtain the personal C and D coefficients before carrying out tests in practice.

The stress state in calcite powder, loaded hydrostatically with permanent strains up to 13% was well described using the same C and D values as discussed in (7).

CONCLUSION

The following conclusions can be drawn from the tests:

1. The thinner the cell is the smaller the coefficient of variation of determining σ_{33}^{A} becomes. This is attributed to the fact that it is easier to produce a homogeneous sand density around the cell when it is thin, and that the changes under load of sand homogeneity and stress state around the cell become small.
2. The behaviour of cell I and II complies well with the theoretical expression while cell III does not. This latter cell, which has a high thickness/diameter ratio, shows pronounced load-history dependency and seems not suitable for measurements in particulate materials.
3. There is a difference, although small, between the C and D values obtained in the tests with the thin cell described here and those obtained in (7). This means that there is a personal factor connected with the use of pressure cells. This is related to the mounting procedure of the cells although the same instruction is given to all operators. This means that an operator should do calibration experiments under controlled conditions in order to determine the personal coefficients C and D.
4. It is possible to measure stress in fine grained materials with good accuracy when using thin pressure cells. Even quite inexperienced operators obtain results with an acceptable coefficient of variation.

REFERENCES

1. Hvorslev, M.J. *The changeable interaction between soils and pressure cells. Tests and reviews at the Waterways Experiment Station.* Tech.rep. S-76-7, U.S. Army Corps of Engineers, Waterways Experiment Station. Vicksburg, MS.. (1976).
2. Askegaard, V. *Stress and strain measurements in solid materials.* Report No. R 92, Dept. of Structural Engineering, Tech. Univ. of Denmark, 1979-
3. Askegaard, V. Applicability of normal and shear stress cells embedded in cohesionless materials. *Experimental Mechanics,* Dec. 1995, pp. 315-321.
4. Askegaard, V. Measurement of pressure between a rigid wall and a compressible medium by means of pressure cells. *Acta Polytechnica Scandinavica,* Ci 11, Copenhagen 1961.
5. Askegaard, V. The pressure cell problem, *Proc. Tagung 88, Silos-Forschung und Praxis.* Karlsruhe 10-11 Octobre 1988, pp. 339-361.
6. Askegaard, V. Consequence of loading history on the measuring error of embedded stress cells. *Transactions of Mechanical Engineering* (Australia), vol. ME 12, No. 3, Sept. 1987, pp. 191-195.
7. Askegaard, V. and Brown, C:J. Influence of personal factor on cell response when mounting embedded pressure cells, *Bulk Solids Handling,* vol. 15, No. 2, April/June 1995, pp. 221-224.

Fifth International Conference on the
BEARING CAPASITY OF
ROADS AND AIRFIELDS

PERFORMANCE PARAMETERS AND TARGET VALUES FOR CONSTRUCTION OF UK ROAD FOUNDATIONS

P. R. FLEMING[1], C. D. F. ROGERS[2] AND M. W. FROST[3].

[1]Lecturer in Geotechnics, Department of Civil and Building Engineering, Loughborough University, Loughborough, Leicestershire, LE11 3TU, UK. Telephone +44 1509 222616. Fax +44 1509 223981, E-mail P.R.Fleming@lboro.ac.uk

[2]Reader in Geotechnical Engineering, Department of Civil and Building Engineering, Loughborough University.

[3]Research Assistant, Department of Civil and Building Engineering, Loughborough University.

ABSTRACT

There is impetus in the UK to move away from empirical design of road foundations and method specification towards analytical design assured by end product testing during construction. Current research at Loughborough University, sponsored by the Highways Agency, is aimed at introducing such a performance based specification. This paper introduces the philosophy behind the research and explains the primary objectives. Selective results are presented from a full-scale field trial construction of a road foundation on a soft subgrade, which was characterised by a variety of field devices and subsequently trafficked. This paper focuses on insitu stiffness modulus (E_R), measured by several devices for comparison. The setting of suitable target values for field E_R is seen as a key factor and the influence of inherent variability in material response a concern. The discrepancy between devices and the problem of long-term stability are discussed and highlighted as a key focus for the remainder of the work.

KEYWORDS: Foundation, Insitu Stiffness, Design, Trafficking, Triaxial.

INTRODUCTION

UK road foundations often consist of an improvement layer of unbound granular materials (capping) compacted onto a soil subgrade, which is either in its natural state or has been treated and/or compacted during the earthworks. This provides the *formation* upon which sub-base is laid and compacted to complete the *foundation,* which aims to provide a suitable platform to construct the upper structural pavement layers. UK road foundations are designed in accordance with HD25/94[1]. The foundation layer thicknesses are designed according to the equilibrium California Bearing Ratio (CBR), which is based on empirical data and experience and aims to ensure the long-term performance of the pavement. The road foundation is then constructed in accordance with the method specification included in the Specification for Highway Works[2]. The use of a recipe specification gives assurance of the performance of closely regulated materials that are known to be adequate.

This approach to design and construction precludes the use of functional soil and granular material performance parameters, which are required for analytical pavement design and for a truly performance based specification for road foundation construction. However, such a specification is desirable to permit the inclusion of alternative materials and/or alternative construction methods. A consortium of Scott Wilson Pavement Engineering Ltd and Loughborough and Nottingham Universities are researching and developing a performance based specification for subgrade and capping materials, funded by the UK Highways Agency. The broad specification philosophy is:

- measure performance parameters of the subgrade and potential capping materials at the site investigation stage,
- incorporate these into a foundation design procedure that provides a stable platform on which to construct and in the long term, and
- measure the performance parameters during construction to assure compliance with the design.

If a specification based on ‘end product’ testing can be implemented and linked to functional design parameters, marginal materials that do not meet the requirements of the current specification could be permitted. This would facilitate the use of more recycled materials and secondary aggregates, thus deriving desirable environmental benefits.

PERFORMANCE PARAMETERS

Road foundations must be designed to meet three critical loading conditions:

- Relatively few (≤ 1000 standard axles) large stress pulses applied directly to the formation surface by construction traffic.
- A small number of very high stresses applied by compaction plant.
- A very large number of small applied stresses caused by trafficking of the completed pavement.

Clearly the first two conditions are applied to the subgrade and compacted capping materials in their state at the time of construction, whereas the third condition is applied to the soil and capping materials in their long-term equilibrium state.

To meet these loading requirements the foundation must possess adequate stiffness (i.e. low resilient deformation) and sufficient strength to prevent permanent deformation. Adequate stiffness is required to allow satisfactory compaction of the overlaying pavement layers, and is achieved using appropriate materials that are well compacted, combined with sufficient thickness to spread the applied transient surface stress over a sufficiently large area to avoid distress to the underlying layers. Equally, the foundation must be able to sustain the applied stress without significant permanent deformation (rutting). If rutting does occur during construction, this can lead to ponding of water at the top of cohesive subgrades. This in turn results in weakening and a reduction in subgrade stiffness in the long term, causing the third loading condition to become more critical. This loss of stiffness can result in high resilient in-service deformations, leading to fatigue cracking of the overlying bound layers.

There are significant problems in measuring these two performance parameters accurately. Both resilient and permanent deformations caused in a foundation are controlled by the nature of the applied loading, in terms of applied stress (axle load), duration of stress pulse (vehicle speed) and direction of vehicle pass. In addition, different soils will behave in different ways to each combination of the above factors. Therefore the tests used to measure performance must match the loading conditions in practice as closely as possible.

Excessive resilient deformation can be guarded against by measuring stiffness *in situ* using dynamic measuring devices. To guard against excessive permanent deformation the foundation layers must possess sufficient shear strength. This can be controlled in practice by limiting the applied stress transmitted to the subgrade, i.e. by providing a capping of sufficient thickness and by ensuring that this capping material is sufficiently well compacted to have achieved an adequate stiffness and internal shear strength. Capping thickness can be designed according to the strength and stiffness of the subgrade, which can be assessed *in situ* during construction by an appropriate test or possibly predicted from laboratory tests. The capping behaviour is more difficult to estimate from laboratory tests but can be controlled by measuring density and stiffness. Direct measurement of subsurface rutting to limit the quantity of permanent deformation induced by trafficking during construction (i.e. rutting at the top of the subgrade and within the capping itself), however, remains a challenge for future research.

TARGET VALUES

Target values for stiffness and shear strength must be determined for construction compliance. It is anticipated that the target values will be specific to the device used to measure them due to variations in their operation. Values of target stiffness have been suggested from other work. For example, a stiffness of 50 to 60 MPa measured on the completed capping and 100 MPa on the completed sub-base have been suggested by the current authors[3]. In Germany, a minimum static plate bearing test modulus (E_{V2}) of 45 MPa is required at the top of formation, although this is negated by a requirement for a thick frost protection layer above the formation. Chaddock and Brown[4] suggested a value of 80 MPa as an acceptable top of capping stiffness, as measured by the Falling Weight Deflectometer with a 450 mm plate and 200 kPa contact stress, whereas a value of 30 MPa as measured by the Lightweight Weight Drop Tester is suggested[5].

Three dynamic stiffness measuring devices have been recommended for detailed evaluation[3] of their ability to confirm the stiffness achieved during construction at the top of the formation:

- The Falling Weight Deflectometer (FWD)
- The TRL Foundation Tester[4] (TFT, developed at Loughborough)
- The Lightweight Drop Tester (LDT)

The TFT and LDT are similar small scale dynamic devices which comprise a 10kg falling mass that impacts via a rubber damper seated on a 300mm bearing plate. The TFT measures the force applied and deflection of the ground (directly) whilst the LDT assumes a constant stress and measures the deflection of the plate. The LDT applies a peak stress of 100 kPa and as a result the TFT and FWD were configured to match this for the field trial described later. The FWD has been included as a benchmark device against which to judge the measurements from the other devices as it appears to be widely accepted by pavement engineers. However it is considered too sophisticated to be proposed in the new specification.

A target composite formation stiffness has been set initially at 50 MPa, based upon field observations and static linear elastic analyses. The achieved formation stiffness depends upon the stiffness of the subgrade, and the thickness and stiffness of the capping layer. Controlled compaction trials are currently being undertaken to refine this target value using the TFT and LDT.

Current UK design guidance[6] requires the formation of ruts in the sub-base, when used as a haul road, to be limited to 40mm to prevent excessive damage to the subgrade beneath. A similar approach for capping is proposed, thus requiring that the formation surface rut be monitored. The proportion of the surface rut that will occur in the subgrade will clearly depend upon the nature of the capping used, although this is difficult to predict and thus the general recommendations given in Table 1 have been suggested to limit rutting in the subgrade to 20mm.

Table 1. Permissible Rutting at the Surface of Capping Materials

Capping thickness, t (mm)	Maximum permissible surface rut (mm)
$t \leq 250$	20
$250 < t \leq 500$	30
$t > 500$	40

FIELD VALIDATION

A series of controlled field trials is being undertaken to validate and/or revise the target values suggested for the performance specification. These will also determine the strength of the relationship between the laboratory parameters used for design and the site compliance tests. To date, one controlled field trial and selective testing on live construction sites have been carried out. The controlled trial site consisted of a reasonably consistent, relatively soft sandy clay subgrade onto which three good quality standard crushed rock capping materials were constructed. Both rutting response to repeated lorry loading and stiffness were measured.

The trial foundations were constructed in twelve 3m long bays. The capping thickness design was based upon laboratory repeated load triaxial test data for undisturbed samples taken during the site investigation. The capping was 450mm deep for nine bays, each material being laid with poor (4 passes), medium (8 passes) and good (16 passes) compaction using a Benford TV 800H twin drum roller. Three bays were constructed as a wedge ranging in thickness from 300 to 600mm and given 16 passes (good compaction) of the same roller. After subgrade exposure, Dynamic Cone Penetrometer (DCP) tests were performed in each bay to provide strength profile measurements *in situ* and undisturbed samples were taken and subjected to repeated load triaxial testing in the laboratory for further correlations. Stiffness was then measured on the exposed subgrade and on the completed capping layer *directly above* the subgrade test positions. Insitu density measurements were also carried out on the completed capping. The road was then trafficked with 1000 passes of a 15 tonne, two axle lorry to examine rutting, which was monitored.

	Row A	*Row B*	*Row C*
Lightweight Drop Tester	x	x	x
TRL Foundation Tester *Lightweight Drop Tester* *Falling Weight Deflectometer*	x	x	x
Falling Weight Deflectometer	x	x	x

Figure 1. Layout of Dynamic Stiffness Test Points for Each Test Bay

The test layout was devised to give 9 locations per bay, to allow each test to be carried out three times on nominally similar freshly compacted capping and such that three sets of tests were carried out at the same point. (Figure 1).

During the course of the trial the weather became very hot causing a firm crust to be formed on the subgrade surface when exposed. Repeat stiffness measurements on the subgrade in the last bays to be covered showed it to have increased in stiffness by 50%.

No more than 30mm of rutting was caused in any bay during trafficking. This excellent foundation performance under trafficking, was primarily attributed to two factors. Firstly, the materials improved during construction due to drying out, rendering the parameters obtained from the site investigation tests no longer relevant at the time of construction. Secondly, the capping material stiffness used in calculating the required layer thickness to give a composite modulus was assumed, from experience, to be 80 MPa. Back-analysis of the FWD stiffness data indicated that the actual capping layer stiffness was in most cases higher than 80 MPa, possibly due to the greater stiffness achievable when compacting onto a stiffer substrate and due to the induction of larger suction stresses as the capping dried. Further research is required to produce a method of estimating site values for capping stiffness.

Subgrade stiffness measurements exhibited relatively high variation with all three devices, despite the fact that the subgrade could be regarded as being reasonably consistent. Figure 2 shows the stiffness readings on the subgrade for Row A. The variability observed was attributed to both natural variations in the subgrade and differences between devices including rate of·loading, transducer type, analysis method and adequacy of plate/ground contact. This variability illustrates the problems of obtaining repeatable measurements of an absolute value of subgrade stiffness for comparison with a target value.

Stiffness readings for the TFT and FWD on top of capping typically exceeded the site target value set for the composite foundation stiffness of 50 MPa. An unusual result is exhibited by Bay 10, though this was probably caused by the presence of backfill to a buried service. The LDT values however typically fell below this target value, yet the excellent performance of the trial road suggests that the target value set for testing by the LDT (only) should be reduced.

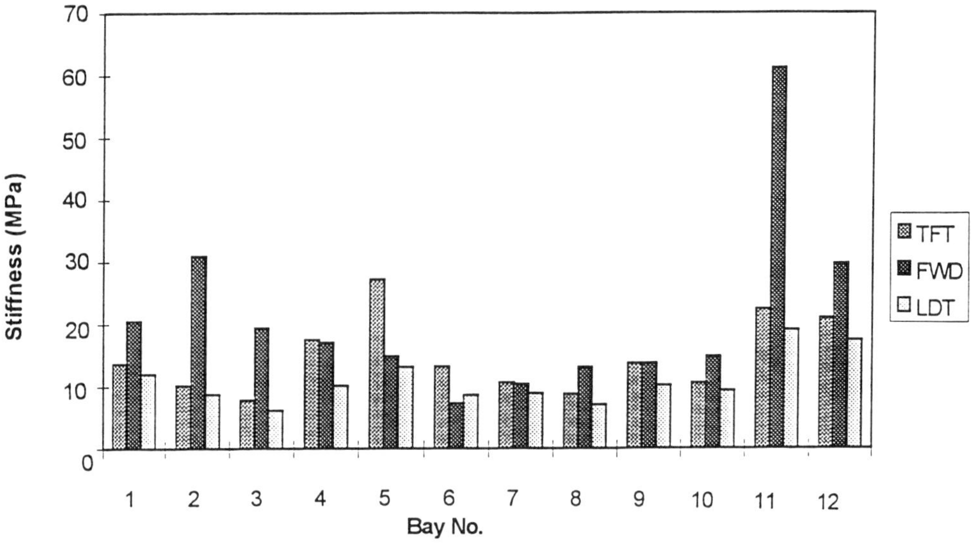

Figure 2. Comparison of Measured Stiffness along Row A on the Subgrade (Three Tests at a Single Point)

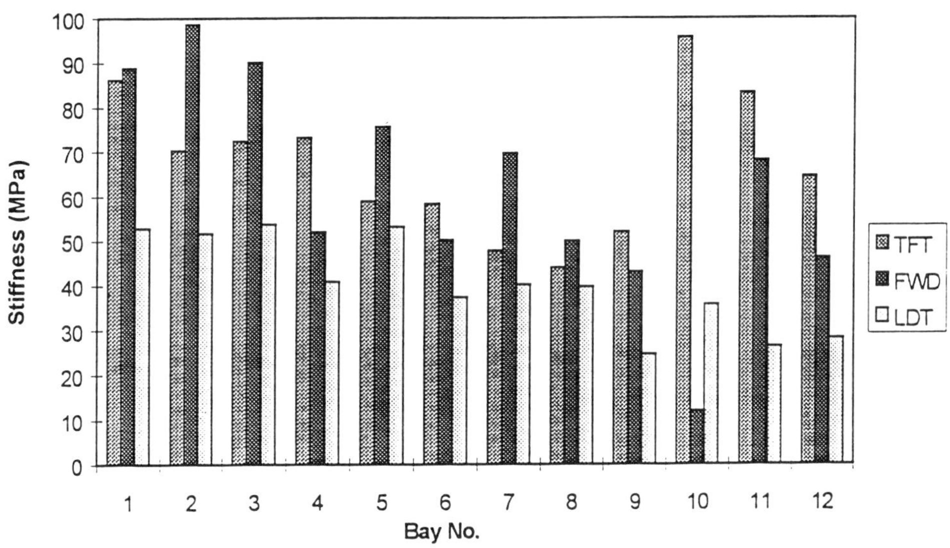

Figure 3. Comparison of Measured Stiffness along Row A on the Surface of the Capping (Three Tests at a Single Point)

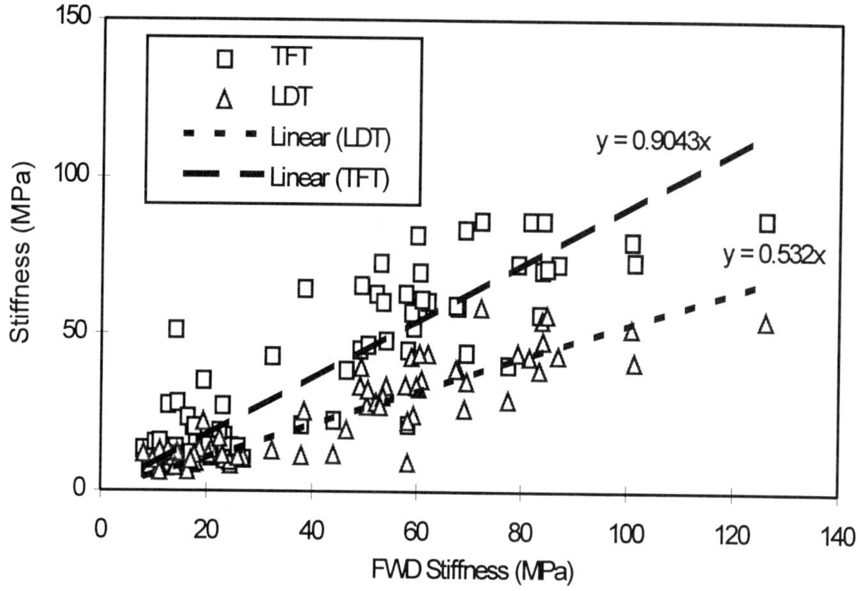

Figure 4. Stiffnesses Measured by the TFT and LDT Compared with that Measured by the FWD on the Capping and Subgrade

The test data from each of the devices, when tested at the same point, were plotted against those from the FWD to seek a correlation between devices. An example is presented in Figure 4, which includes data on both subgrade and capping. It shows that relatively poor correlation to the lines of best fit are obtained (i.e. R<0.8). The LDT showed less scatter than the TFT, although this might be partly due to the data smoothing that occurs during its data processing routine. The TFT measured stiffness values were generally 90% of those measured with the FWD, whereas those with the LDT were approximately 50% of the FWD measurements.

CONCLUSIONS AND RECOMMENDATIONS

A performance based specification for road foundation layers would provide greater assurance of adequate construction and long-term performance. This would allow the use of more marginal capping materials as their in-service performance could be assured. However its success requires accurate measurement of the elastic stiffness of subgrade, capping and sub-base materials, as well as provision to restrict rutting under construction traffic loading. In parallel with the need for suitable site assessment methods, a laboratory test is required to measure the stiffness of subgrade, and ideally also the capping and sub-base materials, under cyclic load and low confining stress, to include the use of measured stiffness in a more analytical design.

The capping thickness designed for the field trial was based upon laboratory triaxial tests on the subgrade. The design was more than adequate in terms of stiffness and resistance to permanent deformation. However, this may have partly been due to weather induced changes in the subgrade, thus confirming that site conditions can improve material performance in the short term but prudent design must always be based on a worst case condition possibly that relating to subgrade equilibrium water content. Further full-scale field trials are currently underway, with efforts focused on establishing lower bound laboratory and field values for stiffness and strength, aiming to design foundations that are close to or at failure for further validation of this work. The fieldwork also showed that significant scatter of the data was produced by all the dynamic stiffness measuring devices, therefore care must be taken in setting target values and field test protocols. Most scatter was observed on the subgrade.

The LDT consistently measured a lower stiffness (approximately 50%) than the FWD or TFT, whereas the TFT measured slightly lower stiffness (approximately 90%) than the FWD. Nevertheless the TFT and LDT are lightweight, portable and simple to use for repeated testing and provide a potentially valuable way forward for stiffness measurement *in situ*.

ACKNOWLEDGEMENTS

The authors are grateful to their colleagues at Nottingham University and Scott Wilson (PE) for their collaboration on this work, and to the Highways Agency for the opportunity to do so.

REFERENCES

1. Design Manual for Roads and Bridges, Volume 7 HD25/94, HMSO London, UK.

2. Manual Of Contract Documents for Highway Works, Volume 1: Specification for Highway Works (MCHW1) August 1993, HMSO, London, UK.

3. Fleming, P. R. and Rogers, C. D. F., Assessment of Pavement Foundations during Construction. Proc of the Institution of Civil Engineers, Transportation, May 1995, Vol 111, pp. 105-115.

4. Chaddock, B.C.J. and Brown A.J. Insitu Tests for Road Foundation Assessment. Proceedings UNBAR4 symposium, Nottingham, 17-19 July 1995, pp 259-266.

5. Nunn, M.E., Brown, A., Weston, D., Nicholls, J.C. Design of Long Life Flexible Pavements for Heavy Traffic, Transport Research Laboratory Report 250, Crowthorne, Berkshire, UK. 1997.

6. Powell, W.D., Potter, J.F., Mayhew, H.C. and Nunn, M.E. The Structural Design of Bituminous Roads, Transport Road Research Laboratory Report LR1132, Crowthorne, Berkshire, UK, 1984.

Fifth International Conference on the
BEARING CAPASITY OF
ROADS AND AIRFIELDS

MECHANICAL RESPONSE OF A GEOSYNTHETIC-REINFORCED PAVEMENT SYSTEM TO CYCLIC LOADING

STEVEN W. PERKINS[1], MUHANNAD ISMEIK[2] AND MACGREGOR L. FOGELSONG[3]

[1]Assistant Professor, Montana State University (MSU), 205 Cobleigh Hall, Bozeman, Montana, 59717, USA, (1)406-994-6105, stevep@ce.montana.edu

[2]Post-Doctoral Research Associate, MSU, 205 Cobleigh Hall, Bozeman, Montana, 59717, USA, (1)406-994-6105, ismeik@ce.montana.edu

[3]Graduate Research Assistant, MSU, 205 Cobleigh Hall, Bozeman, Montana, 59717, USA, (1)406-994-6105, flycast@imt.net

ABSTRACT

A pavement test facility has been developed for the purpose of examining the reinforcement benefit derived by placing geosynthetics (geogrids and geotextiles) in the base course layer of a flexible pavement system. The facility consists of a large concrete box in which a flexible pavement section can be constructed and loaded by a stationary cyclic load applied to a circular plate. The test sections contain an extensive array of stress, strain, moisture content and temperature sensors to assess mechanical behavior during the process of loading. Results from strain measurements in the base course aggregate and in the geosynthetics from test sections containing a geogrid and a geotextile are examined to illustrate and describe mechanisms of reinforcement for these systems. These results indicate certain shear-interface features necessary for effective reinforcement.

INTRODUCTION

The use of geosynthetics (geogrids and geotextiles) for reinforcement of the base course layer in a flexible pavement was first examined in the early 1980's using geotextiles (1, 2) and later with geogrids (3, 4). Experienced gained from studies involving reinforced unsurfaced roads (5) indicated that certain mechanisms of reinforcement thought to be operating in these roads might also be applicable for flexible pavement systems. A description of these mechanisms has been based largely on speculation and engineering judgement. Several studies have provided strain measurements on the geosynthetic to aid in illustrating reinforcement functions (3,6). An adequate understanding of reinforcement mechanisms is necessary to develop a mechanics-based model of behavior. Detailed measurements of the mechanical response of reinforced pavement systems are lacking, which may partially explain the absence of an accepted design procedure for such roadways.

The purpose of this paper is to present experimental data from a laboratory model reinforced flexible pavement system to aid in illustrating the mechanical response and mechanisms of reinforcement. The data presented is part of an on-going study and is limited to four sections completed to-date.

PROPOSED MECHANISMS OF REINFORCEMENT

Historically, the main reinforcement function attributed to geosynthetics in paved roads is what has commonly been referred to as the function of base course lateral restraint. This function was originally described by Bender and Barenberg (7) and later elaborated on by Kinney and Barenberg (5) for geotextile-reinforced unpaved roads. This name may be somewhat misleading in that the function, as envisioned, incorporates mechanisms in addition to preventing lateral movement of the base course aggregate. A more appropriate description might be to describe this reinforcement function, and its associated mechanisms, as a shear-resisting interface.

The reinforcement function of a shear-resisting interface develops through shear interaction of the base course layer with the geosynthetic layer or layers contained in or at the bottom of the base aggregate (Figure 1) and consists of four separate reinforcement mechanisms. Vehicular loads applied to the roadway surface create a lateral spreading motion of the base course aggregate. Tensile lateral strains are created in the base as the material moves down and out away from the applied load. Lateral movement of the base allows for vertical strains to develop leading to a permanent rut in the wheel path.

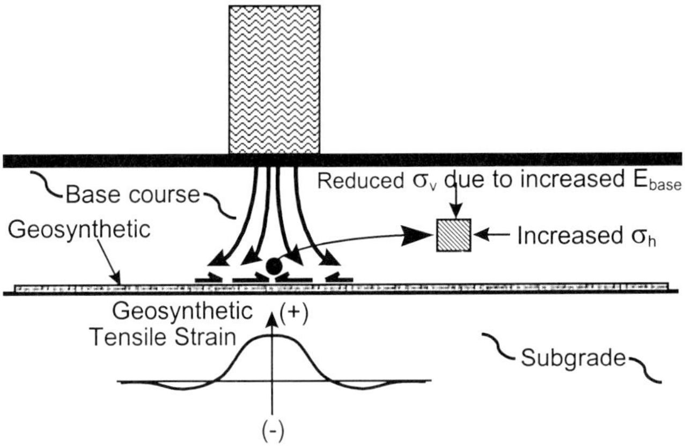

Figure 1. Illustration of shear-resisting interface reinforcement function

Placement of a geosynthetic layer or layers in the base course allows for shear interaction to develop between the aggregate and the geosynthetic as the base attempts to spread laterally. Tensile load is effectively transmitted from the base aggregate to the geosynthetic. Since the geosynthetic is considerably more stiff in tension as compared to the aggregate, far less lateral tensile strain develops in the system. Less lateral strain in the base results in less vertical deformation of the roadway surface. Hence, the first mechanism of reinforcement corresponds to direct prevention of lateral spreading of the base aggregate.

Shear stress developed between the base course aggregate and the geosynthetic provides an increase in lateral stress within the base. This increase in lateral confinement leads to an increase in the mean hydrostatic stress. Granular materials generally exhibit an increase in elastic modulus with increasing mean stress. The second reinforcement mechanism results from an increase in stiffness of the base course aggregate when adequate interaction develops between the base and the geosynthetic. The increased stiffness of this layer results in lower vertical strains in the base. It would also be expected that an increase in modulus of the base would result in lower dynamic, recoverable vertical deformations of the roadway surface, implying that fatigue of the asphalt concrete layer would be reduced.

The presence of a geosynthetic layer in the base can also lead to a change in the state of stress and strain in the subgrade. For layered systems, where a weaker subgrade material lies beneath the base, an increase in modulus of the base layer results in an improved vertical stress distribution on the subgrade. In general, the vertical stress in the base and subgrade directly beneath the applied load should decrease as the base layer stiffness increases. The vertical stress on the subgrade will become more widely distributed, meaning that surface deformation will be less and more uniform. Hence, a third reinforcement mechanism results from an improved vertical stress distribution on the subgrade.

The fourth reinforcement mechanism results from a reduction of shear stress in the subgrade soil. It is expected that shear stress transmitted from the base course to the subgrade would decrease as shearing of the base transmits tensile load to the reinforcement. Less shear stress, coupled with less vertical stress results in a less severe state of loading leading to lower vertical strain in the subgrade.

Prerequisite to realizing the reinforcement mechanisms described above is the development of a strain distribution in the geosynthetic similar to that shown in Figure 1. Haas et al. (3), Miura et al. (6) and Perkins et al. (8) have presented data demonstrating such trends for paved roadways using geogrid reinforcement. Haas et al. has also shown that vertical stress on the subgrade was less when reinforcement was present. Additional measurements of stress and strain in the base and subgrade materials are necessary to experimentally demonstrate these mechanisms. Lateral strain measurements in the base course aggregate can be used to assess the compatibility of strain between the base and the geosynthetic, indicating the effectiveness of the reinforcement. The purpose of this paper is to present data that several of the mechanisms described above and that illustrates the importance of certain geosynthetic characteristics necessary for reinforcement.

PAVEMENT TEST FACILITY

A laboratory scale pavement test facility has been developed for the purpose of evaluating the mechanical response of geosynthetic reinforced pavement systems and the reinforcement mechanisms described above. The facility has been described by Perkins et al. (8). Additional details of the facility are provided below.

Test container and loading arrangement The test container and load apparatus used are shown schematically in Figure 2. A reinforced concrete box with inside dimensions of 2 m by 2 m by 1.5 m in height is used to construct the pavement sections. A pneumatic cylinder provides a peak load of 40 kN to the pavement surface. The load is applied to a 305 mm diameter steel plate resulting in an applied surface pressure of 550 kPa. A waffled rubber pad 4 mm in thickness is placed beneath the steel plate to aid in providing a uniform pressure distribution on the asphalt concrete (AC) surface. A pneumatic binary regulator is used to control the time-history of loading to the plate. The regulator is controlled by a computer and data acquisition software and is set to deliver a load pulse of approximately 0.8 Hz frequency and corresponding to a flat-topped triangular pulse. The I-beam and roller assembly shown in Figure 2 can be used to move the load actuator to different positions within the box. This feature was not used for the tests reported in this paper where the load actuator was centered in the box for all tests.

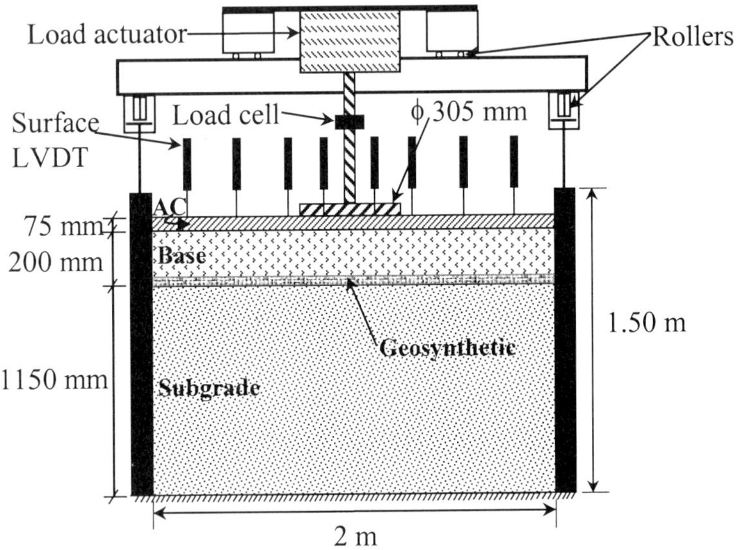

Figure 2. Schematic diagram of pavement test facility

Pavement layer materials Results from four tests are reported in this paper. Two of the tests reported were constructed with a hot-mix asphalt concrete while the other two used a cold-mix asphalt concrete. Hot-mix was used during the months when local hot-mix plants were in operation, with cold-mix being used at other times. All asphalt mixes were fully heated prior to placement. As-constructed properties of the material are given in Table 1. Grain size distributions of the aggregate used in the hot-mix and cold-mix asphalts are given in Figure 3.

A crushed stone aggregate was used for the base course material. A grain size distribution of the material is given in Figure 3 where it is seen that 100 % passes the 19 mm sieve. The material classifies as an A-1-a according to the AASHTO classification system and as a GW according to the Unified Soil Classification System. Specific gravity of the base course is 2.64. As-constructed properties of the base course aggregate are listed in Table 2. The base course was removed and replaced for each test section.

Table 1. As-constructed properties of asphalt concrete

Section	Thickness (mm)	Bulk Density (kN/m^3)	Air Voids (%)	Asphalt Content (%)	Marshall Stability (lb)	Flow
1	79	22.6	6.3	5.7	2040	18
2	77	22.4	6.7	6.2	1375	19
3	75	21.5	12.8	4.7	4650	10
4	75	21.5	12.8	NA	NA	NA

NA=Test data Not Available at the time this paper was prepared.

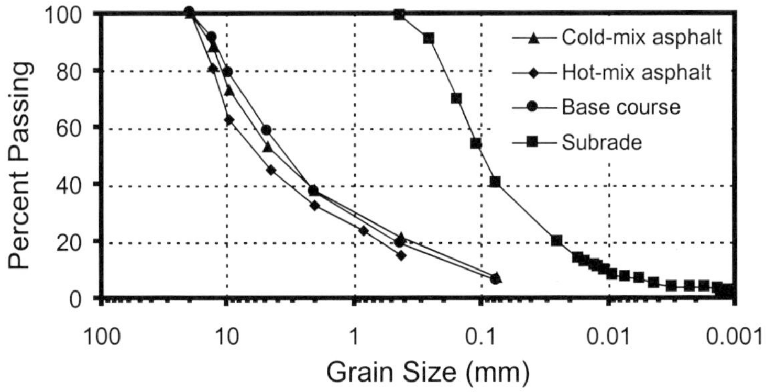

Figure 3. Grain size distribution of the AC aggregate, base course and subgrade

A fine silty sand consisting of the fines trapped in the baghouse of a hot-mix plant was used for the subgrade for the sections reported in this paper. The material has 40 % fines by weight, a liquid limit of 18 % and a plastic limit of nearly the same value, classifying the material as an A-4 or a SM. A grain size distribution of the material is given in Figure 3. Modified Proctor compaction tests indicate a maximum dry density of 18.2 kN/m^3 occurring at a water content of 11.5 %. Specific gravity of the subgrade is 2.68. The subgrade for sections 1 and 2 was replaced prior to the start of the tests. The as-constructed properties of the subgrade are listed in Table 3. The subgrade CBR was determined in situ using a Dynamic Cone Penetrometer. The Dynamic Penetration Index (DPI) of the subgrade soil was determined following construction with the test performed outside the area influenced by the load plate. CBR values were estimated from the DPI using the correlation equation provided in the "User Guide to the Dynamic Cone Penetrometer" (MnROAD, Office of Minnesota Road Research, Minnesota Department of Transportation, USA).

Table 2. As-constructed properties of the base course aggregate

Section	Thickness (mm)	Dry Density (kN/m^3)	Water Content (%)
1	205	20.7	6.1
2	200	20.8	5.5
3	200	20.2	6.2
4	200	20.9	6.3

For sections 3 and 4, the subgrade had not been replaced from the previous tests. For section 3, the subgrade had been used for one other test. For section 4, the subgrade had been used for four other tests. CBR was determined in the manner described above prior to placement of the base layer for that section. Since the test was performed outside the influence of the loaded area, the CBR recorded for the subgrade did not reflect compaction induced in the subgrade from earlier tests.

Table 3. As-constructed properties of the subgrade soil

Section	Dry Density (kN/m^3)	Water Content (%)	CBR
1	17.0	13.8	18
2	17.1	14.0	19
3	18.0	11.5	21
4	18.0	11.4	23

Two polypropylene geosynthetic products were used in the test sections. A biaxial geogrid having an aperture size of 25 by 33 mm was used in section 1 and 3 and a woven geotextile was used in sections 2 and 4. Figure 4 shows results of rapid loading uniaxial tension tests performed on the two materials in their machine and cross-machine directions. The geogrid specimens were loaded to rupture. Loading was discontinued prior to rupture for the geotextile specimens due to limited capacity of the testing machine. The strain rate of loading ranged from 4 to 16 % strain per second. As can be seen from Figure 4, the tensile stiffness characteristics of the two geosynthetics are comparable, with the geotextile stiffness properties bracketing that of the geogrid.

The shear interaction properties of these two materials, however, are more different. Due to the ability for aggregate strike-through and interlocking of the aggregate with the geogrid and the relatively smooth surface texture of the geotextile, shear interaction was expected to be greater with the geogrid. Pull-out tests are currently being conducted to assess these differences. These two materials were selected to provide a contrast between shear-interface characteristics, while the tensile stiffness properties were comparable.

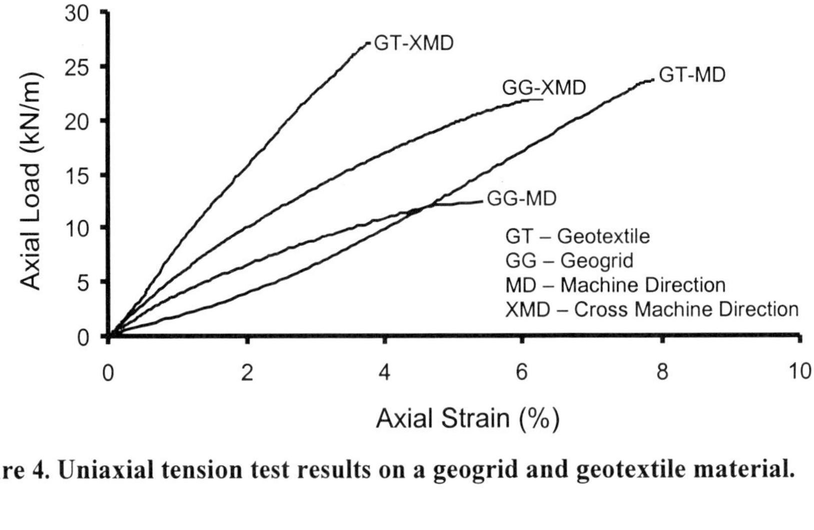

Figure 4. Uniaxial tension test results on a geogrid and geotextile material.

Instrumentation An extensive array of instrumentation was used to quantify the mechanical response of the pavement sections. A load cell was used to measure the cyclic load applied to the pavement surface. As shown in Figure 2, eight LVDT's were used to measure pavement surface deflection. Four H-type strain gages marketed by Dynatest were used to measure tensile strain in the bottom of the AC layer.

Stress cells, also marketed by Dynatest, were used to measure total stress in the base and subgrade materials. Eleven sensors were used in the base and fourteen were used in the subgrade. Strain in the base and subgrade was measured using a LVDT mounted between rectangular end plates measuring 15 by 50 mm and 5 mm thick. The gage length between the end plates was nominally 80 mm. The same number of strain cells was used as the stress cells. Stress and strain cells were oriented in the base and subgrade soils to measure response in the vertical direction and in directions radial and tangential to the center of the applied load. These sensors were placed at different radial distances from the centerline of the applied load to establish the variation of response with radial distance. In the base layer, the instruments were concentrated towards the bottom of the base and above the geosynthetic layer and as close as possible to this layer. In the subgrade, the sensors were concentrated towards the top of the subgrade, however sensors were also placed at various depths throughout the layer. Bonded resistance strain gages were mounted to the geosynthetics at twelve different locations. Similar to other sensors, these gages were oriented to measure strain in the radial and tangential directions. The gages used for the geogrid were approximately 12 mm in length while the gages used for the geotextile were approximately 100 mm in length.

RESULTS

Results presented in this paper concentrate on an examination of the horizontal strain developed in the base course aggregate as compared to the strain developed in the geosynthetic. Four test sections are examined. The materials used and the construction parameters associated with these sections were described in the section on pavement layer materials. Examination of horizontal strain in the base and the geosynthetics illustrates the difference and importance of shear interaction between the base and the geosynthetic.

The strain cells described in the section on instrumentation were used to measure horizontal strain in the base course aggregate in directions radial and tangential to the centerline of the applied load. These sensors were placed at varying radii from the load plate centerline. In sections 1 and 2 where the geosynthetics were placed 40 mm above the base-subgrade interface, the strain cells were placed in a horizontal plane approximately 20 mm above the geosynthetic. In sections 3 and 4 where the geosynthetics were placed between the base and the subgrade, the strain cells were approximately 60 mm above the geosynthetic.

Figure 5 shows results of radial and tangential strain in the base aggregate and on the geogrid from section 3 where extensional strain is taken as positive for both the geosynthetics and the base aggregate. The data points shown in Figure 5 (and the remaining figures) correspond to recorded data from the instrument locations. The trend lines drawn on these figures are best-fit curves that in certain cases have been extrapolated. Extrapolation of these curves is partly based on results described in Perkins et al. (8).

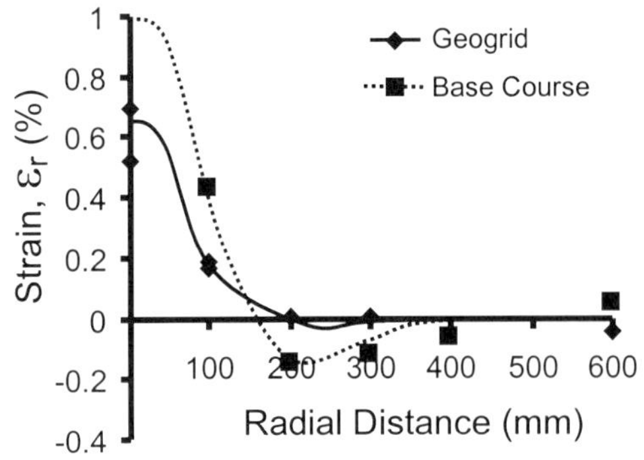

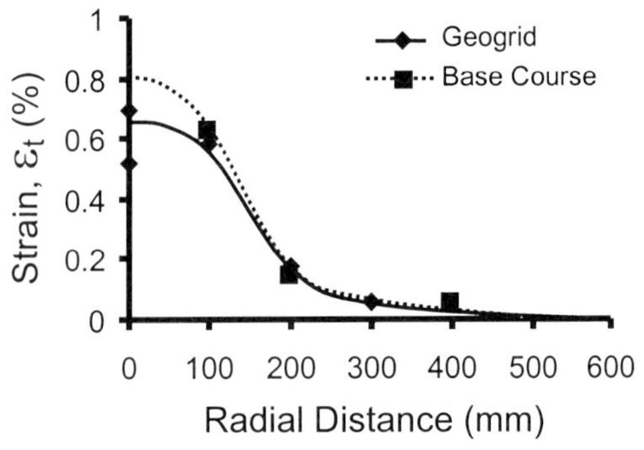

Figure 5. Distribution of permanent radial (ε_r) and tangential (ε_t) strain in the geogrid and base course aggregate in section 3 at 5000 load cycles

The strains presented in Figure 5 correspond to permanent strain (i.e. strain recorded between load pulses when the applied load is zero) immediately after the 5000th load cycle. The 5000th load cycle was chosen because of instances of geosynthetic strain gage fatigue beyond this cycle number. In this and the remaining figures, the trends observed at 5000 load cycles were similar to those at higher load cycle numbers. The data shows a trend of radial strain in the geogrid that is similar to the distribution presented in Figure 1 and which is prerequisite for realizing the reinforcement mechanisms described in the section on proposed mechanisms of reinforcement. The distribution of tangential strain also shows that tensile strain decreases to zero as radius increases. The distribution of radial strain in Figure 5 shows a zone of extension approximately equal to the radius of the load plate, beyond which the material experiences compression. This zone of extension is approximately equal to the zone of tension in the geogrid. The results indicate more extensional strain in the base as compared to the geogrid. In the tangential direction, the compatibility of strain between the geogrid and the base is much better.

Figure 6 shows similar results from section 4 with the geotextile. In this figure, results were also measured after the 5000th load cycle. These results show a similar distribution of radial and tangential strain in the geotextile but approximately 40 % less than that in the geogrid. The radial and tangential strain in the base is greater than that seen in the geotextile and is greater than that measured in section 3. The compatibility of strain between the base and the geotextile is less in section 4 as compared to section 3.

Similar results were seen in sections 1 and 2. Results from radial strain in the base and geogrid from section 1 are compared to radial strain in the base from section 2 in Figure 7. Water vapor intrusion rendered the geotextile strain gages inoperable in section 2, making it impossible to compare strain in the base to that in the geotextile.

Results from the stress cells and the remaining strain cells were consistent with results presented in Figures 5-7 and supported the reinforcement mechanisms described earlier. In particular, for the sections containing the geogrid, radial and tangential stress in the base course beneath the load plate was greater, vertical stress was slightly less vertical strain was less. As an example, Figure 8 shows a comparison of permanent vertical strain in the base course aggregate versus load cycle number for sections 1 and 2 where it is seen that less vertical strain developed in section 1.

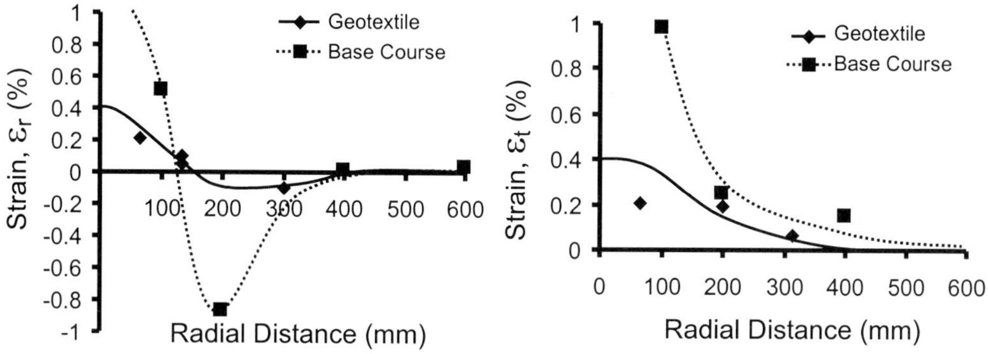

Figure 6. Distribution of permanent radial (ε_r) and tangential (ε_t) strain in the geotextile and base course aggregate in section 4 at 5000 load cycles.

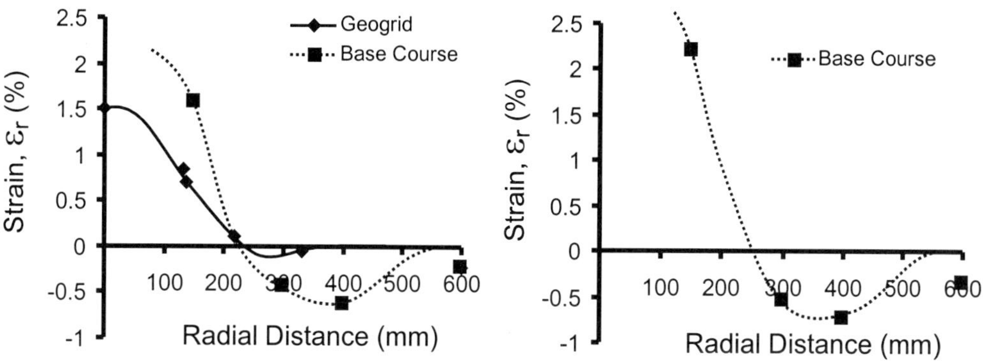

Figure 7. Distribution of peak radial (ε_r) strain in the geogrid and base course aggregate in section 1 at 5000 load cycles and in the base course aggregate in section 2 at 5000 load cycles.

CONCLUSIONS

Reinforcement mechanisms associated with geosynthetics acting as a shear-resisting interface have been described and partially demonstrated through experimental results. An examination of radial and tangential strain developed in the base and in the geosynthetics has demonstrated the development of tensile strain in the geosyntehetic. Compatibility of strain between the geosynthetic and the base aggregate has been observed when proper shear interaction develops. These mechanisms lead to a reduction of lateral and vertical aggregate movement. The purpose of this work is not to make a comparison between geogrid and geotextile products but rather to demonstrate the importance of shear interaction on ensuing reinforcement benefits. One goal of this work is to establish and quantify reinforcement benefit in terms of geosynthetic-aggregate shear interaction characteristics such that these parameters can be established for a host of geosynthetic materials, thereby allowing for reinforcement benefits to be assessed during the design process.

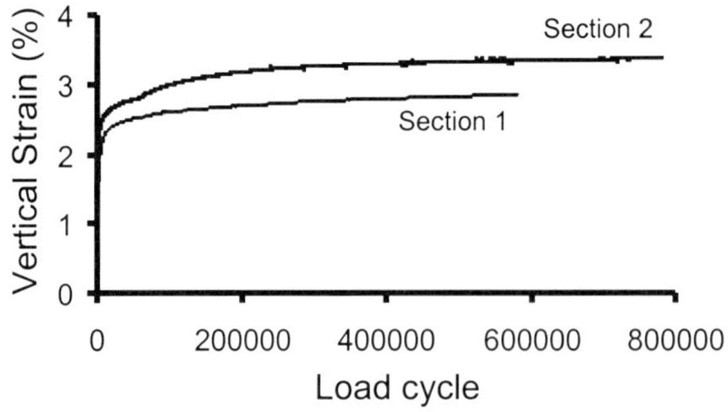

Figure 8. Permanent vertical strain in the base course aggregate versus load cycle for sections 1 and 2.

ACKNOWLGEMENTS

The authors gratefully acknowledge the financial support of the Federal Highway Administration and the Montana Department of Transportation through grant number 8106.

REFERENCES

1. Brown, S.F., Jones, C.P.D. and Brodrick, B.V., "Use of Non-Woven Fabrics in Permanent Road Pavements", *Proceedings of the Institution of Civil Engineers*, London, Part 2, Vol.73, pp. 541-563, 1982
2. Ruddock, E.C., Potter, J.F. and McAvoy, A.R., "A Full Scale Experiment on Granular and Bituminous Road Pavements Laid on Fabrics", *Proceedings of the Second International Conference on Geotextiles*, IFAI, Las Vegas, NV, USA, Vol. 2, pp. 365-370, 1982.
3. Haas R., Wall, J., and Carroll, R.G., "Geogrid Reinforcement of Granular Bases in Flexible Pavements," *Transportation Research Record 1188*, TRB, National Research Council, Washington DC, USA, pp. 19 – 27, 1988.
4. Barksdale, R.D., Brown, S.F. and Chan, F., *Potential Benefits of Geosynthetics in Flexible Pavement Systems*, National Cooperative Highway Research Program Report No. 315 , Transportation Research Board, National Research Council, Washington DC, USA, 56p, 1989.
5. Kinney, T.C. and Barenberg, E.J., "The Strengthening Effect of Geotextiles on Soil-Geotextile-Aggregate Systems", *Proceedings of the Second International Conference on Geotextiles*, IFAI, Las Vegas, NV, USA, Vol. 2, pp. 347-352, 1982.
6. Miura, N., Sakai, A., Taesiri, Y., Yamanouchi, T. and Yasuhara, K. (1990), "Polymer Grid Reinforced Pavement on Soft Clay Grounds", *Geotextiles and Geomembranes*, Vol. 9, pp. 99-123.
7. Bender, D.A. and Barenberg, E.J. (1978), "Design and Behavior of Soil-Fabric-Aggregate Systems", *Transportation Research Record 671*, TRB, National Research Council, Washington, DC, USA, pp. 64-75.
8. Perkins, S.W., Ismeik, M., Fogelsong, M.L., Wang, Y. and Cuelho, E.V. (1998), "Geosynthetic-Reinforced Pavements: Overview and Preliminary Results", To Appear in the *Proceedings of the Sixth International Conference on Geosynthetics*, Atlanta, GA, USA.

Fifth International Conference on the
BEARING CAPACITY OF
ROADS AND AIRFIELDS

STEEL FABRICS AS REINFORCEMENT IN ROAD CONSTRUCTION

SVANTE JOHANSSON[1] OCH LARS ERIK NILSSON[2]

[1]Eng, Swedish National Road Administration, Borlänge, Sweden
[2]Prof, Dr Sc, Gothenburg University, Gothenburg, Sweden

ABSTRACT

The first scientific attempts to use steel reinforcement in road rehabilitation to prevent cracking from frost heave were performed in Norway in the 1960s. The steel fabrics were built into the wearing course. The first trials with steel fabrics built into the unbound base layer were carried out in Finland in the 1970s. A new wave of field trials with steel reinforcement in wearing courses and base courses started in Finland in the late 1980s. During the latest decade, Sweden has further developed steel fabric design and construction techniques (1988-1994) in laboratory experiments (1988-1994) and field trials (1988-1998).

The concept of reinforcing asphalt pavements with steel fabric was based on the following discussion: When the weather is cold, asphalt concrete is hard and brittle. In these conditions, it is very similar to cement concrete. As steel reinforcement increases the strength and decreases the development of cracks in cement concrete, steel reinforcement should give a similar effect in cold asphalt concrete.

Field trials have shown that steel reinforced asphalt pavements exhibit the expected performance. Laboratory experiments show, however, that the united action between asphalt concrete and steel reinforcement is different from the united action between cement concrete and steel reinforcement, as the asphalt is plastic even when it is cold. Experience from the field trials is now so comprehensive that recommendations have been worked out for design and construction of asphalt pavements reinforced with steel fabrics.

Laboratory experiments and field trials have shown that a properly designed and constructed steel fabric reinforced asphalt concrete is an effective rehabilitation method to stop longitudinal cracking from frost heave in paved roads. A few field trials with steel reinforcement in unbound base layers indicate that that method will also counteract cracking from frost heave.

INTRODUCTION

The function of a steel fabric reinforcement in asphalt concrete.
Cement concrete and asphalt concrete have similar structures. Asphalt concrete is a stone skeleton kept together with bitumen. Cement concrete is a stone skeleton kept together with cement. Asphalt is soft during the summer period, but its elastic modulus increases strongly when the temperature falls. The asphalt then becomes hard and brittle. As steel reinforcement increases strength and decreases crack development in cement concrete,

steel reinforcement should give a similar effect in cold asphalt concrete. As cracks from uneven frost heave and low temperature cracks will occur in cold weather, a steel reinforcement should be able to prevent or reduce crack formation.

The united action between asphalt concrete and steel reinforcement is different from the united action between cement concrete and steel reinforcement. In cement concrete reinforced with steel fabrics, the tensile forces are taken up mainly by adhesion between the concrete and steel bars located in the tensile direction. In asphalt concrete reinforced with steel fabrics, the tensile forces are primarily taken up by the steel bars located transversely to the tensile direction. These forces are transferred through welded joints to the steel bars in the tensile direction. This is described under the title "Laboratory Investigations" below. Steel reinforcement in unbound base layers seems to work in the same way.

Steel reinforcement in asphalt is also different from steel reinforcement in cement concrete in several other ways. The transfer of the forces to the steel bars in asphalt concrete varies strongly with the temperature and tensile speed. In cement concrete, the transfer of the forces is independent of the temperature and the tensile speed.

Laboratory Experiments
Tests on cylindrical test bodies have shown how the united action between steel bars and asphalt concrete works. Tests on steel reinforced and unreinforced asphalt slabs have confirmed the mode of action for steel fabrics.

In a few cases, experiments have been performed at different temperatures and at different tensile speeds, although most laboratory tests have been conducted at temperatures close to 0° C.

Field Trials
The first scientific attemps to use steel reinforcement in road rehabilitation were performed in Norway. Field trials were started in the middle of the 1960s by the Norwegian Road Research Laboratories (NRRL) at the Norwegian Road Directorate. The aim of the trials was to examine whether steel fabrics in asphalt pavement could stop longitudinal cracking caused by frost heave in roads. The first trial was performed by Erling Hansen. Further trials were carried out among others by Nymoen. In all, NRRL published 11 reports on the subject between 1968 and 1978. Nymoen and Hansen both published summarizing reports in 1977.

Nymoen writes in his report that the trials show that "...severe longitudinal cracks from frost heave in roads can be stopped by using proper reinforcement with steel fabrics in the new asphalt wearing course as a rehabilitation measure." Hansen makes clear that, in a few trials in which the results were unsatisfactory, the problem was caused by placing the fabrics too close to the road surface.

In Finland, field trials with steel fabrics in road construction have been conducted since the middle of the 1970s. The Finnish trials include steel reinforcement in both asphalt layers and unbound base layers. The aim of the trials was to examine the possibilities of preventing longitudinal cracks resulting from uneven frost heave.

It is said in a first report that reinforcement with steel fabric in unbound layers gave 'positive effects.' No cracks from frost heave had arisen three years after rehabilitation. On a reference road without steel reinforcement, a frost crack returned within two years. The laying of the fabrics worked very well. The laying of asphalt on the fabrics did not function with thin asphalt layers. They attempted to nail the fabrics to the existing pavement to avoid wave formation in the steel fabrics. Trials with 120 mm asphalt concrete worked quite well.

Trials were also carried out with local crack reinforcement (reinforcement with narrow steel fabrics following the crack, in contrast to whole covering reinforcement, which extends from one side of the road to the other). The report does not describe the level on which the fabric is placed, although it does say that local crack reinforcement will not be effective. Cracks will occur at the sides of the steel fabric.

In 1995, Seppo Salmenkaita presented a summary of the Finnish trials "from a practical point of view". He summarizes the results: "The experiments show that reinforcement with steel fabrics in road constructions, properly placed, is an effective and economical method to counteract longitudinal cracks in roads with asphalt pavement...when the fabrics are placed on old pavements it is best first to cover the fabrics with a thin layer of asphalt concrete (thickness 20-30 mm) and then to place the new wearing course on top. If the new wearing course is placed directly on the fabrics the thickness has to be at least 50 mm."

Tore Junes of the Swedish National road Administration (SNRA) has carried out trials in the north of Sweden with steel fabric reinforcement in eight road projects and two cycle paths. The trials were conducted between 1990 and 1993. In the middle and south of Sweden, Svante Johansson, SNRA, has performed five road projects with steel reinforcement. The Swedish field trials are described under the title "Field Studies" below. Characteristic of the Swedish trials is an ambition to find cost-effective solutions for the design and construction of steel fabric reinforced roads.

Figure 1. Cycle path reinforced with steel fabric in 1976. The photo was taken 16 years later. The reinforcement starts where cat Whiskey (black and white) is laying.

Production Trials
During the development of the steel reinforcement technique, there were initially a number of production problems. The design of the steel fabrics depended greatly on the laying process. Among other things, the fabrics must have sufficient torsion stiffness to allow trucks, pavers and other types of traffic to drive on the lain but not covered fabrics.

LABORATORY INVESTIGATIONS

A laboratory test on reinforced asphalt concrete slabs has been carried out at the Swedish Road and Transport Research Institute (VTI) at Linköping, Sweden. See figure 2. The test shows that in a tensile stressed asphalt concrete slab the combined effect of reinforcement and asphalt gives a total tensile stress at break equal to the sum of the stresses in steel reinforcement and in asphalt concrete. In this respect the function of steel reinforced asphalt concrete is better than that of steel reinforced cement concrete. It's due to the asphalt plasticity even when cold. Compare figure 2 and figure 3.

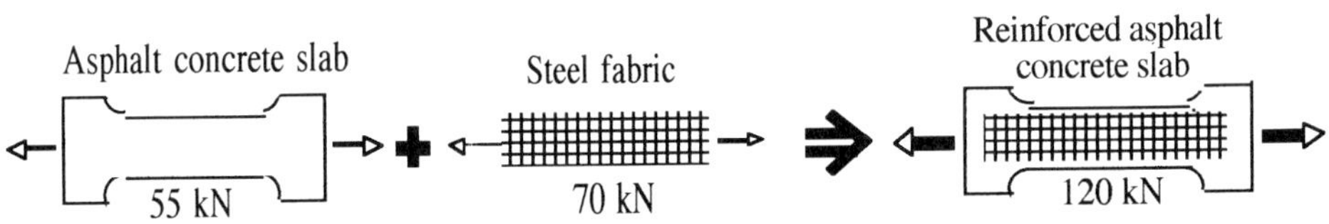

Figure 2. Asphalt concrete slab. Tension capacity. No crack arises. The tension values on the right and left figures are measured values from the first tests at VTI. The tension value in the middle figure is a value recieved from the steel fabric factory. Two slabs were tested. The tension speed was 3mm/s and the temperature was -5°C. Slab thicknesses: 85 - 90 mm. Slab widths: 500 - 560 mm. Total length: 2 m.

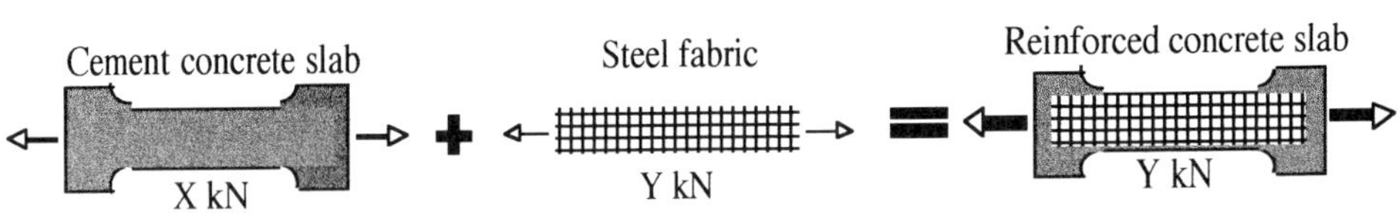

Figure 3. Cement concrete slab. Tension capacity. Tension maximum capacity occurs when the first crack has arised.

In steel reinforced cement concrete the tensile forces are taken up mainly by adhesion between concrete and steel reinforcement with the bars located in the tensile direction. In tensile stressed reinforced asphalt concrete (figure 4) the tensile forces are primarely taken up by bars located transversely to the tensile direction, transferring the forces through welded joints to the tensile direction.

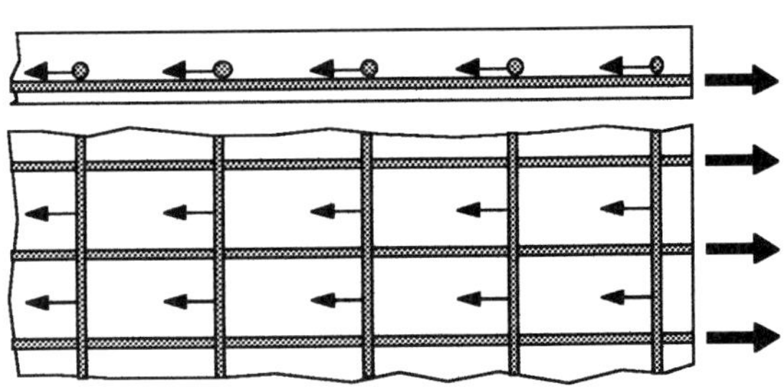

Figure 4. Forces in a frozen asphalt slab reinforced by a steel fabric. Longitudinal (above) and transverse section.

At the same test at VTI different joints between fabrics were tested. The best type of joint for practical use is named 'one edge finger joint' depending on the design. Just one of the two fabrics in the joint has 'fingers'. See figure 15.

Due to a technical accident on a road 'reinforced' by a syntetical geonet the Norvegian Road Research Laboratory in Oslo tested the shearing capacity between an asphalt layer and different kinds of steel fabrics, geonets and geotextiles. The results are shown in figure 5.

The shear stresses in the steel reinforced layer are higher than those in the tack coated layer. This is an important finding. If the shear stresses in the steel reinforced layer had been lower than in the tack coated layer, there would be a risk for obtaining a reduction in the bearing capacity. Laboratory tests not shown in this report instead indicate a not unimportant increase in bearing capacity as a side effect.

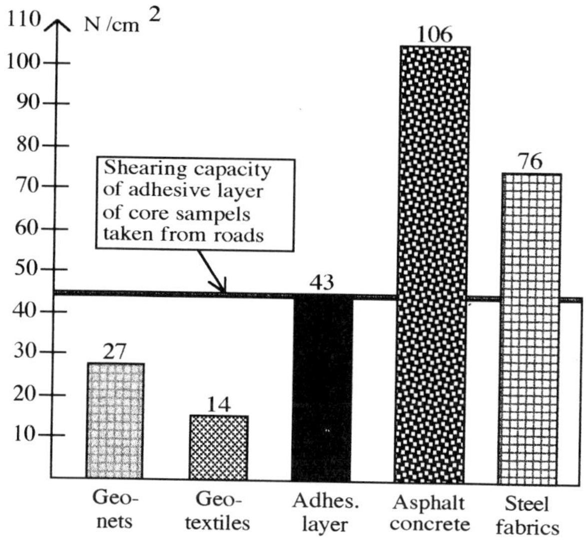

Figure 5: Shearing test results of core samples taken from Norvegian roads. Temperature at testing +20°C. Shearing test of cylinders made at the laboratory gave similar results.

FIELD STUDIES

Steel reinforcement can be carried out as a local crack reinforcement or as a whole covering reinforcement. A whole covering reinforcement can be done with or without a longitudinal mid-joint. All three alternatives have been tested in the Swedish trials. The following is based on the results of the Swedish trials.

Local Crack Reinforcement
Steel fabric reinforcement in bituminous wearing courses can not be carried out as local crack reinforcement on longitudinal cracks from frost heave in roads. The covered cracks will return as new cracks at the sides of the steel fabric. See figures 6-8.

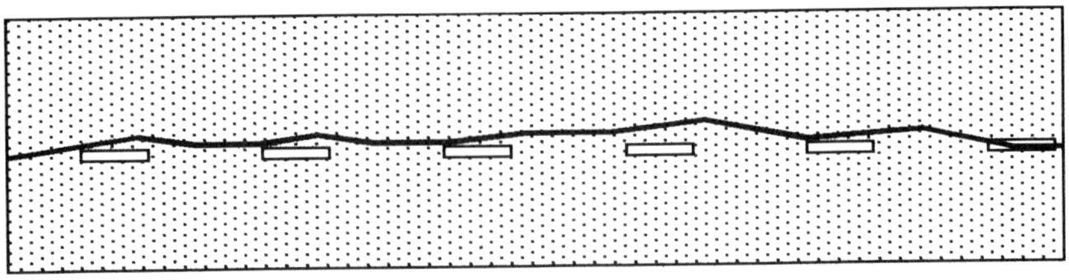

Figure 6. Longitudinal crack from uneven frost heave to be mended.

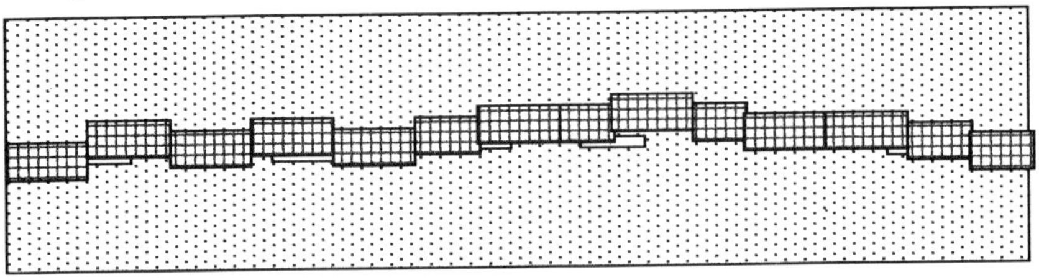

Figure 7. Local crack reinforcement of frost crack. Placement of steel fabric.

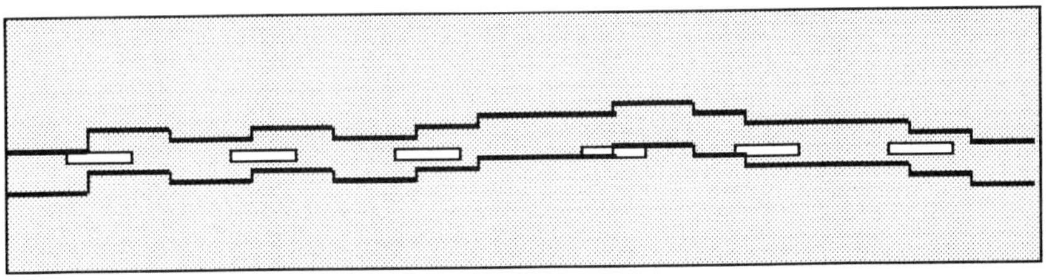

Figure 8. Local crack reinforcement. Crack pattern a few years after the rehabilitation measure.

Whole Covering Steel Fabric Reinforcement with a Longitudinal Mid-joint

Steel reinforcement in bituminous wearing courses can not be carried out as whole covering steel fabric reinforcement with a longitudinal mid-joint on roads with longitudinal cracks from uneven frost heave. The longitudinal crack will return through the wearing course in the longitudinal mid-joint. Neither is it possible - with the joint technique tested thus far - to avoid the crack by placing the joint beside the frost crack. This is illustrated in figures 9-13. The fabrics seem to need to be welded together to be able to stop the joint crack.

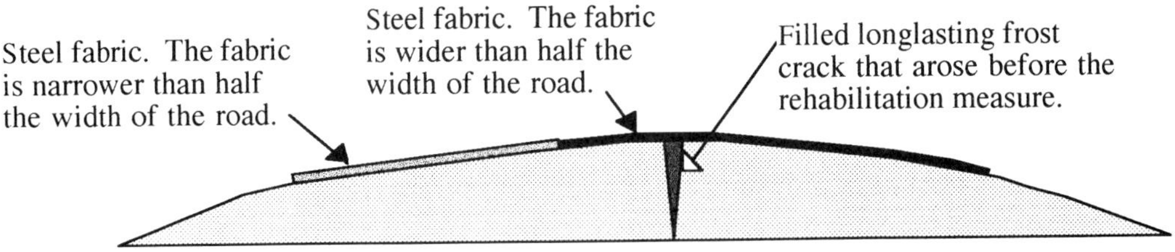

Figure 9. Longitudinal joint placed beside longitudinal frost crack just after placement.

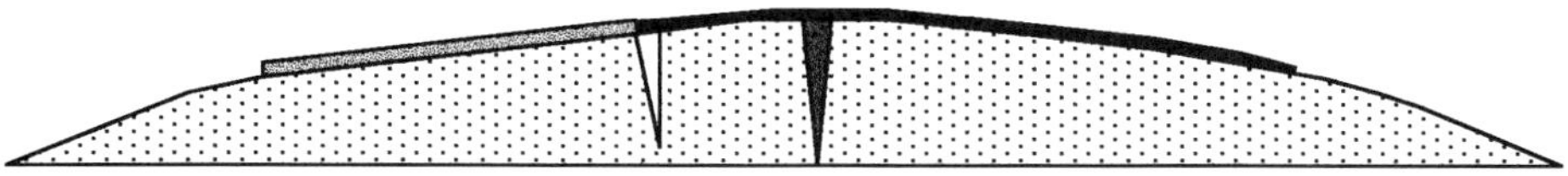

Figure 10. Longitudinal joint placed beside longitudinal frost crack after a few years.

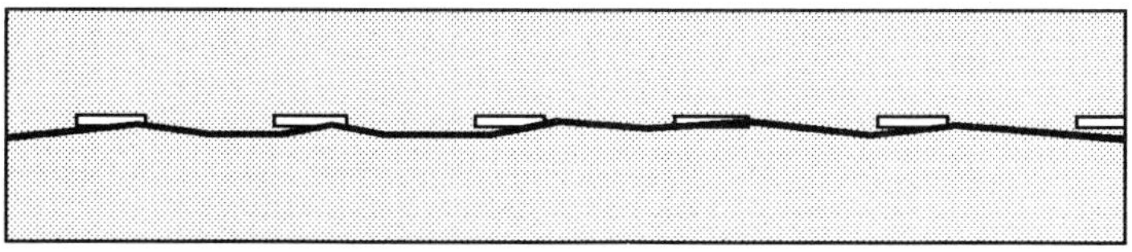

Figure 11. Longitudinal crack to be mended.

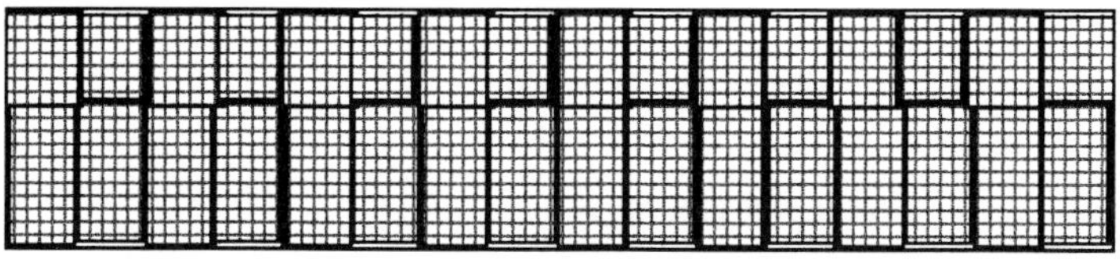

Figure 12. Placement of steel fabrics according to figure 9.

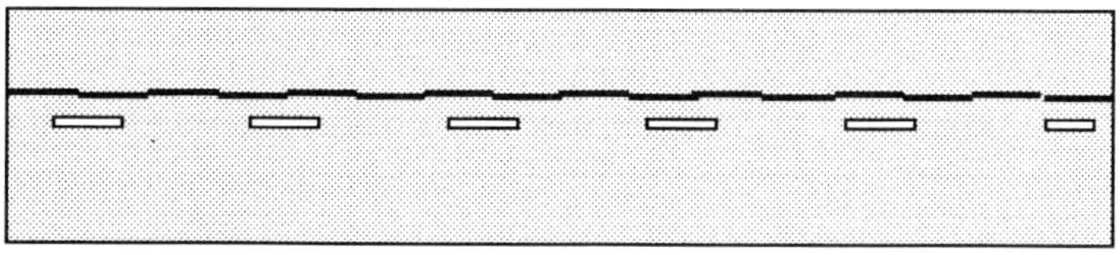

Figure 13. Result. The crack has returned through the wearing course in the longitudinal joint

Whole covering steel fabric reinforcement without mid-joint
Steel reinforcement in bituminous wearing courses can be carried out as a whole covering reinforcement without a longitudinal mid-joint on roads with longitudinal frost cracks.

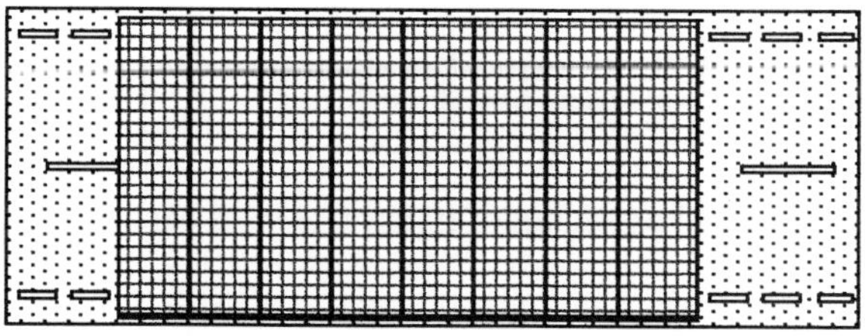

Figure 14. The nets on the road.

Normally a frost crack is longitudinal to the road and situated near the middle of the road. The process causing cracks is developed during early winter. It takes much longer time than the laboratory experiments. The fact that the frost heaving process is very slow makes it impossible to use longitudinal joints. They don't work. Therefore the nets must be oriented as shown in figure 14 i. e. with transversal but not longitudinal joints. If the nets are too long in the road direction they move vertically when the asphalt paver drives on them. In Scandinavia, where the fabrics have 5 mm bars, the maximal length used is 2.30 m. They must run all the way from one side of the road to the other. Otherwise there will be a crack at the end of the nets.

If the asphalt layer on the fabrics is too thin, there may arise small cracks along the transversal joints. In Sweden a minimum thickness of 5 cm of asphalt concrete is recommended. In Norway and Finland thicker layers are frequently used.

Many kinds of antirust fluids are oil products - like asphalt. Therefore it was from the beginning expected that there shouldn't be any rust problems. As a matter of fact it is necessary to spread a tack cote of bitumen emulsion from both the two longitudinal directions to get better adhesion and to avoid rusting.

On horizontal roads without sharp curves the asphalt paver can drive with normal speed on the steel fabrics. The lorries must drive slowly without turning sharply. The fabrics can be placed on the road faster than the paver layes asphalt.

RECOMMENDED DESIGN

Whole covering reinforcement of steel fabrics can be used to stop longitudinal cracks from uneven frost heave on roads. Before putting the reinforcement in place, the following measures must be taken:

- local vertical deformations caused, for instance, by a drum or a rock must be fixed
- extremely low bearing capacity in the outer wheel path on roads without shoulders must be increased.

General requirements for steel fabric reinforcement of wearing courses and base courses to counteract longitudinal crackings caused by uneven frost heave are given below.

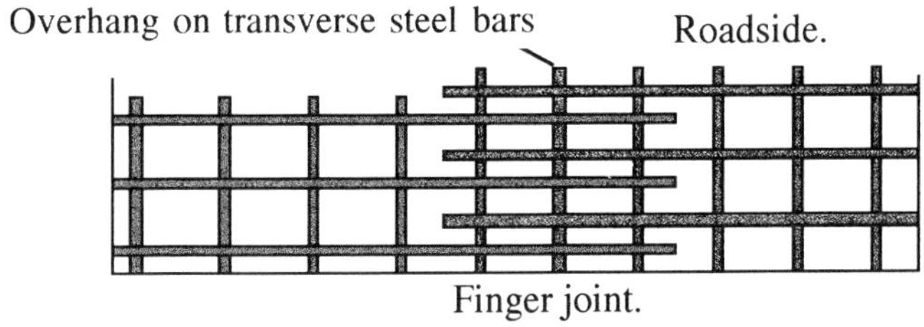

Figure 15. Design of finger joint and overhang of transverse steel bar.

Steel bars. Steel bars in the fabric shall have a steel quality of NPS 500 and a diameter of not less than 5 mm. The transversal bars should have a combined cross-section area of at least 250 mm2/m road length.

Transversal joints. Transversal joints shall be made as finger joints. See figure 15.

Overhang. The overhang of transverse bars shall be a maximum of 30 mm. See figure 15. The overhangs of the longitudinal steel bars shall - if the fabrics will be trafficked before overlay - be carried out so that no damages will arise on car tires.

Tack coating. Tack coating should be carried out with half the amount in each direction of the road.

Traffic regulation. If the steel fabrics shall be trafficked before overlay, the speed limit shall be a maximum of 6 km/h or walking speed. Each road user shall be stopped and informed about the stipulations before crossing.

Special requirements for reinforcement of wearing courses to counteract longitudinal cracks from uneven frost heave are given below. See figure 16.

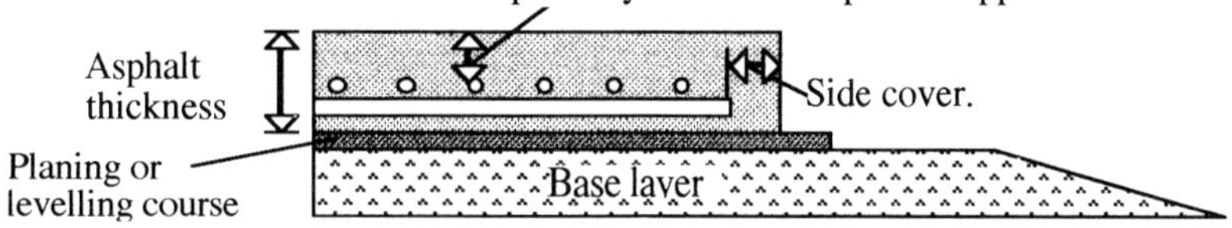

Figur 16 Asphalt thickness, side cover and asphalt layer thickness upon the upper steel bar.

Reinforcement of wearing course:

- The transversal steel bars shall be placed at the bottom and they shall always have contact with existing pavement.
- If contact is lacking between steel bars and the road surface, for instance in the wheel paths, a planing or a levelling course shall be carried out prior to placement of the steel fabrics.

Width of the steel fabrics in wearing courses. Fabrics in wearing courses shall reach from roadside to roadside without a joint. The asphalt paver shall be steered so that a small side cover of asphalt is obtained on the fabrics. The side cover is the distance from the end of a transversal steel bar to the edge of the new wearing course. A side cover of 10-50 mm should be the target.

Reinforcement level in the wearing course. The construction shall be carried out so that the asphalt layer thickness upon the upper steel bar is overall at least 30 mm. It can be estimated that the space under the upper steel bars shall be at least as large as the maximum aggregate size in the wearing course. For an AC16 wearing course and a steel fabric with a steel bar diameter of 5 mm, a minimum asphalt thickness of 50 mm is required.

Special requirements for reinforcement in bound and unbound base layers to counteract longitudinal cracks from uneven frost heave are given below. See figure 17.

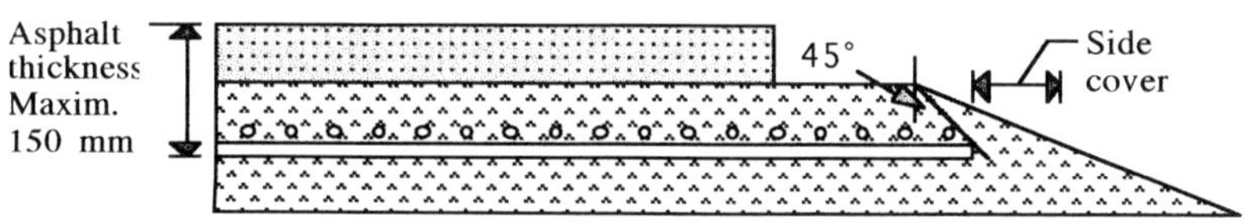

Figur 17 Asphalt thickness and side cover in bound and unbound base layers

Recently a new design method was developed to estimate the effectiveness of a polyester reinforced asphalt overlay.

Nowadays a trend to use composite materials for asphalt reinforcement can be observed. In fact an optimised combination between a well-known reinforcing grid and an ultra-light precoated non woven shows several important advantages as was described in the paper.
The improved and simplified installation and the high bond strength offered by the newly developed composit are important also on the practical point of view.
The tests of the composite reinforced asphalt under cyclic loading indicated a significantly better performance than unreinforced asphalt. The tests also showed that by using a low creep and high modulus raw material for the grid in the composite the service life of the reinforced overlay can be considerably increased.

References

1. Kunst, P.A.J.C.; Deenekamp, Q.; Wattimena, J.S. (1990) Vergleichende Untersuchungen an unbewehrten und HaTelit bewehrten Asphaltprobekörpern, Huesker Synthetic, Germany.
2. Molenaar, A.A.A.; Nods, M.; Design Method for plain and geogrid reinforced overlays on cracked Pavements, Proceedings of the Third International Conference on Reflective Cracking in Pavements, s. 310-320, Maastricht, 1996.
3. Lytton, R.L. (1989) Use of geotextiles for reinforcement and strain relief in asphalt concrete, Geotextiles and Geomembranes, Vol. 8, No. 3.
4. Jacobs, M.M.J. (1995) Crack grows in asphalt mixes, Dissertation Faculty of Civil Engineering, Delft University of Technology, Delft
5. Molenaar A.A.A. (1983) Structural performance and design of flexible pavements and asphalt concrete overlays, Dissertation Faculty of Civil Engineering, Delft University of Technology, Delft
6. Urbanski, Ingenieurbüro für Geotechnik und Baustoffprüfung GmbH & Co.KG, Prüfbericht AsS 25/96/2894, Untersuchungen von Asphaltbohrkernen / Bestimmung des Haftverbundes, 1996.
7. Belgian Road Research Centre (BRRC), Research Report EP 5055, Thermal Cracking Test, 1997.

Fifth International Conference on the
BEARING CAPASITY OF
ROADS AND AIRFIELDS

REINFORCEMENT OF ASPHALT PAVEMENT TO IMPROVE RESISTANCE AGAINST REFLECTION CRACKING

KWANG W. KIM[1], YOUNG S. DOH[2], SUNGBIN LIM[2], XIANG-FAN LI[2] AND S. K. RHEE[3]

[1]Associate Professor, [2]Graduate students
Department of Agricultural and Civil Engineering
Kangwon National University, Chun Chon, 200-701 Korea
[3]Assistant Professor, Department of Civil Engineering,
Kyung Hee University, Yongin 449-701 Korea

ABSTRACT

This study is devised to evaluate specially designed reinforcement techniques against reflection cracking of the typical asphalt pavement overlay in the laboratory. A selected polymer from previous studies was used for asphalt modifier and a fiber, a vinyl and a grid were used as reinforcing materials. Using the reinforced asphalt mixture with the optimum asphalt content from mix design, beam specimens were made for the simulated repeated load testing. A thin interlayer was made using grid and polyethylene film at the bottom of specimen to strengthen the pavement layer against crack by placing it at the bottom of the mould before mixture dump. Simulated-repeated loading was applied to the beam specimen which is placed on a gap (crack) of cement concrete base, using a hydraulic dynamic loading frame. Crack propagation under repeated loading was monitored and effectiveness of the devised crack preventing techniques was evaluated. From the test results, a significant retardation of crack progress was monitored.

INTRODUCTION

Reflective cracking is one of the major distress mechanisms in overlaid asphalt pavement on concrete roads [1]. Reflection crack is a result of vertical loading by traffic and horizontal movement (shrinkage) of base concrete slabs due to drop in temperature. A vertical wheel load, which is simultaneously applied to the asphalt layer on top of an existing crack, creates bending stress at the bottom of asphalt layer. At almost the same time, the wheel load, which is moving across the crack, creates shear stress at perpendicular direction to the bottom face of the layer. The former is responsible for mode I cracking, together with the shrinkage due to temperature drop. The later is responsible for mode II cracking [2]. When the pavement is under stress, the spot at the bottom of asphalt layer on top of an existing crack opening (gap) becomes a crack tip. A high stress concentration is created at this spot and finally a crack initiates propagation from here.

Many studies have been performed in efforts of preventing or retarding progress of reflection cracking through asphalt layer which is paved on top of existing cracks in concrete pavements [3,4,5]. However, no solution for completely preventing this crack has been suggested yet, possibly because of too many variables involved in the cause of reflection cracks. Only retardation is the best way so far developed. One of the possible ways of retarding the crack is to reinforce the material's tensile property to be used in overlay.

Therefore, objective of this study is to evaluate effectiveness of several materials and reinforcing techniques being used for retarding progress of reflection crack in Korea. In order to select the most appropriate material in this purpose, a research is under process by support of Korea Highway Corps. This paper presents methodology and some of the significant results obtained from this work to date.

MATERIALS AND METHODS

Materials

An asphalt (AP-3) with penetration grade of 85-100 which is the most widely used in Korea was used as binder for mixture preparation. Properties of this asphalt are shown in Table 1. The asphalt was modified using a carbon black (CB) and a low-density polyethylene (LDPE). In addition, three reinforcing materials were used in this study and these include a polyethylene fiber (PF), a polyethylene film (vinyl: PV) and a glass grid (GG). Appropriate contents of LDPE, CB and PF were selected as 6% and 10% by weight of asphalt, and 0.35% by weight of total mixture, respectively. Table 2 describes the mixture name and materials used for each mixture.

Table 1. Physical properties of AP-3 asphalt

Material	Penetration (0.1mm)	Ductility (mm)	Flash Point (℃)	Pen. Ratio After TFOT (%)	Asphaltene Content (%)	Solubility In THE (%)	Specific gravity
AP-3	91	150	45.0	60.4	8.0	99.4	1.03

Table 2. Description of various asphalt mixture

Mixture	Description
AP	Normal asphalt mixture
APG	Glass grid (GG) reinforced normal asphalt mixture
APV	Polyethylene vinyl (PV) reinforced normal asphalt mixture
LC	LDPE and CB asphalt mixture
LCG	LDPE, CB and GG reinforced asphalt mixture
LCV	LDPE, CB and PV reinforced asphalt mixture
LCF	LDPE, CB, polyethylene fiber (PF) reinforced asphalt mixture
LCFG	LDPE, CB, PF and GG reinforced asphalt mixture
LCFV	LDPE, CB, PF and PV reinforced asphalt mixture

Thickness of PV and length of PF are 0.2mm and 6mm, respectively. The mesh size and thread diameter of GG are 25mm and 1mm, respectively. LDPE was pre-mixed in asphalt using a high shear mixer together with carbon black. PF was added to the mixture just before mixing. Single layer of PV and GG were place at the bottom of mould before dumping loose hot mix for compaction.

Three sources of aggregate with maximum size of 19mm were examined for use in overlaying asphalt mixtures. A gneiss coarse and fine aggregate, with maximum size 19mm, was selected from the three as the best performing one for a dense-graded asphalt mixture from preliminary study [6]. Table 3 shows gradation of gneiss aggregate together with the specification limit given by Ministry of Construction and Transportation (MCT) of Korea. Table 4 shows physical properties of aggregate.

Table 3. Gradation of gneiss aggregate and dense graded mixture specification (19mm)

Sieve size		25mm	19mm	13mm	#4	#8	#30	#50	#100	#200
Passing (%)	Spec. Limit	100	95～100	75～90	45～65	35～50	18～30	10～21	6～16	4～8
	Gneiss	100	99.1	78.3	52.3	40.1	22.6	17.0	11.6	7.73

Table 4. Physical properties of gneiss aggregate

Classification	Specific gravity	Abrasion (%)	Absorption (%)
Spec. limit	> 2.5	< 35	< 3.0
Gneiss	2.72	18.1	0.7

Marshall mix design was carried out for each mixture combination to determine optimum asphalt content (OAC). The optimum asphalt contents were used in slab specimen mix. Two beam specimens were obtained from saw cutting a slab specimen which was made using a hydraulic roller. The size of the beam specimen was 340mm (length) x 120mm (width) x 50mm (depth).

Methods

Marshall stability, air void, flow, and indirect tensile strength (ITS) were measured for each mixture by standard methods. To simulate an asphalt pavement on top of a crack in a concrete pavement, a beam was overlaid on a discontinued concrete block (depth 120mm), as shown in Figure 1. The beam was bonded using an epoxy on top of the concrete block that has 10mm gap in 2/3 depth from the top and an arbitrarily broken crack at the bottom, as shown in Figure 1.

Cyclic square loads were applied to the top of the beam through a circular loading plate (100mm diameter) at 10Hz using an Instron hydraulic loading system. A rubber pad was used at the interface between loading plate and specimen to simulate tire contact. A maximum load of 5.4kN (550kg) was applied to the specimen to create 686 kPa (7kg/cm^2) or 100psi pressure on top of the specimen. A 0.196kN (20kg) of minimum load was used to keep the loading plate in place during dynamic loading. Test was carried out at 20°C in a temperature-controlled chamber. Figure 2 shows the test system including chamber.

Horizontal expansion of the asphalt mixture beam was measured using an extensometer which was installed at lower level in one side of beam. Vertical crack growth as load cycle accumulating was visually monitored in other side which is pained in white. Test was conducted until the vertical crack length reached 50mm.

Figure 1. Test specimen

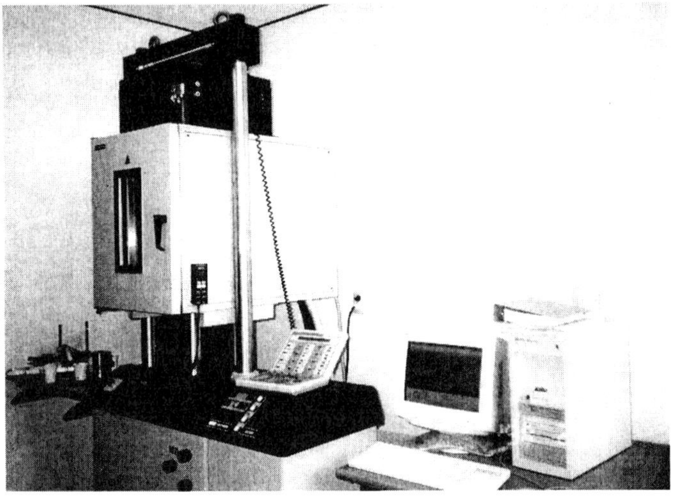

Figure 2. Test system

RESULTS AND DISCUSSIONS

Mixture Properties

Fundamental properties and OCA of each mixture are shown in Table 5. Air voids of three mixtures ranged from 3.42 to 4.16%. Marshall stability of LDPE and CB modified asphalt mixture (LC) was improved approximately 10% than normal asphalt mixture (AP). That of LDPE, CB modified and PF reinforced asphalt mixture (LCF) was improved approximately 22% than AP mixture. Indirect tensile strength (ITS) of LC mixture was not much improved than AP, but ITS of LCF was improved 37% than AP mixture. Therefore fiber showed a significant tensile reinforcing effect on asphalt mixture.

Table 5. Fundamental properties of asphalt mixtures

Property	Mixtures		
	AP	LC	LCF
Optimum asphalt content (%)	4.8	4.5	5.0
Air void (%)	3.24	4.16	3.69
Marshall stability (KN)	15.96	17.46	19.47
Marshall flow (0.1mm)	27	35	35
ITS (Mpa)	7.76×10^6	8.17×10^6	10.65×10^6

AP mixture was used in making APG and APV specimens, LC mixture was used in making LCG and LCV specimens, and LCF was used in making LCFG and LCFV specimens, for cracking resistance test.

Reflection Cracking Resistance

As explained previously, mode I crack propagation test was carried out on two specimens for each mixture, described in Table 2. In the process of test, horizontal deformation (expansion of beam) and vertical crack growth were monitored as the number of cyclic load accumulating. Figure 3 shows a typical horizontal deformation curve and a vertical crack growth curve. Dynamic stability (DS) based on horizontal deformation and number of cycle was measured from recorded data as described in Figure 4. This value represents how many passes (loading) require for the pavement to deform 1 mm horizontally. Vertical crack growth rate was calculated as slope of linear regression line for each specimen.

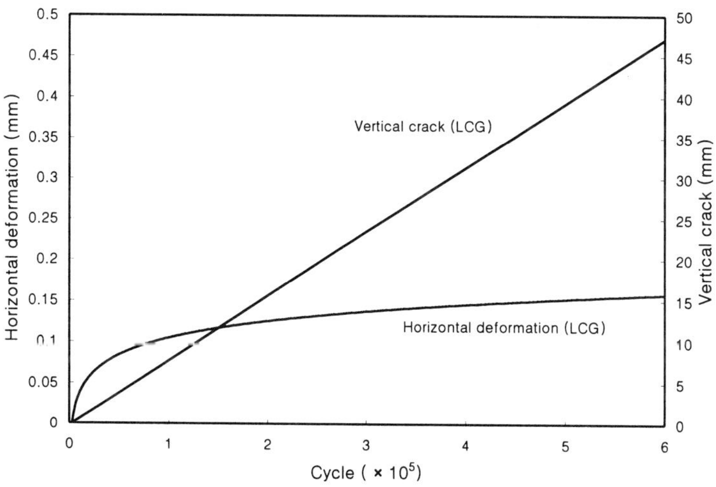

Figure 3. Typical horizontal deformation and vertical crack propagation curve

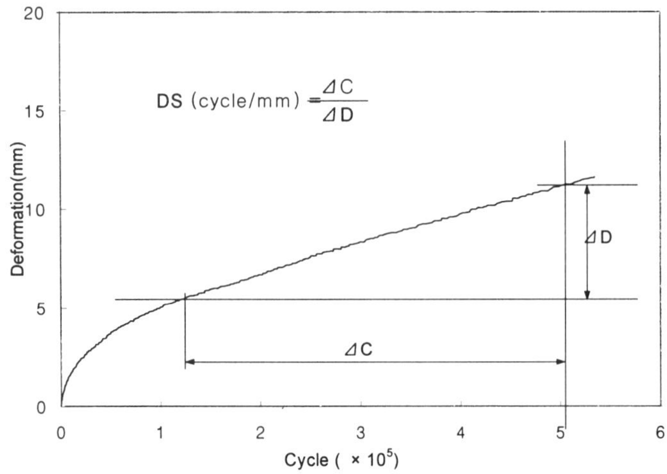

Figure 4. Description of deformation rate

The value in Table 6 is average of two specimens. Fatigue life of normal asphalt mixture (AP) was only 60,000 cycles. When AP mixture was reinforced using only a layer vinyl and grid at the bottom, fatigue life elongated 2.75 times and 8 times, respectively, than AP mixture. The fatigue life increased further when LDPE and CB were used for modification of asphalt binder as shown in fatigue life of LC, LCV and LCG. The fatigue life was extended drastically when 0.3% of fiber (PF) by weight of mixture was added. When fatigue lives of LCF, LCFV and LCFG were compared with normal asphalt mixture (AP), significant increments (16 to 25 times higher) were found. The increment was almost twice when compared with the same combination mixture without fiber (LC, LCV and LCG).

Table 6. Reflection crack propagation test results

Mixture	Fatigue life (cycle)	Final horizontal Displacement(mm)	Horizontal DS (cycle/mm)	Vertical crack growth rate (mm/cycle)
AP	60,000	0.2117	0.81×10^6	6×10^{-4}
APG	480,000	0.4052	8.30×10^6	0.7×10^{-4}
APV	165,000	0.1425	4.30×10^6	2×10^{-4}
LC	495,000	0.1785	7.00×10^6	0.8×10^{-4}
LCG	675,000	0.1549	10.41×10^6	0.7×10^{-4}
LCV	540,000	0.2666	6.80×10^6	0.6×10^{-4}
LCF	980,000	0.3568	13.20×10^6	0.4×10^{-4}
LCFG	1,500,000	0.2683	24.40×10^6	0.1×10^{-4}
LCFV	1,185,000	0.2539	16.40×10^6	0.4×10^{-4}

Therefore, fiber seems to have the most significant effect on fatigue life of asphalt mixtures. Simple reinforcement using vinyl or grid in AP mixture was less efficient than simple modification of asphalt binder using LDPE and CB. Reinforcement using glass grid (GG) was more effective on retarding progress of reflection crack than the reinforcement using vinyl layer. Therefore it was found that reinforcing LDPE and CB modified mixture using fiber and glass grid was the best performing combination.

Figure 5 shows horizontal deformation curves by loading cycle. The horizontal deformation is a result of crack width expansion. In general, horizontal deformation occurs initially within 50,000 loading cycles and then curves were stabilized. Highest deformation was found from simple grid reinforced specimen (APG). However, since horizontal deformations of all specimens were relatively small, comparison of the value seems to give no significant meaning.

Dynamic stability (DS) is measure of loading cycle for horizontal crack to reach 1 mm width. DS of AP mixture was 812,000cycle/mm and that of LCFG was 24,390,000 cycle/mm. This means that, to create 1 mm width of crack, a total of 24×10^6 cycles (or loading pass with 100psi pressure) is required in LCFG mixture. This number is approximately 30 times greater than that of AP mixture.

In visual monitoring of crack, most beam showed thin cracks. The crack of AP mixture initiated at the bottom center of beam. But cracks of other mixture initiated somewhere in the middle of specimen, mostly slightly above the PV or GG layer, and then propagated toword the top. Most beams except for AP, it was very difficult to find a single major crack. In stead, many thin cracks were created around the vertical line of loading. Especially, LCFG beam shows only several hairline cracks (no clearly visible crack) until 10^6 cycle of loading. It was noticed that the gird or vinyl reinforcement at the bottom of polymer-modified and fiber- added mixture prevented for the gap of concrete block to act as a crack tip at which, in general, a major crack initiates.

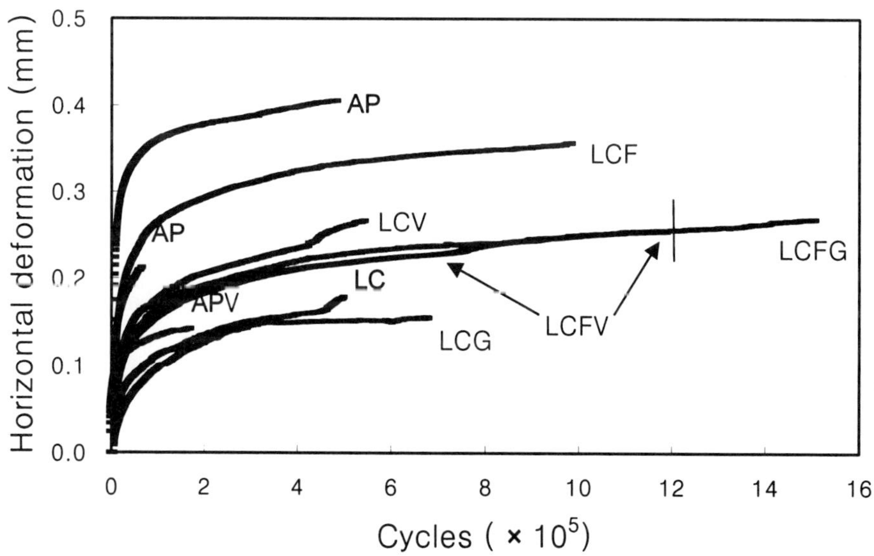

Figure 5. Horizontal deformation curves by loading cycle

CONCLUSIONS

This study evaluated nine different reinforcing techniques, using two binder modifier and three reinforcing materials, for retarding reflection crack propagation in asphalt pavement overlaid on an existing distressed concrete pavement. The two modifiers include a low-density polyethylene (LDPE), carbon back (CB), and three reinforcing materials include polyethylene fiber (PF), polyethylene film (vinyl: PV), and glass grid.

All of the materials used in this study showed a significant effect on improving resistance against crack propagation. However, the most significant feature of the results is that when the fiber added and LDPE and CB modified asphalt mixture was reinforced with the glass grid at the bottom of asphalt layer, almost no visible crack was observed until 1 million cycle loading from the side of beam. Fatigue life and dynamic stability (cycle/mm) of this reinforce combination were 25 times and 30 times, respectively, greater than AP mixture. Vinyl reinforcement at the bottom of asphalt mixture was also effective. But for the grid and vinyl to be more effective, binder modification using LDPE and CB is strongly recommended. Because when the modified asphalt was used in the mixture, the fatigue life of reinforced beam increased drastically.

The methods of reinforcement using grid and vinyl in this study are relatively simple to use in the field without special equipment. The modification of binder is not a new technique. Therefore, the study results can be evaluated in the field without difficulty.

REFERENCES

1. RIGO, J. M., "General Introduction, main conclusions of the 1989 conference on reflective cracking in pavements and future prospects," Proceedings of 2nd international RILEM Conference on Reflective Cracking in Pavements, 1993, pp. 3～20.
2. MOLENAAR, A. A. A., "Evaluation of pavement structure with emphasis on reflective cracking, Proceedings of 2nd international RILEM Conference on Reflective Cracking in Pavements, 1993, pp. 21～48.
3. SOUSA, J. B., SHATNAMI, S. and COX, J. "An approach for investigation reflective fatigue cracking in asphalt-aggregate overlays" Proceedings of 3rd international RILEM Conference on Reflective Cracking in Pavements, 1996, pp. 103～112.
4. LYTTON, R. L. and JAYAWICKRAMA, P., "Reinforcing fiberglass grids for asphalt overlay", Texas Transportation Institute, 1986.
5. JOSEPH, P. E., "Low temperature reflection cracking through asphalt overlays", Ph.D thesis, The University of Waterloo, Ontario, 1987.
6. "A Study on Identifying Methods to Control Reflection Cracks in Overlay" Interim research reports to Korea Highway Corp., 1997.

Fifth International Conference on the
BEARING CAPASITY OF
ROADS AND AIRFIELDS

INVESTIGATION AND ACCELERATED PERFORMANCE TESTING OF ULTRA-THIN FIBER REINFORCED CONCRETE OVERLAY ON ASPHALT AND CONCRETE PAVEMENTS

S.S. Kuo, Ph.D., P.E.
Department of Civil and Environmental Engineering
University of Central Florida
Orlando, FL 32816-4450
USA

Jamshid M. Armaghani, Ph.D., P.E.
Florida Department of Transportation
2006 Northeast Waldo Road
Gainesville, FL 32609
USA

Dave Scherling, P.E.
Department of Transportation
605 Suwannee Street, M.S. 46
Tallahassee, FL 32399- 0450
USA

ABSTRACT

The existing pavement at the New Smyrna Beach Airport parking apron has been severely deteriorated. Alligator cracking in the asphalt, along with transverse, diagonal, and joint cracks in the concrete pavement were exhibited. The Florida Department of Transportation officials have collaborated in designing a new overlay as part of a research project. Proposal for the rehabilitation was the use of ultra-thin fiber reinforced concrete and its studies of the type and optimum content using polyolefin and polypropylene fibers. The UCF Circular Accelerated Test Track facility (UCF-CATT) apparatus was utilized to determine the optimum thickness, joint spacing, durability, and performance.

Laboratory investigation of the material properties included slump cone, compressive, and flexural strength testing. A total of twenty cylinders and eight beams were tested in accordance to ASTM specifications. The results indicated that the addition of fiber had an insignificant effect on the concrete specimen strengths. However, the ability to retard the initial formation of cracks in the concrete was evident because of the fibers residual strength. The fibers were distributed uniformly within the matrix, and ultra-thin fiber reinforced concrete was well bonded to the existing asphalt and concrete pavements, as observed from several specimens cut from the test sections.

A total of eight test sections with various overlay thickness and saw cut spacing were constructed at the UCF-CATT facility. In 157 hours of testing, 160,000 repetitions of 49 kN (11,000 lb) wheel load was applied to the test segments. Results showed that minimal cracks occurred on every test section over the asphalt layer. No cracks were observed on the fibrous concrete over the concrete slab. The width of surface cracks from the fibrous concrete was significantly less when compared to the plain concrete control sections.

INTRODUCTION

The existing pavements (rigid and flexible) at the New Smyrna Beach Airport parking apron has been in service for small private airplanes for over thirty years. Severe alligator cracking appears throughout the asphalt pavement and transverse, diagonal and joint cracks appears along the concrete pavement. In recent years, use of ultra-thin reinforced concrete for overlay of asphalt and concrete pavements has become increasingly popular in pavement rehabilitation (Ramakrishnan, et.al. 1992,1995 and Smith, 1991). Fibers in concrete serve two major functions. First, the fibers may prevent shrinkage cracking of concrete. Second, the fibers can reduce the segregation of the components of the concrete. The permeability of concrete may lessen with the addition of fibers in the mix, thus the corrosion may be reduced. The increase of the Toughness Index of concrete with fibers as additive has been reported by many laboratory experiments. The use of ultra-thin concrete overlay reinforced with polypropylene fibers was first introduced at the St. Louis Airport whitetopping project in 1993 (Mowris, 1995). Subsequently, many rehabilitation's using whitetopping techniques were followed (Speakman, 1996 and Mack, 1997).

The objectives of this research study are to investigate:

1. The type and optimum content of the fiber additive in ultra-thin Portland Cement Concrete (PCC) overlay to extend the pavement service life.
2. The durability and performance testing of ultra-thin PCC overlay using the University of Central Florida (UCF) Circular Accelerated Test Track (CATT) facility.
3. Result from study to be implemented on the New Smyrna Beach Airport Demonstration Project.

Six test sections constructed on the UCF test track over an asphalt consist of:

a. One section was a 3m x 1.2m x 5.1cm control section built with conventional design of PCC overlay with .6m x .6m joint spacing.
b. One section was a 3m x 1.2m x 5.1cm section reinforced with 3M Polyolefin fibers with .6m x .6m joint spacing.
c. One section was a 3m x 1.2m x 5.1cm section reinforced with Fibermesh Polypropylene fibers with .6m x .6m joint spacing.
d. One section was a 3.7m x 1.2m x 8.9cm control section built with conventional design of PCC
   overlay with 1.2m x 1.2m joint spacing.
e. One section was a 3.7m x 1.2m x 8.9cm section reinforced with 3M fibers with 1.2m x 1.2m joint spacing.
f. One section was a 3.7m x 1.2m x 8.9cm section reinforced with Fibermesh fibers with 1.2m x 1.2m joint spacing.

Two test sections were 8.5m x 1.2m x 3.8cm and 8.5m x 1.2m x 2.5cm reinforced with 3M fibers with no joint spacing.

Laboratory testing of specimens were to determine the characteristics of the mix: such as, density, compressive strength, tensile strength. In the performance test the wheel load, velocity, and the number of load applications were recorded and converted in terms of service life of years as traffic volume and aircraft gear load are known. During the course of testing, the pavement conditions were visually inspected and photographed, and the locations and crack patterns were carefully examined and measured. The results from this test study indicate that the ultra-thin fiber reinforced concrete overlay is a very promising technique for the future pavement rehabilitation.

ACCELERATED TESTING OF PAVEMENTS

The purpose of an accelerated test track is to identify, understand, and characterize factors which influence pavement performance in short periods of time. The results from this testing will help to predict future pavement performance for design and evaluation purposes. There are several pavement testing facilities in use for pavement research. The 1985 Organization for Economic Cooperation and Development (OECD) report listed thirty-one full-scale pavement testing facilities built in different countries. The UCF-CATT, see photo 1, is 15.2 meter in diameter to the centerline of a 1.2 meters width concrete pavement that was made to accommodate a dual wheel assembly consisting of a half an axle. This track was built in 1989 by the UCF research team for the project funded by the FDOT.

The operation of this test track is to have three sets of dual truck wheels traveling around a circular path, guided by three radial arms. A 44.7 kW hydraulic motor is mounted to each axle to deliver the power through a planetary gear speed reducer and handle dual-wheel loading up to 133 KN. Power is supplied by a hydraulic pump that is driven by a 164.1 kW diesel engine. This hydraulic transmission provides the speed of rotation for the machine, up to 43 km/h in either a clockwise or a counter-clockwise direction.

Photo 1: UCF Circular Accelerated Test Track.

TEST SECTIONS

Test sections included a 3.81cm layer of asphalt concrete overlaid by two different thicknesses, 5.1cm and 8.9cm of ultra-thin fiber reinforced concrete and a control section of plain Portland Cement Concrete(PCC). Figure 1 shows a plan view of the test track and the testing sections. The ultra-thin fiber reinforced concrete overlays and the control sections were cured for seven days with water sprinklers running twenty-four hours a day. A conventional S-3 asphalt mix was used for the 3.81cm layer under the concrete overlay. The concrete mix design is shown in Table 1, and content of fibers used in the concrete overlay is shown in Table 2.

Table 1. Concrete Mix Design

MATERIAL	QUANTITY
Type III Cement	446.9 kg/m^3 (753 lbs/yd^3)
Sand	789 kg/m^3 (1330 lbs/yd^3)
#89 Rock	917 kg/m^3 (1545 lbs/yd^3)
Air Entraining Admixture (MBVR)	23 ml (0.8 oz)
Set Retarder (LL961R)	436 ml (15.0 oz)
High Range Water Reducer (440N)	3939 ml (135.36 oz)
W/C Ratio	0.39
Air	2.0%
Water = 154 kg/m^3	
Toltal Weight = 2307 kg/m^3 (3889 lbs/ yd^3)	

Table 2. Quantity of Fibers Used

FIBER USED	WEIGHT USED	FIBER CONTENT
Polypropylene	1.8 kg/m^3 (3 lbs/yd^3)	0.2%
Polyolefin	14.8 kg/m^3 (25 lbs/yd^3)	1.63%

MATERIAL PREPARATION AND PLACEMENT

A total of 5.3 m^3 (7 yd^3) of PCC mix was ordered for the project and delivered by Rinker Materials Corporation in two trucks. Truck one with 2.7 m^3 (3 yd^3) of PCC, with a slump test of 12.7cm, poured a control section with no fibers added and two other sections with 3M Polyolefin fibers. An equipment of roller bug, which aligns and buries the fibers horizontally beneath the surface is essential to finish the preparation of the surface for the 3M fiber mix. Truck two delivered 2.7 m^3 (3 ½ yd^3) of concrete mix, with a slump test of 22.9cm for the one control section and two Fibermesh polypropylene fibrous sections. It appeared too much superplasticizer was induced into the mix at the plant.

LABORATORY TEST SAMPLES

A total of twenty 15.2 cm x 30.4 cm cylindrical specimens were cast during the pouring of the test sections. Among the twenty cylinders specimens, ten were of no fiber added, five with 3M fibers added, and five with Fibermesh fibers added. ASTM C31 standards were followed for sample collection. ASTM test procedures, C39 for compressive strength, and C78 for the flexural strength were followed to test all collected samples. Test result of each specimen is tabulated in Table 3.

It is clearly seen from Table 3, that the concrete mix from truck one has a higher compressive strength than that from truck two. The flexural strength of the beam specimens has the same characteristics even though all beam specimens were tested at twenty-eight day curing time. Nevertheless, the compressive strength of the specimens exceeded 41.4 Mpa at seven days to 46.9 Mpa at thirty days and the flexural strength on average of 5.5 Mpa are much greater than that required by FDOT's 27.6 Mpa compressive strength and 4.5 Mpa flexural strength . As expected, the plain PCC cylinder specimens (control sections) broke into pieces at the the compression test, whereas, the fiber reinforced concrete specimens showed cracks yet held together. In beam testing, the plain PCC specimens broke into two halves, however the fiber reinforced concrete specimens showed cracks but remained intact. In order to examine the distribution of polypropylene fibers in the aggregate mix, the beams were forced to break into two halves. Both Fibermesh and 3M fibers were considered to be evenly distributed in the mix.

PERFORMANCE TESTING

The performance testing took place at the UCF - CATT facility. The pavement sections were tested for approximately for 157 hours which resulted in 160,000 load repetitions.. During this testing the temperature varied between 15.5 C and 26.7 C. In addition, the track was constantly sprayed with water at two different locations, thus keeping the entire track wet while the apparatus was in motion.

The examination of the performance of the concrete overlays was undertaken by the visual inspection of the track. This visual examination was documented by photographs, and would include the analysis of all cracking as well as any other modes of failure which is standard to the investigation of the pavement performance. The data collected by this visual method will provide an acceptable method of analyzing the distress of the overlay. As a result, the effectiveness of the different concrete overlay can be evaluated under traffic conditions. Upon the completion of 160,000 load repetitions surface wear on the 3M fibrous sections and the control section of Truck 1 exhibited a slight degree of wear. While, the Fibermesh fibrous and control sections of Truck 2 displayed a relatively extensive surface wear. The severity of the surface wear maybe a result of the slump test. The resulting cracks after the completion of testing are also illustrated in Figure 1.

Table 3. Compressive and Flexural Strengths of Test Specimens at Various Cured Times

Material Type:	Speciman No.	Compressive Strengths, MPa (psi)				Flexural Strength, MPa (psi)
		24 Hours	7 Days	14 Days	28 Days	28 Days
3M Fibers:						
Cylinders:	3M-1					
15.2cm x 30.5cm	3M-2			47(6827)		
(6" x 12")	3M-3		47(6827)			
	3M-4				49(7160)	
	3M-5	30(4300)				
Beams:	3M-B1					
15.2cm x 15.24cm x 61cm	3M-B2					5.7(821)
(6" x 6" x 24")	3M-B3					
Control Section 1:						
Cylinders:	CS1-1	29(4200)				
15.2cm x 30.5cm	CS1-2		48(6937)			
(6" x 12")	CS1-3				53(7690)	
	CS1-4			49(7050)		
	CS1-5					
Beams:	CS1-B1					5.7(823)
Fibermesh Fibers:						
Cylinders:	FM-1	18(2560)				
15.2cm x 30.5cm	FM-2		42(6030)			
(6" x 12")	FM-3				46(6600)	
	FM-4					
	FM-5			42(6108)		
Beams:	FM-B1					
15.2cm x 15.24cm x 61cm	FM-B2					
(6" x 6" x 24")	FM-B3					5.0(730)
Control Section 6:						
Cylinders:	CS6-1		45(6515)			
15.2cm x 30.5cm	CS6-2	17(2400)				
(6" x 12")	CS6-3			46(6719)		
	CS6-4					
	CS6-5				48(6970)	
Beams:	CS6-B1					5.4(790)

LIFE EXPECTANCY ANALYSIS

Since the actual loading subjected to pavement systems varies in magnitude, it is necessary to correlate the loading used in this test to the actual loading conditions expected on the pavement overlays. A method utilized is the equivalent axle load factor (EALF). An equivalent axle load factor defines the damage per pass to a pavement by the axle in question relative to the damage per pass of a standard 80KN (18 kip) single axle load. In this study the concrete overlays were subjected to a 97.9 KN (22 kip) axle load, and the EALF of 2.23 from $(22/18)^4$ apprise for the transformation of these repetitions to an equivalent amount of repetitions of a standard 80KN (18 kip) single axle load (ESAL). Therfore a total of 160,000 applied load repetitions is equivalent to 356,800, 80KN (18 kip) ESAL.

It is then necessary to correlate the damage resulting from the loading applied to the test track to the expected loading conditions of a specific roadway. As an example, a typical medium-traffic highway will be evaluated by a standardized method. First, the actual volume of yearly truck traffic for the roadway must be determined, assuming an average daily traffic (ADT) volume of 7,500 for the subscribed roadway, with an average percent of trucks to by six percent. The average volume of truck traffic, with an AASHTO multlane loading reduction factor of 0.9 to be:

$$7{,}500\,x\,0.06\,x\,0.9\,x\,365\,days = 147{,}825$$

In this project an 48.9 KN (11 kip) half axle wheel load was used for the accelerated testing, which resulted in a 97.9 KN (22 kip) ESAL. As previously determined, a factor of 2.23 EALF must be used to correlate the 97.9 KN (22 kip) axle loading used in the test to the typical 80 KN (18 kip) axle load. In addition, it is evident that every truck wheel does not travel over the same path for any pavement, as a result an assumed occurrence of three was used for this analysis. Hence, implying that every third wheel of trucks traveling on the roadway would cover the same path along the pavement. Allowing the value N to equal the number of track repetitions per year, the test track results can be equated to the typical roadway per year life expectancy.

$$N\,x\,2.23\,x\,3 = 147{,}825$$
$$N = 22{,}096.4$$

The 160,000 repetitions on the track are equivalent to approximately 7.25 years of life.

The primary focus of the performance testing of the different ultra-thin fiber reinforced concrete overlays is to analyze its use to rehabilitate the New Smyrna Beach Airport asphalt and concrete apron. It is necessary to consider the more comprehensive use of this type of overlay for different roadway systems.

However, in considering the loading which the airport runway is subjected to, the maximum of landing gear load is 22.2 KN (5 kips) per wheel, which may be considered as a 44.5 KN (10 kip) axle load. Thus,

$$EALF = \left(\frac{22}{10}\right)^4 = 23$$

The corresponding number of repetitions of a 44.5 KN (10 kip) axle load, which the overlay is subjected to is 11,000,000 repetitions. The160,000 repetitions from 97.9 KN (22 kip) ran on the test sections are equivalent to approximately seventy-five years life for the small aircraft at New Smyrna Beach Airport.

COMPUTER PROGRAM ANALYSIS

The KENSLABS computer program was used to analyze the test sections of pavement concrete overlay. The stresses at different thickness of concrete overlays will be calculated from this program. The program has the capability of analyzing three types of pavement systems:

1. Single rigid layer of PCC slab.
2. Hot mix asphalt (HMA) layer on top of PCC layer.
3. PCC layer on top of a cement treated base.

In this study, the PCC was placed over a HMA layer, similar to case 3 pavement system. Since there is bonding between the asphalt and PCC overlays, the two layers were analyzed as a composite system. The overlay being analyzed has to be divided into discrete rectangular elements in order to carryout the finite element analysis. The section shall be considered a continuous element with a uniform thickness, and each segment shall be divided into 36 sections with 49 nodes. A rectangular contact area produced by each wheel is 14.2cm x 20.6cm (5.6" x 8.1"). The modulus of elasticity of concrete overlay was calculated from: $E = 57{,}000\sqrt{f_c}$

The modulus of elasticity of the asphalt was correlated from the Marshall stability. Table 4 summarizes the stresses in the different overlays.

Table 4. Stresses at the bottom of the Overlays

SECTION	FIBER	THICKNESS cm (in.)	SUBSUM KN (lbs)	SMAX KPa (psi)
6	None	8.9 (3.5)	4.9 (21912)	232 (33.6)
5	3M	8.9 (3.5)	4.9 (21912)	232 (33.6)
4	Fibermesh	8.9 (3.5)	4.9 (21912)	223 (32.4)
3	Fibermesh	5.1 (2.0)	4.9 (21912)	137 (19.9)
2	3M	5.1 (2.0)	4.9 (21912)	143 (20.8)
1	None	5.1 (2.0)	4.9 (21912)	143 (20.7)

SMAX - Maximum stress

SUBSUM = Sum of subgrade reaction

These maximum stresses experienced in the different overlay systems are considered well below the fracture capacity of the concrete. It was previously determined that the modulus of rupture of different concrete specimens ranged form 5.0 MPa (730 psi) to 5.7 MPa (823 psi). The stress ratios only achieved a 2% to 4% range. These small values of stresses developed at the bottom of the overlays, the pavement will not, theoretically, fail in fatigue according the to PCA equation: $Log\left(N_f\right) = f_1 - f_2\left(\frac{\sigma}{S_c}\right)$

CONCLUSION

Based on the results of this research study, the following conclusions can be drawn:

1. Laboratory testing of cylindrical and beam specimens indicated that the addition of fibers in the concrete mix design had an insignificant effect on the compressive and flexural strengths, as compared to the plain concrete. Fibers in the concrete matrix were considered to be quite evenly distributed in the mix as examined from the broken specimens.
2. Several cut specimens from the test sections revealed that the bond between fiber reinforced concrete overlay and asphaltic base is strong. However, the Iowa shearing test was not conducted and the shear strength at the bonding interface is unknown.
3. After application of 356,800 equivalent 80 KN (18 kip) ESAL, it is evident that the cracks in the fiber reinforced concrete test sections were relatively small with respect to the severe loading conditions applied to the test track. The transverse cracking that occurred close to the edge of the section, may be caused by the excessive stress since no steel dowels were

installed at the joints.

4. The most advantageous properties of fiber reinforced concrete overlay are the ability to retard the initial formation of cracks and the ability to restrict the propagation of already formed cracks. This was especially evident in the polyolefin test sections, where cracks that occurred in the early stage of testing were retarded from increasing beyond micro-cracking.
5. The extensive cracking along many of the saw-cut joints may necessitate a re-evaluation of the need for the short joint spacing for ultra-thin fiber reinforced concrete overlay on asphalt. No cracks were observed for the overlay on concrete where no saw-cut joints were implemented.
6. The ultra-thin fiber reinforced concrete overlay can significantly extend the load carrying capacity of asphalt and concrete pavements. It is certainly considered a viable mix design for rehabilitation and restoration of deteriorated pavements.
7. Investigation and performance testing of ultra-thin fiber reinforced concrete overlay with optimum fiber content for an effective life cycle cost should be studied in the future

REFERENCES

Mack, J. "Ultra-Thin Whitetopping the World'. Concrete Pavement Progress, American Concrete Paving Association (ACPA), Vol. 36, No. 2, Spring !997.

Mowris, S. "Whitetopping Restores Air Traffic at the Spirit of St. Louis." Concrete Construction, June 1995.

Ramakrishnan, V., Kakodkar, S. "Evaluation of Non-Metallic Fiber Reinforced Concrete in PCC Pavements and Structures." SDDOT Office of Research, September 1995.

Ramakrishnan, V., Zellars, R. C. "Evaluation and Comparisón of the Physical Properties of Standard Polypropylene Fiber Reinforced Concrete." South Dakota Schools of Mines and Technology, July 1992.

Smith, A. "Colorado Tests Whitetopping for Asphalt Road Rehabilitation." Concrete Construction, January 1991.

Speakman, J., N. Scott III. Ultra-Thin , Fiber Reinforced Overlays for Urban Intersections. Transportation Research Record, No. 1532, TRB, National Research Council, Washington, D.C., 1996.

Fifth International Conference on the
BEARING CAPASITY OF
ROADS AND AIRFIELDS

SURVIVABILITY OF GEOTEXTILES IN ROAD CONSTRUCTION

ARNSTEIN WATN[1], GUDMUND EIKSUND[2] AND ÅSMUND KNUTSON[3]

[1]Senior Scientist, SINTEF Civil and Environmental Engineering, N-7034 Trondheim, Norway
[2]Research Scientist, SINTEF Civil and Environmental Engineering, N-7034 Trondheim, Norway
[3]Head of Division, Norwegian Public Roads Administration, Road Research Laboratory, N-0033 Oslo, Norway

ABSTRACT

SINTEF Civil and Environmental Engineering have performed a research project in two phases on non-woven geotextiles in road constructions. The first phase was a large-scale laboratory test aimed to study the effect of non-woven geotextiles on road deformations at cyclic loading. The second phase was a field test aimed to study the resistance against damage of the geotextiles during construction. The project focused on the correspondence between properties from index tests and the observed behaviour. A clear correspondence was found between the initial tension stiffness of a geotextile and the deformation after cyclic loading. Low correlation was found between observed damage during construction and the evaluation criteria used for classification of geotextiles in Norway. To take into account these findings it is proposed to develop a criterion on survivability based on a combination of deformation energy and stress-strain properties to cover the construction and lifetime requirements. It is proposed to establish a common European research and development project involving producers, public authorities and research organisations.

KEYWORDS: Geotechnical engineering, soil, geotextile, properties.

INTRODUCTION

The criteria for evaluating strength and deformation properties for non-woven geotextiles used in separation and filtration in roads have been discussed for more than twenty years. The first systems for evaluation and classification of geotextiles for separation and filtration in roads were introduced by the Norwegian Road Research Laboratory (NRRL), (Alfheim and Sørli,

1977) (1). Later several systems have been introduced, but generally the classification requirements are mainly empirically based, and to some extent dependent on local conditions and experiences (Forschungsgesellscaft für straβen- und Verkehrswesen, 1994 (2), Rathmayer, 1993 (3), AASHTO, 1990) (4). The evaluation criteria and the index test methods which are used, differ between the systems and a possible co-ordination between the systems have been discussed since their introduction. The European Committee for Standardization (5) has prepared documents for general and specific requirements for different application and function, but without giving any figures.

The Norwegian classification system

Geotextiles for separation and filtration in roads are in Norway divided into four classes dependent on the type of material (maximum grain size) to be used against the geotextile:

Class 1:	Generally not used
Class 2:	Sand and gravel with max. diameter 50 mm
Class 3:	Crushed stone with max. diameter 100 mm
Class 4:	Blasted rock with max. diameter 2/3 of the layer thickness

The classification is based on an evaluation of results from the static puncture tests and the cone drop tests. The tested product will achieve points from the results in the tests referring to each criterion and the classification is then dependent on the total sum of points. For the static puncture test (CBR- test, ISO 12236:1996) the measured force and deformation are used to calculate a corresponding tension (force/mm) and strain (%). The classification criterion is based on the derived tension and strain, the maximum tension, the elongation strain at failure and the tension increase from 20 % to 70% strain (or until strain at failure if less than 70%). The average hole diameter is used as evaluation criterion for the cone drop test (Schalin 1995) (6).

Relevant properties and test methods

There is a clear need for establishing a more fundamental understanding of the required characteristics of the geotextile to fulfil its functions (separation and filtration) in the road. The required properties must reflect the environmental loads imposed on the geotextile during the installation, construction and service lifetime. A theoretical sound correlation between the required properties and the corresponding required parameters found from index tests should be established. A combination of index tests, large scale performance tests, full scale field tests and collection of experiences from the field is believed to be the best way to establish such a correlation.

RESEARCH PROJECT

SINTEF Civil and Environmental Engineering have performed a research project on non-woven geotextiles in road constructions. The NRRL and has participated with observers and supervisors in the project. The project focused on the correspondence between geotextile properties found in index tests and the observed behaviour in laboratory and the field. The first project phase (SINTEF 1996) (7) included index tests and large scale laboratory load test. This part aimed to study the effect of stress-strain properties on non-woven geotextiles on road deformations at cyclic loading. The second phase was a field test (SINTEF 1997) (8) aiming to study the resistance against damage of the geotextiles during the construction. Non-woven geotextiles with different production technology and area weight were used in the research projects.

Laboratory tests

Index tests

The index tests included cone drop tests, static puncture tests and wide width tensile tests. The tests were performed on virgin samples and on samples extracted after the load test. In addition the effect of thermal cycling and stress strain behaviour under frozen conditions were tested. Six different non-woven geotextiles were tested, three corresponding to class 2 and three corresponding to class 3. The geotextiles used in the laboratory tests corresponding to class 3 are listed in Table 1 together with a summary of the results from the laboratory index tests on virgin samples.

.

Table 1 Class 3 geotextiles used in the laboratory test with results from index tests

Ref	Type of product	Area Weight (g/m^2)		Static puncture		Falling cone
		Nom.	Meas.	Max force, N	Displ.at max. force (mm)	Average hole diameter, mm
SNP 3A	Staple fibre, needle punched, polypropylene	190	197.8	2380	57	14
CNP 3B	Continuous filament, needle punched, polypropylene	160	171.5	2252	44	24.2
CTP 3C	Continuous filament, thermally bonded, polypropylene	190	190.8	1970	50.8	19.1

Typical load displacement curves from the static puncture test are shown in Figure 1. Observe the differences in initial stiffness between the different geotextiles. The thermal cycling had no significant effect on the results from the index test measurements.

The resistance against damage and the results from the index tests are used to evaluate the requirements in the classification system. The relevancy of an index test parameter for survivability of the geotextile is studied by correlating the parameter with the *resistance against damage* as defined above. The area weights are also included in the correlation. The results of the correlation are shown in Table 5. The test results from geotextile CTP 4F was not included in the correlation.

Table 4 Resistance against damage

Ref	*Damage* (Sum of hole diameter)	*Resistance against damage* (Average damage)/(damage)
CNP 4A	2793	1.07
SNP 4B	2613	1.15
SNP 4C	3157	0.95
CTP 4D	2655	1.13
SNP 4E	3759	0.80
CTP 4F	-	0.40*)

*) Based on a scaling of the results

Table 5 Correlation between index test results and resistance against damage

Parameter	Correlation
Weight/m^2	0.81
Strength incr. 20-70%	-0.11
Failure strength	0.84
Strain to failure	-0.77
1/(Cone drop hole diam)	-0.26
Number of points	-0.36

The parameters showing best correlation with the resistance against damage is the *push through strength* and the *area weight.* The criteria for *strength increase*, and *the number of points* shows poor correlation. The *strain to failure* and the *cone drop hole diameter* shows a fair negative correlation. The poor correlation for the number of points is remarkable. The low correlation is mainly caused by the fact that the two geotextiles with the most damage have full score based on the criteria in the index test.

The results from the index test do not point out an obvious candidate among the parameters that may explain why CTP 4F should be so severely damaged. In the primary tests the best correlation with the resistance against damage was found for the unit weight and the failure strength. This was not the case for CTP 4F that gives a high score on both unit weight and failure strength. Geotextile CTP 4F has, however, a relatively low value both for *strain to failure* and the *inverse of the cone drop hole diameter.* These low values may partly explain some of the higher degree of damage for the CTP 4F geotextile.

Both CTP 4D and CTP 4F are thermally bonded geotextiles, having a high initial stiffness. As shown in Figure 5, the force-displacement relations from the static puncture test are relatively similar for these to geotextiles compared to the other geotextiles tested. The large difference in degree of damage between CTP 4D and 4F is not reflected by similar differences in the index test results, with a possible exception for the deformation at failure. The damage on CTP 4F is therefore probably caused by material properties not measured in the index tests. A possible explanation may be the properties on the brittleness in the failure or the tear propagation for the geotextile.

CONCLUSIONS AND RECOMMENDATIONS

The project has provided useful information for evaluating relevant properties and requirements for geotextiles to be used for separation and filtration in roads. There are considerable differences in stress strain properties of the geotextiles that is also reflected in the behaviour in the field. Noticeable differences are found in the susceptibility for damage during installation. The criteria used in the existing systems for classification and specification do not seem to reflect properly the behaviour in the field. A revision of the criteria is therefore clearly needed.

The deformation of the geotextiles when subjected to loading, that is, in terms of rutting during installation and construction, is clearly linked to the initial stiffness of the geotextile. A criterion for geotextile survivability is clearly relevant, but has to reflect the behaviour during installation, construction and service lifetime. A criterion for geotextile survivability is suggested based on a combination of requirements for deformation energy and remaining stress and strain till failure. The principle is presented in Figure 8.

The deformation energy related to the installation and construction should be chosen with respect to the type of fill material, construction equipment and type of underground.

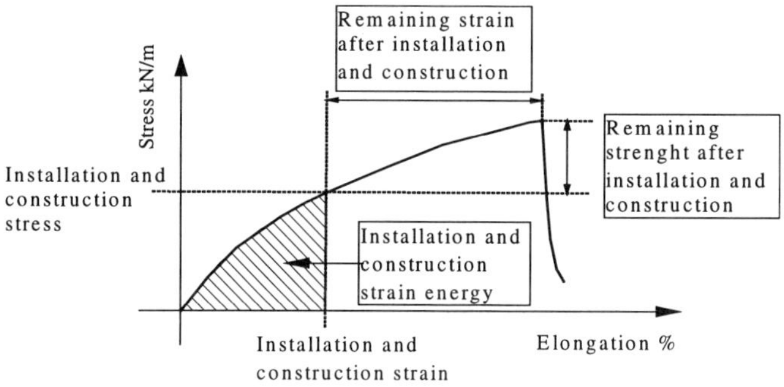

Figure 8 Survivability criterion principle.

The requirements for remaining strength and strain to failure should reflect the expected loads and deformations (settlement) for the service lifetime.

The final criteria should be based on a collection of data from laboratory and field tests correlated with long-term experiences from the field. The field experience should include different type of geotextiles, fill materials, subsoil conditions and construction equipment.

The development of a common European system for specification of geotextiles would be beneficial both for producers and users. Sound and technically verified criteria for stress and strain properties are essential in the development of such a system. We therefore propose to establish a joint research and development project to prepare a common European system for specification of geotextiles for separation and filtration in roads. The project should involve several countries, producers, public authorities and research organisations.

The basis for developing the system should be a combination of theoretical analyses, laboratory testing and collection of long term experience. Some data of long term experiences are available (9), but an updating and systematisation of available information is clearly needed.

Provided such a system can be established, it is likely that this also can be extended to other geosynthetic applications and functions and result in a complete European system for specification of geosynthetics.

ACKNOWLEDGEMENT

The authors would like to thank Du Pont de Nemours (Luxembourg) S.A. for funding the work, and the purchasers of non-woven geotextiles in Norway for providing free samples for testing. We would also like to thank Ingrid Sødergren from Vägverket, Sweden for participating as observer in the field test and for useful contribution to the discussion.

REFERENCES

1. Alfheim S.L., Sørlie A. (1977): " Testing and classification of fabrics for application in road constructions. *Intern. Conf. on the use of fabric in geotechnics* Paris 1977. Vol 2, pp 33-338.

2. Forschungsgesellscaft für straβen- und Verkehrswesen (1994): "Merkblatt für die Anwendung von Geotextilien und Geogittern im Erdbau des Straβaus".

3. Rathmayer, H (1993): "Nonwoven Geotextiles in Road constructions. Quality Requirements- The VTT-GEO Geotextile Specification". *Finnish National Road Administration*, FinnRA Reports 71/1993.

4. AASHTO (1990): "Standard Specifications for Geotextiles". *American Association of State Highway and Transportation Officials,* M288-90, Washington DC.

5. CEN TC 189: "General and specific requirements for geotextiles and geotextile related products used in the construction of roads and other trafficated areas" Preliminary draft.

6. Schalin, L (1995): "A comparison among the Swedish, the Finnish and the Norwegian requirements for separation layers of geotextiles". *Norwegian Road Research Laboratory,* internal report no. 1786 (in Swedish).

7. SINTEF (1996): "Non-Woven Geotextiles in Road Constructions." Report.

8. SINTEF (1997): "Non-woven geotextiles - Field test on damage during installation." Report.

9. Rathmayer H. (1980): "Long-term behaviour of geotextiles installed in road constructions in Finland since 1973. *Väg-och vattenbyggaren* 1980:7/8, pp 33 -36.

Fifth International Conference on the
BEARING CAPASITY OF
ROADS AND AIRFIELDS

AN APPROACH FOR THE DEFINITION OF CONSTRUCTION AND MAINTENANCE STRATEGIES FOR FLEXIBLE PAVEMENTS

JORGE B. SOUSA[1], JORGE C. PAIS[2] AND BERNARD ECKMANN[3]

[1]General Director, CONSULPAV, Taguspark, Tecnologia I, n.26, 2780 Oeiras, Portugal, Fax: +351-1-421-2296, Email: jmbsousa.consulpav@taguspark.pt
[2]Research Engineer, University of Minho, Guimarães, Portugal, Fax: +351-53-510-217, Email jpais@eng.uminho.pt
[3]Research Scientist, NYNAS, N.V., Beliweg 22, 4e Havendok, B-2030 Antwerp, Belgium, bernard@nynas.se

ABSTRACT

Design, Build, Finance and Operate contracts to develop a road network require a re-evaluation of the strategic decisions towards the design and maintenance strategies of flexible pavements.

Probabilistic analysis, based on the variability of pavement material properties and layer thickness as well as on traffic variability is required and needed as a cost effective basis for the many possible design and maintenance strategies.

The NOAH software was used to combine hundreds of possible design scenarios on a compressive basis for the decision making process towards a design criteria. The use of Rosenblueth transformations permitted the reduction of computational efforts and made possible the conclusion of this comprehensive design approach within the time frame of the tender.

The bases for the data variability parameters were actual measurements made in the existing 83 km of freeway. Data from cores and from back-calculation analysis with the FWD in the determination of stiffness layer moduli was used. Permanent deformation (structural) and flexural fatigue failure criteria used was based on SHELL´s criteria. The roughness induced costs, as proposed by the World Bank model, was also used as a part of society cost analysis.

INTRODUCTION

CONSULPAV worked with AUTOESTRADAS DO ATLANTICO (from now on called agency) to propose a strategy to design and maintain 83 Km new highway pavements and 87 Km of existing pavements for a DBFO tender by the Portuguese Highway Authority [1] during 32 years (the tender called for 30 years but is was decided to leave a remaining life of 2 years at the end of the concession to allow some time to rehabilitate the network).

The goal of the process was to provide a strategy of construction and maintenance that would minimize costs to the agency but that would take into consideration user costs. To the extent possible solutions would be found that would minimize both agency and user costs.

The main criteria for evaluation were based on the mechanistic methods proposed by Shell [2] to control fatigue cracking and structural permanent deformation. It was assumed that functional characteristics would be re-establish with a chip seal placed every seven years.

First an evaluation of the most cost effective maintenance strategy was defined based on actual data from 83 km of existing freeway. Then a deterministic analysis was made towards the definition of construction/maintenance alternatives taking into consideration only deterministic concepts. The best solution was evaluated from a probabilistic perspective using the NOAH software [3].

For the two solutions with the least agency costs evaluation of the user costs based on the effects of road roughness was made.

EVALUATION OF MOST COST EFFECTIVE REHABILITATION STRATEGY

Criteria had to be developed to evaluate the most cost effective rehabilitation criteria for all the existing segments of the network. An analysis of the traffic, deflection data, coring and soil analysis led to the determination of the conditions for the existing 83 km of the road network. For all these segments predictions of pavement deterioration had to be made based on the two criteria presented above.

Table 1 – Description of maintenance options considered and associated costs.

NAME OF ACTION	DESCRIPTION OF MAINTENANCE ACTION	COST (PTE/m2)
O5	OVERLAY 5 CM TWO LANES	1270
O10	OVERLAY 10 CM TWO LANES	2540
O15	OVERLAY 15 CM TWO LANES	3810
M5F5	MILL 5 CM + FILL 5 CM ONE LANE	935
M10F10	MILL 10 CM + FILL 10 CM ONE LANE	1870
M12F12	MILL 15 CM + FILL 15 CM ONE LANE	2770
M15F15	MILL 12 CM + FILL 12 CM ONE LANE	2230
M5O10	MILL 5 CM ONE LANE + OVERLAY 10 CM THAT LANE + OVERLAY 5 CM ADJACENT LANE	2840
M5O15	MILL 5 CM ONE LANE + OVERLAY 15 CM THAT LANE + OVERLAY 10 CM ADJACENT LANE	4040

To determined the most cost effective methods for structural rehabilitation three types of scenarios were considered; 1- allow the pavement layer moduli degrade to about 1900 MPa and then place an overlay, 2- allow the pavement deteriorate so that mill and fill would be recommended and, 3 – allow the pavement to reach a level

where mill and overlay would be needed. Based on over 7 years of pavement evaluations performed in Portugal most pavements in need of structural maintenance have bituminous layer moduli that have deteriorated from about 5000 MPa to about 1900 MPa.
It was considered that when an overlay is placed on the right lane for structural reasons an equal thickness needs to be used on the left lane (although due to the lighter traffic the overlay may not be needed). In the mill and fill option only the right lane is affected by this solution. Table 1 presents a summary of all the rehabilitation actions considered. The cost of tack coats between layers has also been considered. The costs considered are variable costs to the contractor.

For each of the pavement sections based on actual *in situ* back calculations analysis the number of ESALS reached by each of the solutions was computed using NOAH. For each solution the relationship between the cost (Portuguese Escudos-PTE) per square meter was related with the number of ESALS obtained by that action. It can be observed in Fig. 1 a typical relationship for 4 sections. For all other road segments identical plots were generated. Although the actual amount of ESALS for each type of actions depended on subgrade conditions, pavement thickness, etc the general trends were identical.

It can be observed that the most cost-effective method to reach the highest number of ESALS for a given cost in the right lane is by overlaying the pavement section. All other methods are less cost effective even taking into consideration that the cost per square meter presented is the cost only of the right lane. It is considered that the placement of the overlay on the left lane is just a "cost" without any "benefit".

Removing pavement layers should only be considered when other requirements, other then structural strength, require it (i.e., curb height, bridge clearance, etc). As such the proposed philosophy for pavement maintenance was to considered that pavements should be overlaid whenever possible and that deterioration of the pavement should only be allowed to the extend that an overlay would still be a feasible option.

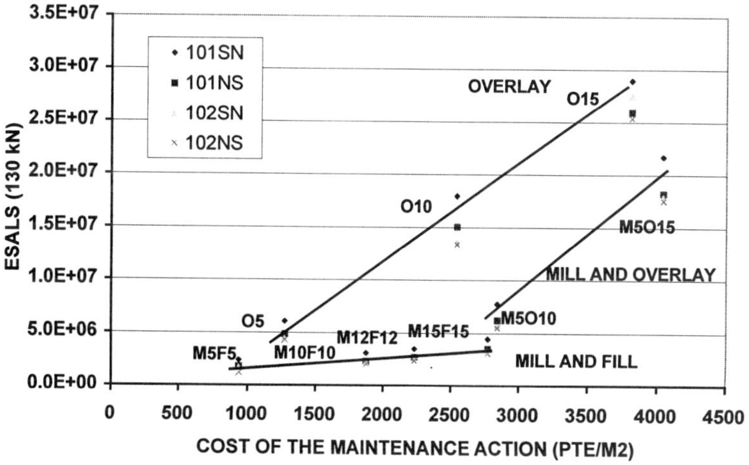

Figure 1 – Typical relationship between cost and type of rehabilitation action and number of ESALS achieved by that action.

EVALUATION OF MOST COST EFFECTIVE CONSTRUCTION/REHABILITATION STRATEGY – DETERMINISTIC EVALUATION

With the understanding that overlays would be the most cost-effective method to maintain the network it was necessary to determined how thick to build the new pavements and consequently how often to overlay during the life of the concession.

Three scenarios were considered:

1- Minimum initial construction costs with as much deferred maintenance as possible. It was considered that the minimum reasonable asphalt concrete layer thickness for construction and rehabilitation was 5 cm. The rational for this solution would be that the agency would minimize the use of its own money and have the maintenance paid by the toll revenue. This strategy was called **Strategy 1**.

2- Initial construction of such an extent that no structural rehabilitation/maintenance actions would be needed over the project life. The rational for evaluating this option was to consider the possibility of borrowing as much money as needed, to minimize user delays and cost, and possible agency costs. This strategy was called **Strategy 3**.

3- An intermediate solution where the pavement would be built with only structural capabilities to resist half of the project was considered. An overlay would then be placed that would last the remainder of the project. This would correspond to the traditional design life considerations where pavements in Portugal are design for 15 or 20 years design life. This strategy was called **Strategy 2**.

In this evaluation the two criteria proposed by the SHELL method for pavement design where used; the limiting tensile strain criteria on the bottom of the asphalt concrete layers and, the maximum compressive strain criteria on the top of the subgrade. It was considered also that at the end of each design life the pavement asphalt layer moduli was half of its initial value. This is a conservative criterion because in most pavements it would be reduced to about 1900 MPa. However because FWD annual evaluation are executed on an annual basis it will be possible to insure that pavement moduli will not degrade below the require value. It was also further assumed that the fatigue failure criteria should not apply for asphalt concrete layers with a moduli degraded to below 2000 MPa. It was also assumed that asphalt concrete layer moduli would not degrade below 500 MPa.

The assumed evolution of the layer moduli function of the life of the pavement can be observed in Table 2. It can be observed that the Strategy 1 a thin pavement was built only with 5 cm thicker bituminous layer. After reaching its design life of 3.4E4 ESALS (failure by fatigue of the asphalt concrete layer) an overlay is placed. However from now on the failure criteria for the overlay is not clearly define. As such two hypotheses were considered; A - one based on subgrade failure and B-another based on fatigue crack propagation. Because after each overlay there is anther interface the life computed based on the tensile strains of each interface was computed. It was considered the failure criteria for hypothesis B based on the fatigue life on the bottom of the layer with at least 2000 MPa.

Table 2 presents for each of the Strategies and hypotheses the estimated moduli and lives based on the above assumptions.

Table 2 – Estimated Moduli at the time of each overlay associated with estimated life after each maintenance action

Strategy 1A and 1B.

INITIAL CONST.	1st OVERLAY	2nd OVERLAY	3rd OVERLAY	4th OVERLAY	5th OVERLAY
					5 cm 40/50 6000 MPa *N=9.9E12*
				5 cm 40/50 6000 MPa >>>> *N=1.7E13*	3000 MPa *N=1.9E32*
			5 cm 40/50 6000 MPa >>>> *N=2.0E13*	3000 MPa >>>> *N=2.2E7*	**2000 MPa** ***N=2.8E6***
		5 cm 40/50 6000 MPa >>>> *N=5.8E11*	3000 MPa >>>> *N=1.4E7*	**2000 MPa>>>>** ***N=1.9E6***	1000 MPa *N=1.1E6*
	5 cm 40/50 6000 MPa >>>> *N=2.5E8*	3000 MPa >>>> *N=3.4E8*	**2000 MPa>>>>** ***N=1.0E6***	1000 MPa >>>> *N=5.9E5*	500 MPa *N=1.4E6*
5 cm 40/50 **6000MPa >>>>** ***N=3.9E4***	**3000 MPa >>>>** ***N=2.0E5***	**2000 MPa >>>>** ***N=1.3E5***	1000 MPa >>>> *N=2.1E5*	500 MPa >>>> *N=4.0E5*	500 MPa
30 cm ABC 200 MPa	200 MPa	200 MPa	200 MPa	200 MPa	200 MPa
Subgrade 100 MPa *N=1.4E5*	100 MPa *N=5.3E5*	100 MPa *N=1.6E6*	100 MPa *N=3.5E6*	100 MPa *N=6.3E6*	100 MPa *N=10.7E6*

Strategy 2A

Initial Construction	1st Overlay
	5 cm 40/50 6000 MPa *N=7.1E9*
5 cm 40/50 6000 MPa >>>> *N=4.4E8*	5 cm 40/50 3000 MPa *N=3.9E13*
13 cm 20/30 10000MPa >>>> *N=7.6E6*	13 cm 20/30 5000 MPa *N=2.6E6*
30 cm ABC 200 MPa	30 cm ABC 200 MPa
Subgrade **100 MPa** ***N=7.1E6***	**Subgrade** **100 MPa** ***N=7.6E6***

Strategy 2B

Initial Construction	1st Overlay
	6 cm 40/50 6000 MPa *N=1.1E11*
5 cm 40/50 6000 MPa >>>> *N=4.4E8*	5 cm 40/50 3000 MPa *N=1.1E12*
13 cm 20/30 10000MPa >>>> *N=7.6E6*	**13 cm 20/30** **5000 MPa** ***N=6.5E6***
30 cm ABC 200 MPa	30 cm ABC 200 MPa
Subgrade **100 MPa** ***N=7.11E6***	Subgrade 100 MPa *N=1.5E7*

Strategy 3

Initial Construction
5 cm 40/50 6000 Mpa *N=8.3E8*
15 cm 20/30 10000MPa *N=20.5E6*
30 cm ABC 200 MPa
Solo Fundação 100 MPa ***N=16.1E6***

ABC – Aggregate Base Course, >>>> - indicates "evolution of the layer moduli towards value in the next cell"
Values in *italic* (*N=*) are estimated number of ESALS (130kN) associated with the distress at each of the corresponding layers. Values in **bold** are adopted (as most likely to occur) limiting conditions after each overlay.

The evolution of the remaining life of the pavement for all these cases is displayed on Fig. 2. It can be noted that for all combinations there is a remaining life at the end of the design period of 32 years.

The determination of the net present value of all these strategies was made considering that every 7 years a chip seal was placed to restore surface conditions. It can be observed in Fig. 3 that for discount rates below 7% the most cost effective solution based on in the agency cost was Strategy 3. For higher discount rates Strategies 2 would be more favorable. Clearly the strategy of deferring maintenance with the least initial cost presents the highest net present value at all discount rates. These analysis deals only with the agency costs. Delay costs for maintenance work and user costs associated with higher IRI values presented in the Strategies 1 and 2 were not considered. Taking them into consideration (as costs to society) would likely point towards the selection of Strategy 3 regardless of the discount rate (within the range considered).

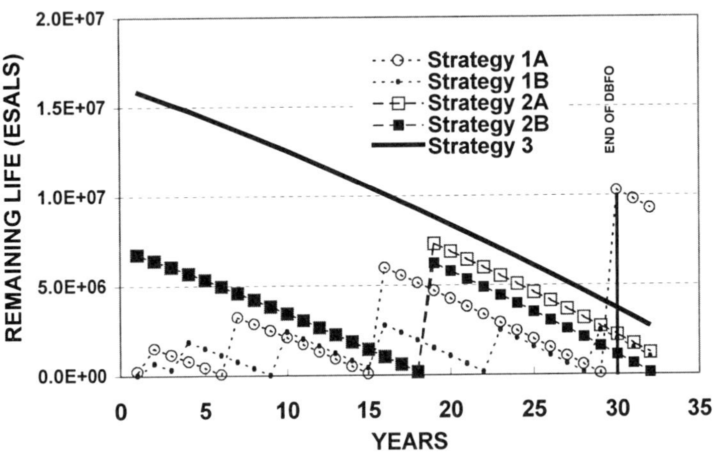

Figure 2– Evolution of remaining life for each of the deterministic strategies considered.

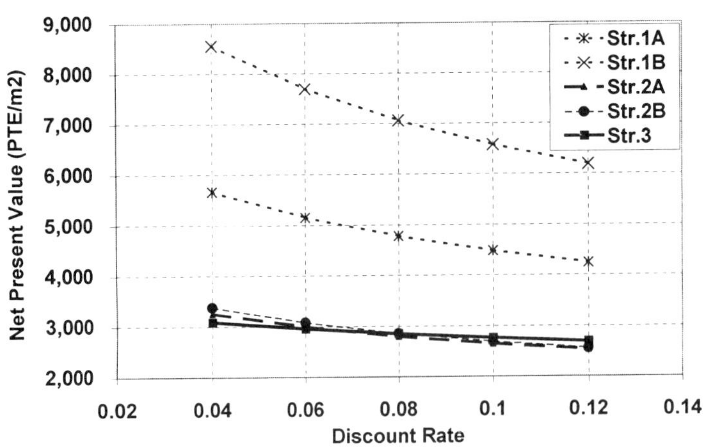

Figure 3 – Net Present Value for each of the deterministic strategies considered.

EVALUATION OF THE MOST COST EFFECTIVE CONSTRUCTION/REHABILITATION STRATEGY USING A PROBABILISTIC EVALUATION

Highway materials have inherent variability. Moduli, pavement layer thickness's and subgrade moduli considered are usually "MEAN" values. It was considered necessary to evaluate to which extent the effect of material variability combined in a probabilistic approach, would affect the conclusions determined above in a deterministic fashion.

From actual values determined from the existing pavement sections, back-calculated stiffness layer moduli, pavement thickness and subgrade moduli were determined from uniform sections. The variability of the data is presented in Fig. 4. For bituminous layers the standard deviation of the stiffness moduli and the thickness was about 10%. The standard deviation for subgrade moduli was about 18%. In Fig. 4 it can also be observed how well a normal distribution with those standard deviation values fits the actual observed data.

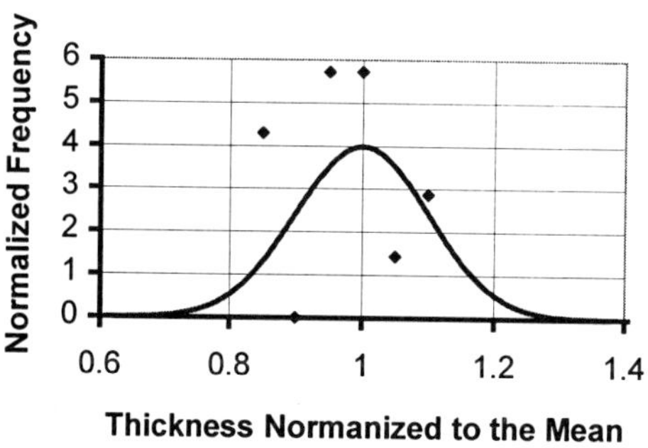

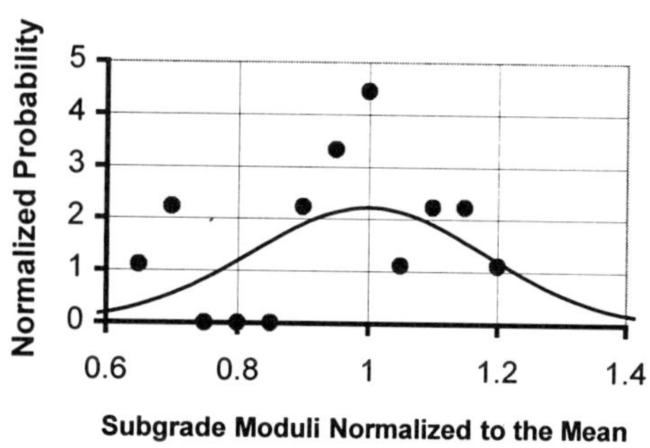

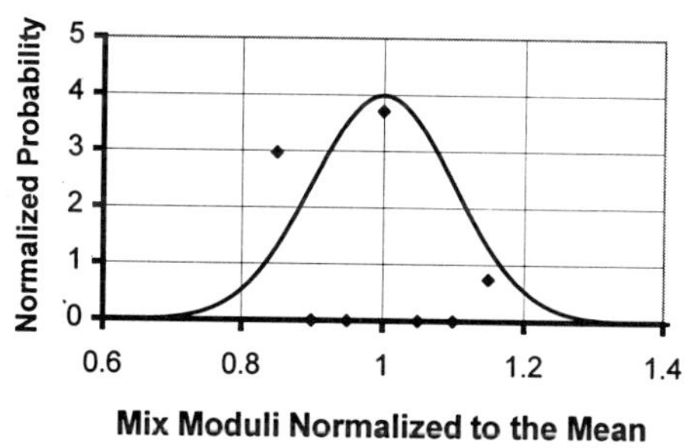

Figure 4 – Normal distribution curves superimposed over actual normalized frequency plots function of the normalized mean of subgrade moduli, bituminous layer thickness moduli and moduli.

It was assumed that the same variability would be present in the new pavements that will have to be constructed. As such the values considered in the deterministic analysis could be considered as corresponding to the mean values expected for those materials. Taking into consideration the actual variability, then there is a probability that the expected life would not be reached, at least in some segments of the pavement, because the layer moduli and layer thickness would be less then expected. This variability would have a stronger effect on thinner pavements given the logarithmic nature of the failure criteria selected. As such two strategies were investigated:

Strategy P1 – For the pavement structure selected from the deterministic analysis (Strategy 3) identify which actions would have to be performed to insure that the pavement would have no structural failure with 90% reliability.

Strategy P2 – Identify which pavement structure would have to be initially built to insure, with 90% reliability, that no structural rehabilitation/maintenance actions would have to be conducted during 32 years.

The NOAH software was used a basic tool to combine hundreds of possible design scenarios on a compressive basis. The use of Rosenblueth transformations permitted the reducion of computational efforts and made possible the conclusion of this compreenshive design approach within the time frame of the tender. The values adopted for each Strategy are presented in Table 3.

Table 3 - Estimated Moduli at the time of each overlay associated with estimated life after each maintenance action taking into account a design relaibility of 90%.

Strategy P1A

Initial Construction	1st Overlay
	6 cm 40/50 6000 MPa
5 cm 40/50 6000 MPa >>>>	5 cm 40/50 3000 MPa
H=15 cm 20/30 STD=1.5 cm E=10000MPa >> STD=1000 MPa *N(90%)=9.2E6*	H=15 cm 20/30 STD=1.5 cm E=5000 MPa STD=500MPa *N(90%)=6.0E6*
30 cm ABC 200 MPa	30 cm ABC 200 MPa
Subgrade E=100 MPa STD=18 MPa *N(90%)=5.6E6*	**Subgrade E=100 MPa STD=18 MPa *N(90%)=11.0E6***

Strategy P1B

Initial Construction	1st Overlay
	7 cm 40/50 6000 MPa
5 cm 40/50 6000 MPa >>>>	5 cm 40/50 3000 MPa
H=15 cm 20/30 STD=1.5 cm E=10000MPa >> STD=1000MPa *N(90%)=9.2E6*	**H=15 cm 20/30 STD=1.5cm E=5000 MPa STD=500 MPa *N(90%)=7.9E6***
30 cm ABC 200 MPa	30 cm ABC 200 MPa
Subgrade E=100 MPa STD=18MPa *N(90%)=5.6E6*	Subgrade E=100 MPa STD=18 MPa *N(90%)=13E6*

Strategy P2

Initial Construction
5 cm 40/50 6000 MPa
E=19 cm 20/30 STD=1.8cm E=10000MPa STD=1000MPa *N(90%)=28.5E6*
30 cm ABC 200 MPa
Subgrade E=100 MPa STD=18 MPa ***N(90%)=14.0E6***

ABC – Aggregate Base Course
STD – Standard Deviation
>>>> - indicates "evolution of the layer moduli towards value in the next cell"
Values in *italic* are estimated number of ESALS (130kN) associated with the distress at each of the corresponding layers.
Values in **bold** are adopted (as most likely to occur) limiting conditions after each overlay or initial construction.

With 90 % reliability it could be inferred that strategy P1 would require an overlay at the end of 14 years. Then either a 6 or 7 cm overlay would have to be placed to insure with 90% reliability that the design life would be sufficient to insure another 18 years of structural life. Strategy P2 would require a 19 cm thick bituminous layer to insure that with 90% reliability the structural design life of 32 years would be achieved.

Figure 5 presents the variation of the remaining life with time for the two strategies.

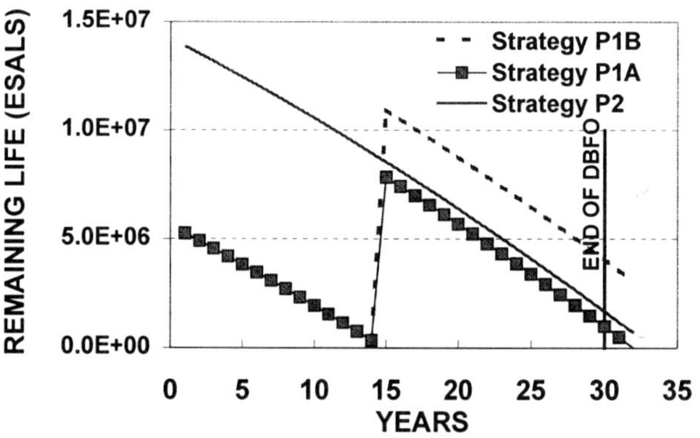

Figure 5 – Variation of remaining life with time for the two probabilities strategies considered to achieve design life with 90% reliability.

The Net Present value of these strategies is presented in Fig. 6. It can be observe that for discount rates below 7% strategy P3 is more cost effective from an agency costing perspective. However no considerations where made for user cost and delay costs. A chip seal treatment was also considered every 7 years to re-establish surface functional requirements.

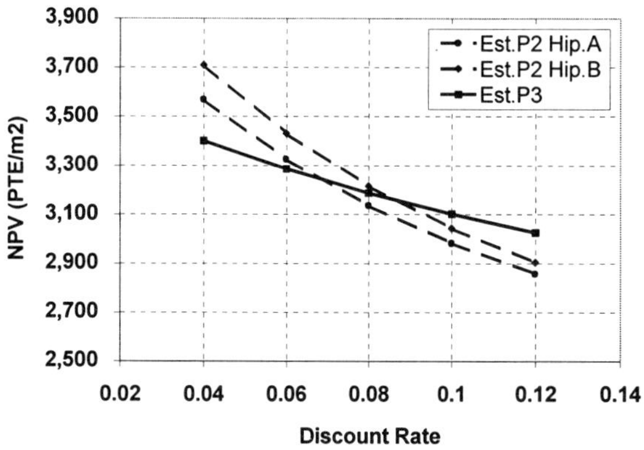

Figure 6 – Variation of Net Present Value for the Probabilistic Strategies over a range of discount rates

INVESTIGATION OF THE COSTS TO SOCIETY OF THE MAINTENANCE STARTEGIES

The evaluations made have only considered the direct costs to the agency. However in the award process the Government should take into consideration the costs to society of the different alternatives. This is a very complex issue. However, a simplified assumption was made considering the costs to the user caused by roughness. The cost model proposed by the World Bank [5] function of the IRI was used assuming the traffic was only composed of VW 1200. This is a truly conservative criterion given that the traffic composition is based on far more expensive vehicles. The evolution of the IRI function of traffic was assumed to be identical to the model proposed by the World Bank between 1.5 and 4. It was assumed that at the end of the design life the pavement would have an IRI value between 4 and 5. The variation of the IRI value between the initial construction until the overlay or end of design life would be an exponential function as presented in Fig. 7.

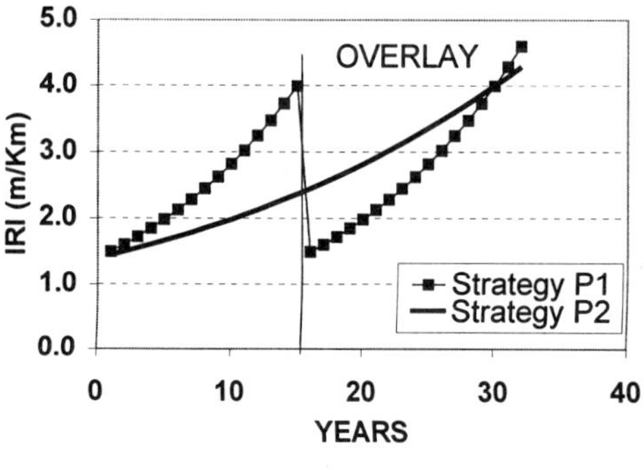

Figure 7 – Assumed variation of IRI between initial construction and overlay or end of design life for the two probabilistic strategies

amount RAP, both at room and at low temperatures, i.e., 22 and 0°C, respectively. Results from the room temperature experiments indicated an increasing trend of fracture toughness as a function of RAP content. A brittle fast fracture occurred at low temperature, and interestingly RAP had no significant affect on the performance of the base asphalt binder. The dynamic response of RAP was evaluated by using the Split Hopkinson Pressure Bar apparatus. The dynamic flow stress was found to increase with RAP content increase like compressive strength under quasi-static conditions. The dynamic flow stress was not affected by incorporating RAP at the low temperature.

INTRODUCTION

For over a century, paved roadways have been constructed using asphalt concrete mixtures in Rhode Island as well as across the United States. However, a major problem still exists in asphalt pavement involving premature distresses and failures, e.g., permanent deformation, fatigue cracking, and low temperature cracking.

Since asphalt binder is a thermoplastic material, its physical property depends on temperature and time (Roberts et al. 1991). In the early 1970s, many highway agencies have recycled the old pavements in the overlay or major reconstruction. Recently, the use of Recycled Asphalt Pavement (RAP) has significantly increased due to the protection of the environment, economy of construction/rehabilitation procedures, and the conservation of materials.

The Strategic Highway Research Program (SHRP) developed the performance-based specification for asphalt binder accompanied by the new system and testing procedures, as a component of "SUPERPAVETM," which stands for Superior Performing Asphalt Pavement. Six new binder testing equipment were recommended to measure the physical properties of modified as well as unmodified asphalt binders that can be related directly to field performance by engineering principles. Among six equipment, the Dynamic Shear Rheometer (DSR) was chosen to evaluate permanent deformation and fatigue cracking resistance characteristics by measuring the properties of asphalt binder at high and intermediate temperatures, respectively. Yet, the Superpave did not include the RAP mixture evaluation. Therefore, an attempt was made to evaluate RAP binder and mixtures utilizing the Superpave tool and to instill confidence to engineers for the use of the RAP binder in asphalt pavement.

Since SHRP performed a limited investigation related to fracture and crack propagation within asphalt mixtures, further mechanical characterization of asphalt binder and mixture is warranted. In the present study, single notched specimens were used for the fracture toughness testing. Besides, a split Hopkinson pressure bar (SHPB) equipment was utilized to characterize the dynamic constitutive behavior.

Specific objectives of the present study were:

1. To investigate the relationship between rheological properties and RAP asphalt binder percentage for evaluation of the permanent deformation resistance characteristics;
2. To investigate quasi-static behavior of asphalt binder containing RAP for the evaluation of permanent deformation resistance characteristics;
3. To investigate the relationship between rheological properties and RAP asphalt binder percentage for evaluation of the fatigue cracking resistance characteristics;
4. To characterize fracture toughness of the asphalt binder containing RAP binder; and
5. To characterize the dynamic response of the asphalt binder containing RAP binder.

PERMANENT DEFORMATION

Permanent deformation or rutting is a type of distress which can be found in the surface of asphalt pavement. Rutting is mostly occurred along the wheel path of the traffic. The surface cross section is no longer in its design position. It causes by many sources (e.g., traffic densification, abrasion, and underlying HMA weakened by moisture damage), and has two principal causes (May and McGennis 1996). Firstly, the rutting is caused by high stress being applied to the subgrade and/or granular subbase layer below the asphalt layer.

Secondly, the rutting results from the asphalt mixture with low shear strength to resist repeated heavy loads. Typically, rutting occurs during the hot summer under high pavement temperatures. Although aggregates play the major roles in causing permanent deformation, the soft asphalt binder can be contributing factor. Since rutting is an accumulation of very small permanent deformation, stiffer asphalt binder and mineral aggregates with high degree of internal friction are needed to increase shear strength.

Rheological Properties by Dynamic Shear Rheometer (DSR) Two base asphalts, i.e., AC-10 and AC-20 typically used in Rhode Island were blended with different amount of RAP binders, i.e., 0, 10, 20, 30, 40, 50, 75 and 100%. DSR tests were performed at Superpave high grade temperatures, i.e., 52, 58, 64, 70, and 76 C in accordance with the procedure of AASHTO TP5. These binders were aged using the rolling thin film oven (RTFO), and DSR tests were performed. Values of G*/sin were determined to evaluate the rutting resistance characteristics of binders for all temperatures.

The RAPs were procured from two asphalt mixing plants (C and L) in Rhode Island. Then, RAP binders were recovered in accordance with the procedure of AASHTO T170-93 at Rhode Island Department of Transportation (RIDOT) laboratory. The base asphalts were procured from a Rhode Island distributor, H. Virgin and RAP asphalt binders were heated at 135°C and 160 C, respectively.

The absolute viscosity was measured at 60 C (140 F) in accordance with the procedure of AASHTO T202-91. The Brookfield rotational viscosity at 135 C was recommended in the Superpave binder specification to determine the handling and pumping characteristics of the asphalt binder at the refinery, terminal, or hot mixing asphalt plant. The test was performed in accordance with the procedure of ASTM D4402. A SC4-21 spindle was used for the base asphalt binders; and SC4-27 was used for the blended asphalt binders.

The viscosity values of Plant C RAP binder were lower than the ones of Plant L RAP binder, and implies that the latter RAP binder is harder than the former. The characteristic of RAP binder is dependent on the age of the RAP. The older the RAP, the harder the RAP binder due to the volatilization and oxidation. However, the age of RAP was not available, because the original sources of the RAPs were not identifiable. The Superpave binder specification requires that the rotational viscosity at 135 C be less than 3 Pa-s. The viscosity values for some RAP binder contents do not meet the requirement. But, the specification states that the viscosity criterion may be violated, if the asphalt supplier warrants that the asphalt can be pumped and mixed at safe temperatures.

The Superpave binder specification requires the rutting factor, G*/sinδ to be a minimum of 1.00 kPa and 2.20 kPa for unaged and RTFO aged binders, respectively. The rutting factor reflects the total resistance of a binder to deform under repeated loading (G*), and the relative energy dissipated into non-recoverable deformation (sin δ) during the loading cycle. A higher value of G*/sinδ implied that the binder behaves more like an elastic material, which is desirable for rutting resistance. Since G*/sinδ for unaged binder requires a minimum of 1.00 kPa to resist tenderness during the lay down process in the field, these values were not considered seriously. Rather, the rutting resistance was evaluated mainly by examining G*/sinδ values of RTFO aged binders, because the aged binder simulated short-term aging, i.e., 1 to 2 years.

As expected, the values of G*/sinδ were increased as the content of RAP binder was increased at all temperatures. It was also observed that the binder with Plant L RAP exhibited the higher G*/sinδ values than the one with Plant C RAP at all corresponding temperatures. It may be noted that the slope for binder with Plant L RAP is stiffer than the one for binder with Plant C RAP.

The high temperature grading of AC-10 used in the present study was 58; and the values of G*/sinδ were evaluated. It was noticed that all values were higher than minimum 2.2 kPa. Similarly, all values at 64 C exhibited higher than the minimum requirement.

Quasi-Static Characterization Experiments with Uniaxial Compression Test The quasi-static experiments were conducted in accordance with ASTM D1074-93 to determine the compressive strength of the binders. The specimen was cylindrical in shape with 4 inches in diameter and 2 inches height.

The specimens were fabricated using the Marshall Compactor (Asphalt Institute 1993). The aggregates used for this study were of size -#30+#50 and were procured from two different contractors, i.e., contractors T and C. The aggregate size was so chosen that the aggregate would not contribute to the strength of the specimen and only binder characteristics can be studied (Kennedy et al. 1982). The virgin asphalt cement used was AC-20. The RAP was obtained from Plant C, and the RAP binder was extracted and recovered. The compaction procedure was in accordance with Kennedy et al. (1982). The heated mix of the binder and aggregates was compacted using the Instron testing machine under a uniaxial load of 6,000 lbs for 15 minutes. The compacted specimen was later ejected and cured overnight.

The specimens were tested in axial compression without lateral support at a uniform rate of vertical displacement of 0.05 in./min-in (1.25 mm/min-mm), i.e., for the present study at 0.1 in/min (2.5 mm/min).

A perceivable increase of over 100% was noticed in both stiffness (E) and σ_u for the 75% RAP and 100% RAP specimen. The addition of 25% RAP to the virgin binder increased the Young's modulus and compressive strength by 26%. The addition of 50% RAP increased the modulus marginally (7%) but increased the compressive strength by 31%. The addition of RAP however, did not cause a noticeable variation in the yield strain of the virgin binder. An aspect to note is the steady increase in the linear range of the stress-strain response when RAP content is increased in the binders.

The mechanism of failure in all the specimens was noticed to be shear dominated. The side view is indicative of shear failure in the specimen, wherein the wedge shaped segments have been removed from the central core. The angle of shear happens to be approximately 45^0. This was observed in all the specimens, irrespective of the RAP content in them. The top view of 100% RAP specimen shows the presence of circumferential cracks and also the formation of different segments along the periphery of the specimen.

FATIGUE CRACKING

Fatigue cracking is the phenomenon of fracture under repeated or fluctuating stress lower than the tensile strength in the asphalt mixture. The intermittent longitudinal cracks are the early sign of the fatigue cracking, usually occurs in the wheel path. Furthermore, the cracks will join and cause more cracks to form. The intermediate stage of fatigue cracking is called "alligator cracking", characterized by transverse cracks connecting the longitudinal cracks. In the severe case, the pothole forms when the aggregates and asphalt binder are no longer bonding. Generally speaking, hot mix asphalt (HMA) must have enough tensile strength to withstand the applied tensile stress at the bottom of the asphalt layer, and to be resilient enough to withstand repeated load applications without cracking.

Rheological Properties and Dynamic Shear Rheometer (DSR) Fatigue cracking typically occurs at normal service temperature about 7 to 10 years later of pavement construction. The pressure aging vessel (PAV) was used to simulate the long-term aging process of asphalt binder by means of pressurized air and elevated temperature. Pressure of 2.10 MPa is applied to age the RTFO residue at temperature of 100 C for 20 hours. Then, the DSR tests were performed on the aged binders to evaluate the fatigue cracking resisting characteristics of asphalt binder with RAP at 19, 22, 25, 28 and 31 C. The PAV aged binder is tested at intermediate temperatures to determine the fatigue cracking parameter, $G^*\sin\delta$.

The maximum limit for the fatigue resistance factor, $G^*\sin\delta$ has been set-up at 5,000 kPa on RTFO and PAV aged binders. The smaller the $G^*\sin\delta$ value, the more elastic the material and the better resistance to fatigue cracking. It was observed that $G^*\sin\delta$ was increased as the amount of RAP content increases. The increase in the $G^*\sin\delta$ value is mainly due to G*. Values of d did not change significantly for different amount of RAP. It may be noted that the AC-10 (PG 58-22) base binder containing 55% C RAP binder content at 22 C or lower temperature did not meet the fatigue resistance criteria. Therefore, it is not desirable to use more than 55% of Plant C RAP in terms of fatigue cracking resistance. Similarly, it is not desirable to use more than 25% of Plant L RAP. It was also observed that the AC-20 (PG 64-22) base binder containing 55% or higher of Plant C RAP did not meet the Superpave binder specification requirement at 25 C, and also the ones with 45% or higher of Plant L RAP did not meet the Superpave binder specification at 25 C. Therefore, it is not desirable to use more than 55% of Plant C RAP and 45% of Plant L RAP.

Fracture Toughness Characterization Fracture toughness (K_{IC}) is the value of the stress intensity at which the crack begins to propagate. In ASTM fracture toughness test, the blank specimen had a diameter of 101.6 mm, and had a width of 76.2 mm (D = 1.35W). The thickness of the specimen was 38.1 mm, as per the relation W/B = 2. For an initial crack length of 34.29 mm, and an a/W ratio of 0.45, the f(a/W) factor was calculated as 8.71.

The asphalt cement and RAP binder were heated to 160^0 C for an hour, till they started to flow. They were further mixed with 550 grams of heated aggregates. The weight of binder as a percentage of the total weight was established by performing preliminary tests and the maximum value of fracture toughness was found to occur at 6% binder content. This mixture was then compacted with 55 blows on either face to account for proper compaction. Finally, the specimen was ejected, cured overnight and machined to the required dimensions. The edge crack was band-sawed and was sharpened up to 1 mm using a diamond saw to provide a sharp crack. The Instron testing machine was used to load the specimen.

In order to study the rate dependency of the virgin asphalt binder, experiments were carried out to evaluate K_{IC} at different rates of loading from an 25.4 mm per minute all the way up to 254 mm per minute. A gradual increase in fracture toughness was noticed with an increasing rate. However, the values stabilized for rates over 152.4 mm per minute. Thus it was decided that further experimentation would be carried out at a displacement rate of 152.4 mm/minute. The maximum average fracture toughness obtained for the virgin binder was 78 MPa m. The error generated in data was within a band of ±10 % and well within tolerable experimental consistencies.

After determining the rate of loading to be used, further experimentation was carried out in a two phase manner. Firstly, the fracture behavior of binders with RAP was studied at room temperature, in order to evaluate fatigue cracking resistance. Secondly, the low temperature fracture characterization was conducted to examine the brittle fast failure, i.e., low-temperature cracking in the upper layers of the pavement.

Two specimens were tested at each increment of 25% RAP content and at the loading rate of 152.4 mm per minute. Also crack sharpening was consistently maintained in all the cases so as to obtain an accurate estimate of fracture toughness.

The trends from the experiments can be represented by a quadratic curve fit. The percentage increase in the fracture toughness values for the 25% and 50% RAP cases were 5% and 6.3 %, respectively when compared to the 0% RAP case. However, there was a significant increase of 71% in the toughness values when 75% RAP was added to the virgin binder. The maximum increase of 118% was noticed in the case of the 100% RAP specimen.

In order to understand this behavior better, the crack propagation in all the specimens was studied. A most common feature is the stable crack propagation leading to final arrest. Also the cracks seem to have propagated in an irregular manner. By a more careful observation of the 0% RAP specimen, it can be seen that the crack actually branched through a distance of 15.24 mm before finally arresting. The crack jump distance increased as the RAP content increased, and is perhaps indicative of reduced arrest toughness. Although stable crack growth was observed in all the experiments, the crack tended to propagate unstably in the 100% RAP specimen. Also the crack opening displacement seems to have decreased with RAP content. These visual observations are indications of plastic zones ahead of the crack front. In this vein it can be concluded that the addition of RAP is decreasing the ductility inherent in the binder and causing it become rather brittle.

Dynamic Characterization Using the Split Hopkinson Pressure Bar (SHPB) The dynamic response of asphalt to compressive stress pulses induced due to traffic loads plays a significant role in the flexible pavement structures.

The SHPB was used to study the dynamic behavior in the present study. The SHPB technique is a well established experimental technique used to study dynamic behavior in a both ductile and brittle materials alike (Lindholm and Yeakley 1968; Nemat-Nasser et al. 1991). A conventional SHPB or the Kolsky Bar consists of a striker bar, an incident bar and a transmitter bar (Kolsky 1949). The specimen under study is sandwiched between the incident and transmitter bar. The striker bar is launched at a predefined velocity towards the incident bar. This impact generates a compressive stress pulse which travels towards the specimen. The amplitude of the stress pulse is a function of the velocity of the striker bar, and its period is approximately equal to twice the travel time of the wave in the striker bar. This wave, upon reaching the incident bar-specimen interface, gets partly reflected back and partly transmitted into the specimen depending on the impedance mismatch and the area mismatch between the specimen and the bar. From one-dimensional wave theory, it has been established that the amplitude of the transmitted pulse is a measure of the stress in the specimen and the amplitude of the reflected pulse is a measure of the strain rate in the specimen. Thus upon integrating the reflected pulse, the strain in the specimen can be determined. The specimen can be subjected to a wide range of strain rates by employing striker bars of various lengths.

The dynamic characterization experiments were conducted by fabricating specimens having a diameter of 43.2 mm (1.729 in.) and a length of 0.125 mm (0.5 in.). The diameter was so selected so as to provide a cross-sectional area mismatch between the pressure bars (diameter 50 mm) and the specimen of 25%, assuming a maximum strain of 25% in the specimen. The specimen constituents were the same as the ones if fracture toughness testing and the specimen was compacted in accordance with the procedure of Kennedy et al. (1982).

Experiments were conducted to determine the dynamic response of virgin and RAP binders. The dynamic flow stress was evaluated at every increment in RAP content and at room temperature (~22^0 C). All the experiments were carried out at a nominal strain rate of 450 /s in order to establish a base for performance comparison. The nominal wave speed in asphalt, was estimated to be 211 m/s.

The flow stress increased as the percentage of RAP was increased. There was a nominal increase of 6% in the values of flow stress obtained for the 25% and 50% RAP specimen when compared to flow stress of 0% RAP. An increase of 12% in the flow stress values was noticed for the 75% RAP specimen. The largest increment of 26% was noticed in the case of the 100% RAP specimen.

The specimen failure was noticed to be shear dominated, much like the one observed in the quasi-static experiments. An examination of the specimen face (facing the incident bar) revealed the presence of circumferential cracks, radial cracks and longitudinal cracks (75% RAP specimen). A more predominant segment formation was noticed in these specimen as compared to specimen tested under static loading conditions. It seems quiet possible that cracking initiated because of circumferential stresses along the periphery of the specimen. As previously mentioned, these stresses may have been caused due to inertial and frictional effects. These reasons may also have been a factor leading to the formation of larger wedge shaped. The side view of the 75% RAP specimen indicates the shear dominant failure in the specimen. This type of failure was typical of all the specimen that failed at room temperature. These visual observations and trends from the dynamic true stress-strain plots are indicators of reduced ductility in the specimen as the RAP percentage increased. The observations from the quasi-static experiments accentuate this argument.

LOW TEMPERATURE CRACKING

Low temperature cracking is considered as the nonload associated cracking. It is caused by the environmental conditions rather than by applied traffic loads. It is characterized by intermittent transverse cracks that occur at the surface of asphalt pavement. Low temperature cracking forms when HMA pavement subjected to high cooling rates and low temperatures develop tensile stresses due to shrinkage. At some point along the pavement, the tensile stress exceeds the fracture strength of the asphalt concrete and transverse cracking develops. The cracking starts at the surface and progresses down with time.

Since HMA mixes have high stiffness at low temperatures, it is very prone to cracking. Mix stiffness at low temperatures is primarily dependent on the property of the asphalt binder because low temperatures do not have a significant effect on aggregates in the mix. Therefore, high asphalt binder stiffness at low temperatures is the predominant cause of this type of cracking. In addition, asphalt binder that are excessively aged, because they are unduly prone to oxidation and/or contained in a mixture constructed with too many air voids, are more prone to low temperature cracking.

Fracture Toughness Test at Low Temperature The methodology adopted in conducing these set of experiments remained much the same as the previous case except the low testing temperature ($0^{o}C$). Thus, the cured specimens were soaked in ice for a period of 24 hours, before testing. Three specimen each were tested at every increment of RAP content and an average of the results was reported. Interestingly, an increase of 30% was noticed in the K_{IC} values from these experiments as compared to that obtained for the 100% RAP specimen at room temperature. The data was fit by a flat trend line which signifies that the addition of RAP does not have any noticeable effect on the performance of the binder at low temperatures.

A brittle fast fracture occurred in all the specimens tested at this temperature. The crack propagation in the 100% RAP specimen is indicative of this fact. Interestingly, stable crack propagation leading to arrest was observed for the 100% RAP specimen at room temperature. The crack opening displacement was also found to be lesser at low temperature than what was observed at room temperature.

In the light of visual observations of crack propagation in the specimens at both temperatures the following can be concluded. The actual numerical values of fracture toughness may be rendered invalid due to the presence of plastic zones at room temperature. This is because fracture toughness is estimated based on the principles of linear elastic fracture mechanics. However, the K_{IC} values obtained at $0^{o}C$ are definitely valid as the binder behaves as a true brittle material at this temperature. Hence, the increasing trend of K_{IC} with RAP content can be considered to be one leading into brittle behavior. More importantly, it is plausible that there is a ductile to brittle transition in the behavior of the binder as the temperature is varied from room $22^{o}C$ to $0^{o}C$.

Dynamic Response of Virgin Asphalt and RAP Binders at Low Temperature Experiments were carried out at low temperature ($0^{o}C$) in order to determine the low temperature performance of RAP binders under dynamic loading conditions. These set of experiments were carried out in as much the same way as the room temperature experiments. The only difference was that the specimen was soaked in an ice bath for a period of 24 hours so as to lower their temperature to $0^{o}C$. These set of experiments were also conducted at a nominal strain rate of 450 /s in order to make a comparison of performance.

There was a marginal increase of 8% in the dynamic flow stress values for the 100% RAP specimen when compared to that of the 0% RAP specimen. The maximum strain (true strain) was also observed to be the same in all the cases. The specimens could not be recovered for postmortem analysis as all of them had fragmented upon impact. This observation is definitely indicative of brittle failure in the specimen.

CONCLUSIONS AND RECOMMENDATIONS

The following conclusion and recommendations can be drawn from the present investigation:

1. The viscosity of asphalt binder was increased as the amount of RAP binder increased. The rotational viscosity of 100% L RAP binder was very high, and it did not meet the limit of 3 Pa-s at 135 C. In this case, the Binder have to be pumped and mixed at higher temperature.
2. The values of G*/sin were increased as the RAP amount was increased. It was also observed that the binder containing Plant L RAP exhibited the higher G*/sin and steeper slope than the one containing Plant C RAP.
3. All values of G*/sin at 58 C for Ac-10 (PG 58-22) base binder met the minimum requirement, i.e., .1.00 and 2.20 kPa for unaged and RTFO aged binders, respectively. It was true for the AC-20 (PG 64-22) base binder.
4. The values of G*sin were increased as the RAP amount was increased, but the increase was due to mainly G*. It was observed that AC-10 (PG 58-22) base binder containing 55% or higher of Plant C RAP did not meet the fatigue cracking resistance criteria, and that the one containing 25% or higher of Plant L RAP did not meet the criteria at 22 C or lower temperatures. It was also observed that AC-20 (PG 64-22) base binder containing 55% or higher of Plant C RAP did not meet the fatigue cracking resistance criteria, and that the one containing 45% or higher of Plant L RAP did not meet the criteria at 25 C.
5. The modification of asphalt binder with RAP significantly affects the Superpave performance related properties of the binder, and generally raised the high temperature grade of the binders. Limited test results with the Superpave technology indicated that the increase in RAP binder appears to enhance the binder's resistance against permanent deformation and fatigue cracking.
6. For the quasi-static loading experiments, it was observed that the compressive strength and stiffness were increased as the amount of RAP increased. The increase in both compressive strength and stiffness reduced the ductility in the specimens as the amount of RAP was increased.
7. The fracture toughness was increased as the amount of RAP increased, at room temperature. However, it was observed that RAP contents had a significant effect on the crack propagation. The crack propagation tended towards instability as the amount of RAP

increases, which increased the crack jump distance in the specimen. In other words, RAP binder addition reduced ductility in a specimen.

8. The fracture toughness values of the binder at low temperature were higher than the ones at room temperature. Again, the crack propagation was seen as a brittle fast fracture in a specimen. However, the addition of RAP content did not have any significant effect on the performance of the binder at low temperature.
9. The dynamic flow stress values of 100% RAP specimen were increased by 25% when compared to the 0% RAP specimen. In the mean time, the dynamic response of the base asphalt binder was not affected by the addition of RAP at low temperature.
10. A comparison of results from the quasi-static and the dynamic experiments indicated an increase in flow stress of many orders of magnitude. Thus, the binder were found to be highly susceptible to rate of loading.
11. It is recommended that further research should be performed at low temperature, e.g. below 0°C to evaluate the low-temperature cracking resistance characteristics of binder with RAP.

REFERENCES

1. Annual book of ASTM standards. (1996). ASTM, Philadelphia, Pa.
2. Brown, R., Kandhal, P., Lee, D. and Lee, K.W. (1996) "Significance of Tests for Highway Materials," ASCE Journal of Materials in Civil Engineering, Vol.8, No.1, pp. 26-40.
3. Kennedy, T., Roberts, F. and Lee, K.W. (1982). "Evaluation of Moisture Susceptibility of Asphalt Mixtures Using the Texas Freeze-Thaw Pedestal Test," Proc. of AAPT. Vol. 51.
4. Kolsky, H. (1949). "An Investigation of the Mechanical Properties of Materials at Very High Rates," Philosophical Transactions of the Royal Society of London, Series B, Vol.62, pp 676-700.
5. Lindholm, U. and Yeakley, L. (1968). "High Strain-Rate Testing: Tension and Compression, "Experimental Mechanics, Vol.8, pp 1-9.
6. May, R. and McGennis, R. (1996) "Superpave Asphalt Mixture Analysis," The Asphalt Institute Research Center, Lexington, KY.
7. Mix-Design Methods for Asphalt Concrete and other hot mix types. (1993). Asphalt Institute MS-2, 6th ed., Lexington, KY.
8. Nemat-Nasser, S., Issacs, J. and Starrett, J. (1991). "Hopkinson Techniques for Dynamic Recovery Experiments," Proceedings of the Royal Society of London, Series A, Vol. 435, pp 371-391.
9. Provisional standards. (1996). Am. Assoc. of State Hwy. and Transp. Officials (AASHTO), Washington, D.C.
10. Roberts, F., Kandhal, P., Brown, R., Lee, D., and Kennedy, T. (1991). "Hot mix asphalt materials, mixture design, and construction." NAPA Education Foundation, Lanham, Md.
11. Standard specifications for transportation materials and methods of sampling and testing. (1993). 16th ed., part II, AASHTO, Washington, D.C.

Fifth International Conference on the
BEARING CAPASITY OF
ROADS AND AIRFIELDS

Recycled Bituminous Planings As Unbound Granular Materials For Road Foundations in the UK

P.R. FLEMING (Lecturer), Dept. of Civil and Building Engineering
Loughborough University, Loughborough, Leicestershire, LE11 3TU, UK
Tel: (0) 1509 222616, email p.r.fleming@lboro.ac.uk

ABSTRACT

Seven million tonnes or more of recycled bituminous bound material is produced annually, and current UK guidance allows its reuse as a capping layer in a road foundation as an unbound layer. However, concern has been expressed over its workability and subsequent performance. Important aspects of its mechanical behaviour are reviewed and discussed. The findings of a short laboratory study are presented from which a comparison of bituminous planings behaviour to that of a pure crushed rock (capping) and to that of a 50/50 blend is made. Results are presented from standard index tests and performance measurements, of strength and stiffness modulus, on larger samples compacted over a clay subgrade. These results demonstrate the relative strength and stiffness gain with increasing compactive effort, the viscoplastic nature of the bituminous planings, and highlight potential construction problems. Recommendations are made with regard to the use of BRP as a capping material and some requirements for effective site control.

Keywords: Foundation, Stiffness, Recycled, Bituminous, Aggregate.

INTRODUCTION

The current production of bituminous road planings, recovered during remedial works, is thought to be around 7 million tonnes per year for the UK. It would apear that it is increasingly being utilised in highway foundations as unbound granular material, and it is known that currently 12 Local Authorities are specifying its use. The potential economic benefits are obvious, and are greatest for insitu reconstruction and widening where minimal haulage is required. The UK demand for aggregates is expected to increase to around 400 million tonnes per year by the year 2000, and the recent introduction of a landfill tax and the likely introduction of a quarried aggregate tax will increase pressure on the use of this and other recycled materials for road and other construction.

The highway designer is faced with the prospect of utilising these recycled bituminous planings (RBP) for which relatively little research is documented, and whilst site specific trials have been undertaken the applicability for its successful use on other sites, for different site conditions and for the long-term is not yet proven. The current UK Department of Transport advice[1], for new motorways and trunk roads, restricts the use of planings to the capping layer only. The capping layer is utilised as a construction expedient for when the subgrade strength (CBR is still specified) is deemed insufficient to adequately place and compact the sub-base layer above. The original capping specification classes, fine (6F1) and coarse (6F2), have been expanded to accomodate this new material (termed 6F3). However, the RBP are treated as equivalent to these conventional material classes and have to comply with the same guidance comprising particle size (grading) limits, 10% fines and achieve a minimum laboratory CBR of 15% for internal stability. The UK operates a method specification, which controls the compacted layer thickness and the type of plant and number of passes required, and as such there is no requirement for any compliance assessment insitu.

The capping layer is often utilised as the works haul road, but is primarily intended as a more stable platform for the upper layer construction. It must also remain stable in the long-term under in-service conditions. Whilst the subgrade CBR is still utilised for design, it is widely recognised[2] that the parameters controlling the behaviour of pavement materials are stiffness modulus and resistance to permanent deformation (strength related), relating to load spreading ability and rutting respectively. Some concerns have been expressed, however, regarding the compactability of planings, resulting in potentially high air voids, and the facility for relatively large permanent deformations under repeated load. Large deformations under construction trafficking and under in-service conditions have been observed, with costly remedial measures. An end product specification may have highlighted this potential problem before work progressed, however.

One major contractor's approach has been to use the BRP blended with good quality crushed rock mixed during the planing process, by spreading crushed rock onto the existing pavement, to avoid further handling. A ratio of 1:1 has been utilised with success on a major motorway widening and reconstruction scheme. It is apparent, however, that the ideal blend will be dependent on the nature of both the BRP and the mixing aggregate. The BRP behaviour will be affected by its constituents including the distribution of particle sizes, the binder content and stiffness and the temperature, both during compaction and subsequent operation. The method of planing will control the particle size distribution. The original binder used will have aged, i.e. oxidised, and consequently stiffened. The BRP moisture content would be expected to be relatively low, possibly reducing workability. The many controlling variables perhaps make the method specification more difficult to apply successfully, and would tend to promote a performance related specification to allow greater flexibility combined with some assurance. However, a greater understanding of the controlling influences on behaviour, and the limits for acceptability, are then required.

FIELD OBSERVATIONS

One recent investigation[3] reported the results of two full-scale trafficking trials of two types of bituminous bound recycled material, in comparison to suitable crushed rock, utilised as sub-base. The bituminous materials comprised a granulated asphalt (GA) produced from waste at source (and blended with some BRP), and bituminous recycled planings (BRP) from a trunk road remediation scheme. The authors' had concluded that, based on the results of permanent deformations under trafficking, the performance was no worse for the recylced materials than the crushed rock, thus suggesting they are suitable for use as sub-base. This constitutes a major piece of work and a brief review is included to highlight several significant factors. The sub-bases (designed at 150mm thick) were laid upon sand and gravel cappings, which were categorised by the Falling Weight Deflectometer (FWD) from which areas of equivalent stiffness (140-180 MPa) were identified for subsequent comparison. The Granulated Asphalt (GA) trial used a crushed limestone sub-base for comparison, and the BRP trial a crushed granite sub-base. Both recycled materials showed a deficiency in fines when graded compared to the specification limits for sub-base. FWD measurements on the completed foundations for the BRP trial showed no improvement over the formation (top of capping) stiffness and in some cases a slight reduction to between 75-125 MPa. This was attributed to rain whilst the capping was exposed, after stiffness assessment, and prior to sub-base construction. At the GA trial, an improvement was generally seen with values in the range 145-200 MPa. Under trafficking, the BRP trial foundations produced surface deformations of 40mm after 100 vehicle passes, interpreted as 400 standard axles (SA), despite sub-base insitu densities measured at 102% of laboratory derived optimums for both the crushed rock and the BRP. The GA trial survived 200 vehicle passes (interpreted as over 1000 SA), with insitu densities measured at 92-95% of the laboratory derived optimums.

These results show the inadequacy of density as a predictor or safeguard of performance. The BRP trial shows the importance of substrate support, especially for relatively thin upper layers, for permanent deformation behaviour. However, excavations into the capping below showed in some instances rutting was within the sub-base alone suggesting internal instability, due perhaps to low stiffness and strength in the layer itself despite the high density insitu. For the GA trial, the capping stiffness was high in relation to recent work[4] on live sites (which typically yielded values of between 50-100MPa). This factor, in combination with a layer thickness of 150mm, provided both good support and thus efficient use of the compactive energy. The lack of potential for yielding in the capping during trafficking was further demonstrated in the GA trial by a significant increase in penetration resistance of the DCP, from measurements both before and after trafficking, showing strength gain under the kneading action of the vehicle rubber tyres. The substantial difference in resistance to permanent deformation behaviour of a low quality aggregate when compacted over a semi-rigid substrate, when compared to a soft clay, has been clearly demonstrated previously[5].

These observations suggest that the performance of the recycled bituminous materials (and the crushed rock) is largely dependent on adequate support during compaction to provide internal stability and to achieve high stiffness. Other work[6] has suggested a maximum compacted layer thickness of 150mm, and indicated that recycled bituminous materials exhibit workability problems. Used as a capping for a low strength subgrade, the support will be relatively poor leading to poorer performance .The current guidance[1], that capping can be constructed in single layers of 250mm thick, will reduce compaction efficiency further. Indeed, it could be argued that the recycled bituminous materials are more suited to sub-base layers whereby a better reaction will be available and the material is then likely to achieve better internal stability and stiffness. However, a strong argument exists for end product testing based on these discussions.

INVESTIGATION OF BRP BEHAVIOUR

In light of these and industry field observations, the following brief investigation was carried out to compare a BRP to a standard quarry won crushed rock capping, and to an equal blend of the two materials. The investigation was carried out in two stages, the first comprising some standard laboratory measurements aimed at classification, workability and strength, comprising particle size distruibution (grading), compactability and CBR. The second phase comprised preparing and compacting the three test materials onto a uniform clay subgrade, in a small test pit out of doors. Measurements were then carried out of resistance to penetration by the DCP, and of stiffness modulus by both the conventional static Plate Bearing Test (300mm diameter) and the more recently available German Dynamic Plate Bearing Test(GDPBT)[4]. The GDPBT provides a transient (18millisecond) load pulse to a 300mm bearing plate giving a peak surface stress of 100kPa, and measures the plate deflection. The three type of material, the crushed rock, the 50:50 blend, and the 100% BRP, are hereafter referred to as 'Rock', 'Blend', and 'BRP' respectively.

Laboratory Tests and Results

The sample preparation and testing procedures carried out were in accordance with the current British Standards. One exception[6] was the measurement of moisture content of the Blend and BRP by oven drying at a temperature of 40-45 degrees centigrade for a period of 48 hours. Mixing of the materials was carried out by hand for adding water and poducing the Blend. The gradings were carried out by wet sieve analysis, and the compaction by vibrating hammer into CBR moulds for subsequent testing. Additionally, the temperature of the samples was measured both during preparation and testing at 16-18 degrees centigrade. Repeat testing was carried out to reduce experimental errors.

The grading results are shown in Figure 1, and the envelope for 6F1 is shown for comparison. It was also visually observed that the BRP particles were, in general, less angular and had a smoother surface texture. The BRP complied with the specification, and had a greater proportion below 10mm than the quarried crushed rock capping. The results of several compactability tests for two materials are presented in Figure 2, with the percentage air voids lines overlaid. These are based upon particle density results of 2.40Mg/m^3 for the BRP and 2.69Mg/m^3 for the crushed rock. The BRP optimum moisture content appears to be slightly higher than for the crushed rock During testing it was found to be impractical to achieve moisture contents greater than this. as excess water drained readily during compaction. Previous work[6] recommended a water content of 2-4 % above optimum for compaction in the field, acknowledging that any excess has to be catered for. This is expected to be difficult to achieve in practice and could be a potential problem for clay subgrades. Releasing significant quantities of water out of the base of the capping, with subsurface drainage not fully operational in most circumstances, could lead to softening and hence promote permanent deformations under traffic. From the compaction tests a target optimum moisture content for all the CBR tests was chosen as 6% (with 5-6% achieved) for the Rock, Blend and BRP, and with several repeat tests. The average values for these tests are presented in Table 1. The data for both the 2.5mm and 5mm penetrations are shown and with the resilient deflection measured during unloading, presented as an Elastic Modulus (by using the standard plate bearing test analysis for an elastic half space). The peak load applied before unloading is also shown for further information.
These results show a significant reduction in strength with increasing content of BRP. The confining effects of the mould appear to influence the increase in CBR from 2.5 to 5mm penetration for the Rock, as might be expected, whilst the Blend and BRP remain relatively constant showing plastic behaviour. According to the UK specification, these results suggest they all have sufficient strength as a capping, or indeed as a sub-base layer. Whilst the inferred stiffness are unrealistic, they indicate a similar trend to the CBR results, and a stiffness ratio for the BRP to Rock of about 4. The Blend results are closer to the BRP than the Rock, perhaps suggesting the BRP dominates the behaviour beyond a simple proportionality relation. However, these results may also be affected by the Rock having a higher proportion of particles in the 10-20mm range.

Test Pit Results

The clay subgrade measures 3.8m by 3.8m in plan, and around 700mm deep, overlying a 300mm granular drainage layer with a water table maintained at around 300mm below the clay surface level. It was characterised over its area, by the DCP, as being of between 4-5% CBR (30-40mm/blow), and of stiffness of 8-10MPa by the GDPBT,and three bays were then marked out. The natural water content of the crushed rock was measured at 4%, slightly below the laboratoy derived, however, the BRP was measured as 2.3% which was much lower than the laboratory derived optimum. However, adding around 5% water to the BRP, or more to allow for losses, was considered impractical to control based on the laboratoty experience and as such was not adjusted. The Blend was thus measured at 3%, as expected. A small vibratory plate compactor was used for compaction. A lower layer thickness of 100mm was chosen, receiving three passes, whilst the upper layer was set at 200mm to further study the compaction efficiency at intervals, by the DCP and GDPBT, and received a total of 18 passes. Final thicknesses were within 5% of these values. After completion of the foundation static plate bearing tests were carried out for comparison to the dynamic plate tests. The results are presented in Table 2.

The DCP results presented are representative of the middle 100mm. The results clearly show an increase in resistance with the number of passes of the compactor, as expected. It would appear, in general, that the Rock achieved internal strength more quickly than the Blend or BRP, and that after 9 passes these latter two behaved similarly. After completion of all the plate testing, repeat DCP measurements were made back through the full construction thickness, and the average layer results were 18mm/blow for the Rock, 26mm/blow for the Blend, and 25mm/blow for the BRP, which correspond to 11%, 7% and 7% respectively. This ranks the Rock most stable, although the values are relatively low in comparison to site experience and very low in comparison to the laboratory values. However, the CBR results inferred from the static plate bearing test (based upon the pressure at 1.25mm deflection on loading) suggests that the difference between the materials is more significant. It further shows some improvement between cycles 1 and 2, and would indicated the Blend and BRP to be of very low strength.

The stiffness results, from the GDPBT, also show a trend with compaction and an asymptote. For the first layer most improvement is seen for the Rock, and allowing for slight variation in the subgrade the Blend and BRP improve by a similar amount. With the addition of the second layer, however, a similar pattern of increasing stiffness is seen for all three bays, and for the Blend and BRP the final results are marginally higher than the Rock. The final stiffness values are 2-3 times those of the subgrade, which would appear appropriate. From extensive field stiffness measurements on live construction sites[4], a relatively consistent correlation of GDPBT stiffness to FWD stiffness was observed to be around 0.6. From this it can be suggeted that FWD stiffness for the three bays would be between 40-48MPa, which is not inconsistent with the field measurements. The static plate results, however, highlight some interesting points regarding the materials' behaviour. This test is a maintained load test, and the readings of deflection for any one pressure increment (four were used) are recorded when they become stable. For granular soils this occurs quickly. However, it was observed for the Blend and the BRP that the time taken to reach a stable plate deflection was at least four times longer than for the Rock. This effect was observed to increase the deflection, compared to that seen immediately upon application of the pressure increment, by around 15% for the Blend and 33% for the BRP. This viscoplastic effect is understood and allowed for when testing bituminous bound surfacing materials. The effects are clearly seen in the static plate results. For the Rock, the stiffness modulus (Eo) is seen to correlate well with that of the GDPBT, assuming a similar correlation to that of the FWD. As expected, the modulus on loading (Ev) increases from the first to second cycle and to around that of Eo showing elastic stability. The Blend and BRP results, however, show relatively lower stiffness and a significant difference between the second cycle loading and unloading moduli, suggesting continuing plastic behaviour. The Blend result for unloading stiffness (Eo) does, however, seem to be significantly influenced by the crushed rock content. Thus, the static plate results give a differing indication for the relative competence of the three materials in comparison to the dynamic test equivalent and is significant for field assessment testing purposes.

The anecdotal site problems of using recycled bituminous materials (RBM) as capping are reinforced by this work and mainly concern the adequate compaction and subsequent performance, both under construction traffic and in-service loading. The construction traffic loading case is the most severe, and depends upon whether the capping forms a haul road for the works, as is often the case in the UK. If so, potential stability problems will quickly be realised. It would appear that, in comparison to a good quality crushed rock, the RBM will require significant watering and water content control during compaction to achieve suitably low air void content, and/or a greater compactive effort to fully develop its strength and stiffness. Field observations have shown that density alone is no guarantee of stability during trafficking, especially for thinner layers, where the subgrade support is vital. This is true for adequate compaction also, and the benefits of greater compactive effort, or thinning of the permissible compacted thickness to below 250mm may not be fullyrealised on a soft subgrade, certainly for the first layer of capping. The deformations under construction haulage vehicles may also be significantly affected by both their speed and the ambient temperature. If the capping is not used as a haul road and the sub-base is placed and compacted, to the current method specification, it is plausible that under the in-service loading condition, further plastic straining could occur leading to loss of support to the upper layers and fatigue. As a result of these influencing factors, and the singnificant risks associated, these materials have been used by blending with proven materials with success, although some long-term monitoring might be prudent if significant thickness have been laid. The results detailed herein showed the effects of a 50:50 blend to be difficult to predict, but performance was significantly enhanced over pure BRP, and may be suitable for many sites where a deficit of BRP to capping material occurs. It is recommended that a field trial be conducted to establish suitability and that field monitoring, by insitu measurement of strength and stiffness, at that time would aim to define performance requirements for the construction as a whole. In this way, the use of relatively simple site assessment equipment as described earlier would assist in the assurance of both uniformity of construction and guard against long-term stability problems. This is the theme of ongoing research at Loughborough.

CONCLUSIONS

Recycled Bituminous Material are currently being used in road foundations as equivalent to traditional capping, and in some cases sub-base, although some concern has been expressed over compactability and adequate performance. Amendments to the method specification are required to assist the introduction new materials.

A laboratory study has shown the relative behaviour of a good quality crushed rock, a 50:50 blend, and pure planings with respect to workability, strength and stiffness. The results ranked the performance of the planings below that of the rock. Additionally, the BRP demonstrated viscoplastic behaviour under loading, andpoor stability when compacted onto a soft subgrade. The stiffness behaviour was also seen to be dependent on rate of loading with implications for site assessment purposes.

ACKNOWLEDGEMENTS

The author is very grateful to Mr Duncan Hill, who carried out the laboratory work as part of his undergraduate project studies.

REFERENCES

1. MCHW, "Specification For Highway Works", HMSO, London 1991.

2. Fleming, P.R. and Rogers, C.D.F.," Assessment of Pavement Foundations during Construction", Transport, Proc. of the I.C.E., Vol. 111(2), May 1995, pp.105-115.

3. Chaddock, BCJ, Coyle, T. and Earland, MG.," The Performanse ofRecycled Bituminous Bound Materials as Unbound Granular Sub-base", UNBAR4, Dawson & Jones, Nottingham University, 1995.

4. Shahid M.A., Thom N.H., Fleming P.R., " In situ Assessment of Road Foundations", The Journal of the IHT & IHIE, Vol 44, No 11 September 1997, pp15-17.

5. Fleming P.R., Rogers C.D.F., " Assessment of Permanent Deformation Characteristics of Four Granular Foundation Materials Insitu", BCRA 4, July 1994, Vol 2, pp 967-988.

6. Berg, F., Milvang-Jensen,O. and Moltved, N.,"Crushed Asphalt as Unbound Base Material", Danish Road Laboratory Report No.69, Danish Road Directorate, 1992.

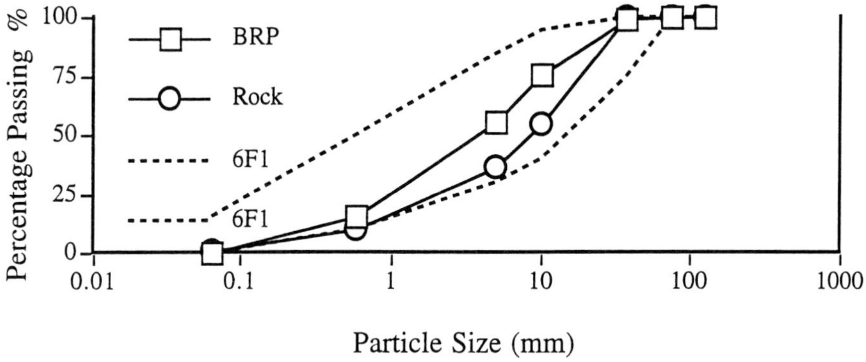

Figure 1. Particle Size Distributions

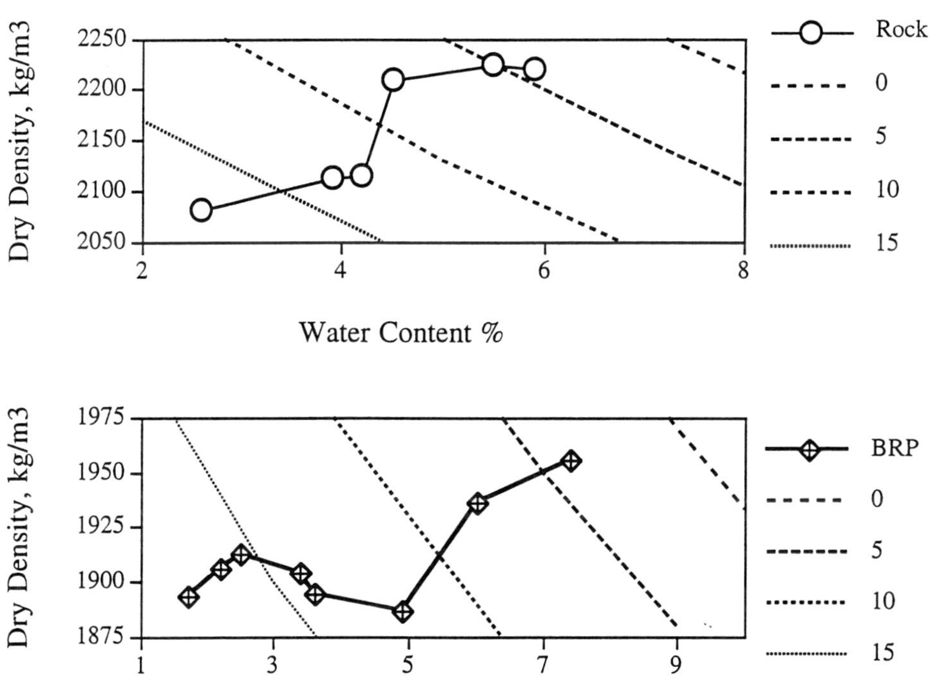

Figure 2. Compactability Results with 0-15% Air Voids Lines

Table 1 Laboratory CBR Results

Material type	CBR (%) at plunger penetration of 2.5mm	CBR (%) at plunger penetration of 5mm	Maximum Force (kN)	Elastic Modulus (MPa)
Rock	182	235	50	1735
Blend	71	80	22	598
BRP	30	30	7	430

Table 2 Test Pit Results

Layer/ No. Passes	Rock Capping		50:50 Blend		BRP	
	DCP mm/blow	GDPBT Ev (MPa)	DCP mm/blow	GDPBT Ev (MPa)	DCP mm/blow	GDPBT Ev (MPa)
Subgrade	30	8	44	10	40	8
Layer 1						
1	--	13	--	12	--	11
2	--	15	--	14	--	13
3	--	17	--	14	--	12
Layer 2						
1	60	17	90	17	105	16
3	53	22	55	23	90	23
6	50	20	50	26	70	30
9	45	26	55	24	60	25
12	30	22	50	27	40	24
18	26	24	35	29	35	27
Static Plate Bearing Test Results (on completed foundation)						
	Load Cycle No.					
	1	2	1	2	1	2
Ev (MPa)	25	38	14	17	8	11
Eo (MPa)	37	37	30	30	15	19
CBR (%)	11	17	4	6	2	3

Note: 1. Static Plate Ev (loading) caculated for 100kPa contact pressure.
2. Static Plate Eo (unloading) calculated from peak pressure.

Fifth International Conference on the
BEARING CAPASITY OF
ROADS AND AIRFIELDS

LABORATORY INVESTIGATION INTO RECYCLING OF CRUMB RUBBER MODIFIED R.A.P.

BYRON E. RUTH[1], MANG TIA[1] AND JAMES C. SETZE[2]

[1]Professor and [2]Graduate Student, University of Florida
Dept. of Civil Engineering, P.O. Box 116580,
Gainesville, Florida 32611-6580, USA

ABSTRACT

An investigation was conducted to evaluate the effects of using RAP containing Crumb Rubber Modifier (CRM). Simulated recycled mixtures containing different types and amounts of CRM were prepared and tested in the Gyratory Testing Machine (GTM). The results of these tests indicated no significant differences in shear strength although a slight increase in air void content occurred with higher CRM contents.

Key words: CRM, RAP, GTM, Recycling

BACKGROUND

Concern over the ability to recycle mixtures containing CRM-RAP and its influence on compactibility initiated this research. Currently the Florida Department of Transportation specifies 12 percent CRM blended into the asphalt using the wet process. This material is specified for open-graded friction courses (FC-2) used on Interstate and primary highways. Also, their specifications limit recycled mixtures to 50 percent RAP but typically the amount is between 20 and 40 percent. Therefore, the CRM contents of the simulated recycled mixtures selected for this investigation were 0, 4, 6, and 9 percent. Complete documentation of this research is presented in the Final Research Report (1).

MATERIALS AND PROCEDURES

The FC-2 RAP gradation was based on the estimated degradation produced by milling. A 50/50 blend of RAP/virgin aggregates were used to meet the gradation requirements for a structural (S-1) mixture. This was a relatively fine-graded mixture having a maximum aggregate size of

19 mm with 53 percent passing the 2.0 mm sieve and 5.5 percent total mineral filler (minus 0.075 mm).

An AC-30 with 12 and 18 percent CRM contents were prepared in a high shear blender. The FC-2 aggregate was heated and mixed with 6.5 percent of the different CRM asphalt blends. Then the hot virgin aggregate and an appropriate amount of AC-30 were added to the RAP to maintain a binder content of 6.5 percent. The hot-mix was compacted and subsequently tested in the Gyratory Testing Machine (GTM) using 18 revolutions for simulation of field compaction and 200 revolutions at 60C to simulate traffic densification. The GTM settings used for evaluation of the mixtures were:

Initial Angle of Gyration	:	3-degrees
Initial Air Roller Pressure	:	62 kPa
Ram Pressure	:	690 kPa

The age-hardening mixture preparation and testing procedures were the same except for that the RAP was heated in a convection oven at 85C for 14 days prior to mixing with virgin aggregate.

EVALUATION OF MIXTURES

Overview Increasing percentages of CRM, rubber particulate size, and simulated age-hardening were evaluated to determine their effect on a recycled mixture. Six samples were compacted for each test condition, two of which were used to determine the Rice density. The remaining four were densified to determine the physical properties of the mix. Analysis of Variance (ANOVA) and Duncan groupings were performed for gyratory shear based on level of compaction densification, the percent of CRM in the binder and the CRM size.

Definition of key terms

- As Compacted: The specimen after 18 revolutions of hot mix compaction in the Gyratory Testing Machine (GTM); Compaction Level 1.
- Initial Densification: The specimen at 5 revolutions at 60C; Compaction Level 2.
- Final Densification: The specimen at 200 revolutions at 60C; Compaction Level 3.

Effects of variables on gyratory shear Table 1 shows the results of ANOVA for the G_s values at all three levels of compaction, as compacted and initial and final densification. The ANOVA uses a linear model with four CRM types, four CRM percentages, three levels of compaction, and their interactions. Examining the individual results of the model, it became clear that all three independent variables, CRM type, CRM percentage and level of compaction were significant. It was also evident that there was interaction between the CRM percent and the compaction level. All other interactions have alpha values greater than 0.05 and are considered insignificant.

Table 1. Analysis of Variance Test for G_s Values

Source	DF	F Value	Alpha
• Linear Model	29	12.18	.0001
• CRM Type	3	3.50	0.018
• CRM Percent	2	5.91	0.004
• Compaction	2	143.77	0.001
• Type/Percent Interaction	4	2.10	0.086*
• Type/Compaction Interaction	6	0.95	0.460*
• Percent/Compaction Interaction	4	3.45	0.010
• Type/Percent/Compaction Interaction	8	1.93	0.064*

* : Results with alpha values of greater than 0.05 are insignificant.

The most significant factor for gyratory shear resistance was the level of compaction. This is supported by Figures 1 to 3 which graphically show the increase in G_s for increasing levels of compaction. To evaluate each independent variable, Duncan groupings were performed. Due to the interaction of the CRM percentage and level of compaction, these Duncans' were run for each cross-referenced condition to remove the interaction.

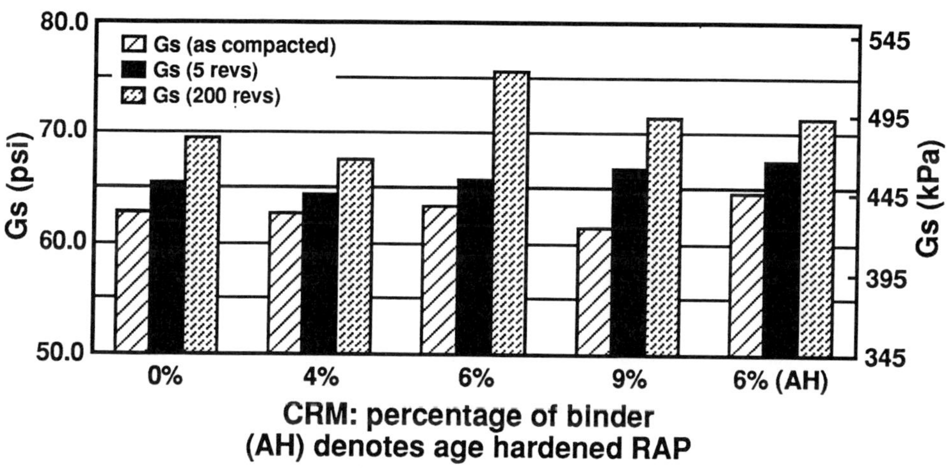

Figure 1. Mean Gyratory Shear Values for S-1 Mixtures with Degraded RAP Containing GF-80 CRM.

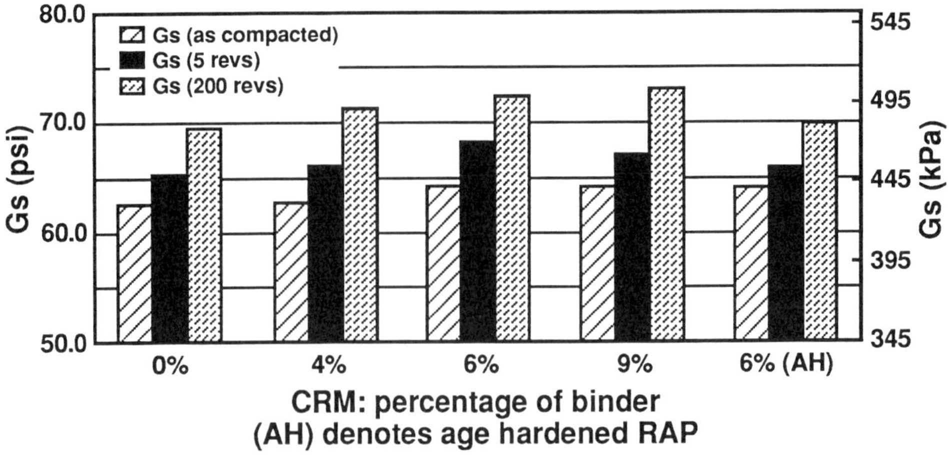

Figure 2. Mean Gyratory Shear Values for S-1 Mixture with Degraded RAP Containing GF-40 CRM.

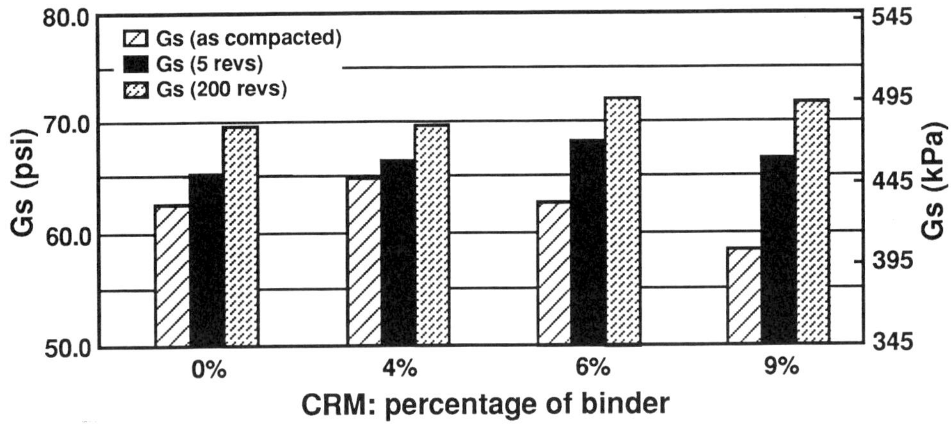

Figure 3. Mean Gyratory Shear Values for S-1 Mixtures Containing Degraded RAP with GF-20 CRM.

As the CRM type did not interact with either the CRM percentage or the level of compaction, a single Duncan grouping, was performed to determine the effects of CRM type on the G_s. Table 2 indicates that there were differences between both the CRM types and the control samples and between the GF-40 and all other CRM types. Looking at the G_s means, however, it was observed that even though the Duncan grouping indicated variation, the difference between the high and low mean was less than 14 kPa, which is within expected experimental error for the GTM. Therefore, it may be concluded that the CRM type did not affect the G_s.

Table 3 presents the effects of CRM percentages on G_s for each level of compaction. Duncan groupings indicated that the percentage of CRM did not affect the G_s at compaction Levels 1 and 2. At final densification, (compaction Level 3), the Duncan grouping given in Table 3 indicated an increase in G_s between the 4 and 6 percent CRM content levels.

Table 2. Duncan Groupings for Rubber Type

Type CRM	Number of Samples	Mean G_s, kPa	Duncan Grouping	
0	24	452		B
GF-20	33	460	A	B
GF-40	36	466	A	
GF-80	36	459	A	B

Table 3. Duncan Groupings for Percent Rubber (CRM)

Level of Compaction	Percent CRM	Number of Samples	Mean G_s (kPa)	Duncan Grouping
Level 1	0	8	432	A
(As-Compacted)	4	12	436	A
	6	11	437	A
	9	12	423	A
Level 2	0	8	450	A
(Initial Densification)	4	12	452	A
	6	11	464	A
	9	12	461	A
Level 3	0	8	479	B
(Final Densification)	4	12	479	B
	6	11	506	A
	9	12	496	A

The results of the Duncan groupings on the level of compaction by CRM percentage are shown in Table 4. It is evident that the compaction level influences the G_s for each CRM percentage. It is also clear that the G_s increased with increasing compaction and when the testing temperature was reduced. The primary difference between compaction Level 1 and 2 is due to changing from a hot-mix temperature (165C) to the densification test temperature of 60C. The density difference between as-compacted (Level 1) and initial densification of 5 revolutions (Level 2) is relatively small and since the G_s is unaffected by CRM content, the difference in G_s can be attributed to binder viscosity.

Air void content effects Figures 4, 5, and 6 are presented to further investigate the significant effects of percent CRM and compaction level as given in Table 1. These figures illustrate the effect that CRM and compaction had on the air void content. In general the air void content varied no more than one percent from the control specimen for both the as-compacted and final densification levels. The air void content appears to increase slightly with increasing percent CRM.

Table 4. Duncan Groupings for Level of Compaction

Percent CRM	Level of Compaction	Number of Samples	Mean G_s (kPa)	Duncan Grouping
4	3	12	479	A
	2	12	452	B
	1	12	436	C
6	3	11	506	A
	2	11	464	B
	1	11	437	C
9	3	12	496	A
	2	12	461	B
	1	12	423	C

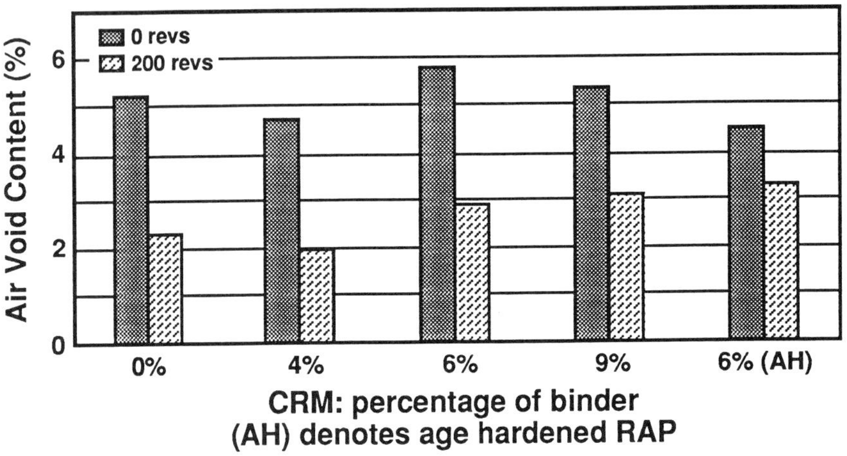

Figure 4. Mean Air Void Contents for S-1 Mix with Degraded RAP Containing GF-80 CRM.

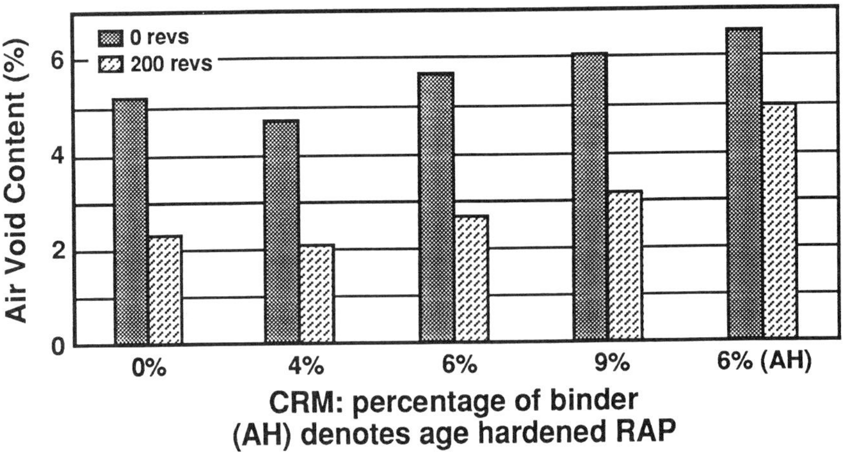

Figure 5. Mean Air Void Contents for S-1 Mix with Degraded RAP Containing GF-40 CRM.

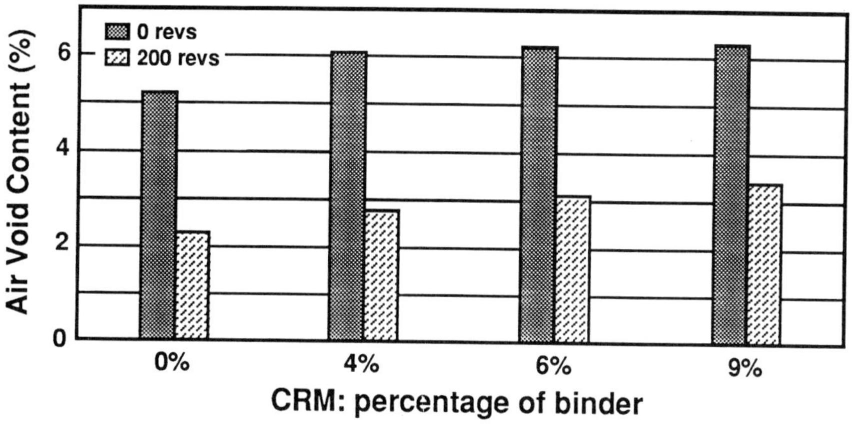

Figure 6. Mean Air Void Contents for S-1 Mix with Degraded RAP Containing GF-20 CRM.

The variation of air void content at different CRM contents is about the same for both as-compacted and densified conditions. Since the G_s was unaffected by CRM content (Table 3) it can be concluded that the air void content variation for the as-compacted (0 rev.) conditions did not affect the G_s values. Consequently, it could be assumed that air void content would not affect the G_s of densified (Compaction Level 2) specimens. However, there was a significant difference in G_s between the 0 to 4 percent and 6 to 9 percent CRM contents for the 200 revolution tests (Table 3). It was apparent that the increase in binder viscosity for CRM contents of 6 percent and more increased the Air Void Content of these densified mixtures (Compaction Level 3).

Age-hardening effects Figures 1 and 2 show the mean values of the 6 percent CRM age-hardened samples in conjunction with non-age-hardened samples. The striking similarity in both graphs is that the age-hardened Gyratory Shear means for GF-40 and GF-80 are very similar to the control samples for as-compacted, initial and final densification conditions. The gyratory shear values for the GF-80 mixtures are about 14 kPa greater than the control samples. The GF-40 shows a smaller increase in G_s, approximately 7 kPa for each condition. Both CRM sizes match so well to the non-age-hardened data that it appears the addition of age-hardened RAP did not affect the shear strength of the mix.

The largest effects of age-hardening are found in the air void contents. Figures 4 and 5 show that the two CRM types exhibit different behavior in both the as-compacted and densified air void content values. With respect to the GF-80 with 6 percent CRM, age-hardening resulted in a substantial decrease of the as-compacted air voids, more than 1.5 percent. Similarly, an air void content difference of 2.3 percent existed for the densified 6 percent GF-40 mixtures shown in Figure 5. The most plausible explanation for these differences was that increased absorption resulted in higher Rice density values which altered the air void content. Figure 7 illustrates the air void contents based upon the Rice density before and after age-hardening. The greater absorption of binder into the aggregate both increased the Rice density and air void content of the recycled S-1 mix containing age-hardened RAP.

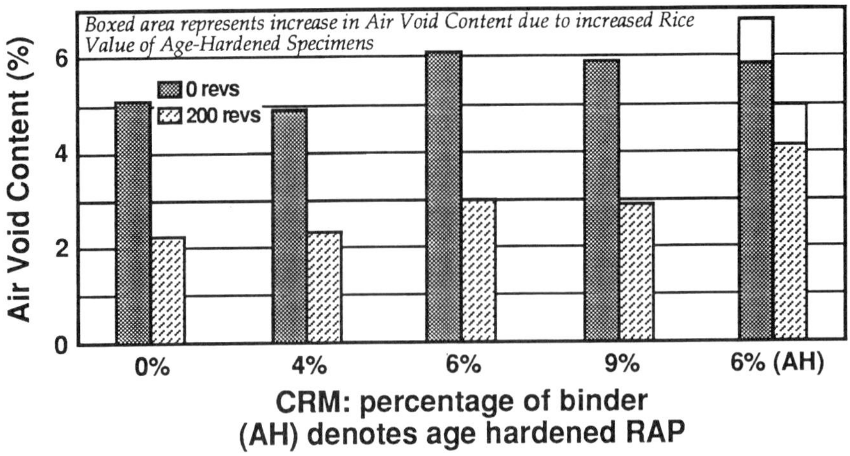

Figure 7. Adjusted Mean Air Void Contents for S-1 Mix with Degraded RAP Containing GF-40.

CONCLUSIONS

The recycled S-1 mixtures formulated with 33 and 50 percent RAP (FC-2) were found to be insensitive to CRM contents as high as nine percent of the total binder in the S-1 mixture. The shear resistance (G_s) and air void contents were essentially unaffected by CRM type (GF-20, GF-40, GF-80). Test parameters for compacted mixtures indicated no statistical difference between mixtures with and without CRM. It was concluded that absorption of the binder into the aggregate and possibly volatile losses were responsible for the increase in maximum density and air void content.

REFERENCES

1. RUTH, B.E., TIA, M., JONSSON, G. and SETZE, J.C. Recycling of asphalt mixtures containing crumb rubber. Final Report - UF Project No. 49104504495-12, Dept. of Civil Engineering, University of Florida, 1997.

Fifth International Conference on the
BEARING CAPASITY OF
ROADS AND AIRFIELDS

RECYCLING OF BITUMINOUS MATERIALS IN HOT ASPHALT PLANTS

M.L. Gallenne and P. Monéron

Laboratoire Central des Ponts et Chaussées

44340 Bouguenais - France

Abstract

Pavement material recycling, though resources-saving, remains a rarely-used technique on account of its cost price and "specific" using conditions. In spite of the French specific characteristics as regards this issue considering the extent of its natural resources, the prospect, in 2002, of the prohibition of demolition materials in dumps necessitates supplying the French authorities with the means to implement the technique in the best conditions.

The paper addresses the problem of hot-recycled demolition asphalt exclusively. The recycling rate achieved, which can be either low (≤ 20 %) or high (≥ 20 %) depending on the sophistication of the study and the equipment used, is discussed according to the type of road works: urban works, regeneration of the whole wearing course (surface layer slippage, cracking, etc.) or partial repair works. The first case requires no specific methodology and few modifications of the classical equipment, whereas the second case requires to respect some rules regarding both material mix-designing and use of the equipment as well as the ensuing environmental restraints. This paper also reports the studies currently in progress to examine stack solids from different type of asphalt recycled in a continuous asphalt mixing plant with a uniflow dryer drum.

The objectives of these studies are to provide construction authorities with reliable recycling procedures suiting all construction site cases, to advise road construction companies in the choice and using conditions of the equipment and, finally, to encourage equipment manufacturers to be more innovative.

1. BITUMINOUS MATERIAL RECYCLING IN FRANCE : STATE - OF - THE - ART

Pavement material recycling, which appears as the best alternative to the ban on asphalt from dumps while placing environmental values on asphalt pavement engineering, must, however, conform to the legislation in force and to the decree of 2 February 98 setting the thresholds for solid and gaseous emissions at the outlet of asphalt mixing plant chimneys, the respect for which requires specific conditions of use of the plants. In this paper, which presents the French recycling method, the accent is put on the adequacy of the equipment to reach both qualitative and environmental objectives.

1.1 Fields of use

Bituminous material recycling appeared in France toward 1980 when maintenance costs became the majority budget heading in road works. Although it is still not very frequent in France, bituminous material recycling is sometimes the only answer to rehabilitation and maintenance problems, because of which the need to eliminate the course in-place, produces asphalt reserves to be recycled. The use of the milling technique (or any other technique to remove materials in-place) is essential in cases where the pavement is seriously damaged or when road works have to be carried out, for which the geometry of the traffic lane must be conserved.
Irreversible damages are surface layer slippage, crazing, extensive cracking and general or partial rutting. Inner city or under structure road works and the partial rehabilitation of a lane, for instance the slow lane of a two-lane dual carriageway road, are examples for which the preservation of the altimetry is vital. In other cases, such as reshaping operations requiring a lot of materials for instance, planing sometimes turns out to be economical. In all cases, asphalt reserves such built up are theoretically homogeneous since coming from a unique site. Nevertheless, it is not always so because the asphalt reserves are also built up through the accumulation of salvage materials of various origin: trenches, rejected asphalt, etc.

1.2 Principle

Salvage asphalt does not have the required characteristics anymore. The proportion of asphalt to be recycled to the total asphalt production (also known as the recycle ratio) will depend on the initial characteristics of the asphalt to be recycled beside the performance desired of the produced asphalt. This performance is no different from that of an asphalt without recycled materials. Generally, the determination of both the type of added binder and the recycle ratio, which requires a preliminary mix-designing study, is necessary to reach the performance desired.

1.3 Preliminary study

When salvage asphalt is of various and varied origin, crushing may be necessary and samples are then taken after the operation. If the asphalt to be recycled results from milling, then the operation must be carried out over a sufficiently long section to allow the adjustment of the device (speed and depth of the cutter principally) and the collection of enough samples. The characterization is, of course, performed on the samples and consists of the particle size analysis, the identification of the asphalt (moisture and binder contents and particle size analysis after stripping of binder) and of the binder (index of penetration, ring and ball temperature, Fraass breaking point, asphaltene content). The recycle ratio is determined from all these elements and from the economical choices of the construction authority. With recycling asphalt of various origin, the recycle ratio is inevitably low (≤ 20 %) considering their probable heterogeneousness.
The accuracy of the characterization is highly dependent on the representativeness of the sampling, the number of samples being in relation with the various origins. The addition binder will be selected according to the recycle ratio, which, when ≥ 20 % is generally regenerant.
From this point, the traditional mix-designing procedure, that is a mix-design elaborated according to level of compaction (gyratory compactor test), water resistance, (Duriez test), rutting and modulus of the produced asphalt, takes over.

1.4 Organization of the work site

Precautions have to be taken for the storage of asphalt to be recycled, the making-up and the implementation of the mixture. As seen in the previous sections, salvage asphalt from a unique site must be set apart from recycling asphalt of various origins, the storage of which should also distinguish between the various provenances. In the first case, the reference board for milling must necessarily be taken before the work starts so as to carry out the study.

The making-up of the mixture, which is a tricky point because of the importance of the asphalt plant configuration on its capacity to produce asphalt with a high recycle ratio, will be explained in detail in the next section. As for implementation, no particular problem arises, apart from the significant workability of the mixture when the recycle ratio is high, which can have repercussions on the deformation property of the material during compaction.

2. IN-PLANT RECYCLING

2.1 Objective

Some rules must be respected when making of a mixture from a significant proportion of recycling asphalt. First, drying and heating of salvage asphalt must be carried out without debasing the binder it contains. When the added binder is a regeneration binder, it must be all the more shielded from the direct flame since it reveals a thermal susceptibility greater than normal. Then, the homogeneousness of the product such produced must be the same than traditional asphalt even if the introduction and the distribution of the constituents are different. Finally, the last constraint concerns gaseous and solid emissions.

2.2 Drying and heating of recycled materials

The two heat exchange principles between asphalt to be recycled and new aggregates are conduction and convection. Conduction is an exchange mode operating through the contact of the warmest aggregates toward the coldest. Here, overheating of aggregates depends on the recycle ratio, on moisture content of recycled materials and on the conduction transfer time.

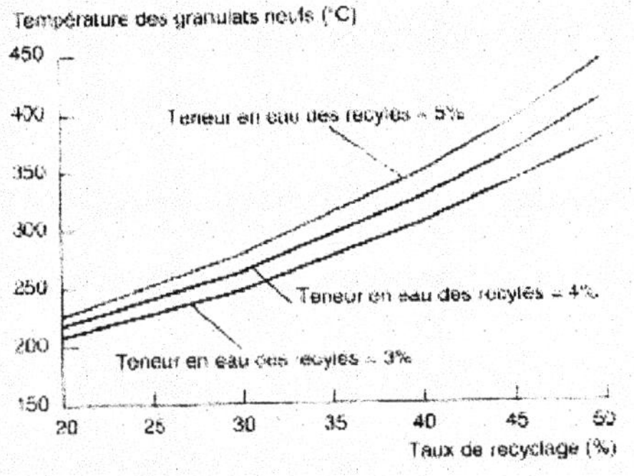

Figure 1 : calculated equilibrium temperature of the dry aggregates according to the recycle ratio and to the moisture content of the recycling asphalt.

Cases	Type of cycle	Type of dryer	Heat exchange	Introduction of the recycled materials
1	Discontinuous	Counterflow	Conduction	Bottom of hot Hot conveyor/elevator
2	Discontinuous	Counterflow	Conduction	In the median part of the dryer (prolongation of the dryer)
3	Discontinuous	Counterflow	Convection + conduction	Specific dryer
4	Continuous	Uniflow	Convection + conduction	In the median part (prolongation of the dryer or additional mixer)
5	Continuous	Counterflow	Conduction	Specific dryer and mixer

Table 1 : different cases according to the type of heat exchange between aggregates, gases and recycling asphalt.

In convection mode, the materials are dried and heated by the flow of hot gas. Two cases are then distinguished, the uniflow dryer when both gases and aggregated circulate in the same direction and the counterflow dryer when they circulate in opposite directions.

Case 1 corresponds to a classical discontinuous asphalt plant, on which no particular adjustments have been made. Conduction is established at the level of the mixer. The evacuation of water vapors may be inadequate when the recyle ratio is too high.
Case 2 enhances conduction through the prolongation of the dryer, which is used to increase slightly the recycle ratio.
Case 3, the latest, having been developed entirely with a view to recycling, has no particular limitation as regards drying.
In case 4, first pieces of equipment were equipped with only one ring known as "the recycling ring" used to introduce the recycled materials in the median part of the drum. Drying and heating areas were afterwards lengthened by 50 %.
Case 5 includes the newest equipment: Retroflux of the Ermont Company, Ultradrum of the Gencor Company, Double Tambour of the Asec Company, MEC plant of the Sim Company. The restrictive factor to good heat exchanges is the heating temperature of the aggregates as shown in figure 5.

2.3 Protection of the binder [1]

The protection concerns both the binder already contained in the asphalt to be recycled and the regeneration binder to be added knowing that it will be all the more susceptible to heat because it is regenerant. The risk of debasing binders is connected with temperatures of gases and new aggregates being too high. The evolution of the temperatures within the uniflow dryer/mixer drum is displayed in figures 2 and 3.

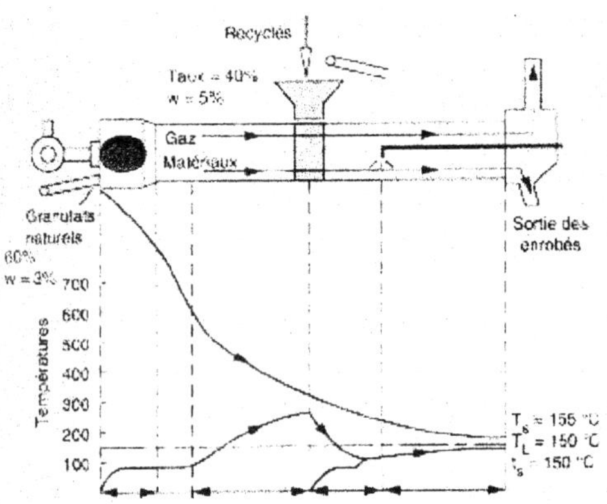

Figure 2 : temperature evolution in a continuous uniflow dryer drum asphalt mixing plant

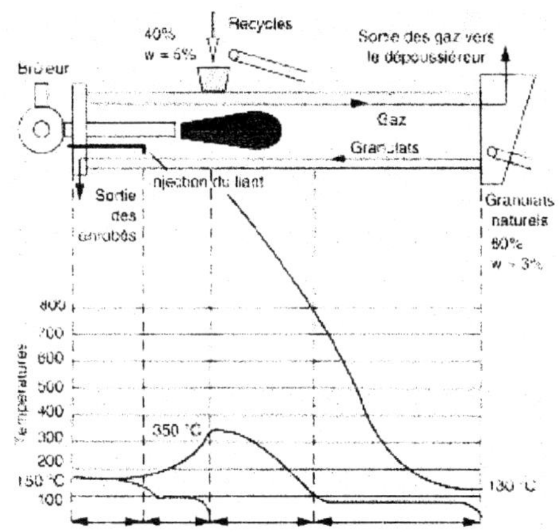

Figure 3 : temperature evolution in a counterflow dryer drum asphalt mixing plant

Risk of recycled asphalt binder
Regarding continuous uniflow asphalt plants (Case 4), several adjustments have been made and the temperature of gases, at the time when recycled materials are introduced, is generally lower than 300°C. Regarding counterflow dryers, the temperature of the new aggregates is approximately 350°C in contact with the recycled materials in view of their moisture content for a recycle ratio of 45 %. Although these temperatures reveal no detectible disorder, quantifying precisely the risk of having the binder of recycled materials debase may be desirable.

Risk of added binder

If a counterflow dryer is used, the added binder is theoretically pumped into materials at 150°C, a temperature at which no risk of debasing of the regeneration binder should be feared. In continuous uniflow asphalt plants, the temperatures of the gases and of the aggregates seam to be about 220°C and 120°C respectively, which, in the case of the gases, can hinder the use of regeneration binders because of their high susceptibility to heat.

2.4 Conveyance and mixing of the product obtained

Conveyance

In discontinuous asphalt mixing plants, dry and hot aggregates are conveyed by a bucket elevator from the dryer to the mixer. When producing recycled asphalt, the conveyed materials are a mixture of dry aggregates and asphalt. If the recycle ratio is too high, the elevator may choke. For Case 3 asphalt plants, the asphalt to be recycled is conveyed by a scraper conveyor.

Mixing

The respect for the homogeneousness of new aggregates and recycling asphalt requires the complete breaking up of the old asphalt before they are mixed with both the new aggregates and the added binder. A satisfactory homogeneousness of aggregates and binders depends on :

- the heating of the recycling asphalt mixture,
- the mixing time of dry aggregates and recycling asphalt,
- the mixing time of this combination with the added binder.

Mixing times must be adapted to the recycle ratio and, although they can be modified on a discontinuous mixer, on a dryer and mixer drum, on the other hand, they evolve with the rotational speed or the inclinaison of the drum only (Cases 4 and 5).

Cases	Type of cycle	Type of dryer	Heating process	Mixing process for aggregates and recycled materials only	Mixing process with the added binder
1	Discontinuous	Counterflow	Mixer	Mixer	Mixer
2	Discontinuous	Counterflow	Dryer drum	Dryer drum	Mixer
3	Discontinuous	Counterflow	Specific dryer	Mixer	Mixer
4	Continuous	Uniflow	Dryer drum	Dryer drum	Mixer drum
5	Continuous	Counterflow	Dryer drum	Dryer drum	Mixer the drum

Table 2 : diversity of asphalt mixing plants, which display different mixing processes.

The mixing efficiencies of a dryer drum, of a mixer drum and of a mixer are difficult to compare.

2.5 Respect of the environment

It concerns here gaseous and solid emissions in the atmosphere.

Gaseous emissions

Gaseous emissions due to HC-vapours are released during the production of recycled asphalt through convective heat exchange as in Cases 3 and 4 in Table 1. Case 3 is not problematic, as long as the recycle ratio remains low, because outgoing gases of the specific recycling asphalt dryer are directly reduced to ashes by the burner of the new aggregate dryer. Regarding Case 4, gaseous emissions are conditioned by the temperatures of the gases, which should be lower than 800°C at the recycling asphalt feed orifice and lower than 400°C at the regeneration binder feed orifice as it is with the new generation equipment, otherwise yellow smoke begins to appear.

When the contact between recycled asphalt and new superheated aggregates can creat a gaseous emission risk.

Dust emissions
The control of stack solid in the atmosphere is dependent upon the capacity of the dust collector to process the fine elements found in the dryer outgoing gases.
In a counterflow dryer, one m^3 of gas contains about 300 g of fines. The quantity is remarkably lower for the uniflow dryer, i.e., 4 g/ m^3. Stack solid is reduced on account of both the protection curtain formed by the aggregates and the introduction of the binder. The particle size is smaller than 500 µm.
There are two frequent types of dust collectors :
- the wet scrubbers,
- the bag filters.

The production of classical asphalt with the bag filter usually does not poses any problem as regards solid emissions whatever the strictness of the legislation which depends on the countries. On the contrary, this is not always true with wet scrubbers. The production of recycled asphalt raised the problem of stack solids containing binder, which can occur with continuous uniflow asphalt plants (Case 4). The risk that the bag filter either chokes or be damaged, increases when the recycle ratio is greater than 35% or when the temperature of the gases is too high (above 200°C).

2.6 Proposals for recycle ratios determined according to the risks for the asphalt plants

The risks related to the different operations as described in sections 2.2 to 2.5 are summarized in Table 3.

Cases	Type of cycle	Type of dryer	Recommended recycle ratio	Risks
1	Discontinuous	Counterflow	15 %	Evacuation of water vapor Choking of the hot elevator
2	Discontinuous	Counterflow	25 %	Choking of the hot elevator
3	Discontinuous	Counterflow	50 %	Gaseous emissions (yellow smoke)
4	Continuous	Uniflow	35 %	Choking of the dust collector Gaseous emissions (yellow smoke)
5	Continuous	Counterflow	50 %	Debasing of the binder Deterioration of the dust collector

Table 3 : a proposal for recycle ratios which allows for such risks

Whatever the recycle ratio selected and lower than 50 %, some technical solutions are available for urban works (use of discontinuous asphalt mixing plants for instance) or other works (use of continuous asphalt mixing plants for instance). However, some solutions remain expensive (Case 3).
The review of the present state of mixing plants functioning and their capacity to recycle asphalt reveals the need for further information to improve our understanding of the relationships between asphalt production, stack solids at the outlet of dryers and atmospheric releases. Chapter 3 gives a detailed account of the studies currently in progress.

3. EFFECT OF ASPHALT RECYCLING ON DRYING AND DUST EXTRACTION OPERATIONS

Stack solids and fine particle processing have been assessed for both drying and dust extraction operations from the comparison of the production of asphalt with and without recycled materials.
The aggregate drying and heating operations within rotational dryer drums are accompanied by stack solids, which must be collected at the level of the dedusting system and reintroduced into the mixture. With counterflow dryer drums, these particles are essentially mineral, whereas, they also contain binder in the case of uniflow dryer drums.

3.1 Experimentation

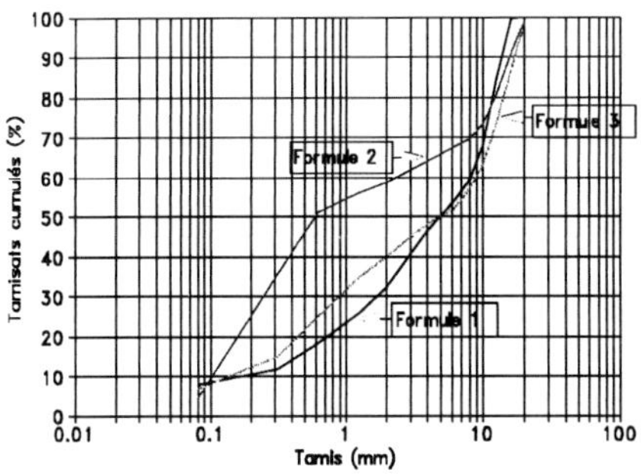

Figure 4: bituminous-bound graded aggregate with recycling size distribution

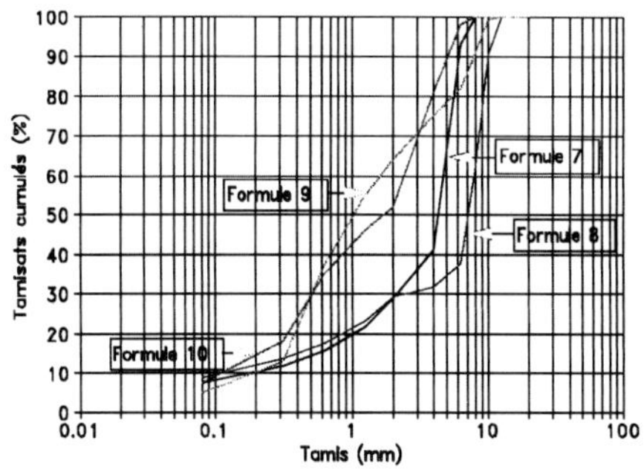

Figure 6 : bituminous concrete without recycling size distribution

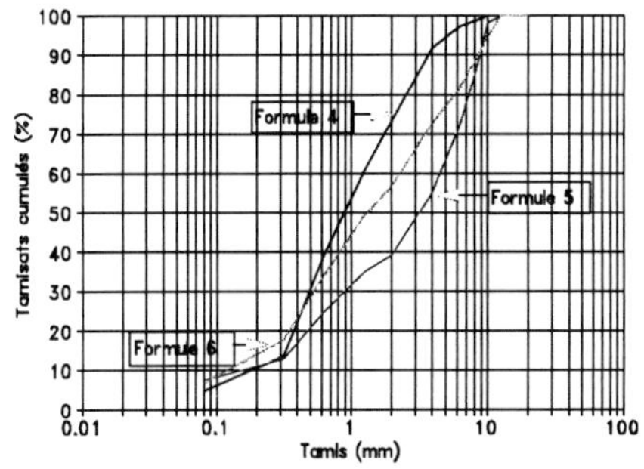

Figure 5 : bituminous concrete with recycling size distribution

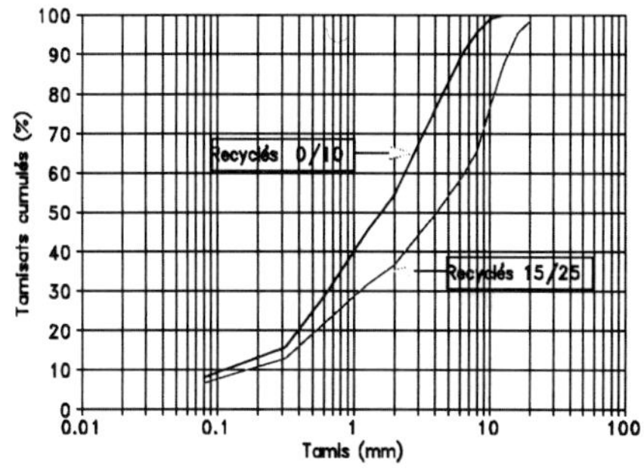

Figure 7 : the two asphalts to be recycled size distribution

The experiments have been carried out on a stationary Ermont dry drum mixing plant of the type TSM 17 XLR. The dryer is uniflow. The plant is equipped with a bag filter, the filtering surface of which is 517 m^2 for a nominal output of 130 T/H, designed to process effectively 45,000 m^3/h of particle-laden gas.

Ten different mix-designs, six of which with recycling procedures, have been experimented (figures 4, 5, 6 and 7). We used two sorts of recycling asphalt : 0/10 and 15/25. The asphalt mixtures produced are 0/10-mm bituminous concrete and 0/20-mm bituminous-bound graded aggregates The binder contents for the recycling asphalt are 4.5 % for 0/10 and 5.0 % for 15/25 respectively.

Mix-designs	1	2	3	4	5	6	7	8	9	10
% of R 0/10	30	15		30	20	20	/	/	/	/
% of R 15/25	/	15	20	/	/	/	/	/	/	/
Rate of flow (t/h)	140	102	114	110	114	112	121	121	118	112
% of global binder	5,9	4,5	3,5	4	5,7	6,5	6,2	5,5	6,5	6,5
% of added binder	4,5	3,1	2,5	2,6	4,8	5,6	/	/	/	/
Velocity of gas flow (m/s)	10,7	13,4	12,1	11,8	11,3	12,3	11	10,3	11	11

Table 4 : parameters of the produced mix-designs

The gas velocity has been measured at the outlet of the dryer with standardized procedures. For all mix-designs, gas samples have been taken at the outlet of both the dryer and the dust collector. They are used to quantify the number and the composition of the particles.

3.2 First results

3.2.1 Size distribution, binder content and quantity of solid particles at the outlet of the dryer

In the dryer drum, gases and aggregates follow the same course, aggregates being directly under the action of smoke gases. The dryer is a rotational drum fitted with buckets, which lift the aggregates then let them fall through the cross section of the drum. In summary, the longitudinal transfer of the aggregates is provided by the rotating movement and the inclinaison of the drum combined to the driving force of the gaseous flow.

The presence of solid particles at the outlet of the dryer is due to the action of the gaseous flow and of its velocity in particular. When falling into the gaseous flow, the displacement of the aggregates is inversely proportional to their radius. The aggregates prone to become stack solids are, thus, the smallest and the maximum particle size of stack solids is approximately 500 µm. In theory, the smallest particles should leave the drum dryer from the very first fall, but fortunately the aggregate curtain and the introduction of binder within the dryer drum favor the trapping of particles liable to become stack solids and thus reducing their quantity in the gases at the outlet of the dryer. The occulting of the drum depends on the aggregate size distribution, the binder content and the rate of the flow.

The main parameters, which have been considered in this experimentation are :

- the velocity of gases in the dryer drum,
- the occulting of the drum by all the aggregates,
- the proportion of particles ≤ 80 µm,
- the proportion of particles ranging between 80 and 500 µm,
- the added binder content,
- the binder content of asphalt to be recycled.

On the other hand, the consideration of the moisture content of fines sands proved impossible.

Although the few results at our disposal cannot be used for a thorough analysis, they teach, however, several things which confirm the intuitive hypotheses ventured on stack solids within dry drum mixers.

Particle size analysis of particles at the outlet of the dryer (Figures 8 and 9)

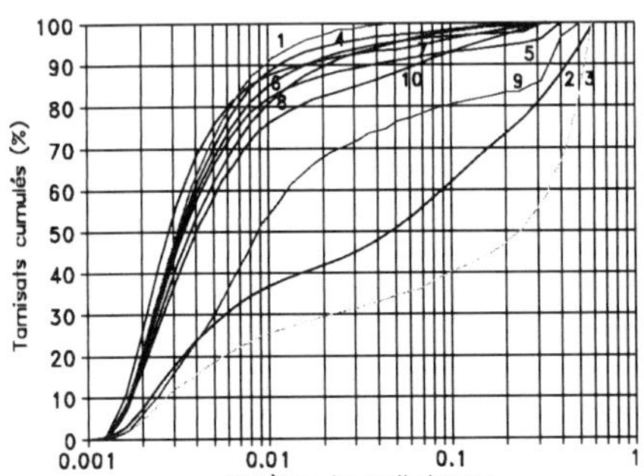

Figure 8 : Size distribution for all mix-designs produced

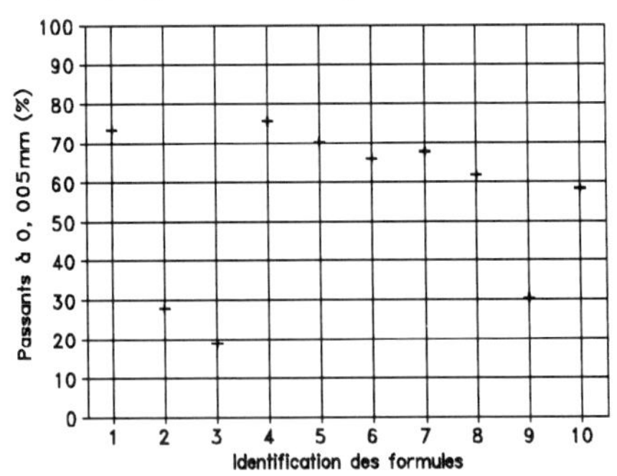

Figure 9 : Percentage of 5 µm sieve undersizes

Particle size analysis is carried out with a laser granulometer equipped with a liquid-carried sample preparation system. Figure 8 shows the particle size distribution achieved with the analysis of all the samples. 5 µm correspond to the radius of particles, which can reach human air cells, its proportion is given in Figure 9.

The mix-designs studied here offer a wide range of the possible asphalt production. The carried particle size analysis shows no difference between the classical production and the production with 0/10 recycled materials. The maximum size is about 400 µm and the proportion of particles smaller than 80 µm is high. As regard 15/25 recycled materials (mix-designs 2 and 3), on the other hand, stack solids occur with particles having a maximum size of 600 µm and a very different size distribution. The proportion of elements smaller than 80 µm is much lower.
In all cases, the percentage of particles smaller than 5 µm is high (20 to 70 % of the cases depending on the mix-designs). So, the filtering capacity of the dust collectors must be around the micron to obtain concentrations in the gases lower than the threshold currently admitted by the legislation.

Binder content of the particles at the outlet of the dryer (Figure 10)

Recycling asphalt is introduced in the median part of the drum, which favors stack solids.

- The presence of binder has been recorded in all the analyzed samples. Stack solids occur over the whole length of the drum, before and after the introduction of the binder.
- For the same mix-design (9 and 6) either produced with traditional methods or with recycling procedures, the binder content of the particles is more important when recycling is considered, an observation, which confirms that the binder content of the particles depends on the recycle ratio.
- Considering all mix-designs together, the binder content of the particles ranges between 0.3 and 5% (from 20 to 800 mg/m3, quantitatively speaking).

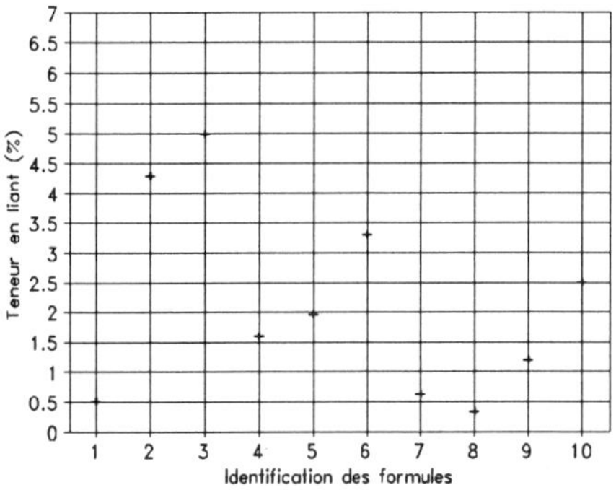

Figure 10 : Binder content of the particles at the outlet of the dryer

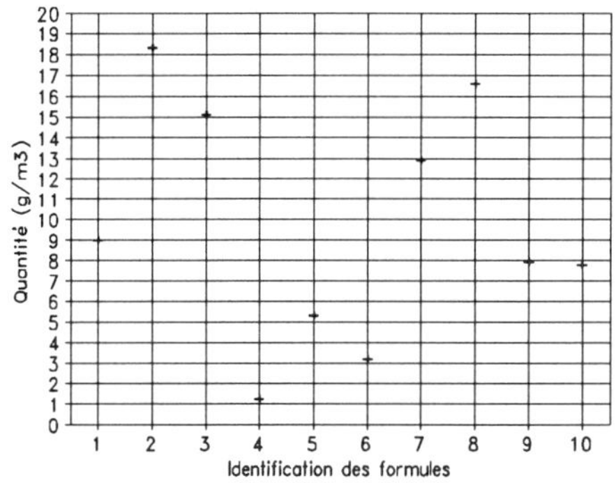

Figure 11 : Quantity of particles at the outlet of the dryer

Quantity of solid particles at the outlet of the dryer (Figure 11)

- The quantity of solid particles is more important with classical production than with recycling (comparison of mix-designs 1 to 6 and 7 to 10), observations, which confirm the self-dedusting capacity of coated particles.
- The quantity of particles is also greater when producing bituminous-bound graded aggregates. This strengthens the observation that the more efficient the occulting by the aggregates is, the fewer stack solids are.
- The quantity of solid particles varies according to aggregates ≤ 315 µm.
- The quantity of solid particles is inversely proportional to the added binder content (mix-design 1, 2, 3, and 7, 8, 9, 10).

Considering all mix-designs together, stack solids range between 1 and 18 g/m^3.

3.2.2. Quantity of the particles released in the atmosphere

All measurements carried out at the outlet of the dust collector give very low concentrations below 10 mg/m^3. These concentrations are lower than the threshold currently admitted by the legislation, figures which demonstrate the collection efficiency of the bag filter.

The future investigations will answer to :

- the ratio recycle influence on the main parameters,
- the wet scrubbers use limits.

CONCLUSION

Hot recycling is a demanding process as regards technique, preliminary studies, selection of adapted equipment and execution of the work. However, the global performance of the road sites resorting to the process is satisfactory [2]. Studies and experiments currently in progress within the network of will soon lead to the formalization of accurate regulation where the application of the technique is concerned. In any case, we must be prepared to use it more regularly in the future since it constitutes one of the best alternatives to the ban on asphalt from dumps in 2002.

REFERENCES

[1] Evolution des centrales Tambour Sécheur Enrobeurs et de leur aptitude à l'utilisation des matériaux recyclés - P. Monéron - Bulletin de Liaison des Ponts et Chaussées - N° 184 - mars - avril 1993 -

[2] Le recyclage des enrobés à chaud en centrale - M. Bertaud - J.P. Lavaud - P. Monéron - Revue Générale des Routes et Aérodromes - N° 712 - novembre 1993 -

EVALUATION OF MODIFIED AND COLD-RECYCLED ASPHALT MIXTURES

KWANG W. KIM[1], KYUDONG JEONG[2], JIYONG LEE[2], SOONJAE LEE[2] AND TAESOON PARK[3]

[1]Associate Professor, [2]Graduate Students,
Department of Agricultural Engineering
Kangwon National University, Chun Chon, 200-701 Korea
[3]Assistant Professor, Department of Civil Engineering
Seoul Polytechnical University, Seoul, 139-743 Korea

ABSTRACT

Recycled asphalt mixture is a good alternative material for asphalt pavement construction. However, there is difficulty in producing high-quality cold-mix recycled materials which have nearly a virgin mixture quality. Therefore, there is a need for improving quality of cold-mix recycled materials, and this study shows the method and result of modifying binder of recycled asphalt mixtures in cold-mix recycle for asphalt-treated base course. An asphalt emulsion, a selected polymer and a rejuvenator was used in this study. A significant improvement in fundamental quality of cold-mix recycled material was achieved from this experimental study.

INTRODUCTION

Every year, a great amount of waste asphalt pavement material occurs and recycle of these waste materials is a hot issue due to increased concern about environmental problems in Korea. Government established a strong regulation for prohibiting dumping waste asphalt materials into a landfill and for recommending recycle [1]. Reclaimed asphalt pavement (RAP) is treated as a valuable resource which can be recycled in the asphalt pavement. Therefore, many recycle techniques are under practice or study to reuse the valuable materials [2,3,4].

However, the recycled asphalt mixtures are not widely accepted for asphalt pavement construction in practice. There are several reasons for this reluctance. Firstly, there is no standard quality manual which specifies acceptance ranges and/or limit of recycled materials for a certain use. Therefore, a conventional standard specification for virgin material is applied to the quality control of recycled materials when these are used for asphalt pavement. However, it is difficult to produce high-quality recycled materials which satisfy the virgin material specification. Secondly, there are serious distresses including permanent deformation and cracking in asphalt pavements. Many people do not assure the quality of recycled materials as the one which has enough durability to overcome the serious damage mechanisms which results from, what they believe, use of lower quality materials. Therefore, they are reluctant in using recycled materials for pavement without evidence that the quality of the recycled material is not inferior to that of virgin one.

This study deals with cold-mix recycle technique for bituminous-treated base course material. Cold-mix process is an environmentally safe and energy saving technique in asphalt pavement industry. Therefore, it is widely used in asphalt mixture for patching, repairing and other purposes [5,6]. However, if cold-mix technique is used in recycle, achieving the quality of the final (recycled) mixture to satisfy a set standard limit is not easy. For example, to use the recycled mixture in the base course, the mixture should satisfy the minimal quality level, which includes Marshall stability over 350kg (3.5kN), air void within 3% to 10% range, and flow within 10 to 40 (×0.1㎜) range. These acceptance criteria are easily achieved by using hot-mix recycle mixture, but is not by cold-mix recycled mixture.

In most cases, the quality of the final mixture depends upon the quality and content of RAP in the recycled mixture. The quality of RAP depends upon service life, condition of the pavement and reclaiming methods which are not usually controllable. However, RAP content in the recycled mixture is an influencing factor. In general, the higher the RAP content, the lower the mixture quality. Therefore, in order to use a higher quantity of the recycled mixture safely in the field, it is necessary to improve quality of the recycled mixture. There can be many methods in improving quality of the cold-mix recycled mixture. The objective of this study is to develop a technique improving a quality by modifying recycled binder using a polymer material when it is recycled, and to evaluate the properties of the modified-recycled mixture.

MATERIALS AND METHODS

Materials

Two reclaimed asphalt pavements were collected from two different road reconstruction sites, and named as RAP-1 and RAP-2. RAP-1 was collected from a urban road in the city of Chun Chon, and RAP-2 was from a country side road near town of Inje. Using a crusher, each RAP was crushed into particles smaller than 25㎜ in size.

An asphalt emulsion (RSC-4) was used for cold mixing. The asphalt content of the emulsion was 45.2% by weight. A rejuvenator (YKS-1) was used as an agent for rejuvenating old asphalt in RAP. Unsaturated polyester resin (UPR) was used together with an initiator for modification of binder in cold-mix asphalt mixture. The physical property of the UPR is shown in Table 1.

Table 1. Physical properties of unsaturated polyester resin (UPR)

Specific gravity	Viscosity (poise)	Styrene content (%)	Acid value
1.138	3.0	40	20

A maximum size of 25㎜ crushed virgin aggregate, crushed sand, and mineral filler were used for a bituminous treated base course mixture (BB-2) based on Korea Ministry of Construction and Transportation specification. Gradation and physical property of aggregates are shown in Tables 2 and 3, respectively.

Table 2. Gradation of virgin aggregates used

Sieve size		40㎜	25㎜	19㎜	10㎜	#4	#8	#30	#50	#100	#200
Passing (%)	25㎜ agg.	100	95.68	63.51	6.28						
	13㎜ agg.			100	96.09	72.11	48.80	22.01	15.08	8.92	3.73
	Fine agg.				100	87.66	64.64	32.22	20.38	10.50	4.98
	Filler							100	94	80	69

Table 3. Physical property of virgin aggregates used.

Specific Gravity	Specification requirement	Coarse Agg.	13㎜ agg.	Fine Agg.	Filler
Saturated surface dry		2.639			
Apparent	> 2.45	2.691	2.689	2.701	3.04

Methods

Aged asphalt in the RAP was recovered by Abson method (ASTM D 1856) [7]. Penetration and absolute viscosity were measured for virgin and recovered asphalts using KS M 2252 and ASTM D 2170, respectively. A certain quantity of RAP and virgin materials were mixed at 60°C and aged for 45 minutes before compacting 75 blows each side using Marshall compactor. Second compaction was applied to the specimen with the same number after aging for 24 hours within the mould in room temperature. The specimen was extruded from the mould after 24 hours of second compaction, and kept at room temperature for another 72 hours before measuring Marshall properties.

A gyratory compactor was also used for the recycled mixture to achieve the similar compaction effort to the field roller compaction. In this case, Superpave level 1 mix design manual was used to determine an appropriate number of gyration [8].

According to the preliminary experiment in Table 4, a relatively good quality was obtained from the mixture with RAP contents below 30%, but a relatively poor quality was resulted from RAP content over 40%. The BB-2 specification limits include air void 3～10%, Marshall stability over 350kg and flow 10～40%. Therefore, the recycled mixture with 40% RAP was intensively studied in this research.

Table 4. Marshall property of recycled asphalt concrete in different RAP content

Recycled mixture with RAP content (%)	Air void (%)	VFA (%)	Marshall Stability (kg)	Flow (1/100cm)
15	7.49	64.06	543	45
30	8.72	60.16	421	31
40	12.03	44.45	266	30
50	11.76	45.28	269	30

RESULTS AND DISCUSSION

Binder Property

Asphalt contents of RAP-1 and RAP-2 were 5.6% and 4.8% respectively. Penetration and absolute viscosity test results of virgin asphalt and two RAPs are shown in the Table 5. According to the results, RAP-1 was found to be much older material and the binder of it was aged much more than RAP-2 as revealed in penetration and viscosity values in the table.

To evaluate the rejuvenating effect of YKS-1, it was added to the recovered binders from RAP-1 and RAP-2 in various contents. Four physical properties were measured for the rejuvenated asphalt binders as shown in Table 6. YKS-1 contents of 15% and 10% by weight of total asphalt binder were needed for the binders from RAP-1 and RAP-2 to reach the target penetration of 60～70, respectively, in recycled mixture.

Table 5. Penetration, viscosity and composition of asphalt

Asphalt	AC content (%)	Penetration (0.1㎜)	Viscosity (ps.)	Saturates (Vol. %)	Aromatics (Vol. %)	Resin (Vol. %)	Asphaltene (Vol. %)
AP-3[1]	-	87	962	9.6	62.4	19.8	5.70
RAP-1	5.6%	17	13,352	4.1	47.1	21.5	27.3
RAP-2	4.8%	32	3,342	-	-	-	-

[1] Recovered from a virgin mixture

Since RAP-1 was found to be more aged from binder analysis, RAP-1 was selected for use in recycled mixture preparation and evaluation. Aggregate was extracted from RAP-1 and gradation and physical properties were evaluated. Table 7 shows physical property of extracted aggregate from RAP-1.

The gradation of blended aggregate (the extracted aggregate and virgin aggregate) in the recycled mixture using 40% RAP is shown in Figure 1. The gradation of extracted aggregate itself from RAP-1 was out of specification limits, but the gradation of blended aggregate fit well within the limits.

Table 6. Relation of physical properties and YKS-1 content in the binders recovered from RAP-1 and RAP-2.

Binder	Property	YKS-1 content (%)					note
		0	10	15	20	25	
RAP-1	Penetration (0.1㎜)	17	41	63	80	99	
	Viscosity (ps.)	13,352	-	2815	1411	936	
	Softening point (°C)	65.8	53.8	49.9	46.0	-	
	Ductility (cm)	20	150+	150+	150+	-	
RAP-2	Penetration (0.1㎜)	32	62	87	120	-	
	Viscosity (ps.)	3,342	-	2,352	1,289	-	
	Softening point (°C)	60.4	50.3	47.4	44.8	-	
	Ductility (cm)	52	150+	150+	150+	-	

Table 7. Specific gravity of the aggregates from RAP-1

Specific gravity	Specification Requirement	Coarse aggregate.	Fine aggregate.	Filler
Saturated surface dry		2.679		
Apparent	>2.45	2.705	2.655	2.696

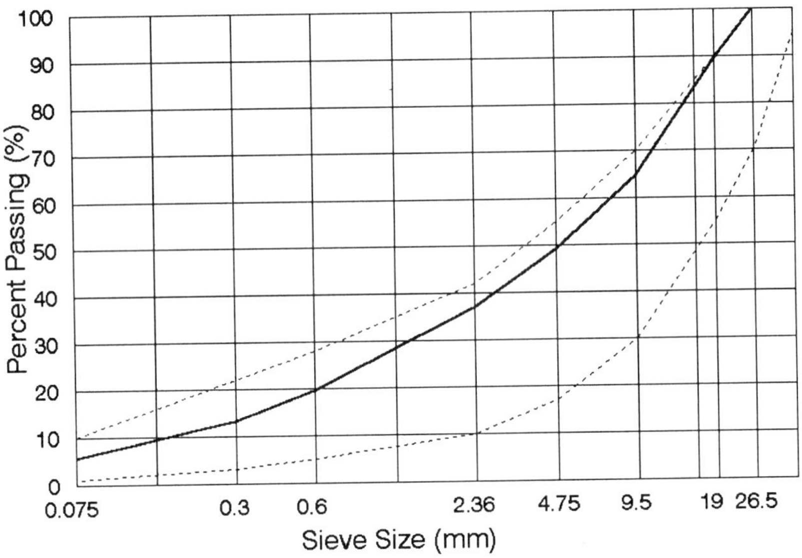

Figure 1. Gradation of blended aggregate.

Mixture Design

Emulsion quantity was calculated based on the method suggested by Asphalt Institute (MS-21) and added to the mixture during mixing process. Mix designs for the recycled mixtures with RAP 40% and various polyester and rejuvenator contents were carried out using asphalt contents of 4.0%, 4.5%, 5.0%. 5.5% and 6.0%. Table 8 shows the results of mix design for different mixtures.

P10R0 mixture containing polyester 10% by weight of asphalt had maximum Marshall stability 498kg, flow 33 and air void 8%. The mixture containing 10% of both polyester and rejuvenator showed maximum stability 366kg, flow 26 and air void 6.92%. In these two mixtures. polyester and rejuvenator combinations were effective on improving stability of recycled mixture.

However, the mixture (P0R10) containing 10% of rejuvenator by weight of asphalt and the mixture (P5R5) containing 5% of both polyester and rejuvenator showed stability below the specification limit for bituminous- treated base course, 350kg. Polyester and rejuvenator combinations in these two mixtures were less effective than previously mentioned two mixtures on improving stability.

Table 8. Mix design results by cold-recycling method for RAP 40%

Polyester (%)	Rejuve. (%)	Mixture name	Asphalt Content (%)	Air Void (%)	Density (g/cm^3)	VFA (%)	Stability (kg)	Flow (1/100cm)
0	0	P0R0	4.0	12.35	2.215	41.25	264	35
			4.5	12.13	2.204	44.45	266	30
			5.0	12.49	2.178	46.05	240	27
			5.5	10.54	2.210	53.02	230	33
			6.0	9.30	2.224	58.40	228	31
10	0	P10R0	4.5	7.99	2.308	55.97	498	33
			5.0	6.94	2.316	62.03	443	30
			5.5	5.13	2.344	71.10	296	32
			6.0	4.09	2.352	77.14	294	24
0	10	P0R10	4.0	10.98	2.250	44.51	332	32
			4.5	10.08	2.255	49.63	311	33
			5.0	8.02	2.289	58.28	330	34
			5.5	6.35	2.314	66.25	305	34
5	5	P5R5	4.0	10.33	2.266	46.21	368	30
			4.5	9.25	2.276	52.01	313	32
			5.0	7.12	2.312	61.35	297	27
			5.5	5.40	2.337	69.95	252	26
10	10	P10R10	4.0	9.56	2.285	48.34	307	31
			4.5	6.92	2.334	59.76	366	26
			5.0	5.88	2.343	66.08	361	32
			5.5	4.57	2.358	73.52	266	28

Table 9 summarized results of the mix design for the mixtures satisfying the base course specification. Optimum asphalt content (OAC) range satisfying all the specification limit was narrowed down to 4.5～5% which is the range of asphalt content satisfying Marshall stability for both mixtures.

In this range, OACs for P10R0 mixture and P10R10 mixture were determined as 4.8% and 4.5%, respectively. Asphalt content at median air void in specification range was importantly considered in determining OAC of the two mixtures.

Table 9. AC range satisfying specification for bituminous-treated base course

Mixture	Marshall stability (>350kg)		Air voids (3~10%)		Flow (10~40)	
	Measured (kg)	AC range (%)	Measured (%)	AC range (%)	Measured (1/100cm)	AC range (%)
P10R0	443~498	4.5~5.0	4.09~7.99	4.5~6.0	24~33	4.5~6.0
P10R10	361~366	4.5~5.0	4.57~9.56	4.0~5.5	26~32	4.0~5.5

Evaluation

At each OAC, specimens were made using Superpave gyratory compactor and Marshall properties were measured. A total of 104 gyration was applied at 600kPa pressure and at 1.25° angle of gyration. From this evaluation, lower air void with higher stability were achieved than those of the Marshall hammer compacted mixture. P10R10 mixture showed better values than P10R0 in the two properties and flow being within the specification range.

Since gyratory compactor is know to simulate field roller better than Marshall compactor, the mixture (P10R10) performing better with gyratory compactor was selected to use in experimental paving using a roller at a test pit. A mix design was carried out on R10P10 mixture using gyratory compacted specimens. Table 10 shows results of mix design. According to this result, OAC was determined as 4.2%, based on the asphalt content range satisfying air void. Table 10 also shows obtained properties when the recycled mixture was made with this OAC using gyratory compactor.

Table 10. Marshall properties of cold-recycled asphalt mixtures which were prepared with a gyratory compactor (RAP 40%, P10R10)

Classifica-tion	Asphalt Content (%)	Air void (%)	Density (g/cm^3)	VFA (%)	Marshall stability (kg)	Flow (1/100cm)	Indirect tensile strength (kg/cm^2)
Mix design	3.5	7.14	2.349	52.99	646	35	-
	4.0	5.39	2.376	63.31	661	40	-
	4.5	3.51	2.405	75.11	584	39	-
	5.0	2.76	2.406	81.01	564	36	-
At OAC	4.2	4.38	2.394	69.19	829	37	3.0

A bituminous-treated base course was experimentally paved at a test pit using the mixture to see if the recycled mixture was compactable in the field condition. Paving was performed during a summer day and compaction was carried twice, one on the day and another on the very next day. Five cores were randomly taken from the pit, and asphalt content and density were measured to evaluate density level (Table 11). Asphalt content and air void were 4.16% and 4.89%, respectively. The asphalt content, 4.16%, was very close to the designed OAC, 4.2%. The density level of the field core was $2.381/2.394 \times 100 = 99\%$ of laboratory compacted specimen. Therefore, the mixture was very well compacted in the pit and field density was much higher than the specification compaction level 96%.

Table 11. Properties of the recycled mixture measured from field core

Mixture	Asphalt content (%)	Air void (%)	Density (g/cm^3)	VMA (%)	VFA (%)	Note
R10P10	4.14	4.90	2.381	14.69	66.61	
		5.09	2.376	14.86	65.72	
		4.21	2.399	14.07	70.08	
	4.18	4.94	2.380	14.72	66.45	
		5.33	2.371	15.07	64.64	
Mean	4.16	4.89	2.381	14.68	66.70	

CONCLUSIONS

The quality of the cold-mix recycled mixture using 40% reclaimed asphalt pavement (RAP) was below the specification limit without modification. Therefore, there was a need for improving the mixture quality. A polymer material was selected to modify the binder in the mix in addition to using a rejuvenator. The roll of rejuvenator is rejuvenating the aged binder in the RAP. Therefore, not much improvement of the recycled mixture quality was observed when rejuvenator was only used. However, the quality of recycled mixture was improved clearly over the specification limit for bituminous- reated base course when an unsaturated polyester resin (UPR) was used as a modifier. This was also proved from the quality control test on cored mixture taken from the experimental paving at a test pit. Therefore, the cold-mix recycled mixture developed in this study could be used satisfactorily for bituminous-treated base course.

Acknowledgement

The authors wish to express their appreciation to the Ministry of Construction and Transportation of Korea for funding of this research.

REFERENCES

1. Kim, K. W. , "Recycling of Waste asphalt pavement materials," Monthly Magazine of Waste, July, 1997, pp. 88～95.
2. Kim, K. W., et al., "Rural Road Pavement Construction using Recycled Concrete and Asphalt Materials," Final Report to Ministry of Agriculture, Kangwon National University, 1996.
3. Kim, K. W. et al., "Development of Modified Recycling Technology for Waste Asphalt Paving Materials," 1st-year Interim report R&D/96-0064, Ministry of Construction and Transport., Kangwon National University, 1997.
4. Kim, K. W., Jeong, K. D., Cho, H. W. and Li, X. F., "Evaluation of Recycled Hot-mix Asphalt Pavement on a Rural Road," Proceedings of annual Conference, Korean Society of Agricultural Engineers, 1997
5. "Asphalt cold-mix recycling," The Asphalt Institute, Manual series, No. 21 (MS-21)., 1983
6. Allen, D. D., "Cold-recycled bituminous concrete using bituminous materials," National cooperative highway research program synthesis of highway practice, 160, 1990.
7. "Recovery of asphalt from solution by Abson method," ASTM D1855, Annual book of ASTM standards, 1989.
8. "Superpave mix design," Asphalt Institute Superpave Series N0. 2(SP-2), 1996.

Fifth International Conference on the
BEARING CAPASITY OF
ROADS AND AIRFIELDS

IN-PLACE ENGINEERING PROPERTIES OF RECYCLED AND STABILIZED PAVEMENT LAYERS

IMRAN SYED[1] AND TOM SCULLION[2]

[1] Graduate Assistant Research and [2] Program Manager
Texas Transportation Institute, The Texas A&M University System, College Station, Texas 77843-3135, U.S.A.

ABSTRACT

Pavement reclamation or base recycling is a rapidly expanding pavement rehabilitation procedure. The existing pavement is pulverized in-place and stabilized prior to resurfacing. Design criteria for designing these pavement layers are not available. This study involved measuring the in-situ strength and visual pavement performance. A structural evaluation was performed on 25 base recycling projects in the Bryan district, Texas, USA. The recycled layers were stabilized with cement or lime and each was 250 mm thick. On the higher volume roadways an unstabilized flexible base (FB) was placed over the stabilized layer followed by a 2 course surface treatment (2 CST). On the other pavements the 2 CST was placed directly on the stabilized layer.

Testing involved the use of Falling Weight Deflectometer, Dynamic Cone Penetrometer, and Ground Penetrating Radar. A correlation was generated between the back calculated layer moduli and the percentage of stabilizer used. Tentative moduli values, to be used in future thickness designs, were also proposed. Of the 25 sections evaluated, 19 were performing well, with high structural strength and no surface distress. No traditional shrinkage cracking of the stabilized bases was found in any section. The only distress found was localized severe longitudinal cracking in several sections. This was related to several factors including a) the high PI subgrade soils, b) the drought of 1996, c) the presence of trees, and d) the strength of the stabilized layer.

In conclusion, base recycling appears to be an effective pavement rehabilitation technique. As the pavements were on average less than three years old at the time of testing, continued long-term monitoring is recommended.

Key Words : Pavement reclamation, base recycling, back calculated moduli values.

INTRODUCTION

The focus of many of the district offices in the Texas Department of Transportation (TxDOT) is the maintenance and rehabilitation of the existing highway network. The rural districts now have large portions of their networks that need major rehabilitation and reconstruction. Many districts are experimenting with innovative rehabilitation strategies for repairing the wide range of pavement types found around the state. One area of major interest is in rehabilitating thin flexible pavements which are near the end of their service lives. It has been estimated that over 70% of the mileage of the Texas network has a thin surface less than 50 mm thick over an unstabilized granular base (*1*). Many kilometers of this thin pavement type highways are being subjected to heavy loads from both agricultural and oil field activities. These roads are generally not structurally adequate to carry these heavy loads and many are exhibiting significant deterioration.

To address this problem in a cost-effective manner several of the districts have actively recycled pavements and stabilized bases in an attempt to correct deterioration and increase the structural capacity. In a one pass operation, the existing surface and base are pulverized, mixed and stabilized to form a lime stabilized base (LSB) or cement stabilized base (CSB). Frequently the reshaped and compacted base is simply sealed with a one or two course surface treatment. For higher volume roads, a new flexible base may be added followed by a thin asphalt surfacing.

Methods Used to Determine Stabilizer Contents for Base Materials
In Texas there are a variety of approaches used by districts to determine the percentage of stabilizer to use when recycling flexible base materials. The most widely used approach is "district experience," in which a fixed percentage of stabilizer is used independent of the base type to be treated. The Bryan district uses an extension of TxDOT method 121-E. Field samples are obtained for each layer to be stabilized. In this method the plasticity index (PI) and percentage of material passing the No. 40 sieve of each layer are used to obtain the percentage stabilizer. If the Bryan district is mixing 150 mm of base with 50 mm of subgrade then an analysis will be performed for each. The final percentage will be the weighted average of the two. For materials with a PI of less than 10, cement would be considered; for all other cases, lime would be used (*1*).
Limited guidance is given in TxDOT's specifications (Red Book 1995) on selecting optimum stabilizer contents. Item 276 on cement treated bases specifies minimum compressive strengths of 5,170 kpa (750 psi) or 3,450 kpa (500 psi). These strength values are high and can only be achieved with relatively high cement contents, usually 5% or above. However, districts using these levels have reported problems with shrinkage cracking (*2*). Most districts are not requiring such high compressive strength values, a common value now used is 1,380 kpa (200 psi).

OBJECTIVES OF STUDY
In-place stabilization is one of the pavement rehabilitation strategies used for low volume farm-to-market roads. This strategy consists of in-place stabilizing the top 250 mm of an existing pavement with either lime or cement and sealing the surface with a surface treatment. The

engineering properties for this stabilized layer generally have not been determined. Currently, engineering judgment is often used to select stabilizer quantities to use for pavement design. The approach followed in this research project is to evaluate projects that have been constructed to determine the in-place engineering properties achieved by stabilization and develop guidelines on stabilizer levels. The objectives of this research are listed below:

i. Dynamic Cone Penetrometer (DCP) test results will be compared with the subgrade modulus as estimated from MODULUS 5.1 (*3*). Thus, a correlation between DCP values and the subgrade modulus will be developed.
ii. A correlation between the stabilizer content and the layer moduli values will be established.
iii. This research will help in the evaluation of completed projects by checking if the construction of the pavement layers conform to the specifications. This shall be achieved by comparing the Falling Weight Deflectometer (FWD) deflection profile after construction with the design deflection profile. The design deflection profile will be determined from the average subgrade modulus and the upper layer moduli values obtained from (ii) above.
iv. A field testing protocol for using Ground Penetrating Radar (GPR) and FWD to evaluate the performance of stabilized bases in terms of resistance to moisture damage and improvements in long-term structural strength will be developed.

TECHNIQUES OF PAVEMENT EVALUATION

Devices that are used to perform non-destructive testing (NDT) on pavements on this project include the FWD and GPR. Deflection measurements from a Dynatest 800, FWD are used to evaluate the structural capacity of insitu pavements by back calculating an elastic moduli of various pavement layers. The analysis of deflection data was conducted using MODULUS. GPR is used to check the consistency, thickness, and entrapped moisture in the pavement layers. A GPR operates by transmitting short pulses of electromagnetic energy from an antenna into the pavement (*4*). These pulses are reflected back to a receiving antenna with the amplitude and arrival time that is related to the electrical properties of the pavement layers. The speed with which the electromagnetic radar wave travels in a particular layer is related to the dielectric constant of that layer.

For many pavement applications, the California Bearing Ratio (CBR) value of a soil is used as a measure of shear strength. The DCP was used for determining the strength profile of these pavements. It penetrates soil layers having CBR strengths in excess of 100 and also measures soil strengths less than a CBR of one.

ENGINEERING ANALYSIS

A total of 25 projects were selected for testing. Details of each of these projects are shown on the next page in Table 1. Fifteen projects were a four-layer system i.e., surface treatment, aggregate base, stabilized sub-base, and subgrade. For four-layer pavements the base was unbound granular aggregate, except for FM 1687, where 1% lime was used to stabilize the base as well. The remaining 10 projects were three-layer pavement systems i.e., surface treatment, stabilized base, and subgrade. Lime was used as the stabilizer in all but two of the projects with the percentage of lime varying between 3% and 10%. FM 3178 and FM 977 used 4% cement as the stabilizer. All the test data is compiled in the report submitted to TxDOT (*5*).

SECTION BY SECTION ANALYSIS

A section by section analysis of each base recycling projects was carried out in the study. In order to evaluate the FWD data and its variation along the project, a **target deflection value** was computed for each project. This was achieved by assigning a target moduli value for each pavement layer, using the average subgrade moduli for the section and the design layer thicknesses. Target deflection value is the acceptable deflection of the pavement based on the engineering properties of the pavement layer for the specified materials and the type of stabilizer. It was based on the level of stabilizer used and was obtained from the graphs of average layer moduli versus stabilizer content presented in the proceeding paragraphs.

Knowing the stabilizer content and the cross-section of the pavement, Figure 4 or 5 (as applicable) is used to calculate the anticipated moduli for the stabilized base and subbase layer. The modulus for the subgrade on each project was obtained from the FWD results. The average subgrade modulus was used in the analysis. Once the layer properties are known, maximum anticipated deflection can be calculated. This is called the **target deflection value**. If the pavement deflection is close to the **target deflection value** the pavement generally has acceptable layer strengths. If the deflections are very much lower than the target value, then the pavement section is very stiff and there is a chance that the pavement would crack. On the other hand if the deflections are far in excess of the target value then the stabilization was not effective and there are chances that the pavement section might deteriorate.

Pavement Rating

A tentative A through F classification has been given to each of the projects as defined below:

A. Structural strength appears sound, no surface distress. No sub-surface problems detected with GPR. No load associated damage anticipated.
B. Structural strength appears sound, however, some surface defects such as longitudinal cracking are present.
C. Some localized areas of weakness or high variability in moduli, little or no visual distress.
D. Some areas of weakness or high variability in moduli, with some defects present, normally longitudinal cracking.
E. Major areas of weakness/little or no stabilization present in some areas -- no surface distress.
F. Major areas of weakness/little or no stabilization in some areas -- with surface

distress.

When reviewing this classification it is critical to keep the following in mind:

1. With stabilized materials there is no guarantee that a low FWD deflection/high layer moduli will translate to a good long-term pavement performance. Therefore, sections with an A rating may not perform better than sections with a C rating. In study 1287 conducted by Texas Transportation Institute (*2*), it was concluded that at the high end of the stabilization scale the opposite was true, the stronger pavement provided poorer long-term performance. This loss in performance was attributed to shrinkage cracks from the stabilized layer. It is of considerable surprise to the research team that although several of the sections included in this study had very stiff bases with moduli in excess of 6,900 Mpa (1,000 ksi), no evidence of shrinkage cracking was found on any of the sections. Regular shrinkage cracking is normally observed as regular transverse cracks.

2. Continued monitoring of these sections will be required to evaluate long-term performance, since the majority of the sections was less than three years old at the time of testing.

3. Variability of the final section is largely related to the variability in the original section. These are low volume roads, and some of these roads initially, had little or no base.

4. With newly constructed flexible base layers, the moduli value anticipated for the base is usually three to four times the subgrade value (*2*). This is of significance in the sections with unstabilized bases only. For example with a subgrade of 69 Mpa (10 ksi) the anticipated unstabilized base moduli value would be 241.5 Mpa (35 ksi) and any increase above this would be attributed to the positive contribution of the stabilizer.

5. The subgrade moduli values obtained under the heavily stabilized subbase layers are inflated. This is because the moduli value obtained from back calculation is dependent upon the stress level seen by the material under FWD loading. With heavily stabilized subbase layers, low subgrade stresses would be transmitted to the subgrade during testing giving an apparent high subgrade modulus value (*2*). If the same subgrade was used on a traditional flexible base pavement substantial lower subgrade values would be back calculated. In general, the subgrade values obtained in this study should not be used for everyday design work.

With these issues in mind the ratings summarized on the following pages provide no guarantees of long-term pavement performance. In general, all we can conclude is that sections with an A or C rating are performing better at this point in time and that sections with a D and F rating appear to have problems. Sections with an E rating are of considerable interest because they indicate no substantial increases in layer strengths but they have little surface distress. One such section is discussed below.

SECTION 12

Road	County	Limits	Date	Structure
FM 111	Burleson	FM 60 to 3.4 Mi E	12-95	LSB (50-50) (3%), & 2 CST

RATING E

The FWD analysis on FM 111 indicates highly variable results. The first half of the section has reasonable deflections and good back calculated moduli values. The second half is considerably weaker. At location 3,965 m (12,999 ft) the measured deflection of 1,092 microns (43 mils) was more than twice the target value of 483 microns (19 mils). This section was stabilized with only 3% lime. The stabilizer content may have been insufficient. Several locations indicated, poorer subgrade strength in the weak areas. The section from 3,660 to 4,880 m (12,000 to 16,000 ft) appears similar to a flexible base section, with high deflection values, indicating little benefit from lime. It is anticipated that this section may deteriorate under load *(1)*.

This is an interesting section as it is built on poor subgrade comprised of highly plastic clays, but, currently there is little or no distress. If this section performs well for several years, it could have a major impact on how we design these base recycling projects in the future. Figure 1 summarizes the target deflection profile for FM 111 *(1)*.

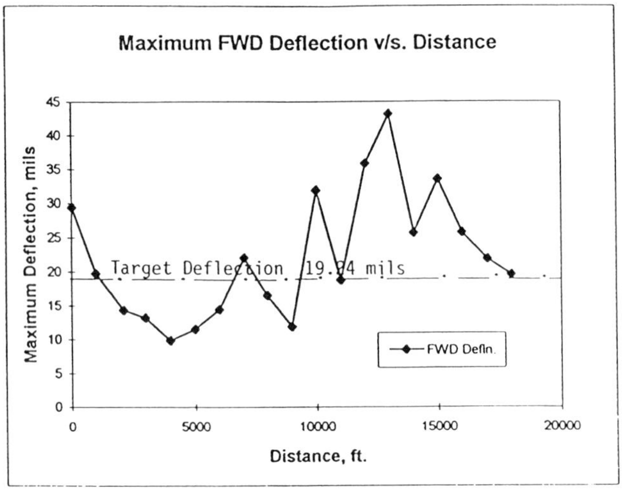

Figure 1. FM 111, Target Deflection Analysis.

FAILURE INVESTIGATION

Of the twenty-five projects evaluated in this study, pavement structural distress was observed in a few projects, viz. FM 542 and FM 246 *(1)*. Unusually high deflections in the pavement structure was also observed in FM 111 *(1)*.

Table 1. Section Location, Construction Date, and Layer Thicknesses

Section No	ROAD	DATE	STRUCTURE	RANKING	E_1	E_2	E_3
1	FM 3090	5-94	LSB (60-40) (3%), 8" FB, & 2 CST	C	21	37	3
2	FM 149	6-94	LSB (60-40) (7%), 8" FB, & 2 CST	B	30	172	2
3	FM 1486	7-94	LSB (60-40) (5%), 8" FB, & 2 CST	C	25	67	3
4	FM 542	6-94	LSB (60-40) (3%), 8" FB, & 2 CST	C	20	82	3
5	FM 542	4-95	LSB (60-40) (5%), 7" FB, & 2 CST	D	45	119	3
6	FM 3478	3-95	LSB (60-40) (6%), 11" FB, & 2 CST	A	29	179	3
7	FM 1362	5-95	LSB (80-20) (4%), 7" FB, & 2 CST	C	16	74	3
8	FM 246	5-95	LSB (40-60) (6%), 9" FB, & 2 CST	C	17	100	2
9	FM 977	7-95	LSB (60-40) (3%), 11" FB, & 2 CST	B	17	80	3
10	FM 27	9-95	LSB (60-40) (5%), 12" FB, & 2 CST	A	11	96	2
11	FM 3411	9-95	LSB (50-50) (6%), 10" FB, & 2 CST	C	21	40	3
12	FM 111	12-95	LSB (50-50) (3%), & 2 CST	E	31	----	2
13	FM 1124	12-95	LSB (50-50) (4%), 5" FB, & 2 CST	C	42	74	3
14	FM 244	5-95	LSB (60-40) (4%), 12" FB, & 2 CST	C	52	64	3
15	FM 2223	9-96	LSB (50-50) (10%), 12" FB, & 2CST	A	22	73	3
16	FM 1687	5-96	LSB (60-40) (XX), XX", & 2 CST	A	47	28	2
17	FM 977	2-96	LSB (30-60) (4%), & 2 CST	A	90	----	3
18	FM 978	4-96	LSB (40-60) (5%), & 2 CST	C	105	----	2
19	FM 1373	7-96	LSB (50-50) (XX), & 2 CST	A	236	----	3
20	FM 3178	8-95	CSB (50-50) (4%), & 2 CST	E	18	----	2
21	FM 977	5-96	CSB (50-50) (4%), & 2 CST	A	72	----	2
22	FM 1935	XXX	LSB (XXX) (4%), & 2 CST	A	54	4	3
23	FM 975	6-96	LSB (60-40) (3%), & 2 CST	C	48	----	2
24	FM 2446	XXX	LSB (XXX) (3.5%), & 2 CST	A	39	----	2
25	FM 2780	XXX	LSB (XXX) (3.5%), & 2 CST	F	73	----	1

E_1, E_2 and E_3: Moduli in MPa for base, subbase and subgrade layers XXX: Not known
---- : 3-layer pavement

In view of the anomalies encountered in these three sites, it was decided to carry out some additional investigations. The purpose of this additional investigation is to analyze the problem of severe longitudinal cracking of the pavement encountered in a few isolated locations. The aim of this effort is to propose a mechanism to explain the cracking and to make recommendations for the prevention of such cracks from occurring in the future. Results of the laboratory testing is awaited at the time of writing this paper *(1)*.

DISCUSSION OF NON-DESTRUCTIVE TESTING DATA

FM 542 : Severe longitudinal cracking was found between two bridge structures (approx. 6 miles south of US 79 at 5508 ft. from start of section). The area of distress is located in a fill zone. The ground water table was measured at 4.5 m from the existing subgrade level. Sub-surface comprises of CLAY with traces of Sand and calcareous nodules. Subgrade modulus is very low in this area, approximately 50% of section average. DCP reading indicated very poor subgrade in cracked areas, noted as heavy clays. Subbase is very strong in this location, 7970 Mpa (1155 ksi). Average stabilized layer is stiff for this section, this could be related to severity of cracking over the heavy clays. Outside the cracked areas, deflections are low and no distresses are present *(1)*.

FM 246 : Some longitudinal cracks were also observed on this road. The ground water table was measured at 2.2 m from the existing subgrade level. Sub-surface comprises of CLAY with traces of Sand and calcareous nodules. The average deflection of 391 microns (15.4 mils), was less than the target deflection of 399 microns (15.7 mils). High deflections in the first 305 m (1000 ft.) and throughout the section attributed to low strength subgrade. The strength of the subbase is low in the first 305 m (1000 ft.), could be "start-up" problems *(1)*.

FM111 : Unusually high deflections were observed in isolated spots along this road. The ground water table was measured at 4.6 m from the existing subgrade level. Sub-surface comprises of CLAY with traces of Sand and calcareous nodules. Highly variable results were obtained from FWD analysis. The first half of the section has reasonable deflections and good back calculated moduli values. The second half is considerably weaker. At location 3962 m (12999 ft.) the measured deflection of 1092 microns (43 mils) was more than twice the target value of 483 micron (19 mils). Only 3% lime was used as a stabilizer which may be insufficient. Several locations indicated, poorer subgrade strength in the weak areas. The section from 3658 m (12000 ft.) to 4877 m (16000 ft.) appears similar to a flexible base section, little benefit from lime. It is anticipated that this section may deteriorate under load. This is an interesting section as it is built on poor subgrade, high PI clays, but, currently there is little or no distress. If this section performs well for several years, it will have a major impact on how we design these base recycling projects in the future *(1)*.

INITIAL CONCLUSION

Triaxial Compression Test was conducted on the cores from the 3 sites. From the results summarized in Table 2, it seems that cracking in the pavement layers is partially due to very stiff pavement layers and the presence of high shrink/swell subgrade soils. Although FM 111 has a few areas with higher deflections, currently there is little or no distress on this pavement structure. FM 542 and FM 246 have achieved a higher compressive strength, but cracks are observed along the pavement surface *(1)*. It would be premature to draw conclusions at this point in time because we are still awaiting the laboratory test results.

TABLE 2

Road	Triaxial Compression Strength	Distress observed
FM 111	1647.6 kPa	High Deflections, no cracking.
FM 246	5864.4 kPa	Longitudinal cracking in few areas.
FM 542	6288.2 kPa	One big longitudinal crack.

CONCLUSIONS AND RECOMMENDATIONS

Correlations between layer moduli, DCP Index, and stabilizer content are presented in this paper. General trend of the correlations, regression analysis performed and suggested design line depends on a number of variable factors. Design charts presented in this paper is an attempt to advance state-of-the-art of performance monitoring and determination of in-place engineering properties of recycled and stabilized layers. Also, to provide some guidelines to the design engineer. With continued performance monitoring of such pavements, these design charts can be modified and a proper design methodology for designing recycled and stabilized pavement layers can be developed. In developing these correlations, expectedly there was a lot of scattering of the data points. Plotting the design line after subtracting one standard deviation, it was observed that this suggested design line generally caters for the lower bound envelope of the data points. Also, it increases the reliability of the prediction.

GROUND PENETRATING RADAR

GPR is a useful tool in the performance evaluation of pavement layers in terms of checks on consistency, moisture trapping and confirmation of layer thickness. In this project, there was no build-up of moisture in stabilized layers. Also, results of the GPR helped in understanding the variability of strengths reported by FWD due to the variation in layer thicknesses.

CORRELATION OF DCP INDEX WITH SUBGRADE MODULI

Procedures presented by the US Army Corps of Engineers *(6)* was used for measuring soil strength and correlating DCP Index with subgrade moduli values. On selected sites the DCP was used to attempt to estimate a modulus profile with depth to compare the back calculated values with those computed from the FWD. Figure 2 illustrates the correlation of DCP Index v/s. Subgrade Moduli for CH (highly plastic clay) soils whereas, figure 3 shows the correlation of DCP Index v/s. Subgrade Moduli for CL (clay of low plasticity) soils. The solid line indicates the regression line or the "best fit" line for the particular distribution. The dotted line indicates the recommended Design Line which was obtained by subtracting one standard deviation from the best fit line *(1)*.

CORRELATION BETWEEN STABILIZER CONTENT AND FWD

Figure 4, Correlation between Stabilizer Content and FWD (3 layer) and Figure 5,

Correlation Stabilizer Content and FWD (4 layer) represents the correlation between the stabilizer content and the layer moduli values as calculated by FWD and MODULUS, based on the field measured stiffnesses of the stabilized pavement layers.

Figure 4 indicates the correlation for a 3 layer pavement system while figure 5 indicates the correlation for a 4 layer pavement system. The solid line in the figures indicate the average line based on the field performance data. Dotted line is the suggested "design line". Design line was obtained by deducting one standard deviation from the average (best fit) line. In figure 5, the correlation line and the design line show no increase in the moduli of the stabilized layer beyond 6% of stabilizer. This is because it has been found in a study (*2*) that stabilizer in excess of 6% does not serve any useful purpose in terms of strength gain (*1*).

RECOMMENDATIONS

This research has demonstrated the effectiveness of non-destructive testing equipment like GPR and FWD in the evaluation of completed projects by checking if the construction of the pavement layers conform to the specifications. The FWD deflection profile after construction has been compared with the design deflection profile. Sure enough, any abnormal behavior in the FWD deflection profile has inherently pointed to some problem in the pavement structure. The design deflection profile was determined from the average subgrade moduli and the upper layer moduli values obtained from the charts presented in this paper. It is clearly evident that a field testing protocol such as using the GPR and FWD can be used to evaluate the performance of stabilized bases in terms of resistance to moisture damage and improvements in long term structural strength of the pavement layers.

Design of recycled and stabilized pavement layers are currently being based on ad hoc procedures. Although these charts would still need refinement, it would definitely be a starting point in the rational design of such pavement layers.

REFERENCES

1. I. M. Syed, and T. Scullion, *Determining In-place Engineering Properties of Recycled Pavement Layers,* Research Report 3903-1, Texas Transportation Institute, Highway Materials Division, College Station, Texas, 1997.

2. D. N. Little, T. Scullion, P. B. V. S. Kota, and J. Bhuiyan, *Identification of the Structural Benefits of Base and Subgrade Stabilization*, Research Report 1287, Texas Transportation Institute, Highway Materials Division, College Station, Texas, 1994.

3. J. Uzan, T. Scullion, C. H. Michalak, M. Paredes, and R. L. Lytton, *A Microcomputer Based Procedure for Back-calculating Layer Moduli from FWD Data*, Research Report 1123-1, Texas Transportation Institute, Highway Materials Division, College Station, Texas, 1988.

4. T. Scullion, Y. Chen, and C. L. Lau, *Colormap - User's Manual with Case Studies*,

Research Report 1341-1, Texas Transportation Institute, Highway Materials Division, College Station, Texas, 1995.

5. I. M. Syed, and T. Scullion, *Determining In-place Engineering Properties of Recycled Pavement Layers,* Research Report 3903-2 (under review of TxDOT), Texas Transportation Institute, Highway Materials Division, College Station, Texas, 1997.

6. S. L. Webster, R. H. Grau, and T. P. Williams, *Description and Application of Dual Mass Dynamic Cone Penetrometer*, Research Report GL-92-3, USAE Waterways Experiment Station, Geotechnical Laboratory, Vicksburg, Mississippi, 1992.

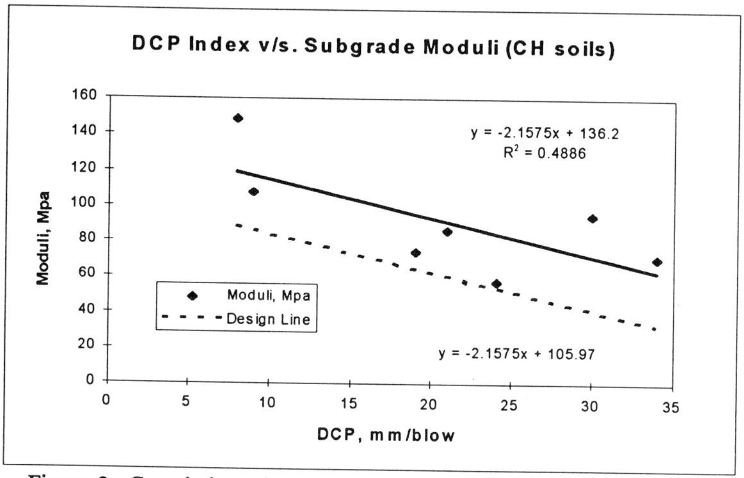

Figure 2. Correlation of DCP Index v/s Subgrade Moduli (CH soils).

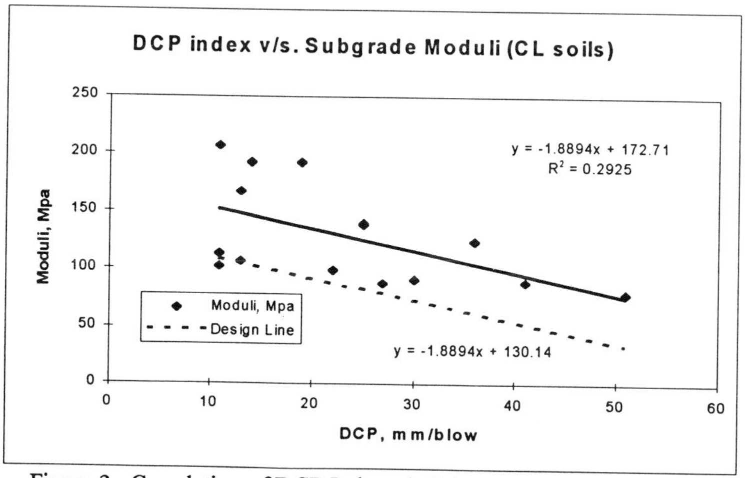

Figure 3. Correlation of DCP Index v/s Subgrade Moduli (CL soils).

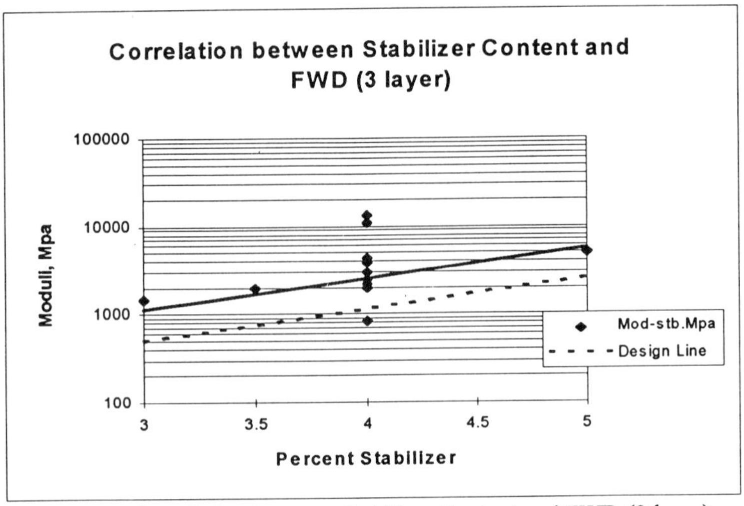

Figure 4. Correlation between Stabilizer Content and FWD (3 layer).

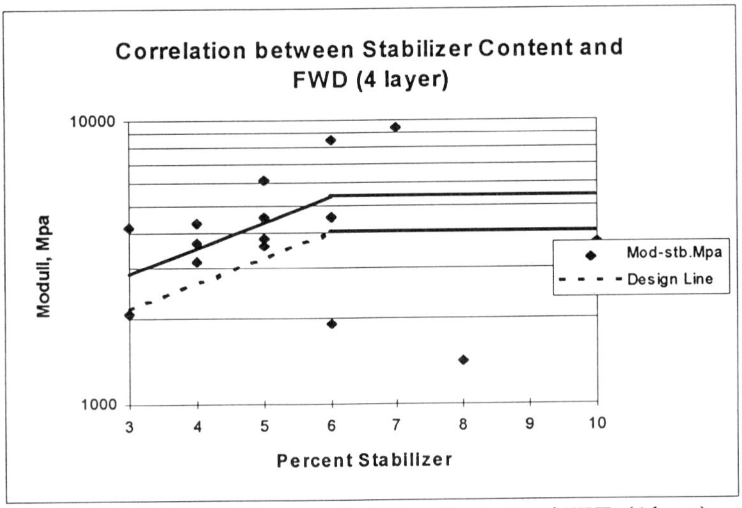

Figure 5. Correlation between Stabilizer Content and FWD (4 layer).

Fifth International Conference on the
BEARING CAPASITY OF
ROADS AND AIRFIELDS

INITIAL RUTTING ON RECONSTRUCTED ROADS AND HOW IT RELATES TO FWD TESTING

CARL A. LENNGREN[1] AND RUNE FREDRIKSSON[2]

[1]Researcher, Vägverket Konsult, S-46228 Vänersborg, Sweden

Fax: +46521279124, E-mail: carl.lenngren@swipnet.se

[2]Researcher, Vägverket Produktion, S-78187 Borlänge, Sweden

Fax: +4624394220, E-mail: rune.fredriksson@vv.se

ABSTRACT

Fatigue and rutting equations related to strain have since long been established in mechanistic overlay design. Rutting is by tradition related to the top of subgrade vertical strain. Less attention was originally given to unbound base and subbase layers. In theory it is possible to use top of base and top of subbase criteria for the design. However, the respective thickness of these layers would certainly influence the proneness for rutting also. In recent years this problem has been addressed and there are now a number of equations from various research institutes throughout the world. Obviously, the initial rutting rate is higher for any newly constructed or reconstructed road and many new models would incorporate this fact. In order to study the feasibility of using models suggested, intense surface characteristic measurements and falling weight deflectometer testing were done on a research road that had been closely monitored for some years prior to a reconstruction. The first test was carried out a month after reconstruction, a second test occurred after the first spring thaw and yet a third test was accomplished about a year after the first test. Thus, the initial rutting and its relation to layer stiffness, stress and strain on the three occasions could be determined. An important aspect is to determine when the initial rutting rate tapers off to a linear increase. As the initial rutting is depending on construction compaction, non-destructive testing in this fashion could be utilized for quality control.

Keywords: Backcalculation, Falling weight deflectometer

INTRODUCTION

By tradition, asphalt concrete road failure is defined as either fatigue failure resulting in cracking or permanent deformation resulting in rutting. The fatigue failure is obvious; all you have to do is to define a certain amount of cracking per area unit, which would count as failure. The rutting failure criterion is by the same token defined as a certain amount of measurable depression on the surface. Perhaps, 20 mm as derived from a 1.2-m long straightedge.

The rutting rate has been related to subgrade strain, i.e. the vertical compression strain on top of the subgrade. This relationship is very confusing to some people, as there is no criterion for the unbound base and subbase layers. It is hard to accept that there should be no permanent deformation closer to the load, albeit placed material should be better off than the subgrade. Many of the traditional criteria were derived from the AASHO road test[1]. What the researchers then actually concluded was that subgrade strain was the best predictor of rutting, not that all rutting occurs in the subgrade. Certainly, since any strain anywhere in the structure can be related to any other strain somewhere else in it, it is plausible to establish such a relation for any given case. The problem, however, is that the subgrade strain is only really valid for the case you established. If something changes, like the type of load and the way it is being applied you must calibrate your criterion for the new situation. Brown mentioned in his keynote lecture at the Eighth International Conference on Asphalt Pavements that "Subgrade strain is a holy cow that should be killed[2]". The present authors agree fully and indeed there are a number of papers in the proceedings from the conference that suggests plausible rutting models[3,4]. They are very important as such as they do address the behavior of granular materials. In addition to rutting models for asphalt concrete, which are usually a Burger's model, one would have a much better understanding of the rutting process for any given structure, not just only an investigated case.

Possibly, due to higher tire ring pressures initial rutting on new and reconstructed roads seems to be a problem that is becoming more and more common. Traditional quality control like coring and checking asphalt concrete layer thickness does not really tackle the problem. The coring is not a non-destructive test limiting its use. Further, the problem may be found a bit deeper than the asphalt layer. Some road authorities have suggested that falling weight deflectometer (FWD) testing could perhaps be used to check various parts of the structure for critical strains. However, as most bearing capacity consulting staffs know, much compaction by traffic does occur during the first year of operation. Therefore, one is reluctant to predict life times based on virgin road tests.

The present paper is not an attempt to verify any complete model. Rather it is addressing the usefulness of such models from a practical standpoint. It looks at parameters that can be assessed with current practices and how to deal with it. Early rut development may not be that easy to determine for instance and by experience we know that the initial rutting rate is very high. We also know that it does vary much possibly due to inadequate compaction control. The paper addresses some of these aspects and what you could expect from early non-destructive testing. The test road presented here seemed to offer a golden opportunity to some insight into the problem.

BACKGROUND

By the efforts of present co-author Rune Fredriksson a 26 km long test road has been thoroughly investigated by laser road surface testers (L-RST) and falling weight deflectometers several times per year since 1993. Rolling deflection meters, ground penetrating radar and other test equipment have also been visiting the site. Thus, plenty of background material was available when parts of the road were reconstructed in 1996. Some parts included complete excavation of the road, insertion of an insulation layer, and a complete backfill with fresh granular material. As it seemed, monitoring rut depth growth on these sections should perhaps enlighten the mechanisms of early rut growth.

The site is situated near town Falun in inland Sweden. It is far from the coast at latitude 61 degrees North. Winters do get cold with an average freezing index of about 1000 degree days Celsius. Traffic is sparse about 2500 AADT of which 11 % are trucks on this two-lane road. The carriageway is 7 - 9 meters wide and the speed limit is set to 90 km/h.

There were three sites that were excavated and filled back with new material. Two 50-meter long sections each on a strong gravel subgrade interspersed with some silt lenses. There was also an 86-meter long section on a clay subgrade. These sections were troubled by uneven frost heave in the past, hence the removal of the in-place material. In order to reduce frost penetrating into the subgrade the new reconstructed sections were outfitted with an insulation layer placed below the new backfill.

The entire test road was measured with the L-RST equipment in September 1996, a little more than a month after the reconstructed parts had been opened for traffic. Within days a falling weight deflectometer was also brought to the site. The testing equipment was back right after the spring thaw in April 1997. An additional L-RST measurement was done yet again in June 1997. Finally, there were both L-RST and FWD tests conducted about a year after the first test in October 1997.

FIELD TESTING

The reconstructed sections are shown in Tables 1 and 2.

Material	**Thickness, [mm]**
Asphalt Concrete Layers	140
Unbound Gravel Base	700
Frost Insulation Layer	50
Moraine Clay Subgrade	Semi-infinite
Annual Average Daily Traffic is 2620	
Road Width is 7.4 m	

Table 1. Reconstructed Road Site 1

Material	Thickness, [mm]
Asphalt Concrete Layers	120
Unbound Gravel Base	500
Frost Insulation Layer	50
Coarse Gravel Subgrade, with some silt pockets	Semi-infinite
Annual Average Daily Traffic is 2540	
Road Width is 8.0 m	

Table 2. Reconstructed Road Sites 2A and 2B

Surface characteristics measurements

As mentioned above a surface characteristics meter had been brought to the site several times per year. In this case the same individual vehicle was used throughout, the VTI (Swedish Road and Transportation Research Institute) Laser Road Surface Tester.

The L-RST system as such consists of mainly three parts. The laser range finder sensors, the accelerometer and the pulse distance recorder make up the data gathering part. The data are processed in real-time by the signal processing card unit into parameters by choice, e.g. rut depth, roughness et cetera. Finally, an off-the-shelf computer stores the information on 3.5-inch diskettes or any other standard computer medium.

There are seventeen optical laser range finders units mounted transversally on a support beam in front of an ordinary van. Some of the sensors pulsate at 16 kHz, others at 32 kHz. The high-frequency sensors are used for crack detection and macrotexture measurements. Needless to say this allows short wavelengths to be recorded. The 32 kHz sampling rate translates to a reading about as frequent as the spot diameter of 0.5 mm when travelling at 25 m/s.

A transversal profile is assessed on the mean averages of the sensors for every decimeter traveled. Then the so-called string method is used to determine the maximum rut depth. I.e. the largest distance between the surface and an imaginary string stretched across the roadway. The left rut depth and the right rut depth are also determined. The average rut depth will be stored every fifth, tenth or twentieth meter traveled. The percentages of section lengths having values over certain preset values are also stored to reflect the variability.

The roughness of the road is sampled in each wheel path. Two accelerometers record the movement of the support beam so its movement can be deducted by integrating the signal twice from the accelerometer. The comparison of the two signals for this particular system is actually done in the velocity mode. Thus, two longitudinal profiles are recorded, one for each wheel-path. Outputs are given as the quarter-car based index IRI (The International Roughness Index) or filtered for a number of different wavelength bands.

Other parameters gathered include crack detection algorithms, cross slope, macrotexture, and megatexture. Megatexture is the intermediate wavelength band between the longer wavelength band related to unevenness and the shorter macrotexture band, about 0.1 - 0.5 meter.

As mentioned the VTI L-RST stores raw data and data treated in real time in five, ten or twenty meter increments. Even if the primary goal here was to study rutting, as the present material would be used for research and since there is no penalty for storing more parameters, it was decided to store all data possible, like roughness, texture and crack detection. Naturally, the distance traveled and reference to on board video recorders are also stored in the data sets. No less than 232 parameters were stored for every twenty meters traveled during the measurements.

Even if it is possible to engage all seventeen range finders to assess the rut depth, only eleven were used for the present study. The reason being that by tradition eleven sensors have been used for over ten years time and secondly that the fall 1996 test was not set up to use more than eleven sensors. The June 1997 and October 1997 test collected by both methods and thus it is possible to compare the two ways of data gathering. As it turned out the number of sensors used for calculating the rut depth did not affect the results, but the seventeen- sensor configuration yielded about half a millimeter higher rut depths. However, when dealing with shallow rutting, it is a good idea to use as many sensors as possible. Miniscule rutting is difficult to detect, and driver variability is reduced the more sensors there are.

All tests were repeated at least once. The crew used was trained for research work, and had over ten years of experience of this type of mission. The repeatability was very good for the sites investigated, except for one occasion at one time. As the faulty run was easy to conclude it was excluded from the present study.

FWD testing

The FWD testing was synchronized length-wise with the L-RST samples. FWD tests were made in the middle of each 20-meter long L-RST cell in the northbound direction. The southbound tests occurred just across the road, i.e. the same sections. Thus, southbound tests were not staggered, which is otherwise a common practice. L-RST cells did not match the southbound tests perfectly as the shift between cells depended on the location of the starting point. The shifts coincide only if the length of the link in question is a multiple of 20m. In the present study ten cells could be used in the northbound direction. Only eight could be utilized in the southbound lane.

All tests were done with three load levels of roughly 40, 50 and 70 kN peak load. The rise time was approximately 24 ms and a dual mass system was used. The load level did not affect the backcalculated moduli much. Hence, one would assume the materials not to be very stress sensitive. In the following a nominal load of 50 kN is discussed throughout, if nothing else is mentioned.

Backcalculation

A linear elastic backcalculation program CLEVERCALC 3.8 was used for determining the E-moduli of the layers at all load levels. Hence, any stress sensitivity prevailing in the materials involved could be assessed. The program will also calculate stresses and strains at various critical points in the structure. Further, by using forward linear elastic programs, stresses and strains could be determined anywhere in the structure.

Normally, one would start this type of analysis with a three-layer model. Bound layers, unbound layers and subgrade. However, as the layer's thickness was known to a high accuracy, a four-layer model was used right away. Thickness values from Tables 1 and 2 were used.

The common option of using a rigid base was opted out in this case as it was found that it did not improve the fit, neither was there any known rock within near the subgrade surface. The fit of the backcalculated deflection basin as compared to the measured one can be expressed as a root mean square error per sensor. As it turned out the four-layer model proved to yield root mean square-values around one percent or less in general. In effect it means that the model adequately describes the deflection basin as measured at the surface. The program solved all of the basins for convergence within a few iterations. In most cases the insulating layer ended up with an E-modulus of about 50 - 80 MPa. It is a surprisingly good result considering the depth and the meager thickness of this layer. Thin layers buried deep down in the structure usually do not contribute much to the bearing capacity; hence they do not show up measuring the deflection at the surface. However, a few stations ended up with unrealistic values of this particular layer and those were recalculated with a fixed value of 80 MPa.

The backcalculated E-moduli for the asphalt concrete did not vary much over time, except for the effect of temperature. Most test basins were near 6000 MPa for a temperature-adjusted modulus at 15 degrees Celsius. The base and subbase modulus varied with the season. Figures 1 and 2 show the variation over time for site 1. The nomenclature 96/2 n refers to the second measurement 1996 in the northbound direction.

The inconsistent values for the spring measurement (97/1) may indicate that the soil could still be frozen or just recently thawed out.

Figure 1. Site 1 Base Modulus

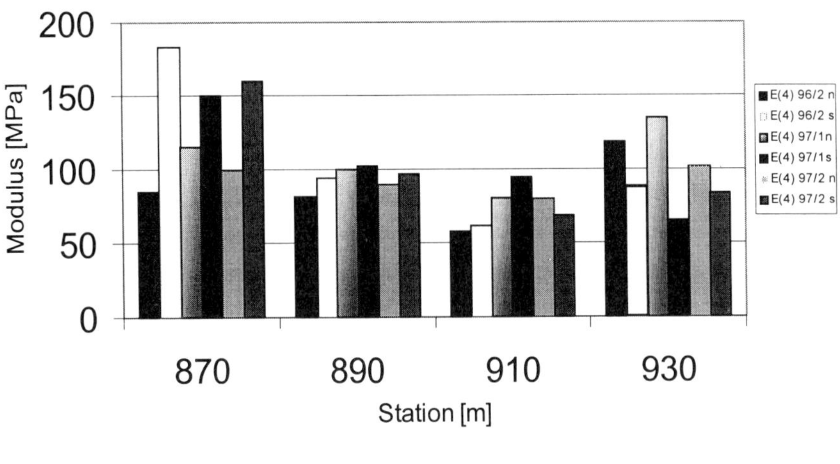

Figure 2. Site 1 Subgrade Modulus

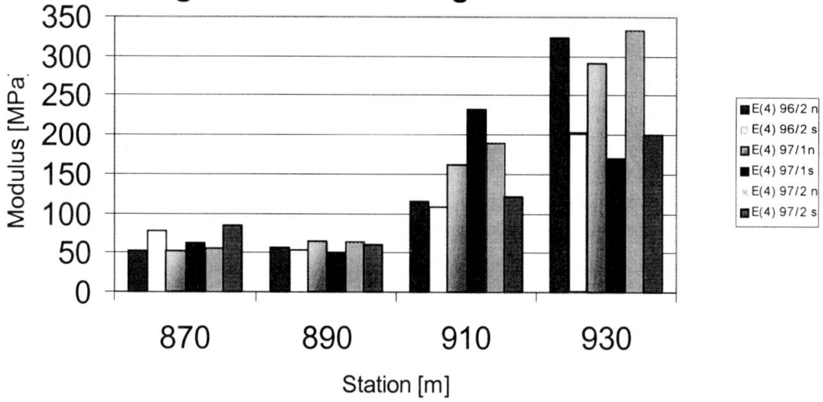

Strains on top of the subgrade were found to be moderate for all three sites. The highest strains backcalculated were around 250 microstrain translating into more than 13 million standard axle loads in the Swedish Design Manual[5]. With a 2 percent traffic growth these sites will experience 1.66 million axles for the next 20-year period, which is considerable less. Thus, subgrade rutting is likely to be very diminutive, as the strains do not exceed 250 microstrain.

RESULTS

Figure 3 shows the rut progression at the four times the three sites have been measured so far. They are divided into the two directions as they do differ significantly.

Figure 3. First year rut growth

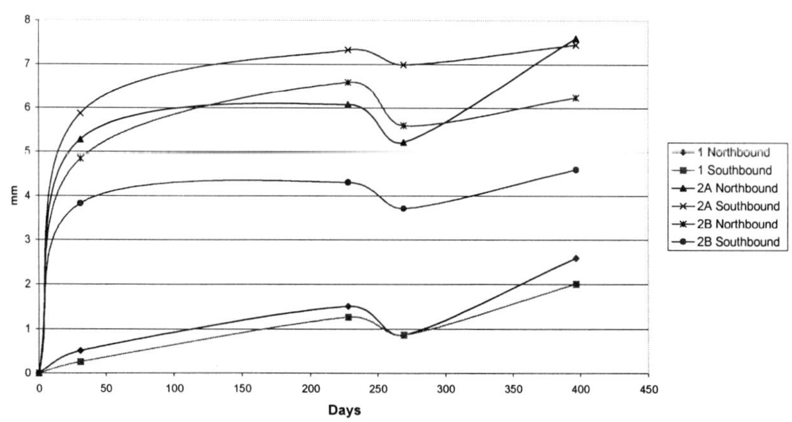

By looking at Figure 3 it is clear that sites 2A and 2B having a very high initial rutting the first month must be considered a failure. Could it be a failure of the insulation layer? It is not quite clear at the present time, but further research should show if this could be the case. Site 1 shows a moderate rut growth rate, which is close to what one would hope to expect. The first month seems to have a rut rate about twice as high as for the rest of the year.

Day 200 is around the time of spring thawing. The reason for the rutting to decrease between April and June is likely due to receding frost heave and a reduction of roughness, which was also assessed from the L-RST data gathering. The summer of 1997 was unusually hot in the area; the accelerated rut growth for two of the sections (1 north and 2A north) may be an effect of asphalt rutting. The backcalculated asphalt strain was significantly higher for these particular stations. It was near 300 microstrain, which is close to the limit 335 stated for this amount of traffic in the Swedish Design Manual. The other stations hovered around 200 – 220 microstrain. The sections exhibiting the extra summer rutting were too few to make this relation statistically significant and were also limited in range. Future coring will tell if these are actually rutted in the asphalt layer.

A regression analysis was performed between backcalculated FWD data as the independent variable and the rutting and rate of rutting as the dependent variable. As there was some variability within each site, 20-m long subsections were compared. A total of 18 cells could be used; two had to be skipped because the southbound L-RST run did not match the FWD test adequately.

The E-modulus and strain proved to be very poor indicators of the rutting. In a previous study of a number of surface characteristics parameters and E-modulus and strain on old pavements due for overlay, a poor to fair correlation between top of subgrade strain and rutting was established[6]. The relationship was stronger for cohesive soils. Assuming that the stress in the unbound layers was better related to the initial rutting it was also investigated in the regression study. Interestingly, it correlated much better than the strain did. The stress taken in the middle of the unbound layer yielded a coefficient of determination of .63, which is a fair relation, considering that some rutting may take place in the asphalt layer also. Further, one might suspect that a full integration of stresses in the entire layer would be a much better predictor as suggested by Bonaquist and Witczak. Future coring, when more data from subsequent years are available will tell if and how much rutting took place in what layer.

Even if a fair relationship was established between unbound base stress and rutting, no other backcalculated parameter correlated better than the subgrade stress and the rutting after one year. (Incidentally, the correlation between the base/subbase stress and the subgrade stress was also rather good, .71.) The subgrade stress-rutting correlation yielded a coefficient of determination of .83. Is this result surprising? Considering that the subgrade stress is a function of how load is being spread in the structure this may not be such a bad predictor of rutting after all. Hence, the subgrade stress will reflect the bearing capacity of all layers above, which is why the correlation was better. Most of the rutting is likely to occur in the unbound base and subbase layers. The regression plot is shown in Figure 4. The standard error of estimate is 1.03, the slope .26 is and the intercept is -1.70.

Since the rut progression is straightforward as seen in Figure 3, the coefficient of determination is also high for the stress vis-a-vis rut depth at the other occasions too. .80, .77 and .80 for the September 96, April 97 and June 97 rut depths respectively.

Figure 4. Subgrade Stress Vs. Rut Depth

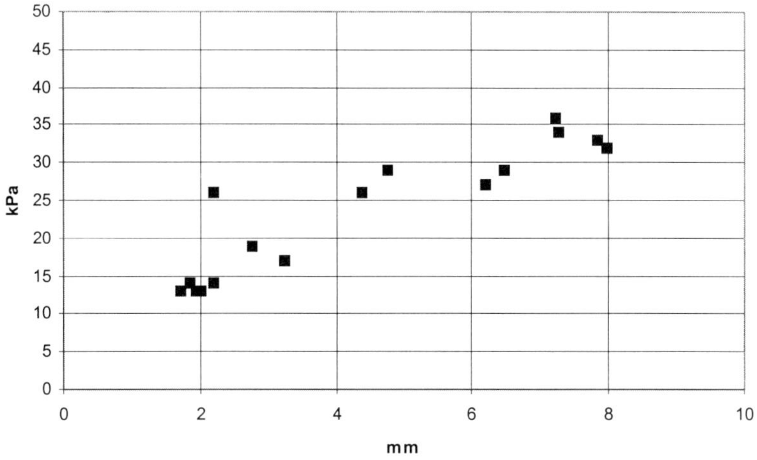

DISCUSSION

The stress on top of the subgrade could serve as an indicator of early rutting development. For the sections investigated it remains to be seen if the rutting rate will change or not. Certainly, rates changed after the first month and some of the seasonal variation may confuse the results. It is now utmost important to follow these sections further, especially during the summer for monitoring asphalt layer rutting. The effects of studded tire wear should also be kept in mind. Even if the traffic volumes are low, the wear should contribute to some of the rutting. As the ruts grow one should also study the shapes of them in order to conclude where in the structure the rutting takes place.

Nunn[7] stated that there is very little structural (subgrade) rutting on very high volume roads in England. The reason being that the ample design for such roads infers very little strain on the subgrade. Just as in the present case the effective strain was far below the limit stated in the design criterion. One may conclude that by limiting the strain on top of any layer rutting will be avoided in that particular layer.

CONCLUSIONS

The initial rutting assessed one year after reconstruction correlated well with the backcalculated stress on top of the subgrade layer one month after the road opened for traffic. The reason being that the stress is a function of the load spreading capabilities of all layers in the structure. As such it can most likely serve as a fair predictor of rutting and an approximate method for newly constructed and reconstructed structures. However, a more constituent model incorporates every layer in the structure and integrates the stresses throughout every layer. The computing resources for such an effort are no longer limited to main frame computers. It remains to be seen though if parameters like shear strength and pore pressures can be easily assessed in the field. If not, complex but accurate models would be of little value.

REFERENCES

1. "The AASHO Road Test History and Description of Project" Highway Research Board. Special Report 61A. Publication 816. 1961.

2. Brown S. Keynote Address, Eighth International Conference on Asphalt Pavements, University of Washington, Seattle, WA, August 10, 1997.

3. Bonaquist, R. F. and Witczak, M. W. "A Comprehensive Constitutive Model for Granular Materials in Flexible Pavement Structures" Proceedings Eighth International Conference on Asphalt Pavements, Vol. 1, pp. 783-802, University of Washington, Seattle, WA, 1997.

4. Ullidtz, P. "Modelling of Granular Materials Using the Discrete Element Method" Proceedings Eighth International Conference on Asphalt Pavements, Vol. 1, pp. 757-769, University of Washington, Seattle, WA, 1997.

5. Swedish Design Manual, "Väg 94 " Del 3. [Swedish] Vägverket, Publikation 1994:23, 1994.

6. Lenngren, C.A. "Relating Bearing Capacity to Pavement Condition" Royal Institute of Technology Dissertation, TRITA VBG 9001 ISSN-0349-5744, 1990.

7. Nunn, M. "Long-Life Flexible Roads" Proceedings Eighth International Conference on Asphalt Pavements, Vol. 1, pp. 3-16, University of Washington, Seattle, WA, 1997.

Fifth International Conference on the
BEARING CAPASITY OF
ROADS AND AIRFIELDS

STRUCTURAL PROPERTIES OF NEW ENGLAND SUBBASE MATERIALS OF FLEXIBLE PAVEMENTS

K. WAYNE LEE[1] AND JEFFREY S. DAVIS[2]

[1]Professor and Dir. of Grad. Studies, Dept. of Civil & Environmental Eng., University of Rhode Island, Kingston, RI 02881, USA, (401)874-2786, lee@egr.uri.edu
[2]Civil Engineer, Engineering Department, Town of Narragansett, Narragansett, RI 02882, USA, (401)782-0669, jdav8136@uriacc.uri.edu

ABSTRACT

An attempt was made to compile a diversity of subbase materials and practices used in New England, USA. Most New England states have specifications for reclaimed materials, but Connecticut (CT) and Massachusetts (MA) are the only states routinely using these in the construction of the subbase layer. CT allows stockpiles of reclaimed materials maintained at inspected private sites and blended there to meet specifications. MA specifies full depth in place milling of the entire pavement structure down to the subgrade.

The present study determined structural properties of various subbase materials with and without reclaimed asphalt pavement (RAP). Granular subbase materials, RAP and in-situ blended materials provided by the New England state highway agencies (SHAs) were tested mainly to determine the resilient modulus (E_{SB}) using the procedure of AASHTO T292-91. It was observed that the majority of granular materials, with 30 percent or more larger size material discarded, have E_{SB} values less than 69,000 kPa (10,000 psi). The moduli of all blends with RAP were higher than the granular material used in the blend. In addition, fundamental properties, coefficient of permeability, resilient modulus and layer coefficients for subbase materials were determined. The layer coefficient for the MA cold in-place recycling (CIR) material was 0.16. It has a permeability in the range of a glacial till (10^{-3} to 10^{-5}). These are excellent characteristics for a subbase material used primarily for load bearing.

Optimum performance characteristics for blends with RAP were also developed. A blend containing up to 50% RAP is recommended for use as the subbase material. However, it was found that all blends have higher moduli and layer coefficients than granular materials. Finally, it was recommended that the procedure of AASHTO TP46 with a 150 mm x 300 mm (6 in. x 12 in.) specimen would be used in the next phase study to determine resilient moduli not to loose the larger size aggregates and to simulate the field condition more closely.

INTRODUCTION

Designs for pavement structures are mainly based upon the strength characteristics of the materials used within the structure. However, it has been observed that materials used in the subbase layer have received the least amount of attention to define strength. Currently, a number of methods exist to characterize the support strength of granular subbase materials. The 1993 AASHTO Guide for Design of Pavement Structures (AASHTO Guide) recommends assigning appropriate layer coefficients based on the resilient modulus value expected (AASHTO 1993). Resilient modulus is defined as deviator stress over recoverable strain measured during specific incremental loading sequences which attempt to recreate vehicular loading conditions on pavement structures. Resilient modulus testing procedures have undergone rapid changes over the last few years. Furthermore, there is no comparative analysis between states for procedures and materials currently being used for the subbase layer. Therefore, the New England Region requires an upgrading and standardization of procedures for the determination of resilient modulus values of subbase materials. In order for the AASHTO Guide recommendations to be implemented regionally, subbase material properties should be characterized. A regional sharing of each state's procedures should be of great benefit for the improvement of road structural designs, since the region shares similar climatic and geological characteristics.

Subbase structural designs today are complicated by State agency's interest in utilizing recycled waste materials blended with existing aggregates. Some States have performed analysis of these materials; however, no history of these analyses are available to aid in future analysis of the various recycled materials in structural subbases. On a regional basis, aggregate types common to more than one State could provide a basis for the development of optimum structural and drainage characteristics for these aggregate types blended with several individual or combined recycled materials.

Therefore, the purposes of the present research were: (1) to compile a database on aggregate properties by aggregate types common to New England; (2) to characterize natural aggregates and recycled material/aggregates blends provided by participating State agencies; (3) to develop optimum performance characteristics for aggregate type and recycled blends which are provided by State agencies; and (4) to recommend appropriate testing for State agencies to develop optimum properties for specific sources of blended materials.

PROPERTIES OF SUBBASE MATERIALS USED IN NEW ENGLAND

Current New England practices on subbase materials and structures have been gathered from participating states in the present study.

Typical Pavement Cross Sections Typical pavement cross sections were surveyed from the previous research project at the University of Rhode Island (URI) (Lee et al. 1996). In order to determine a representative subbase resilient modulus, additional input values were needed for the ELSYM5 computer program (Kooperman et al. 1986). Wheel load and pressure on the pavement were abstracted from a previous research project.(Lee et al. 1990). In any case, the minimum federal single statuary axle limit of 8.7 ton (20 kips) or 2.3 ton (5 kips) per wheel was applied at 689 kPa (100 psi) tire pressure.

STRUCTURAL ANALYSIS OF GRANULAR SUBBASE MATERIALS

The subbase course typically consists of a compacted layer of granular material, or of granular material with RAP.

Experimental Program Although there are many types of material properties for assessing the structural properties of pavement structures, resilient modulus has been adopted as a basis for design in the AASHTO Guide. However, the proposed testing procedure, i.e., AASHTO T274-82, needed to be modified, because most specimens were failed before data collection phase. Therefore, several improved testing procedures were developed, e.g., URI procedure (Kovacs et al. 1991), AASHTO T292-91, and AASHTO T294-92 (AASHTO 1995). The present study was carried out in accordance with the AASHTO T292-91 procedure. This test was performed on the 100 mm x 200 mm (4 in. x 8 in.) cylindrical specimen with a pneumatic actuator. The resilient modulus test was performed with equipment purchased from the H&V Materials Research and Development, Inc., Corvallis, Oregon, July 1989 (H & V 1989).

A compaction permeameter with a 150 mm (6 in.) cylinder diameter was used to perform the test in accordance with AASHTO T215-70.

Fundamental Test Results and Analysis The project began in the fall of 1995. Most of the subbase materials were collected during the winter of 1995 due to the regional nature of this project and the coordination efforts required. These materials were provided from state inspected stock piles rather than ongoing construction projects. The materials were selected by the individual state as representative of subbase materials used for primary state routes (not interstate highways). The majority of the materials tested fall within the band set by each state specification.

Subbase materials were classified in accordance with AASHTO M145-87. Most subbase aggregates were in the A-1-a classification as would be expected. Only the Massachusetts Processed Gravel and the Rhode Island Route 2 materials were classified A-1-b. Both classifications usually contain stone fragments, gravel and sand as their significant constituent materials.

The procedures of AASHTO T180-93 (ASTM D1557-91) were followed to determine the optimum moisture content (OMC) and maximum dry density (γ_{max}).

RESILIENT MODULUS OF GRANULAR SUBBASE MATERIALS

A minimum of 90 percent by weight of the subbase material used to prepare the specimen had a maximum particle size finer than 1/6 the specimen's diameter (100 mm/6 = 17 mm or 4 in./6 = .67 in.). The maximum size of the remaining material was no larger than 1/4 of the specimen's diameter (25 mm or 1 in.). The material larger than maximum size was discarded and noted. This material was prepared at OMC, then was kept in a double plastic bag sealed overnight. This allows the moisture to equilibrate in the sample for even distribution. Prior to testing a fraction of this material was taken to determine moisture content. An additional sample was taken after testing for the same purpose. The two values were averaged, and reported as the representative moisture content.

The previously described material was compacted into a split mold in 5 equal layers with 25 blows per layer using a 2.5 kg (5.5 lb) rammer and 300 mm (12 in.) drop to obtain a specimen of 100 mm (4 in.) diameter and 200 mm (8 in.) height. With this compaction procedure, a 90 percent maximum dry density (AASHTO T-180) was usually achieved.

The subbase specimen was confined in the triaxial chamber using air pressure to achieve the required stress. The triaxial chamber is housed in the walk-in environmental chamber to maintain a temperature of 20^0C (68^0F). A pneumatic actuated loading ram produces the required deviator stress. The pulse duration time (on time) and interval (off time) controls the loading and unloading time. They are metered in terms of percent on the control cabinet. Pulse times of 0.03 to 0.05 seconds are appropriate for full-depth construction with 5 to 12 inches of asphalt concrete and with vehicle speeds of 80 to 96 km/h (50 to 60 mph). Resting time between individual pulses of about 0.7 to 2 seconds is similar to actual conditions in the field. Load (pulse) durations for specimens of cohesive soils of .05 sec. to 0.1 sec. are recommended in AASHTO T292-91. This designation provides no recommended load duration for subbase materials. A load duration of 0.05 was chosen because it produced the optimum load ramp for monitoring with H&V equipment.

Vertical deformation of the subbase specimen is measured by a pair of 100 mm (4 in.) Linear Variable Differential Transformers (LVDTs) attached to the sides of the specimen within the chamber. The resilient modulus of the specimen is calculated and printed by the Resilient Modulus (RM) program.

The result of a single test for granular materials can be presented in a mathematical form that directly incorporates the stress sensitivity of the modulus value in terms of the bulk stress θ (first stress invariant) by $E_{SB} = k_1 \theta^{k2}$. The constants k_1 and k_2 are obtained from regression analysis of the test results and depend on the type of material and physical properties of the specimen during the test (Lee at al. 1997)

It was the URI team's criteria to achieve R^2 values above 70. In some cases the material was not stable enough to provide this result. The material was tested again, and the result was reported if it was reproducible.

The elastic modulus of subbase is highly influenced by the bulk stress calculation. When materials are subjected to higher levels of stress, the moduli value increases accordingly. Resilient moduli of subbase materials were determined with bulk stresses analyzed at mid-depth, and corresponding layer coefficients have been estimated.

Permeability of Granular Subbase Materials In the AASHTO Guide, the effect of drainage is considered by modifying the structural layer coefficient for flexible pavements as a function of (1) the quality of drainage, and (2) the percent of time the pavement structure is exposed to moisture levels approaching saturation. In the present study, the permeability was used to estimate the quality of drainage.

The coefficient of permeability was determined in accordance with AASHTO T215-70. It is a constant head method for the laminar flow of water through granular soils. Materials should not have more than 10 percent passing the 0.075 mm (No. 200) sieve.

The minimum diameter of permeameter should be 8 x max. particle size (8 x 18.75 mm = 150 mm or 8 x .75 in. = 6 in.). A Soiltest Model K-612A compaction permeameter was used. This mold and base permits mechanical compaction at OMC prior to testing. The same soiltest mechanical compactor was used to prepare specimens as was used for the proctor testing. The permeability test results were used to estimate the drainage coefficient for flexible pavement design for New England Subbase materials.

The coefficients of permeability fall within the range of 10^{-3} and 10^{-6}. Clean sand and sand and gravel mixtures are expected to have values from 1 to 10^{-3} and glacial tills have lower values in the 10^{-3} to 10^{-5} range (Lee et al. 1994). The Maine Frenchville subbase material exhibited poor permeability, or practically impervious. When using this material in a pavement structure a drainage layer would be strongly recommended. Although this material has shown a good strength characteristic, it is composed of friable material. It may be noted that fines are generated during compaction and loading which seriously retard drainage.

The New Hampshire sandy gravel produced the highest coefficient of permeability. This material was so uniformly graded that it was difficult to compact and retain in the mold. It had the lowest dry density and a large void ratio. This is what would be expected of a clean course sand with gravel mixture.

Estimation of a Drainage Coefficient A drainage coefficient, m_i is used to reduce or increase the layer coefficient of the subbase layer. Positive drainage within the pavement structure will have a beneficial effect on the life of the pavement structure (AASHTO 1993).

The drainage conditions must be assessed in terms of good, fair, and poor drainage conditions. The method recommended by the FHWA Report TS-80-224 (FHWA 1980) requires the calculation of the time required to drain the base layer to 50% saturation (T_{50}). It is determined for different combinations of permeability (k), length of drainage path (L), effective porosity (n), and slope (S). The present study estimated the drainage coefficient by correlating Table DD.1 of the AASHTO Guide with the permeability of the materials tested in feet per day. The drainage coefficients, m_i were calculated in this fashion, and these m_i values could be refined using specific characteristics for future individual projects.

ANALYSIS OF SUBBASE MATERIALS WITH RECLAIMED MATERIALS

Experimental Program One of the objectives in this study was to develop optimum performance characteristics for recycled material blends. The reclaimed asphalt pavement (RAP) was blended with processed gravel (PG) provided by the Massachusetts Highway Department (MAHWD) in the laboratory at six levels. The results of the PG testing provided the benchmark. The blend levels were 0, 40, 50, 60, 70 and 100 percent milled RAP. The stockpiled RAP which passed a 19 mm (3/4 in.) sieve was procured from the MAHWD. The literature review indicated that the critical pavement performance characteristic is resilient modulus of subbase materials, and this characteristic was examined by blend to determine the optimum amount .

Fundamental Test Results The MAHWD reported that reclaimed base courses contain approximately 1% - 3% asphalt content which is well below the 6% in new pavements. It also indicated that the asphalt cements in reclaimed base courses are old and oxidized and most volatiles have been evaporated. The state of Connecticut has limited the asphalt content of its reclaimed mixture of gravel, RAP and PCC (GRC) to 2%. Asphalt content is beyond the scope of this project, but should be considered as a fundamental property in the future.

Connecticut reclaimed GRC and Massachusetts cold in-place RAP were tested for fundamental properties. The identical procedures for granular subbase materials were followed.

Massachusetts RAP met the state specified gradation limit. Connecticut GRC was found to contain 5% more 3.25 mm (1.5 in) grain size than specified. Both materials had similar mid-range maximum dry density in comparison to the granular materials. The GRC appears to have a high OMC, and this may be attributed to the crushed PCC. The reclaimed materials are classified as A-1-a as good subbase materials.

Resilient Modulus of Reclaimed Subbase Material Blends Connecticut GRC and Massachusetts RAP were tested in addition to laboratory reclaimed subbase material blends. It should be noted that it was difficult to achieve reliable R^2 values when testing reclaimed material. The 50%, 60% and 100% RAP blends were tested a minimum of twice and the higher R^2 test result is provided for this report. This consistent observation of unstable results may be caused by friability of the reclaimed materials.

Resilient moduli were also determined at mid-depth. It was found that the E_{SB} increases with an increase in RAP content. The 50% blend with an E_{SB} of 124 MPa (18 ksi) is at the highest value found in the granular aggregates tested in the present study. Additionally, 124 MPa (18 ksi) is the highest value provided by the AASHO Road Test subbase materials when wet. Massachusetts gravel was the control virgin aggregate used for blending. It exhibited a lower modulus than all RAP blends made in the experimental program. The pure milled asphalt RAP had the highest strength when tested.

Determination of Optimum RAP Amount An experiment was performed with six blends to determine the optimum RAP amount. The pure RAP has by far the highest modulus value. The results were plotted with a second degree regression curve for the final optimization analysis.

A clear trend of increasing strength with increased RAP content has been shown. The 40, 50 and 60% blends provide a minimum increase of 50% higher E_{SB} value than the granular materials. The 60% and 70% blends have an E_{SB} greater than 26,000 which is stiff enough to use for a base layer. RAP with similar modulus values is currently used as a base layer in New England.

It was found that the range of New England granular subbase material moduli are between 47,312 and 123,726 kPa (6,862 and 17,945 psi). Therefore, the blend amount at 123,726 kPa (17,945 psi) was determined as the optimum RAP amount, i.e., 50%.

Permeability of Reclaimed Subbase Materials The coefficient of permeability was determined in accordance with AASHTO T215-70. Reclaimed subbase materials were tested in the same manner as the granular subbase materials.

The Massachusetts RAP was field processed with a CMI RS-500 reclaimer stabilizer. It has a permeability in the range of glacial tills at 10^{-5}. This material provides some poor drainage characteristics.

The Connecticut gravel RAP and concrete exhibited poor permeability, or practically impervious at 10^{-6} cm/sec. When using this material in a pavement structure a drainage layer would be strongly recommended. Although this material has shown a good strength characteristic, it is composed of friable materials. It may be noted that fines are generated during compaction and loading which seriously retard drainage.

It should also be noted that the Connecticut material contains Portland concrete cement rubble. This material may out-perform naturally occurring granular material, because it is lighter and provides the same support

Estimation of a Drainage Coefficient The drainage coefficients for reclaimed subbase materials were assessed by the same method recommended by the FHWA Report TS-80-224 as the granular materials previously discussed. The drainage coefficients were estimated with the permeability of the reclaimed materials tested in feet per day. The m values were estimated but these should be modified according to the specific field project information.

CONCLUSIONS AND RECOMMENDATIONS

The conclusions and recommendations based on the findings and observations of this study have been summarized below.

(1) Typical aggregate types available for structural subbase materials in New England have been identified as follows:

Connecticut:	Bank Run Gravel
Maine:	Frenchville Subbase Sabbatus Subbase
Massachusetts:	Crushed Stone Processed Gravel
New Hampshire:	Sandy Gravel
Rhode Island:	Sandy Gravel
Vermont:	Crushed Stone

(2) The resilient modulus of subbase materials (E_{SB}) was selected as the parameter for structural analysis, and AASHTO T292-91 was initially chosen to determine E_{SB} values for New England subbase materials.

(3) It was noted that the majority of materials with 30 % or more material discarded have E_{SB} values less than 10,000 psi.

(4) Asphalt pavement can be reclaimed in place, eliminating transportation and sorting costs. This saves energy and is usually economically feasible. All utility covers

must be lowered below the level of grinding prior to full depth reclaiming. The layer coefficient for the cold in-place RAP provided by the MAHWD is 0.16. It has a permeability in the range of a glacial till (10^{-3} to 10^{-5}). These are excellent characteristics for a subbase material used primarily for load bearing.

(5) The mixture of gravel, RAP and Portland cement concrete (PCC) from Connecticut has a layer coefficient of 0.10, and shows a better modulus than most of the granular materials. The moduli of all the reclaimed blends tested were higher than the granular material used in the blend.

(6) Layer coefficients for 40, 50 and 60 percent blended materials were between 0.11 and 0.17. The additional layer coefficient strength of a reclaimed material blend of 50 percent RAP is 0.07 higher than the granular material. This increase in support value from a site blended material should make reclaiming worth while.

(7) Milled RAP blended with processed gravel may provide a good reference for material blending. Pure milled asphalt had the highest modulus value. Pure RAP is not likely to be used for a subbase material. The pure RAP may be a better candidate for a base due to its modulus and location in the pavement structure. It also could be used for HMA recycling. A reclaimed aggregate blend containing up to 50% RAP is recommended for use as subbase material.

(8) It was unavoidable to discard significant amounts of material for testing in a 100 mm (4 in.) diameter mold. This would definitely cause the soil fabric to have a greater proportion of fines than specified for construction use. In 1994, AASHTO TP46 was issued in conjunction with Strategic Highway Research Program (SHRP) researchers. Since a grain size of 38.1 mm (1 1/2 in.) is common in subbase materials, AASHTO has increased the specimen size to be more realistic. It is a general agreement among transportation engineers that AASHTO TP46 with a 150 mm x 300 mm (6 in. x 12 in.) specimen should be used whenever the proper equipment is available. Therefore, further testing with the larger mold in accordance with AASHTO TP 46 is recommended to determine structural characteristics of subbase materials which better represent the field conditions.

ACKNOWLEDGMENTS

This research project was funded by the New England Transportation Consortium (NETC). The authors would like to express sincere thanks for NETC's support.

REFERENCES

1. American Association of State Highway and Transportation Officials (AASHTO) (1993). "AASHTO Guide for Design of Pavement Structures." Washington, DC.

2. AASHTO (1995). "Standard Specifications for Transportation Materials and Methods of Sampling and Testing." 17th ed., Washington, DC.

3. Federal Highway Administration (FHWA) (1980). "Highway Subdrainage Design", FHWA Report TS-80-224, U.S. Department of Transportation (DOT), Washington DC.

4. H&V Materials Research and Development (1989). "Resilient Modulus Repeated Load Test System." Corvallis, OR.

5. Kooperman, S., Tiller, G., and Mingtsan, T. (1986). "ELSYM5: Interactive Microcomputer Version." FHWA, U.S. DOT, Washington, DC.

6. Kovacs, W., D., Lee, K, W., Jin, M. (1991). "Seasonal Variation of Soil Resilient Modulus for Rhode Island." Final Report Vols. 1 & 2, The University of Rhode Island (URI), Department of Civil and Environmental Engineering, Prepared for the Rhode Island Department of Transportation (RIDOT), Kingston, RI.

7. Lee, K.W., McEwen, E., Peckham, W. (1990). "Development of a Uniform Truck Management System", Final Research Report Prepared for the New England Transportation Consortium (NETC), Kingston, RI.

8. Lee, K.W., Marcus, A.S., Mao, X.H. (1994), "Determination of Effective Soil Resilient Modulus and Estimation of Layer Coefficients for Unbound Layers of Flexible Pavement in Rhode Island." Department of Civil and Environmental Engineering, URI, Prepared for the RI DOT, Research Report No. 1, Kingston, RI.

9. Lee, K.W. and Mooney, K. (Project Period 9/04/94-5/11/96), "Development of Design Parameters for Pavement Structures in Rhode Island." Department of Civil and Environmental Engineering, URI, Prepared for the RIDOT, Kingston, RI.

10. Lee, K.W., Huston, M., and Davis J. (1997). "Structural Analysis of New England Subbase Materials and Structures." Interim Report to NETC, Kingston, RI.

Fifth International Conference on the
BEARING CAPASITY OF
ROADS AND AIRFIELDS

VERIFICATION OF FIELD PERFORMANCE OF COLD IN-PLACE RECYCLED ASPHALT

KATHERINE LAUTER, M.ENG.
Department of Civil and Environmental Engineering, Carleton University, Canada

A.O. ABD EL HALIM, PhD., P.Eng.
Professor, Department of Civil and Environmental Engineering, Carleton University, Ottawa

ABSTRACT

Thermal cracking is a common pavement distress in Canada and the northern parts of the United States and is responsible for increasing road maintenance and rehabilitation. Cold In-Place Recycling (CIR) is a new rehabilitation method used by the Regional Municipality of Ottawa-Carleton (RMOC) to decrease the rate of occurrence of transverse and reflective thermal cracks. This research involves the collection of field data to validate the in-situ performance of the CIR roads.

This paper gives a description of the CIR process and presents a field research plan which outlines a field data collection program. Data was collected using a pavement management system and a Dynaflect Deflectometer and the data was used to support visual observations.

The results of this research are given to support the RMOC observations. It is recommended that further work be carried out to determine hoe the CIR is able to slow down the phenomenon of transverse cracking.

INTRODUCTION

Roads and highways are vital for the economic development of Canada, and therefore constitute a large part of government expenditures. The continued increase in traffic volume has contributed to the deteriorating conditions observed on all paved roads. In 1989, the cost of road maintenance in Canada totaled $8.4 billion [1]. One of the major problems of the Canadian road and highway network that exhausts a large sum of the maintenance budget is pavement cracking.

Thermal cracking of asphalt pavements is one of the most common stress modes leading to road failures in Canada and in other cold regions. Thermal cracks, which occur when the thermal stresses exceed the tensile strength of the material [2], develop in the top surface of the asphalt layer and propagate downward, partially or entirely, through the layer thickness. Cracks that develop at the bottom of the asphalt layer are typically the result of high load-induced tensile strains, or a crack or discontinuity that exists at the top of the underlying layer. Those resulting from existing cracks are called reflection cracks and commonly occur on sections of roadways that have been rehabilitated by placement of asphalt overlays. The propagation of reflection cracks can be accelerated by freezing shrinkage of the underlayer. Cracks will allow water to penetrate the pavement structure, causing larger cracks and structural damage when subjected to freeze-thaw cycles. The consequence is often premature and costly rehabilitation.

The Regional Municipality of Ottawa-Carleton (RMOC) typifies the cold and variable Canadian climate. Consequently, thermal cracking is a significant pavement distress in this region. The Regional Government of Ottawa-Carleton is responsible for the maintenance and rehabilitation of a road and street network approximately 3000 lane-km in length. They employ several rehabilitation strategies. The more common schemes include partial depth removal of the existing asphalt and placement of a hot mix overlay, microsurfacing, hot mix overlay, and full depth removal of existing asphalt with placement of new hot mix layers. The RMOC has observed that rehabilitation strategies carried out on roads that exhibit a high frequency of thermal cracks do not eliminate the problem. Typically, these rehabilitation methods are susceptible to reflection cracks originating at the underlying thermal cracks in the aged pavement layers. Since 1989, however, the Region has employed a somewhat unconventional strategy for the rehabilitation of roadways with extensive thermal cracking. The new strategy is termed Cold In-Place Recycling (CIR).

Cold In-Place Recycling is a continuous in-place paving process that involves partially milling the existing deteriorated pavement, crushing the milled Reclaimed Asphalt

Pavement (RAP) to a suitable size in a portable crusher screen plant, treating the crushed material with an emulsion, and placing the new material on the roadway with conventional paving and compaction equipment [3]. The CIR layer typically has an open texture, a poor surface durability, and an unknown moisture susceptibility. The layer must then be surface treated or overlayed with a hot mix material.

The CIR material is composed of RAP and a binding agent which is often an emulsion. Virgin aggregate can be added if needed. An asphalt emulsion is a homogeneous mixture of minuscule asphalt droplets that are suspended in water. Once the emulsion contacts the aggregate, the water in the emulsion begins to evaporate. The rate of evaporation is dependent upon the weather and climate conditions. This "breaking" of the emulsion produces asphalt cement which binds the recycled aggregate.

One of the main factors affecting the use of the CIR is the lack of available technical data. At present there are no standard specifications for mix design or quality control. The mechanical properties of the CIR asphalt need to be determined. It is, however, thought of as a good method of reducing thermal cracking by those who use and/or produce it [4].

Visual observations within the RMOC suggest that the CIR material with a Superpave performance graded (PG) overlay is a viable and even preferred option of rehabilitation for roads with significant distress related to thermal cracking. The RMOC has also found that the annual cost for the CIR rehabilitation work, including a 40 mm hot mix overlay, is 55% of the cost of their standard hot mix overlay procedure [4]. Consequently, it is necessary to validate the visual observations with field data for further optimization of field rehabilitation materials and processes. To address the need to quantify the field performance of the CIR material, a field research plan was devised.

FIELD INVESTIGATION

The RMOC uses a pavement management system to collect road inventory and performance data to help efficiently manage the road network [5]. This system collects raw data from the field and calculates performance indices for specific road sections. In addition, a Dynaflect Deflectometer is used to collect deflection data from the roadway which can be used to determine the Structural Adequacy Index (SAI) of the road. The RMOC has observed that the CIR roads tend to perform well in reality, while theoretically their performance indices do not always indicate this to be so.

The field investigation involved the use of the RMOC pavement management system database and historical Dynaflect Deflectometer values to quantify the field observations. Since the CIR rehabilitation is consistent with decreasing transverse and reflective crack production, only the video component of the PMS data library was used to determine the frequency of transverse cracking on selected roads.

Study Roads

Two groups of roads were selected for study sets. The first group included 190 lane-km of roads that had been rehabilitated with the CIR since 1990, while the second group consisted of an equivalent length of roads that had been rehabilitated with the conventional hot mix overlay (CHOL) procedure over the same time period. All selected roads were in rural areas and had similar traffic conditions. In addition, one representative road of each rehabilitation type was selected to assess the pre and post rehabilitation performance of each method with respect to transverse cracking and deflection data.

Pavement Management System

Videos showing the distress on roads in 1994/95 and 1997 were reviewed to determine the transverse crack frequency per lane-kilometre for each study road. These data were used to determine the frequency of transverse cracks per lane kilometre versus age for both the CHOL and the CIR study roads. The pavement condition data for the RMOC is collected using an automated multi-purpose data collection vehicle which uses high resolution video equipment to collect images of the road right-of-way and pavement surface [6]. This data is catalogued on video tapes which comprise the PMS video database. For the purposes of this research, the video tapes were used to count the number of transverse cracks on roads in each group. These values were then converted to the number of transverse cracks per lane kilometre. Linear regression analysis was performed on the yearly average data points.

Dynaflect Deflectometer

The Dynaflect Deflectometer, a non destructive testing device, measures the pavement deflection at five geophones which are located at the centre of the two wheels and, starting at the loading point, are spaced at equal distances of 30.5 cm apart [7].

Since the RMOC typically dynaflects only half of their roads each year, deflection values from both 1996 and 1997 were used. The data from each road were analysed in terms of the average maximum deflection (MD) and the average radius of curvature (R_c). These two parameters were used to compare between the selected study groups to determine any change in the characteristic relative to rehabilitation method.

FIELD RESULTS

Transverse Cracking

As explained previously, the number of transverse cracks per lane kilometer (TC/ln-km) was calculated for groups of CIR and CHOL roads. Figures 1 and 2 show the relationship between transverse crack frequency and pavement age for CIR and CHOL roads respectively. The figures represent the results of the 1997 surface distress survey.

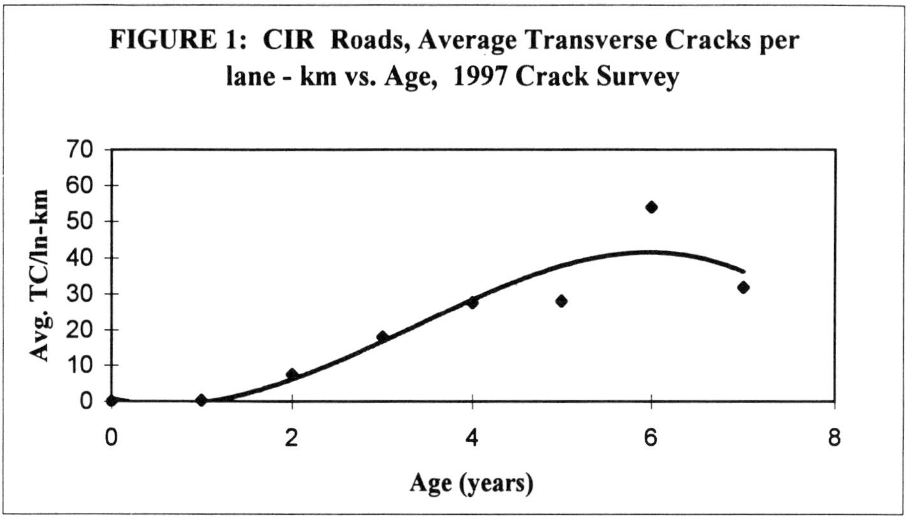

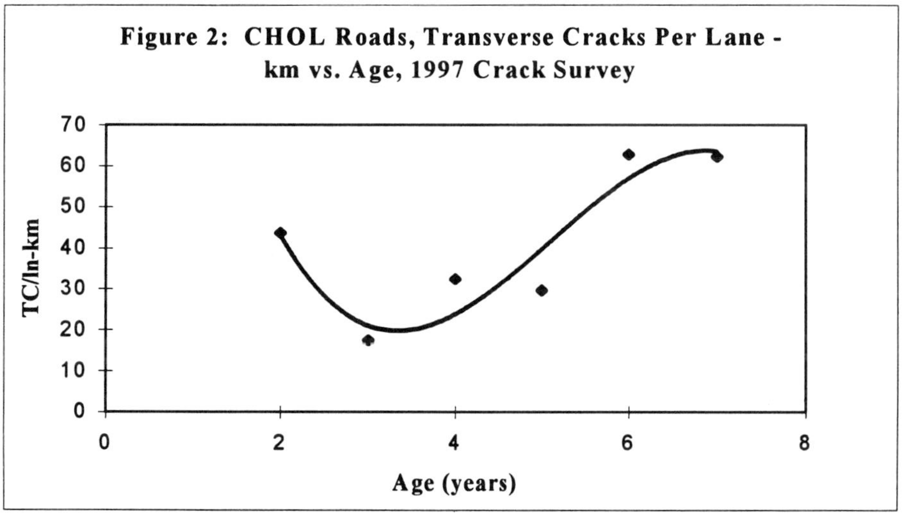

Both figures show that the number of TC/ln-km increases with increasing age. It is evident that cracking on the CIR roads did not begin until the second year following construction while the CHOL roads experienced significant transverse cracking after one year.

These results did not conclusively show which rehabilitation type was superior since the analysis did not account for the different variables, including subgrade type and distress condition before rehabilitation. Consequently two further steps were taken. The groups were divided by subgrade type and analysed, and a representative road from each group was chosen to study the pre and post rehabilitation conditions.

The strength and quality of the pavement subgrade material can affect the frequency of transverse cracks. To determine the influence of the subgrade on the two rehabilitation methods, the CIR and CHOL groups of roads were re-organized according to the subgrade type. Figure 3 shows the average value for the number of cracks per year for each subgrade and rehabilitation type.

Figure 3: Comparison of TC/ln-km/year Between Rehabilitation Method for Selected Subgrade Type

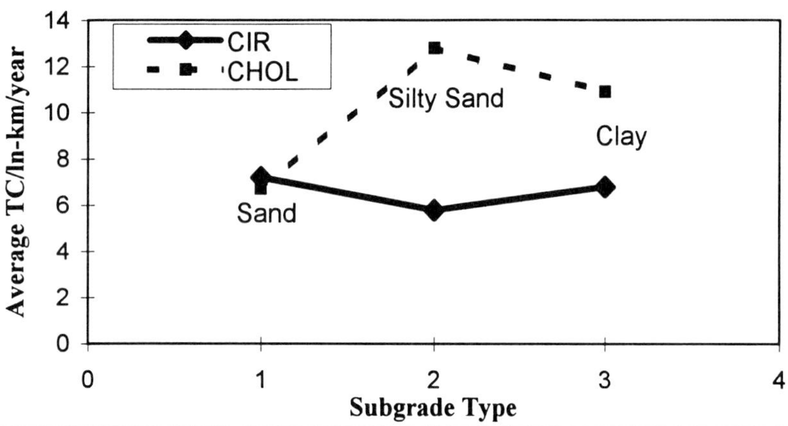

The figure shows that for a sand subgrade, which is typically stable and strong, both the CIR and the CHOL have equivalent crack rates of 7.2 and 6.7, respectively. A significant difference between rehabilitation methods is observed for weaker subgrades, ie. silty sand subgrades. The number of cracks on this weaker subgrade are 5.8 and 12.8 for the CIR and CHOL roads respectively. This observation is further confirmed by the results of the clay subgrade. Clay, the least stable of the subgrades that are examined, resulted in the crack rates of 6.8 and 10.9 for the CIR and CHOL roads. A comparison of these values suggests that on a strong and stable subgrade, like sand, the performance of the pavement is independent of the rehabilitation type. On weaker subgrades, like clay, the performance of the CIR pavement is superior to that of the CHOL layer. In addition, Figure 3 shows that the use of the CIR appears to negate the effect of the subgrade type on the rate of transverse cracking. This observation is very important to pavement designers.

Figure 4 shows two roads constructed on silty sand subgrades and exhibiting identical transverse crack densities in the year before rehabilitation. Two years after rehabilitation, however, the CHOL road shows as much as 40% of the amount of cracks as in 1994, while the CIR road shows less 2% of the same amount of cracks. This suggests that compared to CHOL, the CIR asphalt is less susceptible to reflective cracking, and also may have developed fewer construction cracks produced by rollers during compaction as will be discussed later.

Figure 4: Compare Effect of Rehabilitation Method on Number of TC / ln-km

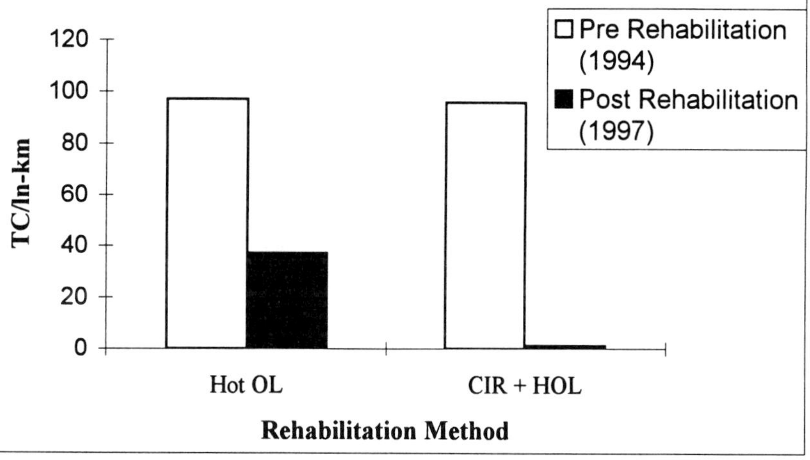

Deflection Measurements

As described earlier, deflection measurements were also used to characterize the field performance of the CIR material. This data was used to calculate the average maximum deflection, and the average radius of curvature. The calculated values for the CIR roads were compared to those of the CHOL roads to determine if surface deflections can be used to evaluate the CIR roads in the same way that they are used to characterize the CHOL roads.

Deflection values for the CIR and CHOL representative road sections for three years prior to rehabilitation and one year following rehabilitation are used. In addition, deflection measurements were taken on the CIR rehabilitated road immediately after placement of the CIR layer and before the hot mix overlay was placed. Measurements were compared to determine the effect of rehabilitation on the deflection characteristics of the road.

Maximum deflection values were plotted by road station for the CIR and CHOL rehabilitated roads, pre and post rehabilitation, as shown in Figures 5 and 6, respectively. At any location, the maximum deflection on the CHOL road is on average two times greater before rehabilitation than after, showing that the addition of the overlay effectively reduced the deflection values of the pavement. Conversely, the maximum deflection for the CIR road was lower before rehabilitation, increased by nearly 50% after replacing the old pavement with a CIR layer, then decreased by 20% once the final hot mix surface course was placed.

Figure 5: Maximum Deflection vs. Location, CIR Road

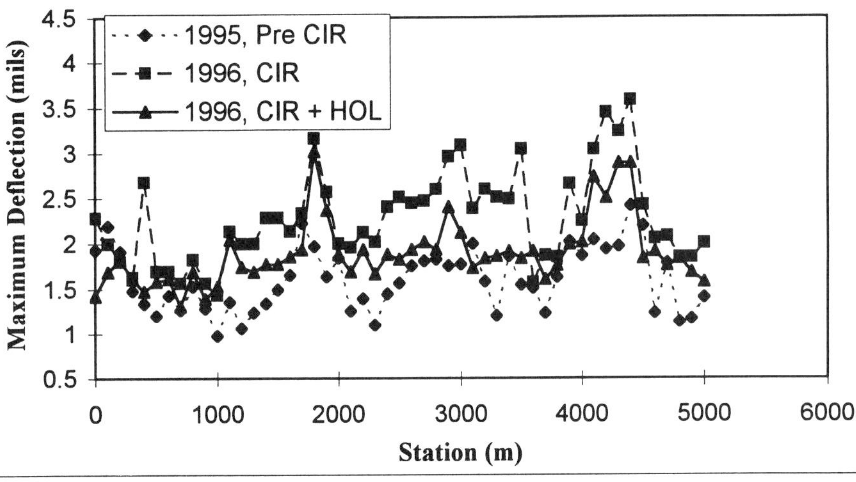

Figure 6: Maximum Deflection vs. Location for RR52, CHOL- Pre(1993) and Post(1997) Rehabilitation

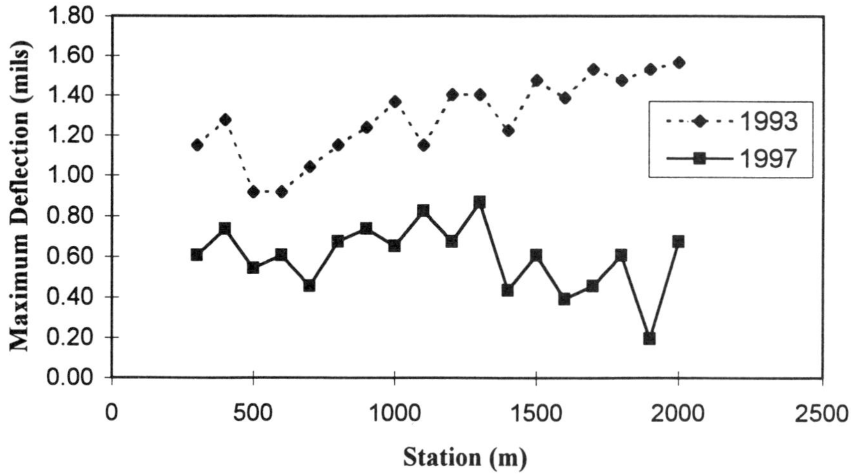

The Radius of Curvature results were quite different. The R_c of the CHOL road increased, on average, by a factor of 1.5 after the rehabilitation was carried out. For the CIR road, the R_c increased, on average, by a factor of nearly 1.5 after the CIR layer was placed, then nearly tripled with the addition of the hot mix surface course.

Field results show that both CIR and CHOL rehabilitation methods will increase the R_c of the material. The maximum deflection will decrease on the CHOL roads, as is expected, but will increase on the CIR roads. Thus, current evaluating methods will not support the use of the CIR for control of thermal cracking.

CONCLUSIONS AND RECOMMENDATIONS

The objectives of this research were to provide field data to support visual observations made by the Regional Municipality of Ottawa-Carleton which indicated that Cold In-Place Recycling produced a material that reduced the frequency of transverse cracking. Field data about transverse cracking were collected for a number of selected CIR and CHOL sites. Time series plots of the transverse crack data were completed for both rehabilitation methods. Dynaflect Deflectometer data were also collected for the same sites. From this data, the radius of curvature and maximum deflection values were computed and analyzed. The effect of subgrade variability on rehabilitation type was also determined using the PMS and Dynaflect data.

The PMS data shows that the number of cracks for both road types increases with age. An examination of the pre and post rehabilitation road condition of one road of each rehabilitation type shows that the CIR material is able to slow down the rate of reflective cracking, while the CHOL asphalt cannot.

The main results of the Dynaflect testing show that the deflection values for the CIR roads are not indicative of actual performance. The Dynaflect data for the CIR pavement shows that the road is in worse shape immediately after rehabilitation than it was before. As this is not true, the Dynaflect should not be used to monitor the CIR rehabilitated roads in the same manner as it is used for the CHOL roads.

Also, observations and analysis of field data concerning the subgrade material show that the CIR and CHOL roads perform the same on strong subgrades, however, the CIR structure is superior to the CHOL structure when placed over a weak subgrade. The CIR structure appears to negate the effect of the subgrade type on the rate of cracking.

The field data collected with the PMS and Dynaflect Deflectometer indicate that the CIR material effectively reduces the rate of thermal and reflective cracking. This supports the visual observations made by the RMOC.

Based on the field investigation that was carried out, the following recommendations are made:

- The Dynaflect Deflectometer should not be used to characterize the CIR material. The deflection values do not have the same meaning as they do for the HMA and therefore must be correlated to the specific material.
- The results show that the CIR appears to negate the effect of subgrade type on the rate of transverse cracking. Further work should be carried out to verify this performance.
- Laboratory tests should be carried out on the CIR material to support the field performance.

ACKNOWLEDGEMENTS

The authors would like to thank Carleton University for their support, and the Regional Municipality of Ottawa-Carleton for the use of their database for analysis.

REFERENCES

1. Roads and Transportation Association of Canada (RTAC), "Canada's roadway infrastructure: selected facts and figures", Ottawa, 1990.

2. Easa, SM., Shalaby, A., Abd El Halim, AO., "Reliability-Based Model For Predicting Pavement Thermal Cracking", Journal of Transportation Engineering, *American Society of Civil Engineers,* Vol. 122, pp.374-380, 1996.
3. Bradbury, A., Kazmierowski, TJ., Cheng, S., Raymond, C., "Cold In-Place Recycling in Ontario - A Case Study", Proceedings, *Canadian Technical Asphalt Association,* pp.153-170, 1991.
4. Croteau, JM., Lee, QS., "Cold In-Place Recycling: Performance and Practices", Presentation at the *XIIIth IRF World Meeting,* Toronto, Ontario, Canada, 1997.
5. Lee, Stephen Q.S., P.Eng., 'Overview of Regional Municipality of Ottawa-Carleton Pavement Management System', June 1997.
6. Lee, Stephen Q.S., VanBarneveld, A., Corbett, M., "Low Temperature Cracking Performance of Superpave and Cold In-Place Recycled Pavements in Ottawa-Carleton", 42nd Canadian Technical Asphalt Association Proceedings, November 1997.
7. Leduc, Captain F.A., "A Prediction Model for Determining the Bearing Capacity of Flexible Airfield Pavements from Heavy Weight Deflectometer Data", Master of Engineering Project, Carleton University, Department of Civil and Environmental Engineering, p.41, September 1994.

Fifth International Conference on the
BEARING CAPASITY OF
ROADS AND AIRFIELDS

ADAPTING PAVEMENT STRUCTURES OF LOWER-RANK ROADS TO TRAFFIC-LOAD REQUIREMENTS FOR MOTORWAYS

MATE SRŠEN[1] and VANJA KEINDL[2]

[1]PhD, Professor, Civil Engineering Institute of Croatia
[2]B.Sc. C.Eng., Civil Engineering Institute of Croatia
HR-10000 Zagreb, Croatia

ABSTRACT

Approximately 400 km of motorways with two lanes in each direction have so far been built in Croatia. According to long-term plans proposed by the Government, it would still be necessary to build as much as 1500 km of motorways in order to provide for a more efficient and safer transport of persons and goods, particularly in the direction of tourist destinations on the Croatian part of the Adriatic. As some sections of future motorways in fact coincide with routes of the existing lower-rank national roads, the issue of adjusting pavement of such roads to the level of roads with the highest level of serviceability (motorways) is considered at the stage of design preparation. This issue is addressed in the case of two sections of the future motorway passing through the continental part of Croatia, i.e. from Zagreb towards the west (Ljubljana) and from Zagreb towards the east (Belgrade). This paper is based on information acquired during pavement studies for these two sections.

Key words: motorways; adapting pavement structure; design methods; condition surveys.

1. INTRODUCTION

The adjustment of pavement structure of a lower-rank road to the traffic load conditions for motorways is a highly complex task requiring answers to a number of questions, such as :

- which methods for analyzing structural condition of existing pavements can be considered reliable?
- which calculation methods can be used to realistically determine the composition and dimensions of layers used for pavement strengthening?

- to what extent the motorway designer should follow the route of the existing road, i.e. in which way it would be possible to resolve structural details caused by the difference between the new and the old alignment (pavement widening and raising)?
- what factors influence selection of materials for the pavement leveling course?

These questions point to the necessity of examining advantages of individual road-construction materials and construction procedures applied in pavement adjustment and strengthening. A major problem is the selection of appropriate material for leveling course as the thickness of such course can be considerable (as much as 50 cm) because it is necessary to adjust the existing pavement cross slope to the cross slope of the new motorway. In this respect, an interesting leveling-course solution is the use of stone material stabilized with the cement/bitumen combination as binder, which gives to this material necessary elasticity and rigidity. This paper is based on concrete experience gained during the study of pavement adjustment issues on two motorway sections in Croatia. The first section runs from Zagreb to Bregana (towards Ljubljana) while the second one starts in Županja and ends in Lipovac (route towards Belgrade)[1].

2. FACTORS INFLUENCING STRUCTURAL CAPACITY

2.1 General

The motorway section Zagreb - Bregana is about fourteen km long (from KM 0+000 to KM 13+660) while the length of the Županja - Lipovac section is approx. thirty kilometers (from KM 438+350 to KM 468+774). In order to take into account differences in pavement thickness and structure as well as the actual condition of the pavement, these sections must be classified on the basis of some appropriate criteria. A quite evident criteria is the original composition of the pavement structure as it is in most cases responsible for current pavement damage. We should also bear in mind all subsequent pavement strengthening activities undertaken during many years of maintenance and upgrading of these road sections. In addition to types and properties of the existing pavement, a highly significant classification criterion consists in analyzing geometrical relationships between the old and new pavement. Two typical cases are:

a) the old and new pavements are inclined in the same direction and
b) the old and new pavements are inclined in different directions.

If the old and new pavements are inclined in the same direction (case a) then the pavement strengthening would consist in simply applying asphalt layers in designed thicknesses. In such a case the pavement adjustment is done by widening the existing pavement, that is by establishing a lateral link between new and old pavement layers while making every attempt to avoid differential settlement. In such cases, it is highly significant to assess as precisely as possible the condition of the existing pavement.

The situation proved much more complex in case b) where pavements of the old and now roads are locally inclined in different directions. The leveling courses then attain as much as 60 cm in thickness (Županja - Lipovac). Therefore, the old pavement can be treated as an improved subgrade, so that the process of assessing condition of pavement as such loses a lot of its significance.

2.2 Pavement condition surveys

2.2.1 Field testing

Pavement deflection measurement

Deflection measurement was performed by means of La Croix deflectograph. Deflections were measured at every 4 meter intervals, at an axial load of 100 kN. The temperature of asphalt surfacing was also recorded and, whenever necessary, the deflection was adjusted to nominal temperature of 20°C. The deflection values were also corrected using the so called seasonal factor, as deflections were not measured in period considered as least favorable from the bearing-capacity standpoint (thawing period). Depending on the method used for pavement strengthening calculation (cf. section 3), deflection values were in some cases correlated with those obtained by means of the Benkelmann's beam. Also depending on the calculation method, measurement results were sometimes analyzed using mathematical-statistics methods (result dispersion, mean value, etc.). Results were then grouped in homogeneous sections.

Borehole drilling and sampling

Boreholes were drilled in order to accurately determine composition of the existing pavement structure, as well as to take samples for laboratory testing. The arrangement and number of boreholes was defined in accordance with the length of individual sections. Boreholes were drilled to the depth of approx. 1 m.

Visual inspection of pavement condition

Visual inspection was performed in order to determine the percentage of cracks, patches and ruts as related to the total pavement surface. In addition, the number of deflections, as well as conditions prevailing at road edges - including roadside ditches and shoulders - were registered. To enable easy reference the results were presented in graphical form. At this stage of inspection, photographs were taken of typical pavement damage as well as of those parts of pavement which are in a relatively good condition. Damage to asphalt surfacing determined through visual inspection was compared to results obtained by laboratory testing of asphalt samples.

Need for more complex investigations

As the modulus of elasticity for individual layers of the existing pavement is an input data in various methods for the design of pavement strengthening (SHELL's program and other methods), it is necessary to obtain in situ values of the modulus of elasticity. The method of geophysical determination of the modulus of elasticity for individual pavement layers is currently being developed in Croatia, but this determination will also be possible in the near future using the first purchased falling weight deflectometer (FWD).

2.2.2 Laboratory testing

Laboratory testing of the loose stone material from pavement structure consisted in defining fine particle content (<0.02 mm) in the sample. The fine particle content and the form of the sieving curve were used to determine probability of volume change. The possibility of estimating volume change in the loose-material layer of the pavement structure is of high significance for determining resistance to freezing action.

Laboratory testing of asphalt samples enabled us to determine the level of compaction for individual layers, and to define the type and quality of the placed asphalt mix. These tests are particularly significant as a guideline in cases when it is necessary to decide whether or not to remove asphalt layers from the existing pavement.

3. DESIGN METHODS FOR PAVEMENT STRENGTHENING AND REHABILITATION

Before designing strengthening thicknesses of the existing pavement, the section under study must be divided into subsections on the basis of criteria that will depend on the design method selected, i.e. on the required input parameters, as described in the following sections.

3.1 Design procedure for pavement strengthening on the motorway section Županja - Lipovac

3.1.1 Pavement strengthening design according to Swiss standards (SNV) and using the method recommended by the American Asphalt Institute (AASHTO) [2, 3]

Knowing that the traffic load forecast amounts to 2.80 x 10^7 of standard axle (80 kN) passes over the design life of the motorway, appropriate strengthening thicknesses were read out from diagrams according to SNV (1977) and from diagrams according to the American Asphalt Institute. It was established that the values from both diagrams almost correspond to one another. Strengthening thicknesses defined (designed) in such a way - expressed in thickness of the asphalt-concrete layer - were then converted, using the structural conversion coefficients, into required thicknesses of the surfacing layer and bituminous base for individual subsections of the motorway.

Values obtained by deflection measurement were adopted as criteria for division into subsections, which in fact determined the designed and the effective thickness of the strengthening.

3.1.2 Pavement strengthening design by means of the SHELL computer program [4]

The following parameters had to be defined to enable data input :

- average monthly air temperatures,
- properties of bitumen in the designed asphalt mix,
- loading period = 0.02 s
- bitumen softening point = 53°C
- penetration index = 0.0°
- thicknesses of the existing pavement layers, and properties of materials contained in individual layers

As the program accepts data input for no more than two basic layers, the following elasticity-modulus combinations were selected :

Old asphalt layers: E=1500 Mpa (the assumption being that the modulus of elasticity has reduced over time due to crack occurrence, and that the old asphalt layers behave as some kind of bedding for the future strengthening), and E = 8000 Mpa (assuming that the asphalt layer compaction occurred over the long-lasting traffic load, which brought about the increase in elasticity modulus). In both cases, moduli of elasticity were read out from the Van der Poel's nomograph.

⇒ All layers situated between asphalt layers and the subgrade were treated as a single layer: E = 150 Mpa, or E =500 Mpa as a limit value.

⇒ Subgrade: E = 50 Mpa.

3.1.3 Overview and analysis of results

Strengthening results obtained by SHELL program, together with required strengthening thicknesses defined on the basis of relevant deflections (Swiss method and AASHTO method) are presented in Table 1. The results are classified by typical motorway sections, taking into account the composition (layer thicknesses) of the existing pavement structure. It can be seen in Table 1 that in every row of columns 6-9 there is one result which approximately corresponds to the result from the corresponding row in column 4. This enables us to conclude that existing-pavement strengthening thicknesses obtained by design according to two different methods generally correspond to one another for assumed properties of materials built into the existing pavement. Nevertheless, this comparison with results obtained by SHELL method must still be considered as no more than orientational.

Table 1: Existing pavement strengthening thicknesses (based on deflections and according to SHELL method)

Motorway section	Pavement thickness (cm)		Strengthening thickness (mm)					
			Based on deflections		According to SHELL method [E (Mpa)]			
(from KM to KM)	Thickness of existing asph. layers	Thickness of the bedding	Design thickness	Effective thickness	$E_{asph.}=15$ $E_{bedd.}=15$	$E_{asph.}=15$ $E_{bedd.}=50$	$E_{asph.}=80$ $E_{bedd.}=15$	$E_{asph.}=80$ $E_{bedd.}=50$
1	2	3	4	5	6	7	8	9
438+350-439+000	12	27	**100**	**120**	160	130	**100**	80
439+000-440+000	16	27	**100**	**120**	140	**100**	60	40
440+000-443+000	12	27	**100**	**120**	160	130	**100**	80
443+000-445+300	12	27	**90**	**100**	160	130	**100**	80
445+300-447+000	16	27	**90**	**100**	140	**100**	60	40
447+000-448+000	12	27	**90**	**100**	160	130	**100**	80
448+000-449+000	23	27	**90**	**100**	**90**	50	20	20
449+000-450+000	17	44	**130**	**160**	**70**	20	20	20
450+000-451+000	12	44	**130**	**160**	**110**	40	60	20
451+000-452+000	17	44	**130**	**160**	**110**	40	20	20
452+000-454+000	12	44	**140**	**170**	**140**	80	60	20
454+000-456+000	7	44	**140**	**170**	**140**	80	110	60
456+000-460+860	12	44	**140**	**170**	**110**	40	60	20

Table 1 shows that the effective strengthening thickness, calculated according to relevant deflections, ranges from 10 to 17 cm of asphalt layers (road base + wearing course), i.e. 13 cm on an average. In order to ensure uniform quality of asphalt layers along the entire route, and taking into account prevailing technical regulations and positive practical experience, it could be appropriate to place the bituminous road base 10 cm in thickness, and the wearing course 4 cm thick, along the entire route.

Considering that asphalt layers used in strengthening procedure are placed over the cement-stabilized stone material of no less than 15 cm in thickness (which is not required for structural reasons but in order to provide for proper geometry of the motorway cross section), it would be quite reasonable to try to keep asphalt strengthening as thin as practicable. However, current experience regarding behavior of asphalt layers placed on the cement-stabilized material, particularly with respect to their resistance to crack reflection, show that an overall asphalt layer thickness of about 15 cm would be highly desirable.

It can be seen from the typical cross section (Figure 1) which is related to the proposed pavements-strengthening solution that differences between the left and right edge of the existing and new pavements sometimes attain as much as 50 cm, i.e. that the so called "wedges" appear in the cross section of the pavement of the future motorway, which have to be filled with an appropriate material. To this date, similar occurrences have successfully been resolved by a leveling course (wedge-shaped fill) made of the cement-stabilized stone material. It is quite obvious that asphalt leveling courses of such thickness would not be an appropriate solution.

In addition, it should be noted that, on the existing pavement, the asphalt-layer thickness sometimes attains as much as 23 cm. In such a case, it would hardly be appropriate to add new asphalt layers of considerable thickness. This is also why it would certainly be advantageous to use thinner asphalt layers "hardened" by an intermediate layer made of cement-stabilized stone material.

The use of cement stabilization is particularly appropriate in cases where there is a horizontal difference in alignment between the old and new route, i.e. where it is necessary to widen the existing pavement, which is in fact the case on the motorway section under study. In zones where pavement widening is required, it will be necessary to place a cement stabilization layer 30 cm in thickness. In this way, an additional bearing capacity is given to the portion of the traffic lane that is not situated directly above the existing pavement. In other words, pavement at such widening consists of the wearing course made of "split mastic asphalt" 35 mm in thickness, bituminous base course 120 mm thick, cement stabilized stone material (30 cm), and a loose layer of crushed stone material whose quality is similar to the material used for the hard shoulder.

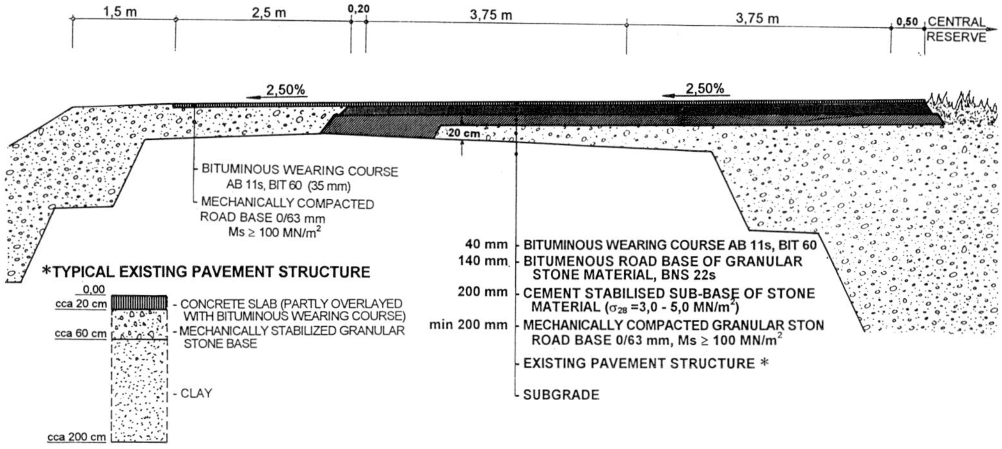

Figure 1 Typical cross section at the Županja - Lipovac motorway

3.2 Design procedure for pavement strengthening on the motorway section Zagreb - Bregana

3.2.1 Pavement strengthening design by means of the SHELL program [4]

Appropriate parameters, fully complying to those presented in 3.1.2, were defined at the stage of data input, so that it is not necessary to mention them once again in this section.

In addition to these parameters, the minimum required thickness was defined for the strengthening layer. This thickness amounts to 20.1 cm of asphalt concrete, which corresponds to the structural strengthening number of 84.42 mm.

3.2.2 Pavement strengthening design by means of the DARWin program [5]

The pavement strengthening design is based on defining difference between structural numbers of the new and existing pavements.
The following assumptions were adopted in the structural number calculation for the new pavement structure :

- number of equivalent 80 kN axles in the design period: 2.82×10^7
- initial pavement serviceability: 4.5
- minimum pavement serviceability: 2.5
- design reliability level: 99.8%
- standard deviation: 0.44
- resilient subgrade modulus: 150 Mpa
- number of pavement-construction stages: 1

The calculation was made using the component-analysis method whereby - based on AASHTO replacement coefficients - the structural number is calculated for the existing pavement structure.
The thickness of pavement strengthening is expressed by means of the minimum structural strengthening number, which amounts to 83 mm.
It is evident that calculation results obtained by SHELL and DARWin methods correspond very much to one another.
Based on AASHTO replacement coefficients, the following dimensions were adopted for strengthening layers:

⇒ asphalt concrete	mm	40 x 0.42 = 16.8 mm
⇒ bitumenized crushed stone	mm	140 x 0.35 = 49.0 mm
⇒ cement-stabilized subbase	mm	200 x 0.20 = 40.0 mm
	mm	380 SN = 105.8 mm

As asphalt layers are in a very poor conditions, the only solution is their full removal. The underlying gravel material can then be considered as subgrade for the new pavement which is why it has to be leveled to the cross slope as defined for the subgrade. A crushed stone leveling layer, 20 cm in thickness, must be placed above the properly leveled subgrade. The top surface of this leveling layer must be adjusted so as to become parallel to the slope of the future pavement.

The described pavement strengthening solution is presented in Figure 2.

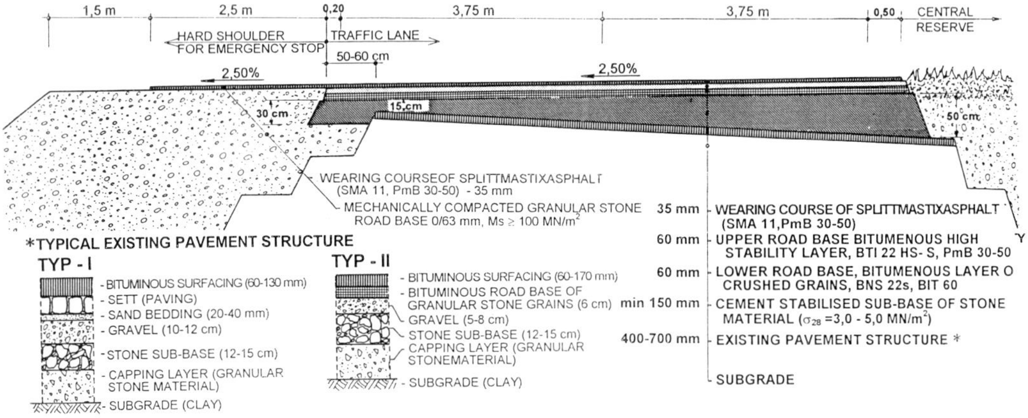

Figure 2: Typical cross section at the Zagreb - Bregana motorway

4. CONCLUSION AND RECOMMENDATION

Conversion of a lower-rank road pavement into a motorway pavement is certainly a highly complex task which, in addition to the use of appropriate design methods, also requires a lot of technical skill and understanding. Appropriate field and laboratory tests must be made in order to determine the structure and condition of the existing pavement, after which pavement strengthening has to be designed on the basis of accepted technical methods. The thickness of strengthening designed in such a way is in most cases treated as a minimum requirement, which needs to be adapted to the specific geometry of the future motorway. It is precisely the geometry of the old road as related to the future motorway (here we primarily deal with cross-sectional differences) that brings about the need to resort to significant thicknesses (up to 60 cm) of the so called leveling layer. It is at this point that comes up the question of selecting an appropriate material for such layers. If a loose stone material is used for leveling course of such thickness, that would result - due to traffic load - in the formation of ruts as well as in other plastic deformations of the pavement. Asphalt layers of such thickness would also be inappropriate because of their poor rigidity. In such situations, Croatian builders have most often been using the layer of a cement-stabilized stone material. However, the high volume and the excessive rigidity of such layer inevitably brings about the appearance of cracks in the cement stabilization, and these cracks then find their way through asphalt layers. This brings us to an obvious need to use the material whose rigidity and elasticity would be situated somewhere in between asphalts and cement-stabilized layers. A material with such properties is currently used in some countries under the name of STABICOL. The leveling-layer realization in accordance with the STABICOL procedure [6] would consist in applying the mixture of cement-stabilized stone material to which bitumen is added, in thicknesses that would depend on technical requirements reached at the design stage. As the leveling course significantly contributes to the overall bearing capacity of the pavement, there is an urgent need to develop an appropriate method to analyze and determine structural characteristics of this layer.

5. ACKNOWLEDGEMENT

This paper is based on a pavement study which has been performed for the Croatian Road Authority. Permission to use the material is gratefully acknowledged.

6. REFERENCES

[1] Civil Engineering Institute of Croatia (IGH). Study of pavement structures for motorway sections Zagreb - Bregana and Županja - Lipovac (sections from KM 0+000.00 to KM 13+660.00 and from KM 438+350.00 to KM 468+774.42), Zagreb (1997) (original in Croatian).

[2] Schweizerische Normenvereinigung (SNV) 640 738, 1977.

[3] AASHTO Guide for Design of Pavement Structures, Washington, D.C., USA (1993).

[4] Shell International Petroleum Company Limited. SPDM-PC User Manual-Shell Pavement Design Method for use on a personal computer (Version 1994, Release 2.0) London, England, February 1994.

[5] AASHTOWARE™ DARWin™ 3.0 Pavement Design and Analysis System - User's Guide, AASHTO, Washington, D.C., USA 1997.

[6] Sršen M. Correspondence between COLAS GmbH, Austria and IGH d.d., Croatia about technical cooperation (1997).

Fifth International Conference on the
BEARING CAPASITY OF
ROADS AND AIRFIELDS

ROAD STRENGTHENING AND THE USE OF PERFORMANCE INDICATORS

PER MAGNE NOSS[1] AND DAG ESE[2]

[1]Research Director, Dr.ing., Rieber & Søn ASA, Nodest Asfalt, N-3002 Drammen, Norway, Fax 47-32848810
[2]Sector Manager, Norwegian Public Road Adm., Sogn og Fjordane, Leikanger, N-5840 Norway, Fax 47-57653986

ABSTRACT

The Norwegian policy for the determination of bearing capacity of roads has been gradually changed the last five years. More weight has been put on interpretation of PMS data and the analysis of measured and expected development of transversal and longitudinal evenness. Sections that show too short service life of the surfacing are then subjected to a detailed study by use of deflection measurements and detailed studies from test pits. An automatic dynamic cone penetrometer (DCP) has been developed to measure in situ stability of well graded base courses. The DCP works well, but has not yet been used on a broad scale.

During the period 1986 through 1990 the Norwegian Public Roads Administration developed a complete pavement management system (PMS). The system incorporates existing subsystems, such as condition surveys and tender preparation, and is supplemented with a project and network-level maintenance and rehabilitation planning system. Today the complete system is interconnected by data processing links and consist of pavement condition surveys, road inventory surveys, road data bank, project planning and network optimisation including performance trend (past history and future predictions) and life cycle cost analysis.

Function contracts have been used in order to optimise alternative methods for road rehabilitation including design of local strengthening and design of new asphalt surfacing. The asphalt contractors are responsible for both the design and the construction, and have to give warranties with regard to pavement performance. The warranties are linked to measured values of rut depth and IRI after 5 years. The allowed maximum values are normally set at such a level that the intervention levels for maintenance of rut depth and road roughness shall not be passed within a period of 10 years.

Key words: Performance indicators, strengthening, performance trend, function contracts.

INTRODUCTION

A major part of the Norwegian road network are low volume roads, and about 85 % of the roads have AADT < 3000. Up to 1985 most roads were designed with base courses of well graded gravel. The climate is wet and especially the upper part of the road pavements are subjected to many freeze/thaw cycles during a normal year. The pavement rehabilitation is normally linked to damages due to lack of bearing capacity, uneven frost heave, weak shoulders etc.

From about 1980 it was observed that the service life of the bituminous surface courses were reduced compared with earlier experience. This reduction has been linked to an observed increase in the maximum tire pressure - up from 0.65 MPa in 1965 to 0.9 MPa in 1975. The increased tire pressure has given higher shear stresses in the base course, and thus a demand for higher stability of the material used. In the recent years we also have had a move from use of dual tires to super single tires which gives an additional increase of the shear stresses in the upper part of the pavement.

The base courses of well graded gravel normally show good performance in situ when they are dry, but will often have a significant reduction in the shear strength when they are subjected to unfavourable moisture and freeze/thaw conditions. Most of the roads have variable drainage conditions, and therefore the pavement performance often vary significantly along a given road. A main challenge has been to design local strengthening in order to get a more homogeneous service life of the asphalt surfacing.

We have not been able to derive good estimates of the in situ stability of granular base courses by use of backcalculated E-modulus from deflection measurements (Falling Weight Deflectometer (FWD) and Dynaflect). Laboratory tests have also confirmed that the resilient modulus of well graded gravel is considerably less influenced by changes of the fines content and the degree of saturation than the CBR_{soaked}. Therefore, the last 5 years there has been a change in the policy for the determination of bearing capacity of roads. More weight has been put on interpretation of PMS data and the analysis of measured and expected development of transversal and longitudinal evenness. Sections that show too short service life of the surfacing are then subjected to a detailed study by use of deflection measurements and detailed studies from test pits. We have also been looking for a good device for in situ measurement of the stability in a well graded base courses. In 1993[1)] we started our first tests with a new developed automatic dynamic cone penetrometer (DCP). Comprehensive field investigations including damage surveying were also performed on low volume roads with gravel base courses. The observed damages were correlated with Dynaflect, FWD and DCP and DCP measurements. A criterion to check the stability of the base course was developed. The automatic DCP works well, but has not yet been used on a broad scale.

PAVEMENT MANAGEMENT SYSTEM

During the period 1986 through 1990 the Norwegian Public Roads Administration[2)] developed a complete pavement management system (PMS). The system incorporates existing subsystems, such as condition surveys and tender preparation, and is supplemented with a project and network-level maintenance and rehabilitation planning system. Today the complete system is interconnected by data processing links and consist of pavement condition surveys, road inventory surveys, road data bank, project planning and network optimisation including performance trend (past history and future predictions) and life cycle cost analysis.

The Norwegian Asphalt Contractors Association[3)] has co-operated with the road authorities in the development of equipment for condition monitoring and in the development of special parts of the PMS system like data processing, performance trends, functional specification linked to pavement condition etc. This co-operation between the Norwegian Public Roads Administration and the Norwegian Asphalt Contractors Association has been the basis for the tests, evaluation and use of different function contracts on asphalt pavements in Norway. The first contracts of this type took place in 1989 and was linked to wear resistance against studded tires.

Pavement Condition Parameters. The main pavement condition parameters measured in Norway are longitudinal evenness, transversal evenness and crossfall. The international roughness index (IRI) which is based on a standardised computation of data describing the longitudinal profile, is used to describe the longitudinal evenness. The rut depth, rut area, and rut width are computed from the measured cross-sectional profile. Friction are only checked on new laid pavements. Due to the wear from studded tires there are normally not any problems with friction. Intervention levels for rut depth and road roughness, so-called maintenance standards, have been established for pavement sections of the road network. These standards, shown below, are not to be exceeded. $IRI_{90/10}$ means that maintenance shall normally be performed when these values are exceeded on 10 % of the section length considered.

Road Roughness (International Roughness Index –IRI(mm/m))

	Road Network Level				Road Section Level	
AADT	Main Roads $IRI_{90/10}$	Other Roads $IRI_{90/10}$	Main Roads $IRI_{50/50}$	Other Roads $IRI_{50/50}$	Main Roads $IRI_{90/10}$	Other Roads $IRI_{90/10}$
<300	4.5	5.5	2.5	3.0	5.0	6.0
301-1500	4.5	5.5	2.5	3.0	5.0	6.0
1501-5000	4.0	5.0	2.0	2.5	4.5	5.5
>5000	3.5	4.5	2.0	2.0	4.0	5.0

	Road Network Level	Road Section Level
Ruth Depth	$RD_{90/10} < 15$ mm	$RD_{90/10} < 15$ mm
	$RD_{50/50} < 11$ mm	
Crossfall:	Design value $\pm$ 1.5%	Design value $\pm$ 1.5%

Equipment for Pavement Condition Monitoring. ALFRED, is a combined unit measuring both cross section, including crossfall, and longitudinal profile. A divided beam 4000 mm long with 17 ultrasonic sensors with a spacing of 250 mm is the main part of the equipment. Effective measuring speed is 0 to 80 km/h.

Cross sectional profiles are measured by the 17 senors from the divided beam with a variable length of 2000 - 2375 mm. The residual individual sensor spacing is 125 - 250 mm. For PMS purpose a beam length of 2000 mm has been standardized. Rut depth is measured for each 1 m along the road. For special purposes the measuring distance can be reduced to 25 cm between transverse profiles. Crossfall and curve radius are computed using signals provided by a rate gyro, an inclinometer and the ultrasonic sensor bar.

There are two ways to measure the longitudinal profile. The 4000 mm beam measures the 17 distances to the road surface at 1 m intervals, and every 1 m length is described by five distances measured four times. This method was used until 1996, but now the longitudinal profile is computed by using signals provided by a high quality laser and different gyros resulting in a very detailed and accurate description of the longitudinal profile.

PC-programs have been developed to take care of quality assurance, graphical presentation, and reporting of automatically collected raw data. In addition to standard rut depth, area, width, and crossfall computations, rules have also been developed to interpret the cross-profile shape. Rut shape combined with traffic volume, lane widths, and so forth are valuable parameters in the PMS decision process.

On basis of the longitudinal profile, PC programs present IRI for 20 m and 100 m intervals. The same profile data are also the basis for 3 m straight edge simulation used in asphalt contracts. Combined straight edge (single irregularities) and IRI data give pavement engineers valuable tools to follow-up different PMS sections through spring thaw.

Pavement performance prediction. Rut depth and roughness (IRI-value) are recorded at least once a year all over the national road network. The 90th percentile for the initial and the latest recorded rut depth and roughness are computed for each PMS section and form the basis for a prediction of the further development of rut depth and roughness.

The development of rut depth (mainly due to wear of studded tires) is considered to follow a straight line, except for the first year when the development of rut depth is accelerated.

Normally the development of roughness is considered to follow the parabola

$$IRI_d = IRI_i + (IRI_t - IRI_i) * (A_d/A_t)^{1.5}$$

where

IRI_d = predicted roughness at a desired future date (IRI value, mm/m)
IRI_i = roughness just after the latest rehabilitation (IRI value, mm/m)

IRI_t = latest recorded roughness (IRI value, mm/m)
A_d = age of the pavement surface at the desired future date (years)
A_t = age of pavement surface when the latest roughness recording was made (years)

Typical curves for the development of ruts and roughness are shown in Figure 1

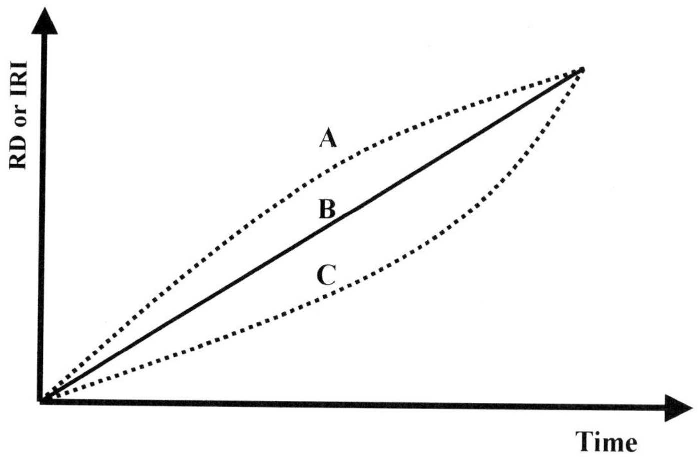

FIGURE 1 Illustration of different trends for development of rut depth (RD) and IRI

Trend-curves of type A are found on some new or strengthened road sections. Such development of ruts or roughness can be linked unsatisfactory compaction of the subgrade, subbase or base course. Retarded settlements will also effect the IRI in the same manner. If the angle of the curve after some years (ref. curve A) is below a given value, then the bearing capacity of this PMS section is said to be sufficient. A new surfacing is expected to have a reasonable functional service life even if the service life of the previous surfacing was short. Therefore, it is very important to study the trend of the curve and not only the total damage of the surfacing.

Curves of type C indicate that the yearly damage factor increases with the age of the surfacing. This is believed to be a normal development of damage for a well constructed road section However, when it comes to development of rut depth our experience show that a straight line gives the best illustration of what happens in the field

Example: Usage of the Trend Curve for Roughness. A new surfacing on a given road section was laid in 1989, and the measurements of roughness the same year gave an initial roughness IRI_{1988}. The yearly damage $factor_{roughness}$ for 1994 is defined as $\Delta IRI_{1994} = IRI_{1994} - IRI_{1993}$. Mean annual damage $factor_{roughness}$ for the period 1988 - 1994 is defined as $\Delta IRI_{Mean} = (IRI_{1994} - IRI_{1988})/6$.

Theoretically the relationship $F = \Delta IRI_{1994}/\Delta IRI_{Mean}$ will tell if we follow curve A or C. $F < 1$ indicates curve A and $F > 1$ indicates curve C. Curves of type C can be found for road sections with good bearing capacity (long functional service life of the surfacing), while on road sections with unsatisfactory bearing capacity the deterioration to larger extent will depend on the combined effect of environment and traffic. For years with unfavourable climatic conditions, high ground water level, large frost heave and critical thawing, the annual damage factors will be much larger than for a year with favourable climatic conditions. Therefore, the trend curves may be buckled and not as smooth as shown in Figure 1.

Today crack propagation is not measured, and the effect of crack propagation is believed to be reflected during the measurements of rut depth and roughness.

The simple way of predicting the pavement performance has the advantage that several site-specific conditions (wear resistance, cross section, axle load limit, traffic conditions, travel speed, and climate) which are difficult to monitor, are reflected in the surface condition recordings. Thus they need not to be monitored separately for this purpose.

When pavement performance predictions are necessary before recordings are available, the user of the PMS program must estimate both the initial and some future values for roughness and rut depth. If the latest rehabilitation was similar to the previous one, the pavement performance during this period might be a good estimate for future performance.

If the latest rehabilitation differs significantly from the previous one (e.g., if it includes strengthening or widening), there will be a problem in making good pavement performance predictions using the PMS program. However, after a year when the first recordings are available, the prediction model can be used.

The main problem with such predicts appears when there are considerable changes in factors that have significant influence on the pavement performance, such as changes in axle load limit, traffic volume or travelling speed.

FUNCTION CONTRACTS

Today the Norwegian asphalt contractors give different warranties linked to the quality of the asphalt mix laid on the road. The quality is determined by registration of development trends of different pavement condition parameters. Some contracts include a special penalty or bonus system linked to the warranty given by the asphalt contractor. The warranty can be linked to the wear resistance against studded tire, the development of ruts on the road and/or the homogeneity of the asphalt mix. In some cases the bonus/penalty is linked to measured condition parameters on new laid asphalt pavements, while in other special road rehabilitation contracts where strengthening of the existing road is a part of the asphalt contract, the asphalt contractors are responsible for design of necessary strengthening and give warranties linked to pavement performance.

STRENGTHENING OF EXISTING ROADS

About 85 % of the Norwegian national road network has AADT < 3000. On these roads the pavement rehabilitation is often linked to damages due to lack of bearing capacity, uneven frost heave, weak shoulders etc. The pavement performance may vary significantly and, therefore, a main challenge has been to design local strengthening in order to get a more homogeneous service life of the asphalt surfacing along the road.

As an attempt to optimise alternative methods for road rehabilitation including design of local strengthening and design of new asphalt surfacing, some special contracts have been developed. The asphalt contractors are responsible for both the design and the construction. In these contracts the asphalt contractors have to give warranties with regard to pavement performance. Normally the warranties are linked to maximum values of measured rut depth, IRI-values and crossfall deviations measured with ALFRED after 5 years. These values are set at such a level that the intervention levels for maintenance of rut depth and road roughness shall not be exceeded within a period of 10 years.

A function contract linked to the strengthening of an existing road in Norway was first attempted in Sogn og Fjordane County in 1991. Analysis of the trend-curves for rut depth and longitudinal evenness for 20 m long road sections showed that the pavement performance varied significantly along a road. Thus, it was need both for local strengthening and also a general strengthening of the whole road in order to get a sufficient and more homogeneous service life of the asphalt surfacing. Special action was taken to reduce uneven frost heave. Critical spots were located by measuring the longitudinal profile early in the spring (start of thawing period with maximum frost heave) and in summer. Special spots with uneven frost heave were located by comparing single irregularities from these two measurements. Trend-curves for measured IRI are shown in Figure 2. The measurements are made in mid summer and represent pavement condition in the non frost period. Expected further development of IRI is shown with the dashed lines.

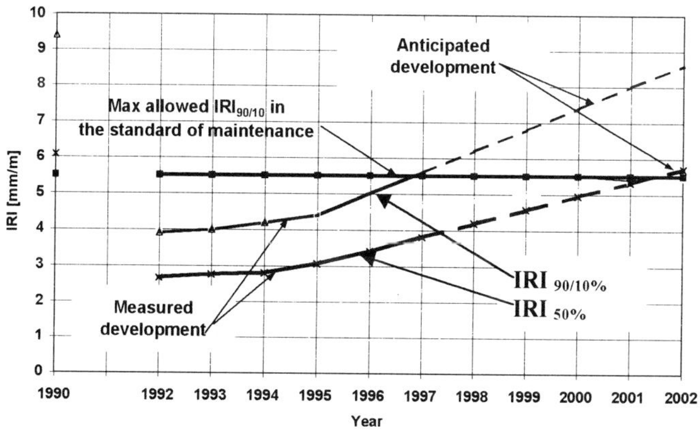

FIGURE 2 Trend-curves for development of road roughness

When the road was strengthening, the pavement was designed to allow for 80 kN axle loads in the spring thaw period and for 100 kN axle loads the rest of the year. However, from January 1995 it was decided on a national level that there should no axle load restriction during the thawing period. This change of permitted axle load limit during the thawing period is reflected by the trend-curves in Figure 2. While the results up to 1994 indicated that the maintenance intervention level should not be reached before about 2001 which was intended, this level was reached in 1997. This indicates that the increase from 80 to 100 kN of the allowed axle load limit during the thawing period on this particular road section has reduced the service life by around 40 %.

Even though the wanted service life was not reached due to change of axle load policy in the example shown, the experience with the use of function contract was good. It should be concluded that we generally have good experience with the projects where the road strengthening has been designed by the contractor as a part of a function contract and where the contractor has given warranty linked to pavement performance. We believe that this is the best way to optimise effective design methods for strengthening of roads. As a result the effect of the designed strengthening is always analysed. The bonus is normally small, and therefore, the competition amongst the contractors will lead to an optimisation of the design and construction procedures for road strengthening.

LITERATURE

1. ESE, DAG; MYRE, JOSTEIN; NOSS, PER MAGNE; VÆRNES, EINAR — "The Use of Dynamic Cone Penetrometer (DCP) for Road Strengthening Design in Norway." BCRA, Minneapolis 1994

2. HAUGØDEGÅRD, TORLEIF; JOHANSEN, JOHNNY M.; BERTELSEN, DAG; GABESTAD, KNUT — "Norwegian Public Road Administration: A Complete Management System in Operation." NPRA, Oslo 1994

3. NOSS, PER MAGNE — "Measurement and Usage of Surface Characteristics of Asphalt Pavements." Eurasphalt-Eurobitume Congress, Strasbourg 1996

Fifth International Conference on the
BEARING CAPASITY OF
ROADS AND AIRFIELDS

EFFECTS OF SLAB SIZE ON AIRPORT PAVEMENT PERFORMANCE ANALYSIS OF FIELD SURVEY DATA

Michel J. Hovan[1], P.E, Ph.D., and Edward Guo[2], Ph.D.

ABSTRACT

The Federal Aviation Administration has conducted an investigation to determine the adequacy of its current policy of allowing 7.7- by 7.7-m (25- by 25-ft) pavement concrete slabs at U.S. airports. The goal of the investigation was to study the performance of rigid airport pavements against slab size. In a previous study [1], results obtained from finite element analysis indicate that larger Portland cement concrete (PCC) slabs (7.7 by 7.7 m) do not perform as well as smaller slabs (smaller than 6.2 by 6.2 m (20 by 20 ft)). In the analysis, it was assumed that maximum bending stress could be used as an indicator of pavement response and as a predictor of pavement performance. The Pavement Condition Index (PCI) was used to measure the performance of airport pavements at 174 U.S. airports. A large quantity of PCI data were obtained and about 27 million m^2 (288 million ft^2) of PCC runway, taxiway, and apron were surveyed. In this paper the effects of PCC pavement slab size on pavement performance are evaluated with measured PCI values.

OVERVIEW OF SURVEYED DATA

Pavement survey data were collected to cover different environmental, traffic, and age condition. The following assumptions were made: (1) pavement sections were designed properly for a design life of 20 years; (2) pavement sections have been properly maintained; (3) Portland cement concrete (PCC) overlaid sections are treated the same as new PCC pavement, and the age is counted from the time of the overlay construction; and (4) all subjective effects of surveyors on the Pavement Condition Index (PCI) results are neglected.

The analysis of the survey data was based on data from 27 million m^2 (288 million ft^2) of pavement from 2820 features of 174 airports. The airports in the survey are located in 23 states and one in Japan. Each type of pavement (runway, taxiway, and apron) was approximately one-third of the total of surveyed pavements.

Figure 1 presents the percentage of pavements by age. Thirty-eight percent of the surveyed pavements are older than 20 years—their design life has been exceeded, but they are still in service. Of the pavements younger than 20 years old, about one-third are between 16 to 20 years. The surveyed pavements aged between 11 to 15 years are the smallest percentage of the age groups (only 10% of total pavements).

---

[1] FAA William J. Hughes Technical Center, Airport Technology Research and Development Branch, AAR-410, Atlantic City International Airport, NJ 08405, (609)485-5552
[2] Galaxy Scientific Corporation, 2500 English Creek Avenue, Pleasantville, NJ 08232, (609)645-0900

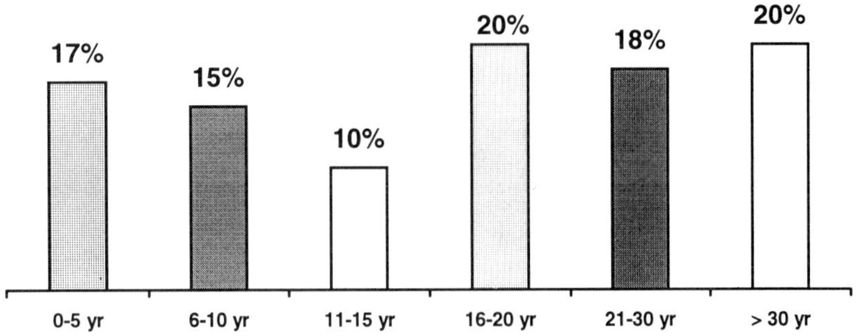

Figure 1 – Pavements Grouped by Age

Figure 2 shows the percentage of pavements involved in this project by the year they were built. Sixteen percent of the pavements were built in the past 10 years, which is less than those built in any of the time periods depicted.

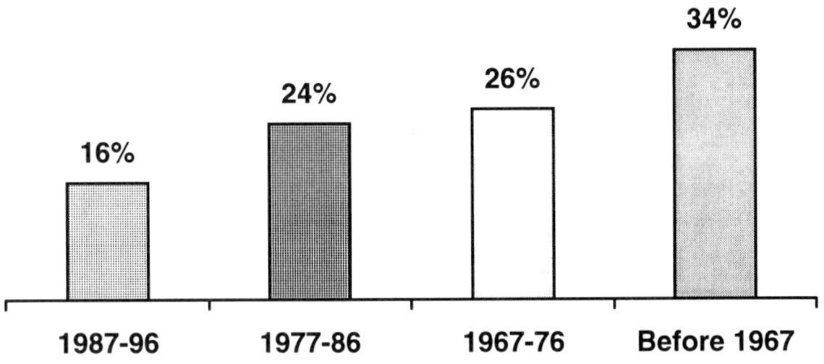

Figure 2 – Pavements Grouped by Year Built

Figure 3 shows the distribution of pavements grouped by PCC average slab size. Fifty-five percent of the total used data come from the slabs smaller than 46 m^2 (500 ft^2). Most of the slabs, in the 46- to 58-m^2 (500- to 625-ft^2) group are 7.7 by 6.2 m (25 by 20 ft) or 7.7 by 7.7 m (25 by 25 ft). Only 5% of the slabs are greater than 58 m^2 (625 ft^2).

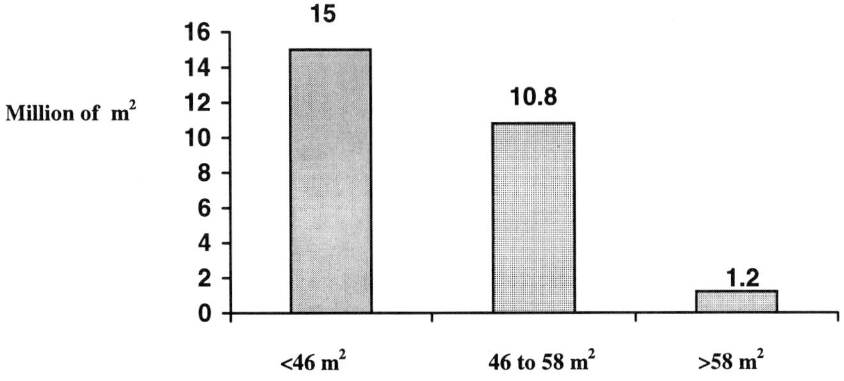

Figure 3 – Pavements Grouped by Slab Size

In addition to using an average area of slabs to define slab size group, the smallest slab length is another way to define the slab group. For example, a 9.29-m^2 (100-ft^2) slab usually has a 3.33 m (10 ft) length for two sides. Similarly, a 11.6-m^2 (125-ft^2) slab is usually sized as 3.33 by 4.16 m (10 by 12.5 ft), a 13.9-m^2 (150-ft^2) slab is 3.33 by 4.95 m (10 by 15 ft), and a 18.5-m^2 (200-ft^2) slab is 3.33 by 6.66 m (10 by 20 ft). Although their areas are different, their shortest side is the same length.

Based on theoretical analysis [2], the maximum total bending stresses caused by aircraft loading and temperature variation are more sensitive to the smallest length than to the area of the slabs. For example, the total bending stress of a 6.1- by 7.7-m (20- by 25-ft) slab would be close to that of a 6.1- by 6.1-m (20- by 20-ft) slab if a single wheel is applied at the 6.1-m (20-ft) edge. Table 1 shows the groups by smallest slab length.

Group	Slabs	Total Area
S1	3.1 x 3.04 m (10 × 10 ft), 3.1 x 3.8 m (10 × 12.5 ft), 3.1 x 4.6 m (10 × 15 ft), 3.1 x 5.3 m (10 × 17.5 ft), 3.1 by 6.1 m (10 × 20 ft)	0.37 million m^2 (3.97 million ft^2)
S2	3.8 x 3.8 m (12.5 × 12.5 ft), 3.8 x 4.56 m (12.5 × 15 ft), 3.8 x 6.1 m (12.5 × 20 ft), 3.8 x 7.70 m (12.5 × 25 ft), 4.6 x 4.6 m (15 × 15 ft), 4.6 by 5.70 m (15 × 18.75 ft), 5.7 x 6.1 m (18.75 × 20 ft), 5.7 x 7.7 m (18.75 × 25 ft)	4.8 million m^2 (51.4 million ft^2)
S3	6.1 x 6.1 m (20 × 20 ft), 6.1 x 7.7 m (20 × 25 ft)	2.2 million m^2 (23.6 million ft^2)
S4	7.7 x 7.7 m (25 × 25 ft), 7.7 x 9.12 m (25 × 30 ft), 7.7 x 12.2 m (25 × 40 ft), 7.7 x 15.2 m (25 × 50 ft)	7.7 million m^2 (82.8 million ft^2)

Table 1 – Pavements Grouped by Smallest Slab Length

The S2 group contains slabs with the smallest side lengths of 3.8 (12.5 ft), to 5.7 m (18.75 ft). The percentages of pavements in the four groups (S1, S2, S3, and S4) are presented in Figure 4.

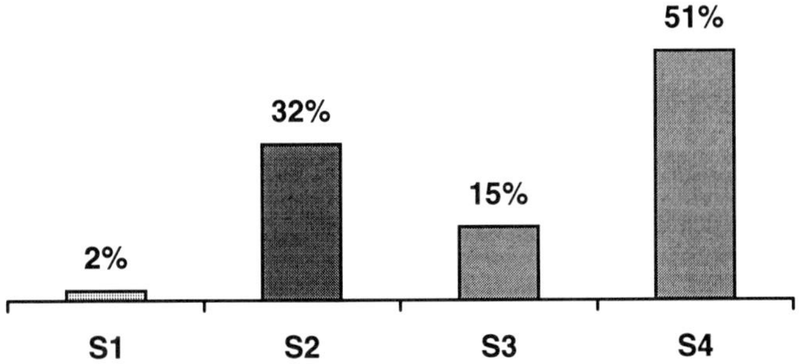

Figure 4 – Pavements Grouped by Smallest Slab Length

VARIATION OF PCI BY PAVEMENT TYPE, AGE, AND PERIOD BUILT

The average PCI of the 0- to 30-year old surveyed pavements grouped by function are given in Figure 5. This Figure indicates that the average runway PCI is higher than that of both the taxiway and the apron.

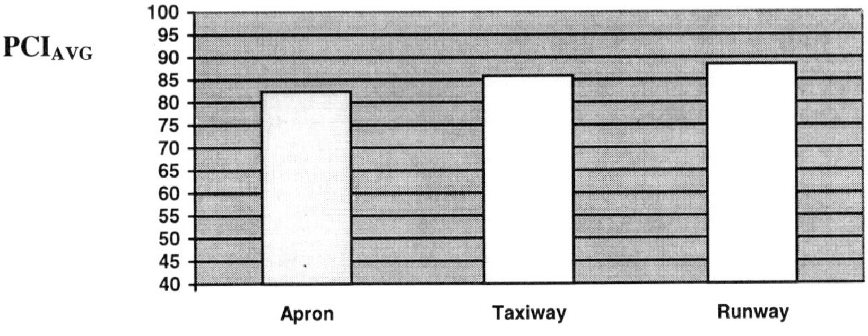

Figure 5 – PCI of Different Types of Pavements

The average PCI values (obtained through tabulation of the collected data) of all surveyed pavements grouped by age are shown in Figure 6. The average PCI decreases as age increases. There is a greater decrease in the PCI value once pavements are older than 30 years. The average PCI as a function of age is presented in Figure 7 for different pavement functions of aprons, taxiways, and runways. The average PCI decreases as age increases except for the 21- to 30-year aprons and for the 16- to 20-year-old runways, the

PCI of a runway generally is higher, followed by the taxiway, except in the 11- to 15-year-old pavements. For pavements in the same age group, the PCI's of aprons are always lower.

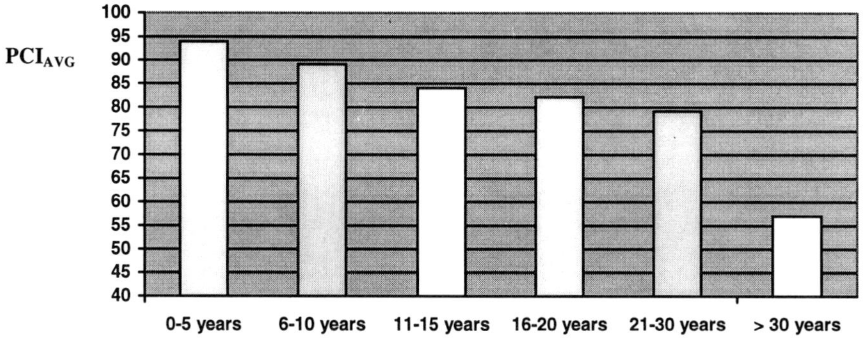

Figure 6 – PCI of Pavements by Age

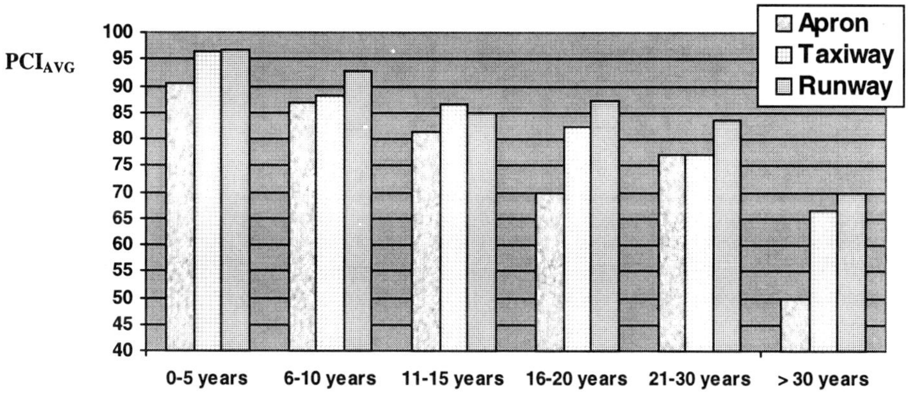

Figure 7 – PCI of Pavements by Age and Functions

GENERAL EFFECTS OF SLAB SIZE ON THE MEASURED PCI

Figures 8 and 9 show the PCI's of pavements grouped by three sizes: smaller than 46 m^2 (500 ft^2), between 46 and 58 m^2 (500 and 625 ft^2), and greater than 58 m^2 (625 ft^2). As indicated in Figure 8, higher PCI values were measured for pavements with smaller slabs. This is consistent for the different types of pavements (apron, taxiway, and runway). Figure 9 also indicates that slabs larger than 58 m^2 (625 ft^2) deteriorate much quicker than the smaller sizes since the average PCI's of the larger slabs are significantly lower.

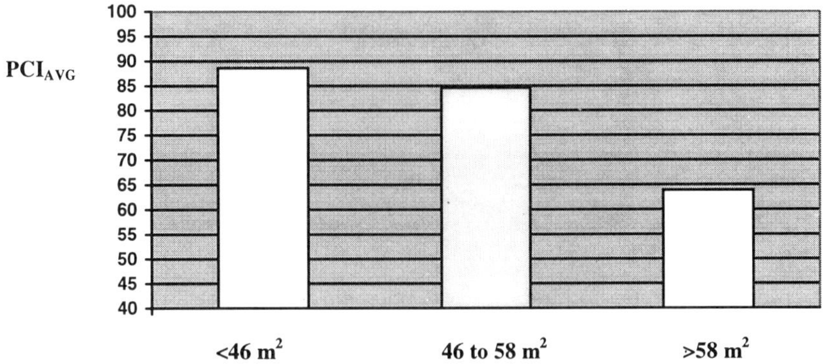

Figure 8 - Pavements by Slab Size

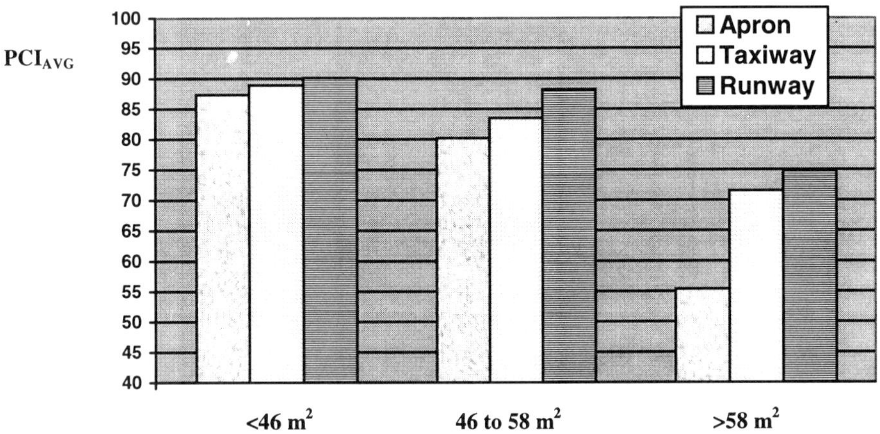

Figure 9 - Pavements by Slab Size and Functions

Figures 10 and 11 present the PCI's of pavements grouped by the smallest slab length as defined in Table 1. Figure 10 shows that the PCI's of larger slabs (S3 and S4) are lower than for the smaller slabs. The PCI's of groups S1 and S2, and S3 and S4 are close to one another. The drop in PCI values occurs between group S2 and group S4. This indicates that the performance of pavements using a shortest slab length smaller than 6.1 m (20 ft) is better than those with larger (shortest) lengths. For the purpose of maintaining pavement performance, the shortest slab size length of 6.1 m (20 ft) can be chosen as a conservative value.

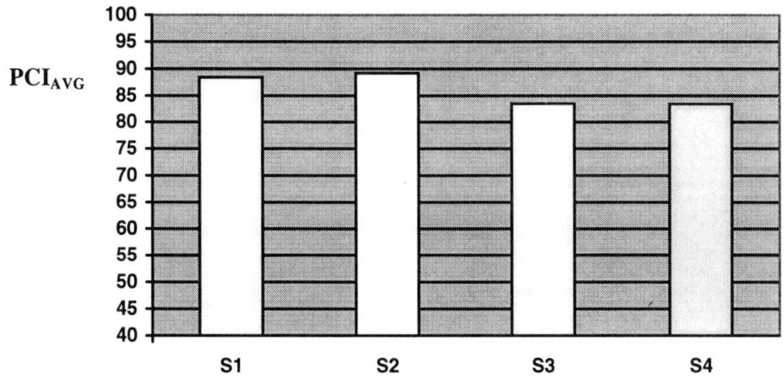

Figure 10 - PCI of Pavements by Slab Side Length

Figure 11 shows that the larger slabs deteriorate more quickly on aprons (S3 plus S4) than runways and taxiways. With sufficient verification, it may be advantageous to avoid using larger slabs for apron design.

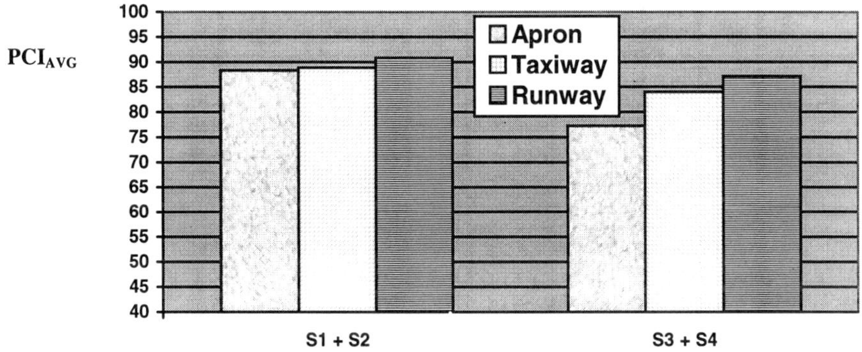

Figure 11 - PCI of Pavements by Slab Length and Pavement Type

EFFECTS EVALUATED BY PCI DISTRIBUTION CURVES.

Many researchers have been interested in evaluating the adequacy of FAA design specification [1]. This specification allows for a maximum slab size of 25 ft, a design life of 20 years, and calls for acceptable performance (but no PCI threshold value is specified). A simple procedure can be used to take advantage of collected pavement PCI data to evaluate one performance component (the PCI value) of pavements designed according to the specification. The following assumptions are used:

- If the surveyed PCI value of a pavement is lower than a threshold PCI value, then consider the pavement to be in need of some rehabilitation.

- To increase the sample size of pavement data around the 20-year mark, pavements between 16 to 23 years old are used to simulate behavior of 19- and 20-year-old pavements.

Different investigators have suggested different PCI threshold values [3, 4]. The threshold PCI value may depend on many factors. The size of the airport and the configuration of the traffic mixture are two important factors that should be considered.

For example, Reference 5 suggests threshold values for runways, taxiways, and aprons to be 93, 89, and 89 respectively. Figure 12 shows PCI distribution curves for runways and taxiways based on the surveyed results in Reference 5. The curves were drawn by using the data of pavements older than 15 years. Because only a few aprons with large areas were older than 15 years, the PCI distribution curve for aprons is not given. Figure 12 indicates that about 94% of runway areas had PCI values higher than their threshold value (93) and only 55% of taxiway areas had PCI values higher than 89 in 1990 when the pavements were surveyed.

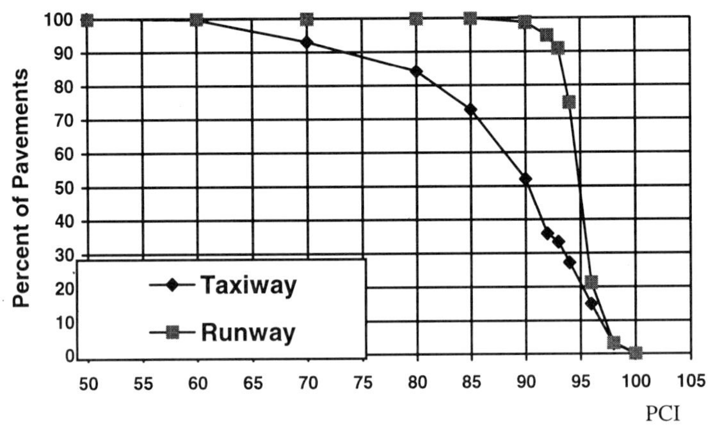

Figure 12 - PCI Distribution Curve for Taxiways and Runways

Figure 13 illustrates PCI curves for aprons, taxiways, and runways for all surveyed pavements from 16 to 23 years old. The survey data were obtained from 174 airports ranging from large hubs to regional airports. If the commonly used (by airport managers) values of 80, 75, and 70 are taken as threshold values for runways, taxiways, and aprons, then 80% of the surveyed runways, 75% of the surveyed taxiways, and 60% of the surveyed aprons would be higher than the threshold values near the end of their design life. The distribution curves of the type in Figure 13 may also be used to investigate the effects of slab size on the PCI's, an indicator of pavement performance.

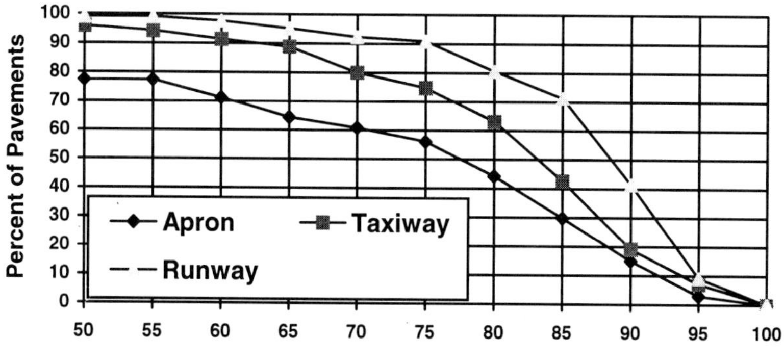

Figure 13 - Distribution of PCI for all Pavements From 16 to 23 Years old

Figure 14 compares the values of PCI's for different pavement functions.

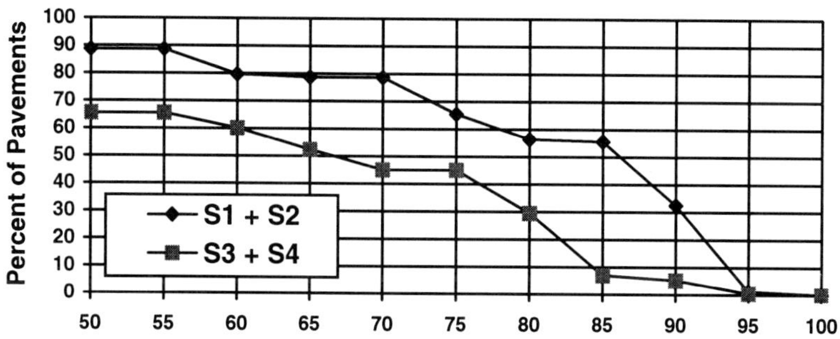

(a) Apron

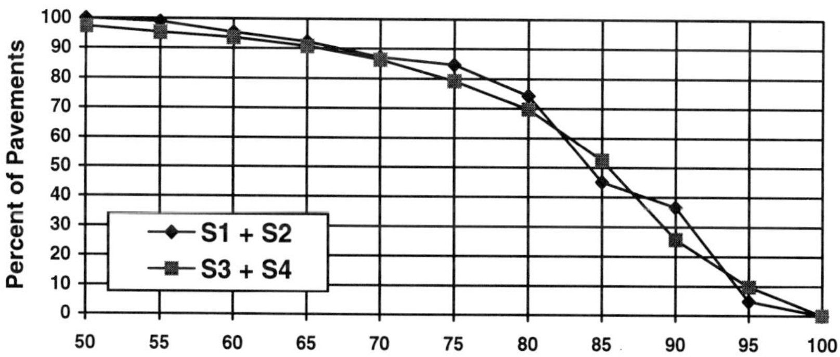

(b) Taxiway

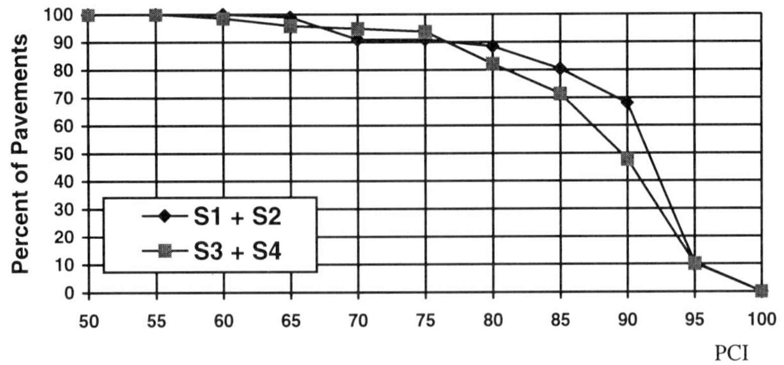

(c) Runway

Figure 14 - PCI Distribution Grouped by Slab Size and Function

Figure 14(a) indicates that smaller slabs (S1 + S2) for apron areas performed significantly better than larger slabs (S3 + S4). If the threshold value of 70 is still assumed, the PCI's of 79% of the surveyed apron areas using smaller slabs were higher than the threshold, whereas only 45% of aprons using larger slabs had a PCI higher than the threshold value. The advantage of using smaller slabs for taxiways and runways does not seem as significant as for aprons although the smaller slabs mostly performed slightly better.

In Figure 15, the PCI's of pavements affected by age and by size are presented. The following findings may be summarized as follows:

- The PCI's of pavements using 46- to 58-m^2 (500- to 625-ft^2) slabs are generally lower than or close to those using smaller slabs. However, the PCI's of pavements using slabs greater than 58 m^2 (625 ft^2) are much lower than the others except for the taxiways and runways that were younger than 10 years old at the time of survey.

- The PCI of all pavements using slabs between 46 to 58 m^2 (500 to 625 ft^2) are mostly lower than those using slabs smaller than 46 m^2 (500 ft^2). However, the differences are not significant before their service life is ended (the age is within 20 years). Namely, the performance of pavements using slabs with smallest slab length equal to or smaller than 25 ft is relatively similar for pavements aged between 0 and 20 years.

- It can been seen that the slab size effects on PCI's for older pavements (between 21 to 30 years) are much greater. Namely, the larger slabs performed worse than smaller slabs for older pavements (20 to 30 years), and this is especially true for apron pavements.

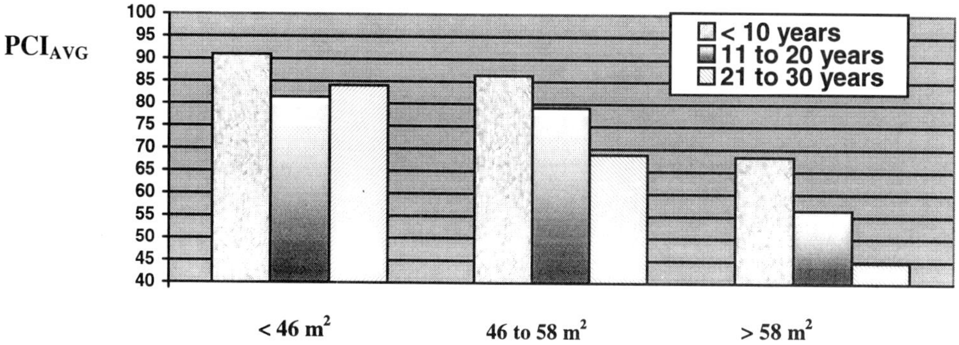

(a) Apron

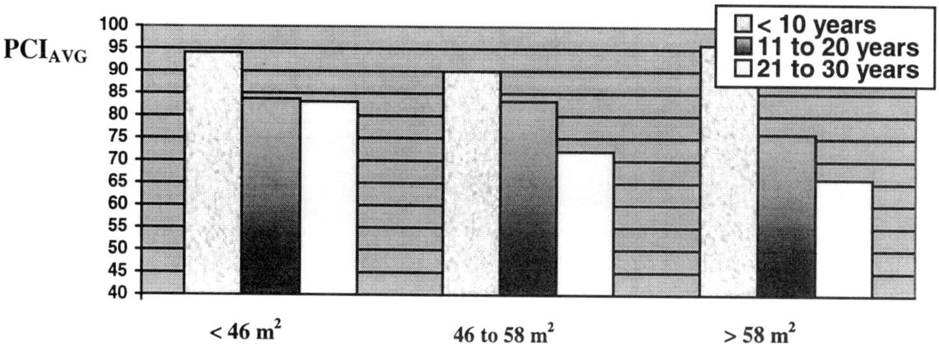

(b) Taxiway

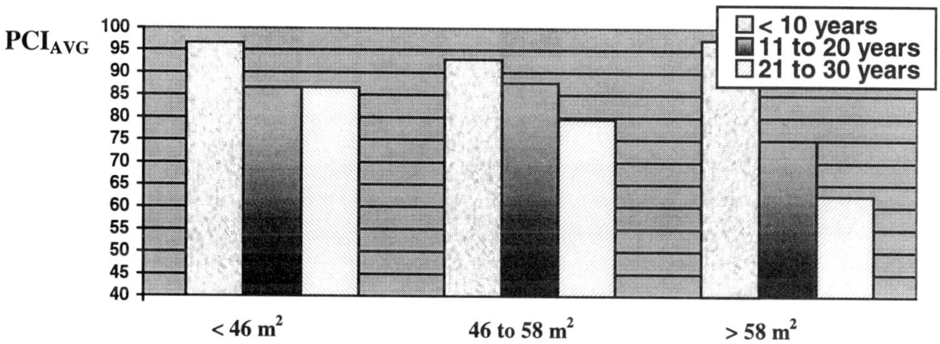

(c) Runway

Figure 15 - PCI of Pavements Affected by Size and by Age

CONCLUSIONS

- The data based on surveyed PCI values suggest that smaller slabs generally perform better.

- In a comparison between pavements in the following two groups: group 1 using slabs equal to or smaller than 7.7 by 7.7 m (25 by 25 ft) and group 2 using slabs larger than 7.7 by 7.7 m (25 by 25 ft), the analyses show that the pavements in group 2, which are not recommended by AC 150/5320-6D, performed worse than the other pavements using smaller slabs. This is true for all functional pavement groups for pavements older than ten years.

- In a comparison between pavements in the following two groups: group 1 using an average slab size smaller than 46 m^2 (500 ft^2) and group 2 using an average slab size between 46 and 58 m^2 (500 to 625 ft^2), the analyses show that the performance of 21- to 30-year old pavements using smaller slabs were significantly better than the performance of larger slabs. However, there was no significant difference between pavement performance in the above two groups for pavements younger than 20 years. This finding indicates that pavements following the joint spacing requirement by AC 150/5320-6D (slabs are not larger than 7.7 by 7.7 m (25 by 25 ft)) perform similarly within 20 years of design life. After 20 years of design life, larger slabs deteriorate more quickly than small slabs.

- In a comparison between pavements in the following two groups: group one's smallest slab lengths smaller than 6.1 m (20 ft), including 3.1, 3.8, 4.6, and 5.8 m (10, 12.5, 15, and 18.75 ft) and group two's smallest slab lengths of 6.1 and 7.7 m (20 and 25 ft), the analysis shows that pavements in group 1 generally performed better than those in group 2. The difference of pavement performance is the largest for aprons in all pavement groups.

- In a study using data analysis of 16- to 23-year-old pavements to investigate pavement performance near the end of the design life, aprons performed significantly worse than taxiways and runways. Taxiways performed worse than runways.

- A drop in pavement performance was found for pavements using a shortest slab side length of 6.1 m (20 ft) or greater. Further studies are recommended to determine the optimal spacing range below 6.1 m (20 ft) for airport pavements.

REFERENCES

1. Airport Pavement Design and Evaluation, FAA Advisory Circular, AC 150/5320-6D, 1988.
2. Guo, E.G., "Effects of Slab Size on Airport Pavement Performance," In Press. FAA Report DOT/FAA/AR-96/117.
3. Eckrose/Green Associates, Inc. "Airport Pavement Inspection by PCI", Survey Handbook, 1996.
4. McQueen, R.D., "Evaluation of Runway and Taxiway Pavements at Atlantic City International Airport, Atlantic City, NJ," June 1993. McQueen and Associates.
5. HLA, PCI, and SDI, "Dallas/Fort Worth International Airport Pavement Management Consulting Service, Final Report," March 1992.

Fifth International Conference on the
BEARING CAPASITY OF
ROADS AND AIRFIELDS

TRIALS TO MINIMISE REFLECTION CRACKING IN UK MILITARY AIRFIELD PAVEMENTS

JOHN POTTER[1] AND JOHN COOK[2]

[1] Project Manager, Transport Research Laboratory, Crowthorne RG45 6AU, United Kingdom. Fax: +44 1344 770686, Email: johnp@h.trl.co.uk
[2]Head of Airfield Pavements Section, Defence Estate Organisation , Ministry of Defence, Sutton Coldfield, B75 7RL, UK. Fax: +44 121 311 3636

ABSTRACT

Jointed concrete runways and taxiways are frequently resurfaced with asphalt to increase the strength, improve the profile or reduce the risk of foreign object damage. The thickness of asphalt applied is often greater than that required for structural purposes in order to reduce the occurrence of reflection cracking above joints and cracks in the underlying concrete.

Trials have been carried out at a number of military airfields in the UK to assess the performance of different anti-reflection cracking systems. Reinforcement grids, modified binders, asphalt inlays and the crack and seat process are being studied. The performance of test sections are being compared with control areas with conventional Marshall asphalt overlay. Three types of grids (steel, glass and polyester), two modified asphalt overlays and two types of guillotine crack inducing equipment for the cracking and seating have been used. The crack and seat approach has been used in main runway and taxiway rehabilitation schemes and a methodology has been developed for inducing the required type of cracking into large unreinforced concrete slabs. The effect of crack spacing, slab thickness and size on the effective stiffness modulus of the cracked concrete, determined from Falling Weight Deflectometer measurements, is being studied.

This paper describes the trials, the measurements made during and after construction and the monitoring of the condition over time.

Key words: reflective cracking, trials, performance, FWD,

INTRODUCTION

For many years, jointed unreinforced pavement quality concrete has been used to construct airfield runways and taxiways at RAF airbases. Historically, for UK military airfield construction, the standard practice has been to omit load transfer dowel bars between adjacent concrete slabs. Although single-slab construction is usually employed, in the 1950's twin slab construction was sometimes specified for use at airfields from which heavy aircraft operated. In the twin-slab construction the joints of the top layer were offset to those of the bottom layer and a separation membrane was usually specified to be laid between the two layers.

In areas where jet blast and fuel spillage are not a major consideration, asphalt overlays provide an economic means of restoring or improving pavement life. However, after a few years in service, reflection cracks generally start to occur in the surface of the asphalt overlay above the joints between the concrete paving slabs due to the thermal movements in the underlying concrete. Reflection cracks also occur in the surface of flexible composite construction which is frequently used on taxiways. This type of construction employs a composite roadbase comprising a continuously laid cement-bound layer under an asphalt roadbase and an asphalt surfacing. Nunn (1) has described crack initiation as occurring in the surface of the asphalt and propagating downwards to the joint in the concrete pavement or to the transverse shrinkage crack in the cement-bound roadbase.

In the early stages of development, reflection cracks may barely be visible and are not considered to be a structural problem. However, when they propagate through the pavement, infiltration of water can weaken the foundation and fine material may be pumped to the surface, resulting in the creation of voids beneath the concrete. Traffic loading exacerbates the situation but of greater concern on airfields is the likelihood of spalling at the cracks and the potential for FOD to aircraft.

In order to inhibit the occurrence of reflection cracking, a greater thickness of asphalt overlay is often applied than is required for structural strengthening. Although this provides an added benefit of better thermal insulation to the concrete, which helps to reduce thermal movements, it increases the cost.

Since 1989, trials have been conducted on runways and taxiways at six military airfields in the UK in order to assess alternative techniques for inhibiting reflection cracking in asphalt overlays on the jointed unreinforced concrete pavements and these trials are described in the following sections of this paper.

METHODS OF INHIBITING REFLECTION CRACKING

The types of anti-reflection cracking treatments being assessed are:

- geogrid reinforcement
- stress absorbing membrane interlayers (SAMIs)
- modified asphalt overlay
- alternative asphalt overlay materials
- crack and seat treatment to the concrete
- asphalt inlay over concrete joints

The conventional overlay material used at UK military airfields is Marshall asphalt with or without a Friction Course surface layer and these are the materials that have been used in most of the trials. A brief description of the methods and the materials used is given below.

The crack and seat method is described in more detail because it has now been used in a major rehabilitation scheme on two taxiways at RAF Lyneham.

Geogrid reinforcement. Reinforcing grids are designed to enhance the tensile strength of an asphalt overlay by absorbing the horizontal tensile stresses above the concrete joints and distributing them over a wider area. As a result, the stress concentrations in the asphalt above the joints, caused when the concrete contracts with reduction in temperature, should be reduced by the presence of the geogrid and hence this should reduce the initiation of reflection cracking.

Three different geogrids have been used in these trials: HaTelit 40/17 which is a polyester grid of 40mm square mesh size, Glasgrid 8501 which is a fibreglass grid with 12.5mm square grid size and Bitufor™ system that incorporates Mesh Track MT1, a galvanised hexagonal steel mesh with dimensions 80mm by 118mm and a series of transverse reinforcing strands, which is bonded to the underlying surface with a specified slurry surfacing.

The recommended installation procedures for the different types of geogrid used in the trials differ in detail. In the Bitufor™ system, the Mesh Track can be placed directly on the concrete or on an asphalt regulating course. The installation methods for both HaTelit and Glasgrid recommend that they are placed between two asphalt layers so that when an overlay is being applied to the concrete slabs an asphalt regulating or base layer is required. The HaTelit is bonded to the asphalt using a specially sprayed adhesive or bitumen emulsion and the Glasgrid is self adhesive. The Glasgrid manufacturer states that, if the asphalt has been milled, it is imperative that a new asphalt regulating layer is used.

Stress absorbing membrane interlayers (SAMIs). SAMIs generally consist of a thick layer of binder which is often modified and spread with single size chippings which are rolled in. They provide a flexible layer which is able to deform horizontally without breaking, in order to allow large movements to take place in the vicinity of the cracks or joints. They also provide a waterproofing role to protect the pavement structure and foundation should the surface of the overlay crack. Some SAMIs are applied only to the cracks thus minimising the cost of materials and installation.

Fibrescreed RC 100, the SAMI used in the trial at RAF Finningley, is a polymer modified rubberised bituminous compound containing chopped fibres, selected granite aggregate and other fillers. It is applied hot, using a screed box to fill the voids in the cracks and joints and form the membrane in one operation.

Modified asphalt. The resistance of asphalt to cracking depends mainly on the binder content and its elastic characteristics although the coefficient of expansion of the aggregate also contributes to the performance. The binder content and its characteristics affect the asphalt's ability to absorb stresses generated at cracks and its self-healing properties as well as its resistance to ageing which causes the asphalt to become brittle with time. These properties are achieved by the use of a viscous binder and high binder content; one way of achieving this is by the use of polymer modified bitumens.

determination of density, stiffness modulus, yield strain, binder penetration and softening point and for visual examination.

Table 1. Design and construction of the trials

Airfield/ /Construction Date	Section	Treatment	Overlay Material	Overlay Thickness (mm)	Existing Construction
Finningley March 1989	T1 T2 T2/3 T3 T4 T4/5 T5	Fibrescreed FC layer MA control DBM layer SBS MA MA control Fibrescreed	MA+FC MA+FC MA+FC MA+FC SBS MA MA MA	120 140 120 120 40 40 40	2-layer JPCP on shale Slab size: 3mx3m HRA on lean concrete
Northolt Aug. 1993 July 1997	S1 S2 S3 S4	Glasgrid Control Bitufor™ Control	MA MA MA MA	100 100 70 70	HRA on JPCP on hoggin JPCP on unbound subbase
Brize Norton July 1994 Aug. 1994	T1 T2 T6 T3 T4 T6	SBR MA EVA MA Control MA Glasgrid HaTelit Control	SBR MA EVA MA MA MA MA MA	40 40 40 100 100 100	 The grids were installed within the overlay
Marham Aug. 1995	S1 S2 S3	Asphalt inlay	UL-M Safepave Slurry	23 21 5	Marshall asphalt on JPCP
Coningsby Sept. 1996	S1 S2 S3 S4	Control C&S (1m) C&S (0.5m) C&S (0.75m)	MA MA MA MA	150-210 150-210 150-210 150-210	2 layer JPCP on unbound granular subbase Slab size 3mx3m
Lyneham December 1996 Sept. 1997		C&S (2m, 1.5m, 1m, 0.75m) C&S (1.2m)	none MA	 170	JPCP on lean concrete Slab size 6mx3m JPCP on lean concrete or asphalt Slab size 6mx6m/6mx3m

JPCP - jointed unreinforced concrete pavement; FC - Friction Course; HRA - hot rolled asphalt, MA - Marshall asphalt, C&S - crack and seat, (crack spacing)

Performance of the test sections. Visual surveys were carried out annually for the first 2 years and then every other year thereafter. Cores were taken after approximately 1 year and 4 years in service. It was interesting to see that the binder penetration of the softer Friction Course with 200pen bitumen had fallen after 4 years to about 60 pen, the same as that of the Marshall asphalt mixed initially with 100 pen bitumen. Similarly, the Marshall asphalt wearing course with modified binder appeared to lose its initial advantage over conventional Marshall asphalt after 4 years because the yield strain of both materials had reduced to similar values. The stiffness modulus of the modified wearing course was similar to the conventional Marshall asphalt when it was first laid but after 4 years it was about 4Gpa compared with 2.5Gpa for the standard material. Although this is beneficial in improving loadspreading, it might not produce a material more resistant to reflection cracking.

The most recent visual survey was carried out in March 1998, nine years after construction. Nearly all the cracking is transverse to the direction of the runway although there is a little longitudinal cracking above the concrete slabs on the runway even though it is 61m wide. The development of transverse cracking with time, in terms of crack length, is shown in Figure 1. The length of cracking is expressed as a percentage of the maximum length of cracking that could occur by taking into account the total length of the transverse joints within the test section across the width of the runway and the values are normalised to take account of the different lengths of the test sections. In terms of the length of cracking both sections T2 and T3, with the Friction Course and the macadam lower layers respectively, showed less cracking than the control bearing in mind that the percentages of cracking are very small. On the Marshall asphalt edge strip only section T2 showed less cracking than the control section.

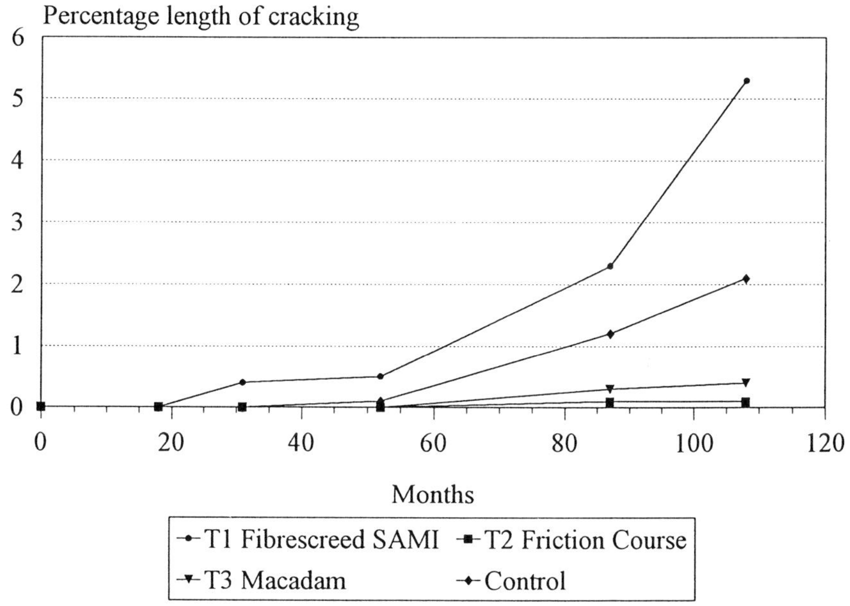

Figure 1. The percentage of transverse reflection cracking by length on the runway at RAF Finningley

The development of transverse reflection cracking on the flexible composite taxiway is shown in Figure 2. The crack length is expressed in terms of a cracking factor defined as the length of transverse cracking (in metres) within the test section divided by the length of the test section (in metres). Figure 2 shows clearly that the SBS modified Marshall asphalt in section T4 and the Fibrescreed SAMI in section T5 demonstrate a significant reduction in reflection cracking compared to the control section with the 40mm conventional overlay of Marshall asphalt. These results demonstrate the importance of applying a SAMI directly above the cracks or joints on the existing pavement.

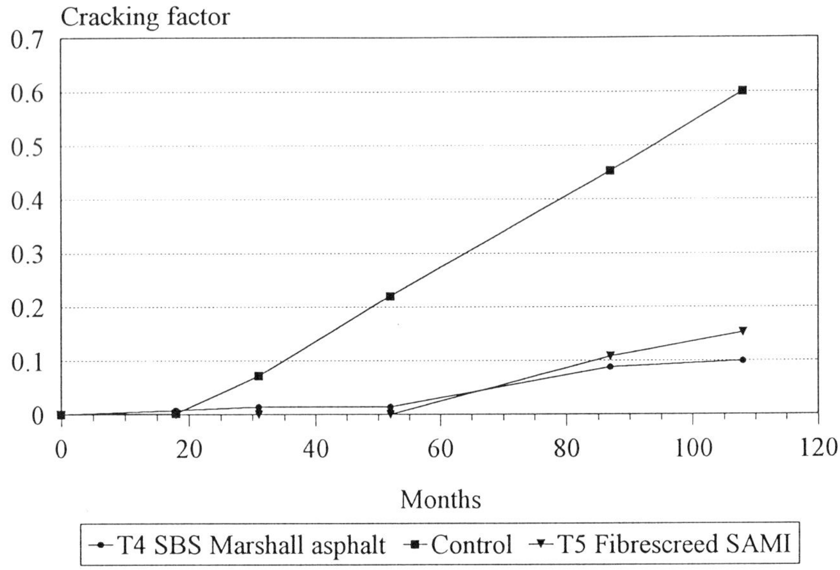

Figure 2. Length of transverse reflection cracking on the taxiway at RAF Finningley

The severity of cracking was also recorded in terms of low, moderate and high depending on crack width and whether a single crack, a crack with branches or a multiple crack was observed. In Figures 1 and 2 the severity of cracking was not taken into account. However, Table 2 shows that for the trial on the taxiway, if the severity of cracking is included then the performance of the Fibrescreed treatment and the SBS modified asphalt shows a further improvement compared to the control section.

Table 2. Reflection cracking after nine years in service

	Length of cracking (m)				
Section	Severity of cracking			Total	Total including Severity
	Low	Moderate	High		
4: SBS MA	3.1	2.4	0.8	6.3	8.3
Control	3.4	10.0	15.9	29.3	50.2
5: Fibrescreed	2.9	4.7	0.0	7.6	10.0

Total including severity = L + 1.5M + 2H

RAF NORTHOLT

Design and construction. Two trials were conducted at RAF Northolt. The first, in June 1993, was part of the resurfacing works on the southern taxiway where the existing construction was 100mm of hot rolled asphalt (HRA) overlay on 200mm of jointed concrete pavement on 150mm of hoggin. Slurry surfacing had been applied to the HRA but extensive and severe reflection cracks were present outlining the 9m x 6m concrete bays. There were also many halving cracks. Two test sections were built; one with Glasgrid and a control section. Along the main length of the taxiway Glasgrid was installed beneath the new Marshall asphalt basecourse and wearing course which were laid to a total thickness of 100mm. Prior to the Glasgrid installation, one side of the taxiway was milled and, on the other side, a thin regulating course was laid to change the crossfall so that over about one half of the width the Glasgrid was installed on a milled surface. Part of these main works formed test section S1. Test section S2 was an adjacent control area on which the Marshall asphalt was laid to the same nominal thickness and no preparatory works were carried out on the existing cracks.

The second trial, on the northwest taxiway, was constructed in August 1997 using the Bitufor™ system as the anti-reflection cracking treatment. Part of the area being overlaid was used as the control section with no Bitufor™ treatment. The existing taxiway consisted predominantly of 9m x 6m concrete bays with some 3m x 3m bays. The concrete was 250mm thick laid on an unbound granular sub-base. The overall thickness of the Marshall asphalt basecourse and wearing course overlay was 70mm but it increased to 110mm where it tied in at a bellmouth with the runway.

Measurements during construction. For both trials, a detailed crack survey was carried out on the original surfaces; for the 1993 trial, the survey was conducted before milling. Cores were taken to determine thicknesses and compressive strength of the concrete. On the 1997 trial, FWD measurements were made to determine load transfer efficiencies between adjacent concrete slabs and also to determine the stiffness characteristics.

Performance of the test sections. For the most recent inspection in March 1998, a visual survey was conducted on the Glasgrid trial together with FWD measurements on the Bitufor™ trial to determine the stiffness moduli and the load transfer efficiency of the concrete joints where fine cracks had occurred in the asphalt.

On the Glasgrid trial of August 1993, considerable transverse reflection cracking has occurred over the half of the test section where the original asphalt surfacing was milled out to change the cross profile. On the other half of the taxiway where a regulating course was laid to increase the cross profile, the cracking was less but significant. A longitudinal crack had formed above the transition between the regulating course and the milled surface. Although the results of the survey have not yet been analysed quantitatively, it is clear that for satisfactory performance, Glasgrid must be installed on a smooth surface.

On the Bitufor™ trial, some fine cracks have occurred in the surface of the asphalt overlay above both longitudinal and transverse joints in the concrete after only 6 months in service. Investigation into the cause of the premature cracking showed that the joint sealant between adjacent concrete slabs had extruded into the overlay possibly producing localised stresses in the asphalt during compaction which could have generated the fine cracks.

RAF BRIZE NORTON

Design and construction. Two trial areas of 40mm of Marshall asphalt wearing course incorporating SBR modifier, section T1, and EVA modifier, section T2, were constructed in July 1994 on taxiway 7 together with a control section T5. The existing asphalt surface of the taxiway was milled to a depth of 40mm to incorporate the Marshall asphalt wearing course inlays of the two test sections and control. Section T1 incorporated 10 per cent of Asphapol 2000 SBR latex by weight of binder and section T2 incorporated 5 per cent of Polybilt 103 EVA by weight of binder and the control section was Marshall asphalt wearing course with 100 pen bitumen. The size of each test section was 50m x 12m.

Two further trial sections of the same area and incorporating asphalt reinforcement grids were constructed in August 1994 on taxiway 6. Section T3 incorporated Glasgrid, similar to that used at RAF Northolt, and section T4, contained HaTelit. A control section, T6, was also constructed without a reinforcement grid. The grids were installed on a newly laid Marshall asphalt basecourse and overlaid with 50mm Marshall asphalt wearing course.

Measurements during construction. Prior to the construction of the test sections, a visual condition survey was carried out on the existing asphalt surfacing showed a pattern of overbanded longitudinal cracking together with irregularly spaced transverse cracks. Cores were extracted from the test sections to measure the properties of the modified Marshall asphalt and to examine bonding of the grids. Bulk density, air voids content, stiffness modulus, fatigue life and resistance to deformation were similar for both modified materials and the control.

Performance of the test sections. The performance of the test sections is being assessed by monitoring crack development. The most recent survey was conducted in March 1998, 3 years 7 months after construction at which reflection cracks mirroring the severe cracks in the original pavement were observed on trial sections T1 and T2 with the modified Marshall asphalt wearing courses and on the control section T5. On the geogrid sections (T3 and T4) and in the control section, T6, no cracking had occurred. However, it should be borne in mind that these sections had a total overlay thickness of 100mm.

RAF MARHAM

Design and construction. The trial formed part of a maintenance scheme on the southern taxiway during August 1995 in which the friction characteristics of the existing construction were being improved by the application of a slurry surfacing and, in two areas, by the use of proprietary thin surfacings. The existing construction comprised 200mm of Marshall asphalt on the original 250mm of jointed unreinforced concrete on 220mm of lean concrete on a hoggin sub-base. An extensive pattern of reflection cracking was present in the Marshall asphalt wearing course, outlining the layout of the original concrete bays. Most of the cracks had been sealed. The most severe cracks were milled out to a depth of about 50mm and a width of 300mm - 400mm and were reinstated with 14mm hot rolled asphalt; elsewhere any crack sealant were burnt off.

Three test sections were constructed each of 50m in length over the full 30m width of the taxiway. In section S3, slurry surfacing was applied which was the standard treatment applied to the remainder of the taxiway. Sections S1 and S2 were proprietary surfacings known as UL-M and Safepave in the UK, laid to a nominal thickness of about 20mm.

Measurements during construction. A visual survey was carried out of the reflection cracking on the original Marshall asphalt and the location of the slot treatments were recorded.

Performance of the test sections. The last visual inspection was conducted in December 1997, sixteen months after the surfaces were laid. The Safepave and UL-M were in good condition and no reflection cracking had occurred but the surface of the slurry surfacing was extensively cracked and the grid pattern of the underlying longitudinal and transverse concrete joints was clearly visible over much of the taxiway. Over the cracks that had been milled then reinstated with asphalt, two parallel cracks were present over the edges of the reinstatement. Clearly, the asphalt inlay approach is not a satisfactory method of inhibiting reflection cracking when a slurry surfacing is the only surface layer.

RAF CONINGSBY

Design and construction. As part of the application of an asphalt overlay on the jointed unreinforced concrete on the northern taxiway, a crack and seat trial was carried out in September 1995. The trial area consisted of a section of the taxiway approximately 450m long by 18m wide. Three test sections were constructed with crack spacings of 1m, 0.75m and 0.5m together with a control section with no special treatment. The overlay was designed to be nominally 150mm of Marshall asphalt but in practice the crossfall was changed to improve the run-off for surface water and this increased the overlay thickness steadily across the taxiway from 150mm to 210mm thus providing the potential for additional information from the trial.

The unreinforced concrete was of two layer construction, nominally 200mm on 150mm laid on a varying thickness of foundation. The upper slabs were approximately 3m square and the lower were 4.5m square. Throughout the pavement, the transverse and longitudinal joints in the upper layer were constructed to overlie the midpoints of the lower-layer slabs.

The cracking was carried out using two Arrows D500 mobile hydraulic hammers working in tandem and an Albaret Orthopactor PF2 pneumatic tyre roller (PTR) weighing 14 tonnes was used for the seating. The Arrows hammer had a specially designed 200mm wide blade mounted on a traversing head that is positioned perpendicularly to the lane to be cracked. The crack pattern was outlined by wetting the surface and observing the pattern as it dried. Cores extracted through the cracks confirmed that the required shape and direction of the cracks was achieved.

Measurements during construction. A visual survey before the start of the works was made so that any subsequent reflection cracking could be related back to the original survey. The longitudinal profile of the concrete was measured along each line of slabs to provide a baseline for assessing the improvement after treatment. A Dynatest 8000 Falling Weight

Deflectometer (FWD) was used to determine the stiffness modulus of the concrete before and after cracking and seating and also the load transfer efficiency across the transverse joints in the upper layer of concrete. The calculation of layer stiffness from FWD data was carried out using the MODULUS computer program, developed by Scullion and Michalak (2). The change in effective stiffness of the concrete after cracking and seating depended on the crack spacing. On average, at 1m spacing, the reduction in stiffness modulus was 45 per cent, at 0.75m spacing, 50 per cent and at 0.5m spacing 65 per cent.

Performance of the test sections. The last visual inspection carried out in February 1998 showed that no reflection cracking had occurred in any section. The longitudinal profile variance measured using the TRL Highspeed Survey Vehicle immediately after overlay and again in February 1998 showed that the longitudinal profile had been improved by the overlay and that no deterioration had occurred in service.

RAF LYNEHAM

Design and construction. Following the successful construction of the crack and seat trial at RAF Coningsby, a trial was carried out at RAF Lyneham prior to using the technique on the northeast (NET) and the southern taxiway (ST). The ability of the crack and seat and overlay approach to inhibit reflection cracking above joints in the concrete had been demonstrated by Potter and Mercer (3) on UK highways. The trial was carried out on an aircraft service platform with 370mm of jointed concrete pavement using the Arrows D500 drophammer.

Subsequently, the NET and the ST were maintained using the crack and seat and overlay technique but this time using a Badger Breaker guillotine with a single blade 2.1m wide. The existing construction of the NET and ST comprised 380mm of jointed unreinforced concrete in bays of 6m x 3m and 6m x 6m laid on a sub-base of lean concrete over part of the taxiways and an asphalt foundation elsewhere. It was necessary to develop a method to crack the 6m x 6m square slabs which involved applying a light impact in the longitudinal direction in the centre of the slab which did not create a crack at that stage. The guillotine then cracked each half of the slab with the blade transverse to the direction of the taxiway at the required crack spacing and this caused the longitudinal halving crack to form from the earlier impact. A Marshall asphalt overlay of 170mm was applied over the cracked and seated taxiways.

Measurements during construction. The trial was monitored by visual inspection of the crack pattern, by coring through the cracks and by FWD measurements to determine effective stiffness moduli at 0.5m intervals along each slab as well as load transfer efficiency between slabs. Crack spacings of between 0.75m and 2m were investigated. The reduction in effective stiffness modulus was greater than that experienced at RAF Coningsby being between 65 and 90 per cent increasing with decrease in crack spacing. However, it was important to ensure that for design purposes, the stiffness modulus exceeded 5 Gpa. This was achieved provided that the crack spacing was greater than 1m.

For the main works, nearly one year later, a detailed visual survey of the concrete surfacing was made before the cracking commenced to provide a datum for future crack surveys. FWD measurements were carried out to ensure that the specified stiffness modulus was achieved after cracking and seating. A crack spacing of 1.2m was used to ensure that the threshold stiffness modulus was achieved for the concrete. Cores were taken through cracks to check that full depth fine cracks were produced and that no shattering of the concrete occurred.

Performance of the test sections. The anti-reflective cracking performance of the crack and seat treatment will be monitored by annual visual inspections.

CONCLUSIONS

Trials have been carried out successfully to evaluate a variety of treatments to inhibit reflection cracking in asphalt overlays over jointed concrete airfield pavements and in asphalt overlays over cracked flexible composite pavements.

Although evaluation is still in progress, some preliminary conclusions can be drawn.

- The application of airfield Friction Course directly onto the jointed concrete pavement before overlaying with Marshall asphalt has reduced the occurrence of reflection cracking after nine years in service.
- A Stress Absorbing Membrane Interlayer applied to a cracked flexible composite pavement reduced the amount of reflection cracking by about 80 per cent and the severity of cracks compared to an equivalent control area after nine years in service.
- An SBS modified Marshall asphalt wearing course overlay applied to a cracked flexible composite pavement reduced the amount of reflection cracking by about 80 per cent and the severity of cracks compared to an equivalent conventional Marshall asphalt after nine years in service.
- After 42 months in service reflection cracks mirroring the crack pattern in the underlying surface had occurred in both SBR modified and EVA modified Marshall asphalt wearing courses.
- The use of a fibreglass reinforcing grid demonstrated the importance of installing it on an asphalt regulating layer and not directly on a milled surface under an asphalt overlay.
- After 42 months in service no reflection cracks had occurred in test sections incorporating a polyester grid and a fibreglass grid under 100mm of Marshall asphalt overlay.
- The crack and seat technique has been applied successfully to jointed unreinforced concrete pavements with bay size between 3m x 3m and 6m x 6m.

FUTURE ACTIVITIES

It is planned to continue a programme of annual monitoring of the trials to measure the development of reflection cracking to compare the performance of various anti-reflection cracking treatments. An assessment will also be made of the extent of the problem of reflection cracking on UK military airfields to develop a structured programme of rehabilitation.

ACKNOWLEDGEMENTS

The work presented in this paper forms part of the research programme of the Defence Estate Organisation and is published by permission of the Director of the Defence Estate Organisation and the Chief Executive of the Transport Research Laboratory.

Copyright Transport Research Laboratory 1998. The information contained herein does not necessarily reflect the views or policies of the Ministry of Defence. Whilst every effort has been made to ensure that the information presented in this paper is relevant, accurate and up-to-date at the time of publication, the Transport Research Laboratory cannot accept any liability for any error or omission.

REFERENCES

1. NUNN, M. E. An investigation of reflection cracking in composite pavements in the United Kingdom. Proceedings of the RILEM conference on reflective cracking in pavements, Liege (1989)

2. SCULLION, T AND MICHALAK, C. MODULUS 4.0 User's manual, Texas Transportation Institute Research Report 1123-4F. Texas Transportation Institute, Texas A&M University, Texas, USA. (1990)

3. POTTER, J. F AND MERCER, J. Performance of the crack and seat method for inhibiting reflection cracking. Proceedings of the third International RILEM conference on reflective cracking in pavements, Maastricht (1996)

Fifth International Conference on the
BEARING CAPASITY OF
ROADS AND AIRFIELDS

EVALUATION, MAINTENANCE, AND REHABILITATION OF PAVEMENTS AT ATLANTA INTERNATIONAL AIRPORT

W. CHARLES GREER, JR.

Senior Vice President
Law Engineering and Environmental Services, Inc.
114 Townpark Drive
Kennesaw, Georgia, 30144, USA
770-499-6649 --- CGREER@LAWCO.COM

ABSTRACT

The Atlanta International Airport is one of the busiest airports in the world. Current aircraft operations are approximately 800,000 per year. Decisions to shut down any of the four runways, three of which are 24 to 29 years old, for reconstruction or major rehabilitation can have a detrimental impact on airport operations throughout the world. This would particularly be the case if the shut down were required due to unexpected failure of the pavements. In 1984, the Department of Aviation for the City of Atlanta implemented a program of periodic investigation and structural evaluation for the aircraft pavements. Approximately every three years, the runways, taxiways, and aprons are tested with the Falling Weight Deflectometer. The Pavement Condition Index procedures along with limited destructive testing which includes cores, soil test borings, and laboratory tests such as resilient modulus of elasticity are also used to evaluate the pavements.

These programs have allowed the airport with the use of very limited strategic maintenance activities to extend the useful lives of three of the runways well beyond their 20 year design life. However, with the approach of the end of the useful life, it has become necessary to carefully evaluate and plan the anticipated reconstruction or major rehabilitation in an orderly manner. The five test programs (1984, 1987, 1990, 1994, and 1997) are presented along with the evaluation procedures, results, and the development of a database for all information including photographs since 1987 along with a five year rehabilitation program.

Key words: condition, maintenance, nondestructive testing, pavements, rehabilitation.

INTRODUCTION

The Atlanta International Airport is one of the busiest airports in the world. Current aircraft operations are approximately 800,000 per year. Decisions to shut down any of the four runways, three of which are 24 to 29 years old, for reconstruction or major rehabilitation can have a detrimental impact on airport operations throughout the world. This would particularly be the case if the shut down were required due to unexpected failure of the pavements. In 1984, the Department of Aviation for the City of Atlanta implemented a program of periodic investigation and structural evaluation for the aircraft pavements. Approximately every three years, the runways, taxiways, and aprons are tested with the Falling Weight Deflectometer. The Pavement Condition Index procedures along with limited destructive testing which includes cores, soil test borings, and laboratory tests such as resilient modulus of elasticity are also used to evaluate the pavements. Photographs at the same locations have been obtained since 1987 during each test program for visual documentation of typical conditions. All of the available information including photographs has been entered into a computer database which will provide the basis for a more complete pavement management system.

BACKGROUND

The Atlanta International Airport has a four parallel runway system. Runways 8R and 8L are located to the north side of the airfield and Runways 9R and 9L are located to the south side of the airfield. Runway 8R was reconstructed in 40 days and 40 nights in 1969. Runway 9R was constructed in 1972. Runway 9L was reconstructed in 1974 and extended in 1984. Runway 8L was constructed in 1984. The basic design for all of the runways is 16 inches of concrete with a design flexural strength (third-point loading) of 650 pounds per square inch (psi). The concrete is underlain by 6 inches of cement treated aggregate (CTA) with a design compressive strength of 750 psi at 7 days. Runway 8R has 6 inches of unbound aggregate in lieu of the CTA. The CTA and unbound aggregate are underlain by 6 inches of soil cement with a design compressive strength of 400 psi at 7 days. The soil-cement is underlain by prepared subgrade. The soil subgrade typically has a modulus of subgrade reaction of 100 to 150 pounds per cubic inch.

The runways are 150 feet wide and range in length from 9,000 feet to approximately 12,000 feet. The longitudinal joint spacing is 25 feet and the transverse joint spacing is 50 or 75 feet. The concrete has temperature reinforcing steel to maintain shrinkage cracks in a tight condition.

Runway 8R was 15 years old in 1984 and was thought to possibly be at or near the end of its service life. The pavement surface exhibited cracking in many locations. The cracks were map type in nature with small spacings of 2-3 inches and interconnected. Cores of the concrete indicated that many of these cracks extended from the surface of the pavement down to depths of only 3 to 5 inches. The cracks appeared to be related to shrinkage rather than loads.

An investigation and structural evaluation was performed utilizing the Falling Weight Deflectometer along with conventional destructive testing to determine the need for rehabilitation or reconstruction. The recommendations from this study were to replace portions of two slabs, repair major cracks along joints, and reevaluate the pavement in 3 years. This was accomplished and the runway is still in service.

Similar testing was performed for the other runways including portions of Runway 8L which was under construction. This 1984 study was the first in a series of what currently has become five programs of evaluation to help the Department of Aviation plan for maintenance and rehabilitation of the pavements at the Atlanta International Airport. The latest program has culminated in a five year plan for the timing of the reconstruction or rehabilitation of the runways and taxiways. In addition, all data including photographs dating back to 1987 have been entered in a computer database to allowe more detailed and faster analysis. This database will also provide the basis for a more complete pavement management system in the future.

INVESTIGATIVE PROCEDURES

Falling Weight Deflectometer Testing: The Falling Weight Deflectometer (FWD) is a device which drops a weight on the pavement surface. The force of the weight is transmitted through a spring system and a hard rubber type pad to the pavement surface. The diameter of the load plate is approximately 11.8 inches. The amount of weight and height of drop can be varied so that the impact force can range up to approximately 30,000 pounds for the devices used in the evaluation programs. The heavy weight FWD (loads up to approximately 55,000 pounds) has not been required for the evaluation of the pavements at the Atlanta International Airport.

The FWD has several transducers on it which allow the measurement of pavement surface deflection when the pavement surface is subjected to loads. One transducer measures the deflection through a small hole at the center of the plate. Six additional transducers are located at offsets up to 6 feet. The transducers allow the "deflection basin" to be determined. The magnitude and shape of the deflection basin along with the material characteristics are used to evaluate the load carrying capacity of the pavement.

The deflection transducers can be placed across joints or cracks to evaluate the load transfer across the joints or cracks in pavements. The FWD tests for the investigation were performed at three basic locations in a test slab. One location was in the interior portion (Center) of the slab. The second location was across the transverse joint (Transverse) midway between the longitudinal joints. The third location was across the longitudinal joint (Longitudinal) midway between the transverse joints. The fourth location, at which tests were performed in some test slabs, was across the transverse joint at a corner (Corner).

Three basic lines of testing were performed down the runways. The center line consisted of test slabs on 100 to 150 foot centers (i.e. every 2nd slab) with test slab locations alternating in the first slab on either side of the centerline. The longitudinal joints for the centerline slab tests were the first longitudinal joints off of the centerline joint. These slabs and longitudinal joints are the ones subjected to most of the aircraft wheel loads. The left line of test slabs was located along the outer two lines of slabs left of the centerline. The test slabs were located in every 4th slab for a spacing between tests of 200 to 300 feet and alternated either side of the outer longitudinal joint. The right line of test slabs was located in a manner similar to that for the left line except on the right side of the runway centerline. These outer lines represent slabs which get few load repetitions except at the exits and taxiways.

Pavement Condition Index: In 1990, 1994, and 1997, the Pavement Condition Index (PCI) of each of the runways was determined in accordance with the procedures outlined by the Federal Aviation Administration [1]. The PCI is numerical indicator of the condition of the pavement. Defects such as cracks, spalls, patches, etc. are quantified as to extent and severity. The defects are then transformed into a PCI value by means of algorithms. The PCI values range from 0 to 100 with 100 being the best condition. The runways were delineated into three sections across the pavement. The center 50 feet were designated as the Center, and the outer 50 feet on each side were designated as Left and Right. The pavement was divided longitudinally at key exits and cross taxiways. For Runway 9L, the extensions were denoted as separate sections.

Pavement Cores: The FWD test results were reviewed in detail so that locations for coring could be selected to represent a wide range of stiffness values. Four-inch diameter cores of the concrete were obtained at various locations by utilizing water cooled diamond bit core drilling procedures. The cores were measured for thickness. They were returned to the laboratory for strength and modulus of elasticity testing. Petrographic examinations were also performed since alkali-silica reactions were found in the cores in the 1990 study. Horizontal cracks were also observed at some locations in the concrete cores at depths of several inches since 1990. Cores of the CTA and soil-cement (4 inch diameter) were obtained and returned to the laboratory for compressive strength and modulus of elasticity tests.

Soil Test and Hand Auger Borings: Soil test borings by mechanical drill rigs and manual hand auger borings were performed at selected locations to determine the consistency of the subgrade soils. Undisturbed samples were also obtained for laboratory testing which included resilient modulus of elasticity.

Laboratory Testing: In the laboratory the concrete cores were trimmed and cut so that strength tests could be performed on the top halves and the bottom halves. The cores were capped where applicable and placed in saturated limewater for at least 40 hours. Split tensile tests were performed on the core sections in accordance with American Society for Testing and Materials (ASTM) Designation C496. In addition, tensile and compressive moduli of elasticity were determined in the split tensile tests by placing strain gauges on the ends of the cores. Modulus of elasticity and compressive strength tests were performed on other core sections in accordance with ASTM Designations C469 and C39, respectively.

Other cores or portions of cores could not be tested due to the location of temperature reinforcement steel and/or the presence of cracks. Several of these remaining core sections were cut and polished for petrographic examination under a stereoscopic optical microscope with magnification up to 100x. The nature of the cracks and cement paste were carefully observed. Alkali-silica reactions between the cement paste and the granitic gneiss aggregates were observed beginning in the 1990 study. The extent of these reactions was evaluated in 1994 and 1997 by use of the "Rapid Identification of Alkali Silica Reaction Products in Concrete", American Association of State Highway and Transportation Officials (AASHTO) T299-93. Selected cores were broken and the "fresh" faces were coated with uranyl acetate and visually observed under shortwave ultraviolet light for evidence of reaction products.

The cores of the CTA and the soil-cement were be tested for compressive strength and modulus of elasticity. Resilient modulus of elasticity tests have been performed on undisturbed samples of the soil subgrade in general accordance with AASHTO P46 or similar procedures in every test program.

Ride Quality Testing: In 1994 and 1997, the ride quality of each runway was measured and evaluated. The profile of each runway was measured with a South Dakota Profilometer. Three ultrasonic sensors were mounted orthogonal to the pavement surface on the bumper in front of the left and right tires and midway between the tires. Four longitudinal profiles were obtained for each runway. Two profiles were along the centerline, the third one was 15 feet left of the centerline, and the fourth was 15 feet right of the centerline. The profiles runs were made at 55 to 60 miles per hour. Profile elevations were obtained at an interval of approximately 2.166 feet. The data were analyzed by APR Consultants, Inc. with their VSWEEP software. The aircraft used for the analysis was a Boeing 727-200 weighing 169,000 pounds. The other parameters for the evaluation were an elevation of 1026 feet, a daytime temperature of 59F, and no headwind.

Impact Echo Testing: In 1997, the impact echo procedure was used in a small area on Runway 9R. The procedure is a nondestructive technique which was used to assess the extent of horizontal cracks within the concrete pavement layers. The tined surface of the concrete was ground smooth but the grooves for surface friction were not ground due to time constraints and the extent of grinding required. The results of the tests were inconclusive and unrepeatable. This was most likely due to the grooves.

Ground Penetrating Radar Testing: In 1997, ground penetrating radar was used for evaluation of horizontal cracks. The device used was manual and capable of obtaining data over a 12 inch wide path while being moved over the concrete surface at a speed of approximately 1 foot per second. The use of this technique was considered experimental for the 1997 program and was limited to the same small area used for the impact echo testing. The procedure showed promise but an exploratory core did not encounter a horizontal crack indicated by the radar technique. More evaluation will be required before it can be used on a production basis.

DISCUSSION OF RESULTS AND ANALYSIS

FWD Tests: The FWD test data for 1997 were analyzed in terms of stiffness which is defined as the load on the plate divided by the center of plate deflection and is expressed in kips/inch. The average center of slab stiffness values for the center line slabs for the four runways ranged from 5375 to 6610 kips/inch. The average longitudinal and corner joint stiffness values were significantly lower and ranged from 2680 to 5218 kips/inch. However, the stiffness values have generally been dropping with time. This trend is illustrated in Figure 1 which presents the frequency diagram (percentage basis) for the longitudinal joint stiffness for Runway 9R for each of the 5 studies. The percentage of test slabs with longitudinal joint stiffness values below 4000 kips per inch has increased form approximately 1 percent in 1984 to 33 percent in 1997. The frequency of test slabs with longitudinal joint stiffness values below 3000 kips per inch is expected to increase significantly in the near future based on the observed trends. Analysis of data for a recently replaced section of Taxiway M at the Atlanta International Airport indicated that 3000 kips per inch is a critical value for longitudinal joint stiffness.

The joint efficiency which is defined as the deflection at Sensor 2 (unloaded slab) divided by the deflection at Sensor 1 (loaded slab) and is expressed as a percentage was also calculated for each test location. The average joint efficiencies for the center line for all four runways in 1997 ranged from 71 to 89 percent for the transverse joints and 61 to 91 percent for the longitudinal and corner joints. The longitudinal joint efficiency values for Runway 9R were typically in excess of 80 percent in the 1997 study which is consistent with the previous evaluations. This indicates that the expansive forces due to the alkali-silica reactivity are likely keeping the joints closed which results in good load transfer across the joints. However, the overall load carrying capacity for Runway 9R is slowly decreasing as shown in Figure 1. Similar, but slower rates of decrease were found for the other runways.

In 1990, Greer [2] presented the details of the structural analysis of Runway 8R based on the 1984 and 1987 data. The data and structural analyses subsequent to that report have indicated that structurally Runway 8R and other three runways will serve very well for several more years. These runways will most likely be considered to have failed when maintenance becomes a nuisance for the owner rather than large scale structural failure. This is particularly true considering that the alkali-silica reactivity in Runway 9R is causing slow degradation of the joints which is confirmed with the decreases in stiffness values at the longitudinal joints. Therefore, the PCI data will be a good indicator of when the runways should be considered for reconstruction or rehabilitation.

Pavement Condition Index: The results of the PCI surveys are summarized in Figure 2. It has been the author's experience that major rehabilitation or reconstruction will be necessary when the PCI values drop below 50. This is particularly the case when numerous sample units have PCI values below 50. It can be seen in Figure 2 that Runway 9R is the one most likely to reach this condition first and that it would be expected to occur in 1999 to 2000. Runway 8R is likely to reach the critical condition around 2001 and Runway 9L which has begun to exhibit significant drops in PCI since 1990 will likely reach the critical condition in

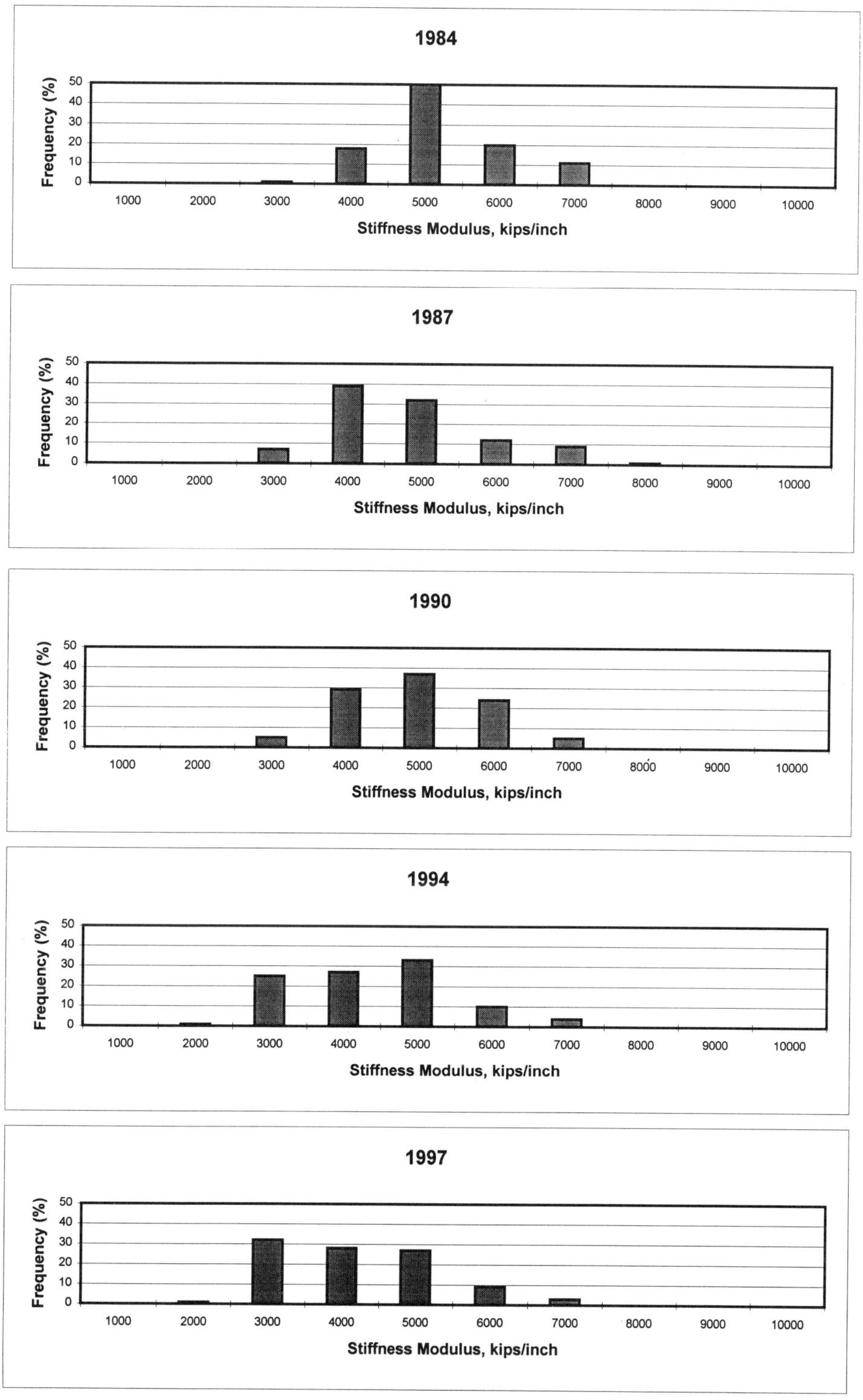

FIGURE 1. STIFFNESS MODULUS FREQUENCIES - LONGITUDINAL JOINTS - CENTERLINE - RUNWAY 9R.

2003. Runway 8L will likely reach a critical PCI condition sometime between 2005 and 2010.

Laboratory Tests: The average split tensile strengths ranged from 420 to 610 psi. These values indicate that the design flexural strength of 650 psi was achieved for the concrete [3,4,5]. The values also indicate that, in general, the flexural strength of the in-place concrete has not significantly increased beyond that which was measured in the acceptance testing at 28 days.

The resilient modulus of elasticity test results for the undisturbed samples of the subgrade ranged from approximately 4,000 psi to 10,000 psi. The majority of the results were between 6,000 and 8,000 psi. These values are consistent with the typical California Bearing Ratio (CBR) values of 5 to 6 for soils at the Atlanta International Airport.

The petrographic examinations of the polished cores indicated a significant amount of microscopic cracking. However, the cracks were relatively tight. Carbonation of the paste was evident in the upper portions of the cores but did not appear to pose a problem. Extensive evidence of alkali-silica reaction products was observed in the cores from Runway 9R. Many of the air voids in the cores were also filled with gel from the reactions. Some amounts of alkali-silica reactions were observed in cores from Runways 8R and 9L. The vertical cracks observable with the unaided eye in most cores generally stopped at depths between 2.5 and 6 inches.

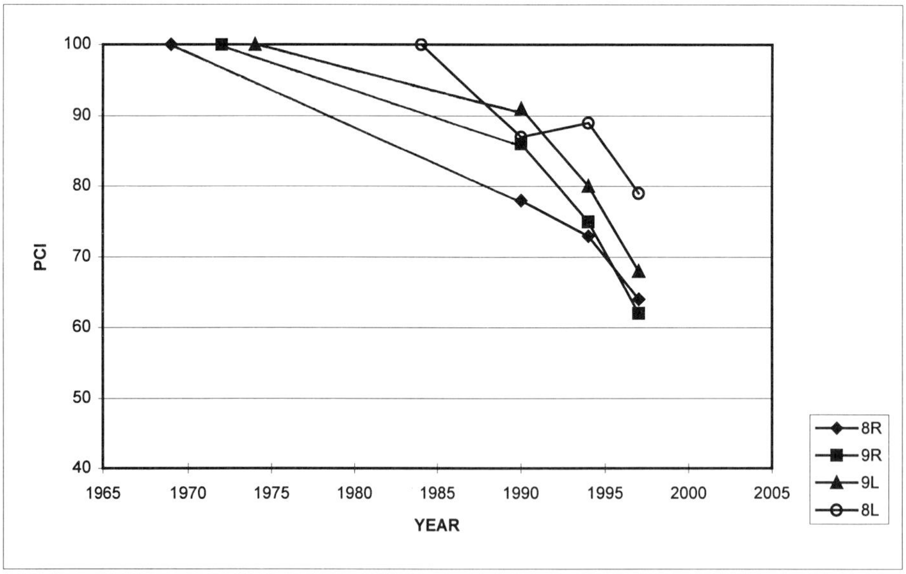

FIGURE 2. PCI VALUES VERSUS TIME FOR ALL FOUR RUNWAYS.

Ride Quality Testing: The Ride Quality Index (RQI) values ranged from 1.0 to 3.1 in 1994 for all four runways. In 1997, the RQI values ranged from 0.9 to 3.4. Runway 9R exhibited the largest detrimental change in RQI from 1994 to 1997. Goldman and Von Gierke [6] have suggested a ride quality comfort criteria of 0.4g to be a reasonable threshold value. When this is transformed into RQI, an RQI value above 4.0 is considered to be indicative of a rough pavement. The RQI data suggest that the runways are slowly getting rougher, but more data will be required in order to better predict the rates of increase in roughness.

ANALYSIS

Support Capability: The support capability available on top of the CTA and unbound aggregate base was analyzed by using the FWD data, the laboratory strength data and Westergaard analysis techniques as previously reported by Greer [2]. A modulus of subgrade reaction of 500 pci or more was found to be available on top of these materials. The analysis of the joint deflections for the FWD data indicate that the transverse joints are performing well. However, the longitudinal joints are not performing as well as the transverse joints.

Stress Analysis: A analysis of stresses in the pavements due to loads was performed in the manner previously reported by Greer [2]. The analysis confirmed that the overall structural capacity of the pavements is very high. Allowable repetitions for all four runways are expected to be in excess of 3 million. Based on the current operations of approaximately 100,000 per year for each end of each runway and estimates of past traffic, the runways are expected to have structural lives of 30 to 40 years.

General: It is clear that three of the four runways are beginning to approach the end of their structural life. However, failure will most likely be defined as occurring when the functional condition becomes a nuisance in terms of timing and cost rather than massive structural failure of the pavements. The PCI procedures which provide a very good measure of the functional condition of the pavements was chosen to establish the time when reconstruction or rehabilitation should be set. Table 1 below presents the timing and actions for maintenance and rehabilitation/reconstruction which were established through the year 2005.

Table 1. Recommended Maintenance and Rehabilitation/Reconstruction.

	YEAR								
RUNWAY	1997	1998	1999	2000	2001	2002	2003	2004	2005
8L	m	m	m	m	m	m	m	m	m
8R	m	m	m	m	**R**	m	m	m	m
9L	m	m	m	m	m	m	**R**	m	m
9R	m	m	**R**	m	m	m	m	m	m

m = Annual routine maintenance (joint repair, joint sealing, rubber removal, etc.).
R = Planned major rehabilitation or reconstruction.

CONCLUSION AND RECOMMENDATIONS

The findings of the evaluation programs have resulted in the development of a multi-year plan for maintenance and reconstruction of the runways at the Atlanta International Airport based on sound engineering principles. The personnel involved with these issues can now determine how the maintenance and reconstruction activities can be implemented with a minimum of disruption to the airline operations and the traveling public throughout the world. The evaluation program will be performed again in 2000. However, it has been recommended that the PCI determinations be made and analyzed on an annual basis in the future.

ACKNOWLEDGEMENTS

The author thanks Mr. Andy Bell, Deputy Commissioner of Aviation and Mr. John Culpepper and Mr. Frank Hayes of Aviation Consulting Engineers, Inc. for their confidence and support throughout the past 25 years. The author also thanks them for the opportunity to be involved with the pavement management, evaluation, and construction process at the Atlanta International Airport.

REFERENCES

1. "Guidelines and Procedures for Maintenance of Airport Pavements", Advisory Circular AC 150/5380-6, U.S. Departemnt of Transportation, Federal Aviation Administration, December, 1982.

2. Greer, W.C., "Structural Evaluation of Runway 8L-26R at Atlanta International Airport", Third International Conference on Bearing Capacity of Roads and Airfields, Trondheim, Norway, 1982.

3. Greer, W.C., "The Estimation of Flexural Strength from other Types of Tests", Special Technical Publication M-5, Law Engineering, Inc., 1982.

4. Packard, R.G., "Design of Concrete Airport Pavement", Portland Cement Association Engineering Bulletin, 1973.

5. Darter, M.I., "Design of Zero-Maintenance Plain Jointed Concrete Pavement, Volume 1 - Development of Design Procedures", Department of Civil Engineering, University of Illinois at Urbana - Champaign, Illinois, 1976.

6. Goldman and Von Gierke, "Effects of Shock and Vibration on Man", Volume III, Chapter 44, <u>Shock and Vibration Handbook</u>, McGraw-Hill Publishers, 1961.

Fifth International Conference on the
BEARING CAPASITY OF
ROADS AND AIRFIELDS

CASE STUDY: DESIGN AND CONSTRUCTION OF BONDED PCC OVERLAY ON A D-CRACKED AIRFIELD PAVEMENT

By ROBERT K. KUMAPLEY[1], LESLIE TITUS-GLOVER[1], AND CURT A. BECKEMEYER[2]

[1]Project Engineer, ERES Consultants, Inc., 505 West University Avenue, Champaign, Illinois, U.S.A 61820-3915.
[2]Division Manager, ERES Consultants, Inc., Champaign, Illinois, U.S.A

ABSTRACT

Concerns about reflective cracking in the overlay and poor bonding between the overlay and the underlying slabs have generally precluded the placement of bonded portland cement concrete (PCC) overlays on significantly distressed pavement, especially one that is severely D-cracked. However, innovative pavement analysis, design, and construction techniques have been employed for successful construction of a bonded PCC overlay on a 40-year-old, severely D-cracked runway. This paper discusses the decision to use this unconventional rehabilitation method, as well as the design and construction techniques that helped to make it a success.

Keywords D-cracking, DEFratio, Nondestructive testing, FWD, PCC overlay.

INTRODUCTION

This paper illustrates use of innovative pavement rehabilitation techniques for successful construction of a bonded concrete overlay in the 30.5-m (100-ft) wide keel area of Runway 13-31 at Forbes Field in Topeka, Kansas. The runway was originally constructed in 1955 and consisted of jointed plain concrete pavement (JPCP) placed directly on a clayey subgrade. This runway is 61 m (200 ft) wide with slab sizes of 7.62 m by 7.62 m (25 ft by 25 ft). In 1975, an asphalt concrete (AC) overlay was placed over the entire runway, and in 1986, an asphalt slurry was placed on the AC overlay (1). Figure 1 shows a typical pavement section that existed prior to rehabilitation.

The Runway 13-31 was rehabilitated in 1995 because the AC overlay exhibited significant deterioration, and this deterioration was magnified by the presence of D-cracking in the upper 100 mm (4 in) of the portland cement concrete (PCC) layer at the JPCP joints.

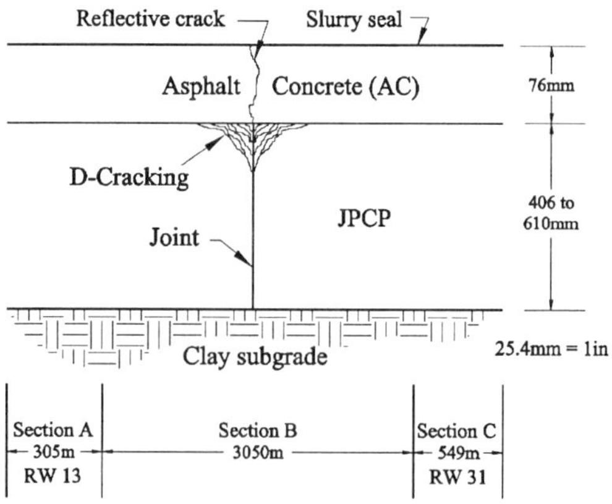

Figure 1. Runway 13-31 pavement section prior to rehabilitation.

Figure 2 shows a JPCP pavement at Forbes Field with D-cracking. As can be seen in figure 2, D-cracking generally occurs at the joints and cracks in concrete pavements, where moisture is most accessible to the concrete (2).

Figure 2. PCC pavement with severe D-cracking at Forbes Field.

RUNWAY 13-31 CONDITION ASSESSMENT

The following paragraphs are summaries of the efforts performed in 1966, 1989, and 1992 to assess the conditions of Runway 13-31. These efforts included a review of previous pavement studies of the runway, pavement coring, falling weight deflectometer (FWD) testing, and a thorough data analysis.

Visual condition surveys In 1966, an investigation of the airfield pavements at Forbes Field revealed that D-cracking was a serious concern on Runway 13-31. In this survey report, it was noted that the concrete pavements placed on a free-draining base exhibited significantly less D-cracking than those pavements, such as Runway 13-31, that were placed directly on the subgrade. In both the 1989 and 1992 surveys, the Pavement Condition Index

(PCI) procedure described in ASTM D5340 (3) was used to assess and rate the surface condition of the runway. The PCI procedure uses a rating scale from 0 to 100, with 100 representing a pavement in excellent condition. A PCI value of 65 to 70 is a typical threshold for triggering rehabilitation on major runways and taxiways. A summary of the PCI results from the 1989 and 1992 condition surveys is provided in table 1.

Table 1. Runway condition assessment results.

Runway section	Average PCI values	
	1989 survey	1992 survey
13 End (Section A)	59 (Good)	44 (Fair)
Interior (Section B)	53 (Fair)	40 (Poor)
31 End (Section C)	52 (Fair)	42 (Fair)

The 1992 PCI survey values clearly indicate the need for major rehabilitation. The major distresses noted on the runway surface were reflective cracking, raveling, and weathering of the AC surface.

Pavement coring Cores were taken from the AC-overlaid PCC pavement to supplement the visual inspection data. The primary purpose of the coring was to assess the condition of the underlying PCC, locate the boundaries of the D-cracking, and quantify the extent of the D-cracking. The coring was necessary because the visual condition surveys only address distress evident on the surface of the pavement and do not provide any direct indication of the level of D-cracking in the underlying PCC layer. The AC overlay tends to mask the underlying pavement conditions.

The PCC cores revealed D-cracking at the joints throughout the entire runway. Inspection of the cores revealed that the D-cracking was generally limited to the upper 100 mm (4 in) of the PCC slabs. D-cracking typically originates at the bottom of a PCC slab and propagates upward with continued freeze-thaw cycles. However, it is believed that local climatic conditions are such that frequent freeze-thaw cycles occur in the top portion of the pavement, and very few cycles occur in the bottom of the slab. Studies at other airfields in Kansas and Oklahoma have revealed similar D-cracking situations (4; 5). Therefore, the D-cracking in Runway 13-31 represented a situation that is potentially unique to that region of the United States.

FWD testing The FWD is a nondestructive impulse loading device that creates a dynamic load by dropping a weight onto a spring connected to a loading plate. Seven geophones measure the resulting pavement deflection. One sensor is located at the center of the loading plate, and the remaining six sensors are spaced at intervals of 305 mm (12 in) away from the load plate. By varying the weight and height from which the weight is dropped, the magnitude of the load can be changed.

Pavement deflection data were collected during the 1992 survey at transverse joints (at midpanel in the main gear path) and center of PCC slabs using a heavyweight FWD. The presence of reflective transverse and longitudinal cracks in the AC layer clearly identified the joint locations in the underlying JPCP. Three target load magnitudes of 156, 200, and 290 kN (35, 45, and 65 kips) were applied to the pavement at each test location. The target loads were selected to cover the range of single wheel loads of expected operational aircraft at Forbes Field.

Estimation of existing pavement layer properties The elastic moduli for the PCC layer (E_{PCC}) and the modulus of subgrade reaction for the clay subgrade (k_{sgd}) were determined through the analysis (backcalculation) of the deflection basins obtained from midslab during the FWD testing. A computer-based backcalculation software, ERESBACK 2.0 (6), was used to estimate the strength parameters for the various layers. The results from the backcalculation efforts indicated a mean elastic PCC modulus (E_{PCC}) of approximately 53,800 MPa (7,800,000 psi) and a mean modulus of subgrade reaction (k) of 70 kPa (250 psi/in). The backcalculation results also showed E_{PCC} for sound concrete is usually between 17,250 and 69,000 MPa (2,500,000 and 10,000,000 psi), and values below 13,800 MPa (2,000,000 psi) typically represent deteriorated PCC. The backcalculation results indicate that the PCC within the interior portions of the slab is sound. However, this result gives no indication of the soundness of the PCC at the joint, where D-cracking has occurred.

Nondestructive D-cracking evaluation The deflections from the joints of the PCC were used to assess the soundness of the PCC at the joints as well as assess the extent of D-cracking in the PCC layer. The ratio of deflections at the slab edge to the deflections at the slab center, DEF_{ratio}, was used to investigate the presence and extent of D-cracking in Runway 13-31. The DEF_{ratio} is computed as follows (7):

$$DEF_{ratio} = \frac{DEF_{joint}}{DEF_{center}} \qquad (1)$$

where

DEF_{ratio} = Ratio of deflections at slab edge and center of slab.
DEF_{joint} = Deflection under the FWD load plate at the joint of slab.
DEF_{center} = Deflection under the FWD load plate at the center of slab.

Studies have shown that DEF_{ratio} values greater than 3.5 are indicative of severe D-cracking. These studies have also shown that DEF_{ratio} values for sound PCC will always be less than 3, even for the worst load transfer efficiency (LTE) case of 0 percent (7; 8). DEF_{ratio} values greater than 3.5 were common throughout Runway 13-31, indicating the presence of severe D-cracking. These findings are consistent with the coring results.

Summary of Runway 13-31 conditions The condition of Runway 13-31 at the time of the 1992 study can be summarized as follows:

- Deteriorated AC surface with the AC layer debonded from the underlying PCC layer in many locations. The deteriorated and debonded AC layer represents a serious foreign object damage (FOD) hazard.
- Severe D-cracking at the transverse and longitudinal joints in the underlying JPCP was prevalent throughout the entire runway.
- The D-cracking was generally confined to the upper 100 mm (4 in) of the PCC layer.

PERFORMANCE OF EXISTING BONDED OVERLAY

Separate evaluations were performed on two sections of apron at Forbes field that had received bonded concrete overlays in 1986 and 1989 (8). The overlay evaluations were conducted to address the following concerns that the design team had with the bonded concrete overlay rehabilitation strategy:

- The condition of the bond at the interface between the overlay and the existing PCC at the time of construction and throughout the design life of the rehabilitated pavement.
- The potential for loss of bond at the interface between the overlay and existing PCC with continued D-cracking.
- The future structural condition of the overlay and the original JPCP.

Unless past performance indicates otherwise, it is expected that D-cracking would continue in the existing PCC. The continued development of D-cracking could reduce or break the bond at the interface between the overlay and the existing PCC. The loss of bond would reduce the structural capabilities of the pavement and greatly reduce the pavement's expected life.

The two sections that were evaluated are located in the Kansas Air National Guard (KANG) aircraft parking apron, west of Taxiway B. The original pavements consisted of 430-mm (17-in) thick JPCP placed directly on a clay subgrade and had D-cracking at the joints, in the top 100 mm (4 in) only. The D-cracking was removed completely by milling prior to the placement of a 230-mm (9-in) bonded JPCP overlay. Figure 4 shows a typical core taken from a joint in the KANG apron test section.

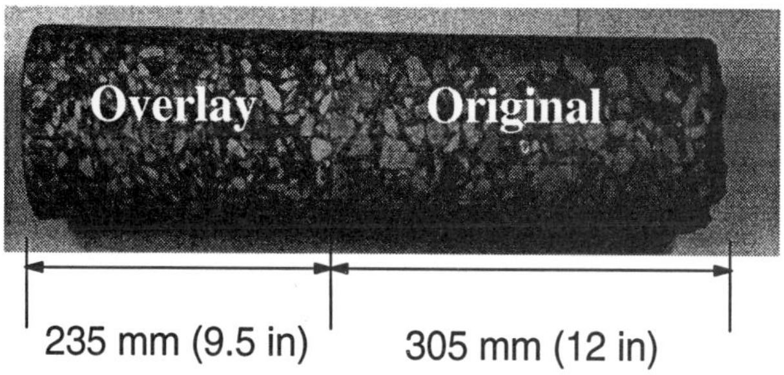

Figure 4. A PCC core from a joint in the KANG apron test section.

The results of the study indicate that, after 6 to 9 years of service, the rehabilitated JPCP is in good structural condition. Significant friction and bonding still exist between the overlay and the original PCC layer of the rehabilitated JPCP. The D-cracking in the existing pavement did not progress as had been expected, and the PCC overlay was providing adequate structural capacity after several years in service. Details on the performance of the rehabilitated D-cracked JPCP were presented by Kumapley and Owusu-Antwi in their paper in *Transportation Research Record 1568* (8).

REHABILITATION ALTERNATIVES

The following rehabilitation options were considered by the design team:

- Reconstruction of the entire runway with either AC or PCC.
- Placement of an unbonded PCC overlay in the 30.5-m (100-ft) wide keel, and application of AC in the outer areas of the runway to match.
- Removal of the AC layer and construction of a new AC overlay.
- Removal of the AC layer, milling of the D-cracked areas of PCC located in the keel area, and construction of a bonded PCC overlay in the 30.5-m (100-ft) wide keel. This option included the placement of a slurry seal on the outer areas of the runway.

Despite traditional thoughts and practices, the results of the pavement evaluation efforts showed that a bonded PCC overlay was a feasible option for rehabilitation of Runway 13-31. Because the D-cracking was limited to the top 100 mm (4 in) of the JPCP slabs, milling the top portion of the PCC would remove the D-cracked material and provide a good condition for placement of a bonded overlay.

BONDED CONCRETE OVERLAY DESIGN

The bonded overlay for Runway 13-31 was designed using mechanistic-empirical pavement analysis techniques, and the design aircraft for the overlay was the C-141. For design purposes, the C-141 was assumed to be operating at its maximum takeoff weight of 1,440 kN (323,000 lb), and the design traffic was assumed to be 100,000 departures.

To conduct the mechanistic design, the finite element software ILLI-SLAB was employed (9). The software was used to model the proposed pavement section, simulate aircraft loadings, and calculate the expected load responses (stresses and strains) within the pavement system. To identify the required overlay design thickness, the following process was used:

1. Calculate the critical stresses that are expected to occur in the pavement as a result of the main gear loading of the design aircraft, the C-141.
2. Estimate the allowable number of departures (passes) that the design aircraft can make until pavement failure occurs.
3. Estimate fatigue damage that will be expected by 100,000 departures of the design aircraft. The estimated fatigue damage is simply the ratio of the expected departures (100,000) to the number of allowable departure calculated in step 2.
4. Add the estimated fatigue damage to the fatigue damage that has already occurred in the existing pavement.
5. Identify the overlay thickness that produced an acceptable level of total fatigue damage (past plus expected) over the design life of the overlay.

To determine the critical stresses in the JPCP, the proposed pavement section and the backcalculated material properties were used to model the existing pavement structure, and the PCC overlay was assumed to have an elastic modulus (E_{PCC}) of 34,500 MPa (5,000,000 psi). Full bond (full friction) was assumed between the existing PCC and the new overlay.

The results of the design checks indicated that a bonded overlay thickness of at least 140 mm (5.5 in) should be able to accommodate 100,000 departures of C-141 aircraft at maximum takeoff weight of 1437 kN (323,000 lb).

SELECTION OF PREFERRED OPTION

Based on the results of a life cycle cost analysis, the bonded PCC overlay option was selected for rehabilitation of the runway. A major advantage of this rehabilitation option on this project was the incorporation of the remaining 305 to 510 mm (12 to 20 in) of existing sound PCC as a structural element in the pavement.

The following rehabilitation process was adopted for construction of the bonded PCC overlay:

1. Mill off existing AC 75 mm (3 in) and top 15 mm (0.5 in) of PCC in the 30.5-m (100-ft) keel section.
2. Mill additional 90 mm of PCC at all joints in the keel.
3. Construct bonded PCC overlay with 140 mm (5.5 in) and 230 mm (9.0 in) at slab interior and joints, respectively.
4. Match existing JPCP joints.

Figure 5 shows the milling detail at the joints.

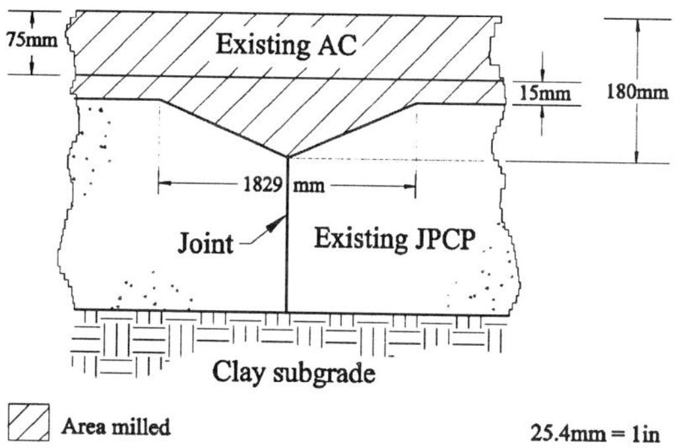

Figure 5. Milling detail at joints.

Because the D-cracking was confined to the upper 100 mm (4.0 in) of the PCC along the pavement joints, not as much milling was required in the interior portions of the slabs. Using a greater milling depth at the joints presented a design and construction challenge; however, the use of this milling technique resulted in significant savings in the cost of removing the D-cracked concrete.

BONDED PCC OVERLAY CONSTRUCTION

Construction of the Runway 13-31 bonded overlay occurred from August to November 1995. In general, the construction process consisted of three major components—milling off the AC surface and the top portion of the PCC surface, preparation of the milled PCC surface, and placement of the PCC overlay. Descriptions of these processes are provided in the paragraphs that follow.

Surface Milling The AC overlay and the top portion of the PCC were removed using cold milling equipment. As noted earlier, the AC and PCC removal was performed over the entire length of the runway, but was confined to the runway keel section (inner 30.5-m [100-ft] width). The pavement milling was performed to the depths shown in figure 5.

Pavement surface preparation After completing the milling operations, the milled pavement surface was carefully prepared to ensure good bonding between the overlay and the milled surface. First, the milled PCC surface was cleaned to remove dust and particulate material. Then, immediately prior to the placement of the overlay, the milled surface received a final cleaning with high-pressure air. There was no grouting or wetting of the milled surface prior to placement of the overlay.

Overlay construction The PCC bonded overlay was constructed using slip-form paving techniques. Figure 6 shows the details of the PCC overlay thickness construction.

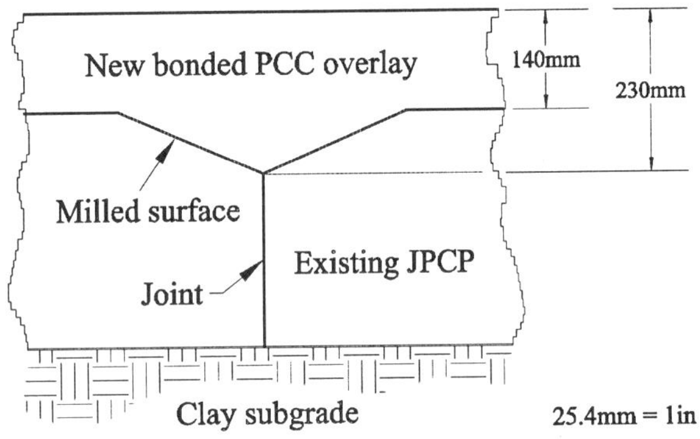

Figure 6. PCC overlay detail at joints.

The paving was performed in the longitudinal direction of the runway, with special care taken at the transverse and longitudinal joints. During the overlay construction, fresh concrete was placed in the milled joint areas immediately ahead of the paving operation. This technique minimized the amount of differential consolidation between the joint and slab interior areas. In addition, transverse finishing of the overlay surface at the existing transverse joints was performed to minimize dips in the overlay at the joints.

During construction, several aspects of the work were monitored closely to ensure good long-term performance of the overlay. The key items that were monitored included:

- Water-to-cement (W/C) ratio of PCC.
- Expected maximum ambient air temperature during the day.

- Sawcut depths and timing.
- Removal of unsound PCC and surface preparation prior to overlay.
- Use of a durable aggregate with a maximum size of 25 mm (1 in).

To achieve the design objective, it was important to develop adequate bonding between the overlay and the existing PCC and to minimize shrinkage in the overlay. To ensure good bonding, the milled surface was kept clean and a concrete mixture with a low W/C ratio was used. The low W/C ratio minimized water build-up at the interface and minimized shrinkage within the overlay. In addition, placement operations were performed during cool temperatures or at night to minimize shrinkage. Shrinkage cracking in the overlay was also controlled by providing contraction joints that matched the locations of the joint in the underlying JPCP.

Performance of the rehabilitated pavement A visual inspection of Runway 13-31 was conducted in January 1998 to assess the performance of the bonded PCC overlay. The inspection revealed that the rehabilitated JPCP was in good condition with no obvious structural problems to date. Although it is premature to draw conclusive inferences about the suitability of bonded PCC overlay for rehabilitation of D-cracked concrete, preliminary results indicate that the technique looks promising, especially where D-cracking is confined to the top portion of the original pavement. The performance of Runway 13-31 will be continually monitored to assess the effectiveness of bonded PCC overlays in the long term.

CONCLUSION

As this paper illustrated, innovative pavement engineering and construction techniques can be cost-effective airfield pavement rehabilitation strategies. In the case of Runway 13-31 at Forbes Field, the desire to go beyond traditional thinking resulted in a high-quality, cost-effective, and long-term runway rehabilitation.

By employing a detailed pavement evaluation and design process, the design team was able to pinpoint the current conditions and performance characteristics of the Runway 13-31 pavements. Using the results of the evaluation process, the design team was then able to develop a cost-effective rehabilitation strategy that specifically addressed the runway's pavement deficiencies. The innovation used in design was then carried into the construction process to provide the airport with a high-quality runway pavement.

The findings in this paper may be useful to pavement engineers considering a bonded PCC overlay for rehabilitation of D-cracked JPCP, especially where D-cracking is confined to the top layer of the original pavement.

ACKNOWLEDGEMENTS

The results presented in this paper are from work conducted for VanDoren Hazard Stallings, Inc. and for the Kansas Air National Guard at Forbes Field, Topeka, Kansas. The authors wish to express their appreciation to Metropolitan Topeka Airport Authority (MTAA), the Kansas Air National Guard (KANG), VanDoren Hazard Stallings, Inc., and ERES Consultants, Inc. The opinions and conclusions expressed in this paper are those of the authors and in no way reflect those of any of the aforementioned agencies.

REFERENCES

1. ERES CONSULTANTS, INC. "Runway 13-31 Study, Forbes Field, Topeka, Kansas," Final Report for Bartlett & West Engineers, Inc., Topeka, Kansas. ERES Project No. 067-S0-2. Champaign, Illinois, 1993.

2. SCHWARTZ, R. D. *D-Cracking of Concrete Pavements.* National Cooperative Highway Research Program Report 134. Transportation Research Record, National Research Council, Washington, D.C., 1987

3. AMERICAN SOCIETY FOR TESTING AND MATERIALS. *Annual Book of ASTM Standards, Road and Paving Materials; Vehicle-Pavement Systems,* Vol. 04.03, 546-593, 1997.

4. ERES CONSULTANTS, INC. "Pavement Rehabilitation Program 300 and 400 Ramps and Taxiways." Final Report for The Boeing Military Airplane Company, Wichita, Kansas. Prepared by ERES Consultants, Inc. Champaign, Illinois, 1985.

5. ERES CONSULTANTS, INC. "Pavement Evaluation for Taxiways J, J1, J2, and J3 at Tulsa International Airport, Tulsa, Oklahoma," Final Report for Tulsa Airports Improvement Trust and Poe and Associates, Inc. ERES Project No. 95-088-S1, Champaign, Illinois, 1996.

6. SMITH , K. D., M.J. WADE, D.G. PESHKIN, L. KHAZANOVICH, H. T. YU, and M. I. Darter. *Performance of Concrete Pavements, Volume II-Evaluation of In-service Concrete Pavements*, Report FHWA-RD-95-110. FHWA, U.S. Department of Transportation, 1996.

7. KUMAPLEY, R.K., OWUSU-ANTWI, E.B., AND TITUS-GLOVER, L. "Nondestructive Technique for Evaluating D-cracking in Airfield Pavements," *Proceedings of the 6th International Purdue Conference on Concrete Pavement, Volume 2*, pp. 139-154, Purdue University, West Lafayette, Indiana, 1997.

8. KUMAPLEY, R.K. AND OWUSU-ANTWI, E.B. "Performance of a Rehabilitated D-Cracked Jointed Plain Concrete Airfield Pavement," *Transportation Research Record No. 1568: Pavement Rehabilitation and Design*, National Research Council, Washington, D.C., pp. 148-154, 1997.

9. KHAZANOVICH, L. *Structural Analysis of Multi-Layered Concrete Pavement Systems*. PhD. Thesis. University of Illinois at Urbana, Illinois, 1994.

Fifth International Conference on the
BEARING CAPASITY OF
ROADS AND AIRFIELDS

USE OF SEGMENTAL BLOCK PAVING ON AIRFIELD APRONS

DR IAN VAN WIJK, JOHN BYRNES AND STEFAANS POOLMAN

Director, Africon Engineering International, Pretoria, RSA
V&V Consulting Engineers, Cape Town, RSA
Under-Secretary, Namibia Dept of Transport, Windhoek, Namibia

ABSTRACT

Sections of the main concrete apron at the Windhoek International Airport had deteriorated badly and required urgent attention. The main type of distress was spalling of joints. The concrete slabs (of 350 mm) were still in a sound structural condition with very few structural cracks. As alternative the conventional repair method of partial depth repair of the spalls and the replacement of the joint seals, overlying the distressed areas with segmental blocks (also called concrete paving blocks or pavers) was considered. The placement of 60 mm segmental blocks with a strength of 35 MPa onto a 25 mm layer of sand on the distressed areas proved to be cost-effective (the construction cost was 60% lower than the conventional cost) and produced the required results.

The segmental blocks increased the maximum deflection but reduced the stresses at the bottom of the concrete slab and increased the allowable coverages up to 10 times. The repaired section has been performing well for almost 10 years. The segmental blocks not only rectified the problems with the spalling, but also increased the bearing capacity and produced more employment opportunities than conventional construction (an important consideration in developing countries).

INTRODUCTION

The apron of Windhoek International Airport was constructed in the 1960's and consisted of jointed unreinforced concrete slabs on an imported natural gravel subbase.

By the late 1980's sections of the apron had deteriorated badly and was in need of repair. The repair was investigated and found to be extremely expensive. Covering the apron with segmental block paving (i.e. concrete paving blocks, pavers) was suggested, investigated and eventually carried out. Segmental blocks have successfully been used in places like the UK[1)] and Australia in heavy duty resurfacing schemes, but not in Southern Africa. Over the next few years the apron was monitored and the solution was found to be extremely successful.

STATE OF APRON

The apron consisted of 3 m by 3 m concrete slabs with an average thickness of 350 mm placed on a 150 mm imported gravel layer. The subgrade materials were of average quality. The apron joint sealant was old and brittle and was pushing out. This resulted in the ingress of incompressible material into the joints preventing expansion. The result was serious spalling.

There were also numerous slabs that showed surface defects as a result of what looked like rain during construction. The top few millimetres of concrete were very weak and were weathering.

The spalling along the joints and cracks was a particular cause for concern. Not only were the spalled bits causing damage to propellers and jet turbines, but were also frequently being picked up and thrown in the direction of the terminal causing damage to vehicles and windows. With aircraft traffic increasing the repair of the apron became urgent.

After an investigation it was revealed that the structural capacity was fine. The slab cracking generally occurred along lines of underground fuel pipes. The subsurface was easily repaired in these places with cementitious grout. The surface appeared to have been cast in rainy conditions and it is doubtful if the correct curing was applied. The exposed concrete was weathered and new surfaces seem to weather quite quickly. This would account for the high rate of spalling along joints and cracks.

The joints had not been maintained on a regular basis resulting in hardening of the jointing material and the joints filling with debris. The cracking and spalling along the joints was so severe and widespread that the repair would have entailed cutting and epoxy repairing all the joints. The time and cost to undertake this repair was far too high.

STANDARD REPAIR METHODS

Cores were taken of the apron and the concrete strength was found to be adequate. Thus the apron did not need replacing. The adequate strength was supported by the types of distress i.e. very few structural cracks.

The standard repair method involved removing the joint sealant, cleaning the joint, cutting the spalled joints into straight lines, epoxy filling the spalled areas and replacing the joint sealant.

The cracked slabs required epoxy injection and joint treatment as described above. The slabs with weak surfaces needed epoxy coverings.

Namibia did not have the expertise and equipment available to undertake the job. This would have had to be brought in from South Africa. The cost of the exercise was prohibitive.

SEGMENTAL PAVING BLOCKS METHOD

It was suggested covering the distressed area (about 100 m by 70 m) with segmental block paving. At first the idea was "laughed at". However, after some discussion and preliminary investigation it was decided to carry out a full investigation.

The two problems that were not immediately solved were the approach to the raised area and the size and quality of the blocks to be used. It was decided to opt for the thinner 60 mm block instead of a 80 mm block because of the good support conditions. The approach to the blocks was made with asphalt at a slope far less than the maximum allowed.

An approximation was made at the increased strength of the pavement. Although this resulted in a higher strength there was no experience to draw upon to determine the behaviour of the compound structure. The existing surface would be less susceptible to temperature changes. The blocks are proven to be self-cleansing, i.e. dirt is removed in the rain.

Three problems were envisaged:

- The blocks must be contained.
- The blocks would cause a height differential of about 85 mm.
- Spilled fuel might accumulate in the sand causing a potential "bomb".

CONSTRUCTION AND COST

A small reinforced ring beam was cast along the entire perimeter to contain the blocks. The beam was anchored to the existing concrete apron with epoxied dowels. Service manholes were raised. Bedding sand was placed, compacted and screeded to a thickness of 25 mm. The blocks were placed on the sand and locked by using a light tap with a rubber mallet. The whole area was compacted with a flat base compactor. The blocks were then treated with an aircraft fuel fuel-resistant bitumen spray. This was to prevent fuel leakage into the sand layer.

Asphaltic concrete (premix) was used to create a ramp for the height differential.
The entire cost of the segmental block paving option was only 60% of the traditional repair method.

An added advantage was the additional use of labour that is very important for developing countries.

POST-CONSTRUCTION TESTING

Falling Weight Deflectometer (FWD) tests were conducted after the repair on the apron on uncovered concrete slabs in a good condition and on concrete slabs that had been overlaid with the 60 mm segmental blocks.

Seventeen tests were done on the uncovered section and fourteen on the overlaid section. The purpose of the testing was to quantify the structural effects of the segmental blocks.

The maximum FWD deflection on the segmental blocks was about 50% higher than that on the concrete slab (123 μm vs 80 μm under a 40 kN impulse load).

The elastic modulus of the concrete slab (of 350 mm) was determined through appropriate back calculation methods. Values ranged between 15 GPa and 40 GPa, with an average of about 25 GPa. The stiffness of the layers below the slab ranged from 80 and 300 MPa, with an average of about 175 MPa. The average modulus of subgrade reaction (k) was about 50 MPa/m.

The combined modulus of the segmental blocks and the bedding sand layer ranged between 1100 MPa and 12000 MPa with an average of about 2500 MPa. The average stiffness of the blocks was calculated as approximately 2500 MPa and that of the sand about 150 MPa. These values are within the expected range of 900 to 7500 MPa for segmental blocks[2)].

The maximum deflections and stresses in the concrete slab were calculated under a Boeing B747 loading. The calculated average deflection on the segmental block sections would be about 0,8 to 0,9 mm, with maximum values in the order of 1,5 mm. This is less than 2 mm and should not create problems. The stresses at the bottom of the concrete slab are 10 to 20% lower under a B747 loading with segmental blocks on the slabs than without segmental blocks. This results in increases of 10 to 20% in the factor of safety. This will increase the number of coverage to failure significantly. An increase of 10% in the factor of safety (F.S.) will increase the number of coverages to failure five fold and an increase in F.S. to 20% will increase the coverages about ten fold[3)].

SUMMARY

The idea is very simple. It worked and it was cheap. The maintenance on the apron now is minimal. This raises the question whether, in the correct circumstances, aprons should not be designed as segmental block paving structures. The repair method is extremely successful, cost effective (about 60% the cost of conventional repair), and will be used again, especially in remote areas. The use of more labour than with the conventional repair makes the use of segmental blocks very attractive in developing countries.

The Department of Transport in Namibia took the initiative for an untried alternative. It resulted in much criticism from certain airlines but also much praise from the engineering community.

The deflections are higher on the apron with the segmental blocks, but this is due to the lower modulus of the blocks and particularly the bedding sand. The moduli of the blocks are within normal ranges. The segmental blocks reduce the stresses at the bottom of the concrete slab by 10 to 20% compared to slabs not overlaid with blocks. This resulted in higher factors of safeties (also 10 to 20% higher) and more coverages up to failure (up to ten fold).

The sections overlaid with segmental blocks are structurally adequate and did increase the structural capacity of the apron significantly, although this was not the prime objective.

REFERENCES

1. BUDGE, C. Field Day for Blocks. NCE Concrete Supplement, July 1995.

2. SHAKEL, B. Design and construction of Interlocking Concrete Block Pavements. Elsener Applied Science, 1990.

3. MAJIDZADEH, K et al. Mechanistic Design of Rigid Pavements. Vol 1 : Development of Design Procedures. DTFH 11 - 9568, 1984, USA.

Fifth International Conference on the
BEARING CAPASITY OF
ROADS AND AIRFIELDS

DESIGN AND CONSTRUCTION OF THE RUNWAY PAVEMENT FOR THE KUALA LUMPUR INTERNATIONAL AIRPORT USING QUASI-MECHANISTIC APPROACH AND NEW MATERIAL

MOHD SABRI HASIM[1] AND AMINUDDIN BIN YAHYA[2]

[1]Resident Engineer, Kuala Lumpur International Airport Berhad ,
Site office, 43900 Sepang, Selangor, MALAYSIA, Fax: 03-8431354
[2]Engineering Manager, Kuala Lumpur International Airport Berhad,
Head Quarters, 50490 Kuala Lumpur, MALAYSIA, Fax: 03-2535671

ABSTRACT

Sandstone deposits exist in abundance at the Kuala Lumpur International Airport (KLIA) site. To save transport cost and expedite the construction period of the first runway, it was imminent to utilise fully the sandstone aggregate quarried at site. The sandstone aggregate was turned into cement treated base, accepted for modern runway construction.

Asphalt aging is predominant in the Malaysian climate. Thicker bitumen film thickness can reduce aging. The use of modified bitumen allows more bitumen to be used without affecting the permanent deformation characteristic of the asphalt. Polymer modified bitumen was used at KLIA to increase durability and improve mechanical properties of the asphalt.

The pavement design was carried out using mechanistic approach which allows new untested materials such as the above, to be used. Local research provided the evidence to accept the material, making the approach a semi verified mechanistic method. Performance based specification was also included to enhance the probability of survival of the pavement.

This paper describes the design and construction experience from the first runway of the KLIA. Description is given on the design aspects where mechanistic interpretation was used incorporating local research findings. The construction of the pavement was carried out using the latest paving technique. Included are the application of pre-cracking and crack relief interlayer to mitigate reflective cracking.

INTRODUCTION

The latest asphalt pavement technology was applied during the design and construction of the first runway pavement at Kuala Lumpur International Airport (KLIA). This technology was used to evaluate the problems strategically, technically and commercially. The challenge is to ensure successful completion of the runway within the given time frame whilst meeting the requirement of International standards and guidelines.

As a result, the runway contain new advance features where local material properties and availability was maximised. The technology was driven to provide strong and safe pavement at the lowest possible cost and shortest period of construction. The first runway pavement is a product of mechanistic interpretation put into reality. This realisation was back up by several years of local data collected by the Public Works Department (JKR) through research and forensic work.

The design and construction experience gained from the exercise open up wider perspective to improve the bearing capacity of roads and airfields pavement using locally available and cheap materials combined with improved technology supported with local research findings.

PAVEMENT DESIGN

Overall design constraint. Structural and functional requirements for runway pavement is guided by the International Civil Aviation Organisation (ICAO). For KLIA, minimum construction cost is required whilst maximising the engineering properties of the material. The pavement is to serve with minimum maintenance and life cycle cost. Maximum safety and durability is required under high service temperature, very heavy rainfall and humidity level. The pavement must be constructed on time, before and ready for use for the next XVI Commonwealth Games in September 1998. Race against time is the biggest challenge.

Technical limitation. The most appropriate pavement design procedure should be a locally verified empirical procedure, which is not available. Empirically developed procedure having similar loading, material composition and environmental conditions are also not available. Empirical local data is only partially available, for a full-depth asphalt pavement structure. However, relying heavily on imported bitumen supply can be uneconomical and risky. Thick asphalt will be slow to construct since it requires many lifts. Granite for asphalt is far from the project site. The added transport cost and time can effect progress and cost of the project.

Sandstone deposit is readily available in abundance in the project site. Sandstone is technically inferior to granite. However, when mixed with small amount of cement it becomes a very good Cement Treated Base (CTB) material. Laboratory tests proved the suitability of the sandstone for construction. The recently completed runway rehabilitation at Sandakan airport, also managed by JKR was a successful evidence of using CTB in runway pavement. Empirical evidence from this work that has performed well for the last five years is used to justify the application of CTB in runway pavement.

Pavement design approach. Having the above constraints and technical limitation, the philosophy of runway pavement design for KLIA has to rely on mechanistic approach. The predicted failure mode and mechanism was addressed carefully based on local research findings. Appropriate solutions were devise to mitigate, retard or eliminate the failure mechanisms.

Aircraft loading. The critical design load from the aircraft fleet operating in Malaysia is the MD11, this design aircraft is having the highest single wheel pressure at departure and most damaging fatigue cycle. The pavement was also checked to accommodate the B747, the new Boeing 777 series and future New Large Aircraft (NLA).

The tyre pressure of the MD11, Boeing 747 and most likely the new B777 is estimated at 1.44 Mpa. These high tyre pressures calls for the asphalt material to have optimum mechanical properties. Resistance to permanent deformation is critical. Investigation on several in-service International Airport in Malaysia reveals the need to focus on permanent deformation.

Design life. A design life of ten years was initially chosen with the provision of upgrading the pavement in the fifth year. At this time pavement evaluation will be carried out to check the true bearing capacity. The translation of B777 and NLA damaging effect should be matured by then for comprehensive upgrading decision. However, the construction cost savings for ten years design life compared to a twenty years design life is small since cheap CTB material will be used. The pavement design life of twenty (20) years was selected.

Predicted mode and mechanism of failure incorporating local factors. Conventionally, the pavement is expected to fail by asphalt fatigue due to repetitive application of the design aircraft load. Structurally induced permanent deformation can also happen due to the load affecting the total pavement structure, predominantly the subgrade.

Research by local scientists found permanent deformation to heavy duty pavement at high stressed areas due to asphalt problems in hot climate (1). Early fatigue cracking due to rapid aging of asphalt is also a mode of failure reported to play a significant role (2). Moisture is another main factor affecting asphalt performance

Cement treated base, the main structural layer is highly prone to cracking. The cracked CTB can cause reflective cracking. Handling reflective cracking using several methods was experimented in a full scale trial (3). This study is the empirical evidence to address this issue.

Structural requirements. The multilayer analysis was used to determine the required component material thickness. The computation for the multilayer model was carried out using the CIRCLY Program, distributed by Mincad System, Australia. The final thickness was checked for compliance to the ICAO Aerodrome design procedure and also the United States Federal Aviation Authority (FAA) circulars.

The cement treated base was designed for a modulus value of 5,000 Mpa. At this value, the stresses and strains under the asphalt material is low and asphalt fatigue failure is not critical. The compressive stresses and strains on the subgrade was used as the control criteria to derive the thickness requirement. The total thickness and material composition was modeled to reduced the compressive strain on the subgrade and avoid subgrade strain failure.

Improvement to the subgrade to achieve higher design soaked CBR of 4.5 % was carried out. The subgrade modulus was estimated at ten times the CBR value. Latest development in pavement design for tropical countries (4) recommended to divide the subgrade into two categories either far from water table or close to the water table. High embankment and far water table reduces the chances of the runway subgrade to be soaked. The assumption of soaked condition is therefore very conservative and safe. The selected pavement structure is shown in Graphic 1.

Graphic 1: Pavement components

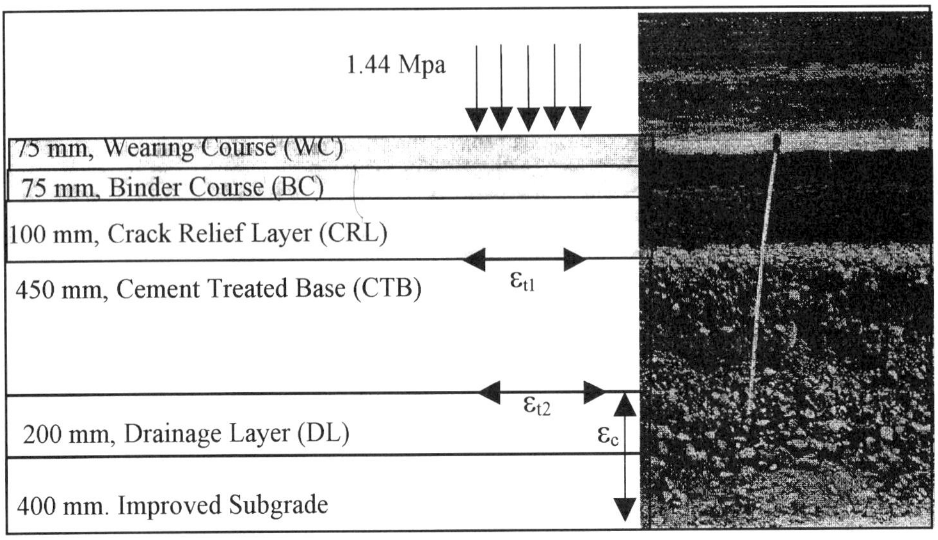

Maximum displacement under the 1.44 Mpa pressure is used for comparative purposes since this value can be verified using the Heavy Weight Deflectometer (HWD). The designed maximum displacement at failure under 1.44 Mpa pressure was estimated at 2.0 mm. Table 1 shows the recorded maximum deflection taken in some airfield pavement in Malaysia. The data indicated safety factor of more than 5 for the KLIA runways. In deflection terms, the strength of KLIA runway is twice of the current in-service main runway at Subang.

Table 1 : Comparative vertical displacement of runway pavement

Airfield	Pavement Layers	Maximum Deflection *
Subang International Airport	500 mm dense asphalt on 100 mm lime stabilise soil.	0.4 - 0.8
Sandakan Airport	250 mm dense asphalt on 300 mm CTB.	0.3 - 0.6
KLIA	150 mm dense asphalt on 100 mm CRL on 450 mm CTB	<0.2 -.0.4

* *Maximum deflection (mm) measured using the Heavy Weight Deflectometer (HWD) @ 1.44 Mpa pressure along the estimated aircraft wheel-path.*

Functional requirements. Several important functional requirements for the above selected type of pavement were identified as shown in Table 2.

Table 2. Functional requirement for the pavement and the solution

Functional requirements	Need statement	Solution
1. Safety against skidding (surface)	High Micro-texture High macro-texture	Granite stone Grooving
2. Anti reflective cracking (internal)	CTB prone to random cracks Reflective cracking	Pre-cracking CRL
3. Durability (CTB and asphalt surface)	Durability of asphalt Durability of CTB	High binder content Polymer Asphalt High cement content
4. Drainability (surface and internal)	2000 mm yearly rainfall >80 % Humidity Thick surface water film thickness Subgrade stability CRL porosity	Grooving CRL vertical drain Slotted drain Subsoil drain Upper drainage layer (CRL) Lower Drainage Layer
5. Regularity (surface)	Wearing course evenness	Multiple lifts Laser Controlled Pavers

These requirements was assessed and resolved base on experimental evidence from the research projects that was carried out by JKR. Anti-reflective cracking and durability are the two additional functional requirements being addressed carefully compare to conventional pavement construction technique.

MATERIALS AND CONSTRUCTION

Aggregates for asphalt. Granite sources are available outside KLIA boundaries within 15 km radius. The sources came from similar geological formation, having similar mechanical, chemical and petrographical properties. The experience and performance of granite in asphalt is well established. The specification for aggregate gradation for asphalt and other layers is shown in Table 3.

Aggregate for CTB and drainage layer. A hill named Hill33 located in the project area contained meta-sandstone deposit. This material was tested and found to be suitable for the CTB and drainage layer application. Comparative relevant properties of the Sandstone compared to conventional granite are also shown in Table 3. The properties compared are Density, Aggregate crushing Value (ACV), Aggregate Impact Value (AIV), Los Angeles Abrasion Value (LAV), Polishing Stone Value (PSV) and Water Absorption (WA).

A quarry was opened and operated by KLIAB nominated contractor to produce and supply Sandstone aggregate for CTB material and underlying layers. Mobile crushers were used to crush the Sandstone. Transporting Hill 33 material to the job site is quick and cheap, a major contribution to the speed and progress of the project. The estimated aggregate cost is half of that supplied from other quarries.

Table 3: Specified aggregate gradation for asphalt and other pavement layers

Sieve size (mm)	DL	CTB	CRL	BC	WC	Comparative Aggregate Properties
75	100					**GRANITE**
50		100	100			Density: 1.7
37,5	74 - 100	90 - 100	75 - 100	100		ACV: 25 %
28.0				90 - 100		AIV: 20 %
25.0					100	LAV: 30 %
20.0	60 - 90	70 - 90	50 - 70	72 - 90	76 - 95	PSV: 49
14.0				58 - 80	64 - 89	WA: 0.5 %
10.0	25 - 75	45 - 70		52 - 75	56 - 81	
5.0	10 - 45	36 - 65	8 - 20	36 - 55	41 - 55	
3.35				30 - 50	32 - 48	**SANDSTONE**
2.36	10 - 20					Density: 1.4
1.18	0 - 10			18 - 32	16 - 31	ACV: 33 %
.600		15 - 30				AIV: 32 %
.425				11 - 25	12 - 16	LAV: 36 %
.150			0 - 5	5 - 14	6 - 10	PSV: <40
.075		5 - 15		1 - 7	3 - 7	WA: 1.8 %
Binder/Cement content		5 - 8 %	3 - 4 %	4 - 6 %	5 - 7 %	

Aggregate for subgrade improvement. The KLIA project area is mostly situated in the Sepang Land System (5). The soil consists of mixed variety on sedimentary and metamorphic rocks. Many of the soil types are very silty and very susceptible to erosion and slipping. For the subgrade improvement, Hill33 Sandstone is suitable. About 2 million tons of additional Hill33 material is used as the fill material for the improved subgrade. In total, about 5 million tons of meta-sandstone was used. Substantial time and cost s was saved in this exercise.

Cement stabilized base. ICAO recommended to used stabilise material for the pavement base layer. The base layer is the main structural or load bearing layer. Laboratory tests indicated 10 N/mm^2 28 days strength mixing Hill33 Sandstone with 90-120 kg/m^3 cement. High strength CTB was specified for structural and durability reasons. Pre-cracking is necessary since the high strength CTB will behave similar to concrete. The pre-cracking will eliminate formation of random cracking. The estimated modulus value for pre-cracked CTB is 5,000 Mpa.

Pre-cracking. Several pre-cracking equipment was tested .The use of kerfing tools attached to steel tandem rollers and dozers was used (See Graphics 2). These equipment were capable of kerfing to a depth of one third of the paved CTB lift. Diluted emulsion was poured in the kerf. The pre-cracking was carried out in straight line at 3 m interval forming a 3 m by 3m pre-cracked CTB slab. Each layer was pre-cracked.

Graphics 2: Graphics of pre-cracking

CTB Placement. Conventional paver was used to pave the CTB layers. To avoid segregation, high powered paver was used. Two lifts of 225 mm each was initially tried but the target strength was not achieved due to difficulty to compact. Three lifts, each 150 mm compacted depth was applied. Compaction was carried out immediately to avoid moisture lost. The third and final layer was sprayed with bituminous emulsion for curing and priming to prepare for asphalt layers.

BITUMINOUS LAYERS

Conventional and Polymer modified bitumen. Conventional bitumen penetration grade 80-100 was used. Polymer is added to the conventional bitumen to turn it into Polymer Modified Bitumen (PMB). To ensure the polymer is added into the bitumen, insitu blending was specified. High shear blending facility was also specified so that wider range of polymer product can be blended. Three types of polymer were accepted as the bitumen modifier. The three polymers were Low Density Polyethylene (LDPE), Ethyl Methyl Acrelyte (EMA) and Styrene Butadiene Styrene (SBS) containing a proprietary resin. The reference standard for the PMB is base on the STRATEGIC HIGHWAY RESEARCH PROGRAM (SHRP) performance grading for bitumen (6). All the selected polymer modified bitumen passes SHRP's PG-76 performance grade. Each PMB has its own merits and disadvantages. The requirement for in-situ blending play a very important role to achieve high quality results.

The specified properties of the PMB are given in Table 4. Relevant properties of the approved product are also shown.

when loaded. At the time of writing a thorough analysis of the HWD test results using back-analysis method is in progress. Full performance analysis and reporting will be carried out at a later stage. Parallel to the design approach and the above evidence it is predicted that the pavement will deteriorate functionally first due to environmental factors.

CONCLUSION AND RECOMMENDATION

The runway pavement described above was designed using mechanistic approach supported with some relevant technical performance data. Form early test data it is shown that the use of untested material is capable of meeting relevant pavement engineering requirements.

The long term and fatigue performance of the pavement is yet to be carefully studied. Long term performance monitoring is needed to organise the verification of the mechanistic approach in a more orderly manner. It is necessary to collect correct performance data directed to verify the approach. The data is vital for the verification and also for future forecast of maintenance needs.

ACKNOWLEDGMENT

The author would like to thank the Director General of JKR, **Dato' Omar Ibrahim** for making the exercise possible using local Engineers expertise and to KLIAB Managing Director, **Dato' Jamilus Husin** for the support and confidence. Special thanks to all those Engineers and Technicians involved in the design, supervision and construction of the asphalt pavement at KLIA. Last but not least thanks to the assistance Director General of JKR, **Dato' Haris Abas** for his direct effort of developing local pavement specialist.

REFERENCES

1. HARUN M.H, JONES C.R., 1992, The performance of polymer modified asphaltic concrete on climbing lanes in Malaysia, Proc. 16 th ARRB Conference, Australia.
2. ISMAIL, H. R, JONES C.R., 1992, The deterioration of bituminous binder in asphaltic concrete on climbing lanes in Malaysia, Proc., 16 th ARRB Conference, Australia.
3. ROLT J, HASIM M.S., HAMEED A.M.A, SUFFIAN Z, 1996, The prediction and treatment of reflection cracking in thin bituminous overlays. Proc. 2 nd Malaysian Road Conference, Kuala Lumpur.
4. TRL, 1993, Overseas Road Note 31, 4 Th. Edition, A guide to structural of bitumen surfaced roads in tropical and sub-tropical climates. Overseas Unit TRL, Crowthorne, Berkshire, U.K.
5. LAWRANCE C.J., Terrain evaluation in West Malaysia Part 2, TRRL Supplementary Report 378, Overseas Unit TRRL, Crowthorne, Berkshire, U.K. 1978.
6. SHRP 1992, Binder Specification, 818 Connecticut Avenue, N.W, Washington , D.C 20006
7. SHRP 1992, Superpave Mix Design System, N.W, Washington , D.C 20006

Fifth International Conference on the
BEARING CAPASITY OF
ROADS AND AIRFIELDS

STRUCTURAL EVALUATION FOR CRACKED AIRPORT ASPHALT PAVEMENTS

JIANJUN YIN[1], YOSHITAKA HACHIYA[2] AND KATSUHISA SATO[3]

[1]Dr. Eng., Guest Researcher, Runways Laboratory, Port and Harbor Research Institute, Ministry of Transport, 1-1, Nagase 3, Yokosuka 239-0826, Japan, Fax: 81-468-44-4471, E-mail: yin@phri.go.jp
[2]Dr. Eng., Chief, Runways Laboratory, Port and Harbor Research Institute, Ministry of Transport, 1-1, Nagase 3, Yokosuka 239-0826, Japan
[3]Professor, Department of Civil and Environmental Engineering, Nagaoka College of Technology, Nishikatagai-cho 888, Nagaoka 940-0817, Japan

ABSTRACT

Two aspects were included: (1) evaluating the seriousness of cracks, and (2) identifying the effectiveness of maintenance and rehabilitation (M&R) for cracks. First, a theoretical analysis of the elastic three-layer pavement model was performed. It was concluded that the pavement deflections D_{45} and D_{250} at cracked and uncracked sections should be used as indices to evaluate the seriousness of cracks. A structural evaluation procedure was proposed based on statistical tests. Then, the effectiveness of sealing, patching and overlay was identified by comparing the FWD deflections before and after the M&R.

Key Words: structural evaluation, asphalt pavement, crack, FWD, M&R

INTRODUCTION

Airport pavement cracks result in higher risks for transportation because of the higher moving speeds. In particular, earthquake induced cracks can be more serious and harmful because they are often wider and accompanying with more severe faults. Therefore, urgent inspection, evaluation and effective repair are needed as soon as the crack appeared. A simple and valid procedure need to be developed to evaluate the seriousness of cracks and to assure the effectiveness of urgent repairs.

The Falling Weight Deflectometer (FWD) is an effective, non-destructive method to inspect pavement bearing capacity. By using FWD deflections with corresponding criteria, the pavement structural condition can be evaluated and the maintenance and rehabilitation (M&R) alternatives can be recommended accordingly. For pavement structural evaluation, Abe et al. performed two years of continuous FWD measurement at more than 300 sites all over Japan. They proposed four FWD deflection indices D_0, D_{150}, D_0-D_{20}, D_0-D_{150} and associated criteria to evaluate the soundness of the overall pavement structure, subgrade, asphalt layer and the pavement layers above subgrade, respectively [1]. To analyze the response of cracked pavement under FWD loading conditions, Uddin investigated the effects of pavement discontinuities by means of three dimensional finite element analysis code ABAQUES, in which the pavement crack was simulated as an uni-directional gap element GAPUNI [2]. Jung et al. used the Boussinesq's theory and Odemark's equivalent layer thickness method to investigate the degree of structural integrity of two layer pavements by calculating the effective moduli of surface course and subgrade [3].

However, the procedure in reference 1) is mainly concerned with road asphalt pavement structures and road FWD test conditions. It is not suitable for evaluating cracked airport asphalt pavements because road asphalt pavement has lower strength, lighter FWD loads and smaller loading plates. The method in reference 2) requires many assumptions, such as the moduli of pavement layers, the friction coefficient at the interface of GAPUNI and the depth of cracks. The accuracy of these assumptions is difficult to verify. The two layer system in reference 3) is far from the reality of airport asphalt pavements. In addition, these procedures rely heavily on the absolute, true FWD deflections. These deflections are difficult to obtain because many factors (e.g., pavement temperature, FWD equipment etc.) influence them. Therefore, a new evaluation procedure was developed, as presented herein.

Two aspects were included in this paper: evaluating the seriousness of earthquake induced cracks in airport asphalt pavements and identifying the effectiveness of the M&R for cracks. First, a theoretical analysis of the elastic, three-layer pavement model was performed. Under FWD loading conditions the pavement deflections D_{45} and D_{250} at cracked and uncracked sections were selected as indices to evaluate the seriousness of cracks. The structural evaluation procedure was proposed based on statistical tests. Its validity was verified by coring tests. Then, by comparing the FWD deflections before and after M&R, the effectiveness of sealing, patching and overlay was identified.

SERIOUSNESS OF CRACKS

Theoretical analysis

Airport asphalt pavement has a thicker structure, higher bearing capacity and functionality than road pavement. Hence, the FWD equipment was designed for a maximum impact load of 200kN, loading plate diameter of 45cm, maximum sensor distance of 250cm and an impulsive force duration of approximately 20-30 milliseconds (ms). Under standard FWD loading condition, the distribution of vertical stress in the pavement structure is shown in

Fig. 1. The accumulated vertical displacements of a pavement structure can be detected by FWD deflection sensors at different distances from the load center.

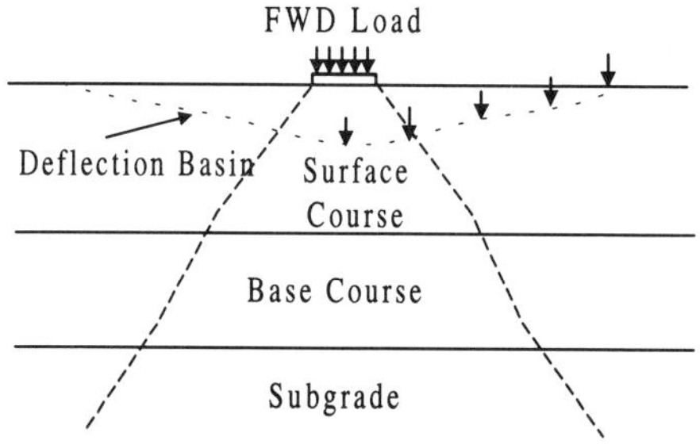

Fig. 1 Distribution of the vertical stress under FWD loading

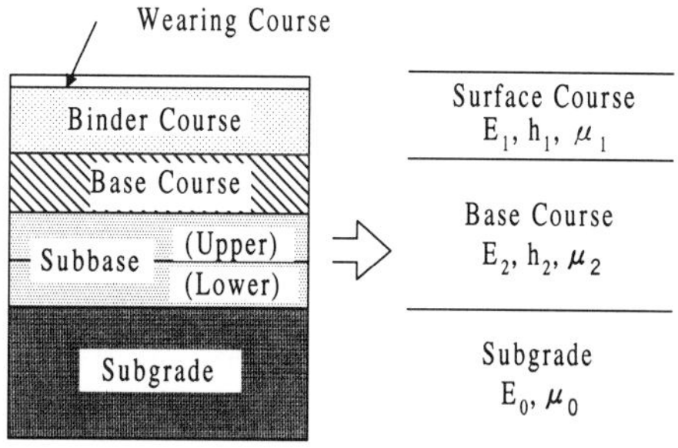

Fig. 2. Three-layer pavement system

Existing research revealed that the deflections far away from the load center are determined by the modulus of the bottom layer, while those at intermediate distances from the load center are determined by the modulus of the middle layer, and those near the load center are determined by the modulus of the top layer [1), 4)]. The theoretical analysis was carried out based on this principle.

The airport asphalt pavement structure was modeled as an elastic, three-layer system with asphalt surface course, base course and subgrade, as shown in **Fig. 2.** Three kinds of asphalt pavements with thin, medium and thick structures were selected for analysis. The thickness, range of elastic modulus and Poisson's ratio of each layer are listed in **Table 1**, which represented the general airport asphalt pavement structures currently used in Japan.

Table 1 Pavement structural parameters

	Surface	Base	Subgrade
Thickness (cm)			
Thin	15	60	
Medium	23.5	90	
Thick	30	120	
Elastic modulus (MPa)	500 - 12000	250 - 3000	50 - 1000
Poisson's ratio	0.30	0.30	0.35

Utilizing the BISAR program and varying the elastic modulus of surface course, base course and subgrade, under 200kN of FWD loading the deflections at load center (D_0), 45cm (D_{45}) and 250cm (D_{250}) away from the load center, respectively, were investigated. Because that

D_0 represents the overall pavement strength and D_{250} mainly represents the strength of subgrade, D_{45} was selected for two reasons: 1) D_{45} mainly represents the accumulated vertical displacements of pavement structures beneath the asphalt layer where the thickness of airport asphalt layers is generally less than 45cm. 2) D_{45} has larger deflection values and hence has relatively higher precision than either D_{60} or D_{90}.

The results of theoretical analysis are shown in **Fig. 3**. It reveals the following relationships. As the elastic modulus of surface course decreases, D_0 increases, whereas D_{45} has relatively smaller increase and D_{250} remains nearly constant. On the contrary, as the elastic modulus of the subgrade decreases, all deflections including D_{250} increase significantly. Moreover, as the elastic modulus of the base course decreases, D_{250} keeps nearly constant, whereas both D_0 and D_{45} increase significantly.

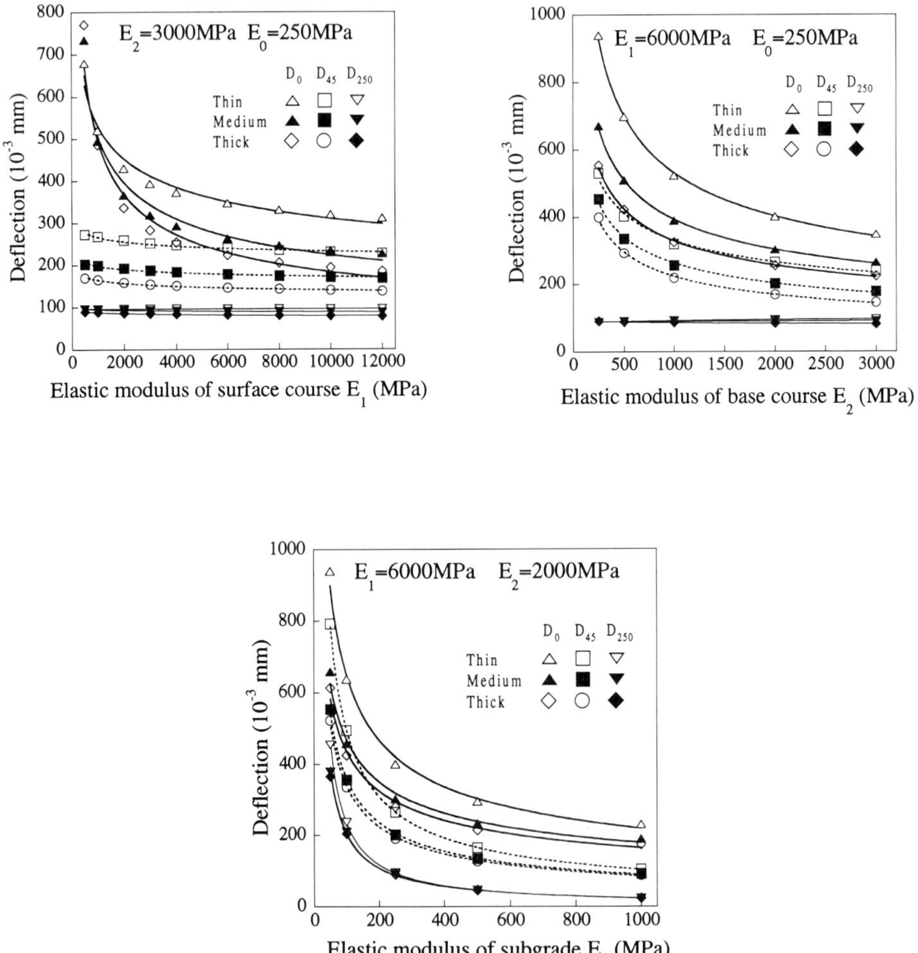

Fig. 3. Relationships between pavement deflections with elastic modulus of each layer

The elastic moduli of cracked pavement layers may appear to be lower than that of the uncracked pavement layers. Therefore, the difference in deflections D_{45} or D_{250} at cracked and uncracked sections certainly results from the lower apparent elastic moduli of the cracked layers. Thus, the seriousness of cracks could be identified qualitatively, and the evaluation procedure was developed as a result.

Evaluation procedure

FWD measurements at two positions must be conducted to evaluate the seriousness of cracks: far away from the crack and near to the crack, as shown in **Fig. 4**. The following procedure is recommended to make the evaluation: 1) Compare D_{250} of the uncracked section with that of the cracked section. If the latter is significantly larger than the former, the elastic modulus must be lower in the subgrade of the cracked section. This indicated that the crack exists in the subgrade. 2) If no significant difference exists between the two D_{250} values, compare D_{45} of the uncracked section with that of the cracked sections. If the latter is significantly larger than the former, the elastic modulus must be lower in the base course of the cracked section. This indicates that the crack exists in the base course. 3) If neither the D_{250} nor the D_{45} show significant difference, the crack only exists in the surface course. 4) The deflections D_{45} and D_{250} of the cracked section on FWD loading side were used for making this evaluation. That is to say, deflection D_{45} is measured at the position near the crack, whereas deflection D_{250} is measured at the position overpass the crack, as shown in **Fig. 4 b)** and **c)**, respectively.

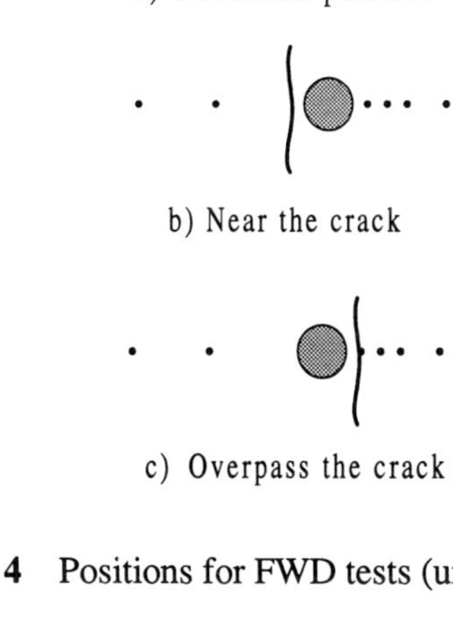

Fig. 4 Positions for FWD tests (unit: in cm)

The evaluation procedure was illustrated by the flow chart in **Fig. 5**.

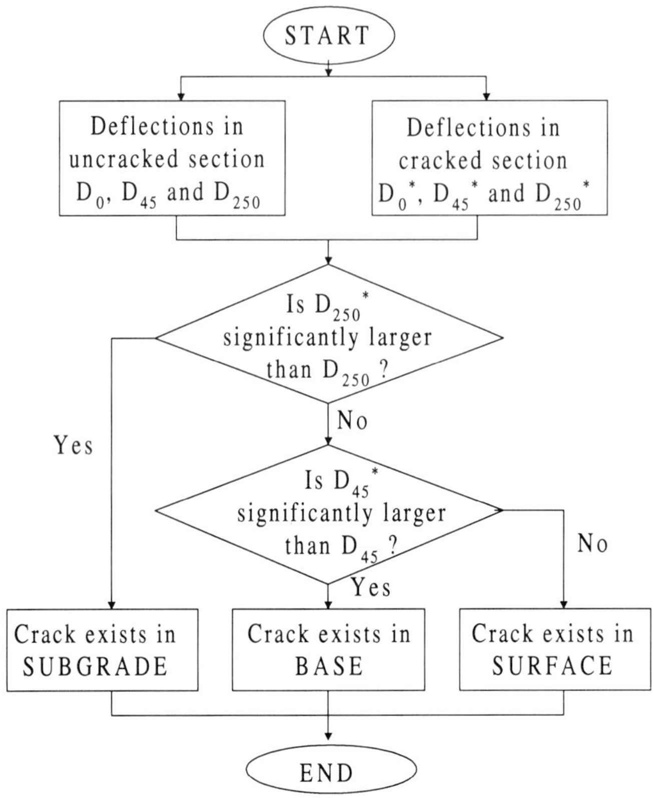

Fig. 5 Flow chart for evaluating the seriousness of crack

A t-statistics test was employed to check the significance of difference between the two deflection values. If the two deflection values, a_1 and a_2, satisfy **Eq. (1)**, the difference is not significant based on the statistical level of α, and vice versa.

$$|a_1 - a_2| \leq t_{\alpha/2} \cdot \sqrt{\frac{s_1^2}{n-1} + \frac{s_2^2}{m-1}} \quad (1)$$

in which,

a_1, a_2 - average deflections;

n, m - number of FWD measurements;

s_1, s_2 - standard deviation of deflections;

$t_{\alpha/2}$ - double side t-distribution value with freedom of m+n-2 and statistical level of α.

The evaluation procedure proposed here was confirmed by coring tests.

M&R EFFECTIVENESS

The effectiveness of three kinds of M&R measures, namely, sealing, patching and overlay, was identified by comparing pavement deflections before and after the M&R. They are: 1) Sealing. Asphalt slurry sealing material was injected into cracks to prevent the pavements from deteriorating by water penetration and the repetitive traffic loads. 2) Patching. After the crack was sealed, a 2m wide band on each side of the wider cracks was patched. The section was first cut 10cm deep and then paved with a 5cm coarse-graded asphalt concrete lower layer and a 5cm dense graded asphalt concrete upper layer. 3) Overlay. After the crack was sealed, the full length of the cracked sections was overlaid with a 4cm coarse-graded asphalt concrete lower layer and a 5cm dense-graded asphalt concrete upper layer.

Before and after the M&R measures, the ratios of deflections at the distance of X, namely, 0, 45 and 250cm away from the load center, in cracked sections to that of the uncracked sections, RD_X, were applied to identify the effectiveness of each M&R measure.

Table 2 illustrates the effectiveness of sealing. It was revealed that, for pavement cracks in either the subgrade or base course, sealing clearly decreases RD_0 and RD_{45}, whereas it shows no significant decrease in RD_{250}. However, for cracks in surface course, sealing appears to have less effect on the deflections.

Table 2 Effectiveness of sealing

Seriousness	Before sealing			After sealing		
of cracks	RD_0	RD_{45}	RD_{250}	RD_0	RD_4 5	RD_{250}
Subgrade	2.20	1.86	1.38	1.24	1.10	1.20
Base	1.60	1.50	0.95	1.31	1.18	0.85
Surface	1.15	1.15	0.81	1.16	1.03	0.94

Table 3 illustrates the effectiveness of patching. It was revealed that patching was effective for pavements with more severe cracks in either the subgrade or the base course.

Table 3 Effectiveness of patching

Seriousness	Before patching			After patching		
of cracks	RD_0	RD_{45}	RD_{250}	RD_0	RD_{45}	RD_{250}
Subgrade	2.36	1.85	1.63	1.18	1.09	1.02
Base	1.94	2.37	0.76	1.20	1.15	1.06
Surface		N/A			N/A	

Note: N/A-- No sections has this kind of M&R.

Table 4 illustrates the effectiveness of overlay. It was revealed that the structural capacity of pavement sections with cracks in the base course can be as good as that of the uncracked sections if properly overlaid.

Table 4 Effectiveness of overlay

Seriousness of cracks	Before overlay RD_0	RD_{45}	RD_{250}	After overlay RD_0	RD_{45}	RD_{250}
Subgrade	N/A			N/A		
Base	2.24	2.00	1.18	1.02	0.96	1.02
Surface	N/A			N/A		

Note: N/A-- No sections has this kind of M&R.

CONCLUSIONS

The seriousness of earthquake induced cracks in airport asphalt pavements can be evaluated by comparing the FWD deflections D_{45} and D_{250} at cracked sections to that of uncracked sections. The t-statistics test is recommended to identify the significance of the differences in deflections.

From the viewpoint of structural capacity, sealing, patching and overlay are effective for repairing cracks, but their usage is different. Sealing is effective for repairing lighter cracks in the subgrade and base course; patching is suitable for repairing more severe cracks; overlay can completely rehabilitate the cracked pavement.

REFERENCES

1. ABE, N. MARUYAMA, T. HIMENO, K. and HAYASHI, M. Structural Evaluation of Pavements based on FWD Deflection Indices. Journal of Civil Engineering, JSCE, No. 460/V-18, pp. 41-48, 1993 (in Japanese).

2. UDDIN, W. ZHANG, D. and FERNANDEZ, F. Finite Element Simulation of Pavement Discontinuities and Dynamic Load Response. Transportation Research Record 1448, TRB, pp. 100-106, 1994.

3. JUNG, F. W. and STOLLE, D. F. E. Nondestructive Testing with Falling Weight Deflectometer on Whole and Broken Asphalt Concrete Pavements. Transportation Research Record 1377, TRB, pp. 183-192, 1992.

4. HUANG, Y. H. Pavement Analysis and Design. Prentice Hall, Englewood Cliffs, New Jersey 07632, 1993.

Fifth International Conference on the
BEARING CAPASITY OF
ROADS AND AIRFIELDS

THE RUNWAY PAVEMENT EVALUATION, REHABILITATION DESIGN AND PERFORMANCE AFTER CONSTRUCTION ON THE "BELGRADE" AIRPORT

MIROSLAV J. STEFANOVIC[1] AND JOVANKA P. DJURAN[2]

Head Engineer, The Highway Institute - Pavement Department,
11000 Belgrade, Yugoslavia, Fax.:+38 466 866
Director, The Highway Institute - Pavement Department,
11000 Belgrade, Yugoslavia, Fax.: +38 466866

ABSTRACT

The runway pavement was three times under rehabilitation since its construction in 1962, and the last one (fourth) was carried out in September 1996. Up to 1996 the pavement structure consisted of two or three layers of asphalt concrete over original jointed cement concrete pavement (not cracked and seated). Since, the first two pavement rehabilitations have been carried out in order to improve the bearing capacity, while the third and fourth rehabilitation have been oriented much more to solve reflection cracking. The data from pavement condition monitoring (deflection measurement by FWD and visual survey on the pavement distress), were subject to analysis in order to carry out pavement evaluation and rehabilitation design for the next service cycle. This paper refers to the pavement condition before and after the last rehabilitation, and some specifics of the pavement rehabilitation design from crack reflection retardation point of view.

Key words: crack, pavement, reflection, rehabilitation.

INTRODUCTION

Total length of the runway pavement is 3500m, and width 45m icluding shoulders. A part of the runway pavement, 3000m long was built in 1962, and a 500m extention was built in 1970 for heavy crafts. Original pavement construction was rigid type - jointed cement concrete pavement (JCCP). The oldest part was reinforced 2-3 times, between 1980 and 1996, with layers of asphalt concrete, and it is the matter of our analysis, and the subject of the present paper. Pavement structure and dates of construction of layers are shown in the following table:

Table 1: Pavement structure in 1996

PAVEMENT LAYERS	PAVEMENT LAYER THICKNESS			
	At runway axis		At 15 m from RW axis	
	Chainage km 0.0-2.1	Chainage km 2.1-3.0	Chainage km 0.0-2.1	Chainage 2.1-3.0
AC overlay 1986(*)	6-8 cm	9 cm	6-6.5 cm	7 cm
AC overlay 1981	7 cm	9 cm	6 cm	3.5 cm
AC overlay 1980	6-8 cm	0	1-1.5 cm	0
JCCP upper layer 1962 lower layer 1962	20-25 cm 10 cm	20 cm 10 cm	20-25 cm 10 cm	20 cm 10 cm
Gravel subbase 1962	35-45 cm	40 cm	40-50 cm	40-50 cm

AC=asphalt concrete; JCCP=jointed cement concrete pavement
(*)=reinforced with polyester mesh in central zone at chainage km 0.90-2.10.

Subgrade layer was built from clay silt , over the present ground of macroporose loess soil.

Untill 1996, we observed indicators of pavement condition such as pavement distress and deflection. Collected data allowed observation of reflection cracking, pavement evaluation with Pavement Condition Index (PCI), prediction of pavement condition, observation of the change of deflections and definition of Back Calculation Moduli of pavement layers. Due to significant pavement distress in the central zone of the runway pavement, a runway pavement evaluation was performed in 1996, as well as rehabilitation design, rehabilitation construction, and construction evaluation. For the purposes of an analysis, the runway pavement was divided by cross section profile in 8 zones (see Figure 1), the width of which (7.5m) matches the width of the slab of the original jointed cement concrete pavement. Zone 1 and 8 refer to shoulders. According to an analysis of available data and indicators of pavement condition, we defined homogeneous sections along each zone of the runway pavement.

PAVEMENT EVALUATION IN 1996

Pavement distress visual survey was carried out continually along each zone. For the purposes of expression of the pavement condition, using Pavement Condition Index - PCI, a longitudinal division of each zone in 30 section units was carried out, where each section is 100m long. A scheme of the runway pavement with appearance of certain types of distress in 1996 is shown in Figure 1. Appearance of alligator cracking was found along reflected longitudinal joints and cracks in wheel paths.

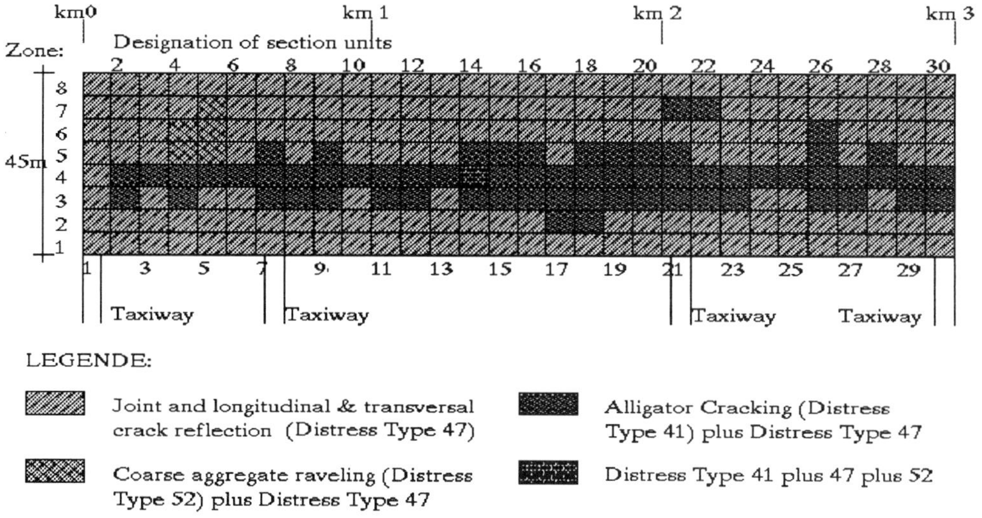

Figure 1: Scheme of RW 12-30 layout with pavement distress appearance in 1996

Pavement condition was evaluated according to the data on pavement distress and expressed through PCI. The worst conditon was found in central zones, especially in the zone 4, as is shown in Figure 2:

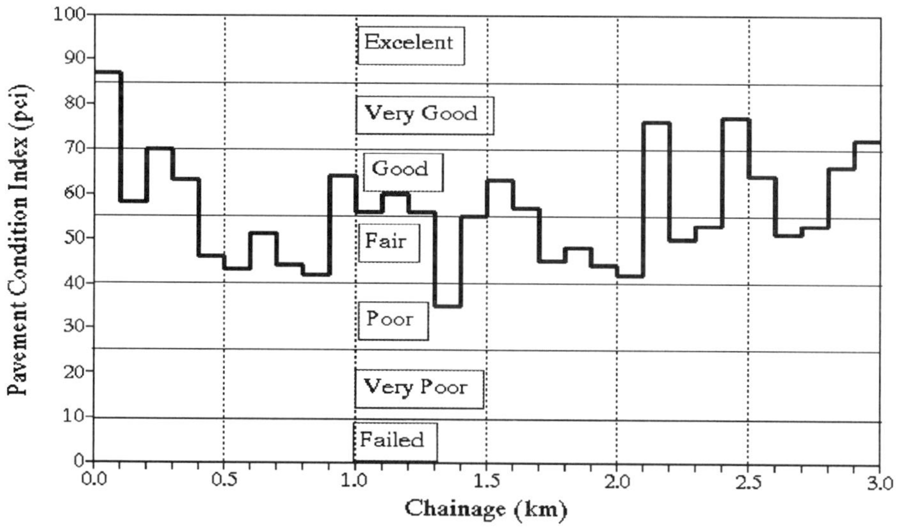

Figure 2: Pavement condition in 1996 - zone 4

According to standards for pavement condition shaped in practical engineering experience ( ), PCI=65 - critical condition, PCI=50 - alarming condition, and PCI=40 minimum acceptable condition, a conclusion was reached that the runway pavement was in need of rehabilitation.

Joints and cracks reflection from each slab of the original jointed cement concrete pavement through asphalt concrete overlay built in 1980/81 and through the overlay built in 1986 were analised. The Figure 3 shows an example of data on joints and cracks reflection in 1996 through the asphalt concrete overlays for the zone 4.

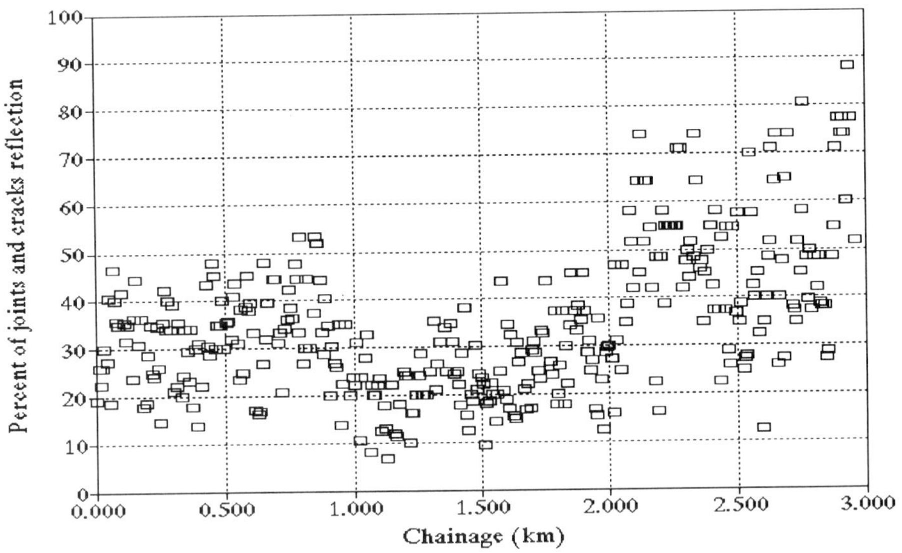

Figure 3: Joints and cracks reflectioin in 1996 - zone 4

Similarly to the example of 1996, shown in the figure above, relatively great range of reflection rate per homogeneous sections was ascertained according to the data from 1986 (before the last overlay), as well as the data from 1989 and 1993 (after the last overlay built in 1986). In order to observe a relatively rough regularity of this phenomenon in the function of time or traffic load, we estimated 85% values for relatively 'homogeneous sections. This way, certain regularity was established: this is shown in Figure 4. Basic data of pavement distress was from 1979, before 1st overlay.

From the figure 4 we can make a conclusion that joints and cracks from the original JCCP reflected more in the sections km 2.1-2.6 and km 2.6-3.0 where the smallest total thickness of asphalt overlay (see Table 1), in spite of the lowest level of the volume of traffic that occurs there. Further analysis established certain general dependence of reflection rate on traffic, thickness of the first overlay and the total thickness of the asphalt concrete overlay (this is not shown in this paper).

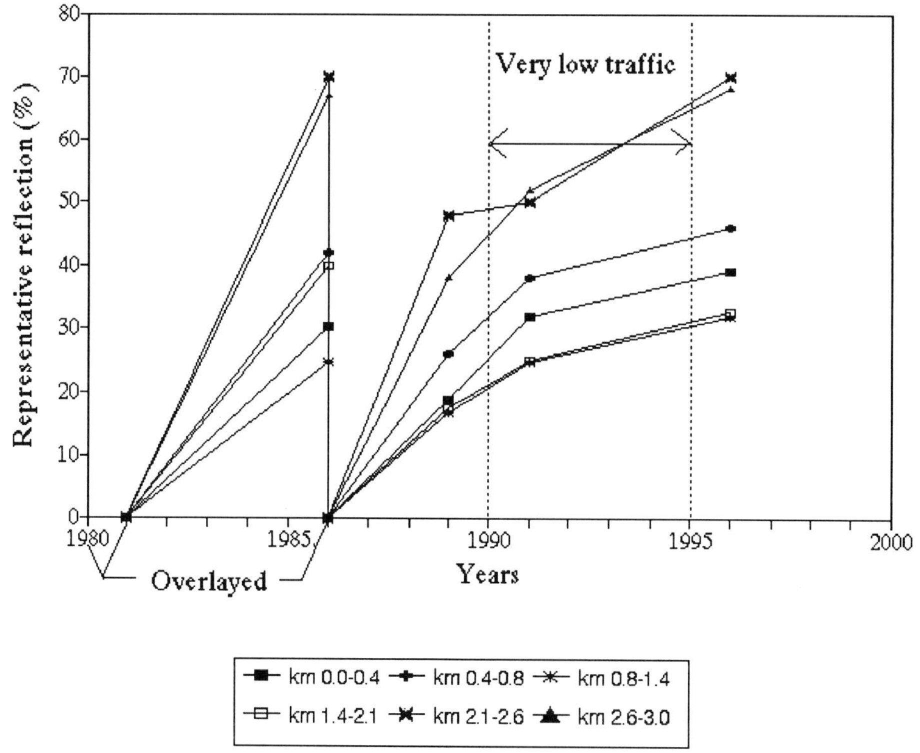

Figure 4: Joints and cracks reflection history - zone 4

Deflections were measured by FWD Kuab 150 (F=50, 100 and 150 kN; circular plate diametar D=45cm) at chosen test locations by longitudinal profile along zone 3-6, and approximately at the position of center of original concrete slabs. Standard configuration of seizmometers was placed at a distance of r=0.0; 0.6 and 0.9m from the centre of the circular plate. At specifically chosen positions, additional measurements were carried out also at a distance r=1.2 and 1.5m. All measurements demonstrated that the largest deflection occurs in the zone 4 - the central zone of the runway pavement. In relation to 1986, an increase in deflection was observed at distances r=0.0; 0.6 and 0.9m, and Surface Curvature Index, and a particularly significant increase of deflection in chainage km 2.1-3.0. This leads to a conclusion that pavement structure suffered deterioration, which is in agreement with pavement distress. Results of the normalized deflection "D_0" (P=150kN; Tac=20°C) from 1986 and 1996 for the zone 4 are shown in Figure 5, and representative deflections and PCI for the zone 4 are shown in Figure 6. Relatively good coincidence between the represntative deflection "D_0" and PCI was observed at the chainage km 0.0-2.2.

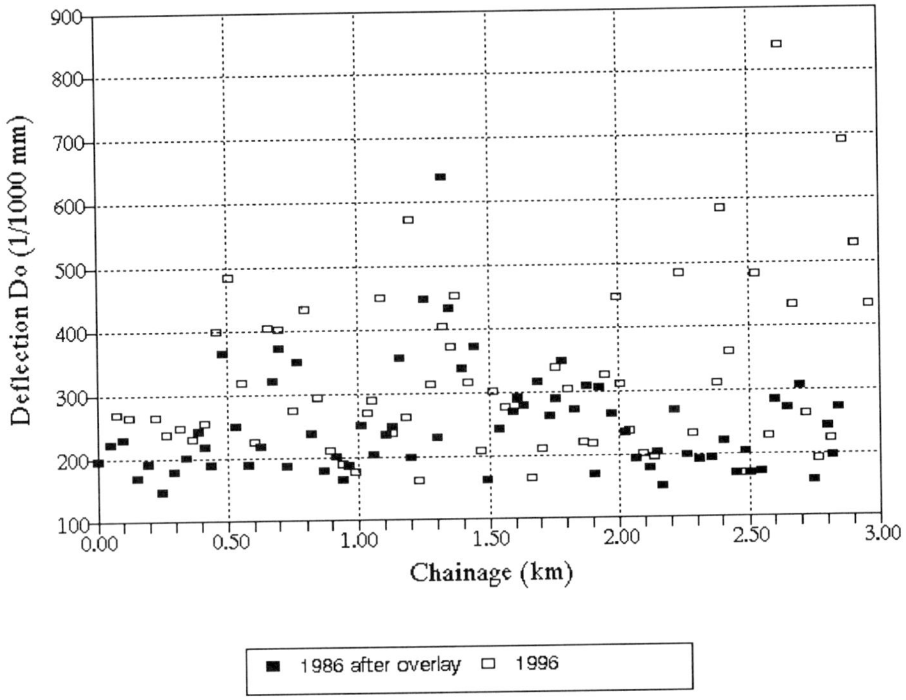

Figure 5: Deflection data from measurements in 1986 and 1996 - zone 4

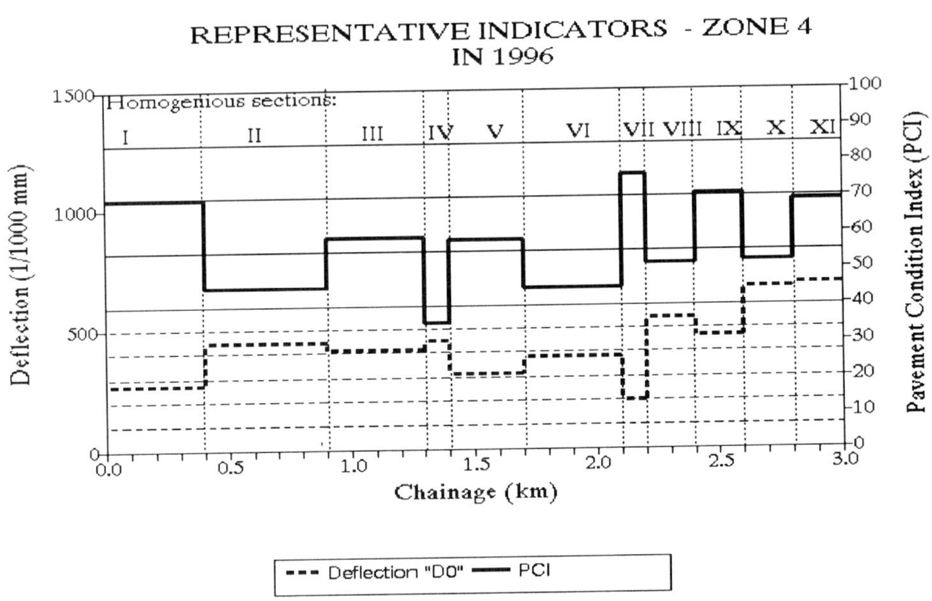

Figure 6: Representative deflection "D_0" and PCI in 1996 - zone 4

The back -calculated layer modulus have been analysed in accordance to deflections and deflection basins measured in 1996 and the available data on the thickness of pavement layers. For the zone 4, a model of three-layered elastic system has been selected for calculation of dynamic moduls "E" for each layer (E1 - of all asphalt layers; E2 - of all portland cement concrete layers; E3- halfspace modul beneath concrete layers). Computer programs WESDEF and MODEF (1) were used. Due to specific pavement structure and pavement distress for certain number of results of the measurement it wasn't possible to find a solution for calculated deflection with acceptable tolerance for the difference between the calculated and the measured deflection basin (RMS<2%). For deflection basins for which the proposed criteria were satisfied, moduls E1 and E3 are shown in Figure 7:

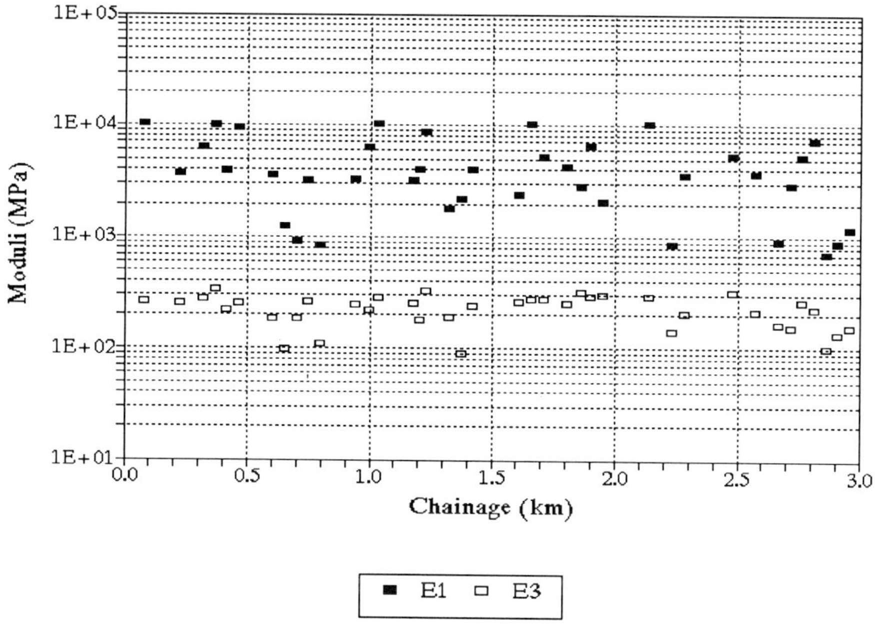

Figure 7: Back calculated modulus E1 and E3 - zone 4

Back calculated modulus E2 (concrete layers) in the zone 4 appeare in a relatively wide range of E2=1000-68950 MPa. This implies that there are locations where this layer completely cracked (E2=1000 MPa), as well as sections where portland cement concrete slabs are divided in large fragments that still behave as slabs (E2>3000 MPa). At the time of deflection measurements the subgrade wasn't in the water-saturated condition, so, during the year lower values of E3 than those shown in Figure 7 can be expected. A combination of moduls in zone 4 that is realistic, and very inconvenient, shows: E1=1000 Mpa; E2=3000 Mpa; E3=100 Mpa (saturated silty clay subgrade).

Crack load transfer efficiency (LT) has been analysed for 22 previously chosen locations of joints and cracks reflections in the zones 3, 4, 5 and 6. A special measurement of deflections was performed in accordance to AASHTO 1993 Guide for pavement design.

Poor load transfer (LT<50%) was ascertained in 32% of the test spots, medium load transfer (LT=50%-70%) was found in 14% of the test spots, and good load transfer was found in 54% of the test spots. Poor load transfer was observed mostly in the chainage km 2.2 - 3.0.

Surface and subsurface drainage system functionality evaluation has been carried out in 1996. According to visual survey, test research on site, and according to mathematical model analise of whole system we made an estimate that the system lacks good functional value, in particular on the right side of the runway pavement and on the chainage km 2.0 - 3.0, where water remains in the system and moistens the pavement subgrade. It has been evaluated that a detailed examination, and most probably reconstruction works, were necessary.

PAVEMENT REHABILITATION

In the preliminary project we analysed a variety of maintenance and rehabiitation strategies, as well as several rehabilitation alternatives in 1996:

* Service life t=>15 years: (a) full and partial reconstruction of central zone;
  (b) reinforcement with portland cement concrete;
  (c) reinforcement with asphalt layers;
* Service life t=5-10 years: (d) reinforcement with asphalt layers
* Service life t<5 years: (e) surface treatment with micro asphalt

According to predictions for the pavement condition and LCCOST analysis which includes predicted cost of rehabilitation and routine maintenance in the forthcoming period, as well as the cost of the airport being closed for the time of rehabilitation construction, we defined a rating of alternatives for economical purposes. According to the available budget in 1996/97 and the rating of rehabilitation alternatives for economical purposes, the Airport authorities made a decision to elaborate the final project for service life of 5-10 years, i.e., for asphalt concrete overlay with up to 10 cm upgrading of vertical alighnement.

Traffic forecast is shown on Table 2. For the purposes of rehabilitation project for the pavement traffic volume forecast was converted to a number of operations of the airplane B-727/200 each year (this is not shown).

Table 2: Traffic forecast

AIRCRAFT TYPE	PERCENTAGE OF TOTAL TRAFFIC	CUMMULATIVE NUMBER OF AIRCRAFT TYPE IN PERIODE 1996/2005
DC-9/M-80	5%	13461
DC-10/MD-11	4%	10769
B-727	10%	26922
B-737	48%	129224
B-747	1%	2692
A-310/A-320	5%	13461

Predicted absolute minimal temperatures for the winter period in asphalt layer at the depth Z=0.635-1.37cm during winter period are T=4.4°C up to -32.2°C. Predicted weighted mean annual air temperature is w-MAAT=13.5°C, and effective annual temperature of the asphalt layer Tyeff=20°C.

In the general approach to final projects of the asphalt overlay, from the point of view of pavement bearing capacity, we applied French method (2). The necessary thickness of the asphalt concrete overlay over the existing pavement is Dac<=6cm for next service life of 10 years. It was checked through additional pavement structure analysis (stress, strain, fatigue, permanent deformation). Nevertheless, through special analysis of crack reflection under method suggested by A.A.A. Molenaar (3, 4) we provided an estimate that this layer isn't sufficient to prevent reflection cracking. According to examples from airports in France, the adopted measure was to slow down reflection cracking by application of an interlayer of dense sand asphalt conctrete with polymer-modified bitumen. Designed pavement rehabilitation comprised:

1. Preparations:
   (a) local correction of pavement profile by milling of asphalt concrete;
   (b) local correction of pavement profile through construction of a leveling layer of asphalt concrete 0/11mm with standard bitumen type Bit.60;
   (c) local repairs of the pavement distress such as alligator cracking at the depth of 6cm with asphalt concreterete 0/11mm with polymer bitumen at the depth of 4cm, or alternative milling at the depth of 6cm asphalt concrete and replacement with new asphalt concrete 0/11mm , Bit.60;
   (d) crack (opening over 10mm) remix in place 30cm wide with fine asphalt concrete 0/11mm with polymer bitumen to the depth of 4cm, or alternative milling to the depth of 6cm of asphalt concrete and replacement with new asphalt concrete 0/11mm with Bit.60;
   (e) channel cutting for the cabling of electrical installations, laying of plastic tubes and covering with 6cm of asphalt concrete 0/11mm;

2. A layer for reflection cracking retardation:
   (f) spraying with an emulsion of polymer-modified bitumen and
   (g) construction of 1.5cm sand asphalt concrete (min. 10% of filler, crushed limestone sand, min. 10% mass content of bitumen, 2% void content) with polymer modified bitumen PmB 80A required properties;

3. Overlay:
   (h) spraying with an amulsion of polymer-modified bitumen and
   (i) construction of 6cm asphalt concrete 0/16mm with standard bitumen Bit.60 (Pen=61-70; Tr&b=53-55°C; IP=min.0; TFOT=max.40%).

For the described solution, a structural analysis was elaborated (stress, strain, fatigue, permanent deformation, reflection). Dynamic modules of elasticity and the fatigue law of asphalt mixtures (Tac=20oC; f=10Hz) are defined analytically and according to the results of previously performed laboratory dynamic tests for similar types of asphalt concrete

0/16mm with Bit.60. (Vb=11%vol.). For asphalt concrete 0/16mm the estimated module is E=6000 MPa, and for the sand asphalt E=1500 MPa. Interlook (poor, medium, good) between fragments of existing bonded pavement layers was evaluated according to an analysis of load transfer efficiency over existing cracks. According to the results of the analysis the estimate was made that the solution satisfies the needs of service life of 9 years from the point of view of bearing capacity as well as for reflection cracking for about 7 years. In the cases of poor interlook between fragments of pavement reflection cracking was predicted after only 1-2 years, i.e., about 30% of existing cracks mostly on the chainage km 2.2-3.0.

Rehabilitation construction was performed in September 1996. According to statistical data on the climate in that period nice and sunny weather was expected. Nevertheless, it was raining every day. Preparations planned as a part of pavement rehabilitation were performed at night, so that in the daytime, between 5 and 22, traffic took place. During the construction of the sand asphalt layer and the overlay of asphalt concrete, the Airport was closed for the traffic. In comparison to projected positions, what has been carried out is the following:

- the cracks were repaired only in the chainage km 0.0-1.0;
- the accomplished thickness of the sand asphalt layer was most often larger than 1.5cm;
- in the asphalt concrete overlay 0/16mm lower compactness than planned was ascertained on few locations;
- in the course of laying the asphalt concrete overlay there occured segregation of the mixture on few locations, while locally this mixture didn't reach the required temperature;
- laying of the asphalt concrete layers was partly carried out in rainy weather on few locations.

An analysis of the completed pavement rehabilitation allowed us to estimate duration of 5-7 years. The following estimate was established for the pavement condition in March of 1997 and February of 1998 according to pavement distress:

- PCI=95-100 March 1997
- PCI=90-100 February 1998

ACKNOWLEDGEMENTS

An approach to pavement rehabilitation such as asphalt concrete ovelayed portland cement concrete should be very detailed and permanent monitoring should be done. Apart from an evaluation from the point of view of distress and bearing capacity, it is indispensable to analyse reflection cracking that can dictate service life of asphalt concrete overlays. It is also very important to evaluate the functional value of the drainage system. A difficulties in calculating moduls of elasticity of the layers according to deflection basins could be expected.

The estimate is that application of a layer of sand asphalt with the purpose of retardation of reflection cracking is effective for the cases of good and medium crack load transfer. In the cases of poor load transfer and when the deflections difference is relatively great, we believe that this layer doesn't have any significant effect.

REFEENCES

Goran Mladenovic: "Computer programa "MODEF"; Proceeding of Coference of compuer application i road engineering, Nis, Yugoslavia, 1996;

SBA & STBA: "Dimensionnement des chaussees, Instuction sur le dimensionnement des chausses d'aerodromes", Paris, 1989;

A.A.A. Molenaar : "Structural performance and design of flexible road constructions and asphal concrete overlays, 1983;

A.A.A. Molenaar: "State of the art of pavement evaluation", The 4th International Conference on the Bearing Capacity of Roas and Airfields, 1994, Minneapolis, Minnesota, U.S.A.

INTRODUCTION

Johannesburg International Airport (JIA), the air transport hub of Southern Africa, is currently handling about 130 000 domestic, regional and international flights per annum. Last year JIA handled 8.2 million passengers surpassing Cairo International Airport (7.8 million), thereby becoming Africa's busiest airport. Opened in 1952 and upgraded in 1972, the airport operates three runways, fifteen taxiways and four aprons (see Figure 1). Both flexible (asphalt) and rigid (concrete) pavement structures have been used on JIA.

As part of the ongoing maintenance and rehabilitation programme of airside pavements, the Airports Company South Africa (ACSA) appointed Stewart Scott Inc. to assess the condition of pavements and, subsequently, to carry out short and long term pavement rehabilitation. The design process comprised the following phases: pavement investigation, air traffic analysis, pavement evaluation, structural capacity analysis, rehabilitation design and comparison of rehabilitation alternatives.

The objectives of the paper are: **a)** to briefly describe the methodology used for the design of long-term flexible pavement rehabilitation of JIA's main runway, *03L/21R* (4,450 km) and taxiway, *Bravo* (1,120 km), and **b)** to detail relevant aspects of this design process. The paper focuses on the development of: constant low strain asphalt fatigue criteria, asphalt sub-layering schemes, and a specific asphalt analysis algorithm for airport pavements. Management and technical factors accounted for a successful design are also described.

The design resulted in various rehabilitation strategies and corresponding alternatives. A 3-stage long-term rehabilitation strategy, flexible enough to accommodate both operational and financial requirements, was adopted by the Airport Authority .

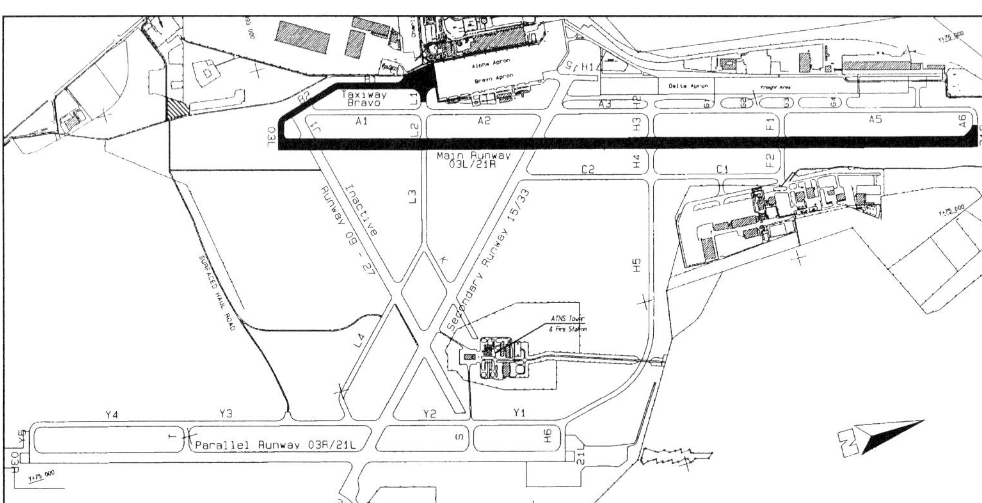

Figure 1: Layout of Johannesburg International Airport (project area shaded in black)

OVERVIEW OF DATA AQUISITION

Pavement investigation The purpose of this phase was to determine the current condition of pavement structures. This included the determination of pavement material characteristics through field and laboratory work. The various aspects of investigation activities are summarised below:

Visual assessments The classification of distress types and the rating of distress were done according to the South African recommendations for centre and edge strips [Africon, 1997].
Riding quality evaluation Riding quality measurements were done using a High Speed Profilometer. A profile of the vertical acceleration experienced by an Airbus A-320 was also recorded. These were undertaken to determine the improvement in riding quality that would be achieved through the pavement rehabilitation.
Ground Penetrating Radar (GPR) measurements A ground penetrating radar, with a 2 GHz antenna, was used to provide continuous data on pavement layer thickness. This proved to be essential information for the backcalculation of layer elastic moduli.
Asphalt sampling Accurate asphalt layer thicknesses were determined by examining the asphalt cores. Also, various types of cracks occurring at both top and bottom of asphalt layers could be investigated. Asphalt slabs were cut from the pavement to obtain beam samples for dynamic stiffness and fatigue testing, at constant low strain.
Deflection measurements Two types of equipment were used for measuring surface deflections: Lacroix Deflectograph (used to assist in the identification of weak pavement areas) and **H**eavy **W**eight **D**eflectometer (HWD - used to determine the layer elastic moduli). The HWD applied loads at various levels to a maximum of 220 kN, equivalent to a fully loaded Boeing B-747 main gear wheel.
As-built Data As-built data provided an important input to the definition of uniform pavement sections, as the runway and taxiway have been built using a large variety of pavement structures. The typical runway pavement structure consisted of ± 300 mm of various asphalt concrete layers, ± 280 mm **w**ater **b**ound **m**acadam (WBM) and/or **t**ar **p**enetrated **m**acadam (TPM), and granular layers.

Air traffic analysis This analysis was undertaken to provide input data on the historical and future pavement loading to be used in the pavement structural capacity analysis and rehabilitation design. As the data that was obtained on JIA's **A**ir **T**raffic **M**ovements (ATM's) was not in the input format required for the pavement analysis, the following inter-related activities had to be done:
Aircraft Spectrum Definition An analysis of the existing and potential future aircraft spectrum was made to select a representative spectrum of aircrafts that would significantly influence the pavement design [Witczak, 1972]. These included the Airbus A-300/320/340, Boeing B-707/727/737/747 and McDonnell Douglas MD-11.
Ultra High Capacity Aircraft (UHCA) Evaluation Due to the analysis period of 20 years, it was decided to evaluate the influence of future use of UHCA's, e.g. A-3XX. Airbus foresee about 1400 UHCA's to be ordered over the analysis period) [FLUG REVUE, 1997].
Historical and Future Traffic Growth Evaluation A study was undertaken to determine the historical growth rates and select future growth rates for the analysis period [Witczak,

1972]. This study looked at various growth scenarios. Due to the recent political changes in South Africa, the growth in the last few years has been ± 20% annually. The average annual growth rate over the design period is however expected to stabilise to ± 6%.
Traffic Distribution and Aircraft Loading The objective of this stage of traffic analysis was to make a realistic estimate of the distribution of future traffic and corresponding aircraft loading over the numerous existing facilities. Past runway utilisation, based on ACSA's detailed ATM records, was used as a guideline for distributing the future traffic. The weighting of aircraft load was done according to the function of pavement [ICAO, 1983].
The air traffic analysis resulted in identifying the ***uniform loading sections*** of both airside elements, characterised by the same ATM's and aircraft loading.

PAVEMENT EVALUATION AND ANALYSIS

This section describes some of the processes that were used to analyse the pavement condition data and presents the results of the analyses. These results were used in the rehabilitation design process.

Software evaluation Due to its flexible pavement design methodology that allows for a variety of material performance models and sub-layering schemes, the Airport Pavement Structural Design System (APSDS) [MINCAD, 1996] was chosen over LEDFAA [FAA, 1995]. The system calculates the cumulative damage induced by the whole traffic spectrum consisting of any combination of aircraft types and load configurations [Wardle and Rodway, 1995]. The Cumulative Damage Factor (CDF) is defined as the number of repetitions of a given material response parameter divided by the "allowable" repetitions of the response parameter that would cause failure. The CDF of a Continuously Graded Asphalt (CGA) layer, based on the horizontal tensile strain at the bottom of layer, is presented in Figure 2. Because the contribution of each aircraft/load configuration is individually considered, there are no systematic errors of the type that can result from simplified design systems based on approximating multi-wheel undercarriages by "equivalent" single wheels. Furthermore, the approach does not involve converting aircraft passes to pavement "coverages". The system also includes the potentially ameliorating effect of aircraft wander via a user defined standard deviation that is appropriate to the particular airside element, e.g. taxiway, runway or parking bay.

Pavement evaluation and analysis In order to perform the structural capacity analysis and rehabilitation design, the taxiway and runway had to be divided in sections having similar functional and structural conditions (***uniform pavement sections***) based on the following:
Preliminary analysis of Lacroix Deflectograph and HWD test results such as abnormal measured deflections or string of measured deflections, correlation between measured deflections and the surface temperature, and the uniformity of magnitude of measured deflections within the same uniform pavement section.
Backcalculation of layer elastic moduli A preliminary backcalculation was done using all the deflection values determined in the previous stage. The possible outliers were identified based on both the error per sensor values and the coefficient of variation per individual layer. The elimination of outliers in the final backcalculation resulted in more realistic

stiffness values. A *t*-distribution of results was assumed [Van As and Joubert, 1996]. The 95th percentile values were determined, thus allowing just 5% of the pavement to fail due to layers weaker than those assumed in the design.

Asphalt sub-layering and temperature correction In order to determine the necessity of sub-layering the asphalt layers in the rehabilitation design process, Stewart Scott requested the South African Council for Scientific and Industrial Research (CSIR) to determine the resilient moduli. 3rd point loading fatigue tests, at constant low strain values, were performed on beams cut out of the recovered asphalt slabs. Furthermore, six cores were extracted from each slab so as to be representative of the complete volume. Two cores were taken from the top, middle and bottom parts, respectively. The resilient modulus values, derived using indirect tensile stiffness and indirect tensile strength tests, clearly indicated the existence of *in-situ* asphalt sub-layering. The following stiffness distribution, as percentages of the 3rd point loading test values for middle sections, was found:

Table 1 : Asphalt Sub-Layering - Stiffness Distribution and Temperature Assumptions

Asphalt Sub-Layer	***Top***	***Middle***	***Bottom***
Stiffness (%)	61	100	37
Field Temperature (°C)	40	30	20

The final backcalculated moduli were considered as representative for the middle part of each slab. Subsequently, the moduli for the other sub-layers (having particular thicknesses corresponding to each considered pavement structure) were determined by multiplying the backcalculated values by the above mentioned coefficients. The resilient modulus tests were carried out at a laboratory temperature of 25°C. In order to simulate the field night conditions, a temperature correction, using the Asphalt Institute approach [Johnson and Baus, 1992] was applied to each sub-layer. The final backcalculated moduli of other layers (WBM, TPM, subgrade, etc) were not subjected to any correction. Also, the Poisson ratios of all materials were kept constant, throughout preliminary and final back calculations.

Comparison with the values found in the literature review The design values of layer moduli and various adjustments for each uniform pavement section were compared to those published [Theyse *et al.*, 1996]. A good agreement was generally found between the theoretical and practical values.

Performance criteria (fatigue curves) investigation Generally, the performance criterion (or fatigue curve) is defined as the relationship between the value of the critical response parameter and the number of load repetitions, sustained until the particular pavement material will fail in a specific mode of failure. The asphalt performance criteria considered for this analysis, from literature review, past experience and CSIR tests, are presented in Figure 3 [Valkering and Stapel, 1992; AA Loudom *et al.*; 1996, Theyse *et al.*, 1996]. It was decided that the failure of WBM, in the given circumstances, would be unlikely. As such, an investigation of existing performance criteria for WBM was unnecessary. The permanent deformation of the selected and subgrade materials, resulting in the deformation of road surface (rutting), was to represent the mode of fatigue for these layers. The fatigue curves defined by South African Mechanistic Design Method were chosen for the subgrade analysis [Theyse *et al.*, 1996].

BCRA '98-Article_Page 6 Diagram 34

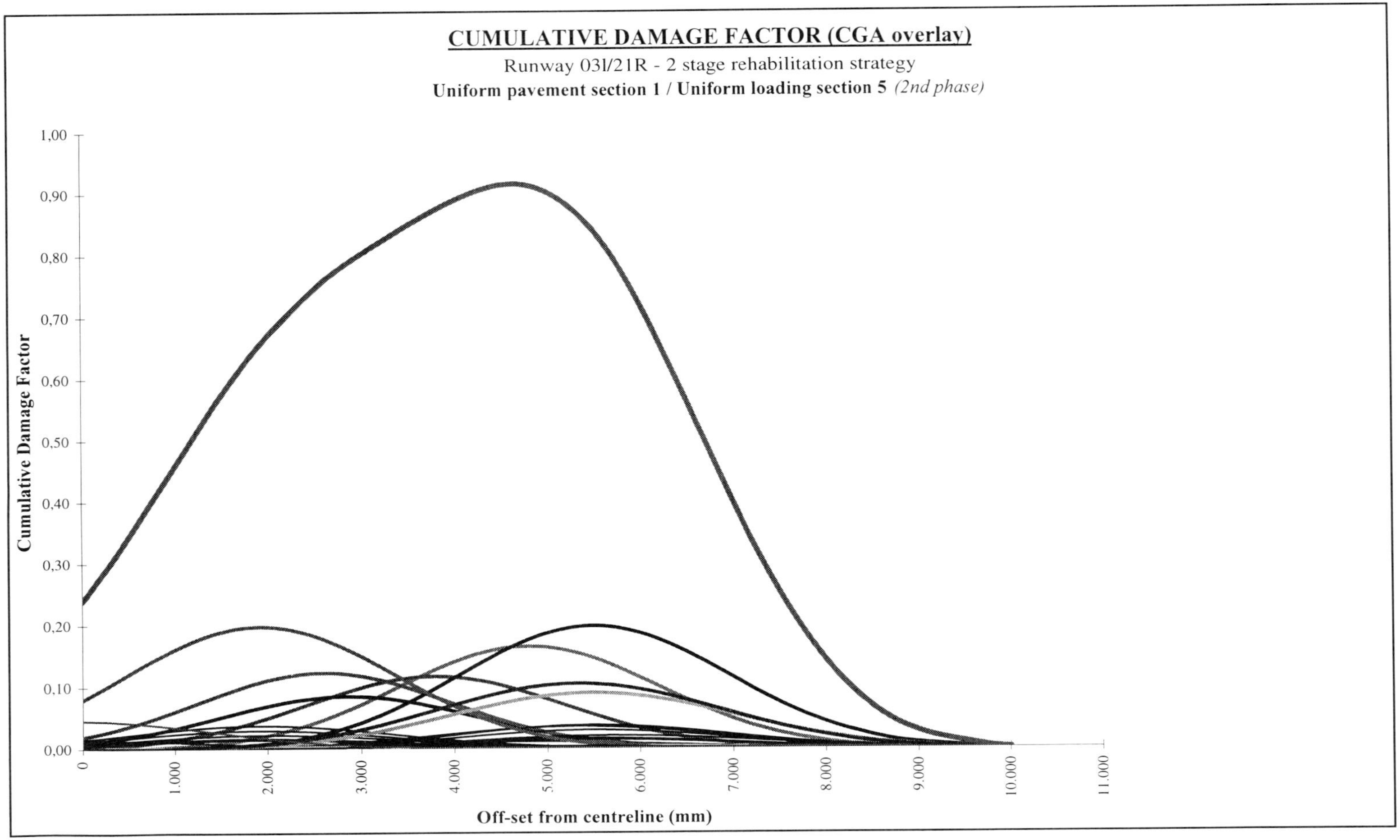

ASPHALT SEQUENTIAL FAILURE ALGORITHM (ASFA)

The necessity of accounting for different asphalt sub-layer performance models within the same layer led to the development of a specific asphalt analysis algorithm for airport pavements, ASFA. The material characteristics were defined for each layer and/or sub-layer in the pavement structure under consideration. The structural analysis would involve a ***linear elastic*** analysis of the multi-layer system (where asphalt sub-layers were considered independent layers). Pavement response was expressed in terms of stresses or strains at critical positions in the pavement structure, in accordance with the material type used in each layer or sub-layer and its failure mode. The pavement response (strains at each predefined off-set), for each aircraft's loading, was afterwards correlated with the specific performance criteria, resulting in the layer/sub-layer's Damage Factors. The sum of all Damage Factors, at predefined off-sets, resulted in the layer/sub-layer's Cumulative Damage Factor (CDF). Where all the CDF's were less than 1.0, the pavement structure was considered acceptable. Where the CDF's exceeded 1.0, material characteristic adjustments were to be applied to the various asphalt layers, existing or added. The process was iterated, for a number of phases, until the last failure of an asphalt layer or until all CDF's of unfailed asphalt layers were less than 1.0. ASFA is briefly described in the following Figure.

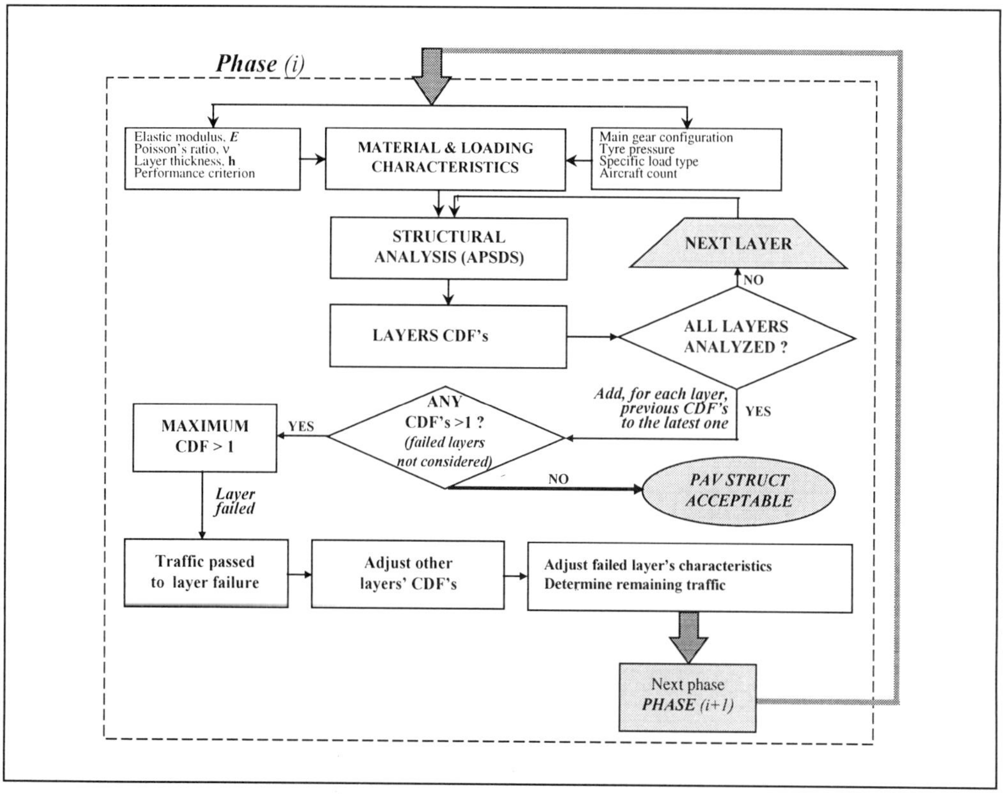

Figure 4: Asphalt Sequential Failure Algorithm (flow chart)

Pavement failure using the CDF method, was checked for two types of failure modes. Firstly, the development of fatigue cracking in the asphalt layers through excessive repetitions of high horizontal strain levels at the bottom of the layer. Secondly, the overloading of the subgrade layer causing high vertical compressive strain and leading to rutting of the subgrade and the upper layers.
Static structural analyses, using APSDS and ASFA, were performed on each uniform design section for every rehabilitation strategy. ASFA allowed a speedy investigation of a large range of layer thicknesses, in terms of both mill & replace and/or overlay alternatives.
For example, a rehabilitation design for 20 years analysis period, using ASFA, resulted in the milling out, at year 0, of the top cracked and/or aged 110 mm of asphalt and replacement with an asphaltic **L**arge **A**ggregate **M**ix for **B**ase**s** (LAMBS) plus the addition of a Continuously Graded Asphalt (CGA) [DOT, 1996]. After 10 years the top CGA would fail. A 75 mm CGA overlay is then added, for the second 10 year period. (APSDS analyses showed that overlays with thicknesses between 50 and 70 mm would reach CDF's greater than 1.0, for this second period.) The 75 mm new overlay reached a CDF of 0.9. The rehabilitation solution was therefore found acceptable as the subgrade CDF also remains less than 1.0 (see Figure 2).
The Boeing B-777's six-wheel main landing gear made a significant impact on the asphalt layer CDF's. This confirmed the necessity of including UHCA's, having similar landing gears with that of B-777 but applying larger loads, in the rehabilitation design process.

STRUCTURAL CAPACITY ANALYSIS

APSDS analyses of the existing pavement structures were done with the view of establishing their structural capacities (remaining life), for the predicted future traffic loading. The existing pavement structures were defined in terms of: layer elastic moduli, Poisson's ratios and thicknesses; existing asphalt sub-layering; and South African Mechanistic Design based performance criteria. The predicted traffic loading was defined for a 20 years design life, including UHCA's. Generally, the remaining life of the existing pavement structures was found to be between 2 and 6 years. At the end of the period, the bottom asphalt sub-layer would fail, resulting in a rapid propagation of tensile cracks through the other asphalt sub-layers. The cracked state of the asphalt layer would inevitably lead to a sub-grade failure and therefore to severe rutting in the whole pavement structure.

REHABILITATION DESIGN

The purpose of rehabilitation measures was to address the following pavement aspects:
Improve pavement structural capacity The pavement structure required strengthening to accommodate the traffic loading in the selected design period and thus prevent fatigue cracking of asphalt layers or rutting of subgrade layers.
Repair existing pavement functional and structural distress and
Eliminate surface drainage problems Surface level corrections were necessary at certain locations on the runway to eliminate water ponding during rain storms.

Management considerations The following general factors influencing the selection of rehabilitation measures from a management viewpoint were considered:
Design period Due to the uncertainty of future changes in the aircraft and air travel industry over very long periods it was decided to limit the analysis period to 20 years. As JIA has become the busiest airport in Africa, the frequency of runway and taxiway rehabilitation work was to be kept to a minimum.
A minimum pavement design period of 10 years and a maximum design period of 20 years were therefore analysed. In addition, at the request of JIA management, a two stage design strategy was also analysed. The concept of this strategy was to mill and replace in-situ asphalt layers at year 0 to a depth that would be sufficient for a 20 year period, but add layers for a 10 year period. A rehabilitation at year 10 would be done, but would consist mainly of applying additional layers with very little, if any, milling work required.
Air traffic distributions during rehabilitation work JIA management had indicated that the runway must stay operational during daylight hours and up to 22:00 at night. Thus, the rehabilitation measures had to be of such a nature that the runway was to be placed in operation every morning at the specified hour.
Condition of the existing pavement The severe surfacing distress of the existing pavement was of such a severity that at least the surfacing layer would require replacing at a short notice. Loose aggregate on the runway and surface unevenness could also cause operational problems.
Funding of the rehabilitation work Three rehabilitation strategies (as described above) plus a holding action were considered to allow flexibility for various budget levels.

Technical considerations The general technical factors considered in the rehabilitation design were as follows:
Selection of rehabilitation design method The rehabilitation design process was entirely based on South African Mechanistic Design Method. A multi-layered system was defined for the pavement structure.
Selection of uniform design sections The ***uniform loading sections*** were assigned to the ***uniform pavement sections*** resulting in 13 ***uniform design sections***.
Selection of rehabilitation materials The selection of pavement rehabilitation materials was done based on both theoretical and practical criteria. The theoretical aspects included the elastic modulus values, performance criteria, thickness, etc. The practical considerations comprised: price, material reliability, site procedures, drainage. Two asphaltic materials had therefore been chosen: LAMBS (selected to replace the milled out asphalt in the existing upper pavement layers) and CGA (selected to be used in the layers that will be added on top of the existing or future asphalt surface)
Restriction on sudden surface level changes Due to aircraft operational requirements, no sudden change in surface level was to be allowed. Consideration had also to be given to the tie-in to the levels of the adjoining taxiways [ICAO, 1983; FAA, 1989].
Surface drainage problems Level corrections were required on some localised areas of the runway to eliminate water ponds which were causing potential aquaplaning problems.
Alternative rehabilitation designs It was decided that alternative rehabilitation designs were to be carried out based on two sets of performance criteria, for each uniform design section. The first alternative, was based on South African Mechanistic Design set of fatigue curves. The CSIR laboratory test results would provide with the second set of fatigue curves.

Rehabilitation design results For each of the two alternative rehabilitation designs, numerous alternatives (e.g. mill & replace, overlay, rework pavement layers, etc.) were analysed, in the framework of a certain strategy, using the APSDS and ASFA. The use of CSIR fatigue curves indicated much better fatigue life for the existing lower asphalt sub-layers.
The selected rehabilitation measures, based on the first design (using SAMDM performance criteria, as nothing similar was found in the literature for the constant low strain fatigue curves developed by CSIR), were as follows:

1-Stage Rehabilitation Strategy (20 years design life)
- Mill existing asphalt to depths between 80 and 110 mm;
- Replace milled asphalt with LAMBS, incorporating modified bitumen binder to achieve higher E moduli (min. 5500 MPa). The depth of milling was to be function of the thickness and strength of the existing upper pavement layers.
- Apply a CGA overlay to the required thickness based on the supporting strength of the layers below, the traffic loading and the design period. The CGA overlay thickness varies between 50 and 250 mm for the 20 year design period.

2-Stage Rehabilitation Strategy (20 years design life)
- Mill existing asphalt to depths between 80 and 110 mm (at year 0).
- Replace milled asphalt with LAMBS, incorporating modified bitumen binder (at year 0).
- Apply a CGA overlay, with thickness varying between 50 and 150 mm (at year 0).
- Apply an additional 100 mm asphalt overlay (at year 10). This 100mm overlay is a conservative value, calculated for the worst areas.

In order to accommodate both operational and financial requirements, the Airport Authority requested the expansion of the 2-stage rehabilitation strategy to a 3-stage one. The following 3-stage strategy, for 20 years design life, was subsequently adopted:
- Mill existing asphalt to depths between 90 and 110 mm (at year 0).
- Replace milled asphalt with LAMBS, 55 or 75 mm, and CGA, 35 mm (at year 0).
- Apply a CGA overlay, thickness to be confirmed based on future passed traffic (at year 2-3).
- Apply an additional asphalt overlay (at year 10).

CONCLUSIONS

The rehabilitation design methodology for airport pavements presented in this paper allowed for fast and effective investigations of various rehabilitation strategies and corresponding alternatives. The development of asphalt low strain fatigue curves, sub-layering schemes, and, subsequently, of an Asphalt Sequential Failure Algorithm optimised both the structural capacity analysis and rehabilitation design process. The necessity of including the future potential use of Ultra High Capacity Aircrafts in the design process was also demonstrated. The flexibility and efficiency of this methodology led the Airport Authority to request the consultant to prepare a long-term, functional and structural, pavement rehabilitation programme for JIA's whole network of airside elements.

REFERENCES

1. AFRICON Manual for the Visual Assessment of Flexible Pavement Distresses, Report, Pretoria, RSA, 1997.
2. Witczak, M.W. Design Analysis-Asphalt Concrete Overlay Requirements for Runway 18-36, Washington National Airport, Report RR72-4, The Asphalt Institute, College Park, Maryland, U.S.A., 1972.
3. FLUG REVUE Datafiles Airbus A3XX-100 & Airbus A3XX, FLUG REVUE On line, July Update, Internet, 1997.
4. International Civil Aviation Organization (ICAO) Aerodrome Design Manual, Part 3: Pavements, 2nd Edition, Montreal, Canada, 1983.
5. MINCAD Systems APSDS-Airport Pavement Structural Design System, User Manual, Richmond, Australia, 1996.
6. Federal Aviation Administration (FAA) Airport Pavement Design for the Boeing 777 Airplane, Advisory Circular No 150/5320-16, U.S. Department of Transportation, Washington, U.S.A., 1995.
7. WARDLE, L.J. and RODWAY, B. An Improved Flexible Airport Pavement Design, Method Airports Pavement Section, 20th World Road Congress (Seminar on Airfield Pavements), Montreal, Canada, 1995.
8. Van AS, S.C. and JOUBERT, H.S. Applied Statistics for Civil Engineers, Civil Engineering Department, Faculty of Engineering, University of Pretoria, RSA, 1996.
9. JOHNSON, A.M. and BAUS, R.L. Alternative Method for Temperature Correction of Backcalculated Equivalent Pavement Moduli, Transportation Research Record 1355, Transportation Research Board, Washington, D.C., U.S.A., 1992.
10. THEYSE, H.L., DE BEER, M. and RUST, F.C. Overview of the South African Mechanistic Pavement Design Analysis Method, Paper No 961294, Transportation Research Board, Washington, D.C., USA, 1996.
11. VALKERING, C.P. and STAPEL, F.D.R. The Shell Pavement Design Method on a Personal Computer, Proceedings of the 5th International Conference on Asphalt Pavements, pp 351-374, Vol I, International Society for Asphalt Pavements, UK, 1992.
12. AA LOUDOM & Partners High Modulus Asphalt Base (LAMBS), Report, Durban, RSA, 1996.
13. Department of Transport (DOT) Construction Materials for the Restoration of Airfield Pavements in South Africa, Report RR93/319, DOT, Pretoria, RSA, 1996.
14. Federal Aviation Administration (FAA) Standards for Specifying Construction of Airports, Advisory Circular No 150/5370-10A, U.S. Department of Transportation, Washington, D.C., U.S.A., 1989.

Fifth International Conference on the
BEARING CAPASITY OF
ROADS AND AIRFIELDS

USE OF SUPERPAVE BINDER SPECIFICATION ON AIRFIELD PAVEMENTS A CASE STUDY

GEIR LANGE

Senior engineer, Civil Aviation Administration, N-0032 Oslo, Norway

ABSTRACT

The new main airport for Oslo at Gardermoen (OSL) is the first airport where specifications according to Superpave has been used. This paper_describes the reasons behind choosing Superpave binder specification instead of ordinary requirement for bitumen. Further more this paper describes how_requirements to a polymer modified binder should be set, what we do today and what we need tomorrow and which part of Superpave we can use in Norway (and the rest of Europe) with regards to our climate. The specification for OSL is discussed and test results from the new runway with an Split Mastic Asphalt with PG 64-34 binder are presented.
This paper summarises our experiences in using Superpave instead of the ordinary requirement for binders on airfield pavements in Norway.

INTRODUCTION

Background

On the 8th of October 1992 the Norwegian Government finally decided that the new main airport in the Oslo region should be located at Gardermoen, 47 km North of the capitol. Exactly 6 years later, October 8th 1998, the new airport at Gardermoen is scheduled to be opened, designed for 12 million passengers in its first year of service - 1999.

This challange could not be handled by the CAA alone. In addition to Gardermoen The Norwegian CAA (CAA) is responsible for 46 of Norway's 55 Airports. To handle the planning and construction of the new airport, CAA established a new company, fully owned by CAA, named Oslo Airport AS (OSL).

The new airport will have approximately 2,2 million m^2 paved surfaces in total, nearly 2,0 million m^2 of these are going to be flexible pavements. On the apron and the four stopways PCC are being used.
Over_the past 10 years CAA has done_extensive work with regard to construction and maintenance on our nearly 10 million square meters of flexible pavements.

Climate

Gardermoen is situated in an area that experiences more severe climate conditions compared with the other airports in Norway, with the exeption of Røros and Spitsbergen. To design an asphalt pavement, or rather choose a binder with correct properties was a challenge taking into account that ordinary Norwegian binder specification enable binder suppliers to deliver a large range of neat binder with different types and amounts of additives, and still meet the requirements set in the National specifications. This paper gives a case study of how the requirements in the tender documents for the asphalt surfacing contract was designed based on experiences from other airports, the existing requirements and why the Superpave level 1 was used in the new requirements.

Figure 1 shows the temperature variations in the asphalt pavement. The pavement temperature on the airfields located in the red spot on the map in figure 1 can vary from -40 °C to +60 °C (58,9 °C registered during August of 1997 at Oslo Airport, Fornebu). The airports located in the dark spots is Spitsbergen the most Northern airport located at the 78th degree latitude, followed by Banak, Bardufoss, Røros, Fagernes and Gardermoen (located at the 58th degree latitude). On these sites neat binder is not sufficient and some kind of polymer have to be used. The white spots have a temperature variation less than 60 °C.

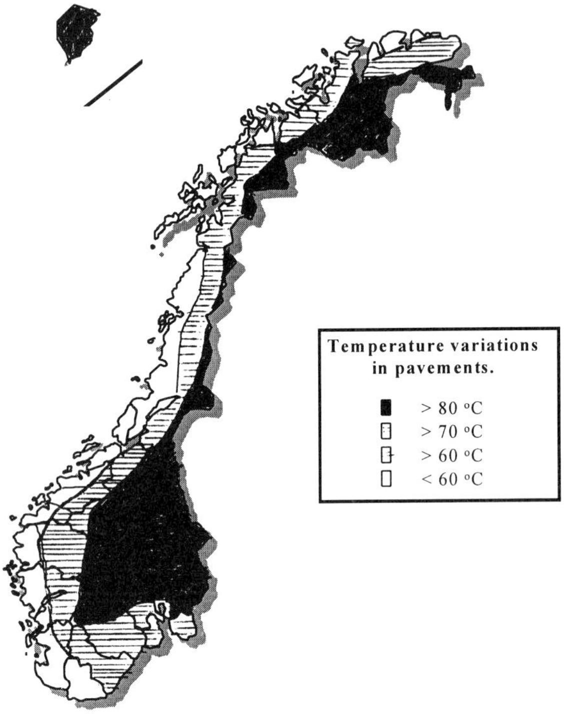

Figure 1 Temperature variations in pavement (1)

R&D project Ageing of Asphalt pavement on airfields

In 1991 CAA started a Research & Development program in order to develop asphalt pavements more resistant to ageing and low temperature cracking that could meet the requirement of a long service life (> 20 years). CAA gave a presentation of this project during the 4th BCRA conference in Minneapolis Minnesota 1994. (2).

In this R&D project a total of 50 samples (450 by 450 mm) were produced in laboratory and included in an accelerated ageing test program in two phases in order to simulate 10 years of field exposure with Norwegian climate. In phases one, a standard asphalt mix in Norway was chosen (Asphalt concrete with 11 mm Chip size) to allow direct mutual comparison between the 22 samples, they were all made of the same raw materials (aggregate, fillers, amount and type of fibres and polymers) In phase two, 9 SMA and 4 porous friction course samples were chosen to investigate and the properties of such mixes. Additionally 6 samples were taken from different airports. In order to allow direct mutual comparison between phase one and two, 5 samples identical with phase one samples were produced.

However three of a total of 41 samples in the two phases were mixed with PmB's from three different bitumen suppliers. When ordering the PmB CAA required that:

- the PmB should meet the Norwegian specification for PmB's,
- penetration grade binder B 250 being used
- SBS polymer the with a 6 % content by weight being used

Sample #. Phase/number	Type of Mix	Binder Pen.grade/amount	Polymer type and content	Binder supplier
1-22	SMA 11	B 250 + 6,5%	SBS - 6%	A
2-18	SMA 11 + cellulose fibre	B 250 + 7,0%	SBS - 6%	B
2-10	SMA 16	B 250 + 6,0%	SBS - 6%	C
2-12	SMA 16 + cellulose fibre	B 250 + 6,0%	SBS - 6%	A
1-14	HMA 11	B 250 + 7,0%	SBS - 6%	C
2-5	HMA 11	B 250 + 7,0%	SBS - 6%	A
2-19	HMA 11	B 250 + 7,0%	SBS - 6%	A?

Figure 2 Job Mix Formula for the 6 different samples

Figure 3 shows the result from this project for these tree different PmB's when all the other parameters in the mix and the climatic exposure were identical.

Sample #	Original binder			Recovered binder			After wetherometer		
	Pen	SP	PI	Pen	SP	PI	Pen	SP	PI
1-22	103		2,7	92		2,7	48		1,5
2-18	110	63,0	4,1	86	63,9	3,4	40	67,3	1,8
2-10	107	>94,8	>8,5	93,7	94,2	>7,9	34	76	2,8
2-12	64	73,5	4,1	56	76	4,1	26	82,7	3,1
1-14	103		2,7	86		2,3	56		1,5
2-05	107	>94,8	8,5	84	91,2	7,3	39	81,7	3,9
2-19	64	73,5	4,1	59	74,3	4,0	36	78	3,2

Figure 3 Penetration at 25 °C, Softening Point and Penetration index for the 7 samples.

As shown in figure 3 the use of the existing specification is not adequate to choose a binder capable of serving a specified climate or temperature range. The requirement is just a list different parameters that have to be reported. On the other hand, if CAA use the existing requirements we can not be sure what kind of PmB we get on the airfields and which low-temperature resistance a specific binder has. Choosing binder A instead of C could probably result in several years shorter service life, or if binder B is chosen it could result in low temperature cracking. It is clear that all the above three PmB's meet the requirement in the existing specification. How is it than possible for the builder to be sure that the binder offered by the contractor is good enough when it meets the requirements set in the specification ? Additional or other requirements have to be set.

On approximately 95% of the Norwegian airports CAA use a penetration grade B 180 binder. On the remaining 5% B180 has been used in the past, but severe low temperature cracking has occurred in the pavement during the first winter.

In order to avoid thermal cracking, polymer modified binder were used on two airports in 1993 and 1995. The binder mix design and binder was chosen based on the experiences gained during the 5 years Research and Development program done by CAA.
Due to severe cracking (each 7-8 m) the runway at Banak Airport had to be rebuilt in 1992. The old asphalt concrete with a B 60 binder was recycled and used as a binder course. The additional binder in this layer and the SMA 11 layer was a PmB equal to type B in table 2. The requirements in the tender paper was 6,5% binder (B 250), adding 6,0% SBS polymer by binder weight. PmB requirement was according to Norwegian specifications.

Binder suppliers have normally claimed that «you get what you ask for». After the R&D project CAA learned that specifying a penetration grade and the type/amount of polymer, do not automatically result in a specific binder property for the mix.

Furthermore when suppliers gives presentations of PmB's, they often refer to penetration and softening point etc. As shown in figure 3 this could not be used as requirements in tender papers, but only to show the effect of different types and amounts of polymers in the same original binder.

The new Oslo Airport Gardermoen
Based on the experiences from the above mentioned R&D program CAA were faced with some basic problems when the requirements in the tender document for the new main airport in Oslo had to be set. This was during the Autumn of 1994 and the Strategic Highway Research Program was just finished, and results from Superpave were starting to be published.
Based on the discussions at several international conferences and seminars and published literature, this could be an alternative for the binder specification at Gardermoen. It was claimed that the binder spesification was valid both for original and polymer modified binders.

Three alternatives were considered:

- use the existing specifications
- use functional requirement for the end product - the asphalt mix
- Use Superpave Binder specifications

LABORATORY INVESTIGATIONS

CAA did not have enough knowledge or experience with this new set of specifications. In order to choose mix design strategies, Ed Harrigan/Advanced Asphalt Technologies L.P. (Virginia) and Von Quintes/Brent Rauhut Engineering Inc. (TEXAS) was engaged to do such study based upon the Superpave mix design procedures (3).

Temperature - performance grade
To choose the Performance Grade temperature data from the site were collected for the period 1957 - 1993. Figure 4 shows the daily maximum and minimum air temperature and the pavement temperatures calculated according to Superpave equations.

Air Temperatures (^oC)	Pavement Temperatures (^oC) According to SHARP
33	47 at 20 mm depth
-35	-29 at surface

Figure 4 Estimated Pavement Temperatures for Gardermoen according to SHARP

Based upon these pavement temperatures the performance grade of PG 52-34 is required by the AASHTO MP1. However in such cases when the pavement experiences heavy wheel loads and high tire pressure, the Superpave recommends that the high temperature grade of the binder should be increased by two grades (each 6^oC), to bolster the asphalt binders contribution to rutting resistance.

This meant that the requirement for Gardermoen should be a PG 64-34.

The minimum pavement surface temperature was calculated from the minimum air temperature. The Superpave equation gives the same result as actual measurements on airfields in Norway. The SHARP equation developed to calculate the 7 days maximum pavement temperature at 20 mm pavement depth are however not adequate for Norwegian conditions (or other countries in northern Europe). Even though Norway is located on high latitudes, oceanic currents makes climatic conditions quite different to the US-north-west for instance. The use of latitude as a parameter will therefor lead to quite different results than what SHPR-originally intended.
During the joint venture project (1) a new equation was developed to calculate the surface temperature based upon the air temperature. The T_{depht} equation is however the same for the 7 days maximum temperature in the pavement. If this has been developed before the requirements were calculated for the Gardermoen project, the temperature T_{20} would have been 55^oC instead of 47^oC (figure 4). By increasing the high temperature with two grades, the final performance grade requirement would have been a PG 70-34.

Original binder
Based on experience gained from comparing different ordinary penetration grade binder supplied to several contracts on Norwegian airfields and the fact that the effect of a polymer (SBS, EVA etc.) depends largely on the original binder, the majority of the different binders supplied in Norway were included in this study.

The anticipated penetration grade binder that could be used was a B180 or B250. Samples of original binder was collected from 4 suppliers in Scandinavia and tested according to (4):

Supplier	Penetration grade	AASHTO MP1 Performance Grade
W1	B 180	PG 52-28
W2	B 250	PG 46-34
X1	B 180	PG 52-28
X2	B 250	PG 52-28
Y1	B 180	PG 52-28
Y2	B 250	PG 52-34
Z1	B 180	PG 52-28
Z2	B 250	PG 52-28
Z3	Soft Bitumen 250	PG 52-28

Figure 5 *Classification of binders in these study*

Polymers - Types and amounts
One of these binders (Y2) was chosen and blended with different types and amounts of polymers

Supplier/ Binder	Penetration grade	Polymer	Performance Grade
Y2A	B 250	None	PG 52-34
Y2B	B 250	6 % SBS	PG 64-34
Y2C	B 250	6 % HDPE	PG 70-28
Y2D	B 250	5 % LDPE	PG 64-28
Y2E	B 250	4 % LDPE	PG 58-34

Figure 6 *Classification of one binder with different polymers (types/amounts)*

Bitumen modification with commercially available polymers is an effective way of raising the high temperature component of the performance grade, but could have a slight negative effect on the low temperature component. Our strategy was to find a bitumen supplied in Norway that met the low temperature grade -34 ^{o}C and thereafter modify it to reach the high temperature grade 64 ^{o}C without raising the low temperature grade (i.e. to -28 ^{o}C). To be able to raise the PG 46-34 binder to a PG 64-34 it would require a large amount of polymers which have an unfavourable effect on the low temperature grade.
In general if we only use a PG 64-34 requirement a very soft binder could be modified to meet the requirement if sufficient amount of polymers are added.

The requirements for Oslo Airport Gardermoen were therefor divided in two parts:

- Original Binder PG 52-34
- Modified Binder PG 64-34

Additives

In order to investigate the adequate binder content in the SMA 11 mix and the effect of adding cellulose fibres to the mix, the Schellenberg SMA Draindown Test at 170 °C were used:

Test Parameters	Draindown %	Test Result
6,5 % Y2B /No fibre's	3,20 %	Fail
6,5 % Y2B / 0,3% fibre's by mass of mix	0,04 %	Pass
6,0 % Y2D /No fibre's	0,85 %	Fail

Figure 7 Draindown Test

Other requirements

The development of the Direct Tension was not finished when this study was done. (By January 1998 there are still no reliable product on the marked known for the CAA). This apparatus is essential when testing PmB's. Therefor additional requirements had to be set for the PmB's. The ductility test was recommended.

The final requirement

The final requirements were as follows:

Performance grade original binder	PG 52-34
Performance grade modified binder	PG 64-34
Binder content	6,2 %
Additives:	
Cellulose fibres	0,3 %
Amine	0,3 - 0,5 %
Ductility after RTFOT/10 °C	min 500 mm.

Figure 8 Final requirement for the Gardermoen binder

The same binder is used in all 3 asphalt layers but the binder content and additives varies.

The suppliers experienced some problems in meeting this requirements even when CAA in advance had tested the binders on the Norwegian market. Especially the m-value requirement was difficult to meet. The contractor used the laboratory at SINTEF to classify the binder and the pre-study was carried out by a US-company.
SINTEF have a Cannon BBR and the Advanced Asphalt Technologies L.P./Brent Rauhut Engineering Inc. used a AST BBR which gives different values (5). We believe this is the main reason why the supplier not meet the requirement with an ordinary PmB and had to use a special binder instead. However this problem is now solved by adjusting the two rheometers.

FIELD STUDIES

Paving started in 1995 at Gardermoen when the taxiways were paved. One of the taxiways has been used as a transportation route during construction of the new runway. After 3 years of heavy traffic the pavement shows no sign of neither rutting nor low temperature cracking.

Figure 9 comparise the test results from the Gardermoen binder with two other PmB's resently used on airfield pavements in Norway.

Requirements		Gardermoen Sample from 1996	Gardermoen Ref.	Banak	Bardufoss
Original binder					
Flash point	Min 230°C	280	255		
Viskosity 135 °C	Max 3 Pa s	1,6	0,57	1,1	0,7
Dynamic shear at 64 °C	Min 1,0 kPa	2,555	1,04	71 °C = 1,0	68,7 °C = 1,0
RTFOT					
Mass loss	Max 1,0%	0,48	0,06	0	0
Dynamic shear at 64 °C	Min 2,2 kPa	6,32	2,2	66,6 °C = 2,2	64,1 °C = 2,2
PAV					
Dynamic shear at 19 °C	Max 5000 kPa	1415,6	880	12 °C = 5000	9,3 °C = 5000
S at -24 °C	Max 300 Mpa	138,9	106	-23 °C = 300	-26,2 °C = 300
m at -24 °C	Min 0,300	0,310	0,327	-21 °C = 0,300	- 26,0 °C = 0,300
Direct tension	Min 1,0%	NA	NA	NA	NA
Performance grade		PG 64-34 ?	PG 64-34	PG 64-28	PG 64-34

Figure 9 Gardermoen bbnder versus B250 + 6% SBS

During the joint venture project (1) the binder used at one of the airports mentioned on page 5, has been classified according to Superpave to meet the requirements of Performance grade PG 64-34.

The same requirements and same asphalt mix was used when the runway at Bardufoss Airport was repaved in 1995. This binder has been classified as a PG 64-28.

CONCLUSIONS AND RECOMMENDATIONS

The Superpave binder requirements may be used on PmB's if the results from the Direct Tension Test is reliable. If not, additional requirements have to be used (SHRP +)

Ductility may not be this additional requirement, but probably elastic recovery.

Having used PG 64-34 at Gardermoen, it shows no sign of neither rutting nor low-temperature cracking after 3 years of heavy traffic (wheel loads 25 tons/tire pressure 1,9 MPa)

Calculations of high temperature range is not adequate in Northern Europe. A new equation is developed during (1) for the high temperature grade. At Gardermoen this means $T_{20} = 55^{o}C$ instead of $T_{20} = 47^{o}C$ as calculated according to Superpave.

ACKNOWLEDGEMENTS

The study carried out by Harrigan/Von Quintes was funded by Oslo Hovedflyplass AS (OSL) and was essential in useing the Superpave binder spesifications on the first airport in the world. A thanks to Arnfinn Erlien and Tore Herland (project managers from OSL for the two «runway» contracts at Gardermoen.

REFERENCES

1. «New Asphalt technologies - use of SHRP in Norwegian climate». Joint Venture projects between STATOIL, SND, CAA, SINTEF, Norwegian Public Road Administration, Binder suppliers and Contractors.
2. «Evaluation of Long-Term Asphalt Pavement Performance on Norwegian Airfields» 4th International Conference on Bearing Capacity of Roads and Airfields Minneapolis Minnesota 1994.
3. «Pavement and Mix Design Strategies for the Oslo Airport Gardermoen», Harrigan/Von Quintes, February 17, 1995
4. AASHTO Standard Specification MP1, Performance graded Asphalt Binder
5. «Direct Tension Test - Current Status», David A. Anderson Presentation held in Trondheim September 1997.

Fifth International Conference on the
BEARING CAPASITY OF
ROADS AND AIRFIELDS

ANALYTICAL OVERLAY DESIGN AND VALIDATION : AN INDIAN CASE STUDY

ANIMESH DAS[1], K. S. REDDY[2] AND B. B. PANDEY[3]

[1] Senior Research Fellow (CSIR), [2] Assistant Professor, [3] Professor,
Civil Engineering Department, Indian Institute of Technology, Kharagpur,
India - 721 302, Fax: +91-03222-55303, Email: animesh@civil.iitkgp.ernet.in

ABSTRACT

Performance studies were conducted on a large number of in-service pavements in India for over ten years. Overlay design methodologies were developed independently by Benkelman beam and Analytical methods. The overlay thicknesses determined by both the methods for the study area under consideration were found to be in agreement. Analytical design method, however, is more suitable for new pavement materials with unknown performance history.

Key Words: Benkelman beam, analytical pavement design, pavement performance, overlay thickness

1 INTRODUCTION

A major research project was launched by the Government of India, in the eighties, for the development of a overlay design method by conventional Benkelman Beam Deflection (BBD) method and design curves were developed after a study of performance of overlays for a period of about ten years. BBD database, pavement composition of different layers, subgrade conditions, axle load spectrum and lateral wander of commercial vehicles on National Highways (NH) of Eastern India were taken into account to develop overlay design charts [1].

Fatigue behaviour and dynamic modulus of asphaltic mixes at different temperatures and resilient modulus of granular material and subgrade soil were determined in the laboratory. Fatigue and rutting criteria based on performance data of a number of in-service National Highways were used for analysis of pavements and an analytical overlay design method was developed for Indian traffic and climatic conditions [2].

The design charts evolved from the two fundamentally different approaches were compared and it was found that overlay thicknesses were by and large in agreement for majority of the locations. The discrepancy in the computed overlay thickness in a few cases could be due to incorrect estimation of elastic modulus of existing bituminous layer. This paper presents methodologies developed for overlay design using Benkelman beam as well as analytical technique based on field performance of pavements in India.

2 BENKELMAN BEAM DESIGN METHOD

2.1 Introduction

The Indian Roads Congress (IRC) guidelines [3] for strengthening of road pavement using BBD technique are based on limited experience gathered in India during the sixties and seventies and has little relevance to the present day traffic. In order to develop a simple and scientific method of overlay design using Benkelman beam, Ministry of Surface Transport, (MOST), Government of India, launched a research scheme R-6 in the East, North, West and South India [4]. IIT Kharagpur was entrusted with the investigations of pavement sections in the Eastern zone [5].

2.2 Test Sections

Several test sections located on the National Highways in the states of West Bengal, Bihar and Orissa were selected for carrying out deflection and performance studies. Fig. 1 shows the National Highway network in Eastern India on which thirteen test sections were selected for periodical deflection tests and condition survey.

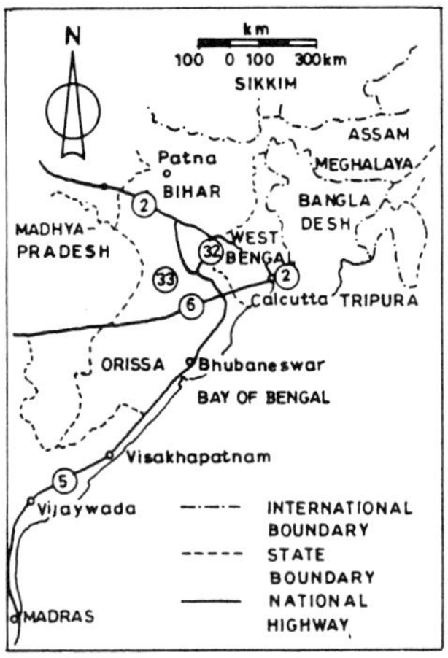

Fig. 1. Relevant National Highway Network in Eastern Zone

All the sections have two-lane two-way carriage ways. Average Annual Rainfall was in the range of 1300-1600mm and the Average Annual Pavement Temperature (AAPT) was about 35°C. Clayey, silty and sandy soils formed the subgrade of different pavements under study. The length of twelve test sections was 300m each while one section was 250m long.

The sections were divided into three sub-sections and different overlay thicknesses were provided on the subsections. The overlay thickness for the mid-subsection was computed as per the IRC guidelines [3] and this design was designated as Standard Design (SD). Other two sections had overlay thickness lower and higher than that of the standard design section and were termed as Under-Design (UD) and Over-Design (OD) sections, respectively.

2.3 Field Studies

2.3.1 Performance Studies and Failure Criteria

On each pavement section BBD measurements were made periodically as per IRC-81 [3]. For each sub-section, mean ($\bar{x}$) and standard deviation (σ) of the rebound deflections were computed. Each lane of the test section was divided into bays of 10m length and the cracked surface area in each bay was noted as percentage of that bay area. Rut depth measurements were made under a 3 m straight edge in each bay. Based on the condition survey data, the section was considered to have reached the end of its service life if the cracked area was 25%. Similarly, if the mean rut depth reached a value of 20mm, the pavement was deemed to have failed in rutting mode.

Pits of 1.0m × 1.0m size were excavated at either end of the test section at a distance of 1.0m from the pavement edge and thickness composition of the pavements were recorded. Field density was determined and the subgrade soil samples were collected for determination of field moisture content and CBR value.

2.3.2 Effect of Climate

Benkelman deflections of pavement depend to a great extent on the degree of subgrade saturation. Overlay design is based on the maximum deflection determined towards the fag end of the monsoon. Ratio of post-monsoon to pre-monsoon deflections was computed for different pavements to estimate the effect of the subgrade moisture on the deflections. Fig. 2 shows variation of deflection ratios for pavements having clay subgrades. It was found that 90% of the deflection ratio values were less than 1.6. The values of seasonal correction factors suggested to adjust the measured deflection values are given in Table-1.

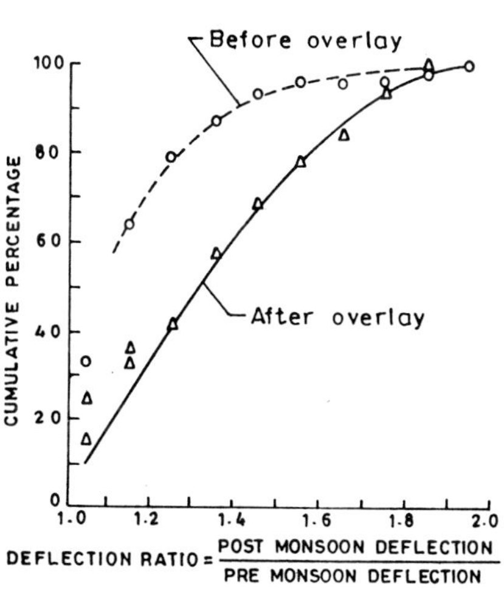

Fig. 2. Deflection Ratio vs. Cumulative Percentage.(East Zone)

Table-1 Suggested Values of Seasonal Correction Factor

Measurements taken in the month of	Seasonal Correction Factor	
	Clayey Soil	Sandy Soil
July, August, September, October	1.00	1.00
November, December, January, February	1.25	1.15
March, April, May, June	1.50	1.25

Deflection studies conducted on thick un-cracked bituminous overlays indicated that temperature correction for rebound deflection is about 0.005 mm / °C. The correction would be additive for pavement temperature lower than 35°C and subtractive for higher temperature.

2.3.3 Axle Load Study

Axle load measurements were made at several sites near R-6 test sections on NH-2 and NH-6 in West Bengal and on NH-5 in Orissa using a portable weigh bridge [6]. The Vehicle Damage Factor (VDF), defined as equivalent number of standard axle loads per commercial vehicle, was established for different National Highways in Eastern zone. Table-2 shows the VDF values for different National Highways in Eastern India.

Table-2 Vehicle Damage Factor

section	VDF
NH-2 (between Burdwan and Calcutta)	3.95
NH-5 (for the stretches between Jharporkharia and Bhubaneswar)	4.26
NH-5 (for the stretches between Bhubaneswar and Tangi)	4.57
NH-6 (between Kharagpur and Calcutta)	4.90
Average for National Highways in Eastern India	4.41

2.3.4 Lateral Placement Characteristics of Commercial Vehicles

A traffic study conducted on National Highways [7] in rural locations, indicated that commercial vehicles generally travel with the mean location of the outer wheel paths positioned about 1.90m away from the pavement edge. The mean location of inner wheel paths for both up and down traffic lies close to the centre line of the two-way two-lane pavements and fifty percent of the wheel paths lie in the central one metre band.

2.4 Development of Allowable Deflection Criteria

For each sub-section, $\bar{x}$, $\bar{x} + \sigma$ and $\bar{x} + 2\sigma$ values were computed and corrected for temperature and seasonal effect. The number of standard axle repetitions to cause failure of the overlays was obtained from traffic volume, VDF and lateral placement characteristics. The relationships obtained between overlay life and initial deflections form the allowable deflection criteria for different levels of reliability are shown in Fig. 3

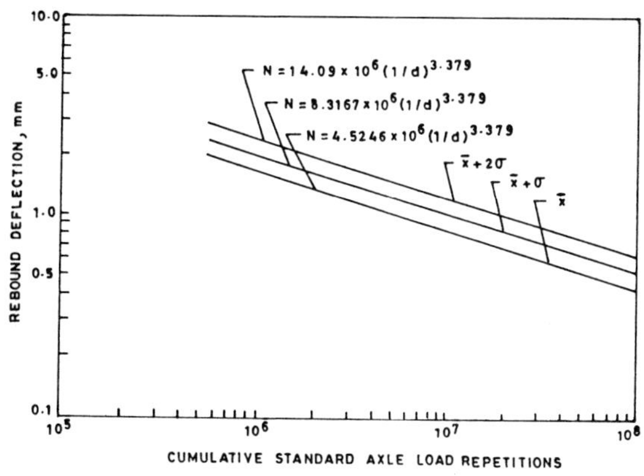

Fig. 3 Allowable Benkelman Beam Deflections vs. Cumulative Standard Axle Load Repetitions for Failure of Pavement

2.5 Development of Overlay Design Charts

Using the allowable deflection criteria overlay design charts are developed for different cumulative standard axle repetitions. Test results of Kingham [8] and R-6 report [5] were used for developing the design charts. A typical overlay design chart for 84% reliability is presented in Fig. 4.

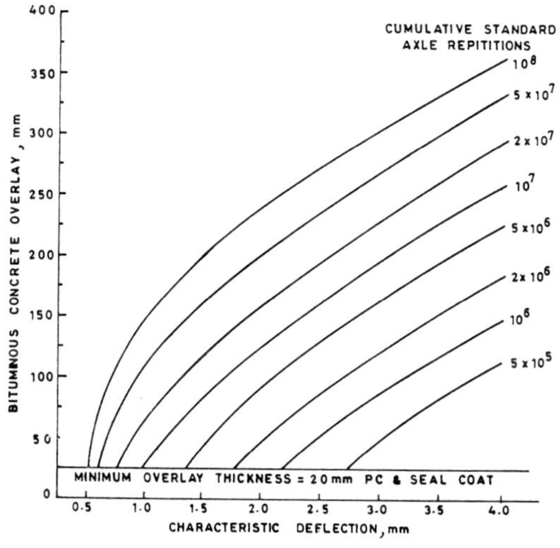

Fig. 4 Overlay Thickness for 84 % Reliability

3 ANALYTICAL DESIGN METHOD

3.1 Introduction

In the analytical design method, the pavement is idealised as a layered structure which behaves elastically under fast moving vehicles and each layer is characterized by its resilient modulus (M_R) and Poisson's ratio (μ). Horizontal tensile strain (ϵ_t) at the bottom of the bituminous layer and vertical compressive strain (ϵ_Z) on the subgrade are identified as the causative factors for fatigue and rutting failures respectively [9, 10]. A computer program **FPAVE** was developed [2] for analysis of stresses in the layered elastic structure. A dual wheel system, each wheel weighing 20.5kN at a tyre pressure of 0.56MPa, was assumed to act on the pavement.

3.2 Characterisation of Pavement Materials

3.2.1 Resilient Modulus

Subgrade: Since M_R values of subgrade soil were available only for a few sections [11], the following equations proposed by Lister and Powell [12] were used for estimation of M_R values in MPa

$$M_R = 10 \times CBR \quad for\ CBR < 5\% \tag{1}$$

$$M_R = 17.6 \times CBR^{0.64} \quad for\ CBR \geq 5\% \tag{2}$$

Granular Base: Base courses of pavements in India are generally made up of Water Bound Macadam (WBM) or Wet Mix Macadam (WMM) [13]. The use of laboratory equations for estimation of M_R value requires iterative finite element analysis for stress computation. Due to limited triaxial test data available [11] for pavement materials, Shell's equation [9] was used to obtain M_R value of the granular layer.

$$\frac{M_{R_{granular}}}{M_{R_{subgrade}}} = 0.2 \times (h)^{0.45} \tag{3}$$

where, $M_{R_{granular}}$ and $M_{R_{subgrade}}$ are resilient modulus values of granular base and subgrade soil respectively, h is the thickness of the granular layer in millimetre

Bituminous Surfacing: Bituminous Concrete (BC), Dense Bituminous Macadam (DBM) and Bituminous Macadam (BM) are the most common bituminous mixes in India [13] that form the surfacing of heavy duty bituminous pavements. BC is used as a wearing course, whereas DBM forms the binder course.

A strain controlled fatigue testing machine was fabricated [14] in the Highway Engineering Laboratory of the Institute to determine the dynamic modulus (E_d) value and fatigue behaviour of the bituminous mixes. Tests were carried out at a different temperatures at a frequency of 10 Hertz. The E_d values are presented in Table-3.

Table-3 Dynamic Modulus Values (MPa) of Bituminous Materials

Mix	Temperature, °C			
type	25	30	35	40
BC 80/100 bitumen	1966	1455	1077	797
BC 60/70 bitumen	2610	2055	1615	1270
BC 30/40 bitumen	4928	3809	2944	2276
DBM 80/100 bitumen	2100	960	600	-
DBM 60/70 bitumen	2815	1520	750	-
DBM 30/40 bitumen	4600	2905	1835	-
BM 80/100 bitumen	-	-	500	-

3.2.2 Poisson's Ratio

Poisson's ratio (μ) values adopted in the present analysis for bituminous mix, granular material and subgrade soil are 0.5, 0.35 and 0.35 respectively.

3.3 Structural Design Criteria

3.3.1 Fatigue Criteria

Laboratory Fatigue Equation: A total of sixty BC beams were tested under four point bending in the fatigue testing machine at temperatures of 20, 25 and 30°C. Tests were done under controlled strain mode and the failure of the specimen was defined as the number of cycles required until the amplitude of applied load for a particular value of central deflection reduced to half of its initial value. The regression equation derived from the laboratory fatigue test is given as

$$N_{lf} = 5.975 \times 10^{-5} [1/\epsilon_t]^{3.565} [1/E_d]^{1.4747} \quad r^2 = 0.74 \tag{4}$$

where, N_{lf} = number of repetitions for laboratory fatigue failure, ϵ_t = tensile strain at the bottom of the bituminous beam, r^2 = co-efficient of determination

Calibration of Fatigue Equation To arrive at the field calibrated fatigue equation for Indian conditions, 120 BC road sections and 160 BM road sections constructed with 80/100 bitumen investigated under research schemes of R-6, R-56 and R-19 sponsored by MOST [4, 2, 15] were analysed and Eq. 4 was modified to obtain the following fatigue equations for BC and BM for 84% reliability.

BC

$$N_f = 1.001 \times 10^{-1} [1/\epsilon_t]^{3.565} [1/E_d]^{1.4747} \tag{5}$$

BM

$$N_f = 2.26 \times 10^{-2} [1/\epsilon_t]^{3.565} [1/E_d]^{1.4747} \tag{6}$$

where, N_f = number of Cumulative Standard Axle (CSA) repetitions to produce 25% surface crack due to flexural fatigue on in-service pavement, ϵ_t = tensile strain at the bottom of the bituminous layer

3.3.2 Rutting Criteria

Considering vertical strain on the subgrade as an index to rutting potential, Reddy and Pandey [16] obtained a rutting criterion based on data from a few sections in Eastern India. Since then considerable amount of data was collected from different parts of the country and the equation given by Reddy and Pandey [16] was modified as

$$N_r = 2.56 \times 10^{-8} [1/\epsilon_Z]^{4.5337} \tag{7}$$

where, N_r= number of CSA repetitions to produce a rutting of 20 mm, ϵ_Z = vertical strain on subgrade

3.4 Development of Overlay Design Charts

A computer program **IITOLAY** was developed using the program **FPAVE** as a subroutine for design of BC overlay. The thickness of existing bituminous and granular layers and the moduli of pavement layers were given as input. The fatigue and rutting strains (ϵ_t and ϵ_Z) for a standard axle load were calculated at the critical points of the bituminous pavement structure and also the allowable strains were obtained for the assumed design period. The thickness of BC overlay was adjusted iteratively so that computed strains are approximately equal to that of the allowable strains. Fig. 5 presents a typical overlay design chart derived from analytical pavement design considerations.

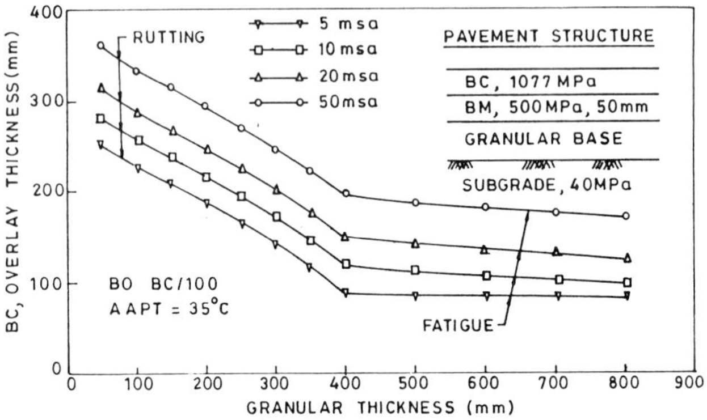

Fig. 5 Bituminous Pavement Overlay Design Chart (Analytical Method)

4 COMPARISON OF OVERLAY DESIGN

For the purpose of comparing the two design methods, a number of pavement section of National Highways under research scheme R-6 were selected. Using the BBD method, the overlay requirement for a design traffic of 20 million standard axles (msa) was determined from Fig. 4. For analytical design, the pavement sections were considered as a four layered structure and thickness of the BC layer was determined in an iterative manner. Elastic modulus of the existing bituminous surfacing obtained from some limited FWD testing [17], was taken as 500MPa. The comparison of overlay requirement by the two methods is given in Fig. 6.

It is found that results from both the methods are in agreement for majority of the cases. The discrepancy in the overlay thickness of the two design methods in a number of cases is due to the fact that a constant modulus value of 500 MPa was assigned to the existing bituminous surfacing, though its value could be different depending upon the degree of damage.

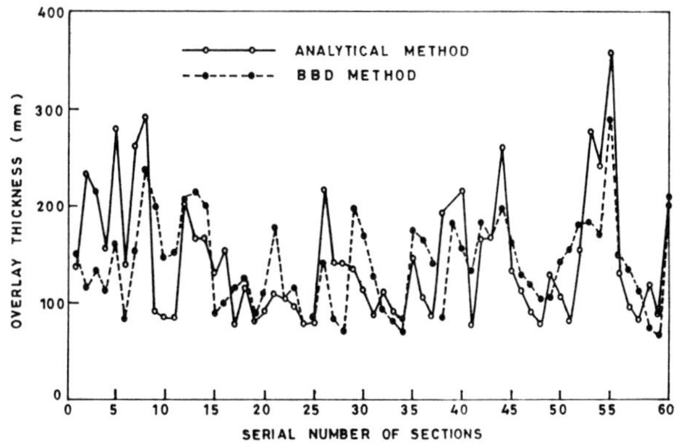

Fig. 6. Comparison of Overlay Thicknesses (BBD Vs. Analytical Method)

5 CONCLUSION

Overlay designs by Benkelman beam and analytical methods developed independently from the same set of performance data of National Highways in Eastern India are found to be in broad agreement. Analytical design method, being more rational, can be used with confidence for overlay design for a wide range of construction practices.

6 ACKNOWLEDGEMENTS

Authors are grateful to the MOST for sponsoring the research schemes R-56 and R-6 (Eastern zone) and making available the R-19 data. The co-operation rendered by the officers of PWD in Bihar, Orissa and West Bengal is greatly acknowledged. The first author acknowledges CSIR, Govt. of India, for sponsoring his research fellowship.

References

[1] REDDY, K. S., ROY, D., DAS A. and PANDEY, B. B. Bituminous overlays and pavement rehabilitation, IRC journal, N. Delhi, No. 56-1, pp. 119-142, 1995.

[2] Draft Final Report, Research Scheme R-56, Development of computer program and design charts for analytical design of flexible pavement, submitted to the MOST, by Civil Engg. Dept., IIT Kharagpur, June, (1997).

[3] IRC: 81-1981. Tentative Guidelines for Strengthening of Flexible Road Pavements using Benkelman Beam Deflection Technique, IRC, (1981).

[4] Draft Final report, Development of methods such as Benkelman beam deflection method for evaluation of structural capacity of existing flexible pavements and also for estimation and design of overlays for strengthening of any weak pavement, research scheme R-6, submitted to MOST, by Central Road Research Institute (CRRI), New Delhi, (1995).

[5] Final report. Eastern Zone. Research Scheme R-6. Phase-II, submitted to the MOST, by Civil Engg. Dept., IIT, Kharagpur, (1993).

[6] REDDY, K. S., ROY, D. and PANDEY, B. B. Axle weights and vehicle damage factors on National Highways in Eastern India, HRB Bulletin, No.50, IRC, N. Delhi, pp67-82, 1994.

[7] REDDY, K. S. and PANDEY, B. B. Lateral Placement of Commercial Vehicles on Highways, HRB Bulletin, IRC, N. Delhi, No. 53, pp63-84, 1995.

[8] KINGHAM, R. I. Development of Asphalt Institute's deflection method for designing asphalt concrete overlays for asphalt pavements, Research Report 69-3, The Asphalt Institute, Maryland, 1969.

[9] Shell Pavement Design Manual– Asphalt Pavements and Overlays for Road Traffic, Shell International Petroleum Company Limited, London, 1978.

[10] Asphalt Overlays and Pavement Rehabilitation (MS-17), The Asphalt Institute, College Park, Maryland, 2nd Edition, 1983.

[11] PANDEY, B. B. and NAIDU, P. K. Elastic modulus of materials in flexible pavement design, HRB Bulletin, No. 50, IRC, N. Delhi, pp21-41.

[12] LISTER, N. W. and POWELL, W. D. Design practice for bituminous pavements in the United Kingdom, Proc. of 6th Int. Conf. on Struct. Des. of Asph. Pavements, Vol I, pp220-231, 1987.

[13] Specification for Road and Bridge Work, 3rd revision, MOST, Govt of India, published by IRC, 1995.

[14] PANDEY, B. B., SINGH, G. G., DASH, N. K. and DAS, A. Development of computer interfaced fatigue test set–up for soil stabilised pavement materials, Indian Geotechnical Journal, N. Delhi, Vol 26(2), April, pp200-211, 1996.

[15] Final Report, research scheme R-19, Pavement performance study : study on existing pavement Sections, Vol.II, submitted to MOST, by CRRI, (1994).

[16] REDDY, K. S. and PANDEY, B. B. Analytical design of flexible pavements, IRC, N. Delhi, Vol. 54-1, pp81-113, 1993.

[17] KUMAR, T. S., DAS, A. and PANDEY, B. B. Evaluation of pavement by Falling Weight Deflectometer. "Role of Civil Engineering for Development of Infrastructure in the 21st Century", ASCE-India Section, Calcutta, Sep, 1997.

Fifth International Conference on the
BEARING CAPASITY OF
ROADS AND AIRFIELDS

STRUCTURAL CAPACITY OF THE SAME-HIMO ROAD PAVEMENT

A.S. IDABAGA
Civil Engineer, Ministry of Works/CML Tanzania

L. MUJJUNGI
Chief Engineer, Ministry of Works/CML Tanzania

C. OVERBY
Senior Engineer, NPRA/NRRL Norway

1. ABSTRACT.

The Non-destructive bearing capacity measurements (deflections and DCP measurements) and Indirect Tensile Strength methods have been utilised during the detailed investigation of the Same-Himo road in the Kilimanjaro region in Tanzania . The road was rehabilitated by the use of a bitumen stabilised base material, utilising local materials of subbase quality. The mentioned methods are among several methods used for the investigation on the defects (cracks) to some of the section of the road.

2. INTRODUCTION

The bearing capacity measurements (deflections and DCP measurements) and the indirect tensile strength may identify any of the investigated sections that have marginal or inadequate structural strength.
The determined pavement structural strength together with the laboratory test results from the cored specimens may identify the causes of the prevailing cracks.

3. LABORATORY INVESTIGATIONS

The laboratory testing of core samples from the bitumen stabilised base has been a very important part of the investigation. The laboratory analyses describe the properties and the condition of the material which is essential for the understanding of pavement behaviour.

3.1 Indirect tensile strength

The indirect tensile strength varies with specimen temperature. To be sure that the specimens are tested at predetermined temperature, the specimen are cured for four hours. Core samples were taken from all 100 m detail-investigated sections. The samples were taken from the outer wheel path and between wheel paths at profiles 20, 50 and 80 m, and from both left and right hand sides of the road.

3.1.1 Testing Procedure.

Indirect tensile strength testing involves applying a static compressive load across the diametrical axis of a cylindrical specimen. The indirect tensile test is carried out using the same loading jack as the marshall apparatus, and applying the load with a rate of 2in/min. (see ASTM D 1559 – 82). Before testing, all the 370 specimens (14 were not tested) were cured in a water bath at 29°C (design temperature) as the result of the test will vary with specimen temperature.

The indirect tensile strength S_t, is calculated using the following formula:

$$S_t = (2.P_{max}) / (\pi.t.D) . 1000 \quad (kPa)$$

Where P_{max} = maximum force (N),
t = specimen height (mm), and
D = specimen diameter (mm).

The elastic modulus is calculated using the following formula:

$$\text{E-modulus} = 6.1\, S_t + 100 \quad (Mpa)$$

where S_t = indirect tensile strength in kPa.

Note the elasticity modulus has not actually been measured, but estimated from indirect tensile strength data. Correlation studies have shown a very good correlation between actual measurements of E-modulus and calculations of E-modulus from indirect tensile strength data.

Obviously the indirect tensile strength and E-modulus will vary depending on other apparent parameters like density (voids), bitumen content, fines content and properties of the binder.

Figure 1 shows a summary of the E-modulus. The sections with cracks are marked with "C" both on the graph and on chainage labels. Each bar represents the average of twelve cores, from profiles 20, 50 and 80 m, outer and between wheel paths and upper and lower core.

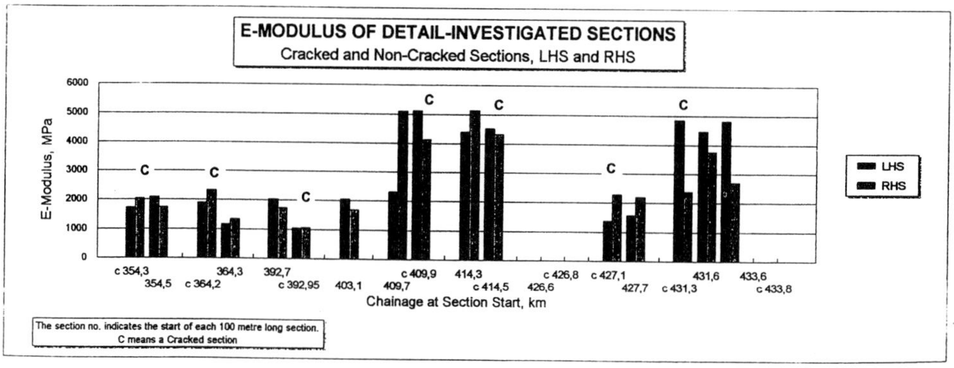

Figure 1 E-modulus for all detail-investigated sections, Cracked and Non-cracked.

The average E-modulus for all samples is 2800 MPa. Three of the "problem areas", A5 (km 409), A6 (km 414) and A9 (km 431), have very high E-modulus, some reaching a maximum of 5000 MPa. The five other "problem areas" have E-modulus between 1000 and 2000 MPa. There is no information from "as built data" that can explain these difference

4. FIELD STUDIES

4.1 Structural capacity of the pavement

The structural capacity of the pavement for the selected sections was measured with the use of non destructive methods which are Benkelman beams and the DCP. The deflection measurements included maximum deflection, radius of curvature and curvature function (CF). Each 100-metre section covered 44 deflection measurements, while the DCP measurements include 24 measurements to a depth of minimum 800 mm.

Deflection measurements were also carried out within the "problem area" sections and at 50 metre intervals for both lanes. Also, a few shorter sections were included as they became interesting during the course of investigation.

4.1.1 Deflection measurements

The Benkelman beam measurements carried out followed the rebound (recovery) concept. Maximum deflection and radius of curvature were measured during the same measuring sequence. The radius of curvature was calculated from a measuring point 200 mm from the point of maximum deflection.

The magnitude of maximum surface deflection is a function of the wheel load, the area of contact between the tyre and the road surface, the speed of the wheel, and the stress-strain characteristics and thickness of the various pavement layers and the subgrade.
The required wheel load accuracy was +/- 50 kilograms and during the test work the wheel load was always within the allowable range of 4030 - 4080 kilograms and the tyre pressure used during the investigation was 590 kN/m^2 (85 psi).

4.1.2 Temperature adjustments

The stiffness of bitumen-bound pavement materials is normally very dependant on temperature, and result in significant changes in the stiffness of the bitumen which will influence the magnitude of deflections. The pavement temperatures were recorded at the following depths: 25mm, 50mm, 75mm, 100mm, 125mm and 150mm during the deflection testing, so that appropriate adjustments could be made if required. The deflection measurements took place each hour from 7.00 am and up to 6.00 pm, measuring in both outer and inner wheel paths.

Beyond a pavement temperature of 32^{0}C the effect of temperature is insignificant, this is valid for both outer and inner wheel paths.

4.1.3 Radius of curvature and curvature function

The following equation has been used in calculations of the radius of curvature:

$$R = \frac{L^2}{2(D_0 - D_L)}$$

where,

R = radius of curvature (m)
L = distance between the point of maximum deflection and the

point where the radius of curvature was determined

D_0 = maximum deflection for test point

D_L = deflection measured at test point when the test load is 200 mm from the point at which the maximum deflection was produced (in direction of travel).

The Curvature Function (CF) of a deflection bowl is given by:

$$CF = D_0 - D_{200}$$

where,

D_0 = maximum deflection for a test point

D_{200} = the deflection measured at the test point when the test load is 200 mm from the point at which the maximum deflection was produced (in direction of travel)

4.1.4 Field data

The deflection measurements are summarised in Tables. Values are also given for maximum deflection and radius of curvature. For estimation of pavement residual life in terms of E80's 90%-ile has been calculated for outer wheel path for both maximum deflection and radius of curvature.

Figure 2 shows the graphical presentation of 90%-ile maximum deflections.

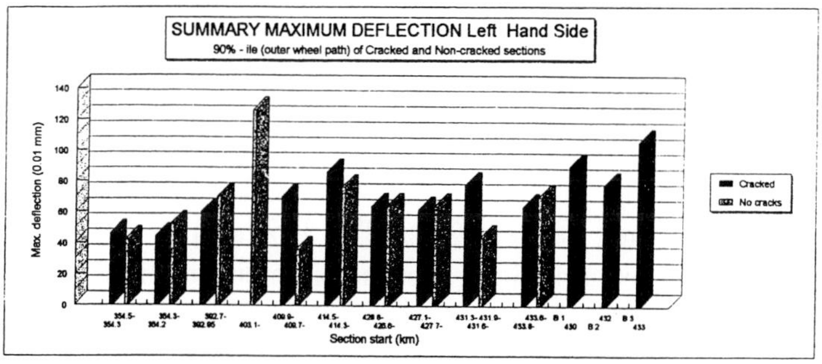

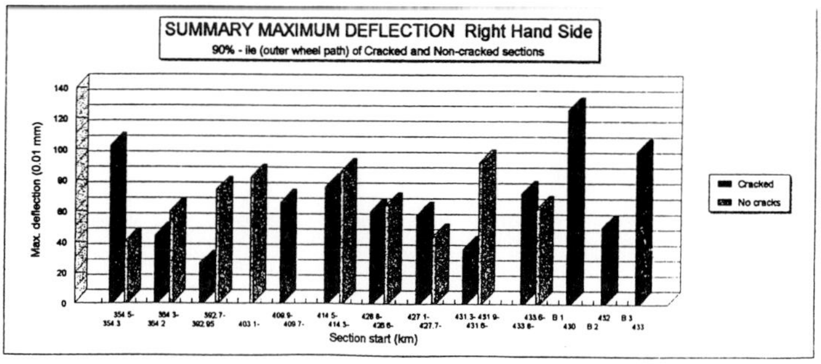

Figure 2 Graphical presentation of 90%-ile maximum deflection all detail-investigated sections.

4.2 Dynamic Cone Penetrometer (DCP)

The DCP instrument measures the penetration per blow into a pavement through all the different pavement layers. Penetration is a function of the in-situ shear strength of the materials.

DCP measurements were carried out only on the 100 metres long sections identified for detailed investigations. Each of the investigated sections was tested as follows:

* Three profiles, profile 20 m, 50 m and 80 m, respectively.

* Each profile was tested at the following locations: centre of shoulder, outer and inner, and between wheel paths, both left and right hand side respectively. In total 24 DCP measurements per section.

* The DCP penetrated all the pavement layers (inclusive the surfacing and the bitumen stabilised base course layer) to a depth of 800 mm.

In calculation of the pavement Structure and Modified Number both the DCP/CBR correlations have been used together with AASHTO strength coefficients for granular materials.

In the calculations of pavement life E80 kN, the 90%-ile values have been used.

4.2.2 Temperature corrections

The temperature reference is the average of the measured temperatures for the following depths: 25, 50, 75, 100 and 125 mm.

A major part of the DCP test work was carried out at an average pavement temperature between 25 - 35^{0}C. However in this study, the DCP (DN/mm) penetration figures for the bitumen stabilised layer have not been temperature corrected, as such would be time consuming and with insignificant impact on the conclusions arrived at.

4.2.2 Field data

Figures 3.1 and 3.2 shows the DCP strength parameters, DN_{800} and BN_{100} across the full pavement width, including the shoulders

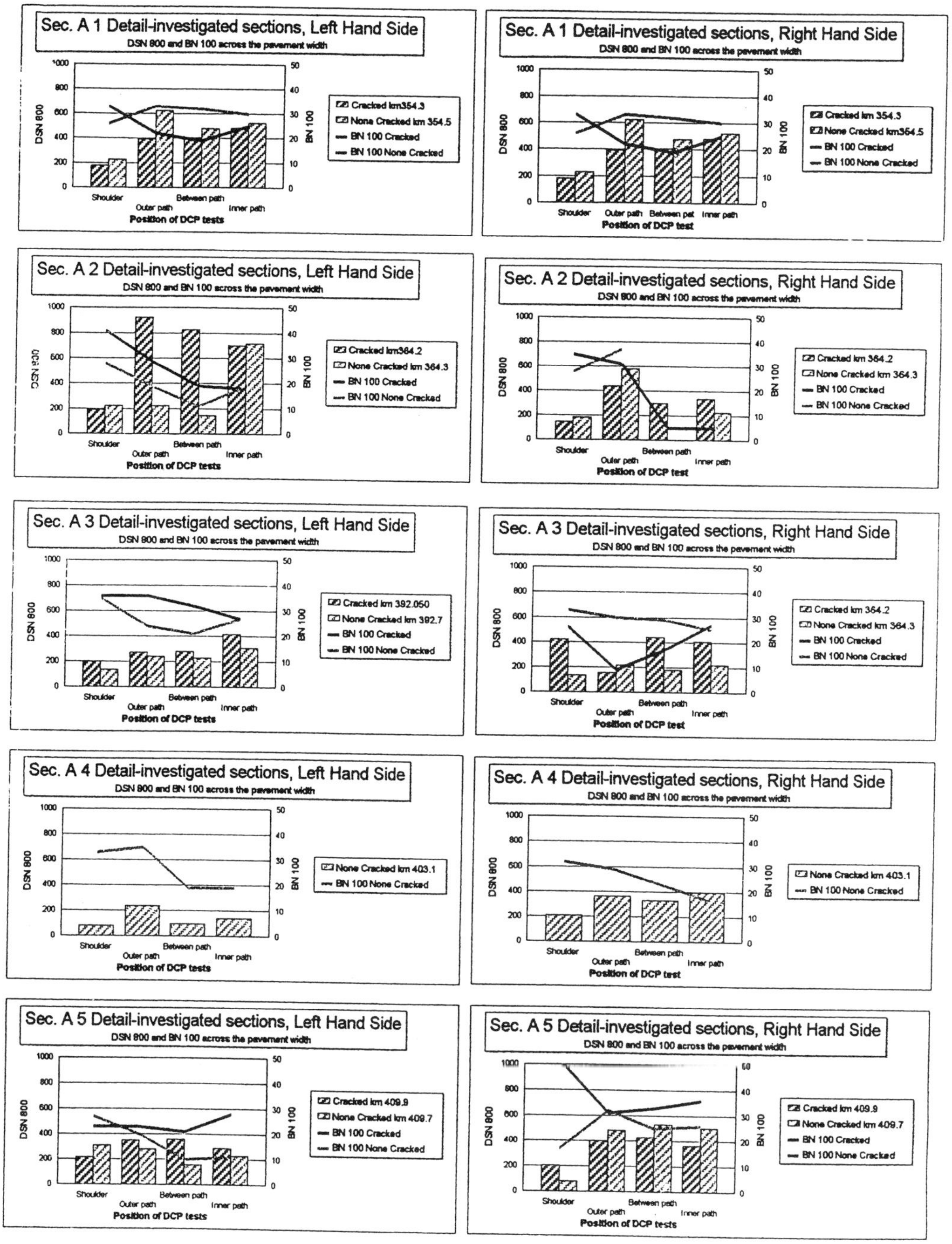

Figure 3.1 DCP strength parameters, DSN_{800} and BN_{10}

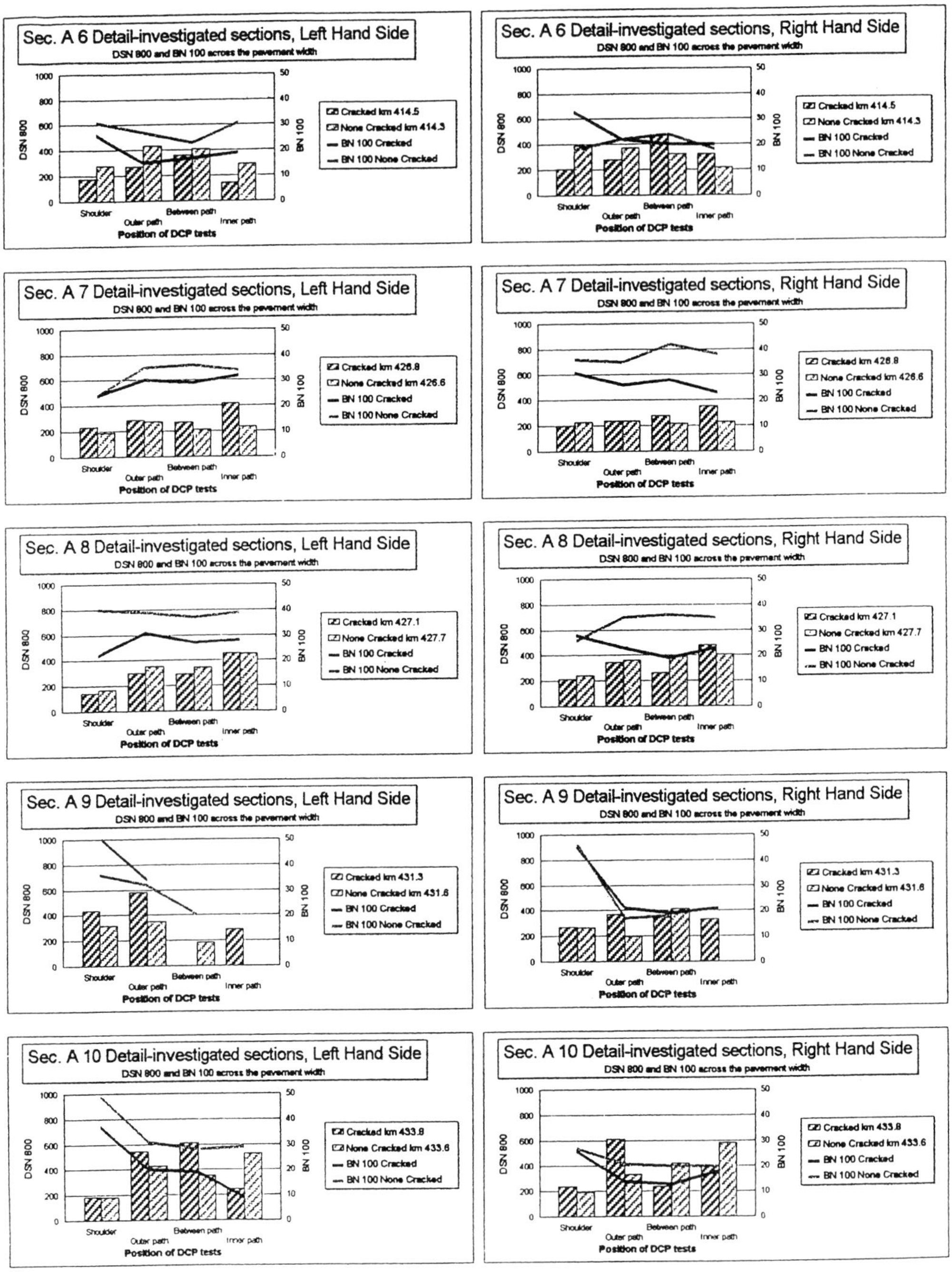

Figure 3.2DCP strength parameters, DSN_{800} and BN_{100}

5. CONCLUSIONS

1. Using maximum deflection and radius of curvature with design charts from the various road authorities bearing capacities in terms of E80 kN repetitions the general bearing capacity of the road is adequate for the design period.

2. Similar conclusions can be quoted when using the DCP_{800} approach and the use of Modified Structural Number, although both these two methods generally indicate a higher service life.

3. Generally the E-modulus derived from the Indirect Tensile Strength are high.

4. However, there seems no relationship between higher/lower Bearing capacity and cracked/non cracked sections.

ACKNOWLEDGEMENTS

The work described in this report is a part of the Institutional cooperation between the Ministry of Works, Communication and Transport, Central Materials Laboratory (MWTC/CML) and the Norwegian Public Roads Administration, Norwegian Road Research Laboratory (NPRA/NRRL) financed by the Norwegian Agency for Development Cooperation (NORAD).

REFERENCES

1. NRRL/NPRA. Site visit report 30th-3rd December 1994, Same-Himo road, Tanzania.

2. Ministry of Works, Communication and Transport/Norconsul International. Draft Final report Same-Himo report, Dar es salaam. January 1993.

3. H.R. Smith and R. Jones. Measurement of Pavement deflection in tropical and sub-tropical climates. TRRL report 935, 1980.

4. Department of Transport and African Consulting Engineers INC. Project report PR 91/241. Pavement rehabilitation Design Based on Pavement Layer Component tests (CBR and DCP). Pretoria, South Africa 1994.

Fifth International Conference on the
BEARING CAPACITY OF
ROADS AND AIRFIELDS

FLEXIBLE PAVEMENT EVALUTION AND STRENGTHETENING IN LITHUANIA

VIRGAUDAS PUODŽIUKAS
Director, Transport and Road Research Institute,
I. Kanto 25, P.O. Box 2082, LT-3000 Kaunas, Lithuania,
Fax: 3707 20 56 19, E-mail: virpuo @tkti.lt

ABSTRACT

The length of flexible pavements in the Lithuanian national highway network is 10 644km. Asphalt concrete is the most widely used pavement type.

Pavement condition data had been used in the past to develop maintenance, rehabilitation and reconstruction programs generally on a project - by - project basis. At present three specific applications for pavement condition data may be identified:

I. To establish priorities. Condition data such as profile, distress and deflection are used to establish the project in need of maintenance and rehabilitation.
II. To establish maintenance and rehabilitation strategies. Data from visual distress surveys are used to develop an action plan which strategy (repair, surface treatment, overlays, recycling, etc.) is the most appropriate for a given pavement condition.
III. To predict pavement performance. Data such as profile, distress, structural evaluation are projected in to the future to estimate the condition of pavements on a network.

Overlays are generaly being used to increase the pavement structural capacity and to improve pavement surface irregularities. In 1996, Transport and Road Research Institute (TRRI) prepared the project: "Pavement strengthening program for the Via Baltica road sections in Lithuania". The Asphalt Institute - Limiting Pavement Surface Deflections method was used to identify continuos pavement sections of uniform performance, to obtain "static" pavement surface deflections and to determine the expected traffic by use of equivalent axle loads (80kN). The 20 defelection measurements have been taken each kilometer.

Evaluation data and the priority of the Via Baltica road sections strengthening in the Lithuania is presented in the report.
Key words: pavement, deflections, strengthening, overlay.

1. INTRODUCTION

1.1 Lithuanian road network

Lithuanian road network consists of 79800 km of roads. Distribution of Lithuanian road network is presented in table 1.

Lithuanian road network **Table 1**

Road type	Cement concrete	Asphalt	Sett paving	Paved total	Gravel	Gravel and soil	Total
State roads							
Main	12.1	1442.5	0.2	1454.8	0.0		1454.8
National	72.6	3259.1	4.8	3336.5	88.5		3425.0
Regional	5.4	5942.2	18.2	5965.8	10275.7		16241.5
Total	90.1	10643.8	23.2	10757.1	10364.2		21121.3
Local roads							
Total		3454				35707	39161
Urban roads							
Total		4547	63	4610	243		4853
Forests' and National parks' roads							
Total		93		93		14566	14659
Total road network							
Total	90	18737	86	18913	10607	50273	79794

Lithuanian Road Administration (LRA) is responsible for 21121 km of state roads.

Main and Regional roads are the most important for the country - 40% of traffic volume and 60% of axial loads are concentrated on these roads.

LRA is facing the more and more complicated problems. It is necessary to make decisions and to spend the insufficient funds in the most effective way.

In 1993 LRA started to create Lithuanian pavement management system (PMS) and to plan pavement rehabilitation based on network prioritization. In the first stage PMS is used mostly for Main and National roads.

1.2 Data collection

Pavement rehabilitation planning requires various data about road and traffic. The data collection system is developed for this purpose.

The most important road condition parameters are roughness and pavement strength. Up to the year 1996 Russian laboratory KP 514 and Benkelman Beam were used for data registration. In autumn 1996 LRA purchased laser profilometer for roughness and rutting measurement and falling weight deflectometer for registration of pavement strength. Roughness measurements cover 2500-3000 km annually, several hundred kilometers are to be examined for the strength. Automatic measurements are carried out by Transport and Road Research Institute (TRRI) in Kaunas.

Road defects represent the other kind of very important parameters. These measurements are performed by local agencies (regions). Road defects are examined visually. 3 regions from 8 are using portable computers Husky Hunter 16 for collection of road defects while other regions are still using manual data registration. Alligator, longitudinal and transverse cracks, potholes and patches are registered.

Traffic is counted in 105 manual posts, 49 of them are permanent. In 1997 11 stationary automatic traffic counters as well as 4 portable counters were installed. Totally 60 counting posts will be organized. Manual counting is performed by local agencies while TRRI is responsible for automatic counting data.

Also data about road alignment, cross section, year and type of repair, layer thickness and material are collected.

All data collected are entered into the LRA information system which consists of several data bases - road data bank, traffic data base, visual inspection's data base, etc. which runs parallel in LRA and TRRI.

1.3 Pavement strengthening program for the Via Baltica road sections in Lithuania

The aim of this work was to determine pavement strength of the selected sections of the main Lithuanian roads joining the Via Baltica road (i.e. A5 Kaunas - Marijampolė - Suvalkai, A8 Panevėžys - Kėdainiai - Cinkiškis and A10 Panevėžys - Pasvalys - Ryga), to give recommendations for strengthening of the "weak" sections and to make the priority list of the sections selected for strengthening.

This Report also defines the problems related to testing of pavement strength in recent years.

2. DESCRIPTION OF THE METHOD USED

To determine the need for the asphalt concrete overlay the Asphalt Institute Deflection Method was used which allows for testing elastic deformations (deflections). The Method is based on the dependency between pavement surface deflection and the number of traffic load repetitions. Limiting deflection values were calculated at which an intensive pavement failure can be observed. Pavement service life could be extended by inceasing its strength and placing asphalt concrete overlays.

The constituent parts of this Method are as follows:

a) deflection measurements,
b) calculation of a representative rebound deflection (RRD),
c) traffic volume forecast,
d) calculation of overlay thickness.

2.1 Deflection Measurements

Based on the Deflection Method the Benkelman Beam - a device for measuring static deflections - should be used. When using another type of measuring devices the correlation of readings is necessary.

2.2 Calculation of Representative Rebound Deflection (RRD)

Representative rebound deflection (RRD) is calculated by the following formula:

$$\mathbf{RRD = (x+2s)c,} \qquad (1)$$

where $$\mathbf{x = (f\Sigma x) / n,} \qquad (2)$$

Σ - sum of deflection values measured in separate points,
n - number of measurements,
f - temperature adjustment factor depending on: a) pavement surface temperature during measurement, b) mean 5-day air temperature before the day of deflection measurement, c) thickness of pavement courses with bitumen binders, d) thickness of unbound base of the road.
s - standard deviation, calculated as:

$$\mathbf{s = (f^2\, \Sigma x^2 - xf\Sigma x) / (n - 1),} \qquad (3)$$

where Σx^2 - sum of square deflection values measured in separate points,
c - critical period adjustment factor, determined from the ratio of deflection values measured during a critical period and another period.
c = 1 if deflection tests are made during a critical period, i.e. during the spring thaw when pavement strength is the lowest due to access of moisture.

2.3 Traffic volume forecast

Traffic volume forecast is established for a given pavement service life taking into account the growth in the number of different type of vehicles and the sum of vehicle impact upon the road pavement. Traffic volume is calculated for one (design) traffic lane and it is expressed as a sum of 80kN equivalent standard axle load (ESA).

2.4 Calculation of asphalt concrete overlay thickness

Thickness of asphalt concrete overlay is calculated by the graph. Overlay thickness depends on the estimated representative rebound deflection (RRD) value and equivalent standard axle load (ESA). This load is a sum of loads generated by the vehicles crossing the cross-section of the road and expressed as a number of 80kN equivalent axle loads.

3. DEFLECTION TESTING RESULTS

3.1 Deflection measurements

To measure pavement surface deflections the device Dina - 3M was used. In order to define the correlation ratio between the Dina - 3M and the Benkelman Beam readings deflection measurements were taken in 50 deflection test points using both devices. The results showed that the ratio between the readings is 1:1. Deflection measurements were taken at the outer wheelpath in each traffic lane every 50 m in a chess-board order. 20 measurements were taken in each road kilometer. 3 measurements were taken on each test point of the road. When measuring deflections the date, measuring time, air temperature and pavement surface temperature are recorded.

Deflection measurements on the road A5 Kaunas -Marijampolė - Suvalkai were taken from March to April 1995. The minimum pavement surface temperature was 0^{0}C, the maximum - 10^{0}C. Deflection measurements on the road A8 Panevėžys - Kėdainiai - Cinkiškis were taken in April 1995. The minimum pavement surface temperature was 3^{0}C, the maximum - 19^{0}C. Deflection measurements on the road A10 Panevėžys - Pasvalys - Ryga were taken at the beggining of April 1995 when the minimum pavement surface temperature was 3^{0}C, the maximum - 10^{0}C.

To identify a critical pavement condition (coefficient c) pavement surface deflections were measured in the check points, also. 4 check points were selected on the road A5 Kaunas - Marijampolė - Suvalkai (km 18,0; 38,5; 52,7 and 71,95), 4 check points as well were selected on the road A8 Panevėžys - Kėdainiai - Cinkiškis (km 10,0; 39,77; 53,05; and 79,4) and 3 check points - on the road A10 Panevėžys - Pasvalys - Ryga - 3 (km 10,0; 45,0; 54,0).

In spring of 1996 deflection measurements in the check points were repeated. In spring of 1995 deflection measurement in the check points were taken from March 2 to April 19. In spring 1996 - from April 3 to May 2. In spring of 1996 pavement deflections were measured only in the check points. The reason is a limited resource of the Dina - 3M sensor.

3.2 Calculation of representative rebound deflection (RRD)

The calculation of RRD was influenced mainly by a critical period adjustment factor "c" which was estimated by analysing the results of two-year deflection measurements in the check points. Overall, the c-values vary in a rather wide range. If we compare not only different routes but also separate sections of the same road the range of "c" variation on the same road is considerably lower.

For example, on road A5 Kaunas - Marijampolė - Suvalkai the maximum c-value is 2,29, the minimum - 1,32, the difference is 1,7 times. On road A8 Panevėžys - Kėdainiai - Cinkiškis the maximum c-value equals to 1.35, the minimum - to 1.10, they differ 1.2 times; while on the road A10 Panevėžys - Pasvalys - Ryga the maximum c-value equals to 1.12, the minimum - to 1.07 and the difference is 1.05 times only. On the whole Via Baltica road the critical period adjusment value "c" differs 2.14 times. Based on the deflection measurements the mean deflection values are as follows: on the road Kaunas - Marijampolė - Suvalkai - c = 1.65, on the road Panevėžys - Kėdainiai - Cinkiškis - c = 1.24 and on the road Panevėžys - Pasvalys - Ryga - c = 1.09. The mean c-value for the whole Via Baltica road is 1.34.

Using different values of the critical period adjustment factor "c", three variants of RRD were calculated:

I. RRD was calculated using the c-values from the nearest check points.

II. RRD was calculated using the maximum c-value (c = 2.29) from the check point located on the road Kaunas - Marijampolė - Suvalkai, km 18.

III. RRD was calculated using the mean c-values.

RRD calculation results are given in Fig. P.2.2.1...P.2.2.3.

3.3 Traffic volume forecast

Traffic volume of a design traffic lane is expressed as a sum of 80kN equivalent standard axle load (ESA). Traffic volume was predicted for the year 2000, 2005, 2010 and 2015 based on an average annual daily traffic (AADT) in 1995.

When calculating ESA the impact of each type of vehicles on the road pavement and dynamics in the growth of vehicles was taken into consideration.

3.4 Determination of pavement strength

Pavement strength is expressed as a modified structural number (SNC) which is calculated by the formula:

$$\mathbf{SNC = 3{,}2\ DEF^{-0{,}63}}, \qquad (4)$$

Fig. P.2.2.1: Road A5 Kaunas - Marijampolė - Suvalkai

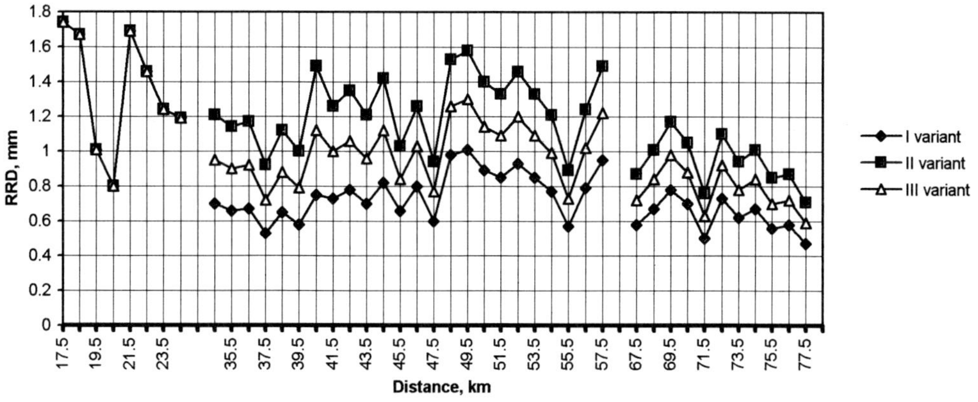

Fig. P.2.2.2: Road A8 Panevėžys - Kėdainiai - Cinkiskis

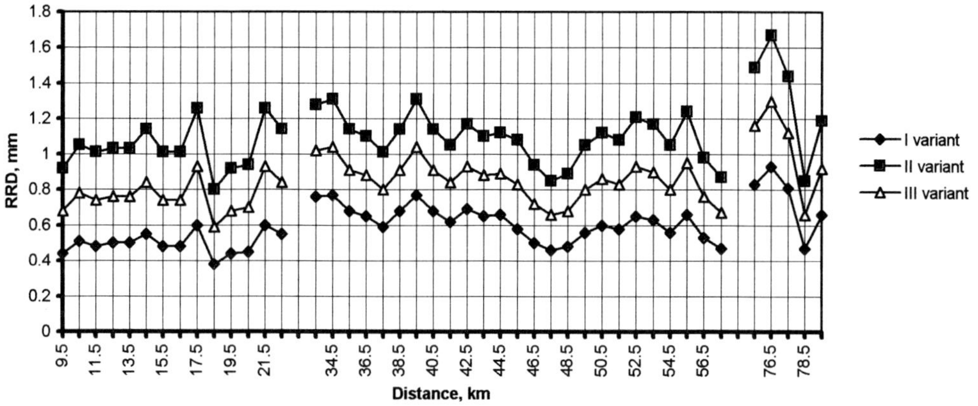

Fig. P.2.2.3: Road A10 Panevėžys - Pasvalys - Ryga

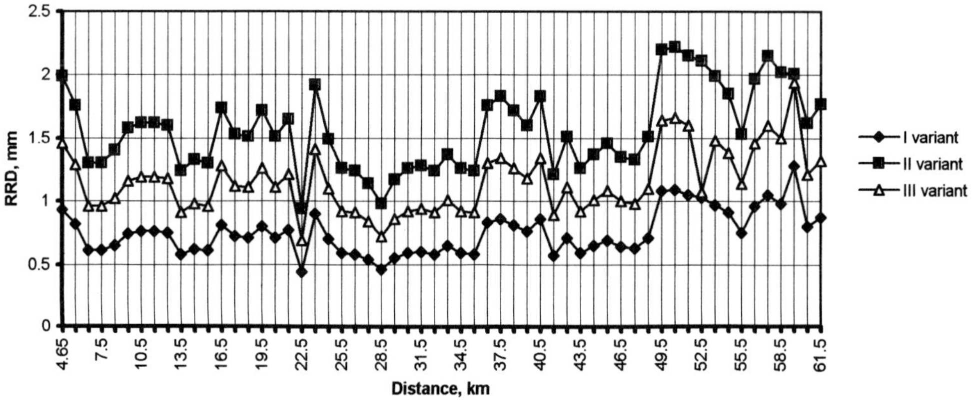

where DEF - representative rebound deflection (RRD) value, calculated according to the above mentioned Method.

Pavement strength expressed as SNC is calculated for three variants depending on the values of RRD.

3.5 Determination of the need for pavement strengthening

The need for asphalt concrete overlay was determined for 1995, 2000, 2005, 2010 and 2015 based on three RRD variants and expressed as thickness of the strengthening layer h (mm).

3.6 Analysis of deflection testing results

The following conclusions could be made after analysis of the three variants characterizing the need for the overlay:

I variant - based on one - year (1995) deflection measurements on the road and two - year (1995 and 1996) deflection measurements in the check points - shows that the pavement strength is too high. The reason for this result is a too short deflection measuring period within which the roadbed and the pavement structure had not reached the most critical condition yet. More reliable results could be obtained when the measuring period is longer (5 years and more).

II variant - when for the calculation of RRD a maximum c-value (c = 2,29) was used - shows that the strength of the existing pavement is too low. According to this variant the majority of the Via Baltica road sections should be strengthened in the nearest future.

III variant. When calculating RRD the mean values of "c" were used by taking the average between the maximum c-value determined during deflection measurements and values in an individual check point. The need for overlay calculated according to this variant shows a real condition of the existing pavement strength. This is confirmed by the data of visual inspection as well.

Based on this variant the road pavement was divided into sections according to the similar strength and the term of strengthening. The rezults are given in Table 2.6.1.

An asphalt concrete mixture of type 0-11 S-V will be used for the overlays.

At the request of Lithuanian Road Administration pavement strengthening is planned to be carried out by a minimal overlay 4 cm thick, except the cases when pavement have to be strengthened by a thicker overlay and without any delay. Already in 1996 the two sections (km 17,0....19,0 and km 21,0...22,0) on the road A5 had to be strengthened by an overlay 5 cm thick.

Table 2.6.1: The need for pavement strengthening on road sections of the same strength

Road A5 Kaunas - Marijampolė - Suvalkai

Section, km		Length	Asphalt overlay thickness, cm									
From...	...to	km	1996	1997	1998	1999	2000	2001	2002	2003	2004	2005
17,0	19,0	2,0	5									
19,0	21,0	2,0				4						
21,0	22,0	1,0	5									
22,0	25,0	3,0		4								
34,0	37,0	3,0					4					
37,0	40,0	3,0								4		
40,0	42,0	2,0				4						
42,0	44,0	2,0					4					
44,0	48,0	4,0							4			
48,0	51,0	3,0			4							
51,0	54,0	3,0				4						
54,0	56,0	2,0									4	
56,0	58,0	2,0				4						
67,0	69,0	2,0									4	
69,0	73,0	4,0						4				
73,0	75,0	2,0										4
75,0	78,0	3,0	Functional layers									

Road A8 Panevėžys - Kėdainiai - Cinkiškis

Section, km		Length	Asphalt overlay thickness, cm									
From...	...to	km	1996	1997	1998	1999	2000	2001	2002	2003	2004	2005
9,00	17,0	8,0	Functional layers									
17,0	18,0	1,0							4			
18,0	20,0	2,0	Functional layers									
20,0	23,0	3,0							4			
33,0	35,0	2,0					4					
35,0	39,0	4,0								4		
39,0	41,0	2,0				4						
41,0	46,0	5,0								4		
46,0	50,0	4,0										4
50,0	56,0	6,0							4			
56,0	58,0	2,0	Functional layers									
75,0	78,0	3,0			4							
78,0	80,0	2,0								4		

Road A10 Panevėžys - Pasvalys - Ryga

Section, km		Length	Asphalt overlay thickness, cm									
From...	...to	km	1996	1997	1998	1999	2000	2001	2002	2003	2004	2005
4,3	6,0	1,7				4						
6,0	9,0	3,0							4			
9,0	13,0	4,0			4							
13,0	16,0	3,0							4			
16,0	22,0	6,0				4						
22,0	23,0	1,0										4
23,0	25,0	2,0				4						
25,0	33,0	8,0									4	
33,0	36,0	3,0						4				
36,0	38,0	2,0			4							
38,0	41,0	3,0				4						
41,0	42,0	1,0									4	
42,0	44,0	2,0						4				
44,0	49,0	5,0					4					
49,0	52,0	3,0	4									
52,0	59,0	7,0			4							

4. CONCLUSIONS AND RECOMMENDATIONS

The Asphalt Institute Method for the determination of the need for pavement strengthening was used 15 - 20 years ago, at present it is not new already. The Method was used to determine the need for the Via Baltica road pavement strengthening because it allowed us to use Dina - 3M for deflection measurements. The characteristic feature of this Method is that the most reliable data are obtained when measuring deflections during the critical condition (the spring thaw) when pavement structure is in the most poor condition.

In order to make a reliable decision on the need for pavement strengthening one should have the deflection measurements data of a number of years (5 years or more). The results of deflection measurements of one or two years are not sufficient.

Since 1997 the Dynatest 8002 FWD Test System is being used in Lithuania. Using the program Elmod based on the deflections measured the elasticity moduli of the four upper pavement courses are calculated and pavement strength is determined which is expressed as a modified structural number SNC. In order to obtain reliable results a five-year deflection testing program was started in 1997 the aim of which is to identify the curves of critical period adjustment factor under Lithuanian conditions.

The use of new technologies would allow us to solve the tasks of pavement strengthening optimization.

5. REFERENCES

1. Lietuvos Respublikos susisiekimo ministerija. Valstybinis Transporto ir kelių tyrimo institutas. Dangų valdymo ir optimizavimo uždavinių formulavimas ir sprendimas. 1994m.

2. Lietuvos Respublikos susisiekimo ministerija. Valstybinis Transporto ir kelių tyrimo institutas. Automobilių, riedančių Lietuvos keliais, svėrimas ir apkrovų analizė. 1994m.

3. Road Deterioration and Maintenance Effects. Models for Planning and Management. William D. O. Paterson. The John Hopkins University Press. 1987.

4. The Highway Design and Maintenance Standards Model. Volume 1: Description of the HDM III Model. Thawat Watanatada and others. The John Hopkins University Press. 1987.

5. Asphalt Institute. Asphalt Overlays for Highway and Street Rehabilitation, Manual Serries No. 17 (MS - 17). 1983.

Fifth International Conference on the
BEARING CAPASITY OF
ROADS AND AIRFIELDS

THE LIVINGSTON COUNTY, IL AGGREGATE BASE STUDY

M.R. THOMPSON[1] AND KEVIN D. HALL[2]

[1]Professor Emeritus, University of Illinois @ Urbana-Champaign
205 N. Mathews, Urbana, IL, 61801, (217) 333-1924, mrthomps@uiuc.edu
[2]Assistant Professor, University of Arkansas
4190 Bell Engr., Fayetteville, AR 72701, (501) 575-7168, kdh3@engr.uark.edu

ABSTRACT

In Illinois, locally available materials frequently have fines (passing 0.075 mm) contents in the upper ranges of gradation specification limits (e.g. 12 to 13% passing 0.075 mm), and occasionally exceed the upper limits. In addition, the fines may be near to or out of the PI specification. High fines/high PI granular materials may weaken after experiencing a moisture increase and/or becoming saturated, leading to premature pavement distress. To address these problems a "low fines" gradation for granular materials and "fines modification" techniques (lime and cement treatment) were proposed. Nine granular base pavement test sections (75 mm asphalt concrete / 300 mm granular base) were constructed (1986/87) in Livingston County, Illinois to investigate field performance of the low-fines and modified-fines granular materials.

Comprehensive laboratory and field tests were conducted on all paving materials prior to construction. Field performance was characterized using surface deflection (FWD) data, rut depth measurements, and periodic visual pavement distress surveys. Through 1995, the sections did not exhibit significant distress. Rut depths ranged from 2 mm to about 6 mm. Since 1995, additional rutting and fatigue cracking have developed. Current rut depths range from 3 mm to 12 mm; several sections exhibit extensive fatigue and some limited surface repairs have been completed. Total traffic on the sections to date is estimated at about 200,000 ESAL. Overall, the granular base materials have performed satisfactorily, indicating the low fines gradation and fines modification techniques represent viable options for improving the performance potential for high fines / high PI granular materials.

Key Words: Flexible pavement, granular base, performance, materials, stabilization

INTRODUCTION

The Illinois Department of Transportation (IDOT) predominantly uses two dense-graded aggregate gradations for granular base/subbase materials; CA-06 (100 % passing 37.5 mm; 43±13% passing 4.75 mm; 4 to 12 percent passing the 0.075 mm sieve) and CA-10 (100% passing 25 mm; 50±10% passing 4.75 mm; 5 to 13 percent passing the 0.075 mm sieve). Frequently, available materials have fines contents in the upper part of the specification ranges and occasionally exceed the upper limits. In addition, aggregates may be near to or out of specification on plasticity index (PI); IDOT requirements for PI are: < 9 for gravel sub-base; < 6 for gravel base; and < 4 for crushed gravel and crushed stone base. *(1)*

When high fines granular subbase/base courses experience a moisture increase, and/or become saturated, they tend to become weak and unstable. Triaxial testing conducted at the U of I on the AASHO Road Test granular base (containing only 6.9 percent passing the 0.075 mm sieve) showed a sixty percent strength loss when the moisture content was increased from optimum (4 to 5 percent) to higher than optimum (6.5 to 7 percent).

To address this problem, IDOT proposed a low fines gradation for granular subbase/base. The gradation (CM-06) is similar to CA-06 but permits only up to four percent passing the 0.075 mm sieve. In addition, "fines modification" techniques using lime and Portland cement were proposed. The "modified fines" may be less plastic, display reduced moisture sensitivity, and show better shear strength and rutting resistance. Preliminary laboratory testing had indicated the modification techniques could produce an acceptable base course. It was decided that the lime and cement "modification techniques" were viable options.

The low fines gradation and the modification procedures represent major departures from current IDOT subbase/base course specifications. Thus, IDOT decided to evaluate the performance of the CM-06 aggregate gradation, lime- and cement-modified aggregates, and CA-06 "control materials" in an experimental aggregate base course project, the Livingston County Test Project (LCTP). A comprehensive report detailing the performance and results from the project is available elsewhere. *(2)*

PROJECT DESCRIPTION

The LCTP is located in Livingston County, approximately four miles south of Pontiac, Illinois on Ocoya Road (FAS Route 1362). Nine test sections, each approximately 1200 feet in length, were constructed in the two-mile project. All test sections consist of 3.0 inches (75 mm) of asphalt concrete (AC) surface over a 12.0 inch (300 mm) base course. Stationing (and Section 1) starts at the western end of the project and continues in an easterly direction. A listing of the granular base materials used in the nine test sections follows.

- *Section 1* Control crushed stone, CA-06 gradation, PI = 4
- *Section 2* Low fines crushed stone, CM-06 gradation, nonplastic
- *Section 3* lime-modified (3 percent hydrated lime - dry weight of aggregate basis) crushed stone, CA-06 gradation, 9.6% passing 0.075 mm, PI = 9 (original) reduced to less than 4 with addition of lime
- *Section 4* Cement-modified (1.5 percent cement; dry weight of aggregate basis) crushed stone, CA-06 gradation, 9.1% passing 0.075 mm
- *Section 5* Control crushed gravel, CA06 gradation, nonplastic
- *Section 6* Low fines crushed gravel, CM-06 gradation, nonplastic

- *Section 7* Control natural gravel, CA-06 gradation, PI = 4
- *Section 8* Partially crushed gravel, CA-06 gradation, PI = 3
- *Section 9* Lime-modified (3 percent lime; dry weight of aggregate basis) natural gravel, CA-06 gradation; 9.5% passing 0.075 mm, PI = 8 to 10 (original) reduced to a range from 2 to 6 with addition of lime

Subgrade Soils

The LCTP includes six major pedologic soil series (all fine-grained) ranging from well to somewhat poorly drained. Subgrade resilient modulus (E_{Ri}) values for the two-mile test section were estimated using March 1986 deflection data collected by IDOT. For surface treatment pavements (the original pavement type prior to construction of the test sections), E_{Ri} can be estimated from the following equation:

$$E_{Ri} = 24.2 - 5.71\, D_3 + 0.35\, D_3^2$$

where:

E_{Ri} - Subgrade resilient modulus, ksi
D_3 - FWD surface deflection (mils) @ 36-inch offset distance, 9-kip FWD loading.

The average E_{Ri} from 44 test points was 4.4 ksi (30 MPa) and the standard deviation was 3.0 ksi (21 MPa). Periodic FWD tests conducted on the sections between 1987 and 1993 indicated "critical" springtime E_{Ri} values generally ranged from 2 to 3 ksi (14 to 21 MPa).

After existing surfacing and base course materials were stripped from the test section roadway and just prior to construction of experimental pavement sections, Dynamic Cone Penetrometer (DCP) tests were conducted. The DCP results were converted to estimated values of California Bearing Ratio (CBR) using the following relation:

$$\text{Log CBR} = 2 - 1.3\, [\text{Log}\, (PR*25.4) - 0.6]$$

where:

CBR - California Bearing Ratio
PR - Penetration rate, inches/blow

CBR values were generally high near the surface and gradually dropped off with depth. Below depths of about five to ten inches (13 to 25 cm), however, the CBR values ranged from 5 to 10. The most striking feature of the *in-situ* CBR values was the variability of results, both within a particular soil series and within a particular Test Section. The very high coefficient of variation for the CBR values indicate a high degree of *in-situ* variability. "Hard and fast" characterizations of the subgrade soils at the site based on CBR are difficult at best. It should be noted that (with regard to "subgrade stability" requirements; CBR > 6) all of the subgrade soils with one exception show "acceptable" CBR values through a depth of 24 to 36 inches (61 to 91 cm).

Traffic

The total two-way average daily traffic (ADT) at the project site was originally estimated as 800 vehicles per day (vpd). Due to industrial activity in the area, the traffic split between lanes was assumed to be 75 percent of the total traffic (600 vpd) travelling westbound, with the remaining 25 percent (200 vpd) assigned to the eastbound lane. Truck traffic was estimated as 25 percent of the total traffic.

Traffic count data from 1988 indicated 900 vpd on the west mile (Section 1 to mid-point of Section 5) and 800-850 vpd (estimated) for the east-mile (mid-point of Section 5 through Section 9). At the time of the LCTP construction, 80-kip (355 kN) trucks were not legal on the LCTP road. However, such trucks were permitted on the west-mile in September of 1991 and on the east-mile in the latter part of 1992. Traffic counts in 1997 indicated total traffic on the west mile to be 1100 vpd, with 4.3 percent single-unit trucks and 18.2 percent multiple-unit trucks. Estimated 80-kN ESAL to date equals about 200,000.

Construction

The 12-inch (305 mm) base course sections were constructed in three 4-inch (102 mm) compacted layers. Each layer (except for the initial 400 feet [122 m] of Sections 1 and 2) was to be compacted to not less than 100 percent of AASHTO T99 ("Standard Proctor") maximum density. The first 400 feet (122 m) of Sections 1 and 2 (Stations 0+19 to 4+19 and 12+30 to 16+30 respectively) were compacted to approximate maximum achievable density as determined by growth curves monitored with nuclear density devices. IDOT field construction records indicate the various Sections were indeed compacted to at least 100% of AASHTO T-99 density. It should be noted that the original CM-06 gravel base could not be compacted in the field; the material was changed to a partially-crushed gravel (CM-06 gradation) to facilitate construction operations.

Sections 2, 3, 4, 9, and the east half of Section 1 were surfaced with 1.75 inches (44 mm) of Class I AC binder before winter weather forced work to shut down in the Fall of 1986. Sections 5, 6, 7, 8, and the west half of Section 1 were left exposed over the winter. The exposed sections were reworked in the Spring of 1987 and surfaced with 1.75 inches (44 mm) of Class I binder. After all sections had been paved with the binder course, 1.25 inches (32 mm) of Class I AC surface course was added to all of the sections.

MATERIALS EVALUATION

Asphalt Concrete

The 3-inch (76 mm) AC surface (AC-10 asphalt cement) was constructed as a 1.75-inch (44 mm) AC binder (4.7% asphalt cement) and a 1.25-inch (32 mm) AC surface course (5.1% asphalt cement). Construction testing records indicated the mixture design asphalt contents and percent passing 0.075 mm sieve requirements were met. IDOT field testing data indicated the average compacted densities of the AC mixtures were 146.3 pcf (23.97 kN/m^3) for the surface and 146.1 pcf (22.94 kN/m^3) for the binder.

Bulk samples of the surface and binder mixtures were taken at the plant site and transported to the University of Illinois Laboratory at Urbana. Resilient modulus and split tensile strength test were conducted. The materials were reheated and Marshall-sized samples compacted. A range of compaction efforts (30 - 50 blows/layer) was utilized to establish the effect of density. Based on the lab testing results (and the average field compacted density), the split tensile strength and resilient moduli (@ 77F [25C]) were estimated as follows:

	Split Tensile Strength		Resilient Modulus	
	psi	(kPa)	ksi	(MPa)
AC Surface	128	(883)	528	(3643)
AC Binder	138	(952)	566	(3905)

Granular Base Materials

Thompson and Smith *(3)* conducted an extensive triaxial testing program with the LCTP granular materials. Repeated-load tests were performed to determine the rapid shear strength, resilient modulus, and the permanent strain accumulation characteristics of the materials. Crushed stone, crushed gravel, and gravel meeting the CA-06 and CM-06 gradations were included in the testing program. Details of the testing program are given elsewhere (3). A brief summary of the report's findings follows.

(1) Friction angles show little variability and ranged from 43 to 46 degrees for near maximum dry density and optimum moisture content conditions (AASHTO T-99).

(2) Cohesion values for the above indicated compaction conditions ranged from 10 to 24 psi (66 to 168 kPa).

(3) Moisture sensitivity, as measured by the RSI (Retained Strength Index), were relatively high and varied from 0.63 to 0.94. It is important to note that the highest fines content was only 10.5 percent.

(4) The resilient modulus parameters "K" and "n" (for the $K\Theta^n$ model) fell within expected ranges. Limited differences were noted among the various granular materials. For a representative bulk stress of 20 psi (138 kPa), resilient moduli (with the exception of Material #8A [partially crushed gravel]) were in the narrow range of 29 to 35 ksi (197 to 244 kPa). It was determined that resilient modulus is not a good property for ranking granular base performance potential.

(5) The rapid shear strength of "conditioned" specimens (Type C testing) was always greater than the "unconditioned" strength. Strength increases ranging from 34% to 217% were produced by conditioning. This is a very significant finding! It is apparent that the shear strength of an "unconditioned" specimen does not represent the strength of an in-service compacted and trafficked granular base material.

(6) Permanent deformation under repeated loading varied considerably. In fact, four of the materials (all gravel products) did not survive the standard conditioning loading (45 psi [311 kPa] deviator stress / 15 psi [104 kPa] confining pressure). A reduced stress state (30 psi [207 kPa]deviator / 15 psi [104 kPa] confining pressure) was used to characterize the permanent deformation behavior of those four materials. It is apparent that permanent deformation under repeated loading provides a more definitive evaluation of granular base materials.

Modified Granular Base Materials

Lime-modified bases were used in Section 3 (crushed stone) and Section 9 (gravel). Moisture-Density-CBR tests (uncured lime-aggregate mixture) were conducted with the crushed stone. Cured compressive strength tests were conducted with the crushed stone and the gravel. The Proctor-sized compression specimens were subjected to an accelerated cure (48-hours @ 120 F [49 C]). Some results of the testing follow.

- The CBR of the lime-modified crushed stone is very sensitive to moisture content. At optimum moisture content (7.6%), the CBR was 38; at a moisture content of 8.3% (less than 1% wet of optimum) the CBR was only 4.
- The laboratory cured (48 hr @ 49C) compressive strengths of the lime-modified granular bases were not high. A compressive strength of 85 psi (587 kPa) was recorded for the crushed stone, while the gravel exhibited a compressive strength of 62 psi (428 kPa).

Clegg Hammer testing performed prior to the placement of the AC binder course indicated the lime- and cement-modified bases achieved higher strengths under field curing conditions. Unconfined compressive strengths estimated from Clegg Hammer data ***(4)*** were approximately 130 psi (896 kPa) for the lime-modified gravel and crushed stone, and approximately 410 psi (2827 kPa) for the cement-modified crushed stone.

PAVEMENT PERFORMANCE

Visible Pavement Distress

Periodic visual pavement distress surveys have been conducted on the LCTP since construction. Cursory "walking surveys" were conducted prior to Fall of 1990. The surveys noted no distress in Sections 1 – 8, with limited longitudinal and transverse cracking noted in Section 9 (lime-modified gravel) in May 1988. By May 1989 Section 9 exhibited two longitudinal edge cracks, six full-width transverse cracks, and three half-lane transverse cracks.

Extensive visual pavement distress surveys were conducted in March 1990, October 1990, July 1991, and April 1993. A final "follow-up" survey was conducted in January 1998. The distress survey conducted by IDOT personnel in March 1990 establishes an initial level of pavement distress to serve as a "baseline" for comparisons with later surveys. The primary distress type recorded in the 1990 IDOT survey was "full-width" (across entire trafficked surface) transverse cracking. Only minor amounts of partial-width and half-lane transverse cracking, longitudinal cracking, and fatigue cracking were recorded. The full-width transverse cracking observed by IDOT personnel reflects early-life thermal cracking of the AC layer. Table 1 summarizes pavement distress observed for the period ending April 30, 1993. Transverse and longitudinal cracking are recorded as "number of cracks". Fatigue cracking is recorded as the number of fatigued locations and the approximate area of the fatigue distress. No distinction is made concerning the relative severity of the fatigue cracking (i.e. "incipient" or "failed"). The primary focus of the follow-up survey in January 1998 was to document fatigue damage to the test sections. A brief discussion of the current state of the pavement distress follows.

Section 1 (Stations 0+19 to 12+30) is the control crushed stone section. A large amount of fatigue cracking is noted, particularly in the westbound lane. Two fatigue-failure areas in the westbound lane have been repaired by removal and replacement of the AC surface. In addition, severe fatigue is noted in the eastbound lane on the west end. This area most likely receives the heaviest traffic loads, both in terms of axle load and volume, of all the sections. A large amount of transverse cracking was noted in the surveys, the majority of which occurs in the eastbound lane.

Section 2 (Stations 12+30 to 24+05) is the "low-fines" crushed stone section. This section contains the most full width transverse cracks recorded in the surveys, most of which were recorded in the original 1990 IDOT survey. The January 1998 survey noted a large amount of fatigue cracking, particularly in the outer wheel path of the westbound lane.

Section 3 (Stations 24+05 to 35+80) is 3% lime-modified crushed stone. The predominate distress features are longitudinal and transverse cracking. Some minor fatigue is also noted. The majority of the distress is recorded in the eastbound lane. Most of the observed pavement distresses were recorded by the 1993 U of I survey.

Section 4 (Stations 35+80 to 47+55) is crushed stone modified by the addition of 1.5% Portland cement. This section exhibits some minor fatigue cracking, most of which was observed in the eastbound lane. Transverse cracking has progressed relatively uniformly from the cracking observed in the original IDOT 1990 survey.

Section 5 (Stations 47+55 to 59+30) is the control crushed gravel section. Only minor amounts of pavement distress are recorded in this section. Most of the pavement distress was observed in the 1993 U of I survey, with some minor fatigue noted in 1998. However, severe fatigue was noted in 1998 at the intersection (Station 52+00) with a north-south road, in the turning movement onto the westbound lane.

Section 6 (Stations 59+30 to 71+05) is the "low fines" crushed gravel section. The primary distress type recorded in this section is transverse cracking. The majority of these cracks were first observed in the U of I 1993 survey. Most of the transverse cracks and observed fatigue (westbound lane) in Section 6 occur in the first 100 feet (30.5 m) of the section.

Section 7 (Stations 71+05 to 82+80) is the control natural gravel section. An area of pavement located at 76+50 that failed in both fatigue and rutting has been patched. Significant areas of minor fatigue cracking were noted in the 1998 survey, primarily in the westbound lane. Transverse cracking, most of which was noted after the 1990 IDOT survey, appears to be progressing relatively uniformly.

Section 8 (Stations 82+80 to 94+55) is a "low-fines" partially crushed gravel. Appreciable amounts of transverse cracking were recorded for this section the majority of which occurs in the westbound lane. The 1998 survey noted appreciable amounts of minor fatigue in both wheel paths of the westbound lane, with some fatigued “spots” in the eastbound lane.

Section 9 (Stations 94+55 to 106+39) is natural gravel modified with 3 percent lime. A relatively large number of full-width transverse cracks were observed in this section, the majority of which were recorded in the 1990 IDOT survey. An appreciable amount of longitudinal cracking was also noted. The 1998 survey noted some isolated spots of low severity fatigue in the eastbound lane. Most of the longitudinal cracks were observed in the westbound lane.

Rutting

Comprehensive rut depth measurements (dip-stick type device at 100-foot intervals in the inner and outer wheel paths) were taken in September 1988, March 1990; May 1993, and September 1997. Average rut depths (including both wheel paths) for each section (EB and WB directions) are presented in Figure 1.

Rut depths range from a minimum of about 3 mm (Section 9 - lime-treated gravel, WB lane) to a maximum of just over 12 mm (Section 1 – control crushed stone, WB lane). Many of the sections have experienced rutting approaching 10 mm. As described earlier, none of the gravel materials survived the *standard* 5000 conditioning triaxial stress repetitions; however they did survive 5000 conditioning stress repetitions of a *reduced* stress state. The rut depth data indicate the gravel sections (Sections 5, 6, 7, and 8) have not rutted to any greater degree than the crushed stone sections. The 3-inch (76 mm) AC surface apparently is providing adequate "protection" (repeated stress states are not excessive) for the gravel base layers.

Some of the disparity in the rut depth data is related to complex truck traffic patterns in the area. The LCTP road services grain, quarry, and hot-mix AC plant facilities, all of which use heavily loaded trucks. The truck loading has undoubtedly contributed to the increased rutting experienced between the 1990/1993 and 1993/1997 rut depth measurements.

SUMMARY OF FINDINGS

Based on the data generated on this project, some summary conclusions may be drawn.

- The IDOT CA-06 gradation is acceptable for compacted aggregate base course construction with crushed stone, gravel, and crushed gravel.

- The proposed CM-06 gradation is acceptable for compacted aggregate base course construction with crushed stone and crushed gravel.

- The proposed CM-06 gradation is not acceptable for compacted aggregate base course construction with gravel (e.g. base layer in Section 9 could not be constructed).

- Lime-modified higher fines content (near 12% specification limit) CA-06 crushed stone and gravel aggregates with higher PI (i.e. Section 9, where PI is out-of-specification) are providing acceptable performance as a compacted aggregate base course.

- Cement-modified higher fines content (near 12% specification limit) CA-06 crushed stone is providing acceptable performance as a compacted aggregate base course.

- The LCTP pavement performance data do not support the establishment of "quantitative rankings" for the various granular base materials. However, all of the granular base materials have performed satisfactorily.

ACKNOWLEDGMENT

This paper is partially based on a report *(2)* prepared in conjunction with IHR-527, *Mechanistic Design for Local Roads*. IHR-527 was part of the Illinois Cooperative Highway and Transportation Research Program sponsored by the Illinois Department of Transportation. The excellent support and cooperation of the Illinois DOT and the Livingston County Highway Department are gratefully acknowledged.

REFERENCES

1. *Flexible Pavement Design for Local Agencies*. Bureau of Local Roads and Streets, Illinois Deptartment of Transportation, Springfield, Illinois. Revised, August, 1988.

2. THOMPSON, M.R. and HALL, K.D., The Livingston County Test Project – An Aggregate Base Study, Civil Engineering Studies, TES No. 87, Illinois Cooperative Highway and Transportation Series No. 253, University of Illinois, July 1993.

3. THOMPSON, M.R. and SMITH, K., Repeated Triaxial Characterization of Granular Materials. Transportation Research Record 1278, Transportation Research Board, Washington DC, pp. 7-17, 1990.

4. OKAMOTO, P.A., BOCK, B.T., and NUSSBAUM, P.J., Nondestructive Tests for Determining Compressive Strength of Cement-Stabilized Soils, Transportation Research Record No. 1295, Transportation Research Board, Washington DC, 1991.

Section	Station From	Station To	Base Material	Cracking Trans.	Cracking Long.	Fatigue # of Areas	Fatigue Total sq. m
1	0.00	12.30	Control Cr. Stone	36	10	12	75
2	12.30	24.05	Low Fines Cr. Stone	33	6	1	2
3	24.05	35.80	Lime-Mod. (3%) Cr. Stone	38	12	1	9
4	35.80	47.55	Cement-Mod. (1.5%) Cr. Stone	30	0	7	42
5	47.55	59.30	Control Cr. Gravel	18	0	1	6
6	59.30	71.05	Low Fines Cr. Gravel	40	0	1	7
7	71.05	82.80	Control Natural Gravel	23	0	3	41
8	82.80	94.55	(Low Fines) Partially Cr. Gravel	40	1	3	15
9	94.55	106.39	Lime-Mod. (3%) Natural Gravel	19	10	0	0
			TOTALS:	277	39	29	197

Table 1. Pavement Distress Summary (through April 1993)

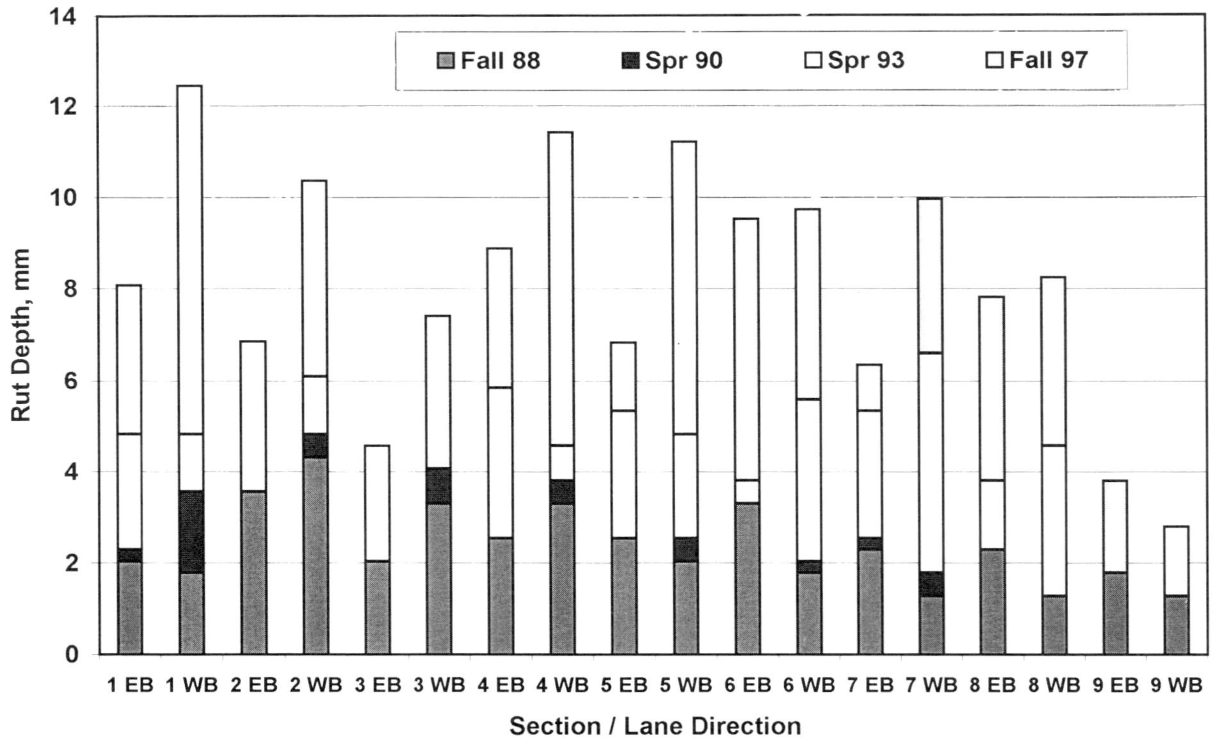

Figure 1. Rut Depths 1990 - 1997.

Fifth International Conference on the
BEARING CAPASITY OF
ROADS AND AIRFIELDS

BACKCALCULATION OF MODULUS ON LOADING TIME DEPENDENT CLAYS

CARL A. LENNGREN

Researcher, Vägverket Konsult, S-46228 Vänersborg, Sweden

Fax: +46521279124, E-mail: carl.lenngren@swipnet.se

ABSTRACT

A freeway in western Sweden was found to deteriorate at a much too high rate. An initial structural bearing capacity evaluation indicated that the design was somewhat on the lean side for the ambient subgrade conditions. To further investigate a falling weight deflectometer designed for use with different load modes was brought to the site. Several tests were done with loads varying from 40 kN up to 68 kN. The load pulse rise time was set to approximately 25 milliseconds. The tests were then repeated with a stiffer buffer configuration, yielding rise times of about 18 milliseconds. Hence, the combined effects of varying load amplitude and frequency could be examined. It was found that the asphalt concrete elastic properties were highly affected by the loading time, which was to be expected and also concluded from other similar tests. The marine clay subgrade elasticity at the present site showed to be very susceptible to the loading time too. Contrary to the asphalt concrete the subgrade stiffness was also much affected by the load. The higher the load the less stiff response from the subgrade. A heavy vehicle moving slowly would be a worst case scenario on this type of pavement. It is obvious that a conversion to equivalent standard axles may not be applicable for this type of subgrade. New testing procedures such as a falling weight deflectometer with even longer load rise times have been suggested. Also, suggestions have been made for the design manual for dealing with this type of clay.

Keywords: Backcalculation, Time-dependent soils, Falling weight deflectometer

OBJECTIVE

The objective of the present study was to investigate if the design parameters used were adequate for the conditions at the site. Further, special attention was to be considered for unusual properties of the subgrade that may alter design practices. Time dependent properties should be investigated in particular.

SCOPE

An advantage using the falling weight deflectometer as a method is the circumstance of having a standardized test. Provided the same type of machine is used the loading time and the shape of the loading curve should be constant over time. Thus, one would mimic the load of a truck traveling at a constant speed, which would be the custom speed on which further pavement overlay design is based. However, what appears to be a benefit does create a problem when different brands of machines are used. If there were a slightest loading time dependency of the material being tested it would be difficult to compare results from different equipment. Problems of this nature have been pointed out in FEHRL seminars held in Delft 1991[1] and Copenhagen 1994[2]. Lukanen[3] addressed how changing the rubber buffers transferring the load to the load plate produced different shapes of the load curve. The present author has previously studied this effect by deliberately altering the loading time by changing the buffer configuration on the same individual machine[4]. The pavements involved were Portland cement concrete, thin layer asphalt concrete and full depth asphalt concrete. In all cases the subgrade consisted mostly of coarse-grained material. In the present case the subgrade consists of clay, which would likely be influenced by the loading time. Thus, it was decided to test the site with an FWD capable of varying the loading time without changing the shape of the loading curve.

FIELD TESTING

Site and location

The site is located approximately 60 km north of Gothenburg on the Swedish West Coast near town Ljungskile. It is situated within a few hundreds of meters away from the sea. Northbound traffic is descending a hill. The actual 800 meter long section investigated slopes down from approximately 17 m to 5 m above the mean sea level. The road has a dual carriageway with two lanes each way. The average annual daily traffic is approximately 10000 vehicles. The truck proportion is about 14 %. The road was opened for traffic in 1991. It was constructed as described in Table 1. The subgrade is rated in the construction records as silty clay.

Material	Thickness, [mm]
Max. 16 mm aggregate size wearing course. Open drain asphalt concrete, B85.	45
Max. 12 mm aggregate size binder course. Dense asphalt concrete, B85.	35
Max. 25 mm aggregate size asphalt bound base.	50
Max. 25 mm aggregate size asphalt bound base.	50
Unbound gravel base.	70
Unbound gravel subbase.	110
Crushed rock subbase, max. aggregate size 200 mm.	540

Table 1. Road as built on the site

Previous Studies

The Swedish Road and Transportation Institute (VTI) brought an FWD to the site in May 1996. All FWD measurements were carried out at a nominal load of 50 kN only, and the rise time used was approximately 24 milliseconds. Along with the results from the non-destructive tests are also visual surveys and some lab tests based on coring reported. It is evident that the pavement has fatigued prematurely, as there are longitudinal cracks in the wheel path of the slow lanes. By regression analysis the strain at the bottom of the asphalt layer was reported to exceed 125 microstrain, which indeed could result in fatigue cracking by the equation quoted in the current Swedish Design Code. The code does take the asphalt concrete seasonal temperature variation into account.

Rutting is measured on annual basis on this category of road. In 1993 after a year and a half under traffic an average rut depth of 7 mm was reported. In 1994 the rutting had increased to 10 mm.

Variable Rise Time Study

A machine (KUAB 150) having the feature of optional rise times of either18 or 24 ms was brought to the site on 22 October 1996. The ambition was to repeat the previous test from May, covering the same wheel paths at the same stations. However, as multiple load levels and two load modes were planned, not all wheel paths in all lanes were tested. Rather, the test was limited to the right wheel path in the slow lane and in between wheel paths in the fast lane in order to get as wide spread of traffic load as possible.

The tests started with the longer load mode yielding a rise time of about 24 ms. Three different load levels were used. After the first 50 kN thump to set the sensors, 40, 50 and 70 kN load levels were used in order to be able to study any effects of non-linearity. When a sufficient number of test were sampled the machine was reconfigured for a shorter load mode. The process involved changing the rubber buffers to a smaller, stiffer size took about half an hour to complete. As asphalt concrete stiffness is very susceptible to changes in temperature, the delay should be as small as possible when comparing the different load modes. Fortunately, the weather did not change much during the day, it was mostly overcast and the air temperature was 5°C when the tests started in the morning. It peaked before noon at 10°C and dropped back to 7°C at the end of the test around 7 p.m. At all times the air and surface temperatures were recorded.

The potential energy from the weights dropped is constant regardless of the loading time. Hence, if the drop height is not changed, the rate of the energy being transferred to the loading plate is higher for the shorter impulse. This will inevitably result in a higher peak load. It is possible to compensate for this effect if the drop height is lowered or if less weight is used. This was not done however when shifting buffers to the short duration impulse load. Hence, the peak loads became higher using the short impulse load. The longer duration yielded peak loads of 41, 51 and 68 kN respectively. The shorter impulse rendered peaks of 52, 66 and 88 kN. Thus, peaks near 51 and 67 kN were coincidental for the two different modes. All load levels are plotted in Figure 1. Note that the area under each set of curves for each load level (low, medium and high) is approximately the same.

Figure 1. Load Pulse Shapes

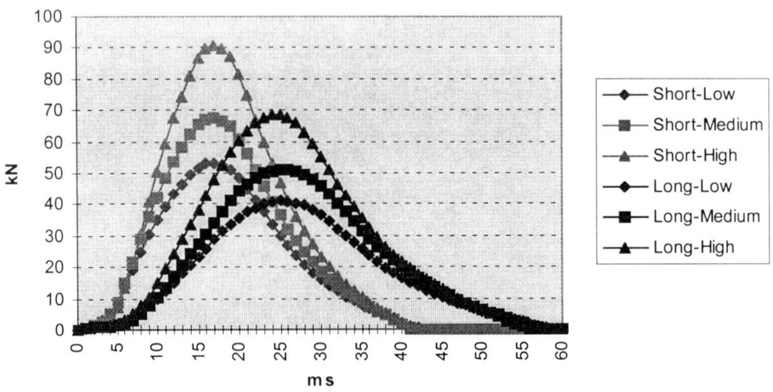

RESULTS

Backcalculation

A linear elastic backcalculation program CLEVERCALC 3.8 was used for assessing the E-moduli of the layers at all load modes and load levels. The idea would be to compare the results from the two nominal load modes as to see if the results would be different. In such cases one would surmise that the loading time does indeed affect the solution. Further, the results should also guide in understanding the magnitude of these effects, even if it may not qualify for determining the parameters describing the non-linear visco-elastic properties that may be.

As is often the case with this type of analysis it is viable to first start out with a three-layer model. Bound layers, unbound layers and subgrade. The common option of using a rigid base was not used in this case as it was found that it did not improve the fit, neither was there any known rock within 10 m from the pavement surface. Modeling the 60 - 70 mm thick unbound base layer separately did not improve the backcalculation either. This is not surprising, a thin layer somewhat down in the structure does not contribute much to the overall bearing capacity, and as such would not be detectable either. The fit of the backcalculated deflection basin as compared to the measured one can be expressed as a root mean square error per sensor. As it turned out the three-layer model proved to yield root mean square-values around one percent or less in general. In effect it means that the model adequately describes the deflection basin as measured at the surface. Tweaking the relationship between the chrushed rock and the subbase did not improve the fit of backcalculated basin either, meaning that the stiffness of these layers was about the same. It is a fair assumption that the difference in stiffness for any coarse granular material is diminutive at these depths. The previous study also indicated that fact.

The thickness of each layer was known from construction records as shown in Table 1. There were also a number of cores of the asphalt layer indicating some variability. Northbound sections at 250 and 800 m seemed to be somewhat thicker and at 500 m thinner than the 180 mm norm. There are suggestions in the literature to backcalculate these variations also, but in the present context no such attempt was made since the main objective was to compare the two different load modes.

As a reminder there are three reasons why a derived backcalculated modulus may differ from what is expected. If the modulus is lower than anticipated, the reason may be:

1. The material is indeed softer than normal.
2. The layer thickness is overrated. An adjacent softer layer is interfering with the result.
3. The material in question is cracked open and appears to be softer in a macro scale.

For case one and two the opposite is true for unusually high moduli. As long as the assumed layer thickness does not deviate more than 10 % from the actual one, the effects on the derived strains are small. An overlay design would not suffer much error from such deviations. In the present study an asphalt thickness of 175 mm was used for the northbound lanes and 180 mm was used for the southbound ones. For both directions an unbound combined base and subbase thickness of 720 mm was used throughout.

The backcalculated moduli are presented in Tables 2 and 3. As is evident from the tables some of the materials involved are certainly affected by the load level. Some of them are also affected by the loading time as well.

When the load increases	E(1)	increases somewhat
When the load increases	E(2)	increases dramatically
When the load increases	E(3)	decreases marginally

When the loading time increases	E(1)	decreases
When the loading time increases	E(2)	decreases somewhat northbound
When the loading time increases	E(2)	decreases dramatically southbound
When the loading time increases	E(3)	decreases dramatically

The asphalt concrete resilient modulus E(1) was recalculated to a nominal temperature of 15°C in the comparison. The actual stiffness northbound was somewhat higher due to the early morning test, which occurred at much colder temperatures. Naturally there are certain uncertainties involved in assessing the right pavement temperature as there may be gradients involved. However, on the particular day of measurement the solar radiation effects were rather limited considering the high northern latitude and the time of the year. In addition to moduli the tables also show the mean pavement surface temperature, strains at the bottom at the asphalt layer and at the top of the unbound layers. Further the fitted solution root mean square values, the center deflection, the impulse stiffness modulus (ISM) and the radius of curvature (30 cm) are shown.

PEAK LOAD [kN]	**SHORT /52**	**SHORT/ 66**	**SHORT/ 88**	**LONG/ 41**	**LONG/ 51**	**LONG/ 68**
E(1) @15°C	5991	6135	6239	5407	5475	5705
E(2)	118	139	153	98	116	122
E(3)	63	61	60	57	55	54
Surf. Temp.	8.9	8.9	8.9	5.9	5.9	5.9
Strain(1)	188	225	290	166	199	253
Strain(2)	567	667	852	505	597	752
Strain(3)	201	251	334	175	218	289
RMS	1.0	.71	.56	1.0	.67	.54
Do	472	578	757	394	485	633
ISM	456	441	429	486	477	469
Radius of curvature	538	455	355	747	616	475

Table 2. Right lane northbound right wheel track averages

PEAK LOAD [kN]	**SHORT/ 52**	**SHORT/ 66**	**SHORT/ 88**	**LONG/ 41**	**LONG/ 51**	**LONG/ 67**
E(1) @15°C	7347	7430	7266	6525	6726	7213
E(2)	182	215	225	106	123	126
E(3)	74	70	71	65	63	62
Surf. Temp.	6.6	6.6	6.6	7.9	7.9	7.9
Strain(1)	139	166	227	141	170	220
Strain(2)	409	482	654	423	501	648
Strain(3)	160	202	270	158	199	261
RMS	.92	.65	.65	1.4	1.3	1.1
Do	357	445	595	367	455	598
ISM	342	338	336	454	450	445
Radius of curvature	801	624	460	858	668	505

Table 3. Left lane southbound between wheel track averages

Stress sensitivity

Figures 2 and 3 show the derived stiffness for the unbound layers E(2) and the subgrade E(3). Each station shows six bars, i.e. the three load levels for each layer, the lowest one to the left. Both figures refer to the northbound test in the right wheel track. Apparently E(2) is indeed stress sensitive as the modulus increase with load for most of the cases. The subgrade stiffness E(3) does not exhibit this behavior at all. On the contrary, the stiffness decreases somewhat, but not dramatically so with the load, i.e. the stress, as would be expected from a cohesive material. The high overburden diminishes this effect as the relative difference in stress states from the varying loads is decreasing with depth. Yet, higher load levels would seemingly show this circumstance more clearly.

Figure 2. Northbound, Right Lane, Right Wheeltrack
Unbound Base and Subgrade Moduli 25ms @ 45, 51 & 67 kN

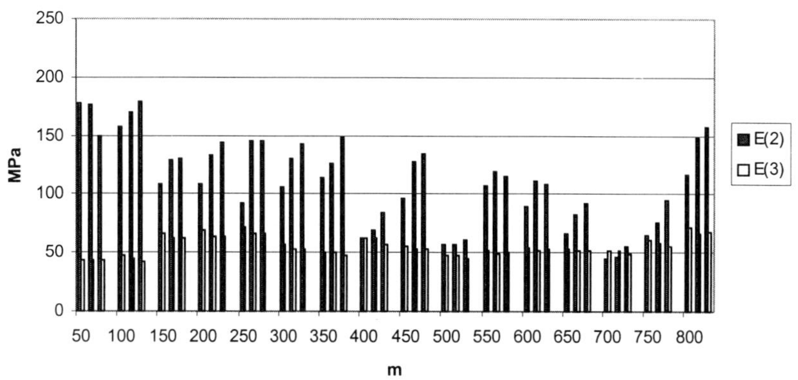

Figure 3. Northbound Right lane, Right Wheeltrack
Unbound Base & Subgrade Moduli, 18 ms @ 52, 66 & 88 kN

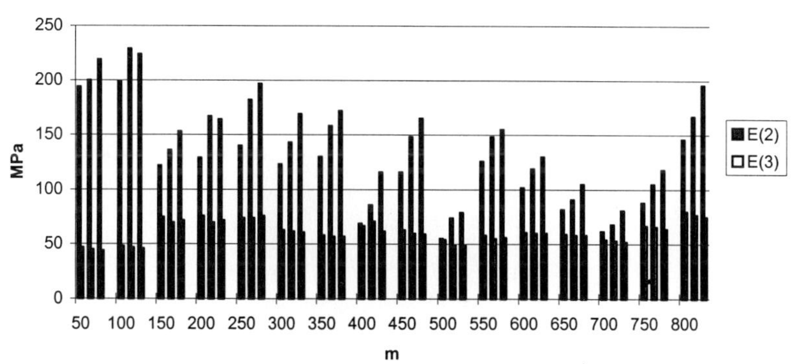

Time dependency
Comparing the load modes, i.e. the two different rise times one does see how large the influence is on the asphalt concrete, which is to be expected. The unbound layers are little affected in the northbound, slow lane, but the subgrade does exhibit an influence of the loading time. Hence, a slow-moving load would induce large movements of the entire structure and most likely would incur all too high strains in the subgrade, but also at the bottom of the asphalt layer as well.

The unbound layers exhibit a slight time-dependency in the right northbound lane. There is no reason to assume that there would be any large deviation between modes. However, in the sparsely truck-trafficked left lane the effect of loading time is significant. Apart from suspecting cohesive material or water being present to a higher degree on this side of the road the most plausible explanation may be lack of compaction. There are very few trucks in the southbound left lane as traffic is going uphill, so the effect of having poorly compacted material in between two time-dependent materials may show up as a frequency dependency. Testing at yet more loading times could perhaps show what is going on. One would certainly like to increase the loading time even more to see whether the effect vanishes or not.

Results from using shorter loading times would most likely be affected by inertia and it would be difficult to discern such effects from the material time dependency properties. The mean modulus for the unbound layers and subgrade are also shown in Figures 4 and 5 for the northbound slow lane and Figures 6 and 7 for the southbound fast lane.

The subgrade does show a consistent time-dependency with loading time. The longer the duration of the load, the softer the material reacts. At a nearby location static plate-loading tests at the time of construction was compared to the two impulse loads used there as well. The plate-loading tests assessed a much lower clay elastic modulus than the FWD did and the comparison showed this effect very clearly. If speeds are even slower than the corresponding long loading time, it could very well explain the rapid evolution of fatigue cracks found here.

Figure 4. Unbound Layers, Slow Lane

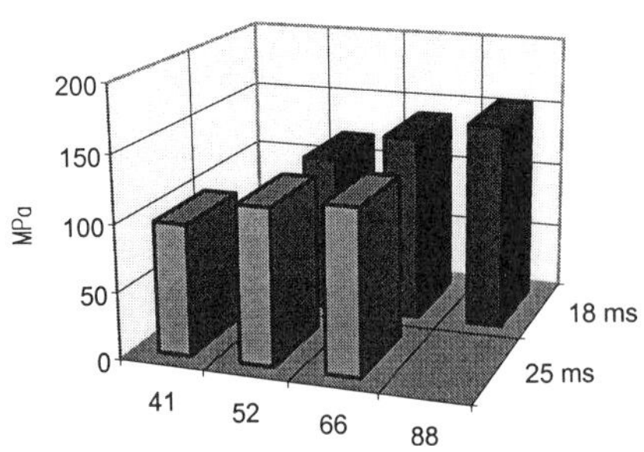

Figure 5. Subgrade Modulus Slow Lane

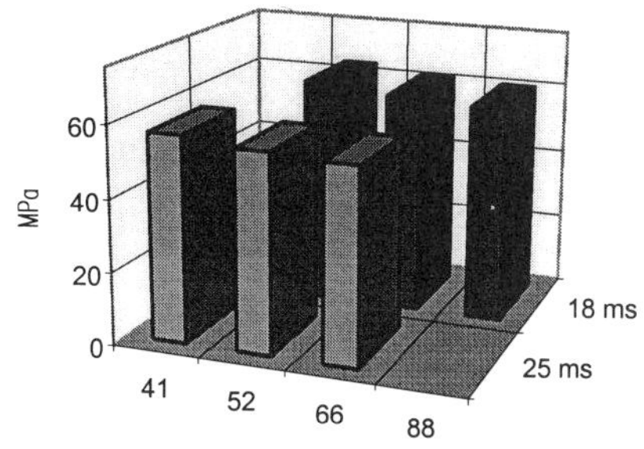

Figure 6. Unbound Layers, Fast Lane

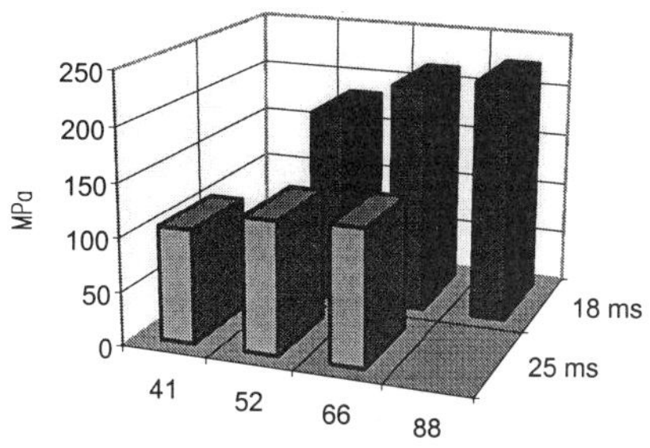

Figure 7. Subgrade Modulus, Fast Lane

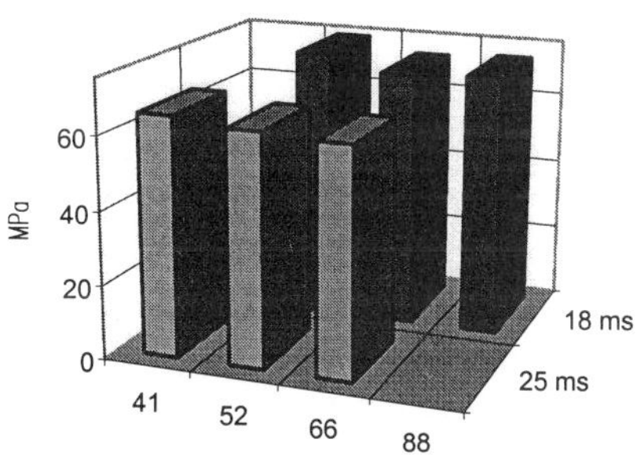

CONCLUSION

The subgrade at the site proved to be very sensitive to the loading time and its stiffness does decrease at longer loading times. The FWD rise times used are corresponding to fairly high truck speeds, especially since the subgrade is far away from the surface. An extrapolation of the times involved could mean a stiffness as low as 20 Mpa for trucks moving less than 50 km/h. Simultaneously the asphalt concrete stiffness will decrease as well, further concentrating the stresses on the subgrade. The resulting deflections would therefore be rather large and even if the asphalt layer is rather thick the resulting strains would be too large and fatigue damage would occur within five years rather than the twenty years it was designed for. The deflections involved for a slow, heavy truck would be as high as 1.2 mm at the surface and about one mm at the top of the subgrade. The corresponding strain at the bottom of the asphalt layer would then be 188 microstrain for a ten-ton standard axle rather than 125 microstrain. Using the Swedish Design Code as a norm this would allow for only 1.5 million passes instead of the design number of eight million standard axles. Expressed in asphalt thickness, the total should be 210 mm rather than the 180 mm designed. Naturally, most trucks do indeed travel faster than 50 km/h though, but there are even worst case scenarios that would be even more damaging. Imagine a 14 ton overload axle on a truck that most likely would go slow due to the excess load. Such an axle would only allow for 214000 passes translating to a fatigue damage 36 times higher than the standard axle.

Thus, certainly there is a concern of speed for pavement design on the present type of clay. There are some new freeway projects planned in the Gothenburg area and it is recommended that the time dependency effects should be thoroughly tested, especially in urban areas where congested traffic may really slow traffic down. The clays involved are not very forgiving to overloads and it may be wise to revise admitting overloads. For construction control and overlay design when these clays are involved, a long rise time FWD is recommended. The manufacturer of the machine used in the study claims that by modifying the equipment up to 50 ms rise time should be possible.

ACKNOWLEDGMENTS

The Regional Road Administration in Gothenburg sponsored the tests and supported the study. Professor emeritus Olle Andersson helped in preparing the manuscript. Their support is much appreciated.

REFERENCES

1. FEHRL-FWD SEMINAR, Full Seminar Report, 11-12 November 1991, Delft, The Netherlands.
2. Third FEHRL FWD Seminar, Seminar Report, Copenhagen, May 17-18 1994
3. Lukanen, Erland O. "Buffer Effects on Falling Weight Deflectometer Loadings and Deflections" Transportation Research Board Annual Meeting, 1992. (Preprint 920298).
4. Lenngren, C.A. "Non-Destructive Testing Utilizing Controlled Variable Rise Time" The 4th International Conference on the Bearing Capacity of Roads and Airfields. Proceedings, Volume 1, Minneapolis 1994. pp. 467-490.

Fifth International Conference on the
BEARING CAPASITY OF
ROADS AND AIRFIELDS

CRUSHED ROCK PAVEMENT STRUCTURE TRIAL SECTIONS

TORKJELL HAUSTVEIT[1], ROAR NÅLSUND[2], TERJE TRENGEREID[3]
JOSTEIN AKSNES[4] and RAGNAR EVENSEN[5]

[1]Public Roads Administration, Oppland County, P.O. box 1010 Skurva, N-2601 Lillehammer, Norway, [2]Norwegian National Rail Administration, Pirsenteret, N-7005 Trondheim, Norway, [3]Public Roads Administration, Hordaland County, P.O. box 3645, N-5033 Fyllingsdalen, Norway, [4]Public Roads Administration, Sør-Trøndelag County, P.O. box 6064 Postterminalen, N-7003 Trondheim, Norway, [5]ViaNova AS, P.O. box 53, N-1312 Slependen, Norway

ABSTRACT

A joint research group on the quality of crushed rock products was established in Norway in 1993. The objective of the research group is to document the functional properties of crushed rock products used in road, railways and airfield pavements. The work of the group is divided into 9 different activities, including construction and analysis of results from field tests. This paper gives a presentation of three field tests.

- Rv 33 Lillo – Opsal in the county of Oppland
- Ev 39 Matre – Nipetjønn in the county of Hordaland
- Ev 39 Bjørkøybekk, Hemnekjølen in the county of Sør-Trøndelag

All the test sites include sections for comparison of the properties of crushed rock and gravel used in the subbase of the road.

The paper also presents results from several FWD-measurements and back calculation of the E-moduli using the program EMODUL. The Norwegian standard for structural design of roads accepts that the E-modulus can be used for adjusting the design layer thickness of the pavement.

For two of the test sites, the analysis concludes that a 10% reduction of the layer thickness can be accepted due to better load distribution properties, when crushed rock is used compared to gravel. In the third test site no significant difference was found between crushed rock and gravel with respect to their load distribution properties. In this test site the E-modulus of the subbase of crushed rock was not lower than what was found in the other two test sites, but the E-moduli of the gravel subbase and the subgrade was higher. All the three test sites will be analysed with respect to the functional properties of the pavements up to 10 years after construction.

Key words: Structural design, crushed rock, trial sections, back calculation.

INTRODUCTION

A joint research group on the quality of crushed rock products was established in Norway in 1993. In the group there are participants from 19 suppliers of rock aggregates, 11 suppliers of equipment for aggregate production and 8 public road agencies. The objective of the research group is to document the functional properties of crushed rock products used in road pavements, railway tracks and airfield pavements. The research activities are planned to continue until 1999.

The activities of the research group include field trial sections at three different sites.

- Rv 33 Lillo – Opsal in the county of Oppland
- Ev 39 Matre – Nipetjønn in the county of Hordaland
- Ev 39 Bjørkøybekk, Hemnekjølen in the county of Sør-Trøndelag

This paper gives a presentation of the field trials and the results at the present time.

For the test sections of Matre – Nipetjønn, the pavement is a reconstruction of an old road with small changes in the horizontal alignment. With respect to structural design and construction work, however, the sections can be regarded as parts of a new road.

The construction of the test sections of Rv 33 in Oppland was done in 1994. The wearing course of stone mastic asphalt was pavedd in 1995. The construction of the test section of Ev 39 in Hordaland was done in 1995 with the binder course as a temporary surface course. The permanent surface course was paved in 1997. The test sections of Ev 39 in Sør-Trøndelag were constructed and completed in 1995.

STRUCTURAL DESIGN OF ROADS IN NORWAY

Since the AASHO Road Test in the early 1960ies, structural design of pavements in Norway has been based upon the principle of equivalent layer thicknesses:

$$SI_{K(n+1)} = \sum_{i=1}^{n} a_i \cdot h_i$$

where $SI_{K(n+1)}$ = the required sum of equivalent thicknesses of the n top layers of a pavement, based on the strength of the material in layer n+1

a_i = the coefficient of load distribution of the material in layer *i*, based on the load distribution properties of the material

h_i = the thickness of layer *i*

The required sum of equivalent thicknesses is a function of several factors:

- the traffic volume, AADT
- the percentage of commercial vehicles
- the annual change in traffic volume
- the design period
- a classification of the materials in the road foundation and the subgrade

Indirectly, the required sum of equivalent thicknesses is a function of the material strength and the sum of 10 tons equivalent single axle load passages over the design period. For new roads the structural design is usually based on a 20 years design period and a legal axle load limit of 10 tons. For trunk roads, however, the thickness of the subbase layer should be based on a 13 tons legal axle load limit.

For many years crushed natural gravel has in Norway been the most used material for basecourses and subbases. This is one of the reasons for using natural gravel as the reference material in structural design of roads. This material is given a coefficient of load distribution, a = 1,0.

The Norwegian Structural Design Manual (4) include SI_K at three different levels of the pavement:

1. at the interface between the base course and the surface course, above which the materials must have a coefficient of load distribution better or equal to 2,5
2. at the interface between the subbase and the base course, above which the materials must have a coefficient of load distribution better or equal to 1,25
3. at the interface between the subgrade and the subbase.

The Design Manual (4) presents a rather comprehensive list of coefficients of load distribution properties based on an identification and classification of the materials to be used in the pavement construction. A small part of this list is presented in table 1.

	Coefficient of load distribution, a
Asphalt surface courses	
Hot mixed, bitumen B60-B180	3,00
Warm mixed, soft bitumen, MB ≥ 6000	1,50
Gravel emulsion	1,75
Base courses	
Dense asphalt base course, hot mix	3,00
Gravel emulsion	1,75
Crushed gravel	1,25
Crushed rock	1,25
Subbase	
Sand, gravel	1,00
Crushed rock	1,00

Table 1. Coefficient of load distribution, examples from ref. (4).

From 1992 it has become possible, within some limitations, to use coefficient of load distributions which differ from the standard ones which are presented in the list. The load distribution property of a material is related to the E-modulus by the equation:

$$a = 0{,}21 \cdot \sqrt[3]{E_{25;10Hz}}$$

where $E_{25;10Hz}$ = the E-modulus at temperature 25°C and frequency 10 Hz
a = the coefficient of load distribution of the material

In a road pavement the properties of the materials will vary over the seasons depending on the climatic conditions. This has to be taken into consideration when the coefficient of load distribution is determined. For mechanical stabilised materials the critical situation is usually related to high water contents during the spring thaw periods. For bituminous materials the dominant factors are the temperature and the time of loading. However, the coefficient of load distribution of these materials is related to standard conditions of 25°C and 10 HZ. Looking at alternative solutions in order to determine the most cost effective pavement, the coefficients of load distribution have great influence on the conclusions.

THE TEST SECTIONS

	Oppland	Hordaland	Sør-Trøndelag
	Rv 33 National Road	Ev 39 Trunk Road	Ev 39 Trunk Road
Total length	1.240 m	400 m	800 m
No of test sections	7	4	8
Traffic volume, AADT	4000	1000	1000
Percentage of heavy vehicles	10 %	15 %	20 %
Start of construction works	1993	1994	1994
Test sections constructed	1994	1995	1995
Permanent wearing course	1995	1997	1995
Mean annual precipitation, mm	600	3.540	1.610
Mean monthly air temp. January	-7,4°C	-2,4°C	-5,8°C
Mean monthly air temp. July	14,8°C	13,6°C	13,6°C

Table 2. Key data of the field test sites

Some key data for the three test sections are presented in table 2. The different test sites are described and discussed more in detail in the paragraphs below.

Oppland, Rv 33

The main objective of the field test in Oppland was to compare three different types of crushed rock with gravel used in the subbase. The test sections are presented in graph 1.

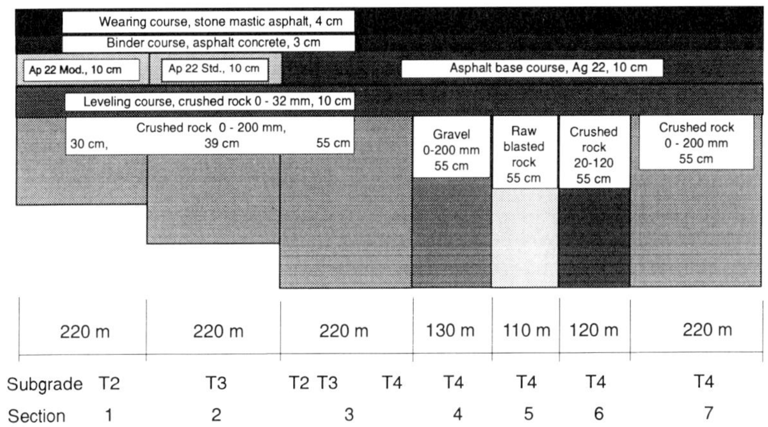

Graph 1. Layout of test site Rv 33 Oppland County

In graph 1 the subgrade is presented by a classification with respect to the frost susceptibility of the materials. The T2 and T3 materials are narrow graded sand and silt susceptible to frost heave, while the T4 materials are moraines highly susceptible to frost heave. In order to ensure good drainage conditions, longitudinal drainage was established along both sides of the road, approximately 2,0 m below the subbase layer.

For the base course three types of hot plant mixed asphalt was used. The Ag 22 is a dense graded mix with 22 maximum aggregate size, 3,9% (by weight) bitumen. In this mix the bitumen content is lower than what is recommended in the standards. The

Ap 22 std is a standard open graded asphalt mix with more than 70% of the aggregate coarser than 11 mm and a bitumen content of 3,0%. The Ap 22 Mod is a modified Ap22 with cellulose fibers and a target value for the bitumen content of 4,5%. However, the test results indicated a bitumen content of 3,3% for this mix.

With the exception of section 4 the materials in the subbase are crushed rock in all test sections. The densely graded material 0 – 200 mm in section 1, 2, 3 and 7 was outside the specification limits with respect the maximum amount of material smaller than 75 μm. The raw blasted rock subbase in section 5 had stones of size up to 40*50*70 cm. The material of 20-120 mm in section 5 was well within the specifications both with respect to mechanical properties and gradation.

Hordaland, Ev 39

The field test Matre - Nipetjønn on trunk road Ev 39 in Hordaland was a reconstruction of an existing road. All test sections are placed on an existing embankment with a total thickness of approximately 100 cm. The major part of the embankment is 40 – 60 cm of moraine susceptible to frost heave, classified as T2, on 50-100 cm of peat. The old wearing course of 10 cm oil gravel was removed before large stone crushed rock 0 – 300 mm was used for reshaping the embankment to an acceptable geometric standard.

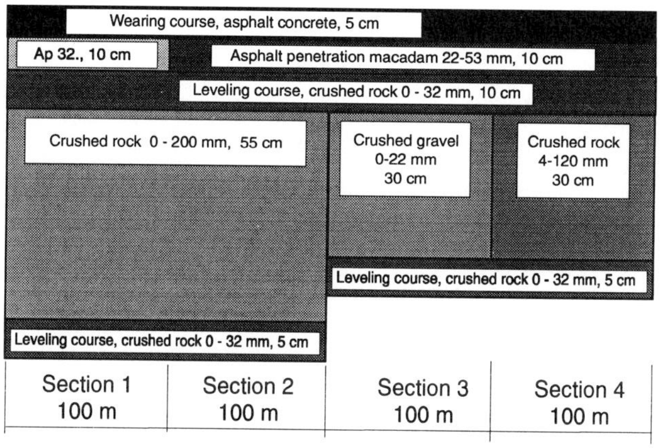

Subgrade: Old road fill of 40 - 60 cm moraine, susceptible to frost heave, T2 on 50 - 100 cm of peat.

Graph 2. Layout of test site Ev 39 Hordaland

Graph 2 shows that the thickness of subbase is greater in section 1 and 2 than in section 3 and 4. This was necessary to obtain a good vertical alignment of the new road. According to the Norwegian standard for structural design of roads (4) the thickness of the base course is smaller than required for all the sections, while the thickness of the subbase is greater than required for section 1 and 2.

The asphalt penetrated macadam is one of the types of base course most used in the county of Hordaland. It is also quite common to accept traffic on this base course a year or two before the surface layer is placed.

Sør-Trøndelag, Ev 39

The test sections in Sør-Trøndelag are built on a relatively large fill of raw blasted rock and moraine. The moraine is classified as susceptible to frost heave, T3, and it is expected that the moraine is soaked with water during the spring thaw periods. The layout of the test site is presented in graph 3.

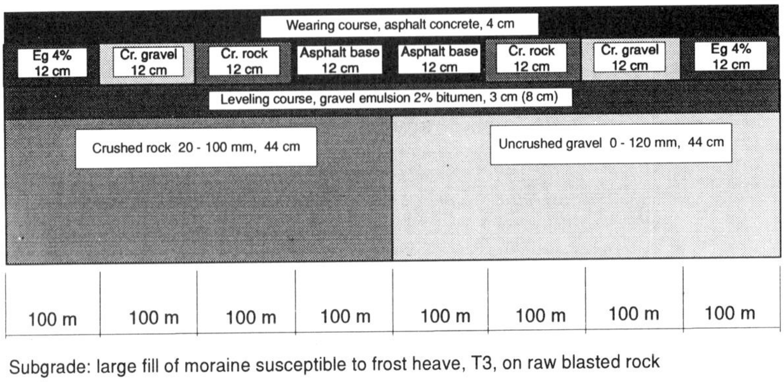

Graph 3. Layout of test sections, Ev 39 Sør-Trøndelag

There are 8 different sections at the test site, each section is 100 meter in length. The field test combines four different materials used as base course and two different materials used as subbase.

Both the gravel and the crushed rock used for subbase in the test sections are of good quality, both with respect to the gradation and the mechanical properties. The surface of the subbase of crushed rock was rather coarse and not well suited for construction traffic. A levelling course of gravel emulsion was therefore laid on the top of the crushed rock subbase. Out of a total thickness of 8 cm, approximately 5 cm penetrated the coarse surface of the subbase, leaving 3 cm as a separate layer of gravel emulsion. On the top of the subbase of crushed gravel it was therefore placed a levelling course of gravel emulsion of a thickness of 3 cm in order to reduce the differences between the test sections.

There were used four different materials for the base course. For all sections the thickness of the base course was 12 cm in spite of the differences in coefficient of load distribution, varying from 3,0 for the hot mixed asphalt to 1,25 for water bound crushed gravel and crushed rock.

COEFFICIENT OF LOAD DISTRIBUTION

On all the test sections several deflection measurements have been made using a falling weight deflectometer. The results from the deflection measurements have been used for back calculation of the E-moduli of the different layers using the program EMODUL developed by SINTEF (5).

In all the calculations the layer thicknesses used are the average, measured thickness of each section.

Rv 33 in Oppland County

Graph 4 presents the E-modulus of the subbase in section 3 – 7, In all the sections with a subbase of crushed rock the E-moduli are significantly greater than the E-modulus of gravel. The differences correspond in average to a 10% higher coefficient of load distribution relative to gravel. There are no significant differences between the crushed rock 0-200 mm with excess of fines in section 3 and 7, raw blasted rock in section 5 and crushed rock 20-120 (with a low content of fines) in section 6.

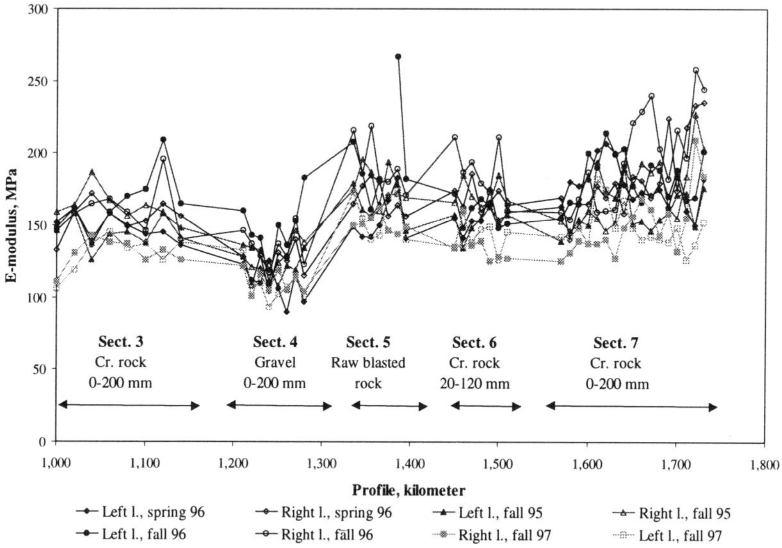

Graph 4. E-modulus of subbase, back calculation of FWD measurements, Rv 33 Oppland County

The E-modulus of the subbase in section 3 is slightly lower than in section 7, in spite of the fact that the subbases are of the same material. The data from the construction gives no explanation of this difference. It can, however, be noticed that low E-modulus of the subbase in section 3 corresponds to a high E-modulus of the subgrade.

Table 3 presents the average E-moduli of asphalt (adjusted to a temperature of 25°C), subbase and subgrade of 8 deflection measurements. As indicated in graph 4, the deflection measurements are done both in left and right lane, the first in the fall of 1995 and the latest in the fall of 1997.

Section	E-modulus, MPa		
	Asphalt (25°C)	Subbase	Subgrade
3. Crushed rock 0-200	3.959	150	172
4. Gravel 0-200	4.112	123	129
5. Large stone crushed rock	3.664	170	155
6. Crushed rock 20-120	3.333	159	138
7. Crushed rock 0-200	4.421	171	154
Average	4.016	164[1)]	150

[1)] Average of section 3, 5,6 and7.

Table 3. Average E-moduli, back calculation of FWD measurements, Rv 33 Oppland

In all the sections of the test site at Rv 33 in Oppland, the E-modulus of asphalt is higher than 3.000 Mpa, which is considered to be a reasonable average for asphalt hot mixes, corresponding to a coefficient of load distribution of 3,0.

Ev 39 Hordaland

Graph 5 presents the E-moduli of the subbase form back calculation of fwd deflection measurements made in the fall 1997. As can be seen from the graph, the results from Hordaland vary more than the results from Rv 33 in Oppland. However, the average E-modulus of crushed stone subbase is higher than the average E-modulus of gravel subbase. The difference correspond to a 17% higher coefficient of load distribution, which is more than the differences from the field test in Oppland.

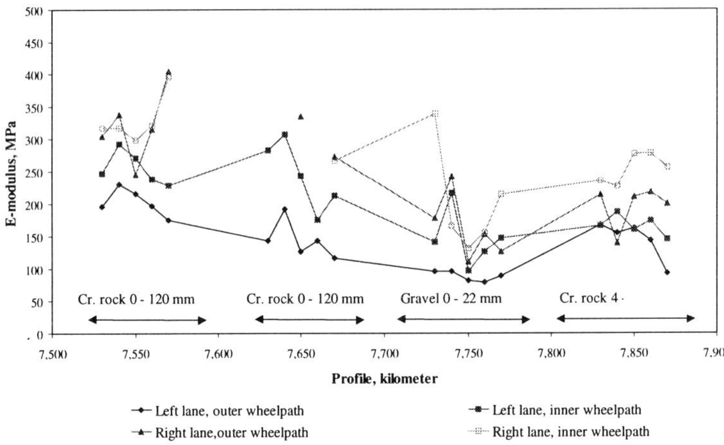

Graph 5. E-modulus of subbase, back calculation of FWD measurements, Ev 39 Hordaland

Section	E-modulus, MPa		
	Asphalt (25°C)[3]	Subbase	Subgrade
1. Ap 32 base course Crushed rock 0-200 subbase	1177	227	199
2. Penetration macadam base course Crushed rock 0-200 subbase	565	264	145
3. Penetration macadam base course Gravel 0-22 subbase	521	149	193
4. Penetration macadam base course Crushed rock 4-120 subbase	792	190	241
Average	623[1]	241[2]	194

[1] Average of section 2, 3 and 4. [2] Average of section 1, 2 and 4. [3] The E-modulus of penetration macadam base course is assumed to be independent of temperature.

Table 4. Average E-moduli, back calculation of FWD measurements, Ev 39 Hordaland

In table 4 the average values of the E-modulus of the base course, subbase and subgrade of the tests in Hordaland are presented. The E-modulus of the base course correspond to a coefficient of load distribution of 2,22 for the coarse graded asphalt hot mix, and 1,79 for the penetration macadam. As for the dense graded base course in the field test in Oppland, the coefficients are slightly higher than the standard coefficient given in ref (4).

Ev 39 in Sør-Trøndelag

Graph 6 presents the E-moduli of the subbase and the subgrade in the field test at Ev 39 in Sør-Trøndelag.

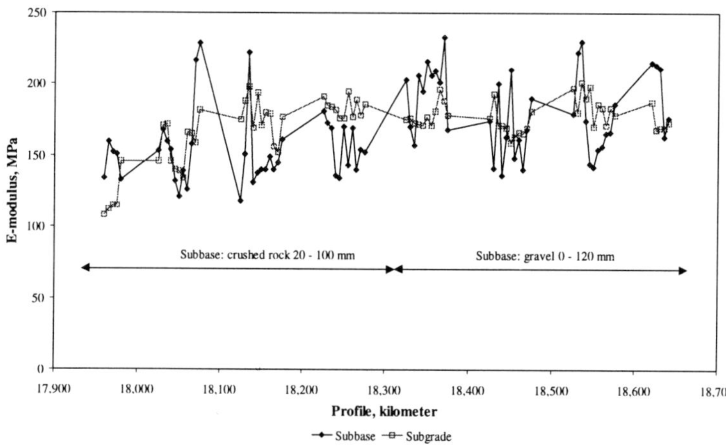

Graph 6. E-modulus of subbase and subgrade, back calculation of FWD measurements, Ev 39 Sør-Trøndelag

When comparing the results from back calculation for subbase layers, the E-modulus of crushed rock in the field test in Sør-Trøndelag has approximately the same value as in Oppland but is lower than the value from Hordaland. The E-modulus of gravel from Sør-Trøndelag is higher than the values from both the other counties.

Sections	E-modulus, MPa		
	Base course	Subbase	Subgrade
Average 1 and 8: Emulsion gravel	120		
Average 2 and 7 Crushed gravel	739		
Average 3 and 6 Crushed rock	720		
Average 4 and 5 Asphalt hot mix	800		
Average 1 - 4 Crushed rock		154	
Average 5 – 8 Gravel		181	
Average 1 – 8 Subgrade			171

1

Table 5. Average E-moduli, back calculation of FWD measurements, Ev 39 Sør-trøndelag

The results from Sør-Trøndelag differ from the results from Oppland and Hordaland with respect to the E-moduli of the base course materials. The E-modulus for base layers of crushed gravel and rock in Sør-Trøndelag are higher than the values obtained for i.e. penetration macadam in Hordaland. This is surprising. Even more surprising is the extremely low E-modulus for bituminous materials in base layers in Sør-Trøndelag, especially for emulsion gravel. Possible explanation can be:

- the FWD measurements were taken at very low temperatures, and the temperature correction formula used to get the E_{25}-modulus, may not be suitable for temperature corrections of this size.
- the emulsion gravel was paved short time before the FWD measurements, and it is quite common that this material increases its strengt after some time due to drainage of excess water.

Results from indirect tensile tests of cores from the pavement performed in 1997, however, show “normal” values for the stiffness of bituminous materials.

CONCLUSIONS AND FUTURE ANALYSIS

The field tests in Oppland and Hordaland indicate that the thickness of subbase can be reduced 10% relative to a subbase of gravel as a result of the differences in the load distribution properties. Relatively seen the coefficient of load distribution for base courses of dense and open graded asphalt hot mix, and of penetration macadam, are in average higher than those presented in the Norwegian standard for structural design of pavements (4). The conclusions form Oppland and Hordaland are, however, not confirmed by the results from Sør-Trøndelag.

Analysis based on back calculation of fwd deflection measurements gives a good background for comparison of alternative materials and layer thicknesses used in a road pavement. However, the preliminary results from the field tests must be checked against analysis of the functional properties of the pavements over several years before the final conclusions can be drawn.

LITTERATURE

1 Forsøksfelt Rv 33, Lillo – Opsal, Oppland. Anleggsrapport
KPG rapport 02, juli 1996 (in Norwegian)

2 Forsøksfelt Rv 01 Matre – Nipertjønn, Hordaland. Anleggsrapport
KPG rapport 08, november 1996 (in Norwegian)

3 Forsøksfelt E 39 Bjørkøybekk på Hemnekjølen, Sør-Trøndelag. Anleggsrapport
KPG rapport 09, januar 1998 (in Norwegian)

4 Vegnormalene, Håndbok 018 ”Vegbygging”
Statens vegvesen, desember 1991 (in Norwegian)

5 Brukerveiledning for EMODUL, versjon 2.00 og 2.01 – Pc-program for etterregning av E-moduler ut fra nedbøyningsmålinger,
Statens vegvesen, Vegdirektoratet, Intern rapport nr 1717, juli 1994
(in Norwegian)

Continuing from page 114

Analysis of these processes identified significant areas of waste or duplication. The financial process indicated considerable resource involved with compensation events arising from the NEC Option C Target Cost form of contract. The target mechanism, while offering an incentive mitigated against improvement in other areas. The original incentive scheme is shown in Figure 2.

Figure 2

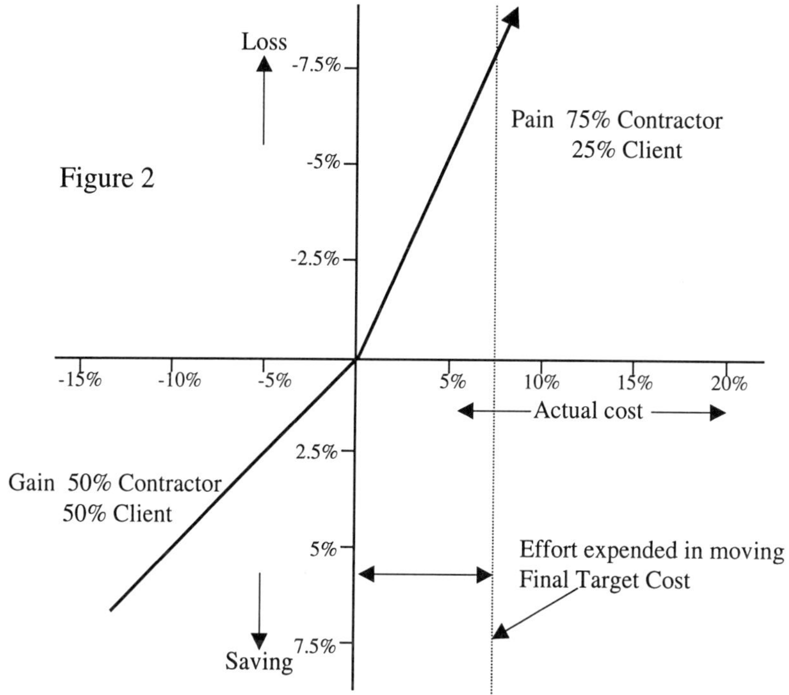

Original Target Cost Mechanism

Normally the design was developed with construction input but the target cost was not fixed until the design package was finalised. There was, therefore, no incentive to bring forward good ideas during design as this would merely lower the target. Once the target was fixed and construction began on site, changes were seen as compensation events and there was considerable effort on both sides to administer these changes. With growing confidence that projects were being managed responsibly, the target cost mechanism was replaced with a cost reimbursable contract attracting a higher fee. This removed any reluctance to bring forward good ideas during the design and focused the whole team on process improvements. The incentive moved from a project by project target to a five-year performance target.

Metrics

The previously in-house design and project management resources were brought into the Pavement Team from April 1997 and the team co-located. At the same time a general manager was introduced with responsibility to the AMEC Southern Civil Engineering Limited's board and BAA for the performance of the team. The team was re-organised into a process/project matrix and focused on achieving BAA's Global project targets and the AMEC business drivers.

BAA Global Project Targets		**AMEC Business Drivers**
Project cost reduction	30%	Better cash flow
Project cost predictability	95%	Increase profit
Project programme reduction	30%	Improved quality
Project programme Predictability	95%	Enhanced reputation
Safety – accident frequency ratios (reportable accidents/100,000hours)	0.5%	Safety improvement

Performance on the cost and programme predictability targets is already operating at close to 100%. The cost predictability graph below contrasts pre-framework outturn to tender costs and framework actual to target costs. The programme predictability graph gives all sectional and project completions and does show where some large savings on programme have been made.

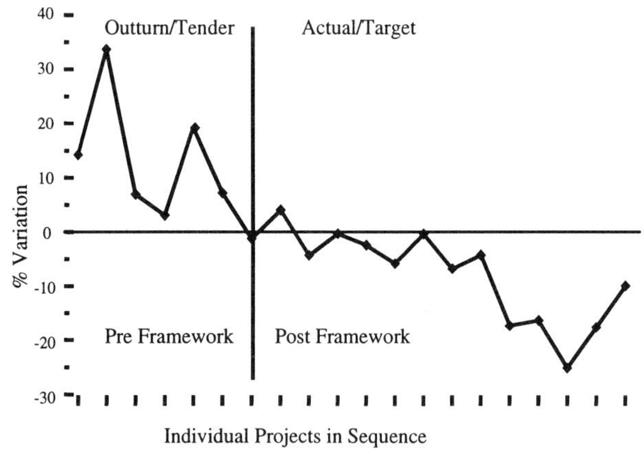

Project Cost Predictability

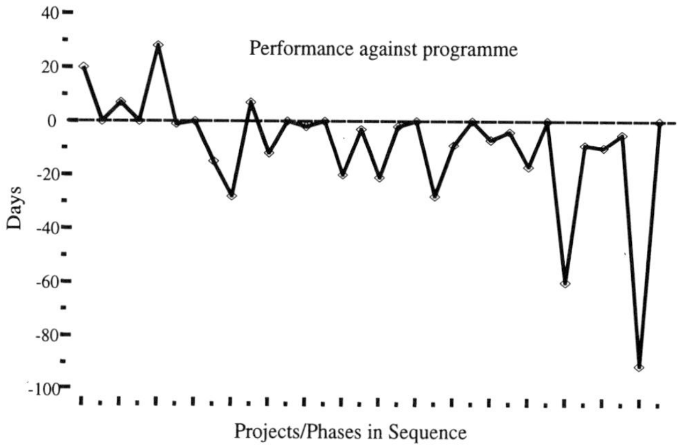

Project Programme Predictability

Project programme reduction for large projects, i.e. those above the EU threshold show savings currently of between 25 – 35%. Having a framework in place and not having to follow individual project EU procurement and tendering procedures typically saves 25% of the overall programme. Further improvement from integrated design, reduced mobilisation and optimisation of operational constraints produce further savings.

Project cost reduction is the raison d'être for developing partnering arrangements and it is necessary to provide hard evidence of the benefits. Initially project costs were split as follows: -

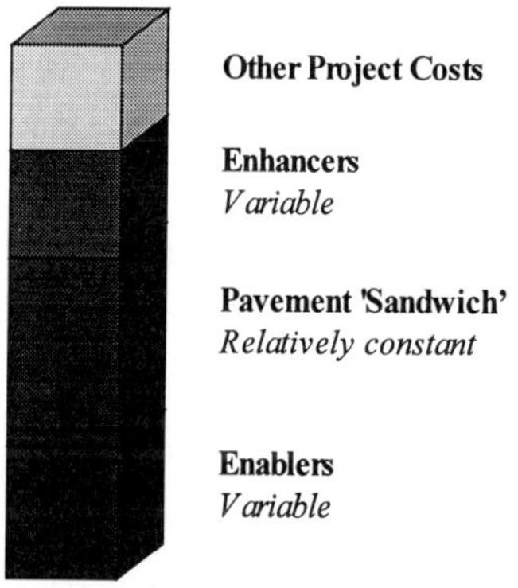

Cost Components

The enablers, enhancers and other project costs tended to be project specific and were, therefore, not easy to compare between projects. The pavement 'sandwich' was much more consistent and was chosen as the initial performance measure.

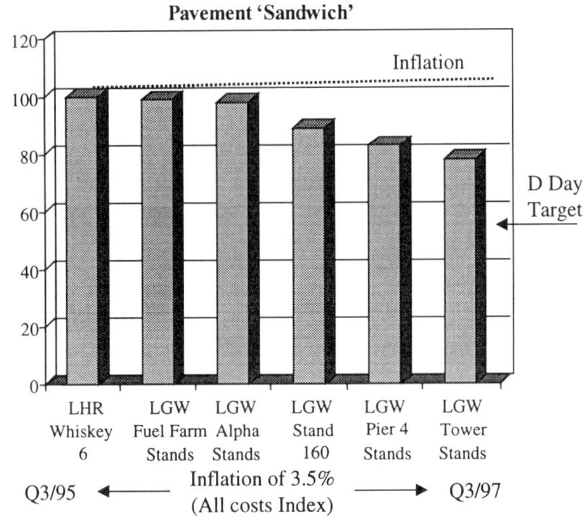

Project Cost Reduction

While these results are meaningful and show a considerable improvement, the pavement sandwich typically only accounts for 35 – 45% of total project costs. Analysis of the project workload revealed a product mix as shown below: -

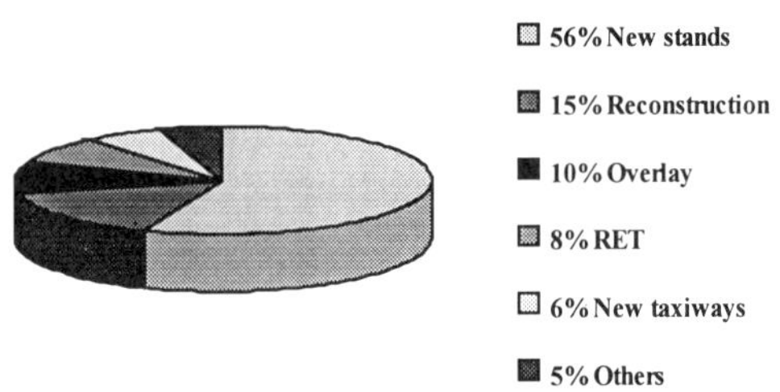

Generic Products by type

The most significant product is new stands which represents over 56% of the total spend. This approach has led to the development of generic products as a benchmark to model project cost reduction. The composition of the generic stand and cost improvement achieved to date is: -

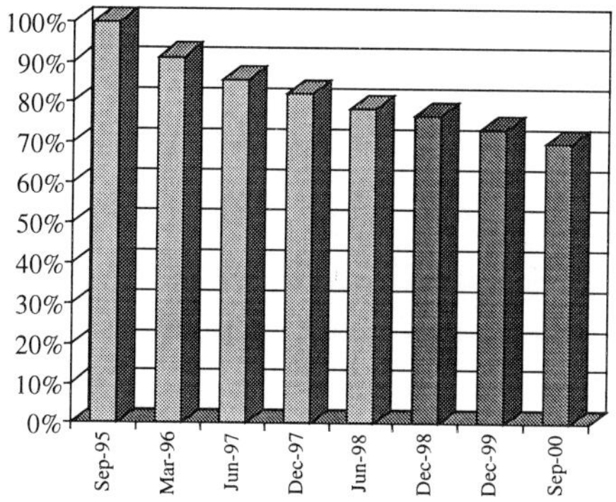

Generic Stand Improvement

10958m²	Pavement
464m	Drainage
252m	Electrical ducts
91m	Fuel main
96m	Firemain
Stand equipment	
Fixed electrical ground power	
Markings	

Generic Stand Components

The model can be used to monitor actual costs and, therefore, improvement to date, but also as a target mechanism for future contracts. Having moved away from an incentive mechanism for individual projects, achievement against this model is used as a key incentive for the framework with AMEC's fee tied to achieving the overall 5-year improvement target.

Whilst the Pavement Team must achieve BAA's global targets, each project has its own Key Performance Indicators that are just as important to its client and customers. A good example is the resurfacing of the Lima cul-de-sac at Heathrow. This cul-de-sac serves Terminal 3 and consequently has many long haul late evening departures and early morning arrivals. Early discussions with Operations established that a working window could only be provided from 22.30 – 04.30 hours each night. This short possession period restricted the effective working period to approximately 4.5 hrs/night and introduced large unproductive costs.

Through detailed discussions with the users and airlines, this window was increased to 05.30 hours but inherent to this was that any failure to hand back on time would severely disrupt aircraft operations for the whole day. The extra hour led to a complete re-programming and the asphalt laying was reduced from 90 to 45 nights and overall costs reduced by 18%. Whilst this was excellent for the client, the handover performance achieved was much more important. Performance like this builds confidence and allows the Team to remove waste that might otherwise be taken for granted.

Moving the Business Forward

The performance of the Pavement Team gives a good indicator of the benefits that can be achieved through long term partnering agreements. Creation of a single co-located delivery team with clear goals and values has provided the environment for success. The Team has invested in Continuous Improvement as the means to take the business forward. Process mapping and activity sampling are used to identify waste in existing processes and then root cause analysis, brainstorming or other CI techniques instigated to identify solutions. Examination of many construction processes reveals productivity typically around 30%.

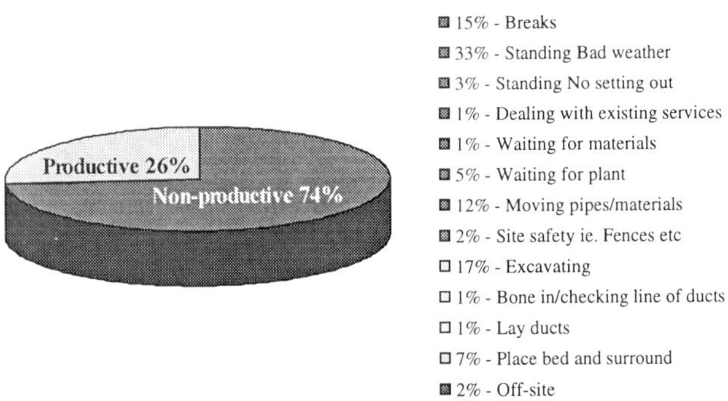

Drainage - Activity Sampling

It must be stressed that this is not the site workers' fault but that current processes only allow them to be that productive. Sustained improvement can only be implemented from removing waste from existing processes.

Taking the Japanese model we need to improve the quality of our products but that has to be measurable quality as defined by our customers. There is a well known paradox – concentrate on quality and productivity will improve, concentrate on productivity and quality reduces. We need to standardise more of our designs and streamline processes to facilitate flow (3). It is also clear that with over 60% of the value of work provided by 2nd and 3rd tier suppliers, the development and involvement of the supply chain becomes a major factor. For partnering to work all stakeholders must share in tangible benefits and be committed to working together to sustain improvement.

Finally and perhaps most importantly, as an industry we need to improve the role of our site workers. The people who work on site all have the ability to improve quality or processes and are our front line. This can be by more involvement in better planning or continuous improvement initiatives. With investment in such as better training and equipment we can all share in the benefit that comes from recovering waste.

References

1. M LATHAM - "Constructing the team". The final report of the Government/Industry Review of procurement and contracting arrangements in the UK. HMSO 1994.

2. J WOMAK, D JONES AND D ROOS – "The Machine that changed the World." Rawson Associates 1990.

3. J WOMAK, D JONES – "Lean Thinking – Banish Waste And Create Wealth In Your Corporation". Simon & Schuster 1996.

Fifth International Conference on the
BEARING CAPASITY OF
ROADS AND AIRFIELDS

USE OF FWD FOR A PAVEMENT STRUCTURE REDESIGN DURING CONSTRUCTION ON VOLCANIC SOIL SUBGRADES

Gabriel D. Berditchevsky and Carlos M. Marcenaro, *Senior Engineers, Louis Berger International, Inc.*
Byron R. Paiz, *Head of the Construction Supervision Division of the Guatemalan Road Authority*

Adress: 7a Avenida y 10a Calle, Zona 13, Módulo 5, Finca La Aurora, Dirección General de Caminos, Ciudad de Guatemala, Guatemala, C.A Fax: + 502-4400780/4400781/4400774

Synopsis

The Guatemalan Ministry of Communications, Transportation, Public Works and Housing (MCTOPV), through the Interamerican Development Bank (IDB), launched an investment of US$ 150 Million in infrastructure, specially roads rehabilitation, during a period of three to four years. For doing so, the IDB loan conditions required that a private company should be hired by the MCTOPV to administrate the financing and technical aspects of the investment projects. The Louis Berger International, Inc. was contracted for doing so, and was engaged also, in reaching the objectives of the investment program settled by the MCTOPV, IDB and the road agency (DGC). Some of the main objectives of the program, were to reduce the average rehabilitation cost per km of execution work, which was about US$ 260 thousand; the contractor commitment to finish the project at the bid proposal time, in contrast with the usually accustomed way in Guatemala that projects extend their construction period even years after its initial time schedule and; lastly, that the hired cost is respected and don't have further increases.

To reach these objectives, the technical-administrative personnel of the MCTOPV-DGC-LBII, proposed a number of actions: first, to enlarge the number of participants to the bids so as to have more competitors an get lower prices; second, to include an evaluative equation that relates the time and the cost proposed by the contractor in the bid so as to obtain the best convenience for the state. Third, the adoption of economic awards and penalties to the advance of the work during the constructive period and to the time offered to finish it (in function of the contractor's work program presented in its offer), and by this to attack the second point. Four, the third objective was handled in coordination with the contractor/supervisors, through the continuous observation of the work progress, the resolution of «in situ» problems and the redesign of the rehabilitation process through the continuous monitoring of the quality of the work by means of the use of the FWD. This last activity was useful too, in adjusting the work advance to the required quality and to the specified time schedule.

The following exposition shows an example of the way a pavement rehabilitation project was redesign during construction several times, with regards to the FWD and roughness monitoring results. These, in order to coup with the original design of recycling volcanic ash material during an extensive and mighty rainy period, which was combined with a relatively short time schedule and a fixed budget. The project was the Rehabilitation of the highway CITO 180, between CA-2 Occidente (Muluá) and Quetzaltenango

1.0 Project Characteristics

1.1 General

The highway Intertroncal Occidente (CITO 180) begins in the highway of the Pacific CA-2 Occidente with a height of 352 Mts. above sea level (asl) in Muluá (Department of Retalhuleu), and it ascends to Quetzaltenango, head of this department with a height of 2,327 Mts. asl.

The project has a longitude of 48.42 Kms.. (Km 177+000 to 125+000) and its alignment go through a relatively high curving line, on the side of steep slopes.
The project topography has characteristic that go from hilly to mountainous. The gradients along the road reach values over 6% in several sections, being the mean of the whole geometrical alignment 4.4%. The allowable speed for light vehicles is limited to an average of 55 to 60 Kms./hour as long as for the heavy vehicles the limit cannot surpass from 30 to 40 Kms./hour.

The actual road was built at the end of the decade of the 50's and its existing structure consists of an asphalt pavement with more than 30 years of use. The surface suffered accumulative deterioration because it has surpassed its useful life, without any or little maintenance.
In 1995 through the Contract Maintenance Program patching work increased and in some sections asphalt concrete overlays were placed.
The existing section has a width of 6 Mts. and shoulders of 0.60 Mts. to 0.80 Mts. to each side, although in most of the project the shoulders are nonexistent. Asphalt concrete layer varies from 7 to 10 Cms. to km 205 and from the 205 to 223 the layer is of 3 Cms.
The base has a variable thickness and its materials vary from silty sand sometimes with cement and gravel, crushed stone, volcanic slag, gravely sand, and from the km. 209 to 223 it has a crushed stone base with thickness varying from 11 to 19 Cms. The subbase has an average of 15 Cms. and it consists mainly of silty sand to sandy silt with certain quantity of gravel. In some sections it doesn't exists, practically between the km 187 and 203 and then between the km 207 and 209.

1.2 Geology of the Surrounding

The road begins in the physiological well know province as Recent Volcanic Slopes, that is characterized by being formed by the hillsides of the mountainous system of the Sierra Madre del Sur by the foot of the volcanic chain. This physiological region is formed by stratums of lava trails, volcanic ash and in some places volcanic mud.

Along the road path, the geological structure of the region varies in the following way:

From Santa Cruz Muluá to San Martín Zapotitlán (Km 177 at the 182), it is conformed by Quaternary Alluviums (Qa)
From San Martín Zapotitlán to the Stay of the Cruz (Km 182 at the 206), it is of the Quaternary (Qv), volcanic rocks, which includes lava trails and volcanic buildings

From Stay of the Cruz to Cantel (Km 206 at the 213), it is of the Tertiary (Tv), with volcanic rocks without dividing, predominantly Mine-Plioceno, it includes tuff, lava trails, and volcanic silts.
From Cantel to the City of Quetzaltenango (Km 217 at the 224) it is of the Quaternary (Qv) revised filled and covered with thick ashy pumice of diverse origin (Pumice type Ignimbrita).
Geologically the region is influenced by the activity of the Volcano Santa María-Santiago.

Still with the shown description, it is observed that in general along the whole path a layer of an ash volcanic layer covers the surface.

1.3 Climate

The climate varies considerably along the project accompanying the strong differences of height that register in its short journey (48 kms.). In Santa Cruz Muluá (350 asl), the annual mean temperature is 30 ºC, with an annual mean precipitation of 3,260 mm, the highest in the Department of Retalhuleu, and a relative humidity of 81%. In the city of Quetzaltenango (2,335 asl) the climate it's colder with an annual mean temperature of 18 ºC. The average annual precipitation is 846 mm and the relative humidity is 82%.

1.4 Traffic

The annual average daily traffic (AADT) in 1996 was of 2945 vehicles/day with 32% consisting of heavy vehicles, while the average rate of annual growth was established in 4%

The traffic projections for this Project gave the following values expressed as ESAL:

Period (years)	Equivalent Single Axle Load (82 kN)-ESAL
10	2,950,000
20	7,434,000

1.5 Geotechnical Profile and Structural Design of the Pavement

The geotechnical profile along the alignment is characterized by the existence of a subgrade formed by fine soil, finding sandy silt to silty sand as well as sandy clays and locally, volcanic slag.
The fine soils (SM, ML – A-2-4, A-4, passing #200: 20-40%) are characterized to possess a very closed humidity-density relationship, that means, the soil is very sensitive to water content changes. With a small increase of the water content above its optimal value its resistance falls very below of its maximum value.
For example, the existing CBRs overcome 10-20% in their dry condition for the sandy silts to silty sands, at a 95% grade AASHTO T180 compaction, and falls to 4-6% with an increase in the water content in 2-4% above the optimum. The CBRs for the clayey soils are around 4-6%.
The plasticity of the soil is generally low (<15% for the clays to <6 to NP for the sandy silts to silty sands) and its limits of consistency vary of LL = 40% (clays) to 15-25% (silts and sands).

1.6 Existing Pavement Deterioration

During the design stage the Consultant studied the superficial condition of the pavement through visual analysis. The state of structural condition was investigated by means of deflection measures with Benkelman Beam. The results of the superficial evaluation are as follows:

Section		Superficial Damages		Other Damages	
From Km	**To Km**	**Qualitative**	**Roughness (IRI assumed)**	**Patching %**	**Cracks %**
177	180	Regular–	4.0	0.70	15.0
180	185	Bad+	5.3	1.20	21.0
185	192	Bad+	5.6	1.50	23.0
192	195	Regular-	4.1	0.80	16.0
195	197	Bad	5.1	1.10	22.0
197	203	Bad+	5.6	1.60	24.0
203	207	Regular-	4.1	0.90	18.0
207	225	Bad	5.2	1.05	22.0
Total	**48 Km.**	**Bad**	**5.1**	**1.15**	**21.1**

A summary of the maximum deflections measured is presented in the Chart 1.

1.7 Rehabilitation Design

Of the alternatives for the pavement rehabilitation studied by the Consultant (on a period of 10 years-service design), the following one was chosen:

a) Road Cross Section: A section having two lanes of 7.2 meters wide and 1.60 meters of shoulders, the whole section should be paved with the asphalt concrete binder.

b) Pavement Structure: Recycling of the asphalt concrete surface layer and existent base in 15 Cms. to be used as subbase, the placement of a 20 cms. crushed stone base layer and on top of it, a hot asphalt concrete layer of 10-12 cms. thickness (5+5).

1.8 The Winning Tender

The project was bid in December/96, and the offers surrendered in January/97 being awarded to Solel Boneh for an amount of Q 55,4 Millions (approximately US $9,0 Millions) and at 300 days calendar time. The contractual procedure finished by the middle of March, and the execution began by the middle of April (real time of beginning), with the first recycling and priming works. The work was developed from the low part of the project (Muluá), where the bank of materials was located, towards the high part. It is important to notice that the rainy season began at the end of April.

2.0 Rehabilitation Redesign

2.1 Phase 1

To corroborate the Consultant's design and to have relative information of the structural and superficial original states of the pavement, a structural and superficial evaluation of the pavement was tried along the whole alignment, the rutting level was measured as well as the roughness an superficial damage.

The equipment used to check the structure properties was a Dynatest FWD and the roughness by an RSP laser type equipment of the company Dynatest (model 5051). TNM, LTD did the measuring work.
The structural evaluation was done with a spacing of 100 meters between each point (the information was evaluated by the program MODULUS), while the roughness was defined in averages sections of 20 meters each one. It is important to notice that this type of equipment was not known in Guatemala before 1997.
The summary of the results is shown in the Chart 1.

Based on those results, the pavement structure was redesign (by the AASHTO Rehabilitation Method) trying to look for an option that favors the constructive procedure, minimizing the work with granular layers so as to improve the construction advance during the rainy season, and offers the pavement the desired structural condition. The new proposed design is presented next, which meant besides of trying to reach the mentioned objectives, a saving of approximately Q 3,2 Millions (US $500,000) to the original cost:

Stationing (Km)	Recycled Subbase (cms.)	Asphalt Concrete Base (cms.)	Asphalt Concrete Course (cms.)
177+000 – 190+000	15	6,5	8 (5+3)
190+000 – 205+000	15	6,0	8 (5+3)
205+000 – 210+000	15	8,0	8 (5+3)
210+000 – 225+000	15	11,0 (6+5)	8 (5+3)

The recycling of the pavement was required so as to rehabilitate the old pavement, i.e., to correct the damages and superficial deformations of the structure (according to the high levels of roughness registered). The work daily goal was to prime all recycled layers before rain starts, and to place the asphalt concrete base immediately after the curing of thc priming (usually after 24 hours). The prime material was a cut back asphalt of an RC-250 type. Also, the contractor was required to build subdrains in the low area of the project, between the station 177+000 to station 190+000, where great quantity of rainwater concentrated and has more areas with clayey subgrade soils.

Because the rainy season began during the recycling process, the advances in work were minimum, making the recovery work complicated because the subgrade got usually saturated as well as the recycled subbase. Because of this picture, it was necessary to adopt an alternative that allows the advance of the work during this station and assures the quality of the finished layers.

2.2 Phase 2

With an advance of 6 Kms., a second structural study (june/97) was carried out by means of the FWD on the worked section in order to corroborate the required goals. In the recycled 6 Kms., the 6.5 cms. thick asphalt concrete base was placed.

In Chart 1, a summary of the results is shown.
The structural number of the section was similar along the section, what indicates a constructive homogeneous procedure. The obtained value varied from 2.3 to 2.4, in this first 6 kilometers. This means an increment of 0.8-0.9 in relation to the original value for the section.
Due to the slow advance of the work and to the uncertainty about the quality of the finished structure, a test section of 250 meters was defined, in which the Contractor tried overlaying the existing structure directly with the asphalt concrete base layer. A visual observation of the procedure, proposed the possibility of its application as a leveling layer
In relation to the results, the Contractor was required to continue and finish the first 12 Kms., some of them were already worked and a test section of 2,5 Kms. would be worked by placing the asphalt concrete base course directly on the existent pavement

2.3 Phase 3

The new results (sept/97) showed the incidence of the rain in the constructive procedure, since the rain was accentuated with the advance of the work. In the first section (to the station 182+000) the benefit of subdrains is observed helped keeping that values of it subgrade elasticity modulus above the one of the following section, which grew in approximately 40%. The advance of almost five months was only of 12 Kms..

The test section of 2,5 Kms. was tried, in which besides the asphalt concrete base, the first concrete asphalt layer (binder) of 5 cms. was placed on top of it.

The new evaluation showed an increase in the structural number of the test section pavement in relation to the original value from 1.6-1.7 to 2.7, result that was satisfying in relation to the previous ones not only because of the structural contribution, but also for the possibility of working during the rainy time, without disturbing the consolidated layers of the pavement.

Based on these results it was opted to continue with the work placing the asphalt concrete base directly on the existing structure, in a thickness based on a new design, as follows:

Stationing (Km)	**Recycled Subbase (cm)**	**Asphalt Concrete Course (cm)**
190+000 – 205+000	6,0	8 (5+3)
205+000 – 210+000	8,0	8 (5+3)
210+000 – 225+000	11,0 (6+5)	8 (5+3)

The asphalt concrete base was designed with a minimum percentage of asphalt cement of 3,5% (finally reached 4,2%), so that it has a similar quality to that of an asphalt concrete surface layer, and by this, to have a thicker asphalt structure that can counteract the structural problems and correct the deformations of the existent pavement. It was request the correction of the roughness through the asphalt layers, using the base as a leveling course.
Previous to the placement of the asphalt concrete leveling course, the current pavement would be treat by milling of 1-2 cms. of the cracked areas (that finally wasn't done), the superficial patching of areas that have a settlement higher than 2 cms. and a deep patching of the areas with a settlement larger than 3 cms. (measured with a 3 meters rule).
The extension of the section of the pavement to the shoulders would be worked with granular material in the areas with an acceptable superficial drainage condition and with subbase material in the areas with a poor drainage condition.
The asphalt cover of this extension would be done as mentioned in section 2.1.
The first 13 Kms. would be in observation and a new evaluation would be carried out in the future, previous to the placement of the superficial asphalt concrete layer, to determine the appropriate solution. Until that date, the subdrain in the lower section would be finished.

2.4 Phase 4

The rainy station concluded by the middle of December/97.
At the beginning of January/98, having the placement of the superficial asphalt concrete course to conclude the work, the appearance of cracks was perceived in the section between the stations 180+000 to 187+500. Adding to this condition, the subdrains were not executed by the Contractor by reasons unaware to the work.

A new evaluation (jan/98) was carried out in those first 13 kilometers (177+000-190+000) to determine the procedure and the actions for the continuation of the work. The summary results are shown in Chart 1.
In relation to these, the following was observed:

1) The registered maximum deflection continues being high in general (730-1300), registering the highest values in the cracked area between the station 180+000 and 188+000. In that area also exists the lowest subgrade elastic modulus. The section between stationing 177+000 and 180+000 presents deflection conditions a little better and its characteristics with a support capacity above the next section (in this area subdrains were built).
2) The roughness registered in that section was between 1.9 and 2.9, which is high so the Contractor was required to take this into account for the placement of the upper layers.

Based on these results a new thickness design was defined for the superficial asphalt concrete course that would replace the original layer of 3 cms., as follows:

Stationing (Km)	177+400 – 180+600	180+600 – 187+600	187+600 – 190+000
Asphalt Concrete Course (cm)	**6**	**6+5**	**5**

3.0 Summary and Conclusions

A last evaluation of the surface and structural conditions was made on 18-April/98. The final results show as follows:

a) The maximum deflections on the structure diminished by more than 50% in the whole alignment, in comparison with the original values.
b) The mean values of the structural number reached the design required ones, but some sections need (177+000-180+000 and 187+000-190+000) an overlay of 3-4 cms. to stand the design values.
c) The average roughness value is 1.85, which is acceptable regarding specifications (< 2.0 m/Km), although there are sections from Quetzaltenango to Muluá (177+000-181+000 and 211+000-216+000) that have average values of 2.2.
d) The use of subdrains in the lower part of the project showed it benefits, in helping the fine soil evacuate the water content from the structure environment.
e) The final time for the rehabilitation work stayed on the limit of the 300 days and the cost of the project was close to Q53 Millions (Q2,5 Millions bellow the bid offer).

A large part of Guatemala lays on volcanic soils of the type found in CITO-180. Many of the projects are under the rehabilitation program, have the pavement conditions as the ones encountered in the CITO-180, and are constructed in a heavy rainy season that goes for more than six month. The behavior of these soils allows it treatment while consolidated and dry so a solution of the kind applied in this project could be an option for keeping the contract time schedule conditions while assuring the required superficial and structural standard of the rehabilitated road.
The use of FWD and RSP equipment as a tool for quality control showed it importance so as to implement a quality control/assurance methodology of the kind used during the construction process to all projects in the program.

4.0 References

1. Zaghloul, S., «Development of performance based quality control specifications for asphalt pavements».
2. Rohde, G., «Determining a Pavement's Structural Number from FWD Testing». Presented at the 73rd TRB Annual Meeting, Washington, D. C., 1994.
3. Cohen, E., Uzan, J.,Rubinstein, I., Cohen, S.,«Development and Implementation of a Rehabilitation System». Presented at the 76th TRB Annual Meeting, Washington, D. C., 1997.
4. «AASHTO Guide for Design of Pavement Structures», Washington, D.C., 1993.

Chart 1: Summary of pavement evaluation results during construction

<table>
<tr><td></td><td colspan="8">**Stationing (Kms)**</td></tr>
<tr><td>**Construction Phase**</td><td>**177+000 – 180+600**</td><td>**180+600 – 182+000**</td><td>**182+000 – 187+500**</td><td>**187+500 – 190+000**</td><td>**190+000 – 192+500**</td><td>**192+500- 205+000**</td><td>**205+000- 210+000**</td><td>**210+000- 225+000**</td></tr>
<tr><td>**Scarifying and Recycling (cms) ([1])**</td><td colspan="4">15</td><td colspan="4"></td></tr>
<tr><td>**Asphalt Concrete Base (cms) ([2])**</td><td colspan="4">6,5</td><td colspan="2">6</td><td>8</td><td>11</td></tr>
<tr><td>**Binder (cms) ([3])**</td><td colspan="8">5</td></tr>
<tr><td>**Asphalt Concrete Surface Course (cms) ([4])+([5])**</td><td>6</td><td colspan="2">6+5</td><td>5</td><td colspan="4">3</td></tr>
<tr><td>**Maximum FWD Deflection (1/1000 mm) (A)**</td><td colspan="8"></td></tr>
<tr><td rowspan="2">18/apr/1998</td><td>560</td><td>450</td><td>430-550</td><td>415-550</td><td>420</td><td>320-430</td><td>470-560</td><td>500-690</td></tr>
<tr><td colspan="8">Final</td></tr>
<tr><td rowspan="2">24/jan/98</td><td>720-830</td><td>860-1085</td><td>840-1085</td><td>670-750</td><td colspan="4" rowspan="2"></td></tr>
<tr><td colspan="4">([1])+([2])+([3])</td></tr>
<tr><td rowspan="2">22/spt/97</td><td colspan="2">719-817</td><td>1050</td><td>859-780</td><td>545-720</td><td colspan="3" rowspan="2"></td></tr>
<tr><td colspan="4">([1])+([2])+([3])</td><td>([2])+([3])</td></tr>
<tr><td rowspan="2">27/jun/97</td><td>1000-760</td><td>910</td><td colspan="6" rowspan="2"></td></tr>
<tr><td colspan="2">([1])+([2])</td></tr>
<tr><td>27/apr/97</td><td></td><td colspan="3">1200</td><td colspan="2">750-900</td><td>900-1100</td><td>1700</td></tr>
<tr><td>2/aug/96 with Benkelman Beam</td><td>450</td><td>550</td><td>700</td><td colspan="2">600</td><td>500</td><td>600</td><td>750-1150</td></tr>
<tr><td>**Subgrade Elasticity Modulus (Kg/cm2) (SD)**</td><td colspan="8"></td></tr>
<tr><td rowspan="2">18/apr/1998</td><td>1250</td><td colspan="2">740</td><td>900</td><td>830</td><td>840-1600</td><td>1500-780</td><td>680</td></tr>
<tr><td colspan="8">Final</td></tr>
<tr><td rowspan="2">24/jan/98</td><td>1300-1100</td><td>661-550</td><td>515</td><td>800-900</td><td colspan="4" rowspan="2"></td></tr>
<tr><td colspan="4">([1])+([2])+([3])</td></tr>
<tr><td rowspan="2">22/spt/97</td><td>1100</td><td>700</td><td>490</td><td>650</td><td>720</td><td colspan="3" rowspan="2"></td></tr>
<tr><td colspan="4">([1])+([2])+([3])</td><td>([2])+([3])</td></tr>
<tr><td rowspan="2">27/jun/97</td><td>1100</td><td>650</td><td colspan="6" rowspan="2"></td></tr>
<tr><td colspan="2">([1])+([2])</td></tr>
<tr><td>27/apr/97</td><td></td><td colspan="2">430</td><td>700</td><td colspan="2">530</td><td>1000</td><td>600</td></tr>
<tr><td>**Structural Number (SD)**</td><td colspan="8"></td></tr>
<tr><td rowspan="2">18/apr/1998</td><td>2.9-3.5</td><td colspan="2">3.3-3.5</td><td>2.9-3.0</td><td colspan="2">3.3-3.6</td><td colspan="2">3.0</td></tr>
<tr><td colspan="8">Final</td></tr>
<tr><td rowspan="2">24/jan/98</td><td>2.2-2.4</td><td>2.1-2.3</td><td>2.3-2.1</td><td>2.5-.2.7</td><td colspan="4" rowspan="2"></td></tr>
<tr><td colspan="4">([1])+([2])+([3])</td></tr>
<tr><td rowspan="2">22/spt/97</td><td>2.4-2.3</td><td>2.4-2.3</td><td colspan="2">2.1-2.2</td><td>2.7</td><td colspan="3" rowspan="2"></td></tr>
<tr><td colspan="4">([1])+([2])+([3])</td><td>([2])+([3])</td></tr>
<tr><td rowspan="2">27/jun/97</td><td colspan="2">2.1-2.3</td><td colspan="6" rowspan="2"></td></tr>
<tr><td colspan="2">([1])+([2])</td></tr>
<tr><td>27/apr/97</td><td></td><td colspan="2">1.5</td><td colspan="3">1.6-1.7</td><td>1.3</td><td>1.2</td></tr>
<tr><td>**Roughness (m/Km) (A)**</td><td colspan="8"></td></tr>
<tr><td rowspan="2">18/apr/98</td><td>1.9-2.2</td><td colspan="6">1.8-1.9</td><td>1.4-2.2</td></tr>
<tr><td colspan="8">Final</td></tr>
<tr><td rowspan="2">24/jan/98</td><td>2.1-3.2</td><td>1.9-3.2</td><td>1.9-2.4</td><td>2.3-2.4</td><td colspan="4" rowspan="2"></td></tr>
<tr><td colspan="4">([1])+([2])+([3])</td></tr>
<tr><td>27.04.97</td><td colspan="4">6.7-7.7</td><td>3.8-4.2</td><td>3.5-6.8</td><td>4.8-5.8</td><td>3.7-6.7</td></tr>
</table>

Note: A = Average Values
SD = Average Values minus Standard Deviation (85%)
([1]), ([2]), ([3]), ([4]), ([5]) = Construction Phases.

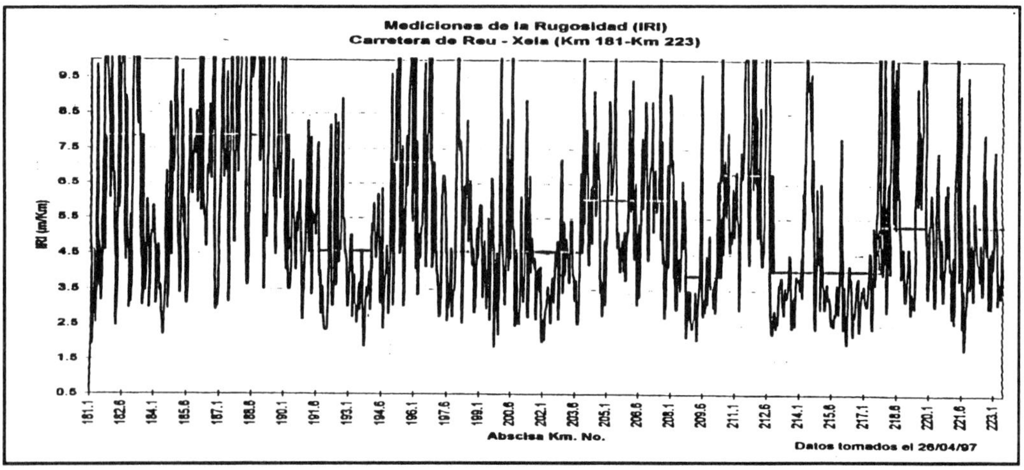
Mediciones de la Rugosidad (IRI)
Carretera de Reu - Xela (Km 181-Km 223)
IRI (m/Km)
Abscisa Km. No.
Datos tomados el 26/04/97

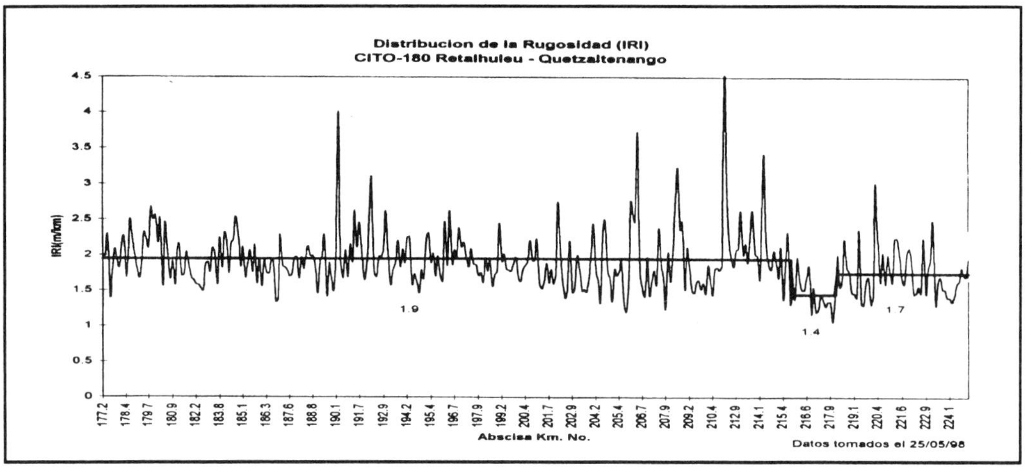
Distribucion de la Rugosidad (IRI)
CITO-180 Retalhuleu - Quetzaltenango
IRI(m/km)
1.9
1.4
1.7
Abscisa Km. No.
Datos tomados el 25/05/98

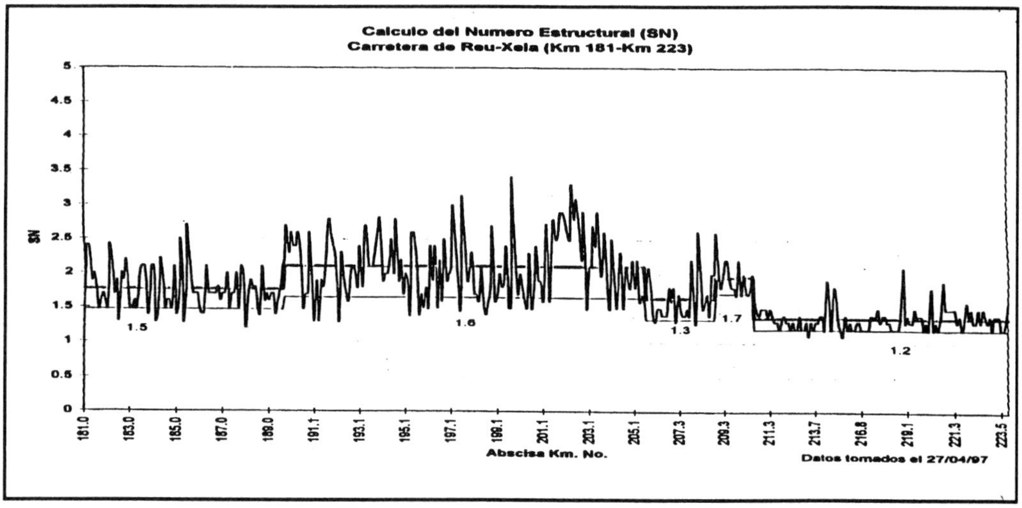

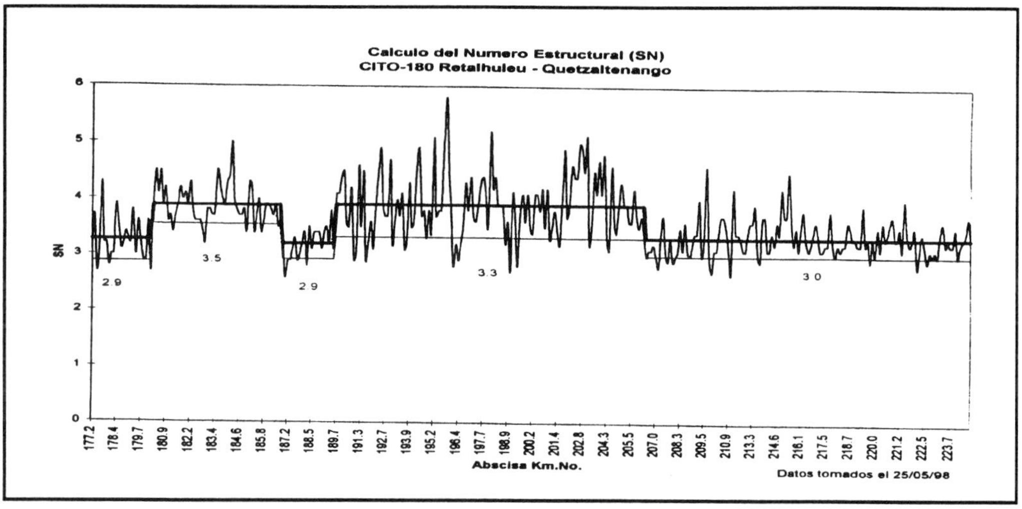

Chart 2: Initial and final roughness and structure number values (Km 177+000-225+000) (According to TNM,ltd Report)

Author Index

INDEX

NOTE A reference is generally given only once, even if the subject may appear several times in the same paper.

A

B

C

D

E

F

G

H

I

J

K

L

M

N

O

P

Q

R

S

T

U

V

W

Z

We contribute to maintain and improve the bearing capasity of roads and airfields by supplying good bitumenous binders of consistent quality

NYNAS

Phønix Contractors

A/S Phønix Contractors was founded in 1907 in Denmark.

PHØNIX
A/S Phønix Contractor

From the humble beginnings manufacturing roofing materials, Phønix expanded to become one of the five largest road contractors in Denmark.
Today Phønix is part of the large Norwegian Rieber & Søn group of companies.
Phønix employs about 1600 skilled and dedicated staff around the world.

Within roads Phønix has particular experience in:

Road Pavements
All sizes of roads, from district roads to major multi-lane highways
All kinds of asphalt products both hot and cold
Phønix has over the years invented and developed much of the equipment and machinery that is widely used among pavement contractors today.

Bridge Pavements
Phønix has applied pavement to some of the largest bridge projects in Denmark.

Airport pavements
Runways, taxiways, aprons

Asphalt products
Products to suit any pavement requirements.
Ongoing research and development

Milling
Recirculating old pavement, stabilising weak foundations

Road markings
Road signs and equipment

Industrial pavements
Specially developed for high static and dynamic loads and resistance against chemicals

Bitumen technology
Phønix has over the years specialised in Bitumen technology, developing new products and techniques for achieving the optimal for each application.
Phønix purchases bitumen directly on the open market and imports it to our own depot and processing facility in Denmark.

Joints
Bituminous joints in concrete structures

Pavement consultancy
Management of maintenance systems
Pavement analysis
Phønix has a major share of the world market for Falling Weight Deflectometers for analysing pavement condition.

Turn-key contracting
Phønix has the capacity to undertake complete projects from planning, through design to execution both for new construction and rehabilitation works.

Projects in Africa

Phønix has executed projects in Zambia, Tanzania, Ghana and Benin. Activities at present are concentrated in Zambia and the focus is on expansion into the neighbouring countries from there.

Phønix has been operating in Zambia and Tanzania since 1991 (in Tanzania through our sister company NODEST) and today we are the largest road contractor in Zambia with jobs around the country, in the Copperbelt district, in Lusaka and in Southern Zambia.
Although it is our intention to grow and expand further in Zambia and the neighbouring countries, we are adamant that this must be a controlled process. Phønix can only maintain the excellent reputation as a serious contractor by focusing first and foremost on continued work of the highest and uncompromised quality.

Address: Vester Allé 1, DK-6600 Vejen, Denmark, Tel.: +45-75 36 11 11, Fax: +45-75 36 42 06, E-mail: Vejadm@phoenix.dk

The best way to give your runway better stability and an extended lifetime

Styrelf®

The successful design of a runway depends on a large number of factors of which the choice of binder represents one of the most important.

Elf Aquitaine group has developed bitumen based products which meet most requirements of a modern airport. In particular, the development of the polymer-modified binder »Styrelf 13 « has made possible tailor made designs to meet the exact needs of each individual airport.

Elf engineers have long experience in designing and developing high quality bitumen products. This expertise is our customers best guarantee for having an airport which will meet the challenges of the ever increasing traffic volume as well as satisfy the needs of a modern aircraft fleet.

For further information please visit our stand in the exhibition area.

Styrelf® 13

elf

Bitumen International

Dynatest

Dynatest is an international, highly specialized group of consulting engineering companies, organized under a parent company known as Dynatest International A/S. Dynatest develops and produces pavement testing, monitoring, and evaluation equipment and software for use by the pavement engineering community worldwide. Dynatest also provides state-of-the-art international pavement consulting engineering services using its equipment, ***methodology*** and software.

Since the early 1970's airports around the world have been using pavement management technology to assist in the management of the pavement infrastructure. Some pavement management systems are based solely on Pavement Condition - visually determined evaluations of the pavement surface. Other systems attempt to integrate pavement condition with structural modelling. Still other systems use structural modelling as the sole management tool.

Each and every pavement is different, specific deterioration modeling for Countries, states or individual pavements, have their limitations. Subgrade and environmental affects should be taken into consideration for each individual section of the pavements. Once this information is known, then an engineer can predict with a greater degree of confidence what will happen to that homogenous section in the future. The engineer knows how and why it will deteriorate. This information is critical in any decision to be made on which rehabilitation option should be chosen. Combining this information with the cost benefit matrix analysis within a **P**avement **M**anagement **S**ystem allows the **R**esident **E**ngineer (R.E.) to make accountable budgetary decisions with a high degree of confidence.

The Performance Economic Rating System (PERS) used within our Road Pavement Management System (over 30 installation world-wide, including Alaska and South Africa DOT) has now been integrated within

The Dynatest – WS Atkins A.I.R.P.O.R.T.S Management System.

Dynatest WS/Atkins

A.I.R.P.O.R.T.S.

A.I.R.P.O.R.T.S. is a state of the art Airfield Pavement Management System being released at the **BCRA'98** Conference.
For a demonstration come and visit us at the Dynatest Stand.

Pavement Consultants

Phønix Pavement Consultants invites you to our in- and outdoor stand in Trondheim

w Falling Weight Deflectometer

ign Software
ndows 95 version)

w Analysis Software

lar Equipment

rs of Experience

There's tough and there's Typar®

No geotextile resists damage during installation better than TYPAR®

Right from the start geotextiles take a lot of punishment. 95% of all damage to them occurs during installation. That's where TYPAR® comes out on top.

Recent independent field trials* demonstrated a direct relationship between the ability of a geotextile to absorb energy (work to break) and failure rate during installation.

The unique TYPAR® process produces a product that combines a high initial modulus (stiffness), with an ability to elongate up to 60% with increased tensile loading before failure. The same field trials concluded that this combination, coupled with a high degree of uniformity, was responsible for the superior performance of TYPAR®. Today over 600 million sq. m is being used in engineering projects around the world. So if you're looking for a geotextile that has shown it can take the knocks and still go the distance, choose TYPAR® - the toughest in the business.

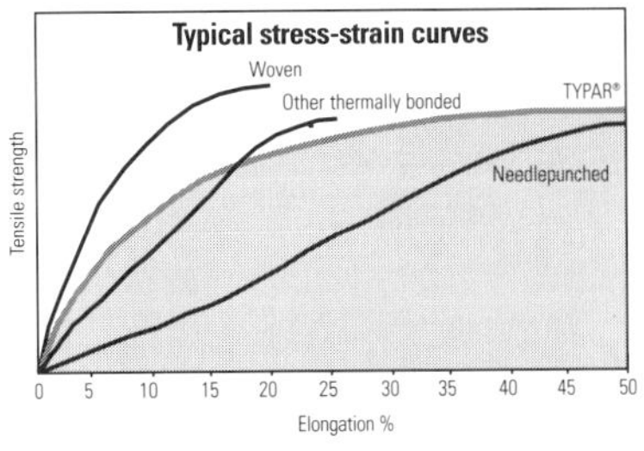

Recommended TYPAR® Styles

Styles	Applications
3337	General purpose separation/filtration/ drainage
3407-2	Separator with greater installation damage resistance
3857	Separator for poorer subgrade

Typar®, DuPont Nonwovens
Fax: 00352 3666 5021
Distributor for Norway:
Nordisk Kartro AS, P.O. Box 54
Holmaveien 20, N-1313 Vøyenenga
Tel.: +47 67 17 36 00, Fax: +47 67 17 36 01

DUPONT®
Typar

® DuPont registered trademark

*SINTEF. Civil and Environmental Engineering Institute. Norway, 1997